PHYSICS:
For Scientists and Engineers

Raymond A. Serway
James Madison University

Cover photo by
Ross Chapple, Hume, Virginia

SAUNDERS GOLDEN SUNBURST SERIES

SAUNDERS COLLEGE PUBLISHING
Philadelphia New York Chicago
San Francisco Montreal Toronto
London Sydney Tokyo Mexico City
Rio de Janeiro Madrid

Address orders to:
383 Madison Avenue
New York, NY 10017

Address editorial correspondence to:
West Washington Square
Philadelphia, PA 19105

This book was set in Laurel by York Graphic Services, Inc.
The editors were John Vondeling, Lloyd Black, Sally Kusch, and Irene Nunes.
The art & design director was Richard L. Moore.
The text design was done by Nancy E. J. Grossman.
The cover design was done by Nancy E. J. Grossman.
The artwork was drawn by Anco/BOSTON.
The production manager was Tom O'Connor.
This book was printed by R. R. Donnelly & Son.

Cover credit: Particle accelerator, © Ross Chapple, Hume, Virginia

To my wife, Elizabeth Ann, and children, Mark, Michele, David and the most recent light in my life, Jennifer Lynn, for their love and understanding.

PHYSICS: FOR SCIENTISTS AND ENGINEERS ISBN 0-03-057903-1

Library of Congress catalog card number 81-53098.

4 039 987

CBS COLLEGE PUBLISHING
Saunders College Publishing
Holt, Rinehart and Winston
The Dryden Press

Physical Data Often Used

Acceleration due to gravity, at the earth's surface	$9.81\ m/s^2$
Average earth-moon distance	3.84×10^8 m
Average earth-sun distance	1.49×10^{11} m
Average radius of the earth	6.37×10^6 m
Density of air	$1.29\ kg/m^3$
Density of water (20°C and 1 atm)	$1.00 \times 10^3\ kg/m^3$
Mass of the earth	5.99×10^{24} kg
Mass of the moon	7.36×10^{22} kg
Mass of the sun	1.99×10^{30} kg
Standard atmospheric pressure	1 atm $= 1.013 \times 10^5$ Pa

Mathematical Symbols

$=$ is equal to	$<$ is less than
$\neq$ is not equal to	$\gg$ is much greater than
$\equiv$ is defined as	$\ll$ is much less than
$\sim$ is proportional to	$\|x\|$ the absolute value of x
$>$ is greater than	$\sum$ the sum of

Preface

This textbook is intended for a two- or three-semester course in introductory physics for students majoring in science or engineering. Most of the text was written over a period of several years while the author taught an introductory physics course at Clarkson College of Technology. During this time, most of the material was classroom-tested, and student critiques were solicited. Many of these comments were taken into consideration for the final version of this text.

The mathematical background of the student taking this course should ideally include one semester of calculus. If that is not possible, the student should be enrolled in a concurrent course in introductory calculus.

The major portion of this book deals with fundamental topics in classical physics: Newtonian Mechanics (Chapters 1–14), Mechanics of Solids and Fluids (Chapter 15), Heat and Thermodynamics (Chapters 16–19), Electricity and Magnetism (Chapters 20–31), and Waves and Optics (Chapters 32–39). Chapter 40 is an introduction to the special theory of relativity and relativistic mechanics. Finally, Chapter 41 covers various topics and concepts in quantum and atomic physics.

My major objective in writing this text has been to provide the student with a clear and logical approach to the basic principles of physics. Emphasis is placed on presenting the concepts and the applications of physics in a precise, but realistic fashion.

It has been my experience that many students struggle through such a course for a variety of reasons such as an inadequate mathematical background. I have tried to keep such students in mind by introducing the calculus slowly. Tutorial remarks are often provided in the text and in problems involving more advanced mathematical techniques. Since many students taking such a course may have had little or no previous training in physics, I have chosen to introduce only a few concepts in each chapter. Furthermore, I have included a large number of illustrative examples which should assist the student in understanding the concepts, and could provide some basis for working out end-of-the-chapter exercises and problems.

It is my view that a textbook should be the student's major "guide" for understanding and learning the material. Furthermore, a textbook should be styled and written for ease in instruction. In order to meet these goals, the book contains the following features:

(1) Most chapters begin with a chapter preview, which includes some introductory remarks on the objectives of that unit.
(2) The text is written in an informal style, which I hope students will find appealing and enjoyable to read.
(3) Many examples of varying difficulty are presented as an aid in understand-

ing the concepts. Many of these worked examples will serve as models for problem-solving. The examples are boxed so as to avoid confusion with the text material.

(4) Many chapters include special topic sections which are intended to expose the student to various contemporary and interesting applications of physics. These topics are closely related to the material covered in that chapter so that the student can appreciate their relevance. The special topics include motion in the presence of resistive forces, energy and the automobile, energy from the tides, rocket propulsion, energy from the wind, thermal pollution, devices such as the oscilloscope, lasers, generators, motors, semiconducting diodes and transistors, filter circuits, power transmission, optical instruments, x-ray diffraction, van Allen belts, and magnetic bottles.

(5) Two introductory chapters are included to "set the stage" for the text and to introduce some basic mathematical tools such as the use of vectors and unit vector notation.

(6) Vector products are introduced later in the text where they are needed in physical applications. The dot product is introduced in Chapter 7, Work and Energy; the cross product is introduced in Chapter 11, which deals with rotational dynamics.

(7) Calculus is introduced gradually, keeping in mind that a course in calculus is often taken concurrently. Several mathematical appendices are included which provide reviews in algebra, geometry, trigonometry, differential calculus and integral calculus.

(8) Questions requiring verbal answers are included at the end of many sections. Some questions provide the student with a means of self-testing the contents of that particular section. Other questions could serve as a basis for classroom discussion. Answers to selected questions can be found in the Student Study Guide.

(9) An extensive set of student exercises and problems is included at the end of each chapter. (The text contains approximately 1300 exercises and 500 problems.) The exercises are straightforward in nature and are intended to test the student's basic understanding of the material. For the convenience of both the student and instructor, the exercises are keyed to the various sections. The problems are generally more challenging and usually involve several concepts. Problems which are especially thought-provoking often include hints. Answers to the odd-numbered exercises and problems are given at the end of the book. In my opinion, assignments should consist of many more exercises than problems. This technique should help in building self-confidence in students.

(10) Marginal notes and comments are used to locate important statements, equations and concepts in the text. Important equations are set in a screened box for review or reference.

(11) Chapter summaries are provided to review the important concepts and relations discussed in that chapter. This feature is especially useful for the student for both problem-solving and self-study.

(12) A number of appendices are included to supplement textual information. In addition to the mathematical reviews, the appendices contain tables of conversion factors, physical data, integrals, derivatives, mathematical symbols, and the SI units of physical quantities.

(13) The international system of units (SI), sometimes called the metric system, is used throughout the book. The "British engineering" system of units is used only to a limited extent in the early chapters on mechanics.

(14) The text contains a generous collection of illustrations and photographs. These are included to clarify and/or expand upon discussions and examples in the text.

As an additional instructional aid to students, a study guide is available which is designed to provide further drill on problem-solving techniques and physical concepts. Most chapters contain a list of objectives, skills necessary for that unit, a review list of important quantities and concepts, a list of equations and their meanings, answers to selected questions from the text, and finally, several programmed exercises which test the students' understanding of concepts and methods of problem solving.

This book is structured in the following sequence: classical mechanics, heat and thermodynamics, electricity and magnetism, matter waves, light waves, optics, relativity and quantum physics. This is a slight departure from the more traditional sequence where matter waves are covered before electricity and magnetism. I have chosen to unify the treatment of waves since this order of topics is typical in a three-semester course. Some instructors may prefer to cover matter waves (Chapters 32, 33, and 34) after completing Chapter 19. Others may prefer to cover waves and optics (Chapters 32–39) before electricity and magnetism. (For this latter order, I suggest that Chapter 35 on electromagnetic waves be covered following the material on electricity and magnetism.) The chapter on relativity was placed near the end of the text since this material is often treated as an introduction to the era of "modern physics." If time permits, instructors may choose to cover Chapter 40 after completing Chapter 14, which concludes the material on newtonian mechanics.

For those instructors teaching a two-semester sequence, some sections and chapters could be deleted without any loss in continuity. I have labeled these in the Table of Contents with asterisks (*). For student enrichment, some of these sections or chapters could be given as reading assignments. Further details regarding optional materials and suggestions to the instructor are given in the instructor's manual.

Those of you who use this textbook might find that certain sections could be further elaborated on or clarified. Any such suggestions, new ideas or criticisms are welcomed and will be taken into consideration for future editions.

ACKNOWLEDGEMENTS

This textbook was written with the assistance of many people. The group of people listed below reviewed parts or all of the manuscript during various phases of its writing. I am grateful for their helpful suggestions, criticisms, and encouragement.

Elmer E. Anderson
University of Alabama

Wallace Arthur
Fairleigh Dickinson University

Richard Barnes
Iowa State University

Marvin Blecher
Virginia Polytechnic Institute and State University

Jerry Faughn
Eastern Kentucky University

William Ingham
James Madison University

Herb Helbig
Clarkson College of Technology

Carl Kocher
Oregon State University

Clem Moses
Canton ATC

Curt Moyer
Clarkson College of Technology

A. Wilson Nolle
The University of Texas at Austin

George Parker
North Carolina State University

Gary Williams
University of California, Los Angeles

Earl Zwicker
Illinois Institute of Technology

I am especially grateful to my colleague J. R. Gordon for writing and solving many of the exercises and problems in this text. I am also grateful to William Ingham who provided an in-depth, line-by-line review of the entire manuscript, and my former colleagues, Herb Helbig and Curt Moyer, who offered many valuable suggestions throughout the development of the text. I am indebted to Lawrence Hmurcik for providing the solutions to all of the exercises and problems, and for his suggestions for improving the quality and accuracy of problem sets. I appreciate the assistance of my son Mark and Walter Curt in proofreading the galleys. I thank

Agatha Brabon, Linda Delosh, Mary Thomas, Georgina Valverde, and Linda Wood for an excellent job in typing various stages of the manuscript. I benefited from valuable discussions with many people including Subash Antani, Gabe Anton, Randall Caton, Don Chodrow, David Kaup, Len Ketelsen, Henry Leap, Frank Moore, Clem Moses, Joe Rudmin, Joe Scaturro, Alex Serway, John Serway, Giorgio Vianson, and Harold Zimmerman. The transformation of the manuscript into the finished text was only possible through the skilled, professional staff at Saunders College Publishing, especially Lloyd Black, developmental editor; Sally Kusch, project editor; and John Vondeling, sponsoring editor and associate publisher. I am most grateful for their hard work and encouragement in completing this project. I also appreciate the efforts and concerns of Joan Garbutt, who patiently worked with the first draft of the manuscript. I would like to acknowledge the contributions of photographs by Lloyd Black, Jim Lehman, and Hugh Strickland.

I am most grateful to the hundreds of students at Clarkson College of Technology and James Madison University who used this text in manuscript form during its development. Many students provided critiques that only a student can give. This feedback was very useful in preparing the final manuscript.

Finally, I am most fortunate to have a family who have continued to share both my joys and sorrows. The completion of this enormous task would not have been possible without their endless love and faith in me.

RAYMOND A. SERWAY
James Madison University
Harrisonburg, VA

To The Student

As I mentioned in the preface, (which I hope you have read), this book contains many features which should be of benefit to you, the student. Therefore, I feel it is appropriate to offer some words of advice which may be useful for understanding the material in the text. These comments are based upon my personal experiences in teaching this course over the last fourteen years.

Maintain a positive attitude towards the subject matter, keeping in mind that physics is the most fundamental of all natural sciences. Other science and engineering courses that follow will use the same physical principles, so it is important that you understand the various concepts, formalisms, and applications discussed in the text.

In order to obtain a thorough understanding of the concepts and principles, you should read the text carefully. Keep in mind that few people are able to absorb the full meaning of scientific writing after one reading. Several readings of the text and your lecture notes may be necessary. Your lectures and laboratory work should supplement the text and should help clarify some of the more difficult material. Memorizing equations, derivations, and definitions presented in the text or in class (in itself) does not necessarily mean you really understand the material. You will increase your level of understanding through a combination of efficient study habits, discussions with other students and instructors, and asking questions when you feel it is necessary. If you are reluctant to ask questions in class, seek private consultations. You will be surprised to find how easily concepts can be learned on a one-to-one basis.

It is important to set up a regular study schedule, preferably on a daily basis. Make sure to stick to the schedule set by your instructor. The lectures will be much more meaningful if you read the corresponding textual material in advance. A good rule of thumb to follow is at least two hours of study for every hour in class. If you are having trouble with the course, seek the advice of the instructor or students who have taken the course. You may find it necessary to seek instruction from more experienced students, and very often review sessions may be offered. In any event, you should try to avoid the practice of delaying study until a few days before an exam. This will often lead to disastrous results. If you feel in need of additional help in understanding the concepts, or in problem-solving, I suggest that you obtain the student study guide as a supplement to the text. The guide, which is keyed to the text, contains statements of chapter objectives, review lists, a review of concepts and equations, answers to selected questions from the text, worked examples, and programmed exercises. The programmed exercises are intended to serve as a self-test of the concepts and your ability to solve problems.

You should make full use of the various features of the text discussed in the preface. For example, marginal notes are useful for reviewing key concepts and definitions, while the appendices provide a review of mathematics and many useful tables. Note that answers to the odd exercises and problems are given at the back of the text. An overview of the entire text is given in the table of contents, while the index will enable you to quickly locate specific material. Footnotes are sometimes used to add notes of interest to the text, or to cite other references on the subject.

R. P. Feynman, Nobel laureate in physics, once said, "You do not know anything until you have practiced." In keeping with this statement, the most important skill you must develop from this course is the ability to solve problems. Your instructor will probably assign 8 or 10 problems per week in this course. You should try to solve as many exercises and problems as possible. Your ability to solve problems will be one of the main tests of your knowledge of physics. It is essential that you understand basic concepts and principles before attempting to solve the problems. It is good practice to try to find alternate solutions to the same problem. For example, many problems in mechanics can be solved using Newton's law of motion, but very often an alternate method using energy considerations is more direct. You should not deceive yourself into thinking you understand the problem after seeing its solution in class. If a problem involves several concepts, be sure to carefully follow a systematic plan in your solution. Always read the problem several times until you are confident you understand the question, and then note the information provided. Your ability to properly interpret the question is an integral part of problem-solving. Finally, decide on the method you feel is applicable to the problem, and proceed with your solution. If you are not successful, it would be wise to reread some portions of the text. Note that exercises are keyed to specific sections in order to simplify the process of obtaining information from the text.

Very often, students fail to recognize the limitations of certain formulas or physical laws in a particular situation. It is very important that you remember the assumptions which underlie various developments. For example, the equations of kinematics in linear motion apply only to a particle moving with constant acceleration. There are many motions for which the acceleration is not constant, such as the motion of an object attached to a spring, or the motion of an object through a resistive medium. In such cases, you must use the more general approach which involves solving the equation of motion.

Physics is a science based upon experimental observations. Therefore I recommend that you try to supplement the text through models and experiments, whenever possible. These home or laboratory experiments can be used to test ideas and models discussed in the text or in class. For example, the common "Slinky" toy is invaluable for demonstrating traveling waves; a piece of string and a ball can be used to investigate pendulum motion; an old pair of Polaroid sunglasses and some discarded lenses and magnifying glass are the components of various experiments in optics; elastic collisions can be demonstrated by studying billiard ball collisions in the pool room, with the addition of a paper-covered table to provide a permanent record of the collisions. The list is endless. When physical models are not available, try to develop "mental" models, and devise thought-provoking experiments to improve your understanding of the concepts or the situation at hand.

It is my hope that you will enjoy reading this text and profit from its content. After you have completed the course, I hope that you will have a good understanding of the ideas of physics, and its application to many real world situations.

Welcome to the exciting world of physics.

Contents Overview

° *These sections are optional.*

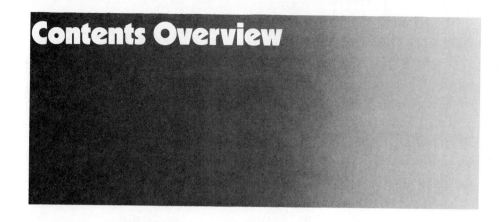

Table of Contents

° *These sections are optional.*

1 Introduction: Physics and Measurement

National Bureau of Standards

Physics is a fundamental science concerned with understanding the natural phenomena that occur in our universe. It is a science based on experimental observations and quantitative measurements. The main objective of the scientific approach is to develop physical theories based on some model and on fundamental laws that will predict the results of some experiments. Fortunately, it is possible to explain the behavior of many physical systems with a limited number of fundamental laws. These fundamental laws are expressed in the language of mathematics, the tool that provides a bridge between theory and experiment.

Whenever a discrepancy arises between theory and experiment, new theories and concepts must be formulated to explain the discrepancy. Many times a theory is satisfactory under limited conditions; a more general theory might be satisfactory without such limitations. A classic example is Newton's laws of motion, which accurately describe the motion of bodies at normal speeds but do not apply to objects moving at speeds comparable to the speed of light. The special theory of relativity developed by Albert Einstein (1879–1955) successfully predicts the motion of objects at speeds approaching the speed of light and hence is a more general theory of motion.

The bulk of this text will be concerned with the study of classical physics, although many contemporary applications will be incorporated in our treatment. Classical physics, developed prior to 1900, includes the theories, concepts, and experiments in three major disciplines: (1) classical mechanics, (2) thermodynamics (heat transfer, temperature, and the behavior of a large number of particles), and (3) electromagnetism (the study of electric and magnetic phenomena and radiation).

Galileo Galilei (1564–1642) made the first significant contributions to classical mechanics through his work on the laws of motion in the presence of constant acceleration. In the same era, Johannes Kepler (1571–1630) used astronomical observations to develop empirical laws for the motion of planetary bodies.

The most important contributions to classical mechanics were provided by Isaac Newton (1642–1727), who developed classical mechanics as a systematic theory and was one of the originators of the calculus as a mathematical tool. Major developments in classical physics continued in the 18th century. However, thermodynamics and electricity and magnetism were not well understood until the latter part of the 19th century, principally because the apparatus for controlled experiments was either too crude or unavailable until then. Although many electric and magnetic phenomena had been studied earlier, it was the work of James Clerk Maxwell (1831–1879) that provided a unified theory of electromagnetism. In this text we will treat the various disciplines of classical physics in separate sections; however, we will see that the disciplines of mechanics and electromagnetism are basic to all the branches of classical and modern physics.

A new era in physics, usually referred to as *modern physics*, began near the end of the 19th century. Modern physics developed mainly because of the discovery that many physical phenomena could not be explained by classical physics. The two most important developments in this modern era were the theories of relativity and quantum mechanics. Einstein's theory of relativity completely revolutionized the traditional concepts of space, time, and energy. Among other things, this theory corrected Newton's laws of motion for describing the motion of objects moving at speeds comparable to the speed of light. The theory of relativity also assumes that the speed of light is the upper limit of the speed of an object or signal and shows the equivalence of mass and energy. The formulation of quantum mechanics by a number of distinguished scientists provided a description of physical phenomena at the atomic level.

Scientists are constantly working at improving our understanding of fundamental laws, and new discoveries are being made every day. In many research areas, there is a great deal of overlap between physics, chemistry, and biology. The many technological developments that we have benefited from in recent times are the result of the efforts of many teams made up of scientists, engineers, and technicians. Some of the most notable recent developments are (1) unmanned space missions and manned moon landings, (2) microcircuitry and high-speed computers, and (3) nuclear energy. The impact of such developments and discoveries on our society has indeed been great, and it is very likely that future discoveries and developments will be exciting, challenging, and of great benefit to humanity.

1.1 STANDARDS OF LENGTH, MASS, AND TIME

The laws of physics are expressed in terms of basic quantities that require a clear definition. For example, such physical quantities as force, velocity, volume, and acceleration can be described in terms of more fundamental quantities which themselves are defined in terms of measurements or comparison with established standards. In mechanics, the three fundamental quantities are length (L), time (T), and mass (M). All other physical quantities are derived from these.

Obviously, if we are to report the results of a measurement to someone who wishes to reproduce this measurement, a standard must be defined. It would be meaningless if a visitor from another planet were to talk to us about a length of 8 "gliches" if we do not know the meaning of the unit "glich." On the other hand, if someone familiar with our system of measurement and weights reports that a wall is 2.0 meters high and our unit of length is defined as 1.0 meter, we then know that the height of the wall is twice our fundamental unit of length. Likewise, if we are told that a person has a mass of 75 kilograms and our unit of mass is defined as 1.0 kilogram, then that person is 75 times as massive as our fundamental unit of mass.[1]

In 1960, an international committee established rules to decide on a set of standards for these fundamental quantities. The system that was established is an adaptation of the metric system, and it is called the *International System* (SI) of units. In this system, the units of mass, length, and time are the kilogram, meter, and second, respectively. Other standard SI units established by the committee are those for temperature (the *kelvin*), electric current (the *ampere*), and luminous intensity (the

[1]The need for assigning numerical values to various physical quantities through experimentation was expressed by Lord Kelvin (William Thomson) as follows: "I often say that when you can measure what you are speaking about, and express it in numbers, you should know something about it, but when you cannot express it in numbers, your knowledge is of a megre and unsatisfactory kind."

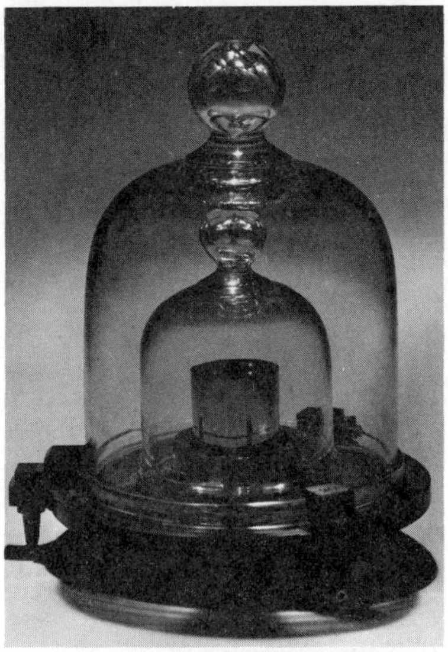

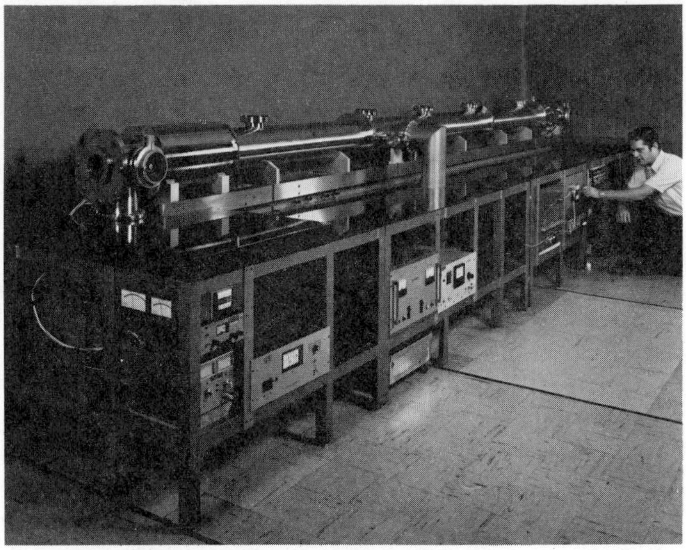

Figure 1.1A *(Left)* The National Standard Kilogram No. 20, an accurate copy of the International Standard Kilogram kept at Sèvres, France, is housed under a double bell jar in a vault at the National Bureau of Standards. (Courtesy National Bureau of Standards, U.S. Dept. of Commerce) *(Right)* The primary frequency standard (atomic clock) at the National Bureau of Standards. When operated as a clock, this device keeps time with an accuracy of about 3 millionths of a second per year. (Courtesy National Bureau of Standards, U.S. Dept. of Commerce)

candela). These six fundamental units are the basic SI units. In the study of mechanics, however, we will be concerned only with the units of mass, length, and time.

The SI unit of mass, the *kilogram,* is defined as the mass of a specific platinum-iridium alloy cylinder kept at the International Bureau of Weights and Measures at Sèvres, France. This mass standard was established in 1901, and there has been no change since that time because platinum-iridium is an unusually stable alloy. The Sèvres cylinder is 3.9 centimeters in diameter and 3.9 centimeters in height. A duplicate is kept at the National Bureau of Standards in Gaithersburg, Md.

Before 1960, the standard for length, the *meter,* was defined as the distance between two lines on a specific platinum-iridium bar stored under controlled conditions. This standard was abandoned for several reasons, a principal one being that the limited accuracy with which the separation between the lines on the bar can be determined does not meet the present requirements of science and technology. Because of the high precision and reproducibility available in optical interferometers, the standard meter was redefined at the 1960 General Conference on Weights and Measures as follows:

> **One meter**—a length equal to 1 650 763.73 wavelengths in vacuum of the radiation corresponding to the transition between the levels $2p_{10}$ and $5d_s$ of the krypton-86 atom.

This spectral line occurs in the red region of the spectrum and represents a standard that can be readily reproduced and measured to one part in 10^9.

Before 1960, the standard of time was defined in terms of the *mean solar day,* which is the length of a day measured throughout the year.[2] Thus, the *mean solar second,* representing the basic unit of time, was originally defined as $(\frac{1}{60})(\frac{1}{60})(\frac{1}{24})$ of a mean solar day. Time that is referenced to the rotation of the earth about its axis is called *universal time.*

In 1967, the second was redefined to take advantage of the high precision that

[2]A solar day is the time interval between successive appearances of the sun overhead.

could be obtained using a device known as an *atomic clock*. In this device, the frequencies associated with certain atomic transitions (which are extremely stable and insensitive to the clock's environment) can be measured to an accuracy of one part in 10^{12}. This is equivalent to an uncertainty of less than one second every 30 000 years. Such frequencies are highly insensitive to changes in the clock's environment. Thus, in 1967 the SI unit of time, the *second*, was redefined using the characteristic frequency of a particular kind of cesium atom as the "reference clock":

> **One second**—the duration of 9 192 631 770 cycles corresponding to the transition between two particular hyperfine levels of the ground state of the cesium-133 atom.

This new standard has the distinct advantage of being "indestructible" and widely reproducible.

The orders of magnitude (approximate values) of various masses, lengths, and time intervals are presented in Tables 1.1 to 1.3. Note the wide range of these quantities.[3] You should study these tables and get a feel for what is meant by a kilogram of mass, for example, or by a time interval of 10^{10} seconds. Systems of units commonly used are the SI or *mks* system, in which the units of mass, length, and time are the kilogram (kg), meter (m), and second (s), respectively; the *cgs* or Gaussian system, in which the units of length, mass, and time are the centimeter (cm), gram (g), and second, respectively; and the British engineering system, in which the units of length, mass, and time are the foot (ft), slug, and second, respectively. Throughout most of this text we will use SI units since they are almost universally

TABLE 1.1 Mass of Various Bodies (Approximate Values)

	MASS (kg)
Sun	2×10^{30}
Earth	6×10^{24}
Moon	7×10^{22}
Shark	1×10^{4}
Human	7×10^{1}
Frog	1×10^{-1}
Mosquito	1×10^{-5}
Hydrogen atom	1.67×10^{-27}
Electron	9.11×10^{-31}

[3] If you are unfamiliar with the use of powers of ten (scientific notation), you should review Section B.1 of the mathematical appendix at the back of this book.

TABLE 1.2 Approximate Values of Some Measured Lengths

	LENGTH (m)
Distance from earth to most remote known normal galaxies	3×10^{26}
Distance from earth to nearest large galaxy (M 31 in Andromeda)	2×10^{22}
Distance from earth to nearest star (Alpha Centauri)	4.3×10^{16}
Mean orbit radius of the earth	1.5×10^{11}
Mean distance from earth to moon	3.8×10^{8}
Mean radius of the earth	6.4×10^{6}
Typical altitude of orbiting earth satellite	2×10^{5}
Length of a football field	9.1×10^{1}
Length of a housefly	5×10^{-3}
Size of smallest dust particles	1×10^{-4}
Size of cells of most living organisms	1×10^{-5}
Diameter of a hydrogen atom	1×10^{-10}
Diameter of an atomic nucleus	1×10^{-14}

TABLE 1.3 Approximate Values of Some Time Intervals

	INTERVAL (s)
Age of the earth	1.3×10^{17}
Average age of a college student	6.3×10^{8}
One day (time for one revolution of earth about its axis)	8.6×10^{4}
Time between normal heartbeats	8×10^{-1}
Period[a] of audible sound waves	1×10^{-3}
Period of typical radio waves	1×10^{-6}
Period of vibration of an atom in a solid	1×10^{-13}
Period of visible light waves	2×10^{-15}
Duration of a nuclear collision	1×10^{-22}

[a] Period is defined as the time interval of one complete vibration.

accepted in science and industry. We will make some limited use of the British engineering units in the study of classical mechanics.

Some of the most frequently used prefixes for the various powers of ten and their abbreviations are listed in Table 1.4. For example, 10^{-3} m is equivalent to 1 millimeter (mm), and 10^3 m is 1 kilometer (km). Likewise, 1 kg is 10^3 g, and 1 megavolt (MV) is 10^6 volts.

TABLE 1.4 Some Prefixes for Powers of Ten

POWER	*PREFIX*	*ABBREVIATION*
10^{-12}	pico	p
10^{-9}	nano	n
10^{-6}	micro	μ
10^{-3}	milli	m
10^{-2}	centi	c
10^{3}	kilo	k
10^{6}	mega	M
10^{9}	giga	G
10^{12}	tera	T

Q1. What types of natural phenomena could serve as alternative time standards?

Q2. The height of a horse is sometimes given in units of "hands." Why do you suppose this is a poor standard of length?

Q3. Express the following quantities using the prefixes given in Table 1.4: (a) 3×10^{-4} m, (b) 5×10^{-5} s, (c) 72×10^2 g.

1.2 DENSITY AND ATOMIC MASS

Any piece of matter tends to resist any change in its motion. This property of matter is called *inertia*. The word *mass* is used to describe the amount of inertia associated with a particular body.

A more fundamental property of any substance is its *density* ρ (Greek letter rho), defined as *mass per unit volume:*

Density

$$\rho \equiv \frac{m}{V} \tag{1.1}$$

For example, aluminum has a density of 2.70 g/cm^3, and lead has a density of 11.3 g/cm^3. Therefore, a piece of aluminum of volume 10 cm^3 has a mass of 27.0 g, while an equivalent volume of lead would have a mass of 113 g. A list of densities for various substances is given in Table 1.5.

TABLE 1.5 Densities of Various Substances

SUBSTANCE	*DENSITY* ρ (g/cm^3)
Gold	19.3
Uranium	18.7
Lead	11.3
Copper	8.93
Iron	7.86
Aluminum	2.70
Magnesium	1.75
Water	1.00
Air	0.0013

The difference in density between aluminum and lead is due, in part, to their different *atomic weights;* the atomic weight of lead is 207 and that of aluminum is 27. However, the ratio of atomic weights, $207/27 = 7.67$, does not correspond to the ratio of densities, $11.3/2.70 = 4.19$. The discrepancy is due to the difference in atomic spacings and atomic arrangements in their crystal structures.

All ordinary matter consists of atoms, and each atom is made up of electrons and a nucleus. Practically all of the mass of an atom is contained in the nucleus, which consists of protons and neutrons. Thus we can understand why the atomic weights of the various elements differ. The mass of a nucleus is measured relative to the mass of an atom of the carbon-12 isotope (carbon has six protons and six neutrons).

The mass of ^{12}C is defined to be 12 atomic mass units (amu), where 1 amu $= 1.66 \times 10^{-27}$ kg. In these units, the proton and neutron have masses of about 1 amu. More precisely,

$$\text{mass of proton} = 1.0073 \text{ amu}$$
$$\text{mass of neutron} = 1.0087 \text{ amu}$$

The mass of the nucleus of ^{27}Al is *approximately* 27 amu. In fact, a more precise calculation shows that the nuclear mass is always slightly *less* than the combined mass of the protons and neutrons making up the nucleus. The processes of nuclear fission and nuclear fusion are based on this mass difference.

One mole of any element (or compound) consists of Avogadro's number, N_0, of molecules of the substance. Avogadro's number is defined so that one mole of carbon-12 atoms has a mass of exactly 12 g. Its value has been found to be $N_0 = 6.02$

$\times\ 10^{23}$ molecules/mole. For example, one mole of aluminum has a mass of 27 g, and one mole of lead has a mass of 207 g. Although the two have different masses, one mole of aluminum contains the same number of atoms as one mole of lead. Since there are 6.02×10^{23} atoms in one mole of any element, the mass per atom is given by

Atomic mass

$$m = \frac{\text{atomic weight}}{N_0} \tag{1.2}$$

For example, the mass of an aluminum atom is

$$m = \frac{27\ \text{g/mole}}{6.02 \times 10^{23}\ \text{atoms/mole}} = 4.5 \times 10^{-23}\ \text{g/atom}$$

Note that 1 amu is equal to $1/N_0$ g.

Example 1.1

A solid cube of aluminum (density 2.7 g/cm^3) has a volume of 0.2 cm^3. How many aluminum atoms are contained in the cube?

Solution: Since the density equals mass per unit volume, the mass of the cube is

$$\rho V = (2.7\ \text{g/cm}^3)(0.2\ \text{cm}^3) = 0.54\ \text{g}.$$

To find the number of atoms, N, we can set up a proportion using the fact that one mole of aluminum (27 g) contains 6.02×10^{23} atoms:

$$\frac{6.02 \times 10^{23}\ \text{atoms}}{27\ \text{g}} = \frac{N}{0.54\ \text{g}}$$

$$N = \frac{(0.54\ \text{g})(6.02 \times 10^{23}\ \text{atoms})}{27\ \text{g}} = 1.2 \times 10^{22}\ \text{atoms}$$

1.3 DIMENSIONAL ANALYSIS

The word *dimension* has a special meaning in physics. It denotes the qualitative nature of a physical quantity. Whether a distance is measured in units of feet or meters or furlongs, it is a distance. We say its dimension is *length.*

The symbols that will be used to specify length, mass, and time are L, M, and T, respectively. We will often use brackets [ ] to denote the dimensions of a physical quantity. For example, in this notation the dimensions of velocity, v, are written $[v] = \text{L/T}$, and the dimensions of area, A, are $[A] = \text{L}^2$. The dimensions of area, volume, velocity, and acceleration are listed in Table 1.6, along with their units in the three common systems. The dimensions of other quantities, such as force and energy, will be described as they are introduced in the text.

In many situations, you may be faced with having to derive or check a specific formula. Although you may have forgotten the details of the derivation, there is a useful and powerful procedure called *dimensional analysis* that can be used to assist in the derivation or to check your final expression. This procedure should be used whenever an equation is not understood and should help minimize the rote memory of equations. Dimensional analysis makes use of the fact that *dimensions can be treated as algebraic quantities.*

To illustrate this procedure, suppose you wish to derive a formula for the distance x traveled by a car in a time t if the car starts from rest and moves with constant acceleration a. In Chapter 3, we will find that the correct expression for this special

TABLE 1.6 Dimensions of Area, Volume, Velocity, and Acceleration

SYSTEM	*AREA* (L^2)	*VOLUME* (L^3)	*VELOCITY* (L/T)	*ACCELERATION* (L/T^2)
SI	m^2	m^3	m/s	m/s^2
cgs	cm^2	cm^3	cm/s	cm/s^2
British engineering	ft^2	ft^3	ft/s	ft/s^2

case is $x = \frac{1}{2} at^2$. The procedure of dimensional analysis is to set up an expression of the form

$$x \sim a^n t^m$$

where n and m are exponents that must be determined and the symbol $\sim$ indicates a proportionality. This relationship is only correct if the dimensions of both sides are the same. Since the dimension of the left side is length, the dimension of the right side must also be length. That is,

$$[a^n t^m] = \mathrm{L}$$

Since the dimensions of acceleration are $\mathrm{L/T^2}$ and the dimension of time is T,

$$(\mathrm{L/T^2})^n \mathrm{T}^m = \mathrm{L}$$

or

$$\mathrm{L}^n \mathrm{T}^{m-2n} = \mathrm{L}$$

Since the exponents of L and T must be the same on both sides, we see that $n = 1$ and $m = 2$. Therefore, we conclude that

$$x \sim at^2$$

This result is off by a factor of 2 from the correct expression, which is $x = \frac{1}{2} at^2$.

Example 1.2

Show that the expression $v = v_0 + at$ is dimensionally correct, where v and v_0 represent velocities, a is acceleration, and t is a time interval.

Solution: Since

$$[v] = [v_0] = \mathrm{L/T}$$

and the dimensions of acceleration are $\mathrm{L/T^2}$, the dimensions of at are

$$[at] = (\mathrm{L/T^2})(\mathrm{T}) = \mathrm{L/T}$$

and the expression is dimensionally correct. On the other hand, if the expression were given as $v = v_0 + at^2$, it would be dimensionally *incorrect*. Try it and see!

Example 1.3

Suppose we are told that the acceleration of a particle moving in a circle of radius r with uniform velocity v is proportional to some power of r, say r^n, and some power of v, say v^m. How can we determine the power of r and v?

Solution: Let us take a to be

$$a = kr^n v^m$$

where k is a dimensionless constant. With the known dimensions of a, r, and v, we see that the dimensional equation must be

$$\mathrm{L/T^2} = \mathrm{L}^n(\mathrm{L/T})^m = \mathrm{L}^{n+m}/\mathrm{T}^m$$

This dimensional equation is balanced under the conditions

$$n + m = 1 \quad \text{and} \quad m = 2$$

Therefore, $n = -1$ and we can write the acceleration

$$a = kr^{-1}v^2 = k\frac{v^2}{r}$$

When we discuss uniform circular motion later, we will see that $k = 1$.

Q4. Does a dimensional analysis give any information on constants of proportionality that may appear in an algebraic expression? Explain.

1.4 CONVERSION OF UNITS

Sometimes it is necessary to convert units from one system to another. Conversion factors between the SI and English systems for units of length are as follows:

1 mile = 1609 m = 1.609 km
1 m = 39.37 in. = 3.281 ft
1 ft = 0.3048 m = 30.48 cm
1 in. = 0.0254 m = 2.540 cm

A more complete list of conversion factors can be found in Appendix A. Units can be treated as algebraic quantities that can cancel each other. For example, suppose we wish to convert 15.0 in. to centimeters. Since 1 in. = 2.54 cm (exactly), we find that

$$15.0\text{ in.} = (15.0\,\cancel{\text{in.}})\left(2.54\,\frac{\text{cm}}{\cancel{\text{in.}}}\right) = 38.1\text{ cm}$$

Example 1.4

The mass of a solid cube is 856 g and each edge has a length of 5.35 cm. Determine the density ρ of the cube in SI units.

Solution: Since 1 g = 10^{-3} kg and 1 cm = 10^{-2} m, the mass, m, and volume, V, in SI units are given by

$$m = 856\,\cancel{\text{g}} \times 10^{-3}\text{ kg}/\cancel{\text{g}} = 0.856\text{ kg}$$

$$V = L^3 = (5.35\,\cancel{\text{cm}} \times 10^{-2}\text{ m}/\cancel{\text{cm}})^3$$

$$= (5.35)^3 \times 10^{-6}\text{ m}^3 = 1.53 \times 10^{-4}\text{ m}^3$$

Therefore

$$\rho = \frac{m}{V} = \frac{0.856\text{ kg}}{1.53 \times 10^{-4}\text{ m}^3} = 5.60 \times 10^3\,\frac{\text{kg}}{\text{m}^3}$$

Q5. Suppose that two quantities A and B have different dimensions. Determine which of the following arithmetic operations *could* be physically meaningful: (a) $A+B$, (b) A/B, (c) $B-A$, (d) AB.

1.5 ORDER-OF-MAGNITUDE CALCULATIONS

It is often useful to compute an approximate answer to a given physical situation even where little information is available. Such results can then be used to determine whether or not a more precise calculation is necessary. These approximations are usually based on certain assumptions, which must be modified if more precision is needed. Thus, we will sometimes refer to an *order of magnitude* of a certain quantity as the power of ten of the number that describes that quantity. Usually, when an order-of-magnitude calculation is made, the results are reliable to within a factor of 10. If a quantity increases in value by three orders of magnitude, this means that its value is increased by a factor of $10^3 = 1000$.

The spirit of attempting order-of-magnitude calculations, sometimes referred to as "guesstimates," is given in the following quotation: "Make an estimate before every calculation, try a simple physical argument . . . before every derivation, guess the answer to every puzzle. Courage: no one else needs to know what the guess is."[4]

Example 1.5

Estimate the number of atoms in 1 cm^3 of a solid.

Solution: From Table 1.2 we note that the diameter of an atom is about 10^{-10} m. Thus, if in our model we assume that the atoms in the solid are solid spheres of this diameter, then the volume of each sphere is about 10^{-30} m^3 (more precisely, volume $= \frac{4\pi}{3}r^3 = \frac{\pi}{6}d^3$, where $r = d/2$). Therefore, since 1 cm^3 = 10^{-6} m^3, the number of atoms in the solid is of the order of $10^{-6}/10^{-30} = 10^{24}$ atoms.

A more precise calculation would require knowledge of the density of the solid and the mass of each atom. However, our estimate agrees with the more precise calculation to within a factor of 10. (For Exercise 16, this same approach should be used.)

Example 1.6

Estimate the number of gallons of gasoline used by all U.S. cars each year.

Solution: Since there are about 200 million people in the United States, an estimate of the number of cars in the country is 40 million (assuming one car and five people per family). We must also estimate that the average distance traveled per year is 10 000 miles. If we assume a gasoline consumption of 20 mi/gal, each car uses about 500 gal/year. Multiplying this by the total number of cars in the United States gives an estimated total consumption of 2×10^{10} gal. This corresponds to a yearly consumer expenditure of over 20 billion dollars! This is probably a low estimate since we haven't accounted for commercial consumption and for such factors as two-car families.

[4] E. Taylor and J.A. Wheeler, *Spacetime Physics*, San Francisco, W.H. Freeman, 1966, p. 60.

Q6. What accuracy is implied in an order-of-magnitude calculation?

Q7. Apply an order-of-magnitude calculation to an everyday situation you might encounter. For example, how far do you walk or drive each day?

1.6 SIGNIFICANT FIGURES

All real measurements of quantities have some degree of inaccuracy. Whenever a physical quantity is measured, both the value and the precision of the measured quantity are important. For example, if observer A measures the speed of an object to be 5.38 m/s to a precision of 1%, the result could be expressed (5.38 ± 0.05) m/s. Therefore the true value must lie between 5.33 m/s and 5.43 m/s. On the other hand, if an independent measurement is made on the same object by observer B, with a precision of only 3%, a value of (5.38 ± 0.16) m/s should be reported. In either case, all three digits in the measured value of 5.38 m/s are significant; however, the last digit is uncertain to some degree. This uncertainty will depend on many factors, such as the quality of the instruments used, experimental technique, and human error. The following rule should be followed when reporting the accuracy of a measurement: *The last figure in the measurement should be the first uncertain figure.* For example, it would be wrong to claim that the speed of an object is 5.384 m/s if the 8 is uncertain.

A few approximate rules regarding arithmetic operations should be mentioned. If two numbers are to be multiplied or divided, the result has the same number of significant figures as the *less accurate* number. For example, if we were to perform the multiplication 3.60×5.387, the result would be 19.4 and not 19.3932. We can only claim three significant figures since the less accurate number, 3.60, contains three significant figures.

When numbers are added or subtracted, the rule is somewhat different. If we wish to compute $115 + 8.35$, the result is 123 and not 123.35. The last significant figure of the result is in the same place relative to the decimal point as the last significant figure of the least accurate number in the sum or difference.

In this text we will assume that, unless stated otherwise, all data given in examples and problems are known to three significant figures, with an accuracy of 1%. For example, a reported length of 6.85 m implies a value (6.85 ± 0.07) m.

Example 1.7

A rectangular plate has a length of (21.3 ± 0.2) cm and a width of (9.80 ± 0.10) cm. Find the area of the plate and the uncertainty in the calculated area.

Solution:

$$\begin{aligned}\text{Area} = lw &= (21.3 \pm 0.2)\ \text{cm} \times (9.80 \pm 0.1)\ \text{cm} \\ &\approx (21.3 \times 9.80 \pm 21.3 \times 0.1 \pm 9.80 \times 0.2)\ \text{cm}^2 \\ &\approx (209 \pm 4)\ \text{cm}^2\end{aligned}$$

Note that the input data were given only to three significant figures, so we cannot claim any more in our result. Furthermore, you should realize that the uncertainty in the product (2%) is approximately equal to the sum of the uncertainties in the length and width (each uncertainty is about 1%).

As a general rule, the relative error in a product or quotient equals the *sum* of the relative errors of the quantities involved. By relative error, we mean the ratio of the absolute error to the quantity itself.[5]

1.7 MATHEMATICAL NOTATION

Many mathematical symbols will be used throughout this book, some of which you are surely aware of, such as the symbol = to denote the equality of two quantities.

[5]This rule is only an approximate one but it is adequate for most purposes. Nevertheless, the result is in reasonable agreement with that obtained from a detailed analysis of error processes.

The symbol $\sim$ is used to denote a proportionality. For example, $y \sim x^2$ means that y is proportional to the square of x.

The symbol $<$ means *less than*, and $>$ means *greater than*. For example, $x > y$ means x is greater than y.

The symbol $\ll$ means *much less than*, and $\gg$ means *much greater than*.

The symbol $\approx$ is used to indicate that two quantities are *approximately equal* to each other.

The symbol $\equiv$ means *is defined as*. This is a stronger statement than a simple $=$.

It is convenient to use a symbol to indicate the change in a quantity. For example, Δx (read delta x) means the *change in the quantity* x. (It does not mean the product of Δ and x). For example, if x_i is the initial position of a particle and x_f is its final position, then the *change in position* is written

$$\Delta x = x_f - x_i$$

We will often have occasion to sum several quantities. A useful abbreviation for representing such a sum is the Greek letter Σ (capital sigma). Suppose we wish to sum a set of five numbers represented by x_1, x_2, x_3, x_4, and x_5. In the abbreviated notation, we would write the sum

$$x_1 + x_2 + x_3 + x_4 + x_5 \equiv \sum_{i=1}^{5} x_i$$

where the subscript i on a particular x represents any one of the numbers in the set. For example, if there are five masses in a system, m_1, m_2, m_3, m_4, and m_5, the *total* mass of the system $M = m_1 + m_2 + m_3 + m_4 + m_5$ could be expressed

$$M = \sum_{i=1}^{5} m_i$$

Finally, the *magnitude* of a quantity x, written $|x|$, is simply the absolute value of that quantity. The magnitude of x is *always positive*, regardless of the sign of x. For example, if $x = -5$, $|x| = 5$; if $x = 8$, $|x| = 8$.

A list of these symbols and their meanings is given in Appendix 1.

1.8 SUMMARY

Mechanical quantities can be expressed in terms of three fundamental quantities, *mass, length*, and *time*, which have the units *kilograms* (kg), *meters* (m), and *seconds* (s), respectively, in the SI system. It is often useful to use the *method of dimensional analysis* to check equations and to assist in deriving expressions.

The *density* of a substance is defined as its *mass per unit volume*. Different substances have different densities mainly because of differences in their atomic masses and crystalline structures.

The number of atoms in one mole of any element or compound is called *Avogadro's number*, N_o, which has the value 6.02×10^{23} atoms/mole.

EXERCISES

Section 1.2 Density and Atomic Mass

1. Calculate the density of a solid cube that measures 5 cm on each side and has a mass of 350 g.

2. A solid sphere is to be made out of copper, which has a density of 8.93 g/cm^3. If the mass of the sphere is to be 475 g, what radius must it have?

3. A hollow cylindrical container has a length of 800 cm and an inner radius of 30 cm. If the cylinder is completely filled with water, what is the mass of the water? Assume 1.0 g/cm^3 as the density of water.

4. Calculate the mass of an atom of (a) helium, (b) iron, and (c) lead. Give your answers in amu and in g. The atomic weights are 4, 56, and 207, respectively, for the atoms given.

5. A small particle of iron in the shape of a cube is observed under a microscope. The edge of the cube is 5×10^{-6} cm long. Find (a) the mass of the cube and (b) the number of iron atoms in the particle. The atomic weight of iron is 56, and its density is 7.86 g/cm^3.

Section 1.3 Dimensional Analysis

6. Show that the expression $x = vt + \frac{1}{2} at^2$ is dimensionally correct, where x is a coordinate and has units of length, v is velocity, a is acceleration, and t is time.

7. The displacement of a particle when moving under uniform acceleration is some function of the time and the acceleration. Suppose we write this displacement $s = ka^m t^n$, where k is a dimensionless constant. Show by dimensional analysis that this expression is satisfied if $m = 1$ and $n = 2$. Can this analysis give the value of k?

8. The square of the speed of an object undergoing a uniform acceleration a is some function of a and the displacement s, according to the expression $v^2 = ka^m s^n$, where k is a dimensionless constant. Show by dimensional analysis that this expression is satisfied if $m = n = 1$.

9. Suppose that the displacement of a particle is related to time according to the expression $s = ct^3$. What are the dimensions of the constant c?

10. (a) One of the fundamental laws of motion states that the acceleration of an object is directly proportional to the resultant force on it and inversely proportional to its mass. From this statement, determine the dimensions of force. (b) The newton (N) is the SI unit of force. According to your results for (a), how can you express a newton using the SI fundamental units of mass, length, and time?

Section 1.4 Conversion of Units

11. Convert the volume 8.50 $in.^3$ to m^3, recalling that 1 in. = 2.54 cm and 1 cm = 10^{-2} m.

12. A rectangular building lot is 100.0 ft by 150.0 ft. Determine the area of this lot in m^2.

13. An object in the shape of a rectangular parallelepiped measures 2.0 in. $\times$ 3.5 in. $\times$ 6.5 in. Determine the volume of the object in m^3.

14. A creature moves at a speed of 5 furlongs per fortnight (not a very common unit of speed). Given that 1 furlong = 220 yards and 1 fortnight = 14 days, determine the speed of the creature in m/s. (The creature is probably a snail.)

15. A solid piece of lead has a mass of 23.94 g and a volume of 2.10 cm^3. From these data, calculate the density of lead in SI units (kg/m^3).

16. Estimate the age of the earth in years using the data in Table 1.3 and the appropriate conversion factors.

17. The proton, which is the nucleus of the hydrogen atom, can be pictured as a sphere of diameter 3×10^{-13} cm having a mass of 1.67×10^{-24} g. Determine the density of the proton in SI units and compare this number with the density of lead, 1.14×10^4 kg/m^3.

18. Using the fact that the speed of light in free space is about 3.00×10^8 m/s, determine how many miles a pulse from a laser beam will travel in one hour.

19. Radio waves are electromagnetic and travel at a speed of about 3.0×10^8 m/s in free space. Use this fact and the data in Table 1.2 to determine the time it would take an electromagnetic pulse to make a round trip from the earth to Alpha Centauri, the star nearest the earth.

20. The mean radius of the earth is 6.37×10^6 m, and that of the moon is 1.74×10^8 cm. From these data calculate (a) the ratio of the earth's surface area to that of the moon and (b) the ratio of the earth's volume to that of the moon. Recall that the surface area of a sphere is $4\pi r^2$ and the volume of a sphere is $\frac{4}{3}\pi r^3$.

21. The mass of a copper atom is 1.06×10^{-22} g, and the density of copper is 8.9 g/cm^3. Determine the number of atoms in 1 cm^3 of copper and compare the result with the estimate in Example 1.4.

22. Aluminum is a very lightweight metal, with a density of 2.7 g/cm^3. What is the weight in pounds of a solid sphere of aluminum of radius 50 cm? The result might surprise you. (Note: a 1-kg mass corresponds to a weight of 2.2 pounds.)

Section 1.5 Order-of-Magnitude Calculations

23. Estimate the total number of times the heart of a human beats in an average lifetime of 70 years. (See Table 1.3 for data.)

24. Estimate the number of ping-pong balls that can be packed into an average-size room (without crushing them).

25. Soft drinks are commonly sold in aluminum containers. Estimate the number of such containers thrown away each year by U.S. consumers. Approximately how many tons of aluminum does this represent?

26. Approximately how many raindrops fall on a 1-acre lot during a 1-in. rainfall?

27. Determine the approximate number of bricks needed to face all four sides of an average-size home.

28. Estimate the number of piano tuners living in New York City. This question was raised by E. Fermi.

Section 1.6 Significant Figures

29. If the length and width of a rectangular plate are measured to be (15.30 ± 0.05) cm and (12.80 ± 0.05) cm, find the area of the plate and the uncertainty in the calculated area.

30. The *radius* of a solid sphere is measured to be (6.50 ± 0.20) cm, and its mass is measured to be (1.85 ± 0.02) kg. Determine the density of the sphere in kg/m^3 and the uncertainty in the density.

31. Calculate (a) the circumference of a circle of radius 3.5 cm and (b) the area of a circle of radius 4.65 cm.

32. Carry out the following arithmetic operations: (a) the sum of the numbers 756, 37.2, 0.83, and 2.5; (b) the product 3.2×3.563; (c) the product $5.6 \times \pi$.

2 Vectors

Lloyd Black

Mathematics is the basic tool used by scientists and engineers to describe the behavior of physical systems. Physical quantities that have both numerical and directional properties are represented by vectors. This chapter is primarily concerned with vector algebra and with some general properties of vectors. The addition and subtraction of vectors will be discussed, together with some common applications to physical situations. Discussion of the product of vectors will be delayed until these operations are needed.[1]

Vectors will be used throughout this text, and it is imperative that you master both their graphical and algebraic properties.

2.1 COORDINATE SYSTEMS AND FRAMES OF REFERENCE

Many aspects of physics deal in some form or other with locations in space. For example, the mathematical description of the motion of an object requires a method for describing the position of the object. Thus, it is perhaps fitting that we first discuss how one describes the position of a point in space. This is accomplished by means of coordinates. A point on a line can be described with one coordinate. A point in a plane is located with two coordinates, and three coordinates are required to locate a point in space.

A coordinate system used to specify locations in space consists of

1. A fixed reference point O, called the origin
2. A set of specified axes or directions
3. Instructions that tell us how to label a point in space relative to the origin and axes.

One convenient coordinate system that we will use frequently is the *cartesian coordinate system,* sometimes called the *rectangular coordinate system.* Such a system in two dimensions is illustrated in Fig. 2.1. An arbitrary point in this system is labeled with the coordinates (x,y). For example, the point P, which has coordinates (5,3), may be reached by first going 5 units along the positive x axis and then 3 units along the positive y axis.[2] Similarly, the point Q has coordinates $(-3,4)$, correspond-

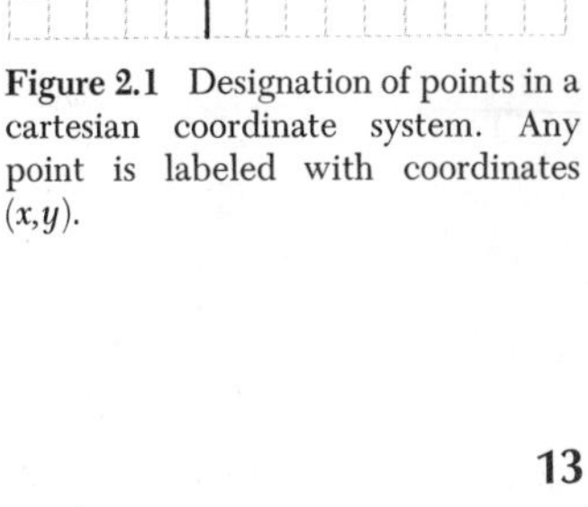

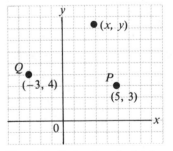

Figure 2.1 Designation of points in a cartesian coordinate system. Any point is labeled with coordinates (x,y).

[1]The dot product is discussed in Section 7.3, and the cross, or vector, product is introduced in Section 11.1.

[2]Positive x is taken to the right of the origin, and positive y is upward from the origin. Negative x is to the left of the origin, and negative y is downward from the origin.

ing to going 3 units in the negative x direction and 4 units in the positive y direction.

Sometimes it is more convenient to represent a point in a plane by its *plane polar coordinates*, (r,θ), as in Fig. 2.2a. In this coordinate system, r is the distance from the origin to the point having cartesian coordinates (x,y) and θ is the angle between r and a fixed axis, usually measured counterclockwise from the positive x axis. From the right triangle in Fig. 2.2b, we see that $\sin\theta = y/r$ and $\cos\theta = x/r$. (A review of trigonometric functions is given in Appendix B.2.) Therefore, starting with plane polar coordinates, one can obtain the cartesian coordinates through the equations

$$x = r\cos\theta \tag{2.1}$$

$$y = r\sin\theta \tag{2.2}$$

Furthermore, it follows that

$$\tan\theta = y/x \tag{2.3}$$

and

$$r = \sqrt{x^2 + y^2} \tag{2.4}$$

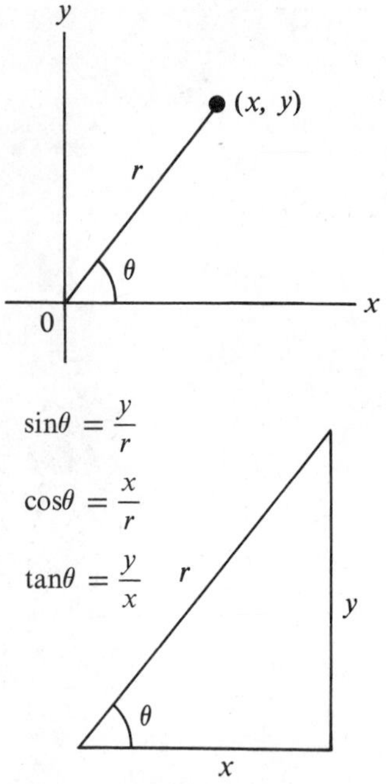

Figure 2.2 (a) The plane polar coordinates of a point are represented by the distance r and the angle θ. (b) The right triangle used to relate (x,y) to (r,θ).

You should note that these expressions relating the coordinates (x,y) to the coordinates (r,θ) apply only when θ is defined as in Fig. 2.2a, where positive θ is an angle measured *counterclockwise* from the positive x axis. If the reference axis for the polar angle θ is chosen to be other than the positive x axis, or the sense of increasing θ is chosen differently, then the corresponding expressions relating the two sets of coordinates will change.

Example 2.1

The cartesian coordinates of a point are given by $(x,y) = (-3.5, -2.5)$ units as in Fig. 2.3. Find the polar coordinates of this point.

Solution:

$$r = \sqrt{x^2 + y^2} = \sqrt{(-3.5)^2 + (-2.5)^2} = 4.3 \text{ units}$$

$$\tan\theta = \frac{y}{x} = \frac{-2.5}{-3.5} = 0.714$$

$$\theta = 216°$$

Note that you must use the signs of x and y to find that θ is in the third quadrant of the coordinate system. That is, $\theta = 216°$ and not 36°.

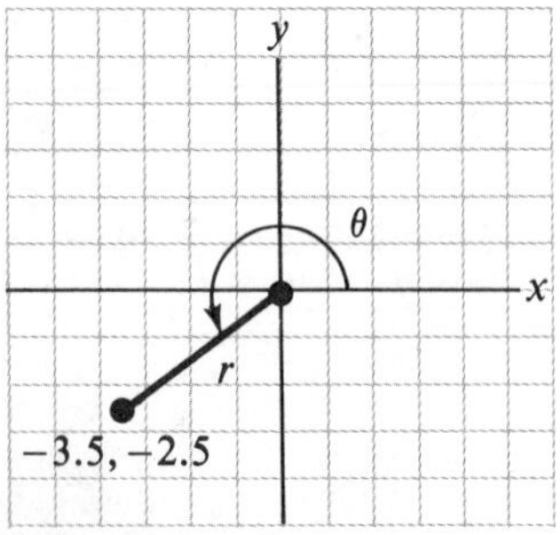

Figure 2.3 (Example 2.1)

2.2 VECTORS AND SCALARS

A *vector* is a physical quantity that is specified by its direction as well as its magnitude. For example, to describe completely the force on an object, we must specify both the direction of the applied force and a number to indicate the force's magnitude. When the motion of an object is described, we must specify both how fast it is moving and the direction of its motion.

A simple example of a vector quantity is the *displacement* of a particle, defined as the *change in the position* of the particle. Suppose the particle moves from some point O to the point P along a straight path, as in Fig. 2.4. We represent this displacement by drawing an arrow from O to P, where the tip of the arrow represents the direction of the displacement and the length of the arrow represents the magnitude of the displacement. If the particle travels along some other path from O to P, such as the broken line in Fig. 2.4, its displacement is still **OP**. The vector displacement along any indirect path from O to P is defined as being equivalent to the displacement for the direct path from O to P. Thus, *the displacement of a particle is completely known if its initial and final coordinates are known.* The path need not be specified. In other words, *the displacement is independent of the path.*

If the particle moves along the x axis from position x_i to position x_f, as in Fig. 2.5, its displacement is given by $x_f - x_i$. As mentioned in Chapter 1, we use the Greek letter delta (Δ) to denote the *change* in a quantity. Therefore, we write the change in the position of the particle (the displacement)

$$\Delta x = x_f - x_i \tag{2.5}$$

Definition of displacement along a line

From this definition, we see that Δx is positive if x_f is greater than x_i and negative if x_f is less than x_i. For example, if a particle changes its position from $x_i = -3$ units to $x_f = 5$ units, its displacement is 8 units.

There are many physical quantities in addition to displacement that are vectors. These include velocity, acceleration, force, and momentum, all of which will be defined in later chapters. In this text, we will use boldface letters, such as **A**, to represent an arbitrary vector. Another common method for vector notation that you should be aware of is to use an arrow over the letter: $\vec{A}$. The magnitude of the vector **A** is written A or, alternatively, $|\mathbf{A}|$. The magnitude of a vector has physical units, such as cm for displacement or m/s for velocity, as discussed in Chapter 1. Vectors combine according to special rules, which will be discussed in later sections.

A *scalar* is a quantity that is completely specified by a number with appropriate units. That is, a scalar has *only* magnitude, and *no* direction. Some physical examples of scalar quantities are mass, density, electric charge, volume, energy, temperature, and time intervals. The rules of ordinary arithmetic are used to manipulate scalar quantities.

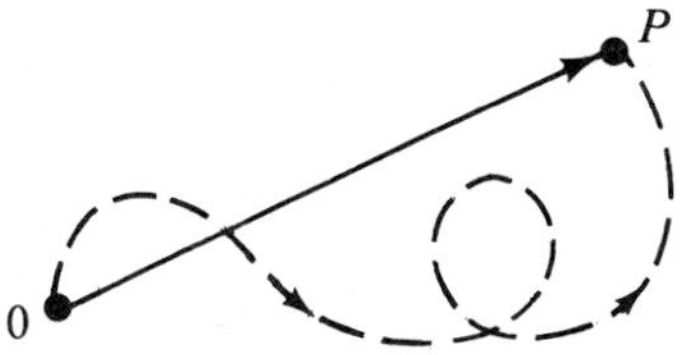

Figure 2.4 As a particle moves from O to P along the broken line, its displacement vector is the arrow drawn from O to P.

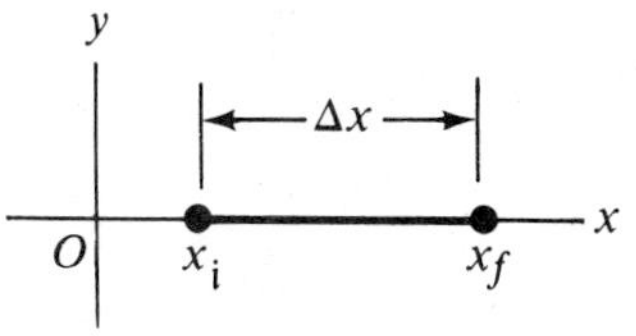

Figure 2.5 A particle moving along the x axis from x_i to x_f undergoes a displacement $\Delta x = x_f - x_i$.

2.3 SOME PROPERTIES OF VECTORS

Equality of Two Vectors. Two vectors **A** and **B** are defined to be equal if they have the same magnitude and point in the same direction. That is, $\mathbf{A} = \mathbf{B}$ only if $A = B$ *and* their directions are the same. For example, all the vectors in Fig. 2.6 are equal even though they have different starting points. This property allows us to translate a vector parallel to itself in a diagram without affecting the vector. In fact, any true vector can be moved parallel to itself.

Addition. When two or more vectors are added together, *all* vectors involved *must* have the same units. For example, it would be meaningless to add a velocity vector to a displacement vector since they are different physical quantities.

The rules for vector sums are conveniently described by geometric methods. To add vector **B** to vector **A**, first draw vector **A** on graph paper and then draw vector **B** with its tail starting from the tip of **A**, as in Fig. 2.7. The *resultant vector* $\mathbf{R} = \mathbf{A} + \mathbf{B}$ is the vector drawn from the tail of **A** to the tip of **B**. This is known as the *triangle method of addition.* An alternative graphical procedure for adding two vectors, known as the *parallelogram rule of addition,* is shown in Fig. 2.8. In this construction, the tails of the two vectors **A** and **B** are together and the resultant vector **R** is the diagonal of a parallelogram formed with **A** and **B** as its sides.

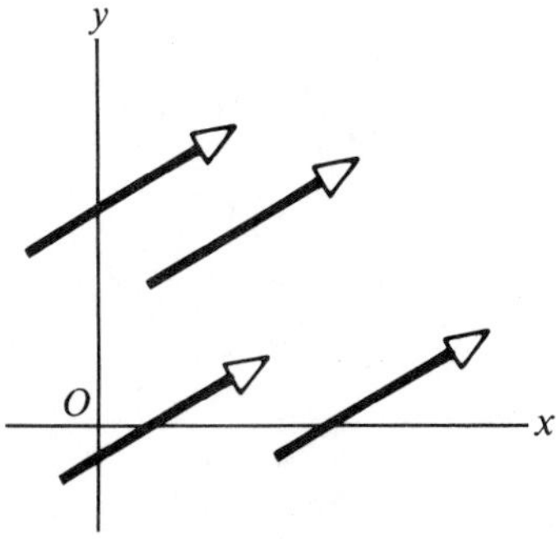

Figure 2.6 A representation of equal vectors.

When two vectors are added, the sum is independent of the order of the addition. This can be seen from the geometric construction in Fig. 2.8 and is known as the *commutative law of addition:*

$$\mathbf{A} + \mathbf{B} = \mathbf{B} + \mathbf{A} \tag{2.6}$$

Commutative law

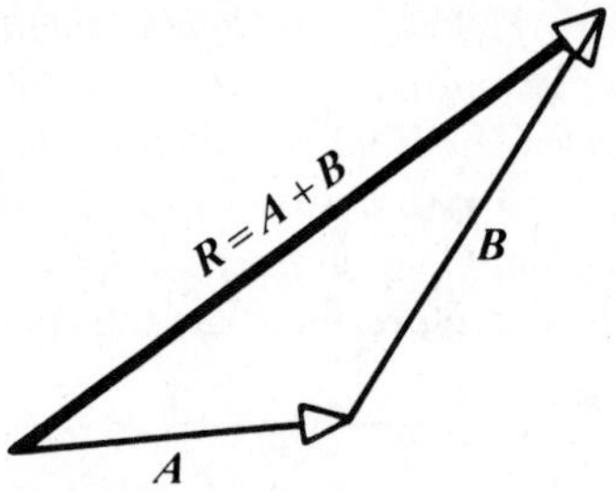

Figure 2.7 When vector $\boldsymbol{A}$ is added to vector $\boldsymbol{B}$, the vector sum $\boldsymbol{R}$ is the vector that runs from the tail of $\boldsymbol{A}$ to the tip of $\boldsymbol{B}$.

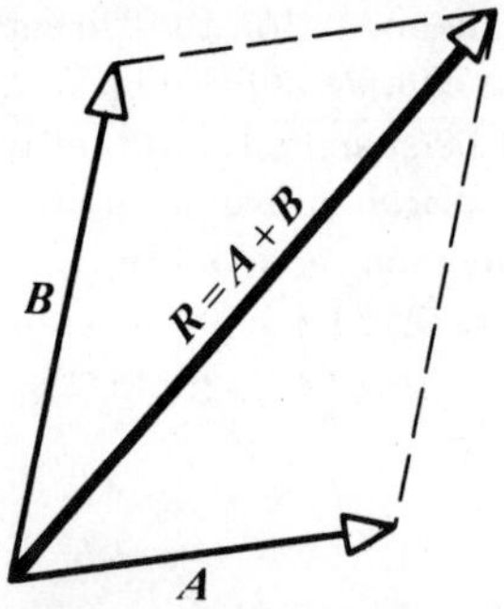

Figure 2.8 This construction shows that $\boldsymbol{A} + \boldsymbol{B} = \boldsymbol{B} + \boldsymbol{A}$. Note that the resultant $\boldsymbol{R}$ is the diagonal of a parallelogram with sides $\boldsymbol{A}$ and $\boldsymbol{B}$.

If three or more vectors are added, their sum is independent of the order in which the individual vectors are added to each other. A geometric proof of this for three vectors is given in Fig. 2.9. This is called the *associative law of addition:*

Associative law

$$\boldsymbol{A} + (\boldsymbol{B} + \boldsymbol{C}) = (\boldsymbol{A} + \boldsymbol{B}) + \boldsymbol{C} \tag{2.7}$$

Thus we conclude that *a vector is a quantity that has both magnitude and direction and also obeys the laws of vector addition* as described in Figs. 2.7 to 2.10.

Geometric constructions can also be used to add more than three vectors. This is shown in Fig. 2.10 for the case of four vectors. The resultant vector sum $\boldsymbol{R} = \boldsymbol{A} + \boldsymbol{B} + \boldsymbol{C} + \boldsymbol{D}$ is *the vector that completes the polygon.* In other words, $\boldsymbol{R}$ is *the vector drawn from the tail of the first vector to the tip of the last vector.* Again, the order of the summation is unimportant.

Negative of a Vector. The negative of the vector $\boldsymbol{A}$ is defined as the vector that when added to $\boldsymbol{A}$ gives zero. That is, $\boldsymbol{A} + (-\boldsymbol{A}) = 0$. The vectors $\boldsymbol{A}$ and $-\boldsymbol{A}$ have the same magnitude but point in opposite directions.

Subtraction of Vectors. The operation of vector subtraction makes use of the definition of the negative of a vector. We define the operation $\boldsymbol{A} - \boldsymbol{B}$ as vector $-\boldsymbol{B}$ added to vector $\boldsymbol{A}$:

$$\boldsymbol{A} - \boldsymbol{B} = \boldsymbol{A} + (-\boldsymbol{B}) \tag{2.8}$$

The geometric construction for subtracting two vectors is shown in Fig. 2.11.

Multiplication of a Vector by a Scalar. If a vector $\boldsymbol{A}$ is multiplied by a positive scalar quantity m, the product $m\boldsymbol{A}$ is a vector that has the same direction as $\boldsymbol{A}$ and

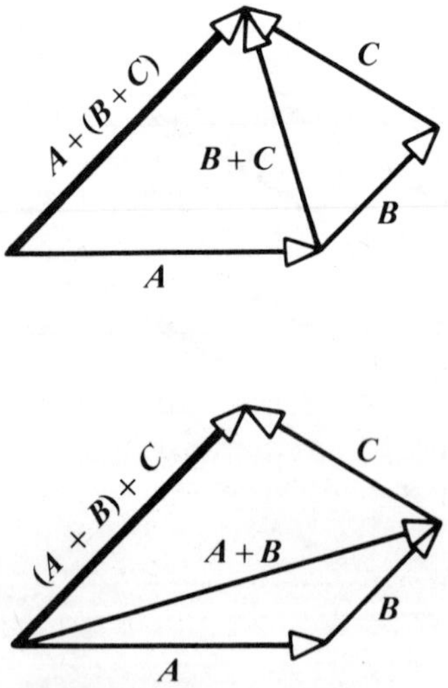

Figure 2.9 Geometric constructions for verifying the associative law of addition.

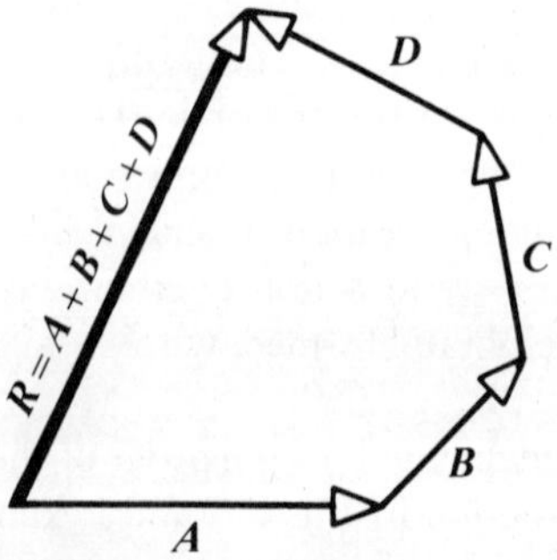

Figure 2.10 Geometric construction for summing four vectors. The resultant vector $\boldsymbol{R}$ completes the polygon.

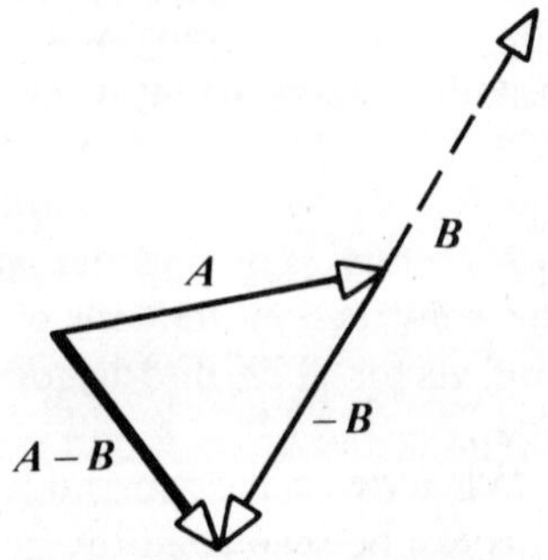

Figure 2.11 This construction shows how to subtract vector $\boldsymbol{B}$ from vector $\boldsymbol{A}$. The vector $-\boldsymbol{B}$ is equal to and opposite the vector $\boldsymbol{B}$.

magnitude mA. If m is a negative scalar quantity, the vector $m\mathbf{A}$ is directed opposite $\mathbf{A}$.

Example 2.2

A car travels 20.0 km due north and then 35.0 km in a direction 60° west of north, as in Fig. 2.12.[3] Find the magnitude and direction of the car's resultant displacement.

Solution: The problem can be solved geometrically using graph paper and a protractor, as shown in Fig. 2.12. The resultant displacement $\mathbf{R}$ is the sum of the two individual displacements $\mathbf{A}$ and $\mathbf{B}$.

An algebraic solution for the magnitude of R can be obtained using the law of cosines from trigonometry as applied to the obtuse triangle (Appendix B.4). Since $\theta = 180° - 60° = 120°$ and $R^2 = A^2 + B^2 - 2AB\cos\theta$, we find that

$$R = \sqrt{A^2 + B^2 - 2\,AB\cos\theta}$$

$$= \sqrt{(20)^2 + (35)^2 - 2(20)(35)\cos 120°} = 48.2 \text{ km}$$

The direction of $\mathbf{R}$ measured from the northerly direction can be obtained from the law of sines from trigonometry:

$$\frac{\sin\beta}{B} = \frac{\sin\theta}{R}$$

$$\sin\beta = \frac{B}{R}\sin\theta = \frac{35}{48.2}\sin 120° = 0.629$$

or

$$\beta \approx 39°$$

Therefore, the resultant displacement of the car is 48.2 km in a direction 39° west of north.

Q1. A book is moved once around the perimeter of a table of dimensions 1 m × 2 m. If the book ends up at its initial position, what is its displacement? What is the distance traveled?

Q2. If $\mathbf{B}$ is added to $\mathbf{A}$, under what condition does the resultant vector have a magnitude equal to $A + B$? Under what conditions is the resultant vector equal to zero?

Q3. Can the magnitude of a particle's displacement be greater than the distance traveled? Explain.

Q4. The magnitudes of two vectors $\mathbf{A}$ and $\mathbf{B}$ are $A = 5$ units and $B = 2$ units. Find the largest and smallest values possible for the resultant vector $\mathbf{R} = \mathbf{A} + \mathbf{B}$.

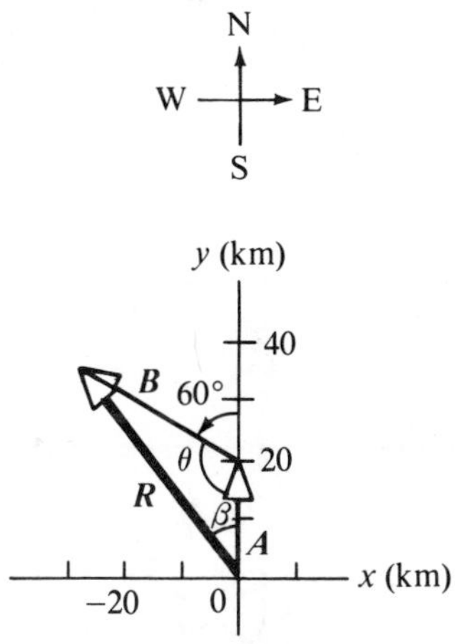

Figure 2.12 (Example 2.2) Graphical method for finding the resultant displacement $\mathbf{R} = \mathbf{A} + \mathbf{B}$.

2.4 COMPONENTS OF A VECTOR AND UNIT VECTORS

The geometric method of adding vectors is not the recommended procedure in situations where high precision is required or in three-dimensional problems. In this section, we describe a method of adding vectors that makes use of the *projections* of a vector along the axes of a rectangular coordinate system. These projections are called the *components* of the vector. Any vector can be completely described by its components.

Consider a vector $\mathbf{A}$ lying in the xy plane and making an arbitrary angle θ with the positive x axis, as in Fig. 2.13. The projection of $\mathbf{A}$ along the x axis, A_x, is called the x component of $\mathbf{A}$, and the projection of $\mathbf{A}$ along the y axis, A_y, is called the y component of $\mathbf{A}$. The components of a vector are not vectors, yet their value depends upon the choice of coordinates. They can be positive or negative numbers with units. From Fig. 2.13 and the definition of the sine and cosine of an angle, we see that $\cos\theta = A_x/A$ and $\sin\theta = A_y/A$. Hence, the rectangular components of $\mathbf{A}$

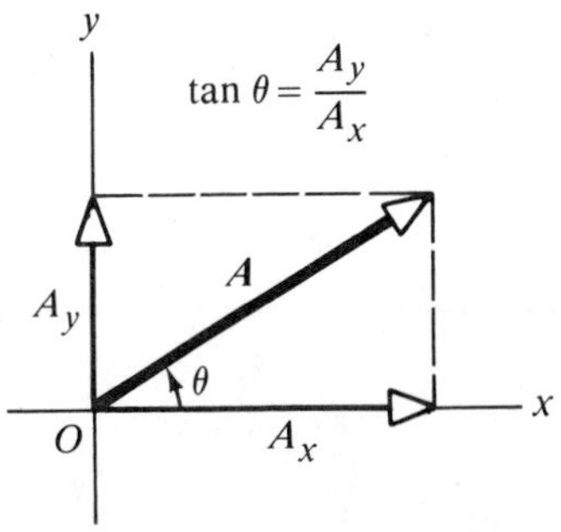

Figure 2.13 Any vector $\mathbf{A}$ lying in the xy plane can be represented by its rectangular components A_x and A_y.

[3]Note that a number such as 20.0 indicates three significant figures (Appendix B.1). From now on, for convenience we will usually write *20* for such a number. Furthermore, for purposes of calculation, we will assume that, unless stated otherwise, all given numbers have three significant figures.

are given by

Components of the vector *A*

$$A_x = A\cos\theta$$

and

$$A_y = A\sin\theta \quad (2.9)$$

These components form two sides of a right triangle the hypotenuse of which has a magnitude A. Thus, it follows that the magnitude of **A** and its direction are related to its rectangular components through the expressions

Magnitude of *A*

$$A = \sqrt{A_x^2 + A_y^2} \quad (2.10)$$

and

Direction of *A*

$$\tan\theta = \frac{A_y}{A_x} \quad (2.11)$$

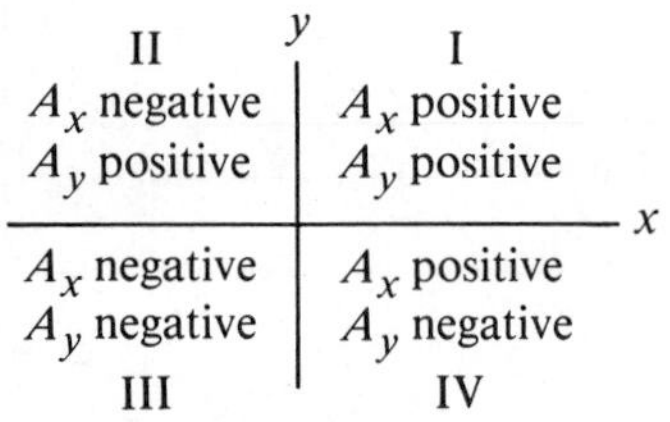

Figure 2.14 The signs of the rectangular components of a vector **A** depend on the quadrant in which the vector is located.

To solve for θ, we can invert this equation and write $\theta = \tan^{-1}(A_y/A_x)$, which is read "$\theta$ equals the angle the tangent of which is the ratio A_y/A_x." *Note that the signs of the rectangular components* A_x *and* A_y *depend on the angle* θ. For example, if $\theta = 120°$, A_x is negative and A_y is positive. On the other hand, if $\theta = 225°$, both A_x and A_y are negative. Figure 2.14 summarizes the signs of the components when **A** lies in the various quadrants.

If you choose reference axes or angle other than what is shown in Fig. 2.13, the components of a vector must be modified accordingly. In many applications it is more convenient to express the components of a vector in a coordinate system having axes that are not horizontal and vertical, but still perpendicular to each other. Suppose a vector ***B*** makes an angle θ with the x' axis defined in Fig. 2.15. The rectangular components of ***B*** along these axes are given by $B_{x'} = B\cos\theta$ and $B_{y'} = B\sin\theta$, as in Eq. 2.9. The magnitude and direction of ***B*** are obtained from expressions equivalent to Eqs. 2.10 and 2.11. Thus, we can express the components of a vector in *any* coordinate system that is convenient for a particular situation.

Another fundamental point concerning vectors is the fact that the components of a vector, such as a coordinate displacement, are different when viewed from different coordinate systems. Furthermore, the components of a vector can change with respect to a fixed coordinate system if the vector changes in magnitude, orientation, or both.

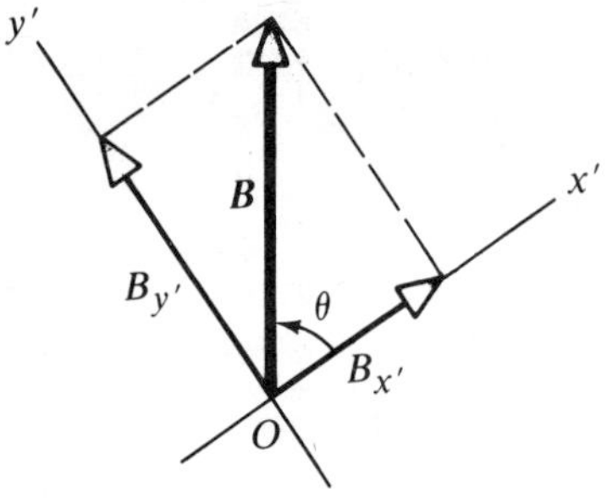

Figure 2.15 The components of a vector ***B*** in a coordinate system that is tilted.

Vector quantities are often expressed in terms of unit vectors. A *unit vector* is *a dimensionless vector of length unity* used to specify a given *direction*. Unit vectors have no other physical significance. They are used simply as a convenience in describing a direction in space. We will use the symbols ***i***, ***j***, and ***k*** to represent unit vectors pointing in the *x*, *y*, and *z* directions, respectively. Thus, the unit vectors ***i***, ***j***, and ***k*** form a set of mutually perpendicular vectors as shown in Fig. 2.16a, where the magnitude of the unit vectors equals unity or where $|\boldsymbol{i}| = |\boldsymbol{j}| = |\boldsymbol{k}| = 1$.

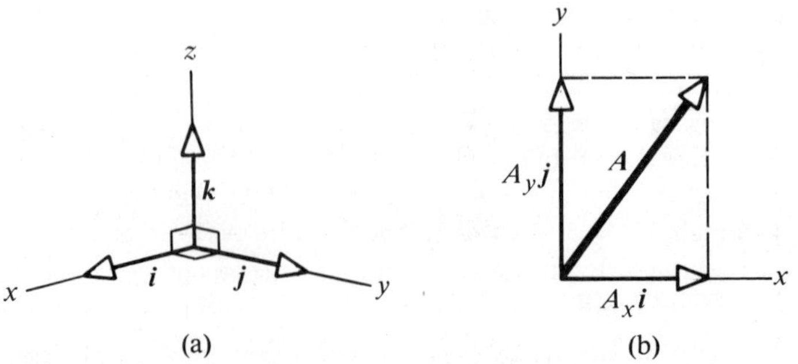

Figure 2.16 (a) The unit vectors ***i***, ***j***, and ***k*** are directed along the *x*, *y*, and *z* axes, respectively. (b) A vector **A** lying in the *xy* plane has an *x* component given by $A_x\boldsymbol{i}$ and a *y* component given by $A_y\boldsymbol{j}$.

Consider a vector $\boldsymbol{A}$ lying in the xy plane, as in Fig. 2.16b. The product of the component A_x and the unit vector $\boldsymbol{i}$ is the vector $A_x\boldsymbol{i}$ parallel to the x axis with magnitude A_x. Likewise, $A_y\boldsymbol{j}$ is a vector of magnitude A_y parallel to the y axis. Thus, the unit-vector form of the vector $\boldsymbol{A}$ is written

$$\boldsymbol{A} = A_x\boldsymbol{i} + A_y\boldsymbol{j} \tag{2.12}$$

Unit-vector form of the vector *A*

Now suppose we wish to add vector $\boldsymbol{B}$ to vector $\boldsymbol{A}$, where $\boldsymbol{B}$ has components B_x and B_y. The procedure for performing this sum is to simply add the x and y components separately. The resultant vector $\boldsymbol{R} = \boldsymbol{A} + \boldsymbol{B}$ is therefore given by

$$\boldsymbol{R} = (A_x + B_x)\boldsymbol{i} + (A_y + B_y)\boldsymbol{j} \tag{2.13}$$

Thus, the rectangular components of the resultant vector are given by

$$\begin{aligned} R_x &= A_x + B_x \\ R_y &= A_y + B_y \end{aligned} \tag{2.14}$$

The magnitude of $\boldsymbol{R}$ and the angle it makes with the x axis can then be obtained from its components using the relationships

$$R = \sqrt{R_x^{\,2} + R_y^{\,2}} = \sqrt{(A_x + B_x)^2 + (A_y + B_y)^2} \tag{2.15}$$

and

$$\tan\theta = \frac{R_y}{R_x} = \frac{A_y + B_y}{A_x + B_x} \tag{2.16}$$

The procedure just described for adding two vectors $\boldsymbol{A}$ and $\boldsymbol{B}$ using the component method can be checked using a geometric construction, as in Fig. 2.17. Again, you must take note of the *signs* of the components when using either the algebraic or the geometric method.

The extension of these methods to three-dimensional vectors is straightforward. If $\boldsymbol{A}$ and $\boldsymbol{B}$ both have x, y, and z components, we express them in the form

$$\boldsymbol{A} = A_x\boldsymbol{i} + A_y\boldsymbol{j} + A_z\boldsymbol{k} \tag{2.17}$$

$$\boldsymbol{B} = B_x\boldsymbol{i} + B_y\boldsymbol{j} + B_z\boldsymbol{k} \tag{2.18}$$

The sum of $\boldsymbol{A}$ and $\boldsymbol{B}$ is given by

$$\boldsymbol{R} = \boldsymbol{A} + \boldsymbol{B} = (A_x + B_x)\boldsymbol{i} + (A_y + B_y)\boldsymbol{j} + (A_z + B_z)\boldsymbol{k} \tag{2.19}$$

Thus, the resultant vector also has a z component given by $R_z = A_z + B_z$. The same procedure can be used to sum up three or more vectors.

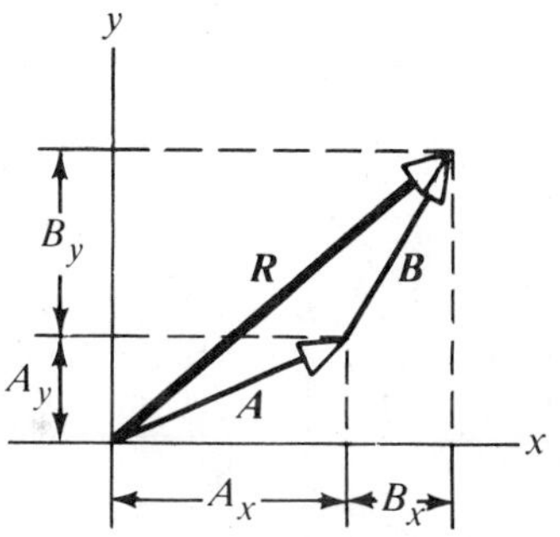

Figure 2.17 Geometric construction showing the relation between the components of the resultant $\boldsymbol{R}$ of two vectors and the individual vector components.

Example 2.3

Find the sum of two vectors $\boldsymbol{A}$ and $\boldsymbol{B}$ lying in the xy plane and given by

$$\boldsymbol{A} = 2\boldsymbol{i} + 2\boldsymbol{j} \qquad \text{and} \qquad \boldsymbol{B} = 2\boldsymbol{i} - 4\boldsymbol{j}$$

Solution: Note that $A_x = 2$, $A_y = 2$, $B_x = 2$, and $B_y = -4$. Therefore, the resultant vector $\boldsymbol{R}$ is given by

$$\boldsymbol{R} = \boldsymbol{A} + \boldsymbol{B} = (2 + 2)\boldsymbol{i} + (2 - 4)\boldsymbol{j} = 4\boldsymbol{i} - 2\boldsymbol{j}$$

or

$$R_x = 4, \; R_y = -2$$

The magnitude of $\boldsymbol{R}$ is given by

$$R = \sqrt{R_x^{\,2} + R_y^{\,2}} = \sqrt{(4)^2 + (-2)^2} = \sqrt{20} = 4.47$$

You should verify that the angle θ that $\boldsymbol{R}$ makes with the x axis is 333°.

Example 2.4

A particle undergoes three consecutive displacements given by $\boldsymbol{d}_1 = (\boldsymbol{i} + 3\boldsymbol{j} - \boldsymbol{k})$ cm, $\boldsymbol{d}_2 = (2\boldsymbol{i} - \boldsymbol{j} - 3\boldsymbol{k})$ cm, and $\boldsymbol{d}_3 = (-\boldsymbol{i} + \boldsymbol{j})$ cm. Find the resultant displacement of the particle.

Solution:

$$\begin{aligned} \boldsymbol{R} &= \boldsymbol{d}_1 + \boldsymbol{d}_2 + \boldsymbol{d}_3 \\ &= (1 + 2 - 1)\boldsymbol{i} + (3 - 1 + 1)\boldsymbol{j} + (-1 - 3 + 0)\boldsymbol{k} \\ &= (2\boldsymbol{i} + 3\boldsymbol{j} - 4\boldsymbol{k}) \text{ cm} \end{aligned}$$

That is, the resultant displacement has components

$R_x = 2$ cm, $R_y = 3$ cm, and $R_z = -4$ cm. Its magnitude is

$$R = \sqrt{R_x{}^2 + R_y{}^2 + R_z{}^2} = \sqrt{(2)^2 + (3)^2 + (-4)^2} = 5.39 \text{ cm}$$

Example 2.5

A hiker begins a trip by first walking 25 km due southeast from her base camp. On the second day, she walks 40 km in a direction 60° north of east, at which point she discovers a forest ranger's tower. (a) Determine the rectangular components of the hiker's displacement on the first and second days.

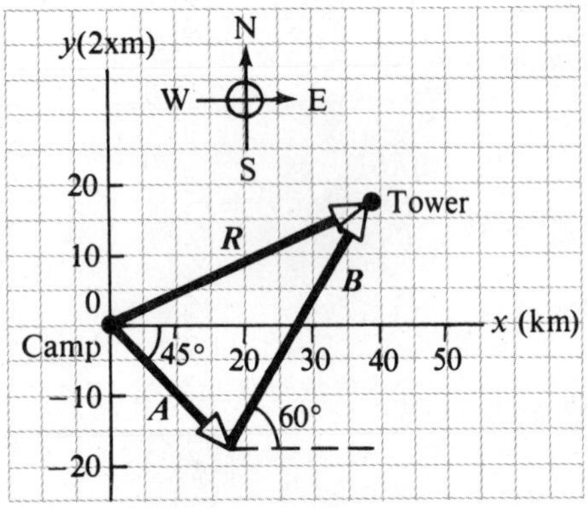

Figure 2.18 (Example 2.5) The total displacement of the hiker is the vector $\boldsymbol{R} = \boldsymbol{A} + \boldsymbol{B}$.

If we denote the displacement vectors on the first and second days by **A** and **B**, respectively, and use the camp as the origin of coordinates, we get the vectors shown in Fig. 2.18. Displacement **A** has a magnitude of 25.0 km and is 45° southeast. Its rectangular components are

$$A_x = A\cos(-45°) = (25 \text{ km})(0.707) = 17.7 \text{ km}$$
$$A_y = A\sin(-45°) = -(25 \text{ km})(0.707) = -17.7 \text{ km}$$

The negative value of A_y indicates that the y coordinate has decreased for this displacement. The signs of A_x and A_y are also evident from Fig. 2.18. The second displacement, **B**, has a magnitude of 40.0 km and is 60° north of east. Its rectangular components are

$$B_x = B\cos 60° = (40 \text{ km})(0.50) = 20.0 \text{ km}$$
$$B_y = B\sin 60° = (40 \text{ km})(0.866) = 34.6 \text{ km}$$

(b) Determine the rectangular components of the hiker's total displacement for the trip.

The resultant displacement for the trip, $\boldsymbol{R} = \boldsymbol{A} + \boldsymbol{B}$, has components given by

$$R_x = A_x + B_x = 17.7 \text{ km} + 20.0 \text{ km} = 37.7 \text{ km}$$
$$R_y = A_y + B_y = -17.7 \text{ km} + 34.6 \text{ km} = 16.9 \text{ km}$$

In unit-vector form, we can write the total displacement $\boldsymbol{R} = (37.7\,\boldsymbol{i} + 16.9\,\boldsymbol{j})$ km.

(c) Determine the magnitude and direction of the total displacement.

The magnitude and direction of the total displacement are given by

$$R = \sqrt{R_x{}^2 + R_y{}^2} = \sqrt{(37.7)^2 + (16.9)^2} = 41.3 \text{ km}$$

and

$$\tan\theta = \frac{R_y}{R_x} = \frac{16.9}{37.7} = 0.448$$

so that

$$\theta = \tan^{-1}(0.448) = 24.1°$$

Therefore, the ranger's tower is 41.3 km and 24.1° north of east from the base camp.

Q5. A vector **A** lies in the xy plane. For what orientations of **A** will both of its rectangular components be negative? For what orientations will its components have opposite signs?

Q6. Can a vector have a component equal to zero and still have a nonzero magnitude? Explain.

Q7. If one of the components of a vector is not zero, can its magnitude be zero? Explain.

Q8. If the component of vector **A** along the direction of vector **B** is zero, what can you conclude about the two vectors?

Q9. If $\boldsymbol{A} = \boldsymbol{B}$, what can you conclude about the components of **A** and **B**?

Q10. Can the magnitude of a vector have a negative value? Explain.

Q11. If $\boldsymbol{A} + \boldsymbol{B} = 0$, what can you say about the components of the two vectors?

2.5 FORCE

Force is an important concept in all branches of physics. If you push or pull an object in a certain direction, you exert a force on that object. The force of gravity exerted on every body on the earth (the weight of the body) is a common force experienced by everyone. Any force on a body is specified completely by its magnitude, direction, and point of application. Force is more fully discussed in Chapter 5; this section simply describes how forces can be treated algebraically. The method of replacing a force by its components is emphasized, since this often simplifies the description of the behavior of a system under the influence of external forces.

The SI unit of force is the newton N, whereas the English unit of force is the more familiar pound (lb). The conversion between the two units is 1 N = 0.2248 lb or 1 lb = 4.448 N.

Suppose a force $\boldsymbol{F}$ acts on an object at the point O at an angle θ to the horizontal, as in Fig. 2.19a. The rectangular components of $\boldsymbol{F}$ are F_x and F_y, where $F_x = F\cos\theta$ and $F_y = F\sin\theta$. The vector sum of the components in Fig. 2.19b is equivalent to the original force $\boldsymbol{F}$. That is, *any force* **F** *can be represented by its rectangular components, provided that the force acting along each component originates at the same point as* **F**.

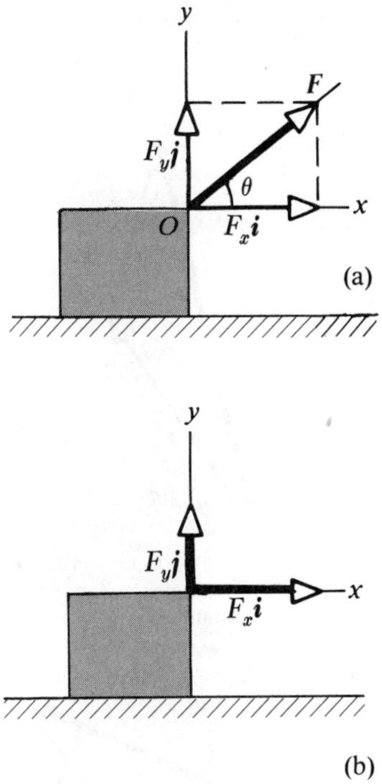

Figure 2.19 (a) The force $\boldsymbol{F}$ acting on the object has components F_x and F_y. (b) The vector sum of the forces $\boldsymbol{F}_x$ and $\boldsymbol{F}_y$ is equivalent to the force $\boldsymbol{F}$ shown in (a).

As a numerical example, consider a single force of magnitude 7 N acting on an object at an angle of 30° to the horizontal, as in Fig. 2.20. We can replace this force by its x and y components as follows: Since $F = 7$ N, its rectangular components are

$$F_x = F\cos\theta = (7\text{ N})(\cos 30°) = 6.06\text{ N}$$

$$F_y = F\sin\theta = (7\text{ N})(\sin 30°) = 3.50\text{ N}$$

Hence, we can express $\boldsymbol{F}$ in unit-vector form:

$$\boldsymbol{F} = F_x\boldsymbol{i} + F_y\boldsymbol{j} = (6.06\,\boldsymbol{i} + 3.50\,\boldsymbol{j})\text{ N}$$

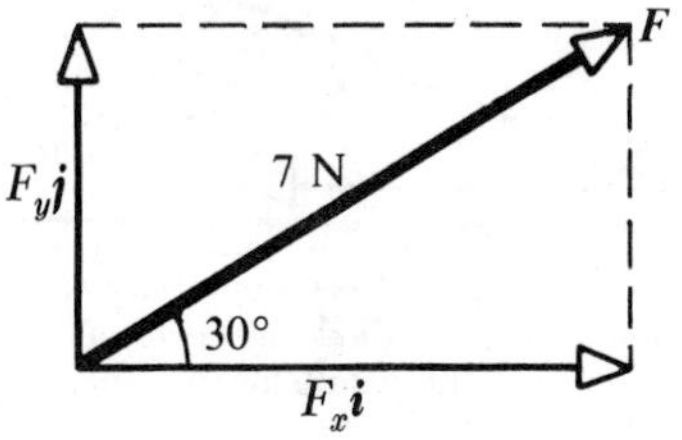

Figure 2.20 The rectangular components of the 7-N force are F_x and F_y.

Now consider two forces acting on an object as in Fig. 2.21a. Suppose you want to find the resultant force on the object, that is, you wish to know what *single* force would be equivalent to the two forces shown. The x and y components of the 12-N force are given by

$$F_{x1} = F_1\cos 60° = (12\text{ N})(0.50) = 6.00\text{ N}$$
$$F_{y1} = F_1\sin 60° = (12\text{ N})(0.866) = 10.4\text{ N}$$

Likewise, the components of the 8-N force are

$$F_{x2} = F_2\cos(105°) = (8\text{ N})(-0.259) = -2.07\text{ N}$$
$$F_{y2} = F_2\sin 105° = (8\text{ N})(0.966) = 7.73\text{ N}$$

Note that the component F_{x2} is negative since it is directed along the negative x axis. We are using the usual sign conventions of analytical geometry, in which x components to the right are positive and those to the left are negative. Likewise, y components upward are positive and those downward are negative. Adding the x and y

---

[4]The newton is defined in Chapter 5.

components gives the components of the resultant force $\boldsymbol{R} = \boldsymbol{F}_1 + \boldsymbol{F}_2$:

$$R_x = F_{x1} + F_{x2} = 6.00\ \text{N} - 2.07\ \text{N} = 3.93\ \text{N}$$
$$R_y = F_{y1} + F_{y2} = 10.4\ \text{N} + 7.73\ \text{N} = 18.1\ \text{N}$$

In unit-vector form, $\boldsymbol{R}$ can be expressed

$$\boldsymbol{R} = (3.93\,\boldsymbol{i} + 18.1\,\boldsymbol{j})\ \text{N}$$

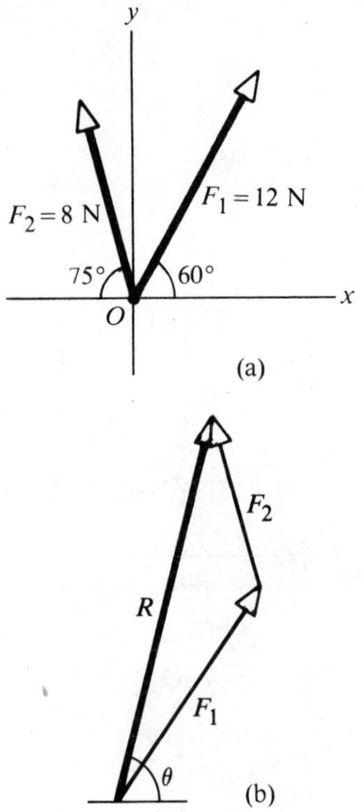

Figure 2.21 (a) Two forces acting on an object at the origin. (b) Graphical method for obtaining the resultant force $\boldsymbol{R}$.

The magnitude and direction of R are given by

$$R = \sqrt{R_x^{\,2} + R_y^{\,2}} = \sqrt{(3.93\ \text{N})^2 + (18.1\ \text{N})^2} = 18.5\ \text{N}$$
$$\theta = \tan^{-1}\frac{R_y}{R_x} = \tan^{-1}\left(\frac{18.1\ \text{N}}{3.93\ \text{N}}\right) = 77.7°$$

You should check these results against the graphical solution shown in Fig. 2.21b.

If you have difficulty keeping track of the various forces and their components, it is suggested that you set up a table similar to the one shown here, which summarizes the above calculations. This procedure is especially useful when dealing with three or more forces.

FORCE	F_x (x component)	F_y (y component)
12 N	6.00 N	10.4 N
8 N	−2.07 N	7.73 N
Resultant R	$R_x = 3.93$ N	$R_y = 18.1$ N

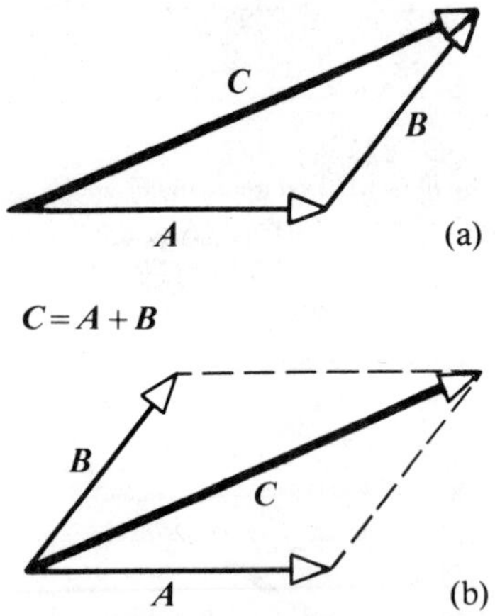

Figure 2.22 (a) Vector addition using the triangle method. (b) Vector addition using the parallelogram rule.

Finally, suppose you wish to determine the magnitude and direction of another force $\boldsymbol{F}$, which when applied to the body will make the resultant force zero. This can easily be calculated by first finding the resultant $\boldsymbol{R}$ of the original forces and then applying the condition that $\boldsymbol{R} + \boldsymbol{F} = 0$, or $\boldsymbol{F} = -\boldsymbol{R}$. That is, $\boldsymbol{F}$ must be equal in magnitude to the resultant of the original forces and in the opposite direction. For instance, the third force $\boldsymbol{F}$ that must be applied to the body given in Fig. 2.21 to make the resultant force zero is given by

$$\boldsymbol{F} = -\boldsymbol{R} = (-3.93\boldsymbol{i} - 18.1\boldsymbol{j})\ \text{N}$$

2.6 SUMMARY

Vectors are quantities that have both magnitude and direction. *Scalars* are quantities that have only magnitude.

Two vectors $\boldsymbol{A}$ and $\boldsymbol{B}$ can be added using either the triangle method or the parallelogram rule. In the triangle method (Fig. 2.22a), the vector $\boldsymbol{C} = \boldsymbol{A} + \boldsymbol{B}$ runs from the tail of $\boldsymbol{A}$ to the tip of $\boldsymbol{B}$. In the parallelogram method (Fig. 2.22b), $\boldsymbol{C}$ is the diagonal of a parallelogram having $\boldsymbol{A}$ and $\boldsymbol{B}$ as its sides.

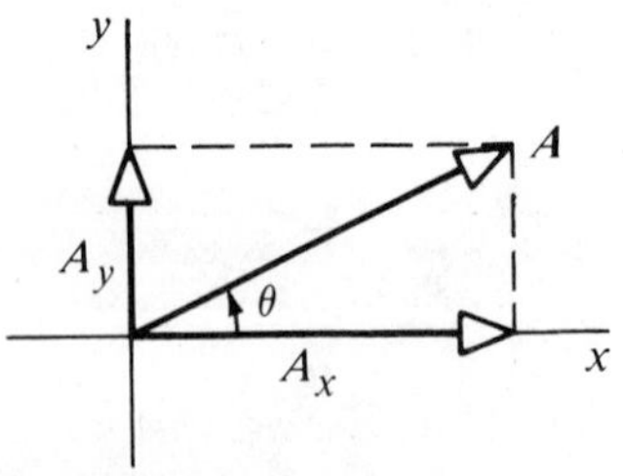

Figure 2.23 The x and y components of a vector $\mathbf{A}$.

The x component of the vector $\boldsymbol{A}$ is equivalent to its projection along the x axis, that is, $A_x = A\cos\theta$, as shown in Fig. 2.23. Likewise, the y component of $\boldsymbol{A}$ is its projection along the y axis, and $A_y = A\sin\theta$.

If a vector $\boldsymbol{A}$ has an x component equal to A_x and a y component equal to A_y, the vector can be expressed in unit-vector form as $\boldsymbol{A} = A_x\boldsymbol{i} + A_y\boldsymbol{j}$. In this notation, $\boldsymbol{i}$ is a unit vector pointing in the positive x direction and $\boldsymbol{j}$ is a unit vector in the positive y direction. Since $\boldsymbol{i}$ and $\boldsymbol{j}$ are unit vectors, $|\boldsymbol{i}| = |\boldsymbol{j}| = 1$.

EXERCISES

Section 2.1 Coordinate Systems and Frames of Reference

1. Two points in the xy plane have cartesian coordinates (2.0, −4.0) and (−3.0, 3.0), where the units are in m. Determine (a) the distance between these points and (b) their polar coordinates.

2. A point in the xy plane has cartesian coordinates (−3.0, 5.0) m. What are the polar coordinates of this point?

3. The polar coordinates of a point are $r = 5.50$ m and $\theta = 240°$. What are the cartesian coordinates of this point?

4. Two points in a plane have polar coordinates (2.50 m, 30°) and (3.80 m, 120°). Determine (a) the cartesian coordinates of these points and (b) the distance between them.

Section 2.2 Vectors and Scalars and Section 2.3 Some Properties of Vectors

5. Vector $\boldsymbol{A}$ is 6 units in length and at an angle of 45° to the x axis. Vector $\boldsymbol{B}$ is 3 units in length and is directed along the positive x axis ($\theta = 0$). Find the resultant vector $\boldsymbol{A} + \boldsymbol{B}$ using (a) graphical methods and (b) the law of cosines.

6. Vector $\boldsymbol{A}$ is 3 units in length and points along the positive x axis. Vector $\boldsymbol{B}$ is 4 units in length and points along the negative y axis. Use graphical methods to find the magnitude and direction of the vectors (a) $\boldsymbol{A} + \boldsymbol{B}$, (b) $\boldsymbol{A} - \boldsymbol{B}$.

7. A pedestrian moves 6 km east and 13 km north. Find the magnitude and direction of the resultant displacement vector using the graphical method.

8. A person walks along a circular path of radius 5 m, around one half of the circle. (a) Find the magnitude of the displacement vector. (b) How far did the person walk? (c) What is the magnitude of the displacement if the circle is completed?

9. A particle undergoes three consecutive displacements such that its *total* displacement is zero. The first displacement is 8 m westward. The second is 13 m northward. Find the magnitude and direction of the third displacement using the graphical method.

10. Each of the displacement vectors $\boldsymbol{A}$ and $\boldsymbol{B}$ shown in Fig. 2.24 has a magnitude of 3 m. Find graphically (a) $\boldsymbol{A} + \boldsymbol{B}$, (b) $\boldsymbol{A} - \boldsymbol{B}$, (c) $\boldsymbol{B} - \boldsymbol{A}$, (d) $\boldsymbol{A} - 2\boldsymbol{B}$.

Section 2.4 Components of a Vector and Unit Vectors

11. Find the x and y components of the vectors $\boldsymbol{A}$ and $\boldsymbol{B}$ shown in Fig. 2.24. Derive an expression for the resultant vector $\boldsymbol{A} + \boldsymbol{B}$ in unit-vector notation.

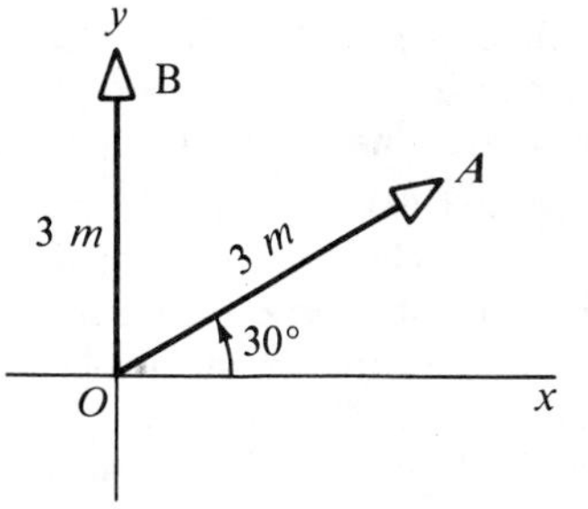

Figure 2.24 (Exercises 10 and 11).

12. A displacement vector $\boldsymbol{A}$ make an angle θ with the positive x axis, as in Fig. 2.13. Find the rectangular components of $\boldsymbol{A}$ for the following values of A and θ: (a) $A = 8$ m, $\theta = 60°$; (b) $A = 6$ ft, $\theta = 120°$; (c) $A = 12$ cm, $\theta = 225°$.

13. Vector $\boldsymbol{A}$ lies in the xy plane. Construct a table of the signs of the x and y components of $\boldsymbol{A}$ when the vector lies in the first, second, third, and fourth quadrants.

14. A displacement vector lying in the xy plane has a magnitude of 50 m and is directed at an angle of 120° to the positive x axis. What are the rectangular components of this vector?

15. A vector has an x component of −25 units and a y component of 40 units. Find the magnitude and direction of this vector.

16. A particle undergoes two displacements. The first has a magnitude of 150 cm and makes an angle of 120° with the positive x axis. The *resultant* displacement has a magnitude of 140 cm and is directed at an angle of 35° to the positive x axis. Find the magnitude and direction of the second displacement.

17. Find the magnitude and direction of the resultant of three displacements having components (3,2) m, (−5,3) m, and (6,1) m.

18. A vector $\boldsymbol{A}$ has a positive x component of 4 units and a negative y component of 2 units. What second vector $\boldsymbol{B}$ when added to $\boldsymbol{A}$ will produce a *resultant* vector three times the magnitude of $\boldsymbol{A}$ directed in the positive y direction?

19. A vector $\boldsymbol{A}$ has a magnitude of 35 units and makes an angle of 37° with the positive x-axis. Describe (a) a vector $\boldsymbol{B}$ that is in the direction opposite $\boldsymbol{A}$ and is one fifth the size of $\boldsymbol{A}$, and (b) a vector $\boldsymbol{C}$ that when added to $\boldsymbol{A}$ will produce a vector twice as long as $\boldsymbol{A}$ pointing in the negative y direction.

20. Vector $\boldsymbol{A}$ has x and y components of -8.7 cm and 15 cm, respectively; vector $\boldsymbol{B}$ has x and y components of 13.2 cm and -6.6 cm, respectively. If $\boldsymbol{A} - \boldsymbol{B} + 3\boldsymbol{C} = 0$, what are the components of $\boldsymbol{C}$?

21. A particle undergoes the following consecutive displacements: 3.5 m south, 8.2 m northeast, and 15.0 m west. What is the resultant displacement?

22. A quarterback takes the ball from the line of scrimmage, runs backward for 10 yards, then sideways parallel to the line of scrimmage for 15 yards. At this point, he throws a 50-yard forward pass straight downfield perpendicular to the line of scrimmage. What is the magnitude of the football's resultant displacement?

23. An airplane flies from city A to city B in a direction due east for 800 miles. In the next part of the trip the airplane flies from city B to city C in a direction 40° north of east for 600 miles. What is the resultant displacement of the airplane between city A and city C?

24. A particle undergoes three consecutive displacements. The first is to the east and has a magnitude of 25 m. The second is to the north and has a magnitude of 42 m. If the resultant displacement has a magnitude of 38 m and is directed at an angle of 30° north of east, what are the magnitude and direction of the third displacement?

25. Two vectors are given by $\boldsymbol{A} = 3\boldsymbol{i} - 2\boldsymbol{j}$ and $\boldsymbol{B} = -\boldsymbol{i} - 4\boldsymbol{j}$. Calculate (a) $\boldsymbol{A} + \boldsymbol{B}$, (b) $\boldsymbol{A} - \boldsymbol{B}$, (c) $|\boldsymbol{A} + \boldsymbol{B}|$, (d) $|\boldsymbol{A} - \boldsymbol{B}|$, (e) the direction of $\boldsymbol{A} + \boldsymbol{B}$ and $\boldsymbol{A} - \boldsymbol{B}$.

26. Three vectors are given by $\boldsymbol{A} = \boldsymbol{i} + 3\boldsymbol{j}$, $\boldsymbol{B} = 2\boldsymbol{i} - \boldsymbol{j}$, and $\boldsymbol{C} = 3\boldsymbol{i} + 5\boldsymbol{j}$. Find (a) the resultant sum of the three vectors and (b) the magnitude and direction of the resultant vector.

27. The vector $\boldsymbol{A}$ has x, y, and z components of 8, 12, and -4 units, respectively. (a) Write a vector expression for $\boldsymbol{A}$ in unit-vector notation. (b) Obtain a unit-vector expression for a vector $\boldsymbol{B}$ one fourth the length of $\boldsymbol{A}$ pointing in the same direction as $\boldsymbol{A}$. (c) Obtain a unit-vector expression for a vector $\boldsymbol{C}$ three times longer than $\boldsymbol{A}$ pointing in the direction opposite $\boldsymbol{A}$.

28. Two vectors are given by $\boldsymbol{A} = -2\boldsymbol{i} + \boldsymbol{j} - 3\boldsymbol{k}$ and $\boldsymbol{B} = 5\boldsymbol{i} + 3\boldsymbol{j} - 2\boldsymbol{k}$. (a) Find a third vector $\boldsymbol{C}$ such that $3\boldsymbol{A} + 2\boldsymbol{B} - \boldsymbol{C} = 0$. (b) What are the magnitudes of $\boldsymbol{A}$, $\boldsymbol{B}$, and $\boldsymbol{C}$?

29. Obtain expressions for the position vectors with polar coordinates (a) 12.8 m, 150°; (b) 3.3 cm, 60°; (c) 22 in., 215°.

30. Vectors $\boldsymbol{A}$ and $\boldsymbol{B}$ have components $A_x = -5.0$ cm, $A_y = 1.1$ cm, $A_z = -3.5$ cm and $B_x = 8.8$ cm, $B_y = -6.3$ cm, $B_z = 9.2$ cm. Determine the components of the vectors (a) $\boldsymbol{A} + \boldsymbol{B}$, (b) $\boldsymbol{B} - \boldsymbol{A}$, (c) $3\boldsymbol{B} + 2\boldsymbol{A}$. (d) Express the vector $\boldsymbol{B} - \boldsymbol{A}$ in unit-vector notation.

31. A vector $\boldsymbol{A}$ has a negative x component 3 units in length and a positive y component 2 units in length. (a) Determine an expression for $\boldsymbol{A}$ in unit-vector notation. (b) Determine the magnitude and direction of $\boldsymbol{A}$. (c) What vector $\boldsymbol{B}$ when added to $\boldsymbol{A}$ gives a resultant vector with no x component and a negative y component 4 units in length?

32. A particle moves in the xy plane from the point (3,0) m to the point (2,2) m. (a) Determine a vector expression for the resultant displacement. (b) What are the magnitude and direction of this displacement vector?

Section 2.5 Force

33. Find the magnitude and direction of a force having x and y components of -5 N and 3 N, respectively.

34. A 40-N force is applied at an angle of 30° to the horizontal. What are the x and y components of this force?

35. Two 25-N forces are applied to an object as shown in Fig. 2.25. Find the magnitude and direction of the resultant force.

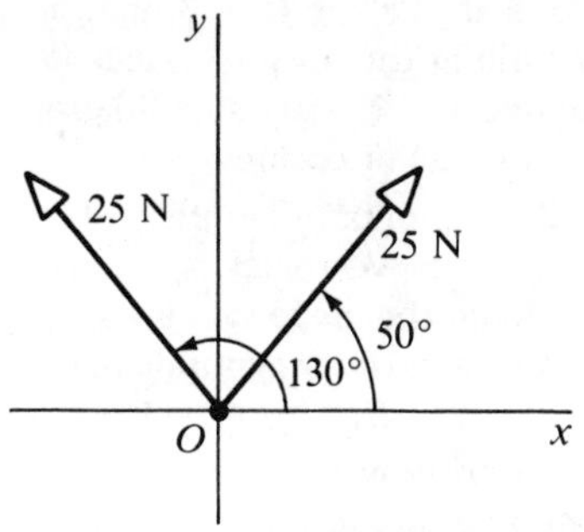

Figure 2.25 (Exercise 35).

36. Three forces are given by $\boldsymbol{F}_1 = 6\boldsymbol{i}$ N, $\boldsymbol{F}_2 = 9\boldsymbol{j}$ N, and $\boldsymbol{F}_3 = (-3\boldsymbol{i} + 4\boldsymbol{j})$ N. (a) Find the magnitude and direction of the resultant force. (b) What force must be added to these three to make the resultant force zero?

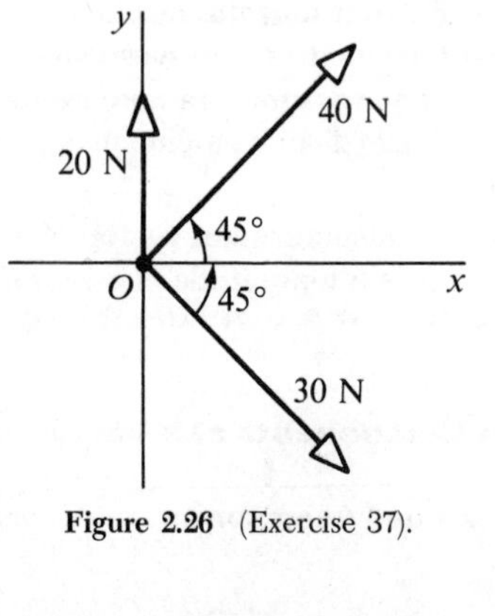

Figure 2.26 (Exercise 37).

37. Three forces act at the point O as shown in Fig. 2.26. Find (a) the x and y components of the resultant force and (b) the magnitude and direction of the resultant force.

PROBLEMS

1. A point P is described by the coordinates (x,y) with respect to the normal cartesian coordinate system shown in Fig. 2.27. Show that (x',y'), the coordinates of this point in the rotated $x'y'$ coordinate system, are related to (x,y) and the rotation angle α by the expressions

$$x' = x\cos\alpha + y\sin\alpha \text{ and } y' = -x\sin\alpha + y\cos\alpha.$$

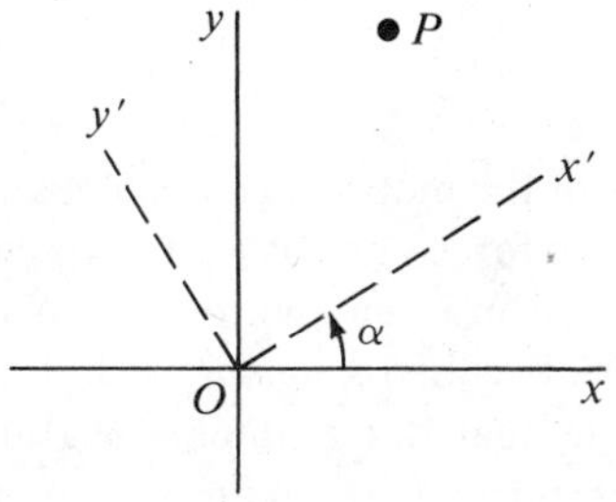

Figure 2.27 (Problem 2).

2. (a) Show that a point lying in the xy plane and having coordinates (x,y) can be described by the position vector $\boldsymbol{r} = x\boldsymbol{i} + y\boldsymbol{j}$. (b) Show that the magnitude of this vector is $r = \sqrt{x^2 + y^2}$. (c) Show that the displacement vector for a particle moving from (x_1,y_1) to (x_2,y_2) is given by $\boldsymbol{d} = (x_2 - x_1)\boldsymbol{i} + (y_2 - y_1)\boldsymbol{j}$. (d) Plot the position vectors $\boldsymbol{r}_1$ and $\boldsymbol{r}_2$ and the displacement vector $\boldsymbol{d}$, and verify by the graphical method that $\boldsymbol{d} = \boldsymbol{r}_2 - \boldsymbol{r}_1$.

3. A particle moves from a point in the xy plane having cartesian coordinates $(-3,-5)$ m to a point with coordinates $(-1,8)$ m. (a) Write vector expressions for the position vectors in unit-vector form for these two points. (b) What is the displacement vector? (See Problem 2 for definition.)

4. A rectangular parallelepiped has dimensions a, b, and c, as in Fig. 2.28. (a) Obtain a vector expression for the face diagonal vector $\boldsymbol{R}_1$. What is the magnitude of this vector? (b) Obtain a vector expression for the body diagonal vector $\boldsymbol{R}_2$. What is the magnitude of this vector?

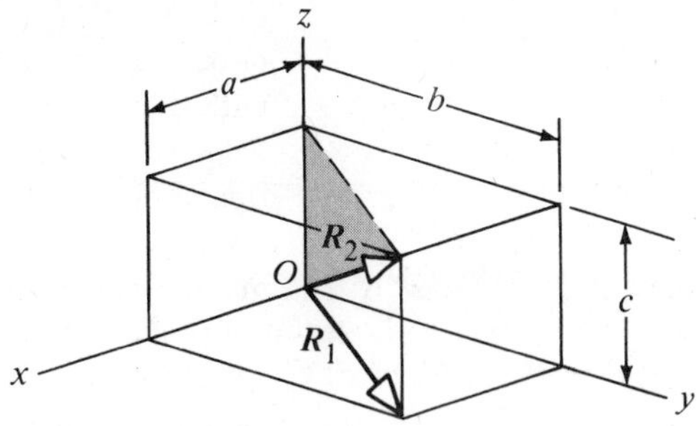

Figure 2.28 (Problem 4).

3 Motion in One Dimension

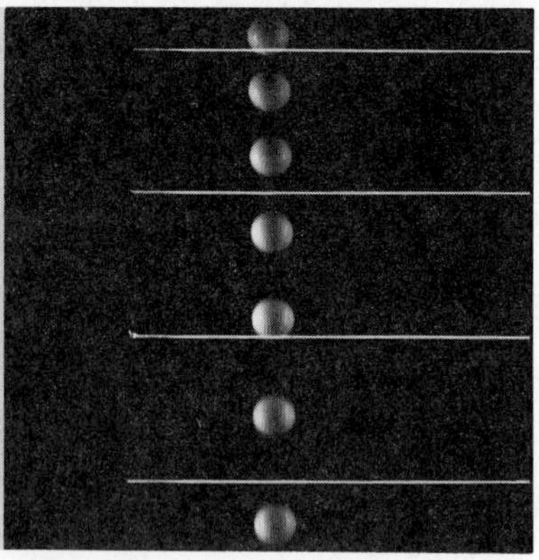

Education Development Center, Newton, Mass.

Mechanics is concerned with the study of the motion of an object and the relation of this motion to such physical concepts as force and mass. It is convenient to first describe motion using the concepts of space and time, without regard to the causes of the motion. This portion of mechanics is called *kinematics*. In this chapter we shall consider motion along a straight line, that is, one-dimensional motion. In the next chapter we shall extend our discussion to two-dimensional motion. Starting with the concept of displacement discussed in the previous chapter, we shall define velocity and acceleration. Using these concepts, we shall proceed to study the motion of objects undergoing constant acceleration. The subject of *dynamics*, which is concerned with the relationships between motion, forces, and the properties of moving objects, will be dealt with in Chapters 5 and 6.

From everyday experience we recognize that motion represents the continuous change in the position of an object. The movement of an object through space may be accompanied by the rotation or vibration of the object. Such motions can be quite complex. However, it is sometimes possible to simplify matters by neglecting the internal motions of the moving object. In many situations, an object can be treated as a *particle* if the only motion being considered is one of translation through space. For example, if we wish to describe the motion of the earth around the sun, we can treat the earth as a particle and obtain reasonable accuracy in a prediction of the earth's orbit. This approximation is justified because the radius of the earth's orbit is large compared with the dimensions of the earth and sun. On the other hand, we could not use the particle description to explain the internal structure of the earth and such phenomena as tides, earthquakes, and volcanic activity. On a much smaller scale, it is possible to explain the pressure exerted by a gas on the walls of a container by treating the gas molecules as particles. However, the particle description of the gas molecules is generally inadequate for understanding those properties of the gas that depend on the internal motions of the gas molecules, namely, rotations and vibrations.

3.1 AVERAGE VELOCITY

The motion of a particle is completely known if its position in space is known at all times. Consider a particle moving along the x axis from point P to point Q. Let its position at point P be x_i at some time t_i, and let its position at point Q be x_f at time t_f. (The indices i and f refer to the initial and final values.) At times other than t_i

and t_f, the position of the particle between these two points may vary as in Fig. 3.1. Such a plot is often called a *position-time graph*. In the time interval $\Delta t = t_f - t_i$, the displacement of the particle is $\Delta x = x_f - x_i$. (Recall that the displacement is defined as the change in the position of the particle, which equals its final minus its initial position value.)

The *average velocity* of the particle, $\bar{v}$, is defined as the ratio of the displacement, Δx, and the time interval, Δt:

$$\bar{v} \equiv \frac{\Delta x}{\Delta t} = \frac{x_f - x_i}{t_f - t_i} \tag{3.1}$$

Average velocity

From this definition, we see that the average velocity has the dimensions of length divided by time, or m/s in SI units and ft/s in British engineering units. The average velocity is independent of the path taken between the points P and Q. This is true because the average velocity is proportional to the displacement, Δx, which in turn depends only on the initial and final coordinates of the particle. It therefore follows that if a particle starts at some point and returns to the same point via any path, its average velocity for this trip is zero, since its displacement along such a path is zero. The displacement should not be confused with the distance traveled, since the distance traveled for any motion is clearly nonzero. Thus, average velocity gives us no details of the motion between points P and Q. (How we evaluate velocity at some instant in time is discussed in the next section.) Finally, note that the average velocity can be positive or negative, depending on the sign of the displacement. (The time interval, Δt, is always positive.) If the coordinate of the particle increases in time (that is, if $x_f > x_i$), then Δx is positive and $\bar{v}$ is positive. On the other hand, if the coordinate decreases in time ($x_f < x_i$), Δx is negative and hence $\bar{v}$ is negative.

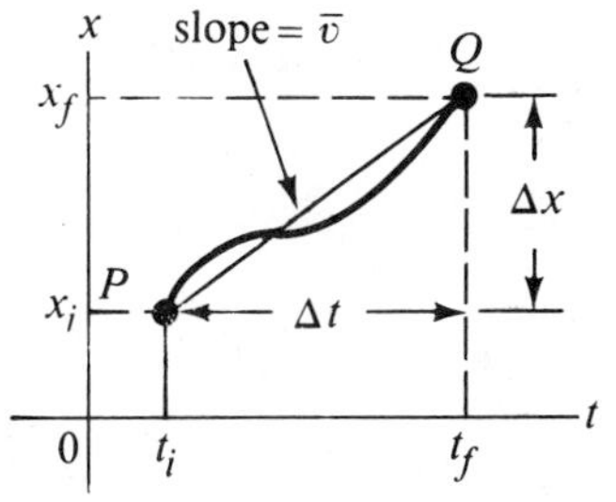

Figure 3.1 Position-time graph for a particle moving along the x axis. The average velocity $\bar{v}$ in the interval $\Delta t = t_f - t_i$ is the slope of the straight line connecting the points P and Q.

The average velocity can also be interpreted geometrically by drawing a straight line between the points P and Q in Fig. 3.1. This line forms the hypotenuse of a triangle of height Δx and base Δt. The slope of this line is the ratio $\Delta x/\Delta t$. Therefore, we see that the *average* velocity of the particle during the time interval t_i to t_f is equal to the slope of the straight line joining the initial and final points on the space-time graph. (The word *slope* will often be used when referring to the graphs of physical data. Regardless of what data are plotted, the word *slope* will represent the ratio of the change in the quantity represented on the vertical axis to the change in the quantity represented on the horizontal axis.)

Example 3.1

A particle moving along the x axis is located at $x_i = 12$ m at $t_i = 1$ s and at $x_f = 4$ m at $t_f = 3$ s. Find its displacement and average velocity during this time interval.

Solution: The displacement is given by

$$\Delta x = x_f - x_i = 4\text{ m} - 12\text{ m} = -8\text{ m}$$

The average velocity is

$$\bar{v} = \frac{\Delta x}{\Delta t} = \frac{x_f - x_i}{t_f - t_i} = \frac{4\text{ m} - 12\text{ m}}{3\text{ s} - 1\text{ s}} = -\frac{8\text{ m}}{2\text{ s}} = -4\text{ m/s}$$

Since the displacement and average velocity are negative for this time interval, we conclude that the particle has moved to the left, toward decreasing values of x.

3.2 INSTANTANEOUS VELOCITY

The velocity of a particle at any instant of time, or at some point on a space-time graph, is called the *instantaneous velocity*. This concept is especially important when the average velocity in different time intervals is *not constant*.

Consider the motion of a particle between the two points P and Q on the space-time graph shown in Fig. 3.2. As the point Q is brought closer and closer to the point

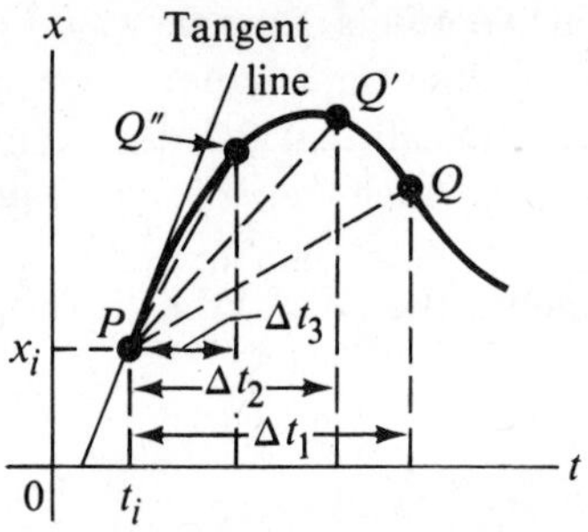

Figure 3.2 Position-time graph for a particle moving along the x axis. As the time intervals starting at t_i get smaller and smaller, the average velocity for that interval approaches the slope of the line tangent at P. The instantaneous velocity at P is defined as the slope of the tangent line at the time t_i.

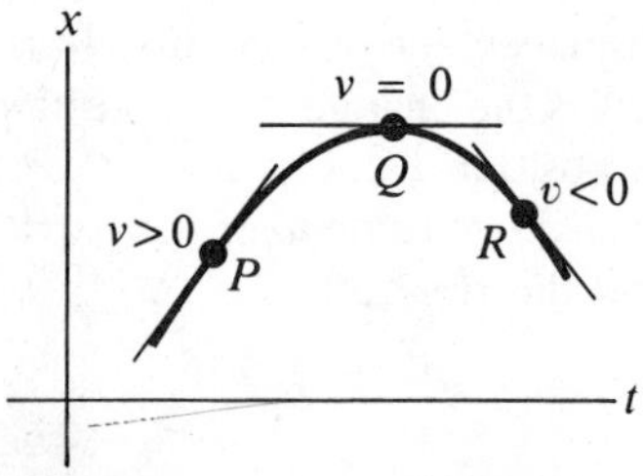

Figure 3.3 In the position-time graph shown here, the velocity is positive at P, where the slope of the tangent line is positive; the velocity is zero at Q, where the slope of the tangent line is zero; and the velocity is negative at R, where the slope of the tangent line is negative.

P, the time intervals (Δt_1, Δt_2, Δt_3, . . .) get progressively smaller. The average velocity for each time interval is the slope of the appropriate dotted line in Fig. 3.2. As the point Q approaches P, the time interval approaches zero, but at the same time the slope of the dotted line approaches that of the line tangent to the curve at the point P. The slope of the line tangent to the curve at P is defined to be the *instantaneous velocity* at the time t_i. In other words, the instantaneous velocity, v, equals the limiting value of the ratio $\Delta x/\Delta t$ as Δt approaches zero[1]:

Definition of instantaneous velocity

$$v \equiv \lim_{\Delta t \to 0} \frac{\Delta x}{\Delta t} \tag{3.2}$$

In the calculus notation, this limit is called the *derivative* of x with respect to t, written dx/dt:

Definition of the derivative

$$v \equiv \lim_{\Delta t \to 0} \frac{\Delta x}{\Delta t} = \frac{dx}{dt} \tag{3.3}$$

The instantaneous velocity can be positive, negative, or zero.

When the slope of the space-time graph is positive, such as at the point P in Fig. 3.3, v is positive. At point R, v is negative since the slope is negative. Finally, the instantaneous velocity is zero at the peak Q (the turning point), where the slope is zero. *From here on, we shall usually use the word* velocity *to designate instantaneous velocity.*

The *instantaneous speed* of a particle is defined as the magnitude of the instantaneous velocity. Hence, by definition, *speed* can never be negative.

---

[1]Note that the displacement, Δx, also approaches zero as Δt approaches zero. However, as Δx and Δt become smaller and smaller, the ratio $\Delta x/\Delta t$ approaches a value equal to the *true* slope of the line tangent to the x versus t curve.

Example 3.2

A particle moves along the x axis. Its x coordinate varies with time according to the expression $x = -4t + 2t^2$, where x is in m and t is in s. The position-time graph for this motion is shown in Fig. 3.4. Note that the particle first moves in the negative x direction for the first second of motion, stops instantaneously at $t = 1$ s, and then heads back in the positive x direction for $t > 1$ s. (a) Determine the displacement of the particle in the time intervals $t = 0$ to $t = 1$ s and $t = 1$ s to $t = 3$ s.

In the first time interval, we can set $t_i = 0$ and $t_f = 1$ s. Since $x = -4t + 2t^2$, we get for the first displacement

$$\begin{aligned}\Delta x_{01} &= x_f - x_i \\ &= [-4(1) + 2(1)^2] - [-4(0) + 2(0)^2] \\ &= -2 \text{ m}\end{aligned}$$

Likewise, in the second time interval we can set $t_i = 1$ s and $t_f = 3$ s. Therefore, the second displacement in this interval is

$$\begin{aligned}\Delta x_{13} &= x_f - x_i \\ &= [-4(3) + 2(3)^2] - [-4(1) + 2(1)^2] \\ &= 8 \text{ m}\end{aligned}$$

Note that these displacements can also be read directly from the position-time graph (Fig. 3.4).

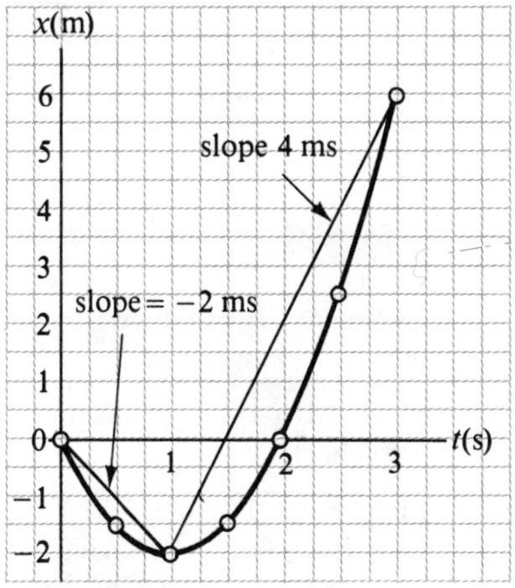

t(s)	x(m)
0	0
0.5	−1.5
1	−2
1.5	−1.5
2	0
2.5	2.5
3	6

Figure 3.4 (Example 3.2) Position-time graph for a particle having an x coordinate that varies in time according to $x = -4t + 2t^2$. Note that $\bar{v}$ is *not* the same as $v = -4 + 4t$.

(b) Calculate the average velocity in the time intervals $t = 0$ to $t = 1$ s and $t = 1$ s to $t = 3$ s.

In the first time interval, $\Delta t = t_f - t_i = 1$ s. Therefore, using Eq. 3.1 and the results from (a) gives

$$\bar{v}_{01} = \frac{\Delta x_{01}}{\Delta t} = \frac{-2 \text{ m}}{1 \text{ s}} = -2 \text{ m/s}$$

Likewise, in the second time interval, $\Delta t = 2$ s; therefore

$$\bar{v}_{13} = \frac{\Delta x_{13}}{\Delta t} = \frac{8 \text{ m}}{2 \text{ s}} = 4 \text{ m/s}$$

These values agree with the slopes of the lines joining these points in Fig. 3.4.

(c) Find the instantaneous velocity of the particle at $t = 2.5$ s.

By measuring the slope of the position-time graph at $t = 2.5$ s, we find that $v = 6$ m/s.[2] (You should show that the velocity is -4 m/s at $t = 0$ and zero at $t = 1$ s.) Do you see any symmetry in the motion? For example, does the speed ever repeat itself?

Example 3.3

The limiting process: The position of a particle moving along the x axis varies in time according to the expression $x = 3t^2$, where x is in m and t is in s. Find the velocity at any time.

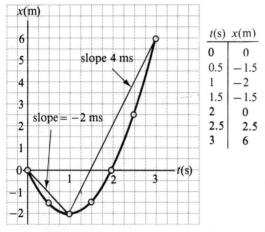

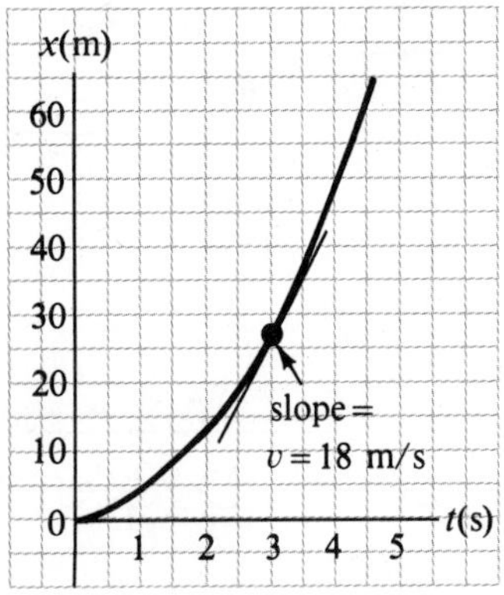

Figure 3.5 (Example 3.3) Position-time graph for a particle having an x coordinate that varies in time according to $x = 3t^2$. Note that the instantaneous velocity at $t = 3$ s equals the slope of the line tangent to the curve at this point.

Solution: The position-time graph for this motion is shown in Fig. 3.5. We can compute the velocity at any time t by using the definition of the instantaneous velocity. If the initial coordinate of the particle at time t is $x_i = 3t^2$, then the coordinate at a later time $t + \Delta t$ is

$$\begin{aligned}x_f &= 3(t + \Delta t)^2 = 3[t^2 + 2t\,\Delta t + (\Delta t)^2] \\ &= 3t^2 + 6t\,\Delta t + 3(\Delta t)^2\end{aligned}$$

Therefore, the displacement in the time interval Δt is

$$\begin{aligned}\Delta x = x_f - x_i &= 3t^2 + 6t\,\Delta t + 3(\Delta t)^2 - 3t^2 \\ &= 6t\,\Delta t + 3(\Delta t)^2\end{aligned}$$

The average velocity in this time interval is

$$\bar{v} = \frac{\Delta x}{\Delta t} = 6t + 3\Delta t$$

To find the instantaneous velocity, we take the limit of this expression as Δt approaches zero. In doing so, we see that the term $3\Delta t$ goes to zero, therefore

$$v = \lim_{\Delta t \to 0} \frac{\Delta x}{\Delta t} = 6t \text{ m/s}$$

Notice that this expression gives us the velocity at *any* general time t. It tells us that v is increasing linearly in time. It is then a straightforward matter to find the velocity at some specific time from the expression $v = 6t$. For example, at $t = 3$ s, the velocity is $v = 6(3) = 18$ m/s. Again, this can be checked from the slope of the graph at $t = 3$ s.

[2] We could also use the rules of differential calculus to find the velocity from the displacement. That is, $v = \frac{dx}{dt} = \frac{d}{dt}(-4t + 2t^2) = 4(-1 + t)$ m/s. Therefore, at $t = 2.5$ s, $v = 4(-1 + 2.5) = 6$ m/s. A review of basic operations in the calculus is provided in Appendix B.

The limiting process can also be examined numerically. For example, we can compute the displacement and average velocity for various time intervals beginning at $t = 3$ s, using the expressions for Δx and $\bar{v}$. The results of such calculations are given in Table 3.1. Notice that as the time intervals get smaller and smaller, the average velocity more nearly approaches the value of the instantaneous velocity at $t = 3$ s, namely, 18 m/s.

TABLE 3.1 Displacement and Average Velocity for Various Time Intervals for the Function $x = 3t^2$

Δt (s)	Δx (m)	$\Delta x/\Delta t$ (m/s)
1.00	21	21
0.50	9.75	19.5
0.25	4.69	18.8
0.10	1.83	18.3
0.05	0.9075	18.15
0.01	0.1803	18.03
0.001	0.018003	18.003

Note: The intervals begin at $t = 3$ s.

Q1. Average velocity and instantaneous velocity are generally different quantities. Can they ever be equal for a specific type of motion? Explain.

Q2. If $\bar{v}$ is nonzero for some time interval Δt, does this mean that the instantaneous velocity is never zero during this interval? Explain.

Q3. If $\bar{v} = 0$ for some time interval Δt and if $v(t)$ is a continuous function, show that the instantaneous velocity must go to zero some time in this interval. (A sketch of x versus t might be useful in your proof.)

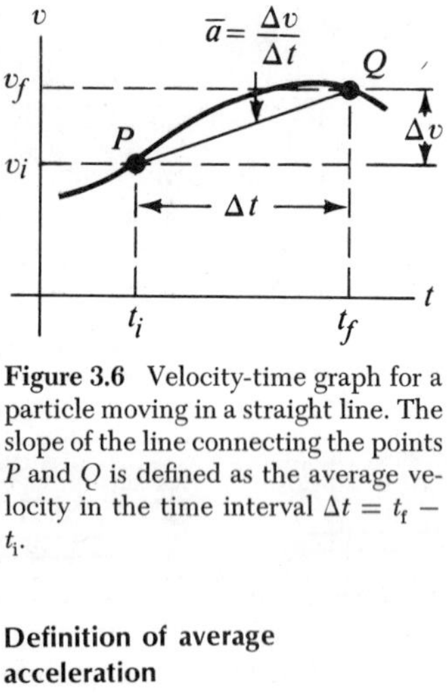

Figure 3.6 Velocity-time graph for a particle moving in a straight line. The slope of the line connecting the points P and Q is defined as the average velocity in the time interval $\Delta t = t_f - t_i$.

3.3 ACCELERATION

When the velocity of a particle changes with time, the particle is said to be *accelerating*. For example, the velocity of a car will increase when you "step on the gas." The car will slow down when you apply the brakes. However, we need a more precise definition of acceleration than this.

Suppose a particle moving along the x axis has a velocity v_i at time t_i and a velocity v_f at time t_f, as in Fig. 3.6. The *average acceleration* of the particle in the time interval $\Delta t = t_f - t_i$ is defined as the ratio $\Delta v/\Delta t$, where $\Delta v = v_f - v_i$ is the *change* in velocity in this time interval:

Definition of average acceleration

$$\bar{a} \equiv \frac{v_f - v_i}{t_f - t_i} = \frac{\Delta v}{\Delta t} \tag{3.4}$$

Acceleration has dimensions of length divided by (time)^2, or L/T^2. Some of the common units of acceleration are meters per second per second (m/s^2) and feet per second per second (ft/s^2).

In some situations, the value of the average acceleration may be different over different time intervals. It is therefore useful to define the *instantaneous acceleration* as the limit of the average acceleration as Δt approaches zero. This concept is analogous to the definition of instantaneous velocity discussed in the previous section. If we imagine that the point Q is brought closer and closer to the point P in Fig. 3.6 and take the limit of the ratio $\Delta v/\Delta t$ as Δt approaches zero, we get the

$$a \equiv \lim_{\Delta t \to 0} \frac{\Delta v}{\Delta t} = \frac{dv}{dt} \qquad (3.5)$$

Definition of instantaneous acceleration

That is, the instantaneous acceleration equals the derivative of the velocity with respect to time, which by definition is the slope of the velocity-time graph. Again you should note that if a is positive, the acceleration is in the positive x direction, whereas negative a implies acceleration in the negative x direction. *From now on we shall use the term* acceleration *to mean instantaneous acceleration.* Average acceleration is seldom used in physics.

Since $v = dx/dt$, the acceleration can also be written

$$a = \frac{dv}{dt} = \frac{d}{dt}\left(\frac{dx}{dt}\right) = \frac{d^2x}{dt^2} \qquad (3.6)$$

That is, the acceleration equals the *second derivative* of the coordinate with respect to time.

Figure 3.7 shows how the acceleration-time curve can be derived from the velocity-time curve. In these sketches, the acceleration at any time is simply the slope of the velocity-time graph at that time. Positive values of the acceleration correspond to those points where the velocity is increasing to the right. The acceleration reaches a maximum at time t_1, when the slope of the velocity-time graph is a maximum. The acceleration then goes to zero at time t_2, when the velocity is a maximum (that is, when the velocity is momentarily not changing and the slope of the v versus t graph is zero). Finally, the acceleration is negative when the velocity to the right is decreasing in time.

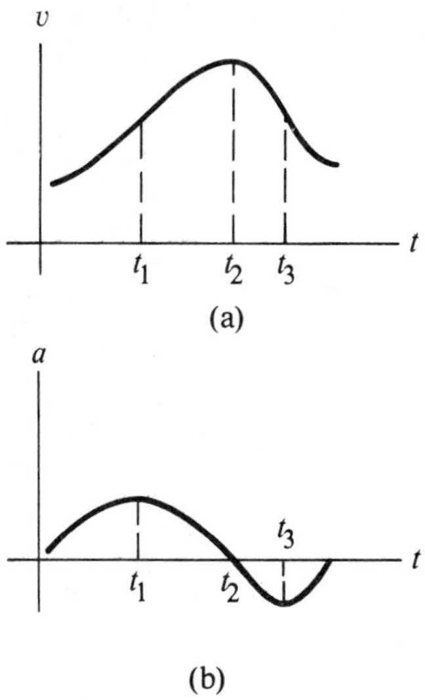

Figure 3.7 The instantaneous acceleration can be obtained from the velocity-time graph (a). At each instant, the acceleration in the a versus t graph (b) equals the slope of the line tangent to the v versus t curve.

Example 3.4

The velocity of a particle moving along the x axis varies in time according to the expression $v = (40 - 5t^2)$ m/s, where t is in s. (a) Find the average acceleration in the time interval $t = 0$ to $t = 2$ s.

The velocity-time graph for this function is given in Fig. 3.8. The velocities at $t_i = 0$ and $t_f = 2$ s are found by substituting these values of t into the expression given for the velocity:

$$v_i = 40 - 5t_i^2 = 40 - 5(0)^2 = 40 \text{ m/s}$$

$$v_f = 40 - 5t_f^2 = 40 - 5(2)^2 = 20 \text{ m/s}$$

Therefore, the average acceleration in the time interval $\Delta t = t_f - t_i = 2$ s is given by

$$\bar{a} = \frac{v_f - v_i}{t_f - t_i} = \frac{(20 - 40) \text{ m/s}}{(2 - 0) \text{ s}} = -10 \text{ m/s}^2$$

The negative sign is consistent with the fact that the slope of the line joining the initial and final points on the velocity-time graph is negative.

(b) Determine the acceleration at $t = 2$ s.

The velocity at time t is given by $v_i = (40 - 5t^2)$ m/s, and the velocity at time $t + \Delta t$ is given by

$$v_f = 40 - 5(t + \Delta t)^2 = 40 - 5t^2 - 10t\,\Delta t - 5(\Delta t)^2$$

Therefore, the change in velocity over the time interval Δt is

$$\Delta v = v_f - v_i = [-10t\,\Delta t - 5(\Delta t)^2] \text{ m/s}$$

Dividing this expression by Δt and taking the limit of the result as Δt approaches zero, we get the acceleration at *any* time t:

$$a = \lim_{\Delta t \to 0} \frac{\Delta v}{\Delta t} = \lim_{\Delta t \to 0} (-10t - 5\,\Delta t) = -10t \text{ m/s}$$

Therefore, at $t = 2$ s, we find that

$$a = -10(2) = -20 \text{ m/s}$$

This result can also be obtained by measuring the slope of the velocity-time graph at $t = 2$ s. Note that the acceleration is not constant in this example. Situations involving constant acceleration will be treated in the next section.

So far we have evaluated the derivatives of a function by starting with the definition of the function and then taking the limit of a specific ratio. Those of you familiar with the calculus should recognize that there are specific rules for taking the derivatives of various functions. These rules, which are listed in Appendix B, enable us to evaluate derivatives quickly.

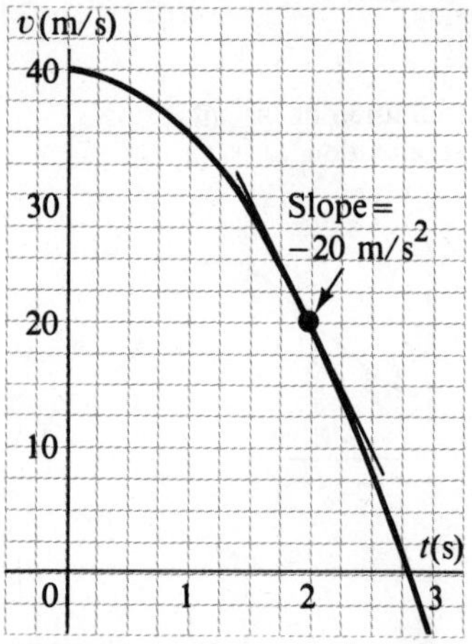

Figure 3.8 (Example 3.4) The velocity-time graph for a particle moving along the x axis according to the relation $v = (40 - 5t^2)$ m/s. Note that the acceleration at $t = 2$ s is equal to the slope of the tangent line at that time.

Suppose x is proportional to some power of t, such as

$$x = At^n$$

where A and n are constants. (This is a very common functional form.) The derivative of x with respect to t is given by

$$\frac{dx}{dt} = nAt^{n-1}$$

Applying this rule to Example 3.3, where $x = 3t^2$, we see that $v = dx/dt = 6t$, in agreement with our result of taking the limit explicitly. Likewise, in Example 3.4, where $v = 40 - 5t^2$, we find that $a = dv/dt = -10t$. (Note that the derivative of any constant is zero.)

Q4. Is it possible to have a situation in which the velocity and acceleration have opposite signs? If so, sketch a velocity-time graph to prove your point.

Q5. If the velocity of a particle is nonzero, can its acceleration ever be zero? Explain.

Q6. If the velocity of a particle is zero, can its acceleration ever be nonzero? Explain.

3.4 ONE-DIMENSIONAL MOTION WITH CONSTANT ACCELERATION

If the acceleration of a particle varies in time, the motion can be complex and difficult to analyze. A very common and simple type of one-dimensional motion occurs when the acceleration is constant, or uniform. In this type of motion, the average acceleration equals the instantaneous acceleration. Consequently, the velocity increases or decreases at the same rate throughout the motion.

If we replace $\bar{a}$ by a in Eq. 3.4, we find that

$$a = \frac{v_f - v_i}{t_f - t_i}$$

For convenience, let $t_i = 0$ and t_f be any arbitrary time t. Also, let $v_i = v_0$ (the initial velocity at $t = 0$) and $v_f = v$ (the velocity at any arbitrary time t). With this notation, we can express the acceleration

$$a = \frac{v - v_0}{t}$$

Velocity as a function of time

$$v = v_0 + at \tag{3.7}$$

This expression enables us to predict the velocity at *any* time t if the initial velocity, acceleration, and elapsed time are known. A graph of velocity versus time for this motion is shown in Fig. 3.9a. The graph is a straight line the slope of which is the acceleration, a, consistent with the fact that $a = dv/dt$ is a constant. From this graph and from Eq. 3.7, we see that the velocity at any time t is the sum of the initial velocity, v_0, and the change in velocity, at. The graph of acceleration versus time (Fig. 3.9b) is a straight line with a slope of zero, since the acceleration is constant. Note that if the acceleration were negative (a decelerating particle), the slope of Fig. 3.9a would be negative.

One of the features of one-dimensional motion with constant acceleration is the fact that, since the velocity varies linearly in time according to Eq. 3.7, we can express the average velocity in any time interval as the arithmetic mean of the

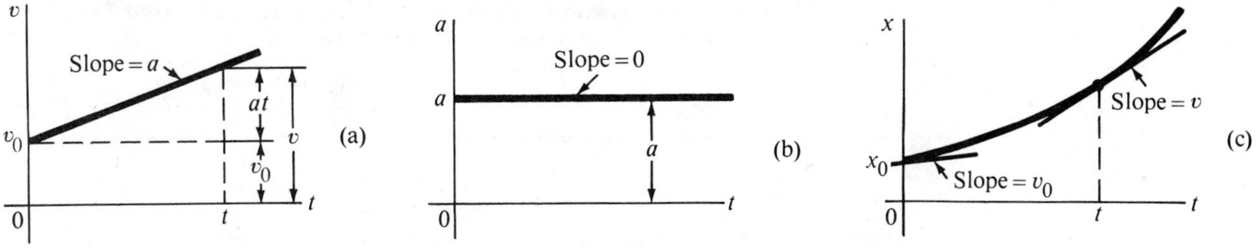

Figure 3.9 A particle moving along the x axis with uniform acceleration a; (a) the velocity-time graph, (b) the acceleration-time graph, and (c) the space-time graph.

initial velocity, v_0, and the final velocity, v:

$$\bar{v} = \frac{v_0 + v}{2} \quad \text{(for constant } a\text{)} \tag{3.8}$$

Note that this expression is only valid when the acceleration is constant, that is, when the velocity varies linearly with time.

We can now use this result and Eq. 3.1 to obtain the displacement as a function of time. Again, we choose $t_i = 0$, at which time the initial position is $x_i = x_0$. This gives

$$\Delta x = \bar{v}\,\Delta t = \left(\frac{v_0 + v}{2}\right)t$$

or

$$x - x_0 = \frac{1}{2}(v + v_0)t \tag{3.9}$$

Displacement as a function of time

We can obtain another useful expression for the displacement by substituting Eq. 3.7 into Eq. 3.9:

$$x - x_0 = \frac{1}{2}(v_0 + v_0 + at)t$$

$$x - x_0 = v_0 t + \frac{1}{2}at^2 \tag{3.10}$$

Finally, we can obtain an expression that does not contain the time by substituting the value of t from Eq. 3.7 into Eq. 3.9. This gives

$$x - x_0 = \frac{1}{2}(v_0 + v)\left(\frac{v - v_0}{a}\right) = \frac{v^2 - {v_0}^2}{2a}$$

$$v^2 = {v_0}^2 + 2a(x - x_0) \tag{3.11}$$

Velocity as a function of displacement

A position-time graph for motion under constant acceleration assuming positive a is shown in Fig. 3.9c. Note that the curve representing Eq. 3.10 is a parabola. The slope of the tangent to this curve at $t = 0$ equals the initial velocity, v_0, and the slope of the tangent line at any time t equals the velocity at that time.

If motion occurs in which the acceleration is *zero,* then we see that

$$\left.\begin{aligned} v &= v_0 \\ x - x_0 &= vt \end{aligned}\right\} \text{when } a = 0$$

That is, when the acceleration is zero, the velocity is a constant and the displacement changes linearly with time.

Equations 3.7 through 3.11 are five *kinematic expressions that may be used to*

TABLE 3.2 Kinematic Equations for Motion in a Straight Line Under Constant Acceleration

EQUATION	*INFORMATION GIVEN BY EQUATION*
$v = v_0 + at$	Velocity as a function of time
$x - x_0 = \frac{1}{2}(v + v_0)t$	Displacement as a function of velocity and time
$x - x_0 = v_0 t + \frac{1}{2}at^2$	Displacement as a function of time
$v^2 = v_0^2 + 2a(x - x_0)$	Velocity as a function of displacement

Note: Motion is along the x axis. At $t = 0$, the position of the particle is x_0 and its velocity is v_0.

solve any problem in one-dimensional motion under constant acceleration. Keep in mind that these relationships were derived from the definition of velocity and acceleration, together with some simple algebraic manipulations and the requirement that the acceleration be constant. It is often convenient to choose the initial position of the particle as the origin of the motion, so that $x_0 = 0$ at $t = 0$. In such a case, the displacement is simply x.

The four kinematic equations that are used most often are listed in Table 3.2 for convenience.

The choice of which kinematic equation or equations you should use in a given situation depends on what is known beforehand. Sometimes it is necessary to use two of these equations to solve for two unknowns, such as the displacement and velocity at some instant. For example, suppose the initial velocity, v_0, and acceleration, a, are given. You can then find (1) the velocity after a time t has elapsed, using $v = v_0 + at$, and (2) the displacement after a time t has elapsed, using $x - x_0 = v_0 t + \frac{1}{2}at^2$. You should recognize that the quantities that vary during the motion are velocity, displacement, and time.

You will get a great deal of practice in the use of these equations by solving a number of exercises and problems. Many times you will discover that there is more than one method for obtaining a solution.

Example 3.5

A certain automobile manufacturer claims that its super-deluxe sportscar will accelerate uniformly from rest to a speed of 87 mi/h in 8 s. (a) Determine the acceleration of the car.

First note that $v_0 = 0$ and the velocity after 8 s is 87 mi/h = 128 ft/s. (It is useful to note that 60 mi/h = 88 ft/s exactly.) Using $v = v_0 + at$, we can find the acceleration:

$$a = \frac{v - v_0}{t} = \frac{128 \text{ ft/s}}{8 \text{ s}} = 16 \text{ ft/s}^2$$

(b) Find the distance the car travels in the first 8 s.

Let the origin be at the original position of the car, so that $x_0 = 0$. Using Eq. 3.9 we find that

$$x = \frac{1}{2}(v_0 + v)t = \frac{1}{2}(128 \text{ ft/s})(8 \text{ s}) = 512 \text{ ft}$$

(c) What is the velocity of the car 10 s after it begins its motion, assuming it continues to accelerate at the rate of 16 ft/s²?

Again, we can use $v = v_0 + at$, with $v_0 = 0$, $t = 10$ s, and $a = 16$ ft/s². This gives

$$v = v_0 + at = 0 + (16 \text{ ft/s}^2)(10 \text{ s}) = 160 \text{ ft/s}$$

which corresponds to 109 mi/h.

Example 3.6

An electron in a cathode ray tube of a TV set enters a region where it accelerates uniformly from a speed of 3×10^4 m/s to a speed of 5×10^6 m/s in a distance of 2 cm. (a) How long is the electron in this region where it accelerates?

Taking the direction of motion to be along the x axis, we can use Eq. 3.9 to find t, since the displacement and velocities are known:

$$x - x_0 = \frac{1}{2}(v_0 + v)t$$

$$t = \frac{2(x - x_0)}{v_0 + v} = \frac{2(2 \times 10^{-2} \text{ m})}{(3 \times 10^4 + 5 \times 10^6) \text{ m/s}}$$
$$= 8 \times 10^{-9} \text{ s}$$

(b) What is the acceleration of the electron in this region?

To find the acceleration, we can use $v = v_0 + at$ and the results from (a):

$$a = \frac{v - v_0}{t} = \frac{(5 \times 10^6 - 3 \times 10^4)\text{ m/s}}{8 \times 10^{-9}\text{ s}}$$

$$= 6.2 \times 10^{14}\text{ m/s}^2$$

We also could have used Eq. 3.11 to obtain the acceleration, since the velocities and displacement are known. Try it! Although a is very large in this example, the acceleration occurs over a very short time interval and is a typical value for such charged particles in acceleration.

Q7. Can the equations of kinematics (Eqs. 3.7 through 3.11) be used in a situation where the acceleration varies in time? Can they be used when the acceleration is zero?

3.5 FREELY FALLING BODIES

It is well known that all objects, when dropped, will fall toward the earth with nearly constant acceleration. There is a legendary story that Galileo Galilei first discovered this fact by observing that two different weights dropped simultaneously from the Leaning Tower of Pisa hit the ground at approximately the same time. Although there is some doubt that this particular experiment was carried out, it is well established that Galileo did perform many systematic experiments on objects moving on inclined planes. Through careful measurements of distances and time intervals, he was able to show that the displacement of an object starting from rest is proportional to the square of the time the object is in motion. This observation is consistent with one of the kinematic equations we derived for motion under constant acceleration (Eq. 3.10). Galileo's achievements in the science of mechanics paved the way for Newton in his development of the laws of motion.

Galileo Galilei (1564–1642)

You might want to try the following experiment. Drop a coin and a crumpled-up piece of paper simultaneously from the same height. If the effects of air friction are negligible, both will experience the same motion and hit the floor at the same time. In the idealized case, where air resistance *is* neglected, such motion is referred to as *free fall.* If this same experiment could be conducted in a good vacuum, where air friction is truly negligible, the paper and coin would fall with the same acceleration, regardless of the shape of the paper. On August 2, 1971, such an experiment was conducted on the moon by astronaut David Scott. He simultaneously released a geologist's hammer and a falcon's feather, and in unison they fell to the lunar surface. This demonstration would have surely pleased Galileo!

Acceleration due to gravity $g = 9.8\text{ m/s}^2$

We shall denote the *acceleration due to gravity* by the symbol **g**. As we shall see in Chapter 6, the magnitude of **g** decreases with increasing altitude. Furthermore, there are slight variations in **g** with latitude. The vector **g** is directed downward toward the center of the earth. At the earth's surface, the magnitude of **g** is approximately 9.8 m/s^2, or 980 cm/s^2, or 32 ft/s^2. Unless stated otherwise, we shall use this value for **g** when doing calculations.

Definition of free fall

When we use the expression *freely falling object,* we do not necessarily refer to an object dropped from rest. A freely falling object is any object moving freely under the influence of gravity, *regardless* of its initial motion. Objects thrown upward or downward and those released from rest are all falling freely once they are released. Furthermore, it is important to recognize that any freely falling object experiences an acceleration directed *downward.* This is true regardless of the initial motion of the object. *An object thrown upward (or downward) will experience the same acceleration as an object released from rest. Once they are in free fall, all objects will have an acceleration downward, equal to the acceleration due to gravity.*

If we neglect air resistance and assume that the gravitational acceleration does not vary with altitude, then the motion of a freely falling body is equivalent to motion in one dimension under constant acceleration. Therefore our kinematic equations for constant acceleration can be applied. We shall take the vertical direction to be the y axis and call y positive upward. With this choice of coordinates, we can replace x by y in Eqs. 3.7, 3.9, 3.10, and 3.11. Furthermore, since positive y is upward, the acceleration is negative (downward) and given by $a = -g$. The negative sign simply indicates that the acceleration is downward. It does not imply that the particle is increasing or decreasing in velocity.[3] With these substitutions, we get the following expressions:

$$v = v_0 - gt \checkmark \tag{3.12}$$

$$y - y_0 = \frac{1}{2}(v + v_0)t \checkmark \tag{3.13}$$

Kinematic equations for a freely falling body

$$y - y_0 = v_0 t - \frac{1}{2} gt^2 \checkmark \tag{3.14}$$

$$v^2 = v_0{}^2 - 2g(y - y_0) \checkmark \tag{3.15}$$

You should note that *the negative sign for the acceleration is already included in these expressions.* Therefore, when using these equations in any free-fall problem, you should simply substitute $g = 9.8$ m/s^2.

Consider the case of a particle thrown vertically upward from the origin with a velocity v_0. In this case, v_0 is *positive* and $y_0 = 0$. Graphs of the displacement and velocity as functions of time are shown in Fig. 3.10. Note that the velocity is initially positive, but decreases in time and goes to zero at the peak of the path. From Eq. 3.12, we see that this occurs at the time $t_1 = v_0/g$. At this time, the displacement has its largest positive value, which can be calculated from Eq. 3.14 with $t = t_1 = v_0/g$. This gives $y_{max} = v_0{}^2/2g$.

At the time $t_2 = 2t_1 = 2v_0/g$, we see from Eq. 3.14 that the displacement is again zero, that is, the particle has returned to its starting point. Furthermore, at time t_2 the velocity is given by $v = -v_0$. (This follows directly from Eq. 3.12.) Hence, there is symmetry in the motion. In other words, both the displacement and the magnitude of the velocity repeat themselves in the time interval $t = 0$ to $t = 2v_0/g$.

In the examples that follow, we shall, for convenience, assume that $y_0 = 0$ at $t = 0$. Notice that this does not affect the solution to the problem. If y_0 is nonzero, then the graph of y versus t (Fig. 3.10a) is simply shifted upward or downward by an amount y_0, while the graph of v versus t (Fig. 3.10b) remains unchanged.

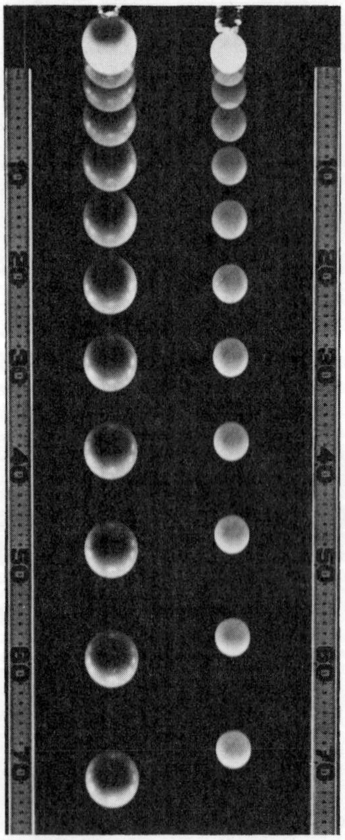

A stroboscopic photograph of two freely falling balls. The time interval between flashes is $\frac{1}{30}$ S and the scale is in cm. Can you determine g from these data?

(a)   (b)

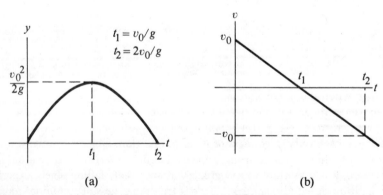

Figure 3.10 Graphs of (a) the displacement versus time and (b) the velocity versus time for a freely falling particle, where y and v are taken to be positive upward. Note the symmetry in the curves about $t = t_1$.

[3]One can also take y positive downward, in which case $a = +g$. The results will be the same, regardless of the convention chosen.

Example 3.7

A golf ball is released from rest from the top of a very tall building. Neglecting air resistance, calculate the coordinate and velocity of the ball after 1, 2, and 3 s.

Solution: We take the origin of coordinates at the starting point of the ball ($y_0 = 0$ at $t = 0$) and note that y is taken positive upward. Since $v_0 = 0$, Eqs. 3.12 and 3.14 become

$$v = -gt = -9.8t$$

$$y = -\frac{1}{2}gt^2 = -4.9t^2$$

where t is in s, v is in m/s, and y is in m. These expressions give the velocity and displacement at any time t after the ball is released. Therefore, at $t = 1$ s we find that

$$v = -9.8(1) = -9.8 \text{ m/s}$$

$$y = -4.9(1)^2 = -4.9 \text{ m}$$

Likewise, at $t = 2$ s, we find that $v = -19.6$ m/s and $y = -19.6$ m. Finally, at $t = 3$ s, $v = -29.4$ m/s and $y = -44.1$ m. The minus signs for v indicate that the velocity vector is directed downward, and the minus signs for y indicate displacements in the negative y direction.

Example 3.8

A stone is thrown from the top of a building with an initial velocity of 20 m/s upward. The building is 50 m high, and the stone just misses the edge of the roof, as in Fig. 3.11. (a) Determine the velocity and coordinate of the stone at any time t while it is in motion.

Again, it is convenient to choose the origin of coordinates at the initial position of the stone, so $y_0 = 0$ at $t = 0$. (If you choose, say, the base of the building as the origin, then $y_0 = 50$ m at $t = 0$.) Since $v_0 = 20$ m/s and $a = -9.8$ m/s^2, Eqs. 3.12 and 3.14 become

$$(1) \qquad v = 20 - 9.8t$$

$$(2) \qquad y = 20t - 4.9t^2$$

where t is in s, v is in m/s, and y is in m. The positions of the stone at various values of t are given in Fig. 3.11. If we let t_1 be the time it takes the stone to reach its maximum height and note that $v = 0$ at maximum height, then (1) gives

$$20 - 9.8t_1 = 0$$

$$t_1 = \frac{20 \text{ m/s}}{9.8 \text{ m/s}^2} = 2.04 \text{ s}$$

Using this value of t in (2) gives the maximum height, y_{max}. That is, at $t = t_1$,

$$y_{max} = (20 - 4.9t_1)t_1 = [20 - 4.9(2.04)]2.04$$
$$= 20.4 \text{ m}$$

One can also calculate y_{max} using Eq. 3.13. Try it!

(b) Find the time it takes the stone to pass by its starting point and the velocity of the stone at this time.

When the stone is at its starting point, its y coordinate is zero. Using (2) with $y = 0$, we obtain the following expression:

$$20t - 4.9t^2 = 0$$

This equation is quadratic in t and has two solutions. Factoring the equation, we get

$$t(20 - 4.9t) = 0$$

One solution is clearly $t = 0$, corresponding to the time the stone starts its motion. The other solution is $t = 4.08$ s, which is the solution we are after. Substituting this value of t into (1), we find that the stone has a velocity of -20 m/s when it passes by its starting point. That is, the velocity of the stone when it arrives at its initial altitude is equal in magnitude to its initial velocity, but opposite in direction. This demonstrates the symmetry of the motion that was discussed earlier.

(c) Find the velocity and position of the stone 5 s after it is thrown.

Since (1) and (2) give the velocity and coordinate at *any* time t, we can obtain the results by setting $t = 5$ s. This gives

$$v = 20 - 9.8(5) = -29 \text{ m/s}$$

$$y = 20(5) - 4.9(5)^2 = -22.5 \text{ m}$$

(d) Find the velocity of the stone just before it hits the ground.

We can use Eq. 3.15 to determine this velocity, noting that the coordinate of the stone just before hitting the ground is $y = -50$ m. This gives

$$v^2 = v_0{}^2 - 2gy = (20 \text{ m/s})^2 - 2\left(9.8 \frac{\text{m}}{\text{s}^2}\right)(-50 \text{ m})$$

$$v^2 = 400 + 980 = 1380 \text{ m}^2/\text{s}^2$$

$$v = \pm 37 \text{ m/s}$$

Since the stone is traveling downward at this time, the physically acceptable solution is $v = -37$ m/s.

(e) Find the total time the stone is in the air.

Using the result from (d) and (1) and calling the total time t_2, we have

$$v = -37 \text{ m/s} = (20 - 9.8t_2) \text{ m/s} \qquad (\text{at } t = t_2)$$

$$t_2 = \frac{(20 + 37) \text{ m/s}}{9.8 \text{ m/s}^2} = 5.8 \text{ s}$$

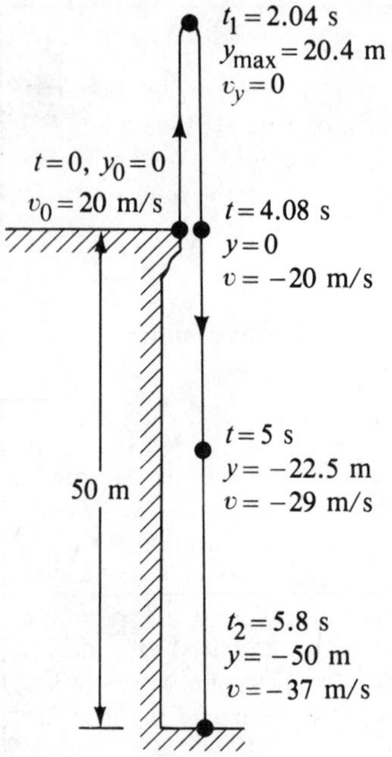

Figure 3.11 (Example 3.8) Position and velocity versus time for a freely falling particle thrown initially upward with a velocity $v_0 = 20$ m/s.

Q8. A ball is thrown vertically upward. What are its velocity and acceleration when it reaches its maximum altitude? What is its acceleration just before it strikes the ground?

Q9. A stone is thrown upward from the top of a building. Does the stone's displacement depend on the location of the origin of the coordinate system? Does the stone's velocity depend on the origin? (Assume that the coordinate system is stationary with respect to the building.) Explain.

Q10. A child throws a marble in the air with an initial velocity v_0. Another child drops a ball at the same instant. Compare the accelerations of the two objects while they are in flight.

3.6 SUMMARY

The *average velocity* of a particle during some time interval is equal to the ratio of the displacement, Δx, and the time interval, Δt:

Average velocity

$$\bar{v} = \frac{\Delta x}{\Delta t} \tag{3.1}$$

The *instantaneous velocity* of a particle is defined as the limit of the ratio $\Delta x/\Delta t$ as Δt approaches zero. By definition, this equals the derivative of x with respect to t, or the time rate of change of the position:

Instantaneous velocity

$$v = \lim_{\Delta t \to 0} \frac{\Delta x}{\Delta t} = \frac{dx}{dt} \tag{3.3}$$

The *speed* of a particle equals the absolute value of the velocity.

The *average acceleration* of a particle during some time interval is defined as the ratio of the change in its velocity, Δv, and the time interval, Δt:

Average acceleration

$$\bar{a} = \frac{\Delta v}{\Delta t} \tag{3.4}$$

The *instantaneous acceleration* is equal to the limit of the ratio $\Delta v/\Delta t$ as $\Delta t \to 0$. By definition, this equals the derivative of v with respect to t, or the time rate of change of the velocity:

Instantaneous acceleration

$$a = \lim_{\Delta t \to 0} \frac{\Delta v}{\Delta t} = \frac{dv}{dt} \tag{3.5}$$

The slope of the tangent to the x versus t curve at any instant equals the instantaneous velocity of the particle. The slope of the tangent to the v versus t curve equals the instantaneous acceleration of the particle. The area under the v versus t curve in any time interval equals the displacement of the particle in that interval.

The *equations of kinematics* for a particle moving along the x axis with uniform acceleration a (constant in magnitude and direction) are

Equations of kinematics

$$v = v_0 + at \quad (3.7)$$

$$x - x_0 = \frac{1}{2}(v_0 + v)t \quad (3.9)$$

$$x - x_0 = v_0 t + \frac{1}{2}at^2 \quad (3.10)$$

$$v^2 = {v_0}^2 + 2a(x - x_0) \quad (3.11)$$

Freely falling body

A body falling freely in the presence of the earth's gravity experiences a gravitational acceleration directed toward the center of the earth. If air friction is neglected, and if the altitude of the motion is small compared with the earth's radius, then one can assume that the acceleration of gravity, g, is constant over the range of motion, where g is equal to 9.8 m/s^2, or 32 ft/s^2. Assuming y positive upward, the acceleration is given by $-g$, and the equations of kinematics for a body in free fall are the same as those given above, with the substitutions $x \rightarrow y$ and $a \rightarrow -g$.

EXERCISES

Section 3.1 Average Velocity

1. A particle moving along the x axis is located initially at $x_i = 2.0$ m. Three minutes later, the particle is at $x_f = -5.0$ m. What is the average velocity of the particle?

2. The displacement over time for a certain particle moving along the x axis is as shown in Fig. 3.12. Find the average velocity in the time intervals (a) 0 to 1 s, (b) 0 to 4 s, (c) 1 s to 5 s, (d) 0 to 5 s.

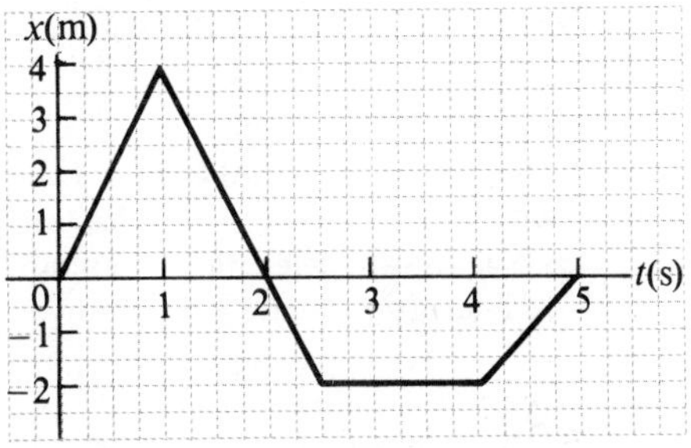

Figure 3.12 (Exercises 2 and 5).

3. A jogger runs in a straight line with an average velocity of 5 m/s for 4 min, and then with an average velocity of 4 m/s for 3 min. (a) What is her total displacement? (b) What is her average velocity during this time?

4. An athlete swims the length of a 50-m pool in 20 s and makes the return trip to the starting position in 22 s. Determine his average velocity in (a) the first half of the swim, (b) the second half of the swim, and (c) the round trip.

Section 3.2 Instantaneous Velocity

5. Find the instantaneous velocity of the particle described by Fig. 3.12 at the following times: (a) $t = 0.5$ s, (b) $t = 2$ s, (c) $t = 3$ s, (d) $t = 4.5$ s.

6. The position of a particle moving along the x axis varies linearly in time according to the expression $x = At + B$, where A and B are constants. (a) What are the dimensions of A and B? (b) Show by calculus and by graphical arguments that the average velocity equals the instantaneous velocity for this situation.

7. The position-time graph of a particle moving along the x axis is as shown in Fig. 3.13. Determine whether the velocity is positive, negative, or zero for the times (a) t_1, (b) t_2, (c) t_3, (d) t_4.

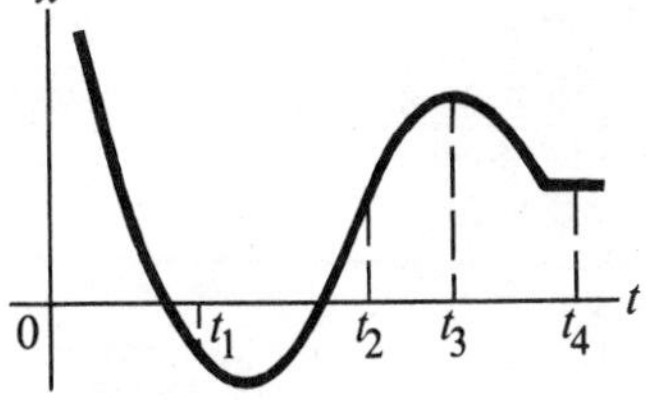

Figure 3.13 (Exercise 7).

8. The position-time graph for a particle moving along the x axis is as shown in Fig. 3.14. (a) Find the average velocity in the time interval $t = 1.5$ s to $t = 4$ s. (b) Determine the instantaneous velocity at $t = 2$ s by measuring the slope of the tangent line shown in the graph. (c) At what value of t is the velocity zero?

9. At $t = 1$ s, a particle moving with constant velocity is located at $x = -3$ m, and at $t = 6$ s, the particle is located at $x = 5$ m. (a) From this information, plot the position as a function of time. (b) Determine the velocity of the particle from the slope of this graph.

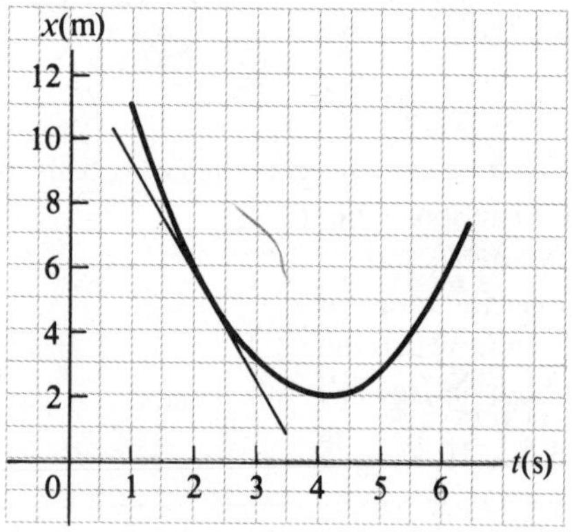

Figure 3.14 (Exercise 8).

Section 3.3 Acceleration

10. A particle moves along the x axis according to the equation $x = 2t + 3t^2$, where x is in m and t is in s. Calculate the instantaneous velocity and instantaneous acceleration at $t = 3$ s.

11. A car traveling in a straight line has a velocity of 30 m/s at some instant. Two seconds later its velocity is 25 m/s. What is its average acceleration in this time interval?

12. The position of a particle moving along the y axis is given by $y = At^3 - Bt$, where A and B are constants, y is in m, and t is in s. (a) What are the dimensions of A and B? (b) Find expressions for the velocity and acceleration as functions of time.

13. A particle moving in a straight line has a velocity of 5 m/s at $t = 0$. Its velocity at $t = 4$ s is 21 m/s. (a) What is its average acceleration in this time interval? (b) Can the average velocity be obtained from the information presented? Explain.

14. The velocity-time graph for an object moving along the x axis is as shown in Fig. 3.15. (a) Plot a graph of the acceleration versus time. (b) Determine the average acceleration of the object in the time intervals $t = 5$ s to $t = 15$ s and $t = 0$ to $t = 20$ s.

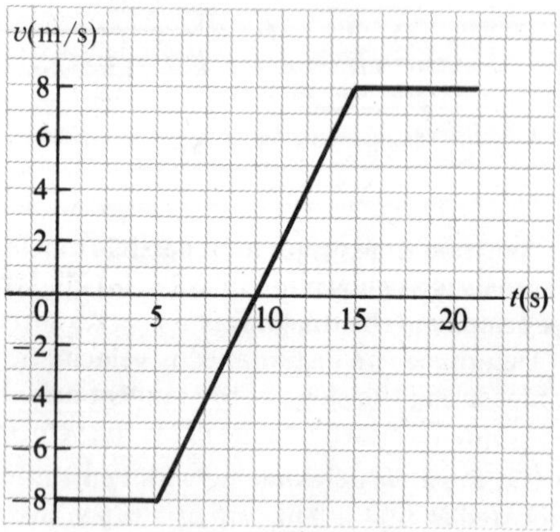

Figure 3.15 (Exercise 14).

15. The velocity of a particle as a function of time is shown in Fig. 3.16. At $t = 0$, the particle is at $x = 0$. (a) Sketch the acceleration as a function of time. (b) Determine the average acceleration of the particle in the time interval $t = 1$ s to $t = 4$ s. (c) Determine the instantaneous acceleration of the particle at $t = 2$ s.

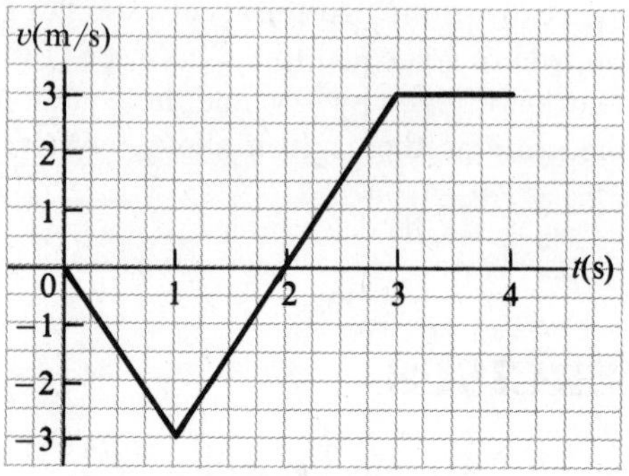

Figure 3.16 (Exercise 15).

16. A particle moves along the x axis according to the equation $x = 2 + 3t - t^2$, where x is in m and t is in s. At $t = 3$ s, find (a) the position of the particle, (b) its velocity, and (c) its acceleration.

17. The velocity of a particle moving along the x axis varies in time according to the relation $v = (15 - 8t)$ m/s. Find (a) the acceleration of the particle, (b) its velocity at $t = 3$ s, and (c) its average velocity in the time interval $t = 0$ to $t = 2$ s.

18. The velocity of a certain particle as a function of time is plotted in Fig. 3.17. (a) Sketch the acceleration as a function of time. (b) Does the particle ever travel with constant acceleration? Explain. (c) Estimate the acceleration of the particle at $t = 6$ s.

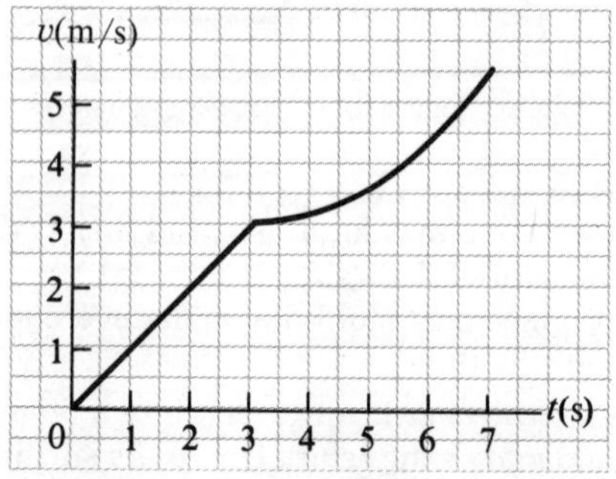

Figure 3.17 (Exercise 18).

Section 3.4 One-Dimensional Motion with Constant Acceleration

19. A particle travels in the positive x direction for 10 s at a constant speed of 50 m/s. It then accelerates uniformly to a speed of 80 m/s in the next 5 s. Find (a) the average acceleration of the particle in the first 10 s, (b) its average acceleration in the interval $t = 10$ s to $t = 15$ s, (c) the total displacement of the particle between $t = 0$ and $t = 15$ s, and (d) its average speed in the interval $t = 10$ s to $t = 15$ s.

20. The position of a particle moving along the x axis varies in time according to the expression $x = 2 + 8t - 2t^2$, where x is in m and t is in s. Find (a) the displacement of the particle in the first 3 s of motion, (b) its acceleration, (c) its initial velocity, (d) the position where the particle comes momentarily to rest, and (e) its average velocity in the first 3 s of motion.

21. A speedboat increases its speed from 50 ft/s to 80 ft/s in a distance of 200 ft. Find (a) the magnitude of its acceleration and (b) the time it takes the boat to travel this distance.

22. A racing car reaches a speed of 50 m/s. At this instant, it decelerates uniformly using a parachute and braking system and comes to rest 5 s later. (a) Determine the deceleration of the car. (b) How far does the car travel after "turning on the brakes"?

23. The acceleration of gravity on the moon is about one sixth as great as on the earth. A stone is thrown vertically upward on the moon, with an initial speed of 20 m/s. (a) How long will the stone remain in motion? (b) What is the maximum height reached by the stone relative to the moon's surface?

24. A particle starts from rest from the top of an inclined plane and slides down with constant acceleration. The inclined plane is 2.0 m long, and it takes 3.0 s for the particle to reach the bottom. Find (a) the acceleration of the particle, (b) its speed at the bottom of the incline, (c) the time it takes the particle to reach the middle of the incline, and (d) its speed at the midpoint.

25. A go-cart travels the first half of a 100-m track with a constant speed of 5 m/s. In the second half of the track, it experiences a mechanical problem and decelerates at 0.2 m/s^2. How long does it take the go-cart to travel the 100-m distance?

26. A car moving at a constant speed of 30 m/s suddenly stalls at the bottom of a hill. The car undergoes a constant deceleration of 2 m/s^2 while ascending the hill. (a) Write equations for the position and the velocity as functions of time, taking $x = 0$ at the bottom of the hill where $v_0 = 30$ m/s. (b) Determine the maximum distance traveled by the car up the hill after stalling.

27. The initial speed of a body is 5.2 m/s. What is its speed after 2.5 s if it (a) accelerates uniformly at 3.0 m/s^2 and (b) accelerates uniformly at -3.0 m/s^2 (that is, it accelerates in the negative x direction)?

28. A body has a velocity of 12 cm/s when its x coordinate is 3 cm. If its x coordinate 2 s later is -5 cm, what is the uniform acceleration of the body?

29. The initial velocity of a body moving along the x axis is -6.0 cm/s when it is located at the origin. If it accelerates uniformly at the rate of 8.0 cm/s^2, what are (a) its coordinate after 2 s and (b) its velocity after 3 s?

30. A proton has an initial velocity of 2.5×10^5 m/s and undergoes a uniform deceleration of 5.0×10^{10} m/s^2. What is its velocity after moving through a distance of 10 cm?

31. An electron has an initial velocity of 3.0×10^5 m/s. If it undergoes an acceleration of 8.0×10^{14} m/s^2, (a) how long will it take to reach a velocity of 5.4×10^5 m/s and (b) how far has it traveled in this time?

32. A railroad car is released from a locomotive on an incline. When the car reaches the bottom of the incline, it has a speed of 30 mi/h, at which point it passes through a retarder track that slows it down. If the retarder track is 30 ft long, what deceleration must it produce to bring the car to rest?

33. A bullet is fired through a board, 10 cm thick, in such a way that the bullet's line of motion is perpendicular to the face of the board. If the initial speed of the bullet is 400 m/s and it emerges from the other side of the board with a speed of 300 m/s, find (a) the deceleration of the bullet as it passes through the board and (b) the total time the bullet is in contact with the board.

34. An electron is accelerated from rest with a constant acceleration of 8×10^{12} m/s^2 to the right. It strikes a plate 4 cm from its starting point. (a) Neglecting gravity, find the final speed of the electron and its time of flight. (b) Treating the effect of gravity as a small perturbation on the electron's motion, determine how far the electron falls during its flight. (Treat the y-direction motion as independent of the x-direction motion.)

35. A hockey player is standing on his skates on a frozen pond when an opposing player skates by with the puck, moving with a uniform speed of 12 m/s. After 3 s, the first player makes up his mind to chase after his opponent. If he accelerates uniformly at 4 m/s^2, (a) how long does it take him to catch the opponent and (b) how far has he traveled in this time? (Assume the player with the puck remains in motion at constant speed.)

36. Until recently, the world's land speed record was held by Colonel John P. Stapp, USAF. On March 19, 1954, he rode a rocket-propelled sled that moved down the track at 632 mi/h. He and the sled were safely brought to rest in 1.4 s. Determine (a) the deceleration he experienced and (b) the distance he traveled during this deceleration.

37. A woman is reported to have fallen 144 ft from the 17th floor of a building, landing on a metal ventilator box, which she crushed to a depth of 18 in. She suffered only minor injuries. Neglecting air resistance, calculate (a) the speed of the woman just before she collided with the ventilator, (b) her deceleration while in contact with the box, and (c) the time it took to crush the box.

Section 3.5 Freely Falling Bodies

38. The *Guinness Book of World Records* lists a man who survived a deceleration of 200g, or 1960 m/s^2. A person seeking to break this record jumps off a cliff 102 m high onto several mattresses having a total thickness of 2 m. (a) What is the velocity of the "record-breaker" just before hitting the mattresses? (b) If the mattresses are crushed to a depth of 0.5 m, what is the record-breaker's deceleration?

39. An object is thrown vertically upward in such a way that it has a speed of 19.6 m/s when it reaches one half its maximum altitude. What are (a) its maximum altitude, (b) its velocity 1 s after it is thrown, and (c) its acceleration when it reaches its maximum altitude?

40. A ball is thrown vertically upward from the ground with an initial speed of 15 m/s. (a) How long does it take the ball to reach its maximum altitude? (b) What is its maximum altitude? (c) Determine the velocity and acceleration of the ball at $t = 2$ s.

41. A ball thrown vertically upward is caught after 3.5 s. Find (a) the initial velocity of the ball and (b) the maximum height it reaches.

42. An object enters the earth's atmosphere with a velocity of 60 mi/h downward when it is 100 miles above the earth. Neglecting the effect of air friction and assuming $g = -32$ ft/s^2, find the velocity of the object just before it hits the earth. (Do you think your result is realistic? Describe what you think really happens in this situation.)

43. A parachutist descending at a speed of 10 m/s drops a camera at an altitude of 50 m. (a) How long does it take the camera to reach the ground? (b) What is the velocity of the camera just before it hits the ground?

44. A ball is thrown vertically upward with an initial speed of 10 m/s. One second later, a stone is thrown vertically upward with an initial speed of 25 m/s. Determine (a) the time it takes the stone to reach the same height as the ball, (b) the velocity of the ball and stone when they are at the same height, and (c) the total time each is in motion before returning to the original height.

PROBLEMS

1. In a recent California driver's handbook, the following table is given listing data on the distance a typical moving vehicle travels for various initial speeds. The *thinking distance* corresponds to the fact that the driver has a finite reaction time, and the *braking distance* is how far the vehicle travels after the brakes are applied. Inspect these data carefully and determine (a) the thinking time, or reaction time, and (b) the deceleration of the vehicle.

SPEED (mi/h)	*THINKING DISTANCE* (ft)	*BRAKING DISTANCE* (ft)	*TOTAL DISTANCE* (ft)
25	27	34	61
35	38	67	105
45	49	110	159
55	60	165	225
65	71	231	302

2. A certain trolley car in San Francisco can stop in 10 s when traveling at maximum speed. On one occasion, the driver sees a dog d m in front of the car and slams on the brakes instantly. The car reaches the dog 8 s later, and the dog jumps off the track just in time. If the car travels 4 m beyond the position of the dog before coming to a stop, how far was the car from the dog? (Hint: You will need three equations.)

3. A rocket is fired vertically upward with an initial velocity of 80 m/s. It accelerates upward at 4 m/s^2 until it reaches an altitude of 1000 m. At that point, its engines fail and the rocket goes into free flight, with acceleration -9.8 m/s^2. (a) How long is the rocket in motion? (b) What is its maximum altitude? (c) What is its velocity just before it collides with the earth? (Hint: Consider the motion while the engine is operating separate from the free-flight motion.)

4. The gasoline consumption of a certain car varies with speed according to the relation mi/gal = $1000/(v + 40)$, where v is in mi/h. The car starts from rest and accelerates to a uniform speed of 60 mi/h in 15 s. If it maintains this speed for 5 miles before uniformly decelerating to rest in 2 s, how much gasoline is consumed by the car?

5. A train travels in time in the following manner. In the first hour, it travels with a speed v, in the next half hour it has a speed $3v$, in the next 90 min it travels with a speed $v/2$, and in the final 2 h it travels with a speed $v/3$, where v is in mi/h. (a) Plot the speed-time graph for this trip. (b) How far does the train travel in this trip? (c) What is the average speed of the train over the entire trip?

6. A "superball" is dropped from a height of 2 m above the ground. On the first bounce the ball reaches a height of 1.85 m, where it is caught. Find the velocity of the ball (a) just as it makes contact with the ground and (b) just as it leaves the ground on the bounce. (c) Neglecting the time the ball spends in contact with the ground, find the total time required for the ball to go from the dropping point to the point where it is caught.

7. The position of a particle traveling along the x axis is given by $x = t^3 - 9t^2 + 6t$, where x is in cm and t is in s. Find (a) the instantaneous velocity of the particle for any time t, (b) the *times* at which the instantaneous velocity is zero, (c) the instantaneous acceleration of the particle at the times found in (b), and (d) the total displacement of the particle in traveling from the first zero to the second zero of the velocity.

8. The coyote, in his relentless attempt to catch the elusive road runner, loses his footing and falls from a sharp cliff, 1500 ft above ground level. After 5 s of free fall the coyote remembers he is wearing his Acme rocket-powered backpack, which he turns on. (a) The coyote comes to the ground with a gentle landing (i.e., zero velocity). Assuming a constant deceleration, find the deceleration of the coyote. (b) Unfortunately for the coyote, he is unable to shut down the rocket as he reaches the ground. Consequently, he is propelled back up into the air. After 5 s the rocket runs out of fuel. Find the maximum height reached by the coyote and his velocity as he reaches the ground for the second time.

9. A young woman named Kathy Kool buys a superdeluxe sportscar that can accelerate at the rate of 16 ft/s^2. She decides to test the car by dragging with another speedster, Stan Speedy. Both start from rest, but experienced Stan leaves 1 s before Kathy. If Stan moves with a constant acceleration of 12 ft/s^2 and Kathy maintains an acceleration of 16 ft/s^2, find (a) the time it takes Kathy to overtake Stan, (b) the distance she travels before she catches him, and (c) the velocities of both cars at the instant she overtakes him.

10. A hockey player takes a slap shot at a puck at rest on the ice. The puck glides over the ice for 10 ft without friction, at which point it runs over rough ice. The puck then decelerates at 20 ft/s^2. If the velocity of the puck is 40 ft/s after traveling 100 ft from the point of impact, (a) what is the average acceleration imparted to the puck as it is struck by the hockey stick? (Assume that the time of contact is 0.01 s.) (b) How far in all does the puck travel before coming to rest? (c) What is the total time the puck is in motion, neglecting contact time?

11. A student stands on the edge of a building 100 ft above the ground and throws a baseball upward with some initial velocity. The ball is blown slightly sideways by a cross wind and then falls to the ground, just missing the building. If the total time for the ball to travel from the student's hand to the ground is 6 s, find (a) the initial velocity of the ball, (b) its final velocity as it hits the ground, and (c) its velocity after 3 s.

12. An ice sled powered by a rocket engine starts from rest on a large frozen lake and accelerates at 40 ft/s^2. After some time t_1 the rocket engine is shut down and the sled moves with constant velocity v for a time t_2. If the total distance traveled by the sled is 17 500 ft and the total time is 90 s, find (a) the times t_1 and t_2 and (b) the velocity v. If at the 17 500-ft mark the sled begins to decelerate at 20 ft/s^2, (c) what is the final position of the sled when it comes to rest and (d) how long does it take to come to rest?

13. A person sees a lightning bolt passing close to an airplane flying off in the distance. The person hears thunder 5 s after seeing the bolt and sees the airplane overhead 10 s after hearing the thunder. If the speed of sound is 1100 ft/s, (a) find the distance the airplane is from the person at the instant of the bolt. (Neglect the time it takes the light to travel from the bolt to the eye.) (b) Assuming the plane travels with a constant speed toward the person, find the velocity of the airplane. (c) Look up the speed of light in air and defend the approximation used in (a).

14. The following relationships represent the displacement x of a particle as a function of time t. The remaining symbols are constants. Find the velocity and acceleration as a function of time by taking appropriate derivatives and give the correct dimensions for the constants A, b, B, and α: (a) $x = Ae^{-bt}$, (b) $x = B \sin \alpha t$.

15. A particle moves along the positive x axis in such a way that its coordinate varies in time according to the expression $x = 4 + 2t - 3t^2$, where x is in m and t is in s. (a) Make a graph of x versus t for the interval $t = 0$ to $t = 2$ s. (b) Determine the initial position and initial velocity of the particle. (c) Determine at what time the particle reaches a *maximum* position coordinate. (Note that at this time, $v = 0$.) (d) Calculate the coordinate, velocity, and acceleration at $t = 2$ s.

4 Motion in Two Dimensions

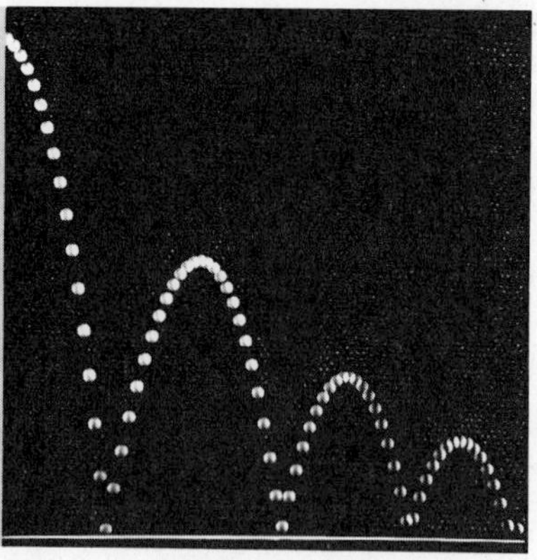

Education Development Center, Newton, Mass.

In this chapter we deal with the kinematics of a particle moving in a plane, or two-dimensional motion. Some common examples of motion in a plane are the motion of projectiles and satellites and the motion of charged particles in uniform electric fields. We begin by showing that velocity and acceleration are vector quantities. As in the case of one-dimensional motion, we shall derive the kinematic equations for two-dimensional motion from the fundamental definitions of displacement, velocity, and acceleration. As special cases of motion in two dimensions, we shall treat motion in a plane with constant acceleration and with uniform circular motion.

4.1 THE DISPLACEMENT, VELOCITY, AND ACCELERATION VECTORS

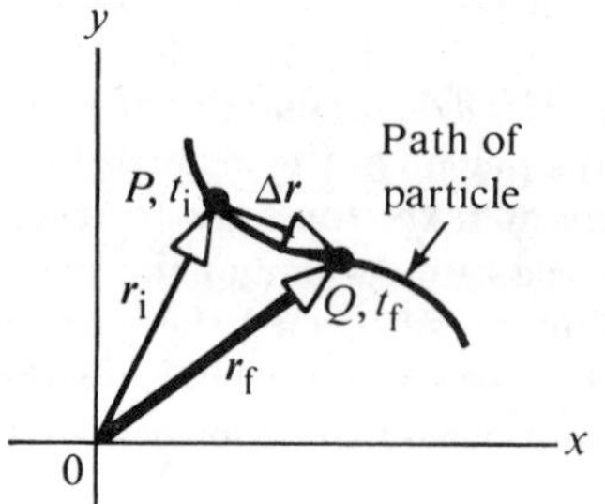

Figure 4.1 A particle moving in the *xy* plane is located with the position vector $\mathbf{r}$ drawn from the origin to the particle. The displacement of the particle as it moves from P to Q in the time interval $\Delta t = t_f - t_i$ is equal to the vector $\Delta\mathbf{r} = \mathbf{r}_f - \mathbf{r}_i$.

In the previous chapter we found that the motion of a particle is completely known if its coordinate is known as a function of time. Now let us extend this idea to the motion of a particle in the *xy* plane. We begin by describing the position of a particle with a *position vector* $\mathbf{r}$, drawn from the origin of some reference frame to the particle located in the *xy* plane, as in Fig. 4.1. At time t_i, the particle is at the point P, and at some later time t_f, the particle is at Q. As the particle moves from P to Q in the time interval $\Delta t = t_f - t_i$, the position vector changes from $\mathbf{r}_i$ to $\mathbf{r}_f$, where the indices i and f refer to initial and final values. Thus, the *displacement vector* for the particle is given by

Definition of the displacement vector

$$\Delta\mathbf{r} \equiv \mathbf{r}_f - \mathbf{r}_i \tag{4.1}$$

The direction of $\Delta\mathbf{r}$ is indicated in Fig. 4.1. Note that the displacement vector equals the difference between the final position vector and the initial position vector. As we see from Fig. 4.1, the magnitude of the displacement vector is less than the distance traveled along the curved path.

We now define the *average velocity* of the particle during the time interval Δt as the ratio of the displacement and the time interval for this displacement:

Average velocity

$$\bar{\mathbf{v}} \equiv \frac{\Delta\mathbf{r}}{\Delta t} \tag{4.2}$$

Since the displacement is a vector and the time interval is a scalar, we conclude that the average velocity is a *vector* quantity directed along $\Delta \boldsymbol{r}$. Note that the average velocity between points P and Q is independent of the path between the two points. This is because the average velocity is proportional to the displacement, which in turn depends only on the initial and final position vectors. As we did in the case of one-dimensional motion, we conclude that if a particle starts its motion at some point and returns to this point via any path, its average velocity is zero for this trip since its displacement is zero.

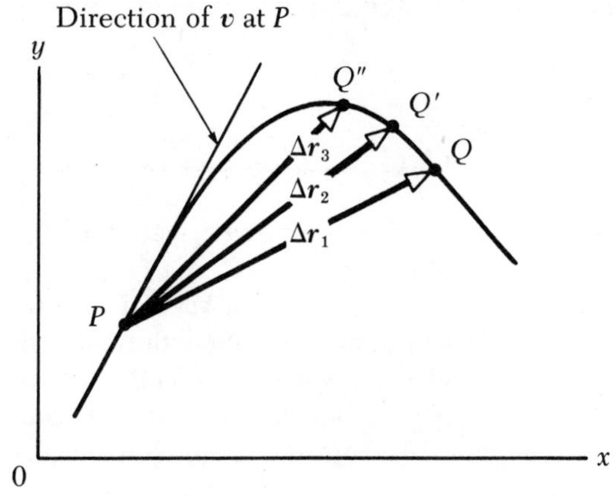

Figure 4.2 As a particle moves between two points, its average velocity is in the direction of the displacement vector $\Delta \boldsymbol{r}$. As the point Q moves closer to P, the direction of $\Delta \boldsymbol{r}$ approaches that of the line tangent to the curve at P. By definition, the instantaneous velocity at P is in the direction of this tangent line.

Consider again the motion of a particle between two points in the xy plane, as in Fig. 4.2. As the time intervals become smaller and smaller, the displacements, $\Delta \boldsymbol{r}_1$, $\Delta \boldsymbol{r}_2$, $\Delta \boldsymbol{r}_3$, . . . , get progressively smaller and the direction of the displacement approaches that of the line tangent to the path at the point P. The *instantaneous velocity*, $\boldsymbol{v}$, is defined as the limit of the average velocity, $\Delta \boldsymbol{r}/\Delta t$, as Δt approaches zero:

Instantaneous velocity

$$\boldsymbol{v} \equiv \lim_{\Delta t \to 0} \frac{\Delta \boldsymbol{r}}{\Delta t} = \frac{d\boldsymbol{r}}{dt} \quad (4.3)$$

That is, the instantaneous velocity equals the derivative of the position vector with respect to time. The direction of the velocity vector is along a line that is tangent to the path of the particle and pointing in the direction of motion. This is illustrated in Fig. 4.3 for two points along the path. The magnitude of the instantaneous velocity vector is called the *speed*.

As the particle moves from P to Q along some path, its instantaneous velocity vector changes from $\boldsymbol{v}_i$ at time t_i to $\boldsymbol{v}_f$ at time t_f (Fig. 4.3). The *average acceleration* of the particle as it moves from P to Q is defined as the ratio of the change in the

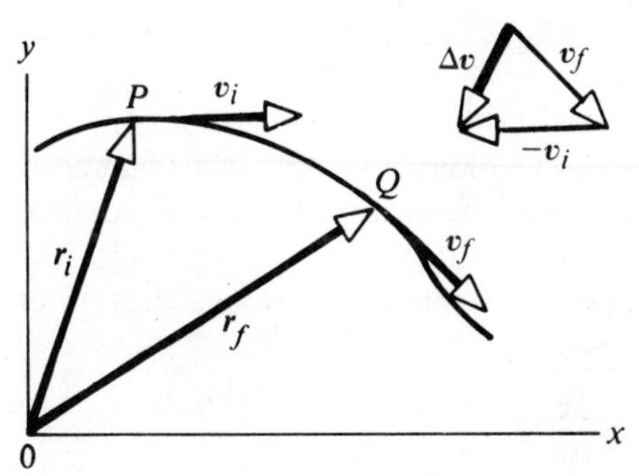

Figure 4.3 The average acceleration vector, $\bar{\boldsymbol{a}}$, for a particle moving from P to Q is in the direction of the change in the velocity, $\Delta \boldsymbol{v} = \boldsymbol{v}_f - \boldsymbol{v}_i$.

instantaneous velocity vector, $\Delta \boldsymbol{v}$, and the elapsed time, Δt:

Average acceleration

$$\bar{\boldsymbol{a}} \equiv \frac{\boldsymbol{v}_f - \boldsymbol{v}_i}{t_f - t_i} = \frac{\Delta \boldsymbol{v}}{\Delta t} \quad (4.4)$$

Since the average acceleration is the ratio of a vector, $\Delta \boldsymbol{v}$, and a scalar, Δt, we conclude that $\bar{\boldsymbol{a}}$ is a vector quantity directed along $\Delta \boldsymbol{v}$. As is indicated in Fig. 4.3, the direction of $\Delta \boldsymbol{v}$ is found by adding the vector $-\boldsymbol{v}_i$ (the reverse of $\boldsymbol{v}_i$) to the vector $\boldsymbol{v}_f$, since by definition $\Delta \boldsymbol{v} = \boldsymbol{v}_f - \boldsymbol{v}_i$.

The *instantaneous acceleration*, $\boldsymbol{a}$, is defined as the limiting value of the ratio $\Delta \boldsymbol{v}/\Delta t$ as Δt approaches zero:

Instantaneous acceleration

$$\boldsymbol{a} \equiv \lim_{\Delta t \to 0} \frac{\Delta \boldsymbol{v}}{\Delta t} = \frac{d\boldsymbol{v}}{dt} \quad (4.5)$$

In other words, the instantaneous acceleration equals the first derivative of the velocity vector with respect to time.

It is important to recognize that a particle can accelerate for several reasons. First, the magnitude of the velocity vector (the speed) may change with time as in one-dimensional motion. Second, a particle accelerates when the direction of the velocity vector changes with time (a curved path) even though its speed is constant. Finally, the acceleration may be due to a change in both the magnitude and the direction of the velocity vector.

Q1. If the average velocity of a particle is zero in some time interval, what can you say about the displacement of the particle for that interval?

Q2. If you know the position vectors of a particle at two points along its path and also know the time it took to get from one point to the other, can you determine the particle's instantaneous velocity? its average velocity? Explain.

Q3. Describe a situation in which the velocity of a particle is perpendicular to the position vector.

Q4. Can a particle accelerate if its speed is constant? Can it accelerate if its velocity is constant? Explain.

Q5. Explain whether or not the following particles have an acceleration: (a) a particle moving in a straight line with constant speed and (b) a particle moving around a curve with constant speed.

Q6. Correct the following statement: "The racing car rounds the turn at a constant velocity of 90 miles per hour."

4.2 MOTION IN TWO DIMENSIONS WITH CONSTANT ACCELERATION

We first consider the motion of a particle in two dimensions with *constant* acceleration. That is, we assume that the magnitude and direction of the acceleration, $\boldsymbol{a}$, remain unchanged during the motion.

Since $\boldsymbol{a}$ is constant, its average value equals its instantaneous value. Thus, from the definition of $\bar{\boldsymbol{a}}$, given by Eq. 4.5, we have

$$\bar{\boldsymbol{a}} = \boldsymbol{a} = \frac{\Delta \boldsymbol{v}}{\Delta t}$$

If we take $\Delta t = t$ (that is, $t_0 = 0$) and recall that the change in velocity is $\Delta \boldsymbol{v} = \boldsymbol{v} - \boldsymbol{v}_0$, then this expression reduces to

Velocity vector as a function of time

$$\boldsymbol{v} = \boldsymbol{v}_0 + \boldsymbol{a}t \tag{4.6}$$

This expression states that the velocity of a particle at some time t equals the *vector* sum of its initial velocity, $\boldsymbol{v}_0$, and the additional velocity $\boldsymbol{a}t$, acquired in the time t as a result of its constant acceleration.

The position vector, $\boldsymbol{r}$, as a function of time for a particle moving with constant acceleration is given by the vector equation

Position vector as a function of time

$$\boldsymbol{r} = \boldsymbol{r}_0 + \boldsymbol{v}_0 t + \frac{1}{2}\boldsymbol{a}t^2 \tag{4.7}$$

This equation says that the displacement vector $\boldsymbol{r} - \boldsymbol{r}_0$ is the vector sum of a displacement $\boldsymbol{v}_0 t$, arising from the initial velocity of the particle, and a displacement $\frac{1}{2}\boldsymbol{a}t^2$, resulting from the uniform acceleration of the particle. Graphical representations of Eqs. 4.6 and 4.7 are shown in Figs. 4.4a and 4.4b. For simplicity in drawing the figure, we have taken $\boldsymbol{r}_0 = 0$ in Fig. 4.4b. That is, we assume that the particle is at the origin at $t = 0$. Note from Fig. 4.4b that $\boldsymbol{r}$ is generally not along the direction of $\boldsymbol{v}_0$ or $\boldsymbol{a}$, since the relation between these quantities is a vector expression. For the same reason, from Fig. 4.4a we see that $\boldsymbol{v}$ is generally not along the direction of $\boldsymbol{v}_0$ or $\boldsymbol{a}$. Finally, if we compare Figs. 4.4a and 4.4b we see that $\boldsymbol{v}$ and $\boldsymbol{r}$ are not in the same direction. This is because $\boldsymbol{v}$ is linear in t, while $\boldsymbol{r}$ is quadratic in t. It is also important to recognize that since Eqs. 4.6 and 4.7 are *vector* expressions having one or more components (in general, three components), we may write the component forms of these expressions along the x and y axes with $\boldsymbol{r}_0 = 0$

$$\boldsymbol{v} = \boldsymbol{v}_0 + \boldsymbol{a}t \quad \begin{cases} v_x = v_{x0} + a_x t & (4.8a) \\ v_y = v_{y0} + a_y t & (4.8b) \end{cases}$$

$$\boldsymbol{r} = \boldsymbol{v}_0 t + \frac{1}{2}\boldsymbol{a}t^2 \begin{cases} x = v_{x0}t + \frac{1}{2}a_x t^2 & (4.9a) \\ y = v_{y0}t + \frac{1}{2}a_y t^2 & (4.9b) \end{cases}$$

These components are illustrated in Fig. 4.4. In other words, two-dimensional motion with constant acceleration is equivalent to two independent motions in the x and y directions with constant accelerations a_x and a_y.

We shall now proceed with a further discussion of Eqs. 4.6 and 4.7. The position vector $\boldsymbol{r}$ for a particle moving in the xy plane can be written

$$\boldsymbol{r} = x\boldsymbol{i} + y\boldsymbol{j}$$

where x, y, and $\boldsymbol{r}$ change with time as the particle moves. From the definition of instantaneous velocity given by Eq. 4.3, we have

$$\boldsymbol{v} = \frac{d\boldsymbol{r}}{dt} = \frac{dx}{dt}\boldsymbol{i} + \frac{dy}{dt}\boldsymbol{j}$$

$$\boldsymbol{v} = v_x\boldsymbol{i} + v_y\boldsymbol{j}$$

Since $\boldsymbol{a}$ is a constant, its components, a_x and a_y, are constants, and we can use Eqs. 4.8a and 4.8b to express $\boldsymbol{v}$:

$$\begin{aligned} \boldsymbol{v} &= (v_{x0} + a_x t)\boldsymbol{i} + (v_{y0} + a_y t)\boldsymbol{j} \\ &= (v_{x0}\boldsymbol{i} + v_{y0}\boldsymbol{j}) + (a_x\boldsymbol{i} + a_y\boldsymbol{j})t \end{aligned}$$

$$\boldsymbol{v} = \boldsymbol{v}_0 + \boldsymbol{a}t \tag{4.6}$$

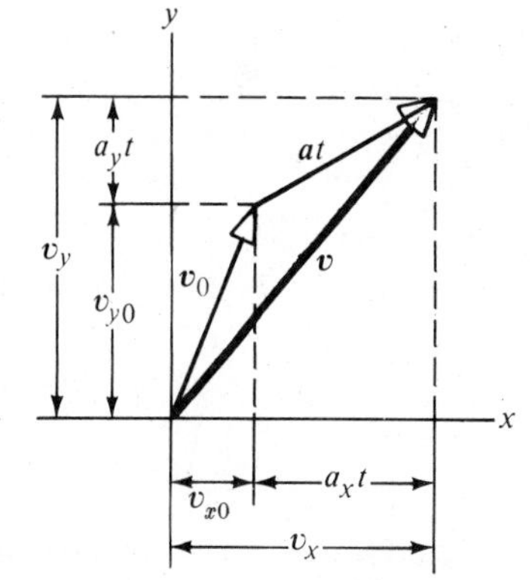

(a)

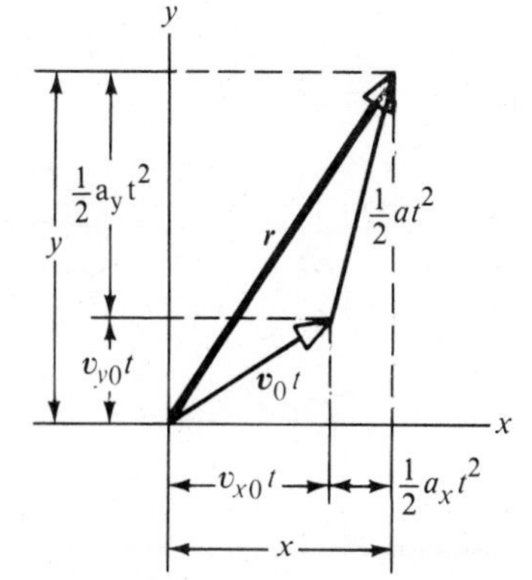

(b)

Figure 4.4 Vector representations and rectangular components of (a) the velocity and (b) the displacement of a particle moving with a uniform acceleration $\boldsymbol{a}$.

Also, since the x and y coordinates versus time are given by Eqs. 4.9a and 4.9b, we can express the position vector $\mathbf{r} = x\mathbf{i} + y\mathbf{j}$ as

$$\mathbf{r} = (v_{x0}t + \frac{1}{2}a_x t^2)\mathbf{i} + (v_{y0}t + \frac{1}{2}a_y t^2)\mathbf{j}$$

$$= (v_{x0}\mathbf{i} + v_{y0}\mathbf{j})t + \frac{1}{2}(a_x\mathbf{i} + a_y\mathbf{j})t^2$$

$$\mathbf{r} = \mathbf{v}_0 t + \frac{1}{2}\mathbf{a}t^2$$

where we have taken $\mathbf{r}_0 = 0$.

Example 4.1

A particle moves in the xy plane with an x component of acceleration only, given by $a_x = 4\text{ m/s}^2$. The particle starts from the origin at $t = 0$ with an initial velocity having an x component of 20 m/s and a y component of -15 m/s. (a) Determine the components of velocity as a function of time and the total velocity vector at any time.

Since $v_{x0} = 20$ m/s and $a_x = 4\text{ m/s}^2$, Eq. 4.8a gives

$$v_x = v_{x0} + a_x t = (20 + 4t)\text{ m/s}$$

Also, since $v_{y0} = -15$ m/s and $a_y = 0$, Eq. 4.8b gives

$$v_y = v_{y0} = -15\text{ m/s}$$

Therefore, using the above results and noting that the velocity vector $\mathbf{v}$ has two components, we get

$$\mathbf{v} = v_x\mathbf{i} + v_y\mathbf{j} = [(20 + 4t)\mathbf{i} - 15\mathbf{j}]\text{ m/s}$$

We could also obtain this result using Eq. 4.6 directly, noting that $\mathbf{a} = 4\mathbf{i}\text{ m/s}^2$ and $\mathbf{v}_0 = (20\mathbf{i} - 15\mathbf{j})$ m/s. Try it!

(b) Calculate the velocity and speed of the particle at $t = 5$ s.

With $t = 5$ s, the result from (a) gives

$$\mathbf{v} = \{[20 + 4(5)]\mathbf{i} - 15\mathbf{j}\}\text{ m/s} = (40\mathbf{i} - 15\mathbf{j})\text{ m/s}$$

That is, at $t = 5$ s, $v_x = 40$ m/s and $v_y = -15$ m/s. The speed is defined as the magnitude of $\mathbf{v}$, or

$$v = |\mathbf{v}| = \sqrt{v_x^2 + v_y^2} = \sqrt{(40)^2 + (-15)^2}\text{ m/s} = 42.7\text{ m/s}$$

(Note that v is larger than v_0. Why?)

The angle θ that $\mathbf{v}$ makes with the x axis can be calculated using the fact that $\tan\theta = v_y/v_x$, or

$$\theta = \tan^{-1}\left(\frac{v_y}{v_x}\right) = \tan^{-1}\left(\frac{-15}{40}\right) = -20.6°$$

(c) Determine the x and y coordinates at any time t and the displacement vector at this time.

Since at $t = 0$, $x_0 = y_0 = 0$, the expressions for the x and y coordinates, Eqs. 4.9a and 4.9b, give

$$x = v_{x0}t + \frac{1}{2}a_x t^2 = (20t + 2t^2)\text{ m}$$

$$y = v_{y0}t = (-15t)\text{ m}$$

Therefore, the displacement vector at any time t is given by

$$\mathbf{r} = x\mathbf{i} + y\mathbf{j} = [(20t + 2t^2)\mathbf{i} - 15t\mathbf{j}]\text{ m}$$

Alternatively, we could obtain $\mathbf{r}$ by applying Eq. 4.7 directly, with $\mathbf{v}_0 = (20\mathbf{i} - 15\mathbf{j})$ m/s and $\mathbf{a} = 4\mathbf{i}\text{ m/s}^2$. Try it!

Thus, for example, at $t = 5$ s, $x = 150$ m and $y = -75$ m, or $\mathbf{r} = (150\mathbf{i} - 75\mathbf{j})$ m. It follows that the distance of the particle from the origin to this point is the magnitude of the displacement, or

$$|\mathbf{r}| = r = \sqrt{(150)^2 + (-75)^2}\text{ m} = 168\text{ m}$$

Note that this is *not* the distance that the particle travels in this time! Can you determine this distance from the available data?

4.3 PROJECTILE MOTION

Assumptions of projectile motion

Anyone who has observed a baseball in motion (or, for that matter, any object thrown in the air) has observed projectile motion. For an arbitrary direction of the initial velocity, the ball moves in a curved path. This very common form of motion is surprisingly simple to analyze if the following three assumptions are made: (1) the acceleration due to gravity, $\mathbf{g}$, is constant over the range of motion and is directed downward,[1] (2) the effect of air resistance is negligible,[2] and (3) the rotation of the

[1]This approximation is reasonable as long as the range of motion is small compared with the radius of the earth (6.4×10^6 m). In effect, this approximation is equivalent to assuming that the earth is flat over the range of motion considered.

[2]This approximation is generally *not* justified, especially at high velocities. In addition, the spin of a projectile, such as a baseball, can give rise to some very interesting effects associated with aerodynamic forces (for example, a curve thrown by a pitcher).

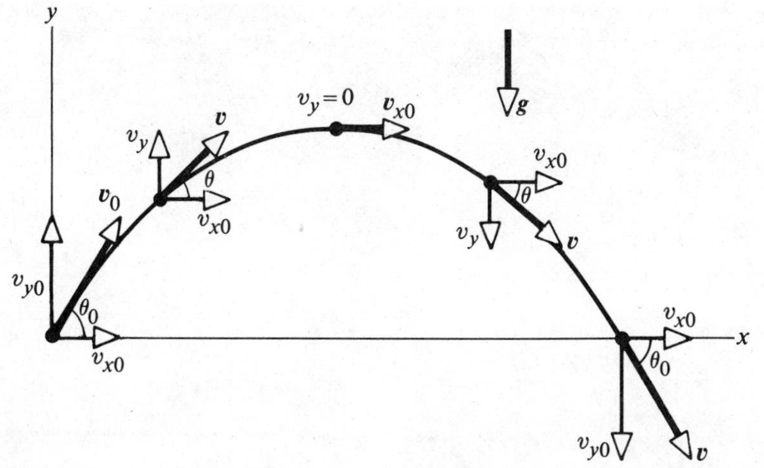

Figure 4.5 The parabolic trajectory of a projectile that leaves the origin with a velocity v_0. Note that the velocity vector, v, changes with time. However, the x component of velocity, v_{x0}, remains constant in time. Also, $v_y = 0$ at the peak.

earth does not affect the motion. With these assumptions, we shall find that the path of a projectile, which we call its *trajectory*, is *always* a parabola. *We shall use these assumptions throughout this chapter.*

If we choose our reference frame such that the y direction is vertical and positive upward, then $a_y = -g$ (as in one-dimensional free fall) and $a_x = 0$ (since air friction is neglected). Furthermore, let us assume that at $t = 0$, the projectile leaves the origin ($x_0 = y_0 = 0$) with a velocity v_0, as in Fig. 4.5. If the vector v_0 makes an angle θ_0 with the horizontal, as in Fig. 4.5, then from the laws of cosines and sines we have

$$\cos\theta_0 = v_{x0}/v_0 \qquad \text{and} \qquad \sin\theta_0 = v_{y0}/v_0$$

Therefore, the initial x and y components of velocity are given by

$$v_{x0} = v_0 \cos\theta_0 \qquad \text{and} \qquad v_{y0} = v_0 \sin\theta_0$$

Substituting these expressions into Eqs. 4.8 and 4.9 with $a_x = 0$ and $a_y = -g$ gives the velocity components and coordinates for the projectile at any time t:

$$v_x = v_{x0} = v_0 \cos\theta_0 = \text{constant} \tag{4.10}$$

Horizontal velocity component

$$v_y = v_{y0} - gt = v_0 \sin\theta_0 - gt \tag{4.11}$$

Vertical velocity component

$$x = v_{x0}t = (v_0 \cos\theta_0)t \tag{4.12}$$

Horizontal position component

$$y = v_{y0}t - \frac{1}{2}gt^2 = (v_0 \sin\theta_0)t - \frac{1}{2}gt^2 \tag{4.13}$$

Vertical position component

From Eq. 4.10 we see that v_x remains constant in time and is equal to the initial x component of velocity, since there is no horizontal component of acceleration. Also, for the y motion we note that v_y and y are identical to the expressions for the freely falling body discussed in Chapter 3. In fact, *all* of the equations of kinematics developed in Chapter 3 are applicable to projectile motion.

If we solve for t in Eq. 4.12 and substitute this expression for t into Eq. 4.13, we find that

$$y = (\tan\theta_0)x - \left(\frac{g}{2{v_0}^2 \cos^2\theta_0}\right)x^2 \tag{4.14}$$

Trajectory of a projectile

which is valid for the angles in the range $0 < \theta_0 < \pi/2$. This is of the form $y = ax - bx^2$, which is the equation of a parabola that passes through the origin. Thus,

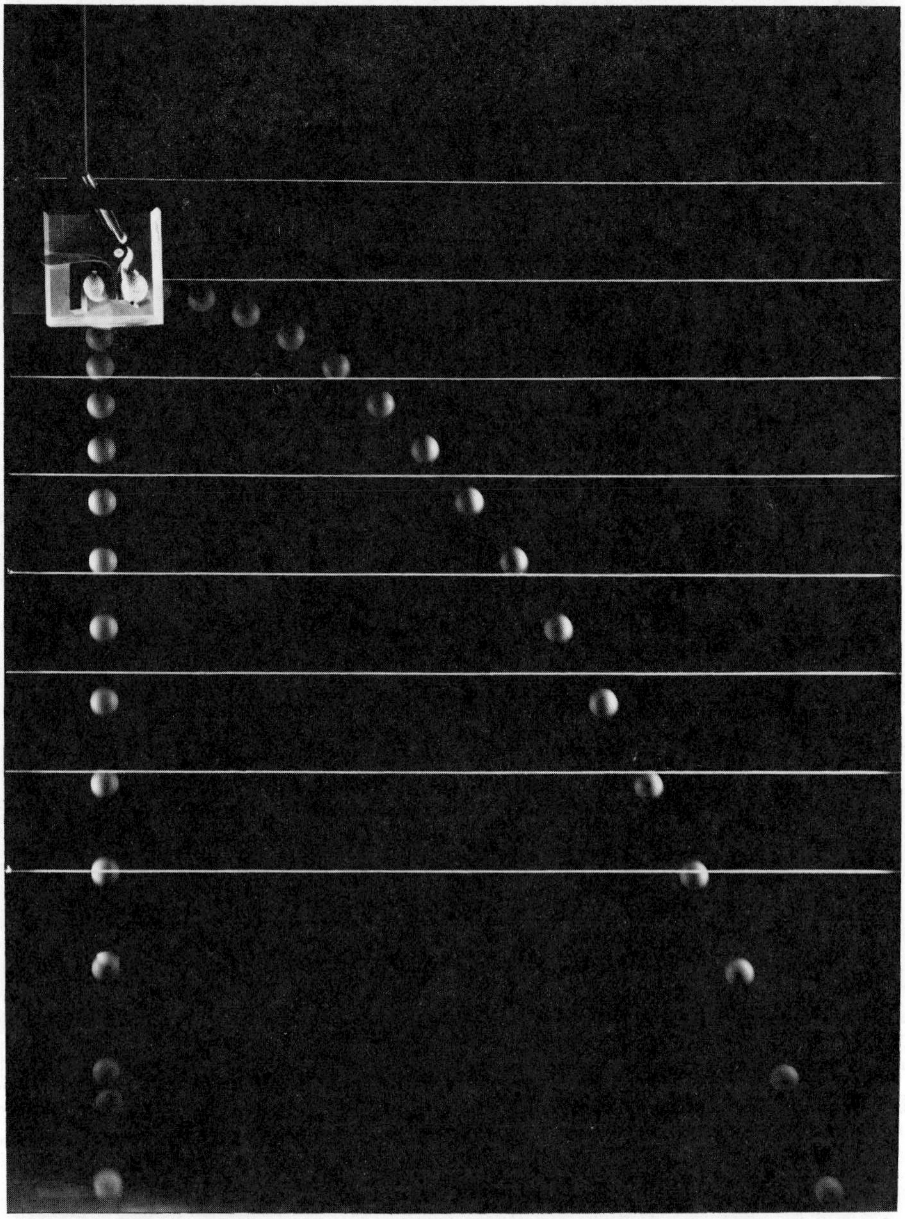

This multiple image photograph of two golf balls released simultaneously illustrates both free fall and projectile motion. The right ball was projected horizontally with an initial velocity of 2 m/s. The light flashes were 1/30 s apart, and the white parallel lines (actually strings) were placed 15¼ cm apart. Can you explain why both balls hit the floor simultaneously?

we have proved that the trajectory of a projectile is a parabola. Note that the trajectory is *completely* specified if v_0 and θ_0 are known.

One can obtain the speed, v, as a function of time for the projectile by noting that Eqs. 4.10 and 4.11 give the x and y components of velocity at any instant. Therefore, by definition, since v is equal to the magnitude of $\boldsymbol{v}$,

Speed

$$v = \sqrt{v_x^{\,2} + v_y^{\,2}} \tag{4.15}$$

Also, since the velocity vector is tangent to the path at any instant, as shown in Fig. 4.5, the angle θ that $\boldsymbol{v}$ makes with the horizontal can be obtained from v_x and v_y

through the expression

$$\tan\theta = \frac{v_y}{v_x} \qquad (4.16)$$

Angle of trajectory

The vector expression for the position vector as a function of time for the projectile follows directly from Eq. 4.7, with $\boldsymbol{a} = \boldsymbol{g}$:

$$\boldsymbol{r} = \boldsymbol{v}_0 t + \frac{1}{2}\boldsymbol{g}t^2$$

This expression is equivalent to Eqs. 4.12 and 4.13 and is plotted in Fig. 4.6. It is interesting to note that the motion can be considered the superposition of the term $\boldsymbol{v}_0 t$, which is the displacement if no acceleration were present, and the term $\frac{1}{2}\boldsymbol{g}t^2$, which arises from the acceleration due to gravity. In other words, if there were no gravitational acceleration, the particle would continue to move along a straight path in the direction of $\boldsymbol{v}_0$. Therefore, the vertical distance, $\frac{1}{2}gt^2$, through which the particle "falls" measured from the straight line is that of a freely falling body. *We conclude that projectile motion is the superposition of two motions: (1) the motion of a freely falling body and (2) uniform motion in the x direction.*

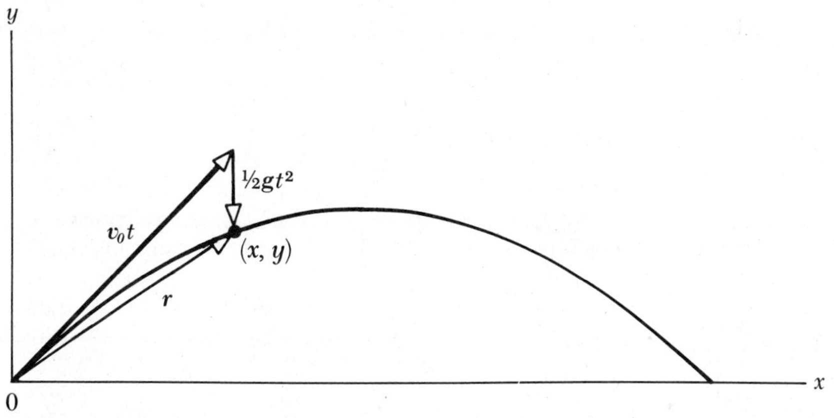

Figure 4.6 The displacement vector, $\boldsymbol{r}$, of a projectile having an initial velocity at the origin of $\boldsymbol{v}_0$. The vector $\boldsymbol{v}_0 t$ would be the displacement of the projectile if gravity were absent, and the vector $\frac{1}{2}\boldsymbol{g}t^2$ is its vertical displacement due to gravity in the time t.

Special Case: Horizontal Range and Maximum Height of a Projectile. Let us assume that a projectile is fired from the origin at $t = 0$ with a positive v_y component, as in Fig. 4.7. There are two special points that are interesting to analyze: the peak with cartesian coordinates labeled $\left(\frac{R}{2}, h\right)$ and the point with coordinates $(R,0)$. The distance R is called the *horizontal range* of the projectile, and h is its *maximum height.* Let us find h and R in terms of v_0, θ_0, and g.

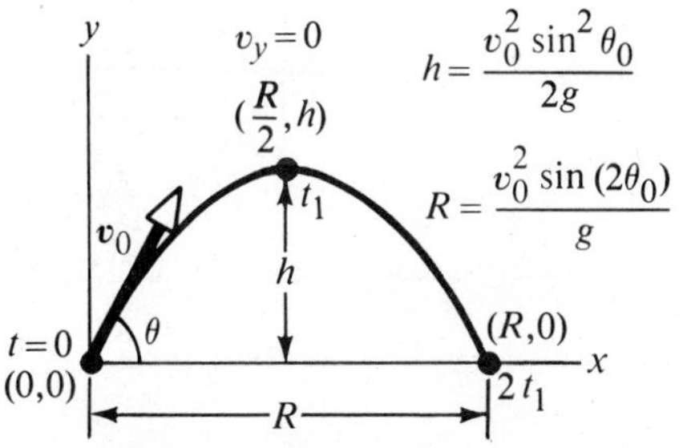

Figure 4.7 A projectile fired from the origin at $t = 0$ with an initial velocity $\boldsymbol{v}_0$. The maximum height of the projectile is h, and its horizontal range is R.

We can determine the maximum height, h, reached by the projectile by noting that at the peak, $v_y = 0$. Therefore, Eq. 4.11 can be used to determine the time t_1 it takes to reach the peak:

$$t_1 = \frac{v_0 \sin\theta_0}{g}$$

Substituting this expression for t_1 into Eq. 4.13 gives h in terms of v_0 and θ_0:

$$h = (v_0 \sin\theta_0)\frac{v_0 \sin\theta_0}{g} - \frac{1}{2}g\left(\frac{v_0 \sin\theta_0}{g}\right)^2$$

Maximum height of projectile

$$h = \frac{v_0^2 \sin^2\theta_0}{2g} \tag{4.17}$$

The range, R, is the horizontal distance traveled in twice the time it takes to reach the peak, that is, in a time $2t_1$. (This can be seen by setting $y = 0$ in Eq. 4.13 and solving the quadratic for t. One solution of this quadratic is $t = 0$, and the second is $t = 2t_1$.) Using Eq. 4.12 and noting that $x = R$ at $t = 2t_1$, we find that

$$R = (v_0 \cos\theta_0)2t_1 = (v_0 \cos\theta_0)\frac{2v_0 \sin\theta_0}{g}$$

$$R = \frac{2v_0^2 \sin\theta_0 \cos\theta_0}{g}$$

Since $\sin 2\theta = 2 \sin\theta \cos\theta$, R can be written in the more compact form

Range of projectile

$$R = \frac{v_0^2 \sin 2\theta_0}{g} \tag{4.18}$$

Keep in mind that Eqs. 4.17 and 4.18 are useful only for calculating h and R if v_0 and θ_0 are known (which means that only $\mathbf{v}_0$ has to be specified). The general expressions given by Eqs. 4.10 through 4.13 are the *most important* results, since they give the coordinates and velocity components of the projectile at *any* time t.

You should note that the maximum value of R from Eq. 4.18 is $R_{max} = v_0^2/g$. This result follows from the fact that the maximum value of $\sin 2\theta_0$ is unity, which occurs when $2\theta_0 = 90°$. Therefore, we see that R is a maximum when $\theta_0 = 45°$, as you would expect if air friction is neglected.

Figure 4.8 illustrates various trajectories for a projectile of a given initial speed. As you

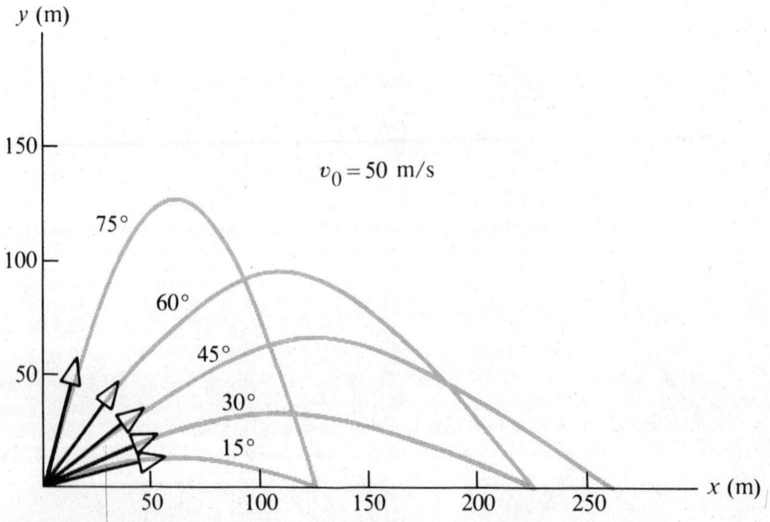

Figure 4.8 A projectile fired from the origin at an initial speed of 50 m/s at various angles of projection. Note that a point along the x axis can be reached at any two complementary values of θ_0.

can see, the range is a maximum for $\theta_0 = 45°$. In addition, for any θ_0 other than 45°, a point with coordinates $(R,0)$ can be reached with *two* complementary values of θ_0, such as 75° and 15°. Of course, the maximum height and time of flight will be different for these two values of θ_0.

Example 4.2

A long-jumper makes a near-record jump of 8.5 m. Assuming the jumper leaves the ground at an angle of 20° to the horizontal, and assuming he is a point mass (a very gross approximation), (a) find the speed at which the jumper leaves the ground.

Since the horizontal range, R, and θ_0 are given, we can use Eq. 4.18 to calculate v_0:

$$R = \frac{v_0^2 \sin 2\theta_0}{g} \quad \text{or} \quad v_0 = \sqrt{\frac{Rg}{\sin 2\theta_0}}$$

$$v_0 = \sqrt{\frac{(8.5\text{ m})(9.8\text{ m/s}^2)}{\sin 40°}} = 11\text{ m/s}$$

(b) Determine the maximum height reached by the jumper.

We can use Eq. 4.17 and the result from (a) to find h:

$$h = \frac{v_0^2 \sin^2\theta_0}{2g} = \frac{(11\text{ m/s})^2(\sin 20°)^2}{2(9.8\text{ m/s}^2)} = 0.78\text{ m}$$

Example 4.3

In a very popular lecture demonstration, a projectile is fired at a falling target in such a way that the projectile leaves the gun at the same time the target is dropped from rest, as in Fig. 4.9. Let us show that if the gun is initially aimed at the target, the projectile will hit the target.[3]

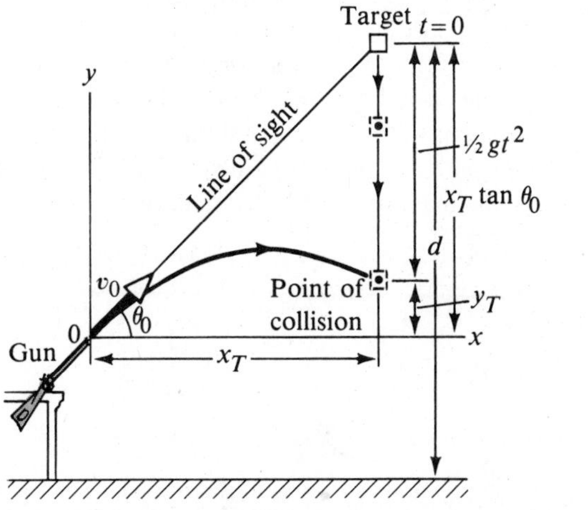

Figure 4.9 (Example 4.3) Schematic diagram of the projectile-and-target demonstration. If the gun is aimed directly at the target and is fired at the same instant the target falls, the projectile will hit the target. Both fall through the same vertical distance in a time t, since both experience the same acceleration, $a_y = -g$.

Solution: We can argue that a collision will result under the conditions stated by noting that both the projectile and the target experience the *same* acceleration, $a_y = -g$, as soon as they are released. Therefore, both will fall through the same vertical distance, $\frac{1}{2}gt^2$, in a time t. First, note from Fig. 4.9 that the initial y coordinate of the target is $x_T \tan\theta_0$ and that it falls through a distance $\frac{1}{2}gt^2$ in a time t. Therefore, the y coordinate of the target as a function of time is

$$y_T = x_T \tan\theta_0 - \frac{1}{2}gt^2$$

Now if we write equations for x and y for the projectile path over time, using Eqs. 4.12 and 4.13 simultaneously, we get

$$y_P = x_P \tan\theta_0 - \frac{1}{2}gt^2$$

Thus, when $x_P = x_T$, we see by comparing the two equations above that $y_P = y_T$ and a collision results.

The result could also be arrived at with vector methods, using expressions for the position vectors for the projectile and target.

You should also note that a collision will *not* always take place. There is the further restriction that a collision will result only when $v_0 \sin\theta_0 \geq \sqrt{gd/2}$, where d is the initial elevation of the target above the *floor*, as in Fig. 4.9. If $v_0 \sin\theta_0$ is less than this value, the projectile will strike the floor before reaching the target.

Example 4.4

A stone is thrown from the top of a building at an angle of 30° to the horizontal, with an initial speed of 20 m/s, as in Fig. 4.10. If the height of the building is 45 m, (a) how long is the stone "in flight"?

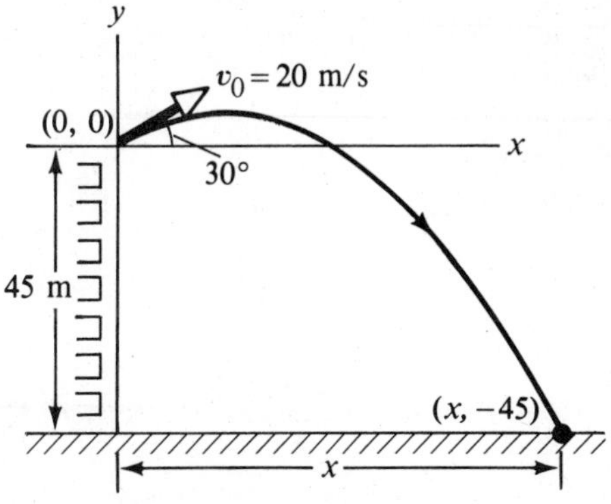

Figure 4.10 (Example 4.4)

[3] In one variation of the demonstration, the target is a tin can held by an electromagnet energized with a small battery. At the instant the projectile leaves the gun, a small switch at the tip of the gun is opened by the moving projectile. This opens the circuit containing the electromagnet, allowing the target to fall.

The initial x and y components of velocity are

$$v_{x0} = v_0 \cos\theta_0 = (20 \text{ m/s})(\cos 30°) = 17.3 \text{ m/s}$$
$$v_{y0} = v_0 \sin\theta_0 = (20 \text{ m/s})(\sin 30°) = 10 \text{ m/s}$$

To find t we can use Eq. 4.13 with $y = -45$ m and $v_{y0} = 10$ m/s (we have chosen the top of the building as the origin as in Fig. 4.10):

$$y = v_{y0}t - \frac{1}{2}gt^2$$
$$-45 = 10t - 4.9t^2$$

Solving the quadratic equation for t gives, for the positive root, $t = 4.22$ s. The negative root has no physical meaning since time intervals are always positive. (Can you think of another way of finding t from the information given?)

(b) Where does the stone strike the ground relative to the building?

We know the time of flight and the x component of the velocity; therefore, from Eq. 4.12 we get

$$x = v_{x0}t = (17.3 \text{ m/s})(4.22 \text{ s}) = 73 \text{ m}$$

(c) What is the speed of the stone just before it strikes the ground?

The y component of the velocity just before the stone strikes the ground can be obtained using Eq. 4.11 with $t = 4.22$ s:

$$v_y = v_{y0} - gt = 10 \text{ m/s} - (9.8 \text{ m/s}^2)(4.22 \text{ s}) = -31.4 \text{ m/s}$$

Since $v_x = v_{x0} = 17.3$ m/s, the required speed is given by

$$v = \sqrt{v_x^2 + v_y^2} = \sqrt{(17.3)^2 + (-31.4)^2} \text{ m/s} = 35.9 \text{ m/s}$$

Q7. Determine which of the following moving objects would exhibit an approximate parabolic trajectory: (a) a ball thrown in an arbitrary direction, (b) a jet airplane, (c) a rocket leaving the launching pad, (d) a rocket with failed engines, (e) a tossed stone moving to the bottom of a pond.

4.4 UNIFORM CIRCULAR MOTION

In this section, we describe the motion of a particle moving in a circle of radius r with a constant speed v as described in Fig. 4.11. Although the particle moves with constant speed, the *direction* of the velocity vector, $\mathbf{v}$, changes with time; therefore the particle accelerates. A common pitfall is to conclude that the acceleration of the particle is zero because its speed is constant. Recall that $\mathbf{a}$ is proportional to the *change in the velocity vector;* therefore, since the direction of $\mathbf{v}$ changes, $\mathbf{a} \neq 0$. The direction of $\mathbf{v}$ is always tangent to the path, but since the speed, v, is constant, there is *no* component of acceleration *tangent* to the path. We shall show that the acceleration vector is perpendicular to the path and *always* points toward the center of the circle, as described in Fig. 4.11b. The magnitude of this radial, or centripetal ("center-seeking"), acceleration is given by

Centripetal acceleration

$$a_r = \frac{v^2}{r} \tag{4.19}$$

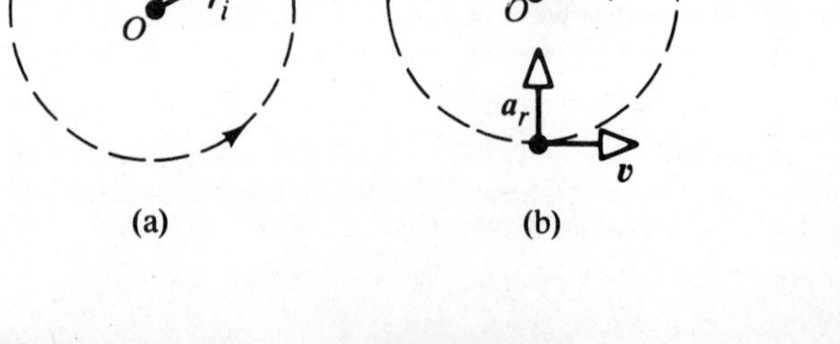

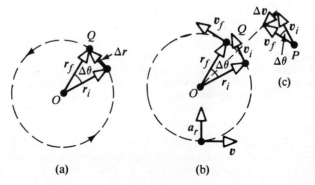

Figure 4.11 Circular motion of a particle moving with a constant speed v. As the particle moves from P to Q, the direction of its velocity vector changes from $\mathbf{v}_i$ to $\mathbf{v}_f$. Note that the direction of the change in velocity, $\Delta\mathbf{v}$ in (c), is toward the center of the circle, which is also the direction of $\mathbf{a}_r$. Also, $\mathbf{v}$ is always tangent to the circle and perpendicular to $\mathbf{a}_r$.

To derive Eq. 4.19, we describe the initial and final positions of the particle by the radius vectors $\boldsymbol{r}_i$ and $\boldsymbol{r}_f$, as shown in Fig. 4.11. The velocities at these two points are labeled $\boldsymbol{v}_i$ and $\boldsymbol{v}_f$. If the time it takes the particle to move from P to Q is Δt, then the displacement of the particle in this time is $\Delta \boldsymbol{r} \approx \boldsymbol{v}\Delta t$, as in Fig. 4.11a.

In Fig. 4.11b we describe the change in direction of the velocity vector as the particle moves from one point on the circle to another point. The change in velocity as the particle moves from P to Q is given by $\Delta \boldsymbol{v} = \boldsymbol{v}_f - \boldsymbol{v}_i$. The average acceleration in the time interval Δt is, by definition, equal to $\Delta \boldsymbol{v}/\Delta t$.

We now draw the vectors $\boldsymbol{v}_i$ and $\boldsymbol{v}_f$ so that they originate at a common point, as in Fig. 4.11c. From this diagram, we see that if Δt is very small (corresponding to a small value of $\Delta\theta$), then the change in velocity is directed approximately toward the center of the circle. Since $\boldsymbol{a}_r = \Delta \boldsymbol{v}/\Delta t$, we conclude that the acceleration vector also points approximately toward the center of the circle.

As the radius vector rotates through a small angle $\Delta\theta$ in the time Δt, the angle between $\boldsymbol{v}_i$ and $\boldsymbol{v}_f$ is also $\Delta\theta$, as shown in Fig. 4.11c. Therefore the triangle OPQ is similar to the triangle formed by $\boldsymbol{v}_i$, $\boldsymbol{v}_f$, and $\Delta \boldsymbol{v}$. From these similar triangles, we can write the ratio

$$\frac{|\Delta \boldsymbol{r}|}{r} = \frac{|\Delta \boldsymbol{v}|}{v}$$

Since $|\Delta \boldsymbol{r}| \approx v\,\Delta t$ and $|\Delta \boldsymbol{v}| \approx a_r\,\Delta t$, this ratio becomes

$$\frac{v\Delta t}{r} \approx \frac{a_r \Delta t}{v}$$

In the limit as $\Delta t \to 0$, this approximate relation becomes exact and we get Eq. 4.19:

$$a_r = \frac{v^2}{r}$$

Thus we conclude that in circular motion *the centripetal acceleration is directed radially inward toward the center of the circle and has a magnitude given by* v^2/r. You should check the dimensions of a_r and show that $[a_r] = \mathrm{L/T^2}$ as required, since it is a true acceleration. We will return to the discussion of circular motion in Section 6.7.

Q8. A student argues that as a satellite orbits the earth in a circular path, it moves with a constant velocity and therefore has no acceleration. The professor claims that the student is wrong since the satellite must have a centripetal acceleration as it moves in its circular orbit. What is wrong with the student's argument?

4.5 TANGENTIAL AND RADIAL ACCELERATION IN CURVILINEAR MOTION

Let us consider the motion of a particle along a curved path where the velocity changes both in direction and in magnitude, as described in Fig. 4.12. In this situation, the velocity of the particle is always tangent to the path; however, the acceleration vector $\boldsymbol{a}$ is now at some angle to the path. As the particle moves along the curved path in Fig. 4.12, we see that the direction of the total acceleration vector, $\boldsymbol{a}$, changes from point to point. This vector can be resolved into two components: a radial component of acceleration, $\boldsymbol{a}_r$, and a tangential component of acceleration,

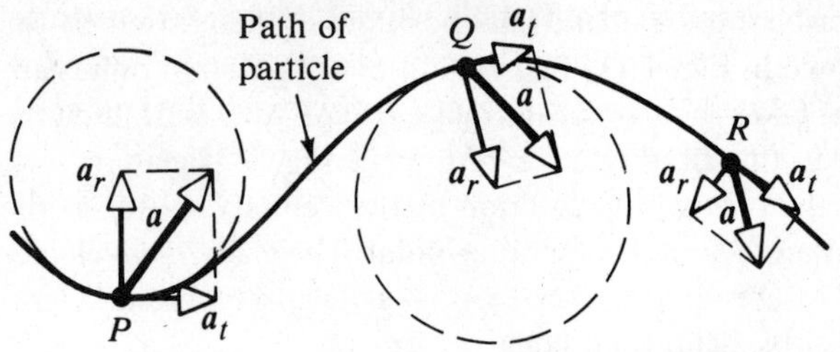

Figure 4.12 The motion of a particle along an arbitrary curved path lying in the xy plane. If the velocity vector $\boldsymbol{v}$ (always tangent to the path) changes in direction and magnitude, the components of the acceleration of the particle are a tangential vector, $\boldsymbol{a}_t$, and a radial vector, $\boldsymbol{a}_r$.

$\boldsymbol{a}_t$. That is, the *total* acceleration vector, $\boldsymbol{a}$, can be written as the vector sum of these components

Total acceleration

$$\boldsymbol{a} = \boldsymbol{a}_r + \boldsymbol{a}_t \tag{4.20}$$

The tangential component of acceleration arises from the change in the speed of the particle, and its magnitude is given by

Tangential acceleration

$$a_t = \frac{dv}{dt} \tag{4.21}$$

The radial component of acceleration is due to the change in direction of the velocity vector and has a magnitude given by

Centripetal acceleration

$$a_r = \frac{v^2}{r} \tag{4.22}$$

where r is the radius of curvature of the path. Since $\boldsymbol{a}_r$ and $\boldsymbol{a}_t$ are the rectangular components of $\boldsymbol{a}$, it follows that $a = \sqrt{a_r{}^2 + a_t{}^2}$. As in the case of uniform circular motion, $\boldsymbol{a}_r$ always points toward the center of curvature, as shown in Fig. 4.12. Also, at a given speed, a_r is large when the radius of curvature is small (as at points P and Q in Fig. 4.12) and small when r is large (such as at point R). The direction of $\boldsymbol{a}_t$ is either in the same direction as $\boldsymbol{v}$ (if v is increasing) or opposite $\boldsymbol{v}$ (if v is decreasing).

Note that in the case of uniform circular motion, where v is constant, $a_t = 0$ and the acceleration is always radial, as we described in Section 4.4. Furthermore, if the direction of $\boldsymbol{v}$ doesn't change, then there is no radial acceleration and the motion is one-dimensional ($a_r = 0$, $a_t \neq 0$).

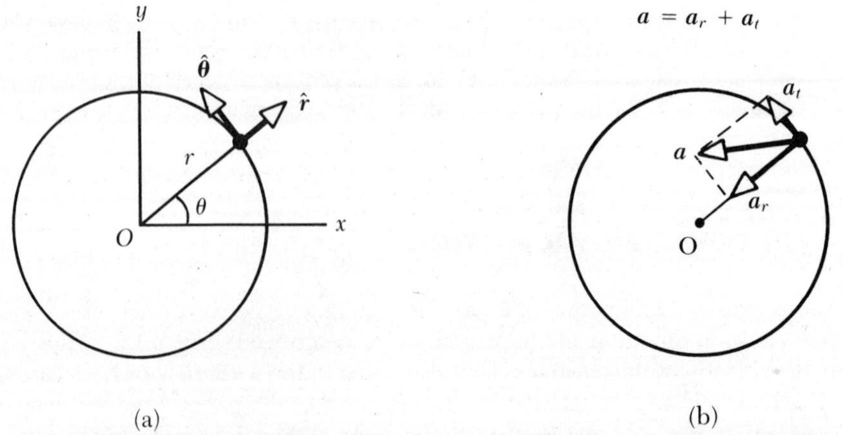

Figure 4.13 (a) Description of the unit vectors $\hat{\boldsymbol{r}}$ and $\hat{\boldsymbol{\theta}}$. (b) The total acceleration $\boldsymbol{a}$ of a particle rotating in a circle consists of a radial component, $\boldsymbol{a}_r$, directed toward the center of rotation, and a tangential component, $\boldsymbol{a}_t$. The component $\boldsymbol{a}_t$ is zero if the speed is constant.

It is convenient to write the acceleration of a particle moving in a circular path in terms of unit vectors. We can do this by defining the unit vectors $\hat{\mathbf{r}}$ and $\hat{\boldsymbol{\theta}}$, where $\hat{\mathbf{r}}$ is *a unit vector directed radially outward along the radius vector,* from the center of curvature, and $\hat{\boldsymbol{\theta}}$ is *a unit vector tangent to the circular path,* as in Fig. 4.13a. The direction of $\hat{\boldsymbol{\theta}}$ is in the direction of increasing θ, where θ is measured counterclockwise from the positive x axis. Note that both $\hat{\mathbf{r}}$ and $\hat{\boldsymbol{\theta}}$ "move along with the particle" and so vary in time relative to a stationary observer. Using this notation, we can express the total acceleration as

$$\mathbf{a} = \mathbf{a}_t + \mathbf{a}_r = \frac{dv}{dt}\hat{\boldsymbol{\theta}} - \frac{v^2}{r}\hat{\mathbf{r}} \tag{4.23}$$

These vectors are described in Fig. 4.13b. The negative sign for $\mathbf{a}_r$ indicates that it is always directed radially inward, *opposite* the unit vector $\hat{\mathbf{r}}$.

Example 4.5

A ball tied to the end of a string 0.5 m in length swings in a vertical circle under the influence of gravity, as in Fig. 4.14. When the string makes an angle of $\theta = 20°$ with the vertical, the ball has a speed of 1.5 m/s. (a) Find the radial component of acceleration at this instant.

Since $v = 1.5$ m/s and $r = 0.5$ m, we find that

$$a_r = \frac{v^2}{r} = \frac{(1.5 \text{ m/s})^2}{0.5 \text{ m}} = 4.5 \text{ m/s}^2$$

(b) When the ball is at an angle θ to the vertical, it has a tangential component of acceleration of magnitude $g \sin\theta$ (the component of $\mathbf{g}$ tangent to the circle). Therefore, at $\theta = 20°$, we find that $a_t = g \sin 20° = 3.4$ m/s². Find the magnitude and direction of the *total* acceleration at $\theta = 20°$.

Since $\mathbf{a} = \mathbf{a}_r + \mathbf{a}_t$, the magnitude of $\mathbf{a}$ at $\theta = 20°$ is given by

$$a = \sqrt{a_r^2 + a_t^2} = \sqrt{(4.5)^2 + (3.4)^2} \text{ m/s}^2 = 5.6 \text{ m/s}^2$$

If ϕ is the angle between $\mathbf{a}$ and the string, then

$$\phi = \tan^{-1}\frac{a_t}{a_r} = \tan^{-1}\left(\frac{3.4 \text{ m/s}^2}{4.5 \text{ m/s}^2}\right) = 37°$$

Note that all of the vectors—$\mathbf{a}$, $\mathbf{a}_t$, and $\mathbf{a}_r$—change in direction *and* magnitude as the ball swings through the circle. When the ball is at its lowest elevation ($\theta = 0$), $a_t = 0$, since there is no tangential component of g at this angle, and a_r is a *maximum,* since v is a maximum. When the ball is at its highest position ($\theta = 180°$), a_t is again zero but a_r is a minimum, since v is a minimum. Finally, in the two horizontal positions, ($\theta = 90°$ and $270°$), $|a_t| = g$ and a_r is somewhere between its minimum and maximum values.

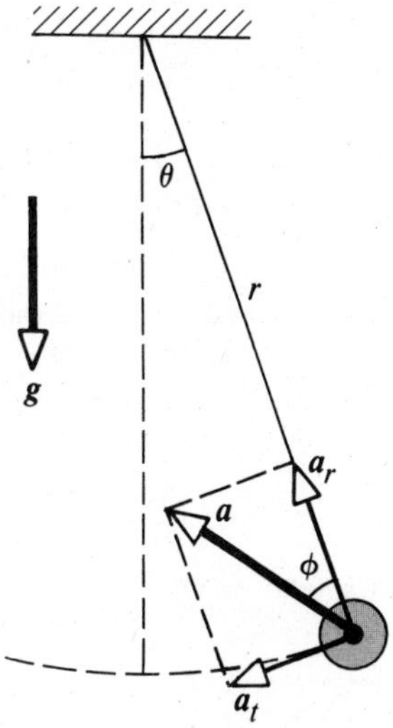

Figure 4.14 (Example 4.5) Circular motion of a ball tied on a string of length r. The ball swings in a vertical plane, and its acceleration, $\mathbf{a}$, has a radial component, $\mathbf{a}_r$, and a tangential component, $\mathbf{a}_t$.

Q9. What is the fundamental difference between the unit vectors $\hat{\mathbf{r}}$ and $\hat{\boldsymbol{\theta}}$ defined in Fig. 4.13 and the unit vectors $\mathbf{i}$ and $\mathbf{j}$?

4.6 RELATIVE VELOCITY AND RELATIVE ACCELERATION

In this section, we describe how observations made by different observers in different frames of reference are related to each other. We shall find that observers in different frames of reference may measure different displacements, velocities, and accelerations for a particle in motion. That is, two observers moving with respect to each other will generally not agree on the outcome of a measurement.

For example, if two cars are moving in the same direction with speeds of 50 mi/h and 60 mi/h, a passenger in the slower car will claim that the speed of the faster car relative to that of the slower car is 10 mi/h. Of course, a stationary observer will measure the speed of the faster car to be 60 mi/h. This simple example demonstrates that velocity measurements differ in different frames of reference.

Next, suppose a person riding on a moving vehicle (observer A) throws a ball straight up in the air according to his frame of reference, as in Fig. 4.15a. According to observer A, the ball will move in a vertical path. On the other hand, a stationary observer (B) will see the path of the ball as a parabola, as illustrated in Fig. 4.15b.

Another simple example is to imagine a package being dropped from an airplane flying parallel to the earth with a constant velocity. An observer on the airplane would describe the motion of the package as a straight line toward the earth. On the other hand, an observer on the ground would view the trajectory of the package as a parabola. Relative to the ground, the package has a vertical component of velocity (resulting from the acceleration of gravity and equal to the velocity measured by the observer in the airplane) *and* a horizontal component of velocity (given to it by the airplane's motion). If the airplane continues to move horizontally with the same velocity, the package will hit the ground directly beneath the airplane (assuming that friction is neglected)!

In a more general situation, consider a particle located at the point P in Fig. 4.16.

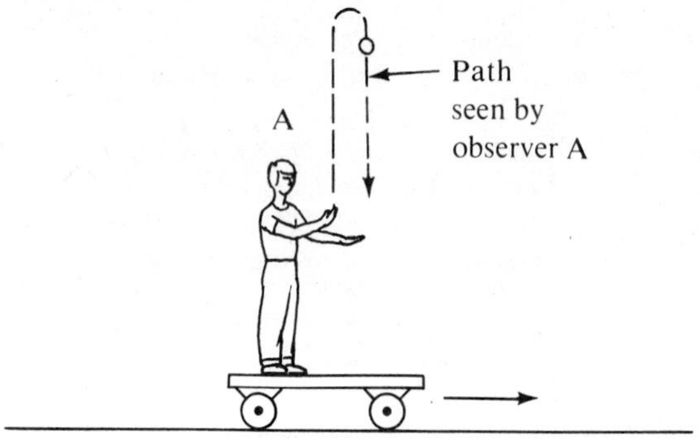

(a)

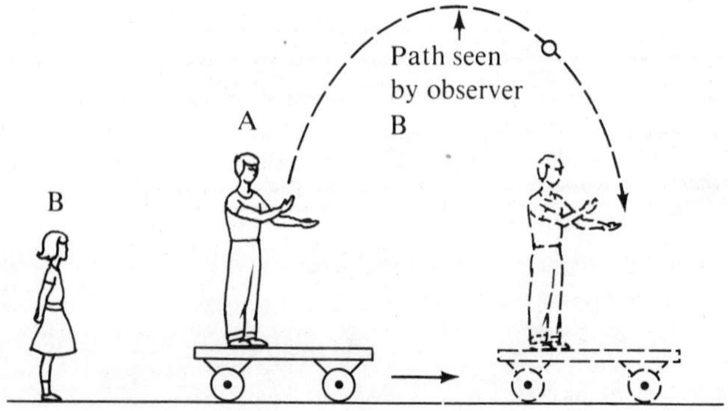

(b)

Figure 4.15 (a) Observer A in a moving vehicle throws a ball upward and sees a straight-line path for the ball. (b) A stationary observer B sees a parabolic path for the same ball.

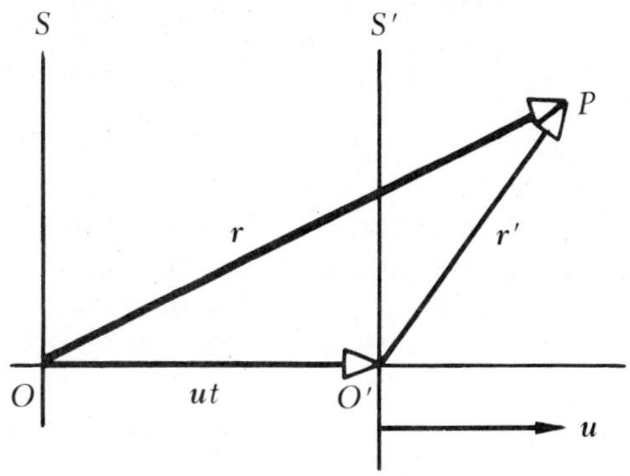

Figure 4.16 A particle located at the point P is described by two observers, one in the fixed frame of reference, S, the other in the frame S′, which moves with a constant velocity $\boldsymbol{u}$ to the right. The vector $\boldsymbol{r}$ is the particle's position vector relative to S, and $\boldsymbol{r}'$ is the position vector relative to S′.

Imagine that the motion of this particle is being described by two observers, one in reference frame S, fixed with respect to the earth, and another in reference frame S′, moving to the right relative to S with a constant velocity $\boldsymbol{u}$. (Relative to an observer in S′, S moves to the left with a velocity $-\boldsymbol{u}$.)

We label the position of the particle with respect to the S frame with the position vector $\boldsymbol{r}$ and label its position relative to the S′ frame with the vector $\boldsymbol{r}'$, at some time t. If the origins of the two reference frames coincide at $t = 0$, then the vectors $\boldsymbol{r}$ and $\boldsymbol{r}'$ are related to each through the expression

$$\boldsymbol{r}' = \mathbf{r} - \boldsymbol{u}t \tag{4.24}$$

Galilean transformation equations

That is, in a time t the S′ frame is displaced to the right by an amount $\boldsymbol{u}t$.

If we differentiate Eq. 4.24 with respect to time and note that $\boldsymbol{u}$ is constant, we get

$$\frac{d\boldsymbol{r}'}{dt} = \frac{d\boldsymbol{r}}{dt} - \boldsymbol{u}$$

$$\boldsymbol{v}' = \boldsymbol{v} - \boldsymbol{u} \tag{4.25}$$

where $\boldsymbol{v}'$ is the velocity of the particle observed in the S′ frame and $\boldsymbol{v}$ is the velocity observed in the S frame. Equations 4.24 and 4.25 are known as *Galilean transformation equations*. They relate the coordinates and velocity of a particle in the earth's reference frame to those measured in a frame of reference in uniform motion with respect to the earth. However, they are *valid only* at particle speeds (relative to both observers) that are small compared with the speed of light ($\approx 3 \times 10^8$ m/s). When the particle speed according to either observer approaches the speed of light, these transformation equations must be replaced by more exact transformation equations, which are used in the special theory of relativity. As it turns out, the relativity transformation equations reduce to the Galilean transformation equations when the particle speed is small compared with the speed of light. We will discuss this in more detail in Chapter 40.

Although observers in the two different reference frames will measure different velocities for the particles, they will measure the *same acceleration* when $\boldsymbol{u}$ is constant. This can be seen by taking the time derivative of Eq. 4.25, which gives

$$\frac{d\boldsymbol{v}'}{dt} = \frac{d\boldsymbol{v}}{dt} - \frac{d\boldsymbol{u}}{dt}$$

But $d\boldsymbol{u}/dt = 0$, since $\boldsymbol{u}$ is constant. Therefore, we conclude that $\boldsymbol{a}' = \boldsymbol{a}$ since $\boldsymbol{a}' = d\boldsymbol{v}'/dt$ and $\boldsymbol{a} = d\boldsymbol{v}/dt$. That is, *the acceleration of the particle measured by an observer in the earth's frame of reference will be the same as that measured by any other observer moving with* constant velocity *with respect to the first observer.*

Example 4.6

A boat heading due north crosses a wide river with a speed of 10 km/h relative to the water. The river has a uniform speed of 5 km/h due east. Determine the velocity of the boat with respect to a stationary ground observer.

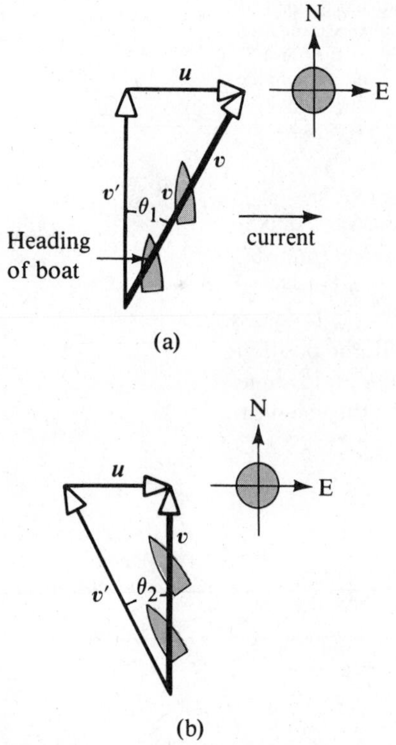

Figure 4.17 (Examples 4.6 and 4.7) (a) If the boat heads north, the motion of the boat relative to the earth is northeast along $\boldsymbol{v}$ when the river flows eastward. (b) If the boat wants to travel north, it must head northwest as shown. In both cases, $\boldsymbol{v} = \boldsymbol{v}' + \boldsymbol{u}$ and the heading of the boat is parallel to $\boldsymbol{v}'$.

Solution: The moving reference frame, S′, is the river, and the observer is in the stationary reference frame, S (the earth). The vectors $\boldsymbol{u}$, $\boldsymbol{v}$, and $\boldsymbol{v}'$ are defined as follows:

$\boldsymbol{u}$ = velocity of the water with respect to the earth
$\boldsymbol{v}$ = velocity of the boat with respect to the earth
$\boldsymbol{v}'$ = velocity of the boat with respect to the water

In this example, $\boldsymbol{u}$ is to the right, $\boldsymbol{v}'$ is straight up, and $\boldsymbol{v}$ is at the angle θ_1, as defined in Fig. 4.17a. Since these three vectors form a right triangle, the speed of the boat with respect to the earth is

$$v = \sqrt{(v')^2 + u^2} = \sqrt{(10)^2 + (5)^2}\ \text{km/h} = 11.2\ \text{km/h}$$

and the direction of $\boldsymbol{v}$ is

$$\theta_1 = \tan^{-1}\left(\frac{u}{v'}\right) = \tan^{-1}\left(\frac{5}{10}\right) = 26.6°$$

Therefore, the boat will be traveling 63.4° north of east with respect to the earth.

Example 4.7

If the boat in Example 4.6 travels with the same speed of 10 km/h relative to the water and wishes to travel due north, as in Fig. 4.17b, in what direction should it head?

Solution: Intuitively, we know that the boat must head upstream. For this example, the vectors $\boldsymbol{u}$, $\boldsymbol{v}$, and $\boldsymbol{v}'$ are oriented as shown in Fig. 4.17b, where $\boldsymbol{v}'$ is now the hypotenuse of the right triangle. Therefore, the boat's speed relative to the earth is

$$v = \sqrt{(v')^2 - u^2} = \sqrt{(10)^2 - (5)^2}\ \text{km/h} = 8.66\ \text{km/h}$$

$$\theta_2 = \tan^{-1}\left(\frac{u}{v}\right) = \tan^{-1}\left(\frac{5}{8.66}\right) = 30°$$

where θ_2 is west of north.

4.7 SUMMARY

If a particle moves with *constant* acceleration $\boldsymbol{a}$ and has a velocity $\boldsymbol{v}_0$ and position $\boldsymbol{r}_0$ at $t = 0$, its velocity and position at some later time t are given by

Velocity vector as a function of time

$$\boldsymbol{v} = \boldsymbol{v}_0 + \boldsymbol{a}t \qquad (4.6)$$

Position vector as a function of time

$$\boldsymbol{r} = \boldsymbol{r}_0 + \boldsymbol{v}_0 t + \frac{1}{2}\boldsymbol{a}t^2 \qquad (4.7)$$

For two-dimensional motion in the xy plane under constant acceleration, these vector expressions are equivalent to two component expressions, one for the motion along x with an acceleration a_x and one for the motion along y with an acceleration a_y.

Projectile motion is two-dimensional motion under constant acceleration, where $a_x = 0$ and $a_y = -g$. In this case, if $x_0 = y_0 = 0$, the components of Eqs. 4.6 and 4.7 reduce to

$$v_x = v_{x0} = \text{constant} \tag{4.10}$$

$$v_y = v_{y0} - gt \tag{4.11}$$

$$x = v_{x0}t \tag{4.12}$$

$$y = v_{y0}t - \frac{1}{2}gt^2 \tag{4.13}$$

Projectile motion equations

where $v_{x0} = v_0 \cos\theta_0$, $v_{y0} = v_0 \sin\theta_0$, v_0 is the initial speed of the projectile, and θ_0 is the angle v_0 makes with the positive x axis. Note that these expressions give the velocity components (and hence the velocity vector) and the coordinates (and hence the position vector) at *any* time t that the projectile is in motion.

The *maximum height, h,* and *horizontal range, R,* are *specific coordinates* in the motion of a projectile:

$$h = \frac{v_0^2 \sin^2\theta_0}{2g} \tag{4.17}$$

Maximum height of a projectile

$$R = \frac{v_0^2 \sin 2\theta_0}{g} \tag{4.18}$$

Horizontal range of a projectile

As you can see from Eqs. 4.10 through 4.13, it is useful to think of projectile motion as the superposition of two motions: (1) uniform motion in the x direction, where v_x remains constant, and (2) motion in the vertical direction, subject to a constant downward acceleration of magnitude $g = 9.8$ m/s^2. Hence, one can analyze the motion in terms of separate horizontal and vertical components of velocity, as in Fig. 4.18.

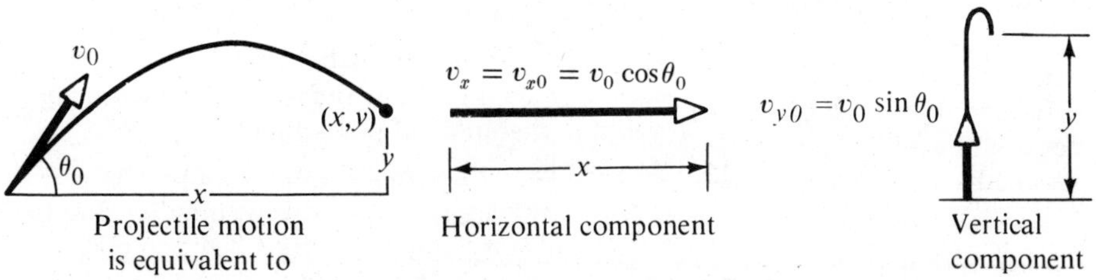

Figure 4.18 Analyzing motion in terms of the horizontal and vertical components of velocity.

A particle moving in a circle of radius r with constant speed v undergoes a centripetal (or radial) acceleration, $\boldsymbol{a}_r$, because the direction of $\boldsymbol{v}$ changes in time. The magnitude of $\boldsymbol{a}_r$ is given by

$$a_r = \frac{v^2}{r} \tag{4.19}$$

Centripetal acceleration

and its direction is always toward the center of the circle.

If a particle moves along a curved path in such a way that the magnitude and direction of $\boldsymbol{v}$ change in time, the particle has an acceleration vector that can be described by two components: (1) a radial component, $\boldsymbol{a}_r$, arising from the change in direction of $\boldsymbol{v}$, and (2) a tangential component, $\boldsymbol{a}_t$, arising from the change in magnitude of $\boldsymbol{v}$. The magnitude of $\boldsymbol{a}_r$ is v^2/r, and the magnitude of $\boldsymbol{a}_t$ is dv/dt.

The velocity of a particle, $\boldsymbol{v}$, measured in a fixed frame of reference, S, is related to the velocity of the same particle, $\boldsymbol{v}'$, measured in a moving frame of reference, S′, by

Galilean velocity transformation

$$\boldsymbol{v}' = \boldsymbol{v} - \boldsymbol{u} \tag{4.25}$$

where $\boldsymbol{u}$ is the velocity of S′ relative to S.

EXERCISES

Section 4.2 Motion in Two Dimensions with Constant Acceleration

1. A particle starts from rest at $t = 0$ at the origin and moves in the xy plane with a constant acceleration of $\boldsymbol{a} = (2\boldsymbol{i} + 4\boldsymbol{j})$ m/s². After a time t has elapsed, determine (a) the x and y components of velocity, (b) the coordinates of the particle, and (c) the speed of the particle.

2. At $t = 0$, a particle moving in the xy plane with constant acceleration has a velocity of $\boldsymbol{v}_0 = (3\boldsymbol{i} - 2\boldsymbol{j})$ m/s at the origin. At $t = 3$ s, its velocity is $\boldsymbol{v} = (9\boldsymbol{i} + 7\boldsymbol{j})$ m/s. Find (a) the acceleration of the particle and (b) its coordinates at any time t.

3. The vector position of a particle varies in time according to the expression $\boldsymbol{r} = (3\boldsymbol{i} - 6t^2\boldsymbol{j})$ m. (a) Find expressions for the velocity and acceleration as functions of time. (b) Determine the particle's position and velocity at $t = 1$ s.

4. A particle initially located at the origin has an acceleration of $\boldsymbol{a} = 3\boldsymbol{j}$ m/s² and an initial velocity of $\boldsymbol{v}_0 = 5\boldsymbol{i}$ m/s. Find (a) the vector position and velocity at any time t and (b) the coordinates and speed of the particle at $t = 2$ s.

5. At $t = 0$ a particle leaves the origin with a velocity of 6 m/s in the positive y direction. Its acceleration is given by $\boldsymbol{a} = (2\boldsymbol{i} - 3\boldsymbol{j})$ m/s². When the particle reaches its *maximum* y coordinate, its y component of velocity is zero. At this instant, find (a) the velocity of the particle and (b) its x and y coordinates.

Section 4.3 Projectile Motion (neglect air resistance in all exercises)

6. A football, kicked at an angle of 50° to the horizontal, travels a horizontal distance of 20 m before hitting the ground. Find (a) the initial speed of the football, (b) the time it is in the air, and (c) the maximum height it reaches.

7. An astronaut on a strange planet finds that she can jump a *maximum* horizontal distance of 30 m if her initial speed is 9 m/s. What is the acceleration of gravity on the planet?

8. It has been said that in his youth George Washington threw a silver dollar across a river. Assuming that the river was 300 m wide, (a) what *minimum initial* speed was necessary to get the coin across the river and (b) how long was the coin in flight?

9. A rifle is aimed horizontally at the center of a large target 150 m away. The initial velocity of the bullet is 450 m/s. Where does the bullet strike the target?

10. A ball is thrown horizontally from the top of a building 35 m high. The ball strikes the ground at a point 80 m from the base of the building. Find (a) the time the ball is in flight, (b) its initial velocity, and (c) the x and y components of velocity just before the ball strikes the ground.

11. A projectile is fired in such a way that its horizontal range is equal to three times its maximum height. What is the angle of projection?

12. Show that the horizontal range of a projectile with a fixed initial speed will be the same for any two complementary angles, such as 30° and 60°.

13. The initial speed of a cannon ball is 200 m/s. If it is fired at a target that is at a horizontal distance of 2 km from the cannon, find (a) the two projected angles that will result in a hit and (b) the total time of flight for each of the two trajectories found in (a).

14. The maximum horizontal distance a certain baseball player is able to hit the ball is 150 m. On one pitch, this player hits the ball in such a way that it has the same initial speed as his maximum-distance hit, but makes an angle of 20° with the horizontal. Where will this ball strike the ground with respect to home plate?

15. A student is able to throw a ball vertically to a maximum height of 40 m. What maximum distance can the student throw the ball in the horizontal direction?

Section 4.4 Uniform Circular Motion

16. Find the acceleration of a particle moving with a constant speed of 6 m/s in a circle 3 m in radius.

17. A particle moves in a circular path 0.4 m in radius with constant speed. If the particle makes five revolutions in each second of its motion, find (a) the speed of the particle and (b) its acceleration.

18. The orbit of the moon about the earth is approximately circular, with a mean radius of 3.84×10^8 m. It takes 27.3 days for the moon to complete one revolution about the earth. Find (a) the mean orbital speed of the moon and (b) its centripetal acceleration.

19. A tire 0.5 m in radius rotates at a constant rate of 200 revolutions per minute. Find the speed and acceleration of a small stone lodged in the tread of the tire (on its outer edge).

20. A hunter uses a stone attached to the end of a rope as a crude weapon in attempting to capture an animal running *away* from him at constant velocity. The stone is swung overhead in a horizontal circle 1.6 m in diameter at the rate of 3 revolutions per second. (a) What is the centripetal acceleration of the stone? (b) What minimum speed must the animal have in order to avoid being struck by the stone after it is released?

Section 4.5 Tangential and Radial Acceleration in Curvilinear Motion

21. A student swings a ball attached to the end of a string 0.5 m in length in a vertical circle. The speed of the ball is 4 m/s at its highest point and 6 m/s at its lowest point. Find the acceleration of the ball at (a) its highest point and (b) its lowest point.

22. A pendulum of length 1 m rotates in a vertical plane (Fig. 4.14). When the pendulum is in the two horizontal positions ($\theta = 90°$ and $\theta = 270°$), its speed is 4 m/s. (a) Find the magnitude of the centripetal acceleration and tangential acceleration for these positions. (b) Draw vector diagrams to determine the direction of the total acceleration for these two positions. (c) Calculate the magnitude and direction of the total acceleration.

23. Figure 4.19 represents the total acceleration of a particle moving clockwise in a circle of radius 3 m at a given instant of time. At this instant of time, find (a) the centripetal acceleration, (b) the speed of the particle, and (c) its tangential acceleration.

24. At some instant of time, a particle moving counterclockwise in a circle of radius 2 m has a speed of 8 m/s and its total acceleration is directed as shown in Fig. 4.20. At this instant, determine (a) the centripetal acceleration of the particle, (b) the tangential acceleration, and (c) the magnitude of the total acceleration.

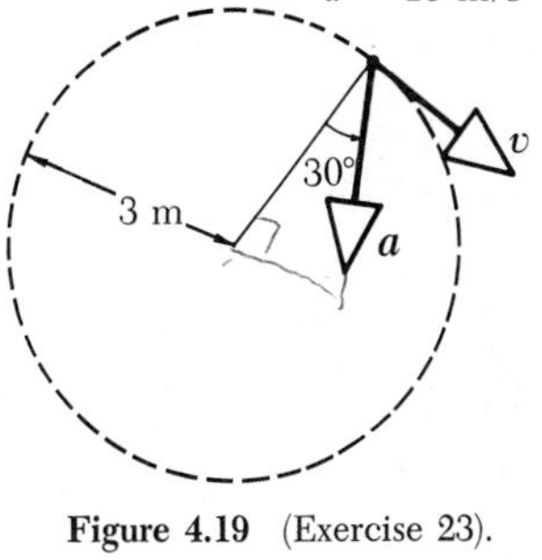

Figure 4.19 (Exercise 23).

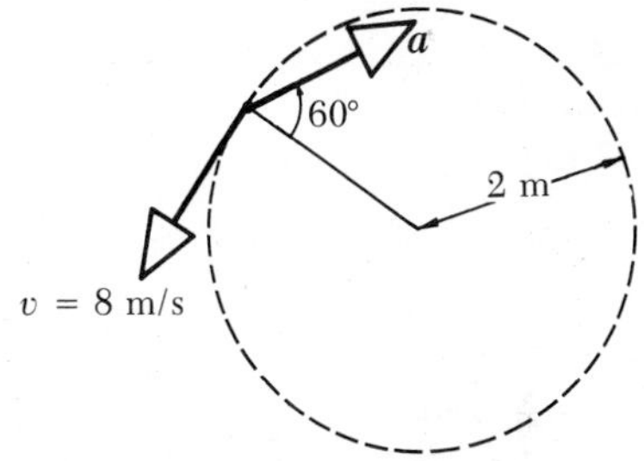

Figure 4.20 (Exercise 24).

25. The speed of a particle moving in a circle 2 m in radius increases at the constant rate of 3 m/s². At some instant, the magnitude of the total acceleration is 5 m/s². At this instant, find (a) the centripetal acceleration of the particle and (b) its speed.

Section 4.6 Relative Velocity and Relative Acceleration

26. A car travels north with a speed of 60 km/h on a straight highway. A truck travels in the opposite direction with a speed of 50 km/h. (a) What is the velocity of the car relative to the truck? (b) What is the velocity of the truck relative to the car?

27. The pilot of an aircraft wishes to fly due west in a wind blowing at 50 km/h toward the south. If the speed of the aircraft in the absence of a wind is 200 km/h, (a) in what direction should the aircraft head and (b) what should its speed be relative to the ground?

28. The pilot of an airplane notes that the compass indicates a heading due west. The airplane's speed relative to the air is 150 km/h. If there is a wind of 30 km/h toward the north, find the velocity of the airplane relative to the ground.

29. Car A travels due west with a speed of 40 km/h. Car B travels due north with a speed of 60 km/h. What is the velocity of car B as seen by the driver of car A?

30. A car travels due east with a speed of 50 km/h. Rain is falling vertically with respect to the earth. The traces of the rain on the side windows of the car make an angle of 60° with the vertical. Find the velocity of the rain with respect to (a) the car and (b) the earth.

31. A river has a steady speed of 0.5 m/s. A student swims upstream a distance of 1 km and returns to the starting point. If the student can swim at a speed of 1.2 m/s in still water, how long does the trip take? Compare this with the time the trip would take if the water were still.

PROBLEMS

1. A rocket is launched at an angle of 53° to the horizontal with an initial speed of 100 m/s. It moves along its initial line of motion with an acceleration of 30 m/s^2 for 3 s. At this time its engines fail and the rocket proceeds to move as a free body. Find (a) the maximum altitude reached by the rocket, (b) its total time of flight, and (c) its horizontal range.

2. A car is parked on a steep incline overlooking the ocean, where the incline makes an angle of 37° with the horizontal. The negligent driver leaves the car in neutral, and the emergency brakes are defective. The car rolls from rest down the incline with a constant acceleration of 4 m/s^2 and travels 50 m to the edge of the cliff. The cliff is 30 m above the ocean. Find (a) the speed of the car when it reaches the cliff and the time it takes to get there, (b) the velocity of the car when it lands in the ocean, (c) the total time the car is in motion, and (d) the position of the car relative to the base of the cliff when the car lands in the ocean.

3. After delivering his toys in the usual manner, Santa decides to have some fun and slide down an icy roof, as in Fig. 4.21. He starts from rest at the top of the roof, which is 8 m in length, and accelerates at the rate of 5 m/s^2. The edge of the roof is 6 m above a soft snowbank, which Santa lands on. Find (a) Santa's velocity components when he reaches the snowbank, (b) the total time he is in motion, and (c) the distance d between the house and the point where he lands in the snow.

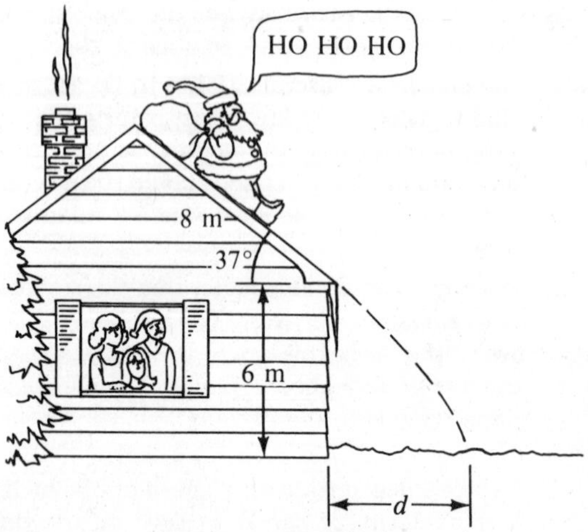

Figure 4.21 (Problem 3).

4. A daredevil is shot out of a cannon at 45° to the horizontal with an initial speed of 25 m/s. A net is located at a horizontal distance of 50 m from the cannon. At what height above the cannon should the net be placed in order to catch the daredevil?

5. The position of a particle moving in the xy plane varies in time according to the equation $\boldsymbol{r} = 3 \cos 2t\boldsymbol{i} + 3 \sin 2t\boldsymbol{j}$, where $\boldsymbol{r}$ is in m and t is in s. (a) Show that the path of the particle is a circle 3 m in radius centered at the origin. (Hint: Let $\theta = 2t$.) (b) Calculate the velocity and acceleration vectors. (c) Show that the acceleration vector always points toward the origin (opposite $\boldsymbol{r}$) and has a magnitude of v^2/r.

6. A student who is able to swim at a speed of 1.5 m/s in still water wishes to cross a river that has a current of velocity 1.2 m/s toward the south. The width of the river is 50 m. (a) If the student starts from the west bank of the river, in what direction should she head in order to swim directly across the river? How long will this trip take? (b) If she heads due east, how long will it take to cross the river? (Note that the student travels farther than 50 m in this case.)

7. A river flows with a uniform velocity v. A person in a motorboat travels 1 km upstream, at which time a log is seen floating by. The person continues to travel upstream for one more hour at the same speed and then returns downstream to the starting point, where the same log is seen again. Find the velocity of the river. (Hint: The time of travel of the boat after it meets the log equals the time of travel of the log.)

8. A skier leaves the ramp of a ski jump with a velocity of 10 m/s, 15° above the horizontal, as in Fig. 4.22. The slope is inclined at 50°, and air resistance is negligible. Find (a) the distance that the jumper lands down the slope and (b) the velocity components just before landing. (How do you think the results might be affected if air resistance were included? Note that jumpers lean forward in the shape of an airfoil to increase their distance. Why does this work?)

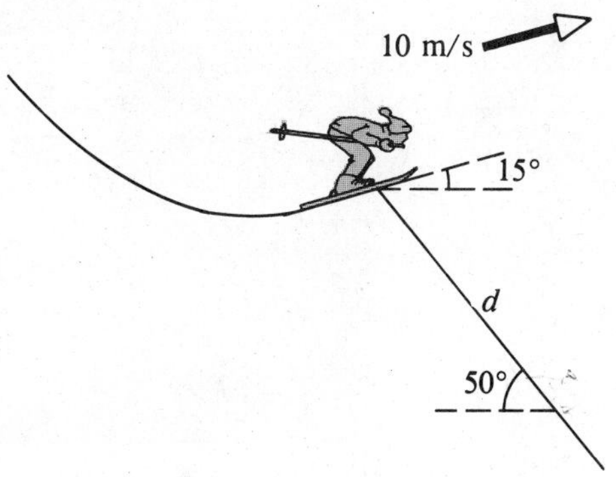

Figure 4.22 (Problem 8).

9. A football is thrown toward a receiver with an initial speed of 20 m/s at an angle of 30° to the horizontal. At that instant, the receiver is 20 m from the quarterback. In what direction and with what constant speed should the receiver run in order to catch the football at the level at which it was thrown?

10. The determined coyote is out once more to try to capture the elusive road runner. The coyote wears a pair of Acme jet-powered roller skates, which provide a constant horizontal acceleration of 15 m/s^2 (Fig. 4.23). The coyote starts off at rest 70 m from the edge of a cliff at the instant the road runner zips by in the direction of the cliff. (a) If the road runner moves with constant speed, determine the minimum speed he must have in order to reach the cliff before the coyote. (b) If the cliff is 100 m above the base of a canyon, determine where the coyote lands in the canyon (assume his skates are still in operation when he is in "flight"). (c) Determine the coyote's velocity components just before he lands in the canyon. (As usual, the road runner is saved by making a sudden turn at the cliff.)

11. A home run in a baseball game is hit in such a way that the ball just clears a wall 21 m high, located 130 m from home plate. The ball is hit at an angle of 35° to the horizontal, and air resistance is negligible. Find (a) the initial speed of the ball, (b) the time it takes the ball to reach the wall, and (c) the velocity components and the speed of the ball when it reaches the wall. (Assume the ball is hit at a height of 1 m above the ground.)

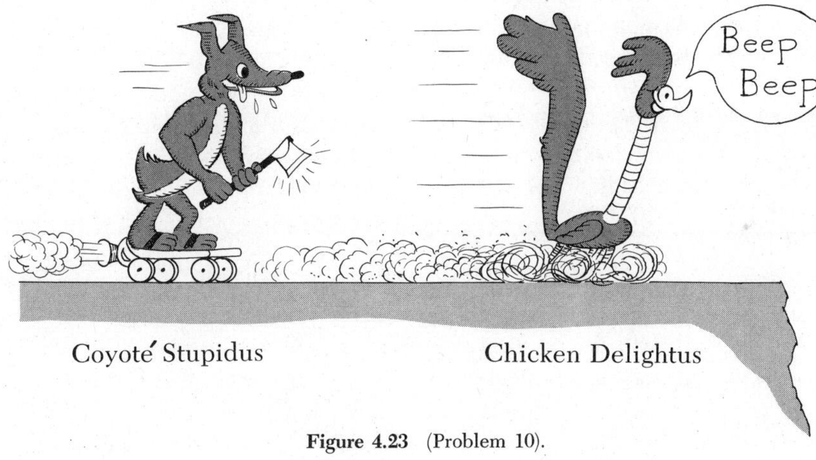

Figure 4.23 (Problem 10).

5 The Laws of Motion

NASA

5.1 INTRODUCTION TO CLASSICAL MECHANICS

In the previous two chapters on kinematics, we described the motion of particles based on the definition of displacement, velocity, and acceleration. However, we would like to be able to answer specific questions related to the causes of motion, such as "What mechanism causes motion?" and "Why do some objects accelerate at a higher rate than others?" In this chapter, we shall describe the change in motion of particles using the concepts of force, mass, and momentum. We shall then discuss the three basic laws of motion, which are based on experimental observations and were formulated nearly three centuries ago by Sir Isaac Newton.

The purpose of classical mechanics is to provide a connection between the change in motion of a body and the external forces acting on it. Keep in mind that classical mechanics deals with objects that are large compared with the dimensions of atoms ($\approx 10^{-10}$ m) and move at speeds that are much less than the speed of light (3×10^8 m/s).

We shall see that it is possible to describe the acceleration of an object in terms of the resultant external force acting on it and the mass of the object. This external force is the result of the interaction of the object with its environment. The mass of an object is a measure of the object's inertia, that is, the tendency of the object to resist an acceleration when a force acts on it.

We shall also discuss *force laws,* which describe the quantitative method of calculating the force on an object if its environment is known. We shall see that although the force laws are rather simple in form, they successfully explain a wide variety of phenomena and experimental observations. These force laws, together with the laws of motion, are the foundations of classical mechanics.

5.2 THE CONCEPT OF FORCE

You are surely familiar with the concept of force from everyday experiences. When you push or pull an object, you exert a force on it. You exert a force when you throw or kick a ball. In these examples, the word *force* is associated with the result of muscular activity and some form of motion. Although forces can cause motion, it does not necessarily follow that force is always correlated with motion. For example, as you sit reading this book, the force of gravity acts on your body, and yet you remain stationary. You can push on a block of stone and not move it.

What force (if any) causes a distant star to drift freely through space? Newton answered such questions by stating that the *change* in motion of an object is caused by forces. If an object moves with uniform motion (constant velocity), its motion does not change; therefore no force is required to maintain the motion. Since only a force can cause a change in motion, we can think of force as that which causes a body to accelerate.

A body accelerates due to an external force

Consider a situation in which several forces act simultaneously on an object. The object will accelerate only if the *net force* acting on it is not equal to zero. We shall often refer to the net force as the *resultant force*, or the *unbalanced force. If the net force is zero, the acceleration is zero and the velocity of the object remains constant. When the velocity of a body is constant or if the body is at rest, it is said to be in equilibrium.*

Definition of equilibrium

A force acting on a body can either deform the body, change its state of motion, or both. For example, if you pull on a coiled spring, as in Fig. 5.1a, the spring elongates from its unstretched position. If the spring is calibrated, the extent of deformation can be used to define a force. On the other hand, if you pull on a cart, as in Fig. 5.1b, the cart could move. Finally, when a football is kicked, as in Fig. 5.1c, the football is both deformed and set in motion. These are all examples of a class of forces called *contact forces*. That is, they represent the result of physical contact between the body and its environment. Other examples of contact forces include the force of a gas on the walls of a container (the result of the collisions of molecules with the walls) and the force of our feet on the floor.

Another class of forces, which do not involve physical contact between a body and its environment but act through empty space, are known as *action-at-a-distance forces*. One good example of such a force is the force of attraction between any two objects, the so-called gravitational force described in Fig. 5.1d. This force keeps objects bound to the earth and gives rise to what we commonly call the *weight* of an object. The planets of our solar system are bound under the action of gravitational forces. Another common example of an action-at-a-distance force is the electric force that one electric charge exerts on another electric charge, as in Fig. 5.1e. These might be an electron and proton forming the hydrogen atom. A third example

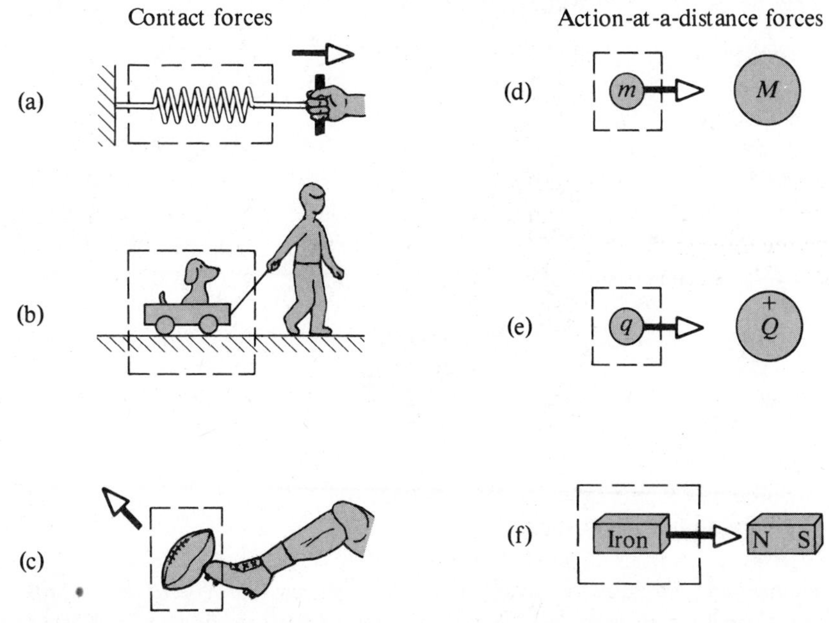

Figure 5.1 Some examples of forces applied to various objects. In each case a force is exerted on the particle or object within the boxed area. The environment external to the boxed area provides the force on the object.

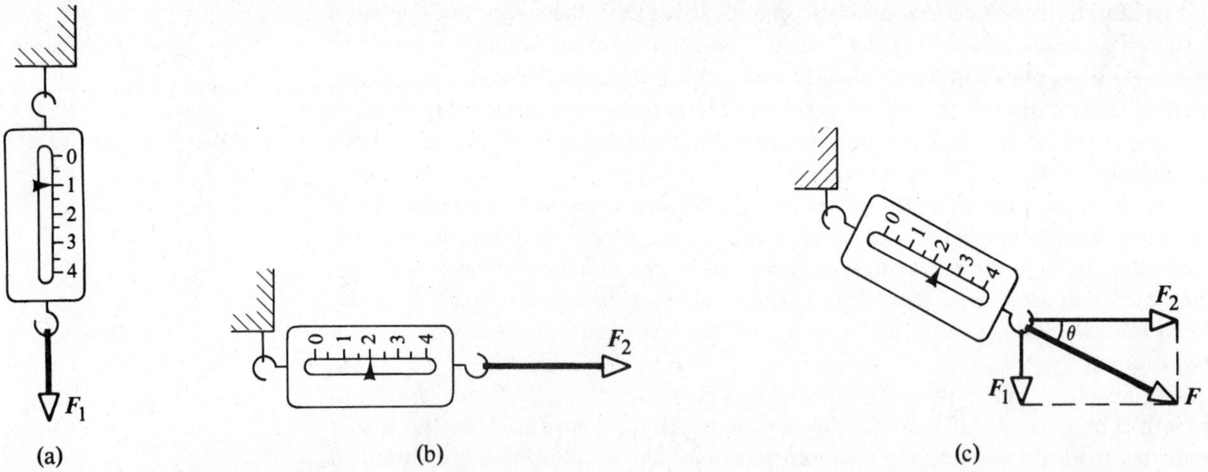

Figure 5.2 The vector nature of a force is tested with a spring scale. (a) The vertical force $\boldsymbol{F}_1$ elongates the spring 1 unit. (b) The horizontal force $\boldsymbol{F}_2$ elongates the spring 2 units. (c) The combination of $\boldsymbol{F}_1$ and $\boldsymbol{F}_2$ elongates the spring $\sqrt{1^2 + 2^2} = \sqrt{5}$ units.

of an action-at-a-distance force is the force that a bar magnet exerts on a piece of iron, as shown in Fig. 5.1f. The forces between atomic nuclei are also action-at-a-distance forces but are very short-range. They are the dominating interaction for particle separations of the order of 10^{-15} m.

Contact forces are actually the result of electromagnetic forces between a large number of atoms in proximity and therefore represent a macroscopic description of action-at-a-distance forces. Nevertheless, in developing models for macroscopic phenomena, it is convenient to use both classifications of forces. However, the only known *fundamental* forces in nature are (1) gravitational attraction between objects because of their masses, (2) electromagnetic forces between charges at rest or in motion, (3) strong nuclear forces between subatomic particles, and (4) weak nuclear forces (the so-called weak interaction), which arise in certain radioactive decay processes. In classical physics, we shall be concerned only with gravitational and electromagnetic forces.

Fundamental forces in nature

It is convenient to use the deformation of a spring to define force. Suppose a force is applied vertically to a spring with a fixed upper end, as in Fig. 5.2a. We can calibrate the spring by defining the unit force, $\boldsymbol{F}_1$, as the force that produces an elongation of 1 cm. If a force $\boldsymbol{F}_2$, applied horizontally as in Fig. 5.2b, produces an elongation of 2 cm, the magnitude of $\boldsymbol{F}_2$ is 2 units. If the two forces $\boldsymbol{F}_1$ and $\boldsymbol{F}_2$ are applied simultaneously, as in Fig. 5.2c, the elongation of the spring is found to be $\sqrt{5} = 2.24$ cm. The single force, $\boldsymbol{F}$, that would produce this same elongation is the vector sum of $\boldsymbol{F}_1$ and $\boldsymbol{F}_2$, as described in Fig. 5.2c. That is, $|\boldsymbol{F}| = \sqrt{F_1{}^2 + F_2{}^2} = \sqrt{5}$ units, and its direction is $\theta = \arctan(-0.5) = -26.6°$. *Since forces are vectors, you must use the rules of vector addition to get the resultant force on a body.* Springs that elongate in proportion to an applied force are said to obey *Hooke's law*. Such springs can be constructed and calibrated to measure unknown forces.

5.3 NEWTON'S FIRST LAW AND INERTIAL FRAMES

Newton's first law

Newton's *first law of motion* states that *a body at rest will remain at rest or a body in uniform motion in a straight line will maintain that motion unless an external resultant force acts on it.* In simpler terms, we can say that *when the resultant force on a body is zero, its acceleration is zero.* That is, when $\Sigma \boldsymbol{F} = 0$, then $\boldsymbol{a} = 0$. From the first law, we conclude that an isolated body (a body that does not interact with its environment) is either at rest or moving with constant velocity.

Suppose an object, such as a book, is lying on a table. Obviously, the book will remain at rest in the absence of any external influences. Now imagine that you push the book with a horizontal force large enough to overcome the force of friction between the book and table. The book can then be set in motion with uniform velocity if the applied force is equal in magnitude to the force of friction and applied in the opposite direction. If the applied force exceeds the force of friction, the book accelerates. If the book is released, it will eventually come to rest since the force of friction retards the motion (or causes a deceleration). Now imagine that the effect of friction is made negligibly small with the use of excellent lubricants. If the book is now set in motion and released, it will move with approximately constant velocity (until it falls off the table).

Isaac Newton (1642–1727)

A better example of uniform motion on a nearly frictionless plane is the motion of a light disk on a column of air (the lubricant), as in Fig. 5.3. If the disk is given an initial velocity, it will coast a great distance before coming to rest. This idea is used in the popular game of air hockey, where the disk makes many collisions with the walls before coming to rest.

Finally, consider a spaceship traveling in space and far removed from any planets or other matter. The spaceship requires some propulsion system to *change* its velocity. However, if the propulsion system is turned off when the spaceship reaches a velocity v, the spaceship will "coast" in space with the same velocity, and the astronauts get a "free ride."

Newton's first law is sometimes called the *law of inertia,* since it applies to objects in an inertial frame of reference. An *inertial frame of reference* is one in which an object will move with constant velocity if left undisturbed. Therefore, *an inertial frame is one with no acceleration.* A reference frame either fixed or moving with constant velocity relative to the distant stars is the best approximation of an inertial frame. The earth is not an inertial frame because of its orbital motion about the sun and rotational motion about its own axis. As the earth travels in its nearly circular orbit about the sun, it experiences a centripetal acceleration of about 4.4×10^{-3} m/s^2 toward the sun. In addition, since the earth rotates about its own axis once every 24 h, a point on the equator experiences an additional centripetal acceleration of 3.37×10^{-2} m/s^2 toward the center of the earth. However, these are small compared with **g** and can often be neglected. In most situations *we shall assume that the earth is an inertial frame.*

Inertial frame

Thus, if an object is in uniform motion (v = constant) an observer in one inertial frame (say, one at rest with respect to the object) will claim that the acceleration and the resultant force on the object are zero. An observer in *any other* inertial

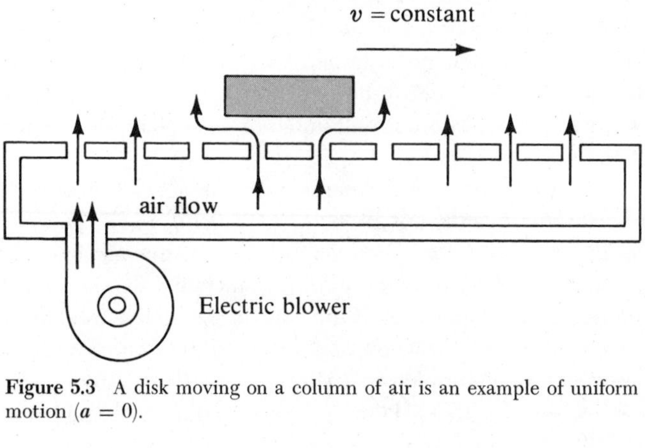

Figure 5.3 A disk moving on a column of air is an example of uniform motion (a = 0).

frame will also find that $\boldsymbol{a} = 0$ and $\boldsymbol{F} = 0$ for the object. According to the first law, a body at rest and one moving with constant velocity are equivalent.

In general, we can state that *inertial frames are those in which Newton's laws of motion are valid.* Unless stated otherwise, we shall usually write the laws of motion with respect to an observer in an inertial frame.

Q1. If a body is at rest, can we conclude that there are no external forces acting on it?

5.4 INERTIAL MASS

Inertia

If you attempt to change the state of motion of any body, the body will resist this change. *The resistance of a body to a change in its state of motion is called* inertia. For instance, consider two large, solid cylinders of equal size, one being balsa wood and the other steel. If you were to push the cylinders along a horizontal, rough surface, it would certainly take more effort to get the steel cylinder rolling. Likewise, once they are in motion, it would require more effort to bring the steel cylinder to rest. Therefore, we say that the steel cylinder has more inertia than the balsa-wood cylinder.

Mass is a term used to quantify inertia. The greater the mass of a body, the less it will accelerate (change its state of motion) under the action of an applied force. For example, if a given force acting on a 3-kg mass produces an acceleration of 4 m/s^2, the same force when applied to a 6-kg mass will produce an acceleration of 2 m/s^2. This idea will be used to obtain a quantitative description of the concept of mass. Before doing so, however, it is important to point out that mass should not be confused with weight. *Mass and weight are two different quantities.* The weight of a body is equal to the force of gravity acting on the body and varies with location. For example, a person who weighs 180 lb on earth weighs only about 30 lb on the moon. On the other hand, the mass of a body is the same everywhere, regardless of location. An object having a mass of 2 kg on earth will also have a mass of 2 kg on the moon.

Mass and weight are different quantities

A quantitative measurement of mass can be made by comparing the accelerations that a given force will produce on different bodies. Suppose a force acting on a body of mass m_1 produces an acceleration $\boldsymbol{a}_1$, and the *same force* acting on a body of mass m_2 produces an acceleration $\boldsymbol{a}_2$. The ratio of the two masses is defined as the *inverse* ratio of the magnitudes of the accelerations produced by the same force:

$$\frac{m_1}{m_2} \equiv \frac{a_2}{a_1} \tag{5.1}$$

If one of these is a standard known mass of, say, 1 kg (the kg is the SI unit of mass), the mass of an unknown can be obtained from acceleration measurements. For example, if the standard 1-kg mass undergoes an acceleration of 3 m/s^2 under the influence of some force, a 2-kg mass will undergo an acceleration of 1.5 m/s^2 under the action of the same force.

We have associated the concept of mass with the acceleration produced by an applied force. *Mass is an inherent property of a body and is independent of the body's surroundings and of the method used to measure the mass.* It is also an experimental fact that *mass is a scalar quantity,* since its value is independent of the direction of motion and independent of the coordinate system used in describing the motion. Finally, *mass is a quantity that obeys the rules of ordinary arithmetic.* That is, several masses can be combined in a simple numerical fashion. For example, if you combine a 3-kg mass with a 5-kg mass, their total mass would be 8 kg. This can be verified experimentally by comparing the acceleration of each object produced by a known force with the acceleration of the combined system using the same force.

5.5 NEWTON'S SECOND LAW

Newton's second law is concerned with the motion of a body that undergoes an acceleration under the action of a nonzero resultant external force. To state the law in its most general form, we first define the *momentum*, $\mathbf{p}$, of a particle as the product of the mass, m, and the velocity, $\mathbf{v}$:

$$\mathbf{p} \equiv m\mathbf{v} \tag{5.2}$$

Definition of momentum

Momentum is a vector quantity that is in the direction of $\mathbf{v}$ and has dimensions of ML/T (kg · m/s in SI units).

Newton's second law of motion states that the time rate of change of momentum of a particle is equal to the resultant external force acting on the particle:

$$\Sigma \mathbf{F} = \frac{d\mathbf{p}}{dt} = \frac{d}{dt}(m\mathbf{v}) \tag{5.3}$$

Newton's second law

This is the most general form of Newton's second law, which is valid in any inertial frame of reference. The notation $\Sigma \mathbf{F}$ represents the *vector sum* of all external forces acting on the particle.

If m is treated as a constant, then Eq. 5.3 can be expressed

$$\Sigma \mathbf{F} = \frac{d}{dt}(m\mathbf{v}) = m\frac{d\mathbf{v}}{dt} \tag{5.4}$$

Since acceleration is defined as $\mathbf{a} = d\mathbf{v}/dt$, Eq. 5.4 can be written

$$\Sigma \mathbf{F} = m\mathbf{a} \tag{5.5}$$

Newton's second law

You should note that Eq. 5.5 is a *vector* expression and hence is equivalent to the following three component equations:

$$\Sigma F_x = ma_x \qquad \Sigma F_y = ma_y \qquad \Sigma F_z = ma_z \tag{5.6}$$

Newton's second law—component form

Therefore, we conclude that the resultant force on a particle is equal to its mass multiplied by its acceleration if the mass is constant.[1] Note that if the resultant force is zero, then $\mathbf{a} = 0$, which corresponds to the equilibrium situation where $\mathbf{v}$ is either constant or zero. Hence, the first law of motion is a special case of the second law.

Units of Force and Mass

The SI unit of force is the *newton*, which is defined as the force that, when acting on a 1-kg mass, produces an acceleration of 1 m/s²:

$$1\ \mathrm{N} \equiv 1\ \mathrm{kg \cdot m/s^2} \tag{5.7}$$

Definition of newton

The unit of force in the cgs system is called the *dyne* and is defined as that force

[1]Equation 5.5 is valid only when the speed of the particle is much less than the speed of light. We will treat the relativistic situation in Chapter 40.

TABLE 5.1 Units of Force, Mass, and Acceleration[a]

SYSTEM OF UNITS	MASS	ACCELERATION	FORCE
SI	kg	m/s^2	$N = kg \cdot m/s^2$
cgs	g	cm/s^2	$dyne = g \cdot cm/s^2$
British engineering	slug	ft/s^2	$lb = slug \cdot ft/s^2$

[a] $1\ N = 10^5\ dyne = 0.225\ lb$.

which, when acting on a 1-g mass, produces an acceleration of 1 cm/s^2:

Definition of dyne

$$1\ \text{dyne} \equiv 1\ \text{g} \cdot \text{cm/s}^2 \tag{5.8}$$

In the British engineering system, the unit of force is the *pound*, defined as the force that, when acting on a 1-slug mass,[2] produces an acceleration of 1 ft/s^2:

Definition of pound

$$1\ \text{lb} = 1\ \text{slug} \cdot \text{ft/s}^2 \tag{5.9}$$

Since 1 kg = 10^3 g and 1 m = 10^2 cm, it follows that 1 N = 10^5 dynes. It is left as an exercise to show that 1 N = 0.225 lb. The units of force, mass, and acceleration are summarized in Table 5.1.

Example 5.1

An object of mass 0.30 kg is placed on a horizontal, frictionless surface. Two forces act on the object as shown in Fig. 5.4. The force $\mathbf{F}_1$ has a magnitude of 5.0 N, and $\mathbf{F}_2$ has a magnitude of 8.0 N. Determine the acceleration of the object.

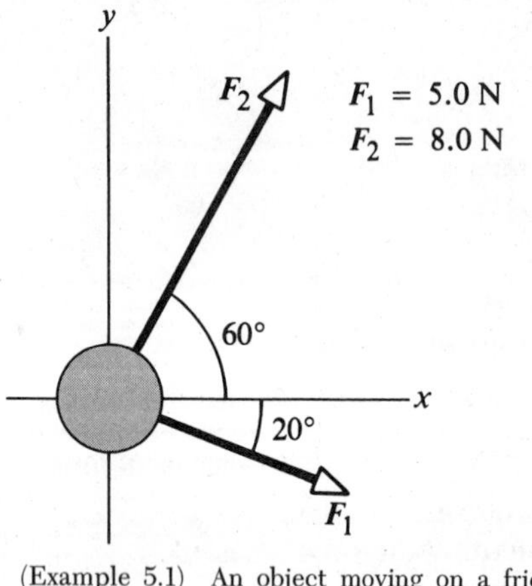

Figure 5.4 (Example 5.1) An object moving on a frictionless surface will accelerate in the direction of the *resultant* force, $\mathbf{F}_1 + \mathbf{F}_2$.

Solution: The resultant force in the x direction is

$$\Sigma F_x = F_{1x} + F_{2x} = F_1 \cos 20° + F_2 \cos 60°$$
$$= 5(0.94) + 8(0.50) = 8.7\ \text{N}$$

The resultant force in the y direction is

$$\Sigma F_y = F_{1y} + F_{2y} = -F_1 \sin 20° + F_2 \sin 60°$$
$$= -5(0.34) + 8(0.87) = 5.2\ \text{N}$$

The resultant force in unit-vector notation is therefore $\mathbf{F} = (8.7\mathbf{i} + 5.2\mathbf{j})$ N. Now, we can use Newton's second law in component form to find the x and y components of acceleration:

$$a_x = \frac{\Sigma F_x}{m} = \frac{8.7\ \text{N}}{0.3\ \text{kg}} = 29\ \text{m/s}^2$$

$$a_y = \frac{\Sigma F_y}{m} = \frac{5.2\ \text{N}}{0.3\ \text{kg}} = 17\ \text{m/s}^2$$

The acceleration has a magnitude of $a = \sqrt{(29)^2 + (17)^2}$ m/s^2 = 34 m/s^2, and its direction is $\theta = \tan^{-1}(17/29) =$ 30° to the positive x axis.

You should show that the object will be in equilibrium ($\mathbf{a} = 0$) if a third force, $\mathbf{F}_3 = (-8.7\mathbf{i} - 5.2\mathbf{j})$ N, is applied to it.

5.6 WEIGHT

We are well aware of the fact that bodies are attracted to the earth. The force exerted by the earth on a body is called the *true weight* of the body, **W**. This force is directed toward the center of the earth.[3]

[2]The *slug* is the *unit of mass* in the British engineering system and is that system's counterpart of the SI *kilogram*. When we speak of going on a diet to lose a few pounds, we really mean that we want to lose a few slugs, that is, we want to reduce our mass. When we lose those few slugs, the force of gravity (pounds) on our reduced mass decreases (since $\mathbf{W} = m\mathbf{g}$) and that is how we "lose a few pounds." Since most of the calculations we shall carry out in our study of classical mechanics will be in SI units, the slug will seldom be used in this text.

[3]This ignores the fact that the mass distribution of the earth is not perfectly spherical.

We have seen that a freely falling body experiences an acceleration g acting toward the center of the earth. Applying Newton's second law to the freely falling body, with $\boldsymbol{a} = \boldsymbol{g}$, gives

Weight

$$\boldsymbol{W} = m\boldsymbol{g} \tag{5.10}$$

Since the weight depends on g, it varies with geographic location. Bodies weigh less at higher altitudes than at sea level. This is because g varies as the inverse square of the distance from the center of the earth. Hence, weight, unlike mass, is not an inherent property of a body. Therefore, you should not confuse mass with weight. For example, if a body has a mass of 70 kg, then the magnitude of its weight in a location where $g = 9.80 \text{ m/s}^2$ is $mg = 686$ N (about 154 lb). At the top of a mountain where $g = 9.76 \text{ m/s}^2$, this weight would be 683 N. This corresponds to a decrease in weight of about 0.4 lb. Therefore, if you want to lose weight without going on a diet, climb a mountain or weigh yourself at 30 000 ft during a jet flight.

Since $W = mg$, we can compare the masses of two bodies by measuring their weights using a spring scale or a chemical balance. That is, the ratio of the weights of two bodies equals the ratio of their masses at a given location.

Q2. Determine your mass in kg and your weight in N.

Q3. How much would you weigh on the moon, where the acceleration of gravity is about one sixth of that on earth?

Q4. How much does an astronaut weigh out in space, far from any planets?

5.7 NEWTON'S THIRD LAW

Newton's third law states that if two bodies interact, the force of body 1 on body 2 is equal to and opposite the force of body 2 on body 1 as described in Fig. 5.5a. That is,

$$\boldsymbol{F}_{12} = -\boldsymbol{F}_{21} \tag{5.11}$$

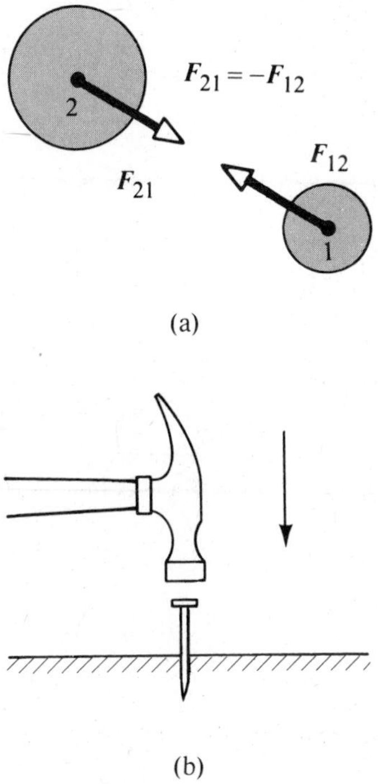

Figure 5.5 Newton's third law. (a) The force of body 1 on body 2 is equal to and opposite the force of body 2 on body 1. (b) The force of the hammer on the nail is equal to and opposite the force of the nail on the hammer.

This is equivalent to stating that *forces always occur in pairs,* or that *a single isolated force cannot exist.* The force that body 1 exerts on body 2 is sometimes called the *action force,* while the force of body 2 on body 1 is called the *reaction force.* Either force can be labeled the action or reaction force. *The action force is equal in magnitude to the reaction force and opposite in direction.* In any case, *the action and reaction forces always act on different objects.*

For example, the force acting on a freely falling projectile is its weight, $\boldsymbol{W} = m\boldsymbol{g}$. This equals the force of the earth on the projectile. The reaction to this force is the force of the projectile on the earth, $\boldsymbol{W}' = -\boldsymbol{W}$. The reaction force, $\boldsymbol{W}'$, must accelerate the earth toward the projectile just as the action force, $\boldsymbol{W}$, accelerates the projectile toward the earth. However, since the earth has such a large mass, the acceleration of the earth due to this reaction force is negligibly small. Another example is shown in Fig. 5.5b. The force of the hammer on the nail is equal to and opposite the force of the nail on the hammer.

You directly experience Newton's third law if you slam your fist against a wall or if you kick a football with your bare foot. You should be able to identify the action and reaction forces in these cases.

The weight of a body, $\boldsymbol{W}$, has been defined as the force the earth exerts on the body. If the body is a book at rest on a table, as in Fig. 5.6a, the reaction force to

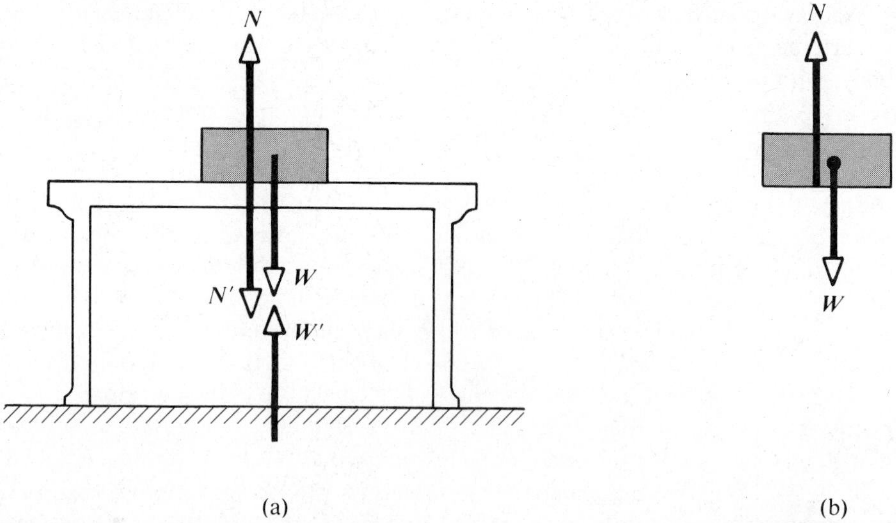

Figure 5.6 When a book is lying on a table, the forces acting on the book are the normal force, **N**, and the force of gravity, **W**, as illustrated in (b). The reaction to **N** is the force of the book on the table **N′**. The reaction to **W** is the force of the book on the earth, **W′**.

W is the force the book exerts on the earth, **W′**. The book does not accelerate since it is held up by the table. The table, therefore, exerts an upward action force, **N**, on the book, called the *normal force*.[3] The normal force balances the weight and provides equilibrium. The reaction to **N** is the force of the book on the table, **N′**. Therefore, we conclude that

Normal force

$$\mathbf{W} = -\mathbf{W}' \qquad \text{and} \qquad \mathbf{N} = -\mathbf{N}'$$

Note that the external forces acting on the book are **W** and **N**, as in Fig. 5.6b. We shall be interested only in such external forces when treating the motion of a body. From the first law, we see that since the book is in equilibrium ($\mathbf{a} = 0$), it follows that $W = N = mg$.

5.8 SOME APPLICATIONS OF NEWTON'S LAWS

In this section we present some simple applications of Newton's laws to bodies that are either in equilibrium ($\mathbf{a} = 0$) or moving linearly under the action of constant external forces. As our model, we shall assume that the bodies behave as particles so that we need not worry about rotational motions. In this section, we shall also neglect the effects of friction for those problems involving motion. This is equivalent to stating that the surfaces are *smooth*. Finally, we shall usually neglect the mass of any ropes involved in a particular problem. In this approximation, the magnitude of the force exerted at any point along the rope, called the *tension*, is the same at all points along the rope.

Tension

When we apply Newton's laws to a body, we shall be interested only in those external forces that act *on the body*. For example, in Fig. 5.6 the only external forces acting on the book are **N** and **W**. The reactions to these forces, **N′** and **W′**, act on the table and on the earth, respectively, and do not appear in Newton's second law as applied to the book.

The normal force, that is, the force of the table on the book, belongs to a general class of forces called *constraint forces*. Another example of a constraint force is the force of a rope on an object when the rope is under tension. Constraint forces place

[3]The word *normal* is used because in the absence of friction, the direction of **N** is always *perpendicular* to the surface.

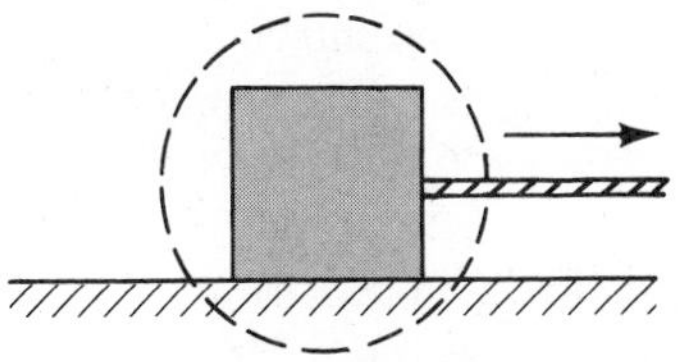

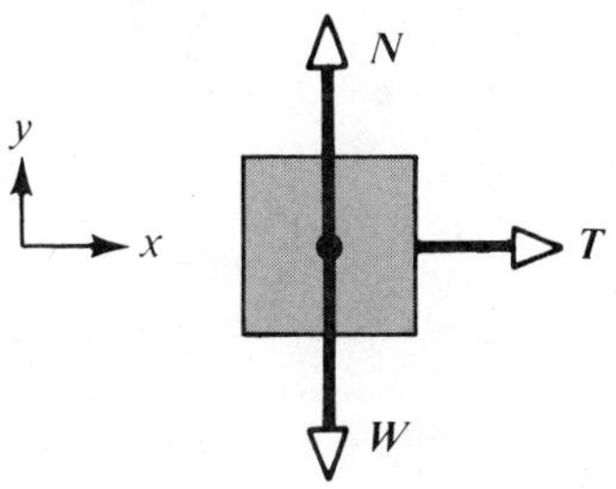

Figure 5.7 (a) A block being pulled to the right on a smooth surface. (b) The free-body diagram that represents the external forces on the block.

limiting conditions on motion. Weight and the normal force play a key role in the following situation.

Consider a block being pulled to the right on the smooth, horizontal surface of a table, as in Fig. 5.7a. Suppose you are asked to find the acceleration of the block and the force of the table on the block. First, note that the horizontal force being applied to the block acts through the string. The force that the string exerts on the block is denoted by the symbol **T**. The magnitude of **T** is called the *tension* in the string. A dotted circle is drawn around the block in Fig. 5.7a to remind you to isolate the block from its surroundings. Since we are interested only in the motion of the block, we must be able to *identify all external forces acting on it.* These are illustrated in Fig. 5.7b. In addition to the force **T**, the force diagram for the block includes the weight, **W**, and the normal force, **N**. As before, **W** corresponds to the force of gravity pulling down on the block and **N** is the upward force of the table on the block. Such a force diagram will be referred to as a *free-body diagram.* The construction of such a diagram is an important step in applying Newton's laws. The *reactions* to the forces we have listed, namely, the force of the string on the hand, the force of the block on the earth, and the force of the block on the table, are not included in the free-body diagram since they act on *other* bodies and not on the block.

Free-body diagrams are important when applying Newton's laws

We are now in a position to apply Newton's second law in component form to the system. The only force acting in the x direction is **T**. Applying $\Sigma F_x = ma_x$ to the horizontal motion gives

$$\Sigma F_x = T = ma_x$$

$$a_x = \frac{T}{m}$$

In this situation, there is no acceleration in the y direction. Applying $\Sigma F_y = ma_y$ with $a_y = 0$ gives

$$N - W = 0 \qquad \text{or} \qquad N = W$$

That is, the normal force is equal to and opposite the weight.

If **T** is a known *and constant* force, then the acceleration, $a_x = T/m$, is also a constant. Hence, the equations of kinematics from Chapter 3 can be used to obtain the displacement, Δx, and velocity, v, as functions of time. Since $a_x = T/m =$ constant, these expressions can be written

$$\Delta x = v_0 t + \frac{1}{2}\left(\frac{T}{m}\right)t^2$$

$$v = v_0 + \left(\frac{T}{m}\right)t$$

where v_0 is the velocity at $t = 0$.

Procedure for applying Newton's laws

The following procedure is recommended for applying Newton's laws:

1. Draw a simple, neat diagram of the system.
2. Isolate the object of interest whose motion is being analyzed. Draw a free-body diagram (or force diagram) for this object, that is, a diagram showing *all external forces acting on the object*. For systems containing more than one object, draw *separate* diagrams for each object. *Do not* include forces that the object exerts on its surroundings.
3. Establish convenient coordinate axes for each body and find the components of the forces along these axes. Now, apply Newton's second law, $\Sigma \mathbf{F} = m\mathbf{a}$, in *component* form. Check your dimensions to make sure that all terms have units of force.
4. Solve the component equations for the unknowns. Remember that you must have as many independent equations as you have unknowns in order to obtain a complete solution.
5. It is a good idea to check the predictions of your solutions for extreme values of the variables. You can often detect errors in your results using this procedure.

Consider a weight W suspended from a *light* rope fastened to the ceiling, as in Fig. 5.8a. The free-body diagram for the weight is shown in Fig. 5.8b, where the forces on it are the weight, $\mathbf{W}$, acting downward, and the force of the rope on the weight, $\mathbf{T}$, acting upward. The force $\mathbf{T}$ is the constraint force in this case. (If we cut the rope, $\mathbf{T} = 0$ and the body goes into free fall.) Note that ropes always *pull* on objects. Try pushing a wagon with a rope.

If we apply the first law to the weight, noting that $\mathbf{a} = 0$, we see that since there are no forces in the x direction, the equation $\Sigma F_x = 0$ provides no helpful information. The condition $\Sigma F_y = 0$ gives

$$\Sigma F_y = T - W = 0 \qquad \text{or} \qquad T = W$$

Note that $\mathbf{T}$ and $\mathbf{W}$ are *not* action-reaction pairs. The reaction to $\mathbf{T}$ is $\mathbf{T}'$, the force exerted on the rope by the weight, as in Fig. 5.8c. The force $\mathbf{T}'$ acts downward and is transmitted to the ceiling. That is, the force of the rope on the ceiling, $\mathbf{T}'$, is *down-*

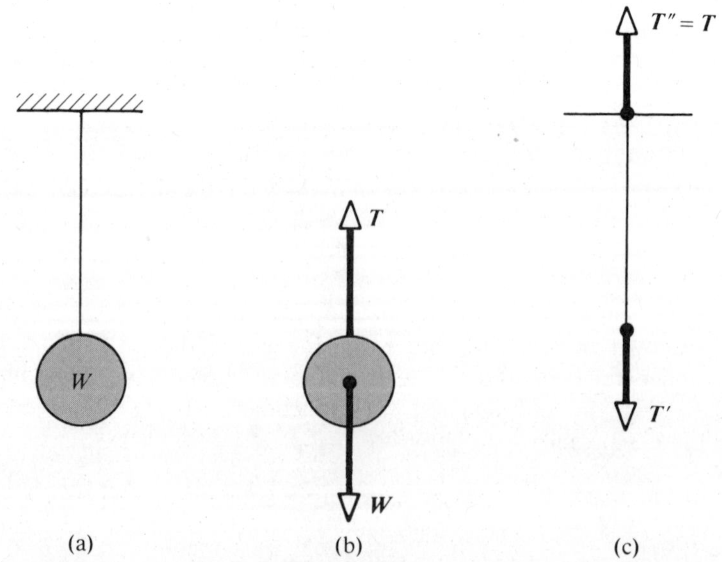

Figure 5.8 (a) A weight W suspended by a light string from a stationary support. (b) The forces acting on the weight are the force of gravity, $\mathbf{W}$, and the tension in the string, $\mathbf{T}$. (c) The forces acting on the string.

ward and equal to W in magnitude. The ceiling exerts an equal and opposite force, $\mathbf{T}'' = \mathbf{T}$, on the rope, as in Fig. 5.8c.

Example 5.2 *A Body at Rest*

A weight W hangs from a rope that is tied to two other ropes that are fastened to the ceiling, as in Fig. 5.9a. The upper ropes make angles of θ and ϕ with the horizontal. Find the tensions in the three ropes.

Solution: First we construct a free-body diagram for the weight, as in Fig. 5.9b. The force of tension in the rope, $\mathbf{T}_3$, supports the weight, so we see that $T_3 = W$. Now, we construct a free-body diagram for the knot, as in Fig. 5.9c. This is a convenient point to choose since *all forces in question act at this point.* We choose the coordinate axes as shown in Fig. 5.9c and resolve the forces into their x and y components as follows:

FORCE	x component	y component
$\mathbf{T}_1$	$-T_1 \cos\theta$	$T_1 \sin\theta$
$\mathbf{T}_2$	$T_2 \cos\phi$	$T_2 \sin\phi$
$\mathbf{T}_3$	0	$-W$

Since the system is stationary, $\mathbf{a} = 0$; therefore, Newton's second law applied to the *knot* gives

$$(1) \quad \Sigma F_x = T_2 \cos\phi - T_1 \cos\theta = 0$$

$$(2) \quad \Sigma F_y = T_1 \sin\theta + T_2 \sin\phi - W = 0$$

That is, from (1) we see that the horizontal components of $\mathbf{T}_1$ and $\mathbf{T}_2$ must be equal in magnitude, and from (2) we see that the sum of the vertical components of $\mathbf{T}_1$ and $\mathbf{T}_2$ must balance the weight. Solving (1) and (2) simultaneously gives

$$(3) \quad T_1 = \frac{W}{\sin\theta + \cos\theta \tan\phi}$$

$$(4) \quad T_2 = \frac{W}{\sin\phi + \cos\phi \tan\theta}$$

As a numerical example, if $W = 10$ N, $\theta = 37°$, and $\phi = 53°$, we find from (3) and (4) that

$$T_1 = \frac{10 \text{ N}}{\sin 37° + \cos 37° \tan 53°} = \frac{10 \text{ N}}{0.60 + 0.80(1.33)} = 6.0 \text{ N}$$

$$T_2 = \frac{10 \text{ N}}{\sin 53° + \cos 53° \tan 37°} = \frac{10 \text{ N}}{0.80 + 0.60(0.75)} = 8.0 \text{ N}$$

Q5. In what situation will $T_1 = T_2$?

Q6. Is it physically possible to have a situation where $\theta = \phi = 0°$?

Example 5.3 *Block on a Smooth Incline*

A block of mass m is placed on a smooth, inclined plane of angle θ, as in Fig. 5.10a. (a) Determine the acceleration of the block after it is released.

The free-body diagram for the block is shown in Fig. 5.10b. The only forces on the block are the normal force, $\mathbf{N}$, acting perpendicular to the plane and the weight, $\mathbf{W}$, acting vertically downward. *It is convenient to choose the coordinate axes with* x *along the incline and* y *perpendicular to it.* Then, we replace the weight vector by a component of magnitude $mg \sin\theta$ along the *positive* x axis and another of magnitude $mg \cos\theta$ in the *negative* y direction. Applying Newton's second law in component form while noting that $a_y = 0$ gives

$$(1) \quad \Sigma F_x = mg \sin\theta = ma_x$$

$$(2) \quad \Sigma F_y = N - mg \cos\theta = 0$$

From (1) we see that the acceleration along the incline is

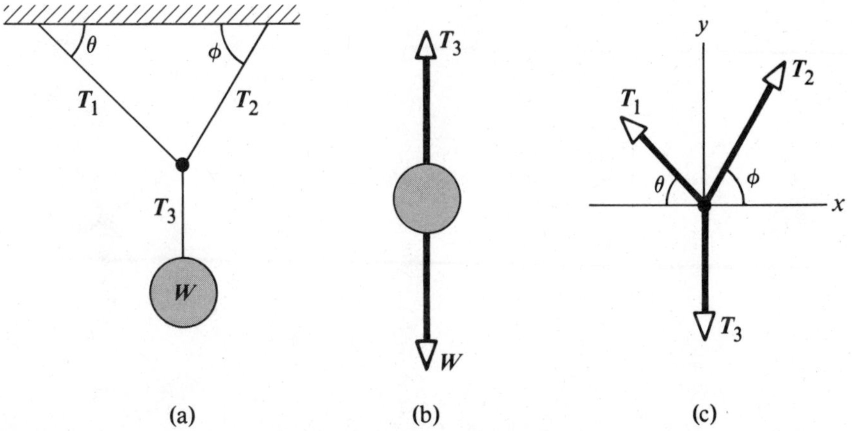

Figure 5.9 (Example 5.2) (a) A weight W suspended from the ceiling. (b) The free-body diagram for the weight. (c) The free-body diagram for the knot.

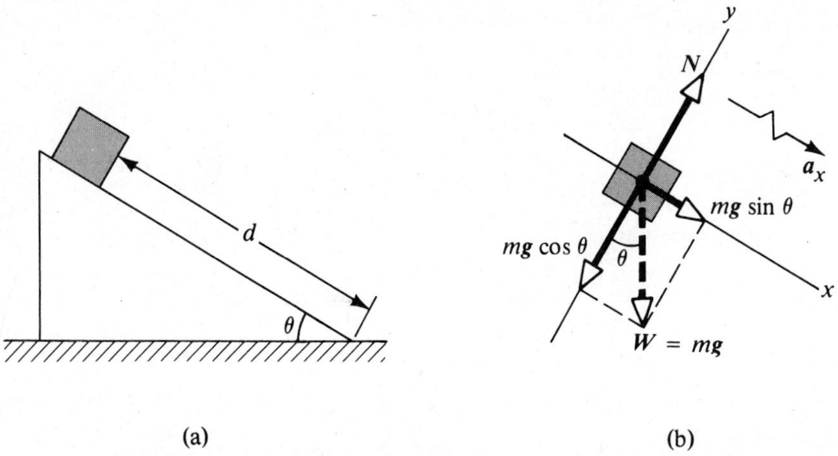

Figure 5.10 (Example 5.3) (a) A block sliding down a smooth incline. (b) The free-body diagram for the block. Note that its acceleration along the incline is g sinθ.

provided by the component of weight down the incline:

$$(3) \qquad a_x = g \sin\theta$$

From (2) we conclude that the component of weight perpendicular to the incline is *balanced* by the normal force, or $N = mg \cos\theta$. The acceleration given by (3) is found to be *independent* of the mass of the block! It depends only on the angle of inclination and on g!

Special cases: We see that when $\theta = 90°$, $a = g$ and $N = 0$. This corresponds to the block in free fall. Also, when $\theta = 0$, $a_x = 0$ and $N = mg$ (its maximum value).

(b) Suppose the block is released from rest at the top, and the distance from the block to the bottom is *d*. How long does it take the block to reach the bottom, and what is its speed just as it gets there?

Since a_x = constant, we can apply the kinematic equation $x - x_0 = v_{x0}t + \frac{1}{2}a_x t^2$ to the block. Since $x - x_0 = d$ and $v_{x0} = 0$, we get

$$d = \frac{1}{2}a_x t^2$$

or

$$(4) \qquad t = \sqrt{\frac{2d}{a_x}} = \sqrt{\frac{2d}{g \sin\theta}}$$

Also, since $v_x^2 = v_{x0}^2 + 2a_x(x - x_0)$ and $v_{x0} = 0$, we find that

$$v_x^2 = 2a_x d$$

$$(5) \qquad v_x = \sqrt{2a_x d} = \sqrt{2gd \sin\theta}$$

Again, *t* and v_x are *independent* of the mass of the block. This suggests a simple method of measuring g using an inclined air track or some other smooth incline. Simply measure the angle of inclination, the distance traveled by the block, and the time it takes to reach the bottom. The value of g can then be calculated from (4) and (5).

Q7. Typical experiments for measuring g using the procedure suggested above yield values of g that are about 5% lower than the true value. How might this be explained?

Example 5.4

Two unequal masses are attached by a light string that passes over a light, frictionless pulley as in Fig. 5.11a. The block of mass m_2 lies on a smooth incline of angle θ. Find the acceleration of the two masses and the tension in the string.

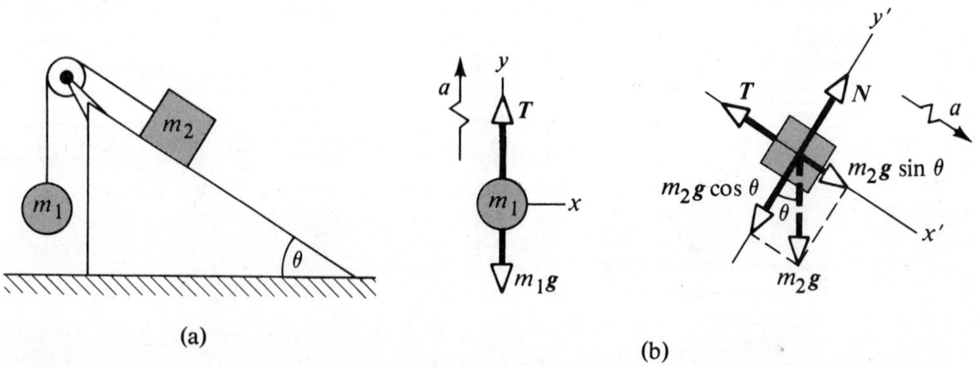

Figure 5.11 (Example 5.4) (a) Two masses connected by a light string over a frictionless pulley. (b) The free-body diagram for m_1. (c) The free-body diagram for m_2 (the incline is smooth).

Solution: Since the two masses are connected by a string (which we assume doesn't stretch), they both have the same acceleration magnitude. The free-body diagrams for the two masses are shown in Figs. 5.11b and 5.11c. Applying Newton's second law in component form to m_1 while *assuming* that $\boldsymbol{a}$ is upward for this mass gives

Equations of motion for m_1:

$$(1) \qquad \Sigma F_x = 0$$

$$(2) \qquad \Sigma F_y = T - m_1 g = m_1 a$$

Note that in order for a to be positive, it is necessary that $T > m_1 g$.

Now, for m_2 it is convenient to choose the positive x' axis along the incline as in Fig. 5.11c. Applying Newton's second law in component form to m_2 gives

Equations of motion for m_2:

$$(3) \qquad \Sigma F_{x'} = m_2 g \sin\theta - T = m_2 a$$

$$(4) \qquad \Sigma F_{y'} = N - m_2 g \cos\theta = 0$$

Expressions (1) and (4) provide no information regarding the acceleration. However, if we solve (2) and (3) simultaneously for the unknowns a and T, we get

$$(5) \qquad a = \frac{m_2 g \sin\theta - m_1 g}{m_1 + m_2}$$

and

$$(6) \qquad T = \frac{m_1 m_2 g(1 + \sin\theta)}{m_1 + m_2}$$

Note that m_2 accelerates down the incline if $m_2 \sin\theta$ exceeds m_1 (that is, if a is positive as we assumed). If m_1 exceeds $m_2 \sin\theta$, the acceleration of m_2 is up the incline and downward for m_1. You should also note that the result for the acceleration, (5), can be interpreted as the resultant unbalanced force on the system divided by the total mass of the system.

Q8. If we take $m_1 = 10$ kg, $m_2 = 5$ kg, and $\theta = 45°$, we find that $a = -4.2$ m/s^2. How do you explain the negative sign? Is it possible for T to be negative?

Example 5.5 *Atwood's Machine*

If two masses hang vertically over a frictionless pulley as in Fig. 5.12a we get what is called *Atwood's machine.*[4] Determine the acceleration of the two masses and the tension in the string.

Solution: In this case, we can get a and T by setting $\theta = 90°$ in (5) and (6) of Example 5.4. Since $\sin 90° = 1$, we get

$$a = \left(\frac{m_2 - m_1}{m_1 + m_2}\right) g$$

$$T = \left(\frac{2m_1 m_2}{m_1 + m_2}\right) g$$

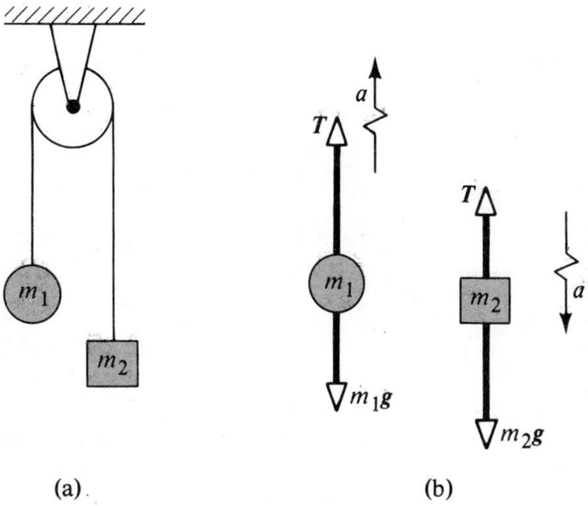

Figure 5.12 (Example 5.5) Atwood's machine. (a) Two masses connected by a light string over a frictionless pulley. (b) Free-body diagrams for m_1 and m_2.

For example, if $m_1 = 2.0$ kg and $m_2 = 4.0$ kg, we find that $a = g/3 = 3.3$ m/s^2 and $T = 26$ N.

Special cases: Note that when $m_1 = m_2$, $a = 0$ and $T = m_1 g = m_2 g$, as we would expect for the static case. Also, if $m_2 \gg m_1$, $a \cong g$ (a freely falling body) and $T \approx 2m_1 g$.

Example 5.6

Weighing a fish in an elevator: A person weighs a fish on a spring scale attached to the ceiling of an elevator, as shown in Fig. 5.13. Show that if the elevator accelerates or decelerates, the spring scale reads a weight different from the true weight of the fish.

Solution: The external forces acting on the fish are its true weight, W, and the upward constraint force, T, exerted on it by the scale. By Newton's third law, T is also the reading of the spring scale. If the elevator is at rest or moving at constant velocity, then the fish is not accelerating and $T = w = mg$ (where $g = 32$ ft/s^2). If the elevator accelerates upward with an acceleration $\boldsymbol{a}$ relative to an observer outside the elevator in an inertial frame (Fig. 5.13a), then the second law applied to the fish of mass m gives the total force F on the fish:

$$(1) \qquad F = T - W = ma \qquad \text{(if } \boldsymbol{a} \text{ is upward)}$$

Likewise, if the elevator accelerates downward as in Fig. 5.13b, Newton's second law applied to the fish becomes

$$(2) \qquad F = T - W = -ma \qquad \text{(if } \boldsymbol{a} \text{ is downward)}$$

Thus, we conclude from (1) that the scale reading, T, is greater than the true weight, W, if $\boldsymbol{a}$ is upward. From (2) we see that T is less than W if $\boldsymbol{a}$ is downward.

[4]You may have occasion to use Atwood's machine in the laboratory to measure the acceleration of gravity.

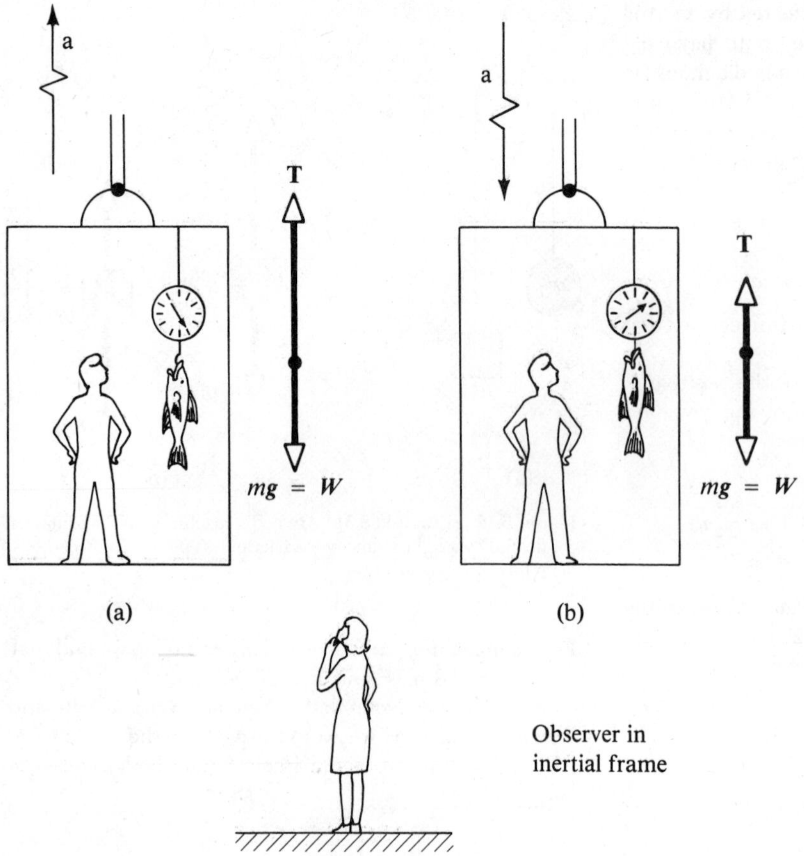

Figure 5.13 (Example 5.6) Apparent weight versus true weight. (a) When the elevator accelerates *upward* the spring scale reads a value *greater* than the true weight. (b) When the elevator accelerates *downward* the spring scale reads a value *less* than the true weight. The spring scale reads the *apparent weight.*

For example, if the true weight of the fish is 10 lb and a is 5 ft/s² upward, then the scale reading is

$$T = ma + mg = mg\left(\frac{a}{g} + 1\right)$$

$$= W\left(\frac{5}{32} + 1\right)$$

$$= 10\left(\frac{37}{32}\right)$$

$$= 11.6 \text{ lb.}$$

If a is 5 ft/s² downward, then

$$T = -ma + mg = mg\left(1 - \frac{a}{g}\right)$$

$$= W\left(1 - \frac{5}{32}\right)$$

$$= 10\left(\frac{27}{32}\right)$$

$$= 8.4 \text{ lb.}$$

Hence, if you buy a fish by the pound in an elevator, make sure the fish is weighed while the elevator is at rest or accelerating downward!

Special cases: If the elevator cable breaks, then the elevator falls freely and $a = -g$. Since $W = mg$, we see from (1) that the apparent weight, T, is zero, that is, the fish appears to be weightless. If the elevator accelerates *downward* with an acceleration *greater* than g, the fish (along with the person in the elevator) will eventually hit the ceiling since its acceleration will still be that of a freely falling body relative to an outside observer.

Q9. The observer in the elevator of Example 5.6 would claim that the "weight" of the fish is T, the scale reading. This is obviously wrong. Why does this observation differ from that of a person outside the elevator at rest with respect to the elevator?

Q10. If you have ever taken a ride in an elevator of a high-rise building, you may have experienced the nauseating sensation of "heaviness" and "lightness" depending on the direction of $\boldsymbol{a}$. Explain these sensations. Are we truly weightless in free fall?

Q11. Why is it that an astronaut in a space capsule orbiting the earth experiences a feeling of weightlessness? Recall that a particle moving in a circular orbit experiences a centripetal acceleration v^2/r. Also, the *only* external force acting on the astronaut is body weight.

5.9 FORCES OF FRICTION

When a body is in motion on a rough surface, or when an object moves through a viscous medium such as air or water, there is resistance to motion because of the interaction of the object with its surroundings. We call such a resistance force a *force of friction*. Forces of friction are very important in our everyday lives. For example, forces of friction allow us to walk or run and are necessary for the motion of wheeled vehicles.

Consider a block on a horizontal table, as in Fig. 5.14a. If we apply an external horizontal force $\boldsymbol{F}$ to the block, acting to the right, the block will remain stationary if $\boldsymbol{F}$ is not too large. The force that keeps the block from moving acts to the left and is the *frictional force*, f. As long as the block is in equilibrium, $f = F$. Since the block

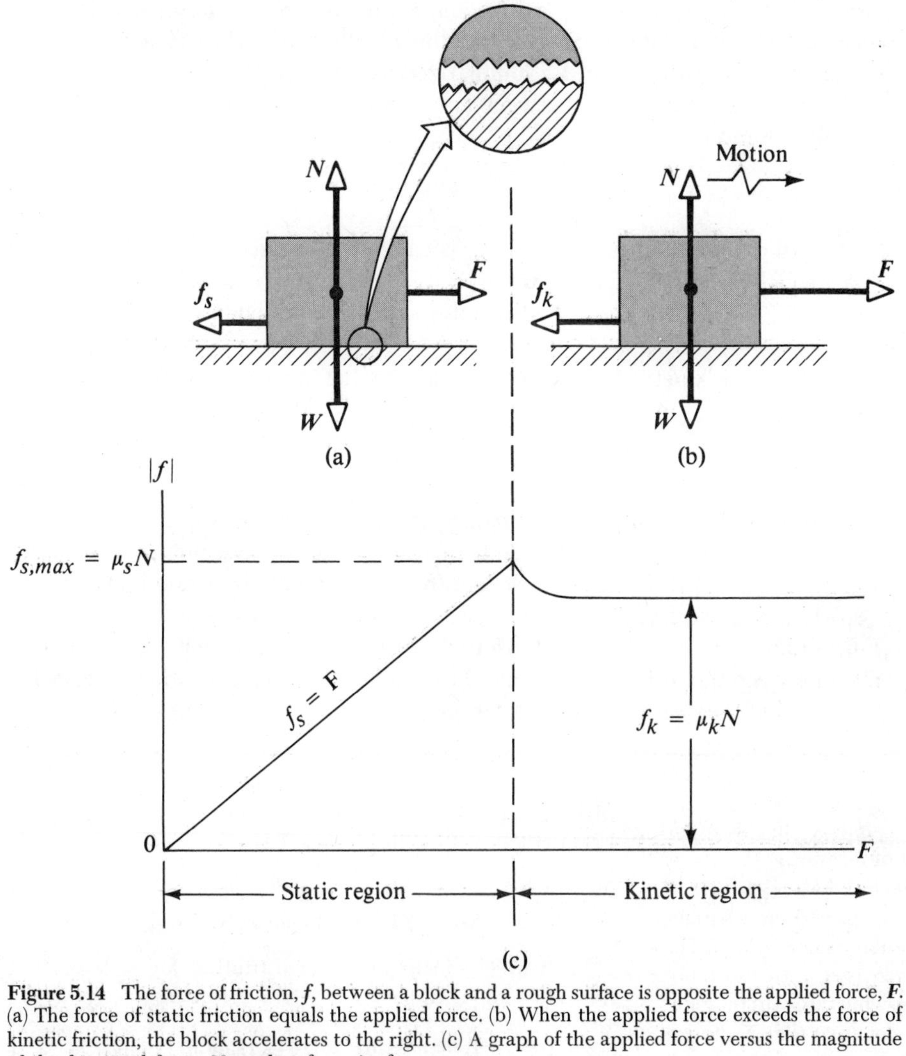

Figure 5.14 The force of friction, f, between a block and a rough surface is opposite the applied force, $\boldsymbol{F}$. (a) The force of static friction equals the applied force. (b) When the applied force exceeds the force of kinetic friction, the block accelerates to the right. (c) A graph of the applied force versus the magnitude of the frictional force. Note that $f_{s,max} > f_k$.

is stationary, we call this frictional force the *force of static friction*, f_s. Experiments show that this force arises from the roughness of the two surfaces, so that contact is made only at a few points, as shown in the "magnified" view of the surfaces in Fig. 5.14a. Actually, the frictional force, when viewed on a microscopic level, is very complicated since it ultimately involves the electrostatic forces between atoms or molecules where the surfaces are in contact.

If we increase the magnitude of $\mathbf{F}$, as in Fig. 5.14b, the block will eventually slip. When the block is on the verge of slipping, f_s is a maximum. When F exceeds $f_{s,max}$ the block moves and accelerates to the right. When the block is in motion, the retarding frictional force becomes *less* than $f_{s,max}$ (Fig. 5.14c). When the block is in motion, we call the retarding force the *force of kinetic friction*, f_k. The unbalanced force in the x direction, $F - f_k$, produces an acceleration to the right. If $F = f_k$ the block moves to the right with constant speed. If the applied force is removed, then the frictional force acting to the left would decelerate the block and eventually bring it to rest.

In a simplified model, we can imagine that the force of kinetic friction is less than $f_{s,max}$ because of the reduction in roughness of the two surfaces when the object is in motion. When the object is stationary, the contact points are said to be *cold-welded*. When the object is in motion, these small welds become ruptured and the frictional force decreases.

Experimentally, one finds that both f_s and f_k are *proportional to the normal force acting on the block* and depend on the roughness of the two surfaces in contact. The experimental observations can be summarized as follows:

1. The force of static friction between any two surfaces in contact is opposite the applied force and can have values given by

Force of static friction

$$f_s \leq \mu_s N \tag{5.12}$$

where the dimensionless constant μ_s is called *the coefficient of static friction*. The equality in Eq. 5.12 holds when the block is on the *verge* of slipping, that is, when $f_{s,max} = \mu_s N$. The inequality holds when the applied force is less than this value.

2. The force of kinetic friction is opposite to the direction of motion and is given by

Force of kinetic friction

$$f_k = \mu_k N \tag{5.13}$$

where μ_k is *the coefficient of kinetic friction*.

3. The values of μ_k and μ_s depend on the nature of the surfaces, but μ_k is generally less than μ_s. Typical values of μ range from around 0.01 for smooth surfaces to 1.5 for rough surfaces.

Furthermore, the coefficients of friction are nearly independent of the area of contact between the surfaces. Although the coefficient of kinetic friction varies with speed, we shall neglect any such variations.

Example 5.7

Experimental determination of μ_s and μ_k: In this example we describe a simple method of measuring the coefficients of friction between an object and a rough surface. Suppose the object is a small block placed on a surface inclined with respect to the horizontal, as in Fig. 5.15. The angle of the inclined plane is increased until the block slips. By measuring the angle θ_c at which this slipping just occurs, we obtain μ_s directly. We note that the only forces acting on the block are its weight, mg, the normal force, $\mathbf{N}$, and the force of static friction, f_s. Taking x parallel to the plane and y perpendicular to the plane, Newton's second law applied to the block gives

Static case: (1) $\quad \Sigma F_x = mg\sin\theta - f_s = 0$

(2) $\quad \Sigma F_y = N - mg\cos\theta = 0$

We can eliminate mg by substituting $mg = N/\cos\theta$ from (2) into (1) to get

(3) $\quad f_s = mg\sin\theta = \left(\dfrac{N}{\cos\theta}\right)\sin\theta = N\tan\theta$

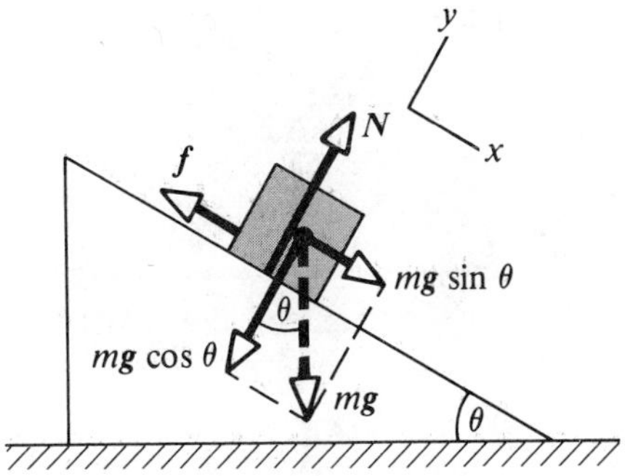

Figure 5.15 (Example 5.7) The external forces acting on a block lying on a rough incline are the weight, *mg*, the normal force, *N*, and the force of friction, *f*. Note that the weight vector is resolved into a component along the incline, $mg\sin\theta$, and a component perpendicular to the incline, $mg\cos\theta$.

When the inclined plane is at the critical angle, θ_c, $f_s = f_{s,max} = \mu_s N$, and so at this angle, (3) becomes

$$\mu_s N = N\tan\theta_c$$
$$\mu_s = \tan\theta_c$$

For example, if we find that the block just slips at $\theta_c = 20°$, then $\mu_s = \tan 20° = 0.364$. Once the block starts to move at $\theta \geq \theta_c$, it will accelerate down the incline and the force of friction is $f_k = \mu_k N$. However, if θ is reduced below θ_c, an angle θ_c' can be found such that the block moves down the incline with constant speed ($a_x = 0$). In this case, using (1) and (2) with f_s replaced by f_k gives

Kinetic case: $\mu_k = \tan\theta_c'$

where $\theta_c' < \theta_c$.

You should try this simple experiment using a coin as the block and a notebook as the inclined plane. Also, you can try taping two coins together to prove that you still get the same critical angles as with one coin.

Example 5.8 *Motion of a Hockey Puck*

A hockey puck is hit with an initial speed of v_0 on a frozen river. If the puck always remains on the ice and moves in a straight line for a distance x before coming to rest, determine the coefficient of kinetic friction between the puck and the ice.

Solution: The forces acting on the puck after it is in motion are shown in Fig. 5.16. If we assume that the force of

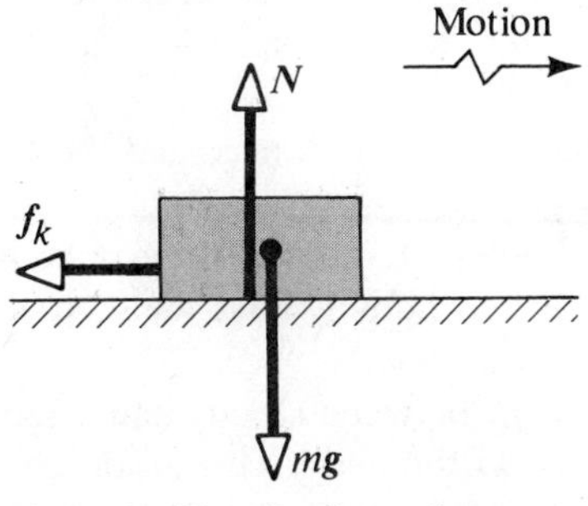

Figure 5.16 (Example 5.8) *After* the puck is given an initial velocity, the external forces acting on it are the weight, *mg*, the normal force, *N*, and the force of kinetic friction, f_k.

friction, f_k, remains constant, then this force produces a uniform deceleration of the puck. Applying Newton's second law to the puck in component form gives

$$(1) \qquad \Sigma F_x = -f_k = ma$$
$$(2) \qquad \Sigma F_y = N - mg = 0 \qquad (a_y = 0)$$

But $f_k = \mu_k N$, and from (2) we see that $N = mg$. Therefore, (1) becomes

$$-\mu_k N = -\mu_k mg = ma$$
$$a = -\mu_k g$$

That is, the acceleration is to the left, corresponding to a deceleration of the puck. Also, the acceleration is independent of the mass of the puck and is *constant* since we are assuming that μ_k remains constant.

Now, since the acceleration is constant, we can use the kinematic equation $v^2 = v_0^2 + 2ax$, with the final speed $v = 0$. This gives

$$v_0^2 + 2ax = v_0^2 - 2\mu_k gx = 0$$
$$\mu_k = \frac{v_0^2}{2gx}$$

In our example, if we take $v_0 = 20$ m/s and $x = 120$ m, we find that

$$\mu_k = \frac{(20\text{ m/s})^2}{2(9.80\text{ m/s}^2)(120\text{ m})} = \frac{400}{2.35 \times 10^3} = 0.170$$

Note that μ_k has no dimensions.

Example 5.9

A block of mass m_1 on a rough, horizontal surface is connected to a second mass m_2 by a light cord over a frictionless pulley as in Fig. 5.17a. A force of magnitude F is applied to mass m_1 as shown. The coefficient of kinetic friction between m_1 and the surface is μ. Determine the acceleration of the masses and the tension in the cord.

Solution: First draw the free-body diagrams of m_1 and m_2 as in Figs. 5.17b and 5.17c. Note that the force $\mathbf{F}$ has components $F_x = F\cos\theta$ and $F_y = F\sin\theta$. Therefore, in this case N is *not* equal to $m_1 g$. Applying Newton's second law to both masses and *assuming* the motion of m_1 is to the right, we get

Motion of m_1: $\Sigma F_x = F\cos\theta - f_k - T = m_1 a$

(1) $\Sigma F_y = N + F\sin\theta - m_1 g = 0$

Motion of m_2: $\Sigma F_x = 0$

(2) $\Sigma F_y = T - m_2 g = m_2 a$

But $f_k = \mu N$, and from (1), $N = m_1 g - F\sin\theta$; therefore

$$(3) \qquad f_k = \mu(m_1 g - F\sin\theta)$$

That is, the frictional force is *reduced* because of the positive y component of $\mathbf{F}$. Substituting (3) and the value of T from (2) into (1) gives

$$F\cos\theta - \mu(m_1 g - F\sin\theta) - m_2(a + g) = m_1 a$$

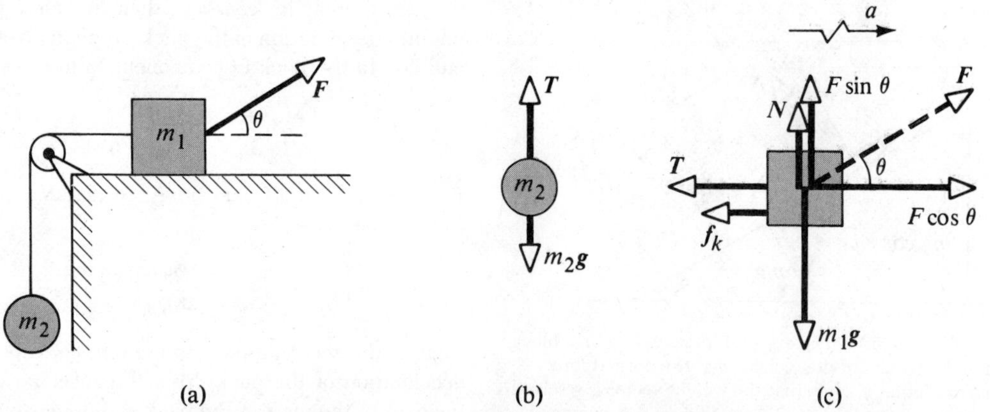

Figure 5.17 (Example 5.9) (a) The external force, F, applied as shown can cause m_1 to accelerate to the right. (b) and (c) The free-body diagrams assuming that m_1 accelerates to the right while m_2 accelerates upward. Note that the force of kinetic friction in this case is given by $f_k = \mu_k N = \mu_k(m_1 g - F\sin\theta)$.

Solving for a, we get

$$(4) \qquad a = \frac{F(\cos\theta + \mu\sin\theta) - g(m_2 + \mu m_1)}{m_1 + m_2}$$

We can find T by substituting this value of a into (2). Note that the acceleration for m_1 can be either to the right or left,[5] depending on the sign of the numerator in (4). If the motion of m_1 is to the *left*, we must reverse the sign of f_k since the frictional force *always opposes* the motion. In this case, the value of a is the same as in (4) with μ replaced by $-\mu$.

Q12. Although the frictional force between two surfaces may decrease as the surfaces are smoothened, the force will again *increase* if the surfaces are made extremely smooth and flat. How do you explain this? (Think about the true origin of friction.)

Q13. Why is it that the frictional force involved in the rolling of one body over another is less than that for sliding motion? (Think about the cold-weld model.)

5.10 SUMMARY

Newton's first law

Newton's first law states that a body at rest will remain at rest or a body in uniform motion in a straight line will maintain that motion unless an external resultant force acts on the body.

Newton's second law

Newton's second law states that the time rate of change of momentum of a body is equal to the resultant force acting on the body. If the mass of the body is constant, the net force equals the product of the mass and its acceleration, or $\Sigma \mathbf{F} = m\mathbf{a}$.

Inertial frame

Newton's first and second laws are valid in an inertial frame of reference. An *inertial frame* is one in which an isolated body does not accelerate.

Mass is a scalar quantity. The mass that appears in Newton's second law is called *inertial mass*.

Weight

The *weight* of a body is equal to the product of its mass and the acceleration of gravity, or $\mathbf{W} = m\mathbf{g}$.

Newton's third law

Newton's third law states that if two bodies interact, the force of body 1 on body 2 is equal to and opposite the force of body 2 on body 1. Thus, an isolated force cannot exist in nature.

Forces of friction

The *maximum force of static friction*, f_s, between a body and a rough surface is proportional to the normal force acting on the body. This maximum force occurs

[5]A close examination of (4) shows that when $\mu m_1 > m_2$, there is a range of values of F for which no motion occurs at a given angle θ.

when the body is on the verge of slipping. In general, $f_s \le \mu_s N$, where μ_s is the *coefficient of static friction.* When a body slides over a rough surface, the *force of kinetic friction,* f_k, is opposite the motion and is also proportional to the normal force. The magnitude of this force is given by $f_k = \mu_k N$, where μ_k is the coefficient of kinetic friction. Usually, $\mu_k < \mu_s$.

More on Free-Body Diagrams

As we have seen throughout this chapter, in order to be successful in applying Newton's second law to a mechanical system you must first be able to recognize all the forces acting on the system. That is, you must be able to construct the correct free-body diagram. The importance of constructing the free-body diagram cannot be overemphasized. In Fig. 5.18 a number of mechanical systems are presented together with their corresponding free-body diagrams. You should examine these carefully and then proceed to construct free-body diagrams for other systems described in the exercises. When a system contains more than one element, it is important that you construct a free-body diagram for *each* element.

As usual, $\boldsymbol{F}$ denotes some applied force, $\mathbf{W} = mg$ is the weight, $\boldsymbol{N}$ denotes a normal force, f is frictional force, and $\boldsymbol{T}$ is the force of tension.

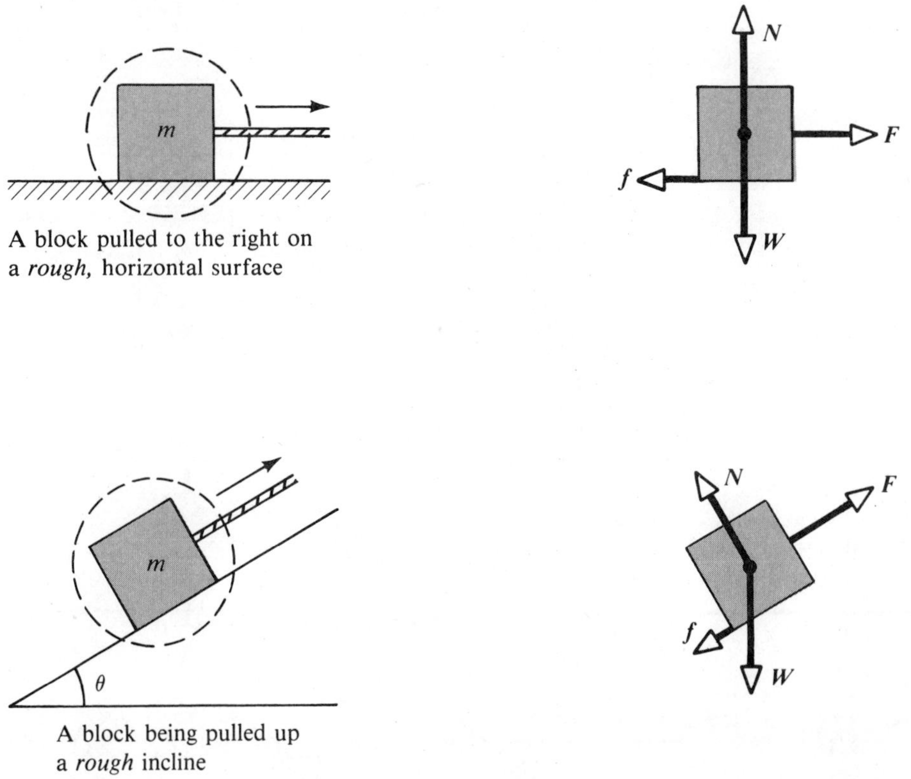

Figure 5.18 Various mechanical configurations (left) and the corresponding free-body diagrams (right).

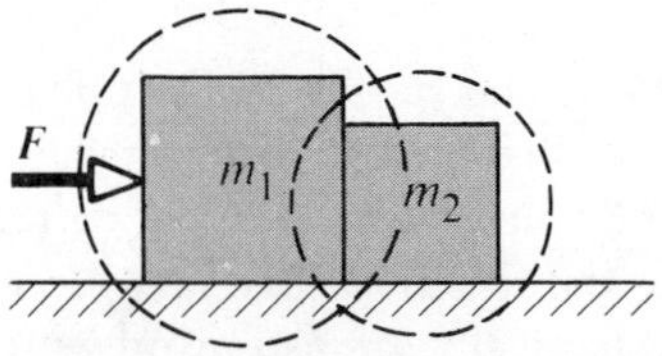

Two blocks in contact, pushed to the right on a smooth surface

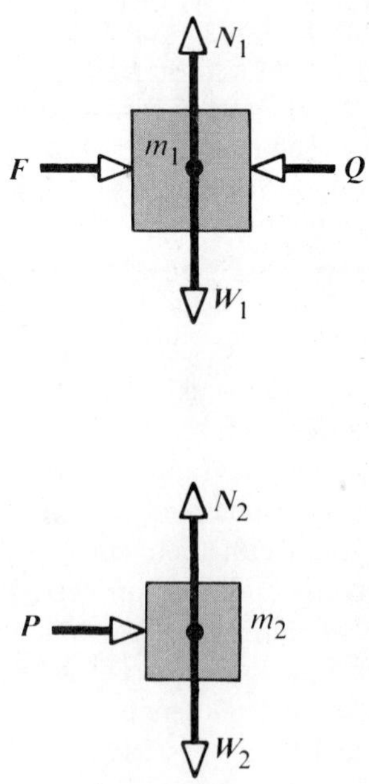

Note: $P = -Q$ since they are an action-reaction pair

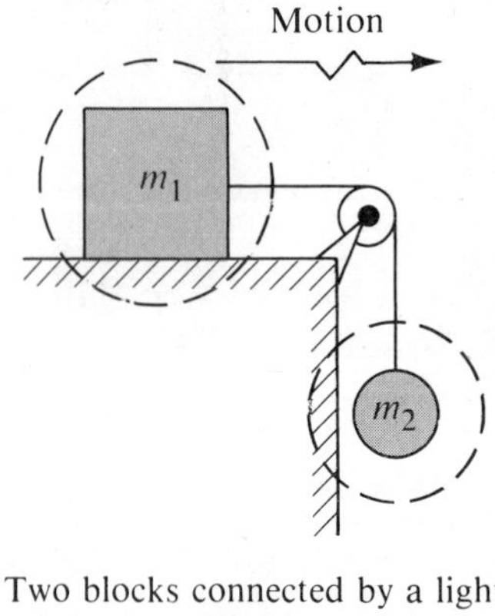

Two blocks connected by a light cord. The surface is rough and the pulley is smooth

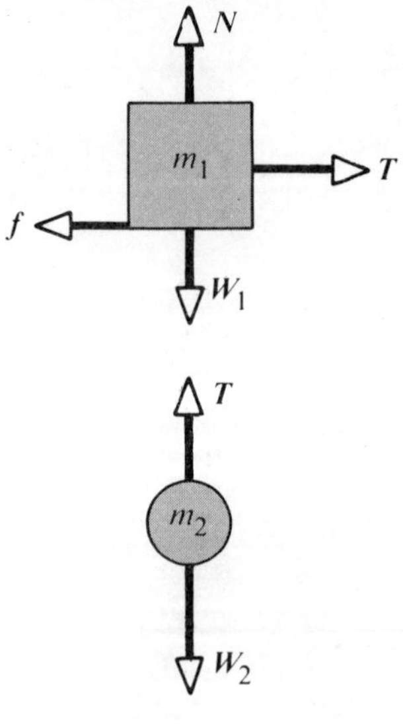

Figure 5.18 (continued)

EXERCISES

Section 5.1 through Section 5.7

1. A force, F, applied to an object of mass m_1 produces an acceleration of 2 m/s^2. The same force applied to a second object of mass m_2 produces an acceleration of 6 m/s^2. (a) What is the value of the ratio m_1/m_2? (b) If m_1 and m_2 are attached, find their acceleration under the action of the force F.

2. An object weighs 25 N at sea level, where $g = 9.8$ m/s^2. What is its weight on planet X, where the acceleration of gravity is 3.5 m/s^2?

3. A person weighs 120 lb. Determine (a) her weight in N and (b) her mass in kg.

4. Verify the following conversions: (a) 1 N = 10^5 dynes, (b) 1 N = 0.225 lb.

5. An object has a mass of 200 g. Find its weight in dynes and in N.

6. A force of 10 N acts on a body of mass 2 kg. What is (a) the acceleration of the body, (b) its weight in N, and (c) its acceleration if the force is doubled?

7. A 6-kg object undergoes an acceleration of 2 m/s^2. (a) What is the magnitude of the resultant force acting on it? (b) If this same force is applied to a 4-kg object, what acceleration will it produce?

8. A 3-kg mass undergoes an acceleration of $\boldsymbol{a} = (2\boldsymbol{i} + 5\boldsymbol{j})$ m/s^2. Find the resultant force, $\boldsymbol{F}$, and its magnitude.

9. Two forces, $\boldsymbol{F}_1$ and $\boldsymbol{F}_2$, act on a 5-kg mass. If $F_1 = 20$ N and $F_2 = 15$ N, find the acceleration in (a) and (b) of Fig. 5.19.

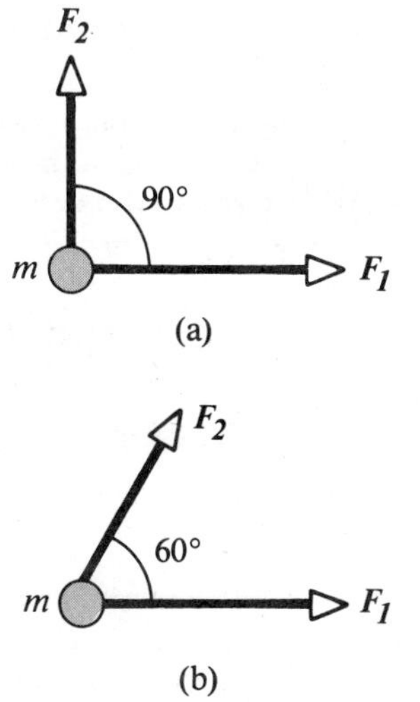

Figure 5.19 (Exercise 9).

10. A 3-kg particle starts from rest and moves a distance of 4 m in 2 s under the action of a single, constant force. Find the magnitude of the force.

11. A 2-kg particle moves along the x axis under the action of a single, constant force. If the particle starts from rest at the origin at $t = 0$ and is observed to have a velocity of $-8.0\boldsymbol{i}$ m/s at $t = 2$ s, what are the magnitude and direction of the force?

12. An electron of mass 9.1×10^{-31} kg has an initial speed of 3.0×10^5 m/s. It travels in a straight line, and its speed increases to 7.0×10^5 m/s in a distance of 5.0 cm. Assuming its acceleration is constant, (a) determine the force on the electron and (b) compare this force with the weight of the electron, which we neglected.

13. A 4-kg object has a velocity of $3\boldsymbol{i}$ m/s at one instant. Eight seconds later, its velocity is $(8\boldsymbol{i} + 10\boldsymbol{j})$ m/s. Assuming the object was subject to a constant net force, find (a) the components of the force and (b) its magnitude.

14. One or more external forces are exerted on each object shown in Fig. 5.1. Clearly identify the reaction to all of these forces. (Note that the reaction forces act on other objects.)

15. A bullet of mass 15 g leaves the barrel of a rifle with a speed of 800 m/s. If the length of the barrel is 75 cm, determine the force that accelerates the bullet, assuming the acceleration is constant. (Note that the actual force is exerted over a shorter time and is therefore greater than this estimate.)

16. A ball is held in a person's hand. (a) Identify all the external forces acting on the ball and the reaction to each of these forces. (b) If the ball is dropped, what force is exerted on it while it is in "flight"? Identify the reaction force in this case. (Neglect air resistance.)

17. A 3-ton truck provides an acceleration of 3 ft/s^2 to a 10-ton trailer. If the truck exerts the same pull on a 15-ton trailer, what acceleration will result?

18. A 15-lb block rests on the floor. (a) What force does the floor exert on the block? (b) If a rope is tied to the block and run over a pulley and the other end attached to a free-hanging 10-lb weight, what is the force of the floor on the 15-lb block? (c) If we replace the 10-lb weight in (b) by a 20-lb weight, what is the force of the floor on the 15-lb block?

Section 5.8 Some Applications of Newton's Laws

19. Find the tension in each cord for the systems described in Fig. 5.20. (Neglect the mass of the cords.)

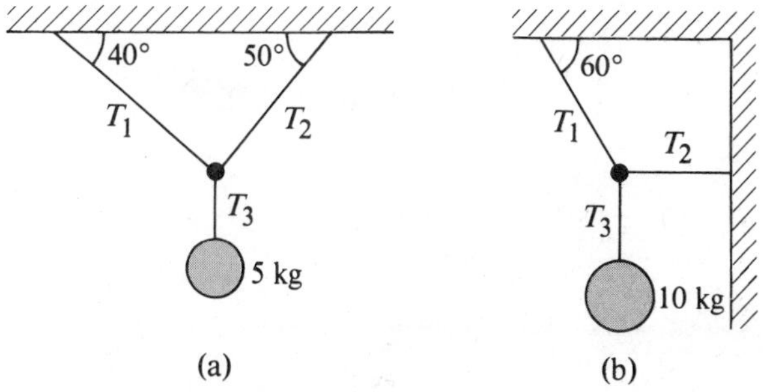

Figure 5.20 (Exercise 19).

20. The systems shown in Fig. 5.21 are in equilibrium. If the spring scales are calibrated in N, what do they read in each case? (Neglect the mass of the pulleys and strings, and assume the incline is smooth.)

21. A 200-lb weight is tied to the middle of a strong rope, and two people pull at opposite ends of the rope in an attempt to lift the weight. (a) What force $\boldsymbol{F}$ must each person apply to suspend the weight as shown in Fig. 5.22? (b) Can they pull in such a way as to make the rope horizontal? Explain.

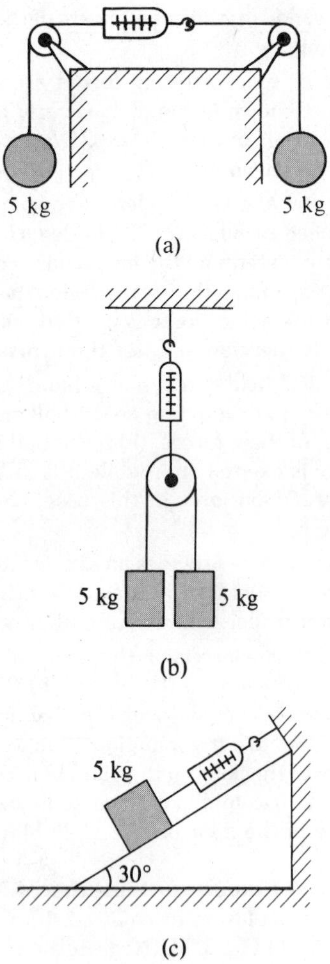

Figure 5.21 (Exercise 20).

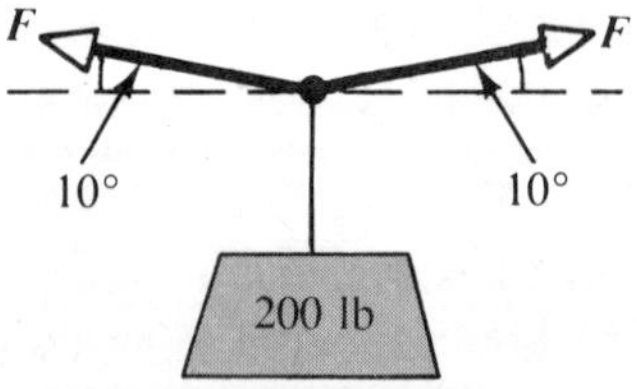

Figure 5.22 (Exercise 21).

22. A block slides down a smooth plane having an inclination of $\theta = 15°$ (Fig. 5.23). If the block starts from rest at the top and the length of the incline is 2 m, find (a) the acceleration of the block and (b) its speed when it reaches the bottom of the incline.

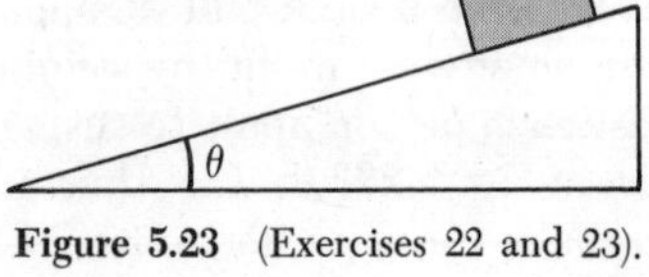

Figure 5.23 (Exercises 22 and 23).

23. A block is given an initial velocity of 5 m/s up a smooth 20° incline (Fig. 5.23). How far up the incline does the block slide before coming to rest?

24. A 50-kg mass hangs from a rope 5 m in length, which is fastened to the ceiling. What horizontal force applied to the mass will deflect it 1 m sideways from the vertical and maintain it in that position?

25. Two masses of 3 kg and 5 kg are connected by a light string that passes over a smooth pulley as in Fig. 5.12. Determine (a) the tension in the string, (b) the acceleration of each mass, and (c) the distance each mass moves in the first second of motion if they start from rest.

26. Two masses are connected by a light string that passes over a smooth pulley as in Fig. 5.11. If the incline is frictionless and if $m_1 = 2$ kg, $m_2 = 6$ kg, and $\theta = 55°$, find (a) the acceleration of the masses, (b) the tension in the string, and (c) the speed of each mass 2 s after they are released from rest.

27. Two masses, m_1 and m_2, situated on a frictionless, horizontal surface are connected by a light string. A force, $\mathbf{F}$, is exerted on one of the masses to the right (Fig. 5.24). Determine the acceleration of the system and the tension, T, in the string.

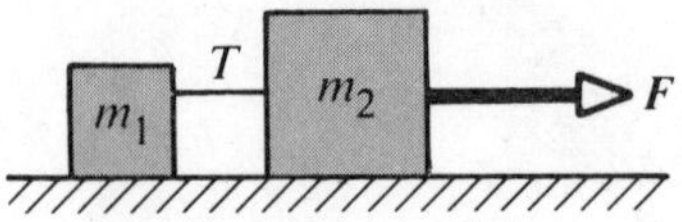

Figure 5.24 (Exercises 27 and 35).

28. The parachute on a race car of weight 8820 N opens at the end of a quarter-mile run when the car is traveling at 55 m/s. What is the total retarding force required to stop the car in a distance of 1000 m in the event of a brake failure?

Section 5.9 Forces of Friction

29. A block moves up a 45° incline with constant speed under the action of a force of 15 N applied *parallel* to the incline. If the coefficient of kinetic friction is 0.3, determine (a) the weight of the block and (b) the minimum force required to allow the block to move *down* the incline at constant speed.

30. The coefficient of static friction between a 4-kg block and a horizontal surface is 0.3. What is the *maximum* horizontal force that can be applied to the block before it slips?

31. A 20-kg block is initially at rest on a rough, horizontal surface. A horizontal force of 75 N is required to set the block in motion. After it is in motion, a horizontal force of 60 N is required to keep the block moving with constant speed. Find the coefficients of static and kinetic friction from this information.

32. A car is traveling at 50 mi/h on a horizontal highway. (a) If the coefficient of friction between the road and tires on a rainy day is 0.1, what is the *minimum* distance in which the car will stop? (b) What is the stopping distance when the surface is dry and $\mu = 0.6$? (c) Why should you avoid "slamming on" your brakes if you want to stop in the shortest distance?

33. A racing car accelerates uniformly from 0 to 80 mi/h in 8 s. The external force that accelerates the car is the frictional force between the tires and the road. If the tires do not spin, determine the *minimum* coefficient of friction between the tires and the road.

34. In a game of shuffleboard, a disk is given an initial speed of 5 m/s. It slides a distance of 8 m before coming to rest. What is the coefficient of kinetic friction between the disk and the surface?

35. Two blocks connected by a light rope are being dragged by a horizontal force $\mathbf{F}$ (Fig. 5.24). Suppose that $F = 50$ N, $m_1 = 10$ kg, $m_2 = 20$ kg, and the coefficient of kinetic friction between each block and the surface is 0.1. (a) Draw a free-body diagram for each block. (b) Determine the tension, T, and the acceleration of the system.

36. A block slides on a *rough* incline. The coefficient of kinetic friction between the block and the plane is μ_k. (a) If the block accelerates *down* the incline, show that the acceleration of the block is given by $a = g(\sin\theta - \mu_k \cos\theta)$. (b) If the block is projected *up* the incline, show that its deceleration is $a = -g(\sin\theta + \mu_k \cos\theta)$.

37. A 3-kg block starts from rest at the top of a 30° incline and slides a distance of 2 m down the incline in 1.5 s. Find (a) the acceleration of the block, (b) the coefficient of kinetic friction between the block and the plane, (c) the frictional force acting on the block, and (d) the speed of the block after it has slid 2 m.

38. In order to determine the coefficients of friction between rubber and various surfaces, a student uses a rubber eraser and an incline. In one experiment the eraser slips down the incline when the angle of inclination is 36° and then moves down the incline with constant speed when the angle is reduced to 30°. From these data, determine the coefficients of static and kinetic friction for this experiment.

39. Two masses are connected by a light string, which passes over a frictionless pulley as in Fig. 5.11. The incline is rough. When $m_1 = 3$ kg, $m_2 = 10$ kg, and $\theta = 60°$, the 10-kg mass accelerates *down* the incline at the rate of 2 m/s^2. Find (a) the tension in the string and (b) the coefficient of kinetic friction between the 10-kg mass and the plane.

40. A box rests on the back of a truck. The coefficient of static friction between the box and the surface is 0.3. (a) When the truck accelerates, what force accelerates the box? (b) Find the *maximum* acceleration the truck can have before the box slides.

41. A block slides down a 30° incline with *constant* acceleration. The block starts from rest at the top and travels 18 m to the bottom, where its speed is 3 m/s. Find (a) the coefficient of kinetic friction between the block and the incline and (b) the acceleration of the block.

PROBLEMS

1. Two blocks of mass 2 kg and 7 kg are connected by a light string that passes over a frictionless pulley (Fig. 5.25). The inclines are smooth. Find (a) the acceleration of each block and (b) the tension in the string.

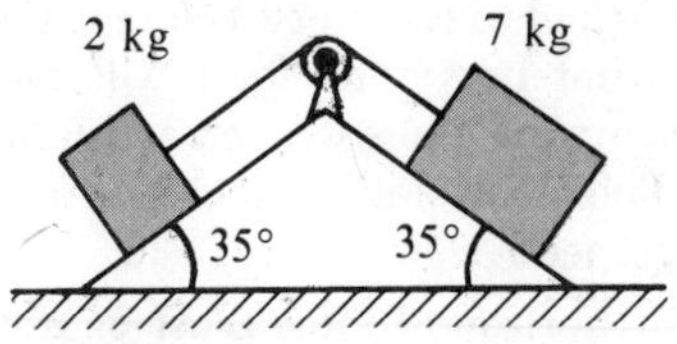

Figure 5.25 (Problems 1 and 2).

2. The system described in Fig. 5.25 is observed to have an acceleration of 1.5 m/s^2 when the inclines are rough. Assume the coefficients of kinetic friction between each block and the inclines are the same. Find (a) the coefficient of kinetic friction and (b) the tension in the string.

3. A 2-kg block is placed on top of a 5-kg block as in Fig. 5.26. The coefficient of kinetic friction between the 5-kg block and the surface is 0.2. A horizontal force $\mathbf{F}$ is applied to the 5-kg block. (a) Draw a free-body diagram for each block. What force accelerates the 2-kg block? (b) Calculate the force necessary to pull both blocks to the right with an acceler-

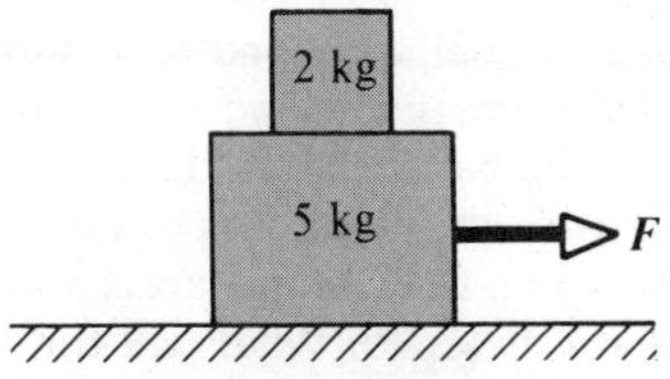

Figure 5.26 (Problem 3).

ation of 3 m/s^2. (c) Find the minimum coefficient of static friction between the blocks such that the 2-kg block does not slip under an acceleration of 3 m/s^2.

4. A car moves with a velocity v_0 down a sloped highway having an angle of inclination θ. The coefficient of friction between the car and the road is μ. The driver applies the brakes at some instant. Assuming that the tires do not skid and that the frictional force is a *maximum*, find (a) the deceleration of the car, (b) the distance the car will move before coming to rest after the brakes are applied, and (c) numerical results for the deceleration and the distance traveled if $v_0 = 60$ mi/h, $\theta = 10°$, and $\mu = 0.6$.

5. In Fig. 5.27, the coefficient of kinetic friction between the 2-kg and 3-kg blocks is 0.3. The horizontal surface and the pulleys are frictionless. (a) Draw free-body diagrams for each block. (b) Determine the acceleration of each block. (c) Find the tension in the strings.

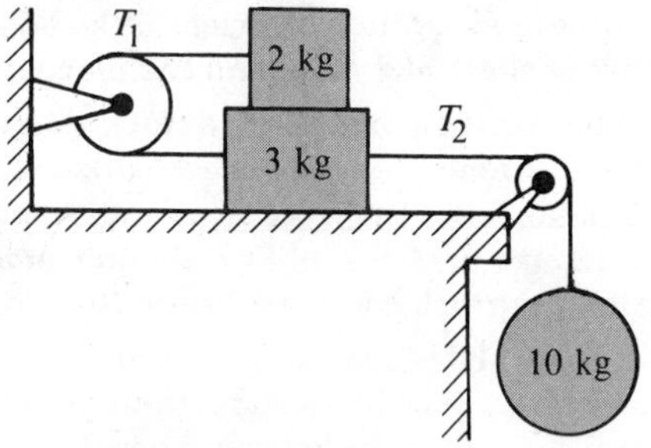

Figure 5.27 (Problem 5).

6. A horizontal force $\boldsymbol{F}$ is applied to a frictionless pulley of mass m_2 as in Fig. 5.28. The horizontal surface is smooth. (a) Show that the acceleration of the block of mass m_1 is *twice* the acceleration of the pulley. Find (b) the acceleration of the pulley and the block and (c) the tension in the string.

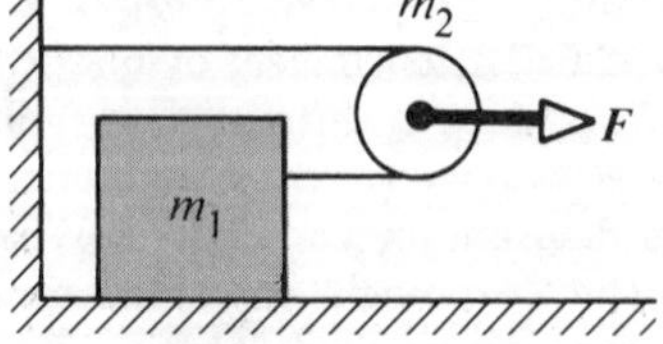

Figure 5.28 (Problem 6).

7. A bowling ball attached to a spring scale is suspended from the ceiling of an elevator as in Fig. 5.13. (The ball replaces the fish!) The scale reads 16 lb when the elevator is at rest. (a) What will the scale read if the elevator accelerates *upward* at the rate of 8 ft/s^2? (b) What will the scale read if the elevator accelerates *downward* at the rate of 8 ft/s^2? (c) If the supporting rope can withstand a maximum tension of 25 lb and the weight of the scale is neglected, what is the maximum acceleration the elevator can have before the rope breaks? (d) If the spring scale weighs 5 lb, which rope breaks first? Why?

8. Three blocks are in contact with each other on a smooth, horizontal surface as in Fig. 5.29. A horizontal force $\boldsymbol{F}$ is applied to m_1. If $m_1 = 2$ kg, $m_2 = 3$ kg, $m_3 = 4$ kg, and $F = 18$ N, find (a) the acceleration of the blocks, (b) the *resultant* force on each block, and (c) the magnitude of the contact forces between the blocks.

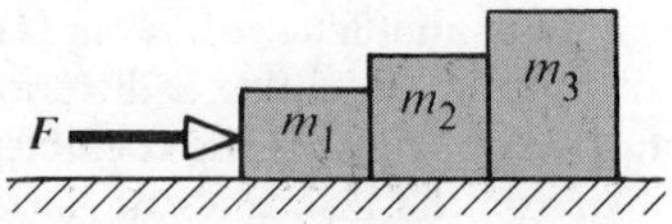

Figure 5.29 (Problems 8 and 9).

9. Repeat Problem 8 given that the coefficient of kinetic friction between the blocks and the surface is 0.1. Use the data given in Problem 8.

10. A 5-kg block is placed on top of a 10-kg block (Fig. 5.30). A horizontal force of 45 N is applied to the 10-kg block, while the 5-kg block is tied to the wall. The coefficient of kinetic friction between the moving surfaces is 0.2. (a) Draw a free-body diagram for each block and identify the action-reaction forces between the blocks. (b) Determine the tension in the string and the acceleration of the 10-kg block.

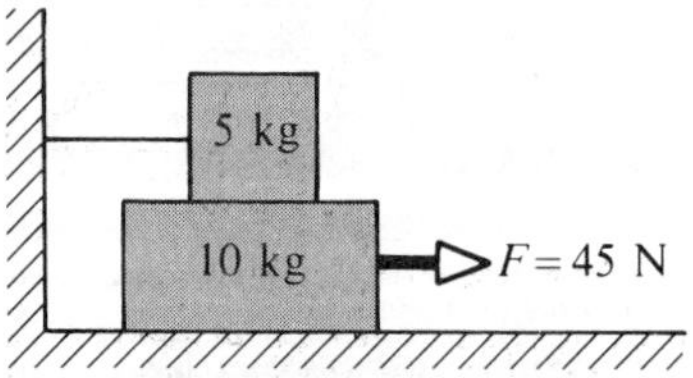

Figure 5.30 (Problem 10).

11. A block of mass m is on a *rough* incline of angle θ. (a) What is the *maximum horizontal* force that can be applied to the block before it slips *up* the plane? (b) What horizontal force will cause the block to move *up* the plane with an acceleration $\boldsymbol{a}$? Take the coefficients of static and kinetic friction to be μ_s and μ_k, respectively.

12. What horizontal force must be applied to the cart shown in Fig. 5.31 in order that the blocks remain *stationary* relative to the cart? Assume all surfaces, wheels, and pulley are frictionless. (Hint: Note that the tension in the string accelerates m_1.)

13. The three blocks in Fig. 5.32 are connected by light strings that pass over frictionless pulleys. The accel-

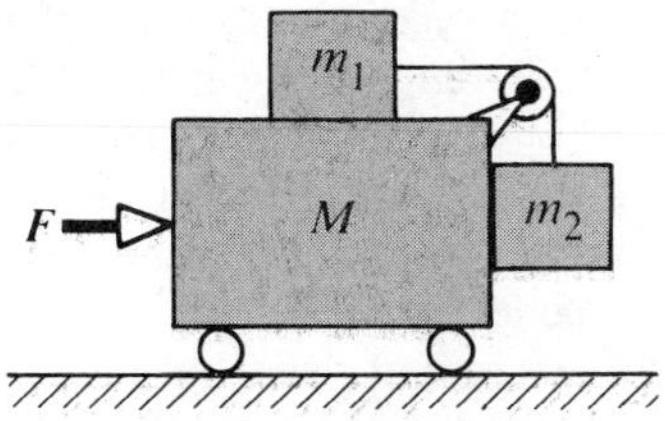

Figure 5.31 (Problem 12).

eration of the system is 2 m/s^2 and the surfaces are rough. Find (a) the tensions in the strings and (b) the coefficient of kinetic friction between blocks and surfaces. (Assume the same μ for both blocks.)

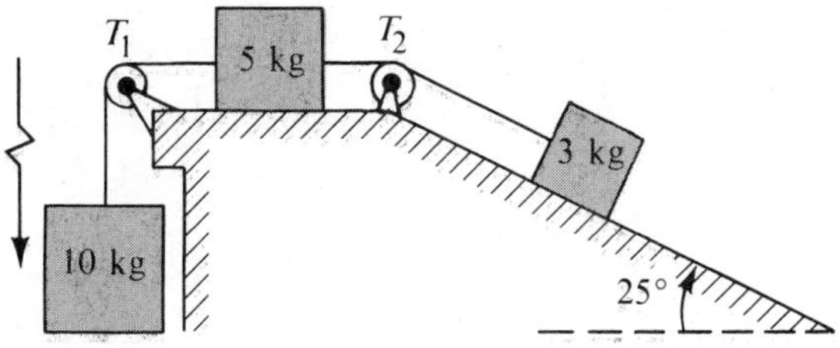

Figure 5.32 (Problem 13).

14. Two blocks on a rough incline are connected by a light string that passes over a frictionless pulley as in Fig. 5.33. Assuming $m_1 > m_2$ and taking the coefficient of kinetic friction for each block to be μ, determine expressions for (a) the acceleration of the blocks and (b) the tension in the string. (Assume that the system is in motion.)

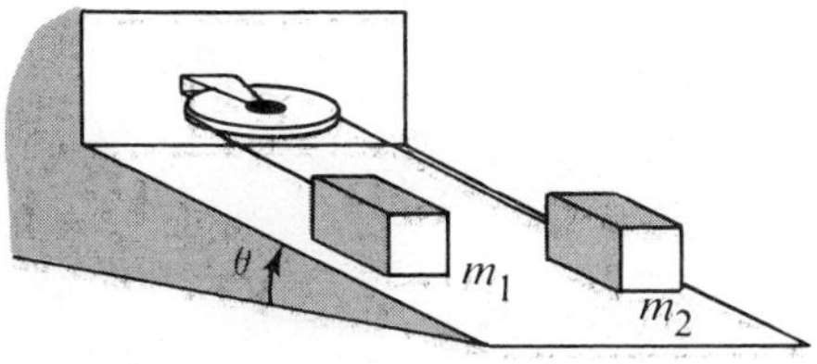

Figure 5.33 (Problem 14).

15. An inventive child named Pat wants to reach an apple in a tree without climbing the tree. Sitting in a chair connected to a rope that passes over a frictionless pulley (Fig. 5.34), Pat pulls on the loose end of the rope with such a force that the spring scale reads 60 lb. Pat's true weight is 64 lb and the chair weighs 32 lb. (a) Draw free-body diagrams for Pat and the chair considered as separate systems, and another diagram for Pat and the chair considered as one system. (b) Show that the acceleration of the system is *upward* and find its magnitude. (c) Find the force that Pat exerts on the chair.

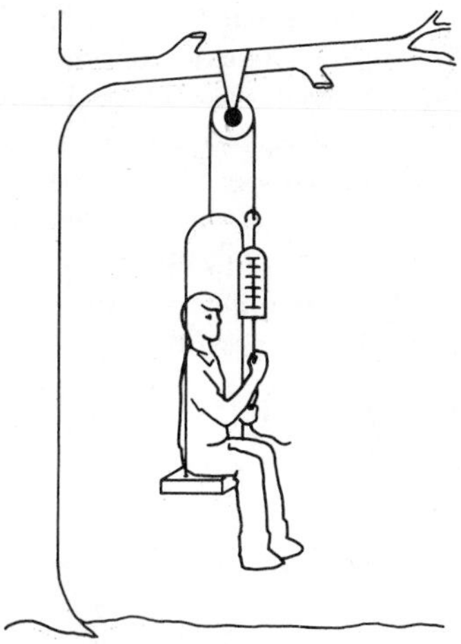

Figure 5.34 (Problem 15).

16. A block of mass m rests on the rough, inclined face of a wedge of mass M as in Fig. 5.35. The wedge is free to move on a frictionless, horizontal surface. A horizontal force $\mathbf{F}$ is applied to the wedge such that the block is *on the verge* of slipping *up* the incline. If the coefficient of static friction between the block and the wedge is μ, find (a) the acceleration of the system and (b) the horizontal force necessary to produce this acceleration.

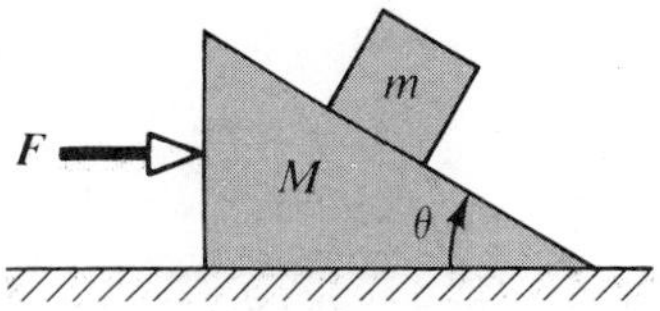

Figure 5.35 (Problem 16).

17. Two blocks are fastened to the top of an elevator as in Fig. 5.36. The elevator accelerates upward at 4 m/s^2. Each rope has a mass of 1 kg. Find the tensions in the ropes at points *A*, *B*, *C*, and *D*.

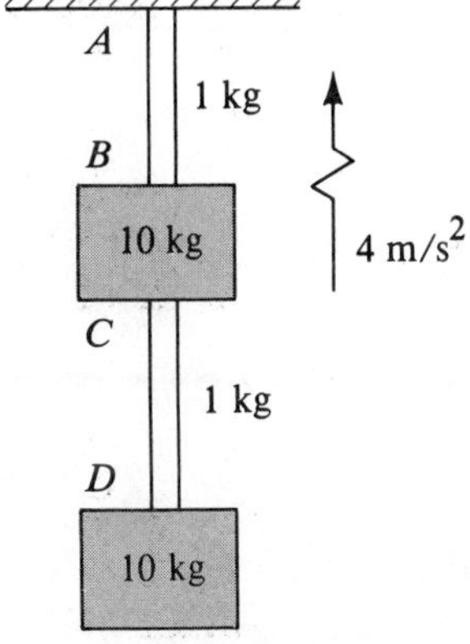

Figure 5.36 (Problem 17).

18. The force on a particle can be obtained from its momentum as a function of time. Find the force on a particle for each case when the momentum measured in kg · m/s varies with time as (a) $\mathbf{p} = (4 + 3t)\mathbf{j}$, (b) $\mathbf{p} = 3t\mathbf{i} + 5t^2\mathbf{j}$, (c) $\mathbf{p} = 4e^{-2t}\mathbf{i}$. (d) If the particle has a mass of 2 kg, find its acceleration at $t = 1$ s for cases (a), (b), and (c).

19. Two masses m and M are attached with strings as shown in Fig. 5.37. If the system is in equilibrium, show that $\tan\theta = 1 + \frac{2M}{m}$.

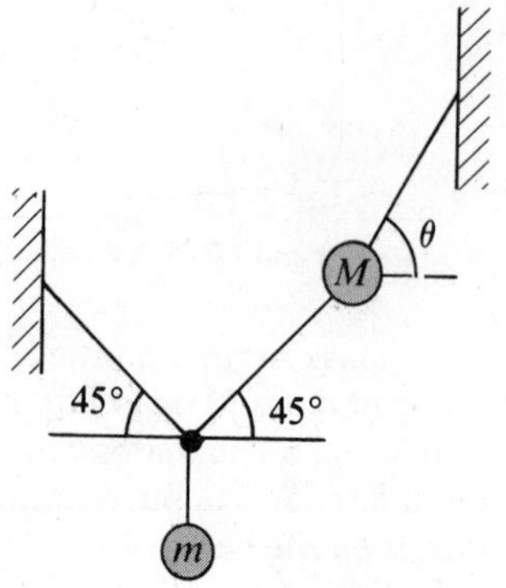

Figure 5.37 (Problem 19).

20. Consider a system consisting of a horse pulling a sled. According to Newton's third law, the force exerted by the horse on the sled is equal to and opposite the force exerted by the sled on the horse. Therefore, one might argue that the system can never move. Explain, using complete force diagrams on the horse and sled, that motion in this system is possible despite Newton's third law. Be sure to identify all of your forces.

21. Find the acceleration of the cart and the mass shown in Fig. 5.38. The pulleys are light and all surfaces are frictionless. What do these results predict in the limits $m_2 \gg m_1$ and $m_1 \gg m_2$?

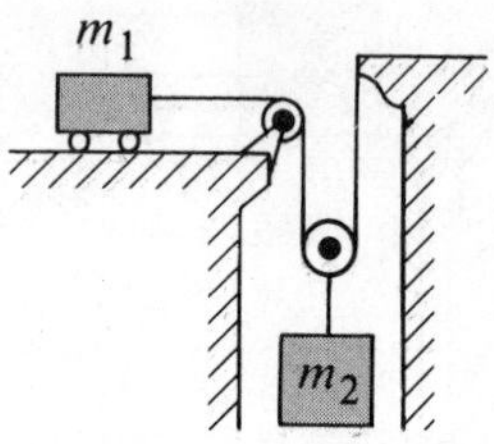

Figure 5.38 (Problem 21).

6 Forces in Nature and More Applications of Newton's Laws

U.S. Air Force Photo

In the previous chapter we introduced Newton's laws of motion and applied them to situations involving linear motion. The only force law discussed quantitatively thus far is the empirical law of friction, based on experimental observation. All observed forces can be attributed to one or more of the following basic interactions: (1) gravitational forces, (2) electromagnetic forces, (3) strong nuclear forces, and (4) weak nuclear forces. In this chapter, we shall briefly describe the basic features of these fundamental forces. We shall show how the universal law of gravity, together with Newton's laws, enables us to understand a variety of familiar motions, such as the motion of satellites. Newton's laws will also be applied to situations involving other types of circular motion. More detailed descriptions of gravitational and electromagnetic forces will be presented later in the text. In the last two sections of this chapter, we shall discuss the motion of an object when observed in an accelerated, or noninertial, frame of reference and the motion of an object through a viscous medium.

6.1 NEWTON'S UNIVERSAL LAW OF GRAVITY

It has been said that Newton was struck on the head by a falling apple while napping under a tree (or some variation of this legend). This supposedly prompted Newton to imagine that perhaps all bodies in the universe are attracted to each other in the same way the apple was attracted to the earth. Newton proceeded to analyze astronomical data on the motion of the moon around the earth. Details of this calculation and further descriptions of planetary motion will be presented in Chapter 14. From the analysis of such data, Newton made the bold statement that the law of force governing the motion of planets has the *same* mathematical form as the force law that attracts a falling apple to the earth.

In 1687 Newton published his work on the universal law of gravity in *Principia*. Newton's law of gravitation states that *every particle in the universe attracts every other particle with a force that is directly proportional to the product of their masses and inversely proportional to the square of the distance between them.* If the particles have masses m_1 and m_2 and are separated by a distance r, the magnitude of this gravitational force is

$$F = G\frac{m_1 m_2}{r^2} \qquad (6.1)$$

Universal law of gravity

where G is a universal constant called the *gravitational constant,* which has been measured experimentally. Its value in SI units is

$$G = (6.673 \pm 0.003) \times 10^{-11} \frac{\mathrm{N \cdot m^2}}{\mathrm{kg^2}} \tag{6.2}$$

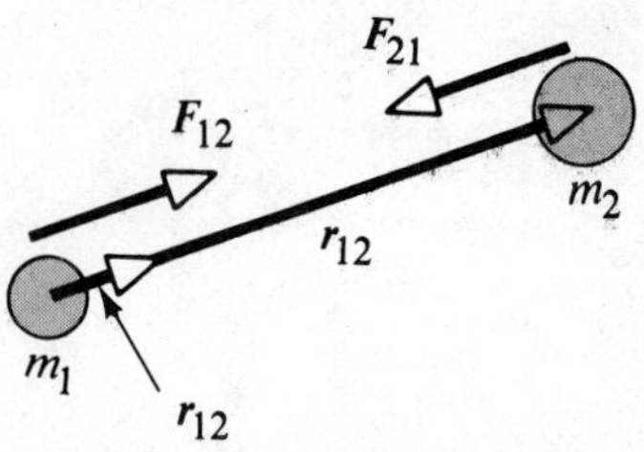

Figure 6.1 The gravitational force between two particles is attractive. The unit vector $\hat{r}_{12}$ is directed from m_1 to m_2. Note that $F_{12} = -F_{21}$.

The force law given by Eq. 6.1 is often referred to as the *inverse-square law,* since it varies as the inverse square of the separation of the particles. We can express this force in vector form by defining a unit vector $\hat{r}_{12}$ (Fig. 6.1). Since this unit vector is in the direction of the displacement vector r_{12} that runs from m_1 to m_2, the force on m_2 due to m_1 is given by

$$F_{21} = -G\frac{m_1 m_2}{r_{12}^2}\hat{r}_{12} \tag{6.3}$$

The minus sign in Eq. 6.3 indicates that m_2 is attracted to m_1, and so the force must be directed toward m_1. Likewise, by Newton's third law the force on m_1 due to m_2, designated F_{12}, is equal in magnitude to F_{21} and in the opposite direction. That is, these forces form an action-reaction pair, and $F_{12} = -F_{21}$.

Properties of the gravitational force

There are several features of the inverse-square law that deserve some attention. The gravitational force is an action-at-a-distance force, which always exists between two particles, regardless of the medium that separates them. The force varies as the inverse square of the distance between the particles and therefore decreases rapidly with increasing separation. The force is proportional to the mass of both particles, as one might intuitively expect.

Another important fact is that *the gravitational force exerted by a finite-size, spherically symmetric mass distribution on a particle outside the sphere is the same as if the entire mass of the sphere were concentrated at its center.* (The proof of this involves the use of integral calculus and is delayed until Chapter 14.) For example, the force on a particle of mass m near the earth's surface has the magnitude

$$F = G\frac{M_e m}{R_e^2}$$

where M_e is the earth's mass and R_e is the earth's radius. This force is directed toward the center of the earth.

6.2 MEASUREMENT OF THE GRAVITATIONAL CONSTANT

The gravitational constant, G, was first measured in an important experiment by Sir Henry Cavendish in 1798. The apparatus he used consists of two small spheres each of mass m fixed to the ends of a light horizontal rod suspended by a fine quartz fiber or thin metal wire, as in Fig. 6.2. Two large spheres each of mass M are then placed near the smaller spheres. The attractive force between the smaller and larger spheres causes the rod to rotate and twist the wire suspension. If the system is oriented as shown in Fig. 6.2, the rod rotates clockwise when viewed from the top. The angle through which the suspended rod rotates is measured by the deflection of a light beam reflected from a mirror attached to the vertical suspension. The moving spot of light is an effective technique for amplifying the motion. The experiment is carefully repeated with different masses at various separations. In addition to providing a value for G, the results show that the force is attractive, proportional to the product mM, and inversely proportional to the square of the distance r.

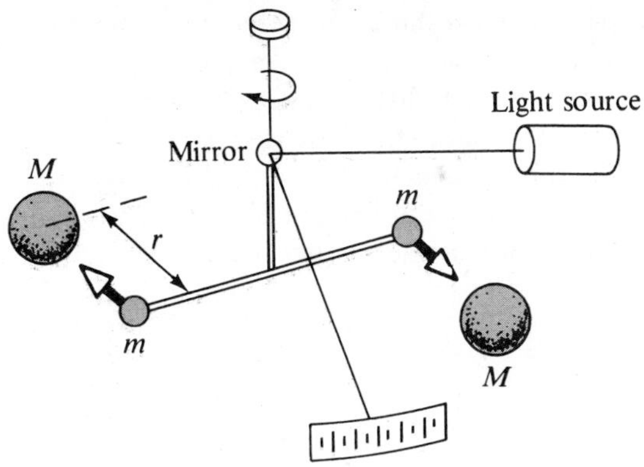

Figure 6.2 Schematic diagram of the Cavendish apparatus for measuring G. The smaller spheres of mass m are attracted to the large spheres of mass M, and the bar rotates through a small angle. A light beam reflected from a mirror on the rotating apparatus measures the angle of rotation.

Example 6.1

Three uniform spheres of mass 2 kg, 4 kg, and 6 kg are placed at the corners of a right triangle as in Fig. 6.3, where the coordinates are in m. Calculate the resultant gravitational force on the 4-kg mass, assuming the spheres are isolated from the rest of the universe.

Solution: First we calculate the individual forces on the 4-kg mass due to the 2-kg and 6-kg masses separately, and then we take a vector sum to get the resultant force on the 4-kg mass.

The force on the 4-kg mass due to the 2-kg mass is upward and given by

$$\boldsymbol{F}_{42} = G\frac{m_4 m_2}{r_{42}^2}\boldsymbol{j} = \left(6.67 \times 10^{-11}\frac{\text{N}\cdot\text{m}^2}{\text{kg}^2}\right)\frac{(4\text{ kg})(2\text{ kg})}{(3\text{ m})^2}\boldsymbol{j}$$
$$= 5.93 \times 10^{-11}\boldsymbol{j}\text{ N}$$

The force on the 4-kg mass due to the 6-kg mass is to the left and given by

$$\boldsymbol{F}_{46} = G\frac{m_4 m_6}{r_{46}^2}(-\boldsymbol{i})$$
$$= \left(-6.67 \times 10^{-11}\frac{\text{N}\cdot\text{m}^2}{\text{kg}^2}\right)\frac{(4\text{ kg})(6\text{ kg})}{(4\text{ m})^2}\boldsymbol{i}$$
$$= -10.0 \times 10^{-11}\boldsymbol{i}\text{ N}$$

Therefore, the resultant force on the 4-kg mass is the vector sum of $\boldsymbol{F}_{42}$ and $\boldsymbol{F}_{46}$:

$$\boldsymbol{F}_4 = \boldsymbol{F}_{42} + \boldsymbol{F}_{46} = (-10.0\boldsymbol{i} + 5.93\boldsymbol{j}) \times 10^{-11}\text{ N}$$

The magnitude of this force is 11.6×10^{-11} N, which is only 2.61×10^{-11} lb! The force makes an angle of 149° with the positive x axis.

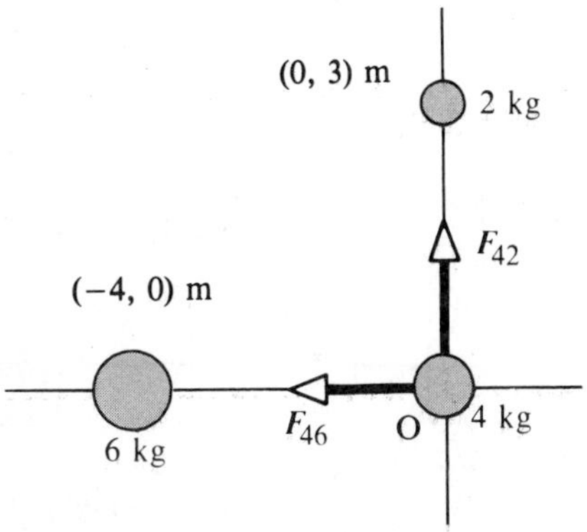

Figure 6.3 (Example 6.1) The *resultant* force on the 4-kg mass is the vector sum $\boldsymbol{F}_{46} + \boldsymbol{F}_{42}$.

Q1. Estimate the gravitational force between you and a person 2 m away from you.

6.3 INERTIAL AND GRAVITATIONAL MASS

In Chapter 3 we described how all objects near the earth fall with the same acceleration if friction is neglected. Although this result is quite familiar, and perhaps obvious by now, it is actually quite remarkable. Although different objects may have different weights, they all fall with the same acceleration. There is no obvious reason that explains why the magnitude of the gravitational force should be proportional to the mass.

Imagine that you perform two experiments on the same object, one with motion in the absence of gravity and another without motion in the presence of gravity. In the first experiment, suppose you have the capability of measuring the force and acceleration for an object moving on a horizontal, frictionless surface. A standard 1-kg mass is at your disposal. A known horizontal force **F**, when applied to the object, produces some acceleration **a**. From Newton's second law, the ratio F/a is the mass of the object, more properly known as the *inertial mass*, m_I:

Inertial mass

$$m_I \equiv \frac{F}{a}$$

Now imagine a second experiment in which you place the same object on a stationary balance. This is, you weigh the object and therefore measure the gravitational force on that object. According to Newton's universal law of gravitation, the force of gravity acting on the object is given by its weight W, where

$$W = G\frac{m_G M_e}{R_e^{\,2}} = m_G g$$

The symbol m_G is used to denote gravitational mass. In a static experiment using a standard 1-kg mass as a test object, you measure the gravitational force. Using these results, the gravitational mass can be determined through the equation

Gravitational mass

$$m_G \equiv \frac{W}{g}$$

Note that the concept of inertia does not enter the picture here since this is a static experiment. The weight, or gravitational force, depends *only* on the properties of the gravitational mass, m_G, and the earth. We now have at our disposal two operational definitions of mass: inertial mass, $m_I = F/a$, and gravitational mass, $m_G = W/g$.

Experiments suggest that inertial mass is equivalent to gravitational mass

A number of early experiments were conducted to test for the equivalence of inertial and gravitational mass. Newton analyzed the motion of pendulums, and his results showed that the motion was independent of the mass or composition of the suspended object. Thus, he concluded that m_I and m_G were equal to an accuracy of one part in 10^3. In 1901, Eötvös determined the equivalence to one part in 10^8. Finally, in 1964 Robert Dicke of Princeton University refined the Eötvös experiment[1] and showed the equivalence of inertial and gravitational mass to three parts in 10^{11}.

These results strongly suggest that inertial mass is *exactly* equal to gravitational mass for all substances. This is surely one of the most amazing discoveries in all of physics! The *principle of equivalence,* which is based on experimental results, is taken to be a fundamental law of nature. In the remainder of this text we shall assume that $m_I = m_G$ and use the symbol m to designate mass.

The equivalence of inertial and gravitational mass can be considered an extraordinary coincidence of nature. There is no experiment that is able to distinguish between acceleration in the laboratory and acceleration due to a gravitational force. Finally, it is interesting to note that the principle of equivalence was taken as a basic assumption by Einstein in developing his general theory of relativity.

---

[1]For further details see Robert Dicke, "The Eötvös Experiment," *Scientific American*, December 1961.

6.4 WEIGHT AND GRAVITATIONAL FORCE

In the previous chapter we defined the weight of a body of mass m as simply mg, where g is the magnitude of the acceleration due to gravity. Now, we are in a position to obtain a more fundamental description of g. Since the force on a freely falling body of mass m near the surface of the earth is given by Eq. 6.1, we can equate mg to this expression to give

Acceleration due to gravity

$$mg = G\frac{M_e m}{R_e^2}$$

$$g = G\frac{M_e}{R_e^2} \qquad (6.4)$$

where M_e is the mass of the earth and R_e is the earth's radius. Since $g = 9.8\ \text{m/s}^2$ at the earth's surface and the radius of the earth is approximately 6.38×10^6 m, we find from Eq. 6.4 that $M_e = 5.98 \times 10^{24}$ kg. Using this result, the average density of the earth is calculated to be

$$\rho_e = \frac{M_e}{V_e} = \frac{M_e}{\frac{4}{3}\pi R_e^3} = \frac{5.98 \times 10^{24}\ \text{kg}}{\frac{4}{3}\pi(6.38 \times 10^6\ \text{m})^3} = 5.50 \times 10^3\ \frac{\text{kg}}{\text{m}^3}$$

Since this value is about twice the density of most rocks at the earth's surface, we conclude that the inner core of the earth has a much higher density.

Now consider a body of mass m a distance h above the earth's surface, or a distance r from the earth's center, where $r = R_e + h$. The magnitude of the gravitational force acting on this mass is given by

$$F = G\frac{M_e m}{r^2} = G\frac{M_e m}{(R_e + h)^2}$$

If the body is in free fall, then $F = mg'$ and we see that g', the acceleration of gravity at the altitude h, is given by

Variation of g with altitude

$$g' = \frac{GM_e}{r^2} = \frac{GM_e}{(R_e + h)^2} \qquad (6.5)$$

Thus, it follows that g' *decreases* with *increasing altitude*. Since the true weight of a body is mg', we see that as $r \to \infty$, the true weight approaches zero.

TABLE 6.1 Acceleration Due to Gravity, g', at Various Altitudes

ALTITUDE h (km)[a]	g' (m/s^2)
1000	7.33
2000	5.68
3000	4.53
4000	3.70
5000	3.08
6000	2.60
7000	2.23
8000	1.93
9000	1.69
10 000	1.49
50 000	0.13

[a]All values are distances above the earth's surface.

Example 6.2 *Value of g with altitude h*

Determine the magnitude of the acceleration of gravity at an altitude of 500 km. By what percentage is the weight of a body reduced at this altitude?

Solution: Using Eq. 6.5 with $h = 500$ km, $R_e = 6.38 \times 10^6$ m, and $M_e = 5.98 \times 10^{24}$ kg gives

$$g' = \frac{GM_e}{(R_e + h)^2}$$

$$= \frac{(6.67 \times 10^{-11}\ \text{N} \cdot \text{m}^2/\text{kg}^2)(5.98 \times 10^{24}\ \text{kg})}{(6.38 \times 10^6 + 0.5 \times 10^6)^2\ \text{m}^2}$$

$$= 8.43\ \text{m/s}^2$$

Since $g'/g = 8.43/9.8 = 0.86$, we conclude that the weight of a body is reduced by about 14 percent at an altitude of 500 km. Values of g' at other altitudes are listed in Table 6.1.

Q2. If you are given the mass and radius of planet X, how would you calculate the acceleration of gravity on the surface of this planet?

Q3. If a hole could be dug to the center of the earth, do you think that the force on a mass m would still obey Eq. 6.1 there? What do you think the force on m would be at the center of the earth? We will return to this point in Chapter 14.

6.5 ELECTROSTATIC FORCES

When two particles are electrically charged and in relative motion, they exert another kind of force on each other called an *electromagnetic force.* It is rather difficult to describe this force when the charges are in motion. We shall provide further details of this force later in the text. However, if the charges are stationary, the force between them is described by an *electrostatic force.* The empirical law used to calculate this force is called *Coulomb's law,* and the force itself is often referred to as the coulombic force. The coulombic force varies as the inverse square of the distance between the charges; in this respect, Coulomb's law is similar to the universal law of gravity. If q_1 and q_2 represent the electric charges on two particles and r is their separation, the electrostatic force between the particles has the magnitude

Electrostatic force

$$F_e = k\frac{q_1 q_2}{r^2} \tag{6.6}$$

where k is a constant, known as the coulomb constant, and the charges in SI units are measured in coulombs (C). The experimental value of k is

$$k = 8.99 \times 10^9 \text{ N} \cdot \text{m}^2/\text{C}^2 \tag{6.7}$$

The smallest amount of charge that has been found in nature is the charge on an electron or proton. This fundamental unit of charge has the value

Charge on an electron or proton

$$e = 1.602 \times 10^{-19} \text{ C} \tag{6.8}$$

One coulomb of charge therefore represents the charge on $\frac{1}{e} \approx 6.2 \times 10^{18}$ electrons or protons.

Properties of electric charge

Electric charge and electrostatic forces are characterized by several important properties:

1. There are two types of electric charges found in nature, which are labeled positive and negative. Charges of the same type *repel* each other, and unlike charges *attract* each other (Fig. 6.4).
2. Charges are scalar quantities and so their values are additive. The net charge on any object has discrete values—it is *quantized.* That is, since e is the fundamental unit of charge, the net charge of an object is Ne, where N is an integer. All elementary charged particles are known to have a charge of $\pm e$. For example, electrons have a charge of $-e$ and protons have a charge of $+e$. Neutrons have no charge.[2]
3. Electric charge is always conserved. In any kind of process, such as a collision event, chemical reaction, or nuclear decay, the total charge of an isolated system does not change.

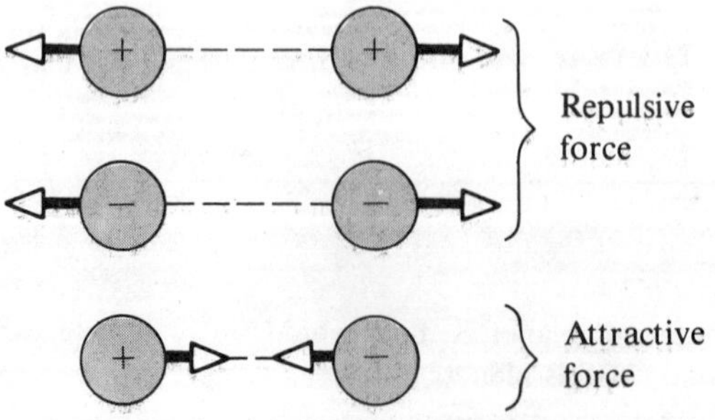

Figure 6.4 The coulomb force between charged particles is repulsive if the particles have like charges and attractive if the particles are oppositely charged.

[2]Recent theories have suggested the possibility of the existence of fractionally charged particles called *quarks,* and there is recent experimental evidence that makes the quark model very compelling.

We have stated that the electric force between two charged elementary particles is much stronger than the gravitational force between them. Let us estimate the relative strength of these interactions for the case of two electrons. The mass of each electron is 9.1×10^{-31} kg, and each has a charge of -1.6×10^{-19} C. Therefore, the ratio of the electric force to the gravitational force at *any* separation r is

$$\frac{F_e}{F_g} = \frac{kq_1q_2/r^2}{Gm_1m_2/r^2} = \frac{kq_1q_2}{Gm_1m_2} \approx \frac{(8.99 \times 10^9)(1.6 \times 10^{-19})^2}{(6.67 \times 10^{-11})(9.1 \times 10^{-31})^2} = 4.2 \times 10^{42}$$

Gravitational forces are therefore negligible at the atomic level.

For macroscopic objects, gravitational forces can be much stronger than electric forces between the objects only because such objects are usually electrically neutral. That is, macroscopic bodies contain approximately as many positive charges as negative charges. However, it is possible to transfer electrons from one object to another such that electric forces can become quite large. For example, if a rubber rod is rubbed with a piece of fur, electrons are transferred from the fur to the rod. The electrified state of the rod is shown by the ability of the rod to attract small bits of paper. Similar experiments can be conducted with other materials. Typical charges formed on objects by rubbing are of the order of microcoulombs ($1\ \mu C = 10^{-6}$ C). However, since there are typically about 10^{24} electrons (more than 10^5 C) in one cm^3 of a solid, one μC of charge represents a *very* small fraction of the total number of electrons. Even so, the force between two such charges can be quite appreciable, as can be seen from the following example.

Example 6.3

Two particles separated by 3.0 cm are positively charged to 1.0 μC. What is the magnitude of the repulsive electric force between them?

Solution: Using Coulomb's law, we find that

$$F_e = k\frac{q_1q_2}{r^2} = \frac{\left(9 \times 10^9 \frac{N \cdot m^2}{C^2}\right)(1 \times 10^{-6}\ C)^2}{(3 \times 10^{-2}\ m)^2}$$

$$\approx 10\ N \approx 2.3\ lb$$

6.6 NUCLEAR FORCES

So far we have discussed only two fundamental forces in nature, the gravitational force and the electrostatic force. Two other fundamental forces are known to exist: the strong nuclear force and the weak nuclear force. The *strong nuclear force* is responsible for the stability of nuclei. In the most naive model, all nuclei consist of a number of positively charged protons and neutral neutrons. (Both protons and neutrons are commonly referred to as *nucleons.*) One would expect that the repulsive coulombic force between the protons would tend to decompose the nuclei. However, the world is made of stable nuclei, such as helium (two protons, two neutrons) and lithium (three protons, four neutrons). Clearly, there must be an attractive force present in nuclei to overcome this strong coulombic repulsion. As we discussed in the previous section, gravitational forces are negligible in comparison to electrostatic forces. It is the strong nuclear force that binds the nucleons together. This strong interaction is independent of charge. That is, the interaction is the same between a pair of protons, a pair of neutrons, and a proton and neutron. For separations of about 10^{-15} m (a typical nuclear dimension), the strong interaction is one to two orders of magnitude stronger than the electrostatic interaction. However, the strong nuclear interaction decreases rapidly with increasing separation and is negligible for separations greater than about 10^{-14} m.

The *weak nuclear force* is another short-range nuclear force that tends to produce instability in certain nuclei. For example, it is responsible for the decomposition of some radioactive nuclei by *beta decay,* a process in which an energetic electron is ejected. The weak interaction is about 12 orders of magnitude weaker than the electrostatic interaction.

6.7 NEWTON'S SECOND LAW APPLIED TO UNIFORM CIRCULAR MOTION

Consider a particle of mass m moving in a circular orbit of radius r, as in Fig. 6.5. In this section, we shall assume that the particle moves with constant speed. Since the velocity vector, $\boldsymbol{v}$, changes its direction during the motion, the particle experiences a *centripetal acceleration directed toward the center of motion,* as described in Chapter 4. This centripetal (or radial) component of acceleration is given by

Centripetal acceleration

$$\boldsymbol{a}_\mathrm{r} = -\frac{v^2}{r}\hat{\boldsymbol{r}} \tag{6.9}$$

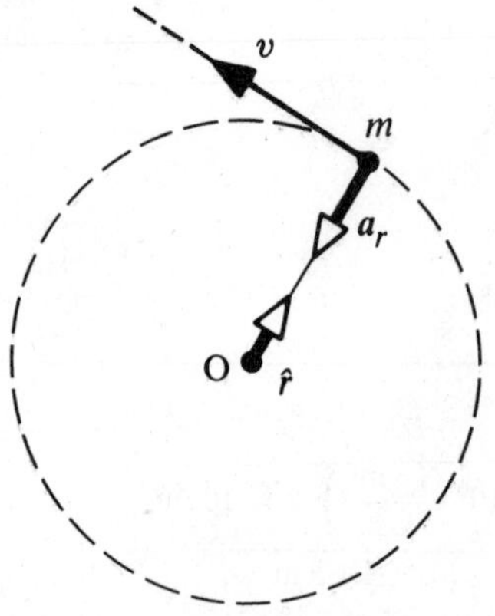

Figure 6.5 A particle of mass m moving in a circle of radius r with constant speed v undergoes a centripetal acceleration, $\boldsymbol{a}_\mathrm{r}$, directed toward the center of rotation.

where $\hat{\boldsymbol{r}}$ is a unit vector directed *radially outward* from the center of motion. Moreover, since v is constant, the tangential component of acceleration is zero, that is, $a_\mathrm{t} = 0$, and therefore $\boldsymbol{a} = \boldsymbol{a}_\mathrm{r} + \boldsymbol{a}_\mathrm{t} = \boldsymbol{a}_\mathrm{r}$. Newton's second law applied to the motion of the particle gives

$$\Sigma \boldsymbol{F} \text{ (along } \hat{\boldsymbol{r}}) = m\boldsymbol{a}_\mathrm{r} = -\frac{mv^2}{r}\hat{\boldsymbol{r}} \tag{6.10}$$

Since there is a centripetal acceleration acting toward the center of rotation, there must be an external centripetal force acting in this direction. If no such force existed, the particle would move along the dotted line shown in Fig. 6.5.

A body can move in circular motion under the influence of such forces as friction, tension (in a cord or spring), a normal force, a gravitational force, or a combination of forces. Let us consider some common examples of uniform circular motion. In each case, be sure to recognize the *external force* (or forces) that constrains the particle to a circular path.

Example 6.4

Satellite motion: A satellite of mass m moves in a circular orbit about the earth at a constant speed v and at an altitude h above the earth's surface, as in Fig. 6.6. (a) Determine the speed of the satellite in terms of G, h, R_e (the radius of the earth), and M_e (the mass of the earth).

Since the *only* external force on the satellite providing the centripetal force is the force of gravity, GmM_e/r^2, acting toward the center of rotation, we conclude that

$$\Sigma \boldsymbol{F}_\mathrm{r} = -G\frac{mM_\mathrm{e}}{r^2}\hat{\boldsymbol{r}} = -\frac{mv^2}{r}\hat{\boldsymbol{r}}$$

$$\frac{GM_\mathrm{e}}{r^2} = \frac{v^2}{r}$$

Since $r = R_\mathrm{e} + h$, we get

$$v = \sqrt{\frac{GM_\mathrm{e}}{r}} = \sqrt{\frac{GM_\mathrm{e}}{R_\mathrm{e} + h}} \tag{6.11}$$

Note that v is independent of the mass of the satellite!

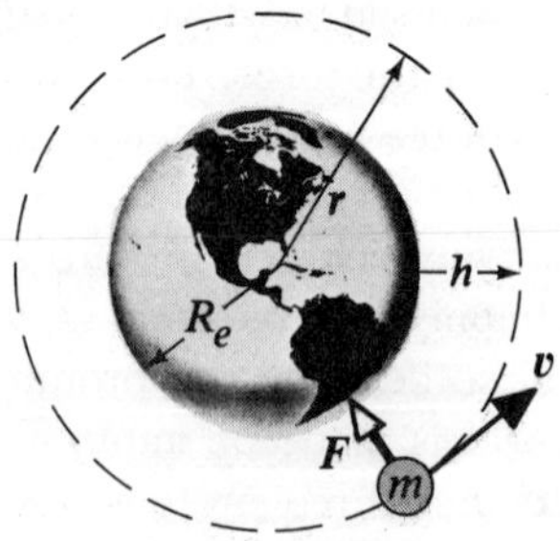

Figure 6.6 (Example 6.4) A satellite of mass m moving in a circular orbit of radius r and with constant speed v around the earth. The centripetal force is provided by the gravitational force between the satellite and the earth.

(b) Determine the satellite's period of revolution, T (the time for one revolution about the earth).

Since the satellite travels a distance of $2\pi r$ (the circumference of the circle) in a time T, we find using Eq. 6.11 that

$$T = \frac{2\pi r}{v} = \frac{2\pi r}{\sqrt{GM_e/r}} = \left(\frac{2\pi}{\sqrt{GM_e}}\right) r^{3/2} \quad (6.12)$$

For example, if we take $h = 1000$ km, then $r = R_e + h = 7.38 \times 10^6$ m, and from Eq. 6.11 we find that $v = 7.35 \times 10^3$ m/s $\approx 16\,400$ mi/h. Also, from Eq. 6.12 we find that $T = 6.31 \times 10^3$ s ≈ 105 min.

The planets move around the sun in approximately circular orbits. The radii of these orbits can be calculated from Eq. 6.12 with M_e replaced by the mass of the sun. The fact that the square of the period is proportional to the cube of the radius of the orbit was first recognized as an empirical relation based on planetary data. We shall return to this topic in Chapter 14.

Q4. Using the ideas developed in Example 6.4, how would you estimate the distance from the earth to the moon?

Q5. How would you explain the fact that planets such as Saturn and Jupiter have periods much greater than one year?

Example 6.5

The conical pendulum: A small body of mass m is suspended from a string of length L. The body revolves in a horizontal circle of radius r with constant speed v, as in Fig. 6.7. Since the string sweeps out the surface of a cone, the system is known as a *conical pendulum*. Find the speed of the body and the period of revolution, T_P.

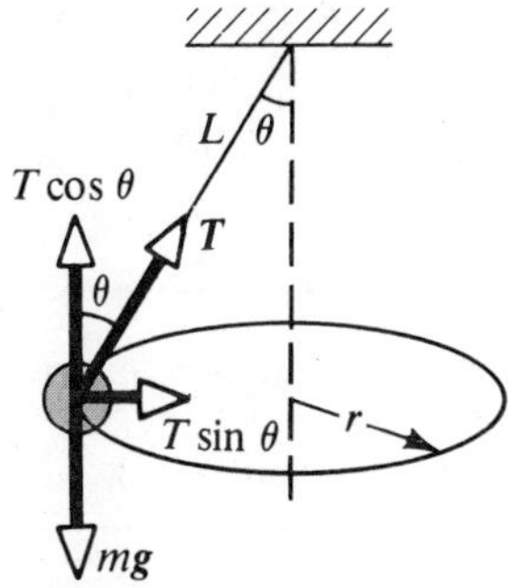

Figure 6.7 (Example 6.5) The conical pendulum and its free-body diagram.

Solution: The free-body diagram for the mass m is shown in Fig. 6.7, where the tension, $\mathbf{T}$, has been resolved into a vertical component, $T\cos\theta$, and a component $T\sin\theta$ acting toward the center of rotation. Since the body does not accelerate in the vertical direction, the vertical component of the tension must balance the weight. Therefore,

$$(1) \qquad T\cos\theta = mg$$

Since the centripetal force in this example is provided by the component $T\sin\theta$, from Newton's second law we get

$$(2) \qquad T\sin\theta = ma_r = \frac{mv^2}{r}$$

By dividing (2) by (1), we eliminate T and find that

$$\tan\theta = \frac{v^2}{rg}$$

But from the geometry, we note that $r = L\sin\theta$, therefore

$$v = \sqrt{rg\tan\theta} = \sqrt{Lg\sin\theta\tan\theta}$$

The period of revolution, T_P, (not to be confused with the tension $\mathbf{T}$) is given by

$$(3) \qquad T_P = \frac{2\pi r}{v} = \frac{2\pi r}{\sqrt{rg\tan\theta}} = 2\pi\sqrt{\frac{L\cos\theta}{g}}$$

The intermediate algebraic steps used in obtaining (3) are left to the reader. Note that T_P is independent of m! If we take $L = 1.0$ m and $\theta = 20°$, we find using (3) that

$$T_P = 2\pi\sqrt{\frac{(1.0\text{ m})(\cos 20°)}{9.8\text{ m/s}^2}} = 1.95\text{ s}$$

Q6. Is it physically possible to have a conical pendulum with $\theta = 90°$?

Example 6.6 *Motion on a Banked Curve*

An engineer wishes to design a curved exit ramp for a tollroad in such a way that a car will not have to rely on friction to round the curve without skidding. At what angle should the ramp be banked?

Solution: On a level track, the centripetal force must be provided by a force of friction between the car and the road. However, if the road is banked at an angle θ, as in Fig. 6.8, the normal force, $\mathbf{N}$, has a component $N\sin\theta$ pointing toward the center of rotation; this component can provide the centripetal force. In the calculation that follows, we assume that only the component $N\sin\theta$ furnishes the centripetal force. Therefore, the banking angle we calculate will be one for which *no* frictional force is

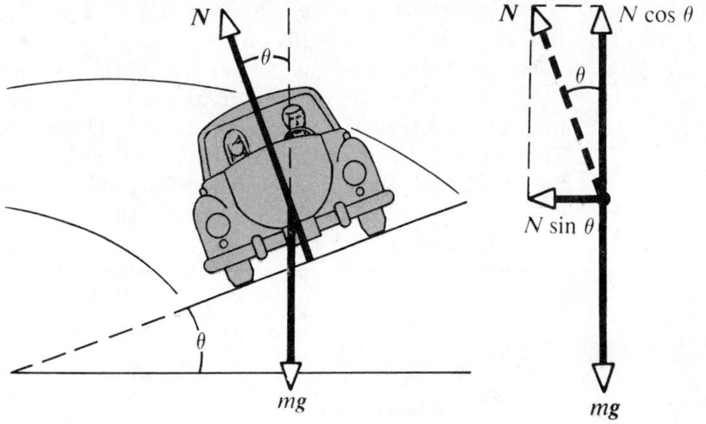

Figure 6.8 (Example 6.6) End view of a car rounding a curve on a road banked at an angle θ to the horizontal. The centripetal force is provided by the horizontal component of the normal force when friction is neglected.

required. Therefore, a car moving at the correct speed can negotiate the curve even on an icy surface. If the speed of the car is v and the radius of curvature is R, then the centripetal force is $N \sin\theta$:

$$(1) \qquad N \sin\theta = \frac{mv^2}{R}$$

Since the vertical component of $\boldsymbol{N}$ is balanced by the weight,

$$(2) \qquad N \cos\theta = mg$$

Dividing (1) by (2) gives

$$\tan\theta = \frac{v^2}{Rg}$$

Therefore, the proper road design depends on the value of R and the expected speed of the vehicles. For example, if $v = 30$ mi/h $= 13.4$ m/s and $R = 50$ m, then $\theta = 20°$. In reality, since the vehicles travel at various speeds, one still has to rely on friction to keep the cars from sliding down the incline if they are going too slow, or to keep them from sliding up the incline if they move faster than the speed for which the road is designed.

Example 6.7 *The Hydrogen Atom*

In the Bohr model of the hydrogen atom, the electron is assumed to move in a circular orbit of radius r about the proton. Since the proton is much more massive than the electron, this assumption is reasonable. The electron and proton are oppositely charged and hence exert a coulombic attractive force given by Eq. 6.6. The centripetal force that maintains the circular motion of the electron is this coulombic force. Determine the orbital speed of the electron if its mass is m_e.

Solution: Since $q_1 = -e$ and $q_2 = +e$, we can apply Eq. 6.6 and Newton's second law to the electron:

$$\boldsymbol{F} = -\frac{ke^2}{r^2}\hat{\boldsymbol{r}} = -m_e\frac{v^2}{r}\hat{\boldsymbol{r}}$$

$$v = \sqrt{\frac{ke^2}{m_e r}}$$

Taking $k = 9 \times 10^9$ N · m^2/C^2, $e = 1.6 \times 10^{-19}$ C, $m_e = 9.1 \times 10^{-31}$ kg, and $r = 0.53 \times 10^{-10}$ m (the radius of the first Bohr orbit), we find that $v = 2.2 \times 10^6$ m/s.

6.8 NONUNIFORM CIRCULAR MOTION

In Chapter 4 we found that if a particle moves with varying speed in a circular path, there is, in addition to the centripetal component of acceleration, a tangential component of magnitude dv/dt. Therefore, the force acting on the particle must also have a tangential and a radial component. That is, since the total acceleration is given by $\boldsymbol{a} = \boldsymbol{a}_r + \boldsymbol{a}_t$, the total force is given by $\boldsymbol{F} = \boldsymbol{F}_r + \boldsymbol{F}_t$. The following example demonstrates this type of motion.

Example 6.8

A small sphere of mass m is attached to the end of a cord of length R, which rotates in a *vertical* circle about a fixed point O, as in Fig. 6.9a. Let us determine the tension in the cord at any instant that the speed of the sphere is v when the cord makes an angle θ with the vertical.

Solution: First we note that the speed is *not* uniform since there is a tangential component of acceleration arising from the weight of the sphere. From the free-body diagram in Fig. 6.9a, we see that the only forces acting on the sphere are the weight, mg, and the constraint force (or tension), $\boldsymbol{T}$. Now we resolve mg into a tangential component, $mg \sin\theta$, and a radial component, $mg \cos\theta$. Applying

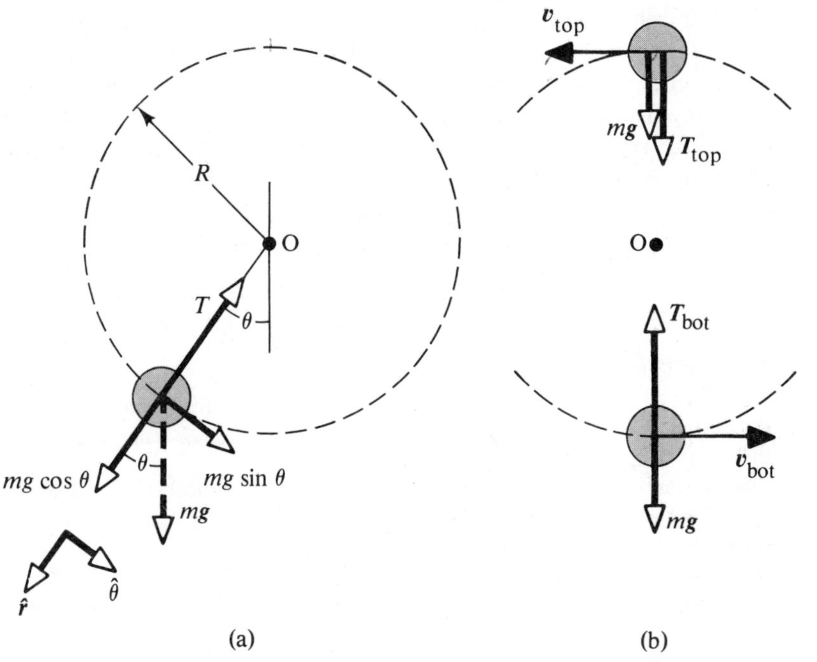

Figure 6.9 (Example 6.8) (a) Forces acting on a mass m connected to a string of length R and rotating in a vertical circle centered at O. (b) Forces acting on m when it is at the top and bottom of the circle. Note that the tension at the bottom is a maximum and the tension at the top is a minimum.

Newton's second law to the tangential motion gives

$$\Sigma F_t = mg\sin\theta = ma_t$$

$$(1) \qquad a_t = g\sin\theta$$

This component causes v to change in time, since $a_t = dv/dt$. Applying Newton's second law to the radial direction and noting that both $\boldsymbol{T}$ and $\boldsymbol{a}_r$ are directed toward O, we get

$$\Sigma F_r = T - mg\cos\theta = \frac{mv^2}{R}$$

$$(2) \qquad T = m\left(\frac{v^2}{R} + g\cos\theta\right)$$

Limiting cases: At the *top* of the path, where $\theta = 180°$, we see from (2) that since $\cos 180° = -1$,

$$T_{top} = m\left(\frac{v_{top}^2}{R} - g\right)$$

This is the *minimum* value of T. Note that at this point $a_t = 0$, and so the acceleration is radial and directed downward, as in Fig. 6.9b.

At the *bottom* of the path, where $\theta = 0$, again from (2) we see that since $\cos 0 = 1$,

$$T_{bot} = m\left(\frac{v_{bot}^2}{R} + g\right)$$

This is the *maximum* value of T. Again, at this point $a_t = 0$, and the acceleration is radial and directed upward.

Q7. Under what condition and at what position could the tension in the cord in Example 6.8 go to zero?

Q8. At what orientation of the system described in Example 6.8 would the cord most likely break if the average speed increased?

6.9 MOTION IN ACCELERATED FRAMES

When Newton's laws of motion were introduced in Chapter 5, we emphasized that the laws are valid when observations are made in an *inertial* frame of reference. In this section, we shall analyze how an observer in a noninertial frame of reference (one that is accelerating) would attempt to apply Newton's second law.

If a particle moves with an acceleration $\boldsymbol{a}$ relative to an observer in an inertial frame, then the inertial observer may use Newton's second law and correctly claim that $\Sigma \boldsymbol{F} = m\boldsymbol{a}$. If an observer in an accelerated frame (the noninertial observer) tries to apply Newton's second law to the motion of the particle, the noninertial observer must introduce *fictitious*, or *pseudo*, forces to make Newton's second law work in that frame. Sometimes, these fictitious forces are referred to as *inertial forces*. These forces "invented" by the noninertial observer *appear* to be real forces

Fictitious or inertial forces

in the noninertial frame. However, we emphasize that these fictitious forces *do not* exist when the motion is observed in an inertial frame. The fictitious forces are used only in a noninertial frame but *do not* represent "real" forces on the body. (By "real" forces, we mean the interaction of the body with its environment.) If the fictitious forces are properly defined in the noninertial frame, then the description of motion in this frame will be equivalent to the description by an inertial observer who considers only real forces. Usually, motions are analyzed using inertial reference frames, but there are cases in which a noninertial frame is more convenient.

Example 6.9

Linear accelerometer: A small sphere of mass m is hung from the ceiling of an accelerating boxcar, as in Fig. 6.10. According to the inertial observer at rest (Fig. 6.10a), the forces on the sphere are the tension $\mathbf{T}$ and the weight $m\mathbf{g}$. The inertial observer concludes that the acceleration of the sphere of mass m is the same as that of the boxcar and that this acceleration is provided by the horizontal component of $\mathbf{T}$. Also, the vertical component of $\mathbf{T}$ balances the weight. Therefore, the inertial observer writes Newton's second law as $\mathbf{T} + m\mathbf{g} = m\mathbf{a}$, which in component form becomes

$$\text{Inertial observer} \begin{cases} (1) & \Sigma F_x = T\sin\theta = ma \\ (2) & \Sigma F_y = T\cos\theta - mg = 0 \end{cases}$$

Thus, by solving (1) and (2) simultaneously, the inertial observer can determine the acceleration of the car through the relation

$$a = g\tan\theta \tag{6.13}$$

Therefore, since the deflection of the string from the vertical serves as a measure of the acceleration of the car, *a simple pendulum can be used as an accelerometer.*

According to the noninertial observer riding in the car, described in Fig. 6.10b, the sphere is at rest and the acceleration is zero. Therefore, the noninertial observer introduces a *fictitious force,* $-m\mathbf{a}$, to balance the horizontal component of $\mathbf{T}$ and claims that the net force on the sphere is *zero!* In this noninertial frame of reference, Newton's second law in component form gives

$$\text{Noninertial observer} \begin{cases} \Sigma F_x' = T\sin\theta - ma = 0 \\ \Sigma F_y' = T\cos\theta - mg = 0 \end{cases}$$

These expressions are equivalent to (1) and (2); therefore the noninertial observer gets the same mathematical results as the inertial observer. However, the physical interpretation of the deflection of the string *differs* in the two frames of reference.

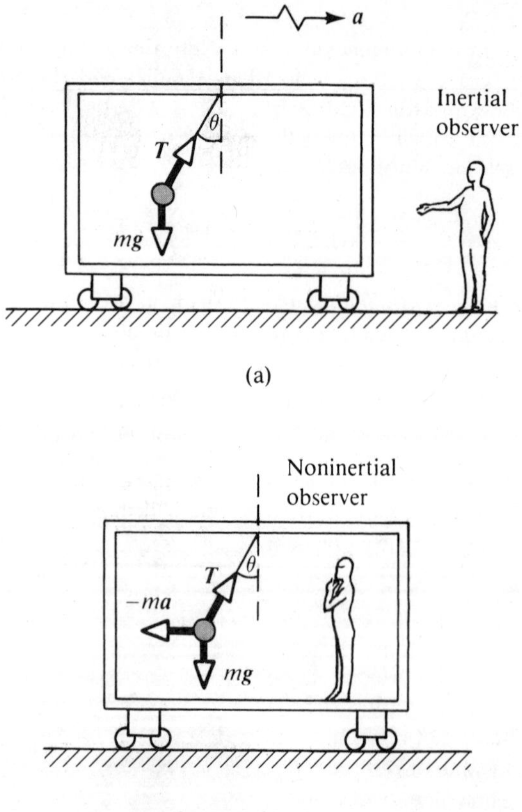

Figure 6.10 (Example 6.9) (a) A ball suspended from the ceiling of a boxcar accelerating to the right is deflected as shown. The inertial observer at rest outside the car claims that the acceleration of the ball is provided by the horizontal component of $\mathbf{T}$. (b) A noninertial observer riding in the car says that the net force on the ball is zero and that the deflection of the string from the vertical is due to a fictitious force, $-m\mathbf{a}$, which balances the horizontal component of $\mathbf{T}$.

Example 6.10

Fictitious force in a rotating system: An observer in a rotating system is another example of a noninertial observer. Suppose a block of mass m lying on a horizontal, frictionless turntable is connected to a string as in Fig. 6.11. According to an inertial observer, if the block rotates uniformly, it undergoes a centripetal acceleration v^2/r, where v is its tangential speed. The inertial observer concludes that this centripetal acceleration is provided by the force of tension in the string, $\mathbf{T}$, and writes Newton's second law $T = mv^2/r$. (The weight vector and the normal force balance and are not included in Fig. 6.11.)

According to a noninertial observer attached to the turntable, the block is at rest. Therefore, in applying Newton's second law, this observer introduces a fictitious *outward* force called the *centrifugal force,* of magnitude mv^2/r. According to the noninertial observer, this "centrifugal" force balances the force of tension and therefore $T - mv^2/r = 0$.

You should be careful when using fictitious forces to describe physical phenomena. Remember that fictitious forces, such as centrifugal force, are used *only* in noninertial, or accelerated, frames of reference. When solving problems, it is generally best to use an inertial frame.

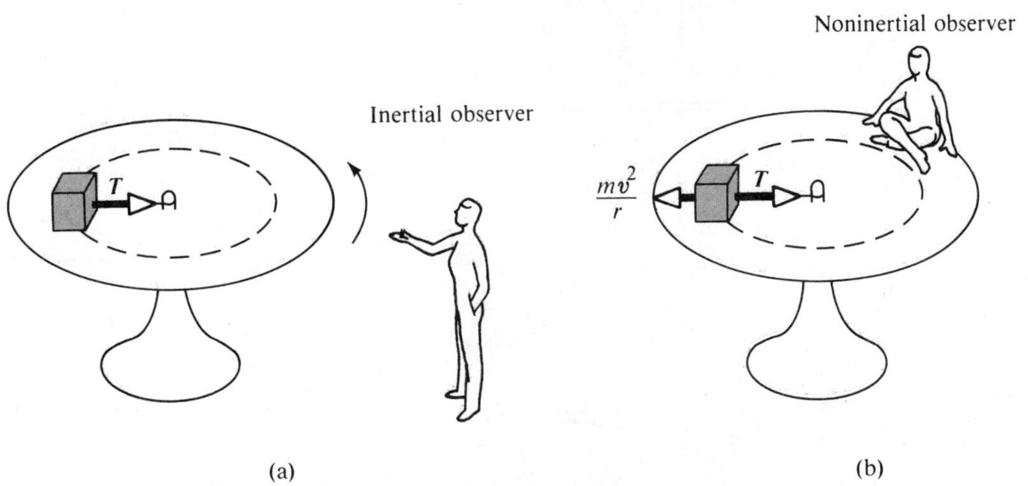

Figure 6.11 (Example 6.10) A block of mass m connected to a string tied to the center of a rotating turntable. (a) The inertial observer claims that the centripetal force is provided by the force of tension, $\boldsymbol{T}$. (b) The noninertial observer claims that the block is not accelerating and therefore introduces a fictitious centrifugal force mv^2/r, which acts outward and balances the tension.

Q9. Because the earth rotates about its axis and about the sun, it is a noninertial frame of reference. Assuming the earth is a uniform sphere, why would the *apparent weight* of an object be greater at the poles than at the equator?

Q10. How would you explain the mysterious force that pushes a rider toward the side of a car as the car rounds a corner?

Q11. When an airplane does a "loop-the-loop" in a vertical plane, at what point would the pilot appear to be heaviest? What is the constraint force acting on the pilot?

Q12. If a pail of water is swung in a vertical circle, under what conditions will no water spill from the pail? What keeps the water from spilling out?

6.10 MOTION IN THE PRESENCE OF RESISTIVE FORCES

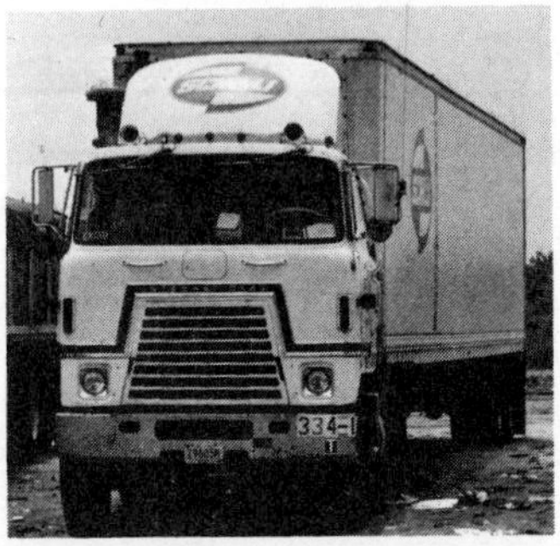

The high cost of fuel has prompted many truck owners to install wind deflectors on their cabs to reduce air drag. (Photo by Lloyd Black)

In the previous chapter we discussed the force of sliding friction, that is, the resistive force on an object moving along a rough, solid surface. Such forces are nearly independent of velocity, and matters are simplified by assuming them to be constant in magnitude. Now let us consider what happens when an object moves through a liquid or gas. In such situations, the medium exerts a resistive force $\boldsymbol{R}$ on the object. The magnitude of this force depends on the velocity of the object, and its direction is always opposite the direction of motion. The magnitude of the resistive force is generally found to increase with increasing velocity. Some examples of such resistive forces are the air resistance to flying airplanes and moving cars and the viscous forces on objects moving through a liquid.

In general, the resistive force can have a complicated velocity dependence. In the following discussions, we will consider two situations. First, we will assume that the

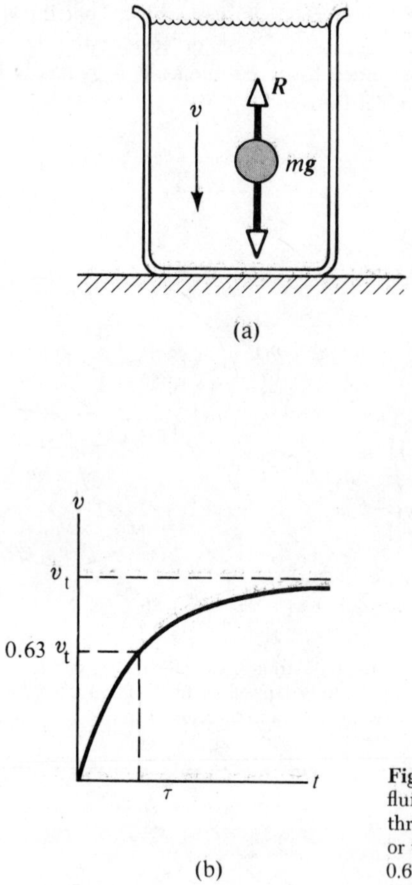

Figure 6.12 (a) A small sphere falling through a viscous fluid. (b) The velocity-time graph for an object falling through a viscous medium. The object reaches a maximum, or terminal, velocity, v_t, and τ is the time it takes to reach $0.63 v_t$.

resistive force is proportional to the velocity. Objects falling through a fluid and very small objects, such as particles of dust moving through air, experience such a force. Second, we will treat situations for which the resistive force is assumed to be proportional to the square of the speed of the object. Large objects, such as a skydiver moving through air in free fall in the presence of gravity, experience such a force.

Resistive Force Proportional to Velocity

When an object moves at low speeds through a viscous medium, it experiences a resistive drag force that is proportional to the velocity of the object. Let us assume that the resistive force, $\mathbf{R}$, has the form

$$\mathbf{R} = -b\mathbf{v} \tag{6.14}$$

where $\mathbf{v}$ is the velocity of the object and b is a constant that depends on the properties of the medium and on the shape and dimensions of the object. If the object is a sphere of radius r, then b is found to be proportional to r.

Consider a sphere of mass m released from rest in a fluid, as in Fig. 6.12a. Assuming the only forces acting on the sphere are the resistive force, $-b\mathbf{v}$, and the weight, $m\mathbf{g}$, let us describe its motion.[3]

[3]There is also a *buoyant* force, which is constant and equal to the weight of the displaced fluid. This will only change the weight of the sphere by a constant factor. We shall discuss such buoyant forces in Chapter 15.

Applying Newton's second law to the vertical motion, choosing the downward direction to be positive, and noting that $\Sigma F_y = mg - bv$, we get

$$mg - bv = m\frac{dv}{dt}$$

where the acceleration is downward. Simplifying the above expression gives

$$\frac{dv}{dt} = g - \frac{b}{m}v \qquad (6.15)$$

Equation 6.15 is called a *differential equation,* and the methods of solving such an equation may not be familiar to you as yet. However, note that initially, when $v = 0$, the resistive force is zero and the acceleration, dv/dt, is simply g. As t increases, the resistive force increases and the acceleration *decreases.* Eventually, the acceleration becomes zero when the resistive force *equals* the weight. At this point, the body continues to move with zero acceleration, and it reaches its *terminal velocity,* v_t. The terminal velocity can be obtained from Eq. 6.15 by setting $a = dv/dt = 0$. This gives

$$mg - bv_t = 0 \qquad \text{or} \qquad v_t = mg/b$$

The expression for v that satisfies Eq. 6.15 with $v = 0$ at $t = 0$ is

$$v = \frac{mg}{b}(1 - e^{-bt/m}) = v_t(1 - e^{-t/\tau}) \qquad (6.16)$$

This function is plotted in Fig. 6.12b. The time $\tau = m/b$ is the time it takes the object to reach 63 percent of its terminal velocity. We can check that Eq. 6.16 is a solution to Eq. 6.15 by direct differentiation:

$$\frac{dv}{dt} = \frac{d}{dt}\left(\frac{mg}{b} - \frac{mg}{b}e^{-bt/m}\right) = -\frac{mg}{b}\frac{d}{dt}e^{-bt/m} = ge^{-bt/m}$$

Substituting this expression and Eq. 6.16 into Eq. 6.15 shows that our solution satisfies the differential equation.

Example 6.11 *Sphere Falling in Oil*

A small sphere of mass 2 g is released from rest in a large cylinder filled with oil. The sphere reaches a terminal velocity of 5 cm/s. Determine the constant τ and the speed of the sphere as a function of time.

Solution: Since the terminal velocity is given by $v_t = mg/b$, the constant b is given by

$$b = \frac{mg}{v_t} = \frac{(2\text{ g})(980\text{ cm/s}^2)}{5\text{ cm/s}} = 392\text{ g/s}$$

Therefore, the time τ is given by

$$\tau = \frac{m}{b} = \frac{2\text{ g}}{392\text{ g/s}} = 5.10 \times 10^{-3}\text{ s}$$

The velocity as a function of time can be calculated using Eq. 6.16:

$$v(t) = v_t(1 - e^{-t/\tau})$$

Since $v_t = 5$ cm/s and $1/\tau = 196\text{ s}^{-1}$, we have

$$v(t) = 5(1 - e^{-196t})\text{ cm/s}$$

Air Drag

We have seen that an object moving through a fluid experiences a resistive drag force. If the object is small and moves at low speeds, the drag force is proportional to the velocity, as we have already discussed. However, for larger objects moving at high speeds through air, such as airplanes, skydivers, and baseballs, the drag force is approximately proportional to the *square* of the speed. In these situations, the magnitude of the drag force can be expressed as

$$R = \frac{1}{2}C\rho Av^2 \qquad (6.17)$$

Drag force for motion in air

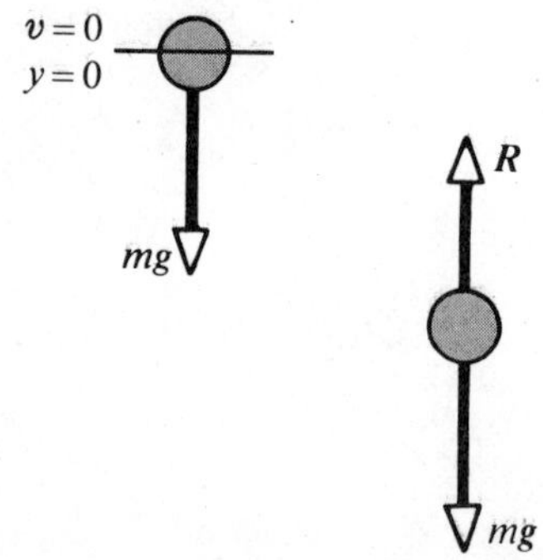

Figure 6.13 An object falling through air experiences a drag force, R, and the force of gravity, mg. The object reaches terminal velocity when the net force is zero, that is, when $R = mg$. Before this occurs, the acceleration varies with speed according to Eq. 6.19.

where ρ is the density of air, A is the cross-sectional area of the falling object measured in a plane perpendicular to its motion, and C is a dimensionless empirical quantity called the *drag coefficient.* The drag coefficient has a value of about 0.5 for spherical objects but can be as high as 1 for irregularly shaped objects.

Consider an airplane in flight experiencing such a drag force. Equation 6.17 shows that the drag force is proportional to the density of air and hence decreases with decreasing air density. Since air density decreases with increasing altitude, the drag force on a jet airplane flying at a given speed must also decrease with increasing altitude. Furthermore, if the plane's speed is doubled, the drag force increases by a factor of 4. In order to maintain constant speed, the propulsive force also increases by a factor of 4 and the power required (force times speed) must increase by a factor of 8.

Now let us analyze the motion of a mass in free fall subject to an upward air drag force given by $R = \frac{1}{2}C\rho Av^2$. Suppose a mass m is released from rest from the position $y = 0$ as in Fig. 6.13. The mass experiences two external forces: the weight, mg, downward and the drag force, $\mathbf{R}$, upward. Hence, the magnitude of the net force is given by

$$F_{\text{net}} = mg - \frac{1}{2}C\rho Av^2 \tag{6.18}$$

Substituting $F_{\text{net}} = ma$ into Eq. 6.18, we find that the mass has a downward acceleration of magnitude

$$a = g - \left(\frac{C\rho A}{2m}\right)v^2 \tag{6.19}$$

Again, we can calculate the terminal velocity, v_t, using the fact that when the weight is balanced by the drag force, the net force is zero and therefore the acceleration is zero. Setting $a = 0$ in Eq. 6.19 gives

$$g - \left(\frac{C\rho A}{2m}\right)v_t^2 = 0$$

Terminal velocity

$$v_t = \sqrt{\frac{2mg}{C\rho A}} \tag{6.20}$$

Using this expression, we can determine how the terminal speed depends on the dimensions of the object. Suppose the object is a sphere of radius r. In this case, $A \sim r^2$ and $m \sim r^3$ (since the mass is proportional to the volume). Therefore, $v_t \sim \sqrt{r}$. That is, as r increases, the terminal speed increases with the square root of the radius.

Table 6.2 lists the terminal speeds for several objects falling through air.

By spreading his arms and legs out from the sides of his body and by keeping the plane of his body parallel to the ground, a skydiver will experience maximum air drag resulting in a specific terminal speed. (U.S. Air Force Photo)

TABLE 6.2 Terminal Speed for Various Objects Falling Through Air

OBJECT	*MASS* (kg)	*AREA* (m²)	v_t (m/s)[a]
Skydiver	75	0.7	60
Baseball (radius 3.66 cm)	0.145	4.2×10^{-3}	33
Golf ball (radius 2.1 cm)	0.046	1.4×10^{-3}	32
Hailstone (radius 0.5 cm)	4.8×10^{-4}	7.9×10^{-5}	14
Raindrop (radius 0.2 cm)	3.4×10^{-5}	1.3×10^{-5}	9

[a]The drag coefficient, C, is assumed to be 0.5 in each case.

Q13. A skydiver in free fall reaches terminal velocity. After the parachute is opened, what parameters change to decrease this terminal velocity?

6.11 SUMMARY

Newton's law of universal gravitation states that every particle in the universe attracts every other particle with a force that is directly proportional to the product of their masses and inversely proportional to the square of the distance between them:

$$F = G\frac{m_1 m_2}{r^2} \tag{6.1}$$

Universal law of gravity

where $G = 6.673 \times 10^{-11}\ \text{N} \cdot \text{m}^2/\text{kg}^2$ is the *gravitational constant.*

For an object above the earth's surface, the acceleration of gravity varies as the inverse square of the distance from the center of the earth. As the separation distance approaches infinity, the acceleration of gravity goes to zero. At the earth's surface, the *acceleration of gravity* is given by

$$g = G\frac{M_e}{R_e^2} \tag{6.4}$$

Acceleration of gravity

where M_e and R_e are the mass and radius of the earth, respectively.

The *electrostatic force* between two charges, q_1 and q_2, separated by a distance r is given by

$$F_e = k\frac{q_1 q_2}{r^2} \tag{6.6}$$

Electrostatic force

This is called *Coulomb's law,* and the constant $k = 8.99 \times 10^9\ \text{N} \cdot \text{m}^2/\text{C}^2$ is the *coulomb constant.* Electrostatic forces are repulsive if the charges have the same sign and are attractive if the charges have opposite signs. Between charged elementary particles, electrostatic forces are much stronger than gravitational forces.

The *strong nuclear interaction* binds the protons and neutrons in atomic nuclei. This interaction is highly attractive in the range 10^{-14} to 10^{-15} m, where it predominates over the coulombic repulsive forces between protons. The interaction is repulsive at very short range. The *weak nuclear force* is an interaction that is short-range and produces instability in certain nuclei.

Newton's second law applied to a particle moving in uniform circular motion states that the net force in the radial direction must equal the product of the mass and the centripetal acceleration:

$$\Sigma \mathbf{F} = m\mathbf{a}_r = -\frac{mv^2}{r}\hat{\mathbf{r}} \tag{6.10}$$

Uniform circular motion

The force that provides the centripetal acceleration could be, for example, the force of gravity (as in satellite motion), the force of friction, or the force of tension (as in a string). A particle moving in nonuniform circular motion has both a centripetal (or radial) acceleration and a nonzero tangential component of acceleration. In the case of a particle rotating in a vertical circle, the tangential acceleration is provided by gravity.

An observer in a noninertial (accelerated) frame of reference must introduce *fictitious forces* when applying Newton's second law in that frame. If these fictitious forces are properly defined, the description of motion in the noninertial frame will be equivalent to that made by an observer in an inertial frame. However, the observers in the two different frames will not agree on the causes of the motion.

Fictitious forces

Terminal velocity

A body moving through a liquid or gas experiences a *resistive force* that is velocity dependent. This resistive force, which opposes the motion, generally increases with velocity. The force depends on the shape of the body and the properties of the medium through which the body is moving. In the limiting case for a falling body, when the resistive force equals the weight ($a = 0$), the body reaches its *terminal velocity*.

EXERCISES

Section 6.1 through Section 6.4

1. Two identical, isolated particles, each of mass 2 kg, are separated by a distance of 30 cm. What is the magnitude of the gravitational force of one particle on the other?

2. A 200-kg mass and a 500-kg mass are separated by a distance of 0.40 m. (a) Find the net gravitational force due to these masses acting on a 50-kg mass placed midway between them. (b) At what position (other than infinitely remote ones) would the 50-kg mass experience a net force of zero?

3. Three 5-kg masses are located at the corners of an equilateral triangle having sides 0.25 m in length. Determine the magnitude and direction of the resultant gravitational force on one of the masses due to the other two masses.

4. Two stars of masses M and $4M$ are separated by a distance d. Determine the location of a point measured from M at which the net force on a third mass would be zero.

5. Four particles are located at the corners of a rectangle as in Fig. 6.14. Determine the x and y components of the resultant force acting on the particle of mass m.

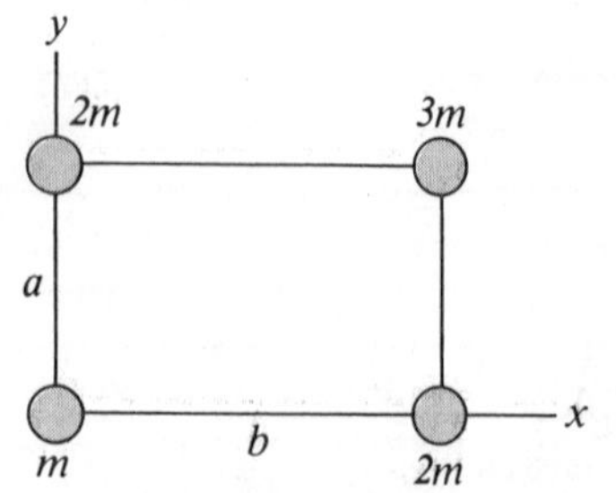

Figure 6.14 (Exercise 5).

6. Calculate the acceleration of gravity at a point that is a distance R_e above the surface of the earth, where R_e is the radius of the earth.

7. Using the data given in Fig. 6.3, determine a vector expression for the resultant force on the 6-kg mass. What is the magnitude of this force?

Section 6.5 Electrostatic Forces

8. Two electrons are separated by a distance of 2×10^{-5} m. What are the magnitude and direction of the electrostatic force on one of the electrons due to the other?

9. The electron and proton of the hydrogen atom are separated by a mean distance of 0.53×10^{-10} m. What is the magnitude of the coulombic force of attraction between the electron and proton for this separation?

10. Three charges, q, $-q$, and $-2q$, are located on the x axis at the positions $x = -d$, 0, and $+d$, respectively, where q is in C and d is in m. Determine the magnitude and direction of the resultant electrostatic force on the charge q.

11. Four equal charges, each with a charge $q = 8\ \mu$C, are located at the corners of a square of sides 0.5 m. Find the resultant electrostatic force on one of the charges due to the other three charges.

12. A charge, $-q$, is located at the origin, and a second charge, $3q$, is placed on the y axis at $y = d$. At what position (other than ∞) would a third charge experience no net force?

13. Two identical uncharged spheres each have a mass of 50 kg. How many electrons would have to be transferred to each sphere (in equal numbers) in order that the electrostatic repulsive force balance their gravitational attraction?

14. Determine the separation between two electrons such that the electrostatic force on either one is equal to its weight on earth. (The mass of an electron is 9.11×10^{-31} kg.)

Section 6.6 Nuclear Forces

15. When two protons are separated by a distance of 10^{-15} m, the strong nuclear attractive force is about 10 times stronger than the electrostatic force between them. Estimate the magnitude of the strong nuclear force for this separation.

Section 6.7 Newton's Second Law Applied to Uniform Circular Motion

16. A 3-kg mass attached to a light string rotates in circular motion on a horizontal, frictionless table.

The radius of the circle is 0.8 m, and the string can support a mass of 25 kg before breaking. What range of speeds can the mass have before the string breaks?

17. A coin is placed 20 cm from the center of a rotating, horizontal turntable. The coin is observed to slip when its speed is 50 cm/s. (a) What provides the centripetal force when the coin is stationary relative to the turntable? (b) What is the coefficient of static friction between the coin and the turntable?

18. What centripetal force is required to keep a 2-kg mass moving in a circle of radius 0.4 m at a speed of 3 m/s?

19. A satellite of mass 600 kg is in a circular orbit about the earth at an altitude equal to the earth's mean radius. Find (a) the satellite's orbital speed, (b) the period of its revolution, and (c) the gravitational force acting on it.

20. A highway curve has a radius of 150 m and is designed for a traffic speed of 40 mi/h (17.9 m/s). (a) If the curve is not banked, determine the minimum coefficient of friction between the car and the road. (b) At what angle should the curve be banked if friction is neglected (Fig. 6.8)?

21. In the Bohr model of the hydrogen atom (Example 6.7), find (a) the centripetal force acting on the electron as it revolves in a circular orbit of radius 0.53×10^{-10} m, (b) the centripetal acceleration of the electron, and (c) the number of revolutions per second made by the electron.

Section 6.8 Nonuniform Circular Motion

22. A pail of water is rotated in a vertical circle of radius 1 m (the approximate length of a person's arm). What is the minimum speed of the pail at the top of the circle if no water is to spill out?

23. A roller-coaster vehicle has a mass of 500 kg when fully loaded with passengers (Fig. 6.15). (a) If the vehicle has a speed of 20 m/s at point A, what is the force of the track on the vehicle at this point? (b) What is the maximum speed the vehicle can have at B in order that it remain on the track?

24. A ball attached to the end of a string 0.8 m in length is rotated in a vertical circle (Fig. 6.9). Determine the minimum speed of the ball at the top of its path if it maintains a circular path. (Note that below this speed, the tension in the string is zero at the top.)

25. A 0.5-kg mass attached to the end of a string swings in a vertical circle of radius $R = 2$ m (Fig. 6.9). When $\theta = 25°$, the speed of the mass is 8 m/s. At this instant, find (a) the tension in the string, (b) the tangential and radial components of acceleration, and (c) the magnitude of the total acceleration.

26. A 40-kg child sits in a conventional swing of length 3 m, supported by two chains. If the child's speed is 6 m/s at the lowest point, find (a) the tension in each chain at the lowest point and (b) the force of the seat on the child at the lowest point. (Neglect the mass of the seat.)

Section 6.9 Motion in Accelerated Frames

27. A ball is suspended from the ceiling of a moving car by a string 25 cm in length. An observer in the car notes that the ball deflects 6 cm from the vertical toward the rear of the car. What is the acceleration of the car?

28. A 0.5-kg object is suspended from the ceiling of an accelerating boxcar as in Fig. 6.10. If $a = 3$ m/s^2, find (a) the angle that the string makes with the vertical and (b) the tension in the string.

29. A 5-kg mass attached to a spring scale rests on a smooth, horizontal surface as in Fig. 6.16. The spring scale, attached to the front end of a boxcar, reads 18 N when the car is in motion. (a) If the spring scale reads zero when the car is at rest, determine the acceleration of the car. (b) What will

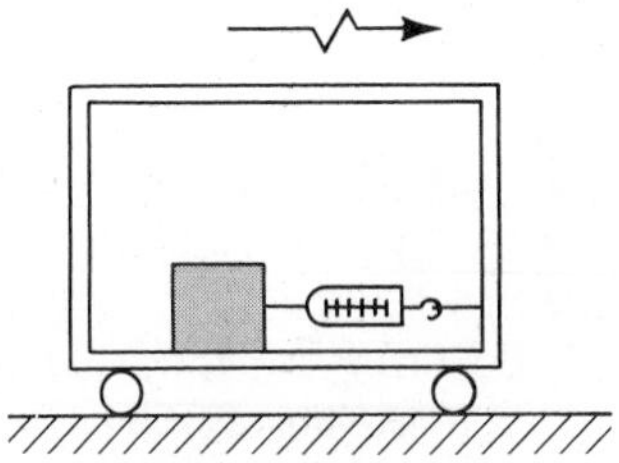

Figure 6.16 (Exercise 29).

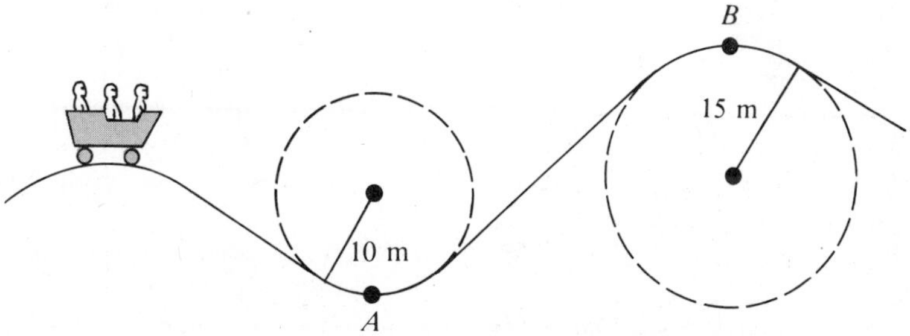

Figure 6.15 (Exercise 23).

the spring scale read if the car moves with constant velocity? (c) Describe the forces on the mass as observed by someone in the car and by someone at rest outside the car.

30. A block is attached to a string, which in turn is connected to a peg at the center of a rotating turntable, as in Fig. 6.11. If the turntable is *rough*, describe the forces on the block as observed by (a) someone on the turntable and (b) an observer at rest relative to the turntable. (c) For a given velocity, does the tension in the string increase, decrease, or remain the same as the turntable is made smoother?

Section 6.10 Motion in the Presence of Resistive Forces

31. A small, spherical bead of mass 3 g is released from rest at $t = 0$ in a bottle of liquid shampoo. The terminal velocity, v_t, is observed to be 2 cm/s. Find (a) the value of the constant b in Eq. 6.16, (b) the time, τ, it takes to reach $0.63v_t$, and (c) the value of the retarding force when the bead reaches terminal velocity.

32. A skydiver of mass 80 kg jumps from a slow-moving aircraft and reaches a terminal speed of 50 m/s. (a) What is the acceleration of the skydiver when her speed is 30 m/s? What is the drag force on the diver when her speed is (b) 50 m/s and (c) 30 m/s?

PROBLEMS

1. A small turtle, appropriately named "Dizzy," is placed on a horizontal, rotating turntable at a distance of 20 cm from its center. Dizzy's mass is 50 g, and the coefficient of static friction between his feet and the turntable is 0.3. Find (a) the *maximum* number of revolutions per second the turntable can have if Dizzy is to remain stationary relative to the turntable and (b) Dizzy's speed and radial acceleration when he is on the verge of slipping.

2. The pilot of an airplane executes a constant-speed loop-the-loop maneuver in a vertical plane. The speed of the airplane is 300 mi/h, and the radius of the circle is 1200 ft. (a) What is the pilot's apparent weight at the lowest point if his true weight is 160 lb? (b) What is his apparent weight at the highest point? (c) Describe how the pilot could experience weightlessness if both the radius and velocity can be varied. (Note that his apparent weight is equal to the force of the seat on his body.)

3. A 4-kg mass is attached to a *horizontal* rod by two strings, as in Fig. 6.17. The strings are under tension when the rod rotates about its axis. If the speed of the mass is constant and equal to 4 m/s, find the tension in the string when the mass is (a) at its lowest point, (b) in the horizontal position, and (c) at its highest point.

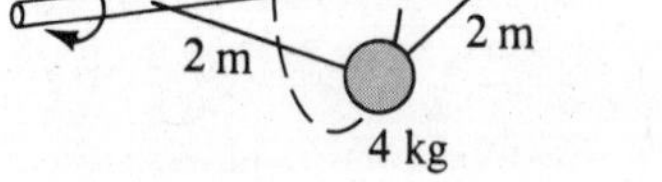

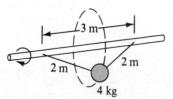

Figure 6.17 (Problems 3 and 4).

4. Suppose the rod in the system shown in Fig. 6.17 is made *vertical* and rotates about this axis. If the mass rotates at a constant speed of 6 m/s in a horizontal plane, determine the tensions in the upper and lower strings.

5. Two small identical spheres each of mass 5 g are suspended from the ceiling by two threads each of length 30 cm (Fig. 6.18). The spheres have the same electric charge, q, and they are in equilibrium when the threads make an angle $\theta = 15°$ with the vertical. (a) What is the charge on each sphere? (b) What is the tension in each thread?

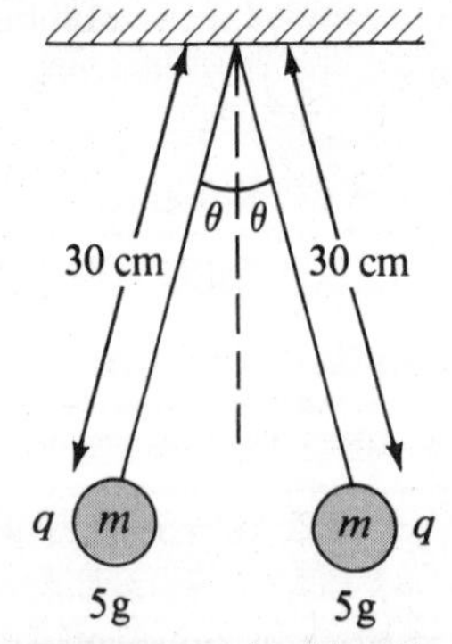

Figure 6.18 (Problem 5).

6. A small sphere of mass m and electric charge q_1 is suspended by a light thread. A second sphere carrying a charge q_2 is placed directly below the first sphere, a distance d away (Fig. 6.19). (a) If both spheres are positively charged, determine the value for d when the tension in the thread goes to zero. (b) If the spheres are oppositely charged, determine the tension in the thread and show that it can never be zero.

7. A car rounds a banked curve as in Fig. 6.8. The radius of curvature of the road is R, the banking angle is θ, and the coefficient of static friction is μ. (a) Determine the *range* of speeds the car can have

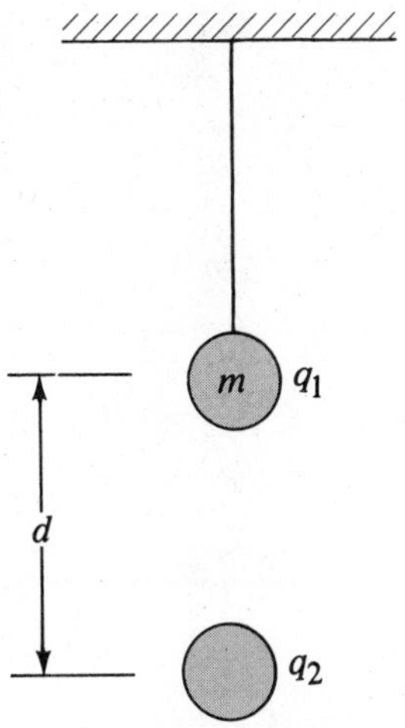

Figure 6.19 (Problem 6).

without slipping up or down the road. (b) Find the minimum value for μ such that the minimum speed is zero. (c) What is the range of speeds possible if $R = 100$ m, $\theta = 10°$, and $\mu = 0.1$ (slippery conditions)?

8. Because of the earth's rotation about its axis, a point on the equator experiences a centripetal acceleration of 0.034 m/s^2, while a point at the poles experience no centripetal acceleration. (a) Show that at the equator the gravitational force on an object (the true weight) must *exceed* the object's apparent weight. (b) What is the apparent weight at the equator and at the poles of a person having a mass of 75 kg? (Assume the earth is a uniform sphere and take $g = 9.800$ m/s^2.)

9. An amusement park ride consists of a large vertical cylinder that spins about its axis fast enough that any person inside is held up against the wall when the floor drops away (Fig. 6.20). The coefficient of static friction between the person and the wall is μ_s, and the radius of the cylinder is R. (a) Show that the *maximum* period of revolution necessary to keep the person from falling is $T = (4\pi^2 R\mu_s/g)^{1/2}$. (b) Obtain a numerical value for T if $R = 4$ m and $\mu_s = 0.4$. How many revolutions per minute does the cylinder make?

10. The acceleration of gravity at an altitude h is given by Eq. 6.5. If $h \ll R_e$, show that the acceleration of gravity at h is given *approximately* by

$$g' \approx g\left(1 - 2\frac{h}{R_e}\right)$$

(Hint: Start with Eq. 6.5, and use the binomial expansion for the denominator.)

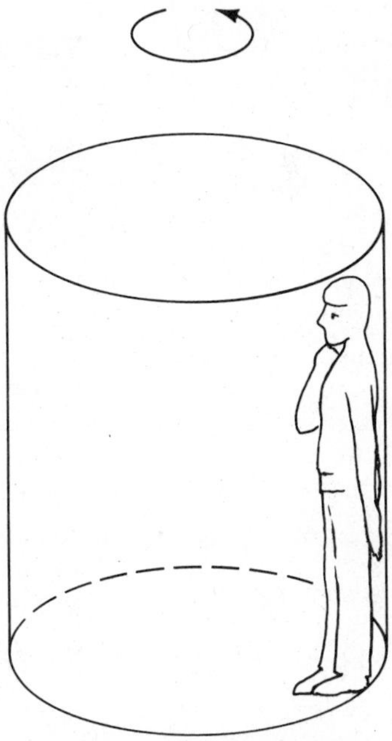

Figure 6.20 (Problem 9).

7 Work and Energy

Dave Coskey, Villanova University

7.1 INTRODUCTION

The concept of energy is perhaps one of the most important physical concepts in both contemporary science and engineering practice. In terms of everyday usage, we think of energy in terms of the cost of fuel for transportation and heating, electricity for lights and appliances, and the foods we consume. However, these ideas do not really define energy. They only tell us that fuels are needed to do a job and that those fuels provide us with something we call *energy*.

Energy can take on various forms, including mechanical energy, electromagnetic energy, chemical energy, thermal (or heat) energy, and nuclear energy. However, the various forms of energy are related to each other through the fact that when energy is transformed from one form to another, *it is always conserved*. This is the point that makes the energy concept so useful. That is, if an isolated system loses energy in some form, then the law of conservation of energy says that the system will gain an equal amount of energy in other forms. For example, when an electric motor is connected to a battery, electrical energy is converted to mechanical energy. It is the transformation of energy from one form to another that provides the basis for unifying the disciplines of physics, chemistry, biology, geology, and astronomy.

In this chapter, we shall be concerned with only the mechanical form of energy. We shall see that the concepts of work and energy can be applied to the dynamics of a mechanical system without resorting to Newton's laws. However, it is important to note that the work-energy concepts are based upon Newton's laws and therefore do not involve any new physical principles.

Although the approach we shall use provides the same results as Newton's laws in describing the motion of a mechanical system, the general ideas of the work-energy concept can be applied to a wide range of phenomena in the fields of electromagnetism and atomic and nuclear physics. In addition, in a complex situation the "energy approach" can often provide a much simpler analysis than the direct application of $\boldsymbol{F} = m\boldsymbol{a}$.

This alternative method of describing motion is especially useful when the force acting on a particle is not constant. In this case, the acceleration is not constant, and we cannot apply the simple kinematic equations we developed in Chapter 3. Often, a particle in nature is subject to a force that varies with the position of the particle. Such forces include gravitational forces and the force exerted on a body attached to a spring. We shall describe techniques for treating such systems with the help of an extremely important development called the *work-energy theorem,* which is the central topic of this chapter. We begin by defining work, a concept that provides a link between the concepts of force and energy. In Chapter 8, we shall discuss the law of conservation of energy and apply it to various problems.

7.2 WORK DONE BY A CONSTANT FORCE

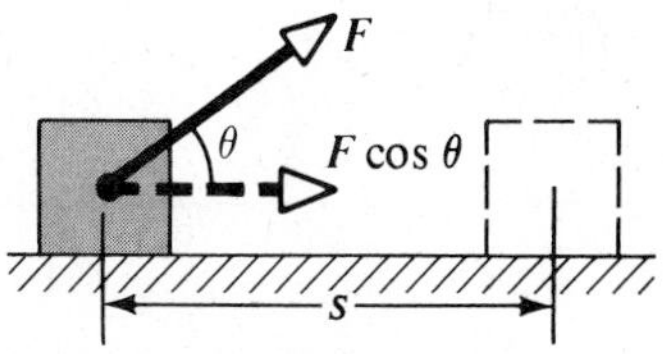

Figure 7.1 If an object undergoes a displacement **s**, the work done by the force **F** is $(F\cos\theta)s$.

Consider an object that undergoes a displacement **s** along a straight line under the action of a constant force **F**, which makes an angle θ with **s**, as in Fig. 7.1. *The work done by the constant force is defined as the product of the component of the force in the direction of the displacement and the magnitude of the displacement.* Since the component of **F** in the direction of **s** is $F\cos\theta$, the work W done by **F** is given by

Work done by a constant force

$$W \equiv (F\cos\theta)s \tag{7.1}$$

According to this definition, work is done by **F** on an object under the following conditions: (1) the object must undergo a displacement and (2) **F** must have a nonzero component in the direction of **s**. From the first condition, we see that a force does no work on an object if the object does not move ($s = 0$). For example, if a person pushes against a brick wall, a force is exerted on the wall but the person does no work since the wall is fixed. However, the person's muscles are contracting in the process so that *internal* energy is being used up.

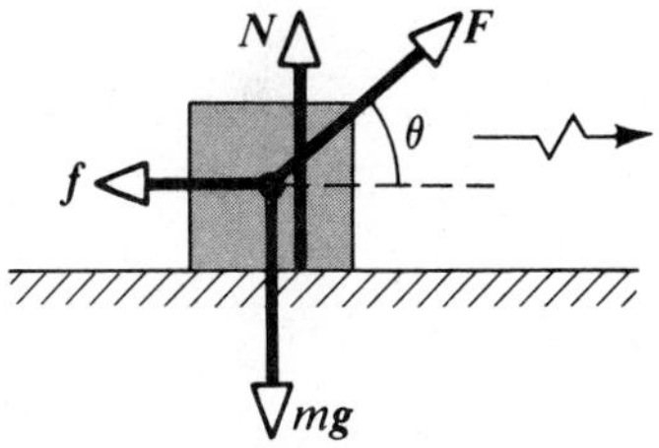

Figure 7.2 When an object is displaced horizontally on a rough surface, the normal force, **N**, and the weight, *m***g**, do *no* work. The work done by **F** is $(F\cos\theta)s$, and the work done by the frictional force is $-fs$.

From the second condition, note that the work done by a force is also zero when the force is perpendicular to the displacement, since $\theta = 90°$ and $\cos 90° = 0$. For example, in Fig. 7.2, both the work done by the normal force and the work done by the force of gravity are zero since both forces are perpendicular to the displacement and have zero components in the direction of **s**.

The sign of the work also depends on the direction of **F** relative to **s**. The work done by the applied force is positive when the component $F\cos\theta$ is in the *same direction* as the displacement. For example, when an object is lifted, the work done by the applied force is positive since the lifting force is upward, that is, in the same direction as the displacement. When the component $F\cos\theta$ is in the direction *opposite* the displacement, W is *negative*. A common example in which W is negative is the work done by a frictional force when a body slides over a rough surface. If the force of sliding friction is **f**, and the body undergoes a linear displacement **s**, the work done by the frictional force is

Work done by a sliding frictional force

$$W_{\mathrm{f}} = -fs \tag{7.2}$$

where the negative sign comes from the fact that $\theta = 180°$ and $\cos 180° = -1$.

Finally, if an applied force **F** acts along the direction of the displacement, then $\theta = 0$, and $\cos 0 = 1$. In this case, Eq. 7.1 gives

Work done when F is along s

$$W = Fs \tag{7.3}$$

Work is a scalar quantity, and its units are force multiplied by length. Therefore, the SI unit of work is the newton · meter (N · m). Another name for the N · m is the joule (J). The units of work in the cgs and British engineering systems are dyne · cm and lb · ft, respectively. These are summarized in Table 7.1. Note that 1 J = 10^7 ergs.

Work is a scalar quantity

Since work is a scalar quantity we can add the individual works done by separate forces to get the *net* work done. For instance, if there are three forces contributing to the work done, there would be three terms in the sum, each corresponding to the work done by a given force. The following example illustrates this point.

TABLE 7.1 Units of Work in the Three Common Systems of Measurement

SYSTEM	*UNIT OF WORK*	*NAME OF COMBINED UNIT*
SI	newton · meter (N · m)	joule (J)
cgs	dyne · centimeter (dyne · cm)	erg
British engineering	pound · foot (lb · ft)	foot · pound (ft · lb)

Example 7.1

A box is dragged across a rough floor by a constant force of magnitude 50 N. The force makes an angle of 37° with the horizontal. A frictional force of 10 N retards the motion, and the box is displaced a distance of 3 m to the right. (a) Calculate the work done by the 50-N force.

From Eq. 7.1 and given that $F = 50$ N, $\theta = 37°$, and $s = 3$ m,

$$W_F = (F\cos\theta)s = (50\ \text{N})(\cos 37°)(3\ \text{m}) = 120\ \text{N}\cdot\text{m} = 120\ \text{J}$$

Note that the vertical component of $\boldsymbol{F}$ does no work.

(b) Calculate the work done by the frictional force.

$$W_f = -fs = (-10\ \text{N})(3\ \text{m}) = -30\ \text{N}\cdot\text{m} = -30\ \text{J}$$

(c) Determine the net work done on the box by all forces acting on it.

Since the normal force, $\boldsymbol{N}$, and the force of gravity, $m\boldsymbol{g}$, are both perpendicular to the displacement, they do no work. Therefore, the net work done on the box is the sum of (a) and (b):

$$W_{\text{net}} = W_F + W_f = 120\ \text{J} - 30\ \text{J} = 90\ \text{J}$$

Later we shall show that the net work done on the body equals the change in kinetic energy, which establishes the physical significance of W_{net}.

Q1. When a particle rotates in a circle, a *centripetal force* acts on it directed toward the center of rotation. Why is it that this force does no work on the particle?

Q2. Explain why the work done by the force of sliding friction is negative when an object undergoes a displacement on a rough surface.

7.3 THE SCALAR PRODUCT OF TWO VECTORS

We have defined work as a *scalar* quantity given by the product of the magnitude of the displacement and the component of the force in the direction of the displacement. It is convenient to express Eq. 7.1 in terms of a *scalar product* of the two vectors $\boldsymbol{F}$ and $\boldsymbol{s}$. We write this scalar product $\boldsymbol{F}\cdot\boldsymbol{s}$. Because of the dot symbol used, the scalar product is often called the *dot product*. Thus, we can express Eq. 7.1 as a scalar product:

Work expressed as a dot product

$$W = \boldsymbol{F}\cdot\boldsymbol{s} = Fs\cos\theta \qquad (7.4)$$

In other words, $\boldsymbol{F}\cdot\boldsymbol{s}$ (read F dot s) is a shorthand notation for $Fs\cos\theta$.

In general, the scalar product of any two vectors **A** *and* **B** *is defined as a scalar quantity equal to the product of the magnitudes of the two vectors and the cosine of the smaller angle between them.* That is, the scalar product (or dot product) of **A** and **B** is defined by the relation

Scalar product of any two vectors *A* and *B*

$$\boldsymbol{A}\cdot\boldsymbol{B} \equiv AB\cos\theta \qquad (7.5)$$

where θ is the smaller angle between **A** and **B**, as in Fig. 7.3, A is the magnitude of **A**, and B is the magnitude of **B**. Note that **A** and **B** need not have the same units.

Note in Fig. 7.3 that $B\cos\theta$ is the projection of **B** onto **A**. Therefore, the definition of $\boldsymbol{A}\cdot\boldsymbol{B}$ as given by Eq. 7.5 can be considered as the product of the magnitude of **A**

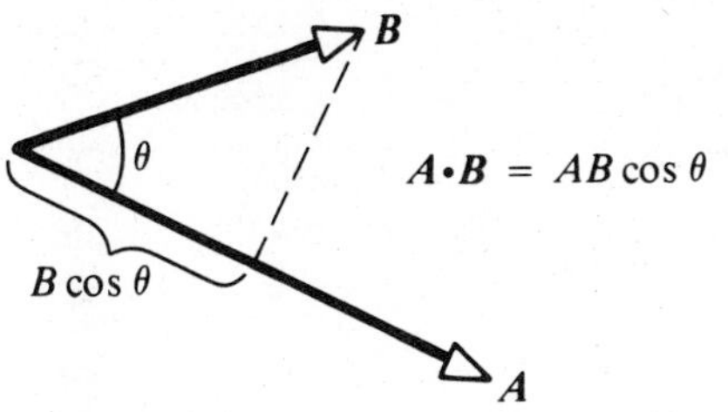

Figure 7.3 The scalar product $\boldsymbol{A} \cdot \boldsymbol{B}$ equals the magnitude of $\boldsymbol{A}$ multiplied by the projection of $\boldsymbol{B}$ onto $\boldsymbol{A}$.

and the projection of $\boldsymbol{B}$ onto $\boldsymbol{A}$.[1] From Eq. 7.5 we also note that the scalar product is *commutative*. That is,

The order of the dot product can be reversed

$$\boldsymbol{A} \cdot \boldsymbol{B} = \boldsymbol{B} \cdot \boldsymbol{A} \tag{7.6}$$

Finally, the scalar product obeys the *distributive law of multiplication*, so that

$$\boldsymbol{A} \cdot (\boldsymbol{B} + \boldsymbol{C}) = \boldsymbol{A} \cdot \boldsymbol{B} + \boldsymbol{A} \cdot \boldsymbol{C} \tag{7.7}$$

The dot product is simple to evaluate from Eq. 7.5 when $\boldsymbol{A}$ is either perpendicular or parallel to $\boldsymbol{B}$. If $\boldsymbol{A}$ is perpendicular to $\boldsymbol{B}$ ($\theta = 90°$), then $\boldsymbol{A} \cdot \boldsymbol{B} = 0$. Also, $\boldsymbol{A} \cdot \boldsymbol{B} = 0$ in the more trivial case when either $\boldsymbol{A}$ or $\boldsymbol{B}$ is zero. If $\boldsymbol{A}$ and $\boldsymbol{B}$ point in the same direction ($\theta = 0°$), then $\boldsymbol{A} \cdot \boldsymbol{B} = AB$. If $\boldsymbol{A}$ and $\boldsymbol{B}$ point in opposite directions ($\theta = 180°$), then $\boldsymbol{A} \cdot \boldsymbol{B} = -AB$. Note that the scalar product is negative when $90° < \theta < 180°$.

The unit vectors $\boldsymbol{i}$, $\boldsymbol{j}$, and $\boldsymbol{k}$, which were defined in Chapter 2, lie in the positive x, y, and z directions, respectively, of a right-handed coordinate system. Therefore, it follows from the definition of $\boldsymbol{A} \cdot \boldsymbol{B}$ that the scalar products of these unit vectors are given by

Dot products of unit vectors

$$\boldsymbol{i} \cdot \boldsymbol{i} = \boldsymbol{j} \cdot \boldsymbol{j} = \boldsymbol{k} \cdot \boldsymbol{k} = 1 \tag{7.8a}$$

$$\boldsymbol{i} \cdot \boldsymbol{j} = \boldsymbol{i} \cdot \boldsymbol{k} = \boldsymbol{j} \cdot \boldsymbol{k} = 0 \tag{7.8b}$$

Since two vectors $\boldsymbol{A}$ and $\boldsymbol{B}$ can be expressed in component form as

$$\boldsymbol{A} = A_x\boldsymbol{i} + A_y\boldsymbol{j} + A_z\boldsymbol{k}$$
$$\boldsymbol{B} = B_x\boldsymbol{i} + B_y\boldsymbol{j} + B_z\boldsymbol{k}$$

using Eqs. 7.8a and 7.8b reduces the scalar product of $\boldsymbol{A}$ and $\boldsymbol{B}$ to

$$\boldsymbol{A} \cdot \boldsymbol{B} = A_xB_x + A_yB_y + A_zB_z \tag{7.9}$$

In the special case where $\boldsymbol{A} = \boldsymbol{B}$, we see that

$$\boldsymbol{A} \cdot \boldsymbol{A} = A_x^2 + A_y^2 + A_z^2 = A^2$$

Example 7.2

The vectors $\boldsymbol{A}$ and $\boldsymbol{B}$ are given by $\boldsymbol{A} = 2\boldsymbol{i} + 3\boldsymbol{j}$ and $\boldsymbol{B} = -\boldsymbol{i} + 2\boldsymbol{j}$. (a) Determine the scalar product $\boldsymbol{A} \cdot \boldsymbol{B}$.

$$\begin{aligned}\boldsymbol{A} \cdot \boldsymbol{B} &= (2\boldsymbol{i} + 3\boldsymbol{j}) \cdot (-\boldsymbol{i} + 2\boldsymbol{j}) \\ &= -2\boldsymbol{i} \cdot \boldsymbol{i} + 2\boldsymbol{i} \cdot 2\boldsymbol{j} - 3\boldsymbol{j} \cdot \boldsymbol{i} + 3\boldsymbol{j} \cdot 2\boldsymbol{j} \\ &= -2 + 6 = 4\end{aligned}$$

where we have used the fact that $\boldsymbol{i} \cdot \boldsymbol{j} = \boldsymbol{j} \cdot \boldsymbol{i} = 0$. The same result is obtained using Eq. 7.9 directly, where $A_x = 2$, $A_y = 3$, $B_x = -1$, and $B_y = 2$.

(b) Find the angle θ between $\boldsymbol{A}$ and $\boldsymbol{B}$.
The magnitudes of $\boldsymbol{A}$ and $\boldsymbol{B}$ are given by

$$A = \sqrt{A_x^2 + A_y^2} = \sqrt{(2)^2 + (3)^2} = \sqrt{13}$$
$$B = \sqrt{B_x^2 + B_y^2} = \sqrt{(-1)^2 + (2)^2} = \sqrt{5}$$

[1]This is equivalent to stating that $\boldsymbol{A} \cdot \boldsymbol{B}$ equals the product of the magnitude of $\boldsymbol{B}$ and the projection of $\boldsymbol{A}$ onto $\boldsymbol{B}$.

Using Eq. 7.5 and the result from (a) gives

$$\cos\theta = \frac{\boldsymbol{A}\cdot\boldsymbol{B}}{AB} = \frac{4}{\sqrt{13}\,\sqrt{5}} = \frac{4}{\sqrt{65}}$$

$$\theta = \cos^{-1}\frac{4}{8.06} = 60.3°$$

Example 7.3

A particle moving in the xy plane undergoes a displacement $\boldsymbol{s} = (2\boldsymbol{i} + 3\boldsymbol{j})$ m while a constant force $\boldsymbol{F} = (5\boldsymbol{i} + 2\boldsymbol{j})$ N acts on the particle. (a) Calculate the magnitude of the displacement and the force.

$$s = \sqrt{x^2 + y^2} = \sqrt{(2)^2 + (3)^2} = \sqrt{13}\text{ m}$$

$$F = \sqrt{F_x^{\,2} + F_y^{\,2}} = \sqrt{(5)^2 + (2)^2} = \sqrt{29}\text{ N}$$

(b) Calculate the work done by the force $\boldsymbol{F}$. Substituting the expressions for $\boldsymbol{F}$ and $\boldsymbol{s}$ into Eq. 7.4 and using Eq. 7.8, we get

$$W = \boldsymbol{F}\cdot\boldsymbol{s} = (5\boldsymbol{i} + 2\boldsymbol{j})\cdot(2\boldsymbol{i} + 3\boldsymbol{j})\text{ N}\cdot\text{m}$$
$$= 5\boldsymbol{i}\cdot 2\boldsymbol{i} + 2\boldsymbol{j}\cdot 3\boldsymbol{j} = 16\text{ N}\cdot\text{m} = 16\text{ J}$$

(c) Calculate the angle θ between $\boldsymbol{F}$ and $\boldsymbol{s}$. Using Eq. 7.5 and the results to (a) and (b) gives

$$\theta = \cos^{-1}\frac{\boldsymbol{F}\cdot\boldsymbol{s}}{Fs} = \cos^{-1}\frac{16}{\sqrt{29}\,\sqrt{13}} = 34.5°$$

Q3. Is there any direction associated with the dot product of two vectors?

Q4. If the dot product of two vectors is positive, does this imply that the vectors must have positive components?

7.4 WORK DONE BY A VARYING FORCE: THE ONE-DIMENSIONAL CASE

Consider an object being displaced along the x axis under the action of a varying force, as in Fig. 7.4. The object is displaced along the x axis from $x = x_i$ to $x = x_f$. In such a situation, we cannot use $W = (F\cos\theta)s$ to calculate the work done by the force, since this relationship applies only when $\boldsymbol{F}$ is constant in magnitude and direction. However, if we imagine that the object undergoes a very small displacement Δx, described in Fig. 7.4a, then the x component of the force, F_x, is approximately constant over this interval and we can express the work done by the force for this small displacement as

$$\Delta W = F_x \Delta x \tag{7.10}$$

Note that this is just the area of the shaded rectangle in Fig. 7.4a. Now, if we imagine that the F_x versus x curve is divided into a large number of such intervals, as in Fig. 7.4a, then the total work done for the displacement from x_i to x_f is approximately equal to the sum of a large number of such terms:

$$W \approx \sum_{x_i}^{x_f} F_x\,\Delta x$$

Work done equals the area under the F versus the x curve

If the displacements are allowed to approach zero, then the number of terms in the sum increases without limit, but the value of the sum approaches a definite value equal to the *true area* under the curve bounded by F_x and the x axis. As you probably have learned in the calculus, this limit of the sum is called an *integral* and is represented by

$$\lim_{\Delta x\to 0}\sum_{x_i}^{x_f} F_x\,\Delta x = \int_{x_i}^{x_f} F_x\,dx$$

The limits on the integral, $x = x_i$ to $x = x_f$, define what is called a *definite integral.* (An *indefinite integral* represents the limit of a sum over a yet-to-be-specified interval. Appendix B.6 gives a brief description of integration.) This definite integral

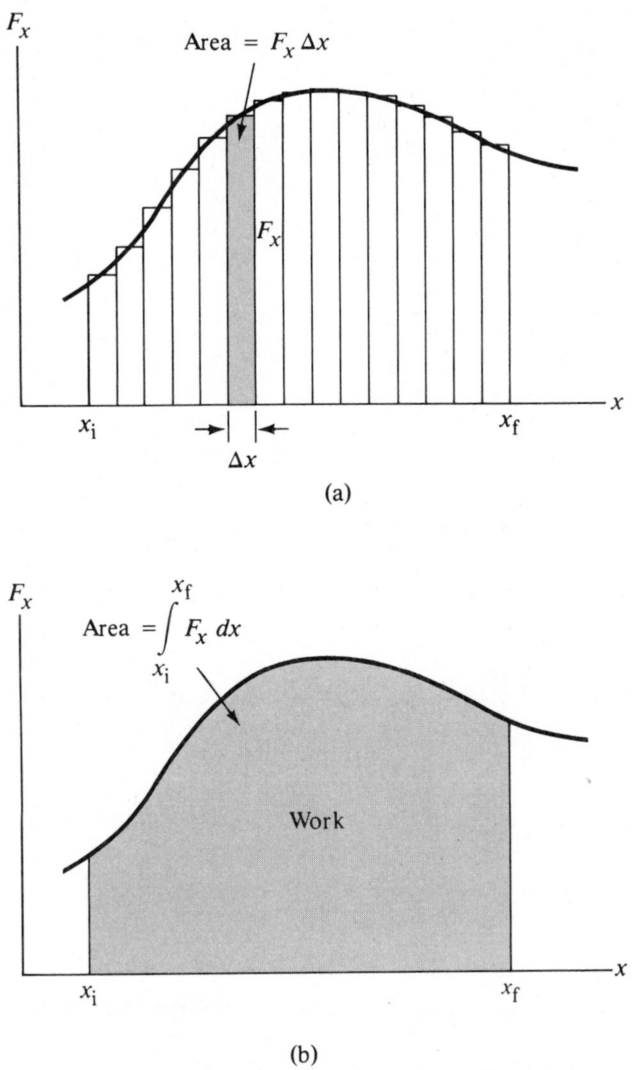

Figure 7.4 (a) The work done by the force F_x for the small displacement Δx is $F_x \Delta x$, which equals the area of the shaded rectangle. The total work done for the displacement x_i to x_f is approximately equal to the sum of the areas of all the rectangles. (b) The work done by the variable force F_x as the particle moves from x_i to x_f is *exactly* equal to the area under this curve.

is numerically equal to the area under the F_x versus x curve between x_i and x_f. Therefore, we can express the work done by F_x for the displacement of the object from x_i to x_f as

$$W = \int_{x_i}^{x_f} F_x \, dx \tag{7.11}$$

Work done by a force

Note that this equation reduces to Eq. 7.1 when $F_x = F\cos\theta$ is constant.

If more than one force acts on the object, the total work done is just the work done by the resultant force. If we express the resultant force in the x direction as ΣF_x (a vector sum), then the *net work* done as the object moves from x_i to x_f is

$$W_{\text{net}} = \int_{x_i}^{x_f} (\Sigma F_x) dx \tag{7.12}$$

Net work done by the resultant force

Example 7.4

A force acting on an object varies with x as shown in Fig. 7.5. Calculate the work done by the force as the object moves from $x = 0$ to $x = 6$ m.

Solution: The work done by the force is equal to the total area under the curve from $x = 0$ to $x = 6$ m. This area is equal to the area of the rectangular section from $x = 0$ to $x = 4$ m plus the area of the triangular section from $x = 4$ m to $x = 6$ m. The area of the rectangle is (4)(5) N · m = 20 J, and the area of the triangle is $\frac{1}{2}$(2)(5) N · m = 5 J. Therefore, the total work done is 25 J.

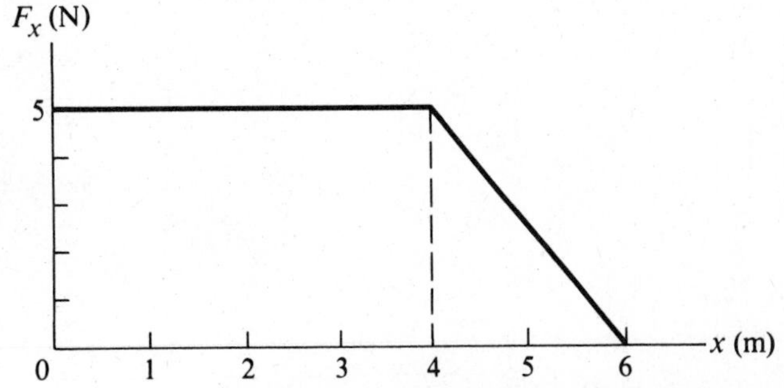

Figure 7.5 (Example 7.4) The force acting on a particle is constant for the first 4 m of motion and then decreases linearly with x from $x = 4$ m to $x = 6$ m. The net work done by this force is the area under this curve.

Work Done by a Spring

A common physical system for which the force varies with position is shown in Fig. 7.6. A body on a horizontal, smooth surface is connected to a helical spring. If the spring is stretched or compressed a small distance from its unstretched, or equilibrium, configuration and then released, the spring will exert a force on the body given by

Spring force (Hooke's law)

$$F_s = -kx \tag{7.13}$$

where x is the displacement of the body from its unstretched ($x = 0$) position and k is a positive constant called the *force constant* of the spring. As we learned in Chapter 5, this force law for springs is known as *Hooke's law*. The value of k is a measure of the stiffness of the spring. Stiff springs have large k values, and soft springs have small k values.

The negative sign in Eq. 7.13 signifies that the force exerted by the spring is always directed *opposite* the displacement. For example, when $x > 0$ as in Fig. 7.6a, the spring force is to the left, or negative. When $x < 0$ as in Fig. 7.6c, the spring force is to the right, or positive. Of course, when $x = 0$ as in Fig. 7.6b, the spring is unstretched and $F_s = 0$. Since the spring force always acts toward the equilibrium position, it is sometimes called a *restoring force*. Once the mass is displaced some distance x_m from equilibrium and then released, it will move from $-x_m$ through zero to $+x_m$. The details of the ensuing oscillating motion will be given in Chapter 13.

Suppose that the block is pushed to the left a distance x_m from equilibrium, as in Fig. 7.6c, and then released. Let us calculate the *work done by the spring* as the body moves from $x_i = -x_m$ to $x_f = 0$. Applying Eq. 7.11, we get

$$W_s = \int_{x_i}^{x_f} F_s\,dx = \int_{-x_m}^{0} (-kx)\,dx = \frac{1}{2}kx_m^2 \tag{7.14a}$$

That is, the work done by the spring is positive since the spring force is in the same direction as the displacement (both are to the right). However, if we consider the work done by the spring as the body moves from $x_i = 0$ to $x_f = x_m$, we find that

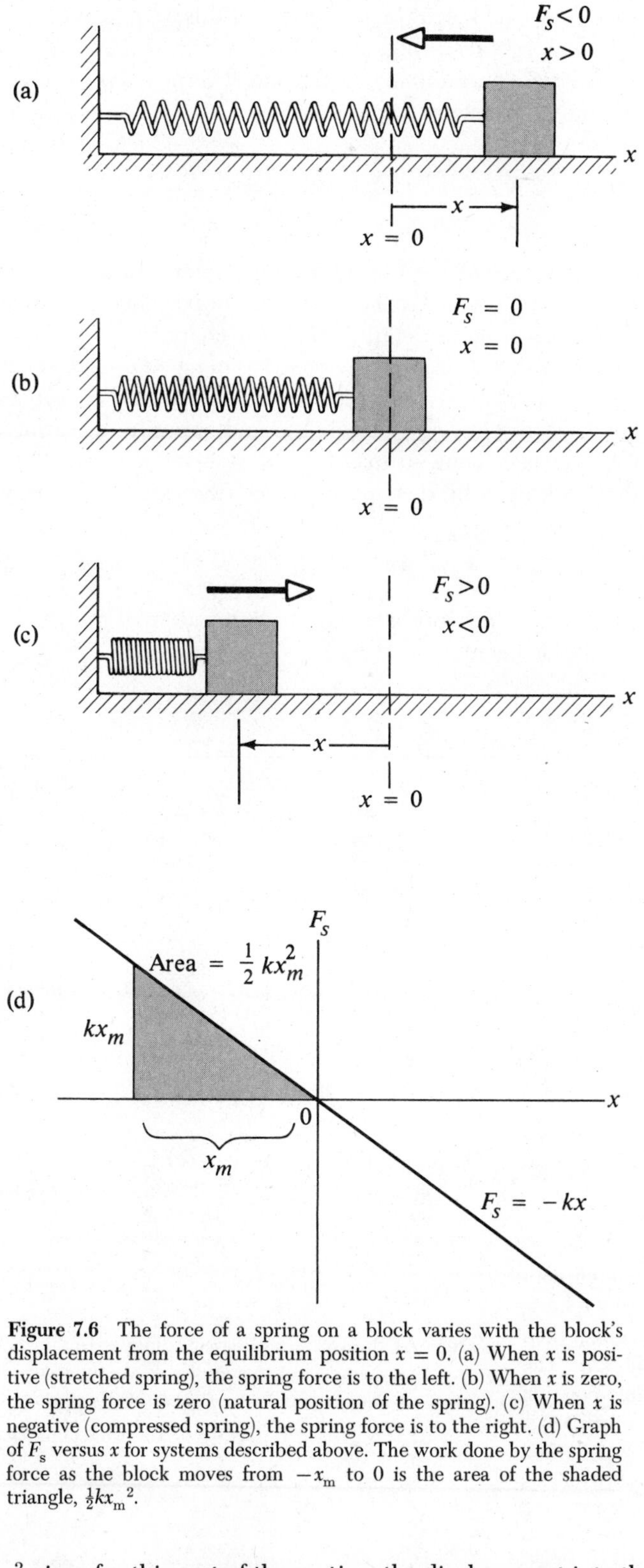

Figure 7.6 The force of a spring on a block varies with the block's displacement from the equilibrium position $x = 0$. (a) When x is positive (stretched spring), the spring force is to the left. (b) When x is zero, the spring force is zero (natural position of the spring). (c) When x is negative (compressed spring), the spring force is to the right. (d) Graph of F_s versus x for systems described above. The work done by the spring force as the block moves from $-x_m$ to 0 is the area of the shaded triangle, $\frac{1}{2}kx_m^2$.

$W_s = -\frac{1}{2}kx_m^2$, since for this part of the motion, the displacement is to the right and the spring force is to the left. Therefore, the *net* work done by the spring as the body moves from $x_i = -x_m$ to $x_f = x_m$ is *zero.*

If we plot F_s versus x as in Fig. 7.6d, we arrive at the same results. Note that the work calculated in Eq. 7.14a is equivalent to the area of the shaded triangle in

Fig. 7.6d, with base x_m and height kx_m. The area of this triangle is $\frac{1}{2}kx_m^2$, which does equal the work done by the spring, Eq. 7.14a.

If the mass undergoes an *arbitrary* displacement from $x = x_i$ to $x = x_f$, the work done by the spring is given by

Work done by a spring

$$W_s = \int_{x_i}^{x_f} (-kx)\,dx = \frac{1}{2}kx_i^2 - \frac{1}{2}kx_f^2 \qquad (7.14b)$$

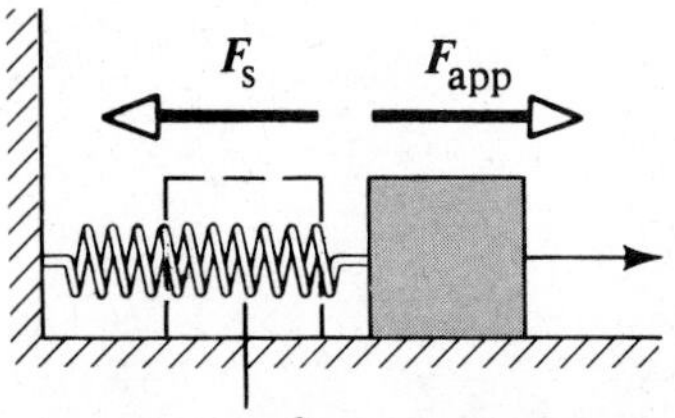

Figure 7.7 A block being pulled to the right on a frictionless surface by a force $\mathbf{F}_{app}$ from $x = 0$ to $x = x_m$. If the process is carried out very slowly, the applied force is equal to and opposite the spring force at all times.

From this equation, we see that the work done is zero for any motion that ends where it began ($x_i = x_f$). We shall make use of this important result in describing the motion of this system in more detail in the next chapter.

Now let us consider the work done by an *external agent* in *very slowly* stretching a spring from $x_i = 0$ to $x_f = x_m$, as in Fig. 7.7. This work can be easily calculated by noting that the *applied force,* $\mathbf{F}_{app}$, is equal to and opposite the spring force, $\mathbf{F}_s$, at any value of the displacement, so that $F_{app} = -(-kx) = kx$. Therefore, the work done by this applied force (the external agent) is given by

$$W_{F_{app}} = \int_0^{x_m} F_{app}\,dx = \int_0^{x_m} kx\,dx = \frac{1}{2}kx_m^2$$

You should note that this work is equal to the negative of the work done by the spring for this displacement.

Example 7.5

A spring with a force constant of 80 N/m is compressed a distance of 3.0 cm from equilibrium on a smooth, horizontal surface, as in Fig. 7.6c. Calculate the work done by the spring as the block moves from $x_i = -3.0$ cm to its unstretched position, $x_f = 0$.

Solution: Using Eq. 7.14a with $x_m = -3.0\text{ cm} = -3 \times 10^{-2}$ m, we get

$$W = \frac{1}{2}kx_m^2 = \frac{1}{2}\left(80\,\frac{\text{N}}{\text{m}}\right)(-3 \times 10^{-2}\text{ m})^2$$
$$= 3.6 \times 10^{-2}\text{ J}$$

Example 7.6 *Measuring k for a Spring*

A common technique used to evaluate the force constant of a spring is described in Fig. 7.8. The spring is hung vertically as shown in Fig. 7.8a. A body of mass m is then attached to the lower end of the spring as in Fig. 7.8b. The spring stretches a distance d from its equilibrium position under the action of the "load" mg. Since the spring force is upward, it must balance the weight mg downward when the system is at rest. In this case, we can apply Hooke's law to give $|\mathbf{F}_s| = kd = mg$, or

$$k = mg/d$$

For example, if a spring is stretched a distance of 2.0 cm by a mass of 0.55 kg, the force constant of the spring is

$$k = \frac{mg}{d} = \frac{(0.55\text{ kg})(9.8\text{ m/s}^2)}{2.0 \times 10^{-2}\text{ m}} = 2.7 \times 10^2\text{ N/m}$$

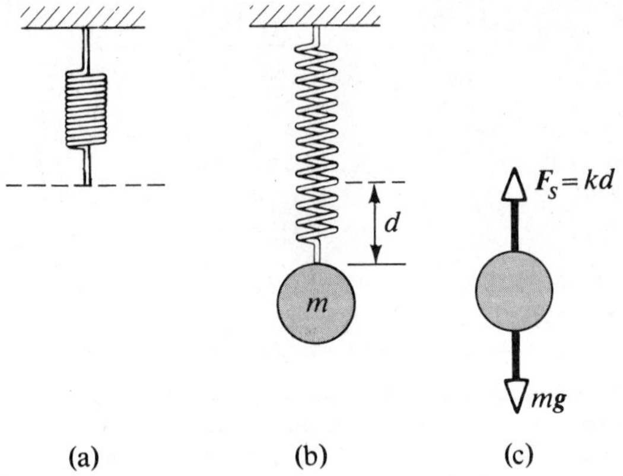

Figure 7.8 (Example 7.6) Determination of the force constant of a helical spring. The elongation d of the spring is due to the weight mg. Since the spring force upward balances the weight, it follows that $k = mg/d$.

Q5. As the load on a spring hung vertically is increased, one would not expect the F_s versus x curve to always remain linear as in Fig. 7.6d. Explain qualitatively what you would expect for this curve as m is increased.

Example 7.7

A sports car on a horizontal surface is pushed by a horizontal force that varies with position according to the graph shown in Fig. 7.9. Determine an approximate value for the total work done in moving the car from $x = 0$ to $x = 20$ m.

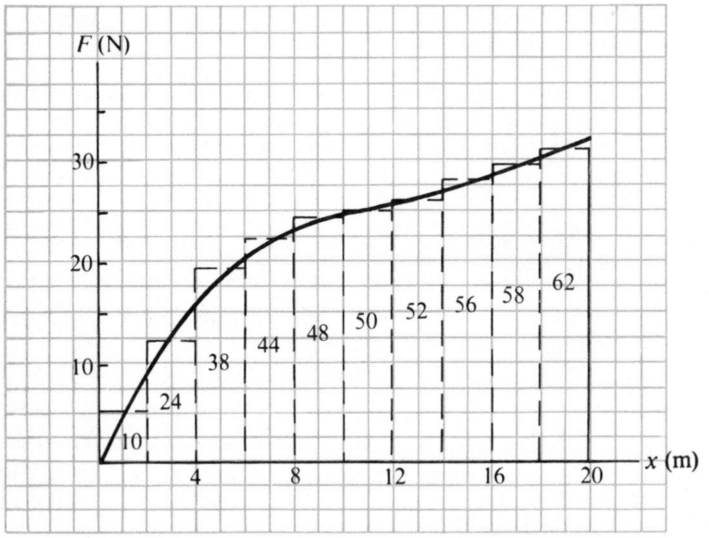

Figure 7.9 (Example 7.7) A graph of force versus position for a car moving along the x axis. The numbers within the rectangles represent the work done (area of rectangle) during that interval.

Solution: We can obtain the result from the graph by dividing the total displacement into many small displacements. For simplicity, we choose to divide the total displacement into ten consecutive displacements, each 2 m in length, as shown in Fig. 7.9. The work done during each small displacement is *approximately* equal to the area of the dotted rectangle. For example, the work done for the first displacement, from $x = 0$ to $x = 2$ m, is the area of the smallest rectangle, (2 m)(5 N) = 10 J; the work done for the second displacement, from $x = 2$ m to $x = 4$ m, is the area of the second rectangle, (2 m)(12 N) = 24 J. Continuing in this fashion, we get the areas indicated in Fig. 7.9, the *sum* of which gives the total work done from $x = 0$ to $x = 20$ m. The result is

$$W_{\text{total}} \approx 442 \text{ J}$$

The accuracy of the result will of course improve as the widths of the intervals are made smaller.

7.5 WORK AND KINETIC ENERGY

In Chapter 5 we found that a particle accelerates when the resultant force on it is not zero. Consider a situation in which a constant force $\boldsymbol{F}_x$ acts on a particle of mass m moving in the x direction. Newton's second law states that $\boldsymbol{F}_x = m\boldsymbol{a}$, where $\boldsymbol{a}$ is constant since $\boldsymbol{F}_x$ is constant. If the particle is displaced from $x_i = 0$ to $x_f = s$, the work done by the force $\boldsymbol{F}_x$ is

$$W = F_x s = (ma_x)s \tag{7.15}$$

However, in Chapter 3 we found that the following relationships are valid when a particle undergoes constant acceleration:

$$s = \frac{1}{2}(v_i + v_f)t \qquad a_x = \frac{v_f - v_i}{t}$$

where v_i is the velocity at $t = 0$ and v_f is the velocity at time t. Substituting these expressions into Eq. 7.15 gives

$$W = m\left(\frac{v_f - v_i}{t}\right)\frac{1}{2}(v_i + v_f)t$$

$$W = \frac{1}{2}mv_f^2 - \frac{1}{2}mv_i^2 \tag{7.16}$$

The product of one half the mass and the square of the speed is defined as the kinetic energy *of the particle.* That is, the kinetic energy, K, of a particle of mass m and speed v is defined as

Kinetic energy is energy associated with the motion of a body

$$K \equiv \frac{1}{2}mv^2 \tag{7.17}$$

This expression is valid in the nonrelativistic limit, that is, when $v \ll c$.

Since the magnitude of the linear momentum of the particle is given by $p = mv$, the kinetic energy is sometimes written

$$K = \frac{1}{2}mv^2 = \frac{(mv)^2}{2m} = \frac{p^2}{2m} \tag{7.18}$$

Kinetic energy is a scalar quantity and has the same units as work. For example, a 1-kg mass moving with a speed of 4.0 m/s has a kinetic energy of 8.0 J. We can think of kinetic energy as energy associated with the motion of a body. It is often convenient to write Eq. 7.16

Work-energy theorem

$$W = K_f - K_i = \Delta K \tag{7.19}$$

Work done equals the change in kinetic energy

That is, *the work done by the constant force* **F** *in displacing a particle equals the change in kinetic energy of the particle.* The change here means the final minus the initial value of the kinetic energy.

Equation 7.19 is an important result known as the *work-energy theorem.* This theorem was derived for the case where the force is constant, but we can show that it is valid even when the force is varying: If the resultant force acting on a body in the x direction is ΣF_x, then Newton's second law states that $\Sigma F_x = ma$. Thus, we can use Eq. 7.12 and express the net work done as

$$W_{net} = \int_{x_i}^{x_f} (\Sigma F_x)\, dx = \int_{x_i}^{x_f} ma\, dx$$

Since the resultant force varies with x, the acceleration and velocity also depend on x. We can now use the following chain rule to evaluate W_{net}:

$$a = \frac{dv}{dt} = \frac{dv}{dx}\frac{dx}{dt} = v\frac{dv}{dx}$$

Substituting this into the expression for W gives

$$W_{net} = \int_{x_i}^{x_f} mv\frac{dv}{dx}\, dx = \int_{v_i}^{v_f} mv\, dv = \frac{1}{2}mv_f^{\,2} - \frac{1}{2}mv_i^{\,2}$$

Note that the limits of the integration were changed because the variable was changed from x to v.

The work-energy theorem given by Eq. 7.19 is also valid in the more general case when the force varies in direction and magnitude while the particle moves along an arbitrary curved path in three dimensions. In this situation, we express the work as

General expression for work done by a force *F*

$$W = \int_i^f \boldsymbol{F} \cdot d\boldsymbol{s} \tag{7.20}$$

where the limits i and f represent the initial and final coordinates of the particle. The integral given by Eq. 7.20 is called a *line integral.* Since the infinitesimal displacement vector can be expressed as $d\boldsymbol{s} = dx\boldsymbol{i} + dy\boldsymbol{j} + dz\boldsymbol{k}$ and since $\boldsymbol{F} = F_x\boldsymbol{i} +$

$F_y \mathbf{j} + F_z \mathbf{k}$, Eq. 7.20 reduces to

$$W = \int_{x_i}^{x_f} F_x\, dx + \int_{y_i}^{y_f} F_y\, dy + \int_{z_i}^{z_f} F_z\, dz \tag{7.21}$$

This is the general expression that is used to calculate the work done by a force when a particle undergoes a displacement from the point with coordinates (x_i, y_i, z_i) to the point with coordinates (x_f, y_f, z_f).[2]

Thus, we conclude that *the net work done on a particle by the resultant force acting on it is equal to the change in the kinetic energy of the particle.* The work-energy theorem also says that the speed of the particle will increase ($K_f > K_i$) if the net work done on it is positive, whereas its speed will decrease ($K_f < K_i$) if the net work done on it is negative. That is, the speed and kinetic energy of a particle will change only if work is done on the particle by some external force. Because of this connection between work and change in kinetic energy, we can also think of the kinetic energy of a body as the work the body can do in coming to rest.

Work can be positive, negative, or zero

Example 7.8

A 6-kg block initially at rest is pulled to the right along a horizontal, smooth surface by a constant, horizontal force of 12 N, as in Fig. 7.10a. Find the speed of the block after it moves a distance of 3 m.

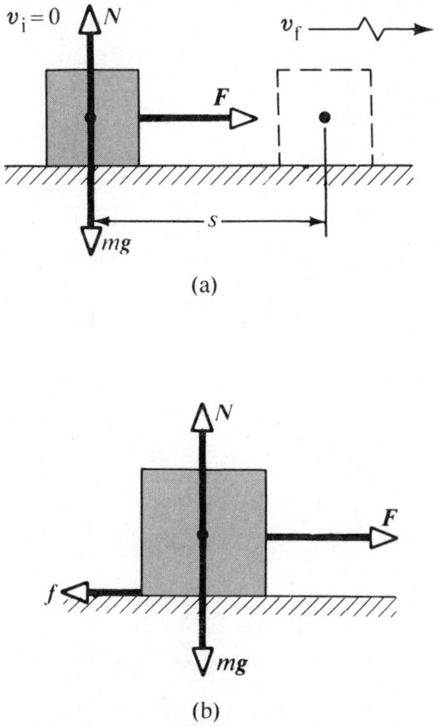

Figure 7.10 (a) Example 7.8. (b) Example 7.9.

Solution: The weight is balanced by the normal force, and neither of these forces does work since the displacement is horizontal. Since there is no friction, the resultant external force is the 12-N force. The work done by this force is

$$W_F = Fs = (12\text{ N})(3\text{ m}) = 36\text{ N}\cdot\text{m} = 36\text{ J}$$

Using the work-energy theorem and noting that the initial kinetic energy is zero, we get

$$W_F = K_f - K_i = \frac{1}{2}mv_f^2 - 0$$

$$v_f^2 = \frac{2W_F}{m} = \frac{2(36\text{ J})}{6\text{ kg}} = 12\,\frac{\text{m}^2}{\text{s}^2}$$

$$v_f = 3.46\,\frac{\text{m}}{\text{s}}$$

As a check of this result, you can also apply the kinematic equation $v_f^2 = v_i^2 + 2as$, where from Newton's second law $a = F/m = 2\text{ m/s}^2$.

Example 7.9

Find the final speed of the block described in Example 7.8 if the surface is rough and the coefficient of kinetic friction is 0.15.

Solution: In this case, we must calculate the net work done on the block, which equals the sum of the work done by the applied 12-N force and the frictional force f, as in Fig. 7.10b. Since the frictional force opposes the displacement, the work this force does is *negative*. The magnitude of the frictional force is given by $f = \mu N = \mu mg$; therefore, the work done by this force is this force multiplied by the displacement (see Eq. 7.2) or

$$W_f = -fs = -\mu mgs = (-0.15)(6)(9.8)(3) = -26.5\text{ J}$$

Therefore, the net work done on the block is

$$W_{\text{net}} = W_F + W_f = 36.0\text{ J} - 26.5\text{ J} = 9.50\text{ J}$$

[2] In the general expression, Eq. 7.21, the component F_x can depend on y and z as well as on x; similarly for F_y and F_z.

Applying the work-energy theorem with $v_i = 0$ gives

$$W_{net} = \frac{1}{2}mv_f^2$$

$$v_f^2 = \frac{2W_{net}}{m} = \frac{19}{6}\frac{m^2}{s^2}$$

$$v_f = 1.78\frac{m}{s}$$

Again, you should check this result using Newton's second law and kinematics. In this case, $\Sigma F_x = F - f = 12 - 8.82 = 3.18$ N, so $a = \Sigma F_x/m = 0.53$ m/s^2.

Example 7.10

A block of mass 1.6 kg is attached to a spring with a force constant of 10^3 N/m, as in Fig. 7.6. The spring is compressed a distance of 2.0 cm and the block is released from rest. (a) Calculate the velocity of the block as it passes through the equilibrium position $x = 0$, if the surface is frictionless.

Following Example 7.5, the work done by the spring with $x_m = -2.0$ cm $= -2 \times 10^{-2}$ m is

$$W_s = \frac{1}{2}kx_m^2 = \frac{1}{2}\left(10^3\frac{N}{m}\right)(-2 \times 10^{-2} \text{ m})^2 = 0.20 \text{ J}$$

Using the work-energy theorem with $v_i = 0$ gives

$$W_s = \frac{1}{2}mv_f^2 - \frac{1}{2}mv_i^2$$

$$0.20 \text{ J} = \frac{1}{2}(1.6 \text{ kg})v_f^2 - 0$$

$$v_f^2 = \frac{0.4 \text{ J}}{1.6 \text{ kg}} = 0.25\frac{m^2}{s^2}$$

$$v_f = 0.50\frac{m}{s}$$

(b) Calculate the velocity of the block as it passes through the equilibrium position if a constant frictional force of 4.0 N retards its motion.

The work done by the frictional force for a displacement of 2×10^{-2} m is given by

$$W_f = -fs = -(4 \text{ N})(2 \times 10^{-2} \text{ m}) = -0.08 \text{ J}$$

The net work done on the block is the work done by the spring plus the work done by friction. From (a), $W_s = 0.20$ J, therefore

$$W_{net} = W_s + W_f = 0.20 \text{ J} - 0.08 \text{ J} = 0.12 \text{ J}$$

Applying the work-energy theorem gives

$$\frac{1}{2}mv_f^2 = W_{net}$$

$$\frac{1}{2}(1.6 \text{ kg})v_f^2 = 0.12 \text{ J}$$

$$v_f^2 = \frac{0.24 \text{ J}}{1.6 \text{ kg}} = 0.15 \text{ m}^2/\text{s}^2$$

$$v_f = 0.39 \text{ m/s}$$

Note that this value for v_f is less than that obtained in the frictionless case. Is this result sensible?

Example 7.11 *Block Pushed Along Incline*

A block of mass m is pushed up a rough incline by a constant force $\boldsymbol{F}$ acting parallel to the incline, as in Fig. 7.11a. The block is displaced a distance d up the incline. (a) Calculate the work done by the force of gravity for this displacement.

The force of gravity is downward but has a component *down* the plane. This is given by $-mg \sin\theta$ if the positive x direction is chosen to be up the plane (Fig. 7.11b). Therefore, the work done by gravity for the displacement d is

$$W_g = (-mg \sin\theta)d = -mgh$$

where $h = d \sin\theta$ is the *vertical* displacement. That is, the work done by gravity equals the force of gravity multiplied by the *vertical displacement.* In the next chapter, we shall show that this result is valid in general for any particle displaced between two points. Furthermore, the result is independent of the path taken between these points.

(b) Calculate the work done by the applied force $\boldsymbol{F}$.

Since $\boldsymbol{F}$ is in the same direction as the displacement, we get

$$W_F = \boldsymbol{F} \cdot \boldsymbol{s} = Fd$$

(c) Find the work done by the force of kinetic friction if the coefficient of friction is μ.

The magnitude of the force of friction is $f = \mu N =$

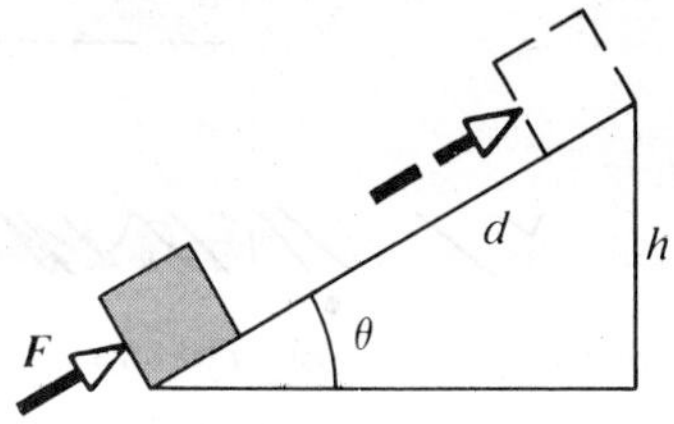

(a)

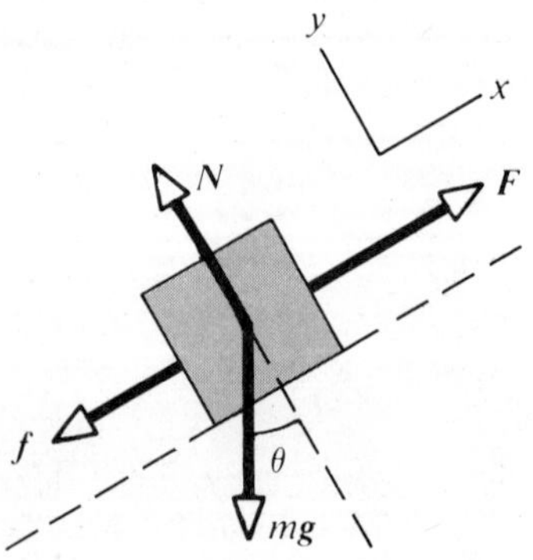

(b)

Figure 7.11 (Example 7.11) The block is pushed up the rough incline by a constant force $\boldsymbol{F}$.

$\mu mg\cos\theta$. Since the direction of this force is *opposite* the direction of the displacement, we find that

$$W_f = -fd = -\mu mgd\cos\theta$$

(d) Find the net work done on the block for this displacement.

Using the results to (a), (b), and (c), we get

$$\begin{aligned} W_{net} &= W_g + W_F + W_f \\ &= -mgd\sin\theta + Fd - \mu mgd\cos\theta \end{aligned}$$

or

$$W_{net} = Fd - mgd(\sin\theta + \mu\cos\theta)$$

For example, if we take $F = 15$ N, $d = 1.0$ m, $\theta = 25°$, $m = 1.5$ kg, and $\mu = 0.30$, we find that

$$\begin{aligned} W_g &= -(mg\sin\theta)d \\ &= -(1.5\text{ kg})\left(9.8\ \frac{\text{m}}{\text{s}^2}\right)(\sin 25°)(1.0\text{ m}) \\ &= -6.2\text{ J} \\ W_F &= Fd = (15\text{ N})(1\text{ m}) = 15\text{ J} \\ W_f &= -\mu mgd\cos\theta \\ &= -(0.30)(1.5\text{ kg})\left(9.8\ \frac{\text{m}}{\text{s}^2}\right)(1.0\text{ m})(\cos 25°) \\ &= -4.0\text{ J} \\ W_{net} &= W_g + W_F + W_f = 4.8\text{ J} \end{aligned}$$

Example 7.12 *Minimum Stopping Distance*

An automobile traveling at 48 km/h can be stopped in a minimum distance of 40 m by applying the brakes. If the same automobile is traveling at 96 km/h, what is the minimum stopping distance?

Solution: We will assume that when the brakes are applied, the car does not skid. To get the minimum stopping distance, d, we take the frictional force f between the tires and road to be a *maximum*. The work done by this frictional force, $-fd$, must equal the change in kinetic energy of the automobile. Since the kinetic energy has a final value of zero and an initial value of $\frac{1}{2}mv^2$, we get

$$\begin{aligned} W_f &= K_f - K_i \\ -fd &= 0 - \frac{1}{2}mv^2 \\ d &= \frac{mv^2}{2f} \end{aligned}$$

If we assume f is the same for the two initial speeds, we can take m and f as constants. Therefore, the ratio of stopping distances is given by

$$\frac{d_2}{d_1} = \left(\frac{v_2}{v_1}\right)^2$$

Taking $v_1 = 48$ km/h, $v_2 = 96$ km/h, and $d_1 = 40$ m gives

$$\frac{d_2}{d_1} = \left(\frac{96}{48}\right)^2 = 4$$

$$d_2 = 4d_1 = 4(40\text{ m}) = 160\text{ m}$$

This shows that *the minimum stopping distance varies as the square of the ratio of speeds.* If the speed is doubled, as it is in this example, the distance increases by a factor of 4.

Q6. Can the kinetic energy of an object have a negative value?

Q7. If the speed of a particle is doubled, what happens to its kinetic energy?

Q8. What can be said about the speed of an object if the net work done on that object is zero?

Q9. Using the work-energy theorem, explain why the force of kinetic friction always has the effect of *reducing* the kinetic energy of a particle.

7.6 POWER

From a practical viewpoint, it is interesting to know not only the work done on an object, but also the rate at which the work is being done. *Power* is defined as *the time rate of doing work.*

If an external force is applied to an object, and if the work done by this force is ΔW in the time interval Δt, then the *average power* during this interval is defined as *the ratio of the work done to the time interval:*

$$\overline{P} = \frac{\Delta W}{\Delta t} \qquad (7.22)$$

Average power

The *instantaneous power, P,* is the limiting value of the average power as Δt approaches zero:

Instantaneous power

$$P = \lim_{\Delta t \to 0} \frac{\Delta W}{\Delta t} = \frac{dW}{dt} \quad (7.23)$$

From Eq. 7.4, we can express the work done by a force $\boldsymbol{F}$ for a displacement $d\boldsymbol{s}$, since $dW = \boldsymbol{F} \cdot d\boldsymbol{s}$. Therefore, the instantaneous power can be written

Instantaneous power

$$P = \frac{dW}{dt} = \boldsymbol{F} \cdot \frac{d\boldsymbol{s}}{dt} = \boldsymbol{F} \cdot \boldsymbol{v} \quad (7.24)$$

where we have used the fact that $\boldsymbol{v} = d\boldsymbol{s}/dt$.

The unit of power in the SI system is J/s, which is also called a *watt* (W):

The watt

$$1 \text{ W} = 1 \text{ J/s} = 1 \text{ kg} \cdot \text{m}^2/\text{s}^3$$

The unit of power in the British engineering system is the horsepower (hp), where

$$1 \text{ hp} = 550 \text{ ft} \cdot \text{lb/s} = 746 \text{ W}$$

A new unit of energy (or work) can now be defined in terms of the unit of power. One kilowatt-hour (kWh) is the energy converted or consumed in 1 h at the constant rate of 1 kW. The numerical value of 1 kWh is

$$1 \text{ kWh} = (10^3 \text{ W})(3600 \text{ s}) = 3.6 \times 10^6 \text{ J}$$

It is important to note that a kWh is a unit of energy, not power. When you pay your electric bill, you are buying energy, and the amount of electricity used is usually in multiples of kWh. For example, an electric bulb rated at 100 W would "consume" 3.6×10^5 J of energy in 1 h.

Although the W and the kWh are commonly used only in electrical applications, they can be used in other scientific areas. For example, an automobile engine can be rated in kW as well as in hp. Likewise, the power consumption of an electrical appliance can be expressed in hp.

Example 7.13

An elevator has a mass of 1000 kg and carries a maximum load of 800 kg. A constant frictional force of 4000 N retards its motion upward, as in Fig. 7.12. (a) What must be the minimum horsepower delivered by the motor to lift the elevator at a constant speed of 3 m/s?

The motor must supply the force $\boldsymbol{T}$ that pulls the elevator upward. From Newton's second law and from the fact that $a = 0$ since v is constant, we get

$$T - f - Mg = 0$$

where M is the *total* mass (elevator plus load), equal to 1800 kg. Therefore,

$$\begin{aligned} T &= f + Mg \\ &= 4 \times 10^3 \text{ N} + (1.8 \times 10^3 \text{ kg})(9.8 \text{ m/s}^2) \\ &= 2.16 \times 10^4 \text{ N} \end{aligned}$$

Using Eq. 7.24 and the fact that $\boldsymbol{T}$ is in the same direction as $\boldsymbol{v}$ gives

$$\begin{aligned} P &= \boldsymbol{T} \cdot \boldsymbol{v} = Tv \\ &= (2.16 \times 10^4 \text{ N})(3 \text{ m/s}) = 6.48 \times 10^4 \text{ W} \\ &= 64.8 \text{ kW} = 86.9 \text{ hp} \end{aligned}$$

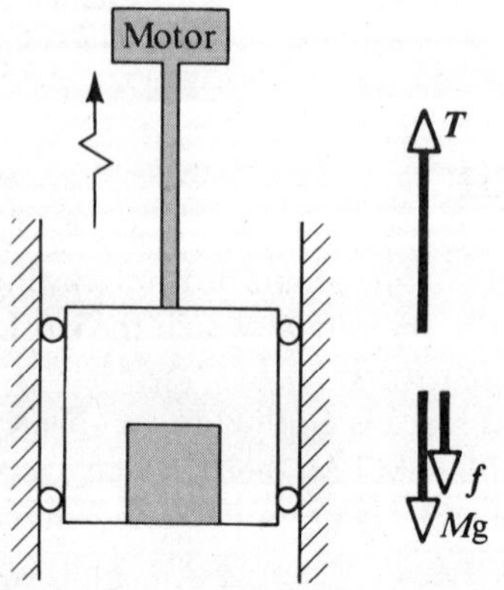

Figure 7.12 (Example 7.13) The motor provides a force $\boldsymbol{T}$ upward on the elevator. A frictional force f and the total weight Mg act downward.

(b) What power must the motor deliver if it is designed to provide an upward acceleration of 1.0 m/s²?

Applying Newton's second law to the elevator gives

$$T - f - Mg = Ma$$

$$T = M(a + g) + f$$
$$= (1.8 \times 10^3 \text{ kg})(1.0 + 9.8) \text{ m/s}^2 + 4 \times 10^3 \text{ N}$$
$$= 2.34 \times 10^4 \text{ N}$$

Therefore, using Eq. 7.24 we get for the required power

$$P = Tv = (2.34 \times 10^4\, v) \text{ W}$$

where v is the instantaneous speed of the elevator in m/s. Hence, the power required increases with increasing speed.

7.7 ENERGY AND THE AUTOMOBILE

Automobiles powered by gasoline engines are known to be very inefficient machines. Even under ideal conditions, less than 15% of the available energy in the fuel is used to power the vehicle. The situation is much worse under stop-and-go driving in the city. The purpose of this section is to use the concepts of energy, power, and forces of friction to analyze some factors that affect automobile fuel consumption.

There are many mechanisms that contribute to the energy losses in a typical automobile.[3] About two thirds of the energy available from the fuel is lost in the engine. Part of this energy ends up in the atmosphere via the exhaust system, and part is used in the engine's cooling system. (The efficiency of engines will be discussed in Chapter 19.) About 10% of the available energy is lost in the automobile's drive-train mechanism; this loss includes friction in the transmission, drive shaft, wheel and axle bearings, and differential. Friction in other moving parts accounts for about 6% of the energy loss, and 4% of the available energy is used to operate fuel and oil pumps and such accessories as power steering, air conditioning, power brakes, and electrical components. Finally, about 14% of the available energy is used to propel the automobile. This energy is used mainly to overcome road friction and air resistance.

Table 7.2 lists the power losses for an automobile with an available fuel power of 136 kW. These data apply to a typical 1450-kg "gas-guzzler" with a gas consumption rate of 6.4 km/liter (15 mi/gal).

TABLE 7.2 Power Losses in a Typical Automobile Assuming a Total Available Power of 136 kW

MECHANISM	*POWER LOSS* (kW)	*POWER LOSS* (%)
Exhaust (heat)	46	33
Cooling system	45	33
Drive train	13	10
Internal friction	8	6
Accessories	5	4
Propulsion of vehicle	19	14

Let us examine the power requirements to overcome road friction and air drag in more detail. The main contribution to road friction is the flexing of the tires. The coefficient of rolling friction, μ, between the tires and the road is about 0.016. For a 1450-kg car, the weight is 14 200 N and the force of rolling friction $\mu N = \mu W =$ 227 N. As the speed of the car increases, there is a small reduction in the normal force as a result of the effect of an airlift as air flows over the top of the car. This causes a slight reduction in the force of rolling friction, f_r, with increasing speed, as shown in Table 7.3.

[3] An excellent article on this subject is the one by G. Waring in *The Physics Teacher*, Vol. 18 (1980), p. 494. The data in Tables 7.2 and 7.3 were taken from this article.

TABLE 7.3 Frictional Forces and Power Requirements for a Typical Car

v (km/h)	N (N)	f_r (N)	f_a (N)	f_t (N)	$P = f_t v$ (kW)
0	14 200	227	0	227	0
32	14 100	226	51	277	2.5
64	13 900	222	204	426	7.6
96.5	13 600	218	465	683	18.3
129	13 200	211	830	1041	37.3
161	12 600	202	1293	1495	66.8

In this table, N is the normal force, f_r is road friction, f_a is air friction, f_t is total friction, and P is the power delivered to the wheels.

Now let us consider the effect of air friction, that is, the drag force that results from air moving past the various surfaces of the car. The drag force associated with air friction for large objects is proportional to the square of the speed (in m/s) (Section 6.10) and may be written

$$f_a = \frac{1}{2} C A \rho v^2 \tag{7.25}$$

where C is the drag coefficient, A is the cross-sectional area of the moving object, and ρ is the density of air. This expression can be used to calculate the values in Table 7.3 using $C = 0.5$, $\rho = 1.293\ \text{kg/m}^3$, and $A \approx 2\ \text{m}^2$.

The magnitude of the total frictional force, f_t, is given by the sum of the rolling friction force and the air drag force:

$$f_t = f_r + f_a \approx \text{constant} + \frac{1}{2} C A \rho v^2 \tag{7.26}$$

At low speeds, road resistance and air drag are comparable, but at high speeds air drag is the predominant resistive force, as shown in Table 7.3. Road friction can be reduced by reducing tire flexing (increase the air pressure slightly above recommended values) and using radial tires. Air drag can be reduced by using a smaller cross-sectional area and streamlining the car. Though driving a car with the windows open does create more air drag resulting in a 3 percent decrease in mpg, driving with the windows closed and the air conditioner running results in a 12 percent decrease in mileage.

The total power needed to maintain a constant speed v equals the product $f_t v$. This must equal the power delivered to the wheels. For example, from Table 7.3 we see that at $v = 96.5$ km/h $= 26.8$ m/s, the required power is

$$P = f_t v = (683\ \text{N})\left(26.8\ \frac{\text{m}}{\text{s}}\right) = 18.3\ \text{kW}$$

This can be broken into two parts: (1) the power needed to overcome road friction, $f_r v$, and (2) the power needed to overcome air drag, $f_a v$. At $v = 26.8$ m/s, these have the values

$$P_r = f_r v = (218\ \text{N})\left(26.8\ \frac{\text{m}}{\text{s}}\right) = 5.8\ \text{kW}$$

$$P_a = f_a v = (465\ \text{N})\left(26.8\ \frac{\text{m}}{\text{s}}\right) = 12.5\ \text{kW}$$

Note that $P = P_r + P_a$.

On the other hand, at $v = 161$ km/h $= 44.7$ m/s, we find that $P_r = 9.0$ kW, $P_a = 57.8$ kW, and $P = 66.8$ kW. This shows the importance of air drag at high speeds.

Example 7.14 *Gas Consumed by Compact Car*

A compact car has a mass of 800 kg, and its efficiency is rated at 14%. (That is, 14% of the available fuel energy is delivered to the wheels.) Find the amount of gasoline used to accelerate the car from rest to 60 mi/h (27 m/s). Use the fact that the energy equivalent of one gallon of gasoline is 1.3×10^8 J.

Solution: The energy required to accelerate the car from rest to a speed v is its kinetic energy, $\frac{1}{2}mv^2$. For this case,

$$E = \frac{1}{2}mv^2 = \frac{1}{2}(800\ \text{kg})\left(27\ \frac{\text{m}}{\text{s}}\right)^2 = 2.9 \times 10^5\ \text{J}$$

If the engine were 100% efficient, each gallon of gasoline would supply an energy 1.3×10^8 J. Since the engine is only 14% efficient, each gallon delivers only $(0.14)(1.3 \times 10^8\ \text{J}) = 1.8 \times 10^7$ J. Hence, the number of gallons used to accelerate the car is

$$\text{Number of gal} = \frac{2.9 \times 10^5}{1.8 \times 10^7}\frac{\text{J}}{\text{J/gal}} = 0.016\ \text{gal}$$

At this rate, a gallon of gas would be used after 62 such accelerations. This demonstrates the severe energy requirements for extreme stop-and-start driving.

Example 7.15 *Power Delivered to Wheels*

Suppose the car described in Example 7.14 has a mileage rating of 35 mi/gal when traveling at 60 mi/h. How much power is delivered to the wheels?

Solution: From the given data, we see that the car consumes $60/35 = 1.7$ gal/h. Using the fact that each gallon is equivalent to 1.3×10^8 J, we find that the total power used is

$$P = \frac{(1.7\ \text{gal})(1.3 \times 10^8\ \text{J/gal})}{1\ \text{h}}$$
$$= \frac{2.2 \times 10^8\ \text{J}}{3.6 \times 10^3\ \text{s}} = 62\ \text{kW}$$

Since 14% of the available power is used to propel the car, we see that the power delivered to the wheels is $(0.14)(62\ \text{kW}) = 8.7$ kW. This is about one half the value obtained for the large 1450-kg car discussed in the text. Mass is clearly an important factor in power-loss mechanisms.

Example 7.16 *Car Accelerating Up a Hill*

Consider a car of mass m accelerating up a hill, as in Fig. 7.13. Assume that the magnitude of the drag force is given by

$$|f| = (218 + 0.70v^2)\ \text{N}$$

where v is the speed in m/s. Calculate the power that the engine must deliver to the wheels.

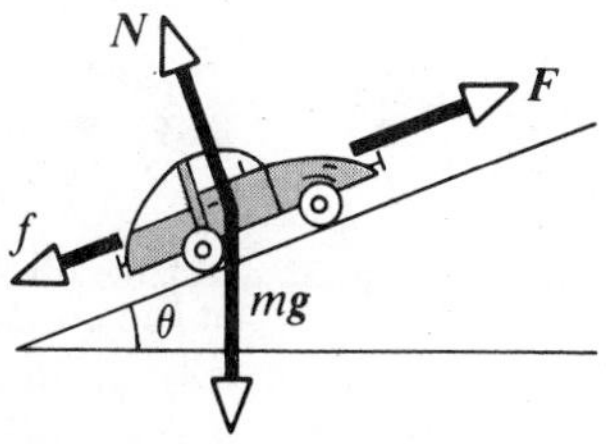

Figure 7.13 (Example 7.16).

Solution: The forces on the car are shown in Fig. 7.13, where $\boldsymbol{F}$ is the force that propels the car and the remaining forces have their usual meaning. Newton's second law applied to the motion along the road surface gives

$$\Sigma F_x = F - |f| - mg\sin\theta = ma$$
$$F = ma + mg\sin\theta + |f|$$
$$= ma + mg\sin\theta + (218 + 0.70v^2)$$

Therefore, the power required for propulsion is

$$P = Fv = mva + mvg\sin\theta + 218v + 0.70v^3$$

In this expression, the term mva represents the power the engine must deliver to accelerate the car. If the car moves at constant speed, this term is zero and the power requirement is reduced. The terms $mvg\sin\theta$ is the power required to overcome the force of gravity as the car moves up the incline. This term would be zero for motion on a horizontal surface. The term $218v$ is the power required to overcome road friction. Finally, the term $0.70v^3$ is the power needed to overcome air drag.

If we take $m = 1450$ kg, $v = 27$ m/s ($= 60$ mi/h), $a = 1\ \text{m/s}^2$, and $\theta = 10°$, the various terms in P are calculated to be

$$mva = (1450\ \text{kg})(27\ \text{m/s})(1\ \text{m/s}^2)$$
$$= 39\ \text{kW} = 52\ \text{hp}$$
$$mvg\sin\theta = (1450\ \text{kg})(27\ \text{m/s})(9.8\ \text{m/s}^2)(\sin 10°)$$
$$= 67\ \text{kW} = 89\ \text{hp}$$
$$218v = 218(27) = 5.9\ \text{kW} = 7.9\ \text{hp}$$
$$0.70v^3 = 0.70(27)^3 = 14\ \text{kW} = 18\ \text{hp}$$

Hence, the total power required is 126 kW, or 167 hp. Note that the power requirements for traveling at *constant* speed on a horizontal surface are only 19.9 kW, or 25.9 hp (the sum of the last two terms). Furthermore, if the mass is halved (as in compact cars), the power required is also reduced by almost the same factor.

Q10. Can the average power ever equal the instantaneous power? Explain.

Q11. In Example 7.13, does the required power increase or decrease as the force of friction is reduced?

Q12. An automobile sales representative claims that a "souped-up" 300-hp engine is a necessary option in a compact car (instead of a conventional 150-hp engine). Suppose you intend to drive the car within speed limits ($\leqslant$55 mi/h) and on flat terrain. How would you counteract this sales pitch?

7.8 SUMMARY

The *work* done by a *constant* force **F** acting on a particle is defined as the product of the component of the force in the direction of the particle's displacement and the magnitude of the displacement. If the force makes an angle θ with the displacement **s**, the work done by **F** is

Work done by a constant force

$$W \equiv Fs\cos\theta \tag{7.1}$$

The *scalar, or dot, product* of any two vectors **A** and **B** is defined by the relationship

Scalar product

$$\mathbf{A}\cdot\mathbf{B} \equiv AB\cos\theta \tag{7.5}$$

where the result is a scalar quantity and θ is the angle between the two vectors. The scalar product obeys the commutative and distributive laws.

The *work* done by a *varying* force acting on a particle moving along the x axis from x_i to x_f is given by

Work done by a varying force

$$W \equiv \int_{x_i}^{x_f} F_x\,dx \tag{7.11}$$

where F_x is the component of force in the x direction. If there are several forces acting on the particle, the net work done by all forces is the sum of the individual work done by each force.

The *kinetic energy* of a particle of mass m moving with a speed v (where v is small compared with the speed of light) is defined as

Kinetic energy

$$K \equiv \frac{1}{2}mv^2 \tag{7.17}$$

The *work-energy theorem* states that the net work done on a particle by external forces equals the change in kinetic energy of the particle:

Work-energy theorem

$$W = K_f - K_i = \frac{1}{2}mv_f^2 - \frac{1}{2}mv_i^2 \tag{7.19}$$

The *instantaneous power* is defined as the time rate of doing work. If an agent applies a force **F** to an object moving with a velocity **v**, the power delivered by that agent is given by

Instantaneous power

$$P = \frac{dW}{dt} = \mathbf{F}\cdot\mathbf{v} \tag{7.24}$$

EXERCISES

Section 7.2 Work Done by a Constant Force

1. How much work is done by a person in raising a 20-kg bucket of water from the bottom of a well that is 30 m deep? Assume the speed of the bucket as it is lifted is constant.

2. A tugboat exerts a constant force of 5000 N on a ship moving at constant speed through a harbor. How much work does the tugboat do on the ship in a distance of 3 km?

3. A 15-kg block is dragged over a rough, horizontal surface by a constant force of 70 N acting at an angle of 25° to the horizontal. The block is displaced 5 m, and the coefficient of kinetic friction is 0.3. Find the work done by (a) the 70-N force, (b) the force of friction, (c) the normal force, and (d) the force of gravity. (e) What is the net work done on the block?

4. A horizontal force of 150 N is used to push a 40-kg box on a rough, horizontal surface through a distance of 6 m. If the box moves at constant speed, find (a) the work done by the 150-N force, (b) the work done by friction, and (c) the coefficient of kinetic friction.

5. A 100-kg sled is dragged by a team of dogs a distance of 2 km over a horizontal surface at a constant velocity. If the coefficient of friction between the sled and the snow is 0.15, find the work done by (a) the team of dogs and (b) the force of friction.

6. Verify the following energy unit conversions: (a) 1 J $= 10^7$ ergs, (b) 1 J = 0.737 ft · lb.

Section 7.3 The Scalar Product of Two Vectors

7. Two vectors are given by $\mathbf{A} = 3\mathbf{i} + 2\mathbf{j}$ and $\mathbf{B} = -\mathbf{i} + 3\mathbf{j}$. Find (a) $\mathbf{A} \cdot \mathbf{B}$ and (b) the angle between $\mathbf{A}$ and $\mathbf{B}$.

8. A vector is given by $\mathbf{A} = -2\mathbf{i} + 3\mathbf{j}$. Find (a) the magnitude of $\mathbf{A}$ and (b) the angle that $\mathbf{A}$ makes with the positive y axis. [In (b), use the definition of the scalar product.]

9. Vector $\mathbf{A}$ has a magnitude of 3 units, and $\mathbf{B}$ has a magnitude of 8 units. The two vectors make an angle of 40° with each other. Find $\mathbf{A} \cdot \mathbf{B}$.

10. A force $\mathbf{F} = (6\mathbf{i} - 2\mathbf{j})$ N acts on a particle that undergoes a displacement $\mathbf{s} = (3\mathbf{i} + \mathbf{j})$ m. Find (a) the work done by the force on the particle and (b) the angle between $\mathbf{F}$ and $\mathbf{s}$.

11. Given two arbitrary vectors $\mathbf{A}$ and $\mathbf{B}$, show that $\mathbf{A} \cdot \mathbf{B} = A_xB_x + A_yB_y + A_zB_z$. (Hint: Write $\mathbf{A}$ and $\mathbf{B}$ in unit vector form and use Eq. 7.8.)

12. Vector $\mathbf{A}$ is 2 units long and points in the positive y direction. Vector $\mathbf{B}$ has a negative x component 5 units long, a positive y component 3 units long, and no z component. Find $\mathbf{A} \cdot \mathbf{B}$ and the angle between the vectors.

13. Using the definition of the scalar product, find the angles between the following pairs of vectors: (a) $\mathbf{A} = 3\mathbf{i} - \mathbf{j}$ and $\mathbf{B} = 2\mathbf{i} + 2\mathbf{j}$, (b) $\mathbf{A} = -\mathbf{i} + 4\mathbf{j}$ and $\mathbf{B} = 2\mathbf{i} + \mathbf{j} + 2\mathbf{k}$, (c) $\mathbf{A} = 2\mathbf{i} + \mathbf{j} + 3\mathbf{k}$ and $\mathbf{B} = -2\mathbf{j} + 2\mathbf{k}$.

14. The scalar product of vectors $\mathbf{A}$ and $\mathbf{B}$ is 6 units. The magnitude of each vector is 4. Find the angle between the vectors.

Section 7.4 Work Done by a Varying Force: The One-Dimensional Case

15. A body is subject to a force F_x that varies with position as in Fig. 7.14. Find the work done by the force on the body as it moves (a) from $x = 0$ to $x = 5$ m, (b) from $x = 5$ m to $x = 10$ m, and (c) from $x = 10$ m to $x = 15$ m. (d) What is the total work done by the force over the distance $x = 0$ to $x = 15$ m?

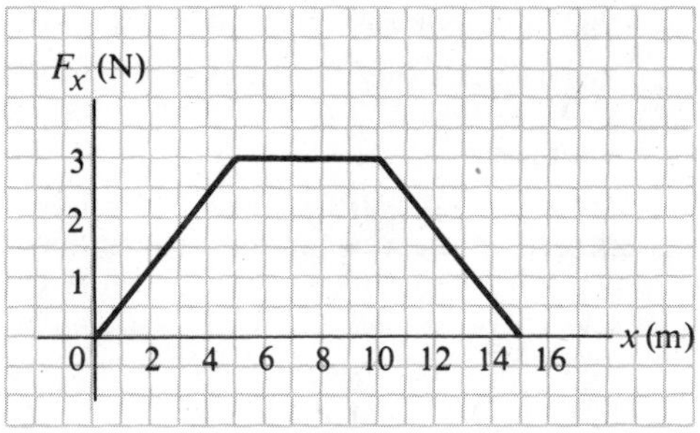

Figure 7.14 (Exercises 15 and 25).

16. The force acting on a particle varies as in Fig. 7.15. Find the work done by the force as the particle moves (a) from $x = 0$ to $x = 8$ m, (b) from $x = 8$ m to $x = 10$ m, and (c) from $x = 0$ to $x = 10$ m.

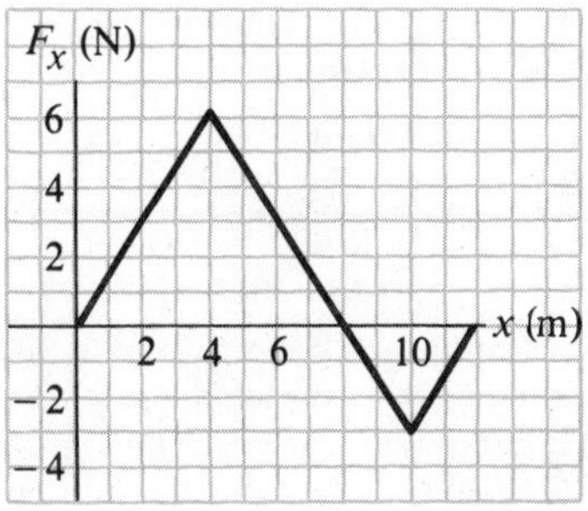

Figure 7.15 (Exercises 16 and 26).

17. The force acting on a particle is given by $F_x = (8x - 16)$ N, where x is in m. (a) Make a plot of this force versus x from $x = 0$ to $x = 3$ m. (b) From your graph, find the net work done by this force as the particle moves from $x = 0$ to $x = 3$ m.

18. When a 3-kg mass is hung vertically on a certain light spring that obeys Hooke's law, the spring stretches 1.5 cm. If the 3-kg mass is removed, (a) how far will the spring stretch if a 1-kg mass is hung on it, and (b) how much work must an external agent do to stretch the same spring 4.0 cm from its unstretched position?

Section 7.5 Work and Kinetic Energy

19. A 0.2-kg ball has a speed of 15 m/s. (a) What is its kinetic energy? (b) If its speed is doubled, what is its kinetic energy?

20. Calculate the kinetic energy of a 1000-kg satellite orbiting the earth at a speed of 7×10^3 m/s.

21. A 3-kg mass has an initial velocity of $\mathbf{v}_0 = (5\mathbf{i} - 3\mathbf{j})$ m/s. (a) What is its kinetic energy at this time? (b) Find the *change* in its kinetic energy if its *velocity changes* to $(8\mathbf{i} + 4\mathbf{j})$ m/s. (Hint: Remember that $v^2 = \mathbf{v} \cdot \mathbf{v}$.)

22. A 0.6-kg particle has a speed of 3 m/s at point A and a speed of 5 m/s at point B. What is its kinetic energy (a) at point A and (b) at point B? (c) What is the total work done on the particle as it moves from A to B?

23. A mechanic pushes a 2000-kg car from rest to a speed of 3 m/s with a constant horizontal force. During this time, the car moves a distance of 30 m. Neglecting friction between the car and the road, determine (a) the work done by the mechanic and (b) the horizontal force exerted on the car.

24. A 40-kg box initially at rest is pushed a distance of 5 m along a rough, horizontal floor with a constant applied force of 130 N. If the coefficient of friction between the box and floor is 0.3, find (a) the work done by the applied force, (b) the work done by friction, (c) the change in kinetic energy of the box, and (d) the final speed of the box.

25. A 4-kg particle is subject to a force that varies with position as shown in Fig. 7.14. The particle starts from rest at $x = 0$. What is the speed of the particle at (a) $x = 5$ m, (b) $x = 10$ m, (c) $x = 15$ m?

26. The force acting on a 6-kg particle varies with position as shown in Fig. 7.15. If its velocity is 2 m/s at $x = 0$, find its speed and kinetic energy at (a) $x = 4$ m, (b) $x = 8$ m, (c) $x = 10$ m.

27. A sled of mass m is given a kick on a frozen pond, imparting to it an initial speed v_0. The coefficient of kinetic friction between the sled and ice is μ_k. (a) Use the work-energy theorem to find the distance the sled moves before coming to rest. (b) Obtain a numerical value for the distance if $v_0 = 5$ m/s and $\mu_k = 0.1$.

28. A 6-kg mass is lifted vertically through a distance of 5 m by a light string with a tension of 80 N. Find (a) the work done by the force of tension, (b) the work done by gravity, and (c) the final speed of the mass if it starts from rest.

29. A 2-kg block is attached to a light spring of force constant 500 N/m as in Fig. 7.6. The block is pulled 5 cm to the right of equilibrium and released from rest. Find the speed of the block as it passes through equilibrium if (a) the horizontal surface is frictionless and (b) if the coefficient of friction between the block and surface is 0.35.

30. A 4-kg block is given an initial speed of 8 m/s at the bottom of a 20° incline. The frictional force that retards its motion is 15 N. (a) If the block is directed *up* the incline, how far will it move before it stops? (b) Will it slide back down the incline?

31. A 3-kg block is moved up a 37° incline under the action of a constant *horizontal* force of 40 N. The coefficient of kinetic friction is 0.1, and the block is displaced 2 m up the incline. Calculate (a) the work done by the 40-N force, (b) the work done by gravity, (c) the work done by friction, and (d) the *change* in kinetic energy of the block. (Note: The applied force is *not* parallel to the incline.)

32. A 4-kg block attached to a string 2 m in length rotates in a circle on a horizontal surface. (a) If the surface is frictionless, identify all the forces on the block and show that the work done by each force is zero for any displacement of the block. (b) If the coefficient of friction between the block and surface is 0.25, find the work done by the force of friction in each revolution of the block.

Section 7.6 Power

33. A certain automobile engine delivers 30 hp (2.24×10^4 W) to its wheels when moving at a constant speed of 27 m/s ($\approx$60 mi/h). What is the resistive force acting on the automobile at that speed?

34. A speedboat requires 130 hp to move at a constant speed of 15 m/s ($\approx$33 mi/h). Calculate the resistive force due to the water at that speed.

35. A 50-kg student climbs a rope 5 m in length and stops at the top. (a) What must her average speed be in order to match the power output of a 200-W light bulb? (b) How much work does she do?

36. A machine lifts a 300-kg crate through a height of 5 m in 8 s. Calculate its power output.

37. A 200-kg crate is pulled along a level surface by an engine. The coefficient of friction between the crate and surface is 0.4. (a) How much power must the engine deliver to move the crate at a constant speed of 5 m/s? (b) How much work is done by the engine in 3 min?

38. A single, constant force $\mathbf{F}$ acts on a particle of mass m. The particle starts at rest at $t = 0$. (a) Show that the instantaneous power delivered by the force at any time t is equal to $\frac{F^2}{m}t$. (b) If $F = 20$ N and $m = 5$ kg, what is the power delivered at $t = 3$ s?

39. A 1500-kg car accelerates uniformly from rest to a speed of 10 m/s in 3 s. Find (a) the work done on the car in this time, (b) the average power delivered by the engine in the first 3 s, and (c) the instantaneous power delivered by the engine at $t = 2$ s.

40. A 65-kg athlete runs a distance of 600 m up a mountain inclined at 20° to the horizontal. He performs this feat in 80 s. Assuming that air resistance is negligible, (a) how much work does he perform and (b) what is his power output during the run?

Section 7.7 Energy and the Automobile

41. Suppose the car described in Table 7.3 has a fuel economy of 6.4 km/liter (15 mi/gal) when traveling at a speed of 60 mi/h. Assuming an available fuel power of 136 kW and an efficiency of 14%, determine the fuel economy of the car if it carries, in addition to the driver, four passengers, each with an average mass of 70 kg.

42. When an air conditioner is added to the car described in Exercise 41, the additional fuel power required to operate the air conditioner is 11 kW. If the fuel economy is 6.4 km/liter without the air conditioner, what is the fuel economy when the air conditioner is operating?

43. The car described in Table 7.3 travels at a constant speed of 129 km/h. At this speed, determine (a) the power needed to overcome air drag, and (b) the total power delivered to the wheels.

44. A passenger car carrying two people has a fuel economy of 25 mi/gal. It travels a distance of 3000 miles. A jet airplane making the same trip with 150 passengers has a fuel economy of 1 mi/gal. Compare the fuel consumed per passenger for the two modes of transportation.

PROBLEMS

1. The direction of an arbitrary vector $\mathbf{A}$ can be completely specified with the angles α, β, and γ that the vector makes with the x, y, and x axes, respectively. If $\mathbf{A} = A_x\mathbf{i} + A_y\mathbf{j} + A_z\mathbf{k}$, (a) find expressions for $\cos\alpha$, $\cos\beta$, and $\cos\gamma$ (these are known as *direction cosines*) and (b) show that these angles satisfy the relation $\cos^2\alpha + \cos^2\beta + \cos^2\gamma = 1$. (Hint: Take the scalar product of $\mathbf{A}$ with $\mathbf{i}$, $\mathbf{j}$, and $\mathbf{k}$ separately.)

2. Three vectors that form a closed triangle satisfy the condition $\mathbf{C} = \mathbf{A} - \mathbf{B}$ (Fig. 7.16). Use this fact and the definition of the scalar product to derive the law of cosines in trigonometry,

$$C^2 = A^2 + B^2 - 2AB\cos\theta.$$

(Hint: Compute the scalar product $\mathbf{C}\cdot\mathbf{C}$.)

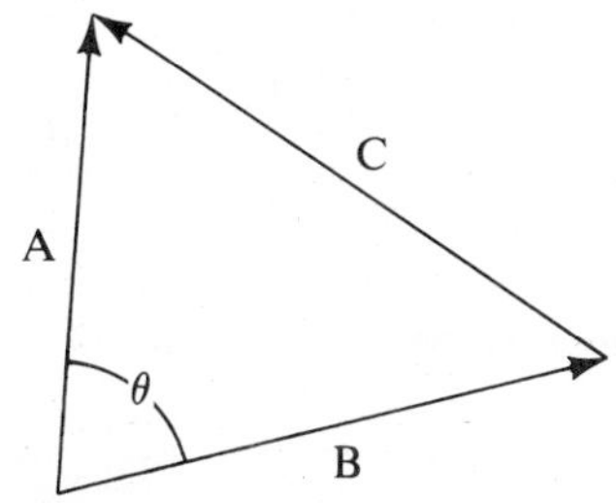

Figure 7.16 (Problem 2).

3. A block of mass m is attached to a light spring of force constant k as in Fig. 7.6. The spring is compressed a distance d from its equilibrium position and released from rest. (a) If the block comes to rest when it first reaches the equilibrium position, what is the coefficient of friction between the block and surface? (b) If the block first comes to rest when the spring is *stretched* a distance of $d/2$ from equilibrium, what is μ?

4. Prove the work-energy theorem, $W = \Delta K$, for a general three-dimensional displacement. (Note that $\mathbf{F} = m\,d\mathbf{v}/dt$ and $d\mathbf{s} = \mathbf{v}\,dt$.)

5. The resultant force acting on a 2-kg particle moving along the x axis varies as $F_x = 3x^2 - 4x + 5$, where x is in m and F_x is in N. (a) Find the net work done on the particle as it moves from $x = 1$ m to $x = 3$ m. (b) If the speed of the particle is 5 m/s at $x = 1$ m, what is its speed at $x = 3$ m?

6. A 4-kg particle moves along the x axis. Its position varies with time according to $x = t + 2t^3$, where x is in m and t is in s. Find (a) the kinetic energy at any time t, (b) the acceleration of the particle and the force acting on it at time t, (c) the power being delivered to the particle at time t, and (d) the work done on the particle in the interval $t = 0$ to $t = 2$ s. (Note that $P = dW/dt$.)

7. A 0.4-kg particle slides on a horizontal, circular track 1.5 m in radius. It is given an initial speed of 8 m/s. After one revolution, its speed drops to 6 m/s because of friction. (a) Find the work done by the force of friction in one revolution. (b) Calculate the coefficient of kinetic friction. (c) What is the total number of revolutions the particle will make before coming to rest?

8. A small sphere of mass m hangs from a string of length L as in Fig. 7.17. A variable horizontal force $\mathbf{F}$ is applied to the mass in such a way that it moves slowly from the vertical position until the string makes an angle θ with the vertical. Assuming the sphere is always in equilibrium, (a) show that $F = mg \tan \theta$. (b) Make use of Eq. 7.20 to show that the work done by the force $\mathbf{F}$ is equal to $mgL(1 - \cos \theta)$. (Hint: Note that $s = L\theta$, and so $ds = L d\theta$.)

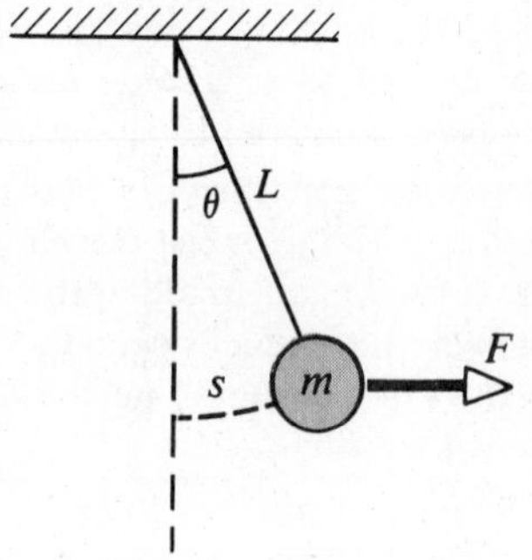

Figure 7.17 (Problem 8).

9. A car of mass m travels with constant speed v on a level road for a distance d. According to actual tests, the drag force is *approximately* given by $\mathbf{f} = -Km\mathbf{v}$, where $K = 0.018\ s^{-1}$. (a) Show that the work done by the engine to overcome the drag force is given by $Kmvd$. (b) Show that the power that the engine must deliver to the wheels to maintain this speed is Kmv^2. (c) Obtain numerical values for the work done and power delivered taking $m = 1500$ kg, $v = 27$ m/s, and $d = 100$ km. (d) If the car has a fuel economy of 15 mi/gal, what is the efficiency of the engine? (In this case, we define efficiency as the work done divided by the energy consumed.)

10. Suppose a car is modeled as a cylinder moving with a speed v, as in Fig. 7.18. In a time Δt, a column of air of mass Δm must be moved a distance $v\,\Delta t$ and hence must be given a kinetic energy $\frac{1}{2}(\Delta m)v^2$. Using this model, show that the power loss due to air resistance is $\frac{1}{2}\rho A v^3$ and the drag force is $\frac{1}{2}\rho A v^2$, where ρ is the density of air.

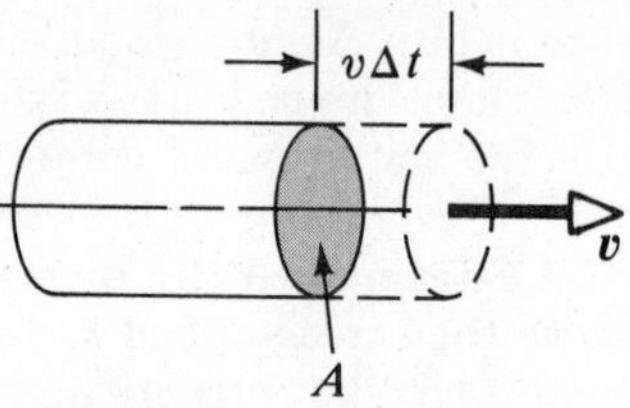

Figure 7.18 (Problem 10).

11. A passenger car of mass 1500 kg accelerates from rest to 97 km/h in 10 s. (a) Find the acceleration of the car. (b) Show that the coefficient of friction between the rear tires and road must be at least 0.55. (c) Determine the limiting frictional force on the car. (d) How much power is delivered by the engine? (Assume that the normal force on each tire is $\frac{1}{4}mg$.)

8 Potential Energy and Conservation of Energy

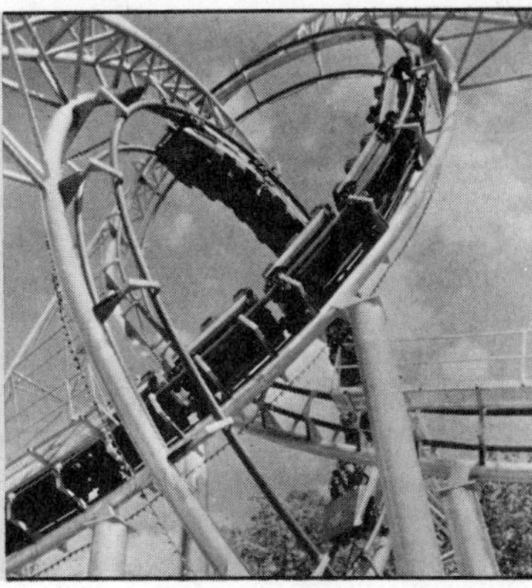

Photo courtesy of Six Flags Great Adventure

In Chapter 7 we introduced the concept of kinetic energy, which is associated with the motion of an object. We found that the kinetic energy of an object can change only if work is done on the object. In this chapter we introduce another form of mechanical energy associated with the position of an object, called *potential energy*. We shall find that the potential energy of a system can be thought of as energy stored in the system that can be converted to kinetic energy. The potential energy concept can be used only when dealing with a special class of forces called *conservative forces*. When only conservative forces, such as gravitational or spring forces, act on a system, the kinetic energy gained (or lost) by the system as its members change their relative positions is compensated by an equal energy lost (or gained) in the form of potential energy. This is known as the *law of conservation of mechanical energy*. A more general energy conservation law applies to an isolated system when all forms of energy and energy transformations are taken into account.

8.1 CONSERVATIVE AND NONCONSERVATIVE FORCES

Conservative Forces

In the previous chapter we found that the work done by the gravitational force acting on a particle equals the weight of the particle multiplied by its vertical displacement, assuming that g is constant over the range of the displacement. As we shall see in Section 8.2, this result is valid for an arbitrary displacement of the particle. That is, the work done by gravity depends only on the initial and final coordinates and is independent of the path taken between these points. When a force exhibits these properties, it is called a *conservative force*. In addition to the gravitational force, other examples of conservative forces are the electrostatic force and the restoring force in a spring. In general, *a force is conservative if the work done by that force acting on a particle moving between two points is independent of the path the particle takes between the points*. That is, the work done on a particle by a conservative force depends only on the initial and final coordinates of the particle. With reference to the *arbitrary* paths shown in Fig. 8.1a, we can write this condition

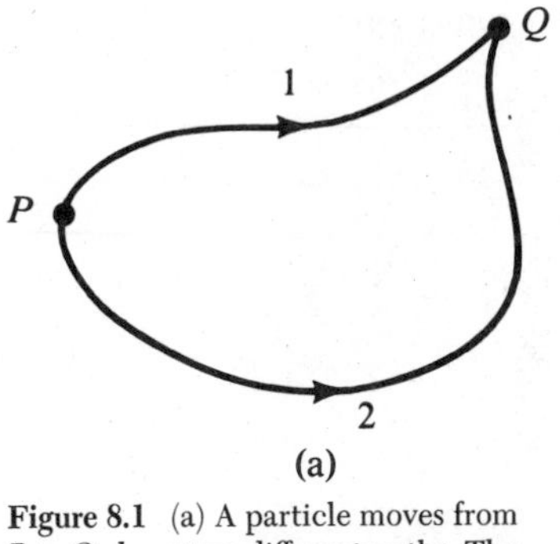

(a)

Figure 8.1 (a) A particle moves from P to Q along two different paths. The work done by a conservative force acting on the particle is the same along each path. If the force is nonconservative, the work done by this force differs along the two paths.

Property of a conservative force

$$W_{PQ}\text{ (along 1)} = W_{PQ}\text{ (along 2)}$$

A conservative force has another property, which can be derived from the above condition. Suppose the particle moves from P to Q along path 1, and then from Q to P along path 2, as in Fig. 8.1b. The work done by a conservative force in the reverse path 2 from Q to P is equal to the *negative* of the work done from P to Q along path 2. Therefore, we can write the original condition of a conservative force

$$W_{PQ} \text{ (along 1)} = -W_{QP} \text{ (along 2)}$$

$$W_{PQ} \text{ (along 1)} + W_{QP} \text{ (along 2)} = 0$$

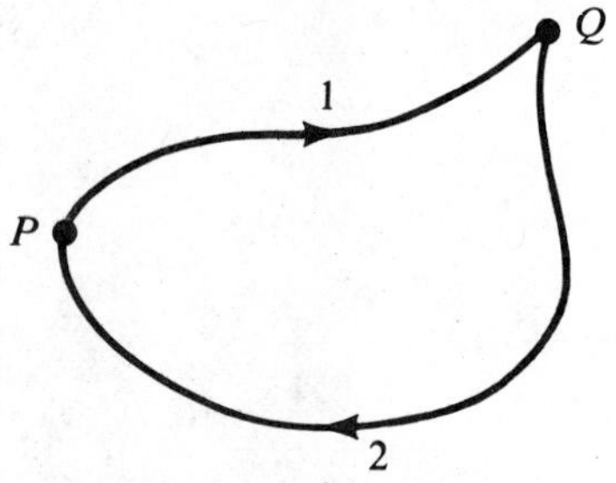

(b)

Figure 8.1 (b) A particle moves from P to Q and then from Q back to P in a closed path.

Hence, a conservative force also has the property that the *total work done by a conservative force on a particle is zero when the particle moves around any closed path and returns to its initial position.*

We can interpret this property of a conservative force in the following manner. The work-energy theorem says that the net work done on a particle displaced between two points equals the change in its kinetic energy. Therefore, if all the forces acting on the particle are conservative, then $W = 0$ for a round trip. This means that the particle will return to its starting point with the same kinetic energy it had when it started its motion.

To illustrate that the force of gravity is conservative, recall that the work done by the gravitational force as a particle of mass m moves between two points of elevation y_i and y_f is given by

Work done by the force of gravity

$$W_g = -mg(y_f - y_i)$$

That is, the work done by the gravitational force mg (in the negative y direction) equals the force multiplied by the displacement in the y direction. From this expression, we first note that W_g depends only on the initial and final y coordinates and is *independent* of the path taken. Furthermore, if y_i and y_f are at the same elevation or if the particle makes a round trip, then $y_i = y_f$ and $W_g = 0$. For example, if a ball is thrown vertically upward with an initial speed v_i, and if air resistance is neglected, the ball must return to the thrower's hand with the same speed (and same kinetic energy) it had at the start of its motion.

Another example of a conservative force is the force of a spring on a block attached to the spring, where the restoring force is given by $F_s = -kx$. In the previous chapter, we found that the work done by the spring on the block is

Work done by the spring force

$$W_s = \frac{1}{2}kx_i^2 - \frac{1}{2}kx_f^2$$

where the initial and final coordinates of the block are measured from the equilibrium position of the block, $x = 0$. We see that W_s again depends only on the initial and final x coordinates. In addition, $W_s = 0$ for a round trip, where $x_i = x_f$.

Nonconservative Forces

A force is nonconservative if the work done by that force on a particle moving between two points depends on the path taken. That is, the work done by a nonconservative force in taking a particle from P to Q in Fig. 8.1a will differ for paths 1 and 2. We can write this

Property of a nonconservative force

$$W_{PQ} \text{ (along 1)} \neq W_{PQ} \text{ (along 2)}$$

Furthermore, from this condition we can show that if a force is nonconservative, the work done by that force on a particle that moves through any closed path is *not necessarily zero.* Since the work done in going from P to Q along path 2 is equal to the negative of the work done in going from Q to P along path 2, it follows from the first condition of a nonconservative force that

$$W_{PQ}\text{ (along 1)} \neq -W_{QP}\text{ (along 2)}$$
$$W_{PQ}\text{ (along 1)} + W_{QP}\text{ (along 2)} \neq 0$$

Property of a nonconservative force

The force of sliding friction is a good example of a nonconservative force. If an object is moved over a rough, horizontal surface between two points along various paths, the work done by the frictional force will certainly depend on the path. The negative work done by the frictional force along any particular path between two points will equal the force of friction multiplied by the length of the path. Various paths involve various amounts of work. The absolute magnitude of the least work done by the frictional force will correspond to a straight-line path between the two points. Furthermore, for a closed path you should note that the total work done by friction is nonzero since the force of friction opposes the motion along the entire path.

As an instructive example, suppose you were to displace a book between two points on a rough, horizontal surface, such as a table. If the book is displaced in a straight line between the two points, the work done by friction is simply $-fd$, where d is the distance between the points. However, if the book is moved along *any other* path between the two points (such as a semicircular path), the work done by friction would be *greater* (in absolute magnitude) than $-fd$. Finally, if the book is moved through any closed path (such as a circle), the work done by friction would clearly be nonzero since the frictional force opposes the motion.

In the example of a ball thrown vertically in the air with an initial speed v_i, careful measurements would show that because of air resistance, the ball would return to the thrower's hand with a speed less than v_i. Consequently, the final kinetic energy is less than the initial kinetic energy. The presence of a nonconservative force has reduced the ability of the system to do work by virtue of its motion. We shall sometimes refer to a nonconservative force as a *dissipative force*. For this reason, frictional forces are often referred to as being dissipative.

8.2 POTENTIAL ENERGY

In the previous section we found that the work done by a conservative force does not depend on the path taken by the particle and is independent of the particle's velocity. The work done is a function only of the particle's initial and final coordinates. For these reasons, we can define a potential energy function U such that the work done equals the decrease in the potential energy. That is, the work done by a conservative force $\mathbf{F}$ as the particle moves along the x axis is[1]

$$W_c = \int_{x_i}^{x_f} F_x\,dx = U_i - U_f = -\Delta U \tag{8.1}$$

That is, *the work done by a conservative force equals the negative of the change in the potential energy associated with that force.* Defining the change in the potential energy as $\Delta U = U_f - U_i$, we can express Eq. 8.1 as

$$\Delta U = U_f - U_i = -\int_{x_i}^{x_f} F_x\,dx \tag{8.2}$$

Change in potential energy

where F_x is the component of $\mathbf{F}$ in the direction of the displacement.

[1]For a general displacement, the work done in two or three dimensions also equals $U_i - U_f$, where $U = U(x,y,z)$. We write this formally $W = \int_i^f \mathbf{F}\cdot d\mathbf{s} = U_i - U_f$.

It is often convenient to establish some particular location, x_i, to be a reference point and to then measure all potential energy differences with respect to this point. With this understanding, we can define the potential energy function as

$$U_f(x) = -\int_{x_i}^{x_f} F_x\, dx + U_i \tag{8.3}$$

Furthermore, the value of U_i is often taken to be zero at some arbitrary reference point. It really doesn't matter what value we assign to U_i, since it only shifts $U_f(x)$ by a constant, and it is only the *change* in potential energy that is physically meaningful. If the conservative force is known as a function of position, we can use Eq. 8.3 to calculate the change in potential energy of a body as it moves from x_i to x_f. It is interesting to note that in the one-dimensional case, a force is *always* conservative if it is a function of position only. This is generally not true for motion involving two- or three-dimensional displacements.

The work done by a nonconservative force does depend on the path as a particle moves from one position to another and could also depend on the particle's velocity or on other quantities. Therefore, the work done is not simply a function of the initial and final coordinates of the particle. We conclude that there is no potential energy function associated with a nonconservative force.

8.3 CONSERVATION OF MECHANICAL ENERGY

Suppose a particle moves along the x axis under the influence of only *one* conservative force, $\boldsymbol{F}_x$. If this is the only force acting on the particle, then the work-energy theorem tells us that the work done by that force equals the increase in kinetic energy of the particle:

$$W_c = \Delta K$$

Since the force is conservative, according to Eq. 8.1 we can write $W_c = -\Delta U$. Hence,

$$\Delta K = -\Delta U$$

$$\Delta K + \Delta U = \Delta(K + U) = 0 \tag{8.4}$$

This is the *law of conservation of mechanical energy,* which can be written in the alternative form

Conservation of mechanical energy

$$K_i + U_i = K_f + U_f \tag{8.5}$$

If we now define the total mechanical energy of the system, E, as the sum of the kinetic energy and potential energy, we can express the conservation of mechanical energy as

$$E_i = E_f \tag{8.6a}$$

where

Total mechanical energy

$$E \equiv K + U \tag{8.6b}$$

The law of conservation of mechanical energy states that the *total mechanical energy of a system remains constant if the only force that does work is a conservative force.* This is equivalent to the statement that if the kinetic energy of a conservative system increases (or decreases) by some amount, the potential energy must decrease (or increase) by the same amount.

If more than one conservative force acts on the system, then there is a potential energy function associated with *each* force. In such a case, we can write the law of conservation of mechanical energy

Conservation of mechanical energy

$$K_i + \Sigma U_i = K_f + \Sigma U_f \tag{8.7}$$

where the number of terms in the sums equals the number of conservative forces present.

8.4 GRAVITATIONAL POTENTIAL ENERGY NEAR THE EARTH'S SURFACE

When a body moves in the presence of the earth's gravity, the gravitational force can do work on the body. In the case of a freely falling body (a projectile), the work done by gravity is a function of the *vertical* displacement of the body. This result is also valid in the more general case where the body undergoes both horizontal and vertical displacement.

Consider a particle being displaced from P to Q along various paths in the presence of a constant gravitational force[2] (Fig. 8.2). The work done along the path PAQ can be broken into two segments. The work done along PA is $-mgh$ (since mg is opposite this displacement), and the work done along AQ is zero (since mg is perpendicular to this path). Hence, $W_{PAQ} = -mgh$. Likewise, the work done along PBQ is also $-mgh$, where $W_{PB} = 0$ and $W_{BQ} = -mgh$. Now consider the general path described by the solid line from P to Q. The curve is broken down into a series of horizontal and vertical steps. There is no work done by the force of gravity along the horizontal steps, since mg is perpendicular to these elements of displacement. Work is done by the force of gravity only along the vertical displacement, where the work done in the nth vertical step is $-mg\,\Delta y_n$. Thus, the total work done by the force of gravity as the particle is displaced upward a distance h is the sum of the work done along each vertical displacement. Summary of all such terms gives

$$W_g = -mg \sum_n \Delta y_n = -mgh$$

Since $h = y_f - y_i$, we can express W_g as

$$W_g = mgy_i - mgy_f \tag{8.8}$$

We conclude that since the work done by the force of gravity is independent of the path, the gravitational force is a conservative force.

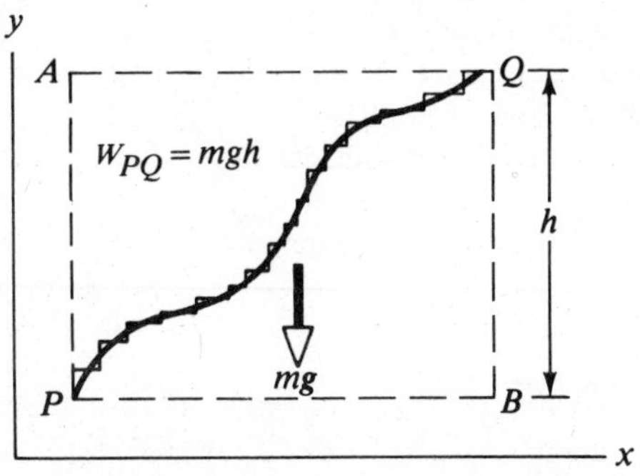

Figure 8.2 A particle that moves between the points P and Q under the influence of gravity can be envisioned as moving along a series of horizontal and vertical steps. The work done by gravity along each horizontal element is zero, and the net work done by gravity is equal to the sum of the works done along the vertical displacement.

[2]The assumption that the force of gravity is constant is a good one as long as the vertical displacement is small compared with the earth's radius.

Since the force of gravity is conservative, we can define a *gravitational potential energy function* U_g as

Gravitational potential energy

$$U_g = mgy \tag{8.9}$$

where we have chosen to take $U_g = 0$ at $y = 0$. Note that this function depends on the choice of origin of coordinates and is valid only when the displacement of the particle in the vertical direction is small compared with the earth's radius. A general expression for the gravitational potential energy will be developed in Chapter 14.

Substituting the definition of U_g (Eq. 8.9) into the expression for the work done by the force of gravity (Eq. 8.8) gives

$$W_g = U_i - U_f = -\Delta U_g \tag{8.10}$$

That is, *the work done by the force of gravity is equal to the initial value of the potential energy minus the final value of the potential energy.* We conclude from Eq. 8.10 that when the displacement is upward, $y_f > y_i$, and therefore $U_i < U_f$ and the work done by gravity is negative. This corresponds to the case where the force of gravity is *opposite* the displacement. When the particle is displaced downward, $y_f < y_i$, and so $U_i > U_f$ and the work done by gravity is positive. In this case, mg is in the *same* direction as the displacement.

The term *potential energy* implies that the particle has the potential, or capability, of gaining kinetic energy when released from some point under the influence of gravity. The choice of the origin of coordinates for measuring U_g is completely arbitrary, since only differences in potential energy are important. However, it is often convenient to choose the surface of the earth as the reference position $y_i = 0$.

If the force of gravity is the *only* force acting on a body, then the total mechanical energy of the body is conserved (Eq. 8.5). Therefore, the law of conservation of mechanical energy for a freely falling body can be written

Conservation of mechanical energy for a freely falling body

$$\frac{1}{2}mv_i^2 + mgy_i = \frac{1}{2}mv_f^2 + mgy_f \tag{8.11}$$

Example 8.1 *Ball in Free Fall*

A ball of mass m is dropped from a height h above the ground as in Fig. 8.3. (a) Determine the speed of the ball when it is at a height y above the ground, neglecting air resistance.

Since the ball is in free fall, the only force acting on it is the gravitational force. Therefore, we can use the law of conservation of mechanical energy. When the ball is released from rest at a height h above the ground, its kinetic energy is $K_i = 0$ and its potential energy is $U_i = mgh$, where the y coordinate is measured from ground level. When the ball is at a distance y above the ground, its kinetic energy is $K_f = \frac{1}{2}mv_f^2$ and its potential energy relative to the ground is $U_f = mgy$. Applying Eq. 8.11, we get

$$K_i + U_i = K_f + U_f$$

$$0 + mgh = \frac{1}{2}mv_f^2 + mgy$$

$$v_f^2 = 2g(h - y)$$

(b) Determine the speed of the ball at y if it were given an initial speed v_i at the initial altitude h.

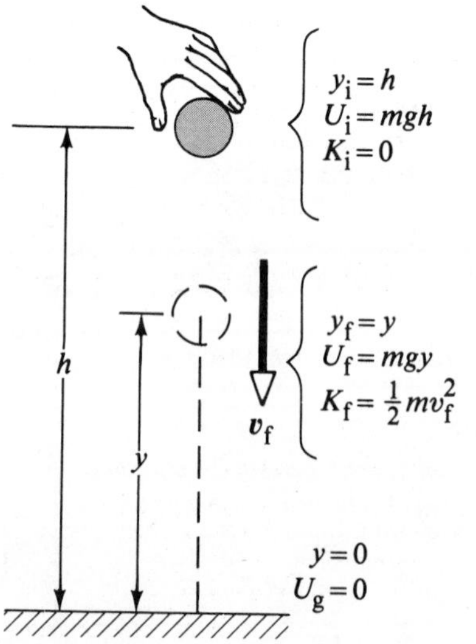

Figure 8.3 (Example 8.1) A ball is dropped from a height h above the floor. Initially, its total energy is its potential energy, equal to mgh relative to the floor. At the elevation y, its energy is the sum of the kinetic and potential energies.

In this case, the initial energy includes kinetic energy equal to $\frac{1}{2}mv_i^2$ and Eq. 8.11 gives

$$\frac{1}{2}mv_i^2 + mgh = \frac{1}{2}mv_f^2 + mgy$$

$$v_f^2 = v_i^2 + 2g(h - y)$$

Note that this result is consistent with an expression from kinematics, $v_y^2 = v_{y0}^2 - 2g(y - y_0)$, where $y_0 = h$. Furthermore, this result is valid even if the initial velocity is at an angle to the horizontal (the projectile situation).

Example 8.2 *The Pendulum*

A pendulum consists of a sphere of mass m attached to a light cord of length l as in Fig. 8.4. The sphere is released from rest when the cord makes an angle θ_0 with the vertical, and the pivot at 0 is frictionless. (a) Find the speed of the sphere when it is at the lowest point, b.

The only force that does work on m is the force of gravity, since the force of tension is always perpendicular to each element of the displacement and hence does no work. Since the force of gravity is a conservative force, the total mechanical energy is conserved, and as the pendulum swings, there is a continuous transfer between potential and kinetic energy. At the instant the pendulum is released, the energy is entirely potential energy. At point b, the pendulum has kinetic energy but has lost some potential energy. At point c, the pendulum has regained its initial potential energy and its kinetic energy is again zero. If we measure the y coordinates from the center of rotation, then $y_a = -l\cos\theta_0$ and $y_b = -l$. Therefore, $U_a = -mgl\cos\theta_0$ and $U_b = -mgl$. Applying the principle of conservation of mechanical energy gives

$$K_a + U_a = K_b + U_b$$

$$0 - mgl\cos\theta_0 = \frac{1}{2}mv_b^2 - mgl$$

$$(1) \qquad v_b = \sqrt{2gl(1 - \cos\theta_0)}$$

(b) What is the tension T in the cord at b?

Since the force of tension does no work, it cannot be determined using the energy method. To find T_b, we can apply Newton's second law to the radial direction. First, recall that the centripetal acceleration of a particle moving in a circle is equal to v^2/r directed toward the center of rotation. Since $r = l$ in this example, we get

$$(2) \qquad \Sigma F_r = T_b - mg = ma_r = mv_b^2/l$$

Substituting (1) into (2) gives for the tension at point b

$$(3) \qquad T_b = mg + 2mg(1 - \cos\theta_0) = mg(3 - 2\cos\theta_0)$$

For example, if $l = 2.0$ m, $\theta_0 = 30°$, and $m = 0.50$ kg, we find from (1) and (3) that

$$v_b = \sqrt{2gl(1 - \cos\theta_0)}$$

$$= \sqrt{2\left(9.8\frac{m}{s^2}\right)(2.0\text{ m})(1 - \cos 30°)}$$

$$= 2.3\text{ m/s}$$

$$T_b = mg(3 - 2\cos\theta_0)$$

$$= (0.50\text{ kg})\left(9.8\frac{m}{s^2}\right)(3 - 2\cos 30°)$$

$$= 6.2\text{ N}$$

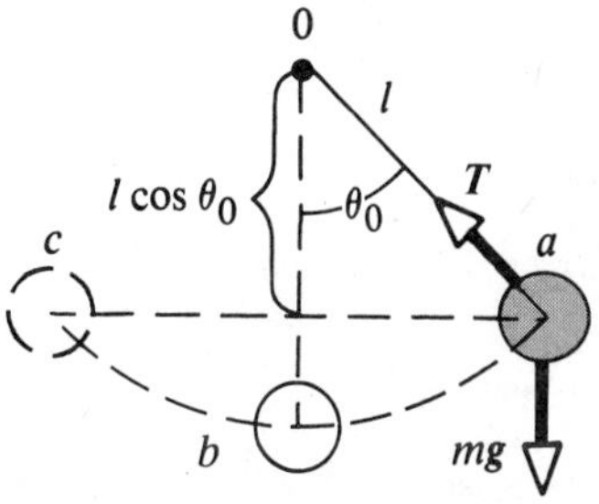

Figure 8.4 (Example 8.2) If the pendulum is released from rest at the angle θ_0, it will never swing above this position during its motion. At the start of the motion, position a, its energy is entirely potential energy. This is transformed into kinetic energy at the lowest elevation, position b.

Q1. A bowling ball is suspended from the ceiling of a lecture hall by a strong cord. The bowling ball is drawn away from its equilibrium position and released from rest at the tip of the demonstrator's nose. If the demonstrator remains stationary, explain why she will not be struck by the ball on its return swing. Would the demonstrator be safe if the ball were given a push from this position?

Q2. Explain why it is possible to have a negative value for the gravitational potential energy.

Q3. A ball is dropped by a person from the top of a building, while another person at the bottom observes its motion. Will these two people agree on the value of the ball's potential energy? on the *change* in potential energy of the ball? on the kinetic energy of the ball?

Q4. When a person runs in a track event at constant velocity, is any work done? (Note that although the runner may move with constant velocity, the legs and arms undergo acceleration.) How does air resistance enter into the picture?

Q5. Our body muscles exert forces when we lift, push, run, jump, etc. Are these forces conservative?

Q6. When nonconservative forces act on a system, does the total mechanical energy remain constant?

8.5 NONCONSERVATIVE FORCES AND THE WORK-ENERGY THEOREM

In realistic physical systems, nonconservative forces, such as friction, are usually present. Therefore, the total mechanical energy is not a constant. However, we can use the work-energy theorem to account for the presence of nonconservative forces. If W_{nc} represents the work done on a particle by all nonconservative forces and W_c is the work done by all conservative forces, we can write the work-energy theorem

$$W_{nc} + W_c = \Delta K$$

Since $W_c = -\Delta U$ (Eq. 8.1), this equation reduces to

$$W_{nc} = \Delta K + \Delta U = (K_f - K_i) + (U_f - U_i) \qquad (8.12)$$

That is, *the work done by all nonconservative forces equals the change in kinetic energy plus the change in potential energy.* Since the total mechanical energy is given by $E = K + U$, we can also express Eq. 8.12 as

Work done by non-conservative forces

$$W_{nc} = (K_f + U_f) - (K_i + U_i) = E_f - E_i \qquad (8.13)$$

That is, *the work done by all nonconservative forces equals the change in the total mechanical energy of the system.* Of course, when there are no nonconservative forces present, it follows that $W_{nc} = 0$ and $E_i = E_f$; that is, the total mechanical energy is conserved.

Example 8.3 *Block Moving on Incline*

A 3-kg block slides down a rough incline 1 m in length as in Fig. 8.5. The block starts from rest at the top and experiences a constant force of friction of magnitude 5 N; the angle of inclination is 30°. (a) Use energy methods to determine the speed of the block when it reaches the bottom of the incline.

Since $v_i = 0$, the initial kinetic energy is zero. If the y coordinate is measured from the bottom of the incline, then $y_i = 0.50$ m. Therefore, the total mechanical energy of the block at the top is potential energy given by

$$E_i = U_i = mgy_i = (3\text{ kg})\left(9.8\frac{\text{m}}{\text{s}^2}\right)(0.50\text{ m}) = 14.7\text{ J}$$

When the block reaches the bottom, its kinetic energy is $\frac{1}{2}mv_f^2$, but its potential energy is *zero* since its elevation is $y_f = 0$. Therefore, the total mechanical energy at the bottom is $E_f = \frac{1}{2}mv_f^2$. However, we cannot say that $E_i = E_f$ in this case. Since there is a nonconservative force that does work W' on the block, namely, the force of friction, $W_{nc} = -fs$, where s is the displacement along the plane. (Recall that the forces normal to the plane do no work on the block since they are perpendicular to the displacement.) Since $f = 5$ N and $s = 1$ m,

$$W_{nc} = -fs = (-5\text{ N})(1\text{ m}) = -5\text{ J}$$

That is, some mechanical energy is lost because of the presence of the retarding force. Applying the work-energy

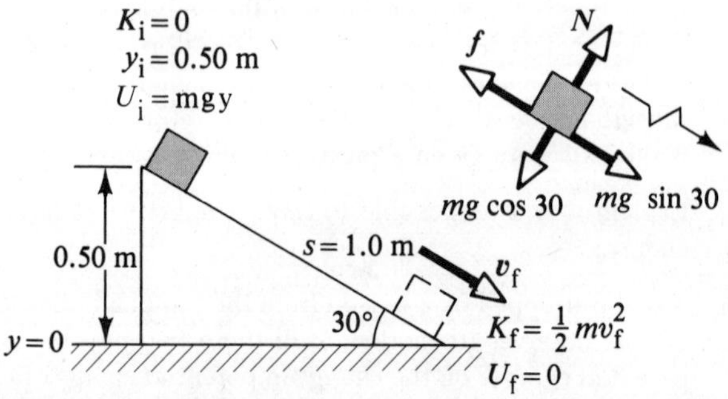

Figure 8.5 (Example 8.3) A block slides down a rough incline under the influence of gravity. Its potential energy decreases while its kinetic energy increases.

theorem in the form of Eq. 8.13 gives

$$W_{nc} = E_f - E_i$$

$$-fs = \frac{1}{2}mv_f^2 - mgy_i$$

$$\frac{1}{2}mv_f^2 = 14.7\text{ J} - 5\text{ J} = 9.7\text{ J}$$

$$v_f^2 = \frac{19.4\text{ J}}{3\text{ kg}} = \frac{6.47\text{ kg}}{\text{kg}}\frac{\text{m}^2}{\text{s}^2}$$

$$v_f = 2.54\text{ m/s}$$

(b) Check the answer to (a) using Newton's second law to first find the acceleration.

Summing the forces along the plane gives

$$mg\sin 30° - f = ma$$

$$a = g\sin 30° - \frac{f}{m} = 9.8(0.5) - \frac{5}{3} = 3.23\frac{\text{m}}{\text{s}^2}$$

Since the acceleration is constant, we can apply the expression $v_f^2 = v_i^2 + 2as$, where $v_i = 0$:

$$v_f^2 = 2as = 2(3.23\text{ m/s}^2)(1\text{ m}) = 6.46\text{ m}^2/\text{s}^2$$

$$v_f = 2.54\text{ m/s}$$

Note that if the plane were frictionless, $W_{nc} = 0$ and we would find that $v_f = 3.13$ m/s and $a = 4.9$ m/s^2.

Example 8.4 *Motion on a Curved Track*

A child of mass m takes a ride on an irregularly curved slide of height h, as in Fig. 8.6. The child starts from rest at the top. (a) Determine the speed of the child at the bottom, assuming there is no friction present.

First, note that the normal force, N, does no work on the child since this force is always perpendicular to each element of the displacement. Furthermore, since there is no friction, $W_{nc} = 0$ and we can apply the law of conservation of mechanical energy. If we measure the y coordinate from the bottom of the slide, then $y_i = h$, $y_f = 0$, and we get

$$K_i + U_i = K_f + U_f$$

$$0 + mgh = \frac{1}{2}mv_f^2 + 0$$

$$v_f = \sqrt{2gh}$$

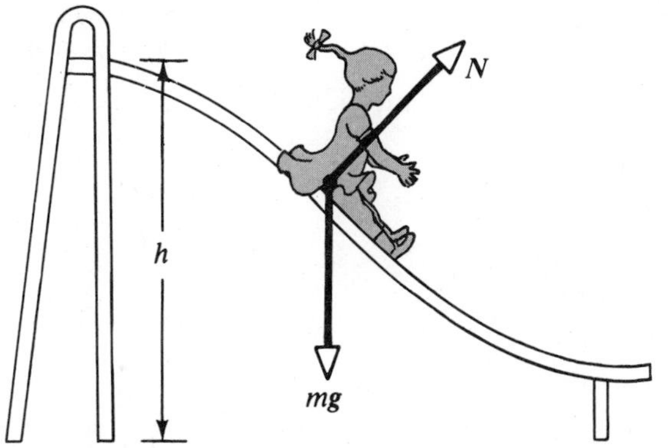

Figure 8.6 (Example 8.4) If the slide is frictionless, the speed of the child at the bottom depends only on the height of the slide and is independent of the shape of the slide.

Note that this result is the same as if the child had fallen vertically through a distance h! For example, if $h = 6$ m, then

$$v_f = \sqrt{2gh} = \sqrt{2\left(9.8\frac{\text{m}}{\text{s}^2}\right)(6\text{ m})} = 10.8\text{ m/s}$$

(b) If there were a frictional force acting on the child, what would be the work done by this force?

In this case, $W_{nc} \neq 0$ and mechanical energy is *not* conserved. We can use Eq. 8.13 to find the work done by friction, assuming the final velocity at the bottom is known:

$$W_{nc} = E_f - E_i = \frac{1}{2}mv_f^2 - mgh$$

For example, if $v_f = 8.0$ m/s, $m = 20$ kg, and $h = 6$ m, we find that

$$W_{nc} = \frac{1}{2}(20\text{ kg})(8.0\text{ m/s})^2 - (20\text{ kg})\left(9.8\frac{\text{m}}{\text{s}^2}\right)(6\text{ m})$$

$$= -536\text{ J}$$

Again, W_{nc} is negative since the *work done by sliding friction is always negative.* Note, however, that because the slide is curved, the normal force changes in magnitude and direction during the motion. Therefore, the frictional force, which is proportional to N, also changes during the motion. Do you think it would be possible to determine μ from these data?

8.6 POTENTIAL ENERGY STORED IN A SPRING

Now let us consider another mechanical system that is conveniently described using the concept of potential, or stored, energy. A block of mass m slides on a frictionless, horizontal surface with constant velocity v_i and collides with a light coiled spring as in Fig. 8.7. The description that follows is greatly simplified by assuming that the spring is very light and therefore its kinetic energy is negligible. The spring exerts a force on the block to the left as the spring is compressed, and eventually the block comes to rest (Fig. 8.7c). The initial energy in the system (block + spring) is the initial kinetic energy of the block. When the block comes to

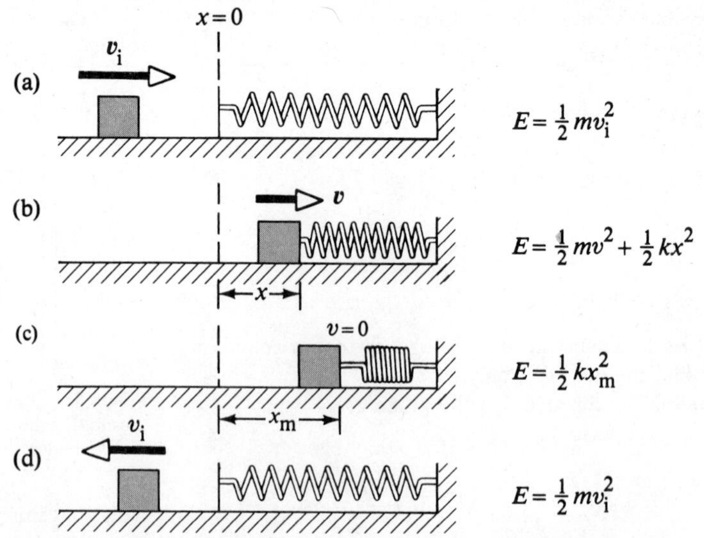

Figure 8.7 A block sliding on a smooth, horizontal surface collides with a light spring. (a) Initially the mechanical energy is all kinetic energy. (b) The mechanical energy is the sum of the kinetic energy of the block and the elastic potential energy stored in the spring. (c) The energy is entirely potential energy. (d) The energy is transformed back to the kinetic energy of the block. Note that the total energy remains constant.

rest after colliding with the spring, the kinetic energy is zero. Since the spring force is conservative and since there are no external forces that can do work on the system, the total mechanical energy of the system must remain constant. Thus, there is a transfer of energy from kinetic energy of the block to potential energy stored in the spring. Eventually, the block moves in the opposite direction and regains all of its initial kinetic energy, as described in Fig. 8.7d.

To describe the potential energy stored in the spring, recall from the previous chapter that the work done by the spring on the block as the block moves from $x = x_i$ to $x = x_f$ is

$$W_s = \frac{1}{2}kx_i^2 - \frac{1}{2}kx_f^2$$

The quantity $\frac{1}{2}kx^2$ is defined as the *elastic potential energy* stored in the spring, denoted by the symbol U_s:

Potential energy stored in a spring

$$U_s = \frac{1}{2}kx^2 \tag{8.14}$$

Note that the elastic potential energy stored in the spring is zero when the spring is unstretched, or undeformed, ($x = 0$). Furthermore, U_s is a *maximum* when the spring has reached its maximum compression (Fig. 8.7c). Finally, U_s is *always* positive since it is proportional to x^2.

The total mechanical energy of the block-spring system can be expressed as

$$E = \frac{1}{2}mv_i^2 + \frac{1}{2}kx_i^2 = \frac{1}{2}mv_f^2 + \frac{1}{2}kx_f^2 \tag{8.15}$$

Applying this expression to the system described in Fig. 8.7 and noting that $x_i = 0$, we get

$$E = \frac{1}{2}mv_i^2 = \frac{1}{2}mv_f^2 + \frac{1}{2}kx_f^2 \tag{8.16}$$

This expression says that for any displacement x_f, when the speed of the block is v_f, the sum of the kinetic and potential energies is equal to a *constant* E, which equals the total energy. In this case, the total energy is the initial kinetic energy of the block.

Now suppose there are nonconservative forces acting on the block-spring system. In this case, we can apply the work-energy theorem in the form of Eq. 8.13, which gives

$$W_{nc} = \left(\frac{1}{2}mv_f^2 + \frac{1}{2}kx_f^2\right) - \left(\frac{1}{2}mv_i^2 + \frac{1}{2}kx_i^2\right) \quad (8.17)$$

That is, the total mechanical energy is not a constant of the motion when nonconservative forces act on the system. Again, if W_{nc} is due to a force of friction, then W_{nc} is *negative* and the final energy is less than the initial energy.

Example 8.5 *Mass-Spring Collision*

A mass of 0.80 kg is given an initial velocity $v_i = 1.2$ m/s to the right and collides with a light spring of force constant $k = 50$ N/m, as in Fig. 8.7. (a) If the surface is frictionless, calculate the initial maximum compression of the spring after the collision.

The total mechanical energy is conserved since $W_{nc} = 0$. Applying Eq. 8.15 to this system with $v_f = 0$ gives

$$\frac{1}{2}mv_i^2 + 0 = 0 + \frac{1}{2}kx_f^2$$

$$x_f = \sqrt{\frac{m}{k}}v_i = \sqrt{\frac{0.8\text{ kg}}{50\text{ N/m}}}(1.2\text{ m/s}) = 0.15\text{ m}$$

(b) If a constant force of friction acts between the block and the surface with $\mu = 0.5$ and if the speed of the block just as it collides with the spring is $v_i = 1.2$ m/s, what is the maximum compression in the spring?

In this case, the mechanical energy of the system is *not* conserved because of the presence of friction, which does negative work on the system. The magnitude of the frictional force is

$$f = \mu N = \mu mg = 0.5(0.80\text{ kg})\left(9.8\ \frac{\text{m}}{\text{s}^2}\right) = 3.9\text{ N}$$

Therefore, the work done by the force of friction as the block is displaced from $x_i = 0$ to $x_f = x$ is

$$W_{nc} = -fx = (-3.9x)\text{ J}$$

Substituting this into Eq. 8.17 gives

$$W_{nc} = \left(0 + \frac{1}{2}kx^2\right) - \left(\frac{1}{2}mv_i^2 + 0\right)$$

$$-3.9x = \frac{50}{2}x^2 - \frac{1}{2}(0.80)(1.2)^2$$

$$25x^2 + 3.9x - 0.58 = 0$$

Solving the quadratic equation for x gives $x = 0.093$ m and $x = -0.25$ m. The physically acceptable root is $x = 0.093$ m $= 9.3$ cm. The negative root is unacceptable since the block must be displaced to the right of the origin after coming to rest. Note that 9.3 cm is *less* than the distance obtained in the frictionless case (a). This result is what we should expect, since the force of friction retards the motion of the system.

Example 8.6

Two blocks are connected by a light string that passes over a frictionless pulley as in Fig. 8.8. The block of mass m_1 lies on a rough surface and is connected to a spring of force constant k. The system is released from rest when the spring is unstretched. If m_2 falls a distance h before coming to rest, calculate the coefficient of kinetic friction between m_1 and the surface.

Solution: In this situation there are two forms of potential energy to consider: the gravitational potential energy and the elastic potential energy stored in the spring. We can write the work-energy theorem

$$(1) \qquad W_{nc} = \Delta K + \Delta U_g + \Delta U_s$$

where ΔU_g is the *change* in the gravitational potential energy and ΔU_s is the *change* in the elastic potential energy of the system. In this situation, $\Delta K = 0$ since the initial and final velocities of the system are zero. Also, W_{nc} is the work done by friction, given by

$$(2) \qquad W_{nc} = -fh = -\mu m_1 gh$$

The change in the gravitational potential energy is associated only with m_2 since the vertical coordinate of m_1 does not change. Therefore, we get

$$(3) \qquad \Delta U_g = U_f - U_i = -m_2 gh$$

where the coordinates have been measured from the low-

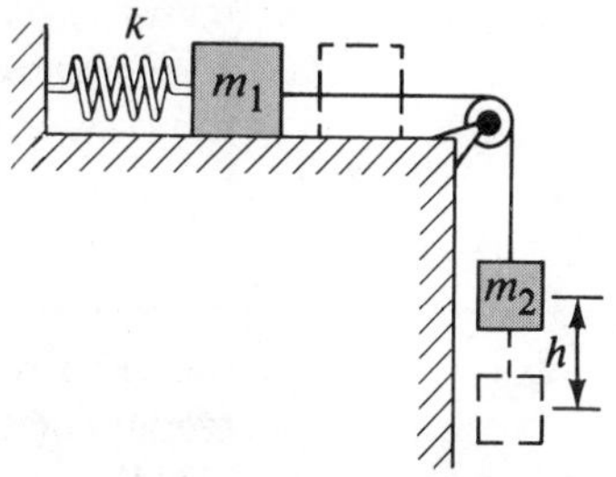

Figure 8.8 (Example 8.6) As the system moves from the highest to the lowest elevation of m_2, the system loses gravitational potential energy but gains elastic potential energy stored in the spring. Some mechanical energy is lost because of the presence of the nonconservative force of friction between m_1 and the surface.

est position of m_2. The change in the elastic potential energy stored in the spring is given by

$$(4) \qquad \Delta U_s = U_f - U_i = \frac{1}{2}kh^2 - 0$$

Substituting (2), (3), and (4) into (1) gives

$$-\mu m_1 gh = -m_2 gh + \frac{1}{2}kh^2$$

$$\mu = \frac{m_2 g - \frac{1}{2}kh}{m_1 g}$$

This represents a possible experimental technique for measuring the coefficient of kinetic friction. For example, if $m_1 = 0.50$ kg, $m_2 = 0.30$ kg, $k = 50$ N/m, and $h = 5.0 \times 10^{-2}$ m, we find that

$$\mu = \frac{(0.30\text{ kg})\left(9.8\,\frac{\text{m}}{\text{s}^2}\right) - \frac{1}{2}\left(50\,\frac{\text{N}}{\text{m}}\right)(5.0 \times 10^{-2}\text{ m})}{(0.50\text{ kg})\left(9.8\,\frac{\text{m}}{\text{s}^2}\right)} = 0.34$$

Q7. If three different conservative forces and one nonconservative force act on a system, how many potential energy terms will appear in the work-energy theorem?

Q8. A block is connected to a spring that is suspended from the ceiling. If the block is set in motion and air resistance is neglected, will the total energy of the system be conserved? How many forms of potential energy are there for this situation?

8.7 RELATIONSHIP BETWEEN CONSERVATIVE FORCES AND POTENTIAL ENERGY

In the previous sections we saw that the concept of potential energy is related to the configuration, or coordinates, of a system. In a few examples, we showed how to obtain the potential energy from a knowledge of the conservative force. (Remember that one can associate a potential energy function only with a conservative force.)

According to Eq. 8.1, the change in the potential energy of a particle under the action of a conservative force equals the negative of the work done. If the system undergoes an infinitesimal displacement, dx, we can express the infinitesimal change in potential energy, dU, as

$$dU = -F_x\, dx$$

Therefore, the conservative force is related to the potential energy function through the relationship

Relation between force and potential energy

$$F_x = -\frac{dU}{dx} \tag{8.18}$$

That is, *the conservative force equals the negative derivative of the potential energy with respect to x.*[*]

We can easily check this relationship for the two examples already discussed. In the case of the deformed spring, $U_s = \frac{1}{2}kx^2$, and therefore

$$F_s = -\frac{dU_s}{dx} = -\frac{d}{dx}\left(\frac{1}{2}kx^2\right) = -kx$$

which corresponds to the restoring force in the spring. Since the gravitational potential energy function is given by $U_g = mgy$, it follows from Eq. 8.18 that $F_g = -mg$.

[*] In a three-dimensional problem, where U depends on x,y,z, the force is related to U through the expression $\mathbf{F} = -\mathbf{i}\,\partial U/\partial x - \mathbf{j}\,\partial U/\partial y - \mathbf{k}\,\partial U/\partial z$, where $\partial/\partial x$, etc., are partial derivatives. In the language of vector calculus, $\mathbf{F}$ is said to equal the negative of the gradient of the scalar quantity $U(x,y,z)$.

We now see that U is an important function, since the conservative force can be derived from it. Furthermore, Eq. 8.18 should clarify the fact that adding a constant to the potential energy is unimportant.

8.8 ENERGY DIAGRAMS AND STABILITY OF EQUILIBRIUM

The qualitative behavior of the motion of a system can often be understood through an analysis of its potential energy curve. Consider the potential energy function for the mass-spring system, given by $U_s = \frac{1}{2}kx^2$. This function is plotted versus x in Fig. 8.9a. The force is related to U through the expression

$$F_s = -\frac{dU_s}{dx} = -kx$$

That is, the force is equal to the negative of the *slope* of the U versus x curve. When the mass is placed at rest at the equilibrium position ($x = 0$), where $F = 0$, it will remain there unless some external force acts on it. If the spring is stretched from equilibrium, x is positive and the slope dU/dx is positive; therefore F_s is negative and the mass accelerates back toward $x = 0$. If the spring is compressed, x is negative and the slope is negative; therefore F_s is positive and again the mass accelerates toward $x = 0$.

From this analysis, we conclude that the $x = 0$ position is one of *stable equilibrium*. That is, any movement away from this position results in a force that is directed back toward $x = 0$. In general, *positions of stable equilibrium correspond to those points for which* U(x) *has a minimum value.*

Stable equilibrium

From Fig. 8.9 we see that if the mass is given an initial displacement x_m and released from rest, its total energy initially is the potential energy stored in the spring, given by $\frac{1}{2}kx_m^2$. As motion commences, the system acquires kinetic energy at the expense of losing an equal amount of potential energy. Since the total energy must remain constant, the mass oscillates between the two points $x = \pm x_m$, called the *turning points*. From an energy viewpoint, the energy of the system cannot exceed $\frac{1}{2}kx_m^2$; therefore the mass must stop at these points and, because of the spring force, accelerate toward $x = 0$.

Another simple mechanical system that has a position of stable equilibrium is that of a ball rolling about in the bottom of a spherical bowl. If the ball is displaced from its lowest position, it will always tend to return to that position when released.

(a)

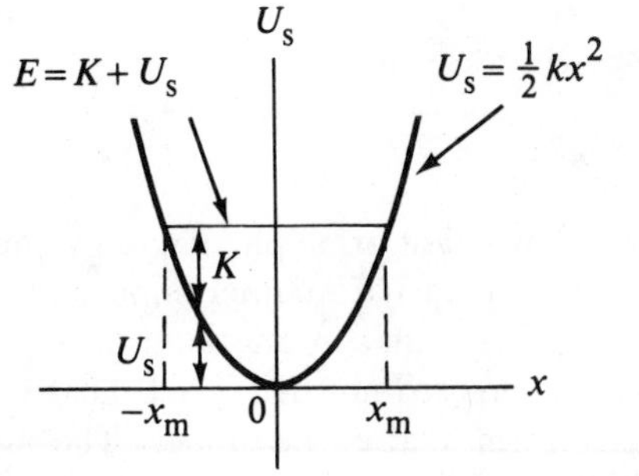

(b)

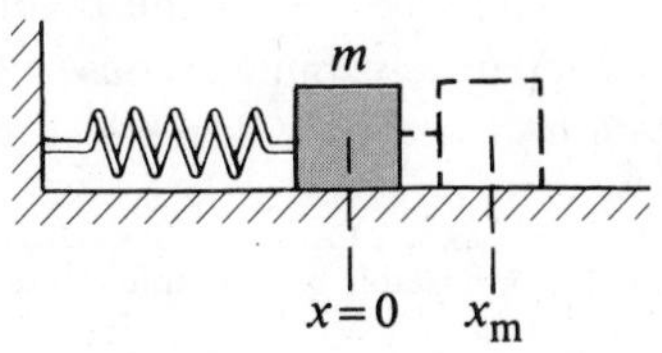

Figure 8.9 (a) The potential energy as a function of x for the block-spring system described in (b). The block oscillates between the turning points, which have the coordinates $x = \pm x_m$. Note that the restoring force of the spring always acts toward $x = 0$, the position of stable equilibrium.

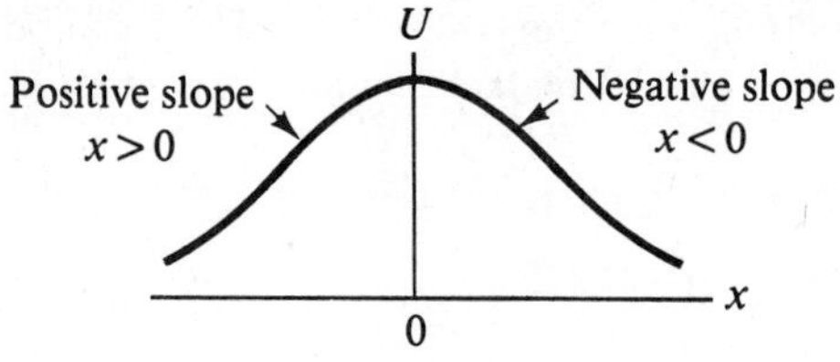

Figure 8.10 A plot of U versus x for a system that has a position of unstable equilibrium, located at $x = 0$. In this case, the force on the system for finite displacements is directed away from $x = 0$.

Unstable equilibrium

Now consider an example where the U versus x curve is as shown in Fig. 8.10. In this case, $F_x = 0$ at $x = 0$, and so the particle is in equilibrium at this point. However, this is a position of *unstable equilibrium* for the following reason. Suppose the particle is displaced to the *right* ($x > 0$). Since the slope is negative for $x > 0$, $F_x = -dU/dx$ is positive and the particle will accelerate away from $x = 0$. Now suppose that the particle is displaced to the left ($x < 0$). In this case, the force is *negative* since the slope is positive for $x < 0$. Therefore, the particle will again accelerate away from the equilibrium position. Therefore, the $x = 0$ position in this situation is called a position of *unstable equilibrium*, since for any displacement from this point, the force pushes the particle farther away from equilibrium. In fact, the force pushes the particle toward a position of lower potential energy. A ball placed on the top of an inverted spherical bowl is obviously in a position of unstable equilibrium. That is, if the ball is displaced slightly from the top and released, it will surely roll off the bowl. In general, *positions of unstable equilibrium correspond to those points for which* U(x) *has a maximum value.*[3]

Neutral equilibrium

Finally, a situation may arise where U is constant over some region, and hence $F = 0$. This is called a position of *neutral equilibrium*. Small displacements from this position produce neither restoring nor disrupting forces.

Q9. Consider a ball fixed to one end of a rigid rod with the other end pivoted on a horizontal axis so that the rod can rotate in a vertical plane. What are the positions of stable and unstable equilibrium?

Q10. A ball rolls on a horizontal surface. Is the ball in stable, unstable, or neutral equilibrium?

Q11. Why is it physically impossible to have a situation where $E - U < 0$?

Q12. What will the curve of U versus x look like if a particle is in a region of neutral equilibrium?

8.9 MASS-ENERGY

Mass-energy equivalence

This chapter has been concerned with the important principle of energy conservation and its application to various physical phenomena. Another important principle, called the *law of conservation of mass*, says that in any kind of ordinary physical or chemical process, matter is neither created nor destroyed. That is, the mass of the system before the process equals the mass after the process. The theory of relativity developed by Einstein in 1905 showed that these concepts of mass and energy must be modified. In particular, one of the consequences of the theory of relativity is that mass and energy are not conserved separately, but are conserved as a single entity called *mass-energy*. That is, *energy and mass are considered to be*

[3]Mathematically, you can test whether an extreme of U is stable or unstable by examining the sign of d^2U/dx^2.

equivalent concepts. Let us pursue this idea further by stating some results of the theory of relativity.[4]

As we stated earlier in the text, the laws of Newtonian mechanics are only valid for particle speeds that are small compared with the speed of light, c ($\approx 3 \times 10^8$ m/s). When the particle speeds are comparable to c, the equations of newtonian mechanics must be replaced by the more general equations predicted by the special theory of relativity. One consequence of the theory of relativity is that the kinetic energy of a particle of mass m moving with a speed v is no longer $\frac{1}{2}mv^2$, but instead is given by

$$K = \frac{mc^2}{\sqrt{1 - \frac{v^2}{c^2}}} - mc^2 \tag{8.19}$$

Relativistic kinetic energy

According to this expression, speeds greater than c are not allowed since K would be imaginary for $v > c$. Furthermore, as $v \to c$, $K \to \infty$. This is consistent with experimental observations, which have shown that no particles travel at speeds greater than c. (That is, c is the ultimate speed.)

From the point of view of the work-energy theorem, v can only approach c, since it would take an infinite amount of work to attain the speed $v = c$.

It is instructive to analyze Eq. 8.19 for the situation where v is small compared with c. In this case, we expect that K should reduce to the newtonian result, $\frac{1}{2}mv^2$. We can check this by using the binomial expansion applied to the radical $\left(1 - \frac{v^2}{c^2}\right)^{-1/2}$, with $v/c \ll 1$. If we let $x = v^2/c^2$, the expansion gives

$$\frac{1}{(1-x)^{1/2}} = 1 + \frac{x}{2} + \frac{3}{8}x^2 + \cdots$$

Making use of this expansion in Eq. 8.19 gives

$$\begin{aligned} K &= mc^2\left(1 + \frac{v^2}{2c^2} + \frac{3}{8}\frac{v^4}{c^4} + \cdots\right) - mc^2 \\ &= \frac{mv^2}{2} + \frac{3}{8}m\frac{v^4}{c^2} + \cdots \\ &\approx \frac{1}{2}mv^2 \qquad \text{for } \frac{v}{c} \ll 1 \end{aligned}$$

Thus, we see that the relativistic kinetic energy expression does indeed agree with the Newtonian expression for speeds that are small compared with c.

It is useful to express Eq. 8.19 in the form

$$K = E - mc^2$$

Rest energy = mc^2

where the constant term mc^2 is an intrinsic energy associated with the particle, called the *rest energy,* and E is the *total energy* of the particle, given by

$$E = \frac{mc^2}{\sqrt{1 - \frac{v^2}{c^2}}} \tag{8.20}$$

Total energy

With this notation, we can express Eq. 8.19 as $E = K + mc^2$. That is, *the total energy equals the sum of the kinetic energy and the rest energy.*

---

[4]Further details on the theory of relativity will be presented in Chapter 40.

Within the framework of this theory, we see that if the total energy is to be conserved, then any change in the rest energy mc^2 (which may be regarded as a kind of internal potential energy) has to be accompanied by a corresponding change in kinetic energy.

Furthermore, if the mass is somehow diminished, it could be converted to energy and do work. As Einstein concluded, "If a body gives off energy E in the form of radiation, its mass diminishes by E/c^2. . . . We are led to the more general conclusion that the mass of a body is a measure of its energy content; if the energy changes by E, the mass changes in the same sense by $E/(9 \times 10^{20})$, the energy being measured in ergs and mass in grams." Later in the same paper he predicts that it may be possible to convert mass to energy using special materials such as radium salts.

Conservation of mass-energy

Since mass can be transformed into energy and energy can be transformed into mass, *the conservation law that governs such processes states that the mass-energy of a system remains constant.* In ordinary experiments, if energy of any form (such as kinetic, potential, or thermal energy) is supplied to an object, the change in mass of the object would be $\Delta m = \Delta E/c^2$. However, since c^2 is so large, the changes in mass in any ordinary mechanical experiment (or chemical reaction) are too small to be detected. On the other hand, the exchange between mass and energy is very important for processes involving nuclei, where fractional changes in mass of about 10^{-3} are readily observed.

The rest energy associated with a small amount of matter is enormous. For example, the rest energy of 1 kg of any substance is calculated to be

$$E_{\text{rest}} = mc^2 = (1\text{ kg})\left(3 \times 10^8 \frac{\text{m}}{\text{s}}\right)^2 = 9 \times 10^{16}\text{ J}$$

This is equivalent to the energy content of about 15 million barrels of crude oil (about one day's consumption in the entire United States)! If we could easily convert this energy to useful work, our energy resources would be unlimited.

In reality, matter is not freely converted to energy. However, in some circumstances, a significant fraction of the rest energy can be converted to work. A good example is the enormous energy released when the uranium-235 nucleus undergoes fission into smaller fragments. This happens because the ^{235}U nucleus is more massive than the sum of the masses of the fission products. The awesome nature of the energy involved in such reactions is demonstrated in the explosion of a nuclear weapon.

Energy can also be converted to rest mass. A good example is a process called *pair production,* in which energy in the form of high-energy radiation called *gamma radiation* can suddenly disappear (under proper conditions) to form an electron and a positron (a positively charged electron). There is also a reverse process, called *pair annihilation,* in which an electron and a positron combine and disappear with the formation of high-energy gamma radiation. The energy of the radiation is exactly equal to the rest energies of the two combining particles plus their kinetic energies.

Example 8.7 *Fission of ^{235}U*

When $^{235}_{92}$U undergoes fission, one reaction is

$$\text{n} + {}^{235}_{92}\text{U} \xrightarrow[\text{fission}]{} {}^{144}_{56}\text{Ba} + {}^{89}_{36}\text{Kr} + 3\text{n}$$

where n denotes a neutron that excites the fission events. The total mass of the constituents on the left side of this equation is somewhat larger than the total mass of those on the right side. When this mass difference is multiplied by c^2, the energy equivalent (energy released) per fission event is about 1.6×10^{-11} J. Using this number, calculate the energy that would be released by 1 kg of ^{235}U, assuming that it all undergoes fission.

Solution: First determine the number of ^{235}U nuclei contained in 1 kg. Since one mole of ^{235}U contains 6.02×10^{23} atoms and the atomic weight of ^{235}U is 235 g/mole, 1 kg (= 1000 g) contains the following number of nuclei:

$$\text{Number of nuclei} = \left(\frac{1000\text{ g}}{235\text{ g/mole}}\right)\left(6.02 \times 10^{23}\,\frac{\text{nuclei}}{\text{mole}}\right)$$
$$= 2.56 \times 10^{24}\text{ nuclei}$$

Therefore, the total energy that would be released if all these nuclei underwent fission is

$$U = (2.56 \times 10^{24})(1.6 \times 10^{-11}\ \text{J}) = 4.1 \times 10^{13}\ \text{J}$$

This is equivalent to the energy content of about 14 000 barrels of crude oil. Comparing this figure with the *total* mass-energy of 1 kg of matter (9×10^{16} J), we see that only about 10^{-3} of the mass-energy is released in the fission process. However, this is a far more efficient form of energy generation than the combustion of a chemical fuel, where only about 10^{-10} of the mass-energy is released.

8.10 CONSERVATION OF ENERGY IN GENERAL

We have seen that the total mechanical energy of a system is conserved when only conservative forces act on the system. Furthermore, we were able to associate a potential energy with each conservative force. In other words, mechanical energy is lost when nonconservative forces, such as friction, are present.

We can generalize the energy conservation principle to include all forces acting on the system, both conservative and nonconservative. In the study of thermodynamics we shall find that mechanical energy can be transformed into thermal energy. For example, when a block slides over a rough surface, the mechanical energy lost is transformed into thermal or heat energy stored in the block, as evidenced by a measurable increase in its temperature. On a submicroscopic scale, we shall see that this internal thermal energy in a solid is associated with vibrations of the atomic constituents about their equilibrium positions. Since this internal atomic motion has kinetic and potential energy, one can say that frictional forces arise fundamentally from conservative atomic forces.[5] Therefore, if we include this increase in the internal energy of the system in our work-energy theorem, the total energy is conserved.

This is just one example of how you can analyze a system and always find that the total energy of an isolated system does not change, as long as you are careful in accounting for all forms of energy. That is, *energy can never be created or destroyed. Energy may be transformed from one form to another, but the total energy of an isolated system is always constant.* From a universal point of view, we can say that the *total energy of the universe is constant.* Therefore, if one part of the universe gains energy in some form, another part must lose an equal amount of energy. No violation of this principle has been found.

Total energy is always conserved

Other examples of energy transformations include the energy carried by sound waves resulting from the collision of two objects, the energy radiated by an accelerating charge in the form of electromagnetic waves (a radio antenna), and the elaborate sequence of energy conversions in a thermonuclear reaction.

In subsequent chapters, we shall see that the energy concept, and especially transformations of energy between various forms, join together the various branches of physics. In other words, one cannot really separate the subjects of mechanics, thermodynamics, and electromagnetism. Finally, from a practical viewpoint, all mechanical and electronic devices rely on some forms of energy transformation.

Q13. Explain the energy transformations that occur during the following athletic events: (a) the pole vault, (b) the shotput, (c) the high jump. What is the source of energy in each case?

Q14. Discuss all the energy transformations that occur during the operation of an automobile.

[5]By labeling friction a nonconservative force, we are able to limit the system we are studying. We have, in effect, avoided the complex problem of describing the dynamics of 10^{23} molecules and their interactions.

8.11 ENERGY FROM THE TIDES

The correlation between the tides and the position of the moon has been known for thousands of years. Newton first reasoned that the two high tides per day are due to the fact that the moon pulls harder on that part of the earth closest to it. The tidal force of the moon produces the greatest effect on the fluid part of the earth, creating a bulge on both the near and far sides of the earth relative to the moon. Thus, there is a high tide on the side of the earth nearest the moon and simultaneously one on the side farthest from the moon (Fig. 8.11). We now know that the sun also produces a tidal force on the oceans, which affects the time and height of the tides. (The solar tidal force is weaker than the lunar tidal force.) The highest high tides and lowest low tides occur when the earth, moon, and sun are aligned so that the sun's tidal force reinforces the moon's tidal force. These are called *spring tides*. On the other hand, when the moon and the sun are at right angles to each other with respect to the earth, their tidal forces do not reinforce each other, resulting in the lower high tides called *neap tides*.

In certain parts of the world, the tidal variations can be as much as 16 m, largely because of the physical nature of the basins holding the water (as opposed to any extra effect of the moon). When these large surges of water occur in narrow channels, energy can be extracted from the water in the following manner. A large dam is constructed in the channel, trapping water on the bay side, as in Fig. 8.12. Large gates are located in the lower portion of the dam, allowing water to flow when they are opened or be trapped when they are closed. The gates are closed at high tide, when the water levels on the bay and ocean sides are equal. At low tides, the ocean water has dropped a distance h and the gates are opened, allowing water to flow out of the bay. The flowing water is used to drive turbines, which generate electricity. After the water levels are again equal at low tide, the gates are closed. At the next high tide, the gates are again opened, allowing water to flow to the bay side and generating more power. At the next high tide, the gates are closed and the cycle is repeated. In this manner, the water flows through the gates four times per day.

We can estimate the power that can be generated in this manner using the follow-

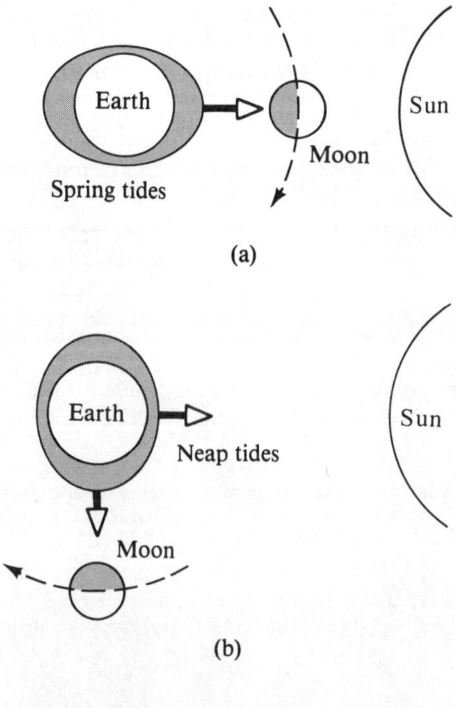

Figure 8.11 Schematic diagram of the tides. (a) The spring tides occur when the earth, moon, and sun are aligned. (b) The neap tides occur when the moon and sun are at right angles relative to the earth. (The figures are not drawn to scale, and the tidal bulges are exaggerated.)

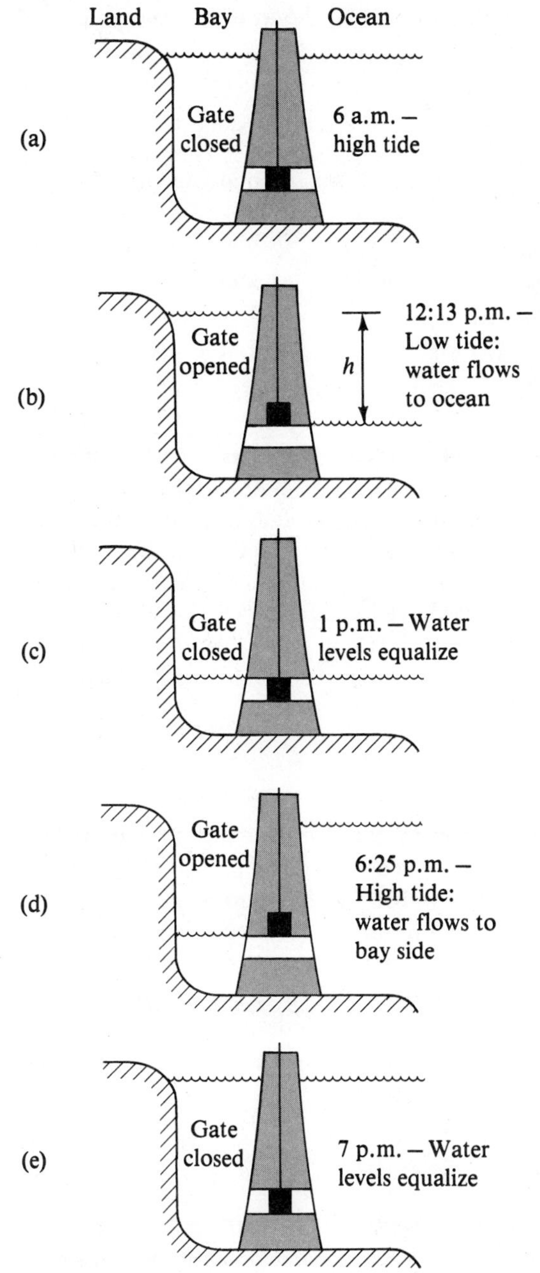

Figure 8.12 Cross-sectional views of a dam used to trap water in a bay during high and low tides. When the gate valve is opened, water flows through the gate and generates electrical power.

ing simple model. Suppose the bay has an area A and the variation between high and low tides is h, as in Fig. 8.13. The center of mass of this volume of water must fall through a distance $h/2$; hence the potential energy of the trapped water is

$$U = mg\frac{h}{2}$$

The mass of the trapped water is $m = \rho V = \rho A h$, where ρ is the density of water; therefore

$$U = \frac{1}{2}\rho A g h^2$$

Since the water flows through the gates four times per day, the energy available

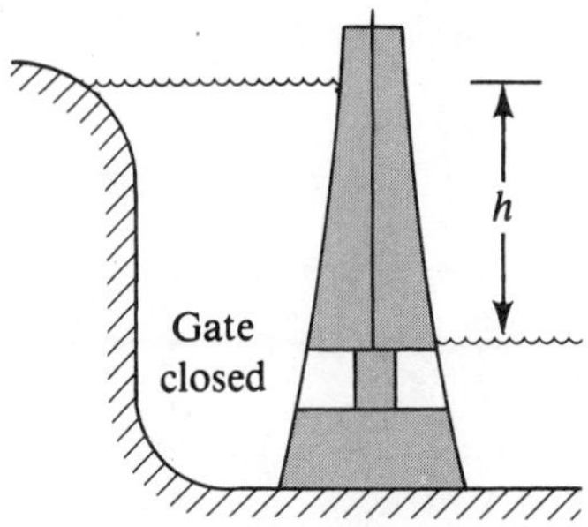

Figure 8.13 The center of gravity of the water trapped in the bay must fall through a distance $h/2$ at low tide before the levels are equalized.

from the tides each day is four times this value, and the available power is

$$P_{\text{max}} = 4\frac{U}{T} = \frac{2\rho Agh^2}{T}$$

where T is one day.

For example, if we take $A = 5 \times 10^7\ \text{m}^2$ (about 2 square miles) and $h = 2$ m, we find that

$$P_{\text{max}} = \frac{2\left(1000\ \frac{\text{kg}}{\text{m}^3}\right)(5 \times 10^7\ \text{m}^2)\left(9.8\ \frac{\text{m}}{\text{s}^2}\right)(2\ \text{m})^2}{(24\ \text{h})\left(3600\ \frac{\text{s}}{\text{h}}\right)}$$

$$= 45 \times 10^6\ \text{W} = 45\ \text{MW}$$

Because of the inefficiency of the electrical generating facilities and other limiting factors, the actual power is 10 to 25% of this value, or 4.5 to 11 MW.

The most successful operating facility using this principle is located on the Rance River in France, where the tides rise as high as 15 m. This facility has the potential of generating around 240 MW, but the average output is about 62 MW. Clearly, tidal power will be useful only in areas that have large tidal variations and convenient natural bays. Some potential sites include the Cook Inlet in Alaska, the San José Gulf in Argentina, and the Passamaquoddy Bay between Maine and Canada.

8.12 SUMMARY

A force *is conservative* if the work done by that force acting on a particle is independent of the path the particle takes between two given points. Alternatively, a force is conservative if the work done by that force is zero when the particle moves through an arbitrary closed path and returns to its initial position. A force that does not meet these criteria is said to be *nonconservative.*

A *potential energy* function U can be associated only with a conservative force. If a conservative force $\boldsymbol{F}$ acts on a particle that moves along the x axis from x_i to x_f, *the change in the potential energy equals the negative of the work done by that force:*

Change in potential energy

$$U_f - U_i = -\int_{x_i}^{x_f} F_x\, dx \tag{8.2}$$

The *law of conservation of mechanical energy* states that if the only force acting on a mechanical system is conservative, the total mechanical energy is conserved:

Conservation of mechanical energy

$$K_i + U_i = K_f + U_f \tag{8.5}$$

The total mechanical energy of a system is defined as the sum of the kinetic energy and potential energy:

Total mechanical energy

$$E \equiv K + U \tag{8.6b}$$

The *gravitational potential energy* of a particle of mass m that is elevated a distance y near the earth's surface is given by

Gravitational potential energy

$$U_g = mgy \tag{8.9}$$

The *work-energy theorem* states that the work done by all nonconservative forces acting on a system equals the change in the total mechanical energy of the system:

$$W_{nc} = E_f - E_i \tag{8.13}$$

Work done by nonconservative forces

The *elastic potential energy* stored in a spring of force constant k is

$$U_s = \frac{1}{2}kx^2 \tag{8.14}$$

Potential energy stored in a spring

If a particle of mass m moves with a speed approaching the speed of light, its kinetic energy is given by

$$K = \frac{mc^2}{\sqrt{1 - v^2/c^2}} - mc^2 \tag{8.19}$$

Relativistic kinetic energy

The total energy of a particle moving at relativistic speeds is

$$E = \frac{mc^2}{\sqrt{1 - v^2/c^2}} \tag{8.20}$$

Total energy

EXERCISES

Section 8.1 Conservative and Nonconservative Forces

1. A 3-kg particle moves from the origin to the position having coordinates $x = 5$ m and $y = 5$ m under the influence of gravity acting in the negative y direction (Fig. 8.14.) Using Eq. 7.21, calculate the work done by gravity in going from O to C along the following paths: (a) OAC, (b) OBC, (c) OC. Your results should all be identical. Why?

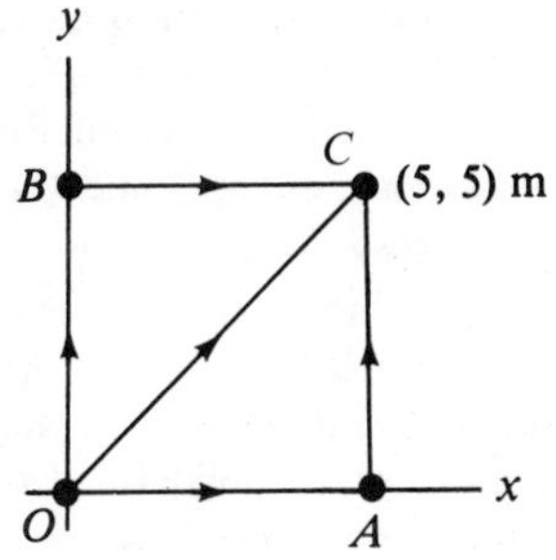

Figure 8.14 (Exercise 1 and Problem 5)

2. (a) Starting with Eq. 7.20 for the definition of work, show that *any constant* force is conservative. (b) As a special case, suppose a particle of mass m is under the influence of force $\boldsymbol{F} = (2\boldsymbol{i} + 5\boldsymbol{j})$ N and moves from O to C in Fig. 8.14. Calculate the work done by $\boldsymbol{F}$ along the three paths OAC, OBC, and OC, and show that they are identical.

3. A particle moves in the xy plane in Fig. 8.14 under the influence of a frictional force that opposes its displacement. If the frictional force has a magnitude of 3 N, calculate the total work done by friction along the following *closed* paths: (a) the path OA followed by the return path AO, (b) the path OA followed by AC and the return path CO, and (c) the path OC followed by the return path CO. (d) Your results for the three closed paths should all be different and nonzero. What is the significance of this?

Section 8.3 Conservation of Mechanical Energy

4. A single conservative force acts on a particle. If its associated potential energy increases by 50 J, find (a) the change in the kinetic energy of the particle, (b) the change in its total energy, and (c) the work done on the particle.

5. A 3-kg particle moves along the x axis under the influence of a single conservative force. If the work done on the particle is 70 J as the particle moves from $x = 2$ m to $x = 5$ m, find (a) the change in the particle's kinetic energy, (b) the change in its potential energy, and (c) its speed at $x = 5$ m if it starts at rest at $x = 2$ m.

6. A single conservative force $F_x = (3x + 5)$ N acts on a 5-kg particle, where x is in m. As the particle moves along the x axis from $x = 1$ m to $x = 4$ m, calculate (a) the work done by this force, (b) the change in the potential energy of the particle, and

(c) its kinetic energy at $x = 4$ m if its speed at $x = 1$ m is 3 m/s.

7. At time t_i, the kinetic energy of a particle is 20 J and its potential energy is 10 J. At some later time t_f, its kinetic energy is 15 J. (a) If only conservative forces act on the particle, what is its potential energy at time t_f? What is its total energy? (b) If the potential energy at time t_f is 5 J, are there any nonconservative forces acting on the particle? Explain.

8. A single constant force $\mathbf{F} = (3\mathbf{i} + 5\mathbf{j})$ N acts on a 4-kg particle. (a) Calculate the work done by this force if the particle moves from the origin to the point with vector position $\mathbf{r} = (2\mathbf{i} - 3\mathbf{j})$ m. Does this result depend on the path? Explain. (b) What is the speed of the particle at $\mathbf{r}$ if its speed at the origin is 4 m/s? (c) What is the change in the potential energy of the particle?

Section 8.4 Gravitational Potential Energy near the Earth's Surface

9. A 2-kg ball hangs at the end of a string 1 m in length from the ceiling of a room. The height of the room is 3 m. What is the gravitational potential energy of the ball relative to (a) the ceiling, (b) the floor, and (c) a point at the same elevation as the ball?

10. A rocket is launched at an angle of 37° to the horizontal from an altitude h with a speed v_0. (a) Use energy methods to find the speed of the rocket when its altitude is $h/2$. (b) Find the x and y components of velocity when the rocket's altitude is $h/2$, using the fact that $v_x = v_{x0} =$ constant (since $a_x = 0$) and the results to (a).

11. A 3-kg mass is attached to a light string of length 1.5 m to form a pendulum (Fig. 8.4). The mass is given an initial speed of 4 m/s at its lowest position. When the string makes an angle of 30° with the vertical, find (a) the *change* in the potential energy of the mass, (b) the speed of the mass, and (c) the tension in the string. (d) What is the maximum height reached by the mass above its lowest position?

12. A 0.4-kg ball is thrown vertically upward with an initial speed of 15 m/s. Assuming its initial potential energy is zero, find its kinetic energy, potential energy, and total mechanical energy (a) at its initial position, (b) when its height is 3 m, and (c) when it reaches the top of its flight. (d) Find its maximum height using the law of conservation of energy.

13. A 0.3-kg ball is thrown into the air and reaches a maximum altitude of 50 m. Taking its initial position as the point of zero potential energy and using energy methods, find (a) its initial speed, (b) its total mechanical energy, and (c) the ratio of its kinetic energy to its potential energy when its altitude is 10 m.

14. A 200-g particle is released from rest at point A along the diameter on the inside of a smooth hemispherical bowl of radius $R = 30$ cm (Fig. 8.15). Calculate (a) its gravitational potential energy at point A relative to point B, (b) its kinetic energy at point B, (c) its speed at point B, and (d) its kinetic energy and potential energy at point C.

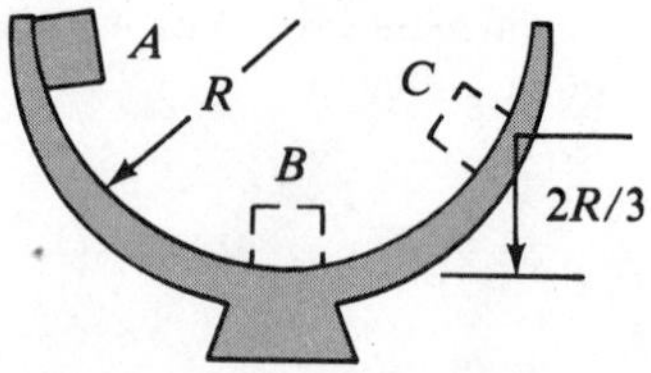

Figure 8.15 (Exercises 14 and 15)

Section 8.5 Nonconservative Forces and the Work-Energy Theorem

15. The particle described in Exercise 14 (Fig. 8.15) is released from point A at rest. The speed of the particle at point B is 1.5 m/s. (a) What is its kinetic energy at B? (b) How much energy is lost as a result of friction as the particle goes from A to B? (c) Is it possible to determine μ from these results in any simple manner? Explain.

16. A 2-kg block is projected up the incline shown in Fig. 8.5 with an initial speed of 3 m/s at the bottom. The coefficient of friction between the block and the incline is 0.7. Find (a) the distance the block will travel up the incline before coming to rest, (b) the total work done by friction while the block is in motion on the incline, and (c) the change in potential energy and change in kinetic energy when the block has traveled 0.3 m up the incline.

17. The total initial mechanical energy of a particle moving along the x axis is 80 J. A frictional force of 6 N is the *only* force acting on the particle. When the total mechanical energy is 30 J, find (a) the distance the particle has traveled, (b) the change in the particle's kinetic energy, and (c) the change in its potential energy.

18. In a given displacement of a particle, its kinetic energy *decreases* by 25 J while its potential energy increases by 10 J. Are there any nonconservative forces acting on the particle? If so, how much work is done by these forces?

19. A child starts from rest at the top of a slide of height $h = 4$ m (Fig. 8.6). (a) What is her speed at the bottom if the incline is frictionless? (b) If she reaches the bottom with a speed of 6 m/s, what percentage of her total energy is lost as a result of friction?

20. A 3-kg particle moving along the x axis has a velocity of $6\mathbf{i}$ m/s when its x coordinate is 3 m. The only force acting on it is a constant retarding force of $-12\mathbf{i}$ N. (a) Find its coordinate when it comes to rest. (b) How much work is done by friction as the particle moves from the origin to the point where it is at rest? (c) What is the change in kinetic energy as the particle moves from the origin to $x = 3$ m?

21. A 0.4-kg bead slides on a curved wire, starting from rest at point A in Fig. 8.16. The segment from A to B is frictionless, and the segment from B to C is rough. (a) Find the speed of the bead at B. (b) If the bead comes to rest at C, find the total work done by friction in going from B to C. (c) What is the net work done by nonconservative forces as the bead moves from A to C?

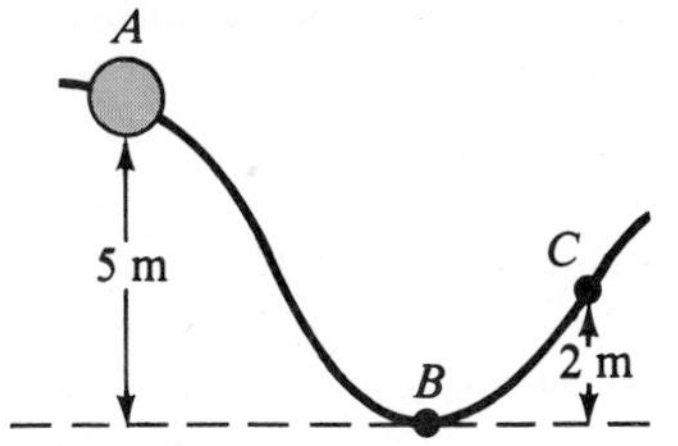

Figure 8.16 (Exercise 21)

22. A 25-kg child on a swing 2 m long is released from rest when the swing supports make an angle of 30° with the vertical. (a) Neglecting friction, find the child's speed at the lowest position. (b) If the speed of the child at the lowest position is 2 m/s, what is the energy loss due to friction?

Section 8.6 Potential Energy Stored in a Spring

23. A spring has a force constant of 400 N/m. How much work must be done on the spring to stretch it (a) 3 cm from its equilibrium position and (b) from $x = 2$ cm to $x = 3$ cm, where $x = 0$ is its equilibrium position? (In the unstretched position, the potential energy is defined to be zero.)

24. A spring has a force constant of 500 N/m. What is the elastic potential energy stored in the spring when (a) it is stretched 4 cm from equilibrium, (b) it is compressed 3 cm from equilibrium, and (c) it is unstretched?

25. A block of mass m is released from rest and slides down a frictionless track of height h above a table (Fig. 8.17). At the bottom of the track, where the surface is horizontal, the block strikes and sticks to a light spring. (a) Find the maximum distance the spring is compressed. (b) Obtain a numerical value for this distance if $m = 0.2$ kg, $h = 1$ m and $k = 490$ N/m.

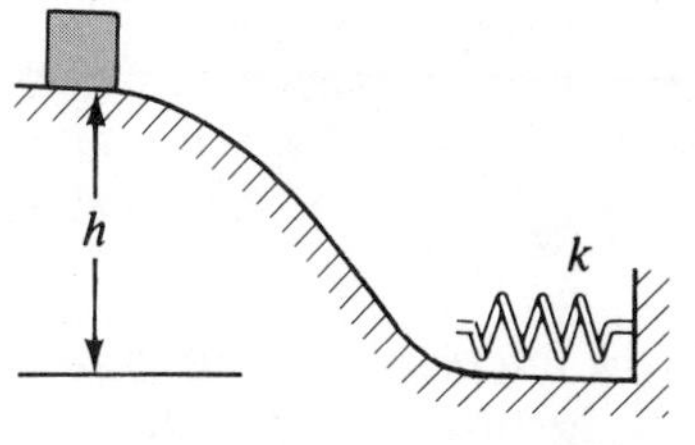

Figure 8.17 (Exercise 25)

26. An 8-kg block travels on a rough, horizontal surface and collides with a spring as in Fig. 8.7. The speed of the block *just before* the collision is 4 m/s. As the block rebounds to the left with the spring uncompressed, its speed as it leaves the spring is 3 m/s. If the coefficient of kinetic friction between the block and surface is 0.4, determine (a) the work done by friction while the block is in contact with the spring and (b) the maximum distance the spring is compressed.

27. A 3-kg mass is fastened to a light spring that passes over a pulley (Fig. 8.18). The pulley is frictionless, and the mass is released from rest when the spring is unstretched. If the mass drops a distance of 10 cm before coming to rest, find (a) the force constant of the spring, and (b) the speed of the mass when it is 5 cm below its starting point.

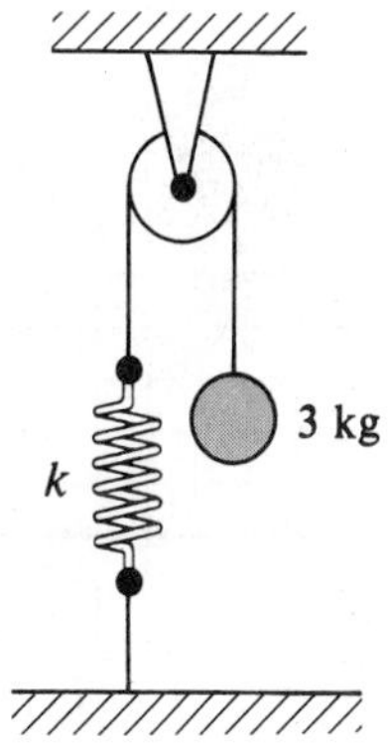

Figure 8.18 (Exercise 27)

28. A child's toy consists of a piece of plastic attached to a spring (Fig. 8.19). The spring is compressed against the floor a distance of 2 cm, and the toy is released. If the mass of the toy is 100 g and it rises to a maximum height of 60 cm, estimate the force constant of the spring.

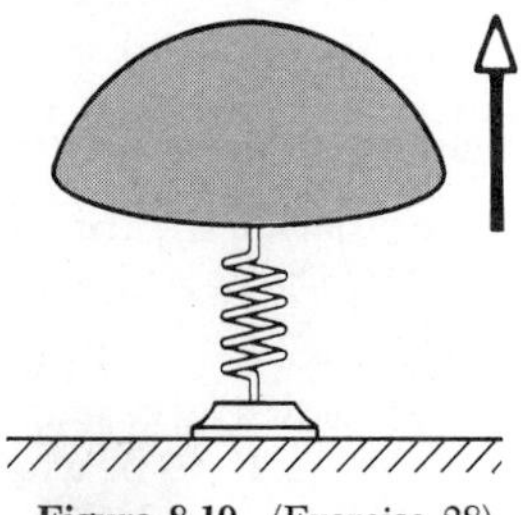

Figure 8.19 (Exercise 28)

Section 8.7 Relationship Between Conservative Forces and Potential Energy

29. The potential energy of a two-particle system separated by a distance r is given by $U(r) = A/r$, where A is a constant. (a) Find the radial force F_r. (b) What two interactions discussed in Chapter 6 can be described with such a potential energy function? What are the signs of A in these cases?

30. The potential energy of a system is given by $U = ax^2 - bx$, where a and b are constants. (a) Find the force F_x associated with this potential energy function. (b) At what value of x is the force zero?

Section 8.8 Energy Diagrams and Stability of Equilibrium

31. Consider the potential energy curve $U(x)$ versus x shown in Fig. 8.20. (a) Determine whether the force F_x is positive, negative, or zero at the various points indicated. (b) Indicate points of stable, unstable, or neutral equilibrium.

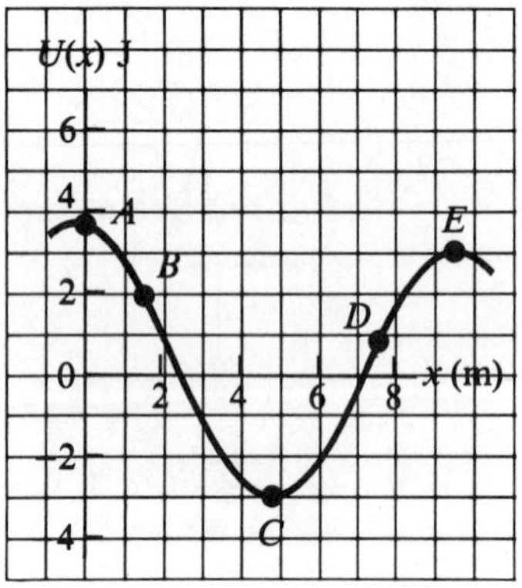

Figure 8.20 (Exercises 31 and 32)

32. With reference to the potential energy curve in Fig. 8.20, make a rough sketch of the F_x versus x curve from $x = 0$ to $x = 8$ m.

33. A right circular cone can be balanced on a horizontal surface in three different ways. Sketch these three equilibrium configurations, and identify them as being positions of stable, unstable, or neutral equilibrium.

Section 8.9 Mass-Energy

34. An electron of mass 9.1×10^{-31} kg is moving with a speed of $0.98c$, where $c = 3 \times 10^8$ m/s. Find (a) its rest energy, (b) its total energy, and (c) its kinetic energy.

35. Calculate the relativistic kinetic energy of a proton (mass $= 1.67 \times 10^{-27}$ kg; velocity $= 1.6 \times 10^8$ m/s). Compare your result with that obtained from the Newtonian expression for kinetic energy.

36. A proton in a high-energy accelerator has a kinetic energy of 2×10^{-8} J. (a) Find the total energy of the proton. (b) Determine the speed of the proton.

37. Calculate the energy released per fission event for the fission reaction described in Example 8.7. Use the following data: the mass of the neutron is 1.008665 amu, the mass of $^{235}_{92}U$ is 235.043915 amu. The mass of $^{89}_{36}Kr$ is 88.916600 amu, and the mass of $^{144}_{56}Ba$ is 143.999923 amu.

38. The mass of the deuteron (the nucleus of heavy hydrogen) is 2.01360 amu, and the masses of its constituents, the proton and neutron, are 1.00731 amu and 1.00867 amu, respectively. Use this information to calculate the binding energy of the deuteron, that is, the minimum energy necessary to "split" the deuteron.

Section 8.11 Energy from the Tides

39. The Bay of Fundy in Canada has an average tidal range of 8 m and an area of 13 000 km^2. What is the average power available from this supply of water assuming an overall efficiency of 25%?

PROBLEMS

1. The masses of the javelin, the discus, and the shot are 0.8 kg, 2.0 kg, and 7.2 kg, respectively, and record throws in the track events using these objects are about 89 m, 69 m, and 21 m, respectively. Neglecting air resistance, (a) calculate the minimum initial kinetic energies that would produce these throws, and (b) estimate the average force exerted on each object during the throw assuming the force acts over a distance of 2 m. (c) Do your results suggest that air resistance is an important factor?

2. A single conservative force acting on a particle varies as $\mathbf{F} = (-Ax + Bx^2)\mathbf{i}$ N, where A and B are constants and x is in m. (a) Calculate the potential energy associated with this force, taking $U = 0$ at $x = 0$. (b) Find the change in potential energy and change in kinetic energy as the particle moves from $x = 2$ m to $x = 3$ m.

3. Prove that the following forces are conservative and find the change in potential energy corresponding to these forces taking $x_i = 0$ and $x_f = x$: (a) $F_x = ax + bx^2$, (b) $F_x = Ae^{\alpha x}$. (a, b, A, and α are all constants.)

4. Find the forces corresponding to the following potential energy functions: (a) K/y, (b) bx^3, (c) e^{-ar}/r. (K, b, and a are all constants.)

5. A force acting on a particle moving in the xy plane is given by $\boldsymbol{F} = (2y\boldsymbol{i} + x^2\boldsymbol{j})$ N, where x and y are in m. The particle moves from the origin to a final position having coordinates $x = 5$ m and $y = 5$ m, as in Fig. 8.14. Calculate the work done by $\boldsymbol{F}$ along (a) OAC, (b) OBC, (c) OC. (d) Is $\boldsymbol{F}$ conservative or nonconservative? Explain.
6. A potential energy function for a system is given by $U(x) = 3x + 4x^2 - x^3$. (a) Determine the force F_x as a function of x. (b) For what values of x is the force equal to zero? (c) Plot $U(x)$ versus x and F_x versus x and indicate points of stable and unstable equilibrium.
7. A 2-kg block situated on a rough incline is connected to a light spring having a force constant of 100 N/m (Fig. 8.21). The block is released from rest when the spring is unstretched and the pulley is frictionless. The block moves 20 cm down the incline before coming to rest. Find the coefficient of kinetic friction between the block and the incline.
8. Suppose the incline is *smooth* for the system described in Problem 7 (Fig. 8.21). The block is released from rest with the spring initially unstretched. (a) How far does it move down the incline before coming to rest? (b) What is the acceleration of the block when it reaches its lowest point? Is the acceleration constant? (c) Describe the energy transformations that occur during the descent of the block.

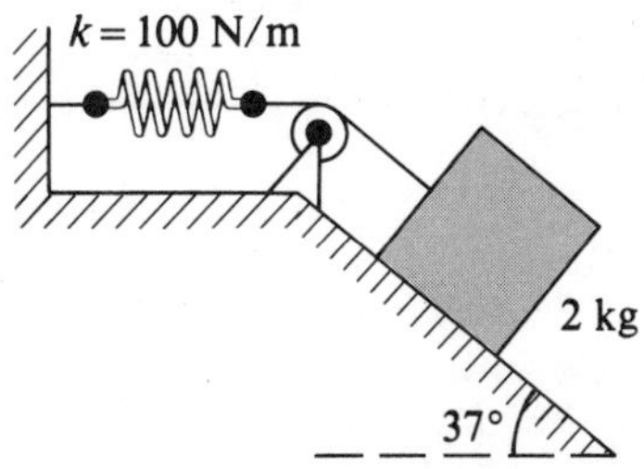

Figure 8.21 (Problems 7 and 8)

9. A 25-kg block is connected to a 30-kg block by a light string that passes over a frictionless pulley. The 30-kg block is connected to a light spring of force constant 200 N/m, as in Fig. 8.22. The spring is unstretched when the system is as shown in the figure, and the incline is smooth. The 25-kg block is pulled a distance of 20 cm down the incline (so that the 30-kg block is 40 cm above the floor) and is released from rest. Find the speed of each block when the 30-kg block is 20 cm above the floor (that is, when the spring is unstretched).
10. An olympic high-jumper whose height is 2 m makes a record leap of 2.3 m over a horizontal bar. Estimate the speed with which he must leave the ground to perform this feat. (Hint: Estimate the position of his center of gravity before jumping, and assume he is in a horizontal position when he reaches the peak of his jump.)

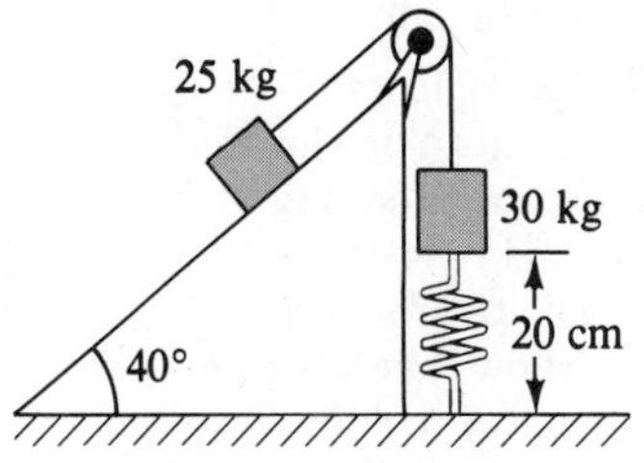

Figure 8.22 (Problem 9)

11. A uniform rope of length L lies on a horizontal, smooth table. Part of the rope, of length d, hangs over the table, and the rope is released from rest (Fig. 8.23). Using energy methods, find (a) the velocity of the rope at the instant all of the rope leaves the table and (b) the time it takes this to occur. (Hint: Note the motion of the center of gravity of the rope.)

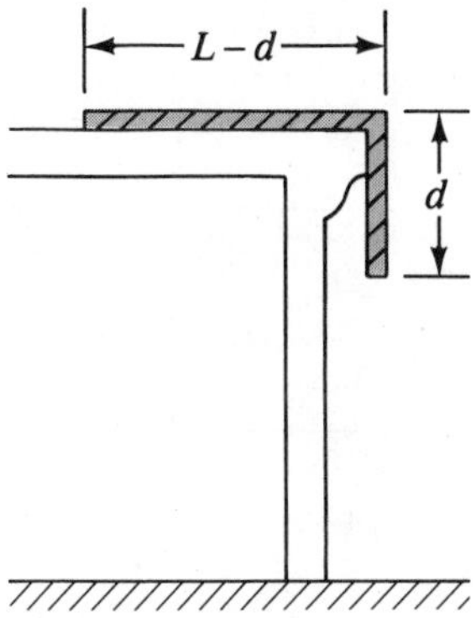

Figure 8.23 (Problem 11)

12. A skier starts at rest from the top of a large hill that is shaped like a hemisphere (Fig. 8.24). Neglecting friction, show that the skier will leave the hill and become "air-borne" at a distance $h = R/3$ below the top of the hill. (Hint: At this point, the normal force goes to zero.)

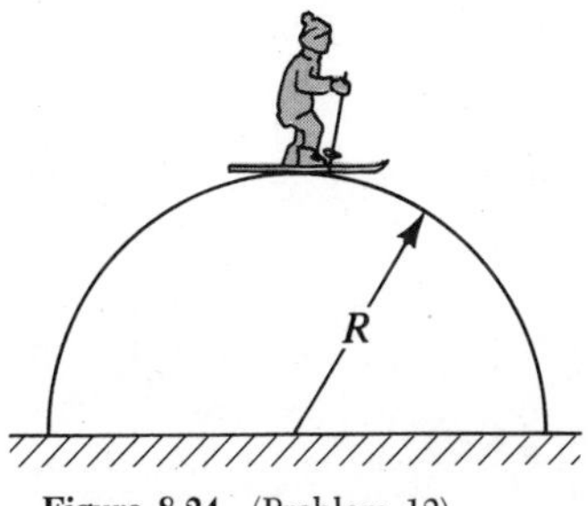

Figure 8.24 (Problem 12)

13. A ball whirls around in a vertical circle at the end of a string. If the ball's total energy remains constant, show that the tension in the string at the bottom is greater than the tension at the top by six times the weight of the ball.

14. A pendulum of length L swings in the vertical plane. The string hits a peg located a distance d below the point of suspension (Fig. 8.25). (a) Show that if the pendulum is released at a height *below* that of the peg, it will return to this height after striking the peg. (b) Show that if the pendulum is released from the horizontal position ($\theta = 90°$) and the pendulum is to swing in a complete circle centered on the peg, then the minimum value of d must be $3L/5$.

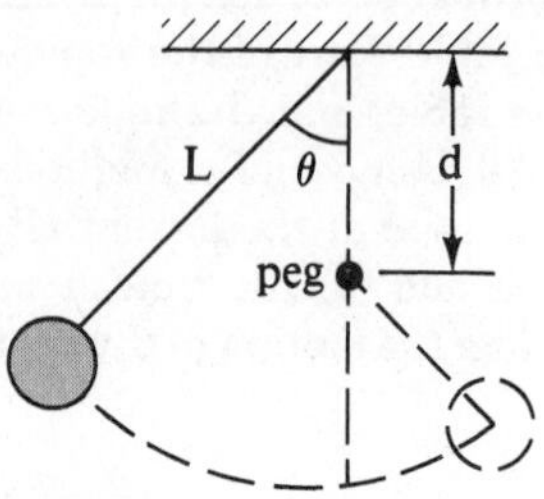

Figure 8.25 (Problem 14)

15. A block of mass m is dropped from rest at a height h directly above the top of a vertical spring having a force constant k. Find the *maximum* distance the spring will be compressed.

16. A frictionless roller coaster is given an initial velocity v_0 at a height h, as in Fig. 8.26. The radius of curvature of the track at point A is R. (a) Find the *maximum* value of v_0 necessary in order that the roller coaster *not* leave the track at A. (b) Using the value of v_0 calculated in (a), determine the value of h' necessary if the roller coaster is to just make it to point B.

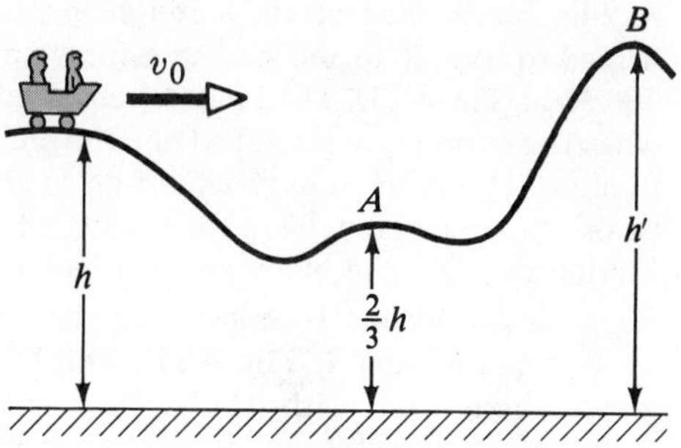

Figure 8.26 (Problem 16)

9 Linear Momentum and Collisions

Harold E. Edgerton, MIT

In this chapter we shall analyze the motion of a system containing many particles. We shall introduce the concept of the linear momentum of the system of particles and show that this momentum is conserved when the system is isolated from its surroundings. The law of momentum conservation is especially useful for treating such problems as the collisions between particles and for analyzing rocket propulsion. The concept of the center of mass of a system of particles will also be introduced. We shall show that the overall motion of a system of interacting particles can be represented by the motion of an equivalent particle located at the center of mass.

9.1 LINEAR MOMENTUM AND IMPULSE

The *linear momentum* of a particle of mass m moving with a velocity $\boldsymbol{v}$ was defined in Chapter 5 to be[1]

Definition of linear momentum of a particle

$$\boldsymbol{p} \equiv m\boldsymbol{v} \tag{9.1}$$

Momentum is a vector quantity since it equals the product of a scalar, m, and a vector, $\boldsymbol{v}$. Its direction is along $\boldsymbol{v}$, and it has dimensions of ML/T. In the SI system, momentum has the units kg · m/s.

If a particle is moving in an arbitrary direction, $\boldsymbol{p}$ will have three components and Eq. 9.1 is equivalent to the component equations given by

$$p_x = mv_x \qquad p_y = mv_y \qquad p_z = mv_z \tag{9.2}$$

We can relate the linear momentum to the force acting on the particle using Newton's second law of motion: *The time rate of change of the momentum of a particle is equal to the resultant force on the particle.* That is,

Newton's second law for a particle

$$\boldsymbol{F} = \frac{d\boldsymbol{p}}{dt} \tag{9.3}$$

From Eq. 9.3 we see that if the resultant force is zero, the momentum of the particle must be constant. In other words, the linear momentum and velocity of a particle are conserved when $\boldsymbol{F} = 0$. Of course, if the particle is *isolated* (that is, if

[1]This expression is nonrelativistic, and so it is valid only when $v \ll c$. For relativistic speeds, $\boldsymbol{p} = m\boldsymbol{v}/(1 - v^2/c^2)^{1/2}$.

it does not interact with its environment), then by necessity, $\mathbf{F} = 0$ and $\mathbf{p}$ remains unchanged. This result can also be obtained directly through the application of Newton's second law in the form $\mathbf{F} = m\, d\mathbf{v}/dt$. That is, when the force is zero, the acceleration of the particle is zero and the velocity remains constant.

Equation 9.3 can be written

$$d\mathbf{p} = \mathbf{F}\, dt \tag{9.4}$$

We can integrate this expression to find the change in the momentum of a particle. If the momentum of the particle changes from $\mathbf{p}_i$ at time t_i to $\mathbf{p}_f$ at time t_f, then integrating Eq. 9.4 gives

$$\Delta\mathbf{p} = \mathbf{p}_f - \mathbf{p}_i = \int_{t_i}^{t_f} \mathbf{F}\, dt \tag{9.5}$$

The quantity on the right side of Eq. 9.5 is called the *impulse* of the force $\mathbf{F}$ for the time interval $\Delta t = t_f - t_i$. Impulse is a vector defined by

Impulse of a force

$$\mathbf{I} = \int_{t_i}^{t_f} \mathbf{F}\, dt = \Delta\mathbf{p} \tag{9.6}$$

Impulse-momentum theorem

Therefore, *the impulse of the force* **F** *equals the change in the momentum of the particle.* This statement, known as the *impulse-momentum theorem,* is equivalent to Newton's second law. In fact, Newton stated the second law in this integral form in his famous publication *Principia.* From this definition, we see that impulse is a vector quantity having a magnitude equal to the area under the force-time curve, as described in Fig. 9.1a. In this figure, it is assumed that the force varies in time in the general manner shown and is nonzero in the time interval $\Delta t = t_f - t_i$. The direction of the impulse vector is the same as the direction of the change in momentum. Impulse has the dimensions of momentum, that is, ML/T. Note that impulse is *not* a

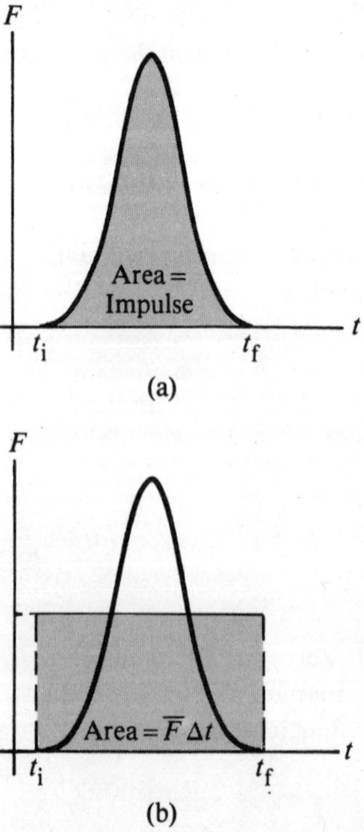

Figure 9.1 (a) A force acting on a particle may vary in time. The impulse is the area under the force versus time curve. (b) The average force (horizontal line) would give the same impulse to the particle in the time Δt as the real time-varying force described in (a).

property of the particle itself, but is a quantity that measures the degree to which an external force changes the momentum of the particle. Therefore, when we say that an impulse is given to a particle, it is implied that momentum is transferred from an external agent to that particle.

Since the force can generally vary in time as in Fig. 9.1a, it is convenient to define a time-averaged force $\bar{\boldsymbol{F}}$, given by

$$\bar{\boldsymbol{F}} = \frac{1}{\Delta t}\int_{t_i}^{t_f} \boldsymbol{F}\, dt \qquad (9.7)$$

where $\Delta t = t_f - t_i$. Therefore, we can express Eq. 9.6 as

$$\boldsymbol{I} = \Delta \boldsymbol{p} = \bar{\boldsymbol{F}}\, \Delta t \qquad (9.8)$$

This average force, described in Fig. 9.1b, can be thought of as the constant force that would give the same impulse to the particle in the time interval Δt as the actual time-varying force gives over this same interval.

In principle, if $\boldsymbol{F}$ is known as a function of time, the impulse can be calculated from Eq. 9.6. The calculation becomes especially simple if the force acting on the particle is constant. In this case, $\bar{\boldsymbol{F}} = \boldsymbol{F}$ and Eq. 9.8 becomes

$$\boldsymbol{I} = \Delta \boldsymbol{p} = \boldsymbol{F}\, \Delta t \qquad (9.9)$$

Impulse when F = constant

In many physical situations, we shall use the so-called *impulse approximation*. In this approximation, *we assume that one of the forces exerted on a particle acts for a short time but is much larger than any other force present.* This approximation is especially useful in treating collisions, where the duration of the collision is very short. When this approximation is made, we refer to the force as an *impulsive force*. For example, when a baseball is struck with a bat, the time of the collision is about 0.01 s, and the average force the bat exerts on the ball in this time is typically several thousand pounds. This is much greater than the force of gravity, and so the impulse approximation is justified. When we use this approximation, it is important to remember that $\boldsymbol{p}_i$ and $\boldsymbol{p}_f$ represent the momenta *immediately* before and after the collision, respectively. Therefore, in the impulse approximation there is very little motion of the particle during the collision.

Impulse approximation

Example 9.1 *Struck Golf Ball*

A golf ball of mass 50 g is struck with a club (Fig. 9.2). The force on the ball varies from zero when contact is made up to some maximum value (where the ball is deformed) back to zero when the ball leaves the club. Thus, the force-time curve is qualitatively described by Fig. 9.1. Assuming that the ball travels a distance of 200 m, (a) estimate the impulse due to the collision.

Neglecting air resistance, we can use the expression for the range of a projectile (Chapter 4) given by

$$R = \frac{v_0^2}{g}\sin 2\theta_0$$

Let us assume that the launch angle is 45°, which provides the maximum range for any given launch speed. The initial velocity of the ball is then estimated to be

$$v_0 = \sqrt{Rg} = \sqrt{(200)(9.8)} = 44 \text{ m/s}$$

Since $v_i = 0$ and $v_f = v_0$ for the ball, the magnitude of the impulse imparted to the ball is

$$I = \Delta p = mv_0 = (50 \times 10^{-3}\text{ kg})\left(44 \frac{\text{m}}{\text{s}}\right) = 2.2 \text{ kg} \cdot \text{m/s}$$

Figure 9.2 A golf ball being struck by a club.

(b) Estimate the time of the collision and the average force on the ball.

From Fig. 9.2, it appears that a reasonable estimate of the distance the ball travels while in contact with the club is the radius of the ball, about 2 cm. The time it takes the club to move this distance (the contact time) is then

$$\Delta t = \frac{\Delta x}{v_0} = \frac{2 \times 10^{-2}\ \text{m}}{44\ \text{m/s}} = 4.5 \times 10^{-4}\ \text{s}$$

Finally, the magnitude of the average force is estimated to be

$$\overline{F} = \frac{I}{\Delta t} = \frac{2.2\ \text{kg}\cdot\text{m/s}}{4.5 \times 10^{-4}\ \text{s}} = 4.9 \times 10^{3}\ \text{N}$$

Note that this force is extremely large compared with the weight (gravity force) of the ball, which is only 0.49 N.

Example 9.2 *The Bouncing Ball*

A ball of mass 100 g is dropped from a height $h = 2$ m above the floor (Fig. 9.3). It rebounds vertically to a height $h' = 1.5$ m after colliding with the floor. (a) Find the momentum of the ball immediately before and after the ball collides with the floor.

Using the energy methods, we can find v_i, the velocity of the ball just before it collides with the floor, through the relationship

$$\frac{1}{2}mv_i^2 = mgh$$

Likewise, v_f, the ball's velocity right after colliding with the floor, is obtained from the energy expression

$$\frac{1}{2}mv_f^2 = mgh'$$

Substituting into these expressions the values $h = 2.0$ m and $h' = 1.5$ m gives

$$v_i = \sqrt{2gh} = \sqrt{(2)(9.8)(2)}\ \text{m/s} = 6.3\ \text{m/s}$$
$$v_f = \sqrt{2gh'} = \sqrt{(2)(9.8)(1.5)}\ \text{m/s} = 5.4\ \text{m/s}$$

Since $m = 0.1$ kg, the vector expressions for the initial and final linear momenta are given by

$$\mathbf{p}_i = m\mathbf{v}_i = -0.63\mathbf{j}\ \text{kg}\cdot\text{m/s}$$
$$\mathbf{p}_f = m\mathbf{v}_f = 0.54\mathbf{j}\ \text{kg}\cdot\text{m/s}$$

(b) Determine the average force exerted by the floor on the ball. Assume the time of the collision is 10^{-2} s (a typical value).

Using Eq. 9.5 and the definition of $\overline{\mathbf{F}}$, we get

$$\Delta\mathbf{p} = \mathbf{p}_f - \mathbf{p}_i = \overline{\mathbf{F}}\,\Delta t$$
$$\overline{\mathbf{F}} = \frac{[0.54\mathbf{j} - (-0.63\mathbf{j})]\ \text{kg}\cdot\text{m/s}}{10^{-2}\ \text{s}} = 1.2 \times 10^{2}\mathbf{j}\ \text{N}$$

Note that this average force is much greater than the force of gravity ($mg \approx 1.0$ N). That is, the impulsive force due to the collision with the floor overwhelms the gravitational force. In this inelastic collision, the energy lost by the ball is transformed into heat, sound, and distortions of the ball and floor.

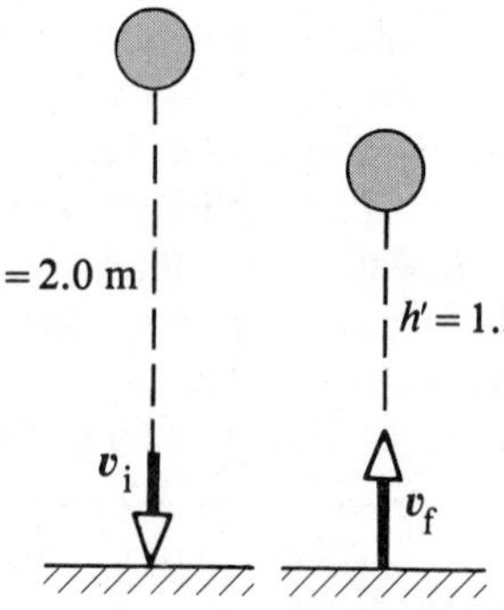

Figure 9.3 (Example 9.2) (a) The ball is dropped from a height h and reaches the floor with a velocity v_i. (b) The ball rebounds from the floor with a velocity v_f and reaches a height h'.

Q1. If the kinetic energy of a particle is zero, what is its linear momentum? If the total energy of a particle is zero, is its linear momentum necessarily zero? Explain.

Q2. If the velocity of a particle is doubled, by what factor is its momentum changed? What happens to its kinetic energy?

Q3. If two particles have equal kinetic energies, are their momenta necessarily equal? Explain.

Q4. Does a large force always produce a larger impulse on a body than a smaller force? Explain.

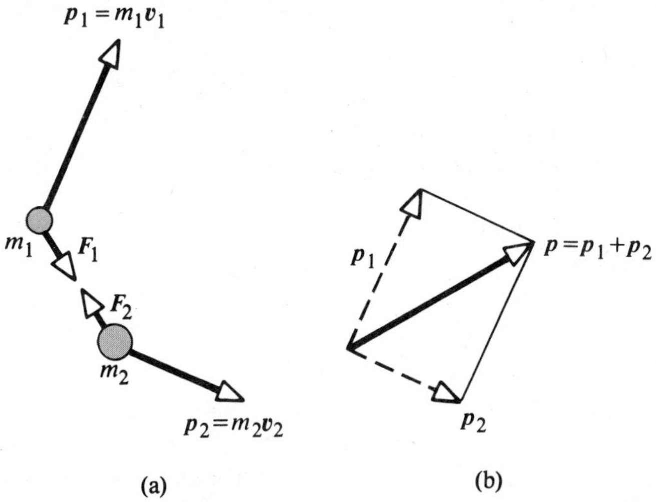

Figure 9.4 (a) At some instant, the momentum of m_1 is $p_1 = m_1v_1$ and the momentum of m_2 is $p_2 = m_2v_2$. If the particles are isolated, $F_1 = -F_2$. (b) The total momentum of the system, P, is equal to the vector sum $p_1 + p_2$.

9.2 CONSERVATION OF LINEAR MOMENTUM FOR A TWO-PARTICLE SYSTEM

Consider two particles that can interact with each other but are isolated from their surroundings (Fig. 9.4). That is, the particles exert forces on each other, but no external forces are present.[2] Suppose that at some time t, the momentum of particle 1 is p_1 and the momentum of particle 2 is p_2. We can apply Newton's second law to each particle and write

$$F_{12} = \frac{dp_1}{dt} \quad \text{and} \quad F_{21} = \frac{dp_2}{dt}$$

where F_{12} is the force on particle 1 due to particle 2 and F_{21} is the force on particle 2 due to particle 1. These forces could be gravitational forces, electrostatic forces, or of some other origin. This really isn't important for the present discussion. However, Newton's third law tells us that F_{12} and F_{21} are equal in magnitude and opposite in direction. That is, they form an action-reaction pair and $F_{12} = -F_{21}$. We can also express this condition as

$$F_{12} + F_{21} = 0$$

or

$$\frac{dp_1}{dt} + \frac{dp_2}{dt} = \frac{d}{dt}(p_1 + p_2) = 0$$

Since the time derivative of the total momentum, $P = p_1 + p_2$, is *zero*, we conclude that the *total* momentum, P, must remain constant, that is,

$$P = p_1 + p_2 = \text{constant} \tag{9.10}$$

The total momentum of an isolated pair of particles is constant

This vector equation is equivalent to three component equations. In other words, Eq. 9.10 in component form says that the total momenta in the x, y, and z directions are all independently conserved, or

[2] A truly isolated system cannot be achieved in the laboratory, since gravitational forces and friction will always be present.

$$P_{ix} = P_{fx} \qquad P_{iy} = P_{fy} \qquad P_{iz} = P_{fz}$$

We can state this law, known as *the conservation of linear momentum,* as follows: *If two particles of masses* m_1 *and* m_2 *form an isolated system, then the total momentum of the system is conserved, regardless of the nature of the force between them* (provided the force obeys Newton's third law). More simply, *whenever two particles collide their total momentum remains constant provided they are isolated.*

Suppose $\mathbf{v}_{1i}$ and $\mathbf{v}_{2i}$ are the initial velocities of particles 1 and 2, and $\mathbf{v}_{1f}$ and $\mathbf{v}_{2f}$ are their velocities at some later time. Applying Eq. 9.10, we can express the conservation of linear momentum of this isolated system in the form

$$m_1\mathbf{v}_{1i} + m_2\mathbf{v}_{2i} = m_1\mathbf{v}_{1f} + m_2\mathbf{v}_{2f} \tag{9.11}$$

Conservation of momentum

$$\mathbf{p}_{1i} + \mathbf{p}_{2i} = \mathbf{p}_{1f} + \mathbf{p}_{2f} \tag{9.12}$$

That is, *the total momentum of the isolated system at all times equals its initial total momentum.* We can also describe the law of conservation of momentum in another way. Since we require that the system be isolated, the only forces acting must be internal to the system (the action-reaction pair). In other words, if there are no external forces present, the total momentum of the system remains constant. Therefore, momentum conservation for an isolated system is an alternative and equivalent statement of Newton's third law.

The law of conservation of momentum is more fundamental than the law of conservation of energy and is considered to be the most important law of mechanics. That is, mechanical energy is only conserved for an isolated system when conservative forces alone act on a system. Momentum is conserved for an isolated two-particle system *regardless* of the nature of the internal forces. In fact, in Section 9.7 we shall show that the law of conservation of linear momentum also applies to an isolated system of n particles.

Example 9.3 *Firing a Cannon*

A 3000-kg cannon rests on a frozen pond as in Fig. 9.5. The cannon is loaded with a 30-kg cannon ball and is fired horizontally. If the cannon recoils to the right with a velocity of 1.8 m/s, what is the final velocity of the cannon ball?

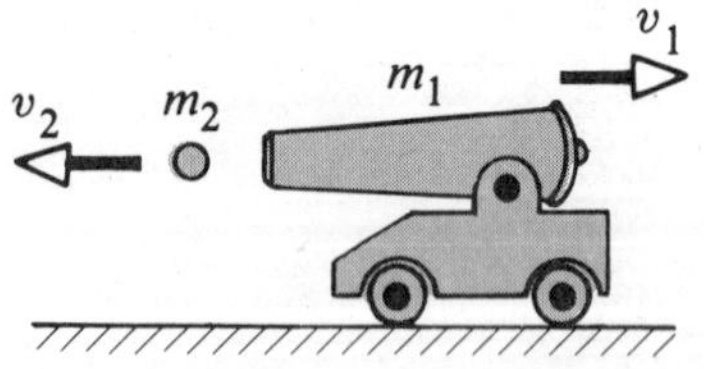

Figure 9.5 (Example 9.3) When the cannon is fired, it recoils to the right.

Solution: In this example, the system consists of the cannon ball and the cannon. The system is not really isolated because of the force of gravity. However, this external force acts in the vertical direction, while the motion of the system is in the horizontal direction. Therefore, momentum is conserved in the x direction since there are no external forces in this direction (assuming the surface is frictionless).

The total momentum of the system before firing is zero. Therefore, the total momentum after firing must be zero, or

$$m_1v_1 + m_2v_2 = 0$$

With $m_1 = 3000$ kg, $v_1 = 1.8$ m/s, and $m_2 = 30$ kg, solving for v_2, the velocity of the cannon ball, gives

$$v_2 = -\frac{m_1}{m_2}v_1 = -\left(\frac{3000\text{ kg}}{30\text{ kg}}\right)1.8\text{ m/s} = -180\text{ m/s}$$

The negative sign for v_2 indicates that the ball is moving to the left after firing, in the direction opposite the movement of the cannon.

Q5. An isolated system is initially at rest. Is it possible for parts of the system to be in motion at some later time? If so, explain how this might occur.

9.3 COLLISIONS

In this section we shall use the law of conservation of momentum to describe what happens when two particles collide with each other. We shall use the term *collision* to represent the event of two particles coming together for a short time, producing impulsive forces on each other. *The impulsive force due to the collision is assumed to be much larger than any external forces present.*

The collision process may be the result of physical contact between two objects, as described in Fig. 9.6a. This is a common observation when two macroscopic objects, such as two billiard balls or a baseball and a bat, collide. The notion of what we mean by a collision must be generalized since "contact" on a submicroscopic scale is ill-defined and meaningless. More accurately, impulsive forces arise from the electrostatic interaction of the electrons in the surface atoms of the two bodies.

To understand this on a more fundamental basis, consider a collision on an atomic scale (Fig. 9.6b), such as the collision of a proton with an alpha particle (the nucleus of the helium atom). Since the two particles are positively charged, they repel each other because of the strong electrostatic force between them at close separations. Such a process is commonly called a *scattering process.*

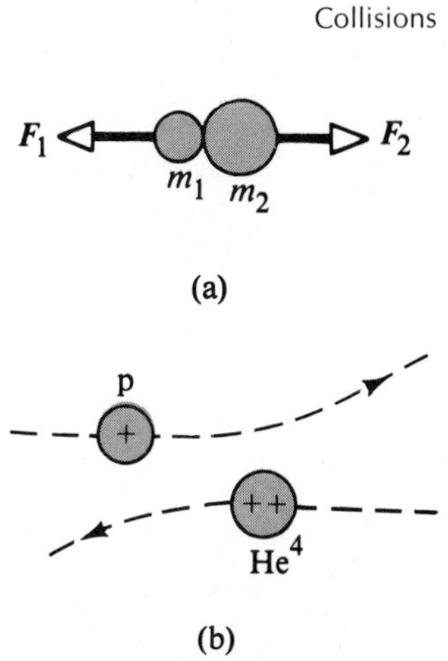

Figure 9.6 (a) The collision between two objects as the result of direct contact. (b) The collision between two charged particles.

When the two particles of masses m_1 and m_2 collide as in Fig. 9.6, the impulse forces may vary in time in a complicated way such as described in Fig. 9.7. If $\boldsymbol{F}_{12}$ is the force on m_1 due to m_2, then the change in momentum of m_1 due to the collision is given by

$$\Delta \boldsymbol{p}_1 = \int_{t_i}^{t_f} \boldsymbol{F}_{12}\, dt$$

Likewise, if $\boldsymbol{F}_{21}$ is the force on m_2 due to m_1, the change in momentum of m_2 is given by

$$\Delta \boldsymbol{p}_2 = \int_{t_i}^{t_f} \boldsymbol{F}_{21}\, dt$$

However, Newton's third law states that the force on m_1 due to m_2 is equal to and opposite the force on m_2 due to m_1, or $\boldsymbol{F}_{12} = -\boldsymbol{F}_{21}$. (This is described graphically in Fig. 9.7.) Hence, we conclude that

$$\Delta \boldsymbol{p}_1 = -\Delta \boldsymbol{p}_2$$

$$\Delta \boldsymbol{p}_1 + \Delta \boldsymbol{p}_2 = 0$$

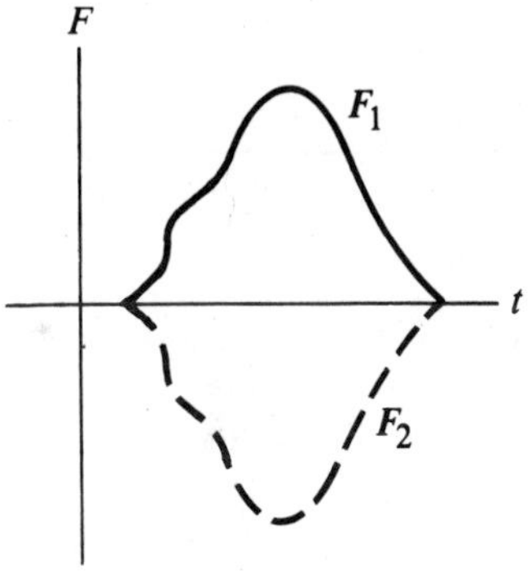

Figure 9.7 The force as a function of time for the two colliding particles described in Fig. 9.6. Note that $\boldsymbol{F}_1 = -\boldsymbol{F}_2$.

Since the total momentum of the system is $\boldsymbol{P} = \boldsymbol{p}_1 + \boldsymbol{p}_2$, we conclude that the *change* in the momentum of the system due to the collision is zero, that is,

$$\boldsymbol{P} = \boldsymbol{p}_1 + \boldsymbol{p}_2 = \text{constant}$$

This is precisely what we expect if there are no external forces acting on the system (Section 9.2). However, the result is also valid if we consider the motion just before and just after the collision. Since the impulsive forces due to the collision are internal, they do not affect the total momentum of the system. Therefore, we conclude that *for any type of collision, the total momentum of the system just before the collision equals the total momentum of the system just after the collision.*

Momentum is conserved for any collision

Whenever a collision occurs between two bodies, we have seen that *the total momentum is always conserved.* However, the total kinetic energy is generally *not* conserved when a collision occurs because some of the kinetic energy is converted into heat and internal elastic potential energy when the bodies are deformed during the collision.

Inelastic collision

We define an *inelastic collision* as a *collision for which the kinetic energy is not conserved. Only momentum is conserved in an inelastic collision.* For a general inelastic collision, we can apply the law of conservation of momentum in the form given by Eq. 9.11. The collision of a rubber ball with a hard surface is inelastic since some of the kinetic energy of the ball is lost when it is deformed. When two objects collide and stick together after the collision, the collision is called *perfectly inelastic.* This is an extreme case of an inelastic collision. For example, if two pieces of putty collide, they stick together and move with some common velocity after the collision. If a meteorite collides with the earth, it becomes buried in the earth and the collision is considered perfectly inelastic. However, not all of the initial kinetic energy is necessarily lost even in a perfectly inelastic collision.

Elastic collision

An *elastic collision* is defined as a *collision in which both momentum and kinetic energy are conserved.* Billiard ball collisions and the collisions of air molecules with the walls of a container at ordinary temperatures are highly elastic. In reality, collisions on a macroscopic scale can only be approximately elastic; however, truly elastic collisions do occur between atomic and subatomic particles.

We summarize the various types of collisions as follows:

Properties of inelastic and elastic collision

1. An *inelastic collision* is one in which only momentum is conserved. Kinetic energy is not conserved in an inelastic collision.
2. A *perfectly inelastic collision* between two objects is an inelastic collision in which the two objects stick together after the collision.
3. An *elastic collision* is one in which both momentum and kinetic energy are conserved.

9.4 COLLISIONS IN ONE DIMENSION

In this section, we treat collisions in one dimension and consider two extreme types of collisions: (1) perfectly inelastic and (2) elastic. The important distinction between these two types of collisions is the fact that *momentum is conserved in both cases, but kinetic energy is conserved only in the case of an elastic collision.*

Perfectly Inelastic Collisions

Consider two particles of masses m_1 and m_2 moving with initial velocities $\boldsymbol{v}_{1i}$ and $\boldsymbol{v}_{2i}$ along a straight line, as in Fig. 9.8. We shall assume that the particles collide "head-on," so that they will be moving along the same line of motion after the collision. If the two particles stick together and move with some common velocity $\boldsymbol{v}_f$ after the collision, then only the linear momentum of the system is conserved. Therefore, we can say that the total momentum before the collision equals the total momentum of the composite system after the collision, that is,

Perfectly inelastic head-on collision

$$m_1 v_{1i} + m_2 v_{2i} = (m_1 + m_2) v_f \tag{9.13}$$

$$v_f = \frac{m_1 v_{1i} + m_2 v_{2i}}{m_1 + m_2} \tag{9.14}$$

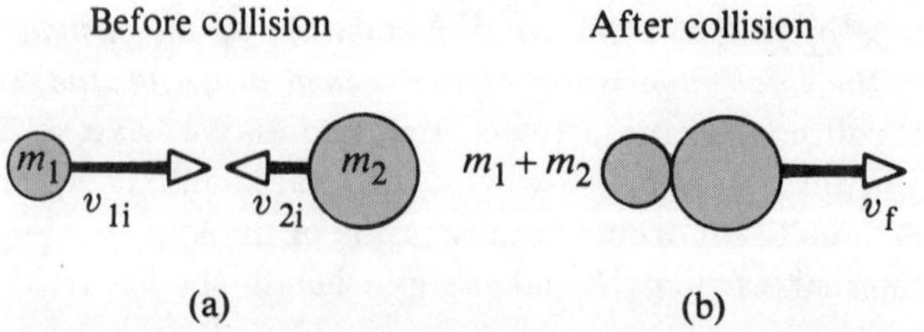

Figure 9.8 Schematic representation of a perfectly inelastic head-on collision between two particles: (a) before the collision and (b) after the collision.

Example 9.4

Two particles collide head-on in a perfectly inelastic fashion as in Fig. 9.8. Suppose $m_1 = 0.5$ kg, $m_2 = 0.25$ kg, $v_{1i} = 4.0$ m/s, and $v_{2i} = -3.0$ m/s. (a) Find the velocity of the composite particle after the collision.

Using Eq. 9.14, we find that

$$v_f = \frac{m_1 v_{1i} + m_2 v_{2i}}{m_1 + m_2}$$
$$= \frac{(0.50 \text{ kg})(4.0 \text{ m/s}) + (0.25 \text{ kg})(-3.0 \text{ m/s})}{(0.50 + 0.25) \text{ kg}}$$
$$= 1.7 \text{ m/s}$$

(b) How much kinetic energy is lost in the collision? The kinetic energy *before* the collision is

$$K_i = \frac{1}{2} m v_{1i}^2 + \frac{1}{2} m_2 v_{2i}^2$$
$$= \frac{1}{2}(0.50 \text{ kg})(4.0 \text{ m/s})^2 + \frac{1}{2}(0.25 \text{ kg})(-3.0 \text{ m/s})^2$$
$$= 5.1 \text{ J}$$

The kinetic energy *after* the collision is

$$K_f = \frac{1}{2}(m_1 + m_2)v_f^2 = \frac{1}{2}(0.75 \text{ kg})(1.7 \text{ m/s})^2 = 1.0 \text{ J}$$

Hence, the *loss* in kinetic energy is

$$K_i - K_f = 4.1 \text{ J}$$

Example 9.5

The ballistic pendulum: The ballistic pendulum (Fig. 9.9) is a system used to measure the velocity of a fast-moving projectile, such as a rifle bullet. The bullet is fired into a large block of wood suspended from some light wires. The bullet is stopped by the block, and the entire system swings through a height h. Since the collision is perfectly inelastic and momentum is conserved, Eq. 9.14 gives the velocity of the system *right after* the collision in the impulse approximation. The kinetic energy *right after* the collision is given by

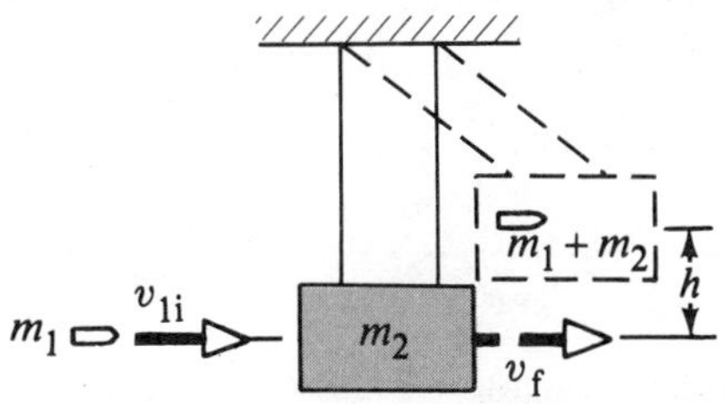

Figure 9.9 (Example 9.5) Diagram of a ballistic pendulum. Note that v_f is the velocity of the system right after the perfectly inelastic collision.

$$(1) \qquad K = \frac{1}{2}(m_1 + m_2)v_f^2$$

With $v_{2i} = 0$, Eq. 9.14 becomes

$$v_f = \frac{m_1 v_{1i}}{m_1 + m_2}$$

Substituting this value of v_f into (1), we get

$$K = \frac{m_1^2 v_{1i}^2}{2(m_1 + m_2)}$$

where v_{1i} is the initial velocity of the bullet. Note that this kinetic energy is *less* than the initial kinetic energy of the bullet. However, *after* the collision, energy is conserved and the kinetic energy at the bottom is transformed into potential energy in the bullet and in the block at the height h; that is,

$$\frac{m_1^2 v_{1i}^2}{2(m_1 + m_2)} = (m_1 + m_2)gh$$
$$v_{1i} = \left(\frac{m_1 + m_2}{m_1}\right)\sqrt{2gh}$$

Hence, it is possible to obtain the initial velocity of the bullet by measuring h and the two masses. For example, if $h = 5$ cm, $m_1 = 5$ g, and $m_2 = 1$ kg, we find that $v_{1i} \approx 200$ m/s. The calculation of the energy lost in this problem is left as an exercise for the student.

Elastic Collisions

Now consider two particles that undergo an elastic head-on collision (Fig. 9.10). In this case, both momentum and kinetic energy are conserved; therefore we can write these conditions

$$m_1 v_{1i} + m_2 v_{2i} = m_1 v_{1f} + m_2 v_{2f}$$
$$\frac{1}{2} m_1 v_{1i}^2 + \frac{1}{2} m_2 v_{2i}^2 = \frac{1}{2} m_1 v_{1f}^2 + \frac{1}{2} m_2 v_{2f}^2$$

Suppose that the masses and the initial velocities of both particles are known. These two equations can be solved for the final velocities in terms of the initial velocities, since there are two equations and two unknowns. Solving for v_{1f} and v_{2f} gives

Elastic collision: relations between final and initial velocities

$$v_{1f} = \left(\frac{m_1 - m_2}{m_1 + m_2}\right)v_{1i} + \left(\frac{2m_2}{m_1 + m_2}\right)v_{2i} \tag{9.15a}$$

$$v_{2f} = \left(\frac{2m_1}{m_1 + m_2}\right)v_{1i} + \left(\frac{m_2 - m_1}{m_1 + m_2}\right)v_{2i} \tag{9.15b}$$

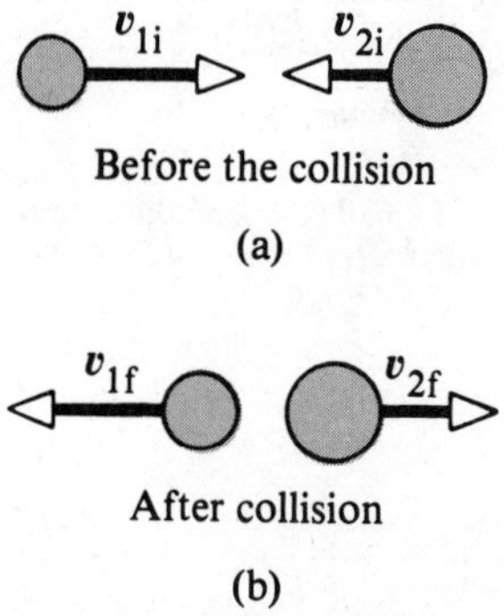

Figure 9.10 Schematic representation of an elastic head-on collision between two particles: (a) before the collision and (b) after the collision.

Note that the appropriate signs for v_{1i} and v_{2i} must be included in Eqs. 9.15a and 9.15b since they are vectors. For example, if m_2 is moving to the left initially, as in Fig. 9.10, then v_{2i} is negative.

Let us consider some special cases: If $m_1 = m_2$, then we see that $v_{1f} = v_{2i}$ and $v_{2f} = v_{1i}$. That is, the particles exchange velocities if they have equal masses. This is what one observes in billiard ball collisions.

If m_2 is initially at rest, $v_{2i} = 0$, and Eqs. 9.15a and 9.15b become

$$v_{1f} = \left(\frac{m_1 - m_2}{m_1 + m_2}\right)v_{1i} \tag{9.16a}$$

$$v_{2f} = \left(\frac{2m_1}{m_1 + m_2}\right)v_{1i} \tag{9.16b}$$

If m_1 is very large compared with m_2, we see from Eqs. 9.16a and 9.16b that $v_{1f} \approx v_{1i}$ and $v_{2f} \approx 2v_{1i}$. That is, when a very heavy particle collides with a very light one initially at rest, the heavy particle continues its motion unaltered after the collision, while the light particle rebounds with a velocity equal to about twice the initial velocity of the heavy particle. An example of such a collision would be the collision of a moving heavy atom, such as uranium, with a light atom, such as hydrogen.

If m_2 is much larger than m_1 and m_2 is initially at rest, then we note from Eqs. 9.16a and 9.16b that $v_{1f} \approx -v_{1i}$ and $v_{2f} \approx 0$. That is, when a very light particle collides with a very heavy particle initially at rest, the light particle will have its velocity reversed, while the heavy particle will remain approximately at rest. For example, imagine what happens when a marble is thrown at a stationary bowling ball.

Example 9.6 *Slowing Down Neutrons by Collisions*

In a nuclear reactor, neutrons are produced when the isotope $^{235}_{92}U$ undergoes fission. These neutrons are moving at high speeds (typically 10^7 m/s) and must be slowed down to about 10^3 m/s. Once the neutrons have slowed down, they have a high probability of producing another fission event and hence a sustained chain reaction. The high-speed neutrons can be slowed down by passing them through a solid or liquid material called a *moderator*. The slowing-down process involves elastic collisions. Let us show that a neutron can lose most of its kinetic energy if it collides elastically with a moderator containing light nuclei, such as deuterium and carbon. Hence, the moderator material is usually heavy water (D_2O) or graphite (which contains carbon nuclei).

Solution: Let us assume that the moderator nucleus of mass m_2 is at rest initially and that the neutron of mass m_1 has an initial velocity v_{1i}. Since momentum and energy are conserved, Egs. 9.16a and 9.16b apply to the head-on collision of a neutron with the moderator nucleus. The initial kinetic energy of the neutron is

$$K_i = \frac{1}{2}m_1 {v_{1i}}^2$$

After the collision, the neutron has a kinetic energy given by $\frac{1}{2}m_1 v_{1f}^2$, where v_{1f} is given by Eq. 9.16a. We can express this energy as

$$K_1 = \frac{1}{2}m_1 {v_{1f}}^2 = \frac{m_1}{2}\left(\frac{m_1 - m_2}{m_1 + m_2}\right)^2 {v_{1i}}^2$$

Therefore, the *fraction* of the total kinetic energy possessed by the neutron *after* the collision is given by

$$(1) \qquad f_1 = \frac{K_1}{K_i} = \left(\frac{m_1 - m_2}{m_1 + m_2}\right)^2$$

From this result, we see that the final kinetic energy of the neutron is small when m_2 is close to m_1 and is zero when $m_1 = m_2$.

We can calculate the kinetic energy of the moderator nucleus after the collision using Eq. 9.16b:

$$K_2 = \frac{1}{2}m_2 v_{2f}^2 = \frac{2m_1^2 m_2}{(m_1 + m_2)^2} v_{1i}^2$$

Hence, the fraction of the total kinetic energy transferred to the moderator nucleus is given by

$$(2) \qquad f_2 = \frac{K_2}{K_i} = \frac{4m_1 m_2}{(m_1 + m_2)^2}$$

Note that since the total energy is conserved, (2) can also be obtained from (1) with the condition that $f_1 + f_2 = 1$, so that $f_2 = 1 - f_1$.

Suppose that heavy water is used for the moderator. Collisions of the neutrons with deuterium nuclei in D_2O ($m_2 = 2m_1$) predict that $f_1 = 1/9$ and $f_2 = 8/9$. That is, 89 percent of the neutron's kinetic energy is transferred to the deuterium nucleus. In practice, the moderator efficiency is reduced because head-on collisions are very unlikely to occur. How would the result differ if graphite were used as the moderator?

Q6. If two objects collide and one is initially at rest, is it possible for both to be at rest after the collision? Is it possible for one to be at rest after the collision? Explain.

Q7. Explain why momentum is conserved when a ball bounces from a floor.

Q8. Is it possible to have a collision in which all of the kinetic energy is lost? If so, cite an example.

Q9. In a perfectly elastic collision between two particles, do both particles have the same kinetic energy after the collision? Explain.

Q10. When a ball rolls down an incline, its momentum increases. Does this imply that momentum is not conserved? Explain.

Q11. Consider an inelastic collision between a car and a large truck. Which vehicle loses more kinetic energy as a result of the collision?

9.5 TWO-DIMENSIONAL COLLISIONS

In the previous section and in Section 9.2, it was shown that the total momentum of a system of two particles is conserved when the system is isolated. For a general collision of two particles, this implies that the total momentum in *each* of the directions x, y, and z is conserved (Eq. 9.12). Thus, for a three-dimensional problem we would get three component equations for the conservation of momentum.

Let us consider a two-dimensional problem in which a particle of mass m_1 collides with a particle of mass m_2, where m_2 is initially at rest (Fig. 9.11). The collision is not head-on, but glancing. The parameter b, defined in Fig. 9.11, is called the *impact parameter*. As you can see from Fig. 9.11, if b is zero, the collision is head-on. After the collision, m_1 moves at an angle θ with respect to the horizontal and m_2 moves at an angle ϕ with respect to the horizontal. Applying the law of conservation of momentum in component form, $P_{xi} = P_{xf}$ and $P_{yi} = P_{yf}$, and noting that $P_y = 0$, we get

$$m_1 v_{1i} = m_1 v_{1f} \cos\theta + m_2 v_{2f} \cos\phi \qquad (9.17a) \quad \text{x-component}$$

$$0 = m_1 v_{1f} \sin\theta - m_2 v_{2f} \sin\phi \qquad (9.17b) \quad \text{y-component}$$

Now let us assume that the collision is elastic, in which case we can also write a third equation for the conservation of kinetic energy, in the form

$$\frac{1}{2}m_1 v_{1i}^2 = \frac{1}{2}m_1 v_{1f}^2 + \frac{1}{2}m_2 v_{2f}^2 \qquad (9.18) \quad \text{Conservation of energy}$$

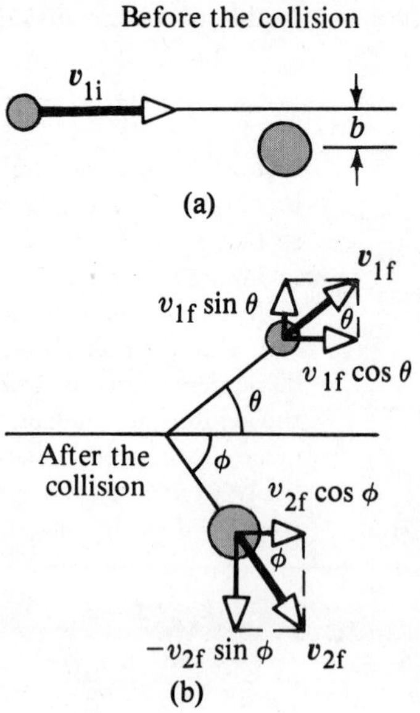

Figure 9.11 Schematic representation of an elastic glancing collision between two particles: (a) before the collision and (b) after the collision. Note that the impact parameter, b, must be greater than zero for a glancing collision.

If we know the initial velocity, v_{1i}, and the masses, we are left with four unknowns. Since we only have three equations, one of the four remaining quantities (v_{1f}, v_{2f}, θ, or ϕ) must be given to determine the motion after the collision from conservation principles alone.

Example 9.7 *Proton–Proton Collision*

A proton collides in a perfectly elastic fashion with another proton initially at rest. The incoming proton has an initial speed of 3.5×10^5 m/s and makes a glancing collision with the second proton, as in Fig. 9.11. (At close separations, the protons exert a repulsive electrostatic force on each other.) After the collision, one proton is observed to move at an angle of 37° to the original direction of motion, and the second deflects at an angle ϕ to the same axis. Find the final speeds of the two protons and the angle ϕ.

Solution: Since $m_1 = m_2$, $\theta = 37°$, and $v_{1i} = 3.5 \times 10^5$ m/s, Eqs. 9.17 and 9.18 become

$$v_{1f} \cos 37° + v_{2f} \cos\phi = 3.5 \times 10^5$$

$$v_{1f} \sin 37° - v_{2f} \sin\phi = 0$$

$$v_{1f}^2 + v_{2f}^2 = (3.5 \times 10^5)^2$$

Solving these three equations with three unknowns simultaneously gives

$$v_{1f} = 2.8 \times 10^5 \text{ m/s} \quad v_{2f} = 2.1 \times 10^5 \text{ m/s} \quad \phi = 53°$$

It is interesting to note that $\theta + \phi = 90°$. This result is *not* accidental. *Whenever two equal masses collide elastically with an impact parameter greater than zero and one of them is initially at rest, their final velocities are* always *at right angles to each other.* The next example illustrates this point in more detail.

Example 9.8

Billiard ball collision: In a game of billiards, the player wishes to "sink" the target ball in the corner pocket, as shown in Fig. 9.12. If the angle to the corner pocket is 35°, at what angle θ is the cue ball deflected? Assume that friction and rotational motion ("English") are unimportant, and assume the collision is elastic.

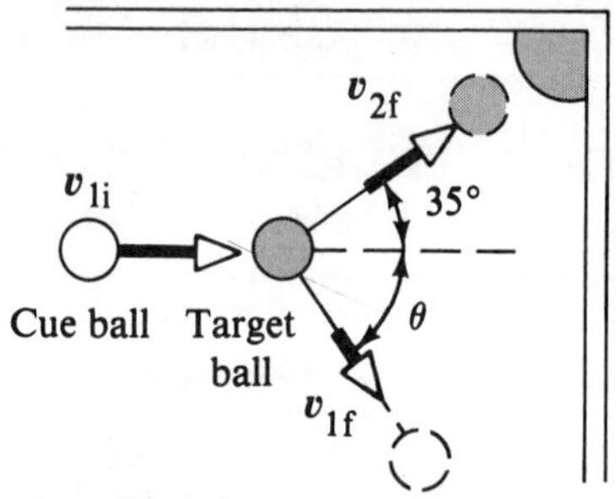

Figure 9.12 (Example 9.8)

Solution: Since the target is initially at rest, $v_{2i} = 0$ and conservation of kinetic energy gives

$$\frac{1}{2}m_1 v_{1i}^2 = \frac{1}{2}m_1 v_{1f}^2 + \frac{1}{2}m_2 v_{2f}^2$$

But $m_1 = m_2$, so that

$$(1) \qquad v_{1i}^2 = v_{1f}^2 + v_{2f}^2$$

Applying conservation of momentum to the two-dimensional collision gives

$$(2) \qquad \mathbf{v}_{1i} = \mathbf{v}_{1f} + \mathbf{v}_{2f}$$

If we square both sides of (2), we get

$$v_{1i}^2 = (\mathbf{v}_{1f} + \mathbf{v}_{2f}) \cdot (\mathbf{v}_{1f} + \mathbf{v}_{2f})$$
$$= v_{1f}^2 + v_{2f}^2 + 2\mathbf{v}_{1f} \cdot \mathbf{v}_{2f}$$

But

$v_{1f} \cdot v_{2f} = v_{1f}v_{2f}\cos(\theta + 35°)$, and so

$$(3) \qquad v_{1i}^2 = v_{1f}^2 + v_{2f}^2 + 2v_{1f}v_{2f}\cos(\theta + 35°)$$

Subtracting (1) from (3) gives

$$2v_{1f}v_{2f}\cos(\theta + 35°) = 0$$

$$\cos(\theta + 35°) = 0$$

$$\theta + 35° = 90° \text{ or } \theta = 55°$$

Again, this shows that whenever two equal masses undergo a glancing elastic collision and one of them is initially at rest, they will move at right angles to each other after the collision.

9.6 THE CENTER OF MASS

In this section we describe the overall motion of a mechanical system in terms of a very special point called the *center of mass* of the system. The mechanical system can be either a system of particles or an extended object. We shall see that the mechanical system moves as if all its mass were concentrated at the center of mass. Furthermore, if the resultant external force on the system is $\boldsymbol{F}$ and the total mass of the system is M, the center of mass moves with an acceleration given by $\boldsymbol{a} = \boldsymbol{F}/M$. That is, the system moves as if the resultant external force were applied to a single particle of mass M located at the center of mass. This result was implicitly assumed in earlier chapters since nearly all examples referred to the motion of extended objects.

Consider a mechanical system consisting of a pair of particles connected by a light, rigid rod (Fig. 9.13). The center of mass is located somewhere on the line joining the particles and is closer to the larger mass. If a single force is applied at some point on the rod closer to the smaller mass, the system will rotate clockwise (Fig. 9.13a). If the force is applied at a point on the rod closer to the larger mass, the system will rotate in the counterclockwise direction (Fig. 9.13b). If the force is applied at the center of mass, the system will not rotate and the rod will move parallel to itself (Fig. 9.13c). Thus, the center of mass can be easily located.

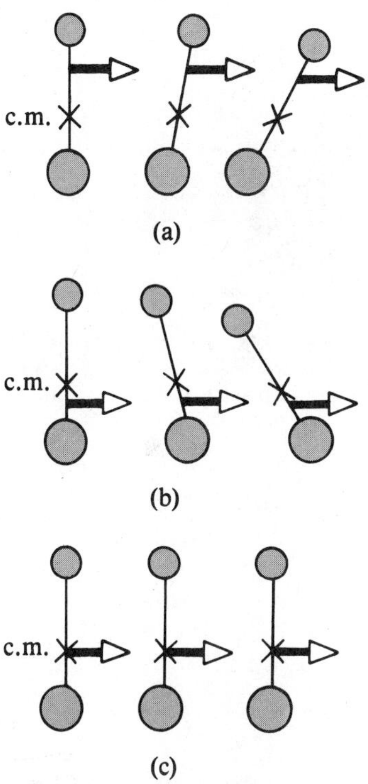

Figure 9.13 Two unequal masses are connected by a light, rigid rod. (a) The system rotates clockwise when a force is applied above the center of mass. (b) The system rotates counterclockwise when a force is applied below the center of mass. (c) The system moves parallel to itself when a force is applied at the center of mass.

The center of mass of the pair of particles described in Fig. 9.14 is located on the x axis and lies somewhere between the particles. The x coordinate of the center of mass in this case is defined to be

$$x_c \equiv \frac{m_1x_1 + m_2x_2}{m_1 + m_2} \qquad (9.19)$$

For example, if $x_1 = 0$, $x_2 = d$, and $m_2 = 2m_1$, we find that $x_c = \frac{2}{3}d$. That is, the center of mass lies closer to the more massive particle. If the two masses are equal, the center of mass lies midway between the particles.

We can extend the center of mass concept to a system of many particles in three dimensions. The x coordinate of the center of mass of n particles is defined to be

x coordinate of the center of mass for a system of particles

$$x_c = \frac{m_1x_1 + m_2x_2 + m_3x_2 + \cdots + m_nx_n}{m_1 + m_2 + m_3 + \cdots + m_n} = \frac{\Sigma m_ix_i}{\Sigma m_i} \qquad (9.20)$$

where x_i is the x coordinate of the ith particle and Σm_i is the *total mass* of the system. For convenience, we shall express the total mass as $M = \Sigma m_i$, where the sum runs over all n particles. The y and z coordinates of the center of mass are similarly defined by the equations

y and z coordinates of the center of mass for a system of particles

$$y_c = \frac{\Sigma m_iy_i}{M} \quad \text{and} \quad z_c = \frac{\Sigma m_iz_i}{M} \qquad (9.21)$$

The center of mass can also be located by its position vector, $\boldsymbol{r}_c$. The rectangular coordinates of this vector are x_c, y_c, and z_c, defined in Eqs. 9.20 and 9.21. Therefore,

$$\begin{aligned} \boldsymbol{r}_c &= x_c\boldsymbol{i} + y_c\boldsymbol{j} + z_c\boldsymbol{k} \\ &= \frac{\Sigma m_i x_i \boldsymbol{i} + \Sigma m_i y_i \boldsymbol{j} + \Sigma m_i z_i \boldsymbol{k}}{M} \end{aligned} \tag{9.22}$$

Vector position of the center of mass for a system of particles

$$\boldsymbol{r}_c = \frac{\Sigma m_i \boldsymbol{r}_i}{M} \tag{9.23}$$

where $\boldsymbol{r}_i$ is the position vector of the ith particle, defined by

$$\boldsymbol{r}_i = x_i\boldsymbol{i} + y_i\boldsymbol{j} + z_i\boldsymbol{k}$$

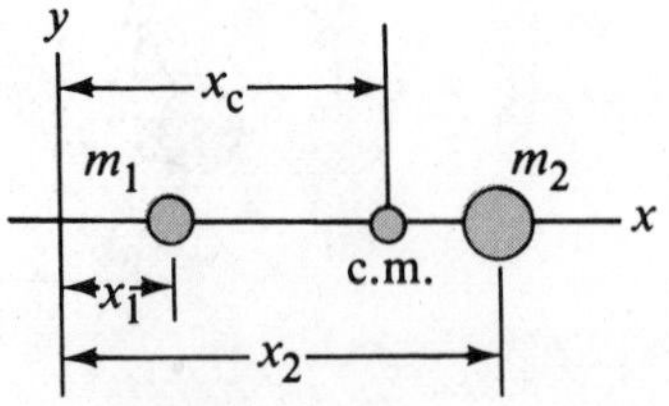

Figure 9.14 The center of mass of two particles on the x axis is located at x_c, a point between the particles, closer to the larger mass.

Although the location of the center of mass for a rigid body is somewhat more cumbersome, the basic ideas we have discussed still apply. We can think of a general rigid body as a system of a large number of particles (Fig. 9.15). The particle separation is very small, and so the body can be considered to have a continuous mass distribution. By dividing the body into elements of mass Δm_i, with coordinates x_i, y_i, z_i, we see that the x coordinate of the center of mass is approximately

$$x_c \approx \frac{\Sigma x_i \,\Delta m_i}{M}$$

with similar expressions for y_c and z_c. If we let the number of elements, n, approach infinity, then x_c will be given precisely. In this limit, we replace the sum by an integral and replace Δm_i by the differential element dm, so that

$$x_c = \lim_{\Delta m_i \to 0} \frac{\Sigma x_i \,\Delta m_i}{M} = \frac{1}{M}\int x\,dm \tag{9.24}$$

Likewise, for y_c and z_c we get

$$y_c = \frac{1}{M}\int y\,dm \quad \text{and} \quad z_c = \frac{1}{M}\int z\,dm \tag{9.25}$$

We can express the vector position of the center of mass of a rigid body in the form

Vector position of the center of mass for a rigid body

$$\boldsymbol{r}_c = \frac{1}{M}\int \boldsymbol{r}\,dm \tag{9.26}$$

where this is equivalent to the three scalar expressions given by Eqs. 9.24 and 9.25.

The center of mass of various homogeneous, symmetric bodies must lie on an axis of symmetry. For example, the center of mass of a homogeneous rod must lie on the rod, midway between its ends. The center of mass of a homogeneous sphere or a homogeneous cube must lie at its geometric center. One can determine the center of mass of an irregularly shaped planar body experimentally by suspending the body

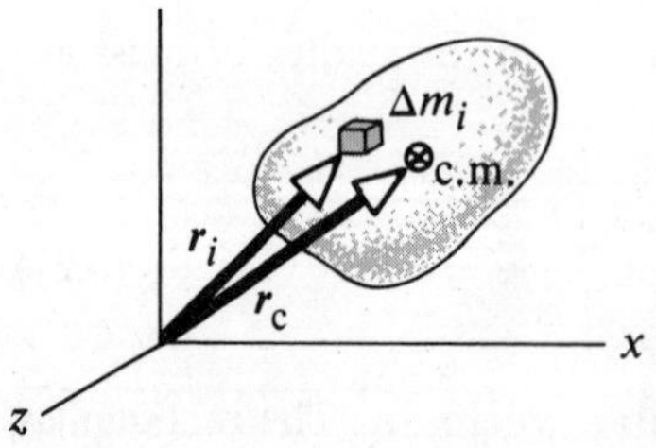

Figure 9.15 A rigid body can be considered a distribution of small elements of mass Δm_i. The center of mass is located at the vector position $\boldsymbol{r}_c$, which has coordinates x_c, y_c, and z_c.

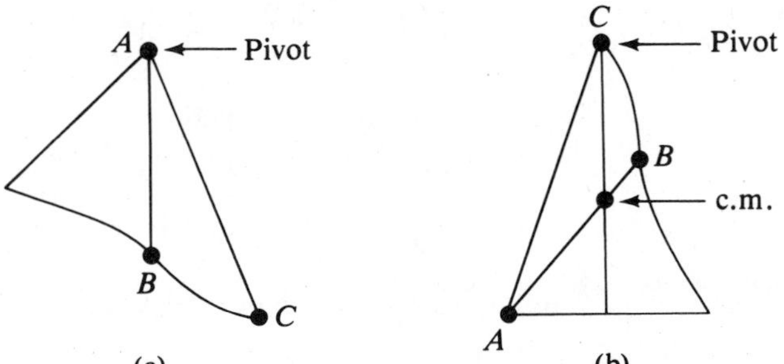

Figure 9.16 An experimental technique for determining the center of mass of an irregular planar object. The object is hung freely from two different pivots, A and C. The intersection of the two vertical lines AB and CD locates the center of mass.

between two different points (Fig. 9.16). The body is first hung from point A, and a vertical line AB is drawn when the body is in equilibrium. The body is then hung from point C, and a second vertical line, CD, is drawn. The center of mass coincides with the intersection of these two lines. In fact, if the body is hung freely from any point, the vertical line through this point must pass through the center of mass.

Since a rigid body is a continuous distribution of mass, each portion is acted upon by the force of gravity. The net effect of all of these forces is equivalent to the effect of a single force, Mg, acting through a special point, called the *center of gravity*. If g is constant over the mass distribution, then the center of gravity coincides with the center of mass. If a rigid body is pivoted at its center of gravity, it will be balanced in any orientation.

Example 9.9

A system consists of three particles located at the corners of a right triangle as in Fig. 9.17. Find the center of mass as measured from the origin.

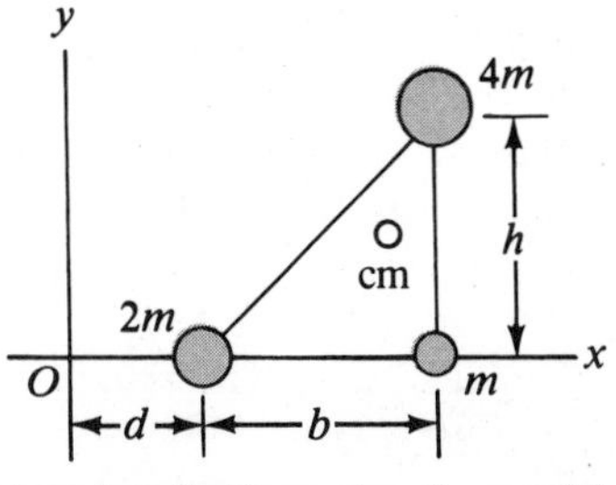

Figure 9.17 (Example 9.9) The center of mass of the three particles is located inside the triangle.

Solution: Using Eq. 9.20 and 9.21 and noting that $z_c = 0$, we get

$$x_c = \frac{\Sigma m_i x_i}{M} = \frac{2md + m(d+b) + 4m(d+b)}{7m}$$

$$= d + \frac{5}{7}b$$

$$y_c = \frac{\Sigma m_i y_i}{M} = \frac{2m(0) + m(0) + 4mh}{7m} = \frac{4}{7}h$$

Therefore, we can express the position vector to the center of mass as

$$\boldsymbol{r}_c = x_c\boldsymbol{i} + y_c\boldsymbol{j} = (d + \frac{5}{7}b)\boldsymbol{i} + \frac{4}{7}h\boldsymbol{j}$$

Example 9.10

(a) Show that the center of mass of a uniform rod of mass M and length L lies midway between its ends (Fig. 9.18).

By symmetry, we see that $y_c = z_c = 0$ if the rod is placed along the x axis. Furthermore, if we call the mass per unit length λ (the linear mass density), then $\lambda = M/L$ for a uniform rod. If we divide the rod into elements of length dx, then the mass of each element is $dm = \lambda\, dx$. Since an arbitrary element is at a distance x from the origin, Eq. 9.24 gives

$$x_c = \frac{1}{M}\int_0^L x\, dm = \frac{1}{M}\int_0^L x\lambda\, dx = \frac{\lambda}{M}\frac{x^2}{2}\Big]_0^L = \frac{\lambda L^2}{2M}$$

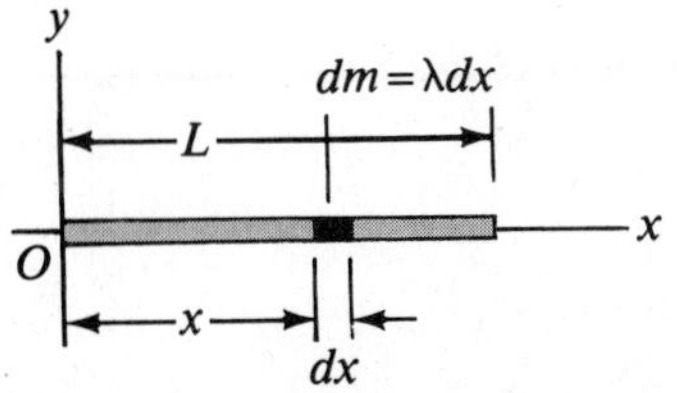

Figure 9.18 (Example 9.10) The center of mass of a uniform rod of length L is located at $x_c = L/2$.

Since $\lambda = M/L$, this reduces to

$$x_c = \frac{L^2}{2M}\left(\frac{M}{L}\right) = \frac{L}{2}$$

(b) Suppose the rod is *nonuniform* and the mass per unit length varies linearly with x according to the expression $\lambda = \alpha x$, where α is a constant. Find the x coordinate of the center of mass as a fraction of L.

In this case, we replace dm by $\lambda\, dx$, where λ is *not* constant. Therefore, x_c is given by

$$x_c = \frac{1}{M}\int_0^L x\, dm = \frac{1}{M}\int_0^L x\lambda\, dx = \frac{\alpha}{M}\int_0^L x^2\, dx = \frac{\alpha L^3}{3M}$$

We can also eliminate α by noting that the total mass of the rod is related to α through the relation

$$M = \int dm = \int_0^L \lambda\, dx = \int_0^L \alpha x\, dx = \frac{\alpha L^2}{2}$$

Substituting this into the expression for x_c gives

$$x_c = \frac{\alpha L^3}{3\alpha L^2/2} = \frac{2}{3}L$$

Q12. Can the center of mass of a body lie outside the body? If so, give examples.

9.7 MOTION OF A SYSTEM OF PARTICLES

We can begin to understand the physical significance and utility of the center of mass concept by taking the time derivative of the position vector of the center of mass, $\boldsymbol{r}_c$, given by Eq. 9.23. Assuming that M remains constant, that is, no particles enter or leave the system, we get the following expression for the *velocity of the center of mass:*

Velocity of the center of mass

$$\boldsymbol{v}_c = \frac{d\boldsymbol{r}_c}{dt} = \frac{1}{M}\Sigma m_i \frac{d\boldsymbol{r}_i}{dt} = \frac{\Sigma m_i \boldsymbol{v}_i}{M} \tag{9.27}$$

where $\boldsymbol{v}_i$ is the velocity of the ith particle. Rearranging Eq. 9.27 gives

Total momentum of a system of particles

$$M\boldsymbol{v}_c = \Sigma m_i \boldsymbol{v}_i = \Sigma \boldsymbol{p}_i = \boldsymbol{P} \tag{9.28}$$

The right side of Eq. 9.28 equals the total momentum of the system. Therefore, we conclude that *the total momentum of the system equals the total mass multiplied by the velocity of the center of mass.* In other words, the total momentum of the system is equal to that of a single particle of mass M moving with a velocity $\boldsymbol{v}_c$.

If we now differentiate Eq. 9.27 with respect to time, we get the *acceleration of the center of mass:*

Acceleration of the center of mass

$$\boldsymbol{a}_c = \frac{d\boldsymbol{v}_c}{dt} = \frac{1}{M}\Sigma m_i \frac{d\boldsymbol{v}_i}{dt} = \frac{1}{M}\Sigma m_i \boldsymbol{a}_i \tag{9.29}$$

Rearranging this expression and using Newton's second law, we get

$$M\boldsymbol{a}_c = \Sigma m_i \boldsymbol{a}_i = \Sigma \boldsymbol{F}_i \tag{9.30}$$

where $\boldsymbol{F}_i$ is the force on particle i.

The forces on any particle in the system may include both external forces (from outside the system) and internal forces (from within the system). However, by Newton's third law, the force of particle 1 on particle 2, for example, is equal to and opposite the force of particle 2 on particle 1. Thus, when we sum over all internal forces in Eq. 9.30, they cancel in pairs and the net force on the system is due *only* to external forces. Thus, we can write Eq. 9.30 in the form

Newton's second law for a system of particles

$$\Sigma \boldsymbol{F}_{\text{ext}} = M\boldsymbol{a}_c = \frac{d\boldsymbol{P}}{dt} \tag{9.31}$$

That is, the resultant external force on the system of particles equals the total mass of the system multiplied by the acceleration of the center of mass. If we compare this to Newton's second law for a single particle, we see that *the center of mass moves like an imaginary particle of mass* M *under the influence of the resultant external force on the system.*

Finally, we see that if the resultant external force is zero, then from Eq. 9.31 it follows that

$$\frac{d\mathbf{P}}{dt} = M\mathbf{a}_c = 0$$

so that

$$\mathbf{P} = M\mathbf{v}_c = \text{constant} \qquad (\text{when } \Sigma\mathbf{F}_{\text{ext}} = 0) \qquad (9.32)$$

That is, *the total linear momentum of a system of particles is conserved if there are no external forces acting on the system.* Therefore, it follows that for an *isolated* system of particles, both the total momentum and velocity of the center of mass are constant in time. This is a generalization to a many-particle system of the law of conservation of momentum that was derived in Section 9.2 for a two-particle system.

Suppose an isolated system consisting of two or more members is at rest. The center of mass of such a system will remain at rest unless acted upon by an external force. For example, consider a system made up of a swimmer and a raft, with the system initially at rest. When the swimmer dives off the raft, the center of mass of the system will remain at rest (if we neglect the friction between raft and water). Furthermore, the momentum of the diver will be equal in magnitude to the momentum of the raft, but opposite in direction.

As another example, suppose an unstable atom initially at rest suddenly decays into two fragments of masses M_1 and M_2, with velocities $\mathbf{v}_1$ and $\mathbf{v}_2$, respectively. (An example of such a radioactive decay is that of the uranium-238 nucleus, which decays into an alpha particle—the helium nucleus—and the thorium-234 nucleus.) Since the total momentum of the system before the decay is zero, the total momentum of the system after the decay must also be zero. Therefore, we see that $M_1\mathbf{v}_1 + M_2\mathbf{v}_2 = 0$. If the velocity of one of the fragments after the decay is known, the recoil velocity of the other fragment can be calculated. Can you explain the origin of the kinetic energy of the fragments?

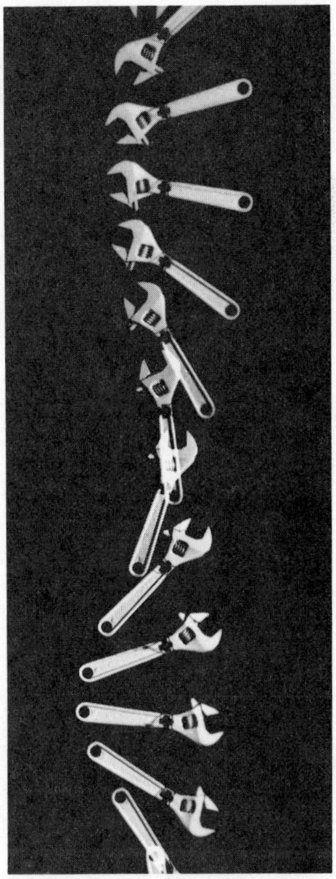

The center of mass of the wrench moves in a straight line as the wrench rotates about this point. (Education Development Center, Newton, Mass.)

Example 9.11 *Exploding Projectile*

A projectile is fired into the air and suddenly explodes into several fragments (Fig. 9.19). What can be said about the motion of the fragments after the collision?

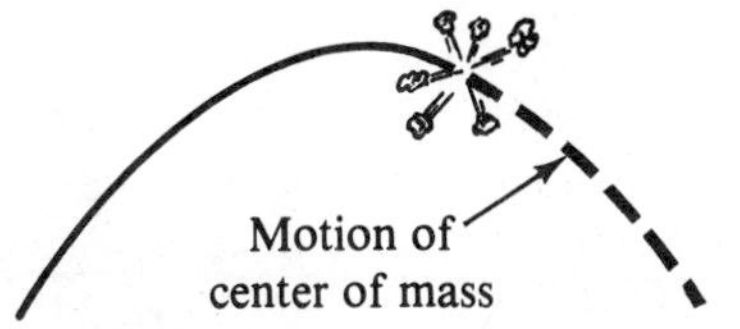

Figure 9.19 (Example 9.11) When a projectile explodes into several fragments, the center of mass of the fragments follows the same parabolic path the projectile would have taken had there been no explosion.

Solution: The only external force on the projectile is the force of gravity. Thus, the projectile follows a parabolic path. If the projectile did not explode, it would continue to move along the parabolic path indicated by the broken line in Fig. 9.19. Since the forces due to the explosion are internal, they do not affect the motion of the center of mass. Thus, after the explosion the center of mass of the fragments follows the *same* parabolic path the projectile would have followed if there had been no explosion.

Example 9.12

A rocket is fired vertically upward. It reaches an altitude of 1000 m and a velocity of 300 m/s. At this instant, the rocket explodes into three equal fragments. One fragment continues to move upward with a speed of 450 m/s right after the explosion. The second fragment has a speed of 240 m/s moving in the easterly direction right after the explosion. (a) What is the velocity of the third fragment right after the explosion?

Let us call the total mass of the rocket M; hence the

mass of each fragment is $M/3$. The total momentum just before the explosion must equal the total momentum of the fragments right after the explosion since the forces of the explosion are internal to the system and cannot affect the total momentum of the system.

Before the explosion: $\boldsymbol{P}_{\mathrm{i}} = M\boldsymbol{v}_0 = 300M\boldsymbol{j}$

After the explosion: $\boldsymbol{P}_{\mathrm{f}} = 240\left(\frac{M}{3}\right)\boldsymbol{i} + 450\left(\frac{M}{3}\right)\boldsymbol{j} + \frac{M}{3}\boldsymbol{v}$

where $\boldsymbol{v}$ is the unknown velocity of the third fragment. Equating these two expressions gives

$$M\frac{\boldsymbol{v}}{3} + 80M\boldsymbol{i} + 150M\boldsymbol{j} = 300M\boldsymbol{j}$$

$$\boldsymbol{v} = (-240\boldsymbol{i} + 450\boldsymbol{j})\ \mathrm{m/s}$$

(b) What is the position of the center of mass relative to the ground 3 s after the explosion? (Assume the rocket engine is nonoperative after the explosion.)

The center of mass of the fragments moves as a freely falling body since the explosion doesn't affect the motion of the center of mass (Example 9.11). If $t = 0$ is the time of the explosion, then $y_0 = 1000$ m and $v_0 = 300$ m/s for the center of mass. Using an expression from kinematics, we get for the y coordinate of the center of mass

$$y_{\mathrm{c}} = y_0 + v_0 t - \frac{1}{2}gt^2 = 1000 + 300t - 4.9t^2$$

Thus, at $t = 3$ s,

$$y_{\mathrm{c}} = [1000 + 300\ (3) - 4.9(3)^2]\ \mathrm{m} \approx 1856\ \mathrm{m}$$

Note that the x coordinate of the center of mass doesn't change. That is, in a given time interval the second fragment moves to the right by the same distance that the third fragment moves to the left.

Q13. A boy stands at one end of a floating raft that is stationary relative to the shore. He then walks to the opposite end of the raft, away from the shore. What happens to the center of mass of the system (boy + raft)? Does the raft move? Explain.

Q14. Three balls are thrown into the air simultaneously. What is the acceleration of their centers of mass while they are in motion?

Q15. Two isolated particles undergo a head-on collision. What is the acceleration of the center of mass after the collision?

Q16. A meter stick is balanced in a horizontal position with the index fingers of the right and left hands. If the two fingers are brought together, the stick remains balanced and the two fingers always meet at the 50-cm mark regardless of their original positions (try it!). Carefully explain this observation.

Q17. A hunter shoots a polar bear on a glacier. How might the hunter, knowing her own weight, be able to *estimate* the weight of the polar bear using a measuring tape and a rope?

9.8 ROCKET PROPULSION

When ordinary vehicles, such as automobiles, boats, and locomotives, are propelled, the driving force for the motion is one of friction. In the case of the automobile, the driving force is the force of the road on the car. A locomotive "pushes" against the tracks; hence the driving force is the force of the tracks on the locomotive. However, a rocket moving in space has no air, tracks, or water to "push" against. Therefore, the source of the propulsion of a rocket must be different. *The operation of a rocket depends upon the law of conservation of momentum as applied to a system of particles, where the system is the rocket plus its ejected fuel.*

The propulsion of a rocket can be understood by first considering the mechanical system consisting of a machine gun mounted on a cart on wheels. As the machine gun is fired, each bullet receives a momentum $m\boldsymbol{v}$ in some direction. For each bullet that is fired, the gun and cart must receive a compensating momentum in the opposite direction (as in Example 9.3). That is, the reaction force of the bullet on the gun accelerates the cart and gun. If there are n bullets fired each second, then the average force on the gun is equal to $\boldsymbol{F}_{\mathrm{av}} = nm\boldsymbol{v}$.

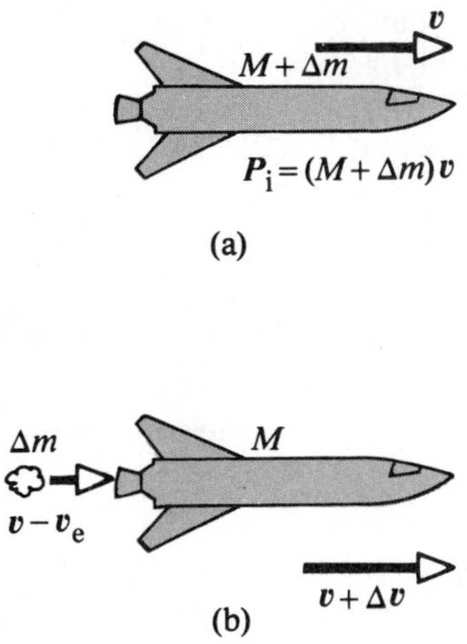

Figure 9.20 Rocket propulsion. (a) The initial mass of the rocket is $M + \Delta m$ at a time t, and its speed is v. (b) At a time $t + \Delta t$, the rocket's mass has reduced to M, and an amount of fuel Δm has been ejected. The rocket's speed increases by an amount Δv.

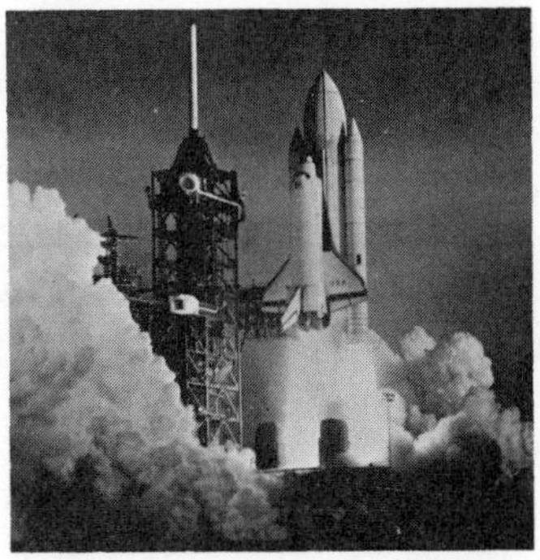

Lift off of the space shuttle Columbia. Massive amounts of thrust are generated by the shuttle's liquid-fueled engines, aided by the two solid fuel boosters. (NASA)

In a similar manner, as a rocket moves in free space (a vacuum), *its momentum changes when some of its mass is released in the form of ejected gases* (Fig. 9.20). *Since the ejected gases acquire some momentum, the rocket receives a compensating momentum in the opposite direction.* Therefore, *the rocket is accelerated as a result of the "push," or thrust, from the exhaust gases.* In free space, the center of mass of the entire system moves uniformly, independent of the propulsion process.

Suppose that at some time t, the momentum of the rocket plus the fuel is $(M + \Delta m)v$ (Fig. 9.20a). At some short time later, Δt, the rocket ejects some fuel of mass Δm and the rocket's speed therefore increases to $v + \Delta v$ (Fig. 9.20b). If the fuel is ejected with a velocity v_e *relative to the rocket,* then the velocity of the fuel relative to a stationary frame of reference is $v - v_e$. Thus, if we equate the total initial momentum of the system to the total final momentum, we get

$$(M + \Delta m)v = M(v + \Delta v) + \Delta m(v - v_e)$$

Simplifying this expression gives

$$M\,\Delta v = v_e\,\Delta m$$

We also could have arrived at this result by considering the system in the center of mass frame of reference. In this frame, the total momentum is zero; therefore if the rocket gains a momentum $M\Delta v$ by ejecting some fuel, the exhaust gases obtain a momentum $v_e\,\Delta m$ in the *opposite* direction, and so $M\,\Delta v - v_e\,\Delta m = 0$. If we now take the limit as Δt goes to zero, then $\Delta v \to dv$ and $\Delta m \to dm$. Furthermore, the increase in the exhaust mass, dm, corresponds to an equal decrease in the rocket mass, so that $dm = -dM$. Note that $dM < 0$. Using this fact, we get

$$M\,dv = -v_e\,dM \qquad (9.33)$$

Integrating this equation, and taking the initial mass of the rocket plus fuel to be M_i and the final mass of the rocket plus its remaining fuel to be M_f, we get

$$\int_{v_i}^{v_f} dv = -v_e \int_{M_i}^{M_f} \frac{dM}{M}$$

$$v_f - v_i = v_e \ln\left(\frac{M_i}{M_f}\right) \qquad (9.34)$$

Expression for rocket propulsion

This is the basic expression of rocket propulsion. First, it tells us that the increase in velocity is proportional to the exhaust velocity, v_e. Therefore, the exhaust velocity should be very high. Second, the increase in velocity is proportional to the logarithm of the ratio M_i/M_f. Therefore, this ratio should be as large as possible, which means that the rocket should carry as much fuel as possible.

The *thrust* on the rocket is the force exerted on the rocket by the ejected exhaust gases. We can obtain an expression for the thrust from Eq. 9.33:

$$\text{Thrust} = M\frac{dv}{dt} = \left|v_e\frac{dM}{dt}\right| \tag{9.35}$$

Here again we see that the thrust increases as the exhaust velocity increases and as the rate of change of mass (burn rate) increases.

Example 9.13

A rocket moving in free space has a speed of 3×10^3 m/s. Its engines are turned on, and fuel is ejected in a direction opposite the rocket's motion at a speed of 5×10^3 m/s relative to the rocket. (a) What is the speed of the rocket once its mass is reduced to one half its mass before ignition?

Applying Eq. 9.34, we get

$$v_f = v_i + v_e \ln\left(\frac{M_i}{M_f}\right)$$

$$= 3 \times 10^3 + 5 \times 10^3 \ln\left(\frac{M_i}{0.5M_i}\right)$$

$$= 6.5 \times 10^3 \text{ m/s}$$

(b) What is the thrust on the rocket if it burns fuel at the rate of 50 kg/s?

$$\text{Thrust} = \left|v_e\frac{dM}{dt}\right| = \left(5 \times 10^3 \frac{\text{m}}{\text{s}}\right)\left(50\frac{\text{kg}}{\text{s}}\right)$$

$$= 2.5 \times 10^5 \text{ N}$$

Q18. Explain the maneuver of decelerating a spacecraft. What other maneuvers are possible?

Q19. An astronaut "walking" in space accidently severs the safety cord attaching him to the spacecraft. What equipment would he need to return to the spacecraft?

Q20. Does the center of mass of a rocket in free space accelerate? Explain.

Q21. Can the speed of a rocket exceed the exhaust velocity of the fuel? Explain.

9.9 SUMMARY

The *linear momentum* of a particle of mass m moving with a velocity $\mathbf{v}$ is defined to be

Linear momentum

$$\mathbf{p} \equiv m\mathbf{v} \tag{9.1}$$

The *impulse of a force* $\mathbf{F}$ on a particle is equal to the change in the momentum of the particle and is given by

Impulse

$$\mathbf{I} = \Delta\mathbf{p} = \int_{t_i}^{t_f} \mathbf{F}\,dt \tag{9.6}$$

Impulsive forces are forces that are very strong compared with other forces on the system, and usually act for a very short time, as in the case of collisions.

The *law of conservation of momentum* for two interacting particles states that if two particles form an isolated system, their total momentum is conserved regardless of the nature of the force between them. Therefore, the total momentum of the system at all times equals its initial total momentum, or

Conservation of momentum

$$\mathbf{p}_{1i} + \mathbf{p}_{2i} = \mathbf{p}_{1f} + \mathbf{p}_{2f} \tag{9.12}$$

When two particles collide, the total momentum of the system before the collision always equals the total momentum after the collision, regardless of the nature of the collision. An *inelastic collision* is a collision for which the mechanical energy is not conserved, but momentum is conserved. A perfectly inelastic collision corresponds to the situation where the colliding bodies stick together after the collision. An *elastic collision* is one in which both momentum and energy are conserved.

Elastic and inelastic collision

In a two- or three-dimensional collision, the components of momentum in each of the three directions (x, y, and z) are conserved independently.

The *vector position of the center of mass of a system of particles* is defined as

$$\boldsymbol{r}_c \equiv \frac{\Sigma m_i \boldsymbol{r}_i}{M} \tag{9.23}$$

Center of mass for a system of particles

where $M = \Sigma m_i$ is the total mass of the system and $\boldsymbol{r}_i$ is the vector position of the ith particle.

The *vector position of the center of mass of a rigid body* can be obtained from the integral expression

$$\boldsymbol{r}_c = \frac{1}{M}\int \boldsymbol{r}\, dm \tag{9.26}$$

Center of mass for a rigid body

The *velocity of the center of mass for a system of particles* is given by

$$\boldsymbol{v}_c = \frac{\Sigma m_i \boldsymbol{v}_i}{M} \tag{9.27}$$

Velocity of the center of mass

The total momentum of a system of particles equals the total mass multiplied by the velocity of the center of mass, that is, $\boldsymbol{P} = M\boldsymbol{v}_c$.

Newton's second law applied to a system of particles is given by

$$\Sigma \boldsymbol{F}_{\text{ext}} = M\boldsymbol{a}_c = \frac{d\boldsymbol{P}}{dt} \tag{9.31}$$

Newton's second law for a system of particles

where $\boldsymbol{a}_c$ is the acceleration of the center of mass and the sum is over all external forces. Therefore, the center of mass moves like an imaginary particle of mass M under the influence of the resultant external force on the system. It follows from Eq. 9.31 that the total momentum of the system is conserved if there are no external forces acting on it.

EXERCISES

Section 9.1 Linear Momentum and Impulse

1. A 3-kg particle has a velocity of $(2\boldsymbol{i} - 4\boldsymbol{j})$ m/s. Find its x and y components of momentum and the magnitude of its total momentum.

2. The momentum of a 1500-kg car is equal to the momentum of a 5000-kg truck traveling at a speed of 25 mi/h. What is the speed of the car?

3. A 1500-kg automobile travels eastward at a speed of 8 m/s. It makes a 90° turn to the north in a time of 3 s and continues with the same speed. Find (a) the impulse delivered to the car as a result of the turn and (b) the average force exerted on the car during the turn.

4. A 0.3-kg ball moving along a straight line has a velocity of $5\boldsymbol{i}$ m/s. It collides with a wall and rebounds with a velocity of $-3\boldsymbol{i}$ m/s. Find (a) the change in its momentum and (b) the average force exerted on the wall if the ball is in contact with the wall for 5×10^{-3} s.

5. The force F_x acting on a 2-kg particle varies in time as shown in Fig. 9.21. Find (a) the impulse of the force, (b) the final velocity of the particle if it is initially at rest, and (c) the final velocity of the particle if it is initially moving along the x axis with a velocity of -2 m/s.

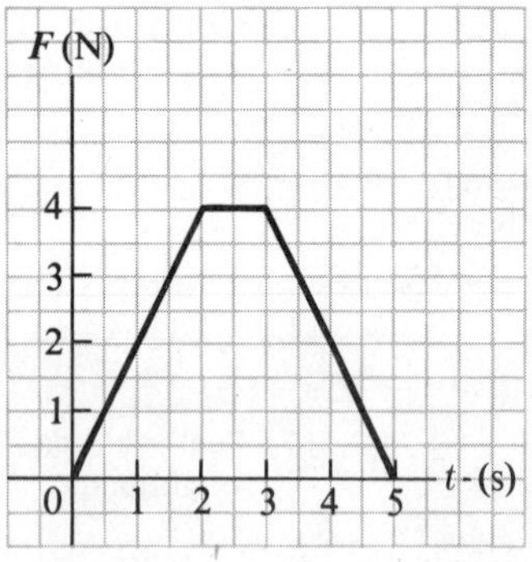

Figure 9.21 (Exercises 5 and 6).

6. Find the average force exerted on the particle described in Fig. 9.21 for the time interval $t_i = 0$ to $t_f = 5$ s.

7. An estimated force-time curve for a baseball struck by a bat is shown in Fig. 9.22. From this curve, determine (a) the impulse delivered to the ball, (b) the average force exerted on the ball, and (c) the peak force exerted on the ball.

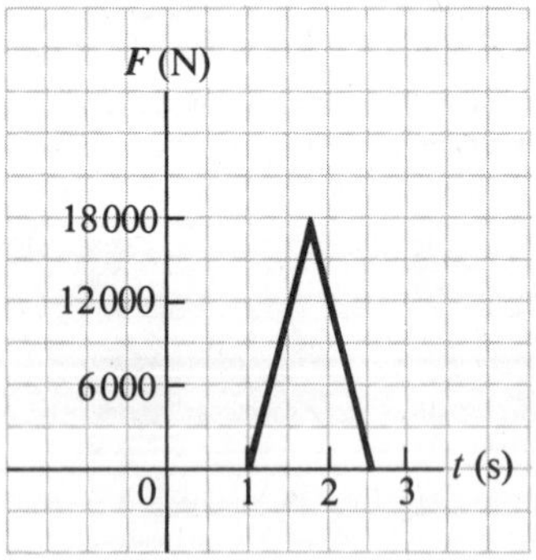

Figure 9.22 (Exercise 7).

8. Calculate the magnitude of the linear momentum for the following cases: (a) a proton of mass 1.67×10^{-27} kg moving with a speed of 5×10^6 m/s; (b) a 15-g bullet moving with a speed of 500 m/s; (c) a 75-kg sprinter running at a speed of 12 m/s, and (d) the earth (mass 5.98×10^{24} kg) moving with an orbital speed of 2.98×10^4 m/s.

9. If the momentum of an object is doubled in magnitude, what happens to its kinetic energy? (b) If the kinetic energy of an object is tripled, what happens to its momentum?

10. A 3-kg particle is initially moving along the y axis with a velocity of 5 m/s. After 5 s, it is moving along the x axis with a velocity of 3 m/s. Find (a) the impulse delivered to the particle and (b) the average force exerted on it in the 5-s interval.

11. A 1.5-kg football is thrown with a speed of 15 m/s. A stationary receiver catches the ball and brings it to rest in 0.02 s. (a) What is the impulse delivered to the ball? (b) What is the average force exerted on the receiver?

12. A single constant force of 80 N accelerates a 5-kg object from a speed of 2 m/s to a speed of 8 m/s. Find (a) the impulse acting on the object in this interval and (b) the time interval over which this impulse is delivered.

13. A 0.16-kg baseball is thrown with a speed of 147 m/s. It is hit straight back at the pitcher with a speed of 160 m/s. (a) What is the impulse delivered to the baseball? (b) Find the average force exerted by the bat on the ball if the ball is in contact with the bat for 2×10^{-3} s. Compare this with the weight of the ball and determine whether or not the impulse approximation is valid in the situation.

Section 9.2 Conservation of Linear Momentum for a Two-Particle System

14. A 40-kg child standing on a frozen pond throws a 2-kg stone to the east with a speed of 5 m/s. Neglecting friction between the child and ice, find the recoil velocity of the child.

15. Two blocks of masses M and $3M$ are placed on a horizontal, frictionless surface. A light spring is attached to one of them, and the blocks are pushed together with the spring between them (Fig. 9.23). A string holding them together is burned, after which the block of mass $3M$ moves to the right with a speed of 2 m/s. What is the speed of the block of mass M? (Assume they are initially at rest.)

16. Consider the cannon described in Example 9.3 and Fig. 9.5. (a) If the cannon's mass is doubled, what is the velocity of the cannon ball after firing? (b) If the cannon is bolted to the ground, the cannon ball is fired with some velocity but the cannon apparently doesn't move. Does this mean momentum is not conserved? Explain.

17. A 65-kg boy and a 40-kg girl, both wearing skates, face each other at rest. The boy pushes the girl, sending her eastward with a speed of 4 m/s. Describe the subsequent motion of the boy. (Neglect friction.)

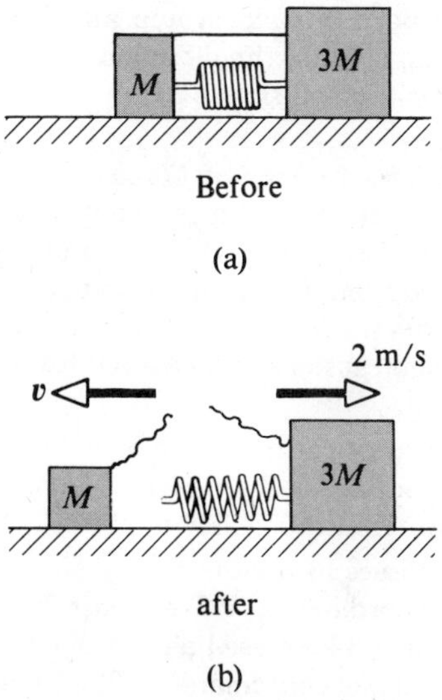

Figure 9.23 (Exercise 15).

Section 9.3 Collisions and Section 9.4 Collisions in One Dimension

18. A 3-kg mass moving initially with a speed of 8 m/s makes a perfectly inelastic head-on collision with a 5-kg mass initially at rest. (a) Find the final velocity of the composite particle. (b) How much energy is lost in the collision?

19. A 2000-kg meteorite has a speed of 80 m/s just before colliding head-on with the earth. Determine the recoil speed of the earth (mass 5.98×10^{24} kg).

20. Consider the ballistic pendulum described in Example 9.5 and shown in Fig. 9.9. (a) Show that the ratio of the kinetic energy after the collision to the kinetic energy before the collision is given by $m_1/(m_1 + m_2)$, where m_1 is the mass of the bullet and m_2 is the mass of the block. (b) If $m_1 = 8$ g and $m_2 = 2$ kg, what percentage of the original energy is left after the inelastic collision? What accounts for the missing energy?

21. An 8-g bullet is fired into a 2.5-kg ballistic pendulum and becomes embedded in it. If the pendulum rises a vertical distance of 6 cm, calculate the initial speed of the bullet.

22. A 90-kg halfback running north with a speed of 9 m/s is tackled by a 120-kg opponent running south with a speed of 3 m/s. If the collision is assumed to be perfectly inelastic and head-on, (a) calculate the velocity of the players just after the tackle and (b) determine the energy lost as a result of the collision. Can you account for the missing energy?

23. A 3-kg sphere makes a perfectly inelastic collision with a second sphere initially at rest. The composite system moves with a speed equal to one third the original speed of the 3-kg sphere. What is the mass of the second sphere?

24. A 1200-kg car traveling initially with a speed of 27 m/s in an easterly direction crashes into the rear end of a 9000-kg truck moving in the same direction at 22 m/s (Fig. 9.24). The velocity of the car right after the collision is 20 m/s to the east. (a) What is the velocity of the truck right after the collision? (b) How much mechanical energy is lost in the collision? How do you account for this loss in energy?

25. A railroad car of mass 2×10^4 kg moving with a speed of 5 m/s collides and couples with three other coupled railroad cars each of the same mass as the single car and moving in the same direction with an initial speed of 2 m/s. (a) What is the speed of the four cars after the collision? (b) How much energy is lost in the collision?

26. Verify Eqs. 9.15a and 9.15b for a perfectly elastic head-on collision.

27. A neutron in a reactor makes an elastic head-on collision with the nucleus of a carbon atom initially at rest. (a) What fraction of the neutron's kinetic energy is transferred to the carbon nucleus? (b) If the initial kinetic energy of the neutron is 1 MeV $= 1.6 \times 10^{-13}$ J, find its final kinetic energy and the kinetic energy of the carbon nucleus after the collision. (The mass of the carbon nucleus is about 12 times the mass of the neutron.)

28. A 3-kg mass moving with a velocity of $8\mathbf{i}$ m/s makes an elastic head-on collision with a 5-kg mass initially at rest. Find (a) the final velocity of each mass and (b) the final kinetic energy of each mass.

29. A 5-g particle moving to the right with a speed of 20 cm/s makes an elastic head-on collision with a 10-g particle initially at rest. Find (a) the final veloc-

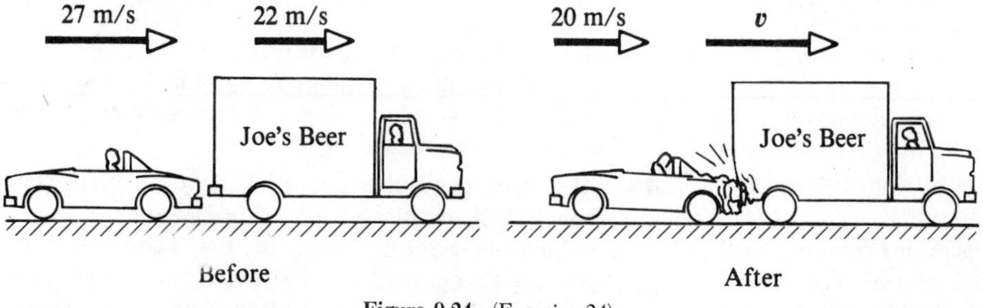

Figure 9.24 (Exercise 24).

ity of each particle and (b) the fraction of the total energy transferred to the 10-g particle.

30. A neutron moving with a velocity of $2 \times 10^6\boldsymbol{i}$ m/s makes a head-on elastic collision with a stationary helium nucleus (the mass of He is 4 amu). Find (a) the final velocity of each particle and (b) the fraction of the initial kinetic energy transferred to the helium nucleus.

31. Two particles of equal mass m collide head-on as shown in Fig. 9.10. Determine which of the following collisions are perfectly elastic for these particles: (a) $v_{1i} = 3$ m/s, $v_{2i} = 0$, $v_{1f} = 0$, $v_{2f} = 2$ m/s; (b) $v_{1i} = 0$, $v_{2i} = -5$ m/s, $v_{1f} = -5$ m/s, $v_{2f} = 0$; (c) $v_{1i} = 4$ m/s, $v_{2i} = -2$ m/s, $v_{1f} = -2$ m/s, $v_{2f} = 4$ m/s.

32. Two billiard balls have velocities of 1.5 m/s and -0.4 m/s before they meet in an elastic head-on collision. What are their final velocities?

Section 9.5 Two-Dimensional Collisions

33. A 200-g cart moves on a horizontal, frictionless surface with a constant speed of 30 cm/s. A 50-g piece of modeling clay is dropped vertically onto the cart. (a) If the clay sticks to the cart, find the final speed of the system. (b) After the collision, the clay has no momentum in the vertical direction. Does this mean that the law of conservation of momentum is violated?

34. A 0.3-kg puck, initially at rest on a horizontal, frictionless surface, is struck by a 0.2-kg puck moving initially along the x axis with a velocity of 2 m/s. After the collision, the 0.2-kg puck has a speed of 1 m/s at an angle of $\theta = 53°$ to the positive x axis (Fig. 9.11). (a) Determine the velocity of the 0.3-kg puck after the collision. (b) Find the fraction of kinetic energy lost in the collision.

35. A 2-kg mass with an initial velocity of $5\boldsymbol{i}$ m/s collides with and sticks to a 3-kg mass with an initial velocity of $-3\boldsymbol{j}$ m/s. Find the final velocity of the composite mass.

36. A bomb initially at rest explodes into three equal fragments. The velocities of two fragments are $(3\boldsymbol{i} + 2\boldsymbol{j})$ m/s and $(-\boldsymbol{i} - 3\boldsymbol{j})$ m/s. Find the velocity of the third fragment.

37. An unstable nucleus of mass 17×10^{-27} kg initially at rest disintegrates into three particles. One of the particles, of mass 5.0×10^{-27} kg, moves along the y axis with a velocity of 6×10^6 m/s. Another particle, of mass 8.4×10^{-27} kg, moves along the x axis with a velocity of 4×10^6 m/s. Find (a) the velocity of the third particle and (b) the total energy given off in the process.

38. A proton moving with a velocity $v_0\boldsymbol{i}$ collides elastically with another proton initially at rest. If both protons have the same speed after the collision, find (a) the speed of each proton after the collision in terms of v_0 and (b) the direction of the velocity vectors after the collision.

Section 9.6 The Center of Mass

39. The mass of the moon is about 0.0123 times the mass of the earth. The earth-moon separation measured from their centers is about 3.84×10^8 m. Determine the location of the center of mass of the earth-moon system as measured from the center of the earth.

40. A 3-kg particle is located on the x axis at $x = -4$ m, and a 5-kg particle is on the x axis at $x = 2$ m. Find the center of mass.

41. Three masses located in the xy plane have the following coordinates: a 2-kg mass has coordinates $(3, -2)$ m; a 3-kg mass has coordinates $(-2, 4)$ m; a 1-kg mass has coordinates $(2, 2)$ m. Find the coordinates of the center of mass.

42. The separation between the hydrogen and chlorine atoms of the HCl molecule is about 1.30×10^{-10} m. Determine the location of the center of mass of the molecule as measured from the hydrogen atom. (Chlorine is 35 times more massive than hydrogen.)

Section 9.7 Motion of a System of Particles

43. A 2-kg particle has a velocity of $(2\boldsymbol{i} - \boldsymbol{j})$ m/s, and a 3-kg particle has a velocity of $(\boldsymbol{i} + 6\boldsymbol{j})$ m/s. Find (a) the velocity of the center of mass and (b) the total momentum of the system.

44. A 5-kg particle moves along the x axis with a velocity of 3 m/s. A 3-kg particle moves along the x axis with a velocity of -2 m/s. Find (a) the velocity of the center of mass and (b) the total momentum of the system.

45. A particle of mass M has an acceleration of 3 m/s^2 in the x direction. A particle of mass $2M$ has an acceleration of 3 m/s^2 in the y direction. Find the acceleration of the center of mass.

46. A 2-kg particle has a velocity of $\boldsymbol{v}_1 = -10t\boldsymbol{j}$ m/s, where t is in s. A 3-kg particle moves with a constant velocity of $\boldsymbol{v}_2 = 4\boldsymbol{i}$ m/s. At $t = 0.5$ s, find (a) the velocity of the center of mass, (b) the acceleration of the center of mass, and (c) the total momentum of the system.

Section 9.8 Rocket Propulsion

47. A rocket engine consumes 75 kg of fuel per second. If the exhaust velocity is 4×10^3 m/s, calculate the thrust on the rocket.

48. The first stage of a Saturn V space vehicle consumes fuel at the rate of 1.5×10^4 kg/s, with an exhaust velocity of 2.6×10^3 m/s. (These are approximate figures.) (a) Calculate the thrust produced by these engines. (b) If the initial mass of the vehicle is 3×10^6 kg, find its *initial* acceleration on the launch pad. [You must include the force of gravity to solve (b).]

49. A rocket moving in free space with its engines off coasts with a speed of 5×10^3 m/s. Its engines are turned on, and at some later time when its mass is reduced to 90% of its initial mass, the speed of the rocket is 6.5×10^3 m/s. Find the exhaust velocity of the ejected fuel, assuming a uniform burn rate.

PROBLEMS

1. Two children in a 90-kg boat are drifting southward with a constant speed of 1.5 m/s. Each child has a mass of 50 kg. What is the velocity of the boat *immediately* after (a) one of the children *falls* off the rear of the boat, (b) one of the children dives off the rear in the northerly direction with a speed of 2 m/s relative to a stationary land observer, and (c) one of the children dives eastward (perpendicular to the boat) with a speed of 3 m/s.

2. A 5-g bullet moving with an initial speed of 400 m/s is fired into and passes through a 1-kg block, as in Fig. 9.25. The block, initially at rest on a frictionless, horizontal surface, is connected to a spring of force constant 900 N/m. If the block moves a distance of 5 cm to the right after impact, find (a) the speed at which the bullet emerges from the block and (b) the energy lost in the collision.

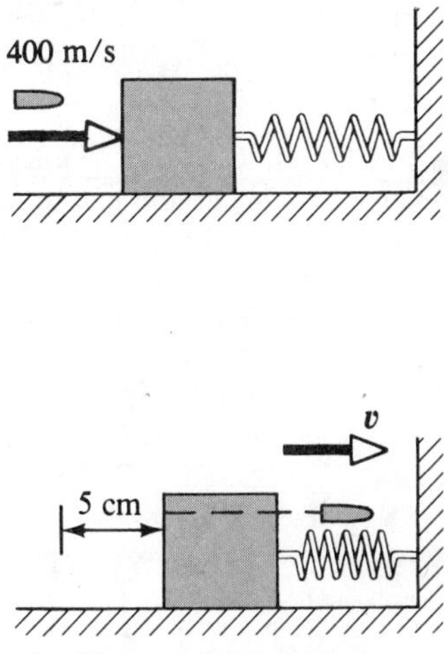

Figure 9.25 (Problem 2).

3. A 6-g bullet is fired into a 2-kg block initially at rest at the edge of a table of height 1 m (Fig. 9.26). The bullet remains in the block, and after impact the block lands 2 m from the bottom of the table. Determine the initial speed of the bullet.

4. The vector position of a 1-g particle moving in the xy plane varies in time according to $\boldsymbol{r}_1 = (3\boldsymbol{i} + 3\boldsymbol{j})t + 2\boldsymbol{j}t^2$. At the same time, the vector position of a 2-g particle moving in the xy plane varies as $\boldsymbol{r}_2 = 3\boldsymbol{i} - 2\boldsymbol{i}t^2 - 6\boldsymbol{j}t$, where t is in s and $\boldsymbol{r}$ is in cm.

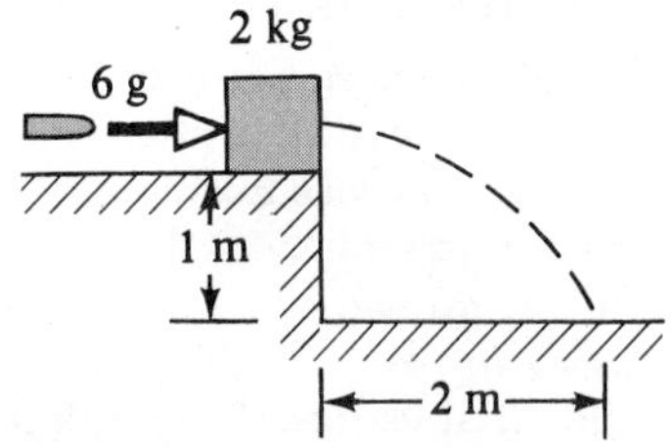

Figure 9.26 (Problem 3).

At $t = 2$ s, determine (a) the vector position of the center of mass, (b) the linear momentum of the system, (c) the velocity of the center of mass, and (d) the acceleration of the center of mass.

5. Two particles each of mass 0.5 kg move in the xy plane. At some instant, their coordinates, velocity components, and acceleration components are as tabulated below. At this instant, find (a) the vector position of the center of mass, (b) the velocity of the center of mass, and (c) the acceleration of the center of mass.

	x (m)	y (m)	v_x (m/s)	v_y (m/s)	a_x (m/s^2)	a_y (m/s^2)
Particle 1	2	3	5	−4	4	0
Particle 2	−2	3	3	8	2	−2

6. A 0.2-kg ball fastened to the end of a string 1.5 m in length to form a pendulum is released in the horizontal position. At the bottom of its swing, the ball collides with a 0.3-kg block initially resting on a frictionless surface (Fig. 9.27). (a) If the collision is elastic, calculate the speed of the ball and of the block just after the collision. (b) If the collision is completely inelastic (they stick), determine the height that the center of mass rises after the collision.

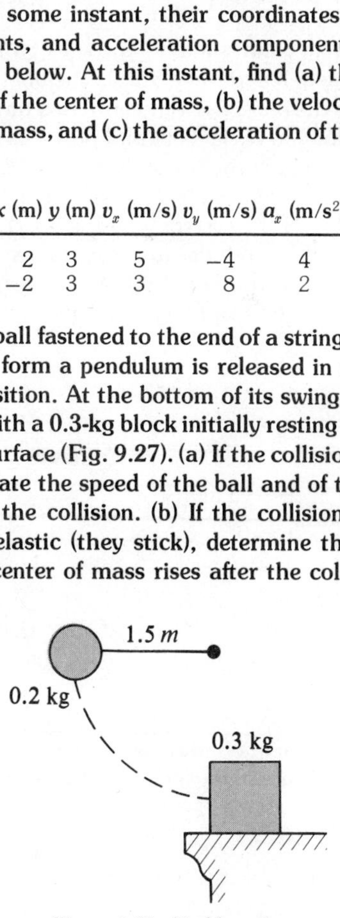

Figure 9.27 (Problem 6).

7. A 60-kg firefighter slides down a pole while a constant frictional force of 300 N retards his motion. A horizontal 20-kg platform is supported by a spring at the bottom of the pole to cushion the fall. The firefighter starts from rest 5 m above the platform, and the spring constant is 2500 N/m. Find (a) the firefighter's speed just before he collides with the platform and (b) the maximum distance the spring will be compressed. (Assume the frictional force acts during the entire motion.)

8. Consider a head-on elastic collision between a moving particle of mass m_1 and an initially stationary particle of mass m_2 (see Example 9.6). (a) Plot f_2, the fraction of energy transferred to m_2, as a function of the ratio m_2/m_1 and show that f_2 reaches a maximum when $m_2/m_1 = 1$. (b) Perform an analytical calculation that verifies that f_2 is a maximum when $m_1 = m_2$.

9. A 1-kg mass moving with an initial speed of 5 m/s collides with and sticks to a 6-kg mass initially at rest. The combined mass then proceeds to collide with and stick to a 2-kg mass also at rest initially. If the collisions are all head-on, find (a) the final speed of the system and (b) the amount of kinetic energy lost.

10. A 40-kg child stands at one end of a 70-kg boat that is 4 m in length (Fig. 9.28). The boat is initially 3 m from the pier. The child notices a turtle on a rock at the far end of the boat and proceeds to walk to that end to catch the turtle. Neglecting friction between the boat and water, (a) describe the subsequent motion of the system (child + boat). (b) Where will the child be *relative to the pier* when he reaches the far end of the boat? (c) Will he catch the turtle? (Assume he can reach out 1 m from the end of the boat.)

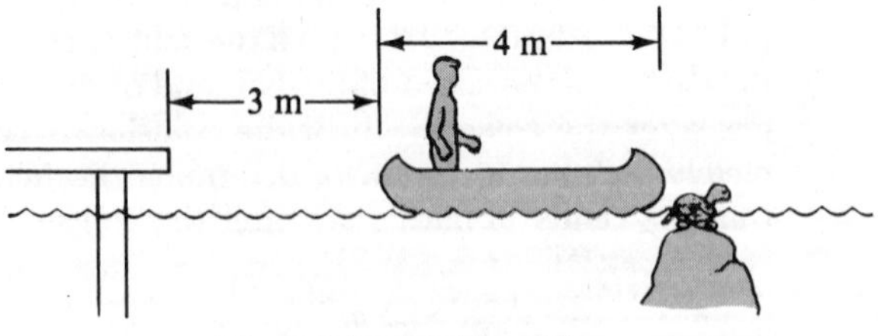

Figure 9.28 (Problem 10).

11. A block of mass M is given an initial velocity v_0 on a rough, horizontal surface. After traveling a distance d, it makes a head-on elastic collision with a block of mass $2M$. How far will the second block move before coming to rest? (Assume the coefficient of friction is the same for each block.)

12. A 7-g bullet is fired into a 1.5-kg ballistic pendulum as in Fig. 9.9. The bullet emerges from the block after the collision with a speed of 200 m/s, and the block rises to a maximum height of 12 cm. Find the initial speed of the bullet.

13. A machine gun held by a soldier fires bullets at the rate of three per second. Each bullet has a mass of 30 g and a speed of 1200 m/s. Find the average force exerted on the soldier.

14. An object of mass M is in the shape of a right triangle with dimensions as shown in Fig. 9.29. Locate the coordinates of the center of mass, assuming the object has a uniform mass per unit area.

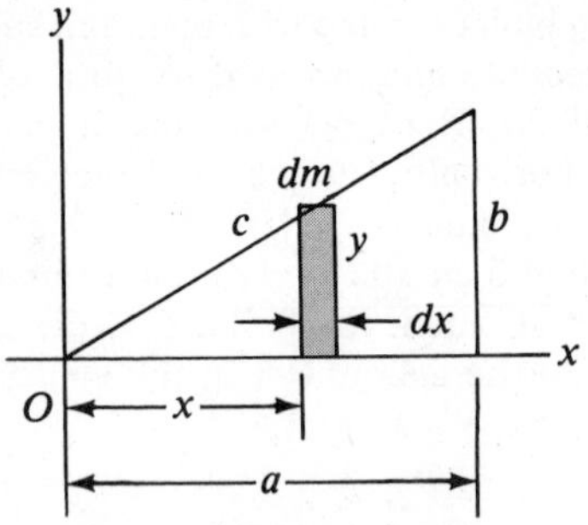

Figure 9.29 (Problem 14).

15. A chain of length L and total mass M is released from rest with its lower end just touching the top of a table, as in Fig. 9.30a. Find the force of the table on the chain after the chain has fallen through a distance x, as in Fig. 9.30b. (Assume each link comes to rest the instant it reaches the table.)

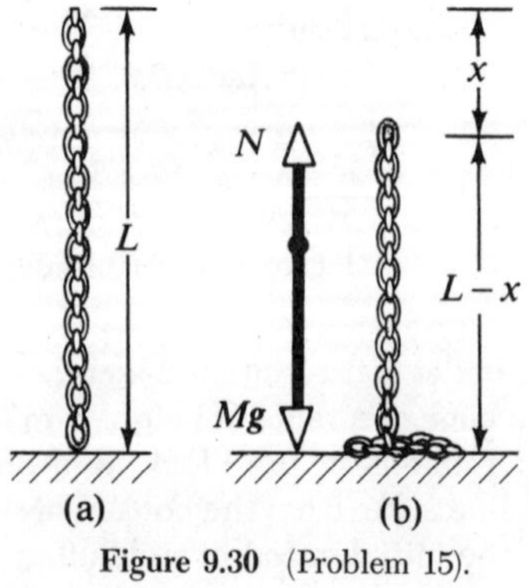

Figure 9.30 (Problem 15).

10 Rotation of a Rigid Body About a Fixed Axis

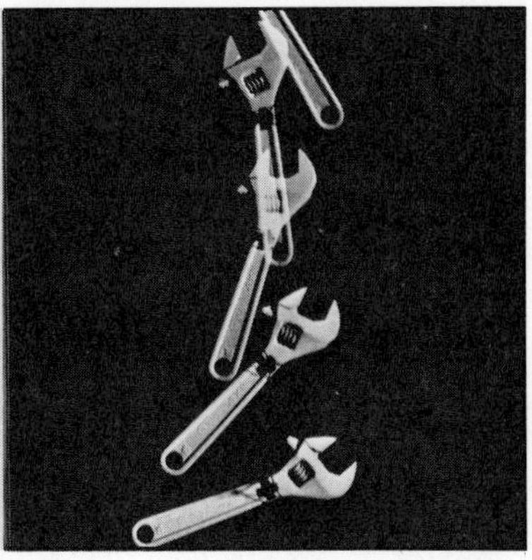

Education Development Center, Newton, Mass.

When an extended body, such as a wheel, rotates about its axis, the motion cannot be analyzed by treating the body as a particle, since at any given time different parts of the body will have different velocities and accelerations. For this reason, it is convenient to consider an extended object as a large number of particles, each with its own velocity and acceleration.

In dealing with the rotation of a body, analysis is greatly simplified by assuming the body to be rigid. A *rigid body* is defined as a body that is nondeformable, or one in which the separations between all pairs of particles in the body remain constant. Needless to say, all real bodies in nature are deformable to some extent; however, our rigid-body model is useful in many situations where deformation is negligible. In this chapter, we shall treat the rotation of a rigid body about a fixed axis, commonly referred to as *pure rotational motion.*

Rigid body

The vector nature of angular quantities, rotations in space, and the concept of angular momentum will be presented in detail in Chapter 11.

10.1 ANGULAR VELOCITY AND ANGULAR ACCELERATION

Figure 10.1 illustrates a planar rigid body of arbitrary shape confined to the xy plane and rotating about a fixed axis through O perpendicular to the plane of the figure. A particle on the body at P is at a fixed distance r from the origin and rotates in a circle of radius r about O. In fact, *every* particle on the body undergoes circular motion about O. It is convenient to represent the position of the point P with its polar coordinates (r,θ). In this representation, the only coordinate that changes in time is the angle θ; r remains constant. (In rectangular coordinates, both x and y vary in time.) As the particle moves along the circle from the positive x axis ($\theta = 0$) to the point P, it moves through an arc length s, which is related to the angular position θ through the relation

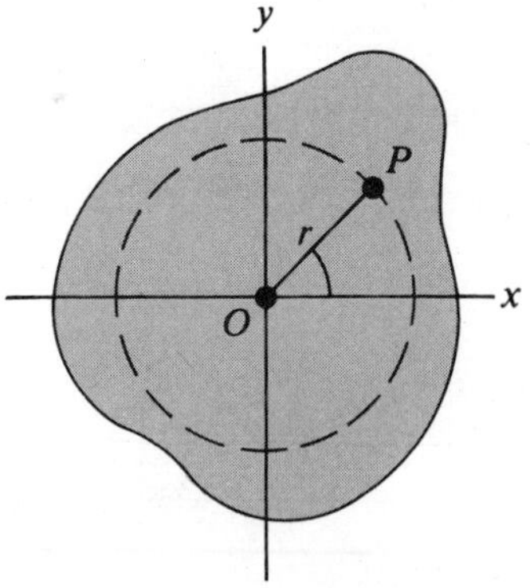

Figure 10.1 Rotation of a rigid body about a fixed axis through O, perpendicular to the plane of the figure (the z axis). Note that a particle at P rotates in a circle of radius r centered at O.

$$s = r\theta \tag{10.1a}$$

$$\theta = s/r \tag{10.1b}$$

Angular position

It is important to make note of the units of θ as expressed by Eq. 10.1b. The angle θ is the ratio of an arc length and the radius of the circle, and hence is a pure number. However, we commonly refer to the unit of θ as a *radian* (rad), where *one rad is the angle subtended by an arc length equal to the radius of the circle.* Since the

Radian

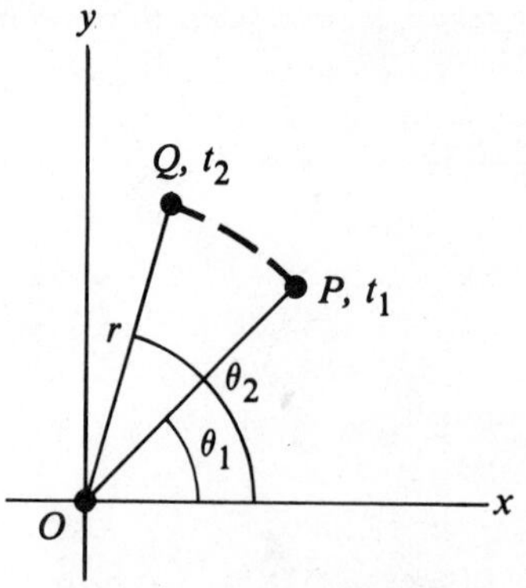

Figure 10.2 A particle on a rotating rigid body moves from P to Q along the arc of a circle. In the time interval $\Delta t = t_2 - t_1$, the radius vector sweeps out an angle $\Delta\theta = \theta_2 - \theta_1$.

circumference of a circle is $2\pi r$, it follows that 360° corresponds to an angle of $2\pi r/r$ rad or 2π rad (one revolution). Hence, 1 rad $= 360°/2\pi \approx 57.3°$. To convert an angle in degrees to an angle in radians, we can use the expression

$$\theta\,(\text{rad}) = \frac{\pi}{180°}\,\theta\,(\text{deg})$$

For example, 60° equals $\pi/3$ rad, and 45° equals $\pi/4$ rad.

As the particle travels from P to Q in Fig. 10.2 in a time Δt, the radius vector sweeps out an angle $\Delta\theta = \theta_2 - \theta_1$, which equals the *angular displacement.* We define the *average angular velocity* $\bar{\omega}$ (omega) as the ratio of this angular displacement to the time interval Δt:

Average angular velocity

$$\bar{\omega} \equiv \frac{\theta_2 - \theta_1}{t_2 - t_1} = \frac{\Delta\theta}{\Delta t} \tag{10.2}$$

In analogy to linear velocity, the *instantaneous angular velocity,* ω, is defined as the limit of the ratio in Eq. 10.2 as Δt approaches zero:

Instantaneous angular velocity

$$\omega \equiv \lim_{\Delta t \to 0} \frac{\Delta\theta}{\Delta t} = \frac{d\theta}{dt} \tag{10.3}$$

Angular velocity has units of rad/s, or s^{-1}, since radians are not dimensional. Let us adopt the convention that the fixed axis of rotation for the rigid body is the z axis, as in Fig. 10.1. We shall take ω to be positive when θ is increasing (counterclockwise motion) and negative when θ is decreasing (clockwise motion).

If the instantaneous angular velocity of a body changes from ω_1 to ω_2 in the time interval Δt, the body has an angular acceleration. The *average angular acceleration* $\bar{\alpha}$ (alpha) of a rotating body is defined as the *ratio of the change in the angular velocity to the time interval* Δt:

Average angular acceleration

$$\bar{\alpha} \equiv \frac{\omega_2 - \omega_1}{t_2 - t_1} = \frac{\Delta\omega}{\Delta t} \tag{10.4}$$

In analogy to linear acceleration, the *instantaneous angular acceleration* is defined as the limit of the ratio $\Delta\omega/\Delta t$ as Δt approaches zero:

Instantaneous angular acceleration

$$\alpha \equiv \lim_{\Delta t \to 0} \frac{\Delta\omega}{\Delta t} = \frac{d\omega}{dt} \tag{10.5}$$

Angular acceleration has units of rad/s^2 or s^{-2}. Note that α is positive when ω is increasing in time and negative when ω is decreasing in time.

For rotation about a fixed axis, we see that every *particle on the rigid body has the* same *angular velocity and the* same *angular acceleration.* That is, the quantities ω and α characterize the entire rigid body. Using these quantities, we can greatly simplify the analysis of rigid-body rotation. Notice that the angular displacement (θ), angular velocity (ω), and angular acceleration (α) are analogous to linear displacement (x), linear velocity (v), and linear acceleration (a), respectively, for the one-dimensional motion discussed in Chapter 3. The variables θ, ω, and α differ dimensionally from the variables x, v, and a, only by a length factor.

We have already indicated how the signs for ω and α are determined; however, we have not specified any direction in space associated with these vector quantities.[1] For rotation about a fixed axis, the only direction in space that uniquely specifies the rotational motion is the direction along the axis of rotation. However, we must also decide on the sense of these quantities, that is, whether they point into or out of the plane of Fig. 10.1.

As we have already mentioned, the direction of $\boldsymbol{\omega}$ is along the axis of rotation, which is the z axis in Fig. 10.1. By convention, we take the direction of $\boldsymbol{\omega}$ to be *out* of the plane of the diagram when the rotation is counterclockwise and *into* the plane of the diagram when the rotation is clockwise. To further illustrate this convention, it is convenient to use the *right-hand rule* shown in Fig. 10.3a. The four fingers of the right hand are wrapped in the direction of the rotation. The extended right thumb points in the direction of $\boldsymbol{\omega}$. Fig. 10.3b illustrates that $\boldsymbol{\omega}$ is also in the direction of advance of a similarly rotating right-handed screw. Finally, the sense of $\boldsymbol{\alpha}$ follows from its definition as $d\boldsymbol{\omega}/dt$. It is the same as $\boldsymbol{\omega}$ if ω is increasing in time and antiparallel to $\boldsymbol{\omega}$ if ω is decreasing in time.

Right-hand rule

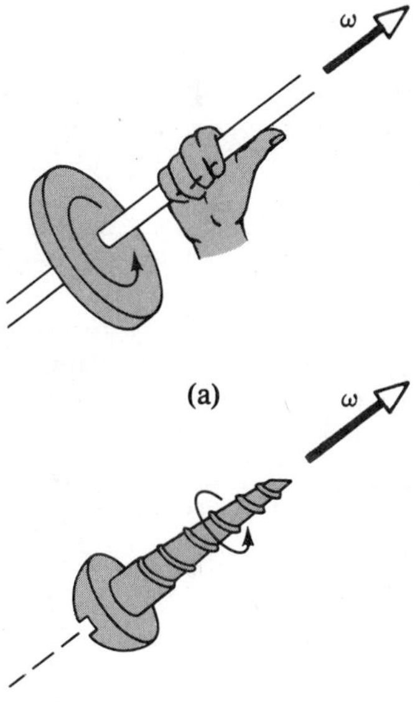

Figure 10.3 (a) The right-hand rule for determining the direction of the angular velocity. (b) The direction of $\boldsymbol{\omega}$ is in the direction of advance of a right-handed screw.

Q1. What is the magnitude of the angular velocity, $\boldsymbol{\omega}$, of the second hand of a clock? What is the direction of $\boldsymbol{\omega}$ as you view a clock hanging vertically? What is the angular acceleration, $\boldsymbol{\alpha}$, of the second hand?

Q2. A wheel rotates counterclockwise in the xy plane. What is the direction of $\boldsymbol{\omega}$? What is the direction of $\boldsymbol{\alpha}$ if the angular velocity is decreasing in time?

10.2 ROTATIONAL KINEMATICS: ROTATIONAL MOTION WITH CONSTANT ANGULAR ACCELERATION

In the study of linear motion, we found that the simplest form of accelerated motion to analyze is motion under constant linear acceleration (Chapter 3). Likewise, for rotational motion about a fixed axis the simplest accelerated motion to analyze is motion under constant angular acceleration. Therefore, we shall next develop kinematic relations for rotational motion under constant angular acceleration. If we write Eq. 10.5 in the form $d\omega = \alpha\, dt$ and let $\omega = \omega_0$ at $t_0 = 0$, we can integrate this expression directly:

Rotational kinematic equations

$$\omega = \omega_0 + \alpha t \qquad (\alpha = \text{constant}) \tag{10.6}$$

Likewise, substituting Eq. 10.6 into Eq. 10.3 and integrating once more (with $\theta = \theta_0$

[1] Although we do not verify it here, the instantaneous angular velocity and instantaneous acceleration are vector quantities, but the corresponding average values are not. This is because angular displacement is not a vector quantity for finite rotations.

at $t_0 = 0$), we get

$$\theta = \theta_0 + \omega_0 t + \frac{1}{2}\alpha t^2 \tag{10.7}$$

Rotational kinematic equations

If we eliminate t from Eqs. 10.6 and 10.7, we get

$$\omega^2 = \omega_0^2 + 2\alpha(\theta - \theta_0) \tag{10.8}$$

Notice that these kinematic expressions for rotational motion under constant angular acceleration are of the *same form* as those for linear motion under constant linear acceleration with the substitutions $x \to \theta$, $v \to \omega$, and $a \to \alpha$. Table 10.1 gives a comparison of the kinematic equations for rotational and linear motion. Furthermore, the expressions are valid for both rigid-body rotation and particle motion about a *fixed* axis.

TABLE 10.1 A Comparison of Kinematic Equations for Rotational and Linear Motion Under Constant Acceleration

Rotational Motion About Fixed Axis with α = Constant Variables: θ and ω	Linear Motion with a = Constant Variables: x and v
$\omega = \omega_0 + \alpha t$	$v = v_0 + at$
$\theta = \theta_0 + \omega_0 t + \frac{1}{2}\alpha t^2$	$x = x_0 + v_0 t + \frac{1}{2} at^2$
$\omega^2 = \omega_0^2 + 2\alpha(\theta - \theta_0)$	$v^2 = v_0^2 + 2a(x - x_0)$

Example 10.1 *Rotating Wheel*

A wheel rotates with a constant angular acceleration of 3.5 rad/s². If the angular velocity of the wheel is 2.0 rad/s at $t_0 = 0$, (a) what angle does the wheel rotate through in 2 s?

$$\theta - \theta_0 = \omega_0 t + \frac{1}{2}\alpha t^2$$

$$= \left(2.0\,\frac{\text{rad}}{\text{s}}\right)(2\text{ s}) + \frac{1}{2}\left(3.5\,\frac{\text{rad}}{\text{s}^2}\right)(2\text{ s})^2$$

$$= 11\text{ rad} = 630° = 1.76\text{ rev}$$

(b) What is the angular velocity at $t = 2$ s?

$$\omega = \omega_0 + \alpha t = 2.0\text{ rad/s} + \left(3.5\,\frac{\text{rad}}{\text{s}^2}\right)(2\text{ s})$$

$$= 9.0\text{ rad/s}$$

We could also obtain this result using Eq. 10.8 and the results of (a). Try it!

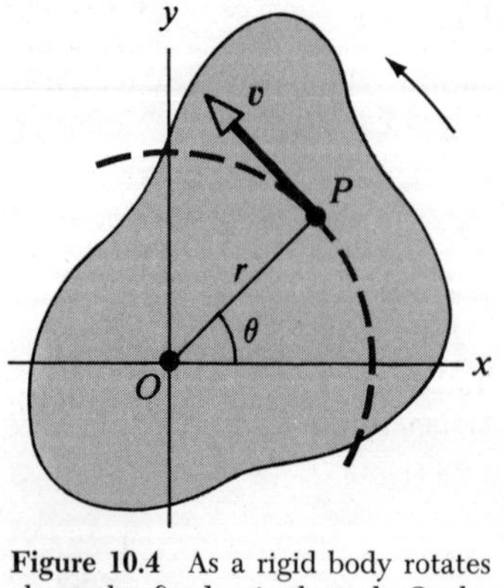

Figure 10.4 As a rigid body rotates about the fixed axis through O, the point P has a linear velocity v, which is always tangent to the circular path of radius r.

Q3. Are the kinematic expressions for θ, ω, and α valid when the angular displacement is measured in degrees instead of in radians?

Q4. A turntable rotates at a constant rate of 45 rotations/min. What is the magnitude of its angular velocity in rad/s? What is its angular acceleration?

10.3 RELATIONSHIPS BETWEEN ANGULAR AND LINEAR QUANTITIES

In this section we shall derive some useful relationships between the angular velocity and acceleration of a rotating rigid body and the linear velocity and acceleration of an arbitrary point in the body. In order to do so, we should keep in mind that when a rigid body rotates about a fixed axis, *every* particle of the body moves in a circle the center of which is the axis of rotation (Fig. 10.4).

We can first relate the angular velocity of the rotating body to the tangential speed, v, of a point P on the body. Since P moves in a circle, the linear velocity vector is always tangent to the circular path, and hence the phrase *tangential velocity*. The magnitude of the tangential velocity of the point P is, by definition, ds/dt, where s is the distance traveled by this point measured along the circular path. Recalling that $s = r\theta$ and noting that r is constant, we get

$$v = \frac{ds}{dt} = r\frac{d\theta}{dt}$$

$$v = r\omega \qquad (10.9)$$

Relationship between linear and angular speed

That is, the tangential speed of a point on a rotating rigid body equals the distance of that point from the axis of rotation multiplied by the angular velocity. Therefore, although every point on the rigid body has the same *angular* velocity, not every point has the same *linear* velocity. In fact, Eq. 10.9 shows that the linear velocity of a point on the rotating body increases as one moves outward from the center of rotation toward the rim, as you would intuitively expect.

We can relate the angular acceleration of the rotating rigid body to the tangential component of the linear acceleration of the point P by taking the time derivative of v:

$$a_t = \frac{dv}{dt} = r\frac{d\omega}{dt}$$

$$a_t = r\alpha \qquad (10.10)$$

Relationship between linear and angular acceleration

That is, the tangential component of the linear acceleration of a point on a rotating rigid body equals the distance of that point from the axis of rotation multiplied by the angular acceleration.

In Chapter 4 we found that a point rotating in a circular path undergoes a centripetal, or radial, acceleration of magnitude v^2/r and directed toward the center of rotation (Fig. 10.5). Since $v = r\omega$ for the point P on the rotating body, we can express this radial component of acceleration as

$$a_r = \frac{v^2}{r} = r\omega^2 \qquad (10.11)$$

Radial, or centripetal, acceleration

Therefore, the magnitude of the *total acceleration* of the point P on the rotating rigid body is given by

$$a = \sqrt{a_t^2 + a_r^2} = \sqrt{r^2\alpha^2 + r^2\omega^4} = r\sqrt{\alpha^2 + \omega^4} \qquad (10.12)$$

Magnitude of total acceleration

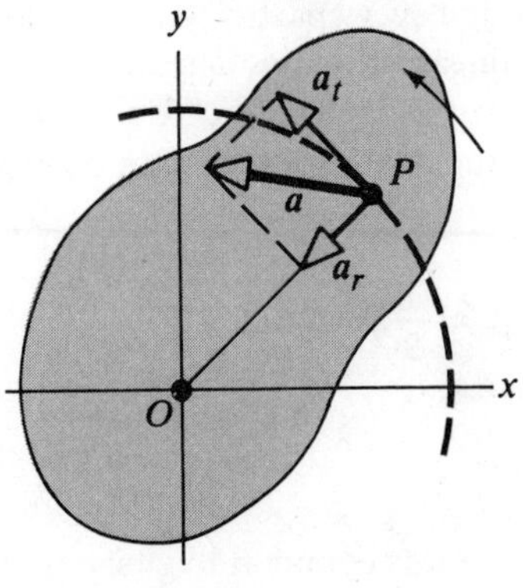

Figure 10.5 As a rigid body rotates about a fixed axis through O, the point P experiences a tangential component of acceleration, $\mathbf{a}_t$, and a centripetal component of acceleration, $\mathbf{a}_r$. The total acceleration of this point is $\mathbf{a} = \mathbf{a}_t + \mathbf{a}_r$.

Example 10.2

The turntable of a record player rotates initially at a rate of 33 revolutions/min and takes 20 s to come to rest. (a) What is the angular acceleration of the turntable, assuming the acceleration is uniform?

Recalling that 1 rev = 2π rad, we see that the initial angular velocity is given by

$$\omega_0 = \left(33 \frac{\text{rev}}{\text{min}}\right)\left(2\pi \frac{\text{rad}}{\text{rev}}\right)\left(\frac{1}{60} \frac{\text{min}}{\text{s}}\right) = 3.46 \frac{\text{rad}}{\text{s}}$$

Using $\omega = \omega_0 + \alpha t$ and the fact that $\omega = 0$ at $t = 20$ s, we get

$$\alpha = -\frac{\omega_0}{t} = -\frac{3.46 \text{ rad/s}}{20 \text{ s}} = -0.173 \frac{\text{rad}}{\text{s}^2}$$

where the negative sign indicates an angular deceleration (ω is decreasing).

(b) How many rotations does the turntable make before coming to rest?

Using Eq. 10.7, we find that the angular displacement in 20 s is

$$\Delta\theta = \theta - \theta_0 = \omega_0 t + \frac{1}{2}\alpha t^2$$

$$= \left[3.46(20) + \frac{1}{2}(-0.173)(20)^2\right] \text{rad} = 34.6 \text{ rad}$$

This corresponds to $34.6/2\pi$ rev, or 5.51 rev.

(c) If the radius of the turntable is 14 cm, what is the initial linear speed of a point on the rim of the turntable?

Using the relation $v = r\omega$ and the value $\omega_0 = 3.46 \frac{\text{rad}}{\text{s}}$ gives

$$v_0 = r\omega_0 = (14 \text{ cm})\left(3.46 \frac{\text{rad}}{\text{s}}\right) = 48.4 \text{ cm/s}$$

(d) What are the magnitudes of the radial and tangential components of the linear acceleration of a point on the rim at $t = 0$?

We can use $a_t = r\alpha$ and $a_r = r\omega^2$, which gives

$$a_t = r\alpha = (14 \text{ cm})\left(0.173 \frac{\text{rad}}{\text{s}^2}\right) = 2.42 \text{ cm/s}^2$$

$$a_r = r\omega_0^2 = (14 \text{ cm})\left(3.46 \frac{\text{rad}}{\text{s}}\right)^2 = 168 \text{ cm/s}^2$$

Q5. When a wheel of radius R rotates about a fixed axis, do all points on the wheel have the same angular velocity? Do they all have the same linear velocity? If the angular velocity is constant and equal to ω_0, describe the linear velocities and linear accelerations of the points at $r = 0$, $r = R/2$, and $r = R$.

10.4 ROTATIONAL KINETIC ENERGY

Let us consider a rigid body as a collection of small particles and let us assume that the body rotates about the fixed z axis with an angular velocity ω (Fig. 10.6). Each particle of the body has some kinetic energy, determined by its mass and velocity. If the mass of the ith particle is m_i and its speed is v_i, the kinetic energy of this particle is

$$K_i = \frac{1}{2} m_i v_i^2$$

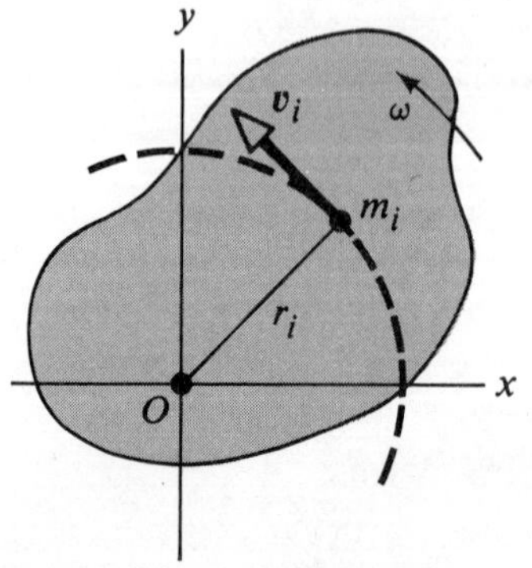

Figure 10.6 A rigid body rotating about the z axis with angular velocity ω. The kinetic energy of the particle of mass m_i is $\frac{1}{2}m_i v_i^2$. The total kinetic energy of the body is $\frac{1}{2}I\omega^2$.

To proceed further, we must recall that although every particle in the rigid body has the same angular velocity, ω, the individual linear velocities depend on the distance r_i from the axis of rotation according to the expression $v_i = r_i\omega$ (Eq. 10.9). The *total* kinetic energy of the rotating rigid body is the sum of the kinetic energies of the individual particles:

$$K = \Sigma K_i = \Sigma \frac{1}{2} m_i v_i^2 = \frac{1}{2} \Sigma m_i r_i^2 \omega^2$$

$$K = \frac{1}{2}(\Sigma m_i r_i^2)\omega^2 \tag{10.13}$$

where we have factored ω^2 from the sum since it is common to every particle. The quantity in parentheses is a property of the rigid body called the *moment of inertia*, *I*:

$$I = \Sigma m_i r_i^2 \quad (10.14)$$

Moment of inertia

Using this notation, we can express the kinetic energy of the rotating rigid body (Eq. 10.13) as

$$K = \frac{1}{2} I\omega^2 \quad (10.15)$$

Kinetic energy of a rotating rigid body

From the definition of moment of inertia, we see that it has dimensions of ML^2 ($kg \cdot m^2$ in SI units and $g \cdot cm^2$ in cgs units). Although we shall commonly refer to the quantity $\frac{1}{2}I\omega^2$ as the *rotational kinetic energy*, you should note that it is not a new form of energy. It is ordinary kinetic energy, since it was derived from a sum over individual kinetic energies of the particles contained in the rigid body. However, the form of the kinetic energy given by Eq. 10.15 is a very convenient one in dealing with rotational motion, providing we know how to calculate I. It is important that you recognize the analogy between kinetic energy associated with linear motion, $\frac{1}{2}mv^2$, and rotational kinetic energy, $\frac{1}{2}I\omega^2$. The quantities I and ω in rotational motion are analogous to m and v in linear motion, respectively. We shall describe how to calculate moments of inertia for rigid bodies in the next section. The following examples illustrate how to calculate moments of inertia and rotational kinetic energy for a distribution of particles.

Example 10.3 *The Oxygen Molecule*

Consider the diatomic molecule oxygen, O_2, which is rotating in the xy plane about the z axis. At room temperature, the "average" separation between the two oxygen atoms is 1.21×10^{-10} m (the atoms are treated as point masses). (a) Calculate the moment of inertia of the molecule about the z axis.

Since the mass of an oxygen atom is 2.77×10^{-26} kg, the moment of inertia about the z axis is

$$I = \Sigma m_i r_i^2 = m\left(\frac{d}{2}\right)^2 + m\left(\frac{d}{2}\right)^2 = \frac{md^2}{2}$$

$$= \left(\frac{2.77 \times 10^{-26}}{2}\ \text{kg}\right)(1.21 \times 10^{-10}\ \text{m})^2$$

$$= 2.03 \times 10^{-46}\ \text{kg} \cdot \text{m}^2$$

(b) If the angular velocity about the z axis is 2.0×10^{12} rad/s, what is the rotational kinetic energy of the molecule?

$$K = \frac{1}{2} I\omega^2$$

$$= \frac{1}{2}(2.03 \times 10^{-46}\ \text{kg} \cdot \text{m}^2)\left(2.0 \times 10^{12}\ \frac{\text{rad}}{\text{s}}\right)^2$$

$$= 4.1 \times 10^{-22}\ \text{J}$$

This is about one order of magnitude smaller than the average kinetic energy associated with the translational motion of the molecule at room temperature (6.2×10^{-21} J).

Example 10.4

Four point masses are fastened to the corners of a frame of negligible mass lying in the xy plane (Fig. 10.7). (a) If the rotation of the system occurs about the y axis with an angular velocity ω, find the moment of inertia about the y axis and the rotational kinetic energy about this axis.

First, note that the particles of mass m that lie on the y axis do not contribute to I_y (that is, $r_i = 0$ for these particles about this axis). Applying Eq. 10.14, we get

$$I_y = \Sigma m_i r_i^2 = Ma^2 + Ma^2 = 2Ma^2$$

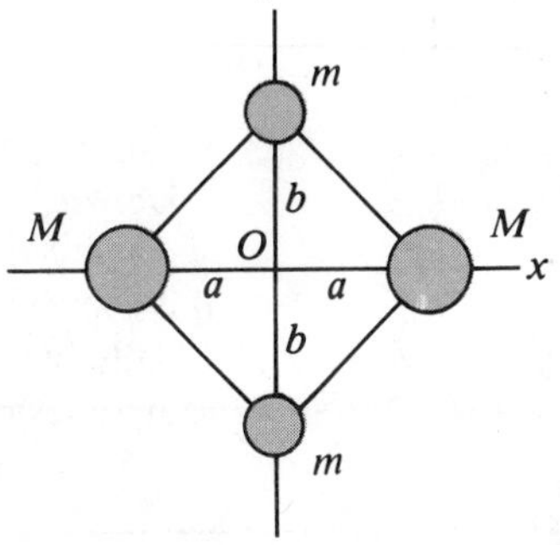

Figure 10.7 (Example 10.4) All particles are at a fixed separation as shown. The moment of inertia depends on the axis about which it is evaluated.

Therefore, the rotational kinetic energy about the y axis is

$$K = \frac{1}{2} I_y \omega^2 = \frac{1}{2}(2Ma^2)\omega^2 = Ma^2\omega^2$$

The fact that the masses m do not enter into this result makes sense, since these particles have no motion about the chosen axis of rotation; hence they have no kinetic energy.

(b) Now suppose the system rotates in the xy plane about an axis through O (the z axis). Calculate the moment of inertia about the z axis and the rotational kinetic energy about this axis.

Since r_i in Eq. 10.14 is the *perpendicular* distance to the axis of rotation, we get

$$I_z = \Sigma m_i r_i^2 = Ma^2 + Ma^2 + mb^2 + mb^2$$
$$= 2Ma^2 + 2mb^2$$

$$K = \frac{1}{2} I_z \omega^2 = \frac{1}{2}(2Ma^2 + 2mb^2)\omega^2 = (Ma^2 + mb^2)\omega^2$$

Comparing the results for (a) and (b), we conclude that the moment of inertia, and therefore the rotational kinetic energy associated with a given angular speed, depend on the axis of rotation. In (b), we would expect the result to include all masses and distances, since all particles are in motion for rotation in the xy plane. Furthermore, the fact that the kinetic energy in (a) is smaller than in (b) indicates that it would take less effort (work) to set the system into rotation about the y axis than about the z axis.

Q6. Suppose $a = b$ and $M > m$ in the system of particles described in Fig. 10.7. About what axis (x, y, or z) does the moment of inertia have the smallest value? the largest value?

10.5 CALCULATION OF MOMENTS OF INERTIA FOR RIGID BODIES

We can evaluate the moment of inertia of a rigid body by imagining that the body is divided into volume elements, each of mass Δm. Now we can use the definition $I = \Sigma r^2 \Delta m$ and take the limit of this sum as $\Delta m \to 0$. In this limit, the sum becomes an integral over the whole body, where r is the perpendicular distance from the axis of rotation to the element Δm. Hence,

Moment of inertia for a rigid body

$$I = \lim_{\Delta m \to 0} \Sigma r^2 \Delta m = \int r^2\, dm \qquad (10.16)$$

To evaluate the moment of inertia using Eq. 10.16, it is necessary to express the element of mass dm in terms of its coordinates. It is common to define a mass density in various forms. For a three-dimensional body, we shall commonly use the *local volume density,* that is, *mass per unit volume.* In this case, we can write

$$\rho = \lim_{\Delta V \to 0} \frac{\Delta m}{\Delta V} = \frac{dm}{dV}$$

$$dm = \rho dV$$

Therefore, the moment of inertia can be expressed in the form

$$I = \int \rho r^2\, dV$$

If the body is homogeneous, then ρ is constant and the integral can be evaluated for a known geometry. If ρ is not constant, then its variation with position must be specified. When dealing with an object in the form of a sheet of uniform thickness t, it is convenient to define a surface density $\sigma = \rho t$, which signifies *mass per unit area.* Finally, when mass is distributed along a uniform rod of cross-sectional area A, we sometimes use linear density, $\lambda = \rho A$, where λ is defined as *mass per unit length.*

Example 10.5 *Uniform Hoop*

Find the moment of inertia of a uniform hoop of mass M and radius R about an axis perpendicular to the plane of the hoop, through its center (Fig. 10.8).

Solution: All elements of mass are at the same distance $r = R$ from the axis. Therefore, applying Eq. 10.16 we get for the moment of inertia about the z axis through O:

$$I_z = \int r^2\, dm = R^2 \int dm = MR^2$$

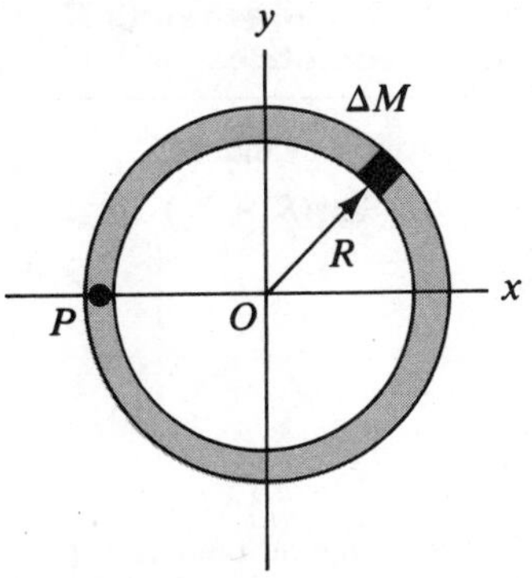

Figure 10.8 (Example 10.5) The mass elements of a uniform hoop are all the same distance from O.

Example 10.6 *Uniform Rigid Rod*

Calculate the moment of inertia of a uniform rigid rod of length L and mass M (Fig. 10.9) about an axis perpendicular to the rod (the y axis) passing through its center of mass.

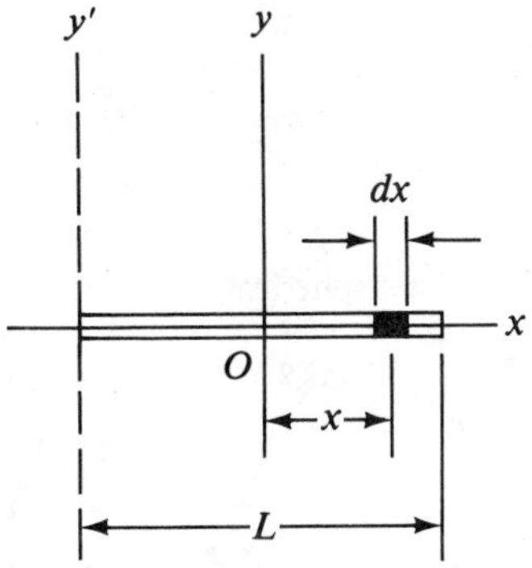

Figure 10.9 (Example 10.6) A uniform rigid rod of length L. The moment of inertia about the y axis is less than that about the y' axis.

Solution: The shaded element of width dx has a mass dm equal to the mass per unit length multiplied by the element of length, dx. That is, $dm = \frac{M}{L}dx$. Substituting this into Eq. 10.16, with $r = x$, we get

$$I_y = \int r^2\,dm = \int_{-L/2}^{L/2} x^2 \frac{M}{L}\,dx = \frac{M}{L}\int_{-L/2}^{L/2} x^2\,dx$$

$$= \frac{M}{L}\left[\frac{x^3}{3}\right]_{-L/2}^{L/2} = \frac{1}{12}ML^2$$

If we were to calculate I about an axis perpendicular to the rod but through one end (the y' axis), the calculation would be similar to the above, with the limits of integration changing to $x = 0$ and $x = L$. It is left as an exercise for the student to show that the result is $\frac{1}{3}ML^2$. Again, we see that I depends on the choice of axis.

Example 10.7 *Uniform Solid Cylinder*

A uniform solid cylinder has a radius R, mass M, and length L. Calculate the moment of inertia of the cylinder about an axis through its center, along its length (the z axis in Fig. 10.10).

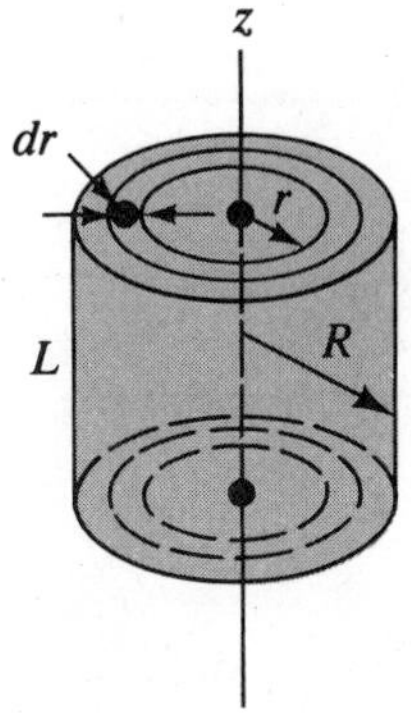

Figure 10.10 (Example 10.7) Calculating I about the z axis for a uniform solid cylinder.

Solution: In this example, it is convenient to divide the cylinder into cylindrical shells of radius r, thickness dr, and length L, as in Fig. 10.10. In this case, cylindrical shells are chosen because one wants all mass elements dm to have a single value for r, which makes the calculation of I more straightforward. The volume of each shell is its cross-sectional area multiplied by the length, or $dV = dA \cdot L = (2\pi r\,dr)L$. If the *mass per unit volume* is ρ, then the mass of this differential volume element is $dm = \rho\,dV = \rho\,2\pi rL\,dr$. Substituting this into Eq. 10.16, we get

$$I_z = \int r^2\,dm = 2\pi\rho L\int_0^R r^3\,dr = \frac{\pi\rho LR^4}{2}$$

However, since the total volume of the cylinder is πR^2L, $\rho = M/V = M/\pi R^2L$. Substituting this into the above result gives

$$I_z = \frac{1}{2}MR^2$$

As we saw in the previous examples, the moments of inertia of rigid bodies with simple geometry (high symmetry) are relatively easy to calculate provided the reference axis coincides with an axis of symmetry. Table 10.2 lists, for some common rigid bodies, moments of inertia about an axis through the center of mass and about an axis parallel to this.[2]

[2]Civil engineers use the moment of inertia concept to characterize the elastic properties (rigidity) of such structures as loaded beams. Hence, it is often useful even in a nonrotational context.

**TABLE 10.2 Moments of Inertia of Homogeneous Rigid Bodies
with Different Geometries**

Hoop or cylindrical shell
$I_c = MR^2$

Hollow cylinder
$I_c = \frac{1}{2}M(R_1^2 + R_2^2)$

Solid cylinder or disk
$I_c = \frac{1}{2}MR^2$

Rectangular plate
$I_c = \frac{1}{12}M(a^2 + b^2)$

Long thin rod
$I_c = \frac{1}{12}ML^2$

Long thin rod
$I = \frac{1}{3}ML^2$

Solid sphere
$I_c = \frac{2}{5}MR^2$

Thin spherical shell
$I_c = \frac{2}{3}MR^2$

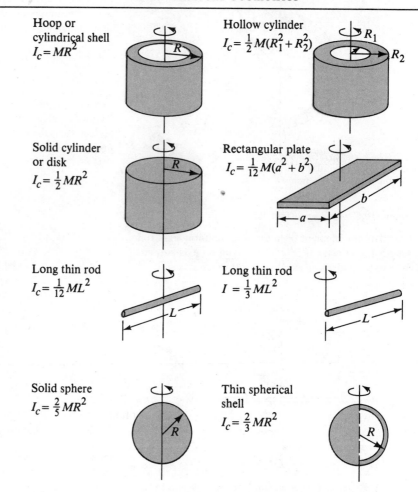

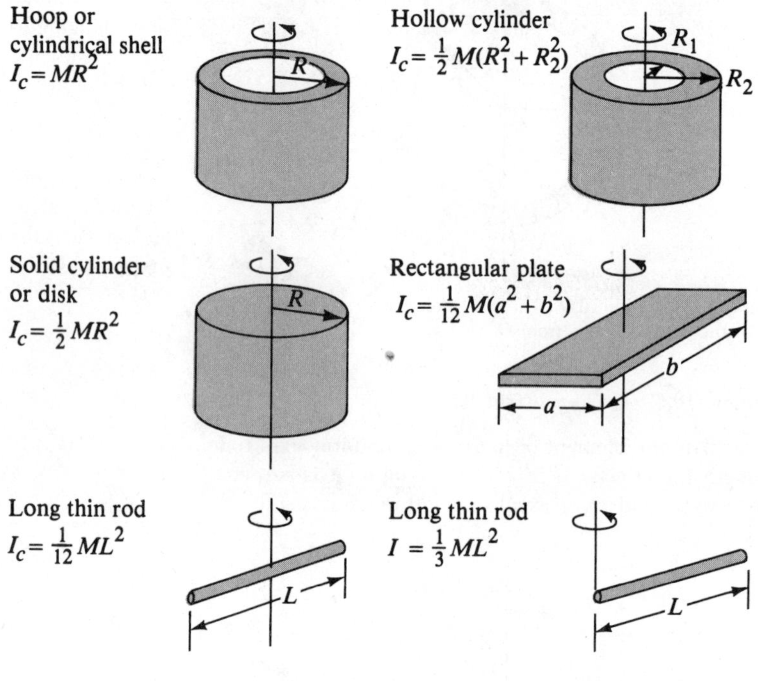

The calculation of moments of inertia about an arbitrary axis can be somewhat cumbersome, even for a highly symmetric body, such as a sphere. In this regard, there is an important theorem, called the *parallel-axis theorem*, that often simplifies the calculation of moments of inertia. Suppose the moment of inertia about any axis through the center of mass is I_c. The parallel-axis theorem states that the moment of inertia about any axis that is *parallel* to and a distance d away from the axis that passes through the center of mass is given by

Parallel-axis theorem

$$I = I_c + Md^2 \tag{10.17}$$

For those interested, a discussion of the parallel-axis theorem follows.

Suppose a body rotates in the xy plane about an axis through O as in Fig. 10.11 and the coordinates of the center of mass are x_c, y_c. Let the element Δm have coordinates x, y relative to the origin. Since this element is at a distance $r = \sqrt{x^2 + y^2}$ from the z axis, the moment of inertia about the z axis through O is

$$I = \int r^2 \, dm = \int (x^2 + y^2) \, dm$$

However, we can relate the coordinates x, y to the coordinates of the center of mass, x_c, y_c, and the coordinates relative to the center of mass, x', y' through the relations $x = x' + x_c$

and $y = y' + y_c$. Therefore,

$$I = \int[(x' + x_c)^2 + (y' + y_c)^2]\,dm$$
$$= \int[(x')^2 + (y')^2]\,dm + 2x_c\int x'\,dm + 2y_c\int y'\,dm + (x_c{}^2 + y_c{}^2)\int dm$$

The first term on the right is, by definition, the moment of inertia about an axis parallel to the z axis, through the center of mass. The second two terms on the right are zero, since by definition of the center of mass $\int x'\,dm = \int y'\,dm = 0$ (x', y' are the coordinates of the mass element relative to the center of mass). Finally, the last term on the right is simply Md^2, since $\int dm = M$ and $d^2 = x_c{}^2 + y_c{}^2$. Therefore, we conclude that

$$I = I_c + Md^2$$

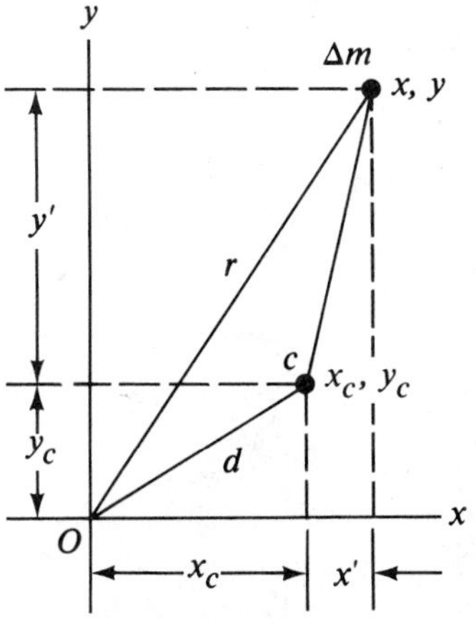

Figure 10.11 The parallel-axis theorem. If the moment of inertia about an axis perpendicular to the figure through the center of mass at c is I_c, then the moment of inertia about the z axis is $I_z = I_c + Md^2$.

Q7. A wheel is in the shape of a hoop as in Fig. 10.8. In two separate experiments, the wheel is rotated from rest to an angular velocity ω. In one experiment, the rotation occurs about the z axis through O; in the other, the rotation occurs about an axis parallel to z through P. Which rotation requires more work?

Q8. Suppose the rod in Fig. 10.9 has a nonuniform mass distribution. In general, would the moment of inertia about the y axis still equal $\frac{1}{12}ML^2$? If not, could the moment of inertia be calculated without knowledge of the manner in which the mass is distributed?

10.6 TORQUE

When a force is properly exerted on a rigid body pivoted about some axis, the body will tend to rotate about that axis. The ability of a force to rotate a body about some axis is measured by a quantity called the *torque* (τ). Consider the wrench pivoted about the axis through O in Fig. 10.12. The applied force $\boldsymbol{F}$ generally can act at an angle ϕ to the horizontal. We define the magnitude of the torque, τ, (Greek letter tau), resulting from the force $\boldsymbol{F}$ by the expression

Definition of torque

$$\tau = rF\sin\phi = Fd \qquad (10.18)$$

Moment arm

It is very important that you recognize that *torque is only defined when a reference point is specified.* The quantity $d = r\sin\phi$, called the *moment arm* (or *lever arm*) of the force $\boldsymbol{F}$, represents the perpendicular distance from the rotation axis to the line of action of $\boldsymbol{F}$. Note that the only component of $\boldsymbol{F}$ that tends to cause a rotation is $F\sin\phi$, the component perpendicular to r. The horizontal component, $F\cos\phi$, passes through O and has no tendency to produce a rotation. If there are two or more forces acting on a rigid body, as in Fig. 10.13, then each has a tendency to produce a rotation about the pivot at O. For example, $\boldsymbol{F}_2$ has a tendency to rotate the body clockwise, and $\boldsymbol{F}_1$ has a tendency to rotate the body counterclockwise. We shall use the convention that the sign of the torque resulting from a force is positive if its turning tendency is counterclockwise and negative if its turning tendency is clockwise. For example, in Fig. 10.13, the torque resulting from $\boldsymbol{F}_1$, which has a moment arm of d_1, is *positive* and equal to $+F_1d_1$; the torque from $\boldsymbol{F}_2$ is *negative* and equal to $-F_2d_2$. Hence, the *net* torque acting on the rigid body about O is

$$\tau_{\text{net}} = \tau_1 + \tau_2 = F_1d_1 - F_2d_2$$

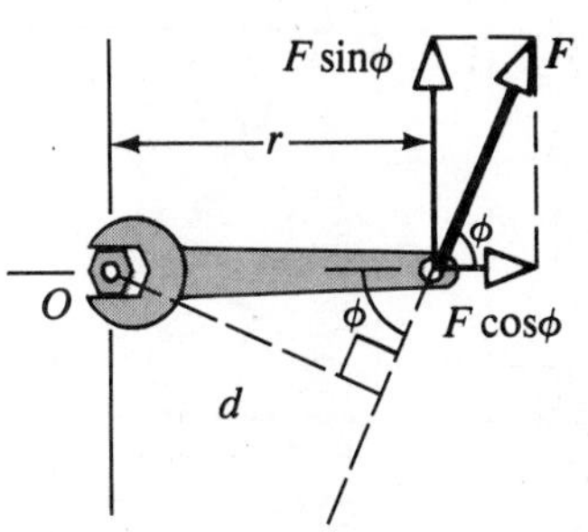

Figure 10.12 The force $\boldsymbol{F}$ has a greater rotating tendency about O as F increases and as the moment arm, d, increases. It is the component $F\sin\phi$ that tends to rotate the system about O.

From the definition of torque, we see that the rotating tendency increases as $\boldsymbol{F}$ increases and as d increases. For example, it is easier to close a door if we push at the doorknob rather than at a point close to the hinge. *Torque should not be confused with force.* Torque has units of force times length, or energy units. In Section

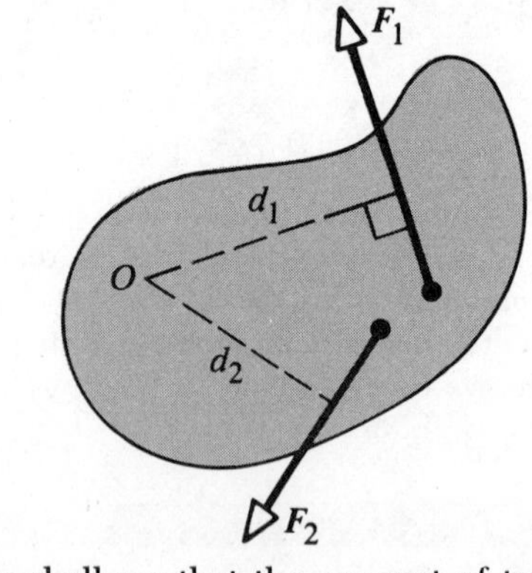

Figure 10.13 The force F_1 tends to rotate the body counterclockwise about O, and F_2 tends to rotate the body clockwise.

10.7 we shall see that the concept of torque is convenient for analyzing the rotational dynamics of a rigid body. The vector nature of torque will be described in detail in the next chapter.

Example 10.8

A solid cylinder is pivoted about a frictionless axle as in Fig. 10.14. A rope wrapped around the outer radius, R_1, exerts a force $\boldsymbol{F}_1$ to the right on the cylinder. A second rope wrapped around another section of radius R_2 exerts a force $\boldsymbol{F}_2$ downward on the cylinder. (a) What is the net torque acting on the cylinder about the z axis through O?

The torque due to $\boldsymbol{F}_1$ is $-R_1F_1$ and is negative because it tends to produce a clockwise rotation. The torque due to $\boldsymbol{F}_2$ is $+R_2F_2$ and is positive because it tends to produce a counterclockwise rotation. Therefore, the net torque is

$$\tau_{\text{net}} = \tau_1 + \tau_2 = R_2F_2 - R_1F_1$$

(b) Suppose $F_1 = 5$ N, $R_1 = 1.0$ m, $F_2 = 6$ N, and $R_2 = 0.5$ m. What is the net torque and which way will the cylinder rotate?

$$\tau_{\text{net}} = (6\text{ N})(0.5\text{ m}) - (5\text{ N})(1.0\text{ m}) = -2\text{ N}\cdot\text{m}$$

Since the net torque is negative, the cylinder will rotate in the clockwise direction.

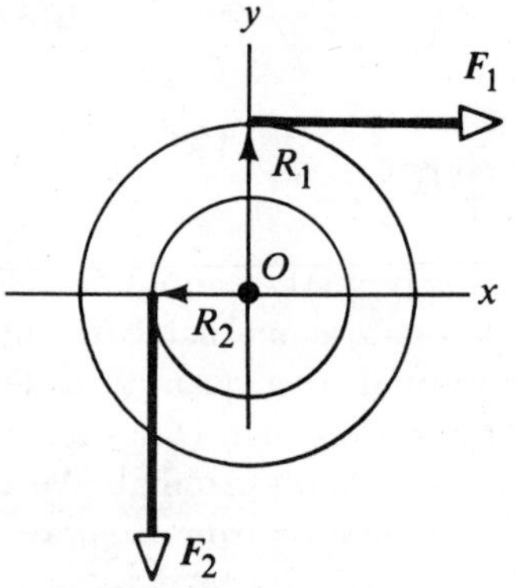

Figure 10.14 (Example 10.8) A solid cylinder pivoted about the z axis through O. The moment arm of $\boldsymbol{F}_1$ is R_1, and the moment arm of $\boldsymbol{F}_2$ is R_2.

Q9. With reference to Fig. 10.14, is it possible to have a situation where the resultant torque on the cylinder due to the two applied forces is zero? Explain.

Q10. Suppose that only two external forces act on a rigid body, and the two forces are equal in magnitude but opposite in direction. Under what condition will the body rotate?

10.7 RELATIONSHIP BETWEEN TORQUE AND ANGULAR ACCELERATION

In this section we shall show that the angular acceleration of a rigid body rotating about a fixed axis is proportional to the net torque acting about that axis. Before discussing the more complex case of rigid-body rotation, it is instructive to first briefly discuss the case of a particle rotating about some fixed point under the influence of an external force. The ideas embodied in this situation will then be extended to the case of a rigid body rotating about a fixed axis.

Consider a particle of mass m rotating in a circle of radius r under the influence of a tangential force $\boldsymbol{F}_t$ as in Fig. 10.15. The tangential force provides a tangential

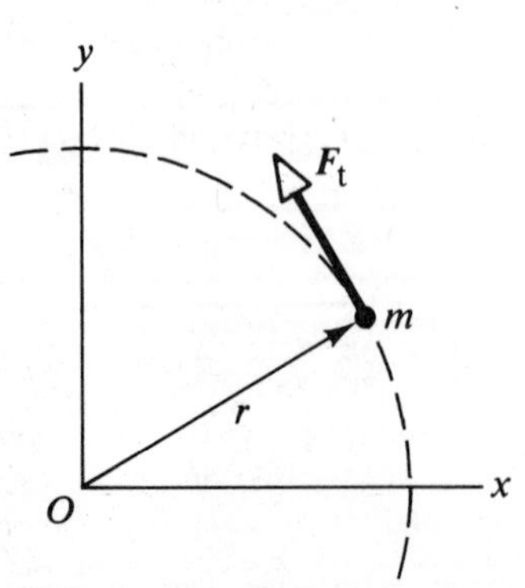

Figure 10.15 A particle rotating in a circle under the influence of a tangential force $\boldsymbol{F}_t$. A centripetal force $\boldsymbol{F}_r$ (not shown) must also be present to maintain the circular motion.

acceleration $\boldsymbol{a}_t$, and

$$F_t = ma_t$$

The torque about the origin due to the force $\boldsymbol{F}_t$ is the product of the magnitude of the force, F_t, and the moment arm of the force:

$$\tau = F_t r = (ma_t)r$$

Since the tangential acceleration is related to the angular acceleration through the relation $a_t = r\alpha$, the torque can be expressed

$$\tau = (mr\alpha)r = (mr^2)\alpha$$

Recall that the quantity mr^2 is the moment of inertia of the rotating mass about the z axis, so that

$$\tau = I\alpha \qquad (10.19)$$

Relationship between torque and angular acceleration

That is, *the torque acting on the particle is proportional to its angular acceleration,* and the proportionality constant is the moment of inertia. It is important to note that $\tau = I\alpha$ is the rotational analogue of Newton's second law of motion, $F = ma$.

Now let us extend this discussion to a rigid body of arbitrary shape rotating about a fixed axis as in Fig. 10.16. The body can be regarded as an infinite number of mass elements dm of infinitesimal size. Each mass element rotates in a circle about the origin, and each has a tangential acceleration $\boldsymbol{a}_t$ produced by a tangential force $d\boldsymbol{F}_t$. For any given element, we know from Newton's second law that

$$dF_t = (dm)a_t$$

The torque $d\tau$ associated with the force $d\boldsymbol{F}_t$ acting about the origin is given by

$$d\tau = r\, dF_t = (r\, dm)a_t$$

Since $a_t = r\alpha$, the expression for $d\tau$ becomes

$$d\tau = (r\, dm)r\alpha = (r^2\, dm)\alpha$$

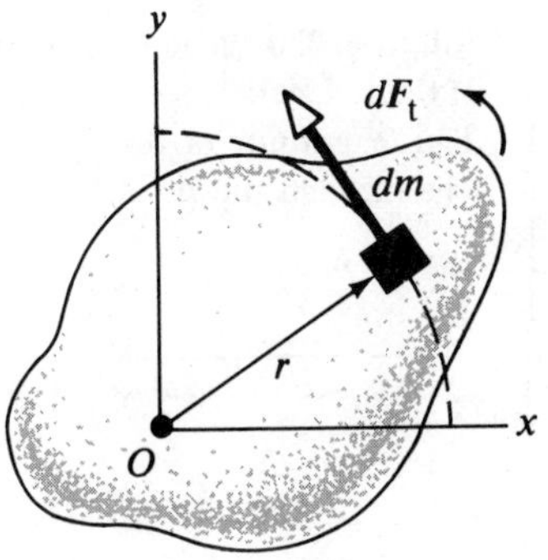

Figure 10.16 A rigid body pivoted about an axis through O. Each mass element dm rotates about O with the same angular acceleration $\boldsymbol{\alpha}$, and the net torque on the body is proportional to $\boldsymbol{\alpha}$.

It is important to recognize that although each point of the rigid body may have a different a_t, all mass elements have the *same* angular acceleration, α. With this in mind, the above expression can be integrated to obtain the net torque about O:

$$\tau_{net} = \int (r^2\, dm)\alpha = \alpha \int r^2\, dm$$

where α can be taken outside the integral since it is common to all mass elements. Since the moment of inertia of the body about the rotation axis through O is defined by $I = \int r^2\, dm$, the expression for τ_{net} becomes

$$\tau_{net} = I\alpha \qquad (10.20)$$

Torque is proportional to angular acceleration

Again we see that the net torque about the rotation axis is proportional to the angular acceleration of the body with the proportionality factor being I, which depends upon the axis of rotation and upon the size and shape of the body.

In view of the complex nature of the system, the important result that $\tau_{net} = I\alpha$ is strikingly simple and in complete agreement with experimental observations. The simplicity of the result lies in the manner in which the motion is described. That is, although each point on the rigid body may not experience the same force, linear acceleration, or linear velocity, *every point on the body has the same angular acceleration and angular velocity at any instant.* Therefore, at any instant the rotating rigid body as a whole is characterized by specific values for angular acceleration, net torque, and angular velocity.

Every point has the same ω and α

Finally, you should note that the result $\tau_{net} = I\alpha$ would also apply if the forces acting on the mass elements had radial components as well as tangential components. This is because the line of action of all radial components must pass through the axis of rotation, and hence would produce *zero* torque about that axis.

Example 10.9 *Rotating Rod*

A uniform rod of length L and mass M is free to rotate about a frictionless pivot at one end, as in Fig. 10.17. The rod is released from rest in the horizontal position. What is the *initial* angular acceleration of the rod and the *initial* linear acceleration of the right end of the rod?

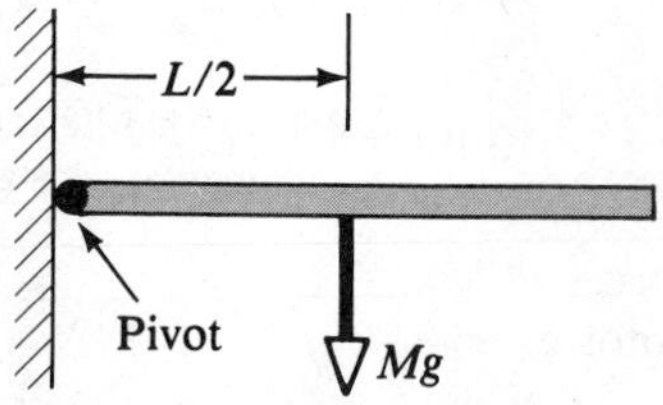

Figure 10.17 (Example 10.9) The uniform rod is pivoted at the left end.

Solution: The weight Mg, located at the geometric center of the rod, acts at its center of mass as shown in Fig. 10.17. The magnitude of the torque due to this force about an axis through the pivot is

$$\tau = \frac{MgL}{2}$$

But $\tau = I\alpha$, where $I = \frac{1}{3}ML^2$ for this axis of rotation. Therefore

$$I\alpha = Mg\frac{L}{2}$$

$$\alpha = \frac{Mg(L/2)}{\frac{1}{3}ML^2} = \frac{3g}{2L}$$

This angular acceleration is common to *all* points on the rod.

To find the linear acceleration of the right end of the rod, we use the relation $a_t = R\alpha$, with $R = L$. This gives

$$a_t = L\alpha = \frac{3}{2}g$$

This result is rather interesting, since $a_t > g$. That is, the end of the rod has an acceleration *greater* than the acceleration due to gravity. Therefore, if a coin were placed at the end of the rod, the end of the rod would fall faster than the coin when released.

Other points on the rod have a linear acceleration less than $\frac{3}{2}g$. For example, the middle of the rod has an acceleration $\frac{3}{4}g$.

10.8 WORK AND ENERGY IN ROTATIONAL MOTION

The description of a rotating rigid body would not be complete without a discussion of the rotational kinetic energy and how its change is related to the work done by external forces.

Again, we shall restrict our discussion to rotation about a fixed axis located in an inertial frame. Furthermore, we shall see that the important relationship $\tau_{net} = I\alpha$ derived in the previous section can also be obtained by considering the rate at which energy is changing with time.

Consider a rigid body pivoted at the point O in Fig. 10.18. Suppose a single external force $\mathbf{F}$ is applied at the point P. The work done by $\mathbf{F}$ as the body rotates through an infinitesimal distance $ds = r\,d\theta$ in a time dt is

$$dW = \mathbf{F} \cdot d\mathbf{s} = (F\sin\phi)r\,d\theta$$

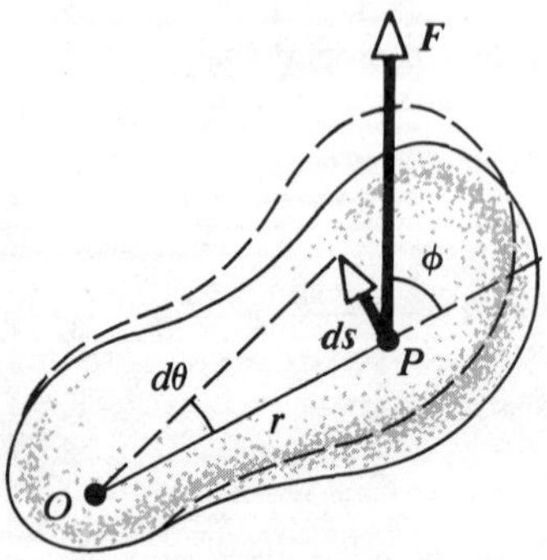

Figure 10.18 A rigid body rotates about an axis through O under the action of an external force $\mathbf{F}$ applied at P.

where $F\sin\phi$ is the tangential component of $\mathbf{F}$, or the component of the force along the displacement. Note from Fig. 10.18 that *the radial component of* **F** *does no work since it is perpendicular to the displacement.*

Since the magnitude of the torque due to $\mathbf{F}$ about the origin was defined as $rF\sin\phi$, we can write the work done for the infinitesimal rotation

$$dW = \tau\,d\theta \tag{10.21}$$

The rate at which work is being done by $\mathbf{F}$ for rotation about the fixed axis is obtained by dividing the left and right sides of Eq. 10.21 by dt:

$$\frac{dW}{dt} = \tau \frac{d\theta}{dt} \tag{10.22}$$

But the quantity dW/dt is, by definition, the instantaneous power, P, delivered by the force. Furthermore, since $d\theta/dt = \omega$, Eq. 10.22 reduces to

Power delivered to a rigid body

$$P = \frac{dW}{dt} = \tau\omega \tag{10.23}$$

This expression is analogous to $P = Fv$ in the case of linear motion, and the expression $dW = \tau\, d\theta$ is analogous to $dW = F_x\, dx$.

The Work-Energy Theorem in Rotational Motion

In linear motion, we found the energy concept, and in particular the work-energy theorem, to be extremely useful in describing the motion of a system. The energy concept can be equally useful in simplifying the analysis of rotational motion. From what we learned of linear motion, we expect that for rotation about a fixed axis, the work done by external forces will equal the change in the rotational kinetic energy. To show that this is in fact the case, let us begin with $\tau = I\alpha$. Using the chain rule from the calculus, we can express the torque as

$$\tau = I\alpha = I\frac{d\omega}{dt} = I\frac{d\omega}{d\theta}\frac{d\theta}{dt} = I\omega\frac{d\omega}{d\theta}$$

Simplifying the above expression and noting that $\tau\, d\theta = dW$, we get

$$\tau\, d\theta = dW = I\omega\, d\omega$$

Integrating this expression and noting that I is a constant, we get for the total work done

Work-energy theorem for rotational motion

$$W = \int_{\theta_0}^{\theta} \tau\, d\theta = \int_{\omega_0}^{\omega} I\omega\, d\omega = \frac{1}{2}I\omega^2 - \frac{1}{2}I\omega_0^{\,2} \tag{10.24}$$

where the angular velocity changes from ω_0 to ω as the angular displacement changes from θ_0 to θ. Note that this expression is analogous to the expression for the work-energy theorem in linear motion with m replaced by I and v replaced by ω. That is, *the net work done by external forces in rotating a rigid body about a fixed axis equals the change in the body's rotational kinetic energy.*

Table 10.3 lists the various equations we have discussed pertaining to rotational motion, together with the analogous expressions for linear motion. The last two equations, involving the concept of angular momentum $\mathbf{L}$, will be discussed in Chapter 11 and are included only for completeness. In all cases, note the similarity between the equations of rotational motion and those of linear motion.

Example 10.10

A wheel of radius R, mass M, and moment of inertia I is mounted on a frictionless, horizontal axle as in Fig. 10.19. A light cord wrapped around the wheel supports a body of mass m. Calculate the linear acceleration of the suspended body, the angular acceleration of the wheel, and the tension in the cord.

Solution: The torque acting on the wheel about its axis of rotation is $\tau = TR$. The weight of the wheel and the normal force of the axle on the wheel pass through the axis of rotation and produce no torque. Since $\tau = I\alpha$, we get

$$\tau = I\alpha = TR$$

$$\alpha = TR/I \tag{1}$$

Now let us apply Newton's second law to the motion of the suspended mass m, making use of the free-body diagram (Fig. 10.19):

$$\Sigma F_y = T - mg = -ma$$

$$(2) \qquad a = \frac{mg - T}{m}$$

We see that the linear acceleration of the suspended mass is equal to the tangential acceleration of a point on the rim of the wheel. Therefore, the angular acceleration of the wheel and this linear acceleration are related by $a = R\alpha$. Using this fact together with (1) and (2) gives

$$a = R\alpha = \frac{TR^2}{I} = \frac{mg - T}{m}$$

$$T = \frac{mg}{1 + \dfrac{mR^2}{I}}$$

Likewise, solving for a and α gives

$$a = \frac{g}{1 + I/mR^2}$$

$$\alpha = \frac{a}{R} = \frac{g}{R + I/mR}$$

For example, if the wheel is a solid disk of $M = 2.0$ kg, $R = 30$ cm, and $I = 0.09$ kg · m², and $m = 0.5$ kg, we find that $T = 3.3$ N, $a = 3.3$ m/s², and $\alpha = 11$ rad/s².

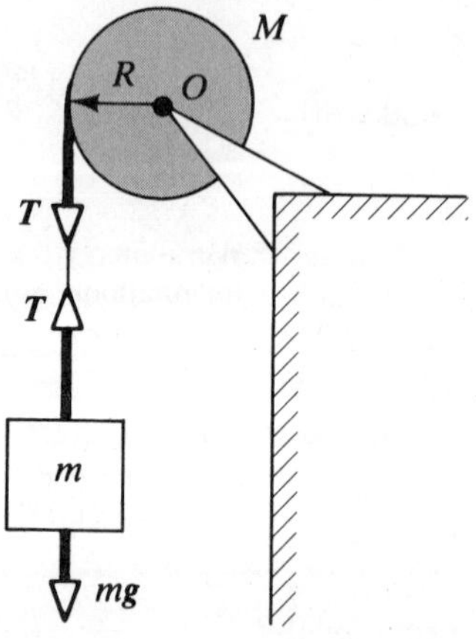

Figure 10.19 (Example 10.10) The cord attached to m is wrapped around the pulley, which produces a torque about the axle through O.

Example 10.11

A uniform rod of length L and mass M is free to rotate on a frictionless pin through one end (Fig. 10.20). The rod is released from rest in the horizontal position. (a) What is the angular velocity of the rod when it is at its lowest position?

The question can be easily answered by considering the mechanical energy of the system. When the rod is in the horizontal position, it has no kinetic energy. Its potential energy relative to the lowest position of its center of mass (O') is $MgL/2$. When it reaches its lowest position, the energy is entirely kinetic energy, $\frac{1}{2}I\omega^2$, where I is the moment of inertia about the pivot. Since $I = \frac{1}{3}ML^2$ (Table 10.2) and since mechanical energy is conserved, we have

$$\frac{1}{2}MgL = \frac{1}{2}I\omega^2 = \frac{1}{2}\left(\frac{1}{3}ML^2\right)\omega^2$$

$$\omega = \sqrt{\frac{3g}{L}}$$

For example, if the rod is a meter stick, we find that $\omega = 5.4$ rad/s.

(b) Determine the linear velocity of the center of mass and the linear velocity of the lowest point on the rod in the vertical position.

$$v_c = r\omega = \frac{L}{2}\omega = \frac{1}{2}\sqrt{3gL}$$

The lowest point on the rod has a velocity equal to $2v_c = \sqrt{3gL}$.

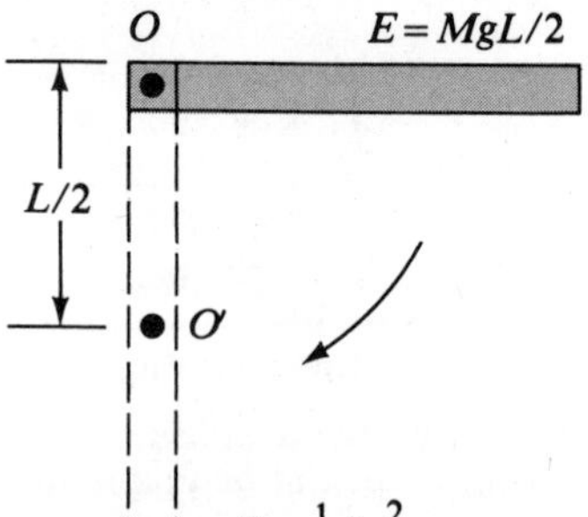

Figure 10.20 (Example 10.11) A uniform rigid rod pivoted at O rotates in a vertical plane under the action of gravity.

Q11. Explain how you might use the apparatus described in Example 10.10 to determine the moment of inertia of the wheel. (Note that if the wheel is not a uniform disk the moment of inertia is not necessarily equal to $\frac{1}{2}MR^2$.)

Q12. Using the results from Example 10.10, how would you calculate the angular velocity of the wheel and the linear velocity of the suspended mass at, say, $t = 2$ s, if the system is released from rest at $t = 0$? Is the relation $v = R\omega$ valid in this situation?

Q13. If a small sphere of mass M were placed at the end of the rod in Fig. 10.20, would the result for ω be greater than, less than, or equal to the value obtained in Example 10.11?

TABLE 10.3 A Comparison of Useful Equations in Rotational and Translational Motion

ROTATIONAL MOTION ABOUT A FIXED AXIS	*LINEAR MOTION*
Angular velocity $\omega = d\theta/dt$	Linear velocity $v = dx/dt$
Angular acceleration $\alpha = d\omega/dt$	Linear acceleration $a = dv/dt$
Resultant torque $\Sigma\tau = I\alpha$	Resultant force $\Sigma F = Ma$
$\alpha = \text{constant} \begin{cases} \omega = \omega_0 + \alpha t \\ \theta - \theta_0 = \omega_0 t + \frac{1}{2}\alpha t^2 \\ \omega^2 = \omega_0{}^2 + 2\alpha(\theta - \theta_0) \end{cases}$	$a = \text{constant} \begin{cases} v = v_0 + at \\ x - x_0 = v_0 t + \frac{1}{2}at^2 \\ v^2 = v_0{}^2 + 2a(x - x_0) \end{cases}$
Work $W = \int_{\theta_0}^{\theta} \tau\, d\theta$	Work $W = \int_{x_0}^{x} F_x\, dx$
Kinetic energy $K = \frac{1}{2}I\omega^2$	Kinetic energy $K = \frac{1}{2}mv^2$
Power $P = \tau\omega$	Power $P = Fv$
Angular momentum $\mathbf{L} = I\boldsymbol{\omega}$	Linear momentum $\mathbf{p} = m\mathbf{v}$
Resultant torque $\boldsymbol{\tau} = d\mathbf{L}/dt$	Resultant force $\mathbf{F} = d\mathbf{p}/dt$

Example 10.12

Consider two masses connected by a string passing over a pulley having a moment of inertia I about its axis of rotation, as in Fig. 10.21. The string does not slip on the pulley, and the system is released from rest. Find the linear velocities of the masses after m_2 falls through a distance h, and the angular velocity of the pulley at this time.

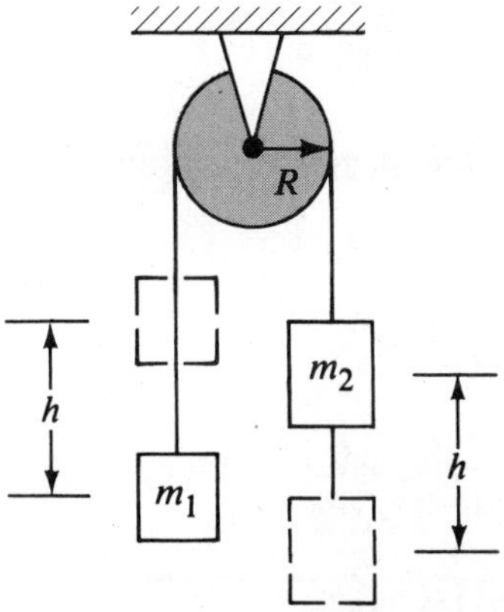

Figure 10.21 (Example 10.12)

Solution: If we neglect friction in the system, then mechanical energy is conserved and we can state that the increase in kinetic energy of the system equals the decrease in potential energy. Since $K_i = 0$ (the system is initially at rest), we have

$$\Delta K = K_f - K_i = \frac{1}{2}m_1v^2 + \frac{1}{2}m_2v^2 + \frac{1}{2}I\omega^2$$

where m_1 and m_2 have a common speed. But $v = R\omega$, so that

$$\Delta K = \frac{1}{2}\left(m_1 + m_2 + \frac{I}{R^2}\right)v^2$$

From Fig. 10.21, we see that m_2 loses potential energy while m_1 gains potential energy. That is, $\Delta U_2 = -m_2gh$ and $\Delta U_1 = m_1gh$. Applying the law of conservation of energy in the form $\Delta K + \Delta U_1 + \Delta U_2 = 0$ gives

$$\frac{1}{2}\left(m_1 + m_2 + \frac{I}{R^2}\right)v^2 + m_1gh - m_2gh = 0$$

$$v = \left[\frac{2(m_2 - m_1)gh}{\left(m_1 + m_2 + \frac{I}{R^2}\right)}\right]^{1/2}$$

Since $v = R\omega$, the angular velocity of the pulley at this instant is given by $\omega = v/R$.

The same result for v can be obtained using $\tau_{net} = I\alpha$ applied to the pulley and Newton's second law applied to m_1 and m_2. The procedure is identical to that used in Example 10.10.

10.9 SUMMARY

The *instantaneous angular velocity* of a particle rotating in a circle or of a rigid body rotating about a fixed axis is given by

$$\omega = \frac{d\theta}{dt} \qquad (10.3)$$

Instantaneous angular velocity

where ω is in rad/s, or s^{-1}.

The *instantaneous angular acceleration* of a rotating body is given by

Instantaneous angular acceleration

$$\alpha = \frac{d\omega}{dt} \tag{10.5}$$

and has units of rad/s^2, or s^{-2}.

When a rigid body rotates about a fixed axis, every part of the body has the same angular velocity and the same angular acceleration. However, different parts of the body, in general, have different linear velocities and linear accelerations.

If a particle or body undergoes rotational motion about a fixed axis under constant angular acceleration α, one can apply equations of kinematics in analogy with kinematic equations for linear motion under constant linear acceleration:

Rotational kinematic equations

$$\omega = \omega_0 + \alpha t \tag{10.6}$$

$$\theta = \theta_0 + \omega_0 t + \frac{1}{2}\alpha t^2 \tag{10.7}$$

$$\omega^2 = \omega_0{}^2 + 2\alpha(\theta - \theta_0) \tag{10.8}$$

When a rigid body rotates about a fixed axis, the angular velocity and angular acceleration are related to the linear velocity and tangential linear acceleration through the relationships

Relationship between linear and angular speed

$$v = r\omega \tag{10.9}$$

Relationship between linear and angular acceleration

$$a_t = r\alpha \tag{10.10}$$

The moment of inertia of a system of particles is given by

Moment of inertia for a system of particles

$$I = \Sigma m_i r_i{}^2 \tag{10.14}$$

If a rigid body rotates about a fixed axis with angular velocity ω, its *kinetic energy* can be written

Kinetic energy of a rotating rigid body

$$K = \frac{1}{2}I\omega^2 \tag{10.15}$$

where I is the moment of inertia about the axis of rotation.

The moment of inertia of a rigid body is given by

Moment of inertia for a rigid body

$$I = \int r^2\, dm \tag{10.16}$$

where r is the distance from the mass element dm to the axis of rotation.

The *torque* associated with a force $\boldsymbol{F}$ acting on a body has a magnitude equal to

Torque

$$\tau = Fd \tag{10.18}$$

where d is the moment arm of the force, which is the perpendicular distance from some origin to the line of action of the force. Torque is a measure of the tendency of the force to rotate the body about some axis.

If a rigid body free to rotate about a fixed axis has a *net external torque* acting on it, the body will undergo an angular acceleration α, where

Net torque

$$\tau_{\text{net}} = I\alpha \tag{10.20}$$

The rate at which work is being done by external forces in rotating a rigid body about a fixed axis, or *the power delivered*, is given by

$$P = \tau\omega \tag{10.23}$$

Power delivered to a rigid body

The net work done by external forces in rotating a rigid body about a fixed axis equals the change in the rotational kinetic energy of the body:

$$W = \frac{1}{2}I\omega^2 - \frac{1}{2}I\omega_0^2 \tag{10.24}$$

Work-energy theorem for rotational motion

This is the work-energy theorem applied to rotational motion.

EXERCISES

Section 10.1 Angular Velocity and Angular Acceleration

1. A particle moves in a circle 1.5 m in radius. Through what angle in radians does it rotate if it moves through an arc length of 2.5 m? What is this angle in degrees?

2. If a particle moving in a circle makes n rev/min, what is its angular velocity in rad/s?

3. A wheel rotates at a constant rate of 3600 rev/min. (a) What is its angular speed? (b) Through what angle (in radians) does it rotate in 1.5 s?

4. Convert the following to degrees: (a) 3.5 rad, (b) 5π rad, (c) 2.2 rev.

Section 10.2 Rotational Kinematics: Rotational Motion with Constant Angular Acceleration

5. A wheel starts from rest and rotates with constant angular acceleration to an angular velocity of 10 rad/s in a time of 2 s. Find (a) the angular acceleration of the wheel and (b) the angle in radians through which it rotates in this time.

6. The turntable of a record player rotates at 33 1/3 rev/min and takes 90 s to come to rest when switched off. Calculate (a) its angular acceleration and (b) the number of revolutions it makes before coming to rest.

7. What is the angular speed in rad/s of (a) the earth in its orbit about the sun and (b) the moon in its orbit about the earth?

8. A wheel rotates in such a way that its angular displacement in a time t is given by $\theta = at^2 + bt^3$, where a and b are constants. Determine equations for (a) the angular speed and (b) the angular acceleration, both as functions of time.

Section 10.3 Relationships Between Angular and Linear Quantities

9. A racing car travels on a circular track of radius 200 m. If the car moves with a constant speed of 80 m/s, find (a) the angular speed of the car and (b) the magnitude and direction of the car's acceleration.

10. The racing car described in Exercise 9 starts from rest and accelerates uniformly to a speed of 80 m/s in 30 s. Find (a) the average angular speed of the car in this interval, (b) the angular acceleration of the car, (c) the magnitude of the car's linear acceleration at $t = 10$ s, and (d) the total distance traveled in the first 30 s.

11. A wheel 4 m in diameter rotates with a constant angular acceleration of 4 rad/s^2. The wheel starts at rest at $t = 0$, and the radius vector at point P on the rim makes an angle of 57.3° with the horizontal at this time. At $t = 2$ s, find (a) the angular speed of the wheel, (b) the linear velocity and acceleration of the point P, and (c) the position of the point P.

12. A cylinder of radius 12 cm starts from rest and rotates about its axis with a constant angular acceleration of 5 rad/s^2. At $t = 3$ s, what is (a) its angular velocity, (b) the linear velocity of a point on its rim, and (c) the radial and tangential components of acceleration of a point on its rim?

13. A disk 6 cm in radius rotates at a constant rate of 1200 rev/min about its axis. Determine (a) the angular speed of the disk, (b) the linear speed at a point 2 cm from its center, (c) the radial acceleration of a point on the rim, and (d) the total distance a point on the rim moves in 2 s.

Section 10.4 Rotational Kinetic Energy

14. A tire of moment of inertia 50 kg · m^2 rotates about a fixed central axis at the rate of 600 rev/min. What is its kinetic energy?

15. The four particles in Fig. 10.22 are connected by light, rigid rods. If the system rotates in the xy plane about the z axis with an angular velocity of 8 rad/s, calculate (a) the moment of inertia of the system about the z axis and (b) the kinetic energy of the system.

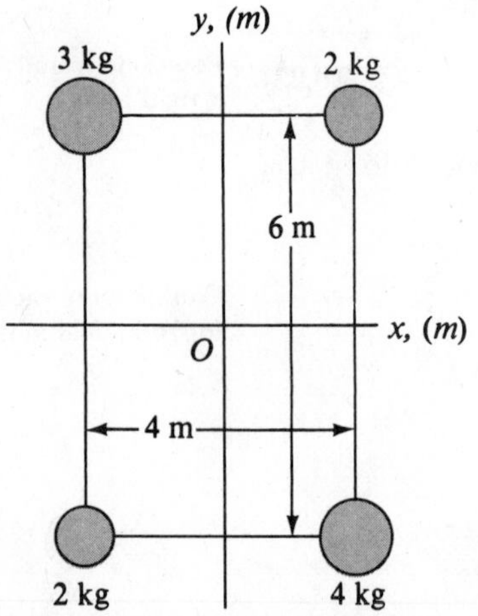

Figure 10.22 (Exercises 15 and 16).

16. The system of particles described in Exercise 15 (Fig. 10.22) rotates about the y axis. Calculate (a) the moment of inertia about the y axis and (b) the work required to take the system from rest to an angular speed of 8 rad/s.

17. Three particles are connected by light, rigid rods lying along the y axis (Fig. 10.23). If the system rotates about the x axis with an angular speed of 2 rad/s, find (a) the moment of inertia about the x axis and the total kinetic energy evaluated from $\frac{1}{2}I\omega^2$ and (b) the linear speed of each particle and the total kinetic energy evaluated from $\Sigma\frac{1}{2}m_i v_i^2$.

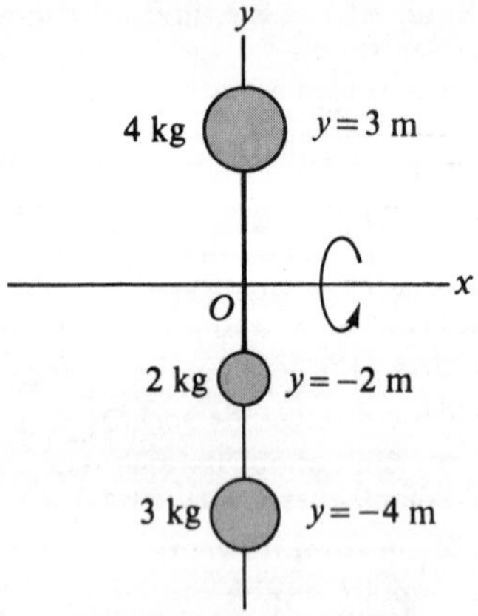

Figure 10.23 (Exercise 17)

Section 10.5 Calculation of Moments of Inertia for Rigid Bodies

18. Following the procedure used in Example 10.6, prove that the moment of inertia about the y' axis of the rigid rod in Fig. 10.9 is $\frac{1}{3}ML^2$.

19. Use the parallel-axis theorem and Table 10.2 to find the moments of inertia of (a) a solid cylinder about an axis parallel to the center of mass axis and passing through the edge of the cylinder and (b) a solid sphere about an axis tangent to the surface of the sphere.

Section 10.6 Torque

20. Calculate the net torque (magnitude and direction) on the beam shown in Fig. 10.24 about (a) an axis through O, perpendicular to the figure and (b) an axis through C, perpendicular to the figure.

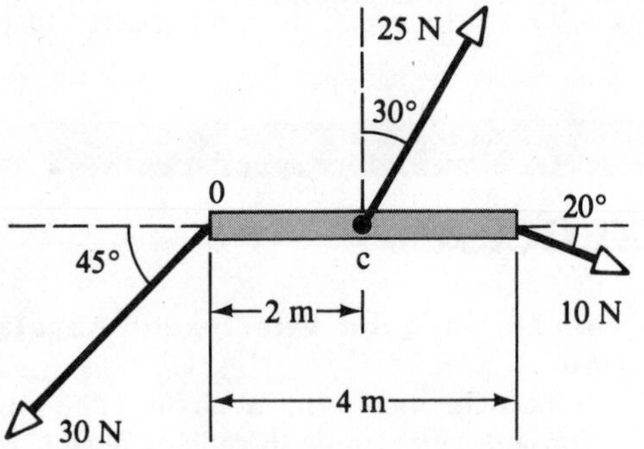

Figure 10.24 (Exercise 20)

21. Find the net torque on the wheel in Fig. 10.25 about the axle through O if a = 5 cm and b = 20 cm.

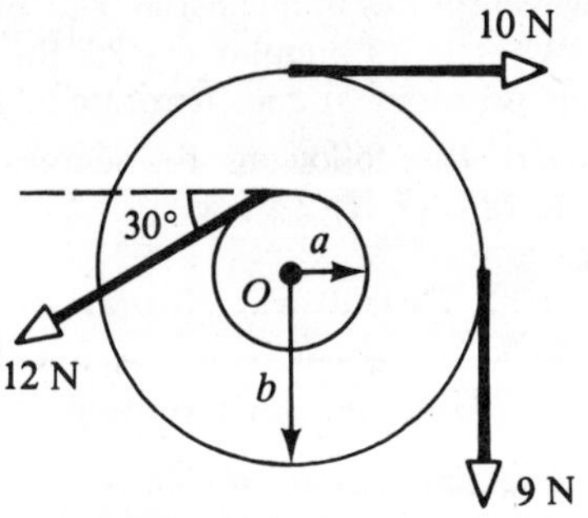

Figure 10.25 (Exercise 21)

Section 10.7 Relationship Between Torque and Angular Acceleration

22. If a motor is to produce a torque of 50 N · m on a wheel rotating at 2400 rev/min, how much power must the motor deliver?

23. The combination of an external force and a frictional force produces a constant total torque of 24 N · m on a wheel rotating about a fixed axis. The external force is applied for 5 s, during which time the angular speed of the wheel increases from 0 to 10 rad/s. The external force is then removed, and the wheel comes to rest in 50 s. Find (a) the moment of inertia of the wheel, (b) the magnitude of the frictional torque, and (c) the total number of revolutions of the wheel.

24. The system described in Example 10.10 (Fig. 10.19) is released from rest. After the mass m has fallen through a distance h, find (a) the linear velocity of the mass m and (b) the angular speed of the wheel.

Section 10.8 Work and Energy in Rotational Motion

25. A wheel 1 m in diameter rotates on a fixed, frictionless, horizontal axle. Its moment of inertia about this axis is 5 kg · m^2. A constant tension of 20 N is maintained on a rope wrapped around the rim of the wheel, so as to cause the wheel to accelerate. If the wheel starts from rest at $t = 0$, find (a) the angular acceleration of the wheel, (b) the wheel's angular speed at $t = 3$ s, (c) the kinetic energy of the wheel at $t = 3$ s, and (d) the length of rope unwound in the first 3 s.

26. (a) A uniform solid disk of radius R and mass M is free to rotate on a frictionless pivot through a point on its rim (Fig. 10.26). If the disk is released from rest in the position shown by the solid line, what is the velocity of its center of mass when it reaches the position indicated by the broken line? (b) What is the speed of the lowest point on the disk in the dotted position? (c) Repeat part (a) if the object is a uniform hoop.

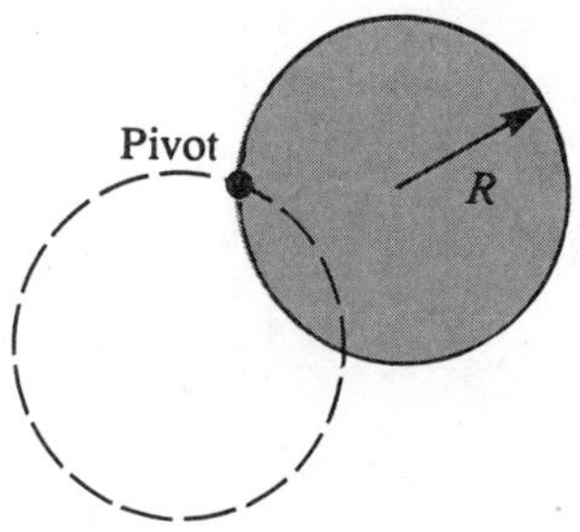

Figure 10.26 (Exercise 26)

27. A 12-kg mass is attached to a cord that is wrapped around a wheel of radius $r = 10$ cm (Fig. 10.27). The acceleration of the mass down the frictionless incline is measured to be 2.0 m/s^2. Assuming the axle of the wheel to be frictionless, determine (a) the tension in the rope, (b) the moment of inertia of the wheel, and (c) the angular speed of the wheel 2 s after it begins rotating, starting from rest.

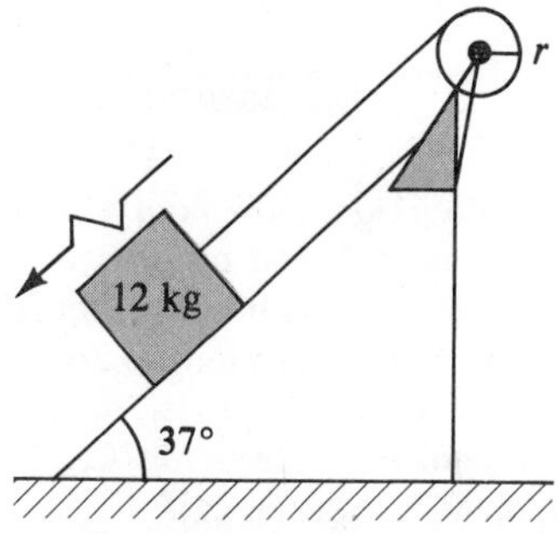

Figure 10.27 (Exercise 27)

PROBLEMS

1. Find by integration the moment of inertia of a hollow cylinder about its symmetry axis. The mass of the cylinder is M, its inner radius is R_1, and its outer radius is R_2. (Check your result against the value given in Table 10.2.)

2. Calculate the moment of inertia of a uniform solid sphere of mass M and radius R about a diameter (see Table 10.2). (Hint: Treat the sphere as a set of disks of various radii, and first obtain an expression for the moment of inertia of one of these disks about the symmetry axis.)

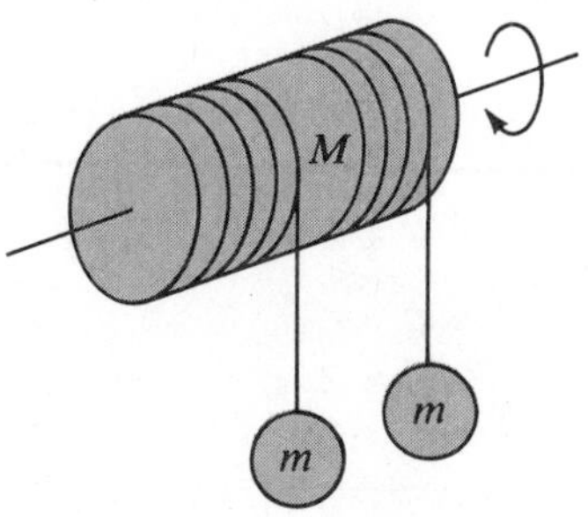

Figure 10.28 (Problem 3)

3. A uniform solid cylinder of mass M and radius R rotates on a horizontal, frictionless axle (Fig. 10.28). Two equal masses hang from light cords wrapped around the cylinder. If the system is released from rest, find (a) the tension in each cord, (b) the acceleration of each mass, and (c) the angular velocity of the cylinder after the masses have fallen a distance h.

4. Suppose the pulley in Fig. 10.21 has a moment of inertia I and radius R. If the cord supporting m_1 and m_2 does not slip, $m_2 > m_1$, and the axle is frictionless, find (a) the acceleration of the masses, (b) the tension supporting m_1 and the tension supporting m_2 (note they are different), and (c) numerical values for T_1, T_2, and a if $I = 5$ kg · m^2, $R = 0.5$ m, $m_1 = 2$ kg, and $m_2 = 5$ kg.

5. A mass m_1 is connected by a light cord to a mass m_2, which slides on a smooth surface (Fig. 10.29). The pulley rotates about a frictionless axle and has a moment of inertia I and radius R. Assuming the cord does not slip on the pulley, find (a) the acceleration of the two masses, (b) the tensions T_1 and T_2, and (c) numerical values for a, T_1, and T_2 if $I = 0.5$ kg · m^2, $R = 0.3$ m, $m_1 = 4$ kg, and $m_2 = 3$ kg. (d) What would your answers be if the inertia of the pulley was neglected?

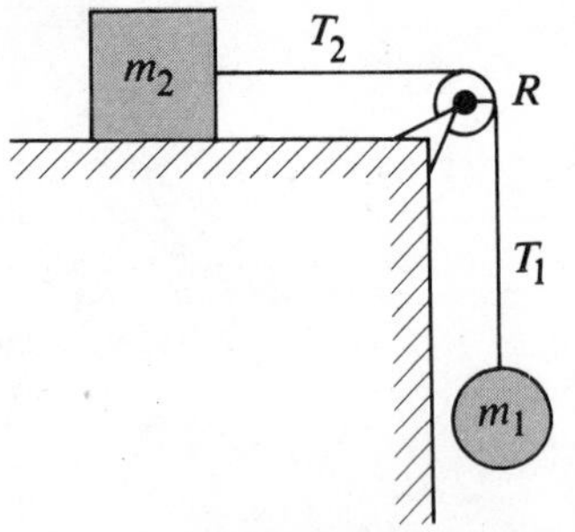

Figure 10.29 (Problem 5)

6. A uniform horizontal plank of mass M and length L is supported at each end by vertical ropes. At the instant one of the ropes breaks, show that (a) the angular acceleration of the plank is $3g/2L$, (b) the acceleration of the center of mass is $3g/4$, and (c) the tension in the supporting rope is $Mg/4$.

7. A long uniform rod of length L and mass M is pivoted about a horizontal, frictionless pin through one end. The rod is released from rest in a vertical position as in Fig. 10.30. At the instant the rod is horizontal, find (a) the angular velocity of the rod, (b) its angular acceleration, (c) the x and y components of the acceleration of its center of mass, and (d) the components of the reaction force at the pivot.

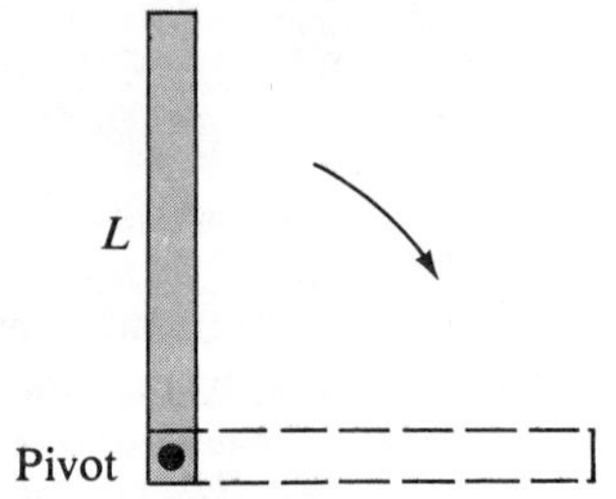

Figure 10.30 (Problem 7)

8. The pulley shown in Fig. 10.31 has a radius R and moment of inertia I. One end of the mass m is connected to a spring of force constant k, and the other end is fastened to a cord wrapped around the pulley. The pulley axle and the incline are smooth. If the pulley is wound counterclockwise so as to stretch the spring a distance d from its *unstretched* position and then released from rest, find (a) the angular velocity of the pulley when the spring is again unstretched and (b) a numerical value for the angular velocity at this point if $I = 1\ \text{kg}\cdot\text{m}^2$, $R = 0.3$ m, $k = 50$ N/m, $m = 0.5$ kg, $d = 0.2$ m, and $\theta = 37°$.

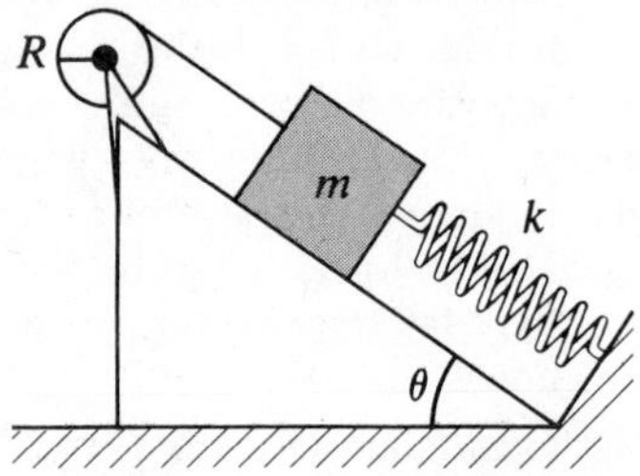

Figure 10.31 (Problem 8)

9. For any given rotational axis, the radius of gyration, K, of a rigid body is defined by the expression $K^2 = I/M$, where M is the total mass of the body and I is the moment of inertia about the given axis. That is, the radius of gyration is equal to the distance of an imaginary point mass M from the axis of rotation such that I for the point mass about that axis is the same as for the rigid body about the same axis. Find the radius of gyration of (a) a solid disk of radius R, (b) a uniform rod of length L, and (c) a solid sphere of radius R, all three rotating about a central axis.

11 Angular Momentum and Torque as Vector Quantities

United Press International Photo

In the previous chapter we learned how to treat the rotation of a rigid body about a fixed axis. This chapter deals in part with the more general case, where the axis of rotation is not fixed in space. We begin by defining a vector product. The vector product is a convenient mathematical tool for expressing such quantities as torque and angular momentum. The central point of this chapter is to develop the concept of the angular momentum of a system of particles, a quantity that plays a key role in rotational dynamics. In analogy to the conservation of linear momentum, we shall find that the angular momentum of any isolated system (an isolated rigid body or any other isolated collection of particles) is always conserved. This conservation law is a special case of the result that the time rate of change of the total angular momentum of any system of particles equals the resultant external torque acting on the system.

11.1 THE VECTOR PRODUCT AND TORQUE

Consider a force $\boldsymbol{F}$ acting on a rigid body at the vector position $\boldsymbol{r}$ (Fig. 11.1). *The origin O is assumed to be in an inertial frame, so that Newton's second law is valid.* The *magnitude* of the torque due to this force relative to the origin is, by definition, equal to $rF \sin\phi$, where ϕ is the angle between $\boldsymbol{r}$ and $\boldsymbol{F}$. The axis about which $\boldsymbol{F}$ would tend to produce rotation is perpendicular to the plane formed by $\boldsymbol{r}$ and $\boldsymbol{F}$. If the force lies in the xy plane as in Fig. 11.1, then the torque, $\boldsymbol{\tau}$ is represented by a vector parallel to the z axis. The force in Fig. 11.1 tends to rotate the body counterclockwise looking down the z axis, so the sense of $\boldsymbol{\tau}$ is toward increasing z, and $\boldsymbol{\tau}$ is in the positive z direction. If we reversed the direction of $\boldsymbol{F}$ in Fig. 11.1, $\boldsymbol{\tau}$ would then be in the negative z direction. The torque involves two vectors, $\boldsymbol{r}$ and $\boldsymbol{F}$, and is in fact defined to be equal to the *vector product*, or *cross product*, of $\boldsymbol{r}$ and $\boldsymbol{F}$:

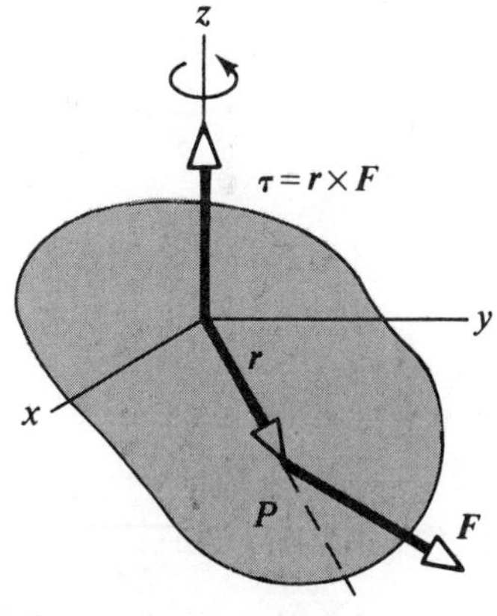

Figure 11.1 The torque vector $\boldsymbol{\tau}$ lies in a direction perpendicular to the plane formed by the position vector $\boldsymbol{r}$ and the applied force $\boldsymbol{F}$.

$$\boldsymbol{\tau} \equiv \boldsymbol{r} \times \boldsymbol{F} \tag{11.1}$$

Definition of torque

We must now give a clear definition of the vector product. Given any two vectors $\boldsymbol{A}$ and $\boldsymbol{B}$, the vector product $\boldsymbol{A} \times \boldsymbol{B}$ is defined as a third vector $\boldsymbol{C}$ the *magnitude* of which is $AB \sin\theta$, where θ is the angle between $\boldsymbol{A}$ and $\boldsymbol{B}$. That is, if $\boldsymbol{C}$ is given by

$$\boldsymbol{C} = \boldsymbol{A} \times \boldsymbol{B} \tag{11.2}$$

then its magnitude is

Magnitude of the cross product

$$C = |\mathbf{C}| = |AB \sin\theta| \tag{11.3}$$

Note that the quantity $AB \sin\theta$ is equal to the area of the parallelogram formed by **A** and **B**, as shown in Fig. 11.2. The *direction* of $\mathbf{A} \times \mathbf{B}$ is perpendicular to the plane formed by **A** and **B**, as in Fig. 11.2, and its sense is determined by the advance of a right-handed screw when turned from **A** to **B** through the *smaller* angle θ. A more convenient rule to use for the direction of $\mathbf{A} \times \mathbf{B}$ is the right-hand rule illustrated in Fig. 11.2. The four fingers of the right hand are pointed along **A** and then "wrapped" into **B** through the angle θ. The direction of the erect right thumb is the direction of $\mathbf{A} \times \mathbf{B}$. Because of the notation, $\mathbf{A} \times \mathbf{B}$ is often read "A cross B"; hence the term *cross product*.

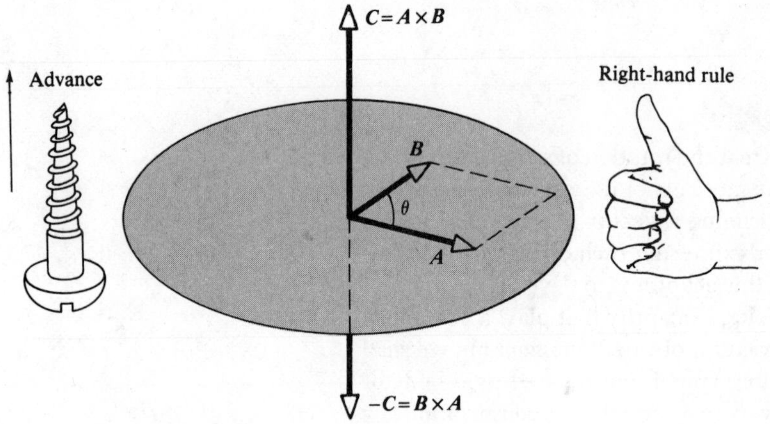

Figure 11.2 The vector product $\mathbf{A} \times \mathbf{B}$ is a third vector **C** having a magnitude $AB \sin\theta$ equal to the area of the parallelogram shown. The direction of **C** is perpendicular to the plane formed by **A** and **B**, and its sense is determined by the right-hand rule.

Some properties of the vector product which follow from its definition are as follows:

Properties of the vector product

1. Unlike the scalar product, the order in which the two vectors are multiplied in a cross product is important, that is,

$$\mathbf{A} \times \mathbf{B} = -\mathbf{B} \times \mathbf{A} \tag{11.4}$$

   Therefore, if you change the order of the cross product, you must change the sign. You could easily verify this relation with the right-hand rule (Fig. 11.2).
2. If **A** is parallel to **B** ($\theta = 0°$ or $180°$), then $\mathbf{A} \times \mathbf{B} = 0$; therefore, it follows that $\mathbf{A} \times \mathbf{A} = 0$.
3. If **A** is perpendicular to **B**, then $|\mathbf{A} \times \mathbf{B}| = AB$.
4. It is also important to note that the vector product obeys the *distributive law*, that is,

$$\mathbf{A} \times (\mathbf{B} + \mathbf{C}) = \mathbf{A} \times \mathbf{B} + \mathbf{A} \times \mathbf{C} \tag{11.5}$$

5. Finally, the derivative of the cross product with respect to some variable such as t is given by

$$\frac{d}{dt}(\mathbf{A} \times \mathbf{B}) = \mathbf{A} \times \frac{d\mathbf{B}}{dt} + \frac{d\mathbf{A}}{dt} \times \mathbf{B} \tag{11.6}$$

where it is important to preserve the multiplicative order of **A** and **B**, in view of Eq. 11.4.

It is left as an exercise to show from Eqs. 11.2 and 11.3 and the definition of unit vectors that the cross products of the rectangular unit vectors $\boldsymbol{i}$, $\boldsymbol{j}$, and $\boldsymbol{k}$ obey the following expressions:

Cross products of unit vectors

$$\boldsymbol{i} \times \boldsymbol{i} = \boldsymbol{j} \times \boldsymbol{j} = \boldsymbol{k} \times \boldsymbol{k} = 0 \tag{11.7a}$$
$$\boldsymbol{i} \times \boldsymbol{j} = -\boldsymbol{j} \times \boldsymbol{i} = \boldsymbol{k} \tag{11.7b}$$
$$\boldsymbol{j} \times \boldsymbol{k} = -\boldsymbol{k} \times \boldsymbol{j} = \boldsymbol{i} \tag{11.7c}$$
$$\boldsymbol{k} \times \boldsymbol{i} = -\boldsymbol{i} \times \boldsymbol{k} = \boldsymbol{j} \tag{11.7d}$$

Note that signs are interchangeable. For example, $\boldsymbol{i} \times (-\boldsymbol{j}) = -\boldsymbol{i} \times \boldsymbol{j} = -\boldsymbol{k}$.

The cross product of *any* two vectors **A** and **B** can be expressed in the following determinant form:

$$\mathbf{A} \times \mathbf{B} = \begin{vmatrix} \boldsymbol{i} & \boldsymbol{j} & \boldsymbol{k} \\ A_x & A_y & A_z \\ B_x & B_y & B_z \end{vmatrix}$$

Expanding this determinant gives the result

$$\mathbf{A} \times \mathbf{B} = (A_yB_z - A_zB_y)\boldsymbol{i} + (A_zB_x - A_xB_z)\boldsymbol{j} + (A_xB_y - A_yB_x)\boldsymbol{k} \tag{11.8}$$

Table 11.1 summarizes some of the properties of the cross product of two vectors.

TABLE 11.1 Some Properties of the Cross Product of Two Vectors

If $\mathbf{C} = \mathbf{A} \times \mathbf{B}$ then $C = |AB \sin\theta|$

$\mathbf{A} \times \mathbf{B} = -\mathbf{B} \times \mathbf{A}$

$\mathbf{A} \times \mathbf{A} = 0$

$\mathbf{A} \times (\mathbf{B} + \mathbf{C}) = \mathbf{A} \times \mathbf{B} + \mathbf{A} \times \mathbf{C}$

$\frac{d}{dt}(\mathbf{A} \times \mathbf{B}) = \mathbf{A} \times \frac{d\mathbf{B}}{dt} + \frac{d\mathbf{A}}{dt} \times \mathbf{B}$

If $\mathbf{A} = A_x\boldsymbol{i} + A_y\boldsymbol{j} + A_z\boldsymbol{k}$ and $\mathbf{B} = B_x\boldsymbol{i} + B_y\boldsymbol{j} + B_z\boldsymbol{k}$,

$$\mathbf{A} \times \mathbf{B} = \begin{vmatrix} \boldsymbol{i} & \boldsymbol{j} & \boldsymbol{k} \\ A_x & A_y & A_z \\ B_x & B_y & B_z \end{vmatrix}$$

or $\mathbf{A} \times \mathbf{B} = (A_yB_z - A_zB_y)\boldsymbol{i} + (A_zB_x - A_xB_z)\boldsymbol{j} + (A_xB_y - A_yB_x)\boldsymbol{k}$

Cross Products of Unit Vectors

$\boldsymbol{i} \times \boldsymbol{i} = \boldsymbol{j} \times \boldsymbol{j} = \boldsymbol{k} \times \boldsymbol{k} = 0$

$\boldsymbol{i} \times \boldsymbol{j} = -\boldsymbol{j} \times \boldsymbol{i} = \boldsymbol{k}$

$\boldsymbol{j} \times \boldsymbol{k} = -\boldsymbol{k} \times \boldsymbol{j} = \boldsymbol{i}$

$\boldsymbol{k} \times \boldsymbol{i} = -\boldsymbol{i} \times \boldsymbol{k} = \boldsymbol{j}$

Example 11.1

Two vectors lying in the *xy* plane are given by $\mathbf{A} = 2\boldsymbol{i} + 3\boldsymbol{j}$ and $\mathbf{B} = -\boldsymbol{i} + 2\boldsymbol{j}$. Find $\mathbf{A} \times \mathbf{B}$, and verify explicitly that $\mathbf{A} \times \mathbf{B} = -\mathbf{B} \times \mathbf{A}$.

Solution: Using Eqs. 11.7a through 11.7d for the cross product of unit vectors gives

$$\mathbf{A} \times \mathbf{B} = (2\boldsymbol{i} + 3\boldsymbol{j}) \times (-\boldsymbol{i} + 2\boldsymbol{j})$$
$$= 2\boldsymbol{i} \times 2\boldsymbol{j} + 3\boldsymbol{j} \times (-\boldsymbol{i}) = 4\boldsymbol{k} + 3\boldsymbol{k} = 7\boldsymbol{k}$$

(We have omitted the terms in $\boldsymbol{i} \times \boldsymbol{i}$ and $\boldsymbol{j} \times \boldsymbol{j}$, which are zero.)

$$\mathbf{B} \times \mathbf{A} = (-\boldsymbol{i} + 2\boldsymbol{j}) \times (2\boldsymbol{i} + 3\boldsymbol{j})$$
$$= -\boldsymbol{i} \times 3\boldsymbol{j} + 2\boldsymbol{j} \times 2\boldsymbol{i} = -3\boldsymbol{k} - 4\boldsymbol{k} = -7\boldsymbol{k}$$

Therefore, $\mathbf{A} \times \mathbf{B} = -\mathbf{B} \times \mathbf{A}$.

As an alternative method for finding $\mathbf{A} \times \mathbf{B}$, we could use Eq. 11.8, with $A_x = 2$, $A_y = 3$, $A_z = 0$ and $B_x = -1$, $B_y = 2$, $B_z = 0$. This gives

$$\mathbf{A} \times \mathbf{B} = (0)\boldsymbol{i} + (0)\boldsymbol{j} + [2 \times 2 - 3 \times (-1)]\boldsymbol{k}$$
$$= 7\boldsymbol{k}$$

Q1. Is it possible to calculate the torque acting on a rigid body without specifying the origin? Is the torque independent of the location of the origin?

Q2. Is the triple product defined by $\boldsymbol{A} \cdot (\boldsymbol{B} \times \boldsymbol{C})$ equal to a scalar or vector quantity? Note that the operation $(\boldsymbol{A} \cdot \boldsymbol{B}) \times \boldsymbol{C}$ has no meaning. Explain.

Q3. In the expression for torque, $\boldsymbol{\tau} = \boldsymbol{r} \times \boldsymbol{F}$, is $\boldsymbol{r}$ equal to the moment arm? Explain.

11.2 ANGULAR MOMENTUM OF A PARTICLE

A particle of mass m, located at the vector position $\boldsymbol{r}$, moves with a velocity $\boldsymbol{v}$ (Fig. 11.3). The instantaneous angular momentum $\boldsymbol{L}$ of the particle relative to the origin O is defined by the cross product of its instantaneous vector position and the instantaneous linear momentum $\boldsymbol{p}$:

Definition of angular momentum of a particle

$$\boldsymbol{L} \equiv \boldsymbol{r} \times \boldsymbol{p} \tag{11.9}$$

The SI units of angular momentum are kg · m^2/s. It is important to note that both the magnitude and direction of $\boldsymbol{L}$ depend on the choice of the origin. The direction of $\boldsymbol{L}$ is perpendicular to the plane formed by $\boldsymbol{r}$ and $\boldsymbol{p}$, and its sense is governed by the right-hand rule. For example, in Fig. 11.3 $\boldsymbol{r}$ and $\boldsymbol{p}$ are assumed to be in the xy plane, so that $\boldsymbol{L}$ points in the z direction. Since $\boldsymbol{p} = m\boldsymbol{v}$, the magnitude of $\boldsymbol{L}$ is given by

$$L = mvr \sin\phi \tag{11.10}$$

where ϕ is the angle between $\boldsymbol{r}$ and $\boldsymbol{p}$. It follows that L is zero when $\boldsymbol{r}$ is parallel to $\boldsymbol{p}$ ($\phi = 0$ or $180°$). In other words, when the particle moves along a line that passes through the origin, it has zero angular momentum with respect to the origin. This is equivalent to stating that it has no tendency to rotate about the origin. On the other hand, if $\boldsymbol{r}$ is perpendicular to $\boldsymbol{p}$ ($\phi = 90°$), then L is a maximum and equal to mrv. In this case, the particle has maximum tendency to rotate about the origin. In fact, at that instant the particle moves exactly as though it were on the rim of a wheel rotating about the origin in a plane defined by $\boldsymbol{r}$ and $\boldsymbol{p}$.

Alternatively, one may note that a particle has nonzero angular momentum about some point if its position vector measured from that point rotates about the point. On the other hand, if the position vector simply increases or decreases in length, the particle moves along a line passing through the origin and therefore has zero angular momentum with respect to that origin.

In the case of the linear motion of a particle, we found that the resultant force on a particle equals the time rate of change of its linear momentum. We shall now show that Newton's second law implies that the resultant torque acting on a particle equals the time rate of change of its angular momentum. Let us start by writing the torque on the particle in the form

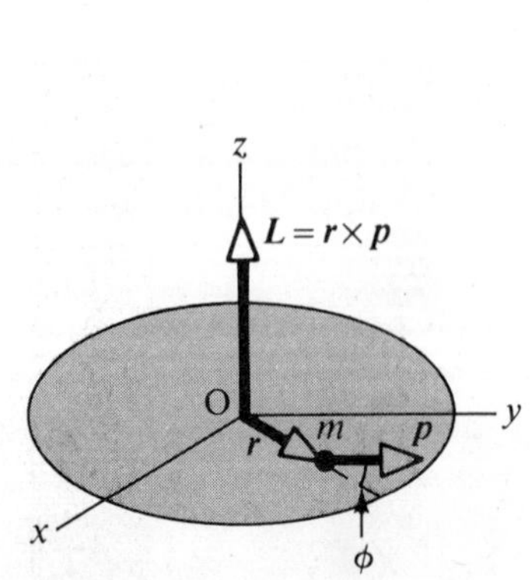

Figure 11.3 The angular momentum $\boldsymbol{L}$ of a particle of mass m and momentum $\boldsymbol{p}$ located at the position $\boldsymbol{r}$ is a vector given by $\boldsymbol{L} = \boldsymbol{r} \times \boldsymbol{p}$. Note that the value of $\boldsymbol{L}$ depends on the origin and is a vector perpendicular to both $\boldsymbol{r}$ and $\boldsymbol{p}$.

$$\boldsymbol{\tau} = \boldsymbol{r} \times \boldsymbol{F} = \boldsymbol{r} \times \frac{d\boldsymbol{p}}{dt} \tag{11.11}$$

where we have used the fact that $\boldsymbol{F} = d\boldsymbol{p}/dt$. Now let us differentiate Eq. 11.9 with respect to time using the rule given by Eq. 11.6.

$$\frac{d\boldsymbol{L}}{dt} = \frac{d}{dt}(\boldsymbol{r} \times \boldsymbol{p}) = \boldsymbol{r} \times \frac{d\boldsymbol{p}}{dt} + \frac{d\boldsymbol{r}}{dt} \times \boldsymbol{p}$$

It is important to adhere to the order of terms since $\boldsymbol{A} \times \boldsymbol{B} = -\boldsymbol{B} \times \boldsymbol{A}$.

The last term on the right in the above equation is zero, since $\boldsymbol{v} = \dfrac{d\boldsymbol{r}}{dt}$ is parallel to

p. Therefore,

$$\frac{d\mathbf{L}}{dt} = \mathbf{r} \times \frac{d\mathbf{p}}{dt} \tag{11.12}$$

Comparing Eqs. 11.11 and 11.12, we see that

$$\boldsymbol{\tau} = \frac{d\mathbf{L}}{dt} \tag{11.13}$$

Torque equals time rate of change of angular momentum

which is the rotational analog of Newton's second law, $\mathbf{F} = d\mathbf{p}/dt$. This result says that *the torque acting on a particle is equal to the time rate of change of its angular momentum.* It is important to note that Eq. 11.13 is valid only if the origins of $\boldsymbol{\tau}$ and $\mathbf{L}$ are the *same*. It is left as an exercise to show that Eq. 11.13 is also valid when there are several forces acting on the particle, in which case $\boldsymbol{\tau}$ is the *net* torque on the particle. *Furthermore, the expression is valid for any origin fixed in an inertial frame.* Of course, the same origin must be used in calculating all torques as well as the angular momentum.

Example 11.2 *Linear Motion*

A particle of mass m moves in the xy plane with a velocity $\mathbf{v}$ along a straight line (Fig. 11.4). What is the magnitude and direction of its angular momentum with respect to the origin O?

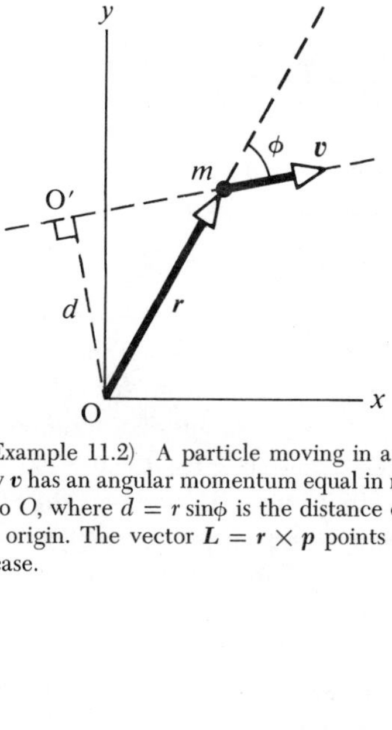

Figure 11.4 (Example 11.2) A particle moving in a straight line with a velocity $\mathbf{v}$ has an angular momentum equal in magnitude to mvd relative to O, where $d = r \sin\phi$ is the distance of closest approach to the origin. The vector $\mathbf{L} = \mathbf{r} \times \mathbf{p}$ points *into* the diagram in this case.

Solution: From the figure we see that the distance of closest approach of the particle from the origin is $d = r \sin\phi$. Therefore, the magnitude of $\mathbf{L}$ is given by

$$L = mvr \sin\phi = mvd$$

The direction of $\mathbf{L}$ from the right-hand rule is *into* the diagram, and so we can write the vector expression $\mathbf{L} = -mvd\mathbf{k}$. Note that the angular momentum relative to the origin O' is *zero*.

Example 11.3 *Circular Motion*

A particle moves in the xy plane in a circular path of radius r, as in Fig. 11.5. (a) Find the magnitude and direction of its angular momentum relative to O when its velocity is $\mathbf{v}$.

Since $\mathbf{r}$ is perpendicular to $\mathbf{v}$, $\phi = 90°$ and the magnitude of $\mathbf{L}$ is simply

$$L = mvr \sin 90° = mvr \qquad (\text{for } \mathbf{r} \text{ perpendicular to } \mathbf{v})$$

The direction of $\mathbf{L}$ is perpendicular to the plane of the circle, and its sense depends on the direction of $\mathbf{v}$. If the sense of the rotation is counterclockwise, as in Fig. 11.5, then by the right-hand rule, the direction of $\mathbf{L} = \mathbf{r} \times \mathbf{p}$ is *out* of the paper. Hence, we can write the vector expression $\mathbf{L} = mvr\mathbf{k}$. On the other hand, if the particle were to move clockwise, $\mathbf{L}$ would point into the paper.

(b) Find an alternative expression for L in terms of the angular velocity, ω.

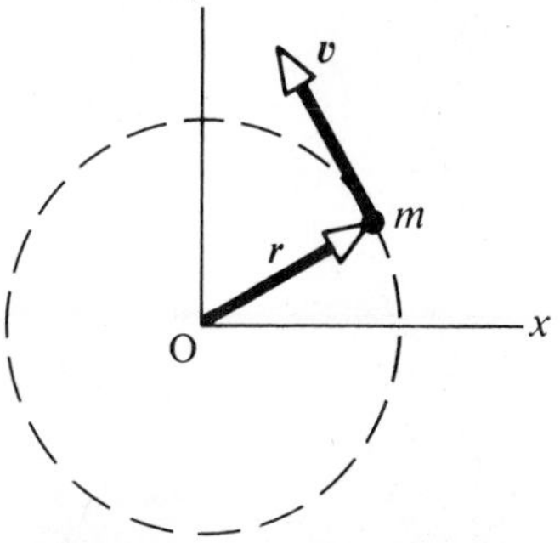

Figure 11.5 (Example 11.3) A particle moving in a circle of radius r has an angular momentum equal in magnitude to mvr relative to the center. The vector $\mathbf{L} = \mathbf{r} \times \mathbf{p}$ points *out* of the diagram.

Since $v = r\omega$ for a particle rotating in a circle, we can express L as

$$L = mvr = mr^2\omega = I\omega$$

where I is the moment of inertia of the particle about the z axis through O. Furthermore, in this case the angular momentum is in the *same* direction as the angular velocity vector, $\boldsymbol{\omega}$ (see Section 10.1), and so we can write $\mathbf{L} = I\boldsymbol{\omega} = I\omega\mathbf{k}$

As a numerical example, a car of mass 1500 kg, moving in a circular race track of radius 50 m with a speed of 40 m/s, has an angular momentum relative to the center of the race track given by

$$L = mvr = (1500 \text{ kg})\left(40 \frac{\text{m}}{\text{s}}\right)(50 \text{ m}) = 3.0 \times 10^6 \text{ kg} \cdot \text{m}^2/\text{s}$$

Q4. If a particle moves in a straight line, is its angular momentum zero with respect to an arbitrary origin? Is its angular momentum zero with respect to any specific origin? Explain.

Q5. If the linear velocity of a particle is constant in time, can its angular momentum vary in time about an arbitrary origin?

Q6. If the torque acting on a particle about an *arbitrary* origin is zero, what can you say about its angular momentum about that origin?

Q7. A particle moves in a straight line, and you are told that the torque acting on it is zero about some unspecified origin. Does this necessarily imply that the net force on the particle is zero? Can you conclude that its velocity is constant? Explain.

Q8. Suppose that the velocity vector of a particle is completely specified. What can you conclude about the *direction* of its angular momentum vector with respect to the direction of motion?

11.3 ANGULAR MOMENTUM AND TORQUE FOR A SYSTEM OF PARTICLES

The total angular momentum, $\mathbf{L}$, of a system of particles about some point is defined as the vector sum of the angular momenta of the individual particles:

$$\mathbf{L} = \mathbf{L}_1 + \mathbf{L}_2 + \cdots + \mathbf{L}_n = \Sigma \mathbf{L}_i$$

where the vector sum is over all of the n particles in the system.

Since the individual momenta of the particles may change in time, the total angular momentum may also vary in time. In fact, from Eqs. 11.11 through 11.13, we find that the time rate of change of the total angular momentum equals the vector sum of *all* torques, including those associated with internal forces between particles and those associated with external forces. However, the net torque associated with internal forces is zero. To understand this, recall that Newton's third law tells us that the internal forces occur in equal and opposite pairs that lie along the line of separation of each pair of particles. Therefore, the torque due to each action-reaction force pair is zero. By summation, we see that the net internal torque vanishes. Finally, we conclude that the total angular momentum can vary with time *only* if there is a net *external* torque on the system, so that we have

$$\Sigma \boldsymbol{\tau}_{\text{ext}} = \Sigma \frac{d\mathbf{L}_i}{dt} = \frac{d}{dt}\Sigma \mathbf{L}_i = \frac{d\mathbf{L}}{dt} \qquad (11.14)$$

That is, *the time rate of change of the total angular momentum of the system about some origin in an inertial frame equals the net external torque acting on the system about that origin.* Note that Eq. 11.14 is the rotational analog of $\mathbf{F}_{\text{ext}} = d\mathbf{p}/dt$ for a system of particles (Chapter 9).

Now consider a symmetric rigid body rotating about a *fixed axis* through the center of mass as in Fig. 11.6. Each particle of the rigid body rotates about the same axis in a circle with an angular velocity $\boldsymbol{\omega}$. The magnitude of the angular momentum of the particle of mass m_i is $m_i v_i r_i$ about the origin O. The direction of $\mathbf{L}_i$ is the same as that of $\boldsymbol{\omega}$. Since $v_i = r_i\omega$, we can express the angular momentum of the ith particle as $m_i r_i^2 \boldsymbol{\omega}$. Summing this over all particles gives the total angular momentum of the rigid body:

Angular momentum of a rigid body about a fixed axis

$$\mathbf{L} = (\Sigma m_i r_i^2)\boldsymbol{\omega} = I\boldsymbol{\omega} \tag{11.15}$$

where I is the moment of inertia of the rigid body about the axis of rotation. It is important to remember that Eq. 11.15 applies only to rotation of a symmetric rigid body about a *fixed axis* through the center of mass.[1] It is also interesting to note from this development that the moment of inertia arises in a natural way as it did when we described the rotational kinetic energy of a rigid body.

Now let us differentiate Eq. 11.15 with respect to time, noting that I is constant for a rigid body:

$$\frac{d\mathbf{L}}{dt} = I\frac{d\boldsymbol{\omega}}{dt} \tag{11.16}$$

Applying Eqs. 11.14 to 11.16 and recalling that the angular acceleration is defined by $\boldsymbol{\alpha} = d\boldsymbol{\omega}/dt$, we obtain the familiar result

$$\Sigma\boldsymbol{\tau}_{\text{ext}} = I\boldsymbol{\alpha} \tag{11.17}$$

That is, *the net external torque acting on a rigid body equals the moment of inertia about the axis of rotation multiplied by its angular acceleration.* Note that the angular acceleration is in the *same* direction as the net external torque vector. For rotation about a fixed axis, recall that $\boldsymbol{\alpha}$ is in the same direction as $\boldsymbol{\omega}$ if ω is increasing in time and in the opposite direction if ω is decreasing in time.

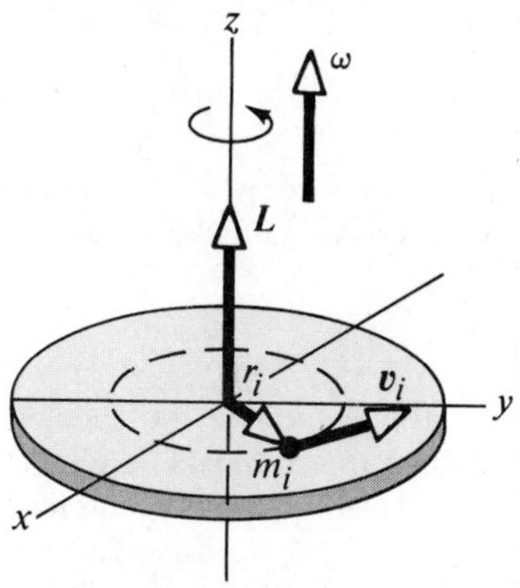

Figure 11.6 When a rigid body rotates about an axis, the angular momentum $\mathbf{L}$ is in the same direction as the angular velocity $\boldsymbol{\omega}$, according to the expression $\mathbf{L} = I\boldsymbol{\omega}$.

Example 11.4 *Rotating Sphere*

A uniform solid sphere of radius $R = 0.50$ m and mass 15 kg rotates in the xy plane about an axis through its center, as in Fig. 11.7. Find its angular momentum when the angular velocity is 3 rad/s.

Solution: The moment of inertia of the sphere about an axis through its center is

$$I = \frac{2}{5}MR^2 = \frac{2}{5}(15\ \text{kg})(0.5\ \text{m})^2 = 1.5\ \text{kg}\cdot\text{m}^2$$

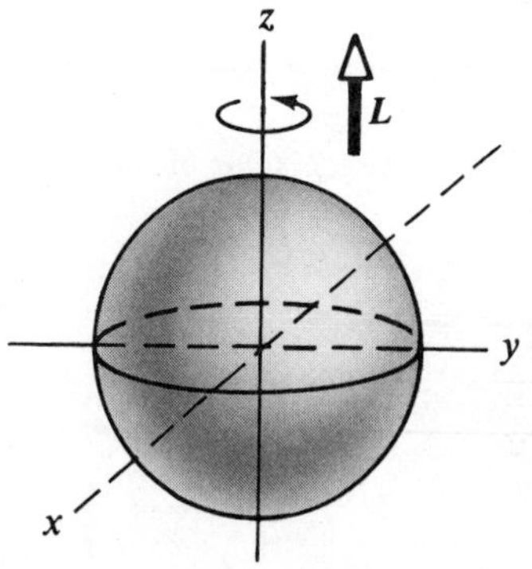

Figure 11.7 (Example 11.4) A sphere that rotates about the z axis in the direction shown has an angular momentum $\mathbf{L}$ in the positive z direction. If the direction of rotation is reversed, $\mathbf{L}$ will point in the negative z direction.

[1] The expression $\mathbf{L} = I\boldsymbol{\omega}$ is not always valid. If a rigid body rotates about an arbitrary axis, $\mathbf{L}$ and $\boldsymbol{\omega}$ will point in different directions. In fact, in this case, the moment of inertia cannot be treated as a scalar. Strictly speaking, $\mathbf{L} = I\boldsymbol{\omega}$ applies only to rigid bodies that rotate about one of three mutually perpendicular axes (called *principal axes*) through the center of mass. This is discussed in more advanced mechanics texts.

Therefore, the magnitude of the angular momentum is

$$L = I\omega = (1.5\ \text{kg}\cdot\text{m}^2)(3\ \text{rad/s}) = 4.5\ \text{kg}\cdot\frac{\text{m}^2}{\text{s}}$$

Example 11.5 *Rotating Rod*

A rigid rod of mass M and length l rotates in a vertical plane about a frictionless pivot through its center (Fig. 11.8). Particles of masses m_1 and m_2 are attached at the ends of the rod. (a) Determine the angular momentum when the angular velocity is ω.

The moment of inertia of the system equals the sum of the moments of inertia of the three components: the rod, m_1, and m_2. Using Table 10.2 we find that the total moment of inertia about the z axis through O is

$$I = \frac{1}{12}Ml^2 + m_1\left(\frac{l}{2}\right)^2 + m_2\left(\frac{l}{2}\right)^2 = \frac{l^2}{4}\left(\frac{M}{3} + m_1 + m_2\right)$$

Therefore, when the angular velocity is ω, the magnitude of the angular momentum is given by

$$L = I\omega = \frac{l^2}{4}\left(\frac{M}{3} + m_1 + m_2\right)\omega$$

(b) Determine the angular acceleration of the system when the rod makes an angle θ with the horizontal.

The torque due to the force m_1g about the pivot is

$$\tau_1 = m_1g\frac{l}{2}\cos\theta \qquad \text{(out of the plane)}$$

The torque due to the force m_2g about the pivot is

$$\tau_2 = -m_2g\frac{l}{2}\cos\theta \qquad \text{(into the plane)}$$

Hence, the net torque about O is

$$\tau_{\text{net}} = \tau_1 + \tau_2 = \frac{1}{2}(m_1 - m_2)gl\cos\theta$$

You should note that τ_{net} is *out* of the plane if $m_1 > m_2$ and is *into* the plane if $m_1 < m_2$. To find α, we use $\tau_{\text{net}} = I\alpha$, where I was obtained in (a). This gives

$$\alpha = \frac{\tau_{\text{net}}}{I} = \frac{2(m_1 - m_2)g\cos\theta}{l\left(\frac{M}{3} + m_1 + m_2\right)}$$

Note that α is zero when θ is $\pi/2$ or $-\pi/2$ (vertical position) and α is a maximum when θ is 0 or π (horizontal position). Furthermore, the angular velocity of the system changes since α varies in time. If $m_1 > m_2$, at what value of θ is ω a maximum? Knowing the angular velocity at some instant, how would you calculate the linear velocity of m_1 and m_2?

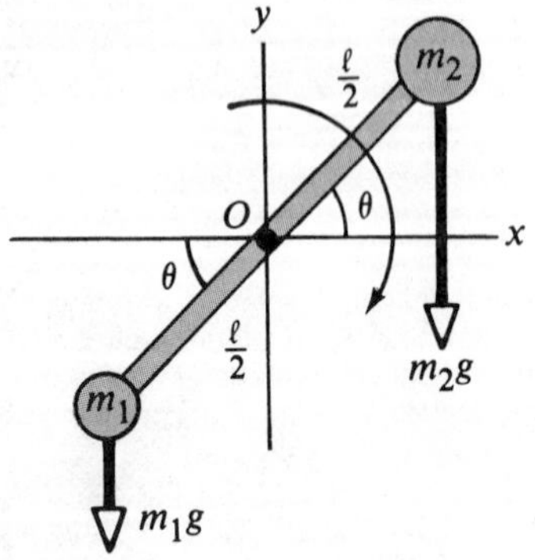

Figure 11.8 (Example 11.5) Since gravitational forces act on the system rotating in a vertical plane, there is in general a net nonzero torque about O when $m_1 \neq m_2$, which in turn produces an angular acceleration according to $\tau_{\text{net}} = I\alpha$.

Example 11.6 *Rotating Disk*

A solid disk of mass M rotates about an axis parallel to the symmetry axis through its center, as in Fig. 11.9. Calculate the angular momentum about the origin O.

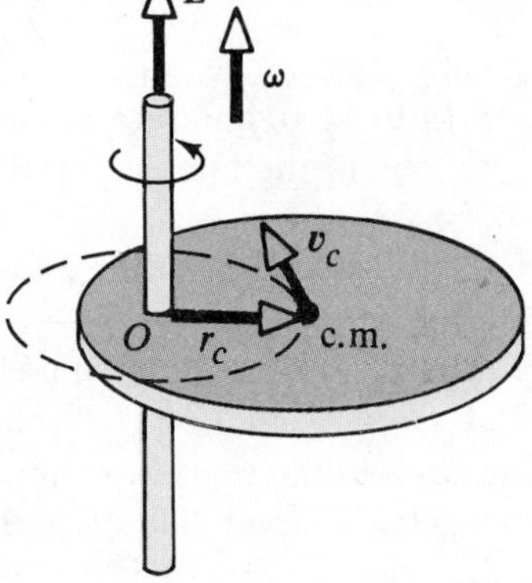

Figure 11.9 (Example 11.6) The angular momentum of a disk about the axis shown is given by $\mathbf{L} = I\boldsymbol{\omega}$, where I is the moment of inertia about the axis of rotation.

Solution: Every particle making up the disk has the same angular velocity, and each has an angular momentum parallel to $\boldsymbol{\omega}$. Therefore, the magnitude of the total angular momentum of the disk according to Eq. 11.15 is

$$L = I\omega$$

where I is the moment of inertia about the axis of rotation through O. According to the parallel-axis theorem, we can express I as

$$I = I_c + Mr^2$$

Hence, the angular momentum can be written in the form

$$L = I\omega = I_c\omega + Mr^2\omega$$

The quantity $I_c\omega$ is the magnitude of the angular momentum about the center of mass. The quantity $I\omega$ is the magnitude of the angular momentum about the axis through O. The distance r is related to the velocity of the center of mass and angular velocity through the relation $v_c = r\omega$, where r is the magnitude of the vector $\mathbf{r}_c$ from O to the center of mass. Hence, we see that the term $Mr^2\omega$ is the magnitude of the vector $\boldsymbol{r}_c \times M\boldsymbol{v}_c$, a vector representing the angular momentum of a particle of mass M moving with a velocity $\boldsymbol{v}_c$. Note that this vector is perpendicular to the plane of rotation and in the same direction as $\boldsymbol{\omega}$. Thus, we can write the angular momentum in the vector form

$$\boldsymbol{L} = I_c\boldsymbol{\omega} + \boldsymbol{r}_c \times M\boldsymbol{v}_c = \boldsymbol{L}_c + \boldsymbol{r}_c \times M\boldsymbol{v}_c$$

The quantity $\boldsymbol{L}_c$ is often called the *spin angular momentum*, since it refers to that part of the angular momentum associated with the spinning motion of the system about the center of mass. The quantity $\boldsymbol{r}_c \times M\boldsymbol{v}_c$ is usually referred to as the *orbital angular momentum*. Although we do not prove it here, the separation of the total angular momentum into spin and orbital parts can be made even for a rotation about an arbitrary axis. In general, we can say that the *angular momentum of a rigid body or system of particles of total mass* M *about any origin* O *equals the sum of the angular momentum about the center of mass plus the angular momentum associated with the motion of the center of mass about* O.

For example, the total angular momentum of the earth consists of a spin part due to rotation about its own axis and an orbital part due to the rotation of the earth about the sun. In a similar fashion, an electron of an atom is said to have both an intrinsic spin angular momentum and an orbital angular momentum due to its motion about the nucleus.

Q9. If the net torque acting on a rigid body is nonzero about some origin, is there any other origin about which the net torque is zero?

Q10. If a system of particles is in motion, is it possible for the total angular momentum to be zero about some origin? Explain.

Q11. A ball is thrown in such a way that it does not spin about its own axis. Does this mean that the angular momentum is zero about an arbitrary origin? Explain.

11.4 CONSERVATION OF ANGULAR MOMENTUM

In Chapter 9 we found that the total linear momentum of a system of particles remains constant when the resultant external force acting on the system is zero. We have an analogous conservation law in rotational motion which states that *the total vector angular momentum of a system is constant if the resultant external torque acting on the system is zero*. This follows directly from Eq. 11.14, where we see that if

$$\Sigma\boldsymbol{\tau}_{\text{ext}} = \frac{d\boldsymbol{L}}{dt} = 0 \tag{11.18a}$$

then

$$\boldsymbol{L} = \text{constant} \tag{11.18b}$$

For a system of particles, we write this conservation law $\Sigma\boldsymbol{L}_i = \text{constant}$. If a body undergoes a redistribution of its mass, then its moment of inertia changes and we express this conservation of angular momentum in the form

$$\boldsymbol{L}_\text{i} = \boldsymbol{L}_\text{f} = \text{constant}$$

If the system is a rigid body rotating about a *fixed* axis, such as the z axis, then we can write $L_z = I\omega$, where L_z is the component of $\boldsymbol{L}$ along the axis of rotation and I is the moment of inertia about this axis. In this case, we can express the conservation of angular momentum as

$$I_\text{i}\omega_\text{i} = I_\text{f}\omega_\text{f} = \text{constant} \tag{11.19}$$

Conservation of angular momentum

This expression is valid for rotations either about a fixed axis or about an axis through the center of mass of the system as long as the axis remains parallel to itself. We only require the net external torque to be zero.

Although we do not prove it here, there is an important theorem concerning the angular momentum relative to the center of mass. This theorem states that *the resultant torque acting on a body about the center of mass equals the time rate of change of angular momentum regardless of the motion of the center of mass.* This theorem applies even if the center of mass is accelerating, provided $\boldsymbol{\tau}$ and $\boldsymbol{L}$ are evaluated relative to the center of mass.

In Eq. 11.19 we have a third conservation law to add to our list. Furthermore, we can now state that the energy, linear momentum, and angular momentum of an isolated system all remain constant.

(United Press International Photo)

There are many examples that demonstrate conservation of angular momentum, some of which should be familiar to you. You may have observed a figure skater undergoing a spin motion in the finale of an act. The angular velocity of the skater increases upon pulling his or her hands and feet close to the body. Neglecting friction between the skater and the ice, we see that there are no external torques on the skater. The change in angular velocity is due to the fact that since angular momentum is conserved, the product $I\omega$ remains constant and a decrease in the moment of inertia of the skater causes an increase in the angular velocity. Similarly, when divers (or acrobats) wish to make several somersaults, they pull their hands and feet close to their bodies in order to rotate at a higher rate. In these cases, the external force due to gravity acts through the center of mass and hence exerts no torque about this point. Therefore, the angular momentum about the center of mass must be conserved, or $I_i\omega_i = I_f\omega_f$. For example, when divers wish to double their angular velocity, they must reduce their moment of inertia to half its initial value.

Example 11.7

A projectile of mass m and velocity v_0 is fired into a solid cylinder of mass M and radius R (Fig. 11.10). The cylinder is initially at rest and is mounted on a fixed horizontal axle that runs through the center of mass. The line of motion of the projectile is perpendicular to the axle and at a distance $d < R$ from the center. Find the angular speed of the system after the projectile becomes imbedded in the cylinder.

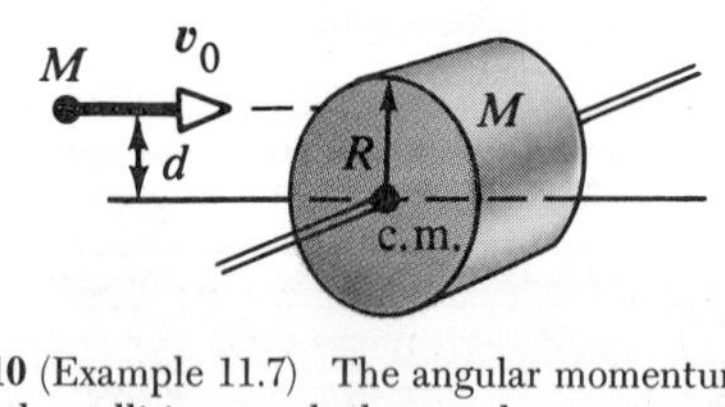

Figure 11.10 (Example 11.7) The angular momentum of the system before the collision equals the angular momentum right after the collision with respect to the center of mass if we neglect the weight of the projectile.

Solution: Let us evaluate the angular momentum of the system (projectile + cylinder) about the axle of the cylinder. About this point, the net external torque on the system is zero if we neglect the force of gravity on the projectile. Hence, the angular momentum of the system is the same before and after the collision.

Before the collision, only the projectile has angular momentum with respect to a point on the axle. The magnitude of this angular momentum is mv_0d, and it is directed along the axle into the paper. After the collision, the total angular momentum of the system is $I\omega$, where I is the total moment of inertia about the axle (projectile + cylinder). Since the total angular momentum is conserved, we get

$$mv_0d = I\omega = \left(\frac{1}{2}MR^2 + md^2\right)\omega$$

$$\omega = \frac{mv_0d}{\frac{1}{2}MR^2 + md^2}$$

This suggests another technique for measuring the velocity of a bullet. You should note that in this totally inelastic collision, mechanical energy is *not* conserved. That is, $\frac{1}{2}I\omega^2 < \frac{1}{2}mv_0^2$. What do you suppose accounts for the energy loss?

Example 11.8 *The Merry-Go-Round*

A horizontal platform in the shape of a circular disk rotates in a horizontal plane about a frictionless vertical axle (Fig. 11.11). The platform has a mass of 100 kg and a radius of 2 m. A student whose mass is 60 kg walks slowly from the rim of the platform toward the center. If the angular velocity of the system is 2 rad/s when the student is at the rim, (a) calculate the angular velocity when the student has reached a point 0.5 m from the center.

Let us call the moment of inertia of the platform I_p and the moment of inertia of the student I_s. Treating the student as a point mass m, we can write the *initial* moment of inertia of the system about the axle of rotation

$$I_i = I_p + I_s = \frac{1}{2}MR^2 + mR^2$$

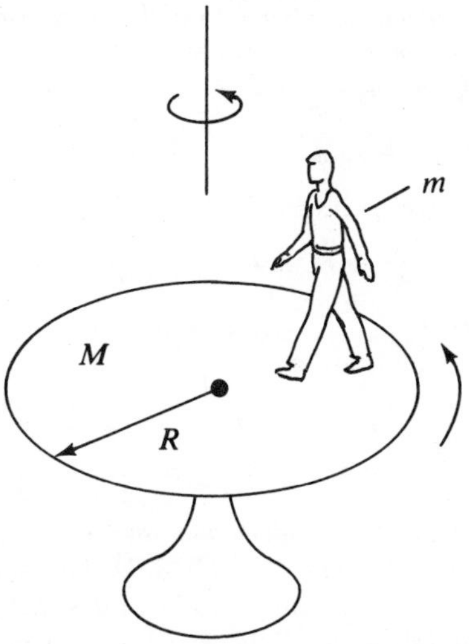

Figure 11.11 (Example 11.8) As the student walks toward the center of the rotating platform, the angular velocity of the system increases since the angular momentum must remain constant.

When the student has walked to the position $r < R$, the moment of inertia of the system *reduces* to

$$I_f = \frac{1}{2}MR^2 + mr^2$$

Since there are no external torques on the system (student + platform) about the axis of rotation, we can apply the law of conservation of angular momentum:

$$I_i\omega_i = I_f\omega_f$$

$$\left(\frac{1}{2}MR^2 + mR^2\right)\omega_i = \left(\frac{1}{2}MR^2 + mr^2\right)\omega_f$$

$$\omega_f = \left(\frac{\frac{1}{2}MR^2 + mR^2}{\frac{1}{2}MR^2 + mr^2}\right)\omega_i$$

Substituting the values given for M, R, m, and ω_i we get

$$\omega_f = \left(\frac{200 + 240}{200 + 15}\right)(2 \text{ rad/s}) = 4.1 \text{ rad/s}$$

(b) Calculate the initial and final kinetic energies of the system.

$$K_i = \frac{1}{2}I_i\omega_i^2 = \frac{1}{2}(440 \text{ kg}\cdot\text{m}^2)\left(2\,\frac{\text{rad}}{\text{s}}\right)^2 = 880 \text{ J}$$

$$K_f = \frac{1}{2}I_f\omega_f^2 = \frac{1}{2}(215 \text{ kg}\cdot\text{m}^2)\left(4.1\,\frac{\text{rad}}{\text{s}}\right)^2 = 1800 \text{ J}$$

Note that the kinetic energy of the system *increases!* Although this result may surprise you, it can be explained as follows: In the process of walking toward the center of the platform, the student had to exert some muscular effort and perform positive work, which in turn is transformed into an increase in kinetic energy of the system. In other words, internal forces within the system did work. Since the student is in a rotating, noninertial frame of reference, he senses an outward "centrifugal" force that varies with r. He must exert a counteracting force to walk inwards, and hence he must perform work, or exert energy. You should show that the gain in kinetic energy can be accounted for using the work-energy theorem.

Example 11.9 *Spinning on a Stool*

A student sits on a pivoted stool while holding a pair of weights, as in Fig. 11.12. The stool is free to rotate about a vertical axis with negligible friction. The student is set in rotating motion with the weights outstretched. Why does the angular velocity of the system increase as the weights are pulled inward?

Figure 11.12 (Example 11.9) What happens to the angular velocity as the student pulls the weights closer to his body?

Solution: The initial angular momentum of the system is $I_i\omega_i$, where I_i refers to the initial moment of inertia of the entire system (student + weights + stool). After the weights are pulled in, the angular momentum of the system is $I_f\omega_f$. Note that $I_f < I_i$ since the weights are now closer to the axis of rotation, reducing the moment of inertia. Since the net external torque on the system is zero, angular momentum is conserved, so $I_i\omega_i = I_f\omega_f$. Therefore, $\omega_f > \omega_i$, or the angular velocity increases. As in the previous example, the increase in kinetic energy of the system arises from the fact that the student must do work in pulling the weights toward the axis of rotation.

Example 11.10 *The Spinning Bicycle Wheel*

In another favorite classroom demonstration, a student holds the axle of a spinning bicycle wheel while seated on a pivoted stool (Fig. 11.13). The student and stool are initially *at rest* while the wheel is spinning in a horizontal plane with an initial angular momentum $\mathbf{L}_0$ pointing upward. Explain what happens if the wheel is inverted about its center by 180°.

Figure 11.13 (Example 11.10) The wheel is initially spinning when the student is at rest. What happens when the wheel is inverted?

Solution: In this situation, the system consists of the student, wheel, and stool. Initially, the total angular momentum of the system is $\mathbf{L}_0$, corresponding to the contribution from the spinning wheel. As the wheel is inverted, a torque is supplied by the student, but this is *internal* to the system. There is *no* external torque acting on the system about the vertical axis. Therefore, *the angular momentum of the system must be conserved.*

Initially, we have

$$\mathbf{L}_{\text{system}} = \mathbf{L}_0 \qquad \text{(upward)}$$

After the wheel is inverted,

$$\mathbf{L}_{\text{system}} = \mathbf{L}_0 = \mathbf{L}_{\text{student+stool}} + \mathbf{L}_{\text{wheel}}$$

In this case, $\mathbf{L}_{\text{wheel}} = -\mathbf{L}_0$ since it is now rotating in the opposite sense. Therefore

$$\mathbf{L}_0 = \mathbf{L}_{\text{student+stool}} - \mathbf{L}_0$$

$$\mathbf{L}_{\text{student+stool}} = 2\mathbf{L}_0$$

This shows that *the student and stool will start to turn, acquiring an angular momentum having a magnitude* twice *that of the spinning wheel and directed upward.*

You should show that if the wheel were in general tilted through an angle θ measured from the vertical axis, the student and stool would acquire an angular momentum of magnitude $L_0(1 - \cos\theta)$. This example illustrates the law of conservation of angular momentum and its vector nature.

Q12. Suppose the student in Fig. 11.12 were to drop the weights to his side rather than pull them inward horizontally. What would account for the increase in the kinetic energy of the system in this situation?

11.5 THE MOTION OF GYROSCOPES AND TOPS

A very unusual and fascinating type of motion that you probably have observed is that of a top spinning about its axis of symmetry as in Fig. 11.14. If the top spins about its axis very rapidly, the axis will rotate about the vertical direction as indicated, thereby sweeping out a cone. The motion of the axis of the top about the vertical, known as *precessional motion,* is usually slow compared with the spin motion of the top. It is quite natural to wonder why the top doesn't fall over. Since the center of mass is not above the pivot point O, there is clearly a net torque acting on the top about O due to the weight force $M\mathbf{g}$. From this description, it is easy to see that the top would certainly fall if it were not spinning. However, because the top is spinning, it has an angular momentum $\mathbf{L}$ directed along its axis of symmetry, and as we shall show, the steady precessional motion is due to the *vector* relationship between the torque on the top and the rate of change of its angular momentum. This is an excellent example of the importance of the directional nature of angular momentum.

The two forces acting on the top are the downward force of gravity, $M\mathbf{g}$, and the normal force, $\mathbf{N}$, acting upward at the pivot point O. The normal force produces no torque about the pivot since its moment arm is zero. However, the force of gravity produces a torque $\boldsymbol{\tau} = \mathbf{r} \times M\mathbf{g}$ about O, where the direction of $\boldsymbol{\tau}$ is perpendicular to the plane formed by $\mathbf{r}$ and $M\mathbf{g}$. By necessity, the vector $\boldsymbol{\tau}$ lies in a horizontal plane perpendicular to the angular momentum vector. The net torque and angular momentum of the body are related through the expression

$$\boldsymbol{\tau} = \frac{d\mathbf{L}}{dt}$$

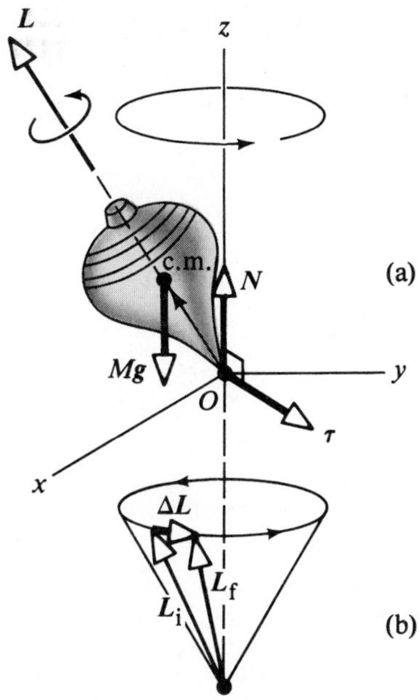

Figure 11.14 Precessional motion of a top spinning about its axis of symmetry. The only external forces acting on the top are the normal force, ***N***, and the force of gravity, ***Mg***. The direction of the angular momentum, ***L***, is along the axis of symmetry.

From this expression, we see that the nonzero torque produces a *change* in angular momentum $d\mathbf{L}$, which is in the same direction as $\boldsymbol{\tau}$. Therefore, like the torque vector, $d\mathbf{L}$ must also be at right angles to $\mathbf{L}$. Figure 11.14b illustrates the resulting precessional motion of the axis of the top. In a time Δt, the change in angular momentum $\Delta\mathbf{L} = \mathbf{L}_f - \mathbf{L}_i = \boldsymbol{\tau}\,\Delta t$. Note that the magnitude of $\mathbf{L}$ doesn't change ($|\mathbf{L}_i| = |\mathbf{L}_f|$). Rather, what is changing is the *direction* of $\mathbf{L}$. Since the change in angular momentum is in the direction of $\boldsymbol{\tau}$, which lies in the xy plane, the top undergoes precessional motion. Thus, the effect of the torque is to deflect the angular momentum of the top in a direction perpendicular to its spin axis. Of course, the torque produces precessional motion only if the top has an initial angular momentum.

We have presented a rather qualitative description of the motion of a top. In general, the motion of such an object is very complex. However, the essential features of the motion can be illustrated by considering the simple gyroscope shown in Fig. 11.15. This device consists of a wheel free to spin about an axle that is pivoted at a distance h from the center of mass of the wheel. If the wheel is given an angular velocity $\boldsymbol{\omega}$ about its axis, the wheel will have a spin angular momentum $\mathbf{L} = I\boldsymbol{\omega}$ directed along the axle as shown. Let us consider the torque acting on the wheel about the pivot O. Again, the force, ***N***, of the support on the axle produces no torque about O. On the other hand, the weight Mg produces a torque of magnitude Mgh about O. The direction of this torque is *perpendicular* to the axle (and perpendicular to $\mathbf{L}$), as described in Fig. 11.15. This torque causes the angular momentum to change in the direction perpendicular to the axle. Hence, the axle moves in the direction of the torque, that is, in the horizontal plane. There is an assumption that we must make in order to simplify the description of the system. The *total* angular momentum of the precessing wheel is actually the sum of the spin angular momentum, $I\boldsymbol{\omega}$, and the angular momentum due to the motion of the center of mass about the pivot. In our treatment, we shall neglect the contribution from the center of mass motion and take the total angular momentum to be just $I\boldsymbol{\omega}$. In practice, this is a good approximation if ω is made very large.

In a time dt the torque due to the weight force adds to the system an *additional* angular momentum equal to $d\mathbf{L} = \boldsymbol{\tau}dt$ in the direction perpendicular to $\mathbf{L}$. This additional angular momentum, $\boldsymbol{\tau}dt$, when added vectorially to the original spin

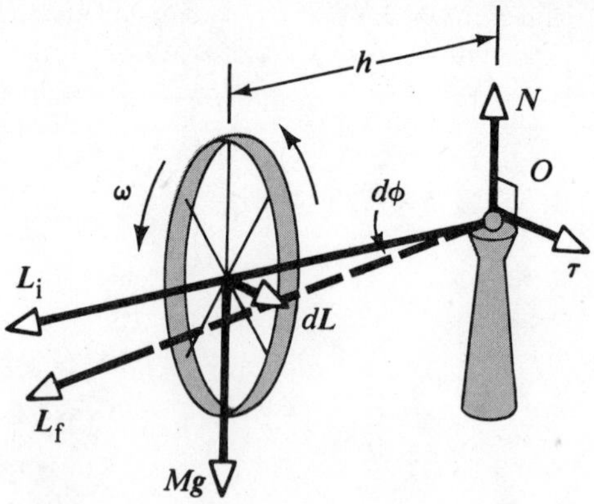

(a)

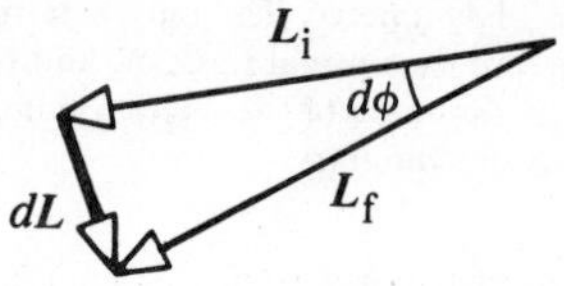

(b)

Figure 11.15 The motion of a simple gyroscope pivoted a distance h from its center of gravity. Note that the weight Mg produces a torque about the pivot that is perpendicular to the axle. This results in a change in angular momentum $d\mathbf{L}$ in the direction perpendicular to the axle. The axle sweeps out an angle $d\phi$ in a time dt.

angular momentum, $I\omega$, *causes a shift in the direction of the total angular momentum.* We can express the magnitude of this change in angular momentum as

$$dL = \tau\, dt = (Mgh)\, dt$$

The vector diagram in Fig. 11.15 shows that in the time dt, the angular momentum vector rotates through an angle $d\phi$, which is also the angle through which the axle rotates. From the vector triangle formed by the vector $\mathbf{L}_i$, $\mathbf{L}_f$, and $d\mathbf{L}$ and from the expression above, we see that

$$d\phi = \frac{dL}{L} = \frac{(Mgh)\, dt}{L}$$

Using $L = I\omega$, we find that the rate at which the axle rotates about the vertical axis is given by

Precessional frequency

$$\omega_p = \frac{d\phi}{dt} = \frac{Mgh}{I\omega} \tag{11.20}$$

The angular frequency ω_p is called the *precessional frequency*. You should note that this result is valid only when $\omega_p \ll \omega$. Otherwise, a much more complicated motion is involved. As you can see from Eq. 11.20, the condition that $\omega_p \ll \omega$ is met when $I\omega$ is large compared with Mgh.

Q13. Why is it easier to keep your balance on a moving bicycle than on a bicycle at rest?

Q14. A scientist at a hotel sought assistance from a bellhop to carry a mysterious suitcase. When the unaware bellhop rounded a corner carrying the suitcase, it suddenly moved away from him for some unknown reason. At this point, the alarmed bellhop dropped the suitcase and ran off. What do you suppose might have been in the suitcase?

11.6 ROLLING MOTION OF A RIGID BODY

In this section we shall treat the motion of a rigid body that is rotating about a moving axis. The general motion of a rigid body in space is very complex. However, we can simplify matters by restricting our discussion to a homogeneous rigid body having a high degree of symmetry, such as a cylinder, sphere, or hoop. Furthermore, we shall assume that the body undergoes rolling motion in a plane.

Consider a cylinder of radius R rolling on a rough, horizontal surface (Fig. 11.16). As the cylinder rotates through an angle θ, its center of mass moves a distance $s = R\theta$. Therefore, the velocity and acceleration of the center of mass for *pure rolling motion* are given by

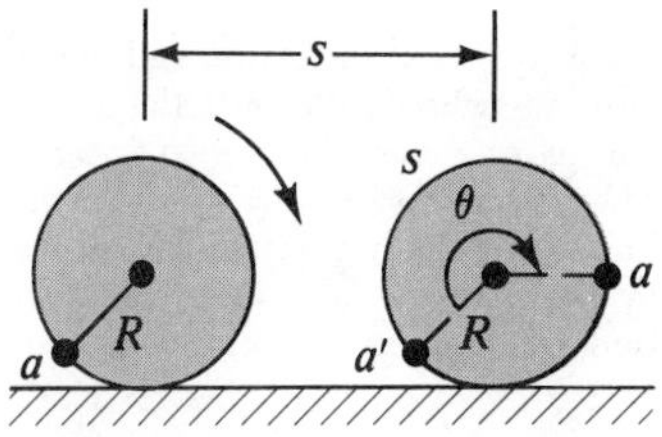

Figure 11.16 For pure rolling motion, as the cylinder rotates through an angle θ, the center of mass moves a distance $s = R\theta$.

$$v_c = \frac{ds}{dt} = R\frac{d\theta}{dt} = R\omega \tag{11.21}$$

$$a_c = \frac{dv_c}{dt} = R\frac{d\omega}{dt} = R\alpha \tag{11.22}$$

The linear velocities of various points on the rolling cylinder are illustrated in Fig. 11.17. Note that the linear velocity of any point is in a direction perpendicular to the line from that point to the contact point. At any instant, the point P is at rest relative to the surface since sliding does not occur. For that reason, the axis through P perpendicular to the diagram is called the *instantaneous axis of rotation.*

A general point on the cylinder, such as Q, has both horizontal and vertical components of velocity. However, the points P and P' and the point at the center of mass are unique and of special interest. By definition, the center of mass moves with a velocity $v_c = R\omega$, whereas the contact point P has zero velocity. Thus, it follows that the point P' must have a velocity equal to $2v_c = 2R\omega$, since all points on the cylinder have the same angular velocity.

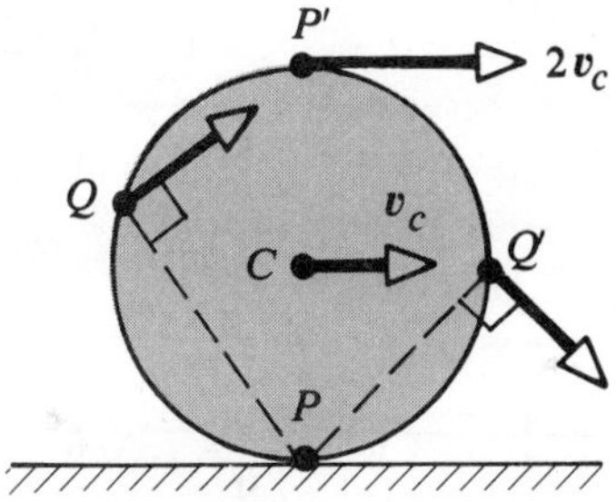

Figure 11.17 All points on a rolling body move in a direction perpendicular to an axis through the contact point P. The center of mass moves with a velocity $\mathbf{v}_c$, while the point P' moves with the velocity $2\mathbf{v}_c$.

We can express the total kinetic energy of the rolling cylinder as

$$K = \frac{1}{2}I_P\omega^2 \tag{11.23}$$

where I_P is the moment of inertia about the axis through P. Applying the parallel-axis theorem, we can substitute $I_P = I_c + MR^2$ into Eq. 11.22 to get

$$K = \frac{1}{2}I_c\omega^2 + \frac{1}{2}MR^2\omega^2$$

$$K = \frac{1}{2}I_c\omega^2 + \frac{1}{2}Mv_c^2 \tag{11.24}$$

Total kinetic energy of a rolling body

where we have used the fact that $v_c = R\omega$.

We can think of Eq. 11.24 as follows: The first term on the right, $\frac{1}{2}I_c\omega^2$, represents the rotational kinetic energy about the center of mass, and the term $\frac{1}{2}Mv_c^2$ represents the kinetic energy the cylinder would have if it were just translating through space without rotating. Thus, we can say that *the total kinetic energy is the sum of a rotational kinetic energy about the center of mass plus the translational kinetic energy of the center of mass.*

We can use energy methods to treat a class of problems concerning the rolling motion of a rigid body down a rough incline. We shall assume that the rigid body in Fig. 11.18 does not slip and is released from rest at the top of the incline. Note that

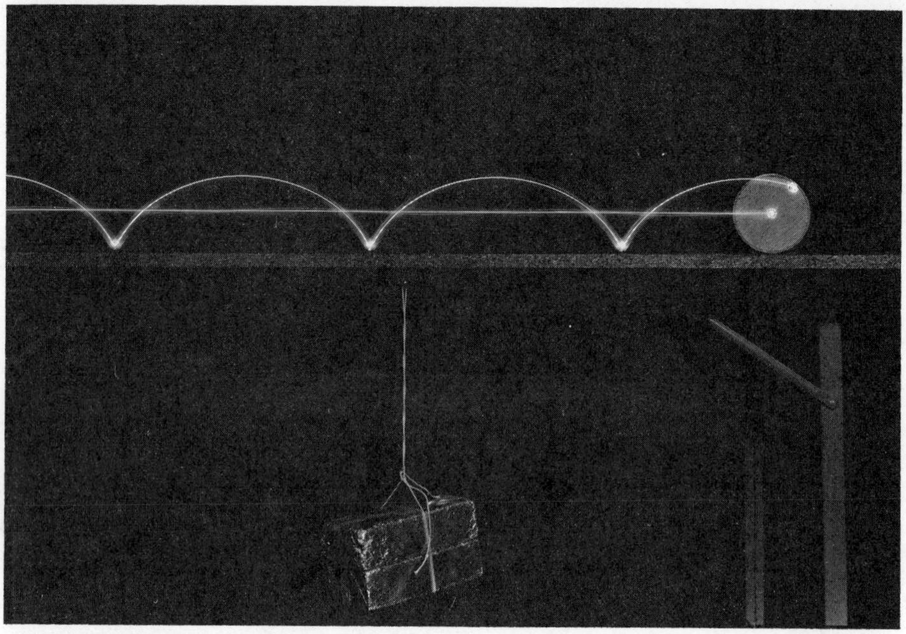

Light sources at the center and rim of a rolling cylinder illustrate the different paths which these points take. The center of mass moves in a straight line, while a point on the rim moves in the path of a cycloid. (Education Development Center, Newton, Mass.)

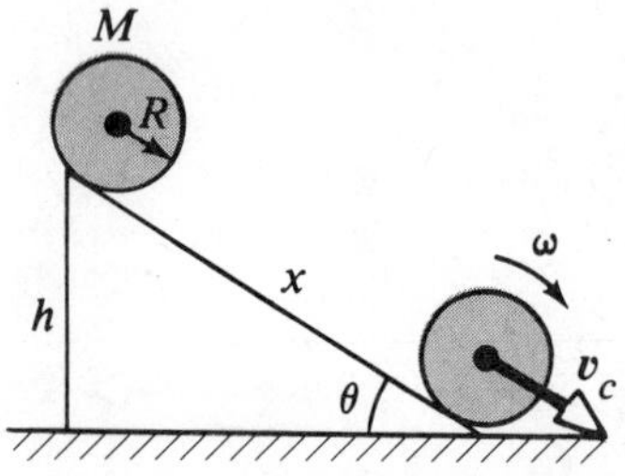

Figure 11.18 A round object rolling down an incline. Mechanical energy is conserved if no slipping occurs.

rolling motion is possible only if a frictional force is present between the object and the incline to produce a net torque about the center of mass. Despite the presence of friction, there is no loss of mechanical energy since the contact point is at rest relative to the surface at any instant. On the other hand, if the rigid body were to slide, mechanical energy would be lost as motion progresses.

Using the fact that $v_c = R\omega$ for pure rolling motion, we can express Eq. 11.24 as

$$K = \frac{1}{2}I_c\left(\frac{v_c}{R}\right)^2 + \frac{1}{2}Mv_c^2$$

Total kinetic energy of a rolling body

$$K = \frac{1}{2}\left(\frac{I_c}{R^2} + M\right)v_c^2 \qquad (11.25)$$

As the rigid body rolls down the incline, it loses potential energy Mgh, where h is the height of the incline. If the body starts from rest at the top, its kinetic energy at the bottom, given by Eq. 11.25, must equal its potential energy at the top. Therefore, the velocity of the center of mass at the bottom can be obtained by equating the two quantities:

$$\frac{1}{2}\left(\frac{I_c}{R^2} + M\right)v_c^2 = Mgh$$

Velocity of center of mass for pure rolling motion

$$v_c = \left(\frac{2gh}{1 + I_c/MR^2}\right)^{1/2} \qquad (11.26)$$

Example 11.11 *Sphere Rolling Down an Incline*

If the rigid body shown in Fig. 11.18 is a solid sphere, calculate the velocity of its center of mass at the bottom and determine the linear acceleration of the center of mass of the sphere.

Solution: For a uniform solid sphere, $I_c = \frac{2}{5}MR^2$, and therefore Eq. 11.26 gives

$$v_c = \left(\frac{2gh}{1 + \frac{2}{5}\frac{MR^2}{MR^2}}\right)^{1/2} = \left(\frac{10}{7}gh\right)^{1/2}$$

The vertical displacement is related to the displacement x along the incline through the relation $h = x\sin\theta$. Hence, we can express the result above as

$$v_c^2 = \frac{10}{7} gx \sin\theta$$

Comparing this with the familiar expression from kinematics, $v_c^2 = 2a_c x$, we see that the acceleration of the center of mass is given by

$$a_c = \frac{5}{7} g \sin\theta$$

The results are quite amazing! Both the velocity and acceleration of the center of mass are *independent* of the mass and radius of the sphere! That is, *all homogeneous solid spheres would experience the same velocity and acceleration on a given incline.* If we repeated the calculations for a hollow sphere, a solid cylinder, or a hoop, we would obtain similar results. The constant factors that appear in the expressions for v_c and a_c depend on the moment of inertia about the center of mass for the specific body. In all cases, the acceleration of the center of mass will be *less* than $g \sin\theta$, the value it would have if the plane were frictionless and no rolling occurred.

Example 11.12

In this example, let us consider the solid sphere rolling down an incline and verify the results of Example 11.11 using dynamic methods. The free-body diagram for the sphere is illustrated in Fig. 11.19.

Solution: Newton's second law applied to the center of mass motion gives

$$(1) \qquad \Sigma F_x = Mg \sin\theta - f = Ma_c$$
$$\Sigma F_y = N - Mg \cos\theta = 0$$

where x is measured downward along the inclined plane. Now let us write an expression for the torque acting on the sphere. A convenient axis to choose is an axis through the center of the sphere, perpendicular to the plane of the figure.[2] Since N and Mg go through this origin, they have zero moment arms and do not contribute to the torque. However, the force of friction produces a torque about this axis equal to fR in the clockwise direction; therefore

$$\tau_c = fR = I_c \alpha$$

Since $I_c = \frac{2}{5}MR^2$ and $\alpha = a_c/R$, we get

$$(2) \qquad f = \frac{I_c \alpha}{R} = \left(\frac{\frac{2}{5}MR^2}{R}\right)\frac{a_c}{R} = \frac{2}{5}Ma_c$$

Substituting (2) into (1) gives

$$a_c = \frac{5}{7} g \sin\theta$$

which agrees with the result of Example 11.11. Note that $a_c < g \sin\theta$ because of the retarding frictional force. It is left as an exercise for you to show that the expression for v_c at the bottom of the incline can be obtained using the expression $v^2 = 2ax$ from kinematics (where $v_0 = 0$).

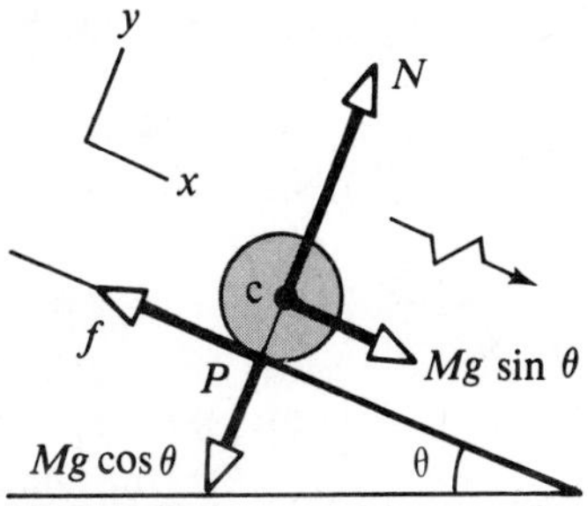

Figure 11.19 (Example 11.12) Free-body diagram for a solid sphere rolling down an incline.

Q15. When a cylinder rolls on a horizontal surface as in Fig. 11.16, are there any points on the cylinder that have only a vertical component of velocity at some instant? If so, where are they?

Q16. Three homogeneous rigid bodies—a solid sphere, a solid cylinder, and a hollow cylinder—are placed at the top of an incline (Fig. 11.20). If they all are released from rest at the same elevation and roll without slipping, which reaches the bottom first? Which reaches last? You should try this at home and note that the result is *independent* of the masses and radii. This is quite amazing!

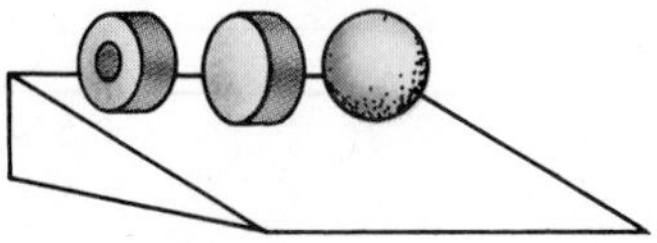

Figure 11.20 (Question 16) Which object wins the race?

11.7 ANGULAR MOMENTUM AS A FUNDAMENTAL QUANTITY

We have seen that the concept of angular momentum is very useful for describing the motion of macroscopic systems. However, the concept is also valid on a submi-

[2]You should note that although the point at the center of mass is not an inertial frame, the expression $\tau_c = I\alpha$ still applies in the center of mass frame.

croscopic scale and has been used extensively in the development of modern theories of atomic, molecular, and nuclear physics. In these developments, it was found that the angular momentum of a system is a *fundamental* quantity. The word *fundamental* in this context implies that angular momentum is an inherent property of atoms, molecules, or their constituents.

In order to explain a variety of experiments on atomic and molecular systems, it is necessary to assign discrete values to the angular momentum. These discrete values are some multiple of a fundamental unit of angular momentum, which equals $\hbar$ (Planck's constant, h, divided by 2π):

$$\text{Fundamental unit of angular momentum} = \hbar = 1.054 \times 10^{-34} \frac{\text{kg} \cdot \text{m}^2}{\text{s}^2}$$

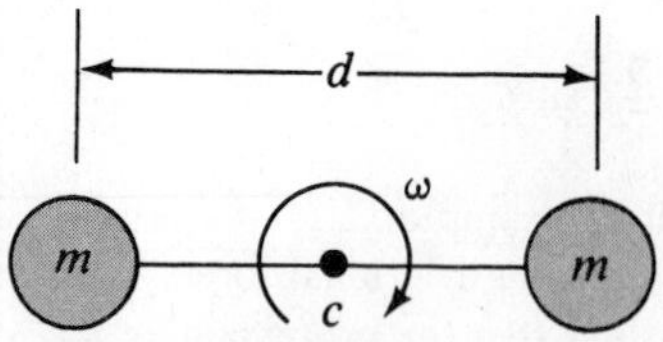

Figure 11.21 The rigid-rotor model of the diatomic molecule. The rotation occurs about the center of mass in the plane of the diagram.

Let us accept this postulate for the time being and show how it can be used to estimate the rotational frequency of a diatomic molecule. Consider the O_2 molecule as a rigid rotor, that is, two atoms separated by a fixed distance d and rotating about the center of mass (Fig. 11.21). Equating the rotational angular momentum to the fundamental unit $\hbar$, we can estimate the lowest rotational frequency:

$$I_c\omega \approx \hbar \qquad \text{or} \qquad \omega \approx \frac{\hbar}{I_c}$$

In Example 10.3, we found that the moment of inertia of the O_2 molecule about this axis of rotation is 1.68×10^{-46} kg · m^2. Therefore,

$$\omega \approx \frac{\hbar}{I_c} = \frac{1.054 \times 10^{-34} \text{ kg} \cdot \text{m}^2/\text{s}}{1.68 \times 10^{-46} \text{ kg} \cdot \text{m}^2} = 6.27 \times 10^{11} \text{ rad/s}$$

This result is in good agreement with measured rotational frequencies. Furthermore, the rotational frequencies are much lower than the vibrational frequencies of the molecule, which are typically of the order of 10^{13} Hz.

This simple example shows that certain classical concepts and mechanical models might be useful in describing some features of atomic and molecular systems. However, a wide variety of phenomena on the submicroscopic scale can be explained only if one assumes discrete values of the angular momentum associated with a particular type of motion.

Historically, the Danish physicist Niels Bohr (1885–1962) was the first to suggest this radical idea in his theory of the hydrogen atom. Strictly classical models were unsuccessful in describing many properties of the hydrogen atom, such as the fact that the atom absorbs and emits radiation at discrete frequencies. Bohr postulated that the electron could only occupy circular orbits about the proton for which the orbital angular momentum was equal to $n\hbar$, where n is an integer. From this rather simple model, one can estimate the rotational frequencies of the electron in the various orbits (Exercise 32).

Although Bohr's theory provided some insight concerning the behavior of matter at the atomic level, it is basically incorrect. Subsequent developments in quantum mechanics from 1924 to 1930 provided models and interpretations that are still accepted. We will discuss this further in Chapter 41.

Later developments in atomic physics indicated that the electron also possesses another kind of angular momentum, called *spin*, which is also an inherent property of the electron. The spin angular momentum is also restricted to discrete values. We shall return to this important property later in the text and discuss its great impact on modern physical science.

11.8 SUMMARY

The *torque* $\boldsymbol{\tau}$ due to a force $\mathbf{F}$ about an origin in an inertial frame is defined to be

$$\boldsymbol{\tau} \equiv \mathbf{r} \times \mathbf{F} \tag{11.1}$$

Torque

Given two vectors $\mathbf{A}$ and $\mathbf{B}$, their *cross product* $\mathbf{A} \times \mathbf{B}$ is a vector $\mathbf{C}$ having a magnitude

$$C = |AB \sin\theta| \tag{11.3}$$

Magnitude of the cross product

where θ is the angle between $\mathbf{A}$ and $\mathbf{B}$. The direction of the vector $\mathbf{C} = \mathbf{A} \times \mathbf{B}$ is perpendicular to the plane formed by $\mathbf{A}$ and $\mathbf{B}$, and its sense is determined by the right-hand rule. Some properties of the cross product include the facts that $\mathbf{A} \times \mathbf{B} = -\mathbf{B} \times \mathbf{A}$ and $\mathbf{A} \times \mathbf{A} = 0$.

The *angular momentum* $\mathbf{L}$ of a particle of linear momentum $\mathbf{p} = m\mathbf{v}$ is given by

$$\mathbf{L} = \mathbf{r} \times \mathbf{p} = m\mathbf{r} \times \mathbf{v} \tag{11.9}$$

Angular momentum of a particle

where $\mathbf{r}$ is the vector position of the particle relative to an origin in an inertial frame. If ϕ is the angle between $\mathbf{r}$ and $\mathbf{p}$, the magnitude of $\mathbf{L}$ is given by

$$L = mvr \sin\phi \tag{11.10}$$

The *net external torque* acting on a particle or rigid body is equal to the time rate of change of its angular momentum:

$$\Sigma\boldsymbol{\tau}_{\text{ext}} = \frac{d\mathbf{L}}{dt} \tag{11.14}$$

The *angular momentum* $\mathbf{L}$ of a symmetric rigid body rotating about a fixed axis of symmetry with angular velocity $\boldsymbol{\omega}$ is given by

$$\mathbf{L} = I\boldsymbol{\omega} \tag{11.15}$$

Angular momentum of a rigid body about a fixed axis

where I is the moment of inertia about the axis of rotation.

The *net external torque* acting on a rigid body equals the product of its moment of inertia about the axis of rotation and its angular acceleration:

$$\Sigma\boldsymbol{\tau}_{\text{ext}} = I\boldsymbol{\alpha} \tag{11.17}$$

If the net external torque acting on a system is zero, the total angular momentum of the system is constant. Applying this *conservation of angular momentum* law to a body whose moment of inertia changes gives

$$I_i\omega_i = I_f\omega_f = \text{constant} \tag{11.19}$$

Conservation of angular momentum

The *total kinetic energy* of a rigid body, such as a cylinder, that is rolling on a rough surface without slipping equals the rotational kinetic energy about its center of mass, $\frac{1}{2}I_c\omega^2$, plus the translational kinetic energy of the center of mass, $\frac{1}{2}Mv_c^2$:

$$K = \frac{1}{2}I_c\omega^2 + \frac{1}{2}Mv_c^2 \tag{11.24}$$

Total kinetic energy of a rolling body

In this expression, v_c is the velocity of the center of mass and $v_c = R\omega$ for pure rolling motion.

EXERCISES

Section 11.1 The Vector Product and Torque

1. Two vectors are given by $\boldsymbol{A} = -3\boldsymbol{i} + \boldsymbol{j}$ and $\boldsymbol{B} = \boldsymbol{i} - 2\boldsymbol{j}$. Find (a) $\boldsymbol{A} \times \boldsymbol{B}$ and (b) the angle between $\boldsymbol{A}$ and $\boldsymbol{B}$.

2. Using the definition of the cross product, verify Eqs. 11.7a through 11.7d for the vector product of unit vectors.

3. Verify Eq. 11.8 for the cross product of any two vectors $\boldsymbol{A}$ and $\boldsymbol{B}$, and show that the cross product may be written in the following determinant form:

$$\boldsymbol{A} \times \boldsymbol{B} = \begin{vmatrix} \boldsymbol{i} & \boldsymbol{j} & \boldsymbol{k} \\ A_x & A_y & A_z \\ B_x & B_y & B_z \end{vmatrix}$$

4. The vectors $\boldsymbol{A}$ and $\boldsymbol{B}$ form two sides of a parallelogram. (a) Show that the area of the parallelogram is given by $|\boldsymbol{A} \times \boldsymbol{B}|$. (b) If $\boldsymbol{A} = (3\boldsymbol{i} + 3\boldsymbol{j})$ m and $\boldsymbol{B} = (\boldsymbol{i} - 2\boldsymbol{j})$ m, find the area of the parallelogram.

5. Find $\boldsymbol{A} \times \boldsymbol{B}$ for the vectors (a) $\boldsymbol{A} = 3\boldsymbol{j}$ and $\boldsymbol{B} = 2\boldsymbol{i} + 2\boldsymbol{j}$, (b) $\boldsymbol{A} = 3\boldsymbol{i} - \boldsymbol{j}$ and $\boldsymbol{B} = 4\boldsymbol{k}$, (c) $\boldsymbol{A} = 3\boldsymbol{j} + \boldsymbol{k}$ and $\boldsymbol{B} = -2\boldsymbol{i}$.

6. Vector $\boldsymbol{A}$ is in the negative y direction, and vector $\boldsymbol{B}$ is in the negative x direction. What is the direction of (a) $\boldsymbol{A} \times \boldsymbol{B}$ and (b) $\boldsymbol{B} \times \boldsymbol{A}$?

7. A particle is located at the vector position $\boldsymbol{r} = (2\boldsymbol{i} + 4\boldsymbol{j})$ m, and the force acting on it is $\boldsymbol{F} = (3\boldsymbol{i} + \boldsymbol{j})$ N. What is the torque about (a) the origin and (b) the point having coordinates (0,6) m?

8. If $|\boldsymbol{A} \times \boldsymbol{B}| = \boldsymbol{A} \cdot \boldsymbol{B}$, what is the angle between $\boldsymbol{A}$ and $\boldsymbol{B}$?

Section 11.2 Angular Momentum of a Particle

9. A particle of mass m moves in a straight line with a constant velocity $\boldsymbol{v} = v\boldsymbol{j}$ along the positive y axis. Determine the angular momentum of the particle (both magnitude and direction) relative to (a) the point having coordinates $(-d,0)$, (b) the point having coordinates $(2d,0)$, and (c) the origin.

10. A particle of mass 0.3 kg moves in the xy plane. At the instant its coordinates are (2,4) m, its velocity is $(3\boldsymbol{i} + 2\boldsymbol{j})$ m/s. At this instant, determine the angular momentum of the particle relative to the origin.

11. A 4-kg particle moves in the xy plane with a constant speed of 2 m/s in the x direction along the line $y = -3$ m. What is its angular momentum relative to (a) the origin and (b) the point $(0,-5)$ m?

12. Two particles move in opposite directions along a straight line (Fig. 11.22). The particle of mass m moves to the right with a speed v while the particle of mass $3m$ moves to the left with a speed v. What is the *total* angular momentum of the system relative to (a) the point A, (b) the point O, and (c) the point B?

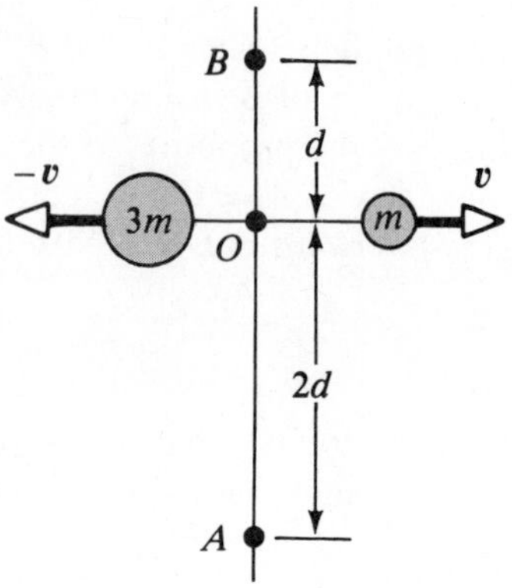

Figure 11.22 (Exercise 12).

13. A light rigid rod 1 m in length rotates in the xy plane about a pivot through the rod's center. Two particles of mass 2 kg and 3 kg are connected to its ends (Fig. 11.23). Determine the angular momentum of the system about the origin at the instant the speed of each particle is 5 m/s.

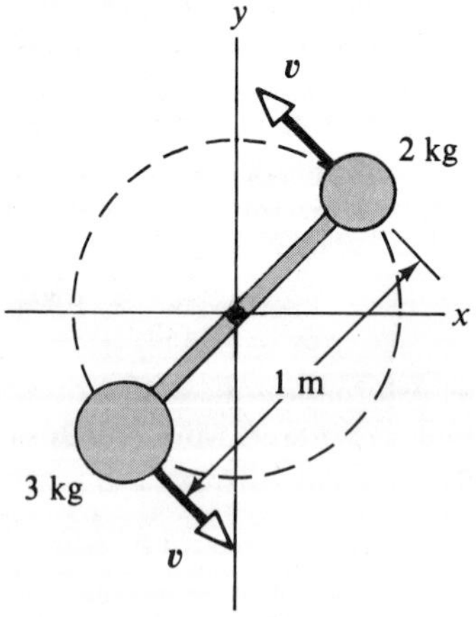

Figure 11.23 (Exercise 13).

14. An airplane of mass 5000 kg flies level to the ground at an altitude of 8 km with a constant speed of 200 m/s relative to the earth. (a) What is the magnitude of the airplane's angular momentum relative to a ground observer who is directly below the airplane? (b) Does this value change as the airplane continues its motion along a straight line?

Section 11.3 Angular Momentum and Torque for a System of Particles

15. A particle of mass m is given a velocity $-v_0\mathbf{j}$ at the point $(-d,0)$ and proceeds to accelerate in the presence of earth's gravity (Fig. 11.24). (a) Find an expression for the angular momentum as a function of time with respect to the origin. (b) Calculate the torque acting on the particle at any time relative to the origin. (c) Using your results to (a) and (b), verify that $\tau = d\mathbf{L}/dt$.

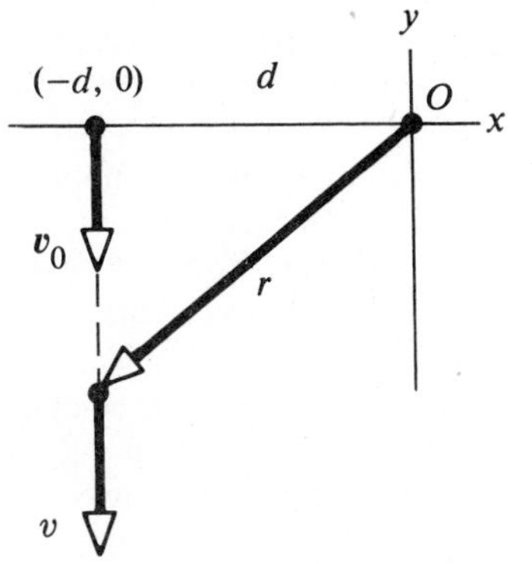

Figure 11.24 (Exercise 15).

16. A uniform solid disk of mass 3 kg and radius 0.2 m rotates about a fixed axis perpendicular to its face (Fig. 11.9). If the angular frequency of rotation is 5 rad/s, calculate the angular momentum of the disk when the axis of rotation (a) passes through its center of mass and (b) passes through a point midway between the center and the rim.

17. A particle of mass 0.3 kg is attached to the 100-cm mark of a meter stick of mass 0.2 kg. The meter stick rotates on a horizontal, smooth table with an angular velocity of 4 rad/s. Calculate the angular momentum of the system if the stick is pivoted about an axis (a) perpendicular to the table through the 50-cm mark and (b) perpendicular to the table through the 0-cm mark.

18. (a) Calculate the angular momentum of the earth due to its spinning motion about its axis. (b) Calculate the angular momentum of the earth due to its orbital motion about the sun and compare this with (a). (Assume the earth is a homogeneous sphere of radius 6.37×10^6 m and mass 5.98×10^{24} kg. Take the earth-sun distance to be 1.49×10^{11} m.)

19. A 3-kg mass is attached to a light cord, which is wound around a pulley (Fig. 10.19). The pulley is a uniform solid cylinder of radius 8 cm and mass 1 kg. (a) What is the net torque on the system about the point O? (b) When the 3-kg mass has a speed v, the pulley has an angular velocity $\omega = v/R$. Determine the total angular momentum of the system about O. (c) Using the fact that $\tau = dL/dt$ and your result from (b), calculate the acceleration of the 3-kg mass.

20. A cylinder with moment of inertia I_1 rotates with angular velocity ω_0 about a vertical, frictionless axle. A second cylinder, with moment of inertia I_2 initially not rotating, drops onto the first cylinder (Fig. 11.25). Since the surfaces are rough, the two eventually reach the same angular velocity ω. (a) Calculate ω. (b) Show that energy is lost in this situation and calculate the ratio of the final to the initial kinetic energy.

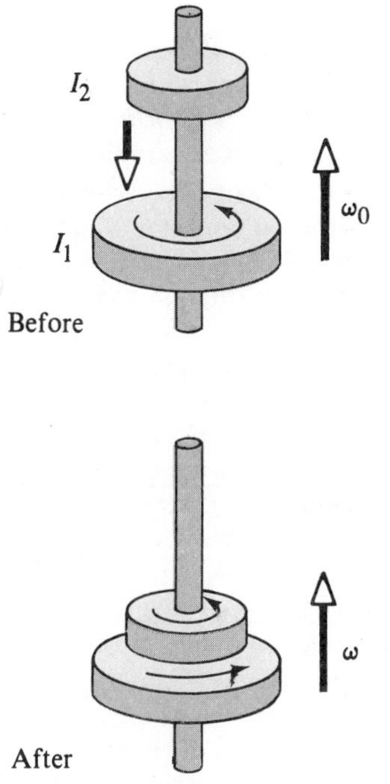

Figure 11.25 (Exercise 20).

21. A uniform solid cylinder of mass 1 kg and radius 25 cm rotates about a fixed vertical, frictionless axle with an angular speed of 10 rad/s. A 0.5-kg piece of putty is dropped vertically onto the cylinder at a point 15 cm from the axle. If the putty sticks to the cylinder, calculate the final angular speed of the system. (Assume the putty is a particle.)

Section 11.4 Conservation of Angular Momentum

22. A particle of mass $m = 10$ g and speed $v_0 = 5$ m/s collides with and sticks to the edge of a uniform solid sphere of mass $M = 1$ kg and radius $R =$ 20 cm (Fig. 11.26). If the sphere is initially at rest and is pivoted about a frictionless axle through O perpendicular to the plane, (a) find the angular velocity of the system after the collision and (b) determine how much energy is lost in the collision.

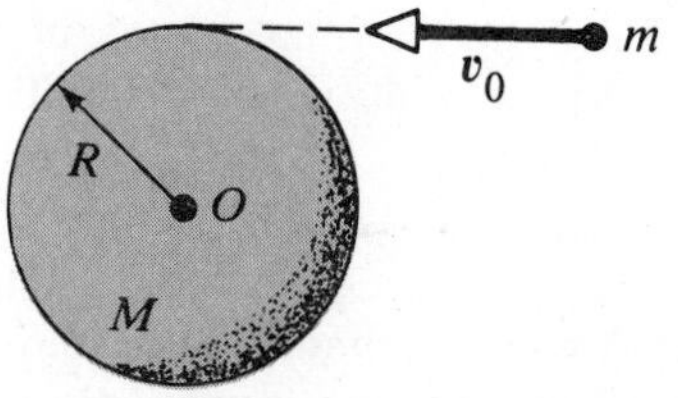

Figure 11.26 (Exercise 22).

23. The student in Fig. 11.12 holds two weights, each of mass 10 kg. When his arms are extended horizontally, the weights are 1 m from the axis of rotation and he rotates with an angular speed of 3 rad/s. The moment of inertia of the student plus the stool is 8 kg · m^2 and is assumed to be constant. If the student pulls the weights horizontally to 0.3 m on the rotation axis, calculate (a) the final angular speed of the system and (b) the change in the mechanical energy of the system.

24. A uniform rod of mass 100 g and length 50 cm rotates in a horizontal plane about a fixed, vertical, frictionless pin through its center. Two small beads, each of mass 30 g, are mounted on the rod such that they are able to slide without friction along its length. Initially the beads are held by catches at positions 10 cm on each side of center, at which time the system rotates at an angular speed of 20 rad/s. Suddenly, the catches are released and the small beads slide outward along the rod. (a) Find the angular speed of the system at the instant the beads reach the ends of the rod. (b) Find the angular speed of the rod after the beads fly off the ends.

25. A woman whose mass is 70 kg stands at the rim of a horizontal turntable having a moment of inertia of 500 kg · m^2 and a radius of 2 m. The system is initially at rest, and the turntable is free to rotate about a frictionless, vertical axle through its center. The woman then starts walking around the rim in a clockwise direction at a constant speed of 1.5 m/s relative to the earth. (a) In what direction and with what angular speed does the turntable rotate? (b) How much work does the woman do to set the system into motion?

26. A bullet of mass 10 g is shot *through* a door initially at rest. The moment of inertia of the door is 4 kg · m^2 about an axis through its hinges. The bullet is fired perpendicular to the door with an initial velocity of 400 m/s, and the angular speed of the door after the collision is 0.3 rad/s. If the bullet passes through the door 0.4 m from the hinge, find (a) the final speed of the bullet and (b) the loss in mechanical energy.

Section 11.6 Rolling Motion of a Rigid Body

27. A spherical shell rolls down a rough incline of height h and angle θ (Fig. 11.18). (a) If the shell is released from rest at the top of the incline, what is the velocity of its center of mass when it reaches the bottom? (b) Calculate the acceleration of its center of mass. Compare your results with those for a solid sphere determined in Example 11.11.

28. A uniform solid disk and a uniform hoop are placed side by side at the top of a rough incline of height h. If they are released from rest and roll without slipping, determine their velocities when they reach the bottom. Which object reaches the bottom first?

29. (a) Determine the acceleration of the center of mass of a uniform solid disk rolling down an incline and compare this acceleration with that of a uniform hoop. (b) What is the minimum coefficient of friction required to maintain pure rolling motion for the disk?

30. A solid sphere has a radius of 0.2 m and a mass of 150 kg. How much work is required to get the sphere rolling with an angular speed of 50 rad/s on a horizontal surface? (Assume the sphere starts from rest and rolls without slipping.)

31. A cylinder of mass 10 kg rolls without slipping on a rough surface. At the instant its center of mass has a speed of 10 m/s, determine (a) the translational kinetic energy of its center of mass, (b) the rotational kinetic energy about its center of mass, and (c) its total kinetic energy.

Section 11.7 Angular Momentum as a Fundamental Quantity

32. In the Bohr model of the hydrogen atom, the electron moves in a circular orbit of radius 0.529 × 10^{-10} m around the proton. Assuming the orbital angular momentum of the electron is equal to $\hbar$, calculate (a) the orbital speed of the electron, (b) the kinetic energy of the electron, and (c) the angular frequency of the electron's motion.

PROBLEMS

1. A particle of mass m is located at the vector position $\boldsymbol{r}$ and has a linear momentum $\boldsymbol{p}$. (a) If $\boldsymbol{r}$ and $\boldsymbol{p}$ both have nonzero x, y, and z components, show that the angular momentum of the particle relative to the origin has components given by $L_x = yp_z - zp_y$, $L_y = zp_x - xp_z$, and $L_z = xp_y - yp_x$. (b) If the particle moves only in the xy plane, prove that $L_x = L_y = 0$ and $L_z \neq 0$.

2. A force $\boldsymbol{F}$ acts on the particle described in Problem 1. (a) Find the components of the torque acting on the particle about the origin when the particle is located at the position $\boldsymbol{r}$ and the force has three components. (b) From this result, show that if the particle moves in the xy plane and the force has only x and y components, the torque (and angular momentum) must be in the z direction.

3. A mass m is attached to a cord passing through a small hole in a frictionless, horizontal surface (Fig. 11.27). The mass is initially orbiting in a circle of radius r_0 with velocity v_0. The cord is then slowly pulled from below, decreasing the radius of the circle to r. (a) What is the velocity of the mass when the radius is r? (b) Find the tension in the cord as a function of r. (c) How much work is done in moving m from r_0 to r? (Note: the tension depends on r!) (d) Obtain numerical values for v, T, and W when $r = 0.1$ m, if $m = 50$ g, $r_0 = 0.3$ m, and $v_0 = 1.5$ m/s.

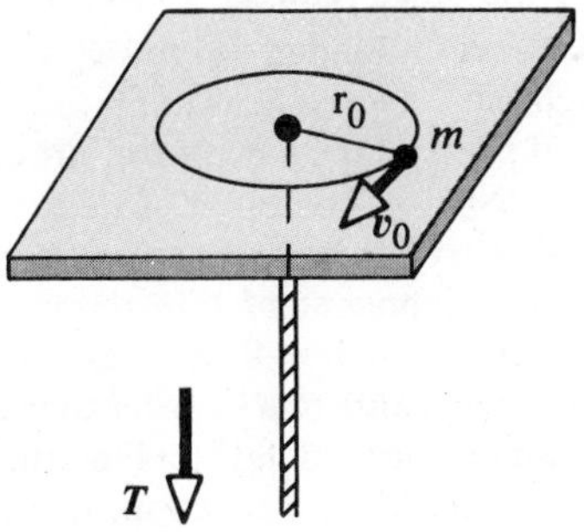

Figure 11.27 (Problem 3).

4. A large, cylindrical roll of tissue paper of initial radius R lies on a long, horizontal surface with the open end of the paper nailed to the surface so that it can unroll easily. The roll is given a *slight* shove ($v_0 \approx 0$) and commences to unroll. (a) Determine the speed of the center of mass of the roll when its radius has diminished to r. (b) Calculate a numerical value for this speed at $r = 1$ mm, assuming $R = 6$ m. (c) What happens to the energy of the system when the paper is completely unrolled? (Hint: Assume the roll has a uniform density and apply energy methods.)

5. A uniform solid sphere of radius r is placed on the inside surface of a hemispherical bowl of radius R. The sphere is released from rest at an angle θ to the vertical and rolls without slipping (Fig. 11.28). Determine the angular speed of the sphere when it reaches the bottom of the bowl.

6. A bowling ball is given an initial speed v_0 on an alley such that it *initially slides without rolling*. The coefficient of friction between the ball and the alley is μ. At the time *pure rolling motion occurs,* show that (a) the velocity of the ball's center of mass is $5v_0/7$

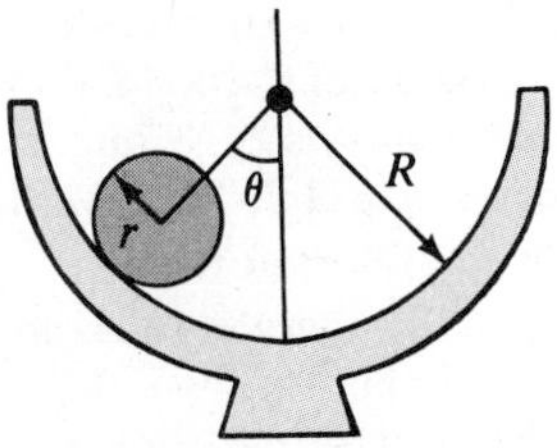

Figure 11.28 (Problem 5).

and (b) the distance it has traveled is $12v_0^2/49\,\mu g$. (Hint: When pure rolling motion occurs, $v_c = R\omega$ and $\alpha = a_c/R$. Since the frictional force provides the deceleration, from Newton's second law it follows that $a_c = -\mu g$.)

7. A constant horizontal force $\boldsymbol{F}$ is applied to a lawn roller in the form of a uniform solid cylinder of radius R and mass M (Fig. 11.29). If the roller rolls without slipping on the horizontal surface, show that (a) the acceleration of the center of mass is $2F/3M$ and (b) the minimum coefficient of friction necessary to prevent slipping is $F/3Mg$. (Hint: Take the torque with respect to the center of mass.)

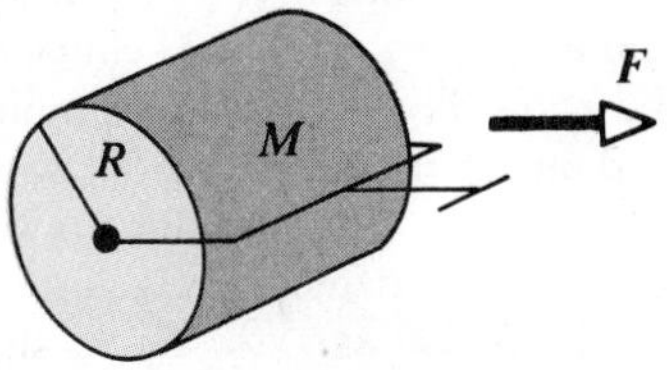

Figure 11.29 (Problem 7).

8. A yoyo is made with a string wound around a uniform disk of radius R and mass M. It is released from rest with the string vertical and its top end tied to a fixed support (Fig. 11.30). As the yoyo de-

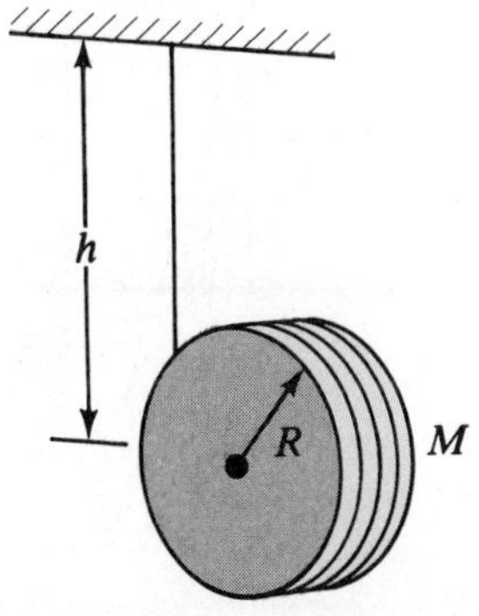

Figure 11.30 (Problem 8).

scends, show that (a) the tension in the string is one third the weight of the yoyo, (b) the acceleration of the center of mass is $2g/3$, and (c) the velocity of the center of mass is $(4gh/3)^{1/2}$. Verify your result to (c) using the energy approach.

9. A small, solid sphere of mass m and radius r rolls without slipping along the track shown in Fig. 11.31. If it starts from rest at the top of the track, (a) what is the minimum value of h (in terms of the radius of the loop R) such that the sphere completes the loop? (b) What are the force components on the sphere at the point P if $h = 3R$?

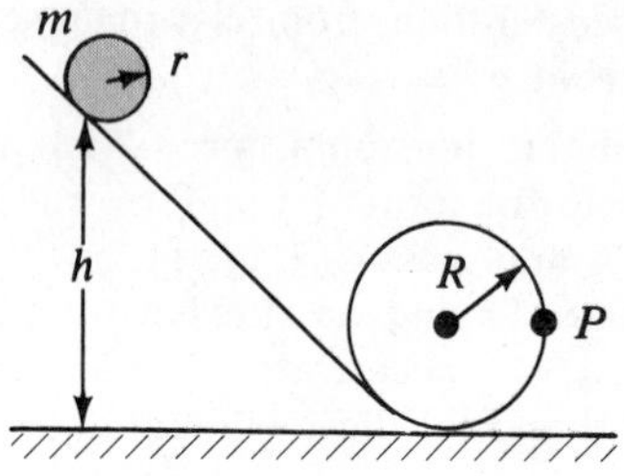

Figure 11.31 (Problem 9).

10. A light rope passes over a light, frictionless pulley. One end is fastened to a bunch of bananas of mass M, and a monkey of mass M clings to the other end of the rope (Fig. 11.32). The monkey climbs the rope in an attempt to reach the bananas. (a) Treating the system as consisting of the monkey, bananas, rope, and pulley, evaluate the net torque about the pulley axis. (b) Using the results to (a), determine the total angular momentum about the pulley axis and describe the motion of the system. Will the monkey reach the bananas?

Figure 11.32 (Problem 10).

11. A spool of wire of mass M and radius R is unwound under a constant force $\mathbf{F}$ (Fig. 11.33). Assuming the spool is a uniform solid cylinder that *doesn't slip*, show that (a) the acceleration of the center of mass is $4\mathbf{F}/3M$ and (b) the force of friction is to the *right* and equal to $\mathbf{F}/3$.

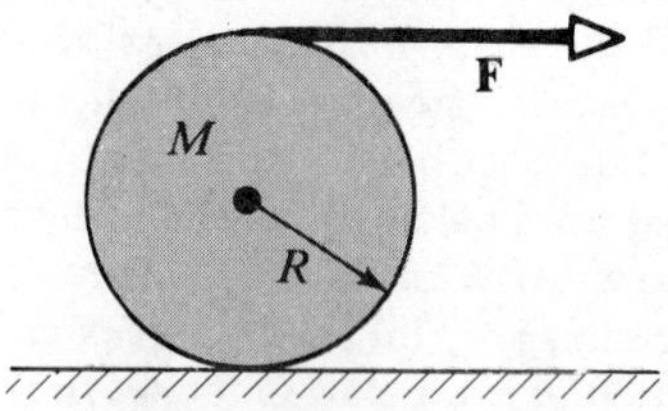

Figure 11.33 (Problem 11).

12. If the cylinder in Fig. 11.33 starts from rest and rolls without slipping, what is the velocity of its center of mass after it has rolled through a distance d? (Assume the force remains constant.)

13. A trailer with loaded weight $\mathbf{W}$ is being pulled by a vehicle with a force $\mathbf{F}$, as in Fig. 11.34. The trailer is loaded such that its center of gravity is located as shown. Neglect the force of rolling friction and assume the trailer has an acceleration a. (a) Find the vertical component of $\mathbf{F}$ in terms of the given parameters. (b) If $a = 2$ m/s^2 and $h = 1.5$ m, what must be the value of d in order that $F_y = 0$ (no vertical load on the vehicle)? (c) Find the value of F_x and F_y if $W = 1500$ N, $d = 0.8$ m, $L = 3$ m, $h = 1.5$ m, and $a = -2$ m/s^2.

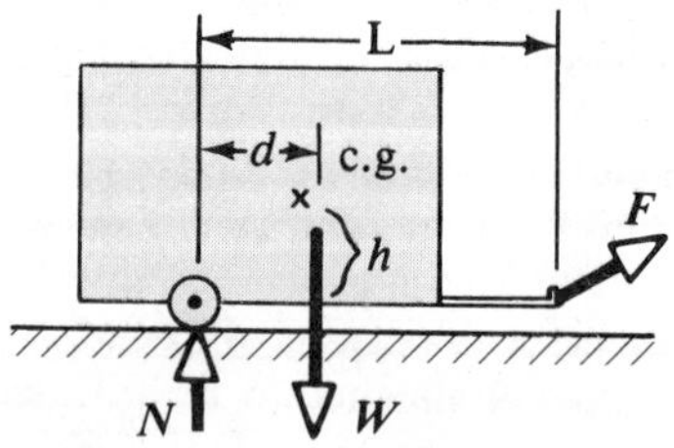

Figure 11.34 (Problem 13).

14. Consider the problem of the solid sphere rolling down an incline as described in Example 11.11. (a) Choose the axis of the origin for the torque equation as the instantaneous axis through the contact point P and show that the acceleration of the center of mass is given by $a_c = \frac{5}{7}g\sin\theta$. (b) Show that the *minimum* coefficient of friction such that the sphere will roll without slipping is given by $f_{min} = \frac{2}{7}\tan\theta$.

15. A uniform solid disk is set into rotation about an axis through its center with an angular velocity ω_0. The rotating disk is lowered to a *rough*, horizontal surface with this angular velocity and released. (a) What is the angular velocity of the disk once pure rolling takes place? (b) Find the fractional loss in kinetic energy from the time the disk is released until pure rolling occurs. (Hint: Angular momentum is conserved about an axis through the point of contact.)

16. Suppose a solid disk of radius R is given an angular velocity ω_0 about an axis through its center and is then lowered to a rough, horizontal surface and released, as in Problem 15. Furthermore, assume that the coefficient of friction between the disk and surface is μ. (a) Show that the *time* it takes pure rolling motion to occur is given by $R\omega_0/3\,\mu g$. (b) Show that the *distance* the disk travels before pure rolling occurs is given by $R^2\omega_0{}^2/18\,\mu g$. (See hint in Problem 6.)

12 Static Equilibrium of a Rigid Body

Carleton Read, Delaware Port Authority

This chapter is concerned with the conditions under which a rigid body is in equilibrium. The term *equilibrium* implies that the body is either at rest or moving with constant velocity. We shall deal with bodies at rest, or bodies in *static equilibrium*. This represents a common situation in engineering practice, and the principles involved are of special interest to civil engineers, architects, and mechanical engineers, who deal with various structural designs, such as bridges and buildings. Those of you who are engineering students will undoubtedly take an intensified course in statics in the future.

In Chapter 5 we stated that one necessary condition for equilibrium is that the net force on an object must be zero. If the object is treated as a single particle, this is the *only* condition that must be satisfied in order that the particle be in an equilibrium state. That is, if the net force on the particle is zero, it will remain at rest (if originally at rest) or move with constant velocity in a straight line (if originally in motion).

The situation with rigid bodies is somewhat more complex because real bodies cannot be treated as particles. A real body has a definite size, shape, and mass distribution. In order that such a body be in equilibrium, the net force on it must be zero *and* the body must have no tendency to rotate. This second condition of equilibrium requires that *the net torque about any origin must be zero.* In order to establish whether or not a body is in equilibrium, we must know the size and shape of the body, the forces acting on different parts of the body, and the points of application of the various forces.

Rigid body

The bodies that will be treated in this chapter are assumed to be rigid. *A rigid body is defined as a body that does not deform under the application of external forces.* That is, all parts of a rigid body remain at a fixed separation with respect to each other when subjected to external forces. In reality, all bodies will deform to some extent under load conditions. Such deformations are usually small and will not affect the conditions of equilibrium. However, deformation is an important consideration in understanding the mechanics of materials, as we will see in Chapter 15.

12.1 THE CONDITIONS OF EQUILIBRIUM OF A RIGID BODY

Consider a single force $\boldsymbol{F}$ acting on a rigid body that is pivoted about an axis through the point O as in Fig. 12.1. The effect of the force on the body depends on its point of application, P. If $\boldsymbol{r}$ is the position vector of this point relative to O, *the torque associated with the force* **F** *about* O *is given by*

$$\boldsymbol{\tau} = \boldsymbol{r} \times \boldsymbol{F} \tag{12.1}$$

Recall that the vector $\boldsymbol{\tau}$ is perpendicular to the plane formed by $\boldsymbol{r}$ and $\boldsymbol{F}$. Furthermore, the sense of $\boldsymbol{\tau}$ is determined by the sense of the rotation that $\boldsymbol{F}$ tends to give to the body. The right-hand rule can be used to determine the direction of $\boldsymbol{\tau}$: Close your hand such that your four fingers wrap in the direction of rotation that $\boldsymbol{F}$ tends to give the body; your thumb will point in the direction of $\boldsymbol{\tau}$. Hence, in Fig. 12.1 $\boldsymbol{\tau}$ is directed *out* of the paper.

As you can see from Fig. 12.1, the tendency of $\boldsymbol{F}$ to make the body rotate about an axis through O depends on the moment arm d (the perpendicular distance to the line of action of the force) as well as on the magnitude of $\boldsymbol{F}$. By definition, the magnitude of $\boldsymbol{\tau}$ is given by Fd.

Now suppose two forces, $\boldsymbol{F}_1$ and $\boldsymbol{F}_2$, act on a rigid body. The two forces will have the same effect on the body only if they have the same magnitude, the same direction, and the same line of action. In other words, *two forces* $\mathbf{F}_1$ *and* $\mathbf{F}_2$ *are equivalent if and only if* $\mathbf{F}_1 = \mathbf{F}_2$ *and if they have the same torque about any given point.* An example of two equal and opposite forces that are *not* equivalent is shown in Fig. 12.2. The force directed toward the right tends to rotate the body clockwise, whereas the force directed toward the left tends to rotate it counterclockwise.

Equivalent forces

If a rigid body is pivoted about an axis through its center of mass, the body can undergo a rotation about this axis if there is a nonzero torque acting on the body. As an example, suppose a rigid body is pivoted about an axis through its center of mass as in Fig. 12.3. Two equal and opposite forces act in the directions shown, such that their lines of action do not pass through the center of mass. Such a pair of forces acting in this manner form what is called a *couple*. Since each force produces the same torque, Fd, about the center of mass, the net torque has a magnitude given by $2Fd$. Clearly, the body will rotate in a clockwise direction and will undergo an angular acceleration about the axis. This is a nonequilibrium situation as far as the rotational motion is concerned. That is, the "unbalanced," or net, torque on the body gives rise to an angular acceleration α according to the relationship $\tau_{\text{net}} = 2Fd = I\alpha$.

Couple

In general, a rigid body will be in rotational equilibrium only if its angular acceleration $\alpha = 0$. Since $\tau_{\text{net}} = I\alpha$ for rotation about a fixed axis, a necessary condition of equilibrium for a rigid body is that *the net torque about any origin must be zero.* We now have *two necessary conditions for equilibrium of a rigid body,* which can be

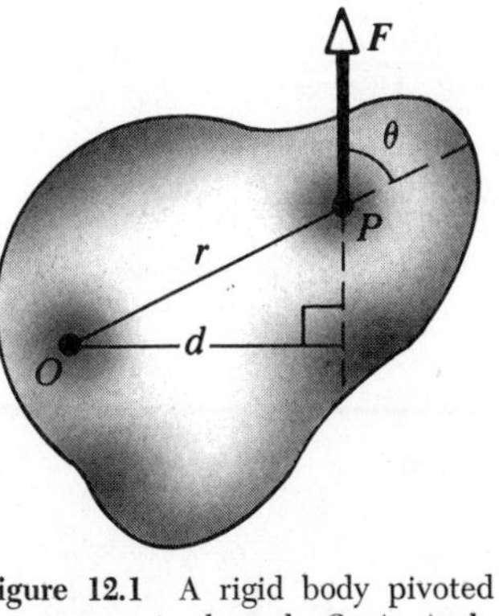

Figure 12.1 A rigid body pivoted about an axis through O. A single force $\boldsymbol{F}$ acts at the point P. The moment arm of $\boldsymbol{F}$ is the perpendicular distance d to the line of action of $\boldsymbol{F}$.

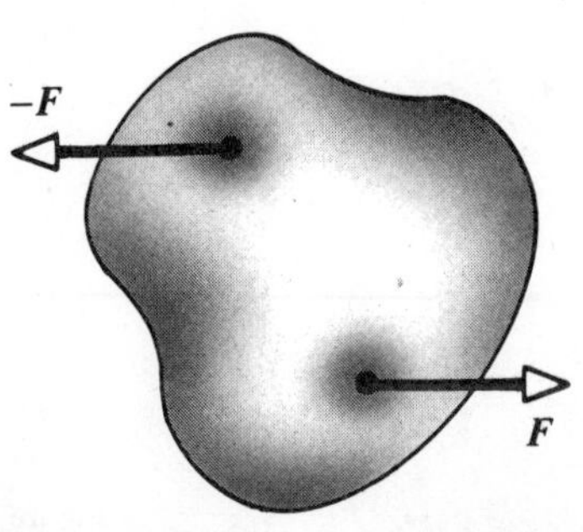

Figure 12.2 The two forces acting on the body are equal in magnitude and opposite in direction, yet the body is not in equilibrium.

Conditions for equilibrium

stated as follows:

1. The resultant external force must equal zero. $$\Sigma F = 0 \qquad (12.2)$$
2. The resultant external torque must be zero about *any* origin. $$\Sigma\tau = 0 \qquad (12.3)$$

The first condition is a statement of *translational equilibrium,* that is, the linear acceleration of the center of mass of the body must be zero when viewed from an inertial reference frame. The second condition is a statement of *rotational equilibrium,* that is, the angular acceleration about any axis must be zero. In the special case of *static equilibrium,* which is the main subject of this chapter, the body is at rest so that it has no linear or angular velocity (that is, $v_c = 0$ and $\omega = 0$).

The two vector expressions given by Eqs. 12.2 and 12.3 are equivalent, in general, to six scalar equations. Three of these come from the first condition of equilibrium, and three follow from the second condition (corresponding to x, y, and z components). Hence, in a complex system involving several forces acting in various directions, you would be faced with solving a set of linear equations with many unknowns. We will restrict our discussion to situations in which all the forces lie in a common plane. Such forces are said to be *coplanar.* We will assume that these forces lie in the xy plane. In this case, we shall have to deal with only *three* scalar equations. Two of these come from balancing the forces in the x and y directions. The third comes from the torque equation, namely, that the net torque about *any* direction perpendicular to the plane must be zero. Hence, these two conditions of equilibrium provide the equations

$$\Sigma F_x = 0 \qquad \Sigma F_y = 0 \qquad \Sigma\tau_z = 0 \qquad (12.4)$$

where the origin of the torque equation is *arbitrary,* as we shall show later.

There are two cases of equilibrium that are often encountered. The first case deals with a rigid body subjected to only two forces, and the second case is concerned with a rigid body subjected to three forces.

Case I. *If a rigid body is subjected to two forces, the body is in equilibrium if and only if the two forces are equal in magnitude and opposite in direction and have the same line of action.* Figure 12.4a shows a situation in which the body is not in equilibrium because the two forces are not along the same line. Note that the torque about any axis, such as one through P, is not zero, which violates the second condition of equilibrium. In Figure 12.4b, the body is in equilibrium because the forces have the same line of action. In this situation, it is easy to see that the net torque about any axis is zero.

Case II. *If a rigid body subjected to three forces is in equilibrium, the lines of action of the three forces must intersect at a common point.* That is, the forces must be

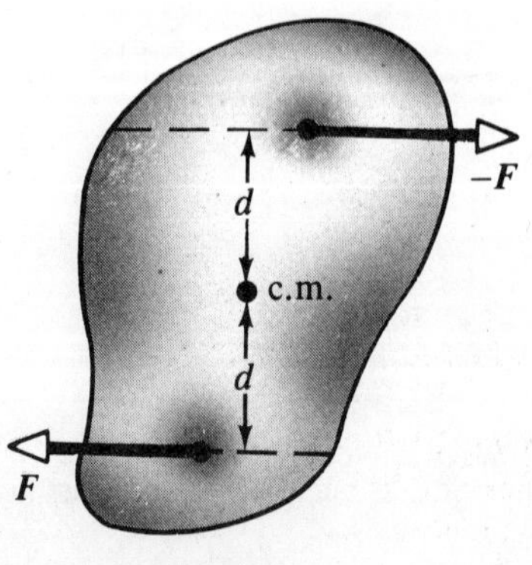

Figure 12.3 Two equal and opposite forces acting on the body form a couple. In this case, the body will rotate clockwise. The net torque on the body about the center of mass is $2Fd$.

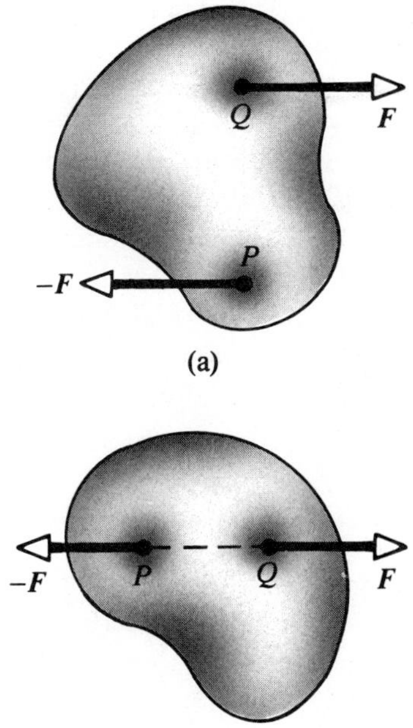

Figure 12.4 (a) The body is not in equilibrium since the two forces do not have the same line of action. (b) The body is in equilibrium since the two forces act along the same line.

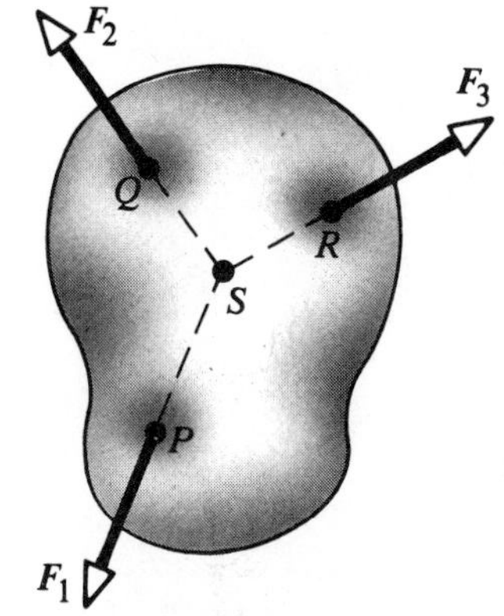

Figure 12.5 If three forces act on a body that is in equilibrium, their lines of action must intersect at a common point S.

concurrent. (One exception to this rule is the situation in which none of the lines of action intersect. In this situation, the forces must be parallel.) Figure 12.5 illustrates this point. The lines of action of the three forces pass through the point S. The conditions of equilibrium require that $\boldsymbol{F}_1 + \boldsymbol{F}_2 + \boldsymbol{F}_3 = 0$ and that the net torque about any axis be zero. Note that as long as the forces are concurrent, the net torque about an axis through S must be zero.

We can easily show that if a body is in translational equilibrium and the net torque is zero with respect to one point, it must be zero about *any* point. Consider a body under the action of several forces such that the resultant force $\Sigma \boldsymbol{F} = \boldsymbol{F}_1 + \boldsymbol{F}_2 + \boldsymbol{F}_3 + \cdots = 0$. Figure 12.6 describes this situation (for clarity, only four forces are shown). The point of application of $\boldsymbol{F}_1$ is specified by the position vector $\boldsymbol{r}_1$. Similarly, the points of application of $\boldsymbol{F}_2, \boldsymbol{F}_3, \ldots$ are specified by $\boldsymbol{r}_2, \boldsymbol{r}_3, \ldots$ (not shown). The net torque about O is

$$\Sigma \tau_O = \boldsymbol{r}_1 \times \boldsymbol{F}_1 + \boldsymbol{r}_2 \times \boldsymbol{F}_2 + \boldsymbol{r}_3 \times \boldsymbol{F}_3 + \cdots$$

Now consider another arbitrary point, O', having a position vector relative to O of $\boldsymbol{r}'$. The point of application of $\boldsymbol{F}_1$ relative to this point is identified by the vector $\boldsymbol{r}_1 - \boldsymbol{r}'$. Likewise, the point of application of $\boldsymbol{F}_2$ relative to O' is $\boldsymbol{r}_2 - \boldsymbol{r}'$, and so forth. Therefore, the torque about O' is

$$\Sigma \tau_{O'} = (\boldsymbol{r}_1 - \boldsymbol{r}') \times \boldsymbol{F}_1 + (\boldsymbol{r}_2 - \boldsymbol{r}') \times \boldsymbol{F}_2 + (\boldsymbol{r}_3 - \boldsymbol{r}') \times \boldsymbol{F}_3 + \cdots$$

$$\Sigma \tau_{O'} = \boldsymbol{r}_1 \times \boldsymbol{F}_1 + \boldsymbol{r}_2 \times \boldsymbol{F}_2 + \boldsymbol{r}_3 \times \boldsymbol{F}_3 + \cdots - \boldsymbol{r}' \times (\boldsymbol{F}_1 + \boldsymbol{F}_2 + \boldsymbol{F}_3 + \cdots)$$

Since the net force is assumed to be zero, the last term in this last expression vanishes and we see that $\Sigma \tau_{O'} = \Sigma \tau_O$. Hence, *if a body is in translational equilibrium and the net torque is zero about one point, it must be zero about any other point.*

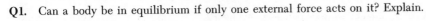

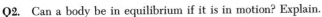

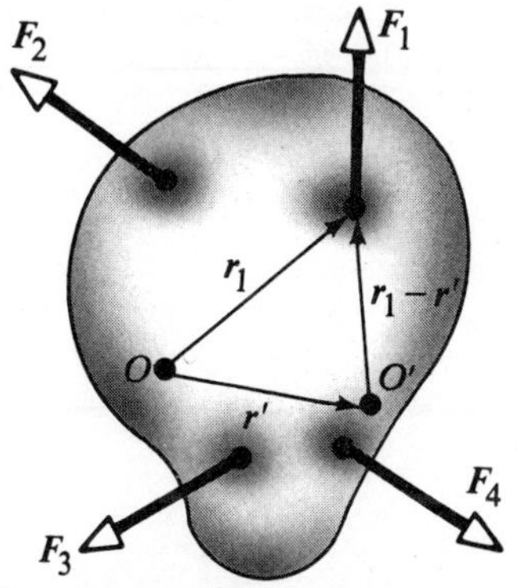

Figure 12.6 Construction for showing that if the net torque about origin O is zero, the net torque about any other origin, such as O', must be zero.

Q1. Can a body be in equilibrium if only one external force acts on it? Explain.

Q2. Can a body be in equilibrium if it is in motion? Explain.

12.2 THE CENTER OF GRAVITY

Whenever we deal with rigid bodies, one of the forces that must be considered is the weight of the body, that is, the force of gravity acting on the body. In order to compute the torque due to the weight force, all of the weight can be considered as being concentrated at a single point called the *center of gravity*. As we shall see, the center of gravity of a body coincides with its center of mass if the body is in a uniform gravitational field.

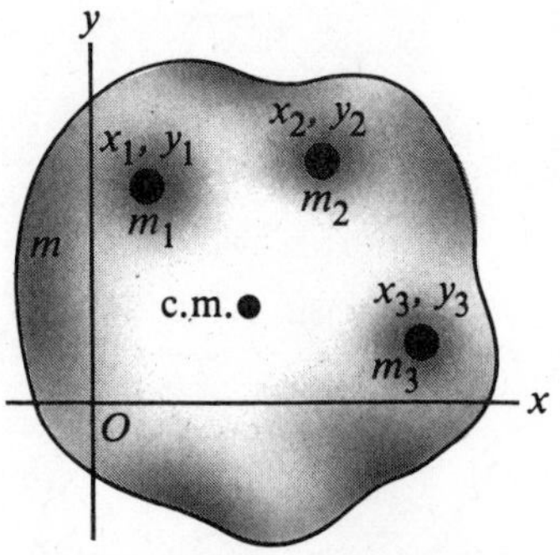

Figure 12.7 A rigid body can be divided into many small particles with specific masses and coordinates. These can be used to locate the center of mass.

Consider a body of arbitrary shape lying in the xy plane, as in Fig. 12.7. Suppose the body is divided into a large number of very small particles of masses m_1, m_2, m_3, . . . having coordinates (x_1,y_1), (x_2,y_2), (x_3,y_3), ... In Chapter 9 we defined the x coordinate of the center of mass of such an object to be

$$x_c = \frac{m_1x_1 + m_2x_2 + m_3x_3 + \cdots}{m_1 + m_2 + m_3 + \cdots} = \frac{\Sigma m_ix_i}{\Sigma m_i}$$

The y coordinate of the center of mass is similar to this, with x_i replaced by y_i.

Let us now examine the situation from another point of view by considering the weight of each part of the body, as in Fig. 12.8. Each particle contributes a torque about the origin equal to its weight multiplied by its moment arm. For example, the torque due to the weight m_1g_1 is $m_1g_1x_1$, and so forth. We wish to locate the one position of the single force **W** (the total weight of the body) whose effect on the rotation of the body is the same as that of the individual particles. This point is called the *center of gravity* of the body. Equating the torque exerted by **W** acting at the center of gravity to the sum of the torques acting on the individual particles gives

$$(m_1g_1 + m_2g_2 + m_3g_3 + \cdots)x_{\text{c.g.}} = m_1g_1x_1 + m_2g_2x_2 + m_3g_3x_3 + \cdots$$

where this expression accounts for the fact that the acceleration of gravity can in general vary over the body. If we assume that g is uniform over the body (as is usually the case), then the g terms in the above expression cancel and we get

$$x_{\text{c.g.}} = \frac{m_1x_1 + m_2x_2 + m_3x_3 + \cdots}{m_1 + m_2 + m_3 + \cdots}$$

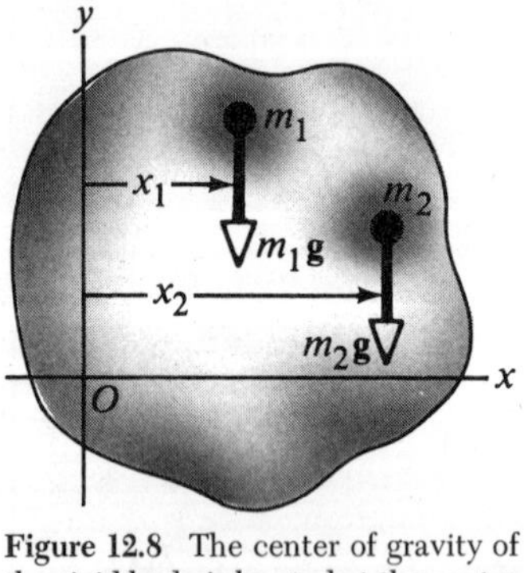

Figure 12.8 The center of gravity of the rigid body is located at the center of mass if the value of g is constant over the body.

In other words, *the center of gravity is located at the center of mass as long as the body is assumed to be in a uniform gravitational field.*

In several examples that will be presented in the next section, we shall be concerned with homogeneous, symmetric bodies for which the center of gravity coincides with the geometric center of the body. In these cases, we shall assume that the center of gravity coincides with the center of mass. Note that a rigid body in a uniform gravitational field can be balanced by a single force equal in magnitude to the weight of the body, as long as the force is directed upward through the center of gravity.

Q3. Locate the center of gravity for the following uniform objects: (a) a sphere, (b) a cube, (c) a cylinder.

Q4. The center of gravity of an object may be located outside the object. Give a few examples for which this is the case.

Q5. You are given an arbitrarily shaped piece of plywood, together with a hammer, nail, and plumb bob. How could you use these items to locate the center of gravity of the plywood? (Hint: Use the nail to suspend the plywood.)

Q6. In order for a chair to be balanced on one leg, where must the center of gravity of the chair be located?

12.3 EXAMPLES OF RIGID BODIES IN STATIC EQUILIBRIUM

In this section we present several examples of rigid bodies in static equilibrium. In working such problems, it is important to first recognize *all* external forces acting on the body being considered. Failure to do so will result in an incorrect analysis. The following procedure is recommended when analyzing a body in equilibrium under the action of several external forces:

Procedure for analyzing a body in equilibrium

1. Make a sketch of the body under consideration.
2. Draw a free-body diagram labeling *all external forces* acting on the body. Try to guess the correct directions for your forces. If you guess wrong, this will show up as a negative sign in your solution for this force. Don't be alarmed. This only indicates that the correct direction of the force is the opposite of what you assumed.
3. Resolve all forces into rectangular components, choosing a convenient frame of reference. Then apply the first condition of equilibrium, which involves balancing forces. Remember to keep track of the *signs* of the various force components.
4. Choose a convenient origin for calculating the net torque on the rigid body. Remember that the choice of the origin for the torque equation is *arbitrary;* therefore choose an origin that will simplify your calculation as much as possible. Becoming adept at this is a matter of practice.
5. The first and second conditions of equilibrium give a set of linear equations with several unknowns. All that is left is to solve them simultaneously for the unknowns in terms of the known quantities.

Example 12.1

The beam balance. A uniform beam of mass M supports two masses m_1 and m_2, as shown in Fig. 12.9. If the knife-edge of the support is under the beam's center of gravity and m_1 is at a distance d from the center, determine the position of m_2 such that the system is balanced.

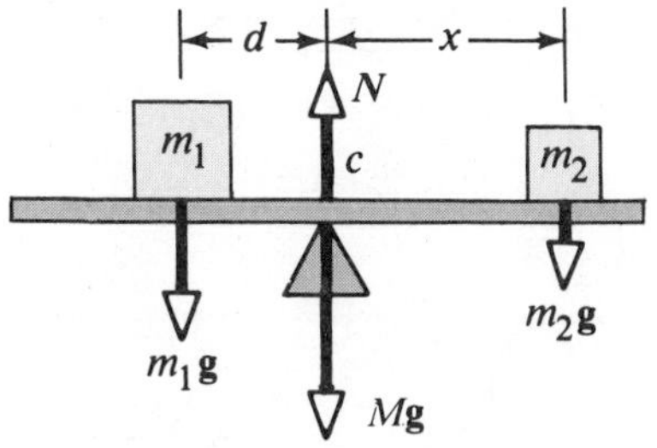

Figure 12.9 (Example 12.1) The beam balance.

Solution: First note that the *external forces acting on the beam* are the weights m_1g, m_2g, and Mg, acting downward, and the force of the knife-edge on the beam, N, acting upward. Obviously, since the forces in the vertical direction must balance, we see that $N = m_1g + m_2g + Mg$. However, this does not tell us the position of m_2. To find this, we must invoke the condition of rotational equilibrium. Taking the center of gravity as the origin of our torque equation, we see that

$$\Sigma\tau_c = m_1gd - m_2gx = 0$$

$$x = \frac{m_1}{m_2}d$$

The problem could also be solved if the knife-edge did not lie under the center of gravity of the beam. What other information would you need in this case?

Example 12.2 *A Problem in Biomechanics*

This is an example from a subject known as *biomechanics*. A weight W is held in the hand with the forearm in the horizontal position, as in Fig. 12.10a. The biceps muscle is located a distance d from the joint, and the weight is a distance l from the joint. Find the upward force that the biceps exerts on the forearm (the ulna) and the downward force of the upper arm (the humerus) acting at the joint. Neglect the weight of the forearm.

Solution: The forces acting on the forearm are equivalent to those acting on a bar of length l, as shown in Fig. 12.10b, where $\mathbf{F}$ is the upward force of the biceps and $\mathbf{R}$ is the downward force at the joint. From the first condition of equilibrium we have

$$F = R + W$$

The sum of the torques about any point must be zero. Taking the joint O as the origin gives

$$Fd - Wl = 0$$

$$F = Wl/d$$

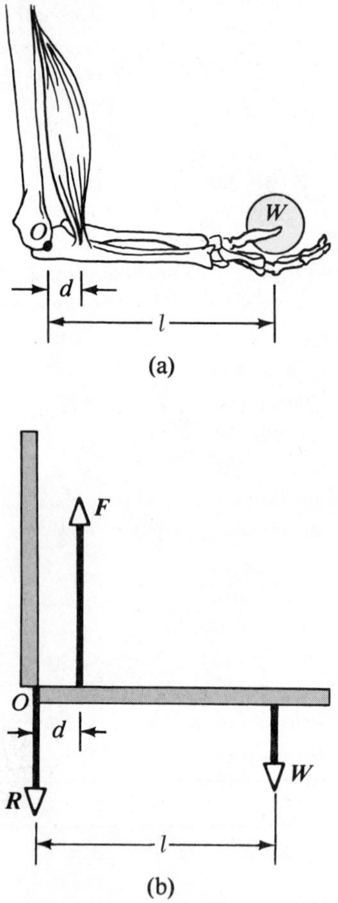

Figure 12.10 (Example 12.2) (a) Forces operative at the elbow joint. (b) The mechanical model for the system described in (a).

Taking $W = 200$ N and using typical values of $d = 5$ cm and $l = 35$ cm, we find that

$$F = \frac{Wl}{d} = (200 \text{ N})\left(\frac{35 \text{ cm}}{5 \text{ cm}}\right) = 1400 \text{ N}$$

$$R = F - W = 1400 \text{ N} - 200 \text{ N} = 1200 \text{ N}$$

These values correspond to $F = 315$ lb and $R = 270$ lb. Hence, the forces at joints and in muscles can be extremely large.

In reality, the biceps makes an angle of about 15° with the vertical, so that $\boldsymbol{F}$ has both a vertical and a horizontal component. If this is included in the model, the results for F and R would change only by about 3%, but there would be an additional horizontal component to the reaction force at the joint of about 375 N. You should analyze this situation in more detail.

Example 12.3

A uniform, horizontal beam of length 8 m and weighing 200 N is pivoted at the wall, with its far end supported by a cable that makes an angle of 53° with the horizontal (Fig. 12.11). If a person weighing 600 N stands 2 m from the wall, find the tension in the cable and the reaction force at the pivot.

Solution: First we must recognize all the external forces acting on the beam. These are the weight of the beam, the force of tension $\boldsymbol{T}$ due to the cable, the reaction force $\boldsymbol{R}$ at the pivot (the direction θ of this force is unknown), and the force of the person on the beam. These are all indicated in the free-body diagram *for the beam* (Fig. 12.11b). Resolving $\boldsymbol{T}$ and $\boldsymbol{R}$ into horizontal and vertical components and applying the first condition of equilibrium, we get

(1) $\quad \Sigma F_x = R\cos\theta - T\cos 53° = 0$

(2) $\quad \Sigma F_y = R\sin\theta + T\sin 53° - 600 - 200 = 0$

Since R, T, and θ are all unknown, we cannot obtain a solution from these expressions alone. (The number of independent equations must equal the number of unknowns.)

Now let us invoke the condition of rotational equilibrium. A convenient origin to choose for our torque equation is the pivot at O, since the reaction force $\boldsymbol{R}$ and the horizontal component of $\boldsymbol{T}$ both have zero moment arms, and hence zero torques, about this pivot. Recalling our convention for the signs of the torques about an origin and noting that the moment arms of the 600-N, 200-N, and $T\sin 53°$ forces are 2 m, 4 m, and 8 m, respectively, we get

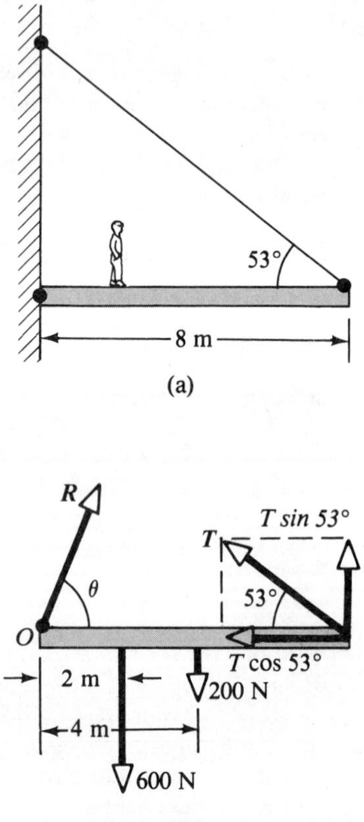

Figure 12.11 (Example 12.3) (a) A uniform beam supported by a cable. (b) The free body diagram for the beam.

(3) $\Sigma\tau_O = (T \sin 53°)8 - 600(2) - 200(4) = 0$

$$T = \frac{2000 \text{ N}}{8 \sin 53°} = 313 \text{ N}$$

Thus, the torque equation with this choice of origin gives us one of the unknowns directly! Substituting this value into (1) and (3) gives

$$R \cos\theta = 188 \text{ N}$$
$$R \sin\theta = 550 \text{ N}$$

Dividing these two equations gives

$$\tan\theta = \frac{550}{188} = 2.93$$
$$\theta = 71.1°$$

Finally,

$$R = \frac{188}{\cos\theta} = \frac{188 \text{ N}}{\cos 71.1°} = 581 \text{ N}$$

If we had selected some other origin for the torque equation, the solution would have been the same. For example, if the origin were chosen at the center of gravity of the beam, the torque equation would involve both T and R. However, this equation, coupled with (1) and (2) could still be solved for the unknown. Try it!

When many forces are involved in a problem of this nature, it is convenient to "keep the books straight" by setting up a table of forces, their moment arms, and their torques. For instance, in the example just given we would construct the following table. Setting the sum of the terms in the last column equal to zero represents the rotational equilibrium condition.

Force Component	*Moment Arm Relative to O (m)*	*Torque About O (N·m)*
$T \sin 37°$	8	$+8T \sin 37°$
$T \cos 37°$	0	0
200 N	4	−4(200)
600 N	2	−2(600)
$R \sin\theta$	0	0
$R \cos\theta$	0	0

Example 12.4 *The Leaning Ladder*

A uniform ladder of length l and weight $W = 50$ N rests against a smooth, vertical wall (Fig. 12.12a). If the coefficient of static friction between the ladder and ground is $\mu = 0.40$, find the *minimum* angle θ_{min} such that the ladder will *not* slip.

Solution: The free-body diagram showing all the external forces acting on the ladder is illustrated in Fig. 12.12b. Note that the reaction force at the ground, $\mathbf{R}$, is the vector sum of a normal force, $\mathbf{N}$, and the force of friction, $\mathbf{f}$. The reaction force at the wall, $\mathbf{P}$, is horizontal, since the wall is smooth. Applying the first condition of equilibrium to the ladder gives

$$\Sigma F_x = f - P = 0$$
$$\Sigma F_y = N - W = 0$$

Since $W = 50$ N, we see from the equation above that $N = W = 50$ N. Furthermore, *when the ladder is on the verge of slipping, the force of friction must be a maximum,* given by $f_{max} = \mu N = 0.40(50) = 20$ N. (Recall that $f_s \le \mu N$.) Thus, at this angle, $P = 20$ N.

To find the value of θ, we must use the second condition of equilibrium. Taking torques about the origin O at the bottom of the ladder, we get

$$\Sigma\tau_O = Pl \sin\theta - W\frac{l}{2}\cos\theta = 0$$

But $P = 20$ N when the ladder is about to slip and $W = 50$ N, so that the expression above gives

$$\tan\theta_{min} = \frac{W}{2P} = \frac{50}{40} = 1.2$$
$$\theta_{min} = 50.2°$$

It is interesting to note that the result does not depend on l!

An alternative approach to analyzing this problem is to consider the intersection O' of the forces $\mathbf{W}$ and $\mathbf{P}$. Since the torque about any origin must be zero, the torque about O' must be zero. This requires that the line of action of $\mathbf{R}$ (the resultant of $\mathbf{N}$ and $\mathbf{f}$) pass through O! That is, since this is a three-force body, the forces must be concurrent. With this condition, one could then obtain the angle ϕ that $\mathbf{R}$ makes with the horizontal (where ϕ is greater than θ), assuming the length of the ladder is known. You should show that in this situation $\tan\phi = 2 \tan\theta$.

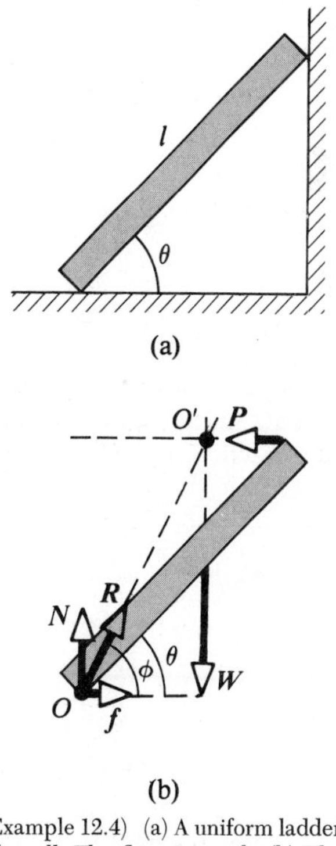

Figure 12.12 (Example 12.4) (a) A uniform ladder at rest, leaning against a smooth wall. The floor is rough. (b) The free-body diagram for the ladder. Note that the forces $\mathbf{R}$, $\mathbf{W}$, and $\mathbf{P}$ pass through a common point O'.

Example 12.5

A cylinder of weight **W** and radius R is to be raised onto a step of height h as shown in Fig. 12.13. A rope is wrapped around the cylinder and pulled horizontally. Assuming the cylinder doesn't slip on the step, find the *minimum* force **F** necessary to raise the cylinder and the reaction force at P.

Solution: When the cylinder is just ready to be raised, the reaction force at Q goes to zero. Hence, at this time there are only three forces on the cylinder, as shown in Fig. 12.13b. From the dotted triangle drawn in Fig. 12.13a, we see that the moment arm d of the weight relative to the point P is given by

$$d = \sqrt{R^2 - (R-h)^2} = \sqrt{2Rh - h^2}$$

The moment arm of **F** relative to P is $2R - h$. Therefore, the net torque acting on the cylinder about P is

$$Wd - F(2R-h) = 0$$
$$W\sqrt{2Rh - h^2} - F(2R-h) = 0$$
$$F = \frac{W\sqrt{2Rh - h^2}}{2R-h}$$

Hence, the second condition of equilibrium was sufficient to obtain the magnitude of **F**. We can determine the components of **N** by using the first condition of equilibrium:

$$\Sigma F_x = F - N\cos\theta = 0$$
$$\Sigma F_y = N\sin\theta - W = 0$$

Dividing gives

$$(1) \qquad \tan\theta = \frac{W}{F}$$

and solving for N gives

$$(2) \qquad N = \sqrt{W^2 + F^2}$$

For example, if we take $W = 500$ N, $h = 0.3$ m, and $R = 0.8$ m, we find that $F = 385$ N, $\theta = 52.4°$, and $N = 631$ N.

The problem can also be solved by noting that since only three forces act on the body, they must be concurrent. Hence, the reaction force **N** must pass through C, which is the point where **F** and **W** intersect. The three forces form the sides of the triangle shown in Fig. 12.13c. Expressions (1) and (2) are obtained directly from this triangle.

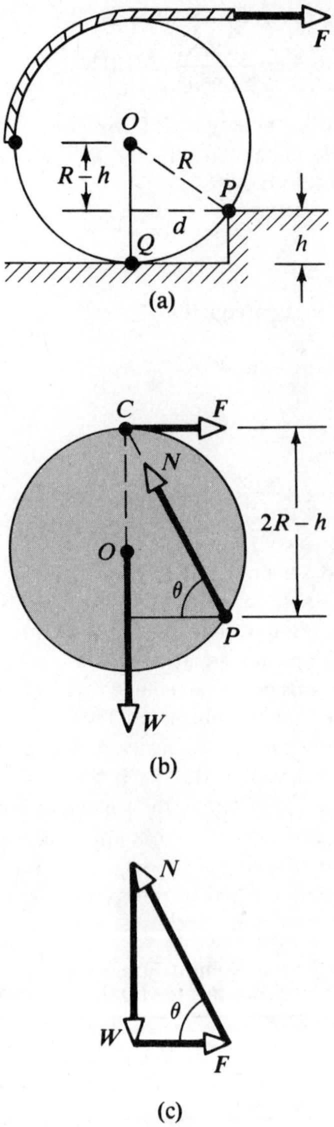

Figure 12.13 (Example 12.5) (a) A cylinder of weight **W** being pulled by a force **F** over a step. (b) The free-body diagram for the cylinder when it is just ready to be raised. (c) The *vector* sum of the three external forces is zero.

12.4 SUMMARY

A rigid body is in *equilibrium* if and only if the following conditions are satisfied: (1) *the resultant external force must be zero* and (2) *the resultant external torque must be zero about* any *origin*. That is,

Conditions for equilibrium

$$\Sigma \mathbf{F} = 0 \qquad (12.2)$$
$$\Sigma \boldsymbol{\tau} = 0 \qquad (12.3)$$

The first condition is the *condition of translational equilibrium,* and the second is the *condition of rotational equilibrium.*

If two forces act on a rigid body, the body is in equilibrium if and only if the forces are equal in magnitude and opposite in direction and have the same line of action.

When three forces act on a rigid body that is in equilibrium, the three forces must be concurrent, that is, their lines of action must intersect at a common point.

The *center of gravity* of a rigid body coincides with the center of mass if the body is in a uniform gravitational field.

EXERCISES

Section 12.1 The Conditions of Equilibrium of a Rigid Body

1. Write the necessary conditions of equilibrium for the body shown in Fig. 12.14. Take the origin of the torque equation at the point O.

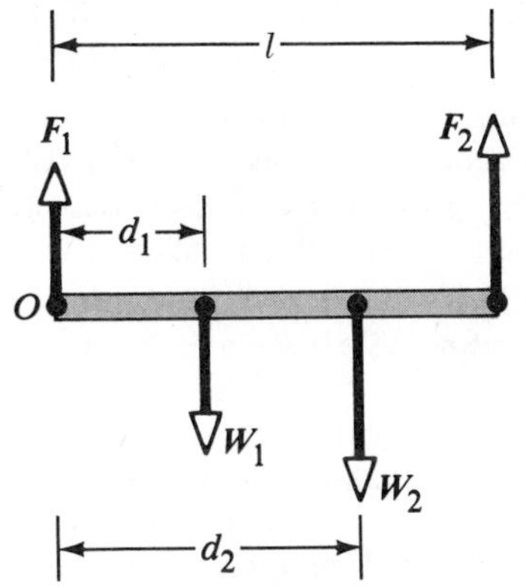

Figure 12.14 (Exercise 1).

2. Write the necessary conditions of equilibrium for the body shown in Fig. 12.15. Take the origin of the torque equation at the point O.

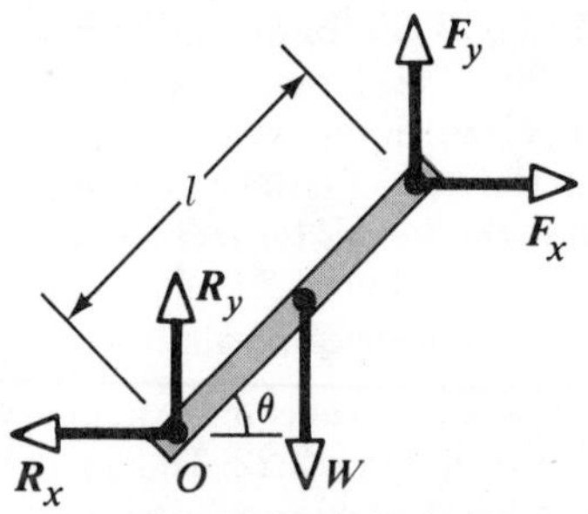

Figure 12.15 (Exercise 2).

3. A uniform beam of weight W and length l has weights W_1 and W_2 at two positions, as in Fig. 12.16. The beam is resting at two points. For what value of x will the beam be balanced at P such that the normal force at O is zero?

4. With reference to Fig. 12.16, find x such that the normal force at O will be one half the normal force at P. Neglect the weight of the beam.

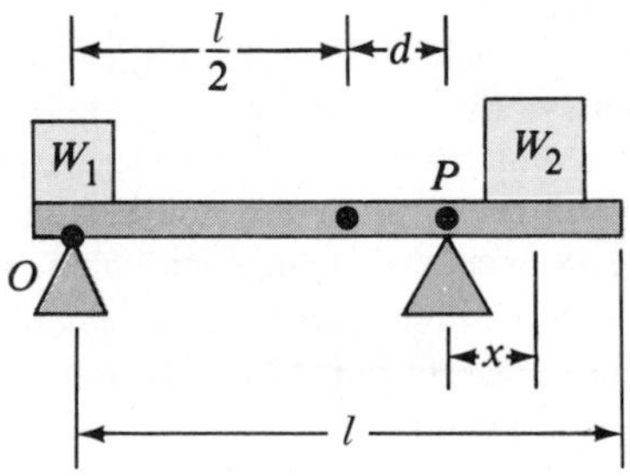

Figure 12.16 (Exercises 3 and 4).

Section 12.2 The Center of Gravity

5. A flat plate in the shape of a letter T is cut with the dimensions shown in Fig. 12.17. Locate the center of gravity. (Hint: Note that the weights of the two rectangular parts are proportional to their volumes.)

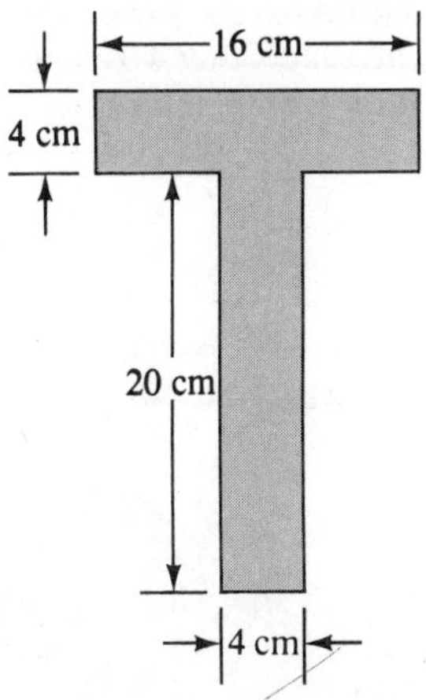

Figure 12.17 (Exercise 5).

6. A carpenter's square has the shape of an L, as in Fig. 12.18. Locate the center of gravity. (See hint in Exercise 5.)

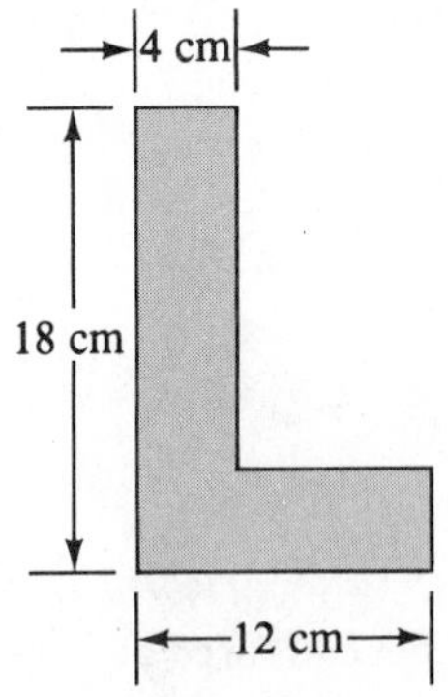

Figure 12.18 (Exercise 6).

Section 12.3 Examples of Rigid Bodies in Static Equilibrium

7. A meter stick supported at the 50-cm mark has masses of 300 g and 200 g hanging from it at the 10-cm and 60-cm marks, respectively. Determine the position at which one would hang a third, 400-g mass to keep the meter stick balanced.
8. Draw free-body diagrams for each of the rigid beams shown in Fig. 12.19. Assume the beams are uniform and have a weight *W*.
9. Repeat Example 12.3 taking the axis of the torque equation through the center of the beam in Fig. 12.11a. Your results for *R*, *T*, and θ should be identical with those obtained in Example 12.3.

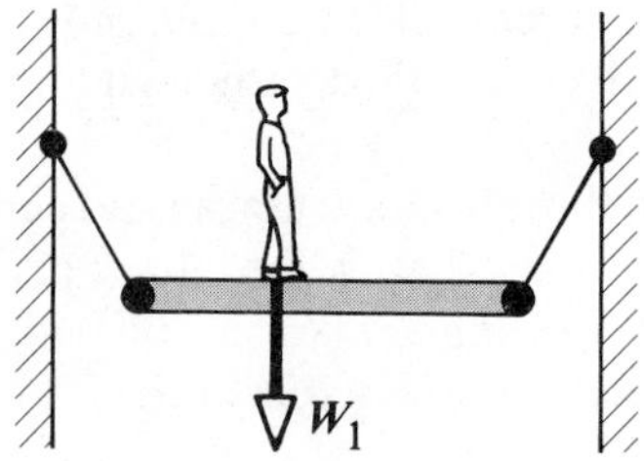

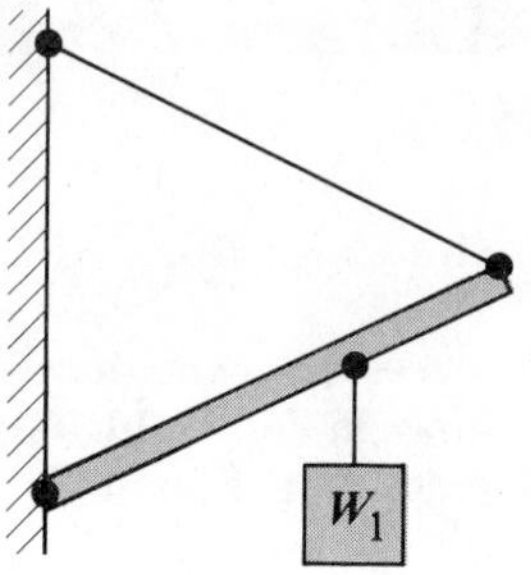

Figure 12.19 (Exercise 8).

10. A uniform plank of length 6 m and mass 30 kg rests horizontally on a scaffold, with 1.5 m of the plank hanging over one end of the scaffold. How far can a painter of mass 70 kg walk on the overhanging part of the plank before it tips?
11. An automobile has a mass of 1600 kg. The distance between the front and rear axles is 3 m. If the normal force on the front tires is 20% larger than the normal force on the rear tires, (a) where is the center of gravity relative to the front axle, and (b) what is the normal force on each tire?

PROBLEMS

1. A uniform beam of length 4 m and mass 10 kg supports a 20-kg mass as in Fig. 12.20. (a) Draw a free-body diagram for the beam. (b) Determine the tension in the supporting wire and the components of the reaction force at the pivot.

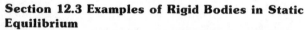

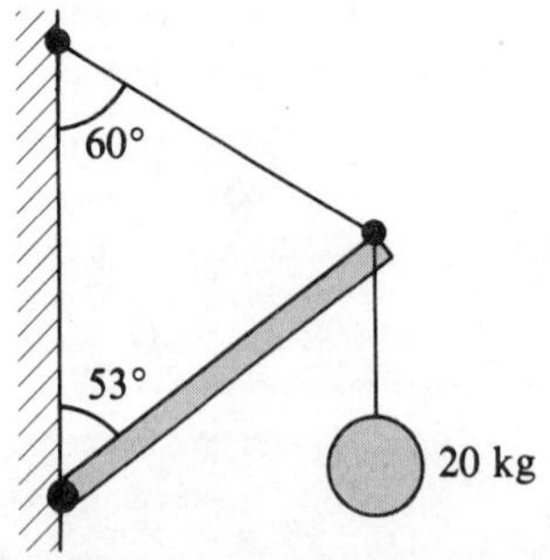

Figure 12.20 (Problem 1).

2. A hungry bear weighing 160 lb walks out on a beam in an attempt to retrieve some "goodies" hanging at the end of the beam (Fig. 12.21). The beam is uniform, weighs 50 lb, and is 20 ft long; the goodies weigh 20 lb. (a) Draw a free-body diagram for the beam. (b) When the bear is at $x = 3$ ft, find the tension in the wire and the components of the reaction force at the hinge. (c) If the wire can withstand a maximum tension of 200 lb, what is the maximum distance the bear can walk before the wire breaks?
3. A 24-lb monkey walks up a 30-lb uniform ladder of length *l*, as in Fig. 12.22. The upper and lower ends of the ladder rest on frictionless surfaces. The lower end of the ladder is fastened to the wall by a horizontal rope that can support a maximum tension of 25 lb. (a) Draw a free-body diagram for the ladder. (b) Find the tension in the rope when the monkey is one third the way up the ladder. (c) Find the maxi-

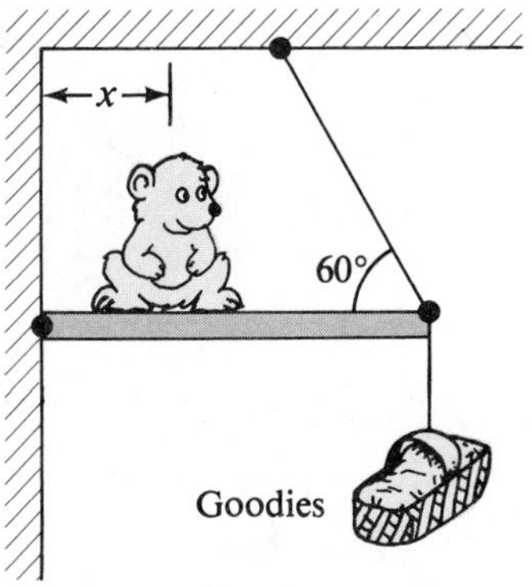

Figure 12.21 (Problem 2).

mum distance *d* the monkey can walk up the ladder before the rope breaks, expressing your answer as a fraction of the length *l*.

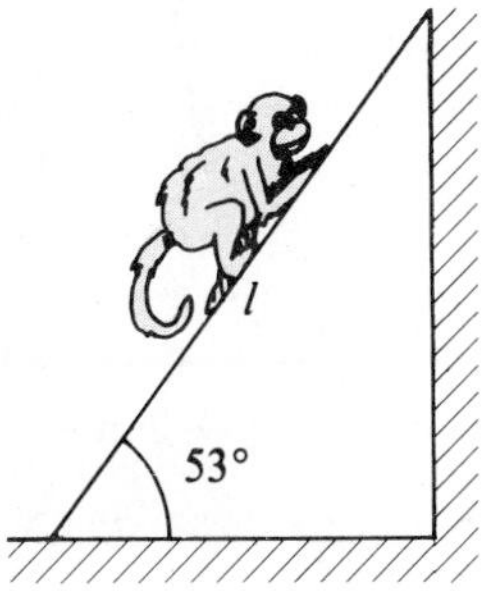

Figure 12.22 (Problem 3).

4. A 300-lb uniform boom is supported by a cable as in Fig. 12.23. The boom is pivoted at the bottom, and a 500-lb weight hangs from its top. Find the tension in the supporting cable and the components of the reaction force on the beam at the hinge.

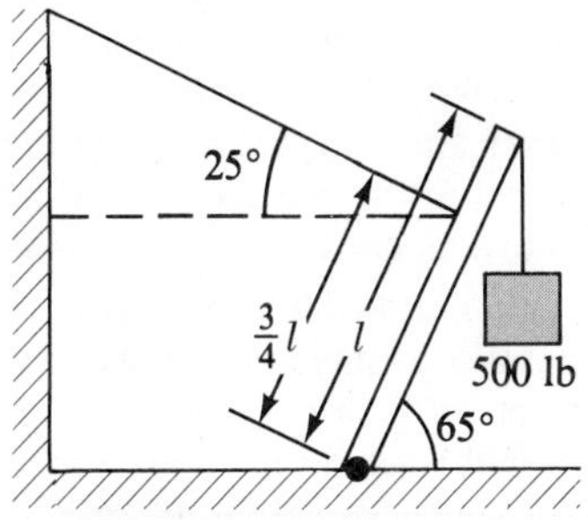

Figure 12.23 (Problem 4).

5. A uniform beam of weight w is inclined at an angle θ to the horizontal with its upper end supported by a horizontal rope tied to a wall and its lower end resting on a rough floor (Fig. 12.24). (a) If the coefficient of static friction between the beam and floor is μ_s, determine an expression for the *maximum* weight W that can be suspended from the top before the beam slips. (b) Determine the magnitude of the reaction force at the floor and the magnitude of the force of the beam on the rope at P in terms of w, W, and μ_s.

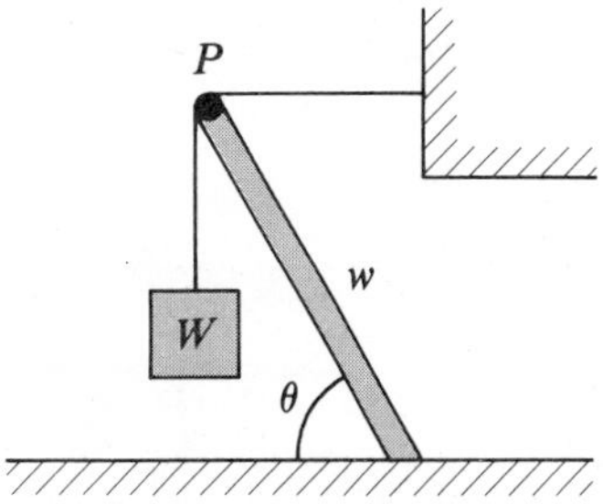

Figure 12.24 (Problem 5).

6. A uniform ladder weighing 50 lb is leaning against a wall (Fig. 12.25). The ladder slips when θ is 60°, where θ is the angle between the ladder and the horizontal. Assuming the coefficients of static friction at the wall and the floor *are the same,* obtain a value for μ_s.

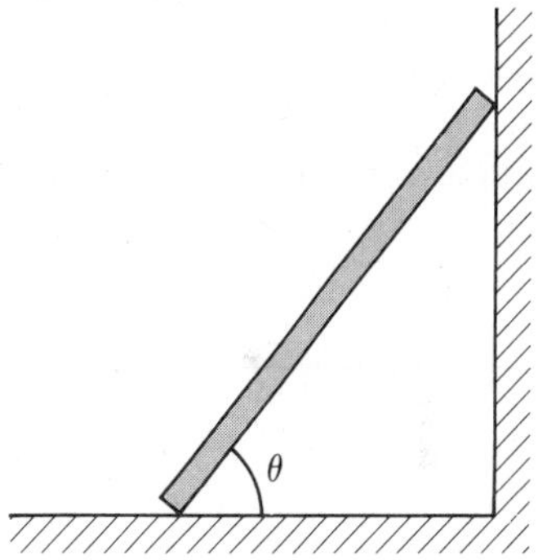

Figure 12.25 (Problem 6).

7. A force **F** acts on a rectangular block weighing 100 lb as in Fig. 12.26. (a) If the block slides with constant speed when $F = 50$ lb and $h = 1$ ft, find the coefficient of sliding friction and the position of the resultant normal force. (b) If $F = 75$ lb, find the value of h for which the block will just begin to tip from a vertical position.

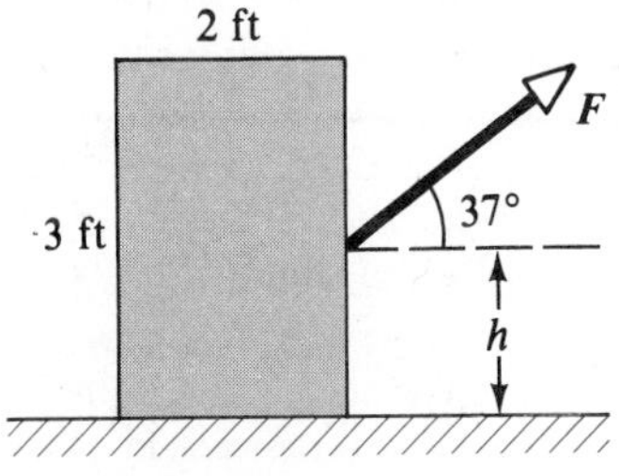

Figure 12.26 (Problem 7).

8. A sign of weight W and width $2l$ hangs from a light, horizontal beam, hinged at the wall and supported by a cable (Fig. 12.27). Determine (a) the tension in the cable and (b) the components of the reaction force at the hinge in terms of W, d, l, and θ.

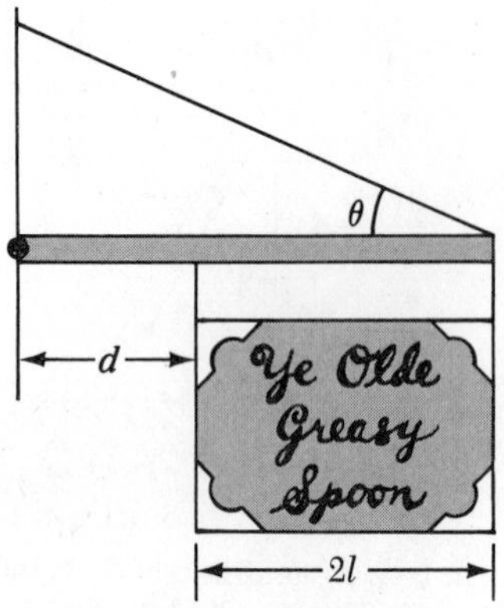

Figure 12.27 (Problem 8).

9. A bridge of length 50 m and mass 8×10^4 kg is supported at each end as in Fig. 12.28. A truck of mass 3×10^4 kg is located 15 m from one end. What are the forces on the bridge at the points of support?

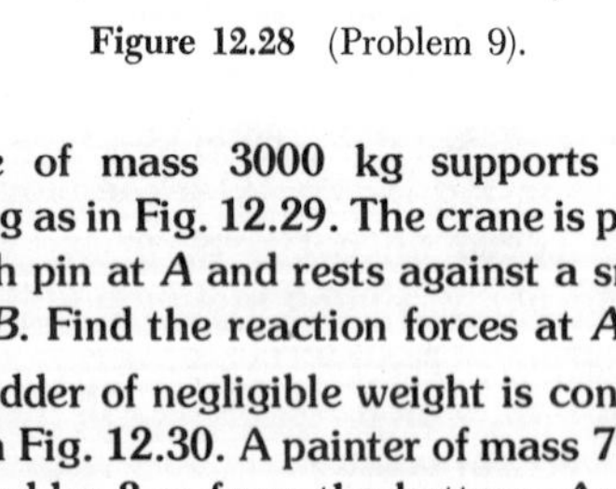

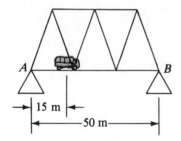

Figure 12.28 (Problem 9).

10. A crane of mass 3000 kg supports a load of 10 000 kg as in Fig. 12.29. The crane is pivoted with a smooth pin at A and rests against a smooth support at B. Find the reaction forces at A and B.

11. A stepladder of negligible weight is constructed as shown in Fig. 12.30. A painter of mass 70 kg stands on the ladder 3 m from the bottom. Assuming the floor is frictionless, find (a) the tension in the horizontal bar connecting the two halves of the ladder, (b) the normal forces at A and B, and (c) the components of the reaction force at the hinge C that the left half of the ladder exerts on the right half. (Hint: Treat each half of the ladder separately.)

12. When a person stands on tiptoe (a strenuous position), the position of the foot is as shown in Fig. 12.31a. The total weight is supported by the force $\boldsymbol{W}$ of the floor on the toe. A mechanical model for the situation is shown in Fig. 12.31b, where $\boldsymbol{T}$ is the tension in the Achilles tendon and $\boldsymbol{R}$ is the force on the foot due to the tibia. Find the values of T and R using the model and dimensions given, with $W = 700$ N.

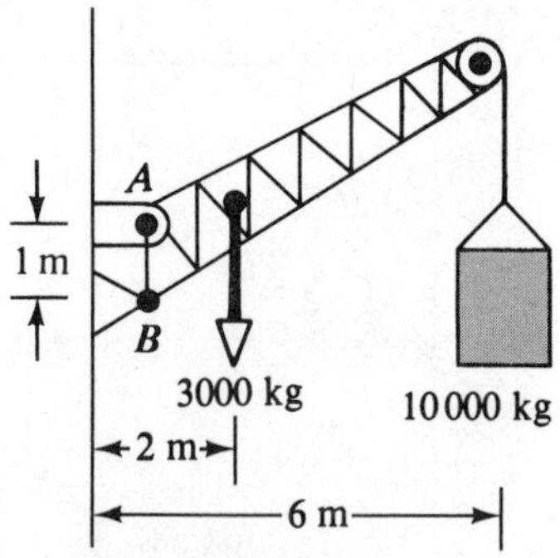

Figure 12.29 (Problem 10).

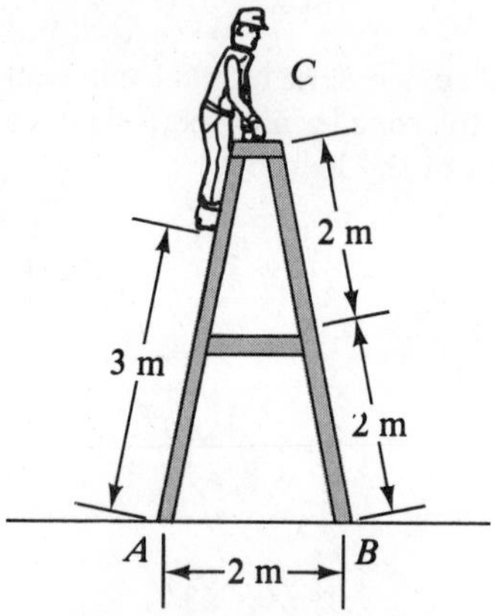

Figure 12.30 (Problem 11).

13. A person bends over and lifts a 200-N weight as in Fig. 12.32a, with the back in the horizontal position. The back muscle attached at a point two thirds up the spine maintains the position of the back, where the angle between the spine and this muscle is 12°. Using the mechanical model shown in Fig. 12.32b and taking the weight of the upper body to be 350 N, find the tension in the back muscle and the compressional force in the spine.

14. The cylinder shown in Fig. 12.33 is held in position by a rope that supplies a force $\boldsymbol{F}$ and by static friction. What is the *minimum* value of μ_s such that the cylinder will remain in equilibrium when $\boldsymbol{F}$ is at the angle θ with the horizontal?

15. A 150-kg mass rests on a 50-kg beam as in Fig. 12.34. The weight is also connected to one end of the beam through a rope and pulley. Assuming the system is in equilibrium, (a) draw free-body diagrams for the weight and beam and (b) find the tension in the rope and the components of the reaction force at the pivot O.

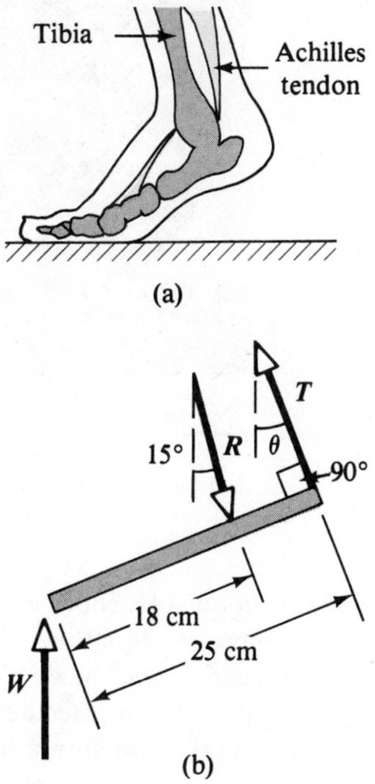

Figure 12.31 (Problem 12).

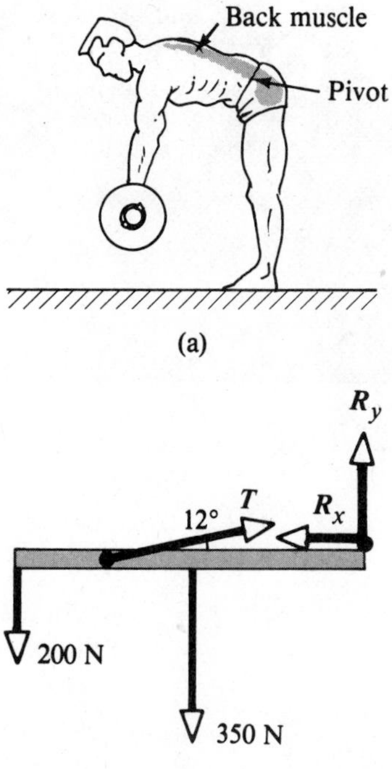

Figure 12.32 (Problem 13).

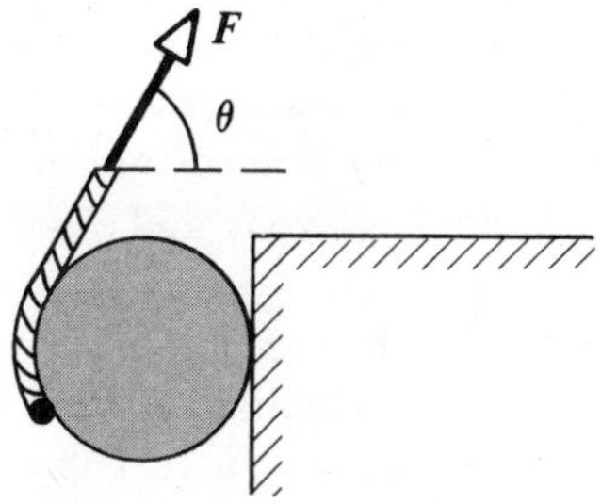

Figure 12.33 (Problem 14).

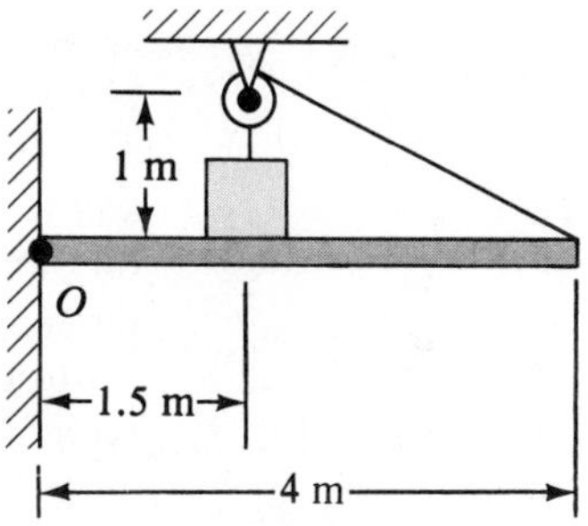

Figure 12.34 (Problem 15).

13 Oscillatory Motion

Griffith Observatory

In previous chapters we pointed out that the motion of a body can be predicted if the external forces acting on it are known. If a force varies in time, the velocity and acceleration of the body will also change with time. A very special kind of motion occurs when the force on a body is proportional to the displacement of the body from equilibrium. If this force always acts toward the equilibrium position of the body, a back-and-forth motion will result about this position. We call such a motion *periodic* or *oscillatory*.

You are most likely familiar with several examples of periodic motion, such as the oscillations of a mass on a spring, the motion of a pendulum, and the vibrations of a stringed musical instrument. However, the number of systems that exhibit oscillatory motion is much more extensive. For example, the molecules in a solid oscillate about their equilibrium positions; electromagnetic waves, such as light waves, radar, and radio waves, are characterized by oscillating electric and magnetic field vectors; and in alternating-current circuits, voltage, current, and electrical charge vary periodically with time.

Most of the material in this chapter deals with *simple harmonic motion*. For this type of motion, an object oscillates between two spatial positions for an indefinite period of time, with no loss in mechanical energy. In real mechanical systems, retarding (or frictional) forces are always present. Such forces reduce the mechanical energy of the system as motion progresses, and the oscillations are said to be *damped*. If an external driving force is applied to overcome this loss in energy, we call the motion a *forced oscillation*.

13.1 SIMPLE HARMONIC MOTION

A particle moving along the x axis is said to exhibit *simple harmonic motion* when its displacement from equilibrium, x, varies in time according to the relationship

Displacement versus time for simple harmonic motion

$$x = A\cos(\omega t + \delta) \qquad (13.1)$$

where A, ω, and δ are constants of the motion. In order to give physical significance to these constants, it is convenient to plot x as a function of t, as in Fig. 13.1. First, we note that A, called the *amplitude* of the motion, is simply the *maximum displacement* of the particle in either the positive or negative x direction. The constant ω is called the *angular frequency* (defined in Eq. 13.4). The constant angle δ is called the

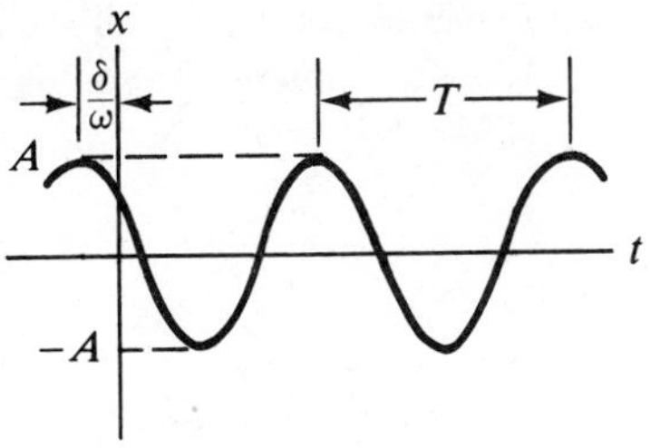

Figure 13.1 Displacement versus time for a particle undergoing simple harmonic motion. The amplitude of the motion is A and the period is T.

phase constant (or phase angle) and along with the amplitude A is determined uniquely by the initial displacement and velocity of the particle. The constant δ tells us what the displacement was at time $t = 0$. The quantity $(\omega t + \delta)$ is called the *phase* of the motion and is useful in comparing the motions of two systems of particles. Note that the function x is periodic and repeats itself when ωt increases by 2π radians.

The time T it takes the particle to go through one full cycle of its motion is called the *period*. That is, the value of x at time t equals the value of x at time $t + T$. We can show that the period of the motion is given by $T = 2\pi/\omega$ by using the fact that the phase increases by 2π radians in a time T:

$$\omega t + \delta + 2\pi = \omega(t + T) + \delta$$

Hence, $\omega T = 2\pi$ or

$$T = \frac{2\pi}{\omega} \tag{13.2}$$

Period

The inverse of the period is called the *frequency* of the motion, f. The frequency represents *the number of oscillations the particle makes per unit time:*

$$f = \frac{1}{T} = \frac{\omega}{2\pi} \tag{13.3}$$

Frequency

The units of f are cycles/s, or hertz (Hz).

Rearranging Eq. 13.3 gives

$$\omega = 2\pi f = \frac{2\pi}{T} \tag{13.4}$$

Angular frequency

The constant ω is called the *angular frequency* and has units of rad/s. We shall discuss the geometric significance of ω in Section 13.4.

We can obtain the velocity of a particle undergoing simple harmonic motion by differentiating Eq. 13.1 with respect to time:

$$v = \frac{dx}{dt} = -A\omega \sin(\omega t + \delta) \tag{13.5}$$

Velocity in simple harmonic motion

The acceleration of the particle is given by dv/dt:

$$a = \frac{dv}{dt} = -A\omega^2 \cos(\omega t + \delta) \tag{13.6}$$

Acceleration in simple harmonic motion

Since $x = A \cos(\omega t + \delta)$, we can express Eq. 13.6 in the form

$$a = -\omega^2 x \tag{13.7}$$

From Eq. 13.5 we see that since the sine and cosine functions oscillate between ± 1, the extreme values of v are equal to $\pm A\omega$. Equation 13.6 tells us that the extreme values of the acceleration are $\pm A\omega^2$. Therefore, the *maximum* values of the velocity and acceleration are given by

Maximum values of velocity and acceleration in simple harmonic motion

$$v_{max} = A\omega \tag{13.8}$$

$$a_{max} = A\omega^2 \tag{13.9}$$

Figure 13.2a represents the displacement versus time for an arbitrary value of the phase constant. The projection of a point moving with uniform circular motion on a reference circle of radius A also moves in sinusoidal fashion. This will be discussed in more detail in Section 13.5.

The velocity and acceleration versus time curves are illustrated in Figs. 13.2b and 13.2c. These curves show that the phase of the velocity differs from the phase of the displacement by $\pi/2$ rad, or 90°. That is, when x is a maximum or a minimum, the velocity is zero. Likewise, when x is zero, the speed is a maximum. Furthermore, note that the phase of the acceleration differs from the phase of the displacement by π radians, or 180°. That is, when x is a maximum, a is a minimum and vice versa.

As we stated earlier, the solution $x = A\cos(\omega t + \delta)$ is a general solution of the equation of motion, where the phase constant δ and the amplitude A must be chosen to meet the initial conditions of the motion. The phase constant is actually impor-

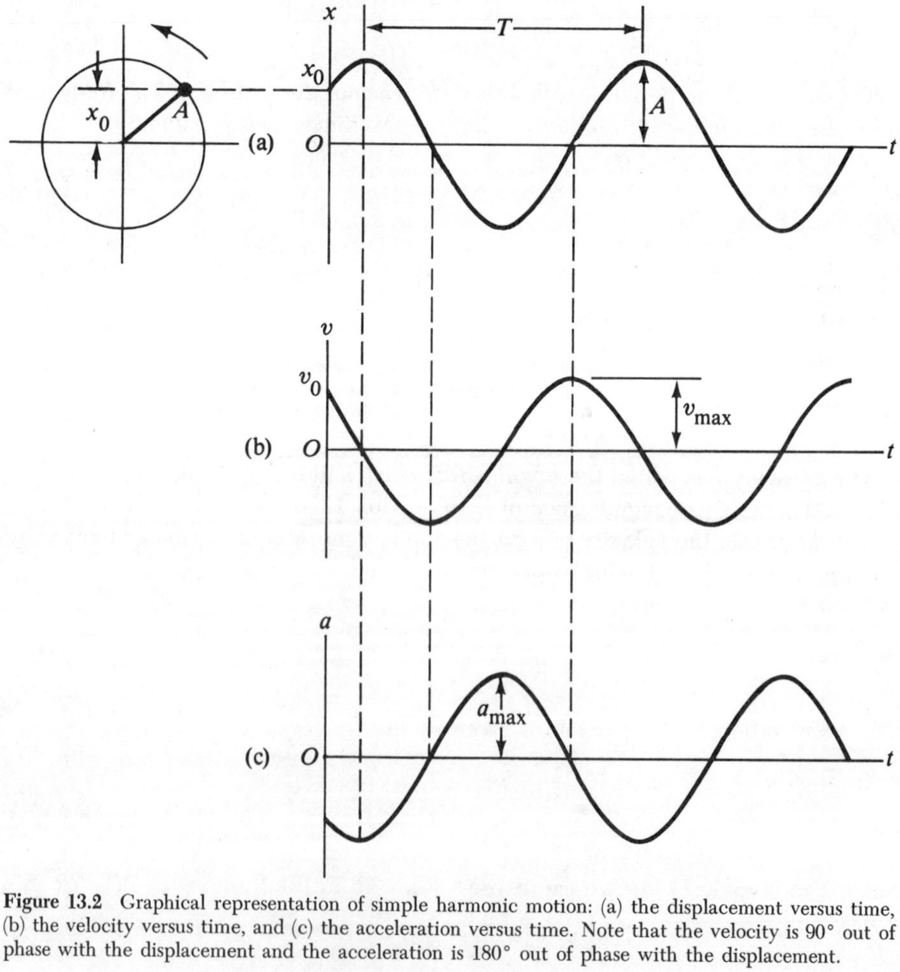

Figure 13.2 Graphical representation of simple harmonic motion: (a) the displacement versus time, (b) the velocity versus time, and (c) the acceleration versus time. Note that the velocity is 90° out of phase with the displacement and the acceleration is 180° out of phase with the displacement.

tant only when comparing the motion of two or more oscillating particles. Suppose that the initial position x_0 and initial velocity v_0 of a single oscillator are given, that is, at $t = 0$, $x = x_0$ and $v = v_0$. Under these conditions, the equations $x = A\cos(\omega t + \delta)$ and $v = -A\omega\sin(\omega t + \delta)$ give

$$x_0 = A\cos\delta \qquad \text{and} \qquad v_0 = -A\omega\sin\delta$$

Dividing these two equations eliminates A, giving

$$\frac{v_0}{x_0} = -\omega\tan\delta$$

The phase angle ϕ and amplitude A can be obtained from the initial conditions

$$\tan\delta = -\frac{v_0}{\omega x_0} \tag{13.10a}$$

Furthermore, if we take the sum $x_0^2 + \left(\frac{v_0}{\omega}\right)^2 = A^2\cos^2\delta + A^2\sin^2\delta$ and solve for A, we find that

$$A = \sqrt{x_0^2 + \left(\frac{v_0}{\omega}\right)^2} \tag{13.10b}$$

Thus, we see that δ and A are known if x_0 and v_0 are specified. We shall treat a few specific cases in the next section.

Properties of simple harmonic motion

We conclude this section by pointing out the following important properties of a particle moving in simple harmonic motion: (1) *The displacement, velocity, and acceleration all vary sinusoidally with time but are not in phase.* (2) *The acceleration of the particle is proportional to the displacement, but in the opposite direction.* (3) *The frequency and the period of motion are independent of the amplitude.*

Example 13.1

A body oscillates with simple harmonic motion along the x axis. Its displacement varies with time according to the equation

$$x = 4.0\cos\left(\pi t + \frac{\pi}{4}\right)$$

where x is in m, t is in s, and the angles in the parentheses are in radians. (a) Determine the amplitude, frequency, and period of the motion.

By comparing this equation with the general relation for simple harmonic motion, $x = A\cos(\omega t + \delta)$, we see that $A = 4.0$ m and $\omega = \pi$ rad/s; therefore $f = \omega/2\pi = \pi/2\pi = 0.50\ \text{s}^{-1}$ and $T = 1/f = 2.0$ s.

(b) Calculate the velocity and acceleration of the body at any time t.

$$v = \frac{dx}{dt} = -4.0\sin\left(\pi t + \frac{\pi}{4}\right)\frac{d}{dt}(\pi t)$$

$$= -4\pi\sin\left(\pi t + \frac{\pi}{4}\right)\ \frac{\text{m}}{\text{s}}$$

$$a = \frac{dv}{dt} = -4\pi\cos\left(\pi t + \frac{\pi}{4}\right)\frac{d}{dt}(\pi t)$$

$$= -4\pi^2\cos\left(\pi t + \frac{\pi}{4}\right)\ \frac{\text{m}}{\text{s}^2}$$

(c) Using the results to (b), determine the position, velocity, and acceleration of the body at $t = 1$ s.

Noting that the angles in the trigonometric functions are in radians, we get at $t = 1$ s

$$x = 4.0\cos\left(\pi + \frac{\pi}{4}\right) = 4.0\cos\left(\frac{5\pi}{4}\right)$$

$$= 4.0(-0.71) = -2.8\ \text{m}$$

$$v = -4\pi\sin\left(\frac{5\pi}{4}\right) = -4\pi(-0.71) = 8.9\ \text{m/s}$$

$$a = -4\pi^2\cos\left(\frac{5\pi}{4}\right) = -4\pi^2(-0.71) = 28\ \text{m/s}^2$$

(d) Determine the maximum speed and maximum acceleration of the body.

By analyzing the general relations for v and a in (b), we note that the maximum values of the sine and cosine functions are unity. Therefore, v oscillates between $\pm 4\pi$ m/s, and a oscillates between $\pm 4\pi^2\ \text{m/s}^2$. Thus, $v_{max} = 4\pi$ m/s and $a_{max} = 4\pi^2\ \text{m/s}^2$. The same results are obtained using $v_{max} = A\omega$ and $a_{max} = A\omega^2$, where $A = 4.0$ m and $\omega = \pi$ rad/s.

(e) Find the displacement of the body between $t = 0$ and $t = 1$ s.

The x coordinate at $t = 0$ is given by

$$x_0 = 4.0\cos\left(0 + \frac{\pi}{4}\right) = 4.0(0.71) = 2.8\ \text{m}$$

In (c), we found that the coordinate at $t = 1$ s was -2.8 m; therefore the displacement between $t = 0$ and $t = 1$ s is

$$\Delta x = x - x_0 = -2.8 \text{ m} - 2.8 \text{ m} = -5.6 \text{ m}$$

Since the particle's velocity changes sign during the first second, the magnitude of Δx is *not* the same as the distance traveled in the first second.

(f) What is the phase of the motion at $t = 2$ s?

The phase is defined as $\omega t + \delta$, where in this case $\omega = \pi$ and $\delta = \pi/4$. Therefore, at $t = 2$ s, we get

$$\text{Phase} = (\omega t + \delta)_{t=2} = \pi(2) + \pi/4 = 9\pi/4 \text{ rad}$$

Q1. What is the total distance traveled by a body executing simple harmonic motion in a time equal to its period if its amplitude is A?

Q2. If the coordinate of a particle varies as $x = -A \cos\omega t$, what is the phase constant δ in Eq. 13.1? At what position does the particle begin its motion?

Q3. Does the displacement of an oscillating particle between $t = 0$ and a later time t necessarily equal the position of the particle at time t? Explain.

Q4. Determine whether or not the following quantities can be in the same direction for a simple harmonic oscillator: (a) displacement and velocity, (b) velocity and acceleration, (c) displacement and acceleration.

Q5. Can the amplitude A and phase constant δ be determined for an oscillator if only the position is specified at $t = 0$? Explain.

13.2 MASS ATTACHED TO A SPRING

In Chapter 7 we introduced the physical system consisting of a mass attached to the end of a spring, where the mass is free to move on a horizontal, frictionless surface (Fig. 13.3). We know from experience that such a system will oscillate back

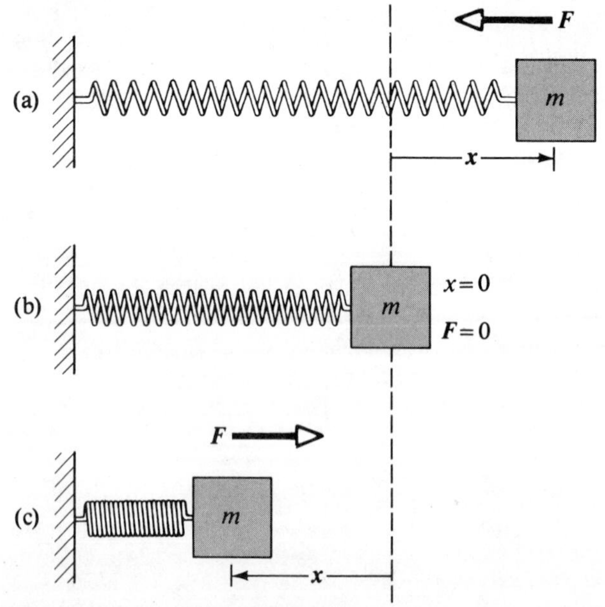

Figure 13.3 A mass attached to a spring on a frictionless surface exhibits simple harmonic motion. (a) When the mass is displaced to the right of equilibrium, the displacement is positive and the acceleration is negative. (b) At the equilibrium position, $x = 0$, the acceleration is zero but the speed is a maximum. (c) When the displacement is negative, the acceleration is positive.

and forth if disturbed from the equilibrium position $x = 0$, where the spring is unstretched. If the surface is frictionless, the mass will exhibit simple harmonic motion. One possible experimental arrangement that clearly demonstrates that such a system exhibits simple harmonic motion is illustrated in Fig. 13.4, in which a mass oscillating vertically on a spring has a marking pen attached to it. While the mass is in motion, a sheet of paper is moved horizontally as shown, and the marking pen traces out a sinusoidal pattern. We can understand this qualitatively by first recalling that when the mass is displaced a small distance x from equilibrium, the spring exerts a force on m given by Hooke's law,

$$F = -kx \tag{13.11}$$

Linear restoring force

where k is the force constant of the spring. We call this a *linear restoring force* since it is linearly proportional to the displacement and is always directed toward the equilibrium position, *opposite* the displacement. That is, when the mass is displaced to the right in Fig. 13.3, x is positive and the restoring force is to the left. When the mass is displaced to the left of $x = 0$, then x is negative and $\mathbf{F}$ is to the right. If we now apply Newton's second law to the motion of m in the x direction, we get

$$F = -kx = ma$$

$$a = -\frac{k}{m}x \tag{13.12}$$

The acceleration of a mass-spring system is proportional to the displacement

That is, *the acceleration is proportional to the displacement of the mass from equilibrium and is in the opposite direction.* If the mass is displaced a maximum distance $x = A$ at some initial time and released from rest, its *initial* acceleration will be $-kA/m$ (that is, it has a maximum negative value). When it passes through the equilibrium position, $x = 0$ and its acceleration is zero. At this instant, its velocity is a maximum. It will then travel to the left of equilibrium and finally reach $x = -A$, at which time its acceleration is kA/m (maximum positive) and its velocity is again zero. Thus, we see that the mass will oscillate between the turning points $x = \pm A$. In one full cycle of its motion, the mass travels a distance $4A$.

We will now describe the motion in a quantitative fashion. This can be accomplished by recalling that $a = dv/dt = d^2x/dt^2$. Thus, we can express Eq. 13.12 as

$$\frac{d^2x}{dt^2} = -\frac{k}{m}x \tag{13.13}$$

Equation of motion for mass-spring system

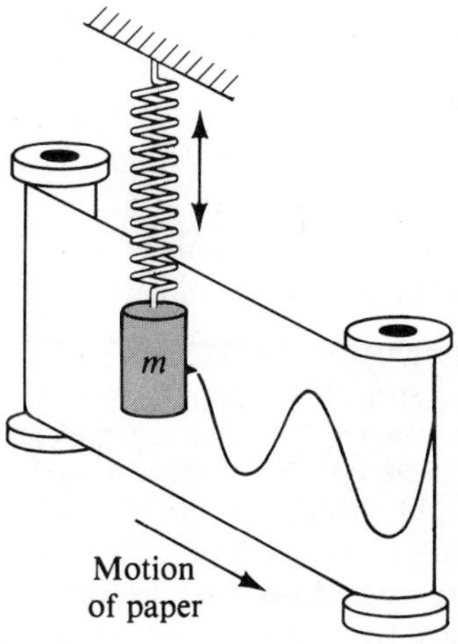

Figure 13.4 An experimental apparatus for demonstrating simple harmonic motion. A pen attached to the oscillating mass traces out a sine wave on the moving chart paper.

If we denote the ratio k/m by the symbol ω^2,

$$\omega^2 = k/m \tag{13.14}$$

then Eq. 13.13 can be written in the form

$$\frac{d^2x}{dt^2} = -\omega^2 x \tag{13.15}$$

What we now require is a solution to Eq. 13.15, that is, a function $x(t)$ that satisfies this second-order differential equation. However, by noting that Eqs. 13.15 and 13.7 are equivalent, we see that the solution must be that of simple harmonic motion:

$$x(t) = A\cos(\omega t + \delta)$$

To see this explicitly, note that if

$$x = A\cos(\omega t + \delta)$$

then

$$\frac{dx}{dt} = -A\frac{d}{dt}\cos(\omega t + \delta) = -A\omega\sin(\omega t + \delta)$$

$$\frac{d^2x}{dt^2} = -A\omega\frac{d}{dt}\sin(\omega t + \delta) = -A\omega^2\cos(\omega t + \delta)$$

Comparing the expressions for x and d^2x/dt^2, we see that $d^2x/dt^2 = -\omega^2 x$ and Eq. 13.15 is satisfied.

The following general statement can be made based on the above discussion: *Whenever the force acting on a particle is linearly proportional to the displacement and in the opposite direction, the particle will exhibit simple harmonic motion.* We shall give additional physical examples in subsequent sections.

Since the period is given by $T = 2\pi/\omega$ and the frequency is the inverse of the period, we can express the period and frequency of the motion for this system as

Period and frequency for mass-spring system

$$T = \frac{2\pi}{\omega} = 2\pi\sqrt{\frac{m}{k}} \tag{13.16}$$

$$f = \frac{1}{T} = \frac{1}{2\pi}\sqrt{\frac{k}{m}} \tag{13.17}$$

That is, the period and frequency depend *only* on the mass and on the force constant of the spring. As we might expect, the frequency is larger for a stiffer spring and decreases with increasing mass.

It is interesting to note that a mass suspended from a vertical spring attached to a fixed support will also exhibit simple harmonic motion. Although there is a gravitational force to consider in this case, the equation of motion still reduces to Eq. 13.15, where the displacement is measured from the equilibrium position of the suspended mass. The proof of this is left as an exercise (Problem 5).

Special Case I. In order to better understand the physical significance of our solution of the equation of motion, let us consider the following special case. Suppose we extend the mass from equilibrium by a distance A and release it from rest from this stretched position, as in Fig. 13.5. We must then require that our solution for $x(t)$ obey the *initial conditions* that at $t = 0$, $x_0 = A$ and $v_0 = 0$. These conditions will be met if we choose $\delta = 0$, giving $x = A\cos\omega t$ as our solution. Note that this is consistent with $x = A\cos(\omega t + \delta)$, where $x_0 = A$ and $\delta = 0$. To check this, we see that the solution $x = A\cos\omega t$ satisfies the condition that $x_0 = A$ at $t = 0$, since $\cos 0 = 1$. Now let us investigate the behavior of the velocity and

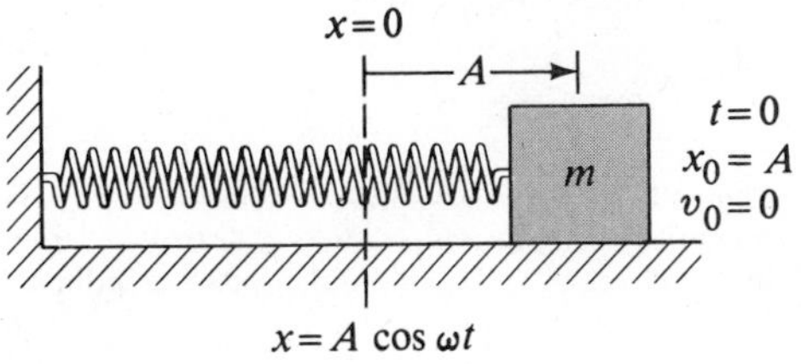

Figure 13.5 A mass-spring system that starts from rest at $x_0 = A$. In this case, $\delta = 0$, and so $x = A\cos\omega t$.

acceleration for this special case. Since $x = A\cos\omega t$

$$v = \frac{dx}{dt} = -A\omega\sin\omega t$$

and

$$a = \frac{dv}{dt} = -A\omega^2\cos\omega t$$

From the velocity expression $v = -A\omega\sin\omega t$, we see that at $t = 0$, $v_0 = 0$, as we require. The expression for the acceleration tells us that at $t = 0$, $a = -A\omega^2$. Physically this makes sense, since the force on the mass is to the left when the displacement is positive. In fact, at this position $F = -kA$ (to the left), and the initial acceleration is $-kA/m$.

We could also use a more formal approach to show that $x = A\cos\omega t$ is the correct solution by using the relation $\tan\delta = -v_0/\omega x_0$ (Eq. 13.10a). Since $v_0 = 0$ at $t = 0$, $\tan\delta = 0$ and so $\delta = 0$.

The displacement, velocity, and acceleration versus time are plotted in Fig. 13.6 for this special case. Note that the acceleration reaches extreme values of $\mp\omega^2 A$ when the displacement has extreme values of $\pm A$. Furthermore, the velocity has extreme values of $\pm A\omega$, which both occur at $x = 0$. Hence, the quantitative solution agrees with our qualitative description of this system.

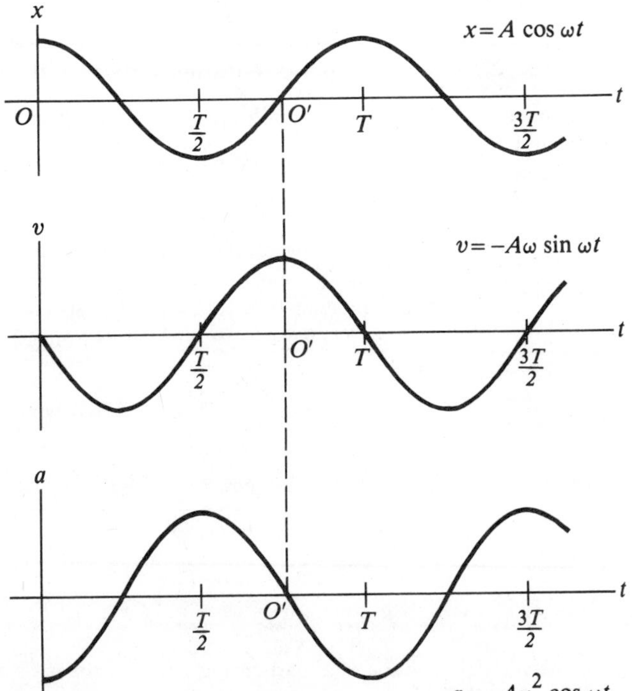

Figure 13.6 Displacement, velocity, and acceleration versus time for a particle undergoing simple harmonic motion under the initial conditions that at $t = 0$, $x_0 = A$ and $v_0 = 0$.

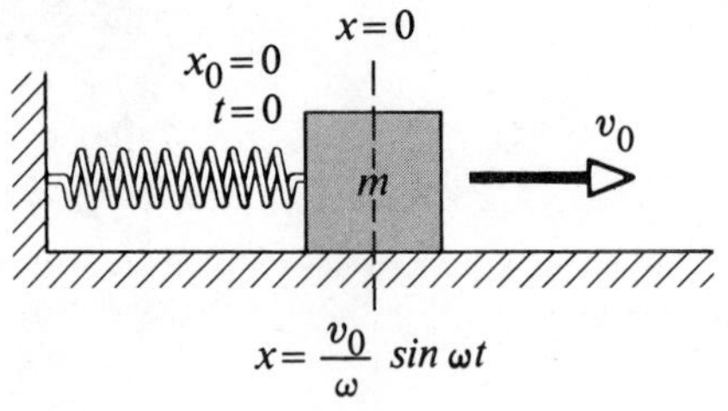

Figure 13.7 The mass-spring system starts its motion at the equilibrium position, $x_0 = 0$ at $t = 0$. If its initial velocity is v_0 to the right, its x coordinate varies as $x = \frac{v_0}{\omega}\sin\omega t$.

Special Case II. Now suppose that the mass starts its motion from the unstretched position so that at $t = 0$, $x_0 = 0$ and $v = v_0$ (Fig. 13.7). Our particular solution must now satisfy these initial conditions. Since the mass is moving toward positive x values at $t = 0$, the solution is expected to be of the form $x = A\sin\omega t$.

Applying $\tan\delta = -v_0/\omega x_0$ and the initial condition that $x_0 = 0$ at $t = 0$ gives $\tan\delta = -\infty$ or $\delta = -\pi/2$. Hence, the solution is $x = A\cos(\omega t - \pi/2)$, which can be written $x = A\sin\omega t$. Furthermore, from Eq. 13.10b we see that $A = v_0/\omega$; therefore we can express our solution as

$$x = \frac{v_0}{\omega}\sin\omega t$$

The velocity and acceleration are given by

$$v = \frac{dx}{dt} = v_0\cos\omega t$$

$$a = \frac{dv}{dt} = -\omega v_0\sin\omega t$$

This is consistent with the fact that the mass always has a maximum speed at $x = 0$, while the force and acceleration are zero at this position. The graphs of these functions versus time in Fig. 13.6 correspond to the origin at O'. What would be the solution for x if the initial motion is to the left in Fig. 13.7?

Example 13.2

A mass of 200 g is connected to a light spring of force constant 5 N/m and is free to oscillate on a horizontal, frictionless surface. If the mass is displaced 5 cm from equilibrium and released from rest, as in Fig. 13.5, (a) find the period of its motion.

First, note that this situation corresponds to Case I, where $x = A\cos\omega t$ and $A = 5 \times 10^{-2}$ m. Therefore,

$$\omega = \sqrt{\frac{k}{m}} = \sqrt{\frac{5\text{ N/m}}{200 \times 10^{-3}\text{ kg}}} = 5\text{ rad/s}$$

Therefore

$$T = \frac{2\pi}{\omega} = \frac{2\pi}{5} = 1.26\text{ s}$$

(b) Determine the maximum speed of the mass.

$$v_{max} = A\omega = (5 \times 10^{-2}\text{ m})(5\text{ rad/s}) = 0.25\text{ m/s}$$

(c) What is the maximum acceleration of the mass?

$$a_{max} = A\omega^2 = (5 \times 10^{-2}\text{ m})(5\text{ rad/s})^2 = 1.25\text{ m/s}^2$$

(d) Express the displacement, speed, and acceleration as functions of time.

The expression $x = A\cos\omega t$ is our special solution for Case I, and so we can use the results from (a), (b), and (c) to get

$$x = A\cos\omega t = (5 \times 10^{-2}\cos 5t)\text{ m}$$

$$v = -A\omega\sin\omega t = (-0.25\sin 5t)\text{ m/s}$$

$$a = -A\omega^2\cos\omega t = (-1.25\cos 5t)\text{ m/s}^2$$

Q6. Describe qualitatively the motion of a mass-spring system if the mass of the spring is not neglected.

Q7. If a mass-spring system is hung vertically and set into oscillation, why does the motion eventually stop?

13.3 ENERGY OF THE SIMPLE HARMONIC OSCILLATOR

Let us examine the mechanical energy of the mass-spring system described in Fig. 13.6. Since the surface is frictionless, we expect that the total mechanical energy is conserved, as was shown in Chapter 8. We can use Eq. 13.5 to express the kinetic energy as

$$K = \frac{1}{2}mv^2 = \frac{1}{2}m\omega^2A^2\sin^2(\omega t + \delta) \quad (13.18)$$

Kinetic energy of a simple harmonic oscillator

The elastic potential energy stored in the spring for any elongation x is given by $\frac{1}{2}kx^2$. Using Eq. 13.1, we get

$$U = \frac{1}{2}kx^2 = \frac{1}{2}kA^2\cos^2(\omega t + \delta) \quad (13.19)$$

Potential energy of a simple harmonic oscillator

We see that K and U are *always* positive quantities. Since $\omega^2 = k/m$, we can express the *total energy* of the simple harmonic oscillator as

$$E = K + U = \frac{1}{2}kA^2[\sin^2(\omega t + \delta) + \cos^2(\omega t + \delta)]$$

But $\sin^2\theta + \cos^2\theta = 1$, where $\theta = \omega t + \delta$, therefore this equation reduces to

$$E = \frac{1}{2}kA^2 \quad (13.20)$$

Total energy of a simple harmonic oscillator

That is, *the energy of a simple harmonic oscillator is a constant of the motion and proportional to the square of the amplitude.* In fact, the total mechanical energy is just equal to the maximum potential energy stored in the spring when $x = \pm A$. At these points, $v = 0$ and there is no kinetic energy. At the equilibrium position, $x = 0$ and $U = 0$, so that the total energy is all in the kinetic energy of the motion. That is, at $x = 0$, $E = \frac{1}{2}mv^2_{max} = \frac{1}{2}m\omega^2A^2$.

Plots of the kinetic and potential energies versus time are shown in Figure 13.8a,

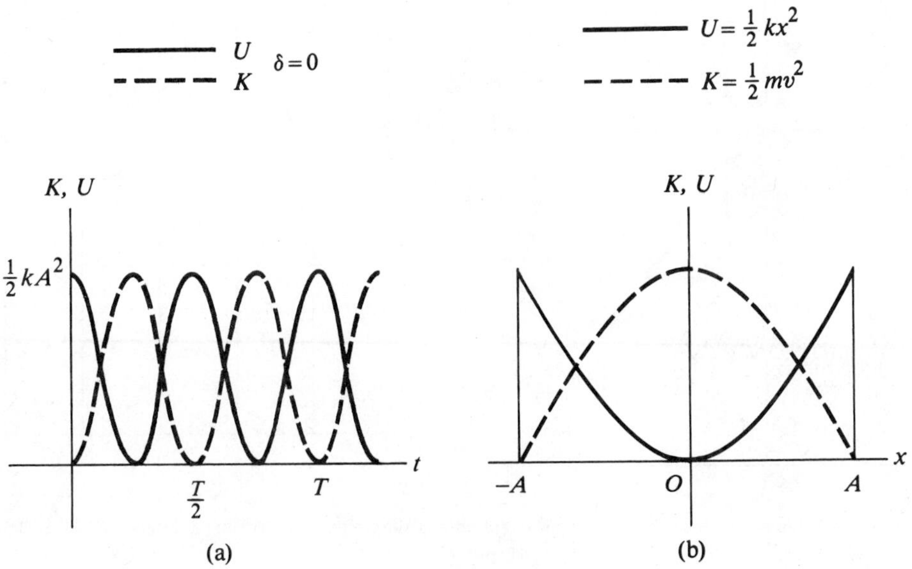

Figure 13.8 (a) Kinetic energy and potential energy versus time for a simple harmonic oscillator with $\delta = 0$. (b) Kinetic energy and potential energy versus displacement for a simple harmonic oscillator. In either plot, note that $K + U =$ constant.

where we have taken $\delta = 0$. Note that both K and U are always positive and their sum at all times is a constant equal to $\frac{1}{2}kA^2$, the total energy of the system. The variations of K and U with displacement are plotted in Fig. 13.8b. You should note that energy is continuously being transferred between potential energy stored in the spring and the kinetic energy of the mass. Figure 13.9 illustrates the position, velocity, acceleration, kinetic energy, and potential energy of the mass-spring system for one full period of the motion. Most of the ideas discussed so far are incorporated in this important figure.

Finally, we can use energy conservation to obtain the velocity for an arbitrary displacement x by expressing the total energy at some arbitrary position as

$$E = K + U = \frac{1}{2}mv^2 + \frac{1}{2}kx^2 = \frac{1}{2}kA^2$$

Velocity as a function of position for a simple harmonic oscillator

$$v = \pm\sqrt{\frac{k}{m}(A^2 - x^2)} \qquad (13.21)$$

Again, this expression substantiates the fact that the speed is a maximum at $x = 0$ and is zero at the turning points, $x = \pm A$.

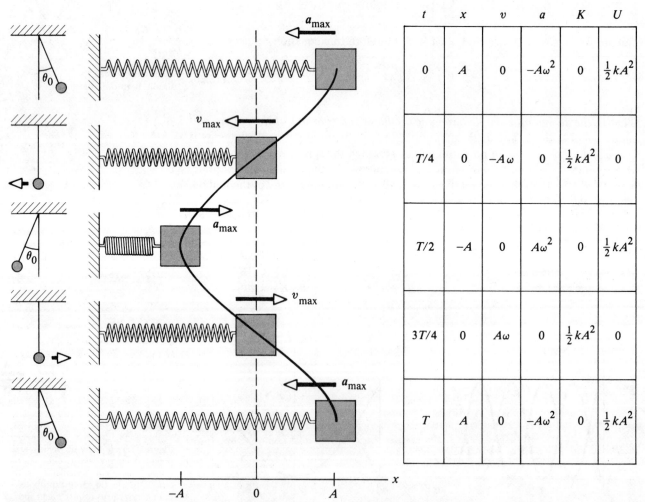

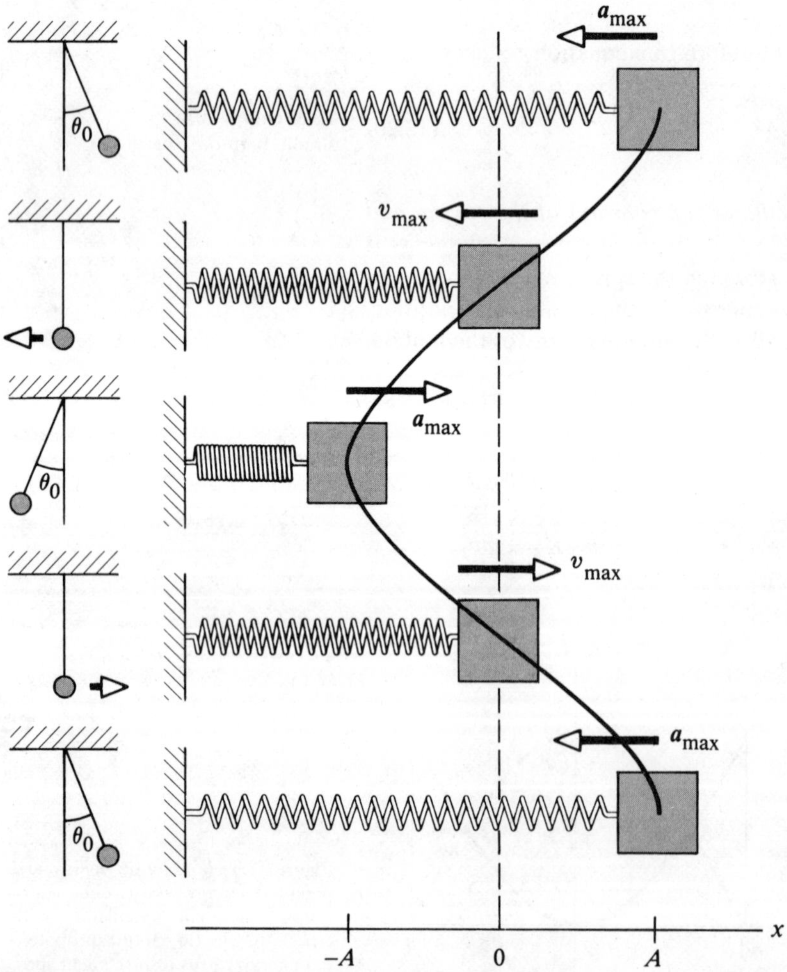

t	x	v	a	K	U
0	A	0	$-A\omega^2$	0	$\frac{1}{2}kA^2$
$T/4$	0	$-A\omega$	0	$\frac{1}{2}kA^2$	0
$T/2$	$-A$	0	$A\omega^2$	0	$\frac{1}{2}kA^2$
$3T/4$	0	$A\omega$	0	$\frac{1}{2}kA^2$	0
T	A	0	$-A\omega^2$	0	$\frac{1}{2}kA^2$

Figure 13.9 Simple harmonic motion for a mass-spring system and its analogy to the motion of a simple pendulum. The parameters in the table refer to the mass-spring system, assuming that at $t = 0$, $x = A$ so that $x = A\cos\omega t$ (Case I).

Example 13.3

A mass of 0.5 kg connected to a light spring of force constant 20 N/m oscillates on a horizontal, frictionless surface. (a) Calculate the total energy of the system and the maximum speed of the mass if the amplitude of the motion is 3 cm.

Using Eq. 13.20, we get

$$E = \frac{1}{2}kA^2 = \frac{1}{2}\left(20\,\frac{\text{N}}{\text{m}}\right)(3 \times 10^{-2}\ \text{m})^2 = 9.0 \times 10^{-3}\ \text{J}$$

When the mass is at $x = 0$, $U = 0$ and $E = \frac{1}{2}mv_{max}^2$; therefore

$$\frac{1}{2}mv_{max}^2 = 9 \times 10^{-3}\ \text{J}$$

$$v_{max} = \sqrt{\frac{18 \times 10^{-3}\ \text{J}}{0.5\ \text{kg}}} = 0.19\ \text{m/s}$$

(b) What is the velocity of the mass when the displacement is equal to 2 cm?

We can apply Eq. 13.21 directly:

$$v = \pm\sqrt{\frac{k}{m}(A^2 - x^2)} = \pm\sqrt{\frac{20}{0.5}(3^2 - 2^2) \times 10^{-4}}$$

$$= \pm 0.14\ \text{m/s}$$

The positive and negative signs indicate that the mass could be moving to the right or left at this instant.

(c) Compute the kinetic and potential energies of the system when the displacement equals 2 cm.

Using the result to (b), we get

$$K = \frac{1}{2}mv^2 = \frac{1}{2}(0.5\ \text{kg})(0.14\ \text{m/s})^2 = 5.0 \times 10^{-3}\ \text{J}$$

$$U = \frac{1}{2}kx^2 = \frac{1}{2}\left(20\,\frac{\text{N}}{\text{m}}\right)(2 \times 10^{-2}\ \text{m})^2 = 4.0 \times 10^{-3}\ \text{J}$$

Note that the sum $K + U$ equals the total energy, E.

Q8. Explain why the kinetic and potential energies of a mass-spring system can never be negative.

Q9. A mass-spring system undergoes simple harmonic motion with an amplitude A. Does the total energy change if the mass is doubled but the amplitude is not changed? Do the kinetic and potential energies depend on the mass? Explain.

13.4 THE PENDULUM

The Simple Pendulum

The simple pendulum is another mechanical system that exhibits periodic, oscillatory motion. It consists of a point mass m suspended by a light string of length L, where the upper end of the string is fixed as in Fig. 13.10. The motion occurs in a vertical plane and is driven by the force of gravity. We shall show that the motion is that of a simple harmonic oscillator, provided the angle θ that the pendulum makes with the vertical is small.

The forces acting on the mass are the tension, T, acting along the string, and the weight mg. The tangential component of the weight, $mg\sin\theta$, always acts toward $\theta = 0$, opposite the displacement. Therefore, the tangential force is a restoring force, and we can write the equation of motion in the tangential direction

$$F_t = -mg\sin\theta = m\frac{d^2s}{dt^2}$$

where s is the displacement measured along the arc and the minus sign indicates that F_t acts toward the equilibrium position. Since $s = L\theta$ and L is constant, this equation reduces to

$$\frac{d^2\theta}{dt^2} = -\frac{g}{L}\sin\theta$$

The right side is proportional to $\sin\theta$, rather than to θ; hence we conclude that the motion is not simple harmonic motion. However, if we assume that θ is *small*, we

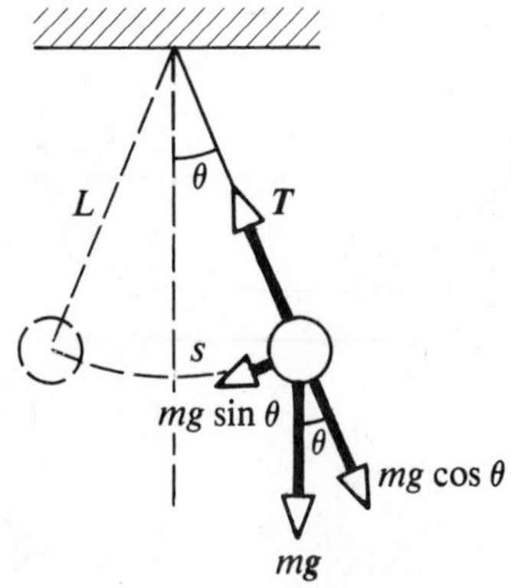

Figure 13.10 The simple pendulum oscillates with simple harmonic motion about the equilibrium position ($\theta = 0$) when θ is small. The restoring force is $mg\sin\theta$, the component of weight tangent to the circle.

can use the approximation $\sin\theta \approx \theta$, where θ is measured in *radians*.[1] Therefore, the equation of motion becomes

Equation of motion for the simple pendulum (small θ)

$$\frac{d^2\theta}{dt^2} = -\frac{g}{L}\theta \tag{13.22}$$

Now we have an expression that is of exactly the same form as Eq. 13.15, and so we conclude that the motion is simple harmonic motion. That is, $\theta = \theta_0 \cos(\omega t + \delta)$, where θ_0 is the *maximum angular displacement* and the angular frequency ω is given by

Angular frequency of motion for the simple pendulum

$$\omega = \sqrt{\frac{g}{L}} \tag{13.23}$$

The period of the motion is

Period of motion for the simple pendulum

$$T = \frac{2\pi}{\omega} = 2\pi\sqrt{\frac{L}{g}} \tag{13.24}$$

In other words, *the period and frequency depend only on the length of the string and the acceleration of gravity*. Since the period is *independent* of the mass, we conclude that *all* simple pendula of equal length oscillate with equal periods.[2] The analogy between the motion of a simple pendulum and the mass-spring system is illustrated in Fig. 13.9.

The simple pendulum is commonly used as a timekeeper. It is also a convenient device for making precise measurements of the acceleration of gravity. Such measurements are important since variations in local values of g can provide information on the location of oil and other valuable underground resources.

Example 13.4

(a) Determine the length of a simple pendulum if you want its period to be 1.00 s.

Using $T = 2\pi\sqrt{L/g}$ and solving for L, we get

$$L = \frac{g}{4\pi^2}T^2 = \left(\frac{9.80}{4\pi^2}\ \text{m/s}^2\right)(1.00\ \text{s})^2 = 0.248\ \text{m}$$

(b) Suppose the pendulum described in (a) is taken to the moon, where the acceleration due to gravity is 1.67 m/s². What would its period be there?

$$T = 2\pi\sqrt{\frac{L}{g_m}} = 2\pi\sqrt{\frac{0.248\ \text{m}}{1.67\ \text{m/s}^2}} = 2.42\ \text{s}$$

The Physical Pendulum

A physical, or compound, pendulum consists of any rigid body suspended from a fixed axis that does not pass through the body's center of mass. The system will oscillate when displaced from its equilibrium position. Consider a rigid body pivoted at a point O that is a distance d from the center of mass (Fig. 13.11). The torque

[1]This approximation can be understood by examining the series expansion for $\sin\theta$, which is $\sin\theta = \theta - \theta^3/3! + \cdots$. For small values of θ, we see that $\sin\theta \approx \theta$. The difference between θ and $\sin\theta$ for $\theta = 15°$ is only about 1%.

[2]The period of oscillation for the simple pendulum with arbitrary amplitude is

$$T = 2\pi\sqrt{\frac{L}{g}}\left(1 + \frac{1}{4}\sin^2\frac{\theta_0}{2} + \frac{9}{64}\sin^4\frac{\theta_0}{2} + \cdots\right)$$

where θ_0 is the maximum angular displacement in radians.

about O is provided by the force of gravity, and its magnitude is $mgd \sin\theta$. Using the fact that $\tau = I\alpha$, where I is the moment of inertia about the axis through O, we get

$$-mgd \sin\theta = I\frac{d^2\theta}{dt^2}$$

The minus sign on the left indicates that the torque about O tends to decrease θ. That is, the force of gravity produces a restoring torque.

If we again assume that θ is small, then the approximation $\sin\theta \approx \theta$ is valid and the equation of motion reduces to

$$\frac{d^2\theta}{dt^2} = -\left(\frac{mgd}{I}\right)\theta = -\omega^2\theta \qquad (13.25)$$

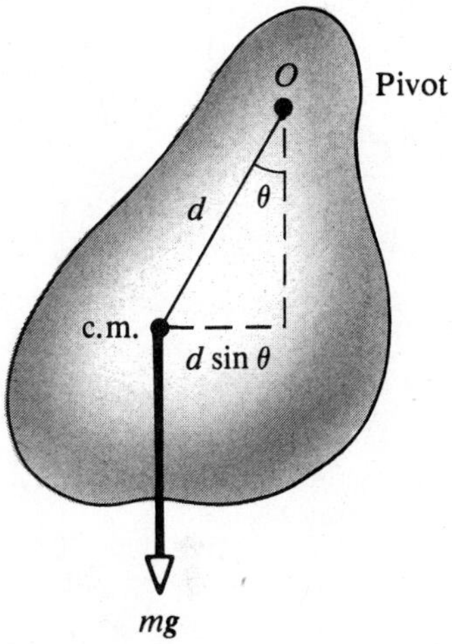

Figure 13.11 The physical pendulum consists of a rigid body pivoted at the point O, and not through the center of mass. At equilibrium, the weight vector passes through O, corresponding to $\theta = 0$. The restoring torque about O when the system is displaced through an angle θ is $mgd \sin\theta$.

Thus, we note that the equation is of the same form as Eq. 13.15, and so the motion is simple harmonic motion. That is, the solution of Eq. 13.25 is $\theta = \theta_0 \cos(\omega t + \delta)$, where θ_0 is the maximum angular displacement and

$$\omega = \sqrt{\frac{mgd}{I}}$$

The period is given by

$$T = \frac{2\pi}{\omega} = 2\pi\sqrt{\frac{I}{mgd}} \qquad (13.26)$$

One can use this result to measure the moment of inertia of a planar rigid body. If the location of the center of mass, and hence of d, are known, the moment of inertia can be obtained through a measurement of the period. Finally, note that Eq. 13.26 reduces to the period of a simple pendulum (Eq. 13.24) when $I = Md^2$, that is, when all the mass is concentrated at the center of mass.

Example 13.5 *A Swinging Rod*

A uniform rod of mass M and length L is pivoted about one end and oscillates in a vertical plane (Fig. 13.12). Find the period of oscillation if the amplitude of the motion is small.

Solution: In Chapter 10 we found that the moment of inertia of a uniform rod about an axis through one end is $\frac{1}{3}ML^2$. The distance d from the pivot to the center of mass is $L/2$. Substituting these quantities into Eq. 13.26 gives

$$T = 2\pi\sqrt{\frac{\frac{1}{3}ML^2}{Mg\frac{L}{2}}} = 2\pi\sqrt{\frac{2L}{3g}}$$

For example, the period of a meter stick ($L = 1$ m) pivoted about one end is calculated to be $T = 1.64$ s. You may want to check this out in the laboratory or at home.

Comment: In one of the early moon landings, an astronaut walking on the moon's surface had a belt hanging from his spacesuit, and the belt oscillated as a compound pendulum. A scientist on earth observed this motion on TV and was able to estimate the acceleration of gravity on the moon from this observation. How do you suppose this calculation was done?

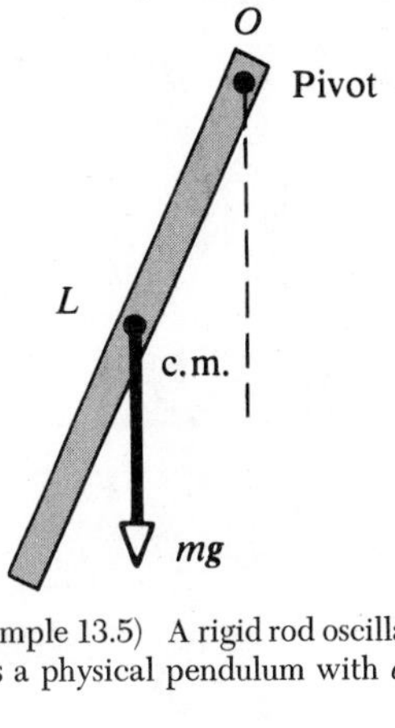

Figure 13.12 (Example 13.5) A rigid rod oscillating about a pivot through one end is a physical pendulum with $d = L/2$ and $I_0 = \frac{1}{3}ML^2$.

Torsional Pendulum

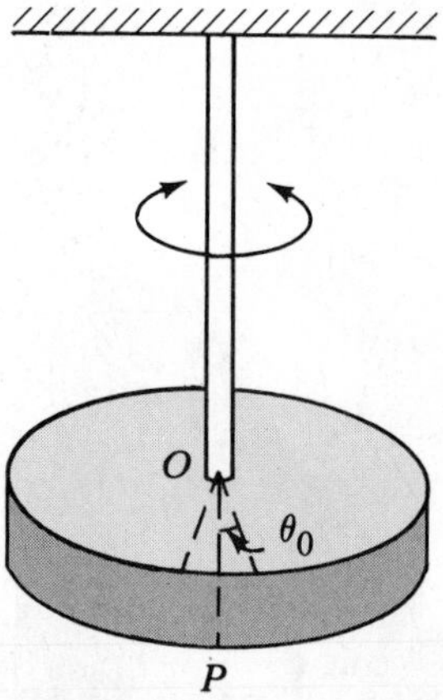

Figure 13.13 A torsional pendulum consists of a rigid body suspended by a wire attached to a rigid support. The body oscillates about the line *OP* with an amplitude θ_0.

Figure 13.13 shows a rigid body suspended by a wire attached at the top to a fixed support. When the body is twisted through some small angle θ, the twisted wire exerts a restoring torque on the body proportional to the angular displacement. That is,

$$\tau = -\kappa\theta$$

where κ (the Greek letter kappa) is called the *torsion constant* of the support wire. The value of κ can be obtained by applying a known torque to twist the wire through a measurable angle θ. Applying Newton's second law for rotational motion gives

$$\tau = -\kappa\theta = I\frac{d^2\theta}{dt^2}$$

$$\frac{d^2\theta}{dt^2} = -\frac{\kappa}{I}\theta \tag{13.27}$$

Again, this is the equation of motion for a simple harmonic oscillator, with $\omega = \sqrt{\kappa/I}$ and a period

$$T = 2\pi\sqrt{\frac{I}{\kappa}} \tag{13.28}$$

This system is called a *torsional pendulum*. The balance wheel of a watch oscillates as a torsional pendulum, energized by the mainspring. Torsional oscillations are also involved in laboratory galvanometers and the Cavendish balance.

Q10. What happens to the period of a simple pendulum if its length is doubled? What happens if the mass that is suspended is doubled?

Q11. A simple pendulum is suspended from the ceiling of a stationary elevator, and the period is determined. Describe the changes, if any, in the period if the elevator (a) accelerates upward, (b) accelerates downward, and (c) moves with constant velocity.

Q12. A simple pendulum undergoes simple harmonic motion when θ is small. Will the motion be *periodic* if θ is large? How does the period of motion change as θ increases?

13.5 COMPARING SIMPLE HARMONIC MOTION WITH UNIFORM CIRCULAR MOTION

Many aspects of simple harmonic motion along a straight line can be better understood and visualized by showing their relationships to uniform circular motion. Consider a particle at point *P* moving in a circle of radius *A* with constant angular velocity ω (Fig. 13.14a). We shall refer to this circle as the *reference circle* for the motion. As the particle rotates, the position vector of the particle rotates about the origin, *O*. At some instant of time, *t*, the angle between *OP* and the *x* axis is $\omega t + \delta$, where δ is the angle that *OP* makes with the *x* axis at $t = 0$. We take this as our reference point for measuring the angular displacement. As the particle rotates on the reference circle, the angle that *OP* makes with the *x* axis *changes* with time. Furthermore, the projection of *P* onto the *x* axis, labeled point *Q*, moves back

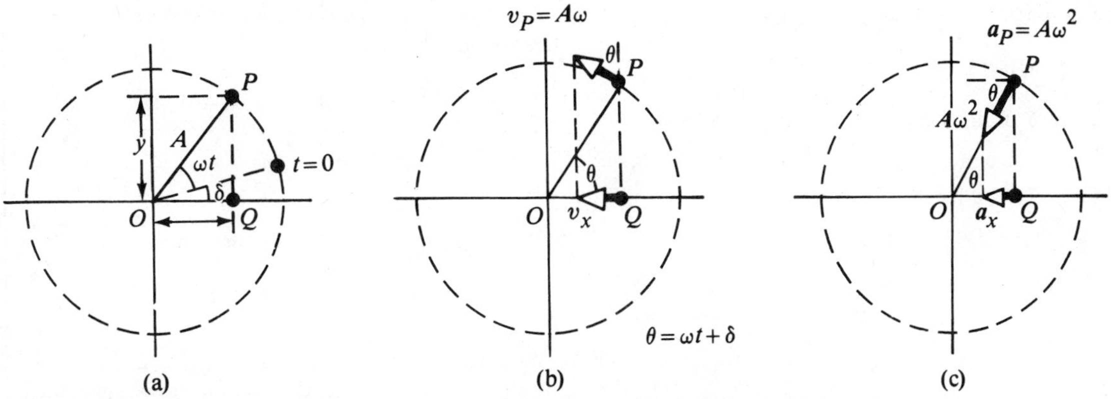

Figure 13.14 Relationship between the uniform circular motion of a point P and the simple harmonic motion of the point Q. A particle at P moves in a circle of radius A with constant angular speed ω. (a) The x components of the points P and Q are equal and vary in time as $x = A\cos(\omega t + \delta)$. (b) The x component of velocity of P equals the velocity of Q. (c) The x component of the acceleration of P equals the acceleration of Q.

and forth along a line parallel to the diameter of the reference circle, between the limits $x = \pm A$.

Note that points P and Q have the *same* x coordinate. From the right triangle OPQ, we see that the x coordinate of P and Q is given by

$$x = A\cos(\omega t + \delta) \tag{13.29}$$

This expression shows that the point Q moves with simple harmonic motion along the x axis. Therefore, we conclude that *simple harmonic motion along a straight line can be represented by the projection of uniform circular motion along a diameter.* By a similar argument, you can see from Fig. 13.14a that the projection of P along the y axis also exhibits simple harmonic motion. Therefore, *uniform circular motion can be considered a combination of two simple harmonic motions,* one along x and one along y, where the two differ in phase by 90°.

The geometric interpretation we have presented shows that the time for one complete revolution of the point P on the reference circle is equal to the period of motion, T, for simple harmonic motion between $x = \pm A$. That is, the angular speed of the point P is the same as the angular frequency, ω, of simple harmonic motion along the x axis. The phase constant δ for simple harmonic motion corresponds to the initial angle that OP makes with the x axis. The radius of the reference circle, A, equals the amplitude of the simple harmonic motion.

Since the relationship between linear and angular velocity for circular motion is $v = r\omega$, the particle moving on the reference circle of radius A has a velocity of magnitude $A\omega$. From the geometry in Fig. 13.14b, we see that the x component of this velocity is given by $-A\omega\sin(\omega t + \delta)$. By definition, the point Q has a velocity given by dx/dt. Differentiating Eq. 13.29 with respect to time, we find that the velocity of Q is the same as the x component of velocity of P.

The acceleration of the point P on the reference circle is directed radially inward toward O and has a magnitude given by $v^2/A = A\omega^2$. From the geometry in Fig. 13.14c, we see that the x component of this acceleration is $-\omega^2 A\cos(\omega t + \delta)$. This also coincides with the acceleration of the projected point Q along the x axis, as you can easily verify from Eq. 13.29.

Example 13.6

A particle rotates counterclockwise in a circle of radius 3.0 m with a constant angular speed of 8 rad/s, as in Fig. 13.14. At $t = 0$, the particle has an x coordinate of 2.0 m. (a) Determine the x coordinate as a function of time.

Since the amplitude of the particle's motion equals the

radius of the circle and $\omega = 8$ rad/s, we have

$$x = A\cos(\omega t + \delta) = 3.0\cos(8t + \delta)$$

We can evaluate δ using the initial condition that $x =$ 2.0 m at $t = 0$:

$$2.0 = 3.0\cos(0 + \delta)$$

$$\delta = \cos^{-1}\left(\frac{2}{3}\right) = 48° = 0.84 \text{ rad}$$

Therefore, the x coordinate versus time is of the form

$$x = 3.0\cos(8t + 0.84) \text{ m}$$

Note that the angles in the cosine function are in radians.

(b) Find the x components of the particle's velocity and acceleration at any time t.

$$v_x = \frac{dx}{dt} = (-3.0)(8)\sin(8t + 0.84)$$

$$= -24\sin(8t + 0.84)\ \frac{\text{m}}{\text{s}}$$

$$a_x = \frac{dv_x}{dt} = (-24)(8)\cos(8t + 0.84)$$

$$= -192\cos(8t + 0.84)\ \frac{\text{m}}{\text{s}^2}$$

From these results, we conclude that $v_{max} = 24$ m/s and $a_{max} = 192$ m/s^2. Note that these values also equal the tangential velocity, $A\omega$, and centripetal acceleration, $A\omega^2$.

13.6 DAMPED OSCILLATIONS

The oscillatory motions we have considered so far have dealt with an ideal system, that is, one that oscillates indefinitely under the action of a linear restoring force. In realistic systems, dissipative forces, such as friction, are present and retard the motion of the system. Consequently, the mechanical energy of the system will diminish in time, and the motion is said to be *damped.*

One common type of retarding force, which we discussed in Chapter 6, is proportional to the velocity and acts in the direction opposite the motion. This is often observed for the motion of an object through a liquid, as shown in Fig. 13.15. Since the retarding force can be expressed as $\mathbf{R} = -b\mathbf{v}$, where b is a constant, and the restoring force is $-kx$, we can write Newton's second law

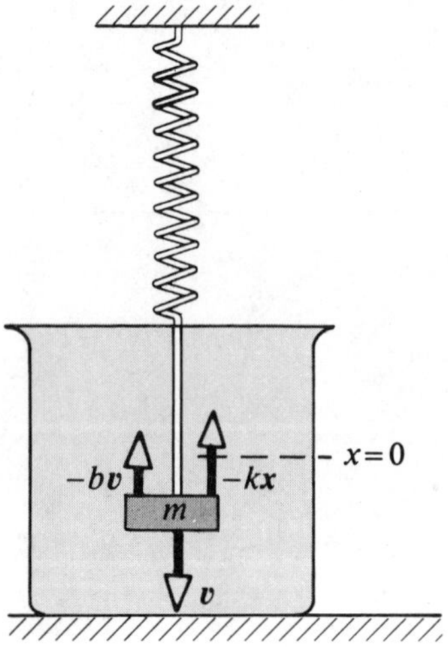

Figure 13.15 One example of a damped oscillator is a body submersed in a fluid. When the body is in motion, a damping force $-bv$ acts in addition to the linear restoring force $-kx$.

$$\Sigma F_x = -kx - bv = ma_x$$

$$-kx - b\frac{dx}{dt} = m\frac{d^2x}{dt^2} \qquad (13.30)$$

The solution of this equation requires mathematics which may not be familiar to you as yet, and so it will simply be stated without proof. When the retarding force is small compared with kx, that is, when b is small, the solution to Eq. 13.30 is

$$x = A\,e^{-\frac{b}{2m}t}\cos(\omega t + \delta) \qquad (13.31)$$

where the frequency of motion is

$$\omega = \sqrt{\frac{k}{m} - \left(\frac{b}{2m}\right)^2} \qquad (13.32)$$

This can be verified by substitution of the solution into Eq. 13.30. Figure 13.16 shows the displacement as a function of time in this case. We see that *when the dissipative force is small compared with the restoring force, the oscillatory character of the motion is preserved but the amplitude of vibration decreases in time,* and the motion will ultimately cease. This is known as an *underdamped oscillator.* The dotted line in Fig. 13.16, which is the *envelope* of the oscillatory curve, represents the exponential factor that appears in Eq. 13.31. This shows that *the amplitude decays exponentially with time.* For motion with a given spring constant and particle mass, the oscillations dampen more rapidly as the maximum value of the dissipative force approaches the maximum value of the restoring force.

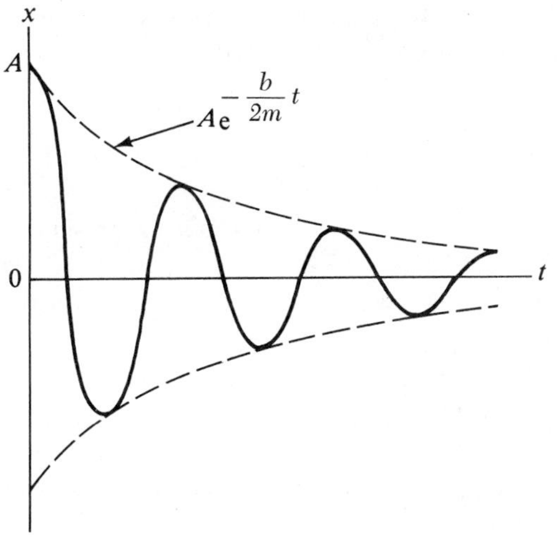

Figure 13.16 Graph of the displacement versus time for an underdamped oscillator. Note the decrease in amplitude with time.

It is convenient to express the frequency of vibration in the form

$$\omega = \sqrt{\omega_0^2 - \left(\frac{b}{2m}\right)^2}$$

where $\omega_0 = \sqrt{k/m}$ represents the frequency of oscillation in the absence of a resistive force (the undamped oscillator). In other words, when $b = 0$, the resistive force is zero and the system oscillates with its natural frequency, ω_0. As the magnitude of the resistive force approaches the value of the restoring force in the spring, the oscillations dampen more rapidly. When b reaches a critical value b_c such that $b_c/2m = \omega_0$, the system does not oscillate and is said to be *critically damped*. In this case, the system returns to equilibrium in an exponential manner with time, as in Fig. 13.17.

If the medium is so viscous that the resistive force is greater than the restoring force, that is, if $b/2m > \omega_0$, the system will be *overdamped*. Again, the displaced system does not oscillate but simply returns to its equilibrium position. As the damping increases, the time it takes the displacement to reach equilibrium also increases, as indicated in Fig. 13.17. In any case, when friction is present, the energy of the oscillator will eventually fall to zero. The loss in mechanical energy dissipates into heat energy of the resistive medium.

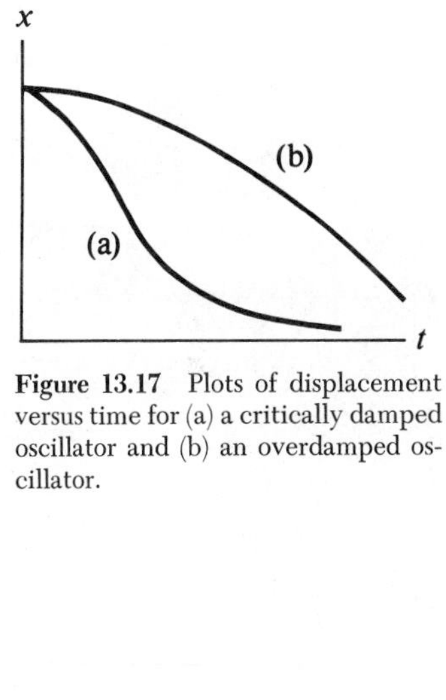

Figure 13.17 Plots of displacement versus time for (a) a critically damped oscillator and (b) an overdamped oscillator.

Q13. Give a few examples of damped oscillations that are commonly observed.

Q14. Will damped oscillations occur for any values of b and k? Explain.

13.7 FORCED OSCILLATIONS

We have seen that the energy of a damped oscillator decreases in time as a result of the dissipative force. It is possible to compensate for this energy loss by applying an external force that does positive work on the system. At any instant, energy can be put into the system by an applied force that acts in the direction of motion of the oscillator. For example, a child on a swing can be kept in motion by appropriately timed "pushes." The amplitude of motion will remain constant if the energy input per cycle of motion exactly equals the energy lost as a result of friction.

A common example of a forced oscillator is a damped oscillator driven by an external force that varies harmonically, such as $F = F_0 \cos\omega t$, where ω is the angular frequency of the force and F_0 is a constant. Adding this driving force to the left side of Eq. 13.30 gives

$$F_0 \cos\omega t - b\frac{dx}{dt} - kx = m\frac{d^2x}{dt^2} \tag{13.33}$$

Again, the solution of this equation is rather lengthy and will not be presented. However, after a sufficiently long period of time, when the energy input per cycle equals the energy lost per cycle, a *steady-state* condition is reached in which the oscillations proceed with constant amplitude. At this time, when the system is in steady state, Eq. 13.33 has the following solution:

$$x = A\cos(\omega t + \delta) \tag{13.34}$$

where

$$A = \frac{F_0/m}{\sqrt{(\omega^2 - {\omega_0}^2)^2 + \left(\frac{b\omega}{m}\right)^2}} \tag{13.35}$$

and where $\omega_0 = \sqrt{k/m}$ is the frequency of the undamped oscillator ($b = 0$).

Equation 13.35 shows that the motion of the forced oscillator is not damped since it is being driven by an external force. That is, the external agent provides the necessary energy to overcome the losses due to the resistive force. Note that the mass oscillates at the frequency of the driving force, ω. Furthermore, *the amplitude is a maximum when the frequency of the driving force equals the natural frequency of oscillation, or when* $\omega = \omega_0$. This is because the denominator of Eq. 13.35 is a minimum when $\omega = \omega_0$, corresponding to a maximum value of A. The frequency ω_0 is called the *resonance frequency* of the system. When the frequency of the driving force equals ω_0, the oscillator is said to be in *resonance* with the driving force.

Physically, the reason for large-amplitude oscillations at the resonance frequency is that energy is being transferred to the system under the most favorable conditions. This can be better understood by taking the first time derivative of x, which gives an expression of the velocity of the oscillator. In doing so, one finds that $v \sim \sin(\omega t + \delta)$. When the applied force is in phase with v, the rate at which work is done on the oscillator by the force $\mathbf{F}$ (or the power) equals Fv. Since the quantity Fv is always positive when $\mathbf{F}$ and $\mathbf{v}$ are in phase, we conclude that *at resonance the applied force is in phase with the velocity and the power transferred to the oscillator is a maximum.*

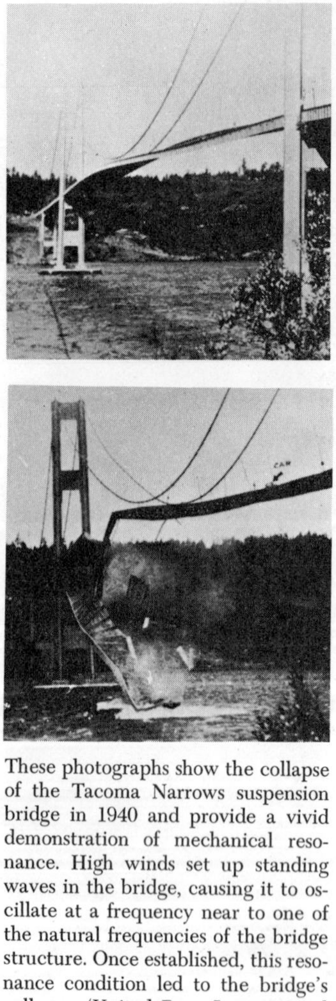

These photographs show the collapse of the Tacoma Narrows suspension bridge in 1940 and provide a vivid demonstration of mechanical resonance. High winds set up standing waves in the bridge, causing it to oscillate at a frequency near to one of the natural frequencies of the bridge structure. Once established, this resonance condition led to the bridge's collapse. (United Press International Photo)

A graph of the amplitude as a function of frequency for the forced oscillator with and without a resistive force is shown in Fig. 13.18. Note that the amplitude increases with decreasing damping ($b \to 0$). Furthermore, the resonance curve is broadened as the damping increases. Under steady-state conditions, and at any driving frequency, the energy transferred into the system equals the energy lost because of the damping force; hence the total energy of the oscillator remains constant. In the absence of a damping force ($b = 0$), we see from Eq. 13.35 that the steady-state amplitude approaches infinity as $\omega \to \omega_0$. In other words, if there are no losses in the system, and we continue to drive an initially motionless oscillator with a sinusoidal force that is in phase with the velocity, the amplitude of motion will build up without limit (Fig. 13.18). This does not occur in practice since some damping will always be present. That is, at resonance the amplitude will be large but finite for small damping.

One experiment that demonstrates a resonance phenomenon is illustrated in Fig. 13.19. Several pendulums of different lengths are suspended from a common beam.

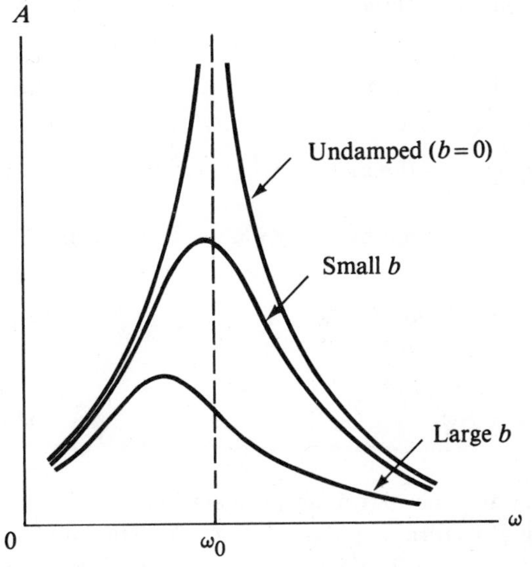

Figure 13.18 Graph of the amplitude versus frequency for a damped oscillator when a periodic driving force is present. When the frequency of the driving force equals the natural frequency, ω_0, resonance occurs. Note that the shape of the resonance curve depends on the size of the damping coefficient, b.

If one of them, such as P, is set in motion, the others will begin to oscillate, since they are coupled by the beam. Of those that are forced into oscillation by this coupling, pendulum Q, whose length is the same as that of P (and hence the two pendula have the same natural frequency), will oscillate with the greatest amplitude.

Later in the text we shall see that the phenomenon of resonance appears in other areas of physics. For example, certain electrical circuits have natural (or resonant) frequencies. A structure such as a bridge has natural frequencies, which can be set into resonance by an appropriate driving force. A striking example of such a structural resonance occurred in 1940, when the Tacoma bridge in Washington was set into torsional oscillation by heavy winds. The amplitude of these oscillations increased steadily until the bridge ultimately was destroyed.

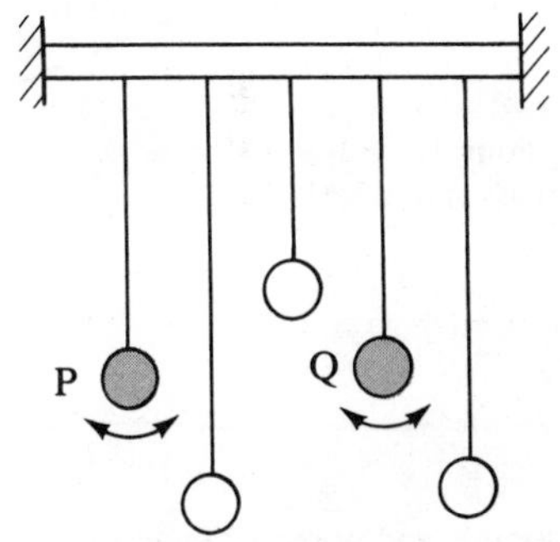

Figure 13.19 If pendulum P is set into oscillation, pendulum Q will eventually oscillate because of the coupling between them and the fact that they have the same natural frequency of vibration.

Q15. Is it possible to have damped oscillations when a system is at resonance? Explain.

Q16. At resonance, what does the phase constant δ equal in Eq. 13.34? (Hint: Compare this with the expression for the driving force, and note that the force must be in phase with the velocity at resonance.)

Q17. A platoon of soldiers marches in step along a road. Why are they ordered to break step when crossing a bridge?

13.8 SUMMARY

The position of a simple harmonic oscillator varies periodically in time according to the relation

$$x = A\cos(\omega t + \delta) \tag{13.1}$$

Displacement versus time for simple harmonic motion

where A is the amplitude of the motion, ω is the angular frequency, and δ is the phase constant. The value of δ depends on the initial position and velocity of the oscillator.

The time for one complete vibration is called the *period* of the motion, defined by

Period

$$T = \frac{2\pi}{\omega} \tag{13.2}$$

The inverse of the period is the *frequency* of the motion, which equals the number of oscillations per second.

The *velocity* and *acceleration* of a simple harmonic oscillator are given by

Velocity in simple harmonic motion

$$v = \frac{dx}{dt} = -A\omega \sin(\omega t + \delta) \tag{13.5}$$

Acceleration in simple harmonic motion

$$a = \frac{dv}{dt} = -A\omega^2 \cos(\omega t + \delta) \tag{13.6}$$

Thus, the maximum velocity is $A\omega$, and the maximum acceleration is $A\omega^2$. The velocity is zero when the oscillator is at its turning points, $x = \pm A$, and the speed is a maximum at the equilibrium position, $x = 0$. The magnitude of the acceleration is a maximum at the turning points and is zero at the equilibrium position.

A mass-spring system exhibits simple harmonic motion on a frictionless surface, with a period given by

Period of motion for mass-spring system

$$T = \frac{2\pi}{\omega} = 2\pi\sqrt{\frac{m}{k}} \tag{13.16}$$

where k is the force constant of the spring and m is the mass attached to the spring.

The kinetic energy and potential energy for a simple harmonic oscillator vary with time and are given by

Kinetic and potential energy of a simple harmonic oscillator

$$K = \frac{1}{2}mv^2 = \frac{1}{2}m\omega^2A^2 \sin^2(\omega t + \delta) \tag{13.18}$$

$$U = \frac{1}{2}kx^2 = \frac{1}{2}kA^2 \cos^2(\omega t + \delta) \tag{13.19}$$

The *total energy* of a simple harmonic oscillator is a constant of the motion and is given by

Total energy of a simple harmonic oscillator

$$E = \frac{1}{2}kA^2 \tag{13.20}$$

The potential energy of a simple harmonic oscillator is a maximum when the particle is at its turning points (maximum displacement from equilibrium) and is zero at the equilibrium position. The kinetic energy is zero at the turning points and is a maximum at the equilibrium position.

A *simple pendulum* of length L exhibits simple harmonic motion for small angular displacements from the vertical, with a *period* given by

Period of motion for a simple pendulum

$$T = 2\pi\sqrt{\frac{L}{g}} \tag{13.24}$$

That is, the period is *independent* of the suspended mass.

A *physical pendulum* exhibits simple harmonic motion about a pivot that does not go through the center of mass. The period of this motion is

$$T = 2\pi\sqrt{\frac{I}{mgd}} \qquad (13.26)$$

Period of motion for a compound pendulum

where I is the moment of inertia about an axis through the pivot and d is the distance from the pivot to the center of mass.

Damped oscillations occur in a system in which a dissipative force acts in addition to the linear restoring force. If such a system is set into motion and then left to itself, the mechanical energy decreases in time because of the presence of the nonconservative damping force. It is possible to compensate for this loss in energy by driving the system with an external periodic force that is in phase with the motion of the system. When the frequency of the driving force matches the natural frequency of the undamped oscillator that starts its motion from rest, energy is continuously transferred to the oscillator and its amplitude increases without limit.

EXERCISES

Section 13.1 Simple Harmonic Motion

1. The displacement of a particle is given by $x = 4\cos(3\pi t + \pi)$, where x is in m and t is in s. Determine (a) the frequency and period of the motion, (b) the amplitude of the motion, (c) the phase constant, and (d) the position of the particle at $t = 0$.

2. For the particle described in Exercise 1, determine (a) the velocity at any time t, (b) the acceleration at any time, (c) the maximum velocity and maximum acceleration, and (d) the velocity and acceleration at $t = 0$.

3. A particle oscillates with simple harmonic motion such that its displacement varies as $x = 5\cos(2t + \pi/6)$, where x is in cm and t is in s. At $t = 0$, find (a) the displacement of the particle, (b) its velocity, and (c) its acceleration. (d) Find the period and amplitude of the motion.

4. A particle moving with simple harmonic motion travels a total distance of 20 cm in each cycle of its motion, and its maximum acceleration is 50 m/s^2. Find (a) the angular frequency of the motion and (b) the maximum speed of the particle.

5. The displacement of a body is given by $x = 8.0\cos(2t + \pi/3)$, where x is in cm and t is in s. Calculate (a) the velocity and acceleration at $t = \pi/2$ s, (b) the maximum speed and the earliest time ($t > 0$) at which the particle has this speed, and (c) the maximum acceleration and the earliest time ($t > 0$) at which the particle has this acceleration.

6. At $t = 0$ a particle moving with simple harmonic motion is at $x_0 = 2$ cm, where its velocity is $v_0 = -24$ cm/s. If the period of its motion is 0.5 s and the frequency is 2 Hz, find (a) the phase constant; (b) the amplitude; (c) the displacement, velocity, and acceleration as functions of time; and (d) the maximum speed and maximum acceleration.

7. A particle moving along the x axis with simple harmonic motion starts from the origin at $t = 0$ and moves toward the right. If the amplitude of its motion is 2 cm and the frequency is 1.5 Hz, (a) show that its displacement is given by $x = 2\sin 3\pi t$ cm. Determine (b) the maximum speed and the earliest time ($t > 0$) at which the particle has this speed, (c) the maximum acceleration and the earliest time ($t > 0$) at which the particle has this acceleration, and (d) the total *distance* traveled between $t = 0$ and $t = 1$ s.

Section 13.2 Mass Attached to a Spring (neglect spring masses)

8. A spring stretches by 3.9 cm when a 10-g mass is hung from it. If a total mass of 25 g attached to this spring oscillates in simple harmonic motion (Fig. 13.2), calculate the period of motion.

9. The frequency of vibration of a mass-spring system is 5 Hz when a 4-g mass is attached to the spring. What is the force constant of the spring?

10. A 1-kg mass attached to a spring of force constant 25 N/m oscillates on a horizontal, frictionless surface. At $t = 0$, the mass is at $x = -3$ cm. (That is, the spring is compressed by 3 cm.) Find (a) the period of its motion, (b) the maximum values of its speed and acceleration, and (c) the displacement, velocity, and acceleration as functions of time.

11. A simple harmonic oscillator takes 12 s to undergo 5 complete vibrations. Find (a) the period of its motion, (b) the frequency in Hz, and (c) the angular frequency in rad/s.

12. A mass-spring system oscillates such that the displacement is given by $x = 0.25 \cos 2\pi t$ m. (a) Find the speed and acceleration of the mass when $x =$ 0.10 m. (b) Determine the maximum speed and maximum acceleration.

13. A 0.5-kg mass attached to a spring of force constant 8 N/m vibrates with simple harmonic motion with an amplitude of 10 cm. Calculate (a) the maximum value of its speed and acceleration, (b) the speed and acceleration when the mass is at $x = 6$ cm from the equilibrium position, and (c) the time it takes the mass to move from $x = 0$ to $x = 8$ cm.

Section 13.3 Energy of the Simple Harmonic Oscillator (neglect spring masses)

14. A 200-g mass is attached to a spring and executes simple harmonic motion with a period of 0.25 s. If the total energy of the system is 2 J, find (a) the force constant of the spring and (b) the amplitude of the motion.

15. A mass-spring system oscillates with an amplitude of 3.5 cm. If the spring constant is 250 N/m and the mass is 0.5 kg, determine (a) the mechanical energy of the system, (b) the maximum speed of the mass, and (c) the maximum acceleration.

16. A simple harmonic oscillator has a total energy E. (a) Determine the kinetic and potential energies when the displacement equals one half the amplitude. (b) For what value of the displacement does the kinetic energy equal the potential energy?

17. The amplitude of a system moving with simple harmonic motion is doubled. Determine the change in (a) the total energy, (b) the maximum velocity, (c) the maximum acceleration, and (d) the period.

18. A mass-spring system of force constant 50 N/m undergoes simple harmonic motion on a horizontal surface with an amplitude of 12 cm. (a) What is the total energy of the system? (b) What is the kinetic energy of the system when the mass is 9 cm from equilibrium? (c) What is its potential energy when $x = 9$ cm?

19. A particle executes simple harmonic motion with an amplitude of 3.0 cm. At what displacement from the midpoint of its motion will its speed equal one half of its maximum speed?

20. Using Eqs. 13.18 and 13.19, plot (a) the kinetic energy versus time and (b) the potential energy versus time for a simple harmonic oscillator. For convenience, take $\delta = 0$. What features do these graphs illustrate?

Section 13.4 The Pendulum

21. Calculate the frequency and period of a simple pendulum of length 10 m.

22. A simple pendulum has a period of 2.50 s. (a) What is its length? (b) What would its period be on the moon?

23. A simple pendulum 2.00 m in length oscillates in a location where $g = 9.80$ m/s^2. How many complete oscillations will it make in 5 min?

24. If the length of a simple pendulum is quadrupled, what happens to (a) its frequency and (b) its period?

25. A simple pendulum has a length of 3.00 m. Determine the *change* in its period if it is taken from a point where $g = 9.80$ m/s^2 to a higher elevation, where $g = 9.79$ m/s^2.

26. A uniform rod is pivoted at one end as in Fig. 13.12. If the rod swings with simple harmonic motion, what must its length be in order that its period be equal to that of a simple pendulum 1 m long?

27. A physical pendulum in the form of a planar body exhibits simple harmonic motion with a frequency of 1.5 Hz. If the pendulum has a mass of 2.2 kg and the pivot is located 0.35 m from the center of mass, determine the moment of inertia of the pendulum.

28. A circular hoop of radius R is hung over a knife edge. Show that its period of oscillation is equal to that of a simple pendulum of length $2R$.

Section 13.6 Damped Oscillations

29. Show that Eq. 13.31 is a solution of Eq. 13.30 provided that $b^2 < 4mk$.

30. Show that the damping constant, b, has units of kg/s.

31. Show that the time rate of change of mechanical energy for a damped, undriven oscillator is given by $dE/dt = -bv^2$ and hence is *always negative.* (Hint: Differentiate the expression for the mechanical energy of an oscillator, $E = \frac{1}{2}mv^2 + \frac{1}{2}kx^2$, and make use of Eq. 13.30.)

Section 13.7 Forced Oscillations

32. For an *undamped* forced oscillator ($b = 0$), show that Eq. 13.34 is a solution of Eq. 13.33, with an amplitude given by Eq. 13.35.

33. A 2-kg mass attached to a spring is driven by an external force $F = 3 \cos 2\pi t$ N. If the force constant of the spring is 20 N/m, determine (a) the period and (b) the amplitude of the motion. (Hint: Assume that there is *no* damping, that is, $b = 0$, and make use of Eq. 13.35.)

34. Calculate the resonant frequencies of the following systems: (a) a 3-kg mass attached to a spring of force constant 240 N/m, (b) a simple pendulum 1.5 m in length.

PROBLEMS

1. A body oscillates with simple harmonic motion according to the equation $x = -7 \cos 2\pi t$ cm. (a) Determine the velocity and acceleration as functions of time. (b) Make a table of x, v, and a versus t for the interval $t = 0$ to $t = 1$ s in steps of 0.1 s. (c) Plot x, v, and a versus time for this interval.
2. A car with bad shock absorbers bounces up and down with a period of 1.5 s after hitting a bump. The car has a mass of 1500 kg and is supported by four springs of equal force constant k. Determine a value for k.
3. A homogeneous rod of length L is pivoted at a distance d above its center of mass (Fig. 13.20). For a small displacement from the vertical equilibrium position, the rod exhibits simple harmonic motion. (a) Find the angular frequency of this motion. (b) If the rod is a meter stick pivoted at the 75-cm mark, what is its period? (The lower end is at 0-cm.)

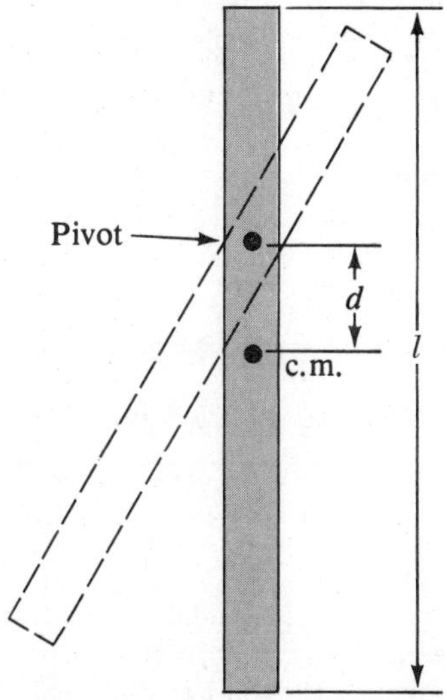

Figure 13.20 (Problem 3).

4. A 50-g mass attached to a spring moves on a horizontal, frictionless surface in simple harmonic motion. Its amplitude is 16 cm, and its period is 4 s. At $t = 0$, the mass is released from rest at $x = 16$ cm, as in Fig. 13.5. Find (a) the displacement as a function of time and its value at $t = 0.5$ s, (b) the magnitude and direction of the force acting on the mass at $t = 0.5$ s, (c) the minimum time required for the mass to reach $x = 8$ cm, (d) the velocity at any time t and the speed at $x = 8$ cm, and (e) the total mechanical energy and force constant of the spring.
5. A mass m is attached to a spring of force constant k hanging vertically (Fig. 13.21). If the mass is released when the spring is *unstretched*, show that (a) the system exhibits simple harmonic motion with a displacement measured from the unstretched position given by $y = \frac{mg}{k}(1 - \cos\omega t)$, where $\omega = \sqrt{k/m}$, and (b) the maximum tension in the spring is $2mg$.

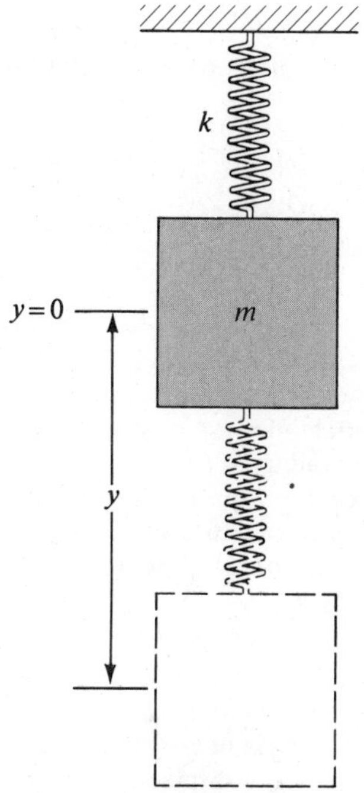

Figure 13.21 (Problem 5).

6. A horizontal platform vibrates with simple harmonic motion in the horizontal direction with a period of 2 s. A body on the platform starts to slide when the amplitude of vibration reaches 0.3 m. Find the coefficient of static friction between the body and the platform.
7. A spherical mass m of radius R is suspended from a light string of length $L - R$ (Fig. 13.22). (a) Determine the moment of inertia for this physical pendulum about the point O using the parallel-axis theorem. (b) Calculate the period for small displacements from equilibrium. (c) If $R \ll L$, show that the period is that of a simple pendulum.
8. A mass m is connected to two rubber bands of length L, each under tension T, as in Fig. 13.23. The mass is displaced by a *small* distance y vertically. Assuming the tension does not change appreciably, show that (a) the restoring force is $-(2T/L)y$ and (b) the system exhibits simple harmonic motion with an angular frequency given by $\omega = \sqrt{2T/mL}$.

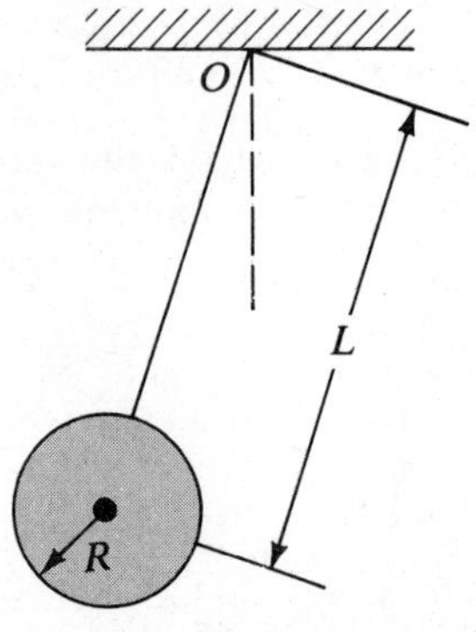

Figure 13.22 (Problem 7).

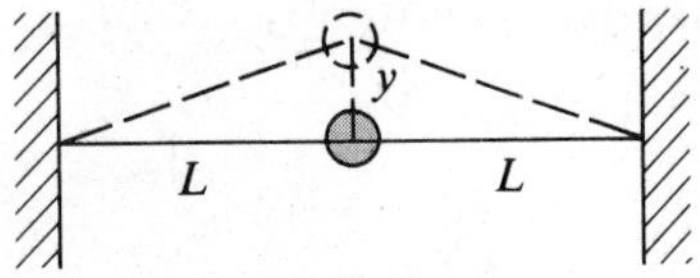

Figure 13.23 (Problem 8).

9. A particle of mass m slides inside a hemispherical bowl of radius R. Show that for small displacements from equilibrium, the particle exhibits simple harmonic motion with an angular frequency equal to that of a simple pendulum of length R. That is, $\omega = \sqrt{g/R}$.

10. A horizontal plank of mass m and length L is pivoted at one end, and the opposite end is attached to a spring of force constant k (Fig. 13.24). The moment of inertia of the plank about the pivot is $\frac{1}{3}mL^2$. If the plank is displaced a *small* angle θ from the horizontal and released, show that it will move with simple harmonic motion with an angular frequency given by $\omega = \sqrt{3k/m}$.

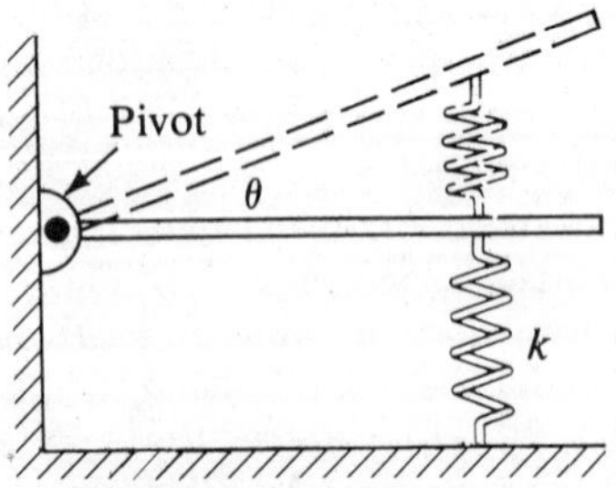

Figure 13.24 (Problem 10).

11. A mass M is attached to the end of a uniform rod of mass m and length l, which is pivoted at the top (Fig. 13.25). (a) Determine the tensions in the rod at the pivot and at the point P when the system is stationary. (b) Calculate the period of oscillation for small displacements from equilibrium, and determine this period for $l = 2$ m. (Hint: Assume the mass M is a point mass, and make use of Eq. 13.26.)

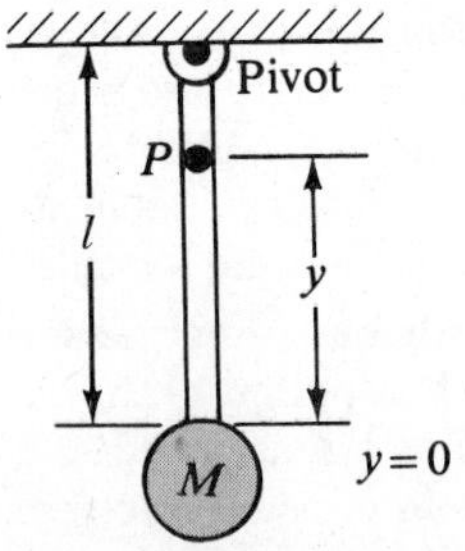

Figure 13.25 (Problem 11).

12. A mass M is connected to a spring of mass m and oscillates in simple harmonic motion on a horizontal, smooth surface (Fig. 13.26). The force constant of the spring is k and the equilibrium length is l. Find (a) the kinetic energy of the system when the mass has a speed v and (b) the period of oscillation. (Hint: Assume that all portions of the spring oscillate in phase and that the velocity of a segment dx is proportional to the distance from the fixed end; that is, $v_x = \frac{x}{l}v$. Also, note that the mass of a segment of the spring is $dm = \frac{m}{l}\,dx$.)

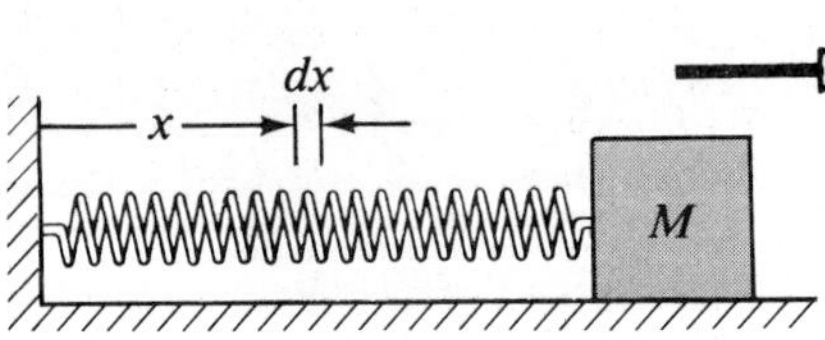

Figure 13.26 (Problem 12).

13. When the simple pendulum illustrated in Fig. 13.27 makes an angle θ with the vertical, its speed is v. (a) Calculate the total mechanical energy of the pendulum as a function of v and θ. (b) Show that when θ is small, the potential energy can be expressed as $\frac{1}{2}mgL\theta^2 = \frac{1}{2}m\omega^2s^2$. (Hint: In part (b), approximate $\cos\theta$ by $\cos\theta \approx 1 - \frac{\theta^2}{2}$.)

14. A mass m is connected to two springs of force constants k_1 and k_2 as in Figs. 13.28a and 13.28b. Show that in each case the mass exhibits simple harmonic motion with periods (a) $T = 2\pi\sqrt{\frac{m(k_1 + k_2)}{k_1k_2}}$ and (b) $T = 2\pi\sqrt{\frac{m}{k_1 + k_2}}$.

15. A pendulum of length L and mass m has a spring of force constant k connected to it at a distance h

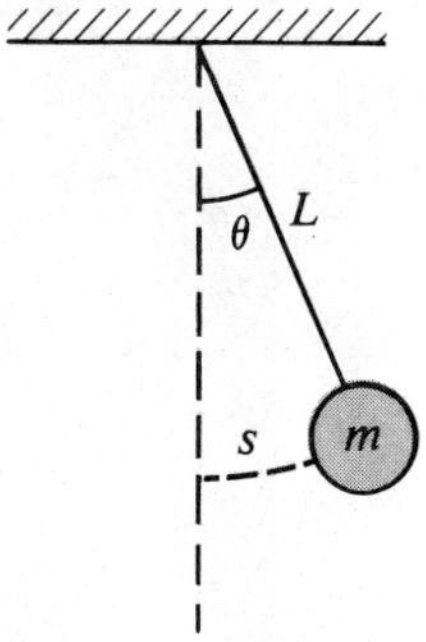

Figure 13.27 (Problem 13).

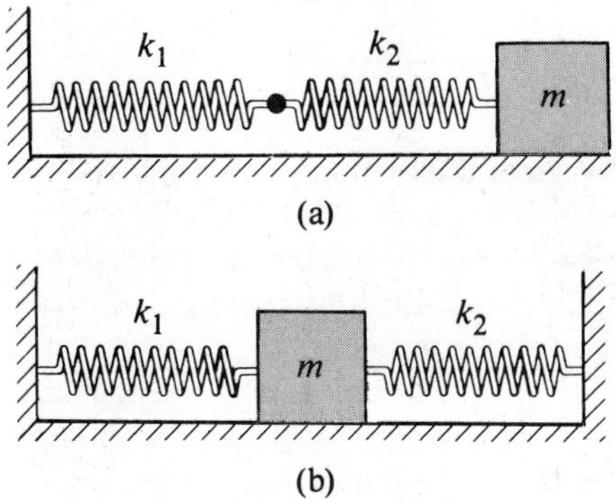

Figure 13.28 (Problem 14).

below its point of suspension (Figure 13.29). Find the frequency of vibration of the system for small values of the amplitude (small θ). (Assume the vertical suspension of length is rigid, but neglect its mass.)

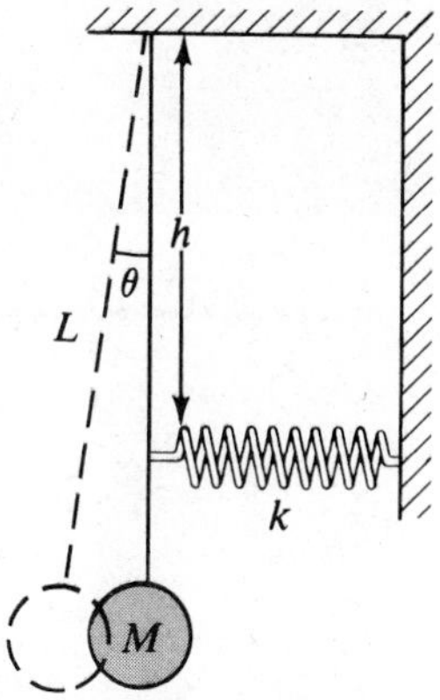

Figure 13.29 (Problem 15).

14 The Law of Universal Gravitation

NASA/JPL

The force of gravity as a fundamental interaction was introduced in Chapter 6. You should recall that the gravitational force on an atomic scale is negligibly small compared with the electrostatic and nuclear forces between elementary particles. However, nuclear forces are very short-range, and electrostatic forces between macroscopic objects are often negligible because of the high degree of charge neutrality in bulk matter. Consequently, massive bodies, such as the planets and the sun, interact primarily via gravitational forces on each other, despite their large separations and the intrinsic weakness of the gravitational force.

Isaac Newton was the first to recognize that the same force that causes objects to fall to the earth is also responsible for the motion of the moon around the earth and the motion of the planets about the sun. In a sense, we can speak of gravity as the "invisible glue" that holds the universe together.

In this chapter we shall study the law of universal gravitation in more detail. Emphasis will be placed on describing the motion of the planets, since astronomical data provide an important test of the validity of the law of universal gravitation. We shall show that the laws of planetary motion developed by Johannes Kepler (1571–1630) follow from the law of universal gravitation and the concept of the conservation of angular momentum. A general expression for the gravitational potential energy will be derived, and the energetics of planetary and satellite motion will be treated. The law of universal gravitation will also be used to determine the force between a particle and an extended body.

14.1 KEPLER'S LAWS

The movements of the planets, stars, and other celestial bodies have been observed by people for thousands of years. In early history, scientists regarded the earth as the center of the universe. This so-called geocentric model was proposed by the Greek astronomer Claudius Ptolemy in the second century A.D. and accepted for the next 1400 years. The Polish astronomer Nicolaus Copernicus (1473–1543) suggested that the earth and the other planets revolved in circular orbits about the sun (the heliocentric hypothesis).

The Danish astronomer Tycho Brahe (1546–1601), who made accurate astronomical measurements over a period of 20 years, provided the basis for the currently accepted model of the solar system. It is interesting to note that these precise observations, made on the planets and 777 stars visible to the naked eye, were carried out with a large sextant and compass, since the telescope had not yet been invented.

The German astronomer Johannes Kepler, Brahe's student, acquired these astronomical data and spent about 16 years trying to deduce a mathematical model for the motion of the planets. After many laborious calculations, he found that Brahe's precise data on the revolution of Mars about the sun provided the answer. Such data are difficult to sort out because the earth is also in motion about the sun. Kepler's analysis first showed that the concept of circular orbits about the sun had to be abandoned. He eventually discovered that the orbit of Mars could be accurately described by an ellipse with the sun at one focus. He then generalized this analysis to include the motion of all planets. The complete analysis is summarized in three statements, known as *Kepler's laws*. These empirical laws applied to the solar system are:

Kepler's laws

1. *All planets move in elliptical orbits with the sun at one of the focal points.*
2. *The radius vector drawn from the sun to any planet sweeps out equal areas in equal time intervals.*
3. *The square of the orbital period of any planet is proportional to the cube of the semi-major axis for the elliptical orbit.*

About 100 years later, Newton demonstrated that these laws were the consequence of a simple force that exists between any two masses. Newton's law of universal gravitation, together with his development of the laws of motion, provides the basis for a full mathematical solution to the motion of planets and satellites. More important, Newton's universal law of gravity correctly describes the gravitational attractive force between *any* two masses.

14.2 THE LAW OF UNIVERSAL GRAVITATION AND THE MOTION OF PLANETS

In Chapter 6 we stated that the gravitational force of attraction between any two particles having masses m_1 and m_2 separated by a distance r has the magnitude

$$F = G\frac{m_1 m_2}{r^2} \tag{14.1}$$

The universal law of gravity

where G is the universal constant of gravity, which has a magnitude in SI units of

$$G = 6.673 \times 10^{-11}\ \text{N} \cdot \text{m}^2/\text{kg}^2 \tag{14.2}$$

In formulating his law of universal gravitation, Newton used the following observation, which suggests that the gravitational force is proportional to the inverse square of the separation. Let us compare the acceleration of the moon in its orbit with the acceleration of an object falling near the earth's surface, such as the legendary apple (Fig. 14.1). Assume that both accelerations have the same cause, namely, the gravitational attraction of the earth. From the inverse-square law, it follows that the acceleration of the moon toward the earth (centripetal acceleration) should be proportional to $1/r_m{}^2$, where r_m is the earth-moon separation. Furthermore, the acceleration of the apple toward the earth should vary as $1/R_e{}^2$, where R_e is the radius of the earth. Using the values $r_m = 3.84 \times 10^8$ m and $R_e = 6.37 \times 10^6$ m, the ratio of the moon's acceleration, a_m, to the apple's acceleration, g, is predicted to be

$$\frac{a_m}{g} = \frac{(1/r_m)^2}{(1/R_e)^2} = \frac{R_e{}^2}{r_m{}^2} = \frac{(6.37 \times 10^6\ \text{m})^2}{(3.84 \times 10^8\ \text{m})^2} = 2.75 \times 10^{-4}$$

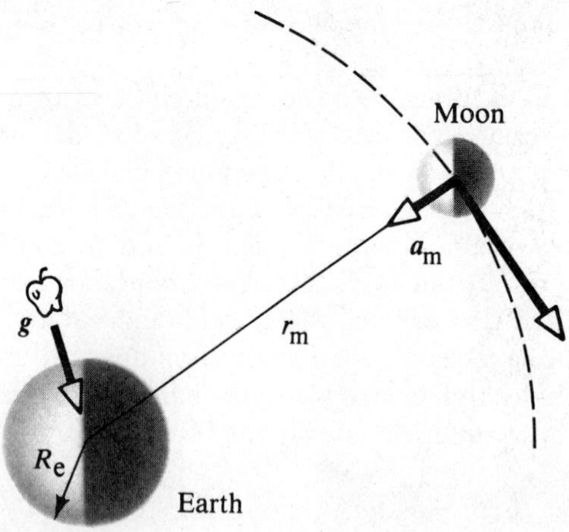

Figure 14.1 As the moon rotates about the earth, the moon experiences a centripetal acceleration a_m directed toward the earth. An object near the earth's surface experiences an acceleration equal to g. (Dimensions are not to scale.)

Therefore

The acceleration of the moon

$$a_m = (2.75 \times 10^{-4})(9.8 \text{ m/s}^2) = 2.7 \times 10^{-3} \text{ m/s}^2$$

The centripetal acceleration of the moon can also be calculated kinematically from a knowledge of its orbital period, $T = 27.32$ days $= 2.36 \times 10^6$ s, and its mean distance from the earth. In a time T, the moon travels a distance $2\pi r_m$, which equals the circumference of its orbit. Therefore, its orbital speed is $2\pi r_m/T$, and its centripetal acceleration is

$$a_m = \frac{v^2}{r_m} = \frac{(2\pi r_m/T)^2}{r_m} = \frac{4\pi^2 r_m}{T^2} = \frac{4\pi^2(3.84 \times 10^8 \text{ m})}{(2.36 \times 10^6 \text{ s})^2} = 2.72 \times 10^{-3} \text{ m/s}^2$$

This agreement provides strong evidence that the inverse-square law of force is correct.

Although these results must have been very encouraging to Newton, he was deeply troubled by an assumption made in the analysis. In order to evaluate the acceleration of an object at the earth's surface, the earth was treated as if its mass were all concentrated at its center. That is, Newton assumed that the earth acts as a particle as far as its influence on an exterior object is concerned. Several years later, and based on his pioneering work in the development of the calculus, Newton proved this point. (The details of the derivation are given in Section 14.8.) For this reason, and because of Newton's inherent shyness, the publication of the theory of gravitation was delayed for about 20 years.

Kepler's Third Law

It is informative to show that Kepler's third law can be predicted from the inverse-square law for circular orbits.[1] Consider a planet of mass M_p moving about the sun of mass M_s in a circular orbit, as in Fig. 14.2. Since the gravitational force on the planet is equal to the centripetal force needed to keep it moving in a circle,

$$\frac{GM_sM_p}{r^2} = \frac{M_pv^2}{r}$$

But the orbital velocity of the planet is simply $2\pi r/T$, where T is its period; therefore the above expression becomes

$$\frac{GM_s}{r^2} = \frac{(2\pi r/T)^2}{r}$$

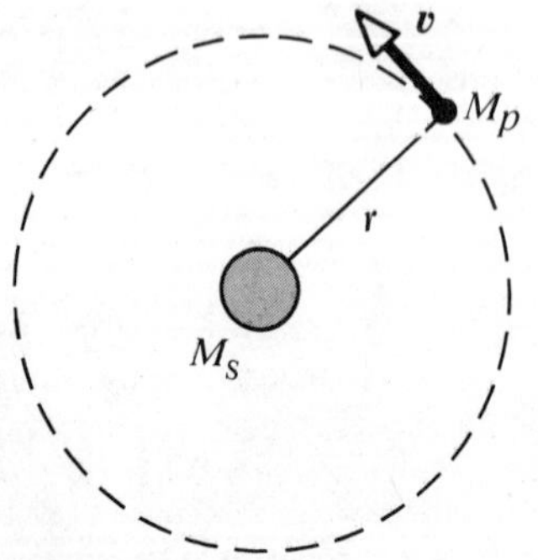

Figure 14.2 A planet of mass M_p moving in a circular orbit about the sun. The orbits of all planets except Mercury and Pluto are nearly circular.

[1]The orbits of all planets except Mercury and Pluto are very close to being circular. For example, the ratio of the semi-minor to the semi-major axis for the earth is $\frac{b}{a} = 0.99986$.

TABLE 14.1 Useful Planetary Data

BODY	*MASS (kg)*	*MEAN RADIUS (m)*	*PERIOD (s)*	*DISTANCE FROM SUN (m)*	$\frac{T^2}{r^3}\left[10^{-19}\left(\frac{s^2}{m^3}\right)\right]$
Mercury	3.18×10^{23}	2.43×10^{6}	7.60×10^{6}	5.79×10^{10}	2.97
Venus	4.88×10^{24}	6.06×10^{6}	1.94×10^{7}	1.08×10^{11}	2.99
Earth	5.98×10^{24}	6.37×10^{6}	3.156×10^{7}	1.496×10^{11}	2.97
Mars	6.42×10^{23}	3.37×10^{6}	5.94×10^{7}	2.28×10^{11}	2.98
Jupiter	1.90×10^{27}	6.99×10^{7}	3.74×10^{8}	7.78×10^{11}	2.97
Saturn	5.68×10^{26}	5.85×10^{7}	9.35×10^{8}	1.43×10^{12}	2.99
Uranus	8.68×10^{25}	2.33×10^{7}	2.64×10^{9}	2.87×10^{12}	2.95
Neptune	1.03×10^{26}	2.21×10^{7}	5.22×10^{9}	4.50×10^{12}	2.99
Pluto	$\approx 1 \times 10^{23}$	$\approx 3 \times 10^{6}$	7.82×10^{9}	5.91×10^{12}	2.96
Moon	7.36×10^{22}	1.74×10^{6}	—	—	—
Sun	1.991×10^{30}	6.96×10^{8}	—	—	—

For a more complete set of data, see, for example, the *Handbook of Chemistry and Physics*, Cleveland, The Chemical Rubber Publishing Co.

$$T^2 = \left(\frac{4\pi^2}{GM_s}\right)r^3 = K_s r^3 \qquad (14.3)$$

Kepler's third law

where K_s is a constant given by

$$K_s = \frac{4\pi^2}{GM_s}$$

This is Kepler's third law. The law is also valid for elliptical orbits if we replace r by the length of the semi-major axis, a (Fig. 14.3). Note that the constant of proportionality, K_s, is independent of the mass of the planet. Therefore, Eq. 14.3 is valid for *any* planet. If we were to consider the orbit of a satellite about the earth, such as the moon, then the constant would have a different value, with the sun's mass replaced by the earth's mass (see Example 6.4). In this case, the proportionality constant equals $4\pi^2/GM_e$.

A collection of useful planetary data is given in Table 14.1. The last column of this table verifies that T^2/r^3 is a constant given by $K_s = 4\pi^2/GM_s = 2.97 \times 10^{-19}\ s^2/m^3$.

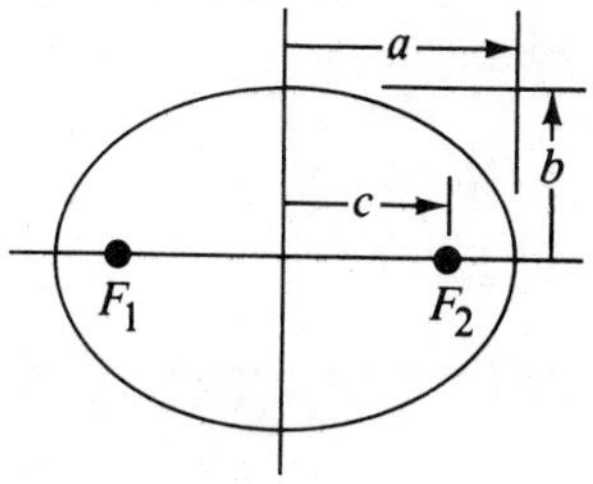

Figure 14.3 Plot of an ellipse. The semi-major axis has a length a, and the semi-minor axis has a length b. The focal points are located at a distance c from the center, and the eccentricity is defined as $e = c/a$.

Example 14.1 *The Mass of the Sun*

Calculate the mass of the sun using the fact that the period of the earth is 3.156×10^7 s and its distance from the sun is 1.496×10^{11} m.

Solution: Using Eq. 14.3, we get

$$M_s = \frac{4\pi^2 r^3}{GT^2} = \frac{4\pi^2(1.496 \times 10^{11}\ m)^3}{\left(6.67 \times 10^{-11}\ \frac{N \cdot m^2}{kg^2}\right)(3.156 \times 10^7\ s)^2}$$

$$= 1.99 \times 10^{30}\ kg$$

Note that the sun is 333 000 times as massive as the earth!

Kepler's Second Law and Conservation of Angular Momentum

Consider a planet (or comet) of mass m moving about the sun in an elliptical orbit (Fig. 14.4). The gravitational force acting on the planet is always along the radius vector, directed toward the sun. Such a force directed toward or away from a fixed point (that is, one that is a function of r only) is called a *central force*.[2] The torque acting on the planet due to this central force is clearly zero since $\mathbf{F}$ is parallel to $\mathbf{r}$.

[2] Another example of a central force is the electrostatic force between two charged particles.

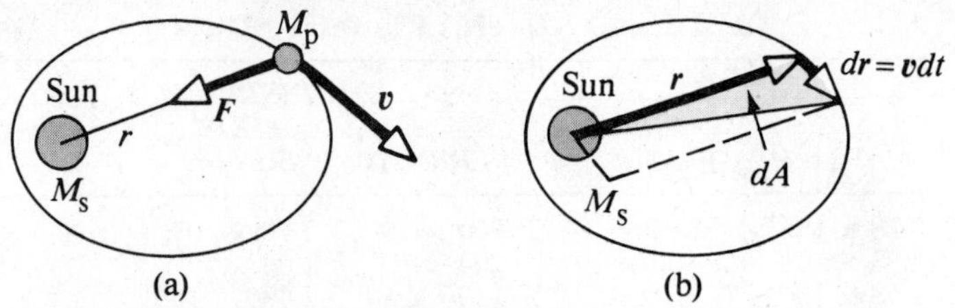

Figure 14.4 (a) The force acting on a planet acts toward the sun, along the radius vector. (b) As a planet orbits the sun, the area swept out by the radius vector in a time dt is equal to one half the area of the parallelogram formed by the vectors $\boldsymbol{r}$ and $d\boldsymbol{r} = \boldsymbol{v}\,dt$.

That is,

$$\boldsymbol{\tau} = \boldsymbol{r} \times \boldsymbol{F} = \boldsymbol{r} \times F(r)\hat{\boldsymbol{r}} = 0$$

Since the torque on a planet is zero, its angular momentum is constant

But recall that the torque equals the time rate of change of angular momentum, or $\boldsymbol{\tau} = d\boldsymbol{L}/dt$. Therefore, since $\boldsymbol{\tau} = 0$. *the angular momentum* $\boldsymbol{L}$ *of the planet is a constant of the motion:*

$$\boldsymbol{L} = \boldsymbol{r} \times \boldsymbol{p} = m\boldsymbol{r} \times \boldsymbol{v} = \text{constant}$$

Since $\boldsymbol{L}$ is a constant of the motion, we see that the planet's motion at any instant is restricted to the plane formed by $\boldsymbol{r}$ and $\boldsymbol{v}$.

We can relate this result to the following geometric consideration. The radius vector $\boldsymbol{r}$ in Fig. 14.4b sweeps out an area dA in a time dt. This area equals one half the area $|\boldsymbol{r} \times d\boldsymbol{r}|$ of the parallelogram formed by the vectors $\boldsymbol{r}$ and $d\boldsymbol{r}$. Since the displacement of the planet in a time dt is given by $d\boldsymbol{r} = \boldsymbol{v}\,dt$, we get

$$dA = \frac{1}{2}|\boldsymbol{r} \times d\boldsymbol{r}| = \frac{1}{2}|\boldsymbol{r} \times \boldsymbol{v}\,dt| = \frac{L}{2m}\,dt$$

Kepler's second law

$$\frac{dA}{dt} = \frac{L}{2m} = \text{constant}$$

where L and m are both constants of the motion. Thus, we conclude that *the radius vector from the sun to any planet sweeps out equal areas in equal times.* It is important to recognize that this result is a consequence of the fact that the force of gravity is a central force, which in turn implies conservation of angular momentum. Therefore, the law applies to *any* situation that involves a central force, whether inverse-square or not.

The inverse-square nature of the force of gravity is not revealed by Kepler's second law. Although we do not prove it here, Kepler's first law is a direct consequence of the fact that the gravitational force varies as $1/r^2$. That is, under an inverse-square force law, the orbits of the planets can be shown to be ellipses with the sun at one focus.

Example 14.2 *Motion in an Elliptical Orbit*

A planet of mass m moves in an elliptical orbit about the sun (Fig. 14.5). The closest and farthest positions of the planet from the sun are called the *perihelion* (indicated by p in Fig. 14.5) and *aphelion* (indicated by a), respectively. If the speed of the planet at p is v_p, what is its speed at a? Assume the distances r_a and r_p are known.

Solution: The angular momentum of the planet relative to the sun is $m\boldsymbol{r} \times \boldsymbol{v}$. At the points a and p, $\boldsymbol{v}$ is perpendicular to $\boldsymbol{r}$. Therefore, the magnitude of the angular momentum at these positions is $L_a = mv_ar_a$ and $L_p = mv_pr_p$. The direction of the angular momentum is out of the plane of the paper. Since angular momentum is conserved, we see that

$$mv_ar_a = mv_pr_p$$

$$v_a = \frac{r_p}{r_a}v_p$$

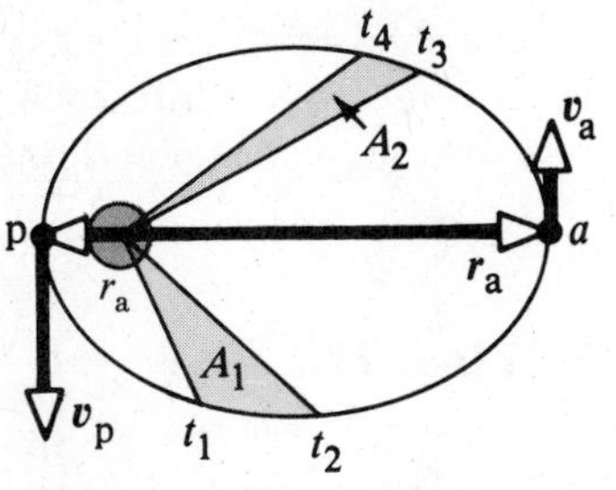

Figure 14.5 (Example 14.2) As a planet moves about the sun in an elliptical orbit, its angular momentum is conserved. Therefore, $mv_ar_a = mv_pr_p$, where the subscripts a and p represent aphelion and perihelion, respectively.

Q1. With reference to Fig. 14.5, consider the area swept out by the radius vector in the time intervals $t_2 - t_1$ and $t_4 - t_3$. Under what condition is A_1 equal to A_2?

Q2. If A_1 equals A_2 in Fig. 14.5, is the average speed of the planet in the time interval $t_2 - t_1$ less than, equal to, or greater than its average speed in the time interval $t_4 - t_3$?

Q3. At what position in its elliptical orbit is the speed of a planet a maximum? At what position is the speed a minimum?

14.3 THE GRAVITATIONAL FIELD

The gravitational force between two masses is an action-at-a-distance type of interaction. That is, the two masses interact even though they are not in contact with each other. An alternative approach in describing the gravitational interaction is to introduce the concept of a *gravitational field*, **g**, at every point in space. When a particle of mass m is placed at a point where the field is **g**, the particle experiences a force $\mathbf{F} = m\mathbf{g}$. In other words, the field **g** exerts a force on the particle. Hence, the gravitational field is defined by

$$\mathbf{g} = \frac{\mathbf{F}}{m}$$

Gravitational field

In other words, the gravitational field at any point equals the gravitational force that a test mass experiences divided by that test mass. Consequently, if **g** is known at some point in space, a test particle of mass m experiences a gravitational force $m\mathbf{g}$ when placed at that point.

As an example, consider an object of mass m near the earth's surface. The gravitational force on the object is directed toward the center of the earth and has a magnitude mg. Thus we see that the gravitational field that the object experiences at some point has a magnitude equal to the acceleration of gravity at that point. Since the gravitational force on the object has a magnitude GM_em/r^2 (where M_e is the mass of the earth), the field **g** at a distance r from the center of the earth is given by

$$\mathbf{g} = \frac{\mathbf{F}}{m} = -\frac{GM_e}{r^2}\hat{\mathbf{r}}$$

This expression is valid at all points *outside* the earth's surface, assuming the earth is spherical. At the earth's surface, where $r = R_e$, **g** has a magnitude of 9.8 m/s².

The field concept is used in many other areas of physics. In fact, the field concept was first introduced by Michael Faraday (1791–1867) in the field of electromagnetism. Later in the text we shall use the field concept to describe electromagnetic interactions. Gravitational, electrical, and magnetic fields are all examples of *vector fields* since a vector is associated with each point in space. On the other hand, a *scalar field* is one in which a scalar quantity is used to describe each point in space. For example, the variation in temperature over a given region can be described by a scalar temperature field.

14.4 GRAVITATIONAL POTENTIAL ENERGY

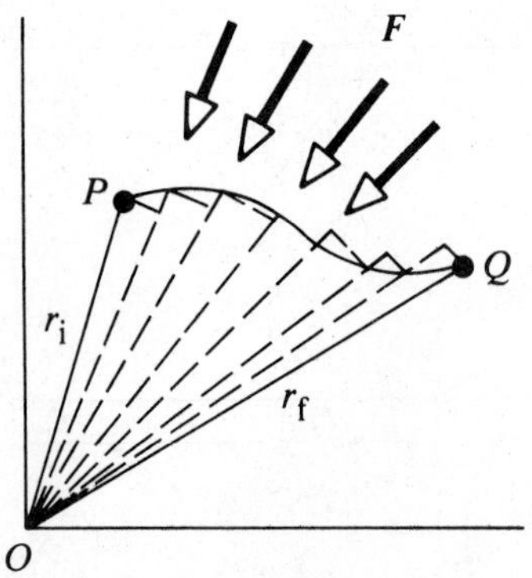

Figure 14.6 A particle moves from P to Q under the action of a central force $\mathbf{F}$, which is in the radial direction. The path is broken into a series of radial and circular segments. Since the work done along the circular segments is zero, the work done is independent of the path.

In Chapter 8 we introduced the concept of gravitational potential energy, that is, the energy associated with the position of a particle. We emphasized the fact that the gravitational potential energy function, $U = mgy$, is valid only when the particle is near the earth's surface. Since the gravitational force between two particles varies as $1/r^2$, we expect that the correct potential energy function will depend on the amount of separation between the particles.

Before we calculate the specific form for the gravitational potential energy, we shall first verify that *the gravitational force is conservative.* In order to establish the conservative nature of the gravitational force, we first note that it is a central force. By definition, a central force is one that depends only on the radial coordinate r, and hence can be represented by $F(r)\hat{\mathbf{r}}$. Such a force acts from some origin and is directed parallel to the radius vector drawn from the origin of the particle under consideration.

Consider a central force acting on a particle moving along the general path P to Q in Fig. 14.6. The central force acts from the point O. This path can be approximated by a series of radial and circular segments. By definition, a central force is always directed along one of the radial segments; therefore the work done along any *radial segment* is given by

$$dW = \mathbf{F} \cdot d\mathbf{r} = F(r)\, dr$$

You should recall that by definition the work done by a force that is perpendicular to the displacement is zero. Hence, the work done along any circular segment is *zero* because $\mathbf{F}$ is perpendicular to the displacement along these segments. Therefore, the total work done by $\mathbf{F}$ is the sum of the contributions along the radial segments:

Work done by a central force

$$W = \int_{r_i}^{r_f} F(r)\, dr \qquad (14.4)$$

where the subscripts i and f refer to the initial and final radial coordinates. This result applies to *any* path from P to Q. Therefore, we conclude that *any central force is conservative.* We are now assured that a potential energy function can be obtained once the form of the central force is specified. You should recall from Chapter 8 that the change in potential energy associated with a given displacement is defined as the negative of the work done by the force during that displacement, or

$$\Delta U = U_f - U_i = -\int_{r_i}^{r_f} F(r)\, dr \qquad (14.5)$$

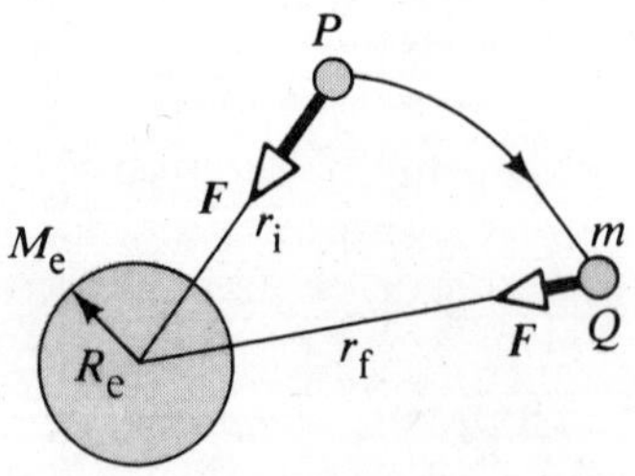

Figure 14.7 As a particle of mass m moves from P to Q above the earth's surface, the potential energy changes according to Eq. 14.7.

We can use this result to evaluate the gravitational potential energy function. Consider a particle of mass m moving between two points P and Q above the earth's surface (Fig. 14.7). The particle is subject to the gravitational force given by Eq. 14.1. We can express the force on m in vector form as

$$\mathbf{F} = -\frac{GM_e m}{r^2}\hat{\mathbf{r}} \qquad (14.6)$$

where $\hat{r}$ is a unit vector directed from the earth to the particle and the negative sign indicates that the force is attractive. Substituting this into Eq. 14.5, we can compute the gravitational potential energy function:

$$U_f - U_i = GM_e m \int_{r_i}^{r_f} \frac{dr}{r^2} = GM_e m \left[-\frac{1}{r} \right]_{r_i}^{r_f}$$

$$U_f - U_i = -GM_e m \left(\frac{1}{r_f} - \frac{1}{r_i} \right) \tag{14.7}$$

The choice of a reference point for the potential energy is completely arbitrary. It is customary to choose the reference point where the force is zero. Taking $U_i = 0$ at $r_i = \infty$, we obtain the important result

Gravitational potential energy $r > R_e$

$$U(r) = -\frac{GM_e m}{r} \tag{14.8}$$

This expression applies to the earth-particle system separated by a distance r, provided that $r > R_e$. The result is not valid for particles moving inside the earth, where $r < R_e$. We shall treat this situation in Section 14.7. Because of our choice of U_i, the function $U(r)$ is always negative (Fig. 14.8).

Although Eq. 14.8 was derived for the particle-earth system, it can be applied to *any* two particles. That is, the gravitational potential energy associated with *any pair* of particles of masses m_1 and m_2 separated by a distance r is given by

Gravitational potential energy for a pair of particles

$$U = -\frac{Gm_1 m_2}{r} \tag{14.9}$$

This expression shows that the gravitational potential energy for any pair of particles varies as $1/r$, whereas the force between them varies as $1/r^2$. Furthermore, the potential energy is *negative* since the force is attractive and we have taken the potential energy as zero when the particle separation is infinity. Since the force between the particles is attractive, we know that an external agent must do positive work to increase the separation between the two particles. The work done by the external agent produces an increase in the potential energy as the two particles are separated. That is, U becomes less negative as r increases. (Note that part of the work done can also produce a change in kinetic energy of the system.) When the two particles are separated by a distance r, an external agent would have to supply an energy *at least* equal to $+Gm_1m_2/r$ in order to separate the particles by an infinite distance. It is convenient to think of the absolute value of the potential energy as the *binding energy* of the system. If the external agent supplies an energy *greater than* the binding energy, Gm_1m_2/r, the additional energy of the system will be in the form of kinetic energy when the particles are at an infinite separation.

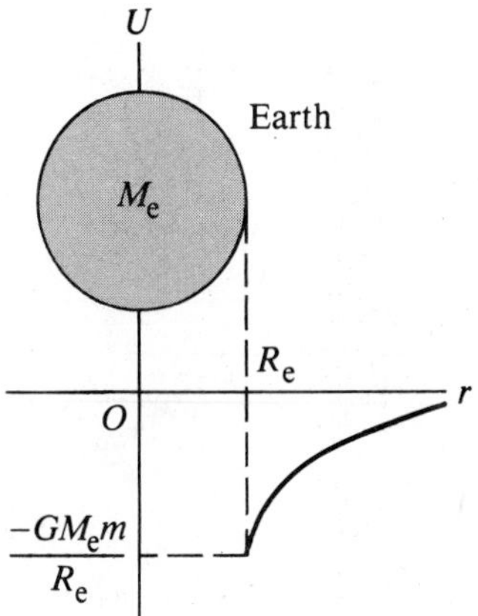

Figure 14.8 Graph of the gravitational potential energy, U, versus r for a particle above the earth's surface. The potential energy goes to zero as r approaches ∞.

We can extend this concept to three or more particles. In this case, the total potential energy of the system is the sum over all *pairs* of particles.[3] Each pair contributes a term of the form given by Eq. 14.9. For example, if the system contains three particles as in Fig. 14.9, we find that

$$U_{\text{total}} = U_{12} + U_{13} + U_{23} = -G\left(\frac{m_1 m_2}{r_{12}} + \frac{m_1 m_3}{r_{13}} + \frac{m_2 m_3}{r_{23}} \right) \tag{14.10}$$

[3] The fact that one can add potential energy terms for all pairs of particles stems from the experimental fact that gravitational forces obey the superposition principle. That is, if $\Sigma \mathbf{F} = \mathbf{F}_{12} + \mathbf{F}_{13} + \mathbf{F}_{23} + \cdots$, then there exists a potential energy term for each interaction $\mathbf{F}_{ij}$.

This represents the total work done by an external agent against the gravitational force in assembling the system from an infinite separation. If the system consists of four particles, there are six terms in the sum, corresponding to the six distinct pairs of interaction forces.

Example 14.3

A particle of mass m is displaced through a small vertical distance Δy near the earth's surface. Let us show that the general expression for the change in gravitational potential energy given by Eq. 14.7 reduces to the familiar relationship $\Delta U = mg\,\Delta y$.

Solution: We can express Eq. 14.7 in the form

$$\Delta U = -GM_e m\left(\frac{1}{r_f} - \frac{1}{r_i}\right) = GM_e m\left(\frac{r_f - r_i}{r_i r_f}\right)$$

If both the initial and the final position of the particle are close to the earth's surface, then $r_f - r_i = \Delta y$ and $r_i r_f \approx R_e^2$. (Recall that r is measured from the center of the earth.) Therefore, the *change* in potential energy becomes

$$\Delta U \approx \frac{GM_e m}{R_e^2}\Delta y = mg\,\Delta y$$

where we have used the fact that $g = GM_e/R_e^2$. Keep in mind that the reference point is arbitrary, since it is the *change* in potential energy that is meaningful.

Q4. If a system consists of five distinct particles, how many terms appear in the expression for the total potential energy?

Q5. Is it possible to calculate the potential energy function associated with a particle and an extended body without knowing the geometry or mass distribution of the extended body?

14.5 ENERGY CONSIDERATIONS IN PLANETARY AND SATELLITE MOTION

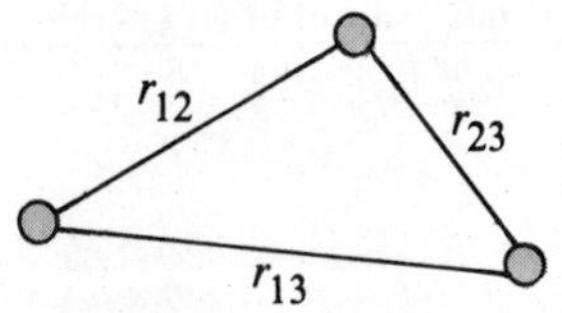

Figure 14.9 Diagram of three interacting particles.

Consider a body of mass m moving with a speed v in the vicinity of a massive body of mass M, where $M \gg m$. The system might be a planet moving around the sun or a satellite in orbit around the earth. If we assume that M is at rest in an inertial reference frame, then the total energy E of the two-body system when the bodies are separated by a distance r is the sum of the kinetic energy of the mass m and the potential energy of the system, given by Eq. 14.9.[4] That is,

$$E = K + U$$

$$E = \frac{1}{2}mv^2 - \frac{GMm}{r} \tag{14.11}$$

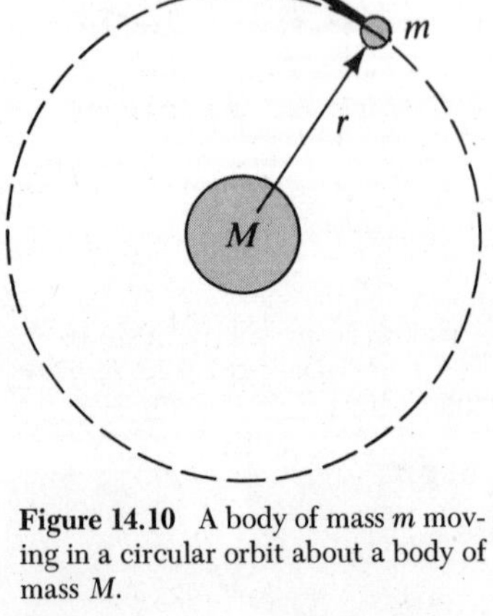

Figure 14.10 A body of mass m moving in a circular orbit about a body of mass M.

Furthermore, the total energy is conserved if we assume the system is isolated. Therefore as the mass m moves from P to Q in Fig. 14.7, the total energy remains constant and Eq. 14.11 gives

$$E = \frac{1}{2}mv_i^2 - \frac{GMm}{r_i} = \frac{1}{2}mv_f^2 - \frac{GMm}{r_f} \tag{14.12}$$

This result shows that E may be positive, negative, or zero, depending on the value of the velocity of the mass m. However, for a bound system, such as the earth and sun, E is necessarily less than zero. We can easily establish that $E < 0$ for the

[4] You might recognize that we have ignored the kinetic energy of the larger mass. To see that this is reasonable, consider an object of mass m falling toward the earth. Since the center of mass of the object-earth system is stationary, it follows that $mv = M_e v_e$. Thus, the earth acquires a kinetic energy equal to $\frac{1}{2}M_e v_e^2 = \frac{1}{2}\frac{m^2}{M_e}v^2 = \frac{m}{M_e}K$, where K is the kinetic energy of the object. Since $M_e \gg m$, the kinetic energy of the earth is negligible.

system consisting of a mass m moving in a circular orbit about a body of mass M, where $M \gg m$ (Fig. 14.10). Newton's second law applied to the body of mass m gives

$$\frac{GMm}{r^2} = \frac{mv^2}{r}$$

Multiplying both sides by r and dividing by 2 gives

$$\frac{1}{2}mv^2 = \frac{GMm}{2r} \tag{14.13}$$

Substituting this into Eq. 14.11, we obtain

$$E = \frac{GMm}{2r} - \frac{GMm}{r}$$

$$E = -\frac{GMm}{2r} \tag{14.14}$$

Total energy for circular orbits

This clearly shows that *the total energy must be negative in the case of circular orbits*. Note that *the kinetic energy is positive and equal to one half the magnitude of the potential energy*. The absolute value of E is also equal to the binding energy of the system.

The total mechanical energy is also negative in the case of elliptical orbits.[5] The expression for E for elliptical orbits is the same as Eq. 14.14 wth r replaced by the semi-major axis length, a. *Both the total energy and the total angular momentum of a planet-sun system are constants of the motion.*

Example 14.4 *Changing the Orbit of a Satellite*

Calculate the work required to move an earth satellite of mass m from a circular orbit of radius $2R_e$ to one of radius $3R_e$.

Solution: Applying Eq. 14.14, we get for the total initial and final energies

$$E_i = -\frac{GM_em}{4R_e} \qquad E_f = -\frac{GM_em}{6R_e}$$

Therefore, the work required to increase the energy of the system is

$$W = E_f - E_i = -\frac{GM_em}{6R_e} - \left(-\frac{GM_em}{4R_e}\right) = \frac{GM_em}{12R_e}$$

For example, if we take $m = 10^3$ kg, we find that $W = 1.04 \times 10^{10}$ J, which is the energy equivalent of 78 gal of gasoline.

If we wish to determine how the energy is distributed after doing work on the system, we find from Eq. 14.13 that the change in kinetic energy is $\Delta K = -GM_em/12R_e$ (it decreases), while the corresponding change in potential energy is $\Delta U = GM_em/6R_e$ (it increases). Thus, the work done is $W = \Delta K + \Delta U = GM_em/12R_e$, as we calculated above. In other words, part of the work done goes into increasing the potential energy and part goes into decreasing the kinetic energy.

Escape Velocity

Suppose an object of mass m is projected vertically upward from the earth's surface with an initial speed v_i, as in Fig. 14.11. We can use energy considerations to find the maximum value of the initial speed such that the object will escape the earth's gravitational field. Equation 14.12 gives the total energy of the object at any point when its velocity and distance from the center of the earth are known. At the

[5] This is shown in more advanced mechanics texts. One can also show that if $E = 0$, the mass would move in a parabolic path, whereas if $E > 0$, its path would be a hyperbola. Nothing in Eq. 14.11 precludes a particle with $E \geq 0$ from reaching infinitely great distances from the gravitating center (that is, the particle's orbit is unbound). Infinitely great distances are energetically forbidden to a particle with $E < 0$, and so its orbit must be bound.

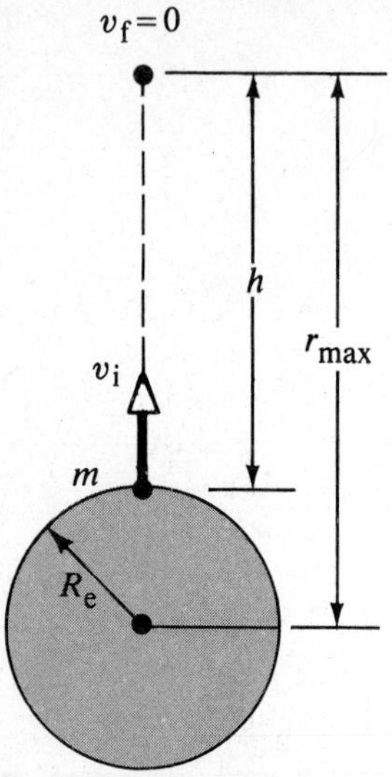

Figure 14.11 An object of mass m projected upward from the earth's surface with an initial speed v_i reaches a maximum altitude h (where $M_e \gg m$).

surface of the earth, where $v = v_i$, $r_i = R_e$. When the object reaches its maximum altitude, $v_f = 0$ and $r_f = r_{max}$. Since the total energy of the system is conserved, substitution of these conditions into Eq. 14.12 gives

$$\frac{1}{2}mv_i^2 - \frac{GM_em}{R_e} = -\frac{GM_em}{r_{max}}$$

Solving for v_i gives

$$v_i^2 = 2GM_e\left(\frac{1}{R_e} - \frac{1}{r_{max}}\right) \tag{14.15}$$

Therefore, if the initial speed is known, this expression can be used to calculate the maximum altitude h, since we know that $h = r_{max} - R_e$.

We are now in a position to calculate the minimum speed the object must have at the earth's surface in order to escape from the influence of the earth's gravitational field. This corresponds to the situation where the object can *just* reach infinity with a final speed of *zero*. Setting $r_{max} = \infty$ in Eq. 14.15 and taking $v_i = v_{esc}$ (the escape velocity), we get

Escape velocity

$$v_{esc} = \sqrt{\frac{2GM_e}{R_e}} \tag{14.16}$$

Note that this expression for v_{esc} is independent of the mass of the object projected from the earth. For example, a spacecraft has the same escape velocity as a molecule. If the object is given an initial speed equal to v_{esc}, its *total* energy is equal to zero. This can be seen by noting that when $r = \infty$, the object's kinetic energy and its potential energy are both zero. If v_i is greater than v_{esc}, the *total* energy will be greater than zero and the object will have some residual kinetic energy at $r = \infty$.

Example 14.5 *Escape Velocity of a Rocket*

Calculate the escape velocity from the earth for a 5000-kg spacecraft, and determine the kinetic energy it must have at the earth's surface in order to escape the earth's field.

Solution: Using Eq. 14.16 with $M_e = 5.98 \times 10^{24}$ kg and $R_e = 6.37 \times 10^6$ m gives

$$v_{esc} = \sqrt{\frac{2GM_e}{R_e}} = \sqrt{\frac{2(6.67 \times 10^{-11})(5.98 \times 10^{24})}{6.37 \times 10^6}}$$

$$= 1.12 \times 10^4 \ \frac{\text{m}}{\text{s}}$$

This corresponds to about 25 000 mi/h.

The kinetic energy of the spacecraft is given by

$$K = \frac{1}{2}mv_{esc}^2 = \frac{1}{2}(5000)(1.12 \times 10^4)^2 = 3.14 \times 10^{11}\ \text{J}$$

Finally, you should note that Eqs. 14.15 and 14.16 can be applied to objects projected vertically from *any* planet. That is, in general, the escape velocity from any planet of mass M and radius R is given by

$$v_{esc} = \sqrt{\frac{2GM}{R}}$$

TABLE 14.2
Escape Velocities for the Planets and the Moon

PLANET	v_{esc} (km/s)
Mercury	4.3
Venus	10.3
Earth	11.2
Moon	2.3
Mars	5.0
Jupiter	60
Saturn	36
Uranus	22
Neptune	24

A list of escape velocities for the planets and the moon is given in Table 14.2. Note that the values vary from 2.3 km/s for the moon to about 60 km/s for Jupiter. These results, together with some ideas from the kinetic theory of gases (Chapter 17), explain why some planets have atmospheres and others do not. As we shall see later, a gas molecule at some temperature has an average velocity that depends on its mass and the temperature. Lighter atoms, such as hydrogen and helium, have a higher average velocity than the heavier species. When the velocity of the lighter atoms is much greater than this average velocity, a significant fraction of the molecules have a chance to escape from the planet. This mechanism also explains why the earth does not retain hydrogen and helium molecules in its atmosphere while much heavier molecules, such as oxygen and nitrogen, do not escape. On the other hand, Jupiter has a very large escape velocity (60 km/s), which enables it to retain hydrogen, the primary constituent of its atmosphere.

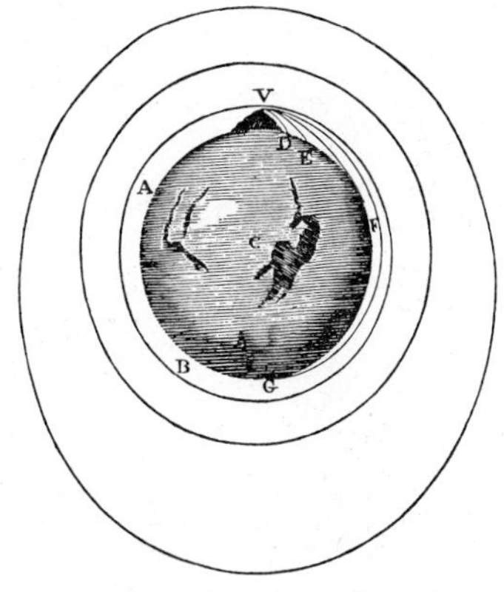

". . . the greater the velocity . . . with which (a stone) is projected, the farther it goes before it falls to the earth. We may therefore suppose the velocity to be so increased, that it would describe an arc of 1, 2, 5, 10, 100, 1000 miles before it arrived at the earth, till at last, exceeding the limits of the earth, it should pass into space without touching."—Newton, *System of the World.*

Q6. Does the escape velocity of a rocket depend on its mass? Explain.

Q7. Compare the energies required to reach the moon for a 10^5-kg spacecraft and a 10^3-kg satellite.

Q8. Explain why it takes more fuel for a spacecraft to travel from the earth to the moon than for the return trip. Estimate the difference.

Q9. Is the magnitude of the potential energy associated with the earth-moon system greater than, less than, or equal to the kinetic energy of the moon relative to the earth?

Q10. Explain carefully why there is no work done on a planet as it moves in a circular orbit around the sun, even though a gravitational force is acting on the planet. What is the *net* work done on a planet during each revolution as it moves around the sun in an elliptical orbit?

14.6 THE GRAVITATIONAL FORCE BETWEEN AN EXTENDED BODY AND A PARTICLE

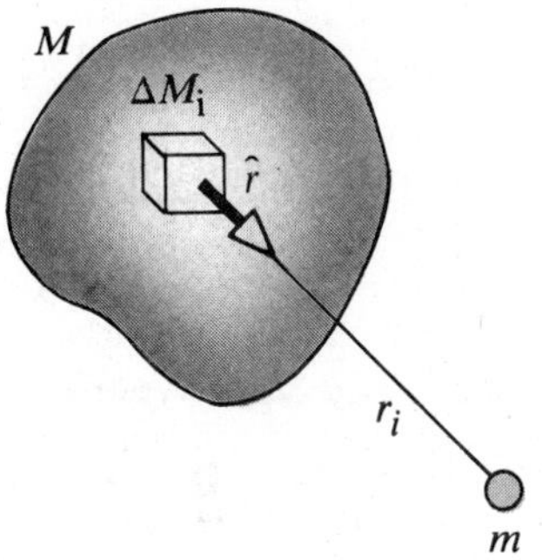

Figure 14.12 A particle of mass m interacting with an extended body of mass M. The potential energy of the system is given by Eq. 14.17. The total force on a particle at P due to an extended body can be obtained by taking a vector sum over all forces due to each segment of the body.

We have emphasized that the law of universal gravitation given by Eq. 14.6 is valid only if the interacting objects are considered as particles. In view of this, how can we calculate the force between a particle and an object having finite dimensions? This is accomplished by treating the *extended* object as a collection of particles and making use of integral calculus. We shall take the approach of first evaluating the potential energy function, from which the force can be calculated.

The potential energy associated with a system consisting of a point mass m and an extended body of mass M is obtained by dividing the body into segments of mass ΔM_i (Fig. 14.12). The potential energy associated with this element and with the particle of mass m is $-Gm\,\Delta M_i/r_i$, where r_i is the distance from the particle to the element ΔM_i. The total potential energy of the system is obtained by taking the sum over all segments as $\Delta M_i \rightarrow 0$. In this limit, we can express U in integral form as

Total potential energy for a particle-extended body system

$$U = -Gm \int \frac{dM}{r} \tag{14.17}$$

Once U has been evaluated, the force can be obtained by taking the negative derivative of this scalar function (see Section 8.6). If the extended body has spherical symmetry, the function U depends only on r and the force is given by $-dU/dr$. We shall treat this situation in Section 14.7. In principle, one can evaluate U for any specified geometry; however, the integration can be cumbersome.

An alternative approach to evaluating the force between a particle and an extended body is to perform a vector sum over all segments of the body. Using the procedure outlined in evaluating U and the law of universal gravitation (Eq. 14.1), the total force on the particle is given by

Total force between a particle and an extended body

$$\boldsymbol{F} = -Gm \int \frac{dM}{r^2} \hat{\boldsymbol{r}} \tag{14.18}$$

where $\hat{\boldsymbol{r}}$ is a unit vector directed from the element dM toward the particle (Fig. 14.12). This procedure is not always recommended, since working with a vector function is more difficult than working with the scalar potential energy function. However, if the geometry is simple, as in the following example, the evaluation of $\boldsymbol{F}$ can be straightforward.

Example 14.6

A homogeneous bar of length L and mass M is at a distance h from a point mass m (Fig. 14.13). Calculate the force on m.

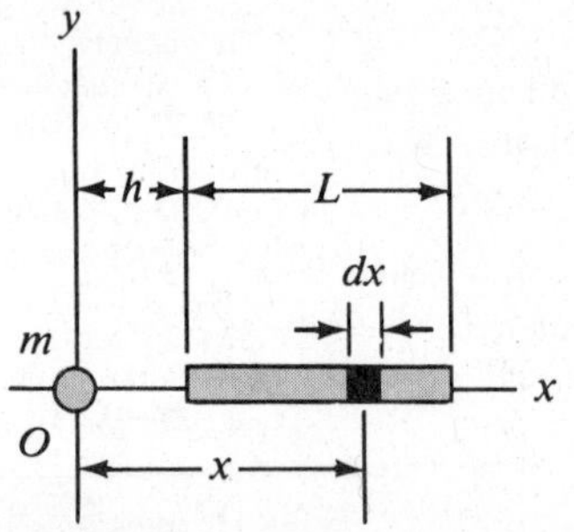

Figure 14.13 (Example 14.6) The force on a particle at the origin due to the bar is to the right. Note that the bar is *not* equivalent to a particle of mass M located at its center of mass.

Solution: The segment of the bar that has a length dx has a mass dM. Since the mass per unit length is a constant, it then follows that the ratio of masses, dM/M, is equal to the ratio of lengths, dx/L, and so $dM = \frac{M}{L}\,dx$. The variable r in Eq. 14.18 is x in our case, and the force on m is to the right; therefore we get

$$\boldsymbol{F} = Gm \int_h^{L+h} \frac{M}{L}\frac{dx}{x^2}\boldsymbol{i}$$

$$\boldsymbol{F} = \frac{GmM}{L}\left[-\frac{1}{x}\right]_h^{L+h}\boldsymbol{i} = \frac{GmM}{h(L+h)}\boldsymbol{i}$$

Note that in the limit $L \to 0$, the force varies as $1/h^2$, which is what is expected for the force between two point masses. Furthermore, if $h \gg L$, the force also varies as $1/h^2$. This can be seen by noting that the denominator of the expression for $\boldsymbol{F}$ can be expressed in the form $h^2\left(1 + \frac{L}{h}\right)$, which is approximately equal to h^2. Thus, when bodies are separated by distances that are large compared with their characteristic dimensions, their shapes become unimportant; they behave like particles.

14.7 GRAVITATIONAL FORCE BETWEEN A PARTICLE AND A SPHERICAL MASS

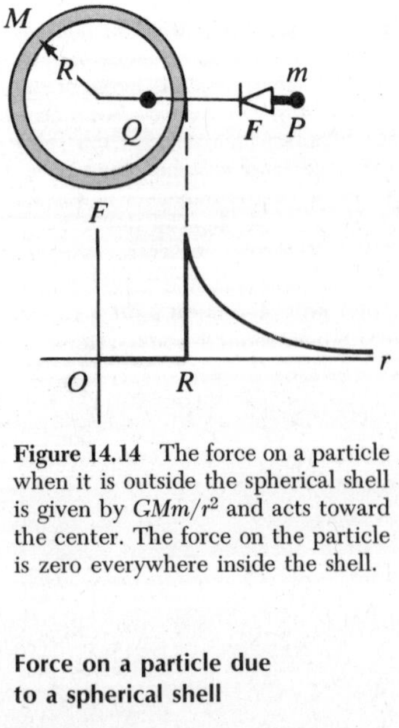

Figure 14.14 The force on a particle when it is outside the spherical shell is given by GMm/r^2 and acts toward the center. The force on the particle is zero everywhere inside the shell.

In this section we shall treat the gravitational force between a particle and a spherically symmetric mass distribution. We have already stated that a large sphere attracts a particle outside it as if the total mass of the sphere were concentrated at its center. This, and other properties of the spherical mass distribution, are proved formally in Section 14.8. Let us describe the nature of the force on a particle when the extended body is either a spherical shell or a solid sphere, and then apply these facts to some interesting systems.

Spherical Shell

1. If a particle of mass m is located *outside* a spherical shell of mass M (say, point P in Fig. 14.14), the spherical shell attracts the particle as though the mass of the shell were concentrated at its center.

2. If the particle is located *inside* the spherical shell (point Q in Fig. 14.14), the force on it is zero. We can express these two important results in the following way:

Force on a particle due to a spherical shell

$$\boldsymbol{F} = -\frac{GMm}{r^2}\hat{\boldsymbol{r}} \quad \text{for } r > R \tag{14.19a}$$

$$F = 0 \qquad \text{for } r < R \qquad (14.19b)$$

The force as a function of the distance r is plotted in Fig. 14.14.

Solid Sphere

3. If a particle of mass m is located *outside* a homogeneous solid sphere of mass M (point P in Fig. 14.15), the sphere attracts the particle as though the mass of the sphere were concentrated at its center. That is, Eq. 14.19a applies in this situation. This follows from case 1 above, since a solid sphere can be considered a collection of concentric spherical shells.

4. If a particle of mass m is located *inside* a homogeneous solid sphere of mass M (point Q in Fig. 14.15), the force on m is due *only* to the mass M' contained within the sphere of radius $r < R$, represented by the dotted line in Fig. 14.15. In other words,

Force on a particle due to a solid sphere

$$\mathbf{F} = -\frac{GmM}{r^2}\hat{\mathbf{r}} \qquad \text{for } r > R \qquad (14.20a)$$

$$\mathbf{F} = -\frac{GmM'}{r^2}\hat{\mathbf{r}} \qquad \text{for } r < R \qquad (14.20b)$$

Since the sphere is assumed to have a uniform density, it follows that the ratio of masses M'/M is equal to the ratio of volumes V'/V, where V is the total volume of the sphere and V' is the volume within the dotted surface. That is,

$$\frac{M'}{M} = \frac{V'}{V} = \frac{\frac{4}{3}\pi r^3}{\frac{4}{3}\pi R^3} = \frac{r^3}{R^3}$$

Solving this equation for M' and substituting the value obtained into Eq. 14.20b, we get

$$\mathbf{F} = -\frac{GmM}{R^3} r\,\hat{\mathbf{r}} \qquad \text{for } r < R \qquad (14.21)$$

That is, the force goes to zero at the center of the sphere, as we would intuitively expect. The force as a function of r is plotted in Fig. 14.15.

5. If a particle is located *inside* a solid sphere having a density ρ that is spherically symmetric but *not* uniform, then M' in Eq. 14.20b is given by an integral of the form $M' = \int \rho\, dV$, where the integration is taken over the volume contained *within* the dotted surface. This integral can be evaluated if the radial variation of ρ is given. The integral is easily evaluated if the mass distribution has spherical symmetry, that is, if ρ is a function of r only. In this case, we take the volume element dV as the volume of a spherical shell of radius r and thickness dr, so that $dV = 4\pi r^2\, dr$. For example, if $\rho(r) = Ar$, where A is a constant, it is left as an exercise to show that $M' = \pi A r^4$. Hence we see from Eq. 14.20b that F is proportional to r^2 in this case and is zero at the center.

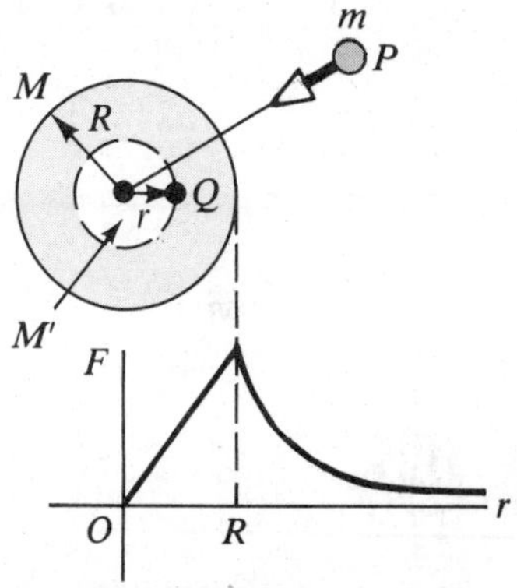

Figure 14.15 The force on a particle when it is outside a uniform solid sphere is given by GMm/r^2 and is directed toward the center. The force on the particle when it is inside such a sphere is proportional to r and goes to zero at the center.

Example 14.7 *A Free Ride*

An object moves in a smooth, straight tunnel dug between two points on the earth's surface (Fig. 14.16). Show that the object moves with simple harmonic motion and find the period of its motion.

Solution: When the object is in the tunnel, the gravitational force on it acts toward the earth's center and is given by Eq. 14.21:

$$F = -\frac{GmM_e}{R_e^{\,3}} r$$

The y component of this force is balanced by the normal force exerted by the tunnel wall, and the x component of

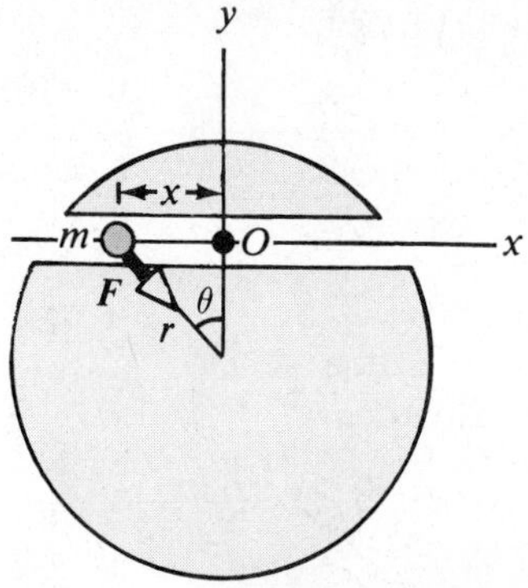

Figure 14.16 A particle moves along a tunnel dug through the earth. The component of the gravitational force $\mathbf{F}$ along the x axis is the driving force for the motion. Note that this component always acts toward the origin O.

the force is given by

$$F_x = -\frac{GmM_e}{R_e^3} r \sin\theta$$

Since the x coordinate of the object is given by $x = r \sin\theta$, we can write F_x in the form

$$F_x = -\frac{GmM_e}{R_e^3} x$$

Applying Newton's second law to the motion along x gives

$$F_x = -\frac{GmM_e}{R_e^3} x = ma$$

$$a = -\frac{GM_e}{R_e^3} x = -\omega^2 x$$

But this is the equation of simple harmonic motion with angular velocity ω (Chapter 13), where

$$\omega = \sqrt{\frac{GM_e}{R_e^3}}$$

The period is calculated using the data in Table 14.1 and the above result:

$$T = \frac{2\pi}{\omega} = 2\pi\sqrt{\frac{R_e^3}{GM_e}} = 2\pi\sqrt{\frac{(6.37 \times 10^6)^3}{(6.67 \times 10^{-11})(5.98 \times 10^{24})}}$$

$$= 5.06 \times 10^6 \text{ s} = 84.3 \text{ min}$$

This period is the same as that of a satellite in a circular orbit just above the earth's surface. Note that the result is *independent* of the length of the tunnel.

It has been proposed to operate a mass-transit system between any two cities using this principle. A one-way trip would take about 42 min. A more precise calculation of the motion must account for the fact that the earth's density is not uniform as we have assumed (Section 6.4). More important, there are many practical problems to consider. For instance, it would be impossible to achieve a frictionless tunnel, and so some auxiliary power source would be required. Can you think of other problems?

Q11. A particle is projected through a small hole into the interior of a large spherical shell. Describe the motion of the particle in the interior of the shell.

Q12. Explain why the force on a particle due to a uniform sphere must be directed toward the center of the sphere. Would this be the case if the mass distribution of the sphere were not spherically symmetric?

Q13. Neglecting the density variation of the earth, what would be the period of a particle moving in a smooth hole dug through the earth's center?

14.8 DERIVATION OF THE GRAVITATIONAL EFFECT OF A SPHERICAL MASS DISTRIBUTION

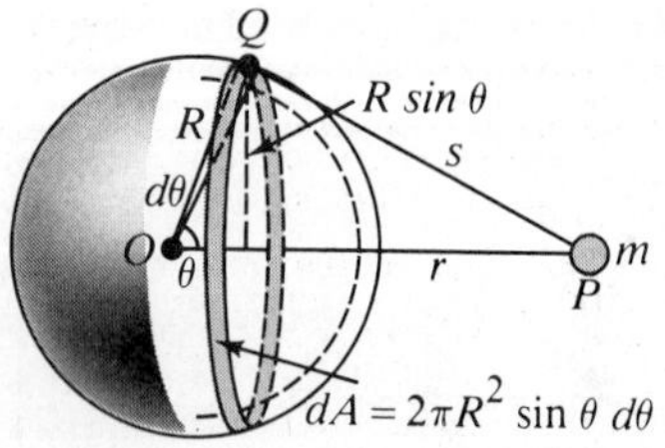

Figure 14.17 Diagram for calculating the gravitational potential energy of a particle interacting with a spherical shell. The shell is divided into circular zones (shaded) for convenience.

The purpose of this section is to prove Eqs. 14.19 and 14.20 using integral calculus. Consider a spherical shell of mass M and radius R with a thickness that is small compared with R (Fig. 14.17). A particle of mass m is placed at a point P, some distance r from the center of the shell. We could calculate the force on m directly, but since this is a vector quantity, a vector sum over all parts of the shell would be required. It is easier to first calculate the potential energy associated with the system (a scalar quantity). Since the mass distribution is spherically symmetric, the potential energy, U, is a function only of the radial distance r, that is, $U(\mathbf{r}) = U(r)$. The force on m can then be obtained from the relation $F_r = -dU/dr$. This is the approach we shall take.

First, let us calculate the mass of a zone of the shell, where the zone is taken perpendicular to the axis OP (the shaded section in Fig. 14.17). Since the width of this zone is $R\, d\theta$ and its radius is $R \sin\theta$, we see that the outer surface area of the zone is $dA = 2\pi R^2 \sin\theta\, d\theta$. The total surface area of the shell is $4\pi R^2$; hence it follows that the mass of the zone is given by

$$dM = \frac{\text{area of zone}}{\text{area of shell}} \times M = \frac{2\pi R^2 \sin\theta\, d\theta}{4\pi R^2} \times M = \frac{1}{2} M \sin\theta\, d\theta$$

Since all parts of the zone are at essentially the same distance s from the point P, from Eq. 14.17 we see that the potential energy associated with this zone and the particle is

$$dU = -\frac{Gm\, dM}{s} = -\frac{GmM}{2}\frac{\sin\theta\, d\theta}{s}$$

The total potential energy of the system is

$$U = -\frac{GmM}{2}\int \frac{\sin\theta\, d\theta}{s} \tag{14.22}$$

We cannot evaluate this integral directly, since it involves two variables, θ and s. However, we can eliminate one of the variables by applying the law of cosines to the triangle OPQ in Fig. 14.17:

$$s^2 = r^2 + R^2 - 2rR\cos\theta$$

Differentiating both sides of this equation with respect to θ and noting that r and R are constants for a particular point P, we get

$$2s\frac{ds}{d\theta} = -2rR(-\sin\theta)$$

$$\sin\theta\, d\theta = \frac{s\, ds}{rR}$$

Substituting this into the integrand of Eq. 14.22 gives

$$U = -\frac{GmM}{2rR}\int_{s_1}^{s_2} ds \tag{14.23}$$

To evaluate U from Eq. 14.23 we must specify the limits of integration. We shall first consider a point P outside the shell, as in Fig. 14.17, and then a point inside the shell.

Outside the Shell

When the particle is outside the shell, where $r > R$, the limits of integration in Eq. 14.23 are $s_1 = r - R$ to $s_2 = r + R$. Therefore,

$$U = -\frac{GMm}{2rR}\int_{r-R}^{r+R} ds = -\frac{GMm}{r} \qquad \text{for } r > R$$

Hence the force on m when it is outside the shell is

$$F_r = -\frac{dU}{dr} = -\frac{d}{dr}\left(-\frac{GMm}{r}\right) = -\frac{GMm}{r^2} \qquad \text{for } r > R$$

This verifies Eq. 14.19a.

Inside the Shell

When the particle is inside the shell, where $r < R$, the limits of integration in Eq. 14.23 are $s_1 = R - r$ to $s_2 = R + r$. Therefore,

$$U = -\frac{GMm}{2rR}\int_{R-r}^{R+r} ds = -\frac{GMm}{R} \qquad \text{for } r < R$$

Since R is a constant, we see that the potential energy is constant for all points within the sphere. Therefore,

$$F_r = -\frac{dU}{dr} = 0 \qquad \text{for } r < R$$

This verifies Eq. 14.19b.

The extension of these ideas to a solid sphere is straightforward, since we can regard a solid sphere as a collection of concentric spherical shells.

14.9 SUMMARY

Kepler's laws

Kepler's laws of planetary motion state that

1. All planets move in elliptical orbits with the sun at one of the focal points.
2. The radius vector drawn from the sun to any planet sweeps out equal areas in equal time intervals.
3. The square of the orbital period of any planet is proportional to the cube of the semi-major axis for the elliptical orbit.

Newton's law of universal gravitation states that the gravitational force of attraction between any two particles of masses m_1 and m_2 separated by a distance r has the magnitude

Universal law of gravity

$$F = G\frac{m_1 m_2}{r^2} \tag{14.1}$$

where G is the universal gravitational constant, equal to $6.673 \times 10^{-11}\ \text{N} \cdot \text{m}^2/\text{kg}^2$.

Kepler's second law is a consequence of the fact that the force of gravity is a *central force,* that is, one that is directed toward a fixed point. This in turn implies that the angular momentum of the planet-sun system is a constant of the motion.

Kepler's third law is consistent with the inverse-square nature of the law of universal gravitation. Newton's second law, together with the force law given by Eq. 14.1, verifies that the period T and radius r of the orbit of a planet about the sun are related by

Kepler's third law

$$T^2 = \left(\frac{4\pi^2}{GM_s}\right) r^3 \tag{14.3}$$

where M_s is the mass of the sun. Most planets have nearly perfect circular orbits about the sun. For elliptical orbits, Eq. 14.3 is valid if r is replaced by the semi-major axis, a.

The gravitational force is a conservative force, and therefore a potential energy function can be defined. *The gravitational potential energy* associated with the two particles separated by a distance r is given by

Gravitational potential energy for a pair of particles

$$U = -\frac{Gm_1 m_2}{r} \tag{14.9}$$

where U is taken to be zero at $r = \infty$. The total potential energy for a system of particles is the sum of energies for all pairs of particles, with each pair given a term of the form given by Eq. 14.9.

If an isolated system consists of a particle of mass m moving with a speed v in the vicinity of a massive body of mass M, the *total energy* of the system is given by

$$E = \frac{1}{2}mv^2 - \frac{GMm}{r} \tag{14.12}$$

That is, the energy is the sum of the kinetic and potential energies. The total energy is a constant of the motion.

If m moves in a circular orbit of radius r about M, where $M \gg m$, *the total energy of the system is*

$$E = -\frac{GMm}{2r} \qquad (14.14)$$

Total energy for circular orbits

The total energy is negative for any bound system, that is, one in which the orbit is closed, such as an elliptical orbit.

The *potential energy* of gravitational attraction between a particle of mass m and an extended body of mass M is given by

$$U = -Gm \int \frac{dM}{r} \qquad (14.17)$$

Total potential energy for a particle-extended body system

where the integral is over the extended body, dM is an infinitesimal mass element of the body, and r is the distance from the particle to the element.

If a particle is outside a uniform spherical shell or solid sphere with a spherically symmetric internal mass distribution, the sphere attracts the particle as though the mass of the sphere were concentrated at the center of the sphere.

If a particle is inside a uniform spherical shell, the gravitational force on the particle is zero.

If a particle is inside a homogeneous solid sphere, the force on the particle acts toward the center of the sphere and is linearly proportional to the distance from the center to the particle.

EXERCISES

Section 14.2 The Law of Universal Gravitation and the Motion of Planets

1. A satellite is in a circular orbit about the earth. (a) Evaluate the constant K that appears in Kepler's third law as applied to this situation. (b) What is the period of the orbit if the satellite is at an altitude of 2×10^6 m?

2. Given that the moon's period about the earth is 27.32 days and the earth-moon distance is 3.84×10^8 m, estimate the mass of the earth. Assume the orbit is circular. Why do you suppose your estimate is high?

3. Using the data in Table 14.1, make a log-log plot of T versus r, and from the slope of this graph verify that T is proportional to $r^{3/2}$ (Kepler's third law).

4. The planet Jupiter has at least 14 satellites. One of them, named Callisto, has a period of 16.75 days and a mean orbital radius of 1.883×10^9 m. From this information, calculate the mass of Jupiter.

5. A satellite of Mars has a period of 459 min. The mass of Mars is 6.42×10^{23} kg. From this information, determine the radius of the satellite's orbit.

6. At its aphelion, the planet Mercury is 6.99×10^{10} km from the sun, and at its perihelion, it is 4.60×10^{10} km from the sun. If its orbital speed is 3.88×10^4 m/s at the aphelion, what is its orbital speed at the perihelion?

7. A satellite is to be sent into orbit about the earth in an equatorial plane such that it will always appear to be stationary relative to an observer on earth. Find the radius of its orbit. (Hint: The satellite must have the same angular velocity as the earth.)

Section 14.4 Gravitational Potential Energy (assume $U = 0$ at $r = \infty$)

8. A satellite of the earth has a mass of 100 kg and is at an altitude of 2×10^6 m. (a) What is the potential energy of the satellite-earth system? (b) What is the magnitude of the force on the satellite?

9. A system consists of three particles, each of mass 5 g, located at the corners of an equilateral triangle with sides of 30 cm. (a) Calculate the potential energy of the system. (b) If the particles are released simultaneously, where will they collide?

10. How much energy is required to move a 1000-kg mass from the earth's surface to an altitude equal to twice the earth's radius?

11. Four particles are positioned at the corners of a square as in Fig. 14.18. Calculate the total potential energy of the system.

Section 14.5 Energy Considerations in Planetary and Satellite Motion

12. A rocket is fired vertically from the earth's surface

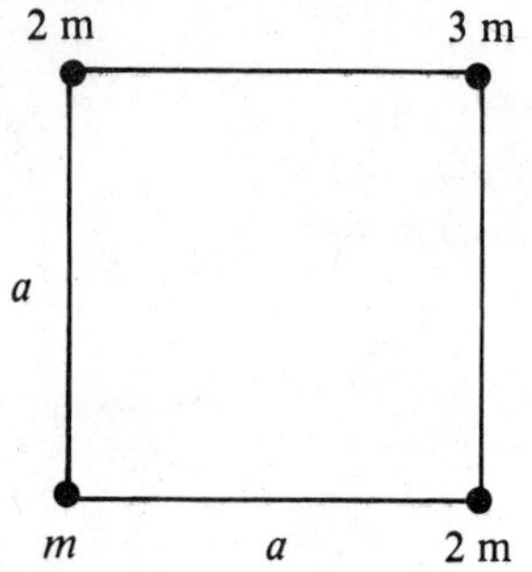

Figure 14.18 (Exercise 11).

and reaches a maximum altitude equal to three earth radii. What was the initial speed of the rocket? (Neglect friction, the earth's rotation, and the earth's orbital motion.)

13. A satellite moves in a circular orbit around a planet. Show that the orbital velocity v and escape velocity of the satellite are related by the expression $v_{esc} = \sqrt{2}v$.

14. Calculate the escape velocity from the moon using the data in Table 14.1.

15. Calculate the escape velocity from Mars using the data in Table 14.1.

16. A spaceship is fired from the earth's surface with an initial speed of 2.0×10^4 m/s. What will its speed be when it is very far from the earth? (Neglect friction.)

17. A 500-kg spaceship is in a circular orbit of radius $2R_e$ about the earth. (a) How much energy is required to transfer the spaceship to a circular orbit of radius $4R_e$? (b) Discuss the change in the potential energy, kinetic energy, and total energy.

18. Two identical spacecrafts, each of mass 1000 kg, travel in free space along the same path. At some instant when their separation is 20 m and each has the *same* velocity, the power is turned off in each vehicle. What are their speeds when they are 2 m apart? (Treat the spacecrafts as particles.)

19. (a) Calculate the minimum energy required to send a 3000-kg spacecraft from the earth to a distant point in space where earth's gravity is negligible. (b) If the journey is to take three weeks, what *average* power would the engines have to supply?

Section 14.6 The Gravitational Force Between an Extended Body and a Particle

20. A uniform rod of mass M is in the shape of a semicircle of radius R (Fig. 14.19). Calculate the force on a point mass m placed at the center of the semicircle.

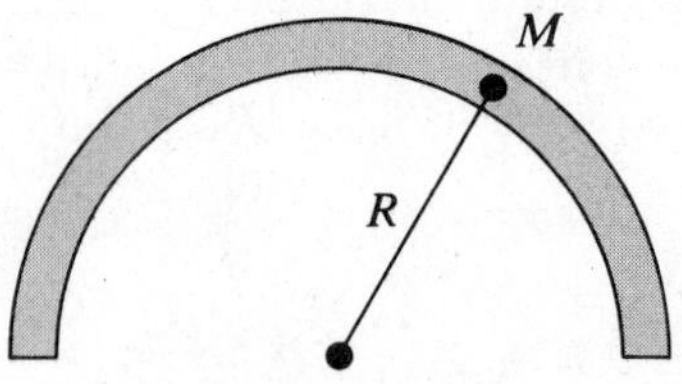

Figure 14.19 (Exercise 20).

21. A *nonuniform* rod of length L is placed along the x axis at a distance h from the origin, as in Fig. 14.13. The mass per unit length, λ, varies according to the expression $\lambda = \lambda_0 + Ax^2$, where λ_0 and A are constants. Find the force on a particle of mass m placed at the origin. (Hint: An element of the rod has a mass $dM = \lambda\ dx$.)

Section 14.7 Gravitational Force Between a Particle and a Spherical Mass

22. A spherical shell has a radius of 0.5 m and mass of 80 kg. Find the force on a particle of mass 50 g placed (a) 0.3 m from the center of the shell and (b) outside the shell 1 m from its center.

23. A uniform solid sphere has a radius of 0.4 m and a mass of 500 kg. Find the magnitude of the force on a particle of mass 50 g located (a) 1.5 m from the center of the sphere, (b) at the surface of the sphere, and (c) 0.2 m from the center of the sphere.

24. A uniform solid sphere of mass m_1 and radius R_1 is inside and concentric with a spherical shell of mass m_2 and radius R_2 (Fig. 14.20). Find the force on a particle of mass m located at (a) $r = a$, (b) $r = b$, (c) $r = c$, where r is measured from the center of the spheres.

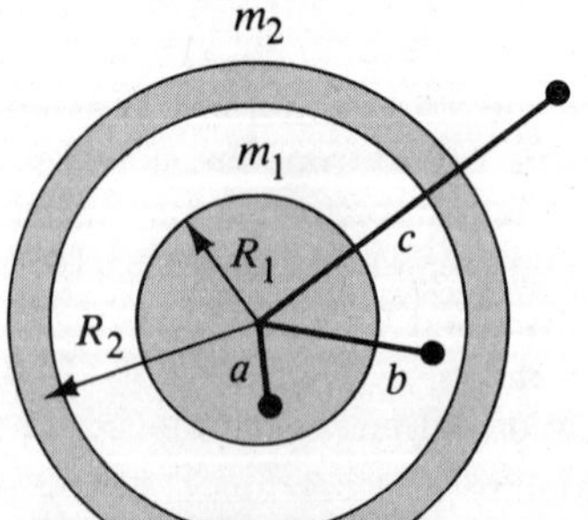

Figure 14.20 (Exercise 24).

PROBLEMS

1. Two astronauts, each of the same mass M, are seated opposite each other in a space station drifting in free space. The room they are in is a cylinder of radius R that rotates about its symmetry axis (Fig. 14.21). (a) What is the minimum angular speed of the cylinder that will keep the astronauts from moving toward each other if they are not strapped in their seats? (b) What angular speed must the cylinder have in order to produce a gravitational force equivalent to that experienced on earth? Obtain a numerical value if $R = 4$ m.

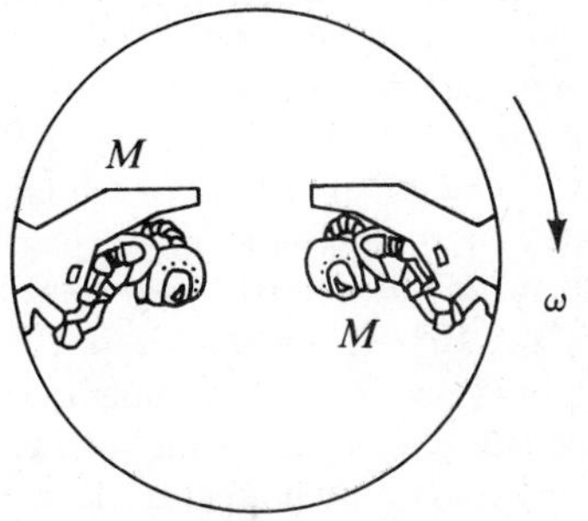

Figure 14.21 (Problem 1).

2. A sphere of mass M and radius R has a *nonuniform* density that varies with r, the distance from its center, according to the expression $\rho = Ar$, for $0 \leq r \leq R$. (a) What is the constant A in terms of M and R? (b) Determine the force on a particle of mass m placed *outside* the sphere. (c) Determine the force on the particle if it is *inside* the sphere. (Hint: See Section 14.7, paragraph 5.)
3. A hypothetical planet of mass M has three moons of equal mass m, each moving in the same circular orbit of radius R (Fig. 14.22). The moons are equally spaced and thus form an equilateral triangle. Find (a) the total potential energy of the system and (b) the orbital speed of each moon such that they maintain this configuration.

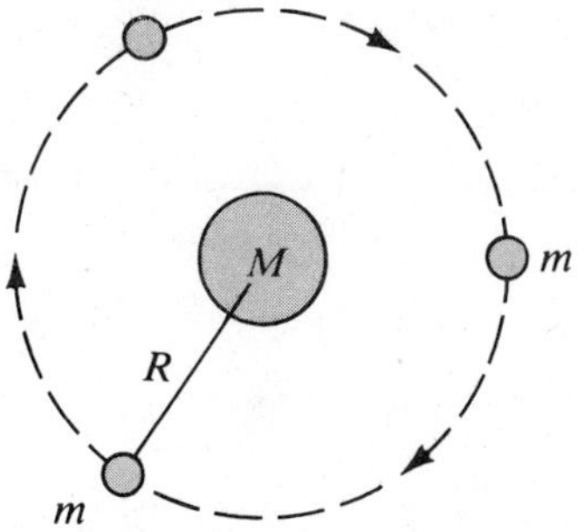

Figure 14.22 (Problem 3).

4. Two stars of masses M and m, separated by a distance d, revolve in circular orbits about their center of mass (Fig. 14.23). Show that each star has a period given by

$$T^2 = \frac{4\pi^2}{G(M + m)} d^3$$

(Hint: Apply Newton's second law to each star, and note that the center of mass condition requires that $Mr_2 = mr_1$, where $r_1 + r_2 = d$.)

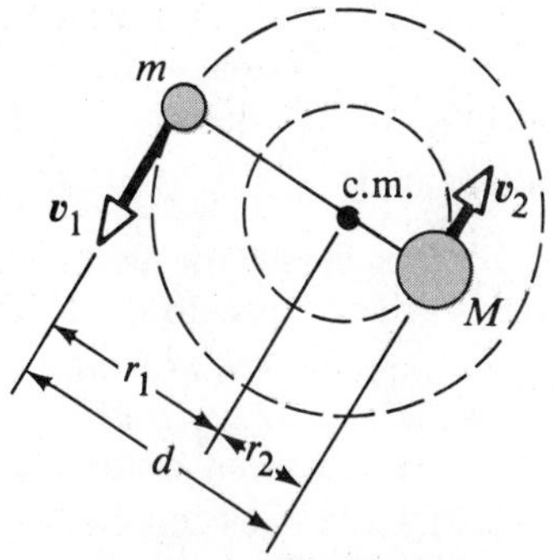

Figure 14.23 (Problem 4).

5. A particle of mass m lies along the symmetry axis of a uniform circular ring of mass M and radius R (Fig. 14.24). (a) Find the force on m if it is at a distance d from the plane of the ring. (b) Show that your result to (a) reduces to what you would intuitively expect (1) when m is at the center of the ring ($d = 0$) and (2) when m is distant from the ring ($d \gg R$).

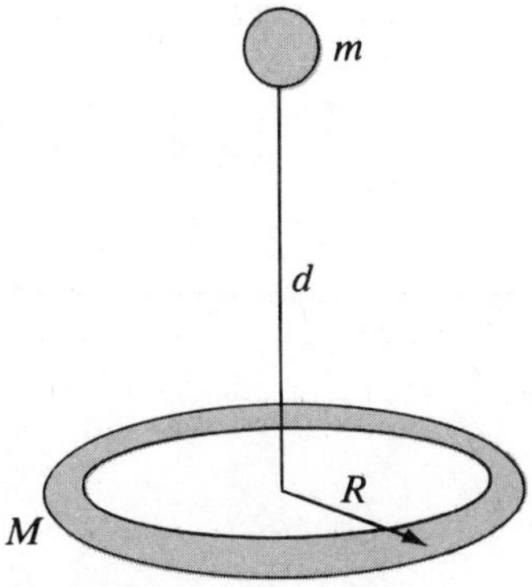

Figure 14.24 (Problem 5).

6. A particle of mass m is located *inside* a uniform solid sphere of radius R and mass M. If the particle is at a distance r from the center of the sphere, (a) show that the gravitational potential energy of the system is given by $U = (GmM/2R^3)r^2 - 3GmM/2R$. (b) How much work is done by the gravitational force in bringing the particle from the surface of the sphere to its center?

7. An object of mass m moves in a smooth straight tunnel of length L dug through a chord of the earth as discussed in Example 14.7 (Fig. 14.16). (a) Determine the effective force constant of the harmonic motion and the amplitude of the motion. (b) Using energy considerations, find the maximum speed of the object. Where does this occur? (c) Obtain a numerical value for the maximum speed if $L = 500$ km.

8. A satellite is in a circular orbit about a planet of radius R. If the altitude of the satellite is h and its period is T, (a) show that the density of the planet is given by $\frac{3\pi}{GT^2}\left(1 + \frac{h}{R}\right)^3$. (b) Calculate the average density of the planet if the period is 200 min and the satellite's orbit is close to the planet's surface.

9. When the Apollo 11 spacecraft orbited the moon, its mass was 9.979×10^3 kg, its period was 119 min, and its mean distance from the moon's center was 1.849×10^6 m. Assuming its orbit was circular and assuming the moon to be a uniform sphere, find (a) the mass of the moon, (b) the orbital speed of the spacecraft, and (c) the minimum energy required for the craft to leave the orbit and escape the moon's gravity.

10. The maximum distance from the earth to the sun (at the aphelion) is 1.521×10^{11} m, and the distance of closest approach (at the perihelion) is 1.471×10^{11} m. If the earth's orbital speed at the perihelion is 3.027×10^4 m/s, determine (a) the earth's orbital speed at the aphelion, (b) the kinetic and potential energy at the perihelion, and (c) the kinetic and potential energy at the aphelion. Is the total energy conserved? (Neglect the effect of the moon and other planets.)

11. Two hypothetical planets of masses m_1 and m_2 and radii r_1 and r_2, respectively, are at rest when they are an infinite distance apart. Because of their gravitational attraction, they head toward each other on a collision course. (a) When their center-to-center separation is d, find the speed of each planet and their *relative* velocity. (b) Find the kinetic energy of each planet *just* before they collide if $m_1 = 2 \times 10^{24}$ kg, $m_2 = 8 \times 10^{24}$ kg, $r_1 = 3 \times 10^6$ m, and $r_2 = 5 \times 10^6$ m. (Hint: Note that both energy and momentum are conserved. Treat the planets as point masses where the distance between their centers is $r_2 - r_1$ and ignore tidal effects.)

12. Using the data in Table 14.1, calculate the total potential energy of the sun-moon-earth system. Assume that the moon and earth are at the same distance from the sun.

15 Mechanics of Solids and Fluids

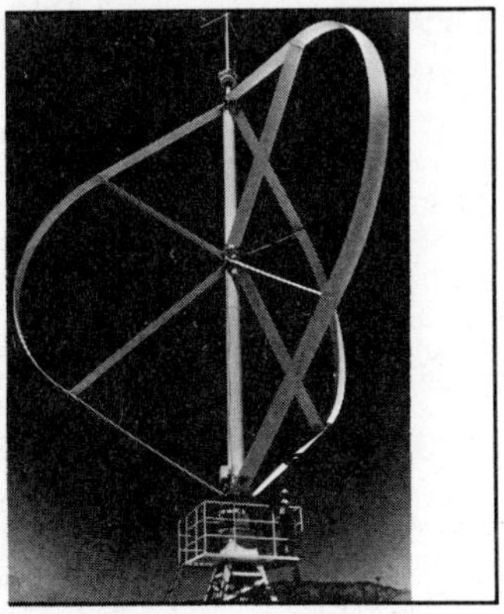

DOE Photo

This chapter deals with the mechanical properties of solids and fluids. This subject, often called *continuum mechanics,* treats the macroscopic behavior of a large number of particles. The deformation of a body under external forces will be used to classify matter as being solid, liquid, or gaseous. Our treatment of solids will assume that they are homogeneous and undergo only small deformations from equilibrium. By *small* we mean that when the deforming forces are removed, the body returns to its original shape. Several elastic constants will be defined, each corresponding to a different type of deformation.

In our treatment of the mechanics of fluids, we shall see that no new physical principles are needed to explain such effects as the buoyant force on a submerged object and the dynamic lift on an airplane wing. First we shall consider a fluid at rest and derive an expression for the pressure of the fluid as a function of its density and depth. We shall find that fluids in motion can be treated (with several important assumptions) using some basic ideas from newtonian mechanics. A very important result of fluid dynamics is Bernoulli's equation, which is a statement of the conservation of energy applied to an ideal fluid.

15.1 STATES OF MATTER

Matter is normally classified as being in one of three states: solid, liquid, or gaseous. Some scientists consider a collection of free, electrically charged particles as the fourth state of matter, referred to as a *plasma.* Everyday experience tells us that a solid has a definite volume and shape, a liquid has a definite volume but no definite shape, and a gas has no definite volume or definite shape. However, these classifications are somewhat artificial. For example, asphalt and glass will tend to flow like a liquid over long periods of time; water can be a solid, liquid, or gas (or combinations of these), depending on the temperature and pressure.

On a microscopic scale all matter consists of some distribution of atoms and molecules, which in turn contain particles that are even more fundamental. The atoms in a solid are located at specific positions in an equilibrium state. They exert strong forces on each other as a result of their proximity. In reality, the atoms vibrate about these equilibrium positions because of thermal agitation. At a sufficiently low temperature, this vibrating motion is slight and the atoms can be considered to be fixed. When an atom is displaced from its equilibrium position, internal atomic forces act so as to restore it to this equilibrium position. Therefore, a displaced atom behaves in much the same way as a vibrating mass-spring system. This manifests itself on a macroscopic scale in the following manner. A solid that is deformed by external forces will tend to return to its original shape and dimensions when the external forces are removed. For this reason, a solid is said to have *elasticity*.

Solids can be classified as being either crystalline or amorphous. A *crystalline* solid is one in which the atoms have an ordered, periodic structure. For example, in the sodium chloride crystal, sodium and chloride atoms occupy alternate corners of a cube. In an *amorphous* solid, such as glass, the atoms are arranged in a disordered fashion.

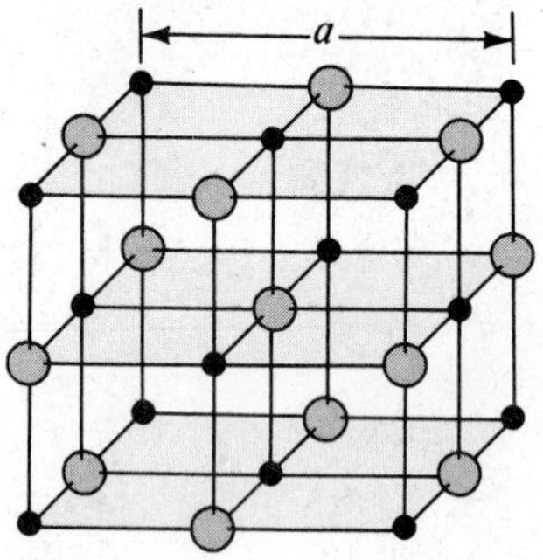

The NaCl structure. The Na^+ and Cl^- ions are at alternate corners of a cube.

In the gaseous state, the molecules are in constant random motion and exert weak forces on each other. The separation distances between the molecules of a gas are, on the average, quite large compared with the dimensions of the molecules. Occasionally the molecules collide with each other, but most of the time they move as nearly free, noninteracting particles. We shall have much more to say about the behavior of gases in subsequent chapters.

In the case of a liquid, the molecules are closely packed, as in a solid. However, the molecular forces in a liquid are not strong enough to keep the molecules in fixed positions. As a result, the molecules wander through the liquid in a random fashion. Since the densities of liquids and solids are similar, the two have the following property in common. When one tries to compress a liquid or a solid, strong repulsive atomic forces act internally to resist the deformation.

15.2 ELASTIC PROPERTIES OF SOLIDS

So far we have assumed that bodies remain undeformed (rigid) when external forces act on them. In reality, all bodies are deformable. That is, it is possible to change the shape or size of a body (or both) through the application of external forces. Although these changes are observed as macroscopic deformations, the internal forces that resist the deformation are due to the short-range electric forces between atoms.

In Chapter 5 we found that the external force required to stretch a spring some distance x from equilibrium is linearly proportional to this distance (Hooke's law). We can generalize Hooke's law to explain the elastic properties of solids in terms of the concepts of stress and strain. *Stress* is a quantity that is proportional to the force causing a deformation; *strain* is a measure of the degree of the deformation. *The generalized Hooke's law states that the stress is proportional to the strain*, and the constant of proportionality depends on the material and on the nature of the deformation. We call this proportionality constant the *elastic modulus*. The elastic modulus is therefore the ratio of the stress to the strain:

$$\text{Elastic modulus} = \frac{\text{stress}}{\text{strain}} \tag{15.1}$$

We shall consider three types of deformation and define an elastic modulus for each. These are (1) the Young's modulus, which measures the resistance of a solid to elongation; (2) the shear modulus, which measures the resistance to motion of the planes of a solid sliding past each other; and (3) the bulk modulus, which measures the resistance that solids or liquids offer to having their volume changed.

In all cases, we shall assume that the elastic substances under consideration are homogeneous and isotropic. That is, the materials have the same elastic properties at all points and in all directions. Of course, this assumption is not generally true. For example, many crystalline substances have anisotropic mechanical properties.

Young's Modulus: Elasticity in Length

Consider a long bar of cross-sectional area A and length l_0 that is clamped at one end (Fig. 15.1a). When an external force $\boldsymbol{F}_l$ is applied longitudinally along the bar (perpendicular to the cross section), internal forces in the bar resist distortion ("stretching"), but the bar attains an equilibrium in which its length is greater and in which the external force is *exactly balanced* by internal forces. In such a situation, the bar is said to be *stressed*. We define the *tensile stress* as the ratio of the internal force (equal to F_l) to the cross-sectional area A. The *tensile strain* in this case is defined as the ratio of the change in length Δl to the original length l_0. Thus, if the solid obeys Hooke's law, we can use Eq. 15.1 to define the *Young's modulus* Y *as*

$$Y \equiv \frac{\text{tensile stress}}{\text{tensile strain}} = \frac{F_l/A}{\Delta l/l_0} \tag{15.2}$$

Definition of Young's modulus

This quantity is typically used to characterize a rod or wire stressed under either tension or compression. Note that since the strain is a dimensionless quantity, Y has units of force per unit area. Typical values are given in Table 15.1. Experiments show that (a) the change in length for a fixed applied force is proportional to the original length and (b) the force necessary to produce a given strain is proportional to the cross-sectional area. Both of these observations are in accord with Eq. 15.2. On an atomic scale, one would expect the force to be proportional to A, since a larger area implies a larger number of interacting atoms in a given layer.

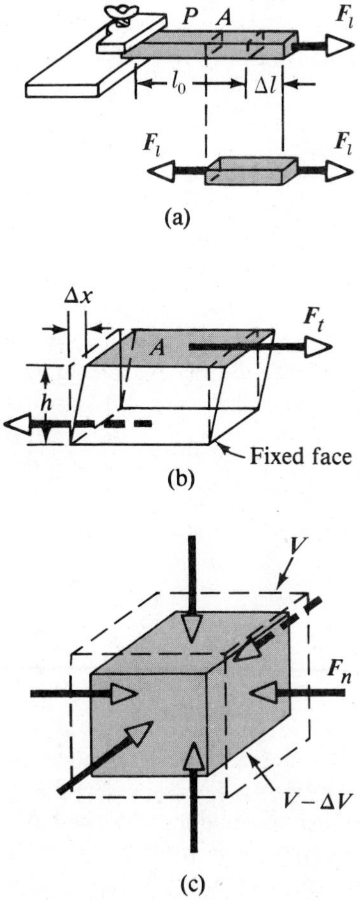

Figure 15.1 Elastic deformations in a solid: (a) a long bar clamped at one end is stretched by an amount $\Delta \ell$ under the action of a force $\boldsymbol{F}_l$; (b) a rectangular parallelopipid is distorted by a force $\boldsymbol{F}_t$ tangent to one of its faces; (c) a solid is compressed on all sides by forces normal to its surfaces.

TABLE 15.1 Elastic Constants

SUBSTANCE	YOUNG'S MODULUS Y (N/m²)	SHEAR MODULUS S (N/m²)	BULK MODULUS B (N/m²)
Aluminum	7.0×10^{10}	2.5×10^{10}	7.0×10^{10}
Brass	9.1×10^{10}	3.5×10^{10}	6.1×10^{10}
Copper	11×10^{10}	4.2×10^{10}	14×10^{10}
Steel	20×10^{10}	8.4×10^{10}	16×10^{10}
Tungsten	35×10^{10}	14×10^{10}	20×10^{10}
Glass	$6.5\text{–}7.8 \times 10^{10}$	$2.6\text{–}3.2 \times 10^{10}$	$5.0\text{–}5.5 \times 10^{10}$
Quartz	5.6×10^{10}	2.6×10^{10}	2.7×10^{10}
Water	—	—	0.21×10^{10}
Mercury	—	—	2.8×10^{10}

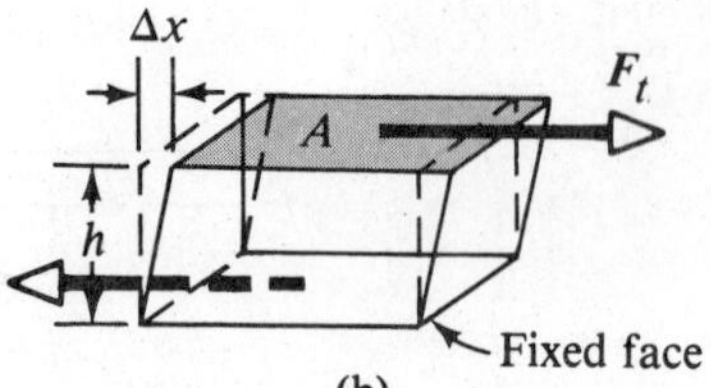

(b)

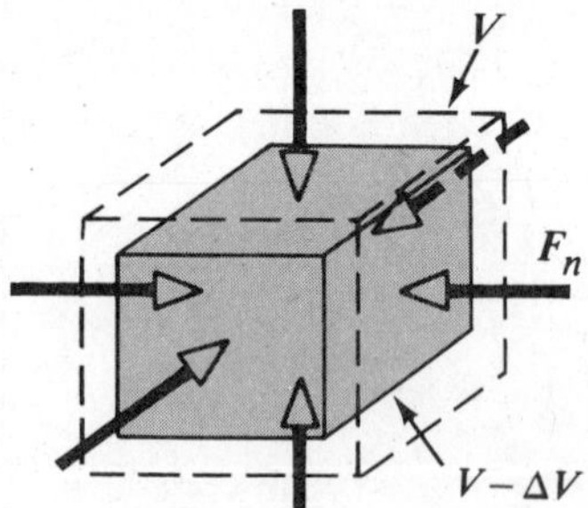

(c)

Figure 15.1 (b) and (c)

It is possible to exceed the *elastic limit* of a substance by applying a sufficiently large stress (Fig. 15.2). At the *yield point,* the stress versus strain curve departs from linearity. A material subjected to a stress beyond the yield point will ordinarily *not* return to its original length when the external force is removed. As the stress is increased even further, the material will ultimately break.

Shear Modulus: Elasticity of Shape

Another type of deformation occurs when a body is subjected to a force F_t tangential to one of its faces of area A, while the opposite face is kept fixed (Fig. 15.1b). If the body is originally a rectangular block, then the cross section becomes a parallelogram under this *shearing stress.* We see that there is no change in volume under this deformation. We define the shearing stress as F_t/A, that is, the ratio of the tangential force to the area A. The *shearing strain* is defined as $\Delta x/h$. In terms of these quantities, the *shear modulus* S is given by

$$S \equiv \frac{\text{shearing stress}}{\text{shearing strain}} = \frac{F_t/A}{\Delta x/h} \tag{15.3}$$

Bulk Modulus: Volume Elasticity

Finally, we can define the *bulk modulus* of a substance, which characterizes its response to uniform squeezing. Suppose that the external forces acting on a body are at right angles to all of its faces (Fig. 15.1c). As we shall see in the next section, this occurs when a body is immersed in a fluid. A body subject to this type of deformation undergoes a change in volume but no change in shape. The *volume stress,* ΔP, is defined as the ratio of the normal force F_n to the area A. When dealing with fluids, we shall refer to the quantity $\Delta P = F_n/A$ as the *pressure.* The *volume strain* is equal to the change in volume, ΔV, divided by the original volume, V. Thus, from Eq. 15.1 we can characterize a volume compression in terms of the *bulk modulus* B, defined as

$$B \equiv \frac{\text{volume stress}}{\text{volume strain}} = -\frac{F_n/A}{\Delta V/V} = -\frac{\Delta P}{\Delta V/V} \tag{15.4}$$

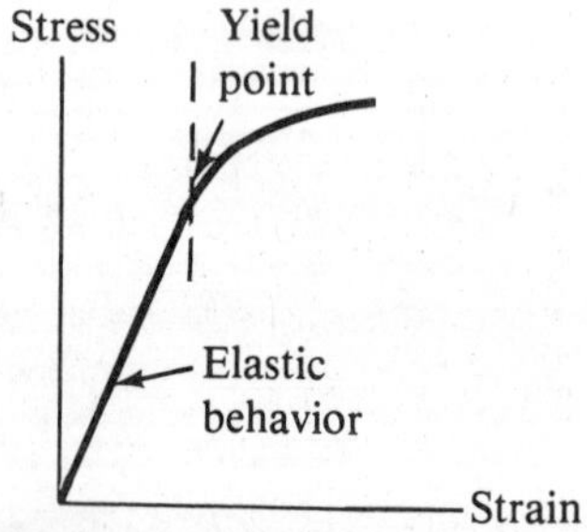

Figure 15.2 Stress versus strain curve for an elastic solid.

Note that the minus sign in Eq. 15.4 is introduced so that B is *always positive.* In other words, an increase in pressure (positive ΔP) causes a decrease in volume (negative ΔV) and vice versa. The reciprocal of the bulk modulus is called the *compressibility* of the material.

Table 15.1 lists typical values of elastic moduli for various substances. Note that both solids and liquids have a bulk modulus. However, there is no shear modulus and no Young's modulus for liquids since a static liquid will not sustain a shearing stress or a tensile stress.

Example 15.1 *Measuring Young's Modulus*

A load of 102 kg is hung on a wire of length 2 m and cross-sectional area 0.1 cm². The wire stretches by 0.22 cm over its unstretched length. Find the tensile stress, tensile strain, and Young's modulus for the wire from this information.

Solution:

$$\text{Tensile stress} = \frac{F_l}{A} = \frac{mg}{A} = \frac{(102\text{ kg})(9.8\text{ m/s}^2)}{0.1 \times 10^{-4}\text{ m}^2}$$

$$= 1.0 \times 10^8\text{ N/m}^2$$

$$\text{Tensile strain} = \frac{\Delta l}{l_0} = \frac{0.22 \times 10^{-2}\text{ m}}{2\text{ m}} = 0.11 \times 10^{-2}$$

$$Y = \frac{\text{tensile stress}}{\text{tensile strain}} = \frac{1.0 \times 10^8\text{ N/m}^2}{0.11 \times 10^{-2}}$$

$$= 9.1 \times 10^{10}\text{ N/m}^2$$

Comparing this value for Y with the values in Table 15.1, we see that the wire is probably brass.

Example 15.2

A square steel plate is 1 m on a side and has a thickness t of 1 cm. The bottom edge is held fixed as in Fig. 15.3. A force $\mathbf{F}_t$ applied along the upper edge produces a displacement $\Delta x = 0.005$ cm. Find the shearing strain, the shearing stress, and the magnitude of the force $\mathbf{F}_t$.

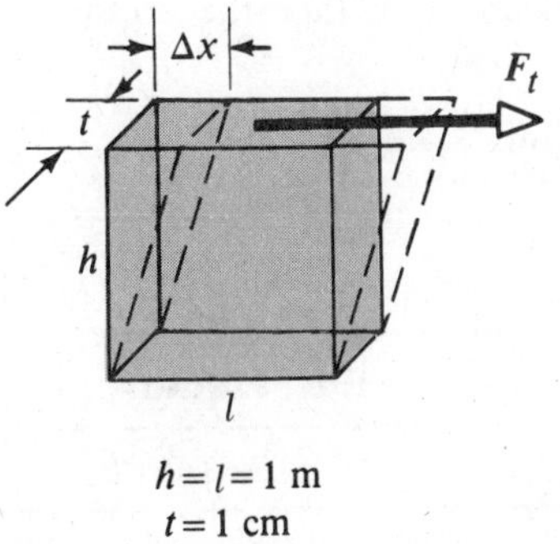

Figure 15.3 (Example 15.2) A shearing strain applied to a steel plate fixed at the bottom edge.

Solution:

$$\text{Shearing strain} = \frac{\Delta x}{h} = \frac{5 \times 10^{-5}\text{ m}}{1\text{ m}} = 5 \times 10^{-5}$$

From Eq. 15.3, the shearing stress is found to be

$$\text{Shearing stress} = \frac{F_t}{A} = S\frac{\Delta x}{h}$$

Since the shear modulus S for steel is equal to 8.4×10^{10} N/m² (Table 15.1), we have

$$\frac{F_t}{A} = \left(8.4 \times 10^{10}\frac{\text{N}}{\text{m}^2}\right)(5 \times 10^{-5}) = 4.2 \times 10^6\frac{\text{N}}{\text{m}^2}$$

The area of one edge is equal to $(1\text{ m})(1 \times 10^{-2}\text{ m}) = 10^{-2}\text{ m}^2$; therefore,

$$F_t = \left(4.2 \times 10^6\frac{\text{N}}{\text{m}^2}\right)(10^{-2}\text{ m}^2) = 4.2 \times 10^4\text{ N}$$

Example 15.3

A solid lead sphere of volume 0.5 m³ is dropped in the ocean to a depth where the water pressure is 2×10^7 N/m². Lead has a bulk modulus of 7.7×10^9 N/m². What is the change in volume of the sphere?

Solution: From the definition of bulk modulus, we have

$$B = -\frac{\Delta P}{\Delta V/V}$$

$$\Delta V = -\frac{V\,\Delta P}{B}$$

In this case, the change in pressure, ΔP, has the value 2×10^7 N/m². (This is large compared to atmospheric pressure, 1.01×10^5 N/m².) Taking $V = 0.5$ m³ and $B = 7.7 \times 10^9$ N/m², we get

$$\Delta V = -\frac{(0.5\text{ m}^3)(2 \times 10^7\text{ N/m}^2)}{7.7 \times 10^9\text{ N/m}^2} = -1.3 \times 10^{-3}\text{ m}^3$$

The negative sign indicates a *decrease* in volume. Using the fact that $V = \frac{4}{3}\pi R^3$, you should show that the corresponding decrease in the radius of the lead sphere is 4.3×10^{-4} m.

Q1. Why do you suppose the increase in length of a wire under a given load is proportional to its length? (Use a microscopic model in your argument.)

Q2. What kind of deformation does a cube of Jello exhibit when it "jiggles"?

15.3 DENSITY AND PRESSURE

The *density* of a homogeneous substance is defined as its mass per unit volume. That is, a substance of mass m and volume V has a density ρ given by

Definition of density

$$\rho \equiv \frac{m}{V} \tag{15.5}$$

The units of density are kg/m^3 in the SI system and g/cm^3 in the cgs system. Table 15.2 lists the densities of various substances. These values vary slightly with temperature, since the volume of a substance is temperature dependent (as we shall see in Chapter 16). Note that the densities of gases under normal conditions are about 1/1000 the densities of solids and liquids. This implies that the average molecular spacing in a gas under these conditions is about ten times greater than in a solid or liquid.

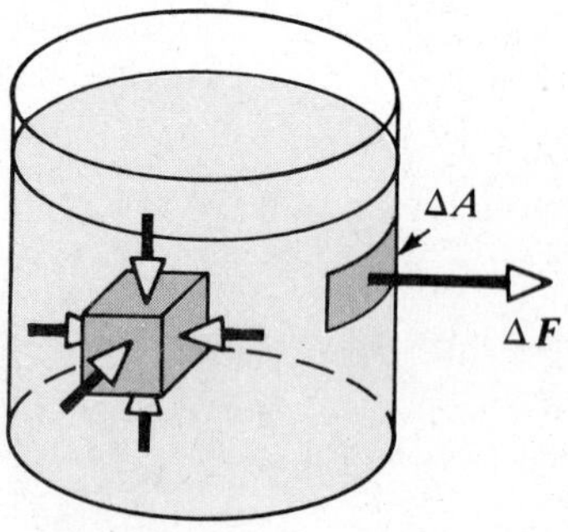

Figure 15.4 The force of the fluid on a submerged object at any point is perpendicular to the surface of the object. The force of the fluid on the walls of the container is perpendicular to the walls at all points.

To define *pressure,* consider an *ideal fluid,* that is, one that does not support a shear stress. The only stress that can exist in such a fluid is one that tends to compress an object. Therefore, *the force exerted by the fluid on an object immersed in it is always perpendicular to the surfaces of the object,* and this force compresses the object (Fig. 15.4).

The pressure at a specific point in a fluid can be measured using the device illustrated in Fig. 15.5. The device consists of an evacuated cylinder containing a light piston connected to a calibrated spring. As the device is submerged, the fluid pressure compresses the spring until the force of the fluid is balanced by the spring force.

If $\mathbf{F}$ is the force on the piston and A is the area of the piston, then the *average pressure* P of the fluid at the submersion level of the device is defined as

Definition of average pressure

$$P \equiv \frac{F}{A} \tag{15.6}$$

The pressure in a fluid is not the same at all points. To define the pressure at a specific point, consider a fluid enclosed as in Fig. 15.4. If the force exerted by the fluid is $\Delta \mathbf{F}$ over a surface element of area ΔA, then the pressure at that point is

Pressure at a point

$$P = \lim_{\Delta A \to 0} \frac{\Delta F}{\Delta A} = \frac{dF}{dA} \tag{15.7}$$

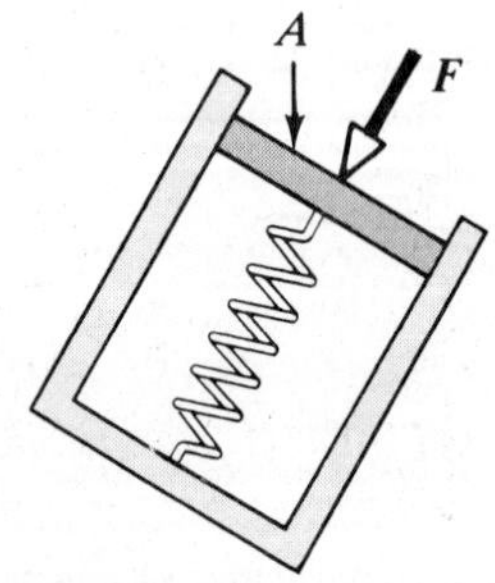

Figure 15.5 A simple device for measuring pressure.

TABLE 15.2 Densities of Some Solids, Liquids, and Gases

SUBSTANCE	ρ (kg/m³)	*SUBSTANCE*	ρ (kg/m³)
Ice	0.917×10^3	Water	1.00×10^3
Aluminum	2.70×10^3	Glycerin	1.26×10^3
Iron	7.86×10^3	Ethyl alcohol	0.806×10^3
Copper	8.92×10^3	Benzene	0.879×10^3
Silver	10.5×10^3	Mercury	13.6×10^3
Lead	11.3×10^3	Air	1.29
Gold	19.3×10^3	Oxygen	1.43
Platinum	21.4×10^3	Hydrogen	8.99×10^{-2}
		Helium	1.79×10^{-1}

The values given are at standard atmospheric pressure and temperature (STP). To convert to g/cm^3, multiply the densities by 10^{-3}.

As we shall see in the next section, the pressure in a fluid in the presence of the force of gravity varies with depth. Therefore, to get the total force on the wall of a container, we would have to integrate Eq. 15.7 over the wall surface.

Since pressure is force per unit area, it has units of N/m^2 in the SI system. Another name for the SI unit of pressure is *pascal* (Pa), where $1\ Pa = 1\ N/m^2$.

15.4 VARIATION OF PRESSURE WITH DEPTH

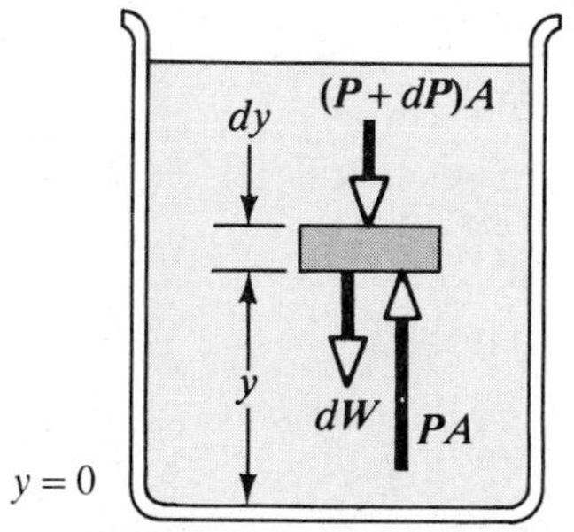

Figure 15.6 The variation of pressure with depth in a fluid. The volume element is at rest, and the forces on it are shown.

Consider a fluid at rest in a container (Fig. 15.6). We first note that *all points at the same depth must be at the same pressure*. If this were not the case, a given element of the fluid would not be in equilibrium. Now let us select a portion of the fluid contained within an imaginary cylinder of cross-sectional area A and height dy. The upward force on the bottom of the cylinder is PA, and the downward force on the top is $(P + dP)A$. The weight of the cylinder, the volume of which is dV, is given by $dW = \rho g\, dV = \rho g A\, dy$, where ρ is the density of the fluid. Since the cylinder is in equilibrium, the forces must add to zero, and so we get

$$\Sigma F_y = PA - (P + dP)A - \rho g A\, dy = 0$$

$$\frac{dP}{dy} = -\rho g \qquad (15.8)$$

From this result, we see that an increase in elevation (positive dy) corresponds to a decrease in pressure (negative dP). If P_1 and P_2 are the pressures at the elevations y_1 and y_2 above the reference level, then integrating Eq. 15.8 gives

$$P_2 - P_1 = -\rho g(y_2 - y_1) \qquad (15.9)$$

If the vessel is open at the top (Fig. 15.7), then the pressure at the depth h can be obtained from Eq. 15.9. Taking atmospheric pressure to be $P_a = P_2$, and noting that the depth $h = y_2 - y_1$, we find that

Pressure at any depth *h*

$$P = P_a + \rho g h \qquad (15.10)$$

where we usually take $P_a \approx 1.01 \times 10^5\ N/m^2$ (14.7 lb/in.²). In other words, the *absolute pressure* P at a depth h below the surface of a liquid open to the atmosphere is *greater* than atmospheric pressure by an amount $\rho g h$. This result also verifies that the *pressure is the same at all points having the same elevation*. Furthermore, the vessel's shape does not affect the pressure.

In view of the fact that the pressure in a fluid depends only upon depth, any increase in pressure at the surface must be transmitted to every point in the fluid. This was first recognized by the French scientist Blaise Pascal (1623–1662) and is called *Pascal's law: Pressure applied to an enclosed fluid is transmitted undiminished to every point of the fluid and the walls of the containing vessel.*

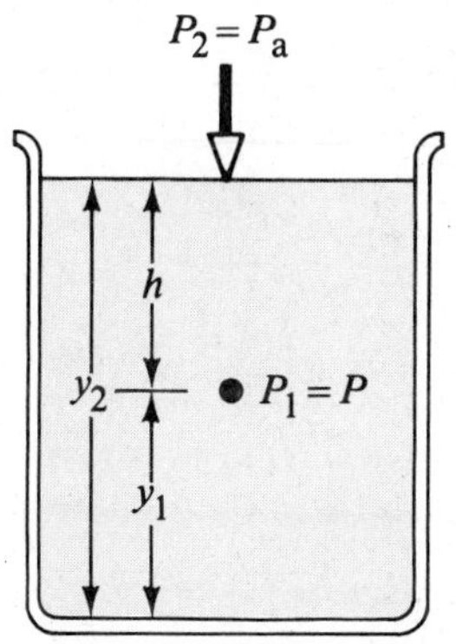

Figure 15.7 The pressure P at a depth h below the surface of a liquid open to the atmosphere is given by $P = P_a + \rho g h$.

An important application of Pascal's law is the hydraulic press illustrated by Fig. 15.8. A force $\boldsymbol{F}_1$ is applied to a small piston of area A_1. The pressure is transmitted through a fluid to a larger piston of area A_2. Since the pressure is the same on both sides, we see that $P = F_1/A_1 = F_2/A_2$. Therefore, the force $\boldsymbol{F}_2$ is larger than $\boldsymbol{F}_1$ by the multiplying factor A_2/A_1. Hydraulic brakes, car lifts, and hydraulic jacks make use of this principle.

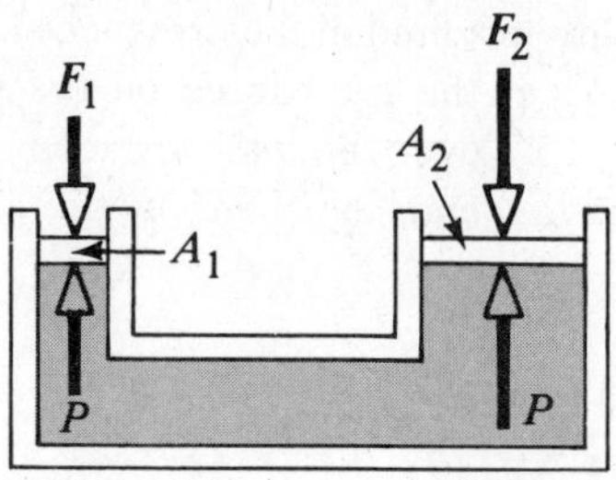

Figure 15.8 Schematic diagram of a hydraulic press. Since the pressure is the same at the left and right sides, a small force $\mathbf{F}_1$ at the left produces a much larger force $\mathbf{F}_2$ at the right.

Example 15.4

(a) Calculate the absolute pressure at an ocean depth of 1000 m. Assume the density of water is 1.0×10^3 kg/m³ and $P_a = 1.01 \times 10^5$ N/m².

$$P = P_a + \rho g h$$
$$= 1.01 \times 10^5 \frac{\text{N}}{\text{m}^2} + (1.0 \times 10^3)(9.8 \times 10^3)\frac{\text{N}}{\text{m}^2}$$
$$P \approx 9.9 \times 10^6 \text{ N/m}^2$$

(b) Calculate the total force that would be exerted on the outside of a circular submarine window of diameter 30 cm at this depth.

$$F = PA = P\pi r^2 = \left(9.9 \times 10^6 \frac{\text{N}}{\text{m}^2}\right)(\pi)(0.15 \text{ m})^2$$
$$\approx 7.0 \times 10^5 \text{ N} = 157\,000 \text{ lb!}$$

To obtain the *net* force on the window, you would have to allow for 1 atm of pressure within the submarine. Obviously, the engineering and construction of windows and vessels that will withstand such enormous pressures are not a trivial matter.

Example 15.5 *The Force on a Dam*

Water is filled to a height H behind a dam of width w (Fig. 15.9). Determine the resultant force on the dam.

Solution: The pressure at the depth h beneath the surface at the shaded portion is

$$P = \rho g h = \rho g(H - y)$$

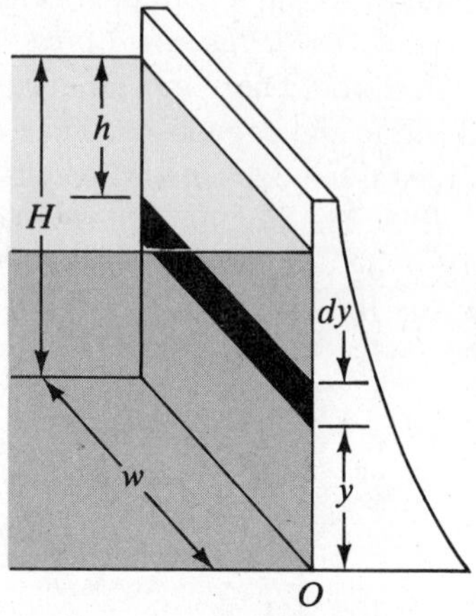

Figure 15.9 (Example 15.5) The total force on a dam must be obtained from the expression $F = \int P\,dA$, where dA is the area of the shaded strip.

(We have left out atmospheric pressure since it acts on both sides of the dam.) Using Eq. 15.7, we find the force on the shaded strip to be

$$dF = P\,dA = \rho g(H - y)w\,dy$$

Therefore, the total force on the dam is

$$F = \int P dA = \int_0^H \rho g(H - y)w\,dy = \frac{1}{2}\rho g w H^2$$

For example, if $H = 30$ m and $w = 100$ m, we find that $F = 4.4 \times 10^8 \text{ N} = 9.9 \times 10^7$ lb!

Q3. Two glass tumblers with *different* shapes and cross-sectional areas are filled to the same level with water. According to Eq. 15.10, the pressure is the same at the bottom of both tumblers. In view of this, why does one tumbler weigh more than the other?

Q4. A flat plate is immersed in a fluid at rest. For what orientation of the plate will the pressure on its flat surface be uniform?

Q5. Since atmospheric pressure is 14.7 lb/in.² and the area of a person's chest is about 200 in.², the force of the atmosphere on one's chest is around 3000 lb! In view of this enormous force, why don't our bodies collapse?

Q6. Describe how you could use an ordinary garden hose filled with water as a makeshift level. Why does it work?

15.5 PRESSURE MEASUREMENTS

One simple device for measuring pressure is the open-tube manometer illustrated in Fig. 15.10a. One end of a U-shaped tube containing a liquid is open to the atmosphere, and the other end is connected to a system of unknown pressure P. The pressure at point B equals $P_a + \rho gh$, where ρ is the density of the fluid. But the pressure at B equals the pressure at A, which is also the unknown pressure P. Therefore, we conclude that

$$P = P_a + \rho gh$$

The pressure P is called the *absolute pressure*, while $P - P_a$ is called the *gauge pressure*. Thus, if the pressure in the system is greater than atmospheric pressure, h is positive. If the pressure is less than atmospheric pressure (a partial vacuum), h is negative.

Another instrument used to measure pressure is the common barometer, invented by Evangelista Torricelli (1608–1647). A long tube is filled with mercury and then inverted into a dish of mercury (Fig. 15.10b). The upper end of the closed tube is nearly a vacuum, and so its pressure can be taken as zero. Therefore, it follows that $P_a = \rho gh$, where ρ is the density of the mercury and h is the height of the mercury column. One atmosphere of pressure is defined to be the pressure equivalent of a column of mercury that is exactly 0.76 m in height at 0°C, with $g = 9.80665\ \text{m/s}^2$. At this temperature, mercury has a density of $13.595 \times 10^3\ \text{kg/m}^3$; therefore

$$P_a = \rho gh = \left(13.595 \times 10^3\ \frac{\text{kg}}{\text{m}^3}\right)\left(9.80665\ \frac{\text{m}}{\text{s}^2}\right)(0.7600\ \text{m}) = 1.013 \times 10^5\ \frac{\text{N}}{\text{m}^2}$$

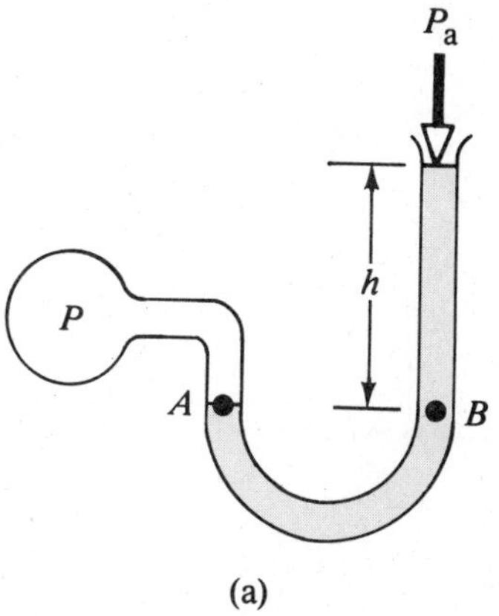

(a)

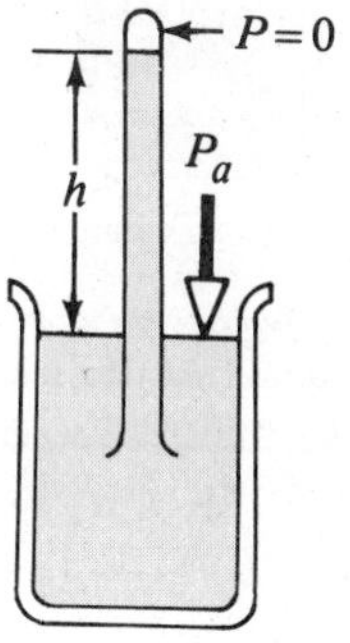

(b)

Figure 15.10 Two devices for measuring pressure: (a) the open-tube manometer; (b) the mercury barometer.

Q7. Pascal made a barometer using water as the working fluid. Why is it impractical to use water?

Q8. If the open end of the U-shaped tube in Fig. 15.10a is evacuated and sealed, what will happen to the height of the column on the right? What will happen to the pressure on the left?

Q9. Explain why a fluid rises in a drinking straw. Why is it easier to sip a soft drink than a malted through a straw?

15.6 BUOYANT FORCES AND ARCHIMEDES' PRINCIPLE

Consider an object having a weight in air of **W**. It is well known that if the object is suspended from a scale, the reading of the scale is less than W when the object is submerged in a fluid (Fig. 15.11a). This is because the fluid exerts an upward force that helps support the object. We call this force *the buoyant force*.

Archimedes' principle

The magnitude of the buoyant force always equals the weight of the fluid displaced by the body. This is known as *Archimedes' principle: A body wholly or partially submerged in a fluid is buoyed up with a force equal to the weight of the fluid displaced by the body.* The buoyant force acts vertically upward through what was the center of gravity of the fluid before the fluid was displaced.

Let us analyze the external forces acting on the submerged object (Fig. 15.11b). These are the upward force of the spring, $\mathbf{F}_s$, the weight **W** acting downward, the upward force of the fluid at the bottom, $\mathbf{F}_1$, and the downward force of the fluid at the top, $\mathbf{F}_2$. In the absence of the fluid, we know that $W = F_s$. But when the body is submerged, the spring scale reads a force less than the weight. Therefore, the force

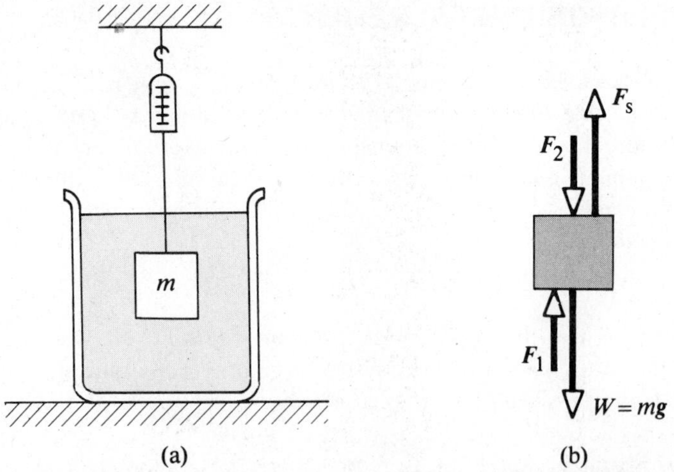

Figure 15.11 (a) The scale reading of an object submerged in a fluid is less than the reading when the object is suspended in air. (b) The external forces acting on the submerged object.

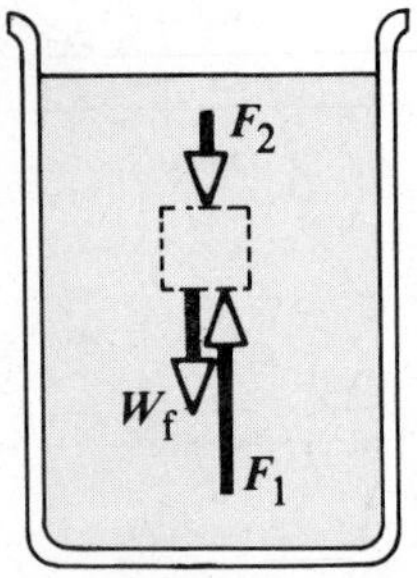

Figure 15.12 The forces acting on a portion of a fluid at rest. The buoyant force, $\mathbf{B}$, equals $\mathbf{F}_1 - \mathbf{F}_2$. Since the fluid is at rest, the buoyant force must balance the weight of this volume of fluid, $\mathbf{W}_f$.

$\mathbf{F}_1$ at the bottom must be greater than the force $\mathbf{F}_2$ at the top. The difference in these two forces is the buoyant force $\mathbf{B} = \mathbf{F}_1 - \mathbf{F}_2$.

We can verify Archimedes' principle in the following manner. Suppose a submerged object is replaced by an equal volume of fluid (the dotted region in Fig. 15.12). Since the fluid in this volume is in equilibrium, the net upward force on it due to the surrounding fluid (the buoyant force) must balance the downward weight of the fluid, $\mathbf{W}_f$, contained in this volume. But the buoyant force, $\mathbf{B} = \mathbf{F}_1 - \mathbf{F}_2$, is the same as that which would act on the original submerged object. Therefore, the buoyant force must equal the weight of the displaced fluid, $\mathbf{W}_f$. This result applies for a submerged object of *any* shape, size, or density.

If the density of the fluid is ρ_f and the volume of the submerged object is V, then the buoyant force is given by

Buoyant force

$$B = W_f = m_f g = \rho_f V g \tag{15.11}$$

where m_f is the mass of the displaced fluid. The weight of the submerged object is $\rho V g$, where ρ is the density of the object. Therefore, if the density of the object is greater than the density of the fluid, the unsupported object will sink. If the density of the object is less than that of the fluid, the unsupported submerged object will accelerate upward and will ultimately float. When a *floating* object is in equilibrium, *part of its volume is submerged.* In this case, the buoyant force equals the weight of the object.

Example 15.5

A piece of aluminum is suspended from a string and then completely immersed in a container of water (Fig. 15.13). The mass of the aluminum is 1 kg and its density is 2.7×10^3 kg/m^3. Calculate the tension in the string before and after the aluminum is immersed in water.

Solution: When the piece of aluminum is suspended in air, the tension in the string, T_1, (the reading on the scale) is equal to the weight mg, assuming that the buoyancy force of air can be neglected:

$$T_1 = mg = (1\text{ kg})\left(9.8\ \frac{\text{m}}{\text{s}^2}\right) = 9.8\text{ N}$$

When the aluminum is immersed in water, it experiences an upward buoyant force $\mathbf{B}$, as in Fig. 15.13b, which reduces the tension in the string. Since the system is in equilibrium,

$$T_2 + B - mg = 0$$

$$T_2 = mg - B = (9.8 - B)\text{ N}$$

In order to calculate B, we must first calculate the volume of the aluminum:

$$V_{Al} = \frac{m}{\rho_{Al}} = \frac{1 \text{ kg}}{2.7 \times 10^3 \text{ kg/m}^3} = 3.7 \times 10^{-4} \text{ m}^3$$

Since the buoyant force equals the weight of the water displaced, we have

$$B = m_w g = \rho_w V_{Al} g$$

$$= \left(1 \times 10^3 \frac{\text{kg}}{\text{m}^3}\right)(3.7 \times 10^{-4} \text{ m}^3)\left(9.8 \frac{\text{m}}{\text{s}^2}\right) = 3.6 \text{ N}$$

Therefore,

$$T_2 = 9.8 - B = 9.8 - 3.6 = 6.2 \text{ N}$$

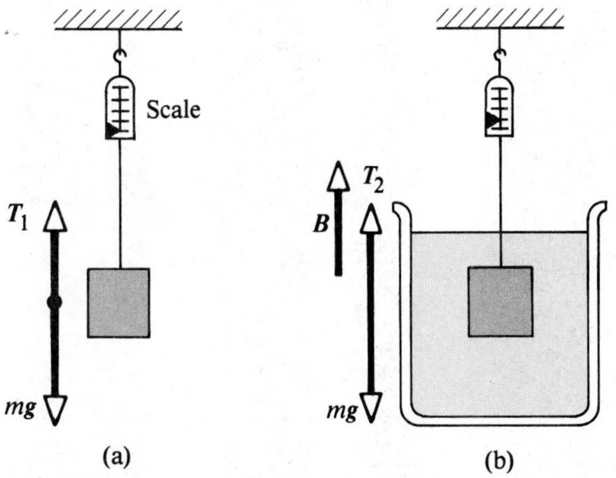

Figure 15.13 (Example 15.5) (a) When the aluminum is suspended in air, the scale reads the true weight, mg (neglecting the buoyancy of air). (b) When the aluminum is immersed in water, the buoyant force, $\mathbf{B}$, reduces the scale reading to $T_2 = mg - B$.

Q10. When an object is immersed in a fluid at rest as in Fig. 15.11, why is the net force on it in the x direction equal to zero?

Q11. A person sits in a boat floating in a small pond. If the person throws a heavy anchor overboard, does the level of the pond rise, fall, or remain the same? (Be careful!)

Q12. (1) Explain why a sealed bottle partially filled with a liquid can float. (2) In order to float more easily, should a swimmer inhale or exhale?

Q13. A piece of wood is partially submerged in a container filled with water. If the container is sealed and pressurized above atmospheric pressure, does the wood rise, fall, or remain at the same level? (Hint: Wood is porous.)

Q14. Lead has a greater density than iron, and both are denser than water. Is the buoyant force on a lead object greater than, less than, or equal to the buoyant force on an iron object of the same dimensions? Explain.

15.7 FLUID DYNAMICS AND BERNOULLI'S EQUATION

The study of fluids in motion, or *fluid dynamics*, is one of the most complex subjects in mechanics. For example, imagine the difficulties one encounters in attempting to describe the motion of a particle in a turbulent river. Fortunately, many features of fluid motion can be understood on the basis of the following four *simplifying assumptions*.

Streamline flow of an ideal fluid

1. The fluid is *nonviscous*. This is equivalent to assuming that the shear modulus is zero. Hence, there is no internal frictional force between adjacent fluid layers.
2. The fluid is *incompressible*. This means that the density of the fluid is constant.
3. The fluid motion is *steady*. Steady-state flow means that the velocity, density, and pressure at each point in the fluid do not change in time.
4. The flow is *irrotational*. This implies that each element of the fluid has zero angular velocity about its center, so that there is no turbulence. Therefore, a small paddle wheel placed anywhere in the fluid will not rotate, and there can be no whirlpools or eddies present in the moving fluid.

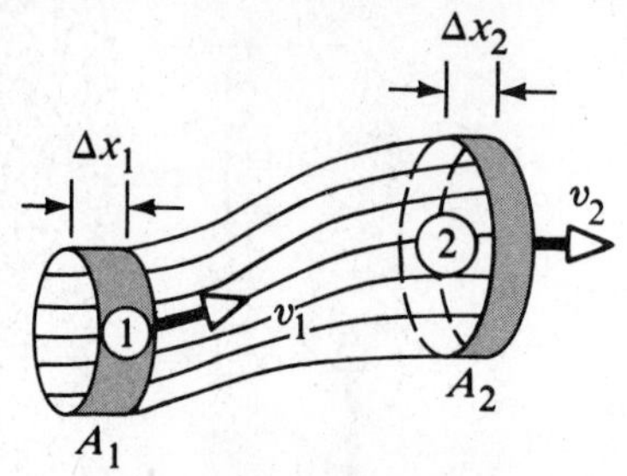

Figure 15.14 A fluid moving with streamline flow through a pipe of varying cross-sectional area. The volume of fluid flowing through A_1 in a time interval Δt must equal the volume flowing through A_2 in the same time interval. Therefore, $A_1v_1 = A_2v_2$.

Figure 15.14 represents an ideal fluid flowing through a pipe of nonuniform size. The particles in the fluid moves along so-called streamlines in steady-state flow. A *streamline* is a line drawn such that at each point the fluid velocity is tangent to the streamline. Every particle starting at the same point follows the same streamline.

In a small time interval Δt, the fluid at the bottom end of the pipe moves a distance $\Delta x_1 = v_1\,\Delta t$. If A_1 is the cross-sectional area in this region, then the mass contained in the shaded region is $\Delta M_1 = \rho_1 A_1\,\Delta x_1 = \rho_1 A_1 v_1\,\Delta t$. Similarly, the fluid that moves through the upper end of the pipe in the time Δt has a mass $\Delta M_2 = \rho_2 A_2 v_2\,\Delta t$. However, since *mass is conserved* and because the flow is steady, the mass that crosses A_1 in a time Δt must equal the mass that crosses A_2 in a time Δt. Therefore $\Delta M_1 = \Delta M_2$, or

$$\rho_1 A_1 v_1 = \rho_2 A_2 v_2 \qquad (15.12)$$

This expression is called the *equation of continuity.*

Since ρ is constant for the steady flow of an *incompressible* fluid, Eq. 15.12 reduces to

The product Av is constant for an incompressible fluid in steady flow

$$A_1 v_1 = A_2 v_2 = \text{constant} \qquad (15.13)$$

That is, *the product of the area and the fluid speed at any point along the pipe is a constant.* Therefore, as one would expect, the speed is high where the tube is constricted and low where the tube is wide. The product Av, which has the dimensions of volume/time, is called the *volume flux,* or flow rate.[1]

Example 15.6

A water hose 2 cm in diameter is used to fill a 20-liter bucket. If it takes 1 min to fill the bucket, what is the speed v at which the water leaves the hose? (Note that 1 liter $= 10^3$ cm^3.)

Solution: The cross-sectional area of the hose is

$$A = \pi\frac{d^2}{4} = \pi\left(\frac{2^2}{4}\right)\text{cm}^2 = \pi\text{ cm}^2$$

According to the data given, the flow rate is equal to 20 liters/min. Equating this to the product Av gives

$$Av = 20\,\frac{\text{liters}}{\text{min}} = \frac{20\times10^3\text{ cm}^3}{60\text{ s}}$$

$$v = \frac{20\times10^3\text{ cm}^3}{(\pi\text{ cm}^2)(60\text{ s})} = 106\text{ cm/s}$$

Q15. Explain why the equation of continuity (Eq. 15.12) would not be valid if the pipe contains a leak somewhere between points 1 and 2 in Fig. 15.14.

[1]The condition Av = constant is equivalent to the fact that the amount of fluid which enters one end of the tube in a given time interval equals the amount of fluid leaving the tube in the same time interval, assuming there are no leaks.

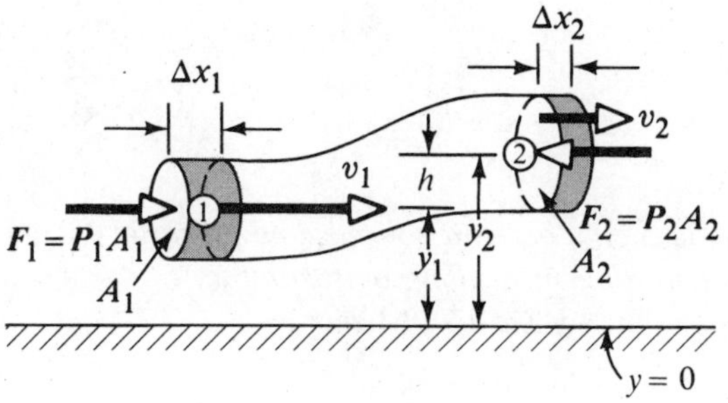

Figure 15.15 A fluid flowing through a constricted pipe with streamline flow. The fluid in shaded region 1 moves to shaded region 2. The volumes of fluid in the two shaded regions are equal.

Bernoulli's Equation

As a fluid moves through a pipe of varying cross section and elevation, the pressure will change along the pipe. In 1738 the Swiss physicist Daniel Bernoulli (1700–1782) first derived a most fundamental expression that relates the pressure to the fluid speed and elevation. As we shall see, this result is a consequence of energy conservation as applied to our ideal fluid.

Again, *we shall assume that the fluid is incompressible and nonviscous and flows in an irrotational, steady-state manner* as described in the previous section. Consider the flow through a nonuniform pipe in a time Δt, as illustrated in Fig. 15.15. The force on the lower end of the fluid is P_1A_1, where P_1 is the pressure at point 1. The work done by this force is $W_1 = F_1\,\Delta x_1 = P_1A_1\,\Delta x_1 = P_1\,\Delta V$, where ΔV is the volume of the lower shaded region. In a similar manner, the work done on the fluid on the upper portion in the time Δt is $W_2 = -P_2A_2\,\Delta x_2 = -P_2\,\Delta V$. (Note that the volume that passes through 1 in a time Δt equals the volume that passes through 2 in the same time interval.) This work is negative since the fluid force opposes the displacement. Thus the net work done by these forces in the time Δt is

$$W = (P_1 - P_2)\,\Delta V$$

Part of this work goes into changing the kinetic energy of the fluid, and part goes into changing the gravitational potential energy. If Δm is the mass passing through 1 and 2 in the time Δt, then the change in kinetic energy is

$$\Delta K = \frac{1}{2}(\Delta m)\,v_2^{\,2} - \frac{1}{2}(\Delta m)\,v_1^{\,2}$$

The change in potential energy is

$$\Delta U = \Delta m g y_2 - \Delta m g y_1$$

Applying the work-energy theorem in the form $W = \Delta K + \Delta U$ (Chapter 8) gives

$$(P_1 - P_2)\,\Delta V = \frac{1}{2}(\Delta m)\,v_2^{\,2} - \frac{1}{2}(\Delta m)\,v_1^{\,2} + \Delta m g y_2 - \Delta m g y_1$$

But $\rho = \Delta m/\Delta V$; therefore if we divide each term by ΔV, the above expression reduces to

$$P_1 - P_2 = \frac{1}{2}\rho v_2^{\,2} - \frac{1}{2}\rho v_1^{\,2} + \rho g y_2 - \rho g y_1$$

Rearranging terms, we get

$$P_1 + \frac{1}{2}\rho v_1^{\,2} + \rho g y_1 = P_2 + \frac{1}{2}\rho v_2^{\,2} + \rho g y_2 \qquad (15.14)$$

This is *Bernoulli's equation* as applied to a nonviscous, incompressible fluid in steady flow. It is often expressed as

Bernoulli's equation

$$P + \frac{1}{2}\rho v^2 + \rho g y = \text{constant} \qquad (15.15)$$

That is, *the sum of the pressure, kinetic energy per unit volume, and potential energy per unit volume has the same value at all points along a streamline.*

When the fluid is at *rest,* $v_1 = v_2 = 0$ and Eq. 15.14 becomes

$$P_1 - P_2 = \rho g(y_2 - y_1) = \rho g h$$

which agrees with Eq. 15.9.

Example 15.7

The *Venturi tube.* The horizontal constricted pipe illustrated in Fig. 15.16, known as a *Venturi tube,* can be used to measure flow velocities in an incompressible fluid. Let us determine the flow velocity at point 2 if the pressure difference $P_1 - P_2$ is known.

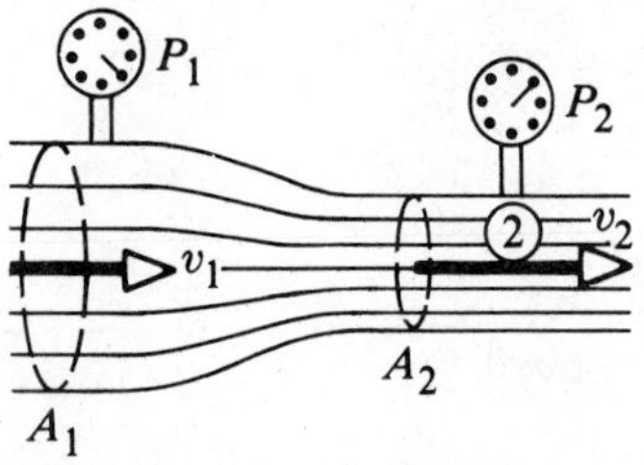

Figure 15.16 (Example 15.7) Schematic diagram of a Venturi tube. The pressure P_1 is greater than the pressure P_2, since $v_1 < v_2$. This device can be used to measure the speed of fluid flow.

Solution: Since the pipe is horizontal, $y_1 = y_2$ and Eq. 15.14 applied to points 1 and 2 gives

$$P_1 + \frac{1}{2}\rho {v_1}^2 = P_2 + \frac{1}{2}\rho {v_2}^2$$

From the equation of continuity (Eq. 15.13), we see that $A_1 v_1 = A_2 v_2$ or

$$v_1 = \frac{A_2}{A_1} v_2$$

Substituting this expression into the previous equation gives

$$P_1 + \frac{1}{2}\rho \left(\frac{A_2}{A_1}\right)^2 {v_2}^2 = P_2 + \frac{1}{2}\rho {v_2}^2$$

$$v_2 = A_1 \sqrt{\frac{2(P_1 - P_2)}{\rho({A_1}^2 - {A_2}^2)}} \qquad (15.16)$$

We can also obtain an expression for v_1 using this result and the continuity equation. Note that since $A_2 < A_1$, it follows that P_1 is *greater* than P_2. In other words, the pressure is *reduced* in the constricted part of the pipe. This result is somewhat analogous to the following situation. Consider a very crowded room, where people are squeezed together. As soon as a door is opened and people begin to exit, the squeezing (pressure) is least near the door where the motion (flow) is greatest.

Example 15.8

Torricelli's law (speed of efflux). A tank containing a liquid of density ρ has a small hole in its side at a distance y_1 from the bottom (Fig. 15.17). The atmosphere above the liquid is maintained at a pressure P. Determine the speed at which the fluid leaves the hole when the liquid level is a distance h above the hole.

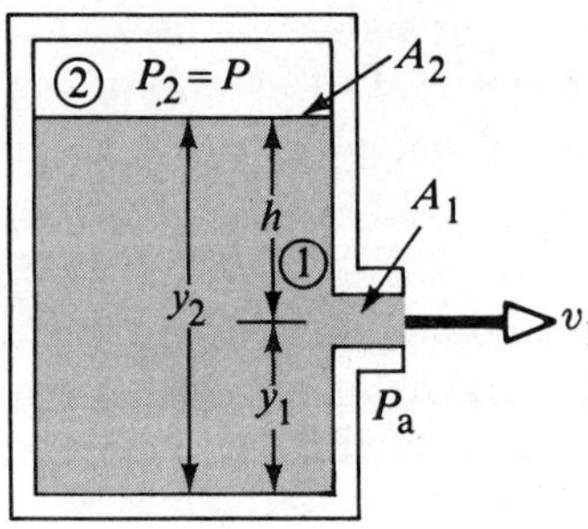

Figure 15.17 (Example 15.8) The speed of efflux, v_1, from the hole in the side of the container is given by Eq. 15.17.

Solution: If we assume the tank is large in cross section compared to the hole ($A_2 \gg A_1$), then the fluid will be approximately at rest at the top, point 2. Applying Bernoulli's equation to points 1 and 2, and noting that at the hole $P_1 = P_a$, we get

$$P_a + \frac{1}{2}\rho v_1^2 + \rho g y_1 = P + \rho g y_2$$

But $y_2 - y_1 = h$, and so this reduces to

Speed of efflux

$$v_1 = \sqrt{\frac{2(P - P_a)}{\rho} + 2gh} \qquad (15.17)$$

If A_1 is the cross-sectional area of the hole, then the flow rate from the hole is given by $A_1 v_1$. When P is large compared with atmospheric pressure (and therefore the term

2*gh* can be neglected), the speed of efflux is mainly a function of *P*. Finally, if the tank is open to the atmosphere, then $P = P_a$ and $v_1 = \sqrt{2gh}$. In other words, the speed of efflux for an open tank is equal to that acquired by a body falling freely through a vertical distance *h*. This is known as *Torricelli's law*.

Q16. Show that each term in Eq. 15.15 has units of energy per unit volume.

Q17. If the pressure in the region above the fluid in Fig. 15.17 is reduced to zero (a vacuum), under what condition will the fluid level remain fixed?

15.8 OTHER APPLICATIONS OF BERNOULLI'S EQUATION

In this section we shall give a qualitative description of some common phenomena that can be explained at least in part by Bernoulli's equation.

Let us examine the "lift" of an airplane wing (Fig. 15.18). We shall assume that the shape of the wing is such that streamline flow is maintained. The air in the region above the wing moves faster than the air below the wing. (Note the difference in the density of streamlines.) As a result, the air pressure above is less than the air pressure below and there is a net upward force, or "lift," on the wing. Of course, the lift depends on several factors, such as the speed of the airplane and the angle between the wing and the horizontal. As this angle increases, turbulent flow above the wing reduces the lift predicted by the Bernoulli effect.

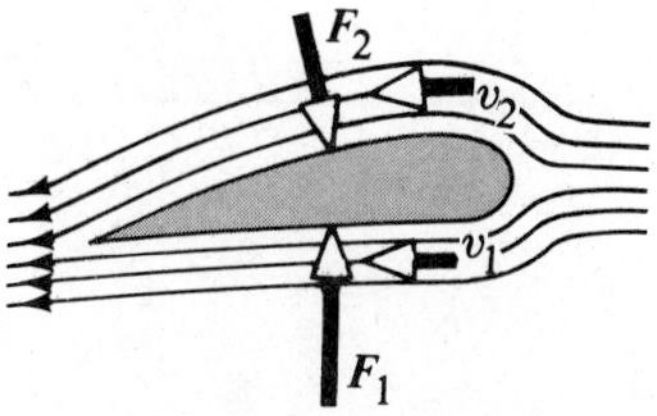

Figure 15.18 Streamline flow around an airplane wing. By Bernoulli's principle, since v_2 is greater than v_1, the pressure at the top is less than the pressure at the bottom, and so there is a net force upward.

The curve of a spinning baseball is another example of dynamic lift (Fig. 15.19). As the ball moves to the right, an observer in the frame of reference of the ball sees the air rushing to the left with a velocity **v**. However, because the ball is spinning as shown, some air is "dragged" with it (that is, its surface is rough). Hence, the air at *B* has a *greater* speed than the air at *A*, as seen by an observer moving with the ball. From Bernoulli's equation, we conclude that the pressure at *A* is greater than the pressure at *B*, and so there is a sideways force that makes the ball curve as shown. A more complete description of the motion must include the details of the viscous drag forces.

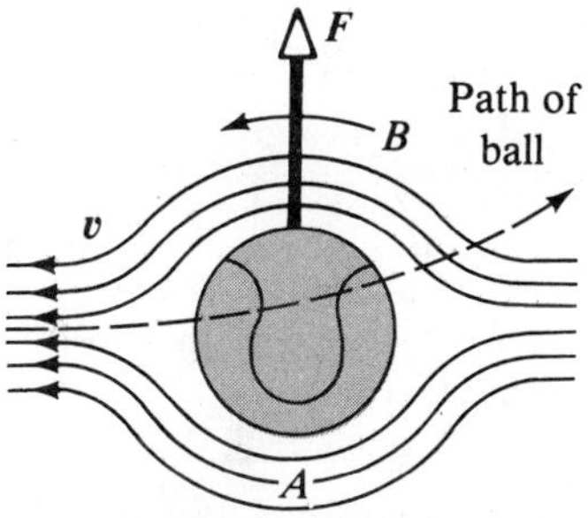

Figure 15.19 Streamline flow around a spinning ball. The ball will curve as shown because of a deflecting force **F**, which arises from the Bernoulli effect.

Finally, a number of devices operate in the manner described in Fig. 15.20. A stream of air passing over an open tube reduces the pressure above the tube, and this reduction in pressure causes the liquid to rise into the air stream. The liquid is then dispersed into a fine spray of droplets. You might recognize that this so-called atomizer is used in perfume bottles and in paint sprayers. The same principle is used in the carburetor of a gasoline engine. In this case, the low-pressure region in the carburetor is produced by air drawn in by the piston from the air filter. The gasoline vaporizes, mixes with the air, and enters the cylinder of the engine for combustion.

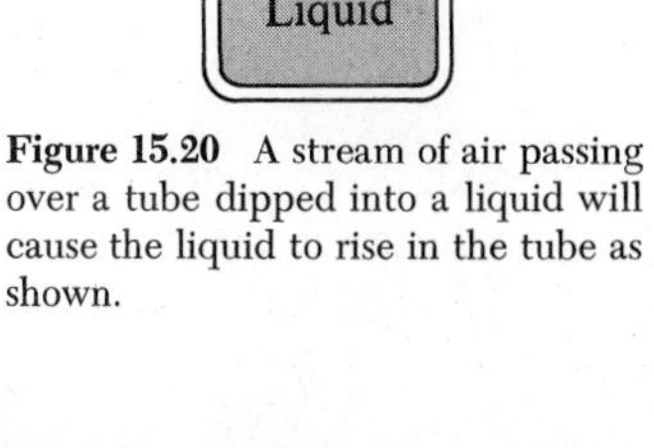

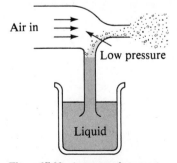

Figure 15.20 A stream of air passing over a tube dipped into a liquid will cause the liquid to rise in the tube as shown.

Q18. Why does a ski-jumper lean forward while in "flight"?

Q19. Why is it easier to throw a curve with a tennis ball than with a baseball?

Q20. Why does a knuckle ball flutter? Can a knuckle ball be effective on a day when there is little air turbulence? Explain.

15.9 ENERGY FROM THE WIND

The wind as a source of energy is not a new concept. In fact, there is some evidence that windmills were used in Babylon and in China as early as 2000 B.C. The kinetic energy carried by the winds originates from solar energy.

Although the wind is a large potential source of energy (about 5 kW per acre in the United States), it has been harnessed only on a small scale. It has been estimated that, on a global scale, the winds account for a total available power of 2×10^{13} W (three times the world energy consumption in 1972). Therefore, if only a few percent of the available power could be harnessed, wind power would represent a significant fraction of our energy needs. As with all indirect energy resources, wind power systems have some disadvantages, which in this case arise mainly from the variability of wind velocities.

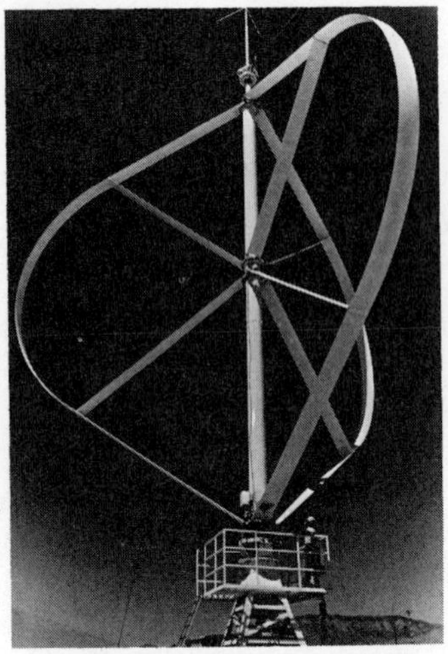

Darrieus wind turbine.

The largest windmill built in the United States was a 1.25-MW generator installed on "Grandpa's Knob" near Rutland, Vermont. The machine's blades were 175 ft in diameter, and the facility operated intermittently between 1941 and 1945. Unfortunately, one of its two main blades broke off as a result of material fatigue and was never repaired. Despite this failure, the windmill was considered a technological success, since wartime needs limited the quality of available materials. Nevertheless, the project was abandoned because costs were not competitive with hydroelectric power. The U.S. Department of Energy is currently planning to develop wind machines capable of generating 1 MW.

We can use some of the ideas developed in this chapter to estimate wind power. Any wind energy machine involves the conversion of the kinetic energy of moving air to mechanical energy, usually a rotating shaft. The kinetic energy per unit volume of a moving column of air is given by

$$\frac{\text{KE}}{\text{volume}} = \frac{1}{2}\rho v^2$$

where ρ is the density of air and v is its speed. The rate of flow of air through a column of cross-sectional area A is Av. (Fig. 15.21). This can be considered as the volume of air crossing the area each second. In the working machine, A is the cross-sectional area of the wind-collecting system, such as a set of rotating propeller blades. Multiplying the kinetic energy per unit volume by the flow rate gives the rate at which energy is transferred, or, in other words, the power:

$$\text{Power} = \frac{\text{KE}}{\text{volume}} \times \frac{\text{volume}}{\text{time}} = \left(\frac{1}{2}\rho v^2\right)(Av) = \frac{1}{2}\rho v^3 A \qquad (15.18)$$

Therefore, the available power per unit area is given by

$$\frac{P}{A} = \frac{1}{2}\rho v^3 \qquad (15.19)$$

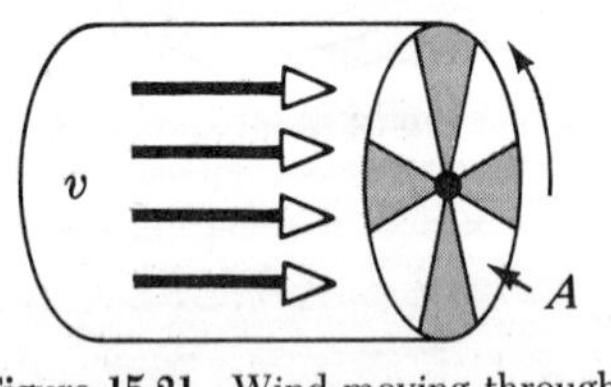

Figure 15.21 Wind moving through a cylindrical column of cross-sectional area A with a speed v.

According to this result, if the moving air column could be brought to rest, a power of $\frac{1}{2}\rho v^3$ would be available for each square meter that is intercepted. For example, if we assume a moderate speed of 12 m/s (27 mi/h) and take $\rho = 1.3$ kg/m³, we find that

$$\frac{P}{A} = \frac{1}{2}\left(1.3\,\frac{\text{kg}}{\text{m}^3}\right)\left(12\,\frac{\text{m}}{\text{s}}\right)^3 \approx 1100\,\frac{\text{W}}{\text{m}^2} = 1.1\,\frac{\text{kW}}{\text{m}^2}$$

Since the power per unit area varies as the cube of the velocity, its value doubles if v increases by only 26%. Conversely, the power output would be halved if the velocity decreased by 26%.

This calculation is based on ideal conditions and assumes that all of the kinetic energy is available for power. In reality, the air stream emerges from the wind generator with some residual velocity, and more refined calculations show that, at best, one can extract only 59.3% of this quantity.[2] The expression for the maximum

[2] For more details, see J. H. Krenz, *Energy Conversion and Utilization*, Boston, Allyn and Bacon, 1976, Chapter 8.

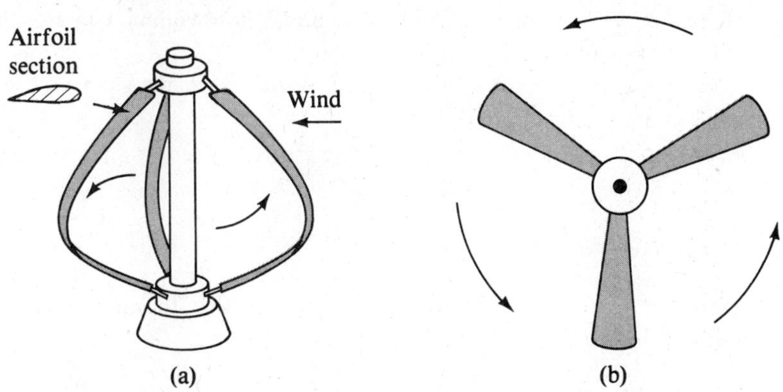

Figure 15.22 (a) A vertical-axis wind generator. (b) A horizontal-axis wind generator.

available power per unit area for the ideal wind generator is found to be

$$\frac{P_{max}}{A} = \frac{8}{27}\rho v^3 \tag{15.20}$$

In a real wind machine, further losses resulting from the nonideal nature of the propeller, gearing, and generator reduce the total available power to around 15% of the value predicted by Eq. 15.19. Sketches of two types of wind turbines are shown in Figure 15.22.

Example 15.9

Calculate the power output of a wind generator having a blade diameter of 80 m, assuming a wind speed of 10 m/s and an overall efficiency of 15%.

Solution: Since the radius of the blade is 40 m, the cross-sectional area of the propellers is given by

$$A = \pi r^2 = \pi(40\text{ m})^2 = 5.0 \times 10^3\text{ m}^2$$

If 100% of the available wind energy could be extracted, the maximum available power would be

$$P_{max} = \frac{1}{2}\rho A v^3 = \frac{1}{2}\left(1.2\,\frac{\text{kg}}{\text{m}^3}\right)(5.0 \times 10^3\text{ m}^2)\left(10\,\frac{\text{m}}{\text{s}}\right)^3$$
$$= 3.0 \times 10^6\text{ W} = 3.0\text{ MW}$$

Since the overall efficiency is 15%, the output power is

$$P = 0.15P_{max} = 0.45\text{ MW}$$

In comparison, a large steam-turbine plant has a power output of about 1 GW. Hence, one would require 2200 such wind generators to equal this output under these conditions. The large number of generators required for reasonable output power is clearly a major disadvantage of wind power. (See Exercise 35.)

15.10 SUMMARY

All forms of matter are normally classified as being in one of three states: solid, liquid, or gaseous.

The *elastic properties* of a solid can be described using the concepts of stress and strain. The *stress* on a solid is a quantity proportional to the force producing a deformation; *strain* is a quantitative measure of the deformation.

Generalizing Hooke's law, one finds that the stress is proportional to the strain, and the constant of proportionality is the *elastic modulus:*

$$\text{Elastic modulus} = \frac{\text{stress}}{\text{strain}} \tag{15.1}$$

Three common types of deformations are (1) the resistance of a solid to elongation under a load, characterized by *Young's modulus, Y,* (2) the resistance of a solid to the motion of planes in the solid sliding past each other, characterized by the *shear modulus*, S, and (3) the resistance of a solid (or liquid) to a volume change, characterized by the *bulk modulus*, *B*.

The *density*, ρ, of a homogeneous substance is defined as its mass per unit volume and has units of kg/m^3 in the SI system:

Density

$$\rho \equiv \frac{m}{V} \tag{15.5}$$

The *pressure*, *P*, in a fluid is the force per unit area that the fluid exerts on an object immersed in the fluid:

Average pressure

$$P = \frac{F}{A} \tag{15.6}$$

In the SI system, pressure has units of N/m^2, and 1 N/m^2 = 1 pascal (Pa).

The pressure in a fluid varies with depth *h* according to the expression

Pressure at any depth *h*

$$P = P_a + \rho g h \tag{15.10}$$

where P_a is atmospheric pressure ($= 1.01 \times 10^5$ N/m^2) and ρ is the density of the fluid.

Pascal's law states that when pressure is applied to an enclosed fluid, the pressure is transmitted undiminished to every point of the fluid and of the walls of the container.

When an object is partially or fully submerged in a fluid, the fluid exerts an upward force on the object called the *buoyant force*. According to Archimedes' principle, the buoyant force is equal to the weight of the fluid displaced by the body.

Various aspects of fluid dynamics (fluids in motion) can be understood by assuming that the fluid is nonviscous and incompressible and that the fluid motion is in a steady state with no turbulence.

Using these assumptions, one obtains two important results regarding fluid flow through a pipe of nonuniform size:

1. The flow rate through the pipe is a constant, which is equivalent to stating that the product of the cross-sectional area, *A*, and the speed, *v*, at any point is a constant. That is,

Equation of continuity

$$A_1 v_1 = A_2 v_2 = \text{constant} \tag{15.13}$$

2. The sum of the pressure, kinetic energy per unit volume, and potential energy per unit volume has the same value at all points along a streamline. That is,

Bernoulli's equation

$$P + \frac{1}{2}\rho v^2 + \rho g y = \text{constant} \tag{15.15}$$

This is known as *Bernoulli's equation* and is fundamental in the study of fluid dynamics.

EXERCISES

Section 15.2 Elastic Properties of Solids

1. A steel wire has a length of 3 m and a cross-sectional area of 0.2 cm^2. Under what load will its length increase by 0.05 cm?

2. A mass of 2 kg is supported by a copper wire of length 4 m and diameter 4 mm. Determine (a) the stress in the wire and (b) the elongation of the wire.

3. A cube of steel 5 cm on an edge is subjected to a shearing force of 2000 N while one face is clamped. Find the shearing strain in the cube.

4. The *elastic limit* of a material is defined as the maximum stress that can be applied to the material before it becomes permanently deformed. If the elastic limit of copper is 1.5×10^8 N/m^2, determine the *minimum* diameter a copper wire can have under a load of 10 kg if its elastic limit is not to be exceeded.

5. If the shear stress in steel exceeds about 4.0×10^8 N/m^2, it ruptures. Determine the shearing force necessary to (a) shear a steel bolt 1 cm in diameter and (b) punch a 1-cm-diameter hole in a steel plate that is 0.5 cm thick.

6. Determine the decrease in volume of a cube of copper 10 cm on an edge if it is subjected to a bulk stress (pressure) of 10^8 N/m^2.

7. What increase in pressure is necessary to decrease the volume of a 6-cm-diameter sphere of mercury by 0.05%? What is the decrease in volume under this pressure?

8. Two wires are made of the same metal, but have different dimensions. Wire 1 is four times longer and twice the diameter of wire 2. If they are both under the same load, compare (a) the stresses in the two wires and (b) the elongations of the two wires.

Section 15.3 Density and Pressure

9. Calculate the mass of a solid iron sphere that has a diameter of 3.0 cm.

10. A solid cube of material 5.0 cm on an edge has a mass of 1.31 kg. What is the material made of, assuming it consists of only one element? (Consult Table 15.2.)

11. Estimate the density of the *nucleus* of an atom. What does this result suggest concerning the structure of matter? (Use the fact that the mass of a proton is 1.67×10^{-27} kg and its radius is about 10^{-15} m.)

Section 15.4 Variation of Pressure with Depth

12. Determine the absolute pressure at the bottom of a lake that is 30 m deep.

13. A rectangular swimming pool has dimensions l = 10 m, w = 5 m, and h = 2 m, where h is the depth. If the pool is completely filled with water, calculate the force exerted *by the water* against (a) the bottom of the pool, (b) the 10-m sides, and (c) the 5-m sides.

14. At what depth in a lake is the absolute pressure equal to three times atmospheric pressure?

15. The spring of the pressure gauge shown in Fig. 15.5 has a force constant of 1000 N/m, and the piston has a diameter of 2 cm. Find the depth in water for which the spring compresses by 0.5 cm.

16. The small piston of a hydraulic lift has a cross-sectional area of 3 cm^2, and the large piston has an area of 200 cm^2 (Fig. 15.8). What force must be applied to the small piston to raise a load of 15 000 N? (In service stations this is usually accomplished with compressed air.)

Section 15.5 Pressure Measurements

17. The U-shaped tube in Fig. 15.10a contains mercury. What is the absolute pressure, P, on the left if h = 20 cm? What is the gauge pressure?

18. If the fluid in the barometer illustrated in Fig. 15.10b is water, what will be the height of the water column in the vertical tube at atmospheric pressure?

19. The open vertical tube in Fig. 15.23 contains two fluids of densities ρ_1 and ρ_2, which do not mix. Show that the pressure at the depth $h_1 + h_2$ is given by $P = P_a + \rho_1 g h_1 + \rho_2 g h_2$.

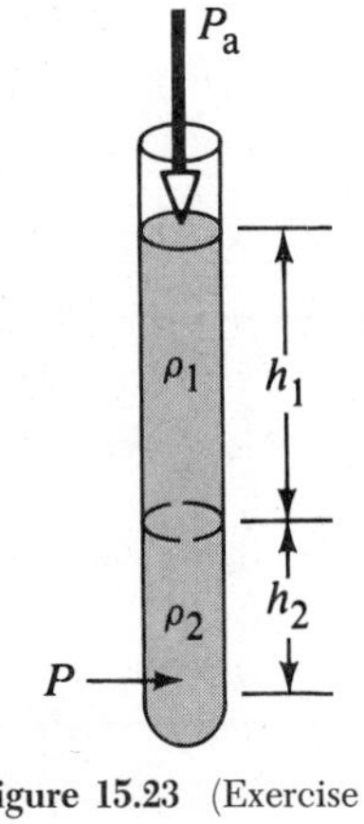

Figure 15.23 (Exercise 19).

Section 15.6 Buoyant Forces and Archimedes' Principle

20. Calculate the buoyant force on a solid object made of copper and having a volume of 0.2 m^3 if it is submerged in water. What is the result if the object is made of steel?

21. Show that only 11% of the total volume of an iceberg is above the water level. (Sea water has a density of 1.03×10^3 kg/m^3, and ice has a density of 0.92×10^3 kg/m^3.)

22. A solid object has a weight of 5.0 N. When it is suspended from a spring scale and submerged in water, the scale reads 3.5 N (Fig. 15.11a). What is the density of the object?

23. A cube of wood 20 cm on a side and of density 0.65 $\times 10^3$ kg/m^3 floats on water. (a) What is the distance from the top of the cube to the water level? (b) How much lead weight has to be placed on top of the cube so that its top is just level with the water?

24. A balloon filled with helium at atmospheric pressure is designed to support a mass M (payload + empty balloon). (a) Show that the volume of the balloon must be *at least* $V = M/(\rho_a - \rho_{He})$, where ρ_a is the density of air and ρ_{He} is the density of helium. (Ignore the volume of the payload.) (b) If $M = 2000$ kg, what radius should the balloon have?

25. A hollow plastic ball has a radius of 5 cm and a mass of 100 g. The ball has a tiny hole at the top through which lead shot can be inserted. How many grams of lead can be inserted into the ball before it sinks in water? (Assume the ball does not leak.)

26. A 10-kg block of metal measuring 12 cm $\times$ 10 cm $\times$ 10 cm is suspended from a scale and immersed in water as in Fig. 15.11. The 12-cm dimension is vertical, and the top of the block is 5 cm from the surface of the water. (a) What are the forces on the top and bottom of the block? (Take $P_a = 1.0130 \times 10^5$ N/m^2.) (b) What is the reading of the spring scale? (c) Show that the buoyant force equals the difference between the forces at the top and bottom of the block.

Section 15.7 Fluid Dynamics and Bernoulli's Equation

27. The rate of flow of water through a horizontal pipe is 2 m^3/min. Determine the velocity of flow at a point where the diameter of the pipe is (a) 10 cm, (b) 5 cm.

28. A large storage tank filled with water develops a small hole in its side at a point 16 m below the water level. If the rate of flow from the leak is 2.5×10^{-3} m^3/min, determine (a) the speed at which the water leaves the hole and (b) the diameter of the hole.

29. Water flows through a constricted pipe at a uniform rate (Fig. 15.15). At one point, where the pressure is 2.5×10^4 Pa, the diameter is 8.0 cm; at another point 0.5 m higher, the pressure is 1.5×10^4 Pa and the diameter is 4.0 cm. (a) Find the speed of flow in the lower and upper sections. (b) Determine the rate of flow through the pipe.

30. Water flows through a horizontal constricted pipe. The pressure is 4.5×10^4 Pa at a point where the speed is 2 m/s and the area is A. Find the speed and pressure at a point where the area is $A/4$.

31. The water supply of a building is fed through a main 6-cm-diameter pipe. A 2-cm-diameter faucet tap located 2 m above the main pipe is observed to fill a 25-liter container in 30 s. (a) What is the speed at which the water leaves the faucet? (b) What is the *gauge pressure* in the 6-cm main pipe? (Assume the faucet is the only "leak" in the building.)

Section 15.8 Other Applications of Bernoulli's Equation

32. An airplane has a mass of 16 000 kg, and each wing has an area of 40 m^2. During level flight, the pressure on the lower wing surface is 7.0×10^4 Pa. Determine the pressure on the upper wing surface.

33. Each wing of an airplane has an area of 25 m^2. If the speed of the air is 50 m/s over the lower wing surface and 65 m/s over the upper wing surface, determine the weight of the airplane. (Assume the plane travels in level flight at constant speed at an elevation where the density of air is 1 kg/m^3. Also assume that all of the lift is provided by the wings.)

Section 15.9 Energy from the Wind

34. Calculate the power output of a windmill having blades 10 m in diameter if the wind speed is 8 m/s. Assume that the efficiency of the system is 20%.

35. According to one rather ambitious plan, it would take 50 000 windmills, each 800 ft in diameter, to obtain an average output of 200 GW. These would be strategically located through the Great Plains, along the Aleutian Islands, and on floating platforms along the Atlantic and Gulf coasts and on the Great Lakes. The annual energy consumption in the United States in 1985 is projected to be 1.3×10^{20} J. What fraction of this could be supplied by the array of windmills?

PROBLEMS

1. One method of measuring the density of a liquid is illustrated in Fig. 15.24. One side of the U-shaped tube is in the liquid being tested; the other side is in water of density ρ_w. When the air is partially removed at the upper part of the tube, show that the density of the liquid on the left is given by $\rho = (h_w/h)\rho_w$.

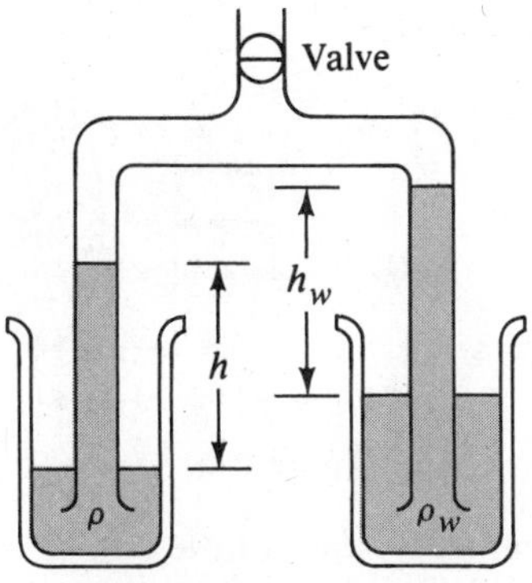

Figure 15.24 (Problem 1).

2. The true weight of a body is its weight when measured in a vacuum where there are no buoyant forces. A body of volume V is weighed in air on a balance using weights of density ρ. If the density of air is ρ_a and the balance reads W', show that the true weight W is given by

$$W = W' + \left(V - \frac{W'}{\rho g}\right)\rho_a g$$

3. A wire of length L, Young's modulus Y, and cross-sectional area A is stretched elastically by an amount ΔL. By Hooke's law, the restoring force is given by $-k\,\Delta L$. (a) Show that the constant k is given by $k = YA/L$. (b) Show that the work done in stretching the wire by an amount ΔL is equal to $\frac{1}{2}\frac{YA}{L}(\Delta L)^2$.

4. One side of the U-shaped tube in Fig. 15.25 is filled with a liquid of density ρ_1 while the other side contains a liquid of density ρ_2. If the liquids do not mix, show that $\rho_2 = (h_1/h_2)\rho_1$.

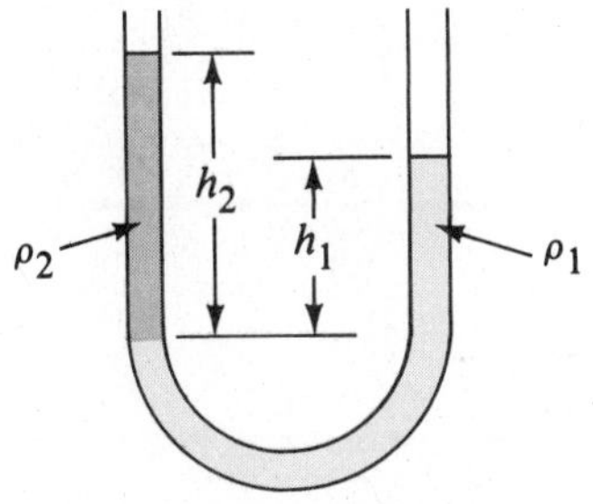

Figure 15.25 (Problem 4).

5. A block of cross-sectional area A, height l, and density ρ is in equilibrium between two fluids of densities ρ_1 and ρ_2 (Fig. 15.26), where $\rho_1 < \rho < \rho_2$. The fluids do not mix. (a) Show that the buoyant force on the block is given by $B = [\rho_1 g y + \rho_2 g(l - y)]A$. (b) Show that the density of the block is equal to $\rho = [\rho_1 y + \rho_2(l - y)]/l$.

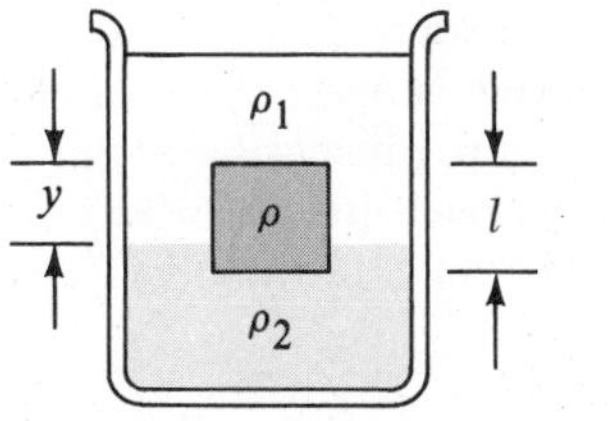

Figure 15.26 (Problem 5).

6. Show that the variation of atmospheric pressure with altitude is given by $P = P_0 e^{-\alpha h}$, where $\alpha = \frac{\rho_0}{P_0} g$, P_0 is atmospheric pressure at some reference level, and ρ_0 is the atmospheric density at this level. Assume that the decrease in atmospheric pressure with increasing altitude is given by Eq. 15.8 and that the density of air is proportional to the pressure.

7. A tank *open to the atmosphere* is filled with liquid, and a leak develops at a distance h below the surface of the liquid (Fig. 15.17). (a) Show that the liquid strikes the floor at a distance $x = 2\sqrt{hy_1}$ from the bottom of the tank. (b) Show that the horizontal distance x is a maximum when the hole is located at $y_1 = h$.

8. A *siphon* is a device that allows a fluid to seemingly defy gravity (Fig. 15.27). The flow must be initiated by a partial vacuum in the tube, as in a drinking straw. (a) Show that the speed of efflux is given by $v = \sqrt{2gh}$. (b) For what values of y will the siphon work? (Incidentally, it has been told that gasoline tastes terrible!)

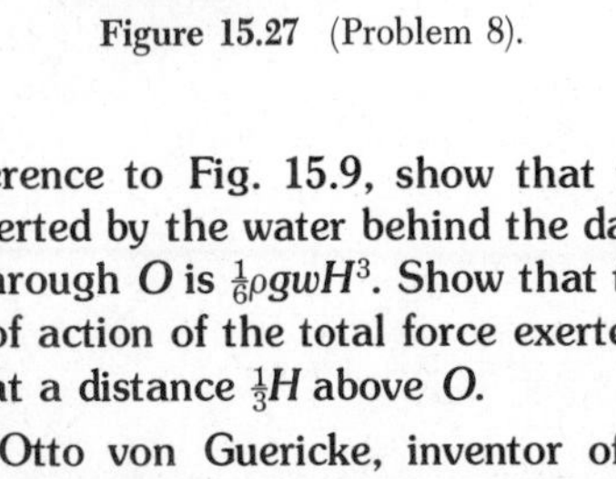

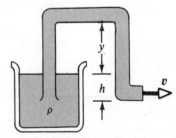

Figure 15.27 (Problem 8).

9. With reference to Fig. 15.9, show that the total torque exerted by the water behind the dam about an axis through O is $\frac{1}{6}\rho g w H^3$. Show that the effective line of action of the total force exerted by the water is at a distance $\frac{1}{3}H$ above O.

10. In 1654 Otto von Guericke, inventor of the air pump, evacuated a sphere made of two brass hemispheres. Two teams of eight horses each *could not pull the hemispheres apart* (Fig. 15.28). (a) Show

that the force F required to pull the evacuated hemispheres apart is $\pi R^2(P_a - P)$, where R is the radius of the hemispheres and P is the pressure inside the hemispheres, which is much less than P_a. (b) Determine the force if $P = 0.1P_a$ and $R = 0.3$ m.

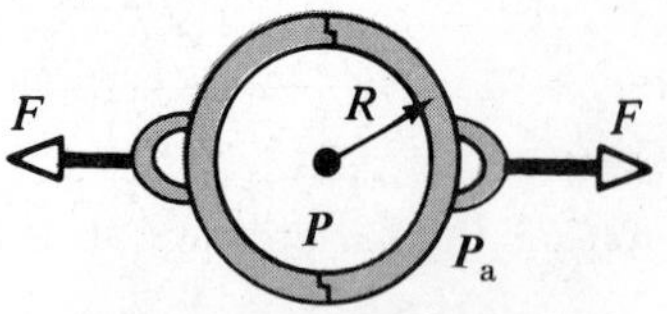

Figure 15.28 (Problem 10).

11. Consider a windmill with blades of cross-sectional area A, as in Fig. 15.29, and assume the mill is facing directly into the wind. (a) If the wind speed is v, show that the kinetic energy of the air that passes

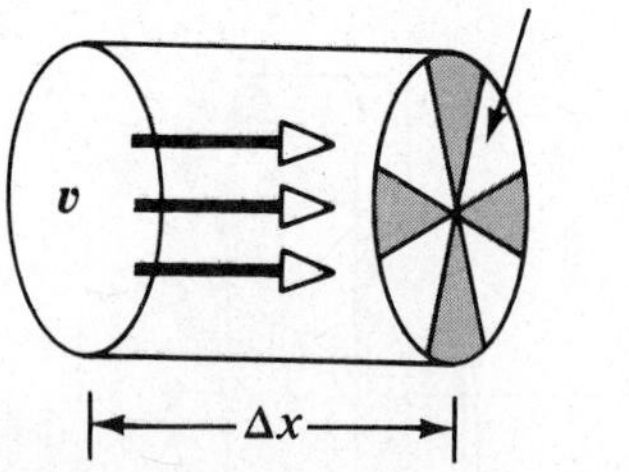

Figure 15.29 (Problem 11).

through the blades in a time Δt is given by $K = \frac{1}{2}\rho A v^3 \Delta t$. (b) What is the maximum available power according to this model? Compare your result with Eq. 15.18.

12. A girl weighing 100 lb sits on a 1 m × 1 m × 0.06 m raft made of solid styrofoam. If the raft *just* supports the girl (that is, the raft is totally submerged), determine the density of the styrofoam.

16 Temperature, Thermal Expansion, and Ideal Gases

Corning Glass Works

The subject of thermal physics deals with phenomena involving energy transfer between bodies at different temperatures. In the study of mechanics such concepts as mass, force, and kinetic energy were carefully defined in order to make the subject quantitative. Likewise, a quantitative description of thermal phenomena requires a careful definition of the concepts of temperature, heat, and internal energy. The science of thermodynamics is concerned with the study of heat flow from a *macroscopic* viewpoint. The laws of thermodynamics provide us with a relationship between heat flow, work, and the internal energy of a system. In practice, suitable observable quantities must be selected to describe the overall behavior of a system. For example, the macroscopic quantities, pressure, volume, and temperature are used to characterize the properties of a gas. Thermal phenomena can also be understood using a *microscopic* approach, which describes what is happening on a microscopic scale. For example, the temperature of a gas is a measure of the average kinetic energy of the gas molecules.

The composition of a body is an important factor when dealing with thermal phenomena. For example, liquids and solids will expand only slightly when heated. On the other hand, a gas will tend to undergo appreciable expansion when heated. Certain substances may melt, boil, burn, or explode, depending on their composition and structure. Thus, the thermal behavior of a substance is closely related to its structure.

It would be far beyond the scope of this book to attempt to present applications of thermodynamics to a wide variety of substances. Instead, we shall examine some rather simple systems, such as a dilute gas and a homogeneous solid. Emphasis will be placed on understanding the key principles of thermodynamics and on providing a basis upon which the thermal behavior of all matter can be understood.

16.1 TEMPERATURE AND THE ZEROTH LAW OF THERMODYNAMICS

When we speak of the temperature of an object, we often associate this concept with the degree of "hotness" or "coldness" of the object when we touch it. Thus, our senses provide us with a qualitative indication of temperature. However, our senses are unreliable and often misleading. For example, if we remove an ice tray and a package of frozen vegetables from the freezer, the ice tray feels colder to the hand even though both are at the same temperature. This is because metal is a better conductor of heat than cardboard. What we need is a reliable and reproducible method for establishing the relative "hotness" or "coldness" of bodies. Scientists have developed various types of thermometers for making such quantitative measurements. Some typical thermometers will be described in the following section.

We are all familiar with the fact that two objects at different initial temperatures will eventually reach some intermediate temperature when placed in contact with each other. For example, a piece of meat placed on a block of ice in a well-insulated container will eventually reach a temperature near 0°C. Likewise, if an ice cube is dropped into a container of warm water, the ice cube will eventually melt and the water's temperature will decrease. If the process takes place in a thermos bottle, the system (water + ice) is approximately isolated from its surroundings.

In order to understand the concept of temperature, it is useful to first define two often used phrases, *thermal contact* and *thermal equilibrium.* Two bodies are in *thermal contact* with each other if energy exchange can occur between them in the absence of macroscopic work done by one on the other. *Thermal equilibrium* is a situation in which two bodies in thermal contact with each other cease to have any net energy exchange. The time it takes the two bodies to reach thermal equilibrium depends on the properties of the bodies and on the pathways available for energy exchange.

Now consider two bodies, A and B, and a third body, C, which will be our thermometer. We wish to determine whether or not A and B are in thermal equilibrium with each other. The thermometer (body C) is first placed in thermal contact with A until thermal equilibrium is reached. At that point, the thermometer's reading will remain constant. The thermometer is then placed in thermal contact with B, and its reading is recorded after thermal equilibrium is reached. If the readings after contact with A and B are the same, then A and B are in thermal equilibrium with each other. We can summarize these results in a statement known as the *zeroth law of thermodynamics (the law of equilibrium): If bodies A and B are separately in thermal equilibrium with a third body, C, then A and B are in thermal equilibrium with each other.* This statement, insignificant and obvious as it may seem, is most fundamental in the field of thermodynamics since it can be used to define temperature. We can think of temperature as the property that determines whether or not a body is in thermal equilibrium with other bodies. That is, *two bodies in thermal equilibrium with each other are at the same temperature.* Conversely, if two bodies have different temperatures, they cannot be in thermal equilibrium with each other.

Q1. Is it possible for two bodies to be in thermal equilibrium if they are not in contact with each other? Explain.

Q2. A piece of copper is dropped into a beaker of water. If the water's temperature rises, what happens to the temperature of the copper? When will the water and copper be in thermal equilibrium?

16.2 THERMOMETERS AND TEMPERATURE SCALES

Thermometers are devices used to define and measure the temperature of a system. A thermometer in thermal equilibrium with a system measures both the temperature of the system and its own temperature. All thermometers make use of the change in some physical property with temperature. (The general name for such properties is *thermometric properties.*) Some of these thermometric properties are (1) the change in volume of a liquid, (2) the change in length of a solid, (3) the change in pressure of a gas at constant volume, (4) the change in volume of a gas at constant pressure, (5) the change in electric resistance of a conductor, and (6) the change in color of a very hot body. A temperature scale can be established for a given substance using one such thermometric property.

The most common thermometer in everyday use consists of a mass of mercury that expands into a glass capillary tube when heated (Fig. 16.1). Thus, the thermometric property in this case is the thermal expansion of the mercury. One can now define any temperature change to be proportional to the change in length of the mercury column. The thermometer can be calibrated by placing it in thermal contact with some natural systems that remain at constant temperature (called a *fixed-point temperature*). One of the fixed-point temperatures normally chosen is that of a mixture of water and ice at atmospheric pressure, which is defined to be zero degrees Celsius, written 0°C. (This was formerly called *degrees centigrade*.) Another convenient fixed point is the temperature of a mixture of water and water vapor (steam) in equilibrium at atmospheric pressure. The temperature of this *steam point* is 100°C. Once the mercury levels have been established at these fixed points, the column is divided into 100 equal segments, each denoting a change in temperature of one Celsius degree.

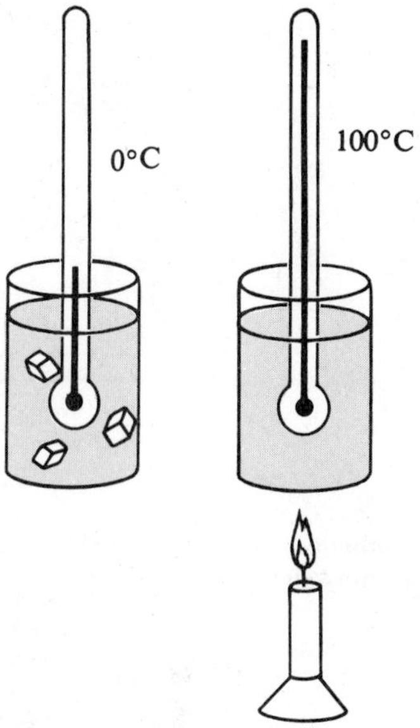

Figure 16.1 Schematic diagram of a mercury thermometer. As a result of thermal expansion, the level of the mercury rises as the mercury is heated from 0°C (the ice point) to 100°C (the steam point).

Other thermometers can be calibrated using these fixed points as reference markers. In all cases, it is assumed that the thermometric property varies *linearly* with temperature. However, different results are usually obtained when different thermometers are used to measure the same temperature. For example, a mercury thermometer and an alcohol thermometer will not agree precisely (except at the calibrated fixed points) because of the different thermal expansion properties of the two liquids.[1] The discrepancies between various thermometers are especially large when the temperatures to be measured are far from the fixed points.

An additional practical problem of any thermometer is its limited temperature range. A mercury thermometer, for example, cannot be used below the freezing point of mercury, which is −39°C. What we need is a universal thermometer whose readings are independent of the thermometric substance. The gas thermometer described in the next section meets this requirement.

16.3 THE CONSTANT-VOLUME GAS THERMOMETER AND THE KELVIN SCALE

The most suitable thermometer devised—one having properties that are nearly independent of the thermometric substance—is the *gas thermometer*. One version of this is the constant-volume gas thermometer shown in Fig. 16.2. The thermometric property in this device is the pressure variation with temperature of a fixed volume of gas. As the gas is heated, its pressure increases and the height of the mercury column in Fig. 16.2 increases. When the gas is cooled, its pressure decreases, hence the column height decreases. Thus, we can define temperature in terms of the concept of pressure discussed in Chapter 15. If the variation of temperature, T, with pressure is assumed to be linear, then

$$T = aP + b \tag{16.1}$$

where a and b are constants. These constants can be determined from two fixed points, such as the ice and steam points described in Section 16.2.

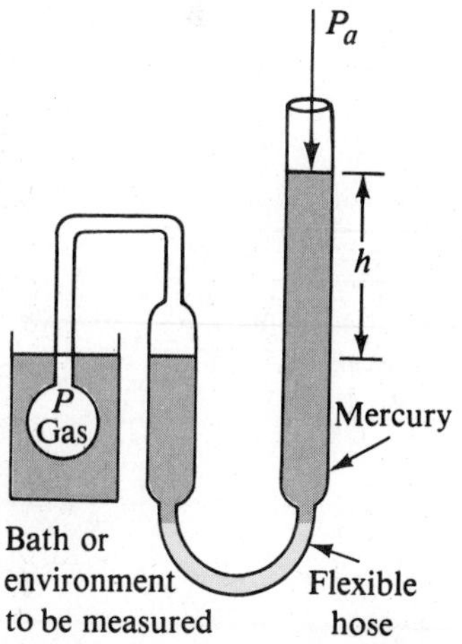

Figure 16.2 A constant-volume gas thermometer measures the pressure of the gas contained in the flask on the left. The volume of gas in the flask is kept constant by raising or lowering the column on the right such that the mercury level on the left remains constant.

Now suppose that temperatures are measured with various gas thermometers containing different gases. Experiments show that the thermometer readings are nearly independent of the type of gas used, so long as the gas pressure is low and the temperature is well above the liquefaction point. The agreement among thermome-

[1]Thermometers that use the same material may also give different readings. This is due in part to difficulties in constructing uniform-bore glass capillary tubes.

ters using various gases improves as the pressure is reduced. This agreement of all gas thermometers at low pressure and high temperature implies that the intercept b appearing in Eq. 16.1 is the same for *all* gases. This fact is illustrated in Fig. 16.3. When the pressure versus temperature curve is extrapolated to very low temperatures, one finds that the pressure is zero when the temperature is $-273.15°C$. This temperature corresponds to the constant b in Eq. 16.1. An extrapolation is necessary since all gases liquify before reaching this temperature.

The triple point of water

Early gas thermometers made use of the ice point and steam point as standard temperatures. However, these points are experimentally difficult to duplicate since they are very sensitive to dissolved impurities in the water. For this reason, a new temperature scale based on a single fixed point with b equal to zero was adopted in 1954 by the International Committee on Weights and Measures. The *triple point of water,* which corresponds to the single temperature and pressure at which water, water vapor, and ice can coexist in equilibrium, was chosen as a convenient and reproducible reference temperature for this new scale. The triple point of water occurs at a temperature of about 0.01°C and a pressure of 4.58 mm Hg. The temperature at the triple point of water on the new scale was arbitrarily set at 273.16 degrees kelvin, abbreviated 273.16 K.[2] This scale is called the *thermodynamic temperature scale* and the SI unit of thermodynamic temperature, the *kelvin,* is defined as *the fraction 1/273.16 of the temperature of the triple point of water.*

Definition of the kelvin

If we take $b = 0$ in Eq. 16.1 and call P_3 the pressure at the triple-point temperature, $T_3 = 273.16$ K, then we see that $a = (273.16\text{ K})/P_3$. Therefore, the temperature at a measured gas pressure P for a constant-volume gas thermometer is defined to be

$$T = \left(\frac{273.16\text{ K}}{P_3}\right)P \qquad \text{(constant } V) \qquad (16.2)$$

As mentioned earlier, one finds experimentally that as the pressure P_3 decreases because the quantity of gas in the thermometer is decreased, the measured value of the temperature approaches the same value for all gases. An example of such a measurement is illustrated in Fig. 16.4, which shows the steam-point temperature measured with a constant-volume gas thermometer using various gases. As P_3 approaches zero, all measurements approach a common value of 373.15 K. Similarly, one finds that the ice-point temperature is 273.15 K.

In the limit of low gas pressures and high temperatures, real gases behave as what is known as an *ideal gas,* which will be discussed in detail in Section 16.6 and Chapter 17. The temperature scale defined in this limit of low gas pressures is called the *ideal-gas temperature,* T, given by

Definition of ideal-gas temperature

$$T \equiv 273.16\text{ K} \lim_{P_3 \to 0} \frac{P}{P_3} \qquad \text{(constant } V) \qquad (16.3)$$

Thus the constant-volume gas thermometer defines a temperature scale that can be reproduced in laboratories throughout the world. Although the scale depends on the properties of a gas, it is independent of which gas is used. In practice, one can use a gas thermometer down to around 1 K using low-pressure helium gas. Helium liquefies below this temperature; other gases liquefy at even higher temperatures.

It would be convenient to have a temperature scale that is independent of the property of any substance. Such a scale is called an *absolute temperature scale,* or *kelvin scale.* Later we shall find that the ideal-gas scale is identical with the absolute

[2]A second fixed point at 0 K is implied by Eq. 16.1. We shall describe the meaning of this point in Chapter 19 when we discuss the second law of thermodynamics.

TABLE 16.1 Fixed-Point Temperatures

FIXED POINT	*TEMPERATURE* (°*C*)	*TEMPERATURE* (*K*)
Triple point of hydrogen	−259.34	13.81
Boiling point of hydrogen at 33360.6 N/m² pressure	−256.108	17.042
Boiling point of hydrogen	−252.87	20.28
Triple point of neon	−246.048	27.102
Triple point of oxygen	−218.789	54.361
Boiling point of oxygen	−182.962	90.188
Triple point of water	0.01	273.16
Boiling point of water	100.00	373.15
Freezing point of tin	231.9681	505.1181
Freezing point of zinc	419.58	692.73
Freezing point of silver	961.93	1235.08
Freezing point of gold	1064.43	1337.58

All values from National Bureau of Standards Special Publication 420, U.S. Department of Commerce, May 1975.

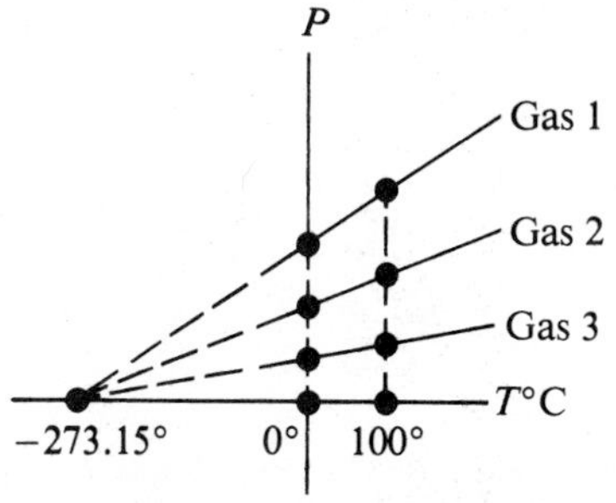

Figure 16.3 Pressure versus temperature for dilute gases. Note that, for all gases, the pressure extrapolates to zero at the unique temperature of −273.15°C.

temperature scale for temperatures above 1 K, where gas thermometers can be used. In anticipation of this, we shall also use the symbol T to denote absolute temperature. The absolute temperature scale will be properly defined when we study the second law of thermodynamics in Chapter 19.

Other methods of thermometry calibrated against gas thermometers have been used to provide various other fixed-point temperatures. The "International Practical Temperature Scale of 1968," which was established by international agreement, is based on measurements in various national standard laboratories. The assigned temperatures of various substances are given in Table 16.1. The platinum resistance thermometer was used to establish all but the last two points in this table. Note that the scale is not defined below 13.81 K.

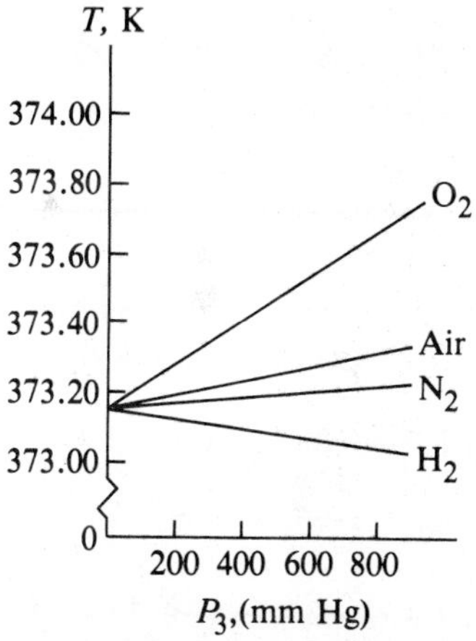

Figure 16.4 The temperature read with a constant-volume gas thermometer versus P_3, the pressure at the steam point of water, for various gases. Note that as the pressure is reduced, the steam-point temperature of water approaches a common value of 373.15 K regardless of which gas is used in the thermometer. Furthermore, the data for helium are nearly independent of pressure, which suggests it behaves like an ideal gas over this range.

16.4 THE CELSIUS, FAHRENHEIT, AND RANKINE TEMPERATURE SCALES[3]

The Celsius temperature, T_C, is shifted from the absolute (or kelvin) temperature T by 273.15°, since by definition the triple point of water (273.16 K) corresponds to 0.01°C. Therefore,

$$T_C = T - 273.15 \qquad (16.4)$$

From this we see that the size of a degree on the kelvin scale is the same as on the Celsius scale. In other words, a temperature difference of 5 Celsius degrees, written 5 C°, is equal to a temperature difference of 5 kelvins. The two scales differ only in the choice of the zero point. Furthermore, the ice point (273.15 K) corresponds to 0.00°C, and the steam point (373.15 K) is equivalent to 100.00°C.

Two other scales used in the United States and in Great Britain are the *Rankine scale* and the *Fahrenheit scale*. The Rankine temperature, T_R, (written °R) is related to the kelvin temperature through the relation

$$T_R = \frac{9}{5}T \qquad (16.5)$$

The Fahrenheit temperature, T_F, (written °F) is shifted from the Rankine tempera-

[3]Named after Anders Celsius (1701–1744), Gabriel Fahrenheit (1686–1736), and William MacQuorn Rankine (1820–1872).

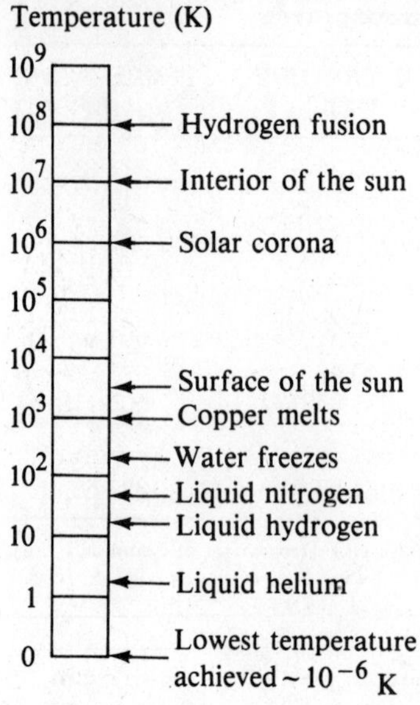

Figure 16.5 Absolute temperatures at which various physical processes take place. Note that the scale is logarithmic.

ture according to the relation

$$T_F = T_R - 459.67 \tag{16.6}$$

Substituting Eqs. 16.4 and 16.5 into Eq. 16.6 we get

$$T_F = \frac{9}{5}T_C + 32\ \mathrm{F}^\circ \tag{16.7}$$

From this expression it follows that the ice point (0.00°C) equals 32°F and the steam point (100.00°C) equals 212°F. Figure 16.5 shows on a logarithmic scale the absolute temperatures for various physical processes and structures.

Example 16.1

An object has a temperature of 50°F. What is its temperature in degrees Celsius and in kelvins?

Solution: Substituting $T_F = 50°\mathrm{F}$ into Eq. 16.7, we get

$$T_C = \frac{5}{9}(T_F - 32) = \frac{5}{9}(50 - 32) = 10°\mathrm{C}$$

Making use of Eq. 16.4, we find that

$$T = T_C + 273.15 = 283.15\ \mathrm{K}$$

Example 16.2

A pan of water is heated from 25°C to 80°C. What is the *change* in its temperature on the kelvin scale and on the Fahrenheit scale?

Solution: From Eq. 16.4, we see that the change in temperature on the Celsius scale equals the change on the kelvin scale. Therefore,

$$\Delta T = \Delta T_C = 80 - 25 = 55\ \mathrm{C}^\circ = 55\ \mathrm{K}$$

From Eq. 16.7, we find that the change in temperature on the Fahrenheit scale is greater than the change on the Celsius scale by the factor 9/5. That is,

$$\Delta T_F = \frac{9}{5}\Delta T_C = \frac{9}{5}(80 - 25) = 99\ \mathrm{F}^\circ$$

In other words, 55 C° = 99 F°, where the notation C° and F° refer to temperature *differences,* not to be confused with actual temperatures, which are written °C and °F.

Q3. In principle, any gas can be used in a gas thermometer. Why is it not possible to use oxygen for temperatures as low as 15 K? What gas would you use? (Look at the data in Table 16.1.)

Other Thermometers

A technique that is often used as a temperature standard in thermometry makes use of a pure platinum wire because its electrical resistance changes with temperature. The *platinum resistance thermometer* is essentially a coil of platinum wire for which the resistance changes by about 0.3% for a temperature change of 1 K. It is commonly used for temperatures ranging from about 14 K to 900 K and can be calibrated to within ±0.0003 K at the triple point of water.

Thermocouple

One of the most useful thermometers for scientific and engineering applications is a device called a *thermocouple*. The thermocouple is essentially a junction formed by two different metals or alloys, labeled A and B in Fig. 16.6. The test junction is placed in the material whose temperature is to be measured, while the opposite ends of the thermocouple wires are maintained at some constant reference temperature (usually in a water-ice mixture) to form two reference junctions. When the reference temperature is different from the temperature of the test junction, an electrical quantity called the *electromotive force* (EMF) is set up in the circuit. The value of this EMF is proportional to the temperature difference and therefore can be used to measure an unknown temperature. An instrument called a *potentiometer* is used to measure the EMF. In practice, one usually uses junctions for which calibration curves are available.

One advantage of the thermocouple is its small mass, which enables it to quickly reach thermal equilibrium with the material being probed. Some common examples of thermocouple junction materials are copper/constantan (an alloy), which is useful over the temperature range of about −180°C to 400°C, and platinum/platinum-10% rhodium, which is useful over the range from about 0°C to 1500°C. Some typical outputs for various thermocouples are given in Fig. 16.7, where the reference junction is at 0°C.

Another thermometer that has extremely high sensitivity is a device called a *thermistor*. This device consists of a small piece of semiconductor material whose

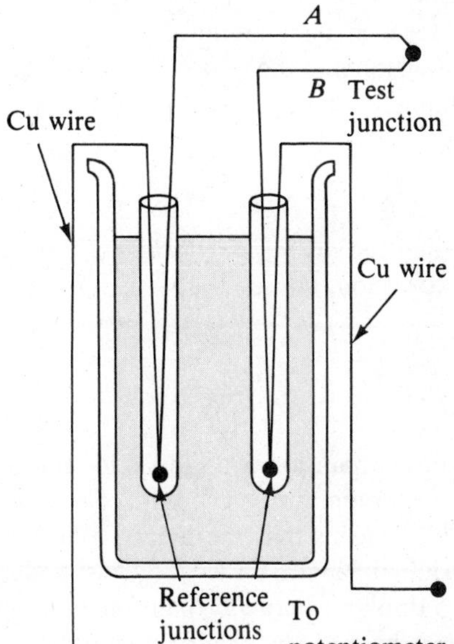

Figure 16.6 Schematic diagram of a thermocouple, which consists of two dissimilar metals, A and B. The reference junction is usually kept at 0°C.

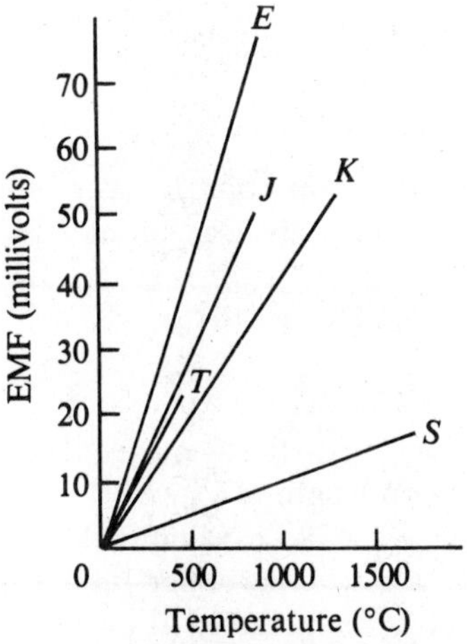

Figure 16.7 Plot of EMF (junction voltage) versus temperature for various thermocouples: E—chromel/constantan; J—iron/constantan; T—copper/constantan; K—chromel/alumel; S—platinum/platinum-10% rhodium.

electrical resistance changes with temperature. Thermistors are usually fabricated from oxides of various metals, such as nickel, manganese, iron, cobalt, and copper, and can be encapsulated in an epoxy. A careful measurement of the resistance serves as an indicator of temperature, with a typical accuracy of ± 0.1 C°. Temperature changes as small as about 10^{-3} C° can be detected with these devices. Most thermistors operate reliably over the temperature range from about -50°C to 100°C. They are often used as clinical thermometers (with digital readout) and in various biological applications.

16.5 THERMAL EXPANSION OF SOLIDS AND LIQUIDS

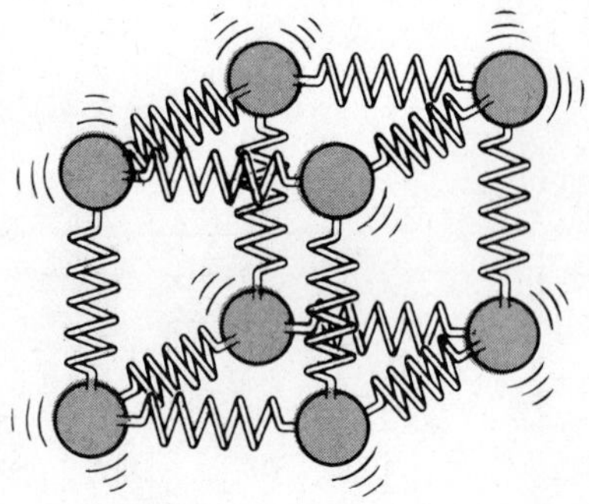

Figure 16.8 A mechanical model of a crystalline solid. The atoms (solid spheres) are imagined to be attached to each other by springs, which reflect the elastic nature of the interatomic forces.

Most bodies expand as their temperature increases. This phenomenon plays an important role in numerous engineering applications. For example, thermal expansion joints must be included in buildings, concrete highways, and bridges to compensate for changes in dimensions with temperature variations.

The overall thermal expansion of a body is a consequence of the change in the average separation between its constituent atoms or molecules. Consider a crystalline solid, which consists of a regular array of atoms held together by electrical forces. We can obtain a mechanical model of these forces by imagining that the atoms are connected by a set of stiff springs as in Fig. 16.8. The interatomic forces are taken to be elastic in nature. At ordinary temperatures, the atoms vibrate about their equilibrium positions with an amplitude of about 10^{-11} m and a frequency of about 10^{13} Hz. The average spacing between the atoms is of the order of 10^{-10} m. As the temperature of the solid increases, the atoms vibrate with larger amplitudes and the average separation between them increases.[4] Consequently, the solid as a whole expands with increasing temperature. If the expansion of an object is sufficiently small compared with its initial dimensions, then the change in any dimension (length, width, or thickness) is, to a good approximation, a linear function of the temperature.

Suppose the linear dimension of a body along some direction is l at some temperature. The length increases by an amount Δl for a change in temperature ΔT. Experiments show that the change in length is proportional to the temperature change and to the original length when ΔT is small enough. Thus, the basic equation for the expansion of a solid is

$$\Delta l = \alpha l \, \Delta T \tag{16.8}$$

where the proportionality constant α is called the *average coefficient of linear expansion* for a given material. From this expression, we see that

Average coefficient of linear expansion

$$\alpha = \frac{1}{l}\frac{\Delta l}{\Delta T} \tag{16.9}$$

In other words, the average coefficient of linear expansion of a solid is the fractional change in length ($\Delta l/l$) per degree change in temperature. The unit of α is deg^{-1}. For example, an α value of 11×10^{-6} (C°)$^{-1}$ means that the length of an object changes by 11 parts per million of its original length for every Celsius degree change in temperature. It may be helpful to think of thermal expansion as an effective magnification or as a photographic enlargement of an object when it is heated.

[4]Strictly speaking, thermal expansion arises from the *asymmetric* nature of the potential energy curve for the atoms in a solid. If the oscillators were truly harmonic, the average atomic separation would not change regardless of the amplitude of vibration.

TABLE 16.2 Expansion Coefficients for Some Materials Near Room Temperature

MATERIAL	*LINEAR EXPANSION COEFFICIENT* α (C°)$^{-1}$	*MATERIAL*	*VOLUME EXPANSION COEFFICIENT* β (C°)$^{-1}$
Aluminum	24×10^{-6}	Alcohol, ethyl	1.12×10^{-4}
Brass and bronze	19×10^{-6}	Benzene	1.24×10^{-4}
Copper	17×10^{-6}	Acetone	1.5×10^{-4}
Glass (ordinary)	9×10^{-6}	Glycerin	4.85×10^{-4}
Glass (pyrex)	3.2×10^{-6}	Mercury	1.82×10^{-4}
Lead	29×10^{-6}	Turpentine	9.0×10^{-4}
Steel	11×10^{-6}	Gasoline	9.6×10^{-4}
Invar (Ni-Fe alloy)	0.9×10^{-6}	Air	3.67×10^{-3}
Concrete	12×10^{-6}	Helium	3.665×10^{-3}

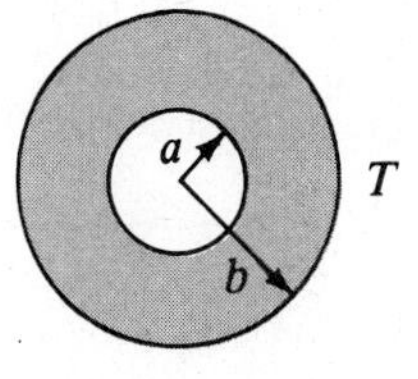

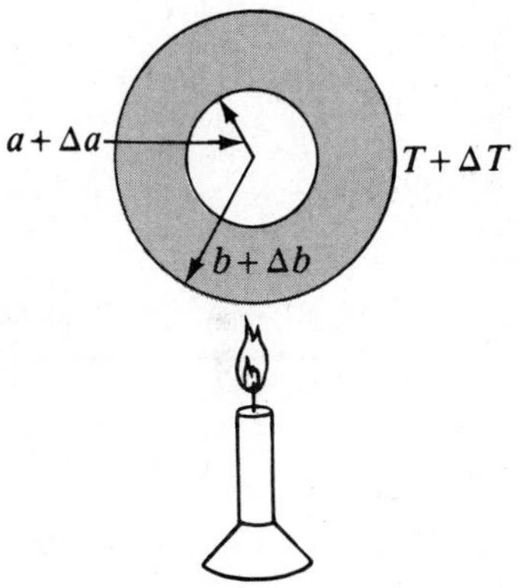

Figure 16.9 Thermal expansion of a homogeneous metal washer. Note that as the washer is heated, all dimensions increase. The expansion is exaggerated.

For example, as a metal washer is heated (Fig. 16.9) all dimensions increase, including the radius of the hole.

The coefficient of linear expansion generally varies with temperature. Usually this temperature variation is negligible on the scale on which most everyday measurements are made. Table 16.2 lists the average coefficient of linear expansion for various materials. Note that for these materials α is positive, indicating an increase in length with increasing temperature. This is not always the case. For example, some single anisotropic crystalline substances, such as calcite ($CaCO_3$), expand along one dimension (positive α) and contract along another (negative α) with increasing temperature.

Since the linear dimensions of body change with temperature,. it follows that the volume of a body also changes as its temperature varies. The change in volume at constant pressure is proportional to the original volume V and to the change in temperature according to the relation

$$\Delta V = \beta V \, \Delta T \tag{16.10}$$

where β is the *average coefficient of volume expansion. For an isotropic solid, the coefficient of volume expansion is three times the linear expansion coefficient,* or $\beta = 3\alpha$. Therefore, Eq. 16.10 can be written

Change in volume of an isotropic solid at constant pressure

$$\Delta V = 3\alpha V \, \Delta T \tag{16.11}$$

The derivation of $\beta = 3\alpha$ is given below.

To show that $\beta = 3\alpha$ for an isotropic solid, consider an object in the shape of a box of dimensions l, w, and h. Its volume at some temperature T is $V = lwh$. If the temperature changes to $T + \Delta T$, its volume changes to $V + \Delta V$, where each dimension changes according to Eq. 16.8. Therefore,

$$\begin{aligned} V + \Delta V &= (l + \Delta l)(h + \Delta h)(w + \Delta w) \\ &= (l + \alpha l \, \Delta T)(h + \alpha h \, \Delta T)(w + \alpha w \, \Delta T) \\ &= lwh(1 + \alpha \, \Delta T)^3 \\ &= V[1 + 3\alpha \, \Delta T + 3(\alpha \, \Delta T)^2 + (\alpha \, \Delta T)^3] \end{aligned}$$

Hence the fractional change in volume is

$$\frac{\Delta V}{V} = 3\alpha \, \Delta T + 3(\alpha \, \Delta T)^2 + (\alpha \, \Delta T)^3$$

Since the product $\alpha \, \Delta T$ is small compared with unity for typical values of ΔT (less than

$\approx$100°C), we can neglect the terms $3(\alpha\,\Delta T)^2$ and $(\alpha\,\Delta T)^3$. In this approximation, we see that

$$\beta = \frac{1}{V}\frac{\Delta V}{\Delta T} = 3\alpha$$

In a similar manner, you should show (Problem 7) that the change in the area of an isotropic solid is given by

Change in area of an isotropic solid

$$\Delta A = 2\alpha A\,\Delta T \tag{16.12}$$

Example 16.3 *Expansion of a Railroad Track*

A steel railroad track has a length of 30 m when the temperature is 0°C. (a) What is its length on a hot day when the temperature is 40°C?

Making use of Table 16.2 and noting that the change in temperature is 40 C°, we find that the increase in length is

$$\Delta l = \alpha l\,\Delta T = [11 \times 10^{-6}\,(\text{C°})^{-1}](30\text{ m})(40\text{ C°}) = 0.013\text{ m}$$

Therefore, its length at 40°C is 30.013 m.

(b) Suppose the ends of the rail are rigidly clamped at 0°C so as to prevent expansion. Calculate the thermal stress set up in the rail if its temperature is raised to 40°C.

From the definition of Young's modulus for a solid (Chapter 15), we have

$$\text{Tensile stress} = \frac{F}{A} = Y\frac{\Delta l}{l}$$

Since Y for steel is 20×10^{10} N/m² we have

$$\frac{F}{A} = \left(20 \times 10^{10}\,\frac{\text{N}}{\text{m}^2}\right)\left(\frac{0.013\text{ m}}{30\text{ m}}\right) = 8.7 \times 10^7\,\frac{\text{N}}{\text{m}^2}$$

Therefore, if the rail has a cross-sectional area of 30 cm², the force of compression in the rail is

$$F = 2.6 \times 10^5\text{ N} \approx 59\,000\text{ lb!}$$

Liquids generally increase in volume with increasing temperature, and have volume expansion coefficients about ten times greater than those of solids (Table 16.2). Water is an exception to this rule, as we can see from its expansion curve, shown in Fig. 16.10. As the temperature increases from 0°C to 4°C, the water contracts. Above 4°C, the water expands with increasing temperature. Therefore, the density of water reaches a *maximum* value of 1000 kg/m³ at 4°C.

We can explain why a pond or lake freezes at the surface from this unusual thermal expansion behavior. As the pond cools, the cooler, denser water at the surface initially flows to the bottom. When the temperature at the surface reaches 4°C, this flow ceases. Consequently, when the surface of the pond is below 4°C,

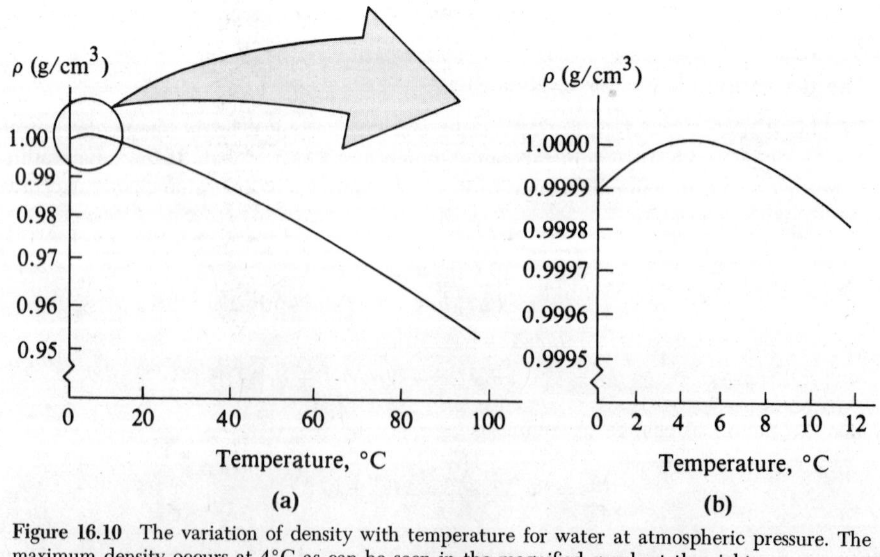

Figure 16.10 The variation of density with temperature for water at atmospheric pressure. The maximum density occurs at 4°C as can be seen in the magnified graph at the right.

equilibrium is reached when the coldest water is at the surface. As the water freezes at the surface, it remains there since ice is less dense than water. The ice continues to build up at the surface, while water near the bottom remains at 4°C. If this did not happen, fish and other forms of marine life would not survive. In fact, if it were not for this peculiarity of water, life as we now know it wouldn't exist!

Q4. Explain why a column of mercury in a thermometer first descends slightly and then rises when placed in hot water.

Q5. Explain why the thermal expansion of a spherical shell made of an isotropic solid is equivalent to that of a solid sphere of the same material.

Q6. A steel wheel bearing is 1 mm smaller in diameter than an axle. How can it be fit onto the axle without removing any material?

16.6 MACROSCOPIC DESCRIPTION OF AN IDEAL GAS

In this section we shall be concerned with the properties of a gas of mass m confined to a container of volume V at a pressure P and temperature T. It would be useful to know how these quantities are related. In general, the equation that interrelates these quantities, called the *equation of state,* is very complicated. However, if the gas is maintained at a very low pressure (or low density), the equation of state is experimentally found to be quite simple. Such a low-density gas is commonly referred to as an *ideal gas.*[5] Most gases at room temperature and atmospheric pressure behave as ideal gases.

It is convenient to express the amount of gas in a given volume in terms of the number of moles, n. By definition, a mole of any substance is that mass of the substance that contains a specific number of molecules called Avogadro's number, N_0. The value of N_0 is approximately 6.022×10^{23} molecules/mole. Avogadro's number is defined to be the number of carbon atoms in 12 g of the isotope carbon-12. The number of moles of a substance is related to its mass m through the expression

$$n = \frac{m}{M} \qquad (16.13)$$

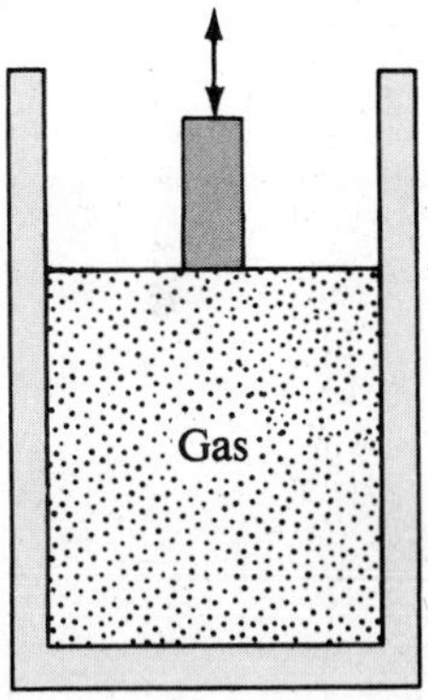

Figure 16.11 An ideal gas contained in a cylinder with a movable piston that allows the volume to be varied. The state of the gas is defined by its pressure, volume, and temperature.

where M is a quantity called the *molecular weight* of the substance, usually expressed in g/mole. For example, the molecular weight of oxygen, O_2, is 32.0 g/mole. Therefore, the mass of one mole of oxygen is 32.0 g.

Now suppose an ideal gas is confined to a cylindrical container whose volume can be varied by means of a movable piston, as in Fig. 16.11. We shall assume that the cylinder does not leak, and hence the mass (or the number of moles) remains constant. For such a system, experiments provide the following information. First, when the gas is kept at a constant temperature, its pressure is inversely proportional to the volume (Boyle's law). Second, when the pressure of the gas is kept constant, the volume is directly proportional to the temperature (the law of Charles and Gay-Lussac). These observations can be summarized by the following *equation of state* for an ideal gas:

Equation of state for an ideal gas

$$PV = nRT \qquad (16.14)$$

[5] To be more specific, the assumption here is that the temperature of the gas is sufficiently high and its pressure sufficiently low that it does not condense into a liquid.

In this expression, R is a constant for a specific gas, which can be determined from experiments, and T is the absolute temperature in kelvins. Experiments on several gases show that as the pressure approaches zero, the quantity PV/nT approaches the same value of R for all gases. For this reason, R is called the *universal gas constant.* In metric units, where pressure is expressed in N/m^2 and volume in m^3, the product PV has units of N · m, or J, and R has the value

The universal gas constant

$$R = 8.31 \text{ J/mole} \cdot \text{K}$$

If the pressure is expressed in atmospheres and the volume in liters (1 liter = 10^3 cm^3 = 10^{-3} m^3), then R has the value

$$R = 0.0821 \text{ liter} \cdot \text{atm/mole} \cdot \text{K}$$

Using this value of R and Eq. 16.13, one finds that the volume occupied by 1 mole of any gas at atmospheric pressure and 0°C (273 K) is 22.4 liters.

The ideal-gas law is often expressed in terms of the total number of molecules, N. Since the total number of molecules equals the product of the number of moles and Avogadro's number, we can write Eq. 16.14

$$PV = nRT = \frac{N}{N_0}RT$$

$$PV = NkT \tag{16.15}$$

where k is called *Boltzmann's constant,* which has the value

Boltzmann's constant

$$k = \frac{R}{N_0} = 1.38 \times 10^{-23} \text{ J/K} \tag{16.16}$$

We have defined an ideal gas as one that obeys the equation of state, $PV = nRT$, under all conditions. In reality, an ideal gas does not exist. However, the concept of an ideal gas is very useful in view of the fact that real gases behave as ideal gases at low pressures. It is common to call quantities such as P, V, and T the *thermodynamic variables* of the system. We note that if the equation of state is known, then one of the variables can always be expressed as some function of the other two thermodynamic variables. That is, given two of the variables, the third can be determined from the equation of state. Other thermodynamic systems are often described with different thermodynamic variables. For example, a wire under tension at constant pressure is described by its length, the tension in the wire, and the temperature.

Example 16.4

An ideal gas occupies a volume of 100 cm^3 at 20°C and a pressure of 10^{-3} atm. Determine the number of moles and the number of molecules of gas in the container.

Solution: The quantities given are volume, pressure, and temperature: $V = 100 \text{ cm}^3 = 0.1$ liter, $P = 10^{-3}$ atm, and $T = 20°\text{C} = 293$ K. Using Eq. 16.14, we get

$$n = \frac{PV}{RT} = \frac{(10^{-3}\text{ atm})(0.1\text{ liter})}{(0.0821\text{ liter} \cdot \text{atm/mole} \cdot \text{K})(293\text{ K})}$$
$$= 4.16 \times 10^{-6} \text{ moles}$$

Note that you must express T as an absolute temperature (K) when using the ideal-gas law. The number of molecules in the container, N, equals the number of moles multiplied by Avogadro's number:

$$N = nN_0$$
$$= (4.16 \times 10^{-6}\text{ mole})\left(6.02 \times 10^{23}\ \frac{\text{molecules}}{\text{mole}}\right)$$
$$= 2.50 \times 10^{18} \text{ molecules}$$

Example 16.5

Pure helium gas is admitted into a cylinder containing a movable piston. The initial volume, pressure, and temperature of the gas are 15 liters, 2 atm, and 300 K. If the volume is decreased to 12 liters and the pressure increased to 3.5 atm, find the final temperature of the gas. (Assume it behaves like an ideal gas.)

Solution: If no gas escapes from the cylinder, the number of moles remains constant; therefore using $PV = nRT$ at the initial and final points gives

$$\frac{P_i V_i}{T_i} = \frac{P_f V_f}{T_f}$$

where i and f refer to the initial and final values. Solving for T_f, we get

$$T_f = \left(\frac{P_f V_f}{P_i V_i}\right) T_i = \frac{(3.5 \text{ atm})(12 \text{ liters})}{(2 \text{ atm})(15 \text{ liters})}(300 \text{ K}) = 420 \text{ K}$$

Q7. Determine the number of grams in one mole of the following gases: (a) hydrogen, (b) helium, and (c) carbon monoxide.

Q8. Two identical cylinders at the same temperature each contain the same kind of gas. If cylinder A contains three times more gas than cylinder B, what can you say about the relative pressures in the cylinders?

Q9. Why is it necessary to use absolute temperature when using the ideal-gas law?

Q10. An inflated rubber balloon filled with air is immersed in a flask of liquid nitrogen that is at 77 K. Describe what happens to the balloon.

16.7 SUMMARY

Two bodies are in *thermal equilibrium* with each other if they have a common temperature.

The *zeroth law of thermodynamics* states that if bodies A and B are separately in thermal equilibrium with a third body, C, then A and B are in thermal equilibrium with each other.

The SI unit of thermodynamic temperature is the *kelvin*, which is defined to be the fraction 1/273.16 of the temperature of the triple point of water.

When a substance is heated, it generally expands. The linear expansion of an object is characterized by an *average expansion coefficient*, α, defined by

$$\alpha = \frac{1}{l}\frac{\Delta l}{\Delta T} \quad (16.9)$$

Average coefficient of linear expansion

where l is the initial length of the object and Δl is the change in length for a temperature change ΔT. The *average volume expansion coefficient*, β, for a homogeneous substance is equal to 3α.

An *ideal gas* is one that obeys the equation of state,

$$PV = nRT \quad (16.14)$$

Equation of state for an ideal gas

where n equals the number of moles of gas, V is its volume, R is the universal gas constant (8.31 J/mole · K), and T is the absolute temperature in kelvins. A real gas behaves approximately as an ideal gas at very low pressures. An ideal gas is used as the working substance in a constant-volume gas thermometer, which defines the absolute temperature scale in kelvins. This absolute temperature T is related to temperatures on the Celsius scale by $T = T_C + 273.15$.

EXERCISES

Section 16.3 The Constant-Volume Gas Thermometer and the Kelvin Scale

1. The pressure in a constant-volume gas thermometer is 0.60 atm at 100°C and 0.25 atm at 0°C. (a) What is the temperature when the pressure is 0.40 atm? (b) What is the pressure at 350°C?

2. A constant-volume gas thermometer registers a pressure of 40 mm Hg when it is at a temperature of 350 K. (a) What is the pressure at the triple point of water? (b) What is the temperature when the pressure reads 2 mm Hg?

3. The gas thermometer shown in Fig. 16.2 reads a pressure of 50 mm Hg at the triple-point temperature. What pressure will it read at (a) the boiling point of water, and (b) the boiling point of sulfur (444.6°C)?

Section 16.4 The Celsius, Fahrenheit, and Rankine Temperature Scales

4. Liquid hydrogen has a boiling point of −252.87°C at atmospheric pressure. Express this temperature in (a) degrees Fahrenheit, (b) degrees Rankine, and (c) kelvins.

5. The boiling point of sulfur is 444.6°C. Express this in (a) degrees Fahrenheit and (b) kelvins.

6. The temperature of one northeastern state varies from 95°F in the summer to −30°F in the winter. Express this range of temperatures in degrees Celsius.

7. The normal human body temperature is 98.6°F. A person with a fever reaches a temperature of 103°F. Express these temperatures in degrees Celsius.

8. A substance is heated from 70°F to 195°F. What is its change in temperature on (a) the Celsius scale and (b) the kelvin scale?

9. Two thermometers are calibrated in degrees Celsius and degrees Fahrenheit. At what temperature are their readings the same?

Section 16.5 Thermal Expansion of Solids and Liquids (use Table 16.2)

10. A copper pipe is 2 m long at 25°C. What is its length at (a) 100°C and (b) 0°C?

11. A structural steel I-beam is 20 m long when installed at 20°C. How much will its length change over the temperature extremes −25°C to 40°C?

12. Calculate the *fractional* change in the volume of an aluminum bar that undergoes a change in temperature of 30 C°. (Note that $\beta = 3\alpha$ for an isotropic substance.)

13. The concrete sections of a certain superhighway are designed to have a length of 30 m. The sections are poured and cured at 10°C. What minimum spacing should the engineer leave *between the sections* to eliminate "buckling" if the concrete is to reach a temperature of 45°C?

14. A steel washer has an inner diameter of 2.000 cm and an outer diameter of 2.500 cm at 20°C. To what temperature must the washer be heated to just fit over a rod that is 2.005 cm in diameter?

15. An automobile fuel tank is filled to the brim with 22 gal of gasoline at −20°C. Immediately afterwards, the vehicle is parked in a garage at 25°C. How much gasoline overflows from the tank as a result of expansion? (Neglect the expansion of the tank.)

16. A metal rod made of some alloy is to be used as a thermometer. At 0°C its length is 30.000 cm and at 100°C its length is 30.050 cm. (a) What is the linear expansion coefficient of the alloy? (b) When the rod is 30.015 cm long, what is the temperature?

17. The active element of a certain laser (light amplifier) is made of a glass rod 20 cm long and 1 cm in diameter. If the temperature of the rod increases by 75 C°, find the increase in (a) its length, (b) its diameter, and (c) its volume. [Take $\alpha = 9 \times 10^{-6}\,(\text{C}°)^{-1}$]

Section 16.6 Macroscopic Description of an Ideal Gas

18. An ideal gas is held in a container at constant volume. Initially, its temperature is 20°C and its pressure is 3 atm. Find the pressure when its temperature is 50°C.

19. The tire of a bicycle is filled with air to a gauge pressure of 50 lb/in.^{2} at 20°C. What is the gauge pressure in the tire on a day when the temperature rises to 35°C? (Assume the volume does not change, and recall that gauge pressure means absolute pressure in the tire minus atmospheric pressure. Furthermore, assume that the atmospheric pressure remains constant.)

20. Show that one mole of any gas at atmospheric pressure (1.01×10^5 N/m^2) and standard temperature (273 K) occupies a volume of 22.4 liters.

21. In modern vacuum systems, pressures as low as 10^{-9} mm Hg are common. Calculate the number of molecules in a 1-m^3 vessel at this pressure if the temperature is 20°C. (Note that 1 atm of pressure corresponds to 760 mm Hg.)

22. Gas is contained in a 3-liter vessel at a temperature of 25°C and a pressure of 5 atm. (a) Determine the number of moles of gas in the vessel. (b) How many molecules are there in the vessel?

23. A cylinder with a movable piston contains gas at a temperature of 27°C, a pressure of 0.2×10^5 Pa, and a volume of 1.5 m^3. What will be its final temperature if the gas is compressed to 0.7 m^3 and the pressure increases to 0.8×10^5 Pa?

24. A gas is heated from 27°C to 127°C while maintained at constant pressure in a vessel whose volume increases. By what factor does the volume change?

25. One mole of oxygen gas is at a pressure of 5 atm and a temperature of 27°C. (a) If the gas is heated at constant volume until the pressure is doubled, what is the final temperature? (b) If the gas is heated such that both the pressure and volume are doubled, what is the final temperature?

26. A cylinder of volume 12 liters contains helium gas at a pressure of 136 atm. How many balloons can be filled with this cylinder at atmospheric pressure if each balloon has a volume of 1 liter?

PROBLEMS

1. Precise temperature measurements are often made using the change in the electrical resistance of a metal or semiconductor with temperature. The resistance varies approximately according to the expression $R = R_0(1 + AT_C)$, where R_0 and A are constants and T_C is the temperature in degrees Celsius. A certain element has a resistance of 50.0 ohms at 0°C and 82.5 ohms at the freezing point of zinc (419.58°C). (a) Determine the constants A and R_0. (b) At what temperature is the resistance equal to 65.5 ohms?

2. A liquid with a coefficient of volume expansion β just fills a spherical shell of volume V at a temperature T (Fig. 16.12). The shell is made of a material that has a coefficient of linear expansion of α. The liquid is free to expand into a capillary of cross-sectional area A at the top. (a) If the temperature increases by ΔT, show that the liquid rises in the capillary by an amount Δh given by $\Delta h = \frac{V}{A}(\beta - 3\alpha)\,\Delta T$. (b) For a typical system, such as a mercury thermometer, why is it a good approximation to neglect the expansion of the shell?

3. A mercury thermometer is constructed as in Fig. 16.12. The capillary tube has a diameter of 0.005 cm, and the bulb has a diameter of 0.30 cm. Neglecting the expansion of the glass, find the change in height of the mercury column for a temperature change of 25 C°.

4. A fluid has a density ρ. (a) Show that the *fractional* change in density for a change in temperature ΔT is given by $\Delta\rho/\rho = -\beta\,\Delta T$. What does the negative sign signify? (b) Water has a maximum density of 1.000 g/cm^3 at 4°C. At 10°C, its density is 0.9997 g/cm^3. What is β for water over this temperature interval?

5. A pendulum clock with a steel suspension system has a period of 1 s at 20°C. If the temperature increases to 25°C, (a) by how much will its period change, and (b) how much time will the clock gain or lose in one week?

6. A steel ball bearing is 2.000 cm in diameter at 20°C. An aluminum plate has a hole in it that is 1.995 cm in diameter at 20°C. What common temperature must they have in order that the ball just squeeze through the hole?

7. The rectangular plate shown in Fig. 16.13 has an area A equal to lw. If the temperature increases by ΔT, show that the increase in area is given by $\Delta A = 2\alpha A\,\Delta T$, where α is the coefficient of linear expansion. What approximation does this expression assume? (Hint: Note that each dimension increases according to $\Delta l = \alpha l\,\Delta T$.)

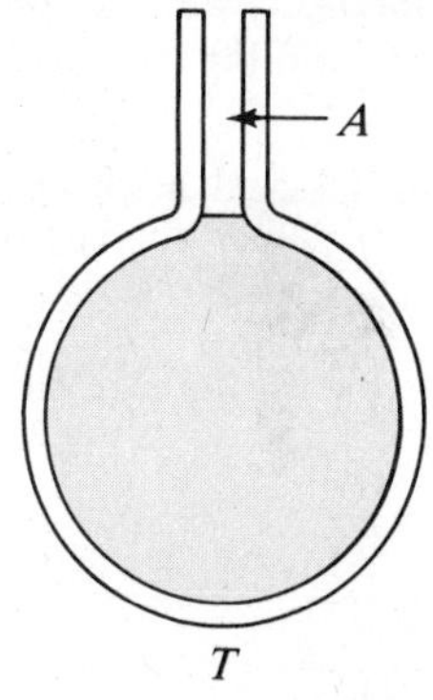

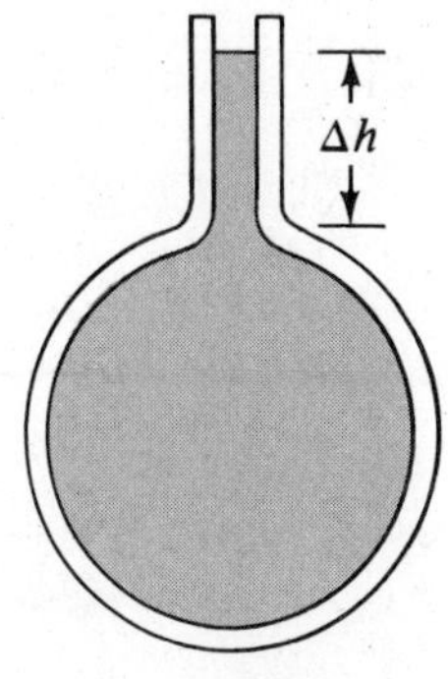

Figure 16.12 (Problem 3).

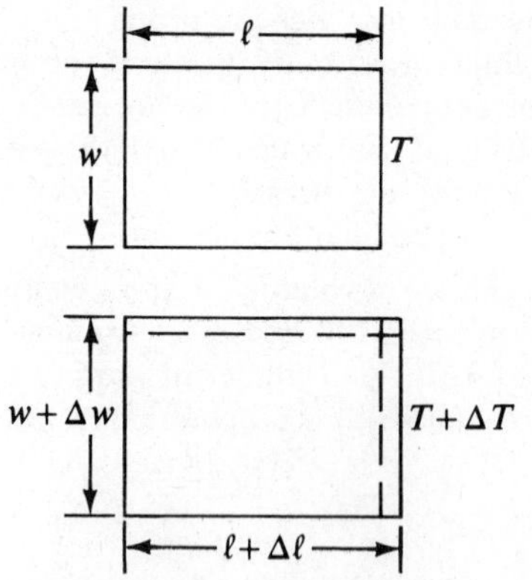

Figure 16.13 (Problem 7).

8. (a) Show that the volume coefficient of thermal expansion for an ideal gas at constant pressure is given by $\beta = 1/T$, where T is the kelvin temperature. Start with the definition of β and use the equation of state, $PV = nRT$. (b) What value does this expression predict for β at 0°C? Compare this with the experimental values for helium and air in Table 16.2.

9. An air bubble originating from a deep-sea diver has a radius of 2 mm at some depth h. When the bubble reaches the surface of the water, it has a radius of 3 mm. Assuming the temperature of the air in the bubble remains constant, determine (a) the depth h of the diver, and (b) the absolute pressure at this depth.

10. Starting with Eq. 16.14, show that the total pressure P in a container filled with a mixture of several different ideal gases is given by $P = P_1 + P_2 + P_3 + \cdots$, where P_1, P_2, etc., are the pressures that each gas would exert if it alone filled the container (or the *partial pressures* of the respective gases). This is known as *Dalton's law of partial pressures.*

11. (a) Show that the density of n moles of a gas occupying a volume V is given by $\rho = nM/V$, where M is the molecular weight. (b) Determine the density of one mole of nitrogen gas at atmospheric pressure and 0°C.

12. A vertical cylinder of cross-sectional area A is fitted with a tight-fitting, frictionless piston of mass m (Fig. 16.14). (a) If there are n moles of an ideal gas in the cylinder at a temperature T, determine the height h at which the piston will be in equilibrium under its own weight. (b) What is the value for h if $n = 3$ moles, $T = 500$ K, $A = 0.05$ m^2, and $m = 5$ kg?

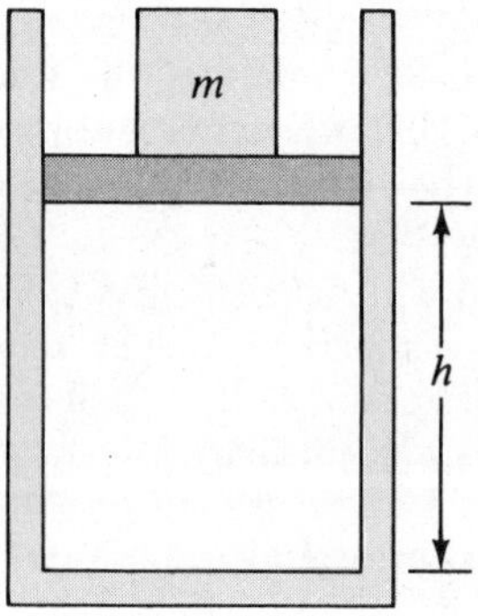

Figure 16.14 (Problem 12).

17 Heat and the First Law of Thermodynamics

Dr. Albert A. Bartlett, University of Colorado, Boulder

It is well known that when two objects at different temperatures are placed in thermal contact with each other, the temperature of the warmer body decreases while the temperature of the cooler body increases. If they are left in contact for some time, they eventually reach a common equilibrium temperature somewhere between the two initial temperatures. When such processes occur, we say that heat is transferred from the warmer to the cooler body. But what is the nature of this heat transfer? Early investigators believed that heat was an invisible, material substance called *caloric*, which was transferred from one body to another. According to this theory, caloric could neither be created nor destroyed. Although the caloric theory was successful in describing heat transfer, it eventually was abandoned when various experiments showed that caloric was in fact not conserved.

The first experimental observation suggesting that caloric was not conserved was made by Benjamin Thompson (1753–1814) at the end of the 18th century. Thompson, an American-born scientist, emigrated to Europe during the Revolutionary War because of his Tory sympathies. Following his appointment as director of the Bavarian arsenal, he was given the title Count Rumford. While supervising the boring of an artillery cannon in Munich, Thompson noticed the great amount of heat generated by the boring tool. The water being used for cooling had to be replaced continuously as it boiled away during the boring process. On the basis of the caloric theory, he reasoned that the ability of the metal filings to retain caloric should decrease as the size of the filings decreased. These heated filings, in turn, presumably transfer caloric to the cooling water, causing it to boil. To his surprise, Thompson discovered that the amount of water boiled away by a blunt boring tool was comparable to the quantity boiled away by a sharper tool for a given turning rate. He then reasoned that if the tool were turned long enough, an almost infinite amount of caloric could be produced from a finite amount of metal filings. For this reason, Thompson rejected the caloric theory and suggested that heat is not a substance, but some form of motion that is transferred from the boring tool to the water. In another experiment, he showed that the heat generated by friction was proportional to the mechanical work done by the boring tool.

Benjamin Thompson (1753–1814.)

There are many other experiments that are at odds with the caloric theory. For example, if you rub two blocks of ice together on a day when the temperature is below 0°C, the blocks will melt. This experiment was first conducted by Sir Humphry Davy (1778–1829). To properly account for this "creation of caloric," we note that mechanical work is done on the system. Thus, we see that the effects of doing mechanical work on a system and of adding heat to it directly, as with a flame, are equivalent. That is, heat and work are both forms of energy.

Although Thompson's observations provided evidence that heat energy is not conserved, it was not until the middle of the 19th century that the modern mechanical model of heat was developed. Before this period, the subjects of heat and mechanics were considered to be two distinct branches of science, and the law of conservation of energy seemed to be a rather specialized result used to describe certain kinds of mechanical systems. After the two disciplines were shown to be intimately related, the law of conservation of energy emerged as a universal law of nature. In this new view, heat is treated as another form of energy that can be transformed into mechanical energy. Experiments performed by James Joule (1818–1889) and others in this period showed that whenever heat is gained or lost by a system during some process, the gain or loss could be accounted for by an equivalent quantity of mechanical work done on the system. Thus, by broadening the concept of energy to include heat as a form of energy, the law of energy conservation was extended.

17.1 HEAT AND THERMAL ENERGY

Definition of heat

The concepts of heat and the internal energy of a substance appear to be synonymous, but there is a subtle distinction between them. The word *heat* should be used only when describing energy transferred from one place to another. That is, *heat flow is an energy transfer that takes place as a consequence of temperature differences only*. On the other hand, *internal energy* is the energy a substance has at some temperature. In the next chapter, we shall show that the energy of an ideal gas is associated with the internal motion of its atoms and molecules. In other words, the internal energy of a gas is essentially its kinetic energy on a microscopic scale; the higher the temperature of the gas, the greater its internal energy. When heat is added to a gas, its internal energy increases and its temperature rises. Its temperature decreases when heat is removed. As an analogy, consider the distinction between work and energy that we discussed in Chapter 7. The work done on (or by) a system is a measure of energy transfer, whereas the mechanical energy (kinetic and/or potential) is a consequence of the motion and coordinates of the system. Thus, when you do work on a system, energy is transferred from you to the system. It makes no sense to talk about the work *of* a system—one can refer only to the *work done on a system* when some process has occurred in which the system has changed in some way. Likewise, it makes no sense to use the term *heat* unless the thermodynamic variables of the system have undergone a change during some process.

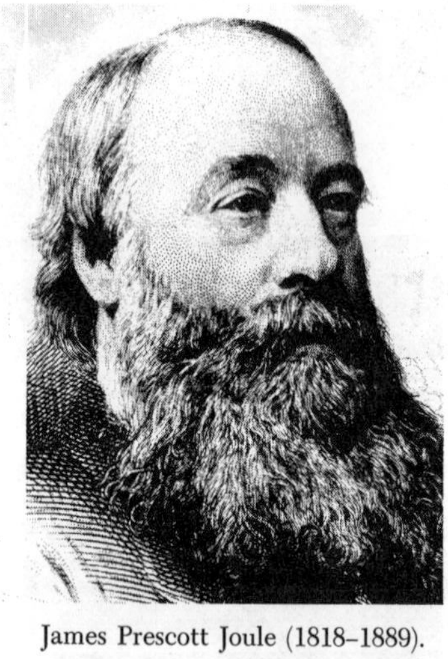

James Prescott Joule (1818–1889).

It is also important to note that energy can be transferred between two systems even when there is no heat flow. For example, when two objects are rubbed together, their internal energy increases since mechanical work is done on them. When an object slides across a surface and comes to rest as a result of friction, its kinetic energy is transformed into internal energy contained in the block and surface. In such cases, the changes in internal energy are measured by corresponding changes in temperature.

17.2 HEAT CAPACITY AND SPECIFIC HEAT

The calorie

It is useful to define a quantity of heat Q in terms of a specific process. The heat unit that is commonly used is the *calorie* (cal), defined as *the amount of heat necessary to raise the temperature of 1 g of water from 14.5°C to 15.5°C.*[1] The *kilocalorie*

[1]Originally, the calorie was defined as the heat necessary to raise the temperature of 1 g of water by 1 C°. However, careful measurements showed that energy depends somewhat on temperature; hence, a more precise definition evolved.

(kcal) is the heat necessary to raise the temperature of 1 kg of water from 14.5°C to 15.5°C (1 kcal $= 10^3$ cal). (Note that the "Calorie," which is used in describing the energy equivalent of foods, is actually a kilocalorie.) The unit of heat in the British engineering system is the *British thermal unit* (Btu), defined as the heat required to raise the temperature of 1 lb of water from 63°F to 64°F. Of course, since we have already recognized that heat is a form of energy, it can be expressed in whatever units happen to be convenient, such as joules, electron-volts, ergs, or foot-pounds. The relationship between the calorie and the mechanical energy unit of joule is found from experiment to be

Mechanical equivalent of heat

$$1 \text{ cal} = 4.186 \text{ J}$$

This result for the so-called *mechanical equivalent of heat* was first established by Joule using an apparatus that will be described in Section 17.5.

The quantity of heat energy required to raise the temperature of a given mass of a substance by some amount varies from one substance to another. For example, the heat required to raise the temperature of 1 g of water by 1 C° is 1 cal, whereas the heat needed to change the temperature of 1 g of carbon by 1 C° is only 0.12 cal. The *heat capacity, C′*, of any substance is defined as *the amount of heat energy needed to raise the temperature of that substance by one Celsius degree*. Therefore, by definition, the heat capacity of 1 g of water is 1 cal/C°, and the heat capacity of 1 g of carbon is 0.12 cal/C°. We shall often refer to a *heat reservoir*, which is considered to be a massive system with a very large heat capacity, such as a lake. The temperature of a heat reservoir is assumed to remain constant during a process. That is, a heat reservoir can exchange heat with another system without itself undergoing any appreciable temperature change.

Heat capacity

In practice, it is often more useful to work with the *specific heat, c*, defined as the heat capacity per unit mass:

Specific heat

$$c = \frac{\text{heat capacity}}{\text{mass}} = \frac{C'}{m} \tag{17.1}$$

The *molar heat capacity, C*, of a substance is defined as the heat capacity per mole:

$$C = \frac{C'}{n} \tag{17.2}$$

Since the number of moles, n, equals the mass, m, divided by the molecular weight, M, we can express the molar heat capacity in the form

Molar heat capacity

$$C = \frac{C'}{n} = \frac{mc}{m/M} = Mc \tag{17.3}$$

Tables found in handbooks usually give the specific heats or the molar heat capacities of substances.

From the definition of heat capacity, we can express the heat energy Q transferred between a system of mass m and its surroundings for a temperature change ΔT as

Heat required to raise the temperature of a substance

$$Q = C' \, \Delta T = mc \, \Delta T \tag{17.4}$$

For example, the heat energy required to raise the temperature of 500 g of water by 3 C° is equal to (500 g)(1 cal/g · C°)(3 C°) = 1500 cal. If the number of moles of

TABLE 17.1 Specific Heat and Molar Heat Capacity for Some Solids at 25°C and Atmospheric Pressure

SUBSTANCE	SPECIFIC HEAT, c_p (cal/g · C°)	MOLAR HEAT CAPACITY (cal/mole · C°)
Aluminum	0.215	5.81
Beryllium	0.436	3.93
Cadmium	0.055	6.18
Copper	0.0924	5.86
Germanium	0.077	5.59
Gold	0.0308	6.07
Iron	0.107	5.98
Lead	0.0305	6.31
Silicon	0.168	4.72
Silver	0.056	6.06

Note: To convert to J/g · K, multiply these values by 4.186.

the system is specified, we can write Q in the form

$$Q = nC\,\Delta T \tag{17.5}$$

Note that when the temperature increases, ΔT and Q are both positive, corresponding to heat flowing into the system. Likewise, when the temperature decreases, ΔT and Q are negative and heat flows out of the system.

Heat capacities of all materials vary somewhat with temperature. If the temperature intervals are not too great, the temperature variation can be ignored and c can be treated as a constant.[2] For example, the specific heat of water (1.00 cal/g · C°) varies by only about 1 percent from 0°C to 100°C at atmospheric pressure. Unless stated otherwise, we shall neglect such variations. When specific heats are measured, one also finds that the amount of heat needed to raise the temperature of a substance depends on other conditions of the measurement. In general, measurements made at constant pressure are different from those made at constant volume. Specific heats measured under conditions of constant pressure are designated c_p, and those measured at constant volume are designated c_v. The difference between the two specific heats for liquids and solids is usually no more than a few percent and is often neglected. Since experimental measurements on solids and liquids are easier to perform under constant-pressure conditions, it is usually c_p that is measured. Table 17.1 gives the specific heat and molar heat capacity of several solid elements. Note that these values are valid at room temperature and atmospheric pressure. Furthermore, these values are considerably less than that of water. Therefore, it takes more heat to raise the temperature of a given mass of water than for most other substances. Large bodies of water will therefore tend to stabilize temperatures in their vicinity, since large heat flows are required to produce significant temperature changes.

Measuring specific heat

One technique for measuring the specific heat of solids or liquids is to simply heat the substance to some known temperature, place it in a vessel containing water of known mass and temperature, and measure the temperature after equilibrium is reached. Since a negligible amount of mechanical work is done in this process, the law of conservation of energy implies that the heat that leaves the warmer body (of

[2] The definitions given by Eq. 17.4 and 17.5 assume that the specific heat does not vary with temperature over the interval ΔT. In general, if c and C vary with temperature over the range T_i to T_f, the correct expression for Q is

$$Q = m\int_{T_i}^{T_f} c\,dT = n\int_{T_i}^{T_f} C\,dT$$

unknown c) must equal the heat that enters the water.[3] Suppose that m_x is the mass of the substance whose specific heat we wish to determine, c_x its specific heat, and T_x its initial temperature. Likewise, let m_w, c_w, and T_w represent the corresponding values for the water. If T is the final equilibrium temperature after everything is mixed, then from Eq. 17.4, we find that the heat gained by the water is $m_w c_w(T - T_w)$, and the heat lost by the substance of unknown c is $-m_x c_x(T - T_x)$. Assuming that the system (water + unknown) does not lose or gain any heat, it follows that the heat gained by the water must equal the heat lost by the unknown (conservation of energy):

$$m_w c_w(T - T_w) = -m_x c_x(T - T_x)$$

Solving for c_x gives

$$c_x = \frac{m_w c_w(T - T_w)}{m_x(T_x - T)} \qquad (17.6)$$

Example 17.1

A 50-g chunk of metal is heated to 200°C and then dropped into a beaker containing 400 g of water initially at 20°C. If the final equilibrium temperature of the mixed system is 22.4°C, find the specific heat of the metal.

Solution: Since the heat lost by the metal equals the heat gained by the water, we can use Eq. 17.6 directly. In our case, $T_x = 200°\text{C}$, $T = 22.4°\text{C}$, $T_w = 20°\text{C}$, $m_x = 50$ g, $m_w = 400$ g, and $c_w = 1$ cal/g · C°. Substituting these values into Eq. 17.6 gives

$$c_x = \frac{(400\text{ g})(1\text{ cal/g}\cdot\text{C}°)(2.4\text{ C}°)}{(50\text{ g})(177.6\text{ C}°)} = 0.108\text{ cal/g}\cdot\text{C}°$$

The metal is most likely iron, as can be seen by comparing this result with the data in Table 17.1.

Q1. Ethyl alcohol has about one half the specific heat of water. If equal masses of alcohol and water in separate beakers are supplied with the same amount of heat, compare the temperature increases of the two liquids.

Q2. Give one reason why coastal regions tend to have a more moderate climate than inland regions.

Q3. A small crucible is taken from a 200°C oven and immersed in a tub full of water at room temperature (often referred to as *quenching*). What is the approximate final equilibrium temperature?

17.3 LATENT HEAT

A substance usually undergoes a change in temperature when heat is transferred between the substance and its surroundings. There are situations, however, where the flow of heat does not result in a change in temperature. This occurs whenever the physical characteristics of the substance change from one form to another, commonly referred to as a *phase change*. Some common phase changes are solid to liquid (melting), liquid to gas (boiling), and the change in crystalline structure of a solid. All such phase changes involve a change in internal energy. The energy required is called the *heat of transformation*.

Consider, for example, the heat required to convert a chunk of ice at −30°C to steam at 100°C. First, heat must be transferred to the ice to change its temperature

[3] For precise measurements, the container for the water should be included in our calculations, since it also gains heat. This would require a knowledge of its mass and composition. However, if the mass of the water is large compared with that of the container, we can neglect the heat gained by the container. Furthermore, precautions must be taken in such measurements to minimize heat transfer from the system to the surroundings.

TABLE 17.2 Latent Heats of Fusion and Vaporization

SUBSTANCE	MELTING POINT (°C)	LATENT HEAT OF FUSION (cal/g)	BOILING POINT (°C)	LATENT HEAT OF VAPORIZATION (cal/g)
Helium	−269.65	1.25	−268.93	4.99
Nitrogen	−209.97	6.09	−195.81	48.0
Oxygen	−218.79	3.30	−182.97	50.9
Ethyl alcohol	−114	24.9	78	204
Water	0.00	79.7	100.00	540
Sulfur	119	9.10	444.60	77.9
Lead	327.3	5.85	1750	208
Aluminum	660	21.5	2450	2720
Silver	960.80	21.1	2193	558
Gold	1063.00	15.4	2660	377
Copper	1083	32.0	1187	1210

Note: To convert latent heats to J/g, multiply the values given by 4.186.

from −30°C to 0°C. When it reaches 0°C, it remains at this temperature while heat is being added until all the ice melts. At this point, further heat transfer will produce a temperature increase of the liquid phase until it reaches 100°C, the boiling point of water. As additional heat is transferred to the boiling water, it remains at 100°C until all the water has changed into the gaseous phase.

The heat required to change the phase of a given mass m of a pure substance is given by

Heat required to change the phase of a substance

$$Q = mL \tag{17.7}$$

where L is called the *latent heat* (hidden heat) of the substance[4] and depends on the nature of the phase change as well as on the properties of the substance. The *latent heat of fusion,* L_f, is used when the phase change is from a solid to a liquid, and the *latent heat of vaporization,* L_v, is the latent heat corresponding to the liquid-to-gas phase change.[5] The latent heat of fusion for water at atmospheric pressure is 79.7 cal/g, and the latent heat of vaporization of water is 540 cal/g. The latent heats of various substances vary considerably, as is seen in Table 17.2.

Phase changes can be described in terms of a rearrangement of molecules when heat is added to or removed from a substance. Consider first the liquid-to-gas phase change. The molecules in the liquid phase are close together, and the intermolecular forces are stronger than in the gas phase, where the molecules are far apart. Therefore, work must be done on the liquid against the attractive intermolecular forces in order to separate the molecules. In the process, the average kinetic energy of the molecules, and hence the temperature, do not change.

Similarly, at the melting point of a solid, we imagine that the amplitude of vibration of the atoms about their equilibrium position becomes large enough to overcome the attractive forces binding them together. The heat energy required to totally melt a given mass of solid is equal to the work required to break the intermolecular bonds and transform the mass from the ordered solid structure to the disordered liquid phase. Since the mean distance between atoms in the gas phase is much larger than in either the liquid or solid phase, we could expect that more work is required to vaporize a given mass of a substance than to melt it. Therefore, it is not

[4] The word *latent* is from the Latin *latere,* meaning *hidden or concealed.*

[5] When a gas cools, it eventually returns to the liquid phase, or *condenses.* The heat per unit mass given up is called the *latent heat of condensation,* which equals the latent heat of vaporization. Likewise, when a liquid cools it eventually solidifies, and the *latent heat of solidification* equals the latent heat of fusion.

surprising that the latent heat of vaporization is much larger than the latent heat of fusion for a given substance.

Example 17.2 *Boiling Liquid Helium*

Liquid helium has a very low boiling point, 4.2 K, and a very low heat of vaporization, 4.99 cal/g (Table 17.2). A constant power of 10 W (1 W = 1 J/s) is transferred to some helium using an immersed electrical heating element. At this rate, how long does it take to boil away 1 kg of liquid helium? (Liquid helium has a density of 0.125 g/cm^3, so that 1 kg corresponds to 8×10^3 cm^3, or 8 liters of liquid.)

Solution: Since $L_v = 4.99$ cal/g for liquid helium, we must supply 4.99×10^3 cal of energy to boil away 1 kg of liquid. The mechanical equivalent of 4.99×10^3 cal is

$$4.99 \times 10^3 \text{ cal} = (4.99 \times 10^3 \text{ cal})\left(4.186 \frac{\text{J}}{\text{cal}}\right)$$

$$= 2.09 \times 10^4 \text{ J}$$

The power supplied to the helium is 10 W = 10 J/s. That is, in 1 s, 10 J of energy are transferred to the helium. Therefore, the time it takes to transfer 2.09×10^4 J is

$$t = \frac{2.09 \times 10^4 \text{ J}}{10 \text{ J/s}} = 2.09 \times 10^3 \text{ s} \approx 35 \text{ min}$$

Since 1 kg of helium corresponds to 8 liters of liquid, this means a "boil off" rate of about 0.23 liters/min. In contrast, 1 kg of liquid nitrogen would boil away in about 3.4 h at the rate of 10 J/s, and 1 kg of water would require about 38 h!

Q4. In a daring lecture demonstration, an instructor dips his wetted fingers into molten lead (327°C) and withdraws them quickly, without getting burned. How is this possible? (Note that this is a dangerous experiment, which you should not attempt.)

17.4 HEAT TRANSFER

In practice, it is important to understand the rate at which heat is transferred between a system and its surroundings and the mechanisms responsible for the heat transfer. You may have used a thermos bottle or some other thermally insulated vessel to store hot coffee for a length of time. The vessel reduces heat transfer between the outside air and the hot coffee. Ultimately, of course, the coffee will reach air temperature since the vessel is not a perfect insulator. In general, there will be no heat transfer between a system and its surroundings when they are at the same temperature.

Heat Conduction

The easiest heat transfer process to describe quantitatively is called *heat conduction*. In this process, the heat transfer can be viewed on an atomic scale as an exchange of kinetic energy between molecules, where the less energetic particles gain energy by colliding with the more energetic particles. Consider a slab of material of thickness Δx and cross-sectional area A with its opposite faces at different temperatures T_1 and T_2, where $T_2 > T_1$ (Fig. 17.1). One finds from experiment that the heat ΔQ transferred in a time Δt flows from the hotter end to the colder end. The rate at which heat flows, $\Delta Q/\Delta t$, is found to be proportional to the cross-sectional area, the temperature difference, and the inverse of the thickness. That is,

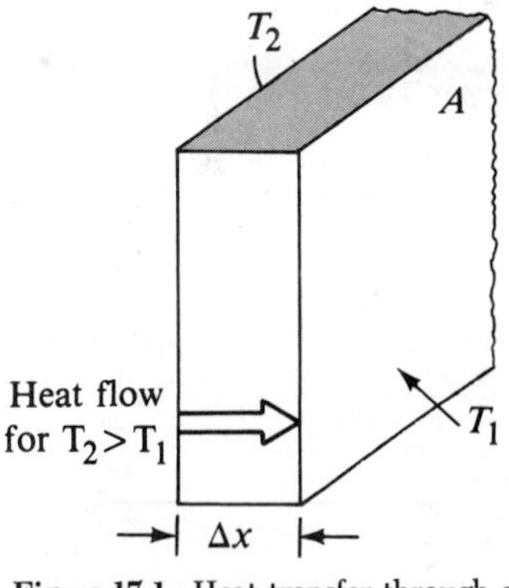

Figure 17.1 Heat transfer through a conducting slab of cross-sectional area A and thickness Δx. The opposite faces are at different temperatures, T_1 and T_2.

$$\frac{\Delta Q}{\Delta t} \sim A \frac{\Delta T}{\Delta x}$$

For a slab of infinitesimal thickness dx and temperature difference dT, we can write the *law of heat conduction*

$$\frac{dQ}{dt} = -kA \frac{dT}{dx} \qquad (17.8)$$

Law of heat conduction

where the proportionality constant k is called the *thermal conductivity* of the material and dT/dx is known as the *temperature gradient* (the variation of temperature with position). The minus sign in Eq. 17.8 denotes that heat flows in the direction of decreasing temperature.

Suppose a substance is in the shape of a long uniform rod of length l, as in Fig. 17.2, and is insulated so that no heat can escape from its surface except at the ends, which are in thermal contact with heat reservoirs having temperatures T_1 and T_2. When a steady state has been reached, the temperature at each point along the rod is constant in time. In this case, the temperature gradient is the same everywhere along the rod and is given by $\frac{dT}{dx} = \frac{T_1 - T_2}{l}$. Thus the heat transferred in a time Δt is

$$\frac{\Delta Q}{\Delta t} = kA\frac{(T_2 - T_1)}{l} \tag{17.9}$$

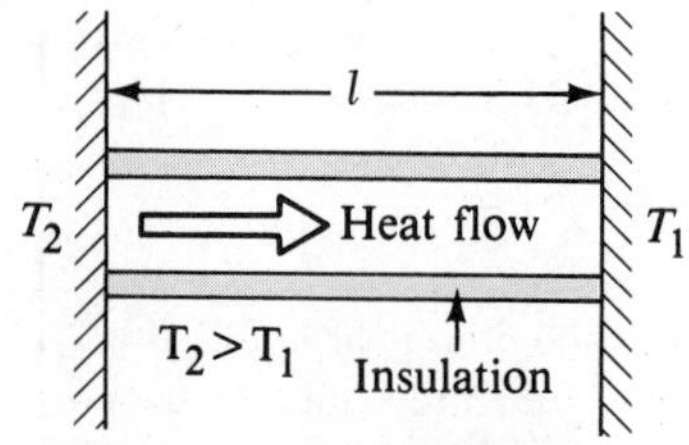

Figure 17.2 Conduction of heat through a uniform, insulated rod of length ℓ. The opposite ends are in thermal contact with heat reservoirs at two different temperatures.

Substances that are good heat conductors have large thermal conductivity values, whereas good thermal insulators have low thermal conductivity values. Table 17.3 lists thermal conductivities for various substances. We see that metals are generally better thermal conductors than nonmetals. This is because in metals there are some electrons more or less free to move through the material and transport energy from one region to another. Nonmetals are poor heat conductors because they do not contain free electrons. It is not surprising that gases are also poor heat conductors in view of their dilute nature.

Melted snow pattern on parking lot indicates the presence of underground steam pipes used to aid snow removal. Heat from the steam is conducted to the pavement from the pipes causing the snow to melt. (Dr. Albert A. Bartlett, University of Colorado, Boulder)

For a compound slab containing several materials of thicknesses $l_1, l_2, \ldots$ and thermal conductivities $k_1, k_2, \ldots$, the rate of heat transfer through the slab at steady state is given by

$$\frac{\Delta Q}{\Delta t} = \frac{A(T_2 - T_1)}{\sum_i (l_1/k_i)} \tag{17.10}$$

where T_1 and T_2 are the temperatures of the outer extremities of the slab (which are held constant) and the summation is over all slabs.

TABLE 17.3 Thermal Conductivities

SUBSTANCE	*THERMAL CONDUCTIVITY k* (cal/s · cm · C°)
Metals (at 25°C)	
Aluminum	0.57
Copper	0.95
Gold	0.75
Iron	0.19
Lead	0.083
Silver	1.02
Gases (at 20°C)	
Air	5.6×10^{-5}
Helium	3.3×10^{-4}
Hydrogen	4.1×10^{-4}
Nitrogen	5.6×10^{-5}
Oxygen	5.7×10^{-5}
Nonmetals (approximate values)	
Glass	2×10^{-3}
Wood	2×10^{-4}
Asbestos	2×10^{-4}
Concrete	2×10^{-3}
Ice	4×10^{-3}
Rubber	5×10^{-4}

Example 17.3

Two slabs of thickness l_1 and l_2 and thermal conductivities k_1 and k_2 are in thermal contact with each other as in Fig. 17.3. The temperatures of their outer surfaces are T_1 and T_2, respectively, and $T_2 > T_1$. Determine the temperature at the interface and the rate of heat transfer through the slabs in the steady-state condition.

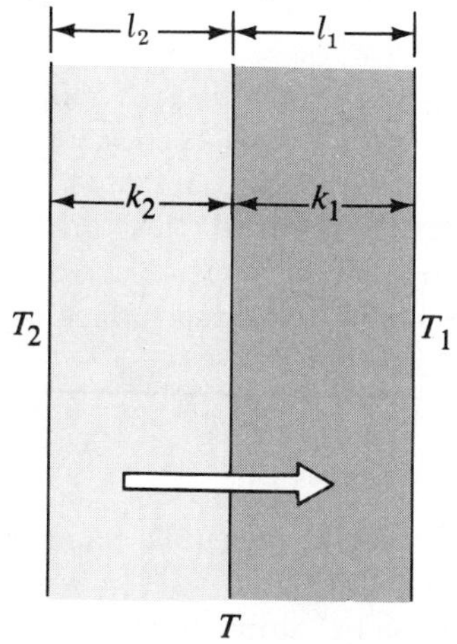

Figure 17.3 Heat transfer by conduction through two slabs in thermal contact with each other. At steady state, the rate of heat transfer through slab 1 equals the rate of heat transfer through slab 2.

Solution: If T is the temperature at the interface, then the rate at which heat is transferred through slab 1 is given by

$$(1) \qquad \frac{\Delta Q_1}{\Delta t} = \frac{k_1 A(T - T_1)}{l_1}$$

Likewise, the rate at which heat is transferred through slab 2 is

$$(2) \qquad \frac{\Delta Q_2}{\Delta t} = \frac{k_2 A(T_2 - T)}{l_2}$$

When a steady state is reached, these two rates must be equal; hence

$$\frac{k_1 A(T - T_1)}{l_1} = \frac{k_2 A(T_2 - T)}{l_2}$$

Solving for T gives

$$(3) \qquad T = \frac{k_1 l_2 T_1 + k_2 l_1 T_2}{k_1 l_2 + k_2 l_1}$$

Substituting (3) into either (1) or (2), we get

$$\frac{\Delta Q}{\Delta t} = \frac{A(T_2 - T_1)}{(l_1/k_1) + (l_2/k_2)}$$

An extension of this model to several slabs leads to Eq. 17.10.

Convection and Radiation

Two other important heat transfer processes, which we shall discuss only briefly, are convection and radiation. *Convection* is heat transfer as the result of the actual movement of a heated substance from one place to another. In some cases, such as in hot-air and hot-water heating systems, the heated substance is forced to move by a fan or pump. This is known as *forced convection*. In *natural*, or *free, convection* the motion is produced as the result of the differences in density between hot and cold regions. Since warmer fluids are generally less dense than cooler fluids, the heated portions will rise according to Archimedes' principle. Convection is the mechanism for the mixing of warm and cool air masses in the atmosphere and hence is a key factor in weather conditions.

The third mechanism of heat transfer is *radiation*. All bodies radiate energy continuously in the form of electromagnetic waves, which we shall discuss in Chapter 35. For example, when we see that the heating element on an electric range is "red hot," we are observing electromagnetic radiation emitted by the hot element. Likewise, the tungsten wire in an incandescent lamp and the surface of the sun also emit radiant energy. Techniques for converting this free solar radiation into useful forms of energy are of current interest.

The rate at which a body emits radiant energy is proportional to the fourth power of its absolute temperature. This is known as *Stefan's law*, often written in the form

$$P = \sigma A e T^4 \qquad (17.11)$$

Stefan's law

where P is the power radiated by the body in W (or J/s), σ is a universal constant equal to 5.6696×10^{-8} W/m^2, A is the surface area of the body in m^2, e is a constant called the *emissivity*, and T is the absolute temperature. The value of e can vary between 0 and 1 depending on the properties of the surface.

A body radiates and also absorbs electromagnetic radiation at rates given by Eq. 17.11. If this were not the case, a body would eventually radiate all of its internal energy and its temperature would reach absolute zero. The energy that the body absorbs comes from the surroundings, which also emit radiant energy. If the body is at a temperature T and its surroundings are at a temperature T_0, the net power gained (or lost) as a result of radiation is given by

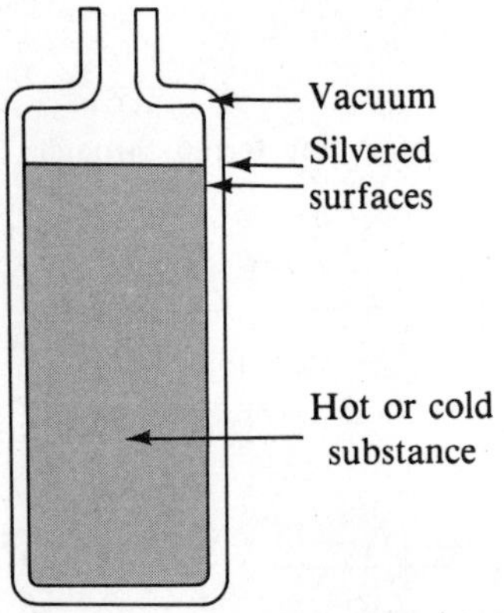

Figure 17.4 Cross-sectional view of a Dewar vessel, used to store hot or cold liquids or other substances.

$$P_{\text{net}} = \sigma A e(T^4 - T_0^4) \tag{17.12}$$

When a body is in equilibrium with its surroundings, it radiates and absorbs energy at the same rate, and so its temperature remains constant. When a body is hotter than its surroundings, it radiates more energy than it absorbs, and so it cools. An *ideal radiator*, or ideal *black body*, ($e = 1$) is one which absorbs all of the energy incident on it (and hence reflects no energy). Therefore, a black body is also a good emitter of radiant energy. Likewise, a highly reflecting surface ($e \approx 0$) is a poor absorber and a poor emitter of radiant energy.

The Dewar

The thermos bottle, called a *Dewar flask*[6] in the scientific community, is a practical example of a container designed to minimize heat losses by conduction, convection, and radiation. Such a container is used to store either cold or hot liquids for long periods of time. The standard construction (Fig. 17.4) consists of a double-walled pyrex vessel with silvered inner walls. The space between the walls is evacuated to minimize heat transfer by conduction and convection. The silvered surfaces minimize heat transfer by radiation by reflecting most of the radiant heat. Very little heat is lost over the neck of the flask since glass is a poor heat conductor. A further reduction in heat loss is obtained by reducing the size of the neck. Dewar flasks are commonly used to store liquid nitrogen (boiling point 77 K) and liquid oxygen (boiling point 90 K).

For other cryogenic liquids, such as liquid helium, which has a very low specific heat (boiling point 4.2 K), it is often necessary to use a double Dewar system in which the Dewar flask containing the liquid is surrounded by a second Dewar flask. The space between the two flasks is filled with liquid nitrogen.

In this photograph of a converted barn we see an alternating pattern of snow-covered and exposed roof. The space between rafters has been well insulated. In contrast, the rafters provide less insulating effect. Snow piles up over the well-insulated spaces while it has melted over the rafters, which allow more heat to escape. (Dr. Albert A. Bartlett, University of Colorado, Boulder)

Q5. In the winter you might notice that some roofs are uniformly covered with snow, while others have regions where the snow has melted. Which houses would you say are better insulated? (See photograph on the left.)

Q6. Why is it possible to hold a lighted match, even when it is burned to within a few millimeters of your fingertips?

Q7. If you wish to cook a piece of meat thoroughly on an open fire, why should you not use a high flame? (Note that carbon is a good thermal insulator.)

Q8. When insulating a wood-frame house, is it better to place the insulation against the cooler outside wall or against the warmer inside wall? (In either case, there is an air barrier to consider.)

Q9. Why is it necessary to store liquid nitrogen or liquid oxygen in vessels equipped with either styrofoam insulation or a double-evacuated wall?

[6]Invented by Sir James Dewar (1842–1923).

Q10. A thermos bottle is constructed with double silvered-glass walls, with the space between them evacuated. Give reasons for the silvered walls and the vacuum jacket.

17.5 THE MECHANICAL EQUIVALENT OF HEAT

When the concept of mechanical energy was introduced in Chapters 7 and 8, we found that whenever friction is present in a mechanical system, some mechanical energy is lost, or is not conserved. Experiments of various sorts show that this lost mechanical energy does not simply disappear, but is transformed into thermal energy. Although this connection between mechanical and thermal energy was first suggested by Thompson's crude cannon boring experiment, it was Joule who first established the equivalence of the two forms of energy.

A schematic diagram of Joule's most famous experiment is shown in Fig. 17.5. The system of interest is the water in a thermally insulated container. Work is done on the water by a rotating paddle wheel, which is driven by weights falling at a constant speed. The water, which is stirred by the paddles, warms up because of the friction between it and the paddles. If the energy lost in the bearings and through the walls is neglected, then the loss in potential energy of the weights equals the work done by the paddle wheel on the water. If the two weights fall through a distance h, the loss in potential energy is $2mgh$, and it is this energy that is used to heat the water. By varying the conditions of the experiment, Joule found that the loss in mechanical energy, $2mgh$, is proportional to the increase in temperature of the water, ΔT, and also proportional to the mass of water used. The proportionality constant (the specific heat of water) was found to be 4.18 J/g · C°. Hence, 4.18 J of mechanical energy will raise the temperature of 1 g of water from 14.5°C to 15.5°C. One calorie is now defined to be *exactly* 4.186 J:

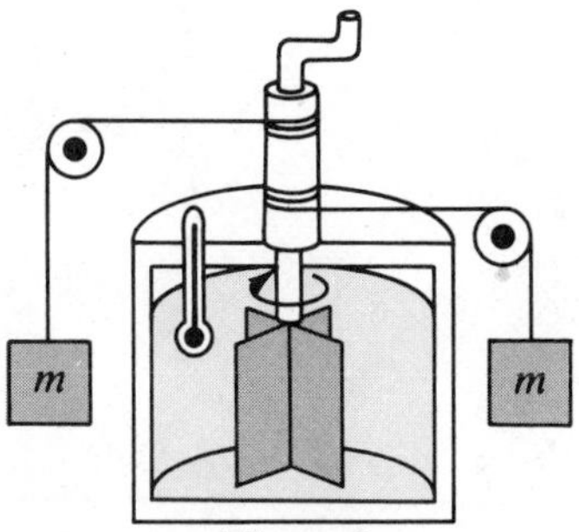

Figure 17.5 An illustration of Joule's experiment for measuring the mechanical equivalent of heat. The falling weights rotate the paddles, causing the temperature of the water to increase.

$$1 \text{ cal} = 4.186 \text{ J} \tag{17.13}$$

Mechanical equivalent of heat

Example 17.4 *Losing Weight the Hard Way*

A student eats a dinner rated at 2000 (food) Calories. He wishes to do an equivalent amount of work in the gymnasium by lifting a 50-kg mass. How many times must he raise the weight to expend this much energy? Assume he raises the weight a distance of 2 m each time and that no work is done when the weight is dropped to the floor.

Solution: Since 1 (food) Calorie $= 10^3$ cal, the work required is 2×10^6 cal. Converting this to J, we have for the total work required

$$W = (2 \times 10^6 \text{ cal})\left(4.186 \frac{\text{J}}{\text{cal}}\right) = 8.37 \times 10^6 \text{ J}$$

The work done in lifting the weight once through a distance h is equal to mgh, and the work done in lifting the weight n times is $nmgh$. Equating this to the total work required gives

$$W = nmgh = 8.37 \times 10^6 \text{ J}$$

Since $m = 50$ kg and $h = 2$ m, we get

$$n = \frac{8.37 \times 10^6 \text{ J}}{(50 \text{ kg})(9.8 \text{ m/s}^2)(2 \text{ m})} = 8.54 \times 10^3 \text{ times}$$

If the student is in good shape and lifts the weight, say, once every 5 s, it would take him about 30 h to perform this feat. Clearly, it is much easier to lose weight by dieting.

17.6 WORK AND HEAT IN THERMODYNAMIC PROCESSES

In the macroscopic approach to thermodynamics we describe the *state* of a system with such variables as pressure, volume, temperature, and internal energy. The number of macroscopic variables needed to characterize a system depends on the nature of the system. For a homogeneous system, such as a gas containing only one type of molecule, usually only two variables are needed, such as pressure and vol-

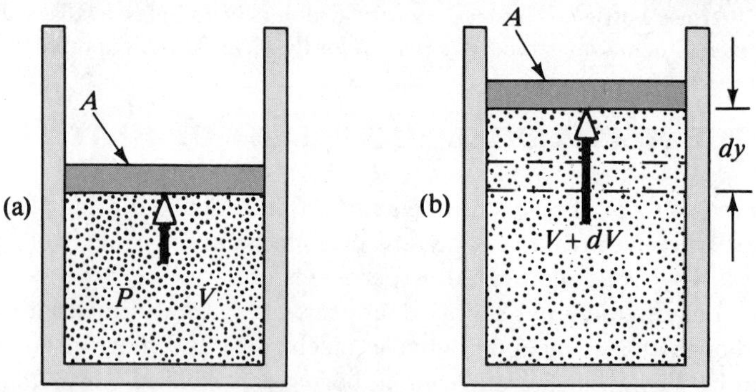

Figure 17.6 Gas contained in a cylinder at a pressure P does work on a moving piston as the system expands from a volume V to a volume $V + dV$.

ume. However, it is important to note that a *macroscopic state* of an isolated system can be specified only if the system is in thermal equilibrium internally. In the case of a gas in a container, internal thermal equilibrium requires that every part of the container be at the same pressure and temperature.

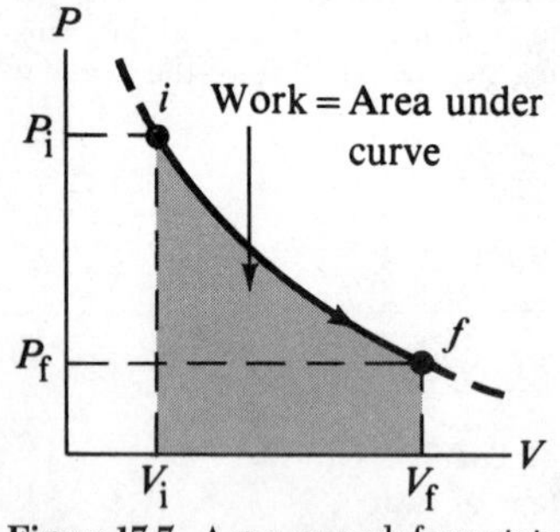

Figure 17.7 A gas expands from state I to state F. The work done by the gas equals the area under the PV curve.

Consider a thermodynamic system such as a gas contained in a cylinder fitted with a movable piston (Fig. 17.6). In equilibrium, the gas occupies a volume V and exerts a uniform pressure P on the cylinder walls and piston. If the piston has a cross-sectional area A, the force exerted by the gas on the piston is $F = PA$. Now let us assume that the gas expands *quasi-statically,* that is, slowly enough to allow the system to move through an (infinite) series of equilibrium states. As the piston moves up a distance dy, the work done by the gas on the piston is

$$dW = F dy = PA\, dy$$

Since $A dy$ is the increase in volume of the gas, dV, we can express the work done as

$$dW = P\, dV \tag{17.14}$$

If the gas expands, as in Fig. 17.6b, then dV is positive and the work done by the gas is positive, whereas if the gas is compressed, dV is negative, indicating that the work done by the gas is negative. (In the latter case, negative work can be interpreted as being work done *on* the system.) Clearly, the work done by the system is zero when the volume remains constant. The total work done by the gas as its volume changes from V_i to V_f is given by the integral of Eq. 17.14:

Work done by a gas

$$W = \int_{V_i}^{V_f} P\, dV \tag{17.15}$$

To evaluate this integral, one must know how the pressure varies during the process. (Note that a *process* is *not* specified merely by giving the initial and final states. That is, a process is a *fully specified* change in state of the system.) In general, the pressure is not constant, but depends on the volume and temperature. If the pressure and volume are known at each step of the process, the states of the gas can then be represented as a curve on a *PV* diagram, as in Fig. 17.7. *The work done in the expansion from the initial state to the final state is the area under the curve in a* PV *diagram.*

Work equals area under the curve in a *PV* diagram

As one can see from Fig. 17.7, the work done in the expansion from the initial state, I, to the final state, F, will depend on the specific path taken between these two states. To illustrate this important point, consider several different paths connecting I and F (Fig. 17.8). In the process described in Fig. 17.8a, the pressure of the

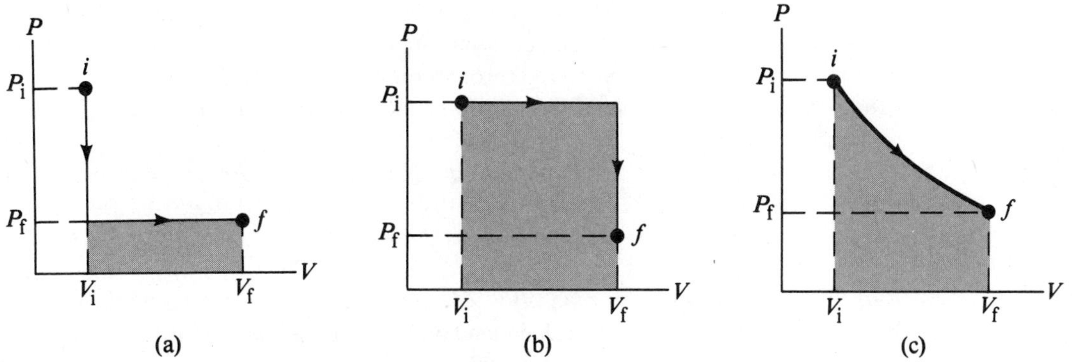

Figure 17.8 The work done by a gas as it is taken from an initial state to a final state depends on the intermediate path between these states.

gas is first reduced from P_i to P_f by cooling at constant volume V_i, and the gas then expands from V_i to V_f at constant pressure P_f. The work done along this path is $P_f(V_f - V_i)$. In Fig. 17.8b, the gas first expands from V_i to V_f at constant pressure P_i, and then its pressure is reduced to P_f at constant volume V_f. The work done along this path is $P_i(V_f - V_i)$, which is greater than that for the process described in Fig. 17.8a. Finally, for the process described in Fig. 17.8c, where both P and V change continuously, the work done has some value intermediate between the values obtained in the first two processes. To evaluate the work in this case, the shape of the PV curve must be known. Therefore, we see that the *work done by a system depends on how the system goes from the initial to the final state.* In other words, *the work done depends on the initial, final, and intermediate states of the system.*

Work done depends on the path between the initial and final states

In a similar manner, the heat transferred into or out of the system is also found to depend on the process. This can be demonstrated by considering the situations described in Fig. 17.9. In each case, the gas has the same initial volume, temperature, and pressure and is assumed to be ideal. In Fig. 17.9a, the gas is in thermal contact with a heat reservoir. If the pressure of the gas is infinitesimally greater than atmospheric pressure, the gas will expand and cause the piston to rise. During this expansion to some final volume V_f, sufficient heat to maintain a constant temperature T_i will be transferred from the reservoir to the gas.

Now consider the thermally insulated system shown in Fig. 17.9b. When the

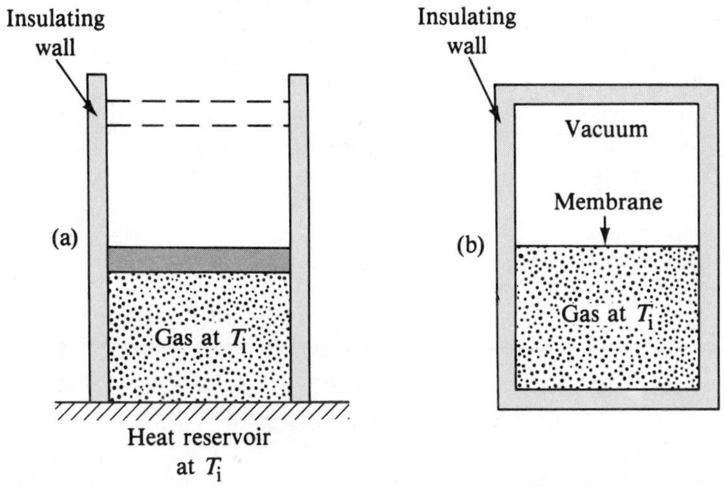

Figure 17.9 (a) A gas at temperature T_i expands slowly by absorbing heat from a reservoir at the same temperature. (b) A gas expands rapidly into an evacuated region by breaking a membrane.

Free expansion of a gas

membrane is broken, the gas expands rapidly into the vacuum until it occupies a volume V_f. In this case, the gas does no work since there is no movable piston. Furthermore, no heat is transferred through the thermally insulated wall, which we call an *adiabatic wall*. This process is often referred to as *adiabatic free expansion*, or simply *free expansion*.

The initial and final states of the ideal gas in Fig. 17.9a are identical to the initial and final states in Fig. 17.9b, but the paths are different. In the first case, heat is transferred slowly to the gas, and the gas does work on the piston. In the second case, no heat is transferred and the work done is zero. Therefore, we conclude that *heat, like work, depends on the initial, final, and intermediate states of the system.* Furthermore, since heat and work depend on the path, neither quantity is independently conserved during a thermodynamic process.

17.7 THE FIRST LAW OF THERMODYNAMICS

When the law of conservation of energy was first introduced in Chapter 8, it was stated that the mechanical energy of a system is conserved in the absence of nonconservative forces, such as friction. That is, the changes in the internal energy of the system were not included in this mechanical model. *The first law of thermodynamics is a generalization of the law of conservation of energy that includes possible changes in internal energy.* It is a universally valid law that can be applied to all kinds of processes. Furthermore, it provides us with a connection between the microscopic and macroscopic worlds.

Change in internal energy

We have seen that energy can be transferred between a system and its surroundings in two ways. One is work done by (or on) the system. This mode of energy exchange results in measurable changes in the macroscopic variables of the system, such as the pressure and volume of a gas. The other is heat transfer, which takes place at the microscopic level and manifests itself by changes in temperature. When energy has been transferred by either or both of these methods, we say that the system has undergone *a change in internal energy*.

To put these ideas on a more quantitative basis, suppose a thermodynamic system undergoes a change from an initial state to a final state in which Q units of heat are absorbed (or removed) and W is the work done by (or on) the system.[7] For example, the system may be a gas whose pressure and volume change from P_i, V_i to P_f, V_f. If the quantity $Q - W$ is calculated for various paths connecting the initial and final equilibrium states (that is, for various *processes*), one finds that $Q - W$ is the same for *all* paths connecting the initial and final states. We conclude that the quantity $Q - W$ is determined completely by the initial and final states of the system, and we call the quantity $Q - W$ the *change in the internal energy of the system.* Although Q and W both depend on the path, the quantity $Q - W$, that is, *the change in internal energy, is independent of the path.* If we represent the internal energy function by the letter U, then the *change* in internal energy, $\Delta U = U_f - U_i$, can be expressed as

First law of thermodynamics

$$\Delta U = U_f - U_i = Q - W \tag{17.16}$$

where all quantities must have the same units of energy. Equation 17.16 is known as *the first law of thermodynamics*. When it is used in this form, we must note that Q is positive when heat *enters* the system and W is positive when work is done *by* the system.

[7] We use the convention that Q is positive if the system absorbs heat and negative if it loses heat. Likewise, the work done is positive if the system does work on the surroundings and negative if work is done on the system.

When a system undergoes an infinitesimal change in state, where a small amount of heat, $đQ$, is transferred and a small amount of work, $đW$, is done, the internal energy also changes by a small amount, dU. Thus, for infinitesimal processes we can express the first law as[8]

First law of thermodynamics for infinitesimal changes

$$dU = đQ - đW \tag{17.17}$$

On a microscopic level, the internal energy of a system includes the kinetic and potential energies of the molecules making up the system. In thermodynamics, we do not concern ourselves with the specific form of the internal energy. We simply use Eq. 17.16 as a definition of the change in internal energy. One can make an analogy here between the potential energy function associated with a body moving under the influence of gravity without friction. The potential energy function is independent of the path, and it is only its change that is of concern. Likewise, the change in internal energy of a thermodynamic system is what matters, since only differences are defined. Any reference state can be chosen for the internal energy since absolute values are not defined.

Now let us look at some special cases. First consider an *isolated system*, that is, one that does not interact with its surroundings. In this case, there is no heat flow and the work done is zero; hence the internal energy remains constant. That is, since $Q = W = 0$, $\Delta U = 0$, and so $U_i = U_f$. We conclude that *the internal energy of an isolated system remains constant.*

Isolated systems

Next consider a process in which the system is taken through a *cyclic process*, that is, one that originates and ends up at the same state. In this case, the change in the internal energy is *zero* and the heat added to the system must equal the work done during the cycle. That is, in a cyclic process,

Cyclic process

$$\Delta U = 0 \qquad \text{and} \qquad Q = W$$

Note that *the net work done per cycle equals the area enclosed by the path representing the process on a* PV *diagram.* As we shall see in Chapter 19, cyclic processes are very important in describing the thermodynamics of *heat engines*, which are devices in which some part of an investment of heat is extracted as mechanical work.

If a process occurs in which the work done is zero, then the change in internal energy equals the heat entering or leaving the system. If heat enters the system, Q is positive and the internal energy increases. For a gas, we can associate this increase in internal energy with an increase in the kinetic energy of the molecules. On the other hand, if a process occurs in which the heat transferred is zero and work is done by the system, then the change in internal energy equals the work done by the system. For example, if a gas is compressed with no heat transferred (by a moving piston, say), the work done is negative and the internal energy again increases. This is because kinetic energy is transferred from the moving piston to the gas molecules.

We have seen that there is really no distinction between heat and work on a microscopic scale. Both can produce a change in the internal energy of a system. Although the macroscopic quantities Q and W are *not* properties of a system, they are related to the internal energy of the system through the first law of thermodynamics. Once a process or path is defined, Q and W can be calculated or measured, and the change in internal energy can be found from the first law. One of the important consequences of the first law is that there is a quantity called *internal energy,* the value of which is determined by the state of the system. The internal energy function is therefore called a *state function.*

---

[8] Note that $đQ$ and $đW$ are not true differential quantities, although dU is a true differential. For further details on this point, see an advanced text in thermodynamics, such as M. W. Zemansky, *Heat and Thermodynamics,* New York, McGraw-Hill, 1968.

17.8 SOME APPLICATIONS OF THE FIRST LAW OF THERMODYNAMICS

In order to apply the first law of thermodynamics to specific systems, it is useful to first define some common thermodynamic processes.

Adiabatic process

An *adiabatic process* is defined as a process for which no heat enters or leaves the system, that is, $Q = 0$. Applying the first law of thermodynamics in this case, we see that

First law for an adiabatic process

$$\Delta U = -W \tag{17.18}$$

An adiabatic process can be achieved either by thermally insulating the system from its surroundings (say, with styrofoam or an evacuated wall) or by performing the process rapidly. From this result, we see that if a gas expands adiabatically, W is positive and so ΔU is negative and the gas cools. Conversely, a gas is heated when it is compressed adiabatically.

Adiabatic processes are very important in engineering practice. Some common examples of adiabatic processes include the expansion of hot gases in an internal combustion engine, the liquefaction of gases in a cooling system, and the compression stroke in a diesel engine.

The *free expansion process* described in Fig. 17.9b is an adiabatic process in which no work is done on or by the gas. Since $Q = 0$ and $W = 0$, we see from the first law that $\Delta U = 0$ for this process. That is, *the initial and final internal energies of a gas are equal in an adiabatic free expansion.* As we shall see in the next chapter, the internal energy of an ideal gas depends only on its temperature. Thus, we would expect no change in temperature during an adiabatic free expansion. This is in accord with experiments performed at low pressures. Careful experiments at high pressures for real gases show a slight decrease in temperature after the expansion.

Isobaric process

A process that occurs at constant pressure is called an *isobaric process.* When such a process occurs, *the heat transferred and the work done are both nonzero.* The work done is simply the pressure multiplied by the change in volume, or $P(V_f - V_i)$.

A process that takes place at constant volume is called an *isovolumetric process.* In such a process, *the work done is clearly zero;* hence from the first law we see that

First law for a constant-volume process

$$\Delta U = Q \tag{17.19}$$

This tells us that *if heat is added to a system kept at constant volume, all of the heat goes into increasing the internal energy of the system.* When a mixture of gasoline vapor and air explodes in the cylinder of an engine, the temperature and pressure rise suddenly because the cylinder volume doesn't change appreciably during the short duration of the explosion.

Isothermal process

A process that occurs at constant temperature is called an *isothermal process,* and a plot of P versus V at constant temperature for an ideal gas yields a hyperbolic curve called an *isotherm.* During an isothermal process, *the internal energy of the system changes as a result of both heat transferred and work done.*

Isothermal Expansion of an Ideal Gas

Suppose an ideal gas is allowed to expand quasi-statically at constant temperature as described by the PV diagram in Fig. 17.10. The curve is a hyperbola that obeys the equation $PV =$ constant. Let us calculate the work done by the gas in the expansion from state I to state F.

The isothermal expansion of the gas can be achieved by placing the gas in good thermal contact with a heat reservoir at the same temperature, as in Fig. 17.9a.

The work done by the gas is given by Eq. 17.15. Since the gas is ideal and the process is quasi-static, we can apply $PV = nRT$ for each point on the path. Therefore, we have

$$W = \int_{V_i}^{V_f} P\,dV = \int_{V_i}^{V_f} \frac{nRT}{V}\,dV$$

But T is constant in this case; therefore it can be removed from the integral. This gives

$$W = nRT\int_{V_i}^{V_f} \frac{dV}{V}$$

To evaluate this integral, we use the fact that $\int \frac{dx}{x} = \ln x$ (Appendix B-5), which gives

$$W = nRT \ln V \Big]_{V_i}^{V_f} = nRT \ln\left(\frac{V_f}{V_i}\right) \tag{17.20}$$

Work done in an isothermal process

Numerically, this work equals the shaded area under the PV curve in Fig. 17.10. If the gas expands isothermally, then $V_f > V_i$ and we see that the work done by the gas is positive, as we would expect. If the gas is compressed isothermally, then $V_f < V_i$ and the work done by the gas is negative. (Negative work here implies that positive work must be done *on* the gas by some external agent to compress it.) In the next chapter we shall find that the internal energy of an ideal gas depends only on temperature. Hence, for an isothermal process $\Delta U = 0$, and from the first law we conclude that the heat given up by the reservoir equals the work done by the gas, or $Q = W$.

Example 17.5

Calculate the work done by 1 mole of an ideal gas that is kept at 0°C in an expansion from 3 liters to 10 liters.

Solution: Substituting these values into Eq. 17.20 gives

$$W = nRT \ln\left(\frac{V_f}{V_i}\right)$$

$$= (1 \text{ mole})(8.31 \text{ J/mole}\cdot\text{K})(273 \text{ K}) \ln\left(\frac{10}{3}\right)$$

$$= 2.73 \times 10^3 \text{ J}$$

The heat that must be supplied to the gas from the reservoir to keep T constant is also 2.73×10^3 J = 653 cal.

The Boiling Process

A liquid of mass m vaporizes at constant pressure P. Its volume in the liquid state is V_ℓ, and its volume in the vapor state is V_v. Let us find the work done in the expansion and the change in internal energy of the system.

Since the expansion takes place at constant pressure, the work done by the system is

$$W = \int_{V_\ell}^{V_v} P\,dV = P\int_{V_\ell}^{V_v} dV = P(V_v - V_\ell)$$

The heat that must be transferred to the liquid to vaporize all of it is equal to $Q = mL_v$, where L_v is the latent heat of vaporization of the liquid. Using the first law and the result above, we get

$$\Delta U = Q - W = mL_v - P(V_v - V_\ell) \tag{17.21}$$

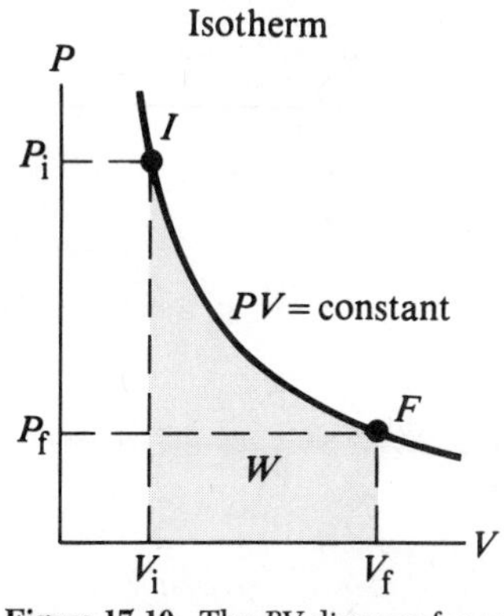

Figure 17.10 The PV diagram for an isothermal expansion of a gas from an initial state to a final state. The curve is a hyperbola.

Example 17.6 *Boiling Water*

One gram of water occupies a volume of 1 cm^3 at atmospheric pressure. When this amount of water is boiled, it becomes 1671 cm^3 of steam. Calculate the change in internal energy for this process.

Solution: Since the heat of vaporization of water is 540 cal/g at atmospheric pressure, the heat required to boil 1 g of water is

$$Q = mL_v = 540 \text{ cal} = 2259 \text{ J}$$

The work done by the system is positive and equal to

$$\begin{aligned} W &= P(V_v - V_l) \\ &= (1.013 \times 10^5 \text{ N/m}^2)[(1671 - 1) \times 10^{-6} \text{ m}^3] \\ &= 169 \text{ J} \end{aligned}$$

Hence, the change in internal energy is given by

$$\Delta U = Q - W = 2259 \text{ J} - 169 \text{ J} = 2090 \text{ J}$$

The internal energy of the system *increases* since ΔU is positive. We see that most of the heat (93%) that is transferred to the liquid goes into increasing the internal energy. Only a small fraction of the heat (7%) goes into external work.

Example 17.7 *Heat Transferred to a Solid*

The internal energy of a solid also increases when heat is transferred to it from its surroundings.

A 1-kg bar of copper is heated at atmospheric pressure. If its temperature increases from 20°C to 50°C, (a) find the work done by the copper.

The change in volume of the copper can be calculated using Eq. 16.11 and the volume expansion coefficient for copper taken from Table 16.2 (remembering that $\beta = 3\alpha$):

$$\begin{aligned} \Delta V &= \beta V \, \Delta T = [5.1 \times 10^{-5} \, (\text{C}°)^{-1}](50°\text{C} - 20°\text{C})V \\ &= 1.5 \times 10^{-3} \, V \end{aligned}$$

But the volume is equal to m/ρ, and the density of copper is 8.92×10^3 kg/m^3. Hence,

$$\begin{aligned} \Delta V &= (1.5 \times 10^{-3})\left(\frac{1 \text{ kg}}{8.92 \times 10^3 \text{ kg/m}^3}\right) \\ &= 1.7 \times 10^{-7} \text{ m}^3 \end{aligned}$$

Since the expansion takes place at constant pressure, the work done is given by

$$\begin{aligned} W = P \, \Delta V &= (1.013 \times 10^5 \text{ N/m}^2)(1.7 \times 10^{-7} \text{ m}^3) \\ &= 1.9 \times 10^{-2} \text{ J} \end{aligned}$$

(b) What quantity of heat is transferred to the copper?

Taking the specific heat of copper from Table 17.1 and using Eq. 17.4, we find that the heat transferred is

$$\begin{aligned} Q &= mc \, \Delta T = (1 \times 10^3 \text{ g})(0.0924 \text{ cal/g} \cdot \text{C}°)(30 \text{ C}°) \\ &= 2.77 \times 10^3 \text{ cal} = 1.16 \times 10^4 \text{ J} \end{aligned}$$

(c) What is the increase in internal energy of the copper?

From the first law of thermodynamics, the increase in internal energy is found to be

$$\Delta U = Q - W = 1.16 \times 10^4 \text{ J}$$

Note that almost *all* of the heat transferred goes into increasing the internal energy. The fraction of heat energy that is used to do work against the atmosphere is only about 10^{-6}! Hence, in the thermal expansion of a solid or a liquid, the small amount of work done is usually ignored.

Q11. When a sealed thermos bottle full of hot coffee is shaken, what are the changes, if any, in (a) the temperature of the coffee and (b) the internal energy of the coffee?

Q12. Using the first law of thermodynamics, explain why the *total* energy of an isolated system is always conserved.

Q13. Is it possible to convert internal energy to mechanical energy? Explain with examples.

17.9 SUMMARY

Heat flow is a form of energy transfer that takes place as a consequence of a temperature difference only. The *internal energy* of a substance is a function of its temperature and generally increases with increasing temperature.

The *calorie* is the amount of heat necessary to raise the temperature of 1 g of water from 14.5°C to 15.5°C. The *mechanical equivalent of heat* is found from experiment to be 1 cal = 4.186 J.

The *heat capacity, C′*, of any substance is defined as the amount of heat energy needed to raise the temperature of the substance by one Celsius degree. The heat required to change the temperature of a substance by ΔT is

$$Q = C' \Delta T = mc \Delta T \tag{17.4}$$

Heat required to raise the temperature of a substance

where m is the mass of the substance and c is its *specific heat,* or heat capacity per unit mass.

The heat required to change the phase of a pure substance of mass m is given by

$$Q = mL \tag{17.7}$$

Latent heat

The parameter L is called the *latent* (hidden) *heat* of the substance and depends on the nature of the phase change and the properties of the substance.

Heat may be transferred by three fundamentally distinct mechanisms: conduction, convection, and radiation. The *conduction* process can be viewed as an exchange of kinetic energy between colliding molecules. The rate at which heat flows by conduction through a slab of area A is given by

$$\frac{dQ}{dt} = -kA\frac{dT}{dx} \tag{17.8}$$

Law of heat conduction

where k is the *thermal conductivity* and $\frac{dT}{dx}$ is the *temperature gradient.*

Convection is a heat transfer process in which the heated substance moves from one place to another.

All bodies radiate and absorb energy in the form of electromagnetic waves. A body that is hotter than its surroundings radiates more energy than it absorbs, whereas a body that is cooler than its surroundings absorbs more energy than it radiates. An *ideal radiator,* or black body, is one that absorbs all energy incident on it; an ideal radiator is also a good emitter of radiation.

A *quasi-static process* is one that proceeds slowly enough to allow the system to always be in a state of equilibrium.

The *work done* by a gas as its volume changes from V_i to V_f is

$$W = \int_{V_i}^{V_f} P\,dV \tag{17.15}$$

Work done by a gas

where P is the pressure, which may vary during the process. In order to evaluate W, the nature of the process must be specified—that is, P and V must be known during each step of the process. Since the work done depends on the initial, final, and intermediate states, it therefore depends on the path taken between the initial and final states.

From *the first law of thermodynamics* we see that when a system undergoes a change from one state to another, the change in its internal energy, ΔU, is given by

$$\Delta U = Q - W \tag{17.16}$$

First law of thermodynamics

where Q is the heat transferred into (or out of) the system and W is the work done by the system. Although Q and W both depend on the path taken from the initial state to the final state, the quantity ΔU is path-independent.

In a *cyclic process* (one that originates and terminates at the same state), $\Delta U = 0$, and therefore $Q = W$. That is, the heat transferred into the system equals the work done during the cycle.

An *adiabatic process* is one in which no heat is transferred between the system and its surroundings ($Q = 0$). In this case, the first law gives $\Delta U = -W$. That is, the internal energy changes as a consequence of work being done by (or on) the system.

In an *adiabatic free expansion* of a gas, $Q = 0$ and $W = 0$, and so $\Delta U = 0$. That is, the internal energy of the gas does not change in such a process.

An *isobaric process* is one that occurs at constant pressure. The work done in such a process is simply $P\,\Delta V$.

An *isothermal process* is one that occurs at constant temperature. The work done by an ideal gas during an isothermal process is

Work done in an isothermal process

$$W = nRT \ln\left(\frac{V_f}{V_i}\right) \tag{17.20}$$

EXERCISES

Section 17.2 Heat Capacity and Specific Heat

1. A 2-kg iron horseshoe initially at 500°C is dropped into a bucket containing 30 kg of water at 20°C. What is the final equilibrium temperature? (Neglect the heat capacity of the container.)

2. Lead pellets, each of mass 1 g, are heated to 200°C. How many pellets must be added to 500 g of water initially at 20°C to make the final equilibrium temperature 25°C? (Neglect the heat capacity of the container.)

3. How many calories of heat are required to raise the temperature of 5 kg of aluminum from 25°C to 50°C?

4. A 50-g piece of copper is at 25°C. If 300 cal of heat is added to the copper, what is its final temperature?

5. What is the final equilibrium temperature when 20 g of milk at 10°C is added to 150 g of coffee at 90°C? (Assume the heat capacities of the two liquids are the same as that of water, and neglect the heat capacity of the container.)

6. A 250-g chunk of aluminum is heated in a furnace and then dropped into a 500-g copper vessel containing 300 g of water. If the temperature of the water rises from 20°C to 35°C, what was the initial temperature of the aluminum?

7. If 100 g of water is contained in a 300-g aluminum vessel at 20°C and an additional 200 g of water at 100°C is poured into the container, what is the final equilibrium temperature of the system?

8. It takes 3.5×10^3 cal to heat 400 g of an unknown substance from 20°C to 35°C. What is the specific heat of the substance?

9. A 50-g ice cube at 0°C is heated until 45 g has become water at 100°C and 5 g has been converted to steam. How much heat was added to do this?

10. How much heat must be added to 10 g of copper at 20°C to completely melt it?

11. One liter of water at 25°C is used to make iced tea. How much ice at 0°C must be added to lower the temperature of the tea to 10°C? (Ice has a specific heat of .50 cal/g°C.)

12. In an insulated vessel, 300 g of ice at 0°C is added to 550 g of water at 16°C. (a) What is the final temperature of the system? (b) How much ice remains? (Ice has a specific heat of .50 cal/g°C.)

13. A 1-kg block of aluminum, initially at 20°C, is dropped into a large vessel of liquid nitrogen, which is boiling at 77 K. Assuming the vessel is thermally insulated from its surroundings, calculate the number of liters of nitrogen that boils away by the time the aluminum reached 77 K. (Note: Nitrogen has a specific heat of 0.21 cal/g · C°, a heat of vaporization of 48 cal/g, and a density of 0.8 g/cm^3.)

14. If 90 g of molten lead at 327.3°C is poured into a 300-g casting made of iron and initially at 20°C, what is the final temperature of the system? (Assume there are no heat losses.)

Section 17.4 Heat Transfer

15. A glass windowpane has an area of 2 m^2 and a thickness of 0.4 cm. If the temperature difference between its faces is 25 C°, how much heat flows through the window per hour?

16. The rod shown in Fig. 17.2 is made of aluminum and has a length of 50 cm and a cross-sectional area of 2 cm^2. One end is maintained at 80°C, and the other end is at 0°C. At steady state, find (a) the temperature gradient, (b) the rate of heat transfer, and (c) the temperature in the rod 15 cm from the cold end.

17. A Thermopane window 5 m^2 in area is constructed of two layers of glass, each 3 mm thick, separated by an air space of 5 mm. If the inside is at 20°C and the outside is at −30°C, what is the heat loss through the window?

18. A styrofoam container in the shape of a box has a surface area of 0.8 m^2 and a thickness of 2 cm. The inside is at 5°C and the outside is at 25°C. If it takes

8 h for 5 kg of ice to melt in the container, determine the thermal conductivity of the styrofoam.

19. A bar of copper is in thermal contact with a bar of aluminum of the same length and area (Fig. 17.11). One end of the compound bar is maintained at 90°C while the opposite end is at 20°C. When the heat flow reaches steady state, find the temperature at their junction.

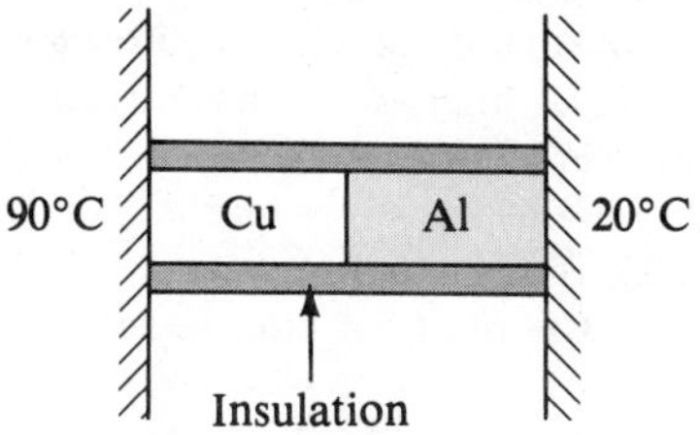

Figure 17.11 (Exercise 19).

20. The tungsten filament of a light bulb has an area of 0.5 cm^2 and is heated to 3000 K. If the emissivity of tungsten is taken to be $e = 0.34$, calculate the power radiated by the bulb.

21. The surface of the sun has a temperature of about 5800 K. Taking the radius of the sun to be 6.96×10^8 m, calculate the total power radiated by the sun. (Assume $e = 1$.)

Section 17.5 The Mechanical Equivalent of Heat

22. Consider Joule's apparatus described in Fig. 17.5. The two masses are 2 kg each, and the tank is filled with 150 kg of water. What is the increase in the temperature of the water after the masses fall through a distance of 1 m?

23. A 1.5-kg copper block is given an initial speed of 3 m/s on a rough, horizontal surface. Because of friction, it finally comes to rest. (a) If 85% of its initial kinetic energy is absorbed by the block in the form of heat, calculate the increase in temperature of the block. (b) What happens to the remaining energy?

24. A 5-g lead bullet traveling with a speed of 275 m/s is stopped by a large tree. If all of its initial kinetic energy is converted to heat in the bullet, find the increase in the temperature of the bullet.

25. A 3-g copper penny at 20°C drops a distance of 30 m to the ground. (a) If 75% of its initial potential energy goes into increasing the internal energy of the penny, determine its final temperature. (b) Does the result depend on the mass of the penny? Explain.

26. Water at the top of Niagara Falls has a temperature of 10°C. If it falls through a distance of 50 m and all of its potential energy goes into heating the water, calculate the temperature of the water at the bottom of the falls.

27. A 75-kg weight-watcher wishes to climb a mountain to work off the equivalent of a large piece of chocolate cake rated at 500 (food) Calories. How high must the person climb?

Section 17.6 Work and Heat in Thermodynamic Processes

28. Using the fact that 1 atm $= 1.013 \times 10^5$ N/m^2, verify the conversion 1 liter · atm = 101.3 J = 24.2 cal.

29. Gas in a container is at a pressure of 2 atm and a volume of 3 m^3. What is the work done by the gas if (a) it expands at constant pressure to twice its initial volume, and (b) it is compressed at constant pressure to one third its initial volume?

30. A gas expands from I to F along three possible paths as indicated in Fig. 17.12. Calculate the work done by the gas along paths IAF, IF, and IBF.

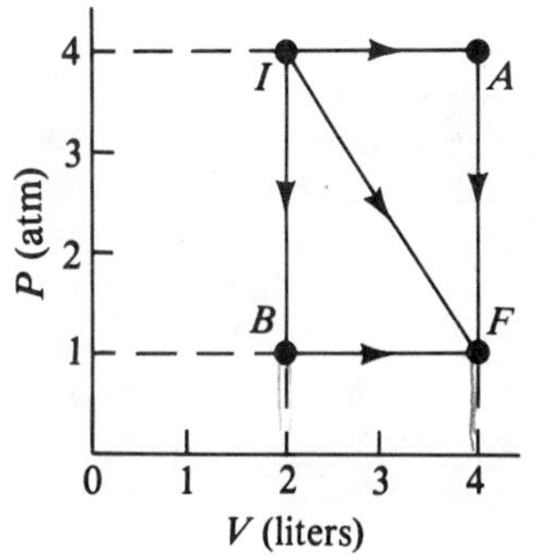

Figure 17.12 (Exercises 30 and 31).

Section 17.7 The First Law of Thermodynamics

31. A gas expands from I to F as in Fig. 17.12. The heat added to the gas is 100 cal when the gas goes from I to F along the diagonal path. (a) What is the change in internal energy of the gas? (b) How much heat must be added to the gas for the indirect path IAF to give the same change in internal energy?

32. A gas is compressed at a constant pressure of 0.3 atm from a volume of 8 liters to a volume of 3 liters. In the process, 400 J of heat energy flows out of the gas. (a) What is the work done by the gas? (b) What is the change in internal energy of the gas?

33. A thermodynamic system undergoes a process in which its internal energy decreases by 300 J. If at the same time, 120 J of work is done on the system, find the heat transferred to or from the system.

34. A gas is taken through the cyclic process described in Fig. 17.13. (a) Find the net heat transferred to the system during one complete cycle. (b) If the cycle is reversed, that is, the process goes along $ACBA$, what is the net heat transferred per cycle?

35. Consider the cyclic process described by Fig. 17.13.

If Q is negative for the process $B \rightarrow C$ and ΔU is negative for the process $C \rightarrow A$, determine the signs of Q, W, and ΔU associated with each process by completing the table in Fig. 17.13.

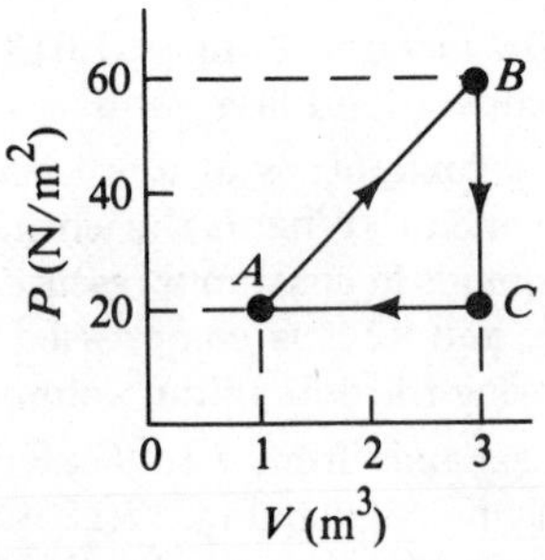

Figure 17.13 (Exercises 34 and 35).

Section 17.8 Some Applications of the First Law of Thermodynamics

36. An ideal gas initially at 300 K undergoes an isobaric expansion at a pressure of 25 N/m^2. If the volume increases from 1 m^3 to 3 m^3 and 80 J of heat is added to the gas, find (a) the change in internal energy of the gas, and (b) its final temperature.

37. Two moles of an ideal gas expands isothermally at 27°C to three times its initial volume. Find (a) the work done by the gas and (b) the heat flow into the system.

38. A 15-g silicon wafer used in a solar cell is heated from 20°C to 150°C at atmospheric pressure. What is the change in its internal energy?

39. One mole of helium gas initially at a temperature of 300 K and pressure of 0.2 atm is compressed isothermally to a pressure of 0.8 atm. Find (a) the final volume of the gas, (b) the work done by the gas, and (c) the heat transferred.

40. One mole of gas initially at a pressure of 2 atm and a volume of 0.3 liters has an internal energy equal to 91 J. In its final state, the pressure is 1.5 atm, the volume is 0.8 liters, and the internal energy equals 182 J. For the three paths *IAF*, *IBF*, and *IF* in Fig. 17.14, calculate (a) the work done by the gas and (b) the net heat transferred in the process.

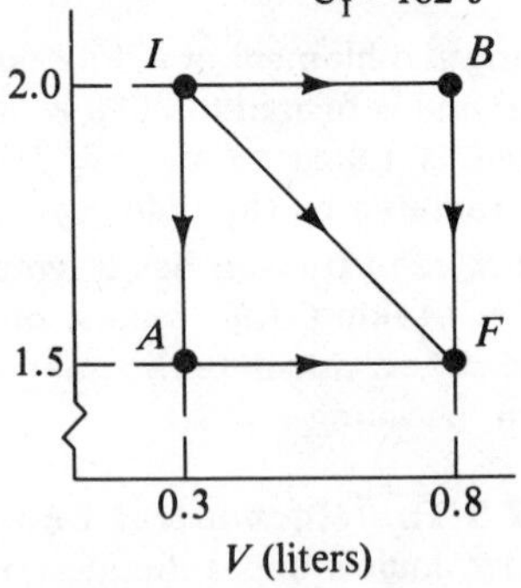

Figure 17.14 (Exercise 40).

PROBLEMS

1. One mole of an ideal gas is contained in a cylinder with a movable piston. The initial pressure, temperature, and volume are P_0, V_0, and T_0, respectively. Find the work done by the gas for the following processes and show the processes in a *PV* diagram: (a) an isobaric compression in which the final volume is one third the initial volume, (b) an isothermal compression in which the final pressure is twice the initial pressure, (c) an isovolumetric process in which the final pressure is twice the initial pressure.

2. An ideal gas initially at pressure P_0, volume V_0, and temperature T_0 is taken through a cycle as described in Fig. 17.15. (a) Find the net work done by the gas per cycle. (b) What is the net heat added to the system per cycle? (c) Obtain a numerical value for the net work done per cycle for one mole of gas initially at 0°C.

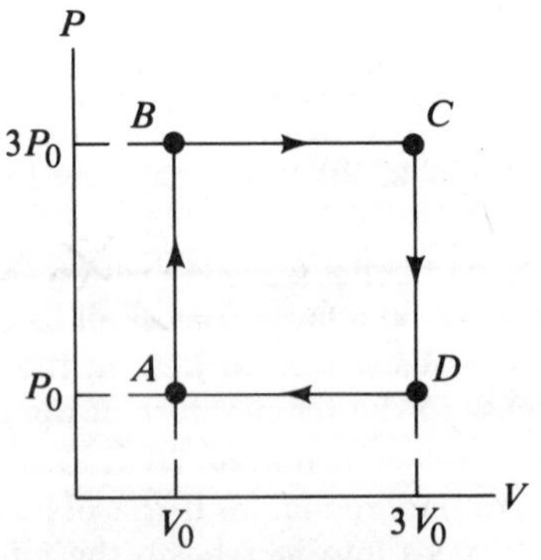

Figure 17.15 (Problem 2).

3. Using the data in Example 17.2 and Table 17.2, calculate the change in internal energy when 1 cm^3 of liquid helium at 4.2 K is converted to helium gas at 273.15 K and atmospheric pressure. (Assume that the molar heat capacity of helium gas is 24.9 J/mole · K, and note that 1 cm^3 of liquid helium is equivalent to 3.1×10^{-2} moles.)

4. A gas expands from a volume of 2 m^3 to a volume of 6 m^3 along two different paths as described in Fig. 17.16. The heat added to the gas along the path *IAF* is equal to 4×10^5 cal. Find (a) the work done by the gas along the path *IAF*, (b) the work done along the path *IF*, (c) the change in internal energy of the gas, and (d) the heat transferred in the process along the path *IF*.

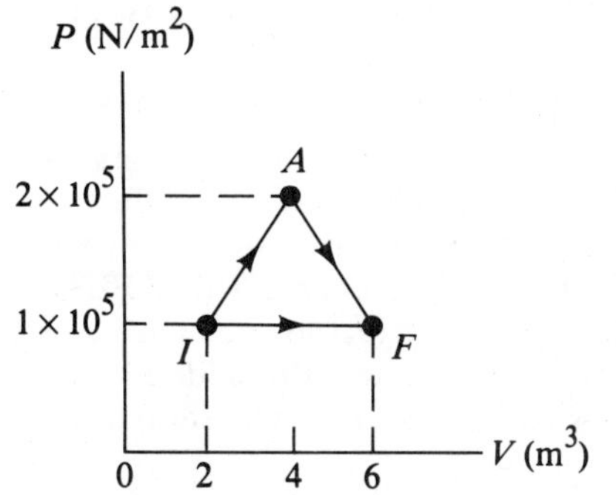

Figure 17.16 (Problem 4).

5. A *flow calorimeter* is an apparatus used to measure the specific heat of a liquid. The technique is to measure the temperature difference between the input and output points of a flowing stream of the liquid while adding heat at a known rate. In one particular experiment, a liquid of density 0.72 g/cm^3 flows through the calorimeter at the rate of 3.5 cm^3/s. At steady state, a temperature difference of 5.8 C° is established between the input and output points when heat is supplied at the rate of 40 J/s. What is the specific heat of the liquid?

6. The inside of a hollow cylinder is maintained at a temperature T_1 while the outside is at a lower temperature, T_2 (Fig. 17.17). The wall of the cylinder has a thermal conductivity k. Neglecting end effects, show that the rate of heat flow from the inner to the outer wall in the radial direction is given by

$$\frac{dQ}{dt} = 2\pi lk \left[\frac{T_1 - T_2}{\ln(b/a)}\right]$$

(Hint: The temperature gradient is given by dT/dr. Note that a radial heat current passes through a concentric cylinder of area $2\pi rl$.)

7. The passenger section of a jet airliner is in the shape of a cylindrical tube of length 30 m and inner radius 2 m. Its walls are lined with a 5-cm thickness of insulating material of thermal conductivity 3×10^{-5} cal/s · cm · C°. The inside is to be maintained at 20°C while the outside is at −40°C. What heating rate is required to maintain this temperature difference? (Use the result from Problem 6.)

8. A thermos bottle in the shape of a cylinder has an inner radius of 4 cm, outer radius of 4.5 cm, and length of 30 cm. The insulating walls have a thermal conductivity equal to 2×10^{-5} cal/s · cm · C°. One liter of hot coffee at 90°C is poured into the bottle. If the outside wall remains at 20°C, how long does it take for the coffee to cool to 50°C? (Neglect end effects and losses by radiation and convection. Use the result from Problem 6 and assume that coffee has the same properties as water.)

9. A "solar cooker" consists of a curved reflecting mirror that focuses sunlight onto the object to be heated (Fig. 17.18). The solar power per unit area reaching the earth at some location is 600 W/m^2, and a small solar cooker has a diameter of 0.5 m. Assuming that 50% of the incident energy is converted into heat energy, how long would it take to evaporate 1 liter of water initially at 20°C? (Neglect the heat capacity of the container.)

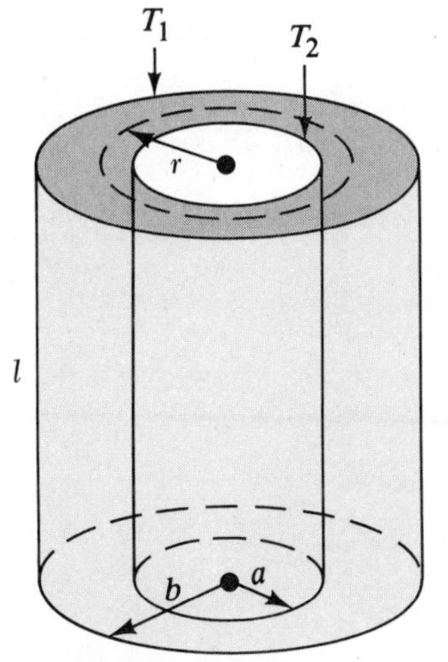

Figure 17.17 (Problem 6).

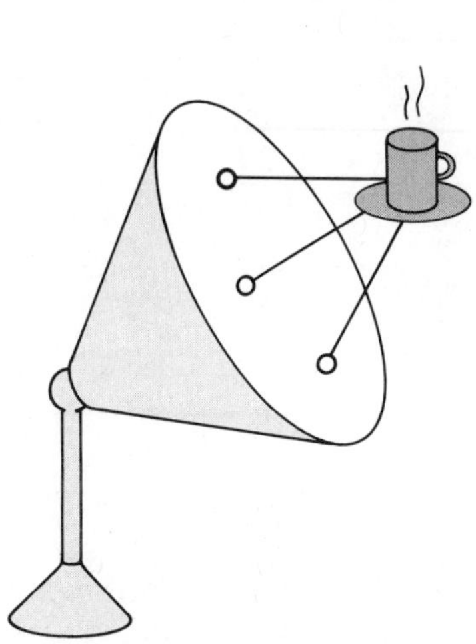

Figure 17.18 (Problem 9).

10. A vessel in the shape of a spherical shell has an inner radius a and outer radius b. The wall has a thermal conductivity k. If the inside is maintained at a temperature T_1 and the outside is at a temperature T_2, show that the rate of heat flow between the surfaces is given by

$$\frac{dQ}{dt} = \frac{4\pi kab}{b - a}\left(T_1 - T_2\right)$$

11. An aluminum rod, 1 m in length and of cross-sectional area 2 cm^2, is inserted vertically into a thermally insulated vessel containing liquid helium at 4.2 K. The rod is initially at 300 K. (a) If one half of the rod is inserted into the helium, how many liters of helium boil off by the time the inserted half cools to 4.2 K? (Assume the upper half does not cool.) (b) If the upper portion of the rod is maintained at 300 K, what is the *approximate* boil-off rate of liquid helium *after* the lower half has reached 4.2 K? (Note that aluminum has a thermal conductivity of 31 J/s · cm · K at 4.2 K, a specific heat of 0.21 cal/g · C°, and a density of 2.7 g/cm^3. See Example 17.2 for data on helium.)

12. What is the minimum heat required to transform 300 g of lead at 20°C to a gas at atmospheric pressure? (Use Tables 17.1 and 17.2.)

13. An aluminum kettle has a circular cross section and is 9 cm in radius and 0.2 cm thick. It is placed on a hotplate and filled with 1 kg of water. If the bottom of the kettle is maintained at 101°C and the inside at 100°C, find (a) the rate of heat flow into the water and (b) the time it takes for all of the water to boil away. (Neglect heat transferred from the sides.)

18 The Kinetic Theory of Gases

Lloyd Black

In the previous chapter we discussed the properties of an ideal gas using such macroscopic variables as pressure, volume, and temperature. We shall now show that such large-scale properties can be explained on a microscopic scale, where matter is treated as a collection of molecules. Newton's laws of motion applied to a collection of particles in a statistical manner provide a reasonable description of thermodynamic processes. In order to keep the mathematics relatively simple, we shall consider only the molecular behavior of gases, where the interactions between molecules are much weaker than in liquids or solids. In the current view of gas behavior, called the *kinetic theory,* gas molecules move about in a random fashion, colliding with the walls of their container and with each other. Perhaps the most important consequence of this theory is that it shows the equivalence between the kinetic energy of molecular motion and the internal energy of the system. Furthermore, the kinetic theory provides us with a physical basis upon which the concept of temperature can be understood.

In the simplest model of a gas, each molecule is considered to be a hard sphere that collides elastically with other molecules or with the container wall. The hard-sphere model assumes that the molecules do not interact with each other except during collisions and that they are not deformed by collisions. This description is adequate only for monatomic gases, where the energy is entirely translational kinetic energy. One must modify the theory for more complex molecules, such as O_2 and CO_2, to include the internal energy associated with rotations and vibrations of the molecules.

18.1 MOLECULAR MODEL FOR THE PRESSURE OF AN IDEAL GAS

We begin this chapter by developing a microscopic model of an ideal gas which shows that the pressure that a gas exerts on the walls of its container is a consequence of the collisions of the gas molecules with the walls. As we shall see, the model is consistent with the macroscopic description of the preceding chapter. The following assumptions will be made:

Assumptions of the molecular model of an ideal gas

1. *The number of molecules is large, and the average separation between them is large* compared with their dimensions. Therefore, the molecules occupy a negligible volume compared with the volume of the container.
2. *The molecules obey Newton's laws of motion, but considered collectively they move in a random fashion.* By random fashion, we mean that the molecules

move in all directions with equal probability and with various speeds. This distribution of velocities does not change in time, despite the collisions between molecules.

3. *The molecules undergo elastic collisions with each other and with the walls of the container.* Thus, the molecules are considered to be structureless (that is, point masses), and in the collisions both kinetic energy and momentum are conserved.
4. *The forces between molecules are negligible except during a collision.* The forces between molecules are short-range, so that the only time the molecules interact with each other is during a collision.
5. *The gas under consideration is a pure gas.* That is, all molecules are identical.

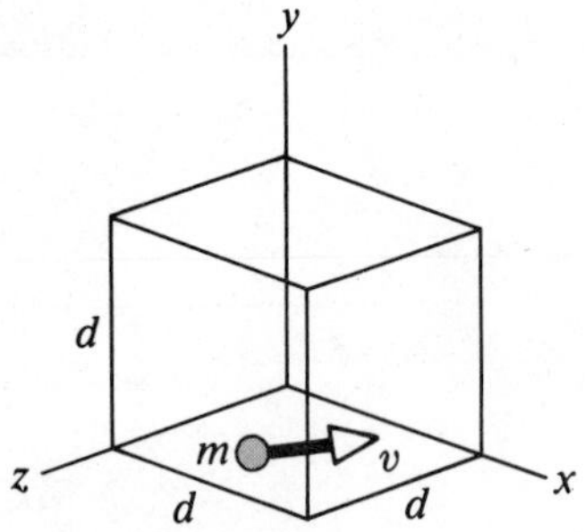

Figure 18.1 A cubical box of sides d containing an ideal gas. The molecule shown moves with velocity v.

Now let us derive an expression for the pressure of an ideal gas consisting of N molecules in a container of volume V. The container is assumed to be in the shape of a cube with edges of length d (Fig. 18.1). Consider the collision of one molecule moving with a velocity v with the right-hand face of the box. The molecule has velocity components v_x, v_y, and v_z. As it collides with the wall elastically, its x component of velocity is reversed, while its y and z components of velocity remained unaltered (Fig. 18.2). Since the x component of momentum of the molecule is mv_x before the collision, and $-mv_x$ afterward, the *change* in momentum of the molecule is given by

$$\Delta p_x = -mv_x - (mv_x) = -2mv_x$$

The momentum delivered to the wall for each collision is $2mv_x$, since the momentum of the system (molecule + container) is conserved. In order that a molecule makes two successive collisions with the same wall, it must travel a distance $2d$ in the x direction in a time Δt. But in a time Δt, the molecule moves a distance $v_x \Delta t$ in the x direction; therefore the time between two successive collisions is $\Delta t = 2d/v_x$. If F is the magnitude of the average force exerted by a molecule on the wall in the time Δt, then from the definition of impulse (which equals change in momentum) we have

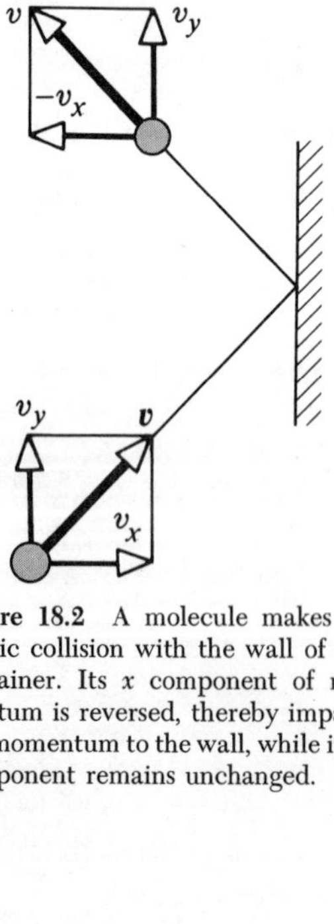

Figure 18.2 A molecule makes an elastic collision with the wall of the container. Its x component of momentum is reversed, thereby imparting momentum to the wall, while its y component remains unchanged.

$$F\Delta t = \Delta p = 2mv_x$$

$$F = \frac{2mv_x}{\Delta t} = \frac{2mv_x}{2d/v_x} = \frac{m{v_x}^2}{d} \qquad (18.1)$$

The total force on the wall is the sum of all such terms for all particles. To get the total pressure on the wall, we divide the total force by the area, d^2:

$$P = \frac{\Sigma F}{A} = \frac{m}{d^3}(v_{x1}^2 + v_{x2}^2 + \cdots)$$

where $v_{x1}, v_{x2}, \ldots$ refer to the x components of velocity for particles 1, 2, etc. Since the average value of v_x^2 is given by

$$\overline{v_x^2} = \frac{v_{x1}^2 + v_{x2}^2 + \cdots}{N}$$

and the volume is given by $V = d^3$, we can express the pressure in the form

$$P = \frac{Nm}{V}\overline{v_x^2} \qquad (18.2)$$

The square of the speed for any one particle is given by

$$v^2 = v_x^2 + v_y^2 + v_z^2$$

Since there is no preferred direction for the molecules, the average values $\overline{v_x^2}$, $\overline{v_y^2}$, and $\overline{v_z^2}$ are equal to each other. Using this fact and the above result, we find that

$$\overline{v_x^2} = \overline{v_y^2} = \overline{v_z^2} = \frac{1}{3}\overline{v^2}$$

Hence, the pressure from Eq. 18.2 can be expressed as

$$P = \frac{1}{3}\frac{Nm}{V}\overline{v^2} \quad (18.3)$$

Pressure and molecular speed

The quantity Nm is the total mass of the molecules, which is equal to nM, where n is the number of moles of the gas and M is its molecular weight in g/mole. Therefore, the pressure can also be expressed in the alternate form

$$P = \frac{1}{3}\frac{nM}{V}\overline{v^2} \quad (18.4)$$

By rearranging Eq. 18.3, we can also express the pressure as

$$P = \frac{2}{3}\frac{N}{V}\left(\frac{1}{2}m\overline{v^2}\right) \quad (18.5)$$

Pressure and molecular kinetic energy

This equation tells us that the pressure is proportional to the number of molecules per unit volume and to the average translational kinetic energy of the molecules.

With this simplified model of an ideal gas, we have arrived at an important result that relates the macroscopic quantities of pressure and volume to a microscopic quantity, average molecular speed. Thus we have a key link between the microscopic world of the gas molecules and the macroscopic world as measured, in this case, with a pressure gauge and meter stick.

In the derivation of this result, note that we have not accounted for collisions between gas molecules. When these collisions are considered, the results do not change since collisions will only affect the momenta of the particles, with no net effect on the walls. This is consistent with one of our initial assumptions, namely, that the distribution of velocities does not change in time. In addition, although our result was derived for a cubical container, it is valid for a container of any shape.

18.2 MOLECULAR INTERPRETATION OF TEMPERATURE

We can obtain some insight into the meaning of temperature by first writing Eq. 18.5 in the more familiar form

$$PV = \frac{2}{3}N\left(\frac{1}{2}m\overline{v^2}\right)$$

Let us now compare this with the empirical equation of state for an ideal gas (Eq. 16.15):

$$PV = NkT$$

Recall that the equation of state is based on experimental facts concerning the macroscopic behavior of gases. Equating the right sides of these expressions, we find that

$$T = \frac{2}{3k}\left(\frac{1}{2}m\overline{v^2}\right) \quad (18.6)$$

Temperature is proportional to average kinetic energy

That is, the absolute temperature of an ideal gas is a measure of the average of the square of the speed of its molecular constituents. Furthermore, since $\frac{1}{2}m\overline{v^2}$ is the average translational kinetic energy per molecule, we see that *temperature is a direct measure of the average molecular kinetic energy.*

By rearranging Eq. 18.6, we can relate the molecular kinetic energy to the temperature:

Average kinetic energy per molecule

$$\frac{1}{2}m\overline{v^2} = \frac{3}{2}kT \tag{18.7}$$

That is, the average translational kinetic energy per molecule is $\frac{3}{2}kT$. Since $\overline{v_x^2} = \frac{1}{3}\overline{v^2}$, it follows that

$$\frac{1}{2}m\overline{v_x^2} = \frac{1}{2}kT \tag{18.8}$$

That is, the average translational kinetic energy per molecule associated with motion in the x direction is $\frac{1}{2}kT$. In a similar manner, for the y and z motions it follows that

Equipartition of energy

$$\frac{1}{2}m\overline{v_y^2} = \frac{1}{2}kT \quad \text{and} \quad \frac{1}{2}m\overline{v_z^2} = \frac{1}{2}kT$$

Thus, each translational degree of freedom contributes an equal amount of energy to the gas, namely, $\frac{1}{2}kT$. A generalization of this result, known as *the theorem of equipartition of energy,* says that *the energy of a system in thermal equilibrium is equally divided among all degrees of freedom.* We shall return to this important point in Section 18.5.

The total translational kinetic energy of N molecules of gas is simply N times the average energy per molecule, which is given by Eq. 18.7:

Total kinetic energy of N molecules

$$E = N\left(\frac{1}{2}m\overline{v^2}\right) = \frac{3}{2}NkT = \frac{3}{2}nRT \tag{18.9}$$

where we have used $k = R/N_0$ for Boltzmann's constant and $n = N/N_0$ for the number of moles of gas.

The square root of $\overline{v^2}$ is called the *root mean square* (rms) *speed* of the molecules. From Eq. 18.7 we get for the rms speed

Root mean square speed

$$v_{\text{rms}} = \sqrt{\overline{v^2}} = \sqrt{\frac{3kT}{m}} = \sqrt{\frac{3RT}{M}} \tag{18.10}$$

The expression for the rms speed shows that at a given temperature, lighter molecules move faster, on the average, than heavier molecules. For example, hydrogen, with a molecular weight of 2 g/mole, moves four times as fast as oxygen, whose molecular weight is 32 g/mole. Note that the rms speed is not the speed at which a gas molecule will move across a room, since it undergoes several billion collisions per second with other molecules under standard conditions. We shall describe this in more detail in Section 18.6.

Table 18.1 lists the rms speeds for various molecules at 20°C.

Example 18.1

A tank of volume 0.3 m^3 contains 2 moles of helium gas at 20°C. Assuming the helium behaves like an ideal gas, (a) find the total internal energy of the system.

Using Eq. 18.9 with $n = 2$ and $T = 293$ K, we get

$$E = \frac{3}{2}nRT = \frac{3}{2}(2\text{ moles})(8.31\text{ J/mole}\cdot\text{K})(293\text{ K})$$

$$= 7.30 \times 10^3\text{ J}$$

(b) What is the average kinetic energy per atom?

From Eq. 18.7, we see that the average kinetic energy

per atom is equal to

$$\frac{1}{2}m\overline{v^2} = \frac{3}{2}kT = \frac{3}{2}(1.38 \times 10^{-23}\ \text{J/K})(293\ \text{K})$$

$$= 6.07 \times 10^{-21}\ \text{J}$$

(c) Determine the rms speed of the atoms.

Since the molecular weight of helium is 4 g/mole $= 4 \times 10^{-3}$ kg/mole, the rms speed is given by

$$v_{\text{rms}} = \sqrt{\frac{3RT}{M}} = \sqrt{\frac{3(8.317\ \text{J/mole}\cdot\text{K})(293\ \text{K})}{4 \times 10^{-3}\ \text{kg/mole}}}$$

$$= 1.35 \times 10^3\ \frac{\text{m}}{\text{s}}$$

The same result is obtained using $v_{\text{rms}} = \sqrt{3kT/m}$. Try it!

TABLE 18.1 Some rms Speeds

GAS	*MOLECULAR WEIGHT* (g/mole)	v_{rms} at 20°C (m/s)*
H_2	2.02	1902
He	4.0	1352
H_2O	18	637
Ne	20.1	603
N_2 and CO	28	511
NO	30	494
CO_2	44	408
SO_2	48	390

*All values calculated using Eq. 17.10.

Q1. Dalton's law of partial pressures, which you have probably learned in chemistry, states: *The total pressure of a mixture of gases is equal to the sum of the partial pressures of gases making up the mixture.* Give a convincing argument of this law based on the kinetic theory of gases.

Q2. One container is filled with helium gas and another with argon gas. If both containers are at the same temperature, which molecules have the higher rms speed?

Q3. If you wished to manufacture an after-shave lotion with a scent that is less "likely to get there before you do," would you use a high- or low-molecular-weight lotion?

18.3 HEAT CAPACITY OF AN IDEAL GAS

We have found that the temperature of a gas is a measure of the average translational kinetic energy of the gas molecules. It is important to note that this kinetic energy is associated with the motion of the center of mass of the molecules. It does not include the energy associated with the internal motion of the molecule, namely, vibrations and rotations about the center of mass. This should not be surprising, since the simple kinetic theory model assumes a structureless molecule.

In view of this, let us first consider the simplest case of an ideal monatomic gas, that is, a gas containing one atom per molecule, such as helium, neon, and argon. Essentially, all of the kinetic energy of such molecules is associated with the motion of their centers of mass. When energy is added to a monatomic gas in a container of fixed volume (by heating, say) all of the added energy goes into increasing the translational kinetic energy of the molecules.[1] There is no other way to store the energy in a monatomic gas. Therefore, from Eq. 18.9 we see that the total internal energy U of N molecules (or n moles) of an ideal monatomic gas is given by

$$U = \frac{3}{2}NkT = \frac{3}{2}nRT \qquad (18.11)$$

Internal energy of an ideal monatomic gas

[1] If the gas is raised to sufficiently high temperatures, the atom can also be excited or even ionized.

If heat is transferred to the system at *constant volume,* the work done by the system is zero. That is, since $\Delta V = 0$, $W = \int P\,dV = 0$. Hence, from the first law of thermodynamics we see that

First law for a constant-volume process

$$Q = \Delta U = \frac{3}{2}nR\,\Delta T \tag{18.12}$$

In other words, all of the heat transferred goes into increasing the internal energy (and temperature) of the system. The constant-volume process from I to F is described in Fig. 18.3, where ΔT is the temperature difference between the two isotherms. Substituting the value for Q given by Eq. 17.5 into Eq. 18.12, we get

$$nC_v\,\Delta T = \frac{3}{2}nR\,\Delta T$$

$$C_v = \frac{3}{2}R \tag{18.13}$$

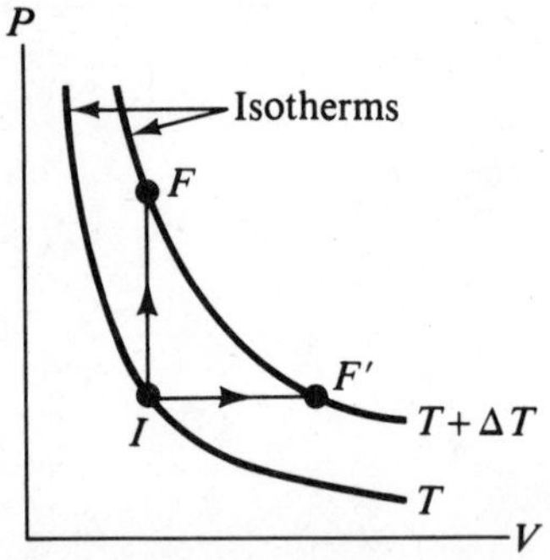

Figure 18.3 Heat is added to an ideal gas in two ways. For the constant-volume path *IF*, the heat added goes into increasing the internal energy of the gas since no work is done. Along the constant-pressure path *IF'*, part of the heat added goes into work done by the gas. Note that the internal energy is constant along any isotherm.

In this notation, C_v is the molar heat capacity of the gas at constant volume. Note that this expression predicts a value of $\frac{3}{2}$R = 2.98 cal/mole · K for all monatomic gases. This is in excellent agreement with measured values of molar heat capacities for such gases as helium and argon over a wide range of temperatures (Table 18.2).

In the limit of differential changes, we can use Eq. 18.12 and the first law of thermodynamics to express the molar heat capacity in the form

$$C_v = \frac{1}{n}\frac{dU}{dT} \tag{18.14}$$

For an ideal monatomic gas, where $U = \frac{3}{2}nRT$, Eq. 18.14 gives $C_v = \frac{3}{2}R$ in agreement with Eq. 18.13.

Now suppose that the gas is taken along the constant-pressure path $I \rightarrow F'$ in Fig. 18.3. Along this path, the temperature again increases by ΔT. The heat that must be transferred to the gas in this process is given by $Q = nC_p\,\Delta T$, where C_p is the molar heat capacity at constant pressure. Since the volume increases in this process, we see that the work done by the gas is $W = P\,\Delta V$. Applying the first law to this process gives

$$\Delta U = Q - W = nC_p\,\Delta T - P\,\Delta V \tag{18.15}$$

TABLE 18.2 Molar Heat Capacities of Various Gases

	MOLAR HEAT CAPACITY (cal/mole · K)			
	C_p	C_v	$C_p - C_v$	$\gamma = C_p/C_v$
		Monatomic Gases		
He	4.97	2.98	1.99	1.67
A	4.97	2.98	1.99	1.67
		Diatomic Gases		
H_2	6.87	4.88	1.99	1.41
N_2	6.95	4.96	1.99	1.40
O_2	7.03	5.04	1.99	1.40
CO	7.01	5.02	1.99	1.40
Cl_2	8.29	6.15	2.14	1.35
		Polyatomic Gases		
CO_2	8.83	6.80	2.03	1.30
SO_2	9.65	7.50	2.15	1.29
H_2O	8.46	6.46	2.00	1.30
CH_4	8.49	6.48	2.01	1.31

Note: All values obtained at 300 K.

In this case, the heat added to the gas is transferred in two forms. Part of it is used to do external work by moving a piston, and the remainder increases the internal energy of the gas. But the change in internal energy for the process $I \rightarrow F'$ is equal to the change for the process $I \rightarrow F$, since U depends only on temperature for an ideal gas and ΔT is the same for each process. In addition, since $PV = nRT$, we note that for a constant-pressure process $P\,\Delta V = nR\,\Delta T$. Substituting this into Eq. 18.15 with $\Delta U = nC_v\,\Delta T$ gives

$$nC_v\,\Delta T = nC_p\,\Delta T - nR\,\Delta T$$

or

$$C_p - C_v = R \tag{18.16}$$

This expression applies to *any* ideal gas. It shows that the molar heat capacity of an ideal gas at constant pressure is greater than the molar heat capacity at constant volume by an amount R, the universal gas constant (1.99 cal/mole · K). This is in good agreement with real gases under standard conditions (Table 18.2).

Since $C_v = \frac{3}{2}R$ for a monatomic ideal gas, Eq. 18.16 predicts a value $C_p = \frac{5}{2}R =$ 4.98 cal/mole · K for the molar heat capacity of a monatomic gas at constant pressure. The ratio of these heat capacities is a dimensionless quantity γ given by

Ratio of heat capacities for an ideal gas.

$$\gamma = \frac{C_p}{C_v} = \frac{\frac{5}{2}R}{\frac{3}{2}R} = \frac{5}{3} = 1.67 \tag{18.17}$$

The values of C_p and γ are in excellent agreement with experimental values for monatomic gases, but in serious disagreement with the values for the more complex gases (Table 18.2). This is not surprising since the value $C_v = \frac{3}{2}R$ was derived for a monatomic ideal gas, and we expect some additional contribution to the specific heat from the internal structure of the more complex molecules. In Section 18.5, we describe the effect of molecular structure on the specific heat of a gas. We shall find that the internal energy and hence the specific heat of a complex gas must include contributions from the rotational and vibrational motions of the molecule.

We have seen that the heat capacities of gases at constant pressure are greater than the heat capacities at constant volume. This difference is a consequence of the fact that in a constant-volume process, no work is done and all of the heat goes into increasing the internal energy (and temperature) of the gas, whereas in a constant-pressure process some of the heat energy is transformed into work done by the gas. In the case of solids and liquids heated at constant pressure, very little work is done since the thermal expression is small (Example 18.7). Consequently C_p and C_v are approximately equal for solids and liquids.

Example 18.2

A cylinder contains 3 moles of helium gas at a temperature of 300 K. (a) How much heat must be transferred to the gas to increase its temperature to 500 K if the gas is heated at constant volume?

For the constant-volume process, the work done is zero. Therefore from Eq. 18.12, we get

$$Q_1 = \frac{3}{2}nR\,\Delta T = nC_v\,\Delta T$$

But $C_v = 2.98$ cal/mole · K for He and $\Delta T = 200$ K; therefore

$$Q_1 = (3\text{ moles})(2.98\text{ cal/mole}\cdot\text{K})(200\text{ K}) = 1.79 \times 10^3\text{ cal}$$

(b) How much heat must be transferred to the gas at constant pressure to raise the temperature to 500 K?

Making use of Table 18.2, we get

$$Q_2 = nC_p\,\Delta T = (3\text{ moles})(4.97\text{ cal/mole}\cdot\text{K})(200\text{ K}) = 2.98 \times 10^3\text{ cal}$$

Note that the difference $Q_2 - Q_1 = 1.19 \times 10^3$ cal is the work done by the gas in this process. Why?

18.4 ADIABATIC PROCESS FOR AN IDEAL GAS

Definition of an adiabatic process

An adiabatic process is one in which there is no heat transfer between the system and its surroundings. In reality, true adiabatic processes cannot occur since a perfect heat insulator between a system and its surroundings does not exist. However, there are processes that are nearly adiabatic. For example, if a gas is compressed (or expands) very rapidly, very little heat flows into (or out of) the system, and so the process is nearly adiabatic. Such processes occur in the cycle of a gasoline engine, which we shall discuss in detail in the next chapter.

It is also possible for a process to be both quasi-static and adiabatic. For example, if a gas that is thermally insulated from its surroundings is allowed to expand slowly against a piston, the process is a quasi-static, adiabatic expansion. In general, *a quasi-static, adiabatic process is one that is slow enough to allow the system to always be near equilibrium, but fast compared with the time it takes the system to exchange heat with its surroundings.*

Suppose that an ideal gas undergoes a *quasi-static, adiabatic* expansion. *At any time during the process, we assume that the gas is in an equilibrium state, so that the equation of state,* PV = nRT, *is valid.* In addition, we shall show that the pressure and volume at any time during the adiabatic process are related by the expression

Relation between P and V for an adiabatic process involving an ideal gas

$$PV^{\gamma} = \text{constant} \tag{18.18}$$

where $\gamma = C_p/C_v$ is assumed to be constant during the process. Thus, we see that all the thermodynamic variables, P, V, and T, change during an adiabatic process.

When a gas expands adiabatically in a thermally insulated cylinder, there is no heat transferred between the gas and its surroundings, and so $Q = 0$. Let us take the change in volume to be ΔV and the change in temperature to be ΔT. The work done by the gas is $W = P\,\Delta V$. Since the internal energy of an ideal gas depends only on temperature, the change in internal energy is $\Delta U = nC_v\,\Delta T$. Hence, the first law of thermodynamics gives

$$\Delta U = nC_v\,\Delta T = -P\,\Delta V$$

From the equation of state of an ideal gas, $PV = nRT$, we see that

$$P\,\Delta V + V\,\Delta P = nR\,\Delta T$$

Eliminating ΔT from these two equations we find that

$$P\,\Delta V + V\,\Delta P = -\frac{R}{C_v}P\,\Delta V$$

Substituting $R = C_p - C_v$ and dividing by PV, we get

$$\frac{\Delta V}{V} + \frac{\Delta P}{P} = -\left(\frac{C_p - C_v}{C_v}\right)\frac{\Delta V}{V} = (1-\gamma)\frac{\Delta V}{V}$$

$$\frac{\Delta P}{P} + \gamma\frac{\Delta V}{V} = 0$$

Taking the limits of differential changes ($\Delta P \rightarrow dP$ and $\Delta V \rightarrow dV$) and integrating, we get

$$\ln P + \gamma \ln V = \ln(\text{constant})$$

which is equivalent to Eq. 18.18:

Adiabatic process

$$PV^{\gamma} = \text{constant}$$

The *PV* diagram for an adiabatic expansion is shown in Fig. 18.4. Note that since $\gamma > 1$, the *PV* curve for the adiabatic expansion is steeper than that for an isothermal expansion. This tells us that the gas cools ($T_f < T_i$) during an adiabatic expansion. Conversely, the temperature increases if the gas is compressed adiabatically. Applying Eq. 18.18 to the initial and final states, we see that

$$P_iV_i^{\gamma} = P_fV_f^{\gamma} \tag{18.19}$$

Using $PV = nRT$, it is left as an exercise to show that Eq. 18.19 can also be expressed as

$$T_iV_i^{\gamma-1} = T_fV_f^{\gamma-1} \tag{18.20}$$

Note that the above analysis is valid only in processes that are slow enough to allow the system to always remain near equilibrium, but fast enough to prevent the system from exchanging heat with its surroundings.

Example 18.3

Air in the cylinder of a diesel engine at 20°C is compressed from an initial pressure of 1 atm and volume of 200 cm³ to a volume of 15 cm³. Assuming that air behaves as an ideal gas ($\gamma = 1.40$) and that the compression is adiabatic, find the final pressure and temperature.

Solution: Using Eq. 18.19, we find that

$$P_f = P_i(V_i/V_f)^{\gamma} = 1 \text{ atm } (200 \text{ cm}^3/15 \text{ cm}^3)^{1.4}$$
$$= 37.6 \text{ atm}$$

Since $PV = nRT$ is always valid during the process and since no gas escapes from the cylinder,

$$\frac{P_iV_i}{T_i} = \frac{P_fV_f}{T_f}$$

$$T_f = \frac{P_fV_f}{P_iV_i}T_i = \frac{(37.6 \text{ atm})(15 \text{ cm}^3)}{(1 \text{ atm})(200 \text{ cm}^3)}(293 \text{ K})$$
$$= 826 \text{ K} = 553°\text{C}$$

18.5 THE EQUIPARTITION OF ENERGY

We have found that model predictions based on specific heat agree quite well with the behavior of monatomic gases, but not with the behavior of complex gases (Table 18.2). Furthermore, the predicted value of the quantity $C_p - C_v = R$ is the same for all gases. This is not surprising, since this difference is the result of the work done by the gas, which is independent of its molecular structure.

In order to explain the variations in C_v and C_p in going from monatomic gases to the more complex gases, let us explain the origin of the specific heat. So far, we have assumed that the sole contribution to the internal energy of a gas is the translational kinetic energy of the molecules. However, the internal energy of a gas actually includes contributions from the translational, vibrational, and rotational motion of the molecules. The rotational and vibrational motions of molecules with structure can be activated by collisions and therefore are "coupled" to the translational motion of the molecules. The branch of physics known as *statistical mechanics* has shown that for a large number of particles obeying newtonian mechanics, the available energy is, on the average, shared equally by each independent mode of energy. Recall that the *equipartition theorem* states that at equilibrium each degree of freedom contributes, on the average, $\frac{1}{2}kT$ of energy per molecule.

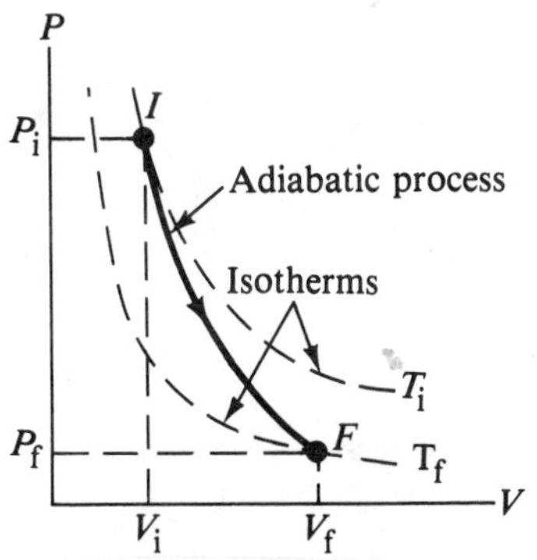

Figure 18.4 The *PV* diagram for an adiabatic expansion. Note that $T_f < T_i$ in this process.

Let us consider a diatomic gas, which we can visualize as a dumbbell-shaped molecule (Fig. 18.5). In this model, the center of mass of the molecule can translate in the *x*, *y*, and *z* directions (Fig. 18.5a). In addition, the molecule can rotate about three mutually perpendicular axes (Fig. 18.5b). We can neglect the rotation about the *y* axis since the moment of inertia and the rotational energy, $\frac{1}{2}I\omega^2$, about this axis are negligible compared with those associated with the *x* and *z* axes. If the two atoms of the molecule are taken to be point masses, then I_y is identically zero. Thus there are five degrees of freedom: three associated with the translational motion and two associated with the rotational motion. Since *each degree of freedom contributes, on the average,* $\frac{1}{2}$kT *of energy per molecule,* the total energy for N molecules is

$$U = 3N\left(\frac{1}{2}kT\right) + 2N\left(\frac{1}{2}kT\right) = \frac{5}{2}NkT = \frac{5}{2}nRT$$

We can use this result and Eq. 18.14 to get the molar heat capacity at constant volume:

$$C_v = \frac{1}{n}\frac{dU}{dT} = \frac{1}{n}\frac{d}{dT}\left(\frac{5}{2}nRT\right) = \frac{5}{2}R$$

From Eqs. 18.16 and 18.17 we find that

$$C_p = C_v + R = \frac{7}{2}R$$

$$\gamma = \frac{C_p}{C_v} = \frac{\frac{7}{2}R}{\frac{5}{2}R} = \frac{7}{5} = 1.40$$

These results agree quite well with most of the data given in Table 18.2 for diatomic molecules. This is rather surprising since we have not yet accounted for the possible vibrations of the molecule. In the vibratory model, the two atoms are joined by an imaginary spring. The vibratory motion adds two more degrees of freedom, corresponding to the kinetic and potential energies associated with vibrations along the length of the molecule. Hence, the equipartition theorem predicts an internal energy of $\frac{7}{2}nRT$ and a higher heat capacity than what is observed. Examination of the experimental data (Table 18.2) suggests that some diatomic molecules, such as H_2 and N_2, do not vibrate at room temperature, and others, such as Cl_2, do. For molecules with more than two atoms, the number of degrees of freedom is even larger and the vibrations are more complex. This results in an even higher predicted heat capacity, which is in qualitative agreement with experiment.

We have seen that the equipartition theorem is successful in explaining some features of the heat capacity of molecules with structure. However, the equipartition theorem does not explain the observed temperature variation in heat capacities. As an example of such a temperature variation, C_v for the hydrogen molecule is $\frac{5}{2}$ R from about 250 K to 750 K and then increases steadily to about $\frac{7}{2}$ R well above 750 K (Fig. 18.6). This suggests that vibrations occur at very high temperatures. At

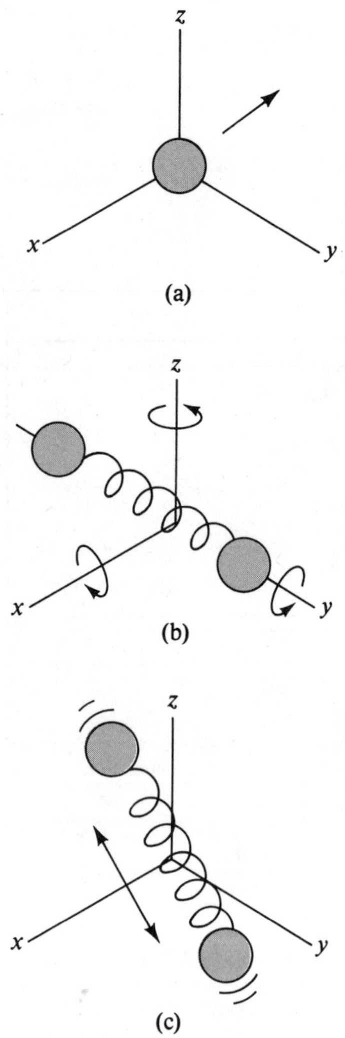

Figure 18.5 Possible motions of a diatomic molecule: (a) translational motion of the center of mass, (b) rotational motion about the various axes, and (c) vibrational motion along the molecular axis.

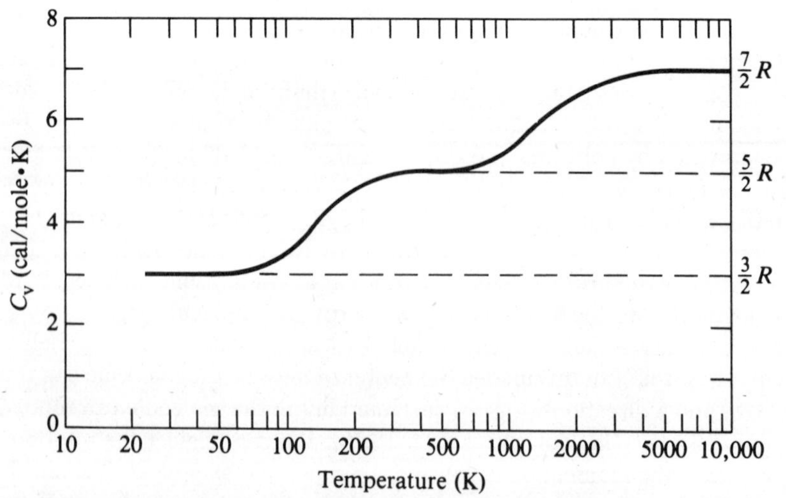

Figure 18.6 The molar heat capacity, C_v, of hydrogen as a function of temperature. The horizontal scale is logarithmic. Note that hydrogen liquefies at 20 K.

temperatures well below 259 K, C_v has a value of about $\frac{3}{2}R$, suggesting that the molecule has only translational energy at low temperatures.

The failure of the equipartition theorem to explain such phenomena is due to the inadequacy of classical mechanics when applied to molecular systems. For a more satisfactory description, it is necessary to use a quantum-mechanical model in which the energy of an individual molecule is quantized. The magnitude of the energy separation between the various vibrational energy levels for a molecule such as H_2 is about ten times as great as the kinetic energy of the molecule, kT, at room temperature. Consequently, collisions between molecules at low temperatures do not provide enough energy to change the vibrational state of the molecule. It is often stated that such degrees of freedom are "frozen out." This explains why the vibrational energy does not contribute to the heat capacities of molecules at low temperatures (below about 1000 K for H_2).

The rotational energy levels are also quantized, but their spacing at ordinary temperatures is small compared with kT. Since the spacing between rotational levels is so small compared with kT, the system behaves classically. However, at sufficiently low temperatures (typically less than 50 K), where kT is small compared with the spacing between rotational levels, intermolecular collisions may not be energetic enough to alter the rotational states. This explains why C_v reduces to $\frac{3}{2}R$ for H_2 in the range from 20 K to about 100 K.

Heat Capacities of Solids

Measurements of heat capacities of solids also show a marked temperature dependence. The heat capacities of solids generally decrease in a nonlinear manner with decreasing temperature and approach zero as the absolute temperature approaches zero. At high temperatures (usually above 500 K), the heat capacities of solids approach the value of about $3R \approx 6$ cal/mole · K, a result known as the *DuLong-Petit law*. The typical data shown in Fig. 18.7 demonstrates the temperature dependence of the heat capacity for two semiconducting solids, silicon and germanium.

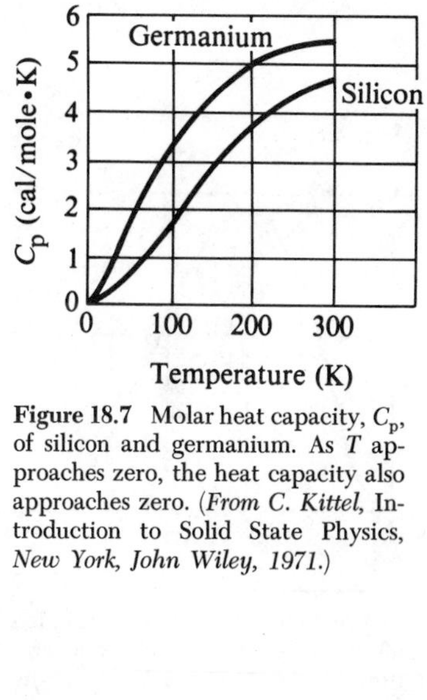

Figure 18.7 Molar heat capacity, C_p, of silicon and germanium. As T approaches zero, the heat capacity also approaches zero. (*From C. Kittel,* Introduction to Solid State Physics, *New York, John Wiley, 1971.*)

The heat capacity of a solid at high temperatures can be explained using the equipartition theorem. For small displacements of an atom from its equilibrium position, each atom executes simple harmonic motion in the x, y, and z directions. The energy associated with vibrational motion in the x direction is

$$E_x = \frac{1}{2}mv_x^2 + \frac{1}{2}kx^2$$

There are analogous expressions for E_y and E_z. Therefore, each atom of the solid has six degrees of freedom. According to the equipartition theorem, this corresponds to an average vibrational energy of $6(\frac{1}{2}kT) = 3\,kT$ per atom. Therefore, the total internal energy of a solid consisting of N atoms is given by

$$U = 3NkT = 3nRT$$

Total internal energy of a solid

From this result, we find that the molar heat capacity $C_v = \dfrac{1}{n}\dfrac{dU}{dT} = 3R$, which agrees with the empirical law of DuLong and Petit. The discrepancies between this model and the experimental data at low temperatures are again due to the inadequacy of classical physics in the microscopic world. One can attribute the decrease in heat capacity with decreasing temperature to a "freezing out" of various vibrational excitations.

18.6 DISTRIBUTION OF MOLECULAR SPEEDS

Thus far we have not concerned ourselves with the fact that not all molecules in a gas have the same speed and energy. Their motion is extremely chaotic. Any individual molecule is colliding with others at the enormous rate of typically a billion times per second. Each collision results in a change in the speed and direction of motion of each of the participant molecules. From Eq. 18.10, we see that average molecular speeds increase with increasing temperature. What we would like to know now is the distribution of molecular speeds. For example, how many molecules of a gas have a speed in the range of, say, 400 to 410 m/s? Intuitively, we expect that the speed distribution depends on temperature. Furthermore, we expect that the distribution peaks in the vicinity of v_{rms}. That is, few molecules are expected to have speeds much less than or much greater than v_{rms}, since these extreme speeds will result only from an unlikely chain of collisions.

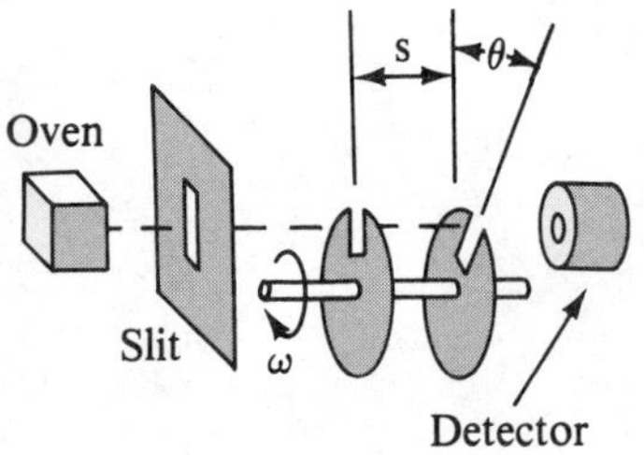

Figure 18.8 A schematic diagram of one apparatus used to measure the speed distribution of gas molecules.

The development of a reliable theory for the speed distribution of a large number of particles appears, at first, to be an almost impossible task. However, in 1860 James Clerk Maxwell (1831–1879) derived an expression that describes the distribution of molecular speeds in a very definite manner. His work, and developments by other scientists shortly thereafter, were highly controversial, since experiments at that time were not capable of directly observing molecules. However, about 60 years later experiments were devised which confirmed Maxwell's predictions.

One experimental arrangement for observing the speed distribution of molecules is illustrated in Fig. 18.8. A substance is vaporized in an oven and forms gas molecules, which are permitted to escape through a hole. The molecules enter an evacuated region and pass through a series of slits to form a collimated beam. The beam is incident on two slotted rotating disks separated by a distance s and displaced from each other by an angle θ. A molecule passing through the first slotted disk will pass through the second slotted disk only if its speed is $v = s\omega/\theta$, where ω is the angular velocity of the disks. Molecules with other speeds will necessarily collide with the second disk and hence will not reach the detector. By varying ω and θ, one can measure the number of molecules in a given range of speeds.

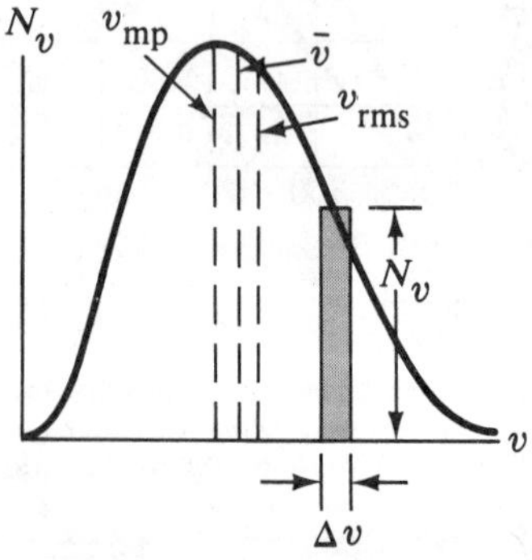

Figure 18.9 The speed distribution of gas molecules at some temperature. The number of molecules in the range Δv is equal to the area of the shaded rectangle, $N_v\,\Delta v$. The function N_v approaches zero as v approaches infinity.

The observed speed distribution of gas molecules in thermal equilibrium is shown in Fig. 18.9. The quantity N_v (which is called the *distribution function*) represents the number of molecules per unit interval of speed.

The number of molecules having a speed in the range from v to $v + \Delta v$ is equal to $N_v\,\Delta v$, represented in Fig. 18.9 by the area of the shaded rectangle. If N is the total number of molecules, then the fraction of molecules with speeds between v and $v + \Delta v$ is equal to $N_v\,\Delta v/N$. This fraction is also equal to the probability that any given molecule has a speed in the range from v to $v + \Delta v$.

The total number of molecules numerically equals the total area under the speed distribution curve. Since the abscissa ranges from $v = 0$ to $v = \infty$ (classically, all molecular speeds are possible), we can express the total number of particles as the sum of the areas of all shaded rectangles. In the limit $\Delta v \rightarrow 0$, this sum is replaced by an integral:

$$N = \lim_{\Delta v \to 0} \left(\sum_{v=0}^{\infty} N_v\,\Delta v \right) = \int_0^{\infty} N_v\,dv \qquad (18.21)$$

The fundamental expression (derived by Maxwell) that describes the most probable distribution of speeds of N gas molecules is given by

Maxwell speed distribution function

$$N_v = 4\pi N \left(\frac{m}{2\pi kT} \right)^{3/2} v^2\, e^{-mv^2/2kT} \qquad (18.22)$$

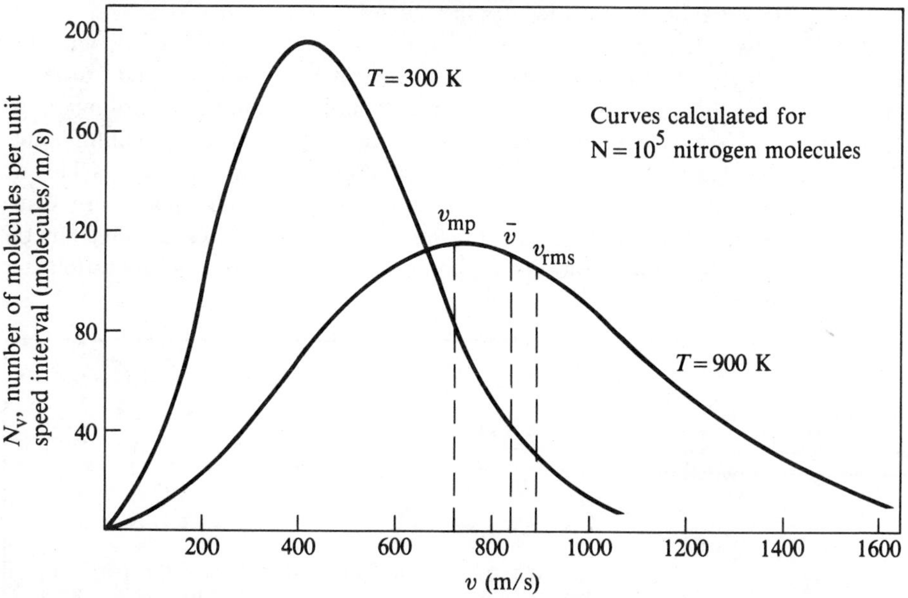

Figure 18.10 The Maxwell speed distribution function for 10^5 nitrogen molecules at temperatures of 300 K and 900 K. The total area under either curve is equal to the total number of molecules, which, in this case, equals 10^5. Note that $v_{rms} > \bar{v} > v_{mp}$.

where m is mass of a gas molecule, k is Boltzmann's constant, and T is the absolute temperature.[2] The function given by Eq. 18.22 satisfies Eq. 18.21. Furthermore, N_v approaches zero in the low- and high-speed limits, as expected. We also note that the speed distribution for a given gas depends only on temperature.

As indicated in Fig. 18.9, the average speed, $\bar{v}$, is somewhat lower than the rms speed. The most probable speed, v_{mp}, is the speed at which the distribution curve reaches a peak. Using Eq. 18.22, one finds that

$$v_{rms} = \sqrt{\overline{v^2}} = \sqrt{3kT/m} = 1.73\sqrt{kT/m} \tag{18.23}$$

rms speed

$$\bar{v} = \sqrt{8kT/\pi m} = 1.60\sqrt{kT/m} \tag{18.24}$$

Average speed

$$v_{mp} = \sqrt{2kT/m} = 1.41\sqrt{kT/m} \tag{18.25}$$

Most probable speed

The details of these calculations are left for the student, but from these equations we see that $v_{rms} > \bar{v} > v_{mp}$.

Figure 18.10 represents specific speed distribution curves for nitrogen molecules. The curves were obtained by using Eq. 18.22 to evaluate the distribution function, N_v, at various speeds and at two temperatures (300 K and 900 K). Note that the curve shifts to the right as T increases, indicating that the average speed increases with increasing temperature, as expected. The asymmetric shape of the curves is due to the fact that the lowest speed possible is zero while the upper classical limit of the speed is infinity. Furthermore, as temperature increases the distribution curve broadens and the range of speeds also increases.

Equation 18.22 shows that the distribution of molecular speeds in a gas depends on mass as well as temperature. At a given temperature, the fraction of particles with speeds exceeding a fixed value increases as the mass decreases. This explains why lighter molecules, such as hydrogen and helium, escape more readily from the earth's atmosphere than heavier molecules, such as nitrogen and oxygen. (See the discussion of escape velocity in Chapter 14. Notice that gas molecules escape even more readily from the moon's surface because its escape velocity is lower.)

---

[2] For the derivation of this expression, see any text on thermodynamics, such as M. W. Zemansky, *Heat and Thermodynamics*, New York, McGraw-Hill, 1968.

The evaporation process

The speed distribution of molecules in a liquid is similar to that shown in Fig. 18.10. The phenomenon of evaporation of a liquid can be understood from this distribution in speeds using the fact that some molecules in the liquid are more energetic than others. Some of the faster-moving molecules in the liquid penetrate the surface and leave the liquid even at temperatures well below the boiling point. The molecules that escape the liquid by evaporation are those that have sufficient energy to overcome the attractive forces of the molecules in the liquid phase. Consequently, the molecules left behind in the liquid phase have a lower average kinetic energy, causing the temperature of the liquid to decrease. Hence evaporation is a cooling process.

Example 18.4

Nine particles have speeds of 5, 8, 12, 12, 12, 14, 14, 17, and 20 m/s. (a) Find the average speed.

The average speed is the sum of the speeds divided by the total number of particles:

$$\bar{v} = \frac{5 + 8 + 12 + 12 + 12 + 14 + 14 + 17 + 20}{9}$$

$$= 12.7 \text{ m/s}$$

(b) What is the rms speed?

The average value of the square of the speed is given by

$$\overline{v^2} = \frac{5^2 + 8^2 + 12^2 + 12^2 + 12^2 + 14^2 + 14^2 + 17^2 + 20^2}{9}$$

$$= 178 \text{ m}^2/\text{s}^2$$

Hence, the rms speed is

$$v_{\text{rms}} = \sqrt{\overline{v^2}} = \sqrt{178 \text{ m}^2/\text{s}^2} = 13.3 \text{ m/s}$$

(c) What is the most probable speed of the particles?

Three of the particles have a speed of 12 m/s, two have a speed of 14 m/s, and the remaining have different speeds. Hence, we see that the most probable speed, v_{mp}, is 12 m/s.

Q4. A gas consists of a mixture of He and N_2 molecules. Do the lighter He molecules travel faster than the N_2 molecules? Explain.

Q5. Although the average speed of gas molecules in thermal equilibrium at some temperature is greater than zero, the average velocity is zero. Explain.

Q6. Why does a fan make you feel cooler on a hot day?

Q7. Alcohol taken internally makes you feel warmer. Yet when it is rubbed on your body, it lowers body temperature. Explain the latter effect.

Q8. A liquid partially fills a container. Explain why the temperature of the liquid decreases when the container is partially evacuated. (Using this technique, it is possible to freeze water at temperatures above 0°C.)

18.7 MEAN FREE PATH

Most of us are familiar with the fact that the strong odor associated with a gas such as ammonia may take several minutes to diffuse through a room. However, since average molecular speeds are typically several hundred meters per second at room temperature, we might expect a time much less than one second. To understand this apparent contradiction, we note that molecules collide with each other, since they are not geometrical points. Therefore, they do not travel from one side of a room to the other in a straight line. Between collisions, the molecules move with constant speed along straight lines.[3] The average distance between collisions is called the *mean free path*. The path of individual molecules is random and resembles that shown in Fig. 18.11. As we would expect from this description, the mean free path is related to the diameter of the molecules and the density of the gas.

[3]Actually, there is a small curvature in the path because of the force of gravity at the earth's surface. However, this effect is small and can be neglected.

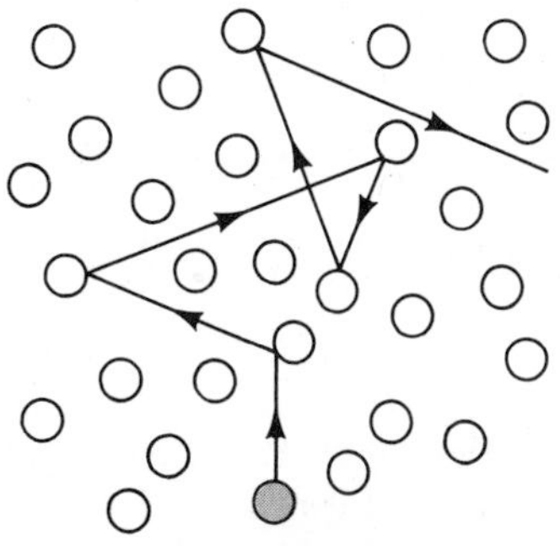

Figure 18.11 A molecule moving through a gas collides with other molecules in a random fashion. This behavior is sometimes referred to as a *random-walk process*. The mean free path increases as the number of molecules per unit volume decreases. Note that the motion is *not* limited to the plane of the paper.

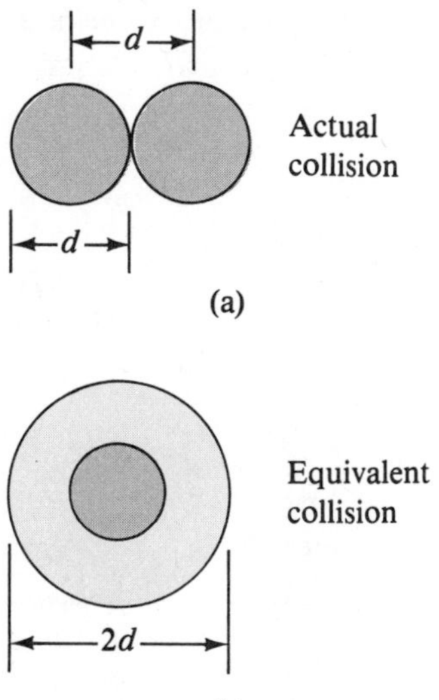

Figure 18.12 (a) Two molecules, each of diameter d, collide if their centers are within a distance d of each other. (b) The collision between the two molecules is equivalent to a point mass colliding with a molecule having an effective diameter of $2d$.

We shall now describe how to estimate the mean free path for a gas molecule. For this calculation we shall assume that the molecules are spheres of diameter d. We see from Fig. 18.12a that no two molecules will collide unless their centers are less than a distance d apart as they approach each other. An equivalent description of the collisions is to imagine that one of the molecules has a diameter $2d$ and the rest are geometrical points (Fig. 18.12b). In a time t, the molecule having the speed that we shall take to be the average speed, $\bar{v}$, will travel a distance $\bar{v}t$. In this same time interval, our molecule with equivalent diameter $2d$ will sweep out a cylinder having a cross-sectional area of πd^2 and a length of $\bar{v}t$ (Fig. 18.13). Hence the volume of the cylinder is $\pi d^2 \bar{v}t$. If n_v is the number of particles per unit volume, then the number of particles in the cylinder is $(\pi d^2 \bar{v}t)n_v$. The molecule of equivalent diameter $2d$ will collide with every particle in this cylinder in the time t. Hence, the number of collisions in the time t is equal to the number of particles in the cylinder, which we found was $(\pi d^2 \bar{v}t)n_v$.

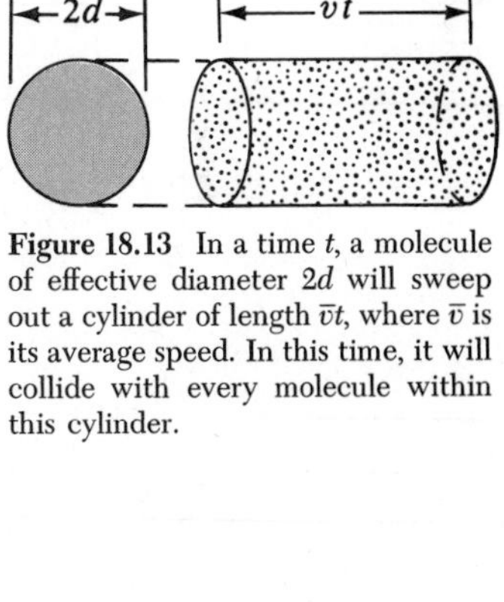

Figure 18.13 In a time t, a molecule of effective diameter $2d$ will sweep out a cylinder of length $\bar{v}t$, where $\bar{v}$ is its average speed. In this time, it will collide with every molecule within this cylinder.

The *mean free path*, l, which is the mean distance between collisions, equals the average distance $\bar{v}t$ traveled in a time t divided by the number of collisions that occurs in the time:

$$l = \frac{\bar{v}t}{(\pi d^2 \bar{v}t)n_v} = \frac{1}{\pi d^2 n_v}$$

Since the number of collisions in a time t is $(\pi d^2 \bar{v}t)n_v$, the number of collisions per unit time, or *collision frequency* f, is given by

$$f = \pi d^2 \bar{v} n_v$$

The inverse of the collision frequency is the average time between collisions, called the *mean free time*.

Our analysis has assumed that particles in the cylinder are stationary. When the motion of the particles is included in the calculation, the correct results are

Mean free path

$$l = \frac{1}{\sqrt{2}\,\pi d^2 n_v} \tag{18.26}$$

Collision frequency

$$f = \sqrt{2}\,\pi d^2 \bar{v} n_v = \frac{\bar{v}}{l} \tag{18.27}$$

Example 18.5

Calculate the mean free path and collision frequency for nitrogen molecules at a temperature of 20°C and a pressure of 1 atm. Assume a molecular diameter of 2×10^{-10} m.

Solution: Assuming the gas is ideal, we can use $PV = NkT$ to obtain the number of molecules per unit volume under these conditions:

$$n_v = \frac{N}{V} = \frac{P}{kT} = \frac{1.01 \times 10^5\ \text{N/m}^2}{(1.38 \times 10^{-23}\ \text{J/K})(293\ \text{K})}$$

$$= 2.50 \times 10^{25}\ \frac{\text{molecules}}{\text{m}^3}$$

Hence, the mean free path is

$$l = \frac{1}{\sqrt{2}\,\pi d^2 n_v}$$

$$= \frac{1}{\sqrt{2}\,\pi(2 \times 10^{-10}\ \text{m})^2 \left(2.50 \times 10^{25}\ \frac{\text{molecules}}{\text{m}^3}\right)}$$

$$= 2.25 \times 10^{-7}\ \text{m}$$

This is about 10^3 times greater than the molecular diameter. Since the average speed of a nitrogen molecule at 20°C is about 511 m/s (Table 18.1), the collision frequency is

$$f = \frac{\bar{v}}{l} = \frac{511\ \text{m/s}}{2.25 \times 10^{-7}\ \text{m}} = 2.27 \times 10^9/\text{s}$$

The molecule collides with other molecules at the average rate of about two billion times each second!

You should note that the mean free path, l, is *not* the same as the average separation between particles. In fact, the average separation, d, between particles is given approximately by $n_v^{-1/3}$. In this example, the average molecular separation is

$$d = \frac{1}{n_v^{1/3}} = \frac{1}{(2.5 \times 10^{25})^{1/3}} = 3.4 \times 10^{-9}\ \text{m}$$

Q9. A vessel containing a fixed volume of gas is cooled. Does the mean free path increase, decrease, or remain constant in the cooling process? What about the collision frequency?

Q10. A gas is compressed at a constant temperature. What happens to the mean free path of the molecules in this process?

18.8 VAN DER WAALS' EQUATION OF STATE

Thus far we have assumed all gases to be ideal, that is, to obey the equation of state, $PV = nRT$. To a very good approximation, real gases behave as ideal gases at ordinary temperatures and pressures. In the kinetic theory derivation of the ideal-gas law, we neglected the volume occupied by the molecules and assumed that intermolecular forces were negligible. Now let us investigate the qualitative behavior of real gases and the conditions under which deviations from ideal-gas behavior are expected.

Consider a gas contained in a cylinder fitted with a movable piston. As noted in Chapter 17, if the temperature is kept constant while the pressure is measured at various volumes, a plot of P versus V yields a hyperbolic curve (an *isotherm*) as predicted by the ideal-gas law (Fig. 18.14).

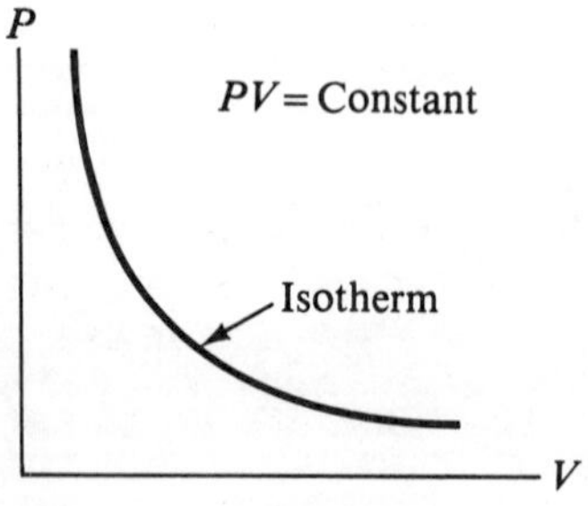

Figure 18.14 The PV diagram of an isothermal process for an ideal gas. In this case, the pressure and volume are related by PV = constant.

Now let us describe what happens to a real gas. Figure 18.15 gives some typical experimental curves taken on a gas at various temperatures. At the higher temperatures, the curves are approximately hyperbolic and the gas behavior is close to ideal. However, as the temperature is lowered, the deviations from the hyperbolic shape are very pronounced.

There are two major reasons for this behavior. First, we must account for the volume occupied by the gas molecules. If V is the volume of the container and b is the volume occupied by the molecules, then $V - b$ is the empty volume available to the gas. The constant b is equal to the number of molecules of gas multiplied by the volume per molecule. As V decreases for a given quantity of gas, the fraction of the volume occupied by the molecules increases.

The second important effect concerns the intermolecular forces when the molecules are close together. At close separations, the molecules attract each other, as we might expect, since gases condense to form liquids. This attractive force reduces the pressure that the molecules exert on the container walls. In other words, a molecule that is on the verge of colliding with the walls is under the influence of attractive forces directed toward the body of the gas. Consequently, the energy of the molecules colliding with the walls is reduced and the resulting pressure is decreased from that of an ideal gas. The net inward force on a molecule near the wall is proportional to the density of molecules, or inversely proportional to the volume. In addition, the pressure at the wall is proportional to the density of molecules. The net pressure is reduced by a factor proportional to the square of the density, which varies as $1/V^2$. Hence, the pressure P is replaced by an effective pressure $P + a/V^2$, where a is a constant.

The two effects just described can now be incorporated into a modified equation of state proposed by J.D. van der Waals (1837–1923) in 1873. For one mole of gas, van der Waals' equation of state is given by

Van der Waals' equation of state

$$\left(P + \frac{a}{V^2}\right)(V - b) = RT \tag{18.28}$$

The constants a and b are empirical and are chosen to provide the best fit to the experimental data for a particular gas.

The experimental curves in Fig. 18.16 for CO_2 are described quite accurately by van der Waals' equation at the higher temperatures (T_3, T_4, and T_5) and outside the shaded regions. Within the shaded region there are major discrepancies. If the van der Waals equation of state is used to predict the PV relationship at a temperature such as T_1, then a nonlinear curve is obtained that is unlike the observed flat portion of the curve in the figure.

The departure from van der Waals equation predictions at the lower temperatures and higher densities is due to the onset of liquefaction. That is, the gas begins to liquefy at the pressure P_c, called the *critical pressure*. In the region within the dotted line below P_c the gas is partially liquefied and the gas vapor and liquid coexist. In the flat portions of the low-temperature isotherms, as the volume is decreased more gas liquefies and the pressure remains constant. At even lower volumes, the gas is completely liquefied. Any further decrease in volume leads to large increases in pressure because liquids are not easily compressed.

It is now realized that a real gas cannot be rigorously described by any simple equation of state, such as Eq. 18.28, because of the complex nature of the intermolecular forces. Nevertheless, the basic concepts involved in Eq. 18.28 are correct. At very low temperatures, the

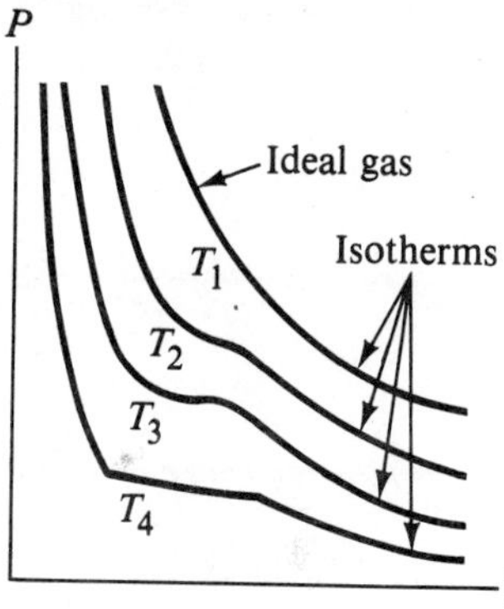

Figure 18.15 Isotherms for a real gas at various temperatures. At higher temperatures, such as T_2, the behavior is nearly ideal. The behavior is not ideal at the lower temperatures.

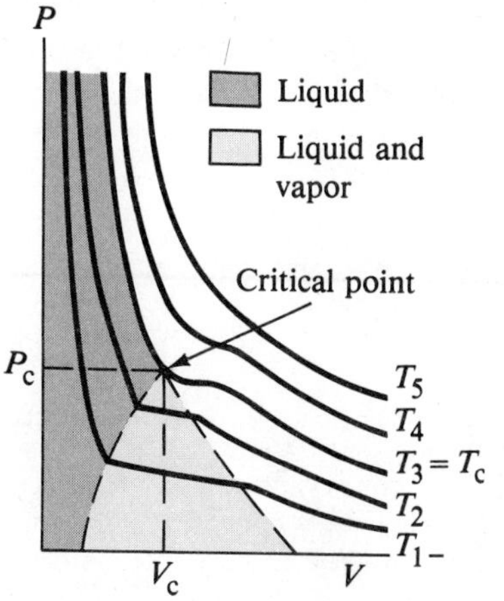

Figure 18.16 Isotherms for CO_2 at various temperatures. Below the critical temperature, T_c, the substance could be in the liquid state, the liquid-vapor equilibrium state, or the gaseous state, depending on the pressure and volume. (*Adapted from K. Mendelssohn,* The Quest for Absolute Zero, *New York, McGraw-Hill, World University Library, 1966.*)

low-energy molecules attract each other and the gas tends to liquefy, or condense. A further pressure increase will accelerate the rate of liquefaction. At the higher temperatures, the average kinetic energy is large enough to overcome the attractive intermolecular forces; hence the molecules do not bind together at the higher temperatures and the gas phase is maintained.

18.9 SUMMARY

The *pressure* of N molecules of an ideal gas contained in a volume V is given by

Pressure and molecular kinetic energy

$$P = \frac{2}{3}\frac{N}{V}\left(\frac{1}{2}m\overline{v^2}\right) \tag{18.5}$$

where $\frac{1}{2}m\overline{v^2}$ is the average kinetic energy per molecule.

The *temperature* of an ideal gas is related to the average kinetic energy per molecule through the expression

Temperature is proportional to kinetic energy

$$T = \frac{2}{3k}\left(\frac{1}{2}m\overline{v^2}\right) \tag{18.6}$$

where k is Boltzmann's constant.

The *average translational kinetic energy per molecule* of a gas is given by

Average kinetic energy per molecule

$$\frac{1}{2}m\overline{v^2} = \frac{3}{2}kT \tag{18.7}$$

Each translational degree of freedom (x, y, or z) has $\frac{1}{2}kT$ of energy associated with it.

The *equipartition of energy theorem* states that the energy of a system in thermal equilibrium is equally divided among all degrees of freedom.

The *total energy* of N molecules (or n moles) of an ideal monatomic gas is given by

Internal energy of an ideal monatomic gas

$$U = \frac{3}{2}NkT = \frac{3}{2}nRT \tag{18.11}$$

The *molar heat capacity* of an ideal monatomic gas at constant volume is $C_v = \frac{3}{2}R$; the molar heat capacity at constant pressure is $C_p = \frac{5}{2}R$. The ratio of heat capacities is $\gamma = C_p/C_v = 5/3$.

An *adiabatic process* is one in which there is no heat transfer between the system and its surroundings.

If an ideal gas undergoes an adiabatic expansion or compression, the first law of thermodynamics together with the equation of state, $PV = nRT$, shows that

Adiabatic process

$$PV^{\gamma} = \text{constant} \tag{18.18}$$

The *most probable speed distribution* of N gas molecules at a temperature T is given by Maxwell's speed distribution function:

Maxwell speed distribution function

$$N_v = 4\pi N\left(\frac{m}{2\pi kT}\right)^{3/2} v^2\, e^{-mv^2/2kT} \tag{18.22}$$

Using this expression, one can find the rms speed, v_{rms}, the average speed, $\overline{v}$, and the most probable speed, v_{mp}:

$$v_{rms} = \sqrt{\frac{3kT}{m}} \quad (18.23) \qquad \bar{v} = \sqrt{\frac{8kT}{\pi m}} \quad (18.24) \qquad v_{mp} = \sqrt{\frac{2kT}{m}} \quad (18.25)$$

The molecules of a gas undergo collisions with each other billions of times each second under standard conditions. If the gas has a volume density n_V and each molecule is assumed to have a diameter d, the average distance between collisions, or *mean free path, l,* is found to be

$$l = \frac{1}{\sqrt{2}\pi d^2 n_V} \qquad (18.26)$$

Mean free path

Furthermore, the number of collisions per second, or *collision frequency, f,* is given by

$$f = \sqrt{2}\pi d^2 \bar{v} n_V = \frac{\bar{v}}{l} \qquad (18.27)$$

Collision frequency

EXERCISES

Section 18.1 Molecular Model for the Pressure of an Ideal Gas

1. Find the average square speed of nitrogen molecules under standard conditions. Recall that 1 mole of any gas occupies a volume of 22.4 liters at standard temperature and 1 atm pressure.

2. Two moles of oxygen gas are confined to a 5-liter vessel at a pressure of 8 atm. Find the average kinetic energy of an oxygen molecule under these conditions. (The mass of an O_2 molecule is 5.34×10^{-26} kg.)

3. In a 1-min interval, a machine gun fires 150 bullets, each of mass 8 g and speed 400 m/s. The bullets strike and become imbedded in a stationary target. If the target has an area of 5 m^2, find the average force and pressure exerted on the target. (Note that these are inelastic collisions.)

4. In a period of 1 s, 5×10^{23} nitrogen molecules strike a wall of area 8 cm^2. If the molecules move with a speed of 300 m/s and strike at an angle of 45° to the normal to the wall, find the pressure exerted on the wall. (The mass of an N_2 molecule is 4.68×10^{-26} kg.)

5. In a 30-s interval, 500 hailstones strike a glass window of area 0.6 m^2 at an angle of 45° to the window surface. Each hailstone has a mass of 5 g and a speed of 8 m/s. If the collisions are assumed to be elastic, find the average force and pressure on the window.

Section 18.2 Molecular Interpretation of Temperature

6. A cylinder contains a mixture of helium and argon gas in equilibrium at a temperature of 150°C. What is the average kinetic energy of each gas molecule?

7. Calculate the root mean square speed of a H_2 molecule at a temperature of 250°C.

8. (a) Determine the temperature at which the rms speed of a He atom equals 500 m/s. (b) What is the rms speed of He on the surface of the sun, where the temperature is 5800 K?

9. Nitrogen molecules have a rms speed of 517 m/s at 300 K. (a) What is the rms speed of nitrogen at 600 K? at 200 K? (b) Construct a graph of v_{rms} versus temperature for helium in intervals of 200 K, over the temperature range 200 K to 2000 K.

10. What is the temperature at which the rms speed of nitrogen molecules equals the rms speed of helium at 20°C?

11. A 5-liter vessel contains nitrogen gas at a temperature of 27°C and a pressure of 3 atm. Find (a) the total translational kinetic energy of the gas molecules and (b) the average kinetic energy per molecule.

Section 18.3 Heat Capacity of an Ideal Gas (use data in Table 18.2)

12. Calculate the change in internal energy of 3 moles of helium gas when its temperature is increased by 2 K.

13. Two moles of oxygen gas are heated from 300 K to 320 K. How much heat is transferred to the gas if the process occurs at (a) constant volume and (b) constant pressure?

14. The total heat capacity, C', of a monatomic gas measured at constant pressure is 14.9 cal/K. Find (a) the number of moles of gas, (b) the total heat capacity at constant volume, and (c) the internal energy of the gas at 350 K. (Recall that $C' = nC$.)

15. In a constant-volume process, 50 cal of heat is transferred to 1 mole of an ideal monatomic gas initially at 300 K. Find (a) the increase in internal energy of the gas, (b) the work done by the gas, and (c) the final temperature of the gas.

16. One mole of hydrogen gas is heated at constant pressure from 300 K to 420 K. Calculate (a) the heat transferred to the gas, (b) the increase in internal energy of the gas, and (c) the work done by the gas.

17. Consider *three* moles of an ideal gas. (a) If the gas is monatomic, find the *total* heat capacity at constant volume and at constant pressure. (b) Repeat (a) for a diatomic gas in which the molecules rotate but do not vibrate.

Section 18.4 Adiabatic Processes for an Ideal Gas

18. Two moles of an ideal gas ($\gamma = 1.40$) expands quasi-statically and adiabatically from a pressure of 5 atm and volume of 12 liters to a final volume of 30 liters. (a) What is the final pressure of the gas? (b) What are the initial and final temperatures?

19. Show that Eq. 18.20 follows from Eq. 18.19 for a quasi-static, adiabatic process. (Note that $PV = nRT$ applies during the process.)

20. An ideal gas ($\gamma = 2$) expands quasi-statically and adiabatically. If the final temperature is one third the initial temperature, (a) by what factor does its volume change? (b) by what factor does its pressure change?

21. One mole of an ideal monatomic gas ($\gamma = 1.67$) initially at 300 K and 1 atm is compressed quasi-statically and adiabatically to one fourth its initial volume. Find its final pressure and temperature.

22. During the compression stroke of a certain gasoline engine, the pressure increases from 1 atm to 20 atm. Assuming that the process is adiabatic and the gas is ideal with $\gamma = 1.40$, (a) by what factor does the volume change and (b) by what factor does the temperature change?

Section 18.5 The Equipartition of Energy

23. If a molecule has f degrees of freedom, show that a gas consisting of such molecules has the following properties: (1) its total internal energy is $fnRT/2$; (2) its molar heat capacity at constant volume is $fR/2$; (3) its molar heat capacity at constant pressure is $(f + 2)R/2$; (4) the ratio $\gamma = C_p/C_v = (f + 2)/f$.

24. Examine the data for polyatomic gases in Table 18.2 and explain why SO_2 has a higher C_v than the other polyatomic gases at 300 K.

25. Inspecting the magnitudes of C_v and C_p for the diatomic and polyatomic gases in Table 18.2, we find that the values increase with increasing molecular mass. Give a qualitative explanation of this observation.

26. Consider 2 moles of an ideal diatomic gas. Find the *total* heat capacity at constant volume and at constant pressure if (a) the molecules rotate but do not vibrate and (b) the molecules rotate and vibrate.

Section 18.6 Distribution of Molecular Speeds

27. Use Fig. 18.10 to *estimate* the number of nitrogen molecules with speeds between 400 m/s and 600 m/s at (a) 300 K and (b) 900 K.

28. Show that the most probable speed of a gas molecule is given by Eq. 18.25. Note that the most probable speed corresponds to the point where the slope of the speed distribution curve, dN_v/dv, is zero.

29. A vessel containing oxygen gas is at a temperature of 400 K. Find (a) the rms speed, (b) the average speed, and (c) the most probable speed of the gas molecules. (The mass of O_2 is 5.31×10^{-26} kg.)

30. Show that the Maxwell speed distribution function given by Eq. 18.22 satisfies Eq. 18.21. (Such a function is said to be *normalized.*)

31. Calculate the most probable speed, average speed, and rms speed for nitrogen gas molecules at 900 K. Compare your results with the values obtained from Fig. 18.10.

32. Using the data in Fig. 18.10, estimate the *fraction* of N_2 molecules that have speeds in the range 1000 m/s to 1200 m/s at 900 K. Note that the total number of molecules is 10^5.

33. At what temperature would the average velocity of helium atoms equal (a) the escape velocity from earth, 1.12×10^4 m/s, and (b) the escape velocity from the moon, 2.37×10^3 m/s? (See Chapter 14 for a discussion of escape velocity, and note that the mass of helium is 6.66×10^{-27} kg.)

34. Fifteen identical particles have the following speeds: one has speed 2 m/s; two have speed 3 m/s; three have speed 5 m/s; four have speed 7 m/s; three have speed 9 m/s; two have speed 12 m/s. Find (a) the average speed, (b) the rms speed, and (c) the most probable speed of these particles.

Section 18.7 Mean Free Path

35. In an ultrahigh vacuum system, the pressure is measured to be 10^{-10} torr (where 1 torr $= 133$ N/m^2). If the gas molecules have a molecular diameter of 3 Å $= 3 \times 10^{-10}$ m and the temperature is 300 K, find (a) the number of molecules in a volume of 1 m^3, (b) the mean free path of the molecules, and (c) the collision frequency, assuming an average speed of 500 m/s.

36. Show that the mean free path for the molecules of an ideal gas is given by

$$l = \frac{kT}{\sqrt{2}\pi d^2 P}$$

where d is the molecular diameter.

37. A cylinder contains 5 moles of oxygen gas at a pressure of 80 atm and temperature of 300 K. Assuming a molecular diameter of 2.5×10^{-10} m, find (a) the number of molecules per unit volume, (b) the mean free path, and (c) the collision frequency.

Section 18.8 Van der Waals' Equation of State (Optional)

38. The constant b that appears in van der Waals' equation of state for oxygen is measured to be 31.8 cm^3/mole. Assuming a spherical shape, estimate the diameter of the molecule.

PROBLEMS

1. A mixture of two gases will diffuse through a filter at rates proportional to their rms speeds. If the molecules of the two gases have masses m_1 and m_2, show that the ratio of their rms speeds (or the ratio of diffusion rates) is given by

$$\frac{(v_1)_{\text{rms}}}{(v_2)_{\text{rms}}} = \sqrt{\frac{m_2}{m_1}}$$

This process is used to obtain uranium enriched with the isotope ^{235}U, which is used in nuclear reactors.

2. A cylinder containing n moles of an ideal gas undergoes a quasi-static, adiabatic process. (a) Starting with the expression $W = \int P\,dV$ and using $PV^\gamma =$ constant, show that the work done is given by

$$W = \frac{1}{\gamma - 1}(P_iV_i - P_fV_f)$$

(b) Starting with the first law in differential form, prove that the work done is also equal to $nC_v(T_i - T_f)$. Show that this result is consistent with the equation in (a).

3. A vessel contains 1 mole of helium gas at a temperature of 300 K. Calculate the approximate number of molecules having speeds in the range from 400 m/s to 410 m/s. (Hint: This number is approximately equal to $N_v\,\Delta v$, where Δv is the range of speeds.)

4. Verify Eqs. 18.23 and 18.24 for the rms and average speed of the molecules of a gas at a temperature T. Note that the average value of v^n is given by

$$\overline{v^n} = \frac{1}{N}\int_0^\infty v^n N_v\,dv$$

and make use of the integrals

$$\int_0^\infty x^3 e^{-ax^2}\,dx = \frac{1}{2a^2}$$

and

$$\int_0^\infty x^4 e^{-ax^2}\,dx = \frac{3}{8a^2}\sqrt{\frac{\pi}{a}}$$

5. Twenty particles, each of mass m and confined to a volume V, have the following speeds: two have speed v; three have speed $2v$; five have speed $3v$; four have speed $4v$; three have speed $5v$; two have speed $6v$; one has speed $7v$. Find (a) the average speed, (b) the rms speed, (c) the most probable speed, (d) the pressure they exert on the walls of the vessel, and (e) the average kinetic energy per particle.

6. A vessel contains 10^4 oxygen molecules at 500 K. (a) Make an accurate graph of the Maxwell speed distribution function, N_v, versus speed with points at speed intervals of 100 m/s. (b) Determine the most probable speed from this graph. (c) Calculate the average and rms speeds for the molecules and label these points on your graph. (d) From the graph, estimate the fraction of molecules with speeds in the range 300 m/s to 600 m/s.

7. The internal energy of a gas consisting of n moles of CO_2 at 300 K is given by $U = anRT + b$, where a and b are constants. (a) From this expression, derive the molar heat capacity at constant volume, C_v. (b) What is C_p for this gas? (c) Use Table 18.2 to obtain a value for the constant a. (d) How many degrees of freedom does the molecule have at this temperature?

8. The compressibility, κ, of a substance is defined as the fractional change in volume of that substance for a given change in pressure:

$$\kappa = -\frac{1}{V}\frac{dV}{dP}$$

(a) Explain why the negative sign in this expression ensures that κ will always be positive. (b) Show that if an ideal gas is compressed *isothermally*, its compressibility is given by $\kappa_1 = 1/P$. (c) Show that if an ideal gas is compressed *adiabatically*, its compressibility is given by $\kappa_2 = 1/\gamma P$. (d) Determine values for κ_1 and κ_2 for a monatomic ideal gas at a pressure of 2 atm.

9. One mole of a gas obeying van der Waals' equation of state is compressed isothermally. At some critical temperature, T_c, the isotherm has a point of zero slope and zero inflection, as in Fig. 18.16. That is, at $T = T_c$, $\frac{\partial P}{\partial V} = 0$ and $\frac{\partial^2 P}{\partial V^2} = 0$. Using Eq. 18.28 and these conditions, show that at the critical point, the pressure, volume, and temperature are given by $P_c = a/27b^2$; $V_c = 3b$, $T_c = 8a/27Rb$.

19 Heat Engines, Entropy, and the Second Law of Thermodynamics

James Madison University

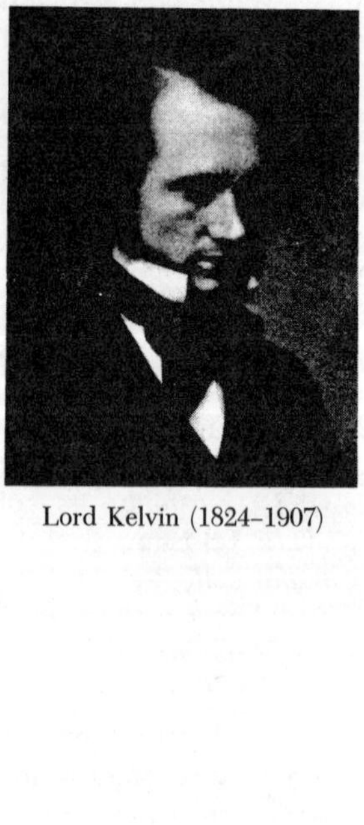

Lord Kelvin (1824–1907)

The first law of thermodynamics is merely the law of conservation of energy generalized to include heat as a form of energy. This law tells us only that an increase in one form of energy must be accompanied by a decrease in some other form of energy. The first law places no restrictions on the types of energy conversions that can occur. Furthermore, it makes no distinction between heat and work. According to the first law, the internal energy of a body may be increased by either adding heat to it or doing work on it. But there is an important difference between heat and work that is not evident from the first law. For example, it is possible to convert work completely into heat but, in practice, it is impossible to convert heat completely into work without changing the surroundings.

The *second law of thermodynamics* establishes which processes in nature do or do not occur. Of all processes permitted by the first law, only certain types of energy conversions can take place. The following are some examples of processes that are consistent with the first law of thermodynamics but proceed in an order governed by the second law of thermodynamics. (1) When two objects at different temperatures are placed in thermal contact with each other, heat flows from the warmer to the cooler object, but never from the cooler to the warmer. (2) Salt dissolves spontaneously in water, but extracting salt from salt water requires some external influence. (3) When a rubber ball is dropped to the ground, it bounces several times and eventually comes to rest. The opposite process does not occur. (4) The oscillations of a pendulum will slowly decrease in amplitude because of collisions with air molecules and friction at the point of suspension. Eventually the pendulum will come to rest. Thus, the initial mechanical energy of the pendulum is converted into thermal energy. The reverse transformation of energy does not occur.

These are all examples of *irreversible* processes, that is, processes that occur naturally in only one direction. None of these processes occur in the opposite temporal order; if they did, they would violate the second law of thermodynamics.[1] That is, the one-way nature of thermodynamic processes in fact *establishes* a direction of time.[2] You may have witnessed the humor of an action film running in reverse, which demonstrates the improbable order of events in a time-reversed world.

The second law of thermodynamics, which can be stated in many equivalent ways, has some very practical applications. From an engineering viewpoint, per-

[1] To be more precise, we should say that the set of events in the time-reversed sense is highly improbable. From this viewpoint, events occur with a vastly higher probability in one direction than in the opposite direction.

[2] See, for example, D. Layzer, "The Arrow of Time," *Scientific American*, December 1975.

haps the most important application is the limited efficiency of heat engines. Simply stated, the second law says that a machine capable of continuously converting thermal energy completely into other forms of energy cannot be constructed.

19.1 HEAT ENGINES AND THE SECOND LAW OF THERMODYNAMICS

A *heat engine* is a device that converts thermal energy into other useful forms of energy, such as mechanical and electrical energy. Power plants generate electricity by converting the potential energy stored in chemical or nuclear fuels into thermal energy. This thermal energy is, in turn, converted into the mechanical energy used to drive an electrical generator. Internal combustion engines and diesel engines, which propel automobiles and aircraft, extract heat from a burning fuel and convert a fraction of this energy into mechanical energy.

All heat engines operate on the same principle. In effect, a heat engine carries a *working substance* through a cyclic process, that is, a process in which the substance eventually returns to its initial state. In the case of a steam engine, the working substance is water. The water is carried through a cycle in which it evaporates into steam in a boiler, the steam expands against a piston and is then condensed with cooling water, and the condensed water finally returns to the boiler. Internal combustion engines use a mixture of fuel and air as the working substance.

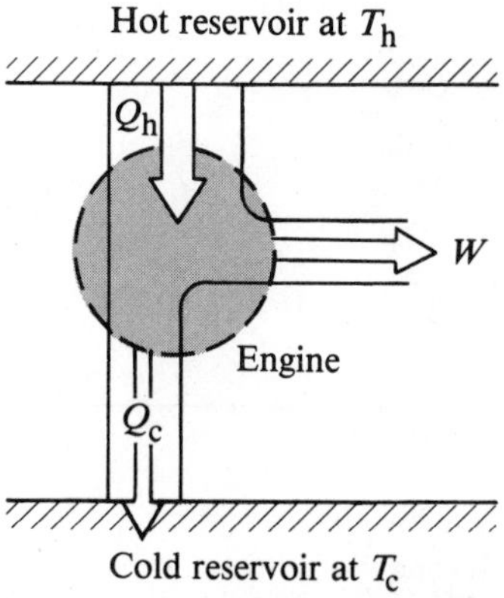

Figure 19.1 Schematic representation of a heat engine. The engine (in the circular area) absorbs heat Q_h from the hot reservoir, expels heat Q_c to the cold reservoir, and does work W.

In the operation of any heat engine, a quantity of heat is extracted from a high-temperature source, some mechanical work is done, and some heat is rejected to a low-temperature medium. It is useful to represent a heat engine schematically as in Fig. 19.1. The engine (represented by the circle at the center of the diagram) absorbs a quantity of heat Q_h from the heat reservoir at a temperature T_h. It does work W and gives up heat Q_c to another heat reservoir at a temperature T_c, where $T_c < T_h$. (The subscripts h and c are used to indicate hot and cold heat reservoirs.) Since the engine is a cyclic device, its initial and final internal energies are equal, so $\Delta U = 0$. Hence, from the first law of thermodynamics we see that the *net work* W *done by the engine equals the net heat flowing into the engine.* As we can see from Fig. 19.1, $Q_{net} = Q_h - Q_c$; therefore

$$W = Q_h - Q_c \tag{19.1}$$

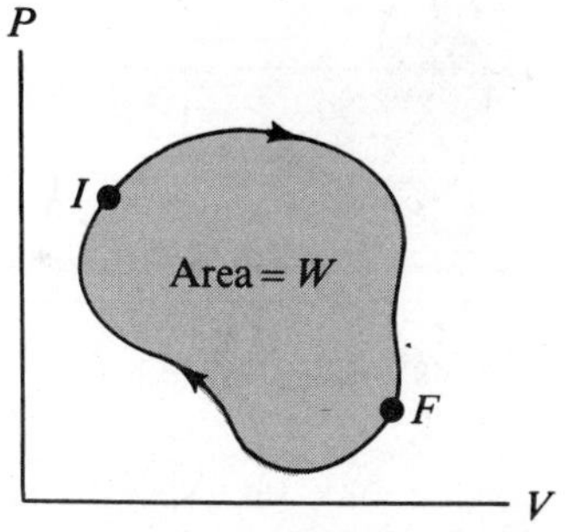

Figure 19.2 The *PV* diagram for an arbitrary cyclic process. The net work done equals the area enclosed by the curve.

where Q_h and Q_c are taken to be positive quantities. If the working substance is a gas, the net work done for a cyclic process is the area enclosed by the curve representing the process in the *PV* plane. This is shown for an arbitrary cyclic process in Fig. 19.2.

The *thermal efficiency, e,* of a heat engine is defined as the ratio of the net work done to the heat absorbed during one cycle:

Thermal efficiency

$$e = \frac{W}{Q_h} = \frac{Q_h - Q_c}{Q_h} = 1 - \frac{Q_c}{Q_h} \tag{19.2}$$

We can think of the efficiency as the ratio of "what you get" (mechanical work) to "what you pay for" (fuel). This result shows that a heat engine has 100% efficiency ($e = 1$) only if $Q_c = 0$, that is, if no heat is rejected to the cold reservoir. In other words, a heat engine with perfect efficiency would have to convert all of the absorbed heat energy Q_h into mechanical work. The second law of thermodynamics says that this is impossible.

In practice, it is found that all heat engines convert only a fraction of the absorbed heat into mechanical work. For example, a good automobile engine has an efficiency of about 20%, and diesel engines have efficiencies ranging from 35% to 40%. On the basis of this fact, the *Kelvin-Planck form* of *the second law of thermodynamics* states the following: *No heat engine, operating in a cycle, can absorb thermal energy from a reservoir and perform an equal amount of work.* This is equivalent to stating that *it is impossible to construct a perpetual-motion machine of the second kind,* that is, a machine that would violate the second law.[3] Figure 19.3 is a schematic diagram of the impossible "perfect" heat engine.

Second law of thermodynamics (Kelvin-Planck)

A refrigerator (or heat pump) is a heat engine running in reverse. This is shown schematically in Fig. 19.4, in which the engine absorbs heat Q_c from the cold reservoir and expels heat Q_h to the hot reservoir. This can be accomplished only if work is done *on* the refrigerator. From the first law, we see that the heat given up to the hot reservoir must equal the sum of the work done and the heat absorbed from the cold reservoir. Therefore, we see that the refrigerator transfers heat from a colder body (the contents of the refrigerator) to a hotter body (the room). In practice, it is desirable to carry out this process with a minimum of work. If it could be accomplished without doing any work, we would have a "perfect" refrigerator (Fig. 19.5). Again, this is in violation of the second law of thermodynamics, which in the form of the *Clausius statement*[4] says the following: *It is impossible to construct a cyclical machine that produces no other effect than to transfer heat continuously from one body to another body at a higher temperature.* In effect, this statement of the second law governs the direction of heat flow between two bodies at different temperatures. Heat will flow from the colder to the hotter body only if work is done on the system. For example, homes are cooled in summer by pumping heat out; the work done on the air conditioner is supplied by the power company.

Second law of thermodynamics (Clausius)

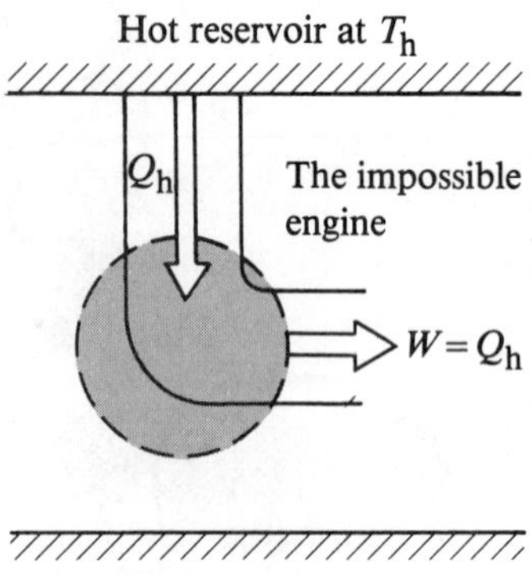

Figure 19.3 Schematic diagram of a heat engine that absorbs heat Q_h from a hot reservoir and does an equivalent amount of work. This perfect engine is impossible to construct.

An efficient refrigerator is one that removes the greatest amount of heat Q_c from the cold reservoir for the least amount of work. The *coefficient of performance* of a refrigerator is defined as the ratio

$$\text{Coefficient of performance} = \frac{Q_c}{W} \tag{19.3}$$

A good refrigerator should have a high coefficient of performance, typically 5 or 6. The impossible (perfect) refrigerator would have an infinite coefficient of performance. In the design of refrigeration systems, thicker and higher-quality insulation (low thermal conductivity) tends to increase the coefficient of performance.

The Clausius and Kelvin-Planck statements of the second law appear, at first sight, to be unrelated. They are, in fact, equivalent in all respects. Although we do not prove it here, one can show that if either statement is false, so is the other.[5]

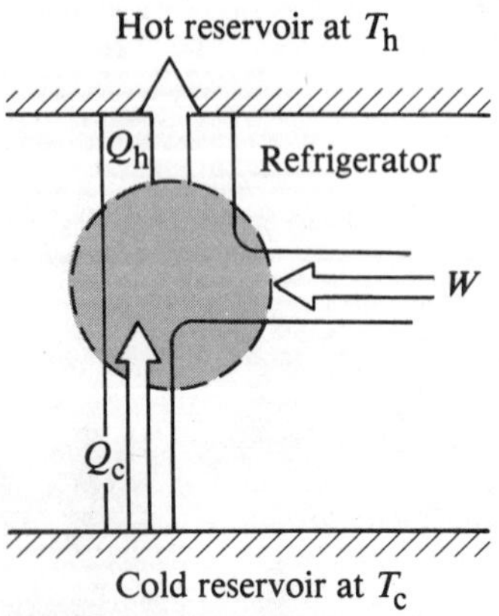

Figure 19.4 Schematic diagram for a refrigerator, which absorbs heat Q_c from the cold reservoir and expels heat Q_h to the hot reservoir. Work W is done *on* the refrigerator.

Q1. What are some factors that affect the efficiency of automobile engines?

Q2. Is it possible to cool a room by leaving the door of a refrigerator open? What happens to the temperature of a room in which an air conditioner is left running on a table in the middle of the room?

19.2 REVERSIBLE AND IRREVERSIBLE PROCESSES

In our introductory remarks we mentioned that real processes have a preferred direction. Heat flows spontaneously from a hot to a cold body when the two are

---

[3] A perpetual-motion machine of the first kind is one that would violate the first law of thermodynamics (energy conservation). This type of machine is also impossible to construct.

[4] First expressed by Rudolf Clausius (1822–1888).

[5] See, for example, F.W. Sears, *Thermodynamics, The Kinetic Theory of Gases, and Statistical Mechanics,* Reading, Mass., Addison-Wesley, 1953, Chapter 7.

placed in contact, but the reverse is accomplished only with some external influence. When a block slides on a rough surface, it eventually comes to rest. The mechanical energy of the block is converted into internal energy of the block and table. Such unidirectional processes are called *irreversible* processes. After any irreversible process occurs, it is impossible to return the system to its original state without affecting its surroundings.

Irreversible process

In general, a process is irreversible if the system and its surroundings cannot be returned to their initial states. Processes that involve the conversion of mechanical energy to internal energy, such as the block sliding on a rough surface, are irreversible. Once the block has come to rest, the internal energy of the block and table cannot be completely converted back into mechanical energy. The process can only be reversed by doing external work, that is, by changing the surroundings.

Reversible process

A process is *reversible* if the system passes from the initial state to the final state through a succession of equilibrium states. If a real process occurs quasi-statically, that is, slowly enough so that each state departs only infinitesimally from equilibrium, it can be considered reversible. For example, we can imagine compressing a gas quasi-statically by dropping some grains of sand onto a frictionless piston (Fig. 19.6). The pressure, volume, and temperature of the gas are well defined during the isothermal compression. The process is made isothermal by placing the gas in thermal contact with a heat reservoir. Some heat is transferred from the gas to the reservoir during the process. Each time a grain of sand is added to the piston, the volume decreases slightly while the pressure increases slightly. Each added grain of sand represents a change to a new equilibrium state. The process can be reversed by slowly removing grains of sand from the piston.

Since a reversible process is defined by a succession of equilibrium states, it can be represented by a line on a *PV* diagram, which establishes the path for the process (Fig. 19.7). Each point on this line represents one of the intermediate equilibrium states. On the other hand, an irreversible process is one that passes from the initial state to the final state through a series of nonequilibrium states. In this case, only the initial and final equilibrium states can be represented on the *PV* diagram. The intermediate, nonequilibrium states may have well-defined volumes, but these states are not characterized by a unique pressure for the entire system. Instead, there are variations in pressure (and temperature) throughout the volume range, and these variations will not persist if left to themselves (i.e., nonequilibrium conditions). For this reason, an irreversible process cannot be represented by a line on a *PV* diagram.

We have stated that a reversible process must take place quasi-statically. In addition, in a reversible process there can be no dissipative effects that produce heat.

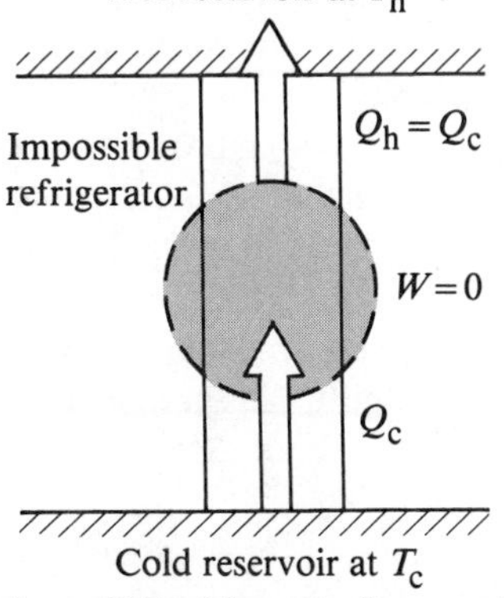

Figure 19.5 Schematic diagram of the impossible refrigerator, that is, one that absorbs heat Q_c from a cold reservoir and expels an equivalent amount of heat to the hot reservoir with $W = 0$.

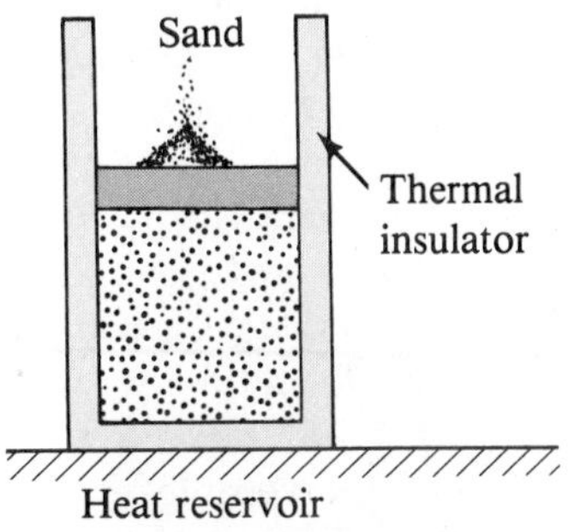

Figure 19.6 A gas in thermal contact with a heat reservoir is compressed slowly by dropping grains of sand onto the piston. The compression is isothermal and reversible.

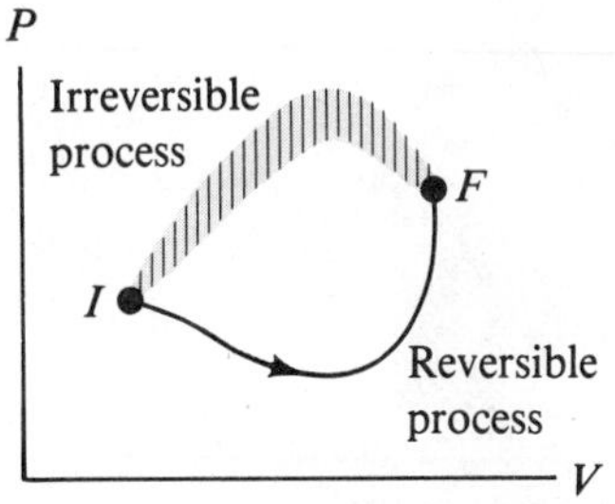

Figure 19.7 A reversible process between the two equilibrium states *I* and *F* can be represented by a line on the *PV* diagram. Each point on this line represents an equilibrium state. An irreversible process passes through a series of nonequilibrium states and cannot be represented by a line on this diagram.

Other effects that tend to disrupt equilibrium, such as heat conduction resulting from a temperature difference, must not be present. In reality, such effects are impossible to eliminate completely, and so it is not surprising that processes in nature are irreversible. Nevertheless, it is possible to approximate reversible processes through carefully controlled procedures. As we shall see in the next section, the concept of reversible processes is especially important in establishing the theoretical limit of the efficiency of a heat engine.

Q3. Describe some common irreversible processes in nature.

Q4. Are unidirectional processes reversible or irreversible? Explain.

Q5. A gas will expand to twice its initial volume at constant temperature if one opens a valve and allows the gas to rush into an evacuated vessel equal in volume to that of the gas container. Explain why the process is irreversible.

19.3 THE CARNOT ENGINE

In 1824 a French engineer named Sadi Carnot (1796–1832) described a working cycle, now called a *Carnot cycle*, that is of great importance from both a practical and a theoretical viewpoint. Using the second law of thermodynamics, he showed that a heat engine operating in this ideal, reversible cycle between two heat reservoirs would be the most efficient engine possible. Such an ideal engine, called a *Carnot engine*, establishes an upper limit on the efficiencies of all engines. That is, the net work done by a working substance taken through the Carnot cycle is the largest possible for a given amount of heat supplied to the substance.

To describe the Carnot cycle, we shall assume that the working substance is an ideal gas contained in a cylinder with a movable piston at one end. The cylinder walls and the piston are thermally nonconducting. Four stages of the Carnot cycle are shown in Fig. 19.8, and the *PV* diagram for the cycle is shown in Fig. 19.9. The Carnot cycle consists of two adiabatic and two isothermal processes, all reversible.

Process involved in the Carnot cycle

1. The process $A \to B$ is an isothermal expansion at temperature T_h, in which the gas is placed in thermal contact with a heat reservoir at temperature T_h (Fig. 19.8a). During the process, the gas absorbs heat Q_h from the reservoir through the base of the cylinder and does work W_{AB} in raising the piston.
2. In the process $B \to C$, the base of the cylinder is replaced by a thermally nonconducting wall and the gas expands adiabatically, that is, no heat enters or leaves the system (Fig. 19.8b). During the process, the temperature falls from T_h to T_c and the gas does work W_{BC} in raising the piston.
3. In the process $C \to D$, the gas is placed in thermal contact with a heat reservoir at temperature T_c (Fig. 19.8c) and is compressed isothermally at temperature T_c. During this time, the gas expels heat Q_c to the reservoir and the work done on the gas by an external agent is W_{CD}.
4. In the final stage, $D \to A$, the base of the cylinder is replaced by a nonconducting wall (Fig. 19.8d) and the gas is compressed adiabatically. The temperature of the gas increases to T_h and the work done on the gas by an external agent is W_{DA}.

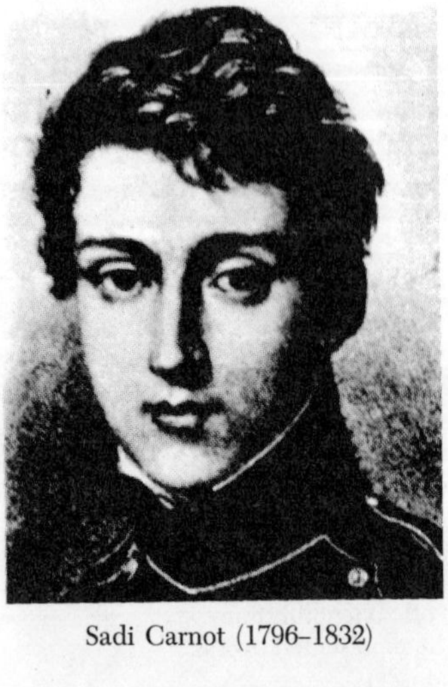

Sadi Carnot (1796–1832)

The net work done in this reversible, cyclic process is equal to the area enclosed by the path *ABCDA* of the *PV* diagram (Fig. 19.9). As we showed in Section 19.1, the net work done in one cycle equals the net heat transferred into the system, $Q_h - Q_c$, since the change in internal energy is zero. Hence, the thermal efficiency of the engine is given by Equation 19.2:

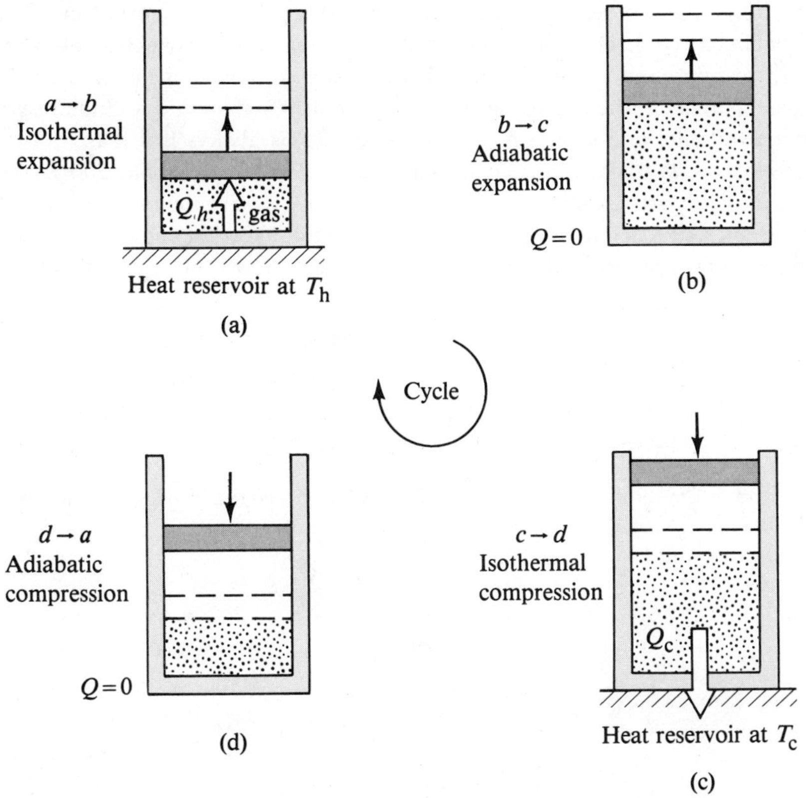

Figure 19.8 The Carnot cycle. In process $A \rightarrow B$, the gas expands isothermally while in contact with a reservoir at T_h. In process $B \rightarrow C$, the gas expands adiabatically ($Q = 0$). In process $C \rightarrow D$, the gas is compressed isothermally while in contact with a reservoir at $T_c < T_h$. In process $D \rightarrow A$, the gas is compressed adiabatically. The upward arrows on the piston indicate sand being removed during the expansions, and the downward arrows indicate the addition of sand during the compressions.

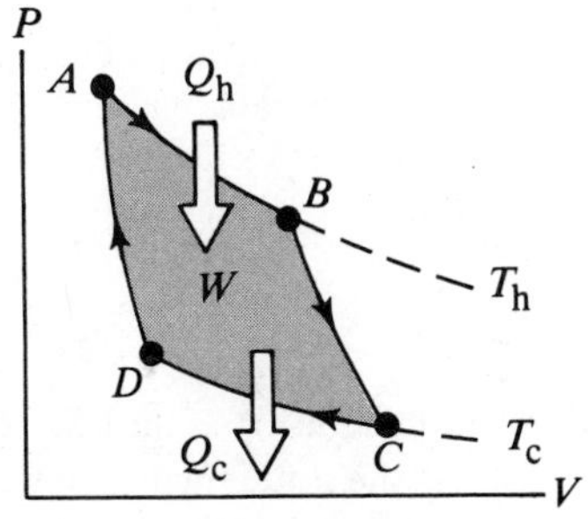

Figure 19.9 The *PV* diagram for the Carnot cycle. The net work done, *W*, equals the net heat received in one cycle, $Q_h - Q_c$. Note that $\Delta U = 0$ for the cycle.

Efficiency of a heat engine

$$e = \frac{W}{Q_h} = 1 - \frac{Q_c}{Q_h}$$

Equation 19.2 gives the efficiency of *any* heat engine operating between two reservoirs. In Example 19.1, we show that for a Carnot cycle, the ratio of heats Q_c/Q_h is given by

Ratio of heats for a Carnot cycle

$$\frac{Q_c}{Q_h} = \frac{T_c}{T_h} \qquad (19.4)$$

Hence, the thermal efficiency of a Carnot engine is given by

Efficiency of a Carnot engine

$$e_c = 1 - \frac{T_c}{T_h} \qquad (19.5)$$

Hence, we see that *all Carnot engines operating between the same two temperatures in a reversible manner have the same efficiency*. It can also be shown that *the efficiency of any reversible engine operating in a cycle between two temperatures is greater than the efficiency of any irreversible (real) engine operating between the same two temperatures.*[6]

[6]See, for example, F.W. Sears, *Thermodynamics, The Kinetic Theory of Gases, and Statistical Mechanics,* Reading, Mass., Addison-Wesley, 1953, Chapter 7.

Equation 19.5 can be applied to any working substance operating in a Carnot cycle between two heat reservoirs. According to this result, the efficiency is zero if $T_c = T_h$, as one would expect. The efficiency increases as T_c is lowered and as T_h increases. However, the efficiency can only be 1 (100%) if $T_c = 0$ K. Such reservoirs are not available, and so the maximum efficiency is always less than 1. In most practical cases, the cold reservoir is near room temperature, about 300 K. Therefore, one usually strives to increase the efficiency by raising the temperature of the hot reservoir. All real engines are less efficient than the Carnot engine since they are subject to such practical difficulties as friction and heat losses by conduction.

Example 19.1 *Efficiency of the Carnot Engine*

Show that the efficiency of a heat engine operating in a Carnot cycle using an ideal gas is given by Eq. 19.5.

Solution: During the isothermal expansion, $A \rightarrow B$ (Fig. 19.8a), the temperature does not change and so the internal energy remains constant. The work done by the gas is given by Eq. 18.15. According to the first law the heat absorbed, Q_h, equals the work done, so that

$$Q_h = W_{AB} = nRT_h \ln \frac{V_B}{V_A}$$

In a similar manner, the heat rejected to the cold reservoir during the isothermal compression $C \rightarrow D$ is given by

$$Q_c = W_{CD} = nRT_c \ln \frac{V_C}{V_D}$$

Dividing these expressions, we find that

$$(1) \qquad \frac{Q_c}{Q_h} = \frac{T_c}{T_h} \frac{\ln(V_C/V_D)}{\ln(V_B/V_A)}$$

We now show that the ratio of the logarithmic quantities is unity by obtaining a relation between the ratio of volumes.

For any quasi-static, adiabatic process, the pressure and volume are related by Eq. 18.18:

$$PV^{\gamma} = \text{constant}$$

During any reversible, quasi-static process, the ideal gas must also obey the equation of state, $PV = nRT$. Substituting this into the above expression to eliminate the pressure, we find that

$$TV^{\gamma-1} = \text{constant}$$

Applying this result to the adiabatic processes $B \rightarrow C$ and $D \rightarrow A$, we find that

$$T_h V_B^{\gamma-1} = T_c V_C^{\gamma-1}$$
$$T_h V_A^{\gamma-1} = T_c V_D^{\gamma-1}$$

Dividing these equations, we obtain

$$(V_B/V_A)^{\gamma-1} = (V_C/V_D)^{\gamma-1}$$

$$(2) \qquad \frac{V_B}{V_A} = \frac{V_C}{V_D}$$

Substituting (2) into (1), we see that the logarithmic terms cancel and we obtain the relation

$$\frac{Q_c}{Q_h} = \frac{T_c}{T_h}$$

Using this result and Eq. 19.2, the thermal efficiency of the Carnot engine is

$$e_c = 1 - \frac{Q_c}{Q_h} = 1 - \frac{T_c}{T_h} = \frac{T_h - T_c}{T_h}$$

Example 19.2

A heat engine operates between two reservoirs at 300 K and 500 K. During each cycle it absorbs 200 cal of heat from the hot reservoir. (a) What is the maximum efficiency of the engine?

The maximum efficiency equals the efficiency of a Carnot engine operating between these two reservoirs, given by Eq. 19.5:

$$e_c = 1 - \frac{T_c}{T_h} = 1 - \frac{300}{500} = 0.40, \text{ or } 40\%$$

(b) Determine the maximum work the engine can perform in each cycle of operation.

Since the efficiency is defined as the ratio W/Q_h, we see that the maximum work the engine can do per cycle is

$$W_{max} = e_c Q_h = 0.40(200 \text{ cal}) = 80 \text{ cal}$$

Any real engine operating between these two reservoirs will have an efficiency less than 40% and will do less than 80 cal of work in each cycle.

Q6. In practical heat engines, which do we have more control of, T_h or T_c? Explain.

Q7. A steam-driven turbine is one major component of an electrical power plant. Why is it advantageous to increase the temperature of the steam as much as possible?

19.4 THE ABSOLUTE TEMPERATURE SCALE

In Chapter 16 we defined temperature scales in terms of changes in certain physical properties with temperature. It is desirable to define a temperature scale that is independent of material properties. The Carnot cycle provides us with the basis for such a temperature scale. Equation 19.4 tells us that the ratio Q_c/Q_h depends *only* on the temperatures of the two heat reservoirs. The ratio of the two temperatures, T_c/T_h, can be obtained by operating a reversible heat engine in a Carnot cycle between these two temperatures and carefully measuring the heats Q_c and Q_h. A temperature scale can be determined with reference to some fixed-point temperatures. The *absolute*, or *kelvin*, temperature scale was defined by choosing 273.16 K as the absolute temperature of the triple point of water.

The temperature of any substance can be obtained in the following manner: (1) take the substance through a Carnot cycle; (2) measure the heat Q absorbed or expelled by the system at some temperature T; (3) measure the heat Q_3 absorbed or expelled by the system when it is at the temperature of the triple point of water. From Eq. 19.4 and this procedure we find that the unknown temperature is given by

$$T = (273.16\text{ K})\frac{Q}{Q_3}$$

As we saw in Example 19.1, the absolute temperature scale is identical to the ideal-gas temperature scale. The absolute temperature scale is independent of the property of the working substance. Therefore it can be applied even at very low temperatures.

In the previous section, we found that the thermal efficiency of any Carnot engine is given by $e_c = 1 - T_c/T_h$. This result shows that a 100% efficient engine is possible only if a temperature of absolute zero is maintained for T_c. If this were possible, any Carnot engine operating between T_h and $T_c = 0$ K would convert all of the absorbed heat into work.[7] Using this idea, Lord Kelvin defined absolute zero as follows: *Absolute zero is the temperature of a reservoir at which a Carnot engine will expel no heat.*

19.5 THE GASOLINE ENGINE

In this section we shall discuss the efficiency of the common gasoline engine. Four successive processes occur in each cycle, as illustrated in Fig. 19.10. During the *intake stroke* of the piston, air that has been mixed with gasoline vapor in the carburetor is drawn into the cylinder. During the *compression stroke,* the intake valve is closed and the air-fuel mixture is compressed approximately adiabatically. At this point a spark ignites the air-fuel mixture, causing a rapid increase in pressure and temperature at nearly constant volume. The burning gases expand and force the piston back, which produces the *power stroke.* Finally, during the *exhaust stroke,* the exhaust valve is opened and the rising piston forces most of the remaining gas out of the cylinder. The cycle is repeated after the exhaust valve is closed and the intake valve is opened.

Processes involved in the Otto cycle

These four processes can be approximated by the *Otto cycle,* a *PV* diagram of which is illustrated in Fig. 19.11.

1. In the process $A \rightarrow B$ (compression stroke), the air-fuel mixture is compressed

[7] Experimentally, it is not possible to reach absolute zero. Temperatures as low as about 10^{-5} K have been achieved with enormous difficulties using a technique called *nuclear demagnetization.* The fact that absolute zero may be approached but never reached is a law of nature known as the *third law of thermodynamics.*

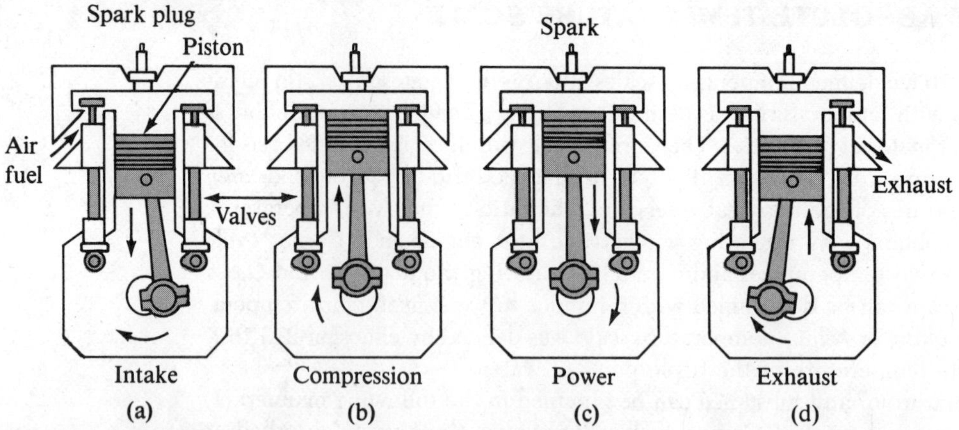

Figure 19.10 The four-stroke cycle of a conventional internal combustion engine. In the intake stroke, air is mixed with fuel. The intake value is then closed, and the air-fuel mixture is compressed by the piston. The mixture is ignited by the spark plug in the power stroke. Finally, the residual gases are exhausted.

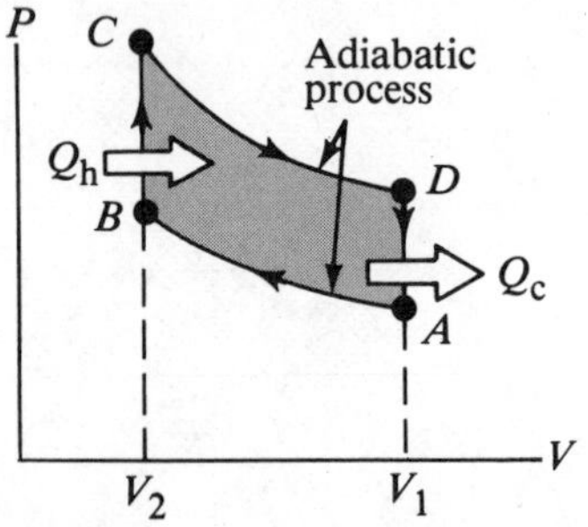

Figure 19.11 The *PV* diagram for the Otto cycle, which approximately represents the processes in the internal combustion engine. No heat is transferred during the adiabatic processes $A \to B$ and $C \to D$.

adiabatically from volume V_1 to volume V_2, and the temperature increases from T_A to T_B. The work done on the gas is the area under the curve *AB*.

2. In the process $B \to C$, combustion occurs and heat Q_h is added to the gas. This is not an inflow of heat but rather a release of heat from the combustion process. During this time the pressure and temperature rise rapidly, but the volume remains approximately constant. No work is done on the gas.
3. In the process $C \to D$ (power stroke), the gas expands adiabatically from V_2 to V_1, causing the temperature to drop from T_C to T_D. The work done by the gas equals the area under the curve *CD*.
4. In the final process, $D \to A$ (exhaust stroke), heat Q_c is extracted from the gas as its pressure decreases at constant volume. (Hot gas is replaced by cool gas.) No work is done during this process.

If the air-fuel mixture is assumed to be an ideal gas, the efficiency of the Otto cycle is shown in Example 19.3 to be

Efficiency of the Otto cycle

$$e = 1 - \frac{1}{(V_1/V_2)^{\gamma-1}} \tag{19.6}$$

where γ is the ratio of the molar heat capacities C_p/C_v and V_1/V_2 is called the *compression ratio*. This expression shows that the efficiency increases with increasing compression ratios. For a typical compression ratio of 8 and $\gamma = 1.4$, a theoretical efficiency of 56% is predicted for an engine operating in the idealized Otto cycle. This is much higher than what is achieved in real engines (15% to 20%) because of such effects as friction, heat loss to the cylinder walls, and incomplete combustion of the air-fuel mixture. Diesel engines have higher efficiencies than gasoline engines because of their higher compression ratios (about 16) and higher combustion temperatures.

Example 19.3 *Efficiency of the Otto Cycle*

Show that the thermal efficiency of an engine operating in an idealized Otto cycle (Fig. 18.11) is given by Eq. 19.6. Treat the working substance as an ideal gas.

Solution: First, let us calculate the work done by the gas during each cycle. No work is done during the processes $B \to C$ and $D \to A$. Work is done on the gas during the adiabatic compression $A \to B$, and work is done by the gas during the adiabatic expansion $C \to D$. The net work done equals the area bounded by the closed curve in Fig. 19.11. Since the change in internal energy is zero for one cycle, we see from the first law that the net work done for each cycle equals the net heat into the system:

$$W = Q_h - Q_c$$

Since the processes $B \to C$ and $D \to A$ take place at constant volume and since the gas is ideal, we find from the definition of heat capacity that

$$Q_h = nC_v(T_C - T_B) \qquad \text{and} \qquad Q_c = nC_v(T_D - T_A)$$

Using these expressions together with Eq. 19.2, we obtain for the thermal efficiency

$$(1) \qquad e = \frac{W}{Q_h} = 1 - \frac{Q_c}{Q_h} = 1 - \frac{T_D - T_A}{T_C - T_B}$$

We can simplify this expression by noting that the processes $A \to B$ and $C \to D$ are adiabatic and hence obey the relation $TV^{\gamma-1}$ = constant. Using this condition, and the facts that $V_A = V_D = V_1$ and $V_B = V_C = V_2$, we find that

$$(2) \qquad \frac{T_D - T_A}{T_C - T_B} = \left[\frac{V_2}{V_1}\right]^{\gamma-1}$$

Substituting (2) into (1) gives for the thermal efficiency

$$(3) \qquad e = 1 - \frac{1}{(V_1/V_2)^{\gamma-1}}$$

This can also be expressed in terms of a ratio of temperatures by noting that since $T_A V_1^{\gamma-1} = T_B V_2^{\gamma-1}$, it follows that

$$\left[\frac{V_2}{V_1}\right]^{\gamma-1} = \frac{T_A}{T_B} = \frac{T_D}{T_C}$$

Therefore (3) becomes

$$(4) \qquad e = 1 - \frac{T_A}{T_B} = 1 - \frac{T_D}{T_C}$$

During this cycle, the lowest temperature is T_A and the highest temperature is T_C. Therefore the efficiency of a Carnot engine operating between reservoirs at these two extreme temperatures $\left(\text{which is given by } e_c = 1 - \frac{T_A}{T_C}\right)$ would be *greater* than the efficiency of the Otto cycle, which is given by (4).

Q8. What factors limit the compression ratio to about 10 for an internal combustion engine? How can one increase the combustion temperature?

19.6 DEGRADATION OF ENERGY

The first law of thermodynamics is a general statement of the conservation of energy. It makes no distinction between the different forms of energy. The second law of thermodynamics says that thermal energy is different from all other forms of energy. Various forms of energy can be converted into thermal energy spontaneously, whereas the reverse transformation is never complete. For example, when a block slides on a table, the force of friction causes the block's kinetic energy to be converted into thermal energy and the block ultimately comes to rest. The reverse energy conversion does not occur. In general, if two kinds of energy, A and B, can be completely converted into each other, we can say that they are *of the same grade.* On the other hand, if form A can be completely converted into form B, but the reverse is never complete, then form A is a higher grade of energy than form B. For example, the kinetic energy of the sliding block is of higher grade than the thermal energy contained in the block and table. Therefore, when high-grade energy is converted into thermal energy, it can never be fully recovered as high-grade energy. This conversion of high-grade energy into thermal energy is referred to as the *degradation of energy.* The energy is said to be degraded because it takes on a form that is less useful for doing work. In other words, in all real processes where heat transfer occurs, the energy available for doing work decreases.

High and low grades of energy

To understand more clearly what we mean by *high-grade* and *low-grade* energy, recall from the previous chapter that thermal energy is actually a measure of the random kinetic energy of the molecules making up a substance. Since the motion of the large number of molecules is chaotic, or disordered, we regard this as a low-grade form of energy. In contrast, the kinetic energy of a macroscopic object, such as a ball, is a high-grade form of energy. It is the result of a highly ordered form of motion since all molecules have a common velocity (apart from their random thermal motions).

In real processes, the disorder in the system increases

When real (irreversible) processes occur, the degree of disorder or chaos in the system increases. For example, consider the isothermal expansion of an ideal gas in a container with a movable piston. As the gas absorbs heat and gradually expands, it maintains a constant temperature by doing work (pushing on the piston). After the expansion, the gas occupies a greater volume than it did originally. The gas molecules become more disordered in that they are not as localized as they were originally. Left by itself, the gas will not become ordered again by giving up its thermal energy to a reservoir. Thus, the flow of heat is in the direction that increases the amount of disorder. In view of these considerations, we can state the second law of thermodynamics as follows: *When an isolated system undergoes a change, the disorder in the system increases.* Furthermore, we can say that the *changes occurring in an isolated system result in a degradation of energy.* Ordered energy is converted into disordered energy.

The measure of the disorder in a system is made quantitative by introducing a quantity called *entropy.* For the moment we can think of entropy as being synonymous with the "degree of disorder" in a system. Simply stated, *an increase in disorder is equivalent to an increase in entropy.* For example, the highly ordered arrangement of atoms in a crystal of sodium chloride has lower entropy than the disordered arrangement of atoms in molten sodium chloride. The vapor phase has even more disorder, and consequently higher entropy. Entropy should not be confused with energy. The total energy of a closed system remains constant, whereas the entropy generally increases, never decreases. In fact, the entropy of a closed system tends to increase toward a maximum value. Entropy is rather abstract and must be defined carefully for every situation. In the next section we shall give a purely thermodynamic definition of entropy.

Q9. Is it possible to construct a heat engine that creates no thermal pollution? Explain.

Q10. Electrical energy (high-grade) can be converted into heat energy (low-grade) with an efficiency of 100%. Why is this number misleading with regard to heating a home? That is, what other factors must be considered in comparing the cost of electric heating versus the cost of hot-air or hot-water heating?

19.7 ENTROPY

The concept of temperature is involved in the zeroth law of thermodynamics, and the concept of internal energy is involved in the first law. Temperature and internal energy are both state functions. That is, they can be used to describe the thermodynamic state of a system. Another state function related to the second law of thermodynamics is the *entropy,* S. In this section we define entropy on a macroscopic scale as it was first expressed by Clausius in 1865.

Consider a quasi-static, reversible process between two equilibrium states. If $đQ_r$ is the heat absorbed or expelled by the system during some small interval of the path, the *change in entropy, dS,* between two equilibrium states is given by the heat transferred $đQ_r$ divided by the absolute temperature, *T,* of the system in this interval. That is,

Clausius definition of change in entropy

$$dS = \frac{đQ_r}{T} \tag{19.7}$$

The subscript r on the $đQ_r$ is used to emphasize that the definition applies only to *reversible* processes. When heat is absorbed by the system, $đQ_r$ is positive and hence the entropy increases. When heat is expelled by the system, $đQ_r$ is negative and the entropy decreases. Note that Eq. 19.7 does not define entropy, but the *change* in entropy. This is consistent with the fact that a change in state always accompanies heat transfer. Hence, the meaningful quantity in describing a process is the *change* in entropy.

To calculate the change in entropy for a finite process, we must recognize that T is generally not constant. If $đQ_r$ is the heat transferred when the system is at a temperature T, then the change in entropy in an arbitrary reversible process between an initial state and a final state is

Changes in entropy for a finite process

$$\Delta S = \int_i^f dS = \int_i^f \frac{đQ_r}{T} \quad \text{(reversible path)} \tag{19.8}$$

Although we do not prove it here, the change in entropy in going from one state to another has the same value for *all* reversible paths connecting the two states.[8] That is, *the change in entropy depends only on the properties of the initial and final equilibrium states.*

In the case of a *reversible, adiabatic* process, no heat is transferred between the system and its surroundings, and therefore $\Delta S = 0$ in this case. Since there is no change in entropy, such a process is often referred to as an *isentropic process.*

Consider the changes in entropy that occur in a Carnot heat engine operating between the temperatures T_c and T_h. In one cycle, the engine absorbs heat Q_h from a hot reservoir at a temperature T_h and rejects heat Q_c to a cold reservoir at a temperature T_c. Thus, the total change in entropy for one cycle is

$$\Delta S = \frac{Q_h}{T_h} - \frac{Q_c}{T_c}$$

where the negative sign in the second term represents the fact that heat Q_c is expelled by the system. In Example 19.1 we showed that for a Carnot cycle,

$$\frac{Q_c}{Q_h} = \frac{T_c}{T_h}$$

Using this result in the previous expression for ΔS, we find that the total change in entropy for a Carnot engine operating in a cycle is *zero.* That is,

Change in entropy for one Carnot cycle is zero

$$\Delta S = 0$$

Now consider a system taken through an arbitrary reversible cycle. Since the entropy function is a state function and hence depends only on the properties of a given equilibrium state, we conclude that $\Delta S = 0$ for *any* reversible cycle. In general, we can write this condition in the mathematical form

$\Delta S = 0$ for any reversible cycle

$$\oint \frac{đQ_r}{T} = 0 \tag{19.9}$$

where the symbol $\oint$ indicates that the integration is over a *closed* path.

Another important property of entropy is the fact that *the entropy of the universe remains constant in a reversible process.* This can be understood by noting that two bodies A and B that interact with each other reversibly must always be in thermal equilibrium with each other. That is, their temperatures must always be equal.

[8] Note that the quantity $đQ_r$ is called an *inexact differential quantity,* whereas $đQ_r/T = dS$ is a perfect differential. This is because heat is not a property of the system, and hence Q is not a state function. Mathematically, we call $1/T$ the *integrating factor* in this case, since the perfect differential $đQ_r/T$ can be integrated.

Therefore, when a small amount of heat $đQ$ is transferred from A to B, the increase in entropy of B is $đQ/T$, while the corresponding change in entropy of A is $-đQ/T$. Thus the total change in entropy of the system (A + B) is zero, and the entropy of the universe is unaffected by the reversible process.[9]

As a special case, we next show how to calculate the change in entropy for an ideal gas that undergoes a quasi-static, reversible process in which heat is absorbed from a reservoir.

Quasi-static, Reversible Process for an Ideal Gas

An ideal gas undergoes a quasi-static, reversible process from an initial state T_i, V_i to a final state T_f, V_f. Let us calculate the change in entropy for this process.

According to the first law, $đQ_r = dU + đW$, where $đW = P\,dV$. For an ideal gas, recall that $dU = nC_v\,dT$ and $P = nRT/V$. Therefore, we can express the heat transferred as

$$đQ_r = dU + P\,dV = nC_v\,dT + nRT\frac{dV}{V} \tag{19.10}$$

We cannot integrate this expression as it stands since the last term contains two variables, T and V. However, if we divide each term by T, we can integrate both terms on the right-hand side:

$$\frac{đQ_r}{T} = nC_v\frac{dT}{T} + nR\frac{dV}{V} \tag{19.11}$$

Assuming that C_v is constant over the interval in question, and integrating Eq. 19.11 from T_i, V_i to T_f, V_f, we get

$$\Delta S = \int_i^f \frac{đQ_r}{T} = nC_v \ln\frac{T_f}{T_i} + nR\ln\frac{V_f}{V_i} \tag{19.12}$$

This expression shows that ΔS depends *only on the initial and final states and is independent of the reversible path.* Furthermore, ΔS can be positive or negative depending on whether the gas absorbs or expels heat during the process. Finally, for a cyclic process ($T_i = T_f$ and $V_i = V_f$), we see that $\Delta S = 0$.

Example 19.4 *Change in Entropy-Melting Process*

A solid substance with a latent heat of fusion L melts at a temperature T_m. Calculate the change in entropy that occurs when m grams of this substance is melted.

Solution: Let us assume that the melting process occurs so slowly that it can be considered a reversible process. In that case the temperature can be considered to be constant and equal to T_m. Making use of Eqs. 19.8 and 17.7, we find that

$$\Delta S = \int \frac{đQ_r}{T} = \frac{1}{T_m}\int đQ = \frac{Q}{T_m} = \frac{mL}{T_m} \tag{19.13}$$

Note that we were able to remove T_m from the integral in this case since the process is isothermal. Also, the quantity Q is the total heat required to melt the substance and is equal to mL (Section 17.3).

As a numerical example, let us calculate the change in entropy when 300 g of lead melts at 327°C (= 600 K). Lead has a latent heat of fusion equal to 5.85 cal/g (Table 17.2):

$$\Delta S = \frac{mL}{T_m} = \frac{(300\text{ g})(5.85\text{ cal/g})}{600\text{ K}} = 2.93\text{ cal/K}$$

[9] Alternatively, we can say that since the universe is, by definition, an isolated system, it never gains or loses heat; hence the change in entropy of the universe is zero for a reversible process.

Q11. What is the change in entropy of a gas that expands quasi-statically and adiabatically from a volume V to a volume $3V$?

Q12. Under what conditions is it possible for the entropy of a system to decrease? Give an example.

19.8 ENTROPY CHANGES IN IRREVERSIBLE PROCESSES

By definition, the change in entropy for a system can be calculated only for reversible paths connecting the initial and final equilibrium states. In order to calculate changes in entropy for real (irreversible) processes, we must first recognize that the entropy function (like internal energy) depends only on the *state* of the system. That is, entropy is a state function. Hence, the change in entropy of a system between any two equilibrium states depends only on the initial and final states. Experimentally one finds that the entropy change is the same for all processes between the initial and final states.[10]

In view of the fact that the entropy of a system depends only on the state of the system, we can now calculate entropy changes for irreversible processes between two equilibrium states. This can be accomplished by devising a reversible process (or series of reversible processes) between the same two equilibrium states and computing $\int dQ_r/T$ for the reversible process. The entropy change for the irreversible process is the same as that of the reversible process between the same two equilibrium states. Let us demonstrate this procedure with a few specific cases.

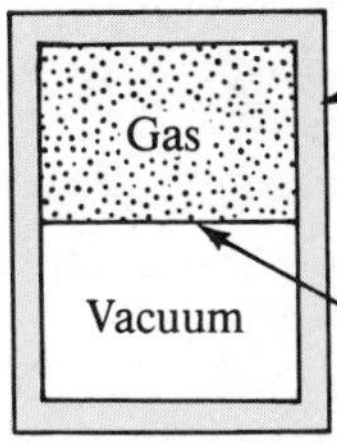

Figure 19.12 Free expansion of a gas. When the partition separating the gas from the evacuated region is ruptured, the gas expands freely and irreversibly so that it occupies a greater final volume. The container is thermally insulated from its surroundings, and so $Q = 0$.

Heat Conduction

Consider the transfer of heat Q from a hot reservoir at temperature T_h to a cold reservoir at temperature T_c. Since the cold reservoir absorbs heat Q, its entropy increases by Q/T_c. At the same time, the hot reservoir loses heat Q and its entropy decreases by Q/T_h. The increase in entropy of the cold reservoir is greater than the decrease in entropy of the hot reservoir since T_c is less than T_h. Therefore, the total change in entropy of the system (universe) is greater than zero:

$$\Delta S_u = \frac{Q}{T_c} - \frac{Q}{T_h} > 0$$

Free Expansion

An ideal gas in an insulated container initially occupies a volume V_i (Fig. 19.12). A partition separating the gas from another evacuated region is suddenly broken so that the gas expands (irreversibly) to a volume V_f. Let us find the change in entropy of the gas and the universe.

The process is clearly neither reversible nor quasi-static. The work done by the gas against the vacuum is zero, and since the walls are insulating, no heat is transferred during the expansion. That is, $W = 0$ and $Q = 0$. Using the first law, we see that the change in internal energy is zero, therefore $U_i = U_f$, where i and f indicate the initial and final equilibrium states. Since the gas is ideal, U depends on temperature only, and so we conclude that $T_i = T_f$.

We cannot use Eq. 19.8 directly to calculate the change in entropy since that equation applies only to reversible processes. In fact, at first sight one might *wrongfully* conclude that $\Delta S = 0$ since there is no heat transferred. To calculate the

[10] It is also possible to show that if this were not the case, the second law of thermodynamics would be violated.

change in entropy, let us imagine a reversible process between the same initial and final equilibrium states. A simple one to choose is an isothermal, reversible expansion in which the gas pushes slowly against a piston. Since T is constant in this process, Eq. 19.8 gives

$$\Delta S = \int \frac{đQ_r}{T} = \frac{1}{T}\int_i^f đQ_r$$

But $\int_i^f đQ_r$ is simply the work done by the gas during the isothermal expansion from V_i to V_f, which is given by Eq. 17.15. Using this result, we find that

Change in entropy during a free expansion

$$\Delta S = nR \ln \frac{V_f}{V_i} \tag{19.14}$$

Since $V_f > V_i$, we conclude that ΔS is positive, and so both the entropy and disorder of the gas (and universe) increase as a result of the irreversible, adiabatic expansion. This result can also be obtained from Eq. 19.12, noting that $T_i = T_f$, and so $dT = 0$.

Example 19.5

Calculate the change in entropy of 2 moles of an ideal gas that undergoes a free expansion to three times its initial volume.

Solution: Using Eq. 19.14 with $n = 2$ and $V_f = 3V_i$, we find that

$$\Delta S = nR \ln \frac{V_f}{V_i} = (3 \text{ moles})(8.31 \text{ J/mole} \cdot \text{K}) \ln 3$$
$$= 27.4 \text{ J/K}$$

Entropy of Mixing

A substance of mass m_1, specific heat c_1, and initial temperature T_1 is mixed with a second substance of mass m_2, specific heat c_2, and initial temperature T_2, where $T_2 > T_1$. (For example, they could both be liquids.) The mixed system is allowed to reach thermal equilibrium. What is the total entropy change for the system?

First, let us calculate the final equilibrium temperature, T_f. Energy conservation requires that the heat lost by one substance equal the heat gained by the other. Since by definition, $Q = mc\,\Delta T$ for each substance, we get $Q_1 = -Q_2$, or

$$m_1c_1\,\Delta T = -m_2c_2\,\Delta T$$
$$m_1c_1(T_f - T_1) = -m_2c_2(T_f - T_2)$$

Solving for T_f gives

$$T_f = \frac{m_1c_1T_1 + m_2c_2T_2}{m_1c_1 + m_2c_2} \tag{19.15}$$

Note that $T_1 < T_f < T_2$, as would be expected.

The mixing process is irreversible since the system goes through a series of nonequilibrium states. During such a transformation, the temperature at any time is not well defined. However, we can imagine that the hot body is slowly cooled to the temperature T_f while the cool body is slowly warmed to T_f by placing them in contact with heat reservoirs at T_f. These would be reversible processes. Applying Eq. 19.8 and noting that $dQ = mcdT$ for an infinitesimal change, we get

$$\Delta S = \int_1 \frac{đQ_1}{T} + \int_2 \frac{đQ_2}{T} = m_1c_1\int_{T_1}^{T_f} \frac{dT}{T} + m_2c_2\int_{T_2}^{T_f} \frac{dT}{T}$$

where we have assumed that the specific heats remain constant. Integrating, we find that

$$\Delta S = m_1 c_1 \ln \frac{T_f}{T_1} + m_2 c_2 \ln \frac{T_f}{T_2} \tag{19.16}$$

Change in entropy for a mixing process

where T_f is given by Eq. 19.15. If Eq. 19.15 is substituted into Eq. 19.16, you can show that one of the terms in Eq. 19.16 will always be positive and the other negative. (The verification of this is left as a problem). However, the positive term will always be larger than the negative term, resulting in a positive value for ΔS. Thus, we conclude that the entropy of the universe (system) increases in this irreversible process.

Example 19.6

One kg of water at 0°C is mixed with an equal mass of water at 100°C. After equilibrium is reached, the mixture has a uniform temperature of 50°C. What is the change in entropy of the system?

Solution: The change in entropy can be calculated from Eq. 19.16 using the values $m_1 = m_2 = 1$ kg, $c_1 = c_2 = 1$ cal/g · K = 4.19 J/g · K, $T_1 = 0°\text{C}$ (= 273 K), $T_2 =$ 100°C (= 373 K), and $T_f = 50°\text{C}$ (= 323 K). Note that you must use absolute temperatures in this calculation.

$$\Delta S = m_1 c_1 \ln \frac{T_f}{T_1} + m_2 c_2 \ln \frac{T_f}{T_2}$$

$$= (10^3 \text{ g})(4.19 \text{ J/g} \cdot \text{K}) \ln \frac{323}{273} + (10^3 \text{ g})(4.19 \text{ J/g} \cdot \text{K}) \ln \frac{323}{373}$$

$$= 705 \text{ J/K} - 603 \text{ J/K} = 102 \text{ J/K}$$

That is, the increase in entropy of the cold water is greater than the decrease in entropy of the warm water as a result of this irreversible mixing process. Consequently, the increase in entropy of the system is 102 J/K.

The cases just described show that the change in entropy of a system is always positive for an irreversible process. In general, the total entropy (and disorder) always increase in irreversible processes. From these considerations, the second law of thermodynamics can be stated as follows: *The total entropy of an isolated system that undergoes a change cannot decrease.* Furthermore, if the process is *irreversible,* the total entropy of an isolated system always *increases.* On the other hand, in a reversible process, the total entropy of an isolated system remains constant. When dealing with interacting bodies that are not isolated, one must be careful to note that the system refers to the bodies *and* their surroundings. When two bodies interact in an irreversible process, the increase in entropy of one part of the system is greater than the decrease in entropy of the other part. Hence, we conclude that *the change in entropy of the universe must be greater than zero for an irreversible process and equal to zero for a reversible process.* Ultimately, the entropy of the universe should reach a maximum value. At this point, the universe will be in a state of uniform temperature and density. All physical, chemical, and biological processes will cease, since a state of perfect disorder implies no energy available for doing work. This gloomy state of affairs is sometimes referred to as an ultimate "heat death" of the universe.

An illustration from Flammarion's novel *La Fin du Monde,* depicting the "heat-death" of the universe.

19.9 ENERGY CONVERSION AND THERMAL POLLUTION

The main source of thermal pollution is waste heat from electrical power plants. In the United States, about 85 percent of the electric power is produced by steam engines, which burn either fossil fuels (coal, oil, or natural gas) or nuclear fuels (uranium-235). The remaining 15 percent of the electric power is generated by water in hydroelectric plants. The overall thermal efficiency of a modern fossil-fuel plant is about 40 percent. The actual efficiencies of any power plant must be lower than the theoretical efficiencies derived from the second law of thermodynamics. One always seeks the highest efficiency possible for two reasons. First, higher efficiency results in lower fuel costs. Second, thermal pollution of the environment is reduced since there is less waste energy in a highly efficient power plant. Since any power plant involves several steps of energy conversion, the inefficiency will accumulate in steps.

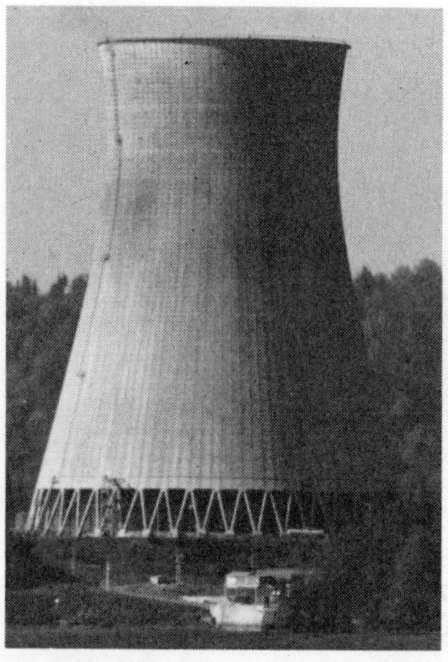

A cooling tower at a reactor site in southern Washington. (From Jonathan Turk and Amos Turk, *Physical Science*, 2nd edition, Philadelphia, Saunders College Publishing, 1981.)

The burning of fossil fuels in an electrical power plant involves three energy-conversion processes: (1) chemical to thermal energy, (2) thermal to mechanical energy, and (3) mechanical to electrical energy. These are indicated schematically in Fig. 19.13.

During the first step, heat energy is transferred from the burning fuel to a water supply, which is converted into steam. In this process about 12 percent of the available energy is lost up the chimney. In the second step, thermal energy in the form of steam at high pressure and temperature passes through a turbine and is converted into mechanical energy. A well-designed turbine has an efficiency of about 47 percent. Steam, which leaves the turbine at a lower pressure, is then condensed into water and gives up heat in the process. Finally, in the third step, the turbine drives an electrical generator of very high efficiency, typically 99 percent. Hence, the overall efficiency is the product of the efficiencies of each step, which for the figures given becomes $(0.88)(0.47)(0.99) = 0.41$, or 41 percent. The thermal energy transferred to the cooling water amounts to about 47 percent of the initial fuel energy.

In the case of nuclear power plants, the steam generated by the nuclear reactor is at a lower temperature than that of a fossil-fuel plant. This is due primarily to material limitations in the reactor. Typical nuclear power plants have an overall efficiency of about 34 percent.

The waste heat from electrical power plants can be disposed of in various ways. The method shown in Fig. 19.13 involves passing water from a river or lake through a condenser and returning it to that source at a higher temperature. This can raise the water temperature of the river or lake by several degrees, which can produce undesirable ecological effects, such as the increased growth of bacteria, undesirable blue-green algae, and pathogenic organisms. Fish and other marine life are also affected since they require oxygen and the percentage of dissolved oxygen in the water decreases with increasing temperature. There is further demand for oxygen in the decomposition of organic matter, which also proceeds at a higher rate as the temperature increases.

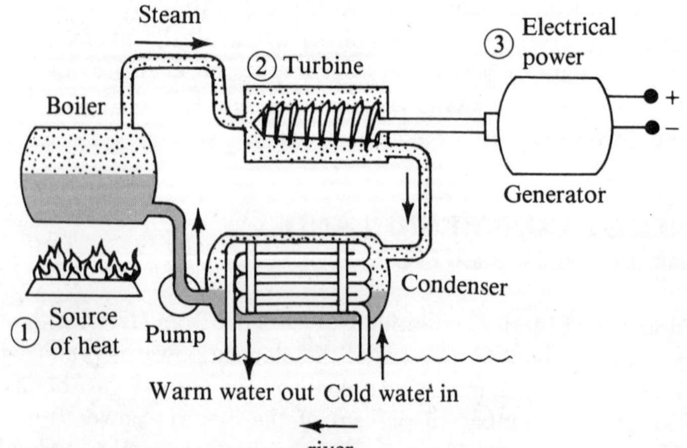

Figure 19.13 Schematic diagram of an electrical power plant.

Cooling towers are also commonly used in disposing of waste heat. These towers usually use the heat to evaporate water, which is then released to the atmosphere. Cooling towers also present environmental problems since evaporated water can cause increased precipitation, fog, and ice. Another type of cooling tower is the dry cooling tower (nonevaporative), which transfers heat to the atmosphere by conduction. However, this type is more expensive and cannot cool to as low a temperature as an evaporative tower.

19.10 SUMMARY

The *first law of thermodynamics* is a generalization of the law of conservation of energy that includes heat transfer in any process.

Real processes proceed in an order governed by the second law of thermodynamics.

A *heat engine* is a device that converts thermal energy into other useful forms of energy. The net work done by a heat engine in carrying a substance through a cyclic process ($\Delta U = 0$) is given by

Work done by a heat engine

$$W = Q_h - Q_c \qquad (19.1)$$

where Q_h is the heat absorbed from a warmer reservoir and Q_c is the heat rejected to a cooler reservoir.

The *thermal efficiency, e,* of a heat engine is defined as the ratio of the net work done to the heat absorbed per cycle:

Thermal efficiency

$$e = \frac{W}{Q_h} = 1 - \frac{Q_c}{Q_h} \qquad (19.2)$$

The *second law of thermodynamics* can be stated in many ways:

Statements of the second law

1. No heat engine operating in a cycle can absorb thermal energy from a reservoir and perform an equal amount of work (Kelvin-Planck statement).
2. A perpetual-motion machine of the second kind is impossible to construct.
3. It is impossible to construct a cyclical machine whose sole effect is to transfer heat continuously from one body to another body at a higher temperature (Clausius statement).

Reversible process

A process is *reversible* if the system passes from the initial to the final state through a succession of equilibrium states. A process can be reversible only if it occurs quasi-statically.

Irreversible process

An *irreversible* process is one in which the system and its surroundings cannot be returned to their initial states. In such a process, the system passes from the initial to the final state through a series of nonequilibrium states.

The *efficiency of a heat engine* operating in the Carnot cycle is given by

Efficiency of a Carnot engine

$$e_c = 1 - \frac{T_c}{T_h} \qquad (19.5)$$

where T_c is the absolute temperature of the cold reservoir and T_h is the absolute temperature of the hot reservoir.

No real heat engine operating (irreversibly) between the temperatures T_c and T_h can be more efficient than an engine operating reversibly in a Carnot cycle between the same two temperatures.

The second law of thermodynamics states that when real (irreversible) processes occur, the degree of disorder in the system increases. When a process occurs in an isolated system, ordered energy is converted into disordered energy. The measure of disorder in a system is called *entropy*, S.

The *change in entropy*, dS, of a system moving quasi-statically between two equilibrium states is given by

Clausius definition of change in entropy

$$dS = \frac{đQ_r}{T} \qquad (19.7)$$

The change in entropy of a system moving reversibly between two equilibrium states is

$$\Delta S = \int_i^f \frac{đQ_r}{T} \qquad (19.8)$$

The value of ΔS is the same for all reversible paths connecting the initial and final states.

The change in entropy for any reversible, cyclic process is zero.

In any reversible process, the entropy of the universe remains constant.

The entropy of a system is a state function, that is, it depends on the state of the system. The change in entropy for a system undergoing a real (irreversible) process between two equilibrium states is the same as that of a reversible process between the same states.

The total entropy of an isolated system always increases in an irreversible process. In general, the total entropy (and disorder) always increases in any irreversible process. Furthermore, the change in entropy of the universe is greater than zero for an irreversible process and is zero for a reversible process.

EXERCISES

Section 19.1 Heat Engines and the Second Law of Thermodynamics

1. A heat engine absorbs 90 cal of heat and performs 25 J of work in each cycle. Find (a) the efficiency of the engine and (b) the heat expelled in each cycle.
2. A heat engine performs 200 J of work in each cycle and has an efficiency of 30 percent. For each cycle of operation, (a) how much heat is absorbed, and (b) how much heat is expelled?
3. An engine absorbs 400 cal from a hot reservoir and expels 250 cal to a cold reservoir in each cycle. (a) What is the efficiency of the engine? (b) How much work is done in each cycle? (c) What is the power output of the engine if each cycle lasts for 0.3 s?
4. A particular engine has a power output of 5 kW and an efficiency of 25 percent. If the engine expels 2000 cal of heat in each cycle, find (a) the heat absorbed in each cycle and (b) the time for each cycle.
5. A refrigerator has a coefficient of performance equal to 5. If the refrigerator absorbs 30 cal of heat from a cold reservoir in each cycle, find (a) the work done in each cycle and (b) the heat expelled to the hot reservoir.
6. In each cycle of its operation, a certain refrigerator absorbs 25 cal from the cold reservoir and expels 32 cal. (a) What is the power required to operate the refrigerator if it works at 60 cycles/s? (b) What is the coefficient of performance of the refrigerator?
7. The heat absorbed by an engine is three times greater than the work it performs. (a) What is its thermal efficiency? (b) What fraction of the heat absorbed is expelled to the cold reservoir?

Section 19.3 The Carnot Engine

8. A heat engine operates between two reservoirs at temperatures of 20°C and 300°C. What is the maximum efficiency possible for this engine?
9. The efficiency of a Carnot engine is 30%. The engine absorbs 200 cal of heat per cycle from a hot reservoir at 500 K. Determine (a) the heat expelled per cycle and (b) the temperature of the cold reservoir.
10. A Carnot engine has a power output of 150 kW. The engine operates between two reservoirs at 20°C and 500°C. (a) How much heat energy is absorbed per hour? (b) How much heat energy is lost per hour?
11. A power plant has been proposed that would make

use of the temperature gradient in the ocean. The system is to operate between 20°C (surface water temperature) and 5°C (water temperature at a depth of about 1 km). (a) What is the maximum efficiency of such a system? (b) If the power output of the plant is 75 MW, how much thermal energy is absorbed per hour? (c) In view of your results to (a), do you think such a system is worthwhile?

12. A heat engine operates in a Carnot cycle between 80°C and 350°C. It absorbs 5×10^3 cal of heat per cycle from the hot reservoir. The duration of each cycle is 1 s. (a) What is the maximum power output of this engine? (b) How much heat does it expel in each cycle?

13. One of the most efficient engines ever built operates between 430°C and 1870°C. Its actual efficiency is 42%. (a) What is its maximum theoretical efficiency? (b) How much power does the engine deliver if it absorbs 3.5×10^4 cal of heat each second?

14. An electrical generating plant has a power output of 500 MW. The plant uses steam at 200°C and exhausts water at 40°C. If the system operates with one half the maximum (Carnot) efficiency, (a) at what rate is heat expelled to the environment? (b) If the waste heat goes into a river whose flow rate is 1.2×10^6 kg/s, what is the rise in temperature of the river?

15. An air conditioner absorbs heat from its cooling coil at 13°C and expels heat to the outside at 30°C. (a) What is the *maximum* coefficient of performance of the air conditioner? (b) If the actual coefficient of performance is one third of the maximum value and if the air conditioner removes 2×10^3 cal of heat energy each second, what power must the motor deliver?

16. A heat pump powered by an electric motor absorbs heat from outside at 5°C and exhausts heat inside in the form of hot air at 40°C. (a) What is the maximum coefficient of performance of the heat pump? (b) If the actual coefficient of performance is 3.2, what *fraction* of the available work (electrical energy) is actually done?

17. An ideal gas is taken through a Carnot cycle. The isothermal expansion occurs at 250°C, and the isothermal compression takes place at 50°C. If the gas absorbs 300 cal of heat during the isothermal expansion, find (a) the heat expelled to the cold reservoir in each cycle, and (b) the net work done by the gas in each cycle.

Section 19.5 The Gasoline Engine

18. A gasoline engine has a compression ratio of 6 and uses a gas with $\gamma = 1.4$. (a) What is the efficiency of the engine if it operates in an idealized Otto cycle? (b) If the actual efficiency is 15%, what fraction of the fuel is wasted as a result of friction and heat losses. (Assume complete combustion of the air-fuel mixture.)

19. A gasoline engine using an ideal diatomic gas ($\gamma = 1.4$) operates between temperature extremes of 300 K and 1500 K. Determine its compression ratio if it has an efficiency of 20%. Compare this efficiency to that of a Carnot engine operating between the same temperatures.

Section 19.7 Entropy

20. One mole of an ideal gas expands isothermally and quasi-statically to twice its initial volume. What is the change in entropy of the gas?

21. One mole of an ideal monatomic gas is heated quasi-statically at constant volume from 300 K to 400 K. What is the change in entropy of the gas?

22. An ice tray contains 500 g of water at 0°C. Calculate the change in entropy of the water as it freezes completely and slowly at 0°C.

23. Calculate the change in entropy of 250 g of water when it is slowly heated from 20°C to 80°C. (Hint: Note that $đQ = mc\,dT$.)

24. One kg of mercury is initially at −100°C. What is its change in entropy when heat is slowly added to raise its temperature to 100°C? (Mercury has a melting temperature of −39°C, a heat of fusion of 2.8 cal/g, and a specific heat of 0.033 cal/g · C°.)

Section 19.8 Entropy Changes in Irreversible Processes

25. If 800 cal of heat flows from a heat reservoir at 500 K to another reservoir at 300 K through a conducting metal rod, find the change in entropy of (a) the hot reservoir, (b) the cold reservoir, (c) the metal rod, and (d) the universe.

26. A glass ampule of volume 150 cm^3 contains 0.15 moles of an ideal gas. The ampule is broken in an evacuated vessel of volume 800 cm^3. In this free expansion of the gas, find the change in entropy of the universe.

27. If 200 g of water at 20°C is mixed with 300 g of water at 75°C, find (a) the final equilibrium temperature of the mixture and (b) the change in entropy of the system.

28. A 2-kg block moving with an initial speed of 5 m/s slides on a rough table and is stopped by the force of friction. Assuming the table and air remain at a temperature of 20°C, calculate the entropy change of the universe.

29. A 70-kg log falls from a height of 25 m into a lake. If the log, the lake, and the air are all at 300 K, find the change in entropy of the universe for this process.

30. A cyclic heat engine operates between two reservoirs at temperatures of 300 K and 500 K. In each cycle, the engine absorbs 700 J of heat from the hot reservoir and does 160 J of work. Find the entropy change in each cycle for (a) each reservoir, (b) the engine, and (c) the universe.

31. A 6-kg block of ice at 0°C is dropped into a lake at 27°C. After the ice has all melted, what is the change in entropy of (a) the ice, (b) the lake, and (c) the universe?

PROBLEMS

1. One mole of an ideal monatomic gas is taken through the cycle shown in Fig. 19.14. The process *AB* is an isothermal expansion. Calculate (a) the net work done by the gas, (b) the heat added to the gas, (c) the heat expelled by the gas, and (d) the efficiency of the cycle.

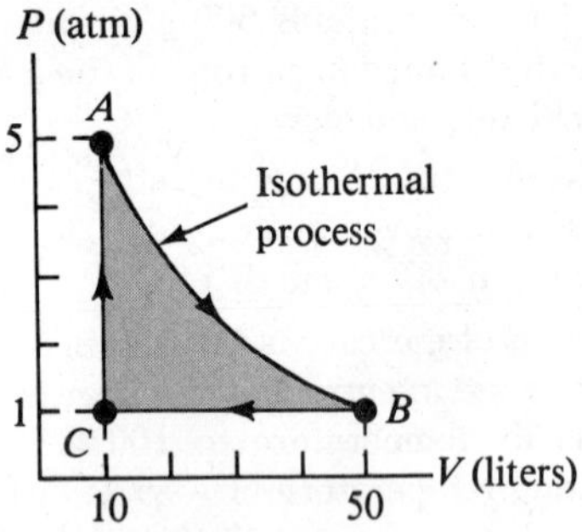

Figure 19.14 (Problem 1).

2. An ideal refrigerator (or heat pump) is equivalent to a Carnot engine running in reverse. That is, heat Q_c is absorbed from a cold reservoir and heat Q_h is rejected to a hot reservoir. (a) Show that the work that must be supplied to run the refrigerator is given by

$$W = \frac{T_h - T_c}{T_c} Q_c$$

(b) Show that the coefficient of performance of the ideal refrigerator is given by

$$\text{C.P.} = \frac{T_c}{T_h - T_c}$$

3. Figure 19.15 represents *n* moles of an ideal gas being taken through a reversible cycle consisting of two isothermal processes at temperatures $3T_0$ and T_0 and two constant-volume processes. For each cycle, determine in terms of *n*, *R*, and T_0 (a) the net heat transferred to the gas and (b) the efficiency of an engine operating in this cycle.

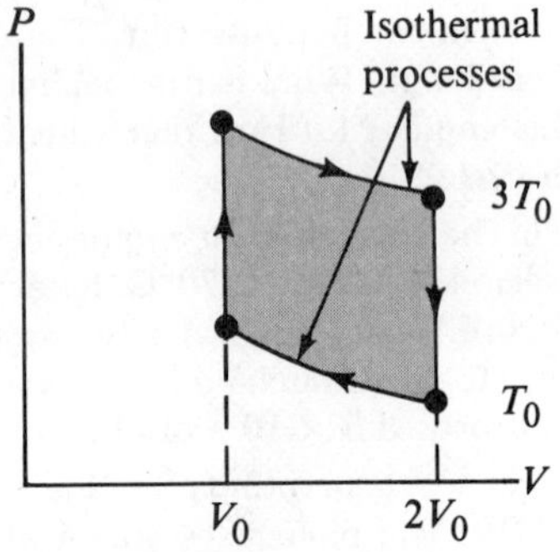

Figure 19.15 (Problem 3).

4. One mole of an ideal gas ($\gamma = 1.4$) is carried through the Carnot cycle described in Fig. 19.9. At point *A*, the pressure is 25 atm and the temperature is 600 K. At point *C*, the pressure is 1 atm and the temperature is 400 K. (a) Determine the pressures and volumes at points *A*, *B*, *C*, and *D*. (b) Calculate the net work done per cycle. (c) Determine the efficiency of an engine operating in this cycle.

5. One mole of a monatomic ideal gas is taken through the reversible cycle shown in Fig. 1916. At point *A*, the pressure, volume, and temperature are P_0, V_0, and T_0, respectively. In terms of *R* and T_0, find (a) the total heat entering the system per cycle, (b) the total heat leaving the system per cycle, (c) the efficiency of an engine operating in this reversible cycle, and (d) the efficiency of an engine operating in a Carnot cycle between the temperature extremes for this process.

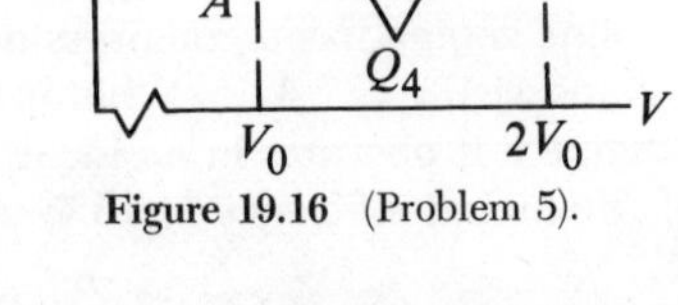

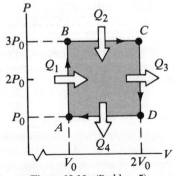

Figure 19.16 (Problem 5).

6. (a) Show that the work done by a system in any reversible cycle is given by $W = \oint TdS$, where the integral is over the closed path corresponding to the cyclic process. (Hint: Make use of the first law, $đQ = dU + PdV$, and the definition of the change in entropy as given by Eq. 19.7.) (b) Construct an entropy-temperature plot for the Carnot cycle described in Fig. 19.9. Identify the process associated with each line on the graph. What is the physical significance of the area enclosed by the cycle on this plot?

7. An idealized Diesel engine operates in a cycle known as the *air-standard Diesel cycle,* shown in Fig. 19.17. Fuel is sprayed into the cylinder at the point of maximum compression, B. Combustion occurs during the expansion $B \rightarrow C$, which is approximated as an isobaric process. The rest of the cycle is the same as in the gasoline engine, described in Fig. 19.11. Show that the efficiency of an engine operating in this idealized Diesel cycle is given by

$$e = 1 - \frac{1}{\gamma}\left(\frac{T_D - T_A}{T_C - T_B}\right)$$

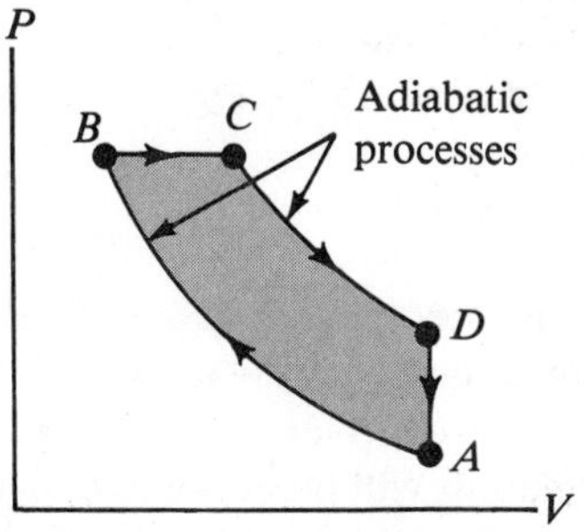

Figure 19.17 (Problem 7).

8. One mole of an ideal gas at an initial pressure of 4 atm and temperature of 300 K is carried through the following reversible cycle: (1) it expands isothermally until its volume is doubled; (2) it is compressed to its original volume at constant pressure; (3) it is compressed isothermally to a pressure of 4 atm; and (4) it expands at constant pressure to its original volume. (a) Make an accurate plot of the cyclic process on a PV diagram. (b) Calculate the work done by the gas per cycle. (c) What is the efficiency of an engine operating in this cycle if 700 J of heat is removed from the gas during the cycle?

9. A system consisting of n moles of an ideal gas undergoes a reversible, *isobaric* process from a volume V_0 to a volume $3V_0$. Calculate the change in entropy of the gas. (Hint: Imagine that the system goes from the initial state to the final state first along an isotherm and then along an adiabatic curve, for which there is no change in entropy.)

10. An electrical power plant has an overall efficiency of 15%. The plant is to deliver 150 MW of power to a city, and its turbines use coal as the fuel. The burning coal produces steam at 190°C, which drives the turbines. This steam is then condensed into water at 25°C by passing it through cooling coils in contact with river water. (a) How many metric tons of coal does the plant consume each day (1 metric ton = 10^3 kg)? (b) What is the total cost of the fuel per year if the delivered price is \$8/metric ton? (c) If the river water is delivered at 20°C, at what minimum rate must it flow over the cooling coils in order that its temperature not exceed 25°C? (Note: the heat of combustion of coal is 7.8×10^3 cal/g.)

11. Consider one mole of an ideal gas which undergoes a quasi-static, reversible process in which the heat capacity C remains constant. Show that the pressure and volume of the gas obey the relation

$$PV^{\gamma'} = \text{constant, where } \gamma' = \frac{C - C_p}{C - C_v}$$

(Hint: Start with the first law of thermodynamics, $đQ = dU + P\,dV$, and use the fact that $đQ = C\,dT$ and $dU = C_v\,dT$. Furthermore, note that $PV = RT$ applies to the gas, so it follows that $P\,dV + V\,dP = R\,dT$.)

20 Electric Fields

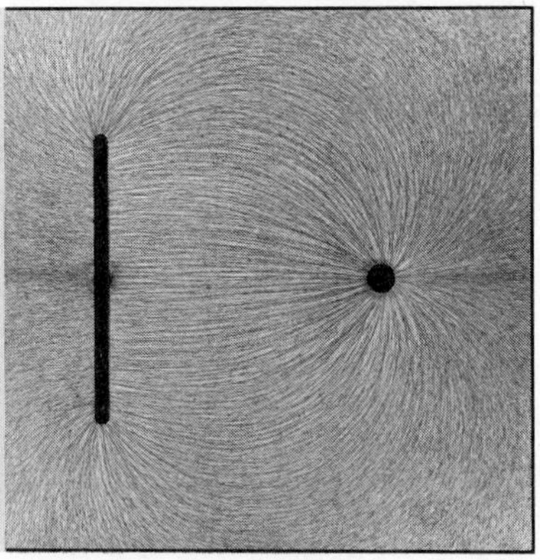

Harold Waage, Princeton University

The electromagnetic force between charged particles is one of the fundamental forces of nature. In this chapter, we begin by describing some of the basic properties of electrostatic forces. We then discuss Coulomb's law, which is the fundamental law of force between any two charged particles. The concept of an electric field associated with a charge distribution is then introduced, and its effect on other charged particles is described. The method for calculating electric fields of a given charge distribution from Coulomb's law is discussed, and several examples are given. Then the motion of a charged particle in a uniform electric field is discussed. We conclude the chapter with a brief discussion of the oscilloscope.

20.1 INTRODUCTION

We now begin the study of that branch of physics known as *electromagnetism*. The science of electromagnetism is concerned with the nature of electric and magnetic phenomena and the connection between them. In the next 12 chapters the fundamental principles of electromagnetism will be presented, together with some applications of interest to scientists and engineers. The laws of electromagnetism play a central role in the understanding of the operation of various electromagnetic devices, such as radio, television, electric motors, computers, radar, and high-energy accelerators.

Observations of electric and magnetic phenomena can be traced back to the ancient Greeks as early as around 800 B.C. It was found, when rubbed, a piece of amber becomes electrified and attracts pieces of straw or feathers. The existence of magnetic forces was known from observations that a naturally occurring stone called *magnetite* (Fe_2O_3) is attracted to iron. (The word *electric* comes from the Greek word for amber, *elektron*. The word *magnetic* comes from the name of the country where magnetite was found, *Magnesia*.) In 1600, William Gilbert (1540–1603) discovered that electrification was not limited to amber, but is a general phenomenon. Other scientists in this period went on to electrify a variety of objects (including chickens and people!). Experiments by Charles Coulomb (1736–1806) confirmed the inverse-square force law for electricity (see Chapter 6). His apparatus, called a *torsion balance*, was equivalent to that used in the Cavendish experiment, with masses replaced by small charged spheres. Apart from these observations, very little else was known about electric and magnetic phenomena. It was not until the 18th century that scientists established that electricity and magnetism were, in fact, related phenomena. In 1820, Hans Oersted (1777–1851) discovered that a compass needle was deflected when placed near a wire carrying an electric current. Experiments by Michael Faraday (1791–1867) showed that when a wire is moved near a magnet (or, equivalently, if a magnet is moved near a wire), an electric current flows in the wire. James Clerk Maxwell (1831–1879) used these observations and other experimental facts as a basis for formulating the laws of electromagnetism as we now understand them.

Maxwell's contributions to the science of electromagnetism were especially significant. The laws he formulated, known as *Maxwell's equations*, are basic to *all* forms of electromagnetic phenomena. They are comparable in importance to Newton's laws of motion and gravitation. For example, Maxwell deduced that waves would be radiated by a wire carrying a current that changes in time. About 20 years later, Heinrich Hertz (1857–1894) verified this prediction by producing electromagnetic waves in the laboratory. This was followed by such practical developments as the radio, or "wireless," invented by Guglielmo Marconi (1874–1937).

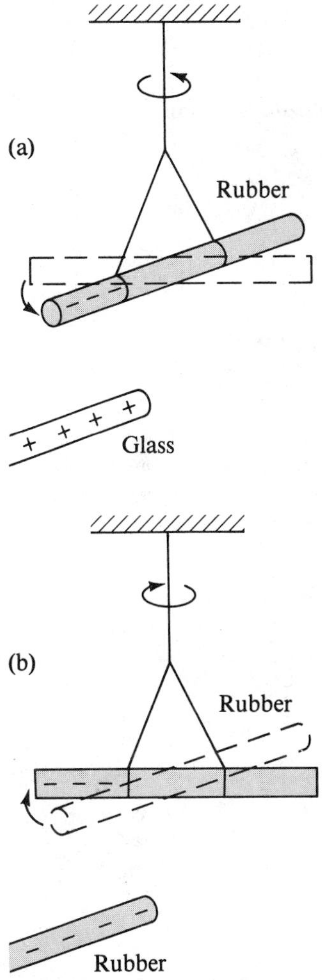

Figure 20.1 (a) A negatively charged rubber rod, suspended by a thread, is attracted to a positively charged glass rod. (b) A negatively charged rubber rod is repelled by another negatively charged rubber rod.

20.2 PROPERTIES OF ELECTRIC CHARGES

A number of simple experiments can be performed to demonstrate the existence of electrical forces and charges. For example, after running a comb through your hair, you will find that the comb will attract bits of paper. The attractive force is often strong enough to suspend the pieces of paper. The same effect occurs with other rubbed materials, such as glass or rubber.

Another simple experiment is to rub an inflated balloon with wool. The balloon will then adhere to the wall or the ceiling of a room. When materials behave in this way, they are said to be *electrified*, or to have become *electrically charged*. You can easily electrify your body by vigorously rubbing your shoes on a wool rug. The charge on your body can be sensed and removed by lightly touching (and startling) a friend. Under the right conditions, a visible spark is seen when you touch. (Experiments such as these work best on a dry day, since an excessive amount of moisture can lead to a leakage of charge from the electrified body to the earth by various conducting paths.)

In a systematic series of rather simple experiments, one finds that there are two kinds of electric charges, which were given the names *positive* and *negative* by Benjamin Franklin (1706–1790). To demonstrate this fact, consider a hard rubber rod that has been rubbed with fur and then suspended by a nonmetallic thread as in Fig. 20.1. When a glass rod that has been rubbed with silk is brought near the rubber rod, the rubber rod will be attracted toward the glass rod. On the other hand, if two charged rubber rods (or two charged glass rods) are brought near each other, as in Fig. 20.1b, the force between them will be repulsive. This observation shows that the rubber and glass are in two different states of electrification. Using the convention suggested by Franklin, the electric charge on the glass rod is called *positive*, and that on the rubber rod is called *negative*. Therefore any charged body that is attracted to a charged rubber rod (or repelled by a charged glass rod) must have a positive charge. Conversely, any charged body that is repelled by a charged rubber rod (or attracted to a charged glass rod) has a negative charge on it. On the basis of these observations, we conclude that *like charges repel one another and unlike charges attract one another.*

Like charges repel
Unlike charges attract

Another important aspect of Franklin's model of electricity is the implication that *electric charge is always conserved.* That is, when one body is rubbed against another, charge is not created in the process. The electrified state is due to a *transfer* of charge from one body to the other. Therefore, one body gains some amount of negative charge while the other gains an equal amount of positive charge. For

Charge is conserved

example, when a glass rod is rubbed with silk, the silk obtains a negative charge that is equal in magnitude to the positive charge on the glass rod. We now know from our understanding of atomic structure *that it is the negatively charged electrons that are transferred* from the glass to the silk in the rubbing process. Likewise, when rubber is rubbed with fur, electrons are transferred from the fur to the rubber, giving the rubber a net negative charge and the fur a net positive charge. This is consistent with the fact that neutral, uncharged matter contains as many positive charges (protons within atomic nuclei) as negative charges (electrons).

Charge is quantized

In 1909, Robert Millikan (1886–1953) discovered that electric charge always occurs as some integral multiple of some fundamental unit of charge, e. In modern terms, the charge q is said to be *quantized*. That is, electric charge exists as discrete "packets." Thus, we can write $q = Ne$, where N is some integer. Other experiments in the same period showed that the electron has a charge $-e$ and the proton has an equal and opposite charge, $+e$. Some elementary particles, such as the neutron, have no charge. A neutral atom must contain as many protons as electrons.

Electrostatic force varies as $1/r^2$

Electric forces between charged objects were measured quantitatively by Coulomb using the torsion balance, which he invented (Fig. 20.2). Using this apparatus, Coulomb confirmed that the electric force between two small charged spheres varies as the inverse square of their separation, that is, $F \sim 1/r^2$. The operating principle of the torsion balance is the same as that of the apparatus used by Cavendish to measure the gravitational constant (Section 6.2), with masses replaced by charged spheres. The electric force between the charged spheres produces a twist in the suspended fiber. Since the restoring torque of the twisted fiber is proportional to the angle through which it rotates, a measurement of this angle provides a quantitative measure of the electric force of attraction or repulsion. If the spheres are charged by rubbing, the electrical force between the spheres is very large compared with the gravitational attraction; hence the gravitational force can be neglected.

From our discussion thus far, we conclude that electric charge has the following important properties: (1) There are two kinds of charges in nature, with the property that unlike charges attract one another and like charges repel one another. (2) The force between charges varies as the inverse square of their separation. (3) Charge is conserved. (4) Charge is quantized.

20.3 INSULATORS AND CONDUCTORS

It is convenient to classify substances in terms of their ability to conduct electrical charge. *Conductors are materials in which electric charges move quite freely, whereas insulators are materials that do not readily transport charge.* Materials such as glass, rubber, and lucite fall into the category of insulators. When such materials are charged by rubbing, only the area that is rubbed becomes charged and there is no tendency for the charge to move into other regions of the material.

Metals are good conductors

In contrast, materials such as copper, aluminum, and silver are good conductors. When such materials are charged in some small region, the charge readily distributes itself over the entire surface of the conductor. If you hold a copper rod in your hand and rub it with wool or fur, it will not attract a small piece of paper. This might suggest that a metal cannot be charged. On the other hand, if you hold the copper rod by a lucite handle and then rub, the rod will remain charged and attract the piece of paper. This is explained by noting that in the first case, the electric charges produced by rubbing will readily move from the copper through your body and finally to earth. In the second case, the insulating lucite handle prevents the flow of charge to earth.

Semiconductors are a third class of materials, and their electrical properties are somewhere between those of insulators and conductors. Silicon and germanium are

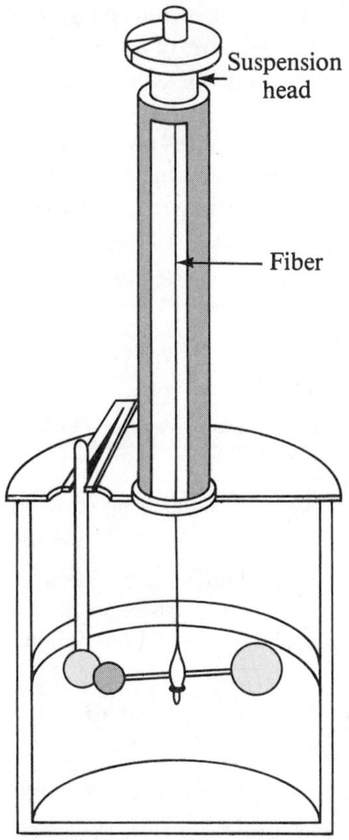

Figure 20.2 Coulomb's torsion balance, which was used to establish the inverse-square law for the electrostatic force between two charges. (*Taken from Coulomb's 1785 memoirs to the French Academy of Sciences.*)

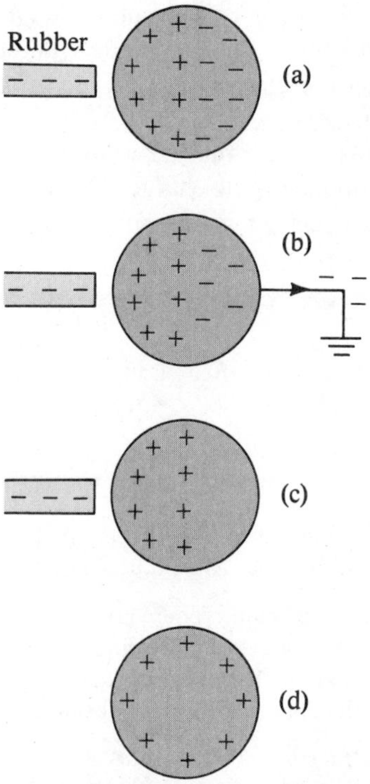

Figure 20.3 Charging a metallic object by induction. (a) The charge on a neutral metallic sphere is redistributed when a charged rubber rod is placed near the sphere. (b) The sphere is grounded, and some of the electrons leave the conductor. (c) The ground connection is removed, and the sphere has a nonuniform positive charge. (d) When the rubber rod is removed, the sphere becomes uniformly charged.

well-known examples of semiconductors commonly used in the fabrication of a variety of electronic devices. The electrical properties of semiconductors can be changed over many orders of magnitude by adding controlled amounts of foreign atoms to the materials.

When a conductor is connected to earth by means of a conducting wire or copper pipe, it is said to be *grounded.* The earth can then be considered an infinite "sink" to which electrons can easily migrate. With this in mind, we can understand how to charge a conductor by a process known as *induction.*

Charging by induction

To understand induction, consider a negatively charged rubber rod brought near a neutral (uncharged) conducting sphere insulated from ground. That is, there is no conducting path to ground (Fig. 20.3a). The region of the sphere nearest the negatively charged rod will obtain an excess of positive charge, while the region of the sphere farthest from the rod will obtain an equal excess of negative charge. (That is, electrons in the part of the sphere nearest the rod migrate to the opposite side of the sphere.) If the same experiment is performed with a conducting wire connected from the sphere to ground (Fig. 20.3b), some of the electrons in the conductor will be repelled to earth. If the wire to ground is then removed (Fig. 20.3c), the conducting sphere will contain an excess of *induced* positive charge. Finally, when the rubber rod is removed from the vicinity of the sphere (Fig. 20.3d), the induced positive charge remains on the ungrounded sphere. Note that the charge remaining on the sphere is uniformly distributed over its surface because of the repulsive forces among the like charges. In the process, the electrified rubber rod loses none of its negative charge.

Thus, we see that charging an object by induction requires no contact with the body inducing the charge. This is in contrast to charging an object by rubbing (that is, charging by *conduction*), which does require contact between the two objects.

Q1. Sparks are often observed (or heard) on a dry day when clothes are removed in the dark. Explain.

Q2. A balloon is negatively charged by rubbing and then clings to a wall. Does this mean that the wall is positively charged? Why does the balloon eventually fall?

Q3. A light, uncharged metal sphere suspended from a thread is attracted to a charged rubber rod. After touching the rod, the sphere is repelled by the rod. Explain.

Q4. A large metal sphere insulated from ground is charged with an electrostatic generator while a person standing on an insulating stool holds the sphere while it is being charged. Why is it safe to do this? Why wouldn't it be safe for another person to touch the sphere after it has been charged?

Q5. If a suspended object A is attracted to object B, which is charged, can we conclude that object A is charged? Explain.

Q6. Why do some clothes cling together and to your body after being removed from a dryer?

20.4 COULOMB'S LAW

In 1785, Coulomb established the fundamental law of electric force between two stationary, charged particles. Experiments show that an electric force has the following properties: (1) The force is inversely proportional to the square of the separation, r, between the two particles and is along the line joining the particles. (2) The force is proportional to the product of the charges q_1 and q_2 on the two particles. (3) The force is attractive if the charges are of opposite sign and repulsive if the charges have the same sign. From these observations, we can express the magnitude of the electric force between the two charges as

Coulomb's law

$$F = k\frac{q_1 q_2}{r^2} \tag{20.1}$$

where k is a constant called the *Coulomb constant*. In his experiments, Coulomb was able to show that the exponent of r was 2 to within an uncertainty of a few percent. Modern experiments have shown that the exponent is 2 to an accuracy of a few parts per *billion*.

The constant k in Eq. 20.1 has a value that depends on the choice of units. The unit of charge in SI units is the coulomb (C). The coulomb is defined in terms of a unit current called the *ampere* (A), where current equals the rate of flow of charge. (The ampere will be defined in Chapter 24.) When the current in a wire is 1 A, the amount of charge that flows past a given point in the wire in 1 s is 1 C. From experiment, we know that the Coulomb constant k in SI units has the value

Coulomb constant

$$k = 8.9875 \times 10^9 \text{ N} \cdot \text{m}^2/\text{C}^2 \tag{20.2}$$

TABLE 20.1 Charge and Mass of the Electron, Proton, and Neutron

PARTICLE	*CHARGE* (C)	*MASS* (kg)
Electron (e)	$-1.6021917 \times 10^{-19}$	9.1095×10^{-31}
Proton (p)	$+1.6021917 \times 10^{-19}$	1.67261×10^{-27}
Neutron (n)	0	1.67492×10^{-27}

To simplify our calculations, we shall use the approximate value

$$k = 9.0 \times 10^9 \text{ N} \cdot \text{m}^2/\text{C}^2 \tag{20.3}$$

The constant k is often written

$$k = \frac{1}{4\pi\varepsilon_0}$$

where the constant ε_0 is known as the *permittivity of free space* and has the value

$$\varepsilon_0 = 8.8542 \times 10^{-12} \text{ C}^2/\text{N} \cdot \text{m}^2 \tag{20.4}$$

The smallest unit of charge known in nature is the charge on an electron or proton.[1] The charge of an electron or proton has a magnitude

Charge on an electron or proton

$$e = 1.6021917 \times 10^{-19} \text{ C} \tag{20.5}$$

Therefore, 1 C of charge is equal to the charge of 6.3×10^{18} electrons (that is, $1/e$). This can be compared with the number of free electrons in 1 cm^3 of copper,[2] which is of the order of 10^{23}. Thus, we see that 1 C is a substantial amount of charge. In typical electrostatic experiments, where a rubber or glass rod is charged by friction, a net charge of the order of 10^{-6} C ($= 1\ \mu$C) is obtained. In other words, only a very small fraction of the total available charge is transferred between the rod and rubbing material.

The charges and masses of the electron, proton, and neutron are given in Table 20.1.

When dealing with Coulomb's force law, you must remember that force is a *vector* quantity and must be treated accordingly. Furthermore, you should note that *Coulomb's law applies exactly only to point charges or particles.* If $\hat{\boldsymbol{r}}$ is taken to be a unit vector from charge q_2 to charge q_1, as in Fig. 20.4, then the electric force on q_1 due to q_2, written $\boldsymbol{F}_{12}$, can be expressed in vector form as

$$\boldsymbol{F}_{12} = k\frac{q_1 q_2}{r^2}\hat{\boldsymbol{r}} \tag{20.6}$$

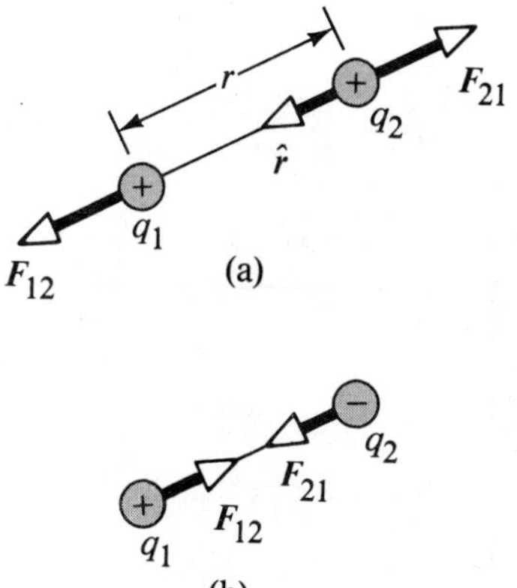

Figure 20.4 Two point charges separated by a distance r exert a force on each other given by Coulomb's law. Note that the force on q_1 is equal to and opposite the force on q_2. (a) When the charges are of the same sign, the force is repulsive. (b) When the charges are of opposite sign, the force is attractive.

Since Coulomb's law obeys Newton's third law, the electric force on q_2 due to q_1 is equal in magnitude to the force on q_1 due to q_2 and in the opposite direction, that is, $\boldsymbol{F}_{21} = -\boldsymbol{F}_{12}$. Finally, from Eq. 20.6 we see that if q_1 and q_2 have the same sign, the product q_1q_2 is positive and the force is repulsive, as in Fig. 20.4a. On the other hand, if q_1 and q_2 are of opposite sign, as in Fig. 20.4b, the product q_1q_2 is negative and the force is attractive.

---

[1] No unit of charge smaller than e has been detected; however, some recent theories have proposed the existence of particles called *quarks* having charges $e/3$ and $2e/3$.

[2] A metal atom, such as copper, contains one or more outer electrons, which are weakly bound to the nucleus. When many atoms combine to form a metal, the so-called free electrons are these outer electrons, which are not bound to any one atom. These electrons move about the metal in a manner similar to gas molecules moving in a container.

The superposition principle

When more than two charges are present, the force between any pair of charges is given by Eq. 20.6. Therefore, the resultant force on any one of them equals the *vector* sum of the forces due to the various individual charges. This principle of *superposition* as applied to electrostatic forces is an experimentally observed fact. For example, if there are four charges, then the resultant force on particle 1 due to particles 2, 3, and 4 is given by

$$\boldsymbol{F}_1 = \boldsymbol{F}_{12} + \boldsymbol{F}_{13} + \boldsymbol{F}_{14}$$

Example 20.1

Consider three point charges located at the corners of a triangle, as in Fig. 20.5, where $q_1 = q_3 = 5\ \mu\text{C}$, $q_2 = -2\ \mu\text{C}$ ($1\ \mu\text{C} = 10^{-6}$ C), and $a = 0.1$ m. Find the resultant force on q_3.

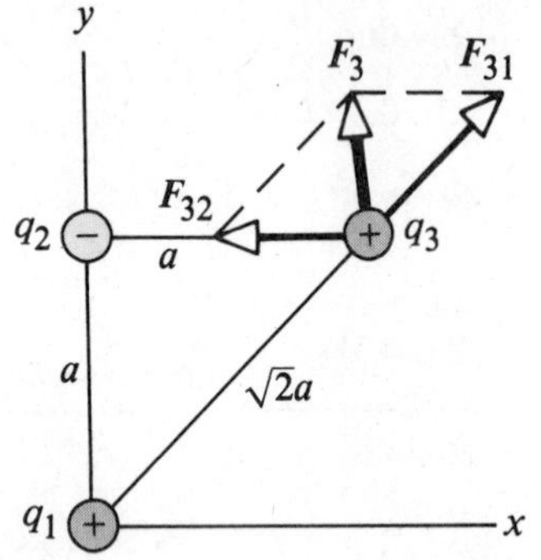

Figure 20.5 (Example 20.1) The force on q_3 due to q_1 is $\boldsymbol{F}_{31}$. The force on q_3 due to q_2 is $\boldsymbol{F}_{32}$. The *total* force, $\boldsymbol{F}_3$, on q_3 is the *vector* sum $\boldsymbol{F}_{31} + \boldsymbol{F}_{32}$.

Solution: First, note the direction of the individual forces on q_3 due to q_1 and q_2. The force on q_3 due to q_2 is attractive since q_2 and q_3 have opposite signs. The force on q_3 due to q_1 is repulsive since they are both positive.

Now let us calculate the magnitude of the forces on q_3. The magnitude of the force on q_3 due to q_2 is given by

$$F_{32} = k\frac{q_3q_2}{a^2}$$

$$= \left(9.0 \times 10^9\ \frac{\text{N}\cdot\text{m}^2}{\text{C}^2}\right)\frac{(5 \times 10^{-6}\ \text{C})(2 \times 10^{-6}\ \text{C})}{(0.1\ \text{m})^2}$$

$$= 9.0\ \text{N}$$

The magnitude of the force on q_3 due to q_1 is given by

$$F_{31} = k\frac{q_3q_1}{(\sqrt{2}a)^2}$$

$$= \left(9.0 \times 10^9\ \frac{\text{N}\cdot\text{m}^2}{\text{C}^2}\right)\frac{(5 \times 10^{-6}\ \text{C})(5 \times 10^{-6}\ \text{C})}{2(0.1\ \text{m})^2}$$

$$= 11\ \text{N}$$

The force $\boldsymbol{F}_{31}$ makes an angle of 45° with the x axis. Therefore, the x and y components of $\boldsymbol{F}_{31}$ are equal, with magnitude given by $F_{31}\cos 45° = 7.9$ N. The force $\boldsymbol{F}_{32}$ is in the negative x direction. Hence, the x and y components of the resultant force on q_3 are given by

$$F_x = F_{31x} + F_{32} = 7.9\ \text{N} - 9.0\ \text{N} = -1.1\ \text{N}$$

$$F_y = F_{31y} = 7.9\ \text{N}$$

We can also express the resultant force on q_3 in unit-vector form as $\boldsymbol{F}_3 = (-1.1\boldsymbol{i} + 7.9\boldsymbol{j})$ N. The magnitude of this vector is given by $\sqrt{(-1.1)^2 + (7.9)^2} = 8.0$ N, and the vector makes an angle of 98° with the x axis.

Example 20.2

Three charges lie along the x axis as in Fig. 20.6. The positive charge $q_1 = 15\ \mu\text{C}$ is at $x = 2$ m, and the positive charge $q_2 = 6\ \mu\text{C}$ is at the origin. Where must a *negative* charge q_3 be placed on the x axis such that the resultant force on it is zero?

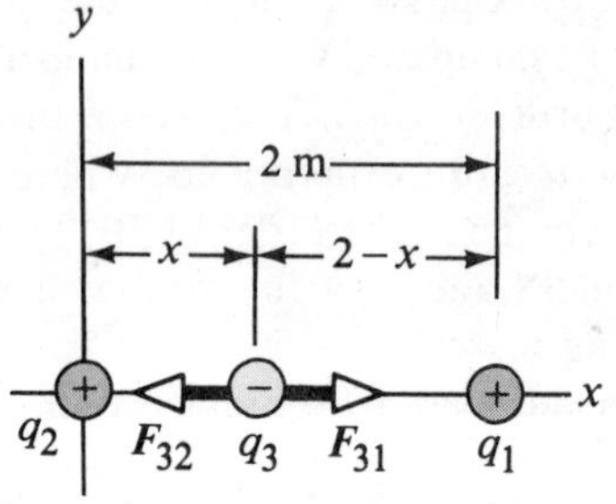

Figure 20.6 (Example 20.2) Three point charges are placed along the x axis. The charge q_3 is negative, whereas q_1 and q_2 are positive. If the net force on q_3 is zero, then the force on q_3 due to q_1 must be equal to and opposite the force on q_3 due to q_2.

Solution: Since q_3 is negative and both q_1 and q_2 are positive, the forces $\boldsymbol{F}_{31}$ and $\boldsymbol{F}_{32}$ are both attractive, as indicated in Fig. 20.6. If we let x be the coordinate of q_3, then the forces $\boldsymbol{F}_{31}$ and $\boldsymbol{F}_{32}$ have magnitudes given by

$$F_{31} = k\frac{q_3q_1}{(2 - x)^2} \quad \text{and} \quad F_{32} = k\frac{q_3q_2}{x^2}$$

If the resultant force on q_3 is zero, then $\boldsymbol{F}_{32}$ must be equal to and opposite $\boldsymbol{F}_{31}$, or

$$k\frac{q_3q_2}{x^2} = k\frac{q_3q_1}{(2 - x)^2}$$

Noting that k and q_3 are common to both sides and solving for x, we find that

$$(2 - x)^2q_2 = x^2q_1$$

$$(4 - 4x + x^2)(6 \times 10^{-6}\ \text{C}) = x^2(15 \times 10^{-6}\ \text{C})$$

Solving this quadratic equation, we find that $x = 0.775$ m. Why is the negative root not acceptable?

Example 20.3

The electron and proton of a hydrogen atom are separated (on the average) by a distance of about 5.3×10^{-11} m. Find the magnitude of the electrical force and the gravitational force between the two particles.

Solution: From Coulomb's law, we find that the attractive electrical force has the magnitude

$$F_e = k\frac{e^2}{r^2} = 9.0 \times 10^9 \frac{\text{N} \cdot \text{m}^2}{\text{C}^2} \frac{(1.6 \times 10^{-19}\ \text{C})^2}{(5.3 \times 10^{-11}\ \text{m})^2}$$

$$= 8.2 \times 10^{-8}\ \text{N}$$

Using Newton's universal law of gravity and Table 20.1, we find that the gravitational force is equal to

$$F_g = G\frac{m_e m_p}{r^2} = \left(6.7 \times 10^{-11} \frac{\text{N} \cdot \text{m}^2}{\text{kg}^2}\right)$$

$$\times \frac{(9.11 \times 10^{-31}\ \text{kg})(1.67 \times 10^{-27}\ \text{kg})}{(5.3 \times 10^{-11}\ \text{m})^2} = 3.6 \times 10^{-47}\ \text{N}$$

The ratio $F_e/F_g \approx 3 \times 10^{39}$. Thus the gravitational force between charged atomic particles is negligible compared with the electrical force.

Q7. Would life be different if the electron were positively charged and the proton were negatively charged? Explain.

Q8. Two charged spheres each of radius a are separated by a distance $r > 2a$. Is the force on either sphere given by Coulomb's law? Explain. (Hint: Refer back to Chapter 14 on gravitation.)

20.5 THE ELECTRIC FIELD

The gravitational field **g** at some point in space was defined in Chapter 14 to be equal to the gravitational force **F** acting on a test mass m_o divided by the test mass. That is, $\mathbf{g} = \mathbf{F}/m_o$. In a similar manner, an electric field at some point in space can be defined in terms of the electric force acting on a test charge q_o placed at that point. To be more precise, *the electric field vector* **E** at some point in space is defined as *the electric force* **F** *acting on a positive test charge placed at that point divided by the magnitude of the test charge* q_o:

$$\mathbf{E} = \frac{\mathbf{F}}{q_o} \qquad (20.7)$$

Definition of electric field

Note that **E** is the field *external* to the test charge—not the field produced by the test charge. The vector **E** has the SI units of newtons per coulomb (N/C). The direction of **E** is in the direction of **F** since we have assumed that **F** acts on a positive test charge. Thus, we can say that *an electric field exists at some point if a test charge at rest placed at that point experiences an electrical force.* Once the electric field is known at some point, the force on *any* charged particle placed at that point can be calculated from Eq. 20.7. Furthermore, the electric field is said to exist at some point (even empty space) regardless of whether or not a test charge is located at that point.

When Eq. 20.7 is applied, we must assume that the test charge q_o is small enough such that it does not disturb the charge distribution responsible for the electric field.[3] For instance, if a vanishingly small test charge q_o is placed near a uniformly

[3]To be more precise, the test charge q_0 should be infinitesimally small to ensure that its presence does not affect the original charge distribution. Therefore, strictly speaking, we should replace Eq. 20.7 by the expression

$$\mathbf{E} = \lim_{q_0 \to 0} \frac{\mathbf{F}}{q_0}$$

It is impossible to follow this prescription strictly in any experiment since no charges smaller in magnitude than e are known to exist. However, as a practical matter, it is almost always possible to select a sufficiently small test charge to obtain any desired degree of accuracy.

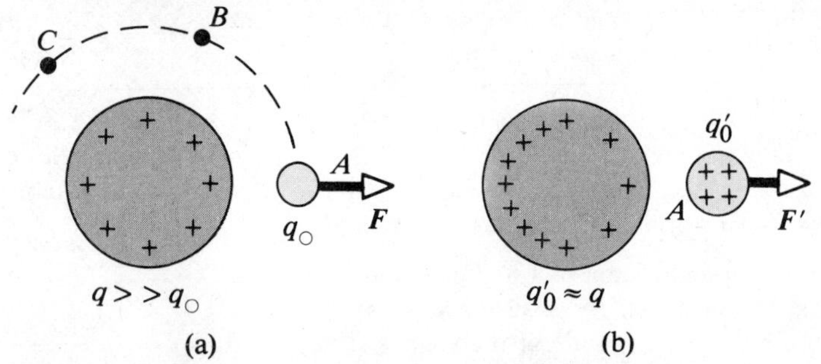

Figure 20.7 (a) When a small test charge q_o is placed near a conducting sphere of charge q (where $q \gg q_o$), the charge on the conducting sphere remains uniform. (b) If the test charge q_o' is of the order of the charge on the sphere, the charge on the sphere is nonuniform.

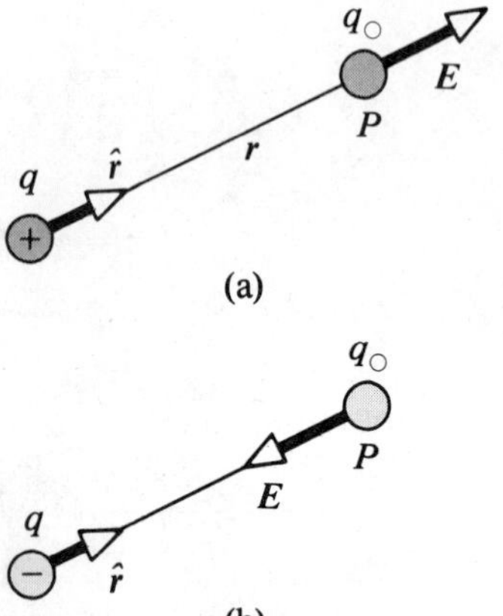

Figure 20.8 A test charge q_o at the point P is at a distance r from a point charge q. (a) If q is positive, the electric field at P points radially *outward* from q. (b) If q is negative, the electric field at P points radially *inward* toward q.

charged metallic sphere as in Fig. 20.7a, the charge on the metallic sphere, which produces the electric field, will remain uniformly distributed. Furthermore, the force $\mathbf{F}$ on the test charge will have the same magnitude at points A, B, and C, which are equidistant from the sphere. If the test charge is large enough ($q_o' \gg q_o$) as in Fig. 20.7b, the charge distribution on the metallic sphere will be redistributed and the ratio of the force to the test charge at point A will be different: ($E'/q_o' \neq E/q_o$). That is, because of this redistribution of charge on the metallic sphere, the electric field at point A set up by the sphere in Fig. 20.7b must be different from that of the field at point A in Fig. 20.7a. Furthermore, the distribution of charge on the sphere will change as the smaller charge is moved from point A to point B or C.

Consider a point charge q located a distance r from a test charge q_o. According to Coulomb's law, the force on the test charge is given by

$$\mathbf{F} = k\frac{qq_o}{r^2}\hat{\mathbf{r}}$$

Since the electric field at the position of the test charge is defined by $\mathbf{E} = \mathbf{F}/q_o$, we find that the electric field *due to the charge* q at the position of q_o is given by

Electric field due to a point charge *q*

$$\mathbf{E} = k\frac{q}{r^2}\hat{\mathbf{r}} \tag{20.8}$$

where $\hat{\mathbf{r}}$ is a unit vector that is directed away from q toward q_o (Fig. 20.8). If q is *positive*, as in Fig. 20.8a, the field is directed radially *outward* from this charge. If q is *negative*, as in Fig. 20.8b, the field is directed *toward* q.

In order to calculate the electric field due to a group of point charges, we first calculate the electric field vectors at the point P individually using Eq. 20.8 and then add them *vectorially*. In other words, *the total electric field due to a group of charges equals the vector sum of the electric fields of all the charges at some point.* This *superposition principle* applied to fields follows directly from the superposition property of electric forces. Thus, the electric field of a group of charges (excluding the test charge q_o) can be expressed as

Electric field due to a group of charges

$$\mathbf{E} = k\sum_i \frac{q_i}{r_i^2}\hat{\mathbf{r}}_i \tag{20.9}$$

where r_i is the distance from the ith charge, q_i, to the point P (the location of the test charge) and $\hat{\mathbf{r}}_i$ is a unit vector directed from q_i toward P.

Example 20.4

Find the electric force on a proton placed in an electric field of 2×10^4 N/C directed along the positive x axis.

Solution: Since the charge on a proton is $+e = +1.6 \times 10^{-19}$ C, the electric force on it is

$$\begin{aligned} \boldsymbol{F} = e\boldsymbol{E} &= (1.6 \times 10^{-19}\text{ C})(2 \times 10^4 \boldsymbol{i}\text{ N/C}) \\ &= 3.2 \times 10^{-15}\boldsymbol{i}\text{ N} \end{aligned}$$

where $\boldsymbol{i}$ is a unit vector in the positive x direction. The weight of the proton $mg = (1.67 \times 10^{-27}\text{ kg})(9.8\text{ m/s}^2) = 1.6 \times 10^{-26}$ N. Hence, the magnitude of the gravitational force is negligible compared with the electric force.

Example 20.5

A charge $q_1 = 7\ \mu$C is located at the origin, and a second charge $q_2 = -5\ \mu$C is located on the x axis 0.3 m from the origin (Fig. 20.9). (a) Find the electric field at the point P with coordinates (0,0.4) m.

First, let us find the magnitudes of the electric fields due to each charge. The fields $\boldsymbol{E}_1$ due to the 7-μC charge and $\boldsymbol{E}_2$ due to the -5-μC charge at P are shown in Fig. 20.9. Their magnitudes are given by

$$\begin{aligned} E_1 &= k\frac{q_1}{r_1^2} = \left(9.0 \times 10^9\ \frac{\text{N} \cdot \text{m}^2}{\text{C}^2}\right)\frac{(7 \times 10^{-6}\text{ C})}{(0.4\text{ m})^2} \\ &= 3.9 \times 10^5\text{ N/C} \end{aligned}$$

$$\begin{aligned} E_2 &= k\frac{|q_2|}{r_2^2} = \left(9.0 \times 10^9\ \frac{\text{N} \cdot \text{m}^2}{\text{C}^2}\right)\frac{(5 \times 10^{-6}\text{ C})}{(0.5\text{ m})^2} \\ &= 1.8 \times 10^5\text{ N/C} \end{aligned}$$

The vector $\boldsymbol{E}_1$ has only a y component. The vector $\boldsymbol{E}_2$ has an x component given by $E_2 \cos\theta = \frac{3}{5}E_2$ and a negative y component given by $-E_2 \sin\theta = -\frac{4}{5}E_2$. Hence, we can express the vectors as

$$\boldsymbol{E}_1 = 3.9 \times 10^5\boldsymbol{j}\text{ N/C}$$
$$\boldsymbol{E}_2 = (1.1 \times 10^5\boldsymbol{i} - 1.4 \times 10^5\boldsymbol{j})\text{ N/C}$$

The resultant field $\boldsymbol{E}$ at P is the superposition of $\boldsymbol{E}_1$ and $\boldsymbol{E}_2$:

$$\boldsymbol{E} = \boldsymbol{E}_1 + \boldsymbol{E}_2 = (1.1 \times 10^5\boldsymbol{i} + 2.5 \times 10^5\boldsymbol{j})\text{ N/C}$$

From this result, we find that $\boldsymbol{E}$ has a magnitude of 2.7×10^5 N/C and makes an angle ϕ of 66° with the positive x axis.

(b) Find the force on a test charge of 2×10^{-8} C placed at P.

From Eq. 20.7 and the results to (a), we get

$$\begin{aligned} \boldsymbol{F} = q_o\boldsymbol{E} &= (2 \times 10^{-8}\text{ C})(1.1 \times 10^5\boldsymbol{i} + 2.5 \times 10^5\boldsymbol{j}) \\ &= (2.2 \times 10^{-3}\boldsymbol{i} + 5.0 \times 10^{-3}\boldsymbol{j})\text{ N} \end{aligned}$$

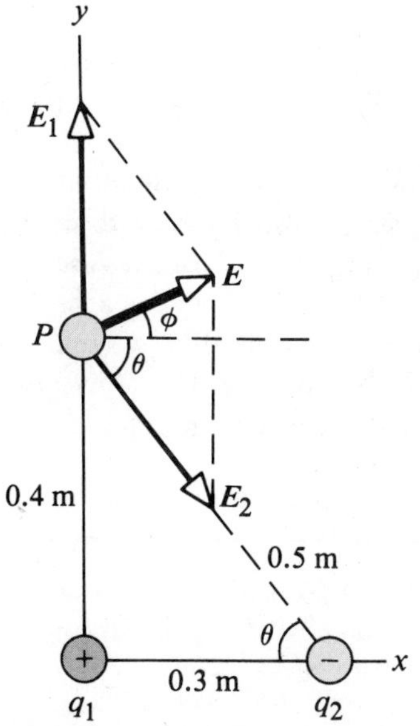

Figure 20.9 (Example 20.5) The total electric field $\boldsymbol{E}$ at P equals the vector sum $\boldsymbol{E}_1 + \boldsymbol{E}_2$, where $\boldsymbol{E}_1$ is the field due to the positive charge q_1 and $\boldsymbol{E}_2$ is the field due to the negative charge q_2.

Example 20.6

Electric field of a dipole: An *electric dipole* consists of a positive charge q and a negative charge $-q$ separated by a distance $2a$, as in Fig. 20.10. Find the electric field $\boldsymbol{E}$ due to these charges along the y axis at the point P, which is a distance y from the origin. Assume that $y \gg a$.

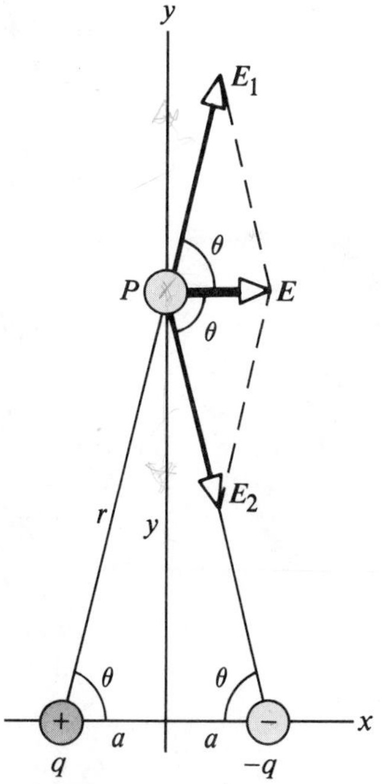

Figure 20.10 (Example 20.6) The total electric field $\boldsymbol{E}$ at P due to two equal and opposite charges (an electric dipole) equals the vector sum $\boldsymbol{E}_1 + \boldsymbol{E}_2$. The field $\boldsymbol{E}_1$ is due to the positive charge q, and $\boldsymbol{E}_2$ is the field due to the negative charge $-q$.

Solution: At P, the fields $\boldsymbol{E}_1$ and $\boldsymbol{E}_2$ due to the two charges are equal in magnitude, since P is equidistant from the two equal and opposite charges. The total field $\boldsymbol{E} = \boldsymbol{E}_1 + \boldsymbol{E}_2$, where the magnitudes of $\mathbf{E}_1$ and $\mathbf{E}_2$ are given by

$$E_1 = E_2 = k\frac{q}{r^2} = k\frac{q}{y^2 + a^2}$$

The y components of $\mathbf{E}_1$ and $\mathbf{E}_2$ cancel each other. The x components are equal since they are both along the x axis. Therefore, $\mathbf{E}$ lies along the x axis and has a magnitude equal to $2E_1 \cos\theta$. From Fig. 20.10 we see that $\cos\theta = a/r = a/(y^2 + a^2)^{1/2}$. Therefore,

$$E = 2E_1 \cos\theta = 2k\frac{q}{(y^2 + a^2)}\frac{a}{(y^2 + a^2)^{1/2}}$$

$$= k\frac{2qa}{(y^2 + a^2)^{3/2}}$$

Using the approximation $y \gg a$, we can neglect a^2 in the denominator and write

$$E \approx k\frac{2qa}{y^3} \qquad (20.10)$$

Thus we see that along the y axis the field of a *dipole* at a distant point varies as $1/r^3$, whereas the more slowly varying field of a *point charge* goes as $1/r^2$. This is because at distant points, the fields of the two equal and opposite charges almost cancel each other. The $1/r^3$ variation in E for the dipole is also obtained for a distant point along the x axis (Problem 6) and for a general distant point. The dipole is a good model of many molecules, such as HCl.

As we shall see in later chapters, neutral atoms and molecules behave as dipoles when placed in an external electric field. Furthermore, many molecules, such as HCl, are permanent dipoles. (HCl is essentially an H^+ ion combined with a Cl^- ion.) The effect of such dipoles on the behavior of materials subjected to electric fields will be discussed in Chapter 22.

Q9. How would you experimentally distinguish an electric field from a gravitational field?

Q10. A "free" electron and "free" proton are placed in an identical electric field. Compare the electric force on each particle. Compare their accelerations.

20.6 ELECTRIC FIELD OF A CONTINUOUS CHARGE DISTRIBUTION

In the previous section, we showed how to calculate the electric field of a point charge using Coulomb's law. The total field of a group of point charges was obtained by taking the vector sum of the individual fields due to all the charges. *This procedure assumes that the superposition principle applies to the electrostatic field.*

A continuous charge distribution

Very often the charges of interest are close together compared with their distances to points of interest. In such situations, the system of charges can be considered to be *continuous*. That is, we imagine that the system of closely spaced charges is equivalent to a total charge that is continuously distributed through a volume or over some surface.

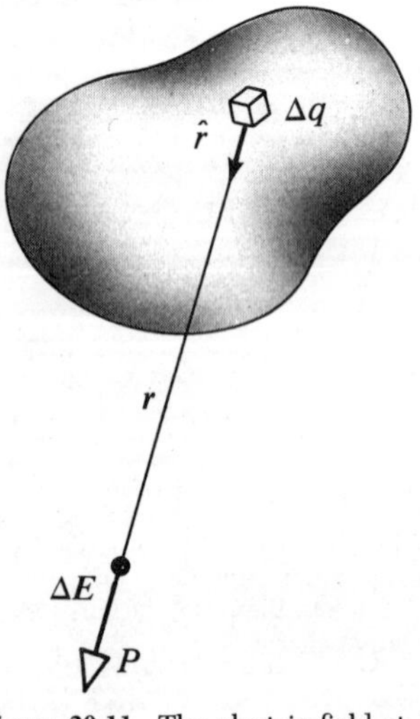

Figure 20.11 The electric field at P due to a continuous charge distribution is the vector sum of the fields due to all the elements Δq of the charge distribution.

To evaluate the electric field of a continuous charge distribution, the following procedure is used. First, we divide the charge distribution into small elements each of which contains a small charge Δq, as in Fig. 20.11. Next, we use Coulomb's law to calculate the electric field due to one of these elements at a point P. Finally, we evaluate the total field at P due to the charge distribution by summing the contributions of all the charge elements (that is, by applying the superposition principle).

The electric field at P due to one element of charge Δq is given by

$$\Delta\mathbf{E} = k\frac{\Delta q}{r^2}\hat{\mathbf{r}}$$

where r is the distance from the element to point P and $\hat{\mathbf{r}}$ is a unit vector directed from the charge element toward P. The total electric field at P due to all elements in the charge distribution is approximately given by

$$\mathbf{E} \approx k\sum_i \frac{\Delta q_i}{r_i^2}\hat{\mathbf{r}}_i$$

where the index i refers to the ith element in the distribution. If the separation between elements in the charge distribution is small compared with the distance to P, the charge distribution can be approximated to be continuous. Therefore, the total field at P in the limit $\Delta q_i \to 0$ becomes

$$\mathbf{E} = k \lim_{\Delta q_i \to 0} \sum_i \frac{\Delta q_i}{r_i^2}\hat{\mathbf{r}}_i = k \int \frac{dq}{r^2}\hat{\mathbf{r}} \tag{20.11}$$

Electric field of a continuous charge distribution

where the integration is a *vector* operation and must be treated with caution. We shall illustrate this type of calculation with several examples. In these examples, we shall assume that the charge is *uniformly* distributed on a line or a surface or throughout some volume. When performing such calculations, it is convenient to use the concept of a charge density along with the following notations:

If a charge Q is uniformly distributed throughout a volume V, the *charge per unit volume*, ρ, is defined by

$$\rho = \frac{Q}{V} \tag{20.12}$$

Volume charge density

where ρ has units of C/m^3.

If a charge Q is uniformly distributed on a surface of area A, the *surface charge density*, σ, is defined by

$$\sigma = \frac{Q}{A} \tag{20.13}$$

Surface charge density

where σ has units of C/m^2.

Finally, if a charge Q is uniformly distributed along a line of length l, the *linear charge density*, λ, is defined by

$$\lambda = \frac{Q}{l} \tag{20.14}$$

Linear charge density

where λ has units of C/m.

If the charge is *nonuniformly* distributed over a volume, surface, or line, we would have to express the charge densities as

$$\rho = \frac{dQ}{dV} \qquad \sigma = \frac{dQ}{dA} \qquad \lambda = \frac{dQ}{dl}$$

where dQ is the amount of charge in a small volume, surface, or length element.

Example 20.7

The electric field due to a charged rod: A rod of length l has a uniform positive charge per unit length λ and a total charge Q. Calculate the electric field at a point P along the axis of the rod, a distance d from one end (Fig. 20.12).

Solution: For this calculation, the rod is taken to be along the x axis. The ratio of Δq, the charge on the segment to Δx, the length of the segment, is equal to the ratio of the total charge to the total length of the rod. That is, $\Delta q/\Delta x = Q/l = \lambda$. Therefore, the charge Δq on the small segment is given by $\Delta q = \lambda\,\Delta x$.

The field $\Delta\mathbf{E}$ due to this segment at the point P is in the negative x direction, and its magnitude is given by[4]

$$\Delta E = k\frac{\Delta q}{x^2} = k\frac{\lambda\,\Delta x}{x^2}$$

Note that each element produces a field in the negative x direction, and so the problem of summing their contribu-

[4]It is important that you understand the procedure being used to carry out integrations such as this. First, you must choose an element whose parts are all equidistant from the point at which the field is being calculated. Next, you must express the charge element Δq in terms of the other variables within the integral (in this example, there is one variable, x.) In examples that have spherical or cylindrical symmetry, the variable will be a radial coordinate.

tions is particularly simple in this case. The total field at P due to all segments of the rod, which are at different distances from P, is given by Eq. 20.11, which in this case becomes

$$E = \int_d^{l+d} k\lambda \frac{dx}{x^2}$$

where the limits on the integral extend from one end of the rod ($x = d$) to the other ($x = l + d$). Since k and λ are constants, they can be removed from the integral. Thus, we find that

$$E = k\lambda \int_d^{l+d} \frac{dx}{x^2} = k\lambda \left[-\frac{1}{x} \right]_d^{l+d}$$

$$= k\lambda \left(\frac{1}{d} - \frac{1}{l+d} \right) = k\lambda \left(\frac{l}{d(l+d)} \right) \quad (20.15)$$

From this result we see that if the point P is *far* from the rod ($d \gg l$), then l in the denominator can be neglected, and $E \approx k\lambda l/d^2 = kQ/d^2$ (where we have used the fact that the total charge $Q = \lambda l$). This is just the form you would expect for a point charge. Therefore, at large distances from the rod, the charge distribution appears to be a point charge of magnitude Q. The use of the limiting technique ($d \to \infty$) is often a good method for checking a theoretical formula.

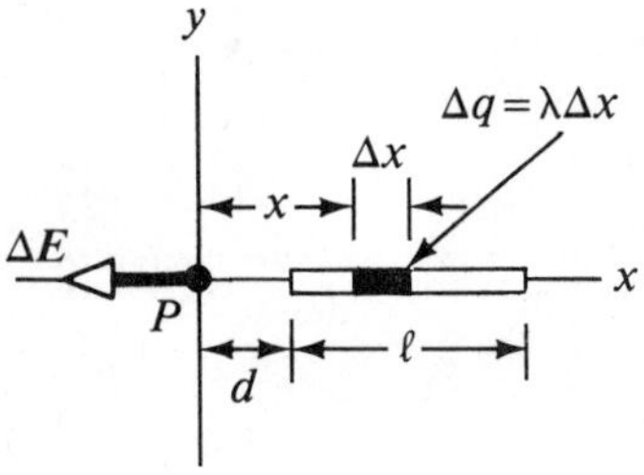

Figure 20.12 (Example 20.7) The electric field at P due to a uniformly charged rod lying along the x axis. The field at P due to the segment of charge Δq is given by $k\,\Delta q/x^2$. The total field at P is the vector sum over all segments of the rod.

Example 20.8

The electric field of a uniform ring of charge: A ring of radius a has a uniform positive charge per unit length, with a total charge Q. Calculate the electric field along the axis of the ring at a point P lying a distance x from the center of the ring (Fig. 20.13a).

Solution: The magnitude of the electric field at P due to the segment of charge Δq is

$$\Delta E = k \frac{\Delta q}{r^2}$$

This field has an x component $\Delta E_x = \Delta E \cos\theta$ along the axis of the ring and a component $\Delta E_\perp$ perpendicular to the axis. But as we see in Fig. 20.13b, the resultant field at P must lie along the x axis since the perpendicular components sum up to zero. That is, the perpendicular component of any element is canceled by the perpendicular component of an element on the opposite side of the ring. Since $r = (x^2 + a^2)^{1/2}$ and $\cos\theta = x/r$, we find that

$$\Delta E_x = \Delta E \cos\theta = \left(k \frac{\Delta q}{r^2} \right) \frac{x}{r} = \frac{kx}{(x^2 + a^2)^{3/2}} \Delta q$$

In this case, all segments of the ring give the *same* contribution to the field at P since they are all equidistant from this point. Thus, we can easily sum over all segments to get the total field at P:

$$E_x = \Sigma \frac{kx}{(x^2 + a^2)^{3/2}} \Delta q = \frac{kx}{(x^2 + a^2)^{3/2}} Q \quad (20.16)$$

Again, note that at large distances from the ring ($x \gg a$), the constant a in the denominator of Eq. 20.16 can be neglected and the result reduces to that of a point charge Q. This result also shows that the field is zero at $x = 0$. Does this surprise you?

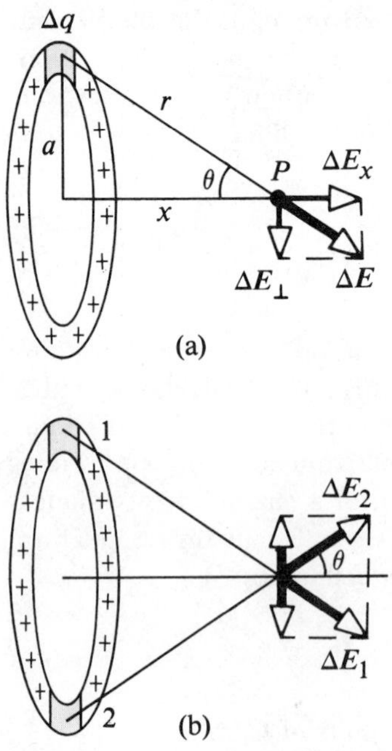

Figure 20.13 (Example 20.8) A uniformly charged ring of radius a. (a) The total electric field at P is along the x axis. (b) Note that the perpendicular component of the electric field at P due to segment 1 is canceled by the perpendicular component due to segment 2, which is opposite segment 1.

Example 20.9

The electric field of a uniformly charged disk: A disk of radius R has a uniform charge per unit area σ. Calculate the electric field along the axis of the disk, a distance x from its center (Fig. 20.14).

Solution: The solution to this problem is straightforward if we consider the disk as a set of concentric rings. We can then make use of Example 20.8, which gives the field of a given ring of radius r, and sum up contributions of all rings making up the disk. By symmetry, the field on an axial point must be parallel to this axis.

The ring of radius r and width dr has an area equal to $2\pi r\,dr$ (Fig. 20.14). The charge dq on this ring is equal to the area of the ring multiplied by the charge per unit area, or $dq = 2\pi\sigma r\,dr$. Using this result in Eq. 20.16 (with a replaced by r) gives for the field due to the ring the expression

$$dE_x = \frac{kx}{(x^2 + r^2)^{3/2}}(2\pi\sigma r\,dr)$$

To get the total field at P, we integrate this expression over the limits $r = 0$ to $r = R$, noting that x is a constant, which gives

$$E_x = kx\pi\sigma \int_0^R \frac{2r\,dr}{(x^2 + r^2)^{3/2}}$$

$$= kx\pi\sigma \int_0^R (x^2 + r^2)^{-3/2}\,d(r^2)$$

$$= kx\pi\sigma \left[\frac{(x^2 + r^2)^{-1/2}}{-1/2}\right]_0^R$$

$$= 2\pi k\sigma \left(1 - \frac{x}{(x^2 + R^2)^{1/2}}\right) \qquad (20.17)$$

It is left as an exercise to show that at large distances from the disk, the electric field approaches that of a point charge of magnitude $Q = \sigma\pi R^2$. The field close to the disk along an axial point can also be obtained from Eq. 20.17 by letting $x \to 0$ (or $R \to \infty$). This gives

$$E_x = 2\pi k\sigma = \frac{\sigma}{2\varepsilon_0} \qquad (20.18)$$

where ε_0 is the permittivity of free space, given by Eq. 20.4.

This result shows that the field close to the surface of the disk is *uniform* since it does not depend on x. As we shall find in the next chapter, the same result is obtained for the field of a uniformly charged infinite sheet.

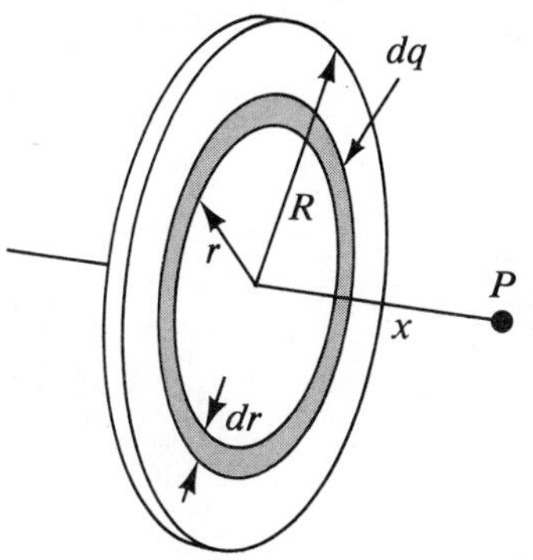

Figure 20.14 (Example 20.9) A uniformly charged disk of radius R. The electric field at an axial point P is directed along this axis, perpendicular to the plane of the disk.

Q11. A negative point charge $-q$ is placed along the axis of a positively charged ring (Fig. 20.13). Describe the motion of the point charge if it is released from rest.

20.7 ELECTRIC FIELD LINES

A convenient aid for visualizing electric field patterns is to draw electric field lines that have the same direction as the electric field vector at any point. These imaginary lines are related to the electric field in any region of space in the following manner: (1) The electric field vector $\mathbf{E}$ is *tangent* to the electric field line at each point. (2) The number of lines per unit area through a surface perpendicular to the lines is proportional to the magnitude of the electric field in a given region. Thus $\mathbf{E}$ is large when the field lines are close together and small when they are far apart.

Properties of electric field lines

These points are illustrated in Fig. 20.15. Note that the density of lines through surface A is greater than the density of lines through surface B. Therefore, the electric field is more intense on surface A than on surface B. Furthermore, the field

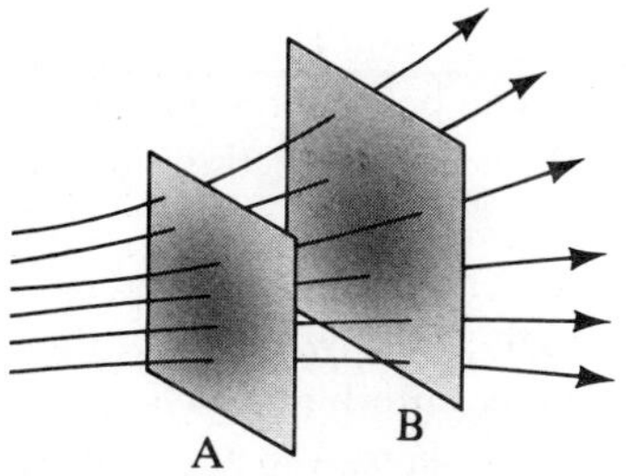

Figure 20.15 Electric field lines penetrating two surfaces. The magnitude of the field is greater on surface A than on surface B.

drawn in Fig. 20.15 is nonuniform since the lines at different locations point in different directions.

The electric field lines for a single positive point charge are shown in Fig. 20.16a. (Note that in our two-dimensional drawings, we show only the field lines that lie in the plane containing the charge.) By convention, the lines are directed *away* from the positive charge, like the spines of a porcupine, since a positive test charge placed in this field would be repelled by the charge q. Similarly, the electric field lines for a single negative point charge are directed toward the charge (Fig. 20.16b). In either case, the lines are along the radial direction and extend all the way to infinity.

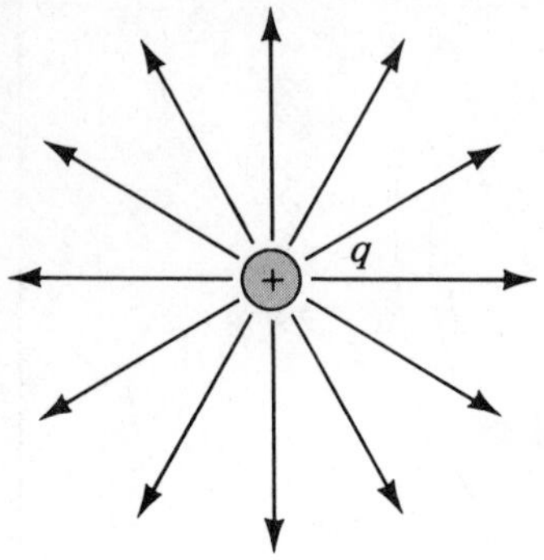

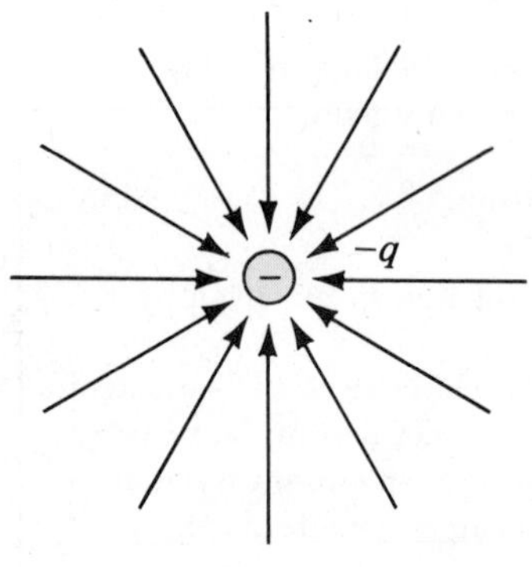

Figure 20.16 The electric field lines for a point charge. (a) For a positive point charge, the lines are radial outward. (b) For a negative point charge, the lines are radial inward. (Note: The figures show only field lines that lie in the plane containing the charge.)

Is this visualization of the electric field in terms of field lines consistent with Coulomb's law? To answer this question, consider an imaginary spherical surface of radius r concentric with the charge. From symmetry, we see that the magnitude of the electric field is the same everywhere on the surface of the sphere. The number of lines, N, that emerge from the charge is equal to the number that penetrate the spherical surface. Hence, the number of lines per unit area on the sphere is $N/4\pi r^2$ (where the surface area of the sphere is $4\pi r^2$). Since E is proportional to the number of lines per unit area, we see that E varies as $1/r^2$. This is consistent with the result obtained from Coulomb's law, that is, $E = kq/r^2$.

It is important to note that electric field lines are not material objects. They are used only to provide us with a qualitative description of the electric field. One problem with this model is the fact that one always draws a finite number of lines from each charge, which makes it appear as if the field were quantized and acted only in a certain direction. The field, in fact, is continuous—existing at every point. Another problem with this model is the danger of getting the wrong impression from a two-dimensional drawing of field lines being used to describe a three-dimensional situation.

Note that since charge is quantized, the number of lines leaving any material object must be 0, $\pm C'e$, $\pm 2C'e$, . . . , where C' is an arbitrary (but fixed) proportionality constant. Once C' is chosen, the number of lines is not arbitrary. For example, if object 1 has charge Q_1 and object 2 has charge Q_2, then the ratio of numbers of lines is $N_2/N_1 = Q_2/Q_1$.

Rules for drawing electric field lines

The rules for drawing electric field lines for any charge distribution are as follows: (1) The lines must begin on positive charges and terminate on negative charges. (2) The number of lines leaving a positive charge or entering a negative charge is proportional to the magnitude of the charge. (3) No two field lines can cross.

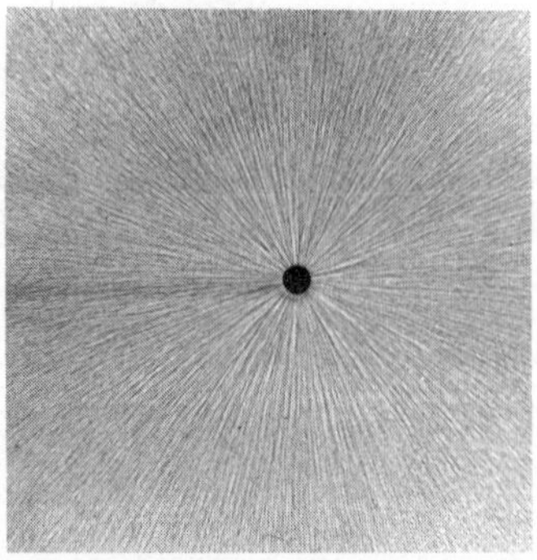

The electric field lines of a single charge. The dark areas are small pieces of thread suspended in oil which align with the electric field produced by a small charged conductor at the center. (Courtesy of Harold M. Waage, Princeton University.)

The electric field lines for two point charges of equal magnitude, but opposite signs (the electric dipole), are shown in Fig. 20.17. In this configuration the number of lines that begin at the positive charge must equal the number that terminate at the negative charge. At points very near the charges, the lines are nearly radial. The high density of lines between the charges indicates a region of strong electric field. The attractive nature of the force between the charges can also be visualized in this model.

Figure 20.18 shows the electric field lines in the vicinity of two equal positive point charges. Again, the lines are nearly radial at points close to either charge. The same number of lines emerge from each charge since the charges are equal in magnitude. At large distances from the charges, the field is approximately equal to that of a single point charge of magnitude $2q$. The bulging out of the electric field lines between the charges indicates the repulsive nature of the electric force between like charges.

Finally, in Figure 20.19 we sketch the electric field lines associated with a positive charge $+2q$ and a negative charge $-q$. In this case, we see that the number of lines leaving the charge $+2q$ is twice the number entering the charge $-q$. Hence only half of the lines that leave the positive charge enter the negative charge. The

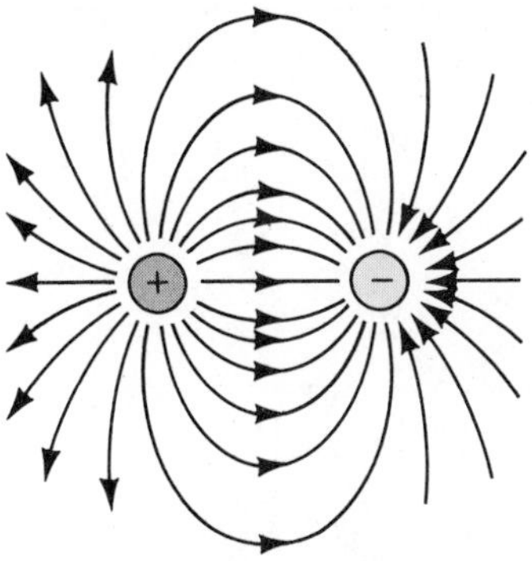

Figure 20.17 The electric field lines for two equal and opposite point charges (an electric dipole). Note that the number of lines that leave the positive charge equals the number that terminate at the negative charge.

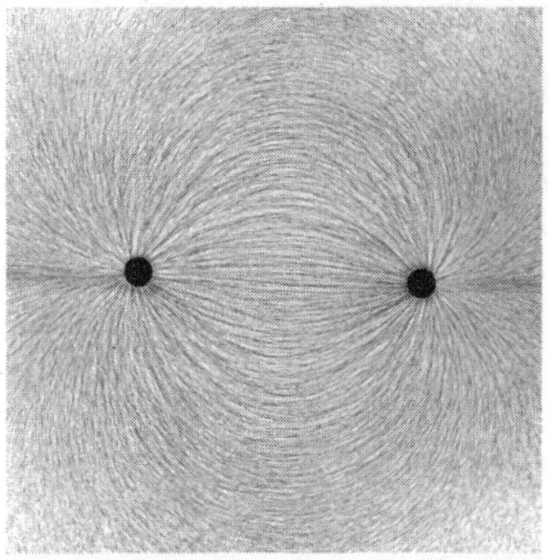

The electric field pattern of a pair of small charged conductors of *opposite* sign (the dipole). The photograph was taken using small pieces of thread suspended in oil which align with the electric field lines. (Courtesy of Harold M. Waage, Princeton University.)

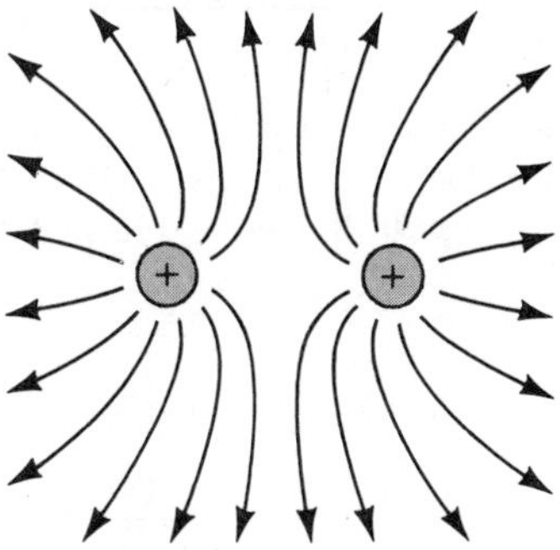

Figure 20.18 The electric field lines for two equal positive point charges.

remaining half beginning on the positive charge leave the system. At large distances from the charges (large compared with the charge separation), the electric field lines are equivalent to those of a single charge $+q$.

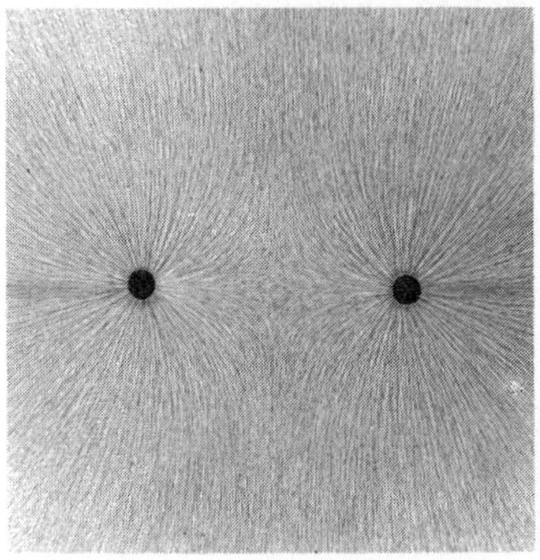

The electric field pattern of a pair of small charged conductors of the *same* sign. The photograph was taken using small pieces of thread suspended in oil which align with the electric field lines. (Courtesy of Harold M. Waage, Princeton University.)

Q12. Explain why electric field lines never cross. (Hint: ***E*** must have a unique direction at all points.)

Q13. A charge $4q$ is at a distance r from a charge $-q$. Compare the number of electric field lines leaving the charge $4q$ with the number entering the charge $-q$.

Q14. For the two equal charges described in Fig. 20.18, at what point (other than ∞) would a third test charge experience no net force?

Q15. In Fig. 20.19, where do the extra lines leaving the charge $+2q$ end up?

20.8 MOTION OF CHARGED PARTICLES IN A UNIFORM ELECTRIC FIELD

In this section we describe the motion of a charged particle in a uniform electric field. As we shall see, the motion is equivalent to that of a projectile moving in a uniform gravitational field. When a particle of charge q is placed in an electric field $\mathbf{E}$, the electric force on the charge is $q\mathbf{E}$. If this is the only force exerted on the charge, then Newton's second law applied to the charge gives

$$\mathbf{F} = q\mathbf{E} = m\mathbf{a}$$

where m is the mass of the charge and we assume that the speed is small compared with the speed of light. The acceleration of the particle is therefore given by

$$\mathbf{a} = \frac{q\mathbf{E}}{m} \tag{20.19}$$

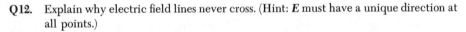

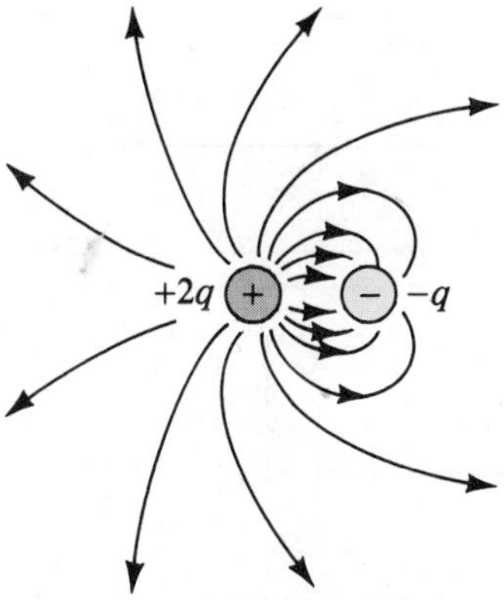

Figure 20.19 The electric field lines for a point charge $+2q$ and a second point charge $-q$. Note that two lines leave the charge $+2q$ for every one that terminates on $-q$.

If ***E*** is uniform (that is, constant in magnitude and direction), we see that the acceleration is a constant of the motion. If the charge is positive, the acceleration will be in the direction of the electric field. If the charge is negative, the acceleration will oppose the electric field.

Example 20.10

A positive point charge q of mass m is released from rest in a uniform electric field $\mathbf{E}$ directed along the x axis as in Fig. 20.20. Describe its motion.

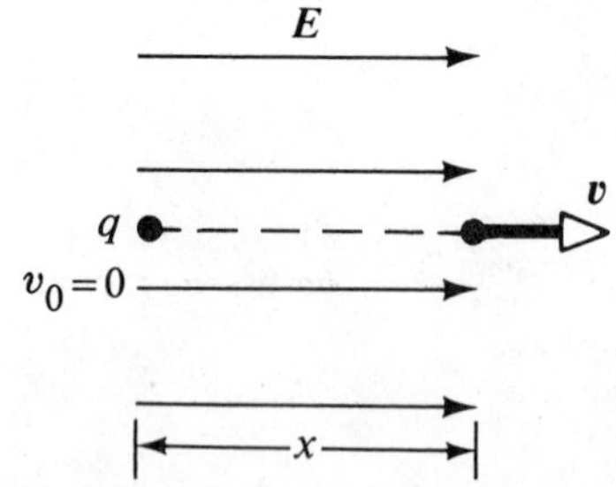

Figure 20.20 (Example 20.10) A positive point charge q in a uniform electric field $\mathbf{E}$ undergoes constant acceleration in the direction of the field.

Solution: The acceleration of the charge is constant and given by qE/m. The motion is simple linear motion along the x axis. Therefore, we can apply the equations of kinematics in one dimension (from Chapter 3):

$$x - x_0 = v_0 t + \frac{1}{2}at^2 \qquad v = v_0 + at$$

$$v^2 = v_0^2 + 2a(x - x_0)$$

Taking $x_0 = 0$ and $v_0 = 0$ gives

$$x = \frac{1}{2}at^2 = \frac{qE}{2m}t^2$$

$$v = at = \frac{qE}{m}t$$

$$v^2 = 2ax = \left(\frac{2qE}{m}\right)x$$

The kinetic energy of the charge after it has moved a distance x is given by

$$K = \frac{1}{2}mv^2 = \frac{1}{2}m\left(\frac{2qE}{m}\right)x = qEx$$

Notice that this result can also be obtained from the work-energy theorem, since the work done by the electric force is $F_e x = qEx$ and $W = \Delta K$.

The electric field in the region between two oppositely charged flat metal plates is approximately uniform, neglecting end effects (Fig. 20.21). Suppose an electron of charge $-e$ is projected horizontally into this field with an initial velocity $v_0\mathbf{i}$. Since the electric field $\mathbf{E}$ is in the positive y direction, the acceleration of the electron is in the negative y direction. That is,

$$\mathbf{a} = -\frac{eE}{m}\mathbf{j} \tag{20.20}$$

Since the acceleration is constant, we can apply the equations of kinematics in two dimensions (from Chapter 4) with $v_{x0} = v_0$ and $v_{y0} = 0$. The components of velocity of the electron after it has been in the electric field a time t are given by

$$v_x = v_0 = \text{constant} \tag{20.21}$$

$$v_y = at = -\frac{eE}{m}t \tag{20.22}$$

Likewise, the coordinates of the electron after a time t in the electric field are given by

$$x = v_0 t \tag{20.23}$$

$$y = \frac{1}{2}at^2 = -\frac{1}{2}\frac{eE}{m}t^2 \tag{20.24}$$

Substituting the value $t = x/v_0$ from Eq. 20.23 into Eq. 20.24, we see that y is proportional to x^2. Hence, the trajectory is a parabola. After the electron leaves the region of uniform electric field, it continues to move in a straight line with a speed $v > v_0$.

Note that we have neglected the gravitational force on the electron. This is a good approximation when dealing with atomic-size particles. For a typical electric field of 10^4 N/C, the ratio of the electric force, eE, to the gravitational force, mg, for the electron is of the order of 10^{14}. The corresponding ratio for a proton is of the order of 10^{11}.

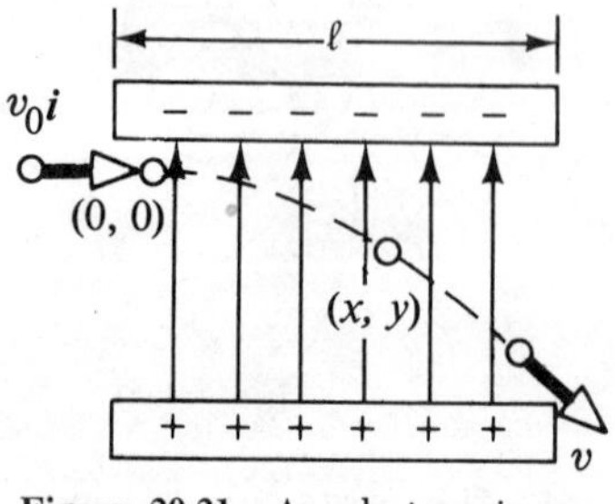

Figure 20.21 An electron is projected horizontally into a uniform electric field produced by two charged plates. The electron undergoes a downward acceleration (opposite $\mathbf{E}$), and its motion is parabolic.

Example 20.11

An electron enters the region of a uniform electric field as in Fig. 20.21, with $v_0 = 3 \times 10^6$ m/s and $E = 200$ N/C. The width of the plates is $l = 0.1$ m. (a) Find the acceleration of the electron while in the electric field.

Since the charge on the electron has a magnitude of 1.60×10^{-19} C and $m = 9.11 \times 10^{-31}$ kg, Eq. 20.20 gives

$$a = -\frac{eE}{m}\mathbf{j} = -\frac{(1.6 \times 10^{-19}\text{ C})\left(200\frac{\text{N}}{\text{C}}\right)}{9.11 \times 10^{-31}\text{ kg}}\mathbf{j}$$
$$= -3.51 \times 10^{13}\mathbf{j}\text{ m/s}^2$$

(b) Find the time it takes the electron to travel through the region of the electric field.

The horizontal distance traveled by the electron while in the electric field is $l = 0.1$ m. Using Eq. 20.23 with $x = l$, we find that the time spent in the electric field is given by

$$t = \frac{l}{v_0} = \frac{0.1\text{ m}}{3 \times 10^6\text{ m/s}} = 3.33 \times 10^{-8}\text{ s}$$

(c) What is the vertical displacement y of the electron while it is in the electric field?

Using Eq. 20.24 and the results from (a) and (b), we find that

$$y = \frac{1}{2}at^2 = -\frac{1}{2}\left(3.51 \times 10^{13}\frac{\text{m}}{\text{s}^2}\right)(3.33 \times 10^{-8}\text{ s})^2$$
$$= -0.0195\text{ m} = -1.95\text{ cm}$$

If the separation between the plates is smaller than this, the electron will strike the positive plate.

(d) Find the speed of the electron as it emerges from the electric field.

The x component of the velocity v_0 remains constant since there is no acceleration in the x direction. The electron's y component of velocity as it emerges from the field can be calculated from Eq. 20.22 with $t = 3.33 \times 10^{-8}$ s:

$$v_y = at = \left(-3.51 \times 10^{13}\frac{\text{m}}{\text{s}^2}\right)(3.33 \times 10^{-8}\text{ s})$$
$$= -1.17 \times 10^6\text{ m/s}$$

Therefore, the speed of the electron as it emerges from the field is given by

$$v = (v_x^2 + v_y^2)^{1/2} = [(3 \times 10^6)^2 + (-1.17 \times 10^6)^2]^{1/2}\text{ m/s}$$
$$= 3.22 \times 10^6\text{ m/s}$$

Q16. If the electron in Fig. 20.21 is projected into the electric field with an arbitrary velocity $\mathbf{v}_0$ (at an angle to $\mathbf{E}$), will its trajectory still be parabolic? Explain.

Q17. In Example 20.11 we found that the speed of the electron increases ($v > v_0$). From this, we conclude that its kinetic energy increases. What is the source of this energy?

20.9 THE OSCILLOSCOPE

The oscilloscope is an electronic instrument that is widely used in making electrical measurements. The main component of the oscilloscope is the cathode ray tube (CRT) (Fig. 20.22). This tube is commonly used to obtain a visual display of elec-

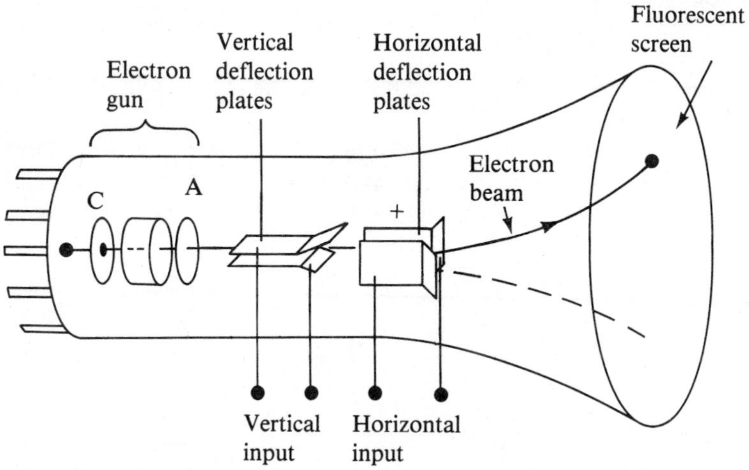

Figure 20.22 Schematic diagram of a cathode ray tube. Electrons leaving the hot cathode C are accelerated to the anode A. The electron gun is also used to focus the beam, and the plates deflect the beam.

tronic information for other applications, including radar systems, television receivers, and computers. The CRT is essentially a vacuum tube in which electrons within the tube are accelerated and deflected under the influence of electric fields. Electrons emitted from a hot filament (the cathode) are accelerated to a series of electrodes located in an assembly referred to as an *electron gun.*

In addition to accelerating the electrons from the cathode C to a positively charged anode A, the electron gun contains an element that focuses the electron beam and one that controls the number of electrons reaching the anode (that is, a brightness control). The details of these elements are not included in Fig. 20.22. The sharply focused electron beam passes through two sets of deflection plates at right angles to each other and finally strikes a screen that is coated with a fluorescent material. As the electrons strike the screen, the atoms of the fluorescent material emit visible light (as well as some x-rays), which produces a visible spot on the screen.

The deflection plates consist of oppositely charged conductors as described in the previous section. The first set of plates produces an electric field that is in the vertical direction (perpendicular to the direction of the electron beam), causing the beam to be deflected in the vertical plane. The second set of plates produces an electric field in the horizontal direction, which deflects the beam in the horizontal plane.[5] Therefore, by varying the magnitude and direction of the fields between these plates, one has control of both the up-and-down and right-to-left motion of the spot on the screen.

In practice, various types of signals with different waveforms can be applied to the horizontal and vertical deflection plates to produce different patterns on the screen of the CRT. The input signal to be examined is usually amplified and fed into the vertical deflection plates of the CRT. A reference signal (either internal or external to the oscilloscope) is applied to the horizontal deflection plates to sweep the beam across the oscilloscope screen.

20.10 SUMMARY

Electric charges have the following important properties:

Properties of electric charges

1. Unlike charges attract one another and like charges repel one another.
2. Electric charge is always conserved.
3. Charge is quantized, that is, it exists in discrete packets that are some integral multiple of the electronic charge.
4. The force between charged particles varies as the inverse square of their separation.

Conductors are materials in which charges move quite freely. Some examples of good conductors are copper, aluminum, and silver. *Insulators* are materials that do not readily transport charge. Some examples are glass, rubber, and wood.

Coulomb's law states that the electrostatic force between two stationary, charged particles separated by a distance r has a magnitude given by

Coulomb's law

$$F = k\frac{q_1 q_2}{r^2} \tag{20.1}$$

---

[5] In some of the larger CRT's and TV picture tubes, a finer focus control is obtained by using magnetic fields as well as electric fields.

where the constant k has the value

$$k = 8.9875 \times 10^9 \frac{\text{N} \cdot \text{m}^2}{\text{C}^2} \qquad (20.2)$$

Coulomb constant

The smallest unit of charge known to exist in nature is the charge on an electron or proton. The magnitude of this charge e is given by

$$e = 1.60207 \times 10^{-19}\ \text{C} \qquad (20.5)$$

Charge on an electron or proton

The *electric field* $\mathbf{E}$ at some point in space is defined as the electric force $\boldsymbol{F}$ that acts on a small positive test charge placed at that point divided by the magnitude of the test charge q_o:

$$\mathbf{E} = \frac{\boldsymbol{F}}{q_o} \qquad (20.7)$$

Definition of electric field

The electric field due to a point charge q at a distance r from the charge is given by

$$\mathbf{E} = k\frac{q}{r^2}\hat{\boldsymbol{r}} \qquad (20.8)$$

Electric field of a point charge q

where $\hat{\boldsymbol{r}}$ is a unit vector directed from the charge to the point in question. The electric field is directed radially outward from a positive charge and is directed *toward* a negative charge.

The *electric field* due to a group of charges can be obtained using the superposition principle. That is, the total electric field equals the *vector sum* of the electric fields of all the charges at some point:

$$\mathbf{E} = k\sum_i \frac{q_i}{r_i^2}\hat{\boldsymbol{r}}_i \qquad (20.9)$$

Electric field of a group of charges

Similarly, the electric field of a continuous charge distribution at some point is given by

$$\mathbf{E} = k\int \frac{dq}{r^2}\hat{\boldsymbol{r}} \qquad (20.11)$$

Electric field of a continuous charge distribution

where dq is the charge on one element of the charge distribution and r is the distance from the element to the point in question.

Electric field lines are useful for describing the electric field in any region of space. The electric field vector $\boldsymbol{E}$ is always tangent to the electric field lines at every point. Furthermore, the number of lines per unit area through a surface perpendicular to the lines is proportional to the magnitude of $\boldsymbol{E}$ in that region.

A charged particle of mass m and charge q moving in an electric field $\boldsymbol{E}$ has an acceleration $\boldsymbol{a}$ given by

$$\boldsymbol{a} = \frac{q\boldsymbol{E}}{m} \qquad (20.19)$$

Acceleration of a charge in an electric field

If the electric field is uniform, the acceleration is constant and the motion of the charge is equivalent to that of a projectile moving in a uniform gravitational field.

EXERCISES

Section 20.4 Coulomb's Law

1. Calculate the net charge on a substance consisting of (a) 5×10^{14} electrons and (b) a combination of 7×10^{13} protons and 4×10^{13} electrons.

2. Two protons in a molecule are separated by 2.5×10^{-10} m. Find the electrostatic force exerted by one proton on the other.

3. A 4.5-μC charge is located 3.2 m from a -2.8-μC charge. Find the electrostatic force exerted by one charge on the other.

4. A 2.2-μC charge is located on the x axis at $x = -1.5$ m, a 5.4-μC charge is located on the x axis at $x = 2.0$ m, and a 3.5-μC charge is located at the origin. Find the net force on the 3.5-μC charge.

5. Three point charges of 8 μC, 3 μC, and -5 μC are located at the corners of an equilateral triangle as in Fig. 20.23. Calculate the net electric force on the 3-μC charge.

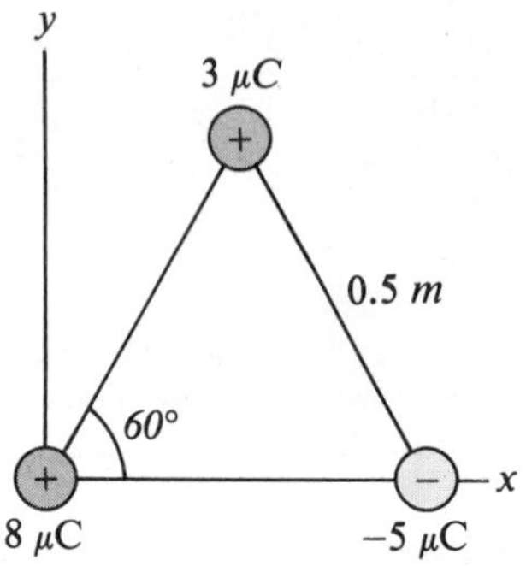

Figure 20.23 (Exercises 5 and 12).

6. A point charge $q_1 = -3.5$ μC is located on the y axis at $y = 0.12$ m, a charge $q_2 = -1.8$ μC is located at the origin, and a charge $q_3 = 2.6$ μC is located on the x axis at $x = -0.12$ m. Find the resultant force on the charge q_1.

7. Four point charges are situated at the corners of a square of sides a as in Fig. 20.24. Find the resultant force on the positive charge q.

8. What are the magnitude and direction of the electric field that will balance the weight of (a) an electron and (b) a proton? (Use the data in Table 20.1.)

9. Three point charges lie along the y axis. A charge $q_1 = -2$ μC is at $y = 2.0$ m, and a charge $q_2 = -3$ μC is at $y = -1.0$ m. Where must a third positive charge, q_3, be placed such that the resultant force on it is zero?

Section 20.5 The Electric Field

10. The electric force on a point charge of 5.0 μC at some point is 3.8×10^{-3} N in the positive x direction. What is the value of the electric field at that point?

11. A point charge of -2.8 μC is located at the origin. Find the electric field (a) on the x axis at $x = 2$ m, (b) on the y axis at $y = -3$ m, (c) at the point with coordinates $x = 1$ m, $y = 1$ m.

12. Three charges are at the corners of an equilateral triangle as in Fig. 20.23. Calculate the electric field intensity at the position of the 8-μC charge due to the 3-μC and -5-μC charges.

13. Two equal point charges each of magnitude 4.0 μC are located on the x axis. One is at $x = 0.8$ m, and the other is at $x = -0.8$ m. (a) Determine the electric field on the y axis at $y = 0.5$ m. (b) Calculate the electric force on a third charge, of -3.0 μC, placed on the y axis at $y = 0.5$ m.

14. Four charges are at the corners of a square as in Fig. 20.24. (a) Find the magnitude and direction of the electric field at the position of the charge $-q$, the coordinates of which are $x = a$, $y = a$. (b) What is the electric force on this charge?

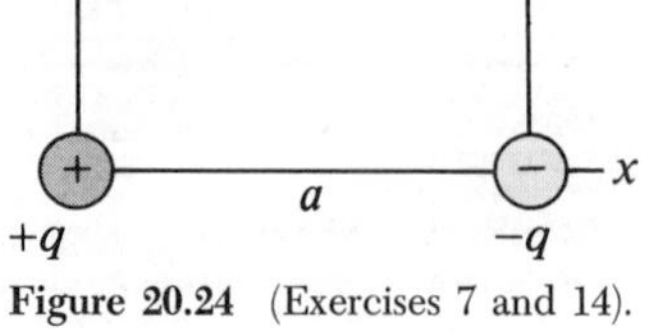

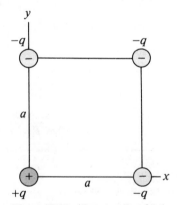

Figure 20.24 (Exercises 7 and 14).

15. Three equal charges q are at the corners of an equilateral triangle of sides a as in Fig. 20.25. (a) At what point (other than ∞) is the electric field zero? (b) What are the magnitude and direction of the electric field at the point P?

16. Find the total electric field along the line of the two charges shown in Fig. 20.26 at the point midway between them.

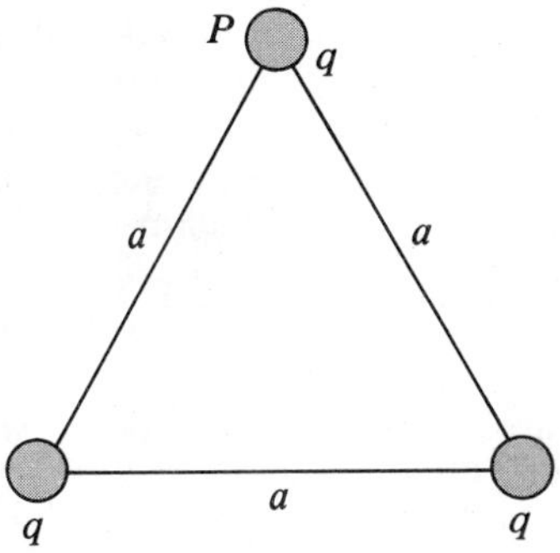

Figure 20.25 (Exercise 15).

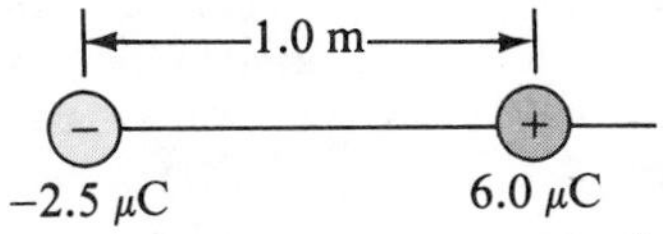

Figure 20.26 (Exercises 16 and 17).

17. In Fig. 20.26, determine the point (other than ∞) at which the total electric field is zero.

18. A charge of $-3\ \mu C$ is located at the origin, and a charge of $-7\ \mu C$ is located along the y axis at $y = 0.2$ m. At what point along the y axis is the electric field zero?

Section 20.6 Electric Field of a Continuous Charge Distribution

19. A rod 10 cm long is uniformly charged and has a total charge of $-50\ \mu C$. Determine the magnitude and direction of the electric field along the axis of the rod, at a point 30 cm from its center.

20. A uniformly charged ring of radius 5 cm has a total charge of 8 μC. Find the electric field on the *axis* of the ring at (a) 2 cm, (b) 4 cm, (c) 8 cm, and (d) 200 cm from the center of the ring.

21. A uniformly charged disk of radius 8 cm carries a charge density of $6 \times 10^{-4}\ C/m^2$. Calculate the electric field on the *axis* of the disk at (a) 2 cm, (b) 6 cm, (c) 20 cm, and (d) 400 cm from the center of the disk.

22. Starting with Eq. 20.13, show that the electric field due to a uniformly charged disk of radius R and total charge Q at distances x that are large compared with R approaches that of a point charge $Q = \sigma\pi R^2$. (Hint: First show that $x/(x^2 + R^2)^{1/2} = (1 + R^2/x^2)^{-1/2}$ and use the binomial expansion $(1 + \delta)^n \approx 1 + n\delta$ when $\delta \ll 1$.)

23. A uniformly charged rod of length 14 cm is bent into the shape of a semicircle as in Fig. 20.27. If the rod has a total charge of $-7.5\ \mu C$, find the magnitude and direction of the electric field at O, the center of the semicircle.

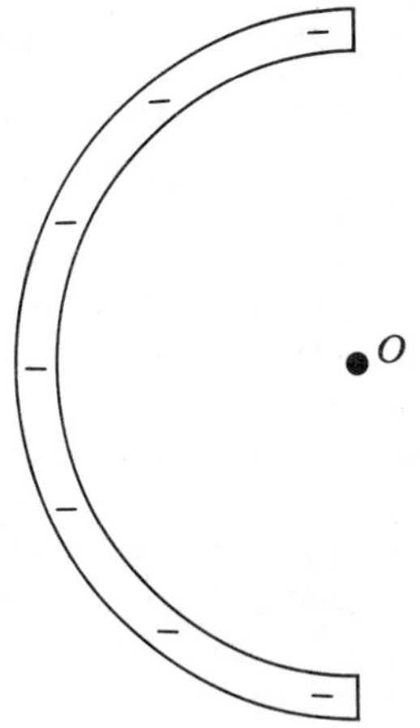

Figure 20.27 (Exercise 23).

Section 20.7 Electric Field Lines

24. Four equal positive point charges are at the corners of a square. Sketch the electric field lines in the plane of the square.

25. Figure 20.28 shows the electric field lines for two point charges separated by a small distance. (a) Determine the ratio q_1/q_2. (b) What are the signs of q_1 and q_2?

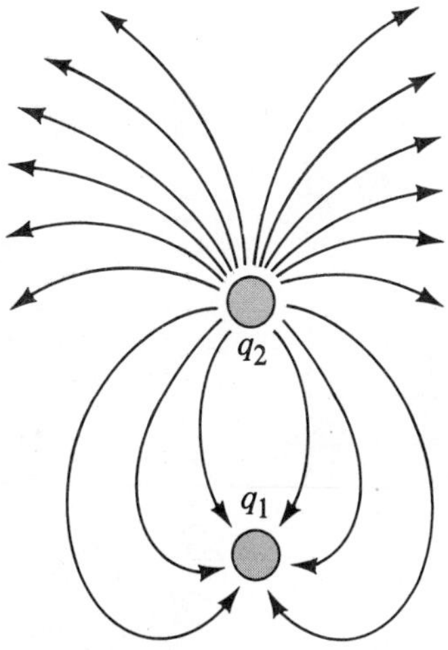

Figure 20.28 (Exercise 25).

26. A positively charged disk has a uniform charge per unit area as described in Example 20.9. Sketch the electric field lines in a plane perpendicular to the plane of the disk passing through its center.

27. A negatively charged rod of finite length has a uniform charge per unit length. Sketch the electric field lines in a plane containing the rod.

Section 20.8 Motion of Charged Particles in a Uniform Electric Field

28. An electron traveling with an initial velocity of $4.0 \times 10^6 \mathbf{i}$ m/s enters a region of a uniform electric field given by $\mathbf{E} = 2.5 \times 10^4 \mathbf{i}$ N/C. (a) Find the acceleration of the electron. (b) Determine the time it takes for the electron to come to rest after it enters the field. (c) How far does the electron move in the electric field before coming to rest?

29. A proton accelerates from rest in a uniform electric field of 500 N/C. At some later time, its speed is 2.50×10^6 m/s (nonrelativistic since v is much less than the speed of light). (a) Find the acceleration of the proton. (b) How long does it take the proton to reach this velocity? (c) How far has it moved in this time? (d) What is its kinetic energy at this time?

30. A proton has an initial velocity of 2.30×10^5 m/s in the x direction. It enters a uniform electric field of 1.50×10^4 N/C in the y direction. (a) Find the time it takes the proton to travel 0.05 m in the x direction. (b) Find the vertical displacement of the proton after it has traveled 0.05 m in the x direction. (c) Determine the components of the proton's velocity after it has traveled 0.05 m in the x direction.

31. An electron is projected at an angle of 37° to the horizontal at an initial speed of 4.50×10^5 m/s, in a region of an electric field $\mathbf{E} = 200\mathbf{j}$ N/C. Find (a) the time it takes the electron to return to its initial height, (b) the maximum height reached by the electron, and (c) its horizontal displacement when it reaches its maximum height.

32. A proton is projected in the x direction into a region of a uniform electric field $\mathbf{E} = -3 \times 10^5 \mathbf{i}$ N/C. The proton travels 4 cm before coming to rest. Determine (a) the acceleration of the proton, (b) its initial speed, and (c) the time it takes the proton to come to rest.

PROBLEMS

1. Four charges are located at the corners of a rectangle as in Fig. 20.29. (a) Find the x and y components of the electric field at the *center* of the rectangle. (b) Determine the electric force on a test charge q_0 placed at the center.

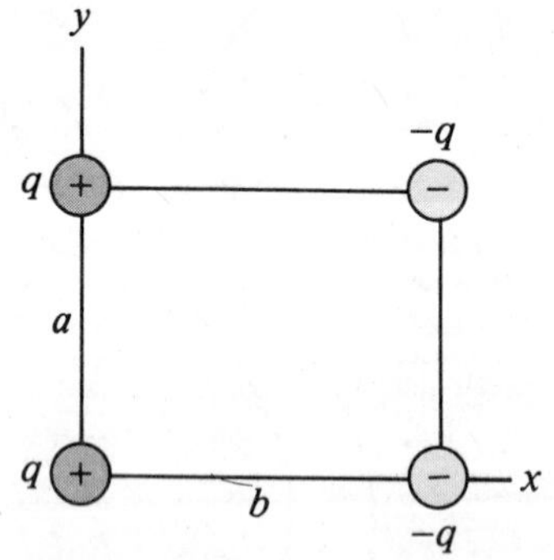

Figure 20.29 (Problems 1 and 2).

2. Calculate the components of the net force on the charge q located at the origin in Fig. 20.29.

3. A charged cork ball of mass 1 g is suspended on a light string in the presence of a uniform electric field as in Fig. 20.30. When $\mathbf{E} = (3\mathbf{i} + 5\mathbf{j}) \times 10^5$ N/C, the ball is in equilibrium at $\theta = 37°$. Find (a) the charge on the ball and (b) the tension in the string.

4. Three charges are located at the corners of an equilateral triangle as in Fig. 20.31. (a) Determine the magnitude and direction of the net electric field at the *center* of the triangle. (b) Find the magnitude and direction of the resultant force on the charge $-q$.

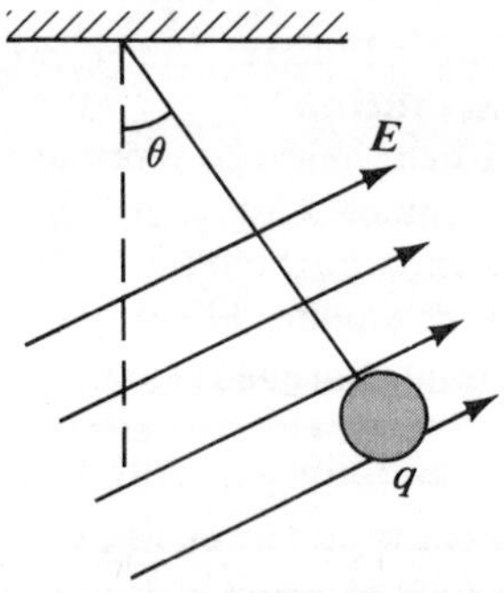

Figure 20.30 (Problem 3).

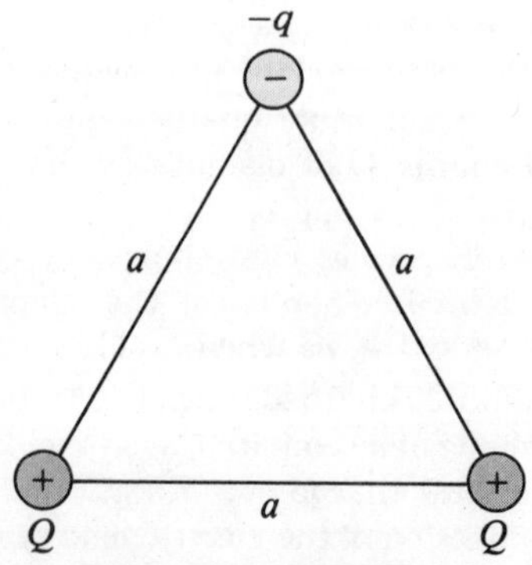

Figure 20.31 (Problem 4).

5. Two small spheres each of mass 2 g are suspended by light strings 10 cm in length (Fig. 20.32). A uniform electric field is applied in the x direction. If the spheres have charges of -5×10^{-8} C and $+5 \times 10^{-8}$ C, determine the electric field intensity that enables the spheres to be in equilibrium at $\theta = 10°$.

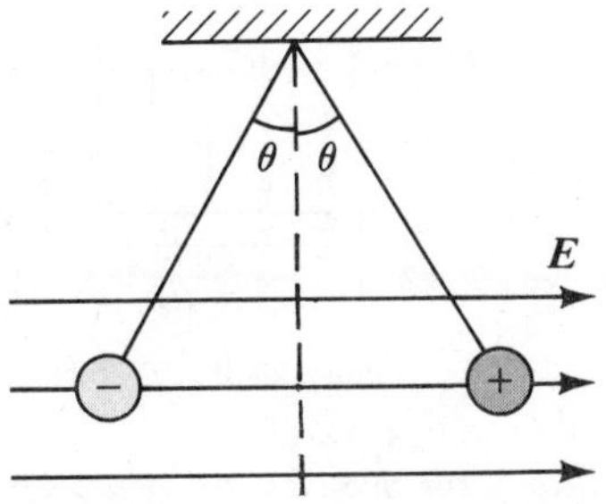

Figure 20.32 (Problem 5).

6. Consider the electric dipole shown in Fig. 20.33. Show that the electric field at a *distant* point along the x axis is given by $E_x = kp/x^3$, where $p = 2qa$ is the dipole moment.

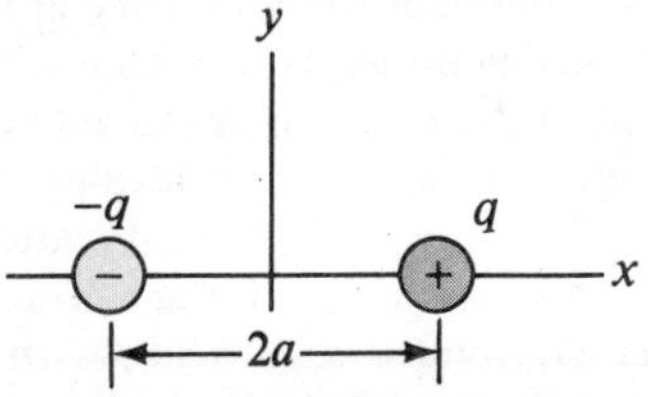

Figure 20.33 (Problem 6).

7. Two equal positive charges q are located on the x axis at $x = a$ and $x = -a$. (a) Show that the field along the y axis is in the y direction and is given by $E_y = 2kqy(y^2 + a^2)^{-3/2}$. (b) Determine the field along the y axis for $y \gg a$, and explain your result. (c) Show that the field is a maximum at $y = \pm a/\sqrt{2}$ (Hint: When E_y is a maximum, $dE_y/dy = 0$.)

8. Three point charges q, $-2q$, and q are located along the x axis as in Fig. 20.34. Show that the electric field at the distant point $P(y \gg a)$ along the y axis is given by $\mathbf{E} = -k\dfrac{2qa^2}{y^4}\mathbf{j}$. This charge distribution, which is essentially two electric dipoles, is called an *electric quadrupole*, and the quantity $2qa^2$ is called the *quadrupole moment*. Note that $\mathbf{E}$ varies as r^{-4} for the quadrupole, compared with variations of r^{-3} for the dipole and r^{-2} for the monopole (a single charge).

9. A uniformly charged rod with charge per unit length λ is bent into the shape of a circular arc of radius R as in Fig. 20.35. The arc subtends an angle 2θ at the center of the circle. Show that the electric field at the center of the circle is in the y direction with a magnitude given by $E_y = 2k\lambda \sin\theta/R$.

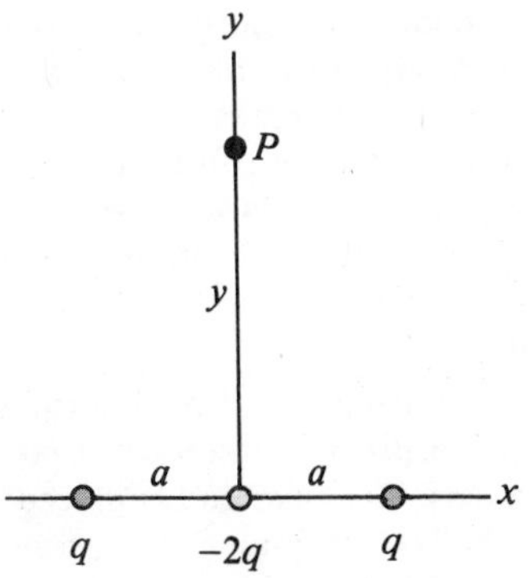

Figure 20.34 (Problem 8).

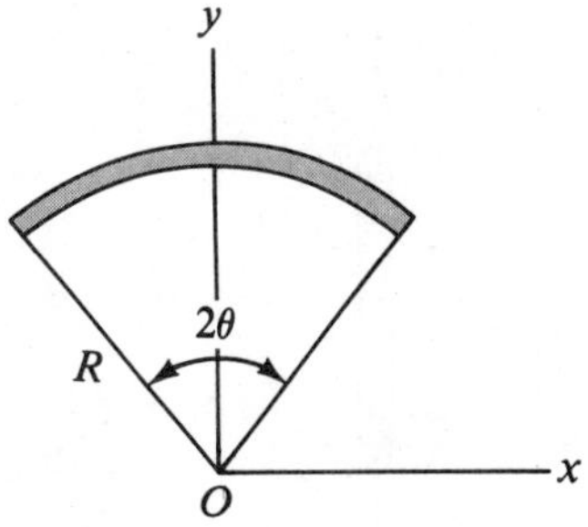

Figure 20.35 (Problem 9).

10. A thin rod of length l and uniform charge per unit length λ lies along the x axis as shown in Fig. 20.36. (a) Show that the electric field at the point P, a distance r from the rod, along the perpendicular bisector has no x component and is given by $E_y = 2k\lambda \sin\theta/r$. (b) Using your result to (a), show that the field of a rod of *infinite* length is given by

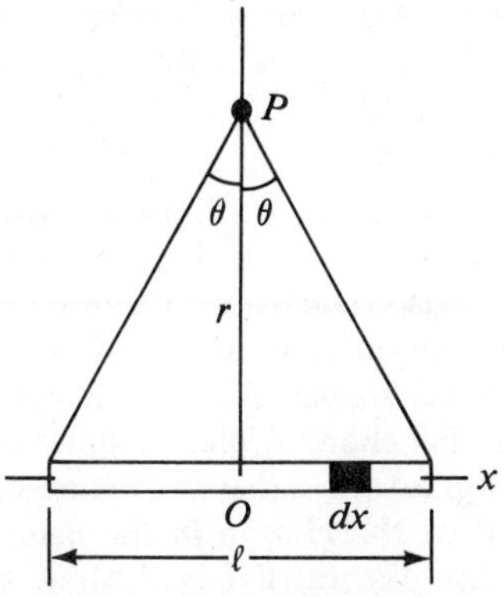

Figure 20.36 (Problem 10).

$E_r = 2k\lambda/r$. (Hint: First calculate the field at P due to an element of length dx, which has a charge $\lambda\ dx$. Then change variables from x to θ using the facts that $x = r\tan\theta$ and $dx = r\sec^2\theta\ d\theta$ and integrate over θ.)

11. A line charge of length l and oriented along the x axis as in Fig. 20.12 has a charge per unit length λ, which varies with x as $\lambda = \lambda_0(x - d)/d$, where d is the distance of the rod from the origin (point P in the figure) and λ_0 is a constant. Find the electric field at the origin. (Hint: An infinitesimal element has a charge $dq = \lambda\ dx$, but note that λ is *not* a constant.)

12. A rod of length l with uniform charge per unit length λ is placed a distance d from the origin along the x axis. A similar rod is placed along the y axis (Fig. 20.37). Determine the net electric field intensity at the origin. (Hint: See Example 20.7.)

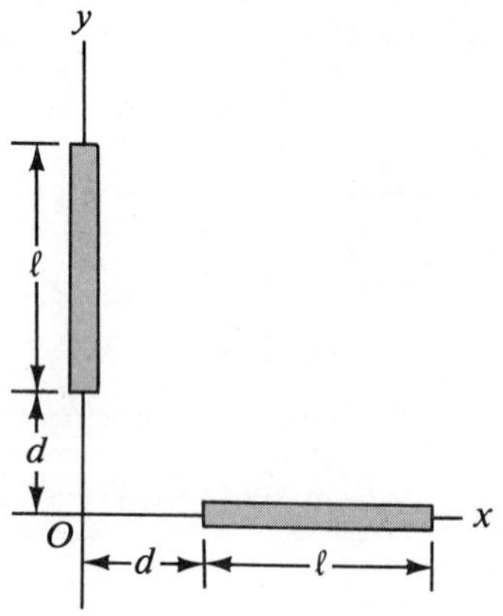

Figure 20.37 (Problem 12).

13. A *negatively* charged particle $-q$ is placed at the center of a uniformly charged ring, where the ring has a total positive charge Q as in Example 20.8. The particle is displaced a *small* distance x along the axis (where $x \ll a$) and released. Show that the particle oscillates with simple harmonic motion along the x axis with a frequency given by $f = \dfrac{1}{2\pi}\left(\dfrac{keQ}{ma^3}\right)^{1/2}$.

14. The *cathode-ray oscilloscope* operates on the following principle. An electron with charge e and mass m is projected with a speed v_0 at right angles to a uniform electric field (Fig. 20.38) and is deflected as shown. A screen is placed a distance L from the charged plates. Ignoring the effects of gravity, (a) show that the equation of the path followed by the charge *in the field* is given by $y = (eE/2mv_0^2)x^2$. (b) If $L \gg d$, show that the charge-to-mass ratio is given by $e/m = hv_0^2/ELd$. (This problem also suggests a means of measuring e/m for other charged particles.)

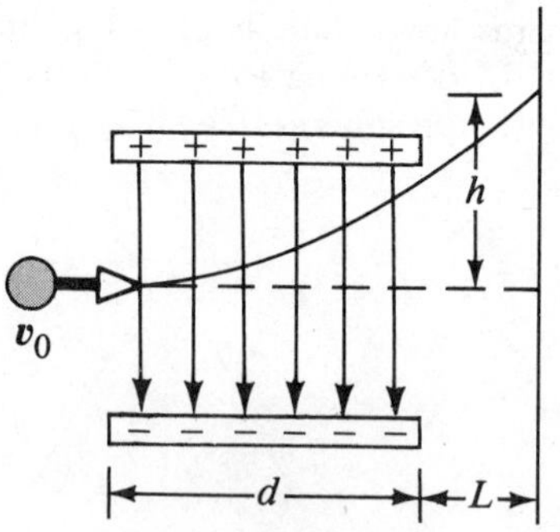

Figure 20.38 (Problem 14).

15. *Millikan's oil drop experiment:* An ingenious technique for determining the charge on an electron was devised by Millikan. The basic apparatus is sketched in Fig. 20.39. Oil is sprayed as fine droplets between two oppositely charged plates, which set up an electric field $\mathbf{E}$. The droplets receive some negative charge by friction with the nozzle of the spraying device. Three forces act on the droplets: the gravitational force $m\mathbf{g}$, the electric force $q\mathbf{E}$, and the buoyant force $\mathbf{B}$. (Recall that according to Archimedes principle, the buoyant force equals the weight of the medium displaced by the oil droplet.) The electric field is varied until the gravitational force is balanced by a combination of the electric force and the buoyant force so that the droplets are either stationary or moving with constant speed. Under this condition, show that the charge q on a droplet is given by

$$q = \frac{(\rho_o - \rho_a)gV}{E}$$

where ρ_o and ρ_a are the densities of oil and air, respectively, and V is the volume of the droplet. (In practice, q is some multiple of the charge on an electron and *many* measurements have to be made to determine e.)

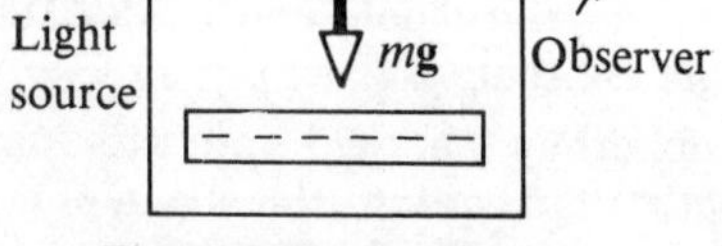

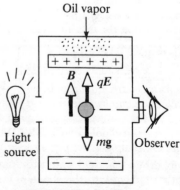

Figure 20.39 (Problem 15).

16. An electric dipole in a uniform electric field is displaced slightly from its equilibrium position, as in Fig. 20.40, where θ is small. The dipole moment is p, and the moment of inertia of the dipole is I. If the dipole is released from this position, show that it exhibits simple harmonic motion with a frequency given by $f = \frac{1}{2\pi}\sqrt{\frac{pE}{I}}$.

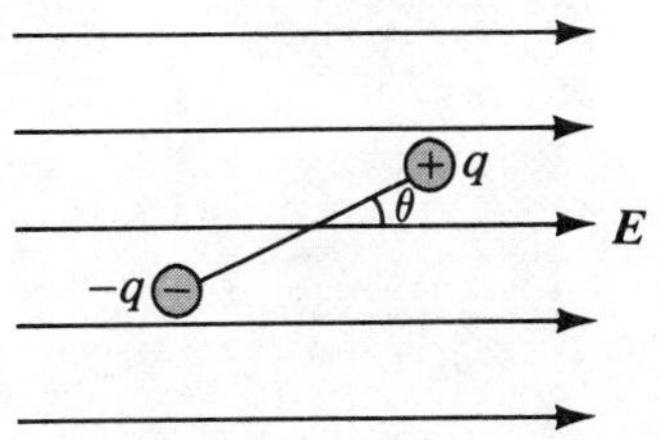

Figure 20.40 (Problem 16).

21 Gauss' Law

General Motors Corporation

In the preceding chapter we showed how to calculate the electric field of a given charge distribution from Coulomb's law. This chapter describes an alternative procedure for calculating electric fields known as *Gauss' law*. This formulation is based on the fact that the fundamental electrostatic force between point charges is an inverse-square law. Although Gauss' law is a consequence of Coulomb's law, it is much more convenient for calculating the electric field of highly symmetric charge distributions. Furthermore, Gauss' law serves as a guide for understanding more complicated problems.

21.1 ELECTRIC FLUX

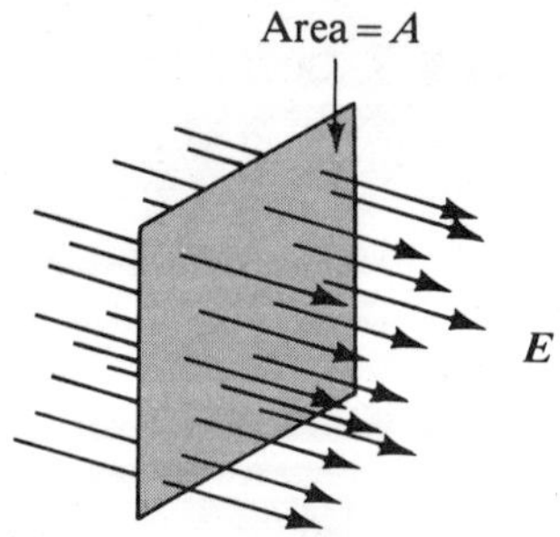

Figure 21.1 Field lines of a uniform electric field penetrating a plane of area A perpendicular to the field. The electric flux, Φ, through this area is equal to EA.

The concept of electric field lines was described qualitatively in the previous chapter. We shall now use the concept of electric flux to put this idea on a quantitative basis. *Electric flux is a measure of the number of electric field lines penetrating some surface.* When the surface being penetrated encloses some net charge, the net number of lines that go through the surface is proportional to the net charge within the surface. The number of lines counted is independent of the shape of the surface enclosing the charge. This is essentially a statement of Gauss' law, which we describe in the next section.

First consider an electric field that is uniform in both magnitude and direction, as in Fig. 21.1. The electric field lines penetrate a rectangular surface of area A, which is perpendicular to the field. Recall that the number of lines per unit area is proportional to the magnitude of the electric field. Therefore, the number of lines penetrating the surface of area A is proportional to the product EA. The product of the electric field strength, E, and a surface area A perpendicular to the field is called the *electric flux*, Φ:

Flux when plane of area A is perpendicular to a uniform field

$$\Phi = EA \tag{21.1}$$

From the SI units of E and A, we see that electric flux has the units of $N \cdot m^2/C$.

If the surface under consideration is not perpendicular to the field, the number of lines (or the flux) through it must be less than that given by Eq. 21.1. This can be easily understood by considering Fig. 21.2, where the surface of area A is at an angle θ to the uniform electric field. Note that the number of lines that cross this area is equal to the number that cross the projected area A', which is perpendicular to the

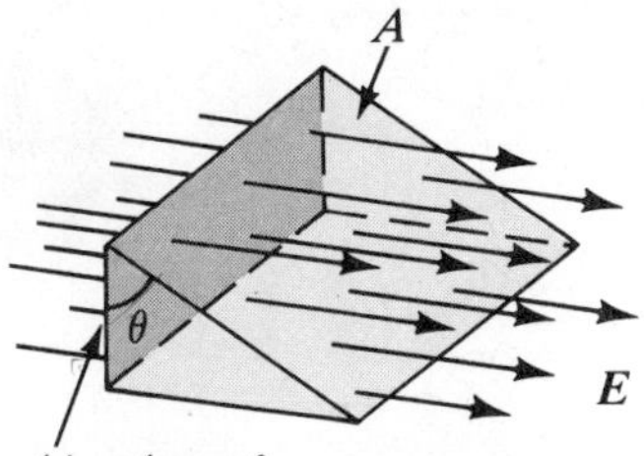

Figure 21.2 Field lines for a uniform electric field through an area A that is at an angle θ to the field. Since the number of lines that go through the shaded area A' is the same as the number that go through A, we conclude that the flux through A' is equal to the flux through A and is given by $\Phi = EA\cos\theta$.

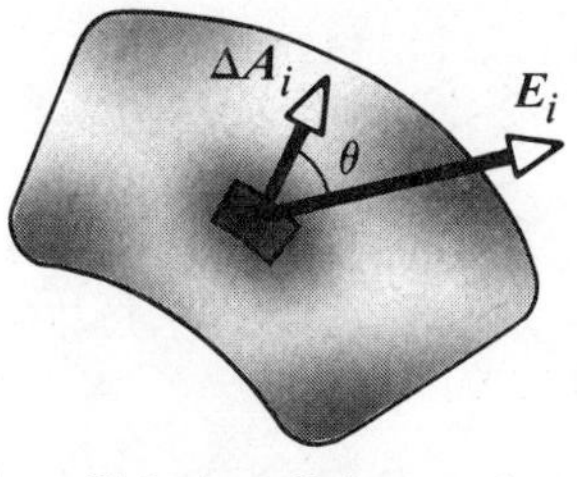

Figure 21.3 A small element of a surface of area ΔA_i. The electric field makes an angle θ with the normal to the surface (the direction of $\Delta \mathbf{A}_i$), and the flux through the element is equal to $E_i\,\Delta A_i\cos\theta$.

field. From Fig. 21.2 we see that the two areas are related by $A' = A\cos\theta$. Since the flux through the area A equals the flux through A', we conclude that the desired flux is given by

Flux when plane of area A is at an angle to a uniform field

$$\Phi = EA\cos\theta \tag{21.2}$$

From this result, we see that the flux through the surface has the maximum value, EA, when the surface is perpendicular to the field (or when the *normal* to the surface is parallel to the field, that is, $\theta = 0°$); the flux is zero when the surface is parallel to the field (or when the normal to the surface is perpendicular to the field, that is, $\theta = 90°$).

In more general situations, the electric field may vary over the surface in question. Therefore, our definition of flux given by Eq. 21.2 has meaning only over a small element of area. Consider a general surface divided up into a large number of small elements, each of area ΔA. The variation in the electric field over the element can be neglected if the element is small enough. It is convenient to define a vector $\Delta\mathbf{A}_i$ whose magnitude represents the area of the ith element and whose direction is *defined* to be *perpendicular* to the surface, as in Fig. 21.3. The electric flux $\Delta\Phi_i$ through this small element is given by

Flux through a small element of area ΔA_i

$$\Delta\Phi_i = E_i\,\Delta A_i\cos\theta = \mathbf{E}_i \cdot \Delta\mathbf{A}_i$$

where we have used the definition of the scalar product of two vectors ($\mathbf{A}\cdot\mathbf{B} = AB\cos\theta$). By summing the contributions of all elements, we obtain the total flux through the surface.[1] If we let the area of each element approach zero, then the number of elements approaches infinity and the sum is replaced by an integral. Therefore *the general definition of electric flux is*

Definition of electric flux

$$\Phi \equiv \lim_{\Delta A_i 0} \Sigma \mathbf{E}_i \cdot \Delta\mathbf{A}_i = \int_{\text{surface}} \mathbf{E}\cdot d\mathbf{A} \tag{21.3}$$

Equation 21.3 is a surface integral, which must be evaluated over the hypothetical surface in question. In general, the value of Φ depends both on the field pattern and on the specified surface.

[1]It is important to note that drawings with field lines have their inaccuracies, since a small area (depending on its location) may happen to have too many or too few penetrating lines. At any rate, it is stressed that the basic definition of electric flux is $\int \mathbf{E}\cdot d\mathbf{A}$.

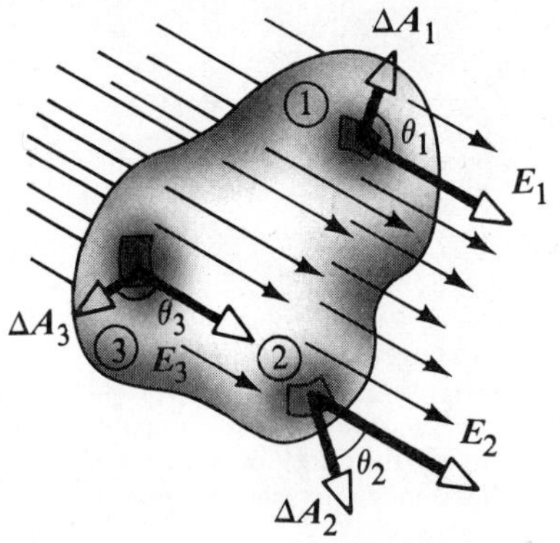

Figure 21.4 A closed surface in an electric field. The area vectors $\Delta \mathbf{A}_i$ are, by convention, normal to the surface and point outward. The flux through an area element can be positive (elements ① and ②) or negative (element ③).

We shall usually be interested in evaluating the flux through a *closed surface*, as shown in Fig. 21.4. (A closed surface is one with no "holes" in it. The surface of a sphere, for example, is a closed surface.) In Fig. 21.4, note that the vectors $\Delta \mathbf{A}_i$ point in different directions for the various surface elements. At each point, these vectors are *normal* to the surface and, by convention, always point *outward*. At the elements labeled ① and ②, $\mathbf{E}$ is outward and $\theta < 90°$; hence the flux $\Delta\Phi = \mathbf{E} \cdot \Delta\mathbf{A}$ through these elements is positive. On the other hand, for elements such as ③, where the field lines are directed into the surface, $\theta > 90°$ and the flux becomes negative with $\cos\theta$. The total, or net, flux through the surface is proportional to the net number of lines penetrating the surface (where the net number means *the number leaving the surface minus the number entering the surface*). If there are more lines leaving the surface than entering, the net flux is positive. If more lines enter than leave the surface, the net flux is negative. Using the symbol $\oint$ to represent an *integral over a closed surface*, we can write the net flux, Φ_c, through a closed surface

Net flux through a closed surface

$$\Phi_c = \oint \mathbf{E} \cdot d\mathbf{A} = \oint E_n \, dA \tag{21.4}$$

where E_n represents the component of the electric field perpendicular, or normal, to the surface and the subscript c denotes a closed surface. Evaluating the net flux through a closed surface could be very cumbersome. However, if the field is normal to the surface at each point and constant in magnitude, the calculation is straightforward. The following example illustrates this point.

Example 21.1 *Flux Through a Cube*

Consider a uniform electric field $\mathbf{E}$ oriented in the x direction. Find the net electric flux through the surface of a cube of edges l oriented as shown in Fig. 21.5. Assume there is no electric charge inside the cube.

Solution: The net flux can be evaluated by summing up the fluxes through each face of the cube. First, note that the flux through *four* of the faces is zero, since $\mathbf{E}$ is perpendicular to $d\mathbf{A}$ on these faces. In particular, the orientation of $d\mathbf{A}$ is perpendicular to $\mathbf{E}$ for the two faces labeled ③ and ④ in Fig. 21.5. Therefore, $\theta = 90°$, so that $\mathbf{E} \cdot d\mathbf{A} = E\,dA\cos 90° = 0$. The fluxes through the planes parallel to the yx plane are also zero for the same reason.

Now consider the faces labeled ① and ②. The net flux through these faces is given by

$$\Phi_c = \int_1 \mathbf{E} \cdot d\mathbf{A} + \int_2 \mathbf{E} \cdot d\mathbf{A}$$

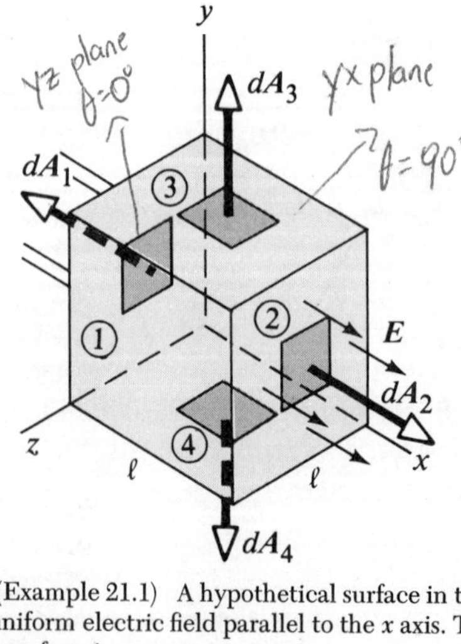

Figure 21.5 (Example 21.1) A hypothetical surface in the shape of a cube in a uniform electric field parallel to the x axis. The net flux through the surface is zero.

For the face labeled ①, $\mathbf{E}$ is constant and inward while $d\mathbf{A}$ is outward ($\theta = 180°$), so that we find that the flux through this face is

$$\int_1 \mathbf{E}\cdot d\mathbf{A} = \int_1 E\,dA\cos 180° = -E\int_1 dA = -EA = -El^2$$

since the area of each face is $A = l^2$.

Likewise, for the face labeled ②, $\mathbf{E}$ is constant and outward and in the same direction as $d\mathbf{A}$ ($\theta = 0°$), so that the flux through this face is

$$\int_2 \mathbf{E}\cdot d\mathbf{A} = \int_2 E\,dA\cos 0° = E\int_2 dA = +EA = El^2$$

Hence, the net flux over all faces is zero, since

$$\Phi_c = -El^2 + El^2 = 0$$

21.2 GAUSS' LAW

In this section we describe a general relation between the net electric flux through a closed surface (often called a *gaussian surface*) and the charge *enclosed* by the surface. This relation, known as *Gauss' law*, is of fundamental importance in the study of electrostatic fields.

First, let us consider a positive point charge q located at the center of a sphere of radius r as in Fig. 21.6. From Coulomb's law we know that the magnitude of the electric field everywhere on the surface of the sphere is $E = kq/r^2$. Furthermore, the field lines are radial outward, and hence are perpendicular (or normal) to the surface at each point. That is, at each point $\mathbf{E}$ is parallel to the vector $\Delta\mathbf{A}_i$ representing the local element of area ΔA_i. Therefore

$$\mathbf{E}\cdot\Delta\mathbf{A}_i = E_n\,\Delta A_i = E\,\Delta A_i$$

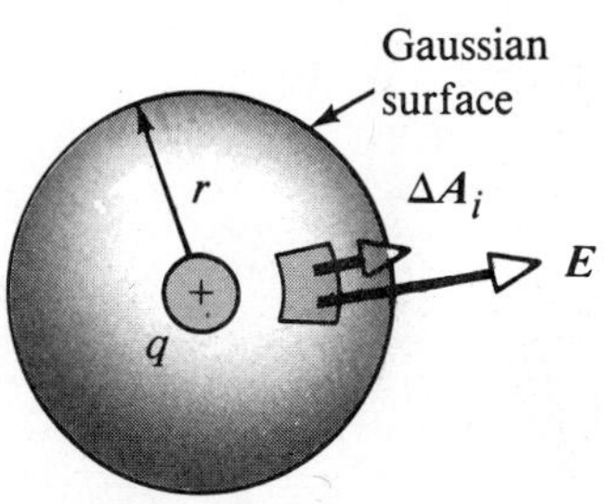

Figure 21.6 A spherical surface of radius r surrounding a point charge q. When the charge is at the center of the sphere, the electric field is normal to the surface and constant in magnitude everywhere on the surface.

and from Eq. 21.4 we find that the net flux through the gaussian surface is given by

$$\Phi_c = \oint E_n\,dA = \oint E\,dA = E\oint dA$$

since E is constant over the surface and given by $E = kq/r^2$. Furthermore, for a spherical gaussian surface, $\oint dA = A = 4\pi r^2$ (the surface area of a sphere). Hence the net flux through the gaussian surface is

$$\Phi_c = \frac{kq}{r^2}(4\pi r^2) = 4\pi kq$$

Recalling that $k = 1/4\pi\varepsilon_0$, we can write this in the form

$$\Phi_c = \frac{q}{\varepsilon_0} \tag{21.5}$$

Flux through a closed surface

Note that this result, which is independent of r, says that the net flux through a spherical gaussian surface is proportional to the charge q inside the surface. The fact that the flux is independent of the radius is a consequence of the inverse-square dependence of the electric field given by Coulomb's law. That is, E varies as $1/r^2$, but the area of the sphere varies as r^2. Their combined effect produces a flux that is independent of r.

Now consider several closed surfaces surrounding a charge q as in Fig. 21.7. Surface S_1 is spherical, whereas surfaces S_2 and S_3 are nonspherical. The flux that passes through surface S_1 has the value q/ε_0. As we discussed in the previous section, the flux is proportional to the number of electric field lines passing through that surface. The construction in Fig. 21.7 shows that the number of electric field lines through the spherical surface S_1 is equal to the number of electric field lines through the nonspherical surfaces S_2 and S_3. Therefore, it is reasonable to conclude that the net flux through any closed surface is independent of the shape of that surface. In

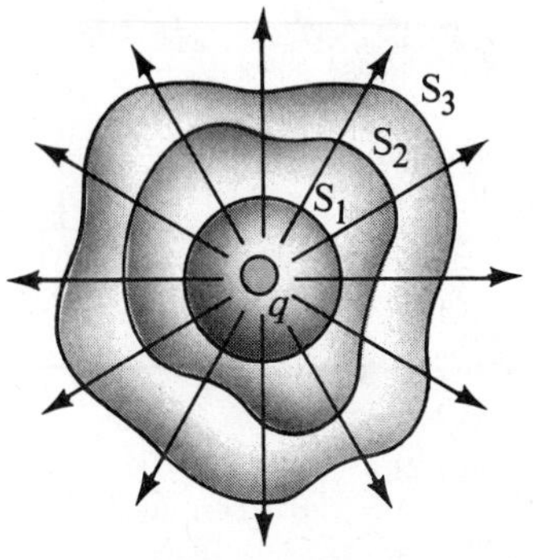

Figure 21.7 Closed surfaces of various shapes surrounding a charge q. Note that the net electric flux through each surface is the same.

fact, *the net flux through any closed surface surrounding a point charge q is given by* q/ε_0.

Now consider a point charge located *outside* a closed surface of arbitrary shape, as in Fig. 21.8. As you can see from this construction, some electric field lines enter the surface, and others leave the surface. However, *the number of electric field lines entering the surface equals the number leaving the surface.* Therefore, we conclude that *the net electric flux through a closed surface that surrounds no charge is zero.* If we apply this result to Example 21.1, we can easily see that the net flux through the cube is zero, since it was assumed there was no charge inside the cube.

The net flux through a closed surface is zero if there is no charge inside

Let us extend these arguments to the generalized case of many point charges, or a continuous distribution of charge. We shall make use of the *superposition principle,* which says that *the electric field due to many charges is the vector sum of the electric fields produced by the individual charges.* That is, we can express the flux through any closed surface as

$$\oint \mathbf{E} \cdot d\mathbf{A} = \oint (\mathbf{E}_1 + \mathbf{E}_2 + \mathbf{E}_3) \cdot d\mathbf{A}$$

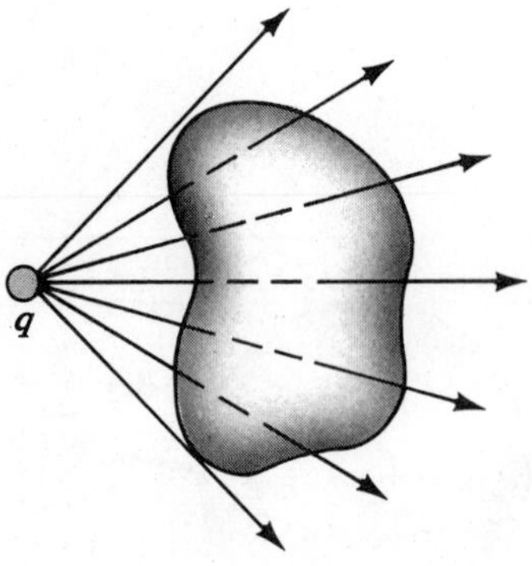

Figure 21.8 A point charge located *outside* a closed surface. In this case, note that the number of lines entering the surface equals the number leaving the surface.

where $\mathbf{E}$ is the total electric field at any point on the surface and $\mathbf{E}_1$, $\mathbf{E}_2$, and $\mathbf{E}_3$ are the fields produced by the individual charges at that point. Consider the system of charges shown in Fig. 21.9. The surface S surrounds only one charge, q_1; hence the net flux through S is q_1/ε_0. The flux through S due to the charges outside it is zero since each electric field line that enters S at one point leaves it at another. The surface S′ surrounds charges q_2 and q_3; hence the net flux through S′ is $(q_2 + q_3)/\varepsilon_0$. Finally, the net flux through surface S″ is zero since there is no charge inside this surface. That is, *all* lines that enter S″ at one point leave S″ at another.

Gauss' law, which is a generalization of the above discussion, states that the net flux through *any* closed surface is given by

Gauss' law

$$\Phi_c = \oint \mathbf{E} \cdot d\mathbf{A} = \frac{q_{in}}{\varepsilon_0} \tag{21.6}$$

where q_{in} represents the net charge inside the gaussian surface and $\mathbf{E}$ represents the electric field at any point on the gaussian surface. In words, Gauss' law states that *the net electric flux through any closed gaussian surface is equal to the net charge inside the surface divided by* ε_0.

When using Eq. 21.6, you should note that although the charge q_{in} is the net charge inside the gaussian surface, the $\mathbf{E}$ that appears in Gauss' law represents the *total electric field,* which includes contributions from charges both inside and outside the gaussian surface. This point is often neglected or misunderstood.

In principle, Gauss' law can always be used to calculate the electric field of a system of charges or a continuous distribution of charge. However, in practice, *the technique is useful only in a limited number of situations where there is a high degree of symmetry.* As we shall see in the next section, *Gauss' law can be used to evaluate the electric field for charge distributions that have spherical, cylindrical, or planar symmetry.* If one carefully chooses the gaussian surface surrounding the charge distribution, the integral in Eq. 21.6 will be easy to evaluate.

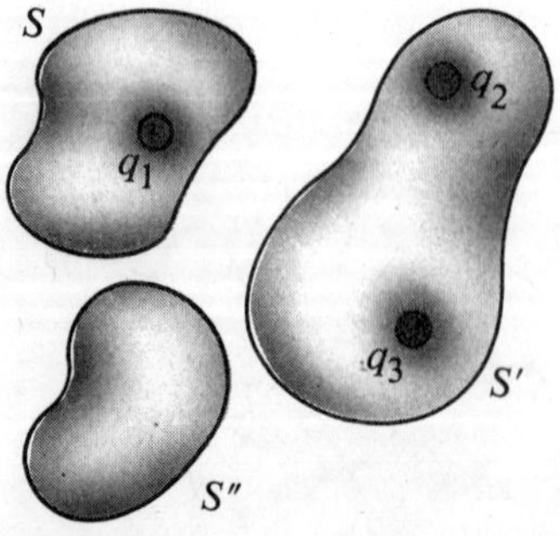

Figure 21.9 The net electric flux through any closed surface depends only on the charge *inside* that surface. The net flux through surface S is q_1/ε_0, the net flux through surface S′ is $(q_2 + q_3)/\varepsilon_0$, and the net flux through surface S″ is zero.

Q1. If the net flux through a gaussian surface is zero, which of the following statements are true? (a) There are no charges inside the surface. (b) The net charge inside the surface is zero. (c) The electric field is zero everywhere on the surface. (d) The number of electric field lines entering the surface equals the number leaving the surface.

Q2. A spherical gaussian surface surrounds a point charge q. Describe what happens to the flux through the surface if (a) the charge is tripled, (b) the volume of the sphere is doubled, (c) the shape of the surface is changed to that of a cube, and (d) the charge is moved to another position inside the surface.

21.3 APPLICATION OF GAUSS' LAW TO CHARGED INSULATORS

In this section we give some examples of how to use Gauss' law to calculate $\mathbf{E}$ for a given charge distribution. It is important to recognize that *Gauss' law is useful only when there is a high degree of symmetry in the charge distribution, as in the case of uniformly charged spheres, long cylinders, and planar sheets.* In such cases, it is possible to find a simple gaussian surface over which the surface integral given by Eq. 21.6 is easily evaluated. *The surface should always be chosen such that it has the same symmetry as that of the charge distribution.* The following examples should clarify this procedure.

Gauss' law is useful for evaluating E when the charge distribution has symmetry

Example 21.2

The electric field due to a point charge: Starting with Gauss' law, calculate the electric field due to an isolated point charge q and show that Coulomb's law follows from this result.

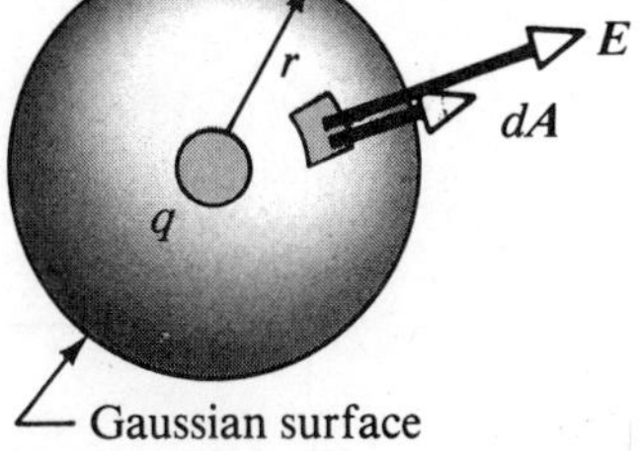

Figure 21.10 (Example 21.2) The point charge q is at the center of the spherical gaussian surface, and $\mathbf{E}$ is parallel to $d\mathbf{A}$ at every point on the surface.

Solution: For this situation we choose a spherical gaussian surface of radius r and centered on the point charge, as in Fig. 21.10. As we saw in the previous chapter, the electric field of a positive point charge is radial outward and is therefore normal to the surface at every point. That is, $\mathbf{E}$ is parallel to $d\mathbf{A}$ at each point, and so $\mathbf{E}\cdot d\mathbf{A} = E\,dA$ and Gauss' law gives

$$\Phi_c = \oint \mathbf{E}\cdot d\mathbf{A} = \oint E\,dA = \frac{q}{\varepsilon_0}$$

By symmetry, E is constant everywhere on the surface, and so it can be removed from the integral. Therefore,

$$\oint E\,dA = E\oint dA = E(4\pi r^2) = \frac{q}{\varepsilon_0}$$

where we have used the fact that the surface area of a sphere is $4\pi r^2$. Hence, the magnitude of the field at a distance r from the charge q is

$$E = \frac{q}{4\pi\varepsilon_0 r^2} = k\frac{q}{r^2}$$

If a second point charge q_0 is placed at a point where the field is E, the electrostatic force on this charge has a magnitude given by

$$F = q_0 E = k\frac{qq_0}{r^2}$$

This, of course, is Coulomb's law. Note that this example is logically circular. It does, however, demonstrate the equivalence of Coulomb's law and Gauss' law.

Example 21.3

A spherically symmetric charge distribution: An insulating sphere of radius a has a uniform charge density ρ and a total positive charge Q (Fig. 21.11). (a) Calculate the electric field intensity at a point *outside* the sphere, that is, for $r > a$.

Since the charge distribution is spherically symmetric, we again select a spherical gaussian surface of radius r, concentric with the sphere, as in Fig. 21.11a. Following the line of reasoning given in Example 21.2, we find that

$$E = k\frac{Q}{r^2} \quad \text{(for } r > a\text{)} \qquad (21.7)$$

Note that this result is identical to that obtained for a point charge. Therefore, we conclude that, for a uniformly charged sphere, the field in the region external to the sphere is *equivalent* to that of a point charge located at the center of the sphere.

(b) Find the electric field intensity at a point *inside* the sphere, that is, for $r < a$.

In this case we select a spherical gaussian surface with radius $r < a$, concentric with the charge distribution (Fig. 21.11b). To apply Gauss' law in this situation, it is important to recognize that the charge q_{in} *within* the

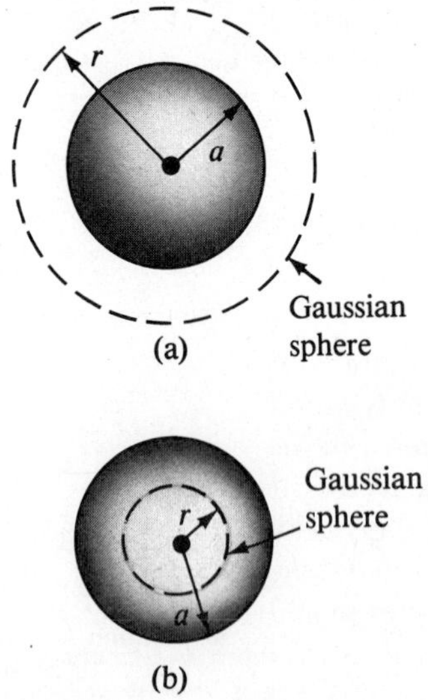

Figure 21.11 (Example 21.3) A uniformly charged insulating sphere of radius a and total charge Q. (a) The field at a point exterior to the sphere is kQ/r^2. (b) The field inside the sphere is due only to the charge *within* the gaussian surface and is given by $(kQ/a^3)r$.

gaussian surface of volume V' is a quantity *less* than the total charge Q. To calculate the charge q_{in}, we use the fact that $q_{\text{in}} = \rho V'$, where ρ is the charge per unit volume and V' is the volume enclosed by the gaussian surface, given by $V' = \frac{4}{3}\pi r^3$ for a sphere. Therefore,

$$q_{\text{in}} = \rho V' = \rho\left(\frac{4}{3}\pi r^3\right)$$

As in Example 21.2, the electric field is constant in magnitude everywhere on the spherical gaussian surface and is normal to the surface at each point. Therefore, Gauss' law in the region $r < a$ gives

$$\oint E\, dA = E\oint dA = E(4\pi r^2) = \frac{q_{\text{in}}}{\varepsilon_0}$$

Solving for E gives

$$E = \frac{q_{\text{in}}}{4\pi\varepsilon_0 r^2} = \frac{\rho \frac{4}{3}\pi r^3}{4\pi\varepsilon_0 r^2} = \frac{\rho}{3\varepsilon_0} r$$

Since by definition $\rho = Q/\frac{4}{3}\pi a^3$, this can be written

$$E = \frac{Qr}{4\pi\varepsilon_0 a^3} = \frac{kQ}{a^3} r \qquad \text{(for } r < a\text{)} \qquad (21.8)$$

Note that this result for E differs from that obtained in (a). It shows that $E \to 0$ as $r \to 0$, as you might have guessed based on the spherical symmetry of the charge distribution. Therefore, the result fortunately eliminates the singularity that would exist at $r = 0$ if E varied as $1/r^2$ inside the sphere. That is, if $E \sim 1/r^2$, the field would be infinite at $r = 0$, which is clearly a physically impossible situation. A plot of E versus r is shown in Fig. 21.12.

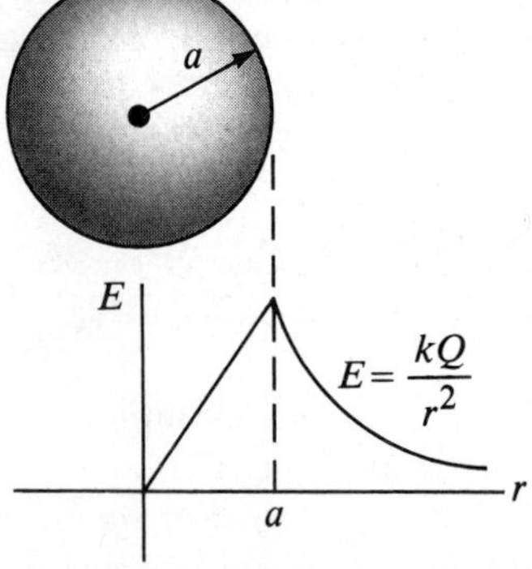

Figure 21.12 (Example 21.3) A plot of E versus r for a uniformly charged insulating sphere of radius a. The field inside the sphere $(r < a)$ varies linearly with r. The field outside the sphere $(r > R)$ is the same as that of a point charge Q located at the origin.

Example 21.4 *The E Field of a Spherical Shell*

A *spherical shell* of radius a has a total charge Q distributed uniformly over its surface (Fig. 21.13). Find the electric field at points inside and outside the shell.

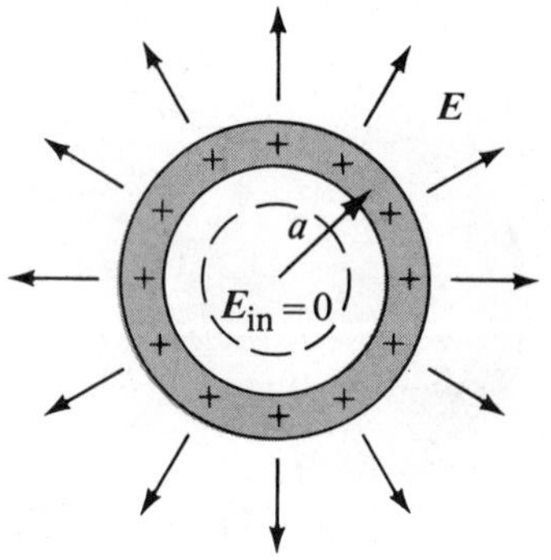

Figure 21.13 (Example 21.4) The electric field inside a uniformly charged spherical shell is *zero*. The field outside is the same as that of a point charge located at the center of the shell.

Solution: The calculation of the field outside the shell is identical to that already carried out for the solid sphere in Example 21.3a. If we construct a spherical gaussian surface of radius $r > a$, concentric with the shell, then the charge inside this surface is Q. Therefore, the field at a point outside the shell is equivalent to that of a point charge Q at the center:

$$E = k\frac{Q}{r^2} \qquad \text{(for } r > a\text{)}$$

The electric field inside the spherical shell is zero. This also follows from Gauss' law applied to a spherical surface of radius $r < a$. Since the net charge inside the surface is zero, and because of the spherical symmetry of the charge distribution, it follows that $E = 0$ in the region $r < a$.

The same results can be obtained using Coulomb's law and integrating over the charge distribution. This calculation is rather complicated and will be omitted. The mathematical steps involved are identical to those given in Section 14.8 for the gravitational field of a spherical shell.

Example 21.5

A cylindrically symmetric charge distribution: Find the electric field at a distance r from a uniform positive line charge of infinite length whose charge per unit length is λ = constant (Fig. 21.14).

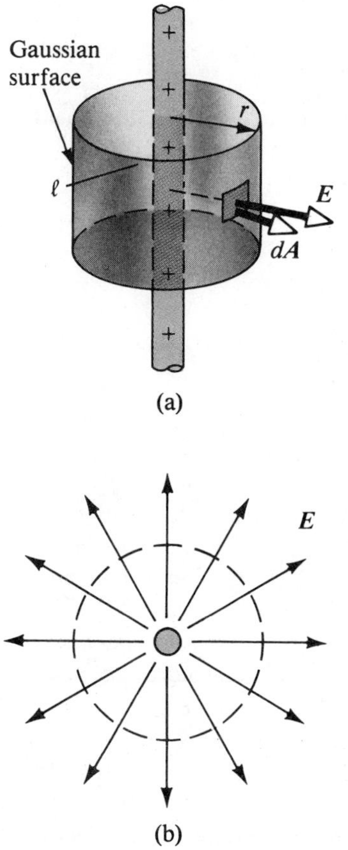

Figure 21.14 (Example 21.5) An infinite line of charge surrounded by a cylindrical gaussian surface concentric with the line charge. The field on the cylinder is constant in magnitude and perpendicular to the surface.

Solution: The symmetry of the charge distribution shows that $\mathbf{E}$ must be perpendicular to the line charge and directed outward as in Fig. 21.14a. The end view of the line charge shown in Fig. 21.14b should help visualize the directions of the electric field lines. In this situation, we select a cylindrical gaussian surface of radius r and length l that is coaxial with the line charge. For the cylindrical part of this surface, $\mathbf{E}$ is constant in magnitude and perpendicular to the surface at each point. Furthermore, the flux through the *ends* of the gaussian cylinder is *zero* since $\mathbf{E}$ is *parallel* to these surfaces.

The total charge inside our gaussian surface is λl, where λ is the charge per unit length and l is the length of the cylinder. Applying Gauss' law and noting the $\mathbf{E}$ is parallel to $d\mathbf{A}$ everywhere on the cylindrical surface, we find that

$$\Phi_c = \oint \mathbf{E} \cdot d\mathbf{A} = E \oint dA = \frac{q_{\text{in}}}{\varepsilon_0} = \frac{\lambda l}{\varepsilon_0}$$

But the area of the cylindrical surface is $A = 2\pi r l$; therefore

$$E(2\pi r l) = \frac{\lambda l}{\varepsilon_0}$$

$$E = \frac{\lambda}{2\pi\varepsilon_0 r} = 2k\frac{\lambda}{r} \tag{21.9}$$

Thus, we see that the field of a cylindrically symmetric charge distribution varies as $1/r$, whereas the field external to a spherically symmetric charge distribution varies as $1/r^2$. Equation 21.9 can also be obtained using Coulomb's law and integration; however, the mathematical techniques necessary for this calculation are more cumbersome.

If the line charge has a finite length, the result for E is *not* the same as that given by Eq. 21.9. For points close to the line charge and far from the ends, Eq. 21.9 gives a good approximation of the actual value of the field. It turns out that Gauss' law is *not useful* for calculating E for a finite line charge. This is because the electric field is no longer constant in magnitude over the surface of the gaussian cylinder. Furthermore, $\mathbf{E}$ is not perpendicular to the cylindrical surface at all points. When there is little symmetry in the charge distribution, as in this situation, it is necessary to calculate $\mathbf{E}$ using Coulomb's law.

Example 21.6

A nonconducting planar sheet of charge: Find the electric field due to a nonconducting, infinite plane with uniform charge per unit area σ.

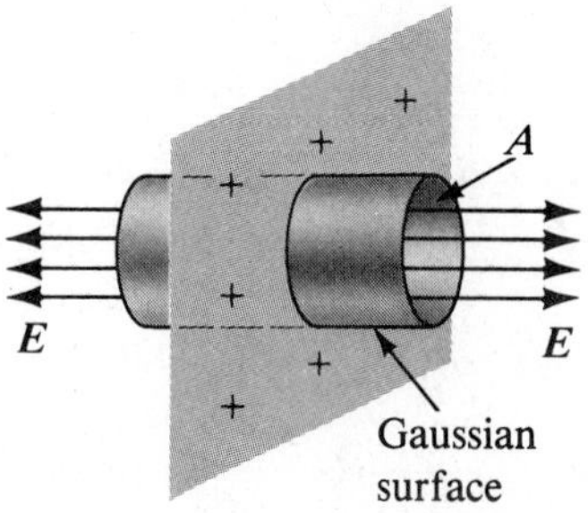

Figure 21.15 (Example 21.6) A cylindrical gaussian surface penetrating an infinite sheet of charge. The flux through each end of the cylinder is EA. There is no flux through the cylinder's surface.

Solution: The symmetry of the situation shows that $\mathbf{E}$ must be perpendicular to the plane and that the direction of $\mathbf{E}$ on one side of the plane must be opposite its direction on the other side, as in Fig. 21.15. It is convenient to choose for our gaussian surface a small cylinder whose axis is perpendicular to the plane and whose ends each have an area A and are equidistant from the plane. Here we see that since $\boldsymbol{E}$ is parallel to the cylindrical surface, there is no flux through this surface. The flux out of *each* end of the cylinder is EA (since $\boldsymbol{E}$ is perpendicular to the ends); hence the *total* flux through our gaussian surface is $2EA$. Noting that the total charge *inside* the surface is σA, we use Gauss' law to get

$$\Phi_c = 2EA = \frac{q_{\text{in}}}{\varepsilon_0} = \frac{\sigma A}{\varepsilon_0}$$

$$E = \frac{\sigma}{2\varepsilon_0} \tag{21.10}$$

Since the distance of the surfaces from the plane does not appear in Eq. 21.10, we conclude that $E = \sigma/2\varepsilon_0$ at *any* distance from the plane. That is, the field is *uniform* everywhere.

Q3. Explain why Gauss' law would *not* be useful in calculating the electric field of (a) an electric dipole, (b) a charged disk, (c) a charged ring, and (d) three point charges at the corners of a triangle.

Q4. If the total charge inside a closed surface is known but the distribution of the charge is unspecified, can you use Gauss' law to find the electric field? Explain.

21.4 CONDUCTORS IN ELECTROSTATIC EQUILIBRIUM

Properties of a conductor in electrostatic equilibrium

A good electrical conductor, such as copper, contains charges (electrons) that are not bound to any atom and are free to move about within the material. When there is no *net* motion of charge within the conductor, the conductor is in electrostatic equilibrium. As we shall see, *a conductor in electrostatic equilibrium* has the following properties: (1) *The electric field is zero everywhere inside the conductor.* (2) *Any excess charge on an isolated conductor must reside entirely on its surface.* (3) *The electric field just outside a charged conductor is perpendicular to the conductor's surface and has a magnitude σ/ε_0, where σ is the charge per unit area at that point.*

The first property can be understood by considering a conducting slab placed in an external field $\boldsymbol{E}$ (Fig. 21.16). In electrostatic equilibrium, the electric field *inside* the conductor must be zero. If this were not the case, the free charges would accelerate under the action of an electric field. Before the external field is applied, the electrons are uniformly distributed throughout the conductor. When the external field is applied, the free electrons accelerate to the left, causing a buildup of negative charge on the left surface (excess electrons) and of positive charge on the right (where electrons have been removed). These charges create their own electric field, which *opposes* the external field. The surface charge density increases until the magnitude of the electric field set up by these charges equals that of the external field, giving a net field of zero *inside* the conductor. In a good conductor, the time it takes the conductor to reach equilibrium is of the order of 10^{-16} s, which for most purposes can be considered instantaneous.

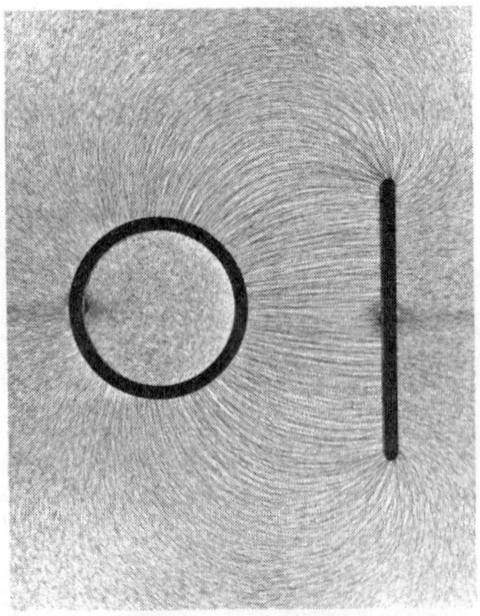

Electric field pattern of a charged conducting plate near an oppositely charged conducting cylinder. Small pieces of thread suspended in oil align with the electric field lines. Note that (1) the electric field lines are perpendicular to the conductors and (2) there are no lines inside the cylinder ($E = 0$). (Courtesy of Harold M. Waage, Princeton University).

We can use Gauss' law to verify the second and third properties of a conductor in electrostatic equilibrium. Figure 21.17 shows an arbitrarily shaped insulated conductor. A gaussian surface is drawn inside the conductor as close to the surface as we wish. As we have just shown, the electric field everywhere inside the conductor is zero when it is in electrostatic equilibrium. Since the electric field is also zero at *every* point on the gaussian surface, we see that the net flux through this surface is zero. From this result and Gauss' law, we conclude that the net charge inside the gaussian surface is zero. Since there can be no net charge inside the gaussian surface (which is arbitrarily close to the conductor's surface), *any net charge on the conduc-*

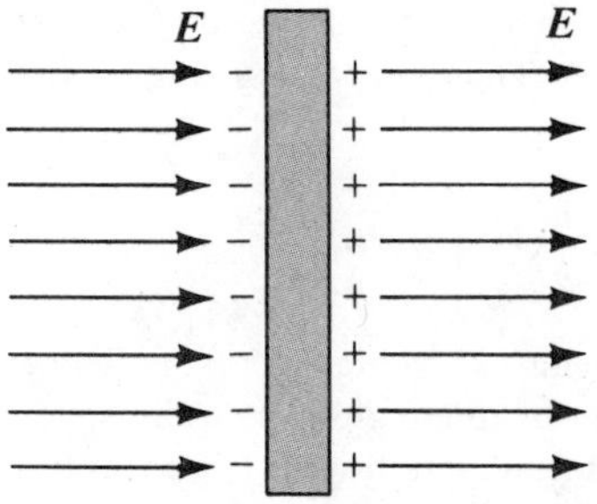

Figure 21.16 A conducting slab in an external electric field **E**. The charges induced on the surfaces of the slab produce an electric field which opposes the external field, giving a resultant field of zero in the conductor.

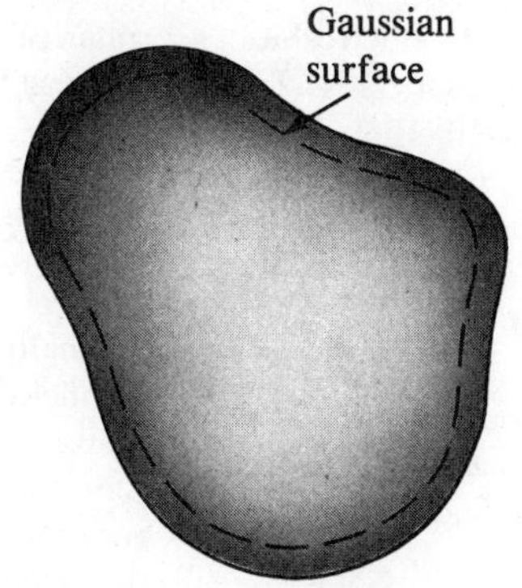

Figure 21.17 An insulated conductor of arbitrary shape. The broken line represents a gaussian surface just inside the conductor.

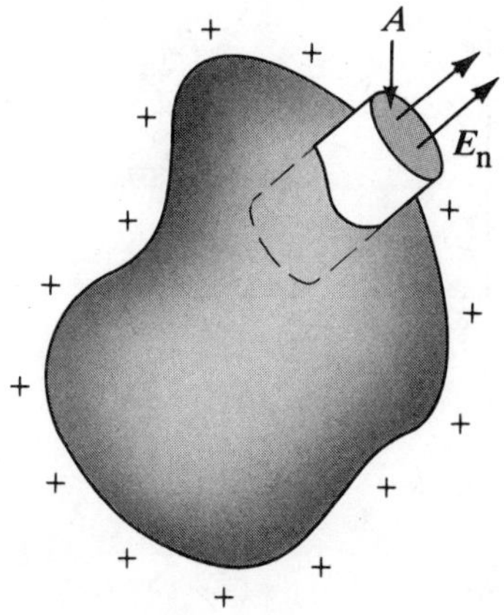

Figure 21.18 A gaussian surface in the shape of a small cylinder is used to calculate the electric field just outside a charged conductor. The flux through the gaussian surface is E_nA. Note that E is zero inside the conductor.

tor must reside on its surface. Gauss' law does *not* tell us how this excess charge is distributed on the surface.

We can use Gauss' law to relate the electric field just outside the surface of a charged conductor in equilibrium. To do this, it is convenient to draw a gaussian surface in the shape of a small cylinder with end faces parallel to the surface (Fig. 21.18). Part of the cylinder is just outside the conductor, and part is inside. There is no flux through the face on the inside of the cylinder since $\mathbf{E} = 0$ inside the conductor. Furthermore, the field is normal to the surface. If $\mathbf{E}$ had a tangential component, the free charges would move along the surface creating surface currents, and the conductor would not be in equilibrium. There is no flux through the cylindrical face of the gaussian surface since $\mathbf{E}$ is tangent to this surface. Hence, the net flux through the gaussian surface is E_nA, where E_n is the electric field just outside the conductor. Applying Gauss' law to this surface gives

$$\Phi_c = \oint E_n\, dA = E_nA = \frac{q_{in}}{\varepsilon_0} = \frac{\sigma A}{\varepsilon_0}$$

We have used the fact that the charge inside the gaussian surface is $q_{in} = \sigma A$, where A is the area of the cylinder's face and σ is the (local) charge per unit area. Solving for E_n gives

$$E_n = \frac{\sigma}{\varepsilon_0} \qquad (21.11)$$

Electric field just outside a charged conductor

Example 21.7

A solid conducting sphere of radius a has a net positive charge $2Q$ (Fig. 21.19). A conducting spherical shell of inner radius b and outer radius c is concentric with the solid sphere and has a *net* charge $-Q$. Using Gauss' law, find the electric field in the regions labeled ①, ②, ③, and ④ and the charge distribution on the spherical shell.

Solution: First note that the charge distribution on both spheres has spherical symmetry, since they are concentric. To determine the electric field at various distances r from the center, we construct spherical gaussian surfaces of radius r.

To find E inside the solid sphere of radius a (region ①), we construct a gaussian surface of radius $r < a$. Since

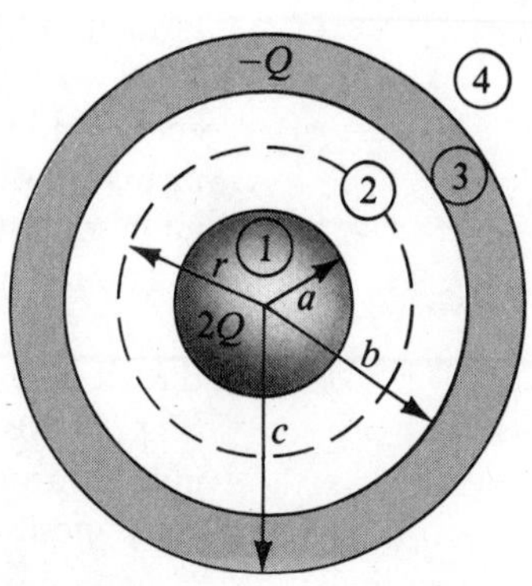

Figure 21.19 (Example 21.7) A solid conducting sphere of radius a and charge $2Q$ surrounded by a conducting spherical shell of charge $-Q$.

there can be no charge inside a conductor in electrostatic equilibrium, we see that $q_{in} = 0$, and so from Gauss' law $E_1 = 0$ for $r < a$. Thus we conclude that the net charge $2Q$ on the solid sphere is distributed on its outer surface.

In region ② between the spheres, where $a < r < b$, we again construct a spherical gaussian surface of radius r and note that the charge inside this surface is $+2Q$ (the charge on the inner sphere). Because of the spherical symmetry, the electric field lines must be radial outward and constant in magnitude on the gaussian surface. Following Example 21.2 and using Gauss' law, we find that

$$E_2A = E_2(4\pi r^2) = \frac{q_{in}}{\varepsilon_0} = \frac{2Q}{\varepsilon_0}$$

$$E_2 = \frac{2Q}{4\pi\varepsilon_0 r^2} = \frac{2kQ}{r^2} \qquad (\text{for } a < r < b)$$

In region ④ outside both spheres, where $r > c$, the spherical gaussian surface surrounds a *total* charge of $q_{in} = 2Q + (-Q) = Q$. Therefore, Gauss' law applied to this surface gives

$$E_4 = \frac{kQ}{r^2} \qquad (\text{for } r > c)$$

Finally, consider region ③, where $b < r < c$. The electric field must be *zero* in this region since the spherical shell is also a conductor in equilibrium. If we construct a Gaussian surface of this radius, we see that q_{in} must be zero since $E_3 = 0$. From this argument, we conclude that the charge on the *inner surface* of the *spherical shell* must be $-2Q$ to cancel the charge $+2Q$ on the solid sphere. (The charge $-2Q$ is induced by the charge $+2Q$ on the solid sphere.) Furthermore, since the net charge on the spherical shell is $-Q$, we conclude that the outer surface of the shell must have a charge equal to $+Q$.

Q5. A person is placed in a large hollow metallic sphere that is insulated from ground. If a large charge is placed on the sphere, will the person be harmed upon touching the inside of the sphere? Explain what will happen if the person also has an initial charge whose sign is opposite that of the charge on the sphere.

Q6. A point charge is placed at the center of an uncharged metallic spherical shell insulated from ground. As the point charge is moved off center, describe what happens to (a) the total induced charge on the shell and (b) the distribution of charge on the interior and exterior surfaces of the shell.

21.5 EXPERIMENTAL PROOF OF GAUSS' LAW AND COULOMB'S LAW

When a net charge is placed on a conductor, the charge distributes itself on the surface in such a way that the electric field inside is zero. Since $\mathbf{E} = 0$ inside a conductor in electrostatic equilibrium, Gauss' law shows that there can be no net charge inside the conductor. We have seen that Gauss' law is a consequence of Coulomb's law (Example 21.2). Hence, it should be possible to test the validity of the inverse-square law of force by attempting to detect a net charge inside a conductor. If a net charge is detected anywhere but on the conductor's surface, Coulomb's law, and hence Gauss' law, are invalid. Many experiments have been performed to show that the net charge on a conductor resides on its surface, including early work by Faraday, Cavendish, and Maxwell. In all reported cases, no electric field could be detected in a closed conductor. The most recent and precise experiments by Williams, Faller, and Hill in 1971 showed that the exponent of r in Coulomb's law is $(2 + \varepsilon)$, where $\varepsilon = (2.7 \pm 3.1) \times 10^{-16}$!

The following experiment can be performed to verify that the net charge on a conductor resides on its surface. A positively charged metal ball at the end of a silk thread is lowered into an uncharged, hollow conductor through a small opening[2] (Fig. 21.20a). The hollow conductor is insulated from ground. The charged ball induces a negative charge on the inner wall of the hollow conductor, leaving an equal positive charge on the outer wall (Fig. 21.20b). The presence of positive

[2]The experiment is often referred to as *Faraday's ice-pail experiment*, since it was first performed by Faraday using an ice pail for the hollow conductor.

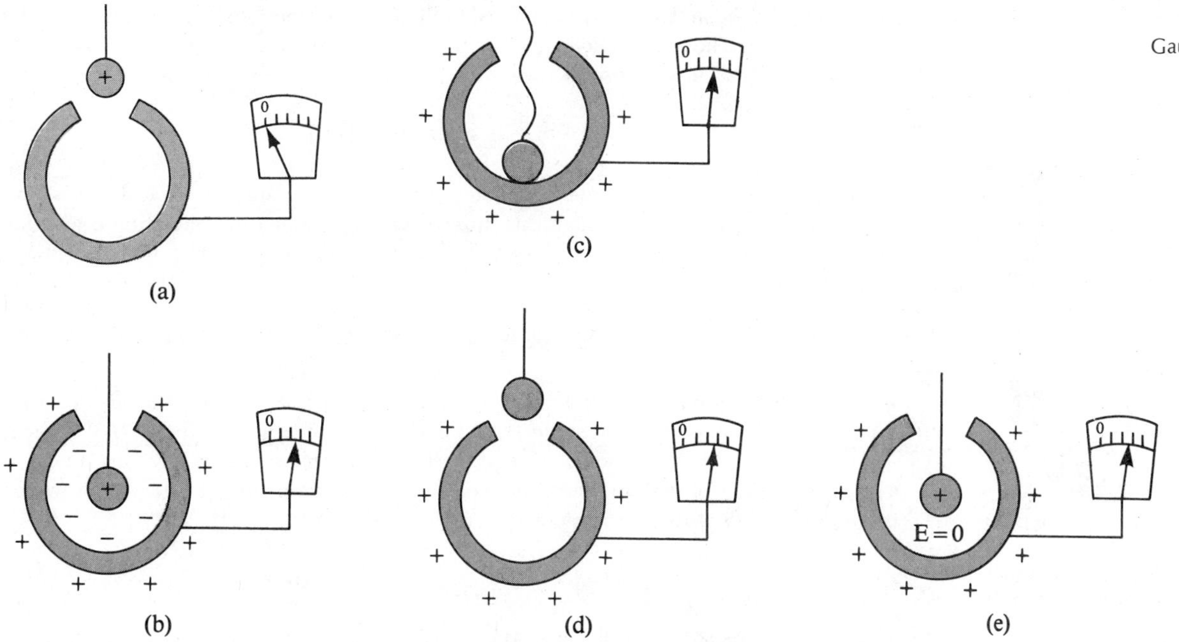

Figure 21.20 An experiment showing that any charge transferred to a conductor resides on its surface in electrostatic equilibrium. The hollow conductor is insulated from ground, and the small metal ball is supported by an insulating thread.

charge on the outer wall is indicated by the deflection of an electrometer (a device used to measure charge). The deflection of the electrometer remains unchanged when the ball touches the inner surface of the hollow conductor (Fig. 21.20c). When the ball is removed, the electrometer reading remains the same and the ball is found to be uncharged (Fig. 21.20d). This shows that *the charge transferred to the hollow conductor resides on its outer surface.* If a small charged metal ball is now lowered into the charged hollow conductor as in Fig. 21.20e, the charged ball will not be attracted to the hollow conductor. This shows that $\boldsymbol{E} = 0$ everywhere inside the hollow conductor. On the other hand, if the small charged ball is placed near the outside of the conductor, the ball will be attracted to the conductor, showing that $\boldsymbol{E} \neq 0$ outside the conductor.

Q7. How would the observations described in Fig. 21.20 differ if the hollow conductor were grounded? How would they differ if the small charged ball were an insulator rather than a conductor?

Q8. What other experiment might be performed on the ball in Fig. 21.20 to show that its charge was transferred to the hollow conductor?

Q9. What would happen to the electrometer reading if the charged ball in Fig. 21.20e touched the inner wall of the conductor? the outer wall?

21.6 DERIVATION OF GAUSS' LAW

One method that can be used to derive Gauss' law involves the concept of the *solid angle*. Consider a spherical surface of radius r containing an area element ΔA. The solid angle $\Delta\Omega$ subtended by this element at the center of the sphere is defined to be

$$\Delta\Omega \equiv \frac{\Delta A}{r^2}$$

From this expression, we see that $\Delta\Omega$ has no dimensions, since ΔA and r^2 both have the dimension of L^2. The unit of a solid angle is called the *steradian*. Since the total surface area of a sphere is $4\pi r^2$, the total solid angle subtended by the sphere at the center is given by

$$\Omega = \frac{4\pi r^2}{r^2} = 4\pi \text{ steradians}$$

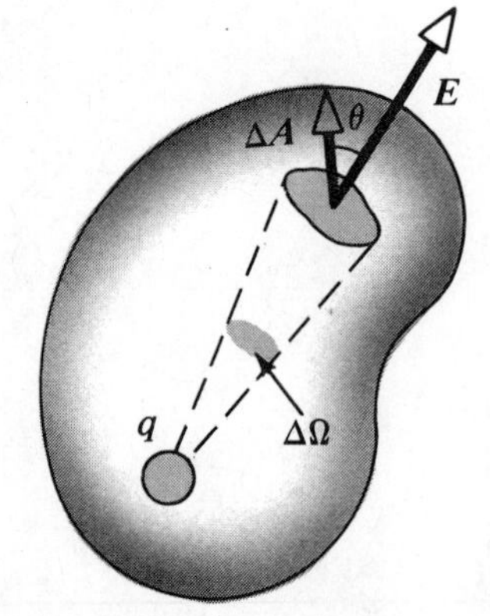

Figure 21.21 A closed surface of arbitrary shape surrounds a point charge q. The net flux through the surface is independent of the shape of the surface.

Now consider a point charge q surrounded by a closed surface of arbitrary shape (Fig. 21.21). The total flux through this surface can be obtained by evaluating $\mathbf{E} \cdot \Delta\mathbf{A}$ for each element of area and summing over all elements of the surface. The flux through the element of area ΔA is

$$\Delta\Phi = \mathbf{E} \cdot \Delta\mathbf{A} = E\cos\theta\, \Delta A = kq\frac{\Delta A \cos\theta}{r^2}$$

where we have used the fact that $E = kQ/r^2$ for a point charge. But the quantity $\Delta A \cos\theta/r^2$ is equal to the solid angle $\Delta\Omega$ subtended at the charge q by the surface element ΔA. From Fig. 21.22 we see that $\Delta\Omega$ is equal to the solid angle subtended by the element of a spherical surface of radius r. Since the total solid angle at a point is 4π steradians, we see that the total flux through the closed surface is

$$\Phi_c = kq\oint\frac{dA\cos\theta}{r^2} = kq\oint d\Omega = 4\pi kq = \frac{q}{\varepsilon_0}$$

Thus we have derived Gauss' law, Eq. 21.6. Note that this result is independent of the shape of the closed surface and independent of the position of the charge within the surface.

21.7 SUMMARY

Electric flux is a measure of the number of electric field lines that penetrate a surface. If the electric field is uniform and makes an angle θ with the normal to the surface, the electric flux through the surface is

Flux of surface in uniform electric field

$$\Phi = EA\cos\theta \qquad (21.2)$$

In general, the electric flux through a surface is defined by the expression

TABLE 21.1 Typical Electric Field Calculations Using Gauss' Law

CHARGE DISTRIBUTION	ELECTRIC FIELD	LOCATION
Insulating sphere of radius R, uniform charge density, and total charge Q	$k\frac{Q}{r_2}$	$r > R$
	$k\frac{Q}{R^3}r$	$r < R$
Thin spherical shell of radius R and total charge Q	$k\frac{Q}{r^2}$	$r > R$
	0	$r < R$
Line charge of infinite length and charge per unit length λ	$2k\frac{\lambda}{r}$	Outside the line charge
Nonconducting, infinite charged plane with charge per unit area σ	$\frac{\sigma}{2\varepsilon_0}$	Everywhere outside the plane
Conductor of surface charge per unit area σ	$\frac{\sigma}{\varepsilon_0}$	Just outside the conductor
	0	Inside the conductor

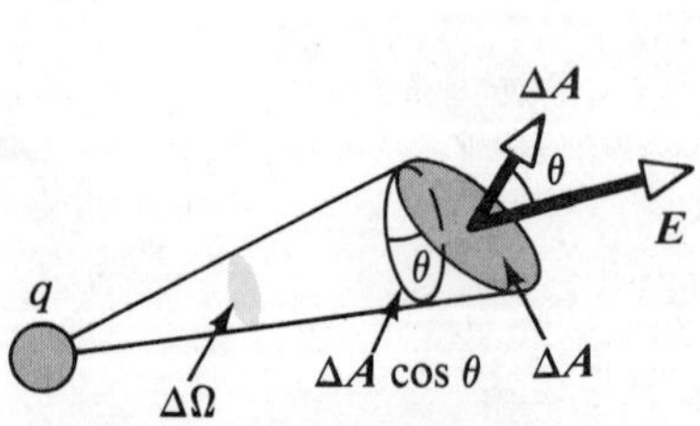

Figure 21.22 The area element ΔA subtends a solid angle $\Delta\Omega = (\Delta A\cos\theta)/r^2$ at the charge q.

$$\Phi = \int_{\text{surface}} \mathbf{E} \cdot d\mathbf{A} \tag{21.3}$$

Definition of electric flux

Gauss' law says that the net electric flux, Φ_c, through any closed gaussian surface is equal to the *net* charge *inside* the surface divided by ε_0:

$$\Phi_c = \oint \mathbf{E} \cdot d\mathbf{A} = \frac{q_{\text{in}}}{\varepsilon_0} \tag{21.6}$$

Gauss' law

Using Gauss' law, one can calculate the electric field due to various symmetric charge distributions. Table 21.1 lists some typical results.

A *conductor in electrostatic equilibrium* has the following properties:

Properties of a conductor in electrostatic equilibrium

1. The electric field is zero everywhere inside the conductor.
2. Any excess charge on an isolated conductor must reside entirely on its surface.
3. The electric field just outside the conductor is perpendicular to its surface and has a magnitude σ/ε_0, where σ is the charge per unit area at that point.

EXERCISES

Section 21.1 Electric Flux

1. An electric field of 5.0×10^4 N/C is applied along the x axis. Calculate the electric flux through a rectangular plane 0.2 m wide and 0.8 m long if (a) the plane is parallel to the yz plane, (b) the plane is parallel to the xy plane, and (c) the plane contains the y axis and its normal makes an angle of 53° with the x axis.

2. A cone with a circular base of radius R stands upright so that its axis is vertical. A uniform electric field $\mathbf{E}$ is applied in the vertical direction. Show that the flux through the cone's surface (not counting its base) is given by $\pi R^2 E$.

Section 21.2 Gauss' Law

3. A point charge of 8 μC is placed at the center of a *spherical* shell of radius 15 cm. What is the total electric flux through (a) the entire surface of the shell and (b) any hemispherical surface of the shell. (c) Do the results depend on the radius? Explain.

4. (a) Two charges of 9 μC and -6 μC are inside a cube of sides 0.3 m. What is the total electric flux through the cube? (b) Repeat (a) if the same two charges are inside a spherical shell of radius 0.3 m.

5. A charge of 120 μC is at the center of a cube of sides 25 cm. (a) Find the total flux through each face of the cube. (b) Find the flux through the whole surface of the cube. (c) Would your answers to (a) or (b) change if the charge were not at the center? Explain.

6. The total electric flux through a closed surface in the shape of a cylinder is 7.50×10^5 N · m²/C. (a) What is the net charge within the cylinder? (b) From the information given, what can you say about the charge within the cylinder? (c) How would your answers to (a) and (b) change if the net flux were -7.50×10^5 N · m²/C?

7. A cube of sides 50 cm is centered at the origin. A point charge of 3 μC is located on the y axis at $y = 100$ cm. (a) Sketch the electric lines for the point charge. (b) What is the net flux through the surface of the cube? (c) Repeat (a) and (b) if a second point charge of 5 μC is located at the center of the cube. (Neglect the lines that go through the edges and corners.)

8. The electric field everywhere on the surface of a hollow sphere of radius 0.2 m is measured to be 6.50×10^3 N/C and points radially toward the center of the sphere. (a) What is the net charge within the sphere's surface? (b) What can you conclude about the nature and distribution of the charge inside the sphere?

9. The electric field on the surface of a hollow sphere is directed radially outward but varies in magnitude over the sphere's surface. A point charge within the sphere has a magnitude Q. (a) What is the sign of Q? (b) What can you say about the position of Q? (c) What is the net flux through the sphere?

Section 21.3 Application of Gauss' Law to Charged Insulators

10. A solid sphere of radius 20 cm has a total charge of 12 μC uniformly distributed throughout its volume. Calculate the electric field intensity at the following distances from the center of the sphere: (a) 0 cm, (b) 5 cm, (c) 20 cm, (d) 50 cm.

11. An inflated balloon in the shape of a sphere of radius 14 cm has a total charge of 8 μC uniformly distributed on its surface. Calculate the electric field intensity at the following distances from the center of the balloon: (a) 13 cm, (b) 14.1 cm, (c) 50 cm.

12. The charge per unit length on a *long*, straight filament is $-70\ \mu C/m$. Find the electric field at the following distances from the filament: (a) 5 cm, (b) 30 cm, (c) 200 cm.

13. A uniformly charged, straight filament 10 m in length has a total positive charge of 5 μC. An uncharged cardboard cylinder 4 cm in length and 13 cm in radius surrounds the filament at its center, with the filament as the axis of the cylinder. Using any reasonable approximations, find (a) the electric field at the surface of the cylinder and (b) the total electric flux through the cylinder.

14. A cylindrical shell of radius 8 cm and length 150 cm has its charge uniformly distributed on its surface. The electric field intensity at a point 12 cm radially outward from its axis (measured from the midpoint of the shell) is 4.8×10^3 N/C. Use approximate relations to find (a) the net charge on the shell and (b) the electric field at a point 4 cm from the axis, measured from the midpoint.

15. A large planar sheet of charge has a charge per unit area of $7.5\ \mu C/m^2$. Find the electric field intensity *just above the surface* of the sheet, measured from its midpoint.

16. Show that the electric field intensity at a point just outside a uniformly charged spherical shell is σ/ε_0, where σ is the charge per unit area on the shell. (Hint: Note that the electric field *inside* the shell is zero.)

Section 21.4 Conductors in Electrostatic Equilibrium

17. A solid copper sphere 25 cm in radius has a total charge of 25 nC. Find the electric field at the following distances measured from the center of the sphere: (a) 24 cm, (b) 26 cm, (c) 85 cm. (d) How would your answers change if the sphere were hollow?

18. A long, straight metal rod has a radius of 3 cm and a charge per unit length of 40 nC/m. Find the electric field at the following distances from the axis of the rod: (a) 2 cm, (b) 4 cm, (c) 75 cm.

19. A square sheet of copper of sides 20 cm is placed in an extended electric field of 3×10^3 N/C directed *perpendicular* to the sheet. Find (a) the charge density of each face of the sheet and (b) the total charge on each face.

20. A conducting sheet 50 cm on a side lies in the *xy* plane. If a total charge of 4×10^{-8} C is placed on the sheet, find (a) the charge density on the sheet, (b) the electric field just above the sheet, and (c) the electric field just below the sheet.

21. A conducting spherical shell of radius 8 cm carries a net charge of $-2\ \mu C$ uniformly distributed on its surface. Find the electric field at points (a) just outside the shell and (b) inside the shell.

22. A hollow (but not necessarily empty) conducting sphere has a uniform charge per unit area of $+\sigma$ on its outer surface and $-\sigma$ on its inner surface. From this information, (a) what can you conclude about the charge in the region *inside* the hollow sphere? (b) What can you say about the electric field just outside the sphere?

23. A *long*, straight wire is surrounded by a hollow metallic cylinder whose axis coincides with that of the wire. The solid wire has a charge per unit length of $+\lambda$, and the hollow cylinder has a *net* charge per unit length of $+2\lambda$. From this information, use Gauss' law to find (a) the charge per unit length on the inner and outer surfaces of the hollow cylinder and (b) the electric field outside the hollow cylinder, a distance *r* from the axis.

PROBLEMS

1. A solid *insulating* sphere of radius *a* has a uniform charge density ρ and a total charge *Q*. Concentric with this sphere is an *uncharged, conducting* hollow sphere whose inner and outer radii are *b* and *c*, as in Fig. 21.23. (a) Find the electric field intensity in the regions $r < a$, $a < r < b$, $b < r < c$, and $r > c$. (b) Determine the induced charge per unit area on the inner and outer surfaces of the hollow sphere.

2. For the configuration shown in Fig. 21.23, suppose that $a = 5$ cm, $b = 20$ cm, and $c = 25$ cm. Furthermore, suppose that the electric field at a point 10 cm from the center is measured to be 3.6×10^3 N/C radially *inward* while the electric field at a point 50 cm from the center is

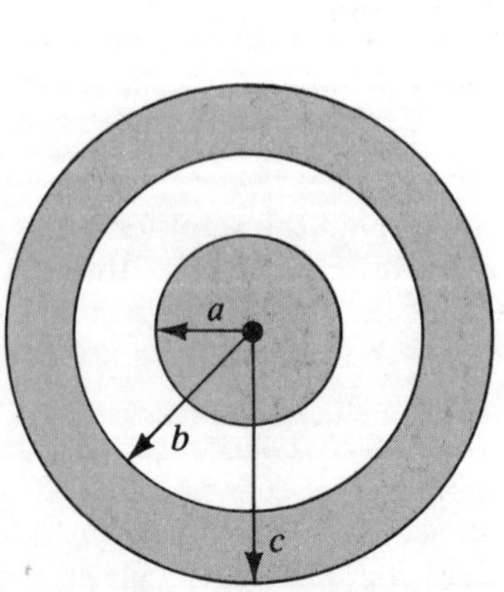

Figure 21.23 (Problems 1 and 2).

2.0×10^2 N/C radially *outward*. From this information, find (a) the charge on the insulating sphere, (b) the net charge on the hollow sphere, and (c) the total charge on the inner and outer *surfaces* of the hollow sphere.

3. A *hollow* insulating sphere has a uniform charge density ρ. Its inner and outer radii are a and b, respectively. Use Gauss' law to find expressions for the electric field in the regions (a) $r < a$, (b) $a < r < b$, (c) $r > b$.

4. A solid insulating sphere of radius R has a *nonuniform* charge density that varies with r according to the expression $\rho = Ar^2$, where A is a constant and r is measured from the center of the sphere. (a) Show that the electric field *outside* the sphere is given by $E = AR^5/5\varepsilon_0 r^2$. (b) Show that the electric field *inside* the sphere is given by $E = Ar^3/5\varepsilon_0$. (Hint: Note that the total charge Q on the sphere is equal to the integral of ρdV, where r extends from 0 to R; also note that the charge q within a radius $r < R$ is *less* than Q. To evaluate the integrals, note that the volume element dV for a spherical shell of radius r and thickness dr is equal to $4\pi r^2 dr$.)

5. Two infinite, nonconducting sheets of charge are parallel to each other as in Fig. 21.24. The sheet on the left has a uniform surface charge density σ, and the one on the right has a uniform charge density $-\sigma$. Calculate the value of the electric field at points (a) to the left of, (b) in between, and (c) to the right of the two sheets.

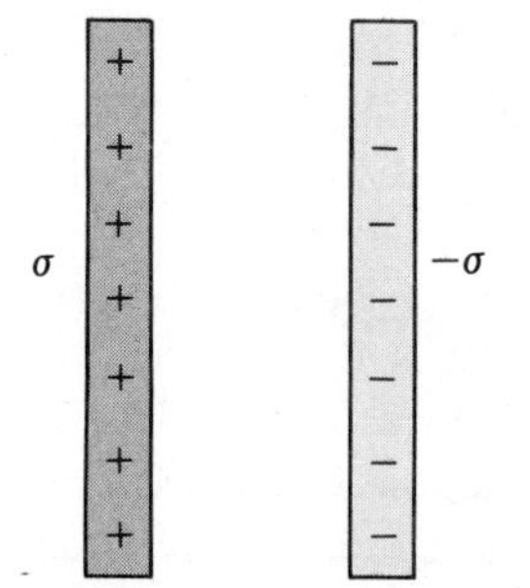

Figure 21.24 (Problems 5 and 6).

6. Repeat the calculations for Problem 5 when both sheets have *positive* uniform charge densities of value σ.

7. A closed surface with dimensions $a = b = 0.4$ m and $c = 0.6$ m is located as shown in Fig. 21.25. The electric field throughout the region is *nonuniform* and given by

$$\mathbf{E} = (3 + 2x^2)\mathbf{i}$$

Calculate the net electric flux leaving the closed surface. What net charge is enclosed by the surface?

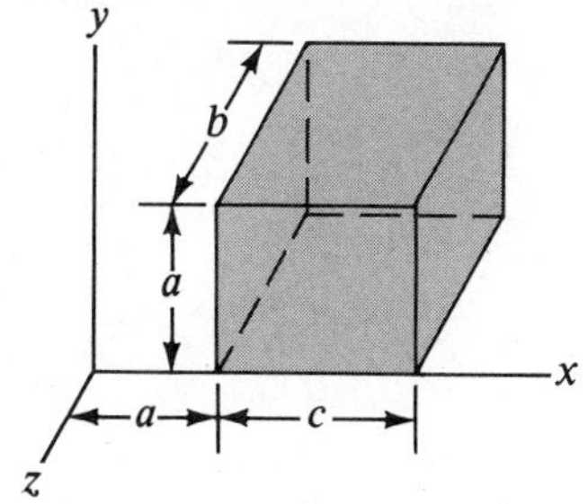

Figure 21.25 (Problem 7).

8. A slab of insulating material (infinite in two of its three dimensions) has a uniform positive charge density ρ as in Fig. 21.26. (a) Show that the electric field a distance x from its center is $E = \rho x/\varepsilon_0$. (b) Suppose an electron of charge e and mass m is placed inside the slab. If it is released from rest at a distance x from the center, show that the electron exhibits simple harmonic motion with a frequency given by $f = \dfrac{1}{2\pi}\sqrt{\dfrac{\rho e}{m\varepsilon_0}}$.

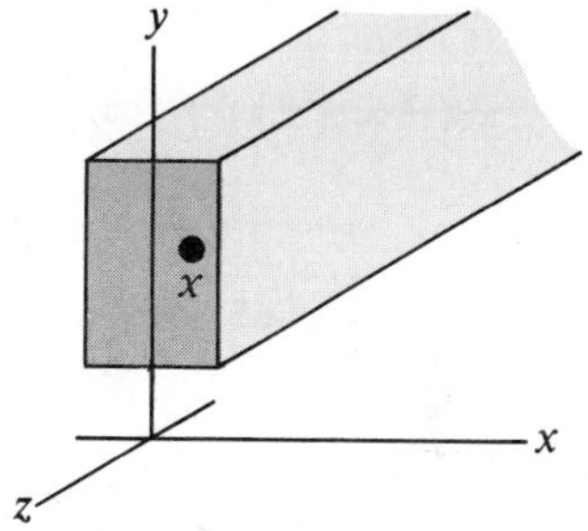

Figure 21.26 (Problem 8).

9. An infinitely long dielectric cylinder of radius R has a volume charge density that varies with the radius as $\rho = \rho_0\left(a - \dfrac{r}{b}\right)$, where ρ_0, a, and b are positive constants and r is the distance from the axis of the cylinder. Use Gauss' law to determine the magnitude of the electric field at radial distances (a) $r < R$ and (b) $r > R$.

10. Consider a dielectric sphere of radius R and having a *uniform* volume charge density ρ. Plot the magnitude of the electric field, E, as a function of the distance from the center of the sphere, r. Let r range over the interval $0 < r < 3R$ and plot E in units of $\dfrac{\rho R}{\varepsilon_0}$.

22 Electric Potential

Edmund Scientific Co.

The concept of potential energy was first introduced in Chapter 8 in connection with such conservative forces as the force of gravity and the elastic force of a spring. By using the law of energy conservation, we were often able to avoid working directly with forces when solving various mechanical problems. In this chapter we shall see that the energy concept is also of great value in the study of electricity. Since the electrostatic force given by Coulomb's law is conservative, one can conveniently describe electrostatic phenomena in terms of an electrical potential energy. This idea enables us to define a scalar quantity called *electric potential.* Because the potential is a scalar function of position, it offers a simpler way of describing electrostatic phenomena than does the electric field. In later chapters we shall see that the concept of the electric potential is of great practical value. In fact, the measured voltage between any two points in an electrical circuit is simply the difference in electric potential between the points.

22.1 POTENTIAL DIFFERENCE AND ELECTRIC POTENTIAL

In Chapter 14, we showed that the gravitational force is conservative. Since the electrostatic force, given by Coulomb's law, is of the same form as the universal law of gravity, it follows that the electrostatic force is also conservative. Therefore, it is possible to define a potential energy function associated with this force.

When a test charge q_0 is placed in an electrostatic field $\mathbf{E}$, the electric force on the test charge is $q_0\mathbf{E}$. The force $q_0\mathbf{E}$ is the vector sum of the individual forces exerted on q_0 by the various charges producing the field $\mathbf{E}$. It follows that the force $q_0\mathbf{E}$ is conservative, since the individual forces governed by Coulomb's law are conservative. The work done by the electric force $q_0\mathbf{E}$ on the test charge for an infinitesimal displacement $d\mathbf{s}$ is given by

$$dW = \mathbf{F} \cdot d\mathbf{s} = q_0\mathbf{E} \cdot d\mathbf{s} \tag{22.1}$$

By definition, the work done by a conservative force equals the negative of the change in potential energy, dU; therefore, we see that

$$dU = -q_0\mathbf{E} \cdot d\mathbf{s} \tag{22.2}$$

For a finite displacement of the test charge between points A and B, *the change* in the potential energy is given by

$$\Delta U = U_B - U_A = -q_0 \int_A^B \mathbf{E} \cdot d\mathbf{s} \tag{22.3}$$

Change in potential energy

The integral in Eq. 22.3 is performed along the path by which q_0 moves from A to B and is called a *path integral,* or *line integral.* Since the force $q_0\mathbf{E}$ is conservative, *this integral does not depend on the path taken between A and B.*

The *potential difference,* $V_B - V_A$, between the points A and B is defined as *the change in potential energy divided by the test charge* q_0:

$$V_B - V_A = \frac{U_B - U_A}{q_0} = -\int_A^B \mathbf{E} \cdot d\mathbf{s} \tag{22.4}$$

Potential difference

Potential difference should not be confused with potential energy. The potential difference is *proportional* to the potential energy, and we see from Eq. 22.4 that the two are related by $\Delta U = q_0\, \Delta V$. Since potential energy is a scalar, electric potential is also a scalar quantity. Note that the change in the potential energy of the charge is the negative of the work done by the electric force. Hence, we see that *the potential difference $V_B - V_A$ equals the work per unit charge that an external agent must perform to move a test charge from A to B without a change in kinetic energy.*

Equation 22.4 defines potential differences only. That is, only *differences* in V are meaningful. The electric potential function is often taken to be zero at some convenient point. We shall usually choose the potential to be zero for a point at infinity (that is, a point infinitely remote from the charges producing the electric field). With this choice, we can say that the *electric potential at an arbitrary point equals the work required per unit charge to bring a test charge from infinity to that point.* Thus, if we take $V_A = 0$ at infinity in Eq. 22.4, then the potential at any point P is given by

$$V_P = -\int_\infty^P \mathbf{E} \cdot d\mathbf{s} \tag{22.5}$$

In reality, V_P represents the potential difference between the point P and a point at infinity. (Equation 22.5 is a special case of Eq. 22.4.)

Since potential difference is a measure of energy per unit charge, the SI units of potential are joules per coulomb, defined to be equal to a unit called the *volt* (V):

$$1\ \text{V} \equiv 1\ \text{J/C}$$

Definition of a volt

That is, 1 J of work must be done to take a 1-C charge through a potential difference of 1 V. Equation 22.4 shows that the potential difference also has units of electric field times distance. From this, it follows that the SI unit of electric field (N/C) can also be expressed as volts per meter:

$$1\ \text{N/C} = 1\ \text{V/m}$$

A unit of energy commonly used in atomic and nuclear physics is the *electron volt,* which is defined as *the energy that an electron (or proton) gains when accelerated through a potential difference of magnitude 1 V.* Since 1 V = 1 J/C and since the fundamental charge is equal to 1.6×10^{-19} C, we see that the electron volt (eV) is related to the joule through the relation

$$1\ \text{eV} = 1.6 \times 10^{-19}\ \text{C} \cdot \text{V} = 1.6 \times 10^{-19}\ \text{J} \tag{22.6}$$

The electron volt

For instance, an electron in the beam of a typical TV picture tube (or cathode ray tube) has a speed of 5×10^7 m/s. This corresponds to a kinetic energy of 1.1×10^{-15} J, which is equivalent to 7.1×10^3 eV. Such an electron has to be accelerated from rest through a potential difference of 7.1 kV to reach this speed.

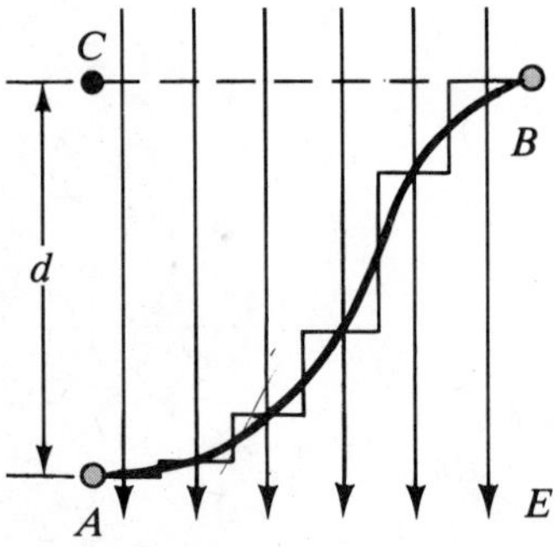

Figure 22.1 The displacement of a charged particle from A to B in the presence of a uniform electric field **E**. If the general curved path between A and B is divided into a succession of small horizontal and vertical displacements, work is done *only* along the vertical steps.

22.2 POTENTIAL DIFFERENCES IN A UNIFORM ELECTRIC FIELD

In this section, we shall describe the potential difference between any two points in a *uniform* electric field. We shall see that the potential difference is independent of the path between these two points; that is, the work done in taking a test charge from point A to point B is the same along all paths. This confirms that a static, uniform electric field is conservative. By definition, a force is conservative if it has this property (see Section 8.1).

Suppose the uniform electric field is in the negative y direction, as in Fig. 22.1. The displacement from A to B can be made by a succession of small horizontal and vertical steps. The displacement vector $d\mathbf{s}$ can be expressed as $d\mathbf{s} = \mathbf{i}\,dx + \mathbf{j}\,dy$. Furthermore, $\mathbf{E} = -E\mathbf{j}$ since the field is in the negative y direction. Therefore, $\mathbf{E} \cdot d\mathbf{s}$, which appears in Eq. 22.4, can be written

$$\mathbf{E} \cdot d\mathbf{s} = -E\mathbf{j} \cdot (\mathbf{i}\,dx + \mathbf{j}\,dy) = -E\,dy$$

Since E is constant and can be removed from the integral of Eq. 22.4, the potential difference between A and B reduces to

$$V_B - V_A = \int_0^d E\,dy = E\int_0^d dy$$

Potential difference in a uniform E field

$$V_B - V_A = Ed \tag{22.7}$$

This result shows that the potential difference between *any* two points in a uniform electric field depends only on the displacement d in the direction *parallel* to **E**. This is because no work is necessary to displace a charged particle in the horizontal steps of Fig. 22.1 since **E** is perpendicular to these displacements. Work is done *only* along the vertical steps.

Using Eqs. 22.4 and 22.7, we find that the change in potential energy of a test charge q_0 as it moves from A to B is given by

$$\Delta U = q_0(V_B - V_A) = q_0 Ed \tag{22.8}$$

Hence, if q_0 is positive, then ΔU is positive. This means that a positive charge would *gain* electric potential energy when it moves in the direction opposite the field.[1] This is analogous to a mass gaining gravitational potential energy when it rises to higher elevations in the presence of gravity. If a positive test charge is released from rest in this field, it experiences an electric force $q_0\mathbf{E}$ in the direction of **E** (downward in Fig. 22.1). Therefore, it accelerates downward, gaining kinetic energy. As it gains kinetic energy, it loses an equal amount of potential energy.

On the other hand, if the test charge q_0 is negative, then ΔU is negative and the situation is reversed. That is, a negative charge displaced in the direction *opposite* **E** *loses* electric potential energy. If a negative charge is released from rest in the field **E**, it accelerates in a direction opposite the electric field.[2]

---

[1]The fact that point B is at a higher potential than point A makes sense from another point of view. An external agent would have to do positive work to move a positive charge from A to B, against the field.

[2]Note that when a charged particle accelerates, it actually loses energy by radiating electromagnetic waves. However, in most cases encountered in this course, there is no *net* loss of energy. Furthermore, the radiation emitted by an accelerating charge is not predicted by the nonrelativistic physics we have studied so far.

Finally, our results show that all points in a plane *perpendicular* to a uniform electric field are at the same potential. This can be seen in Fig. 22.1, where the potential difference $V_B - V_A$ is *equal* to $V_C - V_A$. Therefore, $V_B = V_C$. The name *equipotential surface* is given to any surface consisting of *a set of points having the same potential.* Note that since $\Delta U = q_0 \Delta V$, *no* work is done in moving a test charge between any two points on an equipotential surface. The equipotential surfaces of a uniform electric field consist of a family of planes all *perpendicular* to the field (Fig. 22.1). Equipotential surfaces for fields with other symmetries will be described in later sections.

An equipotential surface

Example 22.1

A 12-V battery is connected between two parallel plates as in Fig. 22.2. The separation between the plates is 0.3 cm, and the electric field is assumed to be uniform. Find the electric field between the plates.

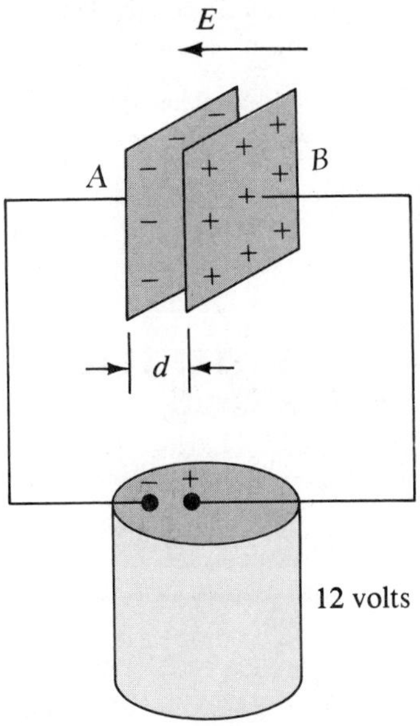

Figure 22.2 (Example 21.1) A 12-V battery connected to two parallel plates. The electric field between the plates has a magnitude given by the potential difference divided by the plate separation d.

Solution: The electric field is directed from the positive plate toward the negative plate. We see that the positive plate (at the right) is at a *higher* potential than the negative plate. Note that the potential difference between plates B and A must equal the potential difference between the battery terminals. This can be understood by noting that all points on a conductor in equilibrium are at the same potential,[3] and hence there is no potential difference between a terminal of the battery and any portion of the plate to which it is connected. Applying Eq. 22.7, we get

$$E = \frac{V_B - V_A}{d} = \frac{12\text{ V}}{0.3 \times 10^{-2}\text{ m}} = 4.0 \times 10^3 \frac{\text{V}}{\text{m}}$$

This configuration, which is called a *parallel-plate capacitor,* will be examined in more detail in the next chapter.

Example 22.2

A proton is released from *rest* in a uniform electric field of 8×10^4 V/m directed along the positive x axis (Fig. 22.3). The proton undergoes a displacement of 0.5 m in the direction of **E.** (a) Find the *change* in the electric potential of the proton as a result of this displacement.

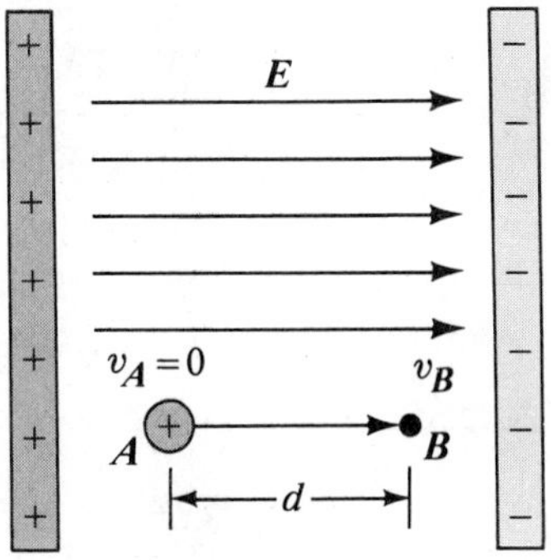

Figure 22.3 (Example 22.2) A proton accelerates from A to B in the direction of the electric field.

Using Eq. 22.4 and noting that the displacement is *in the direction* of the field, we have

$$\Delta V = V_B - V_A = -\int \mathbf{E} \cdot d\mathbf{s} = -\int_0^d E\,dx = -E\int_0^d dx$$

$$= -Ed = -\left(8 \times 10^4 \frac{\text{V}}{\text{m}}\right)(0.5\text{ m}) = -4 \times 10^4\text{ V}$$

Thus, the electric potential of the proton *decreases* as it moves from A to B.

(b) Find the change in potential energy of the proton for this displacement.

$$\Delta U = q_0 \Delta V = e\,\Delta V$$
$$= (1.6 \times 10^{-19}\text{ C})(-4 \times 10^4\text{ V})$$
$$= -6.4 \times 10^{-15}\text{ J}$$

The negative sign here means that the potential energy of the proton decreases as it moves in the direction of **E.** This makes sense since as the proton *accelerates* in the direction

[3]The electric field vanishes within a conductor in electrostatic equilibrium, and so the path integral $\int \mathbf{E} \cdot d\mathbf{s}$ between any two points within the conductor must be zero. A fuller discussion of this point is given in Section 22.6.

of $\mathbf{E}$, it gains kinetic energy and at the same time loses potential energy (mechanical energy is conserved).

(c) Find the speed of the proton after it has been displaced from rest by 0.5 m.

If there are no forces acting on the proton other than the conservative electric force, we can apply the principle of conservation of mechanical energy in the form $\Delta K + \Delta U = 0$; that is, *the decrease in potential energy must be accompanied by an equal increase in kinetic energy*. Noting that the mass of the proton is $m_p = 1.67 \times 10^{-27}$ kg, we have

$$\Delta K + \Delta U = \left(\frac{1}{2}m_p v^2 - 0\right) - 6.4 \times 10^{-15}\ \text{J} = 0$$

$$v^2 = \frac{2(6.4 \times 10^{-15})\ \text{J}}{1.67 \times 10^{-27}\ \text{kg}} = 7.66 \times 10^{12}\ \frac{\text{m}^2}{\text{s}^2}$$

$$v = 2.77 \times 10^6\ \frac{\text{m}}{\text{s}}$$

If an electron was accelerated under the same circumstances, its speed would approach the speed of light and the problem would have to be treated by relativistic mechanics (Chapter 40).

Q1. A negative charge moves in the direction of a uniform electric field. Does its potential energy increase or decrease? Does the electric potential increase or decrease?

Q2. A uniform electric field is parallel to the x axis. In what direction can a charge be displaced in this field without any external work being done on the charge?

Q3. If a proton is released from rest in a uniform electric field, does its electric potential increase or decrease? What about its potential energy? How would your answers differ for an electron?

22.3 ELECTRIC POTENTIAL AND POTENTIAL ENERGY DUE TO POINT CHARGES

An isolated point charge q produces an electric field that is radial. If the charge is positive, the electric field is directed radially outward from the charge, as in Fig. 22.4. Let us calculate the potential difference between the points A and B. As usual, the position of the point charge is taken as the origin of coordinates.

The magnitude of the radial electric field $\mathbf{E}$ at a distance r from the point charge is given by

$$E_r = \frac{kq}{r^2}$$

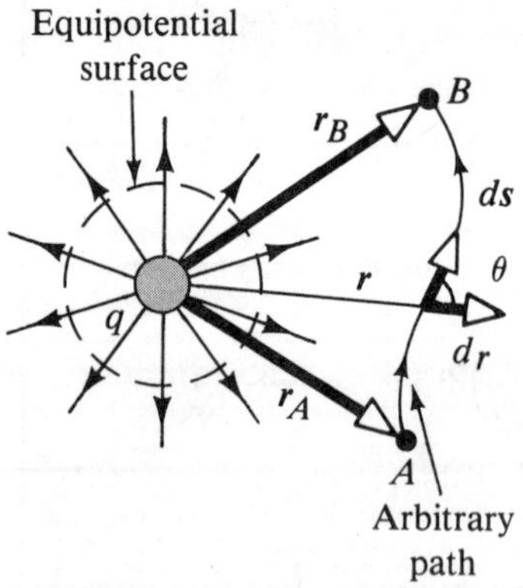

Figure 22.4 The potential difference between points A and B due to a point charge q depends *only* on the initial and final radial coordinates, r_A and r_B, respectively.

where $k = 1/4\pi\varepsilon_0 = 9.0 \times 10^9\ \text{N} \cdot \text{m}^2/\text{C}^2$. The displacement $d\mathbf{s}$ has a radial component dr, which is the *only* component that contributes to $\mathbf{E} \cdot d\mathbf{s}$ since $\mathbf{E}$ is radial. That is, in this case $\mathbf{E} \cdot d\mathbf{s} = E_r\, ds \cos\theta = E_r\, dr$. Hence Eq. 22.4 gives

$$V_B - V_A = -\int E_r\, dr = -kq\int_{r_A}^{r_B} \frac{dr}{r^2} = \frac{kq}{r}\bigg]_{r_A}^{r_B}$$

$$V_B - V_A = kq\left[\frac{1}{r_B} - \frac{1}{r_A}\right] \tag{22.9}$$

The integral $-\int_A^B \mathbf{E} \cdot d\mathbf{s}$ is *independent* of the path between A and B, as it must be. (We had already concluded that the electric field of a point charge is a conservative field, by analogy with the gravitational field of a point mass.) Furthermore, Eq. 22.9 expresses the important result that the potential difference between any two points A and B depends *only* on the *radial* coordinates r_A and r_B. It is customary to choose the reference of potential to be zero at $r_A = \infty$. (This is quite natural since $V \sim$

$1/r_A$, and as $r_A \to \infty$, $V \to 0$.) With this choice, the electric potential due to a point charge at any distance r from the charge is given by

$$V = k\frac{q}{r} \tag{22.10}$$

Potential of a point charge

From this we see that V is constant on a spherical surface of radius r. Hence, we conclude that *the equipotential surfaces (surfaces on which V remains constant) for an isolated point charge consist of a family of spheres concentric with the charge,* as shown in Fig. 22.4. Note that the equipotential surfaces are perpendicular to the lines of electric force, as was the case for a uniform electric field.

The electric potential of two or more point charges is obtained by applying the superposition principle.[4] That is, the total potential at some point P due to several point charges is the sum of the potentials due to the individual charges. For a group of charges, we can write the total potential at P in the form

$$V = k\sum_i \frac{q_i}{r_i} \tag{22.11}$$

The potential of several point charges

where the potential is again taken to be zero at infinity and r_i is the distance from the point P to the charge q_i. Note that the sum in Eq. 22.11 is an *algebraic sum* of scalars rather than a vector sum (which is used to calculate the electric field of a group of charges). Thus, it is much easier to evaluate V than to evaluate $\mathbf{E}$.

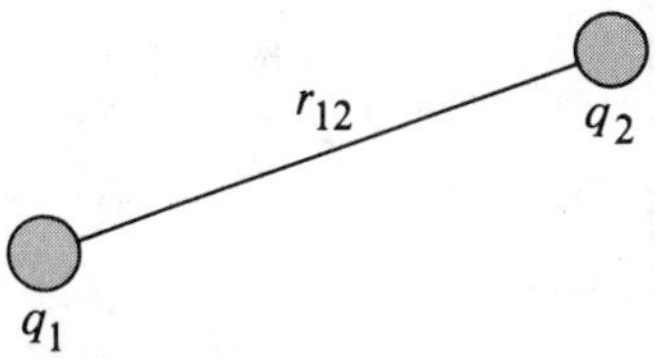

Figure 22.5 If two point charges are separated by a distance r_{12}, the potential energy of the pair of charges is given by kq_1q_2/r_{12}.

We now consider the potential energy of interaction of a system of charged particles. If V_1 is the electric potential due to charge q_1 at a point P, then the work required to bring a second charge, q_2, from infinity to the point P without acceleration is given by q_2V_1. By definition, this work equals the potential energy, U, of the two-particle system when the particles are separated by a distance r_{12} (Fig. 22.5). Therefore, we can express the potential energy as

$$U = q_2V_1 = k\frac{q_1q_2}{r_{12}} \tag{22.12}$$

Electric potential energy of two charges

Note that if the charges are of the same sign, U is positive.[5] This is consistent with the fact that like charges repel, and so positive work must be done *on* the system to bring the two charges near one another. Conversely, if the charges are of opposite sign, the force is attractive and U is negative. This means that negative work must be done to bring the charges into proximity.

If there are more than two charged particles in the system, the total potential energy can be obtained by calculating U for every *pair* of charges and summing the terms algebraically. As an example, the total potential energy of the three charges shown in Fig. 22.6 is given by

$$U = k\left(\frac{q_1q_2}{r_{12}} + \frac{q_1q_3}{r_{13}} + \frac{q_2q_3}{r_{23}}\right) \tag{22.13}$$

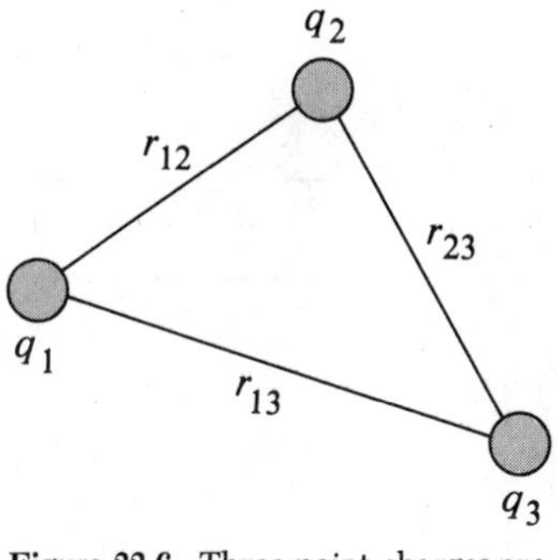

Figure 22.6 Three point charges are fixed at the positions shown. The potential energy of this system of charges is given by Eq. 22.13.

Physically, we can interpret this as follows: Imagine that q_1 is fixed at the position shown in Fig. 22.6, but q_2 and q_3 are at infinity. The work required to bring q_2 from

[4]In doing so, it is important to recognize that noncoincident zero-potential reference points may sometimes be lost.

[5]The expression for the electric potential energy for two point charges, Eq. 22.12, is of the *same* form as the gravitational potential energy of two point masses given by Gm_1m_2/r (Chapter 14). The similarity is not surprising in view of the fact that both are derived from an inverse-square force law.

infinity to its position near q_1 is kq_1q_2/r_{12}, which is the first term in Eq. 22.13. The last two terms in Eq. 22.13 represent the work required to bring q_3 from infinity to its position near q_1 and q_2. (You should show that the result is independent of the order in which the charges are transported.)

Example 22.3

A 5-μC point charge is located at the origin, and a second point charge of $-2\ \mu$C is located on the x axis at the position (3,0) m, as in Fig. 22.7a. (a) If the potential is taken to be zero at infinity, find the total electric potential due to these charges at the point P, whose coordinates are (0,4) m.

The total potential at P due to the two charges is given by

$$V_P = k\left(\frac{q_1}{r_1} + \frac{q_2}{r_2}\right)$$

Since $r_1 = 4$ m and $r_2 = 5$ m, we get

$$V_P = 9 \times 10^9 \frac{\text{N}\cdot\text{m}^2}{\text{C}^2}\left(\frac{5 \times 10^{-6}\text{ C}}{4\text{ m}} - \frac{2 \times 10^{-6}\text{ C}}{5\text{ m}}\right)$$

$$= 7.65 \times 10^3\text{ V}$$

(b) How much work is required to bring a third point charge of 4 μC from infinity to the point P?

$$W = q_3V_P = (4 \times 10^{-6}\text{ C})(7.65 \times 10^3\text{ V})$$

Since 1 V = 1 J/C, W reduces to

$$W = 3.06 \times 10^{-2}\text{ J}$$

(c) Find the *total* potential energy of the system of three charges in the configuration shown in Figure 22.7b.

Using Eq. 22.13, we find that

$$U = k\left(\frac{q_1q_2}{r_{12}} + \frac{q_1q_3}{r_{13}} + \frac{q_2q_3}{r_{23}}\right)$$

$$= 9 \times 10^9\left(\frac{(5 \times 10^{-6})(-2 \times 10^{-6})}{3} + \frac{(5 \times 10^{-6})(4 \times 10^{-6})}{4} + \frac{(-2 \times 10^{-6})(4 \times 10^{-6})}{5}\right)$$

$$= 6.0 \times 10^{-4}\text{ J}$$

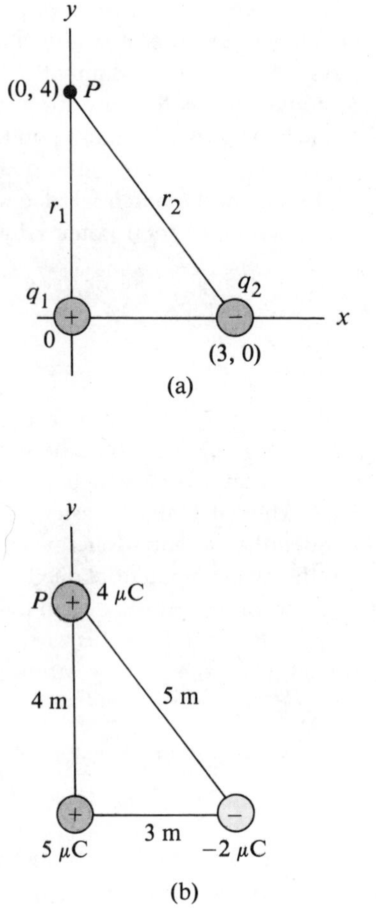

Figure 22.7 (Example 22.3) The electric potential at the point P due to the two point charges q_1 and q_2 is the algebraic sum of the potentials due to the individual charges.

Q4. Give a physical explanation of the fact that the potential energy of a pair of like charges is positive whereas the potential energy of a pair of unlike charges is negative.

Q5. If the electric potential at some point is zero, can you conclude that there are no charges in the vicinity of that point? Explain.

22.4 ELECTRIC POTENTIAL DUE TO CONTINUOUS CHARGE DISTRIBUTIONS

The electric potential due to a continuous charge distribution can be calculated in two ways. If the charge distribution is known, we can start with Eq. 22.10 for the

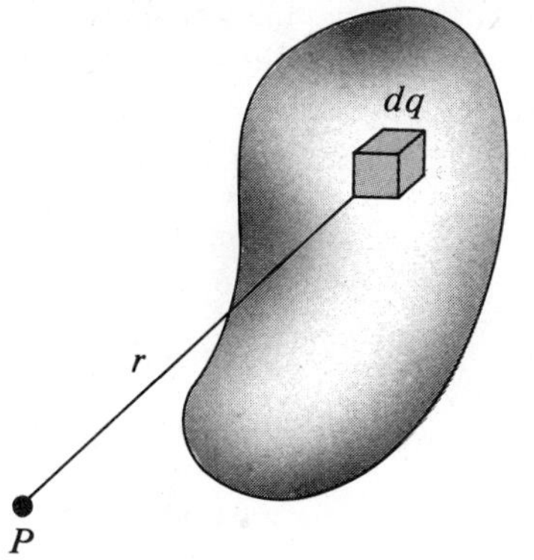

Figure 22.8 The electric potential at the point P due to a continuous charge distribution can be calculated by dividing the charged body into segments of charge dq and summing the potential contributions over all segments.

potential of a point charge. We then consider the potential due to a small charge element dq, treating this element as a point charge (Fig. 22.8). The potential dV at some point P due the charge element dq is given by

$$dV = k\frac{dq}{r} \tag{22.14}$$

where r is the distance from the charge element to the point P. To get the total potential at P, we integrate Eq. 22.14 to include contributions from all elements of the charge distribution. Since each element is, in general, at a different distance from P and since k is a constant, we can express V as

Electric potential due to a continuous charge distribution

$$V = k\int \frac{dq}{r} \tag{22.15}$$

In effect, we have replaced the sum in Eq. 22.11 by an integral. Note that this expression for V uses a particular choice of reference: the potential is taken to be zero for point P located infinitely far from the charge distribution.

The second method for calculating the potential of a continuous charge distribution makes use of Eq. 22.4. This procedure is useful when the electric field is already known from other considerations, such as Gauss' law. If the charge distribution is highly symmetric, we first evaluate **E** at any point using Gauss' law and then substitute the value obtained into Eq. 22.4 to determine the potential difference between any two points. We then choose V to be zero at *any* convenient point. Let us illustrate both methods with several examples.

Example 22.4

Potential due to a uniformly charged ring: Find the electric potential at a point P located on the axis of a uniformly charged ring of radius a and total charge Q. The plane of the ring is chosen perpendicular to the x axis (Fig. 22.9).

Solution: Let us take the point P to be at a distance x from the center of the ring, as in Fig. 22.9. The charge element dq is at a distance equal to $\sqrt{x^2+a^2}$ from the point P. Hence, we can express V as

$$V = k\int \frac{dq}{r} = k\int \frac{dq}{\sqrt{x^2+a^2}}$$

In this case, *each* element dq is at the *same distance* from the point P. Therefore, the term $\sqrt{x^2+a^2}$ can be removed from the integral and V reduces to

$$V = \frac{k}{\sqrt{x^2+a^2}}\int dq = \frac{kQ}{\sqrt{x^2+a^2}} \tag{22.16}$$

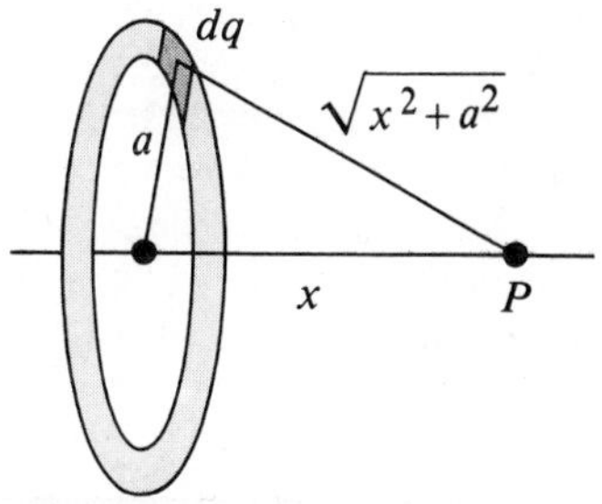

Figure 22.9 (Example 22.4) A uniformly charged ring of radius a, whose plane is perpendicular to the x axis. Any axial point P is at the same distance from all segments of the ring.

The only variable that appears in this expression for V is x. This is not surprising, since our calculation is valid only for points along the x axis, where y and z are both zero. From the symmetry of the situation, we see that along the x axis $\mathbf{E}$ can have only an x component. Therefore, we can use the expression $E_x = -dV/dx$, which we shall derive in Section 22.5, to find the electric field at P:

$$\begin{aligned} E_x &= -\frac{dV}{dx} = -kQ\frac{d}{dx}(x^2 + a^2)^{-1/2} \\ &= -kQ\left(-\frac{1}{2}\right)(x^2 + a^2)^{-3/2}(2x) \\ &= \frac{kQx}{(x^2 + a^2)^{3/2}} \end{aligned} \qquad (22.17)$$

This result agrees with that obtained by direct integration (see Example 20.8). Note that $E_x = 0$ at $x = 0$ (the center of the ring). Could you have guessed this from Coulomb's law?

Example 22.5

Potential of a uniformly charged disk: Find the electric potential along the axis of a uniformly charged disk of radius a and charge per unit area σ (Fig. 22.10).

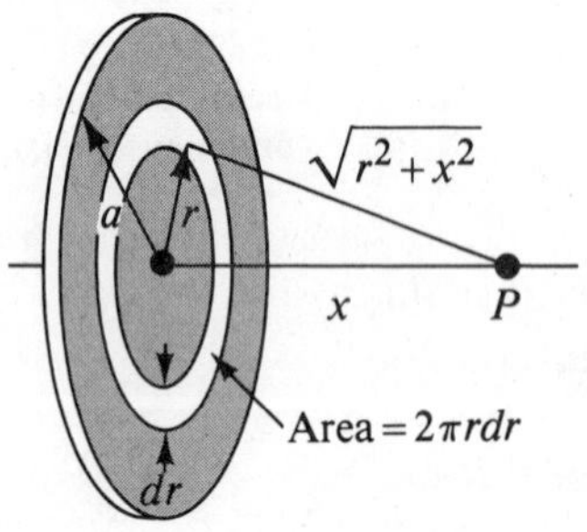

Figure 22.10 (Example 22.5) A uniformly charged disk of radius a, whose plane is perpendicular to the x axis. The calculation of the potential at an axial point P is simplified by dividing the disk into rings of area $2\pi r\, dr$.

Solution: Again we choose the point P to be at a distance x from the center of the disk and take the plane of the disk perpendicular to the x axis. The problem is simplified by dividing the disk into a series of charged rings. The potential of each ring is given by Eq. 22.16 in Example 22.4. Consider one such ring of radius r and width dr, as indicated in Fig. 22.10. The area of the ring is $dA = 2\pi r dr$ (the circumference multiplied by the width), and the charge on the ring is $dq = \sigma dA = \sigma 2\pi r dr$. Hence, the potential at the point P due to this ring is given by

$$dV = \frac{k\, dq}{\sqrt{r^2 + x^2}} = \frac{k\sigma 2\pi r\, dr}{\sqrt{r^2 + x^2}}$$

To find the *total* potential at P, we sum over all rings making up the disk. That is, we integrate dV from $r = 0$ to $r = a$:

$$V = \pi k\sigma \int_0^a \frac{2r\, dr}{\sqrt{r^2 + x^2}} = \pi k\sigma \int_0^a (r^2 + x^2)^{-1/2} 2r\, dr$$

This integral is of the form $u^n\, du$ and has the value $u^{n+1}/n + 1$, where $n = -\frac{1}{2}$ and $u = r^2 + x^2$. This gives the result

$$V = 2\pi k\sigma\,[(x^2 + a^2)^{1/2} - x] \qquad (22.18)$$

As in Example 22.4, we can find the electric field at any axial point by taking the negative of the derivative of V with respect to x. This gives

$$E_x = -\frac{dV}{dx} = 2\pi k\sigma\left(1 - \frac{x}{\sqrt{x^2 + a^2}}\right) \qquad (22.19)$$

The calculation of V and $\mathbf{E}$ for an arbitrary point off the axis is more difficult to perform.

Example 22.6

Potential of a finite line charge: A rod of length l located along the x axis has a uniform charge per unit length and a total charge Q. Find the electric potential at a point P along the y axis at a distance d from the origin (Fig. 22.11).

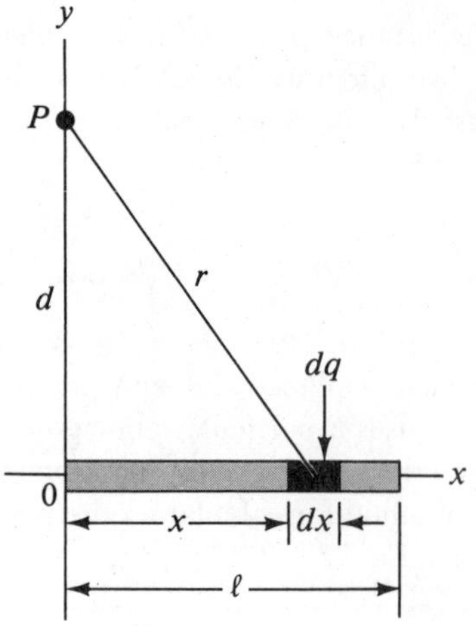

Figure 22.11 (Example 22.6) A uniform line charge of length l located along the x axis. To calculate the potential at P, the line charge is divided into segments each of length dx, having a charge $dq = \lambda\, dx$.

Solution: The element of length dx has a charge dq given by $\lambda\, dx$, where λ is the charge per unit length, Q/l. Since this element is at a distance $r = \sqrt{x^2 + d^2}$ from the point P, we can express the potential at P due to this element as

$$dV = k\frac{dq}{r} = k\frac{\lambda\, dx}{\sqrt{x^2 + d^2}}$$

To get the total potential at P, we integrate this expression over the limits $x = 0$ to $x = l$. Noting that k, λ, and d are constants, we find that

$$V = k\lambda \int_0^l \frac{dx}{\sqrt{x^2 + d^2}} = k\frac{Q}{l}\int_0^l \frac{dx}{\sqrt{x^2 + d^2}}$$

This integral, found in most integral tables, has the value

$$\int \frac{dx}{\sqrt{x^2 + d^2}} = \ln(x + \sqrt{x^2 + d^2})$$

Evaluating V, we find that

$$V = \frac{kQ}{l} \ln\left(\frac{l + \sqrt{l^2 + d^2}}{d}\right) \qquad (22.20)$$

Example 22.7

Potential of a uniformly charged sphere: An insulating solid sphere of radius R has a uniform positive charge density with total charge Q (Fig. 22.12). (a) Find the electric potential at a point *outside* the sphere, that is, for $r > R$. Take the potential to be zero at $r = \infty$.

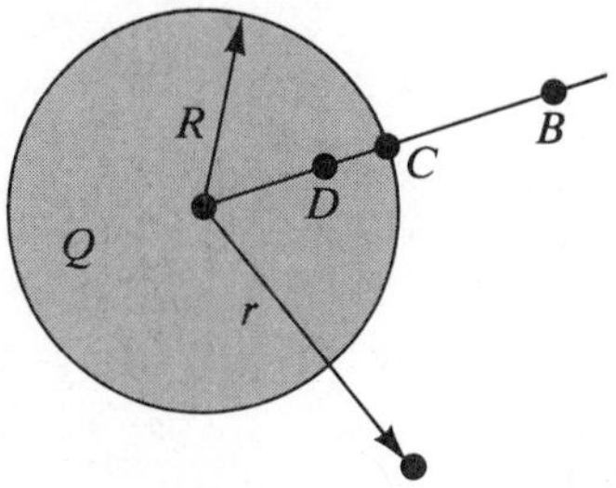

Figure 22.12 (Example 22.7) A uniformly charged insulating sphere of radius R and total charge Q. The electric potential at points B and C is equivalent to that of a point charge Q located at the center of the sphere.

In Example 21.3, we found from Gauss' law that the magnitude of the electric field *outside* a uniformly charged sphere is given by

$$E_r = k\frac{Q}{r^2} \qquad (\text{for } r > R)$$

where the field is directed radially outward when Q is positive. To obtain the potential at an exterior point, such as B in Fig. 22.12, we substitute this expression for E into Eq. 22.5. Since $\mathbf{E} \cdot d\mathbf{s} = E_r\, dr$ in this case, we get

$$V_B = -\int_\infty^r E_r\, dr = -kQ\int_\infty^r \frac{dr}{r^2}$$

$$V_B = k\frac{Q}{r} \qquad (\text{for } r > R)$$

Note that the result is identical to that for the electric potential due to a point charge. Since the potential must be continuous at $r = R$, we can use this expression to obtain the potential at the surface of the sphere. That is, the potential at a point such as C in Fig. 22.12 is given by

$$V_C = k\frac{Q}{R} \qquad (\text{for } r = R)$$

(b) Find the potential at a point *inside* the charged sphere, that is, for $r < R$.

In Example 21.3 we found that the electric field inside a uniformly charged sphere is given by

$$E_r = \frac{kQ}{R^3}r \qquad (\text{for } r < R)$$

We can use this result and Eq. 22.4 to evaluate the potential difference $V_D - V_C$, where D is an interior point:

$$V_D - V_C = -\int_R^r E_r\, dr = -\frac{kQ}{R^3}\int_R^r r\, dr = \frac{kQ}{2R^3}(R^2 - r^2)$$

Substituting $V_C = kQ/R$ into this expression and solving for V_D, we get

$$V_D = \frac{kQ}{2R}\left(3 - \frac{r^2}{R^2}\right) \qquad (\text{for } r < R) \qquad (22.21)$$

At $r = R$, this expression gives a result that agrees with that for the potential at the surface, that is, V_C. A plot of V versus r for this charge distribution is given in Fig. 22.13. Note that although the electric field is *zero* at $r = 0$, the potential is *not zero* at this point, but is given by $V_0 = 3kQ/2R$. This result follows from (Eq. 22.21).

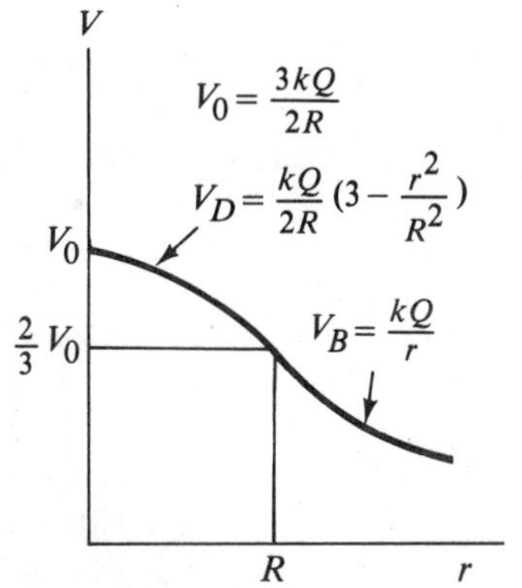

Figure 22.13 (Example 22.7) A plot of the electric potential V versus the distance r from the center of a uniformly charged, insulating sphere of radius R. The curve for V_D inside the sphere is parabolic and joins smoothly with the curve for V_B outside the sphere, which is a hyperbola. The potential has a maximum value V_0 at the center of the sphere.

Q6. Describe the equipotential surfaces for (a) an infinite line of charge and (b) a uniformly charged sphere.

22.5 OBTAINING *E* FROM THE ELECTRIC POTENTIAL

The electric field $\mathbf{E}$ and the potential V are related by Eq. 22.4. Both quantities are determined by a specific charge distribution. We now show how to calculate the

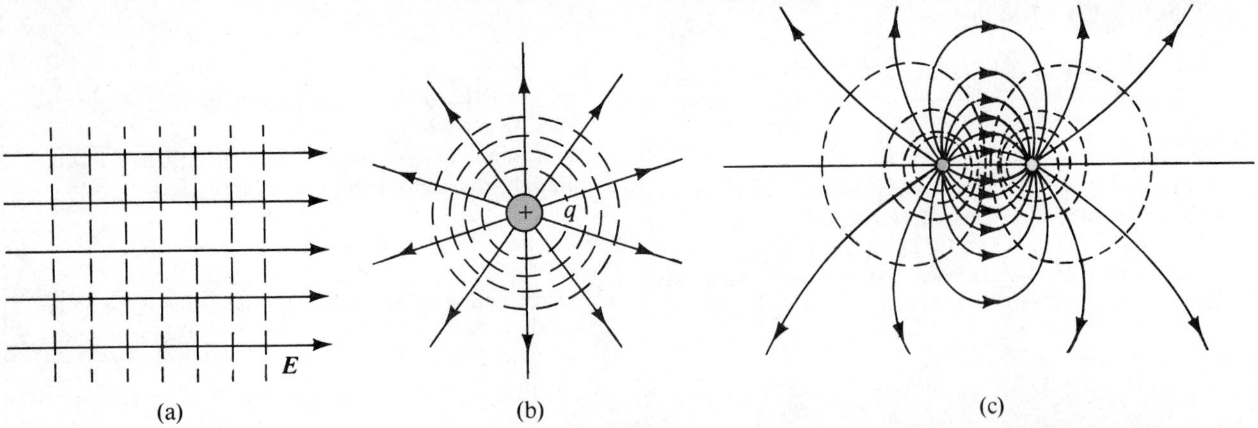

Figure 22.14 Equipotential surfaces (broken lines) and electric field lines (solid lines) for (a) a uniform electric field produced by an infinite sheet of charge, (b) a point charge, and (c) an electric dipole. In all cases, the equipotential surfaces are *perpendicular* to the electric field lines at every point.

electric field if the electric potential is known in a certain region. In many situations, we shall see that the electric field is simply the negative derivative of the electric potential.

From Eq. 22.4, we can express the potential difference dV between two points a distance ds apart as

$$dV = -\mathbf{E} \cdot d\mathbf{s} \tag{22.22}$$

If the electric field has only *one* component, E_x, then $\mathbf{E} \cdot d\mathbf{s} = E_x\, dx$. Therefore, Eq. 22.22 becomes $dV = -E_x\, dx$, or

$$E_x = -\frac{dV}{dx} \tag{22.23}$$

That is, the *electric field is equal to the negative of the derivative of the potential with respect to some coordinate.* Note that the potential change is zero for any displacement perpendicular to the electric field. This is consistent with the notion of equipotential surfaces being perpendicular to the field, as in Fig. 22.14a.

If the charge distribution has *spherical symmetry, where the charge density depends only on the radial distance r,* then the electric field is radial. In this case, $\mathbf{E} \cdot d\mathbf{s} = E_r\, dr$, and so we can express dV in the form $dV = -E_r\, dr$. Therefore,

Equipotential surfaces are always perpendicular to the electric field lines

$$E_r = -\frac{dV}{dr} \tag{22.24}$$

Note that the potential changes only in the radial direction, not in a direction perpendicular to r. Thus V (like E_r) is a function only of r. Again, this is consistent with the idea that *equipotential surfaces are perpendicular to field lines.* In this case the equipotential surfaces are a family of spheres concentric with the spherically symmetric charge distribution (Fig. 22.14b). The equipotential surfaces for the electric dipole are sketched in Fig. 22.14c.

When a test charge is displaced by a vector $d\mathbf{s}$ that lies *within* any equipotential surface, then by definition $dV = -\mathbf{E} \cdot d\mathbf{s} = 0$. This shows that the equipotential surfaces must *always* be *perpendicular* to the electric field lines.

In a general situation, the electric potential is a function of all three spatial coordinates. If $V(\mathbf{r})$ is given in terms of the rectangular coordinates, the electric field components E_x, E_y, and E_z can readily be found from $V(x,y,z)$. The field components are given by

$$E_x = -\frac{\partial V}{\partial x} \qquad E_y = -\frac{\partial V}{\partial y} \qquad E_z = -\frac{\partial V}{\partial z}$$

In these expressions, the derivatives are called *partial derivatives*.[6] What this means is that in the operation $\partial V/\partial x$, *one takes a derivative with respect to x while y and z are held constant.* For example, if $V = 3x^2y + y^2 + yz$, then

$$\frac{\partial V}{\partial x} = \frac{\partial}{\partial x}(3x^2y + y^2 + yz) = \frac{\partial}{\partial x}(3x^2y) = 3y\frac{d}{dx}(x^2) = 6xy$$

Example 22.8

Let us use the potential function for a point charge q to derive the electric field at a distance r from the charge.

Solution: The potential of a point charge is given by Eq. 22.10:

$$V = k\frac{q}{r}$$

Since the potential is a function of r only, it has spherical symmetry and we can apply Eq. 22.24 directly to obtain the electric field:

$$E_r = -\frac{dV}{dr} = -\frac{d}{dr}\left(k\frac{q}{r}\right) = -kq\frac{d}{dr}\left(\frac{1}{r}\right)$$

$$E_r = \frac{kq}{r^2}$$

Thus, the electric field is radial and the result agrees with that obtained using Gauss' law.

Example 22.9

The electric potential of a dipole: An electric dipole consists of two equal and opposite charges separated by a distance $2a$, as in Fig. 22.15. Calculate the electric potential and the electric field at the point P located a distance x from the center of the dipole.

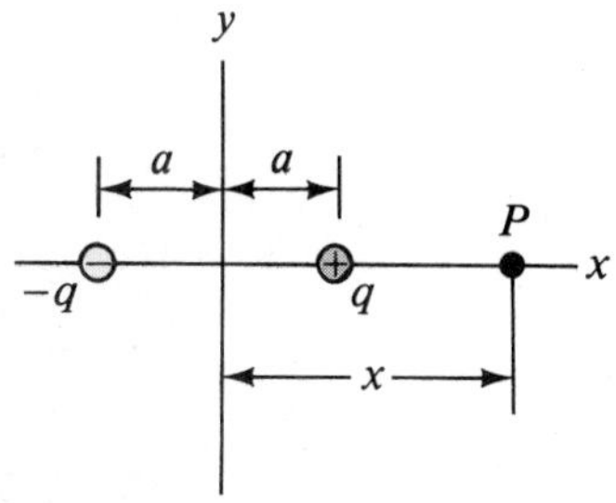

Figure 22.15 (Example 22.9) An electric dipole located on the x axis.

Solution:

$$V = k\Sigma\frac{q_i}{r_i} = k\left(\frac{q}{x-a} - \frac{q}{x+a}\right) = \frac{2kqa}{x^2 - a^2}$$

If the point P is far from the dipole, so that $x \gg a$, then a^2 can be neglected in the term $x^2 - a^2$ and V becomes

$$V \approx \frac{2kqa}{x^2} \qquad (x \gg a)$$

Using Eq. 22.23 and this result, the electric field at P is given by

$$E = -\frac{dV}{dx} = \frac{4kqa}{x^3}$$

Q7. If the potential is constant in a certain region, what is the electric field in that region?

Q8. The electric field inside a hollow, uniformly charged sphere is zero. Does this imply that the potential is zero inside the sphere? Explain.

22.6 POTENTIAL OF A CHARGED CONDUCTOR

In the previous chapter we found that when a conductor in equilibrium carries a net charge, the charge resides on the outer surface of the conductor. Furthermore, we showed that the electric field just outside the surface of a conductor in equilibrium is perpendicular to the surface while the field *inside* the conductor is zero. If

[6]In vector notation, **E** is often written $\mathbf{E} = -\nabla V = -\left(\mathbf{i}\frac{\partial}{\partial x} + \mathbf{j}\frac{\partial}{\partial v} + \mathbf{k}\frac{\partial}{\partial z}\right)V$, where ∇ is called the *gradient operator.*

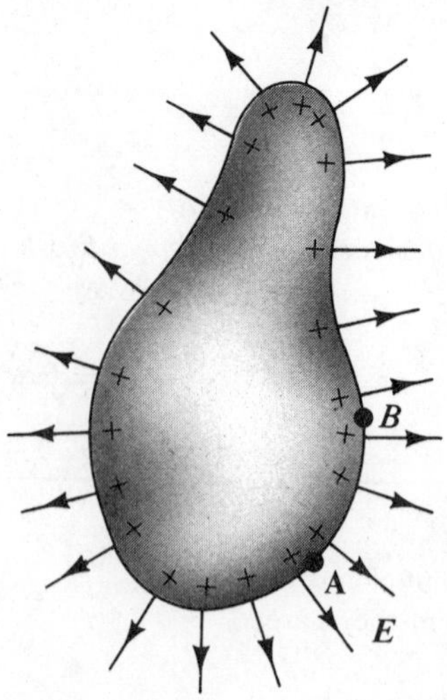

Figure 22.16 An arbitrarily shaped conductor with an excess positive charge. When the conductor is in electrostatic equilibrium, all of the charge resides at the surface, $\mathbf{E} = 0$ inside the conductor, and the electric field just outside the conductor is perpendicular to the surface. The potential is constant inside the conductor and is equal to the potential at the surface. Note that the surface charge density is nonuniform.

the electric field had a component parallel to the surface, this would cause surface charges to move, creating a current and a nonequilibrium situation.

We shall now show that *every point on the surface of a charged conductor in equilibrium is at the same potential.* Consider two points A and B on the surface of a charged conductor, as in Fig. 22.16. Along a surface path connecting these points, **E** is always perpendicular to the displacement $d\mathbf{s}$; therefore $\mathbf{E} \cdot d\mathbf{s} = 0$. Using this result and Eq. 22.4, we conclude that the potential difference between A and B is necessarily zero. That is,

All points on a charged conductor

$$V_B - V_A = -\int_A^B \mathbf{E} \cdot d\mathbf{s} = 0$$

This result applies to *any* two points on the surface. Therefore, V is constant everywhere on the surface of a charged conductor in equilibrium. That is, *the surface of any charged conductor in equilibrium is an equipotential surface.* Furthermore, since the electric field is zero inside the conductor, we conclude that *the potential is constant everywhere inside the conductor and equal to its value at the surface.* Therefore, no work is required to move a test charge from the interior of a charged conductor to its surface. (Note that the potential is *not zero* inside the conductor even though the electric field is zero.)

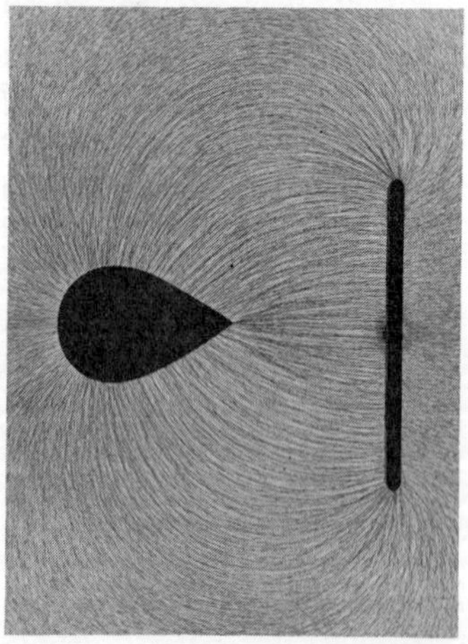

Electric field pattern of a charged conducting plate near an oppositely charged pointed conductor. Small pieces of thread suspended in oil align with the electric field lines. Note that the electric field is most intense near the pointed part of the conductor and at other points where the radius of curvature is small. (Courtesy of Harold M. Waage, Princeton University).

For example, consider a solid metal sphere of radius R and total positive charge Q, as shown in Fig. 22.17a. The electric field outside the charged sphere is given by kQ/r^2 and points radially outward. Following Example 22.7, we see that the potential at the interior and surface of the sphere must be kQ/R relative to infinity. The potential outside the sphere is given by kQ/r. Figure 22.17b is a plot of the potential as a function of r, and Fig. 22.17c shows the variations of the electric field with r.

When a net charge is placed on a spherical conductor, the surface charge density is uniform, as indicated in Fig. 22.17a. However, if the conductor is nonspherical, as in Fig. 22.16, the surface charge density is high where the radius of curvature is small and low where the radius of curvature is large. Since the electric field just outside a charged conductor is proportional to the surface charge density, σ, we see that *the electric field is large near points having a small radius of curvature and reaches very high values at sharp points.*

Figure 22.18 shows the electric field lines around two spherical conductors, one with a net charge Q and one with zero net charge. In this case, the surface charge density is *not* uniform on either conductor. The larger sphere (on the right), with zero net charge, has negative charges induced on its side that faces the charged sphere and positive charge on its side opposite the charged sphere. The broken lines

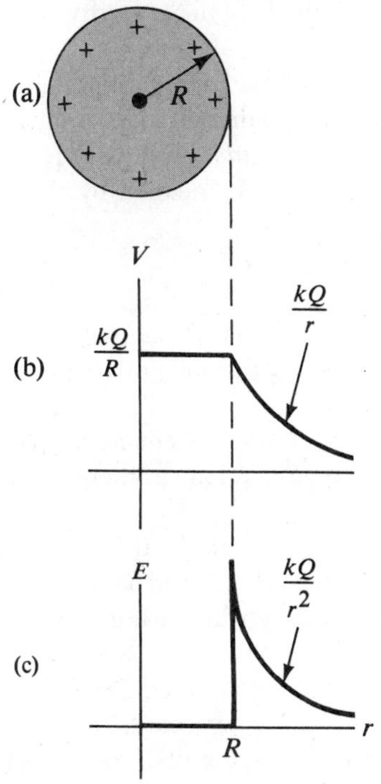

Figure 22.17 (a) The excess charge on a conducting sphere of radius R is uniformly distributed on its surface. (b) The electric potential versus the distance r from the center of the charged conducting sphere. (c) The electric field intensity versus the distance r from the center of the charged conducting sphere.

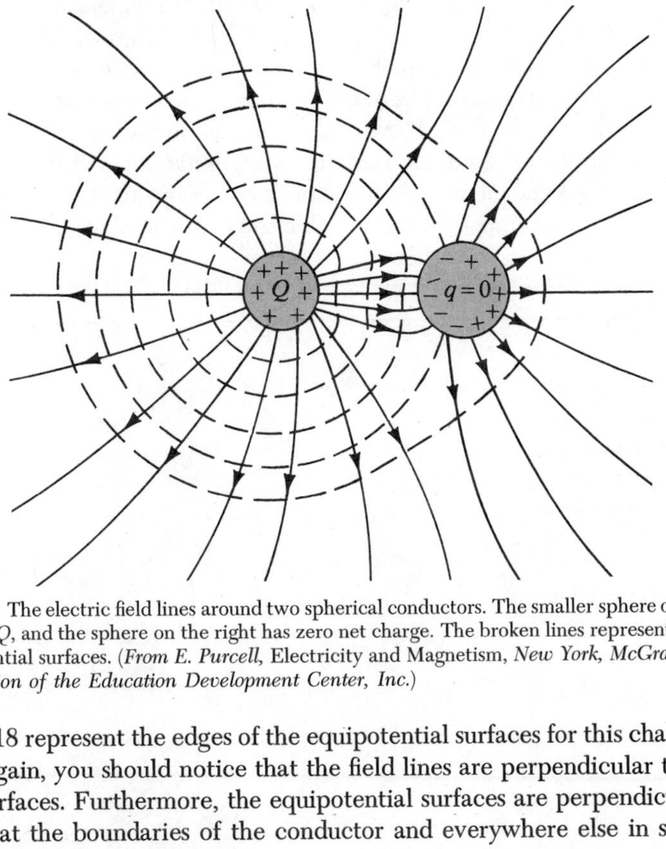

Figure 22.18 The electric field lines around two spherical conductors. The smaller sphere on the left has a net charge Q, and the sphere on the right has zero net charge. The broken lines represent the edges of the equipotential surfaces. (*From E. Purcell,* Electricity and Magnetism, *New York, McGraw Hill, 1965, with permission of the Education Development Center, Inc.*)

in Fig. 22.18 represent the edges of the equipotential surfaces for this charge configuration. Again, you should notice that the field lines are perpendicular to the conducting surfaces. Furthermore, the equipotential surfaces are perpendicular to the field lines at the boundaries of the conductor and everywhere else in space.

A Cavity Within a Conductor

Now consider a conductor of arbitrary shape containing a cavity as in Fig. 22.19. Let us assume there are no charges *inside* the cavity. We shall show that *the electric field inside the cavity must be zero,* regardless of the charge distribution on the *outside* surface of the conductor. Furthermore, the field in the cavity is zero even if an electric field exists outside the conductor.

In order to prove this point, we shall use the fact that every point on the conductor is at the same potential, and therefore any two points A and B on the surface of the cavity must be at the same potential. Now *imagine* that a field $\mathbf{E}$ exists in the cavity, and evaluate the potential difference $V_B - V_A$ defined by the expression

$$V_B - V_A = -\int_A^B \mathbf{E} \cdot d\mathbf{s}$$

We can always find a path between A and B for which $\mathbf{E} \cdot d\mathbf{s}$ is always a positive number (a path along the direction of $\mathbf{E}$), and so the integral must be positive. However, since $V_B - V_A = 0$, the integral must also be zero. This contradiction can be reconciled only if $\mathbf{E} = 0$ inside the cavity. Thus, we conclude that a cavity surrounded by conducting walls is a field-free region as long as there are no charges inside the cavity.

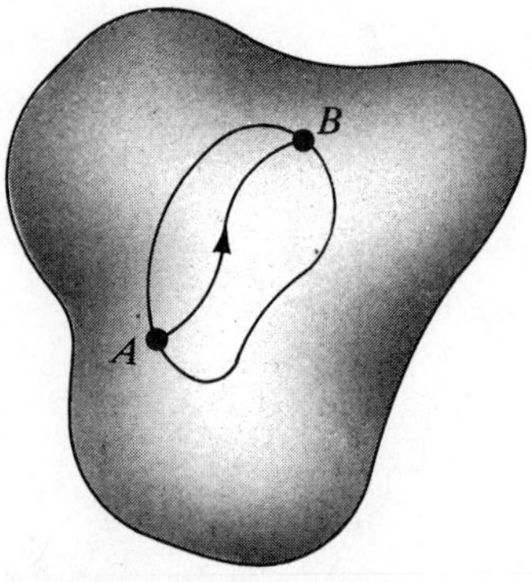

Figure 22.19 A conductor in electrostatic equilibrium containing an empty cavity. The electric field in the cavity is *zero*, regardless of the charge on the conductor.

This result has some interesting applications. For example, it is possible to shield an electronic circuit or even an entire laboratory from external fields by surrounding it with conducting walls. Shielding is often necessary when making highly sensitive electrical measurements.

A phenomenon known as *corona discharge* is often observed near sharp points of a conductor raised to a high potential. This appears as a greenish glow visible to the naked eye. In this process, air becomes a conductor as a result of the ionization of air molecules in regions of high electric fields. At standard temperature and pressure, this discharge occurs at electric field strengths equal to or greater than about 3×10^6 V/m. Since air contains a small number of ions (produced, for example, by cosmic rays), a charged conductor will attract ions of the opposite sign from the air. Near sharp points, where the field is very high, the ions in the air will be accelerated to high velocities. These energetic ions, in turn, collide with other air molecules, producing more ions and an increase in conductivity of the air. The discharge of the conductor is often accompanied by a visible glow surrounding the sharp points.

Example 22.10

Two spherical conductors of radii r_1 and r_2 are separated by a distance much larger than the radius of either sphere. The spheres are connected by a conducting wire as in Fig. 22.20. If the charges on the spheres in equilibrium are q_1 and q_2, respectively, find the ratio of the field strengths at the surfaces of the spheres.

Solution: Since the spheres are connected by a conducting wire, they must both be at the *same* potential V, given by

$$V = k\frac{q_1}{r_1} = k\frac{q_2}{r_2}$$

Therefore, the ratio of charges is

$$(1) \qquad \frac{q_1}{q_2} = \frac{r_1}{r_2}$$

Since the spheres are very far apart, their surfaces are uniformly charged and we can express the electric fields at their surfaces as

$$E_1 = k\frac{q_1}{r_1^2} \quad \text{and} \quad E_2 = k\frac{q_2}{r_2^2}$$

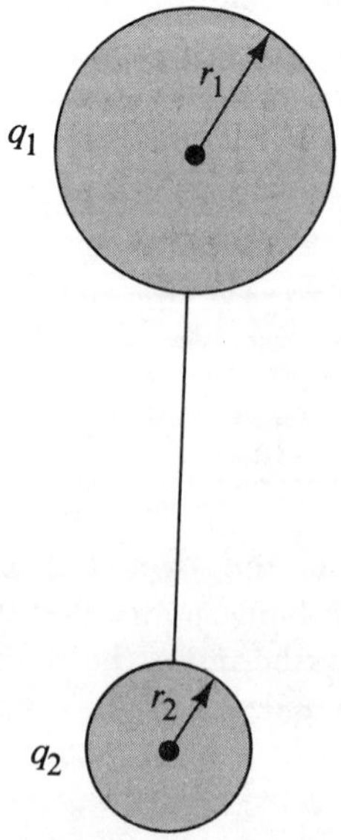

Figure 22.20 (Example 22.10) Two charged spherical conductors connected by a conducting wire. The spheres are at the *same* potential, V.

Taking the ratio of these two fields and making use of (1), we find that

$$\frac{E_1}{E_2} = \frac{r_2}{r_1}$$

Hence, the field is more intense in the vicinity of the smaller sphere.

Q9. Two charged conducting spheres of different radii are connected by a conducting wire as in Fig. 22.20. Which sphere has the greater charge density?

Q10. Explain the origin of the glow that is sometimes observed around the cables of a high-voltage power line.

Q11. Why is it important to avoid sharp edges, or points, on conductors used in high-voltage equipment?

Q12. Why is it safe to stay in an automobile during a severe thunderstorm?

22.7 APPLICATIONS OF ELECTROSTATICS

The principles of electrostatics have been used in various applications, a few of which we shall briefly discuss in this section. Some of the more practical applications include electrostatic precipitators, used to reduce the level of atmospheric pollution, and the xerography process, which has revolutionized imaging process technology. Scientific applications of electrostatic principles include electrostatic generators for accelerating elementary charged particles and the field-ion microscope, which is used to image atoms on the surface of metallic samples.

The Van de Graaf Generator

In the previous chapter we described an experiment that demonstrates a method for transferring charge to a hollow conductor (the Faraday ice-pail experiment). When a charged conductor is placed in contact with the inside of a hollow conductor, all of the charge of the first conductor is transferred to the hollow conductor. In principle, the charge on the hollow conductor and its potential can be increased without limit by repeating the process.

In 1929 Robert J. Van de Graaf used this principle to design and build an electrostatic generator. This type of generator is used extensively in nuclear physics research. The basic idea of the Van de Graaf generator is described in Fig. 22.21. Charge is delivered continuously to a high-voltage electrode on a moving belt of insulating material. The high-voltage electrode is a hollow conductor mounted on an insulating column. The belt is charged at A by means of a corona discharge between comb-like metallic needles and a grounded grid. The needles are maintained at a positive potential of typically 10^4 V. The positive charge on the moving belt is transferred to the high-voltage electrode by a second comb of needles at B. Since the electric field inside the hollow conductor is negligible, the positive charge on the belt easily transfers to the high-voltage electrode, regardless of its potential. In practice, it is possible to increase the potential of the high-voltage electrode until electrical discharge occurs through the air. Since the "breakdown" voltage of air is about 3×10^6 V/m, a sphere 1 m in radius can be raised to a maximum potential of 3×10^6 V. The potential can be increased further by increasing the radius of the hollow conductor and by placing the entire system in a container filled with high-pressure gas.

Van de Graaf generators can produce potential differences as high as 20 million volts. Protons accelerated through such potential differences receive enough energy to initiate nuclear reactions between the protons and various target nuclei.

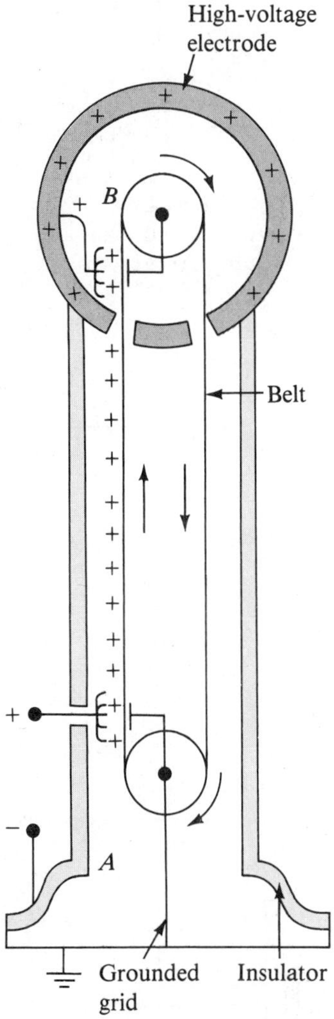

Figure 22.21 Schematic diagram of a Van de Graaf generator. Charge is transferred to the hollow conductor at the top by means of a rotating belt. The charge is deposited on the belt at point A and is transferred to the hollow conductor at point B.

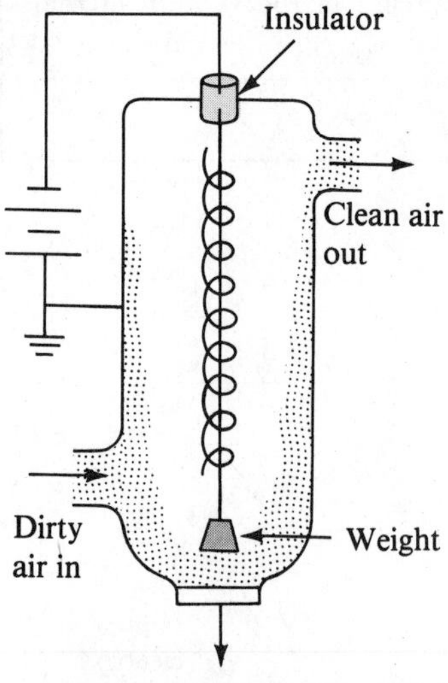

Figure 22.22 Schematic diagram of an electrostatic precipitator. The high voltage maintained on the central wire creates an electrical discharge in the vicinity of the wire.

The Electrostatic Precipitator

One important application of electrical discharge in gases is a device called an *electrostatic precipitator*. This device is used to remove particulate matter from combustion gases, thereby reducing air pollution. They are especially useful in coal-burning power plants and in industrial operations that generate large quantities of smoke. Current systems are able to eliminate more than 99 percent of the ash and dust from the smoke. Figure 22.22 shows the basic idea of the electrostatic precipitator. A high voltage (typically 40 kV to 100 kV) is maintained between a wire running down the center of a duct and the outer wall, which is grounded. The wire is maintained at a negative potential with respect to the walls, and so the electric field is directed toward the wire. The electric field near the wire reaches high enough values to cause a corona discharge around the wire and the formation of positive ions, electrons, and negative ions, such as O_2^-. As the electrons and negative ions are accelerated toward the outer wall by the nonuniform electric field, the dirt particles in the streaming gas become charged by collisions and ion capture. Since most of the charged dirt particles are negative, they are also drawn to the outer wall by the electric field. By periodically shaking the duct, the particles fall loose and are collected at the bottom.

In addition to reducing the level of harmful gases and particulate matter in the atmosphere, the electrostatic precipitator also recovers valuable materials from the stack in the form of metal oxides.

Xerography

The process of xerography is widely used for making photocopies of letters, documents, and other printed materials. The basic idea for the process was developed by Chester Carlson, for which he was granted a patent in 1940. In 1947, the Xerox Corporation launched a full-scale program to develop automated duplicating machines using this process. The huge success of this development is quite evident; today, practically all modern offices and libraries have one or more duplicating machines, and the capabilities of modern machines is on the increase.

Some features of the xerographic process involve simple concepts from electrostatics and optics. However, the one idea that makes the process unique is the use of a photoconductive material to form an image. (A photoconductor is a material that is a poor conductor in the dark but becomes a good electrical conductor when exposed to light.)

The sequence of steps used in the xerographic process is illustrated in Fig. 22.23. First, the surface of a plate or drum is coated with a thin film of the photoconductive material (usually selenium or some compound of selenium), and the photocon-

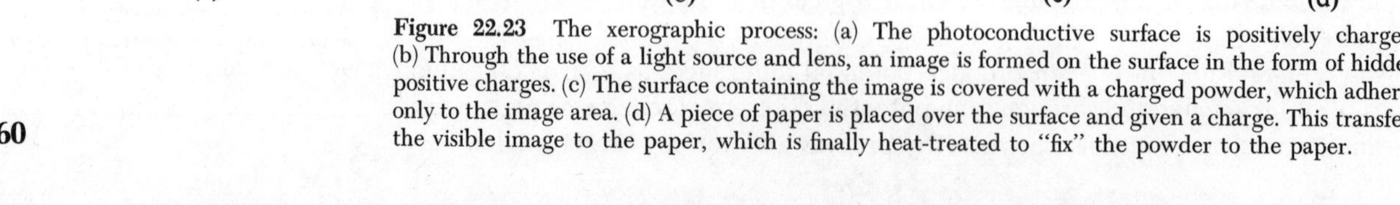

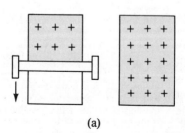

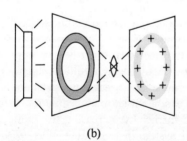

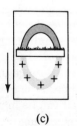

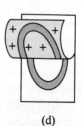

Figure 22.23 The xerographic process: (a) The photoconductive surface is positively charged. (b) Through the use of a light source and lens, an image is formed on the surface in the form of hidden positive charges. (c) The surface containing the image is covered with a charged powder, which adheres only to the image area. (d) A piece of paper is placed over the surface and given a charge. This transfers the visible image to the paper, which is finally heat-treated to "fix" the powder to the paper.

ductive surface is given a positive electrostatic charge in the dark. The page to be copied is then projected onto the charged surface. The photoconducting surface becomes conducting only in areas where light strikes. In these areas, the light produces charge carriers in the photoconductor, which neutralize the positively charged surface. However, the charges remain on those areas of the photoconductor not exposed to light, leaving a hidden image of the object in the form of a positive surface charge distribution.

Next, a negatively charged powder called a *toner* is dusted onto the photoconducting surface. The charged powder adheres only to those areas of the surface that contain the positively charged image. At this point, the image becomes visible. The image is then transferred to the surface of a sheet of positively charged paper.

Finally, the toner material is "fixed" to the surface of the paper through the application of heat. This results in a permanent copy of the original.

The Field-Ion Microscope

In Section 22.6 we pointed out that the electric field intensity can be very high in the vicinity of a sharp point on a charged conductor. A device that makes use of this intense field is the *field-ion microscope*, invented in 1956 by E. W. Mueller of the Pennsylvania State University.

The basic construction of the field-ion microscope is shown in Fig. 22.24. A specimen to be studied is fabricated from a fine wire, and a sharp tip is formed, usually by etching the wire in an acid. Typically, the diameter of the tip is about 10^{-7} m (= 1000 Å). The specimen is placed at the center of an evacuated glass tube containing a fluorescent screen. Next, a small amount of helium is introduced into the vessel. A very high potential difference is applied between the needle and the screen, producing a very intense electric field near the tip of the needle. The helium atoms in the vicinity of this high-field region are ionized by the loss of an electron, which leaves the helium positively charged. The positively charged He^+ ions then accelerate along field lines to the negatively charged fluorescent screen. This results in a pattern on the screen that represents an image of the tip of the specimen.

Under the proper conditions (low specimen temperature and high vacuum), the images of the individual atoms on the surface of the sample are visible, and the atomic arrangement on the surface can be studied. Unfortunately, the high electric fields also set up large mechanical stresses near the tip of the specimen, which limits the application of the technique to strong metallic elements, such as tungsten and rhenium. Figure 22.25 represents a typical field-ion microscope pattern of a platinum crystal.

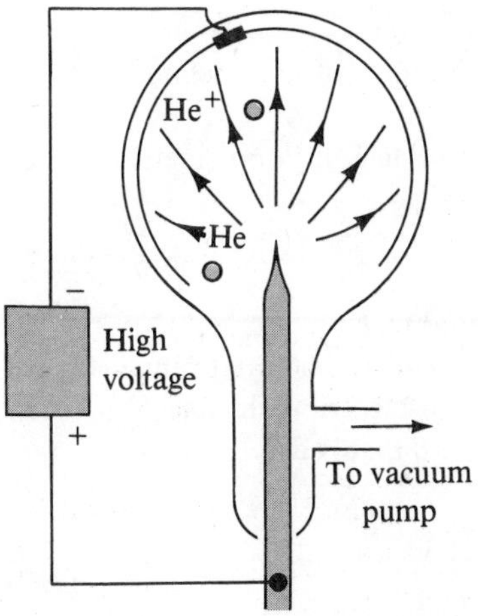

Figure 22.24 Schematic diagram of a field-ion microscope. The electric field is very intense at the tip of the needle-shaped specimen.

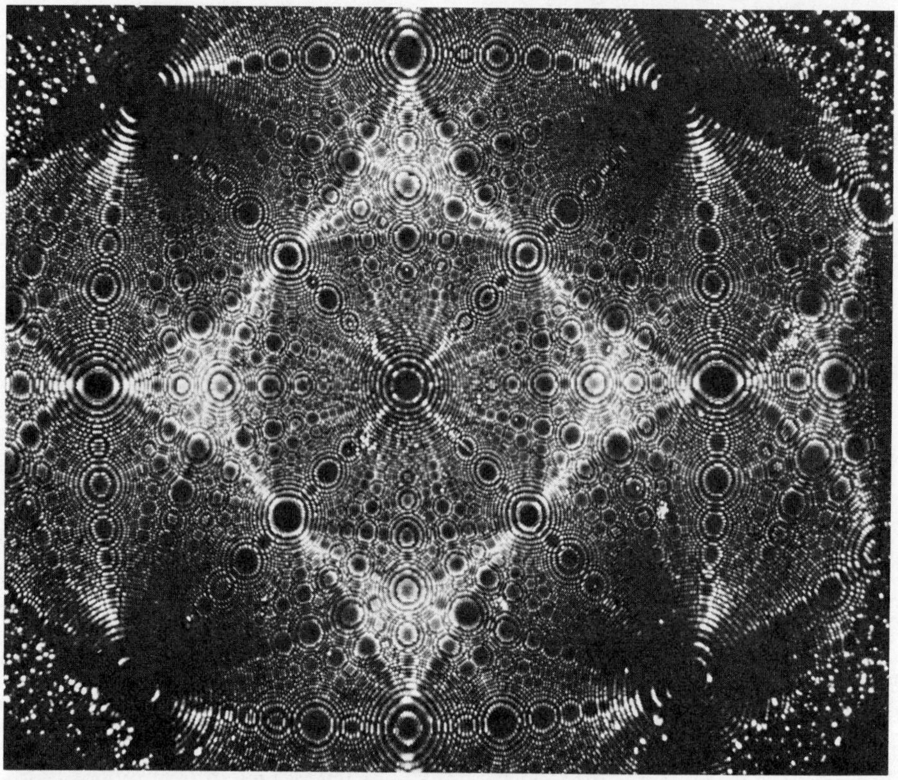

Figure 22.25 Field ion microscope image of the surface of a platinum crystal with a magnification of 1,000,000×. Individual atoms can be seen on surface layers using this technique. (Courtesy of Prof. T. T. Tsong, The Pennsylvania State University.)

22.8 SUMMARY

When a test charge q_0 is moved between points A and B in an electrostatic field $\mathbf{E}$, the *change in the potential energy* is given by

Change in potential energy

$$\Delta U = -q_0 \int_A^B \mathbf{E} \cdot d\mathbf{s} \tag{22.3}$$

The *potential difference* between points A and B in an electrostatic field $\mathbf{E}$ is given by

Potential difference

$$\Delta V = V_B - V_A = -\int_A^B \mathbf{E} \cdot d\mathbf{s} \tag{22.4}$$

where the electric potential V is a scalar and has the units of J/C, defined to be 1 volt (V).

The potential difference between two points A and B in a *uniform* electric field $\mathbf{E}$ is given by

$$V_B - V_A = Ed \tag{22.7}$$

where d is the displacement in the direction *parallel* to $\mathbf{E}$.

Equipotential surfaces are surfaces on which the electric potential remains constant. Equipotential surfaces are *perpendicular* to the electric field lines.

The potential difference between two points A and B that is due to a *point charge* is given by

$$V_B - V_A = kq\left(\frac{1}{r_B} - \frac{1}{r_A}\right) \tag{22.9}$$

Equation 22.9 shows that the potential difference between any two points A and B depends only on the *radial* coordinates r_A and r_B. If the potential is taken to be zero at $r_A = \infty$, the potential of a point charge is given by

$$V = k\frac{q}{r} \tag{22.10}$$

Potential of a point charge

The potential due to a group of point charges is obtained by summing the potentials due to the individual charges. Since V is a scalar, the sum is a simple algebraic operation.

The *potential energy of a pair of point charges* separated by a distance r_{12} is given by

$$U = k\frac{q_1 q_2}{r_{12}} \tag{22.12}$$

Electric potential energy of two charges

This represents the work required to bring the charges from an infinite separation to the separation r_{12}. The potential energy of a distribution of point charges is obtained by summing terms like Eq. 22.12 over all *pairs* of particles.

The *electric potential due to a continuous charge distribution* is given by

$$V = k\int \frac{dq}{r} \tag{22.15}$$

Electric potential due to a continuous charge distribution

If the electric potential is known as a function of coordinates x,y,z, the components of the electric field can be obtained by taking the negative derivative of the potential with respect to the coordinates. For example, the x component of the electric field is given by

$$E_x = -\frac{dV}{dx} \tag{22.23}$$

Every point on the surface of a charged conductor in electrostatic equilibrium is at the same potential. Furthermore, the potential is constant everywhere inside the conductor and equal to its value at the surface. Table 22.1 lists potentials due to several charge distributions.

TABLE 22.1 Potentials Due to Various Charge Distributions

CHARGE DISTRIBUTION	*ELECTRIC POTENTIAL*	*LOCATION*
Uniformly charged ring of radius, a	$V = k\frac{Q}{\sqrt{x^2 + R^2}}$	Along the axis of the ring, a distance x from its center
Uniformly charged disk of radius a	$V = 2\pi k\sigma[(x^2 + a^2)^{1/2} - x]$	Along the axis of the ring, a distance x from its center
Uniformly charged, insulating solid sphere of radius R and total charge Q	$V = k\frac{Q}{r}$	$r \geq R$
	$V = \frac{kQ}{2R}\left(3 - \frac{r^2}{R^2}\right)$	$r < R$
Isolated *conducting* sphere of total charge Q and radius R	$V = k\frac{Q}{R}$	$r \leq R$
	$V = k\frac{Q}{r}$	$r > R$

EXERCISES

Section 22.1 Potential Difference and Electric Potential

1. Concentric spherical surfaces surrounding a point charge at their center are called *equipotential surfaces.* The intersections of these surfaces with a plane through their common center are called *equipotential lines.* How much work is done in moving a charge q a distance s along an arc of an equipotential of circular shape and of radius R?

2. What change in potential energy does a 6-μC charge experience when it is moved between two points for which the potential difference is 40 V? Express the answer in eV.

3. Through what potential difference would one need to accelerate an electron in order for it to achieve a velocity of 60 percent of the velocity of light, starting from rest? ($c = 3.0 \times 10^8$ m/s.)

4. An ion accelerated through a potential difference of 60 V experiences an increase in potential energy of 1.92×10^{-17} J. Calculate the charge on the ion.

5. A positron, when accelerated from rest between two points at a fixed potential difference, acquires a speed of 10 percent of the speed of light. What speed will be achieved by a *proton* if accelerated from rest between the same two points?

Section 22.2 Potential Difference in a Uniform Electric Field

6. How much work is done (by a battery, generator, or other source of electrical energy) in moving Avogadro's number of electrons from an initial point where the electric potential is 6 V to a point where the potential is -10 V? (The potential in each case is measured relative to a common reference point.)

7. A uniform electric field of magnitude 400 V/m is directed in the *negative y* direction in Fig. 22.26. The coordinates of point A are $(-0.4, -0.6)$ m, and those of point B are $(0.5, 0.7)$ m. Calculate the difference in electric potential between A and B using the path ACB.

8. For the situation described in Exercise 7, calculate the change in electric potential while going from point A to point B along the direct path AB. Which point is at the higher potential?

9. Consider two points in an electric field. The potential at point P_1 is $V_1 = -140$ V, and the potential at point P_2 is $V_2 = +260$ V. How much work is done by an external force in moving a charge $q = -12$ μC from P_2 to P_1?

10. The electric field between two charged parallel plates separated by a distance of 2 cm has a uniform value of 1.3×10^4 N/C. Find the potential difference between the two plates. How much energy would be gained by a deuteron in moving from the positive to the negative plate?

11. An electron moving parallel to the x axis has an initial velocity of 5×10^6 m/s at the origin. The velocity of the electron is reduced to 2×10^5 m/s at the point $x = 4$ cm. Calculate the potential difference between the origin and the point $x = 4$ cm. Which point is at the higher potential?

12. A proton moves in a region of a uniform electric field. The proton experiences an increase in kinetic energy of 9×10^{-18} J after being displaced 1 cm in a direction parallel to the field. What is the magnitude of the electric field?

Section 22.3 Electric Potential and Potential Energy Due to Point Charges

13. At what distance from a point charge of 6 μC would the potential equal 2.7×10^4 V?

14. Two point charges are located as shown in Fig. 22.27, where $q_1 = +6$ μC, $q_2 = -4$ μC, $a =$

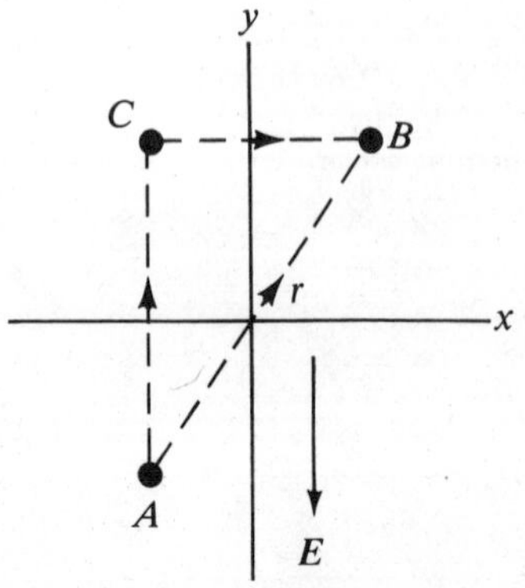

Figure 22.26 (Exercises 7 and 8).

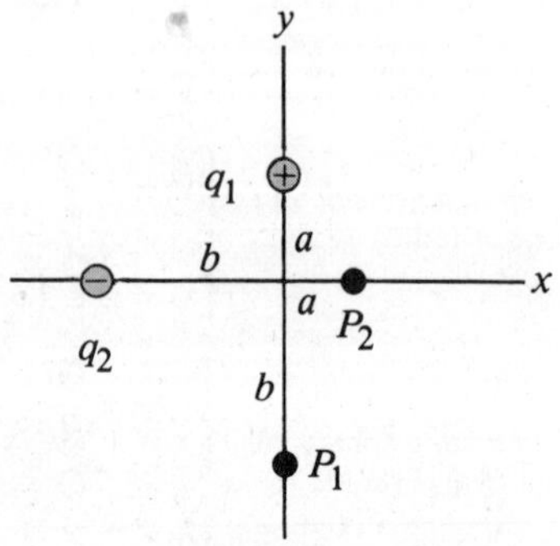

Figure 22.27 (Exercise 14).

0.15 m, and $b = 0.45$ m. Calculate the value of the electrical potential at points P_1 and P_2. Which point is at the higher potential?

15. At a distance r away from a point charge q, the electrical potential is $V = 600$ V and the magnitude of the electric field is $E = 200$ N/C. Determine the value of q and r.

16. A charge $q_1 = -6\ \mu C$ is located at the origin, and a second charge $q_2 = -2\ \mu C$ is located on the x axis at $x = 0.4$ m. Calculate the electric potential energy of this pair of charges.

17. Calculate the energy required to assemble the array of charges shown in Fig. 22.28, where $a = 0.15$ m, $b = 0.25$ m, and $q = 4\ \mu C$.

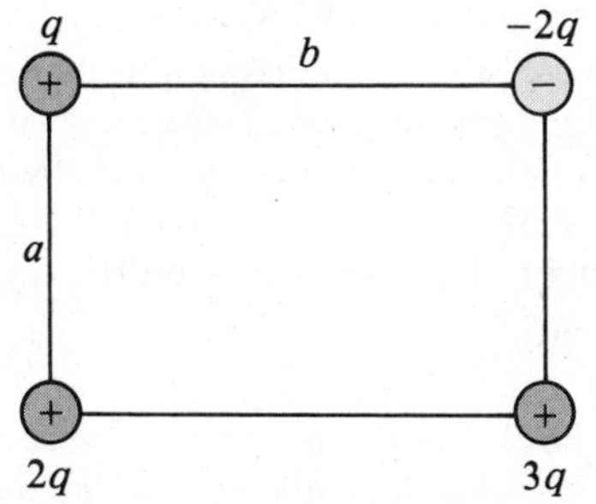

Figure 22.28 (Exercise 17).

18. Four charges are located at the corners of a rectangle as in Fig. 22.29. How much energy would be required to remove the two 4-μC charges to infinity?

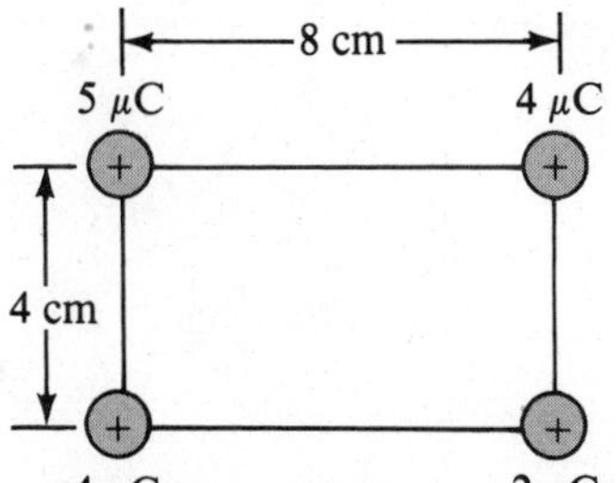

Figure 22.29 (Exercise 18).

19. The three charges shown in Fig. 22.30 are at the vertices of an isosceles triangle. Calculate the electric potential at the *midpoint of the base*, taking $q = 5\ \mu C$.

20. Calculate the value of the electric potential at point P due to the charge configuration shown in Fig. 22.31. Use the values $q_1 = 8\ \mu C$, $q_2 = -12\ \mu C$, $a = 0.1$ m, and $b = 0.15$ m.

Section 22.4 Electric Potential Due to Continuous Charge Distributions

21. Calculate the electric potential at point P on the axis of the annulus shown in Fig. 22.32, which has a uniform surface charge density σ and inner and outer radii a and b, respectively.

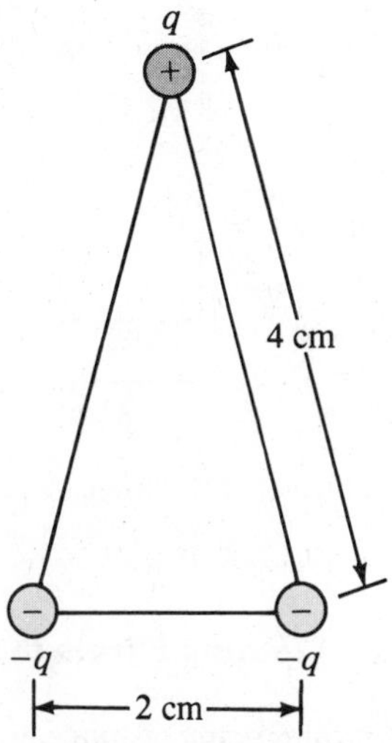

Figure 22.30 (Exercise 19).

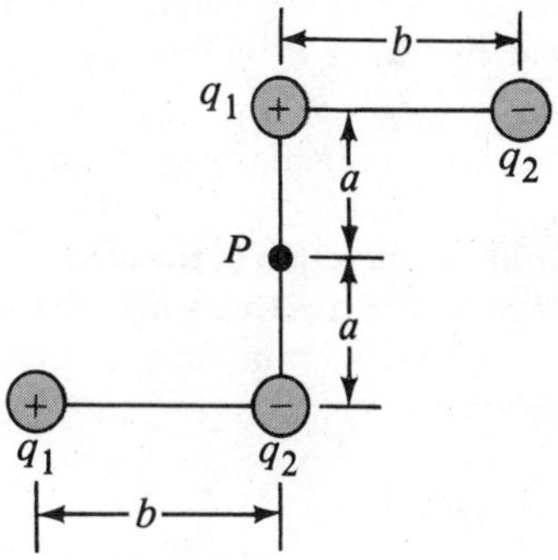

Figure 22.31 (Exercise 20).

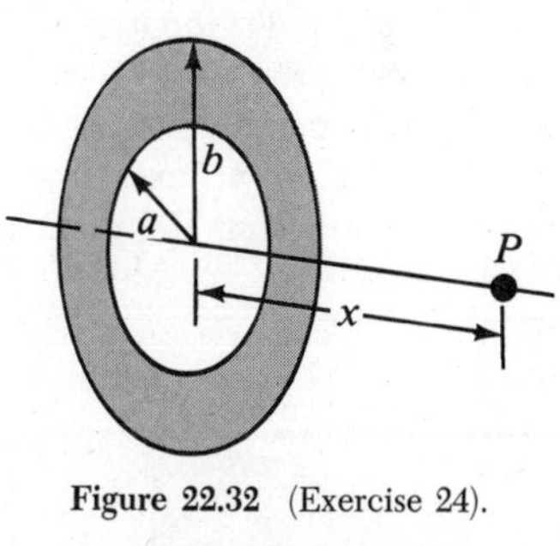

Figure 22.32 (Exercise 24).

22. A rod of length L (Fig. 22.33) lies along the x axis with its left end at the origin and has a *nonuniform* charge density $\lambda = bx$ (where b is a positive constant). (a) What are the units of the constant b? (b) Calculate the electric potential at point A, a distance d from the left end of the rod.

23. For the arrangement described in the previous exercise, calculate the electric potential at point B on the perpendicular bisector of the rod a distance b above the x axis.

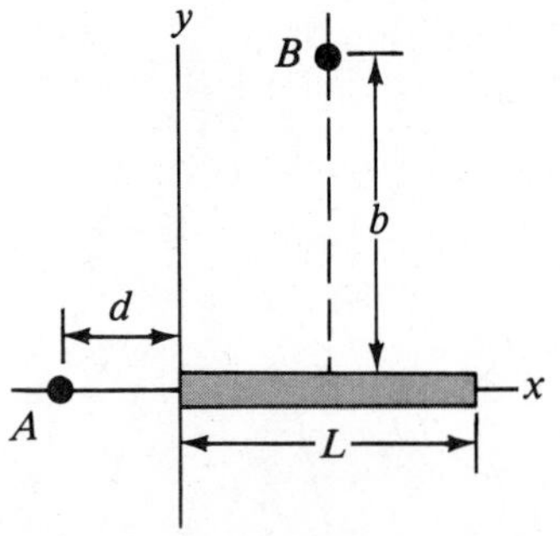

Figure 22.33 (Exercises 25 and 26).

Section 22.5 Obtaining *E* from the Electric Potential

24. Over a certain region of space, the electric potential is given by $V = 3x^2 - 2xy + xyz$. Find the expressions for the x, y, and z components of the electric field over this region. What is the magnitude of the field at the point P, which has coordinates (in meters) $(2, -1, 2)$?

25. The electric potential over a particular region is given by $V = 2x + y^2 - 7$. Determine the angle between the direction of the electric field, $\mathbf{E}$, and the direction of the positive x axis at the point P, which has coordinates (in meters) $(1, 2, 0)$.

26. Two parallel plates are perpendicular to the x axis. The negative plate is in the yz plane and the positive plate is at the point $x = x_0$. The potential at some point between the plates at $x < x_0$ is given by $V(x) = -kx$, where k is a constant. Find an expression for the electric field $\mathbf{E}$ between the two plates.

Section 22.6 Potential of a Charged Conductor

27. How many electrons should be removed from an initially uncharged spherical conductor of radius 0.2 m to produce a potential of 2 kV at the surface?

28. Calculate the surface charge density, σ (in C/m^2), for a solid spherical conductor of radius $R = 0.1$ m if the potential at a distance 0.2 m from the center of the sphere is 800 V.

29. Two spherical conductors of radii r_1 and r_2 are connected by a conducting wire as shown in Fig. 22.20. If $r_1 = 0.3$ m, $r_2 = 0.15$ m, and the field at the surface of the smaller sphere is 500 N/C, calculate the amount of excess charge on the larger sphere.

Section 22.7 Applications of Electrostatics

30. What energy in watts must a Van de Graaf generator deliver if it produces a 100-μA beam of protons at an energy of 12-MeV?

PROBLEMS

1. A square sheet of sides L has a uniform surface charge density σ and is located in the xy plane as in Fig. 22.34. Set up the integral expression necessary to calculate the electric potential at a point P on a line perpendicular to an axis through the center of the sheet. Assume that the point P is a distance d from the sheet.

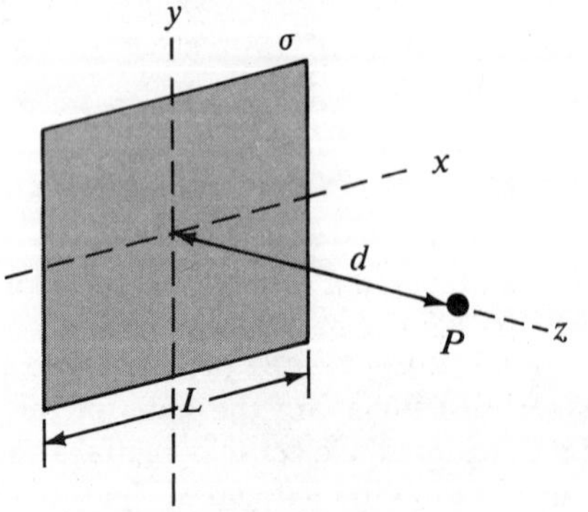

Figure 22.34 (Problem 1).

2. A Geiger-Muller counter is a type of radiation detector that essentially consists of a hollow cylinder (the

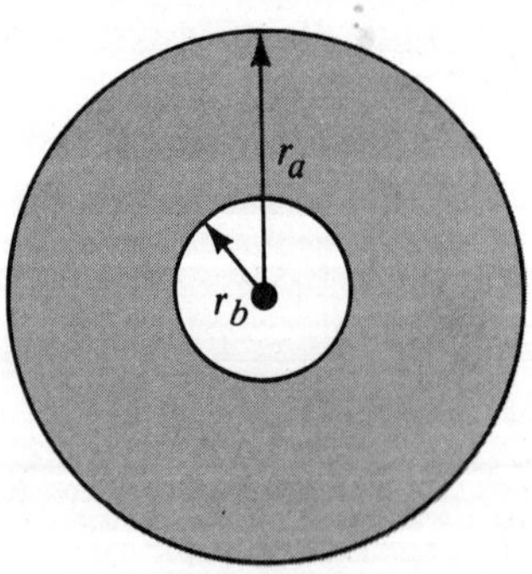

Figure 22.35 (Problem 2).

cathode) of inner radius r_a and a coaxial cylindrical wire (the anode) of radius r_b (Fig. 22.35). The charge per unit length on the anode is λ, while the charge per unit length on the cathode is $-\lambda$. (a) Show that the potential difference in the sensitive region of the detector is given by

$$V = 2k\lambda \ln\left(\frac{r_a}{r_b}\right)$$

(b) Show that the magnitude of the electric field

over that region is given by

$$E = \frac{V}{\ln(r_a/r_b)}\left(\frac{1}{r}\right)$$

where r is the distance from the center of the anode to the point where the field is to be calculated.

3. The Van de Graaf generator shown in Fig. 22.21 has a sphere of radius r at the center of a larger sphere of radius R. For such an arrangement, show that if the charges on the spheres have values of q and Q, respectively, the potential difference between the two spheres will be

$$V_r - V_R = \frac{q}{4\pi\varepsilon_0}\left(\frac{1}{r} - \frac{1}{R}\right)$$

4. It is shown in Example 22.6 that the potential at a point P a distance y above one end of a uniformly charged rod of length l lying along the x axis is given by

$$V = \frac{kQ}{l}\ln\left(\frac{l + \sqrt{l^2 + d^2}}{d}\right)$$

Use this result to derive an expression for the y component of the electric field at the point P.

5. Consider an array of eight equal negative charges located so as to define the corners of a *cube* of edge length $l = 0.15$ m. If each of the eight charges has a charge $q = -6\ \mu$C, determine the potential at the *center* of the cube.

6. A Van de Graaf generator is operating so that the potential difference between the high-voltage electrode and the charging needles (points B and A in Figure 22.21) is 1.5×10^4 V. Calculate the power required to drive the belt (against electrical forces) at an instant when the effective current delivered to the high-voltage electrode is 500 μA.

7. A large rectangle of length $4a$ and width $3a$ contains equal positive charges of magnitude $q_1 = 4\ \mu$C at opposite vertices, as shown in Fig. 22.36. A small rectangle of length $2a$ and width a has equal positive charges $q_2 = 6\ \mu$C located at two vertices as shown in the figure. How much work must be done against electrostatic forces in order to rotate the small rectangle about its long side to the position shown by the broken line? Let $a = 0.1$ m.

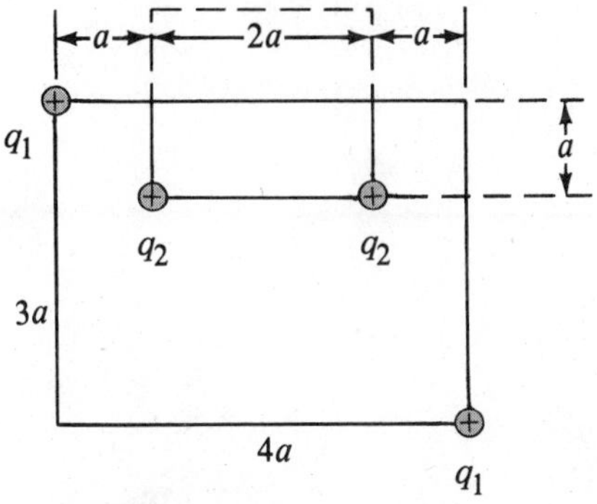

Figure 22.36 (Problem 7).

8. A disk of radius R has a nonuniform surface charge density $\sigma = kr$, where k is a constant and r is measured from the center of the disk (Fig. 22.37). Find (by direct integration) the potential at an axial point P a distance x from the disk.

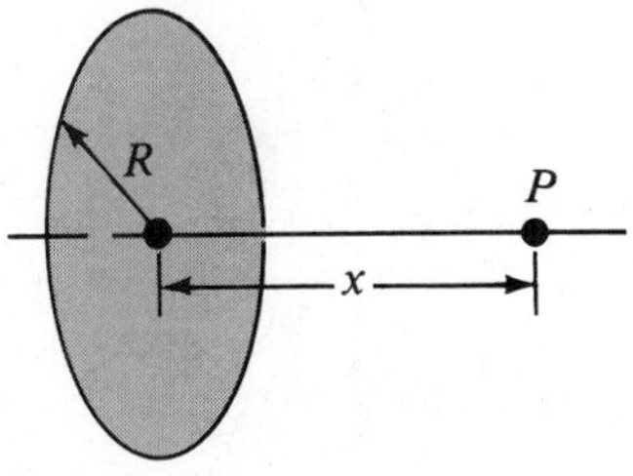

Figure 22.37 (Problem 8).

9. A dipole is located along the y axis as in Fig. 22.38. (a) At a point $P(r,\theta)$, which is far from the dipole ($r \gg a$), the electric potential is given by $V = k\dfrac{p\cos\theta}{r^2}$, where $p = 2qa$. Calculate the radial component of the associated electric field, E_r, and the azimuthal component, E_θ. Do these results seem reasonable for $\theta = 90°$ and $0°$? for $r = 0$? (b) For the dipole arrangement shown, express V in terms of rectangular coordinates using $r = (x^2 + y^2)^{1/2}$ and $\cos\theta = \dfrac{y}{(x^2 + y^2)^{1/2}}$. Using these results, calculate the field components E_x and E_y.

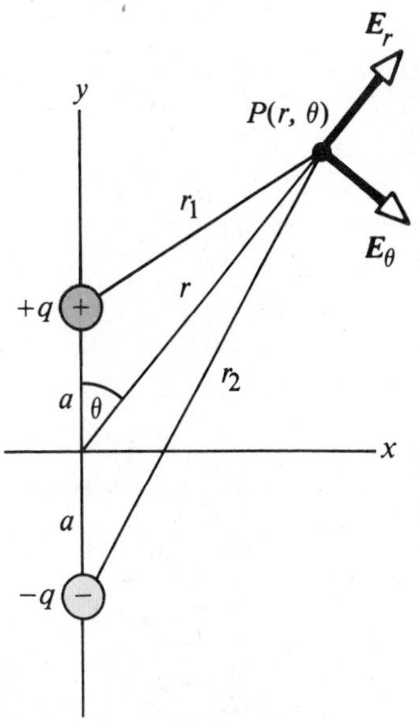

Figure 22.38 (Problem 9).

10. The thin, uniformly charged rod shown in Fig. 22.39 has a length L and a uniform linear charge density λ. Find an expression for the electric potential at point P, a distance b along the positive y axis.

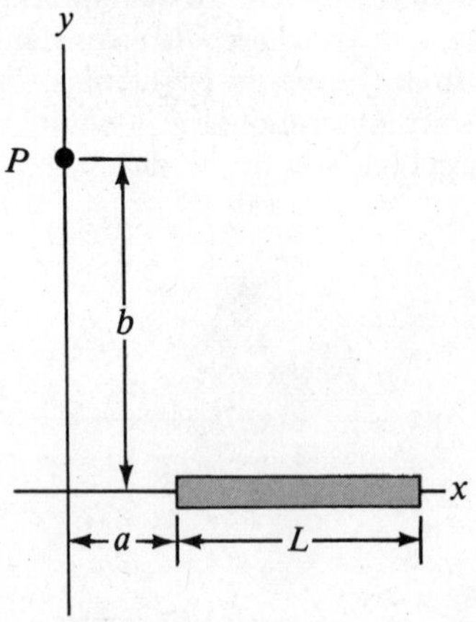

Figure 22.39 (Problem 10).

11. A solid sphere of radius R has a *uniform* charge density ρ and *total* charge Q. Derive an expression for the total electric potential energy of the charged sphere. (Hint: Imagine that the sphere is constructed by adding successive layers of concentric shells of charge $dq = (4\pi r^2\,dr)\rho$ and use $dU = V dq$.)

12. Calculate the work that must be done to charge a spherical shell of radius R to a total charge Q.

13. A spherical drop of water 2 mm in radius has an electric potential of 300 V. (a) What is the charge on the drop? (b) If two such drops of equal charge unite to form a single spherical drop, what is the potential of the resulting drop? (Assume no charge is lost when the two drops unite.)

14. Two point charges of equal magnitude are located along the y axis at equal distances above and below the x axis, as shown in Fig. 22.40. (a) Plot a graph of the potential at points along the x axis over the interval $-3a < x < 3a$. You should plot the potential in units of kQ/a, where k is the Coulomb constant. (b) Let the charge located at $-a$ be *negative* and plot the potential along the y axis over the interval $-4a < y < 4a$.

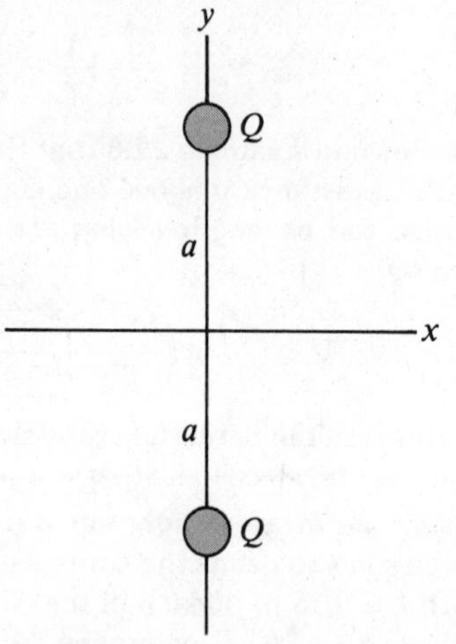

Figure 22.40 (Problem 14).

15. A net positive charge Q is placed on a solid conducting sphere of radius R. Plot a graph of the electric potential V as a function of r, the distance from the center of the sphere. Let r range over the interval $0 < r < 3R$, and plot V in units of kQ/R.

23 Capacitance and Dielectrics

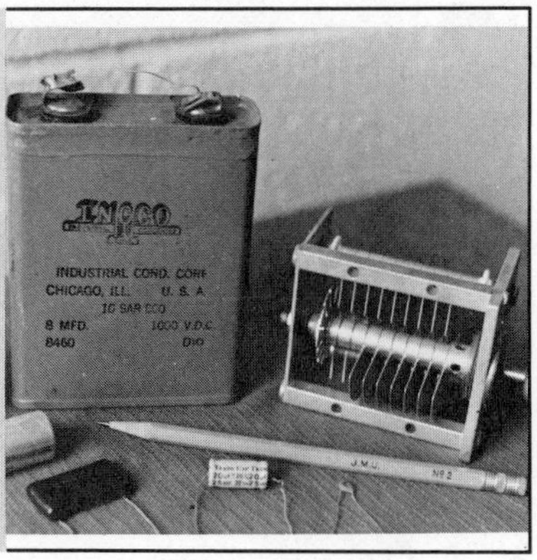

James Madison University

This chapter is concerned with the properties of capacitors, devices that store charge. Capacitors are commonly used in a variety of electrical circuits. For instance, they are used (1) to tune the frequency of radio receivers, (2) as filters in power supplies, (3) to eliminate sparking in automobile ignition systems, and (4) as energy-storing devices in electronic flashing units.

A capacitor basically consists of two conductors carrying equal but opposite charges. The ability of a capacitor to hold charge is measured by a quantity called the *capacitance*. We shall see that the capacitance of a given device depends on its geometry and on the material separating the charged conductors, called a *dielectric*. A dielectric is an insulating material having distinctive electrical properties that can best be understood as a consequence of the properties of atoms.

23.1 DEFINITION OF CAPACITANCE

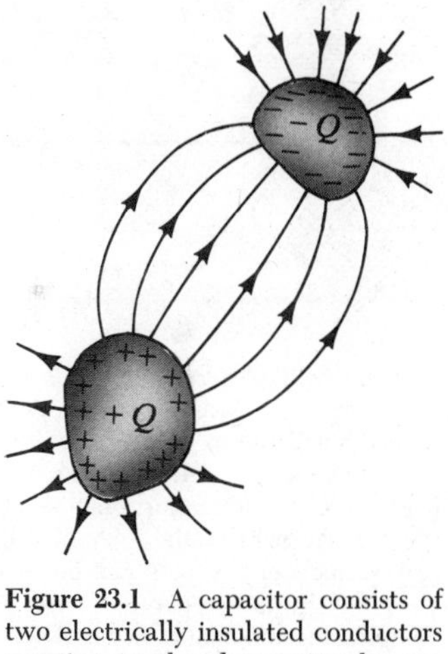

Figure 23.1 A capacitor consists of two electrically insulated conductors carrying equal and opposite charges.

Consider two conductors having a potential difference V between them. Let us assume that the conductors have equal and opposite charges as in Fig. 23.1. This can be accomplished by connecting the two uncharged conductors to the terminals of a battery. Such a combination of charged conductors is a device called a *capacitor*.[1] The potential difference V is found to be proportional to the charge Q on the capacitor. The *capacitance, C,* of a capacitor is defined as the *ratio of the magnitude of the charge on either conductor to the magnitude of the potential difference between them:*

Definition of capacitance

$$C \equiv \frac{Q}{V} \tag{23.1}$$

Note that by definition *capacitance is always a positive quantity*. Furthermore, since the potential difference increases as the stored charge increases, the ratio Q/V is constant for a given capacitor. Therefore, the capacitance of a device is a measure of its ability to store charge and electrical potential energy.

From Eq. 23.1, we see that capacitance has SI units of coulombs per volt. The SI

[1] The proportionality between the potential difference and charge on the conductors can be proved from Coulomb's law or by experiment.

unit of capacitance is the farad (F), in honor of Michael Faraday. That is,

$$[\text{Capacitance}] = 1\text{ F} = 1\text{ C/V}$$

The farad is a very large unit of capacitance. In practice, typical devices have capacitances ranging from microfarads ($1\ \mu\text{F} = 10^{-6}$ F) to picofarads ($1\text{ pF} = 10^{-12}$ F). As a practical note, capacitors are often labeled mF for microfarads and mmF for micromicrofarads (picofarads).

As we shall show in the next section, the capacitance of a device depends on the geometrical arrangement of the conductors. To illustrate this point, let us calculate the capacitance of an isolated spherical conductor of radius R and charge Q. (The second conductor can be taken as a concentric hollow conducting sphere of infinite radius.) Since the potential of the sphere is simply kQ/R (where $V = 0$ at infinity), its capacitance is given by

Capacitance of a charged sphere

$$C = \frac{Q}{V} = \frac{Q}{kQ/R} = \frac{R}{k} = 4\pi\varepsilon_0 R \qquad (23.2)$$

This shows that the capacitance of an isolated charged sphere is proportional to its radius and is independent of both the charge and the potential difference.

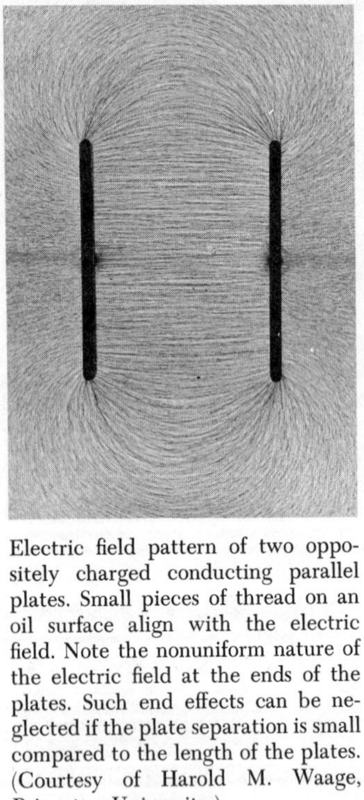

Electric field pattern of two oppositely charged conducting parallel plates. Small pieces of thread on an oil surface align with the electric field. Note the nonuniform nature of the electric field at the ends of the plates. Such end effects can be neglected if the plate separation is small compared to the length of the plates. (Courtesy of Harold M. Waage, Princeton University.)

Q1. What happens to the charge on a capacitor if the potential difference between the conductors is doubled?

23.2 CALCULATION OF CAPACITANCE

The capacitance of a pair of oppositely charged conductors can be calculated in the following manner. A convenient charge of magnitude Q is assumed, and the potential difference is calculated using the techniques described in the previous chapter. One then simply uses $C = Q/V$ to evaluate the capacitance. As you might expect, the calculation is relatively easy to perform if the geometry of the capacitor is simple.

Let us illustrate this with three geometries that we are all familiar with, namely, two parallel plates, two concentric cylinders, and two concentric spheres. In these examples, we shall assume that the charged conductors are separated by a vacuum. The effect of a dielectric material between the conductors will be treated in Section 23.5.

The Parallel-Plate Capacitor

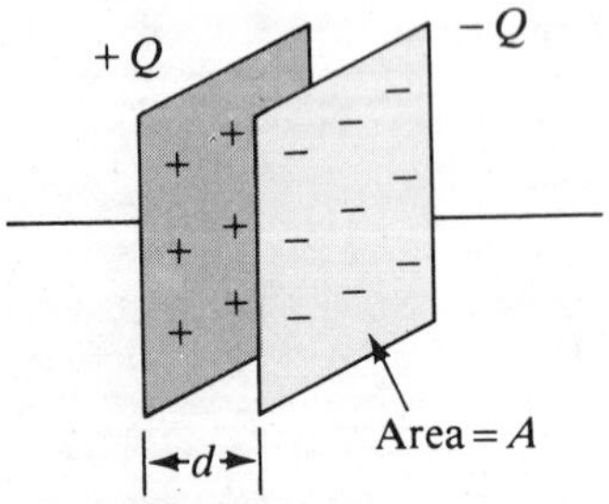

Figure 23.2 A parallel-plate capacitor consists of two parallel plates each of area A, separated by a distance d. The plates carry equal and opposite charges.

Two parallel plates of equal area A are separated by a distance d as in Fig. 23.2. One plate has a charge $+Q$, and the other has a charge $-Q$. The charge per unit area on either plate is $\sigma = Q/A$. If the plates are very close together (compared with their length and width), we can neglect end effects and assume that the electric field is uniform between the plates and zero elsewhere. According to Eq. 21.11, the electric field *between* the plates is given by

$$E = \frac{\sigma}{\varepsilon_0} = \frac{Q}{\varepsilon_0 A}$$

The potential difference *between* the plates equals Ed; therefore

$$V = Ed = \frac{Qd}{\varepsilon_0 A}$$

Substituting this result into Eq. 23.1, we find that the capacitance is given by

$$C = \frac{Q}{V} = \frac{Q}{Qd/\varepsilon_0 A}$$

$$C = \frac{\varepsilon_0 A}{d} \qquad (23.3)$$

Parallel-plate capacitor

That is, *the capacitance of a parallel-plate capacitor is proportional to the area of its plates and inversely proportional to the plate separation.*

Example 23.1

A parallel-plate capacitor has an area of $A = 2\ \text{cm}^2 = 2 \times 10^{-4}\ \text{m}^2$ and a plate separation of $d = 1\ \text{mm} = 10^{-3}\ \text{m}$. Find its capacitance.

Solution: Using Eq. 23.3 gives

$$C = \varepsilon_0 \frac{A}{d} = \left(8.85 \times 10^{-12} \frac{\text{C}^2}{\text{N} \cdot \text{m}^2}\right)\left(\frac{2 \times 10^{-4}\ \text{m}^2}{1 \times 10^{-3}\ \text{m}}\right)$$

$$= 1.77 \times 10^{-12}\ \text{F} = 1.77\ \text{pF}$$

It is left as an exercise to show that the units $\text{C}^2/\text{N} \cdot \text{m}$ equal one farad.

Example 23.2

The cylindrical capacitor: A cylindrical conductor of radius a and charge $+Q$ is concentric with a larger cylindrical shell of radius b and charge $-Q$ (Fig. 23.3a). Find the capacitance of this cylindrical capacitor if its length is l.

Solution: If we assume that l is long compared with a and b, we can neglect end effects. In this case, the field is perpendicular to the axis of the cylinders and is confined to the region between them (Fig. 23.3b). We must first calculate the potential difference between the two cylinders, which is given in general by

$$V_b - V_a = -\int_a^b \mathbf{E} \cdot d\mathbf{s}$$

where $\mathbf{E}$ is the electric field in the region $a < r < b$. In Chapter 21, we showed using Gauss' law that the electric field of a cylinder of charge per unit length λ is given by $2k\lambda/r$. The same result applies here, since the outer cylinder does not contribute to the electric field inside it. Using this result and noting that $\mathbf{E}$ is along r in Fig. 23.3b, we find that

$$V_b - V_a = -\int_a^b E_r\, dr = -2k\lambda \int_a^b \frac{dr}{r} = -2k\lambda \ln\left(\frac{b}{a}\right)$$

Substituting this into Eq. 23.1 and using the fact that $\lambda = Q/l$, we get

$$C = \frac{Q}{V} = \frac{Q}{\dfrac{2kQ}{l} \ln\left(\dfrac{b}{a}\right)} = \frac{l}{2k \ln\left(\dfrac{b}{a}\right)} \qquad (23.4)$$

Note that V is the magnitude of the potential difference given by $2k\lambda \ln(b/a)$, a *positive* quantity. That is, $V = V_a - V_b$ is *positive* since the inner cylinder is at the higher potential.

Our result for C makes sense since it shows that the capacitance is proportional to the length of the cylinders. As you might expect, the capacitance also depends on the radii of the two cylindrical conductors. As an example, a coaxial cable consists of two concentric cylindrical con-

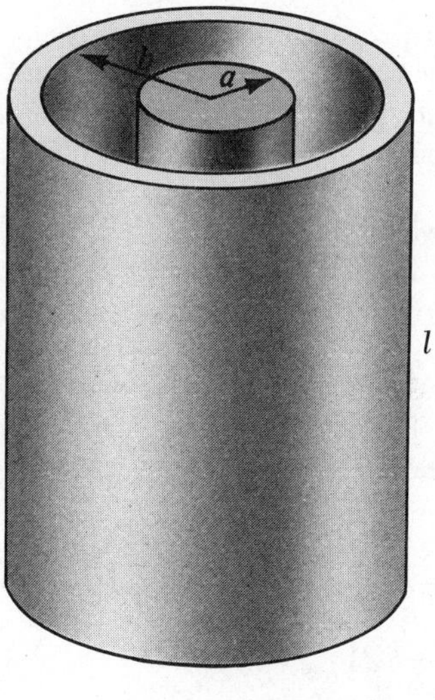

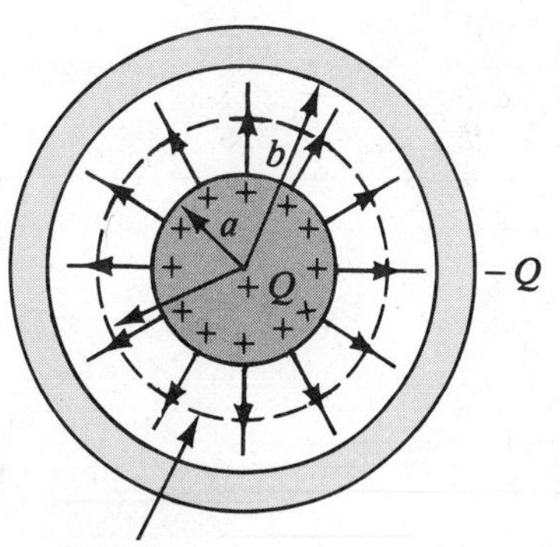

Figure 23.3 (Example 23.2) (a) A cylindrical capacitor consists of a cylindrical conductor of radius a and length l surrounded by a coaxial cylindrical shell of radius b. (b) The end view of a cylindrical capacitor. The broken line represents the end of the cylindrical gaussian surface of radius r and length l.

ductors of radii a and b separated by an insulator. The cable carries currents in opposite directions in the inner and outer conductors. Such a geometry is especially useful for shielding an electrical signal from external influences. From Eq. 23.4, we see that the capacitance per unit length of a coaxial cable is given by

$$\frac{C}{l} = \frac{1}{2k \ln\left(\frac{b}{a}\right)}$$

Example 23.3

The spherical capacitor: A spherical capacitor consists of a spherical conducting shell of radius b and charge $-Q$ that is concentric with a smaller conducting sphere of radius a and charge $+Q$ (Fig. 23.4). Find its capacitance.

Solution: As we showed in Chapter 21, the field outside a spherically symmetric charge distribution is radial and given by kQ/r^2. In this case, this corresponds to the field between the spheres ($a < r < b$). (The field is zero elsewhere.) From Gauss' law we see that only the inner sphere contributes to this field. Thus, the potential difference between the spheres is given by

$$V_b - V_a = -\int_a^b E_r\,dr = -kQ\int_a^b \frac{dr}{r^2} = kQ\left[\frac{1}{r}\right]_a^b$$
$$= kQ\left(\frac{1}{b} - \frac{1}{a}\right)$$

The magnitude of the potential difference is given by

$$V = V_a - V_b = kQ\frac{(b-a)}{ab}$$

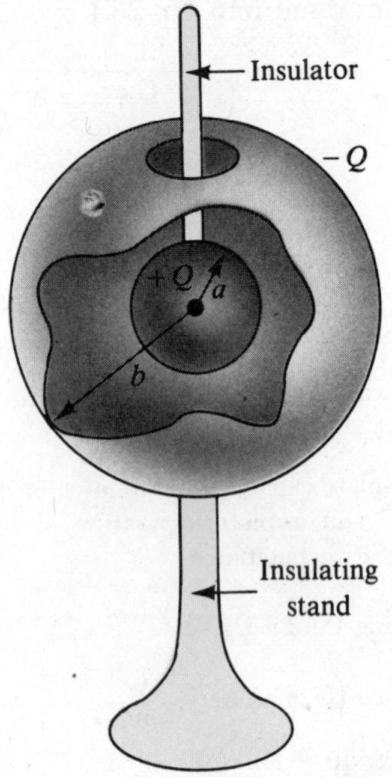

Figure 23.4 (Example 23.3) A spherical capacitor consists of an inner sphere of radius a surrounded by a concentric spherical shell of radius b. The electric field between the spheres is radial outward if the inner sphere is positively charged.

Substituting this into Eq. 23.1, we get

$$C = \frac{Q}{V} = \frac{ab}{k(b-a)} \tag{23.5}$$

Note that if we let the radius b of the outer sphere approach infinity, the capacitance approaches the value $a/k = 4\pi\varepsilon_0 a$. This is the capacitance of an isolated charged sphere, which we obtained earlier (Eq. 23.2).

23.3 COMBINATIONS OF CAPACITORS

Two or more capacitors are often combined in circuits in several ways. The equivalent capacitance of certain combinations can be calculated using methods described in this section. The symbol that is commonly used to represent a capacitor in a circuit is —| |— or sometimes —| (—.

Parallel Combination

Two capacitors connected as shown in Fig. 23.5 are known as a *parallel combination of capacitors.* The left plates of the capacitors are connected by a conducting wire and are therefore at the *same* potential. (Recall from Section 22.6 that all points on a conductor in electrostatic equilibrium are at the same potential.) Likewise, the right plates are common and at a lower potential. In practice, this can be accomplished by connecting points a and b to the positive and negative terminals of a battery, respectively. Since the *potential difference V must be the same across each capacitor,* the charges are given by

$$Q_1 = C_1V \quad \text{and} \quad Q_2 = C_2V$$

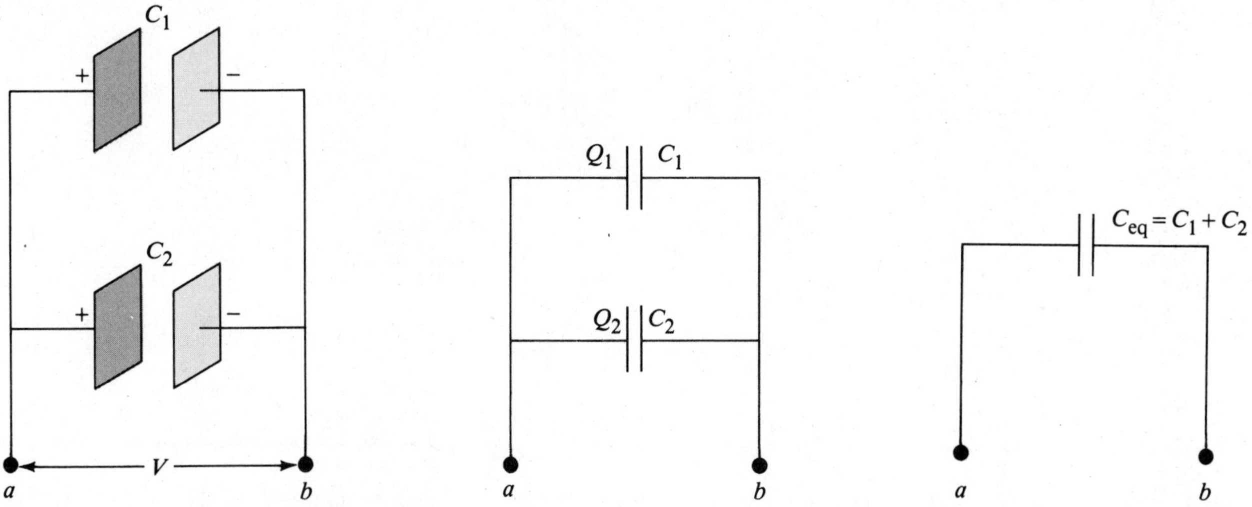

Figure 23.5 A parallel connection of two capacitors. The potential difference is the same across each capacitor, and the equivalent capacitance is $C_{eq} = C_1 + C_2$.

The total charge on both capacitors is given by

$$Q = Q_1 + Q_2 = (C_1 + C_2)V$$

The equivalent capacitance, C_{eq}, of the two capacitors is the ratio of the total charge stored to the potential difference:

$$C_{eq} = \frac{Q}{V} = C_1 + C_2 \tag{23.6}$$

Parallel combination of two capacitors

That is, we can replace C_1 and C_2 by a capacitor of capacitance C_{eq} that will store the same charge Q if the potential difference is V.

If we extend this to three or more capacitors connected in parallel, the equivalent capacitance is given by

$$C_{eq} = C_1 + C_2 + C_3 + \cdots \tag{23.7}$$

Parallel combination of several capacitors

Thus we see that the equivalent capacitance of a parallel combination of capacitors is *larger* than any of the individual capacitances.

Series Combination

Now consider two capacitors connected in series, as illustrated in Fig. 23.6. For this *series combination of capacitors, the capacitors must have the same charge.* This can be understood by noting that whatever charge appears on the right plate of C_1 must come from the left plate of C_2. In other words, the net charge enclosed by the broken line in Fig. 23.6b must be zero, assuming the capacitors are initially uncharged (before a battery is connected between a and b). After the battery is connected, electrons are transferred from the left plate of C_2 to the right plate of C_1, producing a charge of $+Q$ on one plate and $-Q$ on the other.

Since $Q = CV$ can be applied to each capacitor, the potential difference across each is given by

$$V_1 = \frac{Q}{C_1} \quad \text{and} \quad V_2 = \frac{Q}{C_2}$$

But the potential difference across the combination of two capacitors, V, is the sum

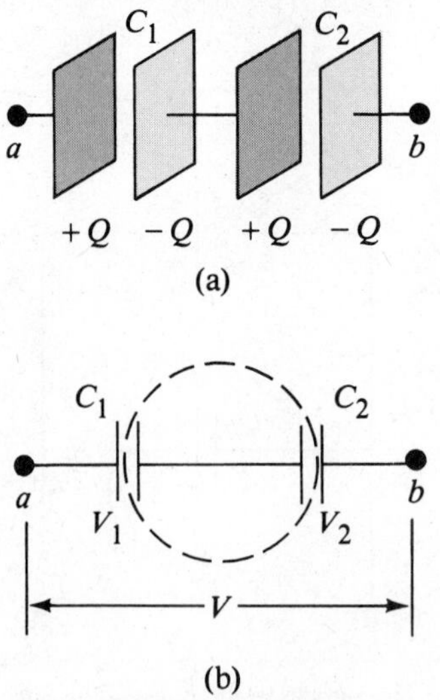

Figure 23.6 A series connection of two capacitors. The charge on each capacitor is the same, and the equivalent capacitance can be calculated from the relation $\frac{1}{C_{eq}} = \frac{1}{C_1} + \frac{1}{C_2}$.

of the potential differences across the individual capacitors:

$$V = V_1 + V_2 = Q\left(\frac{1}{C_1} + \frac{1}{C_2}\right)$$

The equivalent capacitance is $C_{eq} = Q/V$, and so $V = \frac{Q}{C_{eq}}$. Therefore, we see that

Series combination of two capacitors

$$\frac{1}{C_{eq}} = \frac{1}{C_1} + \frac{1}{C_2} \tag{23.8}$$

If three or more capacitors are connected in series, the equivalent capacitance is given by

Series combination

$$\frac{1}{C_{eq}} = \frac{1}{C_1} + \frac{1}{C_2} + \frac{1}{C_3} + \cdots \tag{23.9}$$

This shows that the equivalent capacitance of a series combination is always *less* than any individual capacitance in the combination.

Example 23.4

Find the equivalent capacitance between a and b for the combination of capacitors shown in Fig. 23.7a. All capacitances are in μF.

Solution: Using Eqs. 23.6 and 23.8, we reduce the combination step by step as indicated in the figure. The 1-μF and 3-μF capacitors are in *parallel* and combine according to

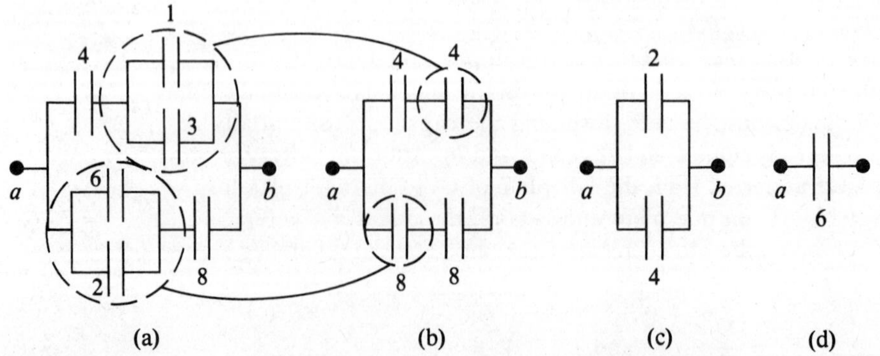

Figure 23.7 (Example 23.4) To find the equivalent combination of the capacitors in (a), the various combinations are reduced in steps as indicated in (b), (c), and (d), using the series and parallel rules described in the text.

$C_{eq} = C_1 + C_2$. Their equivalent capacitance is 4 μF. Likewise, the 2-μF and 6-μF capacitors are also in *parallel* and have an equivalent capacitance of 8 μF. The upper branch in Fig. 23.7b now consists of two 4-μF capacitors in *series,* which combine according to

$$\frac{1}{C_{eq}} = \frac{1}{C_1} + \frac{1}{C_2} = \frac{1}{4} + \frac{1}{4} = \frac{1}{2}$$

$$C_{eq} = 2\ \mu\text{F}$$

Likewise, the lower branch in Fig. 23.7b consists of two 8-μF capacitors in *series,* which give an equivalent capacitance of 4 μF. Finally, the 2-μF and 4-μF capacitors in Fig. 23.7c are in *parallel* and have an equivalent capacitance of 6 μF. Hence, the equivalent capacitance of the circuit is 6 μF.

Q2. The plates of a capacitor are connected to a battery. What happens to the charge on the plates if the connecting wires are removed from the battery? What happens to the charge if the wires are removed from the battery and connected to each other?

Q3. A pair of capacitors are connected in parallel while an identical pair are connected in series. Which pair would be more dangerous to handle after being connected to the same voltage source? Explain.

Q4. Is it always possible to reduce a combination of capacitors to one equivalent capacitor with the rules we have just developed? Explain your answer.

23.4 ENERGY STORED IN A CHARGED CAPACITOR

The process of charging a capacitor involves the transfer of electrical charge from one plate at a lower potential to another plate at a higher potential. Thus we see that work must be done to charge a capacitor.

Consider a parallel-plate capacitor that is initially uncharged, so that the initial potential difference across the plates is zero. Now imagine that the capacitor is connected to a battery and develops a maximum charge Q. We shall assume that the capacitor is charged *slowly* so that the problem can be considered as an electrostatic system. The final potential difference across the capacitor is $V = Q/C$. Since the initial potential difference is zero, the *average* potential difference during the charging process is $V/2 = Q/2C$. From this we might conclude that the work needed to charge the capacitor is $QV/2 = Q^2/2C$. Although this result is correct, a more detailed proof is desirable and is now given.

Suppose that q is the charge on the capacitor at some instant during the charging process. At the same instant, the potential difference across the capacitor is $V = q/C$. The work necessary[2] to transfer an increment of charge dq from the plate of charge $-q$ to the plate of charge q (which is at the higher potential) is given by

$$dW = V\,dq = \frac{q}{C}\,dq$$

Thus, the total work required to charge the capacitor from $q = 0$ to some final charge $q = Q$ is given by

$$W = \int_0^Q \frac{q}{C}\,dq = \frac{1}{2}\frac{Q^2}{C}$$

But the work done in charging the capacitor can be considered as potential energy U stored in the capacitor. Using $Q = CV$, we can express the electrostatic energy

[2]One mechanical analog of this process is the work required to raise a mass through some vertical distance in the presence of gravity.

stored in a charged capacitor in the following alternative forms:

Energy stored in a charged capacitor

$$U = \frac{1}{2}\frac{Q^2}{C} = \frac{1}{2}QV = \frac{1}{2}CV^2 \tag{23.10}$$

This result applies to *any* capacitor, regardless of its geometry. We see that the stored energy increases as C increases and as the potential difference increases. In practice, there is a limit to the maximum energy (or charge) that can be stored. This is because electrical discharge will ultimately occur between the plates of the capacitor at a sufficiently large value of V. For this reason, capacitors are usually labeled with a maximum operating voltage.

The energy stored in a capacitor can be considered as being stored in the electric field created between the plates as the capacitor is charged. This description is reasonable in view of the fact that the electric field is proportional to the charge on the capacitor. For a parallel-plate capacitor, the potential difference is related to the electric field through the relationship $V = Ed$. Furthermore, its capacitance is given by $C = \varepsilon_0 A/d$. Substituting these expressions into Eq. 23.10 gives

Energy stored in a parallel-plate capacitor

$$U = \frac{1}{2}\frac{\varepsilon_0 A}{d}(E^2 d^2) = \frac{1}{2}(\varepsilon_0 Ad)E^2 \tag{23.11}$$

Since the volume of a parallel-plate capacitor is Ad, the *energy per unit volume u*, called the *energy density*, is

Energy density in an electric field

$$u = \frac{U}{Ad} = \frac{1}{2}\varepsilon_0 E^2 \tag{23.12}$$

Although Eq. 23.12 was derived for a parallel-plate capacitor, the expression is generally valid. That is, the *energy density in any electrostatic field is proportional to the square of the electric field intensity at a given point.* (A formal proof of this statement is given in intermediate and advanced courses in electricity and magnetism.)

Example 23.5

Two capacitors C_1 and C_2 (where $C_1 > C_2$) are charged to the same potential difference V_0, but with opposite polarity. The charged capacitors are removed from the battery, and their plates are connected as shown in Fig. 23.8a. The switches S_1 and S_2 are then closed as in Fig. 23.8b. (a) Find the final potential difference between a and b after the switches are closed.

The charges on the capacitors *before* the switches are closed are given by

$$Q_1 = C_1V_0 \qquad \text{and} \qquad Q_2 = -C_2V_0$$

The negative sign for Q_2 is necessary since this capacitor's polarity is *opposite* that of capacitor C_1. After the switches are closed, the charges on the plates redistribute until the

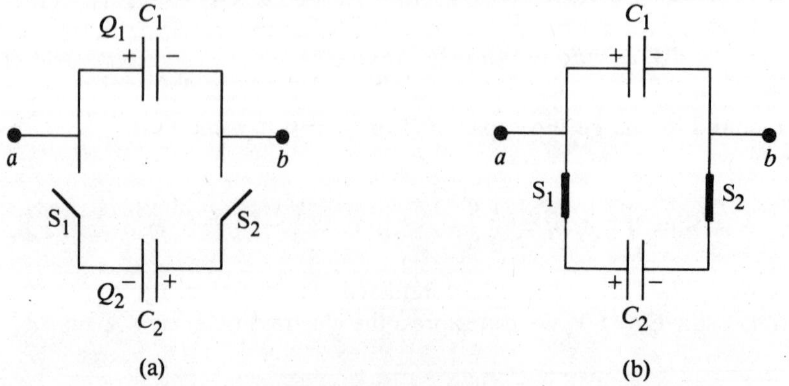

Figure 23.8 (Example 23.5)

total charge Q shared by both capacitors is

$$Q = Q_1 + Q_2 = (C_1 - C_2)V_0$$

Note that the two capacitors are now in *parallel,* and so the final potential difference across each is the *same* and given by

$$V = \frac{Q}{C_1 + C_2} = \left(\frac{C_1 - C_2}{C_1 + C_2}\right)V_0$$

(b) Find the total energy stored in the capacitors before and after the switches are closed.

Before the switches are closed, the total energy stored in the capacitors is given by

$$U_i = \frac{1}{2}C_1V_0^2 + \frac{1}{2}C_2V_0^2 = \frac{1}{2}(C_1 + C_2)V_0^2$$

After the switches are closed and the capacitors have reached an equilibrium charge, the total energy stored in the capacitors is given by

$$U_f = \frac{1}{2}C_1V^2 + \frac{1}{2}C_2V^2 = \frac{1}{2}(C_1 + C_2)V^2$$

$$= \frac{1}{2}(C_1 + C_2)\left(\frac{C_1 - C_2}{C_1 + C_2}\right)^2 V_0^2 = \left(\frac{C_1 - C_2}{C_1 + C_2}\right)^2 U_i$$

Therefore, the ratio of the final to the initial energy stored is

$$\frac{U_f}{U_i} = \left(\frac{C_1 - C_2}{C_1 + C_2}\right)^2$$

This shows that the final energy is *less* than the initial energy. At first, you might think that energy conservation has been violated, but this is not the case since we have assumed that the circuit is ideal. Part of the missing energy appears as heat energy in the connecting wires, which have resistance, and part of the energy is radiated away in the form of electromagnetic waves (Chapter 35).

Q5. If the potential difference across a capacitor is doubled, by what factor does the energy stored change?

Q6. Since the charges on the plates of a parallel-plate capacitor are equal and opposite, they attract each other. Hence, it would take positive work to increase the plate separation. What happens to the external work done in this process?

23.5 CAPACITORS WITH DIELECTRICS

A *dielectric* is a nonconducting material, such as rubber, glass, or paper. When a dielectric material is inserted between the plates of a capacitor, the capacitance generally increases. If the dielectric completely fills the space between the plates, the capacitance increases by a dimensionless factor κ, called the *dielectric constant.*

The following experiment can be performed to illustrate the effect of a dielectric in a capacitor. Consider a parallel-plate capacitor of charge Q_0 and capacitance C_0 in the absence of a dielectric. The potential difference across the capacitor as measured by an electrostatic voltmeter is $V_0 = Q_0/C_0$ (Fig. 23.9a). Notice that the capacitor circuit is *open,* that is, the plates of the capacitor are *not* connected to a battery and charge cannot flow through an ideal voltmeter. (We shall discuss the voltmeter further in Chapter 25.) Hence, there is *no* path by which charge can flow and alter the charge on the capacitor. If a dielectric is now inserted between the plates as in Fig. 23.9b, the voltmeter reading *decreases* by a factor κ to a value V, where

$$V = \frac{V_0}{\kappa}$$

Since $V < V_0$, we see that $\kappa > 1$.

Since the charge Q_0 on the capacitor *does not change,* we conclude that the capacitance must change to the value

$$C = \frac{Q_0}{V} = \frac{Q_0}{V_0/\kappa} = \kappa\frac{Q_0}{V_0}$$

$$C = \kappa C_0 \qquad (22.13)$$

The capacitance of a filled capacitor is greater than that of an empty one by a factor κ.

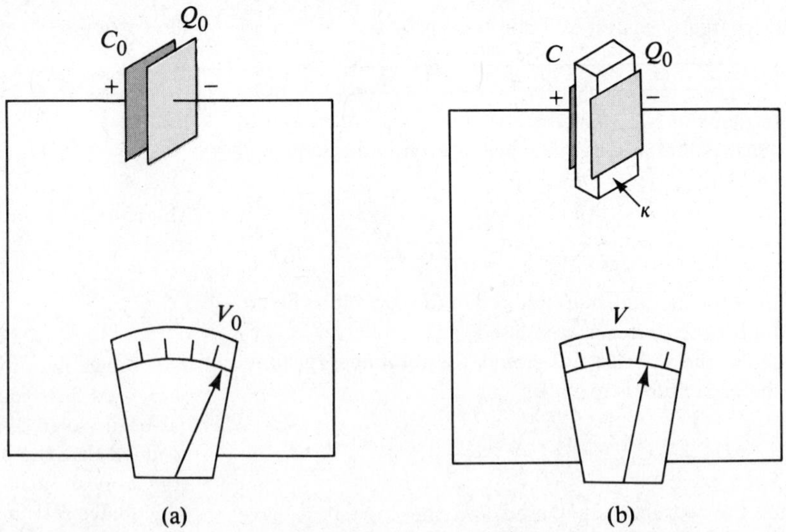

Figure 23.9 When a dielectric is inserted between the plates of a charged capacitor, the charge on the plates remains unchanged, but the potential difference as recorded by an electrostatic voltmeter is reduced from V_0 to $V = V_0/\kappa$. Thus, the capacitance *increases* in the process by the factor κ.

where C_0 is the capacitance in the absence of the dielectric. That is, the capacitance *increases* by the factor κ when the dielectric completely fills the region between the plates.[3] For a parallel-plate capacitor, where $C_0 = \varepsilon_0 A/d$, we can express the capacitance when the capacitor is filled with a dielectric as

$$C = \kappa \frac{\varepsilon_0 A}{d} \tag{23.14}$$

From Eqs. 23.3 and 23.14, it would appear that the capacitance could be made very large by decreasing d, the distance between the plates. In practice, the lowest value of d is limited by the electrical discharge that could occur through the dielectric medium separating the plates. For any given separation d, the maximum voltage that can be applied to a capacitor without causing a discharge depends on the

TABLE 23.1 Dielectric Constants and Dielectric Strengths of Various Materials

MATERIAL	*DIELECTRIC CONSTANT* κ	*DIELECTRIC STRENGTH* (V/m)
Vacuum	1.00000	∞
Air	1.00059	3×10^6
Bakelite	4.9	24×10^6
Fused quartz	3.78	8×10^6
Pyrex glass	5.6	14×10^6
Polystyrene	2.56	24×10^6
Teflon	2.1	60×10^6
Neoprene rubber	6.7	12×10^6
Nylon	3.4	14×10^6
Paper	3.7	16×10^6
Strontium titanate	233	8×10^6
Water	80	—
Silicone oil	2.5	15×10^6

[3] If another experiment is performed in which the dielectric is introduced while the potential difference remains constant by means of a battery, the charge increases to a value $Q = \kappa Q_0$. The additional charge is supplied by the battery and the capacitance still increases by the factor κ.

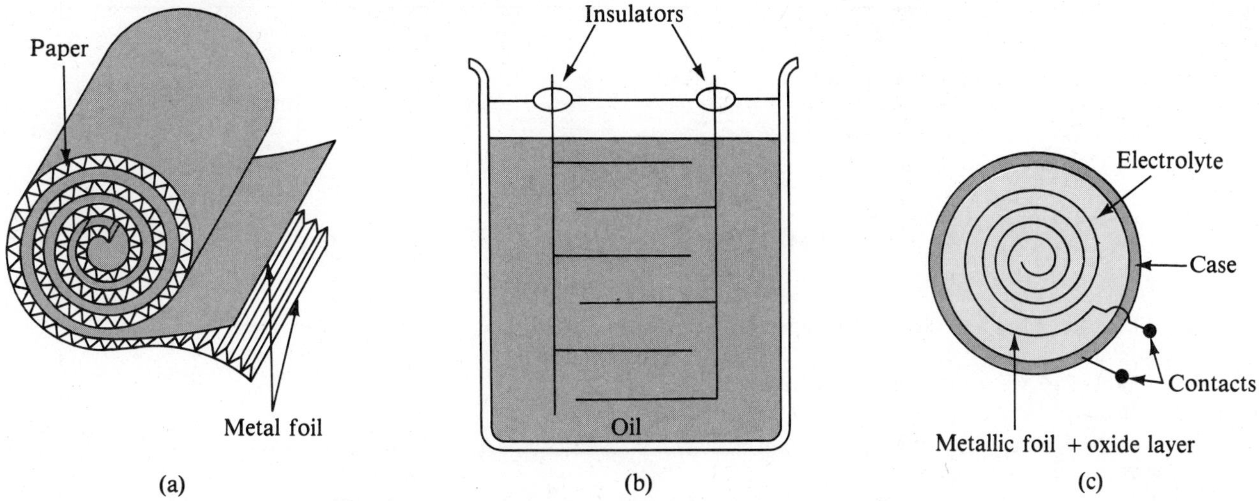

Figure 23.10 Three commercial capacitor designs. (a) A tubular capacitor whose plates are separated by paper and then rolled into a cylinder. (b) A high-voltage capacitor consists of many parallel plates separated by insulating oil. (c) An electrolytic capacitor.

dielectric strength (maximum electric field intensity) of the dielectric, which for air is equal to 3×10^6 V/m. If the field strength in the medium exceeds the dielectric strength, the insulating properties will break down and the medium will begin to conduct. Most insulating materials have dielectric strengths and dielectric constants greater than that of air, as Table 23.1 indicates. Thus, we see that a dielectric increases the capacitance of a capacitor and increases the maximum operating voltage by virtue of its high dielectric strength. In addition, a solid dielectric provides mechanical support between the conducting plates.

Types of Capacitors

Commercial capacitors are often made by spacing two metallic foils with a thin sheet of paraffin-coated paper. These layers are then rolled into the shape of a cylinder to form a small package (Fig. 23.10a). High-voltage capacitors commonly consist of a number of interweaving metal plates immersed in silicone oil (Fig. 23.10b). Small capacitors are often constructed using ceramic materials, such as the titanates, which have high dielectric constants and strength. It is also possible to obtain compact and high-value capacitors with a layer of tantalum (up to 3500 μF) or aluminum oxide (up to 150 000 μF). Variable capacitors (typically 10 to 500 pF) usually consist of two interweaving sets of metal plates, one fixed and the other movable, with air as the dielectric.

An *electrolytic capacitor* is often used to store large amounts of charge at relatively low voltages. This device, shown in Fig. 23.10c, consists of a metallic foil in contact with an electrolyte—a solution that conducts electricity by virtue of the motion of ions. When a voltage is applied between the foil and the electrolyte, a thin layer of metal oxide (an insulator) is formed on the foil, and this layer serves as the dielectric. Very large values of capacitance can be obtained since the dielectric layer is *very thin;* however, the maximum operating voltage is limited. When electrolytic capacitors are used in circuits, the polarity (the plus and minus signs on the device) must be installed properly. If the polarity of the applied voltage is opposite what is intended, the oxide layer will be removed and the capacitor will conduct electricity rather than store charge.

Example 23.6

A parallel-plate capacitor has plates of dimensions 2 cm × 3 cm. The plates are separated by a 1-mm thickness of paper. (a) Find the capacitance of this device.

Since $\kappa = 3.7$ for paper (Table 23.1), we get

$$C = \kappa \frac{\varepsilon_0 A}{d} = 3.7\left(8.85 \times 10^{-12} \frac{\text{C}^2}{\text{N} \cdot \text{m}^2}\right)\left(\frac{6 \times 10^{-4}\ \text{m}^2}{1 \times 10^{-3}\ \text{m}}\right)$$
$$= 19.6 \times 10^{-12}\ \text{F} = 19.6\ \text{pF}$$

(b) What is the maximum charge that can be placed on the capacitor?

From Table 23.1 we see that the dielectric strength of paper is 16×10^6 V/m. Since the thickness of the paper is 1 mm, the maximum voltage that can be applied before breakdown occurs is

$$V_{\text{max}} = E_{\text{max}} d = \left(16 \times 10^6 \frac{\text{V}}{\text{m}}\right)(1 \times 10^{-3}\ \text{m})$$
$$= 16 \times 10^3\ \text{V}$$

Hence, the maximum charge is given by

$$Q_{\text{max}} = CV_{\text{max}} = (19.6 \times 10^{-12}\ \text{F})(16 \times 10^3\ \text{V})$$
$$= 0.31\ \mu\text{C}$$

(c) What is the maximum energy that can be stored in the capacitor?

$$U_{\text{max}} = \frac{1}{2}CV_{\text{max}}^2 = \frac{1}{2}(19.6 \times 10^{-12}\ \text{F})(16 \times 10^3\ \text{V})^2$$
$$= 2.5 \times 10^{-3}\ \text{J}$$

Example 23.7

A parallel-plate capacitor is charged with a battery to a charge Q_0, as in Fig. 23.11a. The battery is then removed, and a slab of dielectric constant κ is inserted between the plates, as in Fig. 23.11b. Find the energy stored in the capacitor before and after the dielectric is inserted.

Solution: The energy stored in the capacitor in the absence of the dielectric is

$$U_0 = \frac{1}{2}C_0 V_0^2$$

Since $V_0 = Q_0/C_0$, this can be expressed as

$$U_0 = \frac{Q_0^2}{2C_0}$$

After the battery is removed and the dielectric is inserted between the plates, the *charge on the capacitor remains the same.* Hence, the energy stored in the presence of the dielectric is given by

$$U = \frac{Q_0^2}{2C}$$

But the capacitance in the presence of the dielectric is given by $C = \kappa C_0$, and so U becomes

$$U = \frac{Q_0^2}{2\kappa C_0} = \frac{U_0}{\kappa}$$

Since $\kappa > 1$, we see that the final energy is *less* than the initial energy by the factor $1/\kappa$. This missing energy can be accounted for by noting that when the dielectric is inserted into the capacitor, it gets pulled into the device. The external agent must do negative work to keep the slab from accelerating. This work is simply the difference $U - U_0$. (Alternatively, the positive work done by the system on the external agent is given by $U_0 - U$.)

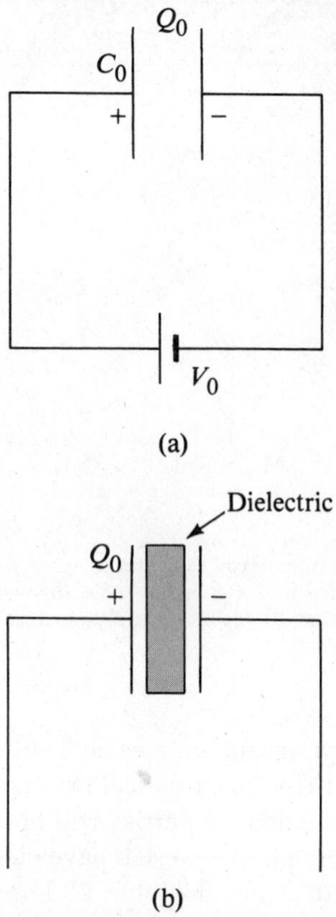

Figure 23.11 (Example 23.7)

Q7. If you want to increase the maximum operating voltage of a parallel-plate capacitor, describe how you could do this for a fixed plate separation.

Q8. Why is it dangerous to touch the terminals of a high-voltage capacitor even after the applied voltage has been turned off? What could be done to make the capacitor safe to handle after the voltage source has been removed?

23.6 ELECTRIC DIPOLE IN AN EXTERNAL ELECTRIC FIELD

The electric dipole, discussed briefly in Example 20.6, consists of two equal and opposite charges separated by a distance $2a$, as in Fig. 23.12. Let us define the *electric dipole moment* of this configuration as the vector $\mathbf{p}$ whose magnitude is $2aq$ (that is, the separation $2a$ multiplied by the charge q).

Electric dipole moment

$$p = 2aq \tag{23.15}$$

The direction of $\mathbf{p}$ is from the negative to the positive charge, as in Fig. 23.12.

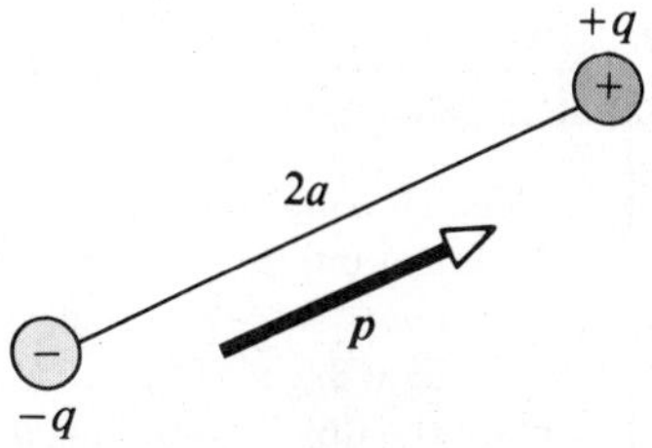

Figure 23.12 An electric dipole consists of two equal and opposite charges separated by a distance $2a$.

Now suppose an electric dipole is placed in a uniform *external* electric field $\mathbf{E}$ as in Fig. 23.13, where the dipole moment makes an angle θ with the field. The forces on the two charges are equal and opposite as shown, each having a magnitude of

$$F = qE$$

Thus, we see that the net force on the dipole is *zero*. However, the two forces produce a net torque on the dipole, and the dipole tends to rotate such that its axis is aligned with the field. The torque due to the force on the positive charge about an axis through O is given by $Fa\sin\theta$, where $a\sin\theta$ is the moment arm of F about O. This force tends to produce a clockwise rotation. Likewise, the torque on the negative charge about O is also $Fa\sin\theta$, and so the net torque about O is given by

$$\tau = 2Fa\sin\theta$$

Since $F = qE$ and $p = 2aq$, we can express τ as

$$\tau = 2aqE\sin\theta = pE\sin\theta \tag{23.16}$$

It is convenient to express the torque in vector form as the cross product of the vectors $\mathbf{p}$ and $\mathbf{E}$:

Torque on an electric dipole in an extended electric field

$$\boldsymbol{\tau} = \mathbf{p} \times \mathbf{E} \tag{23.17}$$

We can also determine the potential energy of an electric dipole as a function of its orientation with respect to the external electric field. In order to do this, you should recognize that work must be done by an external agent to rotate the dipole through a given angle in the field. The work done is then stored as potential energy in the system, that is, the dipole and the external field. The work dW required to rotate the dipole through an angle $d\theta$ is given by $dW = \tau\, d\theta$ (Chapter 10). Since $\tau = pE\sin\theta$, and since the work is transformed into potential energy U, we find that for a rotation from θ_0 to θ,

$$U = \int_{\theta_0}^{\theta} \tau\, d\theta = \int_{\theta_0}^{\theta} pE\sin\theta\, d\theta = pE\int_{\theta_0}^{\theta} \sin\theta\, d\theta$$

$$U = pE[-\cos\theta]_{\theta_0}^{\theta} = pE(\cos\theta_0 - \cos\theta)$$

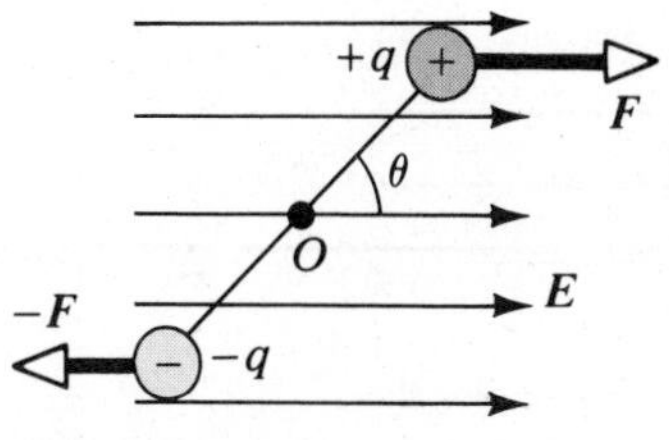

Figure 23.13 An electric dipole in a uniform electric field. The dipole moment $\mathbf{p}$ is at an angle θ with the field, and a torque is exerted on the dipole.

The term involving $\cos\theta_0$ is a constant that depends on the initial orientation of the dipole. It is convenient to choose $\theta_0 = 90°$, so that $\cos\theta_0 = \cos 90° = 0$. In this case, we can express U as

$$U = -pE\cos\theta \tag{23.18}$$

This is equivalent to the dot product of the vectors $\mathbf{p}$ and $\mathbf{E}$:

Potential energy of an electric dipole in an external electric field

$$U = -\mathbf{p} \cdot \mathbf{E} \tag{23.19}$$

Since neutral atoms and molecules contain both positive and negative charges, they behave as electric dipoles when placed in an external electric field. In some cases, the centers of the positive and negative charges coincide (such as atomic hydrogen) and the atom or molecule has no permanent dipole moment. However, an external electric field will tend to separate (or polarize) these charges, and we say that the system experiences an *induced* dipole moment.

There is another class of molecules, called *polar molecules,* in which the positive and negative charges *do not* coincide. These molecules have a permanent electric dipole moment and hence tend to rotate when placed in a uniform external electric field. One example of a polar molecule is HCl, which is essentially an H^+ ion combined with a Cl^- ion. The effect of such dipoles on the behavior of materials subjected to electric fields will be discussed in the next section.

Example 23.8

The HCl molecule has a dipole moment of 3.4×10^{-30} C · m. A sample contains 10^{21} such molecules, whose dipole moments are all oriented in the direction of an electric field of 2.5×10^5 N/C. How much work is required to rotate the dipoles from this orientation ($\theta = 0°$) to one in which all of the moments are perpendicular to the field ($\theta = 90°$)?

Solution: The work required to rotate *one* molecule by 90° is equal to the difference in potential energy between the 90° orientation and the 0° orientation. Using Eq. 23.18 gives

$$\begin{aligned} W &= U_{90} - U_0 = (-pE\cos 90°) - (-pE\cos 0°) \\ &= pE = (3.4 \times 10^{-30}\ \text{C}\cdot\text{m})(2.5 \times 10^5\ \text{N/C}) \\ &= 8.5 \times 10^{-25}\ \text{J} \end{aligned}$$

Since there are 10^{21} molecules in the sample, the *total* work required is given by

$$W_{\text{total}} = (10^{21})(8.5 \times 10^{-25}\ \text{J}) = 8.5 \times 10^{-4}\ \text{J}$$

23.7 AN ATOMIC DESCRIPTION OF DIELECTRICS

In Section 23.5 we found that the potential difference between the plates of a capacitor is reduced by the factor κ when a dielectric is introduced. Since the potential difference between the plates equals the product of the electric field and the separation d, the electric field is also reduced by the factor κ. Thus, if E_0 is the electric field without the dielectric, the field in the presence of a dielectric is

$$E = \frac{E_0}{\kappa} \tag{23.20}$$

This can be understood by noting that a dielectric can be polarized. At the atomic level, a polarized material is one in which the positive and negative charges are slightly separated. If the molecules of the dielectric possess permanent electric dipole moments in the absence of an electric field, they are called *polar molecules* (water is an example). The dipoles are randomly oriented in the absence of an electric field, as shown in Fig. 23.14a. When an external field is applied, a torque is exerted on the dipoles, causing them to be partially aligned with the field, as in Fig. 23.14b. The degree of alignment depends on temperature and on the magnitude of the applied field. In general, the alignment increases with decreasing temperature and with increasing electric field strength. The partially aligned dipoles produce an internal electric field that *opposes* the external field, thereby causing a reduction of the original field.

Polar molecules

If the molecules of the dielectric do not possess a permanent dipole moment, they are called *nonpolar molecules*. In this case, an external electric field produces some charge separation, and the resulting dipole moments are said to be *induced.* These

Nonpolar molecules

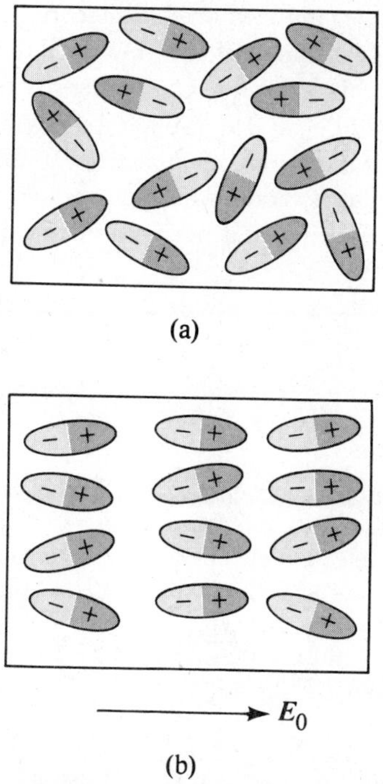

Figure 23.14 (a) Molecules with a permanent dipole moment are randomly oriented in the absence of an external electric field. (b) When an external field is applied, the dipoles are partially aligned with the field.

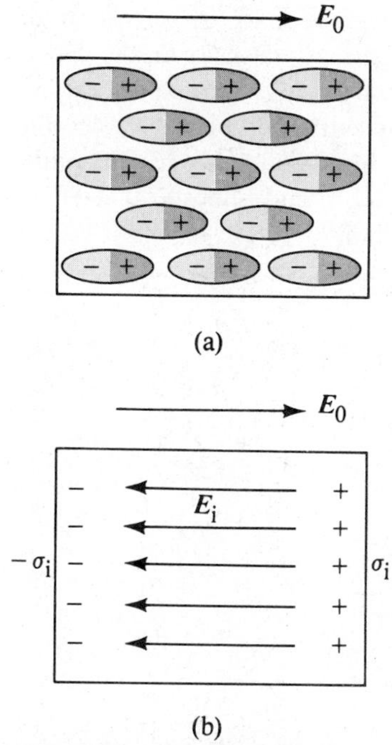

Figure 23.15 (a) When a dielectric is polarized, the molecular dipole moments in the dielectric are aligned with the external field $\mathbf{E}_0$ (the net effect). (b) This polarization causes an induced negative surface charge on one side of the dielectric and an equal positive surface charge on the opposite side. This results in a reduction in the electric field within the dielectric.

induced dipole moments tend to align with the external field, causing a reduction in the internal electric field.

With these ideas in mind, consider a slab of dielectric material in a uniform electric field $\mathbf{E}_0$ as in Figure 23.15a. Positive portions of the molecules are shifted in the direction of the electric field, and negative portions are shifted in the opposite direction. Hence, the applied electric field polarizes the dielectric. The net effect on the dielectric is the formation of an "induced" positive surface charge density σ_i on the right face and an equal negative surface charge density on the left face, as shown in Fig. 23.15b. These induced surface charges on the dielectric give rise to an induced electric field $\mathbf{E}_i$, which *opposes* the external field $\mathbf{E}_0$. Therefore, the net electric field $\mathbf{E}$ in the dielectric has a magnitude given by

$$E = E_0 - E_i \qquad (23.21)$$

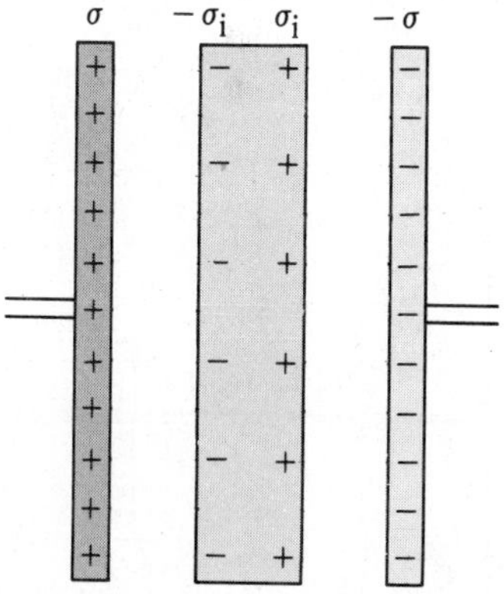

Figure 23.16 Induced charge on a dielectric placed between the plates of a charged capacitor. Note that the induced charge density on the dielectric is *less* than the free charge density on the plates.

In the parallel-plate capacitor shown in Fig. 23.16, the external field E_0 is related to the free charge density σ on the plates through the relation $E_0 = \sigma/\varepsilon_0$. The induced electric field in the dielectric is related to the induced charge density σ_i through the relation $E_i = \sigma_i/\varepsilon_0$. Since $E = E_0/\kappa = \sigma/\kappa\varepsilon_0$, substitution into Eq. 23.21 gives

$$\frac{\sigma}{\kappa\varepsilon_0} = \frac{\sigma}{\varepsilon_0} - \frac{\sigma_i}{\varepsilon_0}$$

$$\sigma_i = \left(\frac{\kappa - 1}{\kappa}\right)\sigma \qquad (23.22)$$

Since $\kappa > 1$, this shows that the charge density σ_i induced on the dielectric is *less* than the free charge density σ on the plates. For instance, if $\kappa = 3$, we see that the induced charge density on the dielectric is two thirds the free charge density on the plates. If there is no dielectric present, $\kappa = 1$ and $\sigma_i = 0$ as expected. However, if the dielectric is replaced by a *conductor,* for which $\kappa = \infty$, then $E \rightarrow 0$ and Eq. 23.22 shows that $\sigma_i \rightarrow \sigma$. That is, the surface charge induced on the conductor will be equal to and opposite that on the plates, resulting in a net field of *zero* in the conductor.

Example 23.9

A parallel-plate capacitor has a capacitance C_0 in the absence of a dielectric. A slab of dielectric material of dielectric constant κ and thickness $\frac{1}{3}d$ is inserted between the plates (Fig. 23.17a). What is the new capacitance when the dielectric is present?

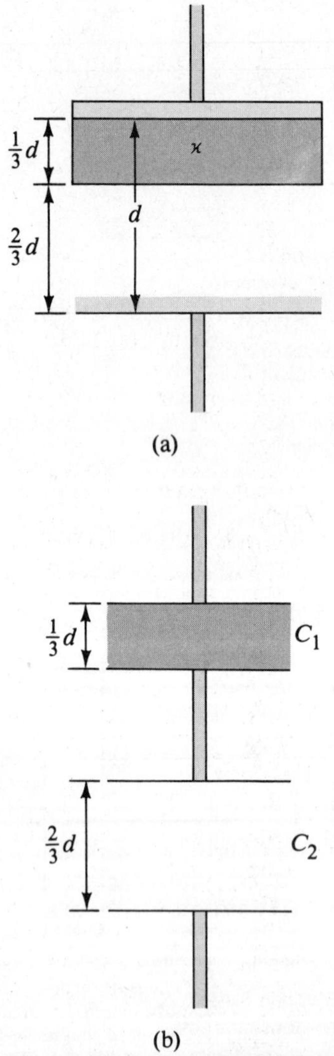

Figure 23.17 (Example 23.9) (a) A parallel-plate capacitor of plate separation d partially filled with a dielectric of thickness $d/3$. (b) The equivalent circuit of the capacitor consists of two capacitors connected in series.

Solution: This capacitor is equivalent to two parallel-plate capacitors of the same area A connected in series, one with a plate separation $d/3$ (dielectric filled) and the other with a plate separation $2d/3$ (Fig. 23.17b). (This step is permissible since there is no potential difference between the lower plate of C_1 and the upper plate of C_2.)[4]

From Eqs. 23.3 and 23.13, the two capacitances are given by

$$C_1 = \frac{\kappa \varepsilon_0 A}{d/3} \quad \text{and} \quad C_2 = \frac{\varepsilon_0 A}{2d/3}$$

Using Eq. 23.8 for two capacitors combined in series, we get

$$\frac{1}{C} = \frac{1}{C_1} + \frac{1}{C_2} = \frac{d/3}{\kappa \varepsilon_0 A} + \frac{2d/3}{\varepsilon_0 A}$$

$$\frac{1}{C} = \frac{d}{3\varepsilon_0 A}\left(\frac{1}{\kappa} + 2\right) = \frac{d}{3\varepsilon_0 A}\left(\frac{1 + 2\kappa}{\kappa}\right)$$

$$C = \left(\frac{3\kappa}{2\kappa + 1}\right)\frac{\varepsilon_0 A}{d}$$

Since the capacitance *without* the dielectric is given by $C_0 = \varepsilon_0 A/d$, we see that

$$C = \left(\frac{3\kappa}{2\kappa + 1}\right)C_0$$

Example 23.10

A parallel-plate capacitor has a plate separation d and plate area A. An uncharged *metal* slab of thickness a is inserted midway between the plates, as shown in Fig. 23.18a. Find the capacitance of the device.

Solution: This problem can be solved by noting that whatever charge appears on one plate of the capacitor must induce an *equal* and *opposite* charge on the metal slab, as shown in Fig. 23.18a. Consequently, the net charge on the metal slab remains zero, and the field inside the slab is zero. Hence, the capacitor is equivalent to two capacitors in *series,* each having a plate separation $(d - a)/2$ as shown in Fig. 23.18b. Using the rule for adding two capac-

[4] You could also imagine placing two thin metallic plates (with a coiled-up conducting wire between them) at the lower surface of the dielectric in Fig. 23.17a and then pulling the assembly out until it becomes like Fig. 23.17b.

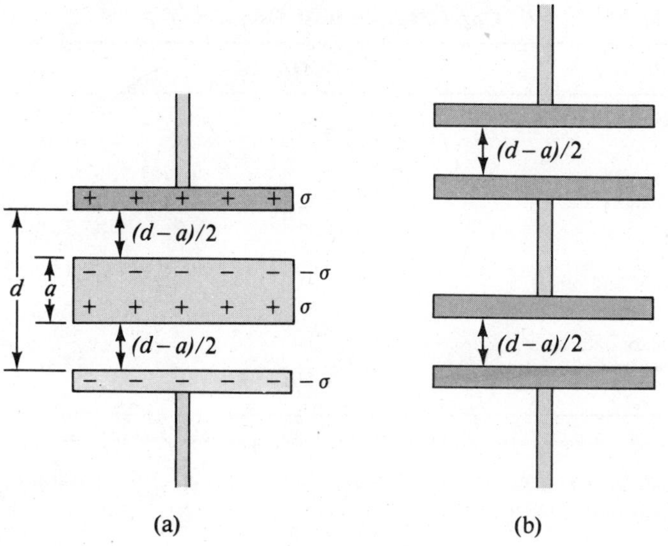

Figure 23.18 (Example 23.10) (a) A parallel-plate capacitor of plate separation d partially filled with a metal slab of thickness a. (b) The equivalent circuit of the device in (a) consists of two capacitors in series, each with a plate separation $(d - a)/2$.

itors in series we get

$$\frac{1}{C} = \frac{1}{C_1} + \frac{1}{C_2} = \frac{1}{\frac{\varepsilon_0 A}{(d-a)/2}} + \frac{1}{\frac{\varepsilon_0 A}{(d-a)/2}}$$

Solving for C gives

$$C = \frac{\varepsilon_0 A}{d - a}$$

Note that C approaches infinity as a approaches d. Why?

Q9. If a dielectric-filled capacitor is heated, how will its capacitance change? (Neglect thermal expansion and assume that the dipole orientations are temperature-dependent.)

Q10. In terms of induced charges, explain why a charged comb attracts small bits of paper.

Q11. If you were asked to design a capacitor where small size and large capacitance were required, what factors would be important in your design?

23.8 SUMMARY

A *capacitor* consists of two equal and oppositely charged conductors with a potential difference V between them. The capacitance C of any capacitor is defined to be the ratio of the magnitude of the charge Q on either conductor to the magnitude of the potential difference V:

$$C \equiv \frac{Q}{V} \tag{23.1}$$

Definition of capacitance

The SI unit of capacitance is coulomb per volt, or farad (F), and 1 F = 1 C/V.

The capacitance of several geometries is summarized in Table 23.2. The formulas apply when the charged conductors are separated by a vacuum.

If two or more capacitors are connected in *parallel*, the potential difference across them must be the same. The equivalent capacitance of a parallel combination of capacitors is given by

$$C_{eq} = C_1 + C_2 + C_3 + \cdots \tag{23.7}$$

Parallel combination

TABLE 23.2 Capacitance and Geometry

GEOMETRY	CAPACITANCE	
Isolated charged sphere of radius R	$C = 4\pi\varepsilon_0 R$	(23.2)
Parallel-plate capacitor of plate area A and plate separation d	$C = \varepsilon_0 \dfrac{A}{d}$	(23.3)
Cylindrical capacitor of length l and inner and outer radii a and b, respectively	$C = \dfrac{l}{2k \ln\left(\dfrac{b}{a}\right)}$	(23.4)
Spherical capacitor with inner and outer radii a and b, respectively	$C = \dfrac{ab}{k(b-a)}$	(23.5)

If two or more capacitors are connected in *series*, the equivalent capacitance of the series combination is given by

Series combination

$$\frac{1}{C_{\text{eq}}} = \frac{1}{C_1} + \frac{1}{C_2} + \frac{1}{C_3} + \cdots \tag{23.9}$$

Work is required to charge a capacitor, since the charging process consists of transferring charges from one conductor at a lower potential to another conductor at a higher potential. The work done in charging the capacitor to a charge Q equals the electrostatic potential energy U stored in the capacitor, where

Energy stored in a charged capacitor

$$U = \frac{Q^2}{2C} = \frac{1}{2}QV = \frac{1}{2}CV^2 \tag{23.10}$$

When a dielectric material is inserted between the plates of a capacitor, the capacitance generally increases by a dimensionless factor κ called the *dielectric constant*. That is,

$$C = \kappa C_0 \tag{23.13}$$

where C_0 is the capacitance in the absence of the dielectric. The increase in capacitance is due to a decrease in the electric field in the presence of the dielectric and to a corresponding decrease in the potential difference between the plates—assuming the charging battery is removed from the circuit before the dielectric is inserted. The decrease in $\mathbf{E}$ arises from an internal electric field produced by aligned dipoles in the dielectric. This internal field produced by the dipoles opposes the original applied field, and this results in a reduction in the net electric field.

An *electric dipole* consists of two equal and opposite charges separated by a distance $2a$. The electric dipole moment $\mathbf{p}$ of this configuration has a magnitude given by

Electric dipole moment

$$p = 2aq \tag{23.15}$$

The *torque* acting on an electric dipole in a uniform electric field $\mathbf{E}$ is given by

Torque on an electric dipole in an extended electric field

$$\boldsymbol{\tau} = \mathbf{p} \times \mathbf{E} \tag{23.17}$$

The *potential energy* of an electric dipole in a uniform external electric field $\mathbf{E}$ is given by

Potential energy of an electric dipole in an external electric field

$$U = -\mathbf{p} \cdot \mathbf{E} \tag{23.19}$$

EXERCISES

Section 23.1 Definition of Capacitance

1. Two parallel wires are suspended in a vacuum. When the potential difference between the two wires is 32 V, each wire has a charge of 95 μC (the two charges are of opposite sign). Calculate the capacitance of the parallel-wire system. Give your answer in μF.

2. An isolated conducting sphere can be considered as one element of a capacitor (the other being a concentric sphere of infinite radius). (a) If the capacitance of this system is 5×10^{-9} F, what is the radius of the sphere? (b) If the potential at the surface of the sphere is 10^4 V, what is the corresponding surface charge density?

3. A parallel-plate capacitor has a capacitance of 6 μF. What charge on each plate will produce a potential difference of 24 V between the plates of the capacitor?

4. Show that the units $C^2/N \cdot m$ equal 1 F.

5. Two conductors insulated from each other are charged by transferring electrons from one conductor to the other. After 2.5×10^{12} electrons have been transferred, the potential difference between the conductors is found to be 16 V. What is the capacitance of the system?

6. The excess charge on each conductor of a simple capacitor is 36 μC. What is the potential difference between the conductors if the capacitance of the system is 6×10^{-2} μF?

Section 23.2 Calculation of Capacitance

7. The plates of a parallel-plate capacitor are separated by 0.1 mm. If the space between the plates is air, what plate area is required to provide a capacitance of 2 pF?

8. An air-filled *cylindrical* capacitor has a capacitance of 10 pF and is 6 cm in length. If the radius of the outside conductor is 1.5 cm, what is the required radius of the inner conductor?

9. An air-filled spherical capacitor is constructed with inner and outer shell radii of 6 and 12 cm, respectively. (a) Calculate the capacitance of the device. (b) What potential difference between the spheres will result in a charge of 1 μC on each conductor?

10. A parallel-plate capacitor has a plate area of 5 cm^2 and a capacitance of 4 μF. What is the plate separation?

11. A cylindrical capacitor has outer and inner conductors whose radii are in the ratio of $b/a = 5/1$. The inner conductor is to be replaced by a wire whose radius is one half of the original inner conductor. By what factor should the length be increased in order to obtain a capacitance equal to that of the original capacitor?

12. An air-filled spherical capacitor has an outer spherical conductor of radius 0.25 m. If the capacitance of the device is to be 1 μF, calculate the required value for the radius of the inner spherical conductor.

13. A capacitor is constructed of interlocking plates as shown in Fig. 23.19 (a cross-sectional view). The separation between adjacent plates is 0.5 cm, and the effective area of overlap of adjacent plates is 6 cm^2. Calculate the capacitance of the unit.

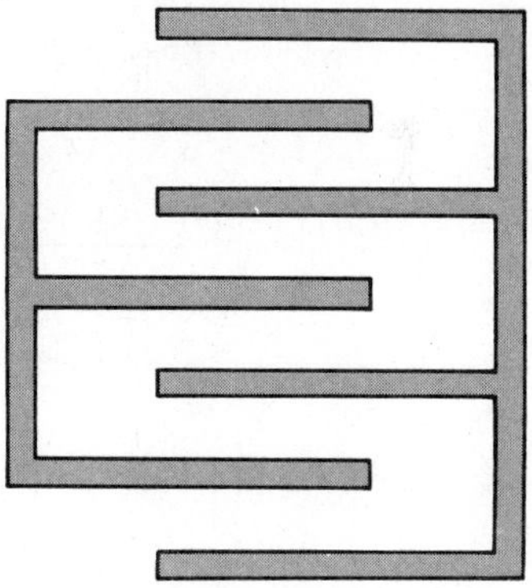

Figure 23.19 (Exercise 13).

Section 23.3 Combinations of Capacitors

14. Two capacitors, $C_1 = 5$ μF and $C_2 = 12$ μF, are connected in parallel. What is the value of the equivalent capacitance of the combination?

15. Calculate the equivalent capacitance of the two capacitors in the previous exercise if they are connected in series.

16. Find the equivalent capacitance between points a and b for the group of capacitors connected as shown in Fig. 23.20 if $C_1 = 3$ μF, $C_2 = 6$ μF, and $C_3 = 2$ μF.

17. For the circuit described in the previous exercise, if the potential between points a and b is 48 V, what charge is stored on the capacitor C_3?

18. Consider the circuit shown in Fig. 23.21, where $C_1 = 6$ μF, $C_2 = 4$ μF, and $V = 22$ V. C_1 is first charged by the closing of switch S_1. Switch S_1 is then opened, and the charged capacitor is connected to the uncharged capacitor by the closing of S_2. Calculate the initial charge acquired by C_1 and the final charge on each of the two capacitors.

19. A 2-μF capacitor charged to 200 V and a 4-μF ca-

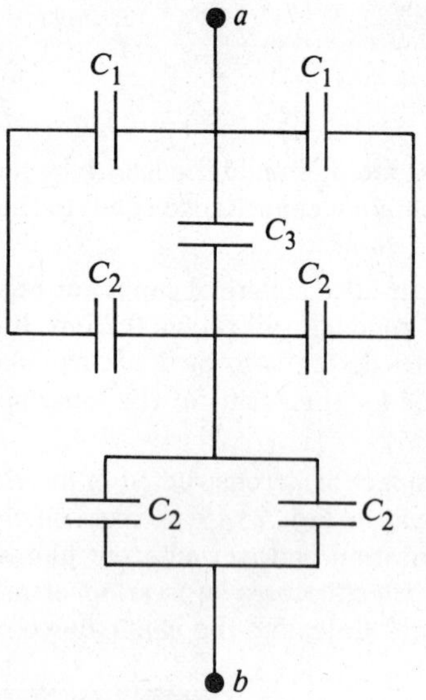

Figure 23.20 (Exercises 16 and 17).

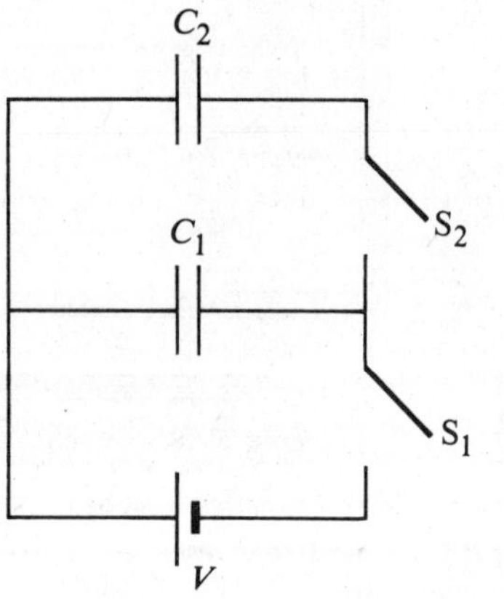

Figure 23.21 (Exercise 18).

pacitor charged to 400 V are connected to each other, with the positive plate of each connected to the negative plate of the other. (a) What is the final value of the charge that resides on each capacitor? (b) What is the potential difference across each capacitor after they have been connected?

20. Consider the group of capacitors shown in Fig. 23.22. (a) Find the equivalent capacitance between

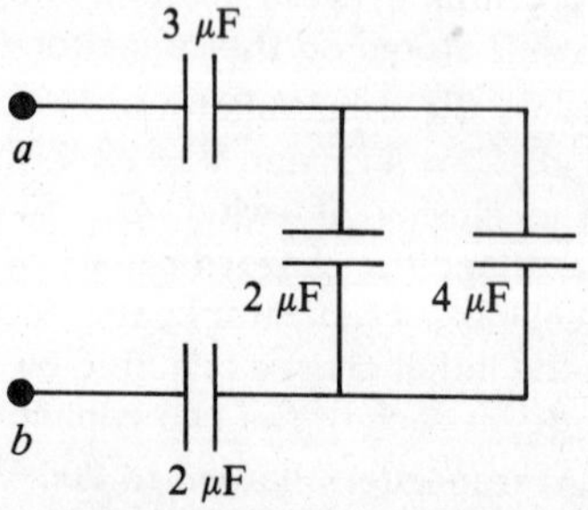

Figure 23.22 (Exercise 20).

points a and b. (b) Determine the charge on each capacitor when the potential difference between a and b is 24 V.

21. Consider the combination of capacitors shown in Fig. 23.23. (a) What is the equivalent capacitance between points a and b? (b) Determine the charge on each capacitor if $V_{ab} = 36$ V.

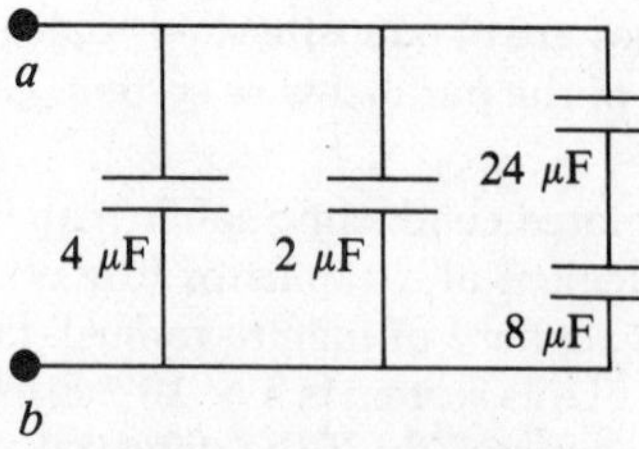

Figure 23.23 (Exercise 21).

22. Four capacitors are connected as shown in Fig. 23.24. (a) Find the equivalent capacitance between points a and b. (b) Calculate the charge on each capacitor if $V_{ab} = 48$ V.

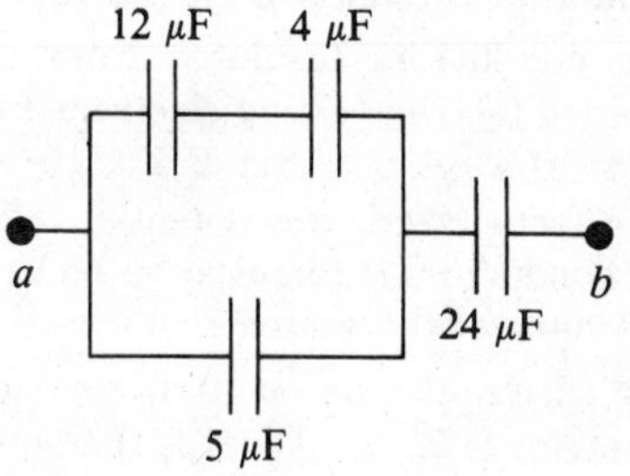

Figure 23.24 (Exercises 22 and 27).

Section 23.4 Energy Stored in a Charged Capacitor

23. A 6-μF parallel-plate capacitor is charged by a 12-V battery. If each plate of the capacitor has an area of 2 cm^2, what is the energy stored in the capacitor? What is the energy density (energy per unit volume) in the electric field of the capacitor?

24. The energy density in a parallel-plate capacitor is given as 2.1×10^{-9} J/m^3. What is the value of the electric field in the region between the plates?

25. Show that the energy associated with a conducting sphere of radius R and charge Q surrounded by a vacuum is given by $U = kQ^2/2R$.

26. A parallel-plate capacitor has a charge Q and plates of area A. Show that the force exerted on each plate by the other is given by $F = Q^2/2\varepsilon_0 A$. (Hint: Use Eq. 23.9 and let $C = \varepsilon_0 A/x$ for an arbitrary plate separation x; then require that the work done in separating the two charged plates be $W = \int F\,dx$.)

27. What total energy is stored in the group of capacitors shown in Fig. 23.24 if $V_{ab} = 48$ V.

28. By what fraction does the energy stored on a charged parallel-plate capacitor change (increase or decrease) when the plate separation is doubled?

29. Calculate the energy stored in a 25-μF capacitor when it is charged to a potential of 120 V.

30. Two capacitors, $C_1 = 16\ \mu$F and $C_2 = 4\ \mu$F, are connected in parallel and charged with a 40-V power supply. (a) Calculate the total energy stored in the two capacitors. (b) What potential difference would be required across the same two capacitors connected in series in order that the combination store the same energy as in (a)?

Section 23.5 Capacitors with Dielectrics and Section 23.7 An Atomic Description of Dielectrics

31. A parallel-plate capacitor has a plate area of 1 cm^2. When the plates are in a vacuum, the capacitance of the device is 2.77 pF. Calculate the value of the capacitance if the space between the plates is filled with nylon. What is the maximum potential difference that can be applied to the plates without causing dielectric breakdown, or discharge?

32. A capacitor is constructed from two square metal plates of side length L and separated by a distance d (Fig. 23.25). One half of the space between the plates (top to bottom) is filled with bakelite ($\kappa =$ 4.9), and the other half is filled with neoprene rubber ($\kappa = 6.7$). Calculate the capacitance of the device, taking $L = 5$ cm and $d = 2$ mm. (Hint: The capacitor can be considered as two capacitors connected in parallel.)

33. A parallel-plate capacitor is to be constructed using paper as a dielectric. If a maximum voltage before breakdown of 6×10^4 V is desired, what thickness of dielectric is needed?

34. A commercial capacitor is constructed as shown in Fig. 23.10a. This particular capacitor is "rolled" from two strips of aluminum separated by two strips of paraffin-coated paper. Each strip of foil and paper is 5 cm wide. The foil is 0.0005 cm thick, and the paper is 0.002 cm thick and has a dielectric constant of 3. What length should the strips be if a capacitor of 2×10^{-8} F is desired? (Use the parallel-plate formula.)

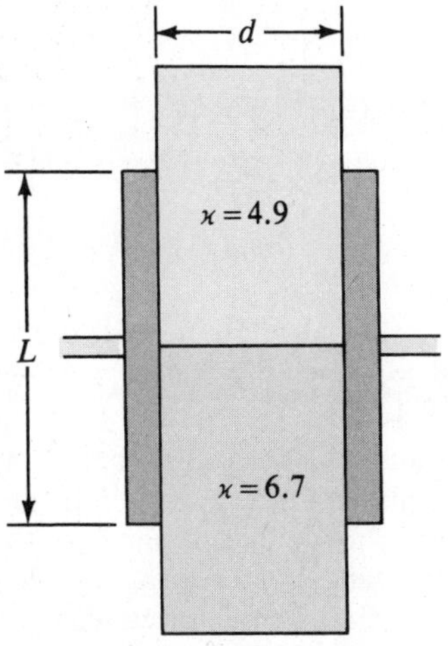

Figure 23.25 (Exercise 32).

35. A capacitor with air between the plates is charged to 100 V and then disconnected from the battery. When a piece of glass is placed between the plates, the voltage across the capacitor drops to 25 V. What is the dielectric constant of the glass? (Assume the glass completely fills the space between the plates.)

PROBLEMS

1. It is possible to obtain large potential differences by first charging a group of capacitors connected in parallel and then activating a switch arrangement that in effect disconnects the capacitors from the charging source and from each other and reconnects them in a *series* arrangement. The group of charged capacitors is then discharged in *series.* What is the maximum potential difference that can be obtained in this manner by using ten capacitors each of 500 μF and a charging source of 800 V?

2. For the system of capacitors shown in Fig. 23.26, find (a) the equivalent capacitance of the system, (b) the potential across each capacitor, (c) the charge on each capacitor, and (d) the total energy stored by the group.

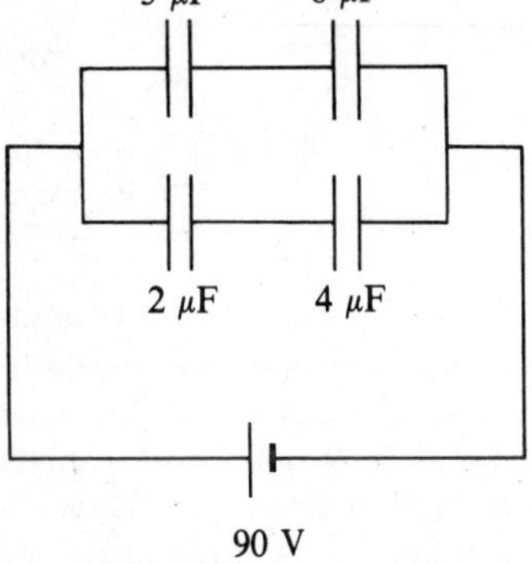

Figure 23.26 (Problem 2).

3. A parallel-plate capacitor is constructed using three different dielectric materials, as shown in Fig. 23.27. (a) Find an expression for the capacitance of the device in terms of the plate area A and d, κ_1, κ_2,

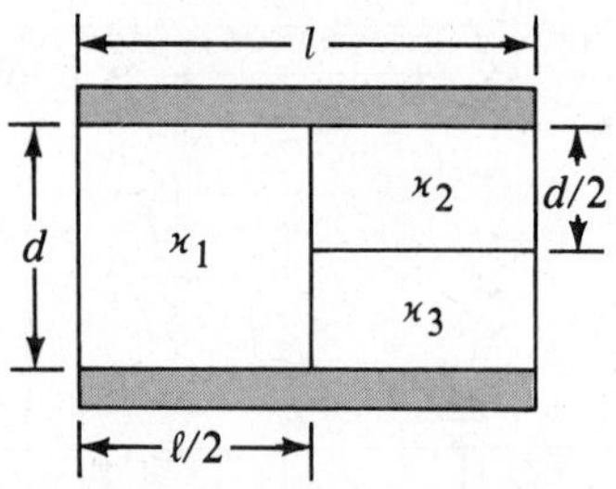

Figure 23.27 (Problem 3).

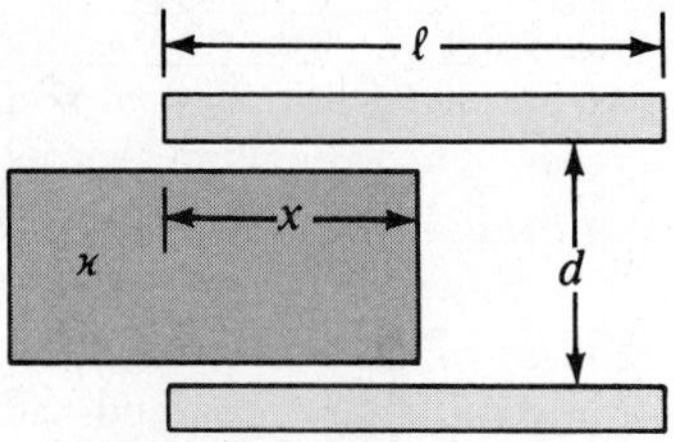

Figure 23.29 (Problem 7).

and κ_3. (b) Calculate the capacitance using the values $A = 1\ \text{cm}^2$, $d = 2$ mm, $\kappa_1 = 4.9$, $\kappa_2 = 5.6$, and $\kappa_3 = 2.1$.

4. A parallel-plate capacitor is to be constructed using Pyrex glass as a dielectric. If the capacitance of the device is to be 0.2 μF and it is to be operated at 6000 V, calculate the minimum required plate area. What is the energy stored in the capacitor at the operating voltage?

5. When a certain air-filled parallel-plate capacitor is connected across a battery, it acquires a charge (on each plate) of 150 μC. While the battery connection is maintained, a dielectric slab is inserted into and fills the region between the plates. This results in the accumulation of an *additional* charge of 200 μC on each plate. What is the dielectric constant of the dielectric slab?

6. The arrangement shown in Fig. 23.28 is known as a *capacitance bridge.* A potential V is applied as shown, and C_1 is adjusted so that the galvanometer G between points b and d reads zero. This "balance" occurs when $C_1 = 4\ \mu$F. If $C_3 = 9\ \mu$F and $C_4 = 12\ \mu$F, calculate the value of C_2.

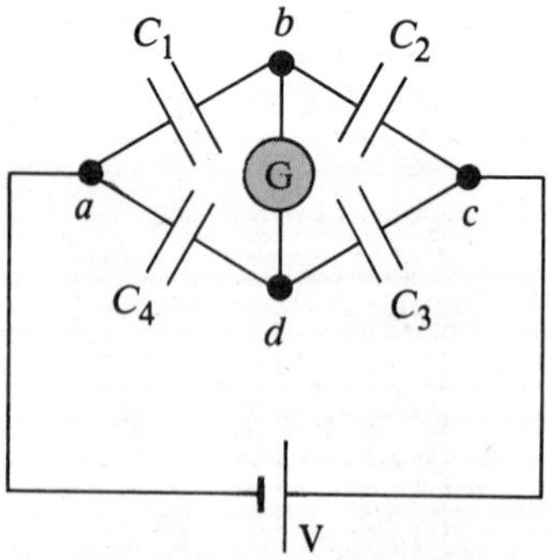

Figure 23.28 (Problem 6).

7. A capacitor is constructed from two square plates of sides l and separation d. A material of dielectric constant κ is inserted a distance x into the capacitor, as in Fig. 23.29. (a) Find the equivalent capacitance of the device. (b) Calculate the energy stored in the capacitor if the potential difference is V. (c) Find the direction and magnitude of the force exerted on the dielectric, assuming a constant potential difference V. Neglect friction and edge effects. (d) Obtain a numerical value for the force assuming $l = 5$ cm, $V = 2000$ V, $d = 2$ mm, and the dielectric is glass ($\kappa = 4.5$). (Hint: The system can be considered as two capacitors connected in *parallel.* Also, recall that $F = -dU/dx$.)

8. An air-dielectric capacitor is formed by two *nonparallel* plates, each of area A. An edge view of the arrangement is shown in Fig. 23.30. Note that the top plate is tilted relative to the bottom plate so that on one edge the plate separation is $d + \Delta d$, while on the other edge it is $d - \Delta d$. Assuming that $\Delta d \ll d$ and that d is small compared with the length of the plate, show that $C = \frac{\varepsilon_0 A}{d}\left[1 + \frac{1}{3}\left(\frac{\Delta d}{d}\right)^2\right]$.

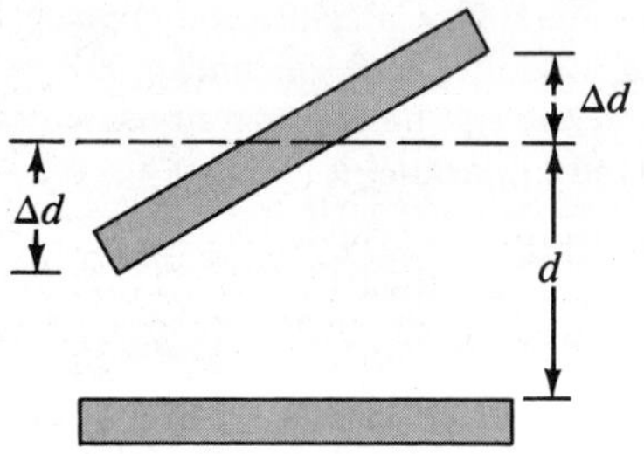

Figure 23.30 (Problem 8).

9. A capacitor $C_1 = 4\ \mu$F is charged to a potential difference of 800 V. The capacitor is then removed from the charging source, and each plate of the charged capacitor is connected to one of the plates of an *uncharged* capacitor $C_2 = 6\ \mu$F. (a) What is the resulting charge on each capacitor? (b) What is the total electrostatic energy associated with the two capacitors *before* and *after* they are connected?

10. Capacitors $C_1 = 4\ \mu$F and $C_2 = 2\ \mu$F are charged as a series combination across a 100-V battery. The two capacitors are disconnected from the battery and from each other. They are then connected positive plate to positive plate and negative plate to negative plate. Calculate the resulting charge on each capacitor.

11. Capacitors $C_1 = 6\ \mu$F and $C_2 = 2\ \mu$F are charged as a parallel combination across a 250-V battery. The capacitors are disconnected from the battery and from each other. They are then connected positive plate to negative plate and negative plate to positive plate. Calculate the resulting charge on each capacitor.

12. A parallel-plate capacitor of plate separation d is charged to a potential difference V_0. A dielectric slab of thickness d and dielectric constant κ is introduced between the plates *while the battery remains connected to the plates.* (a) Show that the ratio of energy stored after the dielectric is introduced to the energy stored in the empty capacitor is given by $U/U_0 = \kappa$. Give a physical explanation for this increase in stored energy. (b) What happens to the charge on the capacitor? (Note that this situation is not the same as Example 23.7, in which the battery was removed from the circuit before introducing the dielectric.)

13. Figure 23.31 shows two capacitors in series. The rigid center section of length b is movable vertically, and the area of each plate is A. Show that the capacitance of the series combination is independent of the position of the center section and is given by $C = \dfrac{A\varepsilon_0}{a - b}$.

14. Calculate the equivalent capacitance between the points a and b in Fig. 23.32. (Hint: Assume a potential difference V between points a and b. Write expressions for V_{ab} in terms of the charges and capacitances for the various possible pathways from a to b, and require conservation of charge for those capacitor plates that are connected to each other.)

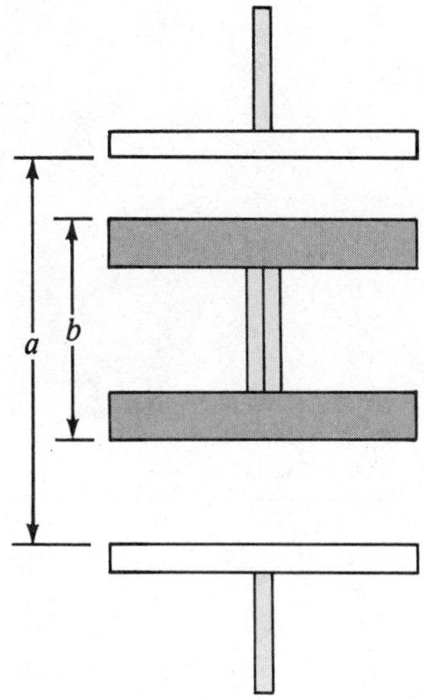

Figure 23.31 (Problem 13).

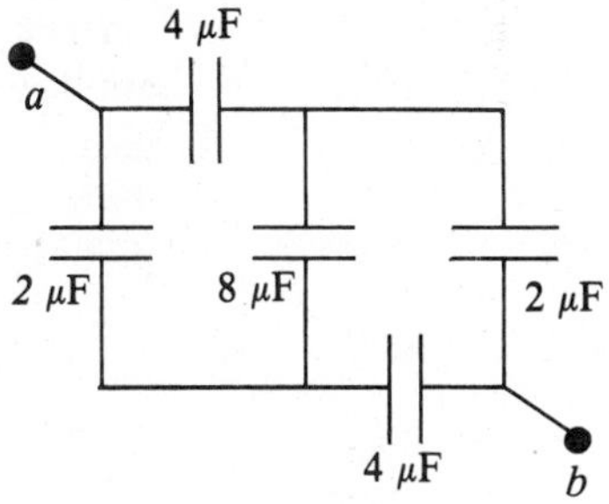

Figure 23.32 (Problem 14).

24 Current and Resistance

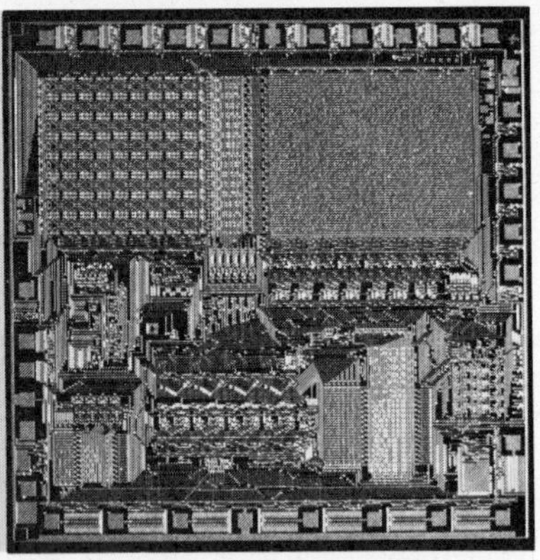

Texas Instruments, Inc.

Thus far our discussion of electrical phenomena has been confined to charges at rest, or electrostatics. We shall now consider situations involving electric charges in motion. The term *electric current,* or simply *current,* is used to describe the rate of flow of charge through some region of space. Most practical applications of electricity deal with electric currents. For example, the battery of a flashlight supplies current to the filament of the bulb when the switch is turned on. A variety of home appliances operate on alternating current. In these common situations, the flow of charge takes place in a conductor, such as a copper wire. However, it is possible for currents to exist outside of a conductor. For instance, a beam of electrons in a TV picture tube constitutes a current.

In this chapter we shall first define current and current density. A microscopic description of current will be given, and some of the factors that contribute to the resistance to the flow of charge in conductors will be discussed. Mechanisms responsible for the electrical resistance of various materials depend on the composition of the material and on temperature. A classical model is used to describe electrical conduction in metals, and some of the limitations of this model are pointed out. Finally, the electrical properties of semiconductors are described qualitatively using a model based on modern concepts.

24.1 ELECTRIC CURRENT AND CURRENT DENSITY

Whenever electric charges move, a current is said to exist. To define current more precisely, consider a surface of area A through which charges are moving as in Fig. 24.1. The current is the rate at which charge flows through this surface. If ΔQ is the net charge that passes through this area in a time interval Δt, the current I is given by

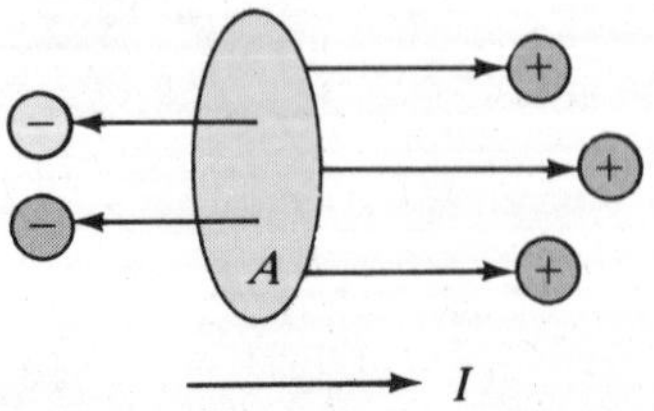

Figure 24.1 Charges in motion through an area A. The time rate of flow of charge through the area is defined as the current I. The direction of the current is in the direction of flow of positive charges.

$$I = \frac{\Delta Q}{\Delta t} \tag{24.1}$$

If the rate at which charge flows varies in time, the current also varies in time and we must write I as the differential limit of the expression above:

Electric current

$$I = \frac{dQ}{dt} \tag{24.2}$$

The SI unit of current is the ampere (A), where

$$1\text{ A} = 1\text{ C/s} \tag{24.3}$$

That is, 1 A of current is equivalent to 1 C of charge passing through the surface in 1 s. In practice, smaller units of current are often used, such as the milliampere ($1\text{ mA} = 10^{-3}\text{ A}$) and the microampere ($1\ \mu\text{A} = 10^{-6}\text{ A}$).

When charges flow through the surface in Fig. 24.1, they can be positive, negative, or both. *It is conventional to choose the direction of the current to be in the direction of flow of positive charge.* In a conductor such as copper, the current is due to the motion of the negatively charged electrons. Therefore, when we speak of current in an ordinary conductor, such as a copper wire, *the direction of the current will be opposite the flow of electrons.* On the other hand, if one considers a beam of positively charged protons in an accelerator, the current is in the direction of motion of the protons. In some cases, the current is the result of the flow of both positive and negative charges. This occurs, for example, in semiconductors, as discussed in Section 24.6. It is common to refer to a moving charge (whether it is positive or negative) as a mobile *charge carrier*. For example, the charge carriers in a metal are electrons.

The direction of the current

It is instructive to relate current to the motion of the charged particles. To illustrate this point, consider the current in a conductor of cross-sectional area A (Fig. 24.2). The volume of an element of the conductor of length Δx (the shaded region in Fig. 24.2) is given $A\,\Delta x$. If n represents the number of mobile charge carriers per unit volume, then the number of mobile charge carriers in the volume element is given by $nA\,\Delta x$. Therefore, the charge ΔQ in this element is given by

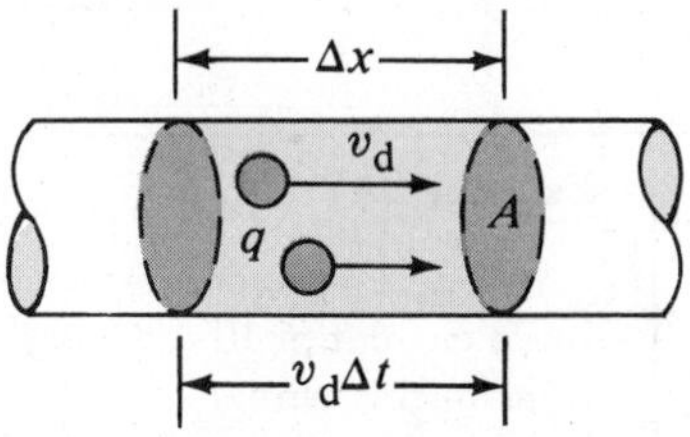

Figure 24.2 A section of a uniform conductor of cross-sectional area A. The charge carriers move with a speed v_d, and the distance they travel in a time Δt is given by $\Delta x = v_d\,\Delta t$. The number of mobile charge carriers in the section of length Δx is given by $nAv_d\,\Delta t$, where n is the number of mobile carriers per unit volume.

$$\Delta Q = (nA\,\Delta x)q$$

where q is the charge on each particle. If the charge carriers move with a speed v_d, the distance they move in a time Δt is given by $\Delta x = v_d\,\Delta t$. Therefore, we can write ΔQ in the form

$$\Delta Q = (nAv_d\,\Delta t)q$$

Using Eq. 24.1, we see that the current in the conductor is given by

$$I = \frac{\Delta Q}{\Delta t} = nqv_dA \tag{24.4}$$

Current in a conductor

The velocity of the charge carriers, $\boldsymbol{v}_d$, is actually an average velocity and is called the *drift velocity*. To understand the meaning of drift velocity, consider a conductor in which the charge carriers are free electrons. In an isolated conductor, these electrons undergo random motion similar to that of gas molecules. When a potential difference is applied across the conductor (say, by means of a battery), an electric field is set up in the conductor, which creates an electric force on the electrons and hence a current. In reality, the electrons do not simply move in straight lines along the conductor. Instead, they undergo repeated collisions with the metal atoms, which results in a complicated zigzag motion (Fig. 24.3). The energy transferred from the electrons to the metal atoms causes an increase in the vibrational energy of the atoms and a corresponding increase in the temperature of the conductor. However, despite the collisions, the electrons move slowly along the conductor (in a direction opposite $\boldsymbol{E}$) with an average velocity called the drift velocity, $\boldsymbol{v}_d$. As we shall see in an example that follows, drift velocities are *much* smaller than the average speed between collisions. We shall discuss this model in more detail in Section 24.3. One can think of the collisions of the electrons within a conductor as being an effective internal friction (or drag force), similar to that

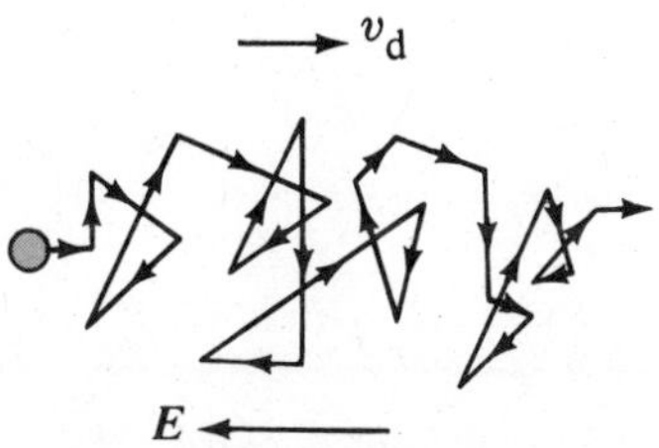

Figure 24.3 A schematic representation of the zigzag motion of a charge carrier in a conductor. The changes in direction are due to collisions with atoms in the conductor. Note that the net motion of electrons is opposite the direction of the electric field.

experienced by the molecules of a liquid flowing through a pipe stuffed with steel wool.

The following quotation is an interesting and amusing description by W.F.G. Swann of electronic conduction in telephone cables.[1]

> Think of the cables which carry the telephone current in the form of electrons. In the absence of the current the electrons are moving in all directions. As many are moving from left to right as are moving from right to left; and the nothingness which is there is composed of two equal and opposite halves, about a million million amperes per square centimeter in one direction, and a million million amperes per square centimeter in the other direction. The telephone current constitutes an upsetting of the balance to the extent of one hundredth of a millionth of an ampere per square centimeter, or about one part in a hundred million million million. Then if this one part in a hundred million million million is at fault by one part in a thousand, we ring up the telephone company and complain that the quality of the speech is faulty.

Example 24.1

A copper wire of cross-sectional area 3×10^{-6} m² carries a current of 10 A. Find the drift velocity of the electrons in this wire.

Solution: We can calculate the density of free electrons by assuming one free electron per copper atom. In that case, the density of free electrons, n, equals the density of atoms, given by $N_0\rho/M$, where N_0 is Avogadro's number, M is the atomic weight of copper, and ρ is its density. That is,

$$n = \frac{N_0\rho}{M}$$

$$= \frac{(6.02 \times 10^{23}\ \text{atoms/mole})(8.95\ \text{g/cm}^3)(1\ \text{electron/atom})}{63.5\ \text{g/mole}}$$

$$= 8.48 \times 10^{22}\ \text{electrons/cm}^3 = 8.48 \times 10^{28}\ \text{electrons/m}^3$$

Using this result and the fact that $I = nqv_dA$, we find that the drift velocity v_d is given by

$$v_d = \frac{I}{neA}$$

$$= \frac{10\ \text{C/s}}{(8.48 \times 10^{28}\ \text{m}^{-3})(1.6 \times 10^{-19}\ \text{C})(3 \times 10^{-6}\ \text{m}^2)}$$

$$= 2.46 \times 10^{-4}\ \text{m/s}$$

This shows that typical drift velocities are very small. For instance, electrons moving with this velocity would take about 68 min to travel 1 m! In view of this low speed, one might wonder why our lights turn on almost instantaneously when a switch is thrown. This apparent discrepancy is resolved by noting that changes in the electric field (that drives the free electrons and produces a current) travel in the conductor with a speed close to the speed of light. The electrons behave much like a nearly incompressible liquid in a pipe. While it may take the liquid a considerable time to traverse the length of the pipe, a flow initiated at one end produces a similar flow at the far end as quickly as a pressure wave can traverse the pipe.

24.2 RESISTANCE AND OHM'S LAW

Charges move in a conductor to produce a current under the action of an electric field inside the conductor. An electric field can exist in the conductor in this case since we are dealing with charges in motion, a *nonelectrostatic* situation. This is in contrast with the situation in which a conductor in *electrostatic equilibrium* (where the charges are at rest) can have no electric field inside.

Consider a conductor of cross-sectional area A carrying a current I. The *current density* J in the conductor is defined to be the current per unit area. Since $I = nqv_dA$, the current density is given by

Current density

$$J \equiv \frac{I}{A} = nqv_d \tag{24.5}$$

[1] W.F.G. Swann, *Physics Today,* June 1951, p. 9.

where J has SI units of A/m^2. In general, the current density is a *vector quantity* in the direction of the drift velocity, $\boldsymbol{v}_d$. That is,

$$J = nq\boldsymbol{v}_d \tag{24.6}$$

From this definition, we see once again that the current density is in the direction of motion of the charges for positive charge carriers and opposite the direction of motion for negative charge carriers.

A current density **J** *and an electric field* **E** *are established in a conductor when a potential difference is maintained across the conductor.* If the potential difference is constant, the current in the conductor will also be constant. Very often, the current density in a conductor is proportional to the electric field in the conductor. That is,

Ohm's law

$$J = \sigma E \tag{24.7}$$

where the constant of proportionality σ is called the *conductivity* of the conductor.[2] Materials that obey Eq. 24.7 are said to follow Ohm's law, named after George Simon Ohm (1787–1854). More specifically, *Ohm's law states that for many materials (including most metals), the ratio of the current density and electric field is a constant, σ, which is independent of the electric field producing the current.* Materials that obey Ohm's law, and hence demonstrate this linear behavior between **E** and *J*, are said to be *ohmic*. Experimentally, one finds that not all materials have this property. Materials that do not obey Ohm's law are said to be *nonohmic*. Ohm's law is *not* a fundamental law of nature, but an empirical relationship valid only for certain materials.

A form of Ohm's law that is more directly useful in practical applications can be obtained by considering a segment of a straight wire of cross-sectional area A and length l, as in Fig. 24.4. A potential difference $V_b - V_a$ is maintained across the wire, creating an electric field in the wire and a current. If the electric field in the wire is assumed to be uniform, the potential difference $V = V_b - V_a$ is related to the electric field through the relationship[3]

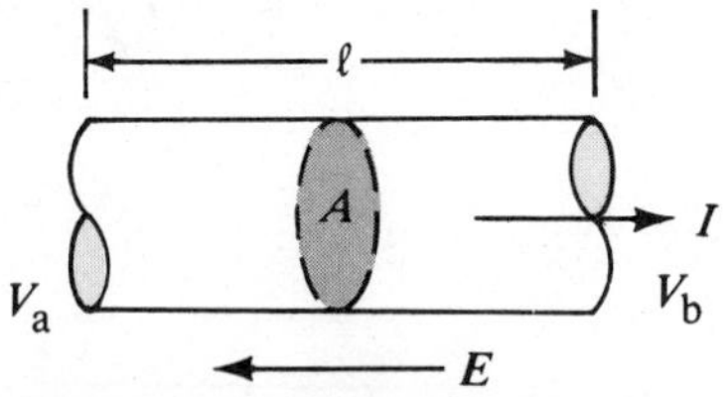

Figure 24.4 A uniform conductor of length l and cross-sectional area A. A potential difference $V_b - V_a$ maintained across the conductor sets up an electric field E in the conductor, and this field produces a current I.

$$V = El$$

Therefore, we can express the magnitude of the current density in the wire as

$$J = \sigma E = \sigma \frac{V}{l}$$

Since $J = I/A$, the potential difference can be written

$$V = \frac{l}{\sigma} J = \left(\frac{l}{\sigma A}\right) I$$

The quantity $l/\sigma A$ is called the *resistance* R of the conductor:

Resistance of a conductor

$$R = \frac{l}{\sigma A} = \frac{V}{I} \tag{24.8}$$

From this result we see that resistance has SI units of volts per ampere. One volt per ampere is defined to be one ohm (Ω):

$$1\,\Omega \equiv 1\,V/A$$

[2]Do not confuse the conductivity σ with the surface charge density, for which the same symbol is used.
[3]This result follows from the definition of potential difference:

$$V_b - V_a = -\int_a^b \mathbf{E} \cdot d\mathbf{s} = E\int_0^l dx = El$$

That is, if a potential difference of 1 V across a conductor causes a current of 1 A, the resistance of the conductor is 1 Ω.

The inverse of the conductivity of a material is called the *resistivity* ρ:

Resistivity

$$\rho \equiv \frac{1}{\sigma} \tag{24.9}$$

Using this definition and Eq. 24.8, the resistance can be expressed as

Resistance of a uniform conductor

$$R = \rho \frac{l}{A} \tag{24.10}$$

where ρ has the units ohm-meters ($\Omega \cdot \text{m}$). Every material has a characteristic resistivity, a parameter that depends only on the properties of the material. On the other hand, as you can see from Eq. 24.10, the resistance of a substance depends on simple geometry as well as on the resistivity of the substance. Good electrical conductors have very low resistivity (or high conductivity), and good insulators have very high resistivity (low conductivity). Table 24.1 gives the resistivities of a variety of materials at 20°C.

TABLE 24.1 Resistivities and Temperature Coefficients of Resistivity for Various Materials

MATERIAL	*RESISTIVITY*[a] ($\Omega \cdot$m)	*TEMPERATURE COEFFICIENT*[a] α [(C°)$^{-1}$]
Silver	1.59×10^{-8}	3.8×10^{-3}
Copper	1.7×10^{-8}	3.9×10^{-3}
Gold	2.44×10^{-8}	3.4×10^{-3}
Aluminum	2.82×10^{-8}	3.9×10^{-3}
Tungsten	5.6×10^{-8}	4.5×10^{-3}
Iron	10×10^{-8}	5.0×10^{-3}
Platinum	11×10^{-8}	3.92×10^{-3}
Lead	22×10^{-8}	3.9×10^{-3}
Nichrome[b]	150×10^{-8}	0.4×10^{-3}
Carbon	3.5×10^{-5}	-0.5×10^{-3}
Germanium	0.46	-48×10^{-3}
Silicon	640	-75×10^{-3}
Glass	10^{10}–10^{14}	
Hard rubber	10^{13}–10^{6}	
Sulfur	10^{15}	
Quartz (fused)	75×10^{16}	

[a]All values at 20°C. [b]A nickel-chromium alloy commonly used in heating elements.

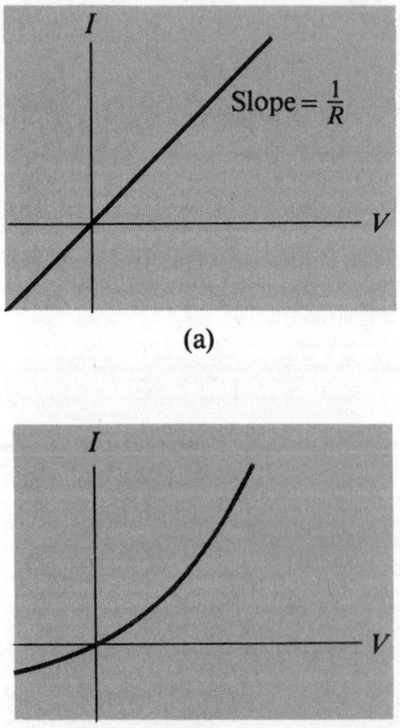

Figure 24.5 (a) The current-voltage curve for an ohmic material. The curve is linear, and the slope gives the resistance of the conductor. (b) A nonlinear current-voltage curve for a semiconducting diode. This device does not obey Ohm's law.

Equation 24.10 shows that the resistance of a given cylindrical conductor is proportional to its length and inversely proportional to its cross-sectional area. Therefore, if the length of a wire is doubled, its resistance doubles. Furthermore, as its cross-sectional area increases, its resistance decreases. The situation is analogous to the flow of a liquid through a pipe. As the length of the pipe is increased, the resistance to liquid flow increases. As its cross-sectional area is increased, the pipe can more readily transport liquid.

Ohmic materials, such as copper, have a linear current-voltage relationship over a large range of applied voltage (Fig. 24.5a). The slope of the I versus V curve in the linear region yields a value for R. Nonohmic materials have a nonlinear current-voltage relationship. One common semiconducting device that has nonlinear I versus V characteristics is the diode (Fig. 24.5b). The effective resistance of this device (inversely proportional to the slope of its I versus V curve) is small for currents in one direction (positive V) and large for currents in the reverse direction (negative

V). In fact, most modern electronic devices, such as transistors, have nonlinear current-voltage relationships; their proper operation depends on the particular way in which they violate Ohm's law. We will discuss these further in Section 24.7.

Example 24.2

(a) Calculate the resistance per unit length of a 22-gauge nichrome wire of radius 0.321 mm.

The cross-sectional area of this wire is

$$A = \pi r^2 = \pi(0.321 \times 10^{-3}\ \text{m})^2 = 3.24 \times 10^{-7}\ \text{m}^2$$

We can use Eq. 24.10 to find the resistance per unit length, noting that the resistivity of nichrome is $1.5 \times 10^{-6}\ \Omega \cdot \text{m}$ (Table 24.1):

$$\frac{R}{l} = \frac{\rho}{A} = \frac{1.5 \times 10^{-6}\ \Omega \cdot \text{m}}{3.24 \times 10^{-7}\ \text{m}^2} = 4.63\ \Omega/\text{m}$$

(b) If a potential difference of 10 V is maintained across a 1-m length of the nichrome wire, what is the current in the wire?

Since a 1-m length of this wire has a resistance of 4.63 Ω, Eq. 24.8 gives

$$I = \frac{V}{R} = \frac{10\ \text{V}}{4.63\ \Omega} = 2.16\ \text{A}$$

(c) Calculate the current density and the electric field in the wire assuming that the current it carries is 2.16 A.

Using Eq. 24.5 we have

$$J = \frac{I}{A} = \frac{2.16\ \text{A}}{3.24 \times 10^{-7}\ \text{m}^2} = 6.67 \times 10^6\ \frac{\text{A}}{\text{m}^2}$$

Since $J = \sigma E$ and $\sigma = \dfrac{1}{\rho}$, the electric field in the wire is given by

$$E = \frac{J}{\sigma} = \rho J$$

$$= (1.5 \times 10^{-6}\ \Omega \cdot \text{m})\left(6.67 \times 10^6\ \frac{\text{A}}{\text{m}^2}\right) = 10.0\ \frac{\text{N}}{\text{C}}$$

Note that copper has a resistivity of $1.7 \times 10^{-8}\ \Omega \cdot \text{m}$, about 10^{-2} that of nichrome. Therefore, a copper wire of the same gauge would have a resistance per unit length of only 0.052 Ω/m. A 1-m length of such copper wire would carry the same current (2.16 A) with an applied voltage of only 0.11 V.

Q1. What is the difference between resistance and resistivity?

Q2. What factors affect the resistance of a conductor?

Q3. Two wires A and B are made of the same metal and have equal lengths, but the resistance of wire A is three times greater than that of wire B. What is the ratio of their cross-sectional areas? How do their radii compare?

Q4. When the voltage across a certain conductor is doubled, the current is observed to increase by a factor of 3. What can you conclude about the conductor?

Q5. We have seen that an electric field must exist inside a conductor that carries a current. How is this possible in view of the fact that in *electrostatics,* we concluded that E must be zero inside a conductor?

Q6. If you were to design an electric heater using nichrome wire as the heating element, what parameters of the wire could you vary to meet a specific power output, such as 1000 W?

24.3 THE RESISTIVITY OF DIFFERENT CONDUCTORS

The resistivity of a conductor depends on a number of factors, one of which is temperature. For most metals, resistivity increases with increasing temperature. The resistivity of a conductor varies in an approximately linear fashion with temperature over a limited temperature range according to the expression

$$\rho = \rho_o[1 + \alpha(T - T_o)] \qquad (24.11)$$

Variation of ρ with temperature

where ρ is the resistivity at some temperature T (in °C), ρ_0 is the resistivity at some reference temperature T_0 (usually taken to be 20°C), and α is called the *temperature coefficient of resistivity*. From Eq. 24.11, we see that the temperature coefficient of resistivity can also be expressed as

Temperature coefficient of resistivity

$$\alpha = \frac{1}{\rho_0}\frac{\Delta\rho}{\Delta T} \tag{24.12}$$

where $\Delta\rho = \rho - \rho_0$ is the change in resistivity in the temperature interval $\Delta T = T - T_0$.

The resistivities and temperature coefficients for various materials are given in Table 24.1. Note the enormous range in resistivities, from very low values for good conductors, such as copper and silver, to very high values for good insulators, such as glass and rubber. An ideal, or "perfect," conductor would have zero resistivity, and an ideal insulator would have infinite resistivity.

Since the resistance of a conductor is proportional to the resistivity according to Eq. 24.10, the temperature variation of the resistance can be written

$$R = R_0[1 + \alpha(T - T_0)] \tag{24.13}$$

Precise temperature measurements are often made using this property, as shown in the following example.

Example 24.3 *A Platinum Resistance Thermometer*

A resistance thermometer made from platinum has a resistance of 50.0 Ω at 20°C. When immersed in a vessel containing melting indium, its resistance increases to 76.8 Ω. From this information, find the melting point of indium. Note that $\alpha = 3.92 \times 10^{-3}\ (\mathrm{C}°)^{-1}$ for platinum.

Solution: Using Eq. 24.13 and solving for ΔT, we get

$$\Delta T = \frac{R - R_0}{\alpha R_0} = \frac{76.8\ \Omega - 50.0\ \Omega}{[3.92 \times 10^{-3}\ (\mathrm{C}°)^{-1}](50.0\ \Omega)}$$

$$= 137\ \mathrm{C}°$$

Since $\Delta T = T - T_0$ and $T_0 = 20°\mathrm{C}$, we find that $T = 157°\mathrm{C}$.

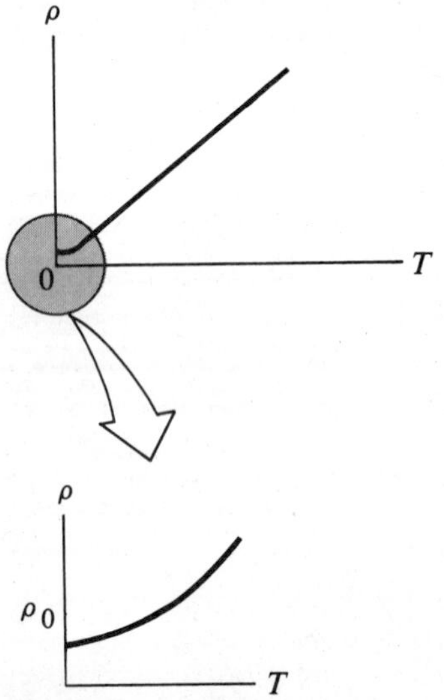

Figure 24.6 Resistivity versus temperature for a normal metal, such as copper. The curve is linear over a wide range of temperatures, and ρ increases with increasing temperature. As T approaches absolute zero (insert), the resistivity approaches a finite value ρ_0.

As mentioned above, many ohmic materials have resistivities that increase linearly with increasing temperature, as shown in Fig. 24.6. In reality, however, there is always a nonlinear region at very low temperatures, and the resistivity usually approaches some finite value near absolute zero (see magnified insert in Fig. 24.6). This residual resistivity near absolute zero is due primarily to collisions of electrons with impurities and imperfections in the metal. In contrast, the high-temperature resistivity (the linear region) is dominated by collisions of electrons with the metal atoms. We shall describe this process in more detail in Section 24.4.

Semiconductors, such as silicon and germanium, have intermediate values of resistivity. The resistivity of semiconductors generally decreases with increasing temperature, corresponding to a negative temperature coefficient of resistivity (Fig. 24.7). This is due to the increase in the density of charge carriers at the higher temperatures. Since the charge carriers in a semiconductor are often associated with impurity atoms, the resistivity is very sensitive to the type and concentration of such impurities. This is described in more detail in Section 24.6. The *thermistor* is a semiconducting thermometer that makes use of the large changes in its resistivity with temperature.

There is a class of metals and compounds whose resistivity goes to *zero* below a certain temperature T_c, called the *critical temperature.* These materials are known as *superconductors*. The resistivity-temperature graph for a superconductor follows that of a normal metal at temperatures above T_c (Fig. 24.8). When the temperature is at or below T_c, the resistivity drops suddenly to zero. This phenomenon was discovered in 1911 by the Dutch physicist H. Kamerlingh Onnes for mercury, which

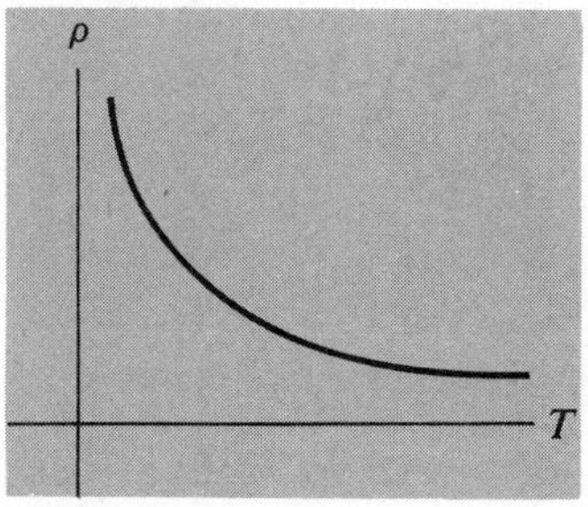

Figure 24.7 Resistivity versus temperature for a pure semiconductor, such as silicon or germanium.

is a superconductor below 4.2 K. Recent measurements have shown that the resistivities of superconductors below T_c are less than $4 \times 10^{-25}\ \Omega \cdot \text{m}$, which is around 10^{17} times smaller than the resistivity of copper!

Today there are thousands of known superconductors, with critical temperatures as high as 23 K for one alloy made of a combination of niobium, aluminum, and germanium. Such common metals as aluminum, tin, lead, zinc, and indium are also superconductors. The critical temperature is sensitive to chemical composition, pressure, and crystalline structure.

One of the truly remarkable features of superconductors is the fact that once a current is set up in the material, the current will persist *without any applied voltage* (since $R = 0$). In fact, steady currents have been observed to persist in superconducting loops for several years with no apparent decay! Another property of superconductors is the fact that a magnetic field is expelled from the interior of the superconductor during the normal-to-superconducting transition.

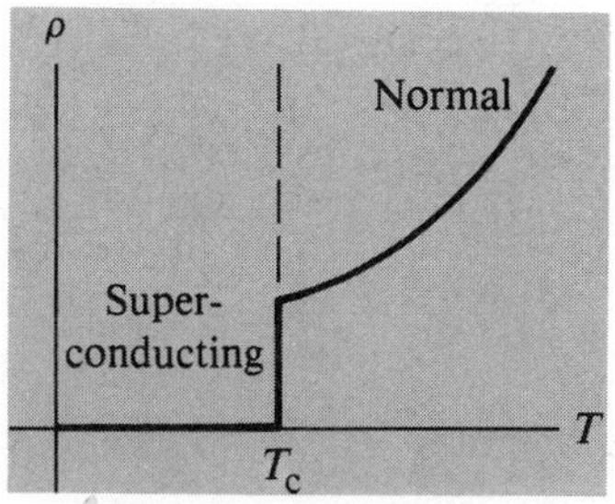

Figure 24.8 Resistivity versus temperature for a pure superconductor, such as mercury, lead, or aluminum. The metal behaves normally down to the critical temperature T_c, below which the resistivity drops to zero.

One important and useful application of superconductivity has been the construction of superconducting magnets, in which the magnetic field intensities are about ten times greater than those of the best normal electromagnets. The idea of using superconducting power lines for transmitting power efficiently is receiving some consideration. Superconducting magnets are also being considered as a means of storing energy. Modern superconducting electronic devices consisting of two thin-film superconductors separated from each other by a thin insulator have been constructed. These devices include magnetometers (a magnetic-field measuring device), ultrafast computer switches, and various microwave devices.

Q7. Why do incandescent lamps usually burn out just after they are switched on?

Q8. What is required in order to maintain a steady current in a conductor?

Q9. Do all conductors obey Ohm's law? Give examples to justify your answer.

Q10. What single experimental requirement makes superconducting devices expensive to operate? In principle, can this limitation be overcome?

24.4 ELECTRICAL ENERGY AND POWER

If a battery is used to establish an electric current in a conductor, there is a continuous transformation of chemical energy stored in the battery to kinetic energy of the charge carriers. This kinetic energy is quickly lost as a result of collisions between the charge carriers and the lattice ions, resulting in an increase in the temperature of the conductor. Therefore, we see that the chemical energy stored in the battery is continuously transformed into thermal energy.

Consider a simple circuit consisting of a battery whose terminals are connected to

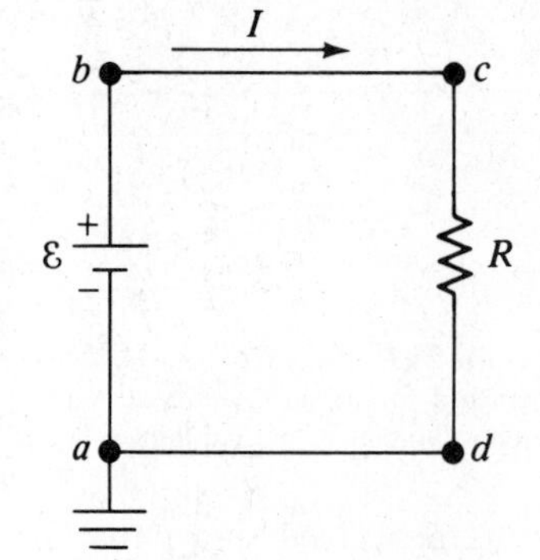

Figure 24.9 A circuit consisting of a battery of emf $\mathcal{E}$ and resistance R. Positive charge flows in the clockwise direction, from the negative to the positive terminal of the battery. Points a and b are grounded.

a resistor R, as shown in Fig. 24.9. The symbol is used to designate a battery (or any other direct current source), and resistors are designated by the symbol. The positive terminal of the battery is at the higher potential. Now imagine following a positive quantity of charge ΔQ moving around the circuit from point a through the battery and resistor and back to a. Point a is a reference point that is grounded (ground symbol), and its potential is taken to be zero. As the charge moves from a to b through the battery, its electrical potential energy *increases* by an amount $V\Delta Q$ (where V is the potential at b) while the chemical potential energy in the battery *decreases* by the same amount. (Recall from Chapter 22 that $\Delta U = q\Delta V$.) However, as the charge moves from c to d through the resistor, it *loses* this electrical potential energy as it undergoes collisions with atoms in the resistor, thereby producing thermal energy. Note that there is no loss in energy for paths bc and da if we neglect the resistance of the interconnecting wires. When the charge returns to point a, it must have the same potential energy (zero) as it had at the start. If this were not the case, the charge would gain energy during each trip around the circuit, which would be a violation of energy conservation.[4] A mechanical system that is somewhat analogous to this circuit is liquid flow through a pipe (Fig. 24.10). The water pump that raises the liquid represents the battery, the liquid flow is the current, and the funnel filled with steel wool represents the resistor. Although such an analogy may be helpful in understanding electrical circuits, you must remember that electrical and fluid properties differ in many ways.

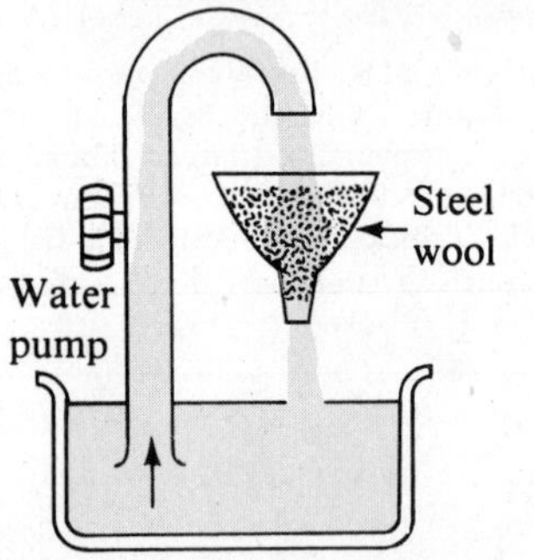

Figure 24.10 A liquid analogy of the simple circuit shown in Fig. 24.9. The pump (battery) lifts water (charge) through some vertical distance. The falling water hits the steel wool (resistance), and the water is recycled.

The rate at which the charge ΔQ *loses* potential energy in going through the resistor is given by

$$\frac{\Delta U}{\Delta t} = \frac{\Delta Q}{\Delta t}V = IV$$

where I is the current in the circuit. Of course, the charge regains this energy when it passes through the battery. Since the rate at which the charge loses energy equals the power P lost in the resistor, we have

Power

$$P = IV \tag{24.14}$$

In this case, the power is supplied to a resistor by a battery. However, Eq. 24.14 can be used to determine the power transferred to *any* device carrying a current I and having a potential difference V between its terminals.

Using Eq. 24.14 and the fact that $V = IR$ for a resistor, we can express the power dissipated in the alternative forms

---

[4]Note that once the current reaches its steady-state value, there is *no* change with time in the kinetic energy associated with the current flow.

$$P = I^2R = \frac{V^2}{R} \qquad (24.15)$$

Power loss in a conductor

When I is in amperes, V in volts, and R in ohms, the SI unit of power is the watt (W). The power lost as heat in a conductor of resistance R is called *joule heat*[5] but is often referred to as *I^2R loss.*

A battery or any other device that provides electrical energy is called a *seat of electromotive force,* usually referred to as an *emf.* (The phrase *electromotive force* is an unfortunate one, since it does not really describe a force but actually refers to a potential difference in volts.) *Neglecting the internal resistance of the battery, the potential difference between points* a *and* b *is equal to the emf* $\mathcal{E}$ *of the battery.* That is, $V = V_b - V_a = \mathcal{E}$, and the current in the circuit is given by $I = V/R = \mathcal{E}/R$. Since $V = \mathcal{E}$, the power supplied by the emf can be expressed as $P = I\mathcal{E}$, which, of course, equals the power lost in the resistor, I^2R.

Example 24.4

An electric heater is constructed by applying a potential difference of 110 V to a nichrome wire of total resistance 8 Ω. Find the current carried by the wire and the power rating of the heater.

Solution: Since $V = IR$, we have

$$I = \frac{V}{R} = \frac{110 \text{ V}}{8\ \Omega} = 13.8 \text{ A}$$

We can find the power rating using $P = I^2R$:

$$P = I^2R = (13.8 \text{ A})^2(8\ \Omega) = 1.52 \text{ kW}$$

If we were to double the applied voltage, the current would double but the power would quadruple.

Example 24.5

A light bulb is rated at 120 V/75 W. That is, its operating voltage is 120 V and it has a power rating of 75 W. The bulb is powered by a 120-V direct-current power supply. Find the current in the bulb and its resistance.

Solution: Since the power rating of the bulb is 75 W and the operating voltage is 120 V, we can use $P = IV$ to find the current:

$$I = \frac{P}{V} = \frac{75 \text{ W}}{120 \text{ V}} = 0.625 \text{ A}$$

Using Ohm's law, $V = IR$, the resistance is calculated to be

$$R = \frac{V}{I} = \frac{120 \text{ V}}{0.625 \text{ A}} = 192\ \Omega$$

What would be the resistance of a 120 V/100 W lamp?

Q11. Two light bulbs both operate from 110 V, but one has a power rating of 25 W and the other of 100 W. Which bulb has the higher resistance? Which bulb carries the greater current?

Q12. Two conductors of the same length and radius are connected across the same potential difference. One conductor has twice the resistance of the other. Which conductor will dissipate more power?

24.5 A MODEL FOR ELECTRICAL CONDUCTION

In this section we describe a classical model of electrical conduction in metals. This model leads to Ohm's law and shows that resistivity can be related to the motion of electrons in metals.

Consider a conductor as a regular array of atoms containing free electrons (sometimes called *conduction,* or *valence,* electrons). These electrons are free to move through the conductor and are approximately equal in number to the number of

[5] It is called *joule heat* even though its dimensions are *energy per unit time,* which are dimensions of power.

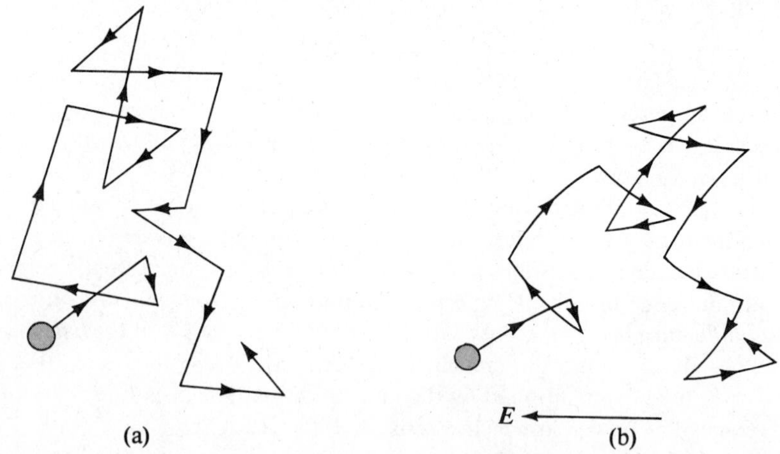

Figure 24.11 (a) A schematic diagram of the random motion of a charge carrier in a conductor in the absence of an electric field. Note that the drift velocity is zero. (b) The motion of a charge carrier in a conductor in the presence of an electric field. Note that the random motion is modified by the field, and the charge carrier has a drift velocity.

atoms. In the absence of an electric field, the free electrons move in random directions through the conductor with average speeds of the order of 10^6 m/s. (These speeds can be properly calculated only if a quantum mechanical description is used.) The situation is similar to the motion of gas molecules confined in a vessel. In fact, some scientists refer to conduction electrons in a metal as an *electron gas*. The conduction electrons are not totally "free" since they are confined to the interior of the conductor and undergo frequent collisions with the array of atoms. These collisions are the predominant mechanism for the resistivity of a metal at normal temperatures. Note that there is no current through a conductor in the absence of an electric field since the *average velocity* of the free electrons is zero. That is, on the average, just as many electrons move in one direction as in the opposite direction, and so there is no net flow of charge.

The situation is modified when an electric field is applied to the metal. In addition to the random thermal motion just described, the free electrons drift slowly in a direction opposite that of the electric field, with an average drift speed v_d, which is much smaller (typically 10^{-4} m/s) than the average speed between collisions (typically 10^6 m/s). Figure 24.11 provides a crude description of the motion of free electrons in a conductor. In the absence of an electric field, there is no net displacement after many collisions (Fig. 24.11a). An electric field $\boldsymbol{E}$ modifies the random motion and causes the electrons to drift in a direction opposite that of $\boldsymbol{E}$ (Fig. 24.11b). The slight curvature in the paths in Fig. 24.11b results from the acceleration of the electrons between collisions, caused by the applied field. One mechanical system somewhat analogous to this situation is a ball rolling down a slightly inclined plane through an array of closely spaced, fixed pegs (Fig. 24.12). The ball represents a conduction electron, the pegs represent the atoms of the solid, and the component of the gravitational force along the incline represents the electric force $e\boldsymbol{E}$.

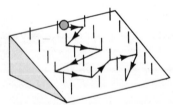

Figure 24.12 A mechanical system somewhat analogous to the motion of charge carriers in the presence of an electric field. The collisions of the ball with the pegs represent the resistance to the ball's motion down the incline.

In our model, we shall assume that the excess energy acquired by the electrons in the electric field is lost to the conductor in the collision process. The energy given up to the atoms in the collisions increases the vibrational energy of the atoms, causing the conductor to heat up. The model also assumes that the motion of an electron after a collision is independent of its motion before the collision.

We are now in a position to obtain an expression for the drift velocity. When a mobile charged particle of mass m and charge q is subjected to an electric field $\boldsymbol{E}$, it experiences a force $q\boldsymbol{E}$. Since $\boldsymbol{F} = m\boldsymbol{a}$, we conclude that the acceleration of the particle is given by

$$a = \frac{qE}{m}$$

This acceleration, which occurs for only a short time between collisions, enables the electron to acquire a small drift velocity. If t is the time since the last collision and $\mathbf{v}_0$ is the initial velocity, then the velocity of the electron after a time t is given by

$$\mathbf{v} = \mathbf{v}_0 + \mathbf{a}t = \mathbf{v}_0 + \frac{q\mathbf{E}}{m}t$$

We now take the average value of $\mathbf{v}$ over all possible times t and all possible values of $\mathbf{v}_0$. If the initial velocities are assumed to be randomly distributed, we see that the average value of $\mathbf{v}_0$ is zero and we get

$$|\bar{\mathbf{v}}| = \frac{qE}{m}t \qquad (24.16)$$

If the *average* time between collisions is τ, then the drift velocity[6] is

$$v_d = \frac{qE}{m}\tau \qquad (24.17)$$

Drift velocity

Substituting this result into Eq. 24.6, we find that the magnitude of the current density is given by

$$J = nqv_d = \frac{nq^2E}{m}\tau \qquad (24.18)$$

Current density

Comparing this expression with Ohm's law, $J = \sigma E$, we obtain the following relationships for the conductivity and resistivity:

$$\sigma = \frac{nq^2\tau}{m} \qquad (24.19)$$

Conductivity

$$\rho = \frac{1}{\sigma} = \frac{m}{nq^2\tau} \qquad (24.20)$$

Resistivity

The average time between collisions is related to the average distance between collisions l (the mean free path) and the average thermal speed $\bar{v}$ through the expression[7]

$$\tau = l/\bar{v} \qquad (24.21)$$

According to this classical model, the conductivity and resistivity do not depend on the electric field. This feature is characteristic of a conductor obeying Ohm's law. The model shows that the conductivity can be calculated from a knowledge of the density of the charge carriers, their charge and mass, and the average time between collisions.

[6]Since the collision process is random, each collision event is *independent* of what happened earlier. This is analogous to the random process of throwing a die. The probability of rolling a particular number on one throw is independent of the result of the previous throw. On the average, it would take six throws to come up with that number, starting at any arbitrary time.

[7]Recall that the thermal speed is the speed a particle has as a consequence of the temperature of *its* surroundings (Chapter 17).

Example 24.6

(a) Using the data and results from Example 24.1 and the classical model of electron conduction, estimate the average time between collisions for electrons in copper at 20°C.

From Eq. 24.20 we see that

$$\tau = \frac{m}{nq^2\rho}$$

where $\rho = 1.7 \times 10^{-8}\ \Omega \cdot \text{m}$ for copper and $n = 8.48 \times 10^{28}$ electrons/m^3 for the wire described in Example 24.1. Substitution of these values into the expression above gives

$$\tau = \frac{(9.11 \times 10^{-31}\ \text{kg})}{(8.48 \times 10^{28}\ \text{m}^{-3})(1.6 \times 10^{-19}\ \text{C})^2(1.7 \times 10^{-8}\ \Omega \cdot \text{m})}$$
$$= 2.5 \times 10^{-14}\ \text{s}$$

(b) Assuming the mean thermal speed for free electrons in copper to be 1.6×10^6 m/s and using the result from (a), calculate the mean free path for electrons in copper.

$$l = \bar{v}\tau = (1.6 \times 10^6\ \text{m/s})(2.5 \times 10^{-14}\ \text{s})$$
$$= 4.0 \times 10^{-8}\ \text{m}$$

which is equivalent to 400 Å (compared with atomic spacings of about 2 Å). Thus, although the time between collisions is very short, the electrons travel about 200 atomic distances before colliding with an atom.

Although this classical model of conduction is consistent with Ohm's law, it is not satisfactory for explaining some important phenomena. For example, classical calculations for $\bar{v}$ using the ideal-gas model are about a factor of 10 smaller than the true values. Furthermore, according to Eqs. 24.18 and 24.19, the temperature variation of the resistivity is predicted to vary as $\bar{v}$, which according to an ideal-gas model (Chapter 18) is proportional to $\sqrt{T}$. This is in disagreement with the linear dependence of resistivity with temperature for pure metals (Fig. 24.6). It is possible to account for such observations only by using a quantum mechanical model, which we shall describe briefly.

According to quantum mechanics, electrons have wavelike properties. If the array of atoms is regularly spaced (that is, periodic), the wavelike character of the electrons makes it possible for them to move freely through the conductor, and a collision with an atom is unlikely. For an idealized conductor, there would be no collisions, the mean free path would be infinite, and the resistivity would be zero. Electron waves are scattered only if the atomic arrangement is irregular (not periodic) as a result of, for example, structural defects or impurities. At low temperatures, the resistivity of metals is dominated by scattering caused by collisions between the electrons and impurities. At high temperatures, the resistivity is dominated by scattering caused by collisions between the electrons and the atoms of the conductor, which are continuously displaced as a result of thermal agitation. The thermal motion of the atoms causes the structure to be irregular (compared with an atomic array at rest), thereby reducing the electron's mean free path.

24.6 CONDUCTION IN SEMICONDUCTORS AND INSULATORS

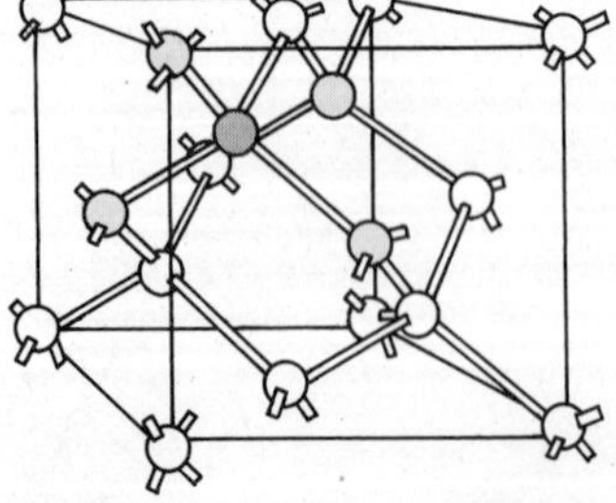

Figure 24.13 Crystalline structure of silicon, germanium, and diamond (carbon). Each atom is in the center of a tetrahedron with four nearest neighbors. The crystalline structure is called a *face-centered lattice.* (*After W. Shockley,* Electrons and Holes in Semiconductors, *New York, Van Nostrand, 1950*)

We have seen that good conductors contain a high density of charge carriers, whereas the density of charge carriers in insulators is nearly zero. Semiconductors are a class of technologically important materials with charge carrier densities intermediate between those of insulators and those of conductors. A full understanding of semiconductors requires a knowledge of quantum mechanics and a treatment beyond the scope of this text. The discussion that follows deals only qualitatively with the mechanisms of conduction in semiconductors.

One way to account for the electrical properties of semiconductors is in terms of the chemical bonding between atoms. Silicon, germanium, and carbon in the form of diamond are materials in which each atom is bound to four nearest-neighbor atoms in a tetrahedral structure (Fig. 24.13). Since an atom in these structures has four valence electrons, every atom shares one electron with each of its nearest neighbors. Therefore, the outer shell of every atom in the solid is completely filled with eight electrons. This sharing of an electron between neighboring atoms is called *covalent bonding*, and a two-dimensional model for this bonding is illustrated in Fig. 24.14. It is difficult to remove an electron from these strong covalent bonds. In fact, it is the high strength of these bonds in crystalline carbon that accounts for the hardness of diamonds. The covalent bonds are somewhat weaker in silicon and germanium since the valence electrons are farther from the nucleus than they are in carbon.

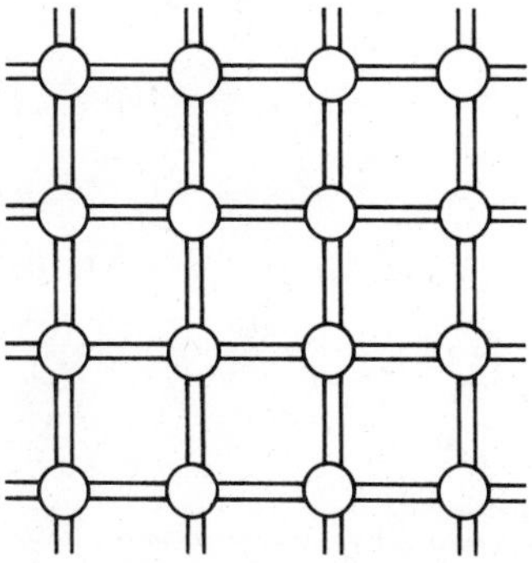

Figure 24.14 A two-dimensional model of a semiconducting lattice. Every atom contains four valence electrons, each of which forms a covalent bond (double lines) with a nearest-neighbor atom.

If the temperature of the material is increased, the atoms vibrate more vigorously about their equilibrium positions. In the process, some electrons acquire sufficient vibrational energy to break out of their covalent bond structure and move independently through the crystal. These electrons then participate in the conduction process. This explains why the resistivity of a pure semiconductor decreases with increasing temperature, corresponding to a negative value of α (Fig. 24.7).

Physicists customarily take a different approach to describe the properties of semiconductors and insulators and use a model that is very useful in dealing with highly ordered crystalline solids. The crystal is viewed as a periodic array of atoms in which symmetry and proximity cause the discrete energy levels of the free atoms to spread out and form energy bands in the solid. Two kinds of energy bands are formed, called *valence bands* and *conduction bands* (Fig. 24.15). In a semiconductor, the valence band is normally filled with electrons and the higher-energy conduction band is normally empty. The valence and conduction bands in a semiconductor are separated by an energy gap E_g, sometimes referred to as the *forbidden energy band* (Fig. 24.15b).

At $T = 0$ K, all electrons are in the valence band and there are no free electrons available for conduction. As the temperature is increased, some electrons "jump" into the conduction band because of thermal excitation and the solid becomes partially conducting. In silicon and germanium, the energy gap is small enough ($\approx$1 eV) to allow an appreciable number of electrons to be thermally excited into the conduction band. (See Table 24.2 for some representative data.) Again, this model explains the decrease in resistivity with increasing temperature for a semiconductor. When an electron moves from the valence band to the conduction band, it leaves behind a vacant crystal site, or so-called *hole*, in the nearly filled valence band. This hole (electron-deficient site) appears as a positive ion of charge $+e$. The hole also acts as a charge carrier in the sense that a valence electron from a nearby bond can transfer into the hole without becoming a free electron. Thus, we see that

TABLE 24.2 Energy Gaps for Various Elements

ELEMENT	E_g (eV)[a]
Carbon	7
Silicon	1.14
Germanium	0.67
Tin	0.1
Copper	0

[a]All values at 20°C.

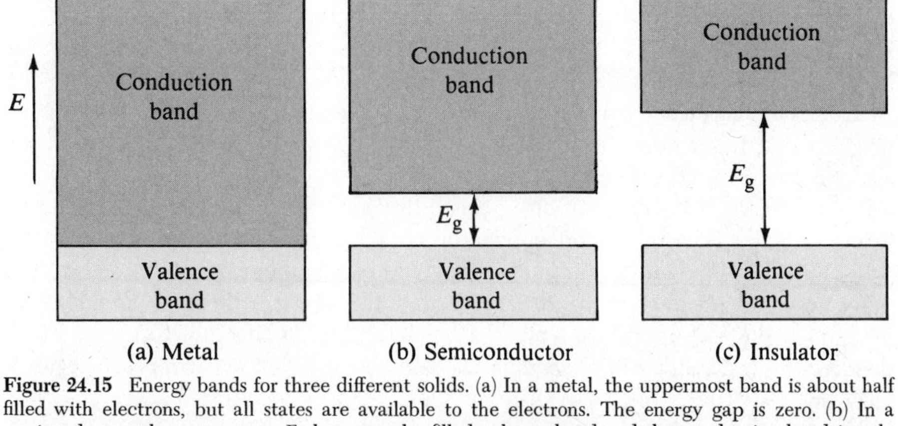

Figure 24.15 Energy bands for three different solids. (a) In a metal, the uppermost band is about half filled with electrons, but all states are available to the electrons. The energy gap is zero. (b) In a semiconductor, the energy gap E_g between the filled valence band and the conduction band is only around 1 eV, and so electrons can be thermally excited into the conduction band. (c) In an insulator, the filled valence and conduction bands are separated by a large energy gap of around 10 eV.

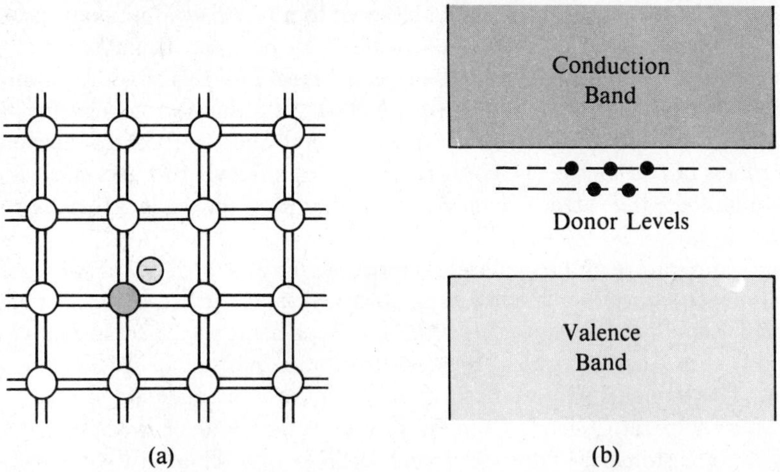

Figure 24.16 (a) Two-dimensional representation of a semiconductor containing a donor atom (dark spot). (b) Energy band diagram of a semiconductor in which the donor levels lie within the forbidden gap, just below the bottom of the conduction band.

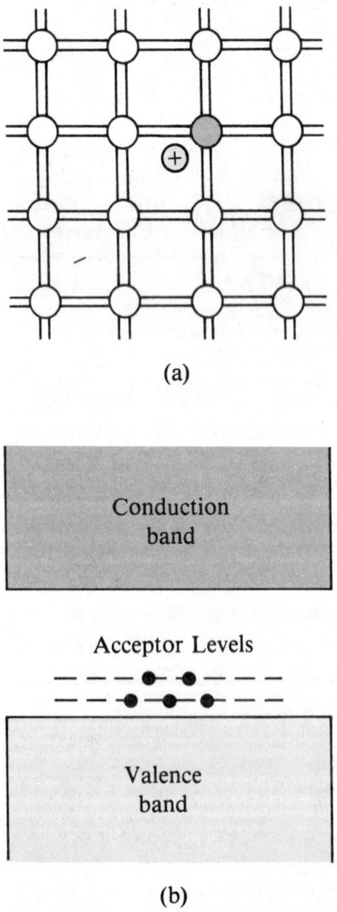

Figure 24.17 (a) Two-dimensional representation of a semiconductor containing an acceptor atom (dark spot). (b) Energy band diagram of a semiconductor in which the acceptor levels lie within the forbidden gap, just above the top of the valence band.

there are both negative and positive charge carriers in a semiconductor. In a pure crystal there are an equal number of conduction electrons and holes. Such combinations of charges are called *electron-hole pairs*, and a pure material that contains such pairs is said to be an *intrinsic semiconductor*. In the presence of an electric field, the hole motion is in the direction of the field and the conduction electrons move opposite the field.

The energy gap for insulators ($\approx$10 eV) is much larger than for semiconductors (Fig. 24.15c). Consequently, very few electrons can be thermally excited into the conduction band at ordinary temperatures; hence the resistivities of insulators are much larger than those of metals and semiconductors. In contrast, the energy gap for a good conductor, such as copper, is zero (Fig. 24.15a). When energy bands are formed in a metal, the uppermost band is only about half filled with electrons.[8] Since the number of states available to the electrons in this band greatly exceeds the number of electrons and since the energy levels are very close together, the electrons can easily change states. At very low temperatures, most of the electrons occupy the valence band, and so the density of charge carriers is small. At room temperature, many electrons are thermally excited into the conduction band, thereby increasing the charge carrier density.

When impurities are added to semiconductors, their band structures and resistivities are modified. The process of adding impurities, called *doping*, is commonly used to obtain the desired conductive properties for devices. For example, when an atom with five valence electrons, such as arsenic, is added to a semiconductor, four valence electrons participate in the covalent bands and one electron is left over (Fig. 24.16a). This extra electron is nearly free and has an energy level in the band diagram that lies within the forbidden gap, just below the conduction band (Fig. 24.16b). Such a pentavalent atom in effect donates an electron to the structure and hence is referred to as a *donor atom*. Since the energy spacing between the donor levels and the bottom of the conduction band is very small (typically, about 0.05 eV), a small amount of thermal energy will cause an electron in these levels to move into the conduction band. (Recall that the average thermal energy of an electron at room temperature is about $kT \approx 0.025$ eV.) Semiconductors doped with donor atoms are called *n-type semiconductors* since the charge carriers are negative electrons.

[8]For some metals, the two allowed bands can overlap.

If a semiconductor is doped with atoms with three valence electrons, such as indium and aluminum, the three electrons form covalent bonds with neighboring atoms, leaving an electron deficiency, or hole, in the fourth bond (Fig. 24.17a). The energy levels of such impurities also lie within the forbidden gap, just above the valence band, as indicated in Fig. 24.17b. Electrons from the valence band have enough thermal energy at room temperature to fill these impurity levels, leaving behind a hole in the valence band. Since a trivalent atom in effect accepts an electron from the valence band, such impurities are referred to as *acceptors*. A semiconductor doped with trivalent (acceptor) impurities is known as a *p-type semiconductor* since the charge carriers are positively charged holes. When conduction is dominated by acceptor or donor impurities, the material is called an *extrinsic semiconductor*. The typical range of doping densities for n- or p-type semiconductors is 10^{13} to 10^{17} cm^{-3}.

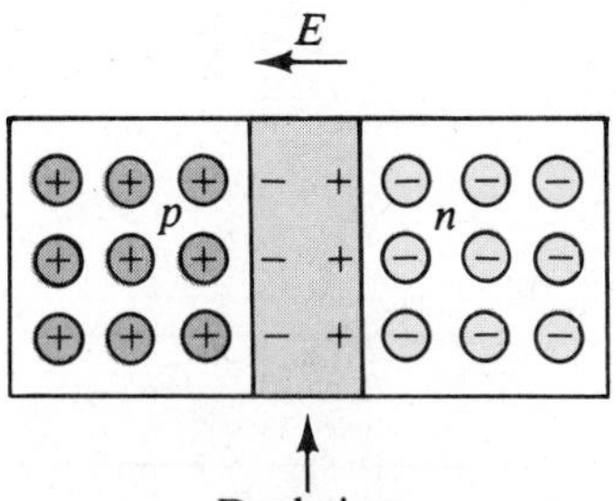

Figure 24.18 A p-n junction diode. The boundary between the p and n regions contains a dipole layer of charge.

24.7 SEMICONDUCTOR DEVICES

The *p-n* Junction

The operation of semiconductor devices, such as diodes, transistors, solar cells, and integrated circuits, is based on the electronic properties of their constituent materials. Such devices are fabricated using a semiconductor that is partly p type and partly n type. The p and n regions are obtained by controlling the number and type of impurity atoms introduced during or following crystal growth.

Consider a semiconducting crystal containing separate, adjoining regions of p-type and n-type materials as in Fig. 24.18. The boundary between these two regions is called a *p-n junction*. When the junction is formed, electrons in the n region tend to diffuse into the p region, while holes in the p region tend to diffuse into the n region. Some electrons entering the p region are captured by vacant crystal sites (holes) in a process called *recombination*. Similarly, holes diffusing into the n region recombine with electrons in that region. This charge transfer process continues until a steady-state condition is reached. At this time, a charged dipole layer called the *depletion layer* is formed over a very narrow region along the boundary; this dipole layer is a region of low conductivity. The electric field set up by the dipole layer acts as a barrier to further diffusion of electrons into the p region and holes into the n region.[9] In other words, the p-n junction consists of a dipole layer that prevents further motion of electrons and holes across the boundary.

The recombination process

The *p-n* Junction Diode

The p-n junction just described acts as a *rectifier*. That is, the junction behaves as a good conductor for an applied voltage of one polarity and as a poor conductor when the applied voltage has the opposite polarity.

Suppose an external battery is connected across the junction, with the p side positive and the n side negative, as in Fig. 24.19a. This polarity for the applied voltage is referred to as a *forward-biased junction*. In this situation, the polarity of the applied voltage is *opposite* that of the dipole layer, a condition that causes a reduction in the field at the junction (or a reduction in the height of the barrier). As a result, free electrons move more readily from the n to the p side of the junction while holes move from the p to the n side. This flow of charge across the junction constitutes a current.

A p-n diode is a rectifier

[9]Under steady-state conditions, the rate at which charge carriers are thermally generated equals the rate at which they recombine.

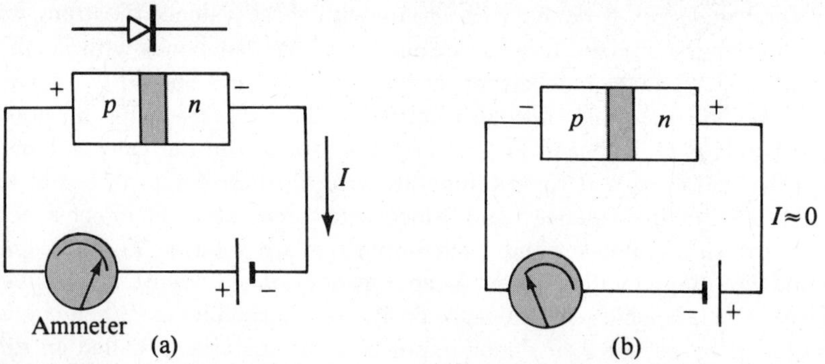

Figure 24.19 (a) A forward-biased *p-n* junction. A large current results from the decrease in the field at the junction. (b) In a reverse-biased junction, the current is approximately zero.

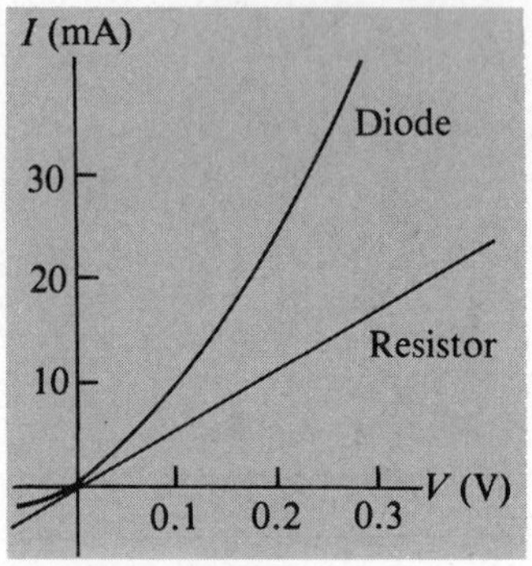

Figure 24.20 Current versus applied voltage for a *p-n* junction diode and a resistor that obeys Ohm's law.

Now consider what happens when the polarity of the battery is reversed, as in Fig. 24.19b, where the p side is negative and the n side is positive. In this case, the junction is said to be *reverse-biased.* Since the applied voltage has the same polarity as that of the dipole layer, the electric field (and the barrier height) at the junction are *enhanced.* This results in a very small current under reverse-bias conditions.

The current-voltage curve of a typical diode is shown in Fig. 24.20, together with the corresponding curve for a resistor obeying Ohm's law. As you can see, the diode has a very small resistance under forward bias and a very large resistance under reverse bias. (Recall that resistance is inversely proportional to the slope of the current-voltage curve.) This is in contrast to the linear I versus V curve of the resistor, for which the resistance is independent of the applied voltage.

Diodes are commonly used to convert alternating current into direct current and to detect radio frequency signals.

The Junction Transistor

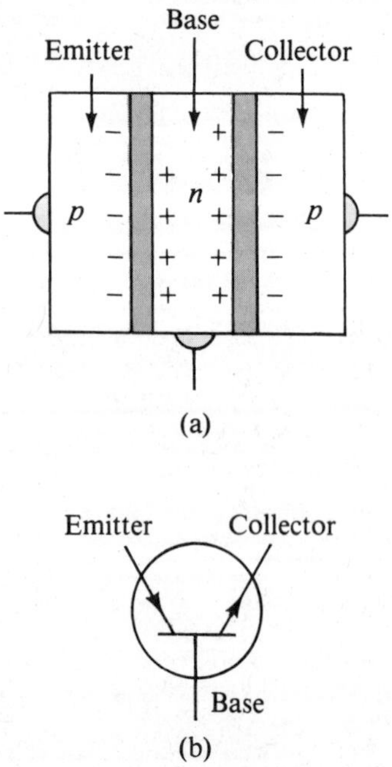

Figure 24.21 (a) The *pnp* transistor consists of an n region (base) sandwiched between two p regions (the emitter and collector). (b) The circuit symbol for the *pnp* transistor.

The development of the transistor by John Bardeen, Walter Brattain, and William Shockley totally revolutionized the world of electronics. For this work, these three men shared a Nobel prize in 1956. By 1960, the transistor had replaced the vacuum tube in nearly every electronic application (with the exception of the cathode ray tube). The advent of the transistor created a multibillion dollar industry, including such popular devices as pocket radios, handheld calculators, computers, television receivers, and electronic games.

The junction transistor (or bipolar transistor) consists of a semiconducting crystal with a very narrow n region sandwiched between two p regions. This configuration is called the *pnp transistor.* Another configuration is the *npn transistor,* which consists of a p region sandwiched between two n regions. Since the operation of the two transistors is essentially the same, we shall describe only the *pnp* transistor.

The structure of the *pnp* transistor, together with its circuit symbol, is shown in Fig. 24.21. The outer regions of the transistor are called the *emitter* (heavily doped) and *collector* (moderately doped). The narrow central region, called the *base,* is lightly doped. Note that the configuration contains *two* junctions. One junction is the interface between the emitter and base regions, and the other is the interface between the base and collector regions.

Suppose a voltage is applied to the transistor when the emitter is at a higher potential than the collector. If we think of the transistor as two diodes back to back, we see that the emitter-base junction is forward-biased and the base-collector junction is reverse-biased. Since the p-type emitter is heavily doped compared with the base region, nearly all of the current consists of holes moving across the emitter-base junction. Most of these holes do not recombine in the base region since the base is very narrow. The holes are finally accelerated across the reverse-biased base-collector junction, resulting in the emitter current, I_e.

Although only a small percentage of holes recombine in the base region, those that do limit the emitter current to a small value. This is due to the accumulation of positive charge in the base region, which increases the height of the barrier that prevents hole diffusion at the emitter-base junction.

In order to increase the current, some of the positive charge on the base must be drawn off; this is accomplished by connecting the base to a second battery as in Fig. 24.22. Those positive charges that are not swept across the collector-base junction leave the base region through this added channel, resulting in a small base current, I_b. A small increase in I_b can significantly lower the potential barrier of the base, resulting in a large increase in the collector current, I_c. In effect, the base current controls the collector current. The collector current is approximately equal to the emitter current, since the base current is very small.

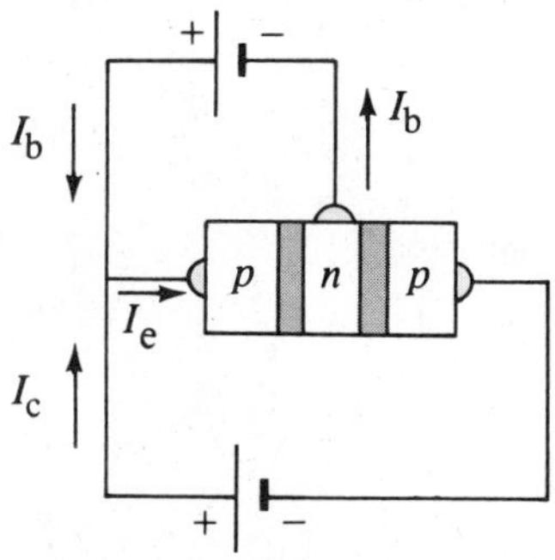

Figure 24.22 A bias voltage applied to the base as shown produces a small base current I_b, which is used to control the collector current I_c.

The *current gain* of the transistor is defined as the ratio of the collector current to the base current:

Current gain

$$\beta = \frac{I_c}{I_b}$$

Values of β for typical transistors range between 10 and 100.

One common application of the transistor is to amplify a small, time-varying signal. A simple amplifier circuit in which the signal is to be amplified is placed in series with the bias voltage on the base. In this configuration, the base current is the sum of a steady current I_b, produced by the bias voltage, and a time-varying current i_b, produced by the input signal. Likewise, the collector current consists of a dc-component I_c and a time-varying component i_c. Assuming the transistor is linear, both components are amplified by the same factor, β.

24.8 SUMMARY

The *electric current* I in a conductor is defined as

Electric current

$$I \equiv \frac{dQ}{dt} \qquad (24.2)$$

where dQ is the charge that passes through a cross section of the conductor in a time dt. The SI unit of current is the ampere (A), where 1 A = 1 C/s.

The current in a conductor is related to the motion of the charge carriers through the relationship

Current in a conductor

$$I = nqv_dA \qquad (24.4)$$

where n is the density of charge carriers, q is their charge, v_d is the drift velocity, and A is the cross-sectional area of the conductor.

The *current density* $\mathbf{J}$ in a conductor is defined as the current per unit area:

$$\mathbf{J} = nq\mathbf{v}_d \qquad (24.6)$$

The current density in a conductor is proportional to the electric field according to the expression

Ohm's law

$$\mathbf{J} = \sigma\mathbf{E} \qquad (24.7)$$

The constant σ is called the *conductivity* of the material. The inverse of σ is called the *resistivity*, ρ. That is, $\rho = 1/\sigma$.

A material is said to obey Ohm's law if its conductivity is independent of the applied field.

The *resistance* R of a conductor is defined as the ratio of the potential difference across the conductor to the current:

Resistance of a conductor

$$R \equiv \frac{V}{I} \tag{24.8}$$

If the resistance is independent of the applied voltage, the conductor obeys Ohm's law.

If the conductor has a uniform cross-sectional area A and a length l, its resistance is given by

Resistance of a uniform conductor

$$R = \frac{l}{\sigma A} = \rho \frac{l}{A} \tag{24.10}$$

The SI unit of resistance is volt per ampere, which is defined to be 1 ohm (Ω). That is, $1\ \Omega = 1\ \text{V/A}$.

The resistivity of a conductor varies with temperature in an approximately linear fashion, that is

Variations of ρ with temperature

$$\rho = \rho_0[1 + \alpha(T - T_0)] \tag{24.11}$$

where α is the temperature coefficient of resistivity and ρ_0 is the resistivity at some reference temperature T_0.

If a potential difference V is maintained across a conductor, the *power,* or rate at which energy is supplied to the conductor, is given by

Power

$$P = IV \tag{24.14}$$

Since $V = IR$ for a device that obeys Ohm's law, we can express the power dissipated in a resistor in the form

Power loss in a conductor

$$P = I^2R = \frac{V^2}{R} \tag{24.15}$$

The electrical energy supplied to a resistor appears in the form of internal energy (thermal energy) in the resistor.

In a classical model of electronic conduction in a metal, the electrons are treated as molecules of a gas. In the absence of an electric field, the average velocity of the electrons is zero. When an electric field is applied, the electrons move (on the average) with a *drift velocity* $\mathbf{v}_d$, which is opposite the electric field. The magnitude of the drift velocity is given by

Drift velocity

$$v_d = \frac{qE}{m}\tau \tag{24.17}$$

where τ is the average time between collisions with the atoms of the metal. The resistivity of the material according to this model is given by

Resistivity

$$\rho = \frac{m}{nq^2\tau} \tag{24.20}$$

where n is the number of free electrons per unit volume.

EXERCISES

Section 24.1 Electric Current and Current Density

1. Calculate the current in the case for which 2×10^{14} electrons pass a given cross section of a conductor each second.

2. The quantity of charge q (in C) passing through a surface of area 1 cm^2 varies with time as $q = 3t^2 - 4t + 2$, where t is in s. (a) What is the instantaneous current through the surface at $t =$ 0.5 s? (b) What is the value of the current density?

3. The current I (in A) in a conductor depends on time as $I = t^2 - 0.5t + 6$, where t is in s. What quantity of charge moves across a section through the conductor during the interval $t = 1$ s to $t = 3$ s?

4. Calculate the drift velocity of the electrons in a conductor that has a cross-sectional area of 8×10^{-6} m^2 and carries a current of 8 A. Take the concentration of free electrons to be 5×10^{28} electrons/m^3.

5. In a particular cathode ray tube, the measured beam current is 60 μA. How many electrons strike the tube screen every 10 s?

6. Figure 24.23 represents a section of a circular conductor of nonuniform diameter carrying a current of 15 A. The radius of cross section A_1 is 0.8 cm. (a) What is the magnitude of the current density across A_1? (b) If the current density across A_2 is one fourth the value across A_1, what is the radius of the conductor at A_2?

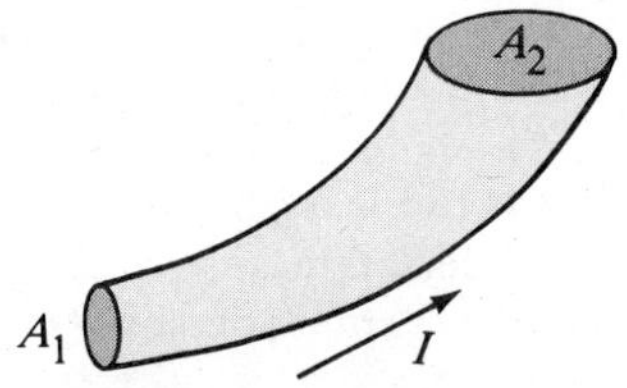

Figure 24.23 (Exercise 6).

7. The current density within a conductor of uniform radius 0.3 cm is 0.35 mA/m^2. In how many seconds will Avogadro's number of electrons pass a given point on the conductor?

Section 24.2 Resistance and Ohm's Law

8. Calculate the resistance at 20°C of a 40-m length of silver wire having a cross-sectional area of 0.4 mm^2.

9. A conducting wire of uniform radius 0.5 cm carries a current of 5 A produced by an electric field of 100 V/m. What is the resistivity of the material?

10. An electric field of 10^4 V/m is applied to a section of fused quartz of uniform cross section. Calculate the resulting current density if the specimen is at a temperature of 20°C.

11. A 40-V potential difference is maintained across a 1-m length of platinum wire that has a cross-sectional area of 0.1 mm^2. What is the current in the wire?

12. What is the resistance of a device that operates with a current of 4 A when the applied voltage is 120 V?

13. Aluminum and copper wires of equal length are found to have the same resistance. What is the ratio of their radii?

14. A potential difference of 12 V is found to produce a current of 0.4 A in a 3.2-m length of wire conductor having a uniform radius of 0.4 cm. (a) Calculate the resistivity of the wire material. (b) What is the resistance of the conductor?

15. A 16.5-m length of wire that is 0.012 cm^2 in cross section has a measured resistance of 0.12 Ω. Calculate the conductivity of the material from which the wire was drawn.

Section 24.3 The Resistivity of Different Conductors

16. If a silver wire has a resistance of 10 Ω at 20°C, what resistance will it have at 40°C? (Neglect any change in length or cross-sectional area due to the change in temperature.)

17. A wire 2 m in length and 0.25 mm^2 in cross-sectional area has a resistance of 43 Ω at 20°C. If the resistance of the wire increases to 43.2 Ω at 32°C, what is the temperature coefficient of resistivity?

18. At 40°C, the resistance of a segment of gold wire is 100 Ω. When the wire is placed in a liquid bath, the resistance decreases to 97 Ω. What is the temperature of the bath?

19. What is the fractional change in the resistance of an iron filament when its temperature changes from 30°C to 45°C?

20. At what temperature will tungsten have a resistivity four times that of copper? (Assume that copper is at room temperature.)

21. Calculate the resistivity of copper from the following data: A potential difference of 1.2 V produces a current of 1.8 A in a 100-m length of copper wire that is 0.18 cm in diameter and at a temperature of 20°C.

22. A segment of nichrome wire is initially at 20°C. Using the data from Table 24.1, calculate the temperature to which the wire must be heated to double its resistance.

Section 24.4 Electrical Energy and Power

23. A 12-V battery is connected to a 100-Ω resistor.

Neglecting the internal resistance of the battery, calculate the power dissipated in the resistor.

24. If a 40-Ω resistor is rated at 100 W (the maximum allowed power), what is the maximum allowed operating voltage?

25. An electric heater operating at full power draws a current of 15 A from a 220-V circuit. (a) What is the resistance of the heater? (b) Assuming constant R, how much current should the heater draw in order to dissipate 1200 W?

26. A current of 8 A is maintained in a 150-Ω resistor for 1 h. Calculate the heat energy developed in the resistor. Give the answer in both J and cal.

27. How much current is being supplied by a 240-V generator delivering 120 kW of power?

28. What is the required resistance of an immersion heater that will increase the temperature of 1.5 kg of water from 10°C to 50°C in 10 min while operating at 110 V?

29. Two conductors made of the same material are connected across a common potential difference. Conductor A has twice the diameter and twice the length of conductor B. What is the ratio of the power delivered to the two conductors?

30. A particular type of automobile storage battery is characterized as "240-ampere-hour, 12 V." What total energy can the battery deliver?

31. In a hydroelectric installation, a turbine delivers 2000 hp to a generator, which in turn converts 90 percent of the mechanical energy into electrical energy. Under these conditions, what current will the generator deliver at a terminal potential difference of 3000 V?

32. Compute the cost per day of operating a lamp that draws 1.2 A from a 110-V line if the cost of electrical energy is \$0.65/kWh.

Section 24.5 A Model for Electrical Conduction

33. Use data from Table 24.1 to calculate the collision mean free path of electrons in copper at a temperature corresponding to an average thermal speed of 1.3×10^6 m/s.

PROBLEMS

1. A resistor is constructed by forming a material of resistivity ρ into the shape of a hollow cylinder of length L and inner and outer radii r_a and r_b, respectively (Fig. 24.24). In use, a potential difference is applied between the ends of the cylinder, producing a current parallel to the axis. (a) Find a general expression for the resistance of such a device in terms of L, ρ, r_a, and r_b. (b) Obtain a numerical value for R when $L = 4$ cm, $r_a = 0.5$ cm, $r_b = 1.2$ cm, and $\rho = 3.5 \times 10^5\ \Omega \cdot \text{m}$.

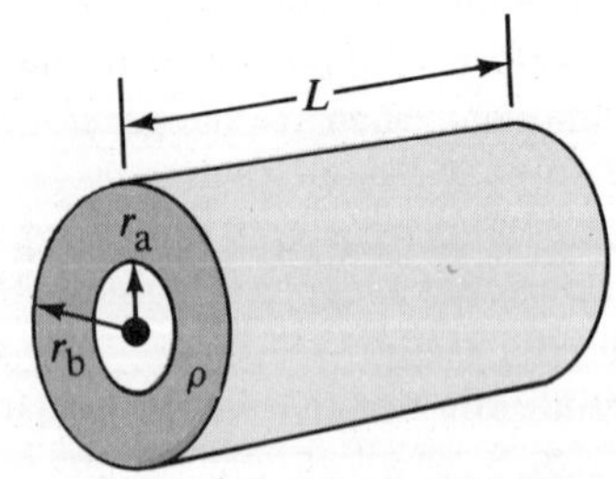

Figure 24.24 (Problems 1 and 2).

2. Consider the device described in Problem 1. Suppose now that the potential difference is applied between the inner and outer surfaces so that the resulting current flows radially outward. (a) Find a general expression for the resistance of the device in terms of L, ρ, r_a, and r_b. (b) Calculate the value of R using the parameter values given in (b) of Problem 1.

3. Two concentric spherical shells with inner and outer radii r_a and r_b, respectively, form a resistive element when the region between the two surfaces contains a material of resistivity ρ. Show that the resistance of the device is given by $R = \dfrac{\rho}{4\pi}\left(\dfrac{1}{r_a} - \dfrac{1}{r_b}\right)$.

4. (a) A 115-g mass of aluminum is formed into a right circular cylinder shaped so that the diameter of the cylinder equals its height. Calculate the resistance between the top and bottom faces of the cylinder at 20°C. (b) Calculate the resistance between opposite faces if the same mass of aluminum is formed into a cube.

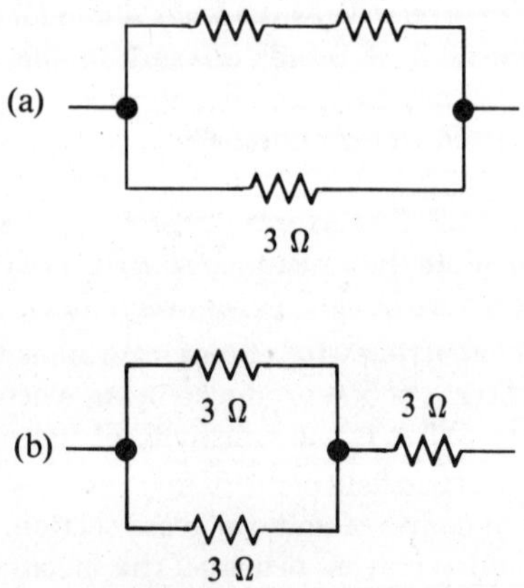

Figure 24.25 (Problem 5).

5. Three resistors, each of value 3 Ω, are arranged in two different arrangements as shown in Fig. 24.25. If the maximum allowable power for each individual resistor is 48 W, calculate the maximum power that can be dissipated by (a) the circuit shown in Fig. 24.25a and (b) the circuit shown in Fig. 24.25b.

6. A material of resistivity ρ is formed into the shape of a truncated cone of altitude h as in Fig. 24.26. The bottom end has a radius b and the top end has a radius a. Assuming a uniform current density through any circular cross section of the cone, show that the resistance between the two ends is $R = \frac{\rho}{\pi}\left(\frac{h}{ab}\right)$.

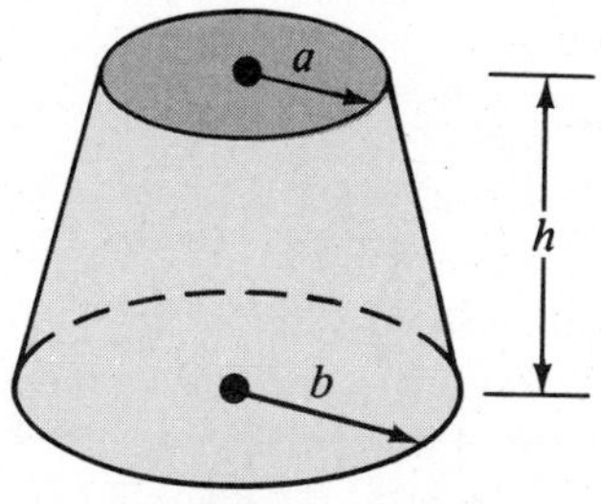

Figure 24.26 (Problem 6).

7. An engineer is in need of a resistor that is to have zero overall temperature coefficient of resistance at 20°C. The design is a composite of right circular cylinders of two materials, as in Fig. 24.27. The ratio of the resistivities of the two materials is $\rho_1/\rho_2 = 3.2$, and the ratio of the lengths of the sections is $l_1/l_2 = 2.6$. The radius r is uniform throughout. Assuming that the temperature of the two sections remains equal, calculate α_1/α_2, the required ratio of temperature coefficients of resistivity of the two materials.

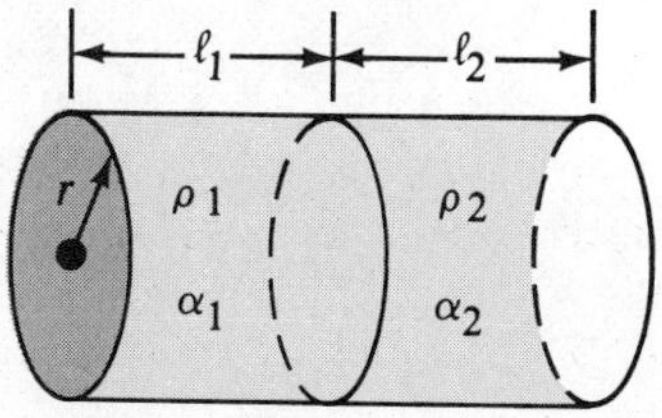

Figure 24.27 (Problem 7).

8. The potential difference across the filament of a lamp is maintained at a constant level while equilibrium temperature is being reached. It is observed that the steady-state current in the lamp is only one tenth of the current drawn by the lamp when it is first turned on. If the temperature coefficient of resistivity for the lamp at 20°C is 0.0045 $(\mathrm{C}°)^{-1}$, and if the resistance increases linearly with increasing temperature, what is the final operating temperature of the filament?

9. A cylindrical tungsten conductor has an initial length L_1 and cross-sectional area A_1. The metal is drawn uniformly to a final length $L_2 = 10L_1$. If the resistance of the conductor at the new length is 75 Ω, what is the initial value of R?

10. (a) A sheet of copper ($\rho = 1.7 \times 10^{-8}\ \Omega \cdot \mathrm{m}$) is 2 mm thick and has surface dimensions of 8 cm × 24 cm. If the long edges are joined to form a hollow tube 24 cm in length, what is the resistance between the ends? (b) What mass of copper would be required to manufacture a spool of copper cable 1500 m in length and having a total resistance of 4.5 Ω?

25 Direct Current Circuits

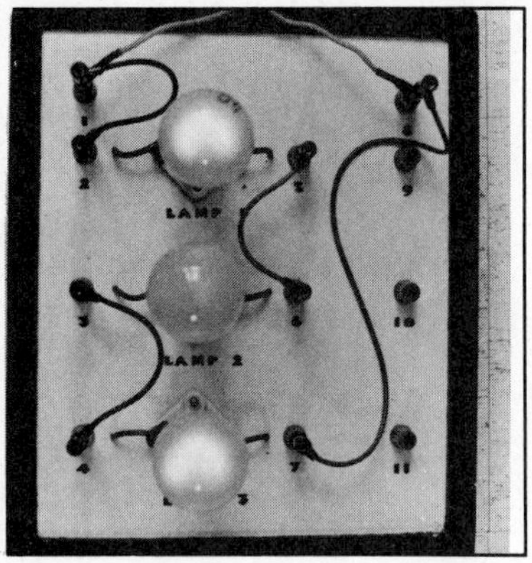

James Madison University

This chapter is concerned with the analysis of some simple circuits whose elements include batteries, resistors, and capacitors in various combinations. The analysis of these circuits is simplified by the use of two rules known as *Kirchhoff's rules*. These rules follow from the laws of conservation of energy and conservation of charge. Most of the circuits analyzed are assumed to be in *steady state,* where the currents are constant in magnitude and direction. In one section we discuss circuits containing resistors and capacitors, for which the current varies with time. Finally, a number of common electrical devices and techniques are described for measuring current, potential differences, resistance, and emfs.

25.1 ELECTROMOTIVE FORCE

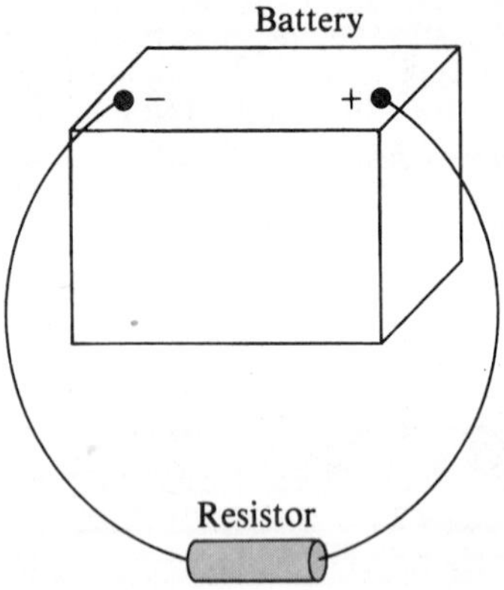

Figure 25.1 A series circuit consisting of a resistor connected to the terminals of a battery.

In the previous chapter we found that a constant current can be maintained in a closed circuit through the use of a source of energy, called an *electromotive force* (abbreviated *emf*). A source of emf is any device (such as a battery or generator) that will *increase* the potential energy of charges circulating in a circuit. One can think of a source of emf as a "charge pump" that forces charges to move in a direction opposite the electrostatic field inside the source. The emf, $\mathcal{E}$, of a source describes the work done per unit charge, and hence the SI unit of emf is the volt.

Consider the circuit shown in Fig. 25.1, consisting of a battery connected to a resistor. We shall assume that the connecting wires have no resistance. The positive terminal of the battery is at a higher potential than the negative terminal. If we were to neglect the internal resistance of the battery itself, then the potential difference across the battery (the terminal voltage) would equal the emf of the battery. However, because a real battery always has some internal resistance r, the terminal voltage is not equal to the emf of the battery. The circuit shown in Fig. 25.1 can be described by the circuit diagram in Fig. 25.2a. The battery within the dotted rectangle is represented by a seat of emf, $\mathcal{E}$, in series with the internal resistance, r. Now imagine a positive charge moving from a to b in Fig. 25.2a. As the charge passes from the negative to the positive terminal of the battery, its potential *increases* by $\mathcal{E}$. However, as it moves through the resistance r, its potential *decreases* by an amount Ir, where I is the current in the circuit. Thus, the terminal voltage of the battery,

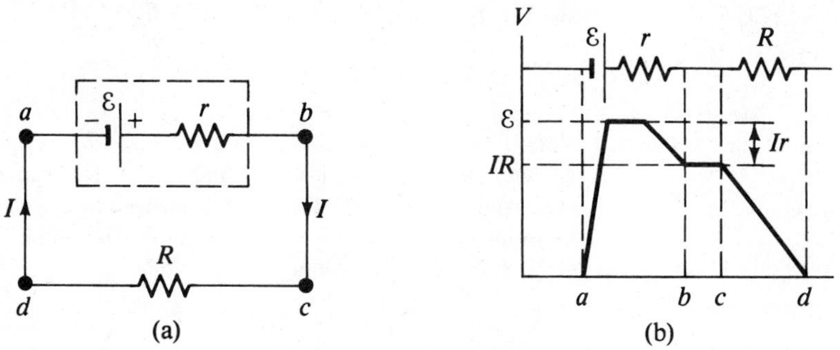

Figure 25.2 (a) Circuit diagram of a source of emf ℰ of internal resistance r connected to an external resistor R. (b) Graphical representation of the changes in potential as the series circuit in (a) is traversed clockwise.

$V = V_b - V_a$, is given by[1]

$$V = \mathcal{E} - Ir \tag{25.1}$$

From this expression, note that ℰ is equivalent to the *open-circuit voltage*, that is, the *terminal voltage when the current is zero.* Figure 25.2b is a graphical representation of the changes in potential as the circuit is traversed in the clockwise direction. By inspecting Fig. 25.2a we see that the terminal voltage V must also equal the potential difference across the external resistance R, often called the *load resistance.* That is, $V = IR$. Combining this with Eq. 25.1, we see that

$$\mathcal{E} = IR + Ir \tag{25.2}$$

Solving for the current gives

$$I = \frac{\mathcal{E}}{R + r} \tag{25.3}$$

This shows that the current in this simple circuit depends on both the resistance external to the battery and the internal resistance. Note also that if the load resistance R is much greater than the internal resistance r, we can neglect r in this analysis. In many circuits we shall ignore this internal resistance.

If we multiply Eq. 25.2 by the current I, the following expression is obtained:

$$I\mathcal{E} = I^2R + I^2r \tag{25.4}$$

This equation tells us that the total power output of the seat of emf, $I\mathcal{E}$, is converted into power dissipated as joule heat in the load resistance, I^2R, *plus* power dissipated in the internal resistance, I^2r. Again, if $r \ll R$, then most of the power delivered by the battery is transferred to the load resistance.

Example 25.1

A battery has an emf of 12 V and an internal resistance of 0.05 Ω. Its terminals are connected to a load resistance of 3 Ω. (a) Find the current in the circuit and the terminal voltage of the battery.

Using Eqs. 25.3 and 25.1, we get

$$I = \frac{\mathcal{E}}{R + r} = \frac{12\ \text{V}}{3.05\ \Omega} = 3.93\ \text{A}$$

$$V = \mathcal{E} - Ir = 12\ \text{V} - (3.93\ \text{A})(0.05\ \Omega) = 11.8\ \text{V}$$

As a check of this result, we can calculate the voltage drop

[1]The terminal voltage in this case is less than the emf by an amount Ir. In some situations, the terminal voltage may *exceed* the emf by an amount Ir. This happens when the current is *opposite* the emf, as in the case of charging a battery with another source of emf.

across the load resistance R. This gives

$$V = IR = (3.93\ \text{A})(3\ \Omega) = 11.8\ \text{V}$$

(b) Calculate the power dissipated in the load resistor, the power dissipated by the internal resistance of the battery, and the power delivered by the battery.

The power dissipated by the load resistor is

$$P_R = I^2R = (3.93\ \text{A})^2(3\ \Omega) = 46.3\ \text{W}$$

The power dissipated by the internal resistance is

$$P_r = I^2r = (3.93\ \text{A})^2(0.05\ \Omega) = 0.8\ \text{W}$$

Hence, the power delivered by the battery is the sum of these quantities, or 47.1 W. This can be checked using the expression $P = I\mathcal{E}$.

Example 25.2

Show that the *maximum* power lost in the load resistance R in Fig. 25.2a occurs when $R = r$, that is, when the load resistance *matches* the internal resistance.

Solution: The power dissipated in the load resistance is equal to I^2R, where I is given by Eq. 25.3:

$$P = I^2R = \frac{\mathcal{E}^2R}{(R + r)^2}$$

When P is plotted versus R as in Fig. 25.3, we find that P reaches a *maximum* value of $\mathcal{E}^2/4r$ at $R = r$. This can also be proved by differentiating P with respect to R, setting the result equal to zero, and solving for R. The details are left as a problem (Problem 12).

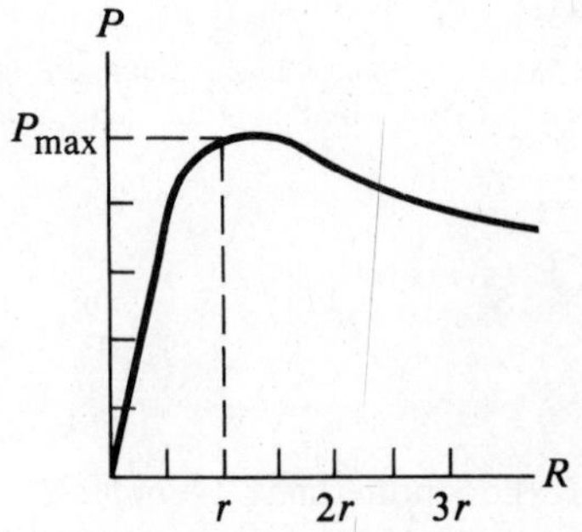

Figure 25.3 Graph of the power P delivered to a load resistor as a function of R. Note that the power into R is a maximum when R equals r, the internal resistance of the battery.

Q1. Under what condition does the potential difference across the terminals of a battery equal its emf? Can the terminal voltage ever exceed the emf? Explain.

25.2 RESISTORS IN SERIES AND IN PARALLEL

For a series connection of resistors, the current is the same in each resistor

When two or more resistors are connected together such that they have only one common point per pair, they are said to be in *series*. Figure 25.4 shows two resistors connected in series. Note *that the current is the same through each resistor since any charge that flows through R_1 must equal the charge that flows through R_2.* Since the potential drop from a to b in Fig. 25.4b equals IR_1 and the potential drop from b to c equals IR_2, the potential drop from a to c is given by

$$V = IR_1 + IR_2 = I(R_1 + R_2)$$

Therefore, we can replace the two resistors in series by a single *equivalent resistance* R_{eq} whose value is the *sum* of the individual resistances:

Two resistors in series

$$R_{eq} = R_1 + R_2 \qquad (25.5)$$

The resistance R_{eq} is equivalent to the series combination $R_1 + R_2$ in the sense that the circuit current is unchanged when R_{eq} replaces $R_1 + R_2$. The equivalent resistance of three or more resistors connected in series is simply

Several resistors in series

$$R_{eq} = R_1 + R_2 + R_3 + \cdots \qquad (25.6)$$

Therefore, the equivalent resistance of a series connection of resistors is always *greater* than any individual resistance.

Note that if the filament of one light bulb in Fig. 25.4 were to break, or "burn

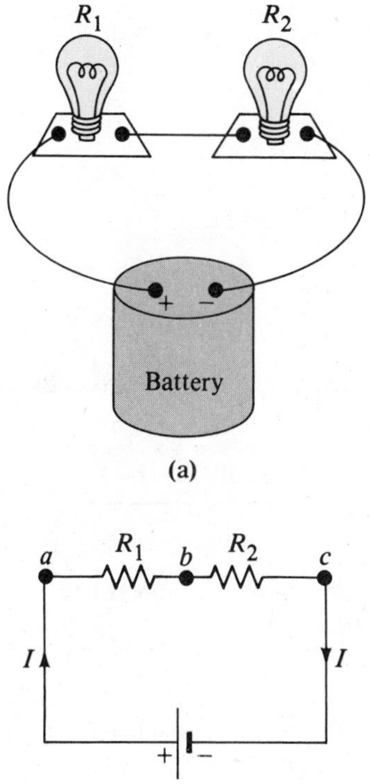

(b)

Figure 25.4 Series connection of two resistors, R_1 and R_2. The current in each resistor is the same, and the equivalent resistance of the combination is given by $R_{eq} = R_1 + R_2$.

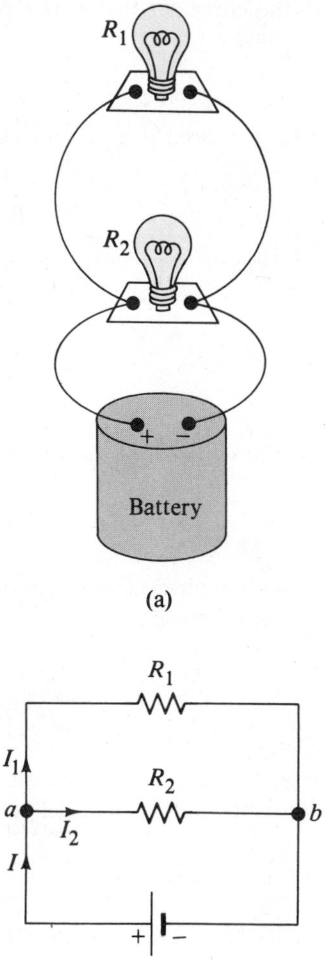

(b)

Figure 25.5 Parallel connection of two resistors, R_1 and R_2. The potential difference across each resistor is the same, and the equivalent resistance of the combination is given by $R_{eq} = R_1R_2/(R_1 + R_2)$.

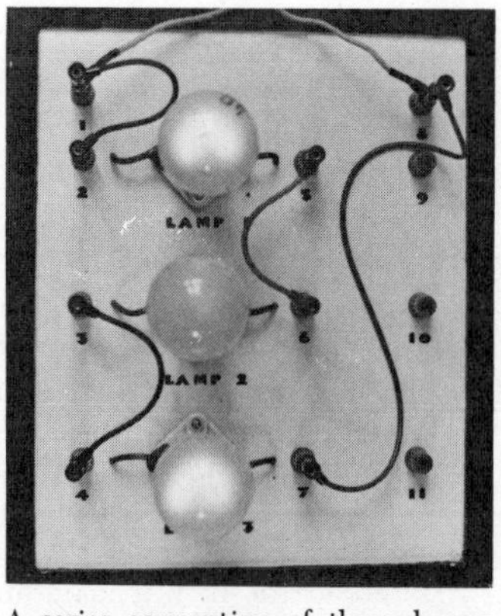

A series connection of three lamps with different power ratings. Why is the illumination of the middle lamp barely visible? How would their intensities differ if they were connected in parallel?

out," the circuit would no longer be complete (an open-circuit condition) and the second bulb would also go out. In many circuits, fuses are used in series with other circuit elements for safety purposes. The conductor in the fuse is designed to melt and open the circuit at some maximum current, the value of which depends on the nature of the circuit. If a fuse is not used, excessive currents could damage circuit elements, overheat wires, and perhaps cause a fire. In modern home construction, circuit breakers are used in place of fuses. When the current in a circuit exceeds some value (typically 15 A), the circuit breaker acts as a switch and opens the circuit.

Now consider two resistors connected in *parallel* as shown in Fig. 25.5. In this case note that *the potential difference across each resistor is the same.* However, the current in each resistor is in general not the same. When the current I reaches point a (called a *junction*), it splits into two parts, I_1 going through R_1 and I_2 going through R_2. If R_1 is greater than R_2, then I_1 will be less than I_2. That is, the charge will tend to take the path of least resistance. Clearly, since charge must be con-

For a parallel connection of resistors, the voltage across each is the same

served, the current I that enters point a must equal the total current leaving this point, $I_1 + I_2$:

$$I = I_1 + I_2$$

Since the potential drop across each resistor must be the *same*, Ohm's law gives

$$I = I_1 + I_2 = \frac{V}{R_1} + \frac{V}{R_2} = V\left(\frac{1}{R_1} + \frac{1}{R_2}\right) = \frac{V}{R_{eq}}$$

From this result, we see that the equivalent resistance of two resistors in parallel is given by

Two resistors in parallel

$$\frac{1}{R_{eq}} = \frac{1}{R_1} + \frac{1}{R_2} \tag{25.7}$$

This can be rearranged to give

$$R_{eq} = \frac{R_1R_2}{R_1 + R_2}$$

An extension of this analysis to three or more resistors in parallel gives the following general expression:

Several resistors in parallel

$$\frac{1}{R_{eq}} = \frac{1}{R_1} + \frac{1}{R_2} + \frac{1}{R_3} + \cdots \tag{25.8}$$

It can be seen from this expression that the equivalent resistance of two or more resistors connected in parallel is always *less* than the smallest resistance in the group.

Household circuits are always wired such that the light bulbs (or appliances, etc.) are connected in parallel, as in Fig. 25.5a. In this manner, each device operates independently of the others, so that if one is switched off, the others remain on.

Example 25.3

Four resistors are connected as shown in Fig. 25.6a. (a) Find the equivalent resistance between a and c.

The circuit can be reduced in steps as shown in Fig. 25.6. The 8-Ω and 4-Ω resistors are in series, and so the equivalent resistance between a and b is 12 Ω (Eq. 25.5). The 6-Ω and 3-Ω resistors are in parallel, and so from Eq. 25.7 we find that the equivalent resistance from b to c is 2 Ω. Hence, the equivalent resistance from a to c is 14 Ω.

(b) What is the current in each resistor if a potential difference of 42 V is maintained between a and c?

The current I in the 8-Ω and 4-Ω resistors is the same since they are in series. Using Ohm's law and the results from (a), we get

$$I = \frac{V_{ac}}{R_{eq}} = \frac{42\ \text{V}}{14\ \Omega} = 3\ \text{A}$$

When this current enters the junction at b, it splits and part of the current goes through the 6-Ω resistor (I_1) and part goes through the 3-Ω resistor (I_2). Since the potential difference across these resistors, V_{bc}, is the *same* (they are in parallel), we see that $6I_1 = 3I_2$, or $I_2 = 2I_1$. Using this result and the fact that $I_1 + I_2 = 3$ A, we find that

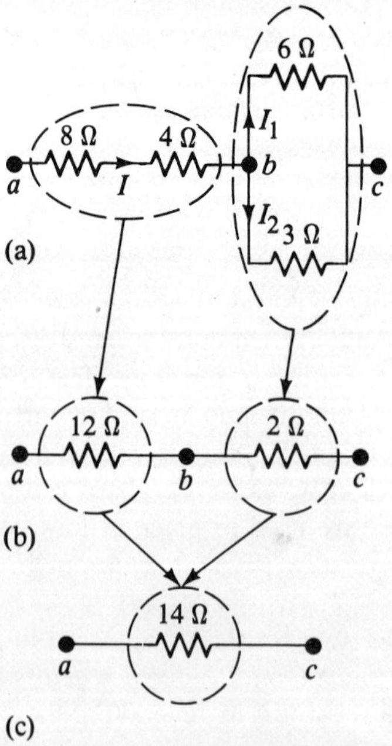

Figure 25.6 (Example 25.3) The equivalent resistance of the four resistors shown in (a) can be reduced in steps to an equivalent 14-Ω resistor.

$I_1 = 1$ A and $I_2 = 2$ A. We could have guessed this from the start by noting that the current through the 3-Ω resistor has to be twice the current through the 6-Ω resistor in view of their relative resistances and the fact that the same voltage is applied to both of them.

As a final check, note that $V_{bc} = 6I_1 = 3I_2 = 6$ V and $V_{ab} = 12I = 36$ V; therefore, $V_{ac} = V_{ab} + V_{bc} = 42$ V, as it must.

Example 25.4

Three resistors are connected in parallel as in Fig. 25.7. A potential difference of 18 V is maintained between points a and b. (a) Find the current in each resistor.

The resistors are in parallel, and the potential difference across each is 18 V. Applying $V = IR$ to each resistor gives

$$I_1 = \frac{V}{R_1} = \frac{18\text{ V}}{3\ \Omega} = 6\text{ A}$$

$$I_2 = \frac{V}{R_2} = \frac{18\text{ V}}{6\ \Omega} = 3\text{ A}$$

$$I_3 = \frac{V}{R_3} = \frac{18\text{ V}}{9\ \Omega} = 2\text{ A}$$

(b) Calculate the power dissipated by each resistor and the total power dissipated by the three resistors.

Applying $P = I^2R$ to each resistor gives

$$3\text{-}\Omega\text{: } P_1 = I_1^2R_1 = (6\text{ A})^2(3\ \Omega) = 108\text{ W}$$
$$6\text{-}\Omega\text{: } P_2 = I_2^2R_2 = (3\text{ A})^2(6\ \Omega) = 54\text{ W}$$
$$9\text{-}\Omega\text{: } P_3 = I_3^2R_3 = (2\text{ A})^2(9\ \Omega) = 36\text{ W}$$

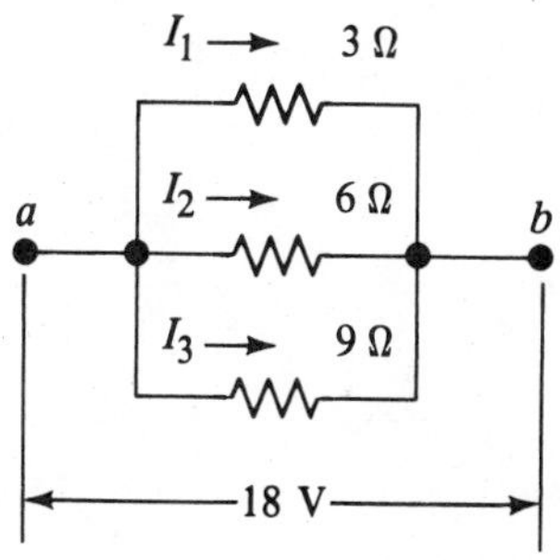

Figure 25.7 (Example 25.4) A parallel combination of three resistors.

This shows that the smallest resistor dissipates the most power since it carries the most current. (Note that you can also use $P = V^2/R$ to find the power dissipated by each resistor.) Summing the three quantities gives a total power of 198 W.

(c) Calculate the equivalent resistance of the three resistors, and from this result find the total power dissipated.

We can use Eq. 25.8 to find R_{eq}:

$$\frac{1}{R_{eq}} = \frac{1}{3} + \frac{1}{6} + \frac{1}{9}$$

$$R_{eq} = \frac{18}{11}\ \Omega$$

Hence, the total power dissipated is

$$P = \frac{V^2}{R_{eq}} = \frac{(18\text{ V})^2}{(18/11)\ \Omega} = 198\text{ W}$$

This agrees with the result obtained in (b), as it must.

Q2. Two different sets of Christmas-tree lights are available. For set A, when one bulb is removed (or burns out), the remaining bulbs remain illuminated. For set B, when one bulb is removed, the remaining bulbs will not operate. Explain the difference in wiring for the two sets of lights.

Q3. An incandescent lamp connected to a 120-V source with a short extension cord will provide more illumination than if it were connected to the same source with a very long extension cord. Explain.

25.3 KIRCHHOFF'S RULES

As we saw in the previous section, simple circuits can be analyzed using Ohm's law and the rules for series and parallel combinations of resistors. Very often it is not possible to reduce a circuit to a single loop. The procedure for analyzing more complex circuits is greatly simplified by the use of two simple rules called *Kirchhoff's rules:*

Statements of Kirchhoff's rules

1. The sum of the currents entering any junction must equal the sum of the currents leaving that junction. (A *junction* is any point in the circuit where the currents can split.)
2. The sum of the potential differences across each element around any closed-circuit loop must be *zero*.

The first rule is a statement of *conservation of charge*. That is, whatever current

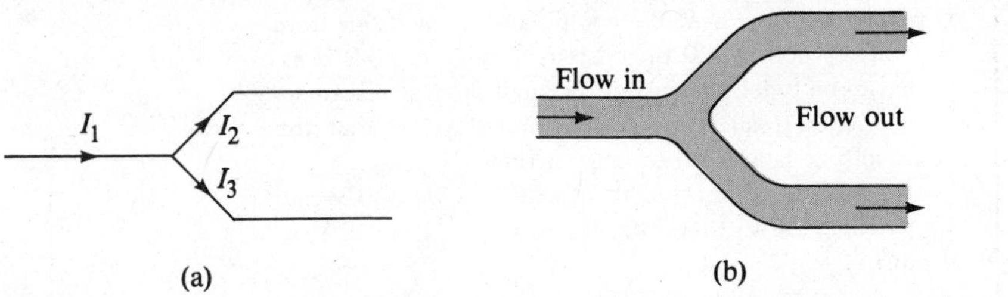

Figure 25.8 A schematic diagram illustrating Kirchhoff's junction rule. Conservation of charge requires that whatever current enters a junction must leave that junction. Therefore, in this case $I_1 = I_2 + I_3$.

enters a given point in a circuit must leave that point, since charge cannot build up at a point. If we apply this rule to the junction shown in Fig. 25.8a, we get

$$I_1 = I_2 + I_3$$

Figure 25.8b represents a mechanical analog to this situation, in which water flows through a branched pipe with no leaks. The flow rate into the pipe equals the total flow rate out of the two branches.

The second rule follows from *conservation of energy*. That is, any charge that moves around *any* closed loop in a circuit (it starts and ends at the same point) must gain as much energy as it loses. Its energy may decrease in the form of a potential drop, $-IR$, across a resistor or as the result of *reversing* a current through a source of emf. In a practical application of the latter case, electrical energy is converted into chemical energy when a battery is charged; similarly, electrical energy may be converted into mechanical energy for operating a motor.

As an aid in applying the second rule, the following points should be noted. These points are summarized in Fig. 25.9.

1. If a resistor is traversed in the direction of the current, the change in potential across the resistor is $-IR$ (Fig. 25.9a).
2. If a resistor is traversed in the direction *opposite* the current, the change in potential across the resistor is $+IR$ (Fig. 25.9b).
3. If a seat of emf is traversed in the direction of the emf (from $-$ to $+$ on the terminals), the change in potential is $+\mathcal{E}$ (Fig. 25.9c).
4. If a seat of emf is traversed in the direction opposite the emf (from $+$ to $-$ on the terminals), the change in potential is $-\mathcal{E}$ (Fig. 25.9d).

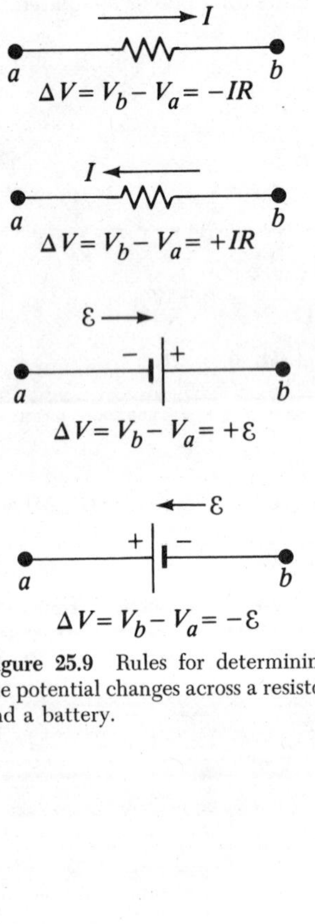

Figure 25.9 Rules for determining the potential changes across a resistor and a battery.

The following examples illustrate the use of Kirchhoff's rules in analyzing circuits. In all cases, it is assumed that the circuits have reached steady-state conditions, that is, the currents in the various branches are constant. If a capacitor is included as an element in one of the branches, *it acts as an open circuit*, that is, the current in the branch containing the capacitor will be zero under steady-state conditions.

When dealing with a multiloop circuit, you should first assign symbols and directions to the currents in the various branches of the circuit. Next, apply Kirchhoff's rules to obtain linear equations containing the unknown currents. It is important to adhere to the assumed directions of currents. The number of independent equations must at least equal the number of unknowns in order to obtain a complete solution. In a particular numerical problem, *a negative answer in one (or more) of the currents indicates that the current is in the direction opposite what was assumed.*

Example 25.5

A single-loop circuit contains two external resistors and two seats of emf as shown in Fig. 25.10. The internal resistances of the batteries have been neglected. (a) Find the current in the circuit.

There are no junctions in this single-loop circuit, and so the current is the same in all elements. Let us assume that the current is in the clockwise direction as shown in Fig. 25.10. Traversing the circuit in the clockwise direction, starting at point a, we see that $a \rightarrow b$ represents a

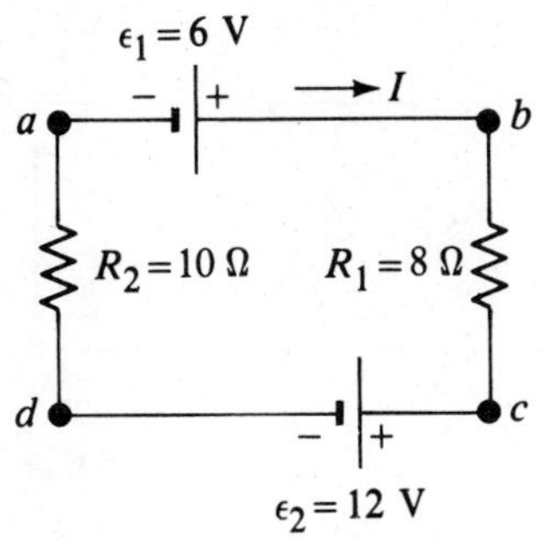

Figure 25.10 (Example 25.5) A series circuit containing two batteries and two resistors, where the polarity of the batteries is reversed.

potential increase of $+\mathcal{E}_1$, $b \to c$ represents a potential decrease of $-IR_1$, $c \to d$ represents a potential decrease of $-\mathcal{E}_2$, and $d \to a$ represents a potential decrease of $-IR_2$. Applying Kirchhoff's second rule gives

$$\sum_i \Delta V_i = 0$$

$$\mathcal{E}_1 - IR_1 - \mathcal{E}_2 - IR_2 = 0$$

Solving for I and using the values given in Fig. 25.10, we get

$$I = \frac{\mathcal{E}_1 - \mathcal{E}_2}{R_1 + R_2} = \frac{6\text{ V} - 12\text{ V}}{8\ \Omega + 10\ \Omega} = -\frac{1}{3}\text{A}$$

The negative sign for I indicated that the current is *opposite* the assumed direction, or *counterclockwise*.

(b) What is the power lost in each resistor?

$$P_1 = I^2R_1 = \left(\frac{1}{3}\text{ A}\right)^2(8\ \Omega) = \frac{8}{9}\text{ W}$$

$$P_2 = I^2R_2 = \left(\frac{1}{3}\text{ A}\right)^2(10\ \Omega) = \frac{10}{9}\text{ W}$$

Hence, the total power lost is $P_1 + P_2 = 2$ W. Note that the 12-V battery delivers power $I\mathcal{E}_2 = 4$ W. Half of this power is delivered to the external resistors. The other half is delivered to the 6-V battery, which is being charged by the 12-V battery. If we had included the internal resistances of the batteries, some of the power would be dissipated as heat in the batteries, so that *less* power would be delivered to the 6-V battery.

Example 25.6

(a) Find the currents I_1, I_2, and I_3 in the circuit shown in Fig. 25.11.

We shall choose the directions of the currents as shown in Fig. 25.11. Applying Kirchhoff's first rule to junction c gives

$$(1) \qquad I_1 + I_2 = I_3$$

There are *three* loops in the circuit, *abcda*, *befcb*, and *aefda* (the outer loop). We need only *two* loop equations to determine the unknown currents. The third loop equation would give no new information. Applying Kirchhoff's second rule to loops *abcda* and *befcb* and traversing these

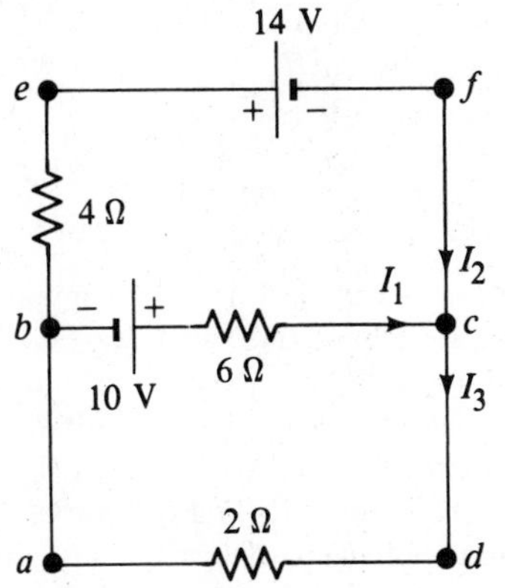

Figure 25.11 (Example 25.6) A circuit containing three loops.

loops in the clockwise direction, we obtain the following expressions:

(2) Loop *abcda:* $10\text{ V} - (6\ \Omega)I_1 - (2\ \Omega)I_3 = 0$

(3) Loop *befcb:* $-14\text{ V} - 10\text{ V} + (6\ \Omega)I_1 - (4\ \Omega)I_2 = 0$

Note that in loop *befcb*, a negative sign is obained when traversing the 6-Ω resistor since the current I_1 is in the opposite direction. A third loop equation for *aefda* gives $-14 = 2I_3 + 4I_2$, which is just the sum of (2) and (3). Expressions (1), (2), and (3) represent three linear, independent equations with three unknowns. We can solve the problem as follows: Substituting (1) into (2) gives

$$10 - 6I_1 - 2(I_1 + I_2) = 0$$

$$(4) \qquad 10 = 8I_1 + 2I_2$$

Dividing each term in (3) by 2 and rearranging the equation gives

$$(5) \qquad -12 = -3I_1 + 2I_2$$

Subtracting (5) from (4) eliminates I_2, giving

$$22 = 11I_1$$

$$I_1 = 2\text{ A}$$

Using this value of I_1 in (5) gives a value for I_2:

$$2I_2 = 3I_1 - 12 = 3(2) - 12 = -6$$

$$I_2 = -3\text{ A}$$

Finally, $I_3 = I_1 + I_2 = -1$ A. Hence, the currents have the values

$$I_1 = 2\text{ A} \qquad I_2 = -3\text{ A} \qquad I_3 = -1\text{ A}$$

The fact that I_2 and I_3 are both negative indicates only that we chose the *wrong* direction for these currents. However, the numerical values are correct.

(b) Find the potential difference between points b and c.

If we traverse the path directly from b to c, we get a potential increase of 10 V across the battery and a potential *decrease* of $6I_1 = 12$ V across the 6-Ω resistor. Therefore, b is at a *higher* potential than c by 2 V. That is, $V_b - V_c = 2$ V. Another way to get this result is to note that the potential difference $V_b - V_c = V_a - V_d$. Since $I_3 = -1$ A, the current I_3 is from a to d and since the potential *drop* across the 2-Ω resistor is 2 V. Therefore, $V_a - V_d = 2$ V.

Example 25.7

The multiloop circuit in Fig. 25.12 contains three resistors, three batteries, and one capacitor. (a) Under steady-state conditions, find the unknown currents.

First note that *the capacitor represents an open circuit, and hence there is no current along path* ghab *under steady-state conditions.* Labeling the currents as shown in Fig. 25.12 and applying Kirchhoff's first rule to junction c, we get

$$(1) \qquad I_1 + I_2 = I_3$$

Kirchhoff's second rule applied to loops *defcd* and *cfgbc* gives

$$(2)\ \text{Loop } defcd: \qquad 4\text{ V} - (3\ \Omega)I_2 - (5\ \Omega)I_3 = 0$$

$$(3)\ \text{Loop } cfgbc: \qquad 8\text{ V} - (5\ \Omega)I_1 + (3\ \Omega)I_2 = 0$$

From (1) we see that $I_1 = I_3 - I_2$, which when substituted into (3) gives

$$(4) \qquad 8\text{ V} - (5\ \Omega)I_3 + (8\ \Omega)I_2 = 0$$

Subtracting (4) from (2), we eliminate I_3 and find

$$I_2 = -\frac{4}{11}\text{ A} = -0.364\text{ A}$$

Since I_2 is negative, we conclude that I_2 is from c to f through the 3-Ω resistor. Using this value of I_2 in (3) and (1) gives the following values for I_1 and I_3:

$$I_1 = 1.38\text{ A} \qquad I_3 = 1.02\text{ A}$$

Under state-steady conditions, the capacitor represents an *open* circuit, and so there is no current in the branch *ghab*.

(b) What is the charge on the capacitor?

We can apply Kirchhoff's second rule to loop *abgha* (or any loop that contains the capacitor) to find the potential difference V_c across the capacitor:

$$-8\text{ V} + V_c - 3\text{ V} = 0$$

$$V_c = 11\text{ V}$$

Since $Q = CV_c$, we find that the charge on the capacitor is equal to

$$Q = (6\ \mu\text{F})(11\text{ V}) = 66\ \mu\text{C}$$

You should check the result for V_c by traversing any other loop, such as the outside loop. Why is the left side of the capacitor positively charged?

Complex networks with many loops and junctions generate large numbers of independent, linear equations and a corresponding large number of unknowns. Such situations can be handled formally using matrix algebra. Computer programs can also be written to solve such large numbers of linear equations.

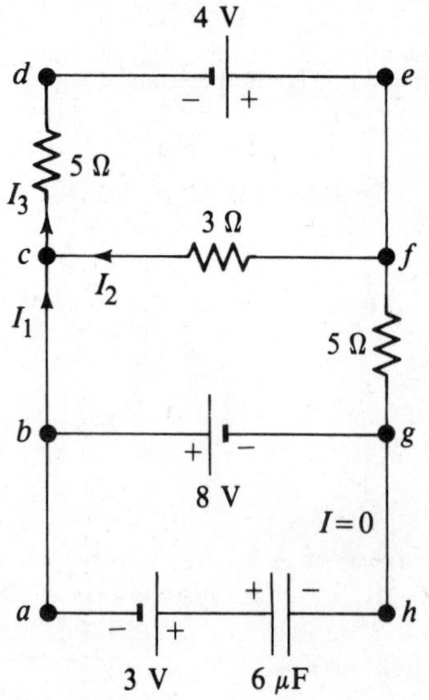

Figure 25.12 (Example 25.7) A multiloop circuit. Note that Kirchhoff's loop equation can be applied to *any* closed loop, including one containing the capacitor.

Q4. With reference to Fig. 25.12, suppose the wire between points g and h is replaced by a 10-Ω resistor. Explain why this change will *not* affect the currents calculated in Example 25.7.

Q5. Is the direction of current through a battery always from negative to positive on the terminals? Explain.

25.4 *RC* CIRCUITS

So far we have been concerned with circuits with constant currents, or so-called *steady-state circuits.* We shall now consider circuits containing capacitors, in which the currents vary in time. When a potential difference is applied across a capacitor, the rate at which it charges up depends on its capacitance and on the resistance in the circuit.

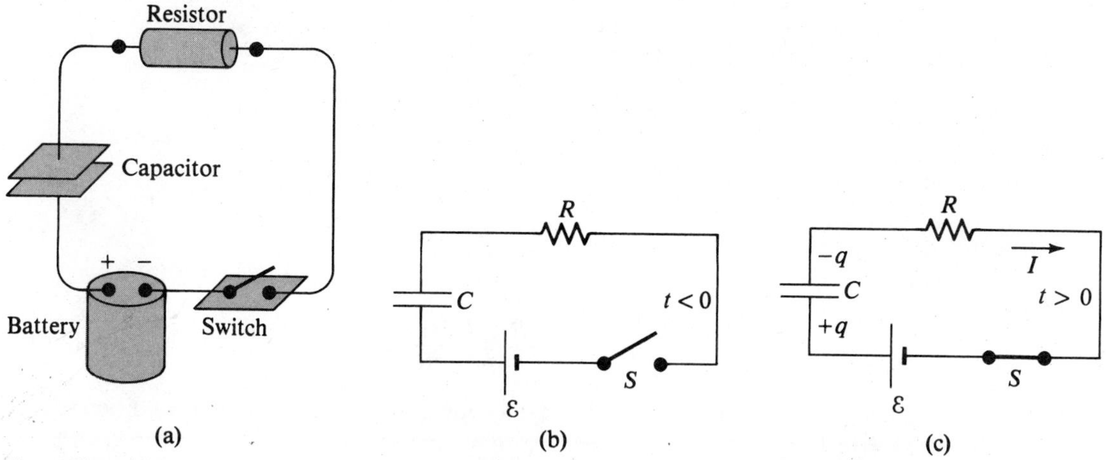

Figure 25.13 (a) A capacitor in series with a resistor, battery, and switch. (b) Circuit diagram representing this system before the switch is closed, $t < 0$. (c) Circuit diagram after the switch is closed, $t > 0$.

Consider the series circuit shown in Fig. 25.13. Let us assume that the capacitor is initially uncharged. There is no current when the switch S is open (Fig. 25.13b). If the switch is closed at $t = 0$, there will be a current through the resistor and the capacitor will begin to charge (Fig. 25.13c). Note that during the charging process, charges do not jump across the plates of the capacitor since the gap between the plates represents an open circuit. Instead, charge is transferred from one plate to the other through the resistor, switch, and battery until the capacitor is fully charged. The value of the maximum charge depends on the emf of the battery. Once the maximum charge is reached, the current in the circuit is zero.

To put this discussion on a quantitative basis, let us apply Kirchhoff's second rule to the circuit *after* the switch is closed. This gives

$$\mathcal{E} - IR - \frac{q}{C} = 0 \qquad (25.9)$$

where IR is the potential drop across the resistor and q/C is the potential drop across the capacitor. Note that q and I are *instantaneous* values of the charge and current, respectively, as the capacitor is being charged.

We can use Eq. 25.9 to find the initial current in the circuit and the maximum charge on the capacitor. At $t = 0$, when the switch is closed, the charge on the capacitor is zero, and from Eq. 25.9 we find that the initial current in the circuit, I_0, is a maximum and equal to

$$I_0 = \frac{\mathcal{E}}{R} \qquad \text{(current at } t = 0\text{)} \qquad (25.10)$$

Maximum current

At this time, *the potential drop is entirely across the resistor.* Later, when the capacitor is charged to its maximum value Q, charges cease to flow, the current in the circuit is zero, and *the potential drop is entirely across the capacitor.* Substituting $I = 0$ into Eq. 25.9 gives the following expression for Q:

$$Q = C\mathcal{E} \qquad \text{(maximum charge)} \qquad (25.11)$$

Maximum charge on the capacitor

To determine analytical expressions for the time dependence of the charge and current, we must solve Eq. 25.9, a single equation containing two variables, q and I. In order to do this, let us differentiate Eq. 25.9 with respect to time. Since $\mathcal{E}$ is a constant, $\frac{d\mathcal{E}}{dt} = 0$ and we get

$$\frac{d}{dt}\left(\mathcal{E} - \frac{q}{C} - IR\right) = 0 - \frac{1}{C}\frac{dq}{dt} - R\frac{dI}{dt} = 0$$

Recalling that $I = \frac{dq}{dt}$, we can express this equation in the form

$$R\frac{dI}{dt} + \frac{I}{C} = 0$$

$$\frac{dI}{I} = -\frac{1}{RC}\,dt \tag{25.12}$$

Since R and C are constants, this can be integrated using the initial condition that at $t = 0$, $I = I_0$:

$$\int_{I_0}^{I} \frac{dI}{I} = -\frac{1}{RC}\int_0^t dt$$

$$\ln\left(\frac{I}{I_0}\right) = -\frac{t}{RC}$$

Current versus time

$$I(t) = I_0\, e^{-t/RC} = \frac{\mathcal{E}}{R}\, e^{-t/RC} \tag{25.13}$$

where $I_0 = \mathcal{E}/R$ is the initial current.

In order to find the charge on the capacitor as a function of time, we can substitute $I = dq/dt$ into 25.13 and integrate once more:

$$\frac{dq}{dt} = \frac{\mathcal{E}}{R}\, e^{-t/RC}$$

$$dq = \frac{\mathcal{E}}{R}\, e^{-t/RC}\, dt$$

We can integrate this expression using the condition that $q = 0$ at $t = 0$:

$$\int_0^q dq = \frac{\mathcal{E}}{R}\int_0^t e^{-t/RC}\, dt$$

In order to integrate the right side of this expression, we use the fact that $\int e^{-ax}\, dx = -\frac{1}{a}\, e^{-ax}$. The result of the integration gives

Charge versus time

$$q(t) = C\mathcal{E}[(1 - e^{-t/RC}] = Q[1 - e^{-t/RC}] \tag{25.14}$$

where $Q = C\mathcal{E}$ is the *maximum* charge on the capacitor.

Plots of Eqs. 25.13 and 25.14 are shown in Figs. 25.14 and 25.15. Note that the charge is zero at $t = 0$ and approaches the maximum value of $C\mathcal{E}$ as $t \rightarrow \infty$. Furthermore, the current has its maximum value of $I_0 = \mathcal{E}/R$ at $t = 0$ and decays exponentially to zero as $t \rightarrow \infty$. The quantity RC, which appears in the exponential of Eqs. 25.13 and 25.14, is called the *time constant*, τ, of the circuit. It represents the time it takes the current to decrease to $1/e$ of its initial value; That is, in a time τ, $I = e^{-1}I_0 = 0.37I_0$. In a time 2τ, $I = e^{-2}I_0 = 0.135I_0$, and so forth. Likewise, in a time τ the charge will increase from zero to $C\mathcal{E}[1 - e^{-1}] = 0.63C\mathcal{E}$.

The following dimensional analysis shows that τ has the unit of time:

$$[\tau] = [\mathrm{RC}] = \left[\frac{\mathrm{V}}{\mathrm{I}} \times \frac{\mathrm{Q}}{\mathrm{V}}\right] = \left[\frac{\mathrm{Q}}{\mathrm{Q/T}}\right] = [\mathrm{T}]$$

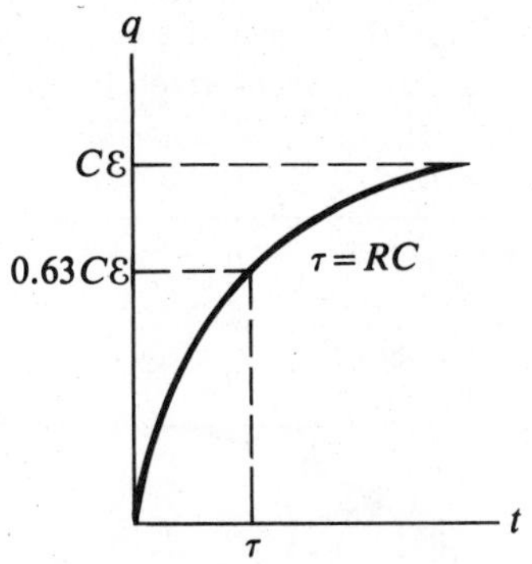

Figure 25.14 Plot of the charge on the capacitor versus time for the circuit shown in Fig. 25.13. After one time constant τ, the charge is 0.63 of the maximum value, $C\mathcal{E}$. The charge approaches its maximum value as t approaches infinity.

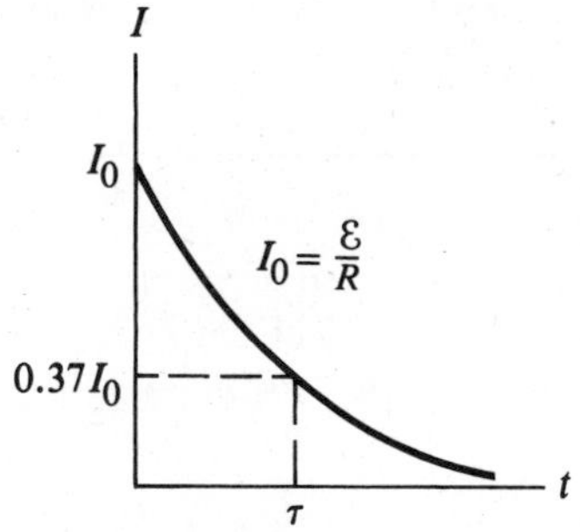

Figure 25.15 Plot of the current versus time for the RC circuit shown in Fig. 25.13. At $t = 0$, the current has its maximum value, $I_0 = \mathcal{E}/R$ and decays to zero exponentially as t approaches infinity. After one time constant τ, the current decreases to 0.37 of its initial value.

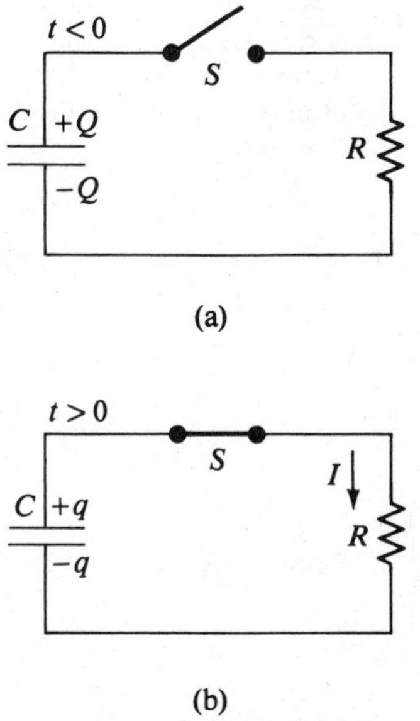

Figure 25.16 (a) A charged capacitor connected to a resistor and a switch, which is opened at $t < 0$. (b) After the switch is closed, a nonsteady current is set up in the direction shown and the charge on the capacitor decreases exponentially with time.

The work done by the battery during the charging process is $Q\mathcal{E} = C\mathcal{E}^2$. After the capacitor is fully charged, the energy stored in the capacitor is $\frac{1}{2}Q\mathcal{E} = \frac{1}{2}C\mathcal{E}^2$, which is just half the work done by the battery. It is left as a problem to show that the remaining half of the energy supplied by the battery goes into joule heat in the resistor (Problem 25.11).

Now consider the circuit in Fig. 25.16, consisting of a capacitor with an initial charge Q, a resistor, and a switch. When the switch is open (Fig. 25.16a), there is a potential difference of Q/C across the capacitor and zero potential difference across the resistor since $I = 0$. If the switch is closed at $t = 0$, the capacitor begins to discharge through the resistor. At some time during the discharge, the current in the circuit is I and the charge on the capacitor is q (Fig. 25.16b). From Kirchhoff's second rule, we see that the potential drop across the resistor, IR, must equal the potential difference across the capacitor, q/C:

$$IR = q/C \tag{25.15}$$

However, the current in the circuit must equal the rate of *decrease* of charge on the capacitor. That is, $I = -dq/dt$, and so Eq. 25.15 becomes

$$-R\frac{dq}{dt} = \frac{q}{C}$$

$$\frac{dq}{q} = -\frac{1}{RC}\,dt \tag{25.16}$$

Integrating this expression using the fact that $q = Q$ at $t = 0$ gives

$$\int_Q^q \frac{dq}{q} = -\frac{1}{RC}\int_0^t dt$$

$$\ln\left(\frac{q}{Q}\right) = -\frac{t}{RC}$$

$$q(t) = Q\,e^{-t/RC} \tag{25.17}$$

Differentiating Eq. 25.17 with respect to time gives the current as a function of time:

$$I = -\frac{dq}{dt} = \frac{Q}{RC}e^{-t/RC} = I_0\,e^{-t/RC} \tag{25.18}$$

Photograph of a large-scale integrating (VLSI) circuit which includes over 65,000 devices on a single chip. (Dimensions are about 1 cm × 1 cm.) This particular chip is an interface processor that communicates with input-output devices. It is part of a 32-bit microprocessor (the APX 432), which integrates about 200,000 transistors and can execute two million instructions per second. (Courtesy Intel Corporation)

where the initial current $I_0 = Q/RC$. Therefore, we see that both the charge on the capacitor and the current decay exponentially at a rate characterized by the time constant $\tau = RC$.

Example 25.8

An uncharged capacitor and a resistor are connected in series to a battery as in Fig. 25.17. If $\mathcal{E} = 12$ V, $C = 5\ \mu$F, and $R = 8 \times 10^5\ \Omega$, find the time constant of the circuit, the maximum charge on the capacitor, the maximum current in the circuit, and the charge and current as a function of time.

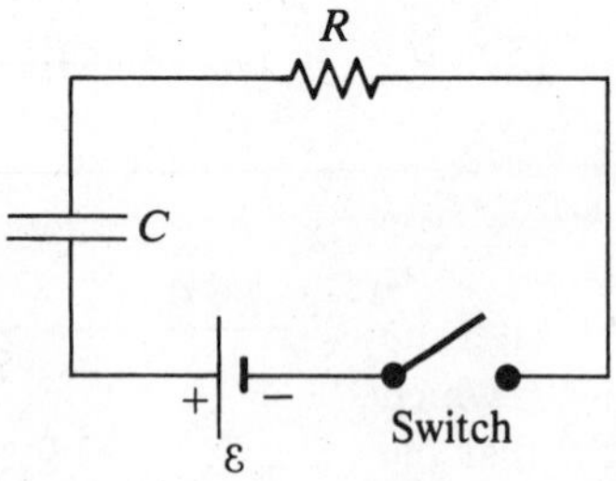

Figure 25.17 (Example 25.8) The switch is closed at $t = 0$.

Solution: The time constant of the circuit is $\tau = RC = (8 \times 10^5\ \Omega)(5 \times 10^{-6}\ \text{F}) = 4$ s. The maximum charge on the capacitor is $Q = C\mathcal{E} = (5 \times 10^{-6}\ \text{F})(12\ \text{V}) = 60\ \mu$C. The maximum current in the circuit is $I_0 = \mathcal{E}/R = (12\ \text{V})/(8 \times 10^5\ \Omega) = 15\ \mu$A. Using these values and Eqs. 25.13 and 25.14, we find that

$$q(t) = 60[1 - e^{-t/4}]\ \mu\text{C}$$

$$I(t) = 15\ e^{-t/4}\ \mu\text{A}$$

Graphs of these functions are given in Fig. 25.18.

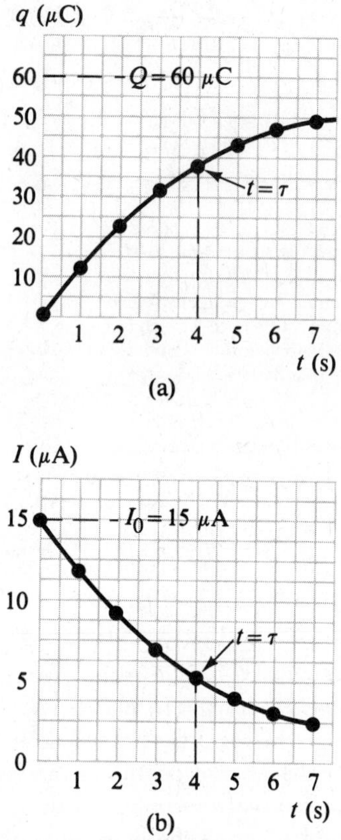

Figure 25.18 (Example 25.8) Plots of (a) charge versus time and (b) current versus time for the *RC* circuit shown in Fig. 25.17, with $\mathcal{E} = 12$ V, $R = 8 \times 10^5\ \Omega$, and $C = 5\ \mu$F.

Q6. With reference to Fig. 25.19, describe what happens to the light bulb after the switch is closed. Assume the capacitor is initially uncharged and assume that the light will illuminate when connected directly across the battery terminals.

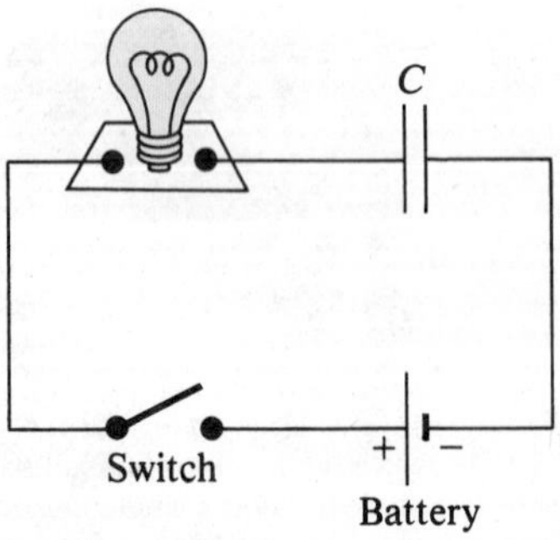

Figure 25.19 (Question 6).

25.5 MEASUREMENTS OF RESISTANCE

From Ohm's law, $V = IR$, it is easy to see that the resistance of a conductor can be obtained if the current through the conductor and the potential difference across it can be measured simultaneously. The current in a circuit can be measured with an ammeter placed in series with the resistor to be measured, as in Fig. 25.20. Ideally, the resistance of the ammeter should be zero so as not to alter the current to be measured. Since any ammeter always has some resistance, its reading will be somewhat less than the true current in the circuit. The potential difference across the resistor can be measured with a voltmeter connected in parallel with the resistor as in Fig. 25.21. An ideal voltmeter should have *infinite* resistance so that no current will pass through it. In practice, the resistance of the voltmeter should be large compared with the resistance to be measured. The effect of the voltmeter on the

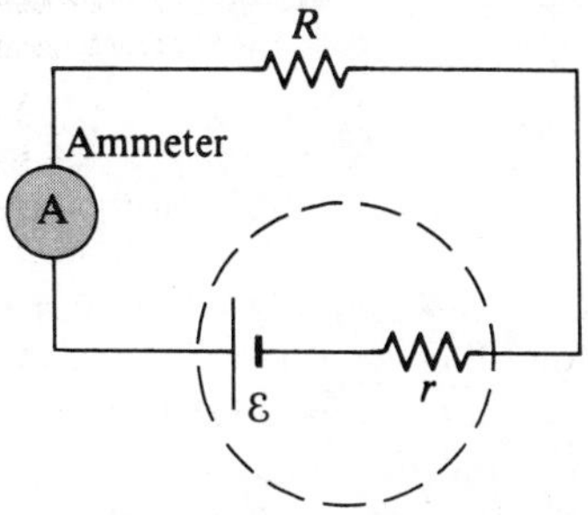

Figure 25.20 The current in a circuit can be measured with an ammeter connected in series with the resistor and battery. An ideal ammeter has zero resistance.

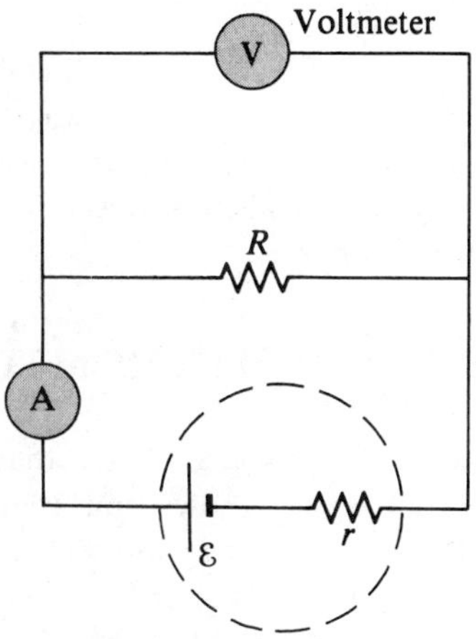

Figure 25.21 The potential difference across a resistor can be measured with a voltmeter connected in parallel with the resistor. An ideal voltmeter has infinite resistance and does not affect the circuit.

circuit is to reduce the resistance of the circuit, causing an increase in the current read by the ammeter and a change in the potential difference being measured.

The ammeter-voltmeter method is not recommended for precise measurements since most meters cannot be read to more than three-digit accuracy. If this method is used, the internal resistance of the voltmeter should be much greater than R and the ammeter's resistance should be much less than R.

Unknown resistances can be accurately measured using a circuit known as a *Wheatstone bridge* (Fig. 25.22). This circuit consists of the unknown resistance, R_x, three known resistors, R_1, R_2, and R_3 (where R_1 is a calibrated variable resistor), a sensitive current-detector called a *galvanometer*,[2] and a source of emf. The principle of its operation is quite simple. The known resistor R_1 is varied until the galvanometer reading is zero, that is, until there is no current from a to b. Under this condition the bridge is said to be balanced.[3] Since the potential at point a equals the potential at point b when the bridge is balanced, the potential difference across R_1 must equal the potential difference across R_x. Likewise, the potential difference across R_2 must equal the potential difference across R_3. From these considerations, we see that

$$(1) \qquad I_1R_1 = I_2R_x$$

$$(2) \qquad I_1R_2 = I_2R_3$$

Dividing (1) by (2) eliminates the currents, and solving for R_x we find

$$R_x = \frac{R_1R_3}{R_2} \tag{25.19}$$

Since R_1, R_2, and R_3 are known quantities, R_x can be calculated. There are a number of similar devices that use the null measurement, such as a capacitance bridge (used to measure unknown capacitances). These devices do not require the use of calibrated meters and can be used with any source of emf.

When very high resistances are to be measured (above $10^5\ \Omega$), the Wheatstone

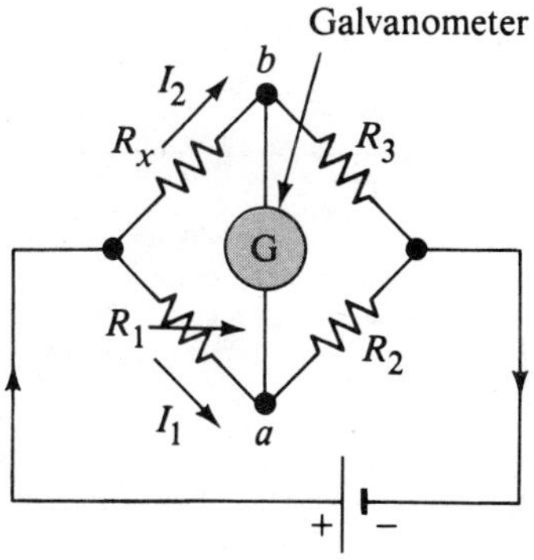

Figure 25.22 Circuit diagram for a Wheatstone bridge. This circuit is often used to measure an unknown resistance R_x in terms of known resistances R_1, R_2, and R_3. When the bridge is balanced, there is no current in the galvanometer.

[2]The details of the galvanometer will be described in Chapter 26.

[3]Measurements in which adjustments are made until some quantity is zero are called *null measurements*.

bridge method becomes difficult for technical reasons. As a result of recent advances in the technology of such solid state devices as the field-effect transistor, modern electronic instruments are capable of measuring resistances as high as $10^{12}\ \Omega$. Such instruments are designed to have an extremely high effective resistance between their input terminals. For example, input resistances of $10^{10}\ \Omega$ are common in most digital multimeters.

25.6 THE POTENTIOMETER

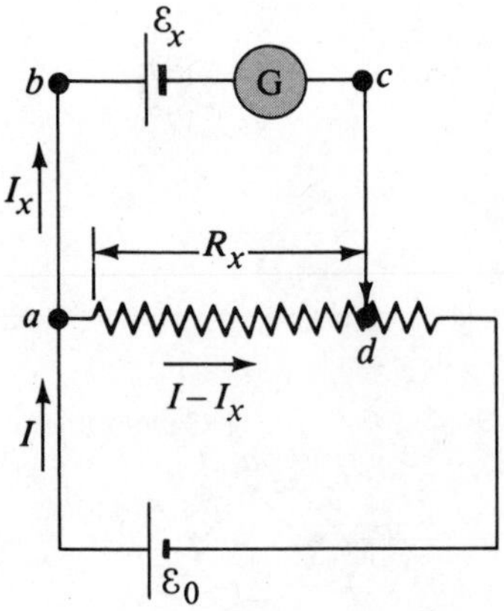

Figure 25.23 Circuit diagram for a potentiometer. The circuit is used to measure an unknown emf $\mathcal{E}_x$ in terms of a known emf $\mathcal{E}_s$, provided by a standard cell.

A *potentiometer* is a circuit that is used to measure an unknown emf, $\mathcal{E}_x$, by comparison with a known emf. Figure 25.23 shows the essential components of the potentiometer. Point d represents a sliding contact used to vary the resistance (and hence the potential difference) between points a and d.

Choosing the currents in the directions shown, we see from Kirchhoff's first rule that the current through the variable resistor is $I - I_x$, where I is the current in the lower branch (through the battery of emf $\mathcal{E}_0$) and I_x is the current in the upper branch. Neglecting the internal resistance of the unknown cell, Kirchhoff's second rule applied to loop $abcd$ gives

$$-\mathcal{E}_x + (I - I_x)R_x = 0$$

where R_x is the resistance between points a and d. The sliding contact at d is now adjusted until the galvanometer reads zero (a balanced circuit). Under this condition, the current in the galvanometer and in the unknown cell is *zero* ($I_x = 0$), and the potential difference between a and d equals the unknown emf, $\mathcal{E}_x$. That is,

$$\mathcal{E}_x = IR_x$$

Next, the cell of unknown emf is replaced by a standard cell of known emf, $\mathcal{E}_s$, and the above procedure is repeated. That is, the moving contact at d is varied until a balance is obtained. If R_s is the resistance between a and d when balance is achieved, then

$$\mathcal{E}_s = IR_s$$

where it is assumed that I remains the same.

Combining this expression with the previous equation, $\mathcal{E}_x = IR_x$, we see that

$$\mathcal{E}_x = \frac{R_x}{R_s}\mathcal{E}_s \tag{25.20}$$

This result shows that the unknown emf can be determined from a knowledge of the standard-cell emf and the ratio of the two resistances.[4]

Q7. Although the internal resistance of the unknown and known emfs was neglected in this treatment, it is really not necessary to make this assumption. Explain why the internal resistances play no role in this measurement.

25.7 HOUSEHOLD WIRING AND ELECTRICAL SAFETY

Household circuits represent a very practical application of some of the ideas we have presented in this chapter concerning circuit analysis. In our world of electrical

[4]In the so-called sliding-wire potentiometer, the variable resistor is a wire of variable length and the ratio of resistances equals the ratio of two lengths of the same wire.

appliances, it is useful to understand the power requirements and limitations of conventional electrical systems and the safety measures that should be practiced to prevent accidents.

In a conventional installation, the utilities company distributes electrical power to individual homes with a pair of power lines. Each user is connected in parallel to these lines, as shown in Fig. 25.24. The potential difference between these wires is about 120 V. The voltage alternates in time, but for the present discussion we shall assume a steady direct current (dc) voltage. (Alternating voltages and currents will be discussed in Chapter 31.) One of the wires is connected to ground, designated by the symbol ⏚, and the potential on the "live" wire oscillates relative to ground.[5] Hence, the 120 V can be considered a time-averaged voltage.

A meter and circuit breaker (or in older installation, a fuse) are connected in series with the wire entering the house as indicated in Fig. 25.24. The circuit breaker is a device that protects against too large a current, which can cause overheating and fires. When the current exceeds some safe value (typically 15 A or 30 A), the circuit breaker disconnects the voltage source from the load. Some circuit breakers make use of the principle of the bimetallic strip discussed in Chapter 16.

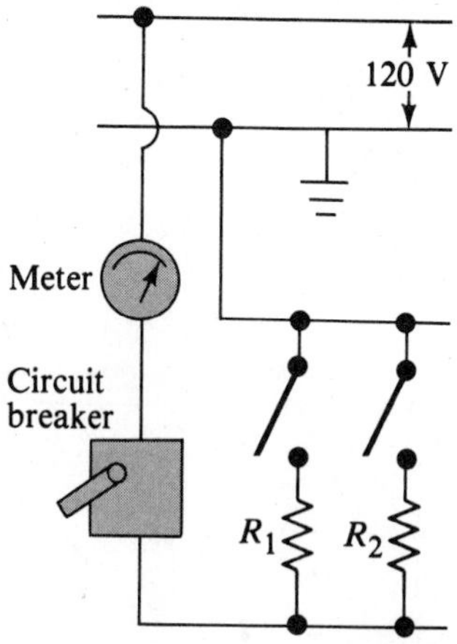

Figure 25.24 Wiring for a household circuit. The resistances R_1 and R_2 represent appliances or other electrical devices.

The wire and circuit breaker are carefully selected to meet the current demands for that circuit. If a circuit is to carry currents as large as 30 A, a heavy wire and appropriate circuit breaker must be selected to handle this current. Other individual household circuits, which are normally used to power lamps and small appliances, often require only 15 A. Therefore, each circuit has its own circuit breaker to accommodate various load conditions.

As an example, consider a circuit in which a toaster, a microwave oven, and a heater are in the same circuit (corresponding to R_1, R_2, . . . in Fig. 25.24). We can calculate the current through each appliance using the expression $P = IV$. The toaster, rated at 1000 W, would draw a current of $1000/120 = 8.33$ A. The microwave oven, rated at 800 W, would draw a current of 6.67 A, and the electric heater, rated at 1300 W, would draw a current of 10.8 A. If the three appliances are operated simultaneously, they will draw a total current of 25.8 A. Therefore, the circuit should be wired to handle at least this much current. In order to accommodate a small additional load, such as a 100-W lamp, a 30-A circuit should be installed. Alternatively, one could operate the toaster and microwave oven on one 15-A circuit and the heater on a separate 15-A circuit.

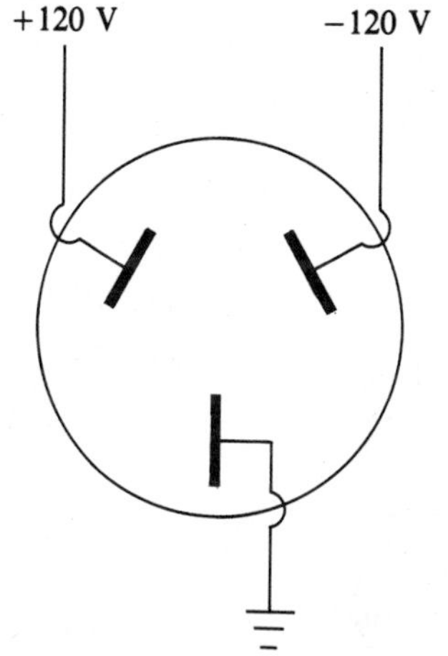

Figure 25.25 Power connections for a 240-V appliance.

Many heavy-duty appliances, such as electric ranges and clothes dryers, require 240 V for their operation. The power company supplies this voltage by providing a third live wire, which is 120 V *below* ground potential (Fig. 25.25). Therefore, the potential difference between this wire and the other live wire (which is 120 V above ground potential) is 240 V. An appliance that operates from a 240-V line requires half the current of one operating from a 120-V line; therefore smaller wires can be used in the higher-voltage circuit without overheating becoming a problem.

ELECTRICAL SAFETY

When the live wire of an electrical outlet is connected directly to ground, the circuit is completed and a short-circuit condition exists. If this were to happen accidentally, a properly operating circuit breaker would "break" the circuit. On the other hand, a person can be electrocuted by touching the live wire (which commonly happens because of a frayed cord or other exposed conductors) while in contact with ground. The ground contact might be made either by the person's touching a water pipe (which is normally at ground potential) or by standing on ground with wet feet, since water is a good electrical conductor. Such situations should be avoided at all costs.

Electrical shock can result in fatal burns, or it can cause the muscles of vital organs, such

[5]The phrase *live wire* is common jargon for a conductor whose potential is above or below ground.

as the heart, to malfunction. The degree of damage to the body depends on the magnitude of the current, the length of time it acts, and the part of the body through which the current passes. Currents of 5 mA or less can cause a sensation of shock, but ordinarily do little or no damage. If the current is larger than about 10 mA, the hand muscles contract and the person may be unable to release the live wire. If a current of about 100 mA passes through the body for only a few seconds, the result could be fatal. Such large currents will paralyze the respiratory muscles and prevent breathing. In some cases, currents of about 1 A through the body may produce serious (and sometimes fatal) burns.

Many 110-V outlets are designed to take a three-pronged power cord. (This feature is required in all new electrical installations.) One of these prongs is the live wire and two are common with ground. The additional ground connection is provided as a safety feature. Many appliances contain a three-pronged 110-V power cord with one of the ground wires connected directly to the casing of the appliance. If the live wire is accidentally shorted to ground (which often occurs when the wire insulation wears off), the current will take the low-resistance path through the appliance to ground. On the other hand, if the casing of the appliance is not properly grounded and a short occurs, anyone in contact with the appliance will experience an electric shock since his or her body will provide a low-resistance path to ground.

Q8. When electricians work with potentially live wires, they often use the backs of their hands or fingers to move wires. Why do you suppose they use this technique?

Q9. Why is it possible for a bird to sit on a high-voltage wire without being electrocuted?

Q10. What procedure would you use to try to save a person who is "frozen" to a live high-voltage wire without endangering your own life?

25.8 SUMMARY

The *emf* of a battery is equal to the voltage across its terminals when the current is zero. That is, the emf is equivalent to the open-circuit voltage of the battery.

The *equivalent resistance* of a set of resistors connected in *series* is given by

Resistors in series

$$R_{eq} = R_1 + R_2 + R_3 + \cdots \tag{25.6}$$

The *equivalent resistance* of a set of resistors connected in *parallel* is given by

Resistors in parallel

$$\frac{1}{R_{eq}} = \frac{1}{R_1} + \frac{1}{R_2} + \frac{1}{R_3} + \cdots \tag{25.8}$$

Kirchhoff's rules

Complex circuits involving more than one loop are conveniently analyzed using two simple rules called *Kirchhoff's rules:*

1. The sum of the currents entering any junction must equal the sum of the currents leaving that junction.
2. The sum of the potential differences across each element around any closed-circuit loop must be *zero.*

The first rule is a statement of *conservation of charge.* The second rule is equivalent to a statement of *conservation of energy.*

When a resistor is traversed in the direction of the current, the change in potential, ΔV, across the resistor is $-IR$. If a resistor is traversed in the direction opposite the current, $\Delta V = +IR$.

If a seat of emf is traversed in the direction of the emf (negative to positive) the change in potential is $+\mathcal{E}$. If it is traversed opposite the emf (positive to negative), the change in potential is $-\mathcal{E}$.

If a capacitor is charged with a battery of emf $\mathcal{E}$ through a resistance R, the current in the circuit and charge on the capacitor vary in time according to the expressions

$$I(t) = \frac{\mathcal{E}}{R} e^{-t/RC} \qquad (25.13)$$

Current versus time

$$q(t) = Q[1 - e^{-t/RC}] \qquad (25.14)$$

Charge versus time

where $Q = C\mathcal{E}$ is the *maximum* charge on the capacitor. The product RC is called the *time constant* of the circuit.

If a charged capacitor is discharged through a resistance R, the charge and current decrease exponentially in time according to the expressions

$$q(t) = Q\,e^{-t/RC} \qquad (25.17)$$

$$I(t) = I_0\,e^{-t/RC} \qquad (25.18)$$

where $I_0 = Q/RC$ is the initial current in the circuit and Q is the initial charge on the capacitor.

A *Wheatstone bridge* is a circuit that can be used to measure an unknown resistance.

A *potentiometer* is a circuit that can be used to measure an unknown emf.

EXERCISES

Section 25.1 Electromotive Force

1. A battery with an emf of 8 V and internal resistance of 0.5 Ω is connected across a load resistor R. If the current in the circuit is 2 A, what is the value of R?

2. What power is dissipated in the internal resistance of the battery in the circuit described in Exercise 1?

3. A certain battery has an open-circuit voltage of 40 V. A load resistance of 10 Ω reduces the terminal voltage to 38 V. What is the value of the internal resistance of the battery?

4. The current in a loop circuit that has a resistance of R_1 is 2 A. The current is reduced to 1.6 A when an additional resistor $R_2 = 3$ Ω is added in series with R_1. What is the value of R_1?

5. If the emf of a battery is 12 V and a current of 50 A is measured when the battery is shorted, what is the internal resistance of the battery?

6. A battery has an emf of 12 V. The terminal voltage of the battery is 10.8 V when it is delivering 18 W of power to an external load resistor R. (a) What is the value of R? (b) What is the internal resistance of the battery?

7. What potential difference will be measured across a 12-Ω load resistor when it is connected across a battery of emf 6 V and internal resistance 0.15 Ω?

Section 25.2 Resistors in Series and in Parallel

8. Two circuit elements with fixed resistances R_1 and R_2 are connected in *series* with a 6-V battery and a switch. The battery has an internal resistance of 5 Ω, $R_1 = 132$ Ω, and $R_2 = 56$ Ω. (a) What is the current through R_1 when the switch is closed? (b) What is the voltage across R_2 when the switch is closed?

9. The components of Exercise 8 are reconnected with R_1 and R_2 in *parallel* across the battery. (a) What is the voltage across R_1 when the switch is closed? (b) What is the current in each resistor?

10. Find the equivalent resistance between points a and b in Fig. 25.26.

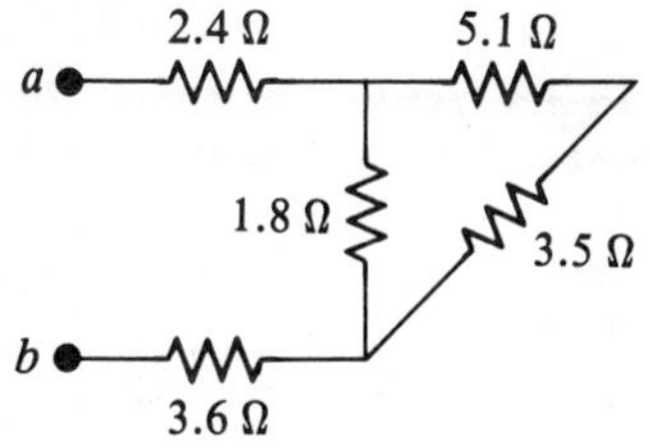

Figure 25.26 (Exercise 10).

11. Find the equivalent resistance between points a and b in Fig. 25.27.

12. A potential difference of 25 V is applied between points a and b in Fig. 25.27. Calculate the current in each resistor.

13. Consider the combination of resistors in Fig. 25.28. (a) Find the resistance between points a and b. (b) If

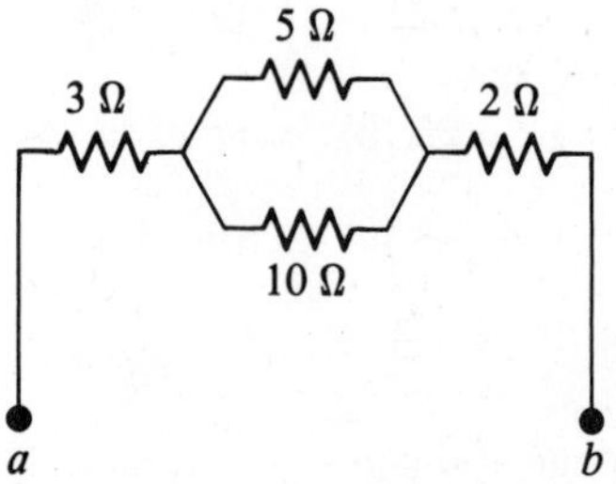

Figure 25.27 (Exercises 11 and 12).

the current in the 5-Ω resistor is 1 A, what is the potential difference between points *a* and *b*?

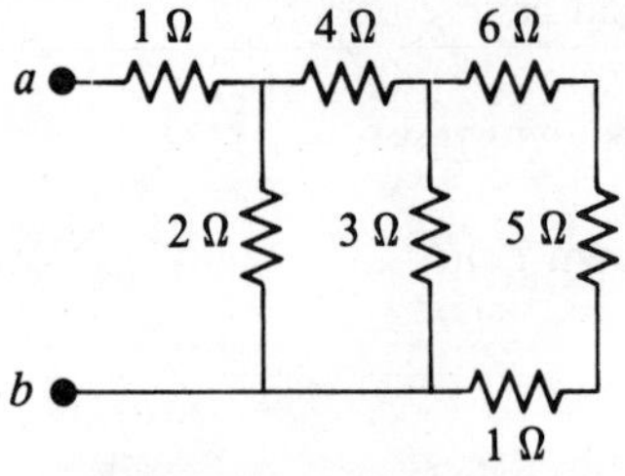

Figure 25.28 (Exercise 13).

14. Consider the circuit shown in Fig. 25.29. Find (a) the current in the 15-Ω resistor and (b) the potential difference between points *a* and *b*.

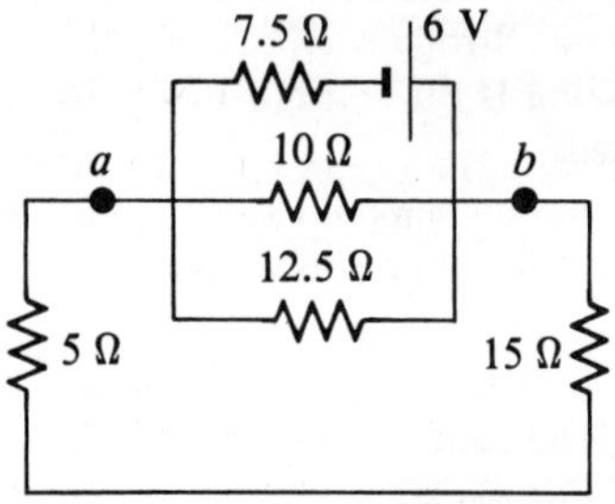

Figure 25.29 (Exercise 14).

Section 25.3 Kirchhoff's Rules (the currents are not necessarily in the directions shown for some circuits.)

15. Find the currents I_1, I_2, and I_3 in the circuit shown in Fig. 25.30.

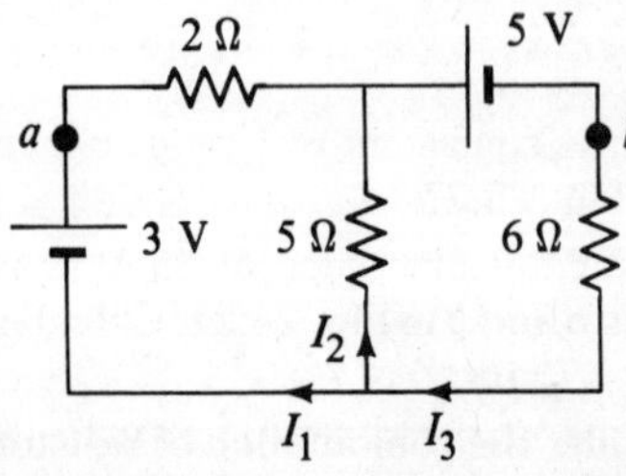

Figure 25.30 (Exercises 15 and 16).

16. Find the potential difference between points *a* and *b* in the circuit in Fig. 25.30.

17. The ammeter in the circuit shown in Fig. 25.31 reads 1 A. Find the currents I_1 and I_2 and the value of $\mathcal{E}$.

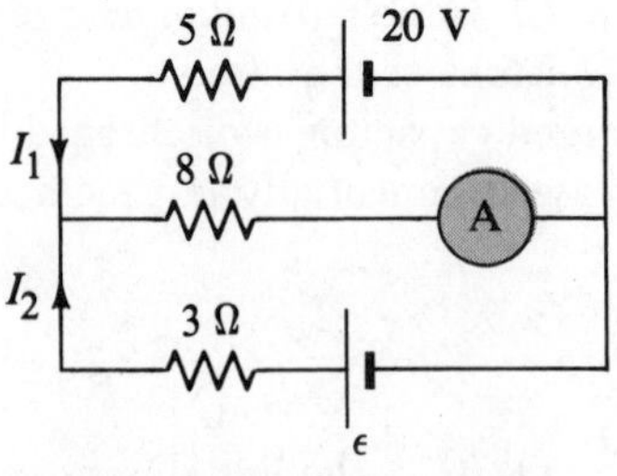

Figure 25.31 (Exercise 17).

18. Consider the circuit shown in Fig. 25.32. Find the value of I_1, I_2, and I_3.

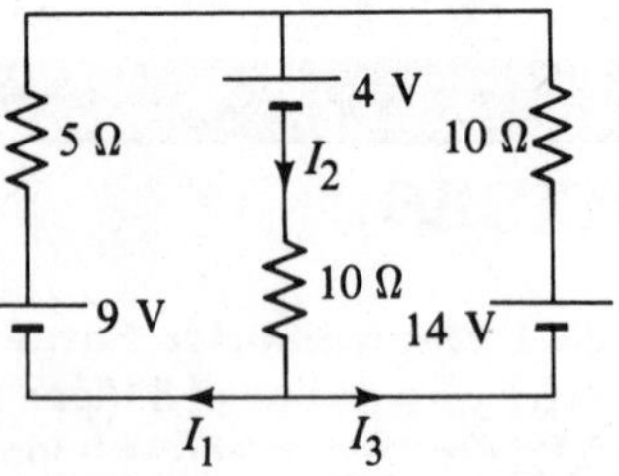

Figure 25.32 (Exercises 18 and 19).

19. (a) Find the value of I_1 and I_3 in the circuit of Fig. 25.32 if the 4-V battery is replaced by a 4-μF capacitor. (b) Determine the charge on the 4-μF capacitor.

20. For the circuit shown in Fig. 25.33, calculate (a) the current in the 2-Ω resistor, and (b) the potential difference between points *a* and *b*.

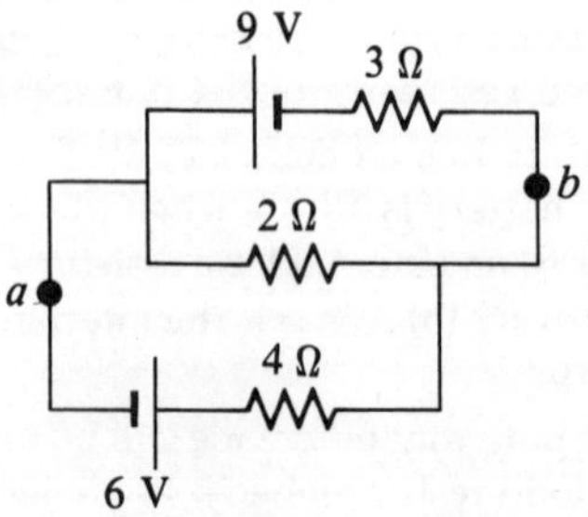

Figure 25.33 (Exercise 20).

21. Determine the value of the current in each of the four resistors shown in the circuit of Fig. 25.34.

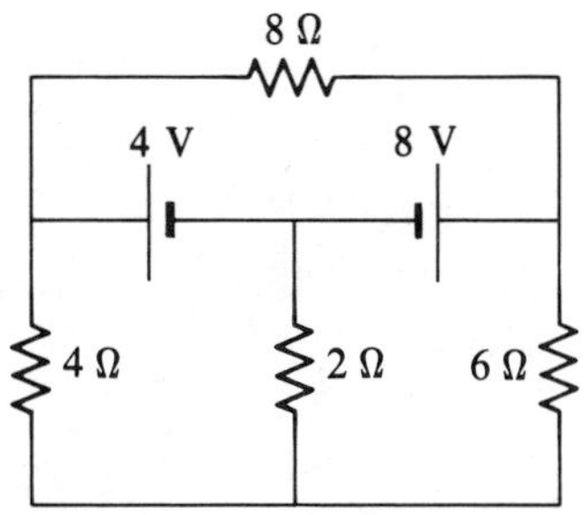

Figure 25.34 (Exercise 21).

22. (a) Calculate the value of R for the circuit shown in Fig. 25.35. (b) Determine the currents in the 6-Ω and 8-Ω resistors.

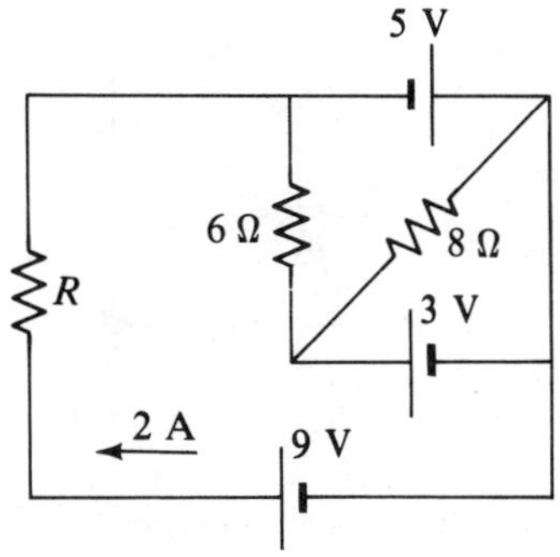

Figure 25.35 (Exercise 22).

Section 25.4 *RC* Circuits

23. A capacitor in an RC circuit is charged to 75% of its maximum value in 1.4 s. What is the time constant of the circuit?

24. Consider a series RC circuit (Fig. 25.13) for which $R = 2 \times 10^6\ \Omega$, $C = 6\ \mu\text{F}$, and $\mathcal{E} = 20$ V. Find (a) the time constant of the circuit, and (b) the *maximum* charge on the capacitor after the switch is closed.

25. The switch in the RC circuit described in Exercise 24 is closed at $t = 0$. Find the current in the resistor R at a time 24 s after the switch is closed.

26. A 3×10^{-3}-μF capacitor with an initial charge of 6.2 μC is discharged through a 1500-Ω resistor. (a) Calculate the current through the resistor 9 μs after the resistor is connected across the terminals of the capacitor. (b) What charge remains on the capacitor after 9 μs? (c) What is the maximum current through the resistor?

27. Consider the capacitor-resistor combination described in Exercise 26. (a) How much energy is stored initially in the charge capacitor? (b) If the capacitor is completely discharged through the resistor, how much energy will be dissipated as heat in the resistor?

28. Dielectric materials used in the manufacture of capacitors are characterized by conductivities that are small but not zero. Therefore, a charged capacitor will slowly lose its charge by "leaking" across the dielectric. If a certain 0.2-μF capacitor leaks charge such that the potential difference decreases to half its initial value in 5 s, what is the equivalent resistance of the dielectric?

Section 25.5 Measurements of Resistance

29. Refer to Fig. 25.21 and show that the ammeter-voltmeter connections shown can be used to obtain a satisfactory calculation for R only when the resistance of the voltmeter is large compared with the resistance being measured.

30. A Wheatstone bridge of the type shown in Fig. 25.22 is used to make a precise measurement of the resistance of a wire connector. The resistor shown in the circuit as R_3 is 1 kΩ. If the bridge is balanced by adjusting R_1 such that $R_1 = 2.5R_2$, what is the resistance of the wire connector, R_x?

31. Consider the case when the Wheatstone bridge shown in Fig. 25.22 is *unbalanced*. Calculate the current through the galvanometer when $R_x = R_3 = 10\ \Omega$, $R_2 = 20\ \Omega$, and $R_1 = 18\ \Omega$. Assume the voltage across the bridge is 55 V.

32. The Wheatstone bridge in Fig. 25.22 is balanced when $R_1 = 15\ \Omega$, $R_2 = 25\ \Omega$, and $R_3 = 40\ \Omega$. Calculate the value of R_x.

Section 25.6 The Potentiometer

33. Consider the potentiometer circuit shown in Fig. 25.23. When a standard cell of emf 1.0186 V is used in the circuit, and the resistance between a and d is 36 Ω, the galvanometer reads zero. When the standard cell is replaced by an unknown emf, the galvanometer reads zero when the resistance is adjusted to 48 Ω. What is the value of the unknown emf?

Section 25.7 Household Wiring and Electrical Safety

34. An electric heater is rated at 1300 W, a toaster is rated at 1000 W, and an electric grill is rated at 1500 W. The three appliances are connected to a common 120-V circuit. (a) How much current does each appliance draw? (b) Is a 30-A circuit sufficient in this situation? Explain.

PROBLEMS

1. The value of a resistor R is to be determined using the ammeter-voltmeter setup shown in Fig. 25.36. The ammeter has a resistance of 0.5 Ω, and the voltmeter has a resistance of 20 000 Ω. Within what range of actual values of R will the measured values be correct to within 5 percent if the measurement is made using the circuit shown in (a) Fig. 25.36a and (b) Fig. 25.36b?

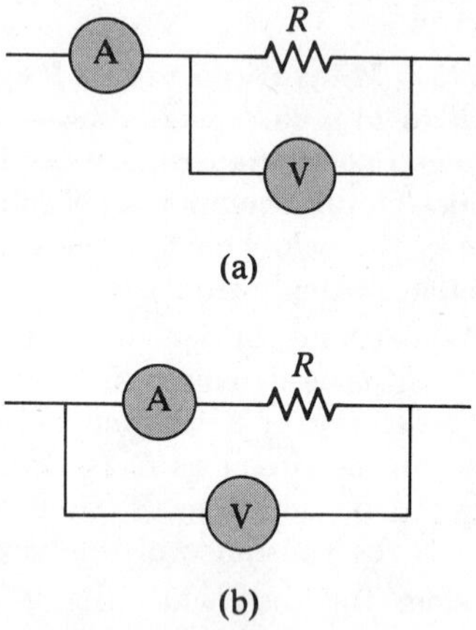

Figure 25.36 (Problem 1).

2. Consider the circuit shown in Fig. 25.37. What are the expected readings of the ammeter and voltmeter?

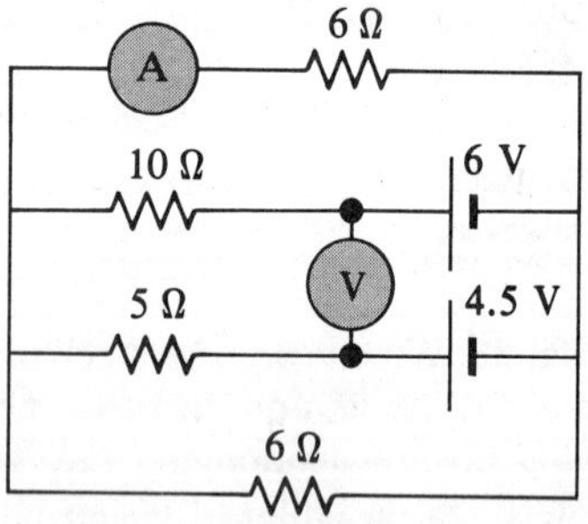

Figure 25.37 (Problem 2).

3. (a) Calculate the current through the 6-V battery in Fig. 25.38. (b) Determine the potential difference between points a and b.

4. Twelve resistors, each of value 1 Ω, are connected so that each is along one edge of a cube, as shown in Fig. 25.39. Find the resistance between the points (a) a and b, (b) a and c, and (c) b and c.

5. Two batteries with emf values of $\mathcal{E}_1$ and $\mathcal{E}_2$ and internal resistances of r_1 and r_2 are connected in parallel across a load resistor R as in Fig. 25.40. (a) Calculate the current delivered to R when $\mathcal{E}_1 = 6$ V, $\mathcal{E}_2 = 4$ V, $r_1 = 0.3$ Ω, $r_2 = 0.1$ Ω, and $R = 10$ Ω.

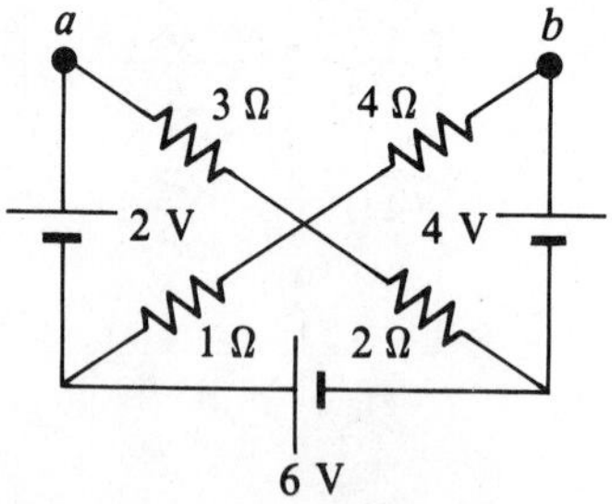

Figure 25.38 (Problem 3).

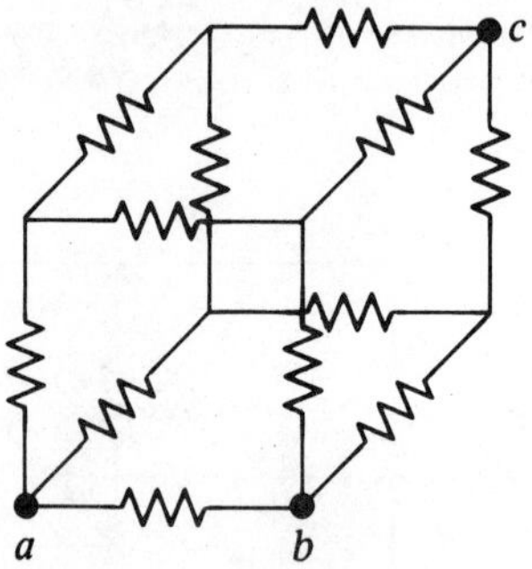

Figure 25.39 (Problem 4).

(b) Show that the "effective" emf of the two batteries in parallel is

$$\mathcal{E}_{\text{eff}} = \frac{r_1 r_2}{r_1 + r_2}\left(\frac{\mathcal{E}_1}{r_1} + \frac{\mathcal{E}_2}{r_2}\right)$$

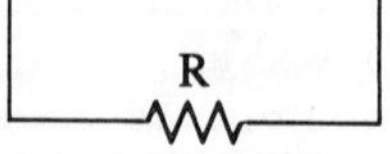

Figure 25.40 (Problem 5).

6. The values of the components in a simple RC circuit (Fig. 25.13) are as follows: $C = 1$ μF, $R = 2 \times 10^6$ Ω, and $\mathcal{E} = 10$ V. At the instant 10 s after the switch in the circuit is closed, calculate (a) the charge on the capacitor, (b) the current in the resistor, (c) the rate at which energy is being stored in the capacitor, and (d) the rate at which energy is being delivered by the battery.

7. A schematic of a device known as an *Ayrton shunt*

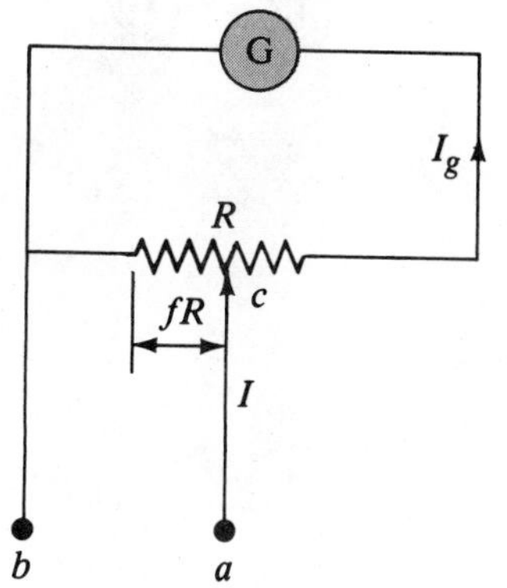

Figure 25.41 (Problem 7).

is shown in Fig. 25.41. The circuit in which the current is to be measured is connected at points a and b. The sliding contact at point c allows the fraction f of the total current that passes through the galvanometer to be varied by selecting $f = 1, 0.1, 0.01, \ldots$. If the resistance of the galvanometer is R_g, show that

$$I_g = \left(\frac{fR}{R + R_g}\right)I$$

8. Consider the circuit shown in Fig. 25.42. Calculate (a) the current in the 4-Ω resistor, (b) the potential difference between points a and b, (c) the terminal potential difference of the 4-V battery, and (d) the thermal energy expended in the 3-Ω resistor during 10 min of operation of the circuit.

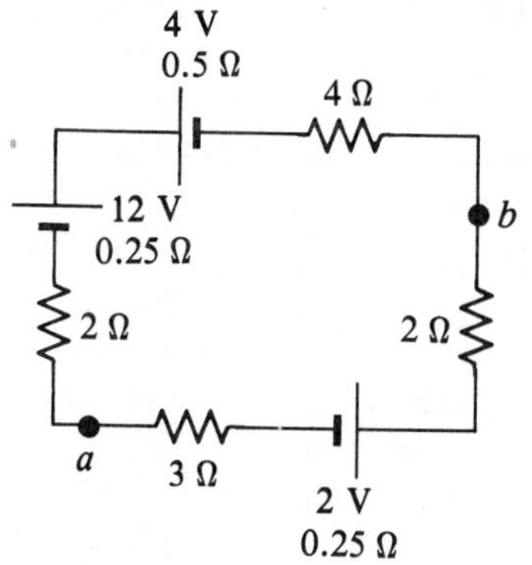

Figure 25.42 (Problem 8).

9. Consider the circuit shown in Fig. 25.43. (a) Calculate the current in the 5-Ω resistor. (b) What power is dissipated by the entire circuit? (c) Determine the potential difference between points a and b. Which point is at the higher potential?

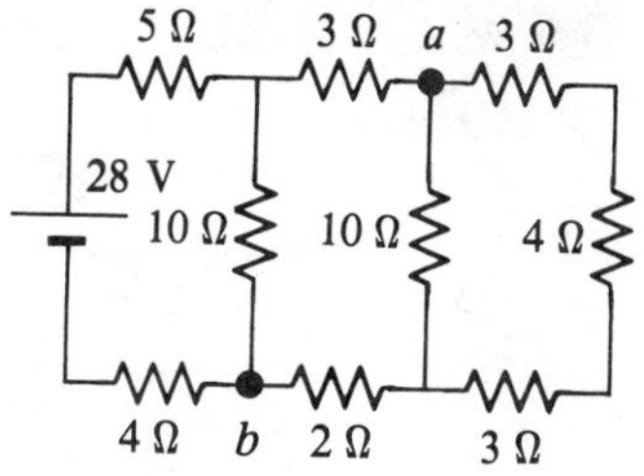

Figure 25.43 (Problem 9).

10. A dc power supply has an open-circuit emf of 40 V and an internal resistance of 2 Ω. It is used to charge two storage batteries connected in series, each having an emf of 6 V and internal resistance of 0.3 Ω. If the charging current is to be 4 A, (a) what additional resistance should be added in series? (b) Find the power lost in the supply, the batteries, and the added series resistance. (c) How much power is converted to chemical energy in the batteries?

11. A battery is used to charge a capacitor through a resistor, as in Fig. 25.13. Show that in the process of charging the capacitor, half of the energy supplied by the battery is dissipated as heat in the resistor and half is stored in the capacitor.

12. A battery has an emf $\mathcal{E}$ and internal resistance r. A variable resistor R is connected across the terminals of the battery. Find the value of R such that (a) the potential difference across the terminals is a maximum, (b) the current in the circuit is a maximum, (c) the power delivered to the resistor is a maximum.

26 Magnetic Fields

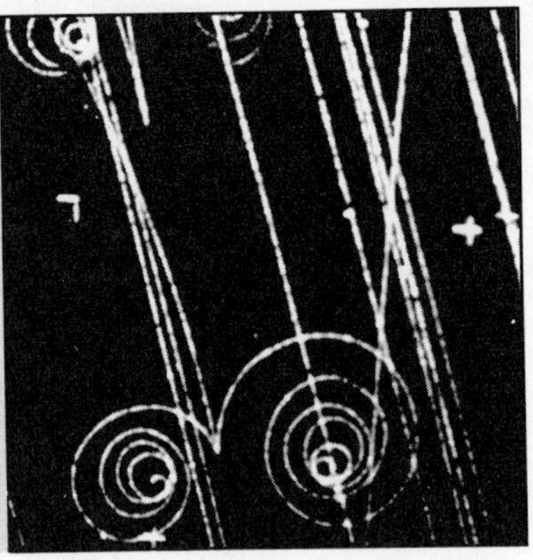

G. Holton, F. J. Rutherford, and F. G. Watson, *Project Physics*, New York, HRW, 1981.

26.1 INTRODUCTION

The behavior of bar magnets is well known to anyone who has studied science. Permanent magnets, which are usually made of alloys containing iron, will attract or repel other magnets. Furthermore, they will attract other bits of iron, which in turn can become magnetized. The list of important technological applications of magnetism is quite extensive. For instance, large electromagnets are used to pick up heavy loads. Magnets are also used in such devices as meters, transformers, motors, and loudspeakers. Magnetic tapes are routinely used in sound recording, TV recording, and computer memories. Intense magnetic fields generated by superconducting magnets are currently being used as a means of containing the plasmas (heated to temperatures of the order of 10^8 K) used in controlled nuclear fusion research.

The phenomenon of magnetism was known to the Greeks as early as 800 B.C. They discovered that certain stones, now called *magnetite* (Fe_3O_4), attract pieces of iron. Legend ascribes the name *magnetite* to the shepherd Magnes, "the nails of whose shoes and the tip of whose staff stuck fast in a magnetic field while he pastured his flocks." In 1269 de Maricourt, using a spherical natural magnet, mapped out the directions taken by a needle when placed at various points on the surface of the sphere. He found that the directions formed lines that encircle the sphere and interact at opposite ends, which he called the *poles* of the magnet. Subsequent experiments showed that every magnet, regardless of its shape, has two poles, called *north* and *south poles*, which exhibit forces on each other in a manner analogous to electrical charges. That is, like poles repel each other and unlike poles attract each other.

In 1600 William Gilbert extended these experiments to a variety of materials. Using the fact that a compass needle orients in preferred directions, he suggested that the earth itself is a large permanent magnet. In 1750 John Michell (1724–1793) used a torsion balance to show that magnetic poles exert attractive or repulsive forces on each other and that these forces vary as the inverse square of their separation. Although the force between two magnetic poles is similar to the force between two electric charges, there is an important difference. Electric charges can be isolated (witness the electron or proton), whereas *magnetic poles cannot be isolated.* That is, *magnetic poles are always found in pairs*. All attempts thus far to detect an isolated magnetic monopole have been unsuccessful. No matter how many times a permanent magnet is cut, each piece will always have a north and a south pole.

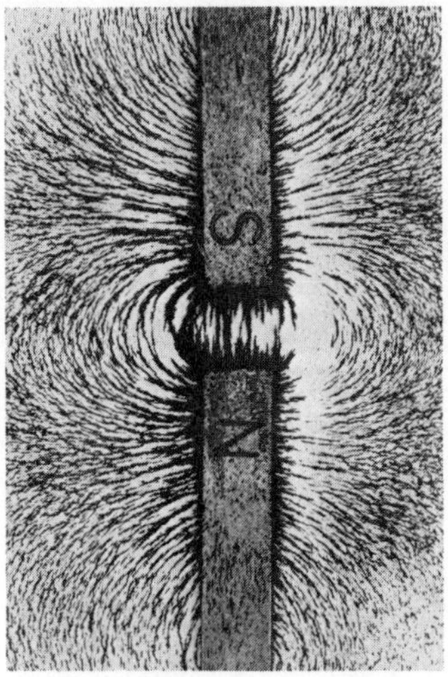

Magnetic field patterns surrounding two bar magnets as displayed with iron filings. This demonstrates the attractive nature of unlike poles. (Courtesy of H. Strickland and J. Lehman, James Madison University)

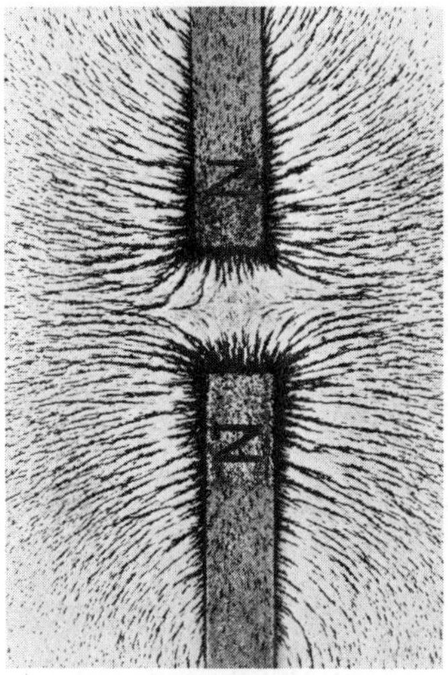

This demonstrates the repulsion between two like poles. (Courtesy of H. Strickland and J. Lehman, James Madison University)

The relationship between magnetism and electricity was discovered in 1819 when, during a lecture demonstration, the Danish scientist Hans Oersted found that an electric current in a wire deflected a nearby compass needle.[1] Shortly thereafter, André Ampère (1775–1836) obtained quantitative laws of magnetic force between current-carrying conductors. He also suggested that electric currents of molecular size are responsible for *all* magnetic phenomena. This idea is the basis for the modern theory of magnetism, which we shall describe in Chapter 30.

In the 1920's, further connections between electricity and magnetism were demonstrated by Faraday and independently by Joseph Henry (1797–1878). They showed that an electric current could be produced in a circuit either by moving a magnet near the circuit or by changing the current in another, nearby circuit. These observations demonstrate that a changing magnetic field produces an electric field. Years later, theoretical work by Maxwell showed that a changing electric field gives rise to a magnetic field.

This chapter examines forces on moving charges and on current-carrying wires in the presence of a magnetic field. The source of the magnetic field itself will be described in Chapter 27.

26.2 DEFINITION AND PROPERTIES OF THE MAGNETIC FIELD

The electric field $\boldsymbol{E}$ at a point in space has been defined as the electric force per unit charge acting on a test charge placed at that point. Similarly, the gravitational field $\boldsymbol{g}$ at a point in space is the gravitational force per unit mass acting on a test mass.

We now define a magnetic field vector $\boldsymbol{B}$ (sometimes called the *magnetic induction* or *magnetic flux density*) at some point in space in terms of a magnetic force that would be exerted on an appropriate test object. Our test object is taken to be a

[1] It is interesting to note that the same discovery was reported in 1802 by an Italian jurist, Gian Dominico Romognosi, but was overlooked, probably because it was published in a newspaper, *Gazetta de Trentino,* rather than in a scholarly journal.

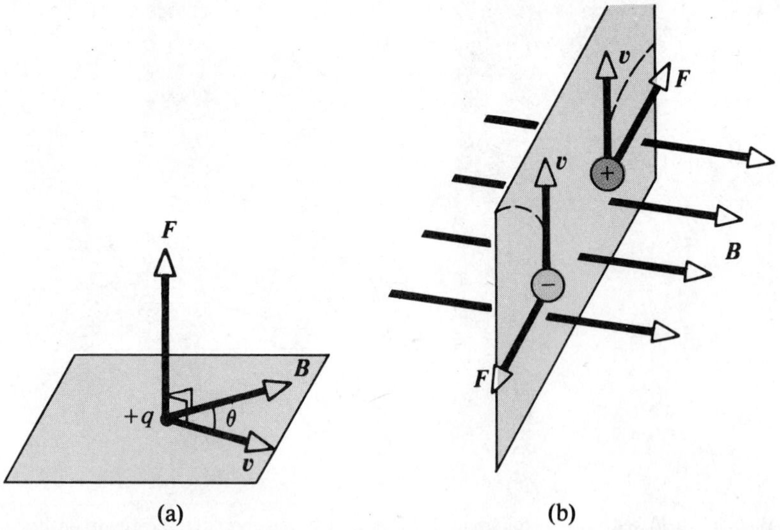

Figure 26.1 The direction of the magnetic force on a charged particle moving with a velocity $\boldsymbol{v}$ in the presence of a magnetic field. (a) When $\boldsymbol{v}$ is at an angle θ to $\boldsymbol{B}$, the magnetic force is perpendicular to both $\boldsymbol{v}$ and $\boldsymbol{B}$. (b) In the presence of a magnetic field, the moving charged particles are deflected as indicated by the dotted lines.

charged particle moving with a velocity $\boldsymbol{v}$. For the time being, let us assume that there are no electric or gravitational fields present in the region of the charge. Experiments on the motion of various charged particles moving in a magnetic field give the following results:

1. The magnetic force is proportional to the charge q and speed v of the particle.
2. The magnitude and direction of the magnetic force depend on the velocity of the particle.
3. When a charged particle moves in a direction *parallel* to the magnetic field vector, the magnetic force $\boldsymbol{F}$ on the charge is *zero*.
4. When the velocity vector makes an angle θ with the magnetic field, the magnetic force acts in a direction perpendicular to both $\boldsymbol{v}$ and $\boldsymbol{B}$; that is, $\boldsymbol{F}$ is perpendicular to the plane formed by $\boldsymbol{v}$ and $\boldsymbol{B}$ (Fig. 26.1a).
5. The magnetic force on a positive charge is in the direction opposite the force on a negative charge moving in the same direction (Fig. 26.1b).
6. If the velocity vector makes an angle θ with the magnetic field, the magnitude of the magnetic force is proportional to $\sin\theta$.

Properties of the magnetic force on a charge moving in a B field

These observations can be summarized by writing the magnetic force in the form

Magnetic force on a charged particle moving in a magnetic field

$$\boldsymbol{F} = q\boldsymbol{v} \times \boldsymbol{B} \tag{26.1}$$

where the direction of the magnetic force is in the direction of $\boldsymbol{v} \times \boldsymbol{B}$, which, by definition of the cross product, is perpendicular to both $\boldsymbol{v}$ and $\boldsymbol{B}$.

Figure 26.2 gives a brief review of the right-hand rule for determining the direction of the cross product $\boldsymbol{v} \times \boldsymbol{B}$. The vector $\boldsymbol{v}$ is turned into the vector $\boldsymbol{B}$ using the four fingers of the right hand, with the palm facing the vector $\boldsymbol{B}$. The thumb then points in the direction of $\boldsymbol{v} \times \boldsymbol{B}$. Since $\boldsymbol{F} = q\boldsymbol{v} \times \boldsymbol{B}$, $\boldsymbol{F}$ is in the direction of $\boldsymbol{v} \times \boldsymbol{B}$ if q is positive (Fig. 26.2a) and in the direction *opposite* $\boldsymbol{v} \times \boldsymbol{B}$ if q is negative (Fig. 26.2b). The magnitude of the magnetic force has the value

$$F = qvB\sin\theta \tag{26.2}$$

where θ is the angle between $\boldsymbol{v}$ and $\boldsymbol{B}$. From this expression, note that F is *zero* when $\boldsymbol{v}$ is parallel to $\boldsymbol{B}$ ($\theta = 0$ or $180°$). Furthermore, the force has its *maximum* value, $F = qvB$, when $\boldsymbol{v}$ is perpendicular to $\boldsymbol{B}$ ($\theta = 90°$).

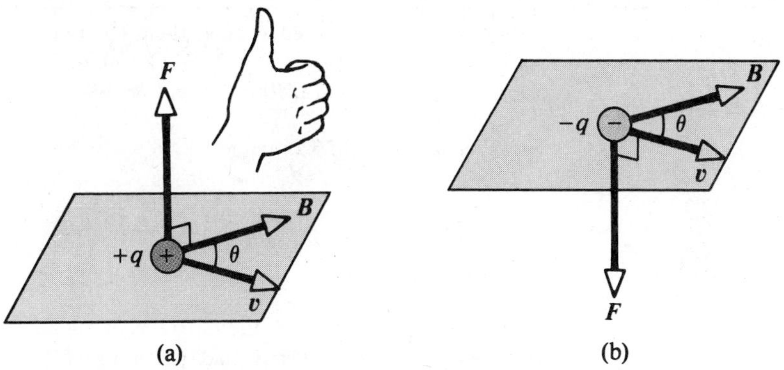

Figure 26.2 The right-hand rule for determining the direction of the magnetic force $\boldsymbol{F}$ acting on a charge q moving with a velocity $\boldsymbol{v}$ in a magnetic field $\boldsymbol{B}$. If q is positive, $\boldsymbol{F}$ is upward in the direction of the thumb. If q is negative, $\boldsymbol{F}$ is downward.

We can regard Eq. 26.1 as an operational definition of the magnetic field at a point in space. That is, the magnetic field is defined in terms of a *sideways* force acting on a moving charged particle. There are several important differences between electric and magnetic forces that should be noted:

Differences between electric and magnetic fields

1. The electric force is always in the direction of the electric field, whereas the magnetic force is perpendicular to the magnetic field.
2. The electric force acts on a charged particle independent of the particle's velocity, whereas the magnetic force acts on a charged particle only when the particle is in motion.
3. The electric force does work in displacing a charged particle, whereas the magnetic force associated with a steady magnetic field does *no* work when a particle is displaced.

This last statement is a consequence of the fact that when a charge moves in a steady magnetic field, the magnetic force is always *perpendicular* to the displacement. That is,

$$\boldsymbol{F} \cdot d\boldsymbol{s} = (\boldsymbol{F} \cdot \boldsymbol{v})\, dt = 0$$

since the magnetic force is a vector perpendicular to $\boldsymbol{v}$. From this property and the work-energy theorem, we conclude that the kinetic energy of a charged particle *cannot* be altered by a magnetic field alone. In other words, when a charge moves with a velocity $\boldsymbol{v}$, *an applied magnetic field can alter the direction of the velocity vector, but it cannot change the speed of the particle.*

A magnetic field cannot change the speed of a particle

The SI unit of the magnetic field is the weber per square meter (Wb/m^2), also called the tesla (T). This unit can be related to the fundamental units by using Eq. 26.1: a 1-coulomb charge moving through a field of 1 tesla with a velocity of 1 m/s perpendicular to the field experiences a force of 1 newton:

$$[\text{B}] = \text{T} = \frac{\text{Wb}}{\text{m}^2} = \frac{\text{N}}{\text{C} \cdot \text{m/s}} = \frac{\text{N}}{\text{A} \cdot \text{m}} \tag{26.3}$$

In practice, the cgs unit for magnetic-field, called the gauss (G), is often used. The gauss is related to the tesla through the conversion

$$1\,\text{T} = 10^4\,\text{G} \tag{26.4}$$

Conventional laboratory magnets can produce magnetic fields as large as about 25 000 G, or 2.5 T. Superconducting magnets that can generate magnetic fields as high as 250 000 G, or 25 T, have been constructed. This can be compared with the earth's magnetic field near its surface, which is about 0.5 G, or 0.5×10^{-4} T.

Example 26.1

A proton moves with a speed of 8×10^6 m/s along the x axis. It enters a region where there is a magntic field of magnitude 2.5 T, directed at an angle of 60° to the x axis and lying in the xy plane (Fig. 26.3). Calculate the initial force and acceleration of the proton.

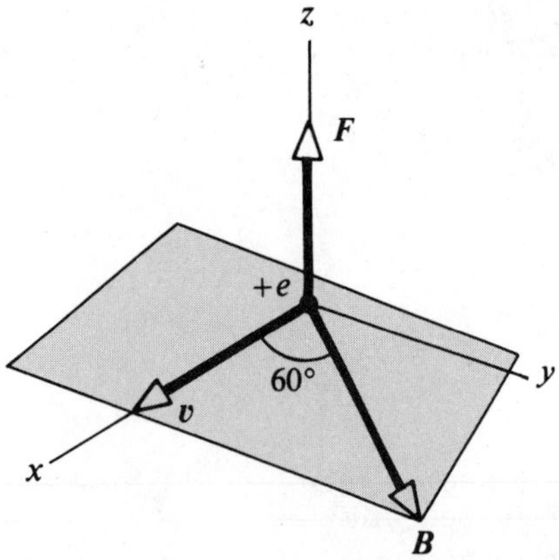

Figure 26.3 (Example 26.1) The magnetic force $\boldsymbol{F}$ on a proton is in the positive z direction when $\boldsymbol{v}$ and $\boldsymbol{B}$ lie in the xy plane.

Solution: From Eq. 26.2, we get

$$\begin{aligned} F &= qvB \sin\theta \\ &= (1.6 \times 10^{-19}\ \text{C})(8 \times 10^6\ \text{m/s})(2.5\ \text{T})(\sin 60°) \\ &= 2.77 \times 10^{-12}\ \text{N} \end{aligned}$$

Note that since $\boldsymbol{v} \times \boldsymbol{B}$ is in the positive z direction and since the charge is positive, the force $\boldsymbol{F}$ is in the positive z direction. You should verify that the units of F in the above calculation reduce to newtons.

Since the mass of the proton is 1.67×10^{-27} kg, its initial acceleration is

$$a = \frac{F}{m} = \frac{2.77 \times 10^{-12}\ \text{N}}{1.67 \times 10^{-27}\ \text{kg}} = 1.66 \times 10^{15}\ \text{m/s}^2$$

in the positive z direction.

Q1. At a given instant, a proton moves in the positive x direction in a region where there is a magnetic field in the negative z direction. What is the direction of the magnetic force? Will the proton continue to move in the positive x direction? Explain.

Q2. Two charged particles are projected into a region where there is a magnetic field perpendicular to their velocities. If the charges are deflected in opposite directions, what can you say about them?

Q3. If a charged particle moves in a straight line through some region of space, can you say that the magnetic field in that region is zero?

Q4. How can the motion of a moving charged particle be used to distinguish between a magnetic field and an electric field? Give a specific example to justify your argument.

Q5. Why does the picture on a TV screen become distorted when a magnet is brought near the screen?

26.3 MAGNETIC FORCE ON A CURRENT-CARRYING CONDUCTOR

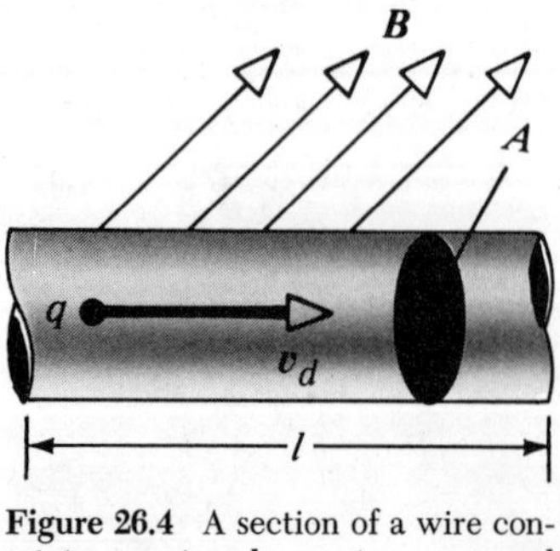

Figure 26.4 A section of a wire containing moving charges in an external magnetic field $\boldsymbol{B}$. The magnetic force on each charge is $q\boldsymbol{v}_d \times \boldsymbol{B}$, and the net force on a straight element is $I\boldsymbol{l} \times \boldsymbol{B}$.

If a wire carrying a current is placed in an external magnetic field, a magnetic force will be exerted on the wire. This follows from the fact that the current represents a collection of many charged particles in motion; hence, the resultant force on the wire is due to the sum of the individual forces on the charged particles. Consider a straight segment of wire of length l and cross-sectional area A, carrying a current I in a uniform *external* magnetic field $\boldsymbol{B}$ as in Fig. 26.4. The magnetic force on a charge q moving with a drift velocity $\boldsymbol{v}_d$ is given by $q\boldsymbol{v}_d \times \boldsymbol{B}$. The force on the charge carriers is transmitted to the "bulk" of the wire through collisions with the atoms making up the wire. To find the total force on the wire, we multiply the force on one charge, $q\boldsymbol{v}_d \times \boldsymbol{B}$, by the number of charges in the segment. Since the volume of the segment is Al, the number of charges in the segment is nAl, where n is the number of charges per unit volume. Hence, the total magnetic force on the wire of

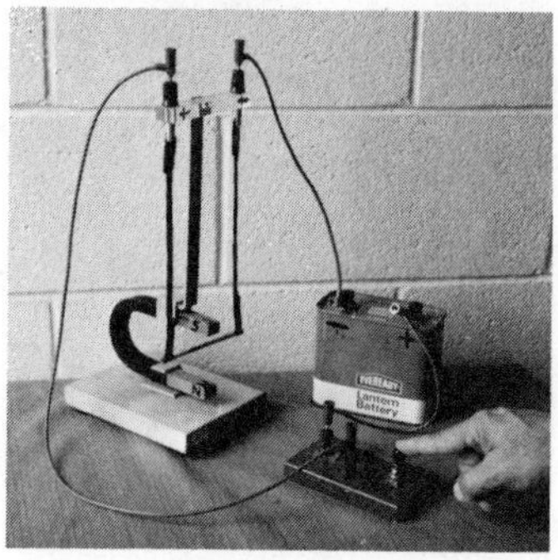

This apparatus demonstrates the force on a current-carrying conductor in an external magnetic field. Why does the bar swing *away* from the magnet after the switch is closed? (Courtesy of J. Lehman, James Madison University)

length l is

$$\boldsymbol{F} = (q\boldsymbol{v}_\mathrm{d} \times \boldsymbol{B})nAl$$

This can be written in a more convenient form by noting that, from Eq. 24.4, the current in the wire is given by $I = nqv_\mathrm{d}A$. Therefore, $\boldsymbol{F}$ can be expressed as

$$\boldsymbol{F} = I\,\boldsymbol{l} \times \boldsymbol{B} \qquad (26.5)$$

Force on a straight wire carrying a current

where $\boldsymbol{l}$ is a vector in the direction of the current I; the magnitude of $\boldsymbol{l}$ equals the length l of the segment. Note that this expression applies only to a straight segment of wire in a uniform external magnetic field. Furthermore, we have neglected the field produced by the current itself. (In fact, the wire cannot produce a force on itself.)

Now consider an arbitrarily shaped wire in an external magnetic field, as in Fig. 26.5. It follows from Eq. 26.5 that the magnetic force on a very small segment $d\boldsymbol{s}$ in the presence of a field $\boldsymbol{B}$ is given by

$$d\boldsymbol{F} = I\, d\boldsymbol{s} \times \boldsymbol{B} \qquad (26.6)$$

Force on a current element

where $d\boldsymbol{F}$ is directed out of the page for the directions assumed in Fig. 26.5. We can consider Eq. 26.6 as an alternative definition of $\boldsymbol{B}$. That is, the field $\boldsymbol{B}$ can be defined in terms of a measurable force on a current element, where the force is a maximum when $\boldsymbol{B}$ is perpendicular to the element and zero when $\boldsymbol{B}$ is parallel to the element.

To get the total force $\boldsymbol{F}$ on the wire, we integrate Eq. 26.6 over the length of the wire:

$$\boldsymbol{F} = I \int_A^C d\boldsymbol{s} \times \boldsymbol{B} \qquad (26.7)$$

Total force on a wire in a magnetic field

In this expression, A and C represent the end points of the wire. Note that when this

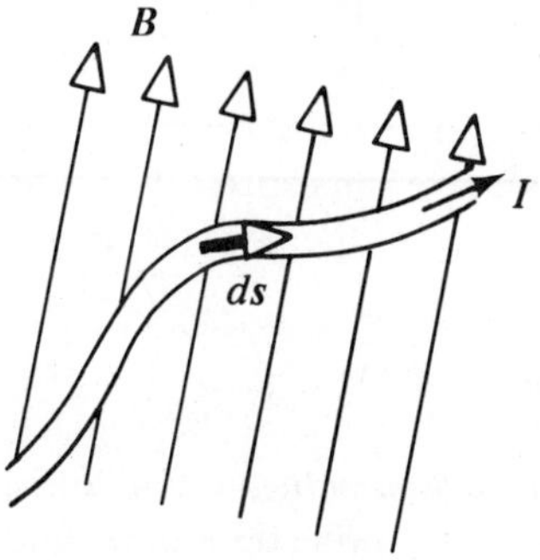

Figure 26.5 A wire of arbitrary shape carrying a current I in an external magnetic field $\boldsymbol{B}$ experiences a magnetic force. The force on any segment $d\boldsymbol{s}$ is given by $I\, d\boldsymbol{s} \times \boldsymbol{B}$.

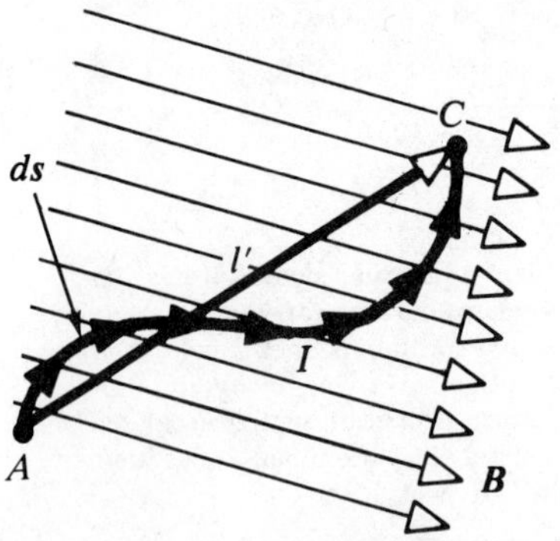

Figure 26.6 A curved current-carrying conductor in a uniform magnetic field. The magnetic force on the conductor is equivalent to the force on a straight segment of length l' running between the ends of the wires, a and b.

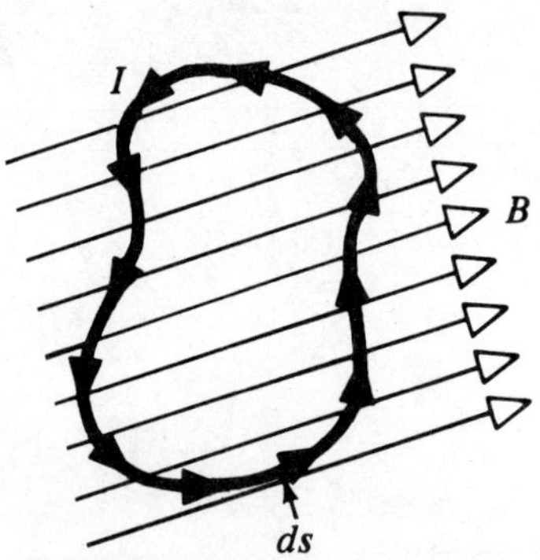

Figure 26.7 A current-carrying loop of arbitrary shape in a uniform magnetic field. The net magnetic force on the loop is 0.

integration is carried out, the magnetic field and the vector $\boldsymbol{ds}$ (that is, the element orientation), may vary at each point.

Now let us consider two special cases involving the application of Eq. 26.7. In both cases, the external magnetic field is taken to be constant in magnitude and direction.

Case I

Consider a curved wire carrying a current I; the wire is located in a uniform external magnetic field $\boldsymbol{B}$ as in Fig. 26.6. Since the field is assumed to be uniform (that is, $\boldsymbol{B}$ has the same value over the region of the conductor), $\boldsymbol{B}$ can be taken outside the integral in Eq. 26.7, and we get

$$\boldsymbol{F} = I\left(\int_A^C \boldsymbol{ds}\right) \times \boldsymbol{B} \tag{26.8}$$

But the quantity $\int_A^C \boldsymbol{ds}$ represents the *vector sum* of all the displacement elements from a to b as described in Fig. 26.6. From the law of addition of many vectors (Chapter 2), the sum equals the vector $\boldsymbol{l}'$, which is directed from a to b. Therefore, Eq. 26.8 reduces to

Force on a wire in a uniform field

$$\boldsymbol{F} = I\boldsymbol{l}' \times \boldsymbol{B} \tag{26.9}$$

Case II

An arbitrarily shaped, closed loop carrying a current I is placed in a uniform external magnetic field $\boldsymbol{B}$ as in Fig. 26.7. Again, we can express the force in the form of Eq. 26.8. In this case, note that the vector sum of the displacement vectors must be taken over the closed loop. That is,

$$\boldsymbol{F} = I(\oint \boldsymbol{ds}) \times \boldsymbol{B}$$

Since the set of displacement vectors forms a *closed polygon* (Fig. 26.7), the vector sum must be *zero*. This follows from the graphical procedure of adding vectors by the *polygon method* (Chapter 2). Since $\oint ds = 0$, we conclude that

$$F = 0 \tag{26.10}$$

The force on a closed loop in a uniform field is zero

That is, *the total magnetic force on any closed current loop in a uniform magnetic field is zero.*

Example 26.2

A wire bent into the shape of a semicircle of radius R forms a closed circuit and carries a current I. The circuit lies in the xy plane, and a uniform magnetic field is present along the positive y axis as in Fig. 26.8. Find the magnetic forces on the straight portion of the wire and on the curved portion.

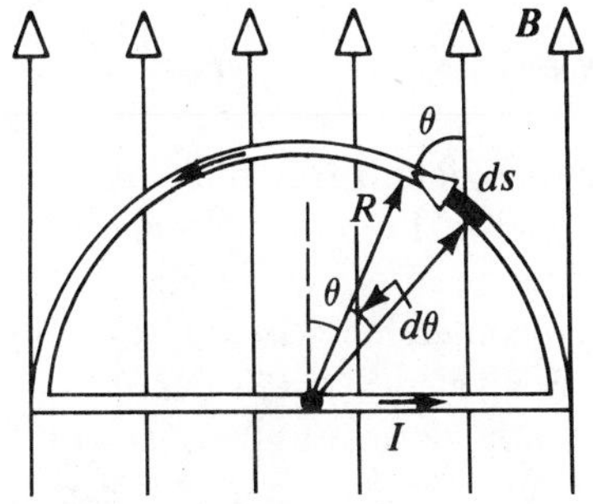

Figure 26.8 (Example 26.2) The net force on a closed current loop in a uniform magnetic field is *zero*. In this case, the force on the straight portion is $2IRB$ and outward, and the force on the curved portion is also $2IRB$ and inward.

Solution: The force on the straight portion of the wire has a magnitude given by $F_1 = IlB = 2IRB$, since $l = 2R$ and the wire is perpendicular to $\mathbf{B}$. The direction of $\mathbf{F}_1$ is *out* of the paper since $\mathbf{l} \times \mathbf{B}$ is outward. (That is, $\mathbf{l}$ is to the right in the direction of the current, and so by the rule of cross products, $\mathbf{l} \times \mathbf{B}$ is outward.)

To find the force on the curved part, we must first write an expression for the force $d\mathbf{F}_2$ on the element ds. If θ is the angle between $\mathbf{B}$ and ds in Fig. 26.8, then the magnitude of $d\mathbf{F}_2$ is given by

$$dF_2 = I|ds \times \mathbf{B}| = IB \sin\theta \, ds$$

where ds is the length of the small element measured along the circular arc. In order to integrate this expression, we must express ds in terms of the variable θ. Since $s = R\theta$, $ds = R\,d\theta$, and the expression for dF_2 can be written

$$dF_2 = IBR \sin\theta \, d\theta$$

To get the *total* force F_2 on the curved portion, we can integrate this expression to account for contributions from *all* elements. Note that the direction of the force on every element is the same: *into* the paper (since $ds \times \mathbf{B}$ is inward). Therefore, the resultant force $\mathbf{F}_2$ on the curved wire must also be *into* the paper. Integrating dF_2 over the limits $\theta = 0$ to $\theta = \pi$ (that is, the entire semicircle) gives

$$F_2 = IRB\int_0^\pi \sin\theta \, d\theta = IRB[-\cos\theta]_0^\pi$$

$$= -IRB(\cos\pi - \cos 0) = -IRB(-1 - 1) = 2IRB$$

This result could have been attained more directly using Case I, discussed in the text (Eq. 26.9). The details are left as an exercise.

Since $F_2 = 2IRB$ and is directed *into* the paper while the force on the straight wire $F_1 = 2IRB$ is *out* of the paper, we see that the *net* force on the closed loop is *zero*. This result is consistent with Case II in the text.

Q6. A current-carrying conductor experiences no magnetic force when placed in a certain manner in a uniform magnetic field. Explain.

26.4 TORQUE ON A CURRENT LOOP IN A UNIFORM MAGNETIC FIELD

We have seen that the net magnetic force on a closed loop carrying a current is zero when the loop is placed in an external uniform magnetic field. Although this is generally true, *there is a net torque on a current loop in an external magnetic field, and this torque tends to rotate the loop.*

To understand this, first consider a rectangular loop carrying a current I in the presence of a uniform magnetic field in the plane of the loop, as in Fig. 26.9a. The forces on the sides of length a are zero since these wires are parallel to the field and hence $ds \times \mathbf{B} = 0$ for these sides. On the other hand, the magnitude of the forces on the sides of length b is given by

$$F_1 = F_2 = IbB$$

where $\boldsymbol{F}_1$ is the force on the left side of the loop and $\boldsymbol{F}_2$ is the force on the right side. The direction of $\boldsymbol{F}_1$ is *out* of the paper in Fig. 26.9 and that of $\boldsymbol{F}_2$ is *into* the paper. An end view of the direction of these forces, shown in Fig. 26.9b, demonstrates that the forces form a couple. Therefore, even though they cancel each other, they tend to rotate the loop. The torque due to this couple about the point O in Fig. 26.9b has a magnitude given by

$$\tau = F_1 \frac{a}{2} + F_2 \frac{a}{2} = IabB$$

where the moment arm about O is $a/2$ for both forces. This, in fact, is the torque about *any* point. But the area of the loop is given by $A = ab$. Hence, the torque can be expressed as

$$\tau = IAB \tag{26.11}$$

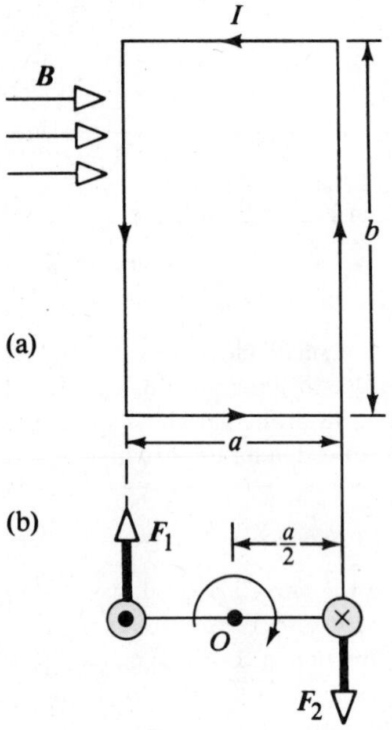

Figure 26.9 (a) A rectangular loop in a uniform magnetic field. There are no forces on the sides of width a parallel to $\boldsymbol{B}$, but there are forces acting on the sides of length b. (b) An end view of the rectangular loop shows that the forces $\boldsymbol{F}_1$ and $\boldsymbol{F}_2$ on the sides of length b create a torque that tends to twist the loop clockwise as shown.

Note that this result is valid only when the field $\boldsymbol{B}$ is in the plane of the loop. The sense of the rotation is clockwise when viewed from the bottom end, as indicated in Fig. 26.9b. If the current were reversed, the forces would reverse their directions and the rotational tendency would be counterclockwise.

Now consider a rectangular loop carrying a current I in a uniform magnetic field. Suppose the field makes an angle θ with the normal to the plane of the loop, as in Fig. 26.10a. For convenience, we shall assume that the field $\boldsymbol{B}$ is perpendicular to the sides of length b. In this case, the magnetic forces $\boldsymbol{F}_3$ and $\boldsymbol{F}_4$ on the sides of length a cancel each other and produce no torque since they pass through a common origin. However, the forces $\boldsymbol{F}_1$ and $\boldsymbol{F}_2$ acting on the sides of length b form a couple and hence produce a torque about *any point*. Referring to the end view shown in Fig. 26.10b, we note that the moment arm of the force $\boldsymbol{F}_1$ about the point O is equal to $\frac{a}{2}\sin\theta$. Likewise, the moment arm of $\boldsymbol{F}_2$ about O is also $\frac{a}{2}\sin\theta$. Since $F_1 =$

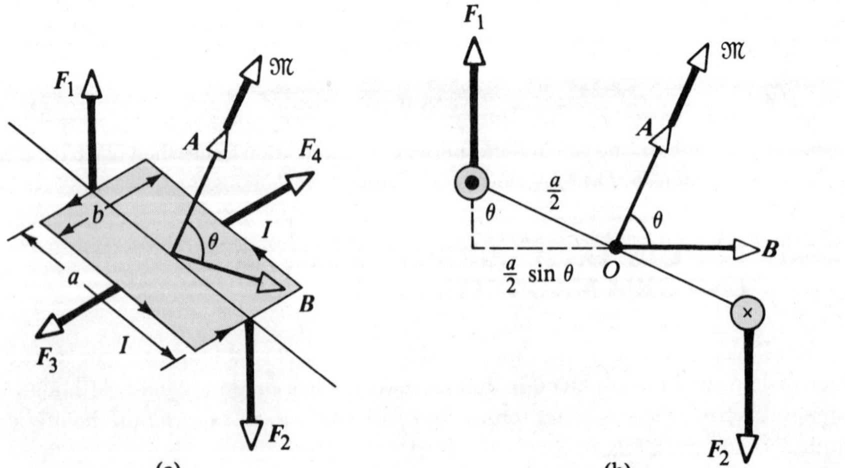

Figure 26.10 (a) A rectangular current loop whose normal makes an angle θ with a uniform magnetic field. The forces on the sides of length a cancel while the forces on the sides of width b create a torque on the loop. (b) An end view of the loop. The magnetic moment $\mathfrak{M}$ is in the direction normal to the plane of the loop.

$F_2 = IbB$, the net torque about O has a magnitude given by

$$\tau = F_1 \frac{a}{2} \sin\theta + F_2 \frac{a}{2} \sin\theta$$
$$= IbB\left(\frac{a}{2} \sin\theta\right) + IbB\left(\frac{a}{2} \sin\theta\right) = IabB \sin\theta$$
$$= IAB \sin\theta$$

where $A = ab$ is the area of the loop. This result shows that the torque has the *maximum* value IAB when the field is parallel to the plane of the loop ($\theta = 90°$) and is *zero* when the field is perpendicular to the plane of the loop ($\theta = 0$). As we see in Fig. 26.10, the loop tends to rotate to smaller values of θ (that is, such that the normal to the plane of the loop rotates toward $\boldsymbol{B}$).

A convenient vector expression for the torque is the following cross-product relationship:

$$\boldsymbol{\tau} = I\mathbf{A} \times \mathbf{B} \tag{26.12}$$

where **A**, a vector perpendicular to the plane of the loop, has a magnitude equal to the area of the loop. The sense of **A** is determined by the right-hand rule as described in Fig. 26.11. By wrapping the four fingers of the right hand in the direction of the current, the thumb points in the direction of **A**. The product $I\mathbf{A}$ is defined to be the *magnetic moment* $\mathfrak{M}$ of the loop. That is,

Magnetic moment of a current loop

$$\mathfrak{M} = I\mathbf{A} \tag{26.13}$$

The SI unit of magnetic moment is ampere-meter^2 ($A \cdot m^2$). Using this definition, the torque can be expressed as

Torque on a current loop

$$\boldsymbol{\tau} = \mathfrak{M} \times \mathbf{B} \tag{26.14}$$

Note that this result is analogous to the torque acting on an electric dipole moment $\boldsymbol{p}$ in the presence of an external electric field $\boldsymbol{E}$, where $\boldsymbol{\tau} = \boldsymbol{p} \times \boldsymbol{E}$ (Section 23.6). If a coil has N turns all of the same dimensions, the magnetic moment and the torque on the coil will clearly be N times greater than in a single loop.

Although the torque was obtained for a particular orientation of $\boldsymbol{B}$ with respect to the loop, the equation $\boldsymbol{\tau} = \mathfrak{M} \times \boldsymbol{B}$ is valid for any orientation. Furthermore, although the torque expression was derived for a rectangular loop, the result is valid for a planar loop of *any* shape.

It is interesting to note the similarity between the rotating tendency of a current loop in an external magnetic field and the motion of a compass needle (or pivoted bar magnet) in such a field. Like the current loop, the compass needle and bar magnet can be regarded as magnetic dipoles. The similarity in their magnetic field lines is described in Fig. 26.12. Note that one face of the current loop behaves as the north pole of a bar magnet while the opposite face behaves as the south pole. The field lines shown in Fig. 26.12 are the patterns due to the bar magnet (Fig. 26.12a) and the current loop (Fig. 26.12b). There is *no* external field present in these diagrams. Furthermore, the diagrams are a simplified, two-dimensional description of the field lines.

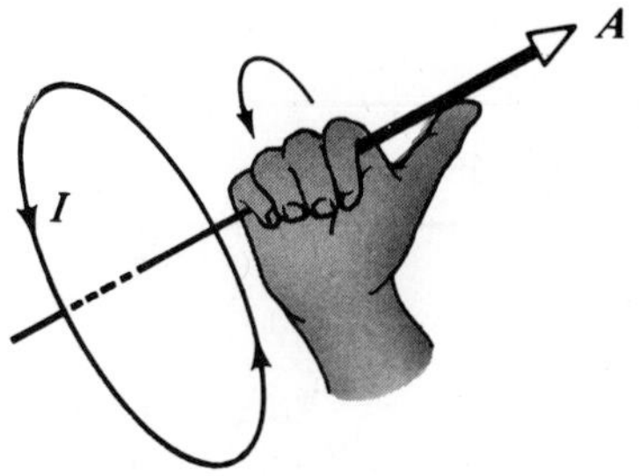

Figure 26.11 The right-hand rule for determining the direction of the magnetic moment $\mathfrak{M}$, which is in the direction of **A**.

The Galvanometer

As we mentioned in the previous chapter, the *galvanometer* is a device used in the construction of both ammeters and voltmeters. The galvanometer operates on the

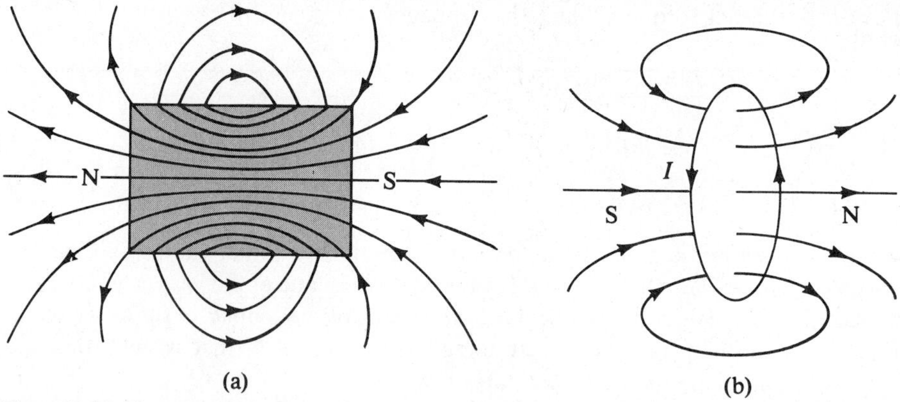

Figure 26.12 The similarity between the magnetic field patterns of (a) a bar magnet and (b) a current loop.

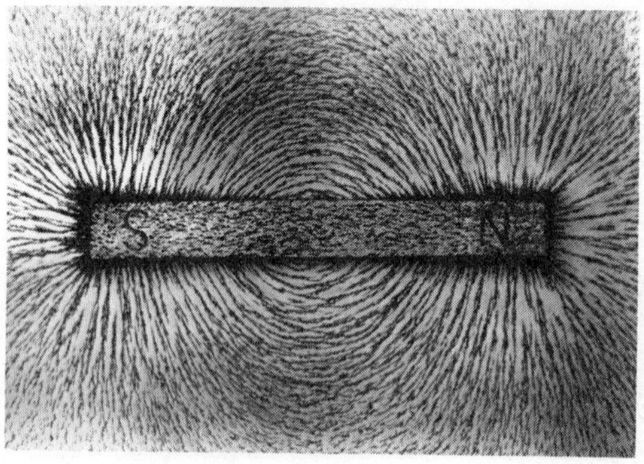

Magnetic field pattern of a bar magnet as displayed by small iron filings on a sheet of paper. (Courtesy of H. Strickland and J. Lehman, James Madison University)

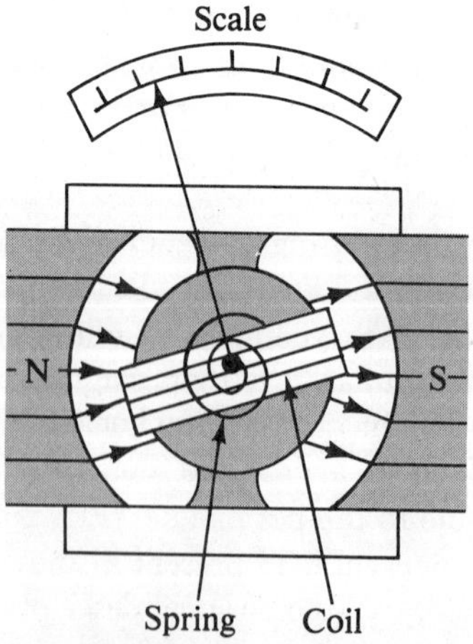

Figure 26.13 The principal components of a pivoted-coil galvanometer. When current passes through the coil, situated in a magnetic field, the magnetic torque causes the coil to twist. The angle through which the coil rotates is proportional to the current through it.

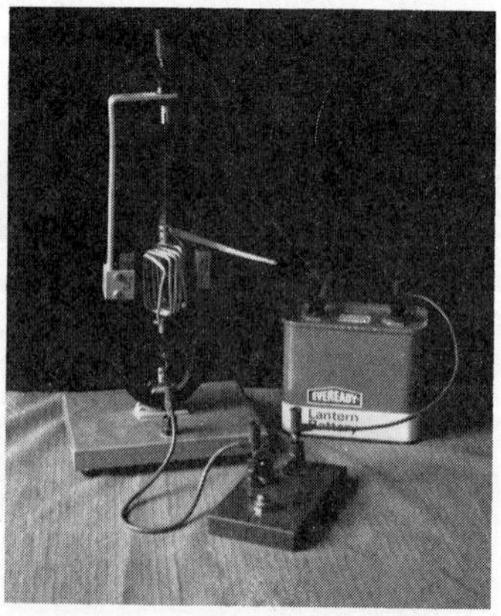

Large-scale model of a galvanometer movement. Why does the coil rotate about the vertical axis after the switch is closed? (Courtesy of J. Lehman, James Madison University)

principle of a torque acting on a current loop in a magnetic field. One type, known as the *pivoted-coil galvanometer*, consists of a coil of wire mounted such that it is free to rotate on a pivot in a radial magnetic field provided by a permanent magnet (Fig. 26.13). The torque experienced by the coil is proportional to the current through it. A restoring torque, provided by a helical spring attached to the pivot, is proportional to the angular displacement of the coil. Hence, the equilibrium deflection angle is also proportional to the current.

In the operation of the galvanometer, the following conditions exist: (1) The motion is strongly damped (Section 13.6) so that the coil rotates until the torque produced by the current loop is balanced by the mechanical restoring torque. (2) Since the magnetic field is radial, the normal to the plane of the coil is always perpendicular to the magnetic field.

Q7. Is it possible to orient a current loop in a uniform magnetic field such that the loop will not tend to rotate? Explain.

Q8. How can a current loop be used to determine the presence of a magnetic field in a given region of space?

26.5 MOTION OF A CHARGED PARTICLE IN A MAGNETIC FIELD

In Section 26.2 we found that the magnetic force acting on a charged particle moving in a magnetic field is always perpendicular to the velocity of the particle. From this property, it follows that *the work done by the magnetic force is zero since the displacement of the charge is always perpendicular to the magnetic force*. Therefore, a static magnetic field changes the *direction* of the velocity but does not affect the speed or kinetic energy of the charged particle.

Consider the special case of a positively charged particle moving in a uniform external magnetic field with its initial velocity vector *perpendicular* to the field. Let us assume that the magnetic field is *into* the page (this is indicated by the crosses in Fig. 26.14). The crosses are used to represent the *tail* of **B**, since **B** is directed *into* the page. Later, we shall use dots to represent the *tip* of a vector directed *out* of the page. In Fig. 26.14 we show that the *particle moves in a circle whose plane is perpendicular to the magnetic field*. This is because the magnetic force **F** is at right angles to **v** and **B** and has a constant magnitude equal to qvB. As the force **F** deflects the particle, the directions of **v** and **F** change continuously, as shown in Fig. 26.14. Therefore the force **F** is a *centripetal force*, which changes only the direction of **v** while the speed remains constant. Note that the sense of the rotation in Fig. 26.14 is counterclockwise for a positive charge. If q were negative, the sense of the rotation would be reversed, or clockwise. Since the resultant force **F** in the radial direction has a magnitude of qvB, we can equate this to the mass m multiplied by the centripetal acceleration v^2/r. From Newton's second law, we find that

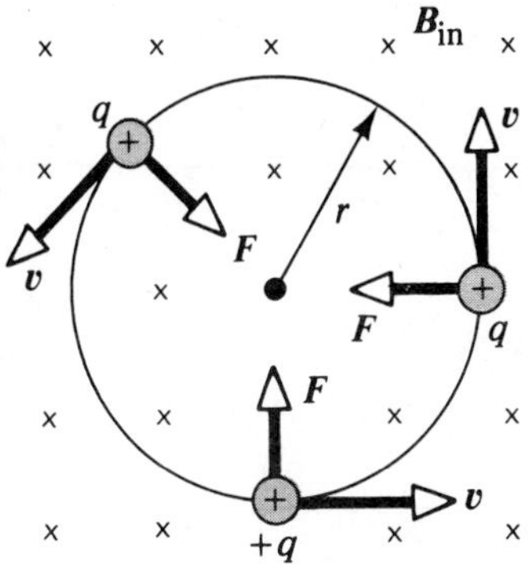

Figure 26.14 When the velocity of a charged particle is perpendicular to a uniform magnetic field, the particle moves in a circle whose plane is perpendicular to **B** (directed inward). The magnetic force **F** on the positive charge is perpendicular to **v** and **B** and is always directed toward the center of the circle. The crosses represent the tail of **B**, which is directed into the page.

$$F = qvB = \frac{mv^2}{r}$$

$$r = \frac{mv}{qB} \qquad (26.15)$$

Radius of the circular orbit

That is, the radius of the path is proportional to the momentum mv of the particle and is inversely proportional to the magnetic field. The angular frequency of the rotating charged particle is given by

Cyclotron frequency

$$\omega = \frac{v}{r} = \frac{qB}{m} \tag{26.16}$$

The period of its motion (the time for one revolution) is equal to the circumference of the circle divided by the speed of the particle:

$$T = \frac{2\pi r}{v} = \frac{2\pi}{\omega} = \frac{2\pi m}{qB} \tag{26.17}$$

These results show that the angular frequency and period of the circular motion do not depend on the speed of the particle or the radius of the orbit. The angular frequency ω is often referred to as the *cyclotron frequency* since charged particles circulate at this frequency in an accelerator called a *cyclotron*, which will be discussed in Section 26.6.

If a charged particle moves in a uniform magnetic field with its velocity at some arbitrary angle to **B**, its path is a helix. For example, if the field is in the x direciton as in Fig. 26.15, there is no component of force in the x direction, and hence $a_x = 0$ and the x component of velocity, v_x, remains constant. On the other hand, the magnetic force $q\mathbf{v} \times \mathbf{B}$ causes the components v_y and v_z to change in time, and the resulting motion is a helix having its axis parallel to the **B** field. The projection of the path onto the yz plane (viewed along the x axis) is a circle. (The projections of the path onto the xy and xz planes are sinusoids!) Equations 26.15 to 26.17 still apply, provided that v is replaced by $v_\perp = \sqrt{v_y^2 + v_z^2}$.

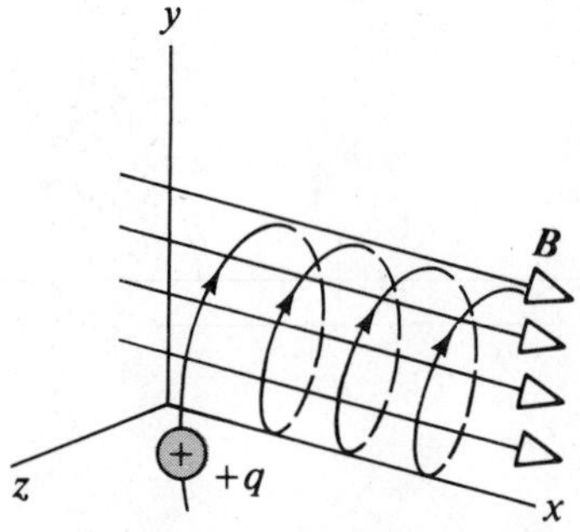

Figure 26.15 A charged particle having a velocity vector that has a component parallel to a uniform magnetic field moves in a helical path.

Example 26.3

A proton is moving in a circular orbit of radius 14 cm when placed in a uniform magnetic field of magnitude 0.35 Wb/m² directed perpendicular to the velocity of the proton. Find the orbital speed of the proton, its angular frequency, and its period of revolution.

Solution: Using Eqs. 26.15 to 26.17 we get

$$v = \frac{rqB}{m} = \frac{(14 \times 10^{-2}\ \text{m})(1.6 \times 10^{-19}\ \text{C})\left(0.35\frac{\text{Wb}}{\text{m}^2}\right)}{1.67 \times 10^{-27}\ \text{kg}}$$

$$= 4.69 \times 10^6 \frac{\text{m}}{\text{s}}$$

$$\omega = \frac{v}{r} = \frac{4.7 \times 10^6\ \text{m/s}}{14 \times 10^{-2}\ \text{m}} = 3.35 \times 10^7\ \text{s}^{-1}$$

$$T = \frac{2\pi}{\omega} = \frac{2\pi}{3.35 \times 10^7\ \text{s}^{-1}} = 1.88 \times 10^{-7}\ \text{s}$$

The bending of an electron beam in an external magnetic field. The apparatus used to take this photograph is part of a system used to measure the ratio e/m. (Courtesy of H. Strickland and J. Lehman, James Madison University)

When charged particles move in *nonuniform* magnetic fields, the motion is rather complex. For instance, in a magnetic field that is strong at the extremities and weak in the middle (Fig. 26.16), the charged particles can oscillate between the end points. This field configuration is known as a *magnetic bottle* since charged particles can be trapped in it. This concept has been used to confine very hot gases ($T \approx 10^6$ K) consisting of electrons and positive ions, known as a *plasma.* Such a plasma-confinement scheme could play a crucial role in achieving a controlled nuclear fusion process, an almost endless source of energy. Unfortunately, the magnetic bottle has its problems. At high particle densities, collisions between the parti-

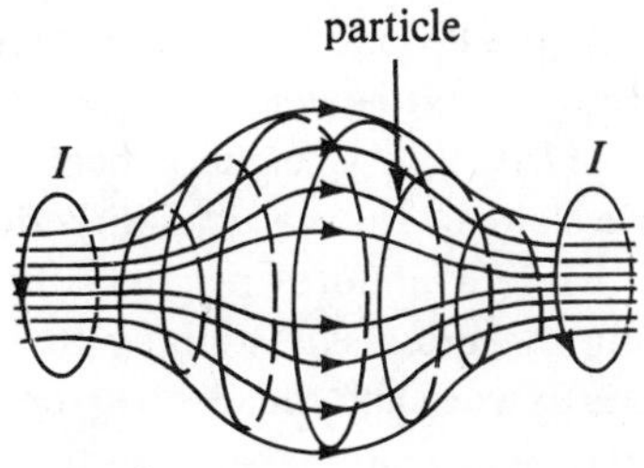

Figure 26.16 A charged particle moving in a nonuniform magnetic field (a magnetic bottle) spirals about the field and oscillates between the end points.

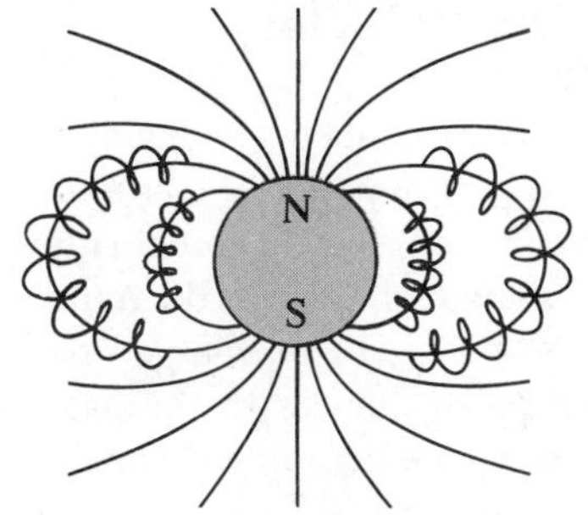

Figure 26.17 The Van Allen belts are made up of charged particles (electrons and protons trapped by the earth's nonuniform magnetic field.

cles cause them to move in offset paths, and eventually they "leak" from the system.

The Van Allen radiation belts consist of charged particles (mostly electrons and protons) surrounding the earth in doughnut-shaped configurations (Fig. 26.17). These radiation belts were discovered in 1958 by James Van Allen using data gathered by the Explorer I satellite. The charged particles, trapped by the earth's nonuniform magnetic field, spiral around the earth's field lines, oscillating between the magnetic poles, where the field is most intense.

The name *cosmic rays* is often used to describe the charged particles, since they originate from the sun, stars, and other bodies in the universe. For instance, protons (ionized hydrogen atoms) are expelled by the sun during large magnetic storms. Fortunately, most of the cosmic rays are carried around the earth's magnetosphere in the solar wind flow, and we are protected from biological damage.

Those charged particles that penetrate into the atmosphere at the poles ionize other atoms, causing them to emit visible radiation. This is the origin of the beautiful Aurora Borealis, or Northern Lights.

Q9. A proton moving horizontally enters a region where there is a uniform magnetic field perpendicular to the proton's velocity, as shown in Fig. 26.18. Describe its subsequent motion. How would an electron behave under the same circumstances?

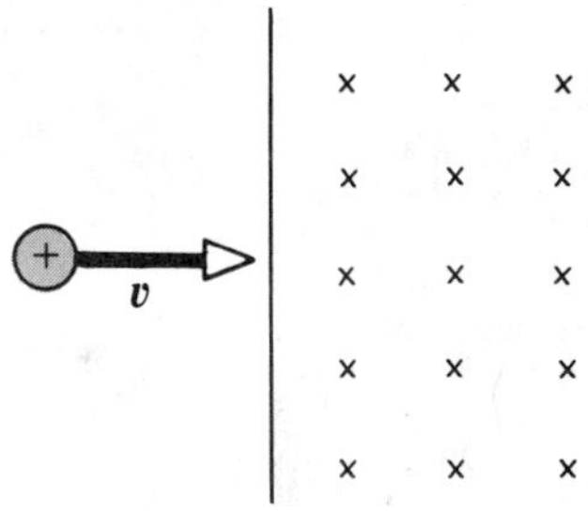

Figure 26.18 (Question 9).

26.6 APPLICATIONS OF THE MOTION OF CHARGED PARTICLES IN A MAGNETIC FIELD

In this section we describe some important devices that involve the motion of charged particles in uniform magnetic fields. For many situations, the charge under consideration will be moving with a velocity $\boldsymbol{v}$ in the presence of both an electric field $\boldsymbol{E}$ and a magnetic field $\boldsymbol{B}$. Therefore, the charge will experience both an electric force $q\boldsymbol{E}$ and a magnetic force $q\boldsymbol{v} \times \boldsymbol{B}$, and so the total force on the charge will be given by

Lorentz force

$$\boldsymbol{F} = q\boldsymbol{E} + q\boldsymbol{v} \times \boldsymbol{B} \tag{26.18}$$

The force described by Eq. 26.18 is known as the *Lorentz force*.

Velocity Selector

In many experiments involving the motion of charged particles, it is important to have a source of particles that move with essentially the same velocity. This can be achieved by applying a combination of an electric field and a magnetic field oriented as shown in Fig. 26.19. A uniform electric field in the vertical direction is provided by a pair of charged parallel plates, while a uniform magnetic field is applied perpendicular to the page (indicated by the crosses). Assuming that q is positive, we see that the magnetic force $q\boldsymbol{v} \times \boldsymbol{B}$ is upward and the electric force $q\boldsymbol{E}$ is downward. If the fields are chosen such that the electric force balances the magnetic force, the particle will move in a straight horizontal line and emerge from the slit. If we equate the upward magnetic force qvB to the downward electric force qE, we find that

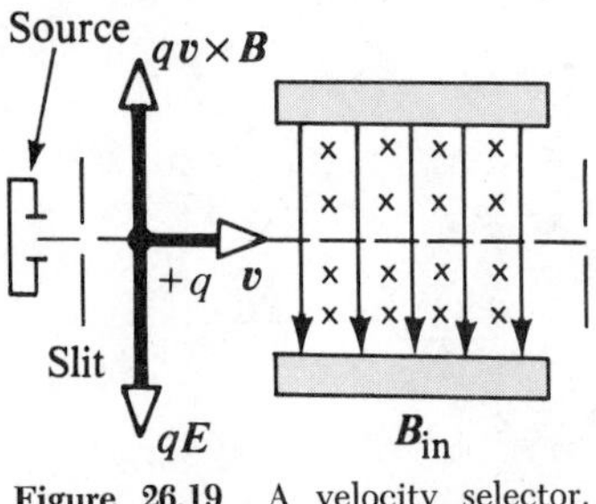

Figure 26.19 A velocity selector. When a positively charged particle is in the presence of both a magnetic field (inward) and an electric field (downward), it experiences both an electric force $q\boldsymbol{E}$ downward and a magnetic force $q\boldsymbol{v} \times \boldsymbol{B}$ upward. If these forces balance each other, the particle moves in a horizontal line.

$$v = \frac{E}{B} \tag{26.19}$$

Hence, only those particles with this velocity will be undeflected. Particles with velocity greater than this will be deflected upward, and those with velocity less than this will be deflected downward.

The Mass Spectrometer

The *mass spectrometer* is an instrument that separates atomic and molecular ions according to their mass-to-charge ratio. In one version, known as the *Bainbridge mass spectrometer,* a beam of ions first passes through a velocity selector and then enters a uniform magnetic field $\boldsymbol{B}_0$ directed into the paper (Fig. 26.20). Upon entering the magnetic field $\boldsymbol{B}_0$, the ions move in a semicircle of radius r before striking a photographic plate at P. From Eq. 26.15, we can express the ratio m/q as

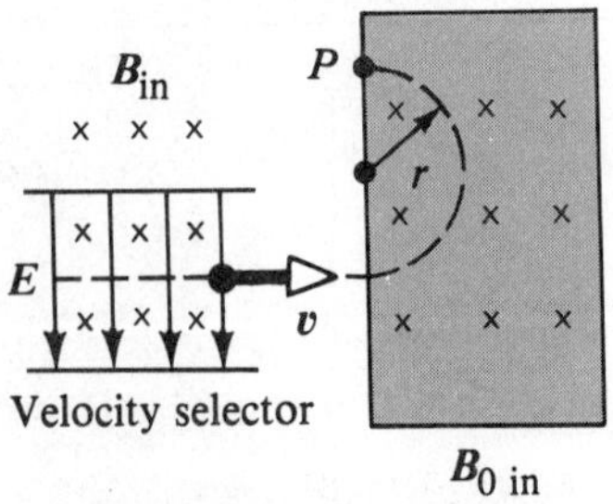

Figure 26.20 A mass spectrometer. Charged particles are first sent through a velocity selector. They then enter a region where the magnetic field $\boldsymbol{B}_0$ (inward) causes positive ions to move in a semicircular path and strike a photographic film at P.

$$\frac{m}{q} = \frac{rB_0}{v} \tag{26.20}$$

Assuming that the magnitude of the magnetic field in the region of the velocity selector is B and using Eq. 26.19, which gives the speed of the particle, we find that

$$\frac{m}{q} = \frac{rB_0B}{E} \tag{26.21}$$

Therefore, one can determine m/q by measuring the radius of curvature and knowing the fields B, B_0, and E. In practice, one usually measures the masses of various isotopes of a given ion with the same charge q. Hence, the mass ratios can be determined even if q is unknown.

A variation of this technique was used by J.J. Thomson (1856–1940) in 1897 to measure e/m for electrons (Fig. 26.21). In this experiment, Thomson showed that the rays of a cathode ray tube could be deflected by both electric and magnetic fields. Furthermore, he was able to show that the rays consisted of charged particles, all having the same charge-to-mass ratio q/m. The results of this crucial experiment represent the discovery of the electron as a fundamental particle of nature.

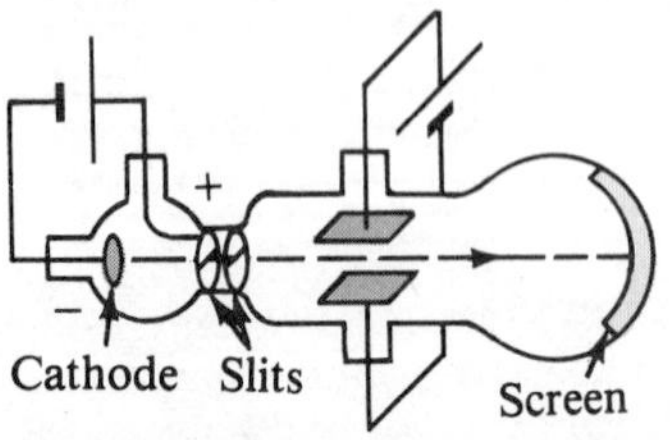

Figure 26.21 Thomson's apparatus for measuring q/m. Electrons are accelerated from the cathode, pass through two slits, and are deflected by both an electric field and a magnetic field (not shown). The deflected beam then strikes a phosphorescent screen.

The Cyclotron

The *cyclotron,* invented in 1934 by E.O. Lawrence and M.S. Livingston, is a machine that can accelerate charged particles to very high velocities. Both electric and magnetic forces play a key role in the operation of the cyclotron. The energetic particles that emerge from the cyclotron are used to bombard other nuclei; this

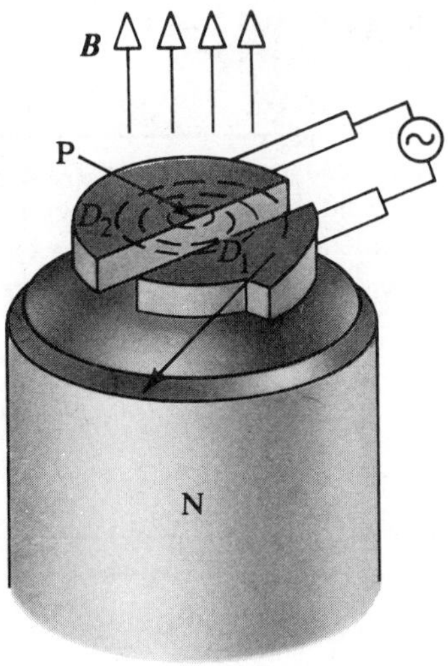

Figure 26.22 The cyclotron consists of an ion source, two dees across which an alternating voltage is applied, and a uniform magnetic field provided by an electromagnet. (The south pole of the magnet is not shown.)

bombardment in turn produces nuclear reactions of interest to researchers. A number of hospitals use cyclotron facilities to produce radioactive substances that can be used in diagnosis and treatment.

A schematic drawing of a cyclotron is shown in Fig. 26.22. Motion of the charges occurs in two semicircular containers, D_1 and D_2, referred to as *dees*. The dees are evacuated in order to minimize energy losses resulting from collisions between the ions and air molecules. A high-frequency alternating voltage is applied to the dees, and a uniform magnetic field provided by an electromagnet is directed perpendicular to the dees. Positive ions released at P near the center of the magnet move in a semicircular path and arrive back at the gap of D_1 in a time $T/2$, where T is the period of revolution, given by Eq. 26.17. The frequency of the applied voltage V is adjusted such that the polarity of the dees is reversed in the same time it takes the ions to complete one half of a revolution. If the phase of the applied voltage is adjusted such that D_2 is at a *lower* potential than D_1 by an amount V, the ion will accelerate across the gap to D_2 and its kinetic energy will increase by an amount qV. The ion then continues to move in D_2 in a semicircular path of larger radius (since its velocity has increased). After a time $T/2$, it again arrives at the gap. By this time, the potential across the dees is reversed (so that D_1 is now negative) and the ion is given another "kick" across the gap. The motion continues such that for each half revolution, the ion gains additional kinetic energy equal to qV. When the radius of its orbit is nearly that of the dees, the energetic ions leave the system through an exit slit as shown in Fig. 26.22.

Bubble chamber photograph. The spiral tracks at the bottom of the photograph are an electron-positron (left and right, respectively) pair formed by a gamma-ray interacting with a hydrogen nucleus. An applied magnetic field causes the electron and the positron to be deflected in opposite directions. The track leaving from the cusp between the two spirals is an additional electron knocked out of a hydrogen atom during this interaction. (G. Holton, F. J. Rutherford, F. G. Watson, *Project Physics*, New York: HRW, 1981.)

It is important to note that the operation of the cyclotron is based on the fact that the time for one revolution is *independent* of the speed (or radius) of the ion.

We can obtain the maximum kinetic energy of the ion when it exits from the cyclotron in terms of the radius R of the dees. From Eq. 26.15 we find that $v = qBR/m$. Hence, the kinetic energy is given by

$$K = \frac{1}{2}mv^2 = \frac{q^2B^2R^2}{2m} \qquad (26.22)$$

When the energy of the ions exceeds about 20 MeV, relativistic effects come into play. (Such effects will be discussed in Chapter 40.) For this reason, the period of the orbit increases and the rotating ions do not remain in phase with the applied voltage. Accelerators have been built which solve this problem by modifying the period of the applied voltage such that it remains in phase with the rotating ion. In 1977, protons were accelerated to 400 GeV (1 GeV = 10^9 eV) in an accelerator in Batavia, Illinois. The system incorporates 954 magnets and has a circumference of 6.3 km (4.1 miles)!

Example 26.4

Calculate the maximum kinetic energy of protons in a cyclotron of radius 0.50 m in a magnetic field of 0.35 T.

Solution: Using Eq. 26.22, we find that

$$K = \frac{q^2B^2R^2}{2m} = \frac{(1.6 \times 10^{-19}\ \text{C})^2(0.35\ \text{T})^2(0.50\ \text{m})^2}{2(1.67 \times 10^{-27}\ \text{kg})}$$

$$= 2.34 \times 10^{-13}\ \text{J} = 1.46\ \text{MeV}$$

where we have used the conversions 1 eV $= 1.6 \times 10^{-19}$ J and 1 MeV $= 10^6$ eV. That is, the kinetic energy acquired by the protons is equivalent to the energy they would gain if they were accelerated through a potential difference of 1.46 MV!

Q10. The *bubble chamber* is a device used for observing tracks of particles that pass through the chamber, which is immersed in a magnetic field. If some of the tracks are spirals and others are straight lines, what can you say about the particles?

26.7 THE HALL EFFECT

In 1879 Edwin Hall discovered that when a current-carrying conductor is placed in a magnetic field, a voltage is generated in a direction perpendicular to both the current and the magnetic field. This observation, known as the *Hall effect*, arises from the deflection of charge carriers to one side of the conductor as a result of the magnetic force experienced by the charge carriers. A proper analysis of experimental data gives information regarding the sign of the charge carriers and their density. The effect also provides a convenient technique for measuring magnetic fields.

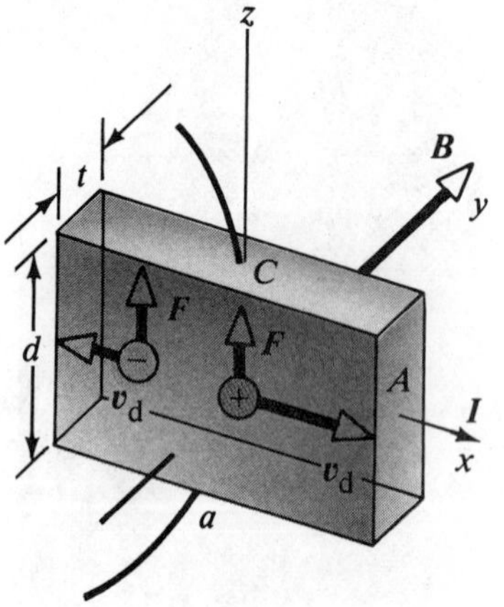

Figure 26.23 To observe the Hall effect, a magnetic field is applied to a current-carrying conductor. When I is in the x direction and $\boldsymbol{B}$ in the y direction as shown, both positive and negative charge carriers are deflected upward in the magnetic field. The Hall voltage is measured between points A and C.

The arrangement for observing the Hall effect consists of a conductor in the form of a flat strip carrying a current I in the x direction as in Fig. 26.23. A uniform magnetic field $\boldsymbol{B}$ is applied in the y direction. If the charge carriers are electrons moving in the negative x direction with a velocity $\boldsymbol{v}_d$, they will experience an *upward* magnetic force $\boldsymbol{F}$. Hence, the electrons will be deflected upward, accumulating at the upper edge and leaving an excess positive charge at the lower edge (Fig. 26.24a). This accumulation of charge at the edges will continue until the

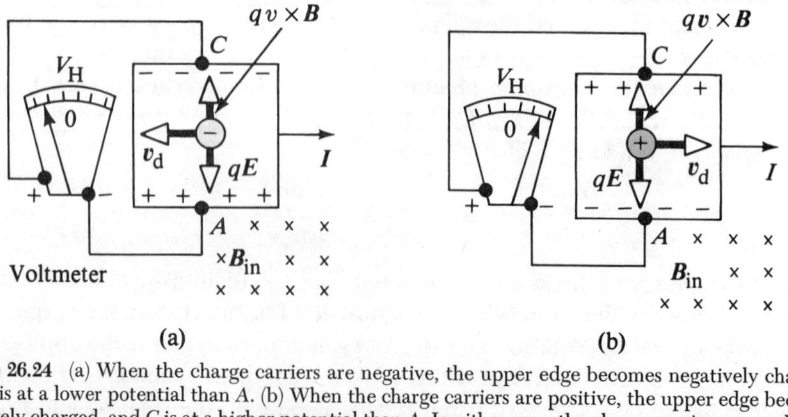

Figure 26.24 (a) When the charge carriers are negative, the upper edge becomes negatively charged, and C is at a lower potential than A. (b) When the charge carriers are positive, the upper edge becomes positively charged, and C is at a higher potential than A. In either case, the charge carriers are no longer deflected when the edges become fully charged.

electrostatic field set up by this charge separation balances the magnetic force on the charge carriers. When this equilibrium condition is reached, the electrons will no longer be deflected upward. A sensitive voltmeter or potentiometer connected across the sample as shown in Fig. 26.24 can be used to measure the potential difference generated across the conductor, known as the *Hall voltage* V_H. If the charge carriers are positive, and hence move in the positive x direction as in Fig. 26.24b, they will also experience an *upward* magnetic force $q\mathbf{v}_d \times \mathbf{B}$. This produces a buildup of positive charge on the upper edge and leaves an excess of negative charge on the lower edge. Hence, the sign of the Hall voltage generated in the sample is opposite the sign of the voltage resulting from the deflection of electrons. The sign of the charge carriers can therefore be determined from a measurement of the polarity of the Hall voltage.

To find an expression for the Hall voltage, first note that the magnetic force on the charge carriers has a magnitude qv_dB. In equilibrium, this force is balanced by the electrostatic force qE_H, where E_H is the electric field due to the charge separation (sometimes referred to as the *Hall field*). Therefore,

$$qv_dB = qE_H$$
$$E_H = v_dB$$

If d is taken to be the width of the conductor, then the Hall voltage V_H measured by the potentiometer is equal to E_Hd, or

$$V_H = E_Hd = v_dBd \qquad (26.23)$$

Thus, we see that the measured Hall voltage gives a value for the drift velocity of the charge carriers if d and B are known.

The number of charge carriers per unit volume, n, can be obained by measuring the current in the sample. From Eq. 24.4, the drift velocity can be expressed as

$$v_d = \frac{I}{nqA} \qquad (26.24)$$

where A is the cross-sectional area of the conductor. Substituting Eq. 26.24 into Eq. 26.23 we obtain

$$V_H = \frac{IBd}{nqA} \qquad (26.25)$$

The Hall voltage

Since A = td, where t is the thickness of the sample, we can also express Eq. 26.25 as

$$V_H = \frac{IB}{nqt} \qquad (26.26)$$

The quantity $1/nq$ is referred to as the *Hall coefficient* R_H. Equation 26.26 shows that a properly calibrated sample can be used to measure the strength of an unknown magnetic field.

Since all quantities appearing in Eq. 26.26 other than nq can be measured, a value for the Hall coefficient is readily obtained. The sign and magnitude of R_H give the sign of the charge carriers and their density. In most metals, the charge carriers are electrons and the charge density determined from Hall effect measurements is in good agreement with calculated values for monovalent metals, such as Li, Na, Cu, and Ag, where n is approximately equal to the number of valence electrons per unit volume. However, this classical model is not valid for metals such as Fe, Bi, and Cd and for semiconductors, such as silicon and germanium. These discrepencies can be explained only by using a model based on the quantum nature of solids.

Example 26.5

A rectangular copper strip 1.5 cm wide and 0.1 cm thick carries a current of 5 A. A 1.2-T magnetic field is applied perpendicular to the strip as in Fig. 26.23. Find the Hall voltage that should be produced.

Solution: If we assume there is one electron per atom available for conduction, then we can take $n = 8.48 \times 10^{28}$ electrons/m^3 (Example 24.1). Substituting this value and the given data into Eq. 26.26 gives

$$V_H = \frac{IB}{nqt}$$

$$= \frac{(5\text{ A})(1.2\text{ T})}{(8.48 \times 10^{28}\text{ m}^{-3})(1.6 \times 10^{-19}\text{ C})(0.1 \times 10^{-2}\text{ m})}$$

$$= 0.442\ \mu\text{V}$$

Hence, the Hall voltage is quite small in good conductors. Note that the width of this sample is not needed in this calculation.

In semiconductors, where n is much smaller than in monovalent metals, one finds a larger Hall voltage since V_H varies as the inverse of n. Current levels of the order of 1 mA are generally used for such materials. Consider a piece of silicon with the same dimensions as the copper strip, with $n = 10^{20}$ electrons/m^3. Taking $B = 1.2$ T and $I = 0.1$ mA, we find that $V_H = 7.5$ mV. Such a voltage is readily measured with a potentiometer.

26.8 SUMMARY

The *magnetic force* that acts on a charge q moving with a velocity $\boldsymbol{v}$ in an external magnetic field $\boldsymbol{B}$ is given by

Magnetic force on a charged particle moving in a magnetic field

$$\boldsymbol{F} = q\boldsymbol{v} \times \boldsymbol{B} \tag{26.1}$$

That is, the magnetic force is in a direction perpendicular both to the velocity of the particle and to the field. The *magnitude* of the magnetic force is given by

$$F = qvB \sin\theta \tag{26.2}$$

where θ is the angle between $\boldsymbol{v}$ and $\boldsymbol{B}$. From this expression, we see that $\boldsymbol{F} = 0$ when $\boldsymbol{v}$ is parallel to (or opposite) $\boldsymbol{B}$. Furthermore, $F = qvB$ when $\boldsymbol{v}$ is perpendicular to $\boldsymbol{B}$.

The SI unit of $\boldsymbol{B}$ is the weber per square meter (Wb/m^2), also called the tesla (T), where

$$[\text{B}] = \text{T} = \frac{\text{Wb}}{\text{m}^2} = \frac{\text{N}}{\text{A} \cdot \text{m}} \tag{26.3}$$

If a straight conductor of length l carries a current I, the force on that current when placed in a uniform *external* magnetic field $\boldsymbol{B}$ is given by

Force on a straight wire carrying a current

$$\boldsymbol{F} = I\boldsymbol{l} \times \boldsymbol{B} \tag{26.5}$$

where the direction of $\boldsymbol{l}$ is in the direction of the current and $|\boldsymbol{l}| = l$.

If an arbitrarily shaped wire carrying a current I is placed in an *external* magnetic field, the force on a very small segment $d\boldsymbol{s}$ is given by

Force on a current element

$$d\boldsymbol{F} = I\, d\boldsymbol{s} \times \boldsymbol{B} \tag{26.6}$$

To determine the total force on the wire, one has to integrate Eq. 26.6, keeping in mind that both $\boldsymbol{B}$ and $d\boldsymbol{s}$ may vary at each point.

The net magnetic force on any *closed* loop carrying a current in a uniform *external* magnetic field is *zero*.

The force on a current-carrying conductor of arbitrary shape in a uniform magnetic field is given by

$$\mathbf{F} = I\mathbf{l}' \times \mathbf{B} \quad (26.9)$$

Force on a wire in a uniform field

where $\mathbf{l}'$ is a vector directed from one end of the conductor to the opposite end.

The *magnetic moment* $\mathfrak{M}$ of a current loop carrying a current I is

$$\mathfrak{M} = I\mathbf{A} \quad (26.13)$$

Magnetic moment of a current loop

where $\mathbf{A}$ is perpendicular to the plane of the loop and $|\mathbf{A}|$ is equal to the area of the loop. The SI unit of $\mathfrak{M}$ is $A \cdot m^2$.

The torque $\boldsymbol{\tau}$ on a current loop when the loop is placed in a uniform *external* magnetic field $\mathbf{B}$ is given by

$$\boldsymbol{\tau} = \mathfrak{M} \times \mathbf{B} \quad (26.14)$$

Torque on a current loop

When a charged particle moves in an external magnetic field, the work done by the magnetic force on the particle is *zero* since the displacement is always *perpendicular* to the direction of the magnetic force. The external magnetic field can alter the direction of the velocity vector, but it cannot change the speed of the particle.

If a charged particle moves in a uniform external magnetic field such that its initial velocity is *perpendicular* to the field, the particle will move in a circle whose plane is *perpendicular* to the magnetic field. The radius r of the circular path is given by

$$r = \frac{mv}{qB} \quad (26.15)$$

where m is the mass of the particle and q is its charge. The angular frequency (cyclotron frequency) of the rotating charged particle is given by

$$\omega = \frac{qB}{m} \quad (26.16)$$

Cyclotron frequency

If a charged particle is moving in the presence of both a magnetic field and an electric field, the total force on the charge is given by the *Lorentz force*,

$$\mathbf{F} = q\mathbf{E} + q\mathbf{v} \times \mathbf{B} \quad (26.18)$$

Lorentz force

That is, the charge experiences both an electric force $q\mathbf{E}$ and a magnetic force $q\mathbf{v} \times \mathbf{B}$.

EXERCISES

Section 26.2 Definition and Properties of the Magnetic Field

1. What force of magnetic origin is experienced by a proton moving north to south with a speed of 7.5×10^6 m/s at a location where the vertical component of the earth's magnetic field is 40 μT directed downward? In what direction is the proton deflected?

2. The magnetic field over a certain region is given by $\mathbf{B} = (2\mathbf{i} - 3\mathbf{j})$ T. An electron moves in the field with a velocity $\mathbf{v} = (\mathbf{i} + 2\mathbf{j} - 3\mathbf{k})$ m/s. Write out in unit-vector notation the force exerted on the electron by the magnetic field.

3. An electron moving along the positive x axis experiences a magnetic deflection in the negative y direction. What is the direction of the magnetic field over this region?

4. A proton moving with a speed of 5×10^7 m/s through a magnetic field of 2 T experiences a magnetic force of magnitude 3×10^{-12} N. What is the angle between the proton's velocity and the field?

5. An electron is projected into a uniform magnetic field given by $\mathbf{B} = (0.2\mathbf{i} + 0.5\mathbf{j})$ T. Find the vector expression for the force on the electron when its velocity is $\mathbf{v} = 5 \times 10^6\mathbf{j}$ m/s.

6. Show that the work done by the magnetic force on a charged particle moving in a magnetic field is zero for any displacement of the particle.

Section 26.3 Magnetic Force on a Current-Carrying Conductor

7. Calculate the magnitude of the force per unit length exerted on a conductor carrying a current of 10 A in a region where a uniform magnetic field has a magnitude of 1.2 T and is directed perpendicular to the conductor.

8. A conductor suspended by two cords as in Fig. 26.25 has a mass per unit length of 0.04 kg/m. What current must exist in the conductor in order for the tension in the supporting wires to be zero if the magnetic field over the region is 3.6 T into the page? What is the required direction for the current?

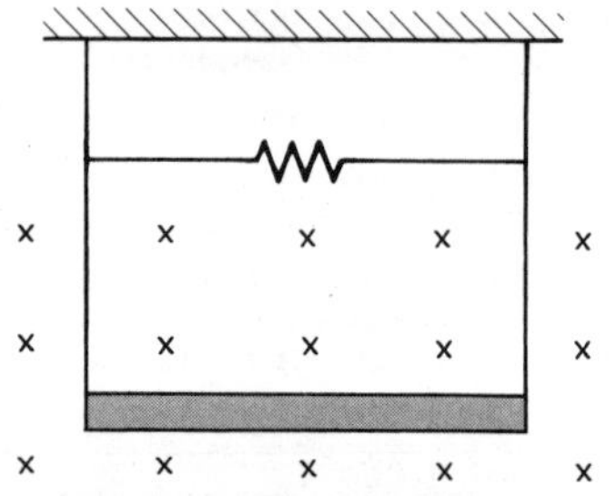

Figure 26.25 (Exercise 8).

9. A current $I = 20$ A is directed along the positive x axis in a conductor that experiences a magnetic force per unit length of 0.12 N/m in the negative y direction. Calculate the magnitude and direction of the magnetic field in the region through which the current passes.

10. A wire 1.2 m in length carries a current of 4 A in a region where a uniform magnetic field has a magnitude of 0.02 T. Calculate the magnitude of the magnetic force on the wire if the angle between the magnetic field and the direction of the current in the wire is (a) 30°, (b) 90°, (c) 180°.

11. The segment of conductor shown in Fig. 26.26 carries a current $I = 0.5$ A. The shorter section is 0.75 m long, and the longer section is 1.5 m long. Determine the magnitude and direction of the magnetic force on the conductor if there is a uniform magnetic field $\mathbf{B} = 0.4\mathbf{i}$ T over the region.

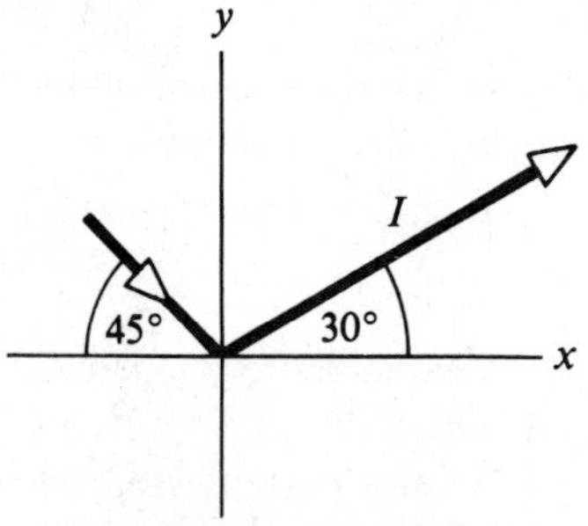

Figure 26.26 (Exercise 11).

12. A rectangular loop with dimensions 10 cm $\times$ 20 cm is suspended by a string, and the lower horizontal section of the loop is immersed in a magnetic field (Fig. 26.27). If a current of 3 A is maintained in the loop in the direction shown, what are the direction and magnitude of the magnetic field required to produce a tension of 4×10^{-2} N in the supporting string? (Neglect the mass of the loop.)

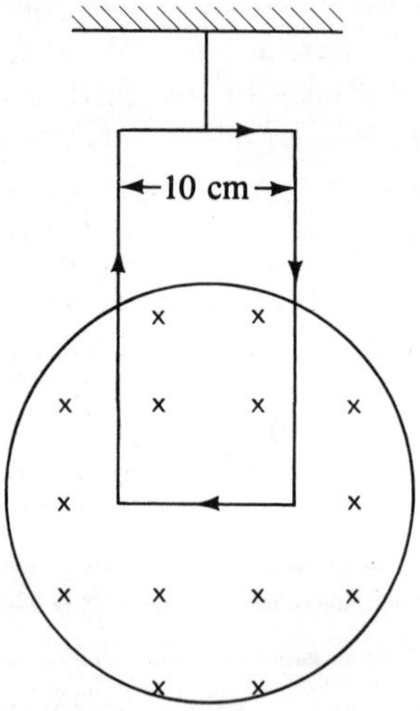

Figure 26.27 (Exercise 12).

Section 26.4 Torque on a Current Loop in a Uniform Magnetic Field

13. A rectangular loop consists of 40 closely wrapped turns and has dimensions 0.25 m by 0.20 m. The loop is hinged along the y axis, and the plane of the coil makes an angle of 45° with the x axis (Fig. 26.28). What is the magnitude of the torque exerted on the loop by a uniform magnetic field of 0.25 T directed along the x axis when the current in the windings has a value of 0.5 A in the direction shown? What is the expected direction of rotation of the loop?

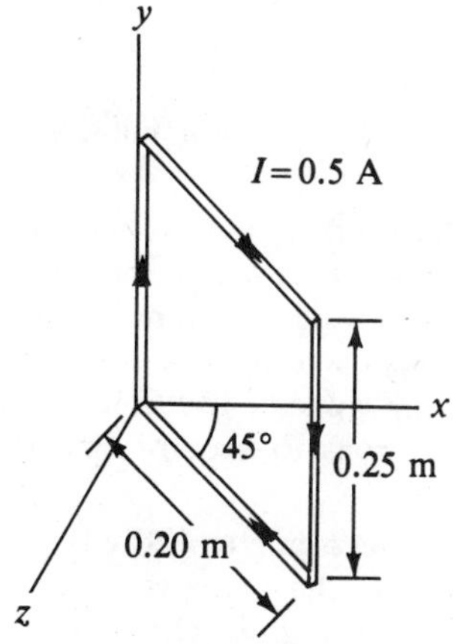

Figure 26.28 (Exercise 13).

14. A current of 6 mA is maintained in a single circular loop of 1 m circumference. An external magnetic field of 0.3 T is directed parallel to the plane of the loop. (a) Calculate the magnetic moment of the current loop. (b) What is the magnitude of the torque exerted on the loop by the magnetic field?

15. A rectangular coil of 150 turns and area 0.12 m^2 is in a uniform magnetic field of 0.15 T. Measurements indicate that the maximum torque exerted on the loop by the field is 6×10^{-4} N · m. (a) Calculate the current in the coil. (b) Would the value found for the required current be different if the 150 turns of wire were used to form a single-turn coil of larger area? Explain.

16. A circular coil of 100 turns has a radius of 0.025 m and carries a current of 0.1 A while in a uniform external magnetic field of 1.5 T. How much work must be done to rotate the coil from a position where the magnetic moment is parallel to the field to a position where the magnetic moment is opposite the field?

Section 26.5 Motion of a Charged Particle in a Magnetic Field

17. Consider a particle of mass m and charge q moving with a velocity $\boldsymbol{v}$. The particle enters a region perpendicular to a magnetic field $\boldsymbol{B}$. Show that while in the region of the magnetic field the kinetic energy of the particle is proportional to the square of the radius of its orbit.

18. A singly charged positive ion has a mass of 2.5×10^{-26} kg. After being accelerated through a potential difference of 250 V, the ion enters a magnetic field of 0.5 T along a direction perpendicular to the direction of the field. Calculate the radius of the path of the ion in the field.

19. What magnetic field would be required to constrain an electron whose energy is 400 eV to a circular path of radius 0.8 m?

20. Calculate the cyclotron frequency ω of a proton in a magnetic field of 3.7 T.

21. A proton, a deuteron, and an alpha particle are accelerated through a common potential difference V. The particles enter a uniform magnetic field $\boldsymbol{B}$ along a direction perpendicular to $\boldsymbol{B}$. The proton moves in a circular path of radius r_p. Find the value of the radii of the orbits of the deuteron, r_d, and the alpha particle, r_α, in terms of r_p.

Section 26.6 Applications of the Motion of Charged Particles in a Magnetic Field

22. What is the required radius of a cyclotron designed to accelerate protons to energies of 15 MeV using a magnetic field of 2 T?

23. An alpha particle with velocity $\boldsymbol{v} = 3 \times 10^5\boldsymbol{i}$ m/s enters a region where the magnetic field has a value $\boldsymbol{B} = 1.2\boldsymbol{k}$ T. Determine the required magnitude and direction of an electric field $\boldsymbol{E}$ that will allow the alpha particle to continue to move along the x axis.

24. A cyclotron designed to accelerate protons is provided with a magnetic field of 2 T and has a radius of 0.4 m. (a) What is the cyclotron frequency? (b) What is the maximum speed acquired by the protons?

25. Consider the mass spectrometer shown schematically in Fig. 26.20. The electric field between the plates of the velocity selector is 800 N/C, and the magnetic field in both the velocity selector and the deflection chamber has a magnitude of 0.8 T. Calculate the radius of the path in the system for a singly charged ion with a mass $m = 1.16 \times 10^{-26}$ kg.

Section 26.7 The Hall Effect

26. A section of conductor 0.15 cm in thickness is used as the experimental specimen in a Hall effect measurement. If a Hall voltage of 60 μV is measured for a current of 15 A in a magnetic field of 1.5 T, calculate the Hall coefficient for the conductor.

27. In an experiment designed to measure the earth's magnetic field using the Hall effect, a copper bar 0.02 m in thickness is positioned along an east-west direction. If a current of 10 A in the conductor results in a measured Hall voltage of 1.8×10^{-12} V, what is the calculated value of the earth's magnetic field? (Assume that $n = 8.48 \times 10^{28}$ electrons/m^3 and that the plane of the bar is rotated to be perpendicular to the direction of $\boldsymbol{B}$.)

PROBLEMS

1. A straight wire of mass 10 g and length 5 cm is suspended from two identical springs which, in turn, form a closed circuit (Fig. 26.29). The springs stretch a distance of 0.5 cm under the weight of the wire. The circuit has a *total* resistance of 12 Ω. When a magnetic field is turned on, directed *out* of the page (indicated by the dots in Fig. 26.29), the springs are observed to stretch an *additional* 0.3 cm. What is the strength of the magnetic field? (The upper portion of the circuit is fixed.)

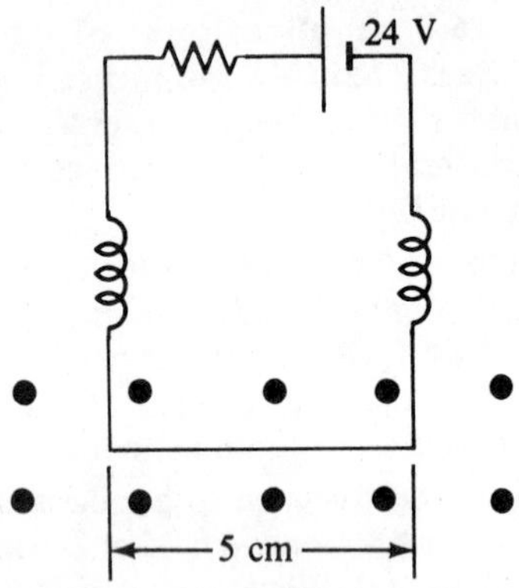

Figure 26.29 (Problem 1).

2. A conducting wire formed into the shape of an M with dimensions as shown in Fig. 26.30 carries a current of 15 A. An external magnetic field $\boldsymbol{B}$ = 2.5 T is directed as shown throughout the region occupied by the conductor. Calculate the magnitude and direction of the net force exerted on the conductor by the magnetic field.

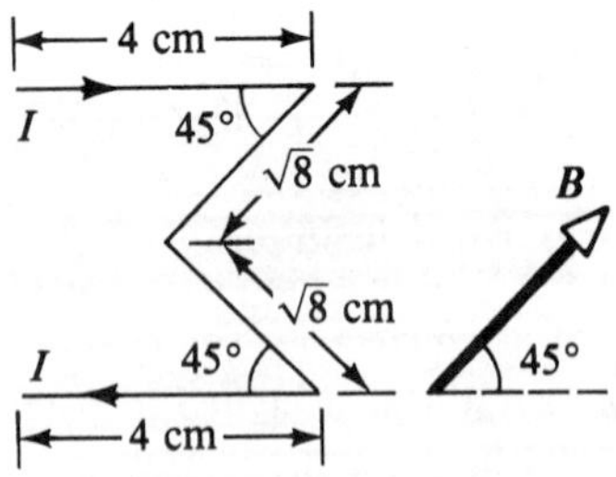

Figure 26.30 (Problem 2).

3. Consider a galvanometer constructed as shown in Fig. 26.13. The coil is wrapped so that the loop is equivalent to a straight wire whose length perpendicular to the field is 0.7 m. The radius of the coil is 1.5 cm, and the radial magnetic field has a value of 0.5 T. The spring, which exerts a restoring torque on the coil, has a characteristic torque constant of 2×10^{-7} N · m/rad (Section 13.4). What value of current will produce a full-scale deflection of $\pi/4$ rad?

4. A cyclotron designed to accelerate deuterons has a magnetic field with a uniform intensity of 1.5 T over a region of radius 0.45 m. If the alternating potential applied between the dees of the cyclotron has a maximum value of 15 kV, what time is required for the deuterons to acquire maximum attainable energy?

5. A mass spectrometer of the Bainbridge type is used to examine the isotopes of uranium. Ions in the beam emerge from the velocity selector with a speed of 3×10^5 m/s and enter a uniform magnetic field of 0.6 T directed perpendicular to the velocity of the ions. What is the distance between the impact points formed on the photographic plate by singly charged ions of ^{235}U and ^{238}U?

6. A uniform magnetic field of 0.15 T is directed along the positive x axis. A positron moving with a speed of 5×10^6 m/s enters the field along a direction that makes an angle of 85° with the x axis (Fig. 26.31). The motion of the particle is expected to be a helix, as described in Section 26.5. Calculate (a) the pitch p and (b) the radius r of the trajectory.

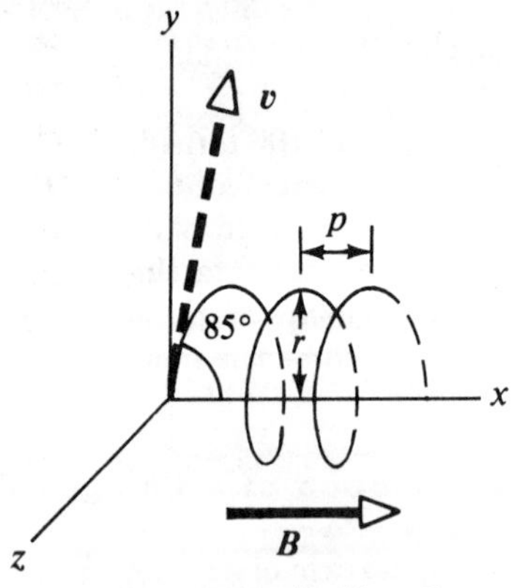

Figure 26.31 (Problem 6).

7. Consider an electron orbiting a proton and maintained in a fixed circular path of radius $R = 5.29 \times 10^{-11}$ m by the coulombic force of mutual attraction. Treating the orbiting charge as a current loop, calculate the resulting torque when the system is in an external magnetic field of 0.4 T directed perpendicular to the magnetic moment of the orbiting electron.

8. A singly charged heavy ion is observed to complete five revolutions in a uniform magnetic field of 5×10^{-2} T in 1.50 ms. Calculate the (approximate) mass of the ion in kg.

9. A conducting wire of circular cross section formed of a material that has a mass density of 2.7 g/cm^3 is placed in a uniform magnetic field with the axis of the wire perpendicular to the direction of the field. A current density of 2.4×10^6 A/m^2 is established in the wire and the magnetic field increased until the magnetic force on the wire just balances the gravitational force. Calculate the value of B when this condition is met.

10. A positive charge $q = 3.2 \times 10^{-19}$ C moves with a velocity $\boldsymbol{v} = (2\boldsymbol{i} + 3\boldsymbol{j} - \boldsymbol{k})$ m/s through a region where both a uniform magnetic field and a uniform electric field exist. (a) Calculate the total force on the moving charge (in unit-vector notation) if $\boldsymbol{B} = (2\boldsymbol{i} + 4\boldsymbol{j} + \boldsymbol{k})$ T and $\boldsymbol{E} = (4\boldsymbol{i} - \boldsymbol{j} - 2\boldsymbol{k})$ V/m. (b) What angle does the force vector make relative to the positive x axis?

11. A rectangular loop of wire carrying a current of 2 A is suspended vertically and attached to the right arm of a balance. After the system is balanced, an external magnetic field is introduced. The field threads the lower end of the loop in a direction perpendicular to the wire. If the width of the loop is 20 cm and it takes 13.5 g of added mass on the left arm to rebalance the system, determine B.

27 Sources of the Magnetic Field

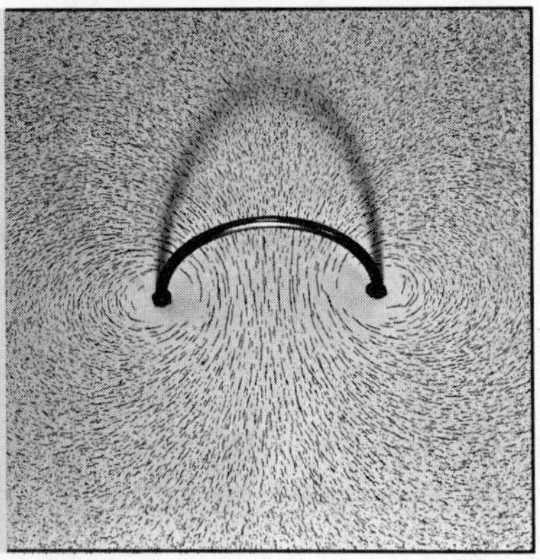

Education Development Center, Newton, Mass.

The preceding chapter treated a class of problems involving the magnetic force on a charged particle moving in a magnetic field. To complete the description of the magnetic interaction, this chapter deals with the origin of the magnetic field, namely, moving charges or electric currents. We begin by showing how to use the law of Biot and Savart to calculate the magnetic field produced at a point by a current element. Using this formalism and the superposition principle, we then calculate the total magnetic field due to a distribution of currents for several geometries. Next, we show how to determine the force between two current-carrying conductors, which leads to the definition of the ampere. Finally, we describe Ampère's law, which is very useful for calculating the magnetic field of highly symmetric configurations carrying steady currents. We apply Ampère's law to determine the magnetic field for several current distributions, including that of a solenoid.

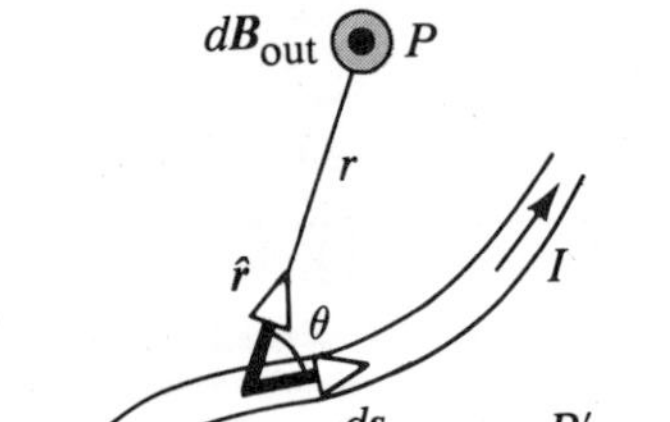

Figure 27.1 The magnetic field $d\mathbf{B}$ at a point P due to a current element ds is given by the Biot-Savart law, Eq. 27.1. The field is out of the paper at P.

27.1 THE BIOT-SAVART LAW

Shortly after Oersted's discovery in 1819 that a compass needle is deflected by a current-carrying conductor, Jean Baptiste Biot and Felix Savart reported that a conductor carrying a steady current produces a force on a magnetic pole. From their experimental results, Biot and Savart were able to arrive at an expression that gives the magnetic field at some point in space in terms of the current that produces the field. The *Biot-Savart law* says that if a wire carries a steady current I, the magnetic field $d\boldsymbol{B}$ at a point P associated with an element $d\boldsymbol{s}$ (Fig. 27.1) has the following properties:

Properties of the magnetic field due to a current element

1. The vector $d\boldsymbol{B}$ is perpendicular both to $d\boldsymbol{s}$ (which is in the direction of the current) and to the unit vector $\hat{\boldsymbol{r}}$ directed from the element to the point P.
2. The magnitude of $d\boldsymbol{B}$ is inversely proportional to r^2, where r is the distance from the element to the point P.
3. The magnitude of $d\boldsymbol{B}$ is proportional to the current and to the length ds of the element.
4. The magnitude of $d\boldsymbol{B}$ is proportional to $\sin\theta$, where θ is the angle between the vectors $d\boldsymbol{s}$ and $\hat{\boldsymbol{r}}$.

The Biot-Savart law can be summarized in the following convenient form:

Biot-Savart law

$$d\boldsymbol{B} = k_{\text{m}} \frac{I\, d\boldsymbol{s} \times \hat{\boldsymbol{r}}}{r^2} \tag{27.1}$$

where k_m is a constant that in SI units is exactly 10^{-7} Wb/A · m. The constant k_m is usually written $\mu_o/4\pi$, where μ_o is another constant, called the *permeability of free space*. That is,

$$\frac{\mu_o}{4\pi} = k_m = 10^{-7}\ \text{Wb/A} \cdot \text{m} \qquad (27.2)$$

Permeability of free space

$$\mu_o = 4\pi k_m = 4\pi \times 10^{-7}\ \text{Wb/A} \cdot \text{m} \qquad (27.3)$$

Hence, the Biot-Savart Law, Eq. 27.1, can also be written

Biot-Savart law

$$d\boldsymbol{B} = \frac{\mu_o}{4\pi} \frac{I\, d\boldsymbol{s} \times \hat{\boldsymbol{r}}}{r^2} \qquad (27.4)$$

It is important to note that the Biot-Savart law gives the magnetic field at a point only for a small element of the conductor. To find the *total* magnetic field $\boldsymbol{B}$ at some point due to a conductor of finite size, we must sum up contributions from all current elements making up the conductor. That is, we must evaluate $\boldsymbol{B}$ by integrating Eq. 27.4:

$$\boldsymbol{B} = \frac{\mu_o I}{4\pi} \int \frac{d\boldsymbol{s} \times \hat{\boldsymbol{r}}}{r^2} \qquad (27.5)$$

where the integral is taken over the entire conductor. This expression must be handled with special care since the integrand is a vector quantity.

There are interesting similarities between the Biot-Savart law of magnetism and Coulomb's law of electrostatics. That is, the current element $I\, d\boldsymbol{s}$ produces a magnetic field, whereas a point charge q produces an electric field. Furthermore, *the magnitude of the magnetic field varies as the inverse square of the distance from the current element,* as does the electric field due to a point charge.

However, the directions of the two fields are quite different. The electric field due to a point charge is radial. In the case of a positive point charge, $\boldsymbol{E}$ is directed from the charge to the field point. On the other hand, the magnetic field due to a current element is perpendicular to both the current element and the radius vector. Hence, if the conductor lies in the plane of the paper, as in Fig. 27.1, $d\boldsymbol{B}$ points *out* of the paper at the point P and into the paper at P'.

The examples that follow illustrate how to use the Biot-Savart law for calculating the magnetic induction of several important geometric arrangements. It is important that you recognize that the magnetic field described in these calculations is *the field due to a given current-carrying conductor.* This is not to be confused with any *external* field that may be applied to the conductor.

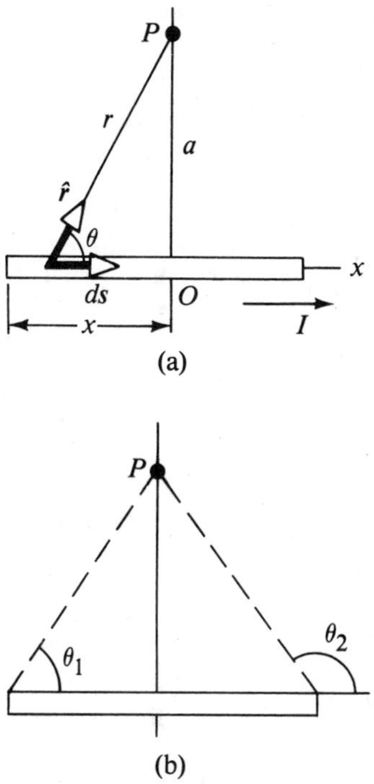

Figure 27.2 (Example 27.1) (a) A straight wire segment carrying a current I. The magnetic field at P due to each element ds is out of the paper, and so the net field is also out of the paper. (b) The limiting angles θ_1 and θ_2 for this geometry.

Example 27.1

Magnetic field of a thin, straight conductor: Consider a thin, straight wire carrying a constant current I and placed along the x axis as in Fig. 27.2. Let us calculate the total magnetic field at the point P located at a distance a from the wire.

Solution: An element ds is at a distance r from P. The direction of the field at P due to this element is out of the paper, since $d\boldsymbol{s} \times \hat{\boldsymbol{r}}$ is out of the paper. In fact, *all* elements give a contribution directly out of the paper at P. Therefore, we have only to determine the magnitude of the field at P. In fact, taking the origin at O and letting P be along the *positive* y axis, with $\boldsymbol{k}$ being a unit vector pointing *out* of the paper, we see that

$$d\boldsymbol{s} \times \hat{\boldsymbol{r}} = \boldsymbol{k}\,|d\boldsymbol{s} \times \hat{\boldsymbol{r}}| = \boldsymbol{k}\,(dx \sin\theta)$$

Substitution into Eq. 27.4 gives $d\boldsymbol{B} = \boldsymbol{k}\, dB$, with

$$(1) \qquad dB = \frac{\mu_o I}{4\pi} \frac{dx \sin\theta}{r^2}$$

In order to integrate this expression, we must somehow relate the variables θ, x, and r. One approach is to express x and r in terms of θ. From the geometry in Fig. 27.2a and some simple differentiation, we obtain the following relationship:

$$(2) \qquad r = \frac{a}{\sin\theta} = a \csc\theta$$

Since $\tan\theta = -a/x$ from the right triangle in Fig. 27.2a,

$$x = -a \cot\theta$$

$$(3) \qquad dx = a \csc^2\theta \, d\theta$$

Substitution of (2) and (3) into (1) gives

$$(4) \qquad dB = \frac{\mu_0 I}{4\pi} \frac{a \csc^2\theta \sin\theta \, d\theta}{a^2 \csc^2\theta} = \frac{\mu_0 I}{4\pi a} \sin\theta \, d\theta$$

Thus, we have reduced the expression to one involving only the variable θ. We can now obtain the total field at P by integrating (4) over all elements subtending angles ranging from θ_1 to θ_2 as defined in Fig. 27.2b. This gives

$$B = \frac{\mu_0 I}{4\pi a} \int_{\theta_1}^{\theta_2} \sin\theta \, d\theta = \frac{\mu_0 I}{4\pi a} (\cos\theta_1 - \cos\theta_2) \quad (27.6)$$

We can apply this result to find the magnetic field of any straight wire if we know the geometry and hence the angles θ_1 and θ_2.

Consider the special case of an infinitely long, straight wire. In this case, $\theta_1 = 0$ and $\theta_2 = \pi$, as can be seen from Fig. 27.2b, for segments ranging from $x = -\infty$ to $x = +\infty$. Since $(\cos\theta_1 - \cos\theta_2) = (\cos 0 - \cos\pi) = 2$, Eq. 27.6 becomes

$$B = \frac{\mu_0 I}{2\pi a} \quad (27.7)$$

A three-dimensional view of the direction of $\boldsymbol{B}$ for a long, straight wire is shown in Fig. 27.3. *The field lines are circles concentric with the wire and are in a plane perpendicular to the wire.* The magnitude of $\boldsymbol{B}$ is constant on any

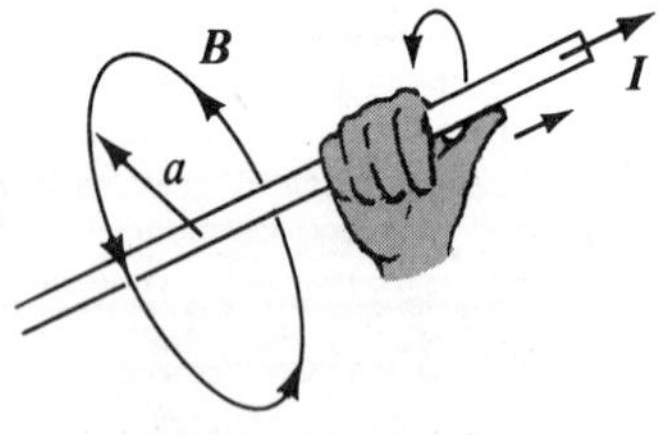

Figure 27.3 The right-hand rule for determining the direction of the magnetic field due to a long, straight wire. Note that the magnetic field lines form circles around the wire.

circle of radius a and is given by Eq. 27.7. A convenient rule for determining the direction of $\boldsymbol{B}$ is to grasp the wire with the right hand, with the thumb along the direction of the current. The four fingers wrap in the direction of the magnetic field.

Our result shows that the magnitude of the magnetic field is proportional to the current and decreases as the distance from the wire increases, as one might intuitively expect. Notice that Eq. 27.7 has the same mathematical form as the expression for the magnitude of the electric field due to a long charged wire (Eq. 21.9).

Example 27.2

A long, straight wire carries a current of 5 A. Find the magnitude of the magnetic field at a distance of 4 cm from the wire.

Solution: Using Eq. 27.7, we find that

$$B = \frac{\mu_0 I}{2\pi a} = \frac{(4\pi \times 10^{-7} \text{ Wb/A} \cdot \text{m})(5 \text{ A})}{2\pi(4 \times 10^{-2} \text{ m})}$$
$$= 2.5 \times 10^{-5} \text{ Wb/m}^2$$

Example 27.3

Calculate the magnetic field at the point O for the closed current loop shown in Fig. 27.4. The loop consists of two straight portions and a circular arc of radius R, which subtends an angle θ at the center of the arc.

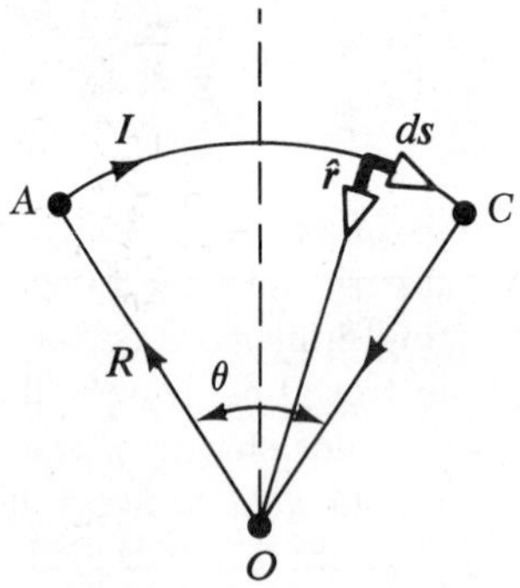

Figure 27.4 (Example 27.3) The magnetic field at O due to the closed current loop is into the paper. Note that the contribution to the field at O due to the straight segments OA and OC is zero.

Solution: First, note that the magnetic field at O due to the straight segments OA and OC is identically *zero*, since $\boldsymbol{ds}$ is parallel to $\hat{\boldsymbol{r}}$ along these paths and therefore $\boldsymbol{ds} \times \hat{\boldsymbol{r}} = 0$. This simplifies the problem because now we need to be concerned only with the magnetic field at O due to the curved portion AC. Note that each element along the path AC is at the same distance R from O, and each gives a contribution $\boldsymbol{dB}$, which is directed into the paper at O. Furthermore, at every point on the path AC, we see that $\boldsymbol{ds}$ is perpendicular to $\hat{\boldsymbol{r}}$, so that $|\boldsymbol{ds} \times \hat{\boldsymbol{r}}| = ds$. Using this information and Eq. 27.4, we get the following expression for the field at O due to the segment ds:

$$dB = \frac{\mu_0 I}{4\pi} \frac{ds}{R^2}$$

Since I and R are constants, we can easily integrate this expression, which gives

$$B = \frac{\mu_0 I}{4\pi R^2} \int ds = \frac{\mu_0 I}{4\pi R^2} s = \frac{\mu_0 I}{4\pi R} \theta \quad (27.8)$$

where we have used the fact that $s = R\theta$, where θ is measured in *radians*. The direction of $\mathbf{B}$ is *into* the paper at O since $d\mathbf{s} \times \hat{\mathbf{r}}$ is into the paper for every segment.

For example, if the loop subtends an angle $\theta = \pi/2$ rad, we find from Eq. 27.8 that $B = \mu_0 I/8R$. On the other hand, if the loop were a *full* circle of radius R, then $\theta = 2\pi$ rad, and we would find that $B = \mu_0 I/2R$ at the center of the loop.

Example 27.4

Magnetic field on the axis of a circular current loop: Consider a circular loop of wire of radius R located in the yz plane and carrying a steady current I, as in Fig. 27.5. Let us calculate the magnetic field at an axial point P a distance x from the center of the loop.

Solution: In this situation, note that any element $d\mathbf{s}$ is perpendicular to $\hat{\mathbf{r}}$. Furthermore, all elements around the loop are at the same distance r from P, where $r^2 = x^2 + R^2$. Hence, the *magnitude* of $d\mathbf{B}$ due to the element $d\mathbf{s}$ is given by

$$dB = \frac{\mu_0 I}{4\pi} \frac{|d\mathbf{s} \times \hat{\mathbf{r}}|}{r^2} = \frac{\mu_0 I}{4\pi} \frac{ds}{(x^2 + R^2)}$$

The direction of the magnetic field $d\mathbf{B}$ due to the element $d\mathbf{s}$ is perpendicular to the plane formed by $\hat{\mathbf{r}}$ and $d\mathbf{s}$, as shown in Fig. 27.5. The vector $d\mathbf{B}$ can be resolved into a component dB_x, along the x axis, and a component $dB_\perp$,

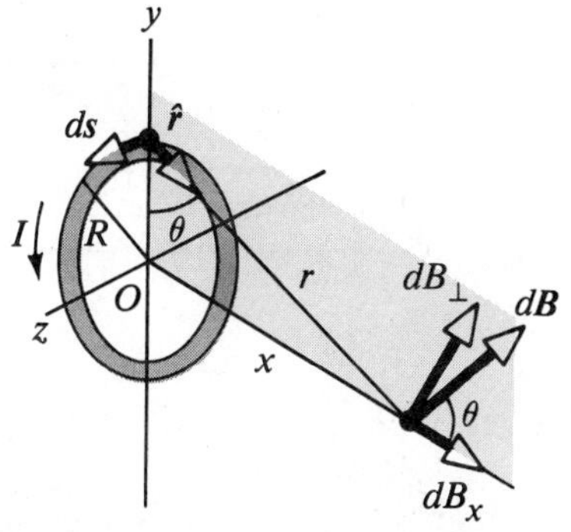

Figure 27.5 (Example 27.4) The geometry for calculating the magnetic field at an axial point P for a current loop. Note that by symmetry the total field $\mathbf{B}$ is along the x axis.

which is perpendicular to the x axis. When the components perpendicular to the x axis are summed over the whole loop, the result is *zero*. That is, by symmetry any element on one side of the loop will set up a perpendicular component that cancels the component set up by an element diametrically opposite it. Therefore, we see that *the resultant field at* P *must be along the* x *axis* and can be found by integrating the components $dB_x = dB \cos\theta$, where this expression is obtained from resolving the vector $d\mathbf{B}$ into its components as shown in Fig. 27.5. That is, $\mathbf{B} = \mathbf{i}B_x$, where B_x is given by

$$B_x = \oint dB \cos\theta = \frac{\mu_0 I}{4\pi} \oint \frac{ds \cos\theta}{x^2 + R^2}$$

where the integral must be taken over the entire loop. Since θ, x, and R are constants for all elements of the loop and since $\cos\theta = R/(x^2 + R^2)^{1/2}$, we get

$$B_x = \frac{\mu_0 IR}{4\pi(x^2 + R^2)^{3/2}} \oint ds = \frac{\mu_0 IR^2}{2(x^2 + R^2)^{3/2}} \qquad (27.9)$$

where we have used the fact that $\oint ds = 2\pi R$ (the circumference of the loop).

To find the magnetic field at the *center* of the loop, we set $x = 0$ in Eq. 27.9. At this special point, this gives

$$B = \frac{\mu_0 I}{2R} \qquad (\text{at } x = 0) \qquad (27.10)$$

which is in agreement with the results of Example 27.3.

It is also interesting to determine the behavior of the magnetic field at large distances from the loop, that is, when x is large compared with R. In this case, we can neglect the term R^2 in the denominator of Eq. 27.9 and get

$$B \approx \frac{\mu_0 IR^2}{2x^3} \qquad (\text{for } x \gg R) \qquad (27.11)$$

Since the magnitude of the magnetic dipole moment $\mathfrak{M}$ of the loop is defined as the product of the current and the area (Eq. 26.13), $\mathfrak{M} = I(\pi R^2)$ and we can express Eq. 27.11 in the form

$$B = \frac{\mu_0}{2\pi} \frac{\mathfrak{M}}{x^3} \qquad (27.12)$$

This result is similar in form to the expression for the electric field due to an electric dipole, $E = kp/y^3$ (Eq. 20.10), where p is the electric dipole moment. The pattern of the magnetic field lines for a circular loop is shown in Fig. 27.6. For clarity, the lines are drawn only for one plane which contains the axis of the loop. The field pattern is axially symmetric.

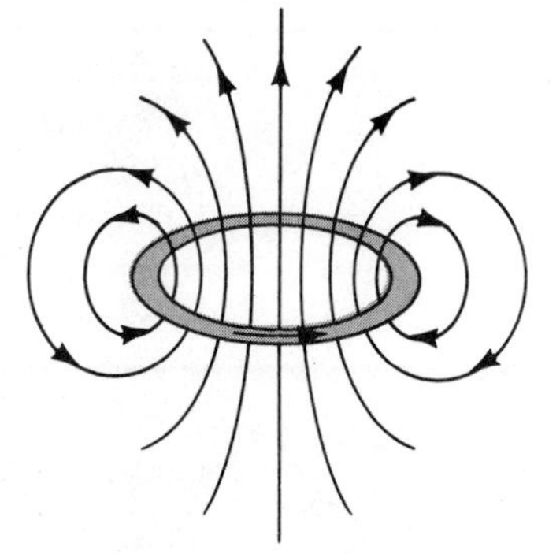

Figure 27.6 Magnetic field lines for a current loop. Far from the loop, the field lines are identical in form to those of an electric dipole.

Q1. Is the magnetic field due to a current loop uniform? Explain.

Q2. A current in a conductor produces a magnetic field which can be calculated using the Biot-Savart law. Since current is defined as the rate of flow of charge, what can you conclude about the magnetic field due to stationary charges? What about moving charges?

27.2 THE MAGNETIC FORCE BETWEEN TWO PARALLEL CONDUCTORS

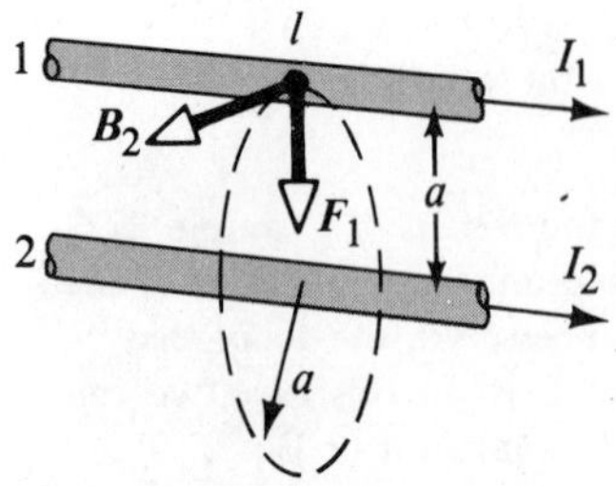

Figure 27.7 Two parallel wires each carrying a steady current exert a force on each other. The field $\boldsymbol{B}_2$ at wire 1 due to wire 2 produces a force on wire 1 given by $F_1 = I_1 lB_2$. The force is attractive if the currents are parallel as shown and repulsive if the currents are antiparallel.

In the previous chapter we described the fact that a magnetic force acts on a current-carrying conductor when the conductor is placed in an external magnetic field. Since a current in a conductor sets up its own magnetic field, it is easy to understand that two current-carrying conductors will exert magnetic forces upon each other. As we shall see, such forces can be used as the basis for defining the ampere and the coulomb. Consider two long, straight, parallel wires separated by a distance a and carrying currents I_1 and I_2 in the same direction, as in Fig. 27.7. We can easily determine the force on one wire due to a magnetic field set up by the other wire. Wire 2, which carries a current I_2, sets up a magnetic field $\boldsymbol{B}_2$ at the position of wire 1. The direction of $\boldsymbol{B}_2$ is *perpendicular* to the wire, as shown in Fig. 27.7. According to Eq. 26.5, the magnetic force on a length l of wire 1 is $\boldsymbol{F}_1 = I_1\boldsymbol{l} \times \boldsymbol{B}_2$. Since $\boldsymbol{l}$ is perpendicular to $\boldsymbol{B}_2$, the magnitude of $\boldsymbol{F}_1$ is given by $F_1 = I_1 lB_2$. Since the field due to wire 1 is given by Eq. 27.7,

$$B_2 = 2k_\mathrm{m}\frac{I_2}{a}$$

we see that

$$F_1 = I_1 lB_2 = I_1 l\left(2k_\mathrm{m}\frac{I_2}{a}\right) = 2k_\mathrm{m}\frac{I_1 I_2}{a}l$$

Therefore, the force per unit length is

Force per unit length between two current carrying parallel wires

$$\frac{F_1}{l} = 2k_\mathrm{m}\frac{I_1 I_2}{a} \tag{27.13}$$

The direction of $\boldsymbol{F}_1$ is downward, toward wire 2, since $\boldsymbol{l} \times \boldsymbol{B}_2$ is downward. If one considers the field set up at wire 2 due to wire 1, the force $\boldsymbol{F}_2$ on wire 2 is found to be equal to and opposite $\boldsymbol{F}_1$. This is what one would expect if Newton's third law of action-reaction is obeyed.[1] When the currents are in opposite directions, the forces are reversed and the wires repel one another. Hence, we find that *parallel conductors carrying currents in the same direction attract each other, whereas parallel conductors carrying currents in opposite directions repel each other.*

The force between two parallel wires each carrying a current is used to define the ampere as follows: *If two long, parallel wires 1 m apart carry the same current and the force per unit length on each wire is 2×10^{-7} N/m, then the current is defined to be 1 A.* The numerical value of 2×10^{-7} N/m is obtained from Eq. 27.13, with $I_1 = I_2 = 1$ A and $a = 1$ m. Therefore, a mechanical measurement can be used to standardize the ampere. For instance, the National Bureau of Standards uses an

[1] Although the total force on wire 1 is equal to and opposite the total force on wire 2, Newton's third law does not apply when one considers two small elements of the wires that are not opposite each other. This apparent violation of Newton's third law and of conservation of momentum is described in more advanced treatments on electricity and magnetism.

instrument called a *current balance* for primary current measurements. These results are then used to standardize other, more conventional instruments, such as ammeters.

The SI unit of charge, the coulomb, can now be defined in terms of the ampere as follows: *If a conductor carries a steady current of 1 A, then the quantity of charge that flows through any cross section in 1 s is 1 C.*

Q3. Two parallel wires carry currents in opposite directions. Describe the nature of the resultant magnetic field due to the two wires at points (a) between the wire and (b) outside the wires in a plane containing the wires.

Q4. Explain why two parallel wires carrying currents in opposite directions repel each other.

Q5. Two wires carrying equal and opposite currents are twisted together in the construction of a circuit. Why does this technique reduce stray magnetic fields?

27.3 AMPÈRE'S LAW

A simple experiment first carried out by Oersted in 1820 clearly demonstrates the fact that a current-carrying conductor produces a magnetic field. In this experiment, several compass needles are placed in a horizontal plane near a long vertical wire, as in Fig. 27.8a. When there is no current in the wire, all compasses in the loop point in the same direction (that of the earth's field), as one would expect. However, when the wire carries a strong, steady current, the compass needles will all deflect in a direction tangent to the circle, as in Fig. 27.8b. These observations show that the direction of $\boldsymbol{B}$ is consistent with the right-hand rule described in Section 27.1. *If the wire is grasped in the right hand with the thumb in the direction of the current, the fingers will wrap (or curl) in the direction of* **B**. When the current is reversed, the compass needles in Fig. 27.8b will also reverse.

Since the compass needles point in the direction of $\boldsymbol{B}$, we conclude that the lines of $\boldsymbol{B}$ form circles about the wire, as we discussed in the previous section. By symmetry, the magnitude of $\boldsymbol{B}$ is the same everywhere on a circular path that is centered on the wire and lying in a plane that is perpendicular to the wire. By varying the current and distance r from the wire, one finds that B is proportional to the current and inversely proportional to the distance from the wire.

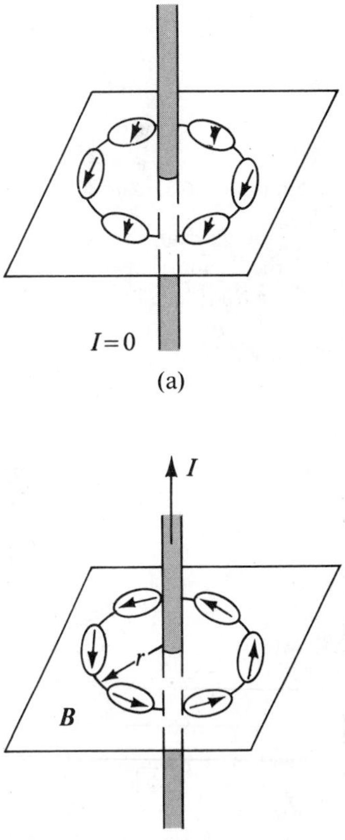

Figure 27.8 (a) When there is no current in the vertical wire, all compass needles point in the same direction. (b) When the wire carries a strong current, the compass needles deflect in a direction tangent to the circle, which is the direction of $\boldsymbol{B}$ due to the current.

Now let us evaluate the product $\boldsymbol{B} \cdot d\boldsymbol{s}$ and sum these products over the closed circular path centered on the wire. Along this path, the vectors $d\boldsymbol{s}$ and $\boldsymbol{B}$ are parallel at each point (Fig. 27.8b), so that $\boldsymbol{B} \cdot d\boldsymbol{s} = B\,ds$. Furthermore, $\boldsymbol{B}$ is constant in magnitude on this circle and given by Eq. 27.7. Therefore, the sum of the products $B\,ds$ over the closed path, which is equivalent to the line integral of $\boldsymbol{B} \cdot d\boldsymbol{s}$, is given by

$$\oint \boldsymbol{B} \cdot d\boldsymbol{s} = B \oint ds = \frac{\mu_0 I}{2\pi r}(2\pi r) = \mu_0 I \qquad (27.14)$$

where $\oint ds = 2\pi r$ is the circumference of the circle.

This result, known as *Ampère's law*, was calculated for the special case of a circular path surrounding a wire. However, the result can be applied in the general case in which an arbitrary closed path is threaded by a *steady current*. That is, *Ampère's law says that the line integral of* **B** · **ds** *around any closed path equals* μ_0I, *where* I *is the total steady current threaded by the path:*

$$\oint \boldsymbol{B} \cdot d\boldsymbol{s} = \mu_0 I \qquad (27.15)$$

Ampère's law

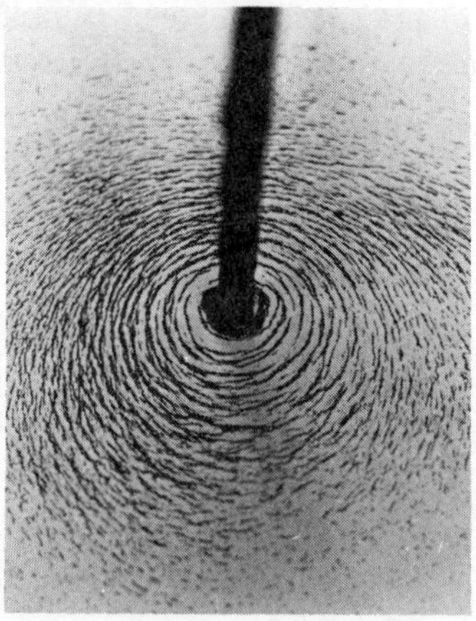

Circular magnetic field lines surrounding a current-carrying conductor as displayed with iron filings. The photograph was taken using 30 parallel wires each carrying a current of $\frac{1}{2}$ A. (Courtesy of J. Lehman and H. Strickland, James Madison University)

It should be recognized that *Ampère's law is valid only for steady currents.* Furthermore, *Ampère's law is useful only for calculating the magnetic field of current configurations with a high degree of symmetry,* just as Gauss' law is useful only for calculating the electric field of highly symmetric charge distributions. The following examples illustrate some symmetric current configurations for which Ampère's law is useful.

Example 27.5 *The **B** Field of a Long Wire*

A long, straight wire of radius a carries a steady current I_0 that is uniformly distributed through the cross section of the wire (Fig. 27.9). Calculate the magnetic field at a distance r from the center of the wire in the regions $r \geqslant a$ and $r < a$.

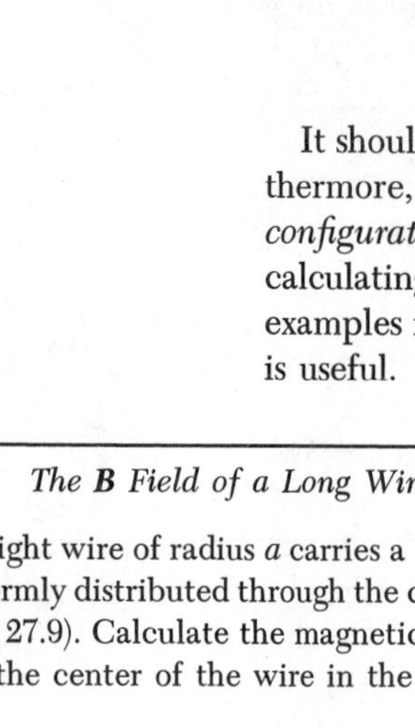

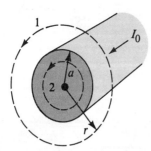

Figure 27.9 (Example 27.5) A long, straight wire of radius a carrying a steady current I_0 uniformly distributed across the wire. The magnetic field at any point can be calculated from Ampère's law using a circular path of radius r, centered on the wire.

Solution: In region 1, where $r \geqslant a$, let us choose a circular path of radius r centered at the wire. From symmetry, we see that $\mathbf{B}$ must be constant in magnitude and parallel to $d\mathbf{s}$ at every point on the path. Since the total current linked by path 1 is I_0, Ampère's law applied to the path gives

$$\oint \mathbf{B} \cdot d\mathbf{s} = B\oint ds = B(2\pi r) = \mu_0 I_0$$

$$B = \frac{\mu_0 I_0}{2\pi r} \quad (\text{for } r \geqslant a) \tag{27.16}$$

which is identical to Eq. 27.7.

Now consider the interior of the wire, that is, region 2, where $r < a$. In this case, note that the current I threaded by the path is *less* than the total current, I_0. Since the current is assumed to be uniform over the cross section of the wire, we see that the fraction of the current threaded by the path of radius $r < a$ must equal the ratio of the area πr^2 enclosed by path 2 and the cross-sectional area πa^2 of the wire.[2] That is,

$$\frac{I}{I_0} = \frac{\pi r^2}{\pi a^2}$$

$$I = \frac{r^2}{a^2} I_0$$

Following the same procedure as for path 1, we can now apply Ampère's law to path 2. This gives

$$\oint \mathbf{B} \cdot d\mathbf{s} = B(2\pi r) = \mu_0 \left(\frac{r^2}{a^2} I_0\right)$$

$$B = \left(\frac{\mu_0 I_0}{2\pi a^2}\right) r \quad (\text{for } r < a) \tag{27.17}$$

The magnetic field versus r for this configuration is sketched in Fig. 27.10. Note that inside the wire, $B \rightarrow 0$ as

[2] Alternatively, the current linked by path 2 must equal the product of the current density, $J = I_0/\pi a^2$, and the area πr^2 enclosed by path 2.

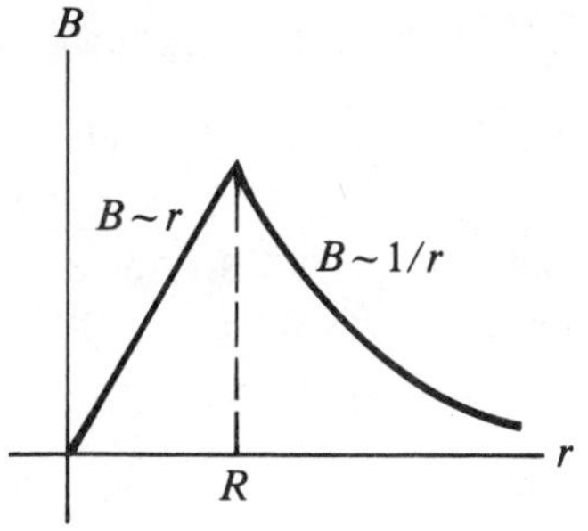

Figure 27.10 A sketch of the magnetic field versus r for the wire described in Example 27.5. The field is proportional to r inside the wire and varies as $1/r$ outside the wire.

$r \to 0$. This result is similar in form to that of the electric field inside a uniformly charged rod.

Example 27.6

The magnetic field of a toroidal coil: The *toroidal coil* consists of N turns of wire wrapped around a doughnut-shaped structure as in Fig. 27.11. Assuming that the turns are closely spaced, calculate the magnetic field *inside* the coil, a distance r from the center.

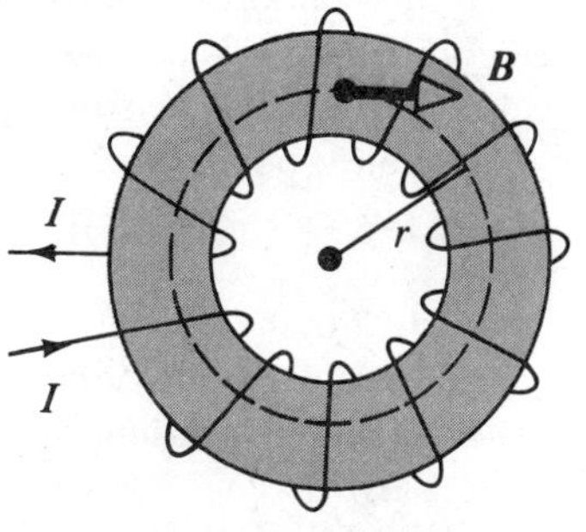

Figure 27.11 (Example 27.6) A toroidal coil consists of many turns of wire wrapped around a doughnut-shaped structure (torus). If the coils are closely spaced, the field inside the toroidal coil is tangent to the dotted circular path and varies as $1/r$, and the exterior field is zero.

Solution: To calculate the field inside the coil, we evaluate the line integral of $\mathbf{B} \cdot d\mathbf{s}$ over a circle of radius r. By symmetry, we see that the magnetic field is constant in magnitude on this path and tangent to it, so that $\mathbf{B} \cdot d\mathbf{s} = B\,ds$. Furthermore, note that the closed path threads N loops of wire, each of which carries a current I. Therefore, the right side of Ampère's law, Eq. 27.15, is $\mu_0 NI$ in this case. Ampère's law applied to this path then gives

$$\oint \mathbf{B} \cdot d\mathbf{s} = B \oint ds = B(2\pi r) = \mu_0 NI$$

$$B = \frac{\mu_0 NI}{2\pi r} \qquad (27.18)$$

This result shows that B varies as $1/r$ and hence is nonuniform within the coil. However, if r is large compared with a, where a is the cross-sectional radius of the toroid, then the field will be approximately uniform inside the coil.

Furthermore, for the ideal toroidal coil, where the turns are closely spaced, the external field is *zero*. This can be seen by noting that the *net* current threaded by any circular path lying outside the toroidal coil is zero (including the region of the "hole in the doughnut"). Therefore, from Ampère's law one finds that $\mathbf{B} = 0$ in the regions exterior to the toroidal coil. In reality, the turns of a toroidal coil form spirals rather than circular loops (the ideal case). As a result, there is always a small field external to the coil.

Example 27.7

Magnetic field of an infinite current sheet: An infinite sheet lying in the yz plane carries a surface current of density $\mathbf{J}_s$. The current is in the y direction, and J_s represents the current per unit length measured along the z axis. Find the magnetic field near the sheet.

Solution: To evaluate the line integral in Ampère's law, let us take a rectangular path around the sheet as in Fig. 27.12. The rectangle has dimensions l and w, where the sides of length l are parallel to the surface. The *net* current through the loop is $J_s l$ (that is, the net current equals the current per unit length multiplied by the length of the rectangle). Hence, applying Ampère's law over the loop and noting that the paths of length w do not contribute to the line integral, we get

$$\oint \mathbf{B} \cdot d\mathbf{s} = \mu_0 I = \mu_0 J_s l$$

$$2Bl = \mu_0 J_s l$$

$$B = \mu_0 \frac{J_s}{2} \qquad (27.19)$$

The result shows that *the magnetic field is independent of the distance from the current sheet.* In fact, the magnetic field is uniform and is everywhere parallel to the plane of the sheet. This is reasonable since we are dealing with an *infinite* sheet of current. The result is analogous to the uniform electric field associated with an infinite sheet of charge. (Example 21.6.)

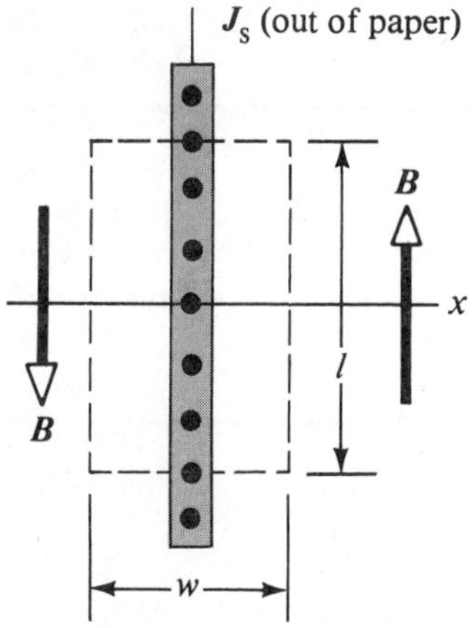

Figure 27.12 A top view of an infinite current sheet lying in the yz plane, where the current is in the y direction. This view shows the direction of $\mathbf{B}$ on both sides of the sheet.

Q6. Is Ampère's law valid for all closed paths surrounding a conductor? Why is it not useful for calculating $\boldsymbol{B}$ for all such paths?

Q7. Compare Ampère's law with the Biot-Savart law. Which is the more general method for calculating $\boldsymbol{B}$ for a current-carrying conductor?

Q8. Is the magnetic field inside a toroidal coil uniform? Explain.

Q9. Describe the similarities between Ampère's law in magnetism and Gauss' law in electrostatics.

Q10. A hollow copper tube carries a current. Why is $\boldsymbol{B} = 0$ inside the tube? Is $\boldsymbol{B}$ nonzero outside the tube?

27.4 THE MAGNETIC FIELD OF A SOLENOID

A *solenoid* is a long wire wound in the form of a helix. With this configuration, one can produce a reasonably uniform magnetic field within a small volume of the solenoid's interior region if the consecutive turns are closely spaced. When the turns are closely spaced, each can be regarded as a circular loop, and the net magnetic field is the vector sum of the fields due to all the turns.

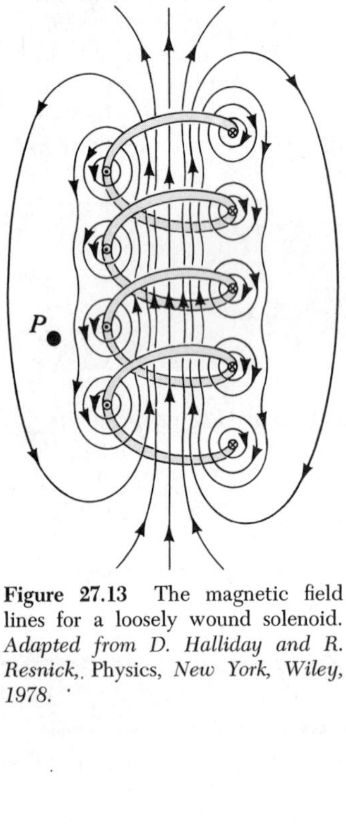

Figure 27.13 The magnetic field lines for a loosely wound solenoid. *Adapted from D. Halliday and R. Resnick,* Physics, *New York, Wiley, 1978.*

Figure 27.13 shows the magnetic field lines of a loosely wound solenoid. Note that the field lines inside the coil are nearly parallel, uniformly distributed, and close together. This indicates that the field inside the solenoid is uniform and strong. The field lines between the turns tend to cancel each other. The field outside the solenoid is both nonuniform and weaker. The field at exterior points, such as *P*, is weak since the field due to current elements on the upper portions tends to cancel the field due to current elements on the lower portions.

If the turns are closely spaced and the solenoid is of finite length, the field lines are as shown in Fig. 27.14. In this case, the field lines diverge from one end and converge at the opposite end. An inspection of this field distribution exterior to the solenoid shows a similarity with the field of a bar magnet. Hence, one end of the solenoid behaves like the north pole of a magnet while the opposite end behaves like the south pole. As the length of the solenoid increases, the field within it becomes more and more uniform. One approaches the case of an *ideal solenoid* when the turns are closely spaced and the length is long compared with the radius. In this case, the field outside the solenoid is weak compared with the field inside the solenoid, and the field inside is uniform over a large volume.

We can use Ampère's law to obtain an expression for the magnetic field inside an ideal solenoid. A cross section of part of our ideal solenoid (Fig. 27.15) carries a current I. Note that for the ideal solenoid, $\boldsymbol{B}$ inside the solenoid is uniform and parallel to the axis and $\boldsymbol{B}$ outside is zero. Consider a rectangular path of length l and width w as shown in Fig. 27.15. We can apply Ampère's law to this path by evaluating $\oint \boldsymbol{B} \cdot d\boldsymbol{s}$ over each of the four sides of the rectangle. The contribution along side 3 is clearly zero, since $\boldsymbol{B} = 0$ in this region. The contributions from sides 2 and 4 are both zero since $\boldsymbol{B}$ is perpendicular to $d\boldsymbol{s}$ along these paths. Side 1, whose length is l, gives a contribution Bl to the integral since $\boldsymbol{B}$ is uniform and parallel to $d\boldsymbol{s}$ along this path. Therefore, the integral over the closed rectangular path has the value

$$\oint \boldsymbol{B} \cdot d\boldsymbol{s} = \int_{\text{path 1}} \boldsymbol{B} \cdot d\boldsymbol{s} = B \int_{\text{path 1}} ds = Bl$$

The right side of Ampère's law involves the *total* current that passes through the area bound by the path of integration. In our case, the total current through the

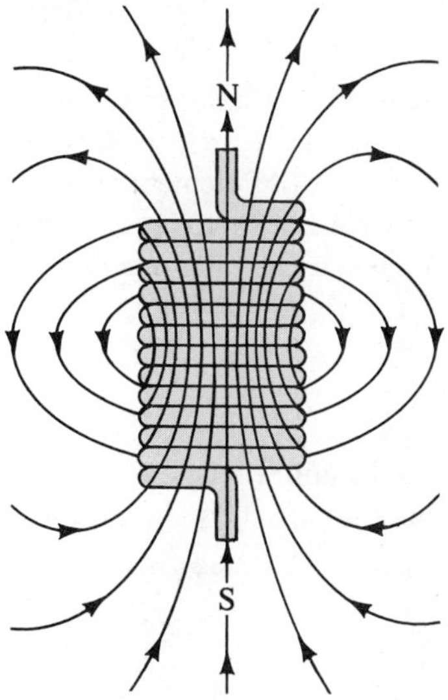

Figure 27.14 Magnetic field lines for a tightly wound solenoid of finite length carrying a steady current I. The field inside the solenoid is nearly uniform and strong. Note that the field lines resemble those of a bar magnet, so that the solenoid effectively has north and south poles.

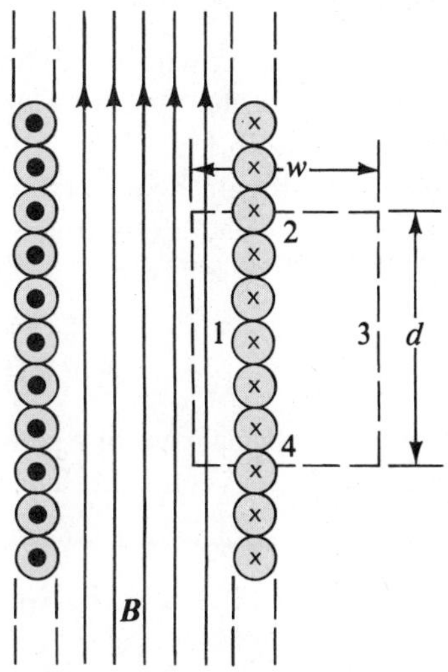

Figure 27.15 A cross-sectional view of a tightly wound solenoid. If the solenoid is long relative to its radius, we can assume that the field inside is uniform and the field outside is zero. Ampère's law applied to the dotted rectangular path can then be used to calculate the field inside the solenoid.

rectangular path equals the current through each turn multiplied by the number of turns. If N is the number of turns in the length l, then the total current through the rectangle equals NI. Therefore, Ampère's law applied to this path gives

$$\oint \mathbf{B} \cdot d\mathbf{s} = Bl = \mu_0 NI$$

$$B = \mu_0 \frac{N}{l} I = \mu_0 nI \qquad (27.20)$$

Magnetic field inside a solenoid

where $n = N/l$ is the number of turns *per unit length* (not to be confused with N).

We also could obtain this result in a simpler manner by reconsidering the magnetic field of a toroidal coil (Example 27.6). If the radius r of the toroidal coil containing N turns is large compared with its cross-sectional radius a, then a short section of the toroidal coil approximates a solenoid with $n = N/2\pi r$. In this limit, we see that Eq. 27.18 derived for the toroidal coil agrees with Eq. 27.20.

Equation 27.20 is valid only for points far from the ends of a very long solenoid. As you might expect, the field near each end is smaller than the value given by Eq. 27.20. At the very end of a long solenoid, the magnitude of the field is about one half that of the field at the center. The field at arbitrary axial points of the solenoid is derived in Section 27.7.

Q11. Why is $\mathbf{B}$ nonzero outside a solenoid? Why is $\mathbf{B} = 0$ outside a toroid? (Note that the lines of $\mathbf{B}$ must form closed paths.)

Q12. Describe the change in the magnetic field inside a solenoid carrying a steady current I if (a) the length of the solenoid is doubled, but the number of turns remains the same and (b) the number of turns is doubled, but the length remains the same.

27.5 MAGNETIC FLUX

The flux associated with a magnetic field is defined in a manner similar to that used to define the electric flux. Consider an element of area dA on an arbitrarily shaped surface, as in Fig. 27.16. If the magnetic field at this element is $\mathbf{B}$, then the magnetic flux through the element is $\mathbf{B} \cdot d\mathbf{A}$, where $d\mathbf{A}$ is a vector perpendicular to the surface whose magnitude equals the area dA. Hence, the total magnetic flux Φ_m through the surface is given by

Magnetic flux

$$\Phi_m = \int \mathbf{B} \cdot d\mathbf{A} \tag{27.21}$$

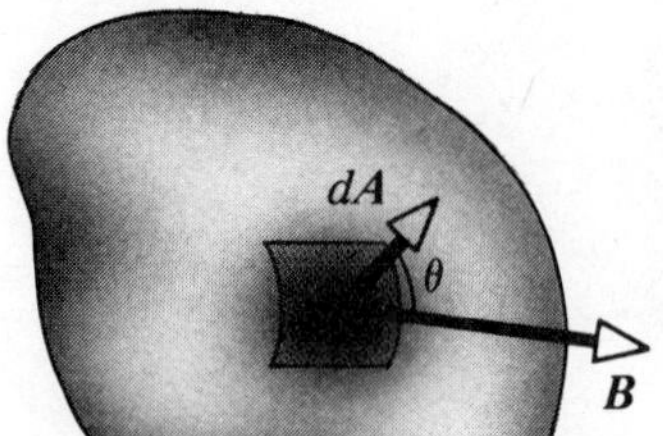

Figure 27.16 The magnetic flux through an area element dA is given by $\mathbf{B} \cdot d\mathbf{A} = BdA\cos\theta$. Note that $d\mathbf{A}$ is perpendicular to the surface.

Consider the special case of a plane of area A and a uniform field $\mathbf{B}$, which makes an angle θ with the vector $d\mathbf{A}$. The magnetic flux through the plane in this case is given by

$$\Phi_m = BA\cos\theta \tag{27.22}$$

If the magnetic field lies in the plane as in Fig. 27.17a, then $\theta = 90°$ and the flux is zero. If the field is perpendicular to the plane as in Fig. 27.17b, then $\theta = 0°$ and the flux is BA (the maximum value).

Since B has units of Wb/m^2, or T, the unit of flux is the weber (Wb), where $1\ Wb = 1\ T \cdot m^2$.

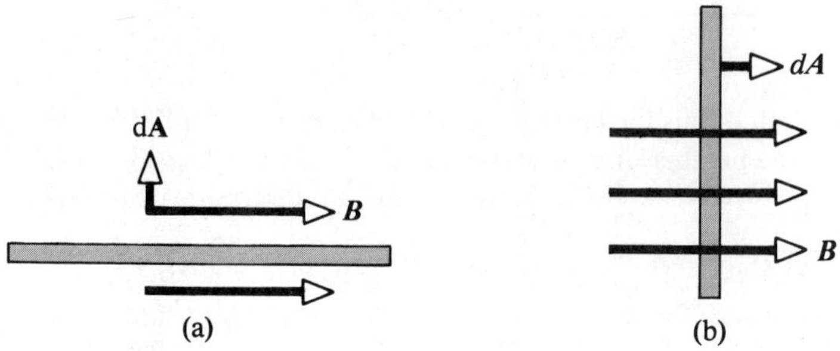

Figure 27.17 (a) The flux is zero when the magnetic field is parallel to the surface of the plane (an edge view). (b) The flux is a maximum when the magnetic field is perpendicular to the plane.

Example 27.8

A rectangular loop of width a and length b is located a distance c from a long wire carrying a current I (Fig. 27.18). The wire is parallel to the long side of the loop. Find the total magnetic flux through the loop.

Solution: From Ampère's law, we found that the magnetic field due to the wire at a distance r from the wire is given by

$$B = \frac{\mu_0 I}{2\pi r}$$

That is, the field *varies* over the loop and is directed *into* the page as shown in Fig. 27.18. Since $\mathbf{B}$ is parallel to $d\mathbf{A}$, we can express the magnetic flux through an area element dA as

$$\Phi_m = \int B\, dA = \int \frac{\mu_0 I}{2\pi r}\, dA$$

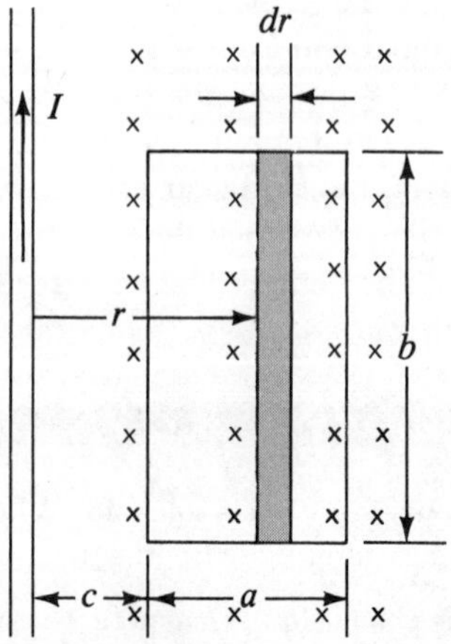

Figure 27.18 (Example 27.8) The magnetic field due to the wire carrying a current I is *not* uniform over the rectangular loop.

Note that since B is not uniform, but depends on r, it cannot be removed from the integral. In order to carry out the integration, we first express the area element (the shaded region in Fig. 27.18) as $dA = b\,dr$. Since r is the only variable that now appears in the integral, the expression for Φ_m becomes

$$\Phi_m = \frac{\mu_0 I}{2\pi} b \int_c^{a+c} \frac{dr}{r} = \frac{\mu_0 I b}{2\pi} \ln r \Big]_c^{a+c}$$

$$= \frac{\mu_0 I b}{2\pi} \ln\left(\frac{a+c}{c}\right)$$

Q13. A planar conducting loop is located in a uniform magnetic field that is directed along the x axis. For what orientation of the loop is the flux a maximum? For what orientation is the flux a minimum?

27.6 GAUSS' LAW IN MAGNETISM

In Chapter 21 we found that the flux of the electric field through a closed surface surrounding a net charge is proportional to that charge (Gauss' law). In other words, the number of electric field lines leaving the surface depends only on the net charge within it. This property is based in part on the fact that electric field lines originate on electric charges.

The situation is quite different for magnetic fields, which are continuous and form closed loops. Magnetic field lines due to currents do not begin or end at any point. The magnetic field lines of the bar magnet in Fig. 27.19 illustrate this point. Note that for any closed surface, the number of lines entering that surface equals the number leaving that surface, and so the net magnetic flux is *zero*. This is in contrast to the case of a surface surrounding one charge of an electric dipole (Fig. 27.20), where the net electric flux is not zero.

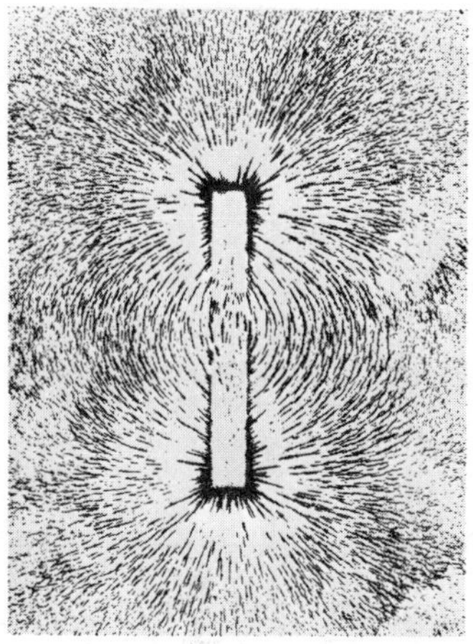

Magnetic field lines of a bar magnet. (Photo, Education Development Center, Newton, Mass.)

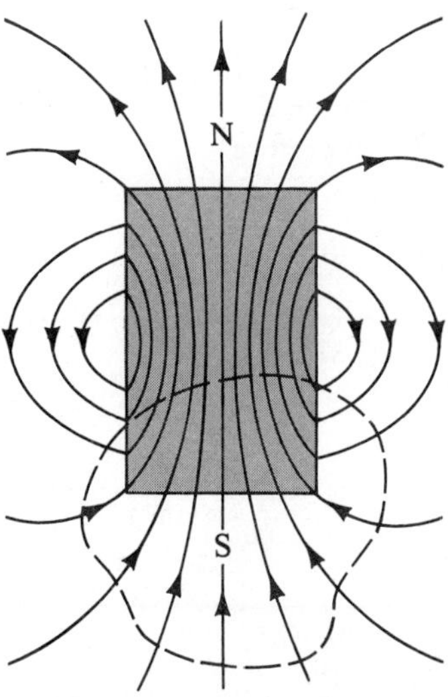

Figure 27.19 The magnetic field lines of a bar magnet form closed loops. Note that the net flux through the closed surface surrounding one of the poles (or any other closed surface) is zero.

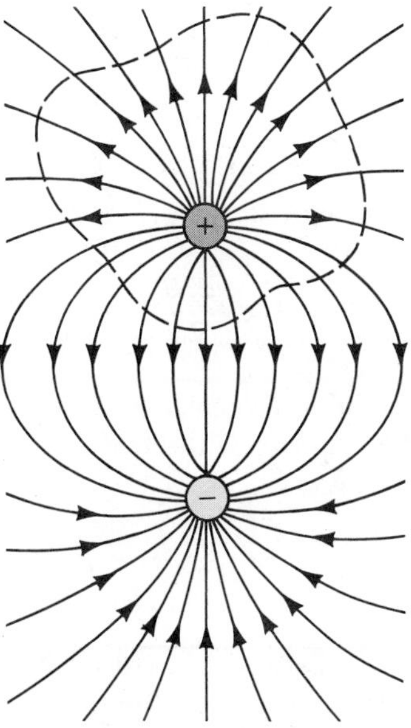

Figure 27.20 The electric field lines of an electric dipole begin on the positive charge and terminate on the negative charge. The electric flux through a closed surface surrounding one of the charges is *not* zero.

Gauss' law in magnetism

Gauss' law in magnetism states that the net magnetic flux through any closed surface is always zero:

$$\oint \mathbf{B} \cdot d\mathbf{A} = 0 \tag{27.23}$$

This statement is based on the experimental fact that *isolated magnetic poles (or monopoles) have not been detected, and perhaps do not even exist.* As we shall see in Chapter 30, the only known sources of magnetic fields are magnetic dipoles (current loops), even in magnetic materials. In fact, all magnetic effects in matter can be explained in terms of magnetic dipole moments (effective current loops) associated with electrons and nuclei.

27.7 THE MAGNETIC FIELD ALONG THE AXIS OF A SOLENOID

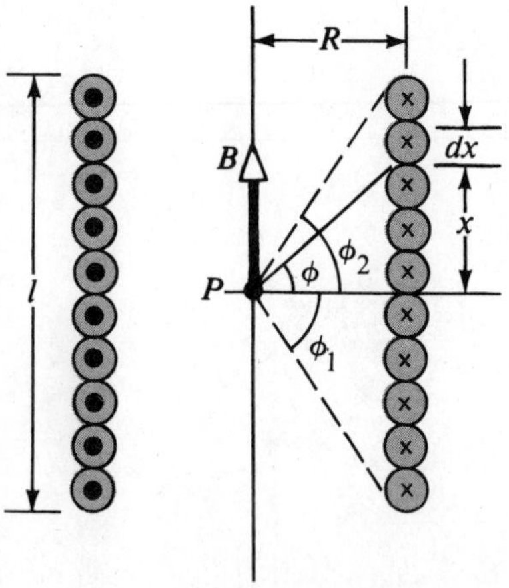

Figure 27.21 The geometry for calculating the magnetic field at an axial point P inside a tightly wound solenoid.

Consider a solenoid of length l and radius R containing N closely spaced turns and carrying a steady current I. Let us determine an expression for the magnetic field at an axial point P inside the solenoid, as indicated in Fig. 27.21.

Perhaps the simplest way to obtain the desired result is to consider the solenoid as a distribution of current loops. The field of any one loop along the axis is given by Eq. 27.9. Hence, the net field in the solenoid is the superposition of fields from all loops. The number of turns in a length dx of the solenoid is $(N/l)\,dx$; therefore the total current in a width dx is given by $I(N/l)\,dx$. Then, using Eq. 27.9, we find that the field at P due to the section dx is given by

$$dB = \frac{\mu_0 R^2}{2(x^2 + R^2)^{3/2}} I\left(\frac{N}{l}\right) dx \tag{27.24}$$

This expression contains the variable x, which can be expressed in terms of the variable ϕ, defined in Fig. 27.21. That is, $x = R\tan\phi$, so that $dx = R\sec^2\phi\, d\phi$. Substituting these expressions into Eq. 27.24 and integrating from ϕ_1 to ϕ_2, we get

$$B = \frac{\mu_0 NI}{2l}\int_{\phi_1}^{\phi_2} \cos\phi\, d\phi = \frac{\mu_0 NI_0}{2l}(\sin\phi_2 - \sin\phi_1) \tag{27.25}$$

If P is at the *midpoint* of the solenoid and if we assume that the solenoid is long compared with R, then $\phi_2 \approx 90°$ and $\phi_1 \approx -90°$; therefore

$$B \approx \frac{\mu_0 NI}{2l}(1 + 1) = \frac{\mu_0 NI}{l} = \mu_0 nI \qquad \text{(at the center)}$$

which is in agreement with our previous result, Eq. 27.20.

If P is a point at the end of a long solenoid (say, the bottom), then $\phi_1 \approx 0°$, $\phi_2 \approx 90°$, and

$$B \approx \frac{\mu_0 NI}{2l}(1 + 0) = \frac{1}{2}\mu_0 nI \qquad \text{(at the ends)}$$

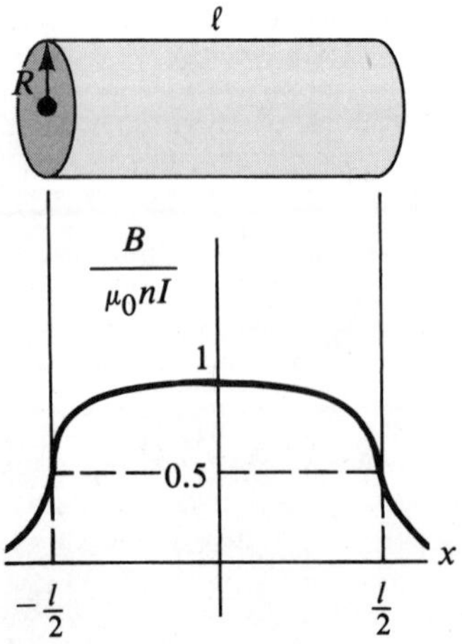

Figure 27.22 A sketch of the magnetic field versus x for a long, tightly wound solenoid. Note that the magnitude of the field at the ends is about one half the value at the center.

This shows that the field at each end of a solenoid approaches *one half* the value at the solenoid's center as the length l approaches infinity.

A sketch of the field at axial points versus x for a solenoid is shown in Fig. 27.22. If the length l is large compared with R, the axial field will be quite uniform over most of the solenoid and the curve will be quite flat except at points near the ends. On the other hand, if l is comparable to R, then the field will have a value somewhat less than $\mu_0 nI$ at the middle and will be uniform only over a small region of the solenoid.

27.8 DISPLACEMENT CURRENT AND THE GENERALIZED AMPÈRE'S LAW

We have seen that charges in motion, or currents, produce a magnetic field. When a current-carrying conductor has high symmetry, we can calculate the magnetic field using Ampère's law, given by Eq. 27.15:

$$\oint \mathbf{B} \cdot d\mathbf{s} = \mu_o I$$

where the line integral is over *any closed path through which the conduction current passes.* If Q is the charge on the capacitor at any instant, the conduction current is defined by

$$I = \frac{dQ}{dt}$$

We shall now show that *Ampère's law in this form is valid only if the conduction is constant in time.* Maxwell recognized this limitation and modified Ampère's law to include all possible situations.

We can understand this problem by considering a capacitor being charged as in Fig. 27.23. The argument given here is equivalent to Maxwell's original reasoning. When the current I changes with time (for example, when an ac voltage source is used), the charge on the plate changes, but *no conduction current passes between the plates.* Now consider the two surfaces S_1 and S_2 bounded by the same path P. Ampère's law says that the line integral of $\mathbf{B} \cdot d\mathbf{s}$ around this path must equal $\mu_o I$, where I is the total current through any surface bounded by the path P.

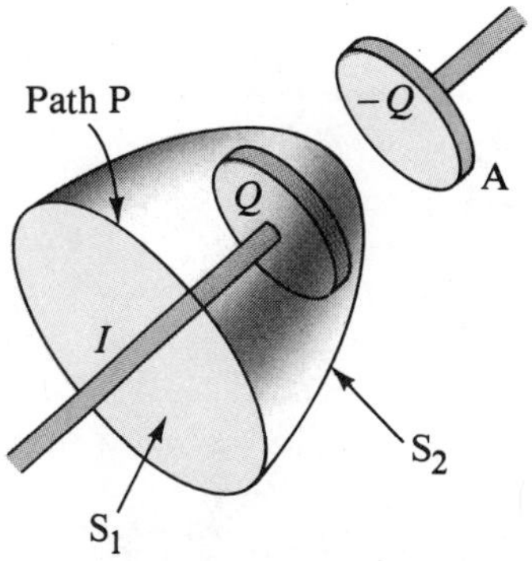

Figure 27.23 The surface S_1 and S_2 are bounded by the same path P. The conduction current passes only through S_1. This leads to a contradictory situation in Ampère's law which is resolved only if one postulates a displacement current through S_2.

When the path P is considered to bound S_1, the result of the integral is $\mu_o I$ since the current passes through S_1. However when the path bounds S_2, the result is *zero* since no conduction current passes through S_2. Thus, we have a contradictory situation which arises from the discontinuity of the current! Maxwell solved this problem by postulating an additional term on the right side of Eq. 27.15, called the *displacement current,* I_d, defined as

Displacement current

$$I_d = \varepsilon_o \frac{d\Phi_e}{dt} \qquad (27.26)$$

Recall that Φ_e is the flux of the electric field, defined as $\Phi_e = \int \mathbf{E} \cdot d\mathbf{A}$.

As the capacitor is being charged (or discharged), *the changing* electric field between the plates may be thought of as a sort of current that bridges the discontinuity in the conduction current. When this expression for the current (Eq. 27.26) is added to the right side of Ampère's law, the difficulty represented by Fig. 27.23 is resolved. No matter what surface bounded by the path P is chosen, some combination of conduction and displacement current will pass through it. With this new term I_d, we can express the generalized form of Ampère's law (sometimes called the *Ampère-Maxwell law*) as[3]

Ampère-Maxwell law

$$\oint \mathbf{B} \cdot d\mathbf{s} = \mu_o(I + I_d) = \mu_o I + \mu_o \varepsilon_o \frac{d\Phi_e}{dt} \qquad (27.27)$$

The meaning of this expression can be understood by referring to Fig. 27.24. The electric flux through S_2 is $\Phi_e = \int \mathbf{E} \cdot d\mathbf{A} = EA$, where A is the area of the plates and E is the uniform electric field strength between the plates. If Q is the charge on the

---

[3]Strictly speaking, this expression is valid only in a vacuum. If a magnetic material is present, one must also include a magnetizing current I_m on the right side of Eq. 27.27 to make Ampère's law fully general. On a microscopic scale, I_m is a current that is as real as the conduction current I.

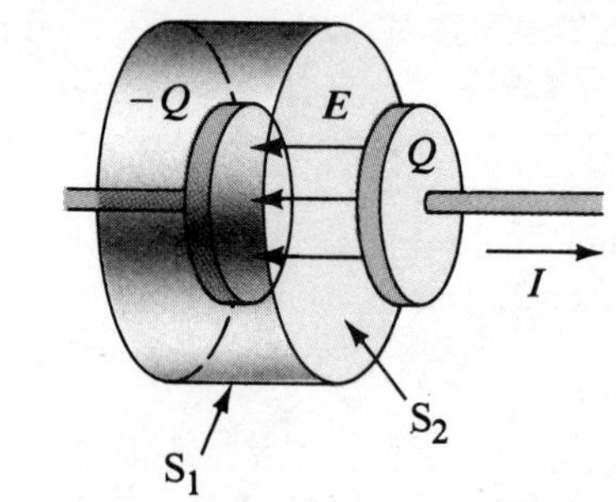

Figure 27.24 The conduction current $I = dQ/dt$ passes through S_1. The displacement current $I_d = \varepsilon_0 d\Phi_e/dt$ passes through S_2. The two currents must be equal for continuity. In general, the total current through any surface bounded by some path is $I + I_d$.

plates at any instant, then one finds that $E = Q/\varepsilon_0 A$ (Section 23.2). Therefore, the electric flux through S_2 is simply

$$\Phi_e = EA = \frac{Q}{\varepsilon_0}$$

Hence, the displacement current I_d through S_2 is

$$I_d = \varepsilon_0 \frac{d\Phi_e}{dt} = \frac{dQ}{dt} \tag{27.28}$$

That is, the displacement current is precisely equal to the conduction current I passing through S_1!

The central point of this formalism is the fact that *magnetic fields are produced both by conduction currents and by changing electric fields.*

Example 27.9

An ac voltage is applied directly across an 8-μF capacitor. The frequency of the source is 3 kHz, and the voltage amplitude is 30 V. Find the displacement current between the plates of the capacitor.

Solution: The angular frequency of the source is given by $\omega = 2\pi f = 2\pi(3 \times 10^3 \text{ Hz}) = 6\pi \times 10^3 \text{ s}^{-1}$. Hence, the voltage across the capacitor in terms of t is

$$V = V_m \sin\omega t = 30 \sin(6\pi \times 10^3 t) \text{ V}$$

We can make use of Eq. 27.28 and of the fact that the charge on the capacitor is given by $Q = CV$ to find the displacement current:

$$I_d = \frac{dQ}{dt} = \frac{d}{dt}(CV) = C\frac{dV}{dt}$$

$$= (8 \times 10^{-6}) \frac{d}{dt}[30 \sin(6\pi \times 10^3 t)]$$

$$= 4.52 \cos(6\pi \times 10^3 t) \text{ A}$$

Hence, the displacement current varies sinusoidally with time and has a *maximum* value of 4.52 A.

27.9 SUMMARY

The *Biot-Savart law* says that the magnetic field $d\mathbf{B}$ at a point P due to a current element $d\mathbf{s}$ carrying a steady current I is

Biot-Savart law

$$d\mathbf{B} = k_m \frac{I\, d\mathbf{s} \times \hat{\mathbf{r}}}{r^2} \tag{27.1}$$

where $k_m = 10^{-7}$ Wb/A · m and r is the distance from the element to the point P. To find the total field at P due to a current-carrying conductor, we must integrate this vector expression over the entire conductor.

The *magnetic field* at a distance r from a long, straight wire carrying a current I

is given by

$$B = \frac{\mu_0 I}{2\pi r} \quad (27.7)$$

Magnetic field of an infinitely long wire

where $\mu_0 = 4\pi \times 10^{-7}$ Wb/A · m is the *permeability of free space.*

The field lines are circles concentric with the wire.

The force per unit length between two parallel wires separated by a distance a and carrying currents I_1 and I_2 has a magnitude given by

$$\frac{F}{l} = 2k_m \frac{I_1 I_2}{a} \quad (27.13)$$

Force per unit length between two wires

The force is attractive if the currents are in the same direction and repulsive if they are in opposite directions.

Ampère's law says that the line integral of $\mathbf{B} \cdot d\mathbf{s}$ around any closed path equals $\mu_0 I$, where I is the total steady current threaded by the path. That is,

$$\oint \mathbf{B} \cdot d\mathbf{s} = \mu_0 I \quad (27.15)$$

Ampère's law

Using Ampère's law, one finds that the fields inside a toroid and solenoid are given by

$$B = \frac{\mu_0 N I}{2\pi r} \quad (27.18)$$

Magnetic field inside a toroid

$$B = \mu_0 \frac{N}{l} I = \mu_0 n I \quad (27.20)$$

Magnetic field inside a solenoid

where N is the total number of turns.

The *magnetic flux* Φ_m through a surface is defined by the surface integral

$$\Phi_m \equiv \int \mathbf{B} \cdot d\mathbf{A} \quad (27.21)$$

Magnetic flux

Gauss' law of magnetism states that the net magnetic flux through any closed surface is zero. That is, isolated magnetic poles (or magnetic monopoles) do not exist.

A *displacement current* I_d arises from a time-varying electric flux and is defined by

$$I_d = \varepsilon_0 \frac{d\Phi_e}{dt} \quad (27.26)$$

Displacement current

The *generalized form of Ampère's law,* which includes the displacement current, is given by

$$\oint \mathbf{B} \cdot d\mathbf{s} = \mu_0 I + \mu_0 \varepsilon_0 \frac{d\Phi_e}{dt} \quad (27.27)$$

Ampère-Maxwell law

This law describes the fact that magnetic fields are produced both by conduction currents and by changing electric fields.

EXERCISES

Section 27.1 The Biot-Savart Law

1. Calculate the magnitude of the magnetic field at a point 10 cm from a long, thin conductor carrying a current of 5 A.

2. A long, thin conductor carries a current of 4 A. At what distance from the conductor is the magnitude of the resulting magnetic field equal to 25 μT?

3. A wire in which there is a current of 2 A is to be formed into a circular loop of one turn. If the required value of the magnetic field at the *center* of the loop is 4 μT, what is the required radius of the loop?

4. At what distance along the *axis* of a circular current loop of radius R is the magnitude of the magnetic field equal to one half the magnitude of the field at the *center* of the loop?

5. Recalling that the current density $J = nqv_d$ (Eq. 24.6), show that the Biot-Savart law can be written

$$d\mathbf{B} = \frac{\mu_0}{4\pi} \frac{q\mathbf{v}_d \times \hat{\mathbf{r}}}{r^2} n \, dV$$

where dV is the volume element of the conductor and the drift velocity $\mathbf{v}_d$ is as defined in Chapter 24.

6. Use the Biot-Savart law to calculate the magnitude of the magnetic field at a point P located at the center of concentric semicircles of radii $a = 5$ cm and $b = 8$ cm (Fig. 27.25) when a current $I = 2$ A is maintained in the loop circuit.

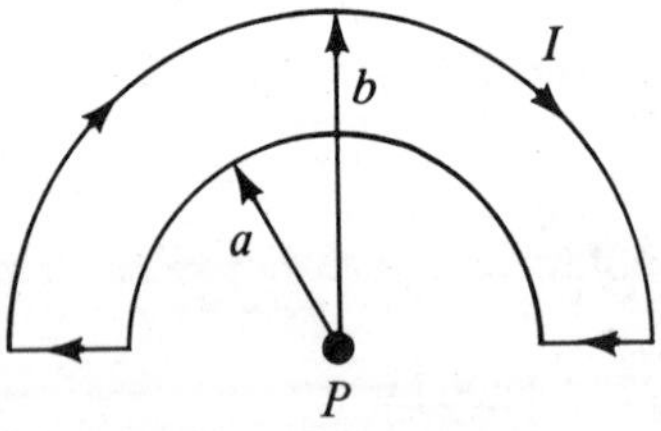

Figure 27.25 (Exercise 6).

7. A segment of wire of total length $4r$ is formed into a shape as shown in Fig. 27.26 and carries a current $I = 4$ A. Find the magnitude and direction of the magnetic field at point P when $r = \pi$ cm.

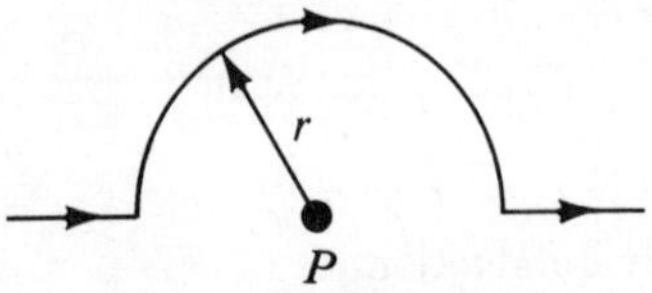

Figure 27.26 (Exercise 7).

8. A closed current path shaped as shown in Fig. 27.27 produces a magnetic field at P, the center of the arc. If the arc subtends an angle of 30° and the total length of wire in the closed path is 1.2 m, what is the magnitude of the field produced at P if the current in the loop is 3 A?

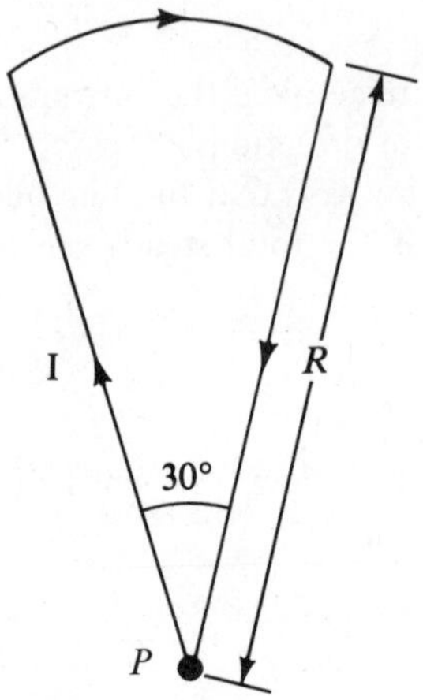

Figure 27.27 (Exercise 8).

9. How many turns should be in a closely wrapped circular coil of radius 0.4 m in order for a current of 3.2 A to produce a magnetic field of 1.61×10^{-4} T at its center?

10. A conductor in the shape of a square of edge length $l = 0.3$ m carries a current $I = 2.5$ A (Fig. 27.28). Calculate the magnitude of the magnetic field produced at the *center* of the square.

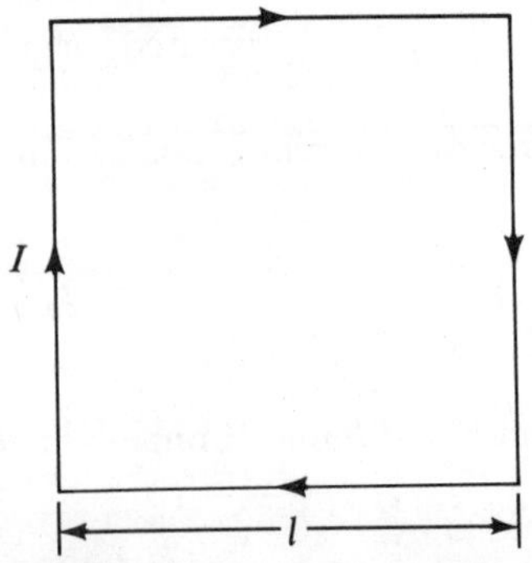

Figure 27.28 (Exercises 10 and 11).

11. If the total length of the conductor in Exercise 10 is formed into a single *circular* turn with the *same* current, what is the value of the magnetic field at the center of the turn?

Section 27.2 The Magnetic Force Between Two Parallel Conductors

12. Two parallel conductors, separated by a distance $a = 0.2$ m, carry currents in the same direction (Fig. 27.29). If $I_1 = 10$ A and $I_2 = 15$ A, what is the force per unit length exerted on each conductor by the other?

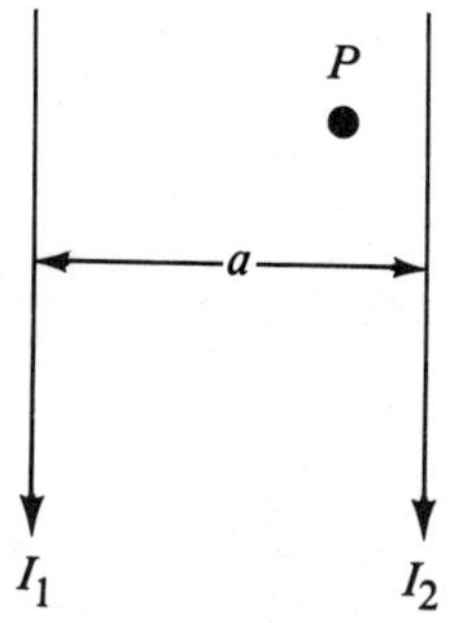

Figure 27.29 (Exercises 12 and 13).

13. For the arrangement of parallel conductors described in Exercise 12 and shown in Fig. 27.29, calculate the magnitude and direction of the magnetic field at point P, located 5 cm to the left of the conductor carrying current I_2.

14. For the arrangement shown in Fig. 27.30, the current in the straight conductor has the value $I_1 = 5$ A and lies in the plane of the rectangular loop, which carries a current $I_2 = 10$ A. The dimensions are $c = 0.1$ m, $a = 0.15$ m, and $l = 0.45$ m. Find the magnitude and direction of the *net force* exerted on the rectangle by the magnetic field of the straight current-carrying conductor.

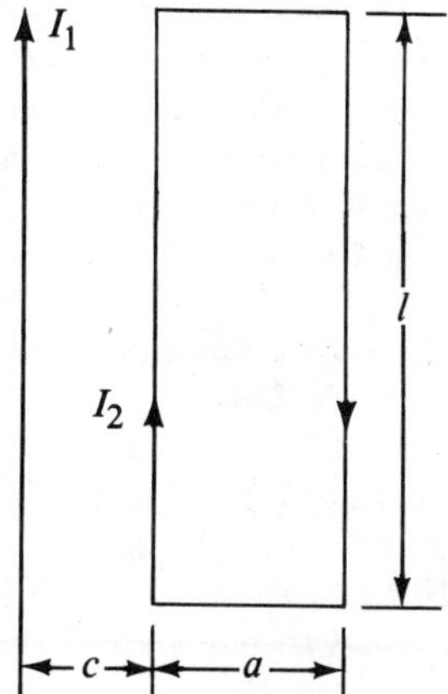

Figure 27.30 (Exercise 14).

15. Four long, parallel conductors carry equal currents $I = 4$ A. An end view of the conductors is shown in Fig. 27.31. The current direction is out of the page at points A and B (indicated by the dots) and into the page at points C and D (indicated by the crosses). Calculate the magnitude and direction of the magnetic field at point P, located at the center of the square of edge length 0.2 m.

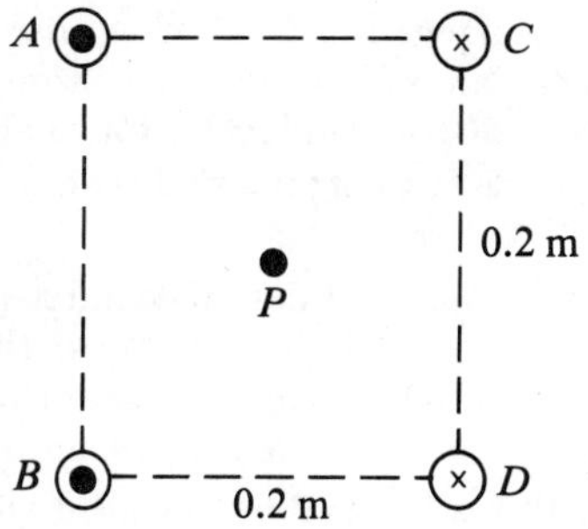

Figure 27.31 (Exercise 15).

16. Two long, parallel conductors carry currents $I_1 = 4$ A and $I_2 = 2$ A, both directed into the page in Fig. 27.32. The conductors are separated by a distance of 10 cm. Determine the magnitude and direction of the resultant magnetic field at point P, located 6 cm from I_1 and 8 cm from I_2.

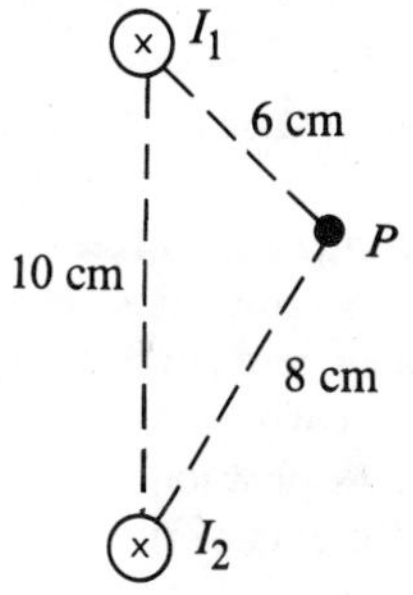

Figure 27.32 (Exercise 16).

Section 27.3 Ampère's Law and Section 27.4 The Magnetic Field of a Solenoid

17. A closely wound solenoid of overall length 0.25 m has a magnetic field $B = 8 \times 10^{-5}$ T due to a current $I = 0.5$ A. How many turns of wire are on the solenoid?

18. What current is required in the windings of a solenoid that has 500 turns uniformly distributed over a length of 0.2 m in order to produce a magnetic field of 1.2×10^{-4} T at the center of the solenoid?

19. A toroidal winding (Fig. 27.11) has a total of 400 turns on a core with inner radius $a = 8$ cm and outer radius $b = 10$ cm. Calculate the magnitude of the magnetic field at a point midway between the inner and outer walls of the core when there is a current of 0.75 A maintained in the windings.

20. A cylindrical conductor of radius $R = 2.5$ cm carries a current $I = 2.5$ A along its length; this current is uniformly distributed throughout the cross section of the conductor. Calculate the magnetic field midway along the radius of the wire (that is, at $r = R/2$).

21. For the conductor described in Exercise 20, find the distance beyond the surface of the conductor at which the magnitude of the magnetic field equals its value at the point midway along the radius.

22. Consider a coaxial arrangement with a wire of radius a along the axis of a thin cylindrical shell of radius b, as in Fig. 27.33. Current is directed *into* the page along the center wire and returns out of the page along the cylinder. If $I = 5$ A, $a = 0.5$ cm, and $b = 1.5$ cm, calculate the magnetic field (a) at point P_1, a distance $r_1 = 1$ cm, and (b) at point P_2, a distance $r_2 = 2$ cm from the center of the wire.

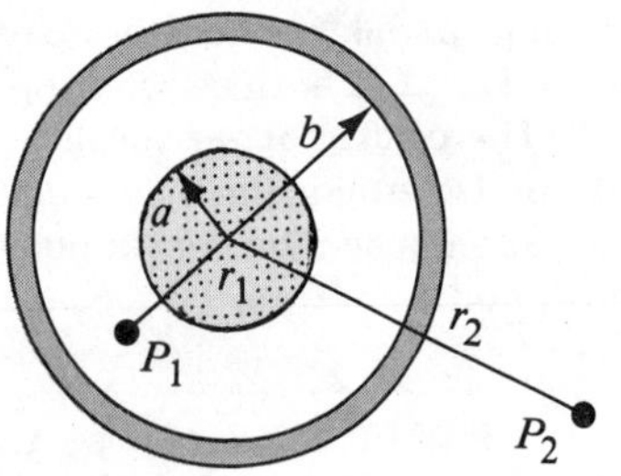

Figure 27.33 (Exercise 22).

Section 27.5 Magnetic Flux

23. A solenoid 4 cm in diameter and 20 cm in length has 250 turns and carries a current of 15 A. Calculate the flux through the surface of a disk of 10-cm radius that is positioned perpendicular to and centered on the axis of the solenoid, as in Fig. 27.34.

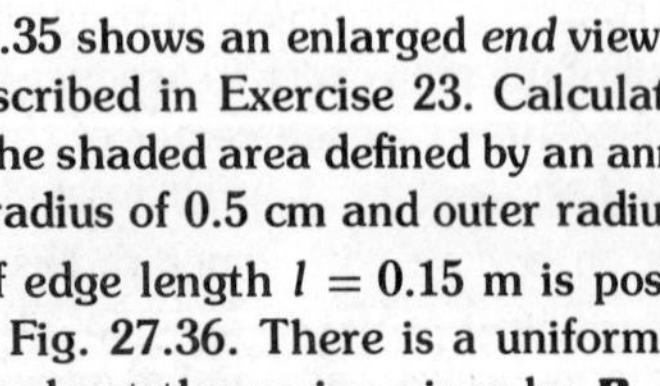

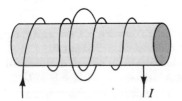

Figure 27.34 (Exercise 23).

24. Figure 27.35 shows an enlarged *end* view of the solenoid described in Exercise 23. Calculate the flux through the shaded area defined by an annulus with an inner radius of 0.5 cm and outer radius of 1 cm.

25. A cube of edge length $l = 0.15$ m is positioned as shown in Fig. 27.36. There is a uniform magnetic field throughout the region given by $\boldsymbol{B} = (6\boldsymbol{i} + 3\boldsymbol{j} + 1.5\boldsymbol{k})$ T. (a) Calculate the flux through the shaded face of the cube. (b) What is the total flux through the six faces of the cube?

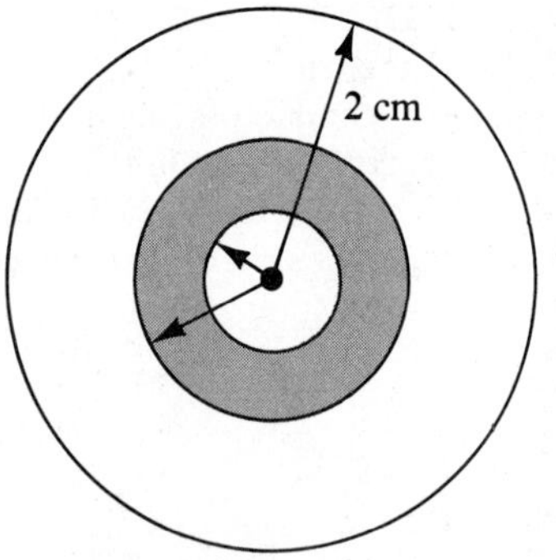

Figure 27.35 (Exercise 24).

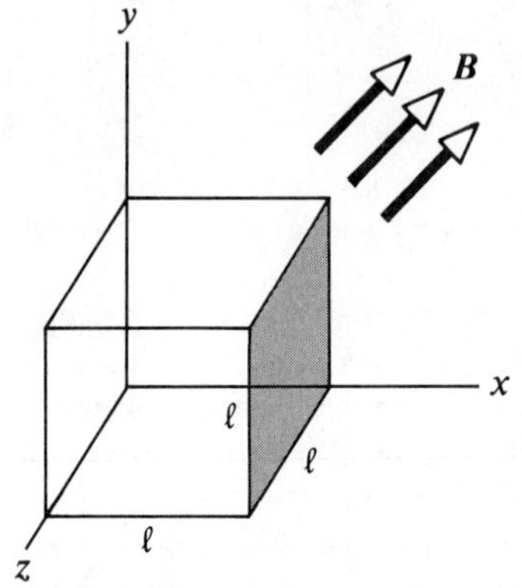

Figure 27.36 (Exercise 25).

Section 27.7 The Magnetic Field Along the Axis of a Solenoid

26. A solenoid has 700 turns, carries a current of 3 A, has a length of 80 cm, and a radius of 4 cm. Calculate the magnetic field along its axis at (a) its center and (b) a point near the end.

27. A solenoid has 400 turns, a length of 50 cm, a radius of 8 cm, and carries a current of 6 A. Calculate the magnetic field at an axial point, a distance of 15 cm from the center (that is, 10 cm from one end).

Section 27.8 Displacement Current and the Generalized Ampère's Law

28. In Example 27.9, we found that the displacement current between the plates of the capacitor varied with time according to the expression $I_d = 4.52 \cos (6\pi \times 10^3 t)$ A. At what time t is the displacement current equal to one half of the maximum value?

29. The applied voltage across the plates of a 3-μF capacitor varies into time according to the expression

$$V_{app} = 6(1 - e^{-t/4}) \text{V}$$

where t is in s. Calculate (a) the displacement current as a function of time and (b) the value of the current at $t = 2$ s.

PROBLEMS

1. It is shown in Example 27.4 that the component of the magnetic field along the axis of a single current loop is given by

$$B_x = \frac{\mu_0 I R^2}{2(x^2 + R^2)^{3/2}}$$

where R is the radius of the loop and x is the distance along the axis from the plane of the loop. When two coaxial coils of the same radius and each with N turns are placed a distance apart equal to their radii, the arrangement is known as a *Helmholtz pair* (Fig. 27.37). (a) Show that at the midpoint between the two coils of a Helmholtz pair, the total field is given by $B = \frac{8\mu_0 NI}{R(5)^{3/2}}$. (b) Show that at a distance $x = R/2$ from a *single* coil, the rate of change of B with respect to x is constant. (Hint: How does $\frac{dB}{dx}$ change with x at the point where $\frac{d^2B}{dx^2} = 0$?)

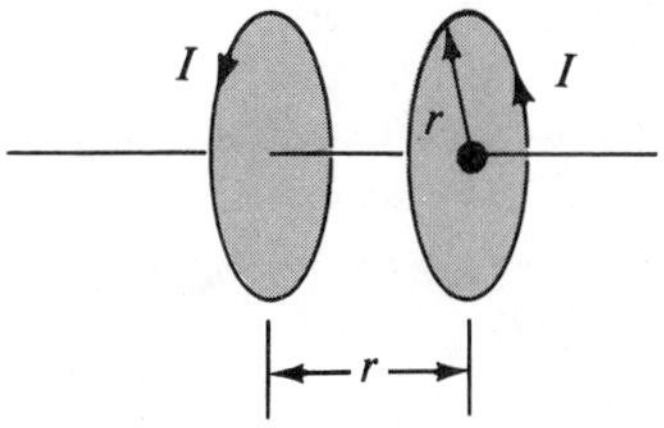

Figure 27.37 (Problem 1).

2. A wire is formed into the shape of a square of edge length L (Fig. 27.38). Show that when the current in the loop is I, the magnetic field at point P a distance x from the center of the square along its axis is given by

$$B = \frac{\mu_0 I L^2}{2\pi\left(x^2 + \frac{L^2}{4}\right)\left(x^2 + \frac{L^2}{2}\right)^{1/2}}$$

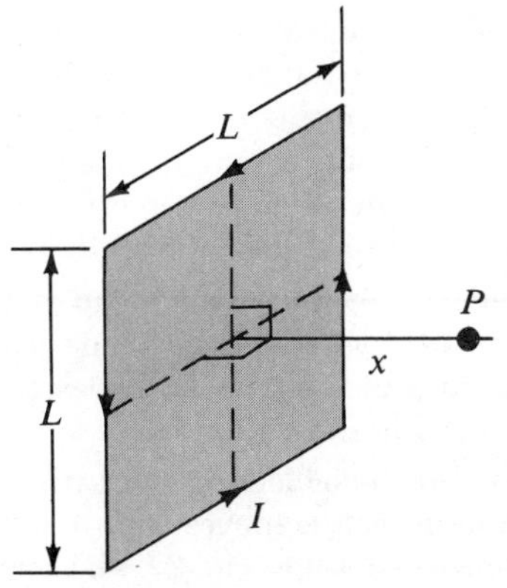

Figure 27.38 (Problem 2).

3. Consider a thin disk of radius R mounted to rotate about the x axis in the yz plane. The disk has a positive uniform surface charge density σ and angular velocity ω. Show that the magnetic field at the center of the disk is given by $B = \frac{1}{2}\mu_0 \sigma \omega R$.

4. A nonconducting ring of radius R is uniformly charged with a total positive charge q. The ring rotates at a constant angular velocity ω about an axis through its center, perpendicular to the plane of the ring. If $R = 0.1$ m, $q = 10\ \mu$C, and $\omega = 20$ rad/s, what is the resulting magnetic field on the axis of the ring a distance of 0.05 m from the center?

5. Two long, parallel conductors are carrying currents in the same direction as in Fig. 27.39. Conductor A carries a current of 100 A and is held firmly in position. Conductor B carries a current I_B and is allowed to slide freely up and down (parallel to A) between a set of nonconducting guides. If the linear density of conductor B is 0.15 g/cm, what value of current I_B will result in equilibrium when the distance between the two conductors is 2 cm?

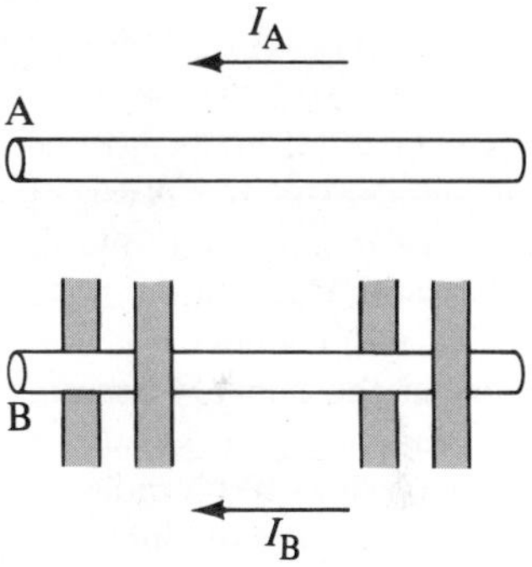

Figure 27.39 (Problem 5).

6. A cylindrical conductor of radius R carries a current I as in Fig. 27.40. The current density J, however, is *not* uniform over the cross section of the conductor but is a function of the radius according to $J = br$, where b is a constant. Find an expression for the magnetic field B (a) at a distance $r_1 < R$ and (b) at a distance $r_2 > R$, measured from the axis.

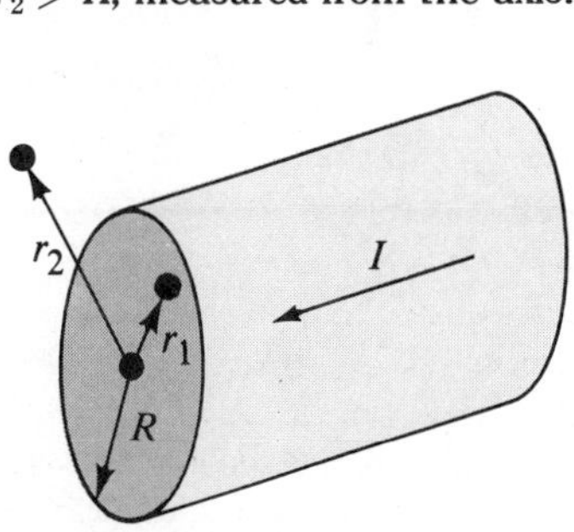

Figure 27.40 (Problem 6).

7. Two parallel conductors carry current in opposite directions as shown in Fig. 27.41. One conductor carries a current of 10 A. Point A is at the *midpoint* between the wires and point C is a distance $d/2$ to the right of the 10-A current. If $d = 10$ cm and I is adjusted so that the magnetic field at C is zero, find (a) the value of the current I and (b) the value of the magnetic field at A.

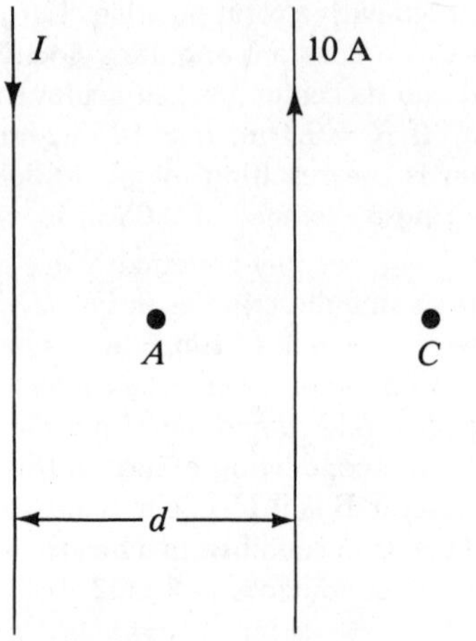

Figure 27.41 (Problem 7).

8. Shown in Fig. 27.42 is the cross section of a nonconducting cylinder that has N wires parallel to the axis of the cylinder and uniformly spaced around the curved surface. The cylinder has a radius R and the current in *each* conductor is I and directed *out* of the plane of the figure. Assuming N to be a very large number and the radius of each wire to be small compared with the radius of the cylinder, find an expression for the magnetic field $\mathbf{B}$ (a) at $r_1 < R$ and (b) at $r_2 > R$. (c) Obtain a numerical value for B when $N = 100$, $R = 5$ cm, $I = 10$ A, and $r_2 = 15$ cm.

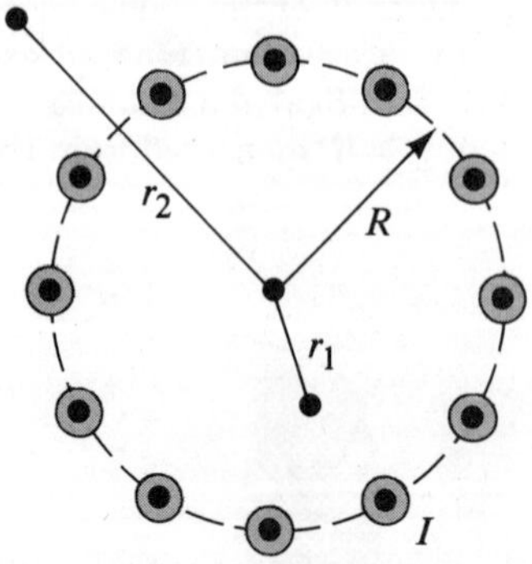

Figure 27.42 (Problem 8).

9. An 8-m length of wire of diameter 3 mm is wrapped in a single layer around a hollow cardboard cylinder of radius 2 cm. Each turn is placed in contact with the adjacent winding. (Assume that adjacent wire turns are electrically insulated from each other, but neglect the thickness of the insulating cover.) A steady current of 5 A is maintained in the wire. (a) Calculate the value of the magnetic field at the center of the solenoidal array. (b) How would the answer to (a) be changed if the total length of wire were used to wrap a *single*-layer solenoid of radius 1 cm?

10. A current I flows around a closed path in the horizontal plane of the circuit shown in Fig. 27.43. The path consists of six arcs with alternating radii r_1 and r_2 connected by radial segments. Each segment of arc subtends an angle of 60° at the common center P, with $r_2/r_1 = 2/3$. This current path produces a magnetic field $\mathbf{B}$ at P. If the path is *modified* so that the ratio $r_2/r_1 = 1/3$, by what factor must I be multiplied in order that the field at P remain the same?

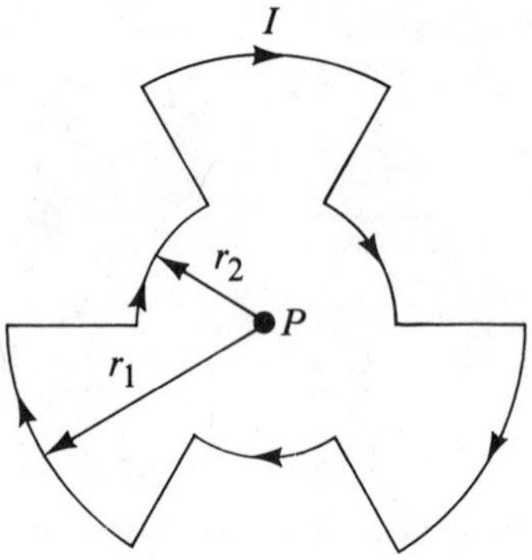

Figure 27.43 (Problem 10).

11. The earth's magnetic field at either pole is about $1\ \text{G} = 1 \times 10^{-4}$ T. Using a model in which you assume that this field is produced by a current loop around the equator, determine the current that would generate such a field. ($R_e = 6.37 \times 10^6$ m.)

12. A straight wire located at the equator is oriented parallel to the earth along the east-west direction. The earth's magnetic field at this point is horizontal and has a magnitude of 3×10^{-5} T. If the mass per unit length of the wire is 5×10^{-3} kg/m, what current must the wire carry in order that the magnetic force balance the weight of the wire?

13. A very long, thin strip of metal of width w carries a current I along its length. Find the magnetic field in the *plane* of the strip (at an external point) a distance b from one edge.

14. A large non-conducting belt with a uniform surface charge density σ moves with a speed v on a set of rollers as shown in Fig. 27.44. Consider a point *just above* the surface of the moving belt. (a) Find an

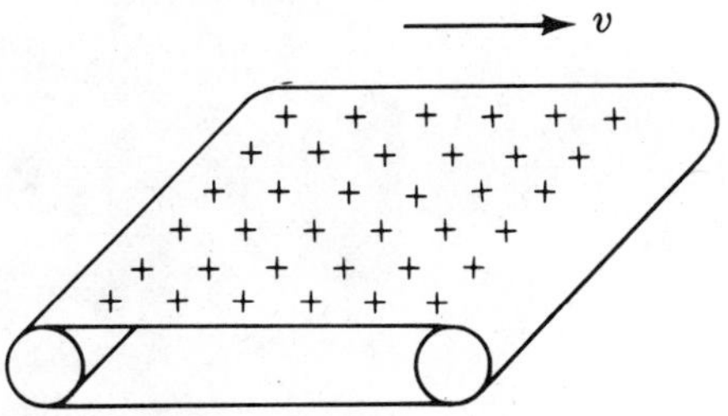

Figure 27.44 (Problem 14).

expression for the magnitude of the magnetic field ***B*** at this point. (b) If the belt is positively charged, what is the direction of ***B***? (Note that the belt may be considered as an infinite sheet.)

15. A wire is bent into the shape shown in Fig. 27.45 (a), and the magnetic field is measured at P_1 when the current in the wire is *I*. The same wire is then formed into the shape shown in Fig. 27.45 (b), and the magnetic field measured at point P_2 when the current is again *I*. If the *total* length of wire is the same in each case, what is the ratio of B_1/B_2?

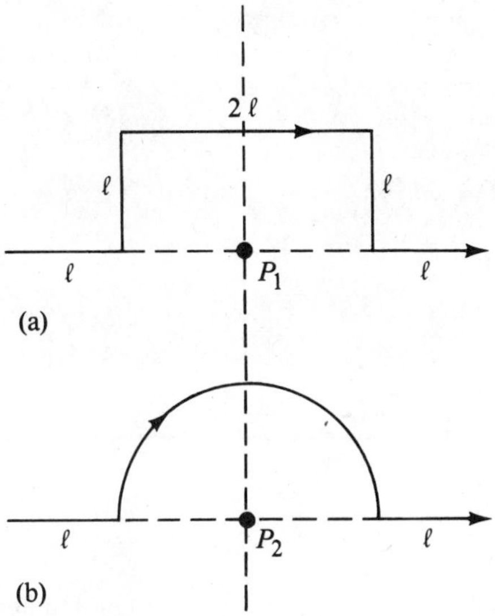

Figure 27.45 (Problem 15).

28 Faraday's Law

Our studies so far have been concerned with the electric field due to stationary charges and the magnetic field of moving charges. This chapter deals with electric fields that originate from changing magnetic fields.

Experiments conducted by Michael Faraday in England and independently by Joseph Henry in the United States in 1831 showed that an electric current could be induced in a circuit by a changing magnetic field. The results of these experiments led to a very basic and important law of electromagnetism known as *Faraday's law of induction*. This law says that the emf induced in a circuit equals the negative of the time rate of change of the magnetic flux through the circuit.

As we shall see, an induced emf can be produced in many ways. For instance, an induced emf and an induced current can be produced in a closed loop of wire when the wire moves through a magnetic field. We shall describe such experiments along with a number of important applications that make use of the phenomenon of electromagnetic induction.

With the treatment of Faraday's law, we complete our introduction to the fundamental laws of electromagnetism. These laws can be summarized in a set of four equations called *Maxwell's equations*. Together with the Lorentz force law, which we shall discuss briefly, they represent a complete theory for describing the interaction of charged objects. Maxwell's equations relate electric and magnetic fields to each other and to their ultimate source, namely, electric charges.

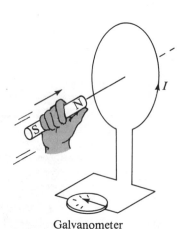

Figure 28.1 When a magnet is moved toward a conducting loop, the galvanometer needle deflects, indicating an induced current.

28.1 FARADAY'S LAW OF INDUCTION

We begin by describing some simple experiments that demonstrate the basic ideas of electromagnetic induction. First, consider a loop of wire connected to a galvanometer as in Fig. 28.1. If a magnet is moved toward the loop, the galvanometer needle will deflect. If the magnet is moved away from the loop, the galvanometer needle will deflect in the opposite direction. If the magnet is held stationary relative to the loop, no deflection is observed. Finally, if the magnet is held stationary and the coil is moved either toward or away from the magnet, the needle will also deflect. From these observations, one concludes that *a current is set up in the circuit as long as there is relative motion between the magnet and the coil.*[1]

[1]The exact magnitude of the current depends on the particular resistance of the circuit, but the existence (and the algebraic sign) do *not*.

These results are quite remarkable in view of the fact that *a current is set up in the circuit even though there are no batteries in the circuit!* We call such a current an *induced current,* which is produced by an *induced emf.*

Now let us consider another apparatus consisting of two *stationary* circuits as shown in Fig. 28.2. Circuit 1 contains a loop connected in series to a resistor, battery, and switch. Circuit 2 is merely a loop with its ends attached to a galvanometer (a low-resistance device). When the switch in circuit 1 is closed, the needle in the galvanometer in circuit 2 *momentarily* deflects, indicating that an induced current is produced in circuit 2. Once the current in circuit 1 reaches its steady-state value, the galvanometer in circuit 2 reads zero. Finally, when the switch in circuit 1 is opened, the galvanometer in circuit 2 shows a momentary deflection in the *opposite* direction. These results show that an induced current and induced emf are produced in circuit 2 as the result of the *change* in current in circuit 1.

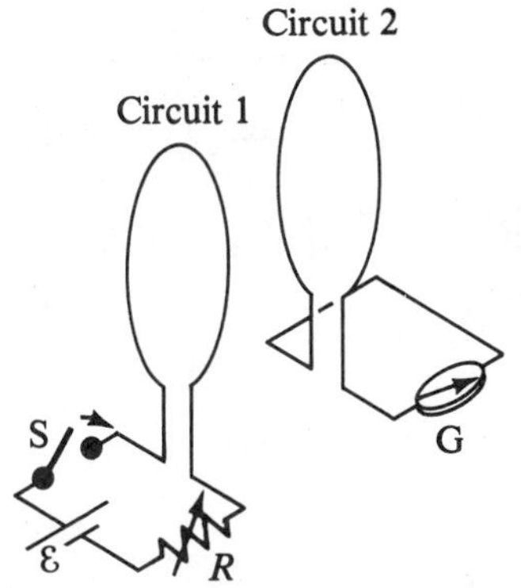

Figure 28.2 When the switch in circuit 1 is closed or opened, the needle of the galvanometer in circuit 2 deflects momentarily.

These two experiments have one thing in common. In both cases, an emf is induced in a circuit when the *magnetic flux* through the circuit *changes with time.* In fact, a general statement that summarizes such experiments involving induced currents and emfs is as follows: *The emf induced in a circuit is directly proportional to the time rate of change of magnetic flux through the circuit.* This statement, known as *Faraday's law of induction,* can be written

$$\mathcal{E} = -\frac{d\Phi_m}{dt} \tag{28.1}$$

Faraday's law

where Φ_m is the magnetic flux threading the circuit (Section 27.5), which can be expressed as

$$\Phi_m = \int \boldsymbol{B} \cdot d\boldsymbol{A} \tag{28.2}$$

The integral given by Eq. 28.2 is taken over the area bounded by the circuit. The meaning of the negative sign in Eq. 28.1 will be discussed in Section 28.3. If the circuit is a coil consisting of N loops all of the same area and if the flux threads all loops, the induced emf is given by

$$\mathcal{E} = -N\frac{d\Phi_m}{dt} \tag{28.3}$$

Suppose the magnetic field is uniform over a loop of area A lying in a plane as in Fig. 28.3. In this case, the flux through the loop is equal to $BA \cos\theta$; hence the induced emf can be expressed as

$$\mathcal{E} = -\frac{d}{dt}(BA \cos\theta) \tag{28.4}$$

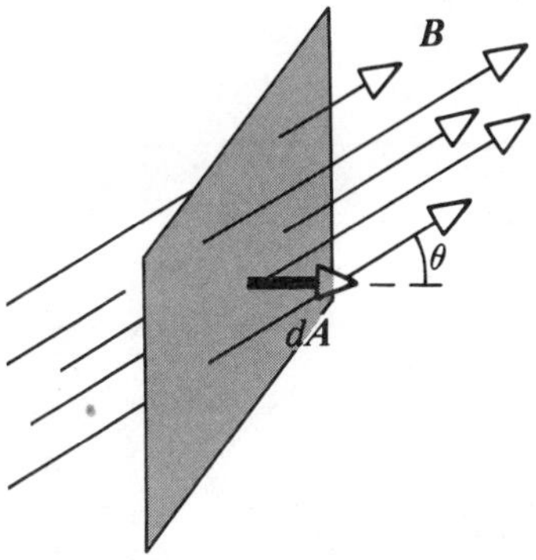

Figure 28.3 A conducting loop of area A in the presence of a uniform magnetic field $\boldsymbol{B}$, which is at an angle θ with the normal to the loop.

From this expression, we see that an emf can be induced in the circuit in several ways: (1) The magnitude of $\boldsymbol{B}$ can vary with time; (2) the area of the circuit can change with time; (3) the angle θ between $\boldsymbol{B}$ and the normal to the plane can change with time; and (4) any combination of these can occur.

The following examples illustrate cases where an emf is induced in a circuit as a result of a time variation of the magnetic field.

Example 28.1

A coil is wrapped with 200 turns of wire on the perimeter of a square frame of sides 18 cm. Each turn has the same area, equal to that of the frame, and the total resistance of the coil is 2 Ω. A uniform magnetic field is turned on perpendicular to the plane of the coil. If the field changes linearly from 0 to 0.5 Wb/m^2 in a time of 0.8 s, (a) find the magnitude of the induced emf in the coil while the field is changing.

The area of the loop is $(0.18 \text{ m})^2 = 0.0324 \text{ m}^2$. The magnetic flux through the loop at $t = 0$ is zero since $B = 0$. At $t = 0.8$ s, the magnetic flux through the loop is $\Phi_m = BA = (0.5 \text{ Wb/m}^2)(0.0324 \text{ m}^2) = 0.0162$ Wb. Therefore, the magnitude of the induced emf is

$$|\mathcal{E}| = \frac{N\,\Delta\Phi_m}{\Delta t} = \frac{200(0.0162 \text{ Wb} - 0 \text{ Wb})}{0.8 \text{ s}} = 4.05 \text{ V}$$

(Note that 1 Wb = 1 V · s.)

(b) What is the magnitude of the induced current in the coil while the field is changing?

Using Ohm's law and the results of (a), we find that the magnitude of the induced current is

$$I = \frac{|\mathcal{E}|}{R} = \frac{4.05 \text{ V}}{2\ \Omega} = 2.03 \text{ A}$$

Example 28.2

A planar loop of wire of area A is placed in a region where the magnetic field is *perpendicular* to the plane. The magnitude of $\boldsymbol{B}$ varies in time according to the expression $B = B_0 e^{-at}$. That is, at $t = 0$ the field is B_0, and at $t > 0$, the field decreases exponentially in time (Fig. 28.4). Find the induced emf in the loop as a function of time.

Solution: Since $\boldsymbol{B}$ is perpendicular to the plane of the loop, the magnetic flux through the loop at time $t > 0$ is given by

$$\Phi_m = BA = AB_0 e^{-at}$$

Noting that the coefficient AB_0 and the parameter a are constants, the induced emf can be calculated from Eq. 28.1:

$$\mathcal{E} = -\frac{d\Phi_m}{dt} = -AB_0 \frac{d}{dt} e^{-at} = aAB_0 e^{-at}$$

That is, the induced emf decays exponentially in time. Note that the maximum emf occurs at $t = 0$, where $\mathcal{E}_{max} = aAB_0$. Why is this true? The plot of $\mathcal{E}$ versus t is similar to the B versus t curve shown in Fig. 28.4.

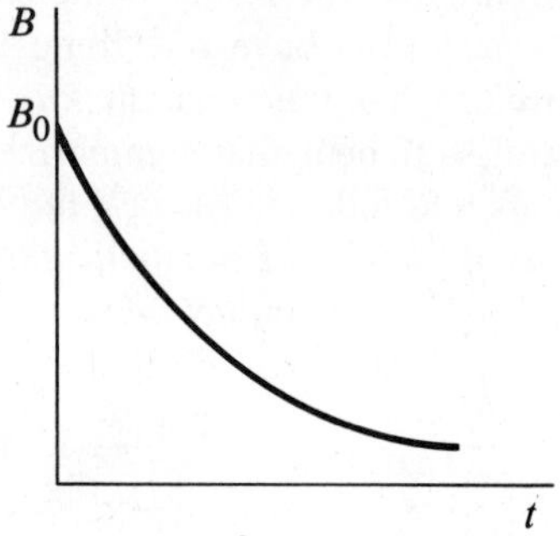

Figure 28.4 (Example 28.2) Exponential decrease of the magnetic field with time. The induced emf and induced current have similar time variations.

Q1. What is the difference between magnetic flux and magnetic field?

Q2. A circular loop is located in a uniform and constant magnetic field. Describe how an emf can be induced in the loop in this situation.

Q3. A loop of wire is placed in a uniform magnetic field. For what orientation of the loop is the magnetic flux a maximum? For what orientation is the flux zero?

28.2 MOTIONAL emf

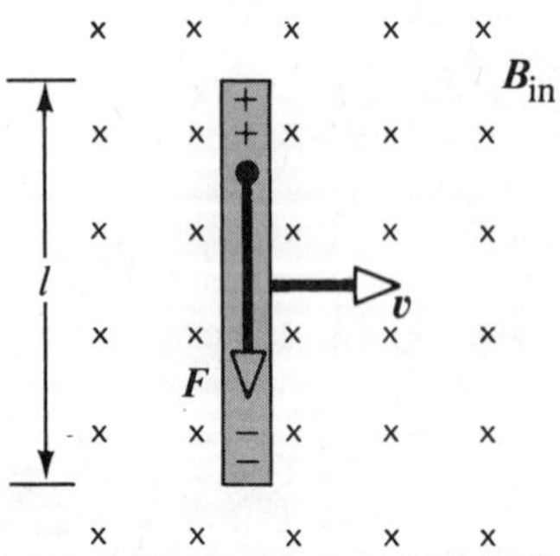

Figure 28.5 A straight conducting bar of length l moving with a velocity $\boldsymbol{v}$ through a uniform magnetic field $\boldsymbol{B}$ directed perpendicular to $\boldsymbol{v}$. An emf equal to Blv is induced between the ends of the bar.

In Examples 28.1 and 28.2, we considered cases in which an emf is produced in a circuit when the magnetic field changes with time. In this section we describe the so-called *motional emf,* which is induced in a conductor moving through a magnetic field.

First, consider a straight conductor of length l moving with constant velocity through a uniform magnetic field directed into the paper as in Fig. 28.5. For simplicity, we shall assume that the conductor is moving perpendicular to the field. The free charges in the conductor will experience a force along the conductor given by $q\boldsymbol{v} \times \boldsymbol{B}$. Under the influence of this force, the electrons will move to the *lower* end and accumulate there, leaving a net positive charge at the upper end. An electric field is therefore produced within the conductor as a result of this charge separation. The charge at the ends builds up until the magnetic force qvB is balanced by the electric force qE. At this point, charge stops flowing and the condition for equilibrium requires that

$$qE = qvB \qquad \text{or} \qquad E = vB$$

This electric field produced in the wire leads to a potential difference V across the ends given by

$$V = El = Blv$$

where the upper end is at a higher potential than the lower end. Thus, a potential difference is maintained as long as there is motion through the field. If the motion is reversed, the polarity of V is also reversed.

A more interesting situation occurs if we now consider what happens when the moving conductor is part of a closed conducting path. This example is particularly useful for illustrating how a changing magnetic flux can cause an induced current in a closed circuit. Consider a circuit consisting of a conducting bar of length l sliding along two fixed parallel conducting rails as in Fig. 28.6a. For simplicity, we shall assume that the moving bar has zero resistance and that the stationary part of the circuit has a resistance R. A uniform and constant magnetic field $\boldsymbol{B}$ is applied perpendicular to the plane of the circuit. As the bar is pulled to the right with a velocity $\boldsymbol{v}$, under the influence of an applied force $\boldsymbol{F}_{app}$, the free charges in the bar experience a magnetic force along the length of the bar. This force, in turn, sets up an induced current since the charges are free to move in a closed conducting path. In this case, the changing magnetic flux through the loop and the corresponding induced emf across the moving bar arise from the change in area of the loop as the bar moves through the magnetic field. As we shall see, if the bar is pulled to the right with a constant velocity, the work done by the applied force is dissipated as joule heat in the circuit's resistive element.

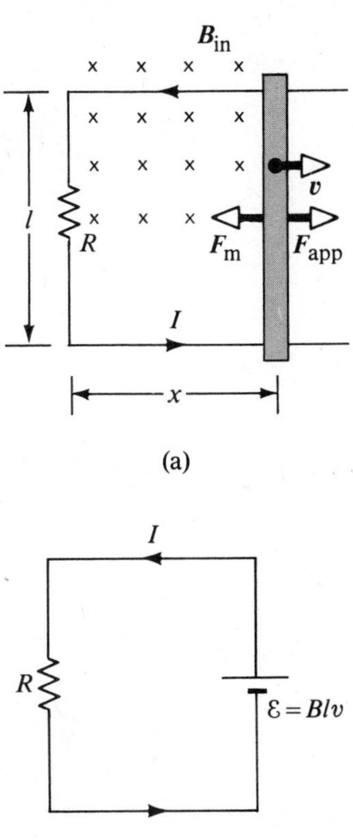

Figure 28.6 (a) A conducting bar sliding with a velocity $\boldsymbol{v}$ along two conducting rails under the action of an applied force $\boldsymbol{F}_{app}$. The magnetic force $\boldsymbol{F}_m$ opposes the motion, and a counterclockwise current is induced in the loop. (b) The equivalent circuit of (a).

Since the area of the circuit at any instant is lx, the external magnetic flux through the circuit is given by

$$\Phi_m = Blx$$

where x is the width of the circuit, which changes with time. Using Faraday's law, we find that the induced emf is

$$\mathcal{E} = -\frac{d\Phi_m}{dt} = -\frac{d}{dt}(Blx) = -Bl\frac{dx}{dt}$$

Motional emf

$$\mathcal{E} = -Blv \tag{28.5}$$

If the resistance of the circuit is R, the magnitude of the induced current is given by

$$I = \frac{|\mathcal{E}|}{R} = \frac{Blv}{R} \tag{28.6}$$

The equivalent circuit diagram for this example is shown in Fig. 28.6b.

Let us examine the system using energy considerations. Since there is no real battery in the circuit, one might wonder about the origin of the induced current and the electrical energy in the system. We can understand this by noting that the external force does work on the conductor, thereby moving charges through a magnetic field. This causes the charges to move with some average drift velocity, and hence a current is established. From the viewpoint of energy conservation, the total work done by the applied force during some time interval should equal the electrical energy that the induced emf supplies in that same period. Furthermore, if the bar moves with constant speed, the work done must equal the energy dissipated as heat in the resistor in this time interval.

As the conductor of length l moves through the uniform magnetic field $\boldsymbol{B}$, it experiences a magnetic force $\boldsymbol{F}_m$ of magnitude IlB (Section 26.3). The direction of this force is opposite the motion of the bar, or to the left in Fig. 28.6a.

If the bar is to move with a *constant* velocity, the applied force must be equal to and opposite the magnetic force, or to the right in Fig. 28.6a. If the magnetic force acted in the direction of motion, it would cause the bar to accelerate once it was in motion, thereby increasing its velocity. This state of affairs would represent a violation of the principle of energy conservation. Using Eq. 28.6 and the fact that $F_{\text{app}} = IlB$, we find that the power delivered by the applied force is

$$P = F_{\text{app}}v = (IlB)v = \frac{B^2l^2v^2}{R} \tag{28.7}$$

This power is equal to the rate at which energy is dissipated in the resistor, I^2R, as we would expect. It is also equal to the power $I\mathcal{E}$ supplied by the induced emf. This example is a clear demonstration of the conversion of mechanical energy into electrical energy (the induced emf) and finally into thermal energy (joule heat).

Example 28.3

A conducting bar of length l rotates with a constant angular velocity ω about a pivot at one end. A uniform magnetic field $\boldsymbol{B}$ is directed perpendicular to the plane of rotation, as in Fig. 28.7. Find the emf induced between the ends of the bar.

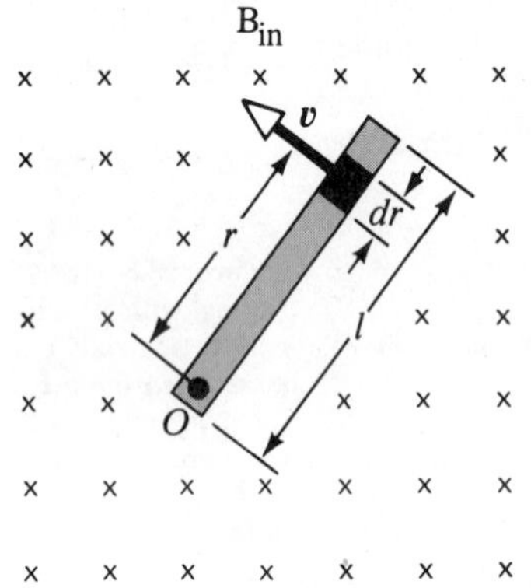

Figure 28.7 (Example 28.3) A conducting bar rotating about a pivot at one end in a uniform magnetic field that is perpendicular to the plane of rotation. An emf is induced across the ends of the bar.

Solution: Consider a segment of the bar of length dr, whose velocity is $\boldsymbol{v}$. According to Eq. 28.5, the emf induced in a conductor of this length moving perpendicular to a field $\boldsymbol{B}$ is given by

$$(1) \qquad d\mathcal{E} = Bv\,dr$$

Each segment of the bar is moving perpendicular to $\boldsymbol{B}$, and so there is an emf generated across each segment; the value of this emf is given by (1). Summing up the emfs induced across all elements, which are in series, gives the total emf between the ends of the bar. That is,

$$\mathcal{E} = \int Bv\,dr$$

In order to integrate this expression, note that the linear speed of an element is related to the angular speed ω through the relationship $v = r\omega$. Therefore, since B and ω are constants, we find that

$$\mathcal{E} = B\int v\,dr = B\omega\int_0^l r\,dr = \frac{1}{2}B\omega l^2$$

Example 28.4

A bar of mass m and length l moves on two frictionless parallel rails in the presence of a uniform magnetic field directed into the paper (Fig. 28.8). The bar is given an initial velocity $\boldsymbol{v}_0$ to the right and is released. Find the velocity of the bar as a function of time.

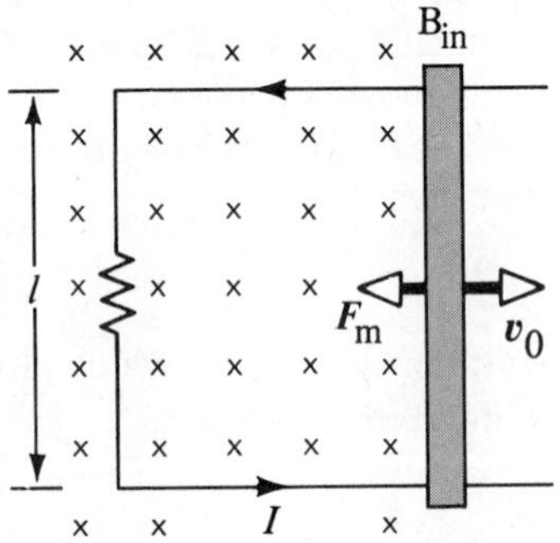

Figure 28.8 (Example 28.4) A conducting bar of length l sliding on two fixed conducting rails is given an initial velocity $\boldsymbol{v}_0$ to the right.

Solution: First note that the induced current is counterclockwise and the magnetic force is $F_{\text{m}} = -IlB$, where the negative sign denotes that the force is to the left and *retards* the motion. This is the *only* horizontal force acting on the bar, and hence Newton's second law applied to motion in the horizontal direction gives

$$F_x = ma = m\frac{dv}{dt} = -IlB$$

Since the induced current is given by Eq. 28.6, $I = Blv/R$, we can write this expression as

$$m\frac{dv}{dt} = -\frac{B^2l^2}{R}v$$

$$\frac{dv}{v} = -\left(\frac{B^2l^2}{mR}\right)dt$$

Integrating this last equation using the initial condition that $v = v_0$ at $t = 0$, we find that

$$\int_{v_0}^{v}\frac{dv}{v} = \frac{-B^2l^2}{mR}\int_0^t dt$$

$$\ln\left(\frac{v}{v_0}\right) = -\left(\frac{B^2l^2}{mR}\right)t = -\frac{t}{\tau}$$

where the constant $\tau = mR/B^2l^2$. From this, we see that the velocity can be expressed in the exponential form

$$v = v_0 e^{-t/\tau}$$

Therefore, the velocity of the bar decreases exponentially with time under the action of the magnetic retarding force. Furthermore, if we substitute this result into Eqs. 28.5 and 28.6, we find that the induced emf and induced current also decrease exponentially with time. That is,

$$I = \frac{Blv}{R} = \frac{Blv_0}{R} e^{-t/\tau}$$

$$\mathcal{E} = IR = Blv_0 e^{-t/\tau}$$

Q4. As the conducting bar in Fig. 28.9 moves to the right, an electric field is set up directed downward. If the bar were moving to the left, explain why the electric field would be upward.

Q5. As the bar in Fig. 28.9 moves perpendicular to the field, is an external force required to keep it moving with constant velocity?

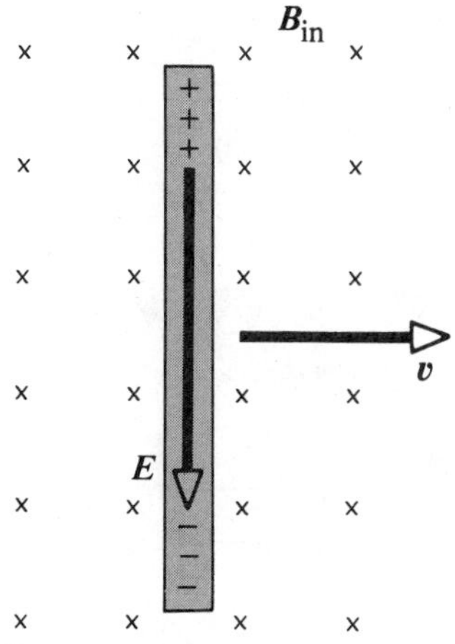

Figure 28.9 (Questions 4 and 5)

28.3 LENZ'S LAW

The direction of the induced emf and induced current can be found from *Lenz's law*,[2] which can be stated as follows: *The induced current and emf are in such a direction as to prevent any change in the net number of lines of flux that pass through the cross-sectional area of the circuit.* That is, the induced current tends to maintain the original flux through the circuit. The interpretation of this statement depends on the circumstances. As we shall see, this law is a consequence of the law of conservation of energy.

Let us return to the example of a bar moving to the right on two parallel rails in the presence of a uniform magnetic field directed into the paper (Fig. 28.10a). As the bar moves to the right, the magnetic flux through the circuit *increases* with time because the area A of the loop increases with time and $\Phi_m = BA$. Lenz's law says that the induced current must be in a direction that *opposes* this change. Since the flux due to the *external* field is increasing *into* the paper, the induced current, if it is to oppose the change, must produce a flux *out* of the paper. Hence, the induced current must be *counterclockwise* when the bar moves to the right to give a counteracting flux out of the paper (recall the right-hand rule).

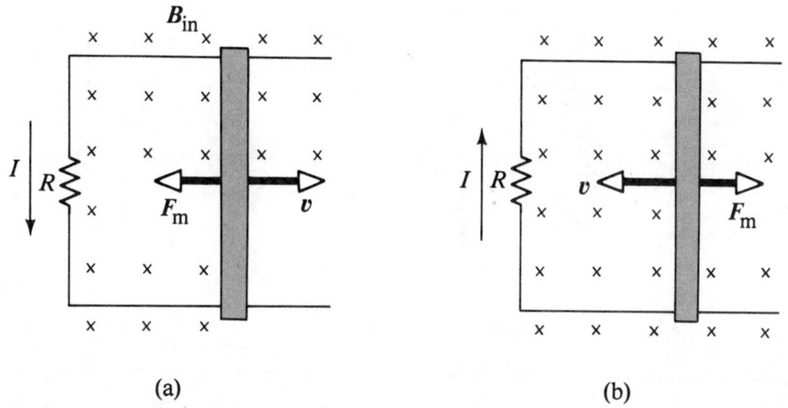

Figure 28.10 (a) As the conducting bar slides on the two fixed conducting rails, the magnetic flux through the loop increases in time. By Lenz's law, the induced current must be *counterclockwise* so as to produce a counteracting flux *out* of the paper. (b) When the bar moves to the left, the induced current must be *clockwise*. Why?

[2]Developed by the German physicist Heinrich Lenz (1804–1865).

On the other hand, if the bar were moving to the *left*, as in Fig. 28.10b, the magnetic flux into the paper would decrease in time, and hence the induced current would have to be *clockwise* to produce a flux into the paper. In either case, we see that *the induced current tends to maintain the original flux through the circuit.*

Let us look at this situation from the viewpoint of energy considerations. As we mentioned in the previous section, the magnetic force $\boldsymbol{F}_m$ on the rod must oppose the motion. If the external force is to the right in Fig. 28.10a, the magnetic force must be to the left. Since $\boldsymbol{B}$ is into the paper, we conclude from the expression $\boldsymbol{F}_m = I\boldsymbol{l} \times \boldsymbol{B}$ that the current in the bar is upward. That is, the induced current is in the direction that produces a magnetic force that *opposes* the external force. If the opposite were true, we can easily see that energy would not be conserved.

Suppose the bar is given a slight push to the right, and we assume the induced current is clockwise. In that case, the magnetic force would be to the right, which would accelerate the rod and increase its velocity. This would cause an increase in the induced current, an increase in the magnetic force, and a continued acceleration of the bar. In effect, this system would acquire energy with zero input energy. This is clearly inconsistent with all experience and with the first law of thermodynamics.

Consider another situation in which a bar magnet is moved toward a stationary conducting loop of wire, as in Fig. 28.11a. As the magnet moves to the right toward the loop, the magnetic flux through the loop increases with time. To counteract this increase in flux to the right, the induced current produces a flux to the *left*, as in Fig. 28.11b; hence the induced current is in the direction shown. Note that the magnetic field lines associated with the induced current oppose the motion of the magnet. Therefore, the left face of the current loop is a north pole, and the right face is a south pole.

On the other hand, if the magnet were moving to the left, the external flux through the loop to the right would *decrease* in time. Hence, the induced current in the loop would be opposite that shown in Fig. 28.11a, so as to produce its own flux to the right. In this case, the left face of the current loop would be a south pole and its right face would be a north pole.

As a final example, consider the system shown in Fig. 28.12, consisting of two adjacent coils wrapped on a common cylindrical core. Circuit 2 contains a battery, a switch, and a resistor, whereas circuit 1 contains only a resistor. When the switch in circuit 2 is open and there is no current, the current in circuit 1 is zero (Fig. 28.12a).

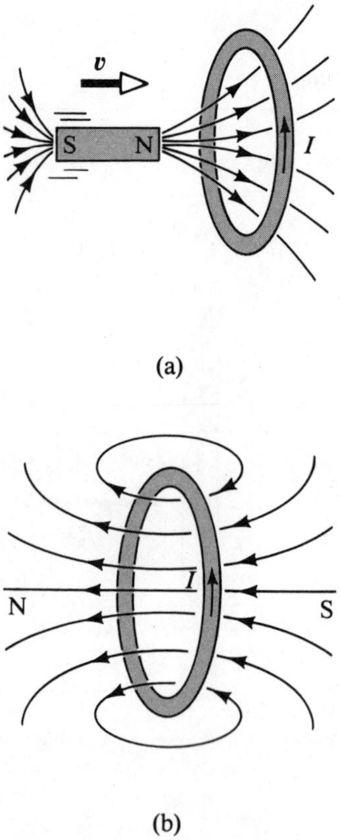

Figure 28.11 (a) When the magnet is moved toward the stationary conducting loop, a current is induced in the direction shown. (b) This induced current produces its own flux to the left to counteract the increasing external flux to the right.

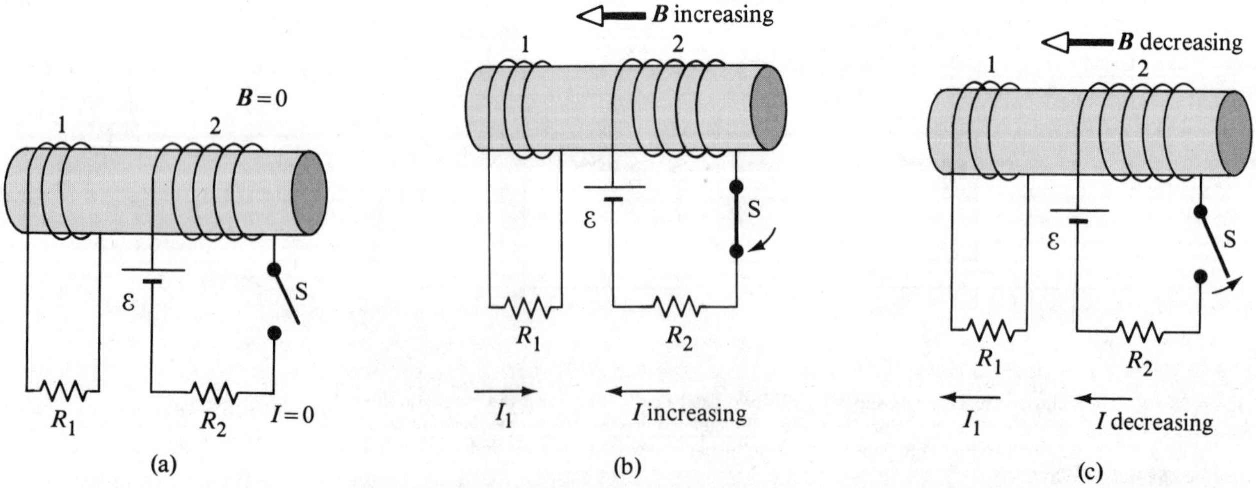

Figure 28.12 Two circuits close to each other. (a) Before the switch in circuit 2 is closed, the current in both circuits is zero. (b) Just after the switch in 2 is closed, a field is produced, which causes a changing flux through 1 and an induced current I_1 whose flux opposes the increase. (c) Just after the switch in 2 is opened, the field decreases in time, again causing a changing flux and an induced current I_1 whose flux opposes the decrease.

When the switch is closed, the current in circuit 2 goes from zero to its maximum value, $\mathcal{E}/R_2$, in some short time. While this current is increasing, the flux due to circuit 2 (which is to the left) is also increasing (Fig. 28.12b). This increasing flux to the left penetrates circuit 1 and induces a current I_1. Because of its direction, I_1 produces a flux within the coil to the right. Once the current in circuit 2 reaches its maximum steady-state value, the change in flux is zero and the induced current I_1 is zero.

Finally, when the switch is opened, the current in circuit 2 decreases toward zero (Fig. 28.12c) and the induced current I_1 momentarily changes its direction, thereby opposing the decrease in the leftward flux.

Example 28.5

A rectangular loop of dimensions l and w and resistance R moves with constant speed v to the right, as in Fig. 28.13a. It continues to move with this speed through a region containing a uniform magnetic field $\boldsymbol{B}$ directed into the paper and extending a distance $3w$. Plot the flux, the induced emf, and the external force acting on the loop as a function of the position of the loop in the field.

Solution: Figure 28.13b shows the flux through the loop as a function of loop position. Before the loop enters the field, the flux is zero. As it enters the field, the flux increases linearly with position. Finally, the flux decreases linearly to zero as the loop leaves the field.

Before the loop enters the field, there is no induced emf since there is no field present (Fig. 28.13c). As the right side of the loop enters the field, the flux inward begins to increase. Hence, according to Lenz's law, the induced current is counterclockwise and the induced emf is given by $-Blv$. This motional emf arises from the magnetic force experienced by charges in the right side of the loop. When the loop is entirely in the field, the *change* in flux is zero, and hence the induced emf vanishes.

From another point of view, the right and left sides of the loop experience magnetic forces that tend to set up currents that cancel one another. As the right side of the loop leaves the field, the flux inward begins to decrease, a clockwise current is induced, and the induced emf is Blv. As soon as the left side leaves the field, the emf drops to zero.

The external force that must act on the loop to maintain this motion is plotted in Fig. 28.13d. When the loop is not in the field, there is no magnetic force on it; hence the external force on it must be zero if v is constant. When the right side of the loop enters the field, the external force necessary to maintain constant speed must be equal to and opposite the magnetic force on that side, given by $F_m = -IlB = -B^2l^2v/R$. When the loop is entirely in the field, the right and left sides experience equal and opposite forces; hence the net force is zero. Finally, as the right side leaves the field, the external force must be equal to and opposite the magnetic force on the left side of the loop. From this analysis, we conclude that power is supplied only when the loop is either entering or leaving the field. Furthermore, this example shows that the induced emf in the loop can be zero even when there is motion through the field! Again, it is emphasized that an emf is induced in the loop *only* when the magnetic flux through the loop *changes* in time.

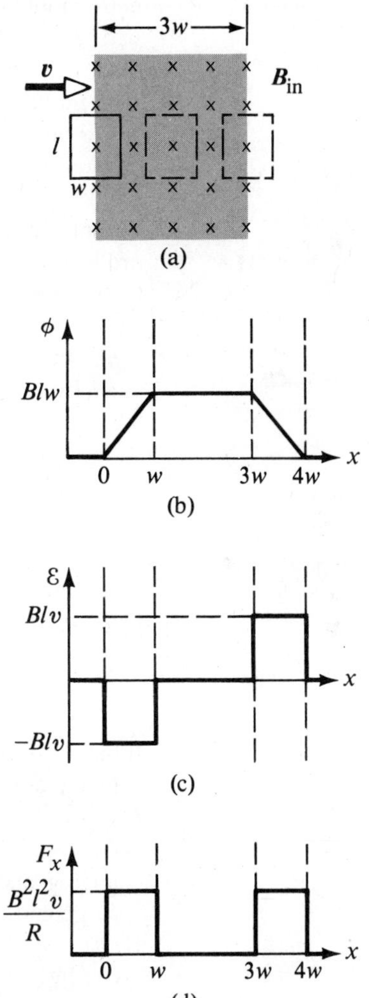

Figure 28.13 (Example 28.5) (a) A conducting rectangular loop of width w and length l moving with a velocity $\boldsymbol{v}$ through a uniform magnetic field extending a distance $3w$. (b) A plot of the flux as a function of the position of the loop. (c) A plot of the induced emf versus the position of the leading edge. (d) A plot of the force versus position such that the velocity of the loop remains constant.

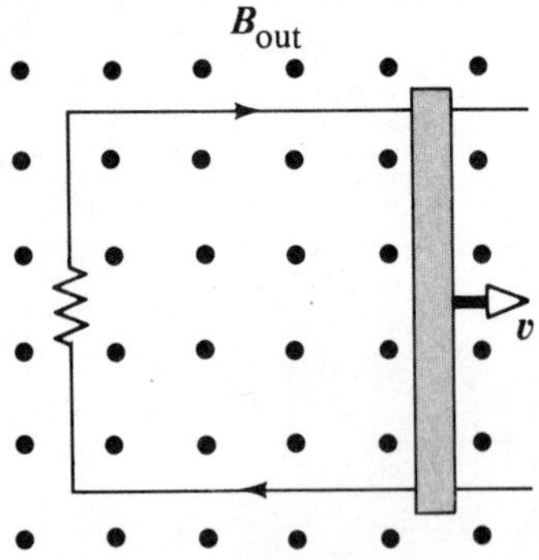

Figure 28.14 (Questions 6 and 7)

Q6. The bar in Fig. 28.14 moves to the right with a velocity v, and the uniform, constant magnetic field is *outward*. Why is the induced current clockwise? If the bar were moving to the left, what would be the direction of the induced current?

Q7. Explain why an external force is necessary to keep the bar in Fig. 28.14 moving with a constant velocity.

28.4 INDUCED emfs AND ELECTRIC FIELDS

We have seen that a changing magnetic flux induces an emf and a current in a conducting loop. We therefore must conclude that *an electric field is created in the conductor as a result of the changing magnetic flux*. In fact, the law of electromagnetic induction shows that *an electric field is always generated by a changing magnetic flux*, even in free space where no charges are present. However, this induced electric field has properties that are quite different from those of an electrostatic field *produced by stationary charges*.

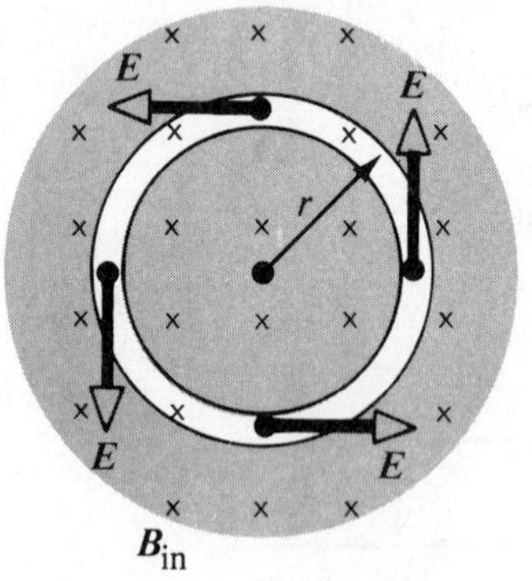

Figure 28.15 A loop of radius r in a uniform magnetic field perpendicular to the plane of the loop. If $\mathbf{B}$ changes in time, an electric field is induced in a direction tangent to the loop.

We can illustrate this point by considering a conducting loop of radius r situated in a uniform magnetic field that is perpendicular to the plane of the loop, as in Fig. 28.15. If the magnetic field changes with time, then Faraday's law tells us that an emf given by $\mathcal{E} = -d\Phi_m/dt$ is induced in the loop. The induced current that is produced implies the presence of an electric field $\mathbf{E}$, which must be tangent to the loop since all points on the loop are equivalent. The work done in moving a test charge q once around the loop is equal to $q\mathcal{E}$. Since the electric force on the charge is qE, the work done by this force in moving the charge once around the loop is given by $qE(2\pi r)$, where $2\pi r$ is the circumference of the loop. These two expressions for the work must be equal; therefore we see that

$$q\mathcal{E} = qE(2\pi r)$$

$$E = \frac{\mathcal{E}}{2\pi r}$$

Using this result, Faraday's law, and the fact that $\Phi_m = BA = \pi r^2 B$ for a circular loop, we find that the electric field can be expressed as

$$E = -\frac{1}{2\pi r}\frac{d\Phi_m}{dt} = -\frac{r}{2}\frac{dB}{dt} \qquad (28.8)$$

If the time variation of the magnetic field is specified, the electric field can easily be calculated from Eq. 28.8. The negative sign indicates that the induced electric field **E** *opposes* the change in the magnetic field. It is important to understand that *this result is also valid in the absence of a conductor*. That is, a free charge placed in a changing magnetic field will also experience the same electric field.

In general, the emf for any closed path can be expressed as the line integral of $\mathbf{E} \cdot d\mathbf{s}$ over that path. Hence, the general form of Faraday's law of induction is given by

Faraday's law in general form

$$\mathcal{E} = \oint \mathbf{E} \cdot d\mathbf{s} = -\frac{d\Phi_m}{dt} \qquad (28.9)$$

It is important to recognize that *the electric field* **E** *that appears in Eq. 28.9 is a nonconservative, time-varying field that is generated by a changing magnetic field.* The field **E** that satisfies Eq. 28.9 could not possibly be an electrostatic field for the following reason. If the field were electrostatic, and hence conservative, the line integral of $\mathbf{E} \cdot d\mathbf{s}$ over a closed loop would be zero, contrary to Eq. 28.9.

Example 28.6

A long solenoid of radius R has n turns per unit length and carries a time-varying current that varies sinusoidally as $I = I_0 \cos\omega t$, where I_0 is the maximum current and ω is the angular frequency of the current source (Fig. 28.16). (a) Determine the electric field outside the solenoid, a distance r from its axis.

First, let us consider an external point and take the path for our line integral to be a circle centered on the solenoid, as in Fig. 28.16. By symmetry we see that the magnitude of **E** is constant on this path and tangent to it. The magnetic flux through this path is given by $BA = B(\pi R^2)$, and hence Eq. 28.9 gives

$$\oint \mathbf{E} \cdot d\mathbf{s} = -\frac{d}{dt}[B(\pi R^2)] = -\pi R^2 \frac{dB}{dt}$$

$$E(2\pi r) = -\pi R^2 \frac{dB}{dt}$$

Since the magnetic field inside a long solenoid is given by Eq. 27.20, $B = \mu_0 n I$, and $I = I_0 \cos\omega t$, we find that

$$E(2\pi r) = -\pi R^2 \mu_0 n I_0 \frac{d}{dt}(\cos\omega t) = \pi R^2 \mu_0 n I_0 \omega \sin\omega t$$

$$E = \frac{\mu_0 n I_0 \omega R^2}{2r} \sin\omega t \qquad (\text{for } r > R)$$

Hence, the electric field varies sinusoidally with time, and its amplitude falls off as $1/r$ outside the solenoid.

(b) What is the electric field inside the solenoid, a distance r from its axis?

For an interior point ($r < R$), the flux threading an integration loop is given by $B(\pi r^2)$. Using the same procedure as in (a), we find that

$$E(2\pi r) = -\pi r^2 \frac{dB}{dt} = \pi r^2 \mu_0 n I_0 \omega \sin\omega t$$

$$E = \frac{\mu_0 n I_0 \omega}{2} r \sin\omega t \qquad (\text{for } r < R)$$

This shows that the amplitude of the electric field *inside* the solenoid increases linearly with r and varies sinusoidally with time.

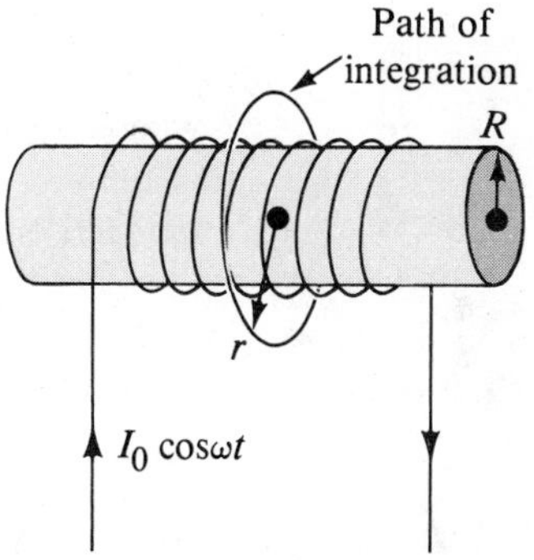

Figure 28.16 (Example 28.6) A long solenoid carrying a time-varying current given by $I = I_0 \cos\omega t$. An electric field is induced both inside and outside the solenoid.

Q8. When a small magnet is moved toward a solenoid, an emf is induced in the coil. However, if the magnet is moved around inside a toroidal coil, there is *no* induced emf. Explain.

28.5 GENERATORS AND MOTORS

Generators and motors are important practical devices that operate on the basis of electromagnetic induction. First, let us consider the *alternating current* (ac) *generator,* a device that converts mechanical energy into electrical energy. In its simplest form, the ac generator consists of a loop of wire rotated by some external means in a magnetic field (Fig. 28.17a). As the loop rotates, the magnetic flux through it changes in time, inducing an emf and, if an external circuit exists, a current. The ends of the wire are connected to slip rings which rotate with the loop. Connections to the load circuit are made using stationary brushes in contact with the slip rings.

To put our discussion of the generator on a quantitative basis, suppose that the loop has N turns (a more practical situation), all of the same area A, and suppose that the loop rotates with a constant angular velocity ω. If θ is the angle between the magnetic field and the normal to the plane of the loop as in Fig. 28.18, then the magnetic flux through the loop at any time t is given by

$$\Phi_m = BA\cos\theta = BA\cos\omega t$$

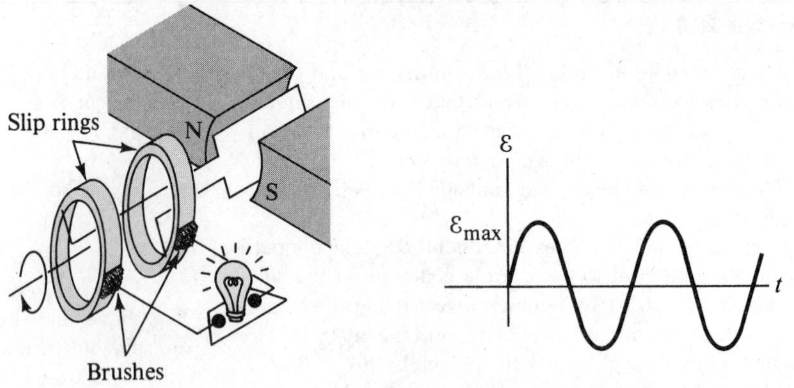

(a) (b)

Figure 28.17 (a) Schematic diagram of an ac generator. An emf is induced in a coil which rotates by some external means in a magnetic field. (b) The alternating emf induced in the loop plotted versus time.

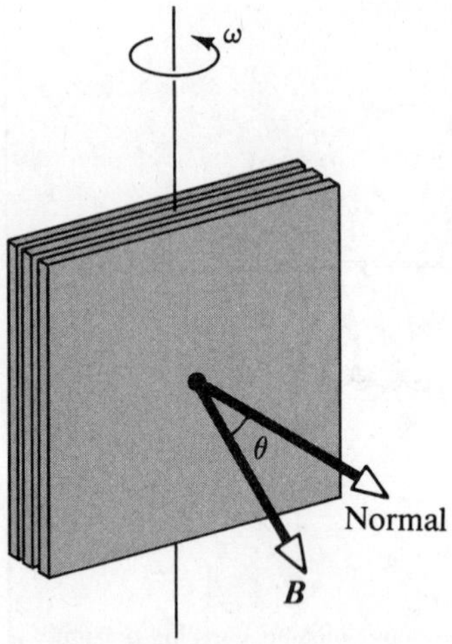

Figure 28.18 A loop of area A containing N turns, rotating with constant angular velocity ω in the presence of a magnetic field. The emf induced in the loop varies sinusoidally in time.

where we have used the relationship between angular displacement and angular velocity, $\theta = \omega t$. (We have set the clock so that $t = 0$ when $\theta = 0$.) Hence, the induced emf in the coil is given by

$$\mathcal{E} = -N\frac{d\Phi_m}{dt} = -NAB\frac{d}{dt}(\cos\omega t) = NAB\omega\,\sin\omega t \tag{28.10}$$

This result shows that the emf varies sinusoidally with time, as plotted in Fig. 28.17b. Note that the maximum emf has the value

$$\mathcal{E}_{max} = NAB\omega \tag{28.11}$$

which occurs when $\omega t = 90°$ or $270°$. In other words, $\mathcal{E} = \mathcal{E}_{max}$ when the magnetic field is in the plane of the coil, and the time rate of change of flux is a maximum. Furthermore, the emf is *zero* when $\omega t = 0$ or $180°$, that is, when $\mathbf{B}$ is perpendicular to the plane of the coil, and the time rate of change of flux is zero.

Example 28.7

An ac generator consists of 8 turns of wire of area $A = 0.09\ m^2$ and total resistance 12 Ω. The loop rotates in a magnetic field $B = 0.5$ T at a constant frequency of 60 Hz. (a) Find the maximum induced emf.

First note that $\omega = 2\pi f = 2\pi(60\ \text{Hz}) = 377\ s^{-1}$. Using Eq. 28.11 with the appropriate numerical values gives

$$\mathcal{E}_{max} = NAB\omega = 8(0.09\ m^2)(0.5\ T)(377\ s^{-1}) = 136\ V$$

(b) What is the maximum induced current?

From Ohm's law and the results to (a), we find that the maximum induced current is

$$I_{max} = \frac{\mathcal{E}_{max}}{R} = \frac{136\ V}{12\ \Omega} = 11.3\ A$$

(c) Determine the time variation of the induced emf and induced current when the output terminals are connected by a low-resistance conductor.

We can use Eq. 28.10 to obtain the time variation of $\mathcal{E}$:

$$\mathcal{E} = \mathcal{E}_{max}\sin\omega t = 136\sin 377t\ V$$

Hence, it follows that the time variation of I is

$$I = I_{max}\sin\omega t = 11.3\sin 377t\ A$$

The *direct current* (dc) *generator* is illustrated in Fig. 28.19a. Such generators are used, for instance, to charge storage batteries used in cars. The components are essentially the same as those of the ac generator, except that the contacts to the rotating loop are made using a split ring, or commutator.

In this configuration, the output voltage always has the same polarity and the current is a pulsating direct current as in Fig. 28.19b. The reason for this can be understood by noting that the contacts to the split ring reverse their roles every half cycle. At the same time, the polarity of the induced emf reverses; hence the polarity of the split ring (which is the same as the polarity of the output voltage) remains the same.

A pulsating dc current is not suitable for most applications. To obtain a more steady dc current, commercial dc generators use many armature coils and commutators distributed so that the sinusoidal pulses from the various coils are out of phase. When these pulses are superimposed, the dc output is almost free of fluctuations.

Motors are devices that convert electrical energy into mechanical energy. Essentially, *a motor is a generator run in reverse.* Instead of generating a current by rotating a loop, a current is supplied to the loop by a battery and the torque acting on the current-carrying loop causes it to rotate.

Useful mechanical work can be done by attaching the rotating armature to some external device. However, as the loop rotates, the changing magnetic flux induces an emf in the loop; this induced emf *always* acts to reduce the current in the loop. If this were not the case, Lenz's law would be violated. The back emf increases in magnitude as the rotational speed of the armature increases. (The phrase *back emf* is used to indicate an emf that tends to reduce the supplied current.) Since the voltage available to supply current equals the difference between the supply voltage and the back emf, the current through the armature coil is limited by the back emf. When a motor is first turned on, there is initially no back emf, and the current is very large since it is limited only by the coil's resistance. As the coils begin to rotate, the induced back emf opposes the externally applied voltage, and the current in the coils is reduced. Under a heavy mechanical load the motor will slow down, which causes the back emf to decrease. This reduction in the back emf increases the current in the coils and therefore the power supplied by the external voltage source. For this reason, the power requirements are greater for starting a motor and for running it under heavy loads. If the motor is allowed to run under no mechanical load, the back emf reduces the current to a value large enough to overcome energy losses by joule heat and friction.

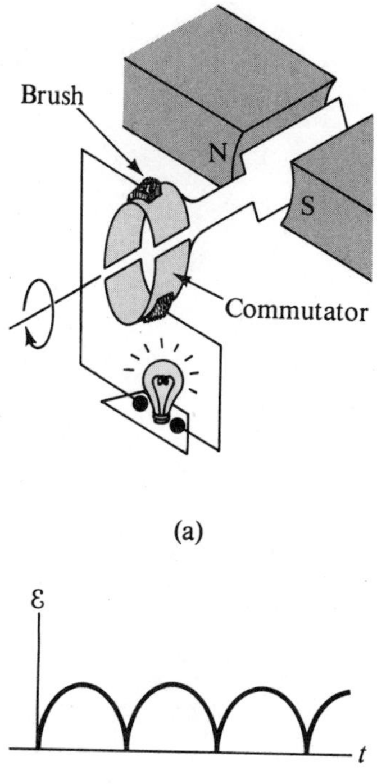

Figure 28.19 (a) Schematic diagram of a dc generator. (b) The emf versus time fluctuates in magnitude but always has the same polarity.

28.6 EDDY CURRENTS

As we have seen, an emf and a current are induced in a circuit by a changing magnetic flux. In the same manner, circulating currents called *eddy currents* are set up in bulk pieces of metal moving through a magnetic field. This can easily be demonstrated by allowing a flat metal plate at the end of a rigid bar to swing as a pendulum through a magnetic field (Fig. 28.20). The metal should be a nonmagnetic material, such as aluminum or copper. As the plate enters the field, the changing flux creates an induced emf in the plate, which in turn causes the free electrons in the metal to move, producing the swirling eddy currents. According to Lenz's law, the direction of the eddy currents must oppose the change that causes them. For this reason, the eddy currents must produce effective magnetic poles on the plate, which are repelled by the poles of the magnet, thus giving rise to a repulsive force that opposes the motion of the pendulum. (If the opposite were true, the pendulum would accelerate and its energy would increase after each swing, in violation of the law of energy conservation.) Alternatively, the retarding force can be "felt" by pulling a metal sheet through the field of a strong magnet.

As indicated in Fig. 28.21, with $\boldsymbol{B}$ into the paper the eddy current is counterclockwise as the swinging plate *enters* the field in position 1. This is because the external flux into the paper is increasing, and hence by Lenz's law the induced current must provide a flux out of the paper. The opposite is true as the plate leaves

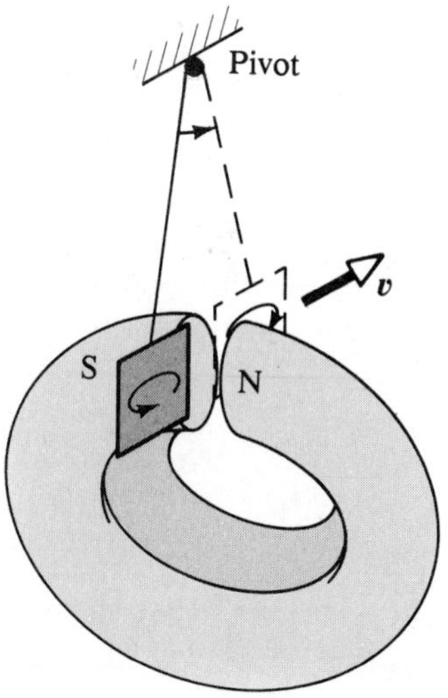

Figure 28.20 An apparatus that demonstrates the formation of eddy currents in a conductor moving through a magnetic field. As the plate enters or leaves the field, the changing flux sets up an induced emf, which causes the eddy currents.

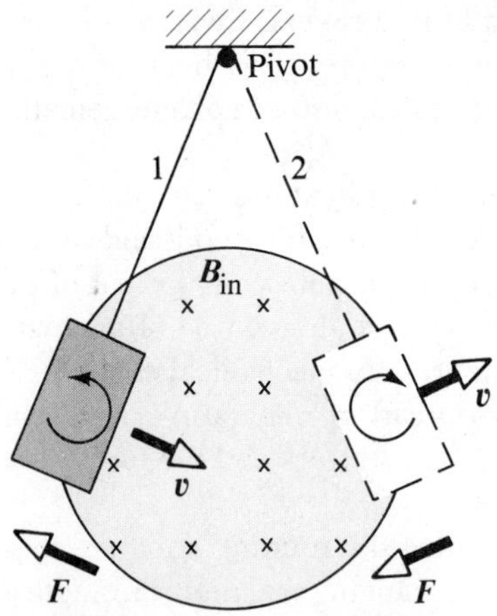

Figure 28.21 As the conducting plate enters the field in position 1, the eddy currents are counterclockwise. However, in position 2, the currents are clockwise. In either case, the plate is repelled by the magnet and eventually comes to rest.

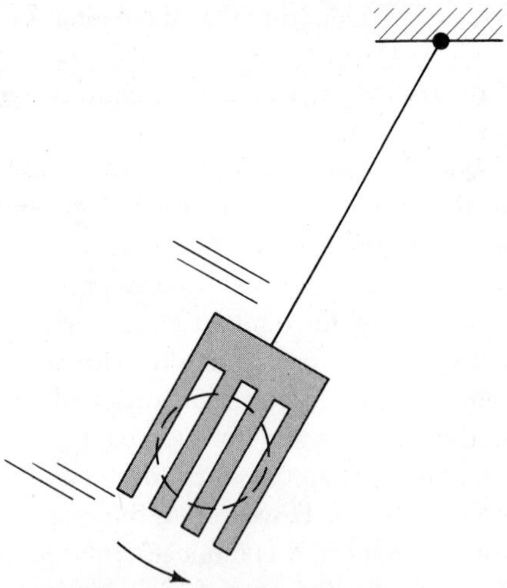

Figure 28.22 When slots are cut in the conducting plate, the eddy currents are reduced and the plate swings more freely through the magnetic field.

the field in position 2, where the current is clockwise. Since the induced eddy current always produces a retarding force $\boldsymbol{F}$ when the plate enters or leaves the field, the swinging plate eventually comes to rest.

If slots are cut in the metal plate as in Fig. 28.22, the eddy currents and the corresponding retarding force are greatly reduced. This can be understood by noting that the cuts in the plate are open circuits for any large current loops that might otherwise be formed.

The braking systems on many subway and rapid transit cars make use of electromagnetic induction and eddy currents. An electromagnet, which can be energized with a current, is positioned near the steel rails. The braking action occurs when a large current is passed through the electromagnet. The relative motion of the magnet and rails induces eddy currents in the rails, and the direction of these currents produces a drag force on the moving vehicle. The loss in mechanical energy of the vehicle is transformed into joule heat. Since the eddy currents decrease steadily in magnitude as the vehicle slows down, the braking effect is quite smooth. Eddy current brakes are also used in some mechanical balances and in various machines.

Eddy currents are often undesirable since they dissipate energy in the form of joule heat. To reduce this energy loss, the moving conducting parts are often laminated, that is, built up in thin layers separated by a nonconducting material such as lacquer or a metal oxide. This layered structure increases the resistance of the possible paths of the eddy currents and effectively confines the currents to individual layers. Such a laminated structure is used in the cores of transformers and motors to minimize eddy currents and thereby increase the efficiency of these devices.

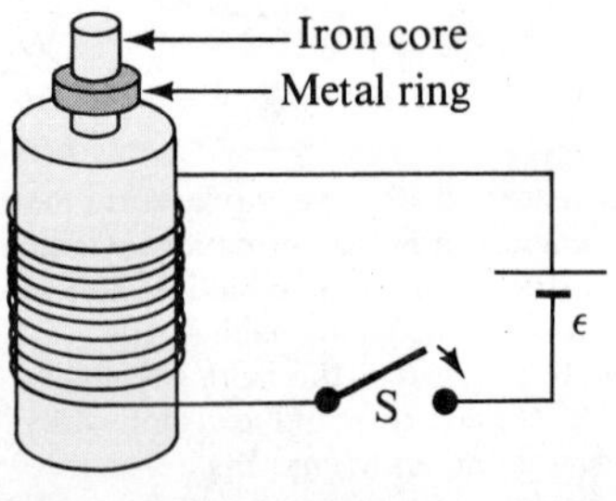

Figure 28.23 (Questions 9 and 10).

Q9. When the switch in the circuit shown in Fig. 28.23 is closed, a current is set up in the coil and the metal ring on the top springs upward. Explain this behavior. (The iron core increases the strength of the coil's magnetic field.)

Q10. If the battery in Fig. 28.23 is replaced by an alternating voltage source, explain why the ring levitates above the coil. If the ring is forced downward, it quickly warms up. Why?

28.7 MAXWELL'S WONDERFUL EQUATIONS

We conclude this chapter by presenting four equations which can be regarded as the basis of all electrical and magnetic phenomena. These equations, known as Maxwell's equations, after James Clerk Maxwell, are as fundamental to electromagnetic phenomena as Newton's laws are to the study of mechanical phenomena. In fact, the theory developed by Maxwell was more far-reaching than even he imagined at the time since it turned out to be in agreement with the special theory of relativity, as Einstein showed in 1905. As we shall see, Maxwell's equations represent laws of electricity and magnetism that have already been discussed. However, the equations have additional important consequences. In Chapter 35 we shall show that these equations predict the existence of electromagnetic waves (traveling patterns of electric and magnetic fields), which travel with a speed $c = 1/\sqrt{\mu_0 \varepsilon_0} \approx 3 \times 10^8$ m/s, the speed of light. Furthermore, the theory shows that such waves are radiated by accelerating charges.

For simplicity, we present Maxwell's equations as applied to *free space,* that is, in the absence of any dielectric or magnetic material. The four equations are:

$$\oint \mathbf{E} \cdot d\mathbf{A} = \frac{Q}{\varepsilon_0} \qquad (28.12)$$

Gauss' law

$$\oint \mathbf{B} \cdot d\mathbf{A} = 0 \qquad (28.13)$$

Gauss' law in magnetism

$$\oint \mathbf{E} \cdot d\mathbf{s} = -\frac{d\Phi_m}{dt} \qquad (28.14)$$

Faraday's law

$$\oint \mathbf{B} \cdot d\mathbf{s} = \mu_0 I + \varepsilon_0 \mu_0 \frac{d\Phi_e}{dt} \qquad (28.15)$$

Ampère-Maxwell law

Let us discuss these equations one at a time. Equation 28.12 is *Gauss' law,* which states that the *total electric flux through any closed surface equals the net charge inside that surface divided by* ε_0. This law relates the electric field to the charge distribution, where electric field lines originate on positive charges and terminate on negative charges. Recall that Gauss' law is equivalent to Coulomb's inverse-square electrostatic force law, which has been confirmed experimentally.

Equation 28.13, which can be considered *Gauss' law in magnetism,* says that *the net magnetic flux through a closed surface is zero.* That is, the number of magnetic field lines that enter a closed volume must equal the number that leave that volume. This implies that magnetic field lines cannot begin or end at any point. If they did, this would mean that isolated magnetic monopoles existed at those points. The fact that isolated magnetic monopoles have not been observed in nature can be taken as a confirmation of Eq. 28.13.

Equation 28.14 is *Faraday's law of induction,* which describes the relationship between an electric field and a changing magnetic flux. This law states that *the line integral of the electric field around any closed path (which equals the emf) equals the rate of change of magnetic flux through any surface area bounded by that path.* One consequence of Faraday's law is the current induced in a conducting loop placed in a time-varying magnetic field.

Equation 28.15 is the generalized form of Ampère's law, which describes a relationship between magnetic and electric fields and electric currents. That is, *the line integral of the magnetic field around any closed path is determined by the sum of the net conduction current through that path and the rate of change of electric flux through any surface bounded by that path.*

Another important consequence of Maxwell's equations is the fact that they can be used to derive the law of charge conservation (Chapter 20).

Once the electric and magnetic fields are known at some point in space, the force on a particle of charge q can be calculated from the expression

$$F = qE + qv \times B \tag{28.16}$$

This is called the *Lorentz force*. Maxwell's equations, together with this force law, give a complete description of all electromagnetic interactions.

It is interesting to note the symmetry of Maxwell's equations. Equations 28.12 and 28.13 are symmetric, apart from the absence of a magnetic monopole term in Eq. 28.13. Furthermore, Eqs. 28.14 and 28.15 are symmetric in that the line integrals of $\boldsymbol{E}$ and $\boldsymbol{B}$ around a closed path are related to the rate of change of magnetic flux and electric flux, respectively. "Maxwell's wonderful equations," as they were called by John R. Pierce[4] are of fundamental importance not only to electronics but to all of science. Heinrich Hertz once wrote, "One cannot escape the feeling that these mathematical formulas have an independent existence and an intelligence of their own, that they are wiser than we are, wiser even than their discoverers, that we get more out of them than we put into them."

28.8 SUMMARY

Faraday's law of induction states that the emf induced in a circuit is directly proportional to the time rate of change of magnetic flux through the circuit. That is,

Faraday's Law

$$\mathcal{E} = -\frac{d\Phi_m}{dt} \tag{28.1}$$

where Φ_m is the magnetic flux, given by

$$\Phi_m = \int \boldsymbol{B} \cdot d\boldsymbol{A}$$

When a conducting bar of length l moves through a magnetic field $\boldsymbol{B}$ with a speed v such that $\boldsymbol{B}$ is perpendicular to the bar, the emf induced in the bar (the so-called *motional emf*) is given by

Motional emf

$$\mathcal{E} = -Blv \tag{28.5}$$

Lenz's law states that the induced current and induced emf in a conductor are in such a direction as to oppose the change that produced them.

A general form of Faraday's law of induction is

Faraday's law in general form

$$\mathcal{E} = \oint \boldsymbol{E} \cdot d\boldsymbol{s} = -\frac{d\Phi_m}{dt} \tag{28.9}$$

where $\boldsymbol{E}$ is a nonconservative, time-varying electric field that is produced by the changing magnetic flux.

When used with the Lorentz force law, $\boldsymbol{F} = q\boldsymbol{E} + q\boldsymbol{v} \times \boldsymbol{B}$, *Maxwell's equations*, given below in integral form, describe *all* electromagnetic phenomena:

Gauss' law (electricity)

$$\oint \boldsymbol{E} \cdot d\boldsymbol{A} = \frac{Q}{\varepsilon_0} \tag{28.12}$$

Gauss' law (magnetism)

$$\oint \boldsymbol{B} \cdot d\boldsymbol{A} = 0 \tag{28.13}$$

Faraday's law

$$\oint \boldsymbol{E} \cdot d\boldsymbol{s} = -\frac{d\Phi_m}{dt} \tag{28.14}$$

Ampère-Maxwell law

$$\oint \boldsymbol{B} \cdot d\boldsymbol{s} = \mu_0 I + \mu_0 \varepsilon_0 \frac{d\Phi_e}{dt} \tag{28.15}$$

[4] John R. Pierce, *Electrons and Waves*, New York, Doubleday Science Study Series, 1964. Chapter 6 is recommended as supplemental reading.

The last two equations are of particular importance for the material discussed in this chapter. Faraday's law describes how an electric field can be induced by a changing magnetic flux. Similarly, the Ampère-Maxwell law describes how a magnetic field can be produced by both a changing electric flux and a conduction current.

EXERCISES

Section 28.1 Faraday's Law of Induction

1. A planar loop of wire consisting of a single turn of cross-sectional area 100 cm^2 is perpendicular to a magnetic field that increases uniformly in magnitude from 0.5 T to 2.5 T in a time of 1.5 s. What is the resulting induced current if the coil has a total resistance of 4 Ω?
2. A coil, formed by wrapping 50 turns of wire in the shape of a square, is positioned in a magnetic field so that the normal to the plane of the coil makes an angle of 30° with the direction of the field. It is observed that if the magnitude of the magnetic field is increased uniformly from 200 μT to 600 μT in 0.4 s, an emf of 80 mV is induced in the coil. What is the total length of the wire?
3. A 20-turn circular coil of radius 5 cm and resistance 0.5 Ω is placed in a magnetic field directed perpendicular to the plane of the coil. The magnitude of the magnetic field varies in time according to the expression $B = 0.02t + 0.05t^2$, where t is in s and B is in T. Calculate the induced emf in the coil at $t = 6$ s.
4. A 50-turn rectangular coil of dimensions 10 cm × 20 cm is "dropped" from a position where $B = 0$ to a new position where $B = 0.5$ T and is directed perpendicular to the plane of the coil. Calculate the resulting average emf induced in the coil if the displacement occurs in 0.2 s.
5. A rectangular loop of area A is placed in a region where the magnetic field is perpendicular to the plane of the loop. The magnitude of the field is allowed to vary in time according to $B = B_0 e^{-t/\tau}$, where B_0 and τ are constants. (a) Use Faraday's law to show that the emf induced in the loop is given by $\mathcal{E} = \frac{AB_0}{\tau} e^{-t/\tau}$. (b) Obtain a numerical value for $\mathcal{E}$ at $t = 6$ s when $A = 0.1$ m^2, $B_0 = 0.3$ T, and $\tau = 3$ s. (c) For the values of A, B_0, and τ given in (b), what is the *maximum* value of $\mathcal{E}$?
6. A tightly wound circular coil has 50 turns, each of radius 0.2 m. A uniform magnetic field is turned on along a direction perpendicular to the plane of the coil. If the field increases linearly from 0 to 0.3 T in a time of 0.3 s, what emf is induced in the windings of the coil?
7. The plane of a rectangular coil of dimensions 10 cm by 8 cm is perpendicular to the direction of a magnetic field **B**. If the coil has 50 turns and a total resistance of 12 Ω, at what rate must the magnitude of **B** change in order to induce a current of 5 mA in the windings of the coil?
8. A square, single-turn coil 0.25 m on a side is placed with its plane perpendicular to a constant magnetic field. An emf of 15 mV is induced in the winding when the area of the coil decreases at a rate of 0.1 m^2/s. What is the magnitude of the magnetic field?

Section 28.2 Motional emf and Section 28.3 Lenz's Law

9. A metal blade spins at a constant rate in the magnetic field of the earth as in Fig. 28.7. The rotation occurs in a region where the component of the earth's magnetic field perpendicular to the plane of rotation is 2.2×10^{-5} T. If the blade is 1.2 m in length and its angular velocity is 15π rad/s, what potential difference is developed between its ends?
10. A 200-turn circular coil of radius 10 cm is located in a uniform magnetic field of 0.8 T such that the plane of the coil is perpendicular to the direction of the field. The coil is rotated at a constant rate (uniform angular velocity) through 90° in a time of 1.5 s, so that the plane of the coil is finally parallel to the direction of the field. (a) Calculate the *average* emf induced in the coil as a result of the rotation. (b) What is the instantaneous value of the emf in the coil at the moment the plane of the coil makes an angle of 45° with the magnetic field?
11. A measured average emf of 20 μV is induced in a small circular coil of 500 turns and 6 cm diameter under the following condition: It is rotated in a uniform magnetic field in 0.05 s from a position where the plane of the coil is parallel to the field to a position where the plane of the coil is at an angle of 45° to the field. What is the value of B within the region where the measurement is made?
12. *Use Lenz's law to answer the following questions concerning the direction of induced currents.* (a) What is the direction of the induced current in resistor R in Fig. 28.24a when the bar magnet is moved to the left? (b) What is the direction of the current induced in the resistor R when the switch S in the circuit of Fig. 28.24b is closed? (c) What is the

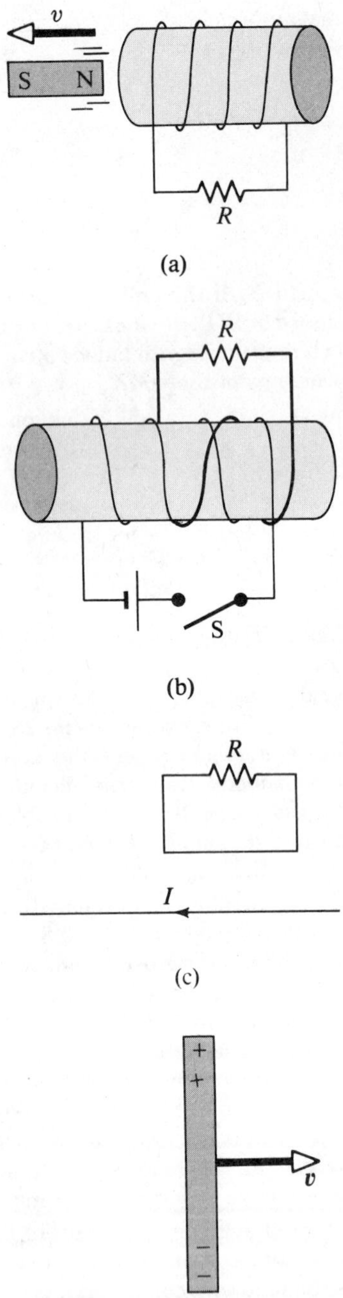

Figure 28.24 (Exercise 12).

direction of the induced current in R when the current I in Fig. 28.24c decreases rapidly to zero? (d) A copper bar is moved to the right while its axis is maintained perpendicular to a magnetic field, as in Fig. 28.24d. If the top of the bar becomes positive relative to the bottom, what is the direction of the magnetic field?

13. Over a region where the *vertical* component of the earth's magnetic field is 40 μT, a 10-m length of wire is held along an east-west direction and moved horizontally to the north with a speed of 30 m/s. Calculate the potential difference between the ends of the wire and determine which end is positive.

14. In the arrangement shown in Fig. 28.25, a conducting bar moves along parallel, frictionless conducting rails connected on one end by a 6-Ω resistor. A 2.5-T magnetic field is directed *into the plane* perpendicular to the movable bar. Let $l = 1.2$ m and neglect the mass of the bar. (a) Calculate the applied force required to move the bar to the right at a *constant* speed of 2 m/s. (b) At what rate is energy dissipated in the resistor?

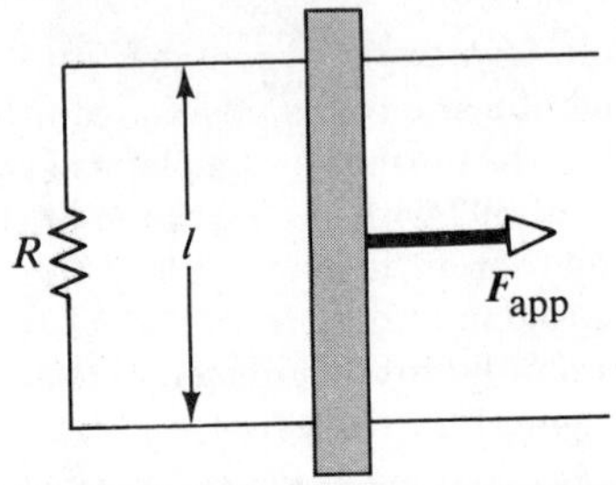

Figure 28.25 (Exercises 14 and 16).

15. A 0.5-kg wire in the shape of a closed rectangle 1 m wide and 2 m long has a total resistance of 5 Ω. The rectangle is allowed to fall through a magnetic field directed perpendicular to the direction of motion of the wire (Fig. 28.26). The rectangle accelerates downward until it acquires a *constant* speed of 10 m/s with the top of the rectangle not yet in that region of the field. Calculate the value of B.

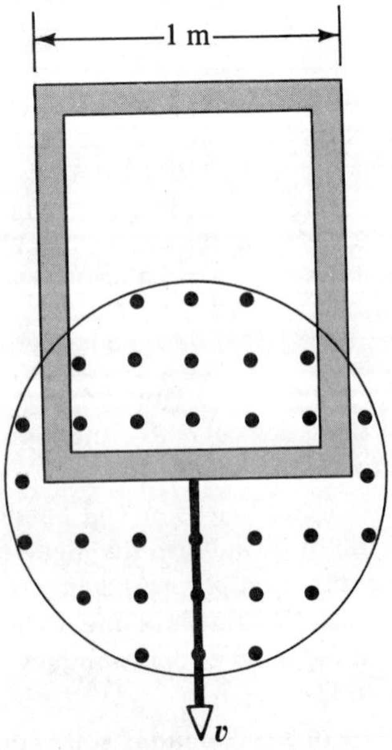

Figure 28.26 (Exercise 15).

16. Consider the arrangement shown in Fig. 28.25. At what speed should the bar be moved to the right in order to produce a current of 0.3 A in the resistor?

17. A small airplane with a wing span of 10 m is flying due north at a speed of 75 m/s over a region where the vertical component of the earth's magnetic field is 0.12 μT. (a) What potential difference is developed between the wing tips? (b) How would the answer to (a) change if the plane were flying due east?

Section 28.4 Induced emfs and Electric Fields

18. A single-turn, circular loop of radius R is coaxial with a long solenoid of radius r and length l and having N turns (Fig. 28.27). The variable resistor is changed so that the solenoid current decreases linearly from 6 A to 1.5 A in 0.2 s. If $R = 0.2$ m, $r = 0.05$ m, $l = 0.8$ m, and $N = 1600$ turns, calculate the induced emf in the circular loop.

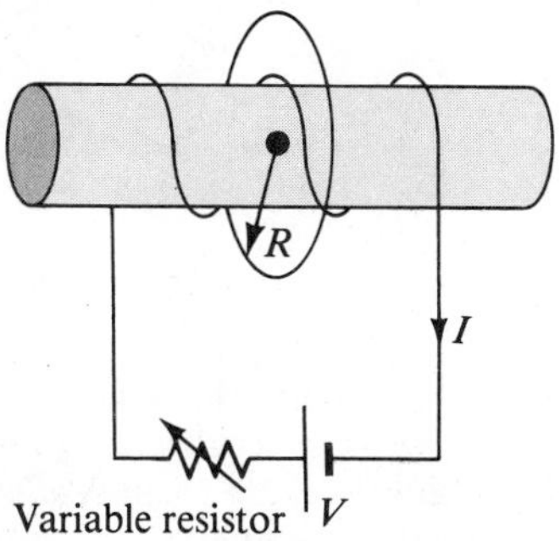

Figure 28.27 (Exercise 18).

19. A magnetic field directed into the page changes with time according to $B = (0.05t^2 + 0.4)$ T, where t is in s. The field has a circular cross-section of radius R = 0.05 m (Fig. 28.28). What are the magnitude and direction of the electric field at point P_1 when $t = 4$ s and $r_1 = 0.04$ m?

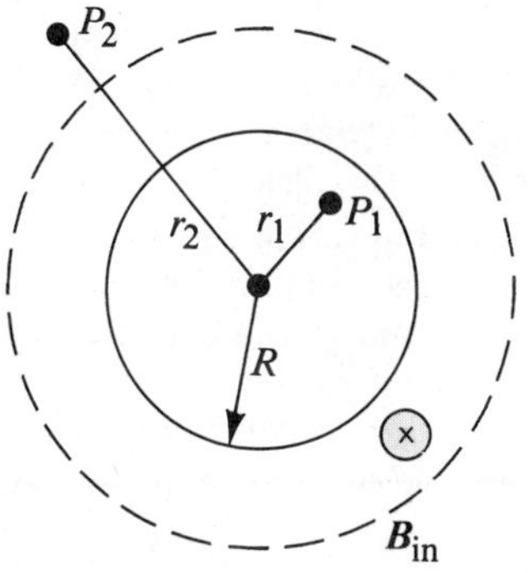

Figure 28.28 (Exercises 19 and 20).

20. For the situation described in Fig. 28.28, $B = (t^3 - 2t^2 + 0.007)$ T, and $r_2 = 0.07$ m. (a) Calculate the magnitude and direction of the force exerted on an electron located at point P_2 when $t = 2$ s. (b) At what time is the magnetic force equal to zero?

21. A solenoid has a radius of 2 cm and has 500 turns/m. The current varies with time according to the expression $I = 2e^{0.5t}$, where I is in A and t is in s. Calculate the electric field at a distance of 5 cm from the axis of the solenoid at $t = 4$s.

Section 28.5 Generators and Motors

22. The coil of a simple ac generator develops a sinusoidal emf of maximum value 72.4 V and frequency 60 Hz. If the coil has dimensions of 10 cm by 20 cm and rotates in a magnetic field of 1.2 T, how many turns are in the winding?

23. A semicircular conductor of radius $R = 0.3$ m is rotated about the axis AC at a constant rate of 120 revolutions per minute (Fig. 28.29). A uniform magnetic field is directed *out of* the plane of rotation and has a magnitude of 1.5 T. (a) Calculate the *maximum* value of the emf induced in the conductor. (b) What is the value of the *average* induced emf for each complete rotation? (c) How would the answers to (a) and (b) change if the uniform field **B** were allowed to extend a distance R above the axis of rotation? (d) Sketch the emf versus time in each case.

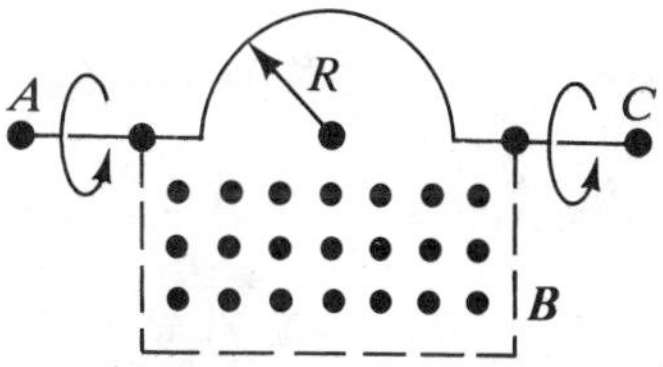

Figure 28.29 (Exercise 23).

24. Consider a 100-turn rectangular coil of cross-sectional area 0.06 m^2 rotating with an angular velocity of 20 rad/s about an axis perpendicular to a magnetic field of 2.5 T. (a) Plot the induced voltage as a function of time over one complete period of rotation. (b) On the same set of axes, plot the voltage induced in the coil for one rotation if the angular velocity is 40 rad/s.

25. A 500-turn circular coil of radius 20 cm is rotating about an axis perpendicular to a magnetic field of 0.01 T. What angular velocity will produce a maximum induced emf of 2/V?

PROBLEMS

1. A long straight wire is parallel to one edge and is in the plane of a single turn rectangular loop as in Fig. 28.30. (a) If the current in the long wire varies in time as $I = I_0 e^{-t/\tau}$, show that the induced emf in the loop is given by

$$\mathcal{E} = \frac{\mu_0 b}{2\pi}\frac{I}{\tau}\ln\left(1 + \frac{a}{d}\right)$$

(b) Calculate the value for the induced emf at $t = 2$ s taking $I_0 = 10$ A, $d = 4$ cm, $a = 6$ cm, $b = 18$ cm and $\tau = 5$ s.

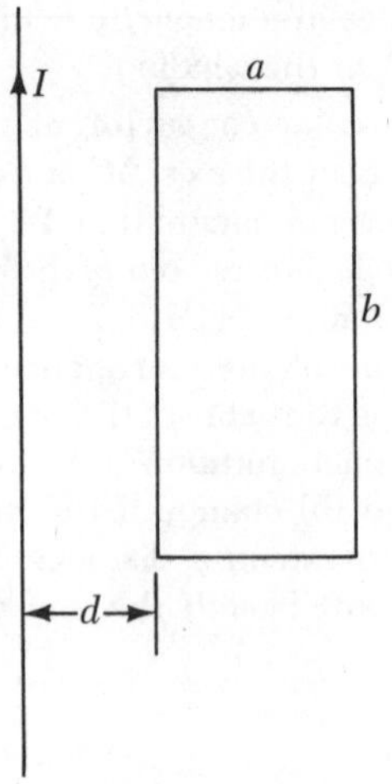

Figure 28.30 (Problem 1).

2. A square loop of wire with edge length $a = 0.5$ m is perpendicular to the earth's magnetic field at a point where $B = 5$ μT, as in Fig. 28.31. The total resistance of the loop and the wires connecting the loop to the galvanometer is 0.3 Ω. If the loop is suddenly collapsed by horizontal forces as shown, what total charge will pass through the galvanometer?

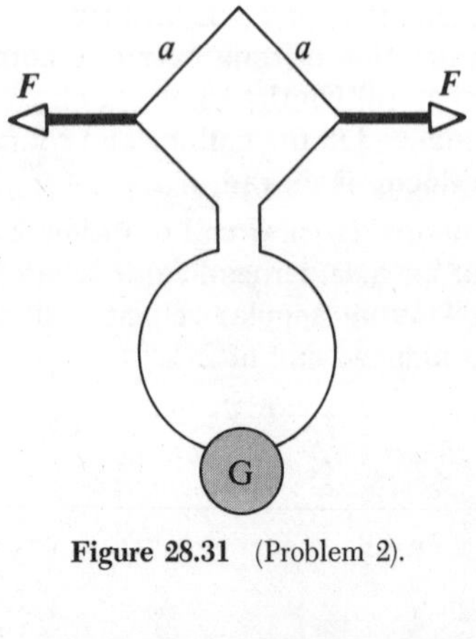

Figure 28.31 (Problem 2).

3. A thin metal strip is allowed to slide down parallel frictionless rails of negligible resistance connected at the bottom end and elevated at an angle θ above the horizontal. A uniform magnetic field $\boldsymbol{B}$ is directed vertically upward throughout the region as shown in Fig. 28.32. The strip has a mass $m = 40$ g, resistance $R = 30$ Ω, and length between the rails $l = 0.2$ m. (a) Derive a general expression for the terminal speed of the strip. (b) Calculate the terminal speed achieved by the strip sliding along the incline if $\theta = 30°$ and B = 2.60 T.

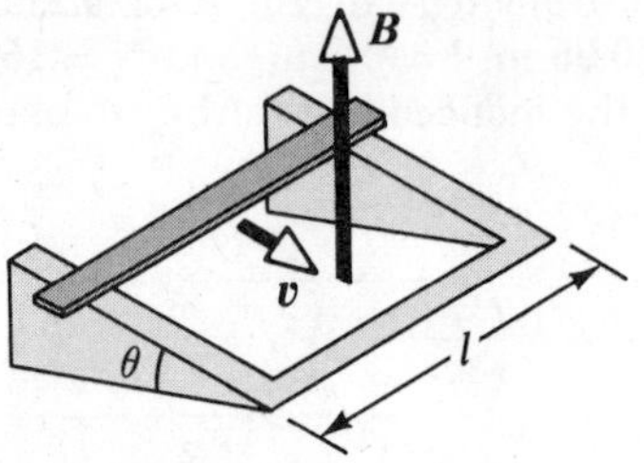

Figure 28.32 (Problem 3).

4. A solenoid wound with 2000 turns/m is supplied with current that varies in time according to $I = 4\sin(120\pi t)$, where I is in A and t is in s. A small coaxial circular coil of 40 turns and radius $r = 5$ cm is located inside the solenoid near its center. (a) Derive an expression that describes the manner in which the emf in the small coil varies in time. (b) At what average rate is energy dissipated in the small coil if the windings have a total resistance of 8 Ω?

5. A conducting disk of radius 0.25 m rotates about an axle through its center with a constant angular speed of $\omega = 10$ rad/s. A uniform magnetic field of 1.2 T acts perpendicular to the plane of the disk (that is, parallel to the axis of rotation). Calculate the potential difference developed between the rim and axle of the disk.

6. A conducting rod of length l moves with velocity $\boldsymbol{v}$ along a direction parallel to a long wire carrying a steady current I. The axis of the rod is maintained perpendicular to the wire with the near end a distance r away, as shown in Fig. 28.33. Show that the emf induced in the rod is given by

$$|\mathcal{E}| = \frac{\mu_0 I}{2\pi} v \ln\left(1 + \frac{l}{r}\right)$$

7. A rectangular coil of N turns, dimensions l and w, and total resistance R rotates with angular velocity ω about the y axis in a region where a uniform magnetic field $\boldsymbol{B}$ is directed along the x axis

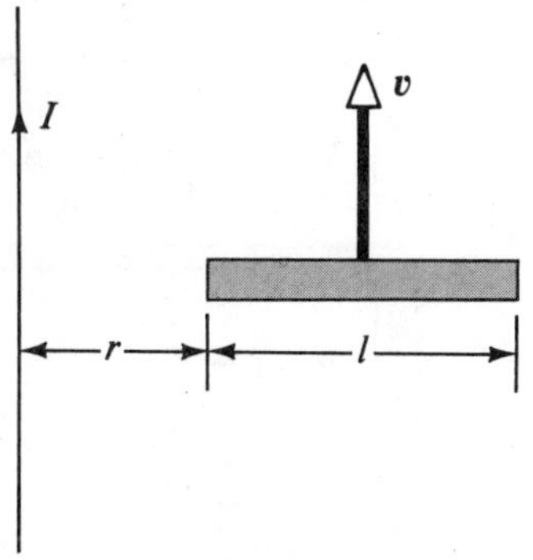

Figure 28.33 (Problem 6).

(Fig. 28.13a). The rotation is initiated so that the plane of the coil is perpendicular to the direction of $\boldsymbol{B}$ at $t = 0$. Let $\omega = 20$ rad/s, $N = 50$ turns, $l = 0.15$ m, $w = 0.25$ m, $R = 12\ \Omega$, and $B = 1.4$ T. Calculate (a) the maximum induced emf in the coil, (b) the maximum rate of change of magnetic flux through the coil, (c) the value of the induced emf at $t = 0.05$ s, and (d) the torque exerted on the loop by the magnetic field at the instant when the emf is a maximum.

8. Consider again the situation described in Example 28.4. Let the length of the bar between the sliding rails be 30 cm, the magnetic field $B = 0.8$ T, the stationary resistor $R = 6\ \Omega$, and the initial velocity $v_0 = 50$ cm/s. If the velocity of the moving bar decreases to $0.4v_0$ in 5 s, what is the value of m_1, the mass of the bar?

9. Figure 28.34 illustrates an arrangement similar to that discussed in Example 28.4, except in this case the bar is pulled horizontally across the set of parallel rails by a string (assumed massless) that passes over an ideal pulley and is attached to a freely suspended mass M. The uniform magnetic field has a magnitude B, the sliding bar has mass m, and the distance between the rails is l. The rails are connected at one end by a load resistor R. Derive an expression that gives the value of the horizontal speed of the bar as a function of time, assuming that the suspended mass was released with the bar at rest at $t = 0$.

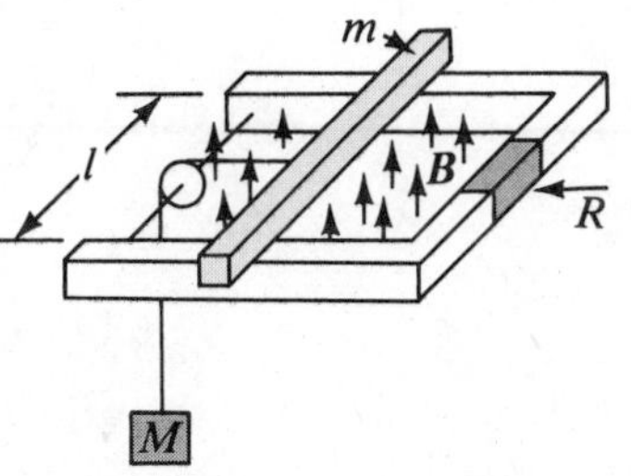

Figure 28.34 (Problem 9).

10. Magnetic field values are often determined by using a device known as a *search coil*. This technique depends on the measurement of the total charge passing through a coil in a time interval during which the magnetic flux linking the windings changes either because of the motion of the coil or because of a change in the value of B. (a) Show that if the flux through the coil changes from Φ_1 to Φ_2, the charge transferred through the coil between t_1 and t_2 will be $Q = N(\Phi_2 - \Phi_1)/R$ where R is the resistance of the coil and associated circuitry (galvanometer). (b) As a specific example, calculate B when a 100-turn coil of resistance 200 Ω and cross-sectional area 40 cm^2 produces the following results: A total charge of 5×10^{-4} C passes through the coil when it is rotated in a uniform field from a position where the plane of the coil is perpendicular to the field into a position where the coil's plane is parallel to the field.

11. A rectangular loop of dimensions l and w moves with a constant velocity $\boldsymbol{v}$ away from a long wire that carries a current I in the plane of the loop (Fig. 28.35). The total resistance of the loop is R. Derive an expression that gives the current in the loop at the instant the near side is a distance r from the wire.

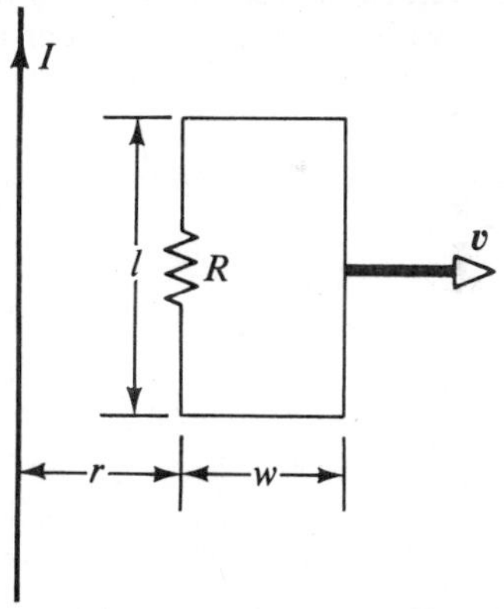

Figure 28.35 (Problem 11).

12. A conducting rod moves with a constant velocity $\boldsymbol{v}$ perpendicular to a long, straight wire carrying a current I as in Fig. 28.36. Show that the emf generated between the ends of the rod is given by

$$|\mathcal{E}| = \frac{\mu_0 v I}{2\pi r} l$$

In this case, note that the emf decreases with increasing r, as you might expect.

13. A horizontal wire is free to slide on the vertical rails of a conducting frame as in Fig. 28.37. The wire has a mass m and length l, and the resistance of the

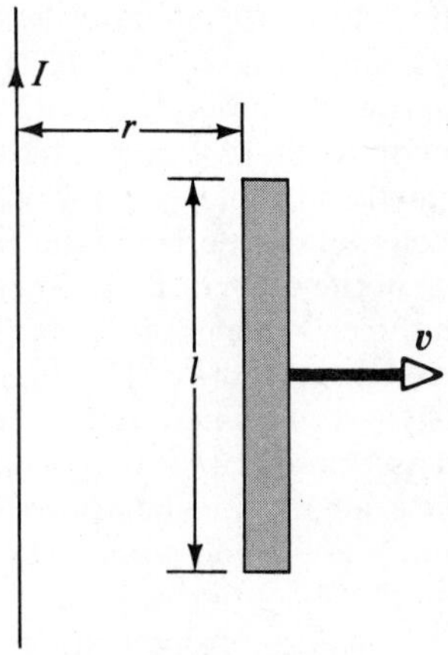

Figure 28.36 (Problem 12).

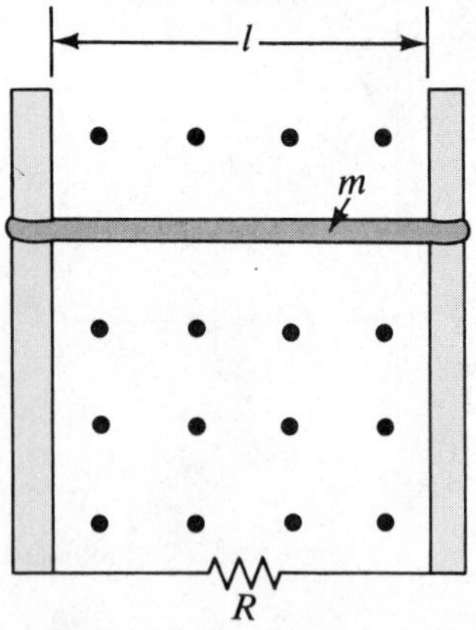

Figure 28.37 (Problem 13).

circuit is R. If a uniform magnetic field is directed perpendicular to the frame, what is the *terminal* velocity of the wire as it falls under the force of gravity? (Neglect mechanical friction.)

14. Shown in Fig. 28.38 is a graph of the induced emf versus time for a coil of N turns rotating with angular velocity ω in a uniform magnetic field directed perpendicular to the axis of rotation of the coil. Copy this sketch (to a larger scale) and on the same set of axes show the graph of $\mathcal{E}(t)$ versus t when (a) the number of turns in the coil is doubled, (b) the angular velocity is doubled, and (c) the angular velocity is doubled while the number of turns in the coil is halved.

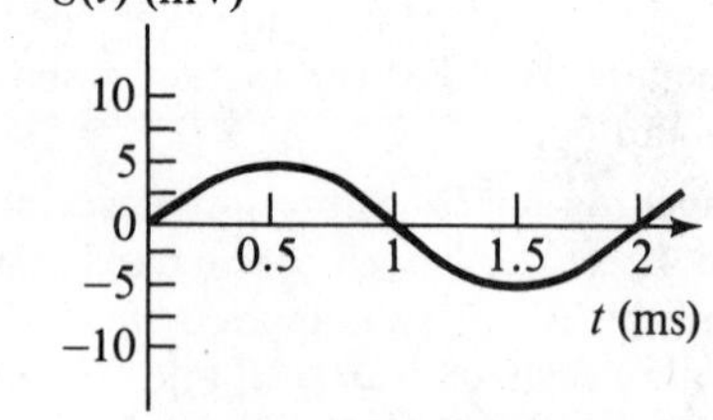

Figure 28.38 (Problem 14).

29 Inductance

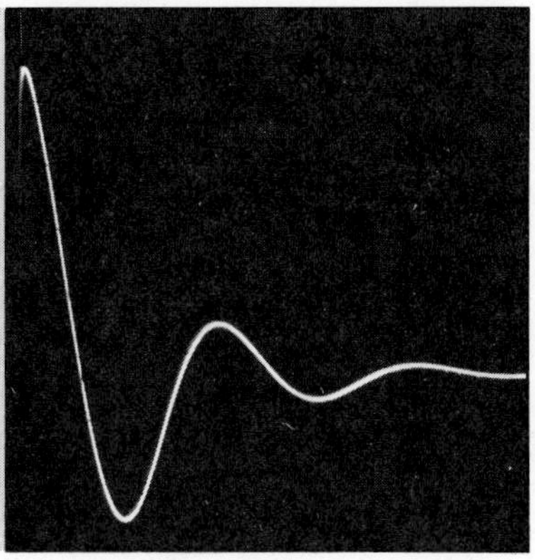

James Madison University

In the previous chapter, we saw that currents and emfs are induced in a circuit when the magnetic flux through the circuit changes with time. This phenomenon of electromagnetic induction has some practical consequences, which we shall describe in this chapter. First, we shall describe an effect known as *self-induction*, in which a time-varying current in a conductor induces an emf in the conductor that opposes the external emf that set up the current. This phenomenon is the basis of the element known as the *inductor*, which plays an important role in circuits with time-varying currents. We shall discuss the concepts of the energy stored in the magnetic field of an inductor and the energy density associated with a magnetic field.

Next, we shall study how an emf can be induced in a circuit as a result of a changing flux produced by an external circuit, which is the basic principle of *mutual induction*. Finally, we shall examine the characteristics of circuits containing inductors, resistors, and capacitors in various combinations. For example, we shall find that in a circuit containing only an inductor and a capacitor, the charge and current both oscillate in a simple harmonic fashion. These oscillations correspond to a continuous transfer of energy between the electric field of the charged capacitor and the magnetic field of the current-carrying inductor.

29.1 SELF-INDUCTANCE

Consider an isolated circuit consisting of a switch, resistor, and source of emf, as in Fig. 29.1. When the switch is closed, the current doesn't immediately jump from zero to its maximum value, $\mathcal{E}/R$. The law of electromagnetic induction (Faraday's law) prevents this from occurring. What really happens is the following. As the current increases with time, the magnetic flux through the loop due to this current also increases with time. This increasing flux induces an emf in the circuit which opposes the change in magnetic flux. By Lenz's law, the induced electric field in the wires must therefore be opposite the direction of flow of the conventional current, and this opposing emf results in a *gradual* increase in the current. For the same reason, when the switch is opened, the current gradually decreases to zero. This effect is called *self-induction* since the changing flux through the circuit arises from the circuit itself. The emf that is set up in this case is called a *self-induced emf*. Later, in Section 29.4, we shall describe a related effect called *mutual induction* in

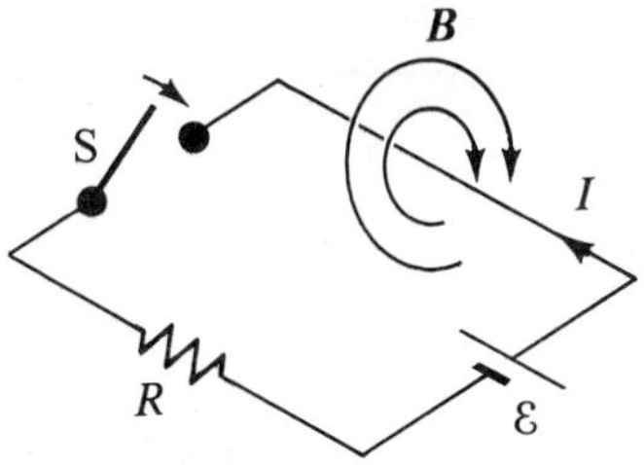

Figure 29.1 After the switch in the circuit is closed, the current produces a magnetic flux through the loop. As the current increases toward its equilibrium value, the flux changes in time and induces an emf in the loop.

which an emf is induced in one circuit as a result of a changing magnetic flux set up by another circuit.

To obtain a quantitative description of self-induction, first note that from Faraday's law the induced emf is given by the negative time rate of change of the magnetic flux. Since the magnetic flux is proportional to the magnetic field, which in turn is proportional to the current in the circuit, *the self-induced emf is always proportional to the time rate of change of the current.* For a closely spaced coil of N turns (a toroidal coil or the ideal solenoid), we find that

Induced emf

$$\mathcal{E} = -N\frac{d\Phi_m}{dt} = -L\frac{dI}{dt} \tag{29.1}$$

where L is a proportionality constant, called the *inductance* of the device, that depends on the geometric features of the circuit. From this expression, we see that the inductance of a coil containing N turns is given by

Inductance of an N-turn coil

$$L = \frac{N\Phi_m}{I} \tag{29.2}$$

Later we shall use this equation to calculate the inductance of some special geometries.

From Eq. 29.1, we can also write the inductance as the ratio

Inductance

$$L = -\frac{\mathcal{E}}{dI/dt} \tag{29.3}$$

This is usually taken to be the defining equation for the inductance of any coil, regardless of its shape, size, or material characteristics. Just as resistance is a measure of the opposition to current, inductance is a measure of the opposition to the *change* in current.

The SI unit of inductance is the henry (H), which, from Eq. 29.3, is seen to be equal to 1 volt-second per ampere:

$$1\text{ H} = 1\frac{\text{V}\cdot\text{s}}{\text{A}}$$

As we shall see, *the inductance of a device depends only on its geometry.* However, the calculation of a device's inductance can be quite difficult for complicated geometries. The examples below involve rather simple situations for which inductances are easily evaluated.

Example 29.1

Inductance of a solenoid: Find the inductance of a uniformly wound solenoid with N turns and length l. Assume that l is long compared with the radius and that the core of the solenoid is air.

Solution: In this case, we can take the interior field to be uniform and given by Eq. 27.20:

$$B = \mu_o nI = \mu_o\frac{N}{l}I$$

where n is the number of turns per unit length, N/l. The flux through each turn is given by

$$\Phi_m = BA = \mu_o\frac{NA}{l}I$$

where A is the cross-sectional area of the solenoid. Using this expression and Eq. 29.2 we find that

$$L = \frac{N\Phi_m}{I} = \frac{\mu_o N^2 A}{l} \tag{29.4}$$

This shows that L depends on geometric factors and is proportional to the square of the number of turns. Since $N = nl$, we can also express the result in the form

$$L = \mu_o\frac{(nl)^2}{l}A = \mu_o n^2 Al = \mu_o n^2 V \tag{29.5}$$

where $V = Al$ is the volume of the solenoid (not to be confused with the abbreviation for volts).

Example 29.2

(a) Calculate the inductance of a solenoid containing 300 turns if the length of the solenoid is 25 cm and its cross-sectional area is 4 cm^2 = 4×10^{-4} m^2.

Using Eq. 29.4 we get

$$L = \frac{\mu_0 N^2 A}{l} = (4\pi \times 10^{-7}\ \text{Wb/A}\cdot\text{m})\frac{(300)^2(4 \times 10^{-4}\ \text{m}^2)}{25 \times 10^{-2}\ \text{m}}$$

$$= 1.81 \times 10^{-4}\ \text{Wb/A} = 0.181\ \text{mH}$$

(b) Calculate the self-induced emf in the solenoid described in (a) if the current through it is *decreasing* at the rate of 50 A/s.

Using Eq. 29.1, and given that $dI/dt = -50$ A/s, we get

$$\mathcal{E} = -L\frac{dI}{dt} = -(1.81 \times 10^{-4}\ \text{H})(-50\ \text{A/s})$$

$$= 9.05\ \text{mV}$$

Q1. A circuit containing a coil, resistor, and battery is in steady state, that is, the current has reached a constant value. Does the coil have an inductance? Does the coil affect the value of the current in the circuit?

Q2. Does the inductance of a coil depend on the current in the coil? What parameters affect the inductance of a coil?

29.2 *RL* CIRCUITS

Circuits that contain coils, such as a solenoid, have a self-inductance that prevents the current from increasing or decreasing instantaneously. A circuit element that has a large inductance is called an *inductor*. The circuit symbol for an inductor is . We shall always assume that the self-inductance of the remainder of the circuit is negligible compared with that of the inductor.

Consider the circuit consisting of a resistor, inductor, and battery shown in Fig. 29.2. The internal resistance of the battery will be neglected. Suppose the switch S is closed at $t = 0$. The current will begin to increase, but the inductor will produce an emf that opposes the increasing current, sometimes referred to as a *back emf*. In other words, the inductor acts like a battery whose polarity is opposite that of the real battery in the circuit. The back emf produced by the inductor is given by

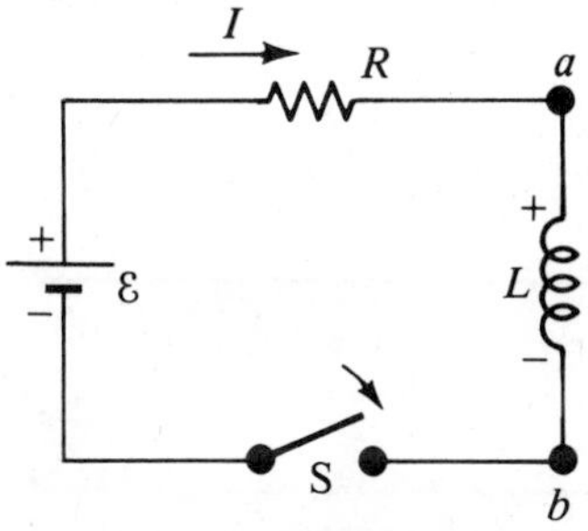

Figure 29.2 A series *RL* circuit. As the current increases toward its maximum value, the inductor produces an emf that opposes the increasing current.

$$\mathcal{E}_L = -L\frac{dI}{dt}$$

Since the current is increasing, dI/dt is positive, and so $\mathcal{E}_L$ is negative. This corresponds to the fact that there is a potential drop in going from a to b across the inductor. For this reason, point a is at a higher potential than point b, as illustrated in Fig. 29.2.

With this in mind, we can apply Kirchhoff's loop equation to this circuit:

$$\mathcal{E} - IR - L\frac{dI}{dt} = 0 \qquad (29.6)$$

where IR is the voltage drop across the resistor. We must now look for a formal solution to this differential equation, which is seen to be similar in form to that of the RC circuit (Section 25.4).

To obtain a mathematical solution of Eq. 29.6, it is convenient to change variables by letting $x = \frac{\mathcal{E}}{R} - I$, so that $dx = -dI$. With these substitutions, Eq. 29.6 can be written

$$x + \frac{L}{R}\frac{dx}{dt} = 0$$

$$\frac{dx}{x} = -\frac{R}{L}\,dt$$

Integrating this last expression gives

$$\ln\frac{x}{x_0} = -\frac{R}{L}t$$

where the integrating constant is taken to be $-\ln x_0$. Taking the antilog of this result gives

$$x = x_0 e^{-Rt/L}$$

Since at $t = 0$, $I = 0$, we note that $x_0 = \mathcal{E}/R$. Hence, the last expression is equivalent to

$$\frac{\mathcal{E}}{R} - I = \frac{\mathcal{E}}{R}e^{-Rt/L}$$

$$I = \frac{\mathcal{E}}{R}(1 - e^{-Rt/L})$$

which represents the solution to Eq. 29.6.

This mathematical solution of Eq. 29.6, which represents the current as a function of time, can also be written:

$$I(t) = \frac{\mathcal{E}}{R}(1 - e^{-Rt/L}) = \frac{\mathcal{E}}{R}(1 - e^{-t/\tau}) \tag{29.7}$$

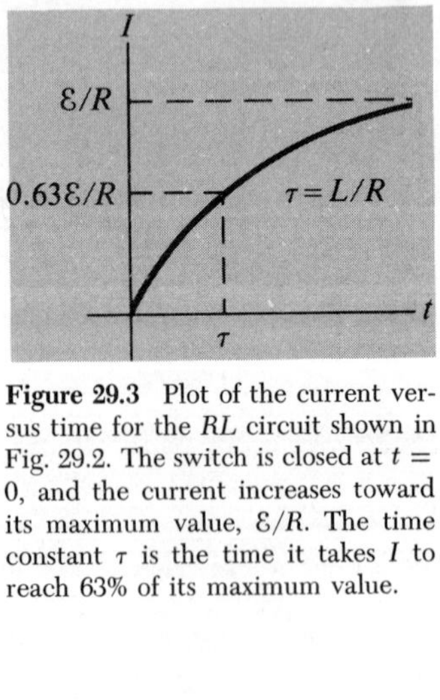

Figure 29.3 Plot of the current versus time for the *RL* circuit shown in Fig. 29.2. The switch is closed at $t = 0$, and the current increases toward its maximum value, $\mathcal{E}/R$. The time constant τ is the time it takes I to reach 63% of its maximum value.

where the constant τ is the *time constant* of the *RL* circuit:

Time constant

$$\tau = L/R \tag{29.8}$$

It is left as an exercise to show that the dimension of τ is time. Physically, τ is the time it takes the current to reach $(1 - e^{-1}) = 0.63$ of its final value, $\mathcal{E}/R$.

Figure 29.3 represents a graph of the current versus time, where $I = 0$ at $t = 0$. Note that the final equilibrium value of the current, which occurs at $t = \infty$, is given by $\mathcal{E}/R$. This can be seen by setting dI/dt equal to zero in Eq. 29.6 (at equilibrium, the change in the current is zero) and solving for the current. Thus, we see that the current rises very fast initially and then gradually approaches the equilibrium value $\mathcal{E}/R$ as $t \to \infty$.

One can show that Eq. 29.7 is a solution of Eq. 29.6 by computing the derivative dI/dt and requiring that $I = 0$ at $t = 0$. Taking the first time derivative of Eq. 29.7, we get

$$\frac{dI}{dt} = \frac{\mathcal{E}}{L}e^{-t/\tau} \tag{29.9}$$

Substitution of this result together with Eq. 29.7 will indeed verify that our solution satisfies Eq. 29.6. That is,

$$\mathcal{E} - IR - L\frac{dI}{dt} = 0$$

$$\mathcal{E} - \frac{\mathcal{E}}{R}(1 - e^{-t/\tau})R - L\left(\frac{\mathcal{E}}{L}e^{-t/\tau}\right) = 0$$

$$\cancel{\mathcal{E}} - \cancel{\mathcal{E}} + \cancel{\mathcal{E}e^{-t/\tau}} - \cancel{\mathcal{E}e^{-t/\tau}} = 0$$

and the solution is verified.

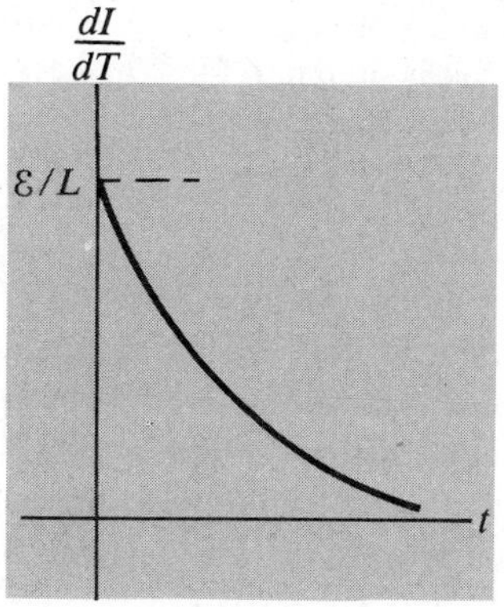

Figure 29.4 Plot of dI/dt versus time for the RL circuit shown in Fig. 29.2. The rate of change of current is a maximum at $t = 0$ when the switch is closed. The rate dI/dt decreases exponentially with time as I increases toward its maximum value.

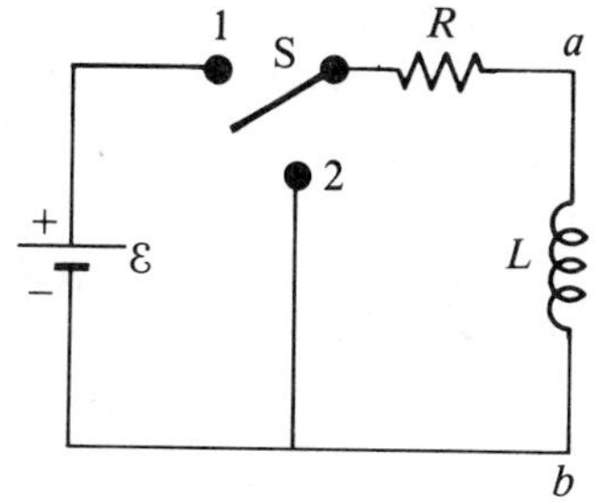

Figure 29.5 An RL circuit. When the switch S is in position 1, the battery is in the circuit. When S is in position 2, the battery is removed from the circuit.

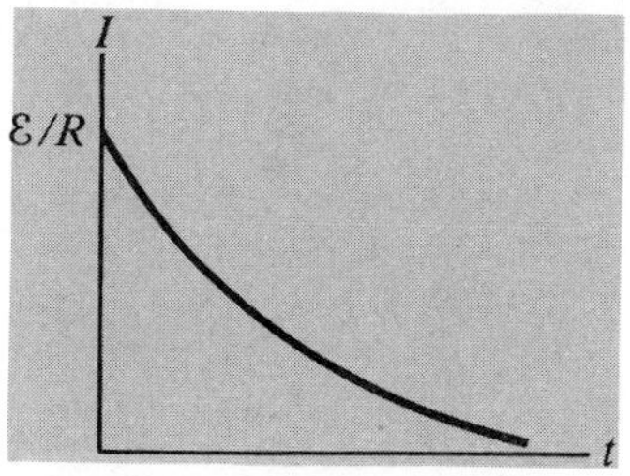

Figure 29.6 Current versus time for the circuit shown in Fig. 29.5 with the switch in position 2. The switch is thrown to position 2 at $t = 0$, at which time the current has its maximum value, $\mathcal{E}/R$, (established with S in position 1 at $t < 0$).

Note that the rate of increase of current, dI/dt, is a *maximum* at $t = 0$ and falls off exponentially to zero as $t \to \infty$ (Fig. 29.4).

Now consider the RL circuit arranged as shown in Fig. 29.5. In this case, the switch can be either in position 1, with the battery in the circuit, or in position 2, where the battery is removed. Suppose the switch is in position 1 for a long enough time to allow the current to reach its equilibrium value, $\mathcal{E}/R$. If the switch is now thrown to position 2 at $t = 0$, we have a circuit with no battery ($\mathcal{E} = 0$). Applying Kirchhoff's circuit law to the loop containing the resistor and inductor with the switch in position 2 gives the expression

$$IR + L\frac{dI}{dt} = 0 \qquad (29.10)$$

It is left as an exercise to show that the solution of this differential equation is

$$I = \frac{\mathcal{E}}{R}e^{-t/\tau} = I_0 e^{-t/\tau} \qquad (29.11)$$

where the current at $t = 0$ is given by $I_0 = \mathcal{E}/R$ and $\tau = L/R$.

The graph of the current versus time (Fig. 29.6) shows that the current is continuously decreasing with time, as one would expect. Furthermore, note that the slope, dI/dt, is always negative and has its maximum value at $t = 0$. The negative slope signifies that $\mathcal{E}_L = -L dI/dt$ is now *positive;* that is, point a is at a lower potential than point b in Fig. 29.5.

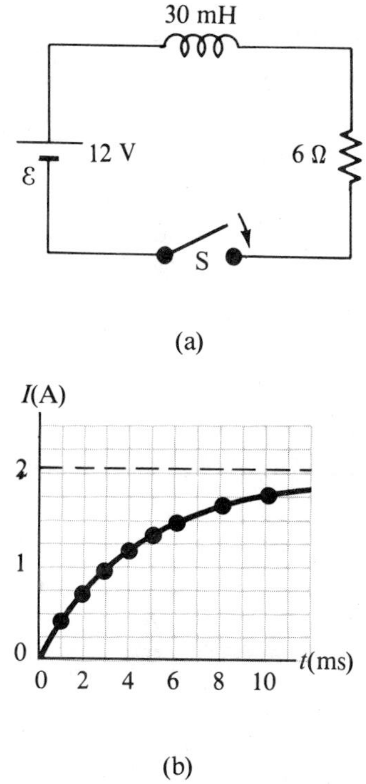

Figure 29.7 (Example 29.4) (a) The switch in this RL circuit is closed at $t = 0$. (b) A graph of the current versus time for the circuit in (a).

Example 29.3

The circuit shown in Fig. 29.7a consists of a 30-mH inductor, a 6-Ω resistor, and a 12-V battery. The switch is closed at $t = 0$. (a) Find the time constant of the circuit.

The time constant is given by Eq. 29.8

$$\tau = \frac{L}{R} = \frac{30 \times 10^{-3}\ \text{H}}{6\ \Omega} = 5\ \text{ms}$$

(b) Calculate the current in the circuit at t = 2 ms.

Using Eq. 29.7 for the current as a function of time (with t and τ in ms), we find that at $t = 2$ ms

$$I = \frac{\mathcal{E}}{R}(1 - e^{-t/\tau}) = \frac{12\ \text{V}}{6\ \Omega}(1 - e^{-0.4}) = 0.659\ \text{A}$$

A plot of Eq. 29.7 for this circuit is given in Fig. 29.7b.

Q3. For the series RL circuit shown in Fig. 29.7a, can the back emf ever be greater than the battery emf? Explain.

Q4. Suppose the switch in the *RL* circuit in Fig. 29.7a has been closed for a long time and is suddenly opened. Does the current instantaneously drop to zero? Why does a spark tend to appear at the switch contacts when the switch is opened?

29.3 ENERGY IN A MAGNETIC FIELD

In the previous section we found that the induced emf set up by an inductor prevents a battery from establishing an instantaneous current. Hence, a battery has to do work against an inductor to create a current. Part of the energy supplied by the battery goes into joule heat dissipated in the resistor, while the remaining energy is stored in the inductor. If we multiply each term in Eq. 29.6 by the current I and rearrange the expression, we get

$$I\mathcal{E} = I^2R + LI\frac{dI}{dt} \tag{29.12}$$

This expression tells us that the rate at which energy is supplied by the battery, $I\mathcal{E}$, equals the sum of the joule heat dissipated in the resistor, I^2R, and the rate at which energy is stored in the inductor, $LIdI/dt$. Thus, Eq. 29.12 is simply an expression of energy conservation. If we let U_{m} denote the energy stored in the inductor at any time, then the rate dU_{m}/dt at which energy is stored in the inductor can be written

$$\frac{dU_{\mathrm{m}}}{dt} = LI\frac{dI}{dt}$$

To find the total energy stored in the inductor, we can rewrite this expression as $dU_{\mathrm{m}} = LI\,dI$ and integrate:

Energy stored in an inductor

$$U_{\mathrm{m}} = \int_0^{U_{\mathrm{m}}} dU_{\mathrm{m}} = \int_0^I LI\,dI = \frac{1}{2}LI^2 \tag{29.13}$$

where L is constant and has been removed from the integral. Equation 29.13 represents the energy stored as magnetic energy in the field of the inductor when the current is I. Note that it is similar in form to the equation for the energy stored in the electric field of a capacitor, $Q^2/2C$. In either case, we see that it takes work to establish a field.

We can also determine the energy per unit volume, or energy density, stored in a magnetic field. For simplicity, consider a solenoid whose inductance is given by Eq. 29.5:

$$L = \mu_0 n^2 V$$

where V is the volume of the solenoid. The magnetic field of a solenoid is given by

$$B = \mu_0 nI$$

Substituting the expression for L and $I = B/\mu_0 n$ into Eq. 29.13 gives

$$U_{\mathrm{m}} = \frac{1}{2}LI^2 = \frac{1}{2}\mu_0 n^2 V\left(\frac{B}{\mu_0 n}\right)^2 = \frac{B^2}{2\mu_0}V \tag{29.14}$$

Hence, the energy stored per unit volume in a magnetic field is given by

Magnetic energy density

$$u_{\mathrm{m}} = \frac{U_{\mathrm{m}}}{V} = \frac{B^2}{2\mu_0} \tag{29.15}$$

Although Eq. 29.15 was derived for the special case of a solenoid, *it is valid for any region of space in which a magnetic field exists.* Note that Eq. 29.15 is similar in form to the equation for the energy per unit volume stored in an electric field, given by $\frac{1}{2}\varepsilon_0 E^2$. In both cases, the energy density is proportional to the *square* of the field strength.

Example 29.4 *The Coaxial Cable*

A long coaxial cable consists of two concentric cylindrical conductors of radii a and b and length l, as in Fig. 29.8. The inner conductor is assumed to be a thin cylindrical shell. Each conductor carries a current I (the outer one being a return path). (a) Calculate the self-inductance L of this cable.

To obtain L, we must know the magnetic flux through any cross section between the two conductors. From Ampère's law (Section 27.3), it is easy to see that the magnetic field *between* the conductors is given by $B = \mu_0 I/2\pi r$. Furthermore, the field is zero outside the conductors and zero inside the inner hollow conductor. The field is zero outside since the *net* current through a circular path surrounding both wires is zero, and hence from Ampère's law, $\oint \mathbf{B} \cdot d\mathbf{s} = 0$. The field is zero inside the inner conductor since it is hollow and there is no current within a radius $r < a$.

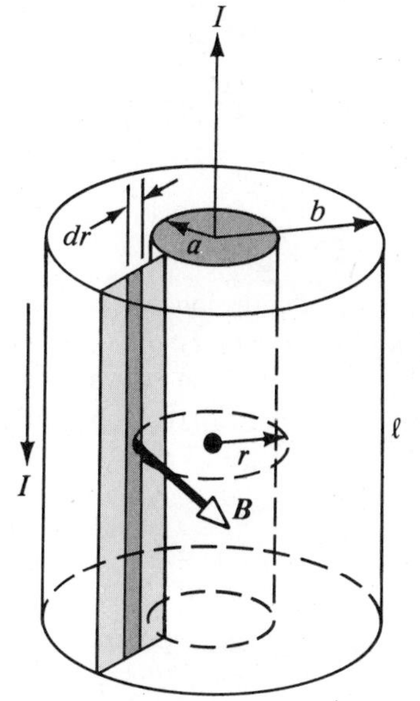

Figure 29.8 (Example 29.5) Section of a long coaxial cable. The inner and outer conductors carry equal and opposite currents.

The magnetic field is *perpendicular* to the shaded rectangular strip of length l and width $b - a$. This is the cross section of interest. Dividing this rectangle into strips of width dr, we see that the area of each strip is $l dr$ and the flux through each strip is $B\,dA = Bl\,dr$. Hence, the *total* flux through any cross section is

$$\Phi_m = \int B\,dA = \int_a^b \frac{\mu_0 I}{2\pi r} l\,dr = \frac{\mu_0 Il}{2\pi}\int_a^b \frac{dr}{r} = \frac{\mu_0 Il}{2\pi}\ln\left(\frac{b}{a}\right)$$

Using this result, we find that the self-inductance of the cable is

$$L = \frac{\Phi_m}{I} = \frac{\mu_0 l}{2\pi}\ln\left(\frac{b}{a}\right)$$

Furthermore, the self-inductance per unit length is given by $(\mu_0/2\pi)\ln(b/a)$.

(b) Calculate the total energy stored in the magnetic field of the cable.

Using Eq. 29.14 and the results to (a), we get

$$U_m = \frac{1}{2}LI^2 = \frac{\mu_0 lI^2}{4\pi}\ln\left(\frac{b}{a}\right)$$

Q5. If the current in an inductor is doubled, by what factor does the stored energy change?

Q6. Discuss the similarities between the energy stored in the electric field of a charged capacitor and the energy stored in the magnetic field of a current-carrying coil.

29.4 MUTUAL INDUCTANCE

Very often the magnetic flux through a circuit varies with time because of varying currents in nearby circuits. This gives rise to an induced emf through a process known as *mutual induction,* so called because it depends on the interaction of two circuits.

Consider two closely wound coils as shown in the cross-sectional view of Fig. 29.9. The current I_1 in coil 1, which has N_1 turns, creates magnetic field lines, some of which pass through coil 2, which has N_2 turns. The corresponding flux through coil 2 produced by coil 1 is represented by Φ_{21}. We define the *mutual*

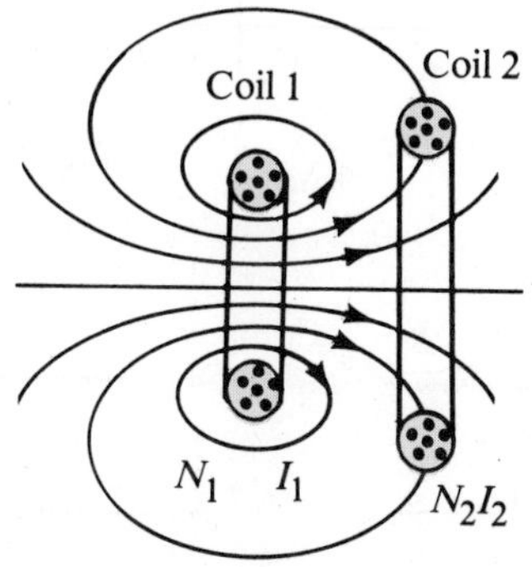

Figure 29.9 A cross-sectional view of two adjacent coils. A current in coil 1 sets up a flux, part of which passes through coil 2.

inductance M_{21} of coil 2 with respect to coil 1 as the ratio of $N_2\Phi_{21}$ and the current I_1:

Definition of mutual inductance

$$M_{21} \equiv \frac{N_2\Phi_{21}}{I_1} \tag{29.16}$$

$$\Phi_{21} = \frac{M_{21}}{N_2} I_1$$

The mutual inductance depends on the geometry of both circuits and on their orientation with respect to one another. Clearly, as the circuit separation increases, the mutual inductance decreases since the flux linking the circuits decreases.

If the current I_1 varies with time, we see from Faraday's law and Eq. 29.16 that the emf induced in coil 2 by coil 1 is given by

$$\mathcal{E}_2 = -N_2 \frac{d\Phi_{21}}{dt} = -M_{21} \frac{dI_1}{dt} \tag{29.17}$$

Similarly, if the current I_2 varies with time, the induced emf in coil 1 due to coil 2 is given by

$$\mathcal{E}_1 = -M_{12} \frac{dI_2}{dt} \tag{29.18}$$

These results are similar in form to the expression for the self-induced emf ($\mathcal{E} = -L\,dI/dt$). *The emf induced by mutual induction in one coil is always proportional to the rate of current change in the other coil.* If the rates at which the currents change with time are equal (that is, if $dI_1/dt = dI_2/dt$), then one finds that $\mathcal{E}_1 = \mathcal{E}_2$. This suggests that $M_{12} = M_{21} = M$, so that Eqs. 29.17 and 29.18 become

$$\mathcal{E}_2 = -M \frac{dI_1}{dt} \quad \text{and} \quad \mathcal{E}_1 = -M \frac{dI_2}{dt} \tag{29.19}$$

The unit of mutual inductance is also the henry.

Example 29.5

A long solenoid of length l has N_1 turns, carries a current I, and has a cross-sectional area A. A second coil containing N_2 turns is wound around the center of the first coil, as in Fig. 29.10. Find the mutual inductance of the system.

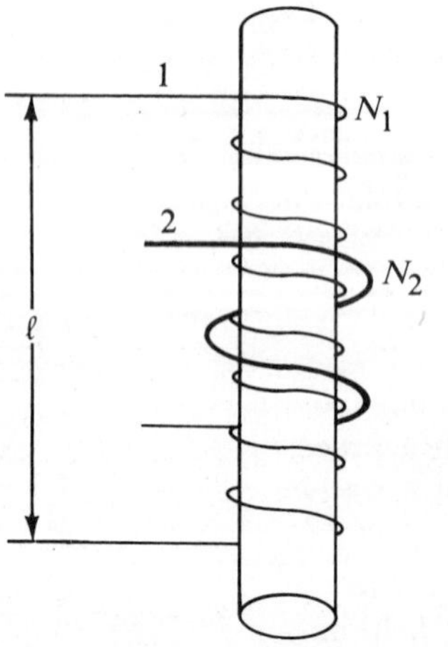

Figure 29.10 (Example 29.6) A small coil of N_2 turns wrapped around the center of a long solenoid of N_1 turns.

Solution: If the solenoid carries a current I_1, the magnetic field at its center is given by

$$B = \frac{\mu_0 N_1 I_1}{l}$$

Since the flux Φ_{21} through coil 2 due to coil 1 is BA, the mutual inductance is

$$M = \frac{N_2\Phi_{21}}{I_1} = \frac{N_2 BA}{I_1} = \mu_0 \frac{N_1 N_2 A}{l}$$

For example, if $N_1 = 500$ turns, $A = 3 \times 10^{-3}$ m^2, $l = 0.5$ m, and $N_2 = 8$ turns, we get

$$M = \frac{(4\pi \times 10^{-7}\ \text{Wb/A}\cdot\text{m})(500)(8)(3 \times 10^{-3}\ \text{m}^2)}{0.5\ \text{m}}$$

$$\approx 30 \times 10^{-6}\ \text{H} = 30\ \mu\text{H}$$

Q7. The centers of two circular loops are separated by a fixed distance. For what relative orientation of the loops will their mutual inductance be a maximum? For what orientation will it be a minimum?

Q8. Two solenoids are connected in series such that each carries the same current at any instant. Is mutual induction present? Explain.

29.5 OSCILLATIONS IN AN *LC* CIRCUIT

When a *charged* capacitor is connected to an inductor as in Fig. 29.11 and the switch is then closed, oscillations will occur in the current and charge on the capacitor. If the resistance of the circuit is neglected, no energy is dissipated as joule heat and the oscillations will persist. In this section we shall neglect the resistance in the circuit.

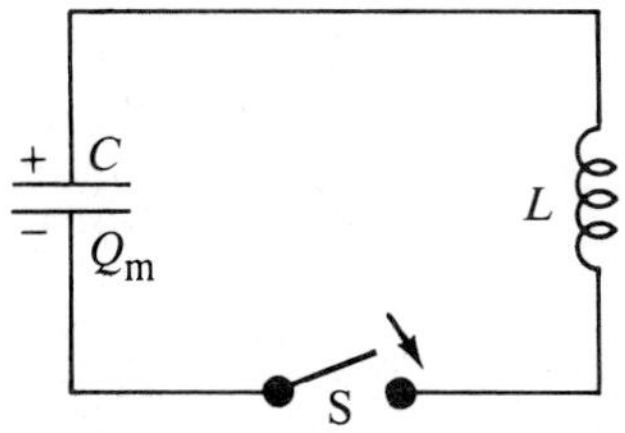

Figure 29.11 A simple LC circuit. The capacitor has an initial charge Q_m, and the switch is closed at $t = 0$.

In the following analysis, let us assume that the capacitor has an initial charge Q_m and that the switch is closed at $t = 0$. It is convenient to describe what happens from an energy viewpoint.

When the capacitor is fully charged, the total energy U in the circuit is stored in the electric field of the capacitor and is equal to $Q_m^2/2C$. At this time, the current is zero and so there is no energy stored in the inductor. As the capacitor begins to discharge, the energy stored in its electric field decreases. At the same time, the current increases and some energy is now stored in the magnetic field of the inductor. Thus, we see that energy is transferred from the electric field of the capacitor to the magnetic field of the inductor. When the capacitor is fully discharged, it stores no energy. At this time, the current reaches its maximum value and all of the energy is now stored in the inductor. The process then repeats in the reverse direction. The energy continues to transfer between the inductor and capacitor indefinitely, corresponding to oscillations in the current and charge.

A graphical description of this energy transfer is shown in Fig. 29.12. The circuit behavior is analogous to the oscillating mass-spring system studied in Chapter 13. The potential energy stored in a stretched spring, $\frac{1}{2}kx^2$, is analogous to the potential energy stored in the capacitor, $Q_m^2/2C$. The kinetic energy of the moving mass, $\frac{1}{2}mv^2$, is analogous to the energy stored in the inductor, $\frac{1}{2}LI^2$, which requires the presence of moving charges. In Fig. 29.12a, all of the energy is stored as potential energy in the capacitor at $t = 0$ (since $I = 0$). In Fig. 29.12b, all of the energy is stored as "kinetic" energy in the inductor, $\frac{1}{2}LI_m^2$, where I_m is the maximum current. At intermediate points, part of the energy is potential energy and part is kinetic energy.

Consider some arbitrary time t after the switch is closed, such that the capacitor has a charge Q and the current is I. At this time, both elements store energy, but the sum of the two energies must equal the total initial energy U stored in the fully charged capacitor at $t = 0$. That is,

$$U = U_C + U_L = \frac{Q^2}{2C} + \frac{1}{2}LI^2 \qquad (29.20)$$

Total energy stored in the LC circuit

Since we have assumed the circuit resistance to be zero, no energy is dissipated as joule heat and hence *the total energy must remain constant in time.* This means that $dU/dt = 0$. Therefore, by differentiating Eq. 29.20 with respect to time while noting that Q and I vary with time, we get

$$\frac{dU}{dt} = \frac{d}{dt}\left(\frac{Q^2}{2C} + \frac{1}{2}LI^2\right) = \frac{Q}{C}\frac{dQ}{dt} + LI\frac{dI}{dt} = 0 \qquad (29.21)$$

The total energy in an LC circuit remains constant

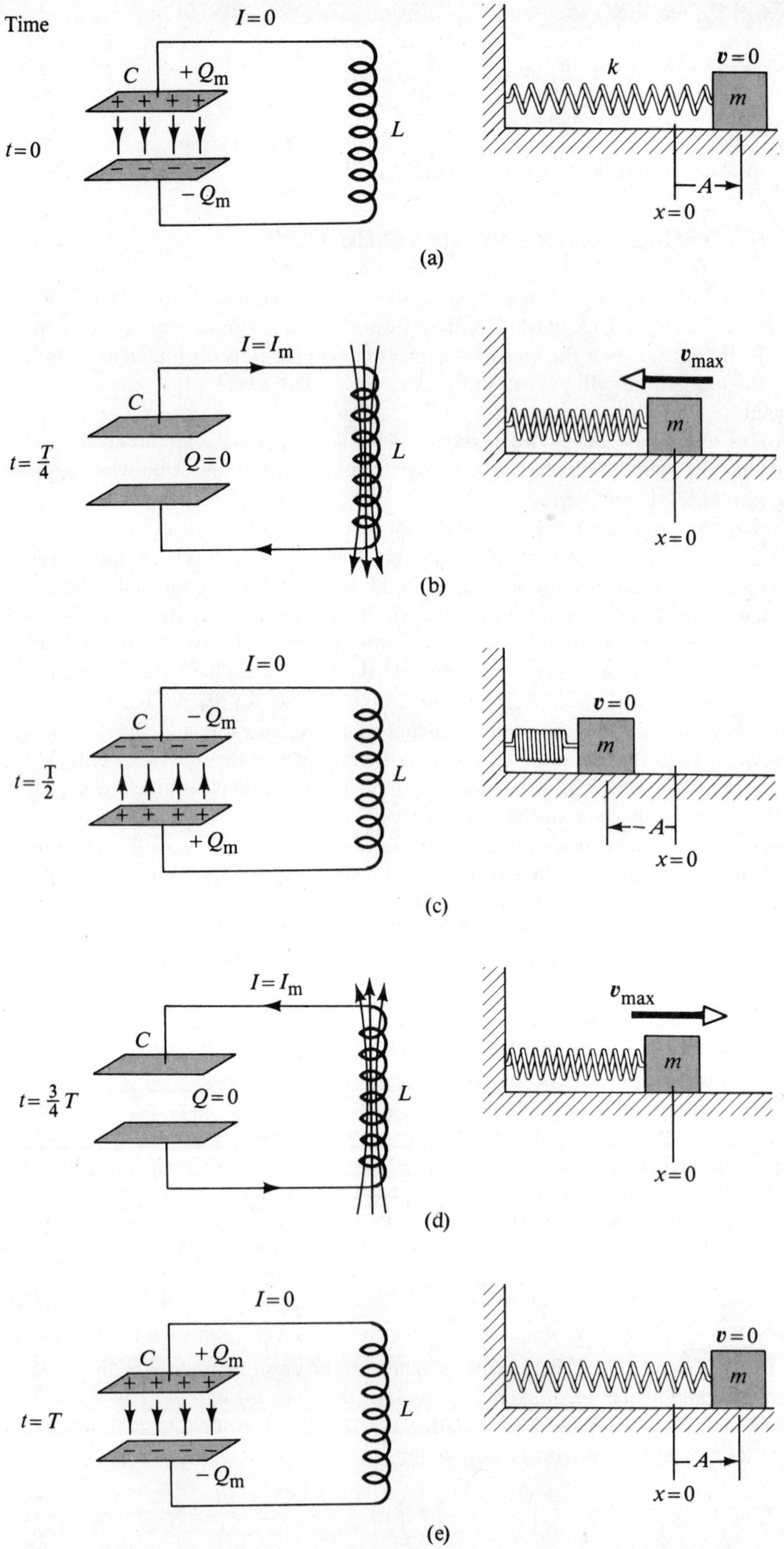

Figure 29.12 Energy transfer in a resistanceless LC circuit. The capacitor has a charge Q_m at $t = 0$ when the switch is closed. The mechanical analog of this circuit, the mass-spring system, is shown at the right.

We can reduce this to a differential equation in one variable by using the relationship between Q and I, namely, $I = dQ/dt$. From this, it follows that $dI/dt = d^2Q/dt^2$. Substitution of these relationships into Eq. 29.21 gives

$$L\frac{d^2Q}{dt^2} + \frac{Q}{C} = 0$$

$$\frac{d^2Q}{dt^2} = -\frac{1}{LC}Q \qquad (29.22)$$

We can solve for the function Q by noting that Eq. 29.22 is of the *same* form as that of the mass-spring system (simple harmonic oscillator) studied in Chapter 13. For this system, the equation of motion is given by

$$\frac{d^2x}{dt^2} = -\frac{k}{m}x = -\omega^2 x$$

where k is the spring constant, m is the mass, and $\omega = \sqrt{k/m}$. The solution of this equation has the general form

$$x = A\cos(\omega t + \delta)$$

where ω is the angular frequency of the simple harmonic motion, A is the amplitude of motion (the maximum value of x), and δ is the phase constant; the values of A and δ depend on the initial conditions. Since Eq. 29.22 is of the same form as the differential equation of the simple harmonic oscillator, its solution is

$$Q = Q_m \cos(\omega t + \delta) \qquad (29.23)$$

Charge versus time for the LC circuit

where Q_m is the maximum charge of the capacitor and the angular frequency ω is given by

$$\omega = \frac{1}{\sqrt{LC}} \qquad (29.24)$$

Angular frequency of oscillation

Note that *the angular frequency of the oscillations depends solely on the inductance and capacitance.*

Since Q varies periodically, the current also varies periodically. This is easily shown by differentiating Eq. 29.23 with respect to time, which gives

$$I = \frac{dQ}{dt} = -\omega Q_m \sin(\omega t + \delta) \qquad (29.25)$$

Current versus time for the LC circuit

To determine the value of the phase angle δ, we examine the initial conditions, which in our situation require that at $t = 0$, $I = 0$ and $Q = Q_m$. Setting $I = 0$ at $t = 0$ in Eq. 29.25 gives

$$0 = -\omega Q_m \sin\delta$$

which shows that $\delta = 0$. This value for δ is also consistent with Eq. 29.23 and the second condition that $Q = Q_m$ at $t = 0$. Therefore, in our case, the time variation of Q and that of I are given by

$$Q = Q_m \cos\omega t \qquad (29.26)$$

$$I = -\omega Q_m \sin\omega t = -I_m \sin\omega t \qquad (29.27)$$

where $I_m = \omega Q_m$ is the *maximum* current in the circuit.

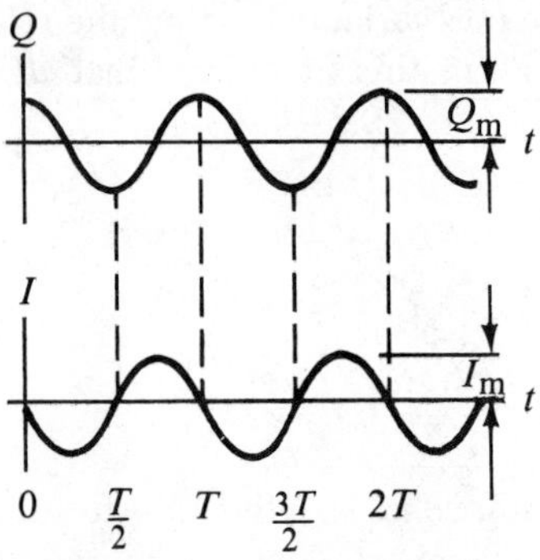

Figure 29.13 Graphs of charge versus time and current versus time for a resistanceless *LC* circuit. Note that Q and I are 90° out of phase with each other.

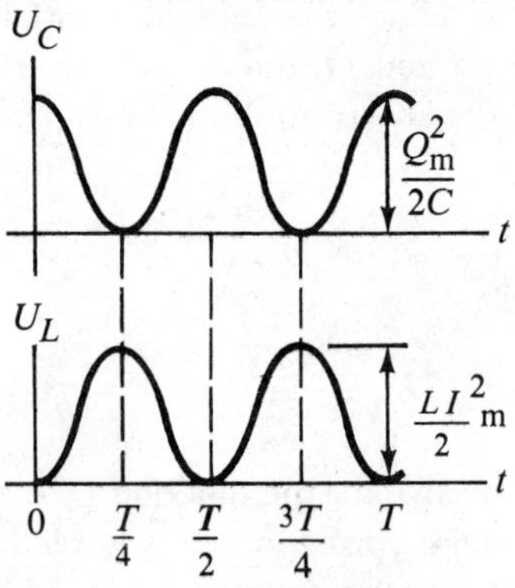

Figure 29.14 Plots of U_C versus t and U_L versus t for a resistanceless *LC* circuit. The sum of the two curves is a constant and equal to the total energy stored in the circuit.

Graphs of Q versus t and I versus t are shown in Fig. 29.13. Note that the charge on the capacitor oscillates between the extreme values Q_m and $-Q_m$, and the current oscillates between I_m and $-I_m$. Furthermore, the current is 90° out of phase with the charge. That is, when the charge reaches an extreme value, the current is zero, and when the charge is zero, the current has an extreme value.

Let us return to the energy of the *LC* circuit. Substituting Eqs. 29.26 and 29.27 in Eq. 29.20, we find that the total energy is given by

Energy of an *LC* circuit

$$U = U_C + U_L = \frac{Q_m^2}{2C}\cos^2\omega t + \frac{LI_m^2}{2}\sin^2\omega t \tag{29.28}$$

This expression contains all of the features that were described qualitatively at the beginning of this section. It shows that the energy of the system continuously oscillates between energy stored in the electric field of the capacitor and energy stored in the magnetic field of the inductor. When the energy stored in the capacitor has its maximum value, $Q_m^2/2C$, the energy stored in the inductor is zero. When the energy stored in the inductor has its maximum value, $\frac{1}{2}LI_m^2$, the energy stored in the capacitor is zero.

Plots of the time variations of U_C and U_L are shown in Fig. 29.14. Note that the sum $U_C + U_L$ is a constant and equal to the total energy, $Q_m^2/2C$. An analytical proof of this is straightforward. Since the maximum energy stored in the capacitor (when $I = 0$) must equal the maximum energy stored in the inductor (when $Q = 0$),

$$\frac{Q_m^2}{2C} = \frac{1}{2}LI_m^2 \tag{29.29}$$

Substitution of this into Eq. 29.28 for the total energy gives

$$U = \frac{Q_m^2}{2C}(\cos^2\omega t + \sin^2\omega t) = \frac{Q_m^2}{2C} \tag{29.30}$$

since $\cos^2\omega t + \sin^2\omega t = 1$.

You should note that the total energy U remains constant *only* if energy losses are neglected. In actual circuits, there will always be some resistance and so energy will be lost in the form of heat. (In fact, even when the energy losses due to wire resistance are neglected, energy will also be lost in the form of electromagnetic waves radiated by the circuit.) In our idealized situation, the oscillations in the circuit persist indefinitely.

Example 29.6

An LC circuit has an inductance of 2.81 mH and a capacitance of 9 pF (Fig. 29.15). The capacitor is initially charged with a 12-V battery when the switch S is open. The battery is then removed from the circuit, and the switch is closed so that the capacitor is shorted across the inductor. (a) Find the frequency of oscillation.

Using Eq. 29.24 gives for the frequency

$$f = \frac{\omega}{2\pi} = \frac{1}{2\pi\sqrt{LC}}$$

$$= \frac{1}{2\pi[(2.81 \times 10^{-3}\text{ H})(9 \times 10^{-12}\text{ F})]^{1/2}} = 10^6\text{ Hz}$$

(b) What are the maximum charge on the capacitor and maximum current in the circuit?

The initial charge on the capacitor equals the maximum charge, and since $C = Q/V$, we get

$$Q_m = CV = (9 \times 10^{-12}\text{ F})(12\text{ V}) = 1.08 \times 10^{-10}\text{ C}$$

From Eq. 29.27, we see that the maximum current is related to the maximum charge:

$$I_m = \omega Q_m = 2\pi f Q_m$$
$$= (2\pi \times 10^6\text{ s}^{-1})(1.08 \times 10^{-10}\text{ C})$$
$$= 6.79 \times 10^{-4}\text{ A}$$

(c) Determine the charge and current as functions of time.

Equations 29.26 and 29.27 give the following expressions for the time variation of Q and I:

$$Q = Q_m \cos\omega t = 1.08 \times 10^{-10} \cos\omega t\text{ C}$$

$$I = -I_m \sin\omega t = -6.79 \times 10^{-4} \sin\omega t\text{ A}$$

where

$$\omega = 2\pi f = 2\pi \times 10^6\text{ rad/s}$$

(d) What is the total energy stored in the circuit?

Using Eq. 29.30 for the total energy gives

$$U = \frac{Q_m^2}{2C} = \frac{(1.08 \times 10^{-10}\text{ C})^2}{2(9 \times 10^{-12}\text{ F})} = 6.48 \times 10^{-10}\text{ J}$$

We could also show that Eq. 29.29 is satisfied:

$$U = \frac{1}{2}LI_m^2 = \frac{1}{2}(2.81 \times 10^{-3}\text{ H})(6.79 \times 10^{-4}\text{ A})^2$$
$$= 6.48 \times 10^{-10}\text{ J}$$

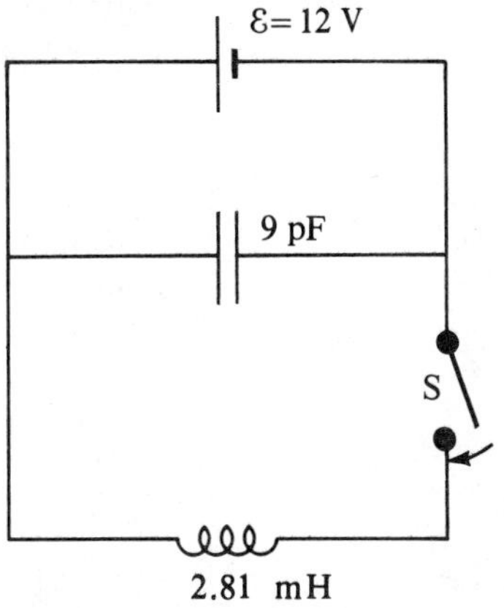

Figure 29.15 (Example 29.7) First the capacitor is fully charged with the switch S open, and then the battery is removed and S is closed.

Q9. In the LC circuit shown in Fig. 29.12, the charge on the capacitor is sometimes zero, even though there is current in the circuit. How is this possible?

Q10. If the resistance of the wires in an LC circuit were not neglected, would the oscillations persist? Explain.

29.6 THE *RLC* CIRCUIT

We now turn our attention to a more realistic circuit consisting of an inductor, a capacitor, and a resistor connected in series, as in Fig. 29.16. We shall assume that the capacitor has an initial charge Q_m before the switch is closed. Once the switch is closed and a current is established, the total energy stored in the circuit at any time is given, as before, by Eq. 29.20. That is, the energy stored in the capacitor is $Q^2/2C$, and the energy stored in the inductor is $\frac{1}{2}LI^2$. However, the total energy is no longer constant, as it was in the LC circuit, because of the presence of a resistor, which dissipates energy as joule heat. Since the rate of energy dissipation through a resistor is I^2R, we have

$$\frac{dU}{dt} = -I^2R \qquad (29.31)$$

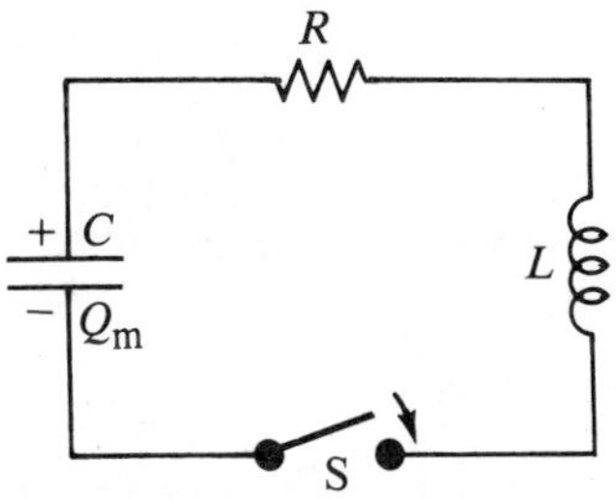

Figure 29.16 A series RLC circuit. The capacitor has a charge Q_m at $t = 0$ when the switch is closed.

where the negative sign signifies that U is *decreasing* in time. Substituting this result into the time derivative of Eq. 29.20 gives

$$LI\frac{dI}{dt} + \frac{Q}{C}\frac{dQ}{dt} = -I^2R \tag{29.32}$$

Using the fact that $I = dQ/dt$ and $dI/dt = d^2Q/dt^2$, and dividing Eq. 29.32 by I, we get

$$L\frac{d^2Q}{dt^2} + R\frac{dQ}{dt} + \frac{Q}{C} = 0 \tag{29.33}$$

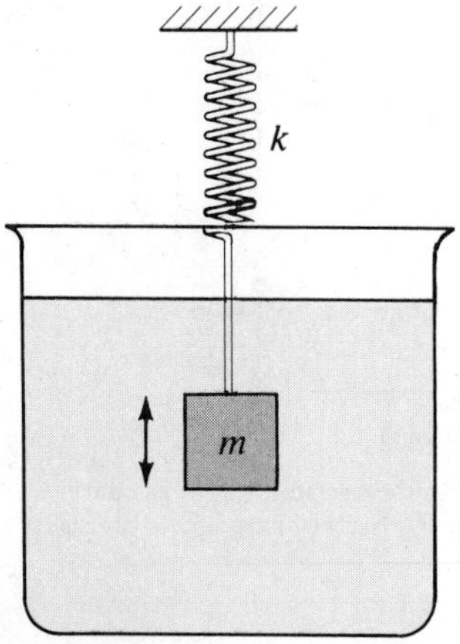

Figure 29.17 A mass-spring system moving in a viscous medium with damped harmonic motion is analogous to an *RLC* circuit.

Note that the *RLC* circuit is analogous to the damped harmonic oscillator discussed in Section 13.6 and illustrated in Fig. 29.17. The equation of motion for this mechanical system is

$$m\frac{d^2x}{dt^2} + b\frac{dx}{dt} + kx = 0 \tag{29.34}$$

Comparing Eqs. 29.33 and 29.34, we see that q corresponds to x, L corresponds to m, R corresponds to the damping constant b, and C corresponds to $1/k$, where k is the force constant of the spring.

The analytical solution of Eq. 29.33 is rather cumbersome and is usually covered in courses dealing with differential equations. Therefore, we shall give only a qualitative description of the circuit behavior.

In the simplest case, when $R = 0$, Eq. 29.33 reduces to that of the simple LC circuit, as expected, and the charge and the current oscillate sinusoidally in time.

Next consider the situation where R is reasonably small. In this case, the solution of Eq. 29.33 is given by

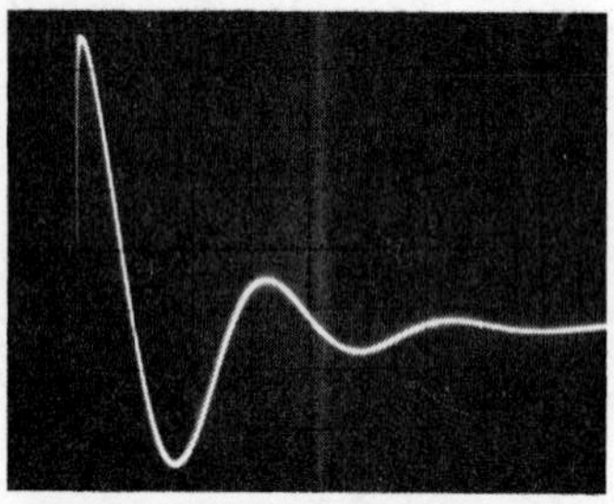

Oscilloscope pattern showing the decay in the oscillations of an *RLC* circuit. The parameters used were $R = 75\Omega$, $L = 10$ mH, $C = 0.19$ μF, and $f = 300$ Hz. (Courtesy of Prof. J. Budmin, James Madison University)

$$Q = Q_m e^{-Rt/2L} \cos\omega_d t \tag{29.35}$$

where

$$\omega_d = \left[\frac{1}{LC} - \left(\frac{R}{2L}\right)^2\right]^{1/2} \tag{29.36}$$

That is, the charge will oscillate with *damped harmonic motion* in analogy with a mass-spring system moving in a viscous medium. From Eq. 29.35, we see that when $R \ll \sqrt{4L/C}$ the frequency ω_d of the damped oscillator will be close to that of the undamped oscillator, $1/\sqrt{LC}$. Since $I = dQ/dt$, it follows that the current will also undergo damped harmonic motion. Plots of the charge and current versus time for the damped oscillator are shown in Fig. 29.18. Note that the maximum values of Q

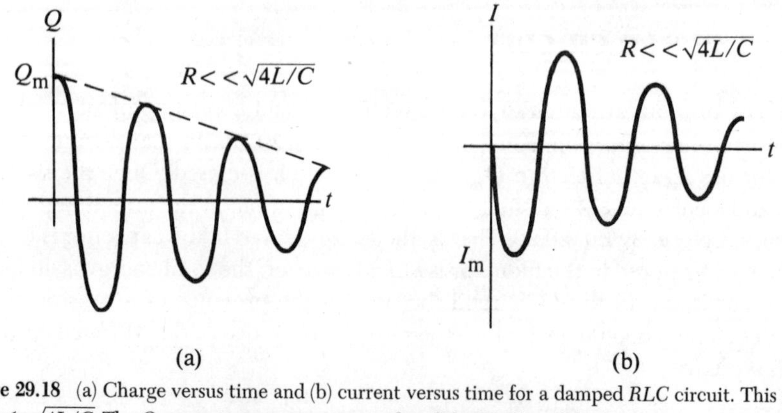

Figure 29.18 (a) Charge versus time and (b) current versus time for a damped *RLC* circuit. This occurs for $R \ll \sqrt{4L/C}$. The Q versus t curve represents a plot of Eq. 29.35, and the I versus t curve is simply the first time derivative of the curve in (a).

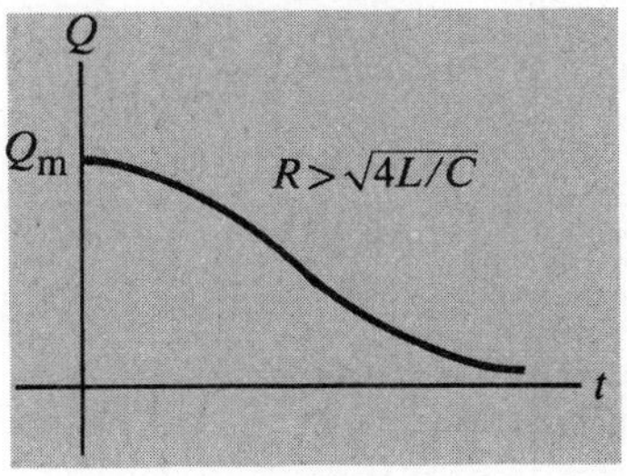

Figure 29.19 Plot of Q versus t for an overdamped RLC circuit, which occurs for values of $R > \sqrt{4L/C}$.

and I decrease after each oscillation, just as the amplitude and maximum velocity of a damped harmonic oscillator decrease in time.

When we consider larger values of R, we find that the oscillations damp out more rapidly; in fact, there exists a critical resistance value R_c above which *no* oscillations occur. The critical value is given by $R_c = \sqrt{4L/C}$. A system with $R = R_c$ is said to be *critically damped*. When R exceeds R_c, the system is said to be *overdamped* (Fig. 29.19).

Q11. Describe the difference between free oscillations in an LC circuit (zero resistance) and damped oscillations in an RLC circuit.

Q12. Give a mechanical analogy of an overdamped RLC circuit for which no oscillations can occur.

29.7 SUMMARY

When the current in a coil changes with time, an emf is induced in the coil according to Faraday's law. The *self-induced emf* is defined by the expression

$$\mathcal{E} = -L\frac{dI}{dt} \qquad (29.1)$$

Induced emf

where L is the *inductance* of the coil. Inductance is a measure of the opposition of a device to a *change* in current. Inductance has the SI unit of henry (H), where $1 \text{ H} = 1 \text{ V}\cdot\text{s/A}$.

The *inductance of any coil,* such as a solenoid or toroid, is given by the expression

$$L = \frac{N\Phi_m}{I} \qquad (29.2)$$

Inductance of an N-turn coil

where Φ_m is the magnetic flux through the coil.

The inductance of a device depends on its geometry. For example, the *inductance of a solenoid* (whose core is a vacuum), as calculated from Eq. 29.2, is given by

$$L = \frac{\mu_o N^2 A}{l} \qquad (29.4)$$

Inductance of a solenoid

where N is the number of turns, A is the cross-sectional area, and l is the length of the solenoid.

If a resistor and inductor are connected in series to a battery of emf $\mathcal{E}$ as shown in Fig. 29.2 and a switch in the circuit is closed at $t = 0$, the current in the circuit

Current in an *RL* circuit

varies in time according to the expression

$$I(t) = \frac{\mathcal{E}}{R}(1 - e^{-t/\tau}) \tag{29.7}$$

where $\tau = L/R$ is the *time constant* of the RL circuit. That is, the current rises to an equilibrium value of $\mathcal{E}/R$ after a time that is long compared to τ.

If the battery is removed from an RL circuit, as in position 2 of Fig. 29.5, the current decays exponentially with time according to the expression

$$I(t) = \frac{\mathcal{E}}{R}e^{-t/\tau} \tag{29.11}$$

where $\mathcal{E}/R$ is the initial current in the circuit.

The *energy stored in the magnetic field of an inductor* carrying a current I is given by

Energy stored in an inductor

$$U_{\mathrm{m}} = \frac{1}{2}LI^2 \tag{29.13}$$

This result is obtained by applying the principle of energy conservation to the RL circuit.

The *energy per unit volume* (or energy density) at a point where the magnetic field is B is given by

Magnetic energy density

$$u_{\mathrm{m}} = \frac{B^2}{2\mu_0} \tag{29.15}$$

That is, the energy density is proportional to the square of the field at that point.

If two coils are close to each other, a changing current in one coil can induce an emf in the other coil. If dI_1/dt is rate of change of current in the first coil, the emf induced in the second is given by

$$\mathcal{E}_2 = -M\frac{dI_1}{dt} \tag{29.19}$$

where M is a constant called the *mutual inductance* of one coil with respect to the other.

If Φ_{21} is the magnetic flux through coil 2 due to the current I_1 in coil 1 and N_2 is the number of turns in coil 2, then the mutual inductance of coil 2 is given by

Mutual inductance

$$M_{21} = \frac{N_2\Phi_{21}}{I_1} \tag{29.16}$$

In an LC circuit with zero resistance, the charge on the capacitor and the current in the circuit vary in time according to the expressions

Charge and current versus time in an *LC* circuit

$$Q = Q_{\mathrm{m}}\cos(\omega t + \delta) \tag{29.23}$$

$$I = -\omega Q_{\mathrm{m}}\sin(\omega t + \delta) \tag{29.25}$$

where Q_{m} is the maximum charge on the capacitor, δ is a phase constant, and ω is the angular frequency of oscillation, given by

$$\omega = \frac{1}{\sqrt{LC}} \tag{29.24}$$

Frequency of oscillation in an LC circuit

The energy in an LC circuit continuously transfers between energy stored in the capacitor and energy stored in the inductor. The *total energy* of the LC circuit at any time t is given by

$$U = U_C + U_L = \frac{Q_m^2}{2C}\cos^2\omega t + \frac{LI_m^2}{2}\sin^2\omega t \tag{29.28}$$

Energy of an LC circuit

where I_m is the maximum current in the circuit. At $t = 0$, all of the energy is stored in the electric field of the capacitor ($U = Q_m^2/2C$). Eventually, all of this energy is transferred to the inductor ($U = LI_m^2/2$). However, the *total* energy remains constant since energy losses are neglected in the ideal LC circuit.

The charge and current in an RLC circuit exhibit a damped harmonic behavior for small values of R. This is analogous to the damped harmonic motion of a mass-spring system in which friction is present.

EXERCISES

Section 29.1 Self-Inductance

1. A "Slinky toy" spring has a radius of 4 cm and an inductance of 275 μH when extended to a length of 1 m. What is the total number of turns in the spring?

2. What is the inductance of a 450-turn solenoid that has a radius of 10 cm and an overall length of 0.75 m?

3. A small air-core solenoid has a length of 3 cm and a radius of 0.4 cm. If the inductance is to be 0.01 mH, how many turns per cm are required?

4. An emf of 24 mV is induced in a 500-turn coil at an instant when the current has a value of 4 A and is changing at a rate of 10 A/s. What is the total magnetic flux through the coil?

5. Calculate the magnetic flux through a 500-turn, 6-mH coil when the current in the coil is 12 mA.

6. Three solenoidal windings of 500, 400, and 300 turns are wrapped at well-spaced positions along a cardboard tube of radius 2 cm. Each winding extends for 10 cm along the cylindrical surface. What is the equivalent inductance of the 1200 turns when the three sets of windings are connected in series?

7. Two coils, A and B, are wound using *equal lengths* of wire. Each coil has the same number of turns per unit length, but coil A has twice as many turns as coil B. What is the ratio of the self-inductance of A to the self-inductance of B? (Note: The radii of the two coils are not equal.)

8. Show that the two expressions for inductance given by

$$L = N\Phi_m/I \quad \text{and} \quad L = -\frac{\mathcal{E}}{dI/dt}$$

have the same units.

9. The current in a 10-H inductor varies in time as $I = 2t^2 - 3t$, where I is in A and t is in s. (a) Calculate the magnitude of the induced emf at $t = 0$ and $t = 3$ s. (b) For what value of t will the induced emf be zero?

10. A solenoidal inductor contains 600 turns, is 20 cm in length, and has a cross-sectional area of 4 cm^2. What uniform rate of decrease of current through the inductor will produce an induced emf of 250 μV?

Section 29.2 *RL* Circuits

11. Verify by direct substitution that the expression for current given in Eq. 29.7 is a solution of Kirchhoff's loop equation for the RL circuit as given by Eq. 29.6.

12. Show that $I = I_0 e^{-t/\tau}$ is a solution of the differential equation

$$IR + L\frac{dI}{dt} = 0,$$

where $\tau = L/R$ and $I_0 = \mathcal{E}/R$ is the value of the current at $t = 0$.

13. Consider the circuit shown in Fig. 29.20, taking $\mathcal{E} = 12$ V, $L = 12$ mH, and $R = 18\ \Omega$. (a) What is the inductive time constant of the circuit? (b) Calculate the current in the circuit at a time 500 μs after the switch S_1 is closed. (c) What is the value of the final steady-state current? (d) How long does it take the current to reach 80% of its maximum value?

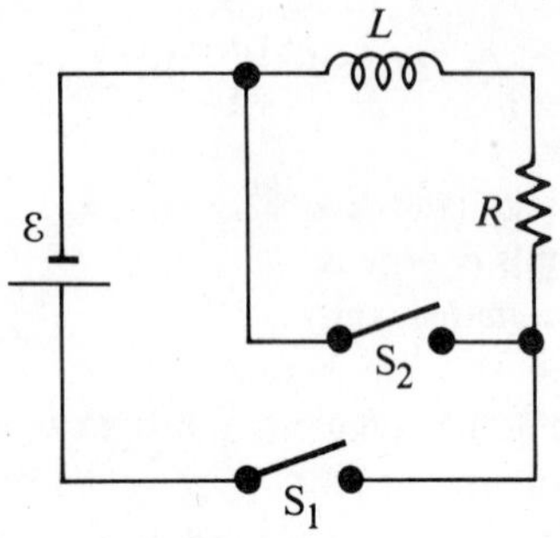

Figure 29.20 (Exercises 13, 14 and 15).

14. Let the following values be assigned to the components in the circuit shown in Fig. 29.20: $\mathcal{E} = 12$ V, $L = 36$ mH, and $R = 24\ \Omega$. (a) Calculate the current in the circuit at a time 1.5 ms after switch S_1 is closed. (b) What is the maximum value of the current in the circuit?

15. Assume that switch S_1 in the circuit of Fig. 29.20 has been closed long enough for the current to reach its *maximum* value. If switch S_1 is now opened and switch S_2 closed at $t = 0$, after what time interval will the current in R be 25% of the maximum value? (Use the numerical values given in Exercise 14.)

16. An inductor with an inductance of 9 H and resistance of 15 Ω is connected across a 60-V battery. (a) What is the *initial* rate of increase of current in the circuit? (b) At what rate is the current changing at $t = 10$ s?

17. In the circuit shown in Fig. 29.2 let $L = 4$ H, $R = 6\ \Omega$, and $\mathcal{E} = 48$ V. What is the self-induced emf, V_{ab}, 0.5 s after the switch is closed?

18. A 6-V battery is connected in series with a resistor and an inductor. The circuit has a time constant of 600 μs, and the maximum current is 300 mA. What is the value of the inductance?

19. Calculate the inductance in an *RL* circuit in which $R = 10\ \Omega$ and the current increases to one fourth its final value in 2 s.

20. For the *RL* circuit shown in Fig. 29.2, use the values for *L*, *R*, and $\mathcal{E}$ given in Exercise 17 to (a) calculate the ratio of the potential difference across the resistor to that across the inductor when $I = 5$ A. (b) Calculate the voltage across the inductor when $I = 8$A.

21. Show that the inductive time constant τ has SI units of seconds.

Section 29.3 Energy in a Magnetic Field

22. Calculate the energy associated with the magnetic field of a 400-turn solenoid in which a current of 2 A produces a flux of 10^{-4} Wb.

23. Consider the circuit shown in Fig. 29.21. What energy is stored in the inductor when the current reaches its final equilibrium value after the switch is closed?

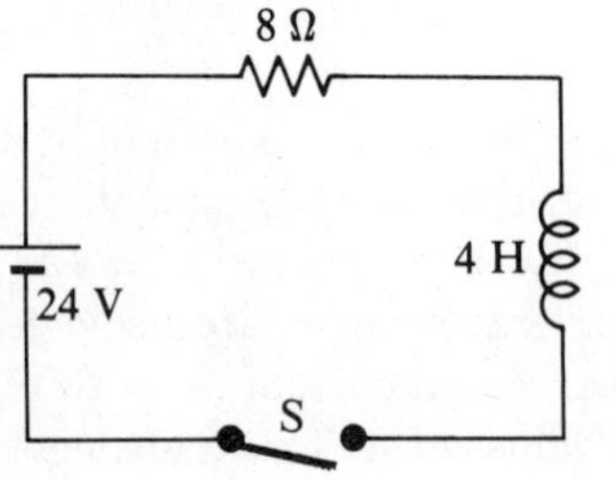

Figure 29.21 (Exercises 23 and 24).

24. The switch in the circuit of Fig. 29.21 is closed at $t = 0$. (a) Calculate the *rate* at which energy is being stored in the inductor after an elapsed time equal to the time constant of the circuit. (b) At what rate is energy being dissipated as joule heat in the resistor at this time? (c) What is the total energy stored in the inductor at this time?

25. An *RL* circuit in which $L = 3$ H and $R = 6\ \Omega$ is connected to a battery with $\mathcal{E} = 24$ V at time $t = 0$. (a) What energy is stored in the inductor when the current in the circuit is 1 A? (b) At what rate is energy being stored in the inductor when $I = 2$ A? (c) What power is being delivered to the circuit by the battery when $I = 1$ A?

26. A long coaxial cable has inner and outer conductor radii of 0.6 cm and 1.2 cm, respectively. (a) Calculate the self-inductance of a 1-m length of the cable. (b) What is the magnetic energy stored in the cable per unit length when the current is 8 A? (c) If the radius of the outside conductor is reduced to 0.8 cm, by what factor must the current be increased in order to maintain a constant value of magnetic energy per unit length of cable?

27. Calculate the magnetic energy density near the center of a closely wound solenoid of 1200 turns/m when the current in the solenoid is 3 A.

28. A uniform *electric* field of magnitude 4×10^5 V/m throughout a cylindrical volume results in a total energy due to the electric field of 1 μJ. What *magnetic* field over this same region will store the same total energy?

29. A battery for which $\mathcal{E} = 12$ V is connected to an *RL* circuit for which $L = 0.5$ H and $R = 4\ \Omega$. When the current has reached one half of its final value, what is the total magnetic energy stored in the inductor?

30. (a) Calculate the energy stored in the magnetic field of the inductor of Exercise 9 at $t = 0.5$ s. (b) At what rate is energy being added to the magnetic field at $t = 0.5$ s?

Section 29.4 Mutual Inductance

31. Two nearby coils, A and B, have a mutual inductance $M = 30$ mH. What is the emf induced in coil A as a function of time when the current in coil B is given by $I = 2 + 3t - t^2$, where I is in A when t is in s?

32. An emf of 96 mV is induced in the windings of a coil when the current in a nearby coil is increasing at the rate of 1.2 A/s. What is the mutual inductance of the two coils?

33. A coil of 20 turns is wound on a long solenoid as shown in Fig. 29.10. The solenoid has a cross-sectional area of 4×10^{-3} m^2 and is wrapped uniformly with 1200 turns per meter of length. Calculate the mutual inductance of the two windings.

34. Two nearby solenoids, A and B, have 300 and 900 turns, respectively. A current of 2 A in coil A produces a flux of 400 μWb at the center of A and a flux of 80 μWb at the center of B. (a) Calculate the mutual inductance of the two solenoids. (b) What is the self-inductance of coil A? (c) What emf will be induced in coil B when the current in coil A increases at the rate of 0.5 A/s?

Section 29.5 Oscillations in an *LC* Circuit and Section 29.6 The *RLC* Circuit

35. An LC circuit of the type shown in Fig. 29.11 has an inductance of 0.63 mH and a capacitance of 10 pF. The capacitor is charged to its maximum value by a 24-V battery. The battery is then removed from the circuit and the capacitor discharged through the inductor. (a) If all resistance in the circuit is neglected, determine the maximum value of the current in the oscillating circuit. (b) At what frequency does the circuit oscillate? (c) What is the maximum energy stored in the magnetic field of the inductor?

36. Calculate the inductance of an LC circuit that oscillates at a frequency of 120 Hz when the capacitance is 8 μF.

37. (a) What capacitance must be combined with an 80-mH inductor in order to achieve a resonant frequency of 200 Hz? (b) What time interval elapses between accumulations of maximum charge of the *same* sign on a given plate of the capacitor?

38. Consider the circuit shown in Fig. 29.16. Let $R = 10\ \Omega$, $L = 1$ mH, and $C = 2\ \mu$F. (a) Calculate the frequency of the damped oscillation of the circuit. (b) What is the value of the critical resistance in the circuit?

39. Consider a series LC circuit ($L = 1.56$ H, $C = 5$ nF). What is the maximum value of a resistor that, if inserted in series with L and C, will allow the circuit to continue to oscillate?

PROBLEMS

1. When the current I in the portion of circuit shown in Fig. 29.22 is 2 A and *increasing* at a rate of 1 A/s, the measured potential difference, $V_{ab} = 8$ V. However, when the current $I = 2$ A and is *decreasing* at a rate of 1 A/s, the measured potential difference $V_{ab} = 4$ V. Calculate the value of L and R.

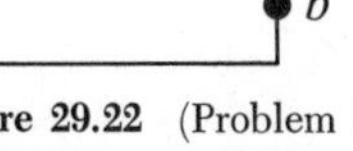

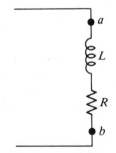

Figure 29.22 (Problem 1).

2. A platinum wire 2.5 mm in diameter is connected in series to a 100-μF capacitor and a 1.2×10^{-3}-μH inductor to form an RLC circuit. The resistivity of platinum is $11 \times 10^{-8}\ \Omega \cdot$m. Calculate the *maximum* length of wire for which the current in the circuit will oscillate.

3. The switch in the circuit of Fig. 29.23 is placed in position 1 at time $t = 0$ and then switched to position 2 at $t = 0.3$ s. The values of the circuit parameters are $\mathcal{E} = 24$ V, $R = 3\ \Omega$, and $L = 1.2$ H. (a) Calculate the current in R at $t = 0.6$ s. (b) For what value of $t > 0.3$ s does the current in R have the same value that it has at $t = 0.15$ s?

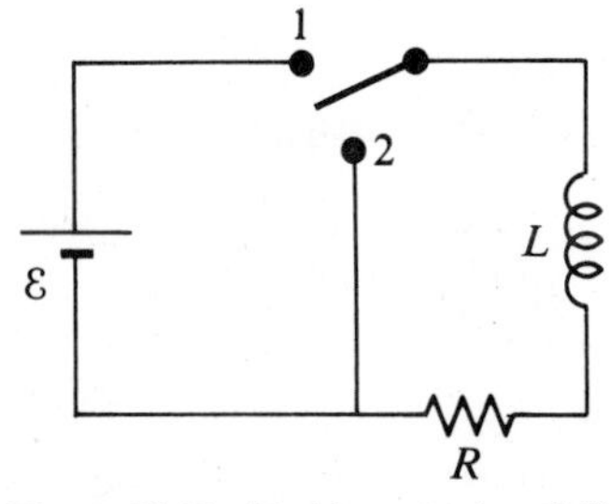

Figure 29.23 (Problems 3, 4, and 5).

4. Let the values of L, R, and $\mathcal{E}$ in the circuit shown in Fig. 29.23 be 3.6 mH, 6 Ω, and 36 V, respectively. Consider the situation when the switch S is placed in position 1 at $t = 0$ and left there indefinitely. Let t_1 equal the time required for the current in R to increase from zero to 9% of its *maximum* value and let t_2 equal the time required for the current in R to increase from 90% to 99% of its maximum value. (a) What is the ratio of t_2 to t_1? (b) At what time will the rate of change of current equal one half of the *initial* rate of change of current? (c) At what time will the potential difference across the resistor have the same magnitude as the potential difference across the inductor? (d) At what time will the power delivered to the inductor equal the power dissipated in the resistor? (e) How is the answer to (d) related to the time constant of the circuit?

5. Assume that the switch in the circuit shown in Fig. 29.23 is initially in position 1. Show that if the switch is thrown from position 1 to position 2, all the energy stored in the magnetic field of the inductor will be dissipated as thermal energy in the resistor.

6. (a) Determine the time constant of the circuit shown in Fig. 29.24. (b) How much energy is stored in the 30-mH inductor when the total energy stored in the circuit is 50% of the maximum possible value? (Neglect mutual inductance between the coils.)

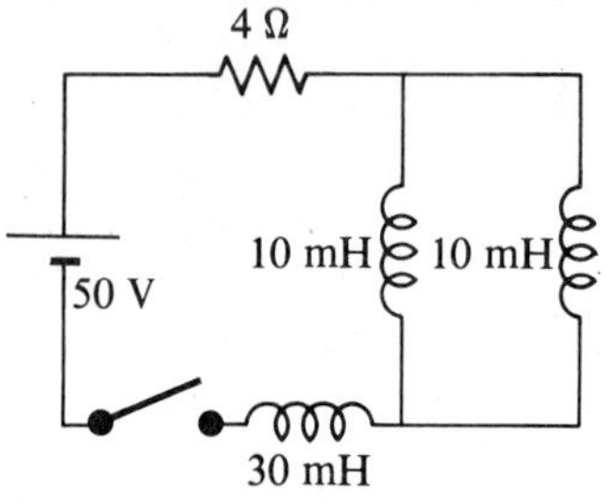

Figure 29.24 (Problem 6).

7. The toroidal coil shown in Fig. 29.25 consists of N turns and has a rectangular cross section. Its inner and outer radii are a and b, respectively. (a) Show that the self inductance of the coil is given by

$$L = \frac{\mu_0 N^2 h}{2\pi} \ln(b/a)$$

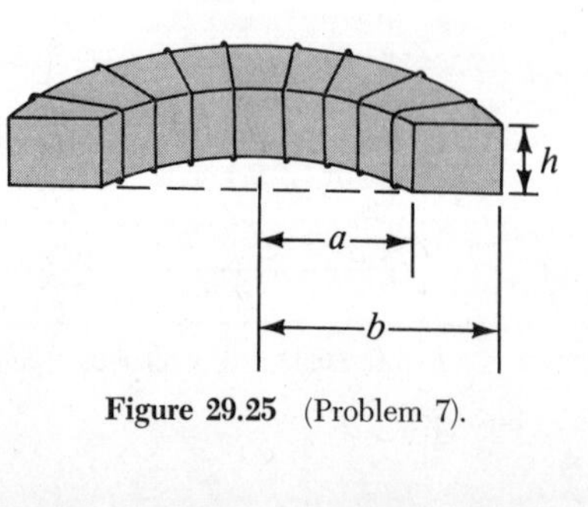

Figure 29.25 (Problem 7).

(b) If $a = 3$ cm, $b = 5$ cm, and $h = 1$ cm, what number of turns will result in an inductance of 0.5 mH?

8. The battery in the circuit shown in Fig. 29.26 has an emf $\mathcal{E} = 24$ V. (a) What current is being delivered by the battery 1 ms after the switch is closed? (b) Determine the potential difference across the 5-Ω resistor 3 ms after the switch is closed. (Neglect the mutual inductance of the coils.)

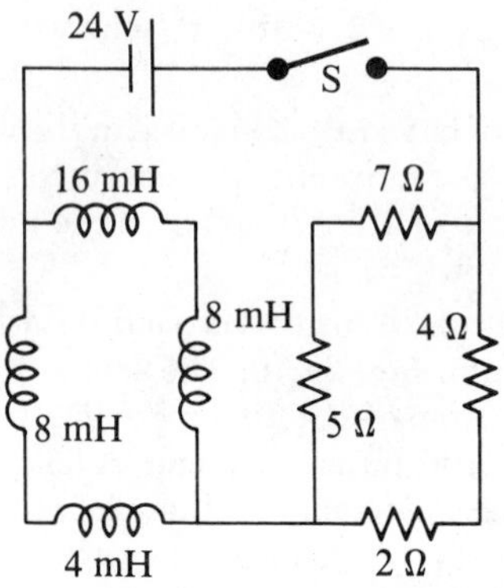

Figure 29.26 (Problem 8).

9. In previous problems, when two or more coils were present in a circuit, it was assumed that they were located so that the flux from one coil did not link the turns of the other coils. Now consider the situation shown in Fig. 29.27, in which two coaxial solenoids are positioned so that some of the flux from one links the windings of the other. (a) Use Kirchhoff's loop theorem to show that the effective inductance of the pair is given by

$$L_{\text{eff}} = L_1 + L_2 \pm 2M$$

where M is the mutual inductance of the two solenoids. (b) Why is it necessary to allow the choice of + and − signs for the $2M$ term?

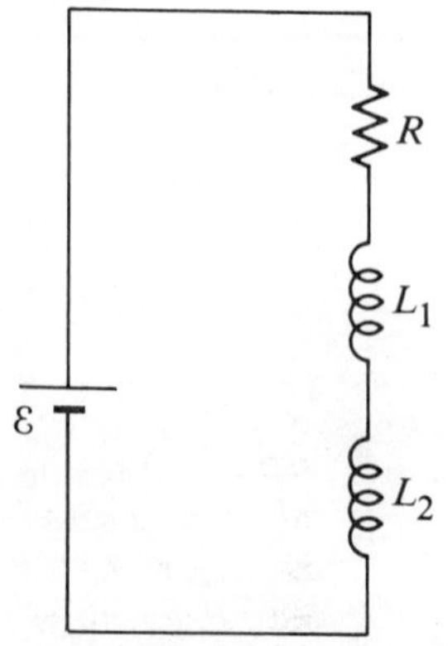

Figure 29.27 (Problem 9).

10. A long wire of radius R carries a steady current I, which is distributed evenly over the cross section of

the conductor. (a) What is the magnetic energy density at $r_1 = R/2$? (b) Calculate the total magnetic energy per unit length stored in the wire. (c) What is the magnetic energy density at $r_2 = 3R$? (d) Calculate the total magnetic energy stored per unit length within a cylindrical shell of inner and outer radii $2R$ and $2.5R$, respectively, when the current-carrying wire is along the axis of the cylindrical shell.

11. An air-core solenoid 0.5 m in length contains 1000 turns and has a cross-sectional area of 1 cm^2. (a) Neglecting end effects, what is the self-inductance? (b) A secondary winding wrapped around the center of the solenoid has 100 turns. What is the mutual inductance? (c) A constant current of 1 A flows in the secondary winding, and the solenoid is connected to a load of $10^3\ \Omega$. The constant current is suddenly stopped. How much charge flows through the load resistor?

12. Two parallel wires, each of radius a, have their centers a distance d apart and carry *equal* currents in *opposite* directions. Neglecting the flux within the wires themselves, calculate the inductance per unit length of such a pair of wires.

13. The current through an inductor L varies in time as shown in Fig. 29.28, where the values of $I(t)$ are in arbitrary units. Sketch a graph of the voltage across the inductor (in arbitrary units) during the time interval 0 to 2 ms.

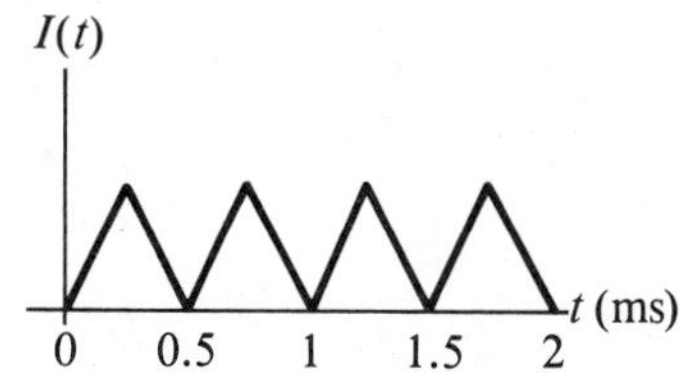

Figure 29.28 (Problem 13).

30 Magnetism in Matter

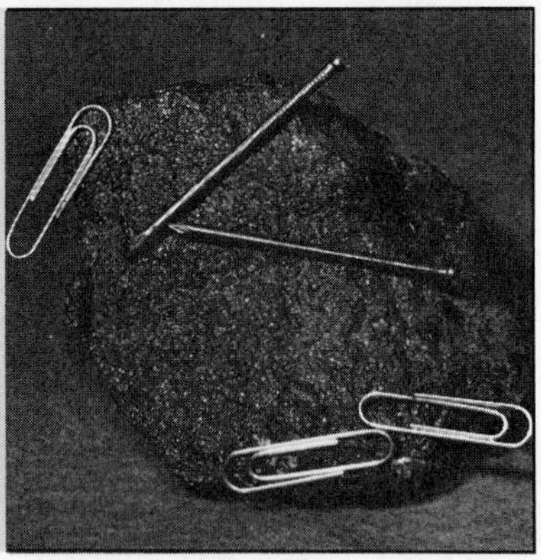

H. L. Levin, *Contemporary Physical Geology*, Saunders College Publishing, 1981

All substances interact with a magnetic field to some extent. Magnetic materials (substances exhibiting strong magnetic interactions) have a wide range of technological importance. Components such as transformers, motors, generators, and inductors commonly use iron cores to change their characteristics for various applications. Communication devices, such as loudspeakers and microphones, often contain magnetic materials. Magnetic tapes are now commonly used for recording sound and television pictures, and thin metallic films of iron alloys or metal oxides are being exploited in computer memories. The operation of such devices and the general nature of matter can be better understood through a study of the basic principles governing the behavior of magnetic substances.

This chapter is concerned with some aspects of the complex processes that occur in magnetic materials. All magnetic effects in matter can be explained on the basis of effective current loops associated with atomic magnetic dipole moments. These atomic magnetic moments can arise both from the orbital motion of the electrons and from an intrinsic, or "built-in," property of the electrons known as *spin*. The mutual forces between these atomic dipoles and the interaction of these dipoles with an external magnetic field are of fundamental importance in describing the behavior of magnetic materials.

The detailed decription of magnetic interactions at the atomic level involves quantum mechanics and is beyond the scope of this text. Therefore, our description of magnetism in matter will be based in part on the experimental fact that the presence of matter in bulk generally modifies the magnetic field produced by currents. For example, when a material is placed inside a toroidal solenoid, the material sets up its own magnetic field, which adds (vectorially) to the field previously present.

We shall describe three categories of materials: paramagnetic, ferromagnetic, and diamagnetic. *Paramagnetic* and *ferromagnetic* materials are those which have atoms with permanent magnetic dipole moments. Substances consisting of atoms with no permanent magnetic dipole moments are called *diamagnetic* substances. For materials whose atoms have permanent magnetic moments, the diamagnetic contribution to the magnetism is usually overshadowed by paramagnetic or ferromagnetic effects.

30.1 THE MAGNETIZATION OF A SUBSTANCE

A ferromagnetic substance such as iron can be magnetized by inserting it into an external magnetic field. In this state of magnetization, the substance acquires a magnetic moment and thus sets up its own magnetic field, which is generally much stronger than the applied magnetic field. All ferromagnetic substances remain magnetized to some extent after the external field is removed. Some materials, such as hard alloy steels, retain their state of magnetization almost indefinitely. These so-called *magnetically hard materials* are used to construct permanent magnets. In contrast, some ferromagnetic materials such as pure iron quickly lose most of their magnetization when the external field is removed. Materials which have this behavior are said to be *magnetically soft.*

In Chapter 27 we found that magnetic fields are generated by electric currents. Similarly, the magnetic moments in a magnetized substance are associated with internal atomic currents. Each atomic current produces a magnetic dipole moment $\mathfrak{M}$. In an unmagnetized substance, these dipoles are randomly oriented, as shown schematically in Fig. 30.1a. When an external magnetic field $\boldsymbol{B}_0$ is applied, the dipoles tend to orient themselves in the direction of the applied field, as in Fig. 30.1b. Hence, the sample acquires a net magnetic moment in the direction of the applied field.

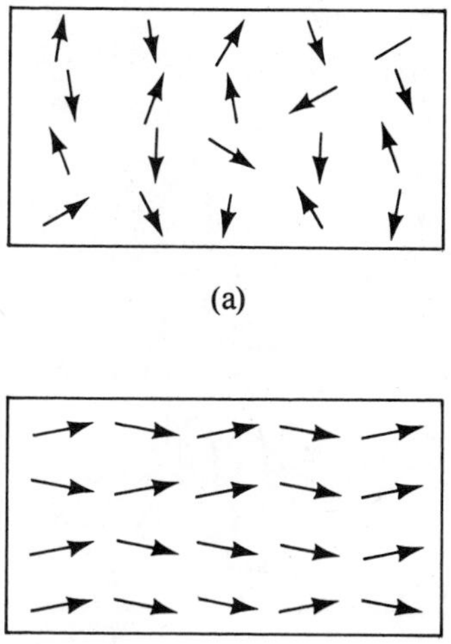

(a)

(b)

Figure 30.1 (a) Random orientation of atomic magnetic dipoles in an unmagnetized substance. (b) When an external field B_0 is applied, the atomic magnetic dipoles tend to align with the field, giving a the sample a net magnetization $\boldsymbol{M}$.

The magnetic state of the material is described by the *magnetization vector* **M**. *The magnitude of* **M** *equals the magnetic moment per unit volume of the sample,* and its direction is that of the sample's magnetic moment. If $\Delta\mathfrak{M}$ is taken to be the magnetic moment contained in a volume element ΔV, then the magnetization vector $\boldsymbol{M}$ associated with this volume element is given by[1]

Magnetization

$$\boldsymbol{M} = \frac{\Delta\mathfrak{M}}{\Delta V} \qquad (30.1)$$

It is convenient to describe the magnetization in terms of an effective surface current that is equivalent to the individual atomic dipoles. Each atomic dipole has a moment given by $\mathfrak{M} = I\,\Delta A$. If an array of dipoles within a volume ΔV are all oriented in the same direction, their current loops all circulate in the same direction, as shown in Fig. 30.2a. Thus, their individual moments point in the same direction, perpendicular to these loops, and produce the net magnetization $\boldsymbol{M}$ (Fig. 30.2b).

Since the individual currents cancel each other at internal boundaries, we can replace the individual current loops by an equivalent surface current I_m, called the *magnetization current* (sometimes called the *amperian current*). Thus, the magnetic effects of a magnetized substance can be represented by an effective current that surrounds the sample. If a volume element of the sample has a cross-sectional area A and length l, then its magnetic moment is given in terms of the magnetization current by $\Delta\mathfrak{M} = I_m A$. Using Eq. 30.1 we can express the magnetization of a uniformly magnetized substance of length l as

$$M = \frac{\Delta\mathfrak{M}}{\Delta V} = \frac{I_m A}{Al} = \frac{I_m}{l} \qquad (30.2)$$

From this result, we see that magnetization has the SI units of amperes per meter. Figure 30.2 shows the correct relative directions of the magnetization and the amperian current.

A typical experimental arrangement used to measure the magnetic properties of matter consists of a toroidal-shaped sample wound with N turns of wire, as in Fig. 30.3. With this geometry, end effects are completely eliminated. When other geometries, such as that of a cylinder, are used, we must account for the demagnet-

[1] In reality, the dipoles in the various volume elements may have *different alignments;* hence $\boldsymbol{M}$ may vary from point to point within the sample. To allow for this possibility, the magnetization should be expressed as the limit of Eq. 30.1 as $\Delta V \to 0$. That is, $\boldsymbol{M} = d\mathfrak{M}/dV$. For the purposes of the present discussion, we shall assume that the magnetization vector is uniform throughout the material.

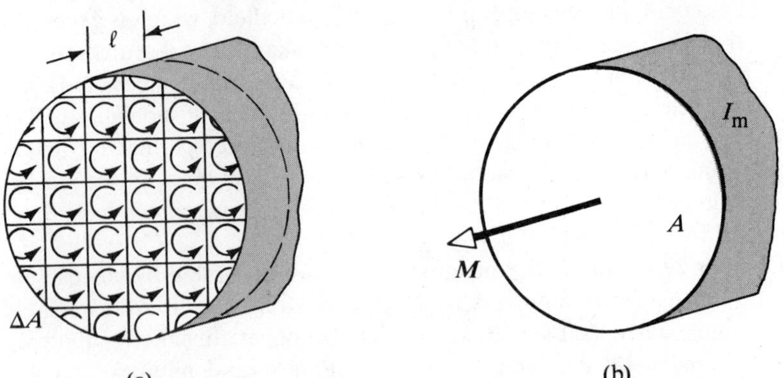

Figure 30.2 (a) Cross section of a magnetized substance, where all the atomic currents circulate in the same direction. (b) The individual atomic currents can be replaced by an effective magnetization current I_m.

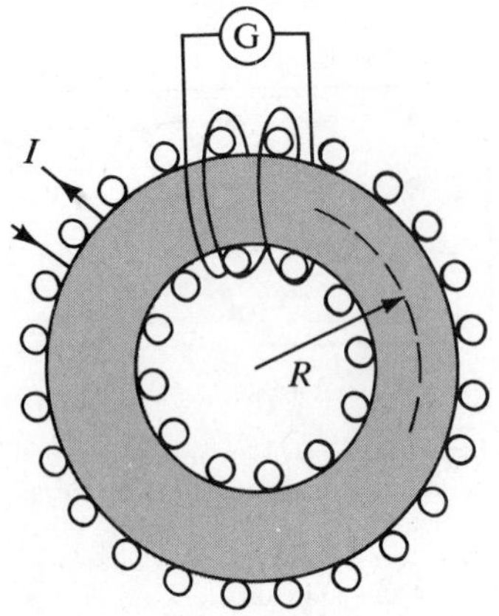

Figure 30.3 Cross section of a toroidal winding arrangement used to measure the magnetic properties of a substance. The material under study fills the core of the toroid, and the secondary circuit containing the galvanometer measures the magnetic flux.

izing effects associated with induced surface dipoles at the ends of the specimen. A secondary coil connected to a galvanometer is used for measuring the magnetic flux BA, where A is the cross-sectional area of the toroid. A fixed current I is passed through the toroidal coil. The magnetic flux through the toroid's cross section is measured first in an empty coil and then in a coil filled with the magnetic substance. The magnetic properties of the substance are then obtained from a comparison of the two measurements.

In Example 27.6, we found that the magnetic induction B_0 inside an *empty* toroid carrying a current I is given by

Magnetic induction in an empty toroidal coil

$$B_0 = \mu_0 \frac{N}{2\pi R} I = \mu_0 n I \tag{30.3}$$

where R is the mean radius of the toroid and n is the number of turns per unit length. When the measurement is repeated with a magnetic core inside the toroid, there are two contributions to the magnetic induction inside the core—one arising from the actual current through the winding, given by Eq. 30.3, and the other from the magnetization associated with the oriented dipoles (or internal currents) in the material. This latter contribution is given by $\mu_0 \boldsymbol{M}$. Hence, the total magnetic induction in the core is given by

Total magnetic induction in the core of a toroid

$$\boldsymbol{B} = \boldsymbol{B}_0 + \mu_0 \boldsymbol{M} \tag{30.4}$$

where the vector notation is necessary since $\boldsymbol{M}$ and $\boldsymbol{B}_0$ are not necessarily in the same direction. If the material is paramagnetic or ferromagnetic, $\boldsymbol{M}$ is in the *same* direction as $\boldsymbol{B}_0$, and so the magnetic induction $\boldsymbol{B}$ is enhanced by the material. For diamagnetic materials, $\boldsymbol{B} < \boldsymbol{B}_0$, since $\boldsymbol{M}$ is *opposite* $\boldsymbol{B}_0$.

It is convenient and traditional to introduce a new field quantity, $\boldsymbol{H}$, called the *magnetic intensity* vector, which is defined by

Magnetic intensity

$$\boldsymbol{H} = \frac{\boldsymbol{B}}{\mu_0} - \boldsymbol{M} \tag{30.5}$$

Solving for $\boldsymbol{B}$ gives

$$\boldsymbol{B} = \mu_0 \boldsymbol{H} + \mu_0 \boldsymbol{M} \tag{30.6}$$

Comparing this with Eq. 30.4, we see that $\boldsymbol{H}$ is proportional to the original external

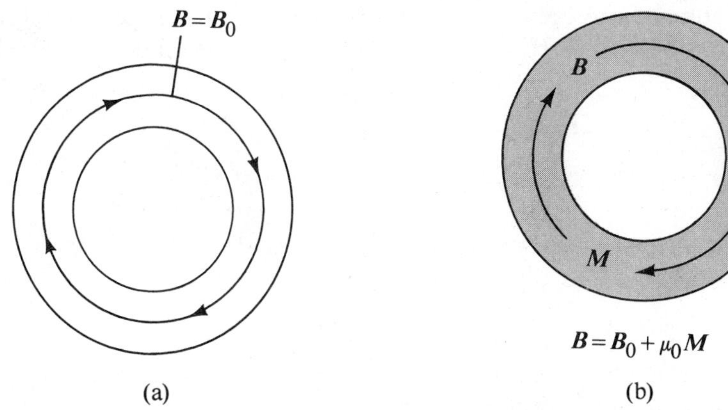

Figure 30.4 Fields in a toroidal coil. (a) In an empty toroid (no magnetic core), the magnetic field is $\mathbf{B}_0$. (b) When the core of the toroid is filled with a magnetic material, the field inside changes as a result of the magnetization $\mathbf{M}$ of the material.

field:

$$\mathbf{B}_0 = \mu_0 \mathbf{H} \tag{30.7}$$

For instance, if we use this result and Eq. 30.3, we find that the magnetic field intensity inside a toroid has a magnitude given by

Magnetic field intensity inside a toroid

$$H = nI \tag{30.8}$$

where H is measured in ampere-turns/meter. Thus, we see that the magnetic flux density $\mathbf{B}$ is composed partly of $\mu_0\mathbf{H}$, which is due to the externally applied current, and partly to $\mu_0\mathbf{M}$, due to the magnetic moments in the magnetized material. A description of the fields in the empty and filled toroid are shown in Fig. 30.4.

For a large class of isotropic materials, specifically, paramagnetic and diamagnetic substancs, the magnetization $\mathbf{M}$ is proportional to the applied field $\mathbf{H}$. In these linear substances, we can write

$$\mathbf{M} = \chi_m \mathbf{H} \tag{30.9}$$

where χ_m is a dimensionless proportionality constant called the *magnetic susceptibility*. This linear relationship *does not* apply to ferromagnetic materials. Typical values for this constant are given in Table 30.1. Note that χ_m is positive for para-

TABLE 30.1 Magnetic Susceptibilities of Some Paramagnetic and Diamagnetic Materials

MATERIAL	χ_m
Aluminum	2.3×10^{-5}
Bismuth	-1.66×10^{-5}
Copper	-0.98×10^{-5}
Gold	-3.6×10^{-5}
Lead	-1.7×10^{-5}
Magnesium	1.2×10^{-5}
Silver	-2.6×10^{-5}
Sodium	-0.24×10^{-5}
Tungsten	6.8×10^{-5}
Water	-0.88×10^{-5}
Hydrogen (1 atm)	-9.9×10^{-9}
Oxygen (1 atm)	2.1×10^{-6}

magnetic materials (where $\mathbf{M}$ is in the same direction as $\mathbf{H}$) and negative for diamagnetic materials ($\mathbf{M}$ opposite $\mathbf{H}$). Substituting Eq. 30.9 for $\mathbf{M}$ into Eq. 30.5 and solving for $\mathbf{B}$, we get

$$\mathbf{B} = \mu_o(\mathbf{H} + \mathbf{M}) = \mu_o(\mathbf{H} + \chi_m \mathbf{H})$$

$$\mathbf{B} = \mu_o(1 + \chi_m)\mathbf{H} = \mu \mathbf{H} \tag{30.10}$$

The constant μ is called the *magnetic permeability* and has the value

Magnetic permeability

$$\mu = \mu_o(1 + \chi_m) \tag{30.11}$$

Materials may also be classified in terms of how their permeability μ compares to μ_o (the permeability of free space) as follows:

Diamagnetic	$\mu < \mu_o$
Paramagnetic	$\mu > \mu_o$
Ferromagnetic	$\mu \gg \mu_o$

Since χ_m is very small for diamagnetic and paramagnetic materials (Table 30.1), μ is very close to μ_o in these cases. For ferromagnetic materials, μ is typically several thousand times larger than μ_o. However, it must again be emphasized that Eq. 30.9 is not valid for all materials under all conditions. For example, permanent magnetic materials such as iron, nickel, and cobalt are nonlinear in that their susceptibility is field-dependent. We shall describe these materials in more detail in Section 30.5.

Example 30.1

A toroidal winding carrying a current of 5 A is wound with 300 turns/m of wire. The core is iron, which has a magnetic permeability of $5000\mu_o$ under the given conditions. Find H, B, and M inside the iron core.

Solution: Using Eqs. 30.8 and 30.10, we get

$$H = nI = \left(300\ \frac{\text{turns}}{\text{m}}\right)(5\ \text{A}) = 1500\ \frac{\text{A}\cdot\text{turns}}{\text{m}}$$

$$B = \mu H = 5000\mu_o H$$
$$= 5000\left(4\pi \times 10^{-7}\ \frac{\text{Wb}}{\text{A}\cdot\text{m}}\right)\left(1500\ \frac{\text{A}\cdot\text{turns}}{\text{m}}\right) = 9.43\ \text{T}$$

Note that B is 5000 times larger than the field in the absence of iron!

Finally, from Eq. 30.5 we find that

$$M = \frac{B}{\mu_o} - H = \frac{5000\mu_o H}{\mu_o} - H$$
$$\approx 5000H = 5000(1500\ \text{A}\cdot\text{turns/m}) = 7.5 \times 10^6\ \text{A/m}$$

where $\mathbf{M}$ is parallel to $\mathbf{H}$.

Example 30.2

Calculate the self-inductance of a toroidal winding whose core is filled with a material of permeability μ. The winding has N turns, and the toroid has a mean radius R.

Solution: The magnetic induction inside the toroid can be calculated using Eqs. 30.8 and 30.10.

$$B = \mu H = \mu \frac{NI}{2\pi R}$$

The total magnetic flux through the cross section of the toroidal coil of area A is

$$\Phi_m = BA = \mu \frac{NIA}{2\pi R}$$

Hence, the self-inductance is

$$L = \frac{N\Phi_m}{I} = \mu\left(\frac{N^2 A}{2\pi R}\right)$$

If the material is ferromagnetic, the inductance increases by a factor μ/μ_o over that of an unfilled toroidal winding.

Q1. What is the difference between magnetic induction and magnetic intensity in a magnetic substance?

Q2. Why is $\mathbf{M} = 0$ in a vacuum? What is the relationship between $\mathbf{B}$ and $\mathbf{H}$ in a vacuum?

30.2 THE MAGNETIC MOMENT OF ATOMS

In this section we describe the magnetic moments associated with individual atoms, which are the source of the magnetization of any substance. We shall use a classical model in which the electrons are assumed to follow circular orbits about the much heavier nucleus. In this model, the circulating electron is viewed as a tiny current loop; the atomic magnetic moment is just the moment associated with this orbital motion. The treatment also shows the relationship between the orbital magnetic moment and the angular momentum of the electron. Although this model has many inherent defects, its predictions are in good agreement with the correct theory from quantum physics.

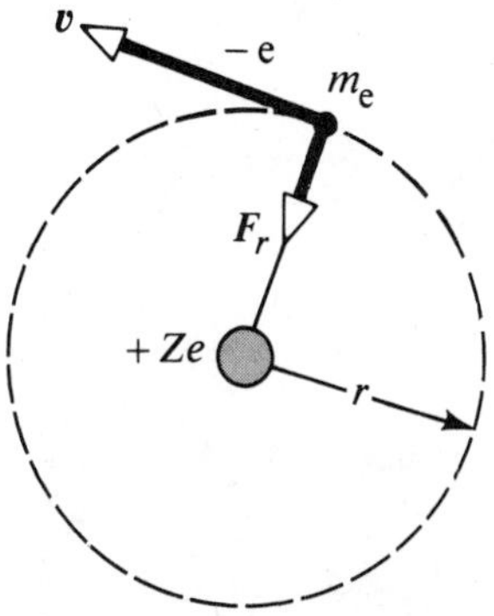

Figure 30.5 An electron moving in a circular orbit around the nucleus of an atom has an orbital angular momentum m_evr out of the page and a magnetic moment $\frac{1}{2}evr$ into the page.

Consider the electron with charge e moving with speed v in a circular orbit of radius r about the nucleus, as in Fig. 30.5. Since the electron travels a distance of $2\pi r$ (the circumference of the circle) in a time T, where T is the time for one revolution, the orbital speed of the electron is

$$v = \frac{2\pi r}{T}$$

The effective current associated with this circulating charge equals the charge divided by the time for one revolution:

$$I = \frac{e}{T} = \frac{ev}{2\pi r}$$

The magnetic moment $\mathcal{M}$ associated with this current loop equals the product IA, where $A = \pi r^2$ is the area of the orbit (see Section 26.4):

$$\mathcal{M} = IA = \left(\frac{ev}{2\pi r}\right)\pi r^2 = \frac{1}{2}evr \tag{30.12}$$

To obtain an expression for v, we can equate the attractive coulombic force between the electron and nucleus (the total charge of the nucleus is $+Ze$, where Z stands for the number of protons) to the required centripetal force. Using Newton's second law and taking m_e as the mass of the electron we get

$$k\frac{Ze^2}{r^2} = m_e a_r = m_e \frac{v^2}{r}$$

$$v = e\left(\frac{kZ}{m_e r}\right)^{1/2} \tag{30.13}$$

where k is Coulomb's constant. Substituting this result into Eq. 30.12, we find that

Orbital magnetic moment

$$\mathcal{M} = \frac{e^2}{2}\left(\frac{kZr}{m_e}\right)^{1/2} \tag{30.14}$$

For instance, to obtain the value of the magnetic moment for the hydrogen atom in its ground state, we take $Z = 1$ and $r = 0.529 \times 10^{-10}$ m, which gives

$$\mathcal{M} = \frac{(1.60 \times 10^{-19}\ \text{C})^2}{2}\left[\left(8.99 \times 10^9 \frac{\text{N}\cdot\text{m}^2}{\text{C}^2}\right)\frac{(0.529 \times 10^{-10}\ \text{m})}{9.11 \times 10^{-31}\ \text{kg}}\right]^{1/2}$$

$$= 9.25 \times 10^{-24}\ \text{A}\cdot\text{m}^2$$

The orbital angular momentum L of the electron is given by

$$L = m_e vr$$

Comparing this with Eq. 30.12, we see that

$$\mathcal{M} = \left(\frac{e}{2m_e}\right)L \tag{30.15}$$

That is, *the magnetic moment of the electron is proportional to its orbital angular momentum.* This classical result is in agreement with the general quantum mechanical treatment of orbital angular momentum. Note that since the electron has a negative charge, the vector $\mathfrak{M}$ is in fact *opposite* the vector **L**. Both vectors are perpendicular to the plane of the orbit.

In atoms containing many electrons, most of the electrons do not contribute to the net magnetic moment. This is because their magnetic moments tend to pair off in such a way that they cancel each other. Hence, only those electrons that are not paired are responsible for the moment.

Another contribution to an atom's magnetic moment is the *spin* of the electron, which is an inherent property (like charge or rest mass). (This classical description of a spinning electron should not be taken literally. The property of spin can be understood only through a quantum mechanical model, the details of which are beyond the level of this text.) The magnetic moment associated with the spin of a free electron, which is called the *Bohr magneton,* has the value

Bohr magneton

$$\mathfrak{M}_s = 9.27 \times 10^{-24}\ \mathrm{A \cdot m^2} \tag{30.16}$$

Classically, we can view the electron as a charged sphere spinning about its axis, as in Fig. 30.6. This spinning motion produces an effective current loop and hence a magnetic moment, which is of the same order of magnitude as that due to the orbital motion. In atoms or ions containing many electrons, the electrons usually pair up with their spins opposed to each other, which results in cancellation of the spin magnetic moments. However, atoms with an odd number of electrons must have at least one "unpaired" electron and a corresponding spin magnetic moment.[2]

The nucleus of an atom can also have a magnetic moment. However, this contribution to the net magnetic moment is about 10^3 times smaller than that due to the electronic moment. This can be understood by referring to Eq. 30.15. Since the nuclear orbital angular momentum is comparable in magnitude to the electronic angular momentum, the nuclear magnetic moment must be smaller than the electronic moment by the ratio $m_e/m_p \approx 5.4 \times 10^{-4}$.

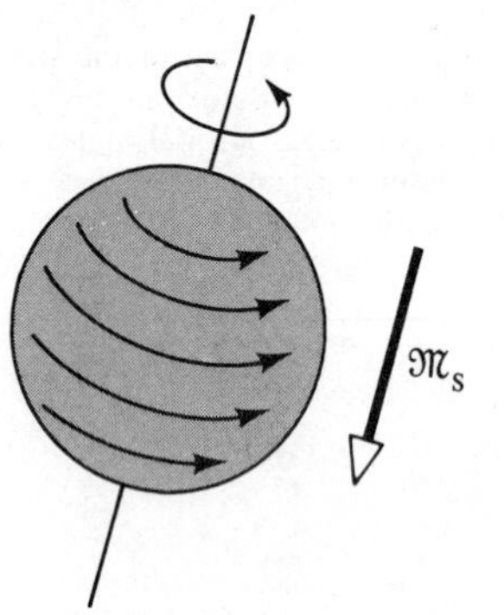

Figure 30.6 Model of a spinning electron. The magnetic moment $\mathfrak{M}_s$ can be viewed as arising from the effective current loops associated with a spinning charged sphere.

[2] Quantum theory shows that the spin angular momentum is quantized in units of $\frac{1}{2}\hbar$, where $\hbar$ (Planck's constant divided by 2π) $= 1.05 \times 10^{-34}$ J · s. The electron's spin is, in fact, $\frac{1}{2}\hbar$. For *any* particle, the *orbital* angular momentum is quantized in *integral multiples* of $\hbar$.

Example 30.3

Estimate the *maximum magnetization* in a long cylinder of iron, assuming there is one unpaired electron spin per atom.

Solution: The maximum magnetization, called the *saturation magnetization,* is obtained when all the magnetic moments in the sample are aligned. If the sample contains n atoms per unit volume, then the saturation magnetization M_s has the value

$$M_s = n\mathfrak{M}$$

where $\mathfrak{M}$ is the magnetic moment per atom. Since the molecular weight of iron is 55 g/mole and its density is 7.9 g/cm^3, the value of n is 8.5×10^{28} atoms/m^3. Assuming each atom contributes one Bohr magneton (due to one unpaired spin) to the magnetic moment, we get

$$M_s = \left(8.5 \times 10^{28}\ \frac{\text{atoms}}{\text{m}^3}\right)\left(9.27 \times 10^{-24}\ \frac{\text{A} \cdot \text{m}^2}{\text{atom}}\right)$$
$$= 7.9 \times 10^5\ \text{A/m}$$

This is about one half the experimentally determined saturation magnetization for annealed iron, which indicates that there are actually *two* unpaired electron spins per atom.

30.3 PARAMAGNETISM

Paramagnetic substances have a positive but small susceptibility ($0 < \chi_m \ll 1$), which is due to the presence of atoms or ions with *permanent* magnetic dipole moments. These dipoles interact only weakly with each other and are randomly oriented in the absence of a magnetic field. When placed in an external magnetic field, the atomic dipoles tend to line up with the field. However, this alignment process must compete with the effects of thermal motion, which tend to randomize the dipole orientations. The degree of alignment, and hence the net magnetization, are proportional to the external field and inversely proportional to the absolute temperature.

Consider a paramagnetic material containing n atoms per unit volume, each with a magnetic moment $\mathfrak{M}$. If *all* the dipoles were aligned, the sample would have a net magnetization equal to $n\mathfrak{M}$. (As mentioned in Example 30.3, this maximum magnetization is often called the *saturation magnetization*, M_s.) Let us assume that in an applied field, the dipole can align either along or opposite the field,[3] as in Fig. 30.7. A dipole aligned with the field has the lowest *potential energy*,[4] given by $U_1 = -\mathfrak{M}B$, whereas a dipole aligned against the field has the highest energy, $U_2 = \mathfrak{M}B$. Hence, the difference in energy between these two orientations is given by $\Delta U = 2\mathfrak{M}B$. This is to be compared with the average thermal energy per atom, $\frac{3}{2}kT$, where k is Boltzmann's constant and T is the absolute temperature. The ratio of these two energies gives a measure of the excess fraction f of moments aligned with the field. Although we shall not derive this result, the fraction f is found to be

$$f = \frac{\mathfrak{M}B}{3kT} \tag{30.17}$$

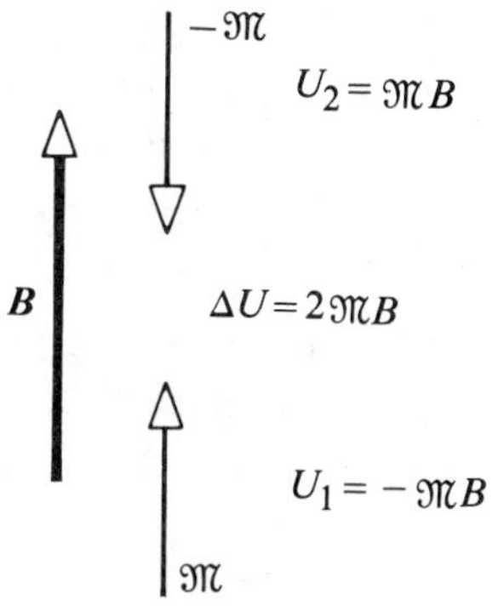

Figure 30.7 When a magnetic moment $\mathfrak{M}$ is aligned parallel to the external field $\boldsymbol{B}$, its potential energy $U_1 = -\mathfrak{M}B$. When $\mathfrak{M}$ is aligned antiparallel to $\boldsymbol{B}$, its potential energy is $U_2 = \mathfrak{M}B$.

That is, as T increases, the excess fraction of moments aligned with the field decreases.

Using this result, we see that the magnetization of a paramagnetic substance containing n paramagnetic atoms per unit volume is given by

$$M = fn\mathfrak{M} = \frac{n\mathfrak{M}^2 B}{3kT} \tag{30.18}$$

Since n, $\mathfrak{M}$, and k are all constants, we can express Eq. 29.17 as

$$M = C\frac{B}{T} \tag{30.19}$$

Curie's law

This is known as *Curie's law*, after its discoverer, Pierre Curie (1859–1906), and the constant C is called *Curie's constant*.[5] Equation 30.19 shows that the magnetization increases with increasing applied field and with decreasing temperature. When $B = 0$, the magnetization is zero, corresponding to a random orientation of dipoles. At very high fields or very low temperatures, the magnetization approaches its maximum, or saturation, value, $M_s = n\mathfrak{M}$, corresponding to a complete alignment of all dipoles. A plot of M versus B/T for the paramagnetic substance chromium potassium alum is shown in Fig. 30.8. The saturation magnetization is reached only at very low temperatures. The maximum value of 3 Bohr magnetons per ion corresponds to the fact that the paramagnetic ion Cr^{3+} has three unpaired electron spins. Note that the curve is approximately linear (and hence obeys Curie's law) at very low values of B/T.

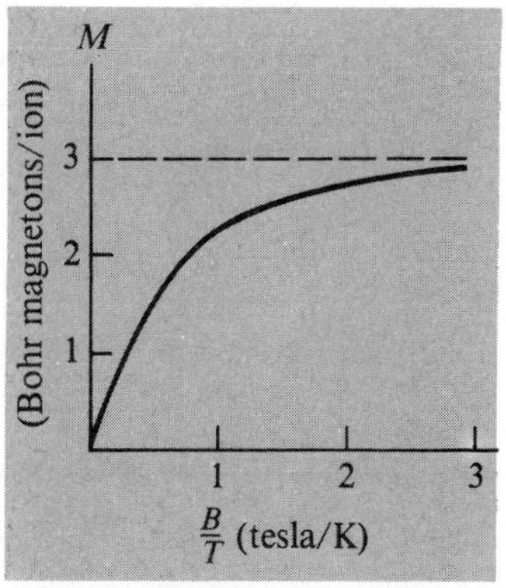

Figure 30.8 A graph of magnetization versus B/T for the paramagnetic substance chromium potassium alum, $CrK(SO_4)_2 \cdot 12H_2O$. The Cr^{3+} ion is the paramagnetic species in this substance.

[3] A correct quantum mechanical description shows that the magnetic dipoles are allowed to have only discrete orientations with respect to the magnetic field. In this case, where the moment is associated with a spin of magnitude $\frac{1}{2}$, the only two possible alignments are along and opposite the field.

[4] In general, the potential energy of a magnetic dipole $\mathfrak{M}$ in an external field $\boldsymbol{B}$ is $U = -\mathfrak{M} \cdot \boldsymbol{B}$.

[5] For a formal proof of this result, see C. Kittel, *Introduction to Solid State Physics*, New York, Wiley, 1971, Chapter 15.

Example 30.4

A paramagnetic substance is at a temperature of 4.2 K. What value of the external field B will produce a magnetization equal to 75% of the maximum magnetization, M_s? (Assume that the magnetic moment per atom has the value $\mathcal{M} = 9.27 \times 10^{-24}\ \text{A} \cdot \text{m}^2$.)

Solution: Since $M_s = n\mathcal{M}$ and we require that $M = 0.75M_s$, Eq. 30.18 gives

$$0.75n\mathcal{M} = \frac{n\mathcal{M}^2 B}{3kT}$$

$$B = 2.25\frac{kT}{\mathcal{M}} = \frac{2.25(1.38 \times 10^{-23}\ \text{J/K})(4.2\ \text{K})}{9.27 \times 10^{-24}\ \text{A} \cdot \text{m}^2}$$

$$= 14.1\ \text{T}$$

This is a very large field, which could be produced with a superconducting magnet. However, if the same calculation is performed for $T = 300$ K (room temperature), a value of 1000 T is obtained for the required magnetic field. This is far greater than any steady field yet produced in the laboratory.

Q3. Explain why some atoms have permanent magnetic dipole moments and others do not.

Q4. What factors can contribute to the total magnetic dipole moment of an atom?

30.4 DIAMAGNETISM

When an external magnetic field is applied to a diamagnetic substance, such as bismuth, the only magnetic moments induced in the substance are aligned *against* the external field. Consequently, the magnetization is *opposite* the external field. Correspondingly, a diamagnetic substance has a *negative* susceptibility. Furthermore, one finds that a diamagnetic sample is *repelled* when placed near the pole of a strong magnet (in contrast with a paramagnetic sample, which is *attracted*). Although the effect of diamagnetism is present in all matter, it is weak compared with paramagnetism and ferromagnetism. Therefore, it can be observed only in those materials whose atoms have no permanent magnetic dipole moment.

We can obtain a qualitative understanding of diamagnetism by applying Lenz's law to the orbital motion of electrons in atoms. Consider two electrons moving in circular orbits with the same speed but in opposite directions, as in Fig. 30.9a. Because their magnetic moments are equal in magnitude and opposite in direction, the two moments cancel each other in the absence of an external field (Fig. 30.9a).

Now consider what happens when an external magnetic field $\boldsymbol{B}$ is applied perpendicular to the plane of the orbits, as in Fig. 30.9b, and increasing into the paper. Lenz's law tells us that currents will be induced in both loops such as to oppose the change in flux. The change in the currents, and corresponding change in the magnetic moments, occur as a result of the change in speed of the electrons (assuming the radii of the orbits are constant[6]). That is, the change in the motion of the electrons is equivalent to an induced current whose magnetic flux opposes the change in flux through the orbit. Since the external magnetic field is increasing into the paper, the *change* in the magnetic moment of *each* charge will be out of the paper, opposite the applied field. Hence, the magnetic moments of the electrons will no longer cancel.

Let us estimate the value of the induced magnetic moment by first determining

[6]For further details, see E. M. Purcell, *Electricity and Magnetism*, McGraw-Hill, 1965, pp. 370–377.

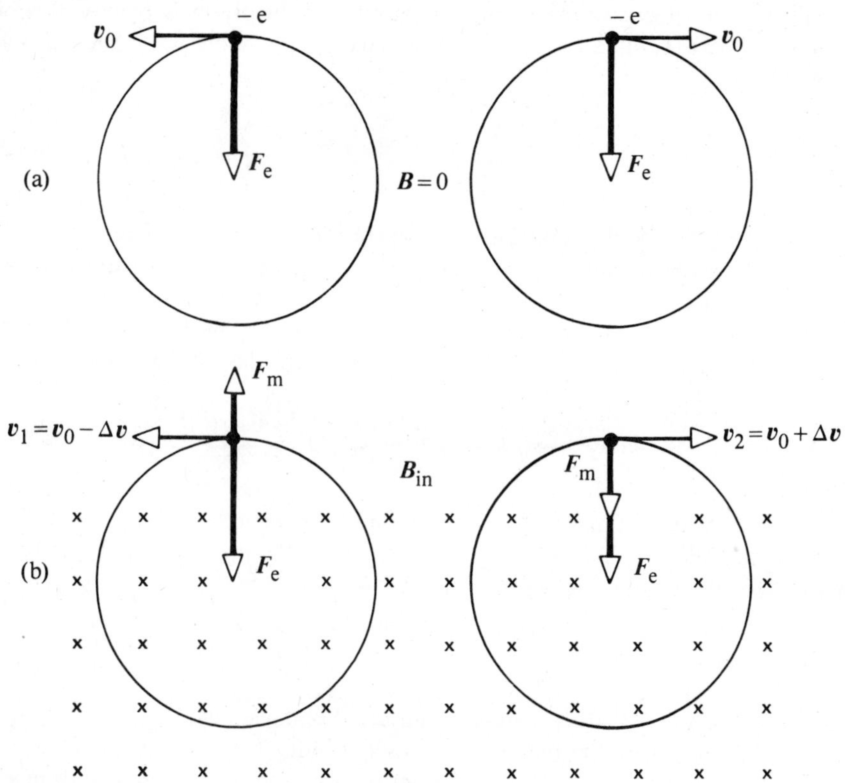

Figure 30.9 (a) Two electrons circulating in opposite directions about a nucleus in the absence of a magnetic field. Each electron experiences a centripetal force $\boldsymbol{F}_e$, which is of electrostatic origin, and each has the same speed v_0. (b) When a magnetic field $\boldsymbol{B}$ is introduced, the centripetal force is modified by an additional magnetic force $\boldsymbol{F}_m$. This decreases the speed of one electron (on the left) and increases the speed of the other (on the right).

the change in speed of the electron as a result of the external magnetic field. In the absence of a magnetic field, each electron moves in a circular orbit under the influence of a centripetal force $\boldsymbol{F}_e$, corresponding to the electrostatic force between the electron and nucleus (Fig. 30.9a). From Newton's second law, we have

$$F_e = \frac{m_e v_0^2}{r} \tag{30.20}$$

After the magnetic field is applied, each electron experiences a magnetic force $-e\boldsymbol{v} \times \boldsymbol{B}$, which also contributes to the centripetal force. The direction of the force $-e\boldsymbol{v} \times \boldsymbol{B}$ is radially outward for the electron on the left in Fig. 30.9b and radially inward for the electron on the right. Since $\boldsymbol{v}$ is perpendicular to $\boldsymbol{B}$, the magnitude of the magnetic force $\boldsymbol{F}_m$ is evB. Hence, we can now write Newton's second law in the form

$$F_e \pm evB = \frac{m_e v^2}{r} \tag{30.21}$$

where v is the speed in the presence of a magnetic field and the two signs refer to the two electrons in Fig. 30.9b. (The minus sign applies to the figure on the left.) Subtracting Eq. 30.20 from Eq. 30.21 and assuming that r remains constant (so that F_e remains constant), we get

$$\pm evB = \frac{m_e v^2}{r} - \frac{m_e v_0^2}{r} = \frac{m_e}{r}(v + v_0)(v - v_0) \tag{30.22}$$

If the change in velocity is very small compared with the original velocity, then we can approximate $v_0 + v \approx 2v$. Since the change in velocity is $\Delta v = v - v_0$, Eq. 30.22 gives

$$\Delta v = \pm \frac{eBr}{2m_e} \tag{30.23}$$

where the negative and positive signs indicate that the electron on the left in Fig. 30.9b slows down while the one on the right speeds up. However, the change in the magnetic moment, $\Delta\mathfrak{M}$, is in the *same* direction for both particles (Fig. 30.10). Since the magnetic moment per electron is given by Eq. 30.12, $\mathfrak{M}_1 = \frac{1}{2}erv$, the change in the magnetic moment for the *two* electrons has a magnitude given by

$$\Delta\mathfrak{M} = 2\left(\frac{1}{2}er\,\Delta v\right) = er\left(\frac{eBr}{2m_e}\right) = \frac{Be^2r^2}{2m_e} \tag{30.24}$$

To estimate the order of magnitude of this induced moment, let us take $B = 5$ T and $r = 10^{-10}$ m. (We use this value of r because this is the approximate radius of an atom.) Using these values, we get

$$\Delta\mathfrak{M} = \frac{5\,\text{T}(1.6 \times 10^{-19}\,\text{C})^2(10^{-10}\,\text{m})^2}{2(9.1 \times 10^{-31}\,\text{kg})} \approx 7 \times 10^{-28}\,\text{A} \cdot \text{m}^2$$

This result is about 10^4 times *smaller* than one Bohr magneton.

We can also estimate the diamagnetic susceptibility. Since the magnetization is given by $M = n\Delta\mathfrak{M}$, where n is the number of atoms per unit volume and since $\boldsymbol{M}$ is opposite $\boldsymbol{H}$, Eq. 30.9 and Eq. 30.24 give

$$\chi_m = -\frac{M}{H} = -\frac{n\,\Delta\mathfrak{M}}{B/\mu_o} = -\frac{\mu_o n e^2 r^2}{2m_e} \tag{30.25}$$

Taking $n = 5 \times 10^{28}$ atoms/m^3 (a typical value for a solid) and $r = 10^{-10}$ m, we find that $\chi_m \approx -8.8 \times 10^{-6}$. This result is in good agreement with the typical values given in Table 30.1. Experimentally, the susceptibility of a diamagnetic substance is almost independent of temperature.

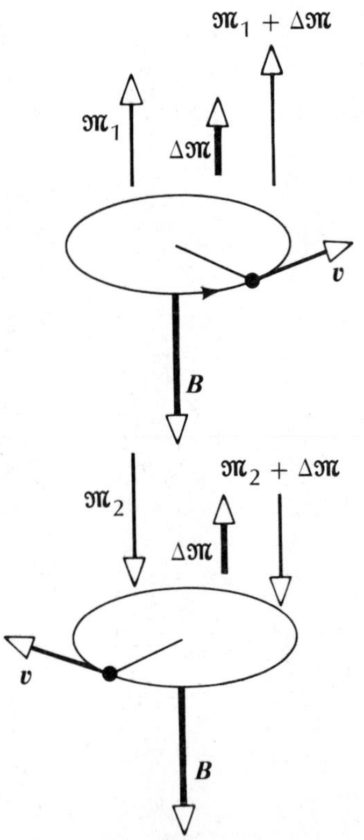

Figure 30.10 Although the two electrons circulate in opposite directions, the *change* in the magnetic moment $\Delta\mathfrak{M}$ is in the *same direction* when a magnetic field is introduced—where $\Delta\mathfrak{M}$ is opposite $\boldsymbol{B}$. Therefore, the two moments *do not* cancel.

Q5. Why is the susceptibility of a diamagnetic substance negative?

Q6. Why can the effect of diamagnetism be neglected in a paramagnetic substance?

30.5 FERROMAGNETISM

Ferromagnetic substances, such as iron, cobalt, and nickel, contain atomic magnetic moments that easily align parallel to each other even in a weak external magnetic field. Furthermore, once the moments are aligned, the magnetization will persist when the external field is removed. This permanent alignment is due to a strong coupling between neighboring moments, which can be understood only by using quantum mechanics. The corresponding forces, which are *electrostatic* in origin, are called *exchange forces.* At sufficiently high temperatures, the thermal energy is large enough to overcome the strong ferromagnetic coupling. Consequently, when the temperature reaches or exceeds a critical temperature, called the *Curie temperature*, the ferromagnetic substance loses its spontaneous magnetization and becomes paramagnetic.

Figure 30.11 is a typical magnetization-temperature curve for a ferromagnet

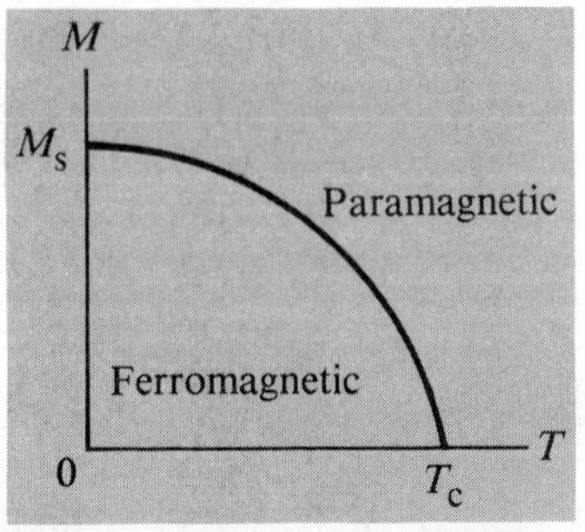

Figure 30.11 Plot of the magnetization versus absolute temperature for a ferromagnetic substance. The magnetic moments are aligned (ordered) below the Curie temperature T_c, where the substance is ferromagnetic. The substance becomes paramagnetic (disordered) above T_c.

TABLE 30.2 Curie Temperature for Several Ferromagnetic Substances

SUBSTANCE	T_c (K)
Iron	1043
Cobalt	1394
Nickel	631
Gadolinium	317
Fe_2O_3	893

(with zero applied field, $\boldsymbol{H} = 0$). Table 30.2 lists the Curie temperatures, T_c, for several ferromagnetic substances. Since the permeability of ferromagnetic substances is very large, the magnetization of a ferromagnet is often several thousand times greater than the applied magnetic intensity, H.

In contrast to paramagnetic substances, the magnetization of ferromagnetic substances is *not* a linear function of the applied field; the susceptibility of a ferromagnetic substance varies as the applied field changes.

Within any ferromagnetic material, there are small, discrete regions called *domains*, in which all the magnetic moments are aligned. The moments are due to ions with unpaired electron spin strongly coupled by an "exchange force." The size of the domains ranges from about 10^{-4} cm to 0.1 cm. In an unmagnetized substance, the domains are randomly oriented. When an external field is applied, the magnetic moments of each domain tend to align with the field, which results in a net magnetization. This effect usually occurs in polycrystalline materials, where individual crystallites form single domains. In large, single-crystal specimens, the boundaries of the domain, called *domain walls,* may either shift or rotate, as in Fig. 30.12. Again, the result is a net magnetization in the direction of the applied field.[7]

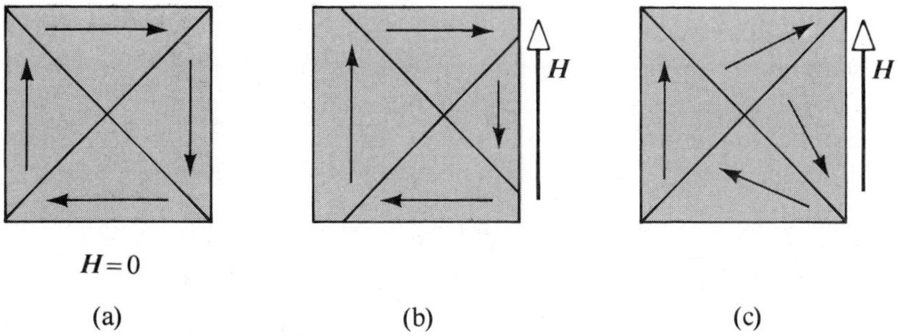

Figure 30.12 Representation of a single-crystal ferromagnetic substance. (a) An unmagnetized sample containing four domains. (b) Magnetization as the result of a shift in domain boundaries. (c) Magnetization due to a rotation of domains.

We shall now describe what happens when a ferromagnetic substance is placed in the field of a toroidal coil. Figure 30.13 shows a plot of the magnetization M versus the applied field H, which can be slowly varied at will. The field H is controlled by the current in the toroidal windings. If the sample is *initially unmagnetized,* an increase in H will cause M to increase from zero until it reaches its saturation value, M_s, at the point P. At this point, all the dipoles are aligned with $\boldsymbol{H}$ and no additional increase in M can be obtained by increasing H. Next, the applied field is reduced in

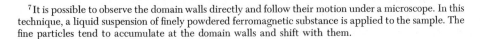

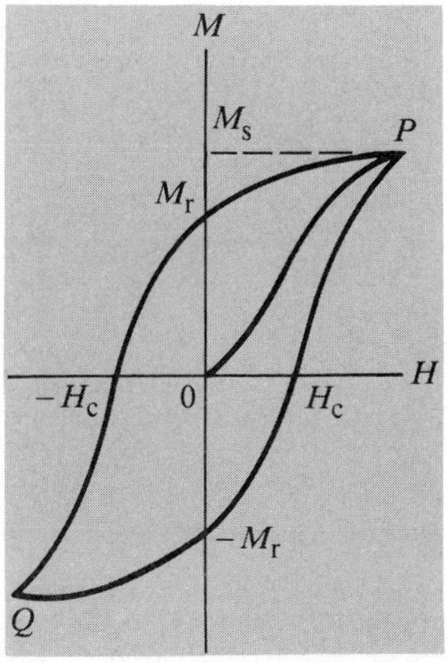

Figure 30.13 The hysteresis loop, M versus H (outer curve), for a ferromagnetic substance. All the moments are aligned at P when M reaches the saturation magnetization, M_s.

[7] It is possible to observe the domain walls directly and follow their motion under a microscope. In this technique, a liquid suspension of finely powdered ferromagnetic substance is applied to the sample. The fine particles tend to accumulate at the domain walls and shift with them.

intensity and the magnetization follows a different path, reducing to a value M_r at $H = 0$. The value M_r, called the *remanent magnetization,* is due to the permanent alignment of a large number of atomic moments in the absence of an external field. If the field $\boldsymbol{H}$ is reversed in direction and increased in strength, the magnetization reduces to zero at a particular value H_c, called the *coercive field.* For even larger negative values of H, the magnetization reverses its direction until the point Q in Fig. 30.13 is reached, where the saturation magnetization becomes $-M_s$. A similar sequence of events occurs as the field strength is decreased to zero and then increased in the original (positive) direction. If the field is increased sufficiently, the curve returns to the point P, where the sample again has its maximum magnetization, M_s.

The effect just described, called *magnetic hysteresis,* shows that the magnetization of a ferromagnetic substance depends on the history of the substance as well as on the strength of the applied field. (The word *hysteresis* literally means *to lag behind.*) One often says that a ferromagnetic substance has a "memory" since it remains magnetized after the external field is turned off. The closed loop in Fig. 30.13 is referred to as a *hysteresis loop.* Its shape and size depend on the properties of the substance and on the strength of the maximum applied field. Very often, the effect of hysteresis is represented by a plot of B versus H, as in Fig. 30.14. Note that the shape is similar to that of the M versus H curve. The hysteresis loop for "hard" ferromagnetic materials (used in permanent magnets) are characterized by a "fat" hysteresis loop and a large *remanent field* B_r, as in Fig. 30.14a. Such materials cannot be easily demagnetized by an external field. This is in contrast to "soft" ferromagnetic materials, such as iron, which have a very narrow hysteresis loop (Fig. 30.14b) and a small remanent field. Such materials are easily magnetized and demagnetized. An ideal soft ferromagnet would exhibit no hysteresis, and hence its remanent field would be zero.

The B versus H curve is useful for another reason. *The area enclosed by the curve represents the change in magnetic energy per unit volume of the sample, which is the energy acquired in the magnetization process.* This energy originates from the source of the external field, that is, the emf in the circuit of the toroidal coil. When the magnetization cycle is repeated, dissipative processes within the material result in a transformation of magnetic energy into internal thermal energy, which raises the temperature of the substance. For this reason, devices subjected to alternating fields (such as transformers) use cores made of soft ferromagnetic substances, which have a "thin" hysteresis loop and a correspondingly small energy loss per cycle.

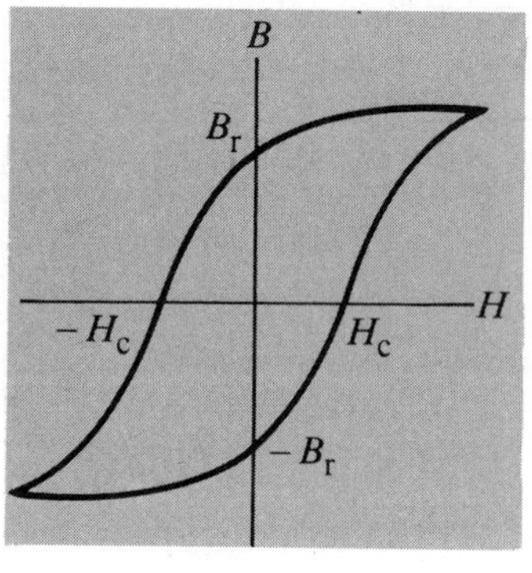

(a)

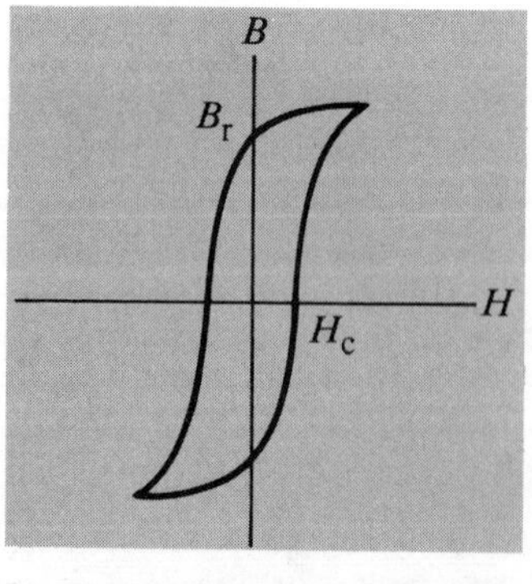

(b)

Figure 30.14 (a) The hysteresis loop (plotted as B versus H) for a hard ferromagnet is characterized by a large remanent field B_r and a large coercive field H_c. (b) The hysteresis loop for a soft ferromagnet has small values of B_r and H_c.

Other Magnetic Substances

It is interesting to note that there are two other classes of magnetic substances, which are related to ferromagnets. These are known as *antiferromagnetic* and *ferrimagnetic* substances. In antiferromagnetic substances, such as chromium and MnF_2, the exchange interaction between atomic dipoles results in an *antiparallel* alignment of adjacent moments, as in Fig. 30.15b. This ordered, antiparallel alignment of dipoles, which occurs at lower temperatures, results in a small net magnetization. Above a critical temperature, called the *Neel temperature,* the ordering disappears and the substance becomes paramagnetic.

Ferrimagnetic substances, such as magnetite (Fe_3O_4), have properties similar to ferromagnets. For example, they can be spontaneously magnetized and show hysteresis. In the ordered state, a ferrimagnetic substance consists of two types of dipoles (with different magnetic moments), which are antialigned as in Fig. 30.15c. This results in a net magnetization that is intermediate between the magnetization of a ferromagnet and that of an antiferromagnet. Ferrimagnetic substances also become paramagnetic (disordered) above the Neel temperature. For Fe_3O_4, which belongs

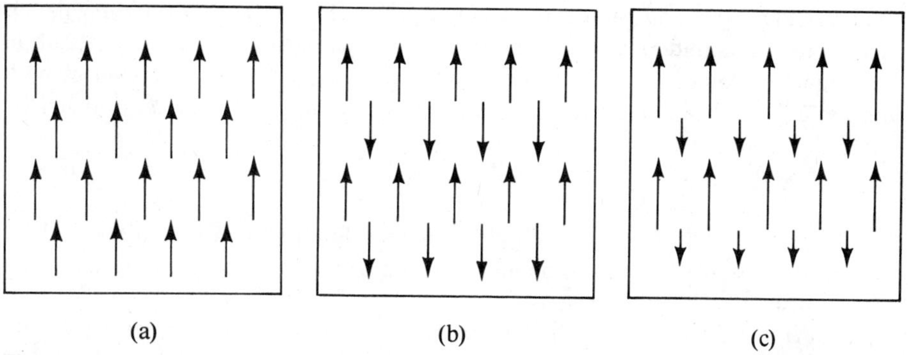

Figure 30.15 (a) In a ferromagnetic substance, all the moments are aligned parallel to each other. (b) In an antiferromagnetic substance, adjacent magnetic moments are antiparallel to each other. (c) In a ferrimagnetic substance, adjacent dipoles having *different* magnetic moments are antiparallel to each other.

to a class of substances known as *ferrites*, the Neel temperature is 858°C. This compound contains *two* ferric ions (Fe^{3+}) for every ferrous ion (Fe^{2+}) and a net moment that results from the Fe^{2+} ion. Ferrites are soft magnetic materials and so exhibit narrow hysteresis curves and small remanent fields. Their high electrical resistivities and high permeabilities make them very suitable for use in high-frequency electronic devices. They are also widely used in computer memory storage.

Q7. Explain the significance of the Curie temperature for a ferromagnetic substance.

Q8. Discuss the difference between ferromagnetic, paramagnetic, and diamagnetic substances.

Q9. What is the difference between hard and soft ferromagnetic materials?

Q10. Explain why it is desirable to use hard ferromagnetic materials to make permanent magnets.

Q11. Why is an ordinary, unmagnetized steel nail attracted to a permanent magnet?

30.6 SUMMARY

The fundamental sources of all magnetic fields are the magnetic dipole moments associated with atoms. The atomic dipole moments can arise both from the orbital motions of the electrons and from an intrinsic property of electrons known as *spin*. To a much smaller degree, the nuclei of some atoms can also contribute to the magnetic moments.

The magnetic properties of substances can be described in terms of their response to an external field. In a broad sense, materials can be described as being *ferromagnetic, paramagnetic,* or *diamagnetic*. The atoms of ferromagnetic and paramagnetic materials have permanent magnetic moments. Diamagnetic materials consist of atoms with no permanent magnetic moments.

When a paramagnetic or ferromagnetic material is placed in an external magnetic field, its dipoles tend to align parallel to the field, and this aligning in turn increases the net field. The increase in the field is quite small in the case of paramagnetic substances. This is because the magnetic dipoles in paramagnetic materials interact weakly with each other and are therefore randomly oriented in the absence of a magnetic field. The dipoles are partially aligned in the presence of an applied field.

The tendency of the moments of a paramagnetic substance to align with an applied field is counteracted by the effect of thermal agitation. The magnetization M (magnetic moment per unit volume) of a paramagnetic substance varies with applied field and temperature approximately according to *Curie's law*, given by

Curie's law

$$M = C\frac{B}{T} \tag{30.19}$$

This law is not valid for very high fields or very low temperatures. At very high fields, M approaches the saturation magnetization, which corresponds to a complete alignment of the dipole moments.

The magnetic moments in a ferromagnetic substance interact strongly with neighboring moments. This results in a high degree of alignment (an ordered system), even in a weak external field, and a correspondingly large increase in the net field. Above a critical temperature, called the *Curie temperature*, a ferromagnetic material becomes paramagnetic. This is because the thermal energy can be sufficiently large to overcome the strong interaction responsible for the parallel alignment of moments. Permanent magnets are made from ferromagnetic materials, for which the magnetic dipoles remain partially aligned with one another in the absence of an applied field.

Although the atoms of a diamagnetic material do not possess a permanent magnetic dipole moment, they are weakly repelled by the poles of a permanent magnet. This is a consequence of Lenz's law. That is, the presence of an external magnetic field alters the orbital motion of the electrons. This alteration results in an induced magnetic moment whose field opposes the external field. Diamagnetism is present in all matter, but it is weak in comparison with paramagnetism and ferromagnetism.

EXERCISES

30.1 The Magnetization of a Substance

1. An iron-core toroid is wrapped with 230 turns of wire per meter of its length. The current in the winding is 6 A. Taking the magnetic permeability of iron to be 5000 μ_o, calculate (a) the magnetic field intensity, (b) the magnetic flux density, and (c) the magnetization.

2. Calculate the magnetic field intensity H of a substance characterized by a magnetization of 1.02×10^6 A · turns/m and a magnetic field of flux density 2.28 T.

3. Show that the product of magnetic field intensity H and magnetic flux density B has SI units of J/m^3.

4. Show that the magnetic susceptibility χ_m is a dimensionless quantity.

5. A specimen of iron has a magnetization of 1.44×10^6 A/m. Each aligned electron in the sample has a magnetic moment of one Bohr magneton. Compute the number of aligned electrons per unit volume.

6. A magnetic field of flux density 1.2 T is to be set up in an iron-core toroid. The toroid has a mean radius of 20 cm, and magnetic permeability of 5000μ_o. (a) What current is required if there are 300 turns of wire in the winding? (b) What is the magnetization under these conditions?

7. A toroid has 300 turns of wire and carries a current of 5 A. The core is made of annealed iron, which has a relative permeability of 400. What is the magnetic field in the toroid at $r = 12$ cm?

8. A toroidal solenoid has an average radius of 12 cm and a cross-sectional area of 2 cm^2. There are 350 turns of wire on the soft iron core, which has a permeability of 800μ_o. Calculate the current necessary to produce a magnetic flux of 4.2×10^{-4} Wb through a cross section of the core.

9. A toroid has an average radius of 18 cm. The current in the coil is 0.4 A. How many turns are required to produce a magnetic intensity of 600 A · turns/m within the toroid?

10. The core material of a certain toroid has a magnetic susceptibility $\chi_m = -0.24 \times 10^{-5}$. The toroid contains 15 turns/cm and carries a current of 5 A. Calculate the magnetization of the core material.

11. A closely wrapped aluminum-core toroid has

10^4 turns/m. (a) What current will result in a magnetization of 1.61 A/m? (b) What is the magnetic flux density in the core?

12. What is the relative permeability of a material that has a magnetic susceptibility of 1.2×10^{-5}?

30.2 The Magnetic Moment of Atoms

13. An electron circulates uniformly in a circular orbit of radius $r = 5.5 \times 10^{-11}$ m. (a) Calculate the effective current necessary to produce a magnetic moment of 9.284×10^{-24} A · m^2. (b) Determine the period of revolution necessary in order to produce the calculated value of effective current.

14. The radius of the electron's orbit in the ground state of hydrogen is $r = 0.529 \times 10^{-10}$ m. (a) Calculate the orbital speed of the electron. (b) Calculate the orbital angular momentum of the electron.

15. Calculate the magnetic moment of an electron that has an angular momentum equal to $h/2\pi = \hbar$ (h = Planck's constant).

16. Cobalt (element no. 27) has a molecular weight of 58.9 and a density of 8.9×10^3 kg/m^3. Assume that each cobalt *atom* has a net magnetic moment of 1.7 Bohr magnetons and calculate the saturation magnetization of cobalt.

17. A total negative charge of 2 μC is distributed uniformly along a thin insulating ring of radius 1 cm. The ring spins with a frequency of 2000 Hz about an axis perpendicular to its plane and through its center. (a) Calculate the magnetic moment due to the rotating charge. (b) Would the answer to (a) be different if all the charge was located at a point on the ring?

18. The earth has a magnetic moment of approximately 8×10^{22} A · m^2. What current in an imaginary one-turn loop around the earth at the equator would cancel out this magnetic moment?

30.3 Paramagnetism

19. A paramagnetic substance has a magnetic moment *per atom* of 8×10^{-24} A · m^2. What magnetic field is required to produce a magnetization that is 0.1% of the saturation magnetization at 300 K?

20. A paramagnetic material has a magnetization of 20% of the saturation value when placed in a magnetic field of 8.9 T. What temperature is required if the magnetic moment per atom is 1.4 Bohr magnetons?

21. A paramagnetic material at a temperature of 9 K has a magnetization equal to 6.7% of the saturation value when placed in a magnetic field of 2.6 T. Calculate the net magnetic moment per atom.

30.4 Diamagnetism

22. Two electrons are moving in opposite directions with the same speed in circular orbits of radius $r = 3 \times 10^{-10}$ m. Calculate the change in the net magnetic moment of the two electrons if a magnetic field of 4 T is applied perpendicular to the common plane of the orbits.

23. An electron moves in a circular orbit of radius $r = 6 \times 10^{-11}$ m. A magnetic field directed perpendicular to the plane of the orbit produces a change in the velocity of the electron (at constant radius) of 6.3 m/s. Calculate the value of B.

PROBLEMS

1. A solenoid 15 cm in length is wrapped uniformly with 350 turns. There is a current of 5 A in the coil. Calculate the values of B and H within the solenoid before and after an iron core is inserted. Assume that the saturation magnetization of iron is 1.72×10^6 A/m.

2. A toroid with 100 turns/m carries a current of 10 A. (a) Calculate the values of B and H within the solenoid before and after an iron core is inserted. Assume that the saturation magnetization of iron is 1.72×10^6 A/m.

3. The density of a specimen of a suspected new element is determined to be 4.15 g/cm^3. The saturation magnetism of the material is found to be 7.6×10^4 A/m, and the measured atomic magnetic moment is 1.2 Bohr magnetons. Calculate the expected value of the atomic weight of the element based on these values.

4. When a certain core material is inserted into a toroidial winding, the magnetic flux density is increased by 80%. If there are 800 turns/m on the toroid and the core magnetization is 5.5×10^6 A/m, what is the current in the windings?

5. A solenoid 80 cm in length has a radius of 1 cm and is wrapped with 250 turns on an aluminum core. Calculate the magnetization in the core when the current in the coil is 5 A.

6. A paramagnetic substance achieves 95% of its saturation magnetization when placed in a magnetic field of 5.2 T at a temperature of 4.2 K. The density of magnetic atoms in the sample is 8×10^{27} atoms/m^3, and the magnetic moment per atom is 5 Bohr magnetons. Calculate the Curie constant for this substance.

7. A toroidal winding filled with a magnetic substance carries a steady current of 4 A. The coil contains a total of 2011 turns, has an average radius of 8 cm, and the core has a cross-sectional area of 1.77 cm^2. (a) What is the magnetic intensity within the core? (b) Calculate the magnetization within the core if the magnetic susceptibility of the core material is 3.38×10^{-4}. (c) Determine the permeability of the core material. (d) Calculate the magnetization current (amperian current) and compare its value to the product NI.

8. Show that the quantity that is the ratio of magnetic dipole moment to angular momentum has SI units of C/kg.

9. Consider an electron moving in a circular orbit of radius r under the influence of simultaneous electric and magnetic forces as shown in Fig. 30.9b. Let the ratio of the magnitudes of the two forces $\frac{F_e}{F_m} = N$ ($N \gg 1$). In general, the magnetic field can be directed into or out of the page. Show that the electron can have *two* possible orbital speeds, given by $v = (N \pm 1)\left(\frac{reB}{m}\right)$.

10. A paramagnetic gas has a magnetic susceptibility of 2.1×10^{-6} at room temperature (300 K). The density of the gas (at 1 atm) is 1.4×10^{-6} kg/m^3. (a) Calculate the average magnetic moment per atom. (b) If a sample of the gas is placed in a magnetic field of 1.5 T, at what temperature will the average thermal energy equal the magnetic energy?

11. Consider a long solenoid of length ℓ containing a core of permeability μ. The core material is magnetized by increasing the current in the coil, so as to produce a changing flux dB/dt. (a) Show that the rate at which work done against the induced emf in the coil is given by

$$\frac{dW}{dt} = I\mathcal{E} = HA\ell\,\frac{dB}{dt}$$

where A is the area of the solenoid. (Hint: Use Faraday's law to find $\mathcal{E}$ and make use of Eq. 30.8.) (b) Use the result of part (a) to show that the total work done in a complete hysteresis cycle equals the area enclosed by the B versus H curve (Fig. 30.14).

31 Alternating Current Circuits

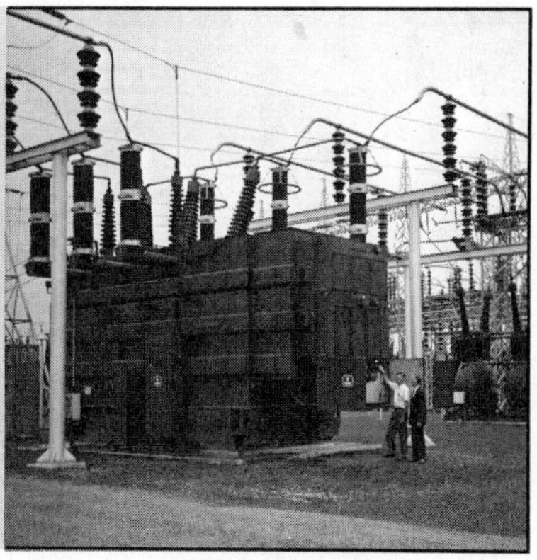

In this chapter, we shall describe the basic principles of simple alternating current (ac) circuits. We shall investigate the characteristics of circuits containing familiar elements and driven by a sinusoidal applied voltage. Our discussion will be limited to analyzing simple series circuits containing a resistor, inductor, and capacitor, both individually and in combination with each other. We shall make use of the fact that these elements respond linearly; that is, the ac current through each element is proportional to the instantaneous ac voltage across the element. We shall find that when the applied voltage of the generator is sinusoidal, the current in each element is also sinusoidal, but not necessarily in phase with the applied voltage. We conclude the chapter with two sections concerning the characteristics of RC filters, transformers, and power transmission.

31.1 RESISTORS IN AN ac CIRCUIT

An ac circuit consists of combinations of circuit elements and a generator, which provides the alternating current. The principles of the ac generator were described in Section 28.5. By rotating a coil in a magnetic field with constant angular velocity ω, a sinusoidal voltage (emf) is induced in the coil; this voltage is given by

$$V = \mathcal{E}_m \sin\omega t \tag{31.1}$$

where $\mathcal{E}_m$ is the *maximum voltage of the ac generator.* The angular frequency ω is given by

$$\omega = 2\pi f = \frac{2\pi}{T}$$

where f is the frequency of the source in Hz and T is the period. Since the output voltage varies sinusoidally with time, we can anticipate that the current will exhibit a similar time variation.

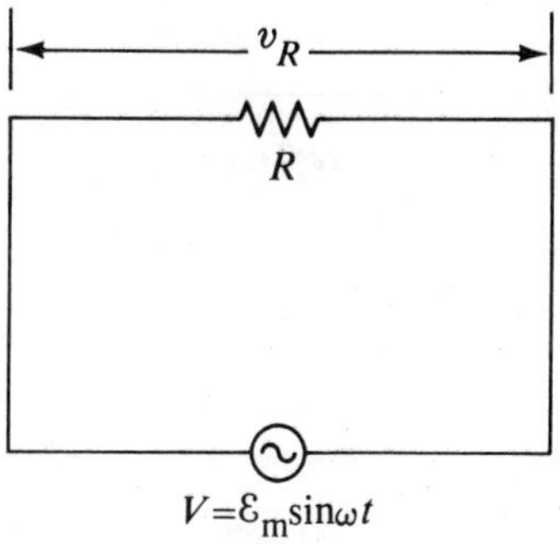

Figure 31.1 A series circuit consisting of a resistor R connected to an ac generator, designated by the symbol —(~)—.

Consider a simple ac circuit consisting of a resistor and an ac generator (designated by the symbol —(~)—), as in Fig. 31.1. From Kirchhoff's circuit equation,

$$V = v_R = \mathcal{E}_m \sin\omega t \tag{31.2}$$

where v_R is the *instantaneous voltage drop across the resistor.* Therefore, the current

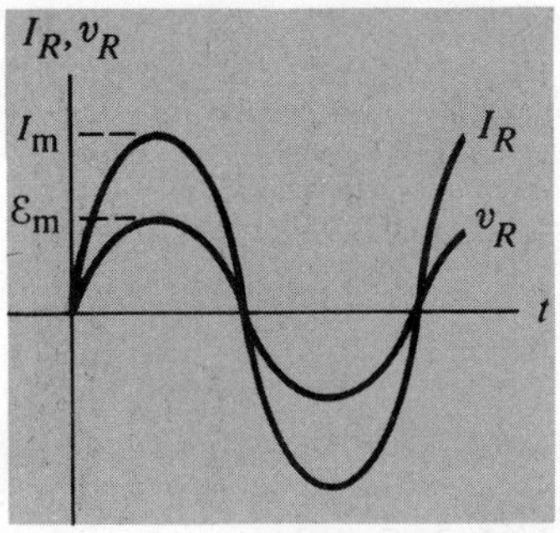

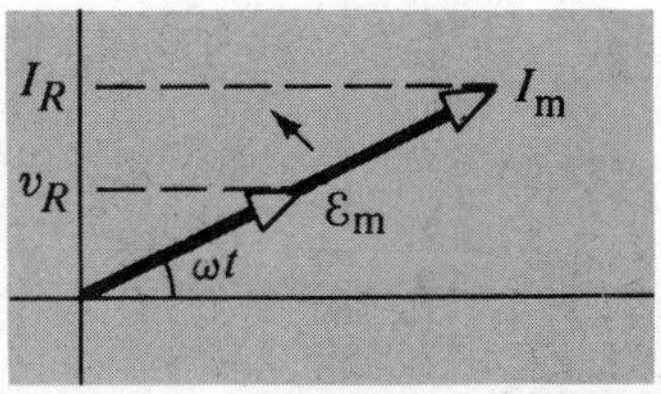

Figure 31.2 (a) Plots of the current and voltage across a resistor as functions of time. The current is in phase with the voltage. (b) A phasor diagram for the resistive circuit, showing that the current is *in phase* with the voltage. The projections of the rotating arrows (phasors) onto the vertical axis represent the instantaneous values v_R and I_R.

is equal to

$$I_R = \frac{V}{R} = \frac{\mathcal{E}_m}{R} \sin\omega t = I_m \sin\omega t \tag{31.3}$$

where I_m is the *maximum current*, given by

Maximum current in a resistor

$$I_m = \frac{\mathcal{E}_m}{R} \tag{31.4}$$

From Eqs. 31.2 and 31.3, we see that the voltage drop across the resistor is

$$v_R = I_m R \sin\omega t \tag{31.5}$$

Since I_R and v_R both vary as $\sin\omega t$, and thus reach their maximum values at the *same time*, they are said to be *in phase*. Graphs of the voltage and current as functions of time (Fig. 31.2a) show that they reach their peak and zero values at the same instant. Very often, we shall use *phasor diagrams* to represent the phase relationship between current and voltage. In these diagrams, alternating quantities such as the current are represented by arrows rotating counterclockwise which corresponds to the *maximum value* of the alternating quantity, and the projection of the phasor onto the vertical axis gives the *instantaneous* value of the alternating quantity. In the case of the single-loop resistive circuit, the current and voltage phasors lie along the same line, as in Fig. 31.2b, since I_R and v_R are *in phase with each other*. Note that although we use arrows to represent phasors, they are not vectors in the ordinary sense.

The current is in phase with the voltage for a resistor

31.2 INDUCTORS IN AN ac CIRCUIT

Now consider an ac circuit consisting only of an inductor connected to the terminals of an ac generator as in Fig. 31.3. If v_L is the *instantaneous voltage drop across the inductor*, then Kirchhoff's rule applied to this circuit gives

$$V = v_L = L\frac{dI}{dt} = \mathcal{E}_m \sin\omega t \tag{31.6}$$

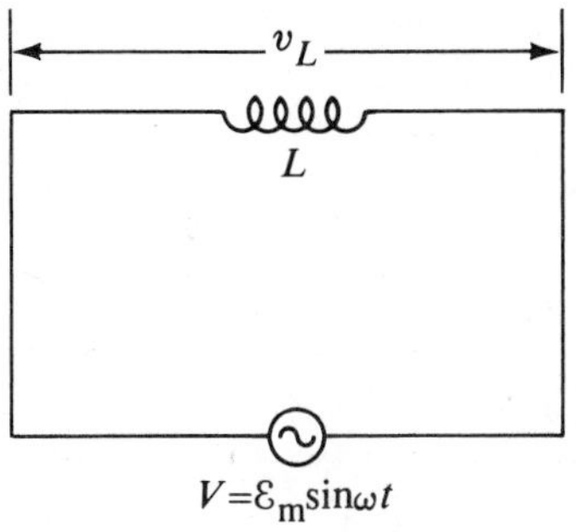

Figure 31.3 A series circuit consisting of an inductor L connected to an ac generator.

Integrating this expression[1] gives the current, as a function of time:

$$I_L = \frac{\mathcal{E}_m}{L}\int \sin\omega t\, dt = -\frac{\mathcal{E}_m}{\omega L}\cos\omega t \tag{31.7}$$

Using the trigonometric identity $\cos\omega t = -\sin(\omega t - \pi/2)$, Eq. 31.7 can also be expressed as

$$I_L = \frac{\mathcal{E}_m}{\omega L}\sin\left(\omega t - \frac{\pi}{2}\right) \tag{31.8}$$

Comparing this result with Eq. 31.6 clearly shows that the current is out of phase with the voltage by $\pi/2$ rad, or 90°. A plot of the voltage and current versus time is given in Fig. 31.4a. Note that the voltage reaches its maximum value at a time that is one quarter of the oscillation period *before* the current reaches its maximum value. The corresponding phasor diagram for this curcuit is shown in Fig. 31.4b. Thus we see that, *for a sinusoidal applied voltage, the current always lags behind the voltage across an inductor by 90°*. This can be understood by noting that since the voltage across the inductor is proportional to dI/dt, the value of v_L is largest when the current is changing most rapidly. Since I versus t is a sinusoidal curve, dI/dt (the slope) is a maximum when the curve goes through zero. This shows that v_L reaches its maximum value when the current is zero.

The current in an inductor lags the voltage by 90°.

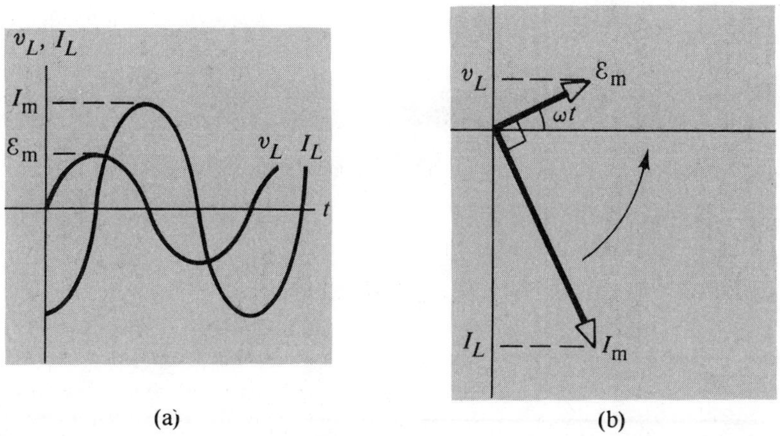

Figure 31.4 (a) Plots of the current and voltage across the inductor as functions of time. The voltage *leads* the current by 90°. (b) The phasor diagram for the inductive circuit. Projections of the phasors onto the vertical axis give the instantaneous values v_L and I_L.

[1]The constant of integration is neglected here since it depends on the initial conditions, which are not important for this situation.

From Eq. 31.7 we see that the maximum current I_m is

Maximum current in an inductor

$$I_m = \frac{\mathcal{E}_m}{\omega L} = \frac{\mathcal{E}_m}{X_L} \tag{31.9}$$

where the quantity X_L, called the *inductive reactance*, is given by

Inductive reactance

$$X_L = \omega L \tag{31.10}$$

The term *reactance* is used so that it is not confused with *resistance*. Recall that I and V are always in phase in a purely resistive circuit, whereas I lags behind V by 90° in a purely inductive circuit.

Using Eqs. 31.6 and 31.9, we find that the instantaneous voltage drop across the inductor can be expressed as

$$v_L = \mathcal{E}_m \sin\omega t = I_m X_L \sin\omega t \tag{31.11}$$

We can think of Eq. 31.11 as Ohm's law for an inductive circuit. It is left as an exercise to show the X_L has the SI unit of ohm.

Note that the reactance of an inductor increases with increasing frequency. This is because at higher frequencies, the current must change more rapidly, which in turn causes an increase in the induced emf associated with a given maximum current.

Example 31.1

In a purely inductive ac circuit (Fig. 31.3), $L = 25$ mH and $\mathcal{E}_m = 150$ V. (a) Find the inductive reactance and maximum current in the circuit if the frequency is 60 Hz.

First, note that $\omega = 2\pi f = 2\pi(60) = 377\ \text{s}^{-1}$. Using Eq. 31.10 gives

$$X_L = \omega L = (377\ \text{s}^{-1})(25 \times 10^{-3}\ \text{H}) = 9.43\ \Omega$$

Substituting this result into Eq. 31.9 gives

$$I_m = \frac{\mathcal{E}_m}{X_L} = \frac{150\ \text{V}}{9.43\ \Omega} = 15.9\ \text{A}$$

(b) Repeat (a) using a frequency of 6 kHz.

For a frequency of 6 kHz, $\omega = 3.77 \times 10^4\ \text{s}^{-1}$ and we find that $X_L = 943\ \Omega$ and $I_m = 0.159$ A.

Q1. What is meant by the statement "the voltage across an inductor leads the current by 90°"?

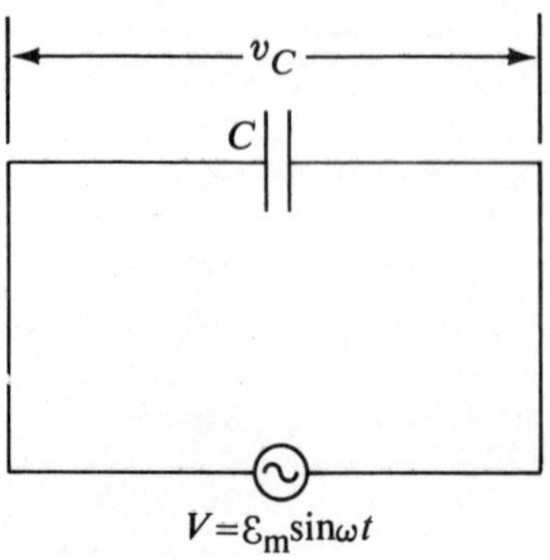

Figure 31.5 A series circuit consisting of a capacitor C connected to an ac generator.

31.3 CAPACITORS IN AN ac CIRCUIT

Figure 31.5 shows an ac circuit consisting of a capacitor connected across the terminals of an ac generator. Kirchhoff's loop rule applied to this circuit gives

$$V = v_C = \mathcal{E}_m \sin\omega t \tag{31.12}$$

where v_C is the *instantaneous voltage drop across the capacitor*. But from the definition of capacitance, $v_C = Q/C$, which when substituted into Eq. 31.12 gives

$$Q = C\mathcal{E}_m \sin\omega t \tag{31.13}$$

Since $I = dQ/dt$, differentiating Eq. 31.13 gives the instantaneous current:

$$I_C = \frac{dQ}{dt} = \omega C \mathcal{E}_m \cos\omega t \tag{31.14}$$

Again, we see that the current is not in phase with the voltage drop across the capacitor, given by Eq. 31.12. Using the trigonometric identity $\cos\omega t = \sin\left(\omega t + \frac{\pi}{2}\right)$, we can express Eq. 31.14 in the alternative form

$$I_C = \omega C \mathcal{E}_m \sin\left(\omega t + \frac{\pi}{2}\right) \tag{31.15}$$

Comparing this expression with Eq. 31.12, we see that the current is 90° out of phase with the voltage across the inductor. A plot of the current and voltage versus time (Fig. 31.6a) shows that the current reaches its maximum value one quarter of a cycle sooner than the voltage reaches its maximum value. The corresponding phasor diagram in Fig. 31.6b also shows that, *for a sinusoidally applied emf, the current always leads the voltage across a capacitor by 90°.*

The current leads the voltage across the capacitor by 90°

From Eq. 31.15, we see that the maximum current in the circuit is given by

$$I_m = \omega C \mathcal{E}_m = \frac{\mathcal{E}_m}{X_C} \tag{31.16}$$

where X_C is called the *capacitive reactance:*

$$X_C = \frac{1}{\omega C} \tag{31.17}$$

Capacitive reactance

Combining Eqs. 31.12 and 31.16, we can express the voltage drop across the capacitor as

$$v_C = I_m X_C \sin\omega t \tag{31.18}$$

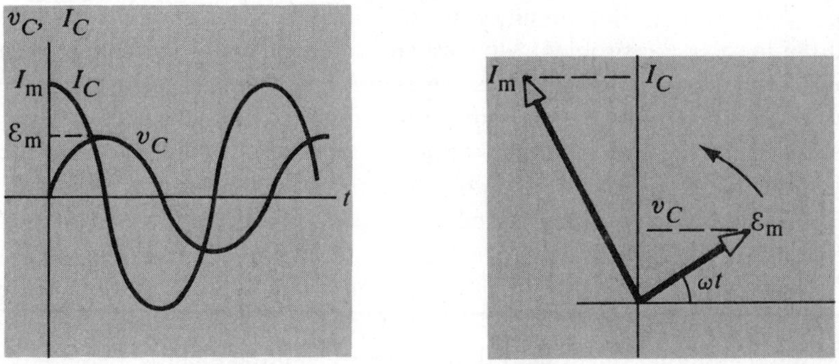

(a) (b)

Figure 31.6 (a) Plots of the current and voltage across the capacitor as functions of time. The voltage *lags behind* the current by 90°. (b) Phasor diagram for the purely capacitive circuit. Projections of the phasors onto the vertical axis gives the instantaneous values v_C and I_C.

The SI unit of X_C is also the ohm. As the frequency of the circuit increases, the current increases but the reactance decreases. For a given maximum applied voltage $\mathcal{E}_m$, the current increases as the frequency increases. As the frequency approaches zero, the capacitive reactance approaches infinity. Therefore, the current approaches zero. This makes sense since the circuit approaches dc conditions as $\omega \rightarrow 0$. Of course, a capacitor passes no current under steady-state dc conditions.

Example 31.2

An 8-μF capacitor is connected to the terminals of an ac generator whose maximum emf is 150 V and whose frequency is 60 Hz. Find the capacitive reactance and the maximum current in the circuit.

Solution: Using Eq. 31.17 and the fact that $\omega = 2\pi f = 377\ \text{s}^{-1}$ gives

$$X_C = \frac{1}{\omega C} = \frac{1}{(377\ \text{s}^{-1})(8 \times 10^{-6}\ \text{F})} = 332\ \Omega$$

Substituting this result into Eq. 31.16, we find that

$$I_m = \frac{\mathcal{E}_m}{X_C} = \frac{150\ \text{V}}{332\ \Omega} = 0.452\ \text{A}$$

Note that if the frequency were doubled, X_C would be halved and the current would be doubled.

Q2. Explain why the reactance of a capacitor decreases with increasing frequency, whereas the reactance of an inductor increases with increasing frequency.

Q3. Why does a capacitor act as a short circuit at high frequencies? Why does it act as an open circuit at low frequencies?

31.4 THE *RLC* SERIES CIRCUIT

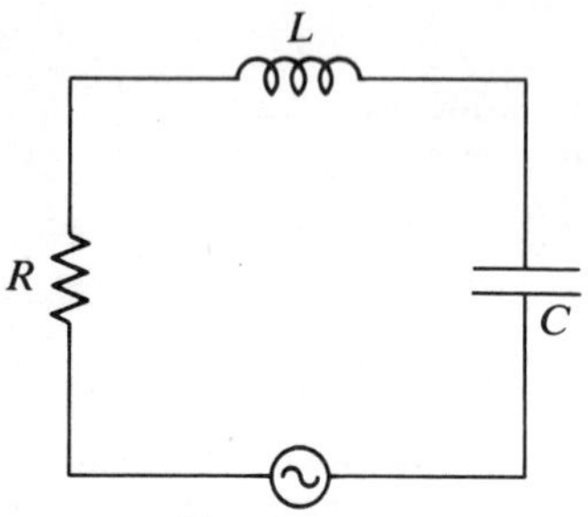

$V = \mathcal{E}_m \sin\omega t$

Figure 31.7 A series circuit consisting of a resistor R, an inductor L, and a capacitor C connected to an ac generator.

We now describe the properties of a series circuit containing a resistor, inductor, capacitor, and ac generator (Fig. 31.7). Let us assume that the sinusoidal current in the circuit has reached a steady-state value. As before, the applied voltage is assumed to vary sinusoidally with time. It is convenient to assume that the applied voltage is given by

$$V = \mathcal{E}_m \sin\omega t$$

while the current varies as

$$I = I_m \sin(\omega t - \phi)$$

The quantity ϕ is called the *phase angle* between the current and the applied voltage. Our aim is to determine ϕ and I_m.

In order to solve this problem, we must construct and analyze the phasor diagram for this circuit. First, note that since the elements are in series, the current everywhere in the circuit must be the same at any instant. That is, *the ac current at all points in a series ac circuit has the same amplitude and phase.* Therefore, as we found in the previous sections, the voltage across each element will have *different* amplitudes and phases, as summarized in Fig. 31.8. In particular, the voltage across the resistor is in phase with the current (Fig. 31.8a), the voltage across the inductor leads the current by 90° (Fig. 31.8b), and the voltage across the capacitor lags behind the current by 90° (Fig. 31.8c). Using these phase relationships, we can express the *instantaneous* voltage drops across the three elements as

$$v_R = I_m R \sin\omega t = V_R \sin\omega t \tag{31.19}$$

$$v_L = I_m X_L \sin\left(\omega t + \frac{\pi}{2}\right) = V_L \cos\omega t \tag{31.20}$$

$$v_C = I_m X_C \sin\left(\omega t - \frac{\pi}{2}\right) = -V_C \cos\omega t \tag{31.21}$$

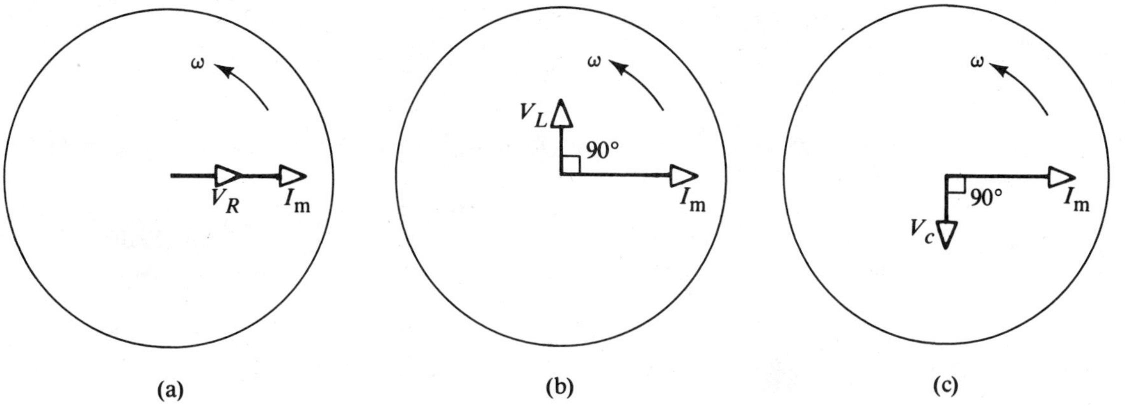

Figure 31.8 Phase relationships between the maximum voltage and current vectors for (a) a resistor, (b) an inductor, and (c) a capacitor.

where V_R, V_L, and V_C are the *maximum* voltages across each element, given by

$$V_R = I_m R \tag{31.22}$$

$$V_L = I_m X_L \tag{31.23}$$

$$V_C = I_m X_C \tag{31.24}$$

At this point, we could proceed by noting that the instantaneous voltage V across the three elements equals the sum

$$V = v_R + v_L + v_C \tag{31.25}$$

Although this analytical approach is correct, it is simpler to obtain the sum by examining the phasor diagram.

Since the current in each element is the same at any instant, we can obtain the resulting phasor diagram by combining the three phasors shown in Fig. 31.8. This gives the diagram shown in Fig. 31.9a, where a single phasor I_m is used to represent the current in each element. To obtain the vector sum of these voltages, it is convenient to redraw the phasor diagram as in Fig. 31.9b. From this diagram, we see that the *vector* sum of the voltage amplitudes V_R, V_L, and V_C equals a phasor whose length is the maximum applied voltage, $\mathcal{E}_m$, where the phasor $\mathcal{E}_m$ makes an angle ϕ with the current phasor, I_m. Note that the voltage phasors V_L and V_C are along the same line, and hence we are able to construct the difference phasor $V_L - V_C$, which is perpendicular to the phasor V_R. From the right triangle in Fig. 31.9b, we see that

$$\mathcal{E}_m = \sqrt{V_R^2 + (V_L - V_C)^2} = \sqrt{(I_m R)^2 + (I_m X_L - I_m X_C)^2}$$

$$\mathcal{E}_m = I_m \sqrt{R^2 + (X_L - X_C)^2} \tag{31.26}$$

where $X_L = \omega L$ and $X_C = 1/\omega C$. Therefore, we can express the maximum current as

$$I_m = \frac{\mathcal{E}_m}{\sqrt{R^2 + (X_L - X_C)^2}}$$

Defining the *impedance* Z of the circuit to be

$$Z \equiv \sqrt{R^2 + (X_L - X_C)^2} \tag{31.27}$$

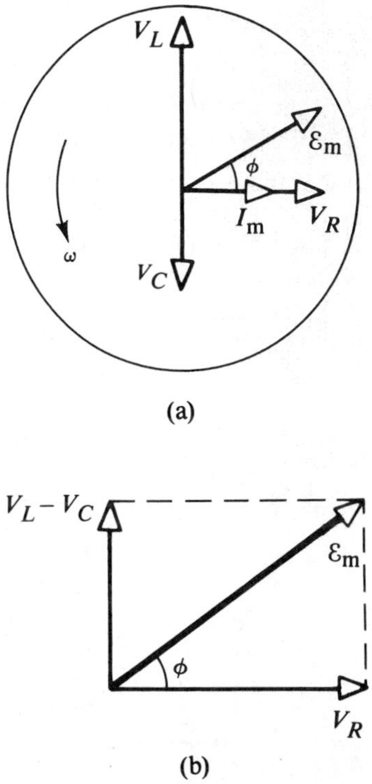

Figure 31.9 (a) The phasor diagram for the series *RLC* circuit shown in Fig. 31.7. Note that the phasor V_R is *in phase* with the current phasor I_m, the phasor V_L *leads* the phasor I_m by 90°, and the phasor V_C *lags behind* the phasor I_m by 90°. The total voltage V_m makes an angle ϕ with the current phasor I_m. (b) Simplified version of the phasor diagram shown in (a).

Impedance

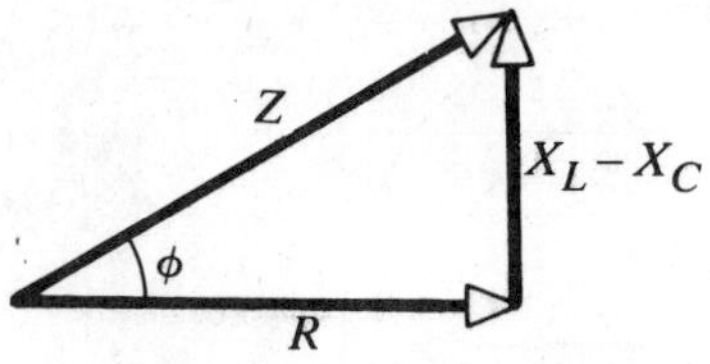

Figure 31.10 The impedance triangle for a series *RLC* circuit, which gives the relationship $Z = \sqrt{R^2 + (X_L - X_C)^2}$.

we can write Eq. 31.26 in the form

$$\mathcal{E}_m = I_m Z \tag{31.28}$$

Impedance also has the SI unit of ohm. We can regard Eq. 31.28 as a generalized Ohm's law applied to an ac circuit.

By removing the common factor I_m from each phasor in Fig. 31.9, we can also construct an impedance triangle, shown in Fig. 31.10. From this phasor diagram, we find that the phase angle ϕ between the current and voltage is given by

Phase angle

$$\tan\phi = \frac{X_L - X_C}{R} \tag{31.29}$$

For example, when $X_L > X_C$ (which occurs at high frequencies), the phase angle is positive, signifying that the current lags behind the applied voltage, as in Fig. 31.9. On the other hand, if $X_L < X_C$, the phase angle is negative, signifying that the current leads the applied voltage. Finally, when $X_L = X_C$, the phase angle is zero. In this case, the ac impedance equals the resistance and the current has its *maximum* value, given by $\mathcal{E}_m/R$. The frequency at which this occurs is called the *resonance frequency*, which will be described further in Section 31.6.

Example 31.3

Analyze a series *RLC* ac circuit for which $R = 250\ \Omega$, $L = 0.6$ H, $C = 3.5\ \mu$F, $\omega = 377\ s^{-1}$, and $\mathcal{E}_m = 150$ V.

Solution: The reactances are given by $X_L = \omega L = 226\ \Omega$ and $X_C = 1/\omega C = 758\ \Omega$. Therefore, the impedance is equal to

$$Z = \sqrt{R^2 + (X_L - X_C)^2}$$
$$= \sqrt{(250\ \Omega)^2 + (226\ \Omega - 758\ \Omega)^2} = 588\ \Omega$$

The maximum current is given by

$$I_m = \frac{\mathcal{E}_m}{Z} = \frac{150\ \text{V}}{588\ \Omega} = 0.255\ \text{A}$$

The phase angle between the current and voltage is

$$\phi = \tan^{-1}\left(\frac{X_L - X_C}{R}\right) = \tan^{-1}\left(\frac{226 - 758}{250}\right)$$
$$= -64.8°$$

Since the circuit is more capacitive than inductive, ϕ is negative and the current leads the applied voltage.

The *maximum* voltages across each element are given by

$$V_R = I_m R = (0.255\ \text{A})(250\ \Omega) = 63.8\ \text{V}$$
$$V_L = I_m X_L = (0.255\ \text{A})(226\ \Omega) = 57.6\ \text{V}$$
$$V_C = I_m X_C = (0.255\ \text{A})(758\ \Omega) = 193\ \text{V}$$

Using Eqs. 31.19, 31.20, and 31.21, and assuming that the applied voltage $V = 150 \sin(\omega t - 64.8°)$, we find that the *instantaneous* voltages across the three elements are given by

$$v_R = 63.8 \sin 377t\ \text{V}$$
$$v_L = 57.6 \cos 377t\ \text{V}$$
$$v_C = -193 \cos 377t\ \text{V}$$

Note that the sum of the three *maximum* voltages $V_R + V_L + V_C = 314$ V, which is much larger than the maximum voltage of the generator, 150 V. The former is a meaningless quantity. This is because when harmonically varying quantities are added, *both their amplitude and their phase* must be taken into account. That is, the voltages must be added in a way that takes account of the different phases. When this is done, Eq. 31.26 is satisfied. You should verify this result.

Q4. Does the phase angle depend on frequency? What is the phase angle when the inductive reactance equals the capacitive reactance?

Q5. In a series *RLC* circuit, what is the possible range of values for the phase angle?

Q6. If the frequency is doubled in a series *RLC* circuit, what happens to the resistance, the inductive reactance, and the capacitive reactance?

31.5 POWER IN AN ac CIRCUIT

When we studied dc circuits in Chapter 25, we defined the power delivered by a battery to an external circuit as the product of the current and the emf of the battery. Likewise, the instantaneous power delivered by an ac generator to any circuit is the product of the generator current and the applied voltage. For the *RLC* circuit shown in Fig. 31.7, we can express the instantaneous power P as

$$\begin{aligned} P = Iv &= I_m \sin(\omega t - \phi)[\mathcal{E}_m \sin\omega t] \\ &= I_m \mathcal{E}_m \sin\omega t \sin(\omega t - \phi) \end{aligned} \tag{31.30}$$

Clearly this result is a complicated function of time and, in itself, is not very useful from a practical viewpoint. What is generally of interest is the average power over one or more cycles. Such an average can be computed by first using the trigonometric identity $\sin(\omega t - \phi) = \sin\omega t \cos\phi - \cos\omega t \sin\phi$. Substituting this into Eq. 31.30 gives

$$P = I_m \mathcal{E}_m \sin^2\omega t \cos\phi - I_m \mathcal{E}_m \sin\omega t \cos\omega t \sin\phi \tag{31.31}$$

We now take the time average of P over one or more cycles, noting that I_m, $\mathcal{E}_m$, ϕ, and ω are all constants. The time average of the first term on the right of Eq. 31.31 involves the average value of $\sin^2\omega t$, which is $\frac{1}{2}$. A graphical description of this result[2] is given in Fig. 31.11. The time average of the second term on the right of Eq. 31.31 is identically zero because $\sin\omega t \cos\omega t = \frac{1}{2}\sin 2\omega t$, whose average value is zero.

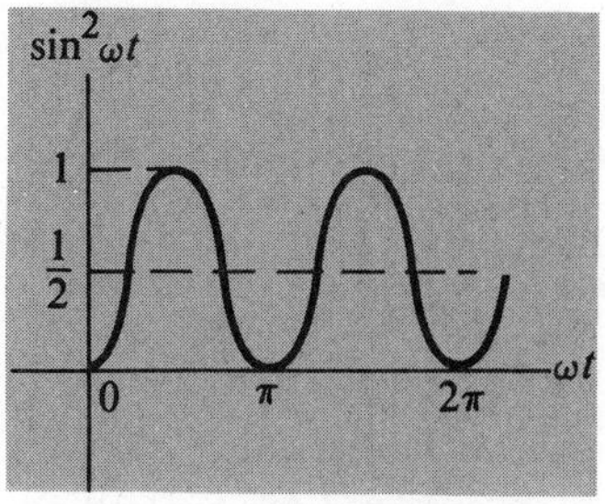

Figure 31.11 A plot of $\sin^2\omega t$ versus ωt, showing that the function is always positive and has an average value of $\frac{1}{2}$ when averaged over one or more cycles.

Therefore, we can express the *average power* P_{av} as

$$P_{av} = \frac{1}{2} I_m \mathcal{E}_m \cos\phi \tag{31.32}$$

When voltages and currents are measured in ac circuits, the instruments used are usually calibrated to read root mean square (rms) values rather than maximum values. (Root mean square values are not to be confused with average values, which are zero for sinusoidal variations of V and I). *The rms value of any sinusoidally varying quantity is simply the maximum value of that quantity divided by* $\sqrt{2}$. This can be understood by considering the square of the instantaneous applied voltage:

$$v^2 = \mathcal{E}_m^2 \sin^2\omega t$$

An average over one or more cycles gives

$$(v^2)_{av} = \mathcal{E}_m^2[\sin^2\omega t]_{av} = \frac{1}{2}\mathcal{E}_m^2$$

Therefore, by definition

$$V_{rms} = \sqrt{(v^2)_{av}} = \frac{\mathcal{E}_m}{\sqrt{2}} \tag{31.33}$$

rms voltage

For example, the usual 120-V household electrical outlet is an rms voltage. The maximum output voltage of the outlet is $\sqrt{2}(120 \text{ V}) = 170 \text{ V}$.

Likewise, the rms current is given by

$$I_{rms} = \sqrt{(I^2)_{av}} = \frac{I_m}{\sqrt{2}} \tag{31.34}$$

rms current

---

[2] Another simple method to prove this result is to make use of the identity $\sin^2\omega t + \cos^2\omega t = 1$. Since the average values of $\sin^2\omega t$ and $\cos^2\omega t$ must be equal, $(\sin^2\omega t)_{av} = (\cos^2\omega t)_{av} = \frac{1}{2}$.

Using these results, we can express the average power as

Average power

$$P_{av} = I_{rms} V_{rms} \cos\phi \qquad (31.35)$$

where the constant $\cos\phi$ is called the *power factor*. By inspecting Fig. 31.9, we note that the maximum voltage drop across the resistor is given by $V_R = \mathcal{E}_m \cos\phi = I_m R$. Using Eqs. 31.33 and 31.34 and the fact that $\cos\phi = I_m R/\mathcal{E}_m$, we find that P_{av} can be expressed as

Average power

$$P_{av} = I_{rms} V_{rms} \cos\phi = I_{rms} \left(\frac{\mathcal{E}_m}{\sqrt{2}}\right) \frac{I_m R}{\mathcal{E}_m} = I_{rms} \frac{I_m R}{\sqrt{2}}$$

$$P_{av} = I_{rms}^2 R \qquad (31.36)$$

In other words, the *average power delivered by the generator is dissipated as heat in the resistor,* just as in the case of a dc circuit. *There is no power loss in an ideal inductor or capacitor.* When the load is purely resistive, then $\phi = 0$, $\cos\phi = 1$, and from Eq. 31.35 we see that $P_{av} = I_{rms} V_{rms}$.

Example 31.4

Calculate the average power delivered to the series *RLC* circuit described in Example 31.3.

Solution: First, we can calculate V_{rms} and I_{rms}:

$$V_{rms} = \frac{\mathcal{E}_m}{\sqrt{2}} = \frac{150\ \text{V}}{\sqrt{2}} = 106\ \text{V}$$

$$I_{rms} = \frac{I_m}{\sqrt{2}} = \frac{\mathcal{E}_m/Z}{\sqrt{2}} = \frac{0.255\ \text{A}}{\sqrt{2}} = 0.180\ \text{A}$$

Since $\phi = -64.8°$, the power factor, $\cos\phi$, is 0.426, and hence the average power is

$$P_{av} = I_{rms} V_{rms} \cos\phi = (0.180\ \text{A})(106\ \text{V})(0.426) = 8.13\ \text{W}$$

The same result can be obtained using Eq. 31.36.

Q7. Energy is delivered to a series *RLC* circuit by a generator. This energy is dissipated as heat in the resistor. What is the source of this energy?

Q8. A particular experiment requires a beam of light of very stable intensity. Why would an ac voltage be unsuitable for powering the light source?

31.6 RESONANCE IN A SERIES *RLC* CIRCUIT

A series *RLC* circuit is said to be in *resonance* when the current has its maximum value. In general, the instantaneous current can be written

$$I = \frac{V}{Z} = \frac{\mathcal{E}_m \sin\omega t}{Z} \qquad (31.37)$$

where Z is the impedance. Substituting Eq. 31.37 into 31.37 gives the relationship

$$I = \frac{\mathcal{E}_m \sin\omega t}{\sqrt{R^2 + (X_L - X_C)^2}} \qquad (31.38)$$

Thus, we see that the current reaches a maximum when $X_L = X_C$, corresponding to $Z = R$. The frequency ω_0 at which this occurs is called the *resonance frequency* of the circuit. To find ω_0, we use the condition $X_L = X_C$, from which we get

$$\omega_0 L = \frac{1}{\omega_0 C}$$

$$\omega_0 = \frac{1}{\sqrt{LC}} \quad (31.39)$$

Resonance frequency

Note that this frequency also corresponds to the natural frequency of oscillation of a resistanceless *LC* circuit (Section 29.5). Therefore, the current in a series *RLC* circuit reaches its *maximum* value when the frequency of the applied voltage matches the natural oscillator frequency, which depends only on L and C. Furthermore, at this frequency the current is in phase with the applied voltage.

A plot of the maximum current versus frequency for a constant applied voltage amplitude is shown in Fig. 31.12. Note that the current amplitude reaches its maximum value at the resonance frequency, ω_0.

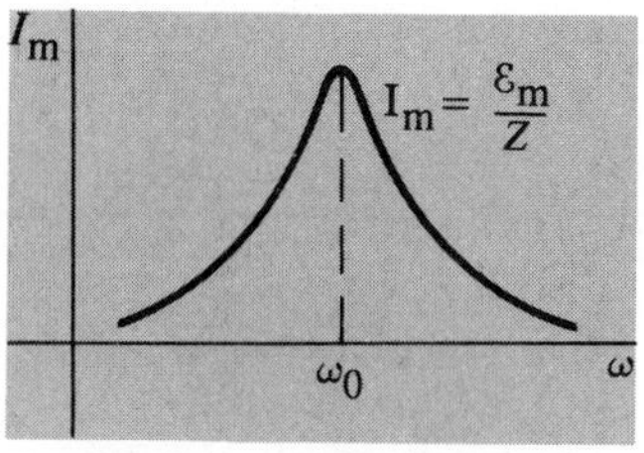

Figure 31.12 Current amplitude in a series *RLC* circuit plotted versus frequency of the generator voltage. Note that the current reaches its maximum value at the resonance frequency, ω_0.

By inspecting Eq. 31.38, one must conclude that the current would become infinite at resonance when $R = 0$. Although the equation predicts this, real circuits always have some resistance, which limits the value of the current. Mechanical systems can also exhibit resonances. For example, when an undamped mass-spring system is driven at its natural frequency of oscillation, its amplitude increases with time, as we discussed in Chapter 13. Large-amplitude mechanical vibrations can be disastrous, as in the case of the Tacoma Narrows Bridge collapse.

It is also interesting to calculate the average power as a function of frequency. Using Eq. 31.36 together with Eq. 31.37, we find that

$$P_{av} = I_{rms}^2 R = \frac{V_{rms}^2}{Z^2} R = \frac{V_{rms}^2 R}{R^2 + (X_L - X_C)^2} \quad (31.40)$$

Since $X_L = \omega L$, $X_C = \dfrac{1}{\omega C}$, and $\omega_0^2 = 1/lC$, the factor $(X_L - X_C)^2$ can be expressed as

$$(X_L - X_C)^2 = \left(\omega L - \frac{1}{\omega C}\right)^2 = \frac{L^2}{\omega^2}(\omega^2 - \omega_0^2)^2$$

Using this result in Eq. 31.40 gives

$$P_{av} = \frac{V_{rms}^2 R \omega^2}{R^2\omega^2 + L^2(\omega^2 - \omega_0^2)^2} \quad (31.41)$$

Power in an *RLC* circuit

This expression shows that at resonance, when $\omega = \omega_0$, the *average power is a maximum* and has the value V_{rms}^2/R. A plot of the average power versus the frequency ω of the applied voltage is shown in Fig. 31.13 for two values of R. As the

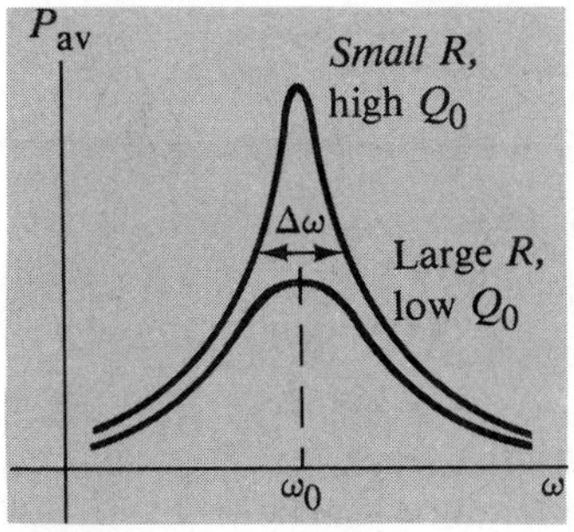

Figure 31.13 Plots of the average power versus frequency for a series *RLC* circuit (see Eq. 31.41). The upper, narrow curve is for a small value of R, and the lower, broad curve is for a large value of R. The width $\Delta\omega$ of each curve is measured between points where the power is half the maximum value, which occurs at the resonance frequency, ω_0.

resistance is made smaller, the curve becomes sharper in the vicinity of the resonance. The sharpness of the curve is usually described by a dimensionless parameter known as the *quality factor*, denoted by Q_o (not to be confused with charge), which is given by the ratio[3]

$$Q_o = \frac{\omega_o}{\Delta\omega} \tag{31.42}$$

where $\Delta\omega$ is the width of the curve measured between the two values of ω for which P_{av} has *half* its maximum value (half-power points). It is left as an exercise to show that the width at the half-power points has the approximate value $\Delta\omega \approx R/L$. Hence,

Quality factor

$$Q_o \approx \frac{\omega_o L}{R} \tag{31.43}$$

That is, Q_o is equal to the ratio of the inductive reactance to the resistance evaluated at the resonance frequency, ω_o.

The curves plotted in Fig. 31.13 show that a high-Q_o circuit responds to a very narrow range of frequencies, whereas a low-Q_o circuit responds to a much broader range of frequencies. Typical values of Q_o in electronic circuits range from 10 to 100.

The receiving circuit of a radio is an important application of a series resonance circuit. The radio is tuned to a particular station (which transmits a specific radio frequency signal) by varying a capacitor, which changes the resonance frequency of the receiving circuit. When the resonance frequency of the circuit matches that of the incoming radio wave, the current in the receiving circuit increases. This signal is then amplified (volume control) and fed to a speaker. Since many signals are often present over a range of frequencies, it is important to design a high-Q_o circuit in order to eliminate unwanted signals. In this manner, stations whose frequencies are near the resonance frequency will give negligibly small signals at the receiver relative to the one that matches the resonance frequency.

Example 31.5

Consider a series *RLC* circuit for which $R = 150\ \Omega$, $L = 20$ mH, $\mathcal{E}_m = 20$ V, and $\omega = 5000\ s^{-1}$. (a) Determine the value of the capacitance for which the current is a maximum.

The current is a maximum at the resonance frequency, ω_o, which should be made to match the "driving" frequency of 5000 s^{-1} in this problem:

$$\omega_o = 5 \times 10^3\ s^{-1} = \frac{1}{\sqrt{LC}}$$

$$C = \frac{1}{(25 \times 10^6\ s^{-2})L}$$

$$= \frac{1}{(25 \times 10^6\ s^{-2})(20 \times 10^{-3}\ H)} = 2.0\ \mu F$$

(b) Calculate the maximum current.

At resonance, $X_L = X_C$ and $Z = R$; therefore

$$I_m = \frac{\mathcal{E}_m}{R} = \frac{20\ V}{150\ \Omega} = 0.133\ A$$

Q9. What is the impedance of an *RLC* circuit at the resonance frequency?

Q10. Consider a series *RLC* circuit in which *R* is an incandescent lamp, *C* is some fixed capacitor, and *L* is a *variable* inductance. The source is 110 V ac. Explain why the lamp glows brightly for some values of *L* and does not glow at all for other values.

[3]The quality factor is also defined as the ratio $2\pi E/\Delta E$, where E is the energy stored per cycle of oscillation and ΔE is the energy lost per cycle of oscillation. One can also define the quality factor for a mechanical system such as a damped oscillator.

31.7 FILTER CIRCUITS

In this section, we give a brief description of RC filters, which are commonly used in ac circuits to modify the time variation of a signal. A filter circuit can be used to smooth out or eliminate a time-varying voltage. For example, radios are usually powered by a 60-Hz ac voltage. The ac voltage is converted to dc using a *rectifier circuit.* After rectification, however, the voltage will still contain a small ac component at 60 Hz (sometimes called *ripple*), which must be filtered. This 60-Hz ripple must be reduced to a value much smaller than the signal to be amplified. Without filtering, the resulting audio signal will include an annoying hum at 60 Hz.

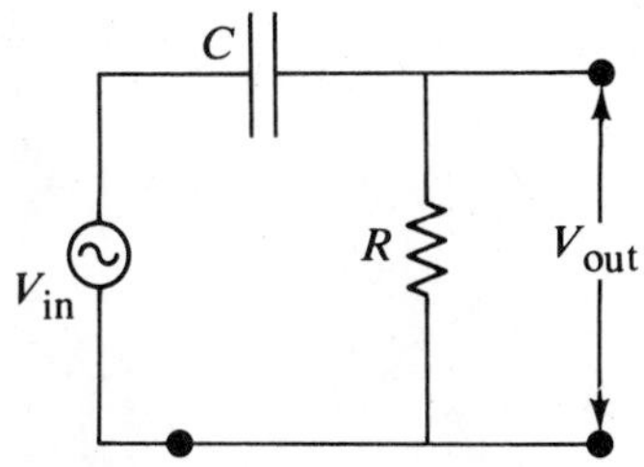

Figure 31.14 A simple RC high-pass filter.

First, consider the simple series RC circuit shown in Fig. 31.14. The input voltage is across the two elements and is represented by $\mathcal{E}_m \sin\omega t$. Since we shall be interested only in maximum values, we can use Eq. 31.28, which shows that the maximum input voltage is related to the maximum current by

$$V_{in} = I_m Z = I_m \sqrt{R^2 + \left(\frac{1}{\omega C}\right)^2}$$

If the voltage across the resistor is considered to be the output voltage, V_{out}, then from Ohm's law the maximum output voltage is given by

$$V_{out} = I_m R$$

Therefore, the ratio of the output voltage to the input voltage is given by

$$\frac{V_{out}}{V_{in}} = \frac{R}{\sqrt{R^2 + \left(\frac{1}{\omega C}\right)^2}} \qquad (31.44)$$

High-pass filter

A plot of Eq. 31.44, given in Fig. 31.15, shows that at low frequencies, V_{out} is small compared with V_{in}, whereas at high frequencies the two voltages are equal. Since the circuit preferentially passes signals of higher frequency while low frequencies are filtered (or attenuated), the circuit is called an *RC high-pass filter.* Physically, the high-pass filter is a result of the "blocking action" of the capacitor to direct current or low frequencies.

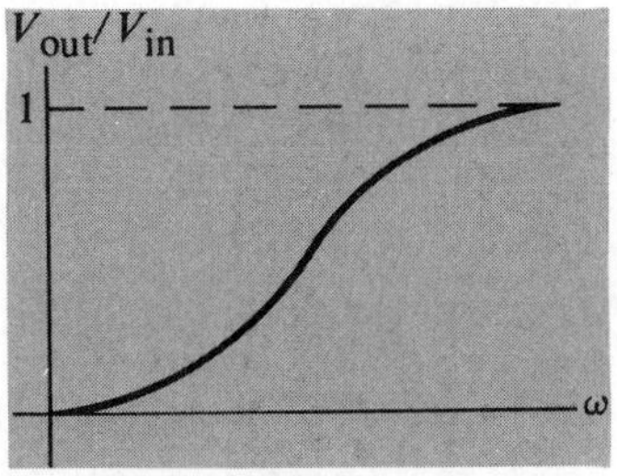

Figure 31.15 Ratio of the output voltage to the input voltage for an RC high-pass filter.

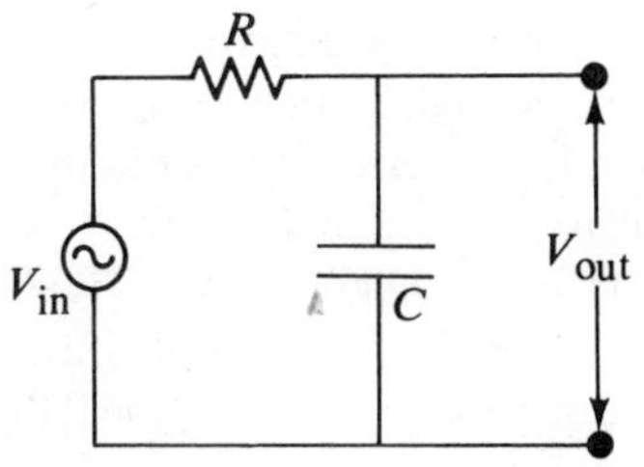

Figure 31.16 A simple RC low-pass filter.

Now consider the RC series circuit shown in Fig. 31.16, where the output voltage is taken across the capacitor. In this case, the maximum voltage equals the voltage across the capacitor. Since the impedance across the capacitor is $X_C = \dfrac{1}{\omega C}$,

$$V_{out} = I_m X_C = \frac{I_m}{\omega C}$$

Low-pass filter

Therefore, the ratio of the output voltage to the input voltage is given by

$$\frac{V_{\text{out}}}{V_{\text{in}}} = \frac{1/\omega C}{\sqrt{R^2 + \left(\frac{1}{\omega C}\right)^2}} \tag{31.45}$$

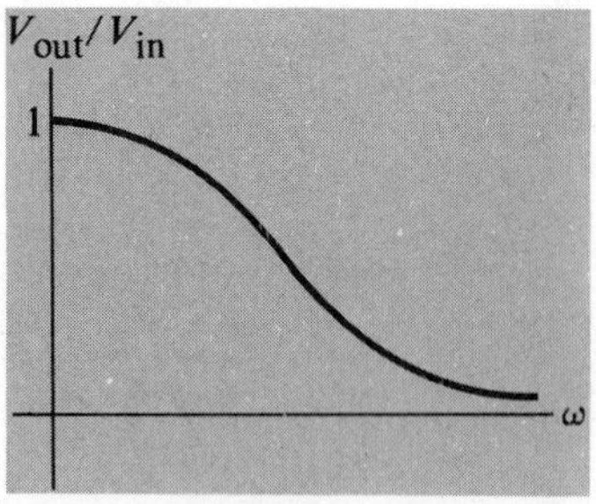

Figure 31.17 Ratio of the output voltage to the input voltage for an *RC* low-pass filter.

This ratio, plotted in Fig. 31.17, shows that in this case the circuit preferentially passes signals of low frequency. Hence, the circuit is called an *RC low-pass filter.*

We have considered only two simple filters. One can also use a series *RL* circuit as a high-pass or low-pass filter. It is also possible to design filters, called *band-pass filters,* that pass only a narrow range of frequencies.

31.8 THE TRANSFORMER AND POWER TRANSMISSION

When electrical power is transmitted over large distances, it is economical to use a high voltage and low current to minimize the I^2R heating loss in the transmission lines. For this reason, 350-kV lines are common, and in many areas even higher-voltage (765 kV) lines are under construction. Such high-voltage transmission systems have met with considerable public resistance because of the potential safety and environmental problems they pose. At the receiving end of such lines, the consumer requires power at a low voltage and high current (for safety and efficiency in design) to operate such things as appliances and motor-driven machines. Therefore, a device is required that will raise (or lower) the ac voltage V and current I without causing appreciable changes in the product IV. The *ac transformer* is a device used for this purpose.

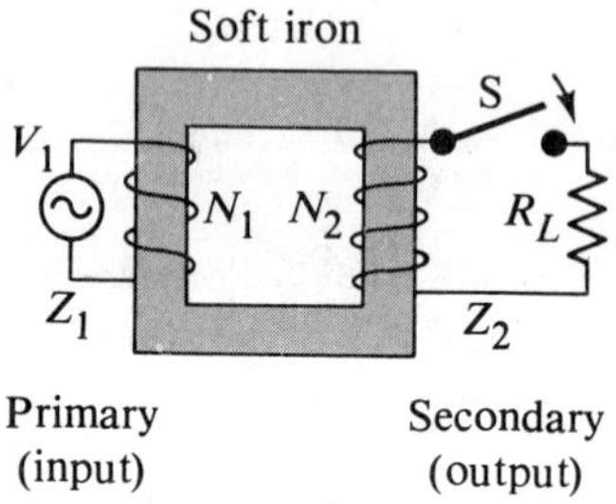

Figure 31.18 An ideal transformer consists of two coils wound on the same soft iron core. An ac voltage V_1 is applied to the primary coil, and the output voltage V_2 is across the load resistance R_L.

In its simplest form, the ac transformer consists of two coils of wire wound around a core of soft iron as in Fig. 31.18. The coil on the left, which is connected to the input ac voltage source and has N_1 turns, is called the *primary* winding (or primary). The coil on the right, consisting of N_2 turns and connected to a load resistor R_L, is called the *secondary.* The purpose of the common iron core is to increase the magnetic flux and to provide a medium in which nearly all the flux through one coil passes through the other coil. Eddy current losses are reduced by using a laminated iron core.[4] Soft iron is used as the core material to reduce hysteresis losses. Joule heat losses due to the finite resistance of the coil wires are usually quite small. Typical transformers have power efficiencies ranging from 90 to 99 percent. In what follows, we shall assume an ideal transformer, for which there are no power losses.

First, let us consider what happens in the primary circuit when the switch in the secondary circuit of Fig. 31.18 is open. If we assume that the resistance of the primary coil is negligible relative to its inductive reactance, then the primary circuit is equivalent to a simple circuit consisting of an inductor connected to an ac generator (described in Section 31.2). Since the current is 90° out of phase with the voltage, the power factor, $\cos\phi$, is zero, and hence the average power delivered from the generator to the primary circuit is zero. Faraday's law tells us that the voltage V_1 across the primary coil is given by

$$V_1 = -N_1 \frac{d\Phi_{\text{m}}}{dt} \tag{31.46}$$

[4] Losses in the core are present even under the condition of no load, that is, when the secondary circuit is open. Most of the power loss in this case is in the form of hysteresis losses as the core is magnetized cyclically.

where Φ_m is the magnetic flux through each turn. If we assume that no flux leaks out of the iron core, then the flux through each turn of the primary equals the flux through each turn of the secondary. Hence, the voltage across the secondary coil is given by

$$V_2 = -N_2 \frac{d\Phi_m}{dt} \tag{31.47}$$

Since $d\Phi_m/dt$ is common to Eqs. 31.46 and 31.47, we find that

$$V_2 = \frac{N_2}{N_1} V_1 \tag{31.48}$$

When N_2 is greater than N_1, the output voltage V_2 exceeds the input voltage V_1. This is referred to as a *step-up transformer*. When N_2 is less than N_1, the output voltage is less than the input voltage, and we speak of a *step-down transformer*.

Since the magnetic flux through the primary and secondary coils must be equal (neglecting leakage of flux lines), this means that

$$N_1 I_1 = N_2 I_2 \tag{31.49}$$

where I_1 and I_2 are the currents in the primary and secondary circuits, respectively, as indicated in Fig. 31.19. Combining this result with Eq. 31.48, we find that

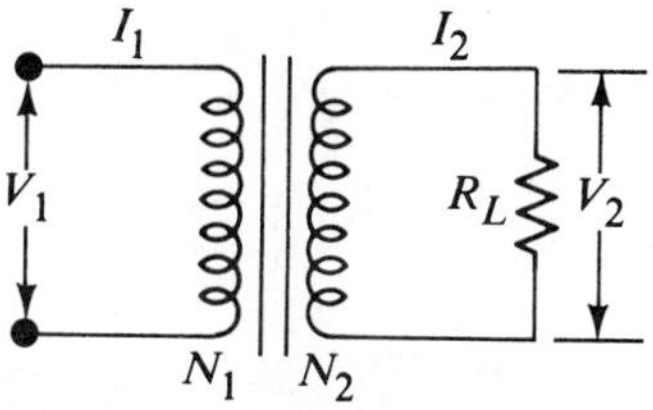

Figure 31.19 Conventional circuit diagram for a transformer.

$$I_1 V_1 = I_2 V_2 \tag{31.50}$$

Physically, this means that the input power equals the output power, a condition corresponding to that found in the so-called *ideal transformer*. Clearly, the value of the load resistance R_L determines the value of the secondary current, since $I_2 = V_2/R_L$. Furthermore, the current in the primary is $I_1 = V_1/R_{eq}$, where R_{eq} is the equivalent resistance of the load resistance R_L when viewed from the primary side, given by

$$R_{eq} = \left(\frac{N_1}{N_2}\right)^2 R_L \tag{31.51}$$

This result can be obtained from Eqs. 31.48 and 31.49. The details are left as a problem. From this analysis, we see that a transformer may be used to match resistances between the primary circuit and the load. In this manner, one can achieve *maximum* power transfer between a given power source and the load resistance.

We can now understand why transformers are useful for transmitting power over long distances. By stepping up the generator voltage, the current in the tansmission line is reduced, thereby reducing I^2R losses. In practice, the voltage is stepped up to around 230 000 V at the generating station, then stepped down to around 20 000 V at a distributing station, and finally stepped down to 110–220 V at the customer's utility poles. The power is supplied by a three-wire cable. In the United States, two of these wires are "hot," with voltages of 110 V with respect to a common ground wire. Most home appliances operating on 110 V are connected in parallel between one of the hot wires and ground. Larger appliances, such as electric stoves and clothes dryers, require 220 V. This is obtained across the two hot wires, which are 180° out of phase so that the voltage difference between them is 220 V.

There is a practical upper limit to the voltages one can use in transmission lines. Excessive voltages could ionize the air surrounding the transmission lines, which could result in a conducting path to ground or to other objects in the vicinity. This, of course, would present a serious hazard to any living creatures. For this reason, a long string of insulators is used to keep high-voltage wires away from their supporting metal towers. Other insulators are used to maintain separation between wires.

Example 31.6

A generator produces 10 A (rms) of current at 400 V. The voltage is stepped up to 4500 V by an ideal transformer and transmitted a long distance through a power line of total resistance 30 Ω. (a) Determine the percentage of power lost when the voltage is stepped up.

From Eq. 31.50, we find that the current in the transmission line is

$$I_2 = \frac{I_1 V_1}{V_2} = \frac{(10\text{ A})(400\text{ V})}{4500\text{ V}} = 0.89\text{ A}$$

Hence, the power lost in the transmission line is

$$P_{\text{lost}} = I_2^2 R = (0.89\text{ A})^2(30\ \Omega) = 24\text{ W}$$

Since the output power of the generator is $P = IV = (10\text{ A})(400\text{ V}) = 4000$ W, we find that the percentage of power lost is

$$\%\text{ power lost} = \left(\frac{24}{4000}\right) \times 100 = 0.6\%$$

(b) What percentage of the original power would be lost in the transmission line if the voltage were not stepped up?

If the voltage were not stepped up, the current in the transmission line would be 10 A and the power lost in the line would be $I^2R = (10\text{ A})^2(30\ \Omega) = 3000$ W. Hence, the percentage of power lost would be

$$\%\text{ power lost} = \frac{3000}{4000} \times 100 = 75\%$$

This example illustrates the advantage of high-voltage transmission lines.

Q11. What is the advantage of transmitting power at high voltages?

Q12. Will a transformer operate if a battery is used for the input voltage across the primary? Explain.

31.9 SUMMARY

If an ac circuit consists of a generator and a resistor, the current in the circuit is in phase with the voltage. That is, the current and voltage reach their maximum values at the same time.

If an ac circuit consists of a generator and an inductor, the current *lags behind* the voltage by 90°. That is, the voltage reaches its maximum value one quarter of a period before the current reaches its maximum value.

If an ac circuit consists of a generator and a capacitor, the current *leads* the voltage by 90°. That is, the current reaches its maximum value one quarter of a period before the voltage reaches its maximum value.

In ac circuits that contain inductors and capacitors, it is useful to define the inductive reactance X_L and capacitive reactance X_C as

Inductive reactance

$$X_L = \omega L \qquad (31.10)$$

Capacitive reactance

$$X_C = \frac{1}{\omega C} \qquad (31.17)$$

where ω is the angular frequency of the ac generator. The SI unit of reactance is the ohm.

The maximum current (or current amplitude) in a series *RLC* circuit is

$$I_m = \frac{\mathcal{E}_m}{\sqrt{R^2 + (X_L - X_C)^2}} \qquad (31.26)$$

where $\mathcal{E}_m$ is the maximum value of the applied voltage.

The quantity in the denominator of Eq. 31.26 is defined as the impedance Z of the circuit, which also has the unit of ohm:

Impedance

$$Z = \sqrt{R^2 + (X_L - X_C)^2} \qquad (31.27)$$

In an RLC series ac circuit, the applied voltage and current are out of phase. The phase angle ϕ between the current and voltage is given by

Phase angle

$$\tan\phi = \frac{X_L - X_C}{R} \qquad (31.29)$$

The sign of ϕ can be positive or negative, depending on whether X_L is greater or less than X_C. The phase angle is zero when $X_L = X_C$.

The average power delivered by the generator in an RLC ac circuit is given by

Average power

$$P_{av} = I_{rms}V_{rms}\cos\phi \qquad (31.35)$$

where I_{rms} and V_{rms} are the root mean square values of the current and voltage, respectively. These are related to their maximum values through the relationships

$$I_{rms} = I_m/\sqrt{2} \qquad \text{and} \qquad V_{rms} = \mathcal{E}_m/\sqrt{2}$$

An equivalent expression for the average power is

Average power

$$P_{av} = I_{rms}^2 R \qquad (31.36)$$

The average power delivered by the generator is dissipated as heat in the resistor. There is no power loss in an ideal inductor or capacitor.

A series RLC circuit is in resonance when the inductive reactance equals the capacitive reactance. When this condition is met, the current given by Eq. 31.26 reaches its maximum value. Setting $X_L = X_C$, one finds that the resonance frequency ω_o of the circuit has the value

Resonance frequency

$$\omega_o = 1/\sqrt{LC} \qquad (31.39)$$

The current in a series RLC circuit reaches its maximum value when the frequency of the generator equals ω_o, that is, when the "driving" frequency matches the resonance frequency.

A transformer is a device designed to raise (or lower) an ac voltage and current without causing an appreciable change in the product IV. In its simplest form, it consists of a primary coil of N_1 turns and a secondary coil of N_2 turns, both wound on a common soft iron core. When a voltage V_1 is applied across the primary, the voltage V_2 across the secondary is given by

$$V_2 = \frac{N_2}{N_1}V_1 \qquad (31.48)$$

In an ideal transformer, the power delivered by the generator must equal the power dissipated in the load. If a load resistor R_L is connected across the secondary coil, this means that

$$I_1V_1 = I_2V_2 = \frac{V_2^2}{R_L} \qquad (31.50)$$

EXERCISES

Section 31.1 Resistors in an ac Circuit

1. In the simple ac circuit shown in Fig. 31.1, let $R = 40\ \Omega$, $\mathcal{E}_m = 120$ V, and the frequency of the generator $f = 60$ Hz. Assume that the voltage across the resistor $V_R = 0$ when $t = 0$. Calculate (a) the maximum current in the resistor and (b) the angular frequency of the generator.

2. Use the values given in Exercise 1 for the circuit of Fig. 31.1 to calculate the current through the resistor at (a) $t = \frac{1}{240}$ s and (b) $t = \frac{1}{180}$ s.

3. In the simple ac circuit shown in Fig. 31.1, $R = 30\ \Omega$. (a) If $V_R = 0.25\mathcal{E}_m$ at $t = 0.002$ s, what is the angular frequency of the generator? (b) What is the next value of t for which V_R will be $0.25\mathcal{E}_m$?

4. The current in the circuit shown in Fig. 31.1 equals 70% of the maximum current at $t = 0.003$ s. What is the frequency f of the generator?

Section 31.2 Inductors in an ac Circuit

5. Show that the inductive reactance X_L has the SI unit of ohm.

6. In a purely inductive ac circuit, as in Fig. 31.3, $\mathcal{E}_m = 120$ V. (a) If the maximum current is 10 A at a frequency of 60 Hz, calculate the inductance L. (b) At what angular frequency ω will the maximum current be reduced to 5 A?

7. For the circuit shown in Fig. 31.3, $\mathcal{E}_m = 90$ V, $\omega = 120\pi$ rad/s, and $L = 100$ mH. Calculate the current in the inductor at $t = 0.002$ s.

8. (a) If $L = 250$ mH and $\mathcal{E}_m = 90$ V in the circuit of Fig. 31.3, at what frequency will the inductive reactance equal 20 Ω? (b) Calculate the maximum value of the current in the circuit at this frequency.

9. What is the inductance of a coil that has an inductive reactance of 40 Ω at an angular frequency of 754 rad/s?

Section 31.3 Capacitors in an ac Circuit

10. Show that the SI unit of capacitance reactance is the ohm.

11. Calculate the capacitive reactance of a 12-μF capacitor when connected to an ac generator having an angular frequency of 180π rad/s.

12. What maximum current will be delivered by an ac generator with $\mathcal{E}_m = 48$ V and $f = 90$ Hz when connected across a 3.7-μF capacitor?

13. A variable-frequency ac generator with $\mathcal{E}_m = 24$ V is connected across a 7.96×10^{-9}-F capacitor. At what frequency should the generator be operated to provide a maximum current of 6 A?

14. The generator in a purely capacitive ac circuit (Fig. 31.5) has an angular frequency of 120π rad/s and $\mathcal{E}_m = 110$ V. If $C = 6\ \mu$F, what is the current in the circuit at $t = \frac{7}{480}$ s?

Section 31.4 The *RLC* Series ac Circuit

15. A series ac circuit contains the following components: $R = 200\ \Omega$, $L = 400$ mH, $C = 5\ \mu$F and a generator with $\mathcal{E}_m = 140$ V operating at 60 Hz. Calculate the (a) inductive reactance, (b) capacitive reactance, (c) impedance, (d) maximum current, and (e) phase angle.

16. A resistor ($R = 900\ \Omega$), a capacitor ($C = 0.25\ \mu$F), and an inductor ($L = 2.5$ H) are connected in series across a 240-Hz ac source for which $\mathcal{E}_m = 140$ V. Calculate the (a) impedance of the circuit, (b) maximum current delivered by the source, and (c) phase angle between the current and voltage. (d) Is the current leading or lagging behind the voltage?

17. A coil with an inductance of 15.3 mH and a resistance of 5 Ω is connected to a *variable*-frequency ac generator. At what frequency will the voltage across the coil lead the current by 60°?

18. A 500-Ω resistor, an inductor, and a capacitor are in series with a generator. When the frequency is adjusted to $500/\pi$ Hz, the inductive reactance is 600 Ω. What is the *minimum* value of capacitance that will result in a circuit impedance of 707 Ω?

19. An ac source with $\mathcal{E}_m = 110$ V and $f = 60$ Hz is connected between points a and d in Fig. 31.20. Calculate the *maximum* voltages between points (a) a and b, (b) b and c, (c) c and d, (d) b and d.

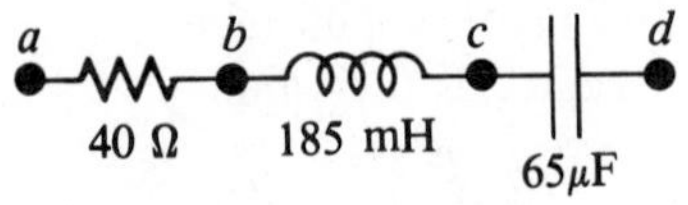

Figure 31.20 (Exercise 19).

20. Draw to scale a phasor diagram showing Z, X_L, X_C, and ϕ for an ac series circuit for which $R = 300\ \Omega$, $C = 11\ \mu$F, $L = 0.2$ H, and $f = 500/\pi$ Hz.

21. An inductor ($L = 530$ mH), a capacitor ($C = 4.43\ \mu$F), and a resistor ($R = 464\ \Omega$) are connected in series. A 60-Hz ac generator produces a maximum current of 310 mA in the circuit. (a) Calculate the required maximum voltage $\mathcal{E}_m$. (b) Determine the angle by which the current in the circuit leads or lags behind the applied voltage.

22. Repeat Exercise 21 for the case where $L = 113$ mH, $C = 53\ \mu$F, $R = 65\ \Omega$, if an applied voltage of frequency 120 Hz produces a maximum current of 2.49 A.

Section 31.5 Power in an ac Circuit

23. Calculate the average power delivered to the series *RLC* circuit described in Exercise 15.

24. Consider the circuit described in Exercise 16. (a) What is the power factor of the circuit? (b) What is the rms current in the circuit? (c) What average power is delivered by the source?

25. The rms terminal voltage of an ac generator is 117 V. The operating frequency is 60 Hz. Write the equation giving the output voltage as a function of time.

26. The average power in a circuit for which the rms current is 8 A is 900 W. Calculate the resistance of the circuit.

27. In a certain series *RLC* circuit, $I_{rms} = 6$ A, $V_{rms} = 240$ V, and the current leads the voltage by 53°. (a) What is the total resistance of the circuit? (b) Calculate the reactance of the circuit ($X_L - X_C$).

28. A series *RLC* circuit has a resistance of 80 Ω and an impedance of 180 Ω. What average power will be delivered to this circuit when $V_{rms} = 120$ V?

Section 31.6 Resonance in a Series RLC Circuit

29. Calculate the resonance frequency of a series *RLC* circuit for which $C = 5$ μF and $L = 80$ mH.

30. The resonant frequency of a series *RLC* circuit is found to be 4.2×10^4 Hz. If the capacitance of the circuit is 300 μF, what is the value of the inductance?

31. A coil of resistance 20 Ω and inductance 10.2 H is in series with a capacitor and a 100-V (rms), 60-Hz source. The current in the circuit is 5 A (rms). (a) Calculate the capacitance in the circuit. (b) What is V_{rms} across the coil?

32. A series *RLC* circuit has $L = 156$ mH, $C = 0.2$ μF, and $R = 88$ Ω. The circuit is driven by a generator with $\mathcal{E}_m = 110$ V and a *variable* angular frequency. Calculate (a) the resonance frequency of the circuit, (b) the quality factor of the circuit, and (c) the two values of ω at which the average power has one half its maximum value.

33. What average power is delivered to the circuit described in Exercise 32 when the frequency is adjusted to 600 Hz?

Section 31.7 Filter Circuits

34. Consider the circuit shown in Fig. 31.14, with $R = 1000$ Ω and $C = 0.01$ μF. Calculate the ratio V_{out}/V_{in} for (a) $\omega = 500$ s^{-1} and (b) $\omega = 5 \times 10^6$ s^{-1}.

35. Assign the values of *R* and *C* given in Exercise 34 to the circuit shown in Fig. 31.16 and calculate V_{out}/V_{in} for (a) $\omega = 50$ s^{-1} and (b) $\omega = 5 \times 10^5$ s^{-1}.

Section 31.8 The Transformer and Power Transmission

36. An ideal transformer has 150 turns on the primary winding and 600 turns on the secondary. If the primary is connected across a 110-V (rms) generator, what is the rms output voltage?

37. A step-up transformer is designed to have an output voltage of 1800 V (rms) when the primary is connected across a 120-V (rms) source. (a) If there are 100 turns on the primary winding, how many turns are required on the secondary? (b) If a load resistor across the secondary draws a current of 0.5 A, what is the current in the primary, assuming ideal conditions?

38. If the transformer in Exercise 37 has an efficiency of 92%, what is the current in the primary when the secondary current is 0.6 A?

39. The primary current of an ideal transformer is 6.5 A when the primary voltage is 96 V. Calculate the voltage across the secondary when a current of 0.8 A is delivered to a load resistor.

PROBLEMS

1. In a series ac circuit, $R = 21$ Ω, $L = 25$ mH, $C = 17$ μF, $\mathcal{E}_m = 150$ V, and $\omega = \frac{2000}{\pi}$ s^{-1}. (a) Calculate the maximum current in the circuit. (b) Determine the maximum voltage across each of the three elements. (c) What is the power factor for the circuit? (d) Show X_L, X_C, R, and ϕ in a phasor diagram for the circuit.

2. A 1000-Ω resistor is connected in series to a 0.6-H inductor and a 2.5-μF capacitor. This *RLC* combination is then connected across a voltage source that varies as $\mathcal{E} = 80 \sin\left(\frac{1000}{\pi}t\right)$ V. Calculate (a) the maximum current, (b) the phase angle, (c) the power factor, (d) V_{rms} across the inductor, and (e) the average power delivered to the circuit.

3. For a certain series *RLC* circuit, $R = 100$ Ω, $I = 2\sqrt{3}\sin(200t - \phi)$ A, $\mathcal{E} = 400\sqrt{3}\sin(200t)$ V, and $X_C = \sqrt{3} \times 10^2$ Ω. Find (a) the impedance, (b) the inductive reactance of the circuit, and (c) the phase angle.

4. A series *RLC* circuit consists of an 8-Ω resistor, a 5-μF capacitor, and a 50-mH inductor. A variable frequency source of 400 V (rms) is applied across the combination. Determine the power delivered to the circuit when the frequency is equal to one half of the resonance frequency.

5. Consider the ideal transformer discussed in Section 31.8. Show that the result stated in Eq. 31.51 can be obtained from the ideal transformer properties stated in Eqs. 31.48 and 31.49.

6. The average power delivered to a series *RLC* circuit at frequency ω (Section 31.6) is given by Eq. 31.41. (a) Show that the maximum current can be written

$$I_m = \omega\mathcal{E}_m[L^2(\omega_o^2 - \omega^2)^2 + (\omega R)^2]^{-1/2}$$

where ω is the operating frequency of the circuit and ω_o is the resonance frequency. (b) Show that the phase angle can be expressed as

$$\phi = \tan^{-1}\left[\frac{L}{R}\left(\frac{\omega_o^2 - \omega^2}{\omega}\right)\right]$$

7. A series *RLC* circuit consists of a 30-Ω resistor, a 133-μF capacitor, and a 159-mH inductor, which also has a resistance of 25 Ω. The voltage source for the circuit is 180 V (rms) at an angular frequency of 120π rad/s. Calculate the rms voltages across (a) the inductor, (b) the capacitor, and (c) the resistor.

8. A series *RLC* circuit has a resonance frequency of 2000/π Hz. When operating at a frequency $\omega > \omega_o$, $X_L = 12\ \Omega$ and $X_C = 8\ \Omega$. Calculate the values of *L and C* for the circuit.

9. Figure 31.21a shows a parallel *RLC* circuit, and the corresponding phasor diagram is given in Fig. 31.21b. The instantaneous voltage (and rms voltage) across each of the three circuit elements is the same, and each is in phase with the current through the resistor. The currents in *C* and *L* lead (or lag behind) the current in the resistor, as shown in Fig. 31.21b. (a) Show that the rms current delivered by the source is given by

$$I_{rms} = V_{rms}\left[\frac{1}{R^2} + \left(\omega C - \frac{1}{\omega L}\right)^2\right]^{1/2}$$

(b) Show that the phase angle ϕ between V_{rms} and I_{rms} is given by

$$\tan\phi = R\left(\frac{1}{X_C} - \frac{1}{X_L}\right)$$

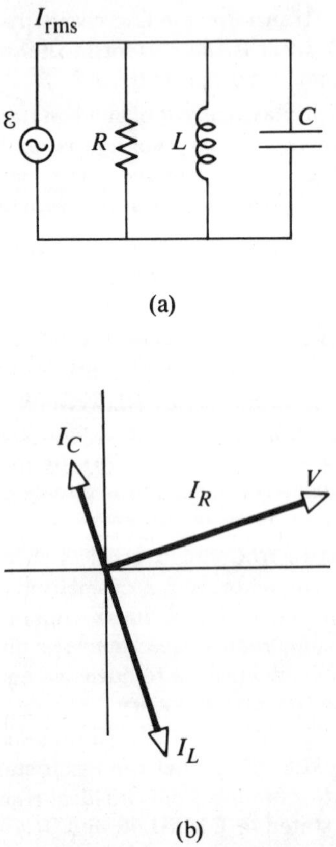

Figure 31.21 (Problem 10).

10. A resistor of 80 Ω and a 200-mH inductor are connected in *parallel* across a 100-V (rms), 60-Hz source. (a) What is the rms current in the resistor? (b) By what angle does the total current lead or lag behind the voltage?

11. An 80-Ω resistor, a 200-mH inductor, and a 0.15-μF capacitor are connected in parallel across a 120-V (rms) source operating at an angular frequency of 374 rad/s. (a) What is the resonant frequency of the circuit? (b) Calculate the rms current in the resistor, inductor, and capacitor. (c) What is the rms current delivered by the source? (d) Is the current leading or lagging behind the voltage? By what angle?

12. A series *RLC* circuit is operating at 2000 Hz. At this frequency, $X_L = X_C = 1884\ \Omega$. The resistance of the circuit is 40 Ω. (a) Prepare a table showing the values of X_L, X_C, and *Z* for *f* = 300, 600, 800, 1000, 1500, 2000, 3000, 4000, 6000, and 10 000 Hz. (b) Plot on the same set of axes X_L, X_C, and *Z* as a function of ln *f*.

13. *Impedance matching:* A transformer may be used to provide maximum power transfer between two ac circuits that have different impedances. (a) Show that the ratio of turns N_1/N_2 needed to meet this condition is given by

$$\frac{N_1}{N_2} = \sqrt{\frac{Z_1}{Z_2}}$$

(b) Suppose you want to use a transformer as an impedance-matching device between an audio amplifier that has an output impedance of 8000 Ω and a speaker that has an input impedance of 8 Ω. What should be the ratio of primary to secondary turns on the transformer?

14. A small transformer is used to supply 6 V ac to a model railroad lighting circuit. The primary has 220 turns and is connected to a standard 110-V, 60-Hz line. Although the resistance of the primary may be neglected, it has an inductance of 150 mH. (a) How many turns are required on the secondary winding? (b) If the transformer is left plugged in, what current will be drawn by the primary when the secondary is open? (c) What power will be drawn by the primary when the secondary is open?

15. A transmission line with a resistance per unit length of $4.5 \times 10^{-4}\ \Omega/\text{m}$ is to be used to transmit 5000 kW of power over a distance of 400 miles (6.44×10^5 m). The terminal voltage of the generator is 4500 V. (a) What is the line loss if a transformer is used to step up the voltage to 500 kV? (b) What fraction of the input power is lost to the line under these circumstances? (c) What difficulties would be encountered by attempting to transmit the 5000 kW of power at the generator voltage of 4500 V?

16. A power transmission line consists of two parallel wires each having a cross-sectional area A. (One serves as a return line, so the affective length is $2l$ when connected to a load.) Show that the power lost as heat in the conductors is given by

$$P_{\text{loss}} = \frac{2\rho l P_L^2}{AV^2 \cos^2\phi}$$

where $\cos\phi$ is the power factor, P_L is the power supplied to the load, and ρ is the resistivity of the conductor. This result shows that power losses may be minimized by minimizing ϕ and increasing the operating voltage V. (Note that V represents an rms value in this problem.)

17. The quality factor of a series RLC circuit is discussed in Section 31.6, with reference to Eq. 31.42, show that $\Delta\omega \approx R/L$ and therefore $Q_0 \approx \omega L/R$.

32 Wave Motion

Jay Freedman

32.1 INTRODUCTION

Most of us first experienced waves as children when we dropped a pebble into a pond. The disturbance created by the pebble excites water waves, which move outward, finally reaching the shore of the pond. If you were to examine carefully the motion of a leaf floating near the disturbance, you would see that it moves up and down and sideways about its original position, but does not undergo any net displacement attributable to the disturbance. That is, the water wave (or disturbance) moves from one place to another, *yet the water is not carried with it.*

An excerpt from a book by Einstein and Infeld gives the following remarks concerning wave phenomena.[1]

> A bit of gossip starting in Washington reaches New York very quickly, even though not a single individual who takes part in spreading it travels between these two cities. There are two quite different motions involved, that of the rumor, Washington to New York, and that of the persons who spread the rumor. The wind, passing over a field of grain, sets up a wave which spreads across the whole field. Here again we must distinguish between the motion of the wave and the motion of the separate plants, which undergo only small oscillations. . . . The particles constituting the medium perform only small vibrations, but the whole motion is that of a progressive wave. The essential new thing here is that for the first time we consider the motion of something which is not matter, but energy propagated through matter.

Water waves represent only one example of a wide variety of physical phenomena that have wavelike characteristics. The world is full of waves: sound waves; mechanical waves, such as a wave on a string; earthquake waves; shock waves generated by supersonic aircraft; and electromagnetic waves, such as visible light, radio waves, television signals, and x-rays. In the present chapter, we shall confine our attention to mechanical waves, that is, waves which travel only in a material substance.

The wave concept is rather abstract. When we observe what we call a water wave, what we see is a rearrangement of the water's surface. Without the water, there would be no wave. A wave traveling on a string would not exist without the

[1] Albert Einstein and Leopold Infeld, *The Evolution of Physics,* New York, Simon and Schuster, 1961. Excerpt from *What is a Wave?*

string. Sound waves travel through air as a result of pressure variations from point to point. In all cases, what we interpret as a wave corresponds to the disturbance of a body or medium. Therefore, we can consider a wave to be the *motion of a disturbance.* The motion of the disturbance (that is, the wave itself, or the state of the medium) is not to be confused with the motion of the particles. The mathematics used to describe wave phenomena is common to all waves. In general, we shall find that mechanical wave motion is described by specifying the positions of all points of the disturbed medium as a function of time.

The mechanical waves discussed in this chapter require (1) some source of disturbance, (2) a medium that can be disturbed, and (3) some physical connection or mechanism through which adjacent portions of the medium can influence each other. We shall find that all waves carry energy and momentum. The amount of energy transmitted through a medium and the mechanism responsible for the transport of energy will differ from case to case. For instance, the power carried by ocean waves during a storm is much greater than the power of sound waves generated by a single human voice.

Wave characteristics

Three physical concepts are important in characterizing waves: the wavelength, the frequency, and the wave velocity. One *wavelength* is the *distance between any two points on a wave that behave identically.* For example, in the case of water waves, the wavelength is the distance between adjacent crests or between adjacent troughs.

Some waves are periodic in nature. The *frequency* of such periodic waves is *the rate at which the disturbance repeats itself.*

Waves travel, or *propagate,* with a specific velocity, which depends on the properties of the medium being disturbed. For instance, sound waves travel through air with a speed of about 344 m/s (781 mi/h), whereas the speed of sound through solids is higher than 344 m/s. A special class of waves which do not require a medium in order to propagate are electromagnetic waves, which travel very swiftly through a vacuum with a speed of about 3×10^8 m/s (186 000 mi/s). We shall discuss electromagnetic waves further in Chapter 35.

32.2 TYPES OF WAVES

One of the simplest ways to demonstrate wave motion is to flip one end of a long rope that is under tension, where the opposite end is fixed, as in Fig. 32.1. Only a portion of the wave is produced in this manner. It consists of a bump (called a pulse) in the rope that travels (to the right in Fig. 32.1) with a definite speed. This type of disturbance is called a *traveling wave.* Figure 32.1 represents four consecutive "snapshots" of the traveling wave. As we shall see later, the speed of the wave depends on the tension in the rope and on the properties of the rope. The rope is the *medium* through which the wave travels. We shall assume that the shape of the wave pulse does not change as it travels along the rope.[2]

Note that, as the wave pulse travels along the rope, *each segment of the rope that is disturbed moves in a direction perpendicular to the wave motion.* Figure 32.2 illustrates this point for one particular segment, labeled *P.* Note that there is no motion of any part of the rope in the direction of the wave. A traveling wave such as this, in which the particles of the disturbed medium move perpendicular to the wave velocity, is called a *transverse wave.*[3]

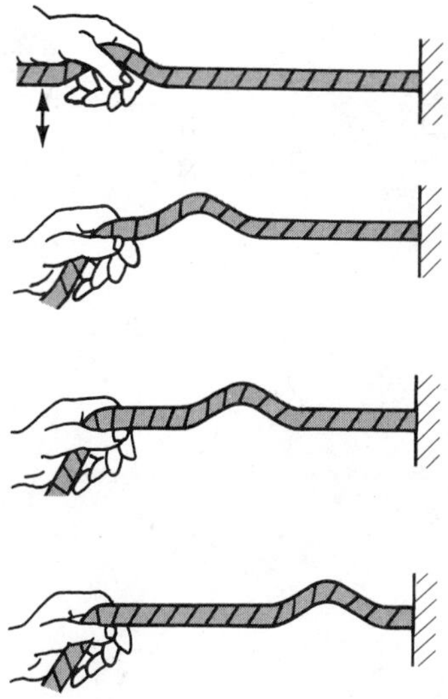

Figure 32.1 A wave pulse traveling down a stretched rope. The shape of the pulse is assumed to remain unchanged as it travels along the rope.

[2] Strictly speaking, the pulse will change its shape and gradually spread out during the motion. This effect is called *dispersion* and is common to all mechanical waves.

[3] Other examples of transverse waves are electromagnetic waves, such as light, radio, and television waves. At a given point in space, the electric and magnetic fields of an electromagnetic wave are perpendicular to the direction of the wave and vary in time as the wave passes. As we shall see later, electromagnetic waves are produced by accelerating charges.

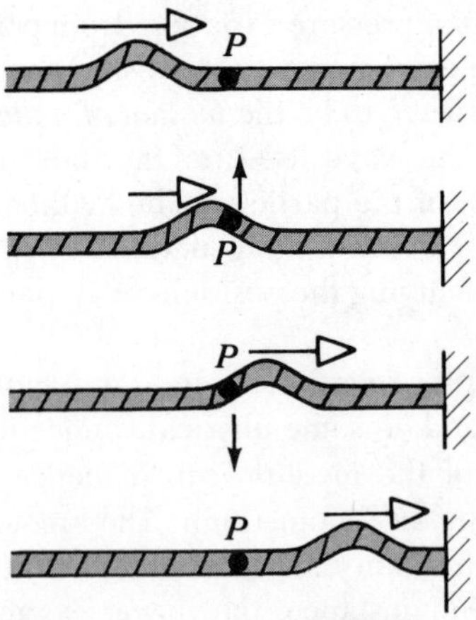

Figure 32.2 A pulse traveling on a stretched rope is a transverse wave. That is, any element P on the rope moves in a direction *perpendicular* to the wave motion.

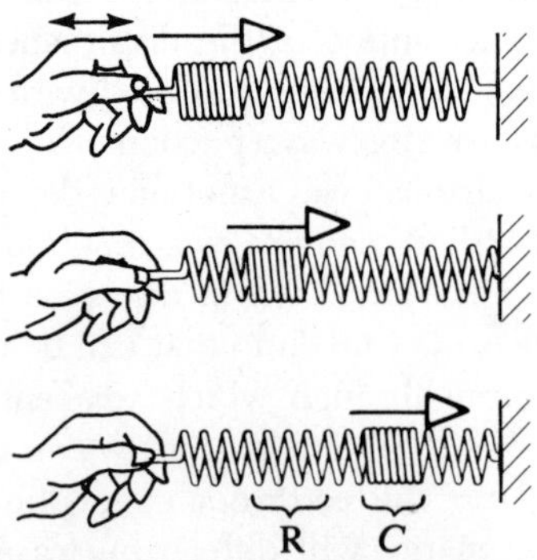

Figure 32.3 A longitudinal pulse along a stretched spring. The disturbance of the medium (the displacement of the coils) is in the direction of the wave motion. For the starting motion described in the text, the compressed region C is followed by an extended region R.

In another class of waves, called *longitudinal waves,* the particles of the medium undergo displacements in a direction *parallel* to the direction of wave motion. Sound waves, which we shall discuss in Chapter 33, are longitudinal waves that result from the disturbance of the medium. The disturbance corresponds to a series of high- and low-pressure regions that travel through air or through any material medium with a certain velocity. A longitudinal pulse can be easily produced in a stretched spring, as in Fig. 32.3. The left end of the spring is given a sudden jerk (consisting of a brief push to the right and equally brief pull to the left) along the length of the spring; this creates a sudden compression of the coils. The compressed region C (pulse) travels along the spring, and so we see that the disturbance is parallel to the wave motion. Region C is followed by a region R, where the coils are extended.[4]

Some waves in nature are neither transverse nor longitudinal, but a combination of the two. Surface water waves are a good example. When a water wave travels on the surface of deep water, water molecules at the surface move in nearly circular paths, as shown in Fig. 32.4, where the water surface is drawn as a series of crests and troughs. Note that the disturbance has both transverse and longitudinal components. As the wave passes, water molecules at the crests move in the direction of the wave, and molecules at the troughs move in the opposite direction. Hence, there is no *net* displacement of a water molecule after the passage of any number of complete waves.

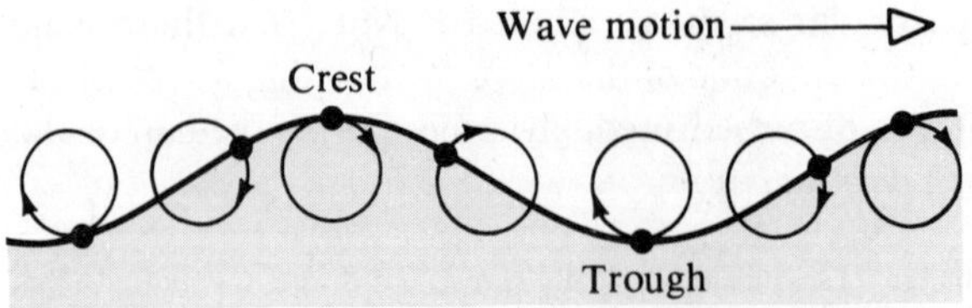

Figure 32.4 Wave motion on the surface of water. The particles at the water's surface move in nearly circular paths. Each particle is displaced horizontally and vertically from its equilibrium position, represented by circles.

[4]In the case of longitudinal pressure waves in a gas, each compressed area is a region of higher-than-average pressure and density, and each stretched area is a region of lower-than-average pressure and density.

Q1. Why is a wave pulse traveling on a string considered a transverse wave?

Q2. How would you set up a longitudinal wave in a stretched spring? Would it be possible to set up a transverse wave in a spring?

32.3 ONE-DIMENSIONAL TRAVELING WAVES

So far we have given only a verbal and graphical description of a traveling wave. Let us now give a mathematical description of a one-dimensional traveling wave. Consider again a wave pulse traveling to the right on a long stretched string with constant speed v, as in Fig. 32.5. The pulse moves along the x axis (the axis of the string), and the transverse displacement of the string is measured with the coordinate y.

Figure 32.5a represents the shape and position of the pulse at time $t = 0$. At this time, the shape of the pulse, whatever it may be, can be represented as $y = f(x)$. That is, y is some definite function of x. The *maximum displacement,* y_m, is called the *amplitude* of the wave. Since the speed of the wave pulse is v, it travels to the right a distance vt in a time t (Fig. 32.5b).

If the shape of the wave pulse doesn't change with time, we can represent the displacement y for all later times measured in a stationary frame with the origin at 0 as

$$y = f(x - vt) \tag{32.1}$$

Wave traveling to the right

Similarly, if the wave pulse travels to the *left*, its displacement is given by

$$y = f(x + vt) \tag{32.2}$$

Wave traveling to the left

The displacement y, sometimes called the *wave function*, depends on the two

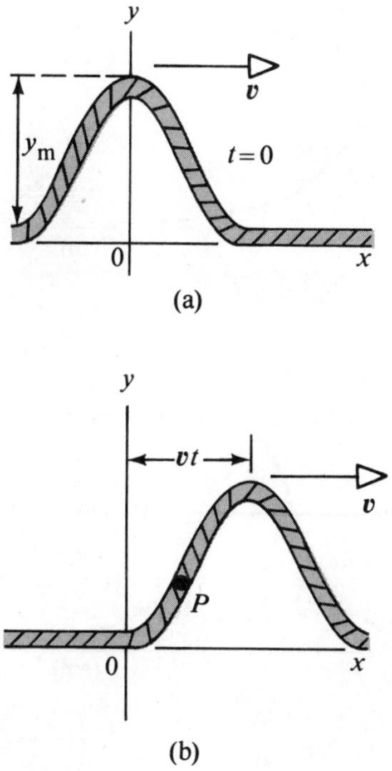

Figure 32.5 A one-dimensional wave pulse traveling to the right with a speed v. (a) At $t = 0$, the shape of the pulse is given by $y = f(x)$. (b) At some later time t, the shape remains unchanged and the vertical displacement is given by $y = f(x - vt)$.

variables x and t. For this reason, it is often written $y(x,t)$, which is read "y as a function of x and t." It is important to understand the meaning of y.

Consider a particular point P on the string, identified by a particular value of its coordinates. As the wave passes the point P, the y coordinate of this point will increase, reach a maximum, and then decrease to zero. Therefore, *the wave function* y *represents the* y *coordinate of any point* P *at any time* t. Furthermore, if t is fixed, then the wave function y as a function of x *defines a curve representing the actual shape of the pulse at this time*. This is equivalent to a "snapshot" of the wave at this time.

For a pulse that moves without changing its shape, the velocity of a wave pulse is the same as the motion of any feature along the pulse profile, such as the crest. To find the velocity of the pulse, we can calculate how far the crest moves in a short time and then divide this distance by the time interval. The crest of the pulse corresponds to that point for which y has its maximum value. In order to follow the motion of the crest, some particular value, say x_0, must be substituted for $x - vt$. (This value x_0 is called the *argument* of the function y.) Regardless of how x and t change individually, we must require that $x - vt = x_0$ in order to stay with the crest. This, therefore, represents the equation of motion of the crest. At $t = 0$, the

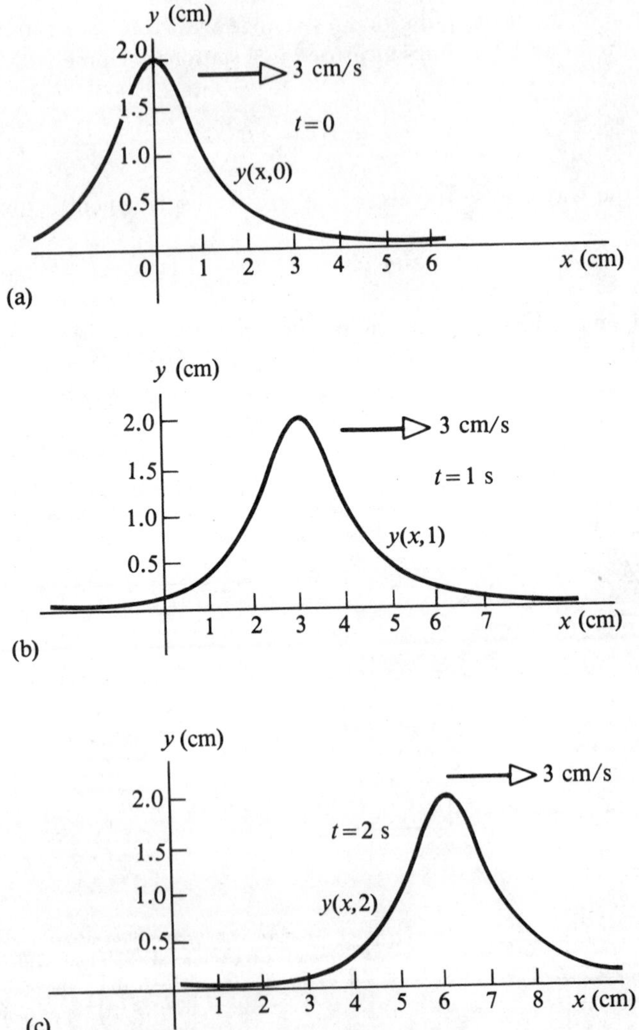

Figure 32.6 (Example 32.1) Graphs of the function $y(x,t) = 2/[(x - 3t)^2 + 1]$ (a) $t = 0$, (b) $t = 1$ s, and (c) $t = 2$ s.

crest is at $x = x_0$; at a time dt later, the crest is at $x = x_0 + v\,dt$. Therefore, the crest has moved a distance $dx = (x_0 + v\,dt) - x_0 = v\,dt$ in a time dt. Clearly, the wave speed, often called the *phase velocity*, is given by

$$v = dx/dt \qquad (32.3)$$

Phase velocity

The wave speed, or phase velocity, must not be confused with the transverse velocity (which is in the y direction) of a particle in the medium.

The following example illustrates how a specific wave function is used to describe the motion of a traveling wave pulse.

Example 32.1

A traveling wave pulse moving to the right along the x axis is represented by the wave function

$$y(x,t) = \frac{2}{(x - 3t)^2 + 1}$$

where x and y are measured in cm and t is in s. Let us plot the waveform at $t = 0$, $t = 1$ s, and $t = 2$ s.

Solution: First, note that this function is of the form $y = f(x - vt)$. By inspection, we see that the speed of the wave is $v = 3$ cm/s. Furthermore, the wave amplitude (the maximum value of y) is given by $y_m = 2$ cm. At times $t = 0$, $t = 1$ s, and $t = 2$ s, the wave function expressions are

$$y(x,0) = \frac{2}{x^2 + 1} \qquad \text{at } t = 0$$

$$y(x,1) = \frac{2}{(x - 3)^2 + 1} \qquad \text{at } t = 1\text{ s}$$

$$y(x,2) = \frac{2}{(x - 6)^2 + 1} \qquad \text{at } t = 2\text{ s}$$

We can now use these expressions to plot the wave function versus x at these times. For example, let us evaluate $y(x,0)$ at $x = 0.5$ cm:

$$y(0.5,0) = \frac{2}{(0.5)^2 + 1} = 1.60\text{ cm}$$

Likewise, $y(1,0) = 1.0$ cm, $y(2,0) = 0.40$ cm, etc. A continuation of this procedure for other values of x yields the waveform shown in Fig. 32.6a. In a similar manner, one obtains the graphs of $y(x,1)$ and $y(x,2)$, shown in Figs. 32.6b and 32.6c, respectively. These snapshots show that the wave pulse moves to the right without changing its shape and has a constant speed of 3 cm/s.

32.4 SUPERPOSITION AND INTERFERENCE OF WAVES

Many interesting wave phenomena in nature cannot be described by a single moving pulse. Instead, one must analyze complex waveforms in terms of a combination of many traveling waves. To analyze such wave combinations, one can make use of the *superposition principle: If two or more traveling waves are moving through a medium, the resultant wave function at any point is the algebraic sum of the wave functions of the individual waves.* This rather striking property is exhibited by many waves in nature. Waves that obey this principle are called *linear waves,* and they are generally characterized by small wave amplitudes. Waves that violate the superposition principle are called *nonlinear waves* and are often characterized by large amplitudes. In this book, we shall deal only with linear waves.

Linear waves obey the superposition principle

One consequence of the superposition principle is the observation that *two traveling waves can pass through each other without being destroyed or even altered.* For instance, when two pebbles are thrown into a pond, the expanding circular surface waves do not destroy each other. In fact, the ripples pass through each other. The complex pattern that is observed can be viewed as two independent sets of expanding circles. Likewise, when sound waves from two sources move through air, they also can pass through each other. The resulting sound one hears at a given point is the resultant of both disturbances.

A simple pictorial representation of the superposition principle is obtained by considering two pulses traveling in opposite directions on a stretched string, as in Fig. 32.7. The wave function for the pulse moving to the right is y_1, and the wave

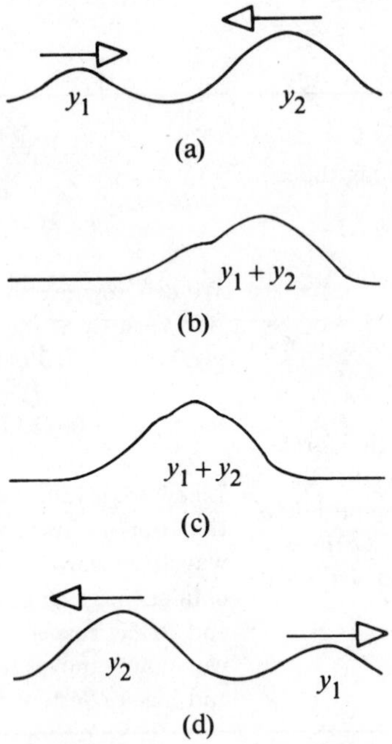

Figure 32.7 Two wave pulses traveling on a stretched string in opposite directions pass through each other. When the pulses overlap, as in (b) and (c), the net displacement of the string equals the sum of the displacements of each pulse. Since the pulses both have positive displacements, we refer to their superposition as *constructive interference.*

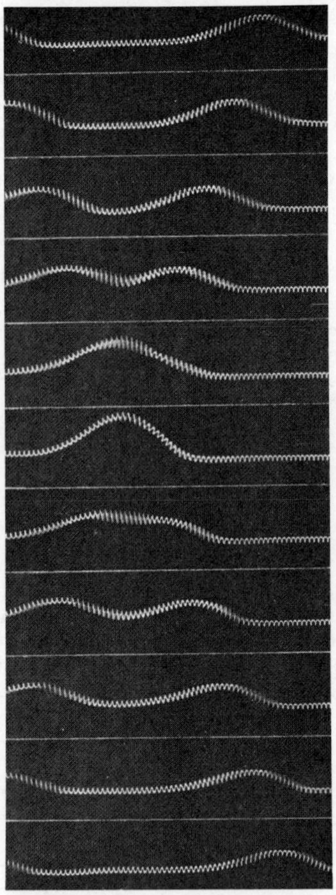

Photograph of superposition of two equal and symmetric pulses traveling in opposite directions on a stretched string. (Photo, Education Development Center, Newton, Mass.)

function for the pulse moving to the left is y_2. The pulses have the same speed, but different shapes. Each pulse is assumed to be symmetric, and both displacements are taken to be positive. When the waves begin to overlap (Fig. 32.7b), the resulting complex waveform is given by $y_1 + y_2$. When the crests of the pulses exactly coincide (Fig. 32.7c), the resulting waveform $y_1 + y_2$ is symmetric. The two pulses finally separate and continue moving in their original directions (Fig. 32.7d). Note that the final waveforms remain unchanged, as if the two pulses never met! The combination of separate waves in the same region of space to produce a resultant wave is called *interference*. For the two pulses shown in Fig. 32.7, the displacements of the individual pulses are in the same direction, and the resultant waveform (when the pulses overlap) exhibits a displacement greater than those of the individual pulses. This type of interference is called *constructive interference.*

Now consider two identical pulses traveling in opposite directions, where one is inverted relative to the other, as in Fig. 32.8. In this case, when the pulses begin to overlap, the resultant waveform is the *arithmetic difference* between the two separate displacements. Again, the two pulses pass through each other as indicated.

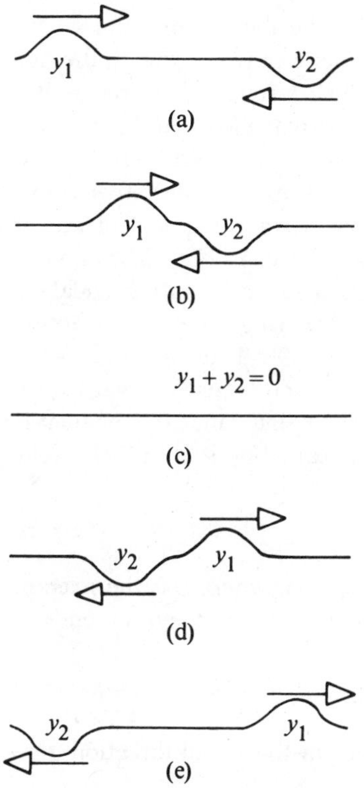

Figure 32.8 Two wave pulses traveling in opposite directions with equal but opposite displacements. When the two overlap, their displacements subtract from each other, corresponding to *destructive interference.* Note that in (c), the displacement is zero for all values of x.

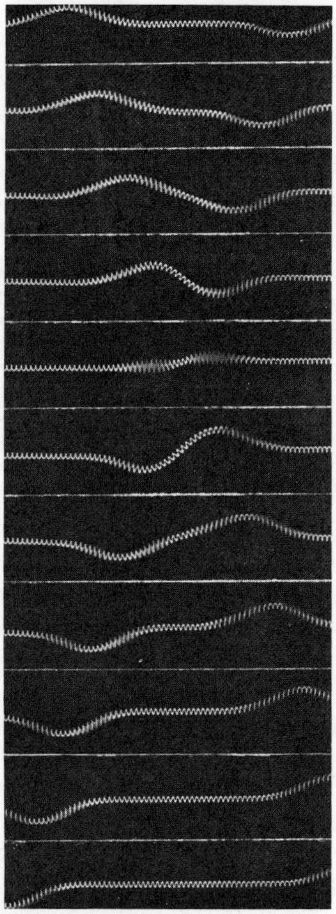

Photograph of superposition of two symmetric pulses traveling in opposite directions, where one is inverted relative to the other. (Photo, Education Development Center, Newton, Mass.)

When the two pulses exactly overlap, they *cancel* each other (assuming the upper positive displacement of the pulse y_1 is equal in magnitude to that of the inverted pulse y_2). At this time, the string is horizontal and the energy associated with the disturbance is contained in the kinetic energy of the string, where the string segments move *vertically.* That is, when the two pulses exactly overlap, the segments of the string on either side of the crossover point are moving vertically, but in opposite directions. When traveling waves cancel each other in this manner, the phenomenon is called *destructive interference.*

32.5 THE VELOCITY OF WAVES ON STRINGS

For linear waves, *the velocity of mechanical waves depends only on the properties of the medium through which the disturbance travels.* In this section, we shall focus our attention on determining the speed of a transverse pulse traveling on a stretched string. If the *tension* in the string is F and its *mass per unit length* is μ, then the wave speed v is given by

Speed of a wave on a stretched string

$$v = \sqrt{F/\mu} \tag{32.4}$$

First, we verify that this expression is dimensionally correct. The dimensions of F are MLT^{-2}, and the dimensions of μ are ML^{-1}. Therefore, the dimensions of F/μ are L^2/T^2; hence the dimensions of $\sqrt{F/\mu}$ are L/T, which are indeed the dimensions of velocity. No other combination of F and μ is dimensionally correct, assuming they are the only variables relevant to the situation.

Now let us use a mechanical analysis to derive the above expression for the speed of a pulse traveling on a stretched string. Consider a pulse moving to the right with a uniform speed v, measured relative to a stationary frame of reference. It is more convenient to choose as our reference frame one that moves along with the pulse with the same speed, so that the pulse appears to be at rest in this frame, as in Fig. 32.9a. This is permitted since Newton's laws are valid in either a stationary frame or one that moves with uniform motion. A *small* segment of the string of length Δs forms the arc of a circle of radius R, as shown in Fig. 32.9a and magnified in Fig. 32.9b. This small segment has a centripetal acceleration equal to v^2/R, which is supplied by the force of tension F in the string. The force F acts on each side of the segment, tangent to the arc, as in Fig. 32.9b. The horizontal components of F cancel, and each vertical component $F \sin\theta$ acts radially inward toward the center of the arc. Hence, the total radial force is $2F \sin\theta$. Since the segment is small, θ is small and we can use the small-angle approximation $\sin\theta \approx \theta$. Therefore, the total radial force can be expressed as

$$F_r = 2F \sin\theta \approx 2F\theta$$

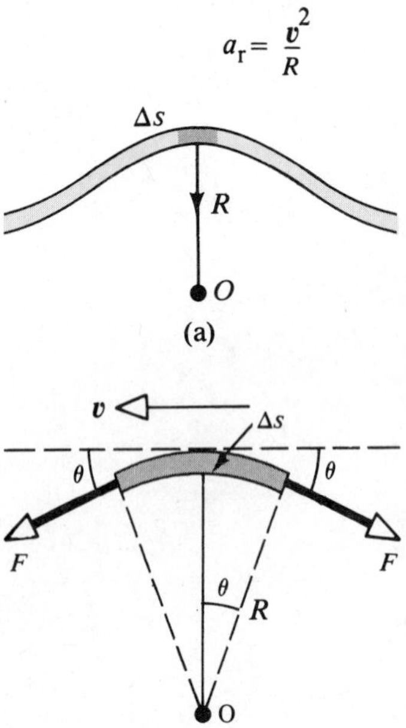

Figure 32.9 (a) To obtain the speed v of a wave on a stretched string, it is convenient to describe the motion of a small segment of the string in a moving frame of reference. (b) The net force on a small segment of length Δs is in the radial direction. The horizontal components of the tension force cancel.

The small segment has a mass given by $m = \mu\Delta s$, where μ is the mass per unit length of the string. Since the segment forms part of a circle and subtends an angle 2θ at the center, $\Delta s = R(2\theta)$, and hence

$$m = \mu\, \Delta s = 2\mu R\theta$$

Applying Newton's second law to this segment in the radial direction gives

$$F_r = mv^2/R \qquad \text{or} \qquad 2F\theta = 2\mu R\theta v^2/R$$

Solving for v gives

$$v = \sqrt{F/\mu}$$

Notice that this derivation is based on the assumption that the pulse height is small relative to the length of the string. Using this assumption, we were able to use the approximation that $\sin\theta \approx \theta$. Furthermore, the model assumes that the tension F is not affected by the presence of the pulse, so that F is the same at all points on the string. Finally, note that this proof does *not* assume any particular shape for the pulse. Therefore, we conclude that a pulse of *any shape* will travel on the string with speed $v = \sqrt{F/\mu}$ without changing its shape.

Example 32.2

A uniform string has a mass of 0.3 kg and a length of 6 m. Tension is maintained in the string by suspending a 2-kg mass from one end (Fig. 32.10). (a) Find the speed of a pulse on this string.

The tension F in the string is equal to the weight of the suspended 2-kg mass multiplied by the gravitational acceleration:

$$F = mg = (2\text{ kg})(9.8\text{ m/s}^2) = 19.6\text{ N}$$

(This calculation of the tension neglects the small mass of the string. Strictly speaking, the string can never be exactly horizontal, and therefore the tension is not uniform.)

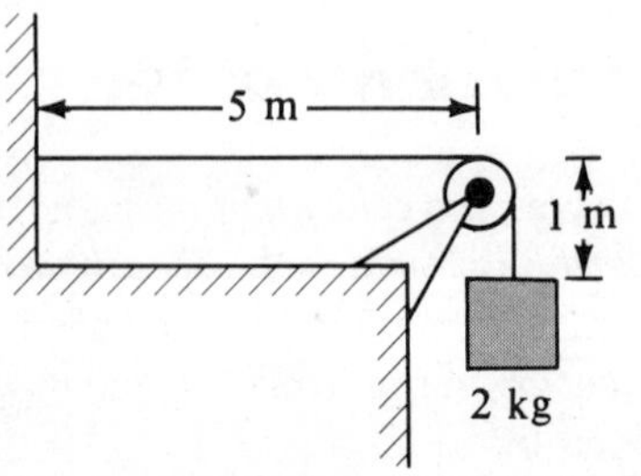

Figure 32.10 (Example 32.2) The tension F in the string is maintained by the suspended mass. The wave speed is calculated using the expression $v = \sqrt{F/\mu}$.

The mass per unit length μ is

$$\mu = \frac{m}{l} = \frac{0.3\ \text{kg}}{6\ \text{m}} = 0.05\ \text{kg/m}$$

Therefore, the wave speed is

$$v = \sqrt{F/\mu} = \sqrt{19.6\ \text{N}/0.05\ \text{kg/m}} = 19.8\ \text{m/s}$$

(b) Determine the time it takes the pulse to travel from the wall to the pulley.

The time it takes the pulse to travel a distance of 5 m is

$$t = \frac{d}{v} = \frac{5\ \text{m}}{19.8\ \text{m/s}} = 0.253\ \text{s}$$

Q3. By what factor would you have to increase the tension in a stretched string in order to double the wave speed?

32.6 REFLECTION AND TRANSMISSION OF WAVES

Whenever a traveling wave reaches a boundary, part or all of the wave will be reflected. For example, consider a pulse traveling on a string fixed at one end (Fig. 32.11). When the pulse reaches the fixed wall, it will be reflected. Since the support attaching the string to the wall is assumed to be rigid, it does not transmit any part of the disturbance to the wall.

Note that the reflected pulse is inverted. This can be explained as follows. When the pulse meets the fixed support, the string produces an upward force on the support. By Newton's third law, the support must then exert an equal and opposite (downward) reaction force on the string. This downward force causes the pulse to invert upon reflection.

Now consider another case where the pulse arrives at the end of a string that is free to move vertically, as in Fig. 32.12. The tension at the free end is maintained by tying the string to a ring of negligible mass that is free to slide vertically on a smooth

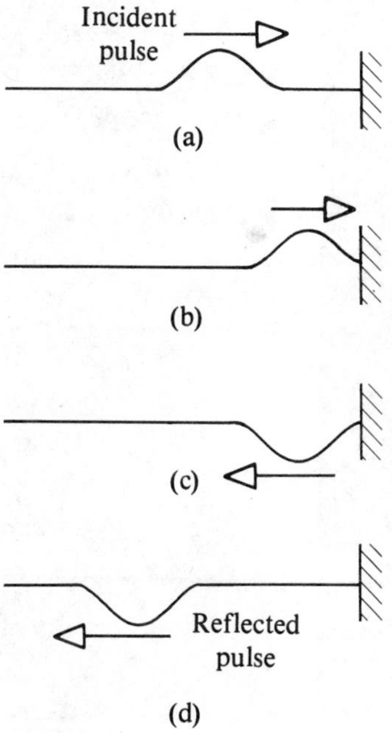

Figure 32.11 The reflection of a traveling wave at the fixed end of a stretched string. Note that the reflected pulse is inverted, but its shape remains the same.

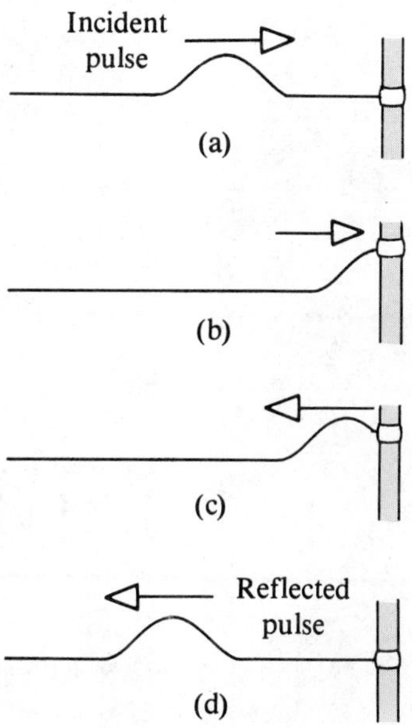

Figure 32.12 The reflection of a traveling wave at the free end of a stretched string. In this case, the reflected pulse is not inverted.

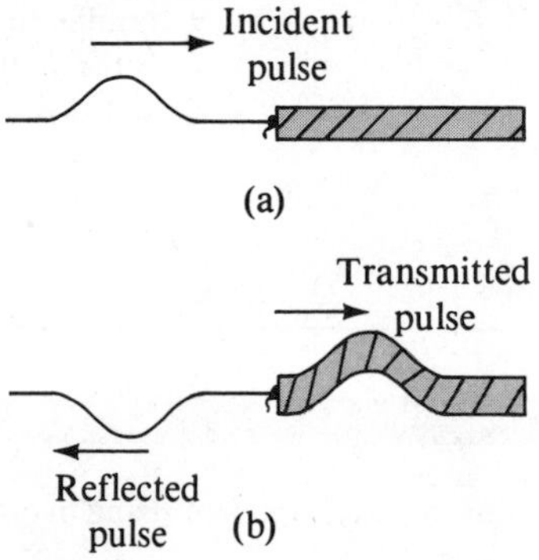

Figure 32.13 (a) A pulse traveling to the right on a light string tied to a heavier string. (b) Part of the incident pulse is reflected (and inverted), and part is transmitted to the heavier string.

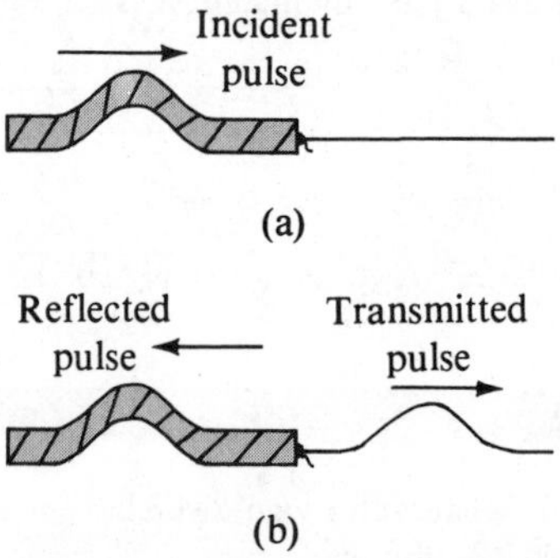

Figure 32.14 (a) A pulse traveling to the right on a heavy string tied to a lighter string. (b) The incident pulse is partially reflected and partially transmitted. In this case, the reflected pulse is not inverted.

post. Again, the pulse will be reflected, but this time its displacement is not inverted. As the pulse reaches the post, it exerts a force on the free end, causing the ring to accelerate upward. In the process, the ring "overshoots" the height of the incoming pulse and is then returned to its original position by the downward component of the tension.

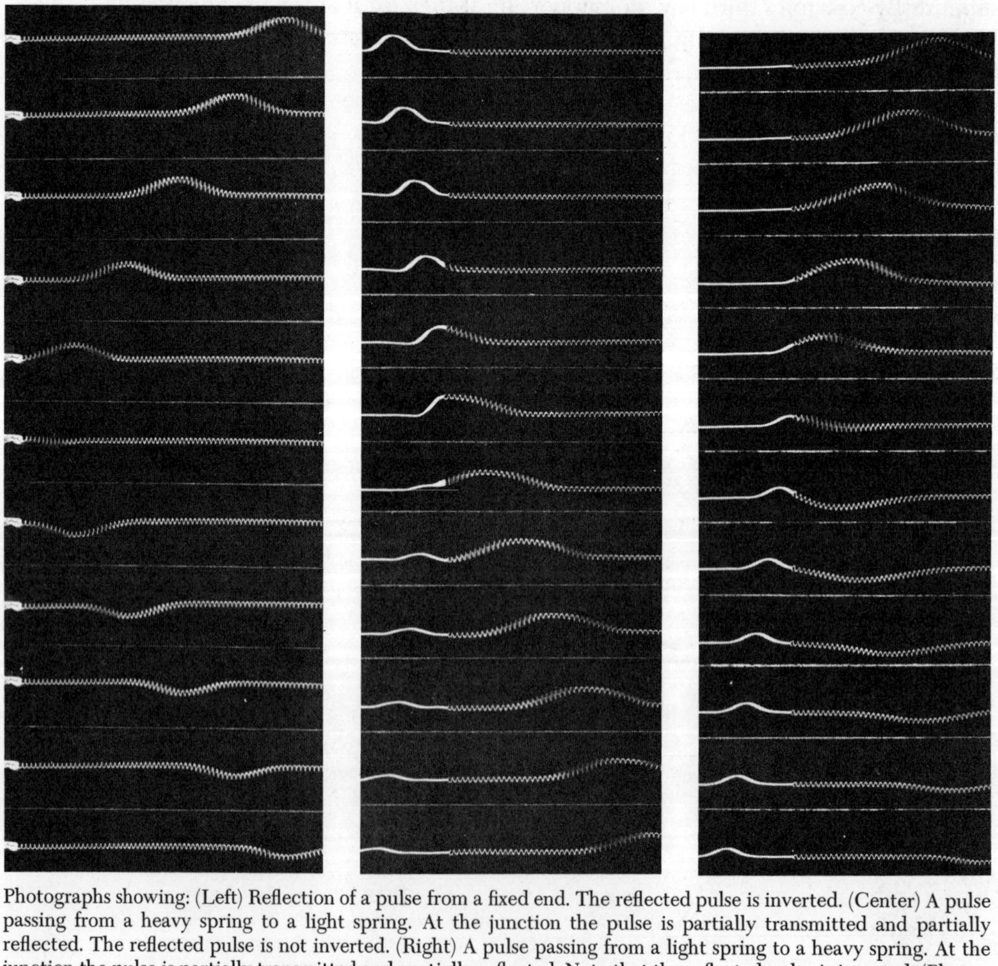

Photographs showing: (Left) Reflection of a pulse from a fixed end. The reflected pulse is inverted. (Center) A pulse passing from a heavy spring to a light spring. At the junction the pulse is partially transmitted and partially reflected. The reflected pulse is not inverted. (Right) A pulse passing from a light spring to a heavy spring. At the junction the pulse is partially transmitted and partially reflected. Note that the reflected pulse is inverted. (Photos, Education Development Center, Newton, Mass.)

Finally, we may have a situation in which the boundary is intermediate between these two extreme cases, that is, one in which the boundary is neither rigid nor free. In this case, part of the incident energy is transmitted and part is reflected. For instance, suppose a light string is attached to a heavier string as in Fig. 32.13. When a pulse traveling on the light string reaches the knot, part of it is reflected and inverted and part of it is transmitted to the heavier string. As one would expect, the reflected pulse has a smaller amplitude than the incident pulse, since part of the incident energy is transferred to the pulse in the heavier string. The inversion in the reflected wave is similar to the behavior of a pulse meeting a rigid boundary, where it is totally reflected.

When a pulse traveling on a heavy string strikes the boundary of a lighter string, as in Fig. 32.14, again part is reflected and part is transmitted. However, in this case, the reflected pulse is not inverted. In either case, the relative heights of the reflected and transmitted pulses depend on the relative densities of the two strings.

If the strings are identical, there is no discontinuity at the boundary, and hence no reflection takes place.

In the previous section, we found that the speed of a wave on a string increases as the density of the string decreases. That is, a pulse travels more slowly on a heavy string than on a light string, if both are under the same tension. The following general rules apply to reflected waves: *When a wave pulse travels from medium A to medium B and $v_A > v_B$ (that is, when B is denser than A), the pulse will be inverted upon reflection. When a wave pulse travels from medium A to medium B and $v_A < v_B$ (A is denser than B), it will not be inverted upon reflection.* Similar rules apply to other kinds of waves.

Q4. When a wave pulse travels on a stretched string, does it always invert upon reflection? Explain.

Q5. Can two pulses traveling in opposite directions on the same string reflect from one another? Explain.

32.7 HARMONIC WAVES

In this section, we introduce an important waveform known as a *harmonic wave*. The shape of a harmonic wave is a sinusoidal curve, as shown in Fig. 32.15. The solid curve represents a snapshot of the traveling harmonic wave at $t = 0$, and the dashed curve represents a snapshot of the wave at some later time t. At $t = 0$, the displacement of the wave can be written

$$y = A \sin\left(\frac{2\pi}{\lambda}x\right) \qquad (32.5)$$

The constant A, called the *amplitude* of the wave, represents the *maximum* value of the displacement. The constant λ, called the *wavelength* of the wave, equals the distance between two successive maxima, which we shall refer to as *crests*, or between any two adjacent points that have the same phase. Thus, we see that the displacement repeats itself when x is increased by any integral multiple of λ. If the wave moves to the right with a phase velocity v, the wave function at some later time t is given by

$$y = A \sin\left[\frac{2\pi}{\lambda}(x - vt)\right] \qquad (32.6)$$

That is, the sine wave moves to the right a distance vt in the time t, as in Fig. 32.15.

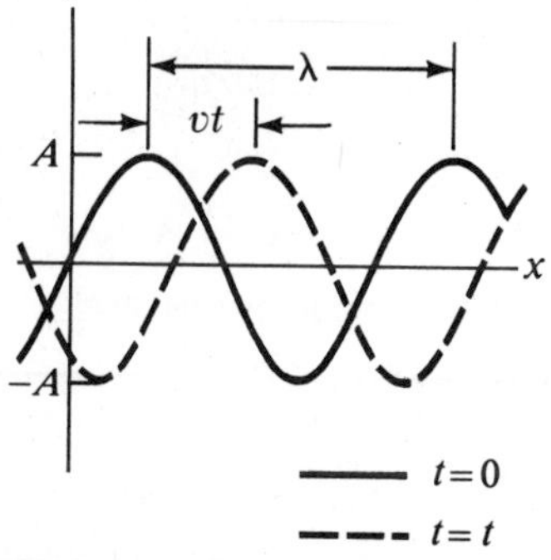

Figure 32.15 A one-dimensional harmonic wave traveling to the right with a speed v. The solid curve represents a snapshot of the wave at $t = 0$, and the dashed curve is a snapshot at some later time t.

Note that the wave function has the form $f(x - vt)$ and represents a wave traveling to the right. If the wave were traveling to the left, the quantity $x - vt$ would be replaced by $x + vt$.

The time it takes the wave to travel a distance of one wavelength is called the *period*, T. Therefore, the speed, wavelength, and period are related by

$$v = \lambda/T \qquad \text{or} \qquad \lambda = vT \tag{32.7}$$

Substituting this into Eq. 32.6, we find that

$$y = A \sin\left[2\pi\left(\frac{x}{\lambda} - \frac{t}{T}\right)\right] \tag{32.8}$$

This form of the wave function clearly shows the *periodic* nature of y. That is, at any given time t (a snapshot of the wave), y has the *same* value at the positions x, $x + \lambda$, $x + 2\lambda$, etc. Furthermore, at any given position x, y has the *same* value at times t, $t + T$, $t + 2T$, etc.

We can express the harmonic wave function in a convenient form by defining two other quantities, called the *wave number* k and the *angular frequency* ω:

Wave number

Angular frequency

$$k = 2\pi/\lambda \tag{32.9}$$

$$\omega = 2\pi/T \tag{32.10}$$

Using these definitions, we see that Eq. 32.8 can be written in the more compact form

Wave function for a harmonic wave

$$y = A \sin(kx - \omega t) \tag{32.11}$$

We shall use this form most frequently.

The frequency f of a harmonic wave equals the number of times a crest (or any other point on the wave) passes a *fixed* point each second. The frequency is related to the period by the relationship

Frequency

$$f = 1/T \tag{32.12}$$

The most common unit for f is s^{-1}, or hertz (Hz). The corresponding unit for T is s.

Using Eqs. 32.9, 32.10, and 32.12, we can express the speed v in the alternative forms

Velocity of a harmonic wave

$$v = \frac{\omega}{k} \tag{32.13}$$

$$v = \lambda f \tag{32.14}$$

The wave function given by Eq. 32.11 assumes that the displacement y is zero at $x = 0$ and $t = 0$. This need not be the case. If the transverse displacement is not zero at $x = 0$ and $t = 0$, we generally express the wave function in the form

General relation for a harmonic wave

$$y = A \sin(kx - \omega t - \phi) \tag{32.15}$$

where ϕ is called the *phase constant*. This constant can be determined from the initial conditions.

Example 32.3

A sinusoidal wave traveling in the positive x direction has an amplitude of 15 cm, a wavelength of 40 cm, and a frequency of 8 Hz. The displacement of the wave at $t = 0$ and $x = 0$ is also 15 cm, as shown in Fig. 32.16. (a) Find

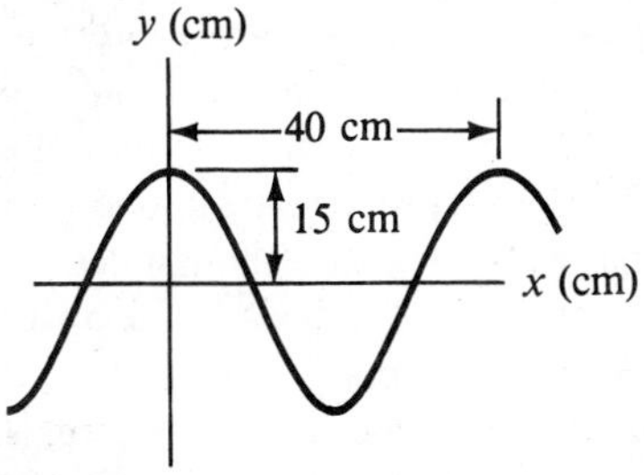

Figure 32.16 (Example 32.3) A harmonic wave of wavelength $\lambda = 40$ cm and amplitude $A = 15$ cm. The wave function can be written in the form $y = A\cos(kx - \omega t)$.

the wave number, period, angular frequency, and speed of the wave.

Using Eqs. 32.9, 32.10, 32.12, and 32.14 and the fact that $\lambda = 40$ cm and $f = 8$ Hz, we find the following:

$$k = 2\pi/\lambda = 2\pi/40 \text{ cm} = 0.157 \text{ cm}^{-1}$$

$$T = 1/f = 1/8 \text{ s}^{-1} = 0.125 \text{ s}$$

$$\omega = 2\pi f = 2\pi(8 \text{ s}^{-1}) = 50.3 \text{ rad/s}$$

$$v = f\lambda = (8 \text{ s}^{-1})(40 \text{ cm}) = 320 \text{ cm/s}$$

(b) Determine the phase constant ϕ, and write a general expression for the wave function.

Since the amplitude $A = 15$ cm and since it is given that $y = 15$ cm at $x = 0$ and $t = 0$, substitution into Eq. 32.15 gives

$$15 = 15\sin(-\phi) \qquad \text{or} \qquad \sin(-\phi) = 1$$

Since $\sin(-\phi) = -\sin\phi$, we see that $\phi = -\pi/2$ rad (or $-90°$). Hence, the wave function is of the form

$$y = A\sin\left(kx - \omega t + \frac{\pi}{2}\right) = A\cos(kx - \omega t)$$

This can be seen by inspection, noting that the cosine function is displaced by 90° from the sine function. Substituting the values for A, k, and ω into this expression gives

$$y = 15\cos(0.157x - 50.3t) \text{ cm}$$

Harmonic Waves on Strings

Harmonic waves can be produced in a very long string by using the scheme shown in Fig. 32.17. One end of the continuous string is connected to a mass-spring system that is suspended vertically. We shall assume an ideal mass-spring system, which oscillates vertically with simple harmonic motion (Chapter 13). As the mass oscillates vertically, a harmonic wave traveling to the right will be set up in the

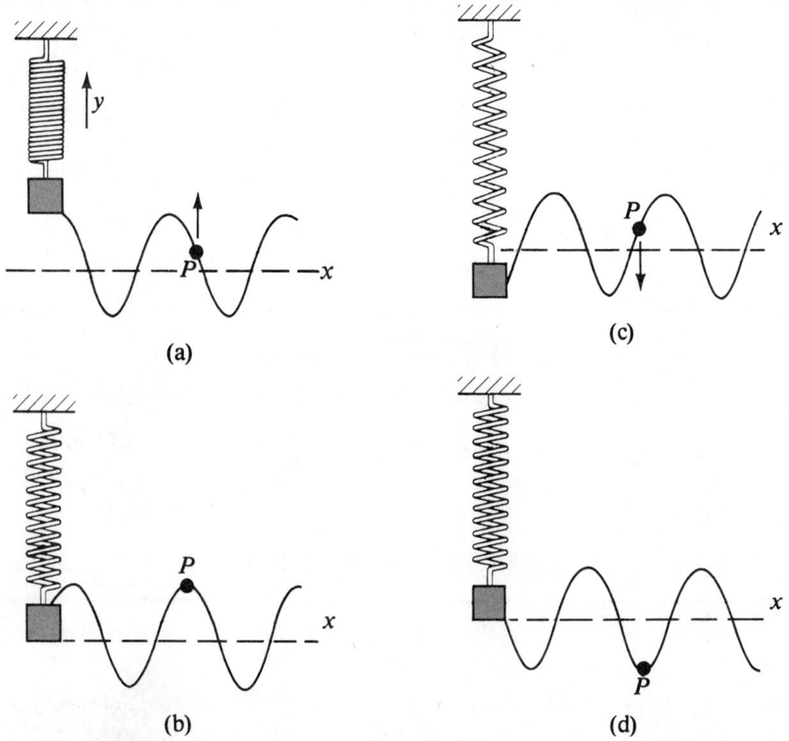

Figure 32.17 One method for producing harmonic waves on a continuous string. The left end of the string is connected to an oscillating mass-spring system that is suspended vertically. Note that every segment, such as P, oscillates with simple harmonic motion in the vertical direction.

string.[5] Figure 32.17 represents snapshots of the waveform at intervals of one quarter of a period. This shows that *each particle of the string such as* P *oscillates vertically in the* y *direction with simple harmonic motion,* assuming the amplitude is small relative to the length of the string. Therefore, every segment of the string can be treated as a simple harmonic oscillator. Note that although each segment oscillates in the y direction, the wave (or disturbance) travels in the x direction with a speed v. Of course, this is the definition of a transverse wave. In this case, the energy carried by the traveling wave is supplied by the mass-spring system. (In reality, the oscillations would gradually decrease in amplitude because of air resistance and the energy delivered to the string.)

If the waveform at $t = 0$ is as described in Fig. 32.17a, then the wave function can be written

$$y = A \sin(kx - \omega t)$$

We can use this expression to describe the motion of any point on the string. The point P (or any other point on the string) moves vertically, and so *its* x *coordinate remains constant*. Therefore, the *transverse velocity,* v_y, (not to be confused with the wave velocity v) and *transverse acceleration,* a_y, are given by

$$v_y = dy/dt]_{x=\text{constant}} = \partial y/\partial t = -A\omega \cos(kx - \omega t) \tag{32.16}$$

$$a_y = dv_y/dt]_{x=\text{constant}} = \partial v_y/\partial t = -A\omega^2 \sin(kx - \omega t) \tag{32.17}$$

The *maximum* values of these quantities are simply the absolute values of the coefficients of the cosine and sine functions:

$$(v_y)_{\text{max}} = A\omega \tag{32.18}$$

$$(a_y)_{\text{max}} = A\omega^2 \tag{32.19}$$

You should recognize that the transverse velocity and transverse acceleration do not reach their maximum values simultaneously. In fact, the transverse velocity reaches its maximum value ($A\omega$) when the displacement $y = 0$, whereas the transverse acceleration reaches its maximum value ($A\omega^2$) when $y = -A$.

Example 32.4

The string shown in Fig. 32.17 is driven at one end at a frequency of 5 Hz. The amplitude of the motion is 12 cm, and the wave speed is 20 m/s. (a) Determine the angular frequency and wave number for this wave, and write an expression for the wave function.

Using Eqs. 32.10, 32.12, and 32.13 gives

$$\omega = 2\pi/T = 2\pi f = 2\pi(5 \text{ Hz}) = 31.4 \text{ rad/s}$$

$$k = \omega/v = \frac{31.4 \text{ rad/s}}{20 \text{ m/s}} = 1.57 \text{ m}^{-1}$$

Since $A = 12$ cm $= 0.12$ m, we have

$$y = A \sin(kx - \omega t) = 0.12 \sin(1.57x - 31.4t) \text{ m}$$

(b) Calculate the *maximum* value for the transverse velocity and transverse acceleration of any point on the string.

Using Eqs. 32.18 and 32.19 and the results to (a) gives

$$(v_y)_{\text{max}} = A\omega = (0.12 \text{ m})(31.4 \text{ s}^{-1}) = 3.77 \text{ m/s}$$

$$(a_y)_{\text{max}} = A\omega^2 = (0.12 \text{ m})(3.14 \text{ s}^{-1})^2 = 118 \text{ m/s}^2$$

Q6. Does the transverse velocity of a segment on a stretched string depend on the wave velocity?

Q7. If you were to periodically shake the end of a stretched rope three times each second, what would be the period of the harmonic waves set up in the string?

[5] In this arrangement, we are assuming that the mass always oscillates in a vertical line. The tension in the string would vary if the mass were allowed to move sideways. Such a motion would make the analysis very complex.

32.8 ENERGY TRANSMITTED BY HARMONIC WAVES ON STRINGS

As waves propagate through a medium, they transport energy and momentum. This is easily demonstrated by hanging a weight on a stretched string and then sending a pulse down the string, as in Fig. 32.18. When the pulse meets the weight, the weight will be momentarily displaced, as in Fig. 32.18b. In the process, energy is transferred to the weight since work must be done in moving it upward.

In this section, we describe the rate at which energy is transported along a string. We shall assume a sinusoidal wave when we calculate the power transferred for this one-dimensional wave. Later, we shall extend these ideas to three-dimensional waves.

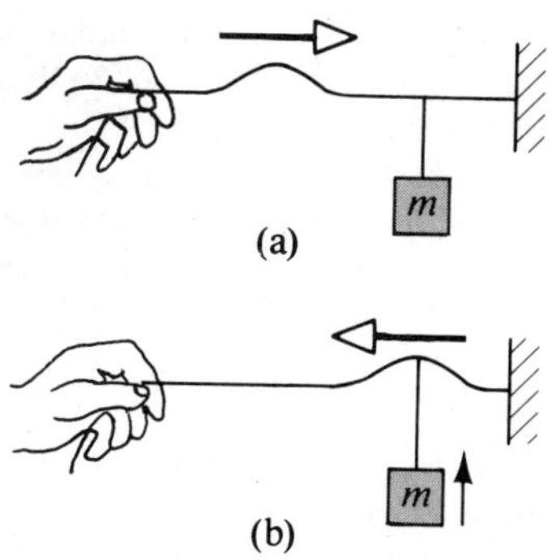

Figure 32.18 A pulse traveling to the right on a stretched string on which a mass has been suspended. (b) Energy and momentum are transmitted to the suspended weight when the pulse arrives.

Consider a harmonic wave traveling on a string (Fig. 32.19). The source of the energy is some external agent at the left end of the string, which does work in producing the oscillations. The points *P*, *Q*, and *R* represent various segments of the string, which move vertically. The wave moves a distance equal to one wavelength, λ, in a time of one period, *T*. To find the power transmitted by the wave, we first calculate the energy contained in one wavelength and divide the result by the period.

First, recall that every point on the string moves vertically with simple harmonic motion. Hence, every segment of equal mass *has the same total energy* (Chapter 13). The energy of segment *P* is entirely potential energy since the segment is momentarily stationary. The energy of segment *Q* is entirely kinetic energy, and segment *R* has both kinetic and potential energy. Consider segment *Q*, which has a *maximum* transverse velocity $(v_y)_{max}$ and mass Δm. The total energy of this segment is given by

$$\Delta E = \frac{1}{2}\Delta m(v_y)^2_{max}$$

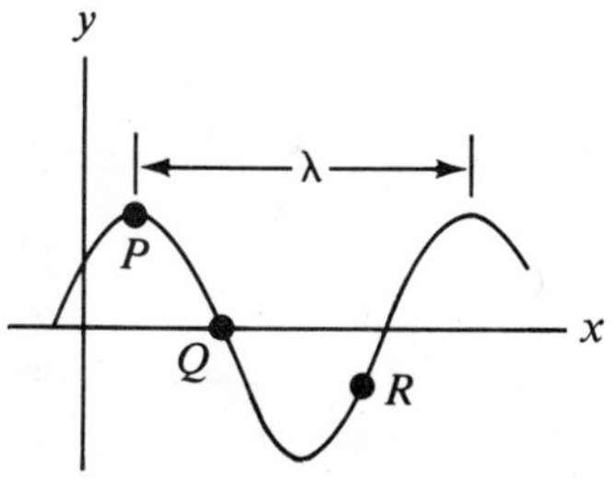

Figure 32.19 A harmonic wave traveling along the *x* axis on a stretched string. Every segment moves vertically, and each has the same total energy. The power transmitted by the wave equals the energy contained in one wavelength divided by the period of the wave.

Since each segment has the same energy, we can obtain the energy contained in one wavelength by summing over all segments. If μ is the mass per unit length, then $\Delta m = \mu\ \Delta\lambda$, where $\Delta\lambda$ is the length of a segment. Therefore, by summing over all segments and using Eq. 32.18, we get the following result for the energy contained in one wavelength:

$$E = \frac{1}{2}\mu\lambda(v_y)^2_{max} = \frac{1}{2}\mu\lambda\omega^2A^2$$

Dividing this expression by the period *T* gives the power, *P*:

$$P = \frac{E}{T} = \frac{1}{2}\mu\frac{\lambda}{T}\omega^2A^2$$

$$P = \frac{1}{2}\mu v\omega^2A^2 \qquad (32.20)$$

Power

This shows that the power transmitted by a harmonic wave on a string is proportional to (a) the wave speed, (b) the square of the frequency, and (c) the square of the amplitude. In fact, *all* harmonic waves have the following general property: *The power transmitted by any harmonic wave is proportional to the square of the frequency and to the square of the amplitude.*

Thus, we see that a wave traveling through a medium corresponds to energy transport through the medium, with no net transfer of matter. An oscillating source provides the energy and produces a harmonic disturbance of the medium. The disturbance is able to propagate through the medium as the result of the interaction between adjacent particles. In order to directly verify Eq. 31.20 experimentally, one would have to design some device at the far end of the string to extract the energy of the wave without producing any reflections.

Example 32.5

A stretched rope having a mass per unit length of $\mu = 5 \times 10^{-2}$ kg/m is under a tension of 80 N. How much power must be supplied to the rope to generate harmonic waves at a frequency of 60 Hz and an amplitude of 6 cm?

Solution: The wave speed on the stretched rope is given by

$$v = \sqrt{T/\mu} = \left(\frac{80 \text{ N}}{5 \times 10^{-2} \text{ kg/m}}\right)^{1/2} = 40 \text{ m/s}$$

Since $f = 60$ Hz, the angular frequency ω of the harmonic waves on the string has the value

$$\omega = 2\pi f = 2\pi(60 \text{ Hz}) = 377 \text{ s}^{-1}$$

Using these values in Eq. 32.20 for the power, with $A = 6 \times 10^{-2}$ m, gives

$$P = \frac{1}{2}\mu v \omega^2 A^2$$

$$= \frac{1}{2}(5 \times 10^{-2} \text{ kg/m})(40 \text{ m/s})(377 \text{ s}^{-1})^2(6 \times 10^{-2} \text{ m})^2$$

$$= 512 \text{ W}$$

Q8. Harmonic waves are generated on a string under constant tension by a vibrating source. If the power delivered to the string is doubled, by what factor does the amplitude change? Does the wave velocity change under these circumstances?

32.9 THE LINEAR WAVE EQUATION

Earlier in this chapter, we introduced the concept of the wave function to represent waves traveling on a string. All wave functions $y(x,t)$ represent solutions of an equation called the *linear wave equation.* This equation gives a complete description of the wave motion, and from it one can derive an expression for the wave velocity. Furthermore, the wave equation is basic to many forms of wave motion. In this section, we shall derive the wave equation as applied to waves on strings.

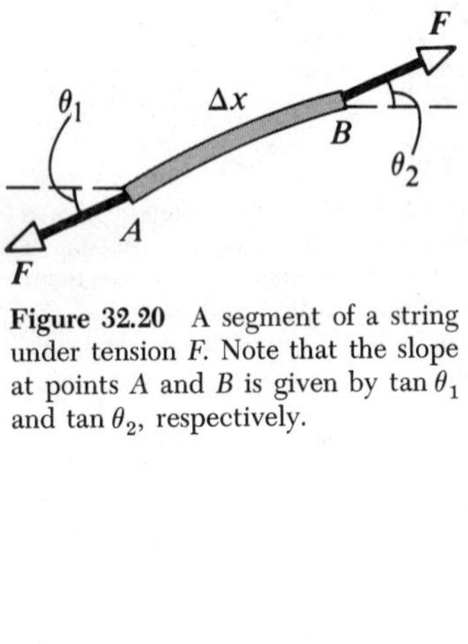

Figure 32.20 A segment of a string under tension F. Note that the slope at points A and B is given by $\tan\theta_1$ and $\tan\theta_2$, respectively.

Consider a small segment of a string of length Δx and tension F, on which a traveling wave is propagating (Fig. 32.20). Let us assume that the ends of the string make small angles θ_1 and θ_2 with the x axis. This is equivalent to the assumption that the vertical displacement of the segment is very small compared with its length.

The net force on the segment in the vertical direction is given by

$$\Sigma F_y = F\sin\theta_2 - F\sin\theta_1 = F(\sin\theta_2 - \sin\theta_1)$$

Since we have assumed that the angles are small, we can use the small-angle approximation $\sin\theta \approx \tan\theta$ and express the net force as

$$\Sigma F_y \approx F(\tan\theta_2 - \tan\theta_1)$$

However, the tangents of the angles at A and B are defined as the slope of the curve at these points. Since the slope of a curve is given by $\partial y/\partial x$, we have[6]

$$\Sigma F_y \approx F[(\partial y/\partial x)_B - (\partial y/\partial x)_A] \quad (32.21)$$

We now apply Newton's second law, $\Sigma F_y = ma_y$, to the segment, where m is the mass of the segment, given by $m = \mu\,\Delta x$. This gives

$$\Sigma F_y = ma_y = \mu\,\Delta x(\partial^2 y/\partial t^2) \quad (32.22)$$

where we have used the fact that $a_y = \partial^2 y/\partial t^2$. Equating Eq. 32.22 to Eq. 32.21 gives

$$\mu\,\Delta x(\partial^2 y/\partial t^2) = F[(\partial y/\partial x)_B - (\partial y/\partial x)_A]$$

$$\frac{\mu}{F}\frac{\partial^2 y}{\partial t^2} = \frac{[(\partial y/\partial x)_B - (\partial y/\partial x)_A]}{\Delta x} \quad (32.23)$$

[6]It is necessary to use partial derivatives because y depends on both x and t.

The right side of Eq. 32.23 can be expressed in a different form if we note that the derivative of any function is defined as

$$\frac{\partial f}{\partial x} = \lim_{\Delta x \to 0} \frac{f(x + \Delta x) - f(x)}{\Delta x}$$

If we associate $f(x + \Delta x)$ with $(\partial y/\partial x)_B$ and $f(x)$ with $(\partial y/\partial x)_A$, we see that in the limit $\Delta x \to 0$, Eq. 32.23 becomes

$$\frac{\mu}{F}\frac{\partial^2 y}{\partial t^2} = \frac{\partial^2 y}{\partial x^2} \qquad (32.24)$$

Linear wave equation

This is the linear wave equation as it applies to waves on a string.

We shall now show that the harmonic wave function represents a solution of this wave equation. If we take the harmonic wave function to be of the form $y(x,t) = A \sin(kx - \omega t)$, the appropriate derivatives are

$$\partial^2 y/\partial t^2 = -\omega^2 A \sin(kx - \omega t)$$
$$\partial^2 y/\partial x^2 = -k^2 A \sin(kx - \omega t)$$

Substituting these expressions into Eq. 32.24 gives

$$k^2 = (\mu/F)\omega^2$$

Using the relation $v = \omega/k$ in the above expression, we see that

$$v^2 = \omega^2/k^2 = F/\mu$$
$$v = \sqrt{F/\mu}$$

This represents another proof of the expression for the wave velocity on a stretched string.

The linear wave equation given by Eq. 32.24 is often written in the form

$$\frac{\partial^2 y}{\partial x^2} = \frac{1}{v^2}\frac{\partial^2 y}{\partial t^2} \qquad (32.25)$$

Linear wave equation in general

This expression applies in general to various types of waves moving through nondispersive media. For waves on strings, y represents the vertical displacement. For sound waves, y corresponds to variations in the pressure or density of a gas. In the case of electromagnetic waves, y corresponds to electric or magnetic field components.

We have shown that the harmonic wave function is one solution of the linear wave equation. Although we do not prove it here, the linear wave equation is satisfied by *any* wave function having the form $y = f(x \pm vt)$. Furthermore, we have seen that the wave equation is a direct consequence of Newton's second law applied to any segment of the string. Similarly, the wave equation in electromagnetism can be derived from the fundamental laws of electricity and magnetism. This will be discussed further in Chapter 35.

32.10 SUMMARY

Transverse wave

A *transverse wave* is a wave in which the particles of the medium move in a direction *perpendicular* to the direction of the wave velocity. An example is a wave on a stretched string.

Longitudinal wave

Longitudinal waves are waves for which the particles of the medium move in a direction *parallel* to the direction of the wave velocity. Sound waves are longitudinal.

Any one-dimensional wave traveling with a speed v in the positive x direction can be represented by a wave function of the form $y = f(x - vt)$. Likewise, a wave traveling in the negative x direction has the form $y = f(x + vt)$. The shape of the wave at any instant (a snapshot of the wave) is obtained by holding t constant.

Superposition principle

The *superposition principle* says that when two or more linear waves move through a medium, the resultant wave function equals the algebraic sum of the individual wave functions. Waves that obey this principle are said to be *linear*. When two waves combine in space, they interfere to produce a resultant wave. The *interference* may be *constructive* (when the individual displacements are in the same direction) or *destructive* (when the displacements are in opposite directions).

The speed of a wave traveling on a stretched string of mass per unit length μ and tension F is

Speed of a wave on a stretched string

$$v = \sqrt{F/\mu} \tag{32.4}$$

When a pulse traveling on a string meets a fixed end, the pulse is reflected and inverted. If the pulse reaches a free end, it is reflected but not inverted.

The wave function for a one-dimensional harmonic wave traveling to the right can be expressed as

Wave function for a harmonic wave

$$y = A\sin[(2\pi/\lambda)(x - vt)] = A\sin(kx - \omega t) \qquad (32.6, 32.11)$$

where A is the amplitude, λ is the wavelength, k is the wave number, and ω is the angular frequency. If T is the period (the time it takes the wave to pass a given point) and f is the frequency, then v, k and ω can be written

Wave number

Angular frequency

$$v = \lambda/T = \lambda f \qquad (32.7, 32.14)$$

$$k = 2\pi/\lambda \tag{32.9}$$

$$\omega = 2\pi/T = 2\pi f \qquad (32.10, 32.12)$$

The *power* transmitted by a harmonic wave on a stretched string is given by

Power

$$P = \frac{1}{2}\mu v\omega^2 A^2 \tag{32.20}$$

The wave function $y(x,t)$ for many kinds of waves satisfies the following linear wave equation:

Linear wave equation in general

$$\frac{\partial^2 y}{\partial x^2} = \frac{1}{v^2}\frac{\partial^2 y}{\partial t^2} \tag{32.25}$$

EXERCISES

Section 32.3 One-Dimensional Traveling Waves

1. A traveling wave pulse moving to the right along the x axis is represented by the following wave function:

$$y(x,t) = \frac{4}{2 + (x - 4t)^2}$$

where x and y are measured in cm and t is in s. Plot the shape of the waveform at $t = 0$, 1, and 2 s.

2. Two wave pulses A and B are moving in *opposite* directions along a stretched string with a speed of 2 cm/s. The amplitude of A is twice the amplitude of B. The pulses are shown in Fig. 32.21 at $t = 0$.

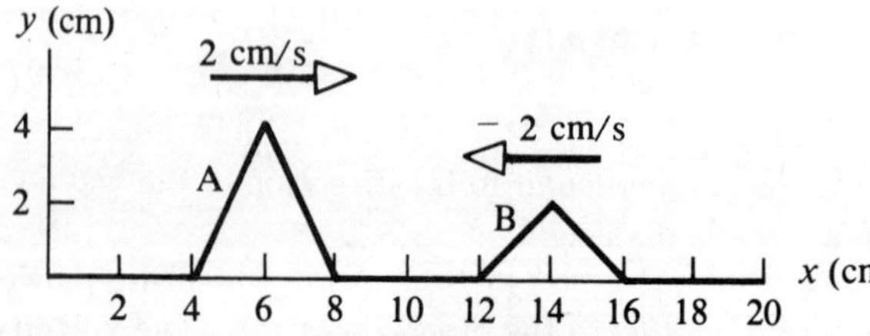

Figure 32.21 (Exercise 2).

Sketch the shape of the string at $t = 1, 1.5, 2, 2.5$, and 3 s.

3. At $t = 0$, a transverse wave pulse in a wire is described by the function

$$y = \frac{6}{x^2 - 3}$$

where x and y are in m. Write the function $y(x,t)$ that describes this wave if it is traveling in the positive x direction with a speed of 4.5 m/s.

Section 32.5 The Velocity of Waves on Strings

4. Transverse waves with a speed of 50 m/s are to be produced in a stretched string. A 5-m length of string with a total mass of 0.06 kg is used. What is the required tension in the string?
5. Calculate the wave speed in the string described in Exercise 4 if the tension in the string is 8 N.
6. The tension in a cord 15 m in length is 20 N. The measured transverse wave speed in the cord is 60 m/s. Calculate the total mass of the cord.
7. Transverse waves travel with a speed of 20 m/s in a string under a tension of 6 N. What tension is required for a wave speed of 30 m/s in the same string?
8. Tension is maintained in a horizontal string as shown in Fig. 32.10. The observed wave speed is 24 m/s when the suspended mass is 3 kg. What is the linear density of the string?

Section 32.7 Harmonic Waves

9. One form of the wave function for a harmonic wave is given by Eq. 32.11. The displacement y is expressed as a function of x and t in terms of the wave number k and the angular frequency ω. Write equivalent equations in which y is shown as a function of x and t in terms of (a) k and v, (b) λ and v, (c) λ and f, and (d) f and v.
10. A harmonic wave train is described by

$$y = 0.15 \sin(0.2x - 30t),$$

where x and y are in m and t is in s. Determine for this wave the (a) amplitude, (b) angular frequency, (c) wave number, (d) wavelength, (e) wave speed, and (f) direction of motion.

11. Determine the quantities (a) through (f) of Exercise 10 when the wave train is described by $y = 0.2 \sin 4\pi(0.4x + t)$. Again x and y are in m and t is in s.
12. In Example 32.3 the harmonic wave was found to be described by $y = 15 \cos(0.157x - 50.3t)$, where x and y are in cm and t is in s. (a) Plot y versus x at $t = 0$ and $t = 0.125$ s. (b) Determine the wave speed from this plot and compare your result with the value found in Example 32.3.
13. (a) Plot y versus t at $x = 0$ for a harmonic wave of the form $y = 15 \cos(0.157x - 50.3t)$, where x and y are in cm and t is in s. (b) Determine the period of vibration from this plot and compare your result with the value found in Example 32.3.
14. (a) Write the expression for y as a function of x and t for a sinusoidal wave traveling along a rope in the positive x direction with the following characteristics: $y_{max} = 8$ cm, $\lambda = 80$ cm, $f = 3$ Hz, and $y(0,t) = y_{max}$ at $t = 0$. (b) Determine the speed and wave number for the wave described in (a).
15. (a) Write the expression for y as a function of x and t for a sinusoidal wave traveling along a rope in the *negative* x direction with the following characteristics: $y_{max} = 8$ cm, $\lambda = 80$ cm, $f = 3$ Hz, and $y(0,t) = 0$ at $t = 0$. (b) Write the expression for y as function of x for the wave in (a) assuming that $y(x,0) = 0$ at the point $x = 10$ cm.
16. Consider a wave in a string described by

$$y = 15 \sin[(\pi/16)(2x - 64t)],$$

where x and y are in cm and t is in s. (a) Calculate the maximum transverse velocity of a point on the string. (b) Calculate the transverse velocity of the point at $x = 6$ cm when $t = 0.25$ s.

17. For the wave described in Exercise 16, calculate (a) the maximum transverse acceleration and (b) the transverse acceleration for the point located at $x = 6$ cm when $t = 0.25$ s.
18. For a certain transverse wave, it is observed that the distance between two successive maxima is 1.2 m. It is also noted that eight crests, or maxima, pass a given point along the direction of travel every 12 s. Calculate the wave speed.
19. A harmonic wave is traveling along a rope. It is observed that the oscillator that generates the wave completes 40 vibrations in 30 s. Also, a given maximum travels 425 cm along the rope in 10 s. What is the wavelength?
20. When a particular wire is vibrating with a frequency of 4 Hz, a transverse wave of wavelength 60 cm is produced. Determine the speed of wave pulses along the wire.

Section 32.8 Energy Transmitted by Harmonic Waves on Strings

21. A stretched rope has a mass of 0.18 kg and a length of 3.6 m. What power must be supplied in order to generate harmonic waves having an amplitude of 0.1 m and a wavelength of 0.5 m and traveling with a speed of 30 m/s?
22. A wire of mass 0.24 kg is 48 m long and under a tension of 60 N. An electric vibrator operating at an angular frequency of 80π rad/s is generating harmonic waves in the wire. The vibrator can supply energy to the wire at a maximum rate of 400 J/s.

What is the maximum amplitude of the wave pulses?

23. Transverse waves are being generated on a rope under *constant tension*. By what factor will the required power be increased or decreased if (a) the length of the rope is doubled and the angular frequency remains constant, (b) the amplitude is doubled and the angular frequency is halved, (c) both the wavelength and the amplitude are doubled, and (d) both the length of the rope and the wavelength are halved?

24. Harmonic waves 5 cm in amplitude are to be transmitted along a string that has a linear density of 4×10^{-2} kg/m. If the maximum power delivered by the source is 300 W and the string is under a tension of 100 N, what is the highest vibrational frequency at which the source can operate?

Section 32.9 The Linear Wave Equation

25. Show that the wave function $y = \ln[A(x - vt)]$ is a solution to Eq. 32.24, where A is a constant.

26. In Section 32.9 it is verified that $y_1 = A \sin(kx - \omega t)$ is a solution to the wave equation. The function $y_2 = B \cos(kx - \omega t)$ describes a wave $\pi/2$ radians out of phase with the first. (a) Determine whether or not $y = A \sin(kx - \omega t) + B \cos(kx - \omega t)$ is a solution to the wave equation. (b) Determine if $y = A(\sin kx)B(\cos \omega t)$ is a solution to the wave equation.

PROBLEMS

1. A traveling wave propagates according to the expression $y = 4.0 \sin(2.0x - 3.0t)$ cm where x is in cm. Determine (a) the amplitude, (b) the wavelength, (c) the frequency, (d) the period, and (e) the direction of travel of the wave.

2. A traveling wave on a string is harmonic, and its transverse displacement is given by

$$y = 3.0 \cos(\pi x - 4\pi t) \text{ cm},$$

where x is in cm. (a) Determine the wavelength and period of the wave. (b) Find the transverse velocity and transverse acceleration at any time t. (c) Calculate the transverse velocity and transverse acceleration at $t = 0$ for a point located at $x = 0.25$ cm. (d) What are the maximum values of the transverse velocity and transverse acceleration?

3. (a) Determine the speed of transverse waves on a stretched string that is under a tension of 80 N if the string has a length of 2 m and a mass of 5 g. (b) Calculate the power required to generate these waves if they have a wavelength of 16 cm and are 4 cm in amplitude.

4. A harmonic traveling wave moving in the positive x direction has an amplitude of 2.0 cm, a wavelength of 4.0 cm, and a frequency of 5 Hz. (a) Determine the speed of the wave and (b) write an expression for the transverse displacement as a function of x and t.

5. Consider the sinusoidal wave of Example 32.3, for which it was determined that

$$y = 15 \cos(0.157x - 50.3t) \text{ cm}.$$

At a given instant, let point A be at the origin and point B be the first point along x that is 60° out of phase with point A. What is the coordinate of point B?

6. A rope of total mass m and length l is suspended vertically. Show that a transverse wave pulse will travel the length of the rope in a time $t = 2\sqrt{l/g}$. (Hint: First find an expression for the velocity at any point a distance x from the lower end of the rope, by considering the tension in the rope as resulting from the weight of the segment below that point.)

7. A transverse wave propagating along the positive x-axis has the following properties: $y_{max} = 6$ cm, $\lambda = 8\pi$ cm, $v = 48$ cm/s, and the displacement of the wave at $t = 0$ and $x = 0$ is -2 cm. Determine the (a) wave number, (b) angular frequency, and (c) phase constant for the wave. (d) What is the first value of t for which the displacement at $x = 0$ will be $+2$ cm? (e) For this initial condition, find the coordinate of the particle on the positive x axis closest to the origin for which $y = 0$.

8. A harmonic wave in a rope is described by the wave function $y = 0.2 \sin[\pi(0.75x - 18t)]$ where x and y are in m and t is in s. This wave is traveling in a rope that has a linear mass density of 0.25 kg/m. If the tension in the rope is provided by an arrangement like the one illustrated in Fig. 32.10, what is the value of the suspended mass?

9. An aluminum wire is clamped at each end under zero tension at room temperature (22°C). The tension in the wire is increased by reducing the temperature. Find the temperature at which the speed of *transverse* waves in the wire will be 100 m/s. (Use the following properties of aluminum: coefficient of linear expansion, $\alpha = 2.2 \times 10^{-5}$ (C°)$^{-1}$; density, $\rho = 2.7 \times 10^3$ kg/m^3; and Young's modulus, $Y = 6.8 \times 10^{11}$ N/m^2.)

10. (a) Show that the speed of longitudinal waves along a spring of force constant k is $v = \sqrt{kl/\mu}$, where l is the unstretched length of the spring and μ is the mass per unit length. (b) A spring of mass 0.4 kg has an unstretched length of 2 m and a force constant of 100 N/m. Using the results to (a), determine the speed of longitudinal waves along this spring.

33 Sound Waves

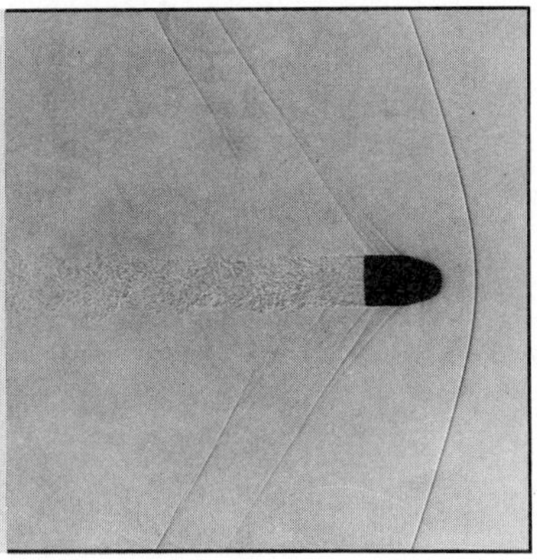

Harold Edgerton, MIT

This chapter deals with the properties of longitudinal waves traveling through various media. Sound waves are the most important example of longitudinal waves. They can travel through any material medium (that is, gases, solids, or liquids) with a speed that depends on the properties of the medium. As sound waves travel through a medium, the particles in the medium vibrate along the direction of motion of the wave. This is in contrast to a transverse wave, where the particle motion is perpendicular to the direction of wave motion. The displacements that occur as a result of sound waves involve the longitudinal displacements of individual molecules from their equilibrium positions. This results in a series of high- and low-pressure regions called *condensations* and *rarefactions*, respectively. If the source of the sound waves, such as the diaphragm of a loudspeaker, vibrates sinusoidally, the pressure variations will also be sinusoidal. We shall find that the mathematical description of harmonic sound waves is identical to that of harmonic string waves discussed in the previous chapter.

There are three categories of longitudinal mechanical waves which cover different ranges of frequency: (1) *Audible waves* are sound waves that lie within the range of sensitivity of the human ear, typically, 20 Hz to 20 000 Hz. They can be generated in a variety of ways, such as by musical instruments, human vocal cords, and loudspeakers. (2) *Infrasonic waves* are longitudinal waves with frequencies below the audible range. Earthquake waves are an example. (3) *Ultrasonic waves* are longitudinal waves with frequencies above the audible range. For example, they can be generated by inducing vibrations in a quartz crystal with an applied alternating electric field. Any device that transforms one form of power into another is called a *transducer*. In addition to the loudspeaker and the quartz crystal, ceramic and magnetic phonograph pickups are common examples of sound transducers. Some transducers can generate ultrasonic waves. Such devices are used in the construction of ultrasonic cleaners and for underwater navigation.

33.1 VELOCITY OF SOUND WAVES

Sound waves are compressional waves traveling through a compressible medium, such as air. The compressed region of air which propagates corresponds to a variation in the normal value of the air pressure. The speed of such compressional waves depends on the compressibility of the medium and on the inertia of the medium. If the compressible medium has a bulk modulus B and an equilibrium density ρ, the speed of sound in that medium is

$$v = \sqrt{B/\rho} \tag{33.1}$$

Recall that the *bulk modulus* (Section 15.4) is defined as the ratio of the change in pressure, ΔP, to the resulting fractional change in volume, $-\Delta V/V$:

Bulk modulus

$$B = -\frac{\Delta P}{\Delta V/V} \tag{33.2}$$

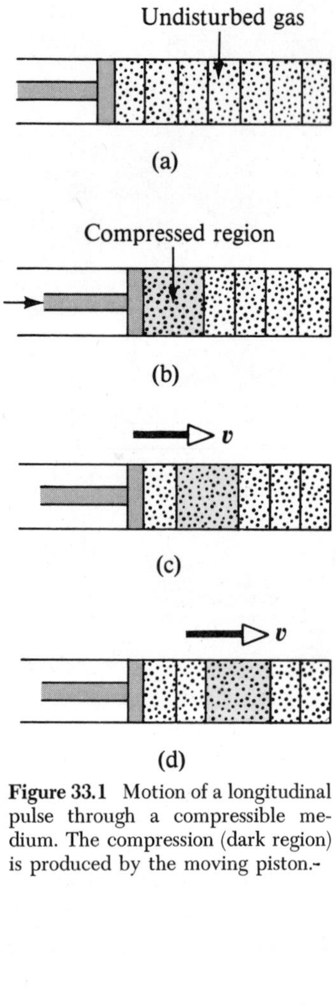

Figure 33.1 Motion of a longitudinal pulse through a compressible medium. The compression (dark region) is produced by the moving piston.

Note that B is always positive, since an increase in pressure (positive ΔP) results in a decrease in volume. Hence, the ratio $\Delta P/\Delta V$ is always negative.

It is interesting to compare Eq. 33.1 with the expression for the speed of transverse waves on a string, $v = \sqrt{T/\mu}$, discussed in the previous chapter. In both cases, the wave speed depends on an elastic property of the medium (B or T) and on an inertial property of the medium (ρ or μ). In fact, the speed of *all mechanical waves* follows an expression of the general form

$$v = \sqrt{\text{elastic property/inertial property}}$$

In order to derive Eq. 33.1, let us first describe pictorially the motion of a longitudinal pulse moving through a long tube containing a compressible gas or liquid (Fig. 33.1). A piston at the left end can be moved to the right to compress the fluid and create the longitudinal pulse. This is a convenient arrangement, since the wave motion is one-dimensional. Before the piston is moved, the medium is undisturbed and of uniform density, as described by the uniformly spaced vertical lines in Fig. 33.1a. When the piston is suddenly pushed to the right (Fig. 33.1b), the medium just in front of it is compressed (represented by the shaded region). The pressure and density in this shaded region are higher than normal. When the piston comes to rest (Fig. 33.1c), the compressed region continues to move to the right, corresponding to a longitudinal pulse traveling down the tube with a speed v. Note that the piston speed does *not* equal v. Furthermore, the compressed region does not "stay with" the piston until it stops.

Let us assume that the equilibrium values of the pressure and density of the medium are P and ρ, respectively, as in Fig. 33.2a. If the piston is pushed to the right with a constant speed u, the distance it moves in a time Δt is equal to $u\Delta t$. Let us assume that the boundary of the compressed region (the leading edge of the longitudinal pulse) moves with a velocity v, which corresponds to the velocity of the disturbance. In the time interval Δt, the wavefront advances a distance $v\,\Delta t$. Furthermore, let us assume that all the fluid in the shaded region moves with the velocity u of the piston. We can now apply the impulse-momentum theorem to this shaded region.

The net force on the compressed region is $A\Delta P$, where ΔP is the *increase* in pressure necessary to compress the fluid and A is the cross-sectional area of the piston. Hence, the impulse imparted to the shaded region in a time Δt is given by

$$\text{Impulse} = F\,\Delta t = (A\Delta P)\,\Delta t$$

Now let us calculate the change in momentum of the mass of fluid set in motion. The mass Δm that is compressed and set in motion equals the density ρ multiplied by the volume $\Delta V = A\,\Delta x$:

$$\Delta m = \rho\,\Delta V = \rho A\,\Delta x = \rho A v\,\Delta t$$

where $\Delta x = v\,\Delta t$ is the length of the fluid set in motion. Since the initial speed of the fluid is zero and the final speed is u, the *change* in momentum is

$$\text{Change in momentum} = (\Delta m)u = (\rho A v\,\Delta t)u$$

Since the impulse acting on a body equals its change in momentum, we see that

$$(A\Delta P)\,\Delta t = (\rho A v\,\Delta t)u$$

$$\Delta P = \rho v u \qquad (33.3)$$

We can obtain another expression for ΔP by using the definition of bulk modulus, given by Eq. 33.2:

$$\Delta P = -B(\Delta V/V) \qquad (33.4)$$

where $\Delta V/V$ is the *fractional* change in volume of the compressed fluid. Since the *original* volume of the compressed fluid is $V = Av\,\Delta t$ and since the change in volume ΔV equals the volume displaced by the piston, where $\Delta V = -Au\,\Delta t$ (the minus sign means the volume has decreased), we see that

$$\frac{\Delta V}{V} = -\frac{Au\,\Delta t}{Av\,\Delta t} = -\frac{u}{v}$$

Substituting this result into Eq. 33.4 gives

$$\Delta P = B\frac{u}{v}$$

Finally, equating this expression to Eq. 33.3, we find that

$$\rho v u = B\frac{u}{v}$$

Rearranging this expression, we arrive at Eq. 33.1:

$$v = \sqrt{B/\rho}$$

Let us now determine the speed of sound waves in various media.

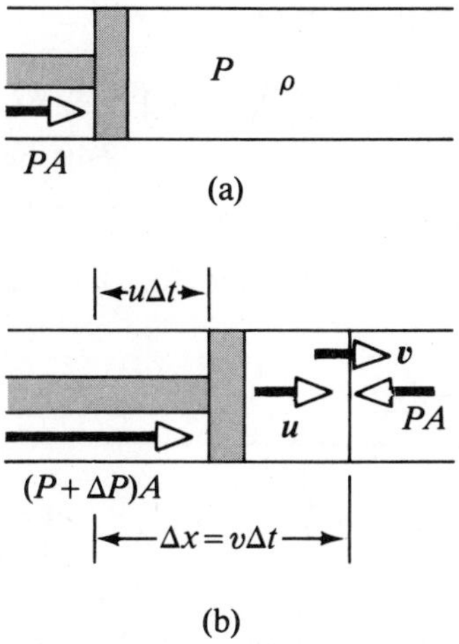

Figure 33.2 A longitudinal wave pulse produced by a piston that is suddenly moved to the right with a speed u. In a time Δt, the wave moves a distance $v\Delta t$, while the piston moves a distance $u\Delta t$.

Sound Waves in a Gas

Most gases are poor heat conductors. Therefore, when a sound wave propagates through a gas, very little heat is transferred between regions of high and low densities. To a good approximation, we can assume that the variations of pressure and volume occur adiabatically, corresponding to no heat transfer between portions of the gas. This is equivalent to assuming that all of the work done in compressing the gas goes into increasing the internal energy of the gas. Furthermore, if the gas is ideal, the pressure and volume during an *adiabatic* process (Section 18.4) are related by the expression

$$PV^{\gamma} = \text{constant} \qquad (33.5)$$

where γ is a constant equal to the ratio of the specific heat at constant pressure to the specific heat at constant volume.

We can use Eq. 33.5 to obtain an expression for the adiabatic bulk modulus. Differentiating Eq. 33.5 gives

$$\gamma P V^{\gamma-1}\,dV + V^{\gamma}\,dP = 0$$

$$dP = -\gamma P\,dV/V$$

Substituting this into Eq. 33.2 (with ΔP and ΔV replaced by dP and dV, respectively) gives

$$B_{\text{adiabatic}} = -\frac{dP}{dV/V} = \gamma P$$

Using this result together with Eq. 33.1, we find that the speed of sound in a gas can be expressed as

Speed of sound in a gas

$$v = \sqrt{\gamma P/\rho} \qquad (33.6)$$

Example 33.1

Calculate the speed of sound in air at atmospheric pressure, taking $P = 1.01 \times 10^5\ \text{N/m}^2$, $\gamma = 1.40$, and $\rho = 1.2\ \text{kg/m}^3$.

Solution: Using Eq. 32.6, we find that

$$v_{\text{air}} = \sqrt{(1.4)(1.01 \times 10^5\ \text{N/m}^2)/1.2\ \text{kg/m}^3} = 343\ \text{m/s}$$

The speed of sound in helium is much greater than this because of the lower density of helium. An interesting and amusing demonstration of this fact is the variation in the human voice when the vocal cavities are partially filled with helium. The demonstrator talks before and after taking a deep breath of helium, an inert gas. The result is a high-pitched voice sounding a bit like that of Donald Duck. The increase in frequency corresponds to an increase in the speed of sound in helium, since frequency is proportional to velocity.

Equation 33.6 can be expressed in another useful form, which uses the equation of state of an ideal gas, $PV = nRT$, or

$$P = nRT/V = \rho RT/M$$

where R is the gas constant, M is the molecular weight, and n is the number of moles of gas. Substituting this expression for P into Eq. 33.6 gives

Speed of sound in a gas

$$v = \sqrt{\gamma RT/M} \tag{33.7}$$

It is interesting to compare this result with the rms speed of molecules in a gas (Eq. 18.10), where $v_{\text{rms}} = \sqrt{3RT/M}$. The two results differ only by the factors γ and 3. It is known that γ lies between 1 and 1.67; hence the two speeds are nearly the same! Since sound waves propagate through air as a result of collisions between gas molecules, one would expect the wave speed to increase as the temperature (and molecular speed) increase.

Example 33.2

Sound waves in a solid bar: If a solid bar is struck at one end with a hammer, a longitudinal pulse will propagate down the bar with a speed

$$v = \sqrt{Y/\rho} \tag{33.8}$$

where Y is the Young's modulus for the material, defined as the longitudinal stress divided by the longitudinal strain (Chapter 15). Find the speed of sound in an aluminum bar.

Solution: Using Eq. 33.8 and the available data for aluminum, $Y = 7.0 \times 10^{10}\ \text{N/m}^2$ and $\rho = 2.7 \times 10^3\ \text{kg/m}^3$, we find that

$$v_{\text{Al}} = \sqrt{\frac{7.0 \times 10^{10}\ \text{N/m}^2}{2.7 \times 10^3\ \text{kg/m}^3}} \approx 5100\ \text{m/s}$$

This is a typical value for the speed of sound in solids. Note that the result is much larger than the speed of sound in gases. This makes sense since the molecules of a solid are close together (in comparison to the molecules of a gas) and hence respond more rapidly to a disturbance.

Example 33.3

Speed of sound in a liquid: Find the speed of sound in water, which has a bulk modulus of about $2.1 \times 10^9\ \text{N/m}^2$ and a density of about $10^3\ \text{kg/m}^3$.

Solution: Using Eq. 33.1, we find that

$$v_{\text{water}} = \sqrt{B/\rho} \approx \sqrt{\frac{2.1 \times 10^9\ \text{N/m}^2}{1 \times 10^3\ \text{kg/m}^3}} = 1500\ \text{m/s}$$

This result is much smaller than that for the speed of sound in aluminum, calculated in the previous example. In general, sound waves travel more slowly in liquids than in solids. This is because liquids are more compressible than solids and hence have a smaller bulk modulus.

The speed of sound in various media is given in Table 33.1.

Q1. Why are sound waves characterized as being longitudinal?

Q2. As a result of a distant explosion, an observer senses a ground tremor and then hears the explosion. Explain.

Q3. Explain why sound travels faster in warm air than in cool air.

TABLE 33.1 Speed of Sound in Various Media

MEDIUM	v (m/s)
Gases	
Air (0°C)	331
Air (100°C)	366
Hydrogen (0°C)	1286
Oxygen (0°C)	317
Helium (0°C)	972
Liquids at 25°C	
Water	1493
Methyl alcohol	1143
Sea water	1533
Solids	
Aluminum	5100
Copper	3560
Iron	5130
Lead	1322
Vulcanized rubber	54

33.2 HARMONIC SOUND WAVES

If the source of a longitudinal wave, such as a vibrating diaphragm, oscillates with simple harmonic motion, the resulting disturbance will also be harmonic. One can produce a one-dimensional harmonic sound wave in a long, narrow tube containing a gas by means of a vibrating piston at one end, as in Fig. 33.3. The darker regions in this figure represent regions where the gas is compressed, and so the density and pressure are *above* their equilibrium values.

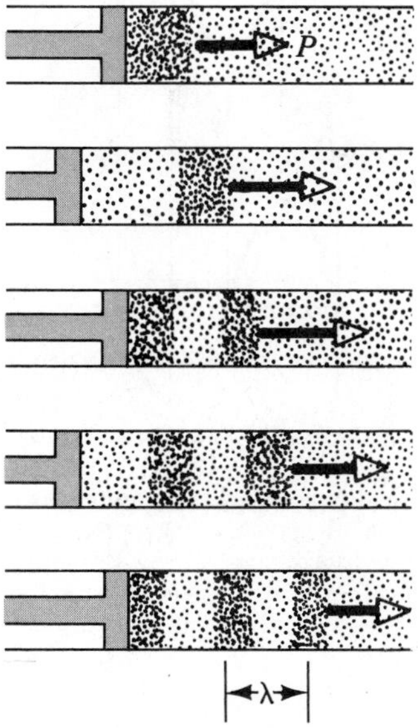

Figure 33.3 A harmonic longitudinal wave propagating down a tube filled with a compressible gas. The source of the wave is a vibrating piston at the left. The high- and low-pressure regions are dark and light, respectively.

A compressed layer is formed at times when the piston is being pushed into the tube. This compressed region, called a *condensation,* moves down the tube as a pulse, continuously compressing the layers in front of it. When the piston is withdrawn from the tube, the gas in front of it expands and the pressure and density in this region fall below their equilibrium values (represented by the lighter regions in Fig. 33.3). These low-pressure regions, called *rarefactions,* also propagate along the tube, following the condensations. Both regions move with a speed equal to the speed of sound in that medium (about 343 m/s in air at 20°C).

As the piston oscillates back and forth, regions of condensation and rarefaction are continuously set up. The distance between two successive condensations (or two successive rarefactions) equals the wavelength, λ. As these regions travel down the tube, any small volume of the medium moves with simple harmonic motion parallel to the direction of the wave. If $s(x,t)$ is the displacement of a small volume element measured from its equilibrium position, we can express this harmonic displacement function as

Harmonic displacement

$$s(x,t) = s_{\mathrm{m}} \cos(kx - \omega t) \tag{33.9}$$

where s_{m} is the *maximum displacement from equilibrium* (the displacement amplitude), k is the wave number, and ω is the angular frequency of the piston. Note that the displacement is along x, the direction of motion of the sound wave, which of course means we are describing a longitudinal wave. The variation in the pressure of the gas, ΔP, measured from its equilibrium value is also harmonic and given by

Pressure variation

$$\Delta P = \Delta P_{\mathrm{m}} \sin(kx - \omega t) \tag{33.10}$$

The derivation of this expression will be given below.

The *pressure amplitude* ΔP_m is the *maximum change in pressure from the equilibrium value.* As we shall show later, the pressure amplitude is proportional to the displacement amplitude, s_m, and is given by

Pressure amplitude

$$\Delta P_m = \rho\omega v s_m \tag{33.11}$$

Thus, we see that a sound wave may be considered as either a displacement wave or a pressure wave. A comparison of Eqs. 33.9 and 33.10 shows that *the pressure wave is 90° out of phase with the displacement wave.* Graphs of these functions are shown in Fig. 33.4. Note that the pressure variation is a maximum when the displacement is zero, whereas the displacement is a maximum when the pressure variation is zero. Since the pressure is proportional to the density, the variation in density from the equilibrium value follows an expression similar to Eq. 33.10.

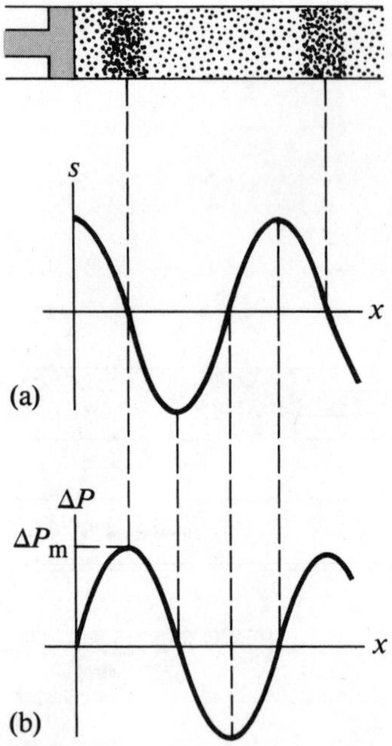

Figure 33.4 (a) Displacement amplitude versus position and (b) pressure amplitude versus position for a harmonic longitudinal wave. Note that the displacement wave is 90° out of phase with the pressure wave.

We shall now give a derivation of Eqs. 32.10 and 32.11. From Eq. 33.4, we see that the pressure variation in a gas is given by

$$\Delta P = -B(\Delta V/V)$$

The volume of a layer of thickness Δx and cross-sectioned area A is $V = A\,\Delta x$. The change in the volume ΔV accompanying the pressure change is equal to $A\,\Delta s$, where Δs is the difference in s between x and $x + \Delta x$. That is, $\Delta s = s(x + \Delta x) - s(x)$. Hence, we can express ΔP as

$$\Delta P = -B\frac{\Delta V}{V} = -B\frac{A\,\Delta s}{A\,\Delta x} = -B\frac{\Delta s}{\Delta x}$$

As Δx approaches zero, the ratio $\Delta s/\Delta x$ becomes $\partial s/\partial x$. (The partial derivative is used here to indicate that we are interested in the variation of s with position at a *fixed* time.) Therefore,

$$\Delta P = -B(\partial s/\partial x)$$

If the displacement is the simple harmonic function given by Eq. 33.9, we find that

$$\Delta P = -B\frac{\partial}{\partial x}[s_m \cos(kx - \omega t)] = Bs_m k \sin(kx - \omega t)$$

Since the bulk modulus is given by $B = \rho v^2$ (Eq. 33.1), the pressure variation reduces to

$$\Delta P = \rho v^2 s_m k \sin(kx - \omega t)$$

Furthermore, from Eq. 32.13, we can write $\omega = kv$, hence ΔP can be expressed as

$$\Delta P = \rho\omega s_m v \sin(kx - \omega t) = \Delta P_m \sin(kx - \omega t)$$

where ΔP_m is the maximum pressure variation, given by Eq. 33.11:

$$\Delta P_m = \rho\omega v s_m$$

Q4. If an alarm clock is placed in a good vacuum and then activated, no sound will be heard. Explain.

Q5. Some sound waves are harmonic, whereas others are not. Give an example of each.

33.3 ENERGY AND INTENSITY OF HARMONIC SOUND WAVES

In the previous chapter, we showed that waves traveling on stretched strings transport energy. The same concepts are now applied to sound waves. Consider a layer of air of mass Δm and width Δx in front of a piston oscillating with a frequency

ω, as in Fig. 33.5. The piston transmits energy to the layer of air.[1] Since the average kinetic energy equals the average potential energy in simple harmonic motion (as was shown in Chapter 13), the average total energy of the mass Δm equals its maximum kinetic energy. Therefore, we can express the average energy of the moving layer of gas as

$$\Delta E = \frac{1}{2}\Delta m\omega^2 s_m^2 = \frac{1}{2}(\rho A\,\Delta x)\omega^2 s_m^2$$

where $A\,\Delta x$ is the volume of the layer. The rate of energy transferred to each layer (or the power) is given by

$$\text{Power} = \frac{\Delta E}{\Delta t} = \frac{1}{2}\rho A\left(\frac{\Delta x}{\Delta t}\right)\omega^2 s_m^2 = \frac{1}{2}\rho A v\omega^2 s_m^2$$

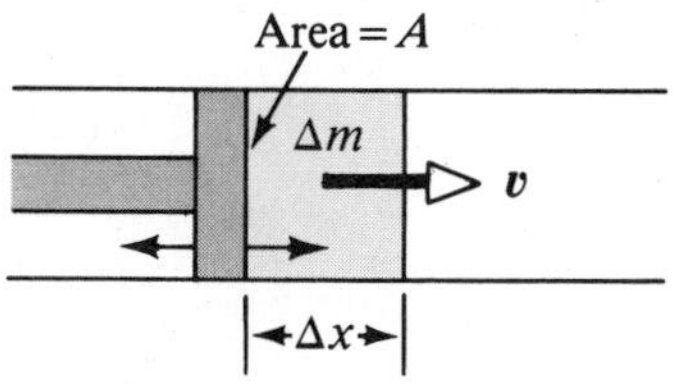

Figure 33.5 An oscillating piston transfers energy to the gas in the tube, causing the layer of width Δx and mass Δm to oscillate with an amplitude s_m.

where $v = \Delta x/\Delta t$ is the velocity of the disturbance to the right. The intensity I of the traveling sound wave is defined as the power per unit area. In this case, the intensity is given by

Intensity of a sound wave

$$I = \frac{\text{power}}{\text{area}} = \frac{1}{2}\rho\omega^2 s_m^2 v \qquad (33.12)$$

Thus, we see that the intensity of the harmonic sound wave is proportional to the square of the amplitude and the square of the frequency (as in the case of a harmonic string wave). This can also be written in terms of the pressure amplitude ΔP_m, using Eq. 33.11, which gives

$$I = \frac{\Delta P_m^2}{2\rho v} \qquad (33.13)$$

Example 33.4

Hearing limitations: The faintest sounds the human ear can detect at a frequency of 1000 Hz correspond to an intensity of about 10^{-12} W/m^2 (the so-called *threshold of hearing*). Likewise, the loudest sounds that the ear can tolerate correspond to an intensity of about 1 W/m^2 (the *threshold of pain*). Determine the pressure amplitudes and maximum displacements associated with these two limits.

Solution: First, consider the faintest sounds. Using Eq. 33.13 and taking $v = 343$ m/s and $\rho = 1.2$ kg/m^3 for air, we get

$$\begin{aligned}\Delta P_m &= (2\rho v I)^{1/2}\\ &= [2(1.2\text{ kg/m}^3)(343\text{ m/s})(10^{-12}\text{ W/m}^2)]^{1/2}\\ &= 2.9\times 10^{-5}\text{ N/m}^2\end{aligned}$$

Since atmospheric pressure is about 10^5 N/m^2, this means the ear can discern pressure fluctuations as small as 3 parts in 10^{10}! The corresponding maximum displacement can be calculated using 33.11, recalling that $\omega = 2\pi f$:

$$\begin{aligned}s_m &= \frac{\Delta P_m}{\rho\omega v} = \frac{2.9\times 10^{-5}\text{ N/m}^2}{(1.2\text{ kg/m}^3)(2\pi\times 10^3\text{ s}^{-1})(343\text{ m/s})}\\ &= 1.1\times 10^{-11}\text{ m}\end{aligned}$$

This is a remarkably small number! If we compare this result for s_m with the diameter of a molecule (about 10^{-10} m), we see that the ear is an extremely sensitive detector of sound waves.

In a similar manner, one finds that the loudest sounds the human ear can tolerate correspond to a pressure amplitude of about 29 N/m^2 and a maximum displacement of 1.1×10^{-5} m. Note that the small pressure amplitudes correspond to fluctuations taking place above and below atmospheric pressure.

Intensity in Decibels

The previous example illustrates the wide range of intensities that the human ear can detect. For this reason, it is convenient to use a logarithmic intensity scale,

[1] Although it is not proved here, the work done by the piston equals the energy carried away by the wave. For a detailed mathematical treatment of this concept, see Frank S. Crawford, Jr., *Waves,* New York, McGraw-Hill, 1968, Berkeley Physics Course, Volume 3, Chapter 4.

Intensity in decibels

where the intensity β is defined by the equation

$$\beta = 10 \log(I/I_0) \tag{33.14}$$

The constant I_0 is the *reference intensity level,* taken to be at the threshold of hearing ($I_0 = 10^{-12}$ W/m^2), and I is the intensity at the level β, where β is measured in decibels[2] (dB). On this scale, the threshold of pain ($I = 1$ W/m^2) corresponds to an intensity level of $\beta = 10 \log(1/10^{-12}) = 10 \log(10^{12}) = 120$ dB. Likewise, the threshold of hearing corresponds to $\beta = 10 \log(1/1) = 0$ dB. Nearby jet airplanes can create intensity levels of 150 dB, and subways and riveting machines have levels of 90 to 100 dB. The electronically amplified sounds heard at rock concerts can be at levels of up to 120 dB, the threshold of pain. Prolonged exposure to such high intensity levels may produce serious damage to the ear. Ear plugs are recommended whenever intensity levels exceed 90 dB. Recent evidence also suggests that "noise pollution" may be a contributing factor to high blood pressure, anxiety, and nervousness. Table 33.2 gives some idea of the sound intensities of various sources.

TABLE 33.2 Decibel Scale Intensity for Some Sources

SOURCE OF SOUND	β (dB)
Nearby jet airplane	150
Jackhammer; machine gun	130
Siren; rock concert	120
Subway; power mower	100
Busy traffic	80
Vacuum cleaner	70
Normal conversation	50
Mosquito buzzing	40
Whisper	30
Rustling leaves	10
Threshold of hearing	0

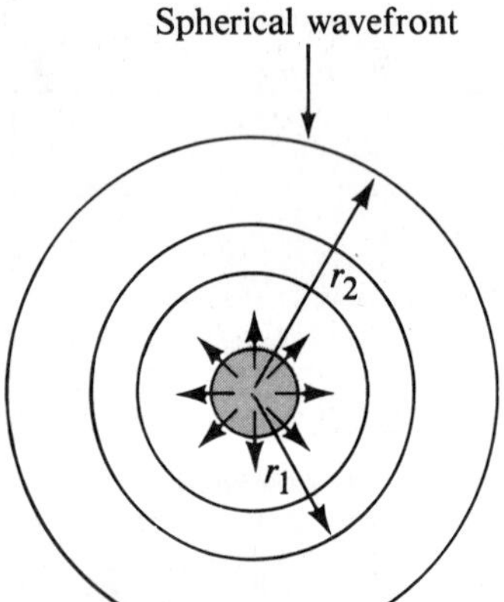

Figure 33.6 A spherical wave propagating radially outward from an oscillating spherical body. The intensity of the spherical wave varies as $1/r^2$.

33.4 SPHERICAL AND PLANAR WAVES

If a spherical body pulsates or oscillates periodically such that its radius varies harmonically with time, a sound wave with spherical wave fronts will be produced (Fig. 33.6). The wave moves outward from the source at a constant speed.

Since all points on the sphere behave in the same way, we conclude that the energy in a spherical wave will propagate equally in all directions. That is, no one direction is preferred over any other. If P_{av} is the average power emitted by the source, then this power at any distance r from the source must be distributed over a spherical surface of area $4\pi r^2$. Hence, the wave intensity at a distance r from the source is

Intensity of a spherical wave

$$I = P_{av}/A = P_{av}/4\pi r^2 \tag{33.15}$$

Since P_{av} is the same through any spherical surface centered at the source, we see that the intensities at distances r_1 and r_2 are given by

$$I_1 = P_{av}/4\pi r_1^2 \quad \text{and} \quad I_2 = P_{av}/4\pi r_2^2$$

Therefore, the ratio of intensities on these two spherical surfaces is

$$\frac{I_1}{I_2} = \frac{r_2^2}{r_1^2}$$

[2]Named after the inventor of the telephone, Alexander Graham Bell (1847–1922).

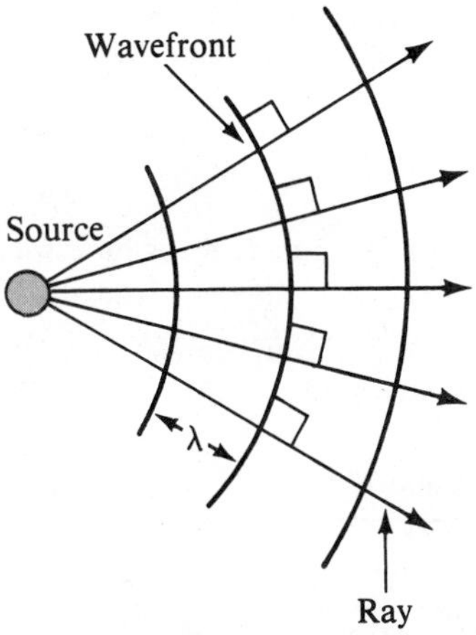

Figure 33.7 Spherical waves emitted by a point source. The circular arcs represent the spherical wavefronts concentric with the source. The rays are radial lines pointing outward from the source, perpendicular to the wavefronts.

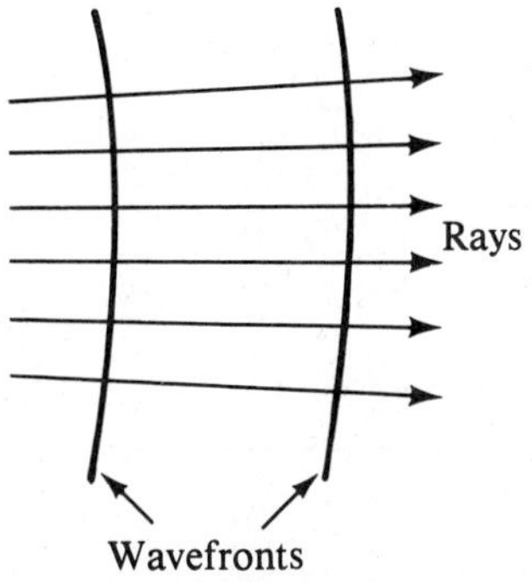

Figure 33.8 At large distances from a point source, the wavefronts are nearly parallel planes and the rays are nearly parallel lines perpendicular to the planes. Hence, a small segment of a spherical wavefront is approximately a planar wave.

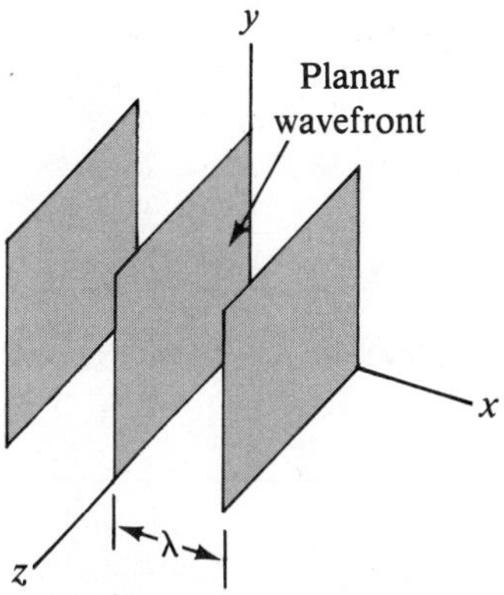

Figure 33.9 Representation of a planar wave moving in the positive x direction. The wavefronts are planes parallel to the yz plane.

In Eq. 33.12 we found that the intensity was also proportional to s_m^2, the square of the wave amplitude. Comparing this result with Eq. 33.15, we conclude that the wave amplitude of a spherical wave must vary as $1/r$. Therefore, we can write the wave function ψ (Greek letter "psi") for an outgoing spherical wave in the form

Wave function for a spherical wave

$$\psi(r,t) = (s_0/r)\sin(kr - \omega t) \tag{33.16}$$

where s_0 is a constant.

It is useful to represent spherical waves by a series of circular arcs concentric with the source, as in Fig. 33.7. Each arc represents a surface over which the phase of the wave is constant. We call such a surface of constant phase a *wavefront*. The distance between adjacent wavefronts equals the wavelength, λ. The radial lines pointing outward from the source are called *rays*.

Wavefront

Now consider a small portion of the wavefronts at *large* distances (large relative to λ) from the source, as in Fig. 33.8. In this case, the rays are nearly parallel and the wavefronts are very close to being planar. Therefore, at distances from the source that are large compared with the wavelength, we can approximate the wavefronts by parallel planes. We call such a wave a *planar wave*. Any small portion of a spherical wave that is far from the source can be considered a planar wave.

Figure 33.9 illustrates a planar wave propagating along the x axis. If x is taken to be the direction of the wave motion (or rays) in Fig. 33.9, then the wavefronts are parallel to the yz plane. In this case, the wave function depends only on x and t and has the form

Planar wave representation

$$\psi(x,t) = s_0 \sin(kx - \omega t) \tag{33.17}$$

That is, the wave function for a planar wave is identical in form to that of a one-dimensional traveling wave. Note that the intensity is the same on successive wavefronts of the planar wave.

Example 33.5

Intensity variations of a point source: A source emits sound waves with a power output of 80 W. Assume the source is a point source. (a) Find the intensity at a distance 3 m from the source.

A point source emits energy in the form of spherical waves (Fig. 33.6). Let P_{av} be the average power output of the source. At a distance r from the source, the power is distributed over the surface area of a sphere, $4\pi r^2$. Therefore, the intensity at a distance r from the source is given by Eq. 33.15. Since $P_{av} = 80$ W and $r = 3$ m, we find that

$$I = \frac{P_{av}}{4\pi r^2} = \frac{80 \text{ W}}{4\pi(3 \text{ m})^2} = 0.71 \text{ W/m}^2$$

which is close to the threshold of pain.

(b) Find the distance at which the sound reduces to a level of 40 dB.

We can find the intensity at the 40-dB level by using Eq. 33.14 with $I_0 = 10^{-12}$ W/m². This gives

$$\log (I/I_0) = 4$$

$$I = 10^4 I_0 = 10^{-8} \text{ W/m}^2$$

Using this value for I in Eq. 33.15 and solving for r, we get

$$r = (P_{av}/4\pi I)^{1/2} = \frac{80 \text{ W}}{(4\pi \times 10^{-8} \text{ W/m}^2)^{1/2}}$$

$$= 2.5 \times 10^4 \text{ m}$$

which equals about 15 miles!

Q6. In Example 33.5, we found that a point source with a power output of 80 W reduces to an intensity level of 40 dB at a distance of about 15 miles. Why do you suppose you cannot normally hear a rock concert going on 15 miles away?

Q7. If the distance from a point source is tripled, by what factor does the intensity decrease?

33.5 THE DOPPLER EFFECT

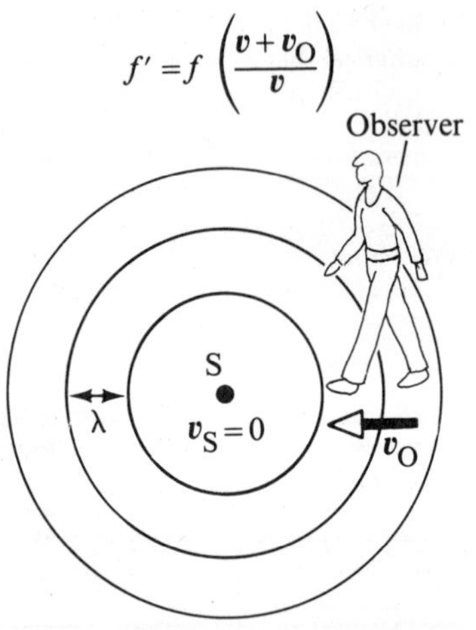

Figure 33.10 An observer O moving with a speed v_O *toward* a stationary point source S hears a frequency f' that is *greater* than the source frequency.

When a car or truck is moving while its horn is blowing, the frequency of the sound you hear is higher as the vehicle approaches you and lower as it moves away from you. This is one example of the Doppler effect.[3] In general, a Doppler effect is experienced whenever there is *relative* motion between the source and the observer. When the source and observer are moving toward each other, the frequency heard by the observer is higher than the frequency of the source. When the source and observer move away from each other, the observer hears a frequency which is lower than the source frequency. Although the Doppler effect is most commonly experienced with sound waves, it is a phenomena common to all harmonic waves. For example, the frequencies of light waves (electromagnetic waves) are also modified by relative motion.

First, let us consider the case where the observer O is moving and the sound source S is stationary. For simplicity, we shall assume that the air is also stationary. Figure 33.10 describes the situation when the observer moves with a speed v_O toward the source (considered as a point source), which is at rest ($v_S = 0$).

We shall take the frequency of the source to be f, the wavelength to be λ, and the velocity of sound to be v. If the observer were also stationary, clearly he or she would detect f wavefronts per second. (That is, when $v_O = 0$ and $v_S = 0$, the observed frequency equals the source frequency.) When the observer travels toward the source, he or she moves a distance $v_O t$ in time of t seconds and in this time detects an *additional* $v_O t/\lambda$ wavefronts. Since the additional number of wavefronts detected *per second* is v_O/λ, the frequency f' heard by the observer is *increased* and given by

$$f' = f + \Delta f = f + v_O/\lambda$$

[3]Named after the Austrian physicist Christian Johann Doppler (1803–1853), who discovered the effect for light waves.

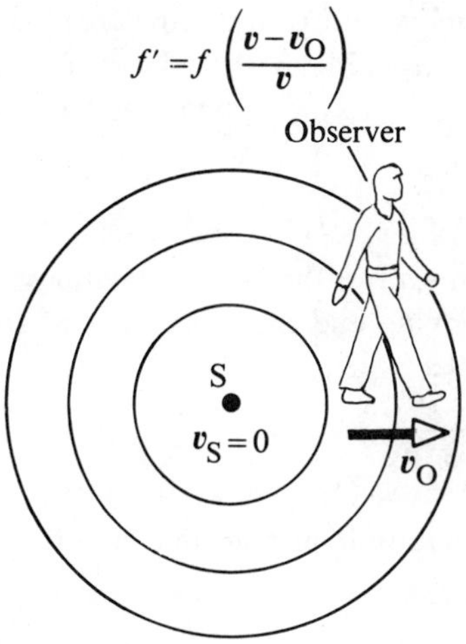

Figure 33.11 An observer O moving with a speed v_O *away* from a stationary source S hears a frequency f' that is *lower* than the source frequency.

Using the fact that $\lambda = v/f$, we see that $v_O/\lambda = (v_O/v)f$, hence f' can be expressed as

$$f' = f\left(\frac{v + v_O}{v}\right) \tag{33.18}$$

Moving *away* from the source, as in Fig. 33.11, an observer detects *fewer* wavefronts per second. Thus, from Eq. 33.18, it follows that the frequency heard by the observer in this case is lowered and given by

$$f' = f\left(\frac{v - v_O}{v}\right) \tag{33.19}$$

In general, when an observer moves with a speed v_O relative to a stationary source, the frequency heard by the observer is

$$f' = f\left(\frac{v \pm v_O}{v}\right) \tag{33.20}$$

Frequency heard by an observer in motion

where the *positive* sign is used when the observer moves *toward* the source and the *negative* sign holds when the observer moves *away* from the source.

Now consider the situation in which the source is in motion and the observer is at rest. If the source moves directly toward observer A in Fig. 33.12, the wavefronts seen by the observer are closer together as a result of the motion of the source in the direction of the outgoing wave. As a result, the wavelength λ' measured by observer A is shorter than the wavelength λ of the source. During each vibration, which lasts for a time T (the period), the source moves a distance $v_ST = v_S/f$ and the wavelength is *shortened* by this amount. Therefore, the observed wavelength λ' is given by

$$\lambda' = \lambda - \Delta\lambda = \lambda - (v_S/f)$$

Since $\lambda = v/f$, the frequency heard by observer A is

$$f' = \frac{v}{\lambda'} = \frac{v}{\lambda - \frac{v_S}{f}} = \frac{v}{\frac{v}{f} - \frac{v_S}{f}}$$

$$f' = f\left(\frac{v}{v - v_S}\right) \tag{33.21}$$

That is, the observed frequency is *increased* when the source moves toward the observer.

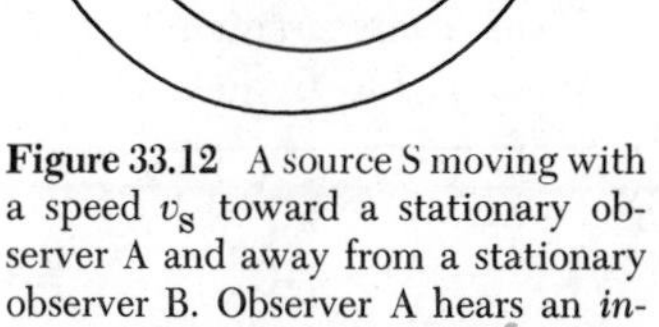

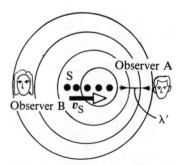

Figure 33.12 A source S moving with a speed v_S toward a stationary observer A and away from a stationary observer B. Observer A hears an *increased* frequency, and observer B hears a *decreased* frequency.

In a similar manner, when the source moves away from an observer B at rest (where observer B is to the left of the source, as in Fig. 33.12), observer B measures a wavelength λ' that is *greater* than λ and hears a *decreased* frequency given by

$$f' = f\left(\frac{v}{v + v_S}\right) \tag{33.22}$$

Combining Eqs. 33.21 and 33.22, we can express the general relationship for the observed frequency when the source is moving and the observer is at rest as

Frequency heard with source in motion

$$f' = f\left(\frac{v}{v \mp v_S}\right) \tag{33.23}$$

Finally, if both the source and the observer are in motion, one finds the following general relationship for the observed frequency:

Frequency heard with observer and source in motion

$$f' = f\left(\frac{v \pm v_O}{v \mp v_S}\right) \tag{33.24}$$

In this expression, the *upper* signs ($+v_O$ and $-v_S$) refer to motion of one *toward* the other, and the lower signs ($-v_O$ and $+v_S$) refer to motion of one *away* from the other.

Example 33.6

A train moving at a speed of 40 m/s sounds its whistle, which has a frequency of 500 Hz. Determine the frequencies heard by a stationary observer as the train approaches and recedes from the observer.

Solution: We can use Eq. 33.21 to get the apparent frequency as the train approaches the observer. Taking $v = 343$ m/s for the speed of sound in air gives

$$f' = f\left(\frac{v}{v - v_S}\right) = (500\text{ Hz})\left(\frac{343\text{ m/s}}{343\text{ m/s} - 40\text{ m/s}}\right)$$

$$= 566\text{ Hz}$$

Likewise, Eq. 33.22 can be used to obtain the frequency heard as the train recedes from the observer:

$$f' = f\left(\frac{v}{v + v_S}\right) = (500\text{ Hz})\left(\frac{343\text{ m/s}}{343\text{ m/s} + 40\text{ m/s}}\right)$$

$$= 448\text{ Hz}$$

Example 33.7

An ambulance travels down a highway at a speed of 75 mi/h. Its siren emits sound at a frequency of 400 Hz. What is the frequency heard by a passenger in a car traveling at 55 mi/h in the opposite direction as the car approaches the ambulance and as the car moves away from the ambulance?

Solution: Let us take the velocity of sound in air to be $v = 343$ m/s and note that 1 mi/h = 0.447 m/s. Therefore, $v_S = 75$ mi/h = 33.5 m/s and $v_O = 55$ mi/h = 24.6 m/s. We can use Eq. 33.24 in both cases. As the ambulance and car approach each other, the observed apparent frequency is

$$f' = f\left(\frac{v + v_O}{v - v_S}\right) = (400\text{ Hz})\left(\frac{343\text{ m/s} + 24.6\text{ m/s}}{343\text{ m/s} - 33.5\text{ m/s}}\right)$$

$$= 475\text{ Hz}$$

Likewise, as they recede from each other, a passenger in the car hears a frequency

$$f' = f\left(\frac{v - v_O}{v + v_S}\right) = (400\text{ Hz})\left(\frac{343\text{ m/s} - 24.6\text{ m/s}}{343\text{ m/s} + 33.5\text{ m/s}}\right)$$

$$= 338\text{ Hz}$$

Note that the *change* in frequency as detected from the car is 475 − 338 = 137 Hz, which is more than 30% of the actual frequency emitted.

Shock Waves

Now let us consider what happens when the source velocity v_S *exceeds* the wave velocity v. This situation is described graphically in Fig. 33.13. The circles represent spherical wavefronts emitted by the source at various times during its motion.

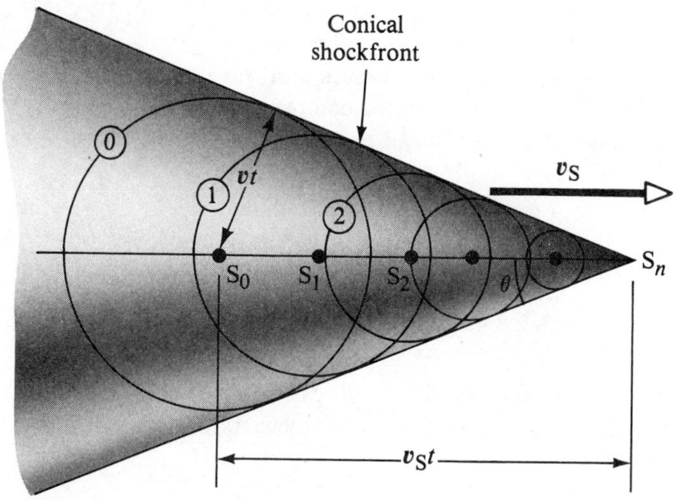

Figure 33.13 Representation of a shock wave produced when a source moves from S_0 to S_n with a speed v_S, which is *greater* than the wave speed v in that medium. The envelope of the wavefronts forms a cone whose apex angle is given by $\sin\theta = v/v_S$.

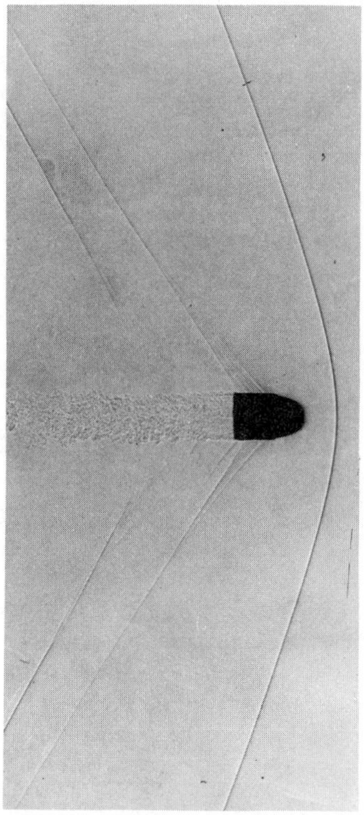

Bullet traveling in air faster than the speed of sound. Note the shape of the shock waves which accompany the bullet.

At $t = 0$, the source is at S_0, and at some later time t, the source is at S_n. In the time t, the wavefront centered at S_0 reaches a radius of vt. In this same interval, the source travels a distance $v_S t$ to S_n. At the instant the source is at S_n, waves are just beginning to be generated and so the wavefront has zero radius at this point. The line drawn from S_n to the wavefront centered on S_0 is tangent to all other wavefronts generated at intermediate times. Thus, we see that the envelope of these waves is a cone whose apex angle θ is given by

$$\sin\theta = v/v_S$$

The ratio v_S/v is referred to as the *mach number*. The conical wavefront produced when $v_S > v$ (supersonic speeds) is known as a *shock wave*. An interesting analogy to shock waves is the V-shaped wavefronts produced by a boat (the bow wave) when the boat's speed exceeds the speed of the surface water waves.

Jet airplanes traveling at supersonic speeds produce shock waves, which are responsible for the loud explosion, or "sonic boom," one hears. The shock wave carries a great deal of energy concentrated on the surface of the cone, with correspondingly large pressure variations. Such shock waves are unpleasant to hear and can cause damage to buildings when aircraft fly supersonically at low altitudes. In fact, an airplane flying at supersonic speeds produces a double boom because two shock fronts are formed, one from the nose of the plane and one from the tail (Fig. 33.14).

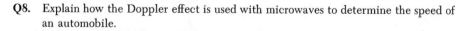

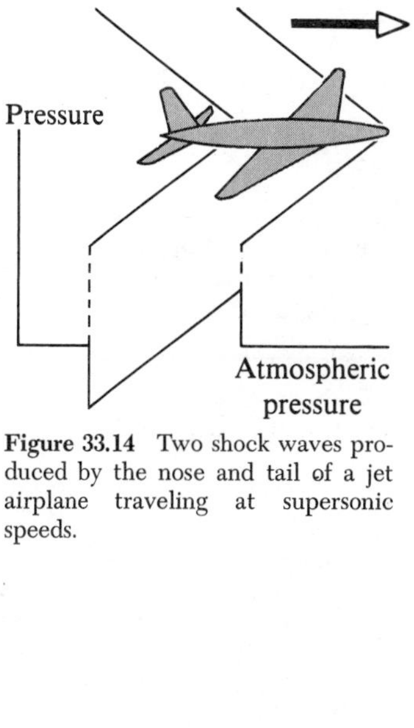

Figure 33.14 Two shock waves produced by the nose and tail of a jet airplane traveling at supersonic speeds.

Q8. Explain how the Doppler effect is used with microwaves to determine the speed of an automobile.

Q9. If you are in a moving vehicle, explain what happens to the frequency of your echo as you move *toward* a canyon wall. What happens to the frequency as you move *away* from the wall?

Q10. Suppose an observer and a source of sound are both at rest, and a strong wind blows toward the observer. Describe the effect of the wind (if any) on (a) the observed wavelength, (b) the observed frequency, and (c) the wave velocity.

33.6 SUMMARY

Sound waves are longitudinal in nature and travel through a compressible medium with a speed that depends on the compressibility and inertia of that medium. The speed of sound in a medium of compressibility B and density ρ is

Speed of sound in a compressible medium

$$v = \sqrt{B/\rho} \tag{33.1}$$

The *speed of sound* in a gas at a pressure P and temperature T is

Speed of sound in a gas

$$v = \sqrt{\gamma P/\rho} = \sqrt{\gamma RT/M} \tag{33.6, 33.7}$$

where M is the molecular weight of the gas and R is the gas constant.

In the case of harmonic sound waves, the variation in pressure from the equilibrium value is given by

Pressure variation

$$\Delta P = \Delta P_{\text{m}} \sin(kx - \omega t) \tag{33.10}$$

where ΔP_{m} is the pressure amplitude. The pressure wave is 90° out of phase with the displacement wave. If the displacement amplitude is s_{m}, ΔP_{m} has the value

Pressure amplitude

$$\Delta P_{\text{m}} = \rho\omega v s_{\text{m}} \tag{33.11}$$

The *intensity of a harmonic sound wave,* which is the power per unit area, is given by

Intensity of a sound wave

$$I = \frac{1}{2}\rho\omega^2 s_{\text{m}}^2 v = \frac{\Delta P_{\text{m}}^2}{2\rho v} \tag{33.13}$$

The *intensity of a spherical wave* produced by a point source is proportional to the average power emitted and inversely proportional to the square of the distance from the source.

The change in frequency heard by an observer whenever there is relative motion between the frequency source and observer is called the *Doppler effect.* If the observer moves with a speed v_{O} and the source is at rest, the observed frequency f' is

Frequency heard by an observer in motion

$$f' = f\left(\frac{v \pm v_{\text{O}}}{v}\right) \tag{33.20}$$

where the positive sign is used when the observer moves toward the source and the negative sign refers to motion away from the source.

If the source moves with a speed v_{S} and the observer is at rest, the observed frequency is

Frequency heard with source in motion

$$f' = f\left(\frac{v}{v \mp v_{\text{S}}}\right) \tag{33.23}$$

where $-v_{\text{S}}$ refers to motion toward the observer and $+v_{\text{S}}$ refers to motion away from the observer.

When the observer and source are both moving, the observed frequency is

Frequency heard with observer and source in motion

$$f' = f\left(\frac{v \pm v_{\text{O}}}{v \mp v_{\text{S}}}\right) \tag{33.24}$$

EXERCISES

Section 33.1 Velocity of Sound Waves

1. Calculate the speed of sound in methane (CH_4) at 288 K, using the values $\gamma = 1.31$, the molecular weight of CH_4, $M = 16$ kg/kmole, and $R = 8314$ J/kmole · K.

2. At what temperature will the speed of sound in methane equal the speed of sound in helium at 288 K? For helium, $\gamma = 1.66$ and $M = 4$ kg/kmole.

3. A worker is at one end of a mile-long section (1.61 km) of iron pipeline when an accidental blast occurs at the other end of the section. The worker receives two sound signals from the blast, one transmitted through the pipe and one through the surrounding air. Use values from Table 33.1 to calculate the elapsed time between the two signals. [Note: first find the speed of sound in air at 300 K and take the speed of sound in iron at that temperature to be 5200 m/s.]

4. A spelunker attempts to determine the depth of a pit in the floor of a cave by dropping a stone into the pit and measuring the time interval between release and the sound of the stone's hitting bottom. If the measured time interval is 10 s, what is the depth of the pit? (Assume a temperature of 15°C.)

5. (a) What are the SI units of bulk modulus as expressed in Eq. 33.2? (b) Show that the SI units of $\sqrt{B/\rho}$ are m/s, as required by Eq. 33.1.

6. The density of aluminum is 2.7×10^3 kg/m^3. Use the value for the speed of sound in aluminum given in Table 33.1 to calculate Young's modulus for this material.

7. A sound wave propagating in air has a frequency of 4000 Hz. Calculate the percent change in wavelength when the wavefront, initially in a region where T = 27°C, enters a region where the air temperature decreases to 10°C.

8. Xenon has a density of 5.9 kg/m^3 at 0°C and 1 atm pressure. Since it is monoatomic, $\gamma = 1.67$. (a) Calculate the speed of sound in xenon at 0°C. (b) What is the bulk modulus of xenon?

Section 33.2 Harmonic Sound Waves

(In this section, use the following values as needed unless otherwise specified: the equilibrium density of air, $\rho = 1.2$ kg/m^3; the velocity of sound in air, $v = 343$ m/s. Also, pressure variations ΔP are measured relative to atmospheric pressure.)

9. Calculate the pressure amplitude of a 2000-Hz sound wave in air if the displacement amplitude is 2×10^{-8} m.

10. A sound wave in air has a pressure amplitude of 4×10^{-3} N/m^2. Calculate the displacement amplitude of the wave at a frequency of 10 kHz.

11. The pressure amplitude corresponding to the threshold of hearing is 2.9×10^{-5} N/m^2. At what frequency will a sound wave in air have this pressure amplitude if the displacement amplitude is 2.8×10^{-10} m?

12. An experimenter wishes to generate in air a sound wave that has a displacement amplitude of 5.5×10^{-6} m. The pressure amplitude is to be limited to 8.4×10^{-1} N/m^2. What is the minimum wavelength the sound wave can have?

13. A sound wave in air has a pressure amplitude of 4 N/m^2 and a frequency of 5000 Hz. $\Delta P = 0$ at the point $x = 0$ when $t = 0$. (a) what is ΔP at $x = 0$ when $t = 2 \times 10^{-4}$ s and (b) what is ΔP at $x = 0.02$ m when $t = 0$?

14. The harmonic displacement of a sound wave is described by $s(x,t) = 0.006 \cos[\pi(5.834x - 2000t)]$, where x is in m and t is in s. (a) What are the frequency, wavelength, and speed of the wave? (b) What is the displacement at the point $x = 0.05$ m when $t = 0$? (c) What is the displacement at $x = 0$ when $t = 3.75 \times 10^{-4}$ s?

15. Consider the sound wave whose harmonic displacement is described in Exercise 14. What is the pressure variation at $x = 0$ when $t = \pi/2\omega$?

16. Write an expression that describes the pressure variation as a function of position and time for a harmonic sound wave in air if $\lambda = 0.1$ m and $\Delta P_m = 0.2$ N/m^2.

17. Write the function that describes the displacement wave corresponding to the pressure wave in Exercise 16.

Section 33.3 Energy and Intensity of Harmonic Sound Waves

18. Calculate the intensity level in dB of a sound wave that has an intensity of 4 μW/m^2.

19. A vacuum cleaner has a measured sound level of 70 dB. What is the intensity of this sound in W/m^2?

20. (a) Calculate the intensity in W/m^2 of the wave described in Exercise 13. (b) Express this intensity in dB.

21. The intensity of a sound wave at a fixed distance from a speaker vibrating at 1000 Hz is 0.6 W/m^2. (a) Determine the intensity if the frequency is increased to 2500 Hz while a *constant* displacement amplitude is maintained. (b) Calculate the intensity if the frequency is reduced to 500 Hz and the displacement amplitude is doubled.

22. Calculate the pressure amplitude corresponding to a sound intensity of 120 dB (a rock concert).

Section 33.4 Spherical and Planar Waves

23. An experiment requires a sound intensity of $1.2\ W/m^2$ at a distance of 4 m from a speaker. What power output is required?
24. A source emits sound waves with a uniform power of 100 W. At what distance will the intensity be just below the threshold of pain, which is $1\ W/m^2$?
25. The sound level at a distance of 3 m from a source is 120 dB. At what distance will the sound level be (a) 100 dB and (b) 10 dB?
26. Spherical waves of wavelength 25 cm are propagating outward from a point source. (a) Compare the wave amplitude at $r = 50$ cm and $r = 200$ cm. (b) Compare the intensity at $r = 50$ cm with the intensity at $r = 100$ cm. (c) Compare the phase of the wave function at a specific time at $r = 50$ cm and $r = 75$ cm.

Section 33.5 The Doppler Effect

27. A commuter train passes a passenger platform at a constant speed of 40 m/s. The train horn is sounded at its characteristic frequency of 320 Hz. (a) What change in frequency is observed by a person on the platform as the train passes? (b) What wavelength does a person on the platform observe as the train approaches?
28. A train is moving parallel to a highway with a constant speed of 20 m/s. A car is traveling in the same direction as the train with a speed of 40 m/s. As the auto overtakes and passes the train, the car horn sounds at a frequency of 510 Hz and the train horn sounds at a frequency of 320 Hz. (a) What frequency does an occupant of the car observe for the train horn just before passing? (b) What frequency does a train passenger observe for the car horn just after passing?
29. A train passenger hears a frequency of 520 Hz as the train approaches a bell on a trackside safety gate; the bell is actually emitting a signal of 500 Hz. What frequency will the passenger hear just after passing the bell?
30. Standing at a crosswalk, you hear a frequency of 510 Hz from the siren on an approaching police car. After the police car passes, the observed frequency of the siren is 430 Hz. Determine the car's speed from these observations.
31. A projectile has a velocity of 725 m/s in air. (a) What is the apex angle of the shock wave associated with the projectile? (b) What is the mach number of the projectile?
32. The Concorde flies at mach 1.5. What is the angle between the direction of propagation of the shock wave and the direction of the plane's velocity?
33. At what speed should a supersonic aircraft fly so that the conical wavefront will have an apex half-angle of 50°?

PROBLEMS

1. (a) The sound level of a jackhammer is measured as 130 dB and that of a siren as 120 dB. Find the ratio of the intensities of the two sound sources. (b) Two sources have measured intensities of $I_1 = 100\ \mu W/m^2$ and $I_2 = 200\ \mu W/m^2$. By how many dB is source 1 lower than source 2?
2. The measured speed of sound in copper is 3560 m/s, and the density of copper is $8.89\ g/cm^3$. Based on this information, by what percent would you expect a block of copper to decrease in volume when subjected to a uniform external (gauge) pressure of 2 atm?
3. Two ships are moving along a line due east. The trailing vessel has a speed relative to a land-based observation point of 64 km/h, and the leading ship has a speed of 45 km/h relative to that station. The two ships are in a region of the ocean where the current is moving uniformly due west at 10 km/h. The trailing ship transmits a sonar signal at a frequency of 1200 Hz. What frequency is monitored by the leading ship? (Use 1520 m/s as the speed of sound in ocean water.)
4. When high-energy, charged particles move through a transparent medium with a velocity greater than the velocity of light in that medium, a shock wave, or bow wave, of light is produced. This phenomenon is called the *Cerenkov effect* and can be observed in the vicinity of the core of a swimming pool reactor due to high-speed electrons move through the water. In a particular case, the Cerenkov radiation produces a wavefront with a cone angle of 53°. Calculate the velocity of the electrons in the water. (Use 2.25×10^8 m/s as the velocity of light in water.)
5. Use Eq. 33.7 to compute the speed of sound in a mixture of 60 percent oxygen and 40 percent nitrogen at 40°C.
6. Consider a longitudinal (compressional) wave of wavelength λ traveling with speed v along the x direction through a medium of density ρ. The *dis-*

placement of the molecules of the medium from their equilibrium position is given by

$$s = s_m \sin(kx - \omega t)$$

Show that the pressure variation in the medium is given by

$$P = -\left(\frac{2\pi\rho v^2}{\lambda} s_m\right) \cos(kx - \omega t)$$

7. In Eq. 33.7 the temperature T must be in degrees kelvin. (a) Starting with this equation, show that the speed of sound in a gas can be expressed as $v = [v_0 + (v_0/546)]t$, where t is the temperature in °C and v_0 is the speed of sound in the gas at 0°C. (Hint: Assume that $t \ll 273$°C and use the expansion $(1 + x)^{1/2} = 1 + \frac{1}{2}x - \frac{1}{8}x^2 + \ldots$ (b) In the case of air, show that this result leads to

$$v = (331 + 0.61t) \text{ m/s}$$

8. Three metal rods are located relative to each other as shown in Fig. 33.15, where $L_1 + L_2 = L_3$. Values of density and Young's modulus for the three materials are $\rho_1 = 2.7 \times 10^3$ kg/m³, $Y_1 = 7 \times 10^{10}$ N/m², $\rho_2 = 11.3 \times 10^3$ kg/m³, $Y_2 = 1.6 \times 10^{10}$ N/m², $\rho_3 = 8.8 \times 10^3$ kg/m³, and $Y_3 = 11 \times 10^{10}$ N/m². (a) If $L_3 = 1.5$ m, what must the ratio L_1/L_2 be if a sound wave is to travel the length of rods 1 *and* 2 in the same time required to travel the length of rod 3? (b) If the frequency of the source is 4000 Hz, determine the phase difference between the wave traveling along rods 1 and 2 and the one traveling along rod 3.

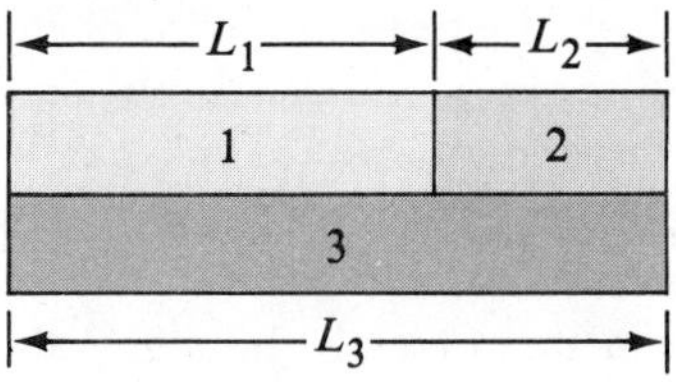

Figure 33.15 (Problem 8).

9. By proper excitation, it is possible to produce both longitudinal and transverse waves in a long metal rod. A particular metal rod is 150 cm long and has a radius of 0.2 cm and a mass of 50.9 g. Young's modulus for the material is 6.8×10^{11} dynes/cm². What must the tension in the rod be if the ratio of the speed of longitudinal waves to the speed of transverse waves is 8?

10. Consider again the situation described in Exercise 3. Show that, in general, the elapsed time interval between arrival of the sound signal through the pipe and through the surrounding air is

$$\Delta t = \frac{l(v_m - v_a)}{v_m v_a}$$

where v_m is the speed of sound in the metal, v_a is the speed of sound in air, and l is the length of the pipe.

11. The gas filling the tube shown in Fig. 33.3 is air at 20°C and at a pressure of 1.5×10^5 N/m². The piston shown is driven at a frequency of 600 Hz. The diameter of the piston is 10 cm, and the amplitude of its motion is 0.1 cm. What power must be supplied to maintain the oscillation of the piston?

12. (a) Use values from Table 32.2 to determine the resultant intensity in dB when a vacuum cleaner and a power mower are operated against a background of busy traffic. (b) In Table 33.2, a buzzing mosquito is rated at 40 dB and normal conversation at 50 dB. How many buzzing mosquitos are required to equal normal conversation in sound intensity?

34 Superposition and Standing Waves

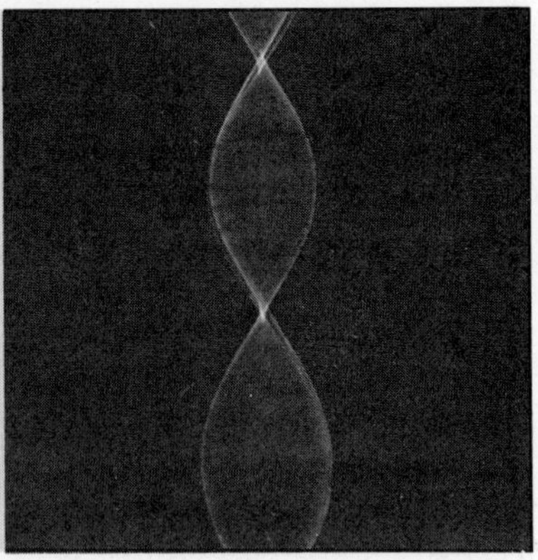

Education Development Center, Newton, Mass.

An important problem that arises when dealing with waves is that of describing the combined effect of two or more waves traveling in the same medium. For instance, what happens to a string when a wave traveling toward its fixed end is reflected back on itself? What is the pressure variation in the air when the instruments of an orchestra play together?

In a linear medium, that is, one in which the restoring force of the medium is proportional to the displacement of the medium, one can apply the *principle of superposition* to obtain the resultant disturbance. This principle can be applied to many types of waves, including waves on strings, sound waves, surface water waves, and electromagnetic waves. The superposition principle states that the actual displacement of any part of the disturbed medium equals the algebraic sum of the displacements caused by the individual waves. We discussed this principle as it applied to wave pulses in Chapter 32. The term *interference* was used to describe the effect produced by combining two waves moving simultaneously through a medium.

This chapter is concerned with the superposition principle as it applies to harmonic waves. If the harmonic waves that combine in a given medium have the same frequency and wavelength, one finds that a stationary pattern, called a *standing wave*, can be produced at certain frequencies. For example, a stretched string fixed at both ends has a discrete set of oscillation patterns, called *modes of vibration*, which depend upon the tension and mass per unit length of the string. These modes of vibration are found in stringed musical instruments. Other musical instruments, such as the organ and flute, make use of the natural frequencies of sound waves in hollow pipes. Such frequencies depend upon the length of the pipe and upon whether one end is open or closed.

We also consider the superposition and interference of waves with different frequencies and wavelengths. When two sound waves with nearly the same frequency interfere, one hears alternate loud and soft sounds called *beats*. The beat frequency corresponds to the rate of alternation between constructive and destructive interference. Finally, we describe how any complex periodic waveform can, in general, be described by a sum of sine and cosine functions.

34.1 SUPERPOSITION AND INTERFERENCE OF HARMONIC WAVES

The superposition principle tells us that when two or more waves move in the same linear medium, the net displacement of the medium (the resultant wave) at any point equals the algebraic sum of the displacements of all the waves. Let us apply this superposition principle to two harmonic waves traveling in the same direction in a medium. If the two waves are traveling to the right and have the same frequency, wavelength, and amplitude but differ in phase, we can express their individual wave functions as

$$y_1 = A_o \sin(kx - \omega t) \qquad \text{and} \qquad y_2 = A_o \sin(kx - \omega t - \phi)$$

Hence, the resultant wave function y is given by

$$y = y_1 + y_2 = A_o\,[\sin(kx - \omega t) + \sin(kx - \omega t - \phi)]$$

In order to simplify this expression, it is convenient to make use of the following trigonometric identity:

$$\sin a + \sin b = 2\cos\left(\frac{a-b}{2}\right)\sin\left(\frac{a+b}{2}\right)$$

If we let $a = kx - \omega t$ and $b = kx - \omega t - \phi$, we find that the resultant wave y reduces to

$$y = \left(2A_o \cos\frac{\phi}{2}\right)\sin\left(kx - \omega t - \frac{\phi}{2}\right) \qquad (34.1)$$

Resultant of two traveling harmonic waves

There are several important features of this result. The resultant wave function y is also harmonic and has the *same* frequency and wavelength as the individual waves. The amplitude of the resultant wave is $2A_o \cos(\phi/2)$, and its phase is equal to $\phi/2$. If the phase constant ϕ equals 0, then $\cos(\phi/2) = \cos 0 = 1$ and the amplitude of the resultant wave is $2A_o$. In other words, the resultant wave is twice as large as either individual wave. In this case, the waves are said to be everywhere *in phase* and thus *interfere constructively*. That is, the crests and troughs of the individual waves occur at the same positions, as is shown by the broken lines in Fig. 34.1a. In general, constructive interference occurs when $\cos(\phi/2) = \pm 1$, or when $\phi = 0, 2\pi, 4\pi, \ldots$. On the other hand, if ϕ is equal to π radians (or any *odd* multiple of π) then $\cos(\phi/2) = \cos(\pi/2) = 0$ and the resultant wave has *zero* amplitude everywhere. In this case, the two waves *interfere destructively*. That is, the crest of one wave coincides with the trough of the second (Fig. 34.1b) and their displacements cancel at

Constructive interference

Destructive interference

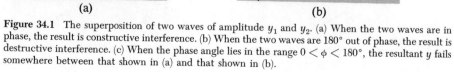

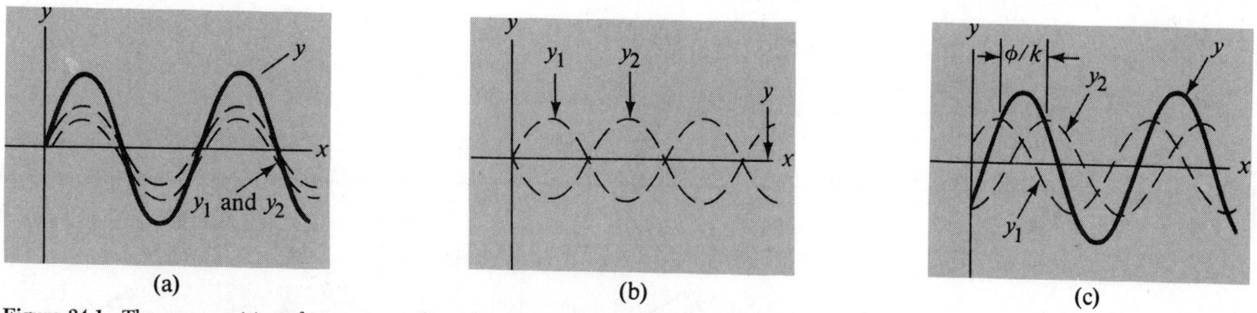

Figure 34.1 The superposition of two waves of amplitude y_1 and y_2. (a) When the two waves are in phase, the result is constructive interference. (b) When the two waves are 180° out of phase, the result is destructive interference. (c) When the phase angle lies in the range $0 < \phi < 180°$, the resultant y fails somewhere between that shown in (a) and that shown in (b).

every point. Finally, when the phase constant has an arbitrary value between 0 and π, as in Fig. 34.1c, the resultant wave has an amplitude whose value is somewhere between 0 and $2A_0$.

Interference of Sound Waves

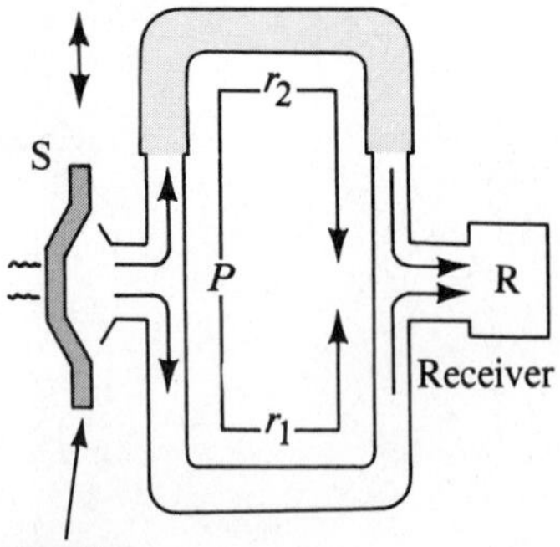

Figure 34.2 An acoustical system for demonstrating interference of sound waves. Sound from the speaker propagates into a tube and splits into two parts at *P*. The two waves, which superimpose at the opposite side, are detected at R. Note that the upper path length, r_2, can be varied by the sliding section.

One simple device for demonstrating interference of sound waves is illustrated in Fig. 34.2. Sound from a loudspeaker S is sent into a tube at *P*, where there is a T-shaped junction. Half the sound intensity travels in one direction and half in the opposite direction. Thus, the sound waves that reach the receiver R at the other side can travel along two different paths. The receiver may be a microphone whose output is amplified and fed into earphones or an oscilloscope. The total distance from the speaker to the receiver is called the *path length, r*. The path length for the lower path is fixed at r_1. Along the upper path, the path length r_2 can be varied by sliding the U-shaped tube, similar to that on a slide trombone. When the difference in the path lengths $\Delta r = |r_2 - r_1|$ is either zero or some integral multiple of the wavelength λ, the two waves reaching the receiver will be in phase and will interfere constructively, as in Fig. 34.1a. For this case, a maximum in the sound intensity will be detected at the receiver. If the path length r_2 is adjusted such that the path difference Δr is $\lambda/2, 3\lambda/2, \ldots, n\lambda/2$ (for n odd), the two waves will be exactly 180° out of phase at the receiver and hence will cancel each other. In this case of completely destructive interference, no sound will be detected at the receiver. This simple experiment is a striking illustration of the phenomenon of interference. In addition, it demonstrates the fact that a phase difference may arise between two waves generated by the same source when they travel along paths of unequal lengths.

It is often useful to express the path difference in terms of the phase difference ϕ between the two waves. Since a phase difference of 2π radians corresponds to a path difference of one wavelength, we arrive at the ratio $2\pi/\lambda = \phi/\Delta r$, or

Relationship between path difference and phase angle

$$\Delta r = \frac{\lambda}{2\pi}\phi \tag{34.2}$$

There are many other examples of interference phenomena in nature. Later, in Chapter 38, we shall describe several interesting interference effects involving light waves.

Example 34.1

Two speakers are driven by the same oscillator at a frequency of 2000 Hz. The speakers are separated by a distance of 3 m, as in Fig. 34.3. A listener is originally at a point *O* located 8 m away along the center line. How far must the listener walk, perpendicular to the center line, before reaching the first minimum in the sound intensity?

Solution: Since the speed of sound in air is 330 m/s and since $f = 2000$ Hz, the wavelength is given by

$$\lambda = \frac{v}{f} = \frac{330 \text{ m/s}}{2000 \text{ Hz}} = 0.165 \text{ m}$$

The first minimum occurs when the two waves reaching *P* are 180° out of phase, or when their path difference, $r_2 - r_1$, equals $\lambda/2$. Therefore, the path difference is

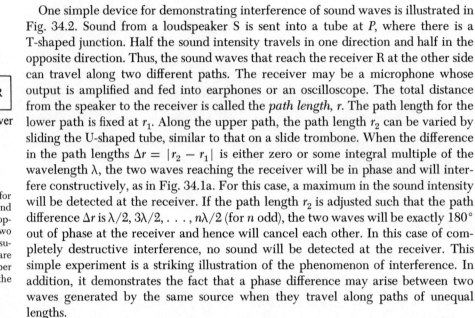

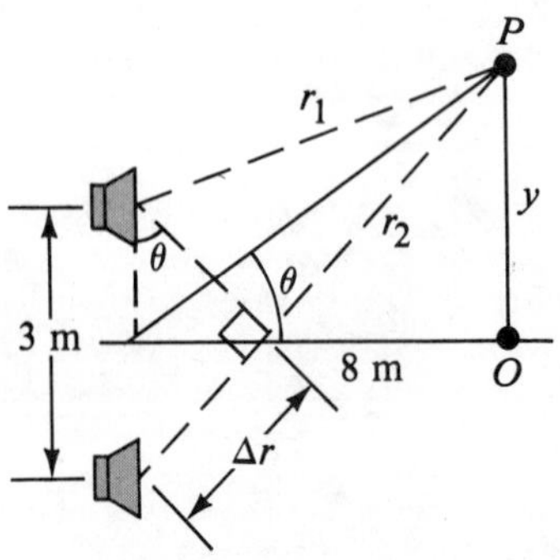

Figure 34.3 (Example 34.1)

given by

$$\Delta r = r_2 - r_1 = \frac{\lambda}{2} = \frac{0.165\text{ m}}{2} = 0.0825\text{ m}$$

From the small right triangle in Fig. 34.3, we see that, to a good approximation, $\sin\theta = \Delta r/3$ for small values of θ, or

$$\sin\theta = \frac{\Delta r}{3} = \frac{0.0825\text{ m}}{3\text{ m}} = 0.0275$$

$$\theta = 1.58°$$

From the large right triangle in Fig. 34.3, we find that $\tan\theta = y/8$, or

$$y = 8\tan\theta = 8\tan 1.58° = 0.22\text{ m}$$

That is, the listener will hear minima in the resultant sound intensity 22 cm to either side of the center line. If the listener remains at these positions, at what other frequencies will minima be heard?

Q1. For certain positions of the movable section in Fig. 34.2, there is no sound detected at the receiver, corresponding to destructive interference. This suggests that perhaps energy is somehow lost! What happens to the energy transmitted by the receiver?

Q2. Does the phenomenon of wave interference apply only to harmonic waves?

Q3. When two waves interfere constructively or destructively, is there any gain or loss in energy? Explain.

34.2 STANDING WAVES

If a stretched string is clamped at both ends, traveling waves will reflect from the fixed ends, creating waves traveling in both directions. The incident and reflected waves will combine according to the superposition principle.

Consider two sinusoidal waves with the same amplitude, frequency, and wavelength, but traveling in *opposite* directions. Their wave functions can be written

$$y_1 = A_0\sin(kx - \omega t) \qquad \text{and} \qquad y_2 = A_0\sin(kx + \omega t)$$

where y_1 represents a wave traveling to the right and y_2 represents a wave traveling to the left. Adding these two functions gives the resultant wave function y:

$$y = y_1 + y_2 = A_0\sin(kx - \omega t) + A_0\sin(kx + \omega t)$$

where $k = 2\pi/\lambda$ and $\omega = 2\pi f$, as usual. Using the trigonometric identity $\sin(a \pm b) = \sin a\cos b \pm \cos a\sin b$, this reduces to

Wave function for a standing wave

$$y = (2A_0\sin kx)\cos\omega t \qquad (34.3)$$

This expression represents the wave function of a *standing wave*. From this result, we see that a standing wave has an angular frequency ω and an amplitude given by $2A_0\sin kx$ (the quantity in the parentheses of Eq. 34.3). That is, every particle of the string vibrates in simple harmonic motion with the same frequency. However, the amplitude of motion of a given particle depends on x. This is in contrast to the situation involving a traveling harmonic wave, in which all particles oscillate with both the same amplitude and the same frequency.

Since the amplitude of the standing wave at any value of x is equal to $2A_0\sin kx$, we see that the *maximum* amplitude has the value $2A_0$. This occurs when the coordinate x satisfies the condition $\sin kx = 1$, or when

$$kx = \frac{\pi}{2}, \frac{3\pi}{2}, \frac{5\pi}{2}, \ldots$$

Since $k = 2\pi/\lambda$, the positions of maximum amplitude, called *antinodes*, are given by

Position of antinodes

$$x = \frac{\lambda}{4}, \frac{3\lambda}{4}, \frac{5\lambda}{4}, \ldots = \frac{n\lambda}{4} \tag{34.4}$$

where $n = 1, 3, 5, \ldots$ Note that *adjacent antinodes are separated by a distance of* $\lambda/2$. Similarly, the standing wave has a *minimum* amplitude of zero when x satisfies the condition $\sin kx = 0$, or when

$$kx = \pi, 2\pi, 3\pi, \ldots$$

giving

Position of nodes

$$x = \frac{\lambda}{2}, \lambda, \frac{3\lambda}{2}, \ldots = \frac{n\lambda}{2} \tag{34.5}$$

where $n = 1, 2, 3, \ldots$. These points of zero amplitude, called *nodes, are also spaced by* $\lambda/2$. The distance between a node and an adjacent antinode is $\lambda/4$.

A graphical description of the standing wave patterns produced at various times by two waves traveling in opposite directions is shown in Fig. 34.4. The upper part of each figure represents the individual traveling waves, and the lower part represents the standing wave patterns. The nodes of the standing wave are labeled N, and the antinodes are labeled A. At $t = 0$ (Fig. 34.4a), the two waves are identical spatially, giving a standing wave of maximum amplitude, $2A_0$. One quarter of a period later, at $t = T/4$ (Fig. 34.4b), the individual waves have moved one quarter of a wavelength (one to the right and the other to the left). At this time, the individual amplitudes are equal and opposite for all values of x, and hence the resultant wave has zero amplitude everywhere. At $t = T/2$ (Fig. 34.4c), the individual waves are again identical spatially, producing a standing wave pattern that is inverted relative to the $t = 0$ pattern.

It is instructive to describe the energy associated with the motion of a standing wave. To illustrate this point, consider a standing wave formed on a stretched string fixed at each end, as in Fig. 34.5. All points on the string oscillate vertically with the same frequency except for the nodes, which are stationary. Furthermore, the various points have different amplitudes of motion. Figure 34.5 represents snapshots of the standing wave at various times over one half of a cycle. Note that since the nodal points are stationary, no energy is transmitted along the string across the center

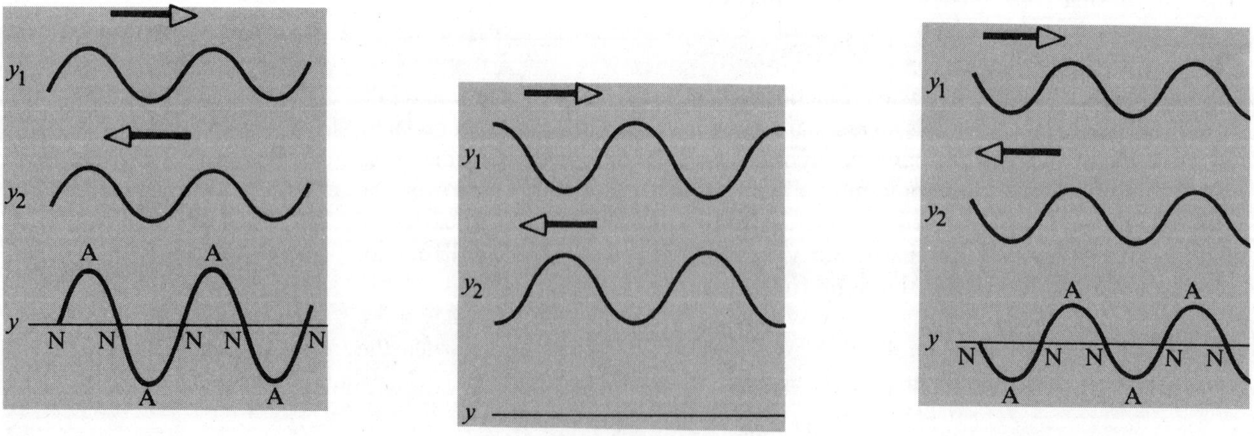

(a) $t = 0$ (b) $t = T/4$ (c) $t = T/2$

Figure 34.4 Standing wave patterns at various times produced by two waves of equal amplitude traveling in *opposite* directions. For the resultant wave y, the nodes (N) are points of zero displacement, and the antinodes (A) are points of maximum displacement.

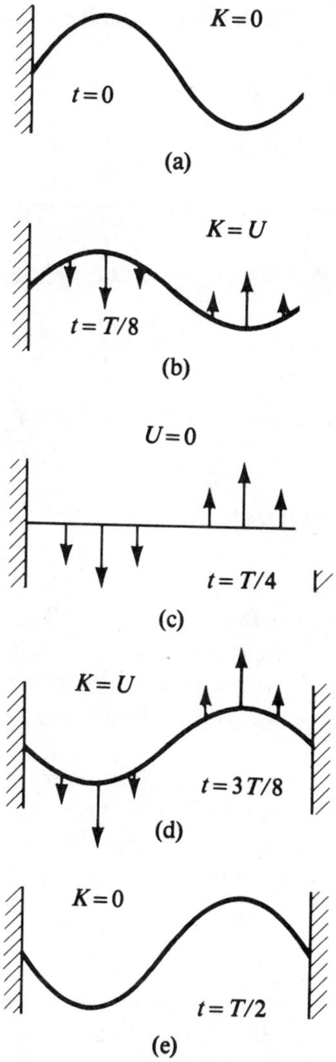

Figure 34.5 A standing wave pattern in a stretched string showing snapshots during one half of a cycle. (a) At $t = 0$, the string is momentarily at rest, and so $K = 0$ and all of the energy is potential energy U associated with the vertical displacements of the string segments (deformation energy). (b) At $t = T/8$, the string is in motion, and the energy is half kinetic and half potential, (c) At $t = T/4$, the string is horizontal (undeformed) and therefore $U = 0$ and all of the energy is kinetic. The motion continues as indicated, and ultimately the initial configuration (a) is repeated.

nodal point. For this reason, standing waves are often called *stationary waves*. Each point on the string executes simple harmonic motion in the vertical direction. That is, one can view the standing wave as a large number of oscillators vibrating parallel to each other. The energy of the vibrating string continuously alternates between elastic potential energy, at which time the string is momentarily stationary (Fig. 34.5a) and kinetic energy, at which time the string is horizontal and the particles have their maximum speed (Fig. 34.5c). The string particles have both potential energy and kinetic energy at intermediate times (Figs. 34.5b and 34.5d).

Example 34.2

Two waves traveling in opposite directions produce a standing wave. The individual wave functions are given by

$$y_1 = 4\sin(3x - 2t)\text{ cm}$$

$$y_2 = 4\sin(3x + 2t)\text{ cm}$$

where x and y are in cm. (a) Find the maximum displacement of the motion at $x = 2.3$ cm.

When the two waves are summed up, the result is a standing wave whose function is given by Eq. 34.3, with $A_o = 4$ cm and $k = 3\text{ cm}^{-1}$:

$$y = (2A_o \sin kx)\cos\omega t = (8\sin 3x)\cos\omega t\text{ cm}$$

Thus, the *maximum* displacement of the motion at $x = 2.3$ cm is given by

$$y_{\max} = 8\sin 3x]_{x=2.3} = 8\sin(6.9\text{ rad}) = 4.63\text{ cm}$$

(b) Find the positions of the nodes and antinodes.

Since $k = 2\pi/\lambda = 3\ \text{cm}^{-1}$, we see that $\lambda = 2\pi/3$ cm. Therefore, from Eq. 34.4 we find that the *antinodes* are located at

$$x = n\left(\frac{\pi}{6}\right)\text{cm} \qquad (n = 1,3,5, \ldots)$$

and from Eq. 34.5 we find that the *nodes* are located at

$$x = n\frac{\lambda}{2} = n\left(\frac{\pi}{3}\right)\text{cm} \qquad (n = 1,2,3, \ldots)$$

Q4. A standing wave is set up on a string as in Fig. 34.5. Explain why no energy is transmitted along the string.

Q5. What is common to *all* points (other than the nodes) on a string supporting a standing wave?

34.3 STANDING WAVES IN A STRING FIXED AT BOTH ENDS

Consider a string of length L that is fixed at both ends, as in Fig. 34.6. Standing waves are set up in the string by a continuous superposition of waves incident on and reflected from the ends. The string has a number of natural patterns of vibration, called *normal modes*. Each of these has a characteristic frequency; the frequencies are easily calculated.

First, note that the ends of the string must be nodes since these points are *fixed*. If the string is displaced at its midpoint and released, the vibration shown in Fig. 34.6b is produced, in which the center of the string is an antinode. For this normal mode, the length of the string equals $\lambda/2$ (the distance between nodes):

$$L = \lambda_1/2 \qquad \text{or} \qquad \lambda_1 = 2L$$

The next normal mode, of wavelength λ_2 (Fig. 34.6c), occurs when the length of the string equals one wavelength, that is, when $\lambda_2 = L$. The third normal mode (Fig. 34.6d) corresponds to the case where the length equals 1.5λ, that is, $\lambda_3 = 2L/3$. Therefore, in general, the wavelengths of the various normal modes can be conveniently expressed as

Wavelengths of normal modes

$$\lambda_n = 2L/n \qquad (n = 1,2,3, \ldots) \tag{34.6}$$

where the index n refers to the nth mode of vibration. The natural frequencies associated with these modes are obtained from the relationship $f = v/\lambda$, where the *wave speed* v *is the same for all frequencies*. Using Eq. 34.6, we find that the frequencies of the normal modes are given by

Frequencies of normal modes

$$f_n = \frac{v}{\lambda_n} = \frac{n}{2L}v \qquad (n = 1,2,3, \ldots) \tag{34.7}$$

Since $v = \sqrt{F/\mu}$, where F is the tension in the string and μ is its mass per unit length, we can also express the natural frequencies of a stretched string as[1]

Normal modes of a stretched string

$$f_n = \frac{n}{2L}\sqrt{F/\mu} \qquad (n = 1,2,3, \ldots) \tag{34.8}$$

The lowest frequency, corresponding to $n = 1$, is called the *fundamental* or the

[1] The laws governing the sound produced by a vibrating string were first published in 1636 by a Franciscan friar, Pére Mersenne, in a treatise entitled "Harmonie Universelle."

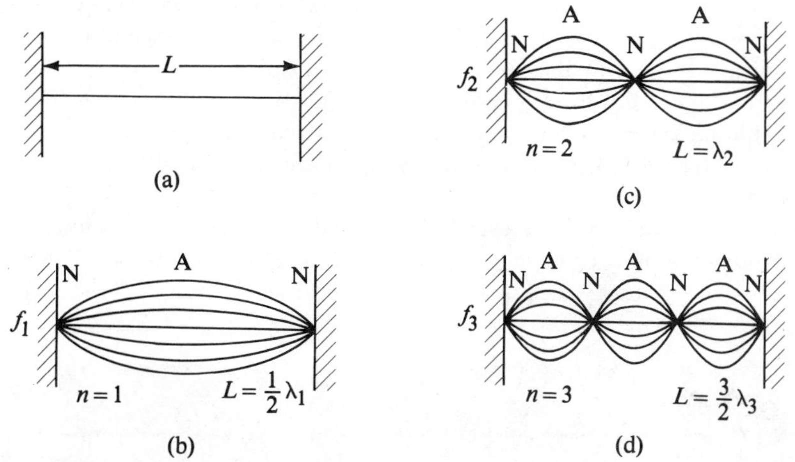

Figure 34.6 Standing waves in a string of length L fixed at both ends. The normal modes of vibration form a harmonic series: (b) the fundamental frequency, or first harmonic; (c) the second harmonic; and (d) the third harmonic.

fundamental frequency, f_1, and is given by

$$f_1 = \frac{1}{2L}\sqrt{F/\mu} \tag{34.9}$$

Clearly, the frequencies of the remaining modes (sometimes called *overtones*) are integral multiples of the fundamental frequency, that is, $2f_1$, $3f_1$, $4f_1$, and so on. These higher natural frequencies, together with the fundamental frequency, are seen to form a *harmonic series.* The fundamental, f_1, is the first harmonic; the frequency $f_2 = 2f_1$ is the second harmonic; the frequency f_n is the nth harmonic. In musical terms, the second harmonic is called the *first overtone,* the third harmonic is the *second overtone,* and so on.

We can obtain the above results in an alternative manner. Since we require that the string be fixed at $x = 0$ and $x = L$, the wave function $y(x,t)$ given by Eq. 34.3 must be *zero* at these points for *all* times. That is, the boundary conditions require that $y(0,t) = 0$ and $y(L,t) = 0$. Since $y = (2A_0 \sin kx) \cos\omega t$, the first condition, $y(0,t) = 0$, is automatically satisfied because $\sin kx = 0$ at $x = 0$. To meet the second condition, $y(L,t) = 0$, we require that $\sin kL = 0$. This condition is satisfied when the angle kL equals an integral multiple of π (180°). Therefore, the allowed vaues of k are[2]

$$k_nL = n\pi \qquad (n = 1,2,3,\ldots) \tag{34.10}$$

Since $k_n = 2\pi/\lambda_n$, we find that

$$(2\pi/\lambda_n)L = n\pi \qquad \text{or} \qquad \lambda_n = 2L/n$$

which is identical to Eq. 34.6.

When a stretched string is distorted such that its initial shape corresponds to any one of its harmonics, after being released it will vibrate at the frequency of that harmonic. However, if the string is struck or bowed, the resulting vibration will include frequencies of various harmonics, including the fundamental. Waves of the "wrong" frequency destroy each other in traveling on a string fixed at both ends. In effect, the string "selects" the normal-mode frequencies when disturbed by a nonharmonic disturbance (which happens, for example, when a guitar string is plucked).

Figure 34.7c shows a stretched string vibrating with its first and second harmonics simultaneously. In this figure, the combined vibration is the superposition of the two

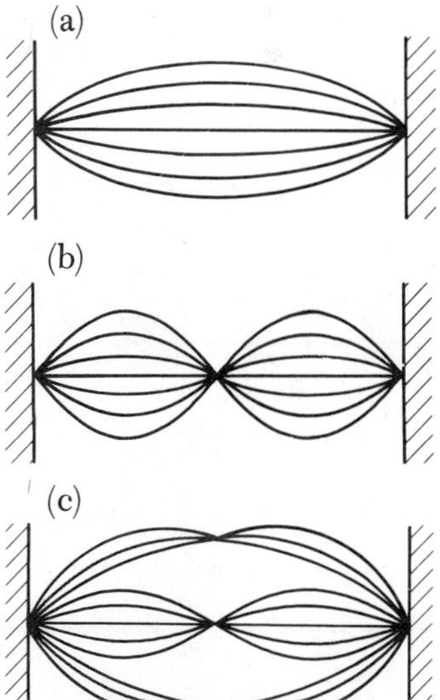

Figure 34.7 Part (c) shows snapshots of a stretched string vibrating in its first harmonic (a) and second harmonic (b) simultaneously.

[2] We exclude $n = 0$ since this corresponds to the trivial case where no wave exists ($k = 0$).

vibrations shown in Figs. 34.7a and 34.7b. The large loop corresponds to the fundamental frequency of vibration, f_1, and the smaller loops correspond to the second harmonic, f_2. In general, the resulting motion, or displacement, can be described by a superposition of the various harmonic wave functions, with different frequencies and amplitudes. Hence, the sound that one hears corresponds to a complex waveform associated with these various modes of vibration. We shall return to this point in Section 34.8.

The frequency, or pitch, of a stringed instrument can be changed either by varying the tension F or by changing the length L. For example, the tension in the strings of guitars and violins is varied by turning pegs located on the neck of the instrument.

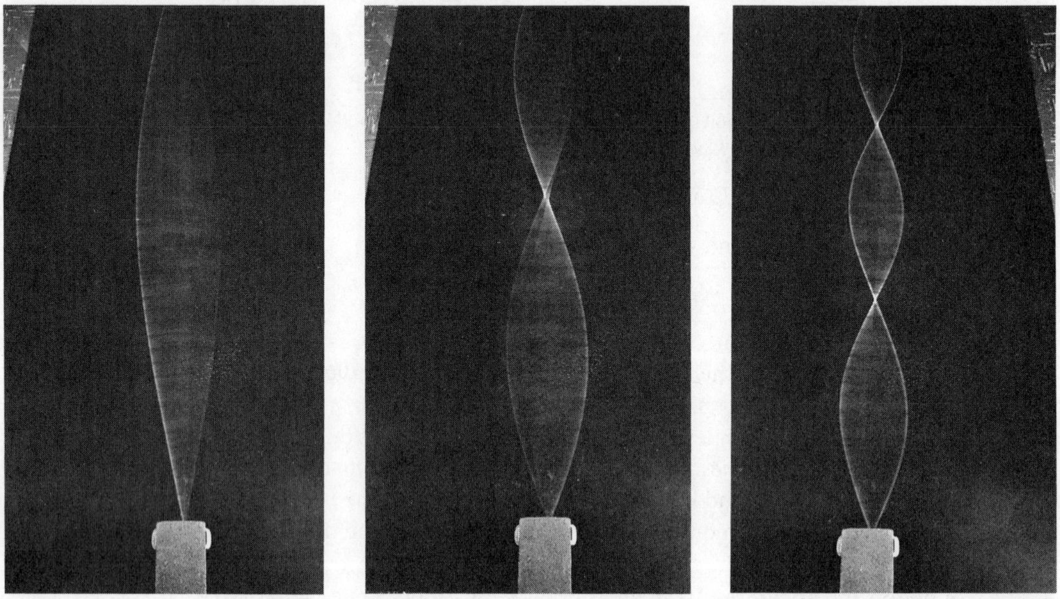

Photographs of standing waves. As one end of the tube is moved from side to side with increasing frequency, patterns with more and more loops are formed; only certain definite frequencies will produce fixed patterns. (Photos, Education Development Center, Newton, Mass.)

As the tension is increased, the frequency of the normal modes increases according to Eq. 34.8. Once the instrument is "tuned," the player varies the pitch by moving his or her fingers along the neck, thereby changing the length of the vibrating portion of the string. As the length is shortened, the pitch increases, since the normal-mode frequencies are inversely proportional to string length.

Example 34.3

The C note of the C-major scale on a piano has a fundamental frequency of 264 Hz, and the A note has a fundamental frequency of 440 Hz. (a) Calculate the frequencies of the first two overtones of the C note.

Since $f_1 = 264$ Hz, we can use Eqs. 34.8 and 34.9 to find the frequencies f_2 and f_3:

$$f_2 = 2f_1 = 528 \text{ Hz}$$
$$f_3 = 3f_1 = 792 \text{ Hz}$$

(b) If the two piano strings for the A and C notes are assumed to have the same mass per unit length and the same length, determine the ratio of tensions in the two strings.

Using Eq. 34.8 for the two strings vibrating at their fundamental frequencies gives

$$f_{1A} = \frac{1}{2L}\sqrt{F_A/\mu} \quad \text{and} \quad f_{1C} = \frac{1}{2L}\sqrt{F_C/\mu}$$

$$f_{1A}/f_{1C} = \sqrt{F_A/F_C}$$
$$F_A/F_C = (f_{1A}/f_{1C})^2 = (440/264)^2 = 2.78$$

(c) While the string densities are, in fact, equal, the A string is 64% as long as the C string. What is the ratio of their tensions?

$$f_{1A}/f_{1C} = (L_C/L_A)\sqrt{F_A/F_C} = (100/64)\sqrt{F_A/F_C}$$
$$F_A/F_C = (0.64)^2(440/264)^2 = 1.14$$

34.4 RESONANCE

We have seen that a system such as a stretched string is capable of oscillating in one or more natural modes of vibration. *If a periodic force is applied to such a system, the resulting amplitude of motion of the system will be larger when the frequency of the applied force is equal or nearly equal to one of the natural frequencies of the system* than when the driving force is applied at some other frequency. We have already discussed this phenomena, known as *resonance,* for mechanical systems and for alternating current circuits.

The corresponding natural frequencies of oscillation of the system are often referred to as *resonant frequencies.* The resonance phenomenon is of great importance in the production of musical sounds. At the atomic level, the electrons and nuclei of atoms and molecules exhibit resonant behavior when exposed to certain frequencies of electromagnetic radiation and applied magnetic fields.

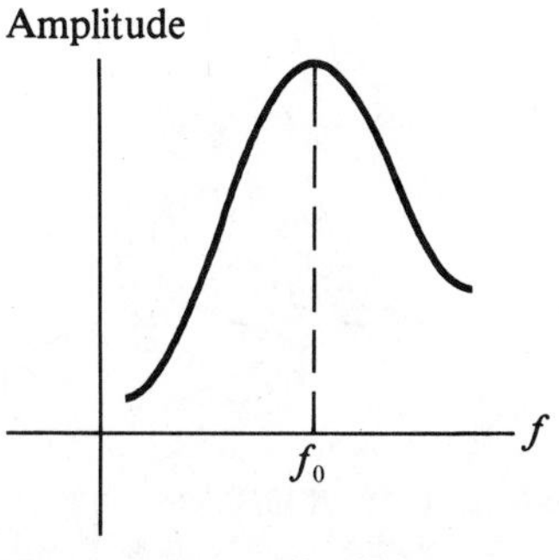

Figure 34.8 Amplitude (response) versus driving frequency for an oscillating system. The amplitude is a maximum at the resonance frequency, f_0.

Whenever a system capable of oscillating is driven by a periodic force, or series of impulses, the resulting amplitude of motion will be large only when the frequency of the driving force is nearly equal to one of the resonant frequencies of the system. Figure 34.8 shows the response of a system to various frequencies, where the peak of the curve represents the resonant frequency, f_0. Note that the amplitude is largest when the frequency of the driving force equals the resonant frequency. When the frequency of the driving force exactly matches one of the resonant frequencies, the amplitude of the motion will be limited by friction in the system. Once maximum amplitude is reached, the work done by the periodic force is used to overcome friction. A system is said to be *weakly damped* when the amount of friction is small, corresponding to a large amplitude of motion when driven at one of its resonant frequencies. The oscillations in such a system will persist for a long time after the driving force is removed. On the other hand, a system with considerable friction, that is, one that is *strongly damped,* will undergo small amplitude oscillations which will decrease rapidly with time once the driving force is removed.

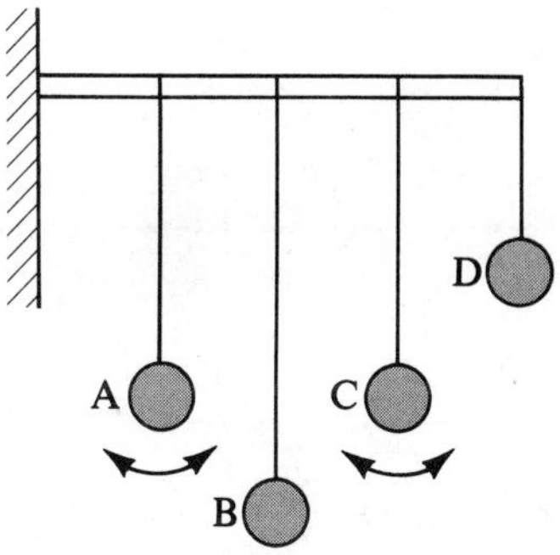

Figure 34.9 If pendulum A is set into oscillation, only pendulum C, whose length is close to the length of A, will eventually oscillate with large amplitude, or resonate.

Examples of Resonance

A playground swing is a pendulum with a natural frequency which depends on its length. Whenever we push a child in a swing with a series of regular impulses, the swing will go higher if the frequency of the periodic force equals the natural frequency of the swing. One can demonstrate a similar effect by suspending several pendula of different lengths from a horizontal support, as in Fig. 34.9. If pendulum A is set into oscillation, the other pendula will soon begin to oscillate as a result of the longitudinal waves transmitted along the beam. However, you will find that those pendula, such as C, whose length is close to the length of A will oscillate with a much larger amplitude than those whose length is much different from the length of A, such as B and D. This is because the natural frequency of C is nearly the same as the driving frequency associated with A.

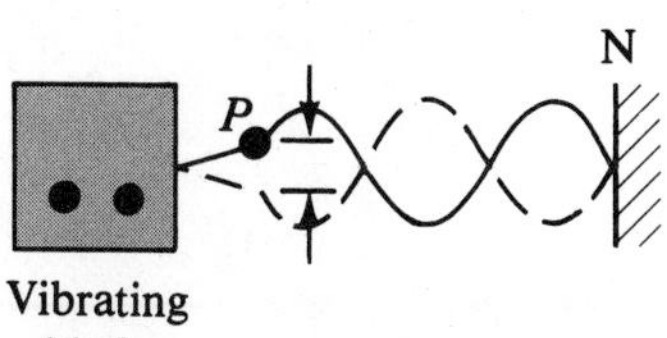

Figure 34.10 Standing waves are set up in a stretched string having one end connected to a vibrating blade when the natural frequencies of the string are nearly the same as those of the vibrating blade.

Next, consider a stretched string fixed at one end and connected at the opposite end to a vibrating blade as in Fig. 34.10. As the blade oscillates, transverse waves sent down the string are reflected from the fixed end. As we found in Section 34.3, the string has natural frequencies of vibration which are determined by its length, tension, and mass per unit length (Eq. 34.8). When the frequency of the vibrating blade equals one of the natural frequencies of the string, standing waves will be produced and the string will vibrate with a large amplitude. In this case, the wave being generated by the vibrating blade is *in phase* with the wave that has been reflected at the fixed end, and so the string absorbs energy from the blade at resonance. Once the amplitude of the standing-wave oscillations reaches a maximum, the energy delivered by the blade and absorbed by the system is lost because of the damping forces. Note that the fixed end is a node, and the point P, which is near the end connected to the vibrating blade, is very nearly a node, since the amplitude of the blade's motion is small compared with that of the string.

As a final example of resonance, consider two identical tuning forks mounted on separate hollow boxes (Fig. 34.11). The hollow boxes augment the sound wave intensity generated by the vibrating tuning forks. If tuning fork A is set into vibration (by someone's striking it, say), tuning fork B will be set into vibration as longitudinal sound waves are received from A. The frequencies of vibration of A and B will be the same, assuming the tuning forks are identical. The energy exchange, or resonance behavior, will not occur if the two have different natural frequencies of vibration. One can test this by offsetting the natural frequency of the receiving fork B by placing a bit of wax on its tip.

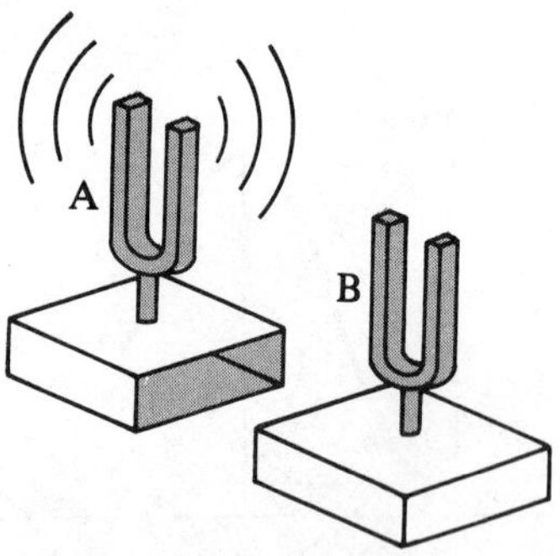

Figure 34.11 If tuning fork A is set into vibration, tuning fork B will eventually vibrate at the same frequency, or resonate, if the two forks are identical.

34.5 STANDING WAVES IN AIR COLUMNS

Standing longitudinal waves can be set up in a tube of air, such as an organ pipe, as the result of interference between longitudinal waves traveling in opposite directions. The phase relationship between the incident wave and the wave reflected from one end depends on whether that end is open or closed. This is analogous to the phase relationships between incident and reflected transverse waves at the ends of a string. *The closed end of an air column is a displacement node,* just as the fixed end of a vibrating string is a displacement node. As a result, at a closed end of a tube of air, the reflected wave is 180° out of phase with the incident wave. Furthermore, since the pressure wave is 90° out of phase with the displacement wave (Section 33.2), *the closed end of an air column corresponds to a pressure antinode* (that is, a point of maximum pressure variation).

If the end of an air column is open to the atmosphere, the air molecules have complete freedom of motion. Therefore, the wave reflected from an open end is nearly in phase with the incident wave when the tube's diameter is small relative to the wavelength of the sound. Consequently, *the open end of an air column is approximately a displacement antinode and a pressure node.*

Strictly speaking, the open end of an air column is not exactly an antinode. When a condensation reaches an open end, it does not reach full expansion until it passes somewhat beyond the end. For a thin-walled tube of circular cross section, this end correction is about $0.6R$, where R is the tube's radius. Hence, the effective length of the tube is somewhat longer than the true length L.

The first three modes of vibration of a pipe open at both ends are shown in Fig. 34.12a. By directing air against an edge at the left, longitudinal standing waves are formed and the pipe resonates at its natural frequencies. All modes of vibration are excited simultaneously (although not with the same amplitude). Note that the ends are displacement antinodes (approximately). In the fundamental mode, the wavelength is twice the length of the pipe, and hence the frequency of the fundamental, f_1, is given by $v/2L$. Similarly, one finds that the frequencies of the overtones are $2f_1, 3f_1, \ldots$ Thus, *in a pipe open at both ends, the natural frequencies of vibration form a harmonic series, that is, the overtones are integral multiples of the fundamental frequency.* Since all harmonics are present, we can express the natural frequencies of vibration as

Natural frequencies of a pipe open at both ends

$$f_n = n\frac{v}{2L} \qquad (n = 1,2,3, \ldots) \tag{34.11}$$

where v is the speed of sound in air.

If a pipe is closed at one end and open at the other, the closed end is a displace-

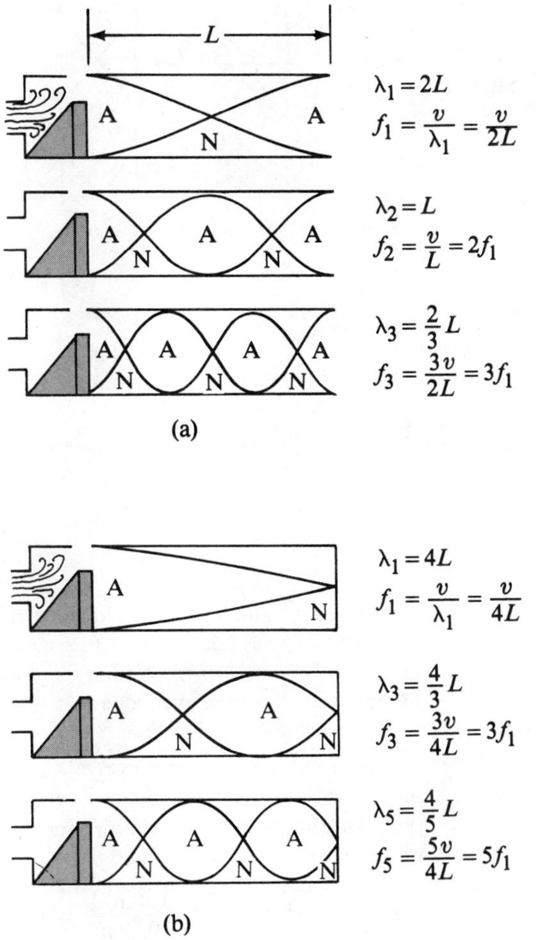

Figure 34.12 (a) Standing longitudinal waves in an organ pipe open at both ends. The natural frequencies which form a harmonic series are $f_1, 2f_1, 3f_1, \ldots$. (b) Standing longitudinal waves in an organ pipe closed at one end. Note that only the *odd* harmonics are present, and so the natural frequencies are $f_1, 3f_1, 5f_1, \ldots$.

ment node (Fig. 34.12b). In this case, the wavelength for the fundamental mode is four times the length of the tube. Hence, the fundamental, f_1, is equal to $v/4L$, and the frequencies of the overtones are equal to $3f_1, 5f_1, \ldots$. That is, *in a pipe closed at one end, only odd harmonics are present,* and these are given by

$$f_n = n\frac{v}{4L} \qquad (n = 1,3,5, \ldots) \qquad (34.12)$$

Natural frequencies of a pipe closed at one end

Example 34.4

A pipe has a length of 2.46 m. (a) Determine the frequencies of the fundamental and the first two overtones if the pipe is open at each end. Take $v = 344$ m/s as the speed of sound in air.

The fundamental frequency of an open pipe is

$$f_1 = \frac{v}{2L} = \frac{344 \text{ m/s}}{2(2.46 \text{ m})} = 70 \text{ Hz}$$

Since all harmonics are present, the first and second overtones are given by $f_2 = 2f_1 = 140$ Hz and $f_3 = 3f_1 = 210$ Hz.

(b) What are the three frequencies determined in (a) if the pipe is closed at one end?

The fundamental frequency of a pipe closed at one end is

$$f_1 = \frac{v}{4L} = \frac{344 \text{ m/s}}{4(2.46 \text{ m})} = 35 \text{ Hz}$$

In this case, only odd harmonics are present, and so the first and second overtones have frequencies given by $f_3 = 3f_1 = 105$ Hz and $f_5 = 5f_1 = 175$ Hz.

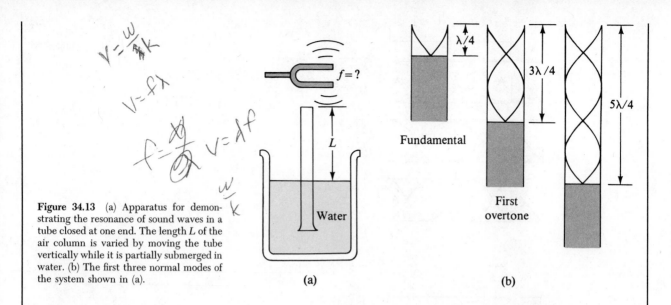

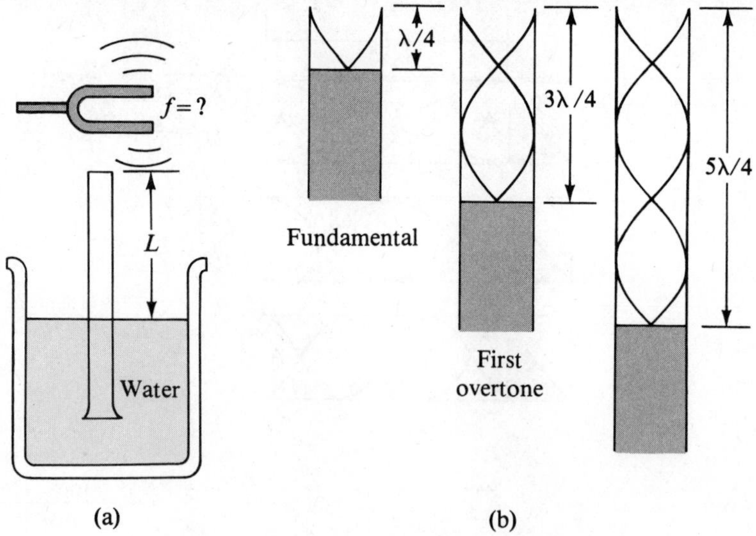

Figure 34.13 (a) Apparatus for demonstrating the resonance of sound waves in a tube closed at one end. The length L of the air column is varied by moving the tube vertically while it is partially submerged in water. (b) The first three normal modes of the system shown in (a).

(c) For the case of the open pipe, how many harmonics are present in the normal human hearing range (20 to 20 000 Hz)?

Since all harmonics are present, $f_n = nf_1$. Hence, the highest frequency corresponds to $n = 20\,000/70 = 285$, so that 285 harmonics are present. Actually, only the first few harmonics will have sufficient amplitude to be heard.

Example 34.5

A simple apparatus for demonstrating resonance in a tube is described in Fig. 34.13a. A long, vertical, open tube is partially submerged in a beaker of water, and a vibrating tuning fork of unknown frequency is placed near the top. The length of the air column, L, is adjusted by moving the tube vertically. The sound waves generated by the fork are reinforced when the length of the column corresponds to one of the resonant frequencies of the tube. The smallest value of L for which a peak occurs in the sound intensity is 9 cm. From this measurement, determine the frequency of the tuning fork and the value of L for the next two resonant modes.

Solution: Since this setup represents a pipe closed at one end, the fundamental has a frequency of $v/4L$ (Fig. 34.13b). Taking $v = 344$ m/s for the speed of sound in air and $L = 0.09$ m, we get

$$f_1 = \frac{v}{4L} = \frac{344 \text{ m/s}}{4(0.09 \text{ m})} = 956 \text{ Hz}$$

From this information about the fundamental mode, we see that the wavelength is given by $\lambda = 4L = 0.36$ m. Since the frequency of the source is constant, we see that the next two resonance modes (Fig. 34.13b) correspond to lengths of $3\lambda/4 = 0.27$ m and $5\lambda/4 = 0.45$ m.

Q6. Some singers claim to be able to shatter a wine glass by maintaining a certain pitch in their voice over a period of several seconds. What mechanism causes the glass to break?

Q7. What limits the amplitude of motion of a real vibrating system that is driven at one of its resonant frequencies?

Q8. If the temperature of the air in organ pipe increases, what happens to the resonance frequencies?

Q9. Explain why your voice seems to sound better than usual when you sing in the shower.

Q10. What is the purpose of the slide on a trombone or the valves on a trumpet?

Q11. Explain why all harmonics are present in an organ pipe open at both ends, but only the odd harmonics are present in a pipe closed at one end.

34.6 STANDING WAVES IN RODS AND PLATES

Standing wave vibrations can also be set up in rods and plates. If a rod is clamped in the middle and stroked at one end, it will undergo longitudinal vibrations as described in Fig. 34.14a. Note that the broken lines in Fig. 34.14 represents *longitudinal* displacements of various parts of the rod. The midpoint is a displacement node since it is fixed by the clamp, whereas the ends are displacement antinodes since they are free to vibrate. This is analogous to vibrations set up in a pipe open at each end. The broken lines in Fig. 34.14a represent the fundamental mode for which the wavelength is $2L$ and the frequency is $v/2L$, where v is the speed of longitudinal waves in the rod. Other modes may be excited by clamping the rod at different points. For example, the second harmonic (Fig. 34.14b) is excited by clamping the rod at a point that is a distance $\lambda/4$ away from one end.

Two-dimensional vibrations can be set up in a flexible membrane stretched over a circular hoop, such as a drumhead. As the membrane is struck at some point, wave pulses traveling toward the fixed boundary are reflected many times. The resulting sound is not melodious, but rather explosive in nature. This is because the vibrating drumhead and the drum's hollow interior produce a disorganized set of waves, which create an unrecognizable note when they reach a listener's ear. This is in contrast to wind and stringed instruments, which produce melodious, recognizable notes.

Some possible normal modes of oscillation of a vibrating, two-dimensional, circular membrane are shown in Fig. 34.15. Note that the nodes are *lines* rather than points, which was the case for a vibrating string. The fixed outer perimeter is one such nodal line. Some other nodal lines are indicated with arrows. The lowest mode of vibration with frequency f_1 (the fundamental) is a symmetric mode with one nodal line, the circumference of the membrane. Note that the other possible modes of vibration are *not* integral multiples of f_1; hence the normal frequencies *do not* form a harmonic series. When a drum is struck, many of these modes are excited simultaneously. However, the higher-frequency modes dampen out more rapidly. With this information, one can understand why the drum is a nonmelodious instrument.

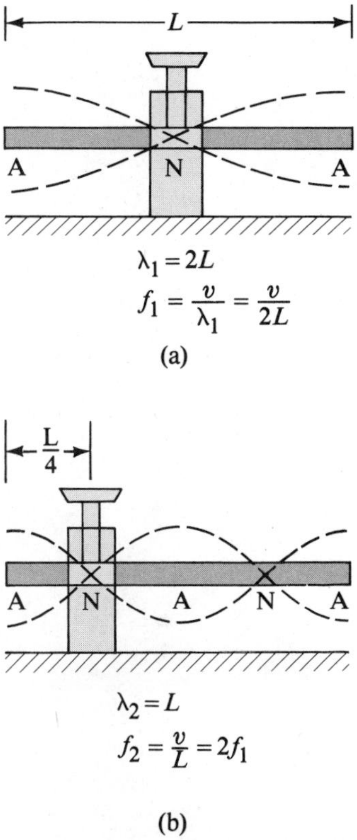

Figure 34.14 Normal longitudinal vibrations of a rod of length L (a) clamped at the middle and (b) clamped at a distance $L/4$ from one end.

34.7 BEATS: INTERFERENCE IN TIME

The interference phenomena we have been dealing with so far involve the superposition of two or more waves with the same frequency traveling in opposite directions. Since the resultant waveform in this case depends on the coordinates of the disturbed medium, we can refer to the phenomenon as *spatial interference.* Standing waves in strings and pipes are common examples of spatial interference.

We now consider another type of interference effect, one that results from the superposition of two waves with slightly *different frequencies* traveling in the *same direction.* In this case, when the two waves are observed at a given point, they are periodically in and out of phase. That is, there is an alternation in time between constructive and destructive interference. Thus, we refer to this phenomenon as *interference in time.* For example, if two tuning forks of slightly different frequencies are struck, one hears a sound of pulsating intensity, called *beats. Beats can therefore be defined as the periodic variation in intensity at a given point due to the superposition of two waves having slightly different frequencies.* The number of beats one hears per second, or *beat frequency,* equals the difference in frequency between the two sources. The maximum beat frequency that the human ear can detect is about 7 beats/s.

Definition of beats

When the beat frequency exceeds this value, it blends indistinguishably with the

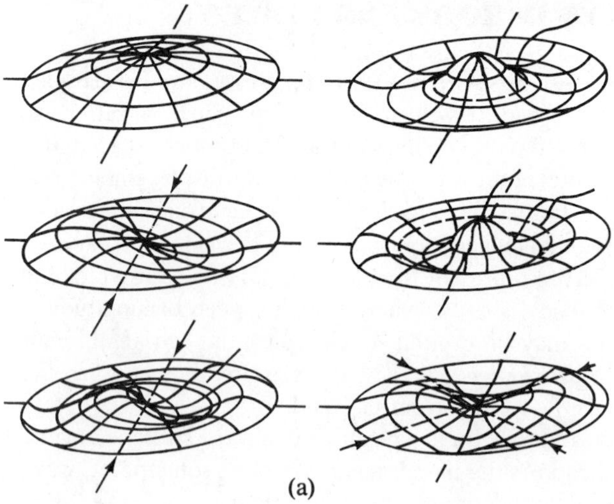

(a)

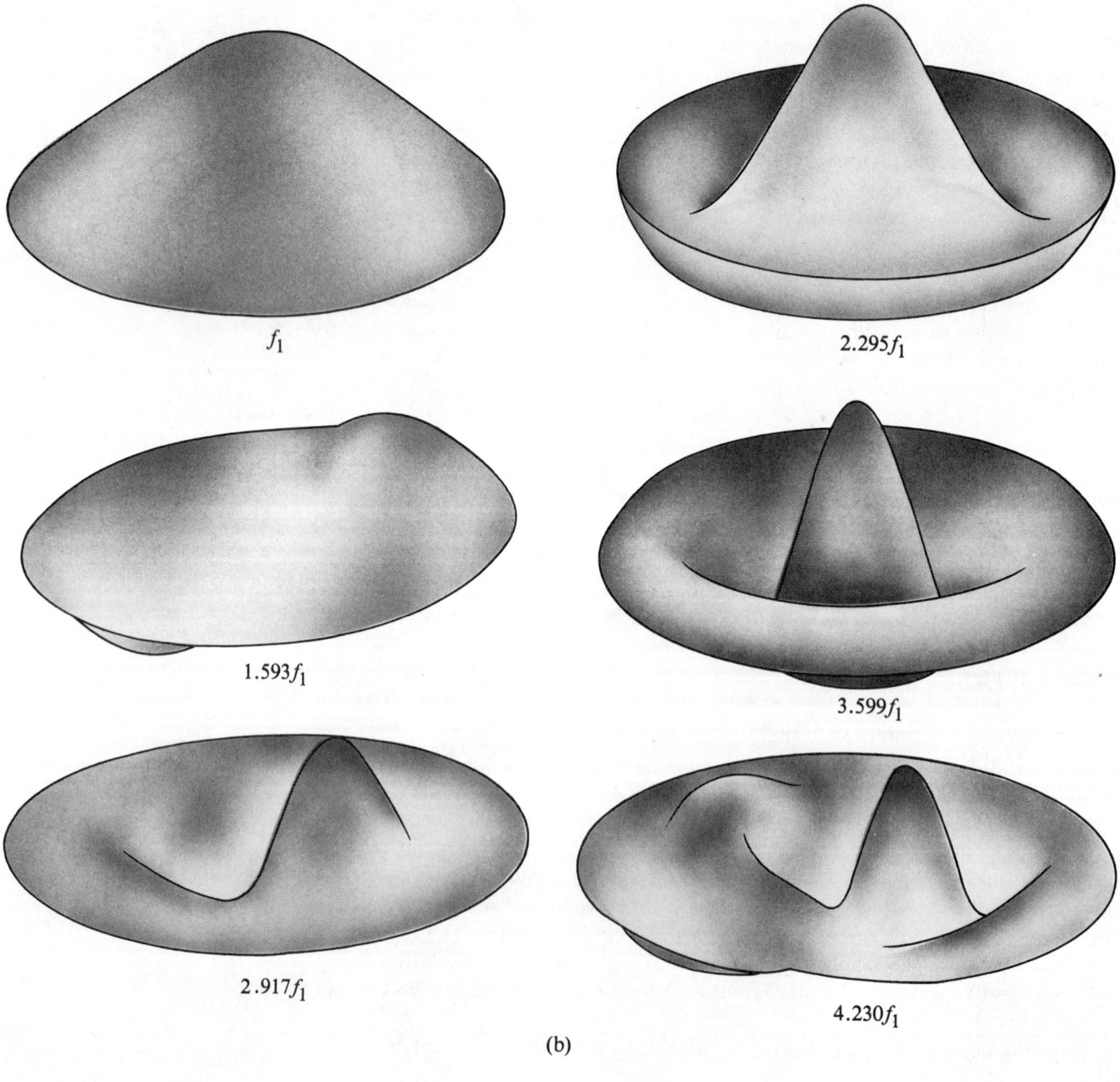

(b)

compound sounds producing the beats. One can use this effect to tune a stringed instrument, such as a piano, by beating a note against a reference tone of known frequency. The string can then be adjusted to equal the frequency of the reference by tightening or loosening it until no beats are heard.

Consider two waves with equal amplitudes traveling through a medium in the *same* direction, but with slightly different frequencies, f_1 and f_2. We can represent the displacement that each wave would produce at a point as

$$y_1 = A_o \cos 2\pi f_1 t \qquad \text{and} \qquad y_2 = A_o \cos 2\pi f_2 t$$

Using the superposition principle, we find that the resultant displacement at that point is given by

$$y = y_1 + y_2 = A_o(\cos 2\pi f_1 t + \cos 2\pi f_2 t)$$

It is convenient to write this in a form that uses the trigonometric identity

$$\cos a + \cos b = 2\cos\left(\frac{a-b}{2}\right)\cos\left(\frac{a+b}{2}\right)$$

Letting $a = 2\pi f_1 t$ and $b = 2\pi f_2 t$, we find that

$$y = 2A_o \cos 2\pi\left(\frac{f_1 - f_2}{2}\right)t \cos 2\pi\left(\frac{f_1 + f_2}{2}\right)t \qquad (34.13)$$

Resultant of two waves of different frequencies but equal amplitude

Graphs demonstrating the individual waveforms as well as the resultant wave are shown in Fig. 34.16. From Eq. 34.13, we see that the resultant vibration at a point

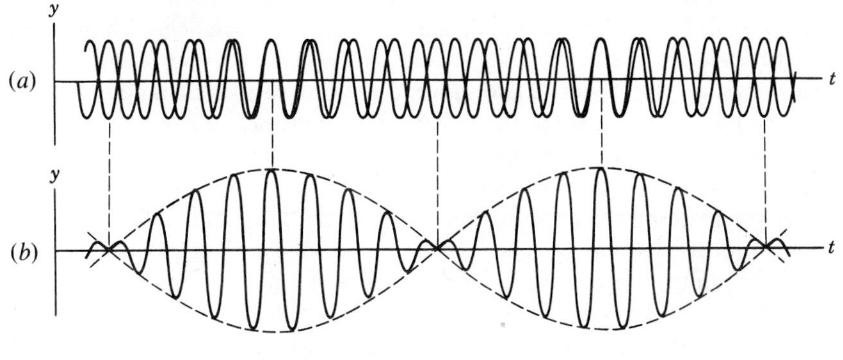

Figure 34.16 Beats are formed by the combination of two waves of slightly different frequencies traveling in the same direction. (a) The individual waves. (b) The combined wave has an amplitude (broken line) that oscillates in time. (*From R. Resnick and D. Halliday,* Physics, *New York, Wiley, 1977; by permission of the publisher*)

has an effective frequency equal to the average frequency, $(f_1 + f_2)/2$, and an amplitude given by

$$A = 2A_o \cos 2\pi\left(\frac{f_1 - f_2}{2}\right)t \qquad (34.14)$$

That is, the *amplitude varies in time* with a frequency given by $(f_1 - f_2)/2$. When f_1 is close to f_2, this amplitude variation is slow, as illustrated by the envelope (broken line) of the resultant waveform in Fig. 34.16b.

Note that a beat, or a maximum in amplitude, will be detected whenever

$$\cos 2\pi\left(\frac{f_1 - f_2}{2}\right)t = \pm 1$$

Figure 34.15 (a) Six normal modes of vibration of a circular membrane (drumhead) fixed at its perimeter. Arrows indicate the nodal lines. (*From P. M. Morse,* Vibration and Sound, *2nd edn., New York, McGraw-Hill, 1948, with permission of the publishers*) (b) Representation of some natural modes of vibration on a circular membrane fixed at its perimeter. Note that the frequencies of vibration *do not* form a harmonic series. (*Courtesy of Ron Hipschman, San Francisco State University*)

That is, there will be *two* maxima in each cycle. Since the amplitude varies with frequency as $(f_1 - f_2)/2$, the number of beats per second, or the beat frequency f_b, is twice this value. That is,

Beat frequency

$$f_b = f_1 - f_2 \tag{34.15}$$

For instance, if two tuning forks vibrate individually at frequencies of 438 Hz and 442 Hz, the resultant sound wave of the combination would have a frequency of 440 Hz (the fundamental of a piano's A note) and a beat frequency of 4 Hz. That is, the listener would hear the 440-Hz sound wave go through an intensity maximum four times every second.

Q12. Explain how a musical instrument such as a piano may be tuned using the phenomenon of beats.

34.8 COMPLEX WAVES

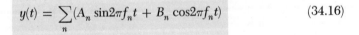

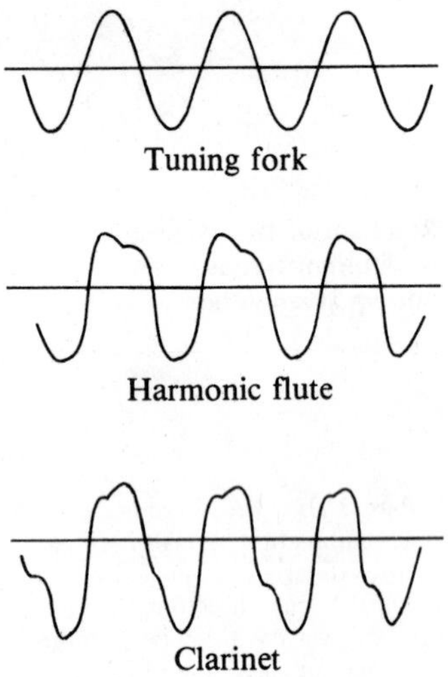

Figure 34.17 Waveform produced by (a) a tuning fork, (b) harmonic flute, and (c) a clarinet, each at approximately the same frequency. *Adapted from C. A. Culver,* Musical Acoustics, *4th ed., New York, McGraw-Hill, 1956, p. 128.*

The sound wave patterns produced by most instruments are very complex. Some characteristic waveforms produced by a tuning fork, a harmonic flute, and a clarinet, each playing the same note, are shown in Fig. 34.17. Although each instrument has its own characteristic pattern, Fig. 34.17 shows that each of the waveforms is periodic in nature. Furthermore, note that the tuning fork produces only one harmonic (the fundamental frequency), whereas the flute and clarinet produce many frequencies, which include the fundamental and various harmonics. Thus, the complex waveforms produced by a violin or clarinet, and the corresponding richness of musical tones, are the result of the superposition of various harmonics. This is in contrast to the drum, in which the overtones do not form a harmonic series.

The problem of analyzing complex waveforms appears at first sight to be a rather formidable task. However, if the waveform is periodic, it can be represented with arbitrary precision by the combination of a sufficiently large number of sinusoidal waves that form a harmonic series. In fact, one can represent any periodic waveform (or function) as a series of sine and cosine terms by using a mathematical technique based on *Fourier's theorem.*[3] The corresponding sum of terms which represents the periodic waveform is called a *Fourier series.*

Let $y(t)$ be any function that is periodic in time with period T, such that $y(t + T) = y(t)$. Fourier's theorem states that this function can be written

Fourier's theorem

$$y(t) = \sum_n (A_n \sin 2\pi f_n t + B_n \cos 2\pi f_n t) \tag{34.16}$$

where the lowest frequency $f_1 = 1/T$.

The higher frequencies are integral multiples of the fundamental, so that $f_n = nf_1$. The coefficients A_n and B_n represent the amplitudes of the various waves. The amplitude of the nth harmonic is proportional to $\sqrt{A_n^2 + B_n^2}$, and its intensity is proportional to $A_n^2 + B_n^2$.

Figure 34.18 represents a harmonic analysis of the waveforms shown in Fig. 34.17. Note the variation of relative intensity with harmonic content for the flute and clarinet. In general, any pleasing music sound (that is, one with good tone quality) contains various harmonics with varying relative intensities.

[3] Developed by Jean Baptiste Joseph Fourier (1786–1830).

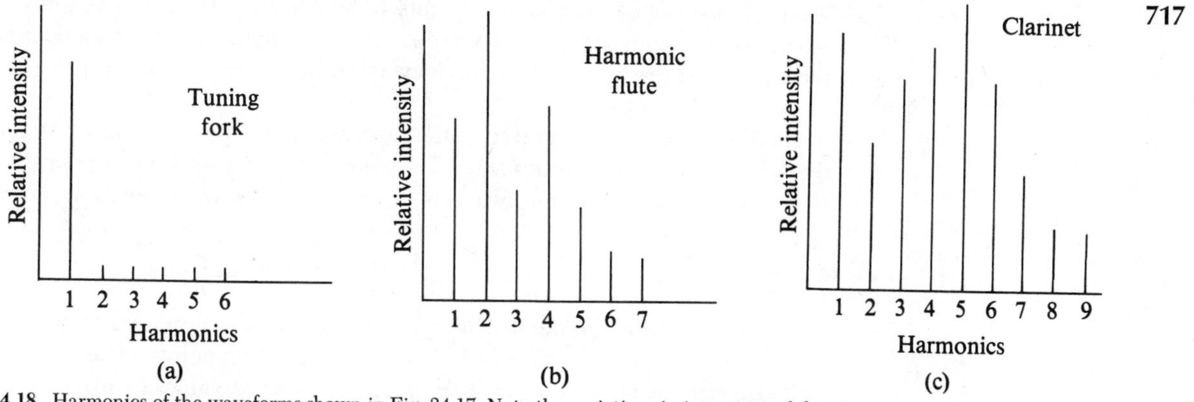

Figure 34.18 Harmonics of the waveforms shown in Fig. 34.17. Note the variations in intensities of the various harmonics. *Adapted from C. A. Culver,* Musical Acoustics, *4th ed., New York, McGraw-Hill, 1956.*

As an example of *harmonic synthesis,* consider the periodic square wave shown in Fig. 34.19. Note that the square wave is synthesized by a series of *odd* harmonics of the fundamental. The series contains only sine functions (that is, $B_n = 0$ for all n). Only the first four odd harmonics and their respective amplitudes are shown. One obtains a better fit to the true waveform by adding more harmonics.

Using modern technology, one can generate musical sounds electronically by mixing any number of harmonics with varying amplitudes. These widely used electronic music synthesizers are able to produce an infinite variety of musical tones and repetitive sequences.

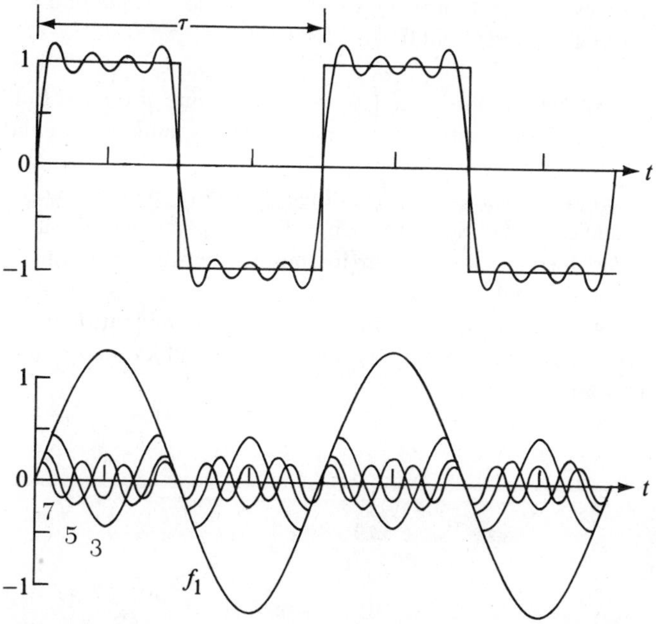

Figure 34.19 Harmonic synthesis of a square wave, which can be represented by the sum of odd harmonics of the fundamental. (*From M. L. Warren,* Introductory Physics, *San Francisco, W. H. Freeman, 1979, p. 178; by permission of the publisher*)

34.9 SUMMARY

When two waves with equal amplitudes and frequencies superimpose, the resultant wave has an amplitude that depends on the phase angle ϕ between the two waves. *Constructive interference* occurs when the two waves are *in phase* everywhere, corresponding to $\phi = 0, 2\pi, 4\pi, \ldots$. *Destructive interference* occurs when the two waves are 180° out of phase everywhere, corresponding to $\phi = \pi, 3\pi, 5\pi, \ldots$.

Standing waves are formed from the superposition of two harmonic waves having the same frequency, amplitude, and wavelength, but traveling in *opposite* directions. The resultant standing wave is described by the wave function

Wave function for a standing wave

$$y = (2A_0 \sin kx) \cos \omega t \qquad (34.3)$$

Hence, its amplitude varies as $\sin kx$. The maximum amplitude points (called *antinodes*) occur at $x = n\pi/2k = n\lambda/4$ (for odd n). The points of zero amplitude (called *nodes*) occur at $x = n\pi/k = n\lambda/2$ (for integral values of n).

One can set up standing waves with specific frequencies in such systems as stretched strings, hollow pipes, rods, and drumheads. The natural frequencies of vibration of a stretched string of length L, fixed at both ends, have frequencies given by

Normal modes of a stretched string

$$f_n = \frac{n}{2L}\sqrt{F/\mu} \qquad (n = 1,2,3, \ldots) \qquad (34.8)$$

where F is the tension in the string and μ is its mass per unit length. The natural frequencies of vibration form a *harmonic series,* that is, $f_1, 2f_1, 3f_1, \ldots$.

The standing wave patterns for longitudinal waves in a hollow pipe depend on whether the ends of the pipe are open or closed. If the pipe is open at both ends, the natural frequencies of vibration form a harmonic series. If one end is closed, only odd harmonics of the fundamental are present.

A system capable of oscillating is said to be *in resonance* with some driving force whenever the frequency of the driving force matches one of the natural frequencies of the system. When the system is resonating, it responds by oscillating with a relatively large amplitude.

The phenomenon of *beats* occurs from the superposition of two waves of slightly different frequencies, traveling in the same direction. For sound waves at a given point, one would hear an alternation in sound intensity with time. Thus, beats correspond to *interference in time.*

Any periodic waveform can be represented by the combination of the sinusoidal waves that form a harmonic series. The process is called *harmonic synthesis* and is based upon *Fourier's theorem.*

EXERCISES

34.1 Superposition and Interference of Harmonic Waves

1. Two harmonic waves are described by

$$y_1 = 5 \sin[\pi(4x - 1200t)]$$
$$y_2 = 5 \sin[\pi(4x - 1200t - 0.25)]$$

where x, y_1, and y_2 are in m and t is in s. (a) What is the amplitude of the resultant wave? (b) What is the frequency of the resultant wave?

2. Two harmonic waves are described by

$$y_1 = 6 \sin\left(\frac{\pi}{15}x - \frac{\pi}{0.005}t\right)$$
$$y_2 = 6 \sin\left(\frac{\pi}{15}x - \frac{\pi}{0.005}t - \phi\right)$$

where x, y_1, and y_2 are in m and t is in s. (a) What is the amplitude of the resultant wave when $\phi = (\pi/6)$ rad? (b) For what value of ϕ will the amplitude of the resultant wave have its maximum value?

3. A harmonic wave is described by

$$y_1 = 8\sin[2\pi(0.1x - 80t)]$$

where y_1 and x are in m and t is in s. Write an expression for a wave that has the same frequency, amplitude, and wavelength as y_1, but when added to y_1 will give a resultant with an amplitude of $8\sqrt{3}$ m.

4. Two speakers are arranged as shown in Fig. 34.3. The distance between the two speakers is 2 m, and they are driven at a frequency of 1500 Hz. An observer is initially at point O located 6 m along the perpendicular bisector of the line joining the two speakers. (a) What distance must the observer move toward P before reaching the first intensity minimum? (Use $v = 330$ m/s.) (b) At what distance from point O toward point P will the observer find a relative maximum in intensity?

5. Two identical sound sources are located along the y axis. Source S_1 is located at (0,0.1) m and source S_2 is located at (0,−0.1) m. The two sources radiate isotropically at a frequency of 1650 Hz and the amplitude of each wave separately is A. A listener is located along the y axis a distance of 5 m from source S_1. (a) What is the phase difference between the sound waves at the position of the listener? (b) What is the amplitude of the resultant wave at the location of the listener? (Use $v = 330$ m/s).

6. Two identical sound sources are located as described in Exercise 5. The frequency of each source is variable. An observer is located at the point (1,0.5) m. (a) What is the lowest frequency that will produce a relative maximum at the location of the observer? (b) What is the lowest frequency that will produce a relative minimum at the observer's location?

7. Two speakers are driven by a common oscillator at 800 Hz and face each other at a distance of 1.25 m. Locate the two points along a line joining the two speakers where relative minima would be expected. (Use $v = 330$ m/s.)

8. For the arrangement shown in Fig. 34.2, let the path length $r_1 = 1.20$ m and the path length $r_2 = 0.80$ m. (a) Calculate the three lowest speaker frequencies that will result in intensity maxima at the receiver. (b) What is the highest frequency within the audible range (20–20 000 Hz) that will result in a minimum at the receiver?

9. Sketch the resultant waveform due to the interference of the two waves y_1 and y_2 in Exercise 2 at $t = 0$ s for (a) $\phi = 0$, (b) $\phi = 90°$, and (c) $\phi = 270°$. Let x range over the interval 0 to 30 m.

34.2 Standing Waves

10. Use the trigonometric identity

$$\sin(a \pm b) = \sin a \cos b \pm \cos a \sin b$$

to show that the resultant of two wave functions each of amplitude A_0, angular frequency ω, and propagation number k and traveling in opposite directions can be written

$$y = (2A_0 \sin kx)\cos\omega t$$

11. Two harmonic waves are described by

$$y_1 = 3\sin\pi(x + 0.6t) \text{ cm}$$
$$y_2 = 3\sin\pi(x - 0.6t) \text{ cm}$$

Determine the *maximum* displacement of the motion at (a) $x = 0.25$ cm, (b) $x = 0.5$ cm, and (c) $x = 1.5$ cm. (d) Find the three smallest values of x corresponding to antinodes.

12. Two harmonic waves traveling in opposite directions interfere to produce a standing wave described by

$$y = 1.5\sin(0.4x)\cos(200t)$$

where x is in m and t is in s. Determine the wavelength, frequency, and speed of the interfering waves.

13. The wave function for a standing wave in a string is given by

$$y = 0.3\sin(0.25x)\cos(120\pi t)$$

where x is in m and t is in s. Determine the wavelength and frequency of the interfering traveling waves.

14. Verify by direct substitution that the wave function for a standing wave given in Eq. 34.3,

$$y = 2A_0 \sin kx \cos\omega t,$$

is a solution of the general linear wave equation, Eq. 32.25:

$$\frac{\partial^2 y}{\partial x^2} = \frac{1}{v^2}\frac{\partial^2 y}{\partial t^2}$$

15. A standing wave is described by the function

$$y = 6\sin(\pi x/2)\cos(100\pi t)$$

where x and y are in m and t is in s. (a) Plot $y(x)$ versus t for $t = 0$, 0.0005 s, 0.001 s, 0.0015 s, and 0.002 s. (b) What is the frequency of the wave? (c) What is the wavelength λ?

16. A standing wave is formed by the interference of the following two traveling waves, each of which has an amplitude $A = \pi$ cm, propagation number $k = (\pi/2)$ cm^{-1}, and angular frequency $\omega = 10\pi$ rad/s. (a) Calculate the distance between the first two antinodes. (b) What is the amplitude of the standing wave at $x = 0.25$ cm?

34.3 Standing Waves in a String Fixed at Both Ends

17. A standing wave is established in a 120-cm-long string fixed at both ends. The string vibrates in four segments when driven at 120 Hz. (a) Determine the wavelength. (b) What is the fundamental frequency?

18. A stretched string is 160 cm long and has a linear density of 0.015 g/cm. What tension in the string will result in a second harmonic of 460 Hz?

19. Consider a tuned guitar string of length L. At what point along the string (fraction of length from one end) should the string be plucked and at what point should the finger be pressed against the fret in order that the first overtone be the most prominent mode of vibration?

20. A stretched string of length L is observed to vibrate in five equal length segments when driven by a 630-Hz oscillator. What oscillator frequency will set up a standing wave such that the string vibrates in three segments?

21. A string with $L = 16$ m and $\mu = 0.015$ g/cm is stretched with a tension of 557 N ($\approx$125 lb). What is the highest harmonic of this string that is within the typical human's audible range (up to 20 000 Hz)?

22. Two pieces of steel wire having identical cross sections have lengths of L and $2L$. The wires are each fixed at both ends and stretched such that the tension in the longer wire is four times greater than that in the shorter wire. If the fundamental frequency in the shorter wire is 60 Hz, what is the frequency of the second harmonic in the longer wire?

23. A stretched string fixed at each end has a mass of 40 g and a length of 8 m. The tension in the string is 49 N. Determine the position of the nodes and antinodes for the second overtone.

34.5 Standing Waves in Air Columns

(In this section, unless otherwise indicated, assume that the velocity of sound in air is 344 m/s.)

24. At a particular instant, the tube in Fig. 34.13a is adjusted so that L, the length above the water surface, is 40 cm. The tuning fork in the figure is replaced by a variable-frequency oscillator that has a frequency range between 20 and 2000 Hz. What are the (a) lowest and (b) highest frequencies within this range that will excite resonant modes in the air column?

25. A tuning fork of frequency 512 Hz is placed near the top of the tube shown in Fig. 34.13a. The water level is lowered so that the length L slowly increases from an initial value of 20 cm. Determine the next two values of L that correspond to resonant modes.

26. Determine the frequency corresponding to the first three harmonics of a 30-cm pipe when it is (a) open at both ends and (b) closed at both ends.

27. Calculate the minimum length for a pipe that has a fundamental frequency of 240 Hz if the pipe is (a) closed at one end and (b) open at both ends.

28. A pipe open at each end has a fundamental frequency of 300 Hz when the velocity of sound in air is 333 m/s. (a) What is the length of the pipe? (b) What is the frequency of the first overtone when the temperature of the air is increased so that the velocity of sound in the pipe is 344 m/s?

29. An organ pipe open at both ends is vibrating in its third harmonic with a frequency of 748 Hz. The length of the pipe is 0.7 m. Determine the speed of sound in air in the pipe.

34.6 Standing Waves in Rods and Plates

30. A 60-cm metal bar which is clamped at one end is struck with a hammer. If the speed of longitudinal (compressional) waves in the bar is 4500 m/s, what is the lowest frequency with which the struck bar will resonate?

31. An aluminum rod is clamped at the one-quarter position and set into longitudinal vibration by a variable-frequency driving source. The lowest frequency that produces resonance is 4400 Hz. The speed of sound in aluminum is 5100 m/s. Determine the length of the rod.

34.7 Beats: Interference in Time

32. Two waves with equal amplitude but with slightly different frequencies are traveling in the same direction through a medium. At a given point the separate displacements are described by

$$y_1 = A_0 \cos\omega_1 t \quad \text{and} \quad y_2 = A_0 \cos\omega_2 t$$

Use the trigonometric identity

$$\cos a + \cos b = 2 \cos\left(\frac{a-b}{2}\right)\cos\left(\frac{a+b}{2}\right)$$

to show that the resultant displacement due to the two waves is given by

$$y = 2A_0\left[\cos\left(\frac{\omega_1 - \omega_2}{2}\right)t\right]\left[\cos\left(\frac{\omega_1 + \omega_2}{2}\right)t\right]$$

PROBLEMS

1. Two speakers are arranged as shown in Fig. 34.3. For this problem, assume that point *O* is 12 m along the center line and the speakers are separated by a distance of 1.5 m. As the listener moves toward point *P* from point *O*, a series of alternating minima and maxima are encountered. The distance between the first minimum and the next maximum is 0.4 m. Using 340 m/s as the speed of sound in air, determine the frequency of the speakers. (Use the approximation $\sin\theta \approx \tan\theta$.)

2. In an arrangement like the one shown in Fig. 34.2, paths r_1 and r_2 are each 1.75 m in length. The top portion of the tube (corresponding to r_2) is filled with air at 0°C (273 K). Air in the lower portion is quickly heated to 200°C (473 K). What is the lowest speaker frequency that will produce an intensity maximum at the receiver? (You may determine the speed of sound in air in different temperatures by using the expression $v = 331(T/273)^{1/2}$ m/s, where T is in K.

3. A light rope 1.5 m in length lies along the *x* axis. It is set into vibration with *one* end fixed at $x = 0$. (a) What is the wavelength of the standing wave corresponding to the fundamental mode? (b) If the rope resonates in its fourth harmonic at a frequency of 320 Hz, what is the speed of transverse waves in the rope? (c) Write an expression for the wave function of the standing wave if the displacement at $x = \lambda/2$ is 4 cm.

4. A variable-length air column as shown in Fig. 34.13a is placed just below a vibrating wire fixed at both ends. The length of the air column is gradually increased from zero until the first position of resonance is observed at $L = 34$ cm. The wire is 120 cm in length and is vibrating in its third harmonic. If the speed of sound in air is 340 m/s, what is the speed of transverse waves in the wire?

5. The frequency of the second harmonic of an "open" pipe (open at both ends) is equal to the frequency of the second harmonic of a "closed" pipe (open at one end). (a) Find the ratio of the length of the closed pipe to the length of the open pipe. (b) If the fundamental frequency of the open pipe is 256 Hz, what is the length of each pipe? (Use $v = 340$ m/s.)

6. Two pipes are each open at one end and are of adjustable length. Each has a fundamental frequency of 480 Hz at 300 K. The air temperature is increased in one pipe to 305 K. (a) If the two pipes are sounded together, what beat frequency will result? (b) By what percent should the length of the 300 K pipe be increased to again match the frequencies? (Use $v = 331(T/273)^{1/2}$ m/s as the speed of sound in air, where T is the air temperature in K.)

7. To maintain a string 1.25 m long under tension in a horizontal position, one end of the string is connected to a vibrating blade and the other end is passed over a pulley and attached to a mass. The mass of the string is 10 g. (a) When the suspended mass is 10 kg, the string vibrates in three equal length segments. Determine the vibration frequency of the blade. (Assume that the point where the string passes over the pulley and the point where it is attached to the blade are both nodes. Also, ignore the contribution to the tension due to the string's mass.) (b) What mass should be attached to the string if it is to vibrate in four equal segments?

8. A speaker at the front of a room and an identical speaker at the rear of the room are being driven by the same oscillator at 456 Hz. A student walks at a uniform rate of 1.5 m/s along the length of the room. How many beats does the student hear per second?

9. Two identical steel wires each fixed at both ends are under equal tension and are vibrating in their third harmonic at 963 Hz. The tension in one wire is increased by 3%. Determine the beat frequency when the two wires now vibrate in their *fundamental* modes.

10. A student located several meters in front of a smooth reflecting wall is holding a board on which a wire is fixed at each end. The wire, vibrating in its third harmonic, is 75 cm long, has a mass of 2.25 g and is under a tension of 400 N. A second student located between the vibrating wire and the wall is moving toward the wall and hears 8.3 beats per second. At what speed does the moving student approach the wall? Use 340 m/s as the speed of sound in air.

11. An air column 2 m in length is open at both ends. The frequency of its *n*th harmonic is 410 Hz and the $(n + 1)$ harmonic frequency is 492 Hz. Determine the speed of sound in air under these conditions.

35 Electromagnetic Waves

California Institute of Technology

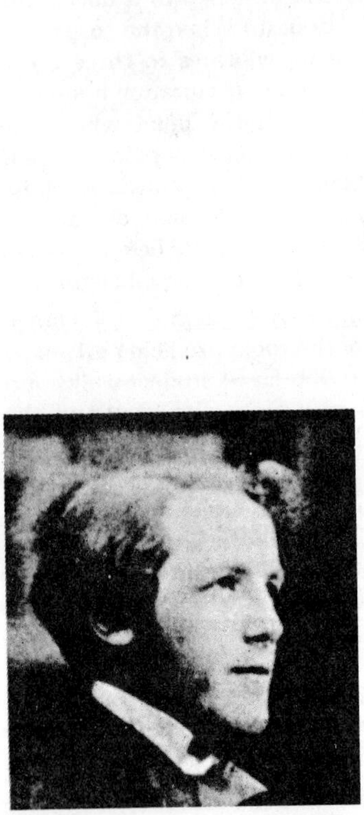

James Clerk Maxwell (1831–1879)

Thus far, the waves we have described have been mechanical waves. Such waves correspond to the disturbance of a medium. By definition, mechanical disturbances such as sound waves, water waves, and waves on a string, require the presence of a medium. This chapter is concerned with the properties of electromagnetic waves which (unlike mechanical waves) can propagate through empty space.

In Section 28.7 we gave a brief description of Maxwell's equations, which form the theoretical basis of all electromagnetic phenomena. The consequences of Maxwell's equations are far reaching and very dramatic for the history of physics. One of Maxwell's equations, the Ampère-Maxwell law, predicts that a time-varying electric field produces a magnetic field just as a time-varying magnetic field produces an electric field (Faraday's law). From this generalization, Maxwell introduced the concept of displacement current, a new source of a magnetic field. Thus, Maxwell's theory provided the final important link between electric and magnetic fields.

Astonishingly, Maxwell's formalism also predicts the existence of electromagnetic waves which propagate through space with the speed of light, given by $c = 1/\sqrt{\mu_0\varepsilon_0}$. This prediction was confirmed experimentally by Hertz, who first generated and detected electromagnetic waves. This discovery has led to many practical communication systems, including radio, television, and radar. On a conceptual level, Maxwell unified the subjects of light and electromagnetism by developing the idea that light is a form of electromagnetic radiation.

Electromagnetic waves are generated by accelerating electric charges. The radiated waves consist of oscillating electric and magnetic fields, which are *at right angles to each other* and also *at right angles to the direction of wave propagation*. Thus, electromagnetic waves are transverse in nature. Maxwell's theory shows that the electric and magnetic field amplitudes, E and B, in an electromagnetic wave are related by $E = cB$. At large distances from the source of the waves, the amplitudes of the oscillating fields diminish with distance, in proportion to $1/r$. The radiated waves can be detected at great distances from the oscillating charges. Furthermore, electromagnetic waves carry energy and momentum and hence exert pressure if they encounter a surface.

Electromagnetic waves cover a wide range of frequencies. For example, radio waves (frequencies of about 10^7 Hz) are electromagnetic waves produced by oscillating charges in a radio tower's transmitting antenna. Light waves are a high-frequency form of electromagnetic radiation (about 10^{14} Hz) produced by oscillating electrons within atomic systems.

35.1 MAXWELL'S EQUATIONS AND HERTZ'S DISCOVERIES

The fundamental laws governing the behavior of electric and magnetic fields are Maxwell's equations, which were discussed in Section 28.7.[1] In this unified theory of electromagnetism, Maxwell showed that electromagnetic waves are a natural consequence of these fundamental laws. Recall that *Maxwell's equations* in free space are given by

$$\oint \mathbf{E} \cdot d\mathbf{A} = \frac{Q}{\varepsilon_0} \tag{35.1}$$

$$\oint \mathbf{B} \cdot d\mathbf{A} = 0 \tag{35.2}$$

$$\oint \mathbf{E} \cdot d\mathbf{s} = -\frac{d\Phi_m}{dt} \tag{35.3}$$

$$\oint \mathbf{B} \cdot d\mathbf{s} = \mu_0 I + \mu_0 \varepsilon_0 \frac{d\Phi_e}{dt} \tag{35.4}$$

As we shall see in the next section, one can combine Eqs. 35.3 and 35.4 and obtain a wave equation for both the electric and the magnetic field. In empty space ($Q = 0, I = 0$), these equations permit a wavelike solution, where the *wave velocity* $(\mu_0\varepsilon_0)^{-1/2}$ *equals the measured speed of light.* This result led Maxwell to the prediction that light waves are, in fact, a form of electromagnetic waves.

Electromagnetic waves were first generated and detected in 1887 by Hertz, using electrical sources. His experimental apparatus is shown schematically in Fig. 35.1. An induction coil is connected to two spherical electrodes with a narrow gap between them (the transmitter). The coil provides short voltage surges to the spheres, charging one positive, the other negative. A spark is generated between the spheres when the voltage between them reaches the breakdown voltage for air. As the air in the gap is ionized, it conducts more readily and the discharge between the spheres becomes oscillatory. From an electrical circuit viewpoint, this is equivalent to an *LC* circuit, where the inductance is that of the loop and the capacitance is due to the spherical electrodes.

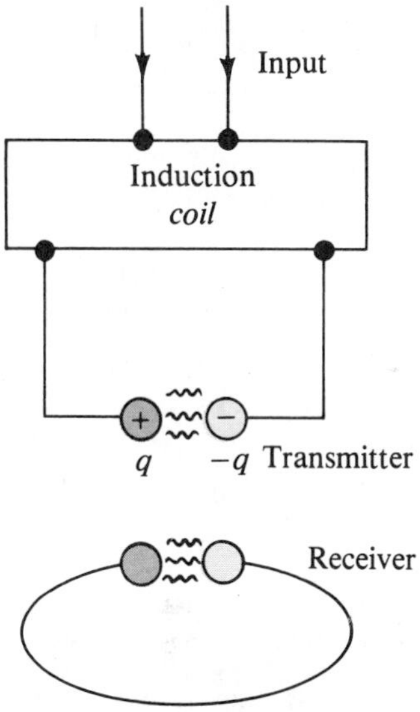

Figure 35.1 Schematic diagram of Hertz's apparatus for generating and detecting electromagnetic waves. The transmitter consists of two spherical electrodes connected to an induction coil, which provides short voltage surges to the spheres, setting up oscillations in the discharge. The receiver is a nearby loop containing a second spark gap.

Since L and C are quite small, the frequency of oscillation is very high, ≈ 100 MHz. (Recall that $\omega = 1/\sqrt{LC}$ for an *LC* circuit.) Electromagnetic waves are radiated at this frequency as a result of the oscillation (and hence acceleration) of free charges in the loop. Hertz was able to detect these waves using a single loop of wire with its own spark gap (the receiver). This loop, placed several meters from the transmitter, has its own effective inductance, capacitance, and natural frequency of oscillation. Sparks were induced across the gap of the receiving electrodes when the frequency of the receiver was adjusted to match that of the transmitter. Thus, Hertz demonstrated that the oscillating current induced in the receiver was produced by electromagnetic waves radiated by the transmitter. Hertz's experiment is analogous to the mechanical phenomenon in which a tuning fork picks up the vibrations from another, identical oscillating tuning fork.

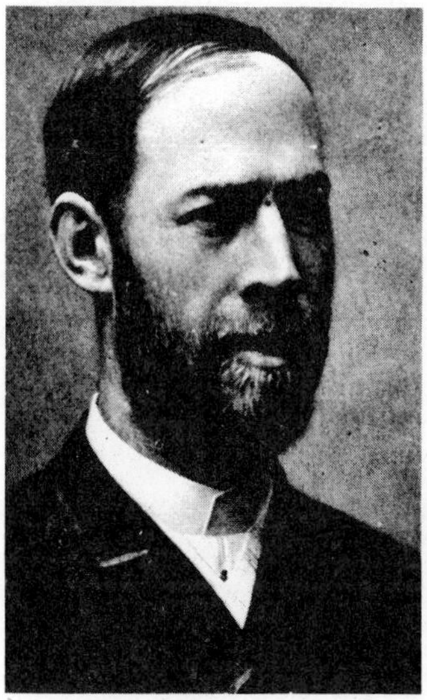

Heinrich Hertz (1857–1894)

In a series of experiments, Hertz also showed that the radiation generated by his spark-gap device exhibited the wave properties of interference, diffraction, reflection, refraction, and polarization, all of which are properties exhibited by light. Thus, it became evident that the radio-frequency waves had properties similar to light waves and differed only in frequency and wavelength.

Perhaps the most convincing experiment performed by Hertz was the measurement of the velocity of the radio-frequency waves, which he determined as follows.

[1]The reader should review Section 28.7 as a background for the material in this chapter.

Radio-frequency waves of known frequency were reflected from a metal sheet and created an interference pattern whose nodal points (where **E** was zero) could be detected. The measured distance between the nodal points allowed determination of the wavelength λ. Using the relation $v = \lambda f$, Hertz found that v was close to 3×10^8 m/s, the known speed of visible light.

35.2 PLANE ELECTROMAGNETIC WAVES

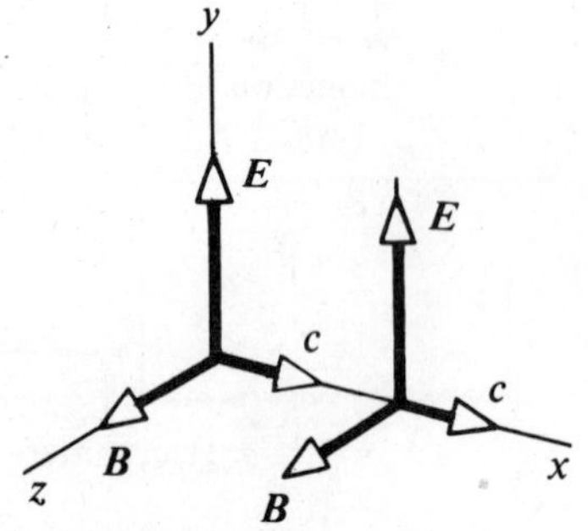

Figure 35.2 A plane polarized electromagnetic wave traveling in the positive x direction. The electric field is along the y direction, and the magnetic field is along the z direction. These fields depend only on x and t.

The properties of electromagnetic waves can be deduced from Maxwell's equations. One approach that can be used to derive such properties would be to solve the second-order differential equation that can be obtained from Maxwell's third and fourth equations. A rigorous mathematical treatment of this sort is beyond the scope of this text. To circumvent this problem, we shall assume that the electric and magnetic vectors have a specific space-time behavior that is consistent with Maxwell's equations.

First, we shall assume that the electromagnetic wave is a *plane wave*, that is, one which travels in one direction. The plane wave we are describing has the following properties. The wave travels in the x direction (the direction of propagation), the electric field **E** is in the y direction, and the magnetic field **B** is in the z direction, as in Fig. 35.2. Waves in which the electric and magnetic fields are restricted to being parallel to certain lines in the yz plane are said to be *linearly polarized waves*.[2] Furthermore, we assume that E and B at any point P depend upon x and t and not upon the y or z coordinates of the point P.

We can relate E and B to each other by using Maxwell's third and fourth equations (Eqs. 35.3 and 35.4). In empty space, where $Q = 0$ and $I = 0$, these equations become

$$\oint \mathbf{E} \cdot d\mathbf{s} = -\frac{d\Phi_{\mathrm{m}}}{dt} \tag{35.5}$$

$$\oint \mathbf{B} \cdot d\mathbf{s} = \varepsilon_0 \mu_0 \frac{d\Phi_{\mathrm{e}}}{dt} \tag{35.6}$$

Using these expressions and the plane wave assumption, one obtains the following differential equations relating E and B. For simplicity of notation, we have dropped the subscripts on the components E_y and B_z:

$$\frac{\partial E}{\partial x} = -\frac{\partial B}{\partial t} \tag{35.7}$$

$$\frac{\partial B}{\partial x} = -\mu_0 \varepsilon_0 \frac{\partial E}{\partial t} \tag{35.8}$$

Note that the derivatives here are partial derivatives. For example, when $\partial E/\partial x$ is evaluated, we assume that t is constant. Likewise, when evaluating $\partial B/\partial t$, x is held constant. Taking the derivative of Eq. 35.7 and combining this with Eq. 35.8, we get

$$\frac{\partial^2 E}{\partial x^2} = -\frac{\partial}{\partial x}(\partial B/\partial t) = -\frac{\partial}{\partial t}(\partial B/\partial x) = -\frac{\partial}{\partial t}(-\mu_0 \varepsilon_0 \partial E/\partial t) \tag{35.9}$$

Wave equations for electromagnetic waves in free space

$$\frac{\partial^2 E}{\partial x^2} = \mu_0 \varepsilon_0 \frac{\partial^2 E}{\partial t^2} \tag{35.10}$$

In the same manner, taking a derivative of Eq. 35.8 and combining this with

[2] Waves with other particular patterns of vibrations of **E** and **B** include *circularly polarized waves*. The most general polarization pattern is *elliptical*.

Eq. 35.10, we get

$$\frac{\partial^2 B}{\partial x^2} = \mu_o \varepsilon_o \frac{\partial^2 B}{\partial t^2} \tag{35.11}$$

Note that Eqs. 35.10 and 35.11 both have the form of the general wave equation,[3] with a speed c given by

$$c = 1/\sqrt{\mu_o \varepsilon_o} \tag{35.12}$$

Substituting the values $\mu_o = 4\pi \times 10^{-7}$ Wb/A · m and $\varepsilon_o = 8.85418 \times 10^{-12}$ C^2/N · m^2 into Eq. 35.12, we find that

The speed of electromagnetic waves

$$c = 2.99792 \times 10^8 \text{ m/s} \tag{35.13}$$

Since this speed is precisely the same as the speed of light in empty space,[4] one is led to believe (correctly) that light is an electromagnetic wave.

The simplest plane wave solution is a sinusoidal wave, for which the field amplitudes E and B vary with x and t according to the expressions

Sinusoidal electric and magnetic fields

$$E = E_m \cos(kx - \omega t) \tag{35.14}$$

$$B = B_m \cos(kx - \omega t) \tag{35.15}$$

where E_m and B_m are the *maximum* values of the fields. The constant $k = 2\pi/\lambda$, where λ is the wavelength, and the angular frequency $\omega = 2\pi f$, where f is the number of cycles per second. The ratio ω/k equals the speed c, since

$$\frac{\omega}{k} = \frac{2\pi f}{2\pi/\lambda} = \lambda f = c$$

Figure 35.3 is a pictorial representation at one instant of a sinusoidal, linearly polarized plane wave moving in the positive x direction.

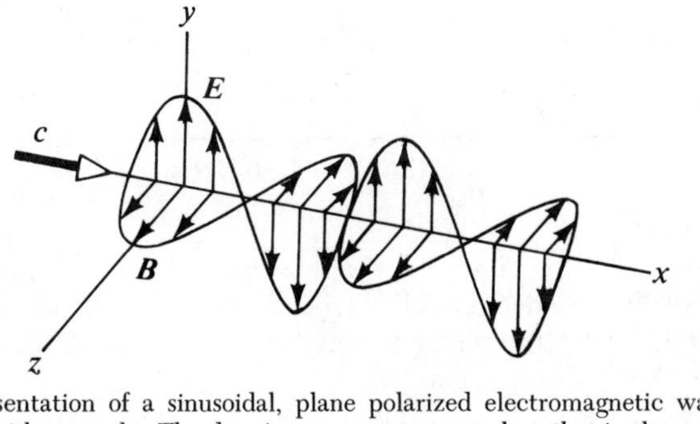

Figure 35.3 Representation of a sinusoidal, plane polarized electromagnetic wave moving in the positive x direction with a speed c. The drawing represents a snapshot, that is, the wave at some instant. Note the sinusoidal variations of $\mathbf{E}$ and $\mathbf{B}$ with x.

Taking partial derivatives of Eqs. 35.14 and 35.15, we find that

$$\frac{\partial E}{\partial x} = -kE_m \sin(kx - \omega t)$$

$$-\frac{\partial B}{\partial t} = \omega B_m \sin(kx - \omega t)$$

[3] The general wave equation is of the form $(\partial^2 f/\partial x^2) = (1/v^2)(\partial^2 f/\partial t^2)$, where v is the speed of the wave and f is the wave amplitude. The wave equation was first introduced in Chapter 32, and it would be useful for the reader to review this material.

[4] In 1972, a measurement of the speed of light in vacuum performed by the National Bureau of Standards using laser light gave a value $c = 2.99792460(6) \times 10^8$ m/s.

Since these must be equal, according to Eq. 35.7, we find that at any instant

$$kE_m = \omega B_m$$

$$\frac{E_m}{B_m} = \frac{\omega}{k} = c$$

The minus sign is ignored here since we are interested only in comparing the amplitudes. Using these results together with Eqs. 34.14 and 34.15, we see that

$$\frac{E_m}{B_m} = \frac{E}{B} = c \qquad (35.16)$$

That is, *at every instant the ratio of the electric field to the magnetic field of an electromagnetic wave equals the speed of light.*

Finally, one should note that electromagnetic waves obey the *superposition principle,* since the differential equations involving E and B are *linear* equations. For example, two waves traveling in opposite directions with the same frequency could be added by simply adding the wave fields algebraically.

Properties of electromagnetic waves

Let us summarize the properties of electromagnetic waves as we have described them:

1. The solutions of Maxwell's third and fourth equations are wavelike, where both E and B satisfy the same wave equation.
2. Electromagnetic waves travel through empty space with the speed of light, $c = 1/\sqrt{\varepsilon_0\mu_0}$.
3. The electric and magnetic field components of plane electromagnetic waves are perpendicular to each other and also perpendicular to the direction of wave propagation. The latter property can be summarized by saying that electromagnetic waves are transverse waves.
4. The relative magnitudes of E and B in empty space are related by $E/B = c$.
5. Electromagnetic waves obey the principle of superposition.

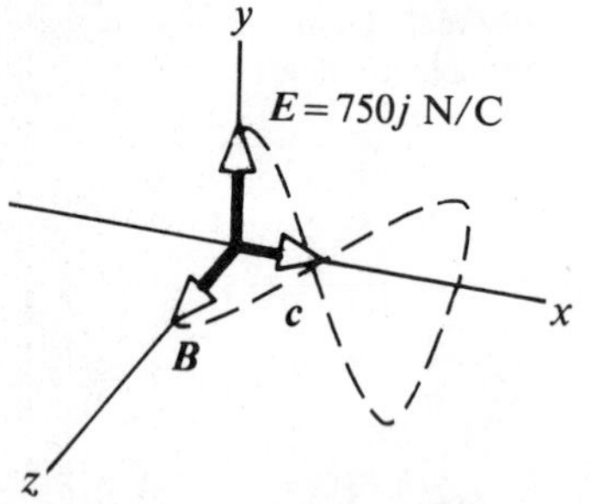

Figure 35.4 (Example 35.1) At some instant, a plane electromagnetic wave moving in the x direction has a maximum electric field of 750 N/C in the positive y direction. The corresponding magnetic field at that point has a magnitude E/c and is in the z direction.

Example 35.1

A plane electromagnetic sinusoidal wave of frequency 40 MHz travels in free space in the x direction, as in Fig. 35.4. At some point and at some instant, the electric field **E** has its *maximum* value of 750 N/C and is along the y axis. (a) Determine the wavelength and period of the wave.

Since $c = \lambda f$ and $f = 40$ MHz $= 4 \times 10^7$ s^{-1}, we get

$$\lambda = \frac{c}{f} = \frac{3 \times 10^8 \text{ m/s}}{4 \times 10^7 \text{ s}^{-1}} = 7.5 \text{ m}$$

The period of the wave T equals the inverse of the frequency, and so

$$T = \frac{1}{f} = \frac{1}{4 \times 10^7 \text{ s}^{-1}} = 2.5 \times 10^{-8} \text{ s}$$

(b) Calculate the magnitude and direction of the magnetic field **B** when **E** $= 750$**j** N/C.

From Eq. 35.16 we see that

$$B_m = \frac{E_m}{c} = \frac{750 \text{ N/C}}{3 \times 10^8 \text{ m/s}} = 2.5 \times 10^{-6} \text{ Wb/m}^2$$

Since **E** and **B** must be perpendicular to each other and both must be perpendicular to the direction of wave propagation (x in this case), we conclude the **B** is in the z direction.

(c) Write expressions for the space-time variation of the electric and magnetic field components for this plane wave.

We can apply Eqs. 35.14 and 35.15 directly:

$$E = E_m \cos(kx - \omega t) = 750 \cos(kx - \omega t) \text{ N/C}$$

$$B = B_m \cos(kx - \omega t)$$
$$= 2.50 \times 10^{-6} \cos(kx - \omega t) \text{ Wb/m}^2$$

where

$$\omega = 2\pi f = 2\pi(4 \times 10^7 \text{ s}^{-1}) = 8\pi \times 10^7 \text{ rad/s}$$

$$k = \frac{2\pi}{\lambda} = \frac{2\pi}{7.5 \text{ m}} = 0.838 \text{ m}^{-1}$$

We shall now give derivations for Eqs. 35.7 and 35.8. To derive Eq. 35.7, we start with Faraday's law, that is, Eq. 35.5:

$$\oint \mathbf{E} \cdot d\mathbf{s} = -\frac{d\Phi m}{dt}$$

Again, let us assume that the electromagnetic plane wave travels in the x direction with the electric field $\mathbf{E}$ in the positive y direction and the magnetic field $\mathbf{B}$ in the positive z direction.

Consider a thin rectangle lying in the xy plane. The dimensions of the rectangle are width dx and height l, as in Fig. 35.5. To apply Eq. 35.5, we must first evaluate the line integral of $\mathbf{E} \cdot d\mathbf{s}$ around this rectangle. The contributions from the top and bottom of this rectangle are zero since $\mathbf{E}$ is perpendicular to $d\mathbf{s}$ for these paths. We can express the electric field on the right side of the rectangle as

$$E(x + dx,t) \approx E(x,t) + \left.\frac{dE}{dx}\right]_{t\ \text{constant}} dx = E(x,t) + \frac{\partial E}{\partial x}\,dx$$

while the field on the left side is simply $E(x,t)$. Therefore, the line integral over this rectangle becomes approximately[5]

$$\oint \mathbf{E} \cdot d\mathbf{s} = E(x + dx,t) \cdot l - E(x,t) \cdot l \approx (\partial E/\partial x)\,dx \cdot l \tag{35.17}$$

Since the magnetic field is in the z direction, the magnetic flux through the rectangle of area $l\,dx$ is approximately

$$\Phi_m = Bl\,dx$$

(This assumes that dx is small compared with the wavelength of the wave.) Taking the time derivative of the flux gives

$$\frac{d\Phi_m}{dt} = l\,dx \left.\frac{dB}{dt}\right]_{x\ \text{constant}} = l\,dx \frac{\partial B}{\partial t} \tag{35.18}$$

Substituting Eqs. 35.17 and 35.18 into Eq. 35.5 gives

$$(\partial E/\partial x)\,dx \cdot l = -l\,dx \frac{\partial B}{\partial t}$$

$$\frac{\partial E}{\partial x} = -\frac{\partial B}{\partial t} \tag{35.19}$$

Thus, we see that Eq. 35.19 is equivalent to Eq. 35.7.

In a similar manner, we can verify Eq. 35.8 by starting with Maxwell's fourth equation in empty space (Eq. 35.6):

$$\oint \mathbf{B} \cdot d\mathbf{s} = \mu_o \varepsilon_o \frac{d\Phi_e}{dt}$$

In this case, we evaluate the line integral of $\mathbf{B} \cdot d\mathbf{s}$ around a rectangle lying in the yz plane and having width dx and length l, as in Fig. 35.6, where the magnetic field is in the z direction. Using the sense of the integration shown, and noting that the magnetic field changes from $B(x,t)$ to $B(x + dx,t)$ over the width dx, we get

$$\oint \mathbf{B} \cdot d\mathbf{s} = B(x,t) \cdot l - B(x + dx,t) \cdot l = -(\partial B/\partial x)\,dx \cdot l \tag{35.20}$$

The electric flux through the rectangle is

$$\Phi_e = El\,dx$$

which when differentiated with respect to time gives

$$\frac{\partial \Phi_e}{\partial t} = l\,dx \frac{\partial E}{\partial t} \tag{35.21}$$

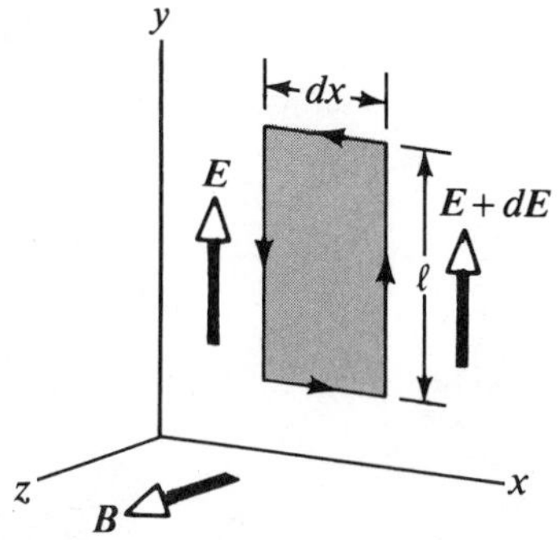

Figure 35.5 As a plane wave passes through a rectangular curve of width dx lying in the xy plane, the electric field along y varies from E to $E + dE$. This spatial variation in $\mathbf{E}$ gives rise to a time-varying magnetic field along the z direction, according to Eq. 35.19.

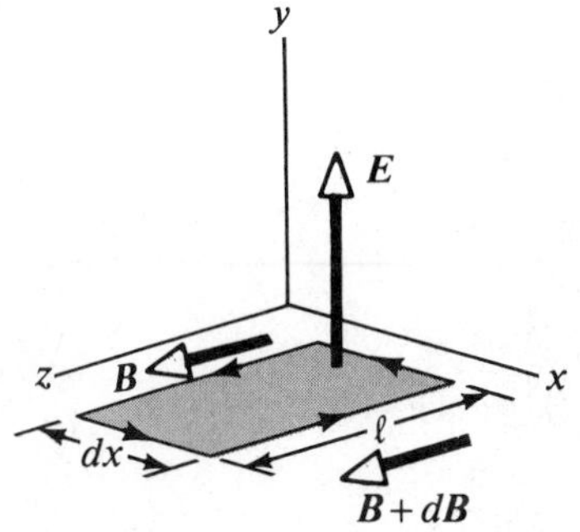

Figure 35.6 As a plane wave passes through a rectangular curve of width dx lying in the xz plane, the magnetic field along z varies from B to $B + dB$. This spatial variation in $\mathbf{B}$ gives rise to a time-varying electric field along the y direction, according to Eq. 35.22.

[5]Since dE/dx means the change in E with x at a given instant t, dE/dx is equivalent to the partial derivative $\partial E/\partial x$. Likewise, dB/dt means the change in B with time at a particular position x, and so we can replace dB/dt by $\partial B/\partial t$.

Substituting Eqs. 35.20 and 35.21 into Eq. 35.6 gives

$$-(\partial B/\partial x)dx \cdot l = \mu_0 \varepsilon_0 l\, dx(\partial E/\partial t)$$

$$\frac{\partial B}{\partial x} = -\mu_0 \varepsilon_0 \frac{\partial E}{\partial t} \tag{35.22}$$

which is equivalent to Eq. 35.8.

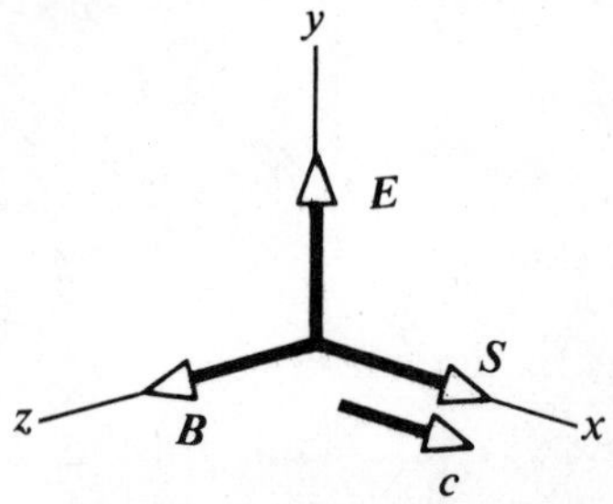

Figure 35.7 The Poynting vector S for a plane electromagnetic wave moving in the x direction is along the direction of propagation.

35.3 ENERGY AND MOMENTUM OF ELECTROMAGNETIC WAVES

Electromagnetic waves carry energy, and as they propagate through space they can transfer energy to objects placed in their path. The rate of flow of energy in an electromagnetic wave is described by a vector **S**, called the *Poynting vector*, defined by the expression

Poynting vector

$$\mathbf{S} \equiv \frac{1}{\mu_0} \mathbf{E} \times \mathbf{B} \tag{35.23}$$

The *magnitude of* **S** *represents the rate at which energy flows through a unit surface area perpendicular to the flow.* The direction of **S** is along the direction of wave propagation (Fig. 35.7). The SI units of the Poynting vector are $\text{J/s} \cdot \text{m}^2 = \text{W/m}^2$. (These are the units S must have since it represents the power per unit area, where the unit area is oriented at right angles to the direction of wave propagation.)

As an example, let us evaluate the magnitude of **S** for a plane electromagnetic wave where $|\mathbf{E} \times \mathbf{B}| = EB$. In this case

Poynting vector for a planar wave

$$S = \frac{EB}{\mu_0} \tag{35.24}$$

Since $B = E/c$, we can also express this as

$$S = \frac{E^2}{\mu_0 c} = \frac{c}{\mu_0} B^2 \tag{35.25}$$

These equations for S *apply at any instant of time.*

What is of more interest for a sinusoidal plane electromagnetic wave is the time average of S taken over one or more cycles, called the *wave intensity*. When this average is taken, one obtains an expression involving the time average of $\cos^2(kx - \omega t)$, which equals $\frac{1}{2}$. Hence, the average value of S (or the intensity of the wave) is

Wave intensity

$$S_{av} = \frac{E_m B_m}{2\mu_0} = \frac{E_m^2}{2\mu_0 c} = \frac{c}{2\mu_0} B_m^2 \tag{35.26}$$

where it is important to note that E_m and B_m represent *maximum* values of the fields.

Recall that the energy per unit volume u_e, the instantaneous energy density associated with an electric field (Section 23.4), is given by

$$u_e = \frac{1}{2}\varepsilon_0 E^2 \tag{23.11}$$

and that the instantaneous energy density u_m associated with a magnetic field (Section 29.3) is given by

$$u_m = \frac{B^2}{2\mu_0} \tag{29.15}$$

Since E and B vary with time for an electromagnetic wave, we see that the energy densities also vary with time. Using the relationships $B = E/c$ and $c = 1/\sqrt{\varepsilon_o\mu_o}$, Eq. 29.15 becomes

$$u_m = \frac{(E/c)^2}{2\mu_o} = \frac{\varepsilon_o\mu_o}{2\mu_o}E^2 = \frac{1}{2}\varepsilon_o E^2$$

Comparing this result with Eq. 23.11, we see that

$$u_m = u_e = \frac{1}{2}\varepsilon_o E^2 = \frac{B^2}{2\mu_o} \qquad (35.27)$$

That is, *for an electromagnetic wave the instantaneous energy density associated with the magnetic field equals the instantaneous energy density associated with the electric field.* Hence, in a given volume the energy is equally shared by the two fields.

The *total instantaneous energy density u* is equal to the sum of the energy densities associated with the electric and magnetic fields:

$$u = u_e + u_m = \varepsilon_o E^2 = \frac{B^2}{\mu_o} \qquad (35.28)$$

Total energy density

When this is averaged over one or more cycles of an electromagnetic wave, we again get a factor of ½. Hence, the total *average* energy per unit volume of an electromagnetic wave is given by

$$u_{av} = \varepsilon_o(E^2)_{av} = \frac{1}{2}\varepsilon_o E_m^2 = \frac{B_m^2}{2\mu_o} \qquad (35.29)$$

Average energy density of an electromagnetic wave

Comparing this result with Eq. 35.26 for the average value of S, we see that

$$S_{av} = cu_{av} \qquad (35.30)$$

In other words, *the intensity of an electromagnetic wave equals the average energy density multiplied by the speed of light.*

Electromagnetic waves transport linear momentum as well as energy. Hence, it follows that pressure (radiation pressure) is exerted on a surface when an electromagnetic wave impinges on it. In what follows, we shall assume that the electromagnetic wave transports a total energy U to a surface in a time t. If the surface *absorbs all* the incident energy U in this time, the total momentum p delivered to this surface is given by

$$p = \frac{U}{c} \quad \text{(complete absorption)} \qquad (35.31)$$

Momentum delivered to an absorbing surface

Furthermore, if the Poynting vector of the wave is S, the *radiation pressure P* (force per unit area) exerted on the perfect absorbing surface is given by

$$P = \frac{S}{c} \qquad (35.32)$$

Radiation pressure exerted on a perfect absorbing surface

We can apply these results to a perfect black body, where *all* of the incident energy is absorbed (none is reflected).

On the other hand, if the surface is a perfect reflector (for example, a mirror with a 100 percent reflecting surface), then the momentum delivered in a time t for normal incidence is *twice* that given by Eq. 35.31, or $2U/c$. That is, a momentum

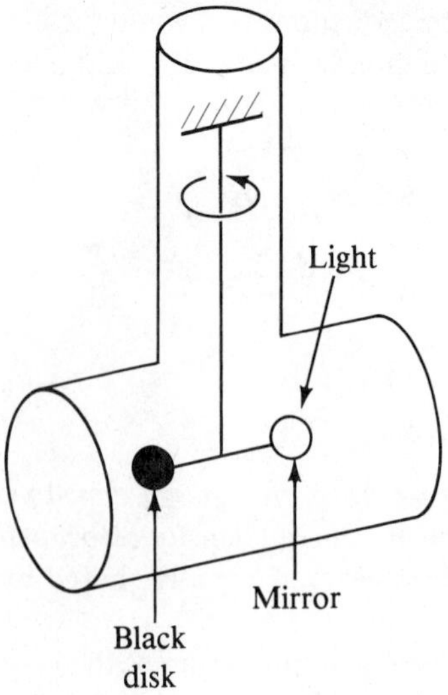

Figure 35.8 An apparatus for measuring the pressure of light. In practice, the system is contained in a high vacuum.

equal to U/c is delivered by the incident wave and U/c is delivered by the reflected wave, in analogy with a ball colliding elastically with a wall. Therefore,

Momentum delivered to a perfectly reflecting surface

$$p = \frac{2U}{c} \quad \text{(complete reflection)} \tag{35.33}$$

The momentum delivered to an arbitrary surface has a value between U/c and $2U/c$, depending on the properties of the surface. Finally, the radiation pressure exerted on a perfect reflecting surface for normal incidence of the wave is given by[6]

Radiation pressure exerted on a perfectly reflecting surface

$$P = \frac{2S}{c} \tag{35.34}$$

Although radiation pressures are very small (about 5×10^{-6} N/m^2 for direct sunlight), they have been measured using torsion balances such as the one shown in Fig. 35.8. Light is allowed to strike either a mirror or a black disk, both of which are suspended from a fine fiber. Light striking the black disk is completely absorbed, and so all of its momentum is transferred to the disk. Light striking the mirror (normal incidence) is totally reflected, hence the momentum transfer is twice as great. The radiation pressure is determined by measuring the angle through which the horizontal portion rotates. The apparatus must be placed in a high vacuum to eliminate the effects of air currents.

[6] For *oblique* incidence, the momentum transferred is $2U\cos\theta/c$ and the pressure is $P = 2S\cos\theta/c$, where θ is the angle between the normal to the surface and the direction of propagation.

Example 35.2

Solar energy: The sun delivers about 1000 W/m^2 of electromagnetic flux to the earth's surface. (a) Calculate the total power that is incident on a roof of dimensions 8 m $\times$ 20 m.

The Poynting vector has a magnitude of $S = 1000$ W/m^2, which represents the power per unit area, or the light intensity. Assuming the radiation is incident *normal* to the roof (sun directly overhead), we get

$$\text{Power} = SA = (1000 \text{ W/m}^2)(8 \times 20 \text{ m}^2) = 1.60 \times 10^5 \text{ W}$$

Note that if this power could *all* be converted into electrical energy, it would provide more than enough power for the average home. Unfortunately, solar energy is not easily harnessed, and the prospects for large-scale conversion are not as "bright" as they may appear from this simple calculation. For example, the conversion efficiency from solar to electrical energy is far less than 100% (typically, 10% for photovoltaic cells). Roof systems for converting solar energy to *thermal* energy have been built with efficiencies of around 50%; however, there are other practical problems with solar energy that must be considered, such as overcast days, geographic location, and energy storage.

(b) Determine the radiation pressure and radiation force on the roof assuming the roof covering is a perfect absorber.

Using Eq. 35.32 with $S = 1000\ \text{W/m}^2$, we find that the radiation pressure is

$$P = \frac{S}{c} = \frac{1000\ \text{W/m}^2}{3 \times 10^8\ \text{m/s}} = 3.33 \times 10^{-6}\ \text{N/m}^2$$

Since pressure equals force per unit area, this corresponds to a radiation force of

$$F = PA = (3.33 \times 10^{-6}\ \text{N/m}^2)(160\ \text{m}^2) = 5.33 \times 10^{-4}\ \text{N}$$

Of course, this "load" is *far* less than the other loads one must contend with on roofs, such as the roof's own weight or a layer of snow.

Example 35.3

A long, straight wire of resistance R, radius a, and length l carries a constant current I as in Fig. 35.9. Calculate the Poynting vector for this wire.

Solution: First, let us find the electric field **E** along the wire. If V is the potential difference across the ends of the wire, then $V = IR$ and

$$E = V/l = IR/l$$

Recall that the magnetic field at the surface of the wire (Example 27.5) is given by

$$B = \mu_o I/2\pi a$$

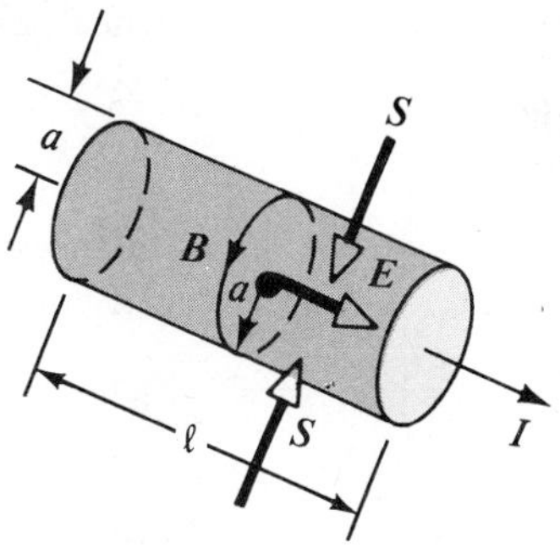

Figure 35.9 (Example 35.3) A wire of length ℓ, resistance R, and radius a carrying a current I. The Poynting vector **S** is directed radially *inward.*

The vectors **E** and **B** are mutually *perpendicular,* as shown in Fig. 35.9, and therefore $|\mathbf{E} \times \mathbf{B}| = EB$. Hence, the Poynting vector **S** is directed radially *inward* and has a magnitude

$$S = \frac{EB}{\mu} = \frac{1}{\mu}\frac{IR}{l}\frac{\mu_0 I}{2\pi a} = \frac{I^2R}{2\pi al} = \frac{I^2R}{A}$$

where $A = 2\pi al$ is the *surface* area of the wire, and the total area through which S passes. From this result, we see that

$$SA = I^2R$$

where SA has units of power (J/s = W). That is, *the rate at which electromagnetic energy flows into the wire,* SA, *equals the rate of energy (or power) dissipated as joule heat,* I^2R.

Q1. For a given incident energy of elecromagnetic wave, why is the radiation pressure on a perfect reflecting surface twice as large as the pressure on a perfect absorbing surface?

Q2. In your own words, describe the physical significance of the Poynting vector.

35.4 RADIATION FROM AN INFINITE CURRENT SHEET

In this section, we shall describe the fields radiated by a conductor carrying a time-varying current. The plane geometry we shall treat reduces the mathematical complexities one would encounter in a lower-symmetry situation, such as an oscillating electric dipole.

Consider an *infinite* conducting sheet lying in the *yz* plane and carrying a *surface current per unit length* J_s in the *y* direction, as in Fig. 35.10. Let us assume that J_s varies sinusoidally with time as

$$J_s = J_o \cos\omega t$$

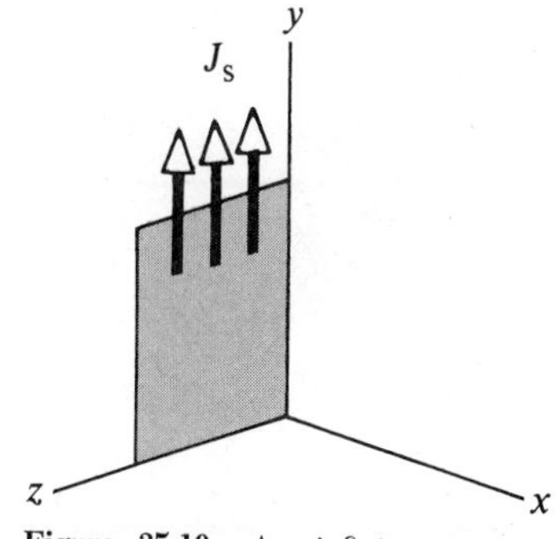

Figure 35.10 An infinite current sheet lying in the *yz* plane. The current is sinusoidal and given by $J_s = J_o \cos\omega t$. The magnetic field is everywhere parallel to the sheet and lies along *z*.

A similar problem for the case of a steady current was treated in Example 27.7, where we found that the magnetic field outside the sheet is everywhere parallel to

Radiated magnetic field

the sheet and lies along the z axis. The magnetic field was found to have a magnitude

$$B_z = -\mu_0 \frac{J_s}{2}$$

In the present situation, where J_s varies with time, this equation for B_z is valid only for distances *close* to the sheet. That is,

$$B_z = -\frac{\mu_0}{2} J_0 \cos\omega t \qquad \text{(for small values of } x\text{)}$$

To obtain the expression for B_z for *arbitrary values* of x, we can investigate the following solution:[7]

$$B_z = -\frac{\mu_0 J_0}{2} \cos(kx - \omega t) \tag{35.35}$$

There are two things to note about this solution, which is unique to the geometry under consideration. First, it agrees with our original solution for small values of x. Second, it satisfies the wave equation as it is expressed in Eq. 35.11. Hence, we conclude that the magnetic field lies along the z axis and is characterized by a transverse traveling wave having an angular frequency ω, wave number $k = 2\pi/\lambda$, and wave speed c.

We can obtain the radiated electric field that accompanies this varying magnetic field by using Eq. 35.16:

Radiated electric field

$$E_y = cB_z = -\frac{\mu_0 J_0 c}{2} \cos(kx - \omega t) \tag{35.36}$$

That is, the electric field is in the y direction, perpendicular to **B**, and has the same space and time dependences.

These expressions for B_z and E_y show that the radiation field of an infinite current sheet carrying a sinusoidal current is a plane electromagnetic wave propagating with a speed c along the x axis, as shown in Fig. 35.11.

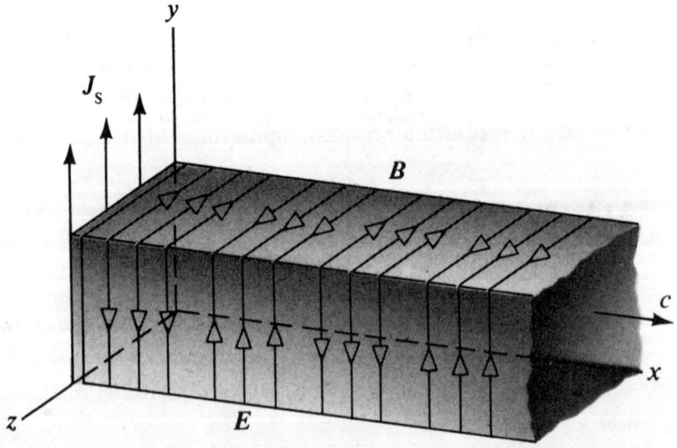

Figure 35.11 Representation of the plane electromagnetic wave radiated by the infinite current sheet lying in the yz plane. Note that **B** is along z, **E** is along y, and the direction of wave motion is along x. Both vectors have a cos $(kx - \omega t)$ behavior.

[7] Note that the solution could also be written in the form $\cos(\omega t - kx)$, which is equivalent to $\cos(kx - \omega t)$. That is, $\cos\theta$ is an even function, which means that $\cos(-\theta) = \cos\theta$.

We can calculate the Poynting vector for this wave by using Eq. 35.24 together with Eqs. 35.35 and 35.36:

$$S = \frac{EB}{\mu_o} = \frac{\mu_o J_o^2 c}{4} \cos^2(kx - \omega t) \tag{35.37}$$

The intensity of the wave, which equals the average value of S, is

$$S_{av} = \frac{\mu_o J_o^2 c}{8} \tag{35.38}$$

The intensity given by Eq. 35.38 represents the average intensity of the outgoing wave on each side of the sheet. The total rate of energy emitted per unit area of the conductor is $2S_{av} = \mu_o J_o^2 c/4$.

Example 35.4

An infinite current sheet lying in the *yz* plane carries a sinusoidal current density that has a maximum value of 5 A/m. (a) Find the *maximum* values of the radiated magnetic field and electric field.

From Eqs. 35.35 and 35.36, we see that the *maximum* values of B_z and E_y are given by

$$B_m = \frac{\mu_o J_o}{2} \quad \text{and} \quad E_m = \frac{\mu_o J_o c}{2}$$

Using the values $\mu_o = 4\pi \times 10^{-7}$ Wb/A · m, $J_o = 5$ A/m, and $c = 3 \times 10^8$ m/s, we get

$$B_m = \frac{(4\pi \times 10^{-7}\ \text{Wb/A}\cdot\text{m})(5\ \text{A/m})}{2} = 3.14 \times 10^{-6}\ \text{Wb/m}^2$$

$$E_m = \frac{(4\pi \times 10^{-7}\ \text{Wb/A}\cdot\text{m})(5\ \text{A/m})(3 \times 10^8\ \text{m/s})}{2}$$

$$= 942\ \text{V/m}$$

(b) What is the average power incident on a second planar surface that is parallel to the sheet and has an area of 3 m²? (Note that the length and width of the plate are both much larger than the wavelength of the light.)

The power per unit area (the average value of the Poynting vector) radiated in each direction by the current sheet is given by Eq. 35.38. Multiplying this by the area of the plane in question gives the incident power:

$$P = \left(\frac{\mu_o J_o^2 c}{8}\right) A$$

$$= \frac{(4\pi \times 10^{-7}\ \text{Wb/A}\cdot\text{m})(5\ \text{A/m})^2(3 \times 10^8\ \text{m/s})}{8}(3\ \text{m}^2)$$

$$= 3.54 \times 10^3\ \text{W}$$

The result is *independent of the distance from the current sheet* since we are dealing with a plane wave.

Q3. Do all current-carrying conductors emit electromagnetic waves? Explain.

35.5 THE PRODUCTION OF ELECTROMAGNETIC WAVES BY AN ANTENNA

Electromagnetic waves arise as a consequence of two effects: (1) a changing magnetic field produces an electric field and (2) a changing electric field produces a magnetic field. Therefore, it is clear that neither stationary charges nor steady currents can produce electromagnetic waves. *The sources of electromagnetic radiation are accelerating electric charges.* Whenever the current through a wire *changes with time,* the wire emits electromagnetic radiation. This, in fact, is the source of radio waves emitted by the antenna of a radio station.

Accelerating charges produce *EM* radiation

First, consider what happens when two conducting rods are connected to the opposite ends of a battery (Fig. 35.12). Before the switch is closed, the current is zero and so there are no fields present (Fig. 35.12a). Just after the switch is closed, charge of opposite signs begins to build up on the rods (Fig. 35.12b), which corresponds to a time-varying current, $I(t)$. The changing charge causes the electric field

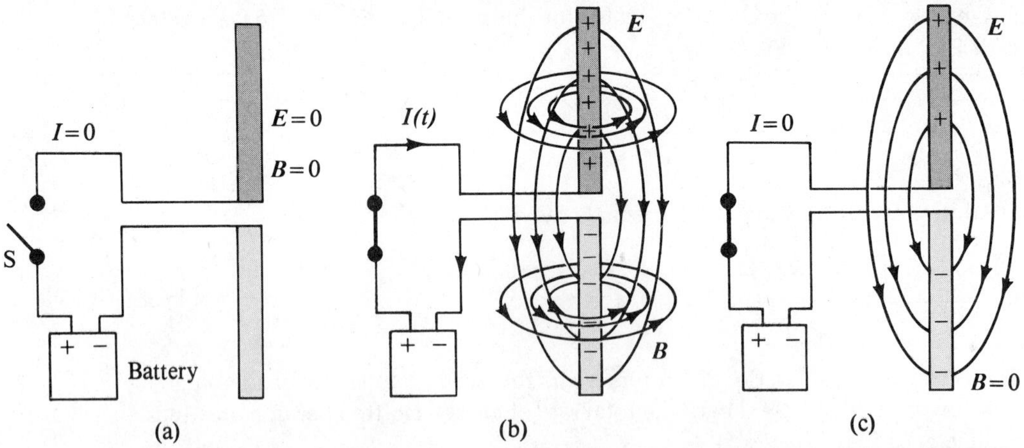

Figure 35.12 A pair of metal rods connected to a battery. (a) When the switch is open and there is no current, the electric and magnetic fields are both zero. (b) After the switch is closed and the rods are being charged (so that a current exists), the rods generate changing electric and magnetic fields. (c) When the rods are fully charged, the current is zero, the electric field is a maximum, and the magnetic field is zero.

to change, which in turn produces a magnetic field around the rods.[8] Finally, when the rods are fully charged, the current is zero and there is no magnetic field (Fig. 35.12c).

Next, let us consider the production of electromagnetic waves by a *half-wave antenna*. In this arrangement, two conducting rods, each one quarter of a wavelength long, are connected to a source of alternating emf (such as an *LC* oscillator), as in Fig. 35.13. The oscillator forces charges to accelerate back and forth between the two rods. Figure 35.13 shows the field configuration at some instant when the current is upward. The electric field lines resemble those of an electric dipole, that is, two equal and opposite charges. Since these charges are continuously oscillating between the two rods, the antenna can be approximated by an oscillating electric dipole. The magnetic field lines form concentric circles about the antenna and are perpendicular to the electric field lines at all points. The magnetic field is zero at all points along the axis of the antenna. Furthermore, **E** and **B** are 90° out of phase in time, that is, **E** at some point reaches its maximum value when **B** is zero and vice versa. This is because when the charges at the ends of the rods are at a maximum, the current is zero.

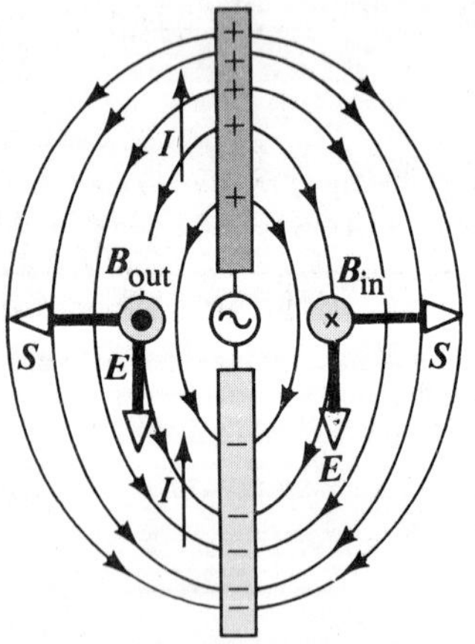

Figure 35.13 A half-wave (dipole) antenna consists of two metal rods connected to an alternating voltage source. The diagram shows **E** and **B** at an instant when the current is upward. Note that the electric field lines resemble those of a dipole.

At the two points shown in Fig. 35.13, Poynting's vector **S** is radially outward. This indicates that energy is flowing away from the antenna at this instant. At later times, the fields and Poynting's vector change direction as the current alternates. Since **E** and **B** are 90° out of phase at points near the dipole, the net energy flow is zero. From this, we might conclude (incorrectly) that no energy is radiated by the dipole.

Since the dipole fields fall off as $1/r^3$ (as in the case of a static dipole), they are not important at large distances from the antenna. However, at these large distances, another effect produces the radiation field. The source of this radiation is the continuous induction of an electric field by a time-varying magnetic field and the induction of a magnetic field by a time-varying electric field. These are predicted by two of Maxwell's equations (Eqs. 35.3 and 35.4). The electric and magnetic fields produced in this manner are in phase with each other and vary as $1/r$. The result is an outward flow of energy at all times.

The electric field lines produced by an oscillating dipole at some instant are

[8] We have neglected the field due to the wires leading to the rods. This is a good approximation if the circuit dimensions are small relative to the length of the rods.

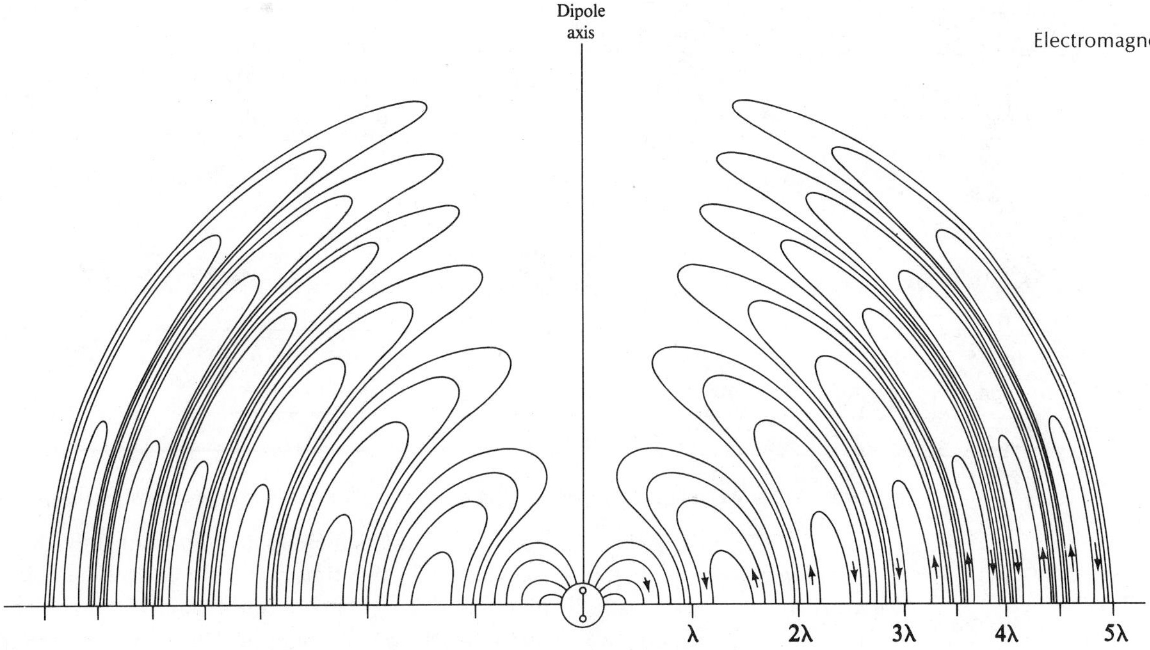

Figure 35.14 Electric field lines surrounding an oscillating dipole at a given instant. The radiation fields propagate outward from the dipole with a speed *c*.

shown in Fig. 35.14. Note that the intensity (and the power radiated) are a maximum in a plane that is perpendicular to the antenna and passing through its midpoint. Furthermore, the power radiated is zero along the antenna's axis. A mathematical solution to Maxwell's equations for the oscillating dipole shows that the intensity of the radiation field varies as $\sin^2\theta/r^2$, where θ is measured from the axis of the antenna. The angular dependence of the radiation intensity (power per unit area) is sketched in Fig. 35.15.

Electromagnetic waves can also induce currents in a *receiving antenna*. The response of a dipole receiving antenna at a given position will be a maximum when its axis is parallel to the electric field at that point and zero when its axis is perpendicular to the electric field.

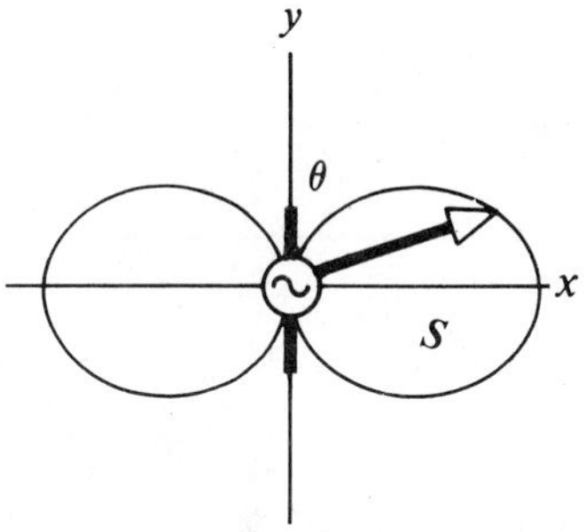

Figure 35.15 Angular dependence of the intensity of radiation produced by an oscillating electric dipole.

Q4. What is the fundamental source of electromagnetic radiation?

Q5. Electrical engineers often speak of the *radiation resistance* of an antenna. What do you suppose they mean by this phrase?

Q6. If a high-frequency current is passed through a solenoid containing a metallic core, the core will heat up by induction. This process also cooks foods in microwave ovens. Explain why the materials heat up in these situations.

Q7. Certain orientations of the receiving antenna on a TV give better reception than others. Furthermore, the best orientation varies from station to station. Explain these observations.

35.6 THE SPECTRUM OF ELECTROMAGNETIC WAVES

We have seen that all electromagnetic waves travel in a vacuum with the speed of light, *c*. These waves transport energy and momentum from some source to a re-

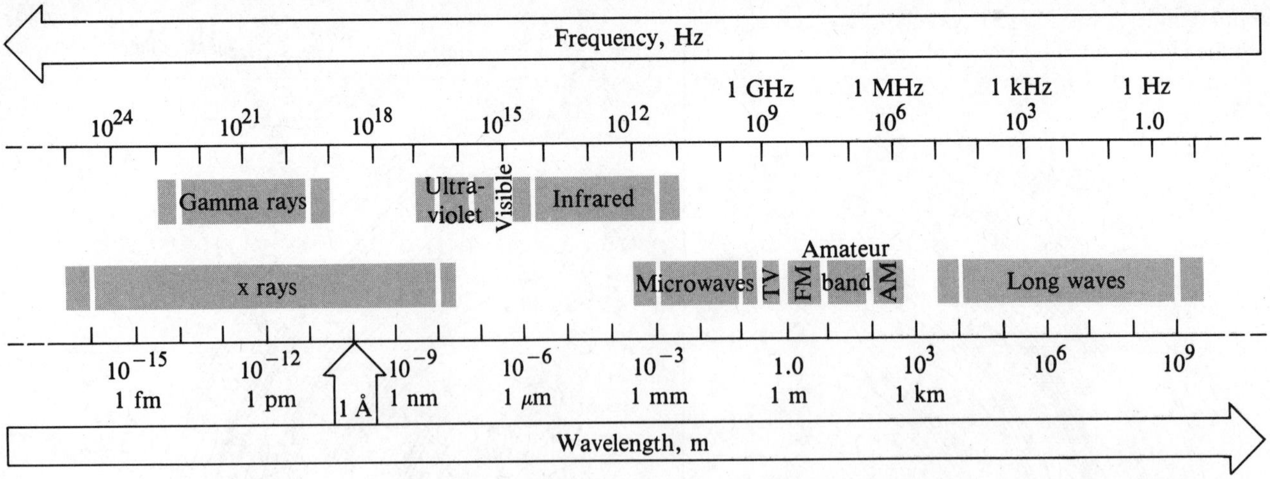

Figure 35.16 The electromagnetic spectrum. Note the overlap between one type of wave and the next.

ceiver. In 1887, Hertz successfully generated and detected the radio-frequency electromagnetic waves predicted by Maxwell.[9] At this time, the only electromagnetic waves recognized were radio waves and visible light. It is now known that other forms of electromagnetic waves exist which are distinguished by their frequency and wavelength.

Since all electromagnetic waves travel through vacuum with a speed c, their frequency f and wavelength λ are related by the important expression

$$c = f\lambda \qquad (35.39)$$

The various types of electromagnetic waves are listed in Fig. 35.16. Note the wide range of frequencies and wavelengths. For instance, a radio wave of frequency 5 MHz (a typical value) would have a wavelength given by

$$\lambda = \frac{c}{f} = \frac{3 \times 10^8 \text{ m/s}}{5 \times 10^6 \text{ s}^{-1}} = 60 \text{ m}$$

Let us give a brief description of these various waves in order of decreasing wavelength. There is no sharp dividing point between one kind of wave and the next. It should be noted that all forms of radiation are produced by accelerating charges.

Radio waves

Radio waves, which were discussed in the previous section, are the result of charges accelerating through conducting wires. They are generated by such electronic devices as LC oscillators and are used in radio and television communication systems.

Microwaves

Microwaves (short-wavelength radio waves), have wavelengths ranging between about 1 mm and 30 cm and are also generated by electronic devices. Because of their short wavelength, they are well suited for the radar systems used in aircraft navigation and for studying the atomic and molecular properties of matter. Microwave ovens represent an interesting domestic application of these waves. In a recent proposal, it was suggested that solar energy could be harnessed by beaming microwaves down to earth from a solar collector in space.[10]

Infrared waves

Infrared waves (sometimes called *heat waves*) have wavelengths ranging from

[9] Following Hertz's discoveries, Marconi succeeded in developing a practical, long-range radio communication system. However, Hertz must be recognized as the true inventor of radio communication.
[10] P. Glaser, "Solar Power from Satellites," *Physics Today,* February, 1977, p. 30.

about 1 mm to the longest wavelength of visible light, 7×10^{-7} m. These waves, produced by hot bodies and molecules, are readily absorbed by most materials. The infrared energy absorbed by a substance appears as heat since the energy agitates the atoms of the body, increasing their vibrational or translational motion, which results in a temperature rise. Infrared radiation has many practical and scientific applications, including physical therapy, infrared photography, and vibrational spectroscopy.

Visible waves

Visible light, the most familiar form of electromagnetic waves, may be defined as that part of the spectrum that the human eye can detect. Light is produced by the rearrangement of electrons in atoms and molecules. The various wavelengths of visible light are classified with colors ranging from violet ($\lambda \approx 4 \times 10^{-7}$ m) to red ($\lambda \approx 7 \times 10^{-7}$ m). The eye's sensitivity is a function of wavelength, the sensitivity being a maximum at a wavelength of about 5.6×10^{-7} m (yellow-green). Light is the basis of the science of optics and optical instruments, which we shall deal with later. The following abbreviations are often used to designate short wavelengths and distances:

$$1 \text{ micrometer } (\mu\text{m}) = 10^{-6} \text{ m}$$
$$1 \text{ nanometer (nm)} = 10^{-9} \text{ m}$$
$$1 \text{ angstrom (Å)} = 10^{-10} \text{ m}$$

Thus, the wavelengths of visible light range from 0.4 to 0.7 μm, or 400 to 700 nm, or 4000 to 7000 Å.

Ultraviolet waves

Ultraviolet light covers wavelengths ranging from about 3.8×10^{-7} (3800 Å) down to 6×10^{-8} m (600 Å). The sun is an important source of ultraviolet light, which is the main cause of suntans. Most of the ultraviolet light from the sun is absorbed by atoms in the upper atmosphere, or stratosphere. This is fortunate since uv light in large quantities produces harmful effects on humans. One important constituent of the stratosphere is ozone (O_3), which results from reactions of oxygen with ultraviolet radiation. This ozone shield converts lethal high-energy ultraviolet radiation into heat, which in turn warms the stratosphere. Recently, there has been a great deal of controversy concerning the possible depletion of the protective ozone layer as a result of the continual use of the freons used in aerosol spray cans and as refrigerants.

x-rays

X-rays are electromagnetic waves with wavelengths in the range of about 10^{-8} m (100 Å) down to 10^{-13} m (10^{-3} Å). The most common source of x-rays is the deceleration of high-energy electrons bombarding a metal target. x-rays are used as a diagnostic tool in medicine and as a treatment for certain forms of cancer. Since x-rays damage or destroy living tissues and organisms, care must be taken to avoid unnecessary exposure or overexposure. x-rays are also used in the study of crystal structure, since x-ray wavelengths are comparable to the atomic separation distances ($\approx$1 Å) in solids.

Gamma rays

Gamma rays are electromagnetic waves emitted by radioactive nuclei (such as ^{60}Co and ^{137}Cs) and during certain nuclear reactions. They have wavelengths ranging from about 10^{-10} m to less than 10^{-14} m. They are highly penetrating and produce serious damage when absorbed by living tissues. Consequently, those working near such dangerous radiation must be protected with heavily absorbing materials, such as thick layers of lead.

35.7 SUMMARY

Electromagnetic waves, which are predicted by Maxwell's equations, have the following properties:

1. The electric and magnetic fields satisfy the following wave equations, which

can be obtained from Maxwell's third and fourth equations:

Wave equations

$$\frac{\partial^2 E}{\partial x^2} = \mu_o \varepsilon_o \frac{\partial^2 E}{\partial t^2} \tag{35.10}$$

$$\frac{\partial^2 B}{\partial x^2} = \mu_o \varepsilon_o \frac{\partial^2 B}{\partial t^2} \tag{35.11}$$

2. Electromagnetic waves travel through a vacuum with the speed of light c, where

The speed of electromagnetic waves

$$c = \frac{1}{\sqrt{\mu_o \varepsilon_o}} = 3.00 \times 10^8 \text{ m/s} \tag{35.12}$$

3. The electric and magnetic fields of an electromagnetic wave are perpendicular to each other and perpendicular to the direction of wave propagation. (Hence, they are transverse waves.)
4. The instantaneous magnitudes of $|\mathbf{E}|$ and $|\mathbf{B}|$ in an electromagnetic wave are related by the expression

$$\frac{E}{B} = c \tag{35.16}$$

5. Electromagnetic waves carry energy. The rate of flow of energy crossing a unit area is described by the Poynting vector **S**, where

Poynting vector

$$\mathbf{S} \equiv \frac{1}{\mu_o} \mathbf{E} \times \mathbf{B} \tag{35.23}$$

6. Electromagnetic waves carry momentum and hence can exert pressure on surfaces. If an electromagnetic wave whose Poynting vector is **S** is completely absorbed by a surface upon which it is normally incident, the radiation pressure on that surface is

$$P = \frac{S}{c} \quad \text{(complete absorption)} \tag{35.31}$$

If the surface totally reflects a normally incident wave, the pressure is doubled.

The electric and magnetic fields of a sinusoidal plane electromagnetic wave propagating in the positive x direction can be written

Sinusoidal electric and magnetic fields

$$E = E_m \cos(kx - \omega t) \tag{35.14}$$

$$B = B_m \cos(kx - \omega t) \tag{35.15}$$

where ω is the angular frequency of the wave and k is the wave number. These equations represent special solutions to the wave equations for **E** and **B**. Since $\omega = 2\omega f$ and $k = 2\pi/\lambda$, where f and λ are the frequency and wavelength, respectively, one finds that

$$\frac{\omega}{k} = \lambda f = c$$

The average value of the Poynting vector for a plane electromagnetic wave has a magnitude given by

$$S_{av} = \frac{E_m B_m}{2\mu_0} = \frac{E_m^2}{2\mu_0 c} = \frac{c}{2\mu_0} B_m^2 \qquad (35.26)$$

Wave intensity

The intensity of a sinusoidal plane electromagnetic wave equals the average value of the Poynting vector taken over one or more cycles.

The fundamental sources of electromagnetic waves are *accelerating electric charges*. For instance, radio waves emitted by an antenna arise from the continuous oscillation (and hence acceleration) of charges within the antenna structure.

The electromagnetic spectrum includes waves covering a broad range of frequencies and wavelengths. The frequency f and wavelength λ of a given wave are related by

$$c = f\lambda \qquad (35.39)$$

EXERCISES

Section 35.2 Plane Electromagnetic Waves

1. Verify that Eq. 35.12 gives c with dimensions of length per unit time.

2. Consider an electromagnetic wave traveling in a medium which has permittivity $\varepsilon = \kappa\varepsilon_o$ and permeability $\mu = \mu_o$. From Eq. 35.12 show that the index of refraction of the material is given by $n = \sqrt{\kappa}$, where κ is the dielectric constant of the medium.

3. (a) Use the relationship $B = \mu_o H$ described in Eq. 30.7 together with the properties of E and B described in Section 35.2 to show that $E/H = \sqrt{\mu_o/\varepsilon_o}$. (b) Calculate the numerical value of this ratio and show that it has SI units of ohms. (The ratio E/H is referred to as the *impedance of free space.*)

4. Figure 35.3 shows a plane electromagnetic sinusoidal wave propagating in the x direction. The wavelength is 75 m, and the electric field vibrates in the xy plane with an amplitude of 35 V/m. Calculate (a) the sinusoidal frequency and (b) the magnitude and direction of $\mathbf{B}$ when the electric field has its maximum value in the negative y direction. (c) Write an expression for B in the form

$$B = B_m \cos(kx - \omega t)$$

with numerical values for B_m, k, and ω.

5. The magnetic field amplitude of an electromagnetic wave is 2×10^{-7} T. Calculate the electric field amplitude if the wave is traveling (a) in free space and (b) in a medium in which the speed of the wave is $0.75c$.

6. An electromagnetic wave in vacuum has an electric field amplitude of 150 V/m. Calculate the amplitude of the corresponding magnetic field.

7. Verify that the following pair of equations for E and B are solutions of Eqs. 35.7 and 35.8:

$$E = \frac{A}{\sqrt{\varepsilon_o\mu_o}} e^{x-ct} \quad \text{and} \quad B = Ae^{x-ct}$$

8. Calculate the maximum value of the magnetic field in a region where the measured maximum value of the electric field is 2 mV/m.

Section 35.3 Energy and Momentum of Electromagnetic Waves

9. An incandescent lamp is radiating isotropically (i.e., identically in all directions) at 15 W. Calculate the maximum values of the electric and magnetic fields at distances of (a) 1 m and (b) 5 m from the source.

10. A helium-neon laser intended for instructional use operates at a typical power of 3.5 mW. (a) Determine the maximum value of the electric field at a point where the cross section of the beam is 8 mm^2. (b) Calculate the electromagnetic energy in a 1-m length of the beam.

11. At what distance from a 30-W isotropic electromagnetic wave power source will $E_m = 10$ V/m?

12. What power must be radiated (isotropically) by a source if the amplitude of the electric field is 20 V/m at a distance of 2 m?

13. A plane electromagnetic wave has an energy flux of 300 W/m^2. A flat, rectangular surface of dimensions 20 cm $\times$ 40 cm is placed perpendicular to the direction of the plane wave. If the surface absorbs half of the energy and reflects half (that is, it is a 50 percent reflecting surface), calculate (a) the total energy absorbed by the surface in a time of 1 min and (b) the momentum absorbed in this time.

14. A disk 0.5 cm in diameter is located 1.5 m from a 150-W light bulb. The surface of the disk is a perfect reflector, and the normal to the plane of the disk makes an angle of 30° with the outward radial direction from the bulb. Calculate the radiation force on the disk.

15. A radio wave transmits 1.5 W/m^2 of power per unit area. A planar surface of area A is perpendicular to the direction of propagation of the wave. Calculate the radiation pressure on the surface if the surface is a perfect absorber.

16. Let the planar surface in Exercise 15 have dimensions of 1.5 m $\times$ 0.8 m. Calculate the momentum delivered to the surface per second.

17. Direct sunlight exerts a typical radiation pressure of 5×10^{-6} N/m^2. Calculate the radiation force on a perfectly reflecting horizontal mirror of dimensions 40 cm $\times$ 80 cm.

18. Determine the momentum per unit volume in a low-power helium-neon laser (3 mW) if the beam diameter is 2 mm.

19. At one location on the earth, the amplitude of the magnetic field due to solar radiation is 2.4 μT. From this value calculate (a) the magnitude of the electric field due to solar radiation, (b) the energy density of the solar component of electromagnetic radiation at this location, and (c) the magnitude of the Poynting vector for the sun's radiation. (d) Compare the value found in (c) to the value of the solar flux given in Example 35.2.

20. A long wire has a radius of 1 mm and a resistance per unit length of 2 Ω/m. Determine the current required if the Poynting vector at the surface of the wire equals 2.68×10^3 W/m^2.

Section 35.4 Radiation from an Infinite Current Sheet

21. A rectangular surface of dimensions 30 cm $\times$ 15 cm is parallel to and 1.2 m from a very large conducting sheet in which there is a sinusoidally varying surface current which has a maximum value of 8 A/m. (a) Calculate the average power incident on the smaller sheet. (b) What power per unit area is radiated by the current-carrying sheet?

22. A large current-carrying sheet is expected to radiate in each direction (normal to the plane of the sheet) at a rate equal to 670 W/m^2 (approximately one half of the solar constant). What maximum value of sinusoidal current density is required?

PROBLEMS

1. Assume that the solar radiation incident on the earth is 1340 W/m^2. (This is the value of the solar flux above the earth's atmosphere.) (a) Calculate the total power radiated by the sun, taking the average earth-sun separation to be 1.49×10^{11} m. (b) Determine the magnitude of the electric and magnetic fields at the earth's surface due to solar radiation.

2. The magnetic field of a linearly-polarized electromagnetic wave is described by the equation

$$B = (1.5 \times 10^{-6}) \sin\left[2\pi\left(\frac{x}{20} - \frac{t \times 10^8}{6.6}\right)\right] T$$

where x is in m and t in s. (a) Calculate the velocity of the wave and (b) write the equation for the associated electric field.

3. Show that the instantaneous value of the energy intensity can be written

$$S = \frac{c}{2}(\varepsilon_0 E^2 + \mu_0 H^2)$$

4. Throughout a region of space, an electromagnetic wave has an intensity of 8×10^{-14} W/m^2. A quarter-wave antenna 3.5 m in length is adjusted so that the axis of the antenna is perpendicular to the direction of propagation of the wave. Calculate the rms voltage between the ends of the antenna. Check your result carefully to be sure it is dimensionally correct.

5. A group of astronauts plan to propel a spaceship by using a "sail" to reflect solar radiation. The sail is totally reflecting, oriented with its plane perpendicular to the direcion to the sun, and 1 km $\times$ 1.5 km in size. What is the maximum acceleration that can be expected for a spaceship of 4 metric tons (4000 kg)? (Use the solar radiation data from Problem 1 and neglect gravitational forces.)

6. Consider a small, spherical particle of radius r located in space a distance R from the sun. (a) Show that the ratio $F_{rad}/F_{grav} \propto 1/r$, where F_{rad} = the force due to solar radiation and F_{grav} = the force of gravitational attraction. (b) The result of (a) means that for a sufficiently small value of r the force exerted on the particle due to solar radiation will exceed the force of gravitational attraction. Calculate the value of r for which the particle will be in equilibrium under the two forces. (Assume that the particle has a perfectly absorbing surface and a mass density of 1.5 g/cm^3. Let the particle be located 1.25 earth orbit diameters from the sun and use 214 W/m^2 as the value of the solar flux at that point.)

7. An astronaut in a spacecraft moving with constant velocity wishes to increase the speed of the craft by using a laser beam attached to the spaceship. The laser beam emits 100 J of electromagnetic energy per pulse, and the laser is pulsed at the rate of 0.2 pulse/s. If the mass of the spaceship plus its contents is 5000 kg, for how long a time must the beam be on in order to increase the speed of the vehicle by 1 m/s in the direction of its initial motion? In what direction should the beam be pointed to achieve this?

8. A community plans to build a facility to convert solar radiation into electrical power. They require 1 MW of power (10^6 W), and the system to be installed has an efficiency of 30 percent (that is, 30 percent of the solar energy incident on the surface is converted to electrical energy). What must be the effective area of a perfectly absorbing surface used in such an installation, assuming a constant energy flux of 1000 W/m^2?

9. A microwave transmitter emits monochromatic electromagnetic waves. The maximum electric field at a distance 1 km from the transmitter is 6.0 V/m. Assuming the transmitter is a point source and neglecting waves reflected from the earth, calculate (a) the maximum magnetic field at this distance and (b) the total power emitted by the transmitter.

10. A thin tungsten filament of length 1 m radiates 60 W of energy in the form of electromagnetic waves. A perfectly absorbing surface in the form of a hollow cylinder of radius 5 cm and length 1 m is placed concentric with the filament. Calculate the radiation pressure acting on the cylinder. (Assume that the radiation is emitted in the radial direction, and neglect end effects.)

11. The torsion balance shown in Fig. 35.8 is used in an experiment to measure radiation pressure. The torque constant (elastic restoring torque) of the suspension fiber is 1×10^{-11} N · m/deg, and the length of the horizontal rod is 6 cm. The beam from a 3-mW helium-neon laser is incident on the black disk, and the mirror disk is completely shielded. Calculate the angle between the *equilibrium* positions of the horizontal bar when the beam is switched from "off" to "on."

12. Monoenergetic x-rays move through a material with a speed of 0.95*c*. The photon flux on a surface perpendicular to the x-ray beam is 10^{13} photons/m^2 · s. (a) Calculate the density of photons (number per unit volume) in the material. (b) If each photon has an energy of 8.88 keV, what is the energy density within the material?

13. A linearly-polarized microwave of wavelength 1.5 cm is directed along the positive *x* axis. The electric field vector has a maximum value of 175 V/m and vibrates in the *xy* plane. (a) Assume that the magnetic field component of the wave can be written $B = B_0 \sin(kx - \omega t)$ and give values for B_0, k, and ω. Also, determine in which plane the magnetic field vector vibrates. (b) Calculate the Poynting vector for this wave. (c) What radiation pressure would this wave exert if directed at normal incidence onto a perfectly reflecting sheet? (d) What acceleration would be imparted to a 500-g sheet (perfectly reflecting and at normal incidence) with dimensions 1 m × 0.75 m?

36 The Nature of Light and the Laws of Geometric Optics

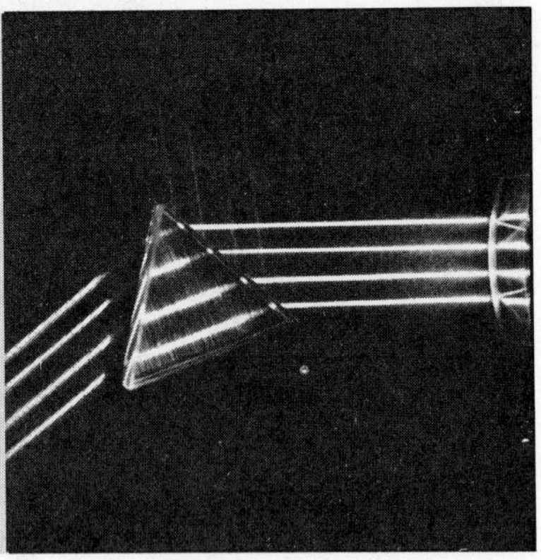

James Madison University

36.1 THE NATURE OF LIGHT

Up until the beginning of the 19th century, light was considered to be a stream of particles (corpuscles) emitted by a light source, which stimulated the sense of sight upon entering the eye. The chief architect of this particle theory of light was, once again, Isaac Newton.[1] With this theory, Newton was able to provide a simple explanation of some known experimental facts concerning the nature of light, namely, the laws of reflection and refraction.

The *law of reflection* states that when light is reflected from a smooth, shiny surface such as a mirror (Fig. 36.1), the angle of incidence, θ_1, equals the angle of reflection, θ_1'. (The angles are conventionally measured from the normal to the surface.) This is explained in the particle theory by treating light as a stream of particles making perfectly elastic collisions with a frictionless surface.

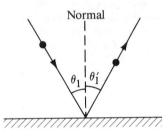

Figure 36.1 Particle description of the law of reflection, $\theta_1 = \theta_1'$. Upon reflection, the component of the velocity perpendicular to the surface is reversed, while the parallel component remains the same.

The law of refraction deals with the manner in which light changes its direction of propagation when it passes from one medium into another. For instance, as a light beam passes from air into glass, as in Fig. 36.2a, the portion of the beam entering the glass is bent *toward* the normal to the surface, so that $\theta_2 < \theta_1$. On the other hand, if the beam passes from glass into air, as in Fig. 36.2b, the beam entering the air bends *away* from the normal. In either case, the light beam transmitted into the second medium is said to be *refracted*, and the angle θ_2 is called the *angle of refraction*. In his particle model, Newton explained this phenomenon by assuming that the light particles entering the glass (as in Fig. 36.2a) are attracted to the glass as they approach the surface, thereby acquiring a change in momentum perpendicular to the surface.

Most scientists accepted Newton's particle theory of light. However, during Newton's lifetime another theory was proposed—one that argued that light might be some sort of wave motion. In 1678, a Dutch physicist and astronomer, Christian Huygens (1629–1695), showed that a wave theory of light could explain the laws of reflection and refraction. The wave theory did not receive immediate acceptance for several reasons. All the waves known at the time (sound, water, etc.) traveled through some sort of medium. On the other hand, light could travel to us from the sun through the vacuum of space. Furthermore, it was argued that if light were

[1] Isaac Newton, *Opticks*, 1704. [The fourth edition (1730) was printed by Dover Publications, New York, 1952.]

some form of wave motion, the waves could bend around obstacles; hence, we should be able to see around corners. It is now known that the bending of light around the edges of an object does indeed occur. This phenomenon, known as *diffraction,* is not easy to observe for light waves because of their short wavelengths. Although experimental evidence for the occurrence of the diffraction of light was discovered by Francesco Grimaldi (1618–1663) around 1660, most scientists rejected the wave theory and adhered to the particle theory for more than a century. This was, for the most part, due to Newton's great reputation as a scientist.

The first clear demonstration of the wave nature of light was provided in 1801 by an experiment of Thomas Young (1773–1829). In his experiment, Young showed that under appropriate conditions, light exhibits an interference behavior. Several years later, a French physicist, Augustin Fresnel (1788–1829), performed a number of detailed experiments dealing with interference and diffraction phenomena. (These topics will be discussed further in Chapters 37 and 38.) In 1850, Jean Foucault (1819–1868) provided further evidence of the inadequacy of the particle theory by showing that the speed of light in liquids is less than in air. (According to the particle model of light, the speed of light would be higher in glasses or in liquids than in air.) Additional developments during the 19th century led to the general acceptance of the wave theory of light.

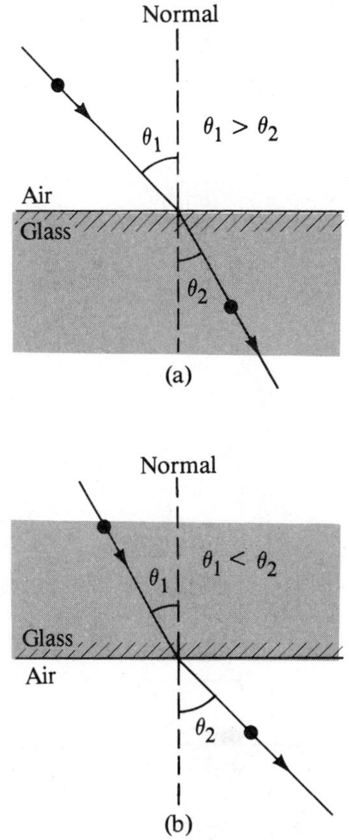

Figure 36.2 Particle description of the law of refraction. (a) When the light beam moves from air into glass, its path is bent toward the normal. (b) When the beam moves from glass into air, its path is bent away from the normal.

The most important development concerning the theory of light was the work of Maxwell, who, in 1873, showed that light was a form of high-frequency electromagnetic waves (Chapter 35). The theory predicts that these waves should have a velocity of about 3×10^8 m/s. Within experimental error, this was equal to the speed of light. As discussed in Chapter 35, Hertz provided experimental confirmation of Maxwell's theory in 1887 by producing and detecting electromagnetic waves. Furthermore, he and other investigators showed that these waves could be reflected, refracted, polarized, and so on. In other words, electromagnetic waves exhibited all the characteristic effects of light waves.

Although the classical theory of electromagnetism was able to explain most known properties of light, subsequent discoveries demonstrated its shortcomings. The most striking experimental fact that is not in accord with this theory is the *photoelectric effect,* also discovered by Hertz. The photoelectric effect is the ejection of electrons from a metal whose surface is exposed to light. A later investigation showed that the kinetic energy of an individual ejected electron is *independent* of the light intensity. An explanation of this phenomenon was proposed by Einstein in 1905, using the concept of quantization developed by Max Planck (1858–1947) in 1900. This model assumes that the energy of a light wave is present in discrete bundles of energy called *photons;* hence, the energy is said to be *quantized.* According to this theory, the energy E of a photon is proportional to the frequency f of the electromagnetic wave, so that

$$E = hf \tag{36.1}$$

Photon energy

where $h = 6.63 \times 10^{-34}$ J · s is Planck's constant. It is important to note that this theory retains some features of both the wave theory and the particle theory of light. The photoelectric effect is the result of energy transfer from a single photon to an electron in the metal.

The energy of the ejected electrons does not change with intensity, but depends only on the frequency of incident light in a manner consistent with Eq. 36.1. Einstein's theory suggested that if the light wave had a frequency below some threshold value (which depends on the type of metal), *no* photoelectric effect would occur regardless of the light intensity. Shortly thereafter, experiments performed by Robert Millikan confirmed this prediction and the Einstein formula given by Eq. 36.1.

In view of these developments, one must regard light as having a dual nature, depending on the system under observation and the circumstances under which it is observed. On one hand, the classical electromagnetic wave theory provides an adequate explanation of light propagation and the effects of interference and diffraction. On the other hand, the photoelectric effect and other experiments involving the interaction of light with matter are best explained by a particle model. Light is light, to be sure. However, the question "Is light a wave or a particle?" is an inappropriate one. The choices it presupposes are too restrictive.

36.2 MEASUREMENTS OF THE SPEED OF LIGHT

Light travels at such a high speed ($c \approx 3 \times 10^8$ m/s) that early attempts to measure its speed were unsuccessful. Galileo attempted to measure c by positioning two observers in towers separated by about one mile. Each observer carried a lantern, with the idea that as soon as observer A received a signal from observer B, he would flash a return signal back to observer B. The velocity could then be obtained knowing the transit time of the light beam between lanterns. The results were inconclusive. Today, we realize that it is impossible to measure c in this manner, since the transit time of the light is very small compared with the reaction time of the observers. We now describe two famous workable methods for determining the speed of light.

Roemer Method

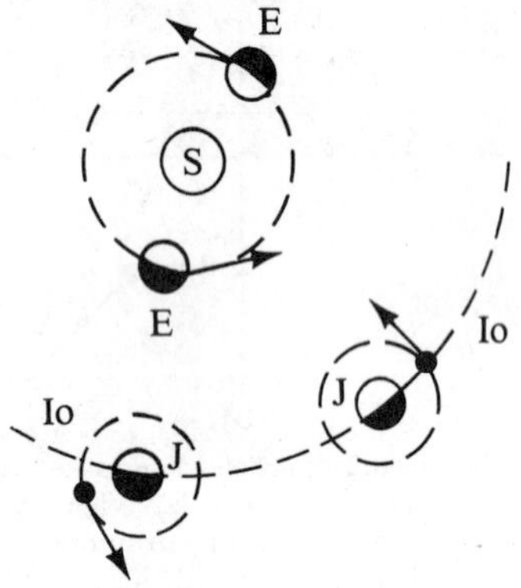

Figure 36.3 Roemer's method for measuring the speed of light.

The first successful estimate of c was made in 1675 by the Danish astronomer Ole Roemer (1644–1710). His technique involved astronomical observations of one of the moons of Jupiter, called Io. At that time, 4 of Jupiter's 14 moons had been discovered, and the period of their orbits was known. Io, the innermost moon, has a period of about 42.5 h. This is measured by observing the eclipse of Io as it passes behind Jupiter (Fig. 36.3). Note that the period of Jupiter is about 12 years, and so as the earth rotates through 180° about the sun, Jupiter turns through an angle of only 15°. Using the orbital motion of Io as a clock, one would expect a constant period in its orbit over long time intervals. However, Roemer observed a systematic variation in Io's period over a long time interval. He found that the observed periods were larger than average when the earth receded from Jupiter and smaller than average when the earth was approaching Jupiter. Roemer attributed this variation in periods to the fact that the distance between the earth and Jupiter was changing during the observations. In a time of 6 months (half the period of the earth), the light from Jupiter has to travel an additional distance equal to the diameter of the earth's orbit. With data existing at that time, he would have estimated the speed of light to be about 2.1×10^8 m/s. The large discrepancy between this value and the currently accepted value of 3.0×10^8 m/s is due to a large error in the assumed diameter of the earth's orbit.

Fizeau's Technique

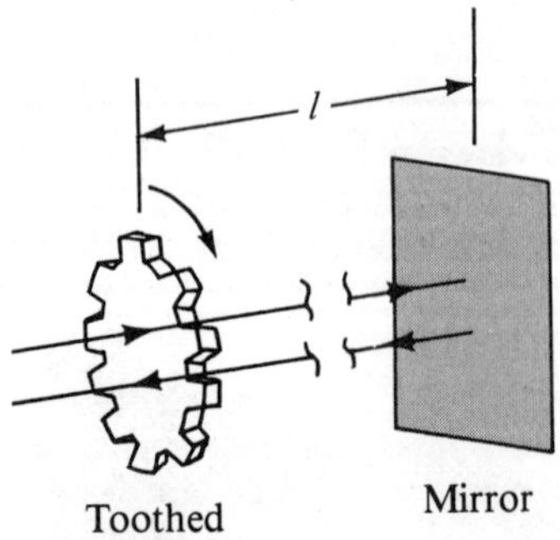

Figure 36.4 Fizeau's method for measuring the speed of light using a rotating, toothed wheel.

The first successful method for measuring c using purely terrestrial techniques was performed in 1849 by Armand H. L. Fizeau (1819–1896). Figure 36.4 represents a simplified diagram of his apparatus.[2] The basic idea is to measure the total time it takes light to travel from some point to a distant mirror and back. If l is the distance between the light source and mirror, and the transit time for one round trip is t, then $c = 2l/t$. To measure the transit time, Fizeau used a rotating toothed

[2]The actual apparatus involved several lenses and mirrors which we have omitted for the sake of clarity. For more details, see F. W. Sears, *Optics*, Reading, Mass., Addison-Wesley, 1949, Chapter 1.

wheel, which effectively "chops" the otherwise continuous beam into a series of light pulses. The rotating wheel has regularly spaced openings along its rim. Therefore, for most speeds, the reflected light is screened by one of the teeth and an observer sees no image of the light source. However, if the angular velocity of the wheel is increased until its speed is high enough to allow the reflected beam to pass through the opening next to the one through which the light first passed, the observer would then see the light source. Knowing the distance l, the number of teeth in the wheel, and the angular velocity of the wheel, Fizeau arrived at a value of $c \approx 3.1 \times 10^8$ m/s. Similar measurements made by subsequent investigators yielded more accurate values for c, approximately 2.9977×10^8 m/s.

A variety of other, more precise measurements have been reported for the determination of c. A recent value obtained using laser techniques[3] is

$$c = 2.997924574(12) \times 10^8 \text{ m/s}$$

The number of significant figures here is certainly impressive. Furthermore, the techniques that have been devised for measuring c are very clever, but often expensive. Now you may ask, "Why bother?" There are several good reasons for measuring c with high precision. One important reason is to test certain modern theories involving fundamental constants. We have seen that the speed of light in a vacuum is related to the permittivity and permeability of free space through the expression $c = 1/\sqrt{\mu_0\epsilon_0}$. Thus, armed with an accurate value for c, scientists are able to provide better tests of the modern theories of electromagnetic interactions.

Christian Huygens (1629–1695).

36.3 HUYGENS' PRINCIPLE

Although the laws of reflection and refraction can be derived from Maxwell's equations, the mathematics is rather complex and hence will not be attempted in this text. A simpler approach for deriving these and other laws of optics is to use a geometric method proposed by Huygens in 1678. Huygens assumed that light is some form of wave motion rather than a stream of particles. He had no knowledge of the nature of the wave or of the electromagnetic character of light. Nevertheless, his simplified wave model is adequate for understanding many practical aspects of the propagation of light.

In effect, Huygens' principle is a geometric construction for determining the position of a new wavefront at some instant from the knowledge of an earlier wavefront. In Huygens' construction, *all points on a given wavefront are taken as point sources for the production of spherical secondary waves, called* wavelets, *which propagate outward with speeds characteristic of waves in that medium. After a time* t *has elapsed, the new position of the wavefront is the surface tangent to these secondary waves.*

Figure 36.5 illustrates two simple examples of Huygens' construction. First, consider the plane wave moving through free space as in Fig. 36.5a. At $t = 0$, the wavefront is indicated by the plane labeled AA'. According to Huygens' construction, each point on this wavefront is considered a point source. (Only a few points on AA' are shown for clarity.) Using these points as sources for the wavelets, we draw circles of radius ct, where c is the speed of light in free space and t is the time of propagation from one wavefront to the next. The plane of tangency to these wavelets is the plane BB', which is parallel to AA'. In a similar manner, Fig. 36.5b shows Huygens' construction for an outgoing spherical wave.

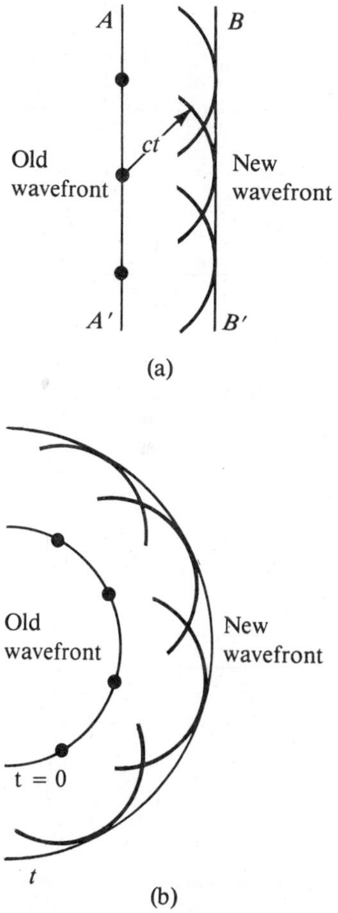

Figure 36.5 Huygens' construction for (a) a plane wave propagating to the right, and (b) a spherical wave.

A rather convincing demonstration of Huygens' principle is obtained with water waves in a shallow tank (called a *ripple tank*), as in Fig. 36.6. Plane waves produced

[3] K. M. Evenson *et al., Phys. Rev. Lett.* **29:** 1346 (1972). For a more complete description of recent measurements of c, see Joseph F. Mulligan, *Am. J. Phys.* *44:* 960 (1976).

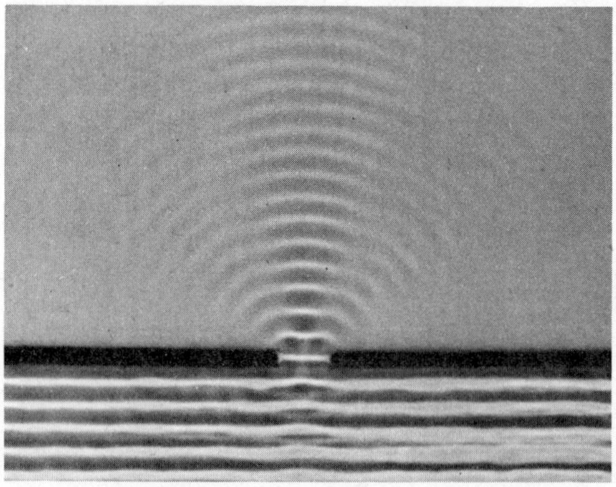

Figure 36.6 Water waves in a ripple tank, which demonstrates Huygens' wavelets. A plane wave is incident on a barrier with a small opening. The opening acts as a source of circular wavelets. (Photo, Education Development Center, Newton, Mass.)

at the left of the slit emerge on the right of the slit as a two-dimensional circular wave propagating outward.

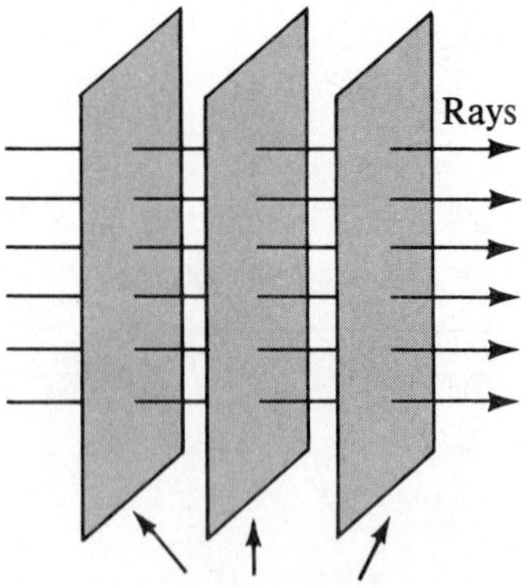

Figure 36.7 A plane wave propagating to the right. Note that the rays, corresponding to the direction of wave motion, are straight lines perpendicular to the wavefronts.

36.4 THE RAY APPROXIMATION IN GEOMETRIC OPTICS

In geometric optics (Chapter 37), we shall use the so-called *ray approximation*. To understand this approximation, first recall that the direction of energy flow of a wave, corresponding to the direction of wave propagation, is called a *ray*. For a given plane wave, the rays are straight lines that are perpendicular to the wavefronts, as illustrated in Fig. 36.7 for a plane wave. In the ray approximation, we assume that the wave moving through a given medium travels in a straight line in the direction of the rays.

If the wave meets a barrier with a circular opening whose diameter is large relative to the wavelength, as in Fig. 36.8a, the wave emerging from the opening continues to move in a straight line (apart from some small edge effects); hence, the ray approximation continues to be valid. On the other hand, if the diameter of the opening in the barrier is of the order of the wavelength, as in Fig. 36.8b, the waves spread out from the opening in all directions. We say that the outgoing wave is noticeably *diffracted*. Finally, if the opening is small relative to the wavelength, the opening can be approximated as a point source of waves (Fig. 36.8c). Thus, the effect of diffraction is more pronounced as the ratio d/λ approaches zero. Similar

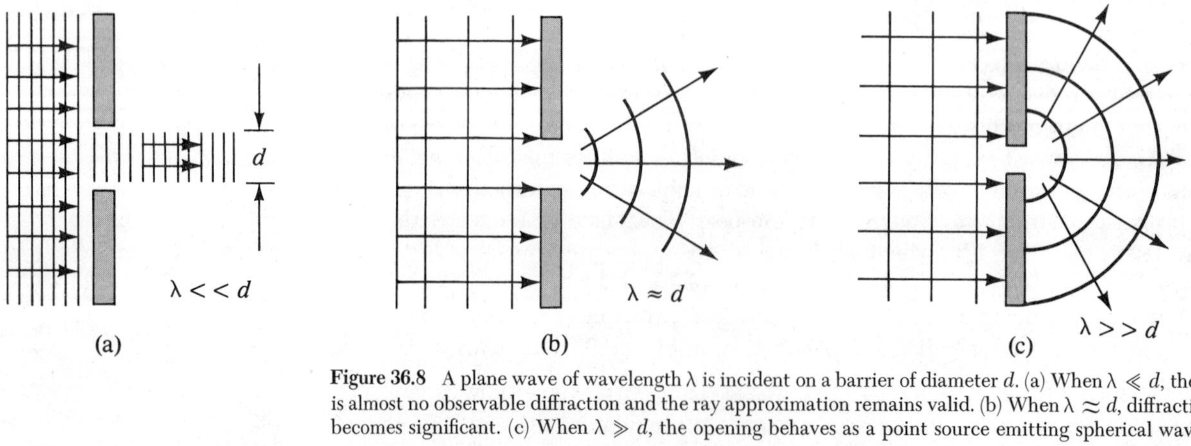

Figure 36.8 A plane wave of wavelength λ is incident on a barrier of diameter d. (a) When $\lambda \ll d$, there is almost no observable diffraction and the ray approximation remains valid. (b) When $\lambda \approx d$, diffraction becomes significant. (c) When $\lambda \gg d$, the opening behaves as a point source emitting spherical waves.

effects are seen when waves encounter an opaque circular object. In this case, when $\lambda \ll d$, the object casts a sharp shadow.

The ray approximation and the assumption that $\lambda \ll d$ will be used here and in Chapter 37, both dealing with geometric optics. This approximation is very good for the study of mirrors, lenses, prisms, and associated optical instruments, such as telescopes, cameras, and eyeglasses. We shall return to the subject of diffraction (where $\lambda \geq d$) in Chapter 39.

Q1. Light of wavelength λ is incident on a slit of width d. Under what conditions is the ray approximation valid? Under what circumstances will the slit produce significant diffraction?

36.5 THE LAWS OF REFLECTION AND REFRACTION AT PLANAR SURFACES

Reflection

When a light ray traveling in a given medium encounters a boundary leading into a second medium, part of the incident ray is reflected back into the first medium and part is transmitted into the second medium. The ray transmitted into the second medium is bent at the boundary and is said to be *refracted.*

When a beam of light consisting of many rays is incident on a smooth, mirrorlike, reflecting surface, as in Fig. 36.9a, each reflected ray will be parallel as indicated. Reflection of light from such a smooth object is called *specular reflection.* On the other hand, if the reflecting surface is rough, as in Fig. 36.9b, the surface will reflect the rays in various directions. Reflection from any rough surface is known as *diffuse reflection.* A surface will behave as a smooth surface as long as the surface variations are small compared with the wavelength of the incident light.

For instance, consider the two types of reflection one can observe from a road's surface while driving a car at night. When the road is dry and rough, light from oncoming vehicles is scattered off the road in different directions, making the road quite visible. On a rainy night, when the road is wet, the road irregularities are filled with water and the surface becomes quite smooth. The resulting specular reflection makes it difficult to see the road clearly. In this book, we shall concern ourselves only with specular reflection. In what follows, the word *reflection* will mean *specular reflection.*

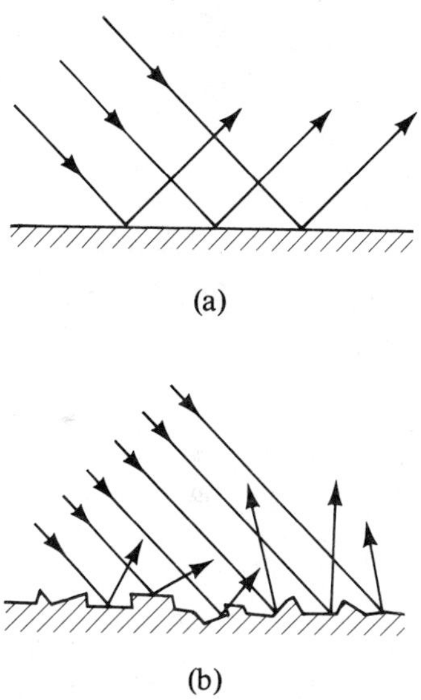

Figure 36.9 Schematic representation of (a) specular reflection, where the reflected rays are parallel, and (b) diffuse reflection, where the reflected rays travel in random directions.

Consider a light ray traveling in air and obliquely incident on a planar, smooth surface (Fig. 36.10). The incident and reflected rays make angles of θ_1 and θ'_1, respectively, with the normal. Experiments show that *the angle of reflection equals the angle of incidence,* that is,

$$\theta'_1 = \theta_1 \tag{36.2}$$

Law of reflection

Refraction

When the incident light ray encounters a boundary leading into another medium as in Fig. 36.10, part of the ray is also transmitted into the second medium. This ray is bent at the boundary and is said to be *refracted.* Furthermore, the refracted ray, the incident ray, and the reflected ray are all coplanar. The angle of refraction, θ_2, depends on the properties of the two media and the angle of incidence through the

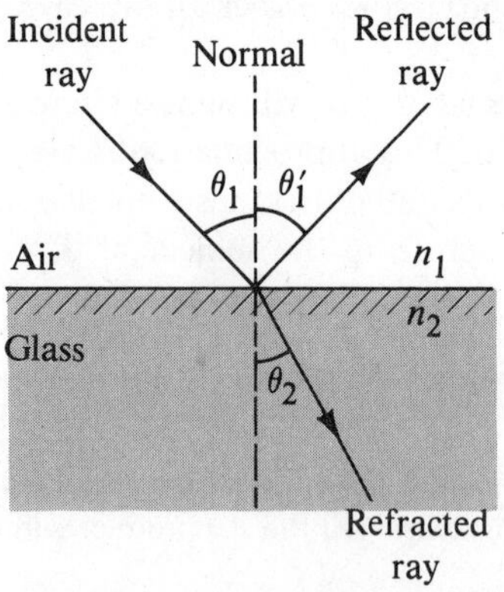

Figure 36.10 A ray obliquely incident on an air-glass interface. The refracted ray is bent toward the normal since $n_2 > n_1$ and $v_2 > v_1$.

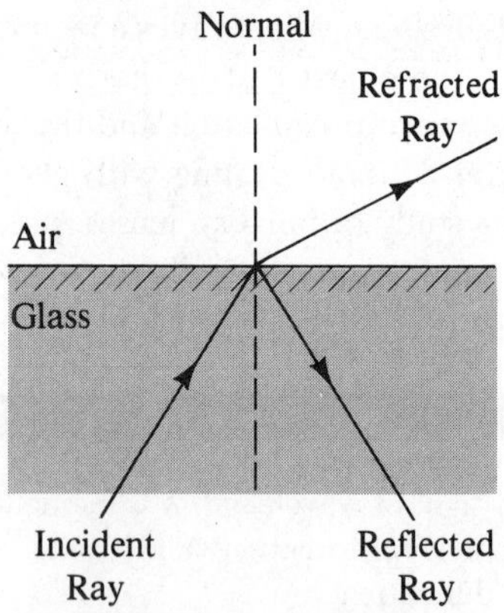

Figure 36.11 The path of a ray traveling from glass into air is the reverse of the path taken when the ray travels from air into glass.

relationship

Snell's law of refraction

$$\frac{\sin\theta_2}{\sin\theta_1} = \frac{v_2}{v_1} = \text{constant} \tag{36.3}$$

where v_1 is the speed of light in medium 1 and v_2 is the speed of light in medium 2. The experimental discovery of this relationship is usually credited to Willebrord Snell (1591–1627) and is therefore known as Snell's law.[4] In Section 36.8, we shall verify these two laws using Huygens' principle.

Laws of reflected and refracted rays

In summary, the three laws governing reflected and refracted rays are

1. The incident ray, the reflected ray, the refracted ray, and the normal to the surface all lie in the same plane.
2. The angle of reflection, θ_1', equals the angle of incidence, θ_1; this is *the law of reflection*.
3. The angle of refraction, θ_2, is related to the angle of incidence, θ_1, through *Snell's law* (Eq. 36.3).

These laws are applicable to the passage of light from any medium to any other medium, for example, when a light ray moves from glass to air (Fig. 36.11). Just as before, the laws of reflection and refraction apply to the reflected and refracted rays. When they are applied, it is found that the path of the ray in going from the glass to air is *reversible*. That is, the path followed by a ray as it moves from *A* to *B* between two media is the same as the path it follows in moving from *B* to *A*. This important result is known as the *reciprocity principle*.

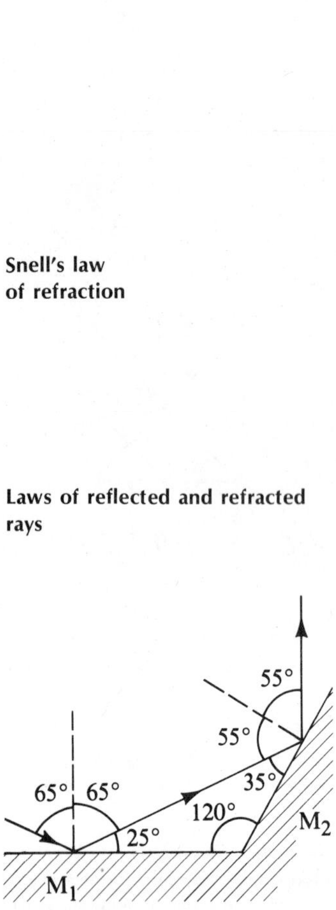

Figure 36.12 (Example 36.1). Mirrors M_1 and M_2 make an angle of 120° with each other.

[4] This law was also deduced from the particle theory of light by René Descartes (1596–1650) and hence is known as *Descartes' law* in France.

Example 36.1

Two mirrors make an angle of 120° with each other, as in Fig. 36.12. A ray is incident on mirror M_1 at an angle of 65° to the normal. Find the direction of the ray after it is reflected from mirror M_2.

Solution: From the law of reflection, we see that the first reflected ray also makes an angle of 65° with the normal. Thus, it follows that this same ray makes an angle of 90° − 65°, or 25°, with the horizontal. From the triangle made by the first reflected ray and the two mirrors, we see that the first reflected ray makes an angle of 35° with M_2 (since the sum of the interior angles of any triangle is 180°). This also means that this ray makes an angle of 55° with the normal to M_2. Hence, from the law of reflection, it follows that the second reflected ray makes an angle of 55° with the normal to M_2. You should be able to show that the second reflected ray makes an angle of 85° with M_1.

Q2. Sound waves have much in common with light waves, including the properties of reflection and refraction. Give examples of such phenomena for sound waves.

Q3. Does a light ray traveling from one medium into another always bend toward the normal as in Fig. 36.10? Explain.

36.6 THE INDEX OF REFRACTION

When light passes from one medium to another, it is refracted because of the difference in the speed of light in the two media. The reason for this variation in speed is rather complex and can be properly explained only with an atomic model. In general, one finds that the speed of light in any material medium is less than the speed of light in a vacuum, where you will recall that $c = 3 \times 10^8$ m/s. Therefore, light travels at its maximum speed in a vacuum. It is convenient to define the *absolute index of refraction, n,* of a medium to be the ratio

$$n \equiv \frac{\text{speed of light in a vacuum}}{\text{speed of light in the medium}} = \frac{c}{v} \qquad (36.4)$$

Index of refraction

From this definition, we see that the index of refraction is a dimensionless number greater than unity, since $v < c$. Furthermore, $n \equiv 1$ for a vacuum. The indices of refraction for various substances measured with respect to a vacuum are listed in Table 36.1.

TABLE 36.1 Index of Refraction for Various Substances Measured with Light of Vacuum Wavelength λ_0 = 589 nm = 5890 Å

SUBSTANCE	*INDEX OF REFRACTION*	*SUBSTANCE*	*INDEX OF REFRACTION*
Solids at 20°C		*Liquids at 20°C*	
Diamond (C)	2.419	Benzene	1.501
Fluorite (CaF_2)	1.434	Carbon disulfide	1.628
Fused quartz (SiO_2)	1.458	Carbon tetrachloride	1.461
Glass, crown	1.52	Ethyl alcohol	1.361
Glass, flint	1.66	Glycerine	1.473
Ice (H_2O)	1.309	Water	1.333
Polystyrene	1.49	*Gases at 0°C, 1 atm*	
Sodium chloride (NaCl)	1.544	Air	1.000293
Zircon	1.923	Carbon dioxide	1.00045

As light travels from one medium to another, the frequency f of the wave does not change. Therefore, since the relation $v = f\lambda$ must be valid in both media, we see that

$$v_1 = f\lambda_1 \qquad \text{and} \qquad v_2 = f\lambda_2$$

where the subscripts refer to the two media. Dividing the two equations and using Eq. 36.4, we see that

$$\frac{\lambda_1}{\lambda_2} = \frac{v_1}{v_2} = \frac{c/n_1}{c/n_2} = \frac{n_2}{n_1} \qquad (36.5)$$

$$\lambda_1 n_1 = \lambda_2 n_2 \qquad (36.6)$$

If medium 1 is a vacuum, then $n_1 = 1$. Hence, it follows from Eq. 36.5 that the

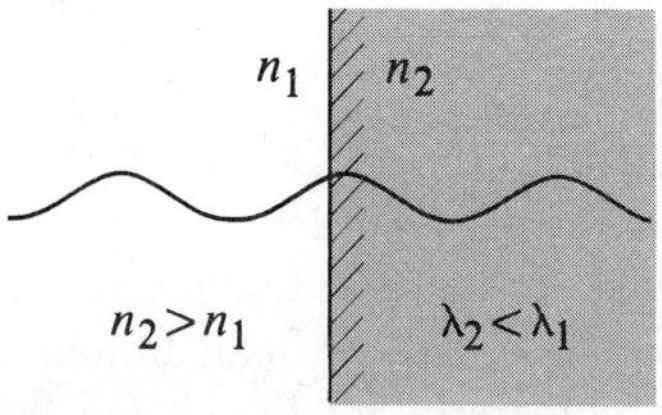

Figure 36.13 Schematic diagram of the *reduction* in wavelength when light travels from a medium of low optical density to one of higher density, that is, $n_2 > n_1$.

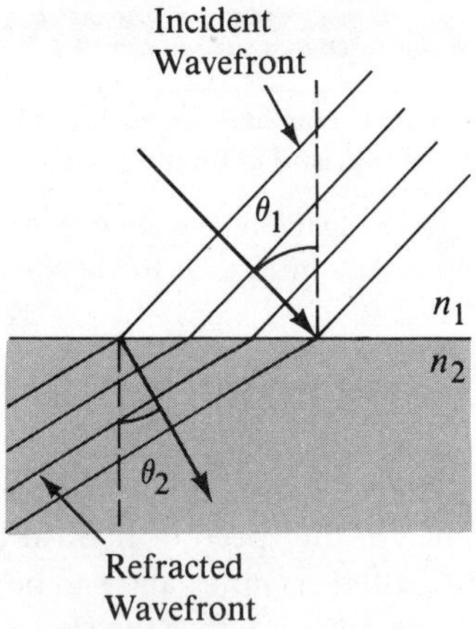

Figure 36.14 Schematic diagram of the law of refraction. Note the matching of the wavefronts at the boundary between the two media.

Index of refraction and wavelength

index of refraction of any medium can be expressed as the ratio

$$n = \frac{\lambda_o}{\lambda_n} \tag{36.7}$$

where λ_o is the wavelength of light in a vacuum and λ_n is the wavelength in the medium whose index of refraction is n. A schematic representation of this reduction in wavelength is shown in Fig. 36.13.

Snell's law of refraction

We are now in a position to express Snell's law in an alternate form that involves the parameters θ_1, θ_2, n_1, and n_2. Substituting Eq. 36.5 into Eq. 36.3, we find that

$$n_1 \sin\theta_1 = n_2 \sin\theta_2 \tag{36.8}$$

This is the most widely used and practical form of Snell's law.

A schematic diagram illustrating the law of refraction is given in Fig. 36.14. In this diagram, the wavelength in medium 1 is greater than the wavelength in medium 2, but the wavefronts are *matched* at the boundary.

Example 36.2

A beam of light of wavelength 550 nm is incident on a slab of transparent material. The incident beam makes an angle of 40° with the normal, and the refracted beam makes an angle of 26° with the normal. (a) Find the index of refraction of the material.

Using Snell's law of refraction (Eq. 36.8) with $\theta_1 = 40°$, $n_1 = 1$, and $\theta_2 = 26°$ gives

$$n_1 \sin\theta_1 = n_2 \sin\theta_2$$

$$n_2 = \frac{n_1 \sin\theta_1}{\sin\theta_2} = \frac{(1)\sin 40°}{\sin 26°} = \frac{0.643}{0.440} = 1.46$$

If we compare this value with the data in Table 36.1, we see that the material is probably fused quartz.

(b) What is the wavelength of light in the material?

The wavelength of light in the medium of refractive index n can be obtained from Eq. 36.7, where $n = 1.46$ in this example:

$$\lambda_n = \frac{\lambda_o}{n} = \frac{550\text{ nm}}{1.46} = 377\text{ nm}$$

Q4. As light travels from one medium to another, does its wavelength change? Does its frequency change? Does its velocity change? Explain.

Q5. A laser beam passing through a nonhomogeneous sugar solution is observed to follow a curved path. Explain.

Q6. A laser beam ($\lambda = 6328$ Å) is incident on a piece of lucite as in the photograph. Part of the beam is reflected and part is refracted. What information can you get from this photograph?

Q7. Suppose blue light were used instead of red light in the experiment shown in the photograph. Would the refracted beam be bent at a larger or smaller angle?

Q8. The level of water in a clear, colorless glass is easily observed with the naked eye. The level of liquid helium in a clear glass vessel is extremely difficult to see with the naked eye. Explain.

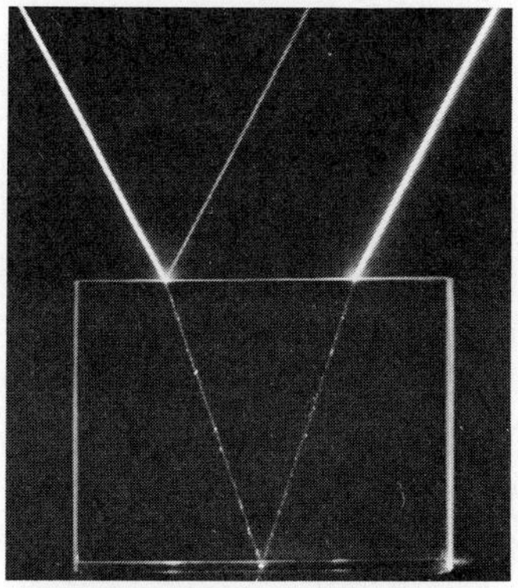

Light from a helium-neon laser beam ($\lambda = 6328$ Å) is incident on a block of lucite. The photograph shows both reflected and refracted rays. Can you identify the incident, reflected and refracted rays? From this photograph, estimate the index of refraction of lucite at this wavelength. (Courtesy of Hugh Strickland and Jim Lehman, James Madison University.)

36.7 DISPERSION AND PRISMS

Another important property of the index of refraction is the fact that n varies with wavelength, as shown by Eq. 36.7. The variation of the index of refraction with wavelength for various materials is shown in Fig. 36.15. This behavior is due to the variation of the speed of light in the material with wavelength (Eq. 36.4). Since n is a function of wavelength, the degree to which a light beam is bent upon entering a substance obliquely depends both on the properties of the substance and on the color (wavelength) of the light. As we see from Fig. 36.15, the index of refraction decreases with increasing wavelength. For instance, blue light will be bent more than red light.

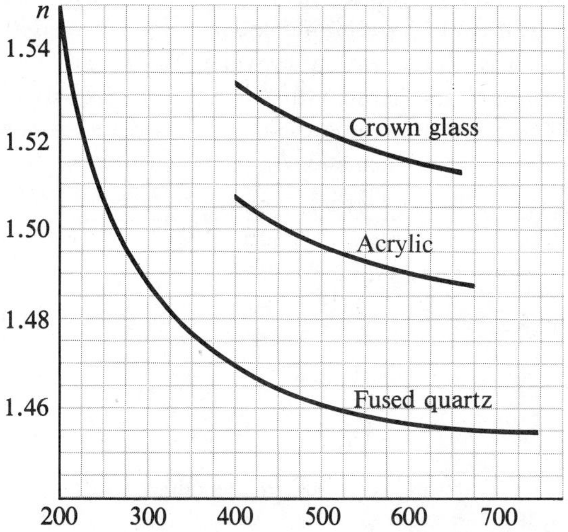

Figure 36.15 Variation of index of refraction with wavelength for several materials.

Dispersion

Any substance in which n varies with wavelength is said to exhibit *dispersion*. To understand the meaning of dispersion, first we must recognize that when light of a single wavelength passes through a piece of glass with nonparallel surfaces as in Fig. 36.16, the emerging light ray experiences a net change of angle δ, called the *angle of deviation*. This particular three-sided geometry is called a *triangular prism*. Now suppose a beam of white light (a combination of all visible wavelengths) is incident on a prism as in Fig. 36.17. As Newton himself showed, the rays that emerge from the second face spread out in a series of colors known as a *spectrum*. These colors, in order of decreasing wavelength, are red, orange, yellow, green, blue, and violet. Clearly, the deviation of light, measured by the angle of deviation δ, depends on the wavelength of a given spectral component. Violet light deviates the most, red light deviates the least, and remaining colors in the visible spectrum fall between these extremes. When light is spread out by a substance such as the prism in Fig. 36.17, the light is said to be *dispersed into a spectrum*.

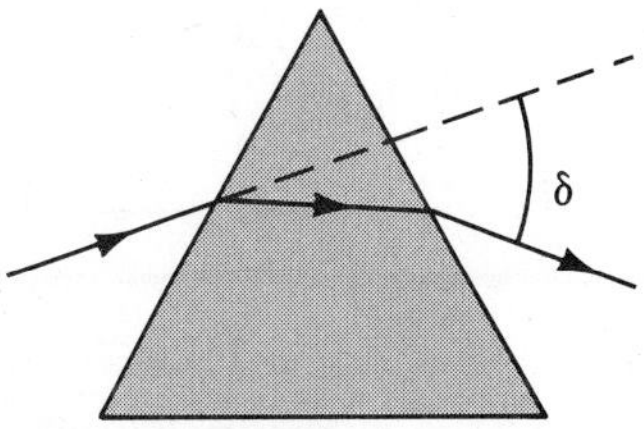

Figure 36.16 A prism refracts a light ray and deviates the light through an angle δ.

A prism is often used in an instrument known as a *prism spectrometer*, the essen-

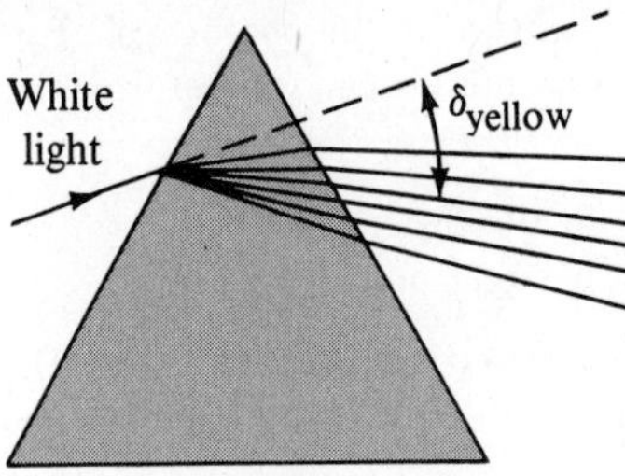

Figure 36.17 Dispersion of white light by a prism. Since n varies with wavelength, the prism disperses the white light into its various spectral components, or colors.

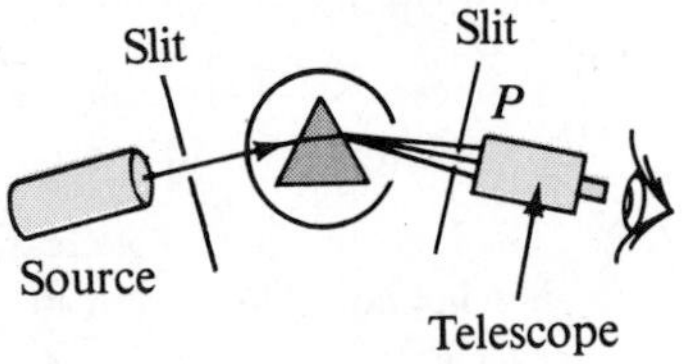

Figure 36.18 Diagram of a prism spectrometer. The various colors in the spectrum are viewed through a telescope.

tial elements of which are shown in Fig. 36.18. The instrument is commonly used to study the spectral characteristics of a light source, such as a sodium vapor lamp. Light from the source is sent through a narrow, adjustable slit and lens to produce a parallel, or collimated, beam. The light then passes through the prism and is dispersed into a spectrum. The refracted light is observed through a telescope. The experimenter sees an image of the slit through the eyepiece of the telescope. The telescope can be moved or the prism can be rotated in order to view the various wavelengths, which have differing angles of deviation.

A similar arrangement is used in another common instrument known as a *prism spectrophotometer*. This type of instrument is used to study the optical absorption characteristics of various substances. The unknown sample under study is placed at a point such as P (Fig. 36.18) so that it is exposed to one wavelength of light. The light intensity passing through the sample is measured with a detector that is particularly sensitive to visible light. In some systems, the sample is exposed to various wavelengths by inserting a rotatable mirror between a fixed prism and the detector.

Example 36.3

A light ray of wavelength 589 nm (produced by a sodium lamp) is incident on a smooth, flat slab of crown glass at an angle of 30° to the normal, as sketched in Fig. 36.19. Find the angle of refraction, θ_2, if medium 1 is air.

Solution: Using Snell's law given by Eq. 36.8, we have

$$\sin\theta_2 = \frac{n_1}{n_2}\sin\theta_1$$

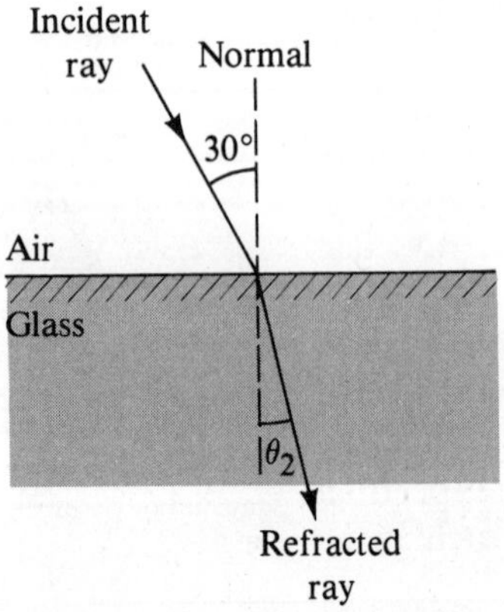

Figure 36.19 (Example 36.3) Refraction of light by glass.

From Table 36.1 we find that $n_1 = 1.00$ for air and $n_2 = 1.52$ for glass. Therefore, we get

$$\sin\theta_2 = \frac{1.00}{1.52}\sin 30° = 0.329$$

$$\theta_2 = \sin^{-1}(0.329) = 19.2°$$

Thus we see that the ray is bent *toward* the normal, as expected. It is left as an exercise to show that if the light ray moved from within the glass toward the glass-air interface at an angle of 30° to the normal, it would be refracted at an angle of 49.5° (that is, *away* from the normal).

Example 36.4

Light of wavelength 589 nm in vacuum passes through a piece of fused quartz ($n = 1.458$). (a) Find the speed of light in quartz.

The speed of light in quartz can be easily obtained using Eq. 36.4:

$$v = \frac{c}{n} = \frac{3 \times 10^8 \text{ m/s}}{1.458} = 2.058 \times 10^8 \text{ m/s}$$

(b) What is the wavelength of this light in quartz?

We can use Eq. 36.7 to calculate the wavelength in quartz, noting that we are given $\lambda_o = 589$ nm $= 589 \times 10^{-9}$ m.

$$\lambda_n = \frac{\lambda_0}{n} = \frac{589 \text{ nm}}{1.458} = 404 \text{ nm}$$

Note that the frequency of the light is the same in vacuum and in quartz. You should be able to show that the frequency is given by $f = 5.09 \times 10^{14}$ Hz.

Example 36.5

A light beam passes from medium 1 to medium 2 through a thick slab of material whose index of refraction is n_2 (Fig. 36.20). Show that the emerging beam is parallel to the incident beam.

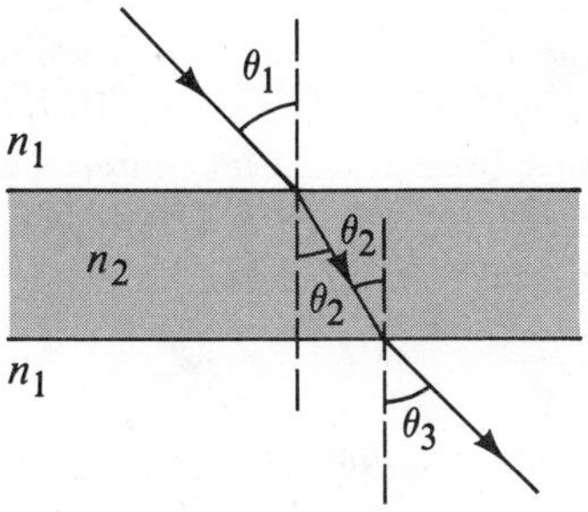

Figure 36.20 (Example 36.5) When light passes through a flat slab of material, the emerging beam is parallel to the incident beam, and therefore $\theta_1 = \theta_3$.

Solution: First, let us apply Snell's law to the upper surface:

$$(1) \qquad \sin\theta_2 = \frac{n_1}{n_2}\sin\theta_1$$

Applying Snell's law to the lower surface gives

$$(2) \qquad \sin\theta_3 = \frac{n_2}{n_1}\sin\theta_2$$

Substituting (1) into (2) gives

$$\sin\theta_3 = \frac{n_2}{n_1}\left(\frac{n_1}{n_2}\sin\theta_1\right) = \sin\theta_1$$

That is, $\theta_3 = \theta_1$, and so the layer does not alter the direction of the beam. It does, however, produce a displacement of the beam. The same result is obtained when light passes through multiple layers of materials.

Example 36.6

Measuring n *using a prism:* A prism is often used to measure the index of refraction of a transparent solid. Although we do not prove it here, one finds that the *minimum angle of deviation*, δ_m, occurs at the angle of incidence θ_1 where the refracted ray inside the prism makes the same angle α with the two prism faces,[5] as in Fig. 36.21. Using the geometry shown, one finds that $\theta_2 = \phi/2$ and

$$\theta_1 = \theta_2 + \alpha = \frac{\phi}{2} + \frac{\delta_m}{2} = \frac{\phi + \delta_m}{2}$$

From Snell's law, we find

$$\sin\theta_1 = n\sin\theta_2$$

$$\sin\left(\frac{\phi + \delta_m}{2}\right) = n\sin(\phi/2)$$

$$n = \frac{\sin\left(\frac{\phi + \delta_m}{2}\right)}{\sin(\phi/2)} \qquad (36.9)$$

Hence, knowing the apex angle ϕ of the prism and measuring δ_m, one can calculate the index of refraction of the prism material. Furthermore, one can use a hollow prism to determine the values of n for various liquids.

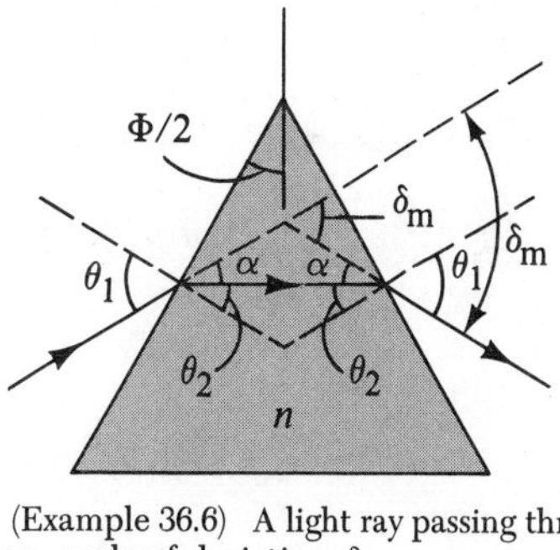

Figure 36.21 (Example 36.6) A light ray passing through a prism at the minimum angle of deviation, δ_m.

[5]For details, see F. A. Jenkins and H. E. White, *Fundamentals of Optics*, New York, McGraw-Hill, 1976, Chapter 2.

36.8 HUYGENS' PRINCIPLE APPLIED TO REFLECTION AND REFRACTION

We can use Huygens' principle to derive the laws of reflection and refraction without concerning ourselves with the details of the mechanism responsible for these phenomena. Figure 36.22 is useful for verifying the law of reflection. The line AA' represents a wavefront of the incident light. As ray 3 travels from A' to C, ray 1 reflects from A and produces a spherical wave of radius AD. Therefore, we see that $A'C = AD$. Meanwhile, the spherical wavelet centered at B has spread only half as far, since ray 2 strikes the surface later than ray 1.

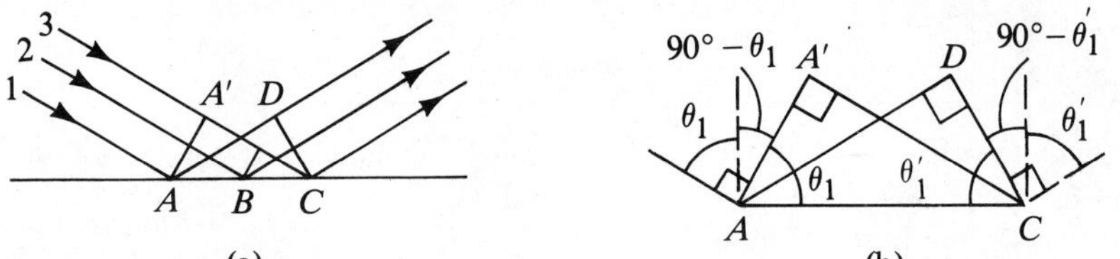

Figure 36.22 (a) Huygens' construction for proving the law of reflection. (b) Note that triangle *ADC* is identical to triangle *AA′C*.

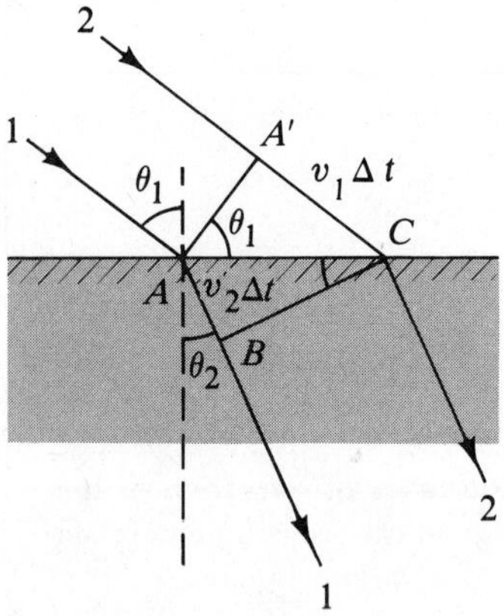

Figure 36.23 Huygens' construction for proving the law of refraction.

Applying Huygens' principle, we find that the *reflected* wavefront is *CD*, a line tangent to all the outgoing spherical wavelets. The rest is geometry. Note that the right triangles *ADC* and *AA′C* are identical, since they have the same hypotenuse and $AD = A'C$. Therefore, we see that $90° - \theta_1' = 90° - \theta_1$, or $\theta_1 = \theta_1'$.

Now let us use Huygens' principle and Fig. 36.23 to derive Snell's law of refraction. Note that in the time interval Δt, ray 1 moves from *A* to *B* while ray 2 moves from *A′* to *C*. The radius of the outgoing spherical wavelet centered at *A* is equal to $v_2\,\Delta t$. The distance *A′C* is equal to $v_1\,\Delta t$. Geometric considerations show that angle *A′AC* equals θ_1, and angle *ACB* equals θ_2. From triangles *AA′C* and *ACB*, we find that

$$\sin\theta_1 = \frac{v_1\,\Delta t}{AC} \quad \text{and} \quad \sin\theta_2 = \frac{v_2\,\Delta t}{AC}$$

Dividing the two equations gives

$$\frac{\sin\theta_1}{\sin\theta_2} = \frac{v_1}{v_2}$$

But from Eq. 36.4, we know that $v_1 = c/n_1$ and $v_2 = c/n_2$. Therefore,

$$\frac{\sin\theta_1}{\sin\theta_2} = \frac{c/n_1}{c/n_2} = \frac{n_2}{n_1}$$

$$n_1 \sin\theta_1 = n_2 \sin\theta_2$$

Another method for deriving the laws of reflection and refraction uses *Fermat's principle*, which states that the path a light beam will follow in traveling between any two points is the path that takes the least time. This principle is discussed further in Section 36.11.

36.9 TOTAL INTERNAL REFLECTION

An interesting phenomenon called *total internal reflection* can occur when light moves from a medium having a high index of refraction to one of a lower index of refraction. Consider a light beam traveling in medium 1 and meeting the interface between medium 1 and medium 2, where $n_1 > n_2$ (Fig. 36.24a). Various possible directions of the beam are indicated by rays 1 through 5. Note that the refracted rays are bent *away* from the normal since $n_1 > n_2$. Clearly, at θ_c, some *critical* value of the angle of incidence, the refracted light ray will move parallel to the interface, so that $\theta_2 = 90°$ (Fig. 36.24b). *For angles of incidence greater than θ_c the beam is entirely reflected at the interface!*

We can use Snell's law to find the critical angle. When $\theta_1 = \theta_c$, $\theta_2 = 90°$, and Snell's law (Eq. 36.8) gives

$$n_1 \sin\theta_c = n_2 \sin 90° = n_2$$

Critical angle for total internal reflection

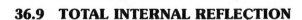

$$\sin\theta_c = \frac{n_2}{n_1} \qquad (36.10)$$

Normal

$\sin \theta_c = \dfrac{n_2}{n_1}$

Normal

(a)

(b)

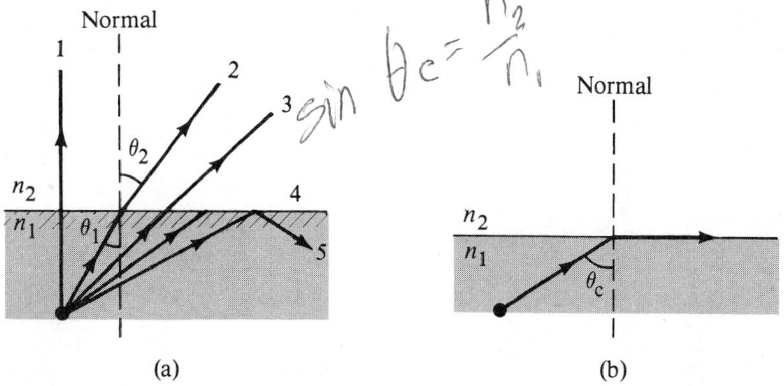

Figure 36.24 (a) A ray traveling from a medium of index n_1 to a medium of index n_2, where $n_1 > n_2$. As the angle of incidence increases, the angle of refraction increases until $\theta_2 \to 90°$ (ray 4). For even larger angles of incidence, total internal reflection occurs (ray 5). (b) The angle of incidence producing an angle of refraction equal to 90° is often called the *critical angle*, θ_c.

(a)

(b)

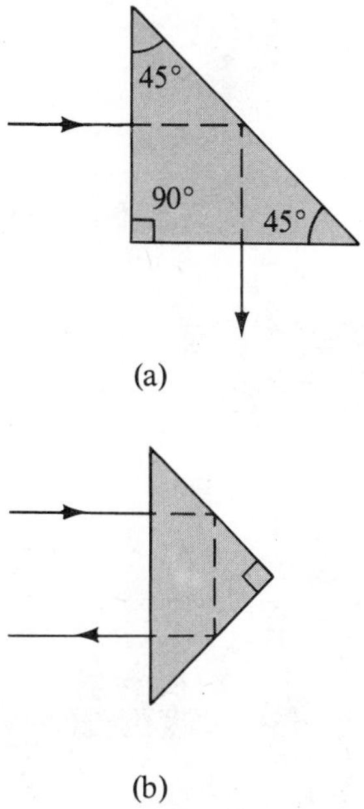

Figure 36.25 Internal reflection in a prism. (a) The ray is deviated by 90°. (b) The direction of the ray is reversed.

where $n_1 > n_2$. Note that this expression can be used only when $n_1 > n_2$. If $n_1 < n_2$, the expression would give $\sin\theta_c > 1$, which is an absurd result since the sine of an angle can never exceed unity.

From Eq. 36.10, we see that θ_c decreases as the ratio n_2/n_1 increases. Hence, the critical angle is small for substances with a large index of refraction, such as diamond, where $n_1 = 2.42$ and $\theta_c = 24°$. For crown glass, $n_1 = 1.52$ and $\theta_c = 41°$. In fact, this property combined with proper shaping causes diamonds and crystal glass to "sparkle."

One can use a prism and the phenomenon of total internal reflection to alter the course of a light beam. Two such possibilities are illustrated in Fig. 36.25. In one case the light beam is deflected by 90° (Fig. 36.25a), whereas in the second case the path of the beam is reversed (Fig. 36.25b).

Example 36.7

Calculate the critical angle for fused quartz, for which the index of refraction is 1.458 at a wavelength of 589 nm.

Solution: Taking $n = 1.00$ for air and using Eq. 36.9, we get

$$\sin\theta_c = \frac{n_2}{n_1} = \frac{1.00}{1.458} = 0.6859$$

$$\theta_c = 43.3°$$

Fiber Optics

Another interesting application of total internal reflection is the use of glass or transparent plastic rods to "pipe" light from one place to another. As indicated in Fig. 36.26, light is confined to travel along the rods, even around gentle curves, as the result of successive internal reflections. The "light pipe" is made flexible with the use of fibers rather than thick rods. If a bundle of parallel fibers is used to construct an optical transmission line, images can be transferred from one point to another.[6] This technique is used in a sizable industry known as *fiber optics*. There is very little light intensity lost in these fibers as a result of reflections on the sides. Any loss in intensity is due essentially to reflections from the two ends and absorption by the fiber material. These devices are particularly useful when one wishes to view an image produced at inaccessible locations. For example, physicians often use this

---

[6]See Narinder S. Kapany, "Fiber Optics," *Scientific American*, November 1960, for further details.

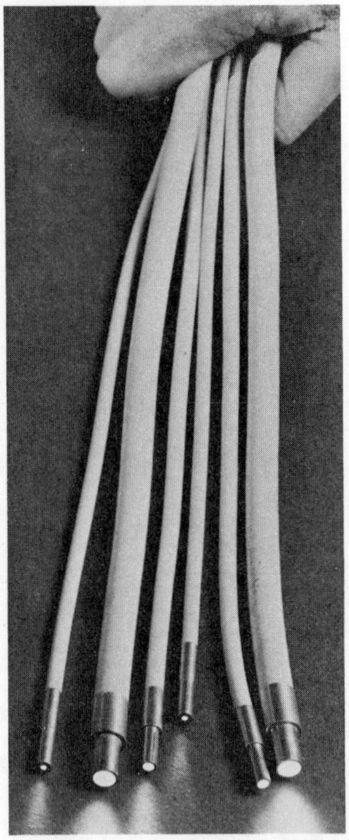

Light pipe, (Edmund Scientific Co.)

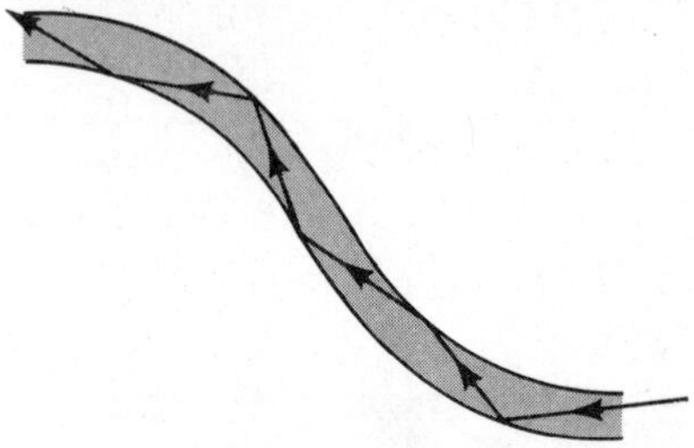

Figure 36.26 Light travels in a curved, transparent rod by multiple internal reflections.

technique to examine various internal organs of the body. Fiber optics are finding increasing use for telecommunications, since they can carry a much higher information density than electrical wires.

Q9. Describe an experiment in which internal reflection is used to determine the index of refraction of a medium.

Q10. Why does a diamond show flashes of color when observed under ordinary white light?

Q11. Explain why a diamond shows more "sparkle" than a glass crystal of the same shape and size.

36.10 LIGHT INTENSITY

So far we have been concerned only with the changes in direction of a light beam upon reflection and refraction at the interface of two media. We now describe how one can calculate the quantity of light reflected or refracted in terms of the energy carried by the incident beam. In general, one finds that the fraction of light reflected and transmitted depends on the indices of refraction of the two media and the angle of incidence. The required expressions can be obtained using Maxwell's equations and appropriate boundary conditions. Since the formalism is rather complex, we shall give only a qualitative description of the factors influencing the relative intensities.[7]

Consider a light beam striking the interface between two media of refractive indices n_1 and n_2, as in Fig. 36.27. Let us assume that the incident light strikes the interface at near-normal incidence, so that θ_1 and θ_2 are very small. If I_o represents the incident light intensity, I_r the reflected intensity, and I_t the transmitted intensity, one finds that the reflected and transmitted intensities are given by

Fresnel's equations

$$I_r = \left(\frac{n_2 - n_1}{n_1 + n_2}\right)^2 I_o \qquad (36.11)$$

$$I_t = \frac{4n_1n_2}{(n_1 + n_2)^2} I_o \qquad (36.12)$$

These expressions, known as *Fresnel's equations*, enable us to calculate I_r and I_t from a knowledge of the indices of refraction and the incident light intensity. When the angle of incidence is not small, one obtains complicated relationships between the intensities, which involve θ_1 and θ_2.

[7]For a more detailed treatment, see F. W. Sears, *Optics*, Reading, Mass., Addison-Wesley, 1949, Chapter 7.

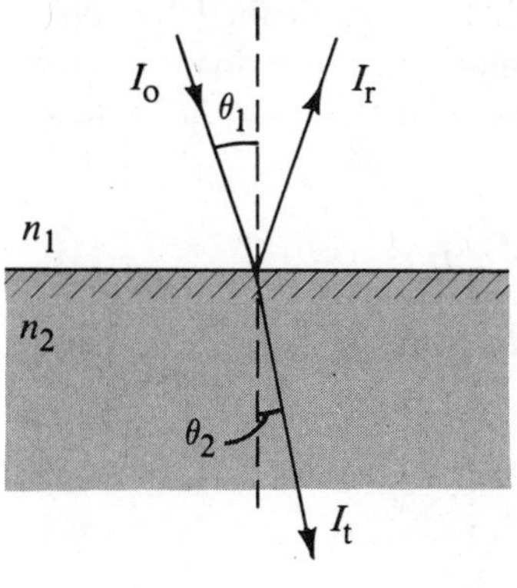

Figure 36.27 The incident light I_o strikes the surface at near-normal incidence. The reflected light intensity is I_r, and the transmitted intensity is I_t.

Note that for normal incidence and when $n_1 = n_2$, we see from Eqs. 36.11 and 36.12 that $I_r = 0$ and $I_t = I_o$. That is, no light is reflected in this case, as we would expect, since, in effect, there is no interface. If $n_2 \gg n_1$, we find that $I_r \approx I_o$ and $I_t \approx 0$. That is, most of the light is reflected when it is incident on a mirror or on any shiny, metallic surface. Furthermore, in the general case, if energy is to be conserved, the sum of the reflected and transmitted intensities must equal the incident intensity, or

$$I_o = I_r + I_t \tag{36.13}$$

It is left as an exercise for you to verify that Eqs. 36.11 and 36.12 are consistent with Eq. 36.13.

Example 36.8 *The Glass-Air Interface*

A light beam of intensity I_o in air is incident normally on a slab of crown glass of index of refraction 1.52. (a) Calculate the reflected and transmitted light intensity.

Using Eqs. 36.11 and 36.12 with $n_2 = 1.52$ and $n_1 = 1.00$, we find that

$$I_r = \left(\frac{n_2 - n_1}{n_1 + n_2}\right)^2 I_o = \left(\frac{1.52 - 1.00}{1.00 + 1.52}\right)^2 I_o = 0.043 I_o$$

$$I_t = \frac{4n_1 n_2}{(n_1 + n_2)^2} I_o = \frac{4(1.00)(1.52)}{(1.00 + 1.52)^2} I_o = 0.957 I_o$$

That is, only 4.3% of the incident light is reflected, and 95.7% is transmitted into the glass. Furthermore, we see that $I_r + I_t = I_o$.

(b) Repeat the calculations for silicon, for which $n = 3.54$.

Using the same procedure as in (a), but with $n_2 = 3.54$, we find that 31% of the light is reflected and 69% is transmitted. Thus, we see that silicon is a better reflecting material than glass.

Absorption of Transmitted Light

In reality, when light travels through a homogeneous, dielectric material, some of the light intensity is *absorbed* by the medium. Consequently, the light intensity decreases with increasing penetration. In air, the absorption is very small, whereas in many liquids and solids, the absorption is very significant. Clearly, a deeply colored object is highly absorbing at certain visible wavelengths, and a transparent material is weakly absorbing in the visible region. The energy absorbed by the material is continuously transferred into thermal energy of atoms and molecules, but the specific absorption mechanism can be quite complicated.

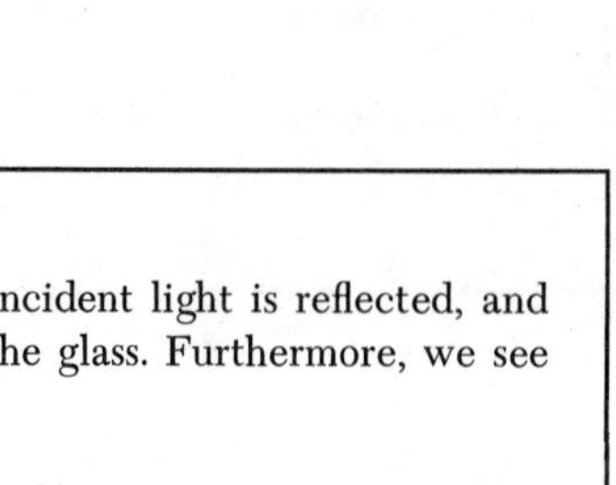

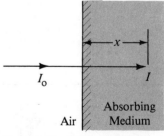

Figure 36.28 Light of intensity I_o is incident on a dielectric material. The intensity at a depth x is reduced to I as a result of absorption by the material.

Suppose a light beam of intensity I_o is incident normally on a material, as in Fig. 36.28. We wish to find the intensity of the transmitted light, I, after it has traveled a distance x into the material. One can obtain a simple expression for I based on the assumptions that the change in light intensity, ΔI, after the light travels a distance Δx is proportional to the light intensity itself and to the distance Δx. That is, $\Delta I = -\alpha I\,\Delta x$, where α is a proportionality constant called the *absorption coefficient* of the medium and the minus sign signifies a decrease in intensity with increasing depth. In the limit of small depths, where $\Delta x \rightarrow dx$ and $\Delta I \rightarrow dI$, we can write this equation in the form

$$\frac{dI}{I} = -\alpha\,dx$$

To solve this familiar equation, we integrate, noting that at the interface, $x = 0$ and $I = I_o$. The result is

$$I = I_o e^{-\alpha x} \qquad (36.14)$$

This equation shows that *the light intensity decreases exponentially with increasing depth.* Although this simple expression has many useful applications, it is important to remember that α is usually a function of the wavelength. For this reason, one must use Eq. 36.14 with care.

Example 36.9 *Light absorption in silicon*

A slab of silicon crystal, used in the fabrication of a solar cell, has an absorption coefficient of about 10^3 cm^{-1} at a particular visible wavelength. At what depth in the crystal will normally incident light be reduced to half its intensity at the surface?

Solution: Using Eq. 36.14, we want to find x such that $I = 0.50I_o$. Therefore,

$$I = I_o e^{-\alpha x} = 0.5I_o$$

Since $\ln e = 1$, we get

$$e^{-\alpha x} = 0.5$$

Taking the natural logarithm of both sides with $\alpha = 10^3$ cm^{-1} gives

$$-\alpha x \ln e = \ln 0.5$$

$$x = \frac{-0.693}{-10^3\ \text{cm}^{-1}} = 6.93 \times 10^{-4}\ \text{cm} = 6.93\ \mu\text{m}$$

Hence, the intensity drops off rapidly in a very short distance. Similarly, one finds that the light intensity drops to 10% of its initial intensity at a depth of about 2.3×10^{-3} cm $= 24\ \mu$m.

36.11 FERMAT'S PRINCIPLE

Statement of Fermat's principle

A general principle that can be used for determining the actual paths of light rays was developed by Pierre de Fermat (1601–1665). *Fermat's principle* states that *when a light ray travels between any two points P and Q, its actual path will be the one that requires the least time.* Fermat's principle is sometimes called the *principle of least time.* An obvious consequence of this principle is that when the rays travel in a single, homogeneous medium, the paths are straight lines because a straight line is the shortest distance between two points. Let us illustrate how to use Fermat's principle to derive the law of refraction.

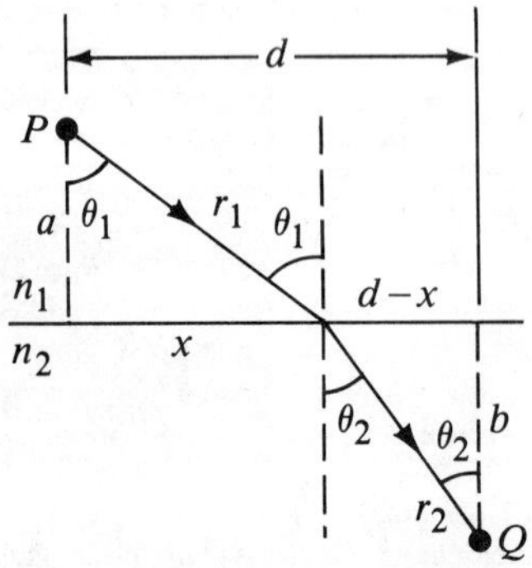

Figure 36.29 Geometry for deriving the law of refraction using Fermat's principle.

Suppose a light ray is to travel from P to Q, where P is in medium 1 and Q is in medium 2 (Fig. 36.29). The points P and Q are at perpendicular distances a and b, respectively, from the interface. The speed of light is c/n_1 in medium 1 and c/n_2 in medium 2. Using the geometry of Fig. 36.29, we see that the time it takes the ray to travel from P to Q is

$$t = \frac{r_1}{v_1} + \frac{r_2}{v_2} = \frac{\sqrt{a^2 + x^2}}{c/n_1} + \frac{\sqrt{b^2 + (d - x)^2}}{c/n_2}$$

We obtain the least time, or the minimum value of t, by taking the derivative of t with respect to x (the variable) and setting the derivative equal to zero. Using this procedure, we get

$$\frac{dt}{dx} = \frac{n_1}{c}\frac{d}{dx}(a^2 + x^2)^{1/2} + \frac{n_2}{c}\frac{d}{dx}[b^2 + (d - x)^2]^{1/2}$$

$$= \frac{n_1}{c}\left(\frac{1}{2}\right)\frac{2x}{(a^2 + x^2)^{1/2}} + \frac{n_2}{c}\left(\frac{1}{2}\right)\frac{2(d - x)(-1)}{[b^2 + (d - x)^2]^{1/2}}$$

$$\frac{dt}{dx} = \frac{n_1 x}{c(a^2 + x^2)^{1/2}} - \frac{n_2(d - x)}{c[b^2 + (d - x)^2]^{1/2}} = 0$$

From Fig. 36.29 and recognizing $\sin\theta_1$ and $\sin\theta_2$ in this equation, we find that

$$n_1 \sin\theta_1 = n_2 \sin\theta_2$$

which is Snell's law of refraction.

It is a simple matter to use a similar procedure to derive the law of reflection. The calculation is left for you to carry out (Problem 7).

36.12 SUMMARY

Huygens' principle states that all points on a wavefront can be taken as point sources for the production of secondary wavelets. At some later time, the new position of the wavefront is the surface tangent to these secondary wavelets.

Huygens' principle

In geometric optics, we use the so-called *ray approximation,* in which we assume that a wave travels through a medium in straight lines in the direction of the rays. Furthermore, we neglect diffraction effects, which is a good approximation as long as the wavelength is short compared with any aperture dimensions.

The basic laws of geometric optics are the *laws of reflection and refraction* for light rays. The law of reflection states that the angle of reflection, θ_1', equals the angle of incidence, θ_1. The law of refraction, or *Snell's Law,* states that

$$n_1 \sin\theta_1 = n_2 \sin\theta_2 \tag{36.8}$$

Snell's law of refraction

where θ_2 is the angle of refraction and n_1 and n_2 are the indices of refraction in the two media.

The absolute index of refraction of a medium, n, is defined by the ratio

$$n \equiv \frac{c}{v} \tag{36.4}$$

Index of refraction

where c is the speed of light in a vacuum and v is the speed of light in the medium. In general, n varies with wavelength and is given by

$$n = \frac{\lambda_0}{\lambda_n} \tag{36.7}$$

Index of refraction and wavelength

where λ_o is the vacuum wavelength and λ_n is the wavelength in the medium.

Total internal reflection can occur when light travels from a medium of high index of refraction to one of lower index of refraction. The minimum angle of incidence, θ_c, for which total reflection occurs at an interface is given by

$$\sin\theta_c = \frac{n_2}{n_1} \quad (\text{where } n_1 > n_2) \tag{36.10}$$

Critical angle for total internal reflection

Fermat's principle Fermat's principle states that when a light ray travels between two points, its path will be the one that requires the least time.

EXERCISES

Section 36.2 Measurements of the Speed of Light

1. In an experiment to measure the speed of light using the apparatus of Fizeau (Fig. 36.4), the distance between light source and mirror was 11.45 km and the wheel had 720 notches. The experimentally determined value of c was 2.998×10^8 m/s. Calculate the minimum angular velocity of the wheel for this experiment.

2. Michelson performed a very careful measurement of the speed of light using an improved version of the technique developed by Fizeau. In one of Michelson's experiments, the toothed wheel was replaced by a wheel with 32 identical mirrors mounted on its perimeter, with the plane of each mirror perpendicular to a radius of the wheel. The total light path was 8 miles in length (obtained by multiple reflections of a light beam within an evacuated tube 1 mile long). For what minimum angular velocity of the mirror would Michelson have calculated the speed of light to be 2.998×10^8 m/s?

3. Experimenters at the National Bureau of Standards have made precise measurements of the speed of light using the property of electromagnetic waves that in vacuum the phase velocity of the waves is $c = \sqrt{1/\mu_0\varepsilon_0}$, where μ_0 (permeability constant) = $4\pi \times 10^{-7}$ $\text{N}\cdot\text{s}^2/\text{C}^2$ and ε_0 (permittivity constant) = 8.854×10^{-12} $\text{C}^2/\text{N}\cdot\text{m}^2$. What value (to four significant figures) does this give for the speed of light in vacuum?

4. As a result of his observations, Roemer concluded that the time interval between successive eclipses of the moon Io by the planet Jupiter increased by 22 min during a 6-month period as the earth moved from a point in its orbit on the side of the sun nearer Jupiter to a position on the side opposite Jupiter (see Fig. 36.3). Using 1.5×10^8 km as the average radius of the earth's orbit about the sun, calculate the speed of light from these data.

Section 36.5 The Laws of Reflection and Refraction at Planar Surfaces and Section 36.6 The Index of Refraction

(Note: In this section if an index of refraction value is not given, refer to Table 36.1. These values correspond to a wavelength of 589 nm [yellow].) However, use $n = 1$ for air and take $c = 3 \times 10^8$ m/s.

5. A light ray in air is incident on a water surface at an angle of 30° *with respect to the normal to the surface.* What is the angle of the refracted ray relative to the normal to the surface?

6. A ray of light in air is incident on a planar surface of fused quartz. The refracted ray makes an angle of 37° with the normal. Calculate the angle of incidence.

7. A light ray initially in water enters a transparent substance at an angle of incidence of 37°, and the transmitted ray is refracted at an angle of 25°. Calculate the speed of light in the transparent material.

8. Light is incident on the interface between air and polystyrene at an angle of 53°. The incident ray, initially traveling in air, is partially transmitted and partially reflected at the surface. What is the angle between the *refracted* and the *reflected* ray?

9. A light source submerged in water sends a beam toward the surface at an angle of incidence of 37°. What is the angle of refraction in air?

10. (a) What is the speed of light in crown glass whose wavelength in vacuum is 589 nm? (b) What thickness of crown glass will equal 100 wavelengths of this light (measured in the glass)?

11. Light of wavelength λ_0 in vacuum has a wavelength of 438 nm in water and a wavelength of 390 nm in benzene. What is the index of refraction of water relative to benzene at the wavelength λ_0?

Section 36.7 Dispersion and Prisms

12. Calculate the index of refraction of an equiangular prism for which the angle of minimum deviation is 37°.

13. A crown glass prism has an apex angle of 15°. What is the angle of minimum deviation of this prism for light of wavelength 525 nm? See Fig. 36.15 for the value of n.

14. Light of wavelength 700 nm is incident on the face of a fused quartz prism at an angle of 75° (with respect to the normal to the surface). The apex angle of the prism is 60°. Use the value of n from Fig. 36.15 and calculate the angle (a) of refraction at this (first) surface, (b) of incidence at the second surface, (c) of refraction at the second surface, and (d) between the incident and emerging rays.

15. Show that if the apex angle ϕ of a prism is small, an approximate value for the angle of minimum deviation can be calculated from $\delta_m = (n - 1)\phi$.

16. For a particular prism and wavelength, $n = 1.62$. Compare the values found for the angle of minimum deviation in this prism when using the approximation of Exercise 15 and the exact form given by Eq. 36.9 when (a) $\phi = 30°$ and (b) $\phi = 10°$.

17. An experimental apparatus includes a prism made of sodium chloride. The angle of minimum deviation for light of wavelength 589 nm is to be 10°. What is the required apex angle of the prism?

Section 36.9 Total Internal Reflection

18. Calculate the critical angle for the following materials when surrounded by air: (a) diamond, (b) flint glass, (c) ice. (Assume that $\lambda = 589$ nm.)
19. A light ray is incident perpendicular to the long face (along the hypotenuse) of a 45°-45°-90° prism surrounded by air, as shown in Fig. 36.25b. Calculate the minimum value of the index of refraction of the prism for which the ray will follow the path shown in the figure.
20. Consider a light ray incident vertically on one of the short faces of a 45°-45°-90° prism, as shown in Fig. 36.25a. Calculate the minimum index of refraction of the prism for which the ray will follow the path shown in the figure if the prism is surrounded by water.

Section 36.10 Light Intensity

21. Verify by direct substitution that Eqs. 36.11 and 36.12 lead to Eq. 36.13.
22. A light ray in air is directed at near-normal incidence onto the surface of a glass block ($n = 1.50$). What percentage of the intensity of the incident beam is reflected?
23. A light beam in air is incident normally onto a block of polystyrene ($n = 1.49$). What fraction of the intensity of the incident beam is transmitted?
24. A light beam in water ($n = 1.33$) is incident normally onto a block of glass ($n = 1.50$). What fraction of the intensity of the incident beam is transmitted? What fraction is reflected?
25. A light ray is incident from air normally onto the surface of a transparent slab. Intensity measurements show that 10 percent of the intensity of the incident beam is reflected. What is the index of refraction of the reflecting material?
26. A monochromatic beam of light of wavelength 500 nm is incident perpendicular to the surface of a thick slab of transparent material. The light entering the surface layer of the material is reduced in intensity to one fourth of its value at the surface at a depth of 1.5 cm. (a) Calculate the absorption coefficient for the material at 500 nm. (b) What thickness of material will reduce the intensity to one half its incident value? (This is sometimes called the *half-value layer*.)
27. Sea water has a fairly weak absorption for red light (700 nm). This wavelength component in sunlight is reduced in intensity to about 1 percent of its initial value at a depth of 30 m. Calculate the absorption coefficient of sea water for the 700-nm wavelength based on the data given here.
28. Use the data given in Example 36.9 to find the depth in a silicon crystal at which light intensity is reduced to 5 percent of its incident value.

PROBLEMS

1. A narrow beam of light in air is incident on a glass block as shown in Fig. 36.20. Let $n_1 = 1$, $n_2 = n$, and $t =$ the thickness of the block. (a) Show that the *lateral deviation* of the light ray as it passes through the glass is given by

$$d = t \sin\theta_1\left(1 - \frac{\cos\theta_1}{\sqrt{n^2 - \sin^2\theta_1}}\right)$$

(b) If the angle of incidence is limited to small values, show that the general result of (a) can be expressed approximately as

$$d \approx t\theta\left(\frac{n-1}{n}\right)$$

where θ is in radians.

2. A light ray is incident on a prism and refracted at the first surface as shown in Fig. 36.30. Let ϕ represent the apex angle of the prism and n its index of refraction. Find in terms of n and ϕ the smallest allowed value of the angle of incidence at the first surface for which the refracted ray will *not* undergo internal reflection at the second surface.

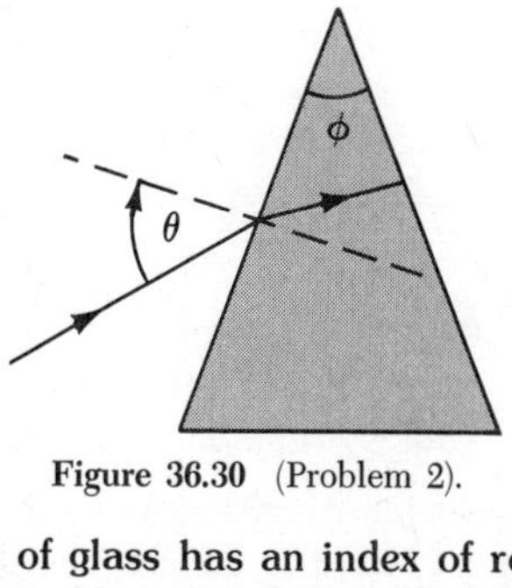

Figure 36.30 (Problem 2).

3. A specimen of glass has an index of refraction of 1.61 for the wavelength corresponding to the prominent bright line in the sodium spectrum. If an equiangular prism is made from this glass, what angle of incidence will result in minimum deviation of the sodium line?
4. (a) Use Eqs. 36.11 and 36.12 to plot I_r/I_o and I_t/I_o (on the same set of axes) as a function of the relative index of refraction, n_1/n_2. Let the ratio n_1/n_2 range over the interval from 0.1 to 10. (b) At what value of n_1/n_2 will the reflected light intensity equal the transmitted light intensity?

5. A light source is 1.5 m below the surface of a swimming pool. Calculate the radius of the circle through which light emerges from the surface of the water.

6. When a polychromatic (multiwavelength) light ray is incident on a prism, the various wavelength components are deviated by different amounts, as shown in Fig. 36.17. This dispersion of the incident light is due to the variation of index of refraction with wavelength (shown for several types of glass in Fig. 36.15). The *dispersive power* of a refracting material is defined by

$$\omega \equiv \frac{n_V - n_R}{n_Y - 1}$$

where n_V, n_R, and n_Y are the respective indices of refraction of three reference wavelengths λ_V, λ_R, and λ_Y. For flint glass, $n_R = 1.644$, $n_Y = 1.650$, and $n_V = 1.665$. For crown glass, $n_R = 1.517$, $n_Y = 1.520$, and $n_V = 1.527$. Calculate the dispersion power of (a) flint glass and (b) crown glass. (c) For small apex angles, the dispersive power can also be written

$$\omega = \frac{\delta_V - \delta_R}{\delta_Y}$$

where the angles of deviation δ are as shown in Fig. 36.31. Show that these two forms for ω are equivalent.

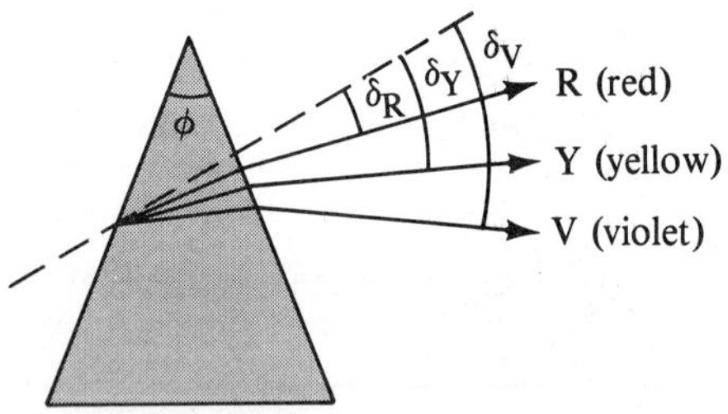

Figure 36.31 (Problem 6).

7. Derive the law of reflection (Eq. 36.2) from Fermat's principle of least time. (See the procedure outlined in Section 36.11 for the derivation of the law of refraction from Fermat's principle.)

8. Uniform planar layers of polystyrene ($n_4 = 1.49$), ice ($n_3 = 1.31$), and flint glass ($n_2 = 1.66$) are surrounded by air ($n_1 = 1$) as shown in Fig. 36.32. A light ray in air is incident on the glass surface at an angle of 33° (with respect to the normal). At each of the four interfaces, a portion of the beam is reflected and the remainder transmitted. Determine the intensity of the emerging beam I_f as a fraction of the incident intensity I_o. (Assume for this problem that there is *no* absorption *within* any of the layers and use Eqs. 36.11 and 36.12.)

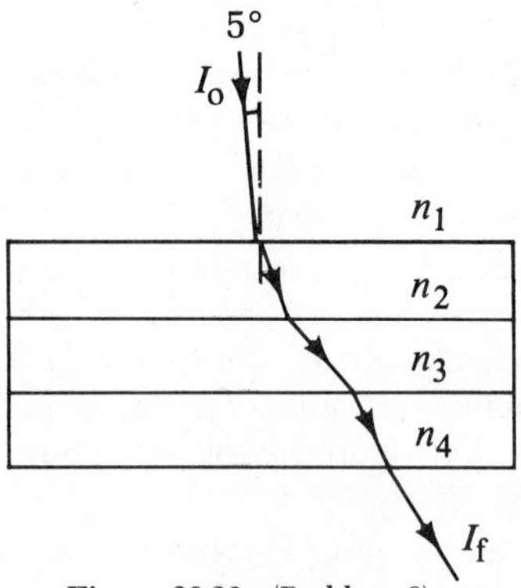

Figure 36.32 (Problem 8).

9. Consider again the situation described in Problem 8. In this case, take into account that the three transparent media have linear absorption coefficients: for glass, $\alpha_2 = 1\ \text{cm}^{-1}$; for ice, $\alpha_3 = 2\ \text{cm}^{-1}$; and for polystyrene, $\alpha_4 = 3\ \text{cm}^{-1}$. Calculate under these conditions the intensity of the emerging beam I_f as a fraction of the intensity of the incident beam I_o, taking into account *both* reflection at the interfaces *between* media and absorption *within* the media layers. Each of the three transparent layers is 2 cm thick.

10. The absorption coefficient for a given material is a function of the wavelength of the incident light. Consider a light ray composed of equal intensities of two wavelengths, λ_1 and λ_2. The light is passing through a slab of transparent material for which $\alpha_1 = 0.346\ \text{cm}^{-1}$ and $\alpha_2 = 0.231\ \text{cm}^{-1}$. (That is, each wavelength has a corresponding absorption coefficient.) The thickness of the slab is adjusted so that 50 percent of the intensity of the λ_1 component is transmitted. Determine the *overall* fraction of the incident beam that is transmitted.

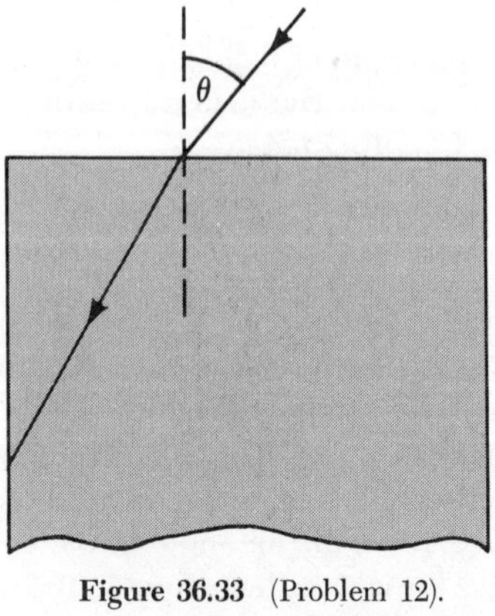

Figure 36.33 (Problem 12).

11. When a monochromatic beam of light passes through a 2.5-cm-thick slab of transparent material, the intensity is reduced by 15 percent. What thickness of this same material will reduce the incident intensity by (a) 30 percent and (b) 99 percent?

12. A light ray of wavelength 589 nm is incident at an angle θ on the top surface of a block of polystyrene, as shown in Fig. 36.33. (a) Find the maximum value of θ for which the refracted ray will undergo *total* internal reflection at the left vertical face of the block. (b) Repeat the calculation for the case in which the polystyrene block is immersed in water. (c) What happens if the block is immersed in carbon disulfide?

37 Geometric Optics

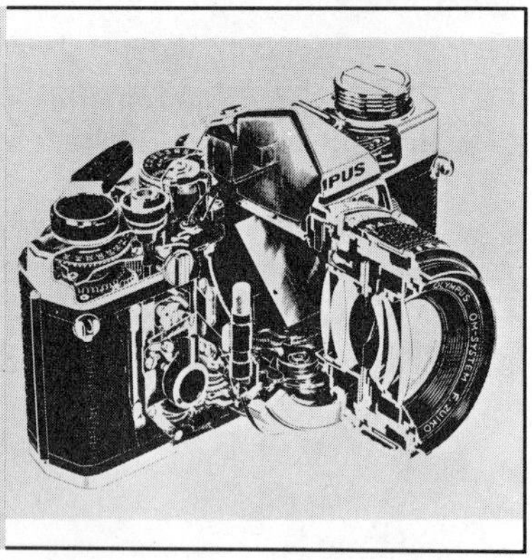

Olympus Camera Corporation

This chapter is concerned with the study of the formation of images when spherical waves fall on planar and spherical surfaces. We shall find that images can be formed by reflection or by refraction. From a practical viewpoint, mirrors and lenses are devices that work on the basis of image formation by reflection and refraction. Such devices, commonly used in optical instruments and systems, will be described in some detail. We shall continue to use the ray approximation and to assume that light travels in straight lines. This corresponds to the field of geometric optics. In subsequent chapters, we shall concern ourselves with interference and diffraction effects, or the field of wave optics.

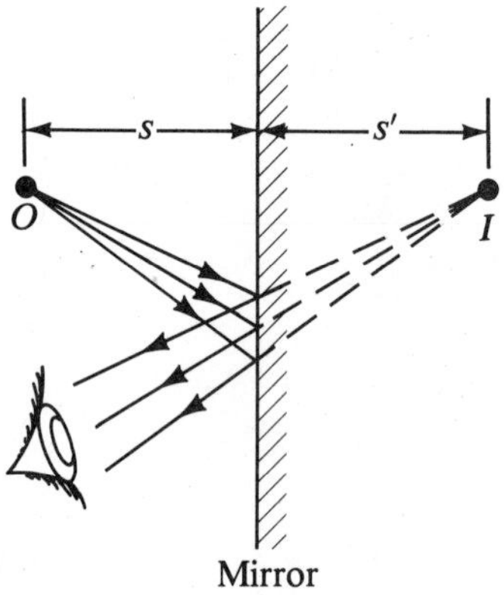

Figure 37.1 An image formed by reflection from a planar mirror. The image point I is located behind the mirror at a distance s', which is equal to object distance s.

37.1 IMAGES FORMED BY PLANAR MIRRORS

When we look into a mirror, what we see is an image of ourselves or some object. The image we see actually corresponds to light originating (or reflected) from the real object, which in turn is reflected from the mirror.

Consider a point source O placed a distance s in front of a planar mirror, as in Fig. 37.1. The spherical waves originating at the point source are represented by rays. The rays reflected from the mirror diverge as shown, but appear to the viewer to come from a point I located behind the mirror. The point I is called the *image* of the object at O. In general, images may be *real* or *virtual*. *A real image is one in which light actually passes through an image point; a virtual image is one in which the light does not really pass through the image point, but appears to diverge from that point.* The images seen in a planar mirror *are always virtual.* Note that real images can be displayed on a screen (as in a movie film), but virtual images cannot be seen on a screen.

Real and virtual images

Using the simple geometric construction shown in Fig. 37.2, we can use rays to locate the image of an object (drawn as a solid arrow) placed to the left of the planar mirror. The horizontal ray starting at P reflects back on itself, while the oblique ray PR is reflected as shown. The two rays seem to originate from P'. A continuation of this procedure for all rays leaving the object results in the virtual image (drawn as a hollow arrow) to the right of the mirror plane. The geometry clearly shows that the triangles PQR and $P'QR$ are congruent, and therefore $PQ = P'Q$. Hence, we conclude that the image formed by an object placed in front of a planar mirror is as far behind the mirror as the object is in front of the mirror. The geometry also shows that the object height h equals the image height h'. Let us define the *lateral magnifi-*

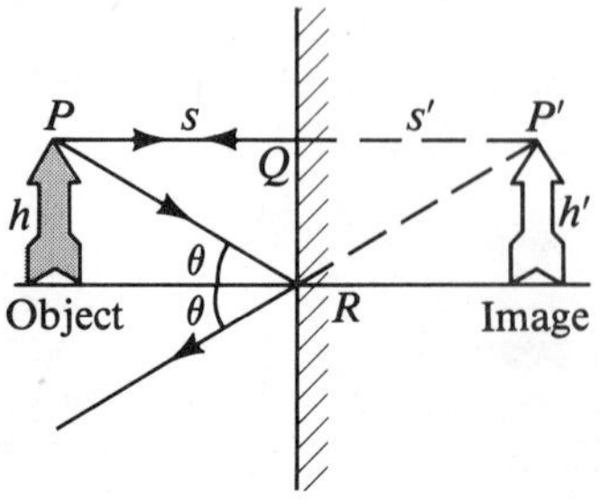

Figure 37.2 Geometric construction used to locate the image of an object placed in front of a planar mirror. Since the triangles PQR and $P'QR$ are congruent, $s = s'$ and $h = h'$.

cation M as follows:

$$M \equiv \frac{\text{image height}}{\text{object height}} = \frac{h'}{h} \qquad (37.1)$$

Lateral magnification

Since M is the ratio of two lengths, it is a form of linear magnification. Note that $M = 1$ for a planar mirror, since $h' = h$ in this case.

The image formed by a planar mirror has one more important property, that of *right-left reversal* between the image and object. That is, a mirror reverses the direction of rays perpendicular to its plane, leaving the direction of those rays parallel to its plane alone. In effect, a mirror changes a right-hand coordinate system into a left-hand one. The mirror's "side-to-side" reversal is merely an artifact of our definition of right and left. This reversal can be seen by standing in front of a mirror and raising your right hand. The image you see raises its left hand. Likewise, your hair appears to be parted on the opposite side (assuming you are not bald) and a mole on your right cheek appears to be on your left.

Thus, we conclude that the image formed by a planar mirror has the following properties: *It is as far behind the mirror as the object is in front; it has unit magnification; it is virtual and erect and has right-left reversal.*

Example 37.1

Two planar mirrors are at right angles to each other, as in Fig. 37.3, and an object is placed as shown. In this situation, multiple images are formed. Locate the positions of these images.

Solution: The image of the object at O in mirror 1 is at I_1, and its image in mirror 2 is at I_2. In addition, a third image is formed at I_3, which will be considered to be the image of I_1 in mirror 2, or, equivalently, the image of I_2 in mirror 1. That is, the image at I_1 (or I_2) serves as the object for I_3. When viewing I_3, as shown in Fig. 37.3, note that the rays reflect twice after leaving the source. The reader should sketch the rays corresponding to viewing the images at I_1 and I_2 and show that the light is reflected only once in these cases.

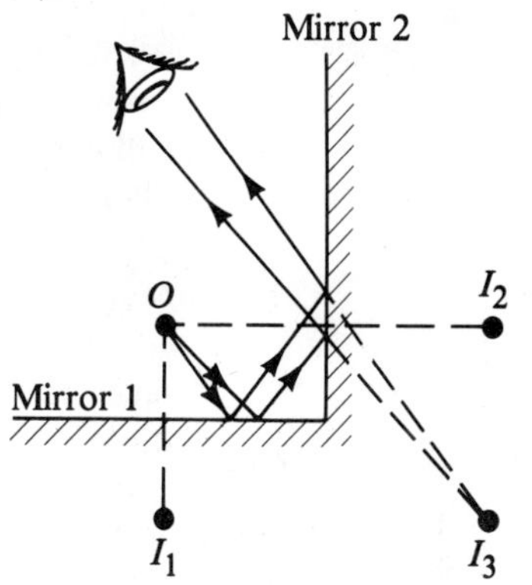

Figure 37.3 (Example 37.1) When an object is placed in front of two perpendicular mirrors as shown, three images are formed.

Q1. A planar mirror reverses left and right, yet images still appear erect. Explain.

Q2. Using a simple ray diagram, as in Fig. 37.2, show that a mirror whose top is at eye level need not be as long as your height in order for you to see your entire body.

37.2 IMAGES FORMED BY SPHERICAL MIRRORS

Concave Mirrors

Let us now consider a spherical mirror of radius of curvature R, whose inner, concave surface is mirrored and faces the incident light, as in Fig. 37.4. This is known as a *concave mirror*. The point C is the *center of curvature* of the mirror, the horizontal line OCV is called the *principal axis* (or *optical axis*), and the point V is the *vertex* of the mirror. A point source is placed at the object point O along the

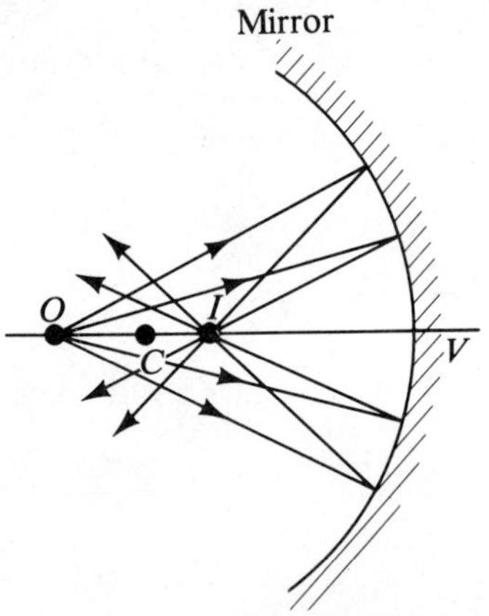

Figure 37.4 A point object placed at O outside the center of curvature of a concave spherical mirror forms a real image at I as shown. If the rays diverging from O are assumed to be paraxial, they all reflect through the same image point.

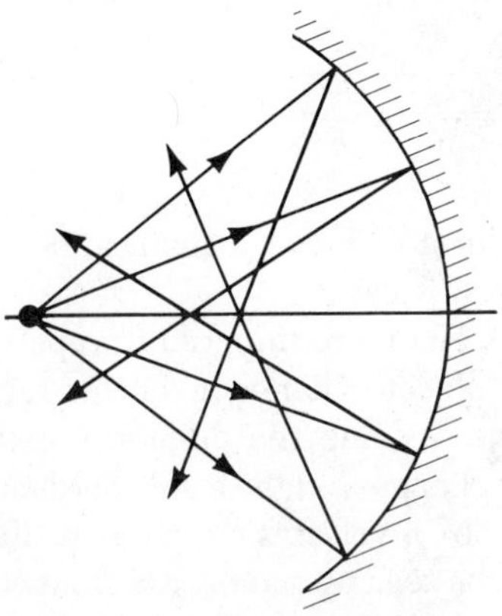

Figure 37.5 Nonparaxial rays reflected from a spherical, concave mirror intersect the optic axis at different points, resulting in a blurred image. This is called spherical aberration.

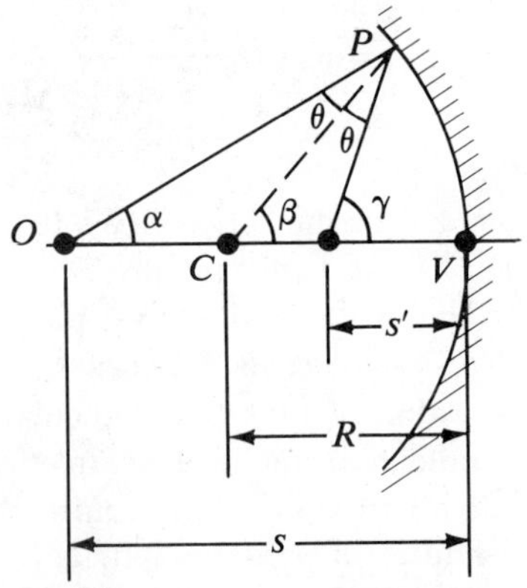

Figure 37.6 Geometry used to derive Eq. 37.3.

principal axis of the mirror, outside the point C. Several diverging rays originating from O are shown. After reflecting from the mirror, these rays converge and meet at the point I, the image point. Since the rays then continue to diverge from I as if there were an object there, *the image is real*. In other words, the image in this case is real, since the light actually passes through the image point.

In what follows, we shall assume that all rays that diverge from the object make a small angle with the principal axis. These rays, called *paraxial rays*, lie close to the principal axis, and all such rays reflect through the image point as in Fig. 37.4. Rays that are nonparaxial would converge to a point on the principal axis other than the image point, producing a blurred image (Fig. 37.5). This effect, called *spherical aberration*, is present to some extent for any spherical mirror. Hence, the assumption of paraxial rays for spherical mirrors is equivalent to neglecting spherical aberration.

We can use the geometry shown in Fig. 37.6 for calculating the image distance s' from a knowledge of the object distance s and radius of curvature R. We shall use the fact that an exterior angle of a triangle equals the sum of the two opposite interior angles. Applying this to triangles OPC and OPI gives

$$\beta = \alpha + \theta \qquad \text{and} \qquad \gamma = \alpha + 2\theta$$

Eliminating θ from the equations, we get

$$\alpha + \gamma = 2\beta \tag{37.2}$$

For sufficiently small angles (the paraxial ray approximation), we can express the angles in radians measure by the appropriate forms[1]

$$\alpha \approx PV/s \qquad \beta \approx PV/R \qquad \gamma \approx PV/s'$$

Substituting these into Eq. 37.2 and noting that the arc length PV is common to all terms, we get

Mirror equation

$$\frac{1}{s} + \frac{1}{s'} = \frac{2}{R} \tag{37.3}$$

This expression is called the *mirror equation*.

[1]Recall that for small values of θ, we can use the approximations $\sin\theta \approx \tan\theta \approx \theta$, where θ is in radians.

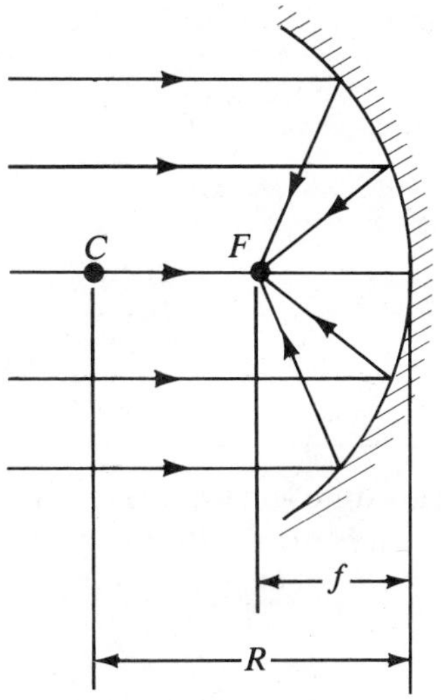

Figure 37.7 Light rays from a distant object ($s = \infty$) reflect from a concave mirror through the focal point F. In this case, the image distance $s' = R/2 = f$, where f is the focal length.

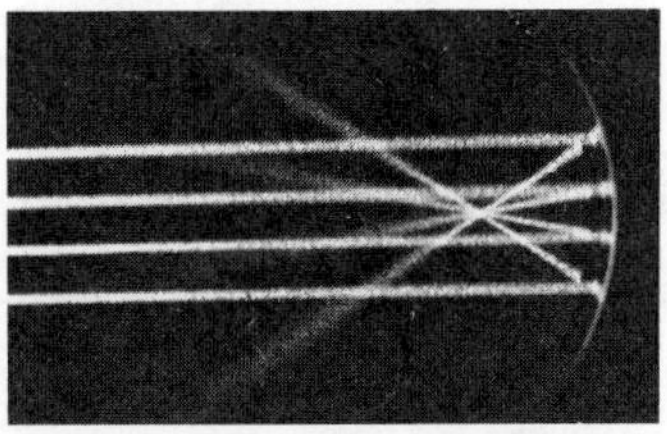

Photograph of the reflection of parallel rays from a concave mirror. (Courtesy of Jim Lehman, James Madison University)

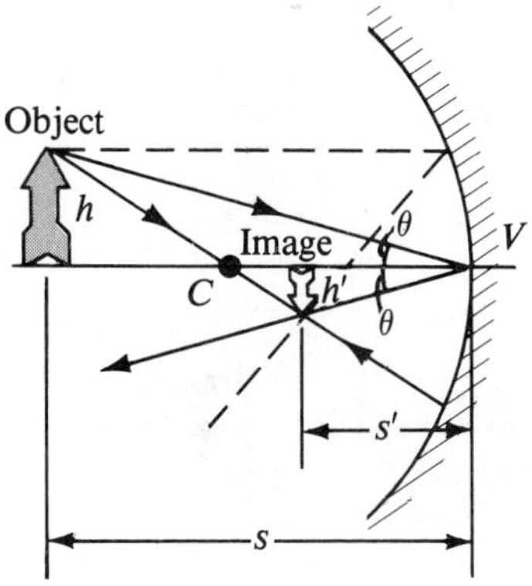

Figure 37.8 Geometry used to compute the lateral magnification of a spherical mirror.

If the object is very far from the mirror, that is, if the object distance s is large compared with R, then $1/s \approx 0$, and we see from Eq. 37.3 that $s' \approx R/2$. That is, the image point is halfway between the center of curvature and the vertex of the mirror, as in Fig. 37.7. Note that the rays are essentially parallel in this figure since the source is assumed to be very far from the mirror. We call the image point in this special case the *focal point, F,* and the image distance the *focal length, f,* where

Focal length

$$f = \frac{R}{2} \tag{37.4}$$

The mirror equation can therefore be expressed in terms of the focal length as

$$\frac{1}{s} + \frac{1}{s'} = \frac{1}{f} \tag{37.5}$$

We can compute the lateral magnification of a spherical mirror by using the geometric construction shown in Fig. 37.8. A ray drawn from the top of the object of height h to the vertex at the angle θ to the horizontal is reflected at the same angle θ (the law of reflection). A second ray drawn from the top of the object through the center of curvature is reflected back on itself. The intersection of these two rays locates the image, which is inverted and has a height h'.

For paraxial rays, any other ray could be used to determine the location of the image, such as the one shown by the broken line in Fig. 37.8.

From the largest triangle in Fig. 37.8, we see that $\tan\theta = h/s$, and the smallest triangle gives $\tan\theta = -h'/s'$. The negative sign signifies that the image is inverted, and so h' is negative. Thus, from Eq. 37.1 and these results, we find that

Magnification of a mirror

$$M = \frac{h'}{h} = -\frac{s'}{s} \tag{37.6}$$

The reader should note the following sign conventions for concave mirrors. The object distance is positive when the object is real, that is, when it lies on the lighted

TABLE 37.1 Sign Convention for Mirrors

s is + if the object is in front of the mirror (real object).
s is − if the object is in back of the mirror (virtual object).

s' is + if the image is in front of the mirror (real image).
s' is − if the image is in back of the mirror (virtual image).

Both f and R are + if the center of curvature is in front of the mirror (concave mirror).
Both f and R are − if the center of curvature is in back of the mirror (convex mirror).

If M is positive, the image is erect.
If M is negative, the image is inverted.

side of the mirror, as in Fig. 37.6. The radius of curvature and focal length are both positive for a concave mirror. The image distance s' is positive if the image is real, as in Fig. 37.6. When the object distance is less than the focal length, s' is negative and a virtual image is formed.

Convex Mirror

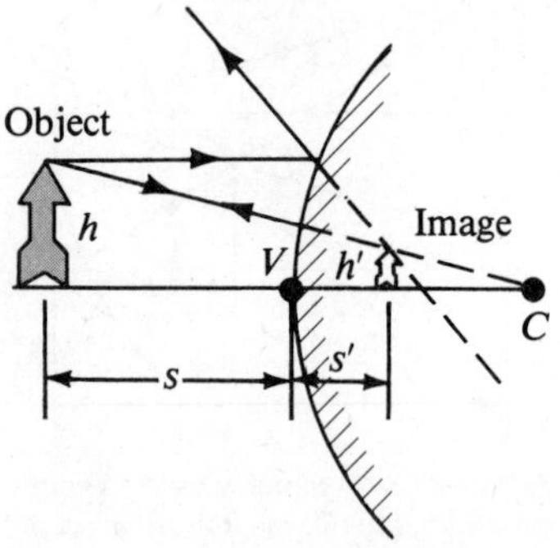

Figure 37.9 The image formed by a spherical, convex mirror is located behind the mirror and hence is virtual.

Figure 37.9 shows the formation of an image from a *convex* mirror, that is, one silvered on the concave surface so that its center of curvature lies beyond the vertex. This is sometimes called a *diverging mirror,* since the rays from any point on the object diverge as if they were coming from a point behind the mirror. Note that the image in Fig. 37.9 is virtual, since it lies behind the mirror, and is erect. Furthermore, as long as the object is real (positive s), the image will always be virtual, erect, and smaller than the object, as illustrated in Fig. 37.9.

We can use Eqs. 37.3, 37.5, and 37.6 for a convex mirror with the following sign conventions. Since the center of curvature is on the virtual side, where there are no light rays, both R and f are taken to be negative. Since s' is negative, as in Eq. 37.9, we see from Eq. 37.6 that the magnification, M, is positive, as it should be for an erect image. The sign rules for using the mirror and magnification equations for concave and convex mirrors are summarized in Table 37.1. Figure 37.10 will be useful in understanding these rules. In the sign convention for mirrors given in Table 37.1, we call the region in which the light rays move the *front side* of the mirror and the other side, where virtual images occur, the *back side.*

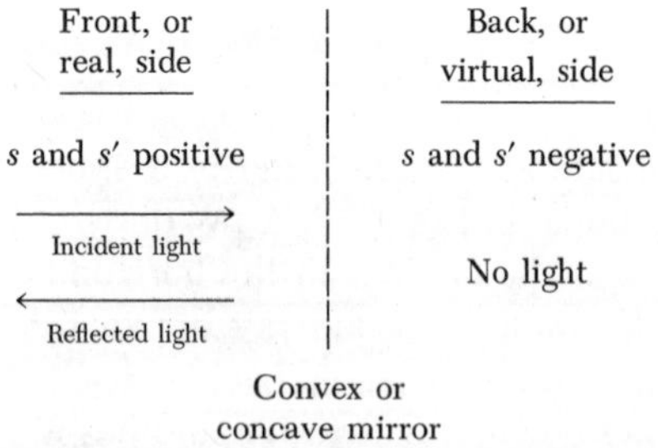

Figure 37.10 Diagram for describing the sign of s and s' for convex or concave mirrors.

Example 37.2

An object of height 5 cm is placed 2 m in front of a concave mirror whose radius of curvature is 40 cm. (a) Determine the position of the image.

Using Eq. 37.3, we have

$$\frac{1}{s} + \frac{1}{s'} = \frac{1}{2\text{ m}} + \frac{1}{s'} = \frac{2}{0.4\text{ m}}$$

$$s' = 0.22\text{ m}$$

Since s' is positive, the image is real.

(b) Determine the magnification and size of the image. The magnification is given by Eq. 37.6:

$$M = \frac{h'}{h} = -\frac{s'}{s} = -\frac{0.22\text{ m}}{2\text{ m}} = -0.11$$

Therefore, $|h'| = 0.11h = 0.11(5\text{ cm}) = 0.55\text{ cm}$. The negative value of M indicates that the image is inverted, as in Fig. 37.8.

Example 37.3

An object 3 cm high is placed 20 cm from a convex mirror having a focal length of 8 cm. Find the position of the final image and its height.

Solution: Since the mirror is convex, its focal length is negative. To find the image position, we use the mirror equation:

$$\frac{1}{s} + \frac{1}{s'} = \frac{1}{f} = -\frac{1}{8\text{ cm}}$$

$$\frac{1}{s'} = -\frac{1}{8\text{ cm}} - \frac{1}{20\text{ cm}}$$

$$s' = -5.71\text{ cm}$$

The negative value of s' indicates that the image is virtual, or behind the mirror, as in Fig. 37.9.

The magnification of the mirror is

$$M = -\frac{s'}{s} = -\frac{-5.71\text{ cm}}{20\text{ cm}} = 0.286$$

Hence, the height of the image is

$$h' = Mh = (0.286)(3\text{ cm}) = 0.858\text{ cm}$$

The image is erect since M is positive.

Q3. Consider a concave spherical mirror with a real object. Is the image always inverted? Is the image always real? Give conditions for your answers.

Q4. Repeat the previous question for a convex spherical mirror.

37.3 RAY DIAGRAMS FOR MIRRORS

The position and size of images formed by mirrors can be conveniently determined by using *ray diagrams*. These graphical constructions tell us the total nature of the image and can be used to check parameters calculated from the mirror and magnifications equations. In these diagrams, one needs to know the position of the object, the location of the center of curvature, and the focal point. In order to locate the image, three rays are then constructed, as shown by the various examples in Fig. 37.11. These rays all start from any object point (chosen to be the top) and are drawn as follows:

1. Ray 1 is drawn from the top of the object (tip of arrow), parallel to the principal axis. After reflection, this ray passes through the focal point, as do all parallel rays.
2. Ray 2 is drawn from the top of the object through the center of curvature. This ray is perpendicular to the mirror and reflects back on itself.
3. Ray 3 is drawn from the top of the object through the focal point. This ray is then reflected back parallel to the principal axis.

These three rays intersect at a point that locates the image. Actually, only two of the three rays are needed to locate the image. The third can serve as a check on the construction. Since sign conventions are often forgotten, the graphical method should be mastered.

In the case of the concave mirror, note what happens as the object is moved closer to the mirror. The real, inverted image in Figs. 37.11a and 37.11b moves to the left as the object approaches the focal point. When the object is at the focal point, the image is infinitely far to the left. However, when the object lies between the focal point and vertex, as in Fig. 37.11c, the image is virtual and erect. Finally, for the convex mirror shown in Fig. 37.11d, the image of a real object is always virtual and erect. In this case, as the object distance increases, the virtual image decreases in size and approaches the focal point as $s \to \infty$. You should construct other diagrams to verify the variation of image position with object position.

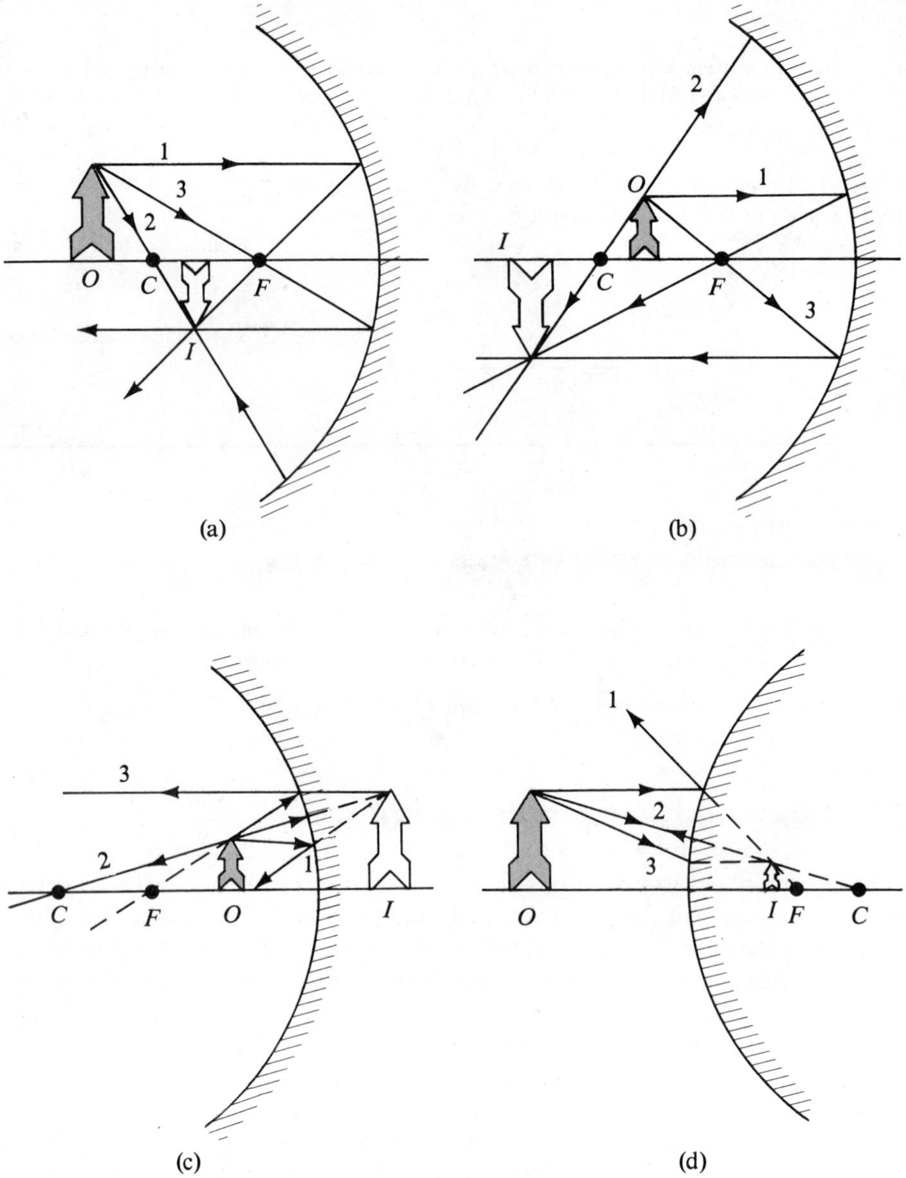

Figure 37.11 Ray diagrams for spherical mirrors. Only two of the three rays shown in each part are needed to locate the image.

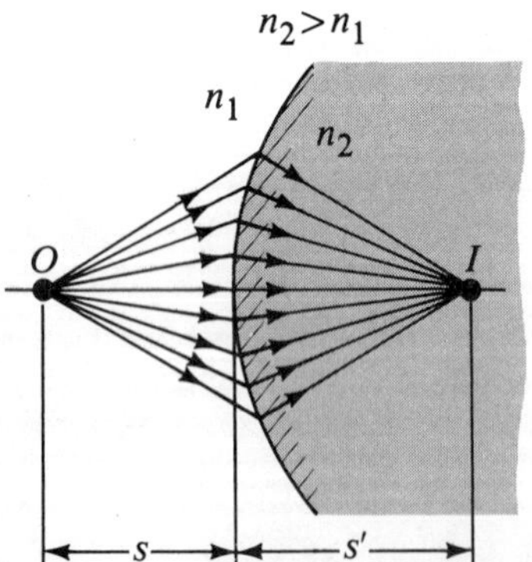

Figure 37.12 Image formed by refraction at a spherical surface. Paraxial rays diverging from a point object at O pass through the same image point I.

37.4 IMAGES FORMED BY REFRACTION

We now describe how images are formed by refraction at a spherical surface. Consider two transparent media with indices of refraction n_1 and n_2, separated by a spherical surface of radius R (Fig. 37.12). Note that all paraxial rays originating at the object point O and refracted at the spherical surface converge to the image point I. Using Snell's law of refraction and the small angle approximation, we find that *the object distance, image distance, and radius of curvature are related by the equation*

Formation of an image by refraction

$$\frac{n_1}{s} + \frac{n_2}{s'} = \frac{n_2 - n_1}{R} \tag{37.7}$$

Furthermore, the magnification of a refracting surface is given by

$$M = \frac{h'}{h} = -\frac{n_1 s'}{n_2 s} \tag{37.8}$$

Magnification of a refracting surface

In order to understand the sign convention for the various parameters, first note that real images are formed on the side of the surface that is *opposite* the side from which the light comes, in contrast to mirrors where real images are formed on the *same* side of the reflecting surface. Therefore, *the sign convention for spherical refracting surfaces is the same as for mirrors, recognizing the change in roles of real and virtual images*. For example, in Fig. 37.12, s, s', and R are all positive.

The sign convention for spherical refracting surfaces is summarized in Table 37.2. The same sign convention will be used for thin lenses, which will be discussed in the next section. As with mirrors, we assume that the front of the refracting surface is the side from which the light approaches the surface.

TABLE 37.2 Sign Convention for Refracting Surfaces

s is + if the object is in front of the surface (real object).
s is − if the object is in back of the surface (virtual object).
s' is + if the image is in back of the surface (real image).
s' is − if the image is in front of the surface (virtual image).
R is + if the center of curvature is in back of the surface.
R is − if the center of curvature is in front of the surface.

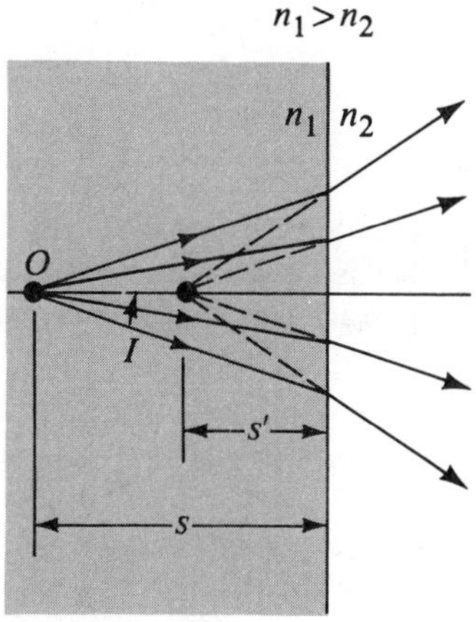

Figure 37.13 The image of a planar refracting surface is virtual, that is, it forms to the left of the refracting surface.

Planar Refracting Surface

If the refracting surface is a *planar surface*, then $R \rightarrow \infty$ and Eq. 37.7 reduces to

$$\frac{n_1}{s} = -\frac{n_2}{s'}$$

$$s' = -\frac{n_2}{n_1} s \tag{37.9}$$

Planar refracting surface

Since the sign of s' is opposite that of s, we conclude that *the image of a planar refractng surface is on the same side of the surface as the object*. This is illustrated in Fig. 37.13 for the situation in which $n_1 > n_2$, where a virtual image forms between the object and surface. The reader should use a ray diagram to show that if $n_1 < n_2$, the image will still be virtual, but will form to the left of the object. Both situations are consistent with Eq. 37.8 and our sign convention.

We shall now derive Eq. 37.7, using the geometric construction shown in Fig. 37.14. Only one refracted ray is needed in this construction. Snell's law applied to this refracted ray gives $n_1 \sin\theta_1 = n_2 \sin\theta_2$. For small angles, $\sin\theta_1 \approx \theta_1$ and $\sin\theta_2 \approx \theta_2$; therefore

$$n_1\theta_1 = n_2\theta_2 \tag{37.10}$$

Again, we make use of the fact that an exterior angle of any triangle equals the sum of the two opposite interior angles. Applying this to triangles OPC and PIC gives

$$\theta_1 = \alpha + \beta \tag{37.11}$$

$$\beta = \theta_2 + \gamma \tag{37.12}$$

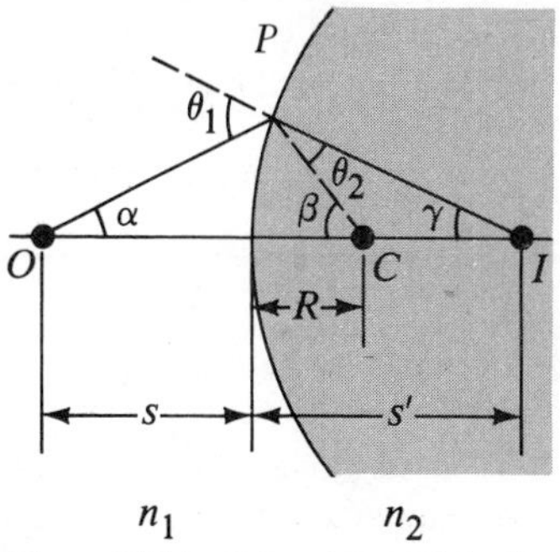

Figure 37.14 Geometry used to derive Eq. 37.7.

Substituting Eq. 37.10 into Eq. 37.12 gives

$$\beta = \frac{n_1}{n_2}\theta_1 + \gamma \tag{37.13}$$

Using Eq. 37.11 in Eq. 37.13 to eliminate θ_1, we obain the expression

$$n_1\alpha + n_2\gamma = (n_2 - n_1)\beta \tag{37.14}$$

Finally, in the small angle approximation $\tan\theta \approx \theta$, we have the relationships $\alpha \approx PV/s$, $\beta \approx PV/R$, and $\gamma \approx PV/s'$. Substituting these into Eq. 37.14 while recognizing that the arc length PV is common to every term gives the final result, Eq. 37.7:

$$\frac{n_1}{s} + \frac{n_2}{s'} = \frac{n_2 - n_1}{R}$$

In a similar manner, one can use a geometric construction to verify the magnification expression given by Eq. 37.8.

Example 37.4

A coin 2 cm in diameter is imbedded in a solid glass ball of radius 30 cm (Fig. 37.15). The index of refraction of the ball is 1.5 and the coin is 20 cm from the surface. Find the position and height of the image.

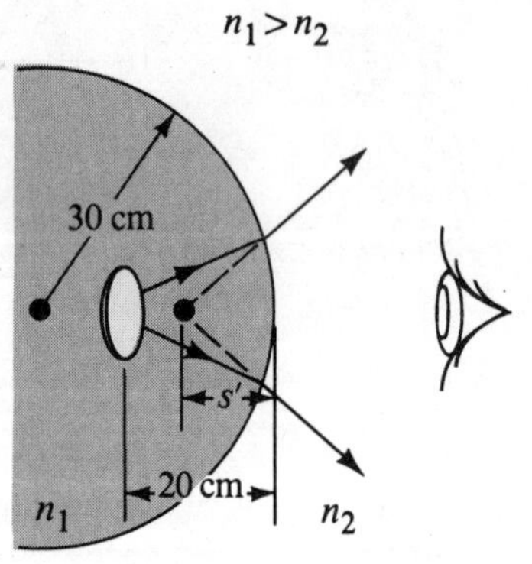

Figure 37.15 (Example 37.4) A coin embedded in a glass ball forms a virtual image between the coin and the glass surface.

Solution: First, note that the rays originating from the object are refracted away from the normal at the surface and diverge outward. Hence, the image is formed in the glass and is virtual. Applying Eq. 37.7 and taking $n_1 = 1.5$, $n_2 = 1$, $s = 20$ cm, and $R = -30$ cm, we get

$$\frac{n_1}{s} + \frac{n_2}{s'} = \frac{n_2 - n_1}{R}$$

$$\frac{1.5}{20\text{ cm}} + \frac{1}{s'} = \frac{1 - 1.5}{-30\text{ cm}}$$

$$s' = -17\text{ cm}$$

The negative sign indicates that the image is in the same medium as the object (the side of incident light), in agreement with our ray diagram. Since the image is in the same medium as the object, it must be a virtual image.

To find the image height, we use Eq. 37.8 for the magnification:

$$M = -\frac{n_1 s'}{n_2 s} = -\frac{1.5(-17\text{ cm})}{1(20\text{ cm})} = \frac{h'}{h}$$

$$h' = (1.28)h = (1.28)(2\text{ cm}) = 2.56\text{ cm}$$

The positive value for M indicates an erect image.

Example 37.5

A small fish is at the bottom of a pond of depth d (Fig. 37.16). What is the *apparent depth* of the fish as viewed from directly overhead?

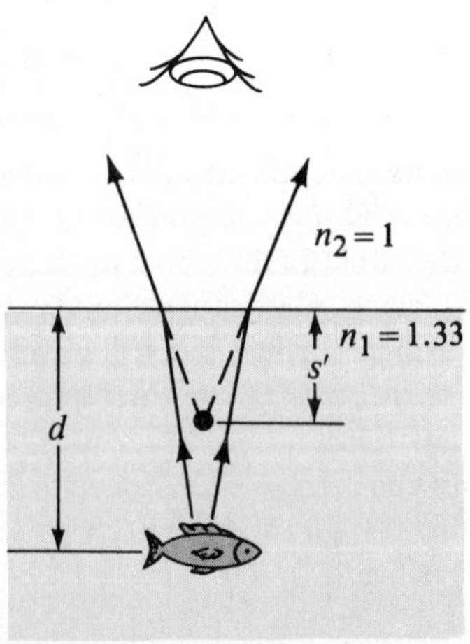

Figure 37.16 (Example 37.5) The apparent depth s' of the fish is less than the true depth d.

Solution: In this example, the refracting surface is a plane, and so $R \to \infty$. Hence, we can use Eq. 37.9 to determine the location of the image. Using the fact that $n_1 = 1.33$ for water and $s = d$ gives

$$s' = -\frac{n_2}{n_1}s = -\frac{1}{1.33}d = -0.75d$$

Again, since s' is negative, the image is virtual, as indicated in Fig. 37.16. The apparent depth is three fourths the actual depth. For instance, if $d = 4$ m, $s' = -3$ m.

Q5. It is well known that distant objects viewed underwater with the naked eye appear blurred and out of focus. On the other hand, the use of goggles provides the swimmer with a clear view of objects. Explain this, using the fact that the indices of refraction of the cornea, water, and air are 1.376, 1.333, and 1.032, respectively.

Q6. Why does a clear stream always appear to be shallower than it actually is?

Q7. A person spearfishing in a boat sees a fish located 3 m from the boat at a depth of 1 m. In order to hit the fish with his spear, should the person aim at the fish, above the fish, or below the fish? Explain.

37.5 THIN LENSES

A *lens* is a transparent medium bounded by two refracting surfaces. Lenses are commonly used to form images by refraction in optical instruments, such as cameras, telescopes, and microscopes. The methods discussed in the previous section will be used here to locate the image position. The essential idea in locating the final image of a lens is to *use the image formed by one refracting surface as the object for the second surface.*

Consider a lens having an index of refraction n and two spherical surfaces of radii of curvature R_1 and R_2, as in Fig. 37.17. An object is placed at point O in front of the first refracting surface at a distance s, which has been chosen for this example so as to produce a virtual image I_1, located to the left of the lens. This image is used as the object for the second surface, of radius R_2, which results in a real image I_2.

Using Eq. 37.7 and assuming $n_1 = 1$, we find that the image formed by the first surface satisfies the equation

$$(1) \qquad \frac{1}{s_1} + \frac{n}{s_1'} = \frac{n-1}{R_1}$$

Now we apply Eq. 37.7 to the second surface, taking $n_1 = n$ and $n_2 = 1$. That is, the object of the second surface (which is the image at I_1) is treated as if it were imbedded in glass. Taking s_2 as the object distance and s_2' as the image distance for the second surface gives

$$(2) \qquad \frac{n}{s_2} + \frac{1}{s_2'} = \frac{1-n}{R_2}$$

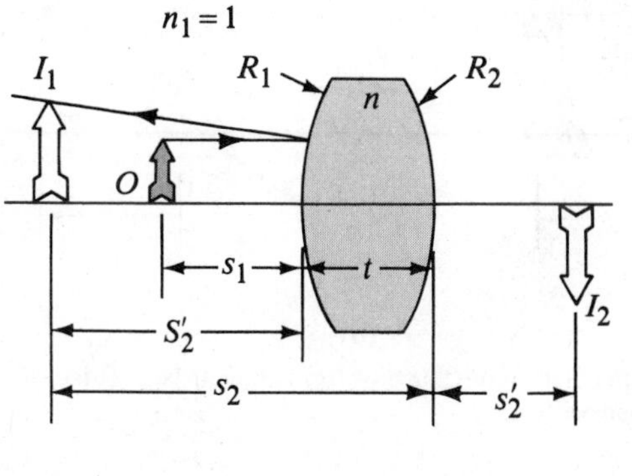

Figure 37.17 To locate the image of a lens, the image at I_1 formed by the first surface is used as the object for the second surface. The final image is at I_2.

But $s_2 = -s_1' + t$, where t is the thickness of the lens. (Remember s_1' is a negative number and s_2 must be positive by our sign convention.) For a thin lens, we can neglect t. In this approximation and from Fig. 37.17, we see that $s_2 = -s_1'$. Hence, (2) becomes

$$(3) \qquad -\frac{n}{s_1'} + \frac{1}{s_2'} = \frac{1-n}{R_2}$$

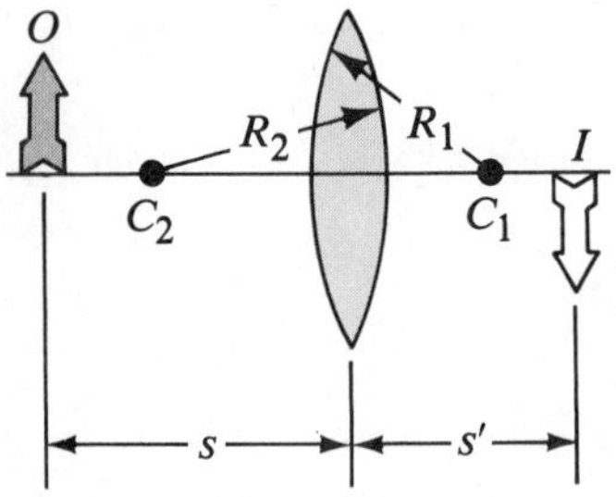

Figure 37.18 The biconvex lens.

Adding (1) and (3), we find that

$$(4) \qquad \frac{1}{s_1} + \frac{1}{s_2'} = (n-1)\left(\frac{1}{R_1} - \frac{1}{R_2}\right)$$

For the thin lens, we can omit the subscripts on s_1 and s_2' in (4) and call the object distance s and the image distance s', as in Fig. 37.18. Hence, we can write (4) in the form

Thin lens formula

$$\frac{1}{s} + \frac{1}{s'} = (n-1)\left(\frac{1}{R_1} - \frac{1}{R_2}\right) \qquad (37.15)$$

This expression relates the image distance s' of a thin lens to the object distance s and to the thin lens properties (index of refraction and radii of curvature). It is valid only for paraxial rays and only when the lens thickness is small relative to the radii R_1 and R_2.

We now define the focal length f of a thin lens as the image distance that corresponds to an infinite object distance, as we did with mirrors. According to this definition and from Eq. 37.15, we see that for $s \to \infty$, $f = s'$; therefore, the focal

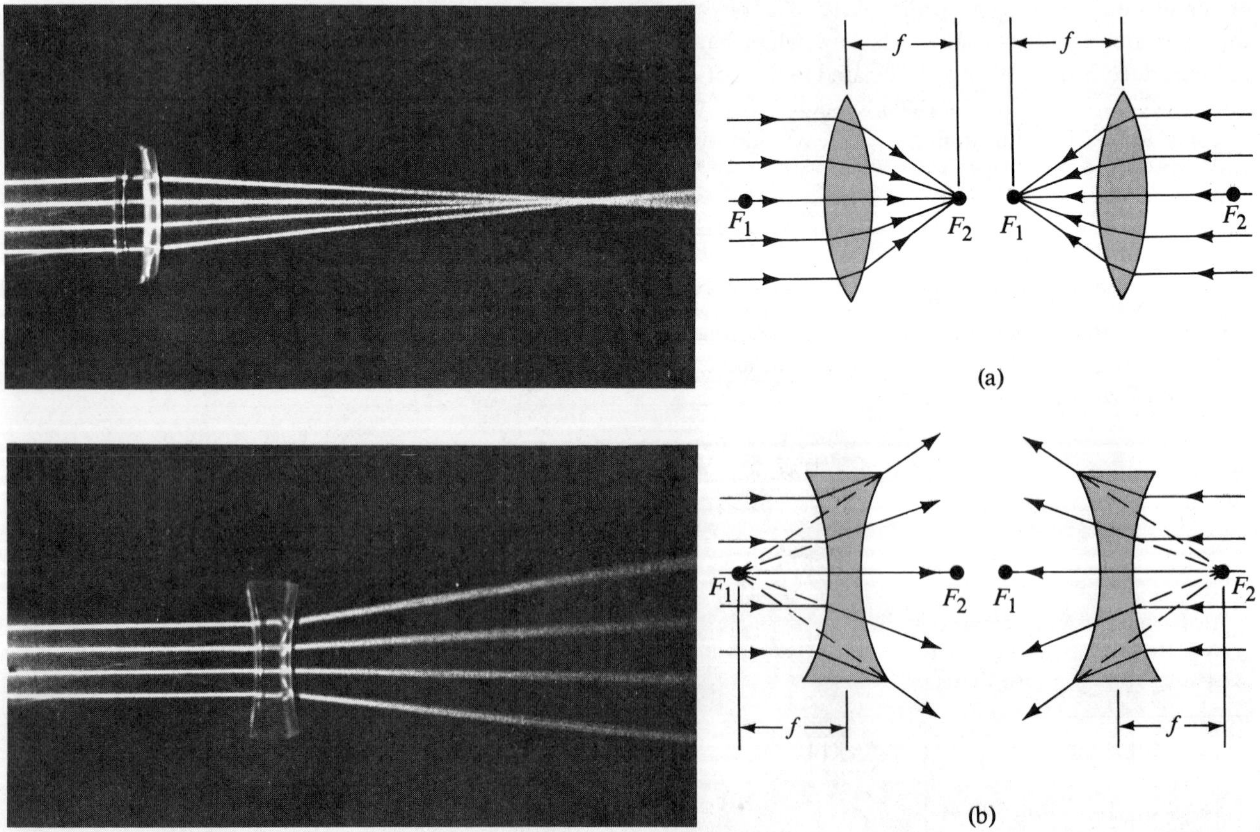

Figure 37.19 (*Left*) Photographs of the effect of converging and diverging lenses on parallel rays. (Courtesy Jim Lehman, James Madison University) (*Right*) The principal focal points of (a) the biconvex lens and (b) the biconcave lens.

length is given by

Lens makers' formula

$$\frac{1}{f} = (n - 1)\left(\frac{1}{R_1} - \frac{1}{R_2}\right) \qquad (37.16)$$

Using this result, we can write Eq. 37.15 in an alternate form identical to Eq. 37.5 for mirrors:

$$\frac{1}{s} + \frac{1}{s'} = \frac{1}{f} \qquad (37.17)$$

Equation 37.16 is called the *lens makers' equation,* since it enables one to calculate f from the known properties of the lens. It can also be used to determine the values of R_1 and R_2 needed for a given index of refraction and desired focal length. A thin lens has *two* focal points, corresponding to incident parallel light rays traveling from the left or right. This is illustrated in Fig. 37.19 for a biconvex lens (converging, positive f) and a biconcave lens (diverging, negative f). Focal point F_1 is sometimes called the *front focal point,* and F_2 is called the *rear focal point.*

Figure 37.20 is useful for obtaining the signs of the quantities appearing in the thin lens equation, Eq. 37.15. It is also applicable to concave or convex refracting

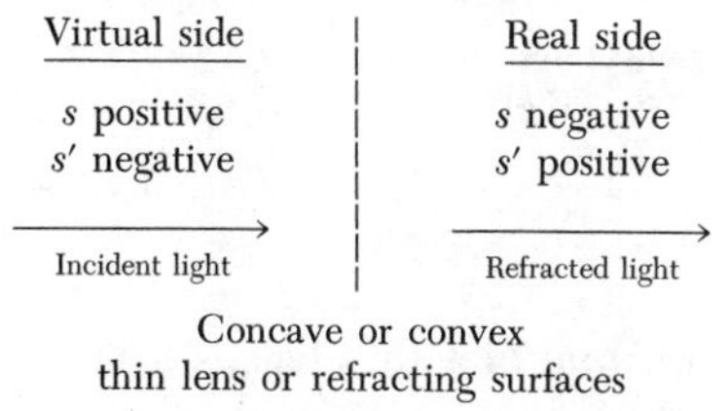

Figure 37.20 Diagram for obtaining the sign of s and s' for a thin lens or a refracting surface.

surfaces. The sign convention for thin lenses (Table 37.3) is the same as for refracting surfaces discussed in the previous section.

TABLE 37.3 Sign Convention for Thin Lenses

s is + if the object is in front of the lens.
s is − if the object is in back of the lens.
s′ is + if the image is in back of the lens.
s′ is − if the image is in front of the lens.
R_1 and R_2 are + if the center of curvature is in back of the lens.
R_1 and R_2 are − if the center of curvature is in front of the lens.

Applying these rules to the *converging* lens in Fig. 37.21a, we see that when $s > f$, the quantities s, s', and R_1 are positive and R_2 is negative. Therefore, in the case of a converging lens, where a real object forms a real image, s, s', and f are all positive. Likewise, for a *diverging* lens (Fig. 37.21b), s and R_2 are positive and s' and R_1 are negative. Thus, f is negative for a diverging lens.

Sketches of various lens shapes are shown in Fig. 37.22. In general, note that a converging lens (positive f) is thicker at the middle than at the edges, whereas a diverging lens (negative f) is thinner at the middle than at the edges.

Consider a single thin lens illuminated by a *real* object, so that $s > 0$. As with

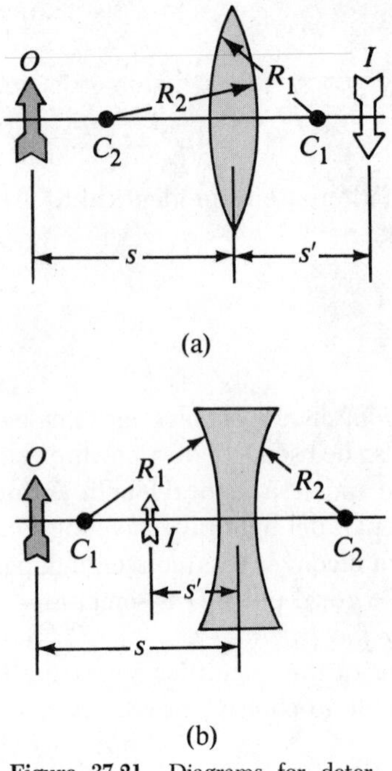

Figure 37.21 Diagrams for determining the signs of the object distances, image distances, and radii of curvature for (a) a biconvex lens and (b) a biconcave lens.

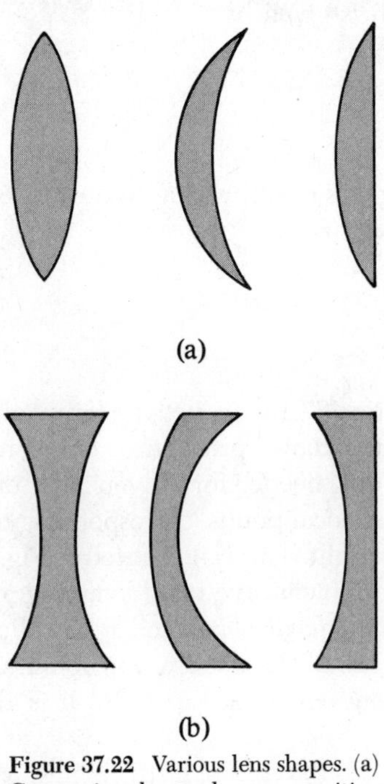

Figure 37.22 Various lens shapes. (a) Converging lenses have a positive focal length and are thickest at the middle. (b) Diverging lenses have a negative focal length and are thickest at the edges.

mirrors, the *lateral magnificaton* of a thin lens is defined as the ratio of the image height h' to the object height h. Since $M = h'/h = -s'/s$, it follows that when M is positive, the image is erect and on the same side of the lens as the object. When M is negative, the image is inverted and on the side of the lens opposite the object.

Ray Diagrams for Thin Lenses

Graphical methods, or ray diagrams, are very convenient for determining the image of a thin lens or a system of lenses. Such constructions should also help clarify the sign conventions that have been discussed. Figure 37.23 illustrates this method for three different single-lens situations. To locate the image, the following two rays are drawn from the top of the object:

1. A ray from O is drawn parallel to the optic axis. After being refracted by the lens, this ray passes through (or appears to come from) one of the focal points.

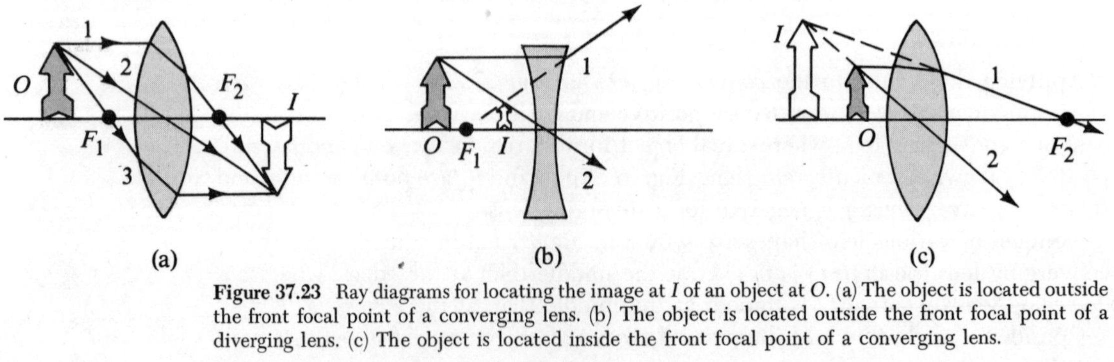

Figure 37.23 Ray diagrams for locating the image at I of an object at O. (a) The object is located outside the front focal point of a converging lens. (b) The object is located outside the front focal point of a diverging lens. (c) The object is located inside the front focal point of a converging lens.

2. A ray from O is drawn through the center of the lens. This ray continues to move in a straight line.

Another useful ray is the one passing through the object and the focal point, as shown in Fig. 37.23a. After being refracted, this ray is parallel to the optic axis. For the converging lens in Fig. 37.23a, where the object is outside the front focal point ($s > f$), the image is real and inverted. On the other hand, when the object is inside the front focal point ($s < f$), as in Fig. 37.23c, the image is virtual, erect, and enlarged. Finally, for the biconcave lens shown in Fig. 37.23b, the image is always virtual and erect.

Example 37.6

A biconvex thin lens is made of glass, which has an index of refraction of $n = 1.52$. Each surface has a radius of curvature equal to 30 cm. An object of height 3 cm is placed 14 cm from the lens, as in Fig. 37.24. Find the focal length of the lens and the position and size of the image.

Solution: We can calculate the focal length using Eq. 37.16 and noting that $R_1 = 30$ cm and $R_2 = -30$ cm. This gives

$$\frac{1}{f} = (1.52 - 1)\left(\frac{1}{30\text{ cm}} - \frac{1}{-30\text{ cm}}\right) = 0.035\text{ cm}^{-1}$$

$$f = 29\text{ cm}$$

Thus, we see that the object is inside the focal point and so the image should be virtual and erect. The ray diagram sketched in Fig. 37.24 illustrates this point.

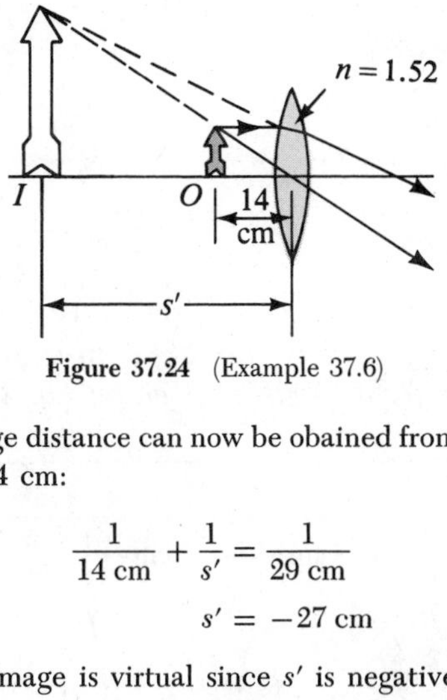

Figure 37.24 (Example 37.6)

The image distance can now be obained from Eq. 37.17, with $s = 14$ cm:

$$\frac{1}{14\text{ cm}} + \frac{1}{s'} = \frac{1}{29\text{ cm}}$$

$$s' = -27\text{ cm}$$

Thus, the image is virtual since s' is negative, in agreement with our ray diagram. The lateral magnification is given by $M = -s'/s = -(-27\text{ cm})/14\text{ cm} = 1.9$. Hence, the image height is $h' = Mh = (1.9)(3\text{ cm}) = 5.7$ cm. The image is enlarged and erect.

Example 37.7

A diverging lens has a focal length of -20 cm (Fig. 37.25). An object 2 cm in height is placed 30 cm in front of the lens. Locate the position and height of the final image.

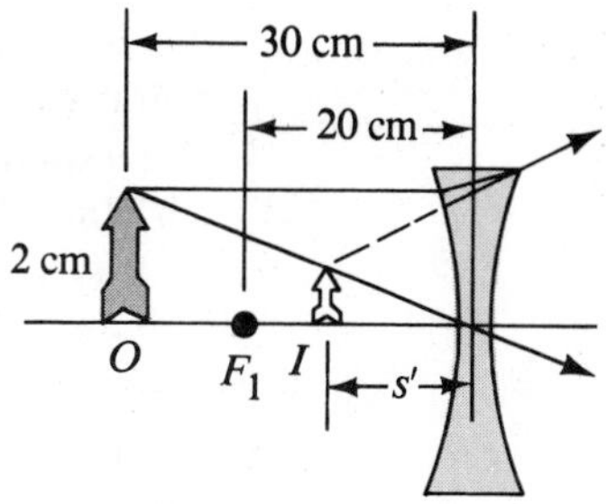

Figure 37.25 (Example 37.7)

Solution: Using the thin lens equation, with $s = 30$ cm and $f = -20$ cm, we get

$$\frac{1}{30\text{ cm}} + \frac{1}{s'} = -\frac{1}{20\text{ cm}}$$

$$s' = -12\text{ cm}$$

Thus, the image is virtual, as indicated in the ray diagram.

The magnification of the lens is given by $M = -s'/s = -(-12\text{ cm})/30\text{ cm} = 0.4$, and so the image size is $h' = Mh = (0.4)(2\text{ cm}) = 0.8$ cm. The image is erect, but reduced in size.

Example 37.8

A converging glass lens ($n = 1.52$) has a focal length of 40 cm in air. Find its focal length when it is immersed in water, which has an index of refraction of 1.33.

Solution: We can use the lens makers' formula (Eq. 37.16) in both cases, noting that R_1 and R_2 remain the same in air and water. In air, we have

$$\frac{1}{f_a} = (n - 1)\left(\frac{1}{R_1} - \frac{1}{R_2}\right)$$

where $n = 1.52$. In water we get

$$\frac{1}{f_w} = (n' - 1)\left(\frac{1}{R_1} - \frac{1}{R_2}\right)$$

where n' is the index of refraction of glass *relative* to water. That is, $n' = 1.52/1.33 = 1.14$. Dividing the two equations gives

$$\frac{f_w}{f_a} = \frac{n - 1}{n' - 1} = \frac{1.52 - 1}{1.14 - 1} = 3.71$$

Since $f_a = 40$ cm, we find that

$$f_w = 3.71 f_a = 3.71(40 \text{ cm}) = 148 \text{ cm}$$

In fact, the focal length of *any* glass lens is increased by the factor $(n - 1)/(n' - 1)$ when immersed in water.

Combination of Thin Lenses

If two thin lenses are used to form an image, the system can be treated in the following manner. First, the image of the first lens is calculated as if the second lens were not present. Next, the image of the first lens is treated as the object of the second lens. The image of the second lens is the final image of the system. If the image of the first lens lies to the right of the second lens, the image is treated as a virtual object for the second lens (that is, s negative). The same procedure can be extended to a system of three or more lenses. The overall magnification of a system of thin lenses equals the *product* of the magnification of the separate lenses.

Now suppose two thin lenses of focal lengths f_1 and f_2 are placed in contact with each other. If s is the object distance for the combination, then application of the thin lens equation to the first lens gives

$$\frac{1}{s} + \frac{1}{s_1'} = \frac{1}{f_1}$$

where s_1' is the image distance for the first lens. Treating this image as the object for the second lens, we see that the object distance for the second lens must be $-s_1'$. Therefore, for the second lens

$$-\frac{1}{s_1'} + \frac{1}{s'} = \frac{1}{f_2}$$

where s' is the final image distance from the second lens. Adding these equations eliminates s_1' and gives

$$\frac{1}{s} + \frac{1}{s'} = \frac{1}{f_1} + \frac{1}{f_2}$$

Focal length of two thin lenses in contact

$$\frac{1}{f} = \frac{1}{f_1} + \frac{1}{f_2} \qquad (37.18)$$

If the two *thin* lenses are in contact with one another, then s' is also the distance of the final image from the first lens. Therefore, *two thin lenses in contact are equivalent to a single thin lens whose focal length is given by Eq. 37.18.*

Example 37.9

Two thin converging lenses of focal lengths 10 cm and 20 cm are separated by 20 cm, as in Fig. 37.26. An object is placed 15 cm in front of the first lens. Find the position of the final image and the magnification of the system.

Solution: First we find the image position for the first lens while neglecting the second lens:

$$\frac{1}{s_1} + \frac{1}{s_1'} = \frac{1}{15 \text{ cm}} + \frac{1}{s_1'} = \frac{1}{10 \text{ cm}}$$

$$s_1' = 30 \text{ cm}$$

where s_1' is measured from the first lens.

Since s_1' is greater than the separation between the two lenses, we see that the image of the first lens lies 10 cm to the *right* of the second lens. We take this as the object distance for the second lenses. That is, we apply the thin

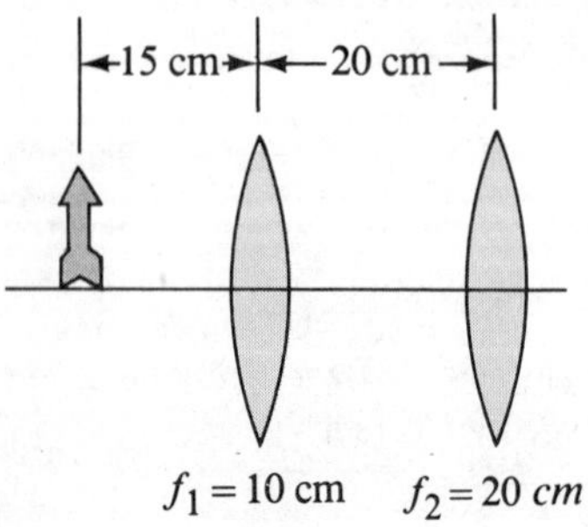

Figure 37.26 (Example 37.9) A combination of two converging lenses.

lens equation to the second lens with $s_2 = -10$ cm, where distances are now measured from the second lens, whose focal length is 20 cm:

$$\frac{1}{s_2} + \frac{1}{s_2'} = \frac{1}{f_2}$$

$$\frac{1}{-10 \text{ cm}} + \frac{1}{s_2'} = \frac{1}{20 \text{ cm}}$$

Solving for s_2' gives $s_2' = (20/3)$ cm. That is, the final image lies (20/3) cm to the *right* of the second lens.

The magnification of each lens separately is given by

$$M_1 = \frac{-s_1'}{s_1} = -\frac{30 \text{ cm}}{15 \text{ cm}} = -2$$

$$M_2 = \frac{-s_2'}{s_2} = -\frac{(20/3) \text{ cm}}{-10 \text{ cm}} = \frac{2}{3}$$

The total magnification M of the two lenses is the product $M_1M_2 = (-2)(2/3) = -4/3$. Hence, the final image is real, inverted, and enlarged.

Q8. Consider the image formed by a thin converging lens. Under what conditions will the image be (a) inverted, (b) erect, (c) real, (d) virtual, (e) larger than the object, and (f) smaller than the object?

Q9. Repeat Question 8 for a thin diverging lens.

Q10. If a cylinder of solid glass or clear plastic is placed above the words LEAD OXIDE and viewed from the side as shown in the photograph, the word "LEAD" appears inverted but the word "OXIDE" does not. Explain.

Can you explain why the word LEAD appears inverted, while the word OXIDE appears upright when viewed through the glass cylinder in the lower photograph? (Courtesy of Jim Lehman, James Madison University)

37.6 LENS ABERRATIONS

One of the basic problems of lenses and lens systems deals with the imperfect quality of the images. The simple theory of mirrors and lenses assumes paraxial rays, which make small angles with the optic axis. In this simple model, all rays leaving a point source focus at a single point, producing a sharp image. Clearly, this is not always the case. The approximations used in this theory are, in part, responsible for the imperfect images in real lenses.

If one wishes to perform a precise analysis of image formation, it is necessary to trace each ray using Snell's law at each refracting surface. This procedure shows that the rays from a point object do *not* focus at a single point. That is, there is no single point image; we say that the image is *blurred*. The departures of the real (imperfect) images from the ideal image predicted by the simple theory are called *aberrations*. A few types of aberrations will now be described.

Spherical Aberrations

Described briefly in Section 37.2, *spherical aberrations* are due to the fact that the focal points of light rays far from the optic axis of a spherical lens (or mirror) are different from those of rays passing through the center. Figure 37.27 illustrates spherical aberration for parallel rays passing through a converging lens. Rays near the middle of the lens have longer focal lengths than rays at the edges. Hence, there is no single focal length for a lens. Many cameras are equipped with an adjustable aperture to reduce spherical aberration when possible. (An *aperture* is an opening that controls the amount of light transmitted through the lens.) Sharper images are produced as the aperture size is reduced, since only the central portion of the lens is exposed to the incident light. At the same time, however, less light is imaged. To compensate for this, a longer exposure time is used on the photographic film.

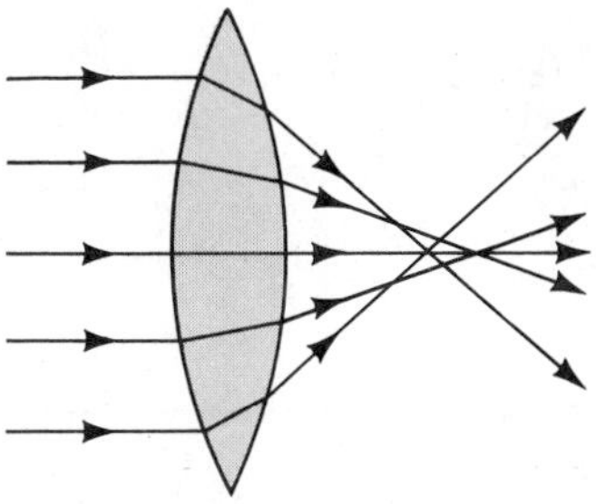
Figure 37.27 Spherical aberration produced by a converging lens. Does a diverging lens produce spherical aberration?

In the case of mirrors, one can eliminate, or at least minimize, spherical aberration by using a parabolic surface rather than a spherical surface. Parabolic surfaces are not often used, however, because they are very expensive to make. Parallel light rays incident on such a surface focus at a common point. Parabolic reflecting surfaces are used in large astronomical telescopes in order to enhance the image intensity. They are also used in flashlights, where a parallel light beam is produced from a small lamp placed at the focus of the surface.

Chromatic Aberrations

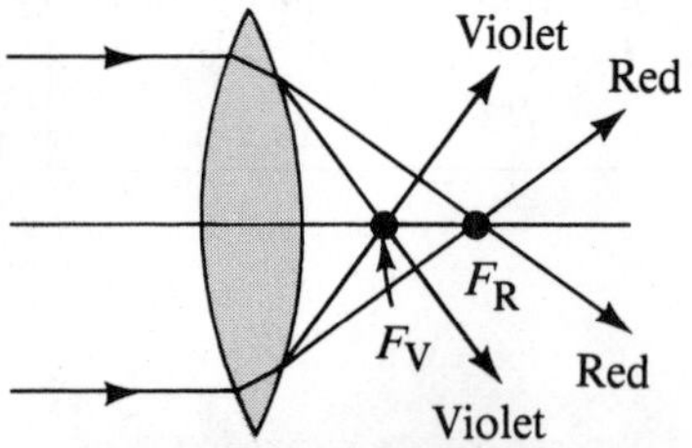

Figure 37.28 Chromatic aberration produced by a converging lens. Rays of different wavelengths focus at different points.

The fact that different wavelengths of light refracted by a lens focus at different points gives rise to *chromatic aberrations*. In the previous chapter, we described how the index of refraction of a material varies with wavelength. When white light passes through a lens, one finds, for example, that violet light rays are refracted more than red light rays (Fig. 37.28). From this, we see that the focal length is larger for red light than for violet light. Other wavelengths (not shown in Fig. 37.28) would have different focal points. The chromatic aberration for a diverging lens is opposite that for a converging lens. Chromatic aberration can be greatly reduced by using a combination of a converging and a diverging lens made from two different types of glass.

Other Aberrations

Several other defects occur as the result of object points being off the optical axis. *Astigmatism* results when a point object off the axis produces two line images at different points. A defect called *coma* is usually found in lenses with large spherical aberration. For this defect, an off-axis object produces a comet-shaped image. *Distortion* in an image exists for an extended object since magnification of off-axis points differs from magnification of those points near the axis. In high-quality optical systems, these defects are minimized by using properly designed, nonspherical surfaces or specific lens combinations.

Q11. Describe two types of aberration common in a spherical lens.

Q12. Explain why a mirror cannot give rise to chromatic aberration.

37.7 THE CAMERA

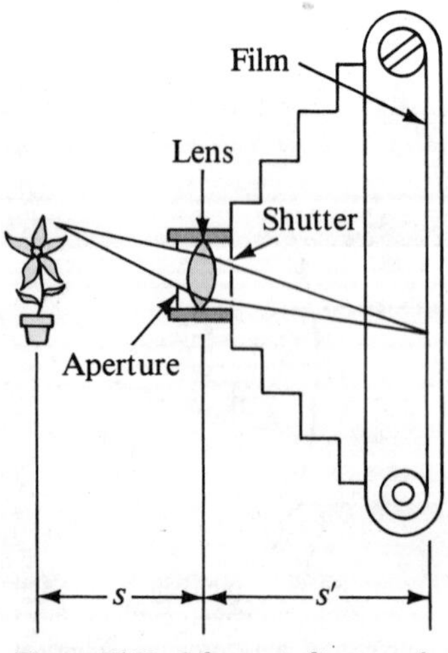

Figure 37.29 Schematic diagram of a simple camera, which contains an aperture, a lens, a shutter, and the film.

The single-lens photographic camera is a simple optical instrument whose essential features are shown in Fig. 37.29. It consists of a light-tight box, a converging lens which produces a real image, and a film at the back to receive the image. Focusing is accomplished by varying the distance between the lens and film with an adjustable bellows or some other mechanical arrangement for moving the lens. For proper focusing, or sharp images, the lens-to-film distance will depend on the object distance as well as on the focal length of the lens. The shutter, located behind the lens, is a mechanical device that is opened for selected time intervals, varying typically from 1 s to (1/1000) s. With this arrangement, one can photograph moving objects by using short exposure times or one can photograph darker scenes (low light intensity) by using longer exposure times. More expensive cameras also have an aperture of adjustable diameter behind the lens to provide further control of the intensity of the light reaching the film. When small aperture diameters are used, only the central portion of the lens receives light, and so the aberration is reduced somewhat.

The brightness (or energy flux) of the image focused on the film depends on the focal length of the lens and on the diameter D of the lens. Clearly, the light intensity I will be proportional to the area of the lens. Since the area is proportional to D^2, we conclude that $I \sim D^2$. Furthermore, the intensity is a measure of the energy received by the film per unit area of the image. Since the area of the image is proportional to $(s')^2$, and $s' \approx f$ (for objects with $s \gg f$), we conclude that the intensity is also proportional to $1/f^2$, so that $I \sim D^2/f^2$. The ratio f/D is defined to be the *f-number* of a lens:

$$f\text{-number} \equiv \frac{f}{D} \tag{37.19}$$

Hence, the intensity of light incident on the film can be expressed as

$$I \sim \frac{1}{(f/D)^2} \sim \frac{1}{(f\text{-number})^2} \tag{37.20}$$

The f-number is a measure of the "light-concentrating" power and determines the "speed" of the lens. A "fast" lens has a small f-number and hence is one with a small focal length and large diameter. Camera lenses are often marked with various f-numbers such as $f/2.8$, $f/4$, $f/5.6$, $f/8$, $f/11$, $f/16$. The various f-numbers are obtained by adjusting the aperture, which effectively changes D. When the f-number is changed by one position (or one "stop"), the light admitted changes by a factor of 2. Likewise, the shutter speeds are changed in steps whose factor is 2. The smallest f-number corresponds to the case where the aperture is wide open and the full lens area is in use. Fast lenses, with f-number as low as about 1.4, are more expensive, since it is more difficult to keep aberrations acceptably small. Cheaper cameras for routine snapshots usually have a fixed focal length and fixed aperture size, with an f-number of about $f/11$.

Example 37.10

The lens of a certain 35-mm camera (where 35 mm is the width of the film strip) has a focal length of 55 mm and a speed of $f/1.8$. The correct exposure time for this speed under certain conditions is known to be (1/500) s. (a) Determine the diameter of the lens.

Fom Eq. 37.19, we find that

$$D = \frac{f}{f\text{-number}} = \frac{55\text{ mm}}{1.8} = 31\text{ mm}$$

(b) Calculate the correct exposure time if the f-number is changed to $f/4$ under the same lighting conditions.

The total light energy received by each part of the image is proportional to the product of the flux and the exposure time. If I is the light intensity reaching the film, then in a time t, the energy received by the film is It. Comparing the two situations, we require that $I_1t_1 = I_2t_2$, where t_1 is the correct exposure time for $f/1.8$ and t_2 is the correct exposure time for some other f-number. Using this result, together with Eq. 37.20, we find that

$$\frac{t_1}{(f_1\text{-number})^2} = \frac{t_2}{(f_2\text{-number})^2}$$

$$t_2 = \left(\frac{f_2\text{-number}}{f_1\text{-number}}\right)^2 t_1 = \left(\frac{4}{1.8}\right)^2\left(\frac{1}{500}\text{ s}\right) \approx \frac{1}{100}\text{ s}$$

That is, as the aperture is reduced in size, the exposure time must increase.

37.8 THE EYE

The eye is a very important optical system which has much in common with the camera. Like the camera, the eye gathers light and produces a sharp image. However, the mechanisms by which the eye controls the amount of light admitted and adjusts to produce correctly focused images are far more complex and intricate than those in the most sophisticated camera. In all respects, the eye is an architectural wonder.

Figure 37.30 shows the essential parts of the eye. The front of the eye is covered by a transparent membrane called the *cornea*. This is followed by a clear liquid region (the *aqueous humor*), a variable aperture (the *iris* and *pupil*), and the *crystal-*

line lens. Most of the refraction occurs in the cornea, since the liquid medium surrounding the lens has an average index of refraction close to that of the lens. The iris, the distinctively colored portion of the eye, is a muscular diaphragm that controls the size of the pupil. It regulates the amount of light entering the eye by dilating the pupil in light of low intensity and contracting the pupil under high-intensity light. The f-stop range of the eye is about $f/2.8$ to $f/16$. Light entering the eye is focused by the cornea-lens system on the back surface of the eye (called the *retina*). The retina contains nerve fibers which branch out into millions of sensing structures called *rods* and *cones*. Optical images received by the retina are transmitted to the brain via the *optic nerve*.

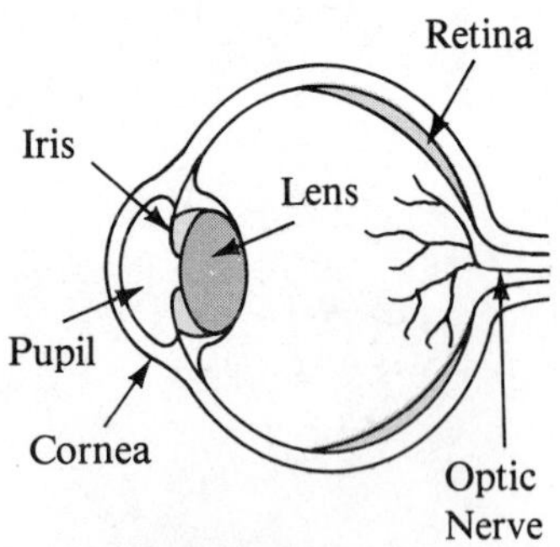

Figure 37.30 Essential parts of the eye. Note the similarity between the eye and the simple camera. Can you correlate the essential parts of the eye with those of the camera?

A distinct image of an object is observed when the image falls on the retina. The eye focuses on a given object by varying the shape of the pliable crystalline lens through an amazing process called *accommodation*. An important component in the accommodation process is the *ciliary muscle* attached to the eye. When the eye is focused on distant objects, the ciliary muscle is relaxed. For an object distance of infinity, the focal length of the eye (the distance between the lens and retina) is about 1.7 cm. The eye is able to focus on nearby objects by tensing the ciliary muscle. This action effectively decreases the focal length by slightly increasing the radius of curvature of the lens, which allows the image to focus on the retina. The lens adjustment in the accommodation process takes place so swiftly that we are not even aware of the change. Again, in this respect, even the finest electronic camera is a toy compared with the eye. It is evident that there is a limit to the accommodation process, since objects that are very close to the eye produce blurred images. The *near point* represents the closest distance for which the lens will produce a sharp image on the retina. This distance usually increases with age and has an average value of around 25 cm.

Although the eye is one of the most remarkable creations in nature, it often does not function perfectly. The eye may have several abnormalities, which can sometimes be corrected with eyeglasses, contact lenses, or surgery. When the relaxed eye produces an image of a distant object *behind* the retina, as in Fig. 37.31a, the abnormality is known as *hyperopia*, and the person is said to be farsighted. The hyperopic eye is shorter than normal. In some cases, the problem is a weak crystalline lens that won't adjust properly. The condition can be corrected with a converging lens, as shown in Fig. 37.31b.

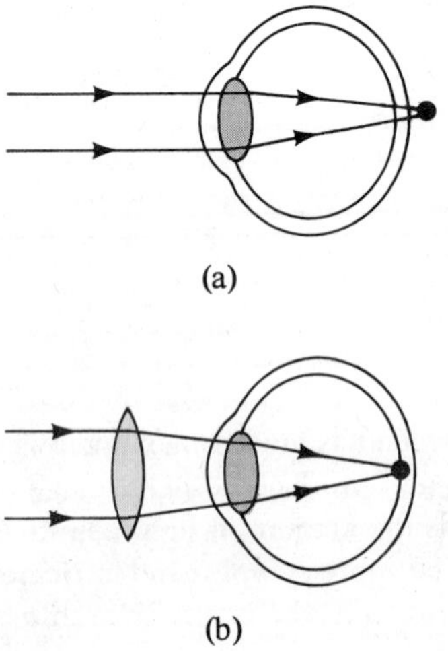

Figure 37.31 (a) A hyperopic eye (farsighted) is slightly shorter than normal, and so the image of a distant object focuses behind the retina. (b) The condition can be corrected with a converging lens.

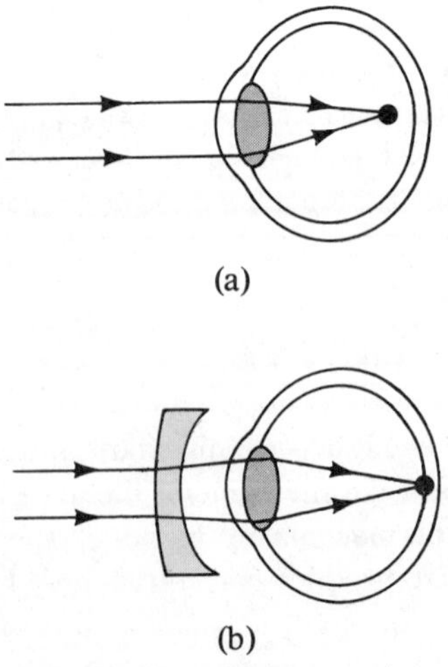

Figure 37.32 (a) A myopic eye (nearsighted) is slightly longer than normal, and so the image of a distant object forms in front of the retina. (b) The condition can be corrected with a diverging lens.

Another condition, known as *myopia*, or nearsightedness, occurs when the eye is longer than normal (Fig. 37.32a). In this case, light from a distant object is focused in *front* of the retina. Nearsightedness can be corrected with a diverging lens, as in Fig. 37.32b.

Beginning with middle age, most people lose some of their accommodation power as a result of a weakening of the ciliary muscles and a hardening of the lens, a condition known as *presbyopia*. This condition can also be corrected with converging lenses. A person may also have an eye defect known as *astigmatism*, in which light from a point source produces a line image on the retina. This defect is caused by an asymmetric shape in the cornea or lens. Astigmatism can be corrected with lenses having different curvatures in two perpendicular directions.

The eye is also subject to several diseases. One disease, which usually occurs in old age, is the formation of *cataracts*, where the lens becomes partially or totally opaque. The only known remedy for cataracts is surgical removal of the lens. Another group of diseases, called *glaucoma*, arise from an abnormal increase in fluid pressure within the eyeball. The pressure increase can lead to a swelling of the lens and to strong myopia. There is a chronic form of glaucoma, in which the pressure increase causes a reduction in blood supply to the retina. This can eventually lead to blindness, since the nerve fibers at the retina eventually die. If the disease is discovered early enough, it can be treated with medicine or surgery.

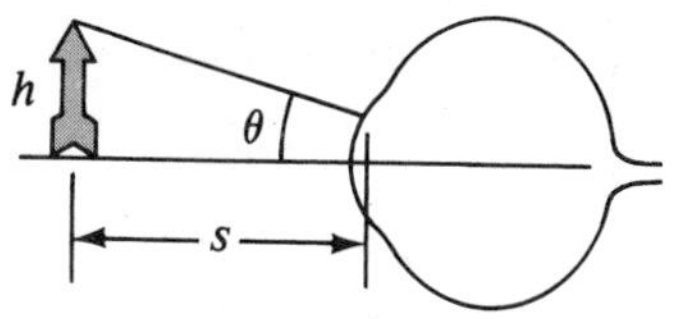

Figure 37.33 An object of height h subtends an angle θ at the eye. As the object moves toward the eye, θ increases and the image height at the retina increases.

Optometrists and ophthalmologists usually prescribe lenses measured in *diopters*. *The power of a lens in diopters equals the inverse of the focal length in meters,* that is, $P = 1/f$. For example, a converging lens whose focal length is $+20$ cm has a power of $+5$ diopters, and a diverging lens whose focal length is -40 cm has a power of -2.5 diopters. From Eq. 37.18, it follows that the power of a combination of two thin lenses placed *in contact* with each other equals the sum of the individual lens powers. For instance, two lenses of power $+2$ diopters and -5 diopters have a combined power of -3 diopters.

37.9 THE SIMPLE MAGNIFIER

A common lens used to increase the apparent size of an object is a converging lens called a *simple magnifier*. Suppose an object of height h is viewed at some distance from the eye, as in Fig. 37.33. Clearly, the size of the image formed at the retina depends on the angle θ subtended by the object at the eye. As the object moves closer to the eye, θ increases and a larger image is observed. However, a normal eye is unable to focus on an object closer than about 25 cm, the near point. Try it! At the near point, θ is a maximum and given by $h/25$.

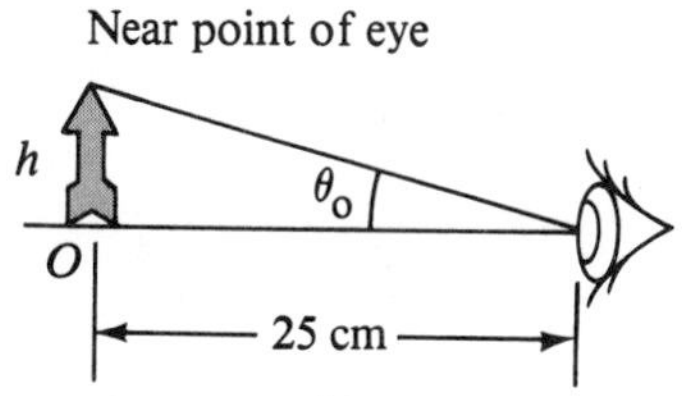

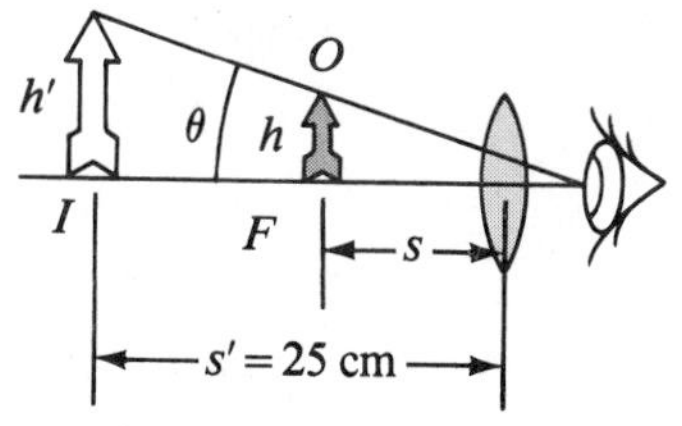

Figure 37.34 (a) An object placed at the near point ($s = 25$ cm) subtends an angle θ_0 at the eye, where $\theta_0 \approx h/25$. (b) An object placed near the focal point of a converging lens produces a magnified image, which subtends an angle $\theta \approx h'/25$ at the eye.

To increase the angular size of an object even more, a converging lens can be placed in front of the eye with the object located just inside the focal point, as in Fig. 37.34. The lens forms a virtual, erect, and enlarged image as shown. Clearly, the lens increases the angular size of the object. As we see from Fig. 37.34, the image on the retina is enlarged by the factor θ/θ_0, where θ_0 is the angle subtended by the object placed at the near point with no lens (Fig. 37.34a) and θ is the angle subtended by the image in the presence of the lens (Fig. 37.34b). We define the *angular magnification*,[2] or *magnifying power*, m of the lens as the ratio θ/θ_0:

Angular magnification

$$m \equiv \frac{\theta}{\theta_0} \tag{37.21}$$

The angular magnification is a maximum when the image is at the near point of the

[2] In general, the angular magnification is defined as the ratio $\tan\theta/\tan\theta_0$. In the small angle approximation, $\tan\theta \approx \theta$ and $\tan\theta_0 \approx \theta_0$, and so $m \approx \theta/\theta_0$. It is left as an exercise to show that in this case, m is also equal to the linear magnification, M.

eye, that is, when $s' = -25$ cm. The object distance s corresponding to this image distance can be calculated from the thin lens formula:

$$\frac{1}{s} + \frac{1}{-25\text{ cm}} = \frac{1}{f}$$

$$s = \frac{25f}{25 + f}$$

From the triangles in Fig. 37.34 and from the small angle approximation, note that $\theta_0 \approx h/25$ and $\theta \approx h/s$, so that Eq. 37.21 becomes

$$m = \frac{\theta}{\theta_0} = \frac{h/25}{h/s} = \frac{25}{s} = \frac{25}{25f/(25 + f)}$$

Magnification with image at near point

$$m = 1 + \frac{25\text{ cm}}{f} \qquad (37.22)$$

The magnification given is the ratio of the angular size seen with the instrument to the angular size that would be seen by examining the object with the naked eye at the near point. In practice, the eye can focus on an image formed anywhere between the near point and infinity. It is left as an exercise to show that the angular magnification is given by (25 cm)/f when the image is viewed at infinity rather than at the near point.

Using a single lens, it is possible to obtain angular magnifications up to about 4× without serious aberration. Magnifications up to about 20× can be achieved by using two lenses to correct for aberrations.

37.10 THE COMPOUND MICROSCOPE AND THE TELESCOPE

In this section, we give a brief description of two important optical instruments, the compound microscope and the astronomical telescope. In its simplest form, each instrument contains two converging lenses. One lens, called the *objective*, forms a real image of the object. The second lens, called the *ocular* (or *eyepiece*) lens, is a simple magnifier for viewing the image formed by the objective.

The Compound Microscope

Figure 37.35 is a schematic diagram of a compound microscope. It consists of an objective lens with a very short focal length, f_o (where $f_o < 1$ cm), and an ocular, or eyepiece, having a focal length f_e of a few cm. The two lenses are separated by a

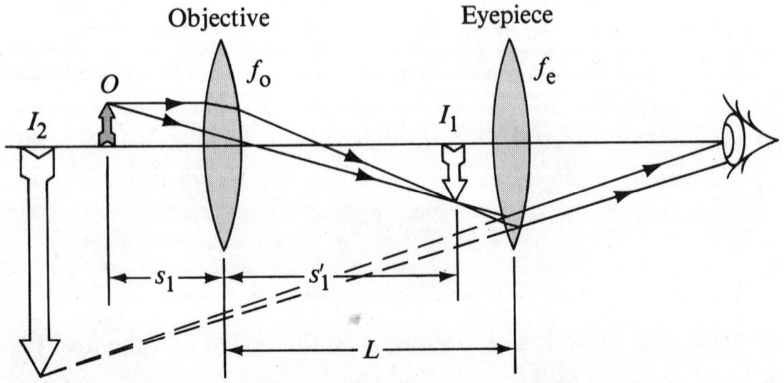

Figure 37.35 Schematic diagram of a compound microscope, which consists of an objective lens and an eyepiece, or ocular, lens.

distance L, where L is much greater than either f_o or f_e. The object, which is placed just outside the focal length of the objective, forms a real, inverted image at I_1, which is at or close to the focal length of the ocular lens. The ocular lens, which serves as a simple magnifier, produces an image of I_1 at I_2, and the image at I_2 is virtual and inverted. The lateral magnification M_1 of the first image is $-s_1'/s_1$. Using the approximations $s_1' \approx L$ and $s_1 \approx f_o$, we see that $M_1 \approx -L/f_o$. Furthermore, the angular magnification of the ocular for an object (corresponding to the image at I_1) placed at the focal point is (25 cm)$/f_e$. Hence, the overall magnification of the compound microscope is the product of the two magnifications:

$$M = M_1 m = -\frac{L}{f_o}\left(\frac{25 \text{ cm}}{f_e}\right) \tag{37.23}$$

Magnification of a compound microscope

The negative sign indicates that the image is inverted.

The Astronomical Telescope

The astronomical telescope sketched in Fig. 37.36 is used to view distant objects. The two lenses are arranged such that the objective lens forms a real, inverted image of the distant object very near the focal point of the ocular lens. Furthermore, the image at I_1 is formed at the rear focal point of the objective, since the object is essentially at infinity. Hence, the two lenses are separated by a distance $f_o + f_e$, which corresponds to the length of the telescope's tube. The eyepiece finally forms an enlarged, inverted image of I_1 at I_2.

The angular magnification is given by θ/θ_o, where θ_o is the angle subtended by the object at the objective and θ is the angle subtended by the final image. From the triangles in Fig. 37.36, and for small angles, we know that $\theta \approx h'/f_e$ and $\theta_o \approx -h'/f_o$. Hence, the angular magnification of the telescope can be expressed as

$$m = \frac{\theta}{\theta_o} = \frac{h'/f_e}{-h'/f_o} = -\frac{f_o}{f_e} \tag{37.24}$$

Angular magnification of a telescope

That is, the angular magnification of a telescope equals the ratio of the objective focal length to that of the eyepiece. Here again, the magnification is the ratio of the angular size seen with the telescope to the angular size seen with the unaided eye. The negative sign in Eq. 37.24 indicates that the final image is inverted.

Astronomical research telescopes used to study very distant objects must have a large diameter in order to gather as much light as possible. It is difficult and expensive to manufacture large lenses for refracting-type telescopes. Another difficulty with large lenses is their large weight, which leads to sagging, an additional source of aberration. These problems can be partially overcome by replacing the objective lens with a reflecting, concave mirror. For example, the reflecting telescope at

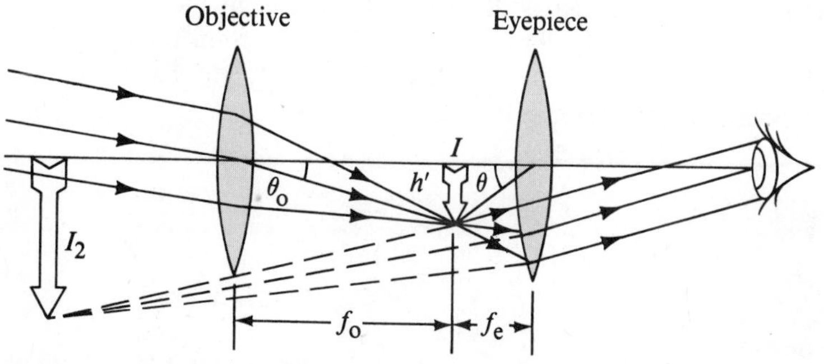

Figure 37.36 Schematic diagram of an astronomical telescope, with the object at infinity.

Mount Palomar Observatory in California has a mirror diameter of 200 in. (5.08 m). This is much larger than the 40-in. (1-m) diameter of the largest refracting telescope, which is located at Yerkes Observatory in Williams Bay, Wisconsin.

37.11 SUMMARY

In the paraxial ray approximation, the object distance s and image distance s' for a spherical mirror of radius R are related by the *mirror equation*

Mirror equation

$$\frac{1}{s}+\frac{1}{s'}=\frac{2}{R}=\frac{1}{f} \qquad (37.3,\ 37.5)$$

where $f = R/2$ is the *focal length* of the mirror.

The *lateral magnification* M of a mirror or lens is defined as the ratio of the image height h' to the object height h:

Magnification of a mirror

$$M=\frac{h'}{h}=-\frac{s'}{s} \qquad (37.6)$$

An image can be formed by refraction from a spherical surface of radius R. The object and image distances for refraction from such a surface are related by

Formation of an image by refraction

$$\frac{n_1}{s}+\frac{n_2}{s'}=\frac{n_2-n_1}{R} \qquad (37.7)$$

where the object is located in the medium of index of refraction n_1 and the image is formed in the medium whose index of refraction is n_2.

The focal length f of a lens is given by

Lens makers' formula

$$\frac{1}{f}=(n-1)\left(\frac{1}{R_1}-\frac{1}{R_2}\right) \qquad (37.16)$$

Converging lenses have positive focal lengths, and *diverging lenses* have negative focal lengths.

For a thin lens, and in the paraxial ray approximation, the object and image distances are related by the expression

Thin lens formula

$$\frac{1}{s}+\frac{1}{s'}=\frac{1}{f} \qquad (37.17)$$

Aberrations are responsible for the formation of imperfect images by lenses and mirrors. *Spherical aberration* is due to the variation in focal points for parallel incident rays that strike the lens at various distances from the optical axis. *Chromatic aberration* arises from the fact that light of different wavelengths focuses at different points when refracted by a lens.

EXERCISES

Section 37.1 Images Formed by Planar Mirrors

1. In a physics laboratory experiment, a torque is applied to a small-diameter wire that is suspended vertically under tensile stress. It is necessary to measure accurately the small angle through which the wire turns as a consequence of the net torque. This is accomplished by attaching a small mirror to the wire and reflecting a beam of light off the mirror and onto a circular scale. Such an arrangement is known as an *optical lever* and is shown from a top view in Fig. 37.37. Show that when the mirror turns through an angle θ, the reflected beam is rotated by an angle 2θ.

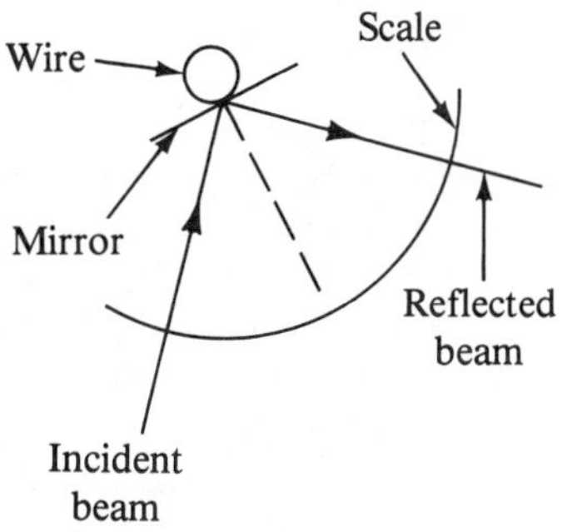

Figure 37.37 (Exercise 1).

2. Two planar mirrors, A and B, are in contact along one edge, and the planes of the two mirrors are at an angle of 45° with respect to each other (Fig. 37.38). A point object is placed at P along the bisector of the angle between the two mirrors. Make a sketch similar to Fig. 37.38 to a suitable scale and locate graphically (a) the image of P in mirror A and the image of P in mirror B. (b) Label the images found in (a) $P_A{}^1$ and $P_B{}^1$, respectively, and locate the image of $P_A{}^1$ in mirror B and the image of $P_B{}^1$ in mirror A. (c) Determine the total number of images for the arrangement described.

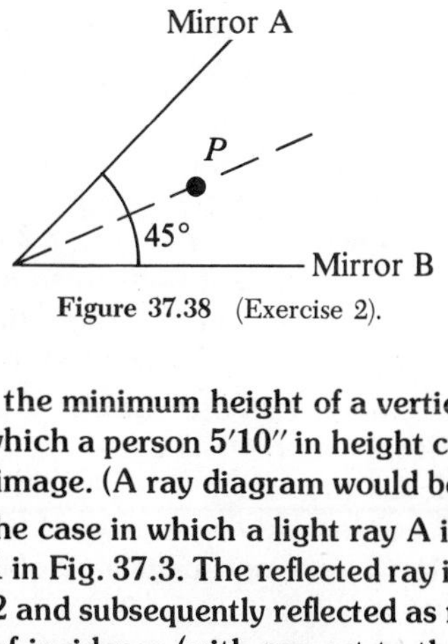

Figure 37.38 (Exercise 2).

3. Determine the minimum height of a vertical planar mirror in which a person 5′10″ in height can see his or her full image. (A ray diagram would be helpful.)

4. Consider the case in which a light ray A is incident on mirror 1 in Fig. 37.3. The reflected ray is incident on mirror 2 and subsequently reflected as ray B. Let the angle of incidence (with respect to the normal) on mirror 1 equal 53° and the point of incidence be located 20 cm from the edge of contact between the two mirrors. Determine the angle between ray A and ray B.

Section 37.2 Images Formed by Spherical Mirrors and Section 37.3 Ray Diagrams for Mirrors

5. A concave mirror has a radius of curvature of 60 cm. Calculate the image position and magnification of an object placed in front of the mirror at distances of (a) 90 cm and (b) 20 cm. (c) Draw ray diagrams to obain the image in each case.

6. The real-image height of a concave mirror is observed to be four times larger than the object height when the object is 30 cm in front of the mirror. (a) What is the radius of curvature of the mirror? (b) Use a ray diagram to locate the image position corresponding to the given object position and the radius of curvature calculated in (a).

7. Calculate the image position and magnification for an object placed (a) 20 cm and (b) 60 cm in front of a convex mirror of focal length 40 cm. (c) Use ray diagrams to locate image positions corresponding to the object positions in (a) and (b).

8. Use a ray diagram to demonstrate that the image of a real object placed in front of a spherical mirror is always virtual and erect when $s < |f|$.

9. A concave mirror has a focal length of 40 cm. Determine the object position for which the resulting image will be erect and four times the size of the object.

10. A convex mirror has a focal length of −20 cm. Determine an object location for which the image will be one half the size of the object.

11. A spherical mirror is to be used to form an image five times the size of an object on a screen located 5 m from the object. (a) Describe the type of mirror required. (b) Where should the mirror be positioned relative to the object?

12. A real object is located at the zero end of a meter stick. A concave mirror located at the 100-cm end of the meter stick forms an image of the object at the 70-cm position. A convex mirror placed at the 60-cm position forms a final image at the 10-cm point. What is the radius of curvature of the convex mirror?

13. A spherical convex mirror has a radius of 40 cm. Determine the position of the virtual image and magnification of the mirror for object distances of (a) 30 cm and (b) 60 cm. (c) Are the images erect or inverted?

Section 37.4 Images Formed by Refraction

14. A glass sphere ($n = 1.50$) of radius 15 cm has a tiny air bubble located 5 cm from the center. The sphere is viewed along a direction parallel to the radius containing the bubble. What is the apparent depth of the bubble below the surface of the sphere?

15. One end of a long glass rod ($n = 1.5$) is formed into the shape of a *convex* surface of radius 6 cm. An object is located in air along the axis of the rod. Find the image positions corresponding to the object at distances of (a) 20 cm, (b) 10 cm, and (c) 3 cm from the end of the rod.

16. Calculate the image positions corresponding to the object positions stated in Exercise 15 if the end of the rod has the shape of a *concave* surface of radius 8 cm.

17. Repeat Exercise 15 if the object is in water surrounding the glass rod instead of in air.

18. A smooth block of ice rests on the floor with one face parallel to the floor. The block has a vertical thickness of 50 cm. Find the location of the image of a pattern in the floor covering as formed by rays that are nearly perpendicular to the block. (Use $n = 1.309$ from Table 36.1.)

19. A flint glass plate ($n = 1.66$) rests on the bottom of an aquarium tank. The plate is 8 cm thick and is covered with water ($n = 1.33$) to a depth of 12 cm. Calculate the apparent thickness of the glass plate as viewed from above the water. (Assume nearly normal incidence.)

20. A glass hemisphere is used as a paperweight with its flat face resting on a stack of papers. The radius of the circular cross section is 4 cm, and the index of refraction of the glass is 1.55. The center of the hemisphere is directly over a letter "O" that is 2.5 mm in height. What is the height of the image of the letter as seen looking along a vertical radius?

Section 37.5 Thin Lenses

21. An object located 32 cm in front of a lens forms an image on a screen 8 cm behind the lens. (a) Find the focal length of the lens. (b) Determine the magnification of the lens. (c) Is the lens converging or diverging?

22. A converging lens has a focal length of 40 cm. Calculate the size of the real image of an object 4 cm in height for the following object distances: (a) 50 cm, (b) 60 cm, (c) 80 cm, (d) 100 cm, (e) 200 cm, (f) ∞.

23. A thin converging lens has a focal length f. Find the object distance if the image is (a) real and twice as large as the object and (b) virtual and twice the size of the object.

24. A convex lens forms a real image of an object at a point located 12 cm to the right of the lens. The object is positioned 50 cm to the left of the lens. (a) Calculate the focal length of the lens. (b) What is the ratio of the height of the image to the height of the object? (c) Is the image erect or inverted? Real or virtual?

25. Construct a ray diagram for the arrangement described in Exercise 24.

26. The left face of a biconvex lens has a radius of curvature of 12 cm, and the right face has a radius of curvature of 18 cm. The index of refraction of the glass is 1.44. (a) Calculate the focal length of the lens. (b) Calculate the focal length if the radii of curvature of the two faces are interchanged.

27. A thin-walled, hollow convex lens is immersed in a tank of water. The hollow lens has $R_1 = 20$ cm and $R_2 = 30$ cm. Calculate the focal length of this "air lens" surrounded by water ($n = 1.33$). Use the derivation of Eq. 37.16 as a guide.

28. A diverging lens is used to form a virtual image of a real object. The object is positioned 80 cm to the left of the lens, and the image is located 40 cm to the left of the lens. (a) Determine the focal length of the lens. (b) If the surfaces of the lens have radii of curvature of magnitude 40 cm and 50 cm, what is the value of the index of refraction of the lens?

29. An object is located 20 cm to the left of a diverging lens of focal length $f = -32$ cm. Determine (a) the location and (b) the magnification of the image.

30. Construct a ray diagram for the arrangement described in Exercise 29.

Section 37.7 The Camera and Section 37.8 The Eye

31. A camera is found to give proper film exposure when it is set at $f/16$ and the shutter is open for (1/32) s. Determine the correct exposure time if a setting of $f/8$ is used. (Assume the lighting conditions are unchanged.)

32. A camera is being used with correct exposure at $f/4$ and a shutter speed of (1/16) s. In order to "stop" a fast-moving subject, the shutter speed is changed to (1/128) s. Find the new f-number setting that should be used to maintain satisfactory exposure.

33. What is the unaided near point for a person required to wear lenses with a power of +1.5 diopters to read at 25 cm?

34. If the aqueous humor of the eye has an index of refraction of 1.34 and the distance from the vertex of the cornea to the retina is 2.2 cm, what is the radius of curvature of the cornea for which distant objects will be focused on the retina? (Assume all refraction occurs in the aqueous humor.)

35. Assume that the camera shown in Fig. 37.29 has a fixed focal length of 6.5 cm and is adjusted to properly focus the image of a distant object. By how much and in what direction must the lens be moved in order to focus the image of an object at a distance of 2 m?

36. Figure 37.31a illustrates the case of a farsighted person who can focus clearly objects that are more distant than 90 cm from the eye. Determine the power of lenses that will enable this person to read comfortably at a normal near point of 25 cm.

37. The eye of a nearsighted person is illustrated in Fig. 37.32a. In this case, the person cannot focus clearly objects that are more distant than 200 cm from the eye. Determine the power of lenses that will enable this person to see distant objects clearly.

Section 37.9 The Simple Magnifier and Section 37.10 The Compound Microscope and the Telescope

38. A philatelist examines a printing detail using a convex lens of focal length 10 cm as a simple magnifier. The lens is held close to the eye, and the lens-to-object distance is adjusted so that the virtual image is formed at the normal near point (25 cm.) Calculate the expected magnification.

39. An astronomical telescope has an objective with focal length 75 cm and an eyepiece with focal length 4 cm. What is the magnifying power of the instrument?

40. The distance between the eyepiece and the objective lens in a compound microscope is 23 cm. The focal length of the eyepiece is 2.5 cm, and the focal length of the objective is 1.2 cm. What is the overall magnification of the microscope? (Assume that the final image is formed 25 cm from the eye.)

41. The desired overall magnification of a compound microscope is 140×. The objective alone produces a lateral magnification of 12×. Determine the required focal length of the eyepiece lens. (Assume that the final image will be 25 cm from the eye.)

PROBLEMS

1. A concave mirror has a radius of 40 cm. (a) Calculate the image distance s' for an arbitrary real object distance s. (b) Obtain values of s' for object distances of 5 cm, 10 cm, 40 cm, and 60 cm. (c) Make a plot of s' versus s using the results of (b).

2. An object placed 10 cm from a concave spherical mirror produces a real image 8 cm from the mirror. If the object is moved to a new position 20 cm from the mirror, what is the position of the image? Is the final image real or virtual?

3. A colored marble is dropped in a large tank filled with benzene ($n = 1.50$). (a) What is the depth of the tank if the apparent depth of the marble when viewed from directly above the tank is 35 cm? (b) If the marble has a diameter of 1.5 cm, what is its apparent diameter when viewed from directly above, outside the tank?

4. Figure 37.39 shows the longitudinal section of a shape formed from glass of index of refraction $n = 1.5$. The ends are hemispheres of radius r and $2r$, and the centers of the hemispherical ends are separated by a distance $4r$. A point object is located in air a distance $0.5r$ from the left end of the glass form. Find the location of the image of the object due to refraction at the two spherical surfaces when $r = 2$ cm.

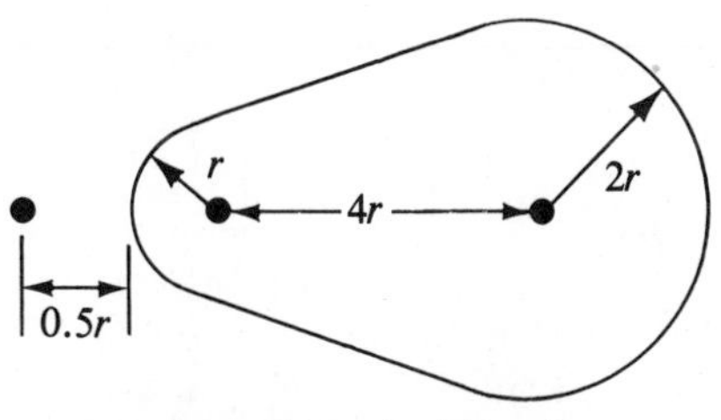

Figure 37.39 (Problem 4).

5. An object 1 cm in height is placed 4 cm to the left of a converging lens of focal length 8 cm. A diverging lens of focal length −16 cm is located 6 cm to the right of the converging lens. Find the position and size of the final image. Is the image inverted or erect? Real or virtual?

6. An object is placed 12 cm to the left of a diverging lens of focal length −6 cm. A converging lens of focal length 12 cm is placed a distance d to the right of the diverging lens. Find the distance d such that the final image is at infinity. Draw a ray diagram for this case.

7. A converging lens has a focal length of 20 cm. Find the position of the image for a real object at distances of (a) 50 cm, (b) 30 cm, (c) 10 cm. (d) Determine the magnification of the lens for these object distances and whether the image is erect or inverted. (e) Draw ray diagrams to locate the images for these object distances.

8. Repeat Problem 7 if the lens is diverging and has a focal length of 15 cm.

9. Find the object distances (in terms of f) of a thin converging lens of focal length f if (a) the image is real and the image distance is four times the focal length and (b) the image is virtual and the image distance is three times the focal length. (c) Calculate the magnification of the lens for cases (a) and (b).

10. A converging lens of focal length 20 cm is separated by 50 cm from a converging lens of focal length 5 cm. (a) Find the final position of the image of an object placed 40 cm in front of the first converging lens. (b) If the height of the object is 2 cm, what is the height of the final image? Is it real or virtual? (c) If the two lenses are placed in contact, what is

the focal length of the combination? (d) Determine the image position of an object placed 5 cm in front of the two lenses in contact.

11. An object is located 36 cm to the left of a biconvex lens of index of refraction 1.5. The left surface of the lens has a radius of curvature of 20 cm. The right surface of the lens is to be shaped so that a real image will be formed 72 cm to the right of the lens. What is the required radius of curvature of the second surface?

12. Since the index of refraction of a material such as glass depends on wavelength, the focal length of a lens made from this material must also depend on the wavelength. This phenomenon (called dispersion) causes the image of an object to be somewhat blurred, and results in what is called chromatic aberration (see Figure 37.28). Suppose that a lens is made from crown glass, whose index of refraction varies with wavelength according to Figure 36.15. Calculate the fractional difference in focal length, $\Delta f/f$, for this lens between wavelengths of (a) $\lambda_1 = 400$ nm and $\lambda_2 = 500$ nm and (b) $\lambda_1 = 400$ nm and $\lambda_3 = 650$ nm. [Assume that $n_1 = 1.532$, $n_2 = 1.522$ and $n_3 = 1.513$.]

38 Interference of Light Waves

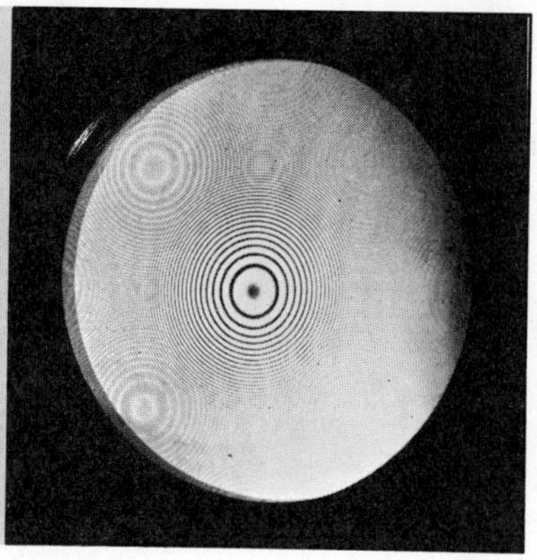

Bausch & Lomb Optical Co.

In the previous chapter on geometric optics, we used the concept of light rays to treat various optical systems. The next two chapters are concerned with the subject of *wave optics,* which deals with the phenomena of interference, diffraction, and polarization of light. These phenomena cannot be adequately explained with ray optics, but we shall describe how the wave theory of light leads to a satisfying description of such phenomena. This chapter is aimed at explaining various types of interference effects associated with light.

38.1 CONDITIONS FOR INTERFERENCE

In Chapter 34, we found that wave interference results from the linear superposition of two waves. If the two waves combine so as to produce a resultant wave intensity that is greater than the separate intensities,[1] this is called *constructive interference.* If the two waves combine so as to produce a resultant wave intensity that is less than the separate intensities, it is called *destructive interference.*

Light waves exhibit two types of interference. One type, discussed in this chapter, arises from the superposition of light waves that either have passed through openings or have undergone only reflection or refraction. The second type of interference, described in the next chapter, arises from the superposition of waves that exhibit diffraction, which is a phenomenon characterizing the ability of waves to bend around obstacles. Fundamentally, all interference associated with light waves arises from the superposition of electric and magnetic field vectors.

Interference effects associated with light waves are not easy to observe because of their short wavelength (about 5×10^{-7} m). In order to observe a sustained interference of light waves, the following conditions must be met: (1) The sources must be *coherent,* that is, they must maintain a constant phase with respect to each other. (2) The sources should be *monochromatic,* that is, of a single wavelength. (3) The *linear superposition principle* should be applicable.

Conditions of interference

Let us elaborate further on the concept of coherent sources. As we have said, two sources (producing two traveling waves) are needed to create interference. When the sources are coherent and maintain a constant relative phase, the resulting interference pattern is stationary at some point in space. This can occur if the two sources each have a definite frequency that does not change with time. On the other

[1] By *intensity* we mean a quantity that is proportional to the local energy density in a wave.

Coherent and noncoherent sources

hand, if the phase difference between the two sources changes with time, the interference pattern also varies with time. Such sources are said to be *noncoherent*. Thus, a *steady interference pattern can be achieved only with two coherent sources.* For example, two equivalent loudspeakers driven by the *same* oscillator will vibrate with equal frequency and will emit coherent sound waves. Similarly, two radio antennas driven by the same oscillator will maintain a constant relative phase. In these cases, the phase of the sources is effectively "locked in" by a common driving force. On the other hand, two independent ordinary light sources are incoherent, since the sources of the waves are randomly vibrating atoms. It is possible to overcome this difficulty by using laser beams. A *laser* is a device that emits a nearly parallel beam of very intense, coherent radiation. A more common method for producing two coherent light sources is to use one monochromatic source to illuminate a screen containing two small apertures. The light emerging from the two apertures is then coherent, since it originates from a common source. The principles behind the operation of the laser will be discussed in Chapter 41.

38.2 YOUNG'S DOUBLE-SLIT EXPERIMENT

The phenomenon of interference of light waves arising from two sources was first demonstrated by Thomas Young in 1801. A schematic diagram of the apparatus used in this experiment is shown in Fig. 38.1. Light is incident on screen A, which is provided with a narrow slit, S_0. The cylindrical waves emerging from this slit arrive at screen B, which contains two narrow, parallel slits, S_1 and S_2. Light emerges from these two slits as cylindrical waves.[2] In effect, slits S_1 and S_2 act as individual light sources that are in phase since they originate from the same cylindrical wavefront. The light from the two slits produces a visible pattern on screen C; the pattern consists of a series of bright and dark parallel bands called *fringes*. The overall light amplitude at a given point on the screen is the result of the superposition of the two wave amplitudes from S_1 and S_2. Two waves that add constructively give a bright fringe, and any two waves that add destructively produce a dark fringe.

We can obain a quantitative description of Young's experiment with the help of

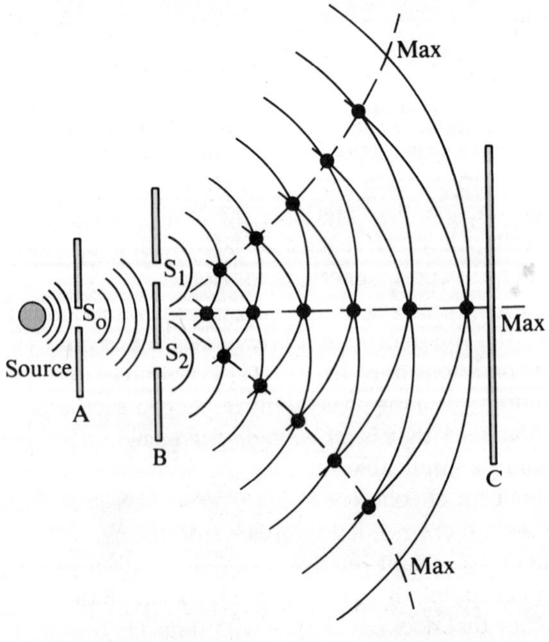

Figure 38.1 Schematic diagram of Young's double-slit experiment. The narrow slits act as sources of cylindrical waves. Slits S_1 and S_2 behave as coherent sources which produce an interference pattern on screen C.

[2] In his original experiment, Young actually used sunlight as the source, and the screens contained pinholes rather than slits.

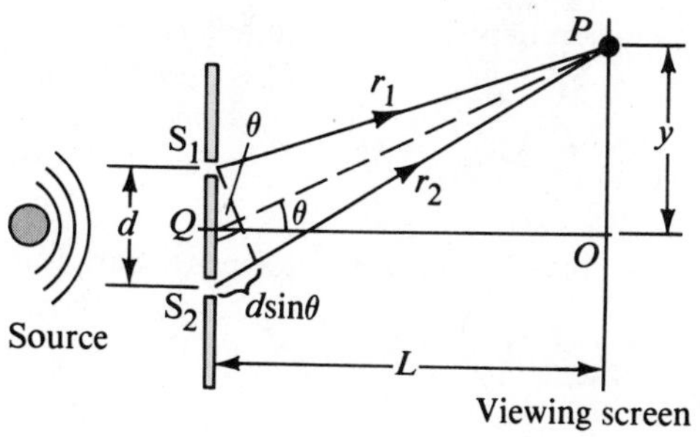

Figure 38.2 Geometric construction for describing Young's double-slit experiment. Note that the path difference between the two rays is $r_2 - r_1 = d\sin\theta$.

Fig. 38.2. Consider a point P on the viewing screen; this point is located a perpendicular distance L from the two identical slits S_1 and S_2, which are separated by a distance d. Let us assume that the source is equidistant from the two slits and is *monochromatic,* that is, emitting light of a single wavelength λ. Under these assumptions, the waves emerging from slits S_1 and S_2 have the same frequency and amplitude and are in phase. The light intensity on the screen at P is the resultant of light coming from both slits. Note that a wave from the lower slit travels farther than a wave from the upper slit by an amount equal to $d\sin\theta$. This distance is called the *path difference,* δ, where

$$\delta = r_2 - r_1 = d\sin\theta \tag{38.1}$$

Path difference

as shown in Fig. 38.2. The value of this path difference will determine whether or not the two waves are in phase when they arrive at P. If the path difference is either *zero* or *some integral multiple of the wavelength,* the two waves are in phase at P and *constructive interference results.* Therefore, the condition for bright fringes, or constructive interference, at P is given by

$$\delta = d\sin\theta = m\lambda \qquad (m = 0, \pm1, \pm2, \pm3 \ldots) \tag{38.2}$$

Condition for bright fringes

The index number m is called the *order number* of the fringe. The central bright fringe at $\theta = 0$ ($m = 0$) is called the *zeroth-order maximum.* The first maximum on either side, when $m = \pm1$, is called the *first-order maximum,* etc.

Similarly, *when the path difference is an odd multiple of* $\lambda/2$, the two waves arriving at P will be opposite in phase and will give rise to *destructive interference.* Therefore, the condition for dark fringes, or destructive interference, at P is given by

$$\delta = d\sin\theta = \left(m + \frac{1}{2}\right)\lambda \qquad (m = 0, \pm1, \pm2 \ldots) \tag{38.3}$$

Condition for dark fringes

It is useful to obtain expressions for the positions of the bright and dark fringes measured vertically from O to P. We shall assume that $L \gg d$ (Fig. 38.2) and consider only points P that are close to O. In this case, θ is small, and so we can use the approximation $\sin\theta \approx \tan\theta$. From the large triangle OPQ in Fig. 38.2, we see that

$$\sin\theta \approx \tan\theta = \frac{y}{L} \tag{38.4}$$

Using this result together with Eq. 38.2, we see that the positions of the bright fringes measured from O are given by

$$y_{\text{bright}} = \frac{\lambda L}{d} m \tag{38.5}$$

Positions of bright fringes

From this expression, we find that the separation between any two adjacent bright fringes is equal to $\lambda L/d$, that is,

$$y_{m+1} - y_m = \frac{\lambda L}{d}(m+1) - \frac{\lambda L}{d}m = \frac{\lambda L}{d} \tag{38.6}$$

Similarly, using Eqs. 38.3 and 38.4, we find that the dark fringes are located at

Positions of dark fringes

$$y_{\text{dark}} = \frac{\lambda L}{d}\left(m + \frac{1}{2}\right) \tag{38.7}$$

This result shows that the separation between adjacent dark fringes is also equal to $\lambda L/d$. Since the quantities L and d are both measurable, we see that the double-slit interference pattern, together with Eq. 38.6, provides a direct determination of the wavelength λ. Young used this technique to make the first measurement of the wavelength of light.

Example 38.1

A screen is separated from a double-slit source by 1.2 m. The distance between the two slits is 0.03 mm. The second-order bright fringe ($m = 2$) is measured to be 4.5 cm from the center line. (a) Determine the wavelength of the light.

We can use Eq. 38.5, with $m = 2$, $y_2 = 4.5 \times 10^{-2}$ m, $L = 1.2$ m, and $d = 3 \times 10^{-5}$ m:

$$\lambda = \frac{dy_2}{mL} = \frac{(3 \times 10^{-5}\text{ m})(4.5 \times 10^{-2}\text{ m})}{2 \times 1.2\text{ m}}$$

$$= 5.6 \times 10^{-7}\text{ m} = 560\text{ nm}$$

(b) Calculate the distance between adjacent bright fringes.

From Eq. 38.6 and the results to (a), we get

$$y_{m+1} - y_m = \frac{\lambda L}{d} = \frac{(5.6 \times 10^{-7}\text{ m})(1.2\text{ m})}{3 \times 10^{-5}\text{ m}}$$

$$= 2.2 \times 10^{-2}\text{ m} = 2.2\text{ cm}$$

Example 38.2

A light source emits light of two wavelengths in the visible region, given by $\lambda = 430$ nm and $\lambda' = 510$ nm. The source is used in a doublet-slit interference experiment in which $L = 1.5$ m and $d = 0.025$ mm. Find the separation between the third-order bright fringes corresponding to these wavelengths.

Solution: Using Eq. 38.5, with $m = 3$ for the third-order bright fringes, we find that the values of the fringe positions are given by

$$y_3 = \frac{\lambda L}{d}m = 3\frac{\lambda L}{d} = 7.74 \times 10^{-2}\text{ m}$$

$$y_3' = \frac{\lambda' L}{d}m = 3\frac{\lambda' L}{d} = 9.18 \times 10^{-2}\text{ m}$$

Hence, the separation between the two fringes is given by

$$\Delta y = y_3' - y_3 = \frac{3(\lambda' - \lambda)}{d}L$$

$$= 1.44 \times 10^{-2}\text{ m} = 1.44\text{ cm}$$

Q1. What is the necessary condition on the path length difference between two waves that interfere (a) constructively and (b) destructively?

Q2. Explain why two distant flashlights will not produce an interference pattern.

Q3. If Young's double-slit experiment were performed under water, how would the observed interference pattern be affected?

38.3 INTENSITY DISTRIBUTION OF THE DOUBLE-SLIT INTERFERENCE PATTERN

We shall now calculate the distribution of light intensity associated with the double-slit interference pattern. Again, suppose that the two slits represent coherent sources of sinusoidal waves. Hence, they have the same angular frequency ω and

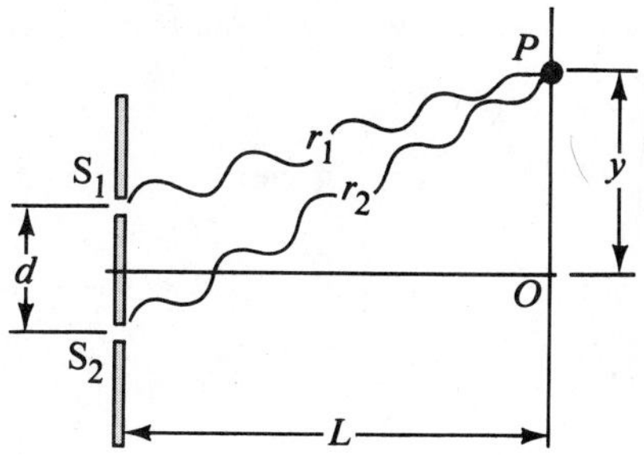

Figure 38.3 Construction for analyzing the double-slit interference pattern. A bright region, or intensity maximum, is observed at O.

a constant phase difference ϕ. The total electric field intensity at the point P on the screen in Fig. 38.3 is the *vector superposition* of the two waves from slits S_1 and S_2. Assuming the two waves have the same amplitude E_0, we can write the electric field intensities at P due to each wave separately as

$$E_1 = E_0 \sin\omega t \qquad \text{and} \qquad E_2 = E_0 \sin(\omega t + \phi) \tag{38.8}$$

Note that although the waves have equal phase at the slits, *their phase difference* ϕ *at* P *depends on the path difference* $\delta = r_2 - r_1 = d \sin\theta$. Since a path difference of λ corresponds to a phase difference of 2π radians (constructive interference), while a path difference of $\lambda/2$ corresponds to a phase difference of π radians (destructive interference), we obtain the ratio

$$\delta/\phi = \lambda/2\pi$$

$$\phi = \frac{2\pi}{\lambda}\delta = \frac{2\pi}{\lambda} d \sin\theta \tag{38.9}$$

Phase difference

This equation gives the precise dependence of the phase difference ϕ on the angle θ.

Using the superposition principle and Eq. 38.8, we can obtain the resultant electric field at the point P:

$$E_P = E_1 + E_2 = E_0[\sin\omega t + \sin(\omega t + \phi)] \tag{38.10}$$

To simplify this expression, we use the following trigonometric identity:

$$\sin A + \sin B = 2 \sin\left(\frac{A + B}{2}\right) \cos\left(\frac{A - B}{2}\right)$$

Taking $A = \omega t + \phi$ and $B = \omega t$, we can write Eq. 38.10 in the form

$$E_P = 2E_0 \cos\left(\frac{\phi}{2}\right) \sin\left(\omega t + \frac{\phi}{2}\right) \tag{38.11}$$

Hence, the electric field at P has the same frequency ω, but its amplitude is multiplied by the factor $2 \cos(\phi/2)$. To check the consistency of this result, note that if $\phi = 0, 2\pi, 4\pi, \ldots$, the amplitude at P is $2E_0$, corresponding to the condition for constructive interference. Referring to Eq. 38.9, we find that our result is consistent with Eq. 38.2. Likewise, if $\phi = \pi, 3\pi, 5\pi, \ldots$, the amplitude at P is zero, which is consistent with Eq. 38.3 for destructive interference.

Finally, to obtain an expression for the light intensity at P, recall *that the intensity of a wave is proportional to the square of its amplitude* (Section 35.3). Using Eq. 38.11, we can therefore express the intensity at P as

$$I \sim E_P^2 = 4E_0^2 \cos^2(\phi/2) \sin^2\left(\omega t + \frac{\phi}{2}\right)$$

Since most light-detecting instruments measure the time average light intensity and the time average value of $\sin^2(\omega t + \phi/2)$ over one cycle is $1/2$, we can write the average intensity at P as

$$I_{av} = I_0 \cos^2(\phi/2) \tag{38.12}$$

where I_0 is the *maximum* possible time average light intensity, given by $I_0 = 2E_0^2$. Substituting Eq. 38.9 into Eq. 38.12, we find that

$$I_{av} = I_0 \cos^2\left(\frac{\pi d \sin\theta}{\lambda}\right) \tag{38.13}$$

Alternatively, since $\sin\theta \approx y/L$ for small values of θ, we can write Eq. 38.13 in the form

$$I_{av} = I_0 \cos^2\left(\frac{\pi d}{\lambda L} y\right) \tag{38.14}$$

Note that constructive interference, which produces intensity maxima, occurs when the quantity $(\pi y d/\lambda L)$ is an integral multiple of 2π, corresponding to $y = (\lambda L/d)m$. This is consistent with Eq. 38.5. A plot of the intensity distribution versus θ is given in Fig. 38.4. Note that the interference pattern consists of equally spaced fringes of equal intensity. However, the result is valid only if the slit-to-screen distance L is large relative to the slit separation, and only for small values of θ.

We have seen that the interference phenomena arising from two sources depend on the relative phase of the waves at a given point. Furthermore, the phase difference at a given point depends on the path difference between the two waves. Finally, it is important to note that the *resultant intensity at a point is proportional to the square of the resultant amplitude.* That is, the intensity is proportional to $|E_1 + E_2|^2$. It would be *incorrect* to calculate the resultant intensity by adding the intensities of the individual waves. This procedure would give a different quantity, namely, $E_1^2 + E_2^2$.

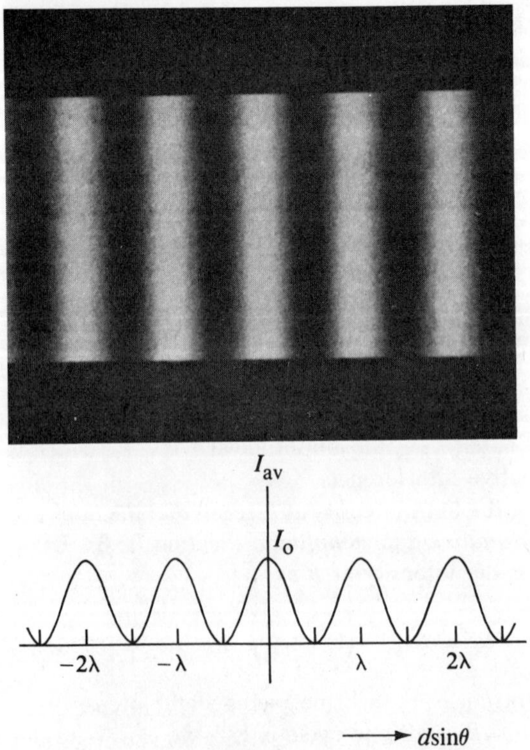

Figure 38.4 Intensity distribution versus $d \sin\theta$ for the double-slit pattern when the screen is far from the two slits ($L \gg d$). (Photo from M. Cagnet, M. Francon, and J. C. Thierr, Atlas of Optical Phenomena, Berlin, Springer-Verlag, 1962)

38.4 PHASOR ADDITION OF WAVES

In the previous section we combined two waves algebraically to obtain the resultant wave amplitude at some point on a screen. Unfortunately, this analytical procedure becomes rather cumbersome when several wave amplitudes have to be added. Since we shall eventually be interested in combining a large number of waves, we now describe a graphical procedure useful for this purpose.

Again, consider a sinusoidal wave whose electric field component is given by

$$E_1 = E_0 \sin\omega t$$

where E_0 is the wave amplitude and ω is the angular frequency. This wave disturbance can be represented graphically with a vector of magnitude E_0, *rotating* about the origin in a counterclockwise direction with an angular frequency ω, as in Fig. 38.5a. Such a rotating vector is called a *phasor* and is commonly used in the field of electrical engineering (see Chapter 30). Note that the phasor makes an angle of ωt with the horizontal axis. The projection of the phasor on the vertical axis represents E_1, the magnitude of the wave disturbance at some time t. Hence, as the phasor rotates in a circle, the projection E oscillates along the vertical axis about the origin.

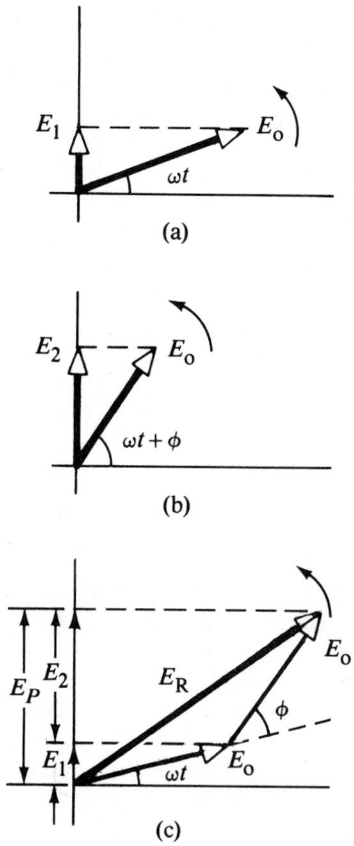

Figure 38.5 (a) Phasor diagram for the wave disturbance $E_1 = E_0 \sin\omega t$. The phasor is a vector of length E_0 rotating counterclockwise. (b) Phasor diagram for the wave $E_2 = E_0 \sin(\theta t + \phi)$. (c) E_R is the resultant phasor formed from the individual phasors shown in (a) and (b).

Now consider a second sinusoidal wave whose electric field is given by

$$E_2 = E_0 \sin(\omega t + \phi)$$

That is, this wave has the same amplitude and frequency as E_1, but its phase is ϕ with respect to the wave E_1. The phasor representing the wave E_2 is given in Fig. 38.5b. The resultant wave, which is the sum of E_1 and E_2, can be obtained graphically by redrawing the phasors end to end, as in Fig. 38.5c, where the tail of the second phasor is placed at the tip of the first phasor. As with vector addition, the resultant phasor E_R runs from the tail of the first phasor to the tip of the second phasor. Furthermore, E_R rotates along with the two individual phasors at the same angular frequency ω. The projection of E_R along the vertical axis equals the sum of the projections of the two phasors. That is, $E_P = E_1 + E_2$.

It is convenient to construct the phasors at $t = 0$ as in Fig. 38.6. From the geometry of the triangle, we see that

$$E_R = E_0 \cos\alpha + E_0 \cos\alpha = 2E_0 \cos\alpha$$

Since the sum of the two opposite interior angles equals the exterior angle ϕ, we see that $\alpha = \phi/2$, so that

$$E_R = 2E_0 \cos(\phi/2)$$

Hence, the projection of the phasor E_R along the vertical axis at any time t is given by

$$E_P = E_R \sin\left(\omega t + \frac{\phi}{2}\right) = 2E_0 \cos(\phi/2) \sin\left(\omega t + \frac{\phi}{2}\right)$$

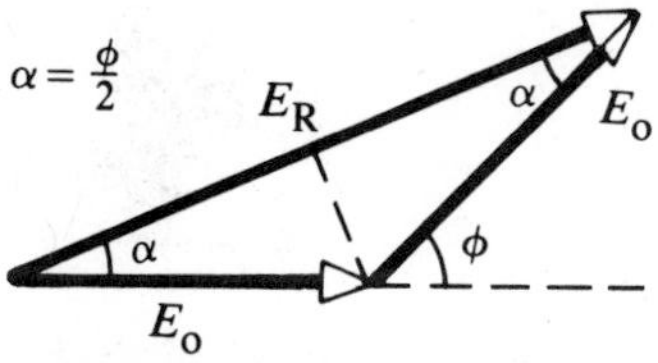

Figure 38.6 Reconstruction of the resultant phasor E_R. From the geometry, note that $\alpha = \phi/2$.

This is consistent with the result obtained algebraically, Eq. 38.11. The resultant phasor has an amplitude $2E_0 \cos(\phi/2)$ and makes an angle of $\phi/2$ with the first phasor. Furthermore, the average intensity at P, which varies as E_P^2, is proportional to $\cos^2(\phi/2)$, as described previously in Eq. 38.12.

We can now describe how to obtain the resultant of several waves which have the same frequency:

1. Draw the phasors representing each wave end to end, as in Fig. 38.7, remembering to maintain the proper phase relationship between waves.
2. The resultant represented by the phasor E_R is the vector sum of the individual phasors. At each instant, the projection of E_R along the vertical axis represents

the time variation of the resultant wave. The phase angle α of the resultant wave is the angle between E_R and the first phasor. From the construction in Fig. 38.7, drawn for four phasors, we see that the phasor of the resultant wave is given by $E_P = E_R \sin(\omega t + \alpha)$.

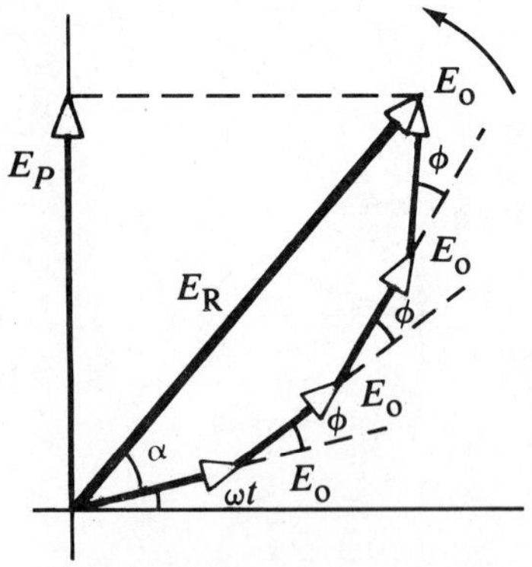

Figure 38.7 The phasor E_R is the resultant of four phasors of equal amplitude E_o. The phase of E_R is α with respect to the first phasor.

Phasor Diagrams for Two Coherent Sources

As an example of the phasor method, consider the interference pattern produced by two coherent sources, which was discussed in the previous section. Figure 38.8 represents the phasor diagrams for various values of the phase difference ϕ, and the corresponding values of the path difference δ, which are obtained using Eq. 38.9.

From Fig. 38.8, we see that the intensity at a point will be a maximum when E_R is a maximum. This occurs at values of ϕ equal to 0, 2π, 4π, etc. Likewise, we see that the intensity at some observation point will be zero when E_R is zero. The first zero-intensity point occurs at $\phi = 180°$, corresponding to $\delta = \lambda/2$, while the other zero points (not shown) occur at $\delta = 3\lambda/2$, $5\lambda/2$, etc. These results are in complete agreement with the analytical procedure described in the previous section.

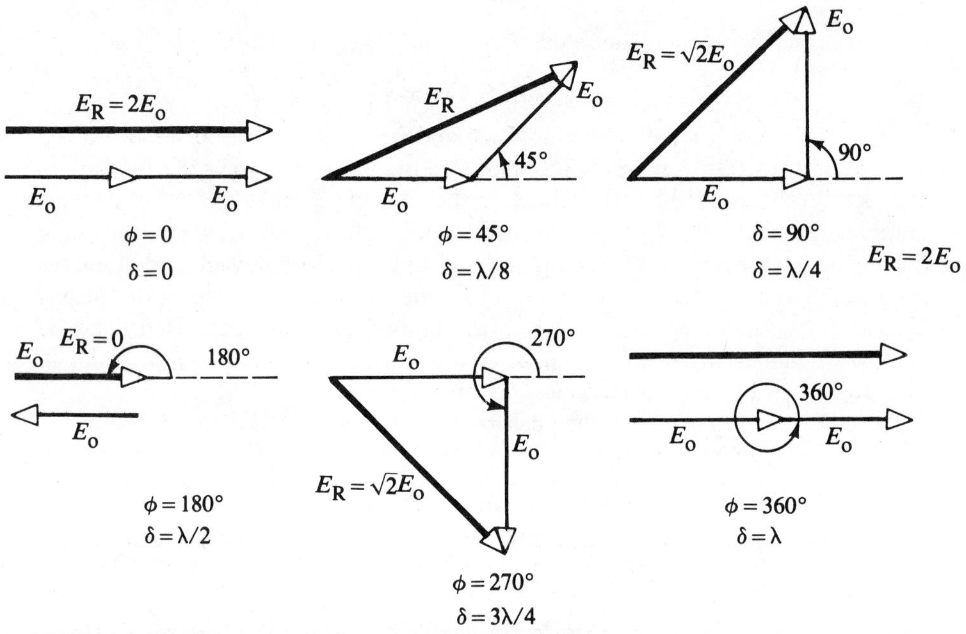

Figure 38.8 Phasor diagrams for the double-slit interference pattern. The resultant phasor E_R is a maximum when $\phi = 0, 2\pi, 4\pi, \ldots$ and is zero when $\phi = \pi, 3\pi, 5\pi, \ldots$.

Three-Slit Interference Pattern

Let us analyze the interference pattern due to three equally spaced slits by using phasor diagrams. A diagram of the slits, shown in Fig. 38.9, shows that the path difference for rays from two adjacent slits is given by $\delta = d \sin\theta$. From Eq. 38.9, recall that the *phase difference* ϕ *between two adjacent rays* is given by

$$\phi = \frac{2\pi}{\lambda}\delta = \frac{2\pi}{\lambda} d \sin\theta$$

From the similar triangles in Fig. 38.9, and for $y \ll L$ (small values of θ), we see that $\delta/d \approx y/L$. Therefore, ϕ can be expressed in terms of y as

$$\phi = \frac{2\pi}{\lambda}\delta \approx \left(\frac{2\pi d}{\lambda L}\right) y$$

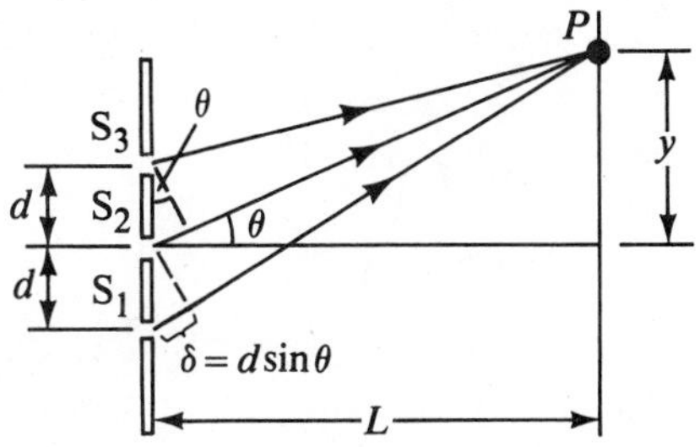

Figure 38.9 Geometry for analyzing the interference pattern of three equally spaced slits. The path difference for rays from two adjacent slits is $d \sin\theta$.

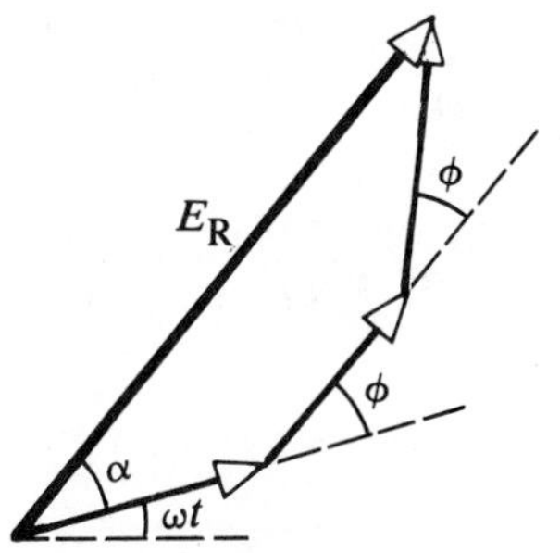

Figure 38.10 Phasor diagram for three equally spaced slits.

This result shows that the phase difference depends on the coordinate y, as in the case of the double-slit interference pattern.

The electric fields at P due to waves from the individual slits can be expressed as

$$E_1 = E_0 \sin\omega t$$
$$E_2 = E_0 \sin(\omega t + \phi)$$
$$E_3 = E_0 \sin(\omega t + 2\phi)$$

Hence, the resultant field at P can be obtained using a phasor diagram as shown in Fig. 38.10.

The phasor diagrams for various specific values of ϕ (or δ) for the three slits are shown in Fig. 38.11. Note that the resultant amplitude at P has a maximum value of $3E_0$ (called the *primary maximum*) when $\phi = 0, \pm 2\pi, \pm 4\pi, \ldots$, similar to the double-slit interference pattern. However, we also find that secondary maxima of amplitude E_0 occur *between* the primary maxima when $\phi = \pm\pi, \pm 3\pi, \ldots$. For these points, the wave from one slit exactly cancels that from another slit (Fig. 38.11d), which gives a net amplitude of E_0, due to the third slit. Finally, total

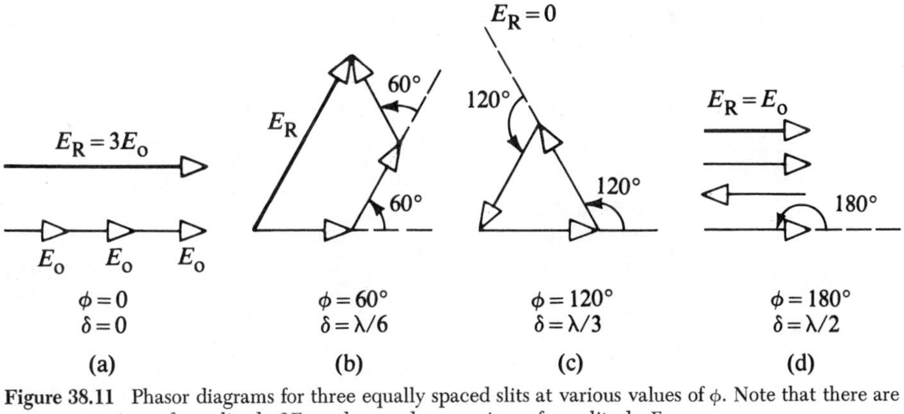

Figure 38.11 Phasor diagrams for three equally spaced slits at various values of ϕ. Note that there are primary maxima of amplitude $3E_0$ and secondary maxima of amplitude E_0.

destructive interference occurs whenever the three phasors form a closed triangle (Fig. 38.11c). These points where $E_R = 0$ correspond to $\phi = \pm 2\pi/3, \pm 4\pi/3, \ldots$. You should construct other phasor diagrams for values of ϕ greater than π. Figure 38.12a is a plot of the resultant amplitude as a function of ϕ (or δ). Since the intensity varies as E_R^2, the intensity distribution reaches maxima and minima at the same positions, as shown in Fig. 38.12b. Note that the primary maxima are nine times more intense than the secondary maxima. The broken line in Fig. 38.12b represents the pattern for the double-slit system for comparison.

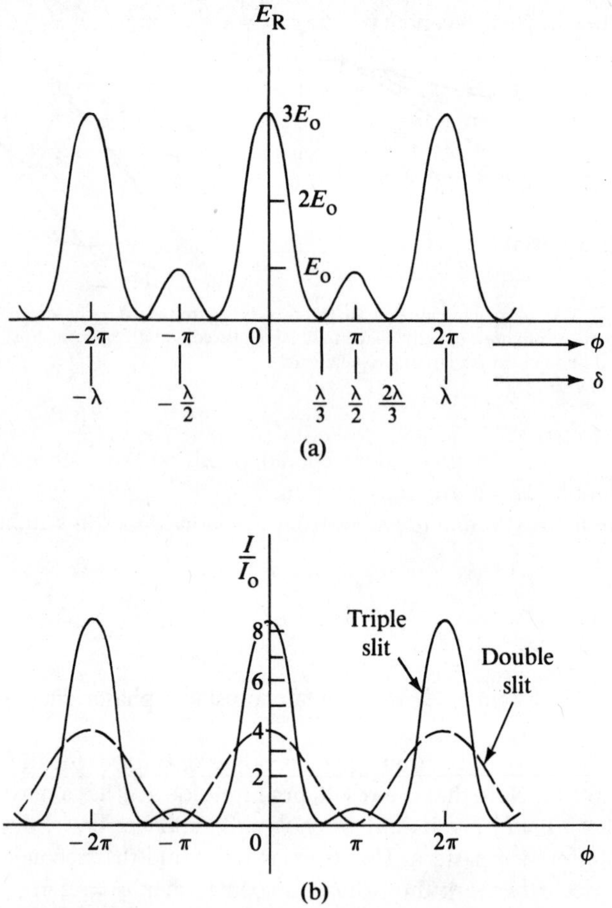

Figure 38.12 (a) Resultant amplitude E_R versus ϕ for the three-slit interference pattern. (b) Intensity distribution for the three-slit interference pattern as compared to that of the double slit.

Interference from *N* Slits

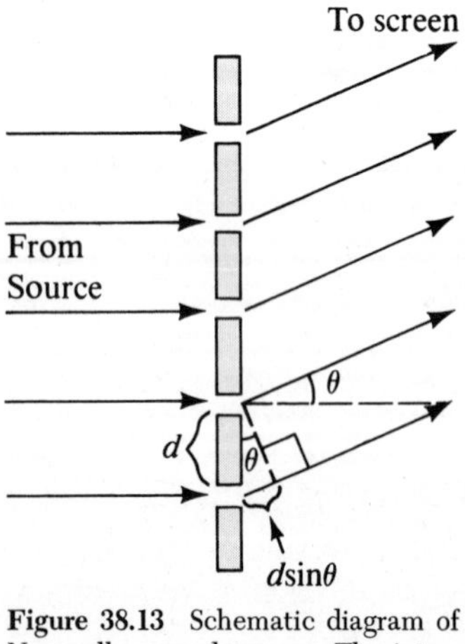

Figure 38.13 Schematic diagram of N equally spaced sources. The interference pattern is observed on a distant screen not shown.

Now suppose there are N equally spaced sources starting off in phase, each separated by a distance d as in Fig. 38.13. The phase difference ϕ between any two adjacent sources as observed on a distant screen at the angle θ is equal to $(2\pi/\lambda)d\sin\theta$. If one constructs a phasor diagram for the N sources (Problem 1), the intensity on the screen is found to be

$$I = I_0\frac{\sin^2(N\phi/2)}{\sin^2(\phi/2)} \tag{38.15}$$

where

$$\phi = \frac{2\pi}{\lambda}d\sin\theta$$

and I_0 is the intensity due to one of the sources.

The interference pattern for $N = 6$ slits is sketched in Fig. 38.14a, and the pattern for a large number of slits is sketched in Fig. 38.14b. Note that the condition for determining the positions of the principal maxima doesn't change as more sources are added. However, as N increases, the principal maxima become narrower and more intense, corresponding to a brighter line on the screen. In fact, the intensity of one of the maxima is a factor of N^2 times the intensity of one of the sources. Furthermore, note that the number of secondary maxima increases as N increases. In fact, the number of such secondary maxima is $N - 2$, and their intensity goes to zero as $N \rightarrow \infty$.

When a large number of sources produce an interference pattern as in Fig. 38.14b, the phenomenon is often referred to as *diffraction*. That is, *diffraction can be viewed as interference arising from the superposition of a large number of waves*. We shall return to this point in the next chapter.

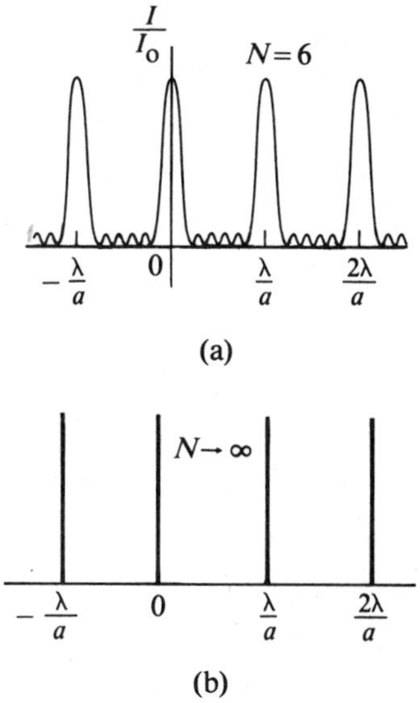

Figure 38.14 (a) Interference pattern for six equally spaced slits. (b) Interference pattern for a large number of equally spaced slits.

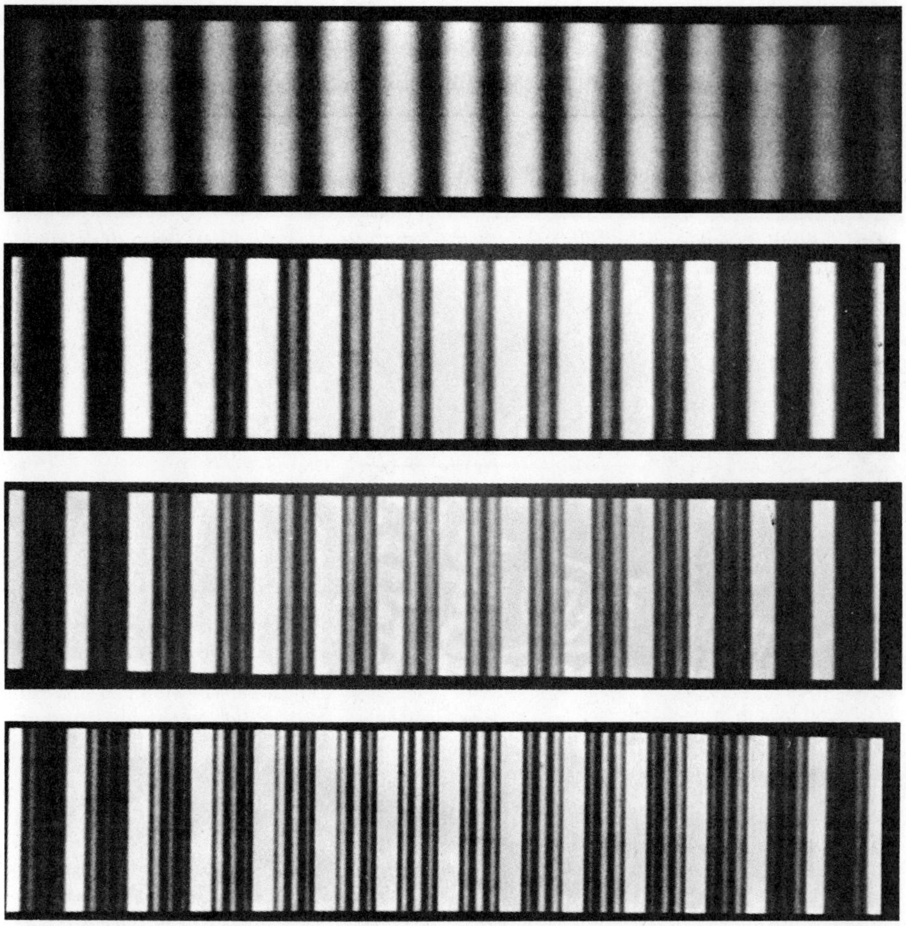

Interference patterns corresponding to (a) two slits, (b) three slits, (c) four slits, and (d) five slits. Note that the number of secondary maxima between the primary maxima is $N - 2$, where N is the number of slits. (From M. Cagnet, M. Francon, and J. C. Thierr, Atlas of Optical Phenomena, Berlin, Springer-Verlag, 1962)

Q4. What is the difference between interference and diffraction?

38.5 CHANGE OF PHASE DUE TO REFLECTION

We have described interference effects produced by two or more coherent light sources. Young's method for producing two coherent light sources involves illuminating a pair of slits with a single source. Another simple, yet ingenious, arrangement for producing an interference pattern with a single light source is known as *Lloyd's mirror*. A light source is placed at S close to a mirror, as illustrated in Fig. 38.15. Waves can reach the viewing point P either by the direct path SP or by

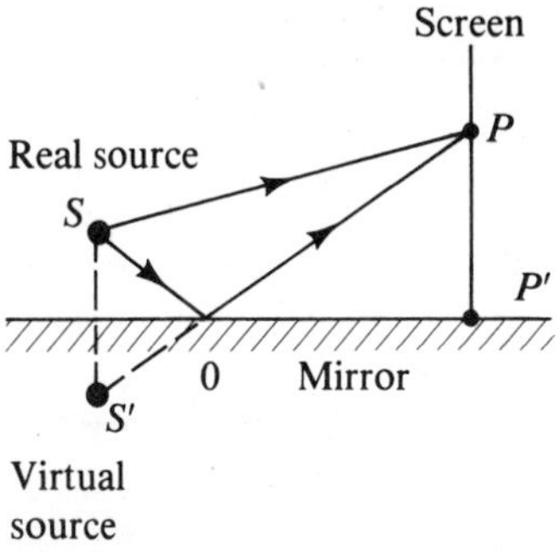

Figure 38.15 Lloyd's mirror. An interference pattern is produced on a screen at P as a result of the combination of the direct ray and the reflected ray. The reflected ray undergoes a phase change of 180°.

the path involving reflection from the mirror. The reflected ray can be treated as a ray originating from a source at S′, located behind the mirror. Note that S′, which is the image of S, can be considered a virtual source. Hence, at observations points far from the source, one would expect an interference pattern due to waves from S and S′ just as is observed for two real coherent sources. An interference pattern is indeed observed. However, the positions of the dark and bright fringes are *reversed* relative to the pattern of two real coherent sources (Young's experiment). This is because the coherent sources at S and S′ differ in phase by 180°. This 180° phase change is produced upon reflection. To illustrate this further, consider the point P', where the mirror meets the screen. This point is equidistant from S and S′. If path difference alone were responsible for the phase difference, one would expect to see a bright fringe at P' (since the path difference is zero for this point), corresponding to the central fringe of the two-slit interference pattern. Instead, one observes a *dark* fringe at P' because of the 180° phase change produced by reflection. In general, *an electromagnetic wave undergoes a phase change of 180° upon reflection from an optically dense medium or from any conducting surface.*

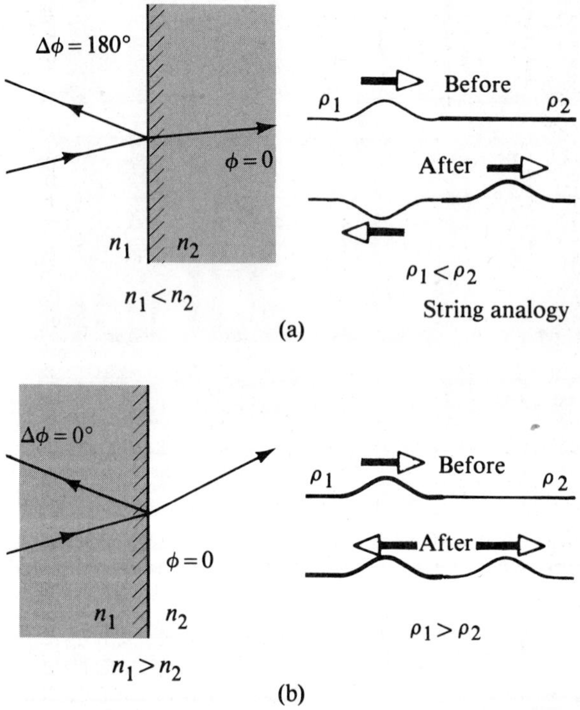

Figure 38.16 (a) A ray reflecting from a medium of higher refractive index undergoes a 180° phase change. The right side shows the analogy with a reflected pulse on a string. (b) A ray reflecting from a medium of lower refractive index undergoes *no* phase change.

It is useful to draw an analogy between reflected light waves and the reflections of a transverse wave on a stretched string when the wave meets the boundary (Section 32.6). The reflected pulse on a string undergoes a phase change of 180° when it is reflected from a denser medium, such as a heavier string. On the other hand, there is no phase change if the pulse reflects from a less dense medium. Similarly, electromagnetic waves undergo a 180° phase change when reflected from a boundary leading to an optically denser medium. There is no phase change when the wave is reflected from a boundary leading to a less dense medium. In either case, the transmitted wave undergoes no phase change. These rules, summarized in Fig. 38.16, can be deduced from Maxwell's equations, but the treatment is beyond the scope of this text.

38.6 INTERFERENCE IN THIN FILMS

Interference effects are commonly observed in thin films, such as thin layers of oil on water and soap bubbles. The various colors that are observed with ordinary white light result from the interference of waves reflected from the opposite surfaces of the film.

Consider a film of uniform thickness t and index of refraction n, as in Fig. 38.17. Let us assume that the light rays traveling in air are nearly normal to the surface. To determine whether the reflected rays interfere constructively or destructively, we must first note the following facts:

1. *A wave that travels from a medium of lower refractive index to one of higher refractive index undergoes a 180° phase change relative to the incident wave.* There is no phase change in the reflected wave if it travels from a medium of higher refractive index to one of lower refractive index.
2. The wavelength of light λ_n in a medium whose refractive index is n (Section 36.6) is given by

$$\lambda_n = \lambda/n \qquad (38.16)$$

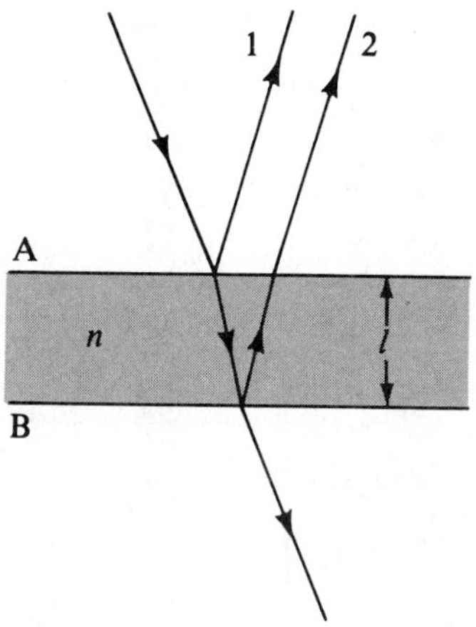

Figure 38.17 Interference observed in light reflected from a thin film is due to a combination of rays reflected from the upper and lower surfaces.

where λ is the wavelength of light in free space.

Let us apply these rules to the film described in Fig. 38.17. According to the first rule, ray 1, which is reflected from the upper surface (A), undergoes a phase change of 180° with respect to the incident wave. On the other hand, ray 2, which is reflected from the lower surface (B), undergoes no phase change with respect to the incident wave. Therefore, ray 1 is 180° out of phase with respect to ray 2, which is equivalent to a path difference of $\lambda_n/2$. Since the path difference between the two rays for normal incidence is twice the film thickness, or $2t$, we can write the condition for constructive interference, or maximum intensity, in the form

$$2t = \left(m + \frac{1}{2}\right)\lambda_n \qquad (m = 0,1,2, \ldots) \qquad (38.17)$$

Note that this condition takes into account two factors: (a) the difference in optical path length for the two rays (the term $m\lambda_n$) and (b) the 180° phase change upon reflection (the term $\lambda_n/2$). Since $\lambda_n = \lambda/n$, we can write Eq. 38.17 in the form

Constructive interference in thin films

$$2nt = \left(m + \frac{1}{2}\right)\lambda \qquad (m = 0,1,2, \ldots) \qquad (38.18)$$

In a similar manner, the condition for destructive interference, or intensity minimum, is

Destructive interference in thin films

$$2nt = m\lambda \qquad (m = 0,1,2, \ldots) \qquad (38.19)$$

Note that these conditions for constructive and destructive interference are valid only when the film is surrounded by a common medium. The surrounding medium may have a refractive index less than or greater than that of the film. In either case, the rays reflected from the two surfaces will be out of phase by 180°. On the other hand, if the film is located between two *different* media, one of lower refractive index and one of higher refractive index, the conditions for constructive and destructive interference are *reversed*. In this case, either there is a phase change of 180° for both ray 1 reflecting from surface A and ray 2 reflecting from surface B or there is no phase change for either ray; hence, the net change in relative phase due to the reflections is *zero*.

Example 38.3

Calculate the thicknesses of a soap bubble film ($n = 1.46$) that will result in constructive interference in the reflected light if the film is illuminated with light whose wavelength in free space is 600 nm.

Solution: The minimum film thickness for constructive interference in the reflected light corresponds to $m = 0$ in Eq. 38.18. This gives $2nt = \lambda/2$, or

$$t = \frac{\lambda}{4n} = \frac{600 \text{ nm}}{4(1.46)} = 103 \text{ nm}$$

From Eq. 38.18, we see that films of thickness $3t$, $5t$, $7t$, etc., will also produce constructive interference.

Example 38.4

Nonreflecting coatings for solar cells: Solar cells are often coated with a transparent thin film such as silicon monoxide (SiO, $n = 1.45$) in order to minimize reflective losses from the surface (Fig. 38.18). A silicon solar cell ($n = 3.5$) is coated with a thin film of SiO for this purpose. Determine the minimum thickness of SiO that will produce the least reflection at a wavelength of 550 nm, which is the center of the visible spectrum.

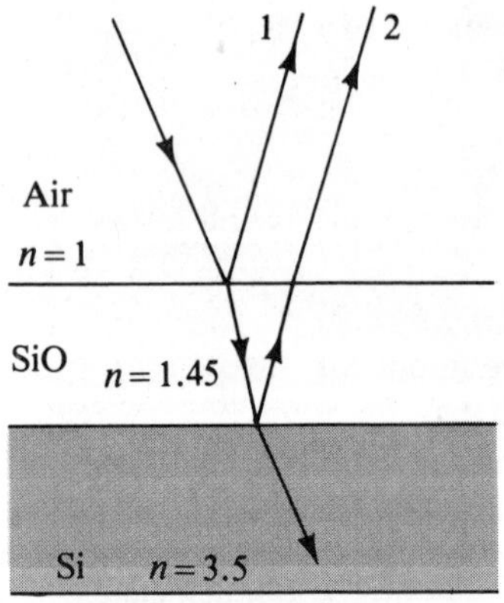

Figure 38.18 (Example 38.4) Reflective losses from a silicon solar cell are minimized by coating it with a thin film of SiO.

Solution: The reflected light is a minimum when rays 1 and 2 meet the condition of destructive interference. Note that *both* rays undergo a 180° phase change upon reflection in this case, one from the upper and one from the lower surface. Hence, the net change in phase is zero and the condition for a reflection *minimum* is given by Eq. 38.19. Since the minimum thickness corresponds to $m = 0$, we get $2nt = \lambda/2$, or

$$t = \frac{\lambda}{4n} = \frac{550 \text{ nm}}{4(1.45)} = 94.8 \text{ nm} = 948 \text{ Å}$$

Typically, such antireflecting coatings reduce the reflective loss from 30% (with no coating) to 10% (with coating). Such coatings will therefore increase the cell's efficiency, since more light will be available to create charge carriers in the silicon cell. Note that in reality, the coating will never be perfectly nonreflecting because the required thickness is wavelength-dependent and the incident light covers a wide range of wavelengths.

Glass lenses used in cameras and other optical instruments are often coated with a transparent thin film, such as magnesium fluoride (MgF_2), to reduce or eliminate unwanted reflections.

Example 38.5

Interference in a wedge-shaped film: A thin wedge-shaped film of refractive index n is illuminated with monochromatic light of wavelength λ as illustrated in Fig. 38.19. Describe the interference pattern observed for this case.

Solution: The interference pattern will be equivalent to that of a thin film of variable thickness surrounded by air. Hence, the pattern will be an alternating series of bright and dark parallel bands. A dark band corresponding to destructive interference appears at the apex O since the upper reflected ray undergoes a 180° phase change while the lower one does not. According to Eq. 38.19, other dark bands appear when $2nt = m\lambda$, so that $t_1 = \lambda/2n$, $t_2 = \lambda/n$, $t_3 = 3\lambda/2n$, etc. Similarly, bright bands will be observed when the thickness satisfies the condition $2nt = (m + \frac{1}{2})\lambda$, corresponding to thicknesses of $\lambda/4n$, $3\lambda/4n$, $5\lambda/4n$, etc. If white light is used to observe the pattern, bands of different colors will be observed at different points, corresponding to the different wavelengths of light.

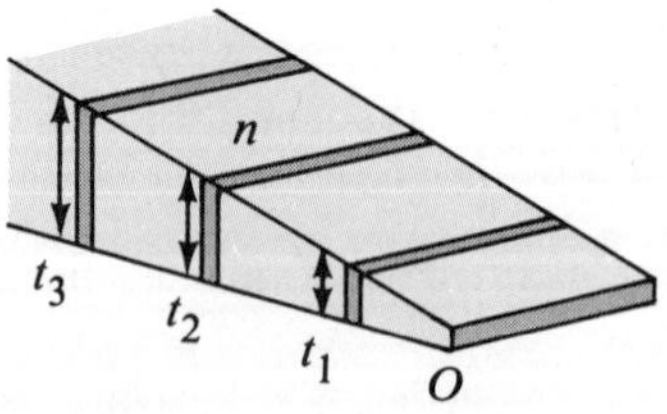

Figure 38.19 (Example 38.5) Interference bands in reflected light can be observed by illuminating a wedge-shaped film with monochromatic light. The dark areas correspond to destructive interference.

Newton's Rings

Another method for observing interference of light waves is to place a planoconvex lens (one having one planar side and one convex side) on top of a planar glass surface as in Fig. 38.20a. With this arrangement, the air film between the glass

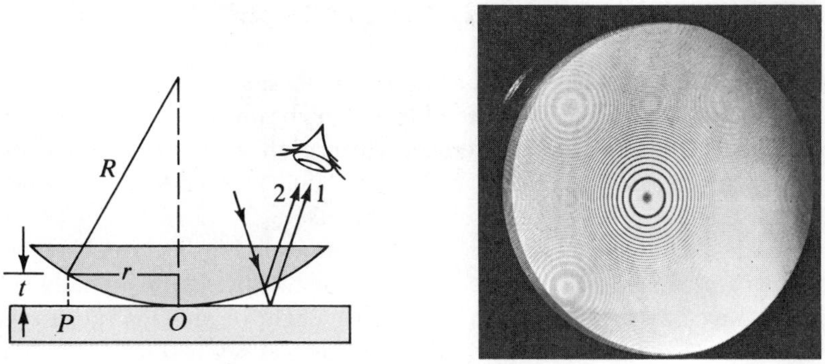

Figure 38.20 (a) The combination of rays reflected from the glass plate and the curved surface of the lens gives rise to an interference pattern known as Newton's rings. (b) Photograph of Newton's rings. (Photo, Bausch & Lomb Optical Co.)

surfaces varies in thickness from zero at the point of contact to some value t at P. If the radius of curvature of the lens R is very large compared with the distance r, and if the system is viewed from above using light of wavelength λ, a series of dark rings is observed. A photograph of such a pattern is shown in Fig. 38.20b. These circular fringes, discovered by Newton, are called *Newton's rings*. Newton's particle model of light could not explain the origin of the rings.

In Fig. 38.20a, the interference effect is due to the combination of ray 1, reflected from the planar glass plate, with ray 2, reflected from the lower part of the lens. Ray 1 undergoes a phase change of 180° upon reflection, since it is reflected from a medium of higher refractive index, whereas ray 2 undergoes no phase change. Hence, the conditions for constructive and destructive interference are given by Eqs. 38.18 and 38.19, respectively, with $n = 1$ since the "film" is air. Here again, one might guess that the contact point O would be bright, corresponding to constructive interference. Instead, the contact point is dark, as seen in Fig. 38.20b, because ray 1, reflected from the planar surface, undergoes a 180° phase change with respect to ray 2. Using the geometry shown in Fig. 38.20a, one can obtain expressions for the radii of the bright and dark bands in terms of the radius of curvature R and wavelength λ. For example, the dark rings have radii given by $r \approx \sqrt{m\lambda R}$. The details are left as a problem for the reader. By measuring the radii of the rings, one can obtain the wavelength, provided R is known. Conversely, if the wavelength is accurately known, the lens maker can use this effect to obtain R.

Q5. In order to observe interference in a thin film, why must the film thickness be thin compared with the wavelength of visible light?

Q6. In the process of evaporation, a soap bubble appears black just before it breaks. Explain this phenomenon in terms of the phase changes that occur upon reflection from the two surfaces.

Q7. If an oil film is observed on water, the film appears brightest at the outer regions, where it is thinnest. From this information, what can you say about the index of refraction of the oil relative to that of water?

Q8. If a soap film on a wire loop is held in air, it appears black in the thinnest regions when observed by reflected light and shows a variety of colors in thicker regions. Explain.

38.7 THE MICHELSON INTERFEROMETER

The *interferometer*, invented by the American physicist A. A. Michelson (1852–1931), is an ingenious device which splits a light beam into two parts and then recombines them to form an interference pattern after they have traveled over different paths. The device can be used for obtaining accurate measurements of wavelengths and for precise length measurements.

A schematic diagram of the interferometer is shown in Fig. 38.21. A beam of light provided by a monochromatic source is split into two rays by a partially silvered mirror M inclined at 45° relative to the incident light beam. One ray is reflected vertically upward toward mirror M_1 while the second ray is transmitted horizontally through M toward mirror M_2. Hence, the two rays travel separate paths l_1 and l_2. After reflecting from mirrors M_1 and M_2, the two rays eventually recombine to produce an interference pattern, which can be viewed through a telescope. The glass plate P, equal in thickness to M, is placed in the path of the horizontal ray in order to equalize the path lengths of the two rays. With this arrangement, each ray will then pass through the same thickness of glass.

The interference condition for the two rays is determined by the difference in the thin optical path lengths. When the two rays are viewed as shown, the image of M_2 is at M_2' parallel to M_1. Hence, M_2' and M_1 form the equivalent of a parallel air film. The effective thickness of the air film is varied by moving mirror M_1 parallel to itself with a finely threaded screw. Under these conditions, the interference pattern is a series of bright and dark circular rings which resemble Newton's rings. If a dark circle appears at the center of the pattern, the two rays interfere destructively. If the mirror M_1 is then moved a distance of $\lambda/4$, the path difference changes by $\lambda/2$ (twice the separation between M_1 and M_2'). The two rays will now interfere constructively, giving a bright circle in the middle. As M_1 is moved an additional distance of $\lambda/4$, a dark circle will appear once again. Thus, we see that successive

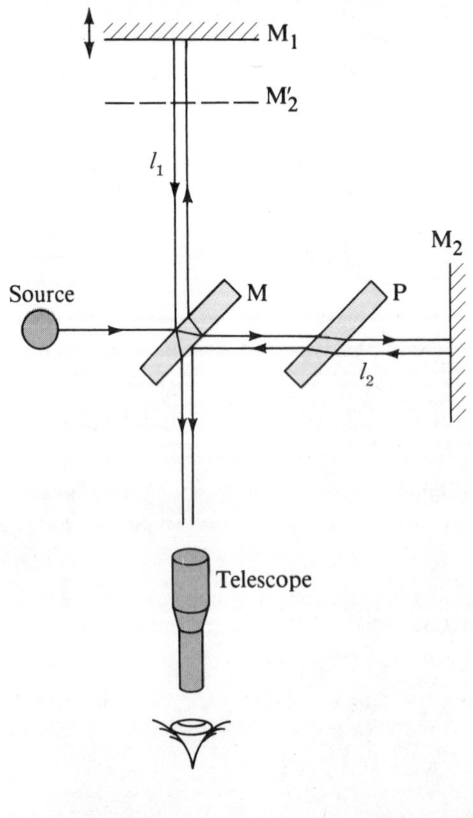

Figure 38.21 Schematic diagram of the Michelson interferometer. A single beam is split into two rays by the half-silvered mirror M. The path difference between the two rays is varied with the adjustable mirror M_1.

dark and bright circles are formed each time M_1 is moved a distance of $\lambda/4$. The wavelength of light is then measured by counting the number of fringe shifts for a given displacement of M_1. Conversely, if the wavelength is accurately known (as with a laser beam), mirror displacements can be measured to within a fraction of the wavelength.

Since the interferometer can accurately measure displacement, it is often used to make highly precise measurements of mechanical components. The fundamental definition of the meter is based on a certain wavelength of light from krypton-86. The interferometer makes such precise wavelength measurements possible.

38.8 SUMMARY

Interference of light waves is the result of the linear superposition of two or more waves at a given point. A sustained interference pattern is observed if (1) the sources are coherent (that is, they maintain a constant relative phase), (2) the sources are monochromatic (of a single wavelength), and (3) the linear superposition principle is applicable.

In Young's double-slit experiment, two slits separated by a distance d are illuminated by a monochromatic light source. An interference pattern consisting of bright and dark fringes is observed on a screen a distance L from the slits. The condition for bright fringes (constructive interference) is given by

Conditions for bright fringes

$$d \sin\theta = m\lambda \qquad (m = 0, \pm 1, \pm 2, \ldots) \qquad (38.2)$$

The condition for dark fringes (destructive interference) is

Conditions for dark fringes

$$d \sin\theta = \left(m + \frac{1}{2}\right)\lambda \qquad (m = 0, \pm 1, \pm 2, \ldots) \qquad (38.3)$$

The index number m is called the *order number* of the fringe.

The *average intensity* of the double-slit interference pattern is given by

$$I_{av} = I_0 \cos^2(\pi d \sin\theta/\lambda) \qquad (38.13)$$

where I_0 is proportional to the *square* of the resultant field amplitude at that point on the screen.

When a series of N slits is illuminated, a diffraction pattern is produced that can be viewed as interference arising from the superposition of a large number of waves. It is convenient to use phasor diagrams to simplify the analysis of interference from three or more equally spaced slits.

An electromagnetic wave undergoes a phase change of 180° upon reflection from an optically dense medium or from any conducting surface.

A wave traveling from a medium of lower refractive index to one of higher refractive index undergoes a 180° phase change relative to the incident wave.

The wavelength of light λ_n in a medium whose refractive index is n is given by

$$\lambda_n = \frac{\lambda}{n} \qquad (38.16)$$

where λ is the wavelength of light in free space.

The condition for constructive interference in a film of thickness t and refractive index n is given by

Constructive interference in thin films

$$2nt = \left(m + \frac{1}{2}\right)\lambda \qquad (m = 0, 1, 2, \ldots) \qquad (38.18)$$

Similarly, the condition for destructive interference is

Destructive interference in thin films

$$2nt = m\lambda \qquad (m = 0,1,2, \ldots) \qquad (38.19)$$

EXERCISES

Section 38.2 Young's Double-Slit Experiment

1. A pair of narrow, parallel slits separated by 0.25 mm are illuminated by the green component from a mercury vapor lamp ($\lambda = 5461$ Å). The interference pattern is observed on a screen located 1.2 m from the plane of the parallel slits. Calculate the distance (a) from the central maximum to the first bright region on either side of the central maximum and (b) between the first and second dark bands in the interference pattern.

2. The slits in a Young's interference apparatus are illuminated with monochromatic light. The third dark band is 9.5 mm from the central maximum. The two slits are 0.15 mm apart, and the screen is 90 cm away from the slits. Calculate the wavelength of the light used.

3. The yellow component of light from a helium discharge tube ($\lambda = 587.5$ nm) is allowed to fall on a plane containing parallel slits that are 0.2 mm apart. A screen is located so that the second bright band in the interference pattern is at a distance equal to 10 slit spacings from the central maximum. What is the distance between the source plane and the screen?

4. One of the bright bands in a Young's interference pattern is located 12 mm from the central maximum. The screen is located 119 cm from the pair of slits that serve as secondary sources. The slits are 0.241 mm apart and are illuminated by the blue light from a hydrogen discharge tube ($\lambda = 486$ nm). How many bright lines are there *between* the central maximum and the 12-mm position?

5. A narrow slit is cut into each of two overlapping opaque squares. The slits are parallel, and the distance between them is adjustable. Monochromatic light of wavelength 6000 Å illuminates the slits, and an interference pattern is formed on a screen 80 cm away. The third dark band is located 1.2 cm from the central bright band. What is the distance between the central bright band and the first bright band on either side of the central band?

6. In a double-slit arrangement as illustrated in Fig. 38.3, $d = 0.15$ mm, $L = 140$ cm, $\lambda = 643$ nm, and $y = 1.8$ cm. (a) What is the path difference δ for the two slits at the point *P*? (b) Express this path difference in terms of the wavelength. (c) Will point *P* correspond to a maximum, a minimum, or an intermediate condition?

7. In deriving Eqs. 38.6 and 38.7, it is assumed (as stated in Eq. 38.4) that $\sin\theta \approx \tan\theta$. The validity of this assumption depends on the requirement that $L \gg d$. In the arrangement of Fig. 38.3, let $L = 40$ cm, $d = 0.5$ mm, and $\lambda = 656.3$ nm (the red line in hydrogen). Under the assumption stated in Eq. 38.4, the ninth-order dark band would be located 4.463 mm on either side of the central maximum. What percent error is introduced by the assumption that $L \gg d$?

8. A third-order maximum is located 4.2 mm above the central maximum in a Young's interference pattern. The distance between the slits equals 200 wavelengths of the incident light. What is the distance between the source plane and the screen?

9. Light of wavelength 546 nm (the intense green line from a mercury discharge tube) produces a Young's interference pattern in which the second-order minimum is along a direction that makes an angle of 18 minutes of arc relative to the direction to the central maximum. What is the distance between the parallel slits?

10. In a double-slit interference experiment, the slits are illuminated with light of wavelength 6800 Å. If the second bright fringe is 3.5 cm from the central line and the slits are 2 m from the observing screen, calculate (a) the slit separation and (b) the position of the second dark fringe.

Section 38.3 Intensity Distribution of the Double-Slit Interference Pattern

11. Make a plot of I/I_0 as a function of θ (see Fig. 38.2) for the interference pattern produced by the arrangement described in Exercise 1. Let θ range over the interval from 0 to 0.3°.

12. In Fig. 38.2, let $L = 1.2$ m and $d = 0.12$ mm and assume that the slit system is illuminated with monochromatic light of wavelength 500 nm. Calculate the phase difference between the two wavefronts arriving at point *P* from S_1 and S_2 when (a) $\theta = 0.5°$ and (b) $y = 5$ mm.

13. For the situation described in Exercise 12, what is the value of θ for which (a) the phase difference will be equal to 0.333 rad and (b) the path difference will be $\lambda/4$?

14. In an arrangement similar to that illustrated in Fig. 38.3, let $L = 140$ cm and $y = 8$ mm. Find the value of the ratio d/λ for which the average intensity at point P will be 60 percent of the maximum intensity.

15. In the arrangement of Fig. 38.3, let $L = 120$ cm and $d = 0.25$ cm. The slits are illuminated with light of wavelength 600 nm. Calculate the distance y above the central maximum for which the average intensity on the screen will be 75 percent of the maximum.

16. Two slits are separated by a distance of 0.18 mm. An interference pattern is formed on a screen 80 cm away by the H_α line in hydrogen ($\lambda = 656.3$ nm). Calculate the fraction of the maximum intensity that would be measured at a point 0.6 cm above the central maximum.

17. In a double-slit interference experiment (Fig. 38.3), $d = 0.2$ mm, $L = 160$ cm, and $y = 1$ mm. What wavelength will result in an average intensity at P that is 36 percent of the maximum?

18. The intensity on the screen at a certain point in a double-slit interference pattern is 64 percent of the maximum value. (a) What is the minimum phase difference (in radians) between sources that will produce this result? (b) Express the phase difference calculated in (a) as a path difference if the wavelength of the incident light is 486.1 nm (H_β line).

19. At a particular location in a Young's interference pattern, the intensity on the screen is 6.4 percent of maximum. (a) Calculate the minimum phase difference in this case. (b) If the wavelength of light is 587.5 nm (from a helium discharge tube), determine the path difference.

Section 38.4 Phasor Addition of Waves

20. The electric fields from three coherent sources are described by $E_1 = E_o \sin\omega t$, $E_2 = E_o \sin(\omega t + \phi)$, and $E_3 = E_o \sin(\omega t + 2\phi)$. Let the resultant field be represented by $E_P = E_R \sin(\omega t + \alpha)$. Use the phasor method to find E_R and α when (a) $\phi = 20°$, (b) $\phi = 60°$, (c) $\phi = 120°$.

21. Repeat Exercise 20 when $\phi = (3\pi/2)$ radians.

22. Use the method of phasors to find the resultant (magnitude and phase angle) of two fields represented by $E_1 = 12 \sin\omega t$ and $E_2 = 18 \sin(\omega t + 60°)$. (Note that in this case the amplitudes of the two fields are unequal.)

23. You are given that $E_1 = 5.77 \sin\omega t$, $E_2 = E_0 \sin(\omega t + \phi)$, and $E_P = E_1 + E_2$. Find ϕ and E_0 such that $E_P = 10 \sin(\omega t + \pi/6)$. (Use the method of phasor addition.)

24. Consider N coherent sources described by $E_1 = E_o \sin(\omega t + \phi)$, $E_2 = E_o \sin(\omega t + 2\phi)$, $E_3 = E_o \sin(\omega t + 3\phi)$, . . . , $E_n = E_o \sin(\omega t + N\phi)$. Find the minimum value of ϕ for which $E_R = E_1 + E_2 + E_3 + \cdots E_n$ will be zero.

25. Sketch a phasor diagram to illustrate the resultant of $E_1 = E_{o1} \sin\omega t$ and $E_2 = E_{o2} \sin(\omega t + \phi)$, where $E_{o2} = 1.5E_{o1}$ and $\pi/6 \le \phi \le \pi/3$. Use the sketch and the law of cosines to show that, for two coherent waves, the resultant *intensity* can be written $I_R = I_1 + I_2 + 2\sqrt{I_1 I_2}\cos\phi$.

26. When illuminated, four equally spaced parallel slits act as multiple coherent sources, each differing in phase from the adjacent one by an angle ϕ. Use a phasor diagram to determine the smallest value of ϕ for which the resultant of the four waves (assumed to be of equal amplitude) will be zero.

Section 38.6 Interference in Thin Films

27. Let the film shown in Fig. 38.17 have an index of refraction of 1.36 and a thickness of 7×10^{-5} cm. A beam of sunlight is incident in air on the top surface of the film. Determine the wavelengths (within the range 4000–7000 Å) that will be strongly reflected by the film.

28. Determine the minimum thickness of a soap film ($n = 1.41$) that will result in constructive interference of (a) the H_α line ($\lambda = 6563$ Å) and (b) the H_γ line ($\lambda = 4340$ Å).

29. A 500-nm-thick oil film in air is illuminated with white light in the direction perpendicular to the film. What wavelengths will be strongly reflected in the range 300–700 nm? (Take $n = 1.46$ for oil.)

30. Repeat Exercise 29 if the oil film is placed on a thick piece of glass ($n = 1.50$).

31. A material having an index of refraction of 1.30 is used to coat a piece of glass ($n = 1.50$). What should be the minimum thickness of this film in order to minimize reflected light at a wavelength of 500 nm?

32. Two rectangular, optically flat glass plates ($n = 1.52$) are in contact along one end and are separated along the other end by a sheet of paper that is 4×10^{-3} cm thick (Fig. 38.22). The top plate is illuminated by monochromatic light ($\lambda = 546.1$ nm). Calculate the number of dark parallel bands crossing the top plate (include the dark band at zero thickness along the edge of contact between the two plates.)

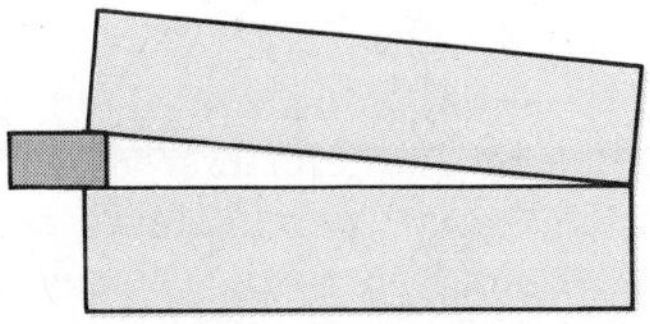

Figure 38.22 (Exercises 32 and 33).

33. An air wedge is formed between two thick glass plates in a manner similar to that described in Exercise 32. Light of wavelength 434 nm is incident vertically on the top plate. In this case, there are 20 bright parallel interference fringes across the top plate. Calculate the thickness of the paper separating the plates.

34. Suppose that the wedge shape shown in Fig. 38.19 has a length (measured along the incline) of 12 cm and that the angle between the two faces is 3.5×10^{-4} rad (2×10^{-2} deg). The wedge has an index of refraction of 1.41 and is illuminated (along the vertical to the top face) by light of wavelength 600 nm. How many bright bands appear in the interference pattern of the reflected light?

Section 38.7 The Michelson Interferometer

35. The mirror on one arm of a Michelson interferometer is displaced a distance Δl. During this displacement, 250 fringe shifts (formation of *successive dark and bright circles*) are counted. The light being used has a wavelength of 632.8 nm. Calculate the displacement Δl.

36. A Michelson interferometer is used to measure an unknown wavelength. The mirror in one arm of the instrument is moved 0.12 mm as 481 *dark* fringes are counted. Determine the wavelength of the light used.

37. Light of wavelength 5505 Å is used to calibrate a Michelson interferometer. By use of a micrometer screw, the platform on which one mirror is mounted is moved 0.18 mm. How many *dark* fringe shifts are counted?

PROBLEMS

1. Consider the case of interference from N coherent sources, each separated by a distance d. Verify that the intensity in the interference pattern is given by Eq. 38.15, where ϕ is the phase difference between adjacent sources.

2. Refer to Fig. 38.20a, which illustrates an arrangement for observing Newton's rings. Assume that n is the index of refraction of the material in the gap between the two glass surfaces. Use the geometry of the figure to show that when $r \ll R$, (a) the radii of the dark fringes are given by $r \approx \sqrt{m\lambda R/n}$ and (b) the radii of the bright fringes are given by $r \approx \sqrt{(m + \frac{1}{2})\lambda R/n}$. (c) When Newton's rings are formed using sodium light ($\lambda = 590$ nm), the diameters of two successive dark rings are 2 mm and 2.23 mm. Calculate the radius of curvature of the convex lens surface.

3. Two sinusoidal vectors of the same amplitude A and frequency ω have a phase difference ϕ. Calculate the resultant amplitude of the two vectors both graphically and analytically if ϕ equals (a) 0, (b) 60°, (c) 90°.

4. Use the method of phasor addition to find the resultant amplitude and phase constant when the following three harmonic functions are combined: $E_1 = \sin(\omega t + \pi/6)$, $E_2 = 3\sin(\omega t + 7\pi/2)$, $E_3 = 6\sin(\omega t + 4\pi/3)$.

5. A fringe pattern is established in the field of view of a Michelson interferometer using light of wavelength 580 nm. A parallel-faced sheet of transparent material 2.5 μm thick is placed in front of one of the mirrors perpendicular to the incident and reflected light beams. An observer counts a fringe shift of six *dark* fringes. What is the index of refraction of the sheet?

6. A glass plate ($n = 1.61$) is covered with a thin uniform layer of oil ($n = 1.2$). A light beam in air of variable wavelength is normally incident onto the oil surface. Observation of the reflected beam shows destructive interference at 500 nm and constructive interference at 750 nm. Calculate the thickness of the oil film from this information.

7. Figure 38.23 illustrates the formation of an interference pattern by the Lloyd's mirror method (see also Section 38.5 and Fig. 38.15). In the case shown here, the actual source S and the virtual source S′ are in a plane 25 cm to the left of the mirror and the screen is a distance $L = 120$ cm to the right of this plane. The source S is a distance $d = 2.5$ mm above the top surface of the mirror (arranged for reflection at glancing incidence), and the light is monochromatic with $\lambda = 620$ nm. (a) Show that, in general, the separation between bright (or dark) fringes on the screen is given by $\Delta y = L\lambda/2d$. (b) Determine the distance of the first bright fringe above the surface of the mirror.

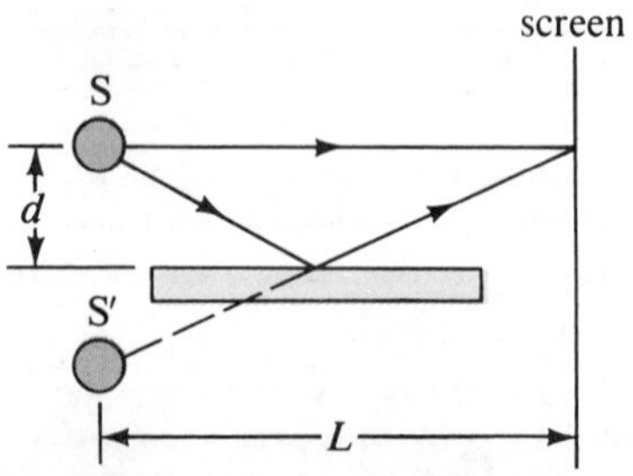

Figure 38.23 (Problem 7).

8. (a) Both sides of a uniform film of index of refraction n and thickness d are in contact with air. For normal incidence of light, an intensity minimum is observed in the reflected light at λ_2 and an intensity maximum is observed at λ_1, where $\lambda_1 > \lambda_2$. If there are no intensity minima observed between λ_1 and λ_2, show that the integer m which appears in Eqs. 38.18 and 38.19 is given by $m = \lambda_1/2(\lambda_1 - \lambda_2)$. (b) Determine the thickness of the film if $n = 1.40$, $\lambda_1 = 500$ nm, and $\lambda_2 = 370$ nm.

9. A piece of transparent material having an index of refraction n is cut into the shape of a wedge as shown in Fig. 38.24. The angle of the wedge is small, and monochromatic light of wavelength λ is normally incident from above. If the height of the wedge is h and the width is l, show that bright fringes occur at the positions $x = \lambda l(m + \frac{1}{2})/2hn$ and dark fringes occur at the positions $x = \lambda lm/2hn$, where $m = 0,1,2, \ldots$ and x is measured as shown.

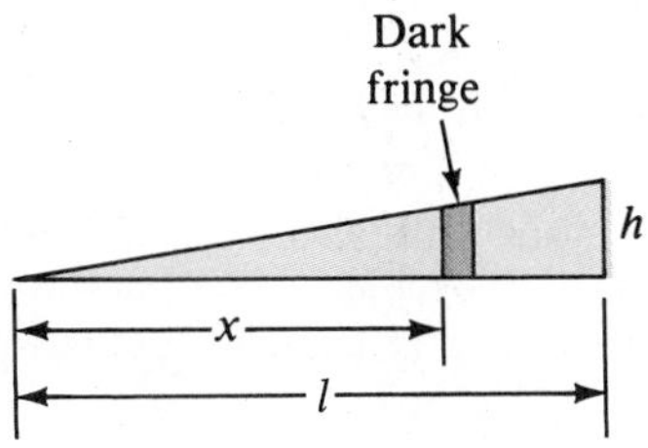

Figure 38.24 (Problem 9).

10. The calculation of Example 38.1 shows that the double-slit arrangement produced fringe separations of 2.2 cm for $\lambda = 560$ nm. Calculate the fringe separation for this same arrangement if the apparatus is submerged in a tank containing a 30% sugar solution ($n = 1.38$).

11. The light source used to illuminate the parallel-slit system illustrated in Fig. 38.2 emits two wavelengths, the longer of which is 700 nm. The fifth dark fringe of the long-wavelength pattern occupies the same position as the fifth bright fringe (not counting the central maximum) of the short-wavelength pattern. Determine the wavelength of the second component.

12. Measurements are made of the intensity distribution in a Young's interference pattern (as illustrated in Fig. 38.4). At a particular value of y (distance from the center of the screen), it is found that $I/I_0 = 0.81$ when light of wavelength 600 nm is used. What wavelength of light should be used to reduce the relative intensity at the same location to 64 percent?

13. The condition for constructive interference by reflection from a thin film in air as developed in Section 38.6 assumes nearly normal incidence. (a) Show that if the light is incident on the film at an angle $\phi_1 \gg 0$ (relative to the normal), then the condition for constructive interference is given by $2nt \cos\theta_2 = (m + 1/2)\lambda$, where θ_2 is the angle of refraction. (b) Calculate the minimum thickness for constructive interference if sodium light ($\lambda = 5.9 \times 10^{-5}$ cm) is incident at an angle of 30° on a film with index of refraction 1.38.

39 Diffraction and Polarization

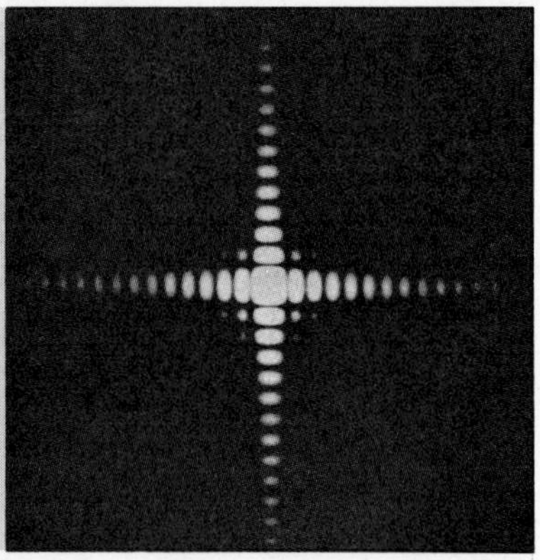

James Madison University

In this chapter, we continue our treatment of physical optics with the discussion of diffraction and polarization phenomena. When light waves pass through a small aperture, an interference pattern is observed rather than a sharp spot of light cast by the aperture. This shows that light spreads in various directions beyond the aperture into regions where a shadow would be expected if light traveled in straight lines. Other waves, such as sound waves and water waves, also have this property of being able to bend around corners. This phenomenon, known as *diffraction*, can be regarded as a consequence of interference from many coherent wave sources. In other words, the phenomena of diffraction and interference are basically equivalent.

In Chapter 35, we discussed the properties of electromagnetic waves and the fact that they are transverse in nature. That is, the electric and magnetic field vectors associated with the wave are perpendicular to the direction of propagation. Under certain conditions, light waves can be plane-polarized. Although ordinary light is usually not polarized, we shall discuss various methods for producing polarized light, such as by using polarizing sheets. The combined evidence of interference and polarization proves that light cannot be composed of classical (newtonian) particles.

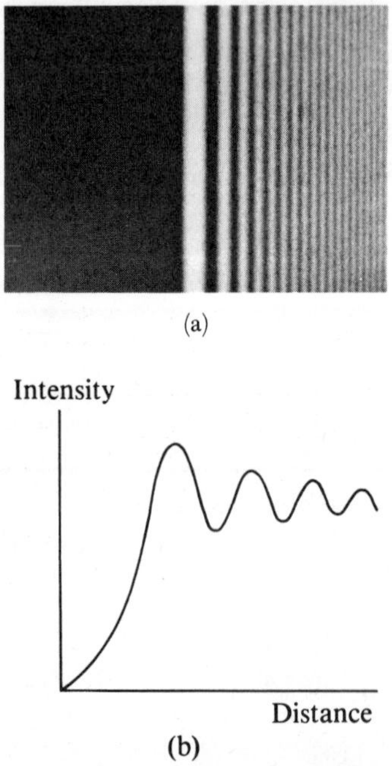

Figure 39.1 (a) Diffraction of light around a straight edge. (From M. Cagnet, M. Francon, and J. C. Thierr, Atlas of Optical Phenomena, Plate 32, Berlin, Springer-Verlag, 1962.) (b) Intensity variation with distance from the edge of the shadow for a straight line.

39.1 INTRODUCTION TO DIFFRACTION

In the previous chapter, we found that light passing through two slits does not produce two distinct bright areas on a screen. Instead, an interference pattern is observed on the screen which shows that the light has deviated from a straight-line path and has entered the otherwise shadowed region. This deviation of light from a straight-line path is called *diffraction*. Diffraction results from the interference of light from many coherent sources. In principle, the intensity of a diffraction pattern at a given point in space can be computed using Huygens' principle, where each point on the wavefront at the source of the pattern is taken to be a point source.

In general, diffraction occurs when waves pass through apertures or around obstacles. As an example of diffraction, if an opaque object is placed between a point source and a screen, the boundary between the shadowed and illuminated regions is not abrupt. A careful inspection of the boundary shows a series of bright and dark bands within the geometric shadow (Fig. 39.1a), which is due to the bending of light around the edge of the obstacle. A plot of intensity versus distance from the edge of the shadow is shown in Fig. 39.1b. Effects of this type were reported in the 17th century by Francesco Grimaldi.

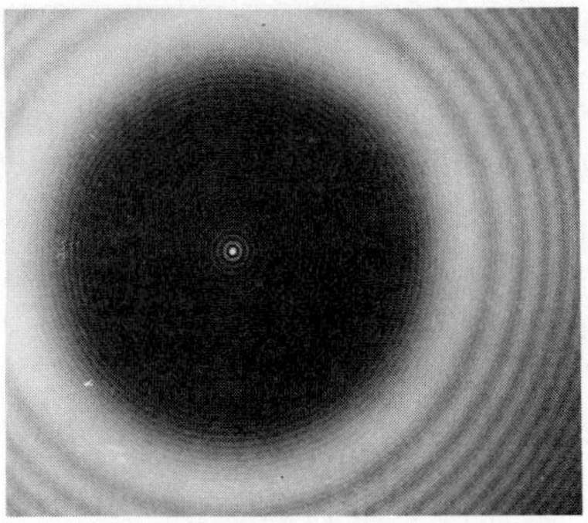

Figure 39.2 Diffraction pattern of a penny, taken with the penny midway between the screen and source. (Courtesy of Prof. P. M. Rinard, from Am. J. Phys. **44**, 70, 1976)

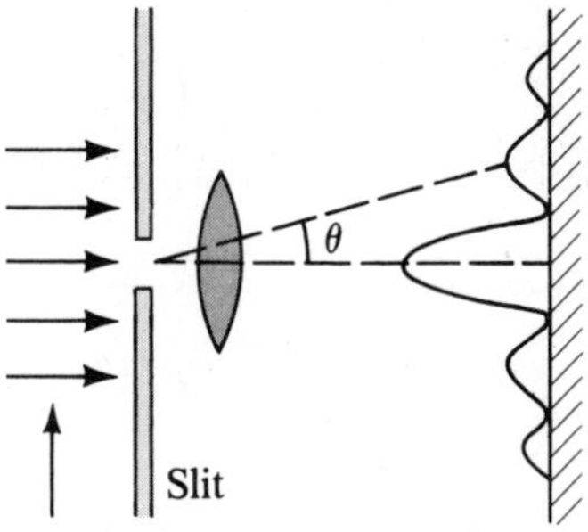

Figure 39.3 A Fraunhofer diffraction pattern of a single slit. The parallel rays are brought into focus on the screen with a converging lens. The pattern consists of a central bright region flanked by much weaker maxima.

One example of diffraction in nature is the occurrence of the "shadow bands" seen on the earth at the time of a total solar eclipse. Diffraction patterns are also observed surrounding the shadows of various opaque objects. For example, Fig. 39.2 shows the shadow of the diffraction pattern of a penny. The pattern exhibits a bright spot at the center, circular fringes near the shadow's edge, and another set of fringes outside the shadow. This pattern was first observed in 1818 by Dominique Arago. The bright spot at the center of the shadow can be explained only through the use of the wave theory of light, which predicts constructive interference at this point. This was certainly a most dramatic experimental proof of the wave nature of light.

Diffraction phenomena are usually classified as being one of two types, which are named after the men who first explained them. The first type, called *Fraunhofer diffraction*, occurs when the rays reaching a point are approximately parallel. This can be achieved experimentally either by placing the observing screen at a large distance from the aperture or by using a converging lens to focus parallel rays on the screen, as in Fig. 39.3. Note that a bright fringe is observed along the axis at $\theta = 0$, with alternating bright and dark fringes on either side of the central bright fringe.

When the observing screen is placed at a *finite* distance from the slit and no lens is used to focus parallel rays, the observed pattern is called a *Fresnel diffraction pattern* (Fig. 39.4). The diffraction patterns shown in Figs. 39.1a and 39.2 are examples of Fresnel diffraction. Another example, shown in Fig. 39.5, is the diffraction pattern of a rectangular aperture. Fresnel diffraction is rather complex to treat quantitatively. Therefore, the following discussions will be restricted to Fraunhofer diffraction.

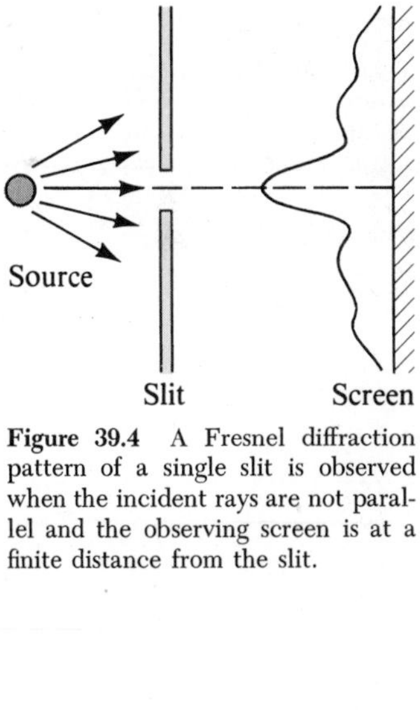

Figure 39.4 A Fresnel diffraction pattern of a single slit is observed when the incident rays are not parallel and the observing screen is at a finite distance from the slit.

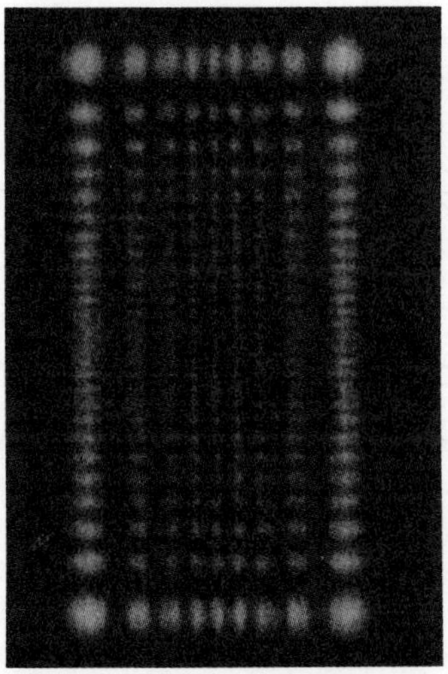

Figure 39.5 Fresnel diffraction pattern for a rectangular aperture. (From M. Cagnet, M. Francon, and J. C. Thierr, Atlas of Optical Phenomena, Plate 34, Berlin, Springer-Verlag, 1962)

Q1. If you place your thumb and index finger very close together and view light passing between them when they are a few cm in front of your eye, dark lines parallel to your thumb and finger will appear. Explain.

Q2. Observe the shadow of your book or some other straight edge when it is held a few inches above a table with a lamp several feet above the book. Why is the shadow of the book somewhat fuzzy at the edges?

Q3. What is the difference between Fraunhofer and Fresnel diffraction?

Q4. Although we can hear around corners, we cannot see around corners. How can you explain this in view of the fact that sound and light are both waves?

39.2 FRAUNHOFER DIFFRACTION OF A SINGLE SLIT

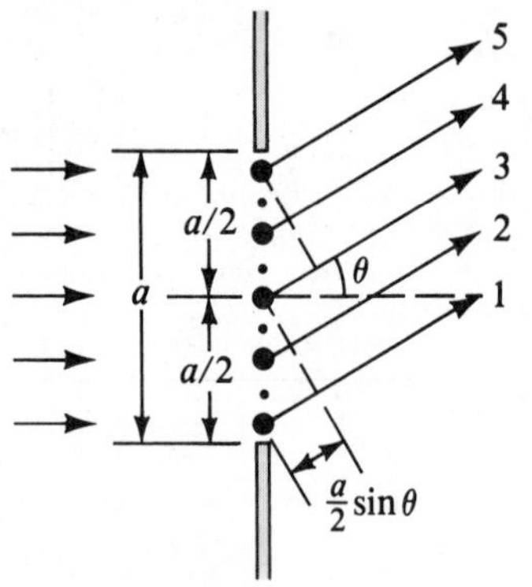

Figure 39.6 Diffraction of light by a narrow slit of width a. Each portion of the slit acts as a point source of waves. The path difference between rays 1 and 3 or between rays 2 and 4 is equal to $(a/2)\ \sin\theta$.

Let us consider Fraunhofer diffraction by a single slit as shown in Fig. 39.6. We can deduce some important features of this problem by examining waves coming from various portions of the slit. According to Huygens' principle, each portion of the slit acts as a source of waves. Hence, light from one portion of the slit can interfere with light from another portion, and, the resultant intensity on the screen will depend on the direction θ.

To analyze the resultant diffraction pattern, it is convenient to divide the slit in two halves as in Fig. 39.6. All the waves that originate from the slit are in phase. Consider waves 1 and 3, which originate from the bottom and center of the slit, respectively. Wave 1 travels farther than wave 3 by an amount equal to the path difference $(a/2)\ \sin\theta$, where a is the width of the slit. Similarly, the path difference between waves 2 and 4 is also equal to $(a/2)\ \sin\theta$. If this path difference is exactly one half of a wavelength (corresponding to a phase difference of 180°), the two waves cancel each other and destructive interference results. This is true, in fact, for any two waves that originate at points separated by half the slit width, since the phase difference between two such points is 180°. Therefore, waves from the upper half of the slit interfere *destructively* with waves from the lower half of the slit when

$$\frac{a}{2}\sin\theta = \frac{\lambda}{2}$$

or when

$$\sin\theta = \frac{\lambda}{a}$$

Similarly, destructive interference results when the path differences $(a/2)\ \sin\theta$ equals λ, $3\lambda/2$, 2λ, $5\lambda/2$, etc. These points occur at progressively larger values of θ. Therefore, the general condition for destructive interference is

Condition for destructive interference

$$\sin\theta = m\frac{\lambda}{a} \qquad (m = \pm1, \pm2, \pm3, \ldots) \tag{39.1}$$

where $|m| \leqslant a/\lambda$.

Equation 39.1 gives the values of θ for which the diffraction pattern has *zero* intensity. However, it tells us nothing about the variation in intensity along the screen. The general features of the intensity distribution along the screen are shown in Fig. 39.7. A broad central bright fringe is observed, flanked by much weaker alternating maxima. The central bright fringe corresponds to those points opposite the slit for which the path difference is zero, or $\theta = 0$. All waves originating from

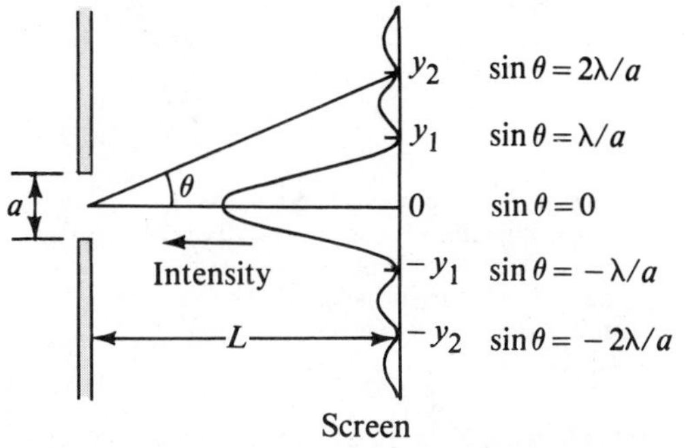

Figure 39.7 Positions of the various minima for the Fraunhofer diffraction pattern of a single slit of width a.

the slit reach this region in phase; hence constructive interference results. The various dark fringes (points of zero intensity) occur at the values of θ that satisfy Eq. 39.1. The positions of the weaker maxima lie approximately halfway between the dark fringes. Note that the central bright fringe is twice as wide as the weaker maxima.

Example 39.1

Light of wavelength 580 nm is incident on a slit of width 0.30 mm. The observing screen is placed 2 m from the slit. Find the positions of the first dark fringes and the width of the central bright fringe.

Solution: The first dark fringes that flank the central bright fringe correspond to $m = \pm 1$ in Eq. 39.1. Hence, we find that

$$\sin\theta = \pm\frac{\lambda}{a} = \pm\frac{5.8 \times 10^{-7}\text{ m}}{0.3 \times 10^{-3}\text{ m}} = \pm 1.93 \times 10^{-3}$$

From the triangle in Fig. 39.7, we see that $\tan\theta = y_1/L$. Since θ is very small, we can use the approximation $\sin\theta \approx \tan\theta$, so that $\sin\theta \approx y_1/L$. Therefore, the positions of the first minima measured from the central axis are given by

$$y_1 \approx L\sin\theta = \pm L\frac{\lambda}{a} = \pm 3.86 \times 10^{-3}\text{ m}$$

The positive and negative signs correspond to the dark fringes on either side of the central bright fringe. Hence, the width of the central bright fringe is $2|y_1| = 7.72 \times 10^{-3}\text{ m} = 7.72$ mm. Note that this value is *much larger* than the width of the slit. However, as the width of the slit is *increased*, the diffraction pattern will *narrow*, corresponding to smaller values of θ. In fact, for large values of a, the various maxima and minima will be so closely spaced that only a large central bright area is observed, which resembles the geometric image of the slit.

Intensity of the Single-Slit Diffraction Pattern

To calculate the intensity at an arbitrary angle θ for the single-slit diffraction pattern, we imagine that the slit consists of a large number of coherent sources as in Fig. 39.8. The total intensity at some point p on the screen is obtained by adding the phasors from each source, as shown in Fig. 39.9. As the number of sources increases, the phasors tend to form the arc of a circle. The total phase difference between the first and last phasors is given by

$$\phi = \frac{2\pi}{\lambda}\delta = \frac{2\pi}{\lambda}a\sin\theta \tag{39.2}$$

Using the method of phasor addition, we see that the resultant amplitude E forms the chord of the circular arc whose radius is R. We can obtain an expression for the resultant amplitude by using the right triangle in Fig. 39.9. This gives

$$\sin\frac{\phi}{2} = \frac{E/2}{R}$$

$$E = 2R\sin\frac{\phi}{2} \tag{39.3}$$

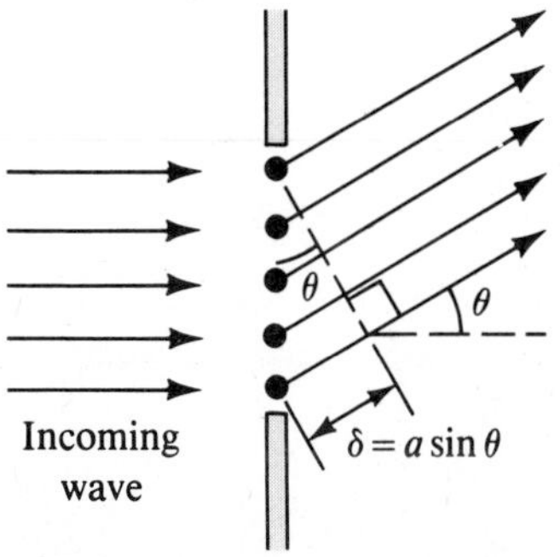

Figure 39.8 A plane wave is incident on a slit of width a. Each point along the slit acts as a source for coherent waves.

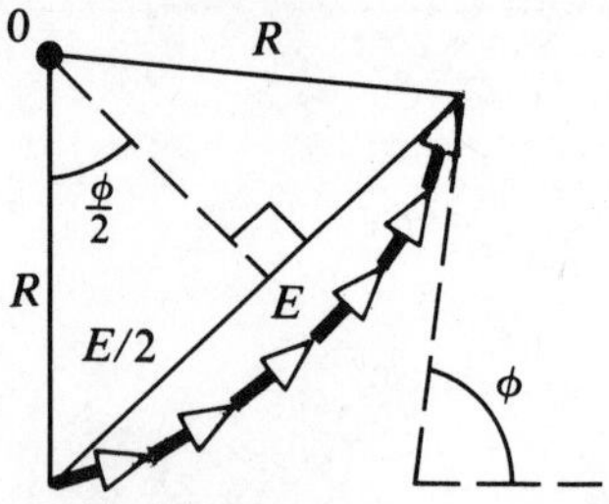

Figure 39.9 Phasor diagram for a large number of coherent sources. Note that the ends of the phasors lie on a circular arc of radius R. The resultant amplitude E equals the length of the chord.

Now, the arc length formed by the phasor equals E_0, the net amplitude observed at $\theta = 0$. Equating this arc length to the product $R\phi$ (where ϕ is in radians), we get

$$R\phi = E_0$$
$$R = E_0/\phi$$

Substituting this result into Eq. 39.3 gives

$$E = E_0\frac{\sin(\phi/2)}{\phi/2}$$

Since the resultant intensity at P is proportional to the *square* of the amplitude E, we find that

Intensity of a single-slit Fraunhofer diffraction pattern

$$I = I_0\left|\frac{\sin(\phi/2)}{\phi/2}\right|^2 \tag{39.4}$$

where $\phi = 2\pi a \sin\theta/\lambda$ and I_0 is the intensity at $\theta = 0$.

A plot of Eq. 39.4 is shown in Fig. 39.10a, and a photograph of a single-slit Fraunhofer diffraction pattern is shown in Fig. 39.10b. Note that most of the light intensity is concentrated in the central bright fringe. Furthermore, according to Eq. 39.4, the various *zeros* in intensity occur when $\phi/2 = m\pi$, which corresponds to

$$\frac{\pi a \sin\theta}{\lambda} = m\pi$$

Condition for intensity minima

$$\sin\theta = m\frac{\lambda}{a}$$

This is in agreement with our earlier result, given by Eq. 39.1.

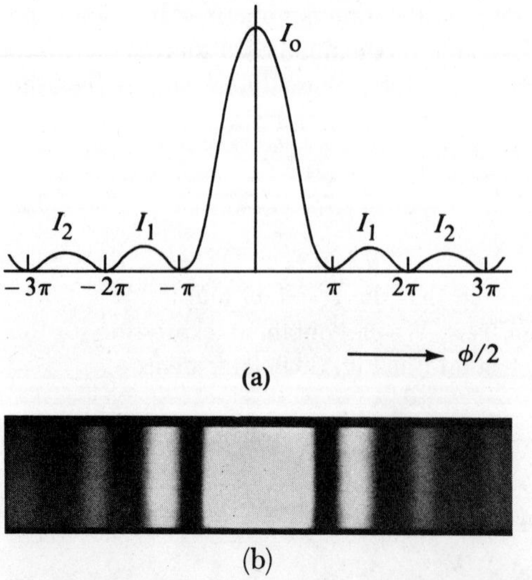

Figure 39.10 (a) A plot of the intensity I versus ϕ for the single-slit Fraunhofer diffraction pattern. (b) Photograph of a single-slit Fraunhofer diffraction pattern. (From M. Cagnet, M. Francon, and J. C. Thierr, Atlas of Optical Phenomena, Plate 18 Berlin, Springer-Verlag, 1962)

Example 39.2

Find the ratio of intensities of the secondary maxima relative to the central maximum for the single-slit Fraunhofer diffraction pattern.

Solution: To a good approximation, we can assume that the secondary maxima lie midway between the zero points. From Fig. 39.10a, we see that this corresponds to $\phi/2$ values of $3\pi/2$, $5\pi/2$, $7\pi/2$, . . . Substituting these into Eq. 39.4 gives for the first two ratios

$$\frac{I_1}{I_0} = \left|\frac{\sin(3\pi/2)}{(3\pi/2)}\right|^2 = \frac{1}{9\pi^2/4} = 0.045$$

$$\frac{I_2}{I_0} = \left|\frac{\sin(5\pi/2)}{5\pi/2}\right|^2 = \frac{1}{25\pi^2/4} = 0.016$$

That is, the secondary maximum that is adjacent to the central maximum has an intensity of 4.5% that of the central bright fringe, and the next secondary maximum has an intensity of 1.6% that of the central bright fringe.

Q5. Describe the change in width of the central maximum of the single-slit diffraction pattern as the width of the slit is made smaller.

39.3 RESOLUTION OF SINGLE-SLIT AND CIRCULAR APERTURES

Suppose that two light sources are far from a narrow slit of width a as in Fig. 39.11. The objects can be considered as two point sources S_1 and S_2, which are *not* coherent. If no diffraction occurred, that is, if geometric optics prevailed, one would observe two distinct bright spots (or images) on the screen at the right. However, because of diffraction effects, each source is imaged as a bright central region flanked by weaker maxima and minima. What is observed on the screen is the sum of the two diffraction patterns.

If the two sources are separated such that their central maxima do not overlap, as in Fig. 39.11a, their images can be distinguished and are said to be *resolved*. On the other hand, if the sources are close together, as in Fig. 39.11b, the two central maxima may overlap. Hence, their images may not be resolved.

To decide when two images are resolved, the following condition is often used: *When the central maximum for one image falls on the first minimum of the other image, the images are said to be just resolved.* This limiting condition of resolution is known as *Rayleigh's criterion*. Figure 39.12 shows the diffraction patterns for three situations. When the objects are far apart, their images are well resolved (Fig. 39.12a). The images are just resolved when their angular separation satisfies Rayleigh's criterion (Fig. 39.12b). Finally, the images are not resolved in

Condition for resolution of two images

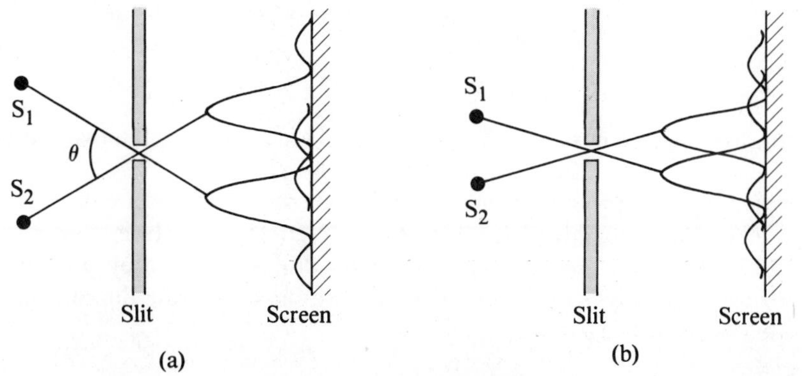

Figure 39.11 Two point sources at some distance from a small aperture each produce a diffraction pattern. (a) The angle subtended by the sources at the aperture is large enough so that the diffraction patterns are distinguishable. (b) The angle subtended by the sources is so small that the diffraction patterns are not distinguishable.

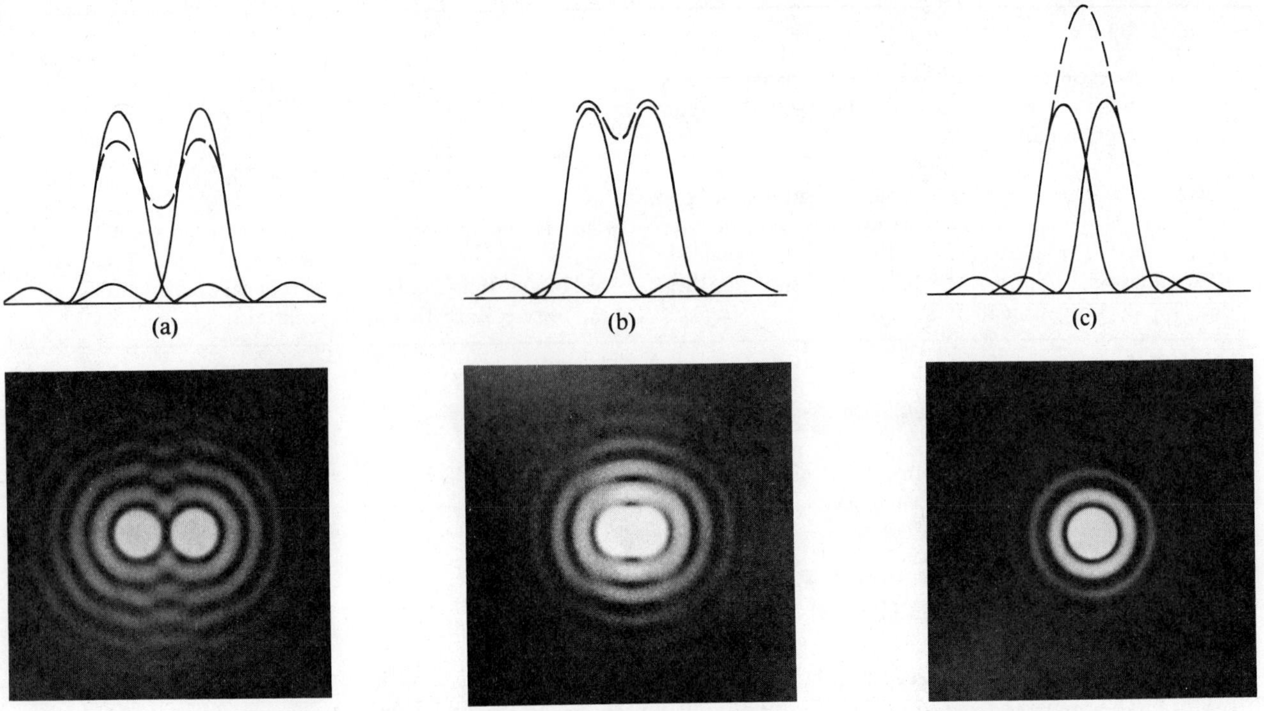

Figure 39.12 The diffraction patterns of two point sources (solid curves) and the resultant pattern (dashed curves) for various angular separations of the sources. (a) The sources are far apart, and the patterns are well resolved. (b) The sources are closer together, and their patterns are just resolved. (c) The sources are so close together that their patterns are not resolved. (From M. Cagnet, M. Francon, and J. C. Thierr, Atlas of Optical Phenomena, Plate 16, Berlin, Springer-Verlag, 1962)

Fig. 39.12c. From this criterion, we can determine the minimum angular separation θ_m subtended by the sources at the slit such that their images will be just resolved. In the previous section, we found that the first minimum in the single-slit diffraction pattern occurs at the angle that satisfied the relationship

$$\sin\theta = \frac{\lambda}{a}$$

According to Rayleigh's criterion, this expression gives the smallest angular separation for which the two images will be resolved. Since $\lambda \ll a$ in most situations, $\sin\theta$ is small and we can use the approximation $\sin\theta \approx \theta$. Therefore, the limiting angle of resolution for a slit is given by

Limiting angle of resolution for a slit

$$\theta_m \approx \frac{\lambda}{a} \tag{39.5}$$

where θ_m is expressed in radians. Hence, the angle subtended by the two sources at the slit must be *greater* than λ/a if they are to be resolved.

Many optical systems, such as microscopes and telescopes, use circular apertures rather than slits. The diffraction pattern of a circular aperture, illustrated in Fig. 39.13, consists of a central circular bright disk surrounded by progressively fainter rings. An analysis of the circular aperture shows that the limiting angle of resolution is given by

Limiting angle of resolution for a circular aperture

$$\theta_m = 1.22\frac{\lambda}{D} \tag{39.6}$$

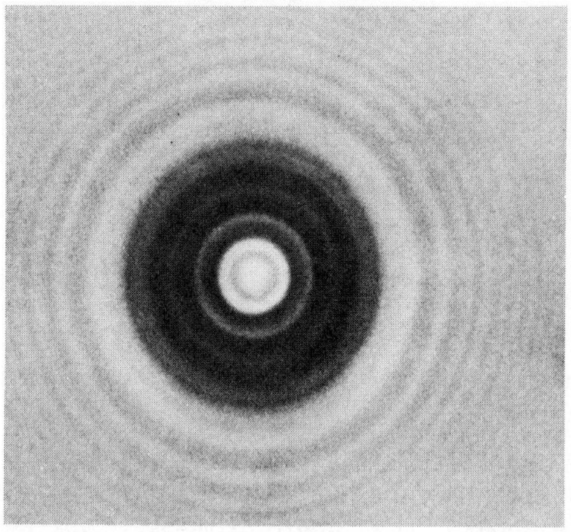

Figure 39.13 The Fresnel diffraction pattern of a circular aperture consists of a central bright disk surrounded by concentric bright and dark rings. (From M. Cagnet, M. Francon, and J. C. Thierr, Atlas of Optical Phenomena, Plate 34, Berlin, Springer-Verlag, 1962)

where D is the diameter of the aperture. Note that Eq. 39.6 is similar to Eq. 39.5 except for the factor 1.22, which arises from a complex mathematical analysis of diffraction from the circular aperture.

Example 39.3

Resolution of a telescope: The telescope at Mt. Palomar has a diameter of 200 in. What is the limiting angle of resolution for this telescope at a wavelength of 600 nm?

Solution: Since $D = 200$ in. $= 5.08$ m and $\lambda = 6 \times 10^{-7}$ m, Eq. 39.6 gives

$$\theta_m = 1.22\frac{\lambda}{D} = 1.22\left(\frac{6 \times 10^{-7}\ \text{m}}{5.08\ \text{m}}\right)$$
$$= 1.44 \times 10^{-7}\ \text{rad} = 0.03\ \text{s of arc}$$

Therefore, any two stars that subtend an angle greater than or equal to this value will be resolved (assuming ideal atmospheric conditions).

It is interesting to compare this to the resolution of a large radiotelescope, such as the world's largest system at Arecibo, Puerto Rico, which has a diameter of 1000 ft (305 m). This telescope detects radio waves at a wavelength of 0.75 m. The corresponding minimum angle of resolution is calculated to be 3×10^{-3} rad (10′19″ of arc), which is more than 10 000 times larger than the calculated minimum angle for the Hale telescope on Mt. Palomar.

It is interesting to note that the Hale telescope can never reach its diffraction limit. Instead, its limiting angle of resolution is always set by atmospheric blurring. This "seeing limit" is usually about 1 s of arc and is never smaller than about 0.1 s of arc. (This is one of the reasons for the current interest in a large space telescope.)

Example 39.4

Resolution of the eye: Calculate the limiting angle of resolution for the eye, assuming a pupil diameter of 2 mm, a wavelength of 500 nm in air, and an index of refraction for the eye equal to 1.33.

Solution: In this example, we can use Eq. 39.6, noting that λ is the wavelength in the medium containing the aperture. Since the wavelength of light in the eye is reduced by the index of refraction of the eye medium, we find that $\lambda = (500\ \text{nm})/1.33 = 376$ nm. Therefore, Eq. 39.6 gives

$$\theta_m = 1.22\frac{\lambda}{D} = 1.22\left(\frac{3.76 \times 10^{-7}\ \text{m}}{2 \times 10^{-3}\ \text{m}}\right)$$
$$= 2.3 \times 10^{-4}\ \text{rad} = 0.013°$$

We can use this result to calculate the minimum separation d between two point sources that the eye can distinguish if they are at a distance L from the observer (Fig. 39.14). Since θ_m is small, we see that

$$\sin\theta_m \approx \theta_m \approx d/L$$
$$d = L\theta_m$$

For example, if the objects are located at a distance of 25 cm from the eye (the near point), then

$$d = (25\ \text{cm})\ (2.3 \times 10^{-4}\ \text{rad}) = 5.8 \times 10^{-3}\ \text{cm}$$

This is approximately equal to the thickness of a human hair.

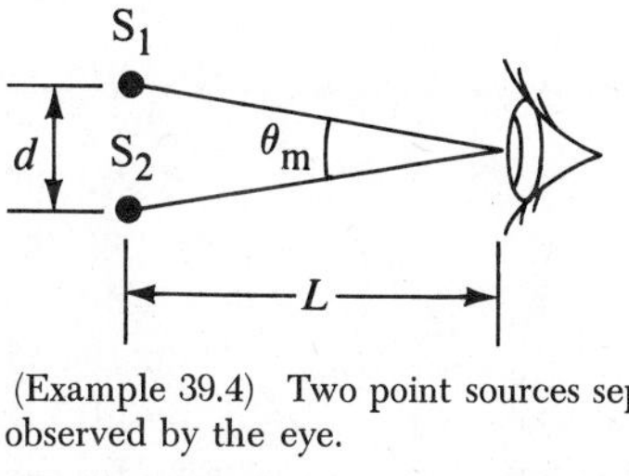

Figure 39.14 (Example 39.4) Two point sources separated by a distance d as observed by the eye.

Q6. Assuming that the headlights of a car are point sources, estimate the maximum distance from an observer to the car at which the headlights are distinguishable from each other.

39.4 THE DIFFRACTION GRATING

The diffraction grating, a very useful device for analyzing light sources, consists of a large number of equally spaced parallel slits. A grating can be made by scratching parallel lines on a glass plate with a precision machining technique. The spaces between each scratch are transparent to the light and hence act as separate slits. A typical grating contains several thousand lines per cm. For example, a grating ruled with 5000 lines/cm has a slit spacing d equal to the inverse of this number, or $d = (1/5000)\text{ cm} = 2 \times 10^{-4}\text{ cm}$.

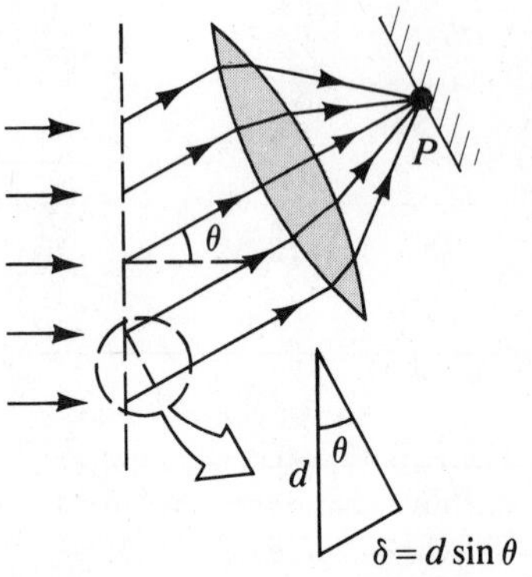

Figure 39.15 Side view of a diffraction grating. The slit separation is d, and the path difference between adjacent slits is $\delta = d \sin\theta$.

A schematic diagram of a section of planar diffraction grating is illustrated in Fig. 39.15. A plane wave is incident from the left, normal to the plane of the grating. A converging lens can be used to bring the rays together at the point P. The intensity of the observed pattern on the screen is the result of the combined effects of interference and diffraction. Each slit produces diffraction, as was described in the previous section. The diffracted beams in turn interfere with each other to produce the final pattern.

Some properties of the interference pattern produced by a diffraction grating were discussed in Section 38.4. Each slit acts as a source of waves, where all waves start at the slits in phase. However, for some arbitrary direction θ measured from the horizontal, the waves must travel *different* path lengths before reaching a particular point P on the screen. From Fig. 39.15, note that the path difference between waves from any two adjacent slits is equal to $d \sin\theta$. If this path difference equals one wavelength or some integral multiple of a wavelength, waves from all slits will be in phase at P and a bright line will be observed. Therefore, the condition for maxima in the interference pattern at the angle θ is

Condition for interference maxima for a grating

$$d \sin\theta = m\lambda \qquad (m = 0,1,2,3, \ldots) \tag{39.7}$$

This expression can be used to calculate the wavelength from a knowledge of the grating spacing and the angle of deviation θ. The integer m represents the *order number* of the diffraction pattern. If the incident radiation contains several wavelengths, each wavelength will deviate through a specific angle. All wavelengths are seen at $\theta = 0$, corresponding to $m = 0$. This is called the *zeroth-order maximum*. The *first-order maximum*, corresponding to $m = 1$, is observed at an angle that satisfies the relationship $\sin\theta = \lambda/d$; the *second-order maximum*, corresponding to $m = 2$, is observed at a larger angle θ, and so on.

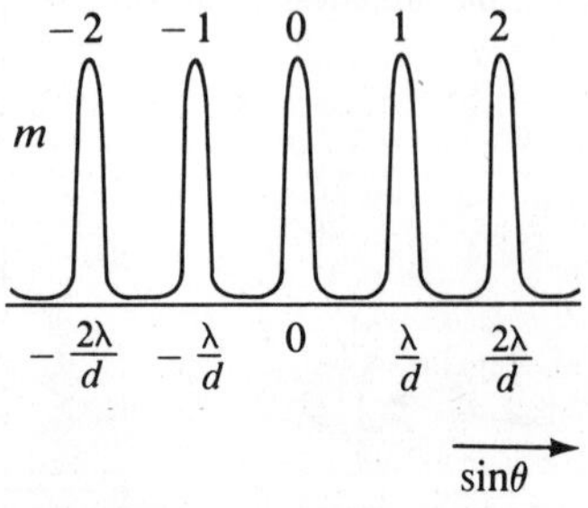

Figure 39.16 The intensity versus $\sin\theta$ for the diffraction grating. The zeroth-, first-, and second-order principal maxima are shown.

A sketch of the intensity distribution for the diffraction grating is shown in Fig. 39.16. If the source contains various wavelengths, a spectrum of lines will be observed at different positions for each order number. Note the sharpness of the principal maxima and the broad range of dark areas. This is in contrast to the broad, bright fringes characteristic of the two-slit interference pattern.

A simple arrangement that can be used to measure various orders of the diffraction pattern is shown in Fig. 39.17. This is a form of a diffraction grating spectrometer. The light to be analyzed passes through a slit, and a parallel beam of light exits from the collimator, which is perpendicular to the grating. The diffracted light leaves the grating at angles that satisfy the grating equation. A telescope is used to view the image of the slit. By measuring the precise angles at which the images of the slit appear for the various orders, the wavelength can be determined.

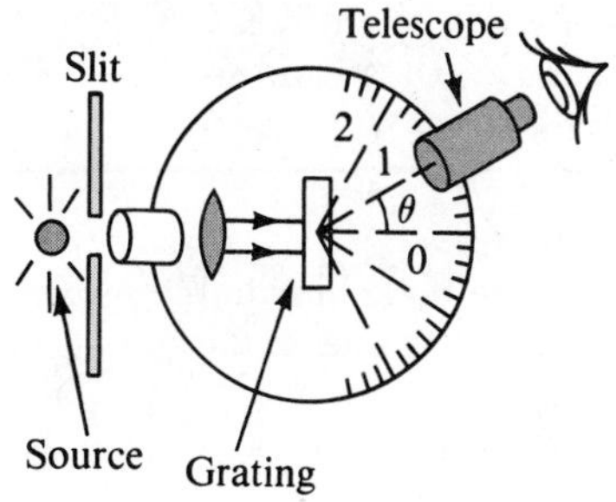

Figure 39.17 Schematic diagram of a diffraction grating spectrometer. The collimated beam incident on the grating is diffracted into the various orders at the angles θ which satisfy the equation $d \sin\theta = m\lambda$, where $m = 0, 1, 2, 3, \ldots$.

Example 39.5

Monochromatic light from a helium-neon laser ($\lambda = 632.8$ nm) is incident normally on a diffraction grating containing 6000 lines/cm. Find the angles at which one would observe the first-order maximum, the second-order maximum, etc.

Solution: First, we must calculate the slit separation, which is equal to the inverse of the number of lines per cm:

$$d = (1/6000)\text{ cm} = 1.667 \times 10^{-4}\text{ cm} = 1667\text{ nm}$$

For the first-order maximum ($m = 1$), we get

$$\sin\theta_1 = \frac{\lambda}{d} = \frac{632.8\text{ nm}}{1667\text{ nm}} = 0.3796$$

$$\theta_1 = 22.31°$$

Likewise, for $m = 2$ we find that

$$\sin\theta_2 = \frac{2\lambda}{d} = \frac{2(632.8\text{ nm})}{1667\text{ nm}} = 0.7592$$

$$\theta_2 = 49.39°$$

However, for $m = 3$ we find that $\sin\theta_3 = 1.139$. Since $\sin\theta$ cannot exceed unity, this does not represent a realistic solution. Hence, only zeroth, first, and second order maxima will be observed for this situation.

Resolving Power of the Diffraction Grating

The diffraction grating is most useful for taking accurate wavelength measurements. Like the prism, the diffraction grating can be used to disperse a spectrum into its components. Of the two devices, the grating is more helpful if one wants to distinguish between two closely spaced wavelengths. We say that the grating spectrometer has a "higher resolution" than a prism spectrometer. If λ_1 and λ_2 are the two nearly equal wavelengths between which the spectrometer can just barely distinguish, the *resolving power R* of the grating is defined as

$$R \equiv \frac{\lambda}{\lambda_2 - \lambda_1} = \frac{\lambda}{\Delta\lambda} \tag{39.8}$$

Resolving power

where $\lambda \approx \lambda_1 \approx \lambda_2$ and $\Delta\lambda = \lambda_2 - \lambda_1$. Thus, we see that a grating with a high resolving power can distinguish small wavelength differences. Furthermore, if N lines of the grating are illuminated, it can be shown that the resolving power in the mth order diffraction equals the product Nm:

$$R = Nm \tag{39.9}$$

Resolving power of a grating

The derivation of Eq. 39.9 is left as a problem (Problem 39.7). Thus, the resolving power increases with increasing order number. Furthermore, R is large for a grating with a large number of illuminated slits. Note that for $m = 0$, $R = 0$, which signifies that *all wavelengths are indistinguishable* for the zeroth-order maximum. However, consider the second-order diffraction pattern ($m = 2$) of a grating that has 5000 rulings illuminated by the light source. The resolving power of such a grating in second-order is $R = 5000 \times 2 = 10\,000$. Therefore, the *minimum* wavelength separation between two spectral lines that can be just resolved, assuming a mean wavelength of 600 nm, is given by $\Delta\lambda = \lambda/R = 6 \times 10^{-2}$ nm = 0.6 Å. For the third-order principal maximum, R = 15 000 and $\Delta\lambda$ = 0.4 Å, and so on.

Example 39.6

Two strong lines in the spectrum of sodium have wavelengths of 589.00 nm and 589.59 nm. (a) What must the resolving power of the grating be in order to distinguish these wavelengths?

$$R = \frac{\lambda}{\Delta\lambda} = \frac{589 \text{ nm}}{589.59 \text{ nm} - 589.00 \text{ nm}} = \frac{589}{0.59} = 998$$

(b) In order to resolve these lines in the second-order spectrum ($m = 2$), how many lines of the grating must be illuminated?

From Eq. 39.9 and the results to (a), we find that

$$N = \frac{R}{m} = \frac{998}{2} = 499 \text{ lines}$$

Q7. A laser beam is incident at a shallow angle on a machinist's ruler that has a finely calibrated scale. The rulings on the scale give rise to a diffraction pattern on a screen. Discuss how you can use this technique to obtain a measure of the wavelength of the laser light.

39.5 DIFFRACTION OF X-RAYS BY CRYSTALS

We have seen that the wavelength of light can be measured with a diffraction grating having a known number of rulings per unit length. In principle, the wavelength of any electromagnetic wave can be determined if a grating of the proper spacing (of the order of λ) is available. X-rays, discovered by W. Roentgen (1845–1923) in 1895, are electromagnetic waves with very short wavelengths (of the order of 1 Å = 10^{-10} m). Obviously, it would be impossible to construct a grating with such a small spacing. However, the atomic spacing in a solid is known to be about 10^{-10} m. In 1913, Max von Laue (1879–1960) suggested that the regular array of atoms in a crystal could act as a three-dimensional diffraction grating for X-rays. Subsequent experiments confirmed this prediction. The diffraction patterns that one observes are rather complicated because of the three-dimensional nature of the crystal. Nevertheless, X-ray diffraction has proved to be an invaluable technique for elucidating crystalline structures and for understanding the structure of matter.[1]

Figure 39.18 is one experimental arrangement for observing X-ray diffraction from a crystal. A collimated beam of X-rays with a continuous range of wavelengths is incident on a crystal, such as one of sodium chloride, for example. The diffracted beams are very intense in certain directions, corresponding to constructive interfer-

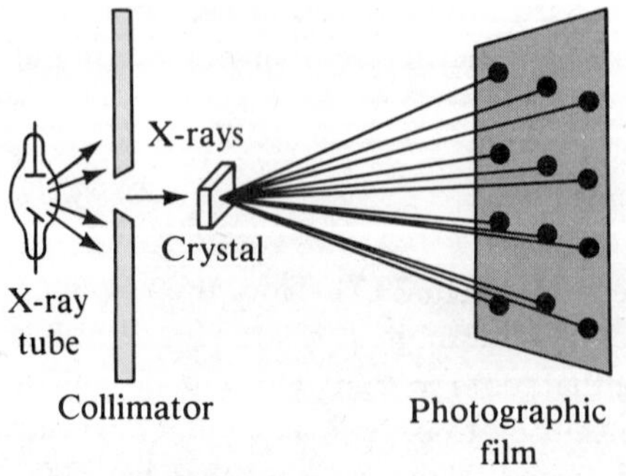

Figure 39.18 Schematic diagram of the technique used to observe the diffraction of X-rays by a single crystal. The array of spots formed on the film by the strongly diffracted beams is called a Laue pattern.

[1]For more details on this subject, see Sir Lawrence Bragg, "X-Ray Crystallography," *Scientific American*, July 1968.

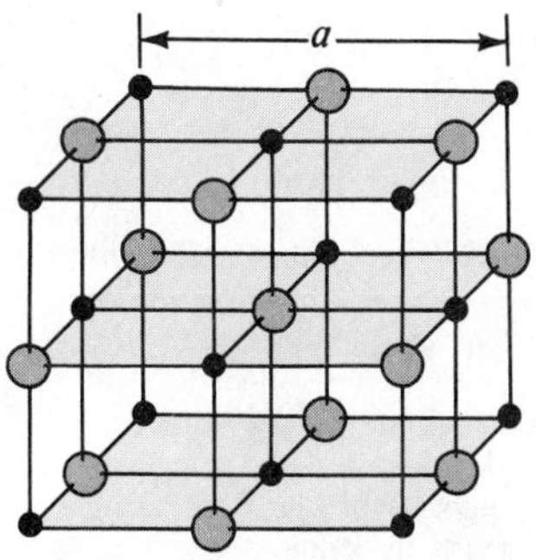

Figure 39.19 A model of the cubic crystalline structure of sodium chloride. The larger spheres represent the Cl^- ions, and the smaller colored spheres represent the Na^+ ions. The length of the cube edge is $a = 0.562737$ nm.

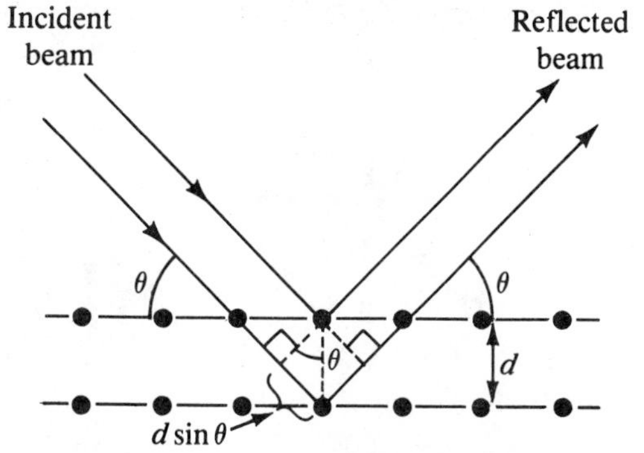

Figure 39.20 A two-dimensional description of the reflection of an X-ray beam from two parallel crystalline planes separated by a distance d. The beam reflected from the lower plane travels farther than the one reflected from the upper plane by an amount equal to $2d \sin\theta$.

ence from waves reflected from layers of atoms in the crystal. The diffracted beams can be detected by a photographic film, and they form an array of spots known as a "Laue pattern." The crystalline structure is deduced by analyzing the positions and intensities of the various spots in the pattern.

The arrangement of atoms in a crystal of NaCl is shown in Fig. 39.19. The smaller, dark spheres represent Na^+ ions and the larger, hollow spheres represent Cl^- ions. Note that the ions are located at the corners of a cube; for this reason, the structure is said to have *cubic symmetry*.

A careful examination of the NaCl structure shows that the ions appear to lie in various planes. The shaded areas in Fig. 39.19 represent one example in which the atoms lie in equally spaced planes. Now suppose an X-ray beam is incident at an angle θ on one of the planes, as in Fig. 39.20. The beam can be reflected from both the upper and the lower plane of atoms. However, the geometric construction in Fig. 39.20 shows that the beam reflected from the lower surface travels farther than the beam reflected from the upper surface. The effective path difference between the two beams is $2d \sin\theta$. The two beams will reinforce each other (constructive interference) when this path difference equals some integral multiple of the wavelength λ. The same is true for reflection from the entire family of parallel planes. Hence, the condition for constructive interference (maxima in the diffracted wave) is given by

$$2d \sin\theta = m\lambda \qquad (m = 1,2,3, \ldots) \qquad (39.10)$$

Bragg's law

This condition is known as *Bragg's law* after W. L. Bragg (1890–1971), who first derived the relationship. If the wavelength and diffraction angle are measured, Eq. 39.10 can be used to calculate the spacing between atomic planes.

39.6 POLARIZATION OF LIGHT WAVES

The wave nature of light has been used to explain the phenomena of interference and diffraction. In Chapter 35 we described the transverse nature of light waves and, in fact, of all electromagnetic waves. Figure 39.21 shows that the electric and magnetic vectors associated with an electromagnetic wave are at right angles to each other and also to the direction of wave propagation. The phenomenon of polarization, which will be described in this section, is firm evidence of the transverse nature of electromagnetic waves.

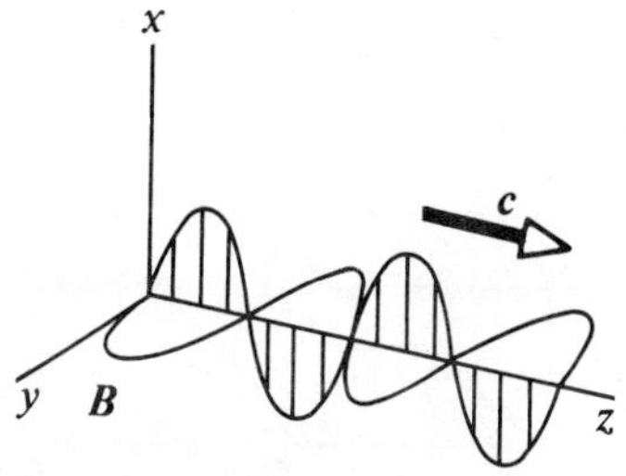

Figure 39.21 Schematic diagram of an electromagnetic wave propagating in the z direction. The electric field vector $\boldsymbol{E}$ vibrates in the xz plane, and the magnetic field vector $\boldsymbol{B}$ vibrates in the yz plane.

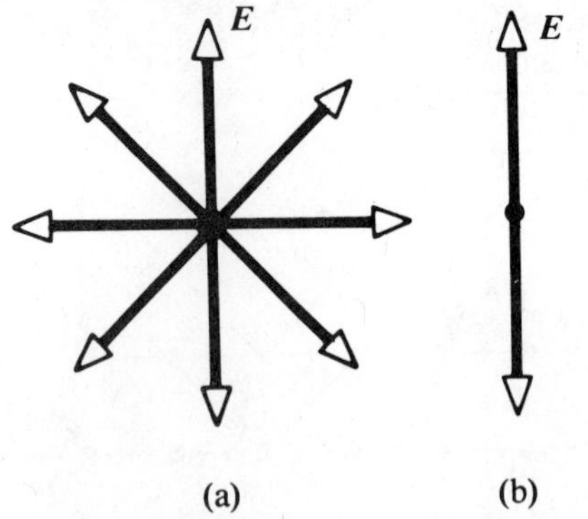

Figure 39.22 (a) An *unpolarized* light beam viewed along the direction of propagation (perpendicular to the page). The transverse electric field vector can vibrate in any direction with equal probability. (b) A linearly polarized light beam with **E** vibrating in the vertical direction.

An ordinary beam of light consists of a large number of waves emitted by the atoms or molecules of the light source. Each atom produces a wave with its own orientation of **E**, as in Fig. 39.21, corresponding to the direction of atomic vibration. The direction of polarization of the electromagnetic wave is defined to be the direction in which **E** is vibrating. However, since all directions of vibration are possible, the resultant electromagnetic wave is a superposition of waves produced by the individual atomic sources. The result is an *unnpolarized* light wave, described in Fig. 39.22a. The direction of wave propagation in this figure is perpendicular to the page. Note that *all* directions of the electric field vector are equally probable all lying in a plane perpendicular to the direction of propagation. At any given point and at some instant of time, there is only one resultant electric field, hence you should not be misled by the meaning of Fig. 39.22a. A wave is said to be *linearly polarized* if only one of these directions of vibration of **E** exists at a particular point, as in Fig. 39.22b. (Sometimes such a wave is described as *plane-polarized*, or simply *polarized*.)

Linearly polarized light

Suppose a light beam traveling in the z direction has an electric field vector that is at an angle θ with the x axis at some instant, as in Fig. 39.23. The vector has components E_x and E_y as shown. Obviously, the light is linearly polarized if one of these components is always zero or if the angle θ remains constant in time. However, if the tip of the vector **E** rotates in a circle with time, the wave is said to be *circularly polarized*. This occurs when the magnitudes of E_x and E_y are *equal*, but differ in phase by 90°. On the other hand, if the magnitudes of E_x and E_y are *not* equal, but differ in phase by 90°, the tip of **E** moves in an ellipse. Such a wave is said to be *elliptically polarized*. Finally, if E_x and E_y are, on the average, equal in magnitude, but have a randomly varying phase difference the light beam is unpolarized.

Circularly polarized light

Elliptically polarized light

It is possible to obtain a linearly polarized beam from an unpolarized beam by removing all waves from the beam except those whose electric field vectors oscillate in a single plane. We shall now discuss four different physical processes for producing polarized light from unpolarized light. These are (1) selective absorption, (2) reflection, (3) double refraction, and (4) scattering.

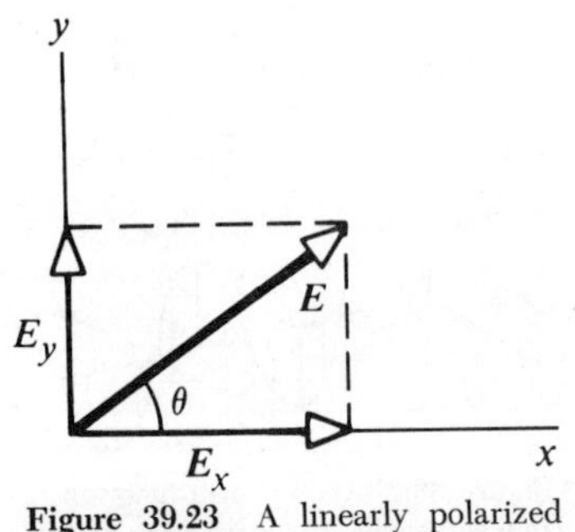

Figure 39.23 A linearly polarized wave with **E** at an angle θ to x has components $E_x = E\cos\theta$ and $E_y = E\sin\theta$.

Polarization by Selective Absorption

The most common technique for obtaining polarized light is to use a material that will transmit waves whose electric field vectors are parallel to a certain direction and will absorb most other directions of polarization. Any substance that has the property of transmitting light with the electric field vector vibrating in only one direction is called a *dichroic substance*. In 1938, E. H. Land discovered a material, which he called *Polaroid*, that polarizes light through selective absorption by oriented molecules. Long-chain hydrocarbon molecules (such as polyvinyl alcohol) in

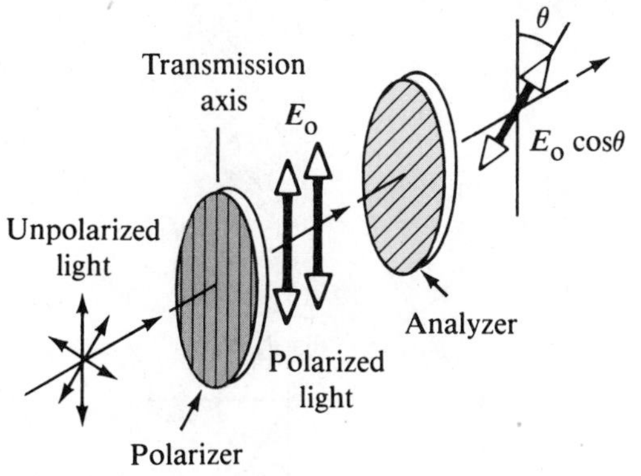

Figure 39.24 Two polarizing sheets whose transmission axes make an angle θ with each other. Only a fraction of the polarized light incident on the analyzer is transmitted.

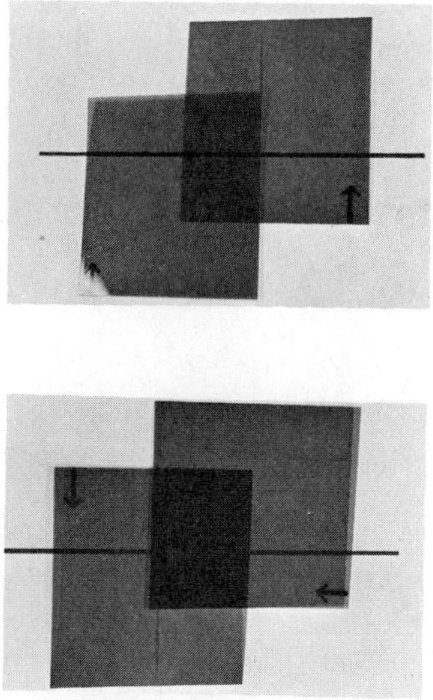

Figure 39.25 Two crossed polarizing sheets. (Courtesy of Jim Lehman, James Madison University)

thin-sheet form are aligned in one direction when the sheet is stretched during fabrication.[2] After the sheet is dipped into a solution containing iodine, the molecules become conducting. However, the conduction takes place primarily along the hydrocarbon chains since the valence electrons of the molecules can move easily only along the chains. As a result, the molecules readily absorb light whose electric field vector is parallel to their length and transmit light whose electric field vector is perpendicular to their length. It is common to refer to the direction perpendicular to the molecular chains as the *transmission axis*. In an ideal polarizer, all light with **E** parallel to the transmission axis is transmitted, and all light with **E** perpendicular to the transmission axis is absorbed.

Figure 39.24 represents an unpolarized light beam incident on the first polarizing sheet, called the *polarizer*, where the transmission axis is indicated by the straight lines on the polarizer. The light that is passing through this sheet is polarized vertically as shown, where the transmitted electric field vector is $\mathbf{E}_0$. A second polarizing sheet, called the *analyzer*, intercepts this beam with its transmission axis at an angle θ to the axis of the polarizer. The component of $\mathbf{E}_0$ perpendicular to the axis of the analyzer is completely absorbed, and the component of $\mathbf{E}_0$ parallel to the axis of the analyzer is transmitted. Since the component of $\mathbf{E}_0$ parallel to the axis of the analyzer is $E_0 \cos\theta$ and since the transmitted intensity varies as the *square* of the transmitted amplitude, we conclude that the transmitted intensity varies as

Light intensity transmitted by two polarizing sheets

$$I = I_0 \cos^2\theta \tag{39.11}$$

where I_0 is the intensity of the polarized wave incident on the analyzer. This expression, known as *Malus's law*,[3] applies to any two polarizing materials whose transmission axes are at an angle θ to each other. From this expression, note that the transmitted intensity is a maximum when the transmission axes are parallel ($\theta = 0$ or $180°$). In addition, the transmitted intensity is zero (complete absorption by the analyzer) when the transmission axes are perpendicular to each other. This variation in transmitted intensity through a pair of polarizing sheets is illustrated in Fig. 39.25.

Polarization by Reflection

Polarized light may also be obtained by the process of reflection. When an unpolarized light beam is reflected from a surface, the reflected light is completely polarized, partially polarized, or unpolarized, depending on the angle of incidence.

[2] An earlier version of a Polaroid film developed by Land consisted of oriented dichroic crystals of quinine sulfate periodide imbedded in a plastic film.

[3] Named after its discoverer, E. L. Malus (1775–1812). Actually, Malus first discovered that reflected light was polarized by viewing it through a calcite ($CaCO_3$) crystal.

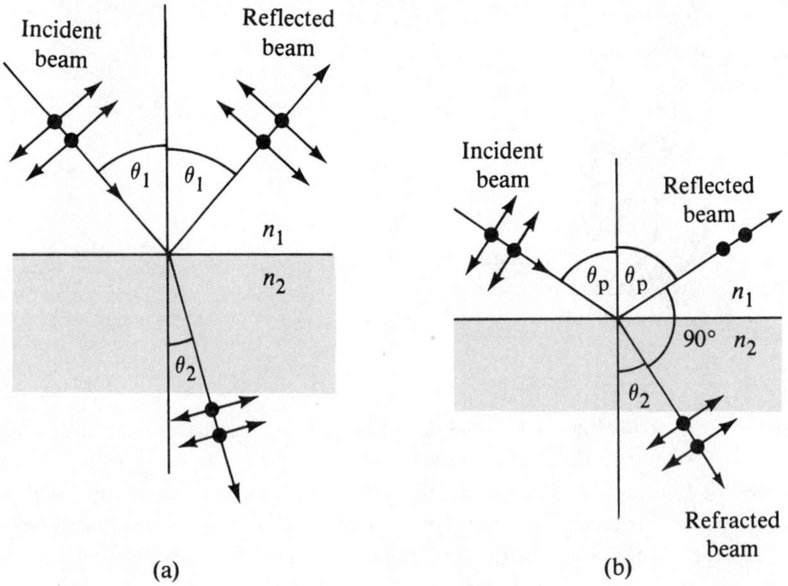

Figure 39.26 (a) When unpolarized light is incident on a reflecting surface, the reflected and refracted beams are partially polarized. (b) The reflected beam is completely polarized when the angle of incidence equals the polarizing angle θ_p, which satisfies the equation $n = \tan\theta_p$.

If the angle of incidence is either 0 or 90° (normal or grazing angles), the reflected beam is unpolarized. However, for intermediate angles of incidence, the reflected light is polarized to some extent.

Suppose an unpolarized light beam is incident on a surface as in Fig. 39.26a. The beam can be described by two electric field components, one parallel to the surface (the dots) and the other perpendicular to the first and to the direction of propagation (the arrows). It is found that the parallel component reflects more strongly than the other component, and this results in a partially polarized beam (Fig. 39.26a). Furthermore, the refracted ray is also partially polarized.

Now suppose the angle of incidence, θ_1, is varied until the angle between the reflected and refracted beams is 90° (Fig. 39.26b). At this particular angle of incidence, the reflected beam is completely polarized with its electric field vector parallel to the surface, while the refracted beam is partially polarized. The angle of incidence at which this occurs is called the *polarizing angle*, θ_p.

The polarizing angle

An expression can be obtained relating the polarizing angle to the index of refraction, n, of the reflecting substance. From Fig. 39.26b, we see that at the polarizing angle, $\theta_p + 90° + \theta_2 = 180°$, so that $\theta_2 = 90° - \theta_p$. Using Snell's law, we have

$$n = \frac{\sin\theta_1}{\sin\theta_2} = \frac{\sin\theta_p}{\sin\theta_2}$$

Since $\sin\theta_2 = \sin(90° - \theta_p) = \cos\theta_p$, the expression for n can be written

Brewster's law

$$n = \frac{\sin\theta_p}{\cos\theta_p} = \tan\theta_p \qquad (39.12)$$

This expression is called *Brewster's law*, and the polarizing angle θ_p is sometimes called *Brewster's angle*, after its discoverer, Sir David Brewster (1781–1868). For example, the Brewster's angle for crown glass ($n = 1.52$) is $\theta_p = \tan^{-1}(1.52) = 56.7°$. Since n varies with wavelength for a given substance, the Brewster's angle is also a function of the wavelength.

Polarization by reflection is a common phenomenon. Sunlight reflected from water, glass, snow, and metallic surfaces is partially polarized. If the surface is horizontal, the electric field vector of the reflected light will have a strong horizontal component. Sunglasses made of polarizing material reduce the glare of reflected light. The transmission axes of the lenses are oriented vertically so as to absorb the strong horizontal component of the reflected light.

Polarization by Double Refraction

When light travels through an isotropic medium, such as glass, it travels with a speed that is the same in all directions. Such isotropic materials are characterized by a single index of refraction. However, in certain crystals, such as calcite and quartz, the speed of light is not the same in all directions. The fundamental reason for this phenomenon is associated with the complex arrangement of the crystalline structures. Such optically anisotropic materials are characterized by *two* indices of refraction. Hence, they are often referred to as *double-refracting*, or *birefringent*, materials.[4]

Calcite is a double-refracting crystal

When an unpolarized beam of light enters a calcite crystal, it splits into two plane-polarized rays which travel with different velocities, corresponding to two different angles of refraction, as in Fig. 39.27. The two rays are polarized in two mutually perpendicular directions, as indicated by the dots and arrows. One ray, called the *ordinary* (O) ray, is characterized by an index of refraction n_O that is the *same* in all directions; hence the ordinary ray has a spherical wavefront. The second ray, called the *extraordinary* (E) ray, travels with different speeds in different directions and hence is characterized by an index of refraction n_E that varies with the direction of propagation.

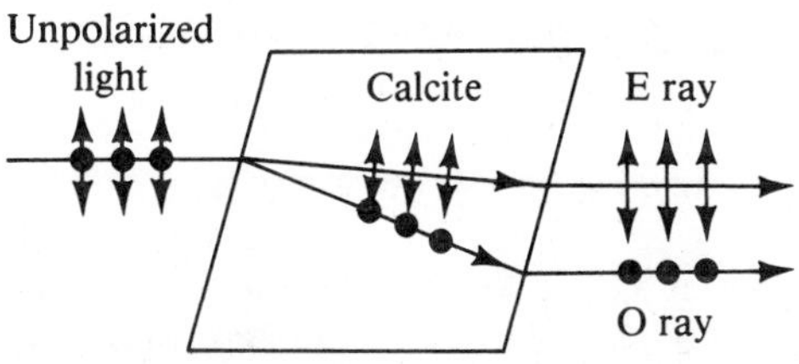

Figure 39.27 When an unpolarized light beam is incident on a calcite crystal, it splits into an ordinary (O) ray and an extraordinary (E) ray. The rays are polarized in mutually perpendicular directions.

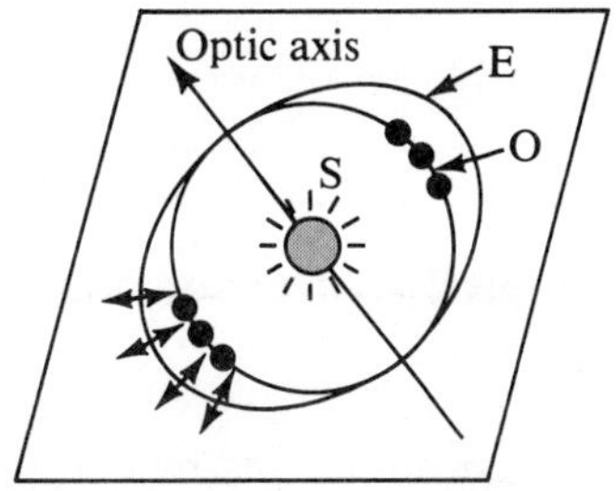

Figure 39.28 A point source S inside a doubly refracting crystal produces a spherical wavefront corresponding to the O ray and an elliptical wavefront corresponding to the E ray. The two waves propagate with the same velocity along the optic axis.

The wavefronts for the extraordinary ray are ellipsoids of revolution. Figure 39.28 illustrates the wavefronts associated with the ordinary and extraordinary rays, assuming a point source within the material. Note that there is one direction, called the *optic axis*, along which the O and E rays have the *same velocity*, corresponding to the direction for which $n_O = n_E$. The difference in velocity for the two rays is a maximum in the direction perpendicular to the optic axis. For example, in calcite $n_O = 1.658$ at a wavelength of 589.3 nm, while n_E varies from 1.658 along the optic axis to 1.486 perpendicular to the optic axis. Values for n_O and n_E for various double-refracting crystals are given in Table 39.1.

[4]For a lucid treatment of this topic, see Elizabeth A. Wood, *Crystals and Light*, New York, Van Nostrand (Momentum), 1964, Chapter 12.

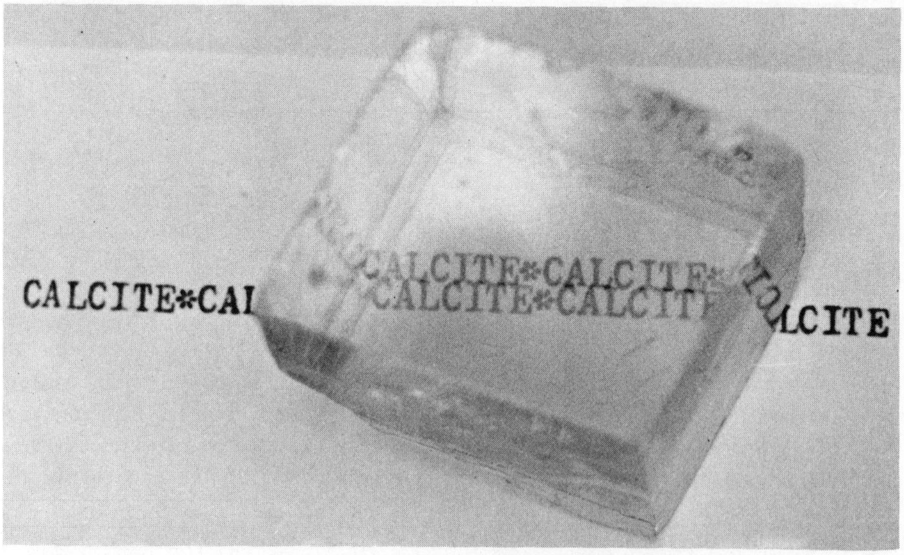

A calcite crystal produces a double image because it is a birefringent (double-refracting) material. (Courtesy of Jim Lehman and Hugh Strickland, James Madison University)

TABLE 39.1 Indices of Refraction for Some Double-Refracting Crystals at a Wavelength of 589.3 nm

Crystal	n_O	n_E	n_O/n_E
Calcite ($CaCO_3$)	1.658	1.486	1.116
Quartz (SiO_2)	1.544	1.553	0.994
Sodium nitrate ($NaNO_3$)	1.587	1.336	1.188
Sodium sulfite ($NaSO_3$)	1.565	1.515	1.033
Zinc chloride ($ZnCl_2$)	1.687	1.713	0.985
Zinc sulfide (Zns)	2.356	2.378	0.991

Polarization by Scattering

When light is incident on a system of particles, such as a gas, the electrons in the medium can absorb and reradiate a part of the light. The process of absorption and reradiation of light by the medium is called *scattering*. It is possible to understand why the scattered light is partially polarized from the following considerations. The incident light wave interacts with the electrons of the medium, causing them to oscillate in response to the oscillating electric vector. Each oscillating electron can be considered a small dipole antenna that radiates light in all directions *except* along its axis of vibration (see Chapter 35). Hence, the scattered light is partially polarized. The resultant wave intensity in the medium equals the sum of the intensities of the incident and scattered waves. A detailed analysis shows that the intensity and degree of polarization of the scattered light depend on the wavelength of the incident light as well as on the location of the observer.

Some phenomena involving the scattering of light in the atmosphere can be understood as follows: When light of various wavelengths λ is incident on air molecules of size d, where $d \ll \lambda$, the relative intensity of the scattered light varies as $1/\lambda^4$. The condition $d \ll \lambda$ is satisfied for scattering from O_2 and N_2 molecules in the atmosphere, whose diameters are about 2 Å. Since light from the sun contains a broad range of wavelengths, shorter wavelengths (blue light) are scattered more efficiently than longer wavelengths (red light). Consequently, the sky appears to be blue. (This also explains why the sky is black in outer space, where there is no atmosphere to scatter the sunlight!)

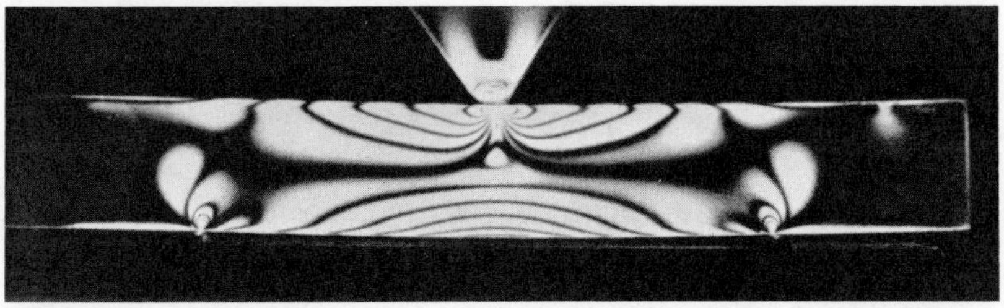

The distribution in strain in a beam shown in a plastic model under stress. The test piece is viewed between two crossed polarizers. The strain pattern disappears when the stress is released. (From M. Cagnet, M. Francon, and J. C. Thierr, Atlas of Physical Phenomena. Plate 40, Berlin, Springer-Verlag, 1962)

The sun appears to be redder at sunset and sunrise for the same reason. The sun's rays must travel a greater distance through the atmosphere at sunrise and sunset, at which times they come in "slantwise." Most of the blue light from the sun is scattered by a large layer of air; hence, the transmitted light reaching the observer is rich in red light.

On a cloudy day, light is also scattered by water droplets whose diameters are much larger than the wavelengths of visible light. Consequently, all wavelengths are scattered with the same efficiency and the clouds appear to be white or gray.

Q8. If a piece of clear cellophane tape is placed between two crossed polarizers, the taped area will transmit more light (and appear lighter) than the nontaped area of the crossed polarizer. Explain.

Q9. Certain sunglasses use a polarizing material to reduce the intensity of light reflected from shiny surfaces, such as water or the hood of a car. What orientation of polarization should the material have in order to be most effective?

Q10. Why is the sky black when viewed from the moon?

39.7 SUMMARY

The phenomenon of *diffraction* arises from the interference of a large number or continuous distribution of coherent sources. Diffraction accounts for the deviation of light from a straight-line path when the light passes through an aperture or around obstacles.

The *Fraunhofer diffraction pattern* produced by a *single slit* of width a on a distant screen consists of a central, bright maximum and alternating bright and dark regions of much lower intensities. The angles θ at which the diffraction pattern has *zero* intensity are given by

$$\sin\theta = m\frac{\lambda}{a} \qquad (m = \pm 1, \pm 2, \pm 3, \ldots) \qquad (39.1)$$

Condition for intensity minima in the single slit differentiation pattern

where $|m| \leq a/\lambda$.

The variation of intensity I with angle θ is given by

$$I = I_0 \left| \frac{\sin(\phi/2)}{\phi/2} \right|^2 \qquad (39.4)$$

Intensity of a single-slit Fraunhofer diffraction pattern

where $\phi = 2\pi a \sin\theta/\lambda$ and I_0 is the intensity at $\theta = 0$.

Rayleigh's criterion, which is a limiting condition of resolution, says that two images formed by an aperture are just distinguishable if the central maximum of the diffraction pattern for one image falls on the first minimum of the other image. The limiting angle of resolution for a slit of width a is given by $\theta_m = \lambda/a$, and the limiting angle of resolution for a circular aperture of diameter D is given by $\theta_m = 1.22\lambda/D$.

A *diffraction grating* consists of a large number of equally spaced, identical slits. The condition for intensity maxima in the interference pattern of a diffraction grating is given by

Condition for intensity maxima for a grating

$$d \sin\theta = m\lambda \qquad (m = 0,1,2,3, \ldots) \qquad (39.7)$$

where d is the spacing between adjacent slits and m is the *order number* of the diffraction pattern. The resolving power of a diffraction grating in the mth order of the diffraction pattern is given by $R = Nm$, where N is the number of rulings in the grating.

Unpolarized light can be polarized by four processes: (1) selective absorption, (2) reflection, (3) double refraction, and (4) scattering.

When polarized light of intensity I_o is incident on a polarizing film, the light transmitted through the film has an intensity equal to $I_o \cos^2\theta$, where θ is the angle between the transmission axis of the polarizer and the electric field vector of the incident light.

In general, light reflected from a dielectric material, such as glass, is partially polarized. However, the reflected light is completely polarized when the angle of incidence is such that the angle between the reflected and refracted beams is 90°. This angle of incidence, called the *polarizing angle* θ_p, satisfies *Brewster's law*, given by

Brewster's law

$$n = \tan\theta_p \qquad (39.12)$$

where n is the index of refraction of the reflecting medium.

EXERCISES

Section 39.2 Fraunhofer Diffraction of a Single Slit

1. In Fig. 39.7, let the slit width $a = 0.5$ mm and assume incident monochromatic light of wavelength 460 nm. (a) Find the value of θ corresponding to the second dark fringe beyond the central maximum. (b) If the observing screen is located 120 cm in front of the slit, what is the distance y from the center of the central maximum to the second dark fringe?

2. A screen is placed 50 cm from a single slit, which is illuminated with light of wavelength 690 nm. If the distance between the first and third minima in the diffraction pattern is 3.0 mm, what is the width of the slit?

3. The second bright fringe in a single-slit diffraction pattern is located 1.4 mm beyond the center of the central maximum. The screen is 80 cm from a slit opening of 0.8 mm. Assuming monochromatic incident light, calculate the wavelength.

4. Light of wavelength 587.5 nm illuminates a single slit 0.75 mm in width. At what distance from the slit should a screen be located if the first minimum in the diffraction pattern is to be 0.85 mm from the center of the screen?

5. Calculate the width of the central maximum for the single-slit diffraction arrangement described in Exercise 4.

6. A Fraunhofer diffraction pattern is produced on a screen 140 cm from a single slit. The distance from the center of the central maximum to the first secondary maximum is $10^4\lambda$. Calculate the slit width.

7. Monochromatic light is incident on a slit of width 0.35 mm. A diffraction pattern is formed on a screen 2 m away. The second dark fringe subtends an angle of 10 min of arc at the center of the slit. Calculate the wavelength of light being used.

8. In Eq. 39.4, let $\phi/2 \equiv \beta$ and show that $I = 0.5I_0$ when $\sin\beta = \beta/\sqrt{2}$.

9. The equation $\sin\beta = \beta/\sqrt{2}$ found in Exercise 8 is known as a *transcendental equation*. One method of solving such an equation is the graphical method. To illustrate this, let $\beta = \phi/2$, $y_1 = \sin\beta$, and $y_2 = \beta/\sqrt{2}$. Plot y_1 and y_2 on the same set of axes over a range from $\beta = 1$ rad to $\beta = \pi/2$ rad. Determine β from the point of intersection of the two curves.

10. A slit of width 1 mm is illuminated with monochromatic light of wavelength 580 nm. A diffraction pattern is formed on a screen 50 cm away. Calculate the intensity on the screen as a fraction of the maximum intensity (I/I_0) at an angular distance of 1.7 min of arc from the center of the diffraction pattern.

11. A diffraction pattern is formed on a screen 120 cm away from a 0.4-mm-wide slit. Monochromatic light of 546.1 nm is used. Calculate the fractional intensity I/I_0 at a point on the screen 4.1 mm from the center of the principal maximum.

12. In the calculation of Example 39.2, it is assumed that the secondary maxima in the diffraction pattern lie midway between points of zero intensity. Verify that this is a reasonable approximation by (a) calculating I/I_0 for $\phi/2 = 1.4303\pi$ and $\phi/2 = 2.4590\pi$. (These are the first two values of $\phi/2$ that satisfy the equation $\tan(\phi/2) = \phi/2$; see Exercise 9.) (b) Compare the results of (a) to the intensity ratios calculated in Example 39.2.

Section 39.3 Resolution of Single-Slit and Circular Apertures

13. In Example 39.4, the limiting angle of resolution of the eye at a wavelength of 500 nm in air is calculated to be $\theta_{min} = 2.3 \times 10^{-4}$ rad. What is the maximum distance from the eye at which two points separated by 1 cm could be resolved?

14. The *resolving power* of a telescope is expressed as $R = 1/\theta_m$, where θ_m is given by Eq. 39.6. Determine the resolving power of a 25-in.-diameter telescope at a wavelength of 550 nm.

15. What is the minimum distance between two points that will permit them to be resolved at 1 km (a) using a terrestrial telescope with a 6.5-cm-diameter objective (assume $\lambda = 550$ nm) and (b) using the unaided eye (assume a pupil diameter of 2.5 mm)?

16. Calculate the angular separation between two points which are just resolvable by a 100-in.-diameter telescope at an average wavelength of 5500 Å.

17. A radar installation operates at a frequency of 9×10^9 Hz and uses an antenna with a diameter of 15 m. Two objects are 150 m apart. At what distance from the antenna would they be at the limit of resolution?

18. Two motorcycles separated laterally by 2 m are approaching an observer who is holding an infrared "snooper scope" (effective $\lambda = 8850$ Å). What aperture diameter is required if the two headlights are to be resolved at a distance of 10 km?

Section 39.4 The Diffraction Grating

19. Collimated light from a hydrogen discharge tube is incident normally on a diffraction grating. The incident light includes four wavelength components: $\lambda_1 = 4101$ Å, $\lambda_2 = 4340$ Å, $\lambda_3 = 4861$ Å, and $\lambda_4 = 6563$ Å. There are 410 lines/mm in the grating. Calculate the angle between (a) λ_1 and λ_4 in the first-order spectrum and (b) λ_1 and λ_3 in the third-order spectrum.

20. The 5015-Å line in helium is observed at an angle of 30° in the second-order spectrum of a diffraction grating. Calculate the angular deviation of the 6678-Å line in helium in the first-order spectrum for the same grating.

21. The 5461-Å line in mercury is measured at an angle of 81° in the third-order spectrum of a diffraction grating. Calculate the number of lines per mm for the grating.

22. Monochromatic light is incident on a grating that is 75 mm wide and ruled with 50 000 lines. The line is imaged at 32.5° in the second-order spectrum. Determine the wavelength of the incident light.

23. A grating with 250 lines/mm is used with an incandescent light source. Assume the visible spectrum to range in wavelength from 400 to 700 nm. In how many orders can one see (a) the entire visible spectrum and (b) the short-wavelength region?

24. A grating is 4 cm wide, and the entire grating is iluminated with monochromatic light of 577 nm. The second-order maximum is formed at an angle of 41.25°. What is the total number of lines in the grating?

25. The full width of a 3-cm-wide grating is illuminated by a sodium discharge tube. The lines in the grating are uniformly spaced at 775 nm. Calculate the angular separation in the first-order spectrum between the two wavelengths forming the sodium doublet ($\lambda_1 = 589.0$ nm and $\lambda_2 = 589.6$ nm).

26. Determine the minimum wavelength difference that can be resolved at 600 nm by the grating described in Exercise 25. (Assume that the full grating width is illuminated.)

27. The H_α line in hydrogen has a wavelength of 6563 Å. This line differs in wavelength from the corresponding spectral line in deuterium (the heavy stable isotope of hydrogen) by 1.8 Å. Determine the minimum number of lines a grating must have in order to resolve these two wavelengths in the first order.

28. (a) Determine the minimum number of lines in a grating that will allow resolution of the sodium doublet in the second order (see Exercise 25 for wavelength values). (b) Calculate the width of the grating if the second-order spectrum is to be formed at an angle of 15°.

Section 39.5 Diffraction of X-Rays by Crystals

29. The dimension labeled a in Fig. 39.19 is the edge length of the unit cell of NaCl. This is twice the distance between adjacent ions, which is the parameter d in Eq. 39.10. Calculate d for the NaCl crystal from the following data: density $\rho = 2.164\ \text{g/cm}^3$, molecular mass $M = 58.45$ g/g · mole, and Avogadro's number $N_o = 6.025 \times 10^{23}$ atoms/g · mole.

30. X-rays of wavelength 1.4 Å are reflected from a NaCl crystal, and the first-order maximum occurs at an angle of 14.4°. (a) What value does this give for the interplane spacing of NaCl? (b) Compare the value found in (a) with the value calculated in Exercise 29.

31. A monochromatic X-ray beam is incident on a NaCl crystal surface. The second-order maximum in the reflected beam is found when the angle between the incident beam and the surface is 20.5°. Determine the wavelength of the X-rays.

32. Monochromatic X-rays of the K_α line of potassium from a nickel target ($\lambda = 1.66$ Å) are incident on a KCl crystal surface. The interplanar distance in KCl is 3.14 Å. At what angle (relative to the surface) should the beam be directed in order that a second-order maximum be observed?

33. A wavelength of 1.29 Å characterizes K_β X-rays from zinc. When a beam of these X-rays is incident on the surface of a crystal whose structure is similar to that of NaCl, a first-order maximum is observed at an angle of 8.15°. Calculate the interplanar spacing based on this information.

34. Show why the Bragg condition expressed by Eq. 39.10 cannot be satisfied in cases where the wavelength is greater than $2d$ (the length of the unit cell).

35. Potassium iodide (KI) has the same crystalline structure as that of NaCl, with $d = 3.53$ Å. A monochromatic X-ray beam shows a diffraction maximum when the angle of incidence is 7.6°. Calculate the X-ray wavelength. (Assume first order.)

Section 39.6 Polarization of Light Waves

36. A light beam is incident on heavy flint glass ($n = 1.65$) at the polarizing angle. Calculate the angle of refraction for the transmitted ray.

37. Light is reflected from a smooth ice surface, and the reflected ray is completely polarized. Determine the angle of incidence. ($n = 1.309$ for ice.)

38. The angle of incidence of a light beam onto a reflecting surface is continuously variable. The reflected ray is found to be completely polarized when the angle of incidence is 48°. What is the index of refraction of the reflecting material?

39. The critical angle for sapphire surrounded by air is 34.4°. Calculate the polarizing angle for sapphire.

40. Three polarizing disks whose planes are parallel are centered on a common axis. The direction of the transmission axis in each case is shown, in Fig. 39.29, relative to the common vertical direction. A plane-polarized beam of light with E_o parallel to the vertical reference direction is incident from the left on the first disk with intensity $I_o = 10$ units (arbitary). Calculate the transmitted intensity I_f when (a) $\theta_1 = 20°$, $\theta_2 = 40°$, and $\theta_3 = 60°$; (b) $\theta_1 = 0°$, $\theta_2 = 30°$, and $\theta_3 = 60°$.

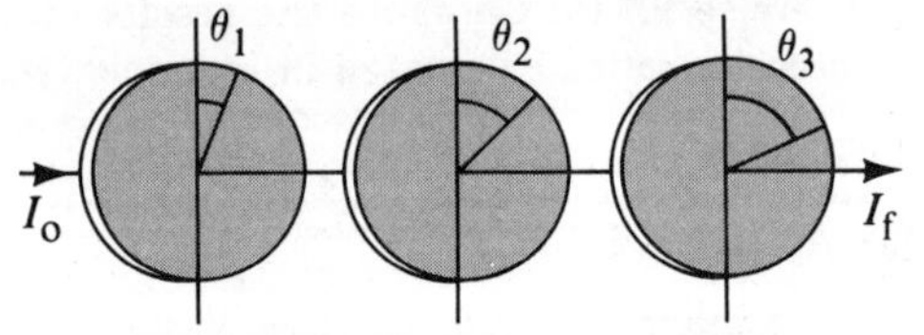

Figure 39.29 (Exercise 40 and Problem 8).

41. Plane-polarized light is incident on a single polarizing disk with the direction of E_o parallel to the direction of the transmission axis. Through what angle should the disk be rotated so that the intensity in the transmitted beam will be reduced by a factor of (a) 3, (b) 5, (c) 10?

42. For a particular transparent medium surrounded by air, show that the critical angle for internal reflection and the polarizing angle are related by $\cot\theta_p = \sin\theta_c$.

PROBLEMS

1. An *unpolarized* beam of light is incident on a stack of polarizing disks. Find the fraction by which the intensity of the transmitted beam is reduced under each of the following conditions. (Note that in each case the angle between the directions of the transmission axes of the first and last disk is 90°.) (a) There are three disks in the stack, and each disk has its transmission axis at an angle of 45° relative to the preceding disk. (b) There are four disks in the stack, and each disk has its transmission axis at an angle of 30° relative to the preceding one. (c) There are seven disks in the stack, and each disk has its axis at an angle of 15° relative to the preceding one. (d) Comment on the different values of I found in (a) (b), and (c).

2. Figure 39.30a is a three-dimensional sketch of the birefringent crystal shown in end view in Fig. 39.27. The dotted lines illustrate how one could cut a thin parallel-faced slab of material from the larger specimen with the optic axis of the crystal parallel to the faces of the plate. A section cut from the crystal in this manner is known as a *retardation plate.* When a beam of light is incident on the plate perpendicular to the direction of the optic axis, as shown in Fig. 39.30b, the O ray and the E ray travel along a

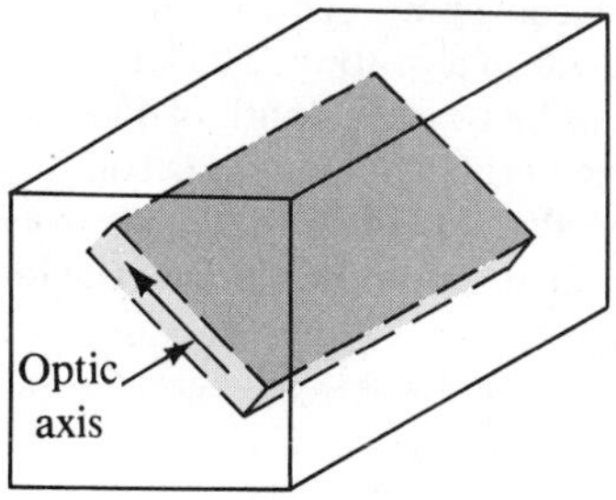

(a)

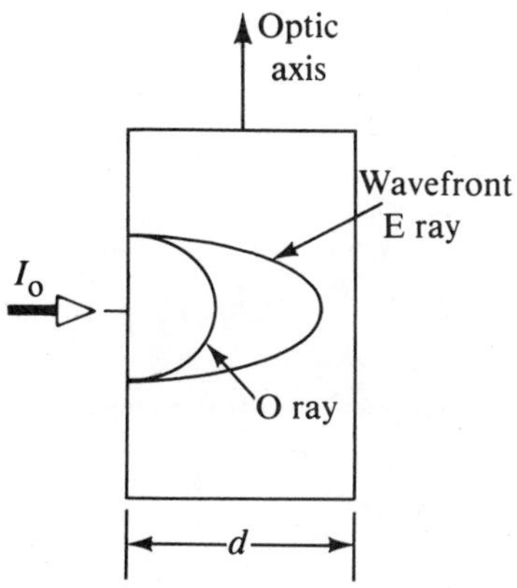

(b)

Figure 39.30 (Problem 2).

single straight line, but with different speeds. (a) Let the thickness of the plate be d and show that the phase difference between the O ray and the E ray in the transmitted beam is given by

$$\theta = \frac{2\pi d}{\lambda_o}|n_O - n_E|$$

where λ_o is the wavelength in air. (Recall that the *optical path length* in a material is the product of the geometric path and the index of refraction.) (b) If in a particular case the incident light has a wavelength of 5500 Å, find the minimum value of d for a quartz plate for which $\theta = \pi/2$. Such a plate is called a *quarter-wave plate.* (Use values of n_O and n_E from Table 39.1.)

3. A diffraction grating of length 4 cm contains 6000 rulings over a width of 2 cm. (a) What is the resolving power of this grating in the first three orders? (b) If two monochromatic waves incident on this grating have a mean wavelength of 400 nm, what is their wavelength separation if they are just resolved in the third order?

4. A single slit of width 0.20 mm is illuminated with light of wavelength 5000 Å. The observing screen is placed 80 cm from the slit. (a) Calculate the width of the central bright fringe and of the secondary maxima of the diffraction pattern. (b) What is the distance between the first and fourth minima?

5. Light of wavelength 500 nm is incident normally on a diffraction grating. If the third-order maximum of the diffraction pattern is observed at an angle of 32°, (a) what is the number of rulings per cm for the grating? (b) Determine the total number of primary maxima that can be observed in this situation.

6. Consider the case of a light beam containing two discrete wavelength components whose difference in wavelength $\Delta\lambda$ is small relative to the mean wavelength λ incident on a diffraction grating. A useful measure of the angular separation of the maxima corresponding to the two wavelengths is the *dispersion D*, given by $D = d\theta/d\lambda$. (The dispersion of a grating should not be confused with its resolving power R, given by Eqs. 39.8 and 39.9.) (a) Starting with Eq. 39.7, show that the dispersion can be written

$$D = \frac{\tan\theta}{\lambda}$$

(b) Calculate the dispersion in the third order for the grating described in Problem 3. Give the answer in units of deg/nm.

7. Derive Eq. 39.9 for the resolving power of a grating, $R = Nm$, where N is the number of lines illuminated and m is the order in the diffraction pattern. Remember that Rayleigh's criterion (Section 39.3) states that two wavelengths will be resolved when the principal maximum for one falls on the first minimum for the other.

8. In Fig. 39.29, suppose that the left and right polarizing disks have their transmission axes perpendicular to each other. Also, let the center disk be rotated on the common axis with an angular velocity ω. Show that if *unpolarized* light is incident on the left disk with an intensity I_0, the intensity of the beam emerging from the right disk will be

$$I = \frac{1}{16} I_0(1 - \cos 4\omega t)$$

This means that the intensity of the emerging beam will be modulated at a rate that is four times the rate of rotation of the center disk. (Hint: Use the trigonometric identities $\cos^2\theta = 1/2(1 + \cos 2\theta)$ and $\sin^2\theta = 1/2(1 - \cos 2\theta)$, and recall that $\theta = \omega t$.)

9. From Eq. 39.4 show that, in the Fraunhofer diffraction pattern of a single slit, the angular width of the central maximum at the point where $I = 0.5I_0$ is $\Delta\theta = 0.884\lambda/a$. (Hint: In Eq. 39.4, let $\phi/2 = \beta$ and solve the resulting transcendental equation graphically.)

10. Sunlight is incident on a diffraction grating which has 2750 lines/cm. The second-order spectrum over the visible range (400–700 nm) is to be limited to 1.75 cm along a screen a distance L from the grating. What is the required value of L?

11. Suppose that the single slit opening in Fig. 39.7 is 6 cm wide and is placed in front of a microwave source operating at a frequency of 7.5 GHz. (a) Calculate the angle subtended by the first minimum in the diffraction pattern. (b) What is the relative intensity I/I_0 at $\theta = 15°$? (c) Consider the case when there are *two* such sources, separated laterally by 20 cm, behind the single slit. What is the maximum distance between the plane of the sources and the slit if the diffraction patterns are to be resolved? (In this case, the approximation that $\sin\theta \approx \tan\theta$ may not be valid because of the relatively small value of the ratio a/λ.)

12. Figure 39.10 shows the relative intensity of a single-slit Fraunhofer diffraction pattern as a function of the parameter $\phi/2 = (\pi a \sin\theta)/\lambda$. Make a plot of the relative intensity I/I_0 as a function of θ, the angle subtended by a point on the screen at the slit, when (a) $\lambda = a$, (b) $\lambda = 0.5a$, (c) $\lambda = 0.1a$, (d) $\lambda = 0.05a$. Let θ range over the interval from 0 to 20° and choose a number of steps appropriate for each case.

13. Light consisting of two wavelength components is incident on a grating. The shorter-wavelength component has a wavelength of 440 nm. The third-order image of this component is coincident with the second-order image of the longer-wavelength component. Determine the value of the longer-wavelength component.

40 Special Theory of Relativity

40.1 INTRODUCTION

Light waves and other forms of electromagnetic radiation travel through free space at the speed $c = 3.00 \times 10^8$ m/s. Numerous experiments suggest that the speed of light is an upper limit for the speeds of particles and mechanical waves.

Most of our everyday experiences and observations deal with objects that move at speeds much less than c. Newtonian mechanics, and the early ideas on space and time, were formulated to describe the motion of such objects. As we saw in the earlier chapters on mechanics, this formalism is very successful for describing a wide range of phenomena. Although newtonian mechanics works very well at low speeds, it fails when applied to particles whose speed approaches c. Experimentally, one can test the predictions of the theory by accelerating an electron through a large electric potential difference. For example, it is possible to accelerate an electron to a speed of $0.99c$ by using a potential difference of several million volts. If the potential difference (and corresponding energy) is increased by a factor of 4, then the speed of the electron should be doubled to $1.98c$, according to newtonian mechanics. However, experiments show that the speed of the electron always remains *less* than c, regardless of the size of the accelerating voltage. Since newtonian mechanics places no upper limit on the speed that a particle can attain, it is contrary to experimental results and is clearly a limited theory.

In 1905, at the age of only 26, Albert Einstein published his *special theory of relativity*. "The relativity theory arose from necessity, from serious and deep contradictions in the old theory from which there seemed no escape. The strength of the new theory lies in the consistency and simplicity with which it solves all these difficulties, using only a few very convincing assumptions. . . ."[1] Although Einstein made many important contributions to science, the theory of relativity alone represents one of the greatest intellectual achievements of the 20th century. Using this theory, one can correctly predict experimental observations over the range of speeds from $v = 0$ to $v \rightarrow c$. Newtonian mechanics, which was accepted for over 200 years, is in fact a specialized case of Einstein's generalized theory. As we shall see, the special theory of relativity is based on two basic postulates, which can be stated as follows:

The postulates of the special theory of relativity

1. The laws of physics must be the same for all observers moving at constant

[1] A. Einstein and L. Infeld, *The Evolution of Physics*, New York, Simon and Schuster, 1961.

velocity with respect to each other. This is equivalent to stating that there is no way to detect absolute, uniform motion. (We presume that there exist privileged frames, known as *inertial frames,* in which *free bodies do not accelerate.*)
2. The speed of light must be the same for *all* inertial observers, *independent of their motion.*

This chapter gives only an introduction to the special theory of relativity, with emphasis on some of the consequences of the theory. These include such phenomena as the slowing down of clocks and the contraction of lengths in moving reference frames as measured by a stationary observer. We shall also discuss the relativistic forms of momentum and energy, terminating the chapter with the famous mass-energy equivalence formula, $E = mc^2$. The interested reader may want to consult a number of excellent books on relativity for more details on the subject.[2]

39.2 THE PRINCIPLE OF RELATIVITY

Initial frame of reference

In order to describe a physical event in analytical form, we must first establish a frame of reference, such as one that is fixed in the laboratory. When Newton's laws of motion were introduced in Chapter 5, we emphasized that these laws are valid in *all* inertial frames of reference. Since an inertial frame of reference is defined as one in which Newton's first law is valid, one can say that *an inertial system is a system in which a free body exhibits no acceleration.* Furthermore, any system moving with constant velocity with respect to an inertial system is also an inertial system. There is no preferred frame. This means that the results of an experiment performed on a vehicle moving with uniform velocity will be the same as those from the same experiment performed in the stationary laboratory. Therefore, *according to the principle of newtonian relativity, the laws of mechanics are the same in all inertial frames of reference.*

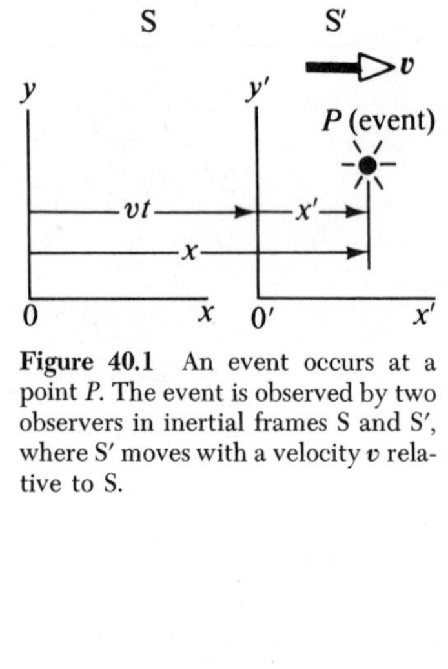

Figure 40.1 An event occurs at a point P. The event is observed by two observers in inertial frames S and S′, where S′ moves with a velocity $\boldsymbol{v}$ relative to S.

Suppose that some physical phenomenon, which we call an *event,* occurs in an inertial system. The event's location and time of occurrence can be specified by the coordinates (x,y,z,t). We would like to be able to transform the space and time coordinates of the event from one inertial system to another moving with uniform relative velocity. This is accomplished by using a so-called *galilean transformation.*

Consider two inertial systems S and S′, as in Fig. 40.1. The system S′ moves with a velocity v along the xx' axes, where v is measured relative to system S. We assume that an event occurs at the point P. The event might be the "explosion" of a flashbulb or a heartbeat. An observer in system S would describe the event with space-time coordinates (x,y,z,t), while an observer in system S′ would use (x',y',z',t') to describe the same event. As we can see from Fig. 40.1, these coordinates are related by the equations

Galilean coordinate transformation

$$\begin{aligned} x' &= x - vt \\ y' &= y \\ z' &= z \\ t' &= t \end{aligned} \tag{40.1}$$

These equations make up what is known as a *galilean transformation of coordinates.* Note that the fourth coordinate, time, is *assumed* to be the same in both inertial

[2]The following books are recommended for more details on the theory of relativity at the introductory level; E. F. Taylor and J. A. Wheeler, *Spacetime Physics,* San Francisco, W. H. Freeman, 1963; R. Resnick, *Introduction to Special Relativity,* New York, Wiley, 1968; A. P. French, *Special Relativity,* New York, Norton, 1968; other suggested readings are selected reprints on "Special Relativity Theory" published by the American Institute of Physics.

systems. That is, within the framework of classical mechanics, clocks are universal, so that the time of an event for an observer in S is the same as the time for the same event in S′. Consequently, the time interval between two successive events should be the same for both observers. Although this assumption may seem obvious, it turns out to be *incorrect* when treating situations in which v is comparable to the speed of light. In fact, this point represents one of the most profound differences between newtonian concepts and the ideas contained in Einstein's theory.

Now suppose two events are separated by a distance Δx in a time Δt as measured by an observer in S. It follows from Eq. 40.1 that the corresponding displacement $\Delta x'$ measured by an observer in S′ is given by $\Delta x' = \Delta x - v\,\Delta t$, where Δx is the displacement measured by an observer in S. Since $\Delta t = \Delta t'$, we find that

$$\frac{\Delta x'}{\Delta t'} = \frac{\Delta x}{\Delta t} - v$$

If the two events correspond to the passage of a moving object (or wave pulse) past two "milestones" in frame S, then $\Delta x/\Delta t$ is the time-averaged velocity in S in the interval and $\Delta x'/\Delta t' = \Delta x'/\Delta t$ is the average velocity in S′. In the limit $\Delta t \rightarrow 0$, this becomes

$$u_x' = u_x - v$$

or

Galilean velocity transformation

$$u_x = u_x' + v \tag{40.2}$$

where u_x' is the velocity of point P relative to S′ and u_x is its velocity relative to S. This is known as the *galilean velocity transformation*. It says that the velocity of an object located at P measured by an observer in the moving frame equals the velocity as measured in the stationary frame minus the velocity of the frame S′. (Since the motion of S′ is along the xx' axes, clearly $u_y' = u_y$ and $u_z' = u_z$. These follow from differentiating Eq. 40.1.)

The galilean transformation equations agree with our everyday experiences. An example is shown in Fig. 40.2. A person in the moving boxcar throws a ball with a speed v in the direction of motion of the boxcar, where v is measured by the person in the boxcar. It follows that the speed of the ball relative to a stationary observer on the ground is $u' + v$. The result is almost obvious. However, Einstein showed that these transformations were incorrect when dealing with objects moving with speeds close to the speed of light. As we shall see later, the galilean transformations are a good approximation only when $v \ll c$.

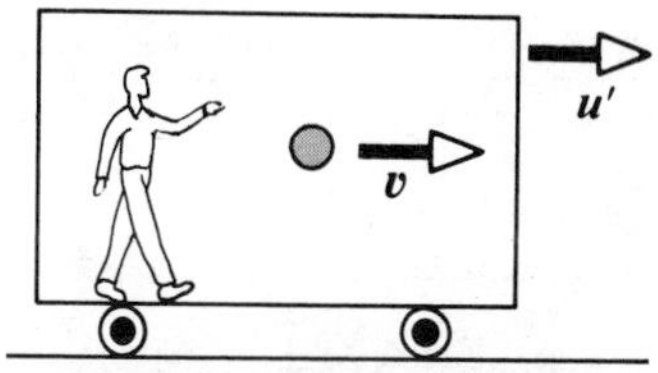

Figure 40.2 A person throws a ball to the right in a moving boxcar. The speed of the ball relative to a stationary observer is $u' + v$, where v is the speed of the ball relative to the person in the boxcar.

40.3 EVIDENCE THAT GALILEAN TRANSFORMATIONS ARE INCORRECT

We have already seen that the galilean transformation is in direct conflict with Einstein's second postulate. What evidence did Einstein have to suspect that the galilean transformation was wrong? The answer lies in the laws of electricity and magnetism, summarized by Maxwell's theory (Chapter 35). Einstein recognized that *Maxwell's equations do not satisfy the galilean transformation.* On the other hand, one might argue that perhaps Maxwell's equations are wrong. However, this is difficult to accept since Maxwell's equations are in total agreement with all known experiments. If we accept that the galilean transformation is incorrect, as suggested by experimental evidence, then newtonian mechanics must be revised, since Newton's laws are known to be consistent with the galilean transformation. (That is, Newton's laws are invariant under a galilean transformation.) In the next section, we shall see how Einstein resolved this conflict with the correct transformation equations.

In Chapter 35 we found that Maxwell's theory predicts the existence of electromagnetic waves which propagate through free space with a speed equal to the speed of light. Maxwell's theory *does not require the presence of a medium* for the wave propagation. This is in contrast to mechanical waves, such as water or sound waves, which do require a medium to support the disturbances. In the 19th century, physicists thought that electromagnetic waves also required a medium through which they could propagate. The medium, called the *ether*, was assumed to be present everywhere, even in free space. Furthermore, the ether had to have the unusual properties of being a massless but rigid medium. Indeed, this is a strange concept. If such a medium existed, then Maxwell's equations should be valid in any frame that was *at rest* relative to the ether. Furthermore, Maxwell's equations would have to be modified in other reference frames moving with respect to the ether frame, called an *absolute frame*. However, all attempts to detect the presence of an ether (the absolute frame) proved futile! The speed of light was always found to be the same in all inertial frames.

The most famous experiment that gave negative results, thus contradicting the ether hypothesis, was performed in 1887 by A. A. Michelson and E. W. Morley (1838–1923).[3] The experiment involved measuring the optical interference pattern produced by two light beams, one traveling along the direction of the earth's orbital motion and the other perpendicular to it. (Details of the interferometer device used in these measurements are given in Section 38.7.) The earth is known to have an orbital speed of $v = 30$ km/s around the sun. Therefore, as the earth moves around the sun, one should be able to detect a change in the measured speed of light. In the Michelson–Morley experiment, the change would be observed as a shift in interference fringes produced by the two light beams. Light traveling in the same direction as the ether should have a velocity $c + v$ with respect to the earth, while light traveling in the direction opposite that of the ether should have a velocity $c - v$. Measurements made over 6-month intervals failed to show any change in the speed of light. The Michelson–Morley experiment was repeated by other researchers under various conditions. Again, the results were negative: no change in the interference pattern could be detected.

The negative results of the Michelson–Morley experiment meant that it was impossible to measure the absolute velocity of the earth with respect to the ether frame. However, it is possible to understand the results within the framework of Einstein's postulates. In other words, we must (1) reject the galilean transformation and look for the correct transformation and (2) recognize that the speed of light is the same for all observers.

Albert Einstein (1879–1955).

40.4 EINSTEIN'S POSTULATES

As mentioned above, Einstein's special theory of relativity is based on two postulates.[4] First, he postulated that the principle of relativity should include *all* the basic laws of physics. In other words, Einstein's principle of relativity states that *the laws of physics are the same in every inertial frame of reference*. In Einstein's own words, ". . . the same laws of electrodynamics and optics will be valid for all frames of reference for which the equations of mechanics hold good." This is, in effect, a

[3] A. A. Michelson and E. W. Morley, *Am. J. Sci.* **134:** 333 (1887).

[4] A. Einstein, "On the Electrodynamics of Moving Bodies, *Ann. Physik* **17:** 891 (1905). For an English translation of this article and other publications by Einstein, see the book by H. Lorentz, A. Einstein, H. Minkowski, and H. Weyl, "The Principle of Relativity," Dover, 1958.

generalization of Newton's principle of relativity, which applies only to the laws of mechanics.

Now consider Einstein's second postulate, which states that *the speed of light has the same value for all observers, independent of the motion of the light source or observer.* Here we are faced with a fundamental problem. We can demonstrate the problem by considering a light pulse sent out by an observer in a boxcar moving with a velocity v (Fig. 40.3). The light pulse has a velocity c relative to the observer in the boxcar. According to the galilean transformation of velocity, the speed of the pulse relative to a stationary ground observer should be $c + v$. This is in obvious contradiction to Einstein's second postulate, which states that the velocity of the light pulse is the same for all observers; that is, both the stationary and moving observers should measure the same velocity for the light pulse, and so $c = c'$. This conclusion seems strange since it contradicts our intuition, or what we often call *common sense.* However, common-sense ideas are based on everyday experiences, which ordinarily do not involve speed-of-light measurements. In order to explain this seemingly strange result, Einstein modified the concepts of space, time, and basic kinematics. In effect, he altered space and time in such a way as to give the result $c = \Delta x'/\Delta t' = \Delta x/\Delta t$ for the speed of light measured by *any* observer located in any inertial frame.

40.5 THE LORENTZ TRANSFORMATION

We have seen that the galilean transformation is not valid when v approaches the speed of light. In this section, we shall derive the correct transformation equations that apply for all speeds in the range of $0 \leq v < c$. This transformation, known as the Lorentz transformation, was developed by H. A. Lorentz (1853–1928) in 1890. However, its real significance in a physical theory was first recognized by Einstein.

We shall derive the Lorentz transformation by considering a rocket moving with a speed v along the xx' axes as in Fig. 40.4a. The frame of the rocket S′ is indicated with the coordinates (x', y', z', t'), while a stationary observer in S uses coordinates

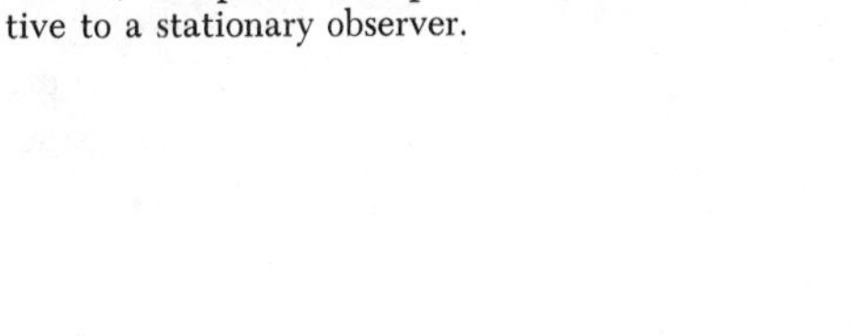

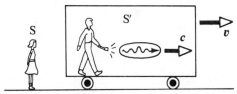

Figure 40.3 A pulse of light is sent out by a person in a moving boxcar. According to the galilean transformation, the speed of the pulse should be $c + v$ relative to a stationary observer.

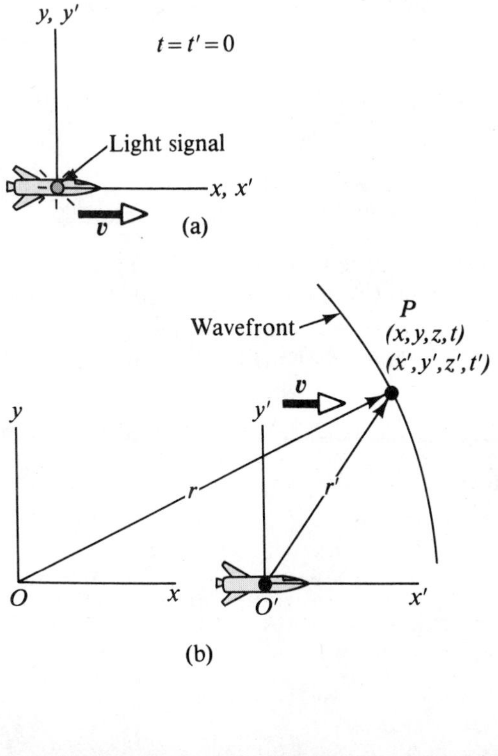

Figure 40.4 A rocket moves with a speed v along the xx' axes. (a) A pulse of light is sent out from the rocket at $t = t' = 0$ when the two systems coincide. (b) Coordinates of some point P on an expanding spherical wavefront as measured by observers in both inertial systems. (This figure is entirely schematic, and you should not be misled by the geometry.)

(x, y, z, t). A flashbulb mounted on the rocket emits a pulse of light at the instant that the origins of the two reference frames coincide.

At the instant the flashbulb goes off and the two origins coincide, we define $t = t' = 0$. The light signal travels as a spherical wave, where the origin of the wavefront is the fixed point O where the flash originated. At some later time, a point such as P on the spherical wavefront is at a distance r from O and a distance r' from O', as shown schematically in Fig. 40.4b. According to Einstein's second postulate, the speed of light should be c for both observers. Hence, the distance to the point P on the wavefront as measured by an observer in S is given by $r = ct$, while the distance to the point P as measured by an observer in S′ is given by $r' = ct'$. That is,

$$r = ct \tag{40.3}$$

$$r' = ct' \tag{40.4}$$

If we accept Einstein's second postulate we must require that the times t and t' taken for the light to reach P be *different*. This is in contrast to the galilean transformation, where $t = t'$.

The radius of a sphere is given by the equation $r^2 = x^2 + y^2 + z^2$ as described by an observer in S. Likewise, the distance r' measured in S′ is given by $(r')^2 = (x')^2 + (y')^2 + (z')^2$. Hence, by squaring Eqs. 40.3 and 40.4, we obtain the following expressions:

$$\text{Observer in S:} \qquad x^2 + y^2 + z^2 = c^2t^2 \tag{40.5}$$

$$\text{Observer in S}': \qquad (x')^2 + (y')^2 + (z')^2 = c^2(t')^2 \tag{40.6}$$

Since the motion of S′ is along the xx' axes, it follows that the y and z coordinates measured in the two frames are always equal. That is, they are unaffected by the motion along x, and therefore $y = y'$ and $z = z'$. Hence, by subtracting Eq. 40.6 from Eq. 40.5, we get

$$x^2 - (x')^2 = c^2t^2 - c^2(t')^2$$

$$x^2 - c^2t^2 = (x')^2 - c^2(t')^2 \tag{40.7}$$

Imposing the condition that $x = vt$ corresponds to $x' = 0$ and using Eq. 40.7, we obtain for x' and t'

$$x' = \frac{x - vt}{\sqrt{1 - v^2/c^2}} \tag{40.8}$$

$$t' = \frac{t - (v/c^2)x}{\sqrt{1 - v^2/c^2}} \tag{40.9}$$

The algebra that gives Eqs. 40.8 and 40.9 is rather tedious and therefore is omitted. However, these equations can be verified by substituting Eqs. 40.8 and 40.9 into the right side of Eq. 40.7.

We now summarize the *Lorentz transformation equations* for transforming from S to S′ as follows:

Lorentz transformation for S → S′

$$x' = \frac{x - vt}{\sqrt{1 - v^2/c^2}} = \gamma(x - vt) \tag{40.10a}$$

$$y' = y \tag{40.10b}$$

$$z' = z \tag{40.10c}$$

$$t' = \frac{t - (v/c^2)x}{\sqrt{1 - v^2/c^2}} = \gamma[t - (v/c^2)x] \tag{40.10d}$$

where the symbol γ (gamma) is defined as

$$\gamma \equiv \frac{1}{\sqrt{1 - v^2/c^2}} \tag{40.11}$$

If we wish to transform coordinates in the S′ frame to coordinates in the S frame, we simply replace v by $-v$ and interchange the primed and unprimed coordinates in Eq. 40.10. The resulting transformation is given by

$$\begin{aligned} x &= \frac{x' + vt'}{\sqrt{1 - v^2/c^2}} = \gamma(x' + vt') \\ y &= y' \\ z &= z' \\ t &= \frac{t' + (v/c^2)x'}{\sqrt{1 - v^2/c^2}} = \gamma[t' + (v/c^2)x'] \end{aligned} \tag{40.12}$$

Inverse Lorentz transformation for S′ → S

Note that in the Lorentz transformation, t depends on t' and x'. Likewise, t' depends on t and x. This is unlike the case of the galilean transformation, in which $t = t'$.

When $v \ll c$, the Lorentz transformation should reduce to the galilean transformation. To check this, note that as $v \to 0$, $v/c^2 \ll 1$ and $v^2/c^2 \ll 1$, so that Eq. 40.10 reduces in this limit to the galilean coordinate transformation equations, given by

$$x' = x - vt \qquad y' = y \qquad z' = z \qquad t' = t$$

Lorentz Velocity Transformation

Let us now derive the Lorentz velocity transformation, which is the relativistic counterpart of the galilean velocity transformation. Suppose that an unaccelerated object is observed in the S′ frame at x_1' at time t_1' and at x_2' at time t_2'. Its speed u_x' measured in S′ is given by

$$u_x' = \frac{x_2' - x_1'}{t_2' - t_1'} = \frac{\Delta x'}{\Delta t'} \tag{40.13}$$

Using Eq. 40.10, we have

$$\Delta x' = \frac{\Delta x - v\,\Delta t}{\sqrt{1 - v^2/c^2}}$$

$$\Delta t' = \frac{\Delta t - (v/c^2)\,\Delta x}{\sqrt{1 - v^2/c^2}}$$

Substituting these into Eq. 40.12 gives

$$u_x' = \frac{\Delta x'}{\Delta t'} = \frac{\Delta x - v\,\Delta t}{\Delta t - (v/c^2)\,\Delta x} = \frac{(\Delta x/\Delta t) - v}{1 - \dfrac{v}{c^2}\dfrac{\Delta x}{\Delta t}}$$

But $\Delta x/\Delta t$ is just the velocity u_x of the object measured in S, and so this expression becomes

$$u_x' = \frac{u_x - v}{1 - u_x v/c^2} \tag{40.14a}$$

Lorentz velocity transformation for S → S′

Similarly, if the object has velocity components along y and z, the components in S′ are given by

$$u'_y = \frac{u_y}{\gamma(1 - u_x v/c^2)} \quad \text{and} \quad u'_z = \frac{u_z}{\gamma(1 - u_x v/c^2)} \tag{40.14b}$$

When u_x and v are both much smaller than c (the nonrelativistic case), we see that the denominator of Eq. 40.14a approaches unity, and so $u'_x \approx u_x - v$. This corresponds to the galilean velocity transformations. In the other extreme, when $u_x = c$, Eq. 40.13 becomes

$$u'_x = \frac{c - v}{1 - cv/c^2} = \frac{c(1 - v/c)}{1 - v/c} = c$$

From this result, we see that an object moving with a speed c relative to an observer in S also has a speed c relative to an observer in S′—*independent* of the relative motion of S and S′. Note that this conclusion is consistent with Einstein's second postulate, namely, that the speed of light must be c with respect to all inertial frames of reference. Furthermore, the speed of an object can never exceed c. That is, the speed of light is the "ultimate" speed. We shall return to this point later when we consider the energy of a particle.

To obtain u_x in terms of u'_x, we replace v by $-v$ in Eq. 40.14a and interchange the roles of u_x and u'_x. This gives

Inverse Lorentz velocity transformation for S′ → S

$$u_x = \frac{u'_x + v}{1 + u'_x v/c^2} \tag{40.15}$$

Example 40.1

Two spaceships A and B are moving in *opposite* directions, as in Fig. 40.5. An observer on the earth measures the speed of A to be $0.75c$ and the speed of B to be $0.85c$. Find the velocity of B with respect to A.

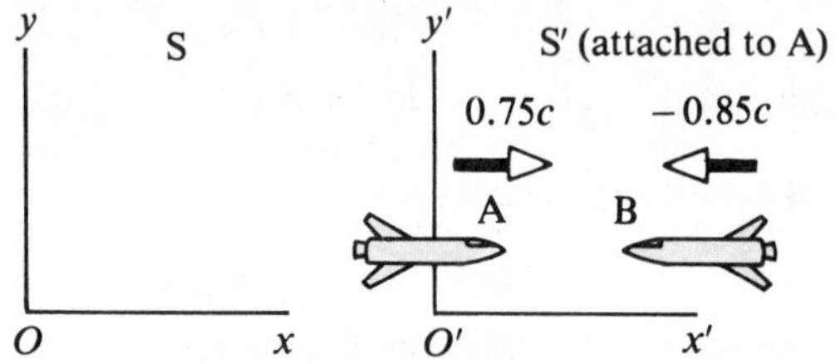

Figure 40.5 (Example 40.1) Two spaceships A and B move in *opposite* directions. The velocity of B relative to A is *less* than c and is obtained by using the relativistic velocity transformation.

Solution: This problem can be solved by taking the S′ frame as being attached to spacecraft A, so that $v = 0.75c$ relative to an observer on the earth (the S frame). Spacecraft B can be considered as an object moving to the left with a velocity $u_x = -0.85c$ relative to the earth observer. Hence, the velocity of B with respect to A can be obtained using Eq. 40.14a:

$$u'_x = \frac{u_x - v}{1 - u_x v/c^2} = \frac{-0.85c - 0.75c}{1 - (-0.85c)(0.75c)/c^2}$$
$$= -0.9771c$$

The negative sign for u'_x indicates that spaceship B is moving in the negative x direction as observed by A. Note that the result is less than c. That is, a body whose speed is less than c in one frame of reference must have a speed less than c in *any other* frame. If the galilean velocity transformation were used in this example, we would find that $u'_x = u_x - v = -0.85c - 0.75c = -1.6c$, which is greater than c.

40.6 CONSEQUENCES OF THE LORENTZ TRANSFORMATION

The Lorentz transformation implies a number of consequences which seem strange at first because they conflict with some of our basic notions. We shall restrict our discussion here to the concepts of length, time, and simultaneity, which are quite different in relativistic mechanics than in newtonian mechanics. For example, we shall see that the distance between two points and the time interval between two events depend on the frame of reference in which they are measured. That is, there is no such thing as absolute length or absolute time in relativity. Furthermore, events at different locations that occur simultaneously in one frame are not simultaneous in another frame.

Time Dilation

Consider a vehicle moving to the right with a speed v as in Fig. 40.6a. A perfectly reflecting mirror is fixed to the ceiling of the vehicle and an observer at O' at rest in this S′ frame holds a flash gun a distance d below the mirror. At some instant, the flash gun goes off and a pulse of light is released. Since the light pulse has a speed c, the time it takes the pulse to travel from the observer to the mirror and back again to the observer is given by

$$\Delta t' = \frac{2d}{c} \tag{40.16}$$

Time measured in S′ frame

Note that this is the time measured by the observer in the moving frame.

Now consider the same set of events as viewed by an observer in the stationary

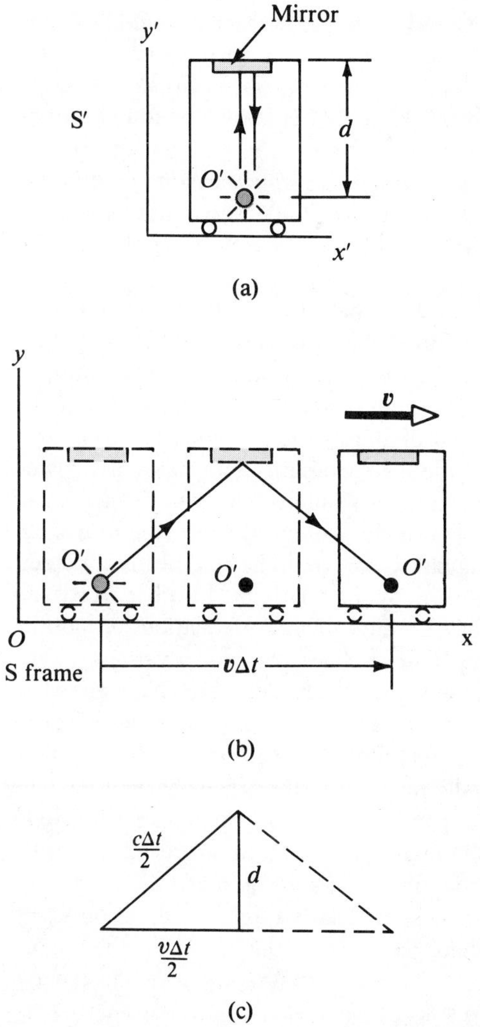

Figure 40.6 (a) A mirror is fixed to a moving boxcar (the S′ frame), and a light pulse leaves O' at rest in the boxcar. (b) Relative to an observer in the S frame, the mirror and O' move with a speed v. Note that the distance the pulse travels is greater than $2d$ as measured in S. (c) This right triangle is useful for calculating the time interval Δt measured in S.

frame S (Fig. 40.6b). According to this observer, the mirror and flash gun are moving to the right with a speed v. Suppose that Δt is time taken for the light pulse to travel from O' to the mirror and back to O' as measured by an observer at O. In this time interval, the flash gun moves a horizontal distance $v\,\Delta t$. Comparing Figs. 40.6a and 40.6b, we see that the light travels farther in the S frame than in the S′ frame. According to Einstein's second postulate, the speed of light must be c as measured by both observers. Therefore, it follows that the time interval Δt measured by an observer in the S frame is *longer* than the time interval $\Delta t'$ measured by an observer in the S′ frame. To obtain a relationship between Δt and $\Delta t'$, it is convenient to use the right triangle shown in Fig. 40.60c. Applying the pythagorean theorem to this triangle gives

$$\left(\frac{c\,\Delta t}{2}\right)^2 = \left(\frac{v\,\Delta t}{2}\right)^2 + d^2$$

Solving for Δt gives

$$\Delta t = \frac{2d}{\sqrt{c^2 - v^2}} = \frac{2d}{c\sqrt{1 - v^2/c^2}} \tag{40.17}$$

Since $\Delta t' = 2d/c$, we can express Eq. 40.17 as

Formula for time dilation

$$\Delta t = \frac{\Delta t'}{\sqrt{1 - v^2/c^2}} = \gamma\,\Delta t' \tag{40.18}$$

This result says that the time interval measured by the observer in S would be *longer* than that measured by the observer in S′ (since $\gamma > 1$). In other words, we must conclude that *according to a stationary observer, a moving clock runs slower by the factor γ than an identical stationary clock.* This effect is known as *time dilation.*

The time interval $\Delta t'$ in Eq. 40.18 is called the *proper time.* In general, proper time is defined as *the time interval between two events as measured by an observer who sees the events occur at the same place.* In our case, the observer at O' measures the proper time. That is, *proper time is always the time measured by an observer moving along with the clock.*

Proper time

In our discussion of time dilation, we have seen that moving clocks slow down by the factor γ. This is true for ordinary mechanical clocks as well as for the light clock just described. In fact, we can generalize these results by stating that *all physical processes, including rates of chemical reactions and biological processes, slow down when in motion.* For example, the heartbeat of an astronaut moving through space would have to keep time with a clock inside the spaceship. Both times are slowed down relative to a stationary clock. The astronaut would not have any sensation of life slowing down in the spaceship. Note that if the biological clock (the heartbeat) did not agree with the time of the moving clock, we would have a means of detecting absolute motion, which violates the principle of relativity. In other words, all clocks depend upon mechanisms that obey Lorentz-invariant equations of motion.

Time dilation is a very real phenomenon that has been verified by various experiments. For example, muons are unstable elementary particles which have one electronic charge and a mass 207 times that of the electron. These unstable particles have a lifetime of only 2.2 μs when measured in their own rest frame. Muons can be produced by the absorption of cosmic radiation high in the atmosphere. If we take 2.2 μs as the average lifetime of a muon, particles moving near the speed of light would travel a distance of only about 600 m and hence could not reach the earth. However, experiments show that a large number of muons do reach the earth. The phenomenon of time dilation explains this effect. Relative to an observer on earth, the muons have a lifetime equal to $\gamma\tau$, where $\tau = 2.2\ \mu s$ is the lifetime in the muon's rest frame. For example, for $v = 0.99c$, $\gamma \approx 7.1$ and $\gamma\tau \approx 16\ \mu s$. Hence, the average

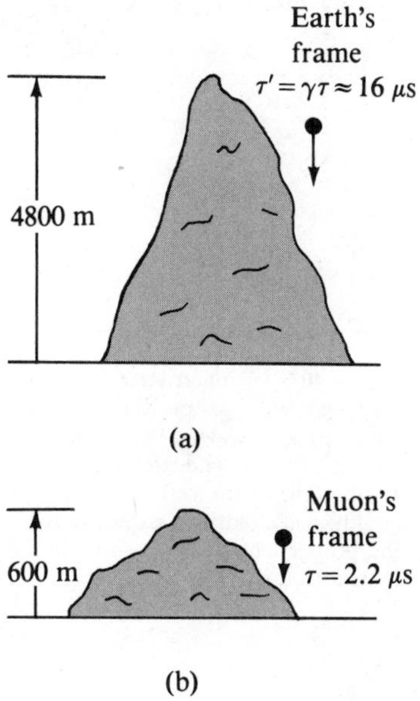

Figure 40.7 (a) Muons traveling with a speed of $0.99c$ travel a distance of about 4800 m as measured by an observer on earth. (b) The muons travel only about 600 m as measured in the muon's reference frame, where its lifetime is about 2.2 μs. Because of time dilation, the muon's lifetime is longer as measured by the observer on earth.

distance traveled as measured by an observer on earth is $v\tau \approx 4800$ m, as indicated in Fig. 40.7a.

The phenomenon of time dilation was also measured by comparing very stable atomic clocks in jet flight with reference clocks on the ground.[5] Time differences of the order of 10^{-9} s were observed, in agreement with the prediction of Einstein's theory.

Example 40.2

The period of a pendulum is measured to be 3 s in the inertial reference frame of the pendulum. What is the period T of the pendulum when measured by an observer moving at a speed of $0.95c$ with respect to the pendulum.

Solution: In this case, the proper time is equal to 3 s. We can use Eq. 40.18 to calculate the period measured by the moving observer. This gives

$$T = \gamma T' = \frac{1}{\sqrt{1 - (0.95c)^2/c^2}} T' = (3.2)(3) = 9.6\ \text{s}$$

Note that the effect of time dilation is reciprocal. That is, an observer in S observes that the clock in S′ is slow, while the observer in S′ observes that the clock that is stationary in S is running slow.

Length Contraction

We have just seen that measured time intervals are not absolute; that is, the time interval between two events depends on the frame of reference in which it is measured. Likewise, the measured distance between two points depends on the frame of reference. *The proper length of an object is defined as the length of the object measured in the reference frame in which the object is at rest.* The length of the object measured in a reference frame in which the object is moving is always less than the proper length. This effect is known as *length contraction.*

Proper length

To understand length contraction quantitatively, consider a meter stick moving to the right with a speed v as in Fig. 40.8. Let us assume that an observer moving along with the stick at the same speed (the S′ frame) measures the length to be $L' =$

[5] J. C. Hafele and R. E. Keating, *Science,* July 14, 1972.

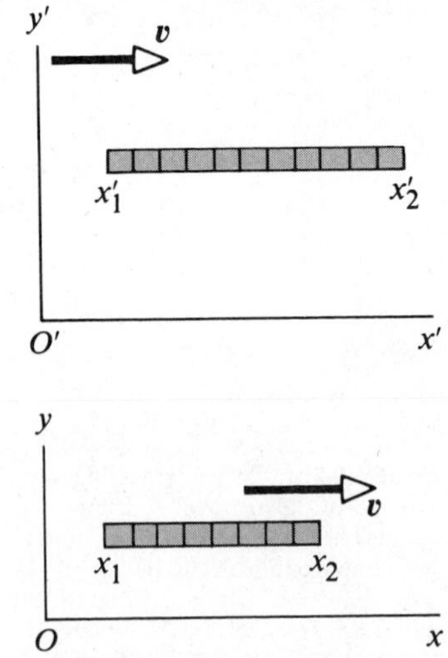

Figure 40.8 A meter stick moves to the right with a speed v. (a) The meter stick as viewed by a frame attached to it. (b) The stick as seen by an observer in the rest frame S. The length measured in S is *shorter* than the proper length L, measured in S′, by the factor $\sqrt{1 - v^2/c^2}$.

$x_2' - x_1'$. From Eq. 40.10 we have

$$x_2' = \gamma(x_2 - vt_2) \qquad \text{and} \qquad x_1' = \gamma(x_1 - vt_1)$$

Subtracting gives

$$(1) \qquad x_2' - x_1' = \gamma(x_2 - x_1) + \gamma v(t_2 - t_1)$$

However, the observer in the stationary frame can make a valid measurement of the rod's length only if he or she measures x_1 and x_2 *at the same instant*. Hence, setting $t_1 = t_2$ in (1) gives

$$x_2' - x_1' = \gamma(x_2 - x_1)$$
$$L' = \gamma L$$

Since L' is the proper length (the length measured in the moving frame), we see that

Length contraction

$$L = \frac{1}{\gamma}L' = \sqrt{1 - v^2/c^2}\,L' \qquad (40.19)$$

According to this result, the length L of the moving stick measured in the S frame is *shorter* than the proper length by the factor $(1 - v^2/c^2)^{1/2}$. You should note that the contraction takes place *only along the direction of motion,* which in this case is x. There are *no* changes in the y and z directions when the object moves in the x direction.

If the observers at O and O' each carried identical sticks oriented along the xx' directions, each would claim that the other's stick was contracted by the same factor, $(1 - v^2/c^2)^{1/2}$. However, a measurement of length is valid only if the end points of the stick are measured at the same instant. Here we arrive at a crucial point. The two observers will not reach the same conclusion with regard to simultaneous events. That is, *events that may be simultaneous for one observer are not simultaneous for another observer who is in motion relative to the first.* We shall return to this important point in the next section.

Example 40.3

A bar 1 m in length and located along the x' axis moves with a speed $0.75c$ with respect to a stationary observer. What is the length of the bar as measured by the stationary observer?

Solution: Using Eq. 40.19 with $L' = 1$ m gives

$$L = \sqrt{1 - v^2/c^2}\,L' = \sqrt{1 - (0.75c)^2/c^2}\,(1\text{ m})$$
$$= 0.66\text{ m}$$

40.7 SIMULTANEITY AND THE RELATIVITY OF TIME

Let us return to the idea of events occurring in a moving frame as viewed from either a stationary frame or the moving frame. Observers in each frame must use clocks to record these events. Either observer can be considered to be at rest while the other is moving. Hence the observer at O in the S frame would claim that the clock in S′ is running slow. Likewise, the observer at O' would claim the clock in S is also running slow. There seems to be an inconsistency in these statements.

In order to understand this fundamental problem, we must consider the manner in which events and time intervals are measured by the two observers. Suppose two events occur at the same place but at different instants in the S′ frame. An observer in the S′ frame would need only *one* clock to measure the time interval $\Delta t'$ since events in this frame occur at the *same* place. On the other hand, the observer in the S frame would require *two* clocks to measure Δt since the events "seen" in the S′ frame occur at *two different* places. As we shall see, it is a fundamental property of time measurements that *two events that are simultaneous in one reference frame are in general not simultaneous in a second frame moving with respect to the first.* That is, *simultaneity is not an absolute concept.*

Simultaneity is not an absolute concept

Einstein devised the following thought experiment to illustrate this point. A boxcar moves with uniform velocity, and two lightning bolts strike the ends of the boxcar, as in Fig. 40.9a, leaving marks on the boxcar and ground at the same instant. The marks left on the boxcar are labeled A' and B', and those on the ground are labeled A and B. An observer at O' moving with the boxcar is midway between A' and B', while a ground observer at O is midway between A and B. The events recorded by the observers are the light signals from the lightning bolts.

If the light signals reach the observer at O at the same time, as indicated in Fig. 40.9b, he or she would rightly conclude that the events at A and B occurred simultaneously. On the other hand, the observer at O', who is moving with the boxcar, observes that the light signal from the front of the boxcar at B' reaches him or her *before* the light signal from the back of the boxcar at A'. Therefore, the observer at O' would conclude that the lightning bolt struck the front of the boxcar *before* it struck the back. This experiment clearly demonstrates that two events

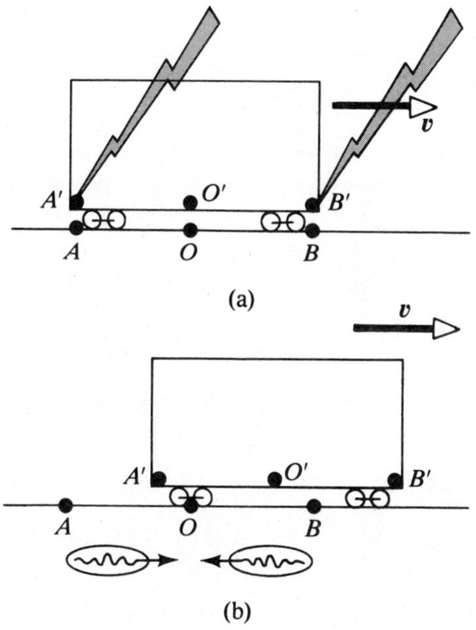

Figure 40.9 Two lightning bolts strike the ends of a moving boxcar. (a) The events appear to be simultaneous relative to the stationary ground observer at O who is midway between the events. (b) The events do not appear to be simultaneous to an observer at O' on the train, who claims that the front of the train is struck *before* the rear of the train.

which appear to be simultaneous to the observer at O *are not* simultaneous as viewed by the observer at O'. In other words, *two events that are simultaneous in one reference frame are not simultaneous in another reference frame that is in motion relative to the first*. This is a special case of the general result that the time interval between two events as measured by one observer is *different* from the time interval measured by another observer moving relative to the first. One can now understand the connection between the relativity of simultaneity and the phenomenon of Lorentz contraction discussed in the previous section.

At this point, the reader might wonder which observer is right concerning the two events. The answer is that *both* observers are correct, since the principle of relativity states that there is no preferred inertial frame of reference. Although the two observers reach different conclusions, both are correct in their own reference frame since *the concept of simultaneity is not absolute*.

Q1. What one measurement will two observers in relative motion *always* agree upon?

Q2. A spaceship in the shape of a sphere moves past an observer on earth with a speed $0.5c$. What shape will the observer see as the spaceship moves past?

Q3. An astronaut moves away from the earth at a speed close to the speed of light. If an observer on earth could make measurements of the astronaut's size and pulse rate, what changes (if any) would he or she measure? Would the astronaut measure any changes?

Q4. Two identically constructed clocks are synchronized. One is put in orbit around the earth while the other remains on earth. Which clock runs slower? When the moving clock returns to earth, will the two clocks still be synchronized?

Q5. Two lasers situated on a moving spacecraft are triggered simultaneously. An observer on the spacecraft claims to see the pulses of light simultaneously. What condition is necessary in order that a stationary observer agree that the two pulses are emitted simultaneously?

Q6. When we say that a moving clock runs slower than a stationary one, does this imply that there is something physically unusual about the moving clock?

Q7. When we speak of time dilation, do we mean that time passes more slowly in moving systems or that it simply appears to do so?

40.8 RELATIVISTIC MOMENTUM

We have seen that the principle of relativity is satisfied if the galilean transformation is replaced by the more general Lorentz transformation. Therefore, in order to properly describe the motion of material particles within the framework of special relativity, we must generalize Newton's laws and the definition of momentum and energy. These generalized definitions of momentum and energy will reduce to the classical (nonrelativistic) definitions for $v \ll c$.

First, recall that the conservation of momentum states that when two bodies collide, the total momentum remains constant, assuming the bodies are isolated (that is, they interact only with each other). Suppose the collision is described in a reference frame S in which the momentum is conserved. If the velocities in a second reference frame S′ are calculated using the Lorentz transformation and the classical definition of momentum, $\mathbf{p} = m\mathbf{u}$, one finds that momentum *is not* conserved in the second reference frame. However, according to the principle of relativity, the momentum must be conserved in all systems. In view of this condition and assuming the Lorentz transformation is correct, we must modify the definition of momentum.

Our definition of relativistic momentum $\boldsymbol{p}$ must satisfy the following conditions:

1. The relativistic momentum must be conserved in all collisions.
2. The relativistic momentum must approach the classical value $m\boldsymbol{u}$ as $\boldsymbol{u} \rightarrow 0$.

The correct relativistic equation for the momentum which satisfies these conditions is given by the expression

Definition of relativistic momentum

$$\boldsymbol{p} \equiv \frac{m\boldsymbol{u}}{\sqrt{1 - u^2/c^2}} = \gamma m\boldsymbol{u} \qquad (40.20)$$

where $\boldsymbol{u}$ is the velocity of the particle. We use the symbol $\boldsymbol{u}$ for particle velocity rather than $\boldsymbol{v}$, which is used for the relative velocity of two reference frames. The proof of this generalized expression for $\boldsymbol{p}$ is beyond the scope of this text. Note that when u is much less than c, the denominator of Eq. 40.20 approaches unity, so that $\boldsymbol{p}$ approaches $m\boldsymbol{u}$. Therefore, the relativistic equation for $\boldsymbol{p}$ reduces to the classical expression when u is small compared with c.

The relativistic force $\boldsymbol{F}$ on a particle whose momentum is $\boldsymbol{p}$ is defined by the expression

$$\boldsymbol{F} \equiv \frac{d\boldsymbol{p}}{dt} \qquad (40.21)$$

where $\boldsymbol{p}$ is given by Eq. 40.20. This expression is identical to the classical statement of Newton's second law, which says that force equals the time rate of change of momentum.

If detailed calculations are carried out, one finds that the acceleration $\boldsymbol{a}$ of a particle decreases under the action of a constant force, in which case $a \simeq (1 - u^2/c^2)^{3/2}$. Furthermore, as the velocity approaches c, the acceleration approaches zero. Hence, it is impossible to accelerate a particle from rest to a speed equal to or greater than c.

Example 40.4

An electron, which has a mass of 9.11×10^{-31} kg, moves with a speed of $0.75c$. Find its relativistic momentum and compare this with the momentum calculated from the classical expression.

Solution: Using Eq. 40.20 with $u = 0.75c$, we have

$$p = \frac{mu}{\sqrt{1 - u^2/c^2}}$$

$$= \frac{(9.11 \times 10^{-31}\ \text{kg})(0.75 \times 3 \times 10^8\ \text{m/s})}{\sqrt{1 - (0.75c)^2/c^2}}$$

$$= 3.10 \times 10^{-22}\ \text{kg} \cdot \text{m/s}$$

The incorrect classical expression would give

$$\text{Momentum} = mu = 2.05 \times 10^{-22}\ \text{kg} \cdot \text{m/s}$$

Hence, the correct relativistic result is 50% greater than the classical result!

40.9 RELATIVISTIC ENERGY

We have seen that the definition of momentum and the laws of motion required generalization to make them compatible with the principle of relativity. This implies that the relation between work and energy must also be modified.

In order to derive the relativistic form of the work-energy relation, let us start with the definition of work done by a force F and make use of the definition of relativistic force, Eq. 40.21. That is,

$$W = \int_{x_1}^{x_2} F\, dx = \int_{x_1}^{x_2} \frac{dp}{dt}\, dx \qquad (40.22)$$

where we have assumed that the force and motion are along the x axis. In order to perform this integration, we make repeated use of the chain rule for derivatives in the following manner:

$$\left(\frac{dp}{dt}\right)dx = \left(\frac{dp}{du}\frac{du}{dt}\right)dx = \frac{dp}{du}\left(\frac{du}{dx}\frac{dx}{dt}\right)dx$$
$$= \frac{dp}{du}u\frac{du}{dx}dx = \frac{dp}{du}u\,du$$

Since p depends on u according to Eq. 40.20, we have

$$\frac{dp}{du} = \frac{d}{du}\frac{mu}{\sqrt{1-u^2/c^2}} = \frac{m}{(1-u^2/c^2)^{3/2}} \tag{40.23}$$

Using these results, we can express the work as

$$W = \int_0^u \frac{dp}{du}u\,du = \int_0^u \frac{mu}{(1-u^2/c^2)^{3/2}}du \tag{40.24}$$

where we have assumed that the particle is accelerated from rest to some final velocity u. Evaluating the integral, we find that

$$W = \frac{mc^2}{\sqrt{1-u^2/c^2}} - mc^2 \tag{40.25}$$

Recall that the work-energy theorem states the work done by a force acting on a particle equals the change in kinetic energy of the particle. Since the initial kinetic energy is zero, we conclude that the work W is equivalent to the relativistic kinetic energy K, that is,

Relativistic kinetic energy

$$K = \frac{mc^2}{\sqrt{1-u^2/c^2}} - mc^2 \tag{40.26}$$

This equation has been confirmed by experiments using high-energy particle accelerators. At low speeds, where $u/c \ll 1$, Eq. 40.26 should reduce to the classical expression $K = \frac{1}{2}mu^2$. We can check this by using the binomial expansion $(1-x^2)^{-1/2} \approx 1 + \frac{1}{2}x + \ldots$ for $x \ll 1$, where the higher-order powers of x are neglected in the expansion. In our case, $x = u/c$, so that

$$\frac{1}{\sqrt{1-u^2/c^2}} = \left(1 - \frac{u^2}{c^2}\right)^{-1/2} \approx 1 + \frac{1}{2}\frac{u^2}{c^2} + \cdots$$

Substituting this into Eq. 40.26 gives

$$K \approx mc^2\left(1 + \frac{1}{2}\frac{u^2}{c^2} + \cdots\right) - mc^2 = \frac{1}{2}mu^2$$

which agrees with the classical result. A graph comparing the relativistic and nonrelativistic expressions is given in Fig. 40.10. Note that in the relativistic case, the particle speed never exceeds c, regardless of the kinetic energy. The two curves are in good agreement when $v \ll c$.

It is useful to write the relativistic kinetic energy in the form

$$K = \gamma mc^2 - mc^2 \tag{40.27}$$

where

$$\gamma = \frac{1}{\sqrt{1-u^2/c^2}}$$

The constant term mc^2, which is independent of the speed, is called the *rest energy*

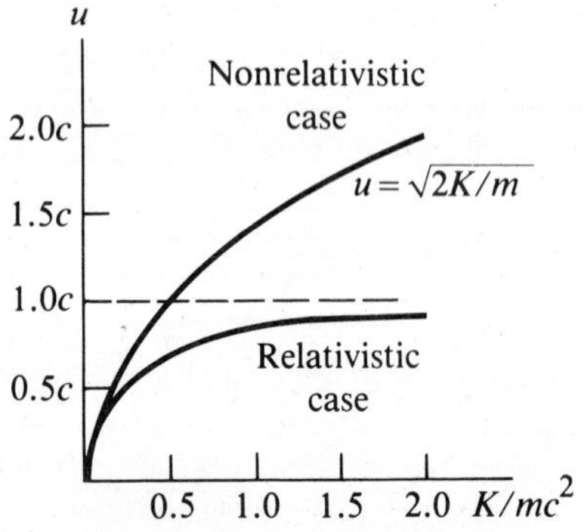

Figure 40.10 A graph comparing relativistic and nonrelativistic kinetic energy. The energies are plotted versus speed. In the relativistic case, v is always *less* than c.

of the particle. The term γmc^2, which depends on the particle speed, is therefore the sum of the kinetic and rest energies. We define γmc^2 to be the *total energy E*, that is,

Definition of total energy

$$E = \gamma mc^2 = K + mc^2 \tag{40.28}$$

Energy-mass equivalence

$$E = \frac{mc^2}{\sqrt{1 - u^2/c^2}} \tag{40.29}$$

This, of course, is Einstein's famous mass-energy equivalence equation. The relation $E = \gamma mc^2$ shows that *mass is a form of energy*. Furthermore, this result shows that a small mass corresponds to an enormous amount of energy. This concept has revolutionized the field of nuclear physics.

In many situations, the momentum or energy of a particle is known rather than its speed. It is therefore useful to have an expression relating the total energy E to the relativistic momentum p. This is accomplished by using the expressions $E = \gamma mc^2$ and $p = \gamma mu$. By squaring these equations and subtracting, we can eliminate u (Exercise 12). The result, after some algebra, is

Energy-momentum relation

$$E^2 = p^2c^2 + (mc^2)^2 \tag{40.30}$$

When the particle is at rest, $p = 0$, and so we see that $E = mc^2$. That is, the total energy equals the rest energy. As we shall discuss in the next chapter, it is well established that there are particles that have zero mass, such as photons (quanta of electromagnetic radiation). If we set $m = 0$ in Eq. 40.30, we see that

$$E = pc \tag{40.31}$$

This equation is an *exact* expression relating energy and momentum for photons and neutrinos, which always travel at the speed of light.

Finally, note that since the mass m of a particle is independent of its motion, m must have the same value in all reference frames. On the other hand, the total energy and momentum of a particle depend on the reference frame in which they are measured, since they both depend on velocity. Since m is a constant, then according to Eq. 40.30 the quantity $E^2 - p^2c^2$ must have the same value in all reference frames. That is, $E^2 - p^2c^2$ is invariant under a Lorentz transformation.

When dealing with electrons or other subatomic particles, it is convenient to express the energy in electron volts (eV), since the particles are usually given this energy by acceleration through a potential difference. Recall that

$$1 \text{ eV} = 1.60 \times 10^{-19} \text{ J}$$

For example, the mass of an electron is 9.11×10^{-31} kg. Hence, the rest energy of the electron is

$$mc^2 = (9.11 \times 10^{-31} \text{ kg})(3.00 \times 10^8 \text{ m/s})^2 = 8.20 \times 10^{-14} \text{ J}$$

Converting this to eV, we have

$$mc^2 = (8.20 \times 10^{-14} \text{ J})(1 \text{ eV}/1.60 \times 10^{-19} \text{ J}) = 0.511 \text{ MeV}$$

where 1 MeV $= 10^6$ eV.

Example 40.5

An electron moves with a speed $u = 0.85c$. Find its total energy and kinetic energy in eV.

Solution: Using the fact that the rest energy of the electron is 0.511 MeV together with Eq. 40.29 gives

$$E = \frac{mc^2}{\sqrt{1 - u^2/c^2}} = \frac{0.511 \text{ MeV}}{\sqrt{1 - (0.85c)^2/c^2}}$$
$$= 1.90(0.511 \text{ MeV}) = 0.970 \text{ MeV}$$

The kinetic energy is obtained by subtracting the rest energy from the total energy:

$$K = E - mc^2 = 0.970 \text{ MeV} - 0.511 \text{ MeV} = 0.459 \text{ MeV}$$

Example 40.6

The total energy of a proton is three times its rest energy. (a) Find the proton's rest energy in eV.

$$\text{Rest energy} = mc^2 = (1.67 \times 10^{-27} \text{ kg})(3 \times 10^8 \text{ m/s})^2$$
$$= (1.50 \times 10^{-10} \text{ J})(1 \text{ eV}/1.60 \times 10^{-19} \text{ J})$$
$$= 938 \text{ MeV}$$

(b) With what speed is the proton moving?

Since the total energy E is three times the rest energy, Eq. 40.29 gives

$$E = 3mc^2 = mc^2/\sqrt{1 - u^2/c^2}$$

$$3 = \frac{1}{\sqrt{1 - u^2/c^2}}$$

Solving for u gives

$$(1 - u^2/c^2) = \frac{1}{9} \quad \text{or} \quad u^2/c^2 = \frac{8}{9}$$

$$u = \frac{\sqrt{8}}{3}c = 2.83 \times 10^8 \text{ m/s}$$

(c) Determine the kinetic energy of the proton in eV.

$$K = E - mc^2 = 3mc^2 - mc^2 = 2mc^2$$

Since $mc^2 = 938$ MeV, $K = 1876$ MeV.

(d) What is the proton's momentum?

We can use Eq. 40.30 to calculate the momentum with $E = 3mc^2$:

$$E^2 = p^2c^2 + (mc^2)^2 = (3mc^2)^2$$
$$p^2c^2 = 9(mc^2)^2 - (mc^2)^2 = 8(mc^2)^2$$
$$p = \sqrt{8}\frac{mc^2}{c} = \sqrt{8}\frac{(938 \text{ MeV})}{c} = 2653 \frac{\text{MeV}}{c}$$

Note that the unit of momentum here is written MeV/c for convenience.

Q8. How would our world differ if the speed of light were only 50 m/s?

Q9. Give a physical argument which shows that it is impossible to accelerate an object of mass m to the speed of light, even with a continuous force acting on it.

Q10. Since mass is a form of energy, can we conclude that a compressed spring has more mass than the same spring when it is not compressed?

40.10   CONFIRMATIONS AND CONSEQUENCES OF RELATIVITY THEORY

The special theory of relativity has been confirmed by a number of experiments. One important experiment concerned with the decay of muons, and time dilation in the muon's reference frame, was discussed in Section 40.6. This section describes further evidence of Einstein's special theory of relativity.

One of the first predictions of relativity that was experimentally confirmed is the variation of momentum with velocity. Experiments were performed as early as 1909 on electrons, which can easily be accelerated to speeds close to c through the use of electric fields. When an energetic electron enters a magnetic field B with its velocity vector perpendicular to B, a magnetic force is exerted on the electron, causing it to move in a circle of radius r (Section 26.5). In this situation, the relativistic momentum of the electron is given by $p = eB/r$. From the relativistic equivalent of Newton's second law, $F = dp/dt$, the variation in momentum with kinetic energy can be checked experimentally. The results of such experiments on electrons and other charged particles are in support of relativistic expressions.

The release of enormous quantities of energy in nuclear fission and fusion processes is a manifestation of the equivalence of mass and energy. The conversion of mass into energy is, of course, the basis of atomic and hydrogen bombs, the most powerful and destructive weapons ever constructed. In fact, all reactions that release energy do so at the expense of mass (including chemical reactions). In a conventional nuclear reactor, the uranium nucleus undergoes fission, a reaction that results in several lighter fragments having considerable kinetic energy. In the case of ^{235}U (the parent nucleus), which undergoes spontaneous fission, the fragments are two lighter nuclei and two neutrons. The total mass of the fragments is *less* than that of the parent nucleus by some amount Δm. The corresponding energy Δmc^2 associated with this mass difference is *exactly* equal to the total kinetic energy of the fragments. This kinetic energy is then used to produce heat and steam for the generation of electrical power.

Next, consider the basic fusion reaction in which two deuterium atoms combine to form one helium atom. This reaction is of major importance in current research and development of controlled-fusion reactors. The decrease in mass which results from the creation of one helium atom from two deuterium atoms is calculated to be $\Delta m = 4.25 \times 10^{-29}$ kg. Hence, the corresponding excess energy which results from one fusion reaction is given by $\Delta mc^2 = 3.83 \times 10^{-12}\ \text{J} = 23.9$ MeV. To appreciate the magnitude of this result, if 1 g of deuterium is converted into helium, the energy released is about 10^{12} J! At the 1981 cost of electrical energy, this would be worth about $30 000.

Example 40.7

The mass of the deuteron, which is the nucleus of "heavy hydrogen," is not equal to the sum of the masses of its constituents, which are the proton and neutron. Calculate this mass difference and determine its energy equivalence.

Solution: Using mass units of amu (defined in Section 1.2), we have

$$m_p = \text{mass of proton} = 1.00783 \text{ amu}$$

$$m_n = \text{mass of neutron} = 1.00866 \text{ amu}$$

$$m_p + m_n = 2.01649$$

Since the mass of the deuteron is 2.01410 amu, we see that the mass difference Δm is 0.00239 amu. By definition, 1 amu $= 1.66 \times 10^{-27}$ kg, and therefore

$$\Delta m = 0.00239 \text{ amu} = 3.97 \times 10^{-30} \text{ kg}$$

Using $E = \Delta mc^2$, we find that

$$\begin{aligned} E &= \Delta mc^2 = (3.97 \times 10^{-30} \text{ kg})(3 \times 10^8 \text{ m/s})^2 \\ &= 3.57 \times 10^{-13} \text{ J} \\ &= 2.23 \text{ MeV} \end{aligned}$$

Therefore, the minimum energy required to separate the proton from the neutron of the deuterium nucleus (the binding energy) is 2.23 MeV.

40.11 SUMMARY

The two basic postulates of the *special theory of relativity* are as follows:

1. All the laws of physics must be the same for all observers moving at constant velocity with respect to each other.
2. In particular, the speed of light must be the same for all inertial observers, independent of their relative motion.

In order to satisfy these postulates, the galilean transformations must be replaced by the Lorentz transformations given by

$$x' = \gamma(x - vt) \tag{40.10a}$$

$$t' = \gamma[t - (v/c^2)x] \tag{40.10b}$$

$$y' = y \tag{40.10c}$$

$$z' = z \tag{40.10d}$$

Lorentz transformation for S → S′

where

$$\gamma = \frac{1}{\sqrt{1 - v^2/c^2}} \tag{40.11}$$

In these equations, it is assumed that the primed system moves with a speed v along the xx' axes.

The relativistic form of the velocity transformation is

Relativistic velocity transformation

$$u'_x = \frac{u_x - v}{1 - u_x v/c^2} \tag{40.14a}$$

where u_x is the speed of an object as measured in the S frame and u'_x is its speed measured in the S′ frame.

Some of the consequences of the special theory of relativity are as follows:

1. Clocks in motion relative to an observer appear to be slowed down by a factor γ. This is known as *time dilation.*
2. Lengths of objects in motion appear to be contracted in the direction of motion.
3. Events that are simultaneous for one observer are not simultaneous for another observer in motion relative to the first.

These three statements can be summarized by saying that duration, length, and simultaneity are not absolute concepts in relativity.

The relativistic expression for the *momentum* of a particle moving with a velocity $\boldsymbol{u}$ is

Relativistic momentum

$$\boldsymbol{p} \equiv \frac{m\boldsymbol{u}}{\sqrt{1 - u^2/c^2}} = \gamma m\boldsymbol{u} \tag{40.20}$$

where

$$\gamma = \frac{1}{\sqrt{1 - u^2/c^2}}$$

The relativistic expression for the *kinetic energy* of a particle is

Relativistic kinetic energy

$$K = \gamma mc^2 - mc^2 \tag{40.27}$$

where mc^2 is called the *rest energy* of the particle.

The total energy E of a particle is related to the mass through the famous energy-mass equivalence expression:

$$E = \gamma mc^2 = \frac{mc^2}{\sqrt{1 - u^2/c^2}} \tag{40.29}$$

Finally, the relativistic momentum is related to the total energy through the equation

Total relativistic energy

$$E^2 = p^2c^2 + (mc^2)^2 \tag{40.30}$$

EXERCISES

Section 40.5 The Lorentz Transformation

1. Show that the Lorentz transformation equations given by Eqs. 40.8 and 40.9 satisfy Eq. 40.7.

2. Two spaceships approach each other, each moving with the *same* speed as measured by an observer on the earth. If their *relative* speed is 0.7c, what is the speed of each spaceship?

3. An observer on earth observes two spacecraft moving in the *same* direction toward the earth. Spacecraft A appears to have a speed of $0.5c$, and spacecraft B appears to have a speed of $0.8c$. What is the speed of spacecraft A measured by an observer in spacecraft B?

4. An electron moves to the right with a speed of $0.90c$ relative to the laboratory frame. A proton moves to the right with a speed of $0.70c$ relative to the electron. Find the speed of the proton relative to the laboratory frame.

Section 40.6 Consequences of the Lorentz Transformation

5. The average lifetime of a pi meson in its own frame of reference is 2.6×10^{-8} s. (This is the proper lifetime.) If the meson moves with a speed of $0.95c$, what is (a) its mean lifetime as measured by an observer on earth and (b) the average distance it travels before decaying, as measured by an observer on earth?

6. An atomic clock is placed in a jet airplane. The clock measures a time interval of 3600 s when the jet moves with a speed of 400 m/s. What corresponding time interval does an identical clock held by an observer on the ground measure? Hint: For $v/c \ll 1$, $\gamma \simeq 1 + v^2/2c^2$.

7. A clock on a moving spacecraft runs 1 s slower per day relative to an identical clock on earth. What is the speed of the spacecraft? (See hint in Exercise 6.)

8. A meter stick moving in a direction parallel to its length appears to be only 75 cm long to an observer. What is the speed of the meter stick relative to the observer?

9. A spacecraft moves at a speed of $0.9c$. If its length is L_0 when measured from in the spacecraft, what is its length measured by a ground observer?

Section 40.8 Relativistic Momentum

10. Calculate the momentum of a proton moving with a speed of (a) $0.01c$, (b) $0.5c$, (c) $0.9c$.

11. An electron has a momentum that is 90% larger than its classical momentum. (a) Find the speed of the electron. (b) How would your result change if the particle were a proton?

Section 40.9 Relativistic Energy

12. Show that the energy-momentum relationship $E^2 = p^2c^2 + (mc^2)^2$ follows from the expressions $E = \gamma mc^2$ and $p = \gamma mu$.

13. A proton moves with a speed of $0.95c$. Calculate its (a) rest energy, (b) total energy, and (c) kinetic energy.

14. An electron has a kinetic energy five times greater than its rest energy. Find (a) its total energy and (b) its speed.

15. Find the speed of a particle whose total energy is 50% greater than its rest energy.

16. A proton in a high-energy accelerator is given a kinetic energy of 50 GeV. Determine the (a) momentum and (b) speed of the proton.

17. Determine the energy required to accelerate an electron from (a) $0.50c$ to $0.75c$ and (b) $0.90c$ to $0.99c$.

Section 40.10 Confirmations and Consequences of Relativity Theory

18. Calculate the binding energy in MeV per nucleon (proton or neutron) in the isotope $^{12}_{6}\text{C}$. Note that the mass of this isotope is exactly 12 amu, and the masses of the proton and neutron are 1.007825 amu and 1.008665 amu, respectively.

19. Consider the decay $^{55}_{24}\text{Cr} \rightarrow {}^{55}_{25}\text{Mn} + \text{e}$, where e is an electron. The ^{55}Cr nucleus has a mass of 54.9279 amu, and the ^{55}Mn nucleus has a mass of 54.9244 amu. (a) Calculate the mass difference in MeV. (b) What is the maximum kinetic energy of the emitted electron?

20. A radium isotope decays by emitting an α particle to a radon isotope according to the scheme $^{226}_{88}\text{Ra} \rightarrow {}^{222}_{86}\text{Rn} + {}^{4}_{2}\text{He}$. The masses of the atoms are 226.0254 (Ra), 222.0175 (Rn), and 4.0026 (He). How much energy is released as the result of this decay?

PROBLEMS

1. An electron has a velocity of $0.75c$. Find the velocity of a proton which has (a) the same kinetic energy as the electron and (b) the same momentum as the electron.

2. The muon is an unstable particle that spontaneously decays into an electron and two neutrinos. If the number of muons at $t = 0$ is N_0, the number at time t is given by $N = N_0 e^{-t/\tau}$ where τ is the mean lifetime, equal to 2.2 μs. Suppose the muons move at a speed of $0.95c$ and there are 5×10^4 muons at $t = 0$. (a) What is the observed lifetime of the muons? (b) How may muons remain after traveling a distance of 3 km?

3. *Doppler effect for light.* If a light source moves with a speed v relative to an observer, there is a shift in the observed frequency analogous to the Doppler effect for sound waves. Show that the observed frequency f_0 is related to the true frequency f through the expression

$$f_0 = \sqrt{\frac{c \pm v_s}{c \mp v_s}} f$$

where the upper signs correspond to the source approaching the observer and the lower signs correspond to the source receding from the observer. [Hint: In the moving frame S′, the period is the proper time interval and is given by $T = 1/f$. Furthermore, the wavelength measured by the observer is $\lambda_0 = (c - v_s)T_0$, where T_0 is the period measured in s.]

4. *The red shift.* A light source recedes from an observer with a speed v_s, which is small compared with c. (a) Show that the fractional shift in the measured wavelength is given by the approximate expression

$$\frac{\Delta\lambda}{\lambda} \approx \frac{v_s}{c}$$

This result is known as the *red shift,* since the visible light is shifted toward the red. (Note that the proper period and measured period are approximately equal in this case.) (b) Spectroscopic measurements of light at $\lambda = 3970$ Å coming from a galaxy in Ursa Major reveals a red shift of 200 Å. What is the recessional speed of the galaxy?

5. A rod of length l_0 moves with a speed v along the horizontal direction. The rod makes an angle of θ_0 with respect to the x' axis as in Fig. 40.8. (a) Show that the length of the rod as measured by a stationary observer is given by $l = l_0[1 - (v^2/c^2)\cos^2\theta_0]^{1/2}$. (b) Show that the angle that the rod makes with the x axis is given by the expression $\tan\theta = \gamma\tan\theta_0$. These results show that the rod is both contracted and rotated. (Take the lower end of the rod to be at the origin of the primed coordinate system.)

6. *Time dilation in an atom.* The atoms in a gas move in random directions with thermal velocities. Because of their motion with respect to a laboratory observer, the frequency at which they radiate is shifted by time dilation. (a) If an atom radiates at a frequency f_0 in its rest frame, show that the observed frequency in the laboratory frame is given by $f = (1 - v^2/c^2)^{1/2} f_0$. (b) If $v \ll c$, show that the fractional change in frequency is given by $\Delta f/f = -v^2/2c^2$. (c) Evaluate the fractional change in frequency for a hydrogen atom at a temperature of 300 K. (Hint: From the kinetic theory of gases, $M\overline{v}^2/2 = 3kT/2$.)

7. *Speed of light in a moving medium.* The motion of a medium such as water influences the speed of light. This effect was first observed by Fizeau in 1851. Consider a light beam passing through a horizontal column of water moving with a speed v. (a) Show that if the beam travels in the same direction as the flow of water, the speed of light measured in the laboratory frame is given by

$$u = \frac{c}{n}\left(\frac{1 + nv/c}{1 + v/nc}\right)$$

where n is the index of refraction of the water. (Hint: Use the velocity transformation relation, Eq. 40.15, and note that the velocity of the water with respect to the *moving* frame is given by c/n.) (b) Show that for $v \ll c$, the expression above is, to a good approximation,

$$u \approx \frac{c}{n} + v - \frac{v}{n^2}$$

8. A cube of density ρ and edge l moves in a direction parallel to one of its edges with a speed v, comparable to c. What is (a) the volume of the cube and (b) the density of the cube as measured by a laboratory observer?

41 Quantum Physics

DOE Photo

In the previous chapter, we discussed the fact that newtonian mechanics must be replaced by Einstein's special theory of relativity when we are dealing with particle velocities comparable to the speed of light. Although many problems were indeed resolved by the theory of relativity in the early part of the 20th century, many experimental and theoretical problems remained unanswered. Attempts to apply the laws of classical physics to explain the behavior of matter on the atomic scale were totally unsuccessful. Various phenomena, such as blackbody radiation, the photoelectric effect, and the emission of sharp spectral lines by atoms in a gas discharge, could not be understood within the framework of classical physics. We shall describe these phenomena because of their importance in subsequent developments.

Another revolution took place in physics between 1900 and 1930. This was the era of a new and more general scheme called *quantum mechanics*. This new approach was highly successful in explaining the behavior of atoms, molecules, and nuclei. Moreover, the quantum theory reduces to classical physics when applied to macroscopic systems. As with relativity, the quantum theory requires a modification of our ideas concerning the physical world.

The basic ideas of quantum theory were first introduced by Max Planck, but most of the subsequent mathematical developments and interpretations were made by a number of distinguished physicists, including Einstein, Bohr, Schrödinger, de Broglie, Heisenberg, Born, and Dirac. Despite the great success of the quantum theory, Einstein frequently played the role of critic, especially with regard to the manner in which the theory was interpreted. In particular, Einstein did not accept Heisenberg's uncertainty principle, which says that it is impossible to obtain a precise measurement of the position and the velocity of a particle simultaneously. According to this principle, it is possible only to predict the *probability* of the future of a system, contrary to the deterministic view held by Einstein.[1]

An extensive study of quantum theory is certainly beyond the scope of this book. This chapter is simply an introduction to the underlying ideas of quantum theory and the wave-particle nature of matter. We shall also discuss some simple applications of quantum theory, including the photoelectric effect, the interpretation of atomic spectra, and the principles of lasers.

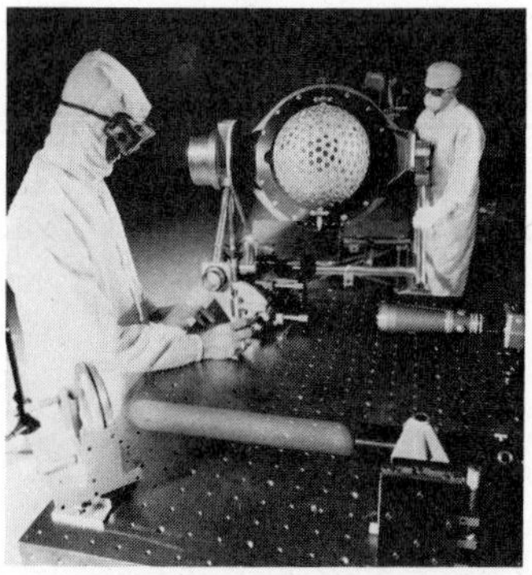

Physicists conduct prelaunch evaluations of the LAGEOS (Laser Geodynamic Satellite) using a low-power laser to test its 426 special reflectors. (NASA)

[1]Einstein's views on the probabilistic nature of quantum theory is brought out in his statement, "God does not play dice with the universe."

41.1 BLACKBODY RADIATION AND PLANCK'S HYPOTHESIS

Max Planck (1858–1947).

Any object, when heated, is known to emit radiation sometimes referred to as *thermal radiation*. The characteristics of this radiation depend on the temperature and properties of the object. At low temperatures, the wavelengths of the thermal radiation are mainly in the infrared region and hence are not observed by the eye. As the temperature of an object is increased, the object eventually will begin to glow red. At sufficiently high temperatures, it appears to be white, as, for example, the glow of a tungsten filament in a light bulb. A careful study of thermal radiation shows that it consists of a continuous distribution of wavelengths that includes the infrared, visible, and ultraviolet portions of the spectrum.

Definition of a blackbody

From a classical viewpoint, thermal radiation originates from accelerated charged particles near the surface of the object. The thermally agitated charges can have a distribution of accelerations, which accounts for the continuous spectrum of radiation emitted by the object. By the end of the 19th century, it had become apparent that the classical theory of thermal radiation was inadequate. The basic problem was concerned with understanding the spectral distribution of radiation emitted by a blackbody. By definition, a *blackbody* is an ideal system that absorbs all radiation incident on it. A good approximation to a blackbody is any heat-resisting material containing a cavity with a very small opening leading to the cavity. The nature of the radiation emitted by a blackbody depends only on the temperature of the cavity walls, which are assumed to be in thermal equilibrium with the radiation. Experimental data for the distribution of energy of blackbody radiation versus wavelength λ are shown in Fig. 41.1 at three different temperatures. Note that the radiated energy varies with wavelength and temperature. Furthermore, as the temperature of the blackbody increases, the total intensity (which equals the area under a given curve) also increases, and the peak of the distribution shifts to shorter wavelengths.

Early attempts to explain these results based on classical theories failed. To describe the radiation spectrum, it is useful to define $I(\lambda,T)\,d\lambda$ to be the power per unit area emitted in the wavelength interval $d\lambda$. The result of a calculation based on a classical model of blackbody radiation known as the *Rayleigh-Jeans law* is

Rayleigh-Jeans law

$$I(\lambda,T) = \frac{2\pi ckT}{\lambda^4} \tag{41.1}$$

where k is Boltzmann's constant. In this classical model of blackbody radiation, the atoms in the cavity walls are treated as a set of oscillators that emit electromagnetic waves at all wavelengths. This model leads to an average energy per oscillator that is proportional to T.

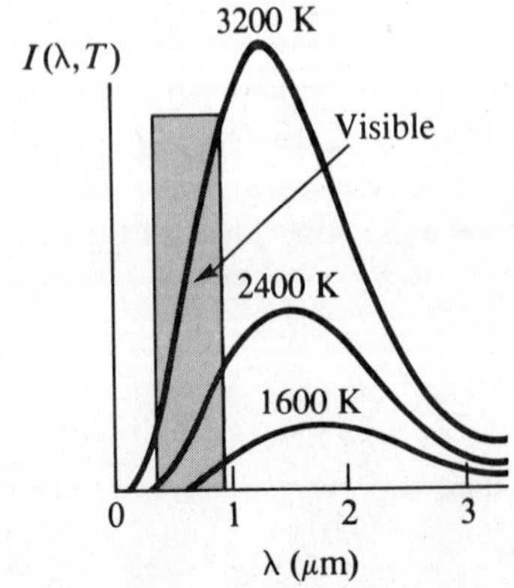

Figure 41.1 Spectral distribution of blackbody radiation versus wavelength at three temperatures. You should note that the total radiation emitted (the area under a curve) increases with increasing temperature. (Adapted from R. B. Leighton, Principles of Modern Physics, 1st edn., New York, McGraw-Hill, 1959; by permission)

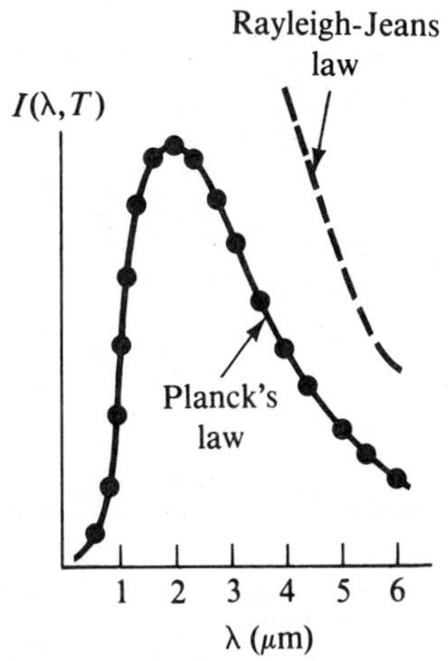

Figure 41.2 Spectral distribution of blackbody radiation versus wavelength. The points are experimental data taken at 1600 K. (Adapted from F. K. Richtmeyer, E. H. Kennard, and J. N. C. Lauritsen, Introduction to Modern Physics, 5th edn., New York, McGraw-Hill, 1955; by permission)

An experimental plot of the blackbody radiation spectrum is shown in Fig. 41.2, together with the theoretical prediction of the Rayleigh-Jeans law. At long wavelengths, Rayleigh-Jeans law is in reasonable agreement with experimental data. However, at short wavelengths there is major disagreement. This can be seen by noting that as λ approaches zero, the function $I(\lambda,T)$ given by Eq. 41.1 approaches infinity; that is, short-wavelength radiation should predominate. This is contrary to the experimental data plotted in Fig. 41.2, which shows that as λ approaches zero, $I(\lambda,T)$ also approaches zero. This contradiction is often called the *ultraviolet catastrophe*. Another major problem with classical theory is that it predicts an *infinite total energy density* since all wavelengths are possible.[2] Physically, an infinite energy in the electromagnetic field is an impossible situation.

In 1900 Max Planck discovered a formula for blackbody radiation that was in complete agreement with experiment at all wavelengths. The empirical function proposed by Planck is given by

$$I(\lambda,T) = \frac{2\pi hc^2}{\lambda^5(e^{hc/\lambda kT} - 1)} \tag{41.2}$$

Planck's radiation law

where h is a constant that can be adjusted to fit the data. The current value of h, known as *Planck's constant*, is given by

$$h = 6.626 \times 10^{-34}\ \mathrm{J \cdot s} \tag{41.3}$$

Planck's constant

You should show that at long wavelengths, Planck's expression, Eq. 41.2, reduces to the Rayleigh-Jeans expression given by Eq. 41.1. Furthermore, at short wavelengths, Planck's law predicts an exponential decrease in $I(\lambda,T)$ with decreasing wavelength, in agreement with experimental results.

In order to derive this formula, Planck made two bold and controversial assumptions concerning the nature of the oscillating atoms of the cavity walls. These assumptions were as follows:

1. The oscillator could have only *discrete* units of energy E_n given by

$$E_n = nhf \tag{41.4}$$

Quantization of energy

where n is a positive integer called a *quantum number* and f is the frequency of the oscillator. The energies of the oscillator are said to be *quantized*, and the allowed energy states are called *quantum states*.

---

[2]The total power per unit area $I = \int_0^\infty I(\lambda, T)\, d\lambda \rightarrow \infty$ when all wavelengths are allowed.

2. The oscillators emit or absorb energy in discrete units of light energy called *quanta* (or *photons*, as they are now called). They do so by "jumping" from one quantum state to another. If the quantum number n changes by one unit, Eq. 41.4 shows that the amount of energy radiated or absorbed by the oscillator equals hf. Hence, the energy of a light quantum corresponding to the energy difference between two adjacent quantum states is given by

$$E = hf \tag{41.5}$$

The oscillator will not radiate or absorb energy if it remains in one of its quantized states.

The key point in this theory is the radical assumption of quantized energy states. This development marked the birth of the quantum theory. At this time, most scientists, including Planck, did not consider the quantum concept to be realistic. Hence, Planck and others continued to search for a more rational explanation of blackbody radiation. However, subsequent developments showed that a theory based on the quantum concept (rather than a classical theory) had to be used to explain a number of phenomena at the atomic level. One of these phenomena, known as the *photoelectric effect*, will be discussed in the next section.

Example 41.1
A 2-kg mass is attached to a massless spring of force constant $k = 25$ N/m. The spring is stretched 0.4 m from its equilibrium position and released. (a) Find the total energy and frequency of oscillation according to classical calculations.

The total energy of the simple harmonic oscillator having an amplitude A is given by $\frac{1}{2}kA^2$. Therefore,

$$E = \tfrac{1}{2}kA^2 = \tfrac{1}{2}(25 \text{ N/m})(0.4 \text{ m})^2 = 2.0 \text{ J}$$

The frequency of oscillation is given by

$$f = \frac{1}{2\pi}\sqrt{\frac{k}{m}} = \frac{1}{2\pi}\sqrt{\frac{25 \text{ N/m}}{2 \text{ kg}}} = 0.56 \text{ Hz}$$

(b) Assuming its energy is quantized, find the quantum number n for the system.

Since the energy is quantized, $E_n = nhf$, and using the result from (a) gives

$$E_n = nhf = n(6.63 \times 10^{-34} \text{ J}\cdot\text{s})(0.56 \text{ Hz}) = 2.0 \text{ J}$$

Therefore

$$n = 5.4 \times 10^{33}$$

We see that n is an extraordinarily large number. In fact, for macroscopic systems such as this, the quantum numbers are so large that the energy difference between adjacent states is very small compared with the total energy of the system. For this reason, one can regard the quantized energy states as a continuum of states for all practical purposes. Therefore, newtonian mechanics gives an adequate description of such macroscopic systems. However, later we shall see that such is not the case for atomic or molecular systems.

Q1. What assumptions were made by Planck in dealing with the problem of blackbody radiation? Discuss the consequences of these assumptions.

41.2 THE PHOTOELECTRIC EFFECT

In the latter part of the 19th century, experiments showed that when light is incident on certain metallic surfaces, electrons are emitted from the surfaces. This phenomenon is known as the *photoelectric effect*, and the emitted electrons are called *photoelectrons*. The first discovery of this phenomenon was made by Hertz, who also first produced the electromagnetic waves predicted by Maxwell. Hertz showed that light must fall on one terminal of a spark gap to allow sparks to pass across the gap. Furthermore, he found that ultraviolet light was more effective in this respect than longer-wavelength light.

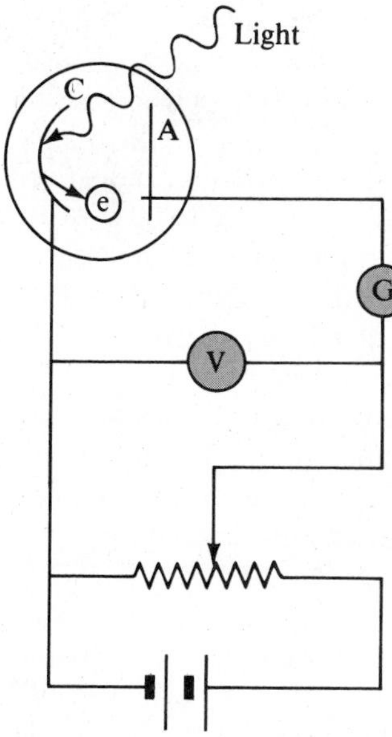

Figure 41.3 Circuit diagram for observing the photoelectric effect. When light strikes the cathode C, photoelectrons are ejected. Electrons collected at the anode A constitute a current in the circuit.

Figure 41.3 is a schematic diagram of an apparatus that can be used to observe this effect. An evacuated tube contains a metal cathode C and a collecting anode A, which are maintained at some potential difference by means of a battery. When the tube is in the dark, the galvanometer G reads zero, indicating that there is no current in the circuit. This corresponds to an open-circuit condition. However, when monochromatic light of the appropriate wavelength shines on the cathode, a current is detected by the galvanometer, indicating a flow of charge from the cathode to the anode. The current associated with this process arises from electrons emitted from the cathode and collected at the anode.

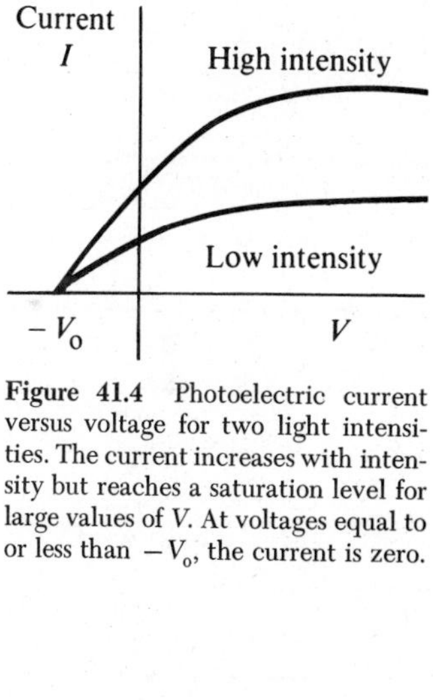

Figure 41.4 Photoelectric current versus voltage for two light intensities. The current increases with intensity but reaches a saturation level for large values of V. At voltages equal to or less than $-V_0$, the current is zero.

A plot of the photoelectric current versus the potential difference V between the anode and cathode is shown in Fig. 41.4 for two different light intensities. Note that for large values of V, the current reaches a maximum value, corresponding to the case where all photoelectrons are collected at the anode. In addition, the current increases as the incident light intensity increases, as you might expect. Finally, when V is negative, the photoelectrons are repelled by the negative anode. Only those electrons having a kinetic energy greater than eV will reach the anode. Furthermore, if V is less than or equal to V_0, called the *stopping potential,* no electrons will reach the anode and the current will be zero. The maximum kinetic energy of the photoelectrons is related to the stopping potential through the relation

Stopping potential

$$K_{\text{max}} = eV_0 \tag{41.6}$$

Several features of the photoelectric effect could not be explained with classical physics and the wave theory of light. The major observations that were not understood are as follows:

Properties of photoelectric emission

1. No electrons are emitted if the incident light frequency falls below some cutoff frequency f_c, which depends on the material being illuminated. For example, in the case of sodium, $f_c = 4.39 \times 10^{11}$ Hz. This is inconsistent with the wave theory, which predicts that the photoelectric effect should occur at any frequency provided the light intensity is high enough.
2. If the light frequency exceeds the cutoff frequency, a photoelectric effect is observed and the number of photoelectrons emitted is proportional to the light intensity. However, the maximum kinetic energy of the photoelectrons is inde-

pendent of light intensity, a fact that cannot be explained using the concepts of classical physics.

3. The maximum kinetic energy of the photoelectrons increases with increasing light frequency.
4. Electrons are emitted from the surface almost instantaneously (less than 10^{-9} s after the surface is illuminated), even at low light intensities. Classically, one would expect that the electrons would require some time to absorb the incident radiation before they acquire enough kinetic energy to escape from the metal.

A successful explanation of the photoelectric effect was given by Einstein in 1905, the same year he published his special theory of relativity. In this paper, for which he received the Nobel prize in 1921, Einstein extended Planck's concept of quantization to the electromagnetic field. He assumed that a light wave (or any electromagnetic wave) of frequency f can be considered a stream of corpuscles, or *photons*, as they are now called. Each photon has an energy E, given by

Energy of a photon

$$E = hf \tag{41.7}$$

where h is Planck's constant. Thus, Einstein viewed light as a stream of particles traveling through space (rather than a wave), where each particle could be absorbed as a unit by an electron. Furthermore, Einstein argued that when an electron in the metal absorbs one of these photons, the maximum kinetic energy acquired by the electron must be hf. However, the electron must also pass through the metal surface and overcome an internal potential barrier, which requires an additional energy ϕ. Hence, in order to conserve energy, the maximum kinetic energy of the ejected photoelectron is given by

Photoelectric effect equation

$$K_{max} = hf - \phi \tag{41.8}$$

That is, the excess energy $hf - \phi$ is the maximum kinetic energy the liberated electron can have outside the surface. The quantity ϕ is known as the *work function* of the substance and is of the order of a few eV for metals. For example, the work function for zinc is about 3.0 eV.

Within the framework of the photon theory of light, one can explain the features of the photoelectric effect that could not be understood using classical concepts. These are briefly described in the order they were introduced earlier:

1. The fact that the photoelectric effect is not observed below a certain cutoff frequency f_c follows from the fact that the maximum kinetic energy equals $hf - \phi$. That is, if the energy of the incoming photon does not exceed the work function ϕ, the electrons will never be ejected from the surface, regardless of the intensity of the beam.
2. The fact that K_{max} is independent of intensity can be understood by using the photon theory with the following argument. If the light intensity is doubled, the number of photons is doubled, which doubles the number of photoelectrons emitted. However, their kinetic energy, which equals $hf - \phi$, depends only on the light frequency and work function, not on the intensity.
3. The fact that K_{max} increased with increasing frequency is easily understood with Einstein's photoelectric effect equation since $K_{max} \sim f$.
4. Finally, the fact that the electrons are emitted almost instantaneously is consistent with the particle view of light, in which the incident energy appears in concentrated packets rather than over a large area (as would be the case in the wave theory).

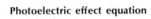

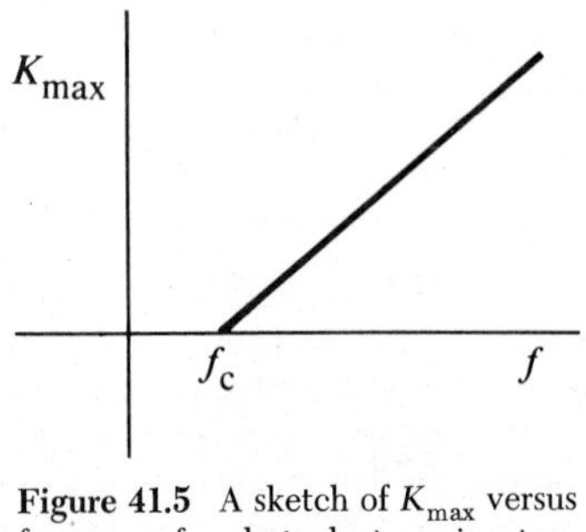

Figure 41.5 A sketch of K_{max} versus frequency for photoelectrons in a typical photoelectric effect experiment. Photons with frequency less than f_c are not sufficiently energetic to eject an electron from the metal.

A final confirmation of Einstein's theory is a test of the prediction of a linear relationship between the frequency f and K_{max}. Indeed, such a linear relationship is observed, as sketched in Fig. 41.5. The slope of such a curve gives a value for h. The intercept on the horizontal axis gives the cutoff frequency f_c, which is related to the work function through the relation $f_c = \phi/h$. This corresponds to a cutoff wavelength of

$$\lambda_c = \frac{c}{f_c} = \frac{c}{\phi/h} = \frac{hc}{\phi} \qquad (41.9)$$

Cutoff wavelength

Wavelengths greater than λ_c for a material with work function ϕ produce no photoelectric effect. Using this technique on various metallic surfaces, Millikan obtained a value for h that was within 0.5% of the value derived from Planck's radiation formula.[3] Millikan was awarded the Nobel prize for this work in 1927. The excellent agreement is strong evidence in support of Einstein's photon concept.

Example 41.2

A sodium surface is illuminated with light of wavelength 300 nm. The work function for sodium metal is 2.46 eV. (a) Find the kinetic energy of the ejected photoelectrons.

The energy of the illuminating light beam is

$$E = hf = \frac{hc}{\lambda} = \frac{(6.63 \times 10^{-34}\ \text{J}\cdot\text{s})(3.00 \times 10^{8}\ \text{m/s})}{300 \times 10^{-9}\ \text{m}}$$

$$= 6.63 \times 10^{-19}\ \text{J}$$

$$= \frac{6.63 \times 10^{-19}\ \text{J}}{1.60 \times 10^{-19}\ \text{J/eV}} = 4.14\ \text{eV}$$

where we have used the conversion 1 eV $= 1.6 \times 10^{-19}$ J. Using Eq. 41.8 gives

$$K_{\text{max}} = hf - \phi = 4.14\ \text{eV} - 2.46\ \text{eV} = 1.68\ \text{eV}$$

(b) Determine the cutoff wavelength for sodium.

The cutoff wavelength can be calculated from Eq. 41.9 after we convert ϕ from eV to J:

$$\phi = 2.46\ \text{eV} = (2.46\ \text{eV})(1.6 \times 10^{-19}\ \text{J/eV})$$

$$= 3.94 \times 10^{-19}\ \text{J}$$

Hence

$$\lambda_c = \frac{hc}{\phi} = \frac{(6.63 \times 10^{-34}\ \text{J}\cdot\text{s})(3.00 \times 10^{8}\ \text{m/s})}{3.94 \times 10^{-19}\ \text{J}}$$

$$= 5.05 \times 10^{-7}\ \text{m} = 505\ \text{nm}$$

This wavelength is in the green region of the visible spectrum.

Q2. If the photoelectric effect is observed for surface A, can you conclude that the effect will also be observed for surface B under the same conditions? Explain.

Q3. Suppose the photoelectric effect occurs in a gaseous target rather than a solid. Will photoelectrons be produced at *all* frequencies of the incident photon? Explain.

41.3 THE COMPTON EFFECT

Further experimental evidence for the photon concept was discovered in 1923 by Arthur H. Compton (1892–1962). His experiment involved the study of the scattering of X-rays by electrons in a carbon target. The intensity of the scattered X-rays was measured as a function of X-ray wavelength and at various scattering angles. Compton observed that the scattered X-rays had a wavelength λ slightly longer than the wavelength of the incident X-rays, λ_0. This change in wavelength, $\Delta\lambda = \lambda - \lambda_0$, called the *Compton shift,* varies with the scattering angle. This result cannot be explained by classical theory. According to a classical model, the X-ray is considered to be an electromagnetic wave of frequency f_0 incident on a material containing electrons. The electromagnetic wave causes the electrons to oscillate and reradiate electromagnetic waves of the same frequency f_0. Hence, the scattered waves are

[3] For actual data and more details on this subject, see R. C. Millikan, *Phys. Rev.* **7**:362 (1916).

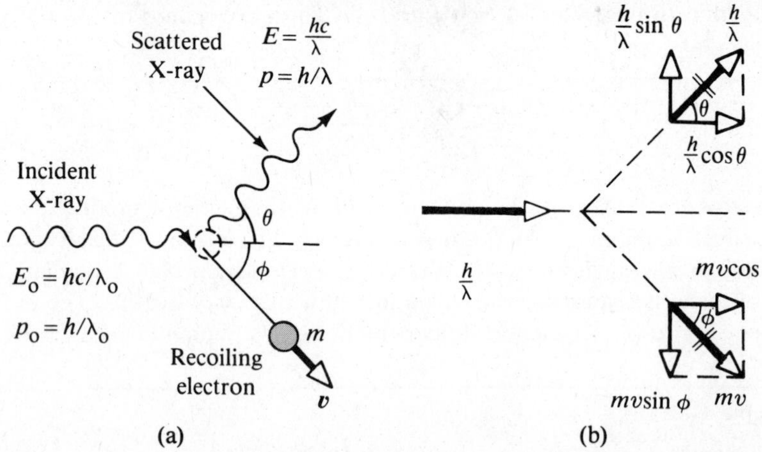

Figure 41.6 (a) Diagram representing Compton scattering of an X-ray by an electron. The scattered X-ray has less energy (is of longer wavelength) than the incident X-ray. The excess energy is taken up by the recoiling electron. (b) Components of momentum before and after the Compton scattering occurs.

predicted to have the same frequency and wavelength as the incident wave, contrary to experimental results.

Compton treated the problem by considering the scattering process as a collision between a photon and an electron. In this model, the photon is treated as a particle of energy:

$$E = hf = hc/\lambda$$

Furthermore, the *rest mass of the photon is taken to be zero;* hence the momentum of the photon can be described by the relationship

$$p = \frac{E}{c} = \frac{hc}{c\lambda} = \frac{h}{\lambda}$$

Figure 41.6a shows the geometry of a collision between a photon of incident wavelength λ_0 and an electron initially at rest. As a result of the collision, part of the photon's energy is transferred to the recoiling electron. Consequently, the energy and frequency of the scattered photon are lowered and its wavelength increases. Assuming that energy and momentum are conserved during the collision, one obtains the following relation for the shift in wavelength of the scattered photon:

Compton shift equation

$$\Delta\lambda = \lambda - \lambda_0 = \frac{h}{mc}(1 - \cos\theta) \quad (41.10)$$

In this expression, known as the *Compton shift equation,* λ is the wavelength of the scattered photon, m is the mass of the electron, and θ is the angle between the directions of the scattered and incident photons. The quantity h/mc, which has dimensions of length, is called the *Compton wavelength.* The Compton wavelength has the numerical value

Compton wavelength

$$\frac{h}{mc} = \frac{6.63 \times 10^{-34}\ \text{J}\cdot\text{s}}{(9.11 \times 10^{-31}\ \text{kg})(3.00 \times 10^{8}\ \text{m/s})} = 2.43 \times 10^{-12}\ \text{m} = 0.0243\ \text{Å}$$

If the scattered photons are observed at $\theta = 90°$, we see from Eq. 41.10 that, in this case, the Compton shift equals the Compton wavelength. Note that the Compton shift depends on the scattering angle θ and not on the wavelength. The experimental results from X-rays scattered from various targets are in excellent agreement with Eq. 41.10, which again confirms the validity of the photon concept.

It is important to recognize the fact that *the Compton effect is purely a quantum phenomenon.* (According to a classical model, it should not occur.) This can be tested formally by letting h approach 0 to see if quantum predictions agree with the laws of classical physics. As you can see from Eq. 41.10, as $h \to 0$, $\Delta\lambda \to 0$, which is in agreement with classical physics.

We can derive the Compton shift expression, Eq. 41.10, by considering the collision between the photon and electron, taking the electron to be initially at rest, as in Fig. 41.6a. Applying conservation of energy to this process, we have

$$\frac{hc}{\lambda_o} = \frac{hc}{\lambda} + K_e$$

where hc/λ_o is the energy of the incident photon, hc/λ is the energy of the scattered photon, and K_e is the kinetic energy of the recoiling electron. Since the electron may recoil at speeds comparable to the speed of light, we must use the relativistic expression for K_e, given by $K_e = \gamma mc^2 - mc^2$ (Eq. 40.27). Therefore,

$$\frac{hc}{\lambda_o} = \frac{hc}{\lambda} + \gamma mc^2 - mc^2 \qquad (41.11)$$

where $\gamma = 1/\sqrt{1 - v^2/c^2}$.

Next, we can apply the law of conservation of momentum to this collision, noting that *both the x and y components of momentum are conserved.* Since the momentum of a photon has a magnitude given by $p = h/\lambda$, and since the relativistic expression for the momentum of the recoiling electron is $p_e = \gamma mv$ (Eq. 40.20), we obtain the following expressions for the x and y components of linear momentum:

$$x \text{ component:} \qquad \frac{h}{\lambda_o} = \frac{h}{\lambda}\cos\theta + \gamma mv \cos\phi \qquad (41.12)$$

$$y \text{ component:} \qquad 0 = \frac{h}{\lambda}\sin\theta - \gamma mv \sin\phi \qquad (41.13)$$

By eliminating v and ϕ from Eqs. 41.11 to 41.13, we obtain a single expression that relates the remaining three variables (λ, λ_o, and θ). After some algebra (Problem 2), one obtains the Compton shift equation:

$$\Delta\lambda = \lambda - \lambda_o = \frac{h}{mc}(1 - \cos\theta)$$

Example 41.3

X-rays of wavelength $\lambda_o = 2.00$ Å are scattered from a block of carbon. The scattered X-rays are observed at an angle of 45° to the incident beam. (a) Calculate the Compton shift and the wavelength of the scattered X-ray at $\theta = 45°$.

Setting $\theta = 45°$ in Eq. 41.10 gives

$$\Delta\lambda = \lambda - \lambda_o = \frac{h}{mc}(1 - \cos\theta)$$

$$= \frac{6.63 \times 10^{-34}\ \text{J}\cdot\text{s}}{(9.11 \times 10^{-31}\ \text{kg})(3.00 \times 10^{8}\ \text{m/s})}(1 - \cos 45°)$$

$$= 7.11 \times 10^{-13}\ \text{m} = 7.11 \times 10^{-4}\ \text{nm} = 0.00711\ \text{Å}$$

Hence, the wavelength of the scattered X-ray at $\theta = 45°$ is $\lambda = 2.00711$ Å

(b) Find the fraction of energy lost by the photon in this collision.

The incident photon energy is $E_0 = hc/\lambda_0$ and the scattered photon energy is $E = hc/\lambda$. Hence, the fraction of energy lost by the photon is

$$\text{Fraction of energy lost} = \frac{E_0 - E}{E_0} = \frac{hc/\lambda_0 - hc/\lambda}{hc/\lambda_0}$$

$$= 1 - \frac{\lambda_0}{\lambda} = \frac{\lambda - \lambda_0}{\lambda} = \frac{\Delta\lambda}{\lambda}$$

$$= \frac{0.00711\ \text{Å}}{2.00711\ \text{Å}} = 0.00354$$

That is, 0.354% of the incident photon energy is lost to the recoiling electron.

Q4. How does the Compton effect differ from the photoelectric effect?

Q5. What assumptions were made by Compton in dealing with the scattering of a photon from an electron?

41.4 ATOMIC SPECTRA

Suppose an evacuated glass tube is filled with an elemental gas, such as neon, helium, or argon. If an electrical discharge is passed through the gas, the tube will emit light whose color is characterized by the properties of the gas. If the emitted light is analyzed with a spectroscope containing a narrow-slit aperture, a series of discrete lines is observed, each line corresponding to a different wavelength, or color. We refer to such a series of lines as a *line spectrum.* The wavelengths contained in a given line spectrum are characteristic of the particular element emitting the light. The simplest line spectrum is observed for atomic hydrogen, and we shall describe this spectrum in some detail. The more complex atoms, such as iron and copper, give completely different line spectra. Since no two elements emit the same line spectrum, this phenomenon represents a practical technique for identifying elements in a given chemical substance.

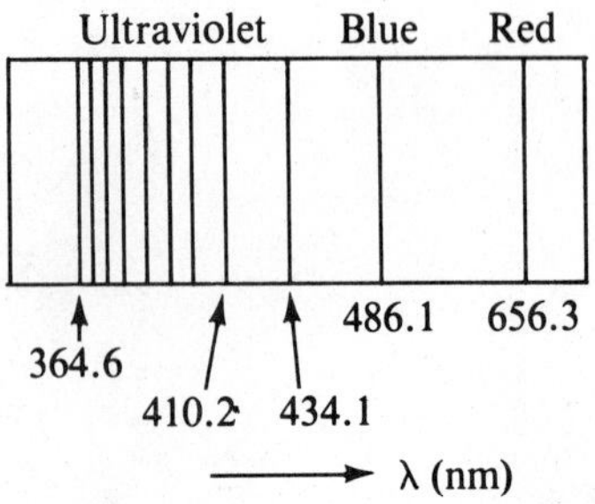

Figure 41.7 A series of spectral lines for atomic hydrogen. The prominent lines labeled are part of the Balmer series.

The line spectrum of hydrogen includes a series of lines in the visible region of the spectrum, shown in Fig. 41.7. Four of the most prominent lines in this region occur at the wavelengths 656.3 nm, 486.1 nm, 434.1 nm, and 410.2 nm.

In 1885, Johann Balmer (1825–1898) found that the wavelengths of these lines can be described by the simple empirical expression

Balmer's formula

$$\frac{1}{\lambda} = R\left(\frac{1}{2^2} - \frac{1}{n^2}\right) \tag{41.14}$$

where n may have integral values of 3, 4, 5, . . . and R is a constant, now called the *Rydberg constant.* If the wavelength λ is in meters, R has the value

Rydberg constant

$$R = 1.0973732 \times 10^7\ \text{m}^{-1} \tag{41.15}$$

The first line in the Balmer series, at 656.3 nm, corresponds to $n = 3$ in Eq. 41.14, the line at 486.1 nm corresponds to $n = 4$, etc. The shortest wavelength in the series occurs at 364.6 nm, corresponding to the series limit, $n = \infty$.

Other line spectra for hydrogen were found following Balmer's discovery. These spectra are called the Lyman, Paschen, and Brackett series after their discoverers. The wavelengths of these series can be described by the following empirical formulas:

Lyman series

$$\frac{1}{\lambda} = R\left(1 - \frac{1}{n^2}\right) \qquad (n = 2,3,4, \ldots) \tag{41.16}$$

Paschen series

$$\frac{1}{\lambda} = R\left(\frac{1}{3^2} - \frac{1}{n^2}\right) \qquad (n = 4,5,6, \ldots) \tag{41.17}$$

Brackett series

$$\frac{1}{\lambda} = R\left(\frac{1}{4^2} - \frac{1}{n^2}\right) \qquad (n = 5,6,7, \ldots) \tag{41.18}$$

Note that the same constant R appears in all series. The same is true for elements other than hydrogen, except that there are slight variations in R from one atom to another. This variation is now known to be due to the effect of different atomic masses. For example, for helium R has the value $1.097223 \times 10^7\ \text{m}^{-1}$.

In addition to emitting light at specific wavelengths, an element can also absorb light at discrete wavelengths. The spectral lines corresponding to this process form what is known as an *absorption spectrum*. Experimentally, one can obtain an absorption spectrum by passing a continuous light beam through a vapor of the element being analyzed. The absorption spectrum consists of a series of dark lines superimposed on the otherwise continuous spectrum emitted by the continuous light source. Experimentally, it is found that each line in the absorption spectrum coincides with a line in the emission spectrum. However, usually not all lines in the emission spectrum of an element are observed in the absorption spectrum. In order to observe some series in an absorption spectrum, it is usually necessary to raise the temperature of the gas to extremely high values. (A detailed explanation of the difference between the emission spectrum and the absorption spectrum is beyond the scope of this book.)

The absorption spectrum of an element has many practical applications. For example, the continuous spectrum of radiation emitted by the sun must pass through the cooler gases of the solar atmosphere and through the earth's atmosphere. The various absorption lines observed in the solar spectrum have been used to identify elements in these gases. The same method is used to analyze the composition of other stars.

41.5 THE BOHR THEORY OF HYDROGEN

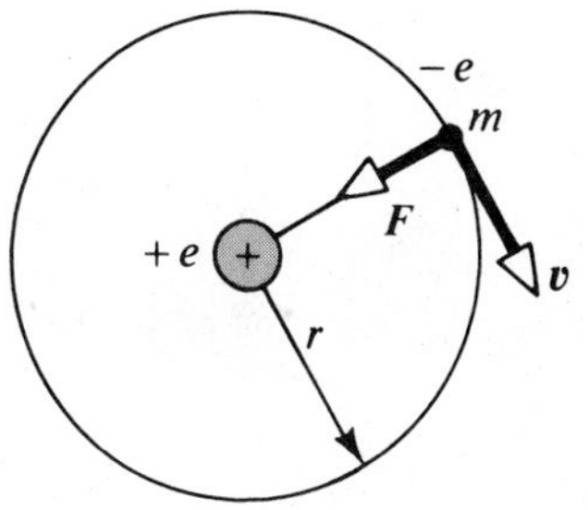

Figure 41.8 Diagram representing Bohr's model of the hydrogen atom. In this model, the orbiting electron is allowed to be only in specific orbits of discrete radii.

At the beginning of the 20th century, scientists were perplexed by the failure of classical physics in explaining the spectral behavior of atomic systems. In 1913, the Danish scientist Niels Bohr (1885–1963) provided the first successful explanation of atomic spectra. His theory contained a combination of ideas from Planck's original quantum theory, Einstein's photon concept, and Rutherford's model of the atom (that of a small, positively charged core, about 10^{-5} Å in diameter, surrounded by electrons in orbits about 1 Å in diameter). Bohr's model of the hydrogen atom contains some classical features as well as some revolutionary postulates that could not be justified within the framework of classical physics. As we shall see, the Bohr model can be applied quite successfully to such hydrogen-like ions as He^{+} and Li^{2+}. However, the theory does not properly describe the spectra of more complex atoms and ions. Bohr's model was eventually abandoned in about 1925, when quantum mechanics was discovered.

Let us describe the Bohr model of the hydrogen atom briefly, keeping in mind that it correctly predicts the hydrogen spectrum, but for the wrong reasons. The basic postulates underlying Bohr's model of the hydrogen atom are as follows:

Postulates of the Bohr theory

1. The electron moves in circular orbits about the nucleus (the planetary model of the atom) under the influence of the coulombic force of attraction between the electron and the positively charged nucleus (Fig. 41.8).
2. The electron can exist only in very specific orbits; hence the states are quantized (Planck's quantum hypothesis). The allowed orbits are those for which the angular momentum of the electron about the nucleus is an integral multiple of $h/2\pi = \hbar$, where h is Planck's constant. That is,

Quantization

$$mvr = \frac{nh}{2\pi} = n\hbar \qquad (n = 1,2,3,\ldots) \qquad (41.19)$$

3. When the electron is in one of its allowed orbits, it does not radiate energy; hence the atom is stable. Such stable orbits are called *stationary states*. (According to classical electromagnetism, the accelerating electron must radiate electromagnetic waves and in doing so would quickly spiral into the nucleus and annihilate the atom.)

4. The atom radiates energy when the electron "jumps" from one allowed stationary state to another. The frequency of the radiation obeys the condition

Energy of an emitted photon

$$hf = E_i - E_f \tag{41.20}$$

where E_i and E_f are the total energies of the initial and final stationary states, respectively. This assumption is equivalent to energy conservation, where the emitted photon of energy hf is involved in the process. Note that this expression for the photon energy agrees with that proposed by Einstein in arriving at his photoelectric effect equation (Eq. 41.8).

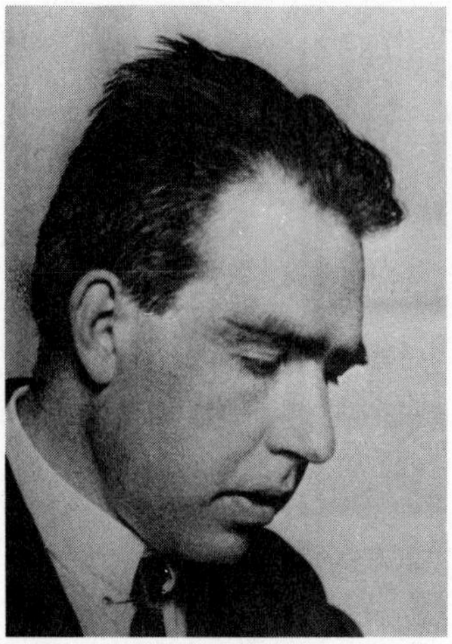

Niels Bohr (1885–1962). (Photo, courtesy of AIP Niels Bohr Library, Margrethe Bohr collection)

Using these assumptions, we shall now calculate the allowed energies of the hydrogen atom, which we can then use to calculate the spectral line positions. We shall use the model described in Fig. 41.8, in which the electron travels in a circular orbit of radius r with an orbital speed v.

Recall that the electrical potential energy of such a system is given by $U = -ke^2/r$, where k is Coulomb's constant. Hence, the total energy of the atom, which is the sum of the kinetic energy and potential energy terms, is given by

$$E = K + U = \tfrac{1}{2}mv^2 - ke^2/r \tag{41.21}$$

Applying Newton's second law to this system, we see that the coulombic attractive force on the electron, ke^2/r^2, must equal ma_r, where $a_r = v^2/r$ is the centripetal acceleration of the electron. That is,

$$\frac{ke^2}{r^2} = \frac{mv^2}{r} \tag{41.22}$$

where $k = 9 \times 10^9 \text{ N} \cdot \text{m}^2/\text{C}^2$ in SI units. Hence, it follows that the kinetic energy is given by

$$\frac{1}{2}mv^2 = \frac{ke^2}{2r} \tag{41.23}$$

We can combine this result with Eq. 41.21 and express the total energy of the atom as simply

Total energy of the hydrogen atom

$$E = -\frac{ke^2}{2r} \tag{41.24}$$

An expression for r is obtained by solving Eqs. 41.19 and 41.22 for v and equating the results:

$$v^2 = \frac{n^2\hbar^2}{m^2r^2} = \frac{ke^2}{mr}$$

Radii of the allowed orbits

$$r = \frac{n^2\hbar^2}{mke^2} \qquad (n = 1,2,3,\ldots) \tag{41.25}$$

That is, the radii of the various orbits have discrete values.

The orbit with the smallest radius, called the *Bohr radius* r_0, corresponds to $n = 1$ and has the value

Bohr radius

$$r_0 = \frac{\hbar^2}{mke^2} \approx 0.529 \text{ Å} \tag{41.26}$$

Substituting Eq. 41.25 into Eq. 41.24 gives the following expression for the energies of the quantum states:

$$E_n = -\frac{mk^2e^4}{2\hbar^2}\left(\frac{1}{n^2}\right) \qquad (n = 1,2,3,\ldots) \tag{41.27}$$

Allowed energy states

Inserting numerical values into Eq. 41.27 gives

$$E_n = -\frac{13.6}{n^2}\text{ eV} \tag{41.28}$$

The *lowest* stationary energy state, or ground state, corresponds to $n = 1$ and has an energy $E_1 = -mk^2e^4/2\hbar^2 = -13.6$ eV. The next state, corresponding to $n = 2$, has an energy $E_2 = \frac{1}{4}E_1 = -3.4$ eV, etc. An energy level diagram showing the energies of these stationary states with corresponding quantum numbers is shown in Fig. 41.9. The uppermost level shown, corresponding to $n \to \infty$, represents the state for which the electron is completely removed from the atom. In this case, $E = 0$ for $r = \infty$. The least energy required to ionize the atom, that is, to completely remove the electron, is called the *ionization* energy. The ionization energy for hydrogen is 13.6 eV.

Ionization energy

Using Eqs. 41.15 and 41.27, together with the fourth Bohr postulate, we find that if the electron jumps from one orbit whose quantum number is n_i to a second orbit whose quantum number is n_f, it emits a photon of frequency f, given by

$$f = \frac{E_i - E_f}{h} = \frac{mk^2e^4}{4\pi\hbar^3}\left(\frac{1}{{n_f}^2} - \frac{1}{{n_i}^2}\right) \tag{41.29}$$

Frequency of an emitted photon

Finally, to compare this result with the empirical formulas for the various spectral series, we use the fact that $\lambda f = c$ and Eq. 41.29 to get

$$\frac{1}{\lambda} = \frac{f}{c} = \frac{mk^2e^4}{4\pi c\hbar^3}\left(\frac{1}{{n_f}^2} - \frac{1}{{n_i}^2}\right) \tag{41.30}$$

Comparing this result with Eq. 41.14, we obtain the following expression for the Rydberg constant:

$$R = \frac{mk^2e^4}{4\pi c\hbar^3} \tag{41.31}$$

If we insert the known values of the various fundamental constants into this expression, the theoretical value for R is found to be in excellent agreement with the spectroscopic value, $R = 1.0973732 \times 10^7\ \text{m}^{-1}$. When Bohr demonstrated this agreement, it was recognized as a major accomplishment of his theory.

In order to compare Eq. 41.30 with spectroscopic data, it is convenient to express it in the form

$$\frac{1}{\lambda} = R\left(\frac{1}{{n_f}^2} - \frac{1}{{n_i}^2}\right) \tag{41.32}$$

Allowed wavelengths in the hydrogen spectrum

We can use this expression to evaluate the wavelengths for the various series in the hydrogen spectrum. For example, in the Balmer series, $n_f = 2$ and $n_i = 3, 4, 5, \ldots$, which corresponds to Eq. 41.14. For the Lyman series, $n_f = 1$ and $n_i = 2, 3, 4, \ldots$. The energy level diagram for hydrogen, shown in Fig. 41.9, indicates the origin of the various spectral lines described earlier. The transitions between levels are repre-

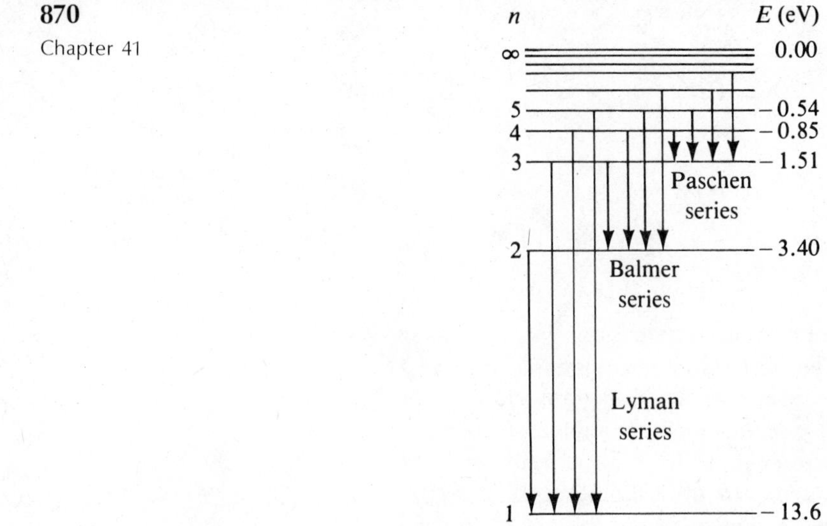

Figure 41.9 Energy level diagram for the hydrogen atom. Some transitions for the Lyman, Balmer, and Paschen series are shown. The quantum numbers are at the left, and the energies of the levels are at the right.

sented by vertical arrows. Note that whenever an electron undergoes a transition from a state designated by n_i to one designated by n_f (where $n_i > n_f$), a photon is emitted whose frequency is given by $(E_i - E_f)/h$. Hence, the Bohr theory successfully predicts the wavelengths of all observed atomic hydrogen spectral lines.

It is possible to obtain a more refined calculation of the energy levels by considering a small correction that depends on nuclear motion. The model we have described assumed that the electron revolves around a massive fixed nucleus of mass M. In reality, the electron and nucleus each revolve around their center of mass, that is, the motion of the nucleus cannot be ignored. The correction for this motion amounts to replacing the electronic mass m by the reduced mass μ, defined as

$$\mu = \frac{mM}{m + M} \tag{41.33}$$

For hydrogen, the correction amounts to only about 1 part in 2000.

If one is dealing with a hydrogen-like ion, such as He^+, Li^{2+}, Be^{3+}, or B^{4+} (one-electron ions), each can be considered as a system of two charges, the electron of mass m and charge $-e$, and the nucleus of mass M and charge $+Ze$, where Z is the atomic number. (Of course, the values of M and μ for these hydrogen-like systems differ from the values for hydrogen.) The radii of the circular orbits for these one-electron ions can only have the values

$$r_n = \frac{n^2\hbar^2}{\mu k Z e^2} \qquad (n = 1,2,3, \ldots) \tag{41.34}$$

which is similar to Eq. 41.25. Furthermore, the allowed energies are given by

Energy levels of hydrogen-like ions

$$E_n = -\frac{\mu k^2 e^4 Z^2}{2\hbar^2 n^2} \qquad (n = 1,2,3, \ldots) \tag{41.35}$$

In addition to the discrete states whose energies are given by Eq. 40.35, there is a continuum of allowed states with energies $E > 0$. An electron can be promoted from the ground state of hydrogen to one of these states by absorbing a photon whose energy is greater than 13.6 eV.

It should be noted that although the Bohr theory correctly predicts the wavelengths of hydrogen and hydrogen-like species, the orbital angular momenta assumed to be integral multiples of $\hbar$ are not the correct values. Quantum mechanics later correctly showed that in a state of energy E_n, possible values of the orbital angular momentum are $\sqrt{l(l+1)}\,\hbar$, where the orbital quantum number $l = 0, 1, 2, \ldots, (n-1)$. Hence, according to quantum mechanics, the electron must have *zero* orbital angular momentum in its ground state (corresponding to $n = 1$).

Bohr's Correspondence Principle

In the previous chapter, we found that newtonian mechanics cannot be used to describe phenomena that occur at speeds approaching the speed of light. Newtonian mechanics is only a special case of relativistic mechanics when $v \ll c$. Similarly, quantum physics is in agreement with classical physics when the quantum numbers are very large. This principle, first postulated by Bohr, is called the *correspondence principle.* For example, consider the hydrogen atom in an orbit for which $n \geq 10\,000$. For such large values of n, the energy differences between adjacent levels approach zero and the levels are nearly continuous. Consequently, the classical model is reasonably accurate. The frequency of light emitted by the atom using this model is the frequency of revolution of the electron in its orbit. Calculations show that for $n \geq 10\,000$, this frequency is different from that predicted by quantum physics by less than 0.015 percent. The details of this calculation are left as a problem.

Example 41.4

The electron in the hydrogen atom makes a transition from the $n = 2$ energy state to the ground state (corresponding to $n = 1$). Find the wavelength and frequency of the emitted photon (Problem 1).

Solution: We can make use of Eq. 41.32 directly to obtain λ, with $n_i = 2$ and $n_f = 1$:

$$\frac{1}{\lambda} = R\left(\frac{1}{{n_f}^2} - \frac{1}{{n_i}^2}\right)$$

$$\frac{1}{\lambda} = R\left(\frac{1}{1} - \frac{1}{2^2}\right) = \frac{3}{4}R$$

$$\lambda = \frac{4}{3R} = \frac{4}{3(1.097 \times 10^7\ \text{m}^{-1})}$$
$$= 1.215 \times 10^{-7}\ \text{m} = 121.5\ \text{nm}$$

This wavelength lies in the ultraviolet region. Since $c = f\lambda$, the frequency of the photon is

$$f = \frac{c}{\lambda} = \frac{3.00 \times 10^8\ \text{m/s}}{1.215 \times 10^{-7}\ \text{m}} = 2.47 \times 10^{15}\ \text{Hz}$$

Example 41.5

The Balmer series for the hydrogen atom corresponds to electronic transitions that terminate in the state of quantum number $n = 2$, as shown in Fig. 41.10. (a) Find the longest-wavelength photon emitted and determine its energy.

The longest-wavelength photon in the Balmer series results from the transition from $n = 3$ to $n = 2$. Using Eq. 41.32 gives

$$\frac{1}{\lambda} = R\left(\frac{1}{{n_f}^2} - \frac{1}{{n_i}^2}\right)$$

$$\frac{1}{\lambda_{max}} = R\left(\frac{1}{2^2} - \frac{1}{3^2}\right) = \frac{5}{36}R$$

$$\lambda_{max} = \frac{36}{5R} = \frac{36}{5(1.097 \times 10^7\ \text{m}^{-1})} = 656.3\ \text{nm}$$

This wavelength is in the red region of the visible spectrum. The energy of this photon is given by

$$E_{photon} = hf = \frac{hc}{\lambda_{max}}$$
$$= \frac{(6.626 \times 10^{-34}\ \text{J}\cdot\text{s})(3.00 \times 10^8\ \text{m/s})}{656.3 \times 10^{-9}\ \text{m}}$$
$$= 3.03 \times 10^{-19}\ \text{J} = 1.89\ \text{eV}$$

We could also obtain the energy of the photon by using the expression $hf = E_3 - E_2$, where E_2 and E_3 are the energy levels of the hydrogen atom, which can be calculated from Eq. 41.28. Note that this is the lowest-energy

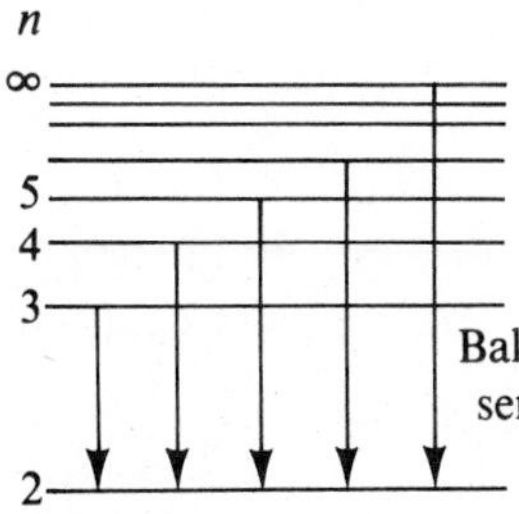

Figure 41.10 (Example 41.5) Transitions responsible for the Balmer series for the hydrogen atom. All transitions terminate in the level $n = 2$.

photon in this series since it involves the smallest energy change.

(b) Find the shortest-wavelength photon emitted and determine its energy.

The shortest-wavelength photon in the Balmer series is emitted when the electron makes a transition from $n = \infty$ to $n = 2$. Therefore

$$\frac{1}{\lambda_{\text{min}}} = R\left(\frac{1}{2^2} - \frac{1}{\infty^2}\right) = \frac{R}{4}$$

$$\lambda_{\text{min}} = \frac{4}{R} = \frac{4}{1.097 \times 10^{-7}\ \text{m}^{-1}} = 364.6\ \text{nm}$$

This wavelength is in the ultraviolet region and corresponds to the series limit. The energy of a photon with this wavelength is

$$E_{\text{photon}} = hc/\lambda_{\text{min}} = 3.41\ \text{eV}$$

This is the maximum-energy photon in this series, since it involves the largest energy change.

Q6. The Bohr theory of the hydrogen atom is based upon several assumptions. Discuss these assumptions and their significance. Do any of these assumptions contradict classical physics?

Q7. Suppose that the electron in the hydrogen atom obeyed classical mechanics rather than quantum mechanics. Why should such a "hypothetical" atom emit a continuous spectrum rather than the observed line spectrum?

Q8. Can the electron in the ground state of hydrogen absorb a photon of energy (a) *less* than 13.6 eV and (b) *greater* than 13.6 eV?

41.6 PHOTONS AND ELECTROMAGNETIC WAVES

In this chapter we have presented very convincing evidence supporting the photon (particle) concept of light. An obvious question that arises at this point is "How can light be considered as a photon when it exhibits wavelike properties?" On the one hand, we describe light in terms of photons with energy hf and momentum h/λ. On the other hand, we must also recognize the fact that light and other electromagnetic waves exhibit interference and diffraction phenomena as discussed in Chapters 38 and 39. Which model is correct? The answer depends on the specific phenomenon being observed. Some experiments can be explained solely on the basis of the photon concept while others are best described with a wave theory. That is, we must accept both theories and therefore admit that the true nature of light is not describable in terms of a single classical concept. However, you should recognize that the same light beam that can eject photoelectrons from a metallic surface can also be diffracted by a grating. In other words, *the photon concept and the wave theory of light complement each other.*

The photon theory and wave theory complement each other

We can perhaps understand why photons are compatible with electromagnetic waves in the following manner. We may suspect that long-wavelength radio waves do not exhibit particle characteristics. Consider, for instance, radio waves at a frequency of 2.5 MHz. The energy of a photon having this frequency is only about 10^{-8} eV. From a practical viewpoint, this energy is too small to be detected as a single photon. A sensitive radio receiver might detect as few as 10^{10} photons/s. Such a high flux of photons would appear, on the average, as a continuous wave. It would be unlikely that any graininess would appear in the detected signal with such a large number of photons reaching the detector each second.

Now consider what happens as we go to higher frequencies, or shorter wavelengths. In the visible region, it is possible to observe both the photon and wave character of light. As we mentioned earlier, a light beam shows interference phenomena and at the same time can produce photoelectrons, which can be explained only by using Einstein's photon concept. At even higher frequencies and correspondingly shorter wavelengths, the momentum and energy of the photon increase. Consequently, the photon description of light begins to dominate over the wave description. For example, an X-ray photon is easily detected as a single event. However, as the wavelength decreases, wave phenomena such as interference and diffraction become more difficult to observe. Since the wavelengths of X-rays are of the order of atomic dimensions, one can use the regular atomic array of a crystal to diffract them. More indirect procedures must be used to detect the wave nature of very high frequency radiation, such as gamma rays.

We have given evidence which shows that all forms of electromagnetic radiation can be described from *two* points of view. In one extreme, electromagnetic waves describe the overall interference pattern formed by a large number of photons. In the other extreme, the photon description is natural when we are dealing with a highly energetic photon of very short wavelength. Hence, light has a *dual* nature; it exhibits both wave and photon characteristics.

Light has a dual nature

A more sophisticated description of what actually occurs uses a probabilistic interpretation known as *wave mechanics*. According to this viewpoint, it is impossile to predict when or where a photon will transfer its energy to a detector. The wave theory will tell us only the probability of such an event occurring. In other words, it is impossible to say precisely where the photon is at any instant. The best one can do is assign some probability of observing the photon at some location within some time interval.

41.7 THE WAVE PROPERTIES OF PARTICLES

In 1924, the French physicist Louis de Broglie (born 1892) wrote a doctoral dissertation in which he proposed that since photons have wave and particle characteristics, perhaps all forms of matter have wave as well as particle properties. This was a highly revolutionary idea which at the time lacked experimental confirmation. However, the suggestion received immediate attention by the scientific community and played an important role in the subsequent development of quantum mechanics.

We have seen that a photon of wavelength λ has a momentum given by $p = h/\lambda$. Hence it follows that the photon wavelength can be specified by its momentum, or $\lambda = h/p$. De Broglie suggested that *material particles of momentum p should also have wavelike properties and a corresponding wavelength.* Since the momentum of a particle of mass m and velocity v is $p = mv$, the *de Broglie wavelength* is

$$\lambda = \frac{h}{p} = \frac{h}{mv} \tag{41.36}$$

De Broglie wavelength

Furthermore in analogy with photons, de Broglie postulated that the frequency of matter waves (that is, waves associated with real particles) obeys the Einstein relation $E = hf$, so that

$$f = \frac{E}{h} \tag{41.37}$$

The dual nature of matter is quite apparent in these two equations. That is, each equation contains both particle concepts (mv and E) and wave concepts (λ and f). Note that the Compton effect confirms the validity of $p = h/\lambda$ for photons, and the photoelectric effect confirms the validity of $E = hf$ for photons. Thus, indirectly, these relations are established experimentally for photons, making the de Broglie hypothesis that much easier to accept.

One important consequence of de Broglie's hypothesis is that he was able to provide an explanation for the quantization of angular momentum postulated by Bohr. De Broglie applied these ideas to the electron in the hydrogen atom and showed that the quantum condition $mvr = n\hbar$ is equivalent to a condition of standing electron waves. In effect, de Broglie required that the circumference of an allowed circular electron orbit must equal an integral number of electron wavelengths, as indicated in Fig. 41.11. That is,

$$2\pi r = n\lambda$$

Since $\lambda = h/mv$, we can write this condition

$$2\pi r = n\frac{h}{mv}$$

$$mvr = n\frac{h}{2\pi} = n\hbar$$

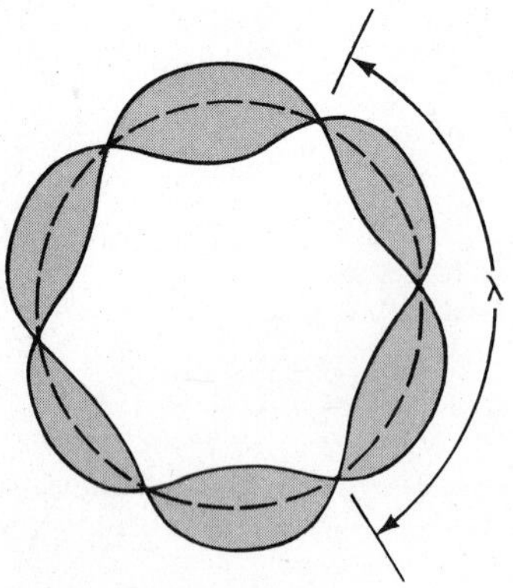

Figure 41.11 Standing waves fit to a circular Bohr orbit. In this particular picture, three wavelengths are fit to the orbit, corresponding to the $n = 3$ energy state of the Bohr theory.

This, of course, is Bohr's postulate of angular momentum quantization.[4] It has been derived by assuming that the electron in the hydrogen atom behaves as a wave and that an integral number of wavelengths must fit each allowed circular orbit. Thus, we see that electron waves that fit the orbits are standing waves because of the boundary conditions imposed. These standing waves have discrete frequencies, corresponding to the allowed wavelengths. If $n\lambda \neq 2\pi r$, a standing-wave pattern can never form a closed circular orbit as in Fig. 41.11.

Since the de Broglie waves have frequencies given by $f = E/h$, we see that the condition of standing waves implies quantized energies. This model is analogous to the case of standing waves on a string fixed at both ends (Chapter 34). The string will vibrate with discrete frequencies (wavelengths) whose values depend on the length of the string and the wave velocity.

De Broglie's proposal that any kind of particle exhibits both wave and particle properties was first regarded as pure speculation. If particles such as electrons had wavelike properties, then under the correct conditions they should exhibit interference phenomena. Three years later, in 1927, C. J. Davisson and L. H. Germer of the United States succeeded in measuring the wavelength of electrons. Their important discovery provided the first experimental confirmation of the matter waves proposed by de Broglie.

It is interesting to point out that the intent of the initial Davisson–Germer experiment was not to confirm the de Broglie hypothesis. In fact, their discovery was made quite by accident, as is often the case. The experiment involved the scattering of electrons from a nickel target in a vacuum. During one experiment, the nickel surface was badly oxidized because of an accidental break in the vacuum system. After the nickel target was heated to remove the oxide coating, subsequent experiments showed that the scattered electrons exhibited intensity maxima and minima at specific angles. The experimenters finally realized that the nickel had formed large crystals upon heating and that the regularly spaced planes of atoms in the crystal served as a diffraction grating for electron matter waves. Thus, an important scientific discovery was made quite by accident.

Shortly thereafter, Davisson and Germer performed more extensive diffraction measurements on electrons scattered from single-crystal targets. Their results showed quite conclusively the wave nature of electrons and confirmed the de Broglie relation $p = h/\lambda$. In the same year, G. P. Thomson of Scotland also observed electron diffraction patterns by passing electrons through very thin gold foils. Diffraction patterns have since been observed for neutrons, protons, and even neutral

[4]Note that de Broglie's analysis still failed to explain the fact that states having an orbital angular momentum quantum number of zero do exist.

atoms. Hence the existence of matter waves has been established in various ways.

The problem of understanding the dual nature of both matter and radiation is conceptually difficult, since the two models seem to contradict each other. This problem as it applies to light was discussed in the previous section. Bohr helped resolve this problem in his *principle of complementarity*, which states that *the wave and particle models of either matter or radiation complement each other.* Neither model can be used exclusively to adequately describe matter or radiation. A complete understanding is obtained only if the two models are combined in a complementary manner.

The principle of complementarity

Example 41.6

A particle of mass 5×10^{-3} g moves with a speed of 8 m/s. Calculate its de Broglie wavelength.

Solution: Using Eq. 41.36 gives

$$\lambda = \frac{h}{mv} = \frac{6.63 \times 10^{-34}\ \text{J}\cdot\text{s}}{(5 \times 10^{-6}\ \text{kg})(8\ \text{m/s})} = 1.66 \times 10^{-29}\ \text{m}$$

Note that this wavelength is much smaller than the size of any available aperture. For example, the diameter of a nucleus is about 10^{-15} m, which is large in comparison with the calculated wavelength. This simple calculation demonstrates why the wave properties of macroscopic objects cannot be observed.

Example 41.7

A particle of charge q and mass m is accelerated from rest through a potential difference V. (a) Find its de Broglie wavelength.

When a charge is accelerated from rest through a potential difference V, its gain in kinetic energy $\frac{1}{2}mv^2$ must equal its loss in potential energy qV, since energy is conserved. That is,

$$\tfrac{1}{2}mv^2 = qV$$

Since $p = mv$, we can express this in the form

$$\frac{p^2}{2m} = qV \qquad \text{or} \qquad p = \sqrt{2mqV}$$

Substituting this expression for p into the de Broglie relation $\lambda = h/p$ gives

$$\lambda = \frac{h}{p} = \frac{h}{\sqrt{2mqV}}$$

(b) Calculate λ if the particle is an electron and $V = 50$ V.

The de Broglie wavelength of the electron accelerated through 50 V is

$$\lambda = \frac{h}{\sqrt{2mqV}}$$
$$= \frac{6.63 \times 10^{-34}\ \text{J}\cdot\text{s}}{\sqrt{2(9.11 \times 10^{-31}\ \text{kg})(1.6 \times 10^{-19}\ \text{C})(50\ \text{V})}}$$
$$= 1.74 \times 10^{-10}\ \text{m} = 1.74\ \text{Å}$$

This wavelength is of the order of atomic dimensions and the spacing between atoms in a solid. Such low-energy electrons are normally used in electron diffraction experiments.

Q9. If matter has a wave nature, why is this wavelike character not observable in our daily experiences?

41.8 THE WAVE FUNCTION

There is a striking similarity between the behavior of light and the behavior of matter. Both have dualistic character in that they behave both as waves and as particles. We can carry the analogy even further. In the case of light waves, we described how a wave theory gives only the probability of finding a photon at a given point within a given time interval. Likewise, matter waves are described by a complex-valued wave function (usually denoted by ψ, the Greek letter psi) whose absolute square $|\psi|^2 = \psi^*\psi$ gives the probability of finding the particle at a given point at some instant.

This interpretation of matter waves was first suggested by Max Born (1882–1970) in 1926. In the same year, Erwin Schrödinger (1887–1961) proposed a wave equation that described the manner in which matter waves change in space and time. The *Schrödinger wave equation* represents a key element in the theory of quantum mechanics. Its role is as important in quantum mechanics as that played by Newton's laws of motion in classical mechanics. Schrödinger's wave equation has been successfully applied to the hydrogen atom and to many other microscopic systems. Its importance in most aspects of modern physics cannot be overemphasized. Since the mathematical complexities of this equation and its solution are beyond the level of this text, we shall not discuss it further.

From the viewpoint of classical mechanics, there is a great conflict in the observation that electrons exhibit both wave properties and particle properties. The mathematical formalism of quantum mechanics resolves this conflict. Each particle is represented by a wave function $\psi(x,y,z,t)$, which is sometimes referred to as a *probability amplitude. The probability of finding the particle within a small volume* dV *near a particular point* (x,y,z) *at time t is given by* $|\psi|^2\,dV$. We use the square of the absolute value of ψ since the wave function can be a complex number.

Experimentally, there is a finite probability of finding a particle at some point and at some instant. The value of the probability must lie between the limits 0 and 1. For example, if the probability is 0.3, this would signify a 30 percent chance of finding the body.

You will most likely be exposed to the mathematical formalism of quantum mechanics in a future course. Although the formalism is self-consistent and predicts experimental observations, one is always faced with many questions regarding physical interpretations. For instance, you might wonder if anything material actually "waves." Actually, there is no medium waving with matter waves as there is in the case of water or sound waves.

The concepts of quantum mechanics, strange as they sometimes may seem, developed from older ideas in classical physics. In fact, if the techniques of quantum mechanics are applied to macroscopic systems rather than atomic systems, the results are essentially identical with those of classical physics. This blending of the two theories occurs when the de Broglie wavelength is small compared with the dimensions of the system. The situation is similar to the agreement between relativistic mechanics and classical mechanics when $v \ll c$. For these reasons, newtonian physics can give an accurate description of a system only if the system's dimensions are large compared with atomic dimensions and only if the system is moving with a speed that is small relative to the speed of light.

41.9 THE UNCERTAINTY PRINCIPLE

If you were to measure the position and velocity of a particle at any instant, you would always be faced with experimental uncertainties in your measurements. According to classical mechanics, there is no fundamental barrier to an ultimate refinement of the apparatus and/or experimental procedures. That is, it would be possible, in principle, to make such measurements with arbitrarily small uncertainty, or with infinite precision. However, since radiation and matter have the wave-particle dual character, it is fundamentally impossible to make *simultaneous* measurements of the particle's position and velocity with infinite precision. This statement, known as the *uncertainty principle,* was first proposed by Werner Heisenberg (1901–1976) in 1927.

Consider a particle moving along the x axis, and suppose that Δx and Δp represent the uncertainties in the measured values of position and momentum, respectively, at some instant. The uncertainty principle says that the product $\Delta x\,\Delta p$ is never less than a number of the order of Planck's constant h. More specifically,

Uncertainty principle

$$\Delta x\,\Delta p \geq \hbar/2 \tag{41.38}$$

That is, *it is physically impossible to measure simultaneously the exact position and exact momentum of a particle.* If Δx is made very small, Δp will be large, and vice versa.

In order to understand the uncertainty principle, let us consider the following experiment. Suppose you wish to measure the position and momentum of an electron. You might be able to perform such an experiment by viewing the electron with a powerful microscope. However, in the process of such a measurement the light beam has to interact with the electron, as shown in Fig. 41.12. Consequently, the photons transfer some energy and momentum to the electron. In order to "see" the electron, at least one photon must be scattered into the microscope. Since the incident photon has momentum h/λ, the *change* in momentum of the electron as the result of the collision is at least h/λ. Therefore, the uncertainty in the measured value of the electron's momentum would be $\Delta p \geq h/\lambda$. In other words, *there is an uncertainty introduced in the momentum of the electron as the result of making a measurement.* There is no way to eliminate this effect. Furthermore, since light also has wave properties, we would expect to be able to determine the position of the electron only to within one wavelength of the light being used, so that $\Delta x \approx \lambda$. Multiplying these two uncertainties gives

$$\Delta x\, \Delta p \geq \lambda \left(\frac{h}{\lambda}\right) = h$$

which agrees with Eq. 41.38 (apart from a small numerical factor).

Heisenberg's uncertainty principle enables us to better understand the dualistic wave-particle nature of both light and matter. We have seen that the wave description is quite the opposite of the particle description. Therefore, if an experiment is designed to reveal the particle character of an electron (such as the photoelectric effect), its wave character will become fuzzy. Likewise, if the experiment is designed to accurately measure the electron's wave properties (such as diffraction from a crystal), its particle character will become fuzzy.

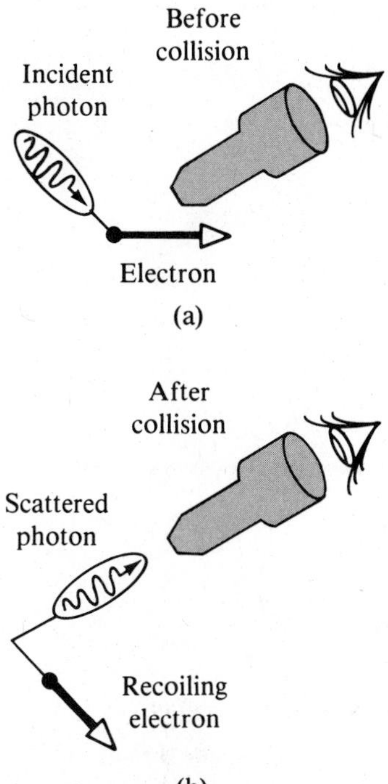

Figure 41.12 A thought experiment for viewing an electron with a powerful microscope. (a) The electron is viewed before colliding with the photon. (b) The electron recoils (is disturbed) as the result of collision with the photon.

Example 41.8

The speed of an electron is measured to be 5.00×10^3 m/s to an accuracy of 0.003%. Find the uncertainty in determining the position of this electron.

Solution: The momentum of the electron is

$$\begin{aligned} p = mv &= (9.1 \times 10^{-31}\text{ kg})(5.00 \times 10^3\text{ m/s}) \\ &= 4.55 \times 10^{-27}\text{ kg}\cdot\text{m/s} \end{aligned}$$

Since the uncertainty in p is 0.003% of this value,

$$\begin{aligned} \Delta p &= (0.00003)p = (0.00003)(4.55 \times 10^{-27}\text{ kg}\cdot\text{m/s}) \\ &= 1.37 \times 10^{-31}\text{ kg}\cdot\text{m/s} \end{aligned}$$

The uncertainty in position can now be calculated by using this value of Δp and Eq. 41.38:

$$\Delta x \Delta p \geq \frac{1}{2}\hbar$$

$$\begin{aligned} \Delta x \geq \frac{\hbar}{2\Delta p} &= \frac{1.05 \times 10^{-34}\text{ J}\cdot\text{s}}{2(1.37 \times 10^{-31}\text{ kg}\cdot\text{m/s})} \\ &= 0.385 \times 10^{-3}\text{ m} = 0.385\text{ mm} \end{aligned}$$

Q10. In what way does Bohr's model of the hydrogen atom violate the uncertainty principle?

41.10 LASERS AND ATOMIC TRANSITIONS

In order to understand how a laser operates, you must first be familiar with some of the basic processes involved in an atomic system. We shall present a qualitative description of such processes before we discuss the principles of the laser.

We have seen that an atom will emit radiation at discrete frequencies, which correspond to the energy separation between the various allowed states. Consider

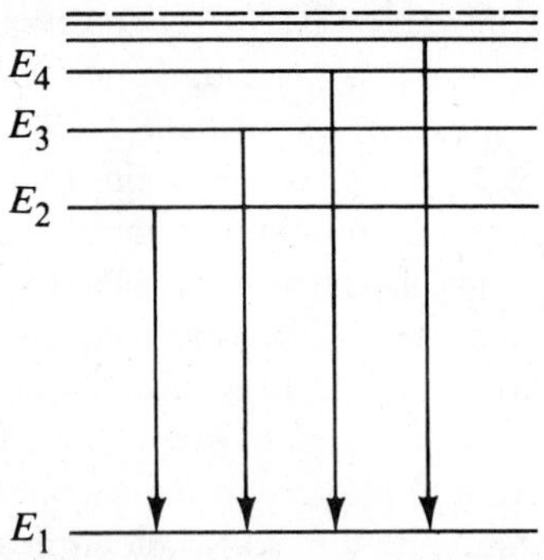

Figure 41.13 Energy level diagram of an atom with various allowed states. The lowest energy state E_1 is the ground state. All others are excited states.

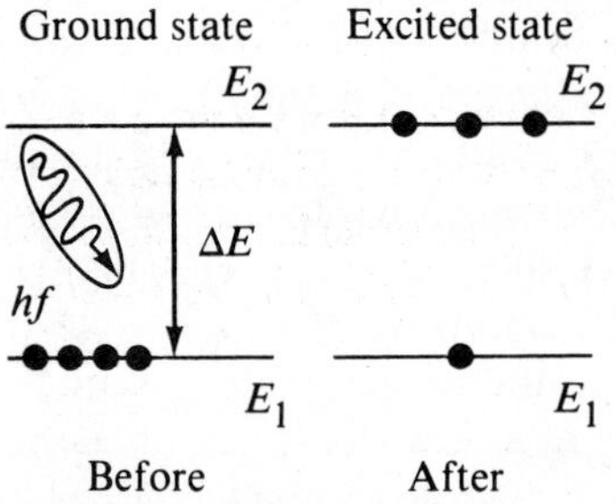

Figure 41.14 Schematic diagram representing the process of *stimulated absorption* of a photon by an atom. The dots represent electrons in the various states.

an atom with many allowed states, labeled E_1, E_2, E_3, . . . , as in Fig. 41.13, where the lowest-lying state, E_1, is called the *ground state*. When light is incident on the atom, only certain photons will interact with the atom. Only those photons whose energy hf matches the energy separation ΔE between two levels can be absorbed. A schematic diagram representing this *stimulated absorption* process is shown in Fig. 41.14. At ordinary temperatures, most of the atoms are in the ground state. If a vapor cell containing many atoms of an element is illuminated by a light beam characterized by a continuous spectrum of photon frequencies, only those photons of energies $(E_2 - E_1)$, $(E_3 - E_1)$, $(E_4 - E_1)$, and so on, can be absorbed. As a result of this absorption process, some atoms are raised to various allowed higher energy levels called *excited states*.

Stimulated absorption

Once an atom is in an excited state, there is a certain probability per unit time that the electron will revert back to a lower energy level by emitting a photon, as described in Fig. 41.15. This process is known as *spontaneous emission*. In typical cases, an atom will remain in an excited state only for about 10^{-8} s.

Spontaneous emission

Finally, there is a third process, which is of major importance in lasers, known as *stimulated emission*. Suppose an atom is in an excited state E_2 as in Fig. 41.16, and a photon of energy $hf = E_2 - E_1$ is incident on it. The incoming photon will increase the probability that the atom will revert to its ground state and thereby emit a second photon of energy hf. This process of speeding up atomic transitions to lower levels is called *stimulated emission*. Note that there are two identical photons that result from this process, corresponding to the incident photon and the emitted

Stimulated emission

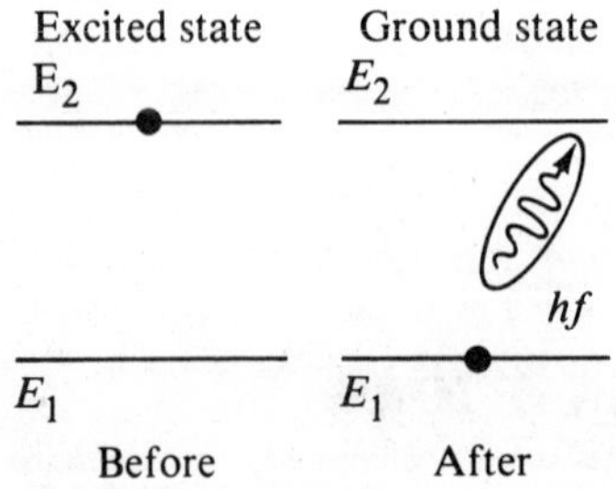

Figure 41.15 Schematic diagram representing the process of *spontaneous emission* of a photon by an atom that is initially in the excited state E_2.

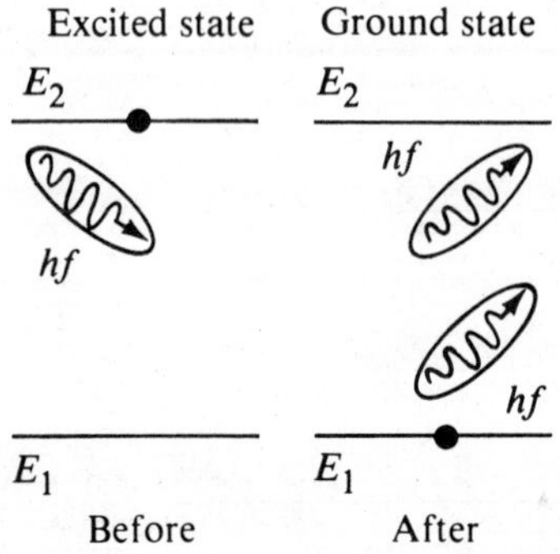

Figure 41.16 Schematic diagram representing the process of *stimulated emission* of a photon by an incoming photon of energy hf. Initially, the atom is in the excited state E_2.

photon. The emitted photon will be exactly *in phase* with the incident photon.[5] These photons can, in turn, stimulate other atoms to emit photons in a chain of similar processes. The many photons produced in this fashion are the source of the intense, coherent light in a laser.

We have described how an incident photon can cause atomic transitions either upward (stimulated absorption) or downward (stimulated emission). Both processes are equally probable. When light is incident on a system of atoms, there is usually a *net* absorption of energy because there are many more atoms in the ground state than in excited states. That is, in a normal situation, there are more atoms in state E_1 ready to absorb photons than there are atoms in states E_2, E_3, . . . ready to emit photons. However, if one can invert the situation so that there are more atoms in an excited state than in a lower state, a condition called *population inversion*, a *net emission* of photons can result. This, in fact, is the fundamental principle involved in the operation of a laser, an acronym for *l*ight *a*mplification by *s*timulated *e*mission of *r*adiation. The amplification process corresponds to a buildup of photons in the system as the result of a chain reaction of such events.

Population inversion

Three conditions must be satisfied in order to achieve laser action:

Condition necessary to achieve laser action

1. There must be an inverted population (that is, more atoms in an excited state than in the ground state).
2. The excited state must be a *metastable state*, which means its lifetime must be long compared with the usually short lifetimes of excited states. In this manner, the process of stimulated emission will occur before spontaneous emission.
3. The emitted photons must be confined within the system long enough to allow them to stimulate further emission from other excited atoms. This is achieved by the use of reflecting mirrors at the ends of the system. One end is made totally reflecting, and the other is slightly transparent to allow the laser beam to escape.

One device that exhibits stimulated emission of radiation is the helium-neon gas laser. The energy level diagram for the neon atom in this system is shown in Fig. 41.17. Neon atoms are excited to state E_3 by collision with excited helium atoms. Stimulated emission occurs as the atoms make a transition to state E_2, and neighboring excited atoms are stimulated, finally resulting in the production of coherent light at a wavelength of 6328 Å. Following this process, the neon atoms revert to the ground state after colliding with the walls.

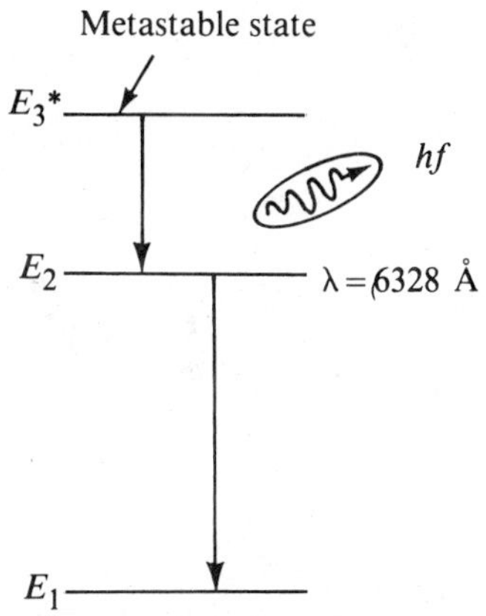

Figure 41.17 Energy level diagram for the neon atom, which emits photons at 6328 Å through stimulated emission. The photon at this wavelength arises from the transition $E_3 \rightarrow E_2$. This is the source of coherent light in the helium-neon gas laser.

Since the development of the first laser in 1960, there has been a tremendous growth in laser technology. Lasers are now available that cover wavelengths in the infrared, visible, and ultraviolet regions. Various active media have been employed (gases, solids, and liquids) to achieve both continuous and pulsed operation, and the number of applications of laser technology is ever increasing. Such applications include the use of laser beams (1) to generate three-dimensional images of objects in a process called *holography*, (2) as a surgical tool for "welding" detached retinas, (3) to perform precision surveying and length measurements, (4) as a potential energy source for inducing fusion reactions, (5) for telephone communication along optical fibers, and (6) for precision cutting of metals and other materials. These and other applications are possible because of the unique characteristics of laser light. In addition to its highly monochromatic and coherent nature, laser light is also highly directional and can be sharply focused to produce extremely high flux densities.

Q11. Explain the difference between laser light and light from an incandescent bulb.

[5]This description accounts for the blending of the wave and particle characteristics of light.

Q12. Explain why laser action could not occur without population inversion between atomic levels.

41.11 SUMMARY

The characteristics of *blackbody radiation* cannot be explained using classical concepts. Planck first introduced the *quantum concept* when he assumed that the atomic oscillators responsible for this radiation existed only in discrete states.

The *photoelectric effect* is a process whereby electrons can be ejected from a metallic surface when light is incident on that surface. Einstein provided a successful explanation of this effect by extending Planck's quantum hypothesis to the electromagnetic field. In this model, light is viewed as a stream of particles called *photons*, each with energy $E = hf$, where f is the frequency and h is Planck's constant. The kinetic energy of the ejected photoelectron is given by

Photoelectric effect equation

$$K_{\text{max}} = hf - \phi \tag{41.8}$$

where ϕ is the work function of the metal.

X-rays from an incident beam are scattered at various angles by electrons in a target such as carbon. In such a scattering event, a shift in wavelength is observed for the scattered X-rays, and the phenomenon is known as the *Compton effect*. Classical physics does not explain this effect. If the X-ray is treated as a photon, conservation of energy and momentum applied to the photon-electron collisions yields the following expression for the Compton shift:

Compton shift equation

$$\Delta\lambda = \frac{h}{mc}(1 - \cos\theta) \tag{41.10}$$

where m is the mass of the electron, c is the speed of light, and θ is the scattering angle.

The Bohr model of the atom is successful in describing the spectra of atomic hydrogen and hydrogen-like ions. One of the basic assumptions of the model is that the electron can exist only in discrete orbits such that the angular momentum mvr is an integral multiple of $h/2\pi = \hbar$. Assuming circular orbits and a simple coulombic attraction between the electron and proton, the energies of the quantum states for hydrogen are calculated to be

Allowed energies of the hydrogen atom

$$E_n = -\frac{mk^2e^4}{2\hbar^2}\left(\frac{1}{n^2}\right) \tag{41.27}$$

where k is Coulomb's constant, e is electronic charge, and n is an integer called a *quantum number*.

If the electron in the hydrogen atom makes a transition from an orbit whose quantum number is n_i to one whose quantum number is n_f, where $n_f < n_i$, a photon is emitted by the atom whose frequency is given by

Frequency of a photon emitted from hydrogen

$$f = \frac{mk^2e^4}{4\pi\hbar^3}\left(\frac{1}{n_f^2} - \frac{1}{n_i^2}\right) \tag{41.29}$$

Using $E = hf = hc/\lambda$, one can calculate the wavelengths of the photons for various transitions in which there is a change in quantum number, $n_i \rightarrow n_f$. The calculated wavelengths are in excellent agreement with observed line spectra.

All matter exhibits both particle and wave character. The dualistic nature of matter was proposed by de Broglie. The *de Broglie wavelength* of any particle of mass m and velocity v is given by

$$\lambda = \frac{h}{p} = \frac{h}{mv} \quad (41.36)$$

De Broglie wavelength

and the frequency of matter waves obeys the Einstein relation $f = E/h$, where E is the total energy of the particle. Subsequent experiments that confirmed the concept of matter waves included the observation of electron diffraction by Davisson and Germer and independently by Thomson.

In the scheme of *quantum mechanics,* a wave function $\psi(x,y,z,t)$ is used to represent the state of a particle. The quantity $|\psi|^2$ gives the probability density for locating the particle near a particular point at a given time. Quantum mechanics is very successful in describing the behavior of atomic and molecular processes.

According to Heisenberg's *uncertainty principle,* it is impossible to simultaneously determine a particle's position and its velocity with infinite precision. More specifically, if Δx is the uncertainty in the measured position and Δp is the uncertainty in momentum, the product $\Delta x\, \Delta p$ is given by

$$\Delta x\, \Delta p \geq \hbar/2 \quad (41.38)$$

Uncertainty principle

Such uncertainties cannot be avoided in any measurement process, since the system being measured must interact with some probe, such as a light beam.

Lasers are devices that work on the principle of stimulated emission of radiation from a system of atoms.

EXERCISES

Section 41.1 Blackbody Radiation and Planck's Hypothesis

1. Calculate the energy of a photon whose frequency is (a) 5×10^{14} Hz, (b) 10 GHz, (c) 30 MHz. Express your answers in eV.
2. Determine the corresponding wavelengths for the photons described in Exercise 1.
3. Show that at *long* wavelengths, Planck's radiation law (Eq. 41.2) reduces to the Rayleigh-Jeans law (Eq. 41.1).
4. Show that at *short* wavelengths or *low* temperatures, Planck's radiation law (Eq. 41.2) predicts an exponential decrease in $I(\lambda,T)$ given by *Wien's radiation law:*

$$I(\lambda,T) = \frac{2\pi hc^2}{\lambda^5} e^{-hc/\lambda kT}$$

5. Consider the mass-spring system described in Example 41.1. If the quantum number n changes by unity, calculate the *fractional* change in energy of the oscillator.

Section 41.2 The Photoelectric Effect

6. The work function for potassium is 2.24 eV. If potassium metal is illuminated with light of wavelength 350 nm, find (a) the maximum kinetic energy of the photoelectrons and (b) the cutoff wavelength.
7. Molybdenum has a work function of 4.2 eV. (a) Find the cutoff wavelength and threshold frequency for the photoelectric effect. (b) Calculate the stopping potential if the incident light has a wavelength of 200 nm.
8. When cesium metal is illuminated with light of wavelength 300 nm, the photoelectrons emitted have a maximum kinetic energy of 2.23 eV. Find (a) the work function of cesium and (b) the stopping potential if the incident light has a wavelength of 400 nm.
9. Consider the metals lithium, iron, and mercury, which have work functions of 2.3 eV, 3.9 eV, and 4.5 eV, respectively. If light of wavelength 300 nm is incident on each of these metals, determine (a) which metals exhibit the photoelectric effect and (b) the maximum kinetic energy for the photoelectron in each case.
10. Light of wavelength 500 nm is incident on a metallic surface. If the stopping potential for the photoelectric effect is 0.45 V, find (a) the maximum energy of

the emitted electrons, (b) the work function, and (c) the cutoff wavelength.

Section 41.3 The Compton Effect

11. Calculate the energy and momentum of a photon of wavelength 500 nm.

12. X-rays of wavelength 2.00 Å are scattered from a block of carbon. If the scattered radiation is detected at 90° to the incident beam, find (a) the Compton shift $\Delta\lambda$ and (b) the kinetic energy imparted to the recoiling electron.

13. X-rays with an energy of 300 keV undergo Compton scattering from a target. If the scattered rays are detected at 30° relative to the incident rays, find (a) the Compton shift at this angle, (b) the energy of the scattered X-ray, and (c) the energy of the recoiling electron.

Section 41.4 Atomic Spectra

14. Calculate the wavelengths of the first three lines in the Balmer series for hydrogen using Eq. 41.14.

15. Calculate the wavelengths of the first three lines in the Lyman series for hydrogen using Eq. 41.16.

Section 41.5 The Bohr Theory of Hydrogen

16. A hydrogen atom initially in its ground state ($n = 1$) absorbs a photon and ends up in the state for which $n = 3$. (a) What is the energy of the absorbed photon? (b) If the atom returns to the ground state, what photon energies could the atom emit?

17. A photon is emitted from a hydrogen atom which undergoes a transition from the state $n = 3$ to the state $n = 2$. Calculate (a) the energy, (b) the wavelength, and (c) the frequency of the emitted photon.

18. Use Eq. 41.25 to calculate the radius of the first, second, and third Bohr orbits of hydrogen.

19. (a) Using Eq. 41.17, calculate the longest and shortest wavelengths for the Paschen series. (b) Determine the photon energies corresponding to these wavelengths.

20. Find the potential energy and kinetic energy of an electron in the ground state of the hydrogen atom.

21. (a) Construct an energy level diagram for the He^+ ion, for which $Z = 2$. (b) What is the ionization energy for He^+?

22. Construct an energy level diagram for the Li^{2+} ion, for which $Z = 3$.

Section 41.7 The Wave Properties of Particles

23. Show that the de Broglie wavelength of an electron accelerated from rest through a potential difference V is given by $\lambda = (12.26/\sqrt{V})$ Å, where V is in volts.

24. Use the results to Exercise 23 to calculate the de Broglie wavelength for an electron with kinetic energy (a) 50 eV and (b) 50 keV.

25. Calculate the de Broglie wavelength of a proton that is accelerated from rest through a potential difference of 10 MV.

26. Find the de Broglie wavelength of a 0.15-kg ball moving with a speed of 20 m/s.

Section 41.9 The Uncertainty Principle

27. A ball of mass 50 g moves with a speed of 30 m/s. If its speed is measured to an accuracy of 0.1%, what is the minimum uncertainty in its position?

28. A proton has a kinetic energy of 1 MeV. If its momentum is measured with an uncertainty of 5%, what is the minimum uncertainty in its position?

PROBLEMS

1. Use Bohr's model of the hydrogen atom to show that when the atom makes a transition from the state n to the state $n - 1$, the frequency of the emitted light is given by

$$f = \frac{2\pi^2 mk^2e^4}{h^3}\left[\frac{2n - 1}{(n - 1)^2n^2}\right]$$

Show that as $n \to \infty$, the expression above varies as $1/n^3$ and reduces to the classical frequency one would expect the atom to emit. (Hint: To calculate the classical frequency, note that the frequency of revolution is $v/2\pi r$, where r is given by Eq. 41.25.) This is an example of the correspondence principle, which requires that the classical and quantum models must agree for large values of n.

2. Derive the formula for the Compton shift (Eq. 41.10) from Eqs. 41.11, 41.12, and 41.13.

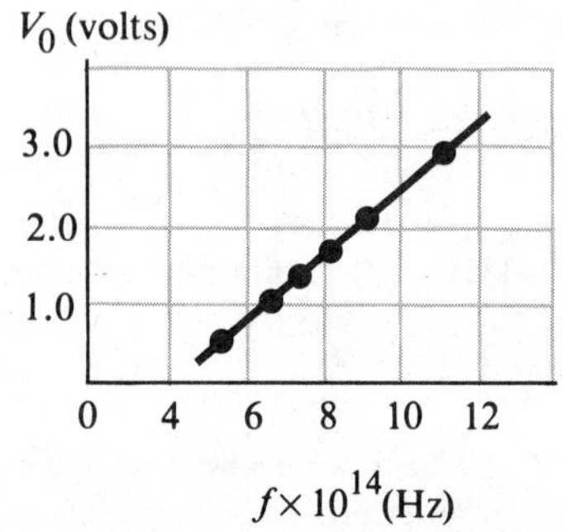

Figure 41.18 (Problem 3).

3. Figure 41.18 shows the stopping potential versus incident photon frequency for the photoelectric effect for sodium. Use these data points to find (a) the work function, (b) the ratio h/e, and (c) the cutoff wavelength. (Data taken from R. A. Millikan, *Phys. Rev.* **7**:362 [1916].)

4. Light from a certain He-Ne laser has a power output of 1.0 mW. The entire beam is incident on a metal target which requires 1.5 eV to remove an electron from its surface. (a) Perform a classical calculation to determine how long it would take the metal to absorb 1.5 eV from the incident beam. (Hint: Assume that the area of an atom is 1 $Å^2$ and first calculate the energy incident on each atom per second.) (b) Compare the (wrong) answer obtained in (a) to the actual response time for photoelectric emission ($\approx 10^{-9}$ s), and discuss the reasons for the large discrepancy.

5. A hydrogen atom is in its ground state ($n = 1$). Using the Bohr theory of the atom, calculate (a) the radius of the orbit, (b) the linear momentum of the electron, (c) the angular momentum of the electron, (d) the kinetic energy, (e) the potential energy, and (f) the total energy.

6. A photon of energy $E = hc/\lambda$ collides with a particle being studied under a powerful microscope. Assuming that the position of the particle being struck by the photon is uncertain by at least the wavelength λ, (a) show that the uncertainty in the time it takes the photon to move one wavelength is $\Delta t \geq \lambda/c$. (b) If the uncertainty in the energy ΔE of the particle is equal to or greater than the energy of the incident photon, show that the relation $\Delta E \Delta t \geq \hbar$ is satisfied. This is another Heisenberg uncertainty relationship, which restricts our ability to know the exact state of a particle.

7. A pulsed laser is constructed with a ruby crystal as the active element. The ruby crystal (Al_2O_3) contains Cr^{3+} impurities, which substitute for the Al^{3+} ions and give the crystal its red color. Laser light is emitted from an "inverted" Cr^{3+}-three-level system. The ruby rod contains a total of 3.0×10^{19} Cr^{3+} ions (a typical value). If the wavelength of the laser light emitted is 694.4 nm, find (a) the energy of one emitted photon, in eV, and (b) the *total* energy available per laser pulse, assuming a total population inversion. (c) Why do you suppose the energy per pulse is much less than this in practice?

8. An electron and a proton each have a thermal kinetic energy of $3kT/2$ at a temperature of 2000 K. Calculate the de Broglie wavelength of each particle. (Recall that k is Boltzmann's constant: $k = 1.38 \times 10^{-23}$ J/K.)

9. Consider the problem of the distribution of blackbody radiation described in Fig. 41.1. Note that as T *increases*, the wavelength λ_{max} at which $I(\lambda,T)$ reaches a maximum shifts toward shorter wavelengths. (a) Show that there is a general relationship between temperature and λ_{max} (known as *Wien's displacement law*), which states that $T\lambda_{max}$ = constant. (b) Obtain a numerical value for this constant. [Hint: Start with Planck's radiation law and note that the slope of $I(\lambda,T)$ versus λ is zero when $\lambda = \lambda_{max}$.]

10. The total power per unit area radiated by a blackbody at a temperature T is given by the area under the $I(\lambda,T)$ versus λ curve, as in Fig. 41.1. (a) Show that this power per unit area is given by

$$\int_0^\infty I(\lambda,T)\, d\lambda = \sigma T^4$$

where $I(\lambda,T)$ is given by Planck's radiation law and σ is a constant independent of T. This result is known as the *Stefan-Boltzmann law* (see Eq. 17.11). To carry out the integration, make the change of variable $x = hc/\lambda kT$ and use the fact that

$$\int_0^\infty \frac{x^3\, dx}{e^x - 1} = \frac{\pi^4}{15}$$

(b) Show that the Stefan-Boltzmann constant σ has the value

$$\sigma = \frac{2\pi^5 k^4}{15c^2h^3} = 5.7 \times 10^{-8}\ \mathrm{W/m^2 \cdot K^4}$$

11. (a) Use the results of Problem 10 to calculate the total power radiated per unit area by a tungsten filament at 3000 K. (Assume that the filament is an ideal radiator.) (b) If the filament is part of a light bulb rated at 75 W, what is the surface area of the filament? (Assume that the main energy loss is due to radiation.)

Appendix A

TABLE A.1. Conversion Factors

LENGTH

	m	cm	km	in.	ft	mi
1 meter	1	10^2	10^{-3}	39.37	3.281	6.214×10^{-4}
1 centimeter	10^{-2}	1	10^{-5}	0.3937	3.281×10^{-2}	6.214×10^{-6}
1 kilometer	10^3	10^5	1	3.937×10^4	3.281×10^3	0.6214
1 inch	2.540×10^{-2}	2.540	2.540×10^{-5}	1	8.333×10^{-2}	1.578×10^{-5}
1 foot	0.3048	30.48	3.048×10^{-4}	12	1	1.894×10^{-4}
1 mile	1609	1.609×10^5	1.609	6.336×10^4	5280	1

MASS

	kg	g	slug	amu
1 kilogram	1	10^3	6.852×10^{-2}	6.024×10^{26}
1 gram	10^{-3}	1	6.852×10^{-5}	6.024×10^{23}
1 slug (lb/g)	14.59	1.459×10^4	1	8.789×10^{27}
1 atomic mass unit	1.660×10^{-27}	1.660×10^{-24}	1.137×10^{-28}	1

TIME

	s	min	h	day	year
1 second	1	1.667×10^{-2}	2.778×10^{-4}	1.157×10^{-5}	3.169×10^{-8}
1 minute	60	1	1.667×10^{-2}	6.994×10^{-4}	1.901×10^{-6}
1 hour	3600	60	1	4.167×10^{-2}	1.141×10^{-4}
1 day	8.640×10^4	1440	24	1	2.738×10^{-3}
1 year (a)	3.156×10^7	5.259×10^5	8.766×10^3	365.2	1

SPEED

	m/s	cm/s	ft/s	mi/h
1 meter/second	1	10^2	3.281	2.237
1 centimeter/second	10^{-2}	1	3.281×10^{-2}	2.237×10^{-2}
1 foot/second	0.3048	30.48	1	0.6818
1 mile/hour	0.4470	44.70	1.467	1

Note: 1 mi/min = 60 mi/h = 88 ft/s.

TABLE A.1 (Continued)

FORCE

	N	dyn	lb
1 newton	1	10^5	0.2248
1 dyne	10^{-5}	1	2.248×10^{-6}
1 pound	4.448	4.448×10^5	1

WORK, ENERGY, HEAT

	J	erg	ft·lb
1 joule	1	10^7	0.7376
1 erg	10^{-7}	1	7.376×10^{-8}
1 ft·lb	1.356	1.356×10^7	1
1 eV	1.602×10^{-19}	1.602×10^{-12}	1.182×10^{-19}
1 cal	4.186	4.186×10^7	3.087
1 Btu	1.055×10^3	1.055×10^{10}	7.779×10^2
1 kWh	3.600×10^6	3.600×10^{13}	2.655×10^6

	eV	cal	Btu	kWh
1 joule	6.242×10^{18}	0.2389	9.481×10^{-4}	2.778×10^{-7}
1 erg	6.242×10^{11}	2.389×10^{-8}	9.481×10^{-11}	2.778×10^{-14}
1 ft·lb	8.464×10^{18}	0.3239	1.285×10^{-3}	3.766×10^{-7}
1 eV	1	3.827×10^{-20}	1.519×10^{-22}	4.450×10^{-26}
1 cal	2.613×10^{19}	1	3.968×10^{-3}	1.163×10^{-6}
1 Btu	6.585×10^{21}	2.520×10^2	1	2.930×10^{-4}
1 kWh	2.247×10^{25}	8.601×10^5	3.413×10^2	1

PRESSURE

	N/m^2	dyn/cm^2	atm
1 newton/meter^2	1	10	9.869×10^{-6}
1 dyne/centimeter^2	10^{-1}	1	9.869×10^{-7}
1 atmosphere	1.013×10^5	1.013×10^6	1
1 centimeter mercury*	1.333×10^3	1.333×10^4	1.316×10^{-2}
1 pound/inch^2	6.895×10^3	6.895×10^4	6.805×10^{-2}
1 pound/foot^2	47.88	4.788×10^2	4.725×10^{-4}

	cmHg	lb/in.^{2}	lb/ft^2
1 newton/meter^2	7.501×10^{-4}	1.450×10^{-4}	2.089×10^{-2}
1 dyne/centimeter^2	7.501×10^{-5}	1.450×10^{-5}	2.089×10^{-3}
1 atmosphere	76	14.70	2.116×10^3
1 centimeter mercury*	1	0.1943	27.85
1 pound/inch^2	5.171	1	144
1 pound/foot^2	3.591×10^{-2}	6.944×10^{-3}	1

*At 0°C and at a location where the acceleration due to gravity has its "standard" value, 9.80665 m/s^2.

TABLE A.2. Symbols, Dimensions, and Units of Physical Quantities

QUANTITY	*COMMON SYMBOL*	*UNIT**	*DIMENSIONS†*	*UNIT IN TERMS OF BASIC SI UNITS*
Acceleration	$\mathbf{a}$	m/s²	L/T²	m/s²
Angle	θ, ϕ	radian		
Angular acceleration	$\boldsymbol{\alpha}$	radian/s²	T^{-2}	s^{-2}
Angular frequency	ω	radian/s	T^{-1}	s^{-1}
Angular momentum	$\mathbf{L}$	kg • m²/s	ML²/T	kg • m²/s
Angular velocity	$\boldsymbol{\omega}$	radian/s	T^{-1}	s^{-1}
Area	A	m²	L²	m²
Atomic number	Z			
Capacitance	C	farad (F)(= C/V)	Q²T²/ML²	A² • s⁴/kg • m²
Charge	q, Q, e	coulomb (C)	Q	A • s
Charge density				
Line	λ	C/m	Q/L	A • s/m
Surface	σ	C/m²	Q/L²	A • s/m²
Volume	ρ	C/m³	Q/L³	A • s/m³
Conductivity	σ	1/Ω • m	Q²T/ML³	A² • s³/kg • m³
Current	I	AMPERE	Q/T	A
Current density	$\mathbf{J}$	A/m²	Q/T²	A/m²
Density	ρ	kg/m³	M/L³	kg/m³
Dielectric constant	κ			
Displacement	$\mathbf{s}$	METER	L	m
Distance	d, h			
Length	l, L			
Electric dipole moment	$\mathbf{p}$	C • m	QL	A • s • m
Electric field	$\mathbf{E}$	V/m	ML/QT²	kg • m/A • s³
Electric flux	Φ	V • m	ML³/QT²	kg • m³/A • s³
Electromotive force	$\mathcal{E}$	volt (V)	ML²/QT²	kg • m²/A • s³
Energy	E, U, K	joule (J)	ML²/T²	kg • m²/s²
Entropy	S	J/K	ML²/T² °K	kg • m²/s² • K
Force	$\mathbf{F}$	newton (N)	ML/T²	kg • m/s²
Frequency	f, ν	hertz (Hz)	T^{-1}	s^{-1}
Heat	Q	joule (J)	ML²/T²	kg • m²/s²
Inductance	L	henry (H)	ML²/Q²	kg • m²/A² • s²
Magnetic dipole moment	$\mathfrak{M}$	N • m/T	QL²/T	A • m²
Magnetic field	$\mathbf{B}$	tesla (T)(= Wb/m²)	M/QT	kg/A • s²
Magnetic flux	Φ_m	weber (Wb)	ML²/QT	kg • m²/A • s²
Mass	m, M	KILOGRAM	M	kg
Molar specific heat	C	J/mole • K		kg • m²/s² • kmole • K
Moment of inertia	I	kg • m²	ML²	kg • m²
Momentum	$\mathbf{p}$	kg • m/s	ML/T	kg • m/s
Period	T	s	T	s
Permeability of space	μ_0	N/A² (= H/m)	ML/Q²T	kg • m/A² • s²
Permittivity of space	ε_0	C²/N • m² (= F/m)	Q²T²/ML³	A² • s⁴/kg • m³
Potential (voltage)	V	volt (V)(= J/C)	ML²/QT²	kg • m²/A • s³
Power	P	watt (W)(= J/s)	ML²/T³	kg • m²/s³
Pressure	P, p	N/m²	M/LT²	kg/m • s²
Resistance	R	ohm (Ω)(= V/A)	ML²/Q²T	kg • m²/A² • s³
Specific heat	c	J/kg • K	L²/T² °K	m²/s² • K
Temperature	T	KELVIN	°K	K
Time	t	SECOND	T	s
Torque	$\boldsymbol{\tau}$	N • m	ML²/T²	kg • m²/s²
Velocity	$\mathbf{v}$	m/s	L/T	m/s
Speed	v			
Volume	V	m³	L³	m³
Wavelength	λ	m	L	m
Work	W	joule (J)(= N • m)	ML²/T²	kg • m²/s²

*The basic SI units are given in upper case letters.
†The symbols M, L, T, and Q denote mass, length, time, and charge, respectively.

TABLE A.3. Atomic Masses of the Naturally Occurring Isotopic Mixtures of the Elements

ELEMENT	*SYMBOL*	*ATOMIC NO.*	*ATOMIC MASS (amu)**	*ELEMENT*	*SYMBOL*	*ATOMIC NO.*	*ATOMIC MASS (amu)**
Actinium	Ac	89	[227]†	Mercury	Hg	80	200.59
Aluminum	Al	13	26.9815	Molybdenum	Mo	42	95.94
Americium	Am	95	[243]	Neodymium	Nd	60	144.24
Antimony	Sb	51	121.75	Neon	Ne	10	20.183
Argon	Ar	18	39.948	Neptunium	Np	93	[237]
Arsenic	As	33	74.9216	Nickel	Ni	28	58.71
Astatine	At	85	[210]	Niobium	Nb	41	92.906
Barium	Ba	56	137.34	Nitrogen	N	7	14.0067
Berkelium	Bk	97	[247]	Nobelium	No	102	[253]
Beryllium	Be	4	9.0122	Osmium	Os	76	190.2
Bismuth	Bi	83	208.980	Oxygen	O	8	15.9994
Boron	B	5	10.811	Palladium	Pd	46	106.4
Bromine	Br	35	79.909	Phosphorus	P	15	30.9738
Cadmium	Cd	48	112.40	Platinum	Pt	78	195.09
Calcium	Ca	20	40.08	Plutonium	Pu	94	[242]
Californium	Cf	98	[249]	Polonium	Po	84	[210]
Carbon	C	6	12.01115	Potassium	K	19	39.102
Cerium	Ce	58	140.12	Praseodymium	Pr	59	140.907
Cesium	Cs	55	132.905	Promethium	Pm	61	[145]
Chlorine	Cl	17	35.453	Protactinium	Pa	91	[231]
Chromium	Cr	24	51.996	Radium	Ra	88	[226.05]
Cobalt	Co	27	58.9332	Radon	Rn	86	[222]
Copper	Cu	29	63.54	Rhenium	Re	75	186.2
Curium	Cm	96	[248]	Rhodium	Rh	45	102.905
Dysprosium	Dy	66	162.50	Rubidium	Rb	37	85.47
Einsteinium	Es	99	[254]	Ruthenium	Ru	44	101.07
Erbium	Er	68	167.26	Samarium	Sm	62	150.35
Europium	Eu	63	151.96	Scandium	Sc	21	44.956
Fermium	Fm	100	[253]	Selenium	Se	34	78.96
Fluorine	F	9	18.9984	Silicon	Si	14	28.086
Francium	Fr	87	[223]	Silver	Ag	47	107.870
Gadolinium	Gd	64	157.25	Sodium	Na	11	22.9898
Gallium	Ga	31	69.72	Strontium	Sr	38	87.62
Germanium	Ge	32	72.59	Sulfur	S	16	32.064
Gold	Au	79	196.967	Tantalum	Ta	73	180.948
Hafnium	Hf	72	178.49	Technetium	Tc	43	[99]
Helium	He	2	4.0026	Tellurium	Te	52	127.60
Holmium	Ho	67	164.930	Terbium	Tb	65	158.924
Hydrogen	H	1	1.00797	Thallium	Tl	81	204.37
Indium	In	49	114.82	Thorium	Th	90	232.038
Iodine	I	53	126.9044	Thulium	Tm	69	168.934
Iridium	Ir	77	192.2	Tin	Sn	50	118.69
Iron	Fe	26	55.847	Titanium	Ti	22	47.90
Krypton	Kr	36	83.80	Tungsten	W	74	183.85
Lanthanum	La	57	138.91	Uranium	U	92	238.03
Lawrencium	Lw	103	[259]	Vanadium	V	23	50.942
Lead	Pb	82	207.19	Xenon	Xe	54	131.30
Lithium	Li	3	6.939	Ytterbium	Yb	70	173.04
Lutetium	Lu	71	174.97	Yttrium	Y	39	88.905
Magnesium	Mg	12	24.312	Zinc	Zn	30	65.37
Manganese	Mn	25	54.9380	Zirconium	Zr	40	91.22
Mendelevium	Md	101	[256]				

*1 amu = 1.6605×10^{-27} kg = $(1/12) \times$ (mass of ^{12}C atom).

†The numbers in brackets are mass numbers of the most stable isotope. These elements are radioactive.

Appendix B
Mathematics Review

These appendices in mathematics are intended as a brief review of operations and methods. Early in this course, you should be totally familiar with basic algebraic techniques, analytic geometry, and trigonometry. The appendices on differential and integral calculus are more detailed and are intended for those students who have difficulties in applying calculus concepts to physical situations.

TABLE B.1. Mathematical Symbols Used in the Text and Their Meaning

SYMBOL	*MEANING*
$=$	is equal to
$\neq$	is not equal to
$\sim$	is proportional to
$>$	is greater than
$<$	is less than
$\gg$ ($\ll$)	is much greater (less) than
$\approx$	is approximately equal to
Δx	the change in x
$\sum_{i=1}^{N} x_i$	the sum of all quantities x_i from $i = 1$ to $i = N$
$\lvert x \rvert$	the magnitude of x (always a positive quantity)
$\Delta x \to 0$	Δx approaches zero
$\frac{dx}{dt}$	the derivative of x with respect to t
$\frac{\partial x}{\partial t}$	the partial derivative of x with respect to t
$\int$	integral

B.1 POWERS OF TEN

You should be familiar with the usage of powers of ten. It is a compact form of writing very large or very small numbers. For example, instead of 10 000, we write 10^4, where the exponent represents the number of zeros; that is, $10^4 = 10 \times 10 \times 10 \times 10 = 10\,000$. Likewise, a small number like 0.0001 can be expressed as 10^{-4}, where the negative exponent indicates that we are dealing with a number less than one. Some other examples of the use of powers of ten are

$$1000 = 10^3 \qquad 0.003 = 3 \times 10^{-3}$$
$$85\,000 = 8.5 \times 10^4 \qquad 0.00085 = 8.5 \times 10^{-4}$$
$$3\,200\,000 = 3.2 \times 10^6 \qquad 0.00002 = 2 \times 10^{-5}$$

If numbers written as powers of ten are *multiplied*, we simply *add* the exponents, maintaining their signs. For example,

$$(3 \times 10^3) \times (5 \times 10^4) = 15 \times 10^7 = 1.5 \times 10^8$$
$$(2 \times 10^5) \times (4 \times 10^{-2}) = 8 \times 10^3$$
$$(5.6 \times 10^4) \times (4.3 \times 10^8) = 24 \times 10^{12}$$

When numbers written as powers of ten are *divided*, we can bring the power of ten from the denominator to the numerator by changing its sign. For example,

$$\frac{8 \times 10^5}{2 \times 10^2} = 4 \times 10^5 \times 10^{-2} = 4 \times 10^3$$
$$\frac{12 \times 10^{-4}}{4 \times 10^{-9}} = 3 \times 10^{-4} \times 10^9 = 3 \times 10^5$$

In general,

$$10^n 10^m = 10^{n+m}$$
$$10^n/10^m = 10^{n-m}$$
$$(10^n)^m = 10^{nm}$$

The various prefixes commonly used for powers of ten are given in Table 1.4.

B.2 ALGEBRA

When algebraic operations are performed, the laws of arithmetic apply. Symbols such as x, y, and z are usually used to represent quantities that are not specified, and symbols such as a, b, and c are used to represent numbers.

You should be familiar with the following operations:

Fractions

$$a\left(\frac{b}{c}\right) = \frac{ab}{c} \qquad \left(\frac{a}{b}\right)\left(\frac{c}{d}\right) = \frac{ac}{bd}$$
$$\frac{(a/b)}{(c/d)} = \frac{ad}{bc} \qquad \frac{a}{b} \pm \frac{c}{d} = \frac{ad \pm cb}{bd}$$

Factoring and combinations

$$ax + bx = x(a + b)$$
$$x^2 - y^2 = (x + y)(x - y)$$
$$x^2 - 2x - 15 = (x + 3)(x - 5)$$

Roots of a quadratic equation

If $ax^2 + bx + c = 0$,

then $x = \dfrac{-b \pm \sqrt{b^2 - 4ac}}{2a}$

Multiplying powers of a given quantity

$$x^n x^m = x^{n+m}$$
$$\frac{x^n}{x^m} = x^{n-m}$$
$$(x^n)^m = x^{nm}$$

Logarithmic functions

$$\begin{cases} \ln = \text{logarithm to base } e \\ \log = \text{logarithm to base } 10 \\ \ln e = 1 \\ \ln e^x = x \\ \ln(xy) = \ln x + \ln y \\ \ln(x/y) = \ln x - \ln y \\ \ln(1/x) = -\ln x \\ \ln x^n = n \ln x \\ \ln a = 2.3026 \log a \\ \log a = 0.43429 \ln a \end{cases}$$

Simultaneous Linear Equations

In order to solve two simultaneous equations involving two unknowns, x and y, we solve one of the equations for x in terms of y and substitute this expression into the other equation.

Example

(1) $5x + y = -8$ (2) $2x - 2y = 4$

Solution: From (2), $x = y + 2$. Substitution of this into (1) gives

$$5(y + 2) + y = -8$$
$$6y = -18$$
$$y = -3$$
$$x = y + 2 = -1$$

Alternate Solution: Multiply (1) by 2 and add the result to (2):

$$10x + 2y = -16$$
$$\underline{2x - 2y = 4}$$
$$12x = -12$$
$$x = -1$$
$$y = x - 2 = -3$$

B.3 GEOMETRY

The *Pythagorean theorem,* which relates the three sides of a right triangle

$$\boxed{c^2 = a^2 + b^2}$$

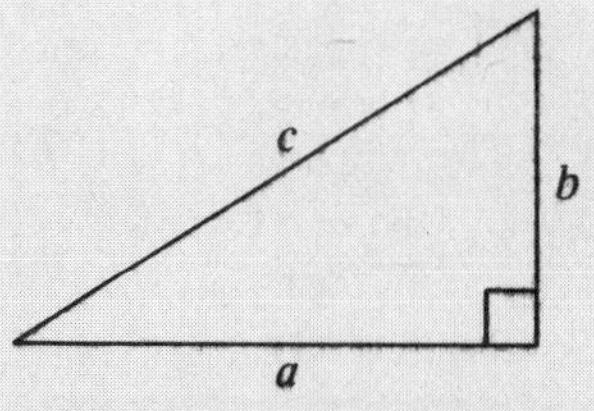

The *distance* d between two points whose coordinates are (x_1,y_1) and (x_2,y_2)

$$\boxed{d = \sqrt{(x_2 - x_1)^2 + (y_2 - y_1)^2}}$$

The *radian measure:* the arc length s is proportional to the radius r for a fixed value of θ (in radians)

$$\boxed{s = r\theta}$$

$$\theta = \frac{s}{r}$$

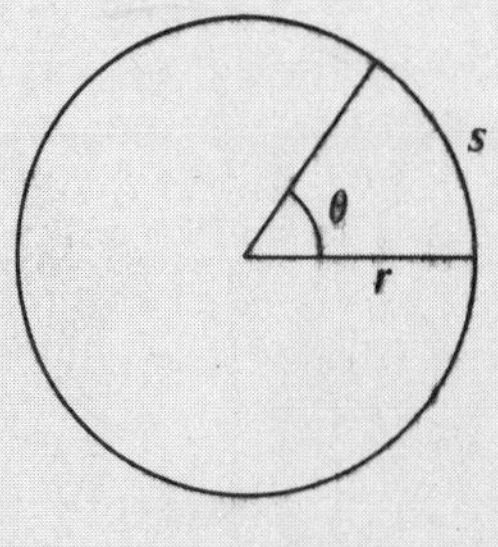

Circumference of a circle

$$\boxed{C = 2\pi r}$$

Area of a circle

$$A = \pi r^2$$

Area of a triangle

$$A = \frac{1}{2}bh$$

Surface area of a sphere

$$A = 4\pi r^2$$

Volume of a sphere

$$V = \frac{4}{3}\pi r^3$$

Volume of a cylinder

$$V = \pi r^2 \ell$$

Equation of a *straight line*

$$y = mx + b$$

$b = y$ intercept
$m =$ slope $= \tan\theta$

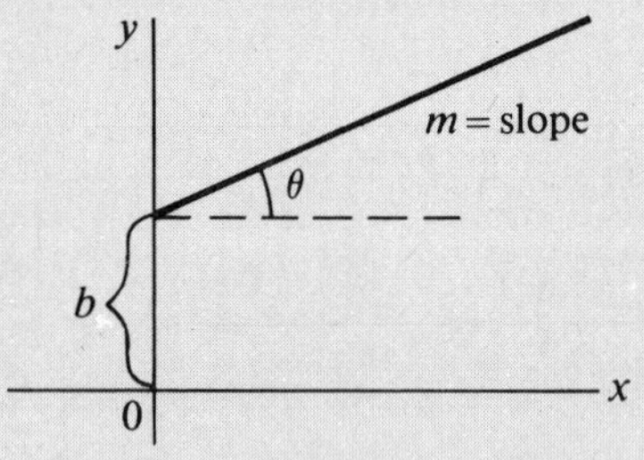

Equation of a *circle* of radius R centered at the origin

$$x^2 + y^2 = R^2$$

Equation of an *ellipse* with the origin at its center

$$\frac{x^2}{a^2} + \frac{y^2}{b^2} = 1$$

$a =$ semi-major axis
$b =$ semi-minor axis

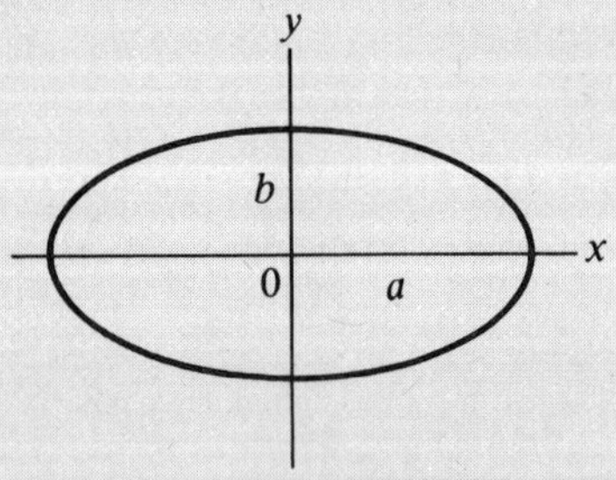

Equation of a *parabola* whose vertex is at $y = b$

$$y = ax^2 + b$$

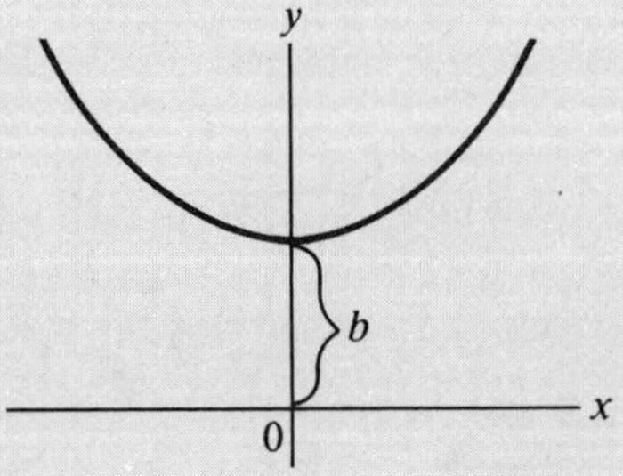

Equation of a *rectangular hyperbola*

$$\boxed{xy = \text{constant}}$$

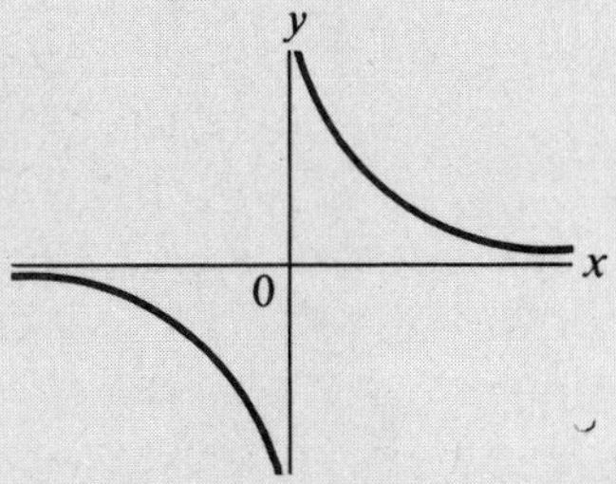

B.4 TRIGONOMETRY

The sine, cosine, and tangent functions in trigonometry are defined in terms of the ratios of the sides of a right triangle:

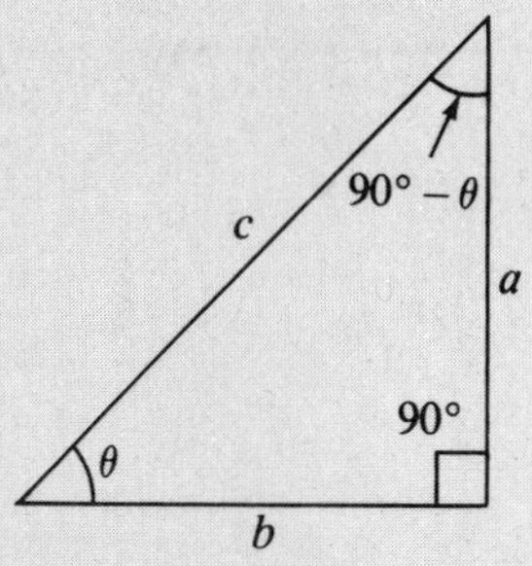

$$\sin\theta = \frac{\text{side opposite } \theta}{\text{hypotenuse}} = \frac{a}{c}$$

$$\cos\theta = \frac{\text{side adjacent } \theta}{\text{hypotenuse}} = \frac{b}{c}$$

$$\tan\theta = \frac{\text{side opposite } \theta}{\text{side adjacent } \theta} = \frac{a}{b}$$

From the above definitions and the Pythagorean theorem, it follows that

$$\sin^2\theta + \cos^2\theta = 1$$

$$\tan\theta = \frac{\sin\theta}{\cos\theta}$$

The cosecant, secant, and cotangent functions are defined by

$$\csc\theta = \frac{1}{\sin\theta} \qquad \sec\theta = \frac{1}{\cos\theta} \qquad \cot\theta = \frac{1}{\tan\theta}$$

The relations at the right follow directly from the right triangle above:

$$\begin{cases} \sin\theta = \cos(90° - \theta) \\ \cos\theta = \sin(90° - \theta) \\ \cot\theta = \tan(90° - \theta) \end{cases}$$

Some properties of trigonometric functions:

$$\begin{cases} \sin(-\theta) = -\sin\theta \\ \cos(-\theta) = \cos\theta \\ \tan(-\theta) = -\tan\theta \end{cases}$$

Relations that apply to *any* triangle:

$$\alpha + \beta + \gamma = 180°$$

Law of cosines
$$\begin{cases} a^2 = b^2 + c^2 - 2bc\cos\alpha \\ b^2 = a^2 + c^2 - 2ac\cos\beta \\ c^2 = a^2 + b^2 - 2ab\cos\gamma \end{cases}$$

Law of sines
$$\left\{ \frac{a}{\sin\alpha} = \frac{b}{\sin\beta} = \frac{c}{\sin\gamma} \right.$$

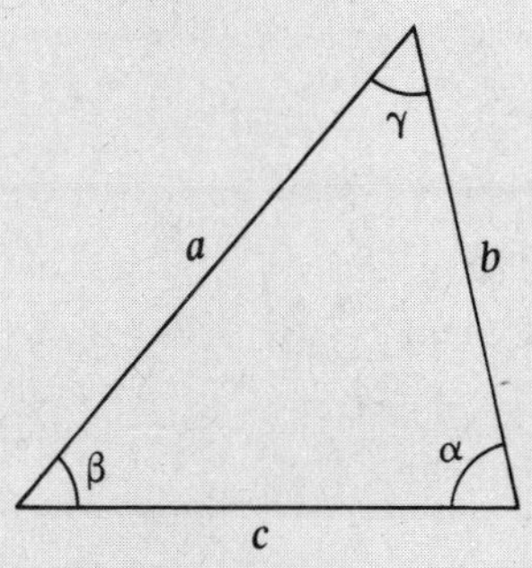

TABLE B.2. Some Trigonometric Identities

$$\sin^2\theta + \cos^2\theta = 1 \qquad \csc^2\theta = 1 + \cot^2\theta$$

$$\sec^2\theta = 1 + \tan^2\theta \qquad \sin^2\frac{\theta}{2} = \frac{1}{2}(1 - \cos\theta)$$

$$\sin 2\theta = 2\sin\theta\cos\theta \qquad \cos^2\frac{\theta}{2} = \frac{1}{2}(1 + \cos\theta)$$

$$\cos 2\theta = \cos^2\theta - \sin^2\theta \qquad 1 - \cos\theta = 2\sin^2\frac{\theta}{2}$$

$$\tan 2\theta = \frac{2\tan\theta}{1 - \tan^2\theta} \qquad \tan\frac{\theta}{2} = \sqrt{\frac{1 - \cos\theta}{1 + \cos\theta}}$$

$$\sin(A \pm B) = \sin A\cos B \pm \cos A\sin B$$

$$\cos(A \pm B) = \cos A\cos B \mp \sin A\sin B$$

$$\sin A \pm \sin B = 2\sin\left[\frac{1}{2}(A \pm B)\right]\cos\left[\frac{1}{2}(A \mp B)\right]$$

$$\cos A + \cos B = 2\cos\left[\frac{1}{2}(A + B)\right]\cos\left[\frac{1}{2}(A - B)\right]$$

$$\cos A - \cos B = 2\sin\left[\frac{1}{2}(A + B)\right]\sin\left[\frac{1}{2}(B - A)\right]$$

TABLE B.3. Series Expansions

$$(a + b)^n = a^n + \frac{n}{1!}a^{n-1}b + \frac{n(n-1)}{2!}a^{n-2}b^2 + \cdots$$

$$(1 + x)^n = 1 + nx + \frac{n(n-1)}{2!}x^2 + \cdots$$

$$e^x = 1 + x + \frac{x^2}{2!} + \frac{x^3}{3!} + \cdots$$

$$\ln(1 \pm x) = \pm x - \frac{1}{2}x^2 \pm \frac{1}{3}x^3 - \cdots$$

$$\left.\begin{aligned} \sin x &= x - \frac{x^3}{3!} + \frac{x^3}{5!} - \cdots \\ \cos x &= 1 - \frac{x^2}{2!} + \frac{x^4}{4!} - \cdots \\ \tan x &= x + \frac{x^3}{3} + \frac{2x^5}{15} + \cdots \qquad |x| < \pi/2 \end{aligned}\right\} \theta \text{ in radians}$$

For $x \ll 1$, the following approximations can be used:

$$(1 + x)^n \approx 1 + nx \qquad \sin x \approx x$$

$$e^x \approx 1 + x \qquad \cos x \approx 1$$

$$\ln(1 \pm x) \approx \pm x \qquad \tan x \approx x$$

B.5 DIFFERENTIAL CALCULUS

In various branches of science, it is sometimes necessary to use the basic tools of calculus, first invented by Newton, to describe physical phenomena. The use of calculus is fundamental in the treatment of various problems in newtonian mechanics, electricity, and magnetism. In this section, we simply state some basic properties and "rules of thumb" that should be a useful review to the student.

First, a *function* must be specified which relates one variable to another (such as coordinate as a function of time). Suppose one of the variables is called y (the dependent variable), the other x (the independent variable). We might have a function relation such as

$$y(x) = ax^3 + bx^2 + cy + d$$

If a, b, c, and d are specified constants, then y can be calculated for any value of x.

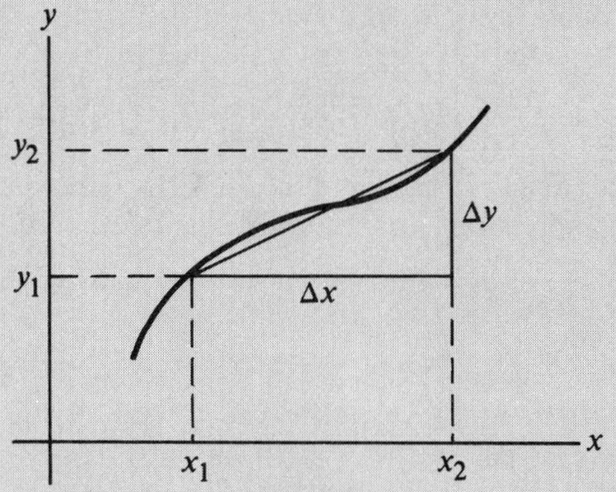

Figure 1

We usually deal with continuous functions, that is, those for which y varies "smoothly" with x.

The *derivative* of y with respect to x is defined as the limit of the slopes of chords drawn between two points on the y versus x curve as Δx approaches zero. Mathematically, we write this definition as

$$\frac{dy}{dx} = \lim_{\Delta x \to 0} \frac{\Delta y}{\Delta x} = \lim_{\Delta x \to 0} \frac{y(x + \Delta x) - y(x)}{\Delta x} \tag{B.1}$$

where Δy and Δx are defined as $\Delta x = x_2 - x_1$ and $\Delta y = y_2 - y_1$ (see Fig. 1).

A useful expression to remember when $y(x) = ax^n$, where a is a *constant* and n is *any* positive or negative number (integer or fraction), is

$$\frac{dy}{dx} = nax^{n-1} \tag{B.2}$$

If $y(x)$ is a polynomial or algebraic function of x, we apply Eq. B.2 to *each* term in the polynomial and take $da/dx = 0$. It is important to note that dy/dx *does not* mean dy divided by dx, but is simply a notation of the limiting process of the derivative as defined by Eq. B.1. In Examples 1 through 4, we evaluate the derivatives of several well-behaved functions.

Example 1

Suppose $y(x)$ (that is, y as a function of x) is given by

$$y(x) = ax^3 + bx + c$$

where a and b are constants. Then it follows that

$$y(x + \Delta x) = a(x + \Delta x)^3 + b(x + \Delta x) + c$$
$$y(x + \Delta x) = a(x^3 + 3x^2\Delta x + 3x\Delta x^2 + \Delta x^3) + b(x + \Delta x) + c$$

so

$$\Delta y = y(x + \Delta x) - y(x) = a(3x^2\Delta x + 3x\Delta x^2 + \Delta x^3) + b\Delta x$$

Substituting this into Eq. B.1 gives

$$\frac{dy}{dx} = \lim_{\Delta x \to 0} \frac{\Delta y}{\Delta x} = \lim_{\Delta x \to 0} [3ax^2 + 3x\Delta x + \Delta x^2] + b$$
$$\frac{dy}{dx} = 3ax^2 + b$$

Example 2

$$y(x) = 8x^5 + 4x^3 + 2x + 7$$

Solution: Applying Eq. B.2 to each term independently, and remembering that d/dx (constant) $= 0$, we have

$$\frac{dy}{dx} = 8(5)x^4 + 4(3)x^2 + 2(1)x^0 + 0$$
$$\frac{dy}{dx} = 40x^4 + 12x^2 + 2$$

Special Properties of the Derivative

A. ***Derivative of the Product of Two Functions.*** If a function y is given by the product of two functions, say, $g(x)$ and $h(x)$, then the derivative of y is defined as

$$\frac{d}{dx}f(x) = \frac{d}{dx}[g(x)h(x)] = g\frac{dh}{dx} + h\frac{dg}{dx} \tag{B.3}$$

B. *Derivative of the Sum of Two Functions.* If a function y is equal to the sum of two functions, then the derivative of the sum is equal to the sum of the derivatives:

$$\frac{d}{dx}f(x) = \frac{d}{dx}[g(x) + h(x)] = \frac{dg}{dx} + \frac{dh}{dx} \tag{B.4}$$

C. *Chain Rule of Differential Calculus.* If $y = f(x)$ and x is a function of some other variable z, then dy/dx can be written as the product of two derivatives:

$$\frac{dy}{dx} = \frac{dy}{dz}\frac{dz}{dx} \tag{B.5}$$

D. *The Second Derivative.* The second derivative of y with respect to x is defined as the derivative of the function dy/dx (or, the derivative of the derivative). It is usually written

$$\frac{d^2y}{dx^2} = \frac{d}{dx}\left(\frac{dy}{dx}\right) \tag{B.6}$$

Example 3

Find the first derivative of $y(x) = x^3/(x + 1)^2$ with respect to x.

Solution: We can rewrite this function as $y(x) = x^3(x + 1)^{-2}$ and apply Eq. B.3 directly:

$$\frac{dy}{dx} = (x + 1)^{-2}\frac{d}{dx}(x^3) + x^3\frac{d}{dx}(x + 1)^{-2}$$

$$= (x + 1)^{-2}\,3x^2 + x^3(-2)(x + 1)^{-3}$$

$$\frac{dy}{dx} = \frac{3x^2}{(x + 1)^2} - \frac{2x^3}{(x + 1)^3}$$

Example 4

A useful formula that follows from Eq. B.3 is the derivative of the quotient of two functions. Show that the expression is given by

$$\frac{d}{dx}\left[\frac{g(x)}{h(x)}\right] = \frac{h\dfrac{dg}{dx} - g\dfrac{dh}{dx}}{h^2}$$

Solution: We can write the quotient as gh^{-1} and then apply Eqs. B.2 and B.3:

$$\frac{d}{dx}\left(\frac{g}{h}\right) = \frac{d}{dx}(gh^{-1}) = g\frac{d}{dx}(h^{-1}) + h^{-1}\frac{d}{dx}(g)$$

$$= -gh^{-2}\frac{dh}{dx} + h^{-1}\frac{dg}{dx}$$

$$= \frac{h\dfrac{dg}{dx} - g\dfrac{dh}{dx}}{h^2}$$

TABLE B.4. Derivatives for Several Functions

$\frac{d}{dx}(a) = 0$	$\frac{d}{dx}(\tan ax) = a\sec^2 ax$
$\frac{d}{dx}(ax^n) = nax^{n-1}$	$\frac{d}{dx}(\cot ax) = -a\csc^2 ax$
$\frac{d}{dx}(e^{ax}) = ae^{ax}$	$\frac{d}{dx}(\sec x) = \tan x \sec x$
$\frac{d}{dx}(\sin ax) = a\cos ax$	$\frac{d}{dx}(\csc x) = -\cot x \csc x$
$\frac{d}{dx}(\cos ax) = -a\sin ax$	$\frac{d}{dx}(\ln ax) = \frac{1}{x}$

Note: The letters a and n are constants.

B.6 INTEGRAL CALCULUS

We can think of integration as the inverse of differentiation. As an example, consider the expression

$$f(x) = \frac{dy}{dx} = 3ax^2 + b$$

which was the result of differentiating the function

$$y(x) = ax^3 + bx + c$$

in Example 1. We can write the first expression $dy = f(x)dx = (3ax^2 + b)dx$ and obtain $y(x)$ by "summing" over all values of x. Mathematically, we write this inverse operation

$$y(x) = \int f(x)dx$$

For the function $f(x)$ given above,

$$y(x) = \int (3ax^2 + b)dx = ax^3 + bx + c$$

where c is a constant of the integration. This type of integral is called an *indefinite integral* since its value depends on the choice of the constant c.

A general *indefinite integral* $I(x)$ is defined as

$$I(x) = \int f(x)dx \tag{B.7}$$

where $f(x)$ is called the *integrand* and $f(x) = \dfrac{dI(x)}{dx}$.

For a *general continuous* function $f(x)$, the integral can be described as the area under the curve bounded by $f(x)$ and the x axis, between two specified values of x, say, x_1 and x_2, as in Fig. 2.

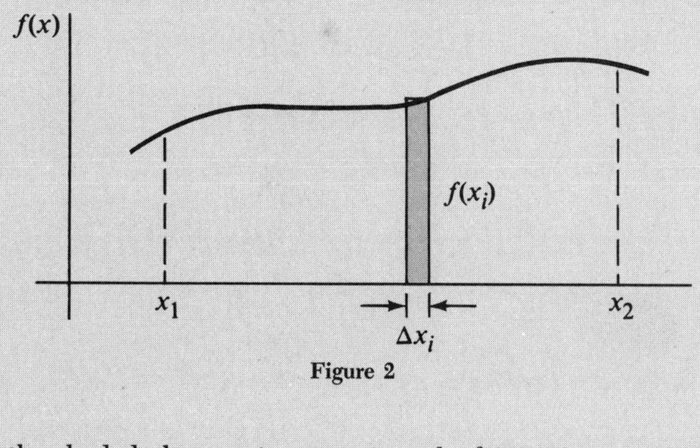

Figure 2

The area of the shaded element is approximately $f_i \Delta x_i$. If we sum all these area elements from x_1 to x_2 and take the limit of this sum as $\Delta x_i \to 0$, we obtain the *true* area under the curve bounded by $f(x)$ and x, between the limits x_1 and x_2:

$$\text{Area} = \lim_{\Delta x \to 0} \sum_i f_i(x)\Delta x_i = \int_{x_1}^{x_2} f(x)dx \tag{B.8}$$

Integrals of the type defined by Eq. B.8 are called *definite integrals.*

One of the common types of integrals that arise in practical situations has the form

$$\int x\,dx = \frac{x^{n+1}}{n+1} + c \qquad (n \neq -1) \tag{B.9}$$

This result is obvious since differentiation of the right-hand side with respect to x gives $f(x) = x^n$ directly. If the limits of the integration are known, this integral becomes a *definite integral* and is written

$$\int_{x_1}^{x_2} x^n dx = \frac{x_2^{\,n+1} - x_1^{\,n+1}}{n+1} \qquad (n \neq -1) \tag{B.10}$$

Examples

1. $\displaystyle \int_0^a x^2 dx = \frac{x^3}{3}\bigg]_0^a = \frac{a^3}{3}$
2. $\displaystyle \int_0^b x^{3/2} dx = \frac{x^{5/2}}{5/2}\bigg]_0^b = \frac{2}{5} b^{5/2}$
3. $\displaystyle \int_3^5 x dx = \frac{x^2}{2}\bigg]_3^5 = \frac{5^2 - 3^2}{2} = 8$

Partial Integration

Sometimes it is useful to apply the method of *partial integration* to evaluate certain integrals. The method uses the property that

$$\int u dv = uv - \int v du \tag{B.11}$$

where u and v are *carefully* chosen so as to reduce a complex integral to a simpler one. In many cases, several reductions have to be made. Consider the example

$$I(x) = \int x^2 e^x dx$$

This can be evaluated by integrating by parts twice. First, if we choose $u = x^2$, $v = e^x$, we get

$$\int x^2 e^x dx = \int x^2 d(e^x) = x^2 e^x - 2\int e^x x dx + c_1$$

Now, in the second term, choose $u = x$, $v = e^x$, which gives

$$\int x^2 e^x dx = x^2 e^x - 2xe^x + 2\int e^x dx + c_1$$

or

$$\int x^2 e^x dx = x^2 e^x - 2xe^x + 2e^x + c_2$$

The Perfect Differential

Another useful method to remember is the use of the *perfect differential.* That is, we should sometimes look for a change of variable such that the differential of the function is the differential of the independent variable appearing in the integrand. For example, consider the integral

$$I(x) = \int \cos^2 x \sin x dx$$

This becomes easy to evaluate if we rewrite the differential as $d(\cos x) = -\sin x dx$. The integral then becomes

$$\int \cos^2 x \sin x dx = -\int \cos^2 x\, d(\cos x)$$

If we now change variables, letting $y = \cos x$, we get

$$\int \cos^2 x \sin x dx = -\int y^2 dy = -\frac{y^3}{3} + c = -\frac{\cos^3 x}{3} + c$$

Table B.5 lists some useful indefinite integrals. A more complete list can be found in various handbooks, such as *The Handbook of Chemistry and Physics,* CRC Press.

TABLE B.5. Some Indefinite Integrals (an arbitrary constant should be added to each of these integrals)

$$\int x^n dx = \frac{x^{n+1}}{n+1} \quad \text{(provided } n \neq -1\text{)}$$

$$\int \frac{dx}{x} = \int x^{-1} dx = \ln x$$

$$\int \frac{dx}{a+bx} = \frac{1}{b}\ln(a+bx)$$

$$\int \frac{dx}{(a+bx)^2} = -\frac{1}{b(a+bx)}$$

$$\int \frac{dx}{a^2+x^2} = \frac{1}{a}\tan^{-1}\frac{x}{a}$$

$$\int \frac{dx}{a^2-x^2} = \frac{1}{2a}\ln\frac{a+x}{a-x} \quad (a^2 - x^2 > 0)$$

$$\int \frac{dx}{x^2-a^2} = \frac{1}{2a}\ln\frac{x-a}{x+a} \quad (x^2 - a^2 > 0)$$

$$\int \frac{x dx}{a^2 \pm x^2} = \pm\tfrac{1}{2}\ln(a^2 \pm x^2)$$

$$\int \frac{dx}{\sqrt{a^2-x^2}} = \sin^{-1}\frac{x}{a} = -\cos^{-1}\frac{x}{a} \quad (a^2 - x^2 > 0)$$

$$\int \frac{dx}{\sqrt{x^2 \pm a^2}} = \ln(x + \sqrt{x^2 \pm a^2})$$

$$\int \frac{x dx}{\sqrt{a^2-x^2}} = -\sqrt{a^2-x^2}$$

$$\int \frac{x dx}{\sqrt{x^2 \pm a^2}} = \sqrt{x^2 \pm a^2}$$

$$\int \sqrt{a^2-x^2}\, dx = \frac{1}{2}\left(x\sqrt{a^2-x^2} + a^2 \sin^{-1}\frac{x}{a}\right)$$

$$\int x\sqrt{a^2-x^2}\, dx = -\tfrac{1}{3}(a^2-x^2)^{3/2}$$

$$\int \sqrt{x^2 \pm a^2}\, dx = \tfrac{1}{2}[x\sqrt{x^2 \pm a^2} \pm a^2 \ln(x + \sqrt{x^2 \pm a^2})]$$

$$\int x\sqrt{x^2 \pm a^2}\, dx = \tfrac{1}{3}(x^2 \pm a^2)^{3/2}$$

$$\int e^{ax} dx = \frac{1}{a} e^{ax}$$

$$\int \ln ax\, dx = (x \ln ax) - x$$

$$\int x e^{ax} dx = \frac{e^{ax}}{a^2}(ax - 1)$$

$$\int \frac{dx}{a + be^{cx}} = \frac{x}{a} - \frac{1}{ac}\ln(a + be^{cx})$$

$$\int \sin ax\, dx = -\frac{1}{a}\cos ax$$

$$\int \cos ax\, dx = \frac{1}{a}\sin ax$$

$$\int \tan ax\, dx = -\frac{1}{a}\ln(\cos ax) = \frac{1}{a}\ln(\sec ax)$$

$$\int \cot ax\, dx = \frac{1}{a}\ln(\sin ax)$$

$$\int \sec ax\, dx = \frac{1}{a}\ln(\sec ax + \tan ax) = \frac{1}{a}\ln\left[\tan\left(\frac{ax}{2} + \frac{\pi}{4}\right)\right]$$

$$\int \csc ax\, dx = \frac{1}{a}\ln(\csc ax - \cot ax) = \frac{1}{a}\ln\left(\tan\frac{ax}{2}\right)$$

$$\int \sin^2 ax\, dx = \frac{x}{2} - \frac{\sin 2ax}{4a}$$

$$\int \cos^2 ax\, dx = \frac{x}{2} + \frac{\sin 2ax}{4a}$$

$$\int \frac{dx}{\sin^2 ax} = -\frac{1}{a}\cot ax$$

$$\int \frac{dx}{\cos^2 ax} = \frac{1}{a}\tan ax$$

$$\int \tan^2 ax\, dx = \frac{1}{a}(\tan ax) - x$$

$$\int \cot^2 ax\, dx = -\frac{1}{a}(\cot ax) - x$$

$$\int \sin^{-1} ax\, dx = x(\sin^{-1} ax) + \frac{\sqrt{1 - a^2x^2}}{a}$$

$$\int \cos^{-1} ax\, dx = x(\cos^{-1} ax) - \frac{\sqrt{1 - a^2x^2}}{a}$$

$$\int \tan^{-1} ax\, dx = x(\tan^{-1} ax) - \frac{1}{2a}\ln(1 + a^2x^2)$$

$$\int \cot^{-1} ax\, dx = x(\cot^{-1} ax) + \frac{1}{2a}\ln(1 + a^2x^2)$$

Appendix C
Calculating the Displacement from the Velocity

The velocity of a particle moving in a straight line can be obtained from a knowledge of its position as a function of time. Mathematically, the velocity equals the derivative of the coordinate with respect to time. It is also possible to find the displacement of a particle if its velocity is known as a function of time. In the calculus, this procedure is referred to as integration, or as the taking of the antiderivative. Graphically, it is equivalent to finding the area under a curve.

Suppose the velocity versus time plot for a particle moving along the x axis is as shown in Fig. 3. Let us divide the time interval $t_f - t_i$ into many small intervals of duration Δt_n. From the definition of average velocity, we see that the displacement during any small interval such as the shaded one in Fig. 3 is given by $\Delta x_n = \overline{v}_n \Delta t_n$, where $\overline{v}_n$ is the average velocity in that interval. Therefore, the displacement during this small interval is simply the area of the shaded rectangle. The total displacement for the interval $t_f - t_i$ is the sum of the area of all the rectangles:

$$\Delta x = \sum \overline{v}_n \Delta t_n$$

where the sum is taken over all the rectangles from t_i to t_f. Now, as each interval is made smaller and smaller, the number of terms in the sum increases and the sum approaches a value equal to the area under the velocity-time graph. Therefore, in

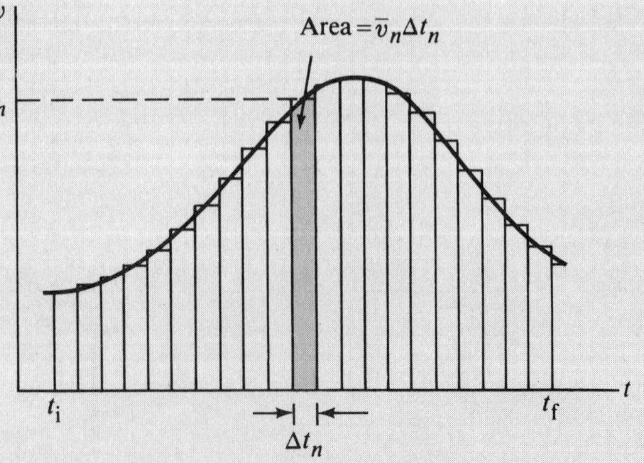

Figure 3

the limit $n \to \infty$, or $\Delta t_n \to 0$, we see that the displacement is given by

$$\Delta x = \lim_{\Delta t_n \to 0} \sum_n v_n \Delta t_n \qquad \text{(C.1)}$$

or

Displacement = area under the velocity-time graph

Note that we have replaced the average velocity $\bar{v}_n$ by the instantaneous velocity v_n in the sum. As you can see from Fig. 3, this approximation is clearly valid in the limit of very small intervals. We conclude that if the velocity-time graph for motion along a straight line is known, the displacement during any time interval can be obtained by measuring the area under the curve.

The limit of the sum in Eq. C.1 is called a *definite integral* and is written

$$\lim_{\Delta t_n \to 0} \sum_n v_n \Delta t_n = \int_{t_i}^{t_f} v(t)dt \qquad \text{(C.2)}$$

where $v(t)$ denotes the velocity at any time t. If the explicit functional form of $v(t)$ is known, the specific integral can be evaluated.

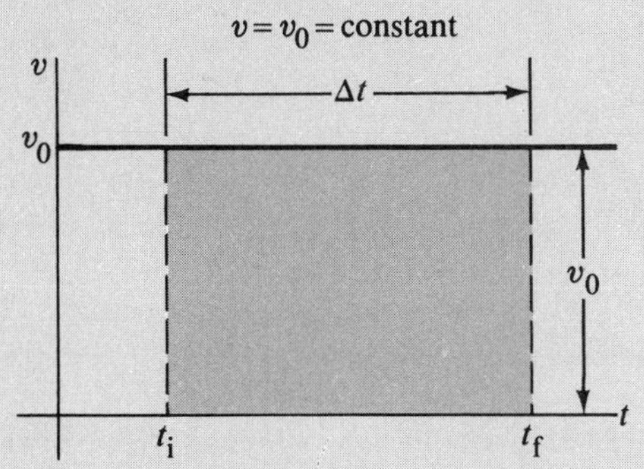

Figure 4

If a particle moves with a constant velocity v_0 as in Fig. 4, its displacement during the time interval Δt is simply the area of the shaded rectangle, that is,

$$\Delta x = v_0 \Delta t \qquad \text{(when } v = v_0 = \text{constant)}$$

As another example, consider a particle moving with a velocity that is proportional to t, as in Fig. 5. Taking $v = at$, where a is the constant of proportionality (the acceleration), we find that the displacement of the particle during the time interval $t = 0$ to $t = t_1$ is the area of the shaded triangle in Fig. 5:

$$\Delta x = \frac{1}{2}(t_1)(at_1) = \frac{1}{2}at_1^2$$

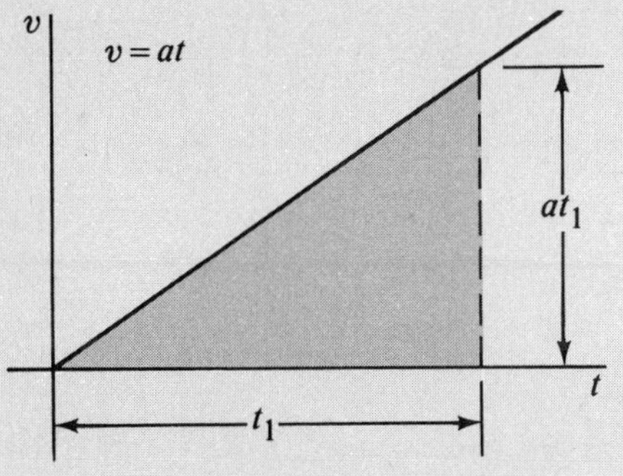

Figure 5

Answers to Odd-Numbered Exercises and Problems

CHAPTER 1

Exercises

1. $2.8\ \text{g/cm}^3$
3. $2.26 \times 10^3\ \text{kg}$
5. (a) $9.83 \times 10^{-16}\ \text{g}$ (b) 1.06×10^7 atoms
7. k cannot be found from this analysis.
9. L/T^3
11. $1.39 \times 10^{-4}\ \text{m}^3$
13. $7.46 \times 10^{-4}\ \text{m}^3$
15. $1.14 \times 10^4\ \text{kg/m}^3$
17. $1.18 \times 10^{17}\ \text{kg/m}^3$
19. $2.87 \times 10^8\ \text{s}$
21. 2.54×10^{22} atoms
23. Estimated at 3×10^9 beats.
25. Assuming two 6-packs per week per family and four people per family, we estimate 30 billion cans per year. Taking the mass of one can as 5 g, we estimate a total mass of 1.5×10^8 kg, corresponding to about 10^5 tons.
27. About 10^4 bricks. Estimating the area of one brick as 3 in. $\times$ 8 in. $= 24\ \text{in.}^2 = 0.17\ \text{ft}^2$, and one wall as having an area of 12 ft $\times$ 30 ft $= 360\ \text{ft}^2$ (for a total area of $1440\ \text{ft}^2$), we estimate a total number of $1440/0.17 \approx 10^4$ bricks.
29. $(195.8 \pm 1.4)\ \text{cm}^2$
31. (a) 22 cm (b) $67.9\ \text{cm}^2$

CHAPTER 2

Exercises

1. (a) 8.6 m (b) (4.5 m, $-63°$) (c) (4.2 m, 135°)
3. $x = -2.75$ m, $y = -4.76$ m
5. (a)

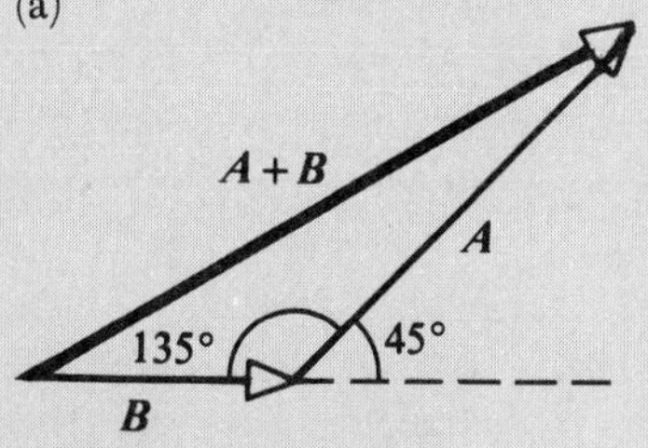

(b) $|\boldsymbol{A} + \boldsymbol{B}| = 8.39$ m

7. (14.3 km, 65.2°)
9.

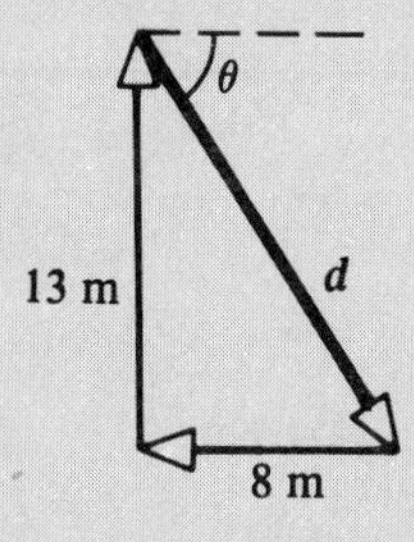

$|\boldsymbol{d}| = \sqrt{8^2 + 13^2} = 15.3$ m
$\theta = -58.4°$

11. $A_x = 2.6$ m, $B_x = 0$, $\boldsymbol{A} + \boldsymbol{B} = (2.6\boldsymbol{i} + 4.5\boldsymbol{j})$ m
$A_y = 1.5$ m, $B_y = 3$ m

13.

Quadrant	I	II	III	IV
x-component	+	−	−	+
y-component	+	+	−	−

15. 47.2 units, $\theta = 122°$
17. (7.21 m, 56.3°)
19. (a) $|\boldsymbol{B}| = 7$ units, $0 = 217°$ (b) $C_x = -28$ units, $C_y = -91$ units
21. 9.2 m west and 2.3 m north, *or* $\boldsymbol{R} = (-9.2\boldsymbol{i} + 2.3\boldsymbol{j})$ m
23. 1260 mi east and 386 mi north, *or* $\boldsymbol{R} = (1260\boldsymbol{i} + 386\boldsymbol{j})$ mi
25. (a) $\boldsymbol{A} + \boldsymbol{B} = 2\boldsymbol{i} - 6\boldsymbol{j}$ (b) $\boldsymbol{A} - \boldsymbol{B} = 4\boldsymbol{i} + 2\boldsymbol{j}$ (c) $|\boldsymbol{A} + \boldsymbol{B}| = 6.32$ (d) $|\boldsymbol{A} - \boldsymbol{B}| = 4.47$
(e) For $\boldsymbol{A} + \boldsymbol{B}$, $\theta = -71.6°$; for $\boldsymbol{A} - \boldsymbol{B}$, $\theta = 26.6°$.
27. (a) $\boldsymbol{A} = 8\boldsymbol{i} + 12\boldsymbol{j} - 4\boldsymbol{k}$ (b) $\boldsymbol{B} = \boldsymbol{A}/4 = 2\boldsymbol{i} + 3\boldsymbol{j} - \boldsymbol{k}$ (c) $\boldsymbol{C} = -3\boldsymbol{A} = -24\boldsymbol{i} - 36\boldsymbol{j} + 12\boldsymbol{k}$
29. (a) $\boldsymbol{r} = (-11.1 + 6.40\boldsymbol{j})$ m (b) $\boldsymbol{r} = (1.65\boldsymbol{i} + 2.86\boldsymbol{j})$ cm (c) $\boldsymbol{r} = (-18.0\boldsymbol{i} - 12.6\boldsymbol{j})$ in.
31. (a) $\boldsymbol{A} = -3\boldsymbol{i} + 2\boldsymbol{j}$ (b) 3.61, 146.3° (c) $\boldsymbol{B} = 3\boldsymbol{i} - 6\boldsymbol{j}$
33. 5.83 N at $\theta = 149°$
35. 38.3 N in the positive y direction
37. (a) $F_x = 49.5$ N, $F_y = 27.1$ N (b) 56.4 N at $\theta = 28.7°$

Problems

3. (a) $\boldsymbol{r}_1 = (-3\boldsymbol{i} - 5\boldsymbol{j})$ m, $\boldsymbol{r}_2 = (-\boldsymbol{i} + 8\boldsymbol{j})$ m (b) $\Delta\boldsymbol{r} = \boldsymbol{r}_2 - \boldsymbol{r}_1 = (2\boldsymbol{i} + 13\boldsymbol{j})$ m

CHAPTER 3

Exercises

1. -3.89×10^{-2} m/s
3. (a) 1.92 km (b) 4.57 m/s
5. (a) 4 m/s (b) −4 m/s (c) zero (d) 2 m/s
7. (a) negative (b) positive (c) zero (d) zero
9. (b) 1.6 m/s
11. -2.5 m/s^2
13. (a) 4 m/s^2 (b) No. The acceleration is not necessarily constant, and so the average velocity cannot be evaluated. If the acceleration were constant, then $\bar{v} = 13$ m/s.
15. (b) 2 m/s^2 (c) 3 m/s^2
17. (a) −8 m/s^2 (b) −9 m/s (c) 7 m/s
19. (a) 0 (b) 6 m/s^2 (c) 825 m (d) 65 m/s
21. (a) 9.75 ft/s^2 (b) 3.08 s
23. (a) 24.5 s (b) 122 m
25. 24 s
27. (a) 12.7 m/s (b) −2.3 m/s
29. (a) 4 cm (b) 18 cm/s
31. (a) 3×10^{-10} s (b) 1.26×10^{-4} m
33. (a) -3.5×10^5 m/s^2 (b) 2.9×10^{-4} s
35. (a) 8.20 s (b) 134 m
37. (a) 96 ft/s (b) -3.07×10^3 ft/s^2, or 96g (c) 3.1×10^{-2} s
39. (a) 39.2 m (b) 17.9 m/s (c) −9.8 m/s^2
41. (a) 17.2 m/s (b) 15.1 m
43. (a) 2.33 s (b) −32.8 m/s

Problems

1. (a) 0.75 s (b) −20 ft/s^2
3. (a) 40.4 s (b) 1735 m (c) −184 m/s
5. (b) $(47/12)v$ (c) $(47/60)v$
7. (a) $(3t^2 - 18t + 6)$ cm/s (b) $(3 \pm \sqrt{7})$ s (c) $-6\sqrt{7}$ cm/s^2, $6\sqrt{7}$ cm/s^2 (d) −74 cm
9. (a) 6.47 s (b) 335 ft (c) $v_j = 104$ ft/s, $v_s = 89.6$ ft/s
11. (a) 79.3 ft/s (b) −113 ft/s (c) −16.7 ft/s
13. (a) 5500 ft (b) 367 ft/s
15. (b) 4 m, 2 m/s (c) 1/3 s (d) −4 m, −10 m/s, −6 m/s^2

CHAPTER 4

Exercises

1. (a) $v_x = 2t$, $v_y = 4t$ (b) $x = t^2$, $y = 2t^2$ (c) $\sqrt{20}t$
3. (a) $\boldsymbol{v} = -12t\boldsymbol{j}$ m/s, $\boldsymbol{a} = -12\boldsymbol{j}$ m/s^2 (b) $\boldsymbol{r} = (3\boldsymbol{i} - 6\boldsymbol{j})$ m, $\boldsymbol{v} = -12\boldsymbol{j}$ m/s
5. (a) $\boldsymbol{v} = 4\boldsymbol{i}$ m/s (b) $x = 4$ m, $y = 6$ m
7. 2.70 m/s^2
9. 54.4 cm below the center of the target
11. 53.1°
13. (a) 14.7°, 75.4° (b) 10.4 s, 39.5 s
15. 80 m

17. (a) 12.6 m/s (b) 395 m/s^2 directed toward the center of rotation
19. $v = 10.5$ m/s, $a = 219$ m/s^2
21. (a) 32 m/s^2 downward (b) 72 m/s^2 upward
23. (a) 13.0 m/s^2 toward the center (b) 6.24 m/s (c) 7.50 m/s^2 along $\boldsymbol{v}$
25. (a) 4 m/s^2 toward the center (b) $\sqrt{8}$ m/s
27. (a) 14.5° north of west (b) 194 km/h
29. 72 km/h, 56.3° north of east
31. 33.6 min (compared with 27.8 min)

Problems

1. (a) 1.53×10^3 m (b) 36.2 s (c) 4.05 km
3. (a) $v_x = 7.14$ m/s, $v_y = -12.1$ m/s (b) $t = 2.47$ s (c) $d = 4.90$ m
5. (b) $v = -6 \sin 2t\boldsymbol{i} + 6 \cos 2t\boldsymbol{j}$ m/s, $a = -12 \cos 2t\boldsymbol{i} - 12 \sin 2t\boldsymbol{j} = -4\boldsymbol{r}$ m/s^2 (c) $\dfrac{v^2}{r} = 12$ $m/s^2 = |\boldsymbol{a}|$
7. 0.139 m/s
9. 7.52 m/s away from the quarterback
11. (a) 41.7 m/s (b) 3.81 s (c) $v_x = 34.2$ m/s, $v_y = -13.4$ m/s, $v = 36.7$ m/s

CHAPTER 5

Exercises

1. (a) 3 (b) 1.5 m/s^2
3. (a) 534 N (b) 54.4 kg
5. 1.96×10^5 dynes, or 1.96 N
7. (a) 12 N (b) 3 m/s^2
9. (a) $(4\boldsymbol{i} + 3\boldsymbol{j})$ m/s^2 (b) $(5.5\boldsymbol{i} + 2.6\boldsymbol{j})$ m/s^2
11. 8 N in the negative x direction
13. (a) $F_x = 2.5$ N, $F_y = 5$ N (b) $F = 5.6$ N
15. 6.4×10^3 N
17. 2 ft/s^2
19. (a) $T_1 = 31.5$ N, $T_2 = 37.5$ N, $T_3 = 49$ N (b) $T_1 = 113$ N, $T_2 = 56.6$ N, $T_3 = 98$ N
21. (a) 576 lb (b) No; F would have to be infinitely large.
23. 3.73 m
25. (a) $T = 36.8$ N (b) $a = 2.45$ m/s^2 (c) 1.23 m
27. $a = \dfrac{F}{m_1 + m_2}$, $T = \dfrac{m_1}{m_1 + m_2}F$
29. (a) 16.3 N (b) 8.07 N
31. $\mu_s = 0.38$, $\mu_k = 0.31$
33. 0.458
35. (b) $T = 16.7$ N, $a = 0.69$ m/s^2
37. (a) 1.78 m/s^2 (b) 0.368 (c) 9.37 N (d) 2.67 m/s
39. (a) 35.4 N (b) 0.601
41. (a) 0.55 (b) 0.25 m/s^2

Problems

1. (a) 3.12 m/s^2 (b) 17.5 N
3. (a) friction between the two blocks (b) 34.7 N (c) 0.306
5. (b) 5.75 m/s^2 (c) $T_1 = 17.4$ N, $T_2 = 40.5$ N
7. (a) 20 lb (b) 12 lb (c) 18 ft/s^2 (d) the upper rope
9. (a) 1.02 m/s^2 (b) 2.04 N, 3.06 N, 4.08 N (c) 14 N between m_1 and m_2, 8 N between m_2 and m_3
11. (a) $mg\left(\dfrac{\sin\theta + \mu_s\cos\theta}{\cos\theta - \mu_s\sin\theta}\right)$ (b) $ma + mg\left(\dfrac{\sin\theta + \mu_k\cos\theta}{\cos\theta - \mu_k\sin\theta}\right)$
13. (a) $T_1 = 78.0$ N, $T_2 = 35.9$ N (b) 0.655
15. (b) 8 ft/s^2 (c) 20 lb
17. $T_A = 304$ N, $T_B = 290$ N, $T_C = 152$ N, $T_D = 138$ N
21. $a_1 = \dfrac{2m_2g}{4m_1 + m_2}$, $a_2 = \dfrac{m_2g}{4m_1 + m_2}$

CHAPTER 6

Exercises

1. 2.96×10^{-9} N
3. 4.62×10^{-8} N toward the center of the triangle
5. $F_x = Gm^2\left[\dfrac{2}{b^2} + \dfrac{3b}{(a^2 + b^2)^{3/2}}\right]$, $F_y = Gm^2\left[\dfrac{2}{a^2} + \dfrac{3a}{(a^2 + b^2)^{3/2}}\right]$
7. $F_6 = (12.6\boldsymbol{i} + 1.92\boldsymbol{j}) \times 10^{-11}$ N, $F_6 = 12.7 \times 10^{-4}$ N
9. 8.20×10^{-8} N
11. $F = 4.41$ N away from the center of the square
13. 2.69×10^{10}
15. 2.3×10^3 N
17. (a) friction (b) 0.128
19. (a) 5.60×10^3 m/s (b) 238 min (c) 1.47×10^3 N
21. (a) 8.20×10^{-8} N (b) 9.01×10^{22} m/s^2 (c) 6.56×10^{15} rev/s
23. (a) 2.49×10^4 N (b) 12.1 m/s

25. (a) 20.4 N (b) $a_t = 4.14\ m/s^2$, $a_r = 32\ m/s^2$ (c) $32.3\ m/s^2$
27. $2.42\ m/s^2$ in the forward direction
29. (a) $3.6\ m/s^2$ to the right (b) zero
31. (a) $1.47\ N \cdot s/m$ (b) 2.03×10^{-3} s (c) 2.94×10^{-2} N

Problems

1. (a) 0.61 rev/s (b) 0.77 m/s, $2.93\ m/s^2$
3. (a) 66.2 N (b) 36.6 N (c) 6.96 N
5. (a) $q = 1.88 \times 10^{-7}$ C (b) 5.07×10^{-2} N
7. (a) $v_{max} = \sqrt{Rg\left(\frac{\tan\theta + \mu}{1 - \mu\tan\theta}\right)}$, $v_{min} = \sqrt{Rg\left(\frac{\tan\theta - \mu}{1 + \mu\tan\theta}\right)}$ (b) $\mu = \tan\theta$ (c) $v_{max} = 16.6$ m/s (37 mi/h), $v_{min} = 8.57$ m/s (19 mi/h)
9. (b) 2.54 s, 23.6 rev/min

CHAPTER 7

Exercises

1. 5.88×10^3 J
3. (a) 317 J (b) −176 J (c) zero (d) zero (e) 141 J
5. (a) 2.94×10^5 J (b) -2.94×10^5 J
7. (a) 3 (b) 74.7°
9. 18.4
13. (a) 63.4° (b) 80.7° (c) 67.8°
15. (a) 7.5 J (b) 15 J (c) 7.5 J (d) 30 J
17. (b) −12 J
19. (a) 22.5 J (b) 90 J
21. (a) 51 J (b) 69 J
23. (a) 9×10^3 J (b) 300 N
25. (a) 1.94 m/s (b) 3.35 m/s (c) 3.87 m/s
27. (a) $v_0^2/2\mu_k g$ (b) 12.8 m
29. (a) 0.791 m/s (b) 0.531 m/s
31. (a) 63.9 J (b) −35.4 J (c) −9.51 J (d) 19.0 J
33. 829 N (186 lb)
35. (a) 0.41 m/s (b) 2.45×10^3 J
37. (a) 3920 W (5.25 hp) (b) 7.06×10^5 J
39. (a) 7.5×10^4 J (b) 2.50×10^4 W (33.5 hp) (c) 3.33×10^4 W (44.7 hp)
41. 6.0 km/liter
43. (a) 29.7 kW (b) 37.3 kW

Problems

1. (a) $\cos\alpha = A_x/A$, $\cos\beta = A_y/A$, $\cos\gamma = A_z/A$, where $A = (A_x^2 + A_y^2 + A_z^2)^{1/2}$
3. (a) $kd/2mg$ (b) $kd/4mg$
5. (a) 20 J (b) 6.71 m/s
7. (a) −5.6 J (b) 0.152 (c) 2.28 rev
9. (c) 7.29×10^7 J 1.97×10^4 W (d) 13.6%
11. (a) $2.7\ m/s^2$ (c) 4.04×10^3 N (d) 146 hp

CHAPTER 8

Exercises

1. (a) $W_{OA} = 0$, $W_{AC} = -147$ J, and so $W_{OAC} = -147$ J (b) $W_{OB} = -147$ J, $W_{BC} = 0$, and so $W_{OBC} = -147$ J (c) $W_{OC} = -147$ J; the gravitational force is conservative.
3. (a) $W_{OAO} = -30$ J (b) $W_{OACO} = -51.2$ J (c) $W_{OCO} = -42.4$ J (d) Friction is a nonconservative force.
5. (a) 70 J (b) −70 J (c) 6.83 m/s
7. (a) 15 J, 30 J (b) Yes. The total energy is not conserved since $E_i = 30$ J and $E_f = 20$ J
9. (a) −19.6 J (b) 39.2 J (c) zero
11. (a) 5.91 J (b) 3.47 m/s (c) 49.6 N (d) 0.816 m
13. (a) 31.3 m/s (b) 147 J (c) 4
15. (a) 0.225 J (b) 0.363 J (c) No. The normal force varies with position, and so the frictional force also varies.
17. (a) 8.33 m (b) −50 J (c) zero
19. (a) 8.85 m/s (b) 54.1%
21. (a) 9.90 m/s (b) −11.8 J (c) −11.8 J
23. (a) 0.180 J (b) 0.100 J
25. (a) $(2\,mgh/k)^{1/2}$ (b) 8.94 cm
27. (a) 588 N/m (b) 0.70 m/s
29. (a) $F_r = A/r^2$ (b) the gravitational force (A negative) and the electrostatic force (A positive or negative)
31. (a) zero at A, C, and E, positive at B, negative at D (b) unstable at A and E, stable at C
33.

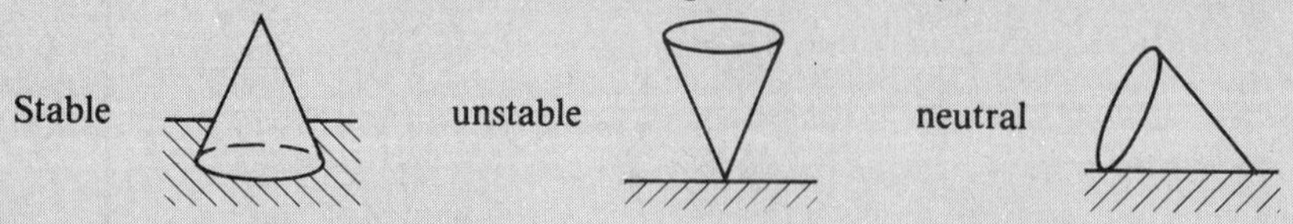

35. 2.74×10^{-11} J, or 171 MeV compared with 2.14×10^{-11} J, or 134 MeV
37. 0.110065 amu, or 103 MeV
39. 47 GW

Problems

1. (a) 3.49 J, 676 J, 741 J (b) 175 N, 338 N, 370 N (c) yes
3. (a) $\Delta U = -\dfrac{ax^2}{2} - \dfrac{bx^3}{3}$ (b) $\Delta U = -\dfrac{A}{\alpha}(1 - e^{\alpha x})$
5. (a) 125 J (b) 50 J (c) 66.7 J (d) nonconservative, since W is path-dependent
7. 0.115
9. 1.07 m/s
11. (a) $v = \sqrt{\dfrac{g}{L}(L^2 - d^2)}$ (b) $t = \sqrt{\dfrac{L}{g}}\ln\left(\dfrac{L + \sqrt{L^2 - d^2}}{d}\right)$
15. $y = \dfrac{mg}{k} + \sqrt{\left(\dfrac{mg}{k}\right)^2 + \dfrac{2mgh}{k}}$

CHAPTER 9

Exercises

1. $p_x = 6$ kg · m/s, $p_y = -12$ kg · m/s, $p = 13.4$ kg · m/s
3. 1.70×10^4 kg · m/s in the northwesterly direction (b) 5.66×10^3 N
5. (a) 12 kg · m/s (b) 6 m/s (c) 4 m/s
7. (a) 1.35×10^4 kg · m/s (b) 9×10^3 N (c) 18×10^3 N
9. (a) quadrupled (b) $\sqrt{3}$ times its initial value
11. (a) 22.5 kg · m/s (b) 1.13×10^3 N
13. (a) 49.1 kg · m/s (b) 2.46×10^4 N (5520 lb)
15. 6 m/s to the left
17. The boy moves westward with a speed of 2.46 m/s.
19. 2.68×10^{-20} m/s
21. 340 m/s
23. 6 kg
25. (a) 2.75 m/s (b) 6.75×10^4 J
27. (a) 0.284, or 28.4% (b) $K_n = 1.15 \times 10^{-13}$ J, $K_c = 0.45 \times 10^{-13}$ J
29. (a) −6.67 cm/s, 13.3 cm/s (b) 8/9
31. (b) and (c) are perfectly elastic
33. (a) 24 cm/s (b) No. The earth recoils by a negligible amount.
35. $\boldsymbol{v} = (2\boldsymbol{i} - 1.8\boldsymbol{j})$ m/s
37. (a) $v_x = -9.3 \times 10^6$ m/s, $v_y = -8.3 \times 10^6$ m/s (b) 4.4×10^{-13} J
39. 4.67×10^6 m (this point lies within the earth)
41. $\left(\dfrac{1}{3}, \dfrac{5}{3}\right)$ m
43. (a) $\boldsymbol{v}_c = (1.4\boldsymbol{i} + 3.2\boldsymbol{j})$ m/s (b) $\boldsymbol{p} = (7\boldsymbol{i} + 16\boldsymbol{j})$ kg · m/s
45. $\boldsymbol{a}_c = (\boldsymbol{i} + 2\boldsymbol{j})$ m/s²
47. 3×10^5 N
49. 1.42×10^4 m/s

Problems

1. (a) 2.04 m/s, south (b) 2.75 m/s to the south (c) 2.30 m/s at an angle 62° south of west
3. 1.48×10^3 m/s
5. (a) $\boldsymbol{r}_c = 3\boldsymbol{j}$ m (b) $\boldsymbol{v}_c = (4\boldsymbol{i} + 2\boldsymbol{j})$ m/s (c) $\boldsymbol{a}_c = (3\boldsymbol{i} - \boldsymbol{j})$ m/s²
7. (a) 6.93 m/s (b) 1.14 m
9. (a) 0.556 m/s (b) 11.1 J
11. $x = \dfrac{2v_0^2}{9\mu g} - \dfrac{4}{9}d$
13. 108 N
15. $\left(\dfrac{3Mg}{L}\right)x$

CHAPTER 10

Exercises

1. 1.67 rad, or 95°
3. (a) 377 rad/s (b) 565 rad
5. (a) 5 rad/s² (b) 10 rad
7. (a) 1.99×10^{-7} rad/s (b) 2.6×10^{-6} rad/s
9. (a) 0.40 rad/s (b) 32 m/s² toward the center
11. (a) 8 rad/s (b) 16 m/s, $a_r = 128$ m/s², $a_t = 8$ m/s² (c) $\theta = 9$ rad
13. (a) 126 rad/s (b) 2.51 m/s (c) 953 m/s² (d) 15.1 m
15. (a) 143 kg · m² (b) 4.58×10^3 J
17. (a) 92 kg · m², 184 J (b) 6 m/s, 4 m/s, 8 m/s, 184 J
19. (a) $\frac{3}{2}MR^2$ (b) $\frac{7}{5}MR^2$
21. 3.2 N · m into the plane
23. (a) 12 kg · m² (b) 2.4 N · m (c) 43.8 rev

25. (a) $2\ \text{rad/s}^2$ (b) 6 rad/s (c) 90 J (d) 4.5 m
27. (a) 46.8 N (b) $0.234\ \text{kg}\cdot\text{m}^2$ (c) 40 rad/s

Problems

3. (a) $\dfrac{Mmg}{M+4m}$ (b) $\dfrac{4mg}{M+4m}$ (c) $\dfrac{1}{R}\sqrt{\dfrac{8mgh}{M+4m}}$
5. (a) $m_1g(m_1+m_2+I/R^2)^{-1}$ (b) $T_2 = m_1m_2g(m_1+m_2+I/R^2)^{-1}$, $T_1 = m_1g(I+m_2R^2)[I+(m_1+m_2)R^2]^{-1}$ (c) $a = 3.12\ \text{m/s}^2$, $T_1 = 26.7$ N, $T_2 = 9.37$ N (d) $a = 5.6\ \text{m/s}^2$, $T_1 = T_2 = 16.8$ N
7. (a) $\omega = \sqrt{3g/l}$ (b) $\alpha = 3g/2l$ (c) $a_x = \frac{3}{2}g$, $a_y = \frac{3}{4}g$ (d) $R_x = \frac{3}{2}Mg$, $R_y = \frac{1}{4}Mg$
9. (a) $0.707R$ (b) $0.289L$ (c) $0.632R$

CHAPTER 11

Exercises

1. (a) $5\mathbf{k}$ (b) 135°
5. (a) $-6\mathbf{k}$ (b) $-4\mathbf{i} - 12\mathbf{j}$ (c) $-2\mathbf{j} + 6\mathbf{k}$
7. (a) $-10\mathbf{k}\ \text{N}\cdot\text{m}$ (b) $8\mathbf{k}\ \text{N}\cdot\text{m}$
9. (a) mvd (out of the plane) (b) $-2mvd$ (into the plane) (c) zero
11. (a) $24\mathbf{k}\ \text{kg}\cdot\text{m}^2/\text{s}$ (b) $-16\mathbf{k}\ \text{kg}\cdot\text{m}^2/\text{s}$
13. $12.5\ \text{kg}\cdot\text{m}^2/\text{s}$ (out of the plane)
15. (a) $\mathbf{L} = md(v_0 + gt)\mathbf{k}$ (b) $\boldsymbol{\tau} = mgd\mathbf{k}$
17. (a) $0.367\ \text{kg}\cdot\text{m}^2/\text{s}$ (b) $1.47\ \text{kg}\cdot\text{m}^2/\text{s}$
19. (a) $0.336\ \text{N}\cdot\text{m}$ (b) $L = 0.28v$ (c) $8.4\ \text{m/s}^2$
21. 7.35 rad/s
23. (a) 8.57 rad/s (b) increases by 234 J (c) The student does work on the system.
25. (a) 0.420 rad/s in the counterclockwise direction (b) 123 J
27. (a) $\left(\frac{6}{5}gh\right)^{1/2}$ (b) $\frac{3}{5}g\sin\theta$
29. (a) $a_c = \frac{2}{3}g\sin\theta$ (disk), $a_c = \frac{1}{2}g\sin\theta$ (hoop) (b) $\frac{1}{3}\tan\theta$
31. (a) 500 J (b) 250 J (c) 750 J

Problems

3. (a) v_0r_0/r (b) $T = (mv_0^2r_0^2)r^{-3}$ (c) $\frac{1}{2}mv_0^2\left(\frac{r_0^2}{r^2} - 1\right)$ (d) 4.5 m/s, 10.1 N, 0.45 J
5. $\omega = \sqrt{\dfrac{10}{7}\dfrac{g}{r^2}(1-\cos\theta)(R-r)}$
9. (a) $2.7(R-r)$ (b) $F_x = -\frac{10}{7}mg\left(\frac{2R+r}{R-r}\right)$ $F_y = -\frac{5}{7}mg$
13. (a) $F_y = \frac{W}{L}\left(d - \frac{ah}{g}\right)$ (b) $F_x = -306$ N, $F_y = 553$ N
15. (a) $\frac{1}{3}\omega_0$ (b) $\frac{2}{3}$

CHAPTER 12

Exercises

1. $F_1 + F_2 - W_1 - W_2 = 0$, $F_2l - W_1d_1 - W_2d_2 = 0$
3. $x = \dfrac{(W_1 + W)\,d + W_1l/2}{W_2}$
5. The y coordinate of the center of mass is 15.3 cm from the bottom. The x coordinate is 8 cm from the left side of the "tee."
7. at the 75-cm mark
11. (a) 1.36 m from the front axle (b) 3560 N on each back tire and 4280 N on each front tire

Problems

1. (b) $T = 213$ N, $R_x = 184$ N, $R_y = 188$ N
3. (b) $T = 17.3$ lb (c) $d = 0.76l$
5. (a) $W = \dfrac{w}{2}\left(\dfrac{2\mu_s\sin\theta - \cos\theta}{\cos\theta - \mu_s\sin\theta}\right)$ (b) $R = (w+W)\sqrt{1+\mu_s^2}$, $F = \sqrt{W^2 + \mu_s^2(w+W)^2}$
7. (a) $\mu_k = 0.57$, $\frac{6}{7}$ ft from the right corner (b) $h = \frac{5}{3}$ ft
9. $N_a = 6.0 \times 10^5$ N, $N_b = 4.8 \times 10^5$ N
11. (a) 133 N (b) $N_A = 429$ N, $N_B = 257$ N (c) $R_x = 133$ N, $R_y = 257$ N
13. $T = 2710$ N, $R_x = 2650$ N
15. (b) $T = 1.07 \times 10^3$ N, $R_x = 991$ N, $R_y = 497$ N

CHAPTER 13

Exercises

1. (a) 1.5 Hz, 0.67 s (b) 4 m (c) π rad (d) -4 m
3. (a) 4.3 cm (b) -5 cm/s (c) -17 cm/s^2 (d) π s, 5 cm
5. (a) -14 cm/s, 16 cm/s^2 (b) 16 cm/s, 1.83s (c) 32 cm/s^2, 1.05 s
7. (b) 6π cm/s, 0.33 s (c) $18\pi^2$, 0.5 s (d) 12 cm
9. 3.95 N/m
11. (a) 2.40 s (b) 0.417 Hz (c) 2.62 rad/s
13. (a) 0.4 m/s, 1.6 m/s^2 (b) 0.32 m/s, -0.96 m/s^2 (c) 0.23 s
15. (a) 0.153 J (b) 0.783 m/s (c) 17.5 m/s^2
17. (a) quadrupled (b) doubled (c) doubled (d) no change
19. 2.6 cm
21. 0.158 Hz, 6.35 s
23. 106
25. increases by 1.78×10^{-3} s
27. 8.5×10^{-2} kg $\cdot$ m^2
33. (a) 1 s (b) 5.09 cm

Problems

1. (a) $v = 14\pi \sin 2\pi t$ cm/s, $a = 28\pi^2 \cos 2\pi t$ cm/s^2

3. (a) $\omega = \left(\dfrac{gd}{d^2 + l^2/12}\right)^{1/2}$ (b) 1.53 s

7. (a) $I = mL^2 + \dfrac{2}{5}mR^2$ (b) $T = 2\pi\sqrt{\dfrac{L}{g}}\left(1 + \dfrac{2}{5}\dfrac{R^2}{L^2}\right)^{1/2}$

11. (a) 2 Mg, $T_p = Mg\left(1 + \dfrac{y}{l}\right)$ (b) $\dfrac{4\pi}{3}\sqrt{\dfrac{2l}{g}} = 2.68$ s

13. (a) $E = \dfrac{1}{2}mv^2 + mgL(1 - \cos\theta)$

15. $\omega = \left(\dfrac{mgL + kh^2}{I}\right)^{1/2}$

CHAPTER 14

Exercises

1. (a) $4\pi^2/GM_e = 9.89 \times 10^{-14}$ s^2/m^3 (b) 127 min
5. 9.37×10^6 m
7. 4.22×10^7 m
9. (a) -1.67×10^{-14} J (b) at the center of the triangle
11. $-20.95\dfrac{Gm^2}{a}$
15. 5.04×10^3 m/s
17. (a) 3.90×10^9 J (b) $|U|$ is halved, K is halved
19. (a) 1.87×10^{11} J (b) 103 kW
21. $Gm\lambda_0 L[d(L + d)]^{-1} + GmAL$ to the right
23. (a) 7.41×10^{-10} N (b) 1.04×10^{-8} N (c) 5.21×10^{-9} N

Problems

1. (a) $(GM/4R^3)^{1/2}$ (b) $(g/R)^{1/2} = 1.57$ rad/s (0.249 rev/s)

3. (a) $U = -\dfrac{3Gm}{R}\left(M + \dfrac{\sqrt{3}}{3}m\right)$ (b) $v = \left(\dfrac{\sqrt{3}Gm}{3R} + \dfrac{GM}{R}\right)^{1/2}$

5. (a) $F = \dfrac{GMmd}{(R^2 + d^2)^{3/2}}$ downward (b) $F = 0$ at the middle and $F \approx \dfrac{GMm}{d^2}$ for $d \gg R$

7. (a) $k = \dfrac{GmM_e}{R_e^3}$, at $\dfrac{L}{2}$ (b) $\dfrac{L}{2}\left(\dfrac{GM_e}{R_e^3}\right)^{1/2}$, at the middle of the tunnel (c) 311 m/s

9. (a) 7.34×10^{22} kg (b) 1.63×10^3 m/s (c) 1.32×10^{10} J

11. (a) $v_1 = m_2\left[\dfrac{2G}{d(m_1 + m_2)}\right]^{1/2}$, $v_2 = m_1\left[\dfrac{2G}{d(m_1 + m_2)}\right]^{1/2}$, $v_{rel} = \left[\dfrac{2G(m_1 + m_2)}{d}\right]^{1/2}$
(b) $K_1 = 1.07 \times 10^{32}$ J, $K_2 = 2.67 \times 10^{31}$ J

CHAPTER 15

Exercises

1. 667 N
3. 9.52×10^{-6}
5. (a) 3.14×10^4 N (7060 lb) (b) 6.28×10^4 N (14 100 lb)
7. 1.40×10^7 N/m^2, 5.65×10^{-8} m^3
9. 0.11 kg

11. 4×10^{17} kg/m^2. Matter contains mostly free space.
13. (a) 9.8×10^5 N (b) 1.96×10^5 N on each side (c) 9.8×10^4 N on each end
15. 1.62 m
17. 1.28×10^5 N/m^2, 2.68×10^4 N/m^2
23. (a) 7 cm (b) 2.8 kg
25. 0.439 kg
27. (a) 4.24 m/s (b) 17.0 m/s
29. (a) 0.83 m/s (lower), 3.3 m/s (upper) (b) 4.15×10^{-3} m^3/s
31. (a) 2.65 m/s (b) 2.31×10^4 N/m^2
33. 4.31×10^4 N
35. 4.9%

CHAPTER 16

Exercises

1. (a) 42.9°C (b) 1.47 atm
3. (a) 68.3 mm Hg (b) 131 mm Hg
5. (a) 832.3°F (b) 717.8 K
7. 37.0°C, 39.4°C
9. −40°C
11. 1.43 cm
13. 1.26 cm
15. 0.95 gal
17. (a) 1.35×10^{-2} cm (b) 6.75×10^{-4} cm (c) 3.18×10^{-2} cm^3
19. 53.3 lb/in.^{2}
21. 3.28×10^{13} molecules
23. 287°C, or 560 K
25. (a) 600 K, or 327°C (b) 1200 K, or 927°C

Problems

1. (a) $R_0 = 50$ ohms, $A = 1.55 \times 10^{-3}\,(\text{C}°)^{-1}$ (b) 200°C
3. 3.28 cm
5. (a) 2.75×10^{-5} s (b) loses 16.6 s each week
9. (a) 24.5 m (b) 3.41×10^5 N/m^2
11. (b) 1.25 kg/m^3

CHAPTER 17

Exercises

1. 23.4°C
3. 2.69×10^4 cal
5. 80.6°C
7. 63.9°C
9. 1.17×10^4 cal
11. 167 g
13. 1.21 liters
15. 9×10^6 cal/h
17. 271 cal/s
19. 63.8°C
21. 3.91×10^{26} W
23. (a) 9.89×10^{-3}C° (b) It goes into heating up the surface.
25. (a) 20.57°C (b) No. The change in potential energy and the heat absorbed are both proportional to the mass; hence the mass cancels.
27. 2.85×10^3 m
29. (a) 6.08×10^5 J (b) -4.05×10^5 J
31. (a) −7.9 J (b) 723 J
33. −420 J (heat leaves the system)
35. $A \rightarrow B(+++)$, $B \rightarrow C(0--)$, $C \rightarrow A(---)$
37. (a) 5.48×10^3 J (b) 5.48×10^3 J
39. (a) 3.08×10^{-2} m^3 (b) -3.46×10^3 J (c) -3.46×10^3 J
41. (a) 101 J (b) 192 J

Problems

1. (a) $-\frac{2}{3}P_0V_0$ (b) $-RT_0\ln 2$ (c) zero
3. 141 J
5. 0.654 cal/g · C°
7. 5.75×10^3 J/s, or 5.75 kW
9. 12.2 h
11. (a) 26.9 liters (b) 8.43 liters/min
13. (a) 725 cal/s (b) 12.4 min

CHAPTER 18

Exercises

1. 2.43×10^5 m^2/s^2
3. $\overline{F} = 8.0$ N, $P = 1.6$ N/m^2
5. $\overline{F} = 0.943$ N, $P = 1.57$ N/m^2
7. 2.54×10^3 m/s
9. (a) 731 m/s at 600 K 422 m/s at 200 K
11. (a) 2.28×10^3 J (b) 6.21×10^{-21} J
13. (a) 202 cal (b) 281 cal

15. (a) 209 J (b) zero (c) 317 K
17. (a) $C_V' = 8.96$ cal/K, $C_P' = 14.9$ cal/K (b) $C_V' = 14.9$ cal/K, $C_P' = 20.9$ cal/K
21. 10.1 atm, 756 K
27. (a) 3.4×10^4 molecules (b) 1.8×10^4 molecules
29. (a) 558 m/s (b) 514 m/s (c) 456 m/s
31. $v_{mp} = 731$ m/s, $\bar{v} = 825$ m/s, $v_{rms} = 895$ m/s
33. (a) 2.37×10^4 K (b) 1.06×10^3 K
35. (a) 3.21×10^{12} molecules (b) 7.78×10^5 m (c) $6.42 \times 10^{-4}\ s^{-1}$
37. (a) 1.96×10^{27} molecules/m^3 (b) 1.84×10^{-9} m (c) $2.42 \times 10^{11}\ s^{-1}$

Problems

3. $N_v \Delta v \approx 1.4 \times 10^{21}$ molecules
5. (a) $3.65v$ (b) $3.99v$ (c) $3v$ (d) $106mv^2/V$ (e) $7.97\,mv^2$
7. (a) $C_V = aR$ (b) $C_P = (a + 1)R$ (c) $a = 3.42$ (d) about 7

CHAPTER 19

Exercises

1. (a) 6.64% (b) 84 cal
3. (a) 37.5% (b) 628 J (c) 2.09 kW
5. (a) 6 cal = 25.1 J (b) 36 cal
7. (a) 33% (b) 2/3
9. (a) 140 cal (b) 350 K
11. (a) 5.1% (b) 1.26×10^{12} cal/h
13. (a) 67.2% (b) 61.5 kW
15. (a) 16.8 (b) 1.49 kW
17. (a) 185 cal (b) 115 cal
19. 1.75, 80% Carnot efficiency
21. 3.59 J/K
23. 46.6 cal/K
25. (a) -6.70 J/K (b) 11.2 J/K (c) zero (d) 4.46 J/K
27. (a) 53°C (b) 7.3 J/K
29. 57.2 J/K
31. (a) 7.33×10^3 J/K (b) -6.67×10^3 J/K (c) 0.66×10^3 J/K

Problems

1. (a) 4.11×10^3 J (b) 1.42×10^4 J (c) 1.01×10^4 J (d) 28.9%
3. (a) $2nRT_0 \ln 2$ (b) 0.273, or 27.3%
5. (a) $10.5RT_0$ (b) $-8.5RT_0$ (c) $4/21 \approx 19\%$ (d) $5/6 \approx 83\%$
9. $nR\left(\frac{\gamma}{\gamma - 1}\right)\ln 3$

CHAPTER 20

Exercises

1. (a) $-80\ \mu C$ (b) $4.8\ \mu C$
3. $|F| = 1.11 \times 10^{-2}$ N. The force is attractive.
5. $\boldsymbol{F} = (0.702\boldsymbol{i} + 0.281\boldsymbol{j})$ N
7. $\vec{F} = 1.35k\frac{q^2}{a^2}(\hat{\imath} + \hat{\jmath})$
9. at $y = 0.652$ m
11. (a) $-6.3 \times 10^3\boldsymbol{i}$ N/C (b) $2.8 \times 10^3\boldsymbol{j}$ N/C (c) $-8.9 \times 10^3(\boldsymbol{i} + \boldsymbol{j})$ N/C
13. (a) $4.3 \times 10^4\boldsymbol{j}$ N/C (b) $-0.128\boldsymbol{j}$ N
15. (a) at the center of the triangle (b) $1.73kq/a^2$ upward
17. 1.82 m to the left of the -2.5-μC charge
19. 5.13×10^6 N/C toward the rod
21. (a) 2.57×10^7 N/C (b) 1.36×10^7 N/C (c) 2.43×10^6 N/C (d) 6.78×10^3 N/C
23. 2.16×10^7 N/C to the left
25. (a) $q_1/q_2 = 1/3$ (b) q_1 is negative; q_2 is positive
27.

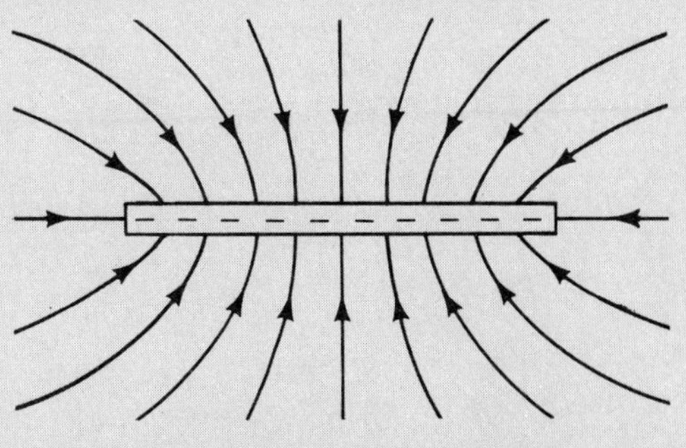

29. (a) 4.79×10^{10} m/s^2 (b) 5.22×10^{-5} s (c) 65.2 m (d) 5.22×10^{-15} J
31. (a) 1.54×10^{-8} s (b) 1.05×10^{-3} m (c) 2.76×10^{-3} m

Problems

1. (a) $E_x = \dfrac{16kqb}{(a^2 + b^2)^{3/2}}$, $E_y = 0$ (b) $F_x = q_0E_x$, $F_y = 0$
3. (a) 1.09×10^{-8} C (b) 5.43×10^{-3} N
5. 4.4×10^5 N/C
7. (b) For $y \gg a$, $E_y = 2kq/y^2$. At large distances, the two charges look like a single charge of magnitude $2q$.
11. $E_x = \dfrac{k\lambda_0}{d}\left[\ln\left(\dfrac{l + d}{d}\right) - \dfrac{l}{l + d}\right]$

CHAPTER 21

Exercises

1. (a) 8×10^3 N·m^2/C (b) zero (c) 4.81×10^3 N·m^2/C
3. (a) 9.04×10^5 N·m^2/C (b) 4.52×10^5 N·m^2/C
5. (a) 2.26×10^6 N·m^2/C (b) 1.36×10^7 N·m^2/C (c) The total flux through all cube faces remains the same, but the flux through one of the faces, as in (a), would depend on the location of the charge.
7. (b) zero (c) 5.65×10^5 N·m^2/C
9. (a) positive (b) It is not at the center of the sphere. (c) Q/ε_0
11. (a) zero (b) 3.62×10^6 N/C (c) 2.88×10^5 N/C
13. (a) 6.92×10^4 N/C (b) 2.26×10^3 N·m^2/C
15. 4.24×10^5 N/C perpendicular to the sheet
17. (a) zero (b) 3.33×10^3 N/C (c) 3.11×10^2 N/C (d) no change
19. (a) 2.26×10^{-8} C/m^2 (b) 1.06 nC
21. (a) 2.81×10^6 N/C directed toward the center (b) since $q_{in} = 0$, $E_{in} = 0$
23. (a) $-\lambda$ on the inner surface, 3λ on the outer surface (b) $E = 3\lambda/2\pi r\varepsilon_0$

Problems

1. (a) For $r < a$, $E = \rho r/3\varepsilon_0$. For $a < r < b$ and for $r > c$, $E = kQ/r^2$. For $b < r < c$, $E = 0$. (b) inner surface $\sigma = -Q/4\pi b^2$, outer surface $\sigma = Q/4\pi c^2$.
3. (a) since $q_{in} = 0$, $E = 0$ (b) $E = \dfrac{\rho}{3\varepsilon_0}\left(\dfrac{r^3 - a^3}{r^2}\right)$ (c) $E = \dfrac{\rho}{3\varepsilon_0}\left(\dfrac{b^3 - a^3}{r^2}\right)$
5. (a) zero (b) σ/ε_0 to the right (c) zero
7. $abc(4a + 2c) = 0.269$ N·m^2/C; $Q = 2.38 \times 10^{-12}$ C
9. (a) $\dfrac{\rho_0 r}{2\varepsilon_0}\left(a - \dfrac{2r}{3b}\right)$ (b) $\dfrac{\rho_0 R^2}{2\varepsilon_0 r}\left(a - \dfrac{2R}{3b}\right)$

CHAPTER 22

Exercises

1. zero
3. 9.2×10^4 V
5. 7.0×10^5 m/s
7. $V_A - V_B = -520$ V
9. 4.8×10^{-3} J
11. -71 V, $x = 0$ is at the higher potential
13. 2.0 m
15. 3 m, 2×10^{-7} C
17. -2.0 J
19. -7.84×10^6 V
21. $V = 2\pi k\sigma(\sqrt{x^2 + b^2} - \sqrt{x^2 + a^2})$
23. $V = \dfrac{kbL}{2}\ln\left[\dfrac{\sqrt{(L/2)^2 + b^2} + L/2}{\sqrt{(L/2)^2 + b^2} - L/2}\right]$
25. 243°
27. 2.78×10^{11} electrons
29. 2.5 nC

Problems

1. $V = k\sigma\displaystyle\int_{-L/2}^{L/2}\int_{-L/2}^{L/2}\frac{dx\,dy}{\sqrt{x^2 + y^2 + d^2}}$
5. -3.33×10^6 V
7. -0.61 J
9. (a) $E_r = \dfrac{2kp\cos\theta}{r^3}$, $E_\theta = \dfrac{kp\sin\theta}{r^3}$ (b) $E_x = \dfrac{3kpxy}{(x^2 + y^2)^{5/2}}$, $E_y = \dfrac{kp(2y^2 - x^2)}{(x^2 + y^2)^{5/2}}$
11. $\dfrac{3}{5}k\dfrac{Q^2}{R}$
13. (a) 66.7 pC (b) 476 V